FEATURES AND BE
Algebra: Concepts and Appl

Student-Friendly	. . . features are included throughout the text.	
	• Lessons focus on **one objective**.	198
	• **Foldables**™ Study Organizers help students actively organize key concepts and create their own review materials.	92
	• Important concepts are described with **Words**, **Symbols**, and **Numbers** as appropriate to help students move from the concrete to the abstract.	71
	• **Your Turn** exercises that follow most Examples give students the opportunity to practice what is presented in the Example before moving on in the lesson.	198
	• **Info-Graphics** help to reduce the reading load by conveying algebraic concepts visually.	58
Reading/Writing/ Communicating in Mathematics	. . . strategies and activities are essential for student success in mathematics.	
	• Frequent **Reading Algebra** notes appear at the point of need.	4, 80, 363
	• **Vocabulary** words are highlighted in yellow.	122
	• Every lesson includes **Communicating Mathematics** exercises that help students use a variety of communication techniques including writing, explaining, listing, and drawing.	61, 130, 319
Intervention and Student Help	. . . are provided throughout the program.	
	• **Prerequisite Skills Review** notes refer the student to help in the **Student Handbook** at the back of the textbook.	142
	• **Look Back** notes point students to the lesson where the needed skill was presented.	206
	• **Homework Help** in the margins of exercise sets link homework exercises to corresponding examples within the lesson. References to the Extra Practice pages for each lesson are also provided.	144
	• **Mixed Review** and **Quiz** exercises are referenced to the skills needed for the homework.	301
	• **Getting Ready** exercises in the Guided Practice review subskills or prerequisite skills needed for the homework.	268
	• Optional **Hands-On Algebra** activities help bridge the gap between the concrete and the abstract.	324
	• A wide variety of Online Study Tools are provided at **www.algconcepts.com**.	49, 180, 265, 361
Test Preparation and Assessment	. . . provide targeted practice for local, state, and national tests.	
	• A **Quiz** appears in the textbook after every two or three lessons with a link to an additional **Self-Check Quiz** at **www.algconcepts.com**.	269
	• Every exercise set concludes with at least one **Standardized Test Practice** problem. Additional Standardized Test Practice problems can also be found at **www.algconcepts.com**.	29, 103, 209
	• Every chapter concludes with a two-page **Preparing for Standardized Tests** lesson.	901–91
Staff Development	. . . features are available to assist new teachers and teachers new to *Algebra: Concepts and Applications.*	
"Sticky Notes" in Chapter 1 provide a "walk-through" of key features. pp. 2–49	• The **Teacher's Handbook** located at the front of the **Teacher's Wraparound Edition** provides detailed information about Student Learning, Assessment, Technology, and Course Planning.	T1–T17
	• The **Resource Manager** located at the beginning of each chapter in the Teacher's Wraparound Edition includes information on Suggested Pacing, Assessment Options, Glencoe Technology, and a lot more.	92a–92d

Teacher's Wraparound Edition

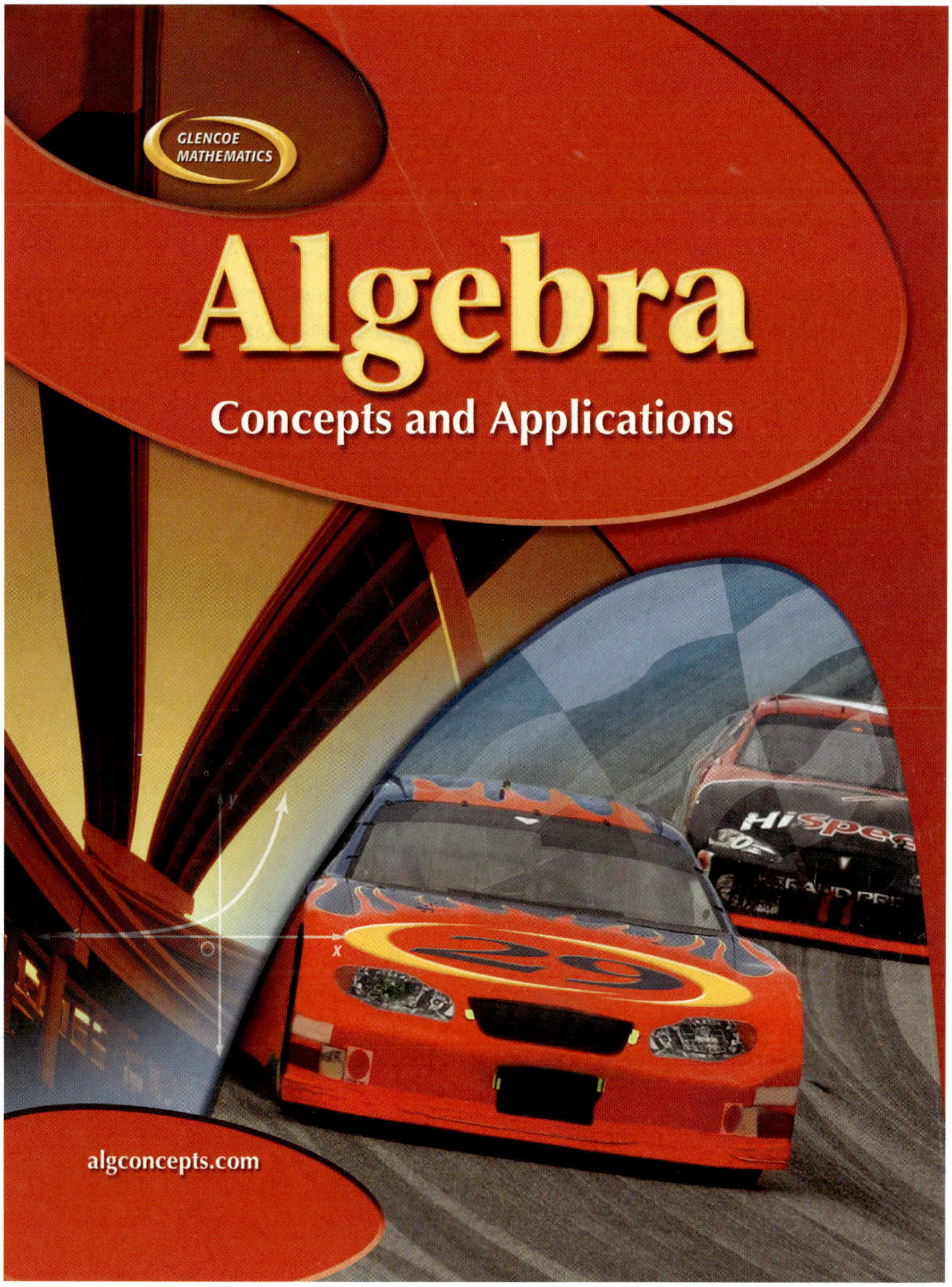

GLENCOE MATHEMATICS

Algebra

Concepts and Applications

algconcepts.com

New York, New York Columbus, Ohio Chicago, Illinois Peoria, Illinois Woodland Hills, California

Student Edition (Also Available in Two Volumes)
StudentWorks Plus CD-ROM
Teacher's Wraparound Edition

Applications

School-to-Workplace Masters
Multimedia Applications CD-ROM
Problem-of-the-Week Cards (online)

Meeting Individual Needs

Chapter Resource Masters
Noteables™ Interactive Study Notebook
Practice Workbook
Prerequisite Skills Workbook
Skills Practice Workbook
Spanish Study Guide and Assessment
Study Guide Workbook
Parent/Student Study Guide (online)

Technology/Multimedia

AlgePASS: Concepts and Applications
Graphing Calculator Masters
Vocabulary PuzzleMaker (online)

Assessment/Evaluation

Diagnostic and Placement Tests
MindJogger Videoquizzes
ExamView® Pro

Manipulatives/Modeling

Overhead Manipulative Resources
Glencoe Mathematics Classroom Manipulative Kit
Glencoe Mathematics Student Manipulative Kit
Hands-On Algebra Masters

Teaching Aids

Answer Key Maker
5-Minute Check Transparencies
Interactive Chalkboard CD-ROM
Solutions Manual
TeacherWorks CD-ROM

The McGraw·Hill Companies

Send all inquiries to:
Glencoe/McGraw-Hill
8787 Orion Place
Columbus, OH 43240-4027

ISBN: 0-07-868170-7 *(Algebra: Concepts and Applications Student Edition)*
ISBN: 0-07-868171-5 *(Algebra: Concepts and Applications Teacher Wraparound Edition)*

Printed in the United States of America.

1 2 3 4 5 6 7 8 9 10 027/055 13 12 11 10 09 08 07 06 05 04

Dear Students, Teachers, and Parents,

Algebra: Concepts and Applications is designed to help you learn algebra and apply it to the real world. Throughout the text, you will be given opportunities to make connections from concrete models to abstract concepts. The real-world photographs and realistic data will help you see algebra in your world. You will also have plenty of opportunities to review and use arithmetic and geometry concepts as you study algebra. And for those of you who love a good debate, you will find plenty of opportunities to communicate your understanding of algebra.

We know that most of you haven't yet decided which careers you would like to pursue, so we've also included a little career guidance. This text offers real examples of how mathematics is used in many types of careers.

You may have to take an end-of-course exam for algebra, a proficiency test for graduation, the SAT, and/or the ACT. When you enter the workforce, you may also have to take job placement tests that include a section on mathematics. All of these tests include algebra problems. Because all Algebra 1 concepts are covered in this text, this program will prepare you for all of those tests.

Each day, as you use ***Algebra: Concepts and Applications,*** you will see the practical value of algebra. You will grow to appreciate how often algebra is used in ways that relate directly to your life. You will have meaningful experiences that will prepare you for the future. If you don't already see the importance of algebra in your life, you soon will!

Sincerely,
The Authors

Jerry Cummins Carol E. Malloy Kay McClain
Yvonne M. Mojica Jack Price

P.S. To help you learn how to use your math book, use the Scavenger Hunt at algconcepts.com. The Scavenger Hunt will help you learn where things are located in each chapter.

Contents in Brief

Authors

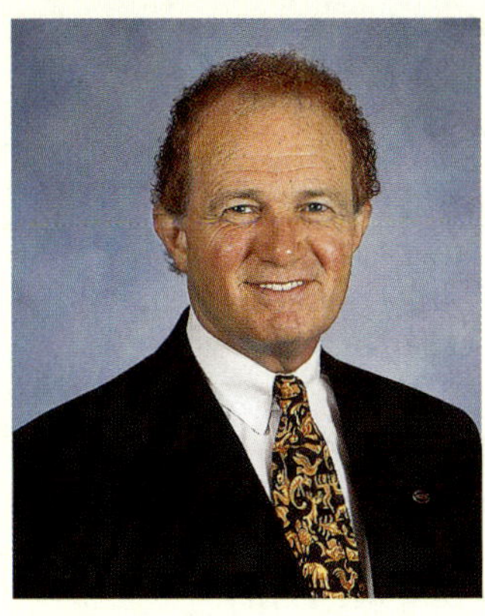

Jerry Cummins
Staff Development Specialist
Bureau of Education and Research
State of Illinois
President, National Council of Supervisors of Mathematics
Western Springs, IL

Carol Malloy
Assistant Professor of Mathematics Education
University of North Carolina at Chapel Hill
Chapel Hill, NC

Kay McClain
Lecturer
George Peabody College
Vanderbilt University
Nashville, TN

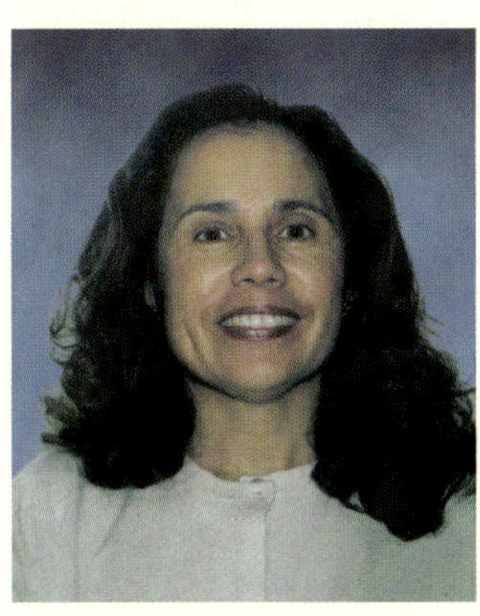

Yvonne Mojica
Mathematics Teacher and Mathematics Department Chairperson
Verdugo Hills High School
Tujunga, CA

Jack Price
Professor, Mathematics Education
California State Polytechnic University
Pomona, CA

Contributing Author
Dinah Zike
Educational Consultant
Dinah-Might Activities, Inc.
San Antonio, TX

Academic Consultants and Teacher Reviewers

Each of the Academic Consultants read all 15 chapters, while each Teacher Reviewer read two chapters. The Consultants and Reviewers gave suggestions for improving the Student Editions and the Teacher's Wraparound Editions.

Academic Consultants

Judith Cubillo
Mathematics Teacher & Department Chairperson
Northgate High School
Walnut Creek, California

Mary C. Enderson
Faculty
Middle Tennessee State University
Murfreesboro, Tennessee

Alan G. Foster
Former Mathematics Teacher & Department Chairperson
Addison Trail High School
Addison, Illinois

Deborah A. Hutchens, Ed.D.
Assistant Principal
Great Bridge Middle School
Chesapeake, Virginia

Nicki Hudson
Mathematics Teacher
West Linn High School
West Linn, Oregon

Daniel Marks, Ph.D.
Associate Professor of Mathematics
Auburn University at Montgomery
Montgomery, Alabama

Donald McGurrin
Senior Administrator for Secondary Mathematics
Wake County Public Schools
Raleigh, North Carolina

C. Vincent Pané, Ed.D.
Associate Professor of Education
Molloy College
Rockville Centre, New York

Gary Shannon, Ph.D.
Professor of Mathematics
California State University
Sacramento, California

Marianne Weber
National Mathematics Consultant
St. Louis, Missouri

Teacher Reviewers

Breta J. Brown
Mathematics Teacher
Hillsboro High School
Hillsboro, North Dakota

Kimberly A. Brown
Mathematics Teacher
McGehee High School
McGehee, Arkansas

Helen Carpini
Mathematics Department Chairperson
Middletown High School
Middletown, Connecticut

Terry Cepaitis
Mathematics Resource Teacher
Anne Arundel County Public Schools
Annapolis, Maryland

Kindra R. Cerfoglio
Mathematics Teacher
Reed High School
Sparks, Nevada

Tom Cook
Mathematics Department Chairperson
Carlisle High School
Carlisle, Pennsylvania

Donna L. Cooper
Mathematics Department Chairperson
Walter E. Stebbins High School
Riverside, Ohio

James D. Crawford
Instructional Coordinator—Mathematics
Manchester Memorial High School
Manchester, New Hampshire

David A. Crine
Mathematics Department Chairperson
Basic High School
Henderson, Nevada

Carol Damiano
Mathematics Teacher
Morton High School
Hammond, Indiana

Douglas D. Dolezal, Ph.D.
Mathematics Educator/
Methods Instructor
Crete High School/Doane
College
Crete, Nebraska

Richard F. Dube
Mathematics Supervisor
Taunton High School
Taunton, Massachusetts

Dianne Foerster
Mathematics Department
Chairperson
Riverview High School
Riverview, Florida

Victoria G. Fortenberry
Mathematics Teacher
Lincoln High School
Tallahassee, Florida

Candace Frewin
Mathematics Teacher
East Lake High School
Tarpon Springs, Florida

Linda Glover
Mathematics Department
Chairperson
Conway High School East
Conway, Arkansas

Karin Sorensen Grandone
Mathematics Department
Supervisor
Bremen District 228
Midlothian, Illinois

R. Emilie Greenwald
Mathematics Teacher
Worthington Kilbourne
High School
Worthington, Ohio

John Scott Griffith
Mathematics Department
Chairperson
New Castle Middle School
New Castle, Indiana

Rebecca M. Gummerson
Mathematics Department
Chairperson
Burley High School
Burley, Idaho

T. L. Watanabe Hall
Mathematics Teacher
Northwood Junior
High School
Kent, Washington

T. B. Harris
Mathematics Teacher
Luray High School
Luray, Virginia

Jerome D. Hayden
Mathematics Department
Chairperson
McLean County Unit District 5
Normal, Illinois

Karlene M. Hubbard
Mathematics Teacher
Rich East High School
Park Forest, Illinois

Joseph Kavanaugh
Academic Head of
Mathematics
Scotia-Glenville Central
School District
Scotia, New York

Ruth C. Keefe
Mathematics Department
Chairperson
Litchfield High School
Litchfield, Connecticut

Roger M. Marchegiano
Supervisor of Mathematics
Bloomfield School District
Bloomfield, New Jersey

Marilyn Martau
Mathematics Teacher (Retired)
Lakewood High School
Lakewood, Ohio

Jane E. Morey
Mathematics Department
Chairperson
Washington High School
Sioux Falls, South Dakota

Grace Clover Mullen
Talent Search Tutor
Dabney Lancaster
Community College
Lexington, Virginia

Laurie D. Newton
Mathematics Teacher
Crossler Middle School
Salem, Oregon

Rinda Olson
Mathematics Department
Chairperson
Skyline High School
Idaho Falls, Idaho

Catherine E. Oppio
Mathematics Teacher
Truckee Meadows
Community College
High School
Reno, Nevada

Peter Pace
Mathematics Teacher
Life Center Academy
Burlington, New Jersey

LaVonne Peterson
Instructional Leader
Minico High School
Rupert, Idaho

Donna H. Preston
Mathematics Teacher
Thomas Worthington High School
Worthington, Ohio

Matthew D. Quibell
Mathematics Teacher
Horseheads Middle School
Horseheads, New York

Linda Rohleder
Mathematics Teacher
Jasper Middle School
Jasper, Indiana

Steve Sachs
Mathematics Department Chairperson
Lawrence North High School
Indianapolis, Indiana

Elizabeth Schopp
Mathematics Department Chairperson
Moberly Senior High School
Moberly, Missouri

Terry Anagnos Serbanos
Mathematics Department Chairperson
Countryside High School
Clearwater, Florida

Eli Shaheen
Mathematics Department Chairperson
Plum Borough School District
Pittsburgh, Pennsylvania

Sue A. Shattuck
Mathematics Teacher
Otsego Middle School
Otsego, Michigan

William Shutters
Mathematics Teacher
Urbandale High School
Urbandale, Iowa

Richard R. Soendlin
Mathematics Department Chairperson
Arsenal Technical High School
Indianapolis, Indiana

Frank J. Sottile, Ed.D.
Supervisor of Mathematics and Computer Science
Scranton School District
Scranton, Pennsylvania

Lana Stahl
K–12 Mathematics Instructional Specialist
Pasadena ISD
Pasadena, Texas

Randall Starewicz
Mathematics Department Chairperson
Lake Central High School
St. John, Indiana

Mary M. Stewart
Secondary Mathematics Curriculum Specialist
Kansas City Kansas Public Schools
Kansas City, Kansas

Dora M. Swart
Mathematics Department Chairperson
W. F. West High School
Chehalis, Washington

Elizabeth Sweeney
Mathematics Teacher
Clayton Midde School
Reno, Nevada

Ruby M. Sylvester
Assistant Principal
High School of Transit Technology
Brooklyn, New York

Lisa Thomson
Mathematics Teacher
Reed High School
Reno/Sparks, Nevada

John Mark Tollett
Mathematics Teacher
Smackover High School
Smackover, Arkansas

Ronald R. Vervaecke
Mathematics Coordinator
Warren Consolidated Schools
Warren, Michigan

Ronald C. Voss
Mathematics Teacher
Sparks High School
Sparks, Nevada

Karen S. Whiteside
Mathematics Teacher
Dunedin High School
Dunedin, Florida

Dale I. Winters
Mathematics Teacher
Worthington Kilbourne High School
Worthington, Ohio

Michael J. Zelch
Mathematics Teacher
Worthington Kilbourne High School
Worthington, Ohio

Table Of Contents

Chapter 1 The Language of Algebra

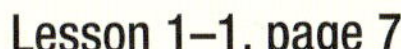

Lesson 1–1, page 7

Chapter 2 Integers

Lesson 2–2, page 61

Lesson 3–4, page 115

Chapter 3 Addition and Subtraction Equations 92

Lesson 4–2, page 150

Chapter 4 Multiplication and Division Equations

Chapter 5 Proportional Reasoning and Probability ... 186

Lesson 5–5, page 217

Lesson 6–3, page 254

Chapter 6 Functions and Graphs

Chapter 7 Linear Equations

Lesson 7–4, page 302

Chapter 8 Powers and Roots

Lesson 8–2, page 345

Investigation, page 411

Chapter 9 Polynomials 380

Lesson 10–3, page 434

Chapter 10 Factoring

Chapter 11 Quadratic and Exponential Functions 456

Lesson 11–2, page 467

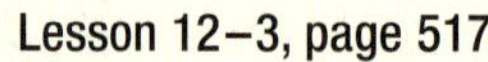

Lesson 12–3, page 517

Chapter 13 Systems of Equations and Inequalities

Lesson 13–4, page 567

Chapter 14 Radical Expressions

Lesson 14–2, page 608

Lesson 15–5, page 666

Chapter 15 Rational Expressions and Equations 636

Preparing for Standardized Test Success

The **Preparing for Standardized Tests** pages at the end of each chapter and at the end of the book have been created to help you get ready for the mathematics portions of your standardized tests. On these pages, you will find strategies for solving problems and test-taking advice to help you maximize your score.

It is important to remember that there are many different standardized tests given by schools and states across the country. Find out as much as you can about your test. Start by asking your teacher and counselor for any information, including practice materials, that may be available to help you prepare.

To help you get ready for these tests, do the Standardized Test Practice question in each lesson. Also review the concepts and techniques contained in the Preparing for Standardized Tests pages at the end of each chapter listed below. This will help you become familiar with the types of math questions that are asked on various standardized tests.

The **Preparing for Standardized Tests** pages in the Student Handbook review common types of problems on standardezed tests and provide practice for each format.

The **Preparing for Standardized Tests** pages are part of a complete test preparation course offered in this text. The test items on these pages were written in the same style as those in state proficiency tests and standardized tests like ACT and SAT. The 15 topics are closely aligned with those tests, the algebra curriculum, and this text. These topics cover all of the types of problems you will see on these tests.

With some practice, studying, and review, you will be ready for standardized test success. Good luck from Glencoe/McGraw-Hill!

Graphing Calculator Quick Reference Guide

Throughout this text, **Graphing Calculator Explorations** have been included so you can use technology to solve problems. These activities use the TI–83 Plus or T1–84 Plus graphing calculator. Graphing calculators have a wide variety of applications and features. If you are just beginning to use a graphing calculator, you will not need to use all of its features. This page is designed to be a quick reference for the features you will need to use as you study from this text.

To darken or lighten the screen:	2nd ▲ or 2nd ▼
To clear an entry:	CLEAR
To get to the home screen:	2nd [QUIT]
To recall an entry:	2nd [ENTRY]
To recall an answer:	2nd [ANS]
To turn the calculator off:	2nd [OFF]

Task	Keystrokes
Using tables	2nd [TABLE]
Use lists	STAT ENTER
Find the mean and median of listed data	STAT ▶ ENTER ENTER
Solve equations	MATH 0
Set the viewing window	WINDOW
Plot points	2nd [DRAW] ▶ ENTER
Enter an equation	Y=
Enter an inequality symbol	2nd [TEST]
Graph an equation	GRAPH
Zoom in	ZOOM 2
Zoom out	ZOOM 3
Graph in the viewing window x: [−10, 10] and y: [−10, 10]	ZOOM 6
Graph an equation with integer coordinates	ZOOM 8
Place a statistical graph in a good viewing window	ZOOM 9
Trace a graph	TRACE
Find the intersection of two graphs	2nd [CALC] 5
Shade an inequality	2nd [DRAW] 7
Enter a program	PRGM ▶ ▶ ENTER

Teacher's Wraparound Edition

TABLE OF CONTENTS

Teacher's Handbook

Chapter Overviews

Student Handbook

"My students need an algebra textbook that has an easy-to-follow lesson format with plenty of examples and exercises."

MOTIVATING STUDENT LEARNING

Each lesson in Glencoe's *Algebra: Concepts and Applications* follows a straightforward format.

- What You'll Learn/ Why It's Important
- Important Concepts Highlighted
- Examples/ Your Turn Exercises
- Check for Understanding
- Exercises

What You'll Learn/ Why It's Important

Students quickly see what concepts are covered in the lesson and why it is beneficial to learn them.

Examples/ Your Turn Exercises

Completely worked-out examples with clear explanations parallel the exercises in the Guided Practice and Practice sections. Your Turn exercises allow students to practice their new skills as they go through the lesson.

Important Concepts Highlighted

Important vocabulary terms are highlighted in yellow, and definitions, rules, and properties are displayed in concept boxes. Many rules and properties are displayed using words, symbols, and models.

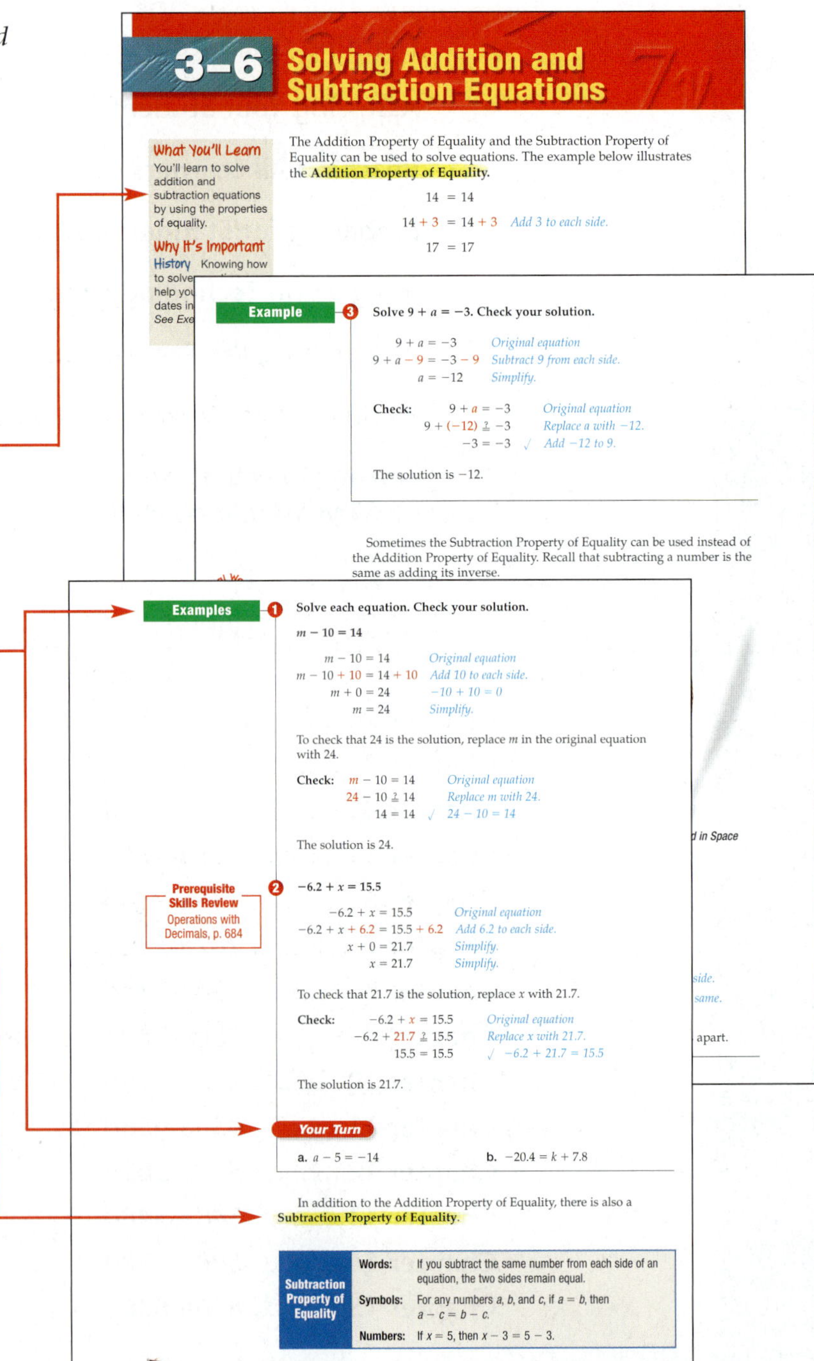

3-6 Solving Addition and Subtraction Equations
What You'll Learn
You'll learn to solve addition and subtraction equations by using the properties of equality.
Why It's Important
The Addition Property of Equality and the Subtraction Property of Equality can be used to solve equations. The example below illustrates the Addition Property of Equality.
14 = 14
14 + 3 = 14 + 3 Add 3 to each side.
17 = 17
Example 3 Solve 9 + a = −3. Check your solution.
9 + a = −3 Original equation
9 + a − 9 = −3 − 9 Subtract 9 from each side.
a = −12 Simplify.
Check: 9 + a = −3 Original equation
9 + (−12) ≟ −3 Replace a with −12.
−3 = −3 ✓ Add −12 to 9.
The solution is −12.
Sometimes the Subtraction Property of Equality can be used instead of the Addition Property of Equality. Recall that subtracting a number is the same as adding its inverse.
Examples 1 Solve each equation. Check your solution.
m − 10 = 14
m − 10 = 14 Original equation
m − 10 + 10 = 14 + 10 Add 10 to each side.
m + 0 = 24 −10 + 10 = 0
m = 24 Simplify.
To check that 24 is the solution, replace m in the original equation with 24.
Check: m − 10 = 14 Original equation
24 − 10 ≟ 14 Replace m with 24.
14 = 14 ✓ 24 − 10 = 14
The solution is 24.
Prerequisite Skills Review
Operations with Decimals, p. 684
2 −6.2 + x = 15.5
−6.2 + x = 15.5 Original equation
−6.2 + x + 6.2 = 15.5 + 6.2 Add 6.2 to each side.
x + 0 = 21.7 Simplify.
x = 21.7 Simplify.
To check that 21.7 is the solution, replace x with 21.7.
Check: −6.2 + x = 15.5 Original equation
−6.2 + 21.7 ≟ 15.5 Replace x with 21.7.
15.5 = 15.5 ✓ −6.2 + 21.7 = 15.5
The solution is 21.7.
Your Turn
a. a − 5 = −14
b. −20.4 = k + 7.8
In addition to the Addition Property of Equality, there is also a Subtraction Property of Equality.
Subtraction Property of Equality
Words: If you subtract the same number from each side of an equation, the two sides remain equal.
Symbols: For any numbers a, b, and c, if a = b, then a − c = b − c.
Numbers: If x = 5, then x − 3 = 5 − 3.
www.algconcepts.com/extra_examples
Lesson 3-6 Solving Addition and Subtraction Equations 123

Check for Understanding

These exercises are designed to be completed with the teacher in class.

- In the **Communicating Mathematics** exercises, students define, describe, and explain mathematical concepts.

  You Decide and **Writing Math** exercises strengthen communication skills. **Getting Ready** exercises review prerequisite skills and subskills.

- Keyed to the examples, the **Guided Practice** exercises present a representative sample of the exercises in the Practice section.

Exercises

These exercises are designed to be completed as homework.

- The **Practice** exercises are separated into A, B, and C sections, indicated only in the Teacher's Wraparound Edition. The Practice exercises generally match the Guided Practice exercises in a 3:1 ratio.

- While completing the **Applications and Problem Solving** exercises, students find numerous opportunities to apply concepts to both real-life and mathematical problem situations.

  Each lesson contains a **Critical Thinking** exercise in which students explain, justify, and prove mathematical relationships.

- The **Mixed Review** is spiraled and cumulative. These exercises comprise about 15% of the total number of exercises in each lesson.

  Each Mixed Review section contains a Standardized Test Practice question, some of which are open-ended.

- There are two **Quizzes** in each chapter.

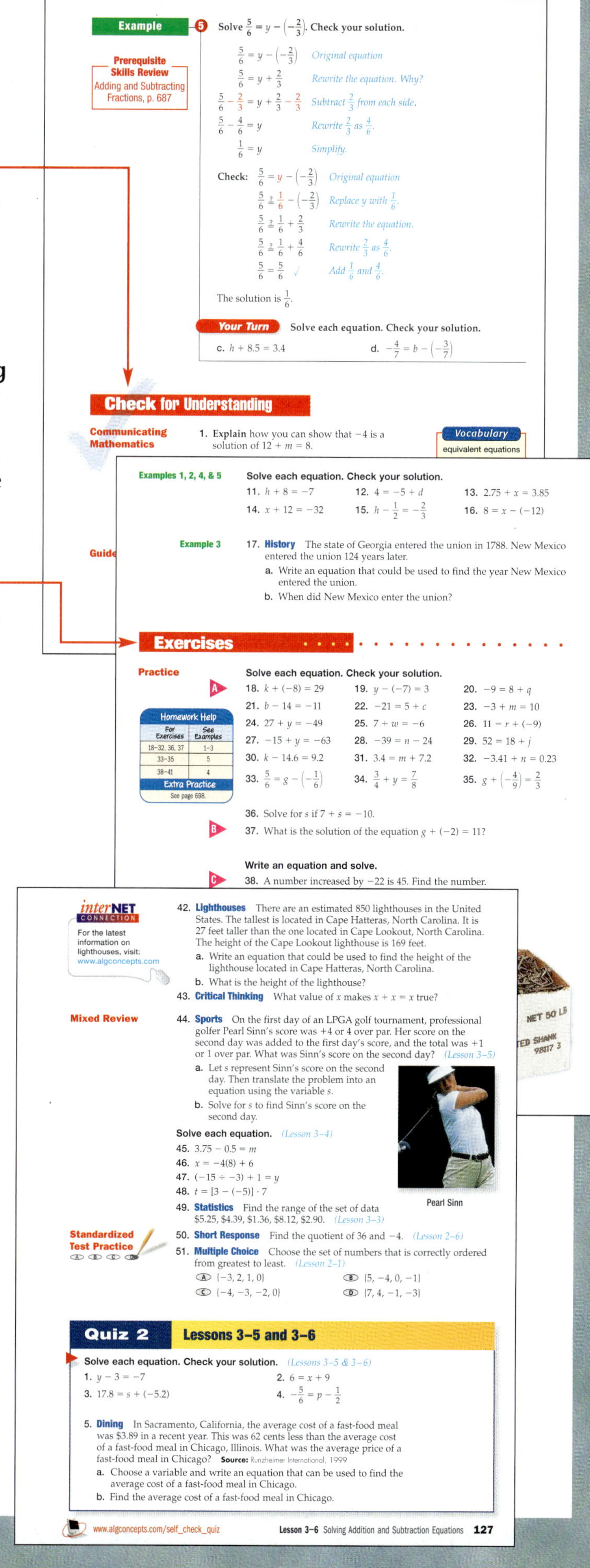

Example 5 Solve $\frac{5}{6} = y - \left(-\frac{2}{3}\right)$. Check your solution.

Prerequisite Skills Review: Adding and Subtracting Fractions, p. 687

$\frac{5}{6} = y - \left(-\frac{2}{3}\right)$ *Original equation*

$\frac{5}{6} = y + \frac{2}{3}$ *Rewrite the equation. Why?*

$\frac{5}{6} - \frac{2}{3} = y + \frac{2}{3} - \frac{2}{3}$ *Subtract $\frac{2}{3}$ from each side.*

$\frac{5}{6} - \frac{4}{6} = y$ *Rewrite $\frac{2}{3}$ as $\frac{4}{6}$.*

$\frac{1}{6} = y$ *Simplify.*

Check: $\frac{5}{6} = y - \left(-\frac{2}{3}\right)$ *Original equation*

$\frac{5}{6} \stackrel{?}{=} \frac{1}{6} - \left(-\frac{2}{3}\right)$ *Replace y with $\frac{1}{6}$.*

$\frac{5}{6} \stackrel{?}{=} \frac{1}{6} + \frac{2}{3}$ *Rewrite the equation.*

$\frac{5}{6} \stackrel{?}{=} \frac{1}{6} + \frac{4}{6}$ *Rewrite $\frac{2}{3}$ as $\frac{4}{6}$.*

$\frac{5}{6} = \frac{5}{6}$ ✓ *Add $\frac{1}{6}$ and $\frac{4}{6}$.*

The solution is $\frac{1}{6}$.

Your Turn Solve each equation. Check your solution.

c. $h + 8.5 = 3.4$ d. $-\frac{4}{7} = b - \left(-\frac{3}{7}\right)$

Check for Understanding

Communicating Mathematics

1. **Explain** how you can show that -4 is a solution of $12 + m = 8$.

Vocabulary: equivalent equations

Guid…

Examples 1, 2, 4, & 5 — Solve each equation. Check your solution.

11. $h + 8 = -7$ 12. $4 = -5 + d$ 13. $2.75 + x = 3.85$
14. $x + 12 = -32$ 15. $h - \frac{1}{2} = -\frac{2}{3}$ 16. $8 = x - (-12)$

Example 3

17. **History** The state of Georgia entered the union in 1788. New Mexico entered the union 124 years later.
    a. Write an equation that could be used to find the year New Mexico entered the union.
    b. When did New Mexico enter the union?

Exercises

Practice — Solve each equation. Check your solution.

Homework Help	
For Exercises	See Examples
18–32, 36, 37	1–3
33–35	5
38–41	4
Extra Practice	See page 698.

A
18. $k + (-8) = 29$ 19. $y - (-7) = 3$ 20. $-9 = 8 + q$
21. $b - 14 = -11$ 22. $-21 = 5 + c$ 23. $-3 + m = 10$
24. $27 + y = -49$ 25. $7 + w = -6$ 26. $11 = r + (-9)$
27. $-15 + y = -63$ 28. $-39 = n - 24$ 29. $52 = 18 + j$
30. $k - 14.6 = 9.2$ 31. $3.4 = m + 7.2$ 32. $-3.41 + n = 0.23$
33. $\frac{5}{6} = g - \left(-\frac{1}{6}\right)$ 34. $\frac{3}{4} + y = \frac{7}{8}$ 35. $g + \left(-\frac{4}{9}\right) = \frac{2}{3}$

36. Solve for s if $7 + s = -10$.

B
37. What is the solution of the equation $g + (-2) = 11$?

Write an equation and solve.

C
38. A number increased by -22 is 45. Find the number.

interNET CONNECTION: For the latest information on lighthouses, visit: www.algconcepts.com

42. **Lighthouses** There are an estimated 850 lighthouses in the United States. The tallest is located in Cape Hatteras, North Carolina. It is 27 feet taller than the one located in Cape Lookout, North Carolina. The height of the Cape Lookout lighthouse is 169 feet.
    a. Write an equation that could be used to find the height of the lighthouse located in Cape Hatteras, North Carolina.
    b. What is the height of the lighthouse?

43. **Critical Thinking** What value of x makes $x + x = x$ true?

Mixed Review

44. **Sports** On the first day of an LPGA golf tournament, professional golfer Pearl Sinn's score was +4 or 4 over par. Her score on the second day was added to the first day's score, and the total was +1 or 1 over par. What was Sinn's score on the second day? (Lesson 3–5)
    a. Let s represent Sinn's score on the second day. Then translate the problem into an equation using the variable s.
    b. Solve for s to find Sinn's score on the second day.

Pearl Sinn

Solve each equation. (Lesson 3–4)

45. $3.75 - 0.5 = m$
46. $x = -4(8) + 6$
47. $(-15 \div -3) + 1 = y$
48. $t = [3 - (-5)] \cdot 7$

49. **Statistics** Find the range of the set of data $5.25, $4.39, $1.36, $8.12, $2.90. (Lesson 3–3)

Standardized Test Practice

50. **Short Response** Find the quotient of 36 and -4. (Lesson 2–6)

51. **Multiple Choice** Choose the set of numbers that is correctly ordered from greatest to least. (Lesson 2–1)
    (A) {−3, 2, 1, 0} (B) {5, −4, 0, −1}
    (C) {−4, −3, −2, 0} (D) {7, 4, −1, −3}

Quiz 2 Lessons 3–5 and 3–6

Solve each equation. Check your solution. (Lessons 3–5 & 3–6)

1. $y - 3 = -7$ 2. $6 = x + 9$
3. $17.8 = s + (-5.2)$ 4. $-\frac{5}{6} = p - \frac{1}{2}$

5. **Dining** In Sacramento, California, the average cost of a fast-food meal was $3.89 in a recent year. This was 62 cents less than the average cost of a fast-food meal in Chicago, Illinois. What was the average price of a fast-food meal in Chicago? Source: Runzheimer International, 1999
   a. Choose a variable and write an equation that can be used to find the average cost of a fast-food meal in Chicago.
   b. Find the average cost of a fast-food meal in Chicago.

www.algconcepts.com/self_check_quiz Lesson 3–6 Solving Addition and Subtraction Equations 127

"I need a complete algebra program for my students who have difficulty reading."

LEARNING GRAPHICALLY

Easy to read

The concise lesson narrative helps students learn about important concepts in a non-threatening, easy-to-read style. New vocabulary words are highlighted in yellow and summarized in the Check for Understanding section. The pages have better readability and are uncluttered.

> A **scale drawing** or **scale model** is used to represent an object that is too large or too small to be drawn or built at actual size. The **scale** is the ratio of a length on the drawing or model to the corresponding length of the real object.
>
> *Vocabulary*
> scale drawing
> scale model
> scale

Reading Algebra

This unique learning tool helps students understand the terminology of algebra, a prerequisite for concept development.

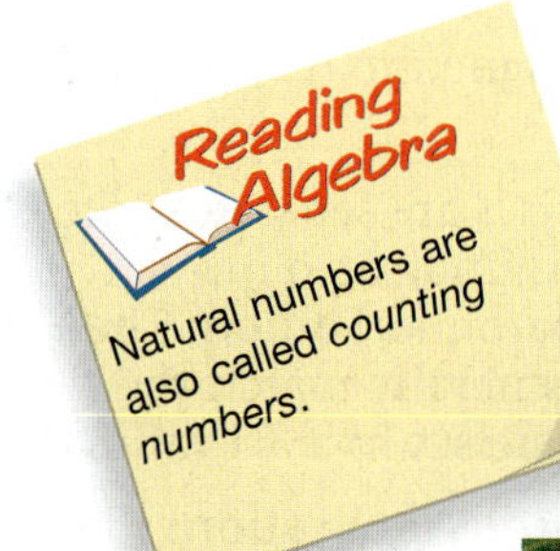

Reading Algebra
Natural numbers are also called counting numbers.

Info-Graphics and Photo-Graphics

Unique info-graphics improve the readability level by conveying algebraic concepts visually. Photo-graphics use photographs paired or overprinted with art to illustrate algebraic concepts.

The greatest common place value for each data item is used to form the *stem*.

Stem	Leaf
1	1 6
2	1 3 9
3	
4	5 5 $\quad 2\|3 = 23$

The *leaves* are formed by the next greatest place value.

In this case, the tens digits are the stems. The ones digits are the leaves. Write the leaves in order from least to greatest.

A *key* is always included. This shows how the digits are related.

Info Graphic

Words, Symbols, and Numbers

This three-pronged approach, in essence a left brain/right brain technique, improves reading comprehension.

Multiplication Property of Equality

Words: If you multiply each side of an equation by the same number, the two sides remain equal.

Symbols: For any numbers a, b, and c, if $a = b$, then $ac = bc$.

Numbers: If $\frac{1}{5}x = 8$, then $5\left(\frac{1}{5}x\right) = 5(8)$.

Understanding and Using the Vocabulary

This vocabulary learning tool is at the end of each chapter. An interactive Study Guide and Review is available on the Internet at **www.algconcepts.com**

State whether each sentence is *true* or *false*. If false, replace the underlined word(s) to make a true statement.

1. In the equation $y = mx + b$, b is the x-intercept of the line.
2. A line that appears to go downhill from left to right has a negative slope.
3. Perpendicular lines have the same slopes.
4. The graph of a linear equation is a straight line.
5. You can write an equation of a line if you know the slope and a point on the line.

Hands-On Algebra

These optional activities give students the opportunity to bridge the gap between the concrete and the abstract.

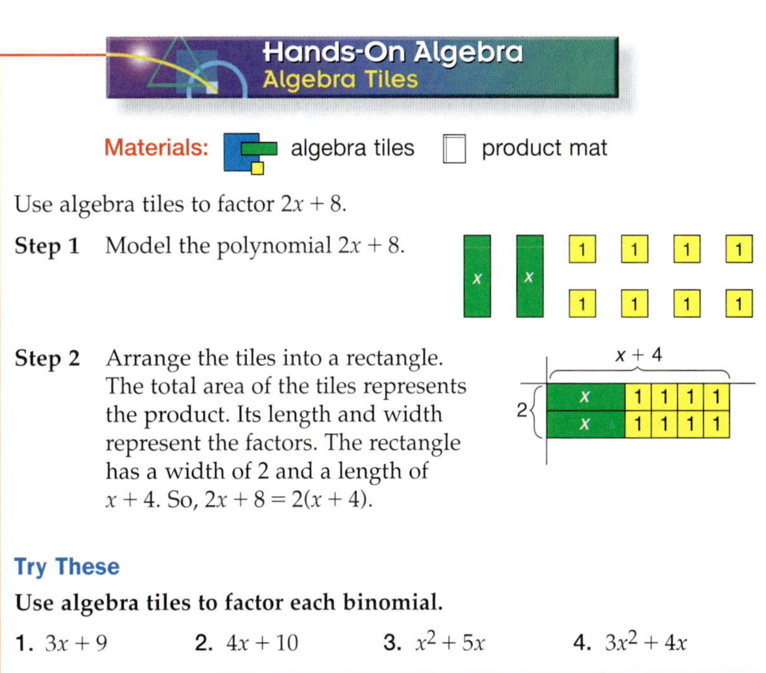
Hands-On Algebra
Algebra Tiles

Materials: algebra tiles product mat

Use algebra tiles to factor $2x + 8$.

Step 1 Model the polynomial $2x + 8$.

Step 2 Arrange the tiles into a rectangle. The total area of the tiles represents the product. Its length and width represent the factors. The rectangle has a width of 2 and a length of $x + 4$. So, $2x + 8 = 2(x + 4)$.

Try These

Use algebra tiles to factor each binomial.

1. $3x + 9$
2. $4x + 10$
3. $x^2 + 5x$
4. $3x^2 + 4x$

Interactive Study Notebook

This note-taking guide reinforces excellent note-taking skills. It includes:

- **Build Your Vocabulary** tool for including mathematical terminology in notes.
- **Review It** activities with links to prior lessons and items to include in **Foldables**™.
- **Bringing It All Together** feature to help students review for the Chapter Test.
- **Are You Ready for the Chapter Test?** feature to allow students to assess their own readiness for the Chapter Test.
- **Teacher Edition** with transparencies to guide note taking. Corresponds to the *Teacher's Wraparound Edition* In-Class Examples and *Interactive Chalkboard CD-ROM*.

ASSESSING YOUR STUDENTS

Algebra: Concepts and Applications helps you assess students' ability to organize information, apply previously learned information, and make conjectures based on gathered data. The following features and components will help you accurately assess each student's achievement.

In the Student Edition...

- **Check Your Readiness** at the beginning of each chapter helps students review prerequisite skills for the chapter.
- **Getting Ready** exercises review prerequisite skills and subskills needed for successful completion of the exercises.
- Every lesson has a **Mixed Review** that includes **Standardized Test Practice**.
- Every chapter has two **Quizzes**.
- Every chapter includes:
  - a **Study Guide and Assessment** that includes vocabulary review, review exercises for each objective, and applications and problem solving,
  - a **Chapter Test**, and
  - **Preparing for Standardized Tests**, a unique two-page testing lesson.
- The **Student Handbook** in the back of the text includes:
  - **Prerequisite Skills Review**,
  - **Extra Practice,**
  - **Mixed Problem Solving,** and
  - **Preparing for Standardized Tests.**

In the Teacher's Wraparound Edition...

- Every lesson includes a **5-Minute Check** that covers the previous lesson or chapter.
- **Error Analysis** in every lesson helps you help your students avoid common errors.
- **Open-Ended Assessment** in every lesson helps students solidify daily learning by modeling, speaking, writing, or acting.

"My students need constant reinforcement and objective assessment."

In the supplementary materials...

The **Chapter Resource Masters** include the following for each chapter.

- 2 multiple-choice tests (Average, Basic)
- 2 free-response tests (Average, Basic)
- 1 open-ended assessment
- 1 mid-chapter test
- 2 quizzes
- 1 cumulative review
- 1 standardized test practice

Also included are 2 semester tests and 1 final test.

Diagnostic and Placement Tests This booklet will help place students in the mathematics course for which they are best prepared.

MindJogger Videoquizzes (VHS and DVD) review each chapter by using a game show format. As students compete on teams, they hear and see each review problem as it is presented and then completely solved.

ExamView® Pro
Use the networkable **ExamView® Pro Testmaker CD-ROM** to:

- Create **multiple versions** of tests.
- Create **modified** tests for **inclusion** students with one mouse click.
- **Edit** existing questions and **add** you own questions.
- Build tests aligned with state standards using built-in **state curriculum correlations.**
- Change **English** tests to **Spanish** with one mouse click and vice versa.

5-Minute Check Transparencies provide a quick review of the previous lesson or chapter. There is one full-color transparency for every lesson. The 5-Minute Check is also printed in the Teacher's Wraparound Edition to make your lesson plans easier to prepare.

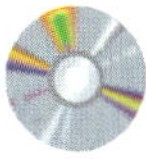

The ***AlgePASS: Concepts and Applications*** **CD-ROM** reviews and reinforces important concepts through a unique Pretest-Tutorial-Guided Practice-Posttest format. Self-paced and easy-to-use, it is also an excellent tool for standardized test preparation.

Alternative Assessment in the Mathematics Classroom, part of the Glencoe Mathematics Professional Series, provides an overview of the latest assessment trends.

"I need an algebra program that reaches all of my students."

REACHING ALL LEARNERS

There are several different learning styles that help us approach and solve problems. Everyone possesses varying degrees of each of these learning styles, but the ways in which they combine and blend are as varied as the personalities of the individuals. Glencoe's *Algebra: Concepts and Applications* provides you with ways to accommodate students with these diverse learning styles.

Learning Style	Description	Where Can I Find This?
verbal/ linguistic	read regularly, write clearly, and easily understand the written word	**Communicating Mathematics** exercises ask students to describe, write, and explain mathematical concepts. Students also express what they have learned in their **Writing Math** exercises.
logical	use numbers, logic, and critical thinking skills	Clearly-written **Examples** present important concepts, and **Critical Thinking** exercises extend those concepts. **Problem-Solving Strategy Workshops** encourage students to practice their logical thinking skills by using various strategies. **You Decide** exercises help students formulate convincing arguments.
visual/ spatial	think in terms of pictures and images	**Info-Graphics** and **Photo-Graphics** illustrate algebraic concepts through the use of photographs and artwork.
auditory/ musical	have "good ears" and can produce rhythms and melodies	Multimedia software, such as the **Multimedia Applications CD-ROM** and **MindJogger Videoquizzes** can be easily incorporated into lessons.
kinesthetic	learn from touch, movement, and manipulating objects	**Hands-On Algebra** activities provide for physical involvement in learning.
interpersonal	understand and work well with other people	Optional **Problem-Solving Strategy Workshops** and **Investigations**, as well as **Hands-On Algebra** activities and **Graphing Calculator Explorations**, allow students to collaborate with others.
intrapersonal	have a realistic understanding of their strengths and weaknesses	**Writing Math** exercises help students personalize mathematics.
naturalist	can distinguish among, classify, and use features of the environment	Interesting lesson openers, **Math In the Workplace** features, as well as **Applications and Problem Solving** exercises show students how mathematics relates to the world around them.

As a mathematics teacher, you may want to assign activities to students that accommodate their strongest learning styles, but frequently ask them to use their weakest learning styles. Additional activities are provided in the bottom margins of the lesson notes in the Teacher's Wraparound Edition. These resources guarantee that your classroom will be a multisensory environment, providing multiple paths for student learning.

GLENCOE'S ASSESSMENT ADVANTAGE

"I want an algebra program that prepares my students for success on Standardized Tests like the SAT and ACT."

PREPARING FOR STANDARDIZED TESTS

Teachers asked for an algebra program that helps to prepare their students for success on state tests, as well as the SAT and the ACT. Glencoe provides an innovative solution.

A Complete Test Prep Course

Two-page lessons at the end of each chapter and an 18-page guide at the back of the book are unique tools for preparing students for success on standardized tests.

- The 15 two-page lessons along with the pages in the Student Handbook can be taught as a test prep course for state and local proficiency tests, the SAT, and the ACT. All standardized test problem types are covered.
- Each Preparing for Standardized Tests lesson covers a specific type of standardized test question and has two completely worked-out examples. The test prep lesson also serves as an additional chapter review.
- Test-Taking Tips and Hints are provided in each lesson.
- Each lesson contains multiple-choice, grid-in, short response, or extended response test questions.
- The **Preparing for Standardized Tests** pages in the Student Handbook review common types of problems on standardized tests and provide practice for each format.

Test-Taking Tip

If there is no figure or diagram, draw one yourself.

Teacher Support

- Each test question is correlated to the chapter where that particular type of question is taught. In this way, intervention can be provided.

Additional Standardized Test Practice Resources

- **Student Edition** A Standardized Test Practice question is in the Mixed Review of each lesson. Some are open-ended. All are correlated to the lesson where the related math skill was taught.
- **Internet** Additional interactive standardized test practice is located at www.algconcepts.com.
- **Workbooks** Glencoe also publishes workbooks that prepare students for success on the SAT, ACT, and state and local proficiency tests.
- **Standardized Test Preparation CD-ROM** (Windows/Macintosh) This CD-ROM contains blocks of test items from state tests, SAT, ACT, TIMSS, and NAEP.

"I want an algebra program that gives my students the opportunity to use available technology tools."

INTEGRATING TECHNOLOGY AND THE INTERNET

Technology, including graphing calculators, graphing software, CD-ROM real-world applications, video, and the Internet, is integrated as a problem-solving tool, a discovery tool, and a review and test prep tool throughout Glencoe's *Algebra: Concepts and Applications.*

Internet Connections throughout the Student Edition refer students to the Glencoe *Algebra: Concepts and Applications* Web site. **www.algconcepts.com**

At this site, students can access links for:

- Problem-Solving Strategy Workshops
- Data Updates
- Career Data
- Investigations
- Chapter Review Activities, and
- Interactive Standardized Test Practice.

The [–] key is used for subtraction. The [(–)] key is used to enter negative integers.

Graphing Calculator Explorations, Technology Tips, and the **Graphing Calculator Tutorial**, all in the Student Edition, give students opportunities to study mathematical concepts using a graphing calculator.

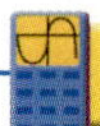

Graphing Calculator Exploration

Solve $x^2 + x - 2 = 0$ by making a table.

Enter the equation in the [Y=] screen. Then press [2nd] [TABLE]. A table with two columns labeled X and Y1 will appear on your screen.

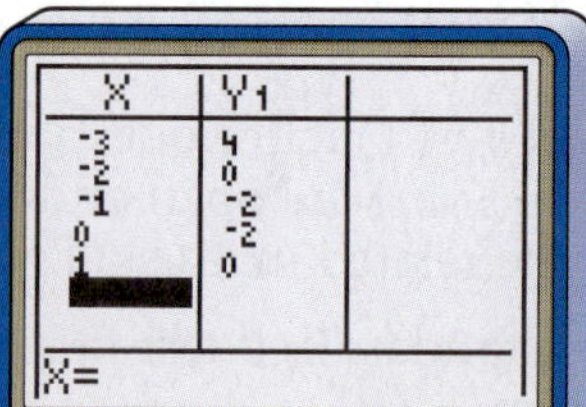

Begin entering values for x. For each x-value entered, a corresponding y-value is calculated using the equation $y = x^2 + x - 2$.

Since $y = 0$ when $x = -2$ and $x = 1$, the roots of the equation are -2 and 1.

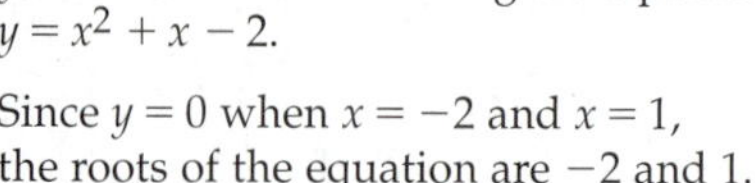

Try These

Make a table to find the roots of each equation.

1. $x^2 - 18x + 81 = 0$
2. $x^2 - 2x - 15 = 0$

Additional resources in the program provide other ways for you to incorporate technology in your teaching and lesson planning.

The **Multimedia Applications CD-ROM** contains 13 activities. Each includes videos of real people using algebra in their careers as well as interactive quizzes and a graphing tool.

Vocabulary PuzzleMaker Software, available online at www.algconcepts.com, improves students' mathematics vocabulary, which results in higher achievement and test scores. There are four types of puzzles.

- crossword
- scramble
- word search using word list
- word search using clues

Students can work on the computer screen or from a printed handout.

The **Graphing Calculator Masters** contain additional graphing calculator activities to be used with either the TI-83 Plus/TI-84 Plus or Casio Algebra FX 2.0 graphing calculators. In addition, Graphing Calculator Explorations for the Casio Algebra FX 2.0 are provided that parallel those for the TI-83 Plus/TI-84 Plus in the Student Edition.

All-in-One Lesson Planner and Resource Center CD-ROM includes a lesson planner and interactive Teacher Edition, so you can customize lesson plans and reproduce classroom resources quickly and easily, from just about anywhere.

Answer Key Maker software allows you to customize answer keys for your assignments from the Student Edition exercises.

Interactive Chalkboard CD-ROM includes fully worked-out examples and the 5-Minute Check Transparencies in a customizable Microsoft® PowerPoint® format.

What's Math Got to Do With It? Real-Life Videos engage students, showing them how math is used in everyday situations.

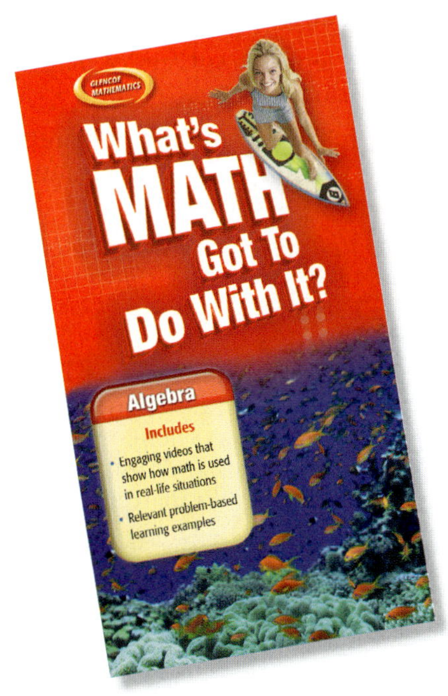

The vast array of technology products for Glencoe's *Algebra: Concepts and Applications* provides students and teachers with all of the tools they need for success in learning and teaching algebra.

"I want an algebra program that was planned with me and my students in mind."

RESEARCHING THE PROGRAM

Glencoe's *Algebra: Concepts and Applications,* as well as Glencoe's entire mathematics series, is the product of ongoing classroom-oriented research that involves students, teachers, curriculum supervisors, administrators, parents, and college-level mathematics educators.

The programs that make up the Glencoe mathematics series are currently being used by millions of students and tens of thousands of teachers. The key reason for the success of these programs in the classroom is the fact that each Glencoe author team is a mix of practicing classroom teachers, curriculum supervisors, and college-level mathematics educators. Glencoe's balanced author teams help ensure that Glencoe mathematics programs are both practical and progressive.

Prior to the publication of Glencoe's *Algebra: Concepts and Applications,* research activities included:

- a review of educational research and recommendations by groups such as NCTM,
- mail surveys of mathematics educators,
- discussion groups involving mathematics teachers, department heads, and supervisors,
- focus groups involving mathematics educators,
- face-to-face interviews with mathematics educators,
- telephone surveys of mathematics educators,
- in-depth analyses of manuscript by a wide range of reviewers and consultants, and
- field tests in which students and teachers used pre-publication manuscript in the classroom.

Feedback from teachers, curriculum supervisors, and even students who currently use Glencoe mathematics programs was also incorporated as Glencoe planned and published this new program. For example, *From the Classroom of...* features, which are printed in the Teacher's Wraparound Edition, are one result of this feedback.

This research makes it possible for Glencoe's authors and editors to publish outstanding instructional resources.

"I need an algebra program that is flexible."

PLANNING YOUR ALGEBRA COURSE

Each chart below contains suggested pacing for two options, core and enhanced, for four 9-week grading periods. The core option covers Chapters 1–13, while the enhanced option covers Chapters 1–15. A suggested pacing chart is provided in the interleaf pages for each chapter of the Teacher's Wraparound Edition.

40- to 50-Minute Class Periods The standard chart is based on 165 teaching days. This allows for teacher flexibility.

Grading Period	Standard Core		Standard Enhanced	
	Chapter	Days	Chapter	Days
1	1 2 3 4-1 to 4-2	13 12 13 3	1 2 3 4-1 to 4-5	12 10 12 8
2	4-3 to 4-7 5 6 7-1 to 7-4	10 13 12 6	4-6 to 4-7 5 6 7	4 12 10 12
3	7-5 to 7-7 8 9 10	7 13 11 11	8 9 10 11	12 9 9 12
4	11 12 13	13 13 13	12 13 14 15	12 12 9 10

Block Scheduling This chart contains suggested pacing for teaching this course using block scheduling in one semester (classes meet every day) or during the entire year (classes meet every other day). A total of 85 days is suggested.

Chapter	Class Periods		Chapter	Class Periods	
	Block Core	Block Enhanced		Block Core	Block Enhanced
1 2 3 4	6.5 5.5 6.5 6.5	5.5 5.5 5.5 5.5	9 10 11 12	5.5 5.5 6.5 6.5	4.5 4.5 5.5 5.5
5 6 7 8	6.5 5.5 6.5 6.5	5.5 5.5 5.5 5.5	13 14 15	6.5 — —	5.5 4.5 5.0

PLANNING YOUR ALGEBRA COURSE (continued)

In our increasingly competitive and technological world, more and more students need to take algebra. However, the abstract concepts of algebra are difficult for some students to grasp. One solution to this problem, is to teach the normal Algebra 1 curriculum in a period of two years. This extended time frame allows students to spend more time on each concept. More importantly, students will have more time to complete hands-on tasks and activities that develop these abstract concepts.

Each chapter's pacing is based on the following.

- 2 days for each lesson
- 2 days for each Investigation
- 4 days for review and testing
- 1 "extra" day

Two-Year Pacing This chart is based on 164 teaching days per year. This time frame allows for special events that may occur during the school day, causing classes to be shortened or omitted, and also for cancellation of school due to weather conditions.

Year One	
Chapter	**Days**
1	21
2	19
3	21
4	21
5	21
6	19
7	21
8*	21
Total	**164**

Year Two	
Chapter	**Days**
Review (Ch. 1–7)	10
8*	21
9	17
10	17
11	21
12	21
13	21
14	17
15	19
Total	**164**

* Some teachers may choose to include Chapter 8 in the first year, and others may choose to include it in the second year.

LESSON OBJECTIVES AND NCTM STANDARDS 2000

Key to NCTM Standards 2000
1 Number & Operations; **2** Algebra; **3** Geometry; **4** Measurement; **5** Data Analysis & Probability; **6** Problem Solving; **7** Reasoning & Proof; **8** Communications; **9** Connections; **10** Representation

Lesson	Lesson Objectives	NCTM Standards 2000
1-1	Translate words into algebraic expressions and equations.	1, 2, 6, 7, 8, 9, 10
1-2	Use the order of operations to evaluate expressions.	1, 6, 8, 9
1-3	Use the commutative and associative properties to simplify expressions.	1, 6, 8, 9
1-4	Use the Distributive Property to evaluate expressions.	1, 6, 8, 9
1-5	Use a four-step plan to solve problems.	1, 6, 7, 8, 9
Investigation	Explore inductive and deductive reasoning.	1, 6, 7, 8, 9
1-6	Collect and organize data using sampling and frequency tables.	1, 5, 6, 7, 8, 9, 10
1-7	Construct and interpret line graphs, histograms, and stem-and-leaf plots.	1, 2, 5, 6, 7, 8, 9, 10
2-1	Graph integers on a number line and compare and order integers.	1, 6, 8, 9, 10
2-2	Graph points on a coordinate plane.	1, 6, 8, 9, 10
2-3	Add integers.	1, 6, 8, 9
2-4	Subtract integers.	1, 6, 8, 9
2-5	Multiply integers.	1, 6, 8, 9
Investigation	Explore matrices.	1, 6, 8, 9
2-6	Divide integers.	1, 6, 8, 9, 10
3-1	Compare and order rational numbers.	1, 6, 8, 9
3-2	Add and subtract rational numbers.	1, 6, 8, 9
3-3	Find the mean, median, mode, and range of a set of data.	1, 5, 6, 8, 9, 10
Investigation	Explore arithmetic sequences.	1, 2, 6, 7, 8, 9, 10
3-4	Determine whether a given number is a solution of an equation.	1, 6, 8, 9
3-5	Solve addition and subtraction equations by using models.	1, 2, 6, 8, 9, 10
3-6	Solve addition and subtraction equations by using the properties of equality.	1, 6, 8, 9
3-7	Solve equations involving absolute value.	1, 6, 8, 9
4-1	Multiply rational numbers.	1, 6, 8, 9
4-2	Use tree diagrams or the Fundamental Counting Principle to count outcomes.	1, 2, 5, 6, 7, 8, 9, 10
Investigation	Explore permutations and combinations.	1, 2, 5, 6, 7, 8, 9
4-3	Divide rational numbers.	1, 6, 8, 9
4-4	Solve multiplication and division equations by using the properties of equality.	1, 6, 8, 9
4-5	Solve equations involving more than one operation.	1, 6, 8, 9
4-6	Solve equations with variables on both sides.	1, 6, 8, 9
4-7	Solve equations with grouping symbols.	1, 6, 8, 9
5-1	Solve proportions.	1, 6, 8, 9
5-2	Solve problems involving scale drawings and models.	1, 2, 3, 4, 6, 8, 9, 10
5-3	Solve problems by using the percent proportion.	1, 6, 8, 9

Lesson	Lesson Objectives	NCTM Standards 2000
5-4	Solve problems by using the percent equation.	1, 6, 8, 9
Investigation	Explore box-and-whisker plots.	1, 2, 6, 7, 8, 9, 10
5-5	Solve problems involving percent of increase or decrease.	1, 2, 6, 8, 9
5-6	Find the probability and odds of a simple event.	1, 5, 6, 8, 9, 10
5-7	Find the probability of mutually exclusive and inclusive events.	1, 5, 6, 8, 9
6-1	Show relations as sets of ordered pairs, in tables, and as graphs.	1, 2, 3, 6, 8, 9, 10
6-2	Solve linear equations for a given domain.	1, 6, 8, 9
6-3	Graph linear relations.	1, 2, 3, 6, 8, 9
6-4	Determine whether a given relation is a function.	1, 6, 8, 9
Investigation	Explore functions.	1, 2, 4, 6, 8, 9, 10
6-5	Solve problems involving direct variations.	1, 2, 6, 8, 9, 10
6-6	Solve problems involving inverse variations.	1, 2, 6, 8, 9, 10
7-1	Find the slope of a line given the coordinates of two points on the line.	1, 3, 6, 8, 9
7-2	Write a linear equation in point-slope form given the coordinates of a point on the line and the slope of the line.	1, 6, 8, 9
7-3	Write a linear equation in slope-intercept form given the slope and y-intercept.	1, 6, 8, 9
7-4	Graph and interpret points on scatter plots.	1, 2, 3, 5, 6, 8, 9, 10
Investigation	Use best-fit lines to make predictions.	1, 2, 3, 4, 6, 7, 8, 9, 10
7-5	Graph linear equations by using the x- and y-intercepts or the slope and y-intercept.	1, 3, 6, 8, 9
7-6	Explore the effects of changing the slopes and y-intercepts of linear functions.	1, 2, 3, 6, 8, 9
7-7	Write an equation of a line that is parallel or perpendicular to the graph of a given equation and that passes through a given point.	1, 3, 6, 8, 9
8-1	Use powers in expressions.	1, 6, 8, 9
8-2	Multiply and divide powers.	1, 6, 8, 9
8-3	Simplify expressions containing negative exponents.	1, 6, 8, 9
8-4	Express numbers in scientific notation.	1, 2, 6, 8, 9, 10
8-5	Simplify radicals by using the Product and Quotient Properties of Square Roots.	1, 6, 8, 9
8-6	Estimate square roots.	1, 6, 8, 9
8-7	Use the Pythagorean Theorem to solve problems.	1, 3, 4, 6, 8, 9
Investigation	Explore the Pythagorean Theorem.	1, 3, 4, 6, 7, 8, 9, 10
9-1	Identify and classify polynomials and find their degree.	1, 2, 6, 8, 9
9-2	Add and subtract polynomials.	1, 6, 8, 9
9-3	Multiply a polynomial by a monomial.	1, 6, 8, 9
9-4	Multiply two binomials.	1, 6, 8, 9
9-5	Develop and use the patterns for $(a + b)^2$, $(a - b)^2$, and $(a + b)(a - b)$.	1, 6, 8, 9
Investigation	Investigate the ratios of areas of different-sized squares.	1, 2, 3, 4, 6, 7, 8, 9
10-1	Find the greatest common factor of a set of numbers or monomials.	1, 6, 8, 9
Investigation	Explore perimeter and area.	1, 2, 3, 6, 7, 8, 9, 10
10-2	Use the GCF and the Distributive Property to factor polynomials.	1, 6, 8, 9
10-3	Factor trinomials of the form $x^2 + bx + c$.	1, 6, 8, 9
10-4	Factor trinomials of the form $ax^2 + bx + c$.	1, 6, 8, 9

Lesson	Lesson Objectives	NCTM Standards 2000
10-5	Recognize and factor the differences of squares and perfect square trinomials.	1, 6, 8, 9
11-1	Graph quadratic functions.	1, 2, 3, 6, 8, 9
11-2	Learn the characteristics of families of parabolas.	1, 2, 3, 6, 8, 9
11-3	Locate the roots of quadratic equations by graphing the related functions.	1, 3, 6, 8, 9
11-4	Solve quadratic equations by factoring and by using the Zero Product Property.	1, 6, 8, 9
11-5	Solve quadratic equations by completing the square.	1, 6, 8, 9
11-6	Solve quadratic equations by using the Quadratic Formula.	1, 6, 8, 9
11-7	Graph exponential functions.	1, 2, 3, 6, 8, 9
Investigation	Explore geometric sequences.	1, 2, 3, 6, 7, 8, 9, 10
12-1	Graph inequalities on a number line.	1, 6, 8, 9
12-2	Solve inequalities involving addition and subtraction.	1, 6, 8, 9
12-3	Solve inequalities involving multiplication and division.	1, 6, 8, 9
12-4	Solve inequalities involving more than one operation.	1, 6, 8, 9
12-5	Solve compound inequalities.	1, 6, 8, 9
12-6	Solve inequalities involving absolute value.	1, 6, 8, 9
12-7	Graph inequalities in the coordinate plane.	1, 3, 6, 8, 9
Investigation	Solve and graph quadratic inequalities.	1, 3, 6, 8, 9, 10
13-1	Solve systems of equations by graphing.	1, 3, 6, 8, 9
13-2	Determine whether a system of equations has one solution, no solution, or infinitely many solutions by graphing.	1, 3, 6, 8, 9, 10
13-3	Solve systems of equations by the substitution method.	1, 6, 8, 9
13-4	Solve systems of equations by the elimination method using addition and subtraction.	1, 6, 8, 9
13-5	Solve systems of equations by the elimination method using multiplication and addition.	1, 6, 8, 9
Investigation	Use matrices to solve systems of equations.	1, 6, 7, 8, 9, 10
13-6	Solve systems of quadratic and linear equations.	1, 6, 8, 9
13-7	Solve systems of inequalities by graphing.	1, 3, 6, 8, 9
14-1	Describe the relationships among sets of numbers.	1, 2, 6, 8, 9, 10
14-2	Find the distance between two points in the coordinate plane.	1, 3, 6, 8, 9
Investigation	Explore the Midpoint Formula.	1, 3, 6, 7, 8, 9
14-3	Simplify radical expressions.	1, 6, 8, 9
14-4	Add and subtract radical expressions.	1, 6, 8, 9
14-5	Solve simple radical equations in which only one radical contains a variable.	1, 6, 8, 9
15-1	Simplify rational expressions.	1, 6, 8, 9
15-2	Multiply and divide rational expressions.	1, 6, 8, 9
15-3	Divide polynomials by binomials.	1, 6, 8, 9
15-4	Add and subtract rational expressions with like denominators.	1, 6, 8, 9
15-5	Add and subtract rational expressions with unlike denominators.	1, 6, 8, 9
15-6	Solve rational equations.	1, 6, 8, 9, 10
Investigation	Explore work problems.	1, 2, 4, 6, 8, 9, 10

CHAPTER 1

Resource Manager

The Language of Algebra

The Instructional Objectives chart lists the objectives for each lesson and correlates those objectives to the NCTM Standards 2000. There is also space for you to reference your state and/or local objectives.

Instructional Objectives

Lesson (pages)	Objectives	NCTM Standards 2000	State/Local Objectives
1-1 (4–7)	Translate words into algebraic expressions and equations.	1, 2, 6, 7, 8, 9, 10	
1-2 (8–13)	Use the order of operations to evaluate expressions.	1, 6, 8, 9	
1-3 (14–18)	Use the commutative and associative properties to simplify expressions.	1, 6, 8, 9	
1-4 (19–23)	Use the Distributive Property to evaluate expressions.	1, 6, 8, 9	
1-5 (24–29)	Use a four-step plan to solve problems.	1, 6, 7, 8, 9	
Investigation (30–31)	Explore inductive and deductive reasoning.	1, 6, 7, 8, 9	
1-6 (32–37)	Collect and organize data using sampling and frequency tables.	1, 5, 6, 7, 8, 9, 10	
1-7 (38–43)	Construct and interpret line graphs, histograms, and stem-and-leaf plots.	1, 2, 5, 6, 7, 8, 9, 10	

Key to NCTM Standards 2000

1 Number & Operations; **2** Algebra; **3** Geometry; **4** Measurement; **5** Data Analysis & Probability; **6** Problem Solving; **7** Reasoning & Proof; **8** Communications; **9** Connections; **10** Representation

Suggested Pacing for standard core, standard enhanced, block core, and block enhanced is provided. The pacing of the course is based on a school year of 165 days.

Suggested Pacing

See page T13 for a complete course-planning calendar.

Standard refers to schedules that provide 45- to 55-minute periods that meet each day.
Block refers to schedules that provide approximately 90-minute periods which may meet every day for one semester or every other day over two semesters.

PACING	DAY 1	DAY 2	DAY 3	DAY 4	DAY 5	DAY 6
Standard Core (Chapters 1–13)	Lesson 1-1	Lesson 1-2		Lesson 1-3	Lesson 1-4	Lesson 1-5
Standard Enhanced (Chapters 1–15)	Lesson 1-1	Lesson 1-2		Lesson 1-3	Lesson 1-4	Lesson 1-5
Block Core (Chapters 1–13)	Lessons 1-1 & 1-2	Lessons 1-3 & 1-4	Lesson 1-5	INV	Lessons 1-6 & 1-7	SG+A
Block Enhanced (Chapters 1–15)	Lessons 1-1 & 1-2	Lessons 1-3 & 1-4	Lesson 1-5 & INV	Lessons 1-6 & 1-7	SG+A	Chapter Test & Lesson 2-1

Instructional Resources

Lesson	Materials and Manipulatives (see below for Glencoe Manipulative Resources)	Blackline Masters (page numbers): Chapter 1 Resource Masters: Study Guide	Practice (Skills & Average)	Reading to Learn Mathematics	Enrichment	Assessment	Hands-On Algebra*	School-to-Workplace*	Graphing Calculator Masters*	5-Minute Check Transparencies
1–1		1	2–3	4	5					1–1
1–2	squares of paper	6	7–8	9	10			1		1–2
1–3	markers (2 different colors)	11	12–13	14	15	47				1–3
1–4		16	17–18	19	20	46	21, 22			1–4
1–5	rectangular box rulers (inch or centimeter) [1, 2, 4] graphing calculator	21	22–23	24	25		23, 24, 25		2	1–5
Investigation	sheets of different-colored paper paper punch three-inch squares of paper two-inch squares of paper									
1–6		26	27–28	29	30					1–6
1–7	straightedge [1, 2, 4] magazines, newspapers, and encyclopedias	31	32–33	34	35	47	26, 27 28		3, 4	1–7
Study Guide & Assessment/ Chapter Test	straightedge [1, 2, 4]					37–45, 48–50				

See page 2c for examples of these instructional materials.

Key to Glencoe Manipulative Resources
[1]Classroom Manipulative Resources [2]Student Manipulative Resources
[3]Overhead Manipulative Resources [4]Hands-On Algebra Masters

The Instructional Resources chart lists all of the blackline masters and transparencies available for each lesson. It also shows the materials and manipulatives your students will need for each lesson.

INV = Investigation *SG+A = Study Guide and Assessment*

DAY 7	DAY 8	DAY 9	DAY 10	DAY 11	DAY 12	DAY 13
Lesson 1–5	INV	Lesson 1–6		Lesson 1–7	SG+A	Chapter Test
Lesson 1–5	INV	Lesson 1–6	Lesson 1–7	SG+A	Chapter Test	
Chapter Test & Lesson 2–1						

Resource Manager

The pages shown on this page are a small sample of the materials available on *TeacherWorks: All-in-One Lesson Planner and Resource Center.*

This CD-ROM includes all of the blackline masters and transparencies available for this program.

It also includes a lesson planner and interactive Teacher Edition, so you can customize lesson plans and reproduce classroom resources quickly and easily, from just about anywhere.

Applications

School-to-Workplace Masters, p. 1

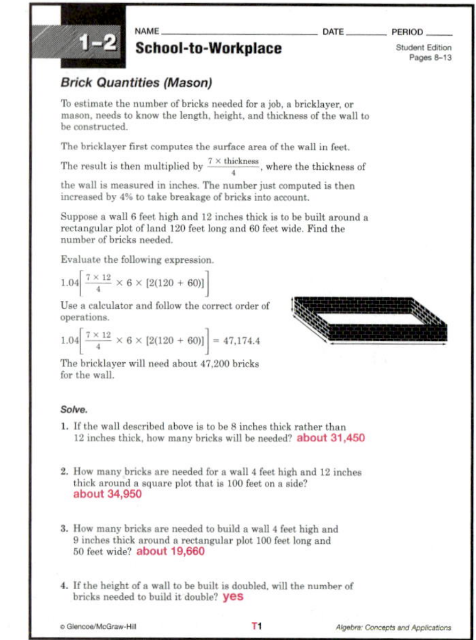

1-2 NAME ______ DATE ______ PERIOD ______

School-to-Workplace

Student Edition Pages 8–13

Brick Quantities (Mason)

To estimate the number of bricks needed for a job, a bricklayer, or mason, needs to know the length, height, and thickness of the wall to be constructed.

The bricklayer first computes the surface area of the wall in feet.

The result is then multiplied by $\frac{7 \times \text{thickness}}{4}$, where the thickness of the wall is measured in inches. The number just computed is then increased by 4% to take breakage of bricks into account.

Suppose a wall 6 feet high and 12 inches thick is to be built around a rectangular plot of land 120 feet long and 60 feet wide. Find the number of bricks needed.

Evaluate the following expression.

$$1.04\left[\frac{7 \times 12}{4} \times 6 \times [2(120 + 60)]\right]$$

Use a calculator and follow the correct order of operations.

$$1.04\left[\frac{7 \times 12}{4} \times 6 \times [2(120 + 60)]\right] = 47{,}174.4$$

The bricklayer will need about 47,200 bricks for the wall.

Solve.

1. If the wall described above is to be 8 inches thick rather than 12 inches thick, how many bricks will be needed? **about 31,450**
2. How many bricks are needed for a wall 4 feet high and 12 inches thick around a square plot that is 100 feet on a side? **about 34,950**
3. How many bricks are needed to build a wall 4 feet high and 9 inches thick around a rectangular plot 100 feet long and 50 feet wide? **about 19,660**
4. If the height of a wall to be built is doubled, will the number of bricks needed to build it double? **yes**

© Glencoe/McGraw-Hill T1 *Algebra: Concepts and Applications*

Manipulatives/Modeling

Hands-On Algebra Masters, pp. 21–28

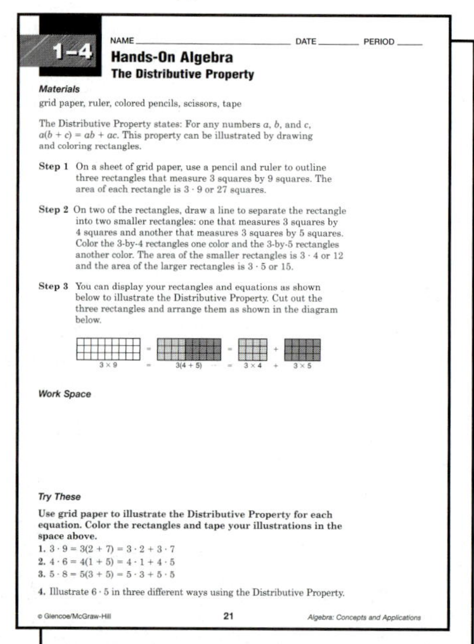

1-4 NAME ______ DATE ______ PERIOD ______

Hands-On Algebra
The Distributive Property

Materials

grid paper, ruler, colored pencils, scissors, tape

The Distributive Property states: For any numbers a, b, and c, $a(b + c) = ab + ac$. This property can be illustrated by drawing and coloring rectangles.

Step 1 On a sheet of grid paper, use a pencil and ruler to outline three rectangles that measure 3 squares by 9 squares. The area of each rectangle is $3 \cdot 9$ or 27 squares.

Step 2 On two of the rectangles, draw a line to separate the rectangle into two smaller rectangles: one that measures 3 squares by 4 squares and another that measures 3 squares by 5 squares. Color the 3-by-4 rectangles one color and the 3-by-5 rectangles another color. The area of the smaller rectangles is $3 \cdot 4$ or 12 and the area of the larger rectangles is $3 \cdot 5$ or 15.

Step 3 You can display your rectangles and equations as shown below to illustrate the Distributive Property. Cut out the three rectangles and arrange them as shown in the diagram below.

$3 \times 9 = 3(4 + 5) = 3 \times 4 + 3 \times 5$

Work Space

Try These

Use grid paper to illustrate the Distributive Property for each equation. Color the rectangles and tape your illustrations in the space above.

1. $3 \cdot 9 = 3(2 + 7) = 3 \cdot 2 + 3 \cdot 7$
2. $4 \cdot 6 = 4(1 + 5) = 4 \cdot 1 + 4 \cdot 5$
3. $5 \cdot 8 = 5(3 + 5) = 5 \cdot 3 + 5 \cdot 5$
4. Illustrate $6 \cdot 5$ in three different ways using the Distributive Property.

© Glencoe/McGraw-Hill 21 *Algebra: Concepts and Applications*

Technology/Multimedia

Graphing Calculator Masters, pp. 2–4

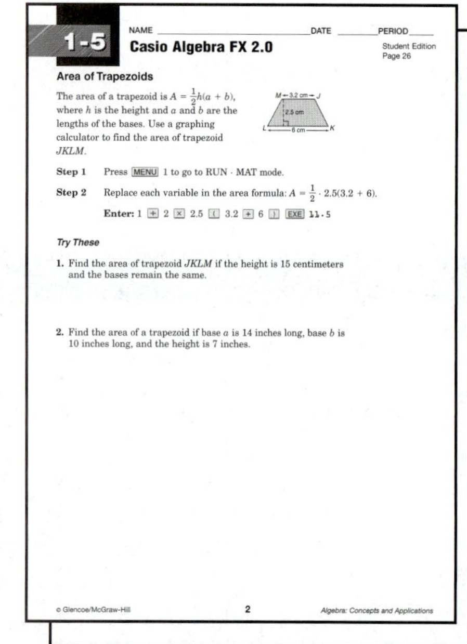

1-5 NAME ______ DATE ______ PERIOD ______

Casio Algebra FX 2.0

Student Edition Page 26

Area of Trapezoids

The area of a trapezoid is $A = \frac{1}{2}h(a + b)$, where h is the height and a and b are the lengths of the bases. Use a graphing calculator to find the area of trapezoid $JKLM$.

Step 1 Press [MENU] 1 to go to RUN · MAT mode.

Step 2 Replace each variable in the area formula: $A = \frac{1}{2} \cdot 2.5(3.2 + 6)$.

Enter: 1 [÷] 2 [×] 2.5 [(] 3.2 [+] 6 [)] [EXE] 11.5

Try These

1. Find the area of trapezoid $JKLM$ if the height is 15 centimeters and the bases remain the same.
2. Find the area of a trapezoid if base a is 14 inches long, base b is 10 inches long, and the height is 7 inches.

© Glencoe/McGraw-Hill 2 *Algebra: Concepts and Applications*

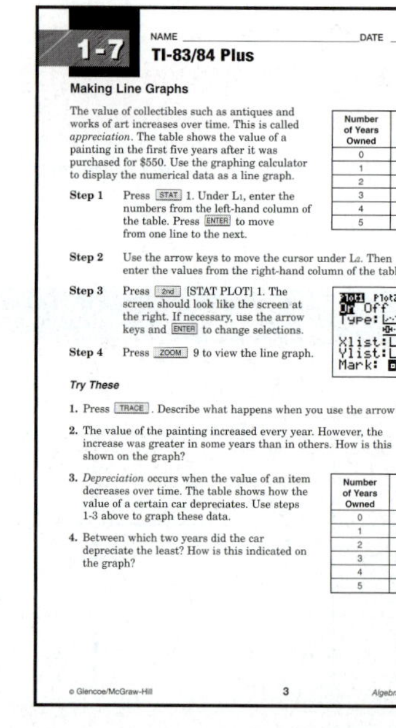

1-7 NAME ______ DATE ______ PERIOD ______

TI-83/84 Plus

Making Line Graphs

The value of collectibles such as antiques and works of art increases over time. This is called *appreciation*. The table shows the value of a painting in the first five years after it was purchased for $550. Use the graphing calculator to display the numerical data as a line graph.

Number of Years Owned	Value (dollars)
0	550
1	600
2	650
3	700
4	770
5	850

Step 1 Press [STAT] 1. Under L_1, enter the numbers from the left-hand column of the table. Press [ENTER] to move from one line to the next.

Step 2 Use the arrow keys to move the cursor under L_2. Then enter the values from the right-hand column of the table.

Step 3 Press [2nd] [STAT PLOT] 1. The screen should look like the screen at the right. If necessary, use the arrow keys and [ENTER] to change selections.

Step 4 Press [ZOOM] 9 to view the line graph.

Try These

1. Press [TRACE]. Describe what happens when you use the arrow keys.
2. The value of the painting increased every year. However, the increase was greater in some years than in others. How is this shown on the graph?
3. *Depreciation* occurs when the value of an item decreases over time. The table shows how the value of a certain car depreciates. Use steps 1-3 above to graph these data.
4. Between which two years did the car depreciate the least? How is this indicated on the graph?

Number of Years Owned	Value (dollars)
0	17,000
1	13,500
2	11,000
3	8000
4	6750
5	4500

© Glencoe/McGraw-Hill 3 *Algebra: Concepts and Applications*

The reduced pages shown give you an overview of the additional resources. Use the full-size pages in blackline master form or in the TeacherWorks lesson planner to supplement your lessons.

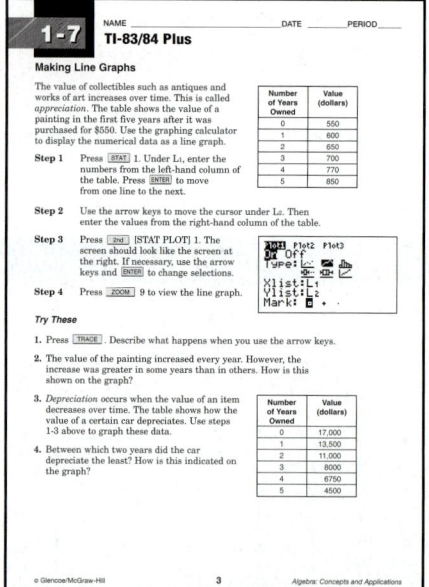

GLENCOE'S ASSESSMENT ADVANTAGE

Assessment Resources

The Assessment Resources chart illustrates the wide variety of assessment materials that are built into the program. Glencoe Instructional Technology provides suggestions on incorporating the use of technology into your daily lessons.

Type	Student Edition	Teacher's Wraparound Edition	Chapter 1 Resource Masters
Ongoing Assessment	Quizzes 1 and 2, pp. 13, 37	5-Minute Check, pp. 4, 8, 14, 19, 24, 32, 38	Mid-Chapter Test, p. 46 Quizzes A and B, p. 47
Mixed Review	Mixed Review, pp. 13, 18, 23, 29, 37, 43 Standardized Test Practice, Chapter 1, pp. 48–49		Cumulative Review, p. 48 Standardized Test Practice, pp. 49–50
Error Analysis	You Decide, pp. 3, 42	Error Analysis, pp. 6, 11, 17, 21, 27, 35, 42	
Standardized Test Prep	Standardized Test Practice, pp. 13, 18, 23, 29, 37, 43 Standardized Test Practice, Chapter 1, pp. 48–49		Standardized Test Practice, pp. 49–50
Open-Ended Assessment	Writing Math, pp. 6, 11 Investigation, pp. 30–31 Portfolio, pp. 3, 31	Modeling: pp. 13, 23 Speaking: pp. 18, 29, 43 Writing: pp. 7, 37	Extended Response Assessment, p. 45
Chapter Assessment	Study Guide and Assessment, pp. 44–46 Chapter Test, p. 47		Multiple-Choice Tests (Forms 1A, 1B), pp. 37–40 Free-Response Tests (Forms 2A, 2B), pp. 41–44

Additional Chapter Resources

Student Edition
Hands-On Algebra, p. 25
Graphing Calculator Exploration, p. 26

Teacher's Classroom Resources
Manipulatives/Modeling
Teacher's Guide for Overhead Manipulative Resources

Meeting Individual Needs
Prerequisite Skills Booklet
Spanish Study Guide and Assessment, pp. 9–15, 105–106

Teaching Aids
Solutions Manual
5-Minute Check Transparencies

Glencoe Technology

Instructional
AlgePASS CD-ROM, Lessons 5, 6
Interactive Chalkboard CD-ROM
StudentWorks Plus CD-ROM
Multimedia Applications CD-ROM, Activity 1
Vocabulary PuzzleMaker (online)

Assessment
ExamView® Pro

Teaching Aids
Answer Key Maker CD-ROM
TeacherWorks CD-ROM

Visit ***www.algconcepts.com*** *for data updates, career information, games, and other interactive activities.*

Mathematics of the Chapter

Students begin the chapter by writing expressions and equations and using the order of operations. A discussion of the Commutative, Associative, and Distributive Properties follows. Students use these properties to simplify and evaluate numerical and algebraic expressions. A four-step plan for problem solving is then presented. Finally, students use sampling to collect data and explore a variety of ways to display data, including frequency tables, line graphs, histograms, and stem-and-leaf plots.

What You'll Learn

Have students read over the lists of key ideas and key vocabulary. Have them make a list of any words with which they are not familiar.

Why It's Important

Point out to students that this is only one of many reasons why each objective is important. Others are provided in each lesson.

CHAPTER

The Language of Algebra

What You'll Learn

Key Ideas

- Write algebraic expressions and equations. *(Lesson 1-1)*
- Simplify and evaluate algebraic expressions. *(Lessons 1-2 to 1-4)*
- Collect, organize, and display data. *(Lessons 1-6 and 1-7)*

Key Vocabulary

algebraic expression *(p. 4)*
data *(p. 32)*
equation *(p. 5)*
variable *(p. 4)*

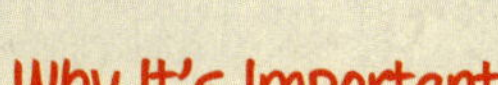

Why It's Important

Biology The feared great white shark is the largest predatory fish in the ocean. But great white sharks are very rare. It is just one of about 368 species of sharks. These sharks range in size from the size of a person's hand to bigger than a bus.

Algebraic expressions and equations are used to model real-life situations. You will write an algebraic expression for the number of shark species in Lesson 1–1.

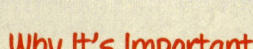

CHAPTER 1 LINKS							
Lesson	1–1	1–2	1–3	1–4	1–5	1–6	1–7
Math in the Workplace	Communication	Business	Construction	Shopping	Savings	Marketing	Research
Applications and Connections	Life Science Biology	Finance Real Estate Sports Gardening Time History Carpentry	Geology Sports Geography	Sports School Retail Health Theater Sound	Science Money Shopping Weather History	Health Science Technology Entertainment Travel Fitness Biology	Travel Physical Science School Weather
Math Integration			Geometry		Geometry	Statistics	Statistics

Study these lessons to improve your skills.

Check Your Readiness

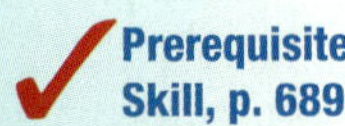

Prerequisite Skill, p. 684

Find each product or quotient.

1. 2×9 **18**	2. 4×7 **28**	3. 9×0 **0**
4. 8×3 **24**	5. 8×8 **64**	6. 12×3 **36**
7. 5×1.2 **6**	8. 12×1.59 **19.08**	9. 160×0.75 **120**
10. $15 \div 3$ **5**	11. $35 \div 5$ **7**	12. $18 \div 3$ **6**
13. $36 \div 6$ **6**	14. $24 \div 2$ **12**	15. $54 \div 9$ **6**
16. $16 \div 2.5$ **6.4**	17. $1.98 \div 2$ **0.99**	18. $150 \div 4.8$ **31.25**

Prerequisite Skill, p. 689

Write each percent as a decimal.

19. 15% **0.15**	20. 30% **0.3**	21. 55% **0.55**
22. 4% **0.04**	23. 100% **1**	24. 12% **0.12**

25. 17.5, 18, 18.5, 19 26. 1.2, 1.8, 2.3, 3.0, 4.8

Prerequisite Skill, p. 691

Write the numbers in each set in order from least to greatest.

25. 19, 17.5, 18, 18.5

26. 2.3, 1.2, 1.8, 4.8, 3.0

27. 32, 30, 18, 29, 14, 20, 26, 21, 37 **14, 18, 20, 21, 26, 29, 30, 32, 37**

28. 3.1, 2.6, 4.9, 2.9, 3.3, 2.8, 3.4, 4.4, 3.0, 2.7
2.6, 2.7, 2.8, 2.9, 3.0, 3.1, 3.3, 3.4, 4.4, 4.9

FOLDABLES™ Study Organizer

Make this Foldable to help you organize your Chapter 1 notes. Begin with four sheets of notebook paper.

❶ **Stack** the sheets with the edges $\frac{3}{4}$ inch apart.

❷ **Fold** up the bottom edges. All tabs should be the same size.

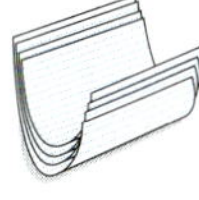

❸ **Staple** along the fold.

❹ **Label** the tabs with topics from the chapter.

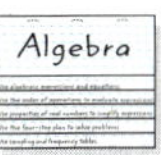

Reading and Writing As you read and study the chapter, use each page to write notes, graphs, and examples for each lesson.

www.algconcepts.com/chapter_readiness

FOLDABLES™ Study Organizer Have each student make a Foldable journal to record important ideas from the chapter or problems that are most difficult. The students can refer to their journals later when reviewing for chapter, unit, or semester tests.

Check Your Readiness

This section provides a review of the basic concepts needed before beginning Chapter 1. Lesson references are included for additional student help.

Use Exercises	To Prepare for Lesson(s)
1–18	1–2, 1–3, 1–4, 1–5
19–24	1–5
25–28	1–7

Quick Review Math Handbook

Glencoe Algebra Lesson	Quick Review Math Handbook Hot Topic
1–1	2–3, 2–5, 6–1, 6–3
1–2	1–2, 1–3, 6–2, 6–3
1–3	1–2, 6–3
1–4	1–2, 7–5
1–5	
Ch. 1 Inv.	2–8, 5–1, 5–2
1–6	2–6, 4–1
1–7	2–1, 4–2, 4–3, 4–4

Noteables™

Interactive Study Notebook with Foldables™

This note-taking guide reinforces excellent note-taking skills. It includes:

- ✓ **Build Your Vocabulary** tool for including mathematical terminology in notes.
- ✓ **Review It** activities with links to prior lessons and items to include in **Foldables™**.
- ✓ **Bringing It All Together** feature to help students review for the Chapter Test.
- ✓ **Are You Ready for the Chapter Test?** feature to allow students to assess their own readiness for the Chapter Test.
- ✓ **Teacher Edition** with transparencies to guide note taking. Corresponds to the *Teacher's Wraparound Edition* In-Class Examples and *Interactive Chalkboard CD-ROM.*

1 FOCUS

5-Minute Check

Find the value of each expression.

1. $\frac{8}{2}$ 4
2. 7 · 6 42
3. 23 + 31 54
4. 17 − 9 8
5. 42 ÷ 6 7

Motivating the Lesson

Hands-On Activity Have a volunteer draw a question mark with a square around it on the board or overhead. Tell students that this symbol represents a number, and the number plus 10 equals 14. Help students discover that the symbol represents the number 4. Now write the equation $x + 10 = 14$ on the board or overhead and help students understand that the variable x, like the symbol, represents an unknown number.

2 TEACH

Teaching Tip When you give examples of words translated into algebraic expressions, use situations involving students in the class, such as their ages in 5 years.

Glencoe's Teacher's Wraparound Edition uses a Four-step Teaching Plan that shows you how to Focus, Teach, Practice/Apply, and Assess each lesson.

1–1 Writing Expressions and Equations

What You'll Learn
You'll learn to translate words into algebraic expressions and equations.

Why It's Important
Communication You can use expressions to represent the cost of your long-distance phone calls. *See Exercise 43.*

Suppose a candy bar costs 45 cents. Then 45 × 2 is the cost of 2 candy bars, 45 × 3 is the cost of 3 candy bars, and so on. Generally, the cost of any number of candy bars is *45 cents times the number of bars*. We can represent this situation with an **algebraic expression**.

$$\underbrace{45}_{\text{45 cents}} \quad \underbrace{\times}_{\text{times}} \quad \underbrace{n}_{\text{the number of bars}}$$

The letter n stands for an unknown number, in this case, candy bars. The unknown n is called a **variable** because its value *varies*. An algebraic expression contains at least one variable and at least one mathematical operation, as shown in the examples below.

$$h \div 3 \qquad 5n + 1 \qquad \frac{r}{t} - 1 \qquad xy \qquad 4 \times a$$

A **numerical expression** contains only numbers and mathematical operations. For example, $6 + 2 \div 1$ is a numerical expression.

Reading Algebra
When one factor in a product is a variable, the multiplication sign is usually omitted. Read $4a$ as *four a*.

Reading Algebra notes help students learn and use the language of algebra.

In an expression involving multiplication, the quantities being multiplied are called **factors**, and the result is the **product**.

$$4 \times 5 \times 8 = 160$$

4, 5, 8: factors; 160: product

To write a multiplication expression such as $4 \times a$, a raised dot or parentheses can be used. A fraction bar can be used to represent division.

$4 \cdot a$, $4(a)$, $(4)(a)$, $4a$ means $4 \times a$

$\frac{t}{2}$ means $t \div 2$

The result of a divison expression is called a **quotient**.

To solve verbal problems in mathematics, you may have to translate words into algebraic expressions. The chart below shows some of the words and phrases used to indicate mathematical operations.

Addition	Subtraction	Multiplication	Division
plus the sum of increased by more than added to the total of	minus the difference of decreased by less than fewer than subtracted from	times the product of multiplied by at of	divided by the quotient of the ratio of per

Resource Manager

Reproducible Masters

Chapter 1 Resource Masters

- *Study Guide*, p. 1
- *Skills Practice*, p. 2
- *Practice*, p. 3
- *Reading to Learn Mathematics*, p. 4
- *Enrichment*, p. 5

Transparencies

5-Minute Check, 1–1

Technology/Multimedia

AlgePASS, Lesson 5

Interactive Chalkboard CD-ROM

Examples
Life Science Link

Write an algebraic expression for each verbal expression.

1 the sum of m and 18

$m + 18$

2 g divided by y

$g \div y$ or $\frac{g}{y}$

Your Turn b. $4 + 8k$ or $8k + 4$

a. 26 decreased by w $26 - w$

b. 4 more than 8 times k

Some kangaroos can travel 30 feet in a single leap.

3 Write a numerical expression to represent the distance a kangaroo can travel if it leaps 4 times.

$30 \cdot 4$ or $30(4)$

4 Write an algebraic expression to represent the distance a kangaroo can travel if it leaps x times.

$30 \cdot x$ or $30x$

You can also translate algebraic expressions into verbal expressions.

Examples

Write a verbal expression for each algebraic expression.

5 $32 - b$

32 less b
b less than 32
the difference of 32 and b
b subtracted from 32
32 decreased by b

6 $(y \div 4) + 9$

y divided by 4, plus 9
the quotient of y and 4, increased by 9
9 added to the ratio of y and 4

Your Turn c–d. Sample answers are given.

c. $15v$ 15 times v

d. $r - \frac{t}{d}$ r minus the quotient of t and d

An **equation** is a mathematical sentence that contains an equals sign (=). Some words used to indicate the equals sign are in the chart at the right. An equation may contain numbers, variables, or algebraic expressions.

Equality	
equals	is equal to
is	is the same as
is equivalent to	is as much as
	is identical to

Examples

Write an equation for each sentence.

7 Three times g equals 21.

$3g = 21$

8 Five more than twice n is 15.

$2n + 5 = 15$

Your Turn

e. A number k divided by 4 is equal to 18. $k \div 4 = 18$ or $\frac{k}{4} = 18$

www.algconcepts.com/extra_examples

In-Class Examples

Examples 1–2

Write an algebraic expression for each verbal expression.

1 the sum of p and 12 $p + 12$

2 the product of k and q kq

Examples 3–4

A python eats 4 pounds of meat each month.

3 Write a numerical expression to represent the amount it eats in 5 months. $4 \cdot 5$ or $4(5)$

4 Write an algebraic expression to represent the amount it eats in d months. $4 \cdot d$ or $4d$

Examples 5–6

Write a verbal expression for each algebraic expression. Sample answers are given.

5 $37 + s$ the sum of 37 and s

6 $5(b - 3)$ 5 times the difference of b and 3

Examples 7–8

Write an equation for each sentence.

7 The quotient of t and 8 equals 20. $t \div 8 = 20$ or $\frac{t}{8} = 20$

8 Seven less than three times g is 31. $3g - 7 = 31$

Study Guide, p. 1

NAME ________ DATE ______ PERIOD _____

Study Guide

Student Edition
Pages 4–7

Writing Expressions and Equations

If the cost of one CD is \$15 and you want to buy three CDs, you know that your total cost will be \$15 × 3, or \$45. The expression 15 × 3 is called a **numerical expression**.

You could use a letter such as n to represent the number of CDs you might buy. Then the expression $15 \times n$, or $15n$, would represent the cost of buying n CDs. The expression $15n$ is called an **algebraic expression** because it contains a variable. A variable such as n is a letter used to represent a number. You can use variables to write verbal expressions as algebraic expressions.

Verbal Expressions	Algebraic Expression
2 plus c c more than 2	$2 + c$
k minus 5 k decreased by 5	$k - 5$
two times the product of 2 and x	$2x$
q divided by the sum of 6 and b the quotient of q and the sum of 6 and b	$\frac{q}{6+b}$

Expressions do not contain equal signs, but tell only which operations to perform. Equations always contain an equal sign.

Example: Write an equation for each sentence.

a. Eight multiplied by 2 equals 16.
$8 \times 2 = 16$

b. Three less than 7 times a number n is 24.
$7n - 3 = 24$

Write an algebraic expression for each verbal expression.

1. the difference of r and 10 $r - 10$
2. increase the product of 3 and a by 1 $3a + 1$

Write a verbal expression for each algebraic expression.

3. $\frac{p}{7}$ p divided by 7
4. $2x + 5$ five more than twice x
5. $\frac{1}{2}(4 + w)$ one-half the sum of 4 and w

Write an equation for each sentence.

6. Seven minus a number z is the same as 15. $7 - z = 15$
7. Ten more than twelve times a number h equals 25. $12h + 10 = 25$

There is a Study Guide, Skills Practice, Reading to Learn Mathematics, and Enrichment Master for every lesson in the Student Edition. Reduced facsimiles of these pages are shown in the Teacher's Wraparound Edition.

In-Class Examples

Examples 9–10

Write a sentence for each equation. **Sample answers are given.**

9 $j + 4 = 21$ **Four more than *j* is equal to 21.**

10 $3z - 12 = 11$ **The product of 3 and *z* decreased by 12 equals 11.**

3 PRACTICE/APPLY

Error Analysis

Watch for students who confuse algebraic expressions and numerical expressions in Exercise 1.

Prevent by explaining that algebraic expressions contain variables, while numerical expressions involve only numbers.

Answers

1. Sample answer: $3 + 7$, $10(6)$, $4 - 1$; $2x$, $5 + a$, gh
2. Sample answer: $5t = 10$, $8 - y = 12$, $4h + 3 = g$

Skills Practice, **p. 2, and *Practice,* p. 3 (shown)**

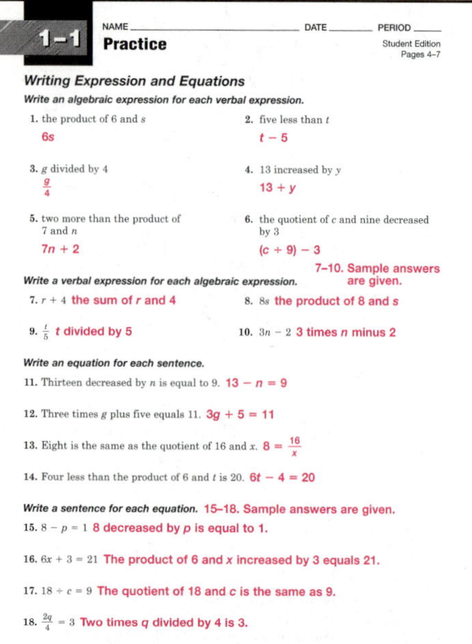

1-1 NAME ______________ DATE ________ PERIOD ______

Practice

Student Edition Pages 4–7

Writing Expression and Equations

Write an algebraic expression for each verbal expression.

1. the product of 6 and s — $6s$
2. five less than t — $t - 5$
3. g divided by 4 — $\frac{g}{4}$
4. 13 increased by y — $13 + y$
5. two more than the product of 7 and n — $7n + 2$
6. the quotient of c and nine decreased by 3 — $(c \div 9) - 3$

Write a verbal expression for each algebraic expression. 7–10. Sample answers are given.

7. $r + 4$ the sum of *r* and 4
8. $8s$ the product of 8 and *s*
9. $\frac{t}{5}$ *t* divided by 5
10. $3n - 2$ 3 times *n* minus 2

Write an equation for each sentence.

11. Thirteen decreased by n is equal to 9. $13 - n = 9$
12. Three times g plus five equals 11. $3g + 5 = 11$
13. Eight is the same as the quotient of 16 and x. $8 = \frac{16}{x}$
14. Four less than the product of 6 and t is 20. $6t - 4 = 20$

Write a sentence for each equation. 15–18. Sample answers are given.

15. $8 - p = 1$ 8 decreased by *p* is equal to 1.
16. $6x + 3 = 21$ The product of 6 and *x* increased by 3 equals 21.
17. $18 \div c = 9$ The quotient of 18 and *c* is the same as 9.
18. $\frac{2q}{4} = 3$ Two times *q* divided by 4 is 3.

Examples

Write a sentence for each equation.

9 $x - 2 = 14$
Two less than x is equivalent to 14.

10 $7y + 6 = 34$
Seven times y increased by 6 is 34.

Your Turn

f. $4b - 5 = 3$ **4 times *b* minus 5 is 3.**

Vocabulary words are highlighted in yellow. (See page 4.) Communicating Math includes a vocabulary list and an exercise relating to it.

Check for Understanding

Communicating Mathematics

1. **Write** three examples of numerical expressions and three examples of algebraic expressions.
2. **Write** three examples of equations.
3. Writing Math Write about a real-life application that can be expressed using an algebraic expression or an equation. **See students' work.**

1–2. See margin.

Vocabulary
algebraic expression
variable
numerical expression
factors
product
quotient
equation

Guided Practice

Examples 1 & 2
4–6. Sample answers are given.

Write an algebraic expression for each verbal expression.

4. t more than s **$s + t$**
5. the product of 7 and m **$7m$**
6. 11 decreased by the quotient of x and 2 **$11 - \frac{x}{2}$**

Examples 5 & 6
7–8. Sample answers are given.

Write a verbal expression for each algebraic expression.

7. $\frac{7}{q}$ **7 divided by *q***
8. $3\ell - 9$ **9 less than the product of 3 and ℓ**

Examples 7 & 8

Write an equation for each sentence.

9. A number m added to 6 equals 17. **$6 + m = 17$**
10. Ten is the same as four times r minus 6. **$10 = 4r - 6$**

Examples 9 & 10

11–12. Sample answers are given.

Write a sentence for each equation.

11. $5 + r = 15$ **11. 5 increased by *r* is 15.**
12. $\frac{4p}{3} = 12$ **12. 4 times *p* divided by 3 is equal to 12.**

Example 4

13. **Biology** The family of great white sharks has w different species. The blue shark family has nine times w plus three different species. Write an algebraic expression to represent the number of species in the blue shark family. **$9w + 3$**

Reteaching Activity

Interpersonal Learners Have pairs of students take turns writing an algebraic expression and showing it to their partner. The partner then recites a verbal expression for the algebraic expression. After several turns by each partner, the students can reverse the activity by reciting a verbal expression for their partner who then writes a corresponding algebraic expression.

Reteaching Activities provide alternative suggestions for presenting the lesson. The Reteaching Activities are keyed to eight commonly-accepted learning styles.

Exercises

Practice

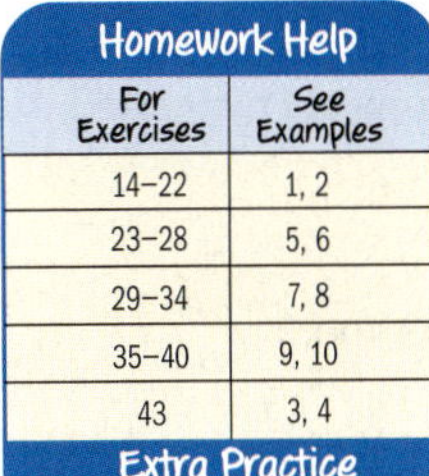

Homework Help

For Exercises	See Examples
14–22	1, 2
23–28	5, 6
29–34	7, 8
35–40	9, 10
43	3, 4

Extra Practice

See page 692.

Write an algebraic expression for each verbal expression.

14. twelve less than y $y - 12$
15. the product of r and s rs
16. the quotient of t and 5 $\frac{t}{5}$
17. three more than five times a
18. p plus the quotient of 9 and 5
19. the difference of 1 and n
20. f divided by y $\frac{f}{y}$
21. ten plus the product of h and 1
22. seven less than the quotient of j and p $\frac{j}{p} - 7$

17. $5a + 3$
18. $p + \frac{9}{5}$
19. $1 - n$

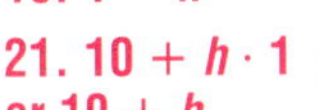

21. $10 + h \cdot 1$ or $10 + h$

23–28. See margin.

Write a verbal expression for each algebraic expression.

23. $9x$
24. $11 + b$
25. $6 - y$
26. $2m + 1$
27. $\frac{3}{r} - 8$
28. $16 - rt$

Write an equation for each sentence.

29. Three plus w equals 15.
30. Five times r equals 7.
31. Two is equal to seven divided by x. $2 = \frac{7}{x}$
32. Five less than the product of two and g equals nine. $2g - 5 = 9$
33. Three minus the product of five and y is the same as two times z.
34. The quotient of 19 and j is equal to the total of a and b and c.

29. $3 + w = 15$
30. $5r = 7$
33. $3 - 5y = 2z$
34. $\frac{19}{j} = a + b + c$

Write a sentence for each equation.

35. $3r = 18$
36. $g + 7 = 3$
37. $h = 10 - i$
38. $6v - 2 = 8$
39. $\frac{t}{4} = 16$
40. $10z + 7 = \frac{6}{r}$

41. Choose a variable and write an equation for *Four times a number minus seven equals the sum of 15 and c and two times the number.*

Applications and Problem Solving

42. **Biology** A smile requires 26 fewer muscles than a frown. Let f represent the number of muscles it takes for a frown and let s represent the number of muscles for a smile. Write an equation to represent the number of muscles a person uses to smile. $s = f - 26$

43. **Communication** A long-distance telephone call costs 20¢ for the first minute plus 10¢ for each additional minute.
    a. Write an expression for the total cost of a call that lasts 15 minutes.
    b. Write an expression for the total cost of a call that lasts m minutes.

44. **Critical Thinking** The ancient Hindus enjoyed number puzzles like the one below. **Source:** *Mathematical History*

    If 4 is added to a certain number, the result divided by 2, that result multiplied by 5, and then 6 subtracted from that result, the answer is 29. Can you find the number?

    a. Choose a variable and write an algebraic equation to represent the puzzle.
    b. Is your answer in part a the only correct way to write the equation? Explain.

35–40. See margin for sample answers.

41. Sample answer: Let x = the variable; $4x - 7 = 15 + c + 2x$

43a. $20 + 10(15 - 1)$

43b. $20 + 10(m - 1)$

44a. $\frac{n + 4}{2} \cdot 5 - 6 = 29$

44b. Sample answer: No, the division symbol and parentheses could be used rather than the fraction bar.

www.algconcepts.com/self_check_quiz

Assignment Guide

Basic: 15–43 odd, 44

Average: 14–40 even, 42–44

4 ASSESS

Open-Ended Assessment

Writing Have students write an algebraic equation involving both a sum and a product. Then have them translate their equation into a sentence.

Answers

For Exercises 23–28 and 35–40, sample answers are given.

23. the product of 9 and x
24. 11 increased by b
25. the difference of 6 and y
26. the product of 2 and m plus 1
27. 8 less than the quotient of 3 and r
28. 16 minus the product of r and t
35. 3 times r is 18.
36. 7 more than g equals 3.
37. h equals 10 minus i.
38. 6 times v minus 2 equals 8.
39. The quotient of t and 4 is 16.
40. The product of 10 and z plus 7 is equivalent to 6 divided by r.

***Enrichment*, p. 5**

1–1

NAME ______________ DATE ________ PERIOD ______

Enrichment

Student Edition Pages 4–7

Perfect, Excessive, Defective, and Amicable Numbers

A **perfect number** is the sum of all of its factors except itself. Here is an example.

$$28 = 1 + 2 + 4 + 7 + 14$$

There are very few perfect numbers. Most numbers are either *defective* or *excessive.*

An **excessive number** is greater than the sum of all of its factors except itself.

A **defective number** is less than this sum.

Two numbers are **amicable** if the sum of the factors of the first number, except for the number itself, equals the second number, and vice versa.

Solve each problem.

1. Write the perfect numbers between 0 and 31.
   6, 28
2. Write the excessive numbers between 0 and 31. 2, 3, 4, 5, 7, 8, 9, 10, 11, 13, 14, 15, 16, 17, 19, 21, 22, 23, 25, 26, 27, 29
3. Write the defective numbers between 0 and 31.
   12, 18, 20, 24, 30
4. Show that 8128 is a perfect number. $8128 = 1 + 2 + 4 + 8 + 16 + 32 + 64 + 127 + 254 + 508 + 1016 + 2032 + 4064$
5. The sum of the reciprocals of all the factors of a perfect number (including the number itself) equals 2. Show that this is true for the first two perfect numbers.
   $\frac{1}{1} + \frac{1}{2} + \frac{1}{3} + \frac{1}{6} = \frac{12}{6} = 2$ $\frac{1}{1} + \frac{1}{2} + \frac{1}{4} + \frac{1}{7} + \frac{1}{14} + \frac{1}{28} = \frac{56}{28} = 2$
6. More than 1000 pairs of amicable numbers have been found. One member of the first pair is 220. Find the other member.
   284
7. One member of the second pair of amicable numbers is 2620. Find the other member.
   2924
8. The Greek mathematician Euclid proved that the expression $2^{n-1}(2^n - 1)$ equals a perfect number if the expression inside the parentheses is prime. Use Euclid's expression with n equal to 19 to find the seventh perfect number.
   $2^{18}(2^{19} - 1) = 137,438,691,32$

Extra Credit

There are two variables, a and b, in the equation $6 + a = 10 - b$. If the value of a is 3, what is the value of b? 1

Lesson 1–2

1 FOCUS

5-Minute Check
Lesson 1–1

Write an algebraic expression for each verbal expression.

1. the product of 7 and n **$7n$**
2. 5 decreased by g **$5 - g$**
3. the sum of b and 19 **$b + 19$**

Write a verbal expression for each algebraic expression. **Sample answers are given.**

4. $(f + 3) \div 2$ **the sum of *f* and 3, divided by 2**
5. $12h$ **12 times *h***

Motivating the Lesson

Hands-On Activity Provide groups of 3–4 students with 20 paper squares of equal size. On the board or overhead, draw 3 groups of 6 objects. Draw a fourth group with 2 objects. Ask the students to model your drawings using their squares. Below your drawings, write the equality $6 + 6 + 6 + 2 = 3 \cdot 6 + 2$. Students should recognize that the expression on the left represents the total number of objects. Focus students' attention on just the expression $3 \cdot 6 + 2$. Ask them to imagine that someone comes into the classroom and evaluates the expression by adding 6 and 2 to get 8, then multiplies 8 by 3 and concludes that there are 24 objects in all. Ask students why the answer 24 is incorrect. Use students' suggestions to begin an introduction of the order of operations.

2 TEACH

Teaching Tip After discussing the paragraph above Example 1, point out that students can confirm Method 1 is correct by using addition to determine how many points Trisha scored: $9 + 9 + 9 + 9 + 9 + 4 = 49$.

1–2 Order of Operations

What You'll Learn
You'll learn to use the order of operations to evaluate expressions.

Why It's Important
Business
Businesspeople use the order of operations to determine the cost of renting a car.
See Exercise 18.

Some expressions have more than one operation. The value of the expression depends on the order in which the operations are evaluated. What is the value of $9 \cdot 5 + 4$?

Method 1		Method 2	
$9 \cdot 5 + 4 = 45 + 4$	*Multiply 9 and 5.*	$9 \cdot 5 + 4 = 9 \cdot 9$	*Add 5 and 4.*
$= 49$	*Add 45 and 4.*	$= 81$	*Multiply 9 and 9.*

Is the answer 49 or 81? The values are different because we multiplied and added in different orders in the two methods. To find the correct value of the expression, follow the **order of operations**.

Order of Operations

1. Find the values of expressions inside grouping symbols, such as parentheses ( ), brackets [ ], and as indicated by fraction bars.
2. Do all multiplications and/or divisions from left to right.
3. Do all additions and/or subtractions from left to right.

According to the order of operations, do multiplication and then addition. So, the value of the expression in Method 1 is correct. The value of the expression is 49.

Examples

Examples illustrate all of the concepts taught in the lesson and closely mirror the progression of Guided Practice and Exercises. Your Turn exercises help you check students' understanding.

Find the value of each expression.

1 $38 - 5 \cdot 6$

$38 - 5 \cdot 6 = 38 - 30$ *Multiply 5 and 6.*
$= 8$ *Subtract 30 from 38.*

2 $\dfrac{4 \times 9}{26 - 8}$

$\dfrac{4 \times 9}{26 - 8} = \dfrac{36}{18}$ *Evaluate the numerator and the denominator separately.*
$= 2$ *Divide 36 by 18.*

Your Turn

a. $7 \cdot 4 + 7 \cdot 3$ **49** b. $12 \div 3 \cdot 5 - 4$ **16** c. $\dfrac{6 + 12}{5(3) - 13}$ **9**

Resource Manager

Reproducible Masters
Chapter 1 Resource Masters
- *Study Guide,* p. 6
- *Skills Practice,* p. 7
- *Practice,* p. 8
- *Reading to Learn Mathematics,* p. 9
- *Enrichment,* p. 10

School-to-Workplace, p. 1

Transparencies
5-Minute Check, 1–2

Technology/Multimedia
AlgePASS, Lesson 6
Interactive Chalkboard CD-ROM

Example 3 — **Finance Link**

The order of operations is useful in solving problems in everyday life.

As a 16-year old, Trent Eisenberg ran his own consulting company called *F1 Computer*. Suppose he charged a flat fee of \$50, plus \$25 per hour. One day he worked 2 hours for one customer and the next day he worked 3 hours for the same customer. Find the value of the expression $50 + 25(2 + 3)$ to find the total amount of money he earned.

Source: *Scholastic Math*

$50 + 25(2 + 3) = 50 + 25(5)$ *Do the operation in parentheses first.*

$= 50 + 125$ *Multiply 25 and 5.*

$= 175$ *Add 50 and 125.*

Trent earned \$175.

In algebra, statements that are true for any number are called **properties**. Four properties of equality are listed in the table below.

Property of Equality	Symbols	Numbers
Substitution	If $a = b$, then a may be replaced by b.	If $9 + 2 = 11$, then $9 + 2$ may be replaced by 11.
Reflexive	$a = a$	$21 = 21$
Symmetric	If $a = b$, then $b = a$.	If $10 = 4 + 6$, then $4 + 6 = 10$.
Transitive	If $a = b$ and $b = c$, then $a = c$.	If $3 + 5 = 8$ and $8 = 2(4)$, then $3 + 5 = 2(4)$.

Examples

Name the property of equality shown by each statement.

4 **If $9 + 3 = 12$, then $12 = 9 + 3$.**

Symmetric Property of Equality

5 **If $z = 8$, then $z \div 4 = 8 \div 4$.**

Substitution Property of Equality *z is replaced by 8.*

Your Turn

d. $7 - c = 7 - c$ **Reflexive**

e. If $10 - 3 = 4 + 3$ and $4 + 3 = 7$, then $10 - 3 = 7$. **Transitive**

www.algconcepts.com/extra_examples

Internet addresses refer students to the Glencoe Algebra: Concepts and Applications Web site for extra examples and a self-check quiz for each lesson.

In-Class Examples

Examples 1–2

Find the value of each expression.

1 $14 + 10 \div 2$ **19**

2 $4 \times (6 + 7)$ **52**

Example 3

Refer to Example 3 in the Student Edition. Suppose that one day Trent worked 1 hour for each of 2 new customers. Find the value of the expression $2(50) + 2(25)$ to find the total amount of money he earned. **\$150**

Teaching Tip As you discuss the Substitution Property, point out that it was used three times in Example 3: when $2 + 3$ was replaced by 5 in the first step, when 25(5) was replaced by 125 in the second step, and when $50 + 125$ was replaced by 175 in the final step.

In-Class Examples

Examples 4–5

Name the property of equality shown by each statement.

4 If $k = 7$, then $k + 3 = 7 + 3$. **Substitution**

5 If $a + 4 = 9$, then $9 = a + 4$. **Symmetric**

An In-Class Example included for every example in the Student Edition exactly parallels the examples in the text.

In-Class Examples

Example 6

Find the value of $[25 + 8(12 - 11)] \div 11$. Identify the property used in each step.

$[25 + 8(12 - 11)] \div 11$
$= [25 + 8(1)] \div 11$ **Substitut.**
$= (25 + 8) \div 11$ **Mult. Ident.**
$= 33 \div 11$ **Substitution**
$= 3$ **Substitution**

Examples 7–8

Evaluate each expression if $x = 4$ and $y = 3$.

7 $xy + 8$ **20**

8 $(2y + 10) \div x$ **4**

These properties of numbers may help to find the value of expressions.

Property	Words	Symbols	Numbers
Additive Identity	When 0 is added to any number *a*, the sum is *a*.	For any number *a*, $a + 0 = 0 + a = a$.	$45 + 0 = 45$ $0 + 6 = 6$ *0 is the identity.*
Multiplicative Identity	When a number *a* is multiplied by 1, the product is *a*.	For any number *a*, $a \cdot 1 = 1 \cdot a = a$.	$12 \cdot 1 = 12$ $1 \cdot 5 = 5$ *1 is the identity.*
Multiplicative Property of Zero	If 0 is a factor, the product is 0.	For any number *a*, $a \cdot 0 = 0 \cdot a = 0$.	$7 \cdot 0 = 0$ $0 \cdot 23 = 0$

When two or more sets of grouping symbols are used, simplify within the innermost grouping symbols first.

Example

6 **Find the value of $5[3 - (6 \div 2)] + 14$. Identify the properties used.**

$5[3 - (6 \div 2)] + 14$
$= 5[3 - 3] + 14$ *Substitution Property of Equality*
$= 5(0) + 14$ *Substitution Property of Equality*
$= 0 + 14$ *Multiplicative Property of Zero*
$= 14$ *Additive Identity*

Your Turn **f.** $(22 - 15) \div 7 \cdot 9$ **9** **g.** $8 \div 4 \cdot 6(5 - 4)$ **12**

You can also apply the properties of numbers to find the value of an algebraic expression. This is called **evaluating** an expression. Replace the variables with known values and then use the order of operations.

Examples

Evaluate each expression if $a = 9$ and $b = 1$.

7 $7 + \left(\frac{a}{b} - 9\right)$

$7 + \left(\frac{a}{b} - 9\right) = 7 + \left(\frac{9}{1} - 9\right)$ *Replace a with 9 and b with 1.*
$= 7 + (9 - 9)$ *Substitution Property of Equality*
$= 7 + 0$ *Substitution Property of Equality*
$= 7$ *Additive Identity*

8 $(a + 4) - 3 \cdot b$

$(a + 4) - 3 \cdot b = (9 + 4) - 3 \cdot 1$ *Replace a with 9 and b with 1.*
$= (13) - 3 \cdot 1$ *Substitution Property of Equality*
$= 13 - 3$ *Multiplicative Identity*
$= 10$ *Substitution Property of Equality*

Your Turn **Evaluate each expression if $m = 8$ and $p = 2$.**

h. $6 \cdot p - m \div p$ **8** **i.** $[m + 2(3 + p)] \div 2$ **9**

Check for Understanding

Communicating Mathematics

1. **Name** two of the three types of grouping symbols discussed in this lesson.
2. **Translate** the verbal expression *six plus twelve divided by three* and *the sum of six and twelve divided by three* into numerical expressions. Use grouping symbols. Evaluate the expressions and explain why they are different.
3. Writing Math Label a section of your math journal "Toolbox." Record all properties given in this course, beginning with this lesson. See students' work.

Vocabulary
order of operations
properties
evaluating

1. parentheses, brackets, fraction bar

2. $6 + 12 \div 3 = 10$; $(6 + 12) \div 3 = 6$; Parentheses change the order of operations.

Guided Practice

Getting Ready State which operation to perform first.

Sample: $3 + 2 \cdot 4$ **Solution:** Multiply 2 and 4.

4. $8 \div 4 \cdot 2$ Divide 8 by 4.
5. $12 - 6 \cdot 2$ Multiply 6 and 2.
6. $5(7 + 7)$ Add 7 and 7.
7. $(10 - 4) \div 3$ Subtract 4 from 10.

Examples 1–3 Find the value of each expression.

8. $7 \cdot 4 + 3$ 31
9. $4(1 + 5) \div 8$ 3
10. $18 \div [3(11 - 8)]$ 2

Examples 4 & 5 Name the property of equality shown by each statement.

11. If $5 + 2n = 5 + 3$ and $5 + 3 = 2 \cdot 4$, then $5 + 2n = 2 \cdot 4$. Transitive
12. If $\frac{y}{2} = 19$, then $19 = \frac{y}{2}$. Symmetric

Example 6 Find the value of each expression. Identify the property used in each step. 13–14. See margin for properties.

13. $8(4 - 8 \div 2)$ 0
14. $5(2) \cdot (15 \div 15)$ 10

Examples 7 & 8 Evaluate each algebraic expression if $q = 4$ and $r = 1$.

15. $4(q - 2r)$ 8
16. $\frac{7q}{r + 3}$ 7
17. $r + \frac{q}{2} \cdot 6$ 13

Examples 7 & 8

18. **Car Rental** The cost to rent a car is given by the expression $25d + 0.10m$, where d is the number of days and m is the number of miles. If Teresa rents the car for five days and drives 300 miles, what is the cost? $155

Exercises

Practice

Find the value of each expression. 22. 40 23. 40

19. $36 \div 4 + 5$ 14
20. $16 - 4 \cdot 4$ 0
21. $4 + 7 \cdot 2 + 8$ 26
22. $42 - (24 \div 2) + 10$
23. $42 - 24 \div (2 + 10)$
24. $24 \div 12 \div 2 \cdot 5$ 5
25. $\frac{7(3 + 6)}{3}$ 21
26. $\frac{4(8 - 2)}{2 \times 2}$ 6
27. $38 - [3(9 + 1)]$ 8

3 PRACTICE/APPLY

Error Analysis

Watch for students who get $\frac{18}{25}$ as the answer to Exercise 10 because they simplify $3(11 - 8)$ as $33 - 8$. ***Prevent by*** stressing that when there are multiple sets of grouping symbols, such as parentheses, students should simplify the expression inside the innermost pair of grouping symbols first and work outward.

Assignment Guide

Basic: 19–53 odd, 54–64
Average: 20–50 even, 51–64
All: Quiz 1, 1–5

Check for Understanding exercises are meant to be completed in class. Getting Ready exercises help students practice prerequisite or subskills for the lesson concepts. The remainder of the Guided Practice exercises are representative of the Practice exercises.

***Study Guide*, p. 6**

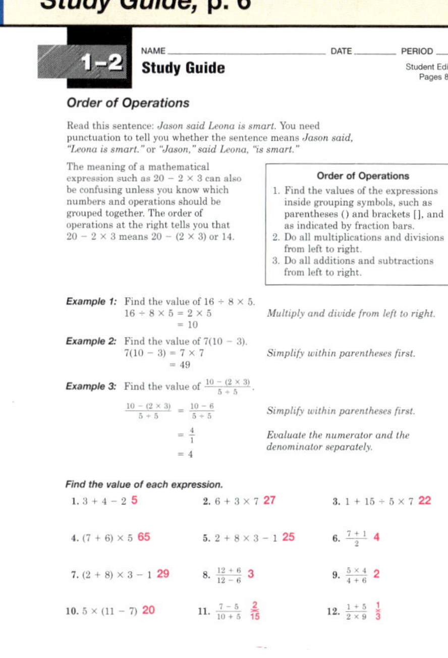

1–2 **Study Guide**

NAME ________ DATE ______ PERIOD ______

Student Edition Pages 8–13

Order of Operations

Read this sentence: *Jason said Leona is smart.* You need punctuation to tell you whether the sentence means *Jason said, "Leona is smart." or "Jason," said Leona, "is smart."*

The meaning of a mathematical expression such as $20 - 2 \times 3$ can also be confusing unless you know which numbers and operations should be grouped together. The order of operations at the right tells you that $20 - 2 \times 3$ means $20 - (2 \times 3)$ or 14.

Order of Operations
1. Find the values of the expressions inside grouping symbols, such as parentheses () and brackets [], and as indicated by fraction bars.
2. Do all multiplications and divisions from left to right.
3. Do all additions and subtractions from left to right.

Example 1: Find the value of $16 \div 8 \times 5$.
$16 \div 8 \times 5 = 2 \times 5$ *Multiply and divide from left to right.*
$= 10$

Example 2: Find the value of $7(10 - 3)$.
$7(10 - 3) = 7 \times 7$ *Simplify within parentheses first.*
$= 49$

Example 3: Find the value of $\frac{10 - (2 \times 3)}{5 + 5}$.
$\frac{10 - (2 \times 3)}{5 + 5} = \frac{10 - 6}{5 + 5}$ *Simplify within parentheses first.*
$= \frac{4}{1}$ *Evaluate the numerator and the denominator separately.*
$= 4$

Find the value of each expression.

1. $3 + 4 - 2$ 5
2. $6 + 3 \times 7$ 27
3. $1 + 15 \div 5 \times 7$ 22
4. $(7 + 6) \times 5$ 65
5. $2 + 8 \times 3 - 1$ 25
6. $\frac{7 + 1}{2}$ 4
7. $(2 + 8) \times 3 - 1$ 29
8. $\frac{12 + 6}{12 - 6}$ 3
9. $\frac{5 \times 4}{4 + 6}$ 2
10. $5 \times (11 - 7)$ 20
11. $\frac{7 - 5}{10 + 5}$ $\frac{2}{15}$
12. $\frac{1 + 5}{2 \times 9}$ $\frac{1}{3}$

Reteaching Activity

Auditory/Musical Learners Have students write a poem or a verse for a rap or rock song about the order of operations.

Answers

13. $8(4 - 8 \div 2)$
$= 8(4 - 4)$ Substitution
$= 8(0)$ Substitution
$= 0$ Multiplicative Property of Zero

14. $5(2) \cdot (15 \div 15)$
$= 10 \cdot 1$ Substitution
$= 10$ Multiplicative Identity

Answers

34. $7(10 - 1 \cdot 3)$
$= 7(10 - 3)$ Multiplicative Identity
$= 7(7)$ Substitution
$= 49$ Substitution

35. $8(9 - 3 \cdot 2)$
$= 8(9 - 6)$ Substitution
$= 8(3)$ Substitution
$= 24$ Substitution

36. $19 - 15 \div 5 \cdot 2$
$= 19 - 3 \cdot 2$ Substitution
$= 19 - 6$ Substitution
$= 13$ Substitution

37. $10(6 - 5) - (20 \div 2)$
$= 10(1) - 10$ Substitution
$= 10 - 10$ Multiplicative Identity
$= 0$ Substitution

38. $\dfrac{9 \cdot 9 - 1}{3(1 + 2) - 1}$
$= \dfrac{81 - 1}{3(3) - 1}$ Substitution
$= \dfrac{80}{9 - 1}$ Substitution
$= \dfrac{80}{8}$ Substitution
$= 10$ Substitution

39. $6(12 - 48 \div 4) + 7 \cdot 1$
$= 6(12 - 12) + 7$ Substitution
$= 6(0) + 7$ Substitution
$= 0 + 7$ Multiplicative Property of Zero
$= 7$ Additive Identity

***Skills Practice*, p. 7, *and Practice*, p. 8 (shown)**

1-2 Practice

NAME ______________ DATE ________ PERIOD _____

Student Edition Pages 8–13

Order of Operations

Find the value of each expression.

1. $16 \div 4 - 3$ **1**	2. $6 + 9 \cdot 2$ **24**
3. $3(8 - 4) \div 2$ **6**	4. $6 \cdot 2 \div 3 + 1$ **5**
5. $21 \div [7(12 - 9)]$ **1**	6. $\frac{7 + 5}{3 \cdot 2}$ **2**

Name the property of equality shown by each statement.

7. $4 + d = 4 + d$ **Reflexive**

8. If $\frac{y}{3} = 9$ and $y = 27$, then $\frac{27}{3} = 9$. **Substitution**

9. If $3c + 1 = 7$, then $7 = 3c + 1$. **Symmetric**

10. If $8 - n = 3 + 1$ and $3 + 1 = 2 \cdot 2$, then $8 - n = 2 \cdot 2$. **Transitive**

Find the value of each expression. Identify the property used in each step.

11. $6(9 - 27 \div 3)$
$= 6(9 - 9)$ Substitution
$= 6(0)$ Substitution
$= 0$ Mult. Prop. of Zero

12. $4(16 \div 16) + 3$
$= 4(1) + 3$ Substitution
$= 4 + 3$ Multiplicative Identity
$= 7$ Substitution

13. $5 + (3 - 6 \div 2)$
$= 5 + (3 - 3)$ Substitution
$= 5 + (0)$ Substitution
$= 5$ Additive Identity

14. $8 \div 2 \cdot 7(9 - 8)$
$= 8 \div 2 \cdot 7(1)$ Substitution
$= 4 \cdot 7(1)$ Substitution
$= 4 \cdot 7$ Multiplicative Identity
$= 28$ Substitution

Evaluate each algebraic expression if s = 5 and t = 3.

15. $3(2s - t)$ **21**	16. $\frac{4s}{t - 1}$ **10**
17. $s + 3t - 8$ **6**	18. $s - \frac{t}{3} \cdot 5$ **0**
19. $(s + t) - 2 \cdot 3$ **2**	20. $3s - 4t + 2$ **5**

Homework Help

For Exercises	See Examples
19–27	1–3
28–33	4, 5
34–39	6
40–49, 51, 53	7, 8

Extra Practice
See page 692.

Name the property of equality shown by each statement.

28. If $x + 3 = 5$ and $x = 2$, then $2 + 3 = 5$. **Substitution**

29. $8t - 1 = 8t - 1$ **Reflexive**

30. If $6 = 3 + 3$, then $3 + 3 = 6$. **Symmetric**

31. $\frac{20 - 2}{9} = \frac{18}{9}$ **Substitution**

32. If $4 \cdot (7 - 7) = 4 \cdot 0$ and $4 \cdot 0 = 0$, then $4 \cdot (7 - 7) = 0$. **Transitive**

33. $a + 1 = 15x$ and $15x = 30$, so $a + 1 = 30$. **Transitive**

Find the value of each expression. Identify the property used in each step. 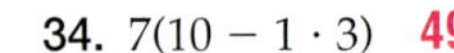 **34–39. See margin for properties.**

B

34. $7(10 - 1 \cdot 3)$ **49**	35. $8(9 - 3 \cdot 2)$ **24**
36. $19 - 15 \div 5 \cdot 2$ **13**	37. $10(6 - 5) - (20 \div 2)$ **0**
38. $\frac{9 \cdot 9 - 1}{3(1 + 2) - 1}$ **10**	39. $6(12 - 48 \div 4) + 7 \cdot 1$ **7**

Evaluate each algebraic expression if $j = 5$ and $s = 2$.

40. $7j - 3s$ **29**	41. $j(3s + 4)$ **50**	42. $j + 5s - 7$ **8**
43. $\frac{9 \cdot 4 + 5 \cdot s}{7 - j}$ **23**	44. $\frac{14 + s}{2(j - 1)}$ **2**	45. $\frac{4js}{s - 1}$ **40**
46. $50 \div js + 6$ **11**	47. $(3s - j)(5s - j)$ **5**	
48. $[3j - s(4 + s)] \div 3$ **1**	49. $2[16 - (j - s)]$ **26**	

C

50. a. Write an algebraic expression for *nine added to the quantity three times the difference of a and b.* **$3(a - b) + 9$**
b. Let $a = 4$ and $b = 1$. Evaluate the expression in part a. **18**

Applications and Problem Solving

51. **Real Estate** The Phams own a \$150,000 home in Rochester, New York, and plan to move to San Diego, California. How much will a similar home in San Diego cost? Evaluate the expression $150{,}000 \div a \times b$ for $a = 79$ and $b = 164$ to find the answer to the nearest dollar.
Source: *USA TODAY* **\$311,392**

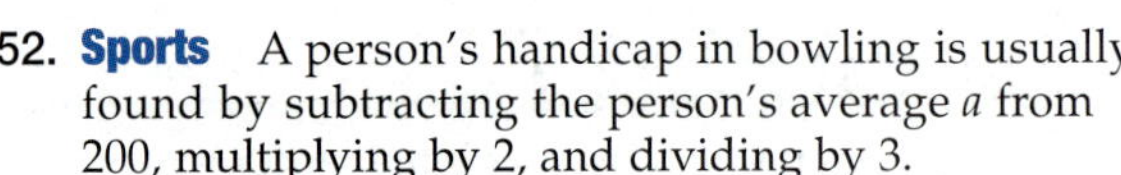

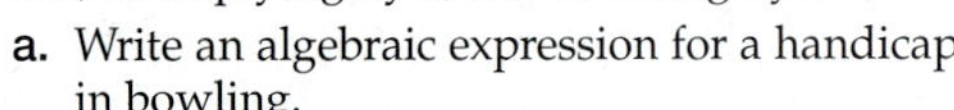

52. **Sports** A person's handicap in bowling is usually found by subtracting the person's average a from 200, multiplying by 2, and dividing by 3.
a. Write an algebraic expression for a handicap in bowling.
b. Find a person's handicap whose average is 170. **20**

52a. $\frac{2(200 - a)}{3}$

53. **Gardening** Mr. Martin is building a fence around a rectangular garden, as shown at the left. Evaluate the expression $2\ell + 2w$, where ℓ represents the length and w represents the width, to find how much fencing he needs. **44 feet**

54b. No; $1 < 2$ is true, but $2 < 1$ is false.

54. **Critical Thinking** The symbol < means "is less than." Are the following properties of equality true for statements containing this symbol? Give examples to explain. a. No; $5 < 5$ is false.

    a. Reflexive b. Symmetric c. Transitive

Mixed Review

54c. Yes; if $6 < 7$ and $7 < 8$, then $6 < 8$ is true.

55–57. Sample answers are given.

57. The quotient of 25 and n is 5.

61. $b + 10 - 1 = 18$

Write a sentence for each equation. *(Lesson 1–1)*

55. $x + 8 = 12$ Eight more than x is 12.
56. $2y = 16$ Two times y is 16.
57. $25 \div n = 5$

Write an equation for each sentence. *(Lesson 1–1)*

58. Six more than g is 22. $g + 6 = 22$
59. Three times c equals 27. $3c = 27$
60. Two is the same as the quotient of 8 and x. $2 = 8 \div x$ or $2 = \frac{8}{x}$
61. b increased by 10 and then decreased by 1 is equivalent to 18.

62. **Time** How many seconds are there in a day? *(Lesson 1–1)*
    a. Write an expression to answer this question. $60 \cdot 60 \cdot 24$
    b. Evaluate the expression. 86,400 s

Standardized Test Practice

Ⓐ Ⓑ Ⓒ Ⓓ

63. **Extended Response** Lincoln's Gettysburg Address began "Four score and seven years ago, . . ." *(Lesson 1–1)* a. $4(20) + 7$
    a. A score is 20. Write a numerical expression for the phrase.
    b. Evaluate the expression to find the number of years. 87

64. **Multiple Choice** At the movie theater, the price for an adult ticket a is \$1.50 less than two times the price of a student ticket s. Choose the algebraic expression that represents the price of an adult ticket in terms of the price of a student ticket. *(Lesson 1–2)* D

    Ⓐ $1.50 - 2s$ Ⓑ $2(s - 1.50)$
    Ⓒ $2s + 2(1.50)$ Ⓓ $2s - 1.50$

Quiz 1 Lessons 1–1 and 1–2

1. Write an algebraic expression for *five less than the product of two and v.* *(Lesson 1–1)* $2v - 5$
2. Write an equation for *the sum of nine and y equals 16.* *(Lesson 1–1)* $9 + y = 16$

Evaluate each algebraic expression if $j = 5$ and $s = 2$. *(Lesson 1–2)*

3. $3(j - s) + 4$ 13
4. $[(11 - j \div 1) + 8] \cdot s$ 28
5. **Carpentry** Ana Martinez is putting molding around the ceiling of her family room. The room measures 12 feet by 16 feet. Evaluate the expression $2\ell + 2w$, where ℓ is the length and w is the width, to find how much molding Ana needs. *(Lesson 1–2)* 56 ft

www.algconcepts.com/self_check_quiz

4 ASSESS

Open-Ended Assessment

Modeling Have students model one of the properties of equality shown in the lesson using manipulatives.

Quiz 1

The Quiz provides students with a brief review of the concepts and skills in Lessons 1–1 and 1–2. Lesson numbers are given to the right of the exercises or instruction lines so students can review concepts not yet mastered.

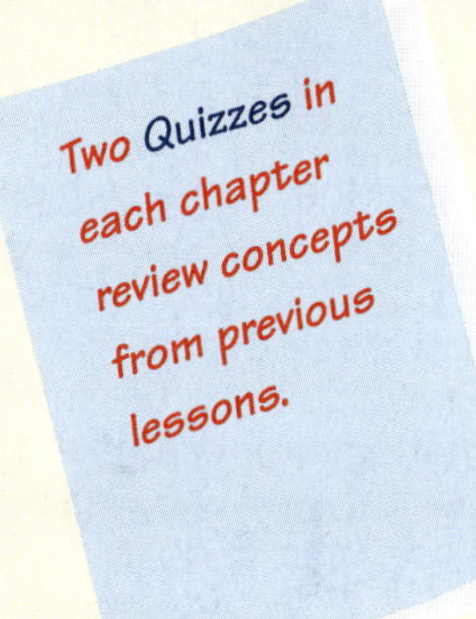

Extra Credit

Using grouping symbols, the operation symbols +, −, ×, and ÷, and the digits 2, 3, and 4, write five expressions whose values are 1, 2, 3, 4, and 5, respectively. Sample answers: $2 + 3 - 4 = 1$; $(4 - 3) \times 2 = 2$; $4 + 2 - 3 = 3$; $(3 - 2) \times 4 = 4$; $(3 - 2) + 4 = 5$

Enrichment, p. 10

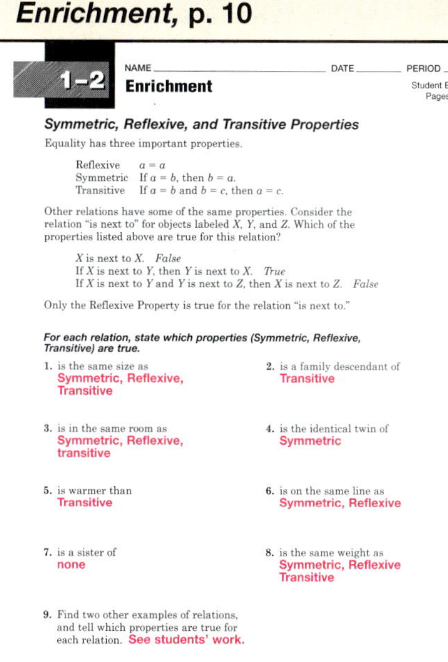
1–2 NAME ______ DATE ______ PERIOD ______

Enrichment

Student Edition Pages 8–13

Symmetric, Reflexive, and Transitive Properties

Equality has three important properties.

Reflexive $a = a$
Symmetric If $a = b$, then $b = a$.
Transitive If $a = b$ and $b = c$, then $a = c$.

Other relations have some of the same properties. Consider the relation "is next to" for objects labeled X, Y, and Z. Which of the properties listed above are true for this relation?

X is next to X. *False*
If X is next to Y, then Y is next to X. *True*
If X is next to Y and Y is next to Z, then X is next to Z. *False*

Only the Reflexive Property is true for the relation "is next to."

For each relation, state which properties (Symmetric, Reflexive, Transitive) are true.

1. is the same size as Symmetric, Reflexive, Transitive
2. is a family descendant of Transitive
3. is in the same room as Symmetric, Reflexive, transitive
4. is the identical twin of Symmetric
5. is warmer than Transitive
6. is on the same line as Symmetric, Reflexive
7. is a sister of none
8. is the same weight as Symmetric, Reflexive Transitive
9. Find two other examples of relations, and tell which properties are true for each relation. See students' work.

Lesson 1–3

1 FOCUS

5-Minute Check
Lesson 1–2

Find the value of each expression.

1. $50 - (20 \div 4)$ **45**

2. $3 + 2 \cdot 8$ **19**

3. $[4(7 + 3) + 2] \div 7$ **6**

Evaluate each algebraic expression if $m = 4$ and $n = 5$.

4. $6m + n$ **29**

5. $\frac{m + 1}{n}$ **1**

Motivating the Lesson

Real-World Connection Tell students that one way to think of the Commutative Property of Addition is that the order in which you add two numbers does not affect their sum. Ask students to think of pairs of real-world activities for which the order they are done matters. For example, a person must get their hair wet before they can shampoo it.

2 TEACH

Teaching Tip As you introduce the Associative Property of Addition, show students how looking for combinations of numbers whose sum is 10 can make simplifying numerical expressions easier.

The Resource Manager provides a snapshot of the four-page Resource Manager at the beginning of the chapter. It lists all of the resources available for this lesson.

1–3 Commutative and Associative Properties

What You'll Learn

You'll learn to use the commutative and associative properties to simplify expressions.

Why It's Important

Construction Lumber yards use the commutative and associative properties to determine the amount of wood to order. *See Exercise 25.*

The **Commutative Property of Addition** states that the sum of two numbers does not depend on the order in which they are added. In the example below, adding 35 and 50 in either order does not change the sum.

$$35 + 50 = 85$$
$$50 + 35 = 85$$

This example illustrates the Commutative Property of Addition.

Commutative Property of Addition		
	Words:	The order in which two numbers are added does not change their sum.
	Symbols:	For any numbers a and b, $a + b = b + a$.
	Numbers:	$5 + 7 = 7 + 5$

Likewise, the order in which you multiply numbers does not matter.

Commutative Property of Multiplication		
	Words:	The order in which two numbers are multiplied does not change their product.
	Symbols:	For any numbers a and b, $a \cdot b = b \cdot a$.
	Numbers:	$3 \cdot 10 = 10 \cdot 3$

Concept boxes highlight definitions, formulas, and other important ideas. Multiple representations—words, symbols, numbers, models—reach students of all learning styles.

Some expressions are easier to evaluate if you group or *associate* certain numbers. Look at the expression below.

$16 + 7 + 3 = 16 + (7 + 3)$ *Group 7 and 3.*

$= 16 + 10$ *Add 7 and 3.*

$= 26$ *Add 16 and 10.*

This is an application of the **Associative Property of Addition**.

Associative Property of Addition		
	Words:	The way in which three numbers are grouped when they are added does not change their sum.
	Symbols:	For any numbers a, b, and c, $(a + b) + c = a + (b + c)$.
	Numbers:	$(24 + 8) + 2 = 24 + (8 + 2)$

Resource Manager

Reproducible Masters

Chapter 1 Resource Masters
- *Study Guide,* p. 11
- *Skills Practice,* p. 12
- *Practice,* p. 13
- *Reading to Learn Mathematics,* p. 14
- *Enrichment,* p. 15
- *Assessment,* p. 47

Transparencies

5-Minute Check, 1–3

Technology/Multimedia

Interactive Chalkboard CD-ROM

The Associative Property also holds true for multiplication.

Associative Property of Multiplication		
	Words:	The way in which three numbers are grouped when they are multiplied does not change their product.
	Symbols:	For any numbers a, b, and c, $(a \cdot b) \cdot c = a \cdot (b \cdot c)$.
	Numbers:	$(9 \cdot 4) \cdot 25 = 9 \cdot (4 \cdot 25)$

Examples

Name the property shown by each statement.

1 $4 \cdot 11 \cdot 2 = 11 \cdot 4 \cdot 2$ Commutative Property of Multiplication

2 $(n + 12) + 5 = n + (12 + 5)$ Associative Property of Addition

Your Turn a. Assoc. (×) b. Comm. (+)

a. $(5 \cdot 4) \cdot 3 = 5 \cdot (4 \cdot 3)$ b. $16 + t + 1 = 16 + 1 + t$

You can use the Commutative and Associative Properties to simplify and evaluate algebraic expressions. To **simplify** an expression, eliminate all parentheses first and then add, subtract, multiply, or divide.

Example

Simplify the expression $15 + (3x + 8)$. Identify the properties used in each step.

$15 + (3x + 8) = 15 + (8 + 3x)$ *Commutative Property of Addition*
$= (15 + 8) + 3x$ *Associative Property of Addition*
$= 23 + 3x$ *Substitution Property*

c. $7 + 6 + 9 + 2a$; Commutative (+)
d. $x \cdot (5 \cdot 20)$; Associative (×), Commutative (×)

Your Turn

Simplify each expression. Identify the properties used in each step.

c. $7 + 2a + 6 + 9$ **$22 + 2a$** d. $(x \cdot 5) \cdot 20$ **$100x$**

Example Geometry Link

4 **The volume of a box can be found using the expression $\ell \times w \times h$, where ℓ is the length, w is the width, and h is the height. Find the volume of a box whose length is 30 inches, width is 6 inches, and height is 5 inches.**

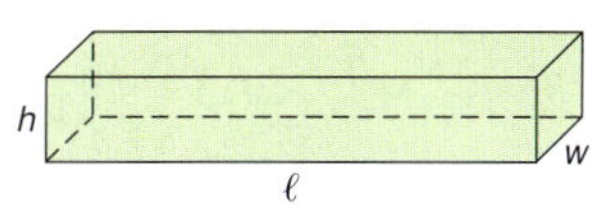

$\ell \times w \times h = 30 \times 6 \times 5$ *Replace ℓ with 30, w with 6, and h with 5.*
$= 30 \times (6 \times 5)$ *Associative Property of Multiplication*
$= 30 \times 30$ *Substitution Property*
$= 900$ *Substitution Property*

The volume of the box is 900 cubic inches.

www.algconcepts.com/extra_examples

Teaching Tip After discussing the Associative Property of Multiplication, write the words *commute* and *associate* on the board or overhead. Then write these definitions.

commute—to travel back and forth
associate—to join or connect together; to combine

Stress that commute implies movement, while associate implies grouping. Refocus students' attention on the Commutative Properties and point out that the numbers "move" because their order changes. Then focus students' attention on the Associative Properties and point out that the numbers remain in the same order even though they have been regrouped.

In-Class Examples

Examples 1–2

Name the property shown by each statement.

1 $8 + (3 + 4) = (3 + 4) + 8$ **Commutative Property of Addition**

2 $7 \cdot (8 \cdot k) = (7 \cdot 8) \cdot k$ **Associative Property of Multiplication**

Example 3

Simplify the expression $(4 \cdot m) \cdot 9$. Identify the properties used in each step.

$(4 \cdot m) \cdot 9$
$= (m \cdot 4) \cdot 9$ **Commutative (×)**
$= m \cdot (4 \cdot 9)$ **Associative (×)**
$= m \cdot 36$ **Substitution**
$= 36m$ **Commutative (×)**

Example 4

Find the volume of a box whose length is 20 inches, width is 12 inches, and height is 3 inches. **720 in^3**

Teaching Tip As you discuss the Closure Property of Whole Numbers, multiply and add several pairs of whole numbers on the board or overhead to show that the sums and products are also whole numbers. Ask students to try to think of any exceptions to the Closure Property of Whole Numbers.

In-Class Example

Example 5

State whether the statement *Subtraction of whole numbers is commutative* is *true* or *false*. If false, provide a counterexample.

false; sample counterexample: $7 - 5 \neq 5 - 7$

Answer

e. **false; sample counterexample:**

$$(12 - 6) - 2 \stackrel{?}{=} 12 - (6 - 2)$$
$$6 - 2 \stackrel{?}{=} 12 - 4$$
$$4 \neq 8$$

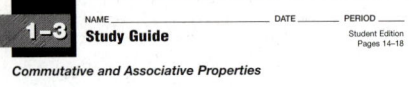

Whole numbers are the numbers 0, 1, 2, 3, 4, and so on. When you add whole numbers, the sum is always a whole number. Likewise, when you multiply whole numbers, the product is a whole number. This is an example of the **Closure Property**. We say that the whole numbers are *closed* under addition and multiplication.

Closure Property of Whole Numbers		
	Words:	Because the sum or product of two whole numbers is also a whole number, the set of whole numbers is closed under addition and multiplication.
	Numbers:	$2 + 5 = 7$, and 7 is a whole number. $2 \cdot 5 = 10$, and 10 is a whole number.

Are the whole numbers closed under division? Study these examples.

$2 \div 1 = 2$ *whole number*

$28 \div 4 = 7$ *whole number*

$5 \div 3 = \frac{5}{3}$ *fraction*

It is impossible to list every possible division expression to prove that the Closure Property holds true. However, we can easily show that the statement is false by finding one **counterexample**. A counterexample is an example that shows the statement is not true. Consider $5 \div 3$ or $\frac{5}{3}$. While 5 and 3 are whole numbers, $\frac{5}{3}$ is not. So, the statement *The whole numbers are closed under division* is false.

Example 5

State whether the statement *Division of whole numbers is commutative* is *true* or *false*. If false, provide a counterexample.

Write two division expressions using the Commutative Property and check to see whether they are equal.

$6 \div 3 \stackrel{?}{=} 3 \div 6$ *Evaluate each expression separately.*

$2 \neq \frac{1}{2}$ *$6 \div 3 = 2$ and $3 \div 6 = \frac{1}{2}$*

We found a counterexample, so the statement is false. Division of whole numbers is *not* commutative.

Your Turn **e. See margin.**

e. State whether the statement *Subtraction of whole numbers is associative* is *true* or *false*. If false, provide a counterexample.

***Study Guide*, p. 11**

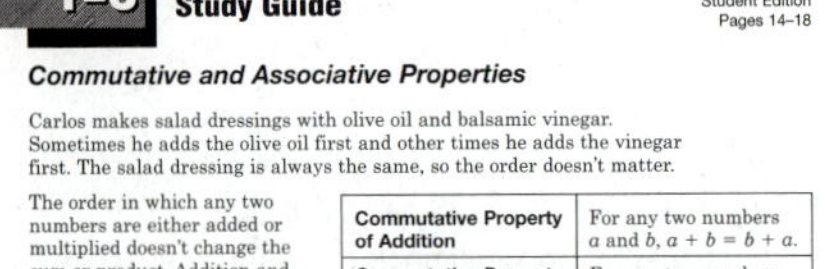

1-3 NAME ________ DATE ______ PERIOD _____

Study Guide — Student Edition Pages 14–18

Commutative and Associative Properties

Carlos makes salad dressings with olive oil and balsamic vinegar. Sometimes he adds the olive oil first and other times he adds the vinegar first. The salad dressing is always the same, so the order doesn't matter.

The order in which any two numbers are either added or multiplied doesn't change the sum or product. Addition and multiplication are said to be *commutative*.

Commutative Property of Addition	For any two numbers a and b, $a + b = b + a$.
Commutative Property of Multiplication	For any two numbers a and b, $a \cdot b = b \cdot a$.

Example 1: $3x + 4y + 5x$
$= 3x + 5x + 4y$ *Commutative (+)*
$= 8x + 4y$

Example 2: $5 \times 11 \times 2$
$= 5 \times 2 \times 11$ *Commutative (×)*
$= 10 \times 11$
$= 110$

You can also regroup numbers when you are adding or multiplying without changing the sum or product. Addition and multiplication are said to be *associative*.

Associative Property of Addition	For any numbers a, b, and c, $(a + b) + c = a + (b + c)$.
Associative Property of Multiplication	For any numbers a, b, and c, $(a \cdot b) \cdot c = a \cdot (b \cdot c)$.

Example 3: $(10 + 5) + 8$
$= 10 + (5 + 8)$ *Associative (+)*
$= 23$

Example 4: $(100 \cdot 4) \cdot 5$
$= 100 \cdot (4 \cdot 5)$ *Associative (×)*
$= 2000$

Simplify each expression.

1. $(d + 7) + 3$ **$d + 10$**
2. $2x + 7 + 5x$ **$7x + 7$**
3. $2 \times 7k \times 5$ **$70k$**
4. $(4a + 2b) + (a + b)$ **$5a + 3b$**
5. $7 \cdot y \cdot 3$ **$21y$**
6. $(8 \times m) \times 4$ **$32m$**

Name the property shown by each statement.

7. $29 + b = b + 29$ **Commutative Property of Addition**
8. $2(4 \cdot 6) = (2 \cdot 4)6$ **Associative Property of Multiplication**
9. $(3 + 21) + 7 = (3 + 7) + 21$ **Associative Property of Addition**
10. $42 \cdot 8 = 8 \cdot 42$ **Commutative Property of Multiplication**

Reteaching Activity

Visual/Spatial Learners Have students make a poster displaying the Commutative and Associative Properties using markers of two different colors. Instruct them to write the two Commutative Properties in one color and the two Associative Properties in the second color. Then have students evaluate some numerical expressions, underlining each step in the color of the property that justifies the step.

Check for Understanding

Communicating Mathematics

1. **Describe** what is meant by the statement *The whole numbers are closed under multiplication.*
   1. For any whole numbers that are multiplied, the product is a whole number.
2. **Write** an equation that illustrates the Commutative Property of Addition. Sample answer: $x + 6 = 6 + x$
3. **You Decide** Abeque says that the expression $(7 \cdot 2) + 5$ equals $7 \cdot (2 + 5)$ because of the Associative Properties of Addition and Multiplication. Jessie disagrees with her. Who is correct? Explain. See margin.

Vocabulary
simplify
whole numbers
counterexample

Guided Practice

Name the property shown by each statement.

Examples 1 & 2

4. $27 + 59 = 59 + 27$ Commutative (+)
5. $(8 + 7) + 3 = 8 + (7 + 3)$ Associative (+)

Example 3

Simplify each expression. Identify the properties used in each step.

6. $(n \cdot 2) \cdot 10$ $n \cdot (2 \cdot 10)$; $20n$
7. $(3 + p + 47)(7 - 6)$ $(3 + 47 + p)(7 - 6)$; $50 + p$

6. Associative (×)
7. Commutative (+)

Example 5

8. State whether the statement *Whole numbers are closed under subtraction* is *true* or *false*. If false, provide a counterexample. False; $2 - 3 = -1$ and -1 is not a whole number.

Example 4

Active volcano in Hawaii

9. **Geology** The table shows the number of volcanoes in the United States and Mexico.
   a. Find the total number of volcanoes in these two countries mentally. 188
   b. Describe the properties you used to add the numbers. See margin.

Location	Number of Volcanoes
U.S. Mainland	69
Alaska	80
Hawaii	8
Mexico	31

Source: *Kids Discover Volcanoes*

10. Assoc. (×) 11. Comm. (+) 12. Comm. (×) 13. Comm. (×) 14. Assoc. (+) 15. Assoc. (×) 16. Assoc. (+) 17. Assoc. (×) 18. Comm. (+) 19. Assoc. (+)

Exercises

Practice

A

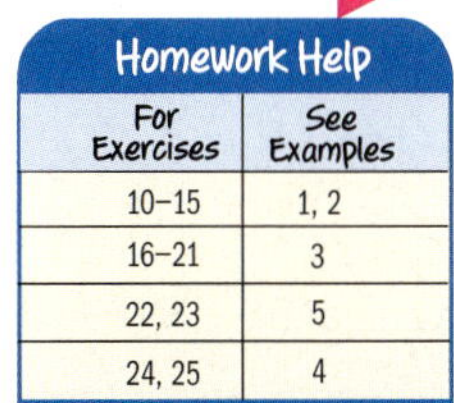
Homework Help

For Exercises	See Examples
10–15	1, 2
16–21	3
22, 23	5
24, 25	4

Extra Practice
See page 692.

Name the property shown by each statement.

10. $(9 \cdot 5) \cdot 20 = 9 \cdot (5 \cdot 20)$
11. $a + 14 = 14 + a$
12. $r \cdot s = s \cdot r$
13. $4 \times 15 \times 25 = 4 \times 25 \times 15$
14. $(7 + 5) + 5 = 7 + (5 + 5)$
15. $c \cdot (d \cdot 10) = (c \cdot d) \cdot 10$

Simplify each expression. Identify the properties used in each step.

16. $h + 1 + 9$ $h + (1 + 9)$; $h + 10$
17. $(r \cdot 30) \cdot 5$ $r \cdot (30 \cdot 5)$; $150r$
18. $17 + k + 23$ $17 + 23 + k$; $40 + k$
19. $6 + (3 + y)$ $(6 + 3) + y$; $9 + y$
20. $2 \cdot (19p)$ $(2 \cdot 19) \cdot p$; $38p$ Assoc. (×)
21. $2 \cdot j \cdot 7$ $2 \cdot 7 \cdot j$; $14j$ Comm. (×)

3 PRACTICE/APPLY

Error Analysis

Watch for students who confuse the Commutative and Associative Properties in Exercises 4–7.
Prevent by pointing out that when either Associative Property is used, the order of the addends or factors does not change but the parentheses move; and when either Commutative Property is used, the order of the addends or factors does change.

Error Analysis alerts you to students' common mistakes and provides strategies for helping them avoid those mistakes. The Assignment Guide provides suggestions for the exercises that are appropriate for a basic course or an average course.

Assignment Guide

Basic: 11–25 odd, 27–33
Average: 10–22 even, 24–33

***Skills Practice*, p. 12, and *Practice*, p. 13 (shown)**

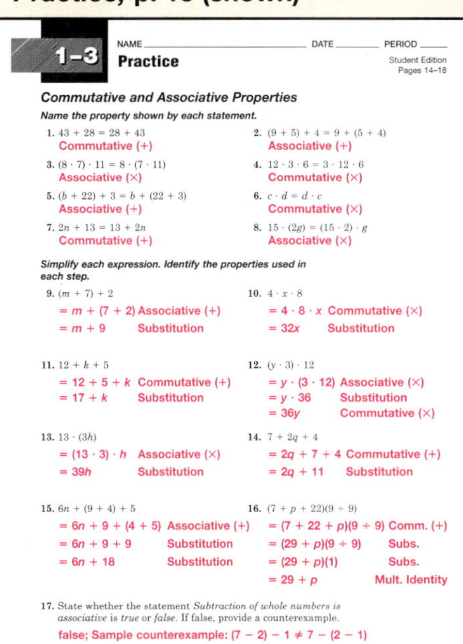
NAME ______ DATE ______ PERIOD ______

1–3 Practice

Student Edition Pages 14–18

Commutative and Associative Properties

Name the property shown by each statement.

1. $43 + 28 = 28 + 43$ Commutative (+)
2. $(9 + 5) + 4 = 9 + (5 + 4)$ Associative (+)
3. $(8 \cdot 7) \cdot 11 = 8 \cdot (7 \cdot 11)$ Associative (×)
4. $12 \cdot 3 \cdot 6 = 3 \cdot 12 \cdot 6$ Commutative (×)
5. $(b + 22) + 3 = b + (22 + 3)$ Associative (+)
6. $c \cdot d = d \cdot c$ Commutative (×)
7. $2n + 13 = 13 + 2n$ Commutative (+)
8. $15 \cdot (2g) = (15 \cdot 2) \cdot g$ Associative (×)

Simplify each expression. Identify the properties used in each step.

9. $(m + 7) + 2$
   $= m + (7 + 2)$ Associative (+)
   $= m + 9$ Substitution
10. $4 \cdot x \cdot 8$
    $= 4 \cdot 8 \cdot x$ Commutative (×)
    $= 32x$ Substitution
11. $12 + k + 5$
    $= 12 + 5 + k$ Commutative (+)
    $= 17 + k$ Substitution
12. $(y \cdot 3) \cdot 12$
    $= y \cdot (3 \cdot 12)$ Associative (×)
    $= y \cdot 36$ Substitution
    $= 36y$ Commutative (×)
13. $13 \cdot (3h)$
    $= (13 \cdot 3) \cdot h$ Associative (×)
    $= 39h$ Substitution
14. $7 + 2q + 4$
    $= 2q + 7 + 4$ Commutative (+)
    $= 2q + 11$ Substitution
15. $6n + (9 + 4) + 5$
    $= 6n + 9 + (4 + 5)$ Associative (+)
    $= 6n + 9 + 9$ Substitution
    $= 6n + 18$ Substitution
16. $(7 + p + 22)(9 \div 9)$
    $= (7 + 22 + p)(9 \div 9)$ Comm. (+)
    $= (29 + p)(9 \div 9)$ Subs.
    $= (29 + p)(1)$ Subs.
    $= 29 + p$ Mult. Identity
17. State whether the statement *Subtraction of whole numbers is associative* is *true* or *false*. If false, provide a counterexample.
    false; Sample counterexample: $(7 - 2) - 1 \neq 7 - (2 - 1)$

Answers

3. Jessie; the Associative Property can be applied to addition or to multiplication, but not to both at the same time.

9b. Use the Commutative Property of Addition to change the order of the numbers to $69 + 31 + 80 + 8$. Then use the Substitution Property, adding 69 and 31 first.

4 ASSESS

Open-Ended Assessment

Speaking Ask students to state the Commutative or Associative Property of Addition and give an example in which they would use the property.

Chapter 1, Quiz A (Lessons 1–1 through 1–3) is available in the *Chapter 1 Resource Masters*, p. 47.

Answers

22. false; sample counterexample:
$5 - 1 \stackrel{?}{=} 1 - 5$
$4 \neq -4$

23. false; sample counterexample:
$(4 \div 2) \div 2 \stackrel{?}{=} 4 \div (2 \div 2)$
$2 \div 2 \stackrel{?}{=} 4 \div 1$
$1 \neq 4$

27. Sample answer:
$(8 - 5) - 3 \stackrel{?}{=} 8 - (5 - 3)$
$3 - 3 \stackrel{?}{=} 8 - 2$
$0 \neq 6$

***Enrichment*, p. 15**

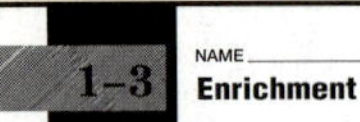

NAME ______________________ DATE ________ PERIOD ______

Enrichment

Student Edition Pages 14–18

Valid and Faulty Arguments

Consider the statements at the right. What conclusions can you make?

(1) Boots is a cat.
(2) Boots is purring.
(3) A cat purrs if it is happy.

From statements 1 and 3, it is correct to conclude that Boots purrs if it is happy. However, it is faulty to conclude from only statements 2 and 3 that Boots is happy. The if-then form of statement 3 is *If a cat is happy, then it purrs.*

Advertisers often use faulty logic in subtle ways to help sell their products. By studying the arguments, you can decide whether the argument is valid or faulty.

Decide if each argument is valid or faulty.

1. (1) If you buy Tuff Cote luggage, it will survive airline travel.
(2) Justin buys Tuff Cote luggage.
Conclusion: Justin's luggage will survive airline travel. **valid**

2. (1) If you buy Tuff Cote luggage, it will survive airline travel.
(2) Justin's luggage survived airline travel.
Conclusion: Justin has Tuff Cote luggage. **faulty**

3. (1) If you use Clear Line long distance service, you will have clear reception.
(2) Anna has clear long distance reception.
Conclusion: Anna uses Clear Line long distance service. **faulty**

4. (1) If you read the book *Beautiful Braids*, you will be able to make beautiful braids easily.
(2) Nancy read the book *Beautiful Braids*.
Conclusion: Nancy can make beautiful braids easily. **valid**

5. (1) If you buy a word processor, you will be able to write letters faster.
(2) Tania bought a word processor.
Conclusion: Tania will be able to write letters faster. **valid**

6. (1) Great swimmers wear AquaLine swimwear.
(2) Gina wears AquaLine swimwear.
Conclusion: Gina is a great swimmer. **faulty**

7. Write an example of faulty logic that you have seen in an advertisement.
See students' work.

State whether each statement is *true* or *false*. If false, provide a counterexample. 22–23. See margin.

22. Subtraction of whole numbers is commutative.
23. Division of whole numbers is associative.

Applications and Problem Solving

24. **Sports** The table shows the point values for different plays in football. The expression below represents the total possible points for a team in a game.

$$6t + 1x + 3f + 2c + 2s$$

If a team scores 3 touchdowns, 2 extra points, 2 field goals, and 2 safeties, how many total points are scored? **30 points**

Type of Score	Number of Points
touchdown, t	6
extra point, x	1
two-point conversion, c	2
field goal, f	3
safety, s	2

25. **Construction** Lumber mills sell wood to lumberyards in *board feet*. The expression shown below represents the number of board feet in a stack of wood.

$$\frac{\text{inches thick} \times \text{inches wide} \times \text{feet long}}{12}$$

Find the number of board feet if the stack of wood is 10 inches thick, 12 inches wide, and 10 feet long.
100 board feet

26. **Geography** The Chattahoochee and Savannah rivers form natural boundaries for the state of Georgia.

a. Write an expression to approximate the total length of Georgia's borders using the map at the right.

b. Evaluate the expression that you wrote in part a. Identify any properties that you used.

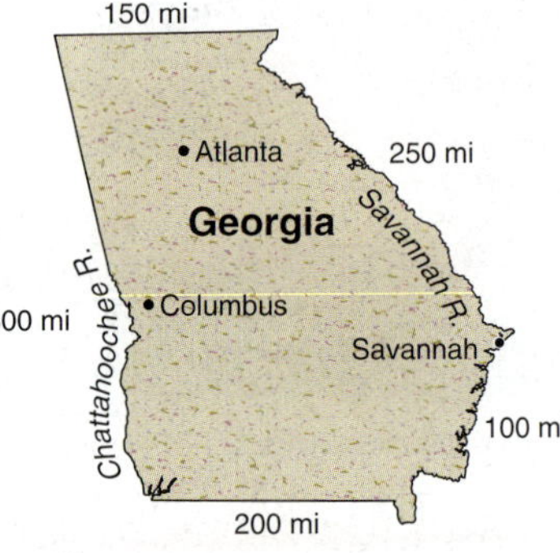

26a. $300 + 150 + 250 + 100 + 200$

26b. 1000 mi; Sample answer: Comm. (+), Assoc. (+), and Substitution

27. **Critical Thinking** Use a counterexample to show that subtraction of whole numbers is not associative. **See margin.**

Mixed Review

Find the value of each expression. *(Lesson 1–2)*

28. $16 \div 2 \cdot 5 \times 3$ **120**
29. $48 \div [2(3 + 1)]$ **6**
30. $25 - \frac{1}{3}(18 - 9)$ **22**

Evaluate each expression if $a = 4$ and $b = 11$. *(Lesson 1–2)*

31. $196 \div [a(b - a)]$ **7**
32. $\frac{ab}{a - 2}$ **22**

Standardized Test Practice

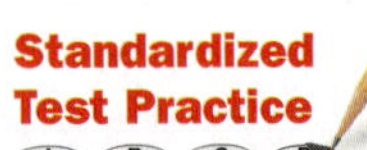

33. **Multiple Choice** Which of the following is the value of $3t - 5q(r + 1)$, if $q = 2$, $r = 0$, and $t = 11$? *(Lesson 1–2)* **A**

Ⓐ 23 Ⓑ 52 Ⓒ 53 Ⓓ 33

www.algconcepts.com/self_check_quiz

Extra Credit

Simplify the expression $4 + 3 \cdot a + 6 \cdot 2$. Identify the properties used in each step.

$4 + 3 \cdot a + 6 \cdot 2$	$= 4 + 3a + 12$	*Substitution Property*
	$= 4 + 12 + 3a$	*Commutative Property of Addition*
	$= (4 + 12) + 3a$	*Associative Property of Addition*
	$= 16 + 3a$	*Substitution Property*

1–4 Distributive Property

What You'll Learn
You'll learn to use the Distributive Property to evaluate expressions.

Why It's Important
Shopping Cashiers use the Distributive Property when they total customers' groceries.
See Exercise 40.

The **Distributive Property** can be applied to simplify expressions. For example, the expression $2 \times (128 + 12)$ can be solved using two different methods.

Method 1

$2 \times (128 + 12) = 2(140)$ *First, add.*
$= 280$ *Then, multiply.*

Method 2

$2 \times (128 + 12) = (2 \times 128) + (2 \times 12)$ *First, distribute.*
$= 256 + 24$ *Multiply.*
$= 280$ *Add.*

The Distributive Property is used in Method 2. Using both methods, the value of the expression is 280.

Distributive Property		
	Symbols:	For any numbers a, b, and c, $a(b + c) = ab + ac$ and $a(b - c) = ab - ac$.
	Numbers:	$2(5 + 3) = (2 \cdot 5) + (2 \cdot 3)$ $2(5 - 3) = (2 \cdot 5) - (2 \cdot 3)$

In the expression $a(b + c)$, it does not matter whether a is placed to the left or to the right of the expression in parentheses. So, $(b + c)a = ba + ca$ and $(b - c)a = ba - ca$.

Examples

Simplify each expression.

1 $3(x + 7)$

$3(x + 7) = (3 \cdot x) + (3 \cdot 7)$ *Distributive Property*
$= 3x + 21$ *Substitution Property*

2 $5(2n + 8)$

$5(2n + 8) = (5 \cdot 2n) + (5 \cdot 8)$ *Distributive Property*
$= 10n + 40$ *Substitution Property*

Your Turn

a. $6(a + b)$ **$6a + 6b$**

b. $(1 + 3t)9$ **$9 + 27t$**

Resource Manager

Reproducible Masters
Chapter 1 Resource Masters
- *Study Guide,* p. 16
- *Skills Practice,* p. 17
- *Practice,* p. 18
- *Reading to Learn Mathematics,* p. 19
- *Enrichment,* p. 20
- *Assessment,* p. 46

Hands-On Algebra, pp. 21–22

Transparencies
5-Minute Check, 1–4

Technology/Multimedia
Interactive Chalkboard CD-ROM

Lesson 1–4

1 FOCUS

5-Minute Check
Lesson 1–3

Name the property shown by each statement.

1. $6 \cdot b \cdot 2 = 2 \cdot 6 \cdot b$ **Commutative Property of Multiplication**
2. $17 + (2 + f) = (17 + 2) + f$ **Associative Property of Addition**

Simplify each expression.

3. $12 + 4k + 3$ **$4k + 15$**
4. $5 \cdot (2 \cdot m)$ **$10m$**
5. Find the volume of the box.

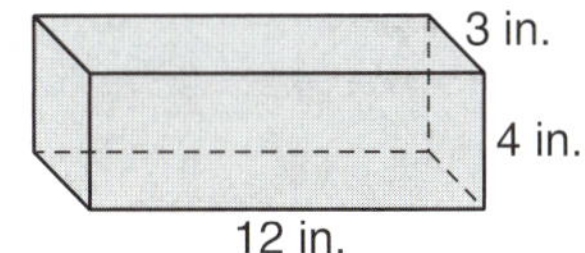

144 in^3

Motivating the Lesson

Real-World Connection Ask students how they could determine the total number of eggs in four cartons if each carton holds one dozen eggs. When multiplication is suggested, multiply 12×4 on the board or overhead by multiplying 2×4 and then 10×4 and adding the two products. Students should recognize this method of multiplying. Point out that this method uses the Distributive Property.

2 TEACH

In-Class Examples
Examples 1–2

Simplify each expression.

1 $5(2 + m)$ **$10 + 5m$**
2 $3(4x + 2)$ **$12x + 6$**

Teaching Tip As you discuss the definition of the *coefficient* of a term, draw students' attention to the *Reading Algebra* feature on page 20. Stress that any variable term such as *x*, *n*, or *b* has a coefficient of 1.

Teaching Tip Before presenting Example 3, write several expressions on the board or overhead, some that are in simplest form and some that are not. Make sure students understand how to distinguish between expressions that are already in simplest form and those that can be simplified.

In-Class Examples

Examples 3–4

Simplify each expression.

3 $8p - 5p$ **$3p$**

4 $10k + 6m - 5k + 2m$ **$5k + 8m$**

A **term** is a number, variable, or product or quotient of numbers and variables.

Examples of Terms		Not Terms	
7	7 is a number.	$7 + x$	$7 + x$ is the sum of two terms.
t	t is a variable.	$8rs + 7y + 6$	$8rs + 7y + 6$ is the sum of three terms.
$5x$	$5x$ is a product.	$x - y$	$x - y$ is the difference of two terms.

The numerical part of a term that contains a variable is called the **coefficient**. For example, the coefficient of $2a$ is 2. **Like terms** are terms that contain the same variables, such as $2a$ and $5a$ or $7xy$ and $3xy$.

Reading Algebra

1 is the *coefficient* of x because $x = 1x$.

Consider the expression $5b + 3b + x + 12x$.

- There are four terms.
- The like terms are $5b$ and $3b$, x and $12x$.
- The coefficients are shown in the table.

Term	Coefficient
$5b$	5
$3b$	3
x	1
$12x$	12

The Distributive Property allows us to combine like terms. If $a(b + c) = ab + ac$, then $ab + ac = a(b + c)$ by the Symmetric Property of Equality.

$2n + 7n = (2 + 7)n$ *Distributive Property*

$= 9n$ *Substitution Property*

The expressions $2n + 7n$ and $9n$ are called **equivalent expressions** because their values are the same for any value of n. An algebraic expression is in **simplest form** when it has no like terms and no parentheses.

Examples

Simplify each expression.

3 $4x + 9x$

$4x + 9x = (4 + 9)x$ *Distributive Property*

$= 13x$ *Substitution Property*

4 $a + 7b + 3a - 2b$

$a + 7b + 3a - 2b = a + 3a + 7b - 2b$ *Commutative Property (+)*

$= (a + 3a) + (7b - 2b)$ *Associative Property (+)*

$= (1 + 3)a + (7 - 2)b$ *Distributive Property*

$= 4a + 5b$ *Substitution Property*

Your Turn

c. $5st + 2st$ **$7st$**

d. $6 + y + 3z + 4y$ **$6 + 5y + 3z$**

www.algconcepts.com/extra_examples

You can use the Distributive Property to solve problems in different, and possibly simpler, ways.

Real World

Example 5

Sports Link

Write an equation representing the area A of a soccer field given its width w and length ℓ as shown in the diagram. Then simplify the expression and find the area if w is 54 yards and ℓ is 60 yards.

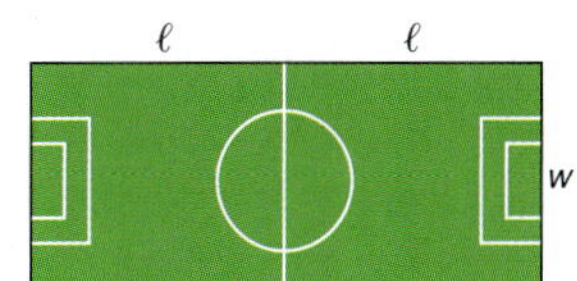

A game on a soccer field

Method 1

$A = w(\ell + \ell)$ *Multiply the total length by the width.*

$= 54(60 + 60)$ *Replace w with 54 and ℓ with 60.*

$= 54(120)$ *Substitution Property*

$= 6480$ *Substitution Property*

Method 2

$A = w\ell + w\ell$ *Add the areas of the smaller rectangles.*

$= 54(60) + 54(60)$ *Replace w with 54 and ℓ with 60.*

$= 3240 + 3240$ *Substitution Property*

$= 6480$ *Substitution Property*

Using either method, the area of the soccer field is 6480 square yards.

Check for Understanding

Communicating Mathematics

1. **Write** an algebraic expression with five terms. One term should have a coefficient of three. Also, include two pairs of like terms.
2. **Explain** why $3xy$ is a term but $3x + y$ is not a term.
3. **Determine** which two expressions are equivalent. Explain how you determined your answer.

   a. $20n + 3p$ b. $16n + p - 4n + 2p$
   c. $16n + 4p + 4n$ d. $12p + 3n$
   e. $12n + 3p$ f. $20n - p - 16n + 2p$

Vocabulary

term
coefficient
like terms
equivalent expressions
simplest form

1. Sample answer: $1 + 2x + 3x + ab + 5ab$
2. The plus sign indicates the sum of two terms.

Guided Practice

3. b, e; Write all the expressions in simplest form and find two that are the same.

Getting Ready Name the like terms in each list of terms.

Sample: $3c$, a, ab, 5, $2c$ **Solution:** $3c$, $2c$

4. $5m, 2n, 7n$ **$2n, 7n$**
5. $8, 8p, 9p, 9q$ **$8p, 9p$**
6. $4h, 10gh, 8, 2h$ **$4h, 2h$**
7. $6b, 6bc, bc$ **$6bc, bc$**

In-Class Example

Example 5

One wing of a building contains three rooms that are identical in size. The floor of each room is 20 meters in length and width. Find the total floor area of the wing. **1200 m²**

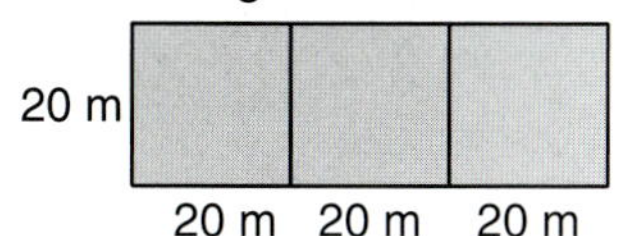

3 PRACTICE/APPLY

Error Analysis

Watch for students who think expressions d and e in Exercise 3 are equivalent.

Prevent by stressing that n and p are not the same. Choose two students whose names begin with N and P. (Use students who are in the class if possible.) Suggest students read $12p + 3n$ using the two names, such as "twelve Pauls and three Natalies." Then have a volunteer read the expression $12n + 3p$ using the same name substitutions. Suggest that students substitute the names of people or objects for the variables in algebraic expressions to help them avoid any confusion.

***Study Guide*, p. 16**

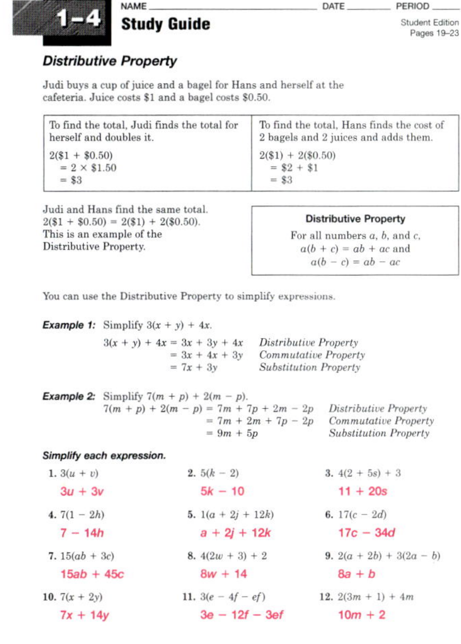

NAME ______ DATE ______ PERIOD ______

1–4 Study Guide

Student Edition Pages 19–23

Distributive Property

Judi buys a cup of juice and a bagel for Hans and herself at the cafeteria. Juice costs $1 and a bagel costs $0.50.

To find the total, Judi finds the total for herself and doubles it.	To find the total, Hans finds the cost of 2 bagels and 2 juices and adds them.
2($1 + $0.50) = 2 × $1.50 = $3	2($1) + 2($0.50) = $2 + $1 = $3

Judi and Hans find the same total. 2($1 + $0.50) = 2($1) + 2($0.50). This is an example of the Distributive Property.

Distributive Property

For all numbers a, b, and c, $a(b + c) = ab + ac$ and $a(b - c) = ab - ac$

You can use the Distributive Property to simplify expressions.

Example 1: Simplify $3(x + y) + 4x$.

$3(x + y) + 4x = 3x + 3y + 4x$ *Distributive Property*

$= 3x + 4x + 3y$ *Commutative Property*

$= 7x + 3y$ *Substitution Property*

Example 2: Simplify $7(m + p) + 2(m - p)$.

$7(m + p) + 2(m - p) = 7m + 7p + 2m - 2p$ *Distributive Property*

$= 7m + 2m + 7p - 2p$ *Commutative Property*

$= 9m + 5p$ *Substitution Property*

Simplify each expression.

1. $3(u + v)$ $3u + 3v$
2. $5(k - 2)$ $5k - 10$
3. $4(2 + 5s) + 3$ $11 + 20s$
4. $7(1 - 2h)$ $7 - 14h$
5. $1(a + 2j + 12k)$ $a + 2j + 12k$
6. $17(c - 2d)$ $17c - 34d$
7. $15(ab + 3c)$ $15ab + 45c$
8. $4(2w + 3) + 2$ $8w + 14$
9. $2(a + 2b) + 3(2a - b)$ $8a + b$
10. $7(x + 2y)$ $7x + 14y$
11. $3(e - 4f - ef)$ $3e - 12f - 3ef$
12. $2(3m + 1) + 4m$ $10m + 2$

Reteaching Activity

Kinesthetic Learners Write large capital letters G and L on sheets of paper, one letter to each sheet. Make a total of four copies of the letter G and three copies of the letter L. Randomly hand out the sheets to seven students. Have these students stand in a line (in any order) in front of the class displaying their sheets. On the board or overhead, write the expression they model, such as 2G + L + 2G + 2L. Rearrange the students several times, having students write each expression that is modeled. Stress that for every arrangement, the simplest form of the resulting expression is 4G + 3L.

Assignment Guide

Basic: 15–41 odd, 42–50
Average: 16–36 even, 37–50

Answer

33. $5(2n + 3r) + 4n + 3(r + 2)$
$= 10n + 15r + 4n + 3r + 6$ Distributive Property
$= 10n + 4n + 15r + 3r + 6$ Commutative (+)
$= 14n + 18r + 6$ Substitution

Every effort is made to show the Answers to exercises (1) on the reduced Student Edition page, or (2) in the margin of the Teacher's Wraparound Edition.

Skills Practice, **p. 17, and** ***Practice,*** **p. 18 (shown)**

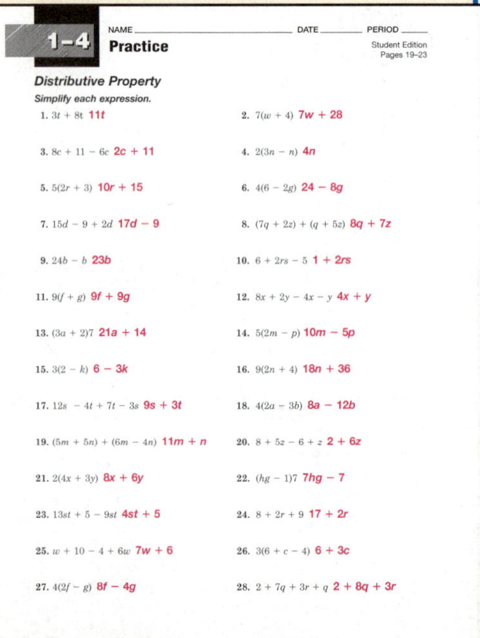

NAME ______________ DATE ________ PERIOD ______

1–4 Practice

Student Edition Pages 19–23

Distributive Property

Simplify each expression.

1. $3t + 8t$ **11t**
2. $7(w + 4)$ **7w + 28**
3. $8c + 11 - 6c$ **2c + 11**
4. $2(3n - n)$ **4n**
5. $5(2r + 3)$ **10r + 15**
6. $4(6 - 2g)$ **24 − 8g**
7. $15d - 9 + 2d$ **17d − 9**
8. $(7q + 2z) + (q + 5z)$ **8q + 7z**
9. $24b - b$ **23b**
10. $6 + 2rs - 5$ **1 + 2rs**
11. $9(f + g)$ **9f + 9g**
12. $8x + 2y - 4x - y$ **4x + y**
13. $(3a + 2)7$ **21a + 14**
14. $5(2m - p)$ **10m − 5p**
15. $3(2 - k)$ **6 − 3k**
16. $9(2n + 4)$ **18n + 36**
17. $12s - 4t + 7t - 3s$ **9s + 3t**
18. $4(2a - 3b)$ **8a − 12b**
19. $(5m + 5n) + (6m - 4n)$ **11m + n**
20. $8 + 5z - 6 + z$ **2 + 6z**
21. $2(4x + 3y)$ **8x + 6y**
22. $(hg - 1)7$ **7hg − 7**
23. $13st + 5 - 9st$ **4st + 5**
24. $8 + 2r + 9$ **17 + 2r**
25. $w + 10 - 4 + 6w$ **7w + 6**
26. $3(6 + c - 4)$ **6 + 3c**
27. $4(2f - g)$ **8f − 4g**
28. $2 + 7q + 3r + q$ **2 + 8q + 3r**

Examples 1–4 **Simplify each expression.**

8. $5x + 9x$ **14x**
9. $4y + 2 - 3y$ **y + 2**
10. $2(5g + 3g)$ **16g**
11. $3(4 - 6m)$ **12 − 18m**
12. $8(2s + 7)$ **16s + 56**
13. $(3a + 5t) + (4a + 2t)$ **7a + 7t**

Example 5

14. **School** Every student at Miller High School must wear a uniform. Suppose shirts or blouses cost \$18 and skirts or pants cost \$25.
   a. If 250 students buy a uniform consisting of a shirt or blouse and a skirt or pants, write an expression representing the total cost. **14a. 250(18 + 25)**
   b. Find the total cost. **\$10,750**

Exercises

Practice 

Homework Help	
For Exercises	See Examples
15–33	1, 4
39, 41	5
Extra Practice	
See page 693.	

Simplify each expression.

15. $16f + 5f$ **21f**
16. $9a + 6a$ **15a**
17. $4r - r$ **3r**
18. $3 + 7 - 2st$ **10 − 2st**
19. $4g - 2g + 6$ **2g + 6**
20. $5a + 7a + 8b + 5b$ **12a + 13b**
21. $14x + 6y - y - 8x$ **6x + 5y**
22. $3(2n + 10)$ **6n + 30**
23. $3(5am - 4)$ **15am − 12**
24. $6(5q + 3w - 2w)$ **30q + 6w**
25. $4a + 7b + (3a - 2b)$ **7a + 5b**
26. $13x - 1 - 8x + 6$ **5x + 5**

B

27. $2y + y + y$ **4y**
28. $bp + 25bp + p$ **26bp + p**
29. $2(15xy + 8xy)$ **46xy**
30. $5(n + 2r) + 3n$ **8n + 10r**
31. $(r + 2s)3 - 2s$ **3r + 4s**
32. $3(2v + 5m) + 2(3v - 2m)$ **12v + 11m**

C

33. Write $5(2n + 3r) + 4n + 3(r + 2)$ in simplest form. Indicate the property that justifies each step. **33. 14n + 18r + 6; See margin for properties.**

34. Is the statement $2 + (s \cdot t) = (2 + s) \cdot (2 + t)$ *true* or *false*? Find values for s and t to show that the statement may be true. Otherwise, find a counterexample to show that the statement is false. **34. false; sample counterexample: $2 + (1 \cdot 3) \neq (2 + 1) \cdot (2 + 3)$**

35. What is the value of $6y$ decreased by the quantity $2y$ plus 1 if y is equivalent to 3? **11**

36. What is the sum of $14xy$, xy, and $5xy$ if x equals 1 and y equals 4? **80**

Applications and Problem Solving

37. **Sports** Rich bought two baseballs for \$4 each and two basketballs for \$22 each. What is the total cost? Use the Distributive Property to solve the problem in two different ways. **\$52**

38. **Retail** Marie and Mark work at a local department store. Each earns \$6.25 per hour. Maria works 24 hours per week, and Mark works 32 hours per week. How much do the two of them earn together each week? **\$350**

39. **Health** If an adult male's height is h inches over five feet, his approximate normal weight is given by the expression $6.2(20 + h)$.
   a. What should the normal weight of a 5'9" male be? **179.8 lb**
   b. How many more pounds should the normal weight of a man that is 6'2" tall be than a man that is 5'9" tall? **31 lb**

40. **Shopping** Luanda went to the grocery store and bought the items in the table below. **a. $4(0.99 + 0.49 + 2.29) + 2(3.29 + 2.69)$**

Item	Cost per Item	Quantity
can of soup	\$0.99	4
can of corn	\$0.49	4
bag of apples	\$2.29	4
box of crackers	\$3.29	2
jar of jelly	\$2.69	2

a. Use the Distributive Property to write an expression representing the cost of the items.

b. Find the change Luanda will receive if she gives the clerk \$30. **\$2.96**

41. **Theater** A school's drama club is creating a stage backdrop with a city theme for a performance. The students sketched a model of buildings as shown at the right.

a. How many square feet of cardboard will they need to make the buildings? Use the expression ℓw, where ℓ is the length and w is the width of each rectangle, to find the area of each rectangle. Then add to find the total area. **96 ft^2**

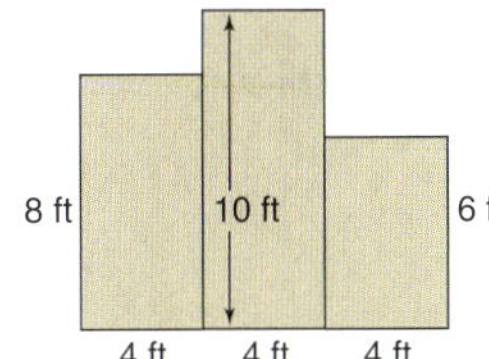

b. Show how to use the Distributive Property as another method in finding the total area of the buildings. **$4(8 + 10 + 6)$ or 96**

42. **Critical Thinking** Use the Distributive Property to write an expression that is equivalent to $3ax + 6ay$.
$3a(x + 2y)$ or $3(ax + 2ay)$ or $a(3x + 6y)$

Mixed Review

Name the property shown by each statement. *(Lesson 1–3)*

43. $8(2 \cdot 6) = (2 \cdot 6)8$ **Comm. (×)**

44. $(7 + 4) + 3 = 7 + (4 + 3)$ **Assoc. (+)**

45. If $19 - 3 = 16$, then $16 = 19 - 3$. *(Lesson 1–2)* **Symmetric**

Find the value of each expression. *(Lesson 1–2)*

46. $8 + 6 \div 2 + 2$ **13**

47. $3(6 - 32 \div 8)$ **6**

48. Write an equation for the sentence *Eighteen decreased by d is equal to f.* *(Lesson 1–1)* **$18 - d = f$**

Standardized Test Practice

49. **Extended Response** Some toys are 30 decibels louder than jets during takeoff. Suppose jets produce d decibels of noise during takeoff. *(Lesson 1–1)*

a. Write an expression to represent the number of decibels produced by the loud toys. **$d + 30$**

b. Evaluate the expression if $d = 140$. **170**

50. **Multiple Choice** Which of the following is an algebraic expression for *six times a number decreased by 17?* *(Lesson 1–1)* **B**

(A) $6n + 17$  (B) $6n - 17$  (C) $17 + 6n$  (D) $17 - 6n$

www.algconcepts.com/self_check_quiz

4 ASSESS

Open-Ended Assessment

Modeling Have students use two different kinds of manipulatives, such as tiles in two colors, to model $x + 2y$. Then have them model $2(x + 2y)$. Finally, have them rearrange their models into two groups by kind (or color) and write the simplified expression that is modeled.

Mid-Chapter Test (Lessons 1–1 through 1–4) is available in the *Chapter 1 Resource Masters*, p. 46.

Extra Practice for each lesson is provided on pages 692–723.

***Enrichment*, p. 20**

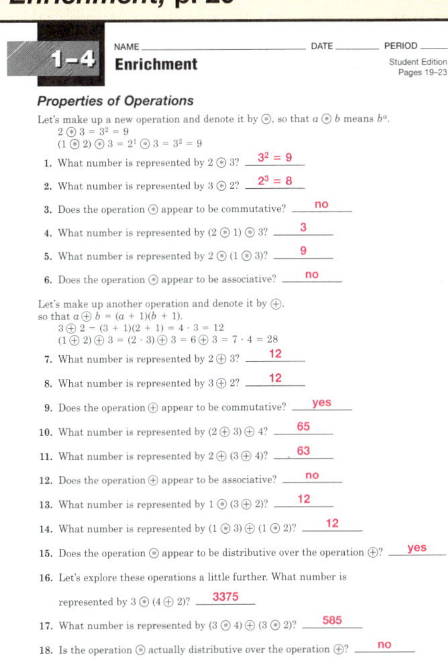

1–4 NAME ______ DATE ______ PERIOD ______

Enrichment

Student Edition Pages 19–23

Properties of Operations

Let's make up a new operation and denote it by ⊛, so that $a ⊛ b$ means b^a.

$2 ⊛ 3 = 3^2 = 9$

$(1 ⊛ 2) ⊛ 3 = 2^1 ⊛ 3 = 3^2 = 9$

1. What number is represented by $2 ⊛ 3$? $3^2 = 9$
2. What number is represented by $3 ⊛ 2$? $2^3 = 8$
3. Does the operation ⊛ appear to be commutative? no
4. What number is represented by $(2 ⊛ 1) ⊛ 3$? 3
5. What number is represented by $2 ⊛ (1 ⊛ 3)$? 9
6. Does the operation ⊛ appear to be associative? no

Let's make up another operation and denote it by ⊕, so that $a ⊕ b = (a + 1)(b + 1)$.

$3 ⊕ 2 = (3 + 1)(2 + 1) = 4 \cdot 3 = 12$

$(1 ⊕ 2) ⊕ 3 = (2 \cdot 3) ⊕ 3 = 6 ⊕ 3 = 7 \cdot 4 = 28$

7. What number is represented by $2 ⊕ 3$? 12
8. What number is represented by $3 ⊕ 2$? 12
9. Does the operation ⊕ appear to be commutative? yes
10. What number is represented by $(2 ⊕ 3) ⊕ 4$? 65
11. What number is represented by $2 ⊕ (3 ⊕ 4)$? 63
12. Does the operation ⊕ appear to be associative? no
13. What number is represented by $1 ⊛ (3 ⊕ 2)$? 12
14. What number is represented by $(1 ⊛ 3) ⊕ (1 ⊛ 2)$? 12
15. Does the operation ⊛ appear to be distributive over the operation ⊕? yes
16. Let's explore these operations a little further. What number is represented by $3 ⊛ (4 ⊕ 2)$? 3375
17. What number is represented by $(3 ⊛ 4) ⊕ (3 ⊛ 2)$? 585
18. Is the operation ⊛ actually distributive over the operation ⊕? no

Extra Credit

Simplify $5a(b + d) + 7[2(ab + bc)]$.
$19ab + 5ad + 14bc$

An Extra Credit exercise is included for every lesson.

Lesson 1-5

1 FOCUS

5-Minute Check
Lesson 1–4

Simplify each expression.

1. $4(b + 3)$ **$4b + 12$**
2. $2(3k + 7)$ **$6k + 14$**
3. $8m + 11m$ **$19m$**

The figure shows a tiled hallway.

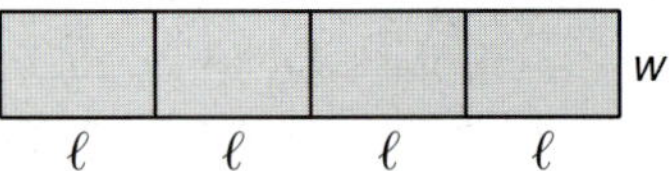

4. Write an equation representing the area A of the hallway given its width w and the length ℓ as shown in the figure. **Sample answer: $A = 4\ell w$**
5. Find the floor area of the hallway if w is 2 yards and ℓ is 3 yards. **24 yd^2**

The 5-Minute Check acts as a bridge to previous lessons. It is also a great way to start each day's lesson.

Motivating the Lesson

Real-World Connection Ask students if they are currently saving money for a special purchase. Ask them to share their savings plan and to discuss how to determine when they will reach their goal. Alternatively, have students imagine they are saving for a vacation. Ask them to explain how they would calculate when they had saved enough money.

1–5 A Plan for Problem Solving

What You'll Learn
You'll learn to use a four-step plan to solve problems.

Why It's Important
Savings The four-step plan is useful for finding the amount of interest earned in a bank account.
See Exercise 18.

In mathematics, solving problems is an important activity. Any problem can be solved using a problem-solving plan like the one below.

1. **Explore** Read the problem carefully. Identify the information that is given and determine what you need to find.
2. **Plan** Select a strategy for solving the problem. Some strategies are shown at the right. If possible, estimate what you think the answer should be before solving the problem.
3. **Solve** Use your strategy to solve the problem. You may have to choose a variable for the unknown, and then write an expression. Be sure to answer the question.
4. **Examine** Check your answer. Does it make sense? Is it reasonably close to your estimate?

Problem-Solving Strategies
- Look for a pattern.
- Draw a diagram.
- Make a table.
- Work backward.
- Use an equation or formula.
- Make a graph.
- Guess and check.

One important problem-solving strategy is *using an equation*. An equation that states a rule for the ralationship between quantities is called a **formula**.

Money in a bank account earns *interest*. You find simple interest by using the formula $I = prt$.

I = interest
p = principal, or amount deposited
r = interest rate, written as a decimal
t = time in years

Example 
Savings Link

Suppose you deposit \$220 into an account that pays 3% simple interest. How much money would you have in the account after five years?

Explore What do you know?
- The amount of money deposited is \$220.
- The interest rate is 3% or 0.03.
- The time is 5 years.

Resource Manager

Reproducible Masters
Chapter 1 Resource Masters
- *Study Guide,* p. 21
- *Skills Practice,* p. 22
- *Practice,* p. 23
- *Reading to Learn Mathematics,* p. 24
- *Enrichment,* p. 25

Graphing Calculator, p. 2
Hands-On Algebra, pp. 23–25

Transparencies
5-Minute Check, 1–5

Technology/Multimedia
Interactive Chalkboard CD-ROM

Data Update For the lastest information on bank interest rates, visit: www.algconcepts.com

Prerequisite Skills Review
Decimals and Percents, p. 689

The Prerequisite Skills Review section on pages 684–691 allows students to refresh arithmetic skills that may be needed.

What do you need to find?

- the amount of money, including interest, at the end of five years

Plan What is the best strategy to use?

Use the formula $I = prt$ and substitute the known values. Add this amount to the original deposit.

Estimate: 1% of \$220 is \$2.20. So, 3% of \$220 is about $3 \times \$2$ or \$6 per year. This will be \$30 in five years. You should have approximately 220 + 30 or \$250 in five years.

Solve $I = prt$ *Interest Formula*

$I = 220 \cdot 0.03 \cdot 5$ or 33 $p = 220, r = 0.03, \text{ and } t = 5$

You will earn \$33 in interest, so the total amount after five years is \$220 + \$33 or \$253.

Examine Is your answer close to your estimate?

Yes, \$253 is close to \$250, so the answer is reasonable.

Your Turn 84.2°F

Science Use $F = 1.8C + 32$ to change degrees Celsius C to degrees Fahrenheit F. Find the temperature in degrees Fahrenheit if it is 29°C.

Another important problem-solving strategy is *using a model*. In the activity below, you will use a model to find a formula for the surface area of a rectangular box.

Hands-On Algebra

Materials:  rectangular box
 ruler

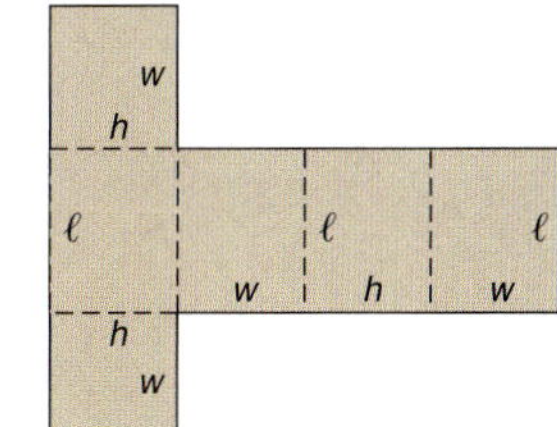

Step 1 Label the edges of a rectangular box ℓ, w, or h to represent the length, width, and height of the box.

Step 2 Take the box apart so that it lies flat on the table with the labels face up.

Try These 1. $wh, wh, \ell h, \ell h, \ell w, \ell w$ 2. $S = 2wh + 2\ell h + 2\ell w$

1. Find the area of each rectangular side of the box in terms of the variables ℓ, w, and h. Be sure to include the top or lid.
2. The sum of the areas is equal to S, the total surface area of a rectangular solid. Express this as a formula in simplest form.
3. Measure the lengths of the sides in centimeters or inches to find the values of ℓ, w, and h. **3–4. See students' work.**
4. Use the formula in Exercise 2 to find the total surface area of your box.

www.algconcepts.com/extra_examples

2 TEACH

Teaching Tip As you discuss Example 1, remind students how to convert 3% to its decimal equivalent 0.03. Show students that if they use $r = 3$ instead of 0.03, the result is $\ell = 3300$ or \$3300 in interest. Make sure all students understand why this answer is not reasonable.

In-Class Example

Example 1

Suppose you deposit \$350 into an account that pays 2% interest. How much money would you have in the account after five years? **\$385**

Hands-On Algebra activities use manipulatives and models to help students learn key concepts.

Hands-On Algebra

Cooperative Learning If students are using a box with a separate lid, instruct them to also label the lid in Step 1. In Step 2, reassure students that their flattened boxes do not have to match exactly the T-shape design shown on page 25.

Additional Hands-On Algebra activities using the problem-solving strategies *use a model* and *make a table* are available in the *Hands-On Algebra Masters*, pp. 24–25.

Hands-On Algebra Masters, p. 23

There are teacher notes for every Hands-On Algebra activity in the Student Edition. These also refer you to the Hands-On Algebra Masters, where you will find a master for the activity.

Teaching Tip As you discuss Example 2, point out that the table is organized first by the number of dimes, beginning with the greatest number possible. For each possible number of dimes, the numbers of nickels are then arranged from greatest to least. Finally, the number of pennies can then be determined for each combination of dimes and nickels. By making an organized list in this way, you can be certain that you are not leaving out or duplicating any combinations of the coins.

In-Class Example

Example 2

How many ways can you make 50¢ using quarters, dimes, and nickels? **10 ways**

Example 2 **Money Link**

How many ways can you make 25¢ using dimes, nickels, and pennies?

Explore A quarter is worth 25¢. How many ways can you make 25¢ without using a quarter?

Plan Make a chart listing every possible combination.

Solve

Coin	Number											
Dimes	2	2	1	1	1	1	0	0	0	0	0	0
Nickels	1	0	3	2	1	0	5	4	3	2	1	0
Pennies	0	5	0	5	10	15	0	5	10	15	20	25

There are 12 ways to make 25¢.

Examine Check that each combination totals 25¢ and that there are no other possible combinations. *The solution checks.*

You can use a graphing calculator to solve problems involving formulas.

Graphing Calculator Tutorial See pp. 750–753.

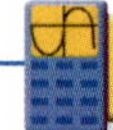

Graphing Calculator Exploration

The area of a trapezoid is $A = \frac{1}{2}h(a + b)$, where h is the height and a and b are the lengths of the bases. Use a graphing calculator to find the area of trapezoid $JKLM$.

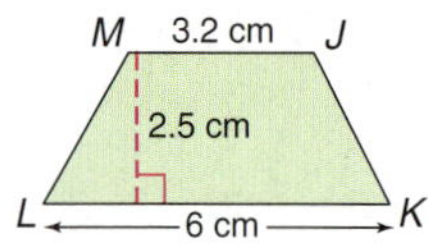

$A = \frac{1}{2} \cdot 2.5(3.2 + 6)$ *Replace each variable with its value.*

Enter: 1 [÷] 2 [×] 2.5 [(] 3.2 [+] 6 [)] [ENTER] *11.5*

Try These

1. Find the area of trapezoid $JKLM$ if the height is 15 centimeters and the bases remain the same. **69 cm²**
2. Find the area of a trapezoid if base a is 14 inches long, base b is 10 inches long, and the height is 7 inches. **84 in²**

The chart below summarizes the properties of numbers. The properties are useful when you are solving problems.

The following properties are true for any numbers *a*, *b*, and *c*.		
Property	**Addition**	**Multiplication**
Commutative	$a + b = b + a$	$ab = ba$
Associative	$(a + b) + c = a + (b + c)$	$(ab)c = a(bc)$
Identity	$a + 0 = 0 + a = a$ 0 is the identity.	$a \cdot 1 = 1 \cdot a = a$ 1 is the identity.
Zero		$a \cdot 0 = 0 \cdot a = 0$
Distributive	$a(b + c) = ab + ac$ and $a(b - c) = ab - ac$	
Substitution	If $a = b$, then a may be substituted for b.	

Graphing Calculator Exploration

Point out to students that no multiplication symbol is needed between 2.5 and the left parenthesis in the keystroke sequence for calculating the area of the given trapezoid. Emphasize that when no operation symbol is used between a number and an expression in parentheses, the calculator assumes that multiplication is the intended operation.

Graphing Calculator Explorations empower students with the ability to use technology to solve problems. The teacher notes clarify or extend the activities in the Student Edition.

Check for Understanding

Communicating Mathematics

1. **List** three reasons for "looking back" when examining the answer to a problem.
2. **Write** a problem in which you need to find the surface area of a rectangular solid. Then solve the problem. **See students' work.**

1. See margin.

Vocabulary
formula

Guided Practice

Getting Ready For each situation, answer the related questions.

Phoebe and Hai picked fourteen pints of raspberries in three hours. Hai picked five more pints than Phoebe.

Samples:	**Solutions:**
How many pints were picked in all?	14
How long did Phoebe and Hai work?	3 hours
Who picked more pints?	Hai
If Phoebe picked x pints, how many did Hai pick?	$x + 5$

Carlos bought 2 more rock CDs than jazz CDs and 3 fewer country CDs than rock CDs. He bought eight CDs, including 1 classical CD.

3. Did Carlos buy more country than rock? **no**
4. Which type of CD did he buy the most of? **rock**
5. If he bought n jazz CDs, how many rock CDs did he buy? $n + 2$

Example 1

6. **Geometry** The perimeter P of a rectangle is the sum of two times the length ℓ and two times the width w.

6a. $P = 2\ell + 2w$

   a. Write a formula for the perimeter of a rectangle.
   b. What is the perimeter of the rectangle? **40 cm**

Example 2

7. **Money** Nate has $267 in bills. None of the bills is greater than $10. He has eleven $10 bills. He has seven fewer $5 bills than $1 bills.
   a. How many $5 and $1 bills does he have? **25 $5 bills, 32 $1 bills**
   b. Describe the problem-solving strategy that you used to solve this problem. **See margin.**

Example 2

8. **Shopping** Two cans of vegetables together cost $1.08. One of them costs 10¢ more than the other.
   a. Would 2 cans of the less expensive vegetable cost more or less than $1.08? **less**
   b. How much would it cost to buy 3 cans of each? **$3.24**

Reteaching Activity

Logical Learners Refer students to Example 2. Point out that they can choose to use any problem-solving strategy in the *Plan and Solve* steps. An alternate strategy might be to draw a diagram like the one below. Draw this diagram on the board or overhead and have students copy and complete it for making change for a quarter. Have them compare their finished diagram to the columns of entries in the table on page 26.

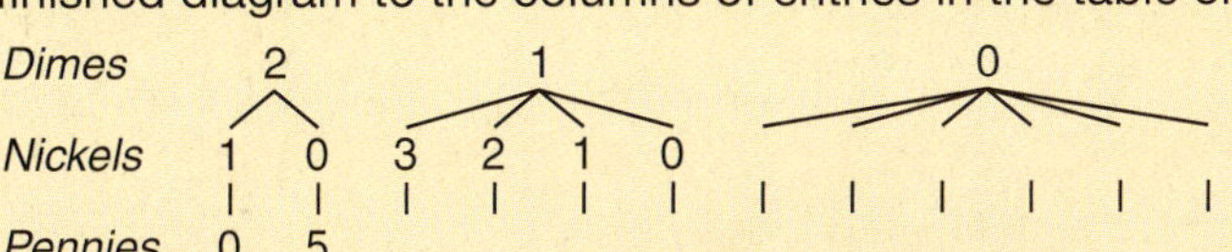

3 PRACTICE/APPLY

Error Analysis

Watch for students who think Carlos bought 2 rock CDs in Exercises 3–5.
Prevent by having students read the directions aloud slowly. Point out that the directions say "2 more rock CDs than jazz CDs." Make sure students understand how this is different from saying Carlos bought "2 rock CDs."

Answers

1. Sample answer: You should look back to determine whether the answer fits the problem, whether there are other possible answers, or whether there might be a better way to solve the problem.

7b. Sample answer: Use an equation. Let x represent the number of $1 bills. Then $x - 7$ represents the number of $5 bills. Write and solve the equation $11(10) + 1x + 5(x - 7) = 267$.

***Study Guide*, p. 21**

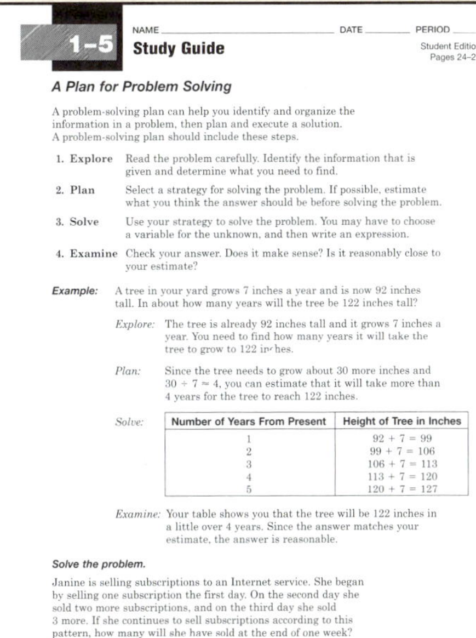

NAME ______ DATE ______ PERIOD ______

1–5 Study Guide

Student Edition
Pages 24–29

A Plan for Problem Solving

A problem-solving plan can help you identify and organize the information in a problem, then plan and execute a solution. A problem-solving plan should include these steps.

1. **Explore** Read the problem carefully. Identify the information that is given and determine what you need to find.
2. **Plan** Select a strategy for solving the problem. If possible, estimate what you think the answer should be before solving the problem.
3. **Solve** Use your strategy to solve the problem. You may have to choose a variable for the unknown, and then write an expression.
4. **Examine** Check your answer. Does it make sense? Is it reasonably close to your estimate?

Example: A tree in your yard grows 7 inches a year and is now 92 inches tall. In about how many years will the tree be 122 inches tall?

Explore: The tree is already 92 inches tall and it grows 7 inches a year. You need to find how many years it will take the tree to grow to 122 inches.

Plan: Since the tree needs to grow about 30 more inches and $30 \div 7 \approx 4$, you can estimate that it will take more than 4 years for the tree to reach 122 inches.

Solve:

Number of Years From Present	Height of Tree in Inches
1	$92 + 7 = 99$
2	$99 + 7 = 106$
3	$106 + 7 = 113$
4	$113 + 7 = 120$
5	$120 + 7 = 127$

Examine: Your table shows you that the tree will be 122 inches in a little over 4 years. Since the answer matches your estimate, the answer is reasonable.

Solve the problem.

Janine is selling subscriptions to an Internet service. She began by selling one subscription the first day. On the second day she sold two more subscriptions, and on the third day she sold 3 more. If she continues to sell subscriptions according to this pattern, how many will she have sold at the end of one week?
28 subscriptions

Assignment Guide

Basic: 9–17 odd, 19–25
Average: 10, 12, 14–25

Answers

9. Sample answer: Let m represent the age of the mother. Then $m - 24$ represents Craig's age. Solve the equation $m + (m - 24) = 56$.

13a. Sample answer:

Person	1	2	3	4	5	6
Number of Handshakes	5	4	3	2	1	0

Exercises

Practice

A

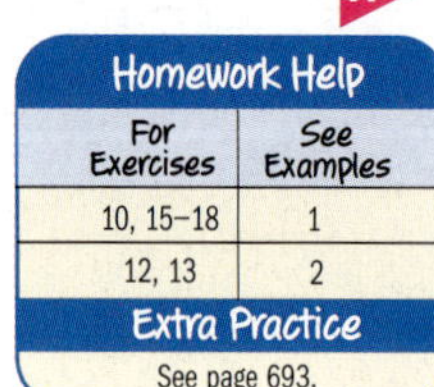

Homework Help	
For Exercises	See Examples
10, 15–18	1
12, 13	2
Extra Practice	
See page 693.	

Solve each problem. Use any strategy.

9. Craig is 24 years younger than his mother. Together their ages total 56 years. How old is each person? Explain how you found your answer.
10. Moira has $500 in the bank at an annual interest rate of 4%. How much money will she have in her account after two years? **$540**
11. Joanne has 20 books on crafts and cooking. She has 6 more cookbooks than craft books. How many of each does she have?
12. How many ways are there to make 20¢ using dimes, nickels, and pennies? **9 ways**
13. Six Explorer Scouts from different packs met for the first time. They all shook hands with each other when they met. **a. See margin.**
    a. Make a chart or draw a diagram to represent the problem.
    b. How many handshakes were there in all? **15**
    c. The number of handshakes h can also be found by using the formula $h = \frac{p(p-1)}{2}$, where p represents the number of people. How many handshakes would there be among 12 people? **66**
    d. Which strategy would you prefer to use to solve the problem: make a table, draw a diagram, or use a formula? **See students' work.**

9. Craig, 16; mother, 40; see margin for explanation.
11. 7 craft books, 13 cookbooks

Applications and Problem Solving

14. **Savings** The table shows the cost of leasing a car for 36 months. Which option is a better deal? Explain. **Option B; although the monthly payments including tax are more, the overall cost is less.**

Type of Fee	Option A Cost ($)	Option B Cost ($)
monthly payment	99	168
monthly tax	6	7
bank fee	495	0
down payment	1956	0
license plates	75	0

15. **Weather** Meteorologists can predict when a storm will hit their area by examining the travel time of the storm system. To do this, they use the following formula.

distance from storm (miles) ÷ speed of storm (miles per hour) = travel time of storm (hours)

At 4:00 P.M. a storm is heading toward the coast at a speed of 30 miles per hour. The storm is about 150 miles from the coast. What time will the storm hit the coast? **9:00 P.M.**

A storm blows in.

***Skills Practice,* p. 22, and *Practice,* p. 23 (shown)**

1–5 NAME ________ DATE ______ PERIOD ____

Practice — Student Edition Pages 24–29

A Plan for Problem Solving

Solve each problem. Use any strategy.

1. Tara read 19 science fiction and mystery novels in 6 months. She read 3 more science fiction novels than mystery novels. How many novels of each type did she read? **8 mystery novels, 11 science fiction novels**
2. Gasoline costs $1.21 per gallon, tax included. Jaime paid $10.89 for the gasoline he put in his car. How many gallons of gasoline did he buy? **9 gal**
3. A coin-operated telephone at a mall requires 40 cents for a local call. It takes quarters, dimes, and nickels and does not give change. How many combinations of coins could be used to make a local call? **7 combinations**
4. Together, Jason and Tyler did 147 sit-ups for the physical fitness test in gym. Jason did 11 fewer sit-ups than Tyler. How many sit-ups did each person do? **Jason, 68; Tyler, 79**
5. The perimeter P of a square can be found by using the formula $P = 4s$, where s is the length of a side of the square. What is the perimeter of a square with sides of length 19 cm? **76 cm**
6. Mrs. Hernandez wants to put a picture of each of her 3 grandchildren on a shelf above her desk. In how many ways can she line up the pictures? **6 ways**
7. Leona is 12 years old, and her sister Vicki is 2 years old. How old will each of them be when Leona is twice as old as Vicki? **Leona, 20; Vicki, 10**
8. Gunther paid for 6 CDs at a special 2-for-1 sale. The CDs that he got at the sale brought the total number of CDs in his collection to 42. How many CDs did he have before the sale? **30 CDs**
9. Phil, Ron, and Felix live along a straight country road. Phil lives 3 miles from Ron and 4 miles from Felix. Felix lives closer to Ron than he does to Phil. How far from Ron does Felix live? **1 mi**
10. Gere has 3 times as many shirts with print patterns as he does shirts in solid colors. He has a total of 16 shirts. How many shirts in print patterns does he have? **12 shirts**

16. **Geometry** The area of a triangle is one-half times the product of the base b and the height h.

a. Write a formula for the area of a triangle.

b. Find the area of the triangle. **$27\ \text{ft}^2$**

a. $A = \frac{1}{2}bh$

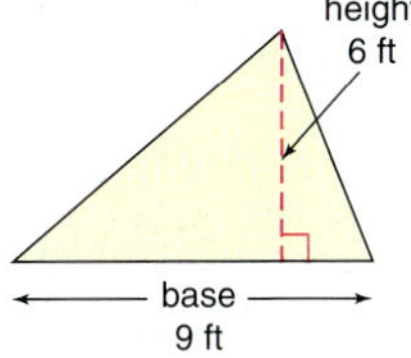

17. **History** Distance traveled d equals the product of rate r and time t.

a. Write the formula for distance. **$d = rt$**

b. In 1936, the *Douglas DC-3* became the first commercial airliner to transport passengers. It flew nonstop from New York to Chicago at an average of 190 miles per hour and the flight lasted approximately 3.7 hours. Find the distance it flew. **703 mi**

18. **Savings** Refer to the table at the right.

Saving Just 50 Cents a School Day . . .

If you start saving in this grade . . .	. . . you'll have this amount at high school graduation*
6th grade	\$630
7th grade	\$540
8th grade	\$450
9th grade	\$360

*Based on 180 school days in each year × 50 cents a day = \$90 a school year.

Source: *Zillions*

a. Suppose you saved 50¢ each school day (180 days) while you were in eighth grade. How much would you have at the end of the school year? **\$90**

b. Suppose you deposited your "school year savings" in an account with a 4% annual interest rate. How much money would you have in your account after a year? **\$93.60**

c. Suppose you saved the 50¢ each school day (180 days) while you were a freshman in high school. Find the sum of this amount and the money already in your account from the eighth grade. **\$183.60**

d. How much money would you have after the second year? **\$190.94**

18e. \$487.47; This is \$37.47 more than if you don't put the money in a bank account.

e. Repeat steps c and d for your junior and senior years. How much money would you have in your account by the time you graduated from high school? Compare this amount to that listed in the table.

19. **Critical Thinking** Refer to Exercise 13. In a meeting, there were exactly 190 handshakes. How many people were at the meeting? **20 people**

Mixed Review

Simplify each expression. *(Lesson 1–4)*

20. $21x - 10x$ **$11x$**
21. $5b + 3b$ **$8b$**
22. $3(x + 2y)$ **$3x + 6y$**
23. $9a + 15(a + 3)$ **$24a + 45$**

Standardized Test Practice

24. **Short Response** State the property shown by $4(ab) = (4a)b$. *(Lesson 1–3)* **Assoc. (×)**

25. **Multiple Choice** The top of a volleyball net is 7 feet 11 inches from the floor. The bottom of the net is 4 feet 8 inches from the floor. How wide is the volleyball net? *(Lesson 1–3)* **A**

Ⓐ 3 feet 3 inches Ⓑ 2.31 feet
Ⓒ 7 feet 1 inch Ⓓ 6 feet

www.algconcepts.com/self_check_quiz

4 ASSESS

Open-Ended Assessment

Speaking Ask students to explain how to use the formula $I = prt$.

Enrichment, p. 25

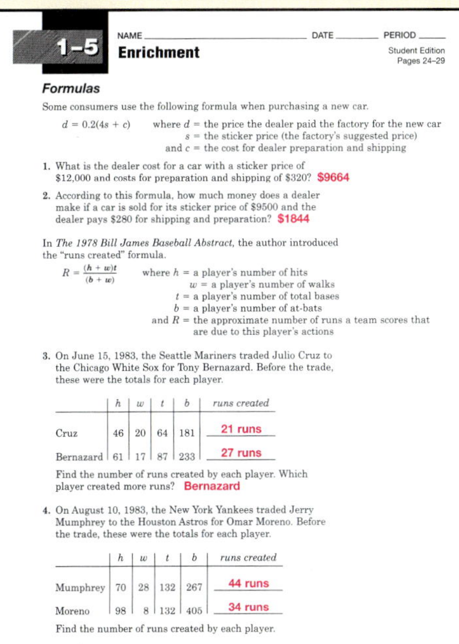

1–5 NAME ______ DATE ____ PERIOD ____

Enrichment Student Edition Pages 24–29

Formulas

Some consumers use the following formula when purchasing a new car.

$d = 0.2(4s + c)$ where d = the price the dealer paid the factory for the new car
s = the sticker price (the factory's suggested price)
and c = the cost for dealer preparation and shipping

1. What is the dealer cost for a car with a sticker price of \$12,000 and costs for preparation and shipping of \$320? **\$9664**

2. According to this formula, how much money does a dealer make if a car is sold for its sticker price of \$9500 and the dealer pays \$280 for shipping and preparation? **\$1844**

In *The 1978 Bill James Baseball Abstract*, the author introduced the "runs created" formula.

$R = \frac{(h + w)t}{(b + w)}$ where h = a player's number of hits
w = a player's number of walks
t = a player's number of total bases
b = a player's number of at-bats
and R = the approximate number of runs a team scores that are due to this player's actions

3. On June 15, 1983, the Seattle Mariners traded Julio Cruz to the Chicago White Sox for Tony Bernazard. Before the trade, these were the totals for each player.

	h	w	t	b	*runs created*
Cruz	46	20	64	181	**21 runs**
Bernazard	61	17	87	233	**27 runs**

Find the number of runs created by each player. Which player created more runs? **Bernazard**

4. On August 10, 1983, the New York Yankees traded Jerry Mumphrey to the Houston Astros for Omar Moreno. Before the trade, these were the totals for each player.

	h	w	t	b	*runs created*
Mumphrey	70	28	132	267	**44 runs**
Moreno	98	8	132	405	**34 runs**

Find the number of runs created by each player.

Extra Credit

Suppose you deposit a sum of money into a bank account paying 5% annually. After five years, you have \$250. How much did you start with? **\$200**

The Standardized Test Practice exercises in each lesson were created to closely parallel those on actual state proficiency tests and college entrance exams.

Investigation

PREPARE

This optional investigation is designed to be completed by groups of 2–3 students over 1–2 days.

Objective

Students model triangular, square, and pentagonal numbers using inductive reasoning. They explore conditional statements using deductive reasoning. Students present their findings by writing a paragraph.

Mathematical Overview

This investigation utilizes the following concepts:

- logical reasoning
- patterns

Suggested Time Management	
Investigation	30–45 min
Extension: Gathering Data	30–45 min
Extension: Summarizing Data	20–30 min

Motivating the Lesson

Have students model the first triangular number using a dot of any color. Then have them add to this model using two dots of a second color to model the second triangular number. Have them add three dots of a third color, and then four dots of a fourth color to model the third and fourth triangular numbers, respectively. Having them use different colors of dots in this manner will help them see the pattern in the sequence of triangular numbers.

The Investigations are optional lessons that provide an opportunity for exploration and cooperative learning. These lessons often involve hands-on learning.

Chapter 1 Investigation

Materials

 5 sheets of different-colored paper

 paper punch

 4 three-inch square slips of paper

 4 two-inch squares of paper

Logical Reasoning

Mathematicians use logical reasoning to discover new ideas and solve problems. *Inductive* and *deductive* reasoning are two forms of reasoning. Let's investigate them to find out how they differ.

Investigate

1. Use the colored paper and the paper punch to make at least 25 dots of each color. You will use these dots to explore patterns.
   a. *Triangular numbers* are represented by the number of dots needed to form different-sized triangles. Use your colored dots to form the first four triangular numbers shown below. **See students' work.**

1b. See margin for fifth triangular number; 15, 21, 28, 36, 45.

   b. Draw the fifth triangular number. Do you see a pattern? Use the pattern to write the next five triangular numbers.
   c. In Step 1b, you used **inductive reasoning**, where a conclusion is made based on a pattern or past events.
2. **Deductive reasoning** is the process of using facts, rules, definitions, or properties in a logical order. You use deductive reasoning to reach valid conclusions.

   If-then statements, called **conditionals**, are commonly used in deductive reasoning. Consider the following conditional.

   If I visit the island of Kauai, then I am in Hawaii.

   The portion of the sentence following *if* is called the **hypothesis**, and the part following *then* is called the **conclusion**. This conditional is true since Kauai is a Hawaiian island.

   a. Use the following information to reach a valid conclusion.
   Conditional: If I visit the island of Kauai, then I am in Hawaii.
   Given: I visit the island of Kauai. **I am in Hawaii.**

Cooperative Learning

This investigation offers an excellent opportunity for using cooperative groups. For more information on cooperative learning strategies and group management, see *Cooperative Learning in the Mathematics Classroom,* one of the titles in the Glencoe Mathematics Professional Series.

Answer

1b.

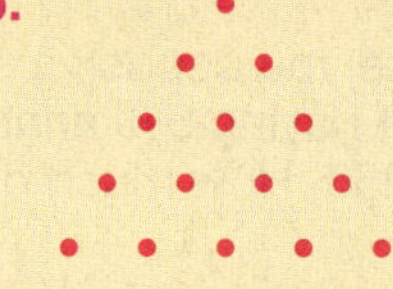

The island of Kauai

b. On a two-inch square, write the hypothesis "I visit the island of Kauai." On a three-inch square, write the conclusion "I am in Hawaii." Place the two-inch square inside the three-inch square.

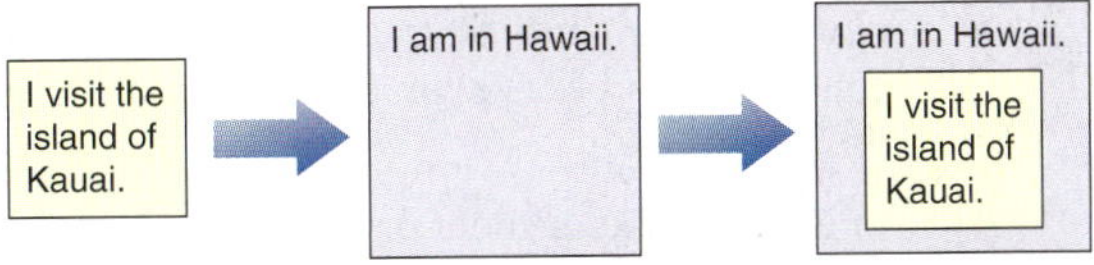

c. Place your pencil on the given statement, "I visit the island of Kauai." Since the pencil is also contained within the square with the conclusion, "I am in Hawaii," the conclusion is valid.

d. Repeat Step 2c, but exchange the hypothesis and conclusion. The new conditional is *If I am in Hawaii, then I visit the island of Kauai.* You can place your pencil in any region marked by the ×'s in the diagram. You may be in Hawaii, visiting Maui, not the island of Kauai. You cannot reach the conclusion using the conditional and the given information.

I am in Hawaii.
I visit the island of Kauai.
X X X

Extending the Investigation

In this extension, you will continue to investigate inductive and deductive reasoning.

1. The first four *square numbers* are 1, 4, 9, and 16. Use the colored dots to make the first five square numbers. Use inductive reasoning to list the first ten square numbers.
2. The first four *pentagonal numbers* are 1, 5, 12, and 22. Use the colored dots to make the first five pentagonal numbers. Use inductive reasoning to list the first ten pentagonal numbers. **1–3. See margin.**
3. For each problem, identify the hypothesis and conclusion. Then use squares as shown above to determine whether a valid conclusion can be made from the conditional and given information.
   a. Conditional: If the living organism is a grizzly bear, then it is a mammal.
      Given: The living organism is a grizzly bear.
   b. Conditional: If Aislyn is in the Sears Tower, then she is in Chicago.
      Given: Aislyn is in Chicago.
4. Write a paragraph explaining the difference between inductive and deductive reasoning. Include an example of each type of reasoning. **See students' work.**

Presenting Your Investigation

Here are some ideas to help you present your conclusions to the class.

- Make a poster showing the triangular, square, and pentagonal numbers.
- Include a description of the patterns you observed.

Investigation For more information on logical reasoning, visit: www.algconcepts.com

MANAGE

Teaching Tip Students may need help building the models for square and pentagonal numbers. Stress that students need to model only the first five numbers of each type. After modeling the first five numbers, students should be able to predict the next five numbers in the sequence.

Working in Groups If students work in groups, they can take turns creating the models for the various numbers. When trying to reach valid conclusions using the paper square Venn diagrams, the group members can work together to validate their conclusions.

Working as a Class Assign half of the class to work on Exercise 1 and the other half to work on Exercise 2. Each group can then present their findings and teach the other students about their topic. After the presentations, hold a class discussion about the differences between inductive and deductive reasoning.

ASSESS

Students' number models should be formed correctly and their sequence of numbers should be correct. Students should be able to explain how inductive and deductive reasoning differ, and give examples of each.

PORTFOLIO Students should add their paragraphs to their portfolios at this time.

Answers

1. 1, 4, 9, 16, 25, 36, 49, 64, 81, 100

1 4 9 16 25

2. 1, 5, 12, 22, 35, 51, 70, 92, 117, 145

1 5 12 22 35

Answers

3a. Hypothesis: the living organism is a grizzly bear
Conclusion: it is a mammal
Valid conclusion: The living organism is a mammal.

3b. Hypothesis: Aislyn is in the Sears Tower
Conclusion: she is in Chicago
No valid conclusion can be made.

Lesson 1–6

1 FOCUS

5-Minute Check
Lesson 1–5

Solve each problem. Use any strategy.

1. Travis is 2 years older than his sister Mickie. Together their ages total 106. How old is each person? **Travis 54, Mickie 52**
2. Mack has $300 in the bank at an annual interest rate of 3%. How much money will Mack have in the bank in four years? **$336**
3. How many ways are there to make 75¢ using dimes and nickels? **8 ways**

The table shows the cost of living in an apartment for one year.

Expense	Apt. A Cost ($)	Apt. B Cost ($)
monthly rent	495	515
average electricity per month	65	23
average gas per month	0	15

4. Which apartment has the lower monthly rent? **apartment A**
5. Overall, which apartment is cheaper to live in? **apartment B**

Motivating the Lesson

Real-World Connection Ask students if they have ever participated in a survey or a poll. Have those who have participated report what they were asked, where they were surveyed, and what kinds of people were being chosen for the survey. If students can remember, ask them whether the survey questions were asked in a way that made students feel more inclined to answer in a certain way.

1–6 Collecting Data

What You'll Learn
You'll learn to collect and organize data using sampling and frequency tables.

Why It's Important
Marketing
Businesses use surveys to collect data in order to test new ideas.
See Example 3.

Researchers gather **data**, or information, using **experiments** and **observational studies.** In an experiment, a test is made under controlled conditions. Subjects in an observational study are studied without action by the investigator.

In either an experiment or an observational study, **sampling** is a convenient way to gather data to make predictions about a **population.** A **sample** is a small group that is used to represent a much larger population. Three important characteristics of a good sample are listed below.

Sampling Criteria	A good sample is: • representative of the larger population, • selected at random, and • large enough to provide accurate data.

A survey can be biased and give false results if these criteria are not followed. Note that there is no given number to make the sample large enough. You must consider each survey individually to see if it is based on a good sample.

Example 1 Health Link

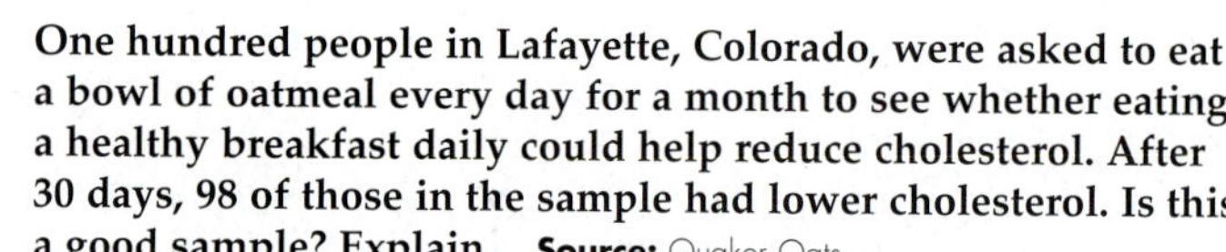

One hundred people in Lafayette, Colorado, were asked to eat a bowl of oatmeal every day for a month to see whether eating a healthy breakfast daily could help reduce cholesterol. After 30 days, 98 of those in the sample had lower cholesterol. Is this a good sample? Explain. **Source:** Quaker Oats

If the people were randomly chosen, then this is a good sample. Also, the sample appears to be large enough to be representative of the population. For example, the results of two or three people would not have been enough to make any conclusions.

Your Turn

Determine whether each is a good sample. Explain.

a. Two hundred students at a school basketball game are surveyed to find the students' favorite sport. **a. No; many of those surveyed would prefer basketball.**

b. Every other person leaving a supermarket is asked to name their favorite soap. **Yes; this would provide a random sample.**

Resource Manager

Reproducible Masters
Chapter 1 Resource Masters
- *Study Guide,* p. 26
- *Skills Practice,* p. 27
- *Practice,* p. 28
- *Reading to Learn Mathematics,* p. 29
- *Enrichment,* p. 30

Transparencies
5-Minute Check, 1–6

Technology/Multimedia
Interactive Chalkboard CD-ROM

After the survey is complete, the gathered data is organized into different types of tables and charts. One way to organize data is by using a **frequency table**. In a frequency table, you use **tally marks** to record and display the frequency of events.

Example 2 Science Link

In an experiment, students "charged" balloons by rubbing them with wool. Then the students placed the balloons on a wall and counted the number of seconds they remained. The class results are shown in the chart at the right. Make a frequency table to organize the data.

Static Electricity Time (s)				
15	52	26	22	25
26	29	33	36	20
43	21	30	39	34
35	27	29	42	35
16	18	21	21	40

Step 1 Make a table with three columns: Time (s), Tally, and Frequency. Add a title.

Step 2 It is sometimes helpful to use *intervals* so there are fewer categories. In this case, we are using intervals of size 10.

Step 3 Use tally marks to record the times in each interval.

Step 4 Count the tally marks in each row and record this number in the Frequency column.

Static Electricity

Time (s)	Tally	Frequency
15–24	𝍸 III	8
25–34	𝍸 IIII	9
35–44	𝍸 II	7
45–54	I	1

Your Turn

c. Make a frequency table to organize the data in the chart at the right. **See margin.**

Noon Temperature (°C)					
32	30	18	29	20	14
21	32	36	15	19	10
16	22	25	30	26	21

In Example 2, suppose the science teacher wanted to know how many balloons stayed on the wall *no more than 44 seconds*. To answer this question, use a **cumulative frequency table** in which the frequencies are accumulated for each item.

Static Electricity

Time (s)	Frequency	Cumulative Frequency
15–24	8	8
25–34	9	17
35–44	7	**24**
45–54	1	25

From the cumulative frequency table, we see that 24 balloons stayed on the wall for 44 seconds or less. Or, 24 balloons stayed on the wall for no more than 44 seconds.

www.algconcepts.com/extra_examples

Answer

Your Turn

c. Sample answer:

Noon Temperature

Temperature (°C)	Tally	Frequency
5–14	II	2
15–24	𝍸 III	8
25–34	𝍸 II	7
35–44	I	1

2 TEACH

In-Class Example

Example 1

One hundred cable-television subscribers are surveyed to find how much time the average American spends reading. Is this a good sample? Explain. **No; many of those surveyed would prefer watching television to reading.**

Teaching Tip Before presenting Example 2, instruct students on how to use tally marks. Point out that the first four marks are vertical lines and that the fifth line crosses the first four. Stress that the groups of tally marks that have been crossed can be counted by 5s.

Teaching Tip Consider actually conducting the experiment discussed in Example 2. Students can compile their own data on the board or overhead and then compare their frequency tables with the table in Example 2.

In-Class Example

Example 2

Make a frequency table to organize the data in the chart below. Use intervals of 5.

Record High Temperatures for Selected U.S. States (°C)				
44	38	53	49	57
48	41	43	41	43
45	38	48	47	47
48	49	46	46	41

Sample answer:

Record High Temperatures for Selected U.S. States

Temp (°C)	Tally	Frequency
35–39	II	2
40–44	𝍸 I	6
45–49	𝍸 𝍸	10
50–54	I	1
55–59	I	1

In-Class Example

Example 3

The owners of a bookstore specializing in travel books are looking for a new location. They counted the number of people who passed by the proposed location during one afternoon. The frequency table below shows the results of their sampling.

Age of People	Tally	Frequency
under 13	𝍸	5
teens	𝍸 III	8
20s	𝍸 𝍸 𝍸 𝍸 𝍸	25
30s	𝍸 𝍸 𝍸 𝍸 𝍸 𝍸 II	32
40s	𝍸 𝍸 𝍸 𝍸 𝍸 𝍸 𝍸 I	36
50s	𝍸 𝍸 𝍸 III	18
60s	𝍸 𝍸 I	11

a. Which group of people passed by the location most frequently? **adults in their 40s**

b. Is this a good location for the bookstore? Explain. **Yes, people in their 40s are likely to be interested in travel books.**

Answers

1. **A frequency table shows the number of times a single event occurs. A cumulative frequency table shows the number of times an event occurs plus the previous events.**
2. **Sample answer: A sample might be biased if there are too few items in the sample or if the sample is not random.**

Once you have summarized data in a frequency table or in a cumulative frequency table, you can analyze the information and make conclusions.

Real World **Example 3** Marketing Link

Owners of a restaurant are looking for a new location. They counted the number of people who passed by the proposed location one day during lunchtime. The frequency table at the right shows the results of their sampling.

Age of People	Tally	Frequency
under 13	𝍸 II	7
teens	𝍸 𝍸	10
20s	𝍸 𝍸 𝍸 III	18
30s	𝍸 𝍸 𝍸 𝍸 𝍸 𝍸 𝍸 𝍸 II	42
40s	𝍸 𝍸 𝍸 𝍸 𝍸 𝍸 𝍸 I	36
50s	𝍸 𝍸 𝍸 IIII	19
60s	𝍸 𝍸 I	11

The right location?

A. Which two groups of people passed by the location most frequently?
adults in their 30s and 40s

B. If the restaurant is an ice cream shop aimed at teens during their lunchtimes, is this a good location for the restaurant? Explain.
Since very few teens pass by the location compared to adults, the owners should probably look for another location.

Check for Understanding

Communicating Mathematics

1–2. See margin.

1. **Explain** the difference between a frequency table and a cumulative frequency table.
2. **List** some examples of how a survey might be biased.

Vocabulary
data
experiments
observational studies
sampling
population
sample
frequency table
tally marks
cumulative frequency table

Guided Practice

Example 1

3. No, the sample is not large enough.

Determine whether each is a good sample. Explain.

3. Four people out of 500 are randomly chosen at a senior assembly and surveyed to find the percent of seniors who drive to school.
4. Six hundred randomly chosen pea seeds are used to determine whether wrinkled seeds or round seeds are the more common type of seed. **Yes, the sample is large enough and it is randomly chosen.**

Reteaching Activity

Verbal/Linguistic Learners Have students give an oral explanation of how to use a frequency table to organize data. Then have them explain how to create a cumulative frequency table for the same data.

Refer to the chart at the right.

Number of Soccer Goals Scored This Season		
1	2	5
1	6	2
6	8	4
2	4	5
5	1	3
4	7	2
2	6	4

Example 2 5. Make a frequency table to organize the data. **See margin.**

Example 3 6. What number of goals was scored most frequently? **2 goals**

Example 3 7. How many times did the team score 8 goals? **1 time**

Example 3 8. How many more times did the soccer team score six goals than three goals? **2 times**

Examples 2 & 3 9. **Technology** When lines of cars get too long at some traffic lights, computers override the signals to turn the lights green and allow the cars to move. A cycle is the number of seconds it takes a light to change from red back to red. The frequency table below shows different traffic light cycles during one afternoon.

Cycle (s)	Tally	Frequency
80	卌 卌 卌 卌 卌 卌 III	33
90	卌 卌 卌 卌 卌 卌 卌 卌 II	42
100	卌 卌 卌 卌 卌 卌 卌 卌 卌 卌 卌 卌	60
110	卌 卌 卌 卌 卌	25

a. Which cycle occurred the most? **100 s**

b. Make a cumulative frequency table of the data. **See margin.**

c. If the standard cycle for a traffic light is 100 seconds, how many times during this period was the cycle less than the standard? **75 times**

Exercises

Practice

Homework Help

For Exercises	See Examples
10–15, 20, 21	1
16–18	2, 3
22	3

Extra Practice

See page 693.

Determine whether each is a good sample. Describe what caused the bias in each poor sample. Explain. **10–15. See margin.**

10. Thirty people standing in a movie line are asked to name their favorite actor.

11. Police stop every fifth car at a sobriety checkpoint.

12. Every other household in a neighborhood of 240 homes is surveyed to determine how many people in the area recycle.

13. Every other household in a neighborhood of 20 homes is surveyed to determine the country's favorite presidential candidate.

14. Every third student on a class roster is surveyed to determine the average number of hours students in the class spend on a computer.

15. All people leaving a sporting goods store are asked to name their favorite golfer.

3 PRACTICE/APPLY

Error Analysis

Watch for students who cannot decide what constitutes a good sample in Exercises 10–15.
Prevent by reviewing the sampling criteria on page 32. Stress that the people being sampled should represent the population being studied. For example, if the survey is about average Americans, the survey participants should be representative of all Americans.

Assignment Guide

Basic: 11–23 odd, 24–28
Average: 10–18 even, 20–28
All: Quiz 2, 1–5

Answer

5.

Number of Goals	Tally	Frequency
1	III	3
2	卌	5
3	I	1
4	IIII	4
5	III	3
6	III	3
7	I	1
8	I	1

***Study Guide*, p. 26**

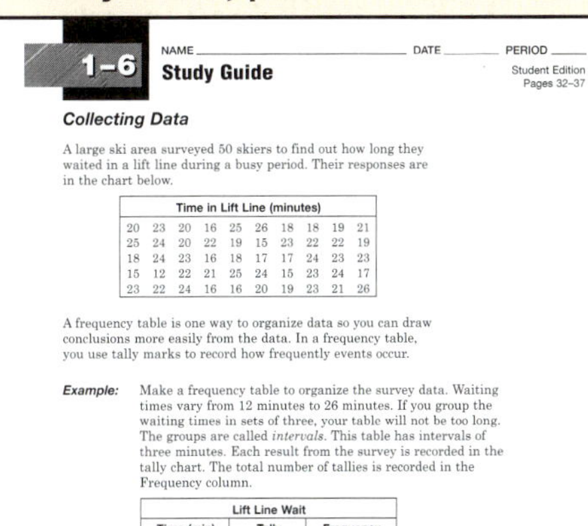

1–6 NAME ______ DATE ______ PERIOD ______

Study Guide Student Edition Pages 32–37

Collecting Data

A large ski area surveyed 50 skiers to find out how long they waited in a lift line during a busy period. Their responses are in the chart below.

Time in Lift Line (minutes)									
20	23	20	16	25	26	18	18	19	21
25	24	20	22	19	15	23	22	22	19
18	24	23	16	18	17	17	24	23	23
15	12	22	21	25	24	15	23	24	17
23	22	24	16	16	20	19	23	21	26

A frequency table is one way to organize data so you can draw conclusions more easily from the data. In a frequency table, you use tally marks to record how frequently events occur.

Example: Make a frequency table to organize the survey data. Waiting times vary from 12 minutes to 26 minutes. If you group the waiting times in sets of three, your table will not be too long. The groups are called *intervals*. This table has intervals of three minutes. Each result from the survey is recorded in the tally chart. The total number of tallies is recorded in the Frequency column.

Lift Line Wait		
Time (min)	Tally	Frequency
12–14	I	1
15–17	卌 卌	10
18–20	卌 卌 II	12
21–23	卌 卌 卌 I	16
24–26	卌 卌 I	11

Make a frequency table to organize the data in the table. Use intervals of $4.

Profits ($)					
92	95	94	91	100	101
90	90	92	105	101	102
99	94	100	95	93	92
102	103	91	97	105	92

Profits ($)	Tally	Frequency
90–93	卌 IIII	9
94–97	卌	5
98–101	卌	5
102–105	卌	5

Answers

9b.

Cycle (s)	Frequency	Cumulative Frequency
80	33	33
90	42	75
100	60	135
110	25	160

10. No; people might be in line because the movie has a favorite star.

11. Yes; the sample is random and representative of drivers in the area.

12. Yes; the sample is random and representative of people in the neighborhood.

13. No; the sample is not representative of the entire country.

14. Yes; the sample is representative of the whole class.

15. No; people might prefer a golfer who is from the same area.

Teaching Tip In Exercise 18c, make sure students count the games that had zero home runs.

Answers

16a.

Type of Pizza	Tally	Frequency
Cheese	卌 卌 III	13
Pepperoni	卌 III	8
Sausage	卌 I	6
Vegetable	III	3

17a.

Quiz Score	Tally	Frequency
6	II	2
7	IIII	4
8	卌 卌 I	11
9	卌 IIII	9
10	IIII	4

17c. Rather than adding all of the numbers, this formula uses multiplication to find the sum. Then the sum is divided by the number of scores, 30.

18a.

Number of Home Runs	Frequency	Cumulative Frequency
0	3	3
1	7	10
2	5	15
3	5	20
4	2	22
5	2	24
6	1	25

Skills Practice, p. 27, and *Practice*, p. 28 (shown)

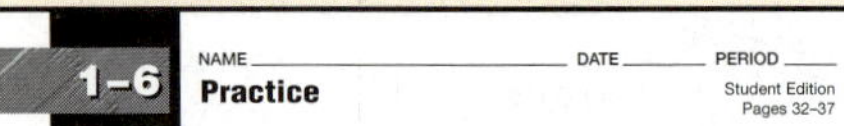

1–6 NAME ________________ DATE ________ PERIOD _____

Practice

Student Edition Pages 32–37

Collecting Data

Determine whether each is a good sample. Describe what caused the bias in each poor sample. Explain.

1. Every third person leaving a music store is asked to name the type of music they prefer. **Yes; the sample is random, appears to be large enough, and music stores sell all types of music.**
2. One hundred students at Cary High School are randomly chosen to find the percentage of people who vote in national elections. **No; students at a high school are not old enough to vote in national elections.**
3. Two out of 25 students chosen at random in a cafeteria lunch line are surveyed to find whether students prefer sandwiches or pizza for lunch. **No; the sample is not large enough.**

Refer to the following chart.

Favorite Leisure Activity									
S	R	C	C	S	R	R	C	S	C
M	S	C	C	C	M	C	C	S	R
S	S	R	M	M	C	M	S	C	R

C = computer games, M = movies, R = reading, S = sports

Favorite Leisure Activity		
Type	Tally	Frequency
Computer Games	卌 卌 I	11
Movies	卌	5
Reading	卌 I	6
Sports	卌 III	8

4. Make a frequency table to organize the data.
5. What is the most popular leisure activity? **computer games**
6. How many more people chose sports over reading? **2**
7. Does the information in the frequency table support the claim that most people do not get enough exercise? Explain. **Sample answer: Yes; 22 out of 30 people preferred leisure activities that involve sitting. However, sports are not the only form of exercise.**

Refer to the following chart.

Number of Breakfasts Eaten Per School Week									
0	5	3	2	0	2	1	3	4	2
5	1	3	2	1	3	1	3	4	1
0	2	3	5	5	2	3	4	1	3

Break. per School Week		
Number	Tally	Frequency
0	III	3
1	卌 I	6
2	卌 I	6
3	卌 III	8
4	III	3
5	IIII	4

8. Make a frequency table to organize the data.
9. How many students eat breakfast fewer than 3 times per week? **15**
10. Should the school consider a campaign to encourage more students to eat breakfast at school? Explain. **Yes; only 4 students out of 30 eat breakfast every school day.**

16. Refer to the chart at the right.
    a. Make a frequency table to organize the data. **See margin.**
    b. How many fewer sausage pizzas were ordered than cheese pizzas? **7**
    c. Suppose x mushroom pizzas were also ordered. Write an expression representing the total number of mushroom, vegetable, and pepperoni pizzas ordered. Write the expression in simplest form. **$x + 11$**

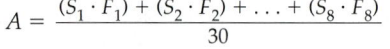

Pizzas Ordered				
C	C	C	P	C
C	P	P	V	S
S	P	C	P	C
S	S	C	P	P
C	V	P	C	C
V	S	S	C	C

C = cheese, P = pepperoni, S = sausage, V = vegetable

17. Refer to the chart at the right.
    a. Make a frequency table to organize the data. **See margin.**
    b. What was the most common score? **8**
    c. Suppose each S represents the score and each F represents the frequency for that score. Explain why the formula below determines the class *average* A for this quiz.

$$A = \frac{(S_1 \cdot F_1) + (S_2 \cdot F_2) + \ldots + (S_8 \cdot F_8)}{30}$$

See margin.

    d. Find the class average for the quiz. **8.3**

Quiz Scores (out of 10 points)				
9	8	8	9	8
8	10	6	7	8
10	8	8	9	9
7	9	7	8	10
9	6	7	8	9
9	10	8	9	8

Reading Algebra

Read S_1 as *S sub 1*. The 1 is called a *subscript*.

C

18. Refer to the chart at the right.
    a. Make a cumulative frequency table to organize the data. **See margin.**
    b. In how many games were there at least three home runs? **10 games**
    c. In how many games were there no more than four home runs? **22 games**

Number of Home Runs in a Game per Month				
2	2	3	5	1
0	1	2	6	3
3	1	1	4	5
2	3	4	3	0
1	0	1	1	2

19. Why do you suppose a coffee and bagel shop would want to locate where a lot of people walk past the store between 7:00 A.M. and 10:30 A.M.? **That is a typical breakfast time and they want the business.**

20. The technician takes a sample because it is representative of the entire supply.

Applications and Problem Solving

20. **Health** When you have a blood test taken for your health, why does the technician only take a few vials of your blood? Use the terms you learned in this lesson to explain your answer.

21. **Marketing** A new cola drink is out on the market. Name three places where the cola company could set up taste tests to determine interest in the drink. **Sample answers: a stadium, restaurant, school campus**

Have students use the frequency table for Example 3 on page 34 as a basis for this activity. Have them work with a family member to estimate the ages of the 30 closest neighbors to their home and make their own frequency table. Based on their results, have students decide what kind of community activities would be the most popular in their neighborhood.

Family Activities provide an opportunity for students to include their family members in the math that they are studying.

22a. Sample answer: a comedy, since more people chose that type as their favorite more than any other

22. **Entertainment** The frequency table at the right shows students' favorite types of movies in one class.

Favorite Type of Movie		
Movie	**Tally**	**Frequency**
Adventure	𝍸 II	7
Comedy	𝍸 IIII	9
Horror	IIII	4
Drama	III	3

a. Suppose you invite students in this class to a party. What type of movie would you show? Explain.

b. In another class, three times more students favored drama, and two fewer favored comedy. Write an expression to find the total number of people in that class who favored drama and comedy. Then find the number. **3(3) + (9 − 2); 16 people**

23. **Critical Thinking** Suppose someone takes a phone survey from a large random sample of people. Do you think that the wording of a question or the surveyor's tone of voice can affect the responses and cause biased results? Explain. **See students' work.**

Mixed Review

24. An adult bus ticket and a child's bus ticket together cost $2.40. The adult fare is twice the child's fare. What is the adult's fare? Use any strategy to solve the problem. *(Lesson 1–5)* **$1.60**

25. **Travel** What distance can a car travel in 5 hours at a constant rate of 55 miles per hour? Use a diagram or the formula $d = rt$ to solve the problem. *(Lesson 1–5)* **275 miles**

26. Simplify the expression $16a + 21a + 30b - 7b$. *(Lesson 1–4)* **$37a + 23b$**

Standardized Test Practice

27. **Short Response** Write a verbal expression for $x + 9$. *(Lesson 1–1)* **nine more than x**

28. **Short Response** Write an algebraic expression for *4 times n less 3*. *(Lesson 1–1)* **$4n - 3$**

1. $11 + 6 + 2a$, Comm. (+); $17 + 2a$ 2. $(4 \cdot 8) \cdot t$, Assoc. (×); $32t$

Quiz 2 Lessons 1–3 through 1–6

Simplify each expression. Identify the properties used in each step. *(Lesson 1–3)*

1. $11 + 2a + 6$

2. $4 \cdot (8t)$

3. **Health** Your optimum exercise heart rate per minute is given by the expression $0.7(220 - a)$, where a is your age. Use your age for a and find your optimum exercise heart rate. *(Lesson 1–4)* **See students' work.**

4. **Fitness** Lorena runs for 30 minutes each day. Find the distance she runs if she averages 660 feet per minute. Use the formula $d = rt$. *(Lesson 1–5)* **19,800 ft** **5. Blue; see margin for table.**

5. **Biology** Make a frequency table to organize the data in the chart at the right. Which eye color occurs the least? *(Lesson 1–6)*

Eye Color				
H	B	B	G	U
H	G	G	B	B
B	G	U	G	H
H	U	B	B	G
H	B	U	B	B

B = brown, U = blue, G = green, H = hazel

www.algconcepts.com/self_check_quiz

4 ASSESS

Open-Ended Assessment

Writing Have students write a description of a frequency table and an explanation of what it shows.

Quiz 2

The Quiz provides students with a brief review of the concepts and skills in Lessons 1–3 through 1–6. Lesson numbers are given to the right of the exercises or instruction lines so students can review concepts not yet mastered.

Answer

Quiz 2

5.

Eye Color		
Color	**Tally**	**Frequency**
brown	𝍸 𝍸	10
blue	IIII	4
green	𝍸 I	6
hazel	𝍸	5

***Enrichment,* p. 30**

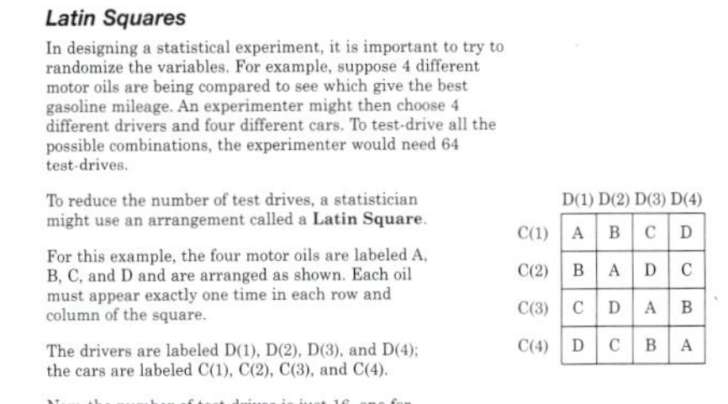

1–6 NAME ______ DATE ______ PERIOD ______

Enrichment

Student Edition Pages 32–37

Latin Squares

In designing a statistical experiment, it is important to try to randomize the variables. For example, suppose 4 different motor oils are being compared to see which give the best gasoline mileage. An experimenter might then choose 4 different drivers and four different cars. To test-drive all the possible combinations, the experimenter would need 64 test-drives.

To reduce the number of test drives, a statistician might use an arrangement called a **Latin Square**.

For this example, the four motor oils are labeled A, B, C, and D and are arranged as shown. Each oil must appear exactly one time in each row and column of the square.

The drivers are labeled D(1), D(2), D(3), and D(4); the cars are labeled C(1), C(2), C(3), and C(4).

Now, the number of test-drives is just 16, one for each cell of the Latin Square.

	D(1)	D(2)	D(3)	D(4)
C(1)	A	B	C	D
C(2)	B	A	D	C
C(3)	C	D	A	B
C(4)	D	C	B	A

Create two 4-by-4 Latin Squares that are different from the example. Sample answers are given.

1.

	D(1)	D(2)	D(3)	D(4)
C(1)	B	C	A	D
C(2)	D	A	C	B
C(3)	C	D	B	A
C(4)	A	B	D	C

2.

	D(1)	D(2)	D(3)	D(4)
C(1)	D	C	B	A
C(2)	A	B	C	D
C(3)	B	D	A	C
C(4)	C	A	D	B

Make three different 3-by-3 Latin Squares.

3.

	D(1)	D(2)	D(3)
C(1)	A	B	C
C(2)	B	C	A
C(3)	C	A	B

4.

	D(1)	D(2)	D(3)
C(1)	A	C	B
C(2)	C	B	A
C(3)	B	A	C

5.

	D(1)	D(2)	D(3)
C(1)	B	C	A
C(2)	C	A	B
C(3)	A	B	C

Extra Credit

Thirty-two students ran laps around the athletic track. The table shows how many students completed each number of laps. How many students completed 3 laps but could not complete a fourth lap? **7 students**

Number of Laps Completed		
Laps	**Tally**	**Cumulative Frequency**
2	𝍸 𝍸 𝍸 𝍸 𝍸 𝍸 II	32
3	𝍸 𝍸 𝍸 𝍸	52
4	𝍸 𝍸 III	65
5	IIII	69
6	I	70

1 FOCUS

5-Minute Check
Lesson 1-6

The frequency table shows the ages of the visitors entering a theme park during the first minute after it opened.

Age	Tally	Frequency
under 13	𝍸𝍸𝍸𝍸 𝍸	25
teens	𝍸𝍸𝍸𝍸 𝍸𝍸 II	32
20s	𝍸𝍸𝍸𝍸 𝍸𝍸𝍸 I	36
30s	𝍸𝍸𝍸 III	18
40s	𝍸𝍸 I	11
50s	𝍸	5
60s	II	2

1. How many people under age 30 entered the park in the first minute? **93 people**
2. Compare the number of people in their 20s to the number of people 30 and older. **They are the same.**
3. Every other teen leaving a mall is surveyed to determine the popularity of a new ice cream flavor among teens. Is this a good sample? Explain. **Yes; the teens are chosen randomly.**

Motivating the Lesson

Real-World Connection Have students look through magazines, encyclopedias, newspapers, and so on for graphs. Discuss the features of the graphs they find.

TECHNOLOGY

An alternative technology option using a graphing calculator is available for teaching this lesson.

1-7 Displaying and Interpreting Data

What You'll Learn

You'll learn to construct and interpret line graphs, histograms, and stem-and-leaf plots.

Why It's Important

Research

Researchers collect data and use graphs to help them make predictions.

See Exercises 4–6.

Graphs are a good way to display and analyze data. The graph at the right is a **line graph**. It shows trends or changes over time. There are no holes in the graph and every point on the graph has meaning. To construct a line graph, include the following items.

1. a title
2. a label on each axis describing the variable that it represents
3. equal intervals on each axis

Note that the graph at the right contains all three items.

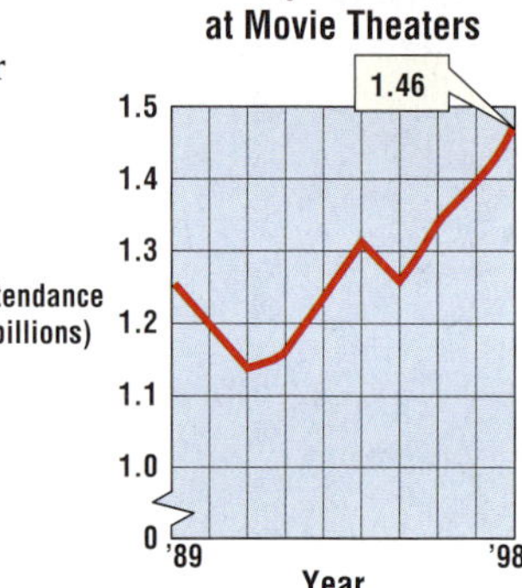

Source: *USA TODAY*

Real World

Example 1

Travel Link

The number of annual visitors to the Grand Canyon is given in the table at the right. Construct a line graph of the data. Then use the graph to predict the number of annual visitors to the Grand Canyon in the year 2010.

Year	Grand Canyon Visitors (millions)
1960	1.2
1970	2.3
1980	2.6
1990	3.8
2000	4.8

Source: National Park Service

Step 1 Draw a horizontal axis and a vertical axis and label them as shown below. Include a title.

Step 2 Plot the points.

Step 3 Draw a line by connecting the points.

You can see from the graph that the general trend is that the number of visitors to the Grand Canyon increases steadily every ten years. A good prediction for the year 2010 might be about 6 or 6.5 million people.

Grand Canyon Visitors

Number of People (millions): 0, 1, 2, 3, 4, 5, 6, 7

Year: 1960, 1970, 1980, 1990, 2000, 2010

Grand Canyon

Resource Manager

Reproducible Masters

Chapter 1 Resource Masters

- *Study Guide,* p. 31
- *Skills Practice,* p. 32
- *Practice,* p. 33
- *Reading to Learn Mathematics,* p. 34
- *Enrichment,* p. 35
- *Assessment,* p. 47

Graphing Calculator, pp. 3–4

Hands-On Algebra, pp. 26–28

Transparencies

5-Minute Check, 1–7

Technology/Multimedia

Interactive Chalkboard CD-ROM

Your Turn

a. The table at the right shows the approximate U.S. consumption of bottled water per person. Construct a line graph of the data. Then use it to predict the amount of bottled water each person will drink in the year 2005. **See margin, page 40.**

Year	Bottled Water (gallons)
1991	9
1993	10.5
1995	12
1997	14
1999	17
2001	19.5

Source: International Bottled Water Association

Another type of graph that is used to display data is a **histogram**. A histogram uses data from a frequency table and displays it over equal intervals. To make a histogram, include the same three items as the line graph: title, axes labels, and equal intervals. In a histogram, all bars should be the same width with no space between them.

Example 2
Physical Science Link

The frequency table is from Example 2 in Lesson 1–6. It shows the various time intervals that "charged" balloons remained stuck to the wall. Construct a histogram of the data.

Static Electricity		
Time (s)	**Tally**	**Frequency**
15–24	𝍸 III	8
25–34	𝍸 IIII	9
35–44	𝍸 II	7
45–54	I	1

Step 1 Draw a horizontal axis and a vertical axis and label them as shown below. Include the title.

Step 2 Label equal intervals given in the frequency table on the horizontal axis. Label equal intervals of 1 on the vertical axis.

Step 3 For each time interval, draw a bar whose height is given by the frequency.

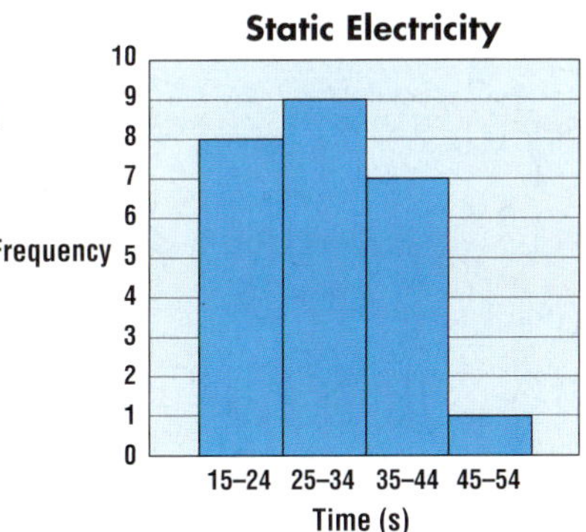

The histogram gives a better visual display of the data than the frequency table. In Lesson 1–6, we used cumulative frequency tables to organize data. Likewise, we can construct **cumulative frequency histograms**.

www.algconcepts.com/extra_examples

2 TEACH

In-Class Examples

Example 1

Construct a line graph of the data given in the table. Use the graph to predict the percent of the labor force in farming in the year 2010.

Percent of the Labor Force in Farming	
Year	**Percent**
1940	17
1950	12
1960	6
1970	3
1980	2
1990	2

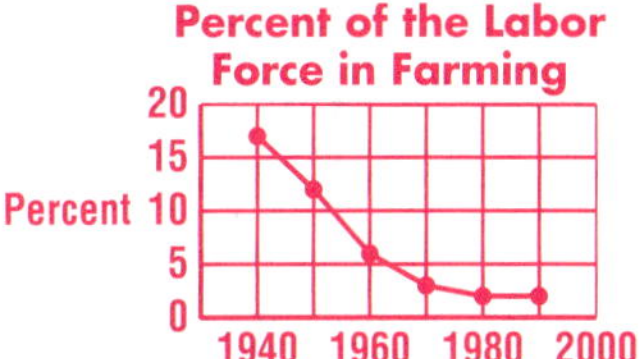

A good prediction might be between 1 and 2 percent.

Example 2

The table shows the number of people in different age groups who entered a new store during the first hour of its grand opening. Construct a histogram of the data.

Age	Tally	Frequency
1–10	𝍸	5
11–20	𝍸 III	8
21–30	𝍸 𝍸 𝍸 𝍸 𝍸	25
31–40	𝍸 𝍸 𝍸 𝍸 𝍸 𝍸 II	32
41–50	𝍸 𝍸 𝍸 𝍸 𝍸 𝍸 𝍸 I	36
51–60	𝍸 𝍸 𝍸 III	18
61–70	𝍸 𝍸 I	11

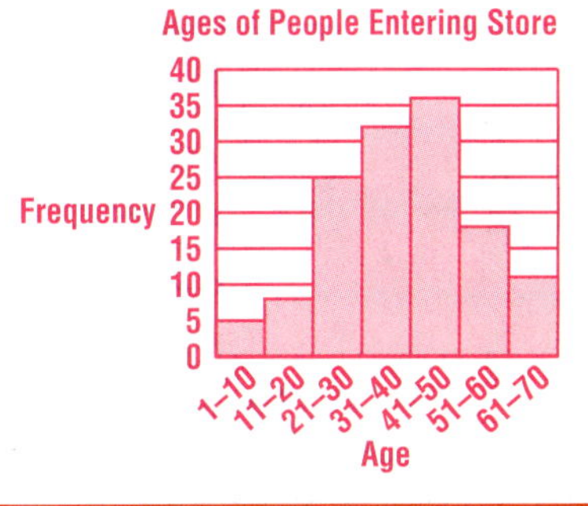

Inclusion Strategies

Assign volunteers to assist students with visual impairments during this lesson. Have the assistants point to the figures and describe them to help each student interpret the graphs and tables. Also, enlarging the figures in this lesson may be helpful to some visually-impaired students.

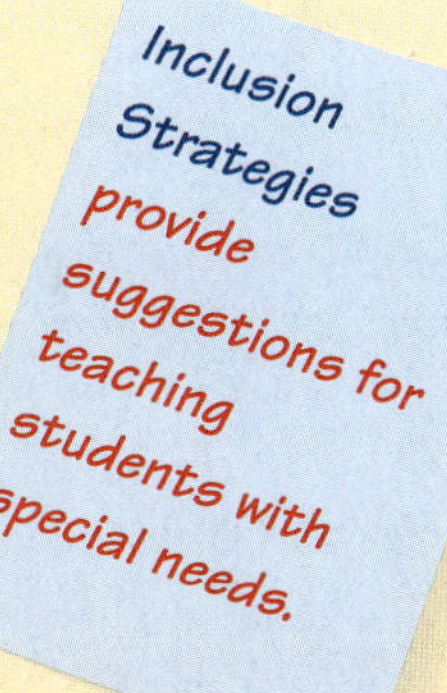

Inclusion Strategies provide suggestions for teaching students with special needs.

In-Class Example

Example 3

Construct a cumulative frequency histogram of the data given in In-Class Example 2 on page 39.

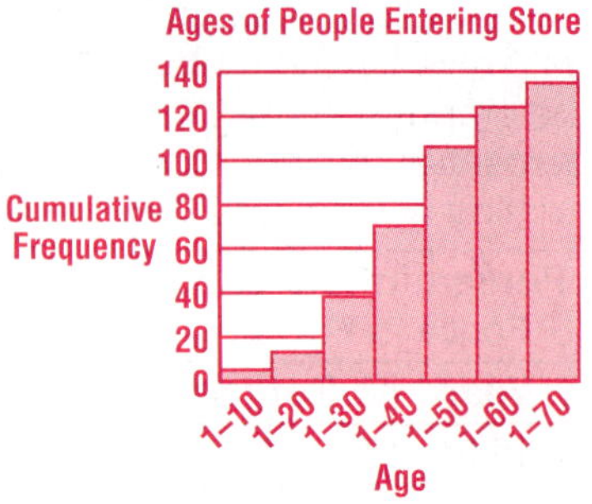

Teaching Tip When introducing stem-and-leaf plots after Example 3, stress that the leaves are always single-digit values.

Answers

a. **U.S. Consumption of Bottled Water**

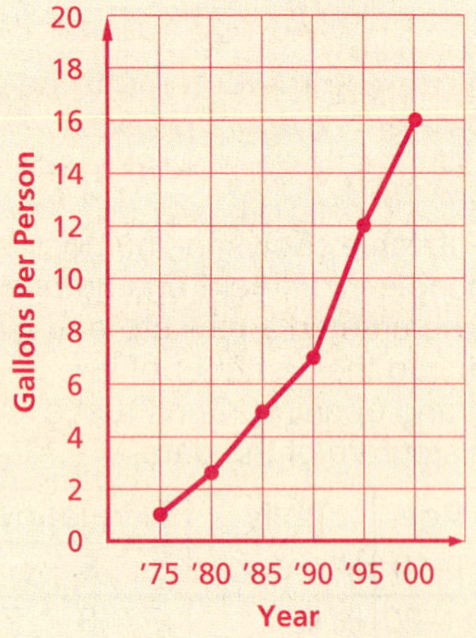

about 24.5 gallons in 2005

b. **Static Electricity**

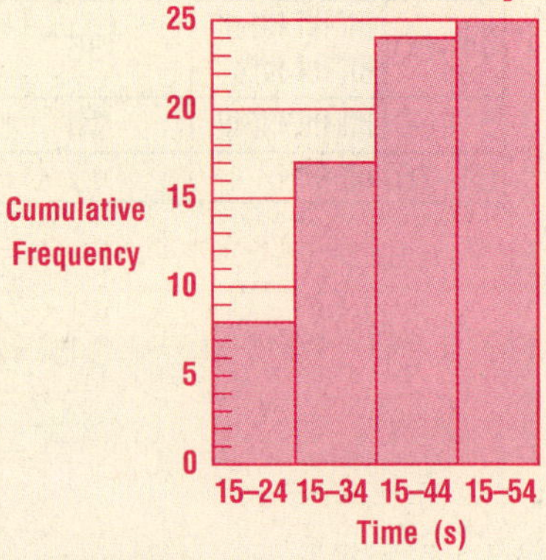

Example 3 **The ages of people who participated in a recent survey are shown in the table at the right. Construct a cumulative frequency histogram to display the data.**

Survey		
Age	Tally	Frequency
1–10	𝍸 III	8
11–20	𝍸 IIII	9
21–30	𝍸	5
31–40	𝍸 𝍸 II	12
41–50	𝍸 III	8

First, make a cumulative frequency table. Then construct a histogram using the cumulative frequencies for the bar heights. Remember to label the axes and include the title.

Survey		
Age	Frequency	Cumulative Frequency
1–10	8	8
11–20	9	17
21–30	5	22
31–40	12	34
41–50	8	42

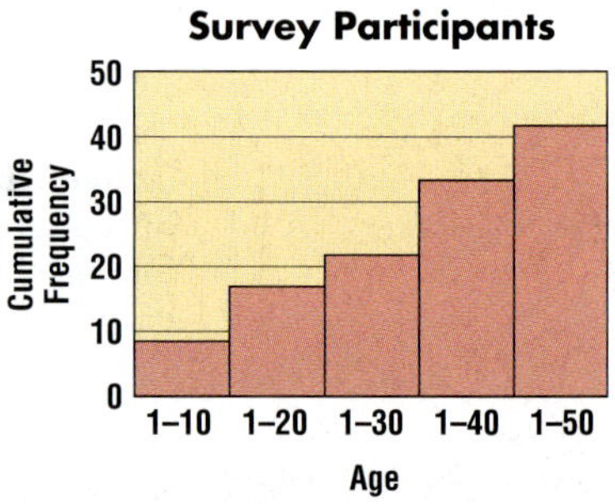

Your Turn See margin.

b. Construct a cumulative frequency histogram of the data in Example 2.

Another way to display data is a **stem-and-leaf plot**.

Info-Graphics use graphic organizers to illustrate algebraic concepts.

Stem	Leaf
1	1 6
2	1 3 9
3	
4	5 5 *2|3 = 23*

The greatest common place value for each data item is used to form the *stem*.

The *leaves* are formed by the next greatest place value.

In this case, the tens digits are the stems. The ones digits are the leaves. Write the leaves in order from least to greatest.

A *key* is always included. This shows how the digits are related.

Info Graphic

In the stem-and-leaf plot at the right, the data are represented by three-digit numbers. In this case, use the digits in the first two place values to form the stems. For example, the values for 102, 108, 114, 115, 125, 127, 131, and 139 are shown in the stem-and-leaf plot at the right.

Stem	Leaf
10	2 8
11	4 5
12	5 7
13	1 9 *11|5 = 115*

Example 4 School Link

The table shows the class results on a 50-question test. Make a stem-and-leaf plot of the grades.

Class Scores					
29	37	48	40	17	34
28	43	37	35	49	29
13	29	42	45	37	46

The tens digits are the stems, so the stems are 1, 2, 3, and 4. The ones digits are the leaves.

Stem	Leaf
1	7 3
2	9 8 9 9
3	7 4 7 5 7
4	8 0 3 9 2 5 6

3|7 = 37

Now arrange the leaves in numerical order to make the results easier to observe and analyze.

Stem	Leaf
1	3 7
2	8 9 9 9
3	4 5 7 7 7
4	0 2 3 5 6 8 9

3|7 = 37

What were the highest and lowest scores?
49 and 13

Which score occurred most frequently?
29 and 37, three times each

How many students received a score of 35 or better?
11 students

Your Turn

c. Make a stem-and-leaf plot of the quiz grades below. **See margin.**
54, 55, 60, 42, 41, 75, 50, 68, 62, 54, 70, 50

Check for Understanding

Communicating Mathematics

1. Line graphs usually show change over time, while histograms compare quantities of similar nature.

1. **Explain** the differences between the use of line graphs and histograms.
2. **Identify** each essential part of a correctly drawn line graph or histogram. **2–3. See margin.**
3. **You Decide** Marcia says that a histogram works as well as a line graph to show trends over time. Manuel says that a histogram shows intervals, not trends. Who is correct? Explain.

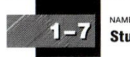

Vocabulary
line graph
histogram
cumulative frequency histograms
stem-and-leaf plot

In-Class Example

Example 4

The table shows the record high temperatures for several states. Make a stem-and-leaf plot of the temperatures.

Record High Temperatures for Several U.S. States (°C)			
44	38	53	49
57	48	41	43
41	43	45	38
48	47	47	48

Stem	Leaf
3	8 8
4	1 1 3 3 4 5 7 7 8 8 8 9
5	3 7

3|8 = 38

Answers

Your Turn

c.

Stem	Leaf
4	1 2
5	0 0 4 4 5
6	0 2 8
7	0 5

6|2 = 62

Exercises

2. A correct graph has a title, horizontal and vertical axes labeled, and equally spaced units on both axes.
3. Manuel; line graphs are usually used to show trends over time. Histograms are usually used to show frequency of items in sets of data.

***Study Guide*, p. 31**

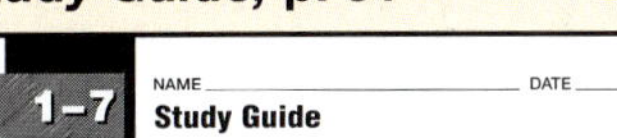

Displaying and Interpreting Data

Data can be easier to analyze if they are presented in the form of a graph. There are many ways to graph data, such as line graphs, histograms, and stem-and-leaf plots. A histogram is a graph of the data in a frequency table.

Example: The frequency table shows the amount of time skiers waited in a lift line. Construct a histogram for the data.

Lift Line Wait		
Time (min)	Tally	Frequency
12–14	|	1
15–17	卌 卌	10
18–20	卌 卌 ||	12
21–23	卌 卌 卌 |	16
24–26	卌 卌 |	11

- The horizontal axis displays the time intervals from the table.
- The vertical axis displays equal intervals of 1.
- For each time interval, draw a bar. The height of the bar is equal to its frequency.
- Label the two axes and title the histogram.

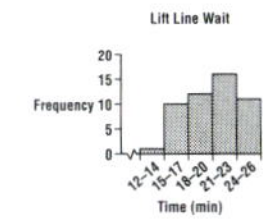

Make a histogram of the data in the frequency table.

Students Visiting Museum		
Ages	Tally	Frequency
8–9	卌 |	6
10–11	卌 卌	10
12–13	卌 卌 卌	15
14–15	卌 卌 ||	12
16–17	||||	4

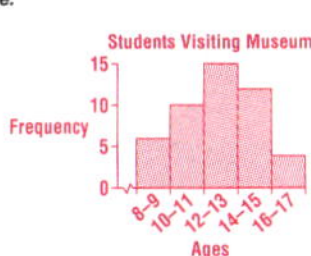

You Decide exercises help students address common errors before they occur.

Reteaching Activity

Intrapersonal Learners Have students write a journal entry describing what concepts they found easiest and most difficult in this lesson.

3 PRACTICE/APPLY

Error Analysis

Watch for students who feel they do not have sufficient information to answer Exercise 8.
Prevent by referring students to the histogram. Ask them how many leaves were 8 to 9 centimeters long. Have them record their answer. Then ask them how many leaves were 10 to 11 centimeters long. Again have them record their answer. Lead students to recognize that they can add the numbers of leaves in each interval to find the total.

Assignment Guide

Basic: 13–23 odd, 24–28
Average: 12–18 even, 19–28

***Skills Practice,* p. 32, and *Practice,* p. 33 (shown)**

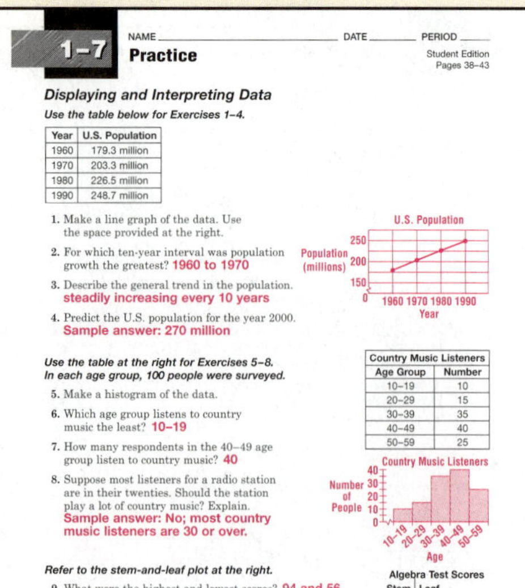

1–7 NAME ______ DATE ______ PERIOD ______
Practice
Student Edition Pages 38–43

Displaying and Interpreting Data

Use the table below for Exercises 1–4.

Year	U.S. Population
1960	179.3 million
1970	203.3 million
1980	226.5 million
1990	248.7 million

1. Make a line graph of the data. Use the space provided at the right.
2. For which ten-year interval was population growth the greatest? 1960 to 1970
3. Describe the general trend in the population. steadily increasing every 10 years
4. Predict the U.S. population for the year 2000. Sample answer: 270 million

Use the table at the right for Exercises 5–8. In each age group, 100 people were surveyed.

Country Music Listeners	
Age Group	Number
10–19	10
20–29	15
30–39	35
40–49	40
50–59	25

5. Make a histogram of the data.
6. Which age group listens to country music the least? 10–19
7. How many respondents in the 40–49 age group listen to country music? 40
8. Suppose most listeners for a radio station are in their twenties. Should the station play a lot of country music? Explain. Sample answer: No; most country music listeners are 30 or over.

Refer to the stem-and-leaf plot at the right.

9. What were the highest and lowest scores? 94 and 56
10. Which test score occurred most frequently? 83
11. In which 10-point interval did most of the students score? 70–80
12. How many students scored 75 or better? 14
13. How many students received a score less than 75? 10

Algebra Test Scores

Stem	Leaf
5	6 7 7 8
6	1 4 9
7	3 3 4 5 5 7 8
8	1 3 3 3 6 9
9	0 1 2 4

7 | 5 = 75

Guided Practice

The table at the right shows the percent of homes in California with internet access.

Year	Percent of Homes On-Line
1997	19
1998	25
1999	28
2000*	31

Source: Pacific Telesis
*estimated

Example 1
4. Make a line graph of the data. **4. See margin.**
5. Between which two years was the growth of on-line access the greatest? **5. 1997–1998**
6. Predict the percent of homes with on-line access in the year 2001. **Sample answer: 34%**

Refer to the histogram at the right.

Example 2
7. Determine the length of most maple leaves. **14–15 cm**

Example 2
8. How many leaves were sampled? **300**

Example 3
9. Construct a cumulative frequency histogram of the data. **See Solutions Manual.**

Example 3
10. How many of the leaves were no more than 15 centimeters long? **240**

Leaf Lengths in a Maple Tree Population

Example 4
11. **Weather** The stem-and-leaf plot at the right shows the daily high temperatures in McComb, Mississippi, in March.
Source: *The Weather Underground*
 a. What was the highest temperature? **81°F**
 b. On how many days was the high temperature in the 70s? **12 days.**
 c. What temperature occurred most frequently? **63°F**

Daily High Temperatures (°F)

Stem	Leaf
4	8
5	4 9 9
6	3 3 3 3 4 4 8 8 8
7	0 0 2 2 2 3 3 5 5 9 9 9
8	1 1

4 | 8 = 48°F

***inter*NET CONNECTION**
Data Update For the latest information on weather forecasts, visit: www.algconcepts.com

Exercises

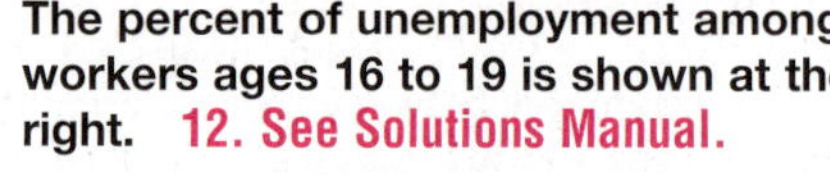

Practice

Homework Help

For Exercises	See Examples
12–14	1
15–18	2
19–23	4

Extra Practice
See page 694.

The percent of unemployment among workers ages 16 to 19 is shown at the right. **12. See Solutions Manual.**

Year	Percent of Working Teens Unemployed
1992	17
1993	19
1994	18
1995	17.5
1996	17
1997	15
1998	13

Source: U.S. Labor Dept.

12. Make a line graph of the data.
13. When was unemployment at its highest? **1993**
14. Describe the general trend in unemployment among teens ages 16 to 19. **It was decreasing.**

Answer

4.

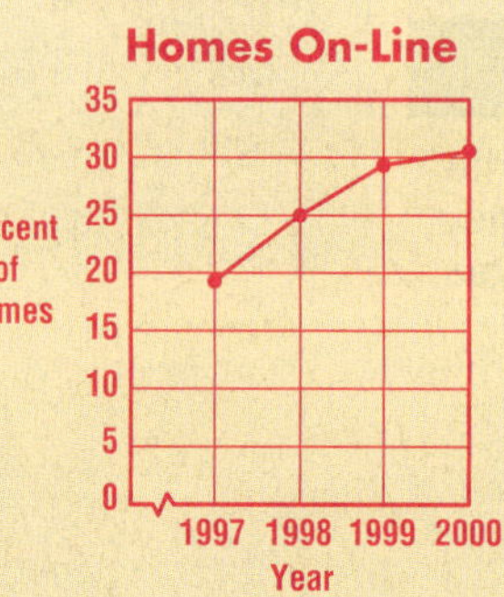

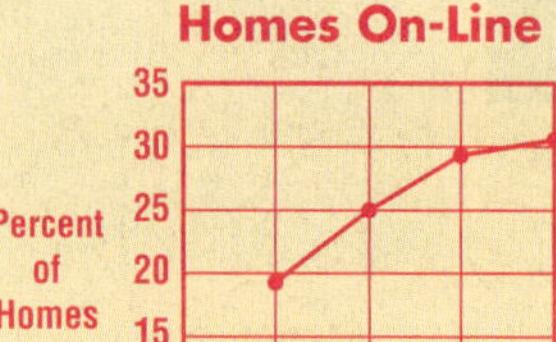

From the Classroom of ...

Helen Carpini
Middletown High School
Middletown, Connecticut

Use Exercises 7 through 10 as an activity lab. Students can make and graph their own sample by collecting and measuring leaves.

From the Classroom of... features contain teaching suggestions from teachers who are creatively teaching Algebra in their classrooms.

In a survey, men and women were asked how long they were willing to stay on hold when calling a customer service representative about a product they purchased. The results are shown in the table at the right.

How Long on Hold?		
Time (min)	Percent of Men	Percent of Women
0–1	28	18
2–3	32	36
4–5	23	27
6–7	7	10
8+	10	9

Source: Bruskin/Goldring for Inference

15. Make a histogram showing the men's responses. **15–18. See margin.**
16. Make a histogram showing the women's responses.
17. How do your histograms compare?
18. Who do you think would hang up the phone sooner, men or women?

The stem-and-leaf plot at the right gives the number of catches of the NFL's leading pass receiver for the first 39 seasons.

Stem	Leaf
6	0 1 2 6 7
7	1 1 1 2 3 3 3 5 7 8
8	2 5 8 8 9
9	0 0 1 2 2 2 3 5
10	0 0 0 1 4 6 8 8
11	2
12	2 3 *12 | 3 = 123*

19. What was the greatest number of catches during a season? **123**
20. How many seasons are represented? **39**
21. What number of catches occurred most frequently? **71, 73, 92, and 100 occur three times.**
22. How many leading pass receivers have at least 90 catches? **19**
23. **Critical Thinking** *Back-to-back stem-and-leaf plots* are used to compare two sets of data. The back-to-back stem-and-leaf plot below compares the performance of two algebra classes on their first test. Which class do you believe did better on the test? Why do you think so? **See margin.**

First Period	Stem	Second Period
9 8	5	
9 9 8 7	6	7 8
7 7 6 5 3	7	2 4 4 4 5 5 8
5 4 1	8	1 3 4 5 7
8 8 6 5 2	9	0 1 1 3 8
		8 | 9 = 89

Mixed Review

24. No; it is much colder in Alaska.

Determine whether each is a good sample. *(Lesson 1–6)*

24. A survey is taken in Alaska to determine how much money an average family in the United States spends on heating their home.
25. In a survey, every third name in the phone book is called and the person answering is interviewed. **Yes; it is a random sample.**
26. Write a formula for the perimeter P of a square with side s in simplest form. *(Lesson 1–5)* $P = 4s$

Standardized Test Practice

27. **Short Response** Write $5x + 3(x - y)$ in simplest form. *(Lesson 1–4)* $8x - 3y$
28. **Multiple Choice** Evaluate $12 \cdot 6 + 3 \cdot 2 - 8$. *(Lesson 1–2)* **C**

   (A) 142 (B) −144 (C) 70 (D) 120

www.algconcepts.com/self_check_quiz

4 ASSESS

Open-Ended Assessment

Speaking Have students explain when they would use a line graph to display data and when they would use a histogram.

Chapter 1, Quiz B (Lessons 1–4 through 1–7) is available in the *Chapter 1 Resource Masters*, p. 11.

Answers

15.

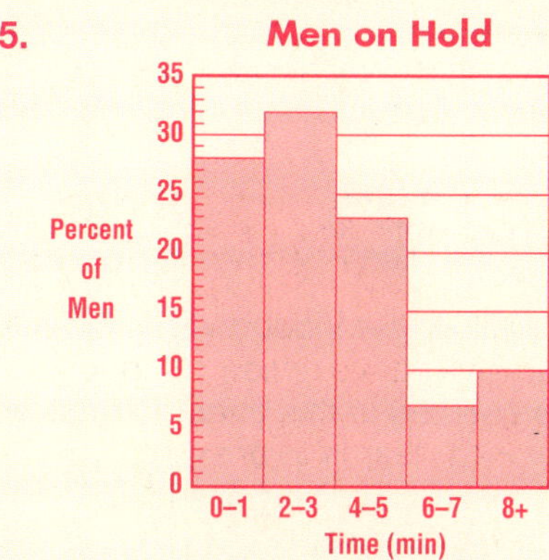

16.

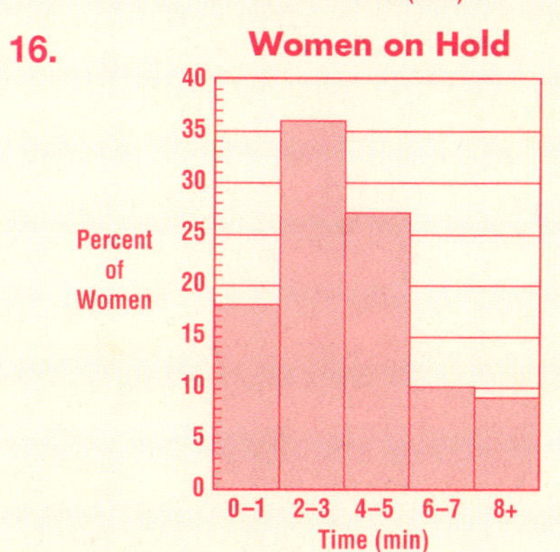

***Enrichment*, p. 35**

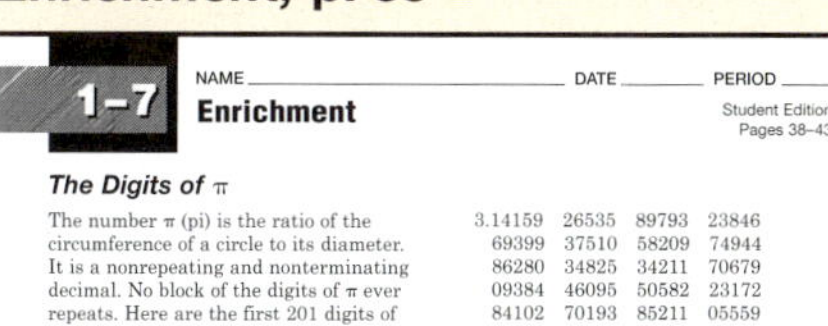

NAME ________ DATE ______ PERIOD ______

1-7 Enrichment

Student Edition Pages 38–43

The Digits of π

The number π (pi) is the ratio of the circumference of a circle to its diameter. It is a nonrepeating and nonterminating decimal. No block of the digits of π ever repeats. Here are the first 201 digits of π including 200 digits that follow the decimal point.

3.14159 26535 89793 23846
69399 37510 58209 74944
86280 34825 34211 70679
09384 46095 50582 23172
84102 70193 85211 05559

26433 83279 50288 41971
59230 78164 06286 20899
82148 08651 32823 06647
53594 08128 34111 74502
64462 29489 54930 38196

Solve each problem.

1. If each of the digits appeared with equal frequency, how many times would each digit appear in the first 200 places following the decimal point? **20**
2. Complete this frequency distribution table for the first 200 digits of π that follow the decimal point.

Digit	Frequency (Tally Marks)	Frequency (Number)	Cumulative Frequency
0	𝍸 𝍸 𝍸 ||||	19	19
1	𝍸 𝍸 𝍸 𝍸	20	39
2	𝍸 𝍸 𝍸 𝍸 ||||	24	63
3	𝍸 𝍸 𝍸 𝍸	20	83
4	𝍸 𝍸 𝍸 𝍸 ||	22	105
5	𝍸 𝍸 𝍸 𝍸	20	125
6	𝍸 𝍸 𝍸 |	16	141
7	𝍸 𝍸 ||	12	153
8	𝍸 𝍸 𝍸 𝍸 ||||	24	177
9	𝍸 𝍸 𝍸 𝍸 |||	23	200

3. Explain how the cumulative frequency column can be used to check a project like this one. **The last number should be 200, the number of items being counted.**
4. Which digit(s) appears most often? **2 and 8**
5. Which digit(s) appears least often? **7**

Extra Credit

A histogram shows the number of people in different age ranges from teens to 70s visiting a bookstore, with the bar for teens on the left. Every bar is shorter than the bar to its left. What can you deduce about the bookstore? **Sample answer: The bookstore attracts more teens than other any age group.**

Answers

17. **Sample answer: The histograms show that a greater percent of men than women are willing to stay on hold for 0–1 minute.**
18. **Sample answer: Women, since a smaller percent are willing to stay on hold for 0–1 minute.**
23. **Sample answer: More students in second period scored 80 or more, so second period did better.**

Study Guide and Assessment

Understanding and Using the Vocabulary

This section provides a listing of the new terms, properties, and phrases that were introduced in this chapter. The exercises check students' understanding of the terms by using a variety of verbal formats including matching, completion, and true/false.

Glossary A complete glossary of terms appears on pages 762–783.

MindJogger Videoquizzes

MindJogger Videoquizzes provide an alternative review of concepts presented in this chapter. Students work in teams to answer questions, gaining points for correct answers.

The Study Guide and Assessment guides students through studying the vocabulary and concepts of the chapter. It includes examples and practice problems for the key concepts from each lesson.

CHAPTER 1 Study Guide and Assessment

Understanding and Using the Vocabulary

After completing this chapter, you should be able to define each term, property, or phrase and give an example of each.

interNET CONNECTION **Review Activities**
For more review activities, visit:
www.algconcepts.com

Algebra

algebraic expression *(p. 4)*
coefficient *(p. 20)*
equation *(p. 5)*
equivalent expressions *(p. 20)*
evaluating *(p. 10)*
factors *(p. 4)*
formula *(p. 24)*
like terms *(p. 20)*
numerical expression *(p. 4)*
order of operations *(p. 8)*
product *(p. 4)*
quotient *(p. 4)*
simplest form *(p. 20)*
simplify *(p. 15)*
term *(p. 20)*
variable *(p. 4)*
whole numbers *(p. 16)*

Statistics

cumulative frequency histogram *(p. 39)*
cumulative frequency table *(p. 33)*
data *(p. 32)*
experiment *(p. 32)*
frequency table *(p. 32)*
histogram *(p. 39)*
line graph *(p. 38)*
observational study *(p. 32)*
population *(p. 32)*
sample *(p. 32)*
sampling *(p. 32)*
stem-and-leaf plot *(p. 40)*
tally marks *(p. 32)*

Logic

conclusion *(p. 30)*
conditional *(p. 30)*
counterexample *(p. 16)*
deductive reasoning *(p. 30)*
hypothesis *(p. 30)*
if-then statement *(p. 30)*
inductive reasoning *(p. 30)*

Choose the correct term to complete each sentence.

1. A (coefficient, term) is a number, a variable, or a product or quotient of numbers and variables.
2. The result of two numbers multiplied together is the (factor, product).
3. A(n) (numerical expression, algebraic expression) contains variables.
4. According to the (order of operations, like terms), you do multiplication before addition.
5. A (counterexample, hypothesis) shows that a statement is not always true.
6. Some examples of (like terms, whole numbers) are $2x$, $10x$, and $-6x$.
7. A (sample, variable) is a group used to represent a much larger population.
8. Any sentence that contains an equals sign is a(n) (equation, formula).
9. Using (sampling, frequency tables) is a way to organize data.
10. A (histogram, stem-and-leaf plot) makes it easier to identify specific data items.

Skills and Concepts

Objectives and Examples

- **Lesson 1–1** Translate words into algebraic expressions and equations.

Write an algebraic expression for *7 decreased by the quantity x divided by 2.*

$7 - (x \div 2)$ or $7 - \frac{x}{2}$

Review Exercises

Write an algebraic expression.

11. the product of 5 and n **$5n$**
12. the sum of 2 and three times x **$2 + 3x$**

Write an equation for each sentence.

13. Six less than two times y equals 14. **13. $2y - 6 = 14$**
14. The quotient of 20 and x is 4. **$\frac{20}{x} = 4$**

www. algconcepts.com/vocabulary_review

Resource Manager

Reproducible Masters
Chapter 1 Resource Masters
- *Assessment*, pp. 37–44, 48–50

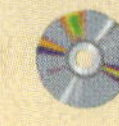

Technology/Multimedia
MindJogger Videoquizzes
ExamView® Pro

Objectives and Examples

- **Lesson 1–2** Use the order of operations to evaluate expressions.

$2 \cdot 7 + 2 \cdot 3 = 14 + 6$ *Multiply.*
$= 20$ *Add.*

- **Lesson 1–3** Use the commutative and associative properties to simplify expressions.

Name the property shown by $3 + x + 2 = 3 + 2 + x$. Then simplify.

$3 + x + 2 = 3 + 2 + x$ *Commutative (+)*
$= 5 + x$ *Substitution*

- **Lesson 1–4** Use the Distributive Property to evaluate expressions.

$5b + 3(b + 2) = 5b + 3 \cdot b + 3 \cdot 2$
$= 5b + 3b + 6$ *Multiply.*
$= (5 + 3)b + 6$ *Distribute.*
$= 8b + 6$ *Simplify.*

- **Lesson 1–5** Use the four-step plan to solve problems.

Explore What do you know? What are you trying to find?

Plan How will you go about solving this? What problem-solving strategy could you use?

Solve Carry out your plan. Does it work? Do you need another plan? If necessary, choose a variable for an unknown and write an expression.

Examine Check your answer. Does it make sense? Is it reasonably close to your estimate?

Review Exercises

Find the value of each expression.

15. $3 + 8 \div 2$ **7**
16. $12 \div 4 + 15 \cdot 3$ **48**
17. $29 - 3(9 - 4)$ **14**
18. $4(11 + 7) - 9 \cdot 8$ **0**
19. Find the value of $3ac - b$ if $a = 6, b = 9$, and $c = 1$. **9**

Name the property shown by each statement. Then simplify.

20. $6 + (7 + b) = (6 + 7) + b$ **Assoc. (+), 13 + *b***
21. $2 \cdot c \cdot 10 = 2 \cdot 10 \cdot c$ **Comm. (×), 20*c***
22. $9 \cdot (5 \cdot f) = (9 \cdot 5) \cdot f$ **Assoc. (×), 45*f***
23. $x(5 + 4) = (5 + 4)x$ **Comm. (×), 9*x***
24. $3 + a + 8 = 3 + 8 + a$ **Comm. (+), 11 + *a***
25. $(g + 1) + 2 = g + (1 + 2)$ **Assoc. (+), *g* + 3**

Simplify each expression.

26. $4(8 + y)$ **32 + 4*y***
27. $7(v - 1)$ **7*v* − 7**
28. $10x + x$ **11*x***
29. $h(2 + a)$ **2*h* + *ah***
30. $5z + 2z - 6$ **7*z* − 6**
31. $10 + 3(4 - d)$ **22 − 3*d***

Use the four-step plan to solve each problem.

32. Finance Mr. Rockwell deposited $1000 in an account that pays 2% interest. How much money would he have in the account after ten years? **$1200**

33. School Jamal is typing a three-page report with approximately 400 words per page for school. He thought he could finish typing the report in 2 hours. After $1\frac{1}{2}$ hours, he had finished 2 pages. **a. 1200**

a. How many words are in his paper?

b. About how many words had Jamal typed in $1\frac{1}{2}$ hours? **800**

Skills and Concepts

The **Objectives and Examples** section reviews the skills and concepts of the chapter and shows completely worked examples.

The **Review Exercises** provide practice for the corresponding objectives.

ExamView® Pro

Use the networkable ***ExamView® Pro Testmaker CD-ROM*** to:

- Create **multiple versions** of tests.
- Create **modified** tests for **inclusion** students with one mouse click.
- **Edit** existing questions and **add** your own questions.
- Build tests aligned with state standards using built-in **state curriculum correlations**.
- Change **English** tests to **Spanish** with one mouse click and vice versa.

Mixed Problem Solving
See pages 724–731.

Objectives and Examples

- **Lesson 1–6** Collect and organize data using sampling and frequency tables.

Make a frequency table for the data {1, 4, 3, 4, 0, 2, 3, 1, 0, 2, 0, 2, 0, 4, 0, 0, 4, 1, 2}.

Number	Tally	Frequency
0	~~IIII~~ I	6
1	III	3
2	IIII	4
3	II	2
4	IIII	4

Review Exercises

Use the frequency table at the left to answer each question. 36. 4

34. How many numbers are in the sample? 19

35. Which number occurs most frequently? 0

36. How many times does the number 2 occur?

37. Make a cumulative frequency table from the data. See margin.

38. How many times does a number less than 2 occur? 9

39. How many times does a number greater than or equal to 2 occur? 10

- **Lesson 1–7** Construct and interpret line graphs, histograms, and stem-and-leaf plots.

Construct a histogram for the data {10, 10, 10, 10, 11, 11, 12, 13, 14, 14, 14, 15, 15}.

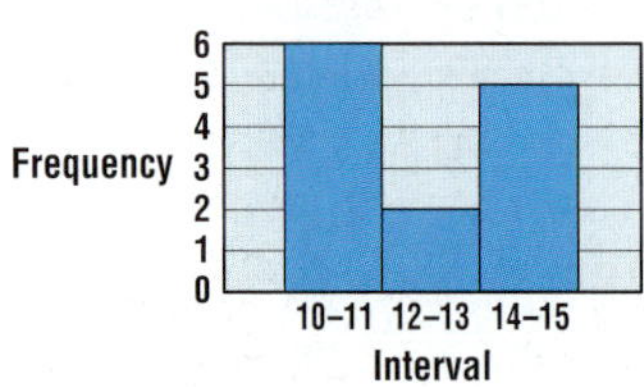

Use the histogram at the left to answer each question.

40. How large is each interval? 2

41. Which interval has the most data? 10–11

42. How many numbers have a value greater than 11? 7

43. Make a cumulative histogram from the data. See margin.

44. How many numbers are in the sample? 13

45. How many numbers have a value less than 14? 8

Applications and Problem Solving

46. Geometry Write an equation to represent the perimeter P of the figure below. Then solve for P if $x = 9$ and $y = 5$. *(Lesson 1–5)* $P = 6x + 2y$; $P = 64$ cm

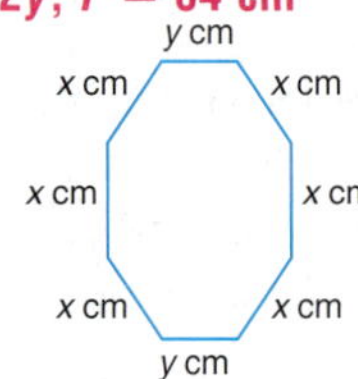

47. Testing The stem-and-leaf plot below shows the scores from a driver's test. *(Lesson 1–7)*

a. What were the highest and lowest scores?

b. Which score occurred most frequently? 88

c. How many people received a score of 76 or better? 10

a. 100; 62

Stem	Leaf
6	2 8
7	4 5 5 6
8	0 4 8 8 8
9	2 4 7
10	0

$7|4 = 74$

Applications and Problem Solving

This section provides additional practice in solving real-world problems that involve the concepts of this chapter.

Answers

37.

Number	Cumulative Frequency
0	6
1	9
2	13
3	15
4	19

43.

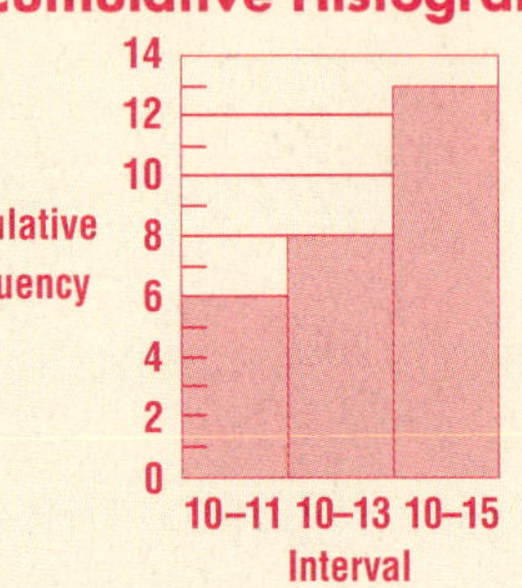

Assessment, pp. 39–40

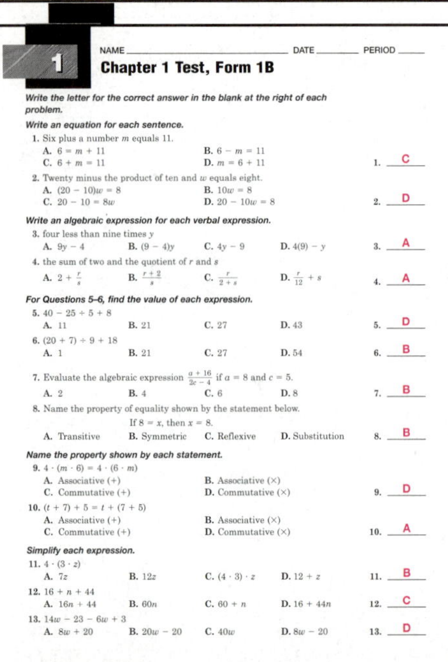

NAME ______ DATE ______ PERIOD ______

Chapter 1 Test, Form 1B

Write the letter for the correct answer in the blank at the right of each problem.

Write an equation for each sentence.

1. Six plus a number m equals 11.
   A. $6 = m + 11$ B. $6 - m = 11$ C. $6 + m = 11$ D. $m = 6 + 11$ — 1. C
2. Twenty minus the product of ten and w equals eight.
   A. $(20 - 10)w = 8$ B. $10w = 8$ C. $20 - 10 = 8w$ D. $20 - 10w = 8$ — 2. D

Write an algebraic expression for each verbal expression.

3. four less than nine times y
   A. $9y - 4$ B. $(9 - 4)y$ C. $4y - 9$ D. $4(9) - y$ — 3. A
4. the sum of two and the quotient of r and s
   A. $2 + \frac{r}{s}$ B. $\frac{r+2}{s}$ C. $\frac{r}{2+s}$ D. $\frac{r}{12} + s$ — 4. A

For Questions 5–6, find the value of each expression.

5. $40 - 25 \div 5 + 8$
   A. 11 B. 21 C. 27 D. 43 — 5. D
6. $(20 + 7) \div 9 + 18$
   A. 1 B. 21 C. 27 D. 54 — 6. B
7. Evaluate the algebraic expression $\frac{a + 16}{2c - 4}$ if $a = 8$ and $c = 5$.
   A. 2 B. 4 C. 6 D. 8 — 7. B
8. Name the property of equality shown by the statement below.
   If $8 = x$, then $x = 8$.
   A. Transitive B. Symmetric C. Reflexive D. Substitution — 8. B

Name the property shown by each statement.

9. $4 \cdot (m \cdot 6) = 4 \cdot (6 \cdot m)$
   A. Associative (+) B. Associative (×) C. Commutative (+) D. Commutative (×) — 9. D
10. $(t + 7) + 5 = t + (7 + 5)$
   A. Associative (+) B. Associative (×) C. Commutative (+) D. Commutative (×) — 10. A

Simplify each expression.

11. $4 \cdot (3 \cdot z)$
   A. $7z$ B. $12z$ C. $(4 \cdot 3) \cdot z$ D. $12 + z$ — 11. B
12. $16 + n + 44$
   A. $16n + 44$ B. $60n$ C. $60 + n$ D. $16 + 44n$ — 12. C
13. $14w - 23 - 6w + 3$
   A. $8w + 20$ B. $20w - 20$ C. $40w$ D. $8w - 20$ — 13. D

Assessment

Four forms of Chapter 1 Test are available in the *Chapter 1 Resource Masters.*

Chapter 1 Test, Form 1B, is shown at the left. Chapter 1 Test, Form 2B, is shown on the next page.

	Form of Test		Level
1A	Multiple Choice	pp. 37–38	Average
1B	Multiple Choice	pp. 39–40	Basic
2A	Free Response	pp. 41–42	Average
2B	Free Response	pp. 43–44	Basic

CHAPTER 1 Test

1. **Explain** why we use the order of operations in mathematics. **See margin.**
2. **List** three like terms with the variable k. **Sample answer: $\frac{k}{2}$, k, $6k$**

Write an algebraic expression for each verbal expression.

3. x increased by 12 **$x + 12$**
4. the quotient of 5 and y **$5 \div y$ or $\frac{5}{y}$**
5. 1 less than 8 times p **$8p - 1$**

Use the order of operations to find the value of each expression.

6. $13 + 4 \cdot 5$ **33**
7. $12 + 6 \div 3 - 4$ **10**
8. $3(8 + 2) - 7$ **23**

Evaluate each expression if $h = 8$, $j = 3$, and $k = 2$.

9. $k(4 + j) + 6$ **20**
10. $\frac{h + k}{h - j}$ **2**

Name the property shown by each statement.

11. If $11 = 7 + x$, then $7 + x = 11$. **Symmetric**
12. $28 \cdot 1 = 28$ **Mult. Identity**
13. $(r \cdot 9) \cdot 3 = r \cdot (9 \cdot 3)$ **Assoc. (×)**
14. $10 + b = b + 10$ **Comm. (+)**
15. $6(m + 2) = 6 \cdot m + 6 \cdot 2$ **Distributive**

Simplify each expression.

16. $n + 5n$ **$6n$**
17. $6x - 4x + 9y - 4y$ **$2x + 5y$**
18. $4(2s + 8t - 1)$ **$8s + 32t - 4$**

19. **Sports** Danny stayed late after every basketball practice to shoot 5 free throws. The chart shows how many free throws he made out of 5 for each night of practice.
    a. Make a frequency table to organize the data. **See margin.**
    b. If Danny has basketball practice 5 days a week, how many weeks did he stay late, shooting free throws? **4 weeks**
    c. What number of free throws did he make most often? **3 free-throws**
    d. How many times did he not make any free throws? **2 times**
    e. How many times did he make all 5 free throws? **1 time**

Free Throws (out of 5)			
1	4	0	2
1	3	3	2
4	3	3	2
5	3	2	2
1	0	4	3

20. **Communication** The line graph shows the growth in sales of prepaid calling cards. **a. 1996–1997 or 1997–1998**
    a. Between which two years was growth in sales the greatest?
    b. Predict the number of sales for the year 2002. **Sample answer: $6.1 billion**

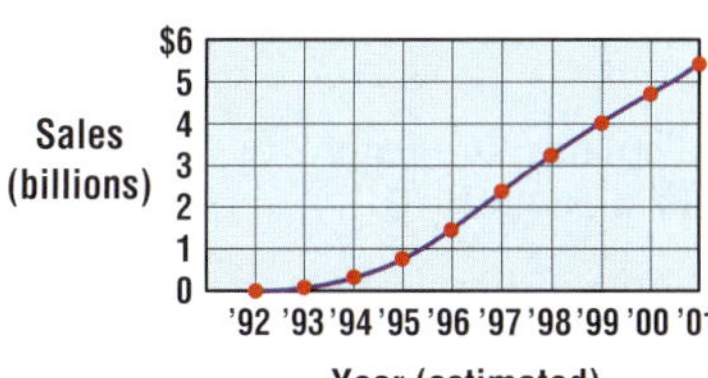

Source: Atlantic ACM

Chapter Test Bonus Question

Add grouping symbols to make the following equation correct.

$6 + 7 \times 2 - 10 \div 5 = 24$

$(6 + 7) \times 2 - (10 \div 5) = 24$

Chapter Test

Answers

1. We use the order of operations so that everyone gets the same answer for a given expression or equation.

19a.

Free Throws	Tally	Frequency
0	||	2
1	|||	3
2	~~||||~~	5
3	~~||||~~ |	6
4	|||	3
5	|	1

The Chapter Resource Masters provide 2 multiple-choice, 2 free-response, 1 open-ended, and 1 mid-chapter test, along with 1 cumulative review, 1 standardized test practice, and 2 quizzes. Reduced facsimiles of many of these pages are shown throughout.

***Assessment*, pp. 43–44**

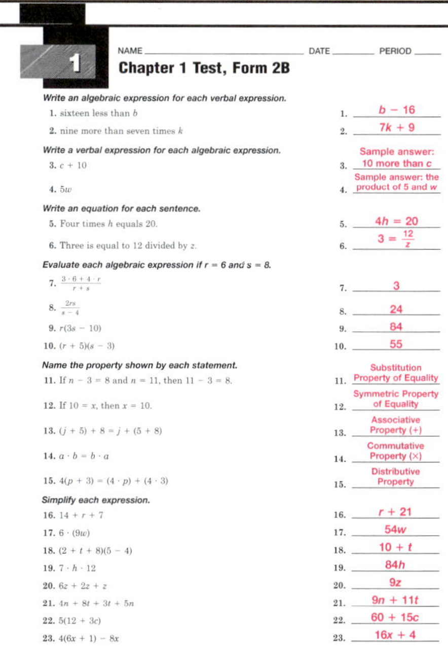

NAME ________ DATE ______ PERIOD ____

1 Chapter 1 Test, Form 2B

Write an algebraic expression for each verbal expression.

1. sixteen less than b — 1. $b - 16$
2. nine more than seven times k — 2. $7k + 9$

Write a verbal expression for each algebraic expression.

3. $c + 10$ — 3. Sample answer: 10 more than c
4. $5w$ — 4. Sample answer: the product of 5 and w

Write an equation for each sentence.

5. Four times h equals 20. — 5. $4h = 20$
6. Three is equal to 12 divided by z. — 6. $3 = \frac{12}{z}$

Evaluate each algebraic expression if $r = 6$ and $s = 8$.

7. $\frac{3 \cdot 6 + 4 \cdot r}{r + s}$ — 7. 3
8. $\frac{2rs}{s - 4}$ — 8. 24
9. $r(3s - 10)$ — 9. 84
10. $(r + 5)(s - 3)$ — 10. 55

Name the property shown by each statement.

11. If $n - 3 = 8$ and $n = 11$, then $11 - 3 = 8$. — 11. Substitution Property of Equality
12. If $10 = x$, then $x = 10$. — 12. Symmetric Property of Equality
13. $(j + 5) + 8 = j + (5 + 8)$ — 13. Associative Property (+)
14. $a \cdot b = b \cdot a$ — 14. Commutative Property (×)
15. $4(p + 3) = (4 \cdot p) + (4 \cdot 3)$ — 15. Distributive Property

Simplify each expression.

16. $14 + r + 7$ — 16. $r + 21$
17. $6 \cdot (9w)$ — 17. $54w$
18. $(2 + t + 8)(5 - 4)$ — 18. $10 + t$
19. $7 \cdot h \cdot 12$ — 19. $84h$
20. $6z + 2z + z$ — 20. $9z$
21. $4n + 8t + 3t + 5n$ — 21. $9n + 11t$
22. $5(12 + 3c)$ — 22. $60 + 15c$
23. $4(6x + 1) - 8x$ — 23. $16x + 4$

Preparing for Standardized Tests

Pages 48–49 are part of a complete test preparation course that is described in detail on page T9 of the Teacher's Handbook. The test items on these pages were written in the same style as those in state proficiency tests and standardized tests like ACT and SAT.

Diagnosis and Prescription

Each of the 10 test questions on page 49 is cross-referenced to the lesson where that SAT or ACT skill is covered. If students miss a particular type of problem, you can have them study that skill.

(See chart at the bottom of page 49. Note that SPT = State Proficiency Test, SAT = Scholastic Assessment Test, and ACT = American College Test.)

The items on the Preparing for Standardized Tests pages were created to closely parallel those on state proficiency tests and college entrance exams, like PSAT, ACT, and SAT.

Assessment, p. 48

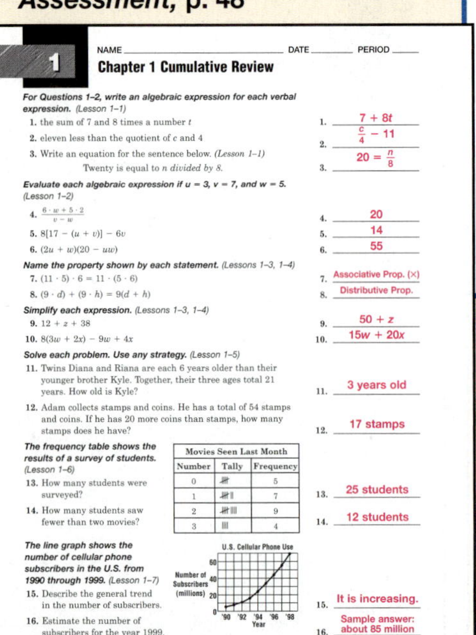

NAME ________ DATE ______ PERIOD ______

1 Chapter 1 Cumulative Review

For Questions 1–2, write an algebraic expression for each verbal expression. *(Lesson 1–1)*

1. the sum of 7 and 8 times a number t — 1. $7 + 8t$
2. eleven less than the quotient of c and 4 — 2. $\frac{c}{4} - 11$
3. Write an equation for the sentence below. *(Lesson 1–1)* Twenty is equal to n *divided by 8.* — 3. $20 = \frac{n}{8}$

Evaluate each algebraic expression if $u = 3$, $v = 7$, and $w = 5$. *(Lesson 1–2)*

4. $\frac{6 \cdot w + 5 \cdot 2}{v - w}$ — 4. 20
5. $8[17 - (u + v)] - 6v$ — 5. 14
6. $(2u + w)(20 - uw)$ — 6. 55

Name the property shown by each statement. *(Lessons 1–3, 1–4)*

7. $(11 \cdot 5) \cdot 6 = 11 \cdot (5 \cdot 6)$ — 7. Associative Prop. (×)
8. $(9 \cdot d) + (9 \cdot h) = 9(d + h)$ — 8. Distributive Prop.

Simplify each expression. *(Lessons 1–3, 1–4)*

9. $12 + z + 38$ — 9. $50 + z$
10. $8(3w + 2x) - 9w + 4x$ — 10. $15w + 20x$

Solve each problem. Use any strategy. *(Lesson 1–5)*

11. Twins Diana and Riana are each 6 years older than their younger brother Kyle. Together, their three ages total 21 years. How old is Kyle? — 11. 3 years old
12. Adam collects stamps and coins. He has a total of 54 stamps and coins. If he has 20 more coins than stamps, how many stamps does he have? — 12. 17 stamps

The frequency table shows the results of a survey of students. *(Lesson 1–6)*

Movies Seen Last Month		
Number	Tally	Frequency
0	𝍸	5
1	𝍸 ‖	7
2	𝍸 ||||	9
3						4

13. How many students were surveyed? — 13. 25 students
14. How many students saw fewer than two movies? — 14. 12 students

The line graph shows the number of cellular phone subscribers in the U.S. from 1990 through 1999. *(Lesson 1–7)*

15. Describe the general trend in the number of subscribers. — 15. It is increasing.
16. Estimate the number of subscribers for the year 1999. — 16. Sample answer: about 85 million

CHAPTER 1 Preparing for Standardized Tests

Number Concept Problems

Standardized tests include many questions written with realistic settings. Read each question carefully. Be sure you understand the situation and what the question asks.

A calculator can help, but you can often find the answer faster with a pencil and your own math skills. Since standardized tests are timed, you will want to find the correct answers as quickly as possible.

Test-Taking Tip

Translate words into arithmetic symbols.

is	=
of	×
per	÷

Example 1

Mrs. Lopez estimates that $\frac{2}{3}$ of the families in her neighborhood will participate in the annual garage sale. If there are 225 families in her neighborhood, how many families does she expect to participate?

Ⓐ 75 Ⓑ 150
Ⓒ 175 Ⓓ 220

Hint Estimate the answer before making any calculations.

Solution First estimate. Since $\frac{2}{3}$ is greater than $\frac{1}{2}$, more than one half of the 225 families will participate. One half of 225 is about 112. So, choice A is not possible.

The word *of* ($\frac{2}{3}$ *of* the families) tells you to use multiplication.

$\frac{2}{3} \cdot 225 = \frac{2(225)}{3}$ *Multiply.*

$= \frac{2(\overset{75}{\cancel{225}})}{\underset{1}{\cancel{3}}}$ *Simplify.*

$= 2(75)$ or 150 *Multiply.*

If you use your calculator, multiply 2 by 225, and then divide the answer by 3.

Two-thirds of the 225 families is 150 families. So, the answer is B.

Example 2

Jan drove 144 miles between 10:00 A.M. and 12:40 P.M. What was her average speed in miles per hour?

Hint Pay attention to the units of measure.

Solution From 10:00 A.M. to 12:40 P.M. is 2 hours and 40 minutes. You need time in *hours*. Convert minutes to hours.

2 hours 40 minutes $= 2\frac{40}{60}$ hours or $2\frac{2}{3}$ hours

$\frac{144}{2\frac{2}{3}} = \frac{144}{\frac{8}{3}}$ *Divide the miles by time. Rename $2\frac{2}{3}$ as $\frac{8}{3}$.*

$= 144 \cdot \frac{3}{8}$ *Multiply by the reciprocal of $\frac{8}{3}$.*

$= \overset{18}{\cancel{144}} \cdot \frac{3}{\underset{1}{\cancel{8}}}$ *Divide by the GCF, 8.*

$= 18(3)$ or 54 *Multiply.*

The answer is 54 miles per hour. Record it on the grid.

- Start with the *left* column.
- Write the answer in the boxes at the top. Write one digit in each column.
- Mark the corresponding oval in each column.
- *Never* grid a mixed number; change it to a fraction or a decimal.

Resource Manager

Reproducible Masters

Chapter 1 Resource Masters

- *Assessment and Evaluation*, pp. 48–50

Preparing for Standardized Tests
For test-taking strategies and more practice, see pages 732–749.

After you work each problem, record your answer on the answer sheet provided or on a sheet of paper.

Multiple Choice

1. Grant purchased a shirt for \$29.95 and 2 pairs of socks for \$2.95 a pair. The sales tax on these purchases was \$2.42. What was the total amount Grant spent? *(Prerequisite Skill)* **C**

   (A) \$35.32 (B) \$35.85
   (C) \$38.27 (D) \$39.37

2. Ariel adds $\frac{1}{2}$ cup of flour to a bowl that already has $3\frac{2}{3}$ cups of flour. How many total cups of flour will be in the bowl? *(Prerequisite Skill)* **B**

   (A) $7\frac{1}{3}$ (B) $4\frac{1}{6}$ (C) 4 (D) $3\frac{1}{6}$

3. The number 1134 is divisible by all of the following except— *(Basic Skill)* **D**

   (A) 3. (B) 6. (C) 9. (D) 12. (E) 14.

4. Dr. Hewson has 758 milliliters of a solution to use for a class lab experiment. She divides the solution evenly among 32 students. If 22 milliliters are left after the experiment, how much of the solution did she give each student? *(Lesson 1–5)* **A**

   (A) 23.0 mL (B) 24.2 mL
   (C) 33.0 mL (D) 35.9 mL

5. For shipping, a company charges \$2.75 in addition to \$1.25 for each \$10 ordered. Which equation represents the cost c for shipping an order worth \$50? *(Lesson 1–1)* **C**

   (A) $\frac{c}{50} = 2.75 + 1.25$
   (B) $c + 2.75 = 1.25 + \frac{50}{10}$
   (C) $c = 2.75 + 1.25\left(\frac{50}{10}\right)$
   (D) $c = \frac{50}{10}(2.75) + 1.25$

6. Baseballs are packed one dozen per box. There are 208 baseballs to be packed. How many more baseballs will be needed to fill the last, partially filled box? *(Lesson 1–5)* **C**

   (A) 0 (B) 4 (C) 8 (D) 12 (E) 18

7. Franco is making a casserole. The recipe uses 8 cups of macaroni and serves 12 people. How many cups of macaroni does the recipe use per person? *(Lesson 1–5)* **A**

   (A) $\frac{2}{3}$ (B) 1
   (C) $\frac{1}{2}$ (D) $\frac{1}{3}$

8. Use the commutative and associative properties to compute the product. *(Lesson 1–3)* **C**

   $$2 \cdot 4 \cdot 2.5 \cdot 15 \cdot 5 \cdot 10$$

   (A) 1500 (B) 12,000
   (C) 15,000 (D) 120,000

Grid In

9. The daily newspaper always follows a particular format. Each even-numbered page contains 6 articles, and each odd-numbered page contains 7 articles. If today's paper has 36 pages, how many articles does it contain? *(Lesson 1–5)* **234**

Extended Response

10. The average annual snowfall in Denver, Colorado, is 59.8 inches. How many feet of snow can Denver residents expect in the next 4 years? *(Lesson 1–5)*

    Part A List the operations you use to solve this problem. Calculate the answer to the nearest hundredth of a foot. Show your work. **See margin.**

    Part B Round your answer to the nearest foot. **20 ft**

www.algconcepts.com/standardized_test

A bubble-in answer sheet for these practice problems is available on page A1 of the *Chapter 1 Resource Masters.*

Additional Practice

Additional test practice questions are available in the *Chapter 1 Resource Masters*, pp. 49–50.

Answer

10A. Sample answer: Use division to change from inches to feet. Use multiplication to find the total snowfall in 4 years. $59.8 \div 12 \times 4 = 19.93$ ft

***Assessment*, pp. 49–50**

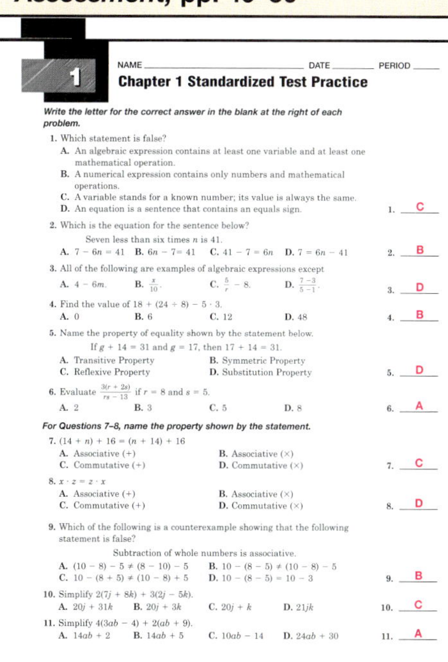

NAME ______ DATE ______ PERIOD ______

Chapter 1 Standardized Test Practice

Write the letter for the correct answer in the blank at the right of each problem.

1. Which statement is false?
   A. An algebraic expression contains at least one variable and at least one mathematical operation.
   B. A numerical expression contains only numbers and mathematical operations.
   C. A variable stands for a known number; its value is always the same.
   D. An equation is a sentence that contains an equals sign. 1. __C__
2. Which is the equation for the sentence below?
   Seven less than six times n is 41.
   A. $7 - 6n = 41$ B. $6n - 7 = 41$ C. $41 - 7 = 6n$ D. $7 = 6n - 41$ 2. __B__
3. All of the following are examples of algebraic expressions except
   A. $4 - 6m$. B. $\frac{x}{10}$. C. $\frac{5}{r} - 8$. D. $\frac{7-3}{5-1}$. 3. __D__
4. Find the value of $18 + (24 \div 8) - 5 \cdot 3$.
   A. 0 B. 6 C. 12 D. 48 4. __B__
5. Name the property of equality shown by the statement below.
   If $g + 14 = 31$ and $g = 17$, then $17 + 14 = 31$.
   A. Transitive Property B. Symmetric Property
   C. Reflexive Property D. Substitution Property 5. __D__
6. Evaluate $\frac{3(r + 2s)}{rs - 13}$ if $r = 8$ and $s = 5$.
   A. 2 B. 3 C. 5 D. 8 6. __A__

For Questions 7–8, name the property shown by the statement.

7. $(14 + n) + 16 = (n + 14) + 16$
   A. Associative (+) B. Associative (×)
   C. Commutative (+) D. Commutative (×) 7. __C__
8. $x \cdot z = z \cdot x$
   A. Associative (+) B. Associative (×)
   C. Commutative (+) D. Commutative (×) 8. __D__
9. Which of the following is a counterexample showing that the following statement is false?
   Subtraction of whole numbers is associative.
   A. $(10 - 8) - 5 \neq (8 - 10) - 5$ B. $10 - (8 - 5) \neq (10 - 8) - 5$
   C. $10 - (8 + 5) \neq (10 - 8) + 5$ D. $10 - (8 - 5) = 10 - 3$ 9. __B__
10. Simplify $2(7j + 8k) + 3(2j - 5k)$.
    A. $20j + 31k$ B. $20j + 3k$ C. $20j + k$ D. $21jk$ 10. __C__
11. Simplify $4(3ab - 4) + 2(ab + 9)$.
    A. $14ab + 2$ B. $14ab + 5$ C. $10ab - 14$ D. $24ab + 30$ 11. __A__

	Chapter 1 Number Concept Problems	
Ex. 1	fraction word problem	SPT
Ex. 2	time word problem	SAT
1	decimal word problem	SPT
2	fraction word problem	SPT
3	divisibility	ACT
4	word problem	SPT
5	write an equation	SPT
6	word problem	ACT
7	fraction word problem	SPT
8	properties	SPT
9	counting	SAT
10	word problem	SPT

CHAPTER 2

Resource Manager

Integers

Instructional Objectives

Lesson (pages)	**Objectives**	**NCTM Standards 2000**	**State/Local Objectives**
2–1 (52–57)	Graph integers on a number line and compare and order integers.	1, 6, 8, 9, 10	
2–2 (58–63)	Graph points on a coordinate plane.	1, 6, 8, 9, 10	
2–3 (64–69)	Add integers.	1, 6, 8, 9	
2–4 (70–74)	Subtract integers.	1, 6, 8, 9	
2–5 (75–79)	Multiply integers.	1, 6, 8, 9	
Investigation (80–81)	Explore matrices.	1, 6, 8, 9, 10	
2–6 (82–85)	Divide integers.	1, 6, 8, 9	

Key to NCTM Standards 2000
1 Number & Operations; **2** Algebra; **3** Geometry; **4** Measurement; **5** Data Analysis & Probability; **6** Problem Solving; **7** Reasoning & Proof; **8** Communications; **9** Connections; **10** Representation

Suggested Pacing *See page T13 for a complete course-planning calendar.*

Standard refers to schedules that provide 45- to 55-minute periods that meet each day.
Block refers to schedules that provide approximately 90-minute periods which may meet every day for one semester or every other day over two semesters.

PACING	DAY 1	DAY 2	DAY 3	DAY 4	DAY 5	DAY 6
Standard Core (Chapters 1–13)	Lesson 2–1		Lesson 2–2		Lesson 2–3	
Standard Enhanced (Chapters 1–15)	Lesson 2–1		Lesson 2–2	Lesson 2–3	Lesson 2–4	Lesson 2–5
Block Core (Chapters 1–13)	Chapter 1 Test & Lesson 2–1	Lessons 2–2 & 2–3	Lessons 2–4 & 2–5	INV & Lesson 2–6	SG+A	Chapter Test & Lesson 3–1
Block Enhanced (Chapters 1–15)	Chapter 1 Test & Lesson 2–1	Lessons 2–2 & 2–3	Lessons 2–4 & 2–5	INV & Lesson 2–6	SG+A	Chapter Test & Lesson 3–1

Instructional Resources

Lesson	Materials and Manipulatives (see below for Glencoe Manipulative Resources)	Blackline Masters (page numbers): Chapter 2 Resource Masters: Study Guide	Practice (Skills & Average)	Reading to Learn Mathematics	Enrichment	Assessment	Hands-On Algebra*	School-to-Workplace*	Graphing Calculator Masters*	5-Minute Check Transparencies
2–1	thermometer decks of playing cards	51	52–53	54	55					2–1
2–2	grid paper [1, 4] graphing calculator straightedge [1, 2] atlas or city map photocopy of a map colored pencils or pens masking tape	56	57–58	59	60	91	33		5–7	2–2
2–3	algebra tiles [1, 2, 3, 4] equation/product mat [1, 2, 4] straightedge [1, 2]	61	62–63	64	65	90	34, 35	2		2–3
2–4	algebra tiles [1, 2, 3, 4] equation/product mat [1, 2, 4]	66	67–68	69	70		36			2–4
2–5	straightedge [1, 2] counters [1, 2, 3, 4] algebra tiles [1, 2, 3, 4]	71	72–73	74	75		37, 38			2–5
Investigation	calculator self-adhesive notes									
2–6	algebra tiles [1, 2, 3, 4]	76	77–78	79	80	91				2–6
Study Guide & Assessment/ Chapter Test	straightedge [1, 2] grid paper [1, 4]					81–89 92–94				

**See page 50c for examples of these instructional materials.*

Key to Glencoe Manipulative Resources

[1]Classroom Manipulative Resources [2]Student Manipulative Resources [3]Overhead Manipulative Resources [4]Hands-On Algebra Masters

INV = Investigation SG+A = Study Guide and Assessment

DAY 7	DAY 8	DAY 9	DAY 10	DAY 11	DAY 12	DAY 13
Lesson 2–4	Lesson 2–5	INV	Lesson 2–6	SG+A	Chapter Test	
INV	Lesson 2–6	SG+A	Chapter Test			

Resource Manager

The pages shown on this page are a small sample of the materials available on *TeacherWorks: All-in-One Lesson Planner and Resource Center.*

This CD-ROM includes all of the blackline masters and transparencies available for this program.

It also includes a lesson planner and interactive Teacher Edition, so you can customize lesson plans and reproduce classroom resources quickly and easily, from just about anywhere.

Applications

School-to-Workplace Masters, p. 2

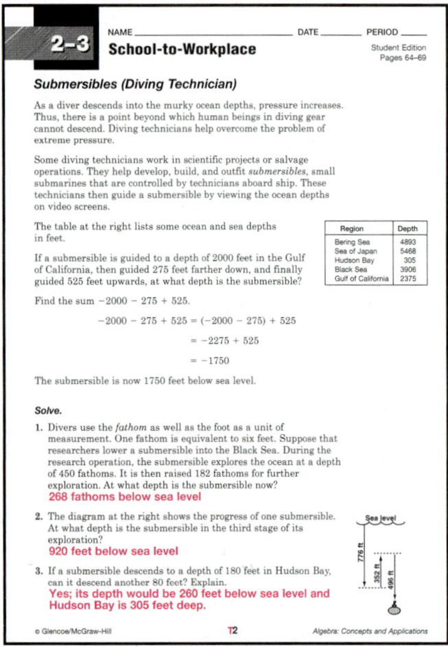

2-3 NAME ______ DATE ______ PERIOD ______

School-to-Workplace

Student Edition Pages 64–69

Submersibles (Diving Technician)

As a diver descends into the murky ocean depths, pressure increases. Thus, there is a point beyond which human beings in diving gear cannot descend. Diving technicians help overcome the problem of extreme pressure.

Some diving technicians work in scientific projects or salvage operations. They help develop, build, and outfit *submersibles*, small submarines that are controlled by technicians aboard ship. These technicians then guide a submersible by viewing the ocean depths on video screens.

The table at the right lists some ocean and sea depths in feet.

Region	Depth
Bering Sea	4893
Sea of Japan	5468
Hudson Bay	305
Black Sea	3906
Gulf of California	2375

If a submersible is guided to a depth of 2000 feet in the Gulf of California, then guided 275 feet farther down, and finally guided 525 feet upwards, at what depth is the submersible?

Find the sum $-2000 - 275 + 525$.

$$-2000 - 275 + 525 = (-2000 - 275) + 525$$
$$= -2275 + 525$$
$$= -1750$$

The submersible is now 1750 feet below sea level.

Solve.

1. Divers use the *fathom* as well as the foot as a unit of measurement. One fathom is equivalent to six feet. Suppose that researchers lower a submersible into the Black Sea. During the research operation, the submersible explores the ocean at a depth of 450 fathoms. It is then raised 182 fathoms for further exploration. At what depth is the submersible now?
268 fathoms below sea level
2. The diagram at the right shows the progress of one submersible. At what depth is the submersible in the third stage of its exploration?
920 feet below sea level
3. If a submersible descends to a depth of 180 feet in Hudson Bay, can it descend another 80 feet? Explain.
Yes; its depth would be 260 feet below sea level and Hudson Bay is 305 feet deep.

© Glencoe/McGraw-Hill T2 *Algebra: Concepts and Applications*

Manipulatives/Modeling

Hands-On Algebra Masters, pp. 33–38

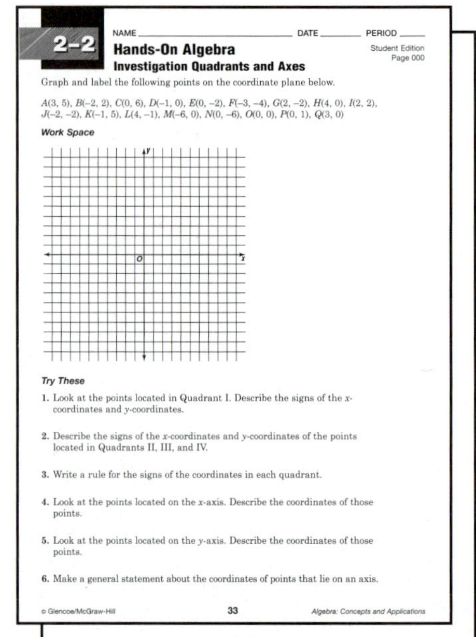

2-2 NAME ______ DATE ______ PERIOD ______

Hands-On Algebra
Investigation Quadrants and Axes

Student Edition Page 000

Graph and label the following points on the coordinate plane below.

$A(3, 5)$, $B(-2, 2)$, $C(0, 6)$, $D(-1, 0)$, $E(0, -2)$, $F(-3, -4)$, $G(2, -2)$, $H(4, 0)$, $I(2, 2)$, $J(-2, -2)$, $K(-1, 5)$, $L(4, -1)$, $M(-6, 0)$, $N(0, -6)$, $O(0, 0)$, $P(0, 1)$, $Q(3, 0)$

Work Space

Try These

1. Look at the points located in Quadrant I. Describe the signs of the x-coordinates and y-coordinates.
2. Describe the signs of the x-coordinates and y-coordinates of the points located in Quadrants II, III, and IV.
3. Write a rule for the signs of the coordinates in each quadrant.
4. Look at the points located on the x-axis. Describe the coordinates of those points.
5. Look at the points located on the y-axis. Describe the coordinates of those points.
6. Make a general statement about the coordinates of points that lie on an axis.

© Glencoe/McGraw-Hill 33 *Algebra: Concepts and Applications*

Technology/Multimedia

Graphing Calculator Masters, pp. 5–7

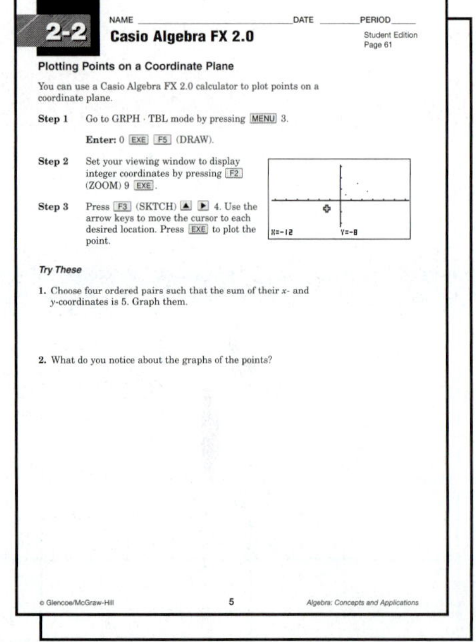

2-2 NAME ______ DATE ______ PERIOD ______

Casio Algebra FX 2.0

Student Edition Page 61

Plotting Points on a Coordinate Plane

You can use a Casio Algebra FX 2.0 calculator to plot points on a coordinate plane.

Step 1 Go to GRPH · TBL mode by pressing MENU 3.

Enter: 0 EXE F5 (DRAW).

Step 2 Set your viewing window to display integer coordinates by pressing F2 (ZOOM) 9 EXE.

Step 3 Press F3 (SKTCH) ▲ ▶ 4. Use the arrow keys to move the cursor to each desired location. Press EXE to plot the point.

Try These

1. Choose four ordered pairs such that the sum of their x- and y-coordinates is 5. Graph them.
2. What do you notice about the graphs of the points?

© Glencoe/McGraw-Hill 5 *Algebra: Concepts and Applications*

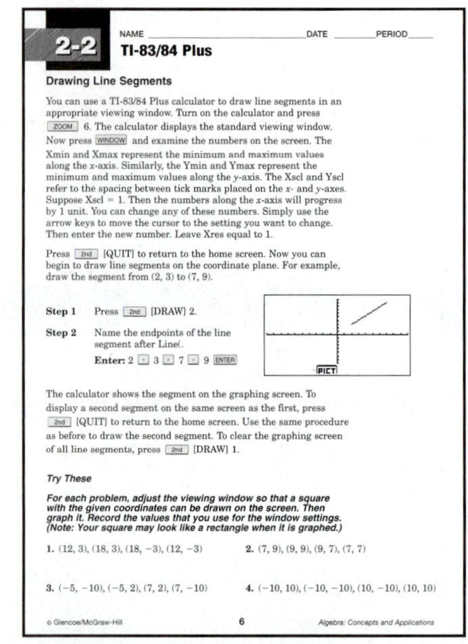

2-2 NAME ______ DATE ______ PERIOD ______

TI-83/84 Plus

Drawing Line Segments

You can use a TI-83/84 Plus calculator to draw line segments in an appropriate viewing window. Turn on the calculator and press ZOOM 6. The calculator displays the standard viewing window. Now press WINDOW and examine the numbers on the screen. The Xmin and Xmax represent the minimum and maximum values along the x-axis. Similarly, the Ymin and Ymax represent the minimum and maximum values along the y-axis. The Xscl and Yscl refer to the spacing between tick marks placed on the x- and y-axes. Suppose Xscl = 1. Then the numbers along the x-axis will progress by 1 unit. You can change any of these numbers. Simply use the arrow keys to move the cursor to the setting you want to change. Then enter the new number. Leave Xres equal to 1.

Press 2nd [QUIT] to return to the home screen. Now you can begin to draw line segments on the coordinate plane. For example, draw the segment from (2, 3) to (7, 9).

Step 1 Press 2nd [DRAW] 2.

Step 2 Name the endpoints of the line segment after Line(.

Enter: 2 , 3 , 7 , 9 ENTER

The calculator shows the segment on the graphing screen. To display a second segment on the same screen as the first, press 2nd [QUIT] to return to the home screen. Use the same procedure as before to draw the second segment. To clear the graphing screen of all line segments, press 2nd [DRAW] 1.

Try These

For each problem, adjust the viewing window so that a square with the given coordinates can be drawn on the screen. Then graph it. Record the values that you use for the window settings. (Note: Your square may look like a rectangle when it is graphed.)

1. (12, 3), (18, 3), (18, −3), (12, −3)
2. (7, 9), (9, 9), (9, 7), (7, 7)
3. (−5, −10), (−5, 2), (7, 2), (7, −10)
4. (−10, 10), (−10, −10), (10, −10), (10, 10)

© Glencoe/McGraw-Hill 6 *Algebra: Concepts and Applications*

GLENCOE'S ASSESSMENT ADVANTAGE

Assessment Resources

Type	Student Edition	Teacher's Wraparound Edition	Chapter 2 Resource Masters
Ongoing Assessment	Quizzes 1 and 2, pp. 63, 79	5-Minute Check, pp. 52, 58, 64, 70, 75, 82	Mid-Chapter Test, p. 90 Quizzes A and B, p. 91
Mixed Review	Mixed Review, pp. 57, 63, 69, 74, 79, 85 Standardized Test Practice, Chapters 1–2, pp. 90–91		Cumulative Review, p. 92 Standardized Test Practice, pp. 93–94
Error Analysis	You Decide, pp. 55, 84	Error Analysis, pp. 56, 61, 68, 73, 78, 84	
Standardized Test Prep	Standardized Test Practice, pp. 57, 63, 69, 74, 79, 85 Standardized Test Practice, Chapters 1–2, pp. 90–91		Standardized Test Practice, pp. 93–94
Open-Ended Assessment	Writing Math, pp. 61, 68 Investigation, pp. 80–81 Portfolio, pp. 51, 81	Modeling: pp. 57, 69 Speaking: pp. 63, 85 Writing: pp. 74, 79	Extended Response Assessment, p. 89
Chapter Assessment	Study Guide and Assessment, pp. 86–88 Chapter Test, p. 89		Multiple-Choice Tests (Forms 1A, 1B), pp. 81–84 Free-Response Tests (Forms 2A, 2B), pp. 85–88

Additional Chapter Resources

Student Edition

Hands-On Algebra, p. 66
Graphing Calculator Exploration, p. 61

Teacher's Classroom Resources

Manipulatives/Modeling
Teacher's Guide for Overhead Manipulative Resources

Meeting Individual Needs
Prerequisite Skills Booklet
Spanish Study Guide and Assessment, pp. 16–21, 107–108

Teaching Aids
Solutions Manual
5-Minute Check Transparencies

Glencoe Technology

Instructional

AlgePASS CD-ROM, Lessons 7, 8
Interactive Chalkboard CD-ROM
StudentWorks Plus CD-ROM
Multimedia Applications CD-ROM, Activity 2
Vocabulary PuzzleMaker (online)

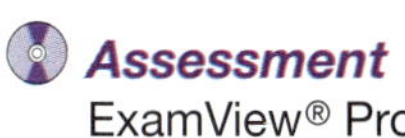

Assessment

ExamView® Pro

Teaching Aids

Answer Key Maker CD-ROM
TeacherWorks CD-ROM

Visit ***www.algconcepts.com***
for data updates, career information, games, and other interactive activities.

Mathematics of the Chapter

This chapter provides students with an in-depth study of integers. Students will begin by comparing and ordering integers, and graphing integers on numbers lines. Students then graph points on coordinate planes. The remainder of the chapter is spent discussing operations with integers, where students add, subtract, multiply, and divide integers.

What You'll Learn

Have students read over the lists of key ideas and key vocabulary. Have them make a list of any words with which they are not familiar.

Why It's Important

Point out to students that this is only one of many reasons why each objective is important. Others are provided in each lesson.

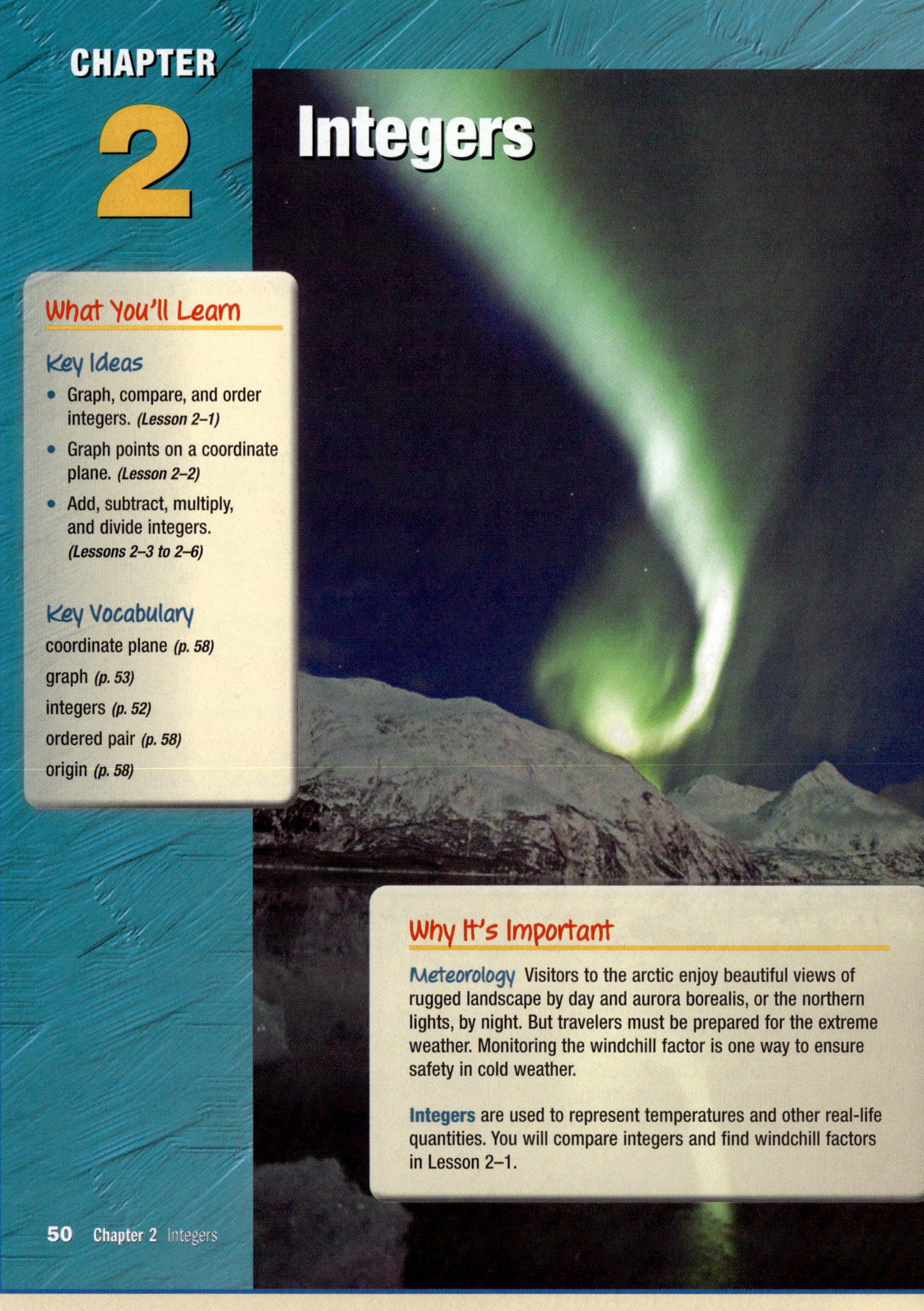

CHAPTER 2 Integers

What You'll Learn

Key Ideas

- Graph, compare, and order integers. *(Lesson 2–1)*
- Graph points on a coordinate plane. *(Lesson 2–2)*
- Add, subtract, multiply, and divide integers. *(Lessons 2–3 to 2–6)*

Key Vocabulary

coordinate plane *(p. 58)*
graph *(p. 53)*
integers *(p. 52)*
ordered pair *(p. 58)*
origin *(p. 58)*

Why It's Important

Meteorology Visitors to the arctic enjoy beautiful views of rugged landscape by day and aurora borealis, or the northern lights, by night. But travelers must be prepared for the extreme weather. Monitoring the windchill factor is one way to ensure safety in cold weather.

Integers are used to represent temperatures and other real-life quantities. You will compare integers and find windchill factors in Lesson 2–1.

50 Chapter 2 Integers

CHAPTER 2 LINKS						
Lesson	**2–1**	**2–2**	**2–3**	**2–4**	**2–5**	**2–6**
Math in the Workplace	Meteorology	Meteorology	Banking	Budgeting	Health	Farming
Applications and Connections	Population History	Biology Entertainment Geography	Games Golf Marketing	Population Meteorology Communications	Patterns Oceanography Chemistry Meteorology	Media Business Animals Energy Swimming
Math Integration		Geometry	Geometry		Geometry	

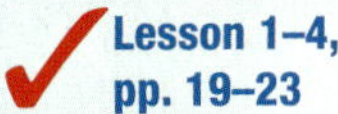

Study these lessons to improve your skills.

✓Check Your Readiness

Lesson 1–4, pp. 19–23

Simplify each expression.

1. $6x + 4x$ **10x**
2. $15a + 2a$ **17a**
3. $11d + 8d$ **19d**
4. $22t + 9t$ **31t**
5. $8y - 2y$ **6y**
6. $14b - 6b$ **8b**
7. $18r - 12r$ **6r**
8. $24c - 11c$ **13c**
9. $32p - 7p$ **25p**

Lesson 1–2, pp. 8–13

Find the value of each expression if $a = 2$, $b = 7$, $c = 6$, and $d = 12$.

10. $a + b$ **9**
11. $d + b$ **19**
12. $d - c$ **6**
13. $b - c$ **1**
14. $5a$ **10**
15. $9c$ **54**
16. $c \div a$ **3**
17. $\frac{d}{c}$ **2**
18. $3a + b$ **13**
19. $4d - c$ **42**
20. $9a + 2c$ **30**
21. $\frac{3d}{c}$ **6**

Lesson 1–3, pp. 14–18

Simplify each expression.

22. $5(3v)$ **15v**
23. $4(8q)$ **32q**
24. $6(6a)$ **36a**
25. $3(12x)$ **36x**
26. $7(6r)$ **42r**
27. $13(2c)$ **26c**
28. $4a(3b)$ **12ab**
29. $9s(4t)$ **36st**
30. $8p(8q)$ **64pq**

Make this Foldable to help you organize your Chapter 2 notes. Begin with four sheets of plain $8\frac{1}{2}$" by 11" paper.

❶ **Stack** the sheets with the edges $\frac{3}{4}$ inch apart.

❷ **Fold** up the bottom edges. All tabs should be the same size.

❸ **Staple** along the fold.

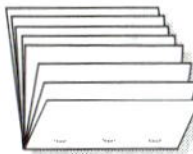

❹ **Label** the tabs as shown.

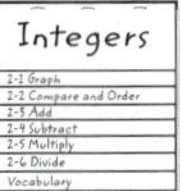

Reading and Writing As you read and study the chapter, use each page to write notes and examples for each lesson.

www.algconcepts.com/chapter_readiness

Check Your Readiness

This section provides a review of the basic concepts needed before beginning Chapter 2. Lesson references are included for additional student help.

Use Exercises	To Prepare for Lesson(s)
1–9	2–3, 2–4
10–21	2–3, 2–4, 2–5, 2–6
22–30	2–5

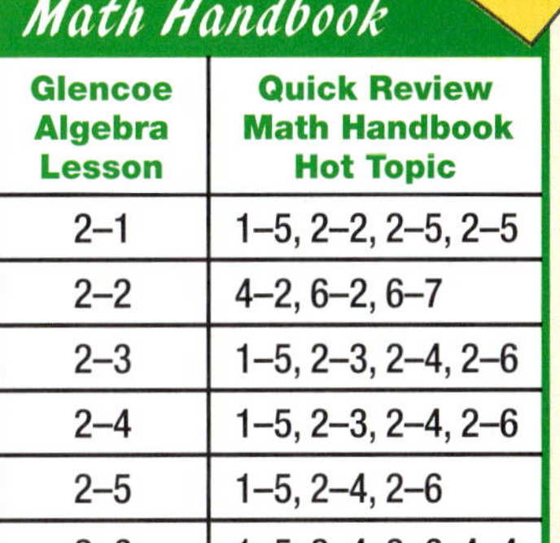

Glencoe Algebra Lesson	Quick Review Math Handbook Hot Topic
2–1	1–5, 2–2, 2–5, 2–5
2–2	4–2, 6–2, 6–7
2–3	1–5, 2–3, 2–4, 2–6
2–4	1–5, 2–3, 2–4, 2–6
2–5	1–5, 2–4, 2–6
2–6	1–5, 2–4, 2–6, 4–4

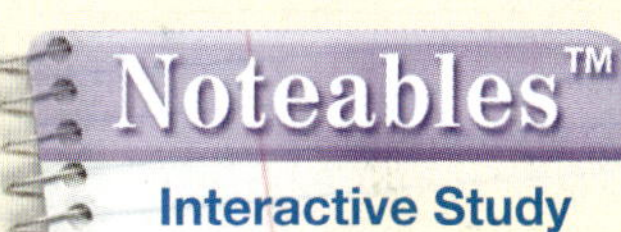

Interactive Study Notebook with Foldables™

This note-taking guide reinforces excellent note-taking skills. It includes:

- ✓ **Build Your Vocabulary** tool for including mathematical terminology in notes.
- ✓ **Review It** activities with links to prior lessons and items to include in **Foldables™**.
- ✓ **Bringing It All Together** feature to help students review for the Chapter Test.
- ✓ **Are You Ready for the Chapter Test?** feature to allow students to assess their own readiness for the Chapter Test.
- ✓ **Teacher Edition** with transparencies to guide note taking. Corresponds to the *Teacher's Wraparound Edition* In-Class Examples and *Interactive Chalkboard CD-ROM.*

FOLDABLES™ Study Organizer Use this Foldable chart for student writing about integers: graphing; ordering and comparing; adding, subtracting, multiplying, and dividing. Have students summarize the chapter by recording rules for operations on integers. Ask students to make up a real-world question for each of the main ideas of the chapter.

Lesson 2–1

1 FOCUS

5-Minute Check
Chapter 1

Simplify each expression.

1. $5(a + 2)$ **$5a + 10$**
2. $9 + 3b + 4b$ **$9 + 7b$**
3. $6 \cdot (9 \cdot c)$ **$54c$**

The frequency table shows the home ZIP code of the first 100 visitors to a new shopping mall on a certain day.

ZIP Code	Tally	Frequency
27605	卌 卌 卌 卌 |	21
27606	卌 卌 卌 卌 卌 卌 ||	32
27607	卌 卌 卌 ||||	19
27608	|||	3
27609	卌	5
27610	卌 卌 ||	12
27611	卌 |||	8

4. From which two ZIP codes did the fewest people come? **27608, 27609**
5. Suppose the ZIP code of the mall is one of those listed in the table. Which one do you think it is? Explain. **27606; The greatest number of people who came to the mall gave their home ZIP code as 27606.**

Motivating the Lesson

Real-World Connection Show students a thermometer that has both positive and negative temperatures on its scale. Help students understand that a thermometer is like a vertical number line.

2–1 Graphing Integers on a Number Line

What You'll Learn
You'll learn to graph integers on a number line and to compare and order integers.

Why It's Important
Meteorology Weather forecasters use integers to determine wind chill. *See Exercise 49.*

There are many ways to represent numbers. One way to represent numbers is with a **number line**. The number line also shows the order of numbers; 2 is to the left of 3, so 2 is smaller than three.

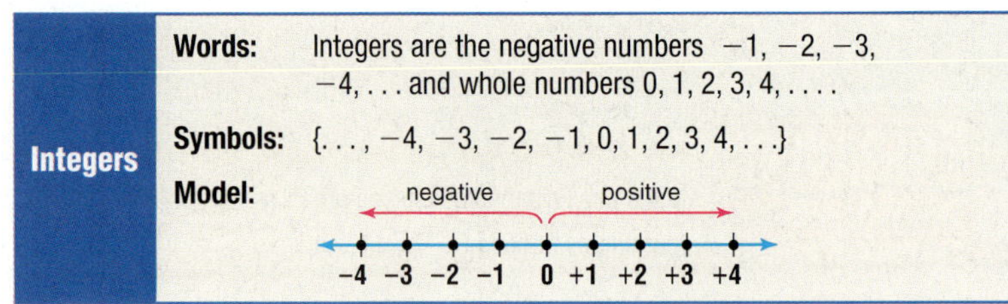

This number line represents the set of whole numbers.
0 1 2 3 4 5 6 7
A number line is drawn by choosing a starting position on a line and marking off equal distances from that point.
The arrowhead indicates that the line and the set of numbers continue indefinitely.

A **negative number** is a number less than zero. To include negative numbers on a number line, extend the line to the left of zero and mark off equal distances. Negative whole numbers are members of the set of **integers**. So, integers can also be represented on a number line.

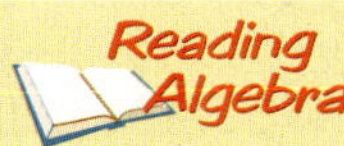

Reading Algebra

Read -3 as *negative 3*. Read $+4$ as *positive 4*. Positive integers usually are written without the $+$ sign. So, $+4$ and 4 are the same number.

Integers

Words: Integers are the negative numbers $-1, -2, -3, -4, \ldots$ and whole numbers $0, 1, 2, 3, 4, \ldots$.

Symbols: $\{\ldots, -4, -3, -2, -1, 0, 1, 2, 3, 4, \ldots\}$

Model: negative ← → positive

−4 −3 −2 −1 0 +1 +2 +3 +4

Zero is neither negative nor positive.

Sets of numbers can also be represented by **Venn diagrams**.

natural numbers	1, 2, 3, 4, . . .
whole numbers	0, 1, 2, 3, . . .
integers	. . . −3, −2, −1, 0, 1, 2, 3, . . .

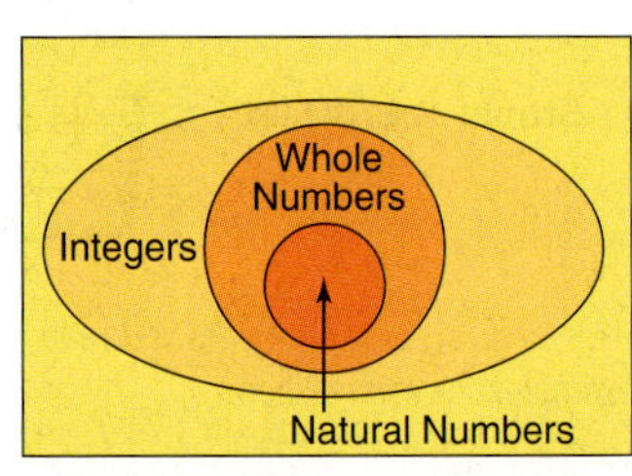

Resource Manager

Reproducible Masters
Chapter 2 Resource Masters
- *Study Guide*, p. 51
- *Skills Practice*, p. 52
- *Practice*, p. 53
- *Reading to Learn Mathematics*, p. 54
- *Enrichment*, p. 55

Transparencies
5-Minute Check, 2–1

Technology/Multimedia
Interactive Chalkboard CD-ROM

The Venn diagram shows that every natural number is also a whole number. Natural numbers are a *subset* of whole numbers. Similarly, whole numbers are a subset of integers.

To **graph** a set of integers, locate the points named by those numbers on a number line and place a dot on the number line. The number that corresponds to a point is called the **coordinate** of that point.

Examples

1 **Name the coordinates of *A*, *B*, and *C*.**

The coordinate of *A* is −4, *B* is 2, and *C* is −1.

2 **Graph points *X*, *Y*, and *Z* on a number line if *X* has coordinate 4, *Y* has coordinate 0, and *Z* has coordinate −3.**

Find each number on a number line. Place a dot on the mark above the number. Then write the letter above the dot.

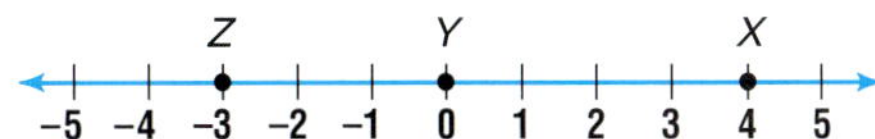

Your Turn **b. See margin.**

a. Name the coordinates of *D*, *E*, and *F*. **−2, 0, 5**

b. Graph points *M*, *N*, and *P* on a number line if *M* has coordinate −3, *N* has coordinate −4, and *P* has coordinate 1.

The numbers on a number line increase as you move to the right and decrease as you move to the left. When graphing two integers on a number line, the number to the right is always greater.

The symbols < and > always point to the lesser number.

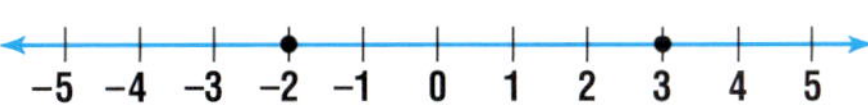

Words:	3 is greater than −2.	**Words:**	−2 is less than 3.
Symbols:	$3 > -2$	**Symbols:**	$-2 < 3$

www.algconcepts.com/extra_examples

2 TEACH

Teaching Tip Sketch the Venn diagram given on page 53 on the board or overhead. Make your sketch as large as possible. Have students come up and write sample numbers in the correct region of the Venn diagram.

Teaching Tip As you discuss the sets of numbers represented in the Venn diagram, point out that the natural numbers are sometimes called the *counting numbers.*

Teaching Tip When discussing the term *subset,* you might want to point out that the set of real numbers can be divided into two completely separate subsets, the *rational numbers* and the *irrational numbers.* Stress that a number is either rational or irrational, but not both.

In-Class Examples

Example 1

Name the coordinates of *G*, *H*, and *J*. **−3, 4, 1**

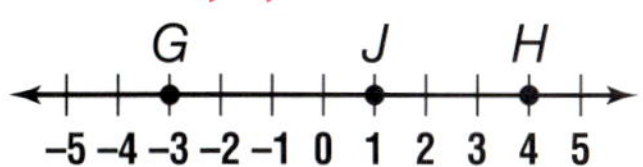

Example 2

Graph points *K*, *L*, and *M* on a number line if *K* has coordinate −4, *L* has coordinate 2, and *M* has coordinate −1.

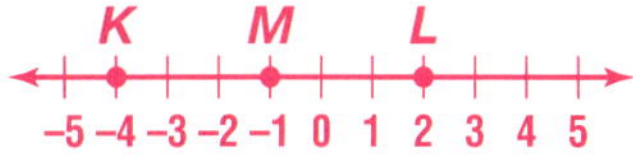

Answer

Your Turn

b.

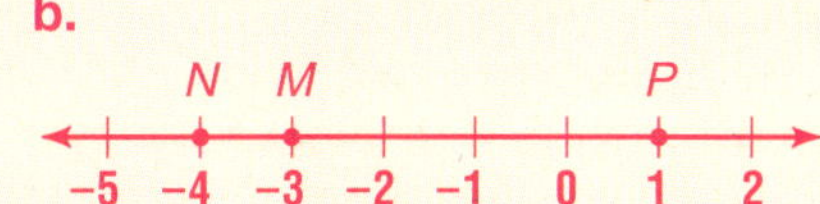

From the Classroom of ...

Dr. Deborah Hutchens
Great Bridge Middle School
Chesapeake, Virginia

The application at the beginning of the lesson creates an opportunity to do an interdisciplinary unit on geography, meteorology, or water conservation.

In-Class Examples

Examples 3–4

Replace each ● with < or > to make a true sentence.

3 3 ● −4 >

4 −1 ● −3 >

Example 5

The table shows the high temperature each day for one week in January in a midwestern city. Order the temperatures from greatest to least.

Day of the Week	Temperature (°C)
Monday	5
Tuesday	−1
Wednesday	8
Thursday	−3
Friday	0
Saturday	−7
Sunday	−2

8°, 5°, 0°, −1°, −2°, −3°, −7°

Examples

Replace each ● with < or > to make a true sentence.

3 4 ● −1

4 is to the right of −1 on the number line. So, 4 > −1.

4 −5 ● −3

−5 is to the left of −3 on the number line. So, −5 < −3.

Your Turn

c. −1 ● −2 > d. 2 ● −2 > e. 0 ● 1 <

Integers are used to compare numbers in many everyday applications.

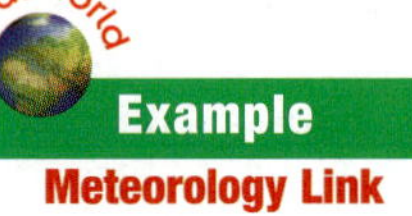

Example Meteorology Link

5 **The table shows the average high temperatures for January in selected cities. Order the temperatures from least to greatest.**

City	Temperature (°C)
Boston, MA	2
Chicago, IL	−2
Detroit, MI	−1
Juneau, AK	3
New York, NY	4
St. Louis, MO	−3

Graph each integer on a number line. Use the first letter of each city name to label the points.

Write the integers as they appear on the number line from left to right.

−3°, −2°, −1°, 2°, 3°, 4° are in order from least to greatest.

Looking at the graphs of 4 and −4 on a number line, you can see that they are the same number of units from 0. We say that they have the same **absolute value**.

Reading Algebra

The symbol for absolute value is two vertical bars on either side of the number. Read $|-4| = 4$ as *the absolute value of negative 4 is 4.*

Absolute Value

Words: The absolute value of a number is the distance it is from 0 on the number line.

Model:

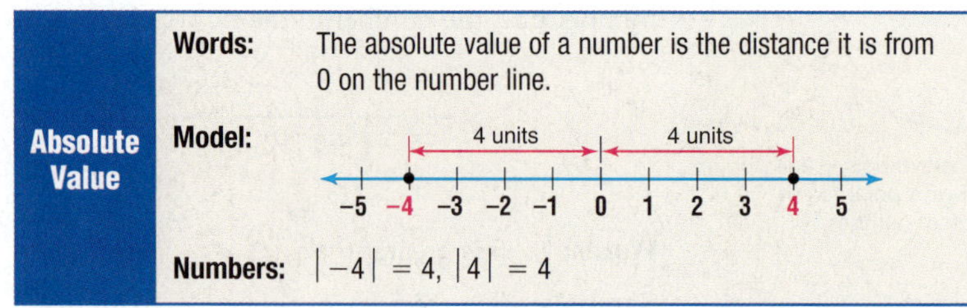

Numbers: $|-4| = 4$, $|4| = 4$

Examples

Evaluate each expression.

$|-3|$

$|-3| = 3$ *The graph of −3 is 3 units away from 0.*

$|-5| - |2|$

$|-5| - |2| = 5 - 2$ *The absolute value of −5 is 5.*

$= 3$ *The absolute value of 2 is 2.*

Your Turn

f. $|9|$ **9** g. $|-2| + |-6|$ **8** h. $|15| - |-4|$ **11**

Check for Understanding

Communicating Mathematics

1. **Describe** a situation in the real world where negative integers are used.

1. Sample answer: temperature, elevation

2. **Draw** a number line from −6 to 6. Graph two points whose coordinates have the same absolute value. **See margin.**

3. **You Decide** Tiffany says that 0 is a negative number. Ramon says that 0 is a positive number. Who is correct? Explain. **Neither; 0 is neither negative nor positive.**

Vocabulary

number line
negative number
integer
Venn diagram
natural number
graph
coordinate
absolute value

Guided Practice

Getting Ready Write an integer for each situation.

Sample: 10 feet below sea level **Solution:** −10

4. 4 degrees above zero **+4**
5. a loss of 6 pounds **−6**
6. 3 inches less rain than normal **−3**
7. a salary increase of $150 **+150**

Example 1 Name the coordinates of each point.

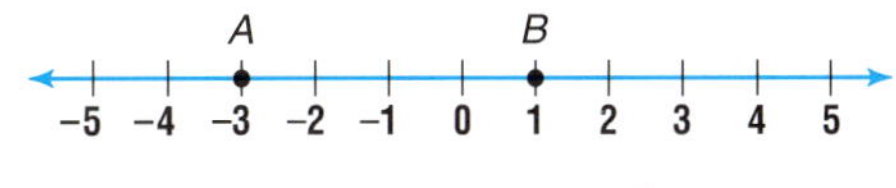

8. A **−3**
9. B **1**

Example 2 Graph each set of numbers on a number line.

10–11. See margin.

10. $\{-3, 1, 4\}$
11. $\{5, 0, -4\}$

Reteaching Activity

Auditory/Musical Learners Provide pairs of students with a deck of playing cards. Instruct them to designate the black cards as positive numbers and the red cards as negative numbers, with aces considered 1, jacks as 11, queens as 12, and kings as 13. Have each student draw a card and then ask the partners to decide which card represents the greater integer.

In-Class Examples

Examples 6–7

Evaluate each expression.

6 $|-1|$ **1**

7 $|4| + |-5|$ **9**

Answers

2. Sample answer:

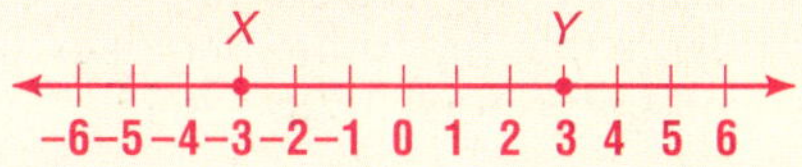

X at −3, Y at 3 on a number line from −6 to 6.

10. Number line from −5 to 5 with points at −3, 1, 4.

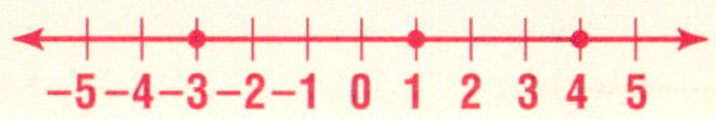

11. Number line from −5 to 5 with points at −4, 0, 5.

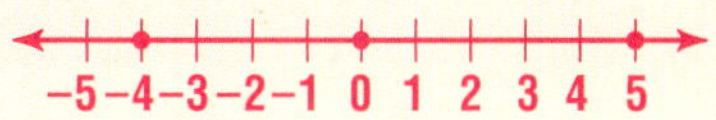

Study Guide, p. 51

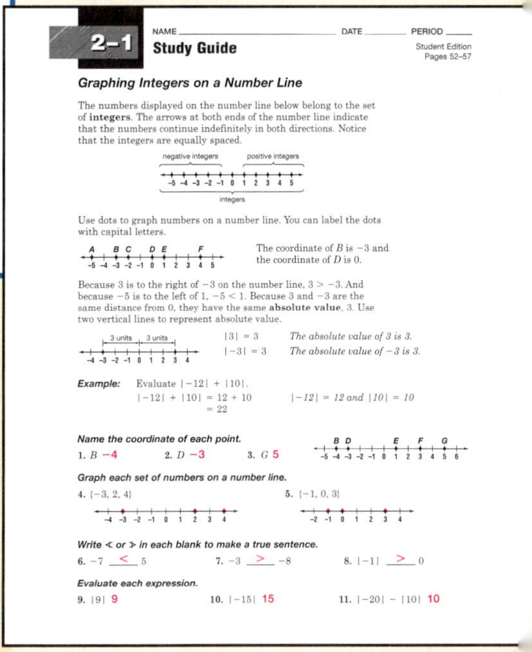

NAME ______ DATE ______ PERIOD ______

Study Guide

Student Edition Pages 52–57

Graphing Integers on a Number Line

The numbers displayed on the number line below belong to the set of **integers**. The arrows at both ends of the number line indicate that the numbers continue indefinitely in both directions. Notice that the integers are equally spaced.

negative integers positive integers
−5 −4 −3 −2 −1 0 1 2 3 4 5
integers

Use dots to graph numbers on a number line. You can label the dots with capital letters.

A B C D E F
−5 −4 −3 −2 −1 0 1 2 3 4 5

The coordinate of B is −3 and the coordinate of D is 0.

Because 3 is to the right of −3 on the number line, $3 > -3$. And because −5 is to the left of 1, $-5 < 1$. Because 3 and −3 are the same distance from 0, they have the same **absolute value**, 3. Use two vertical lines to represent absolute value.

3 units 3 units
−4 −3 −2 −1 0 1 2 3 4

$|3| = 3$ *The absolute value of 3 is 3.*

$|-3| = 3$ *The absolute value of −3 is 3.*

Example: Evaluate $|-12| + |10|$.

$|-12| + |10| = 12 + 10$ *$|-12| = 12$ and $|10| = 10$*

$= 22$

Name the coordinate of each point.

B D E F G
−5 −4 −3 −2 −1 0 1 2 3 4 5 6

1. B **−4** 2. D **−3** 3. G **5**

Graph each set of numbers on a number line.

4. $\{-3, 2, 4\}$
−4 −3 −2 −1 0 1 2 3 4

5. $\{-1, 0, 3\}$
−2 −1 0 1 2 3 4

Write < or > in each blank to make a true sentence.

6. −7 __<__ 5 7. −3 __>__ −8 8. $|-1|$ __>__ 0

Evaluate each expression.

9. $|9|$ **9** 10. $|-15|$ **15** 11. $|-20| - |10|$ **10**

3 PRACTICE/APPLY

Error Analysis

Watch for students who write that $-8 > -5$ in Exercise 12 because they compare the two negative integers in the same way they would compare the positive integers 8 and 5.

Prevent by pointing out to students that the "more negative" an integer is does not mean it is greater, but rather that it is further to the left of 0 on a number line. Stress that the number further to the left on a number line is always less than the other number.

Assignment Guide
Basic: 19–49 odd, 50–60
Average: 18–46 even, 48–60

Skills Practice, **p. 52, and** ***Practice,*** **p. 53 (shown)**

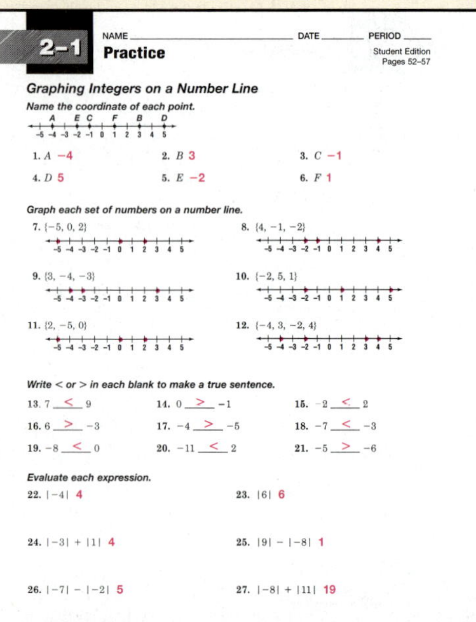

NAME ______ DATE ______ PERIOD ______

2–1 Practice

Student Edition Pages 52–57

Graphing Integers on a Number Line

Name the coordinate of each point.

(Number line from −5 to 5 with points A, E, C, F, B, D)

1. A −4   2. B 3   3. C −1

4. D 5   5. E −2   6. F 1

Graph each set of numbers on a number line.

7. {−5, 0, 2}   8. {4, −1, −2}

9. {3, −4, −3}   10. {−2, 5, 1}

11. {2, −5, 0}   12. {−4, 3, −2, 4}

Write < or > in each blank to make a true sentence.

13. 7 _<_ 9   14. 0 _>_ −1   15. −2 _<_ 2

16. 6 _>_ −3   17. −4 _>_ −5   18. −7 _<_ −3

19. −8 _<_ 0   20. −11 _<_ 2   21. −5 _>_ −6

Evaluate each expression.

22. $|-4|$ 4   23. $|6|$ 6

24. $|-3| + |1|$ 4   25. $|9| - |-8|$ 1

26. $|-7| - |-2|$ 5   27. $|-8| + |11|$ 19

Examples 3 & 4 Replace each ● with < or > to make a true sentence.

12. -8 ● -5 **<**   13. -4 ● 2 **<**   14. 9 ● -7 **>**

Examples 6 & 7 Evaluate each expression.

15. $|-8| + |-2|$ **10**   16. $|-7| - |4|$ **3**

Example 5

17. **Meteorology** The table gives the record low temperatures for each month at the Grand Canyon Airport in Arizona. Order the temperatures from least to greatest.

Month	J	F	M	A	M	J	J	A	S	O	N	D
Temperature (°F)	−22	−17	−7	9	10	26	35	35	22	13	−1	−14

Source: *The Weather Almanac* **−22, −17, −14, −7, −1, 9, 10, 13, 22, 26, 35**

Exercises

Practice  Name the coordinate of each point.

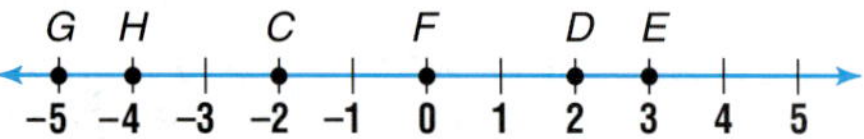

For Exercises	See Examples
Homework Help	
18–23	1
24–29, 46, 47	2
30–38	3, 4
39–45	6, 7
48, 49	5

Extra Practice See page 694.

18. C **−2**   19. D **2**   20. E **3**

21. F **0**   22. G **−5**   23. H **−4**

Graph each set of numbers on a number line. **24–29. See margin.**

24. {2, 3, 5}   25. {−1, −3, 4}   26. {−2, 4, 0}

27. {−3, −2, 1}   28. {−2, −1, 0, 1}   29. {−4, −3, −2, −1}

Replace each ● with < or > to make a true sentence.

30. 4 ● -4 **>**   31. 0 ● -2 **>**   32. -2 ● -1 **<**

33. 2 ● -3 **>**   34. -10 ● 1 **<**   35. -15 ● -10 **<**

36. -5 ● $|-5|$ **<**   37. $|4|$ ● -4 **>**   38. $|-6|$ ● $|-3|$ **>**

Evaluate each expression.

39. $|-6|$ **6**   40. $|10|$ **10**   41. $|-5| - |3|$ **2**

42. $|-7| + |-2|$ **9**   43. $|14| - |-5|$ **9**   44. $|-13| + |-17|$

45. Is $|5| = |-5|$ *sometimes, always* or *never* true? Explain.

46. Order −3, −4, 0, 1, −5, and 3 from least to greatest. **−5, −4, −3, 0, 1, 3**

47. Order −25, 78, −36, 14, and −14 from greatest to least.

44. 30

45. Always; $|5| = 5$ and $|-5| = 5$, so $|5| = |-5|$.

47. 78, 14, −14, −25, −36

Answers

24. 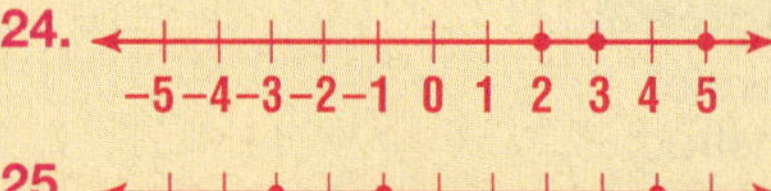

25. 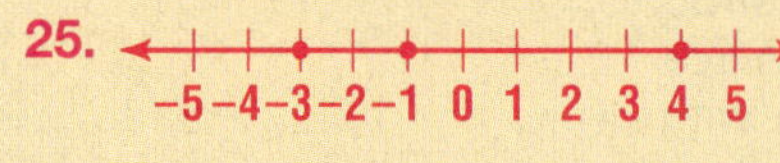

26. (number line from −5 to 5 with points at −2, 0, 4)

27. 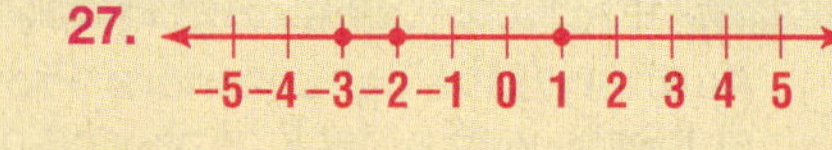

28. 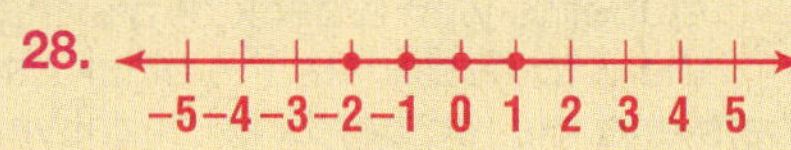

29. 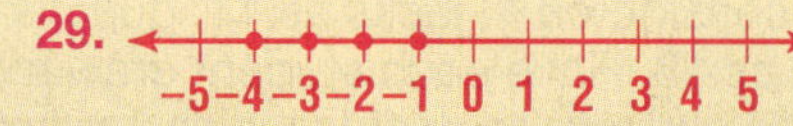

Applications and Problem Solving

48. **Population** In 2002, the population of Texas was 16 million greater than the average of all 50 state populations. The population of Arkansas was 3 million less than the average state population. Write an integer for each situation. **16,000,000; −3,000,000**

49. **Meteorology** *Windchill factor* is an estimate of the cooling effect the wind has on a person in cold weather.

Windchill Factor (°F)

Wind Speed (mph)	Actual Temperature (F°)						
	30	20	10	0	−10	−20	−30
0	30	20	10	0	−10	−20	−30
5	27	16	6	−5	−15	−26	−36
10	16	4	−9	−21	−33	−46	−58
15	9	−5	−18	−36	−45	−58	−72

Source: BMFA, 1999

a. Find the windchill factor when the actual temperature is 0° with a wind speed of 15 mph. **−36°**

b. What is the windchill factor when the actual temperature is −20° with a wind speed of 5 mph? **−26°**

c. Which is less: the windchill factor in part a or part b? **−36°**

50. **Critical Thinking** Determine whether each statement is *true* or *false*. If *false*, give a counterexample.

a. Every integer is a whole number. **false, −1**

b. Every whole number is an integer. **true**

Mixed Review

History Refer to the table for Exercises 51–53.

Heights of United States Presidents					
Height (in.)	63–65	66–68	69–71	72–74	75–77
Number	1	9	14	18	1

51. Make a histogram of the data. *(Lesson 1–7)* **51–52. See margin.**

52. Make a cumulative frequency table for the data. *(Lesson 1–6)*

53. How many presidents were at least six feet tall? *(Lesson 1–6)* **19 presidents**

Simplify each expression. *(Lesson 1–4)*

54. $5x + 6x$ **$11x$**
55. $9a - 3a$ **$6a$**
56. $9x - x + 7x$ **$15x$**
57. $3m + 2n + 4m$ **$7m + 2n$**
58. $3r + 2s + 2r + s$ **$5r + 3s$**
59. $5x + 12y - y + 2x$ **$7x + 11y$**

Standardized Test Practice

60. **Multiple Choice** You have two more sisters than brothers. If you have s sisters, which equation could be used to find b, the number of brothers you have? *(Lesson 1–1)* **B**

Ⓐ $s = b - 2$ Ⓑ $s - 2 = b$ Ⓒ $b = s + 2$ Ⓓ $b = 2s$

www.algconcepts.com/self_check_quiz

Extra Credit

Replace the ● with < or > to make a true sentence.

$|-5| + |3|$ ● $|-3| - |5|$ **>**

4 ASSESS

Open-Ended Assessment

Modeling Have students use a number line model to explain how to decide which of two integers is greater.

Answers

51.

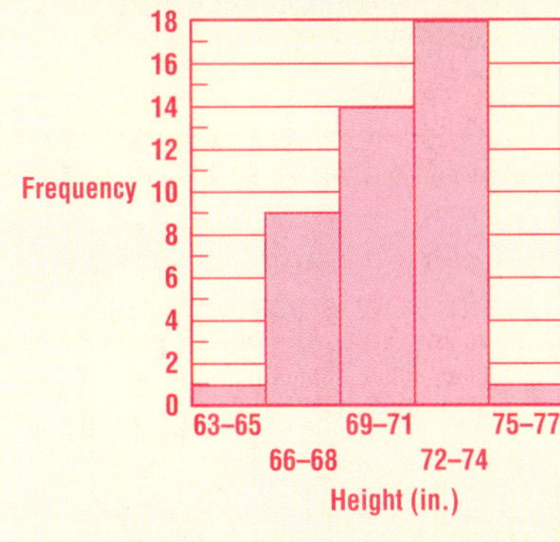

52.

Heights of U.S. Presidents	
Height (in.)	Cumulative Frequency
63–65	1
66–68	10
69–71	24
72–74	42
75–77	43

Enrichment, p. 55

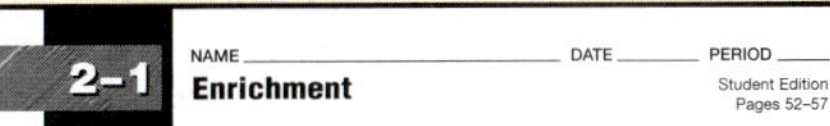

Venn Diagrams

A type of drawing called a **Venn diagram** can be useful in explaining conditional statements. A Venn diagram uses circles to represent sets of objects.

Consider the statement "All rabbits have long ears." To make a Venn diagram for this statement, a large circle is drawn to represent all animals with long ears. Then a smaller circle is drawn inside the first to represent all rabbits. The Venn diagram shows that every rabbit is included in the group of long-eared animals.

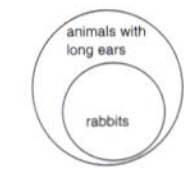

The set of rabbits is called a **subset** of the set of long-eared animals.

The Venn diagram can also explain how to write the statement, "All rabbits have long ears," in if-then form. Every rabbit is in the group of long-eared animals, so if an animal is a rabbit, then it has long ears.

For each statement, draw a Venn diagram. The write the sentence in if-then form.

1. Every dog has long hair.

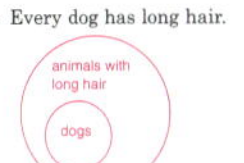

If an animal is a dog, then it has long hair.

2. All rational numbers are real.

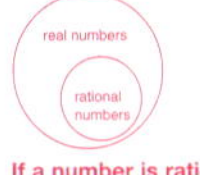

If a number is rational, then it is real.

3. People who live in Iowa like corn.

If a person lives in Iowa, then the person likes corn.

4. Staff members are allowed in the faculty lounge.

If a person is a staff member, then the person is allowed in the faculty cafeteria.

2-2 The Coordinate Plane

1 FOCUS

5-Minute Check
Lesson 2-1

1. Name the coordinates of *A*, *B*, and *C*. **−1, 3, −3**

Replace ● with < or > to make a true statement.

2. −5 ● −1 <
3. −2 ● 2 <
4. 0 ● −4 >
5. Evaluate |−2| + |−1|. **3**

Motivating the Lesson

Hands-On Activity Have students locate several cities in an atlas or find several streets on a city map. Have them explore how each page is divided into grids and how each square is designated by a letter and a number.

2 TEACH

Teaching Tip Before you introduce the lesson content, make sure students know the locations of north, south, east, and west in your classroom. Students from some areas, such as Hawaii, may be more familiar with describing direction in terms of toward and away from mountains rather than compass directions.

MODELING

An alternative hands-on option is available for teaching this lesson.

What You'll Learn
You'll learn to graph points on a coordinate plane.

Why It's Important
Meteorology Weather forecasters at the National Hurricane Center use a coordinate system to track hurricanes. *See Exercise 36.*

In mathematics, you locate a point on a **coordinate system** that is similar to a grid. The coordinate system is formed by the intersection of two number lines that meet at right angles at their zero points.

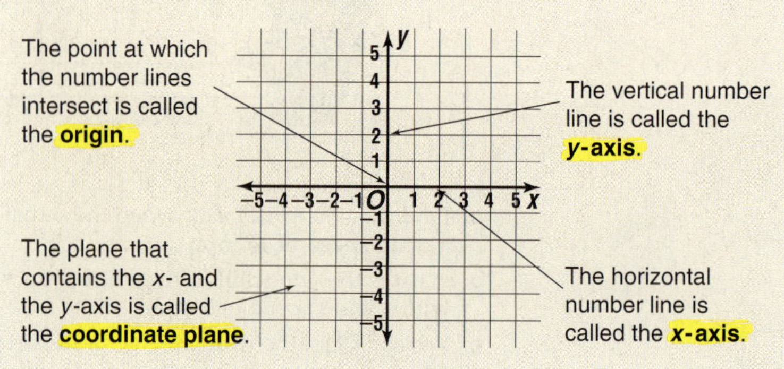

The point at which the number lines intersect is called the **origin**.
The vertical number line is called the ***y*-axis**.
The plane that contains the *x*- and the *y*-axis is called the **coordinate plane**.
The horizontal number line is called the ***x*-axis**.
y
x
O

The directions *east* and *north* tell you how to locate a point on a map. In mathematics, an **ordered pair** of numbers is used to locate any point on a coordinate plane.

The first number in an ordered pair is called the ***x*-coordinate**. It corresponds to a number on the *x*-axis. The second number is called the ***y*-coordinate**. It corresponds to a number on the *y*-axis.

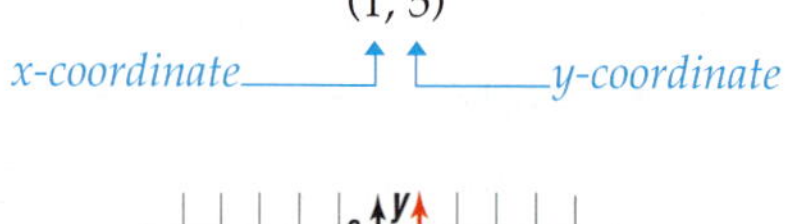

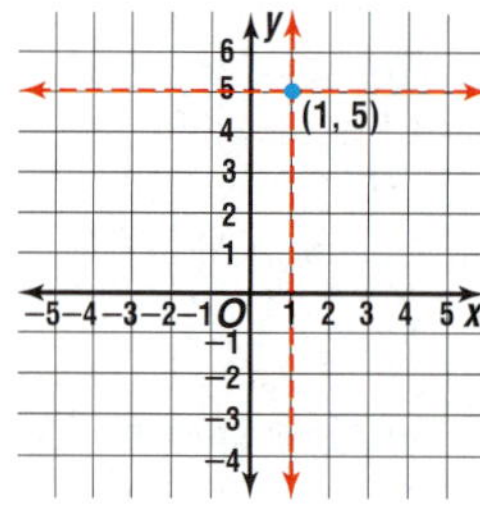

Notice that (1, 5) and (5, 1) are not the same points on the coordinate system.

Resource Manager

Reproducible Masters

Chapter 2 Resource Masters
- *Study Guide,* p. 56
- *Skills Practice,* p. 57
- *Practice,* p. 58
- *Reading to Learn Mathematics,* p. 59
- *Enrichment,* p. 60
- *Assessment,* p. 91

Hands-On Algebra, p. 33
Graphing Calculator, pp. 5–7

Transparencies
5-Minute Check, 2–2

Technology/Multimedia
AlgePASS, Lesson 7
Interactive Chalkboard CD-ROM

Examples

Reading Algebra

The x-coordinate indicates the number of units to move left or right of the origin. The y-coordinate indicates the number of units to move up or down from the origin.

Write the ordered pair that names each point.

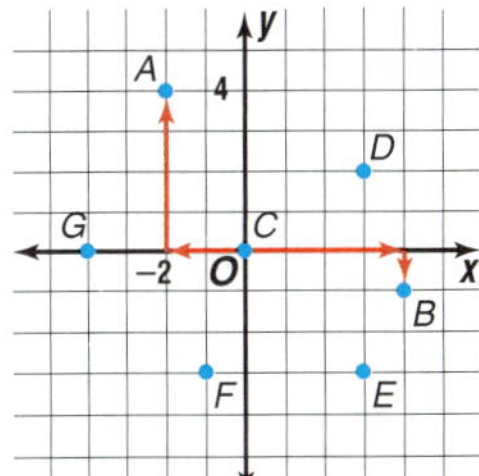

1 A

- Start at the origin. Move left on the x-axis to find the x-coordinate of point A. The x-coordinate is -2.
- Move up along the grid lines to find the y-coordinate. The y-coordinate is 4.

The ordered pair for point A is $(-2, 4)$.

2 B

The x-coordinate is 4, and the y-coordinate is -1.
The ordered pair for point B is $(4, -1)$.

3 C

Point C is the origin. The ordered pair for the origin is $(0, 0)$.

Your Turn

a. D **(3, 2)** b. E **(3, −3)** c. F **(−1, −3)** d. G **(−4, 0)**

A point can be named by both a letter and its ordered pair. For example, $P(2, 3)$ means point P has an x-coordinate of 2 and a y-coordinate of 3. To graph an ordered pair on a coordinate plane, draw a dot at the point that corresponds to the ordered pair. This is called *plotting* the point.

Example

You can assume that each unit on the axes represents 1 unit.

4 **Graph $P(2, 3)$ on a coordinate plane.**

- Start at the origin, O.
- The x-coordinate is 2. So, move 2 units to the right.
- The y-coordinate is 3. Move 3 units up and draw a dot.
- Label the dot with the letter P.

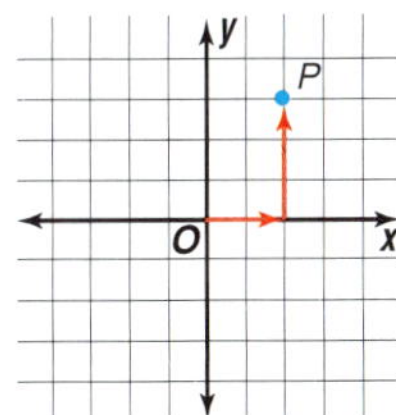

www.algconcepts.com/extra_examples

Teaching Tip As you begin Example 1, have students place one of their fingers at the origin. Have them slide their finger along the x-axis until it is directly under point A and note the value of x at this location. Now instruct them to slide their finger up to point A and then slide it sideways on the grid line until they reach the y-axis. Have them note the value of y at this location. Point out that these values of x and y are the coordinates of point A.

In-Class Examples

Examples 1–3

Write the ordered pair that names each point.

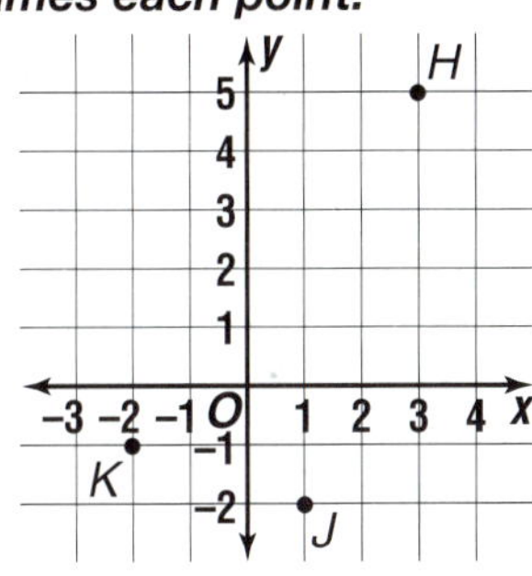

1 H **(3, 5)**

2 J **(1, −2)**

3 K **(−2, −1)**

Example 4

Graph $V(2, 4)$ on a coordinate plane.

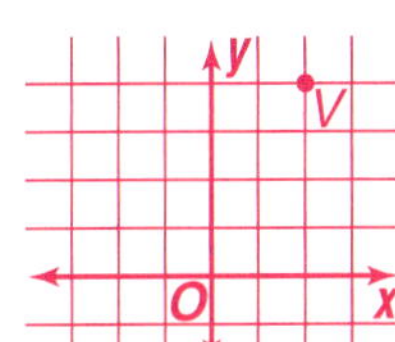

Family Activity

Have students take home a library book containing a map or provide students with a photocopy of a map. Students can show a family member how the map squares are named by a letter-number pair, such as A5. Encourage students to play a hide-and-seek map game in which two people take turns giving the partner clues on the map, such as “a town in square B3 with 8 letters and beginning with K.”

In-Class Example

Example 5

Graph $W(-4, -1)$ on a coordinate plane.

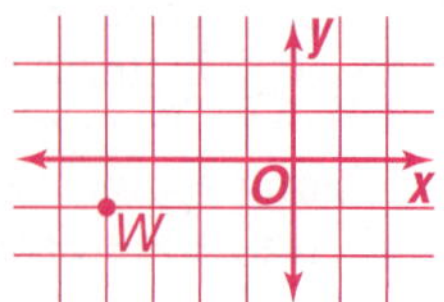

Teaching Tip When discussing the figure showing the four quadrants of a coordinate plane, point out that any point on one (or both) axes is *not* located in any of the quadrants.

In-Class Examples

Examples 6–7

Name the quadrant in which each point is located.

6 $F(0, 1)$ **none**

7 $G(3, -1)$ **IV**

Example 8

The first zeppelin flown in 1900 flew at a speed of 18 mph. Let x represent the number of hours. Then $18x$ represents the total distance traveled in x hours. Evaluate the expression to find the distances traveled in 1, 2, and 3 hours. Then graph the ordered pairs (time, distance).

Time (hours)	Distance (miles)
x	$18x$
1	18
2	36
3	54

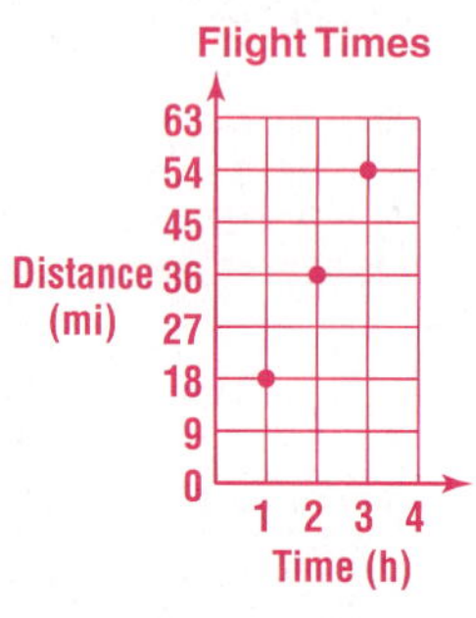

Example 5 Graph $Q(-3, 0)$ on a coordinate plane.

- Start at the origin, O.
- The x-coordinate is -3. So, move 3 units to the left.
- The y-coordinate is 0. So the dot is placed on the axis.

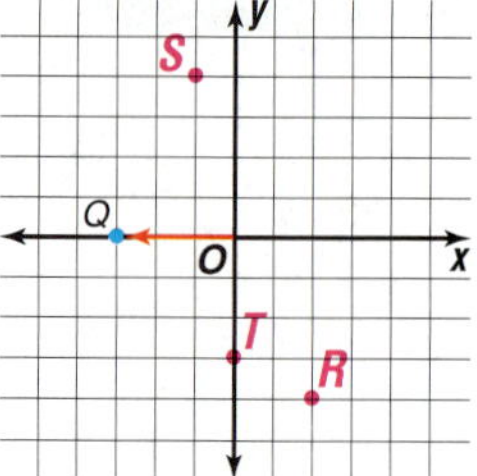

Your Turn

Graph each point on a coordinate plane. **e–g. See above.**

e. $R(2, -4)$ **f.** $S(-1, 4)$ **g.** $T(0, -3)$

The x-axis and the y-axis separate the coordinate plane into four regions, called **quadrants**. The quadrants are numbered as shown at the right. Note that the axes are not located in any of the quadrants.

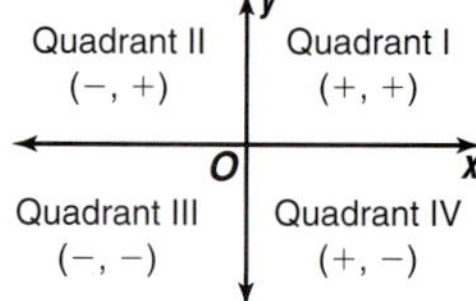

Examples

Name the quadrant in which each point is located.

6 $A(5, -4)$

The x-coordinate is positive, and the y coordinate is negative. So, point A is located in Quadrant IV.

7 $B(2, 0)$

Point B lies on the x-axis. It is not located in a quadrant.

Your Turn

h. $C(-2, -7)$ **III** **i.** $D(-4, 9)$ **II** **j.** $E(0, -3)$ **none**

You can use ordered pairs to show how data are related.

Example 8 **Biology Link**

Dolphins can swim at 30 mph over long distances. Let x represent the number of hours. Then, $30x$ represents the total distance traveled. Evaluate the expression to find the distances traveled in 1, 2, and 3 hours. Then graph the ordered pairs (time, distance).

The data are graphed in the first quadrant because both values are positive.

Time (hours)	Distance (miles)
x	$30x$
1	30
2	60
3	90

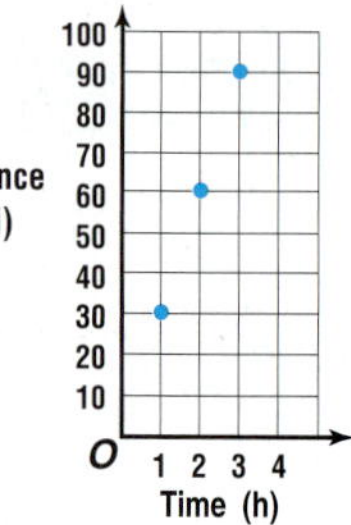

Graphing Calculator Exploration

Refer to the Graphing Calculator Exploration on page 61. To have the calculator display "user-friendly" coordinates (integers, decimals with only one place value), the student must use appropriate settings for the viewing window. The ranges that result in Step 1 are Xmin = -47, Xmax = 47, Ymin = -31, and Ymax = 31. Stress that these settings will result in integer coordinates at the bottom of the graphing screen. Students can also explore what happens when these window settings are multiplied by 10, 100, 1000, or some other power of 10.

Graphing Calculator Tutorial
See pp. 750–753.

Graphing Calculator Exploration

You can plot points on a graphing calculator.

Step 1 Press [ZOOM] 8 [ENTER] to display a coordinate grid.

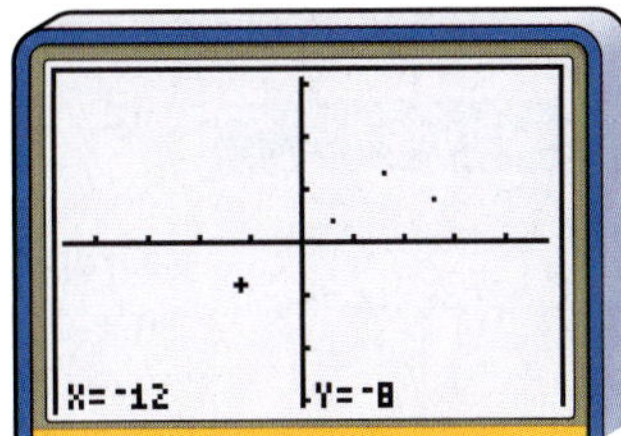

Step 2 Press [2nd] [DRAW] [▶] [ENTER]. Use the arrow keys to move the cursor to each desired location. Press [ENTER] to plot the point.

Try These

1. Choose four ordered pairs such that the sum of their x- and y-coordinates is 5. Graph them. **See students' work.**
2. What do you notice about the graphs of the points?

2. They form a straight line.

Check for Understanding

Communicating Mathematics

1. **Explain** how to graph $(-5, 1)$ on a coordinate plane. **See margin.**
2. **Name** an ordered pair whose graph satisfies each condition. **Sample answers given.**
   a. located in Quadrant IV **(3, −6)**
   b. not located in any quadrant **(0, −2)**
3. *Writing Math* Draw a coordinate system and label the origin, x-axis, y-axis, and quadrants. **See margin.**

Vocabulary
coordinate system
y-axis
x-axis
origin
coordinate plane
ordered pair
x-coordinate
y-coordinate
quadrant

Guided Practice

Examples 1–3

Write the ordered pair that names each point.

4. F **(−2, 4)**
5. G **(3, −2)**

Examples 4 & 5

Graph each point on a coordinate plane.

6. $R(-5, 2)$
7. $S(0, -2)$ **See graph.**

Examples 6 & 7

Name the quadrant in which each point is located.

8. $D(-9, 1)$ **II**
9. $E(0, -6)$ **none**

Example 8

10. **Biology** A tortoise is one of the slowest animals on land. It travels at an average speed of only 20 feet per minute.
    a. Find the distance traveled in 2, 4, and 6 minutes.
    b. Graph the ordered pairs (time, distance). **See margin.**
    a. 40 ft, 80 ft, 120 ft

3 PRACTICE/APPLY

Error Analysis

Watch for students who have the coordinates switched in their ordered pairs for Exercises 4–5 because they find and record the y-coordinate first.

Prevent by having students choose two colors to represent the x- and y-axes. Have them choose colors that they feel have a relative order. For example, students might choose red and green because red comes before green in a traffic signal. Then have students draw a coordinate plane using their colors for the axes. Have them practice finding the coordinates of points on their coordinate planes, being sure to find the x-coordinate first.

Answers

1. Start at the origin. Move 5 units to the left. Then move 1 unit up and draw a dot.

3.

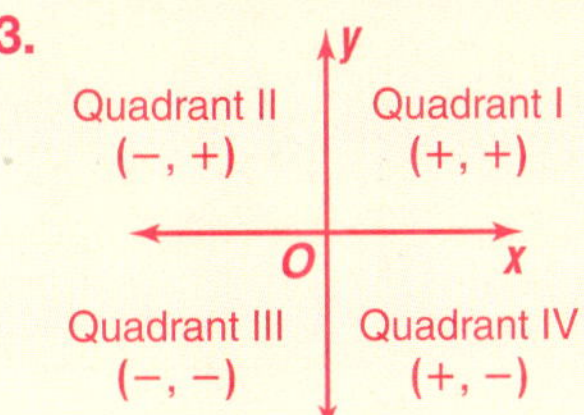

***Study Guide*, p. 56**

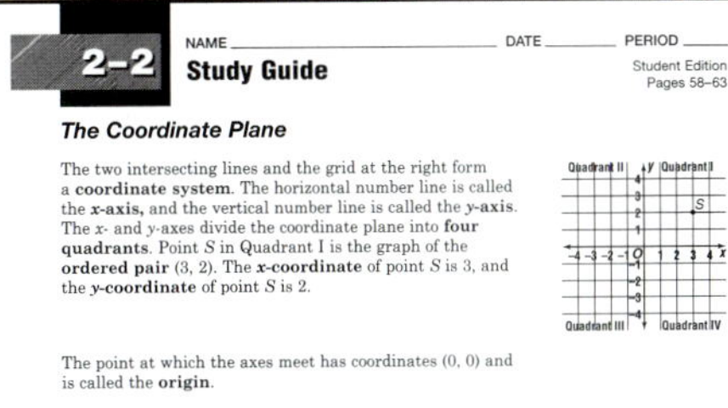

2–2 NAME ______ DATE ______ PERIOD ______

Study Guide

Student Edition Pages 58–63

The Coordinate Plane

The two intersecting lines and the grid at the right form a **coordinate system**. The horizontal number line is called the **x-axis,** and the vertical number line is called the **y-axis**. The x- and y-axes divide the coordinate plane into **four quadrants**. Point S in Quadrant I is the graph of the **ordered pair** (3, 2). The **x-coordinate** of point S is 3, and the **y-coordinate** of point S is 2.

The point at which the axes meet has coordinates (0, 0) and is called the **origin**.

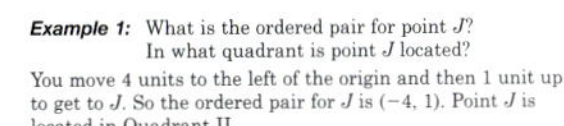

Example 1: What is the ordered pair for point J? In what quadrant is point J located?

You move 4 units to the left of the origin and then 1 unit up to get to J. So the ordered pair for J is $(-4, 1)$. Point J is located in Quadrant II.

Example 2: Graph $M(-2, -4)$ on the coordinate plane. Start at the origin. Move left on the x-axis to −2 and then down 4 units. Draw a dot here and label it M.

Write the ordered pair that names each point.

1. P (1, 3)
2. Q (−4, −3)
3. R (0, 2)
4. T (−4, 0)

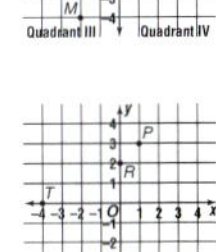

Graph each point on the coordinate plane. Name the quadrant, if any, in which each point is located.

5. $A(5, -1)$ IV
6. $B(-3, 0)$ none
7. $C(-3, 1)$ II
8. $D(0, 1)$ none
9. $E(3, 3)$ I
10. $F(-1, -2)$ III

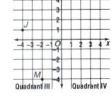

Reteaching Activity

Kinesthetic Learners Use masking tape to make a large coordinate plane on the floor of the school gym or in an unused area of a school parking lot. Direct each student to stand at the location for a specific pair of coordinates as you read them.

Answer

10b.

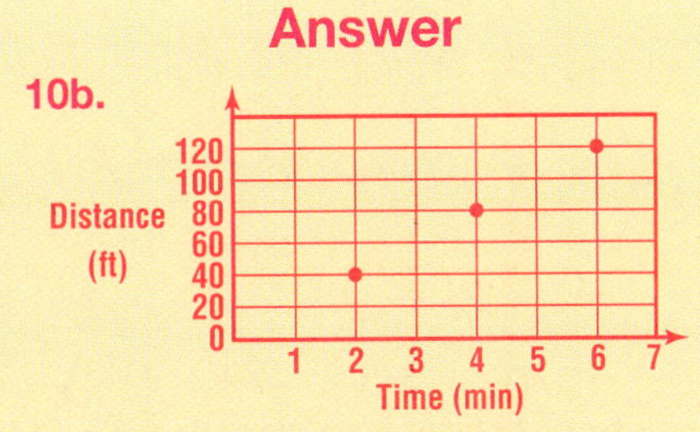

Assignment Guide

Basic: 11–37 odd, 38–45
Average: 12–32 even, 34–45
All: Quiz 1, 1–10

Answers

30. Sample answer:

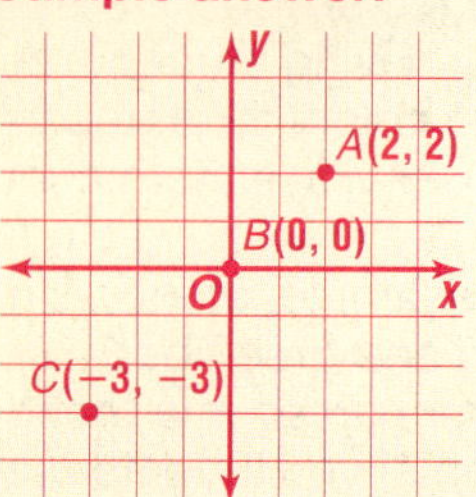

The points appear to be in a straight, diagonal line.

34.

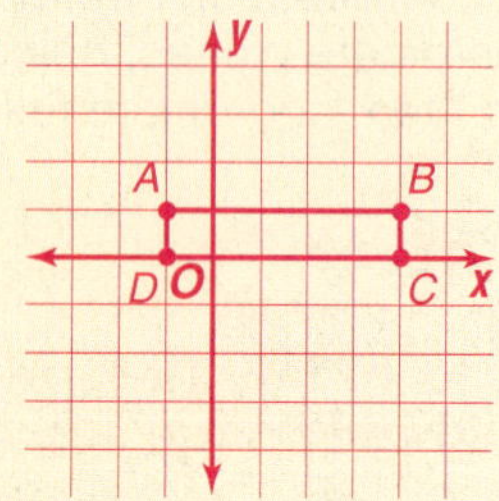

35b–c.

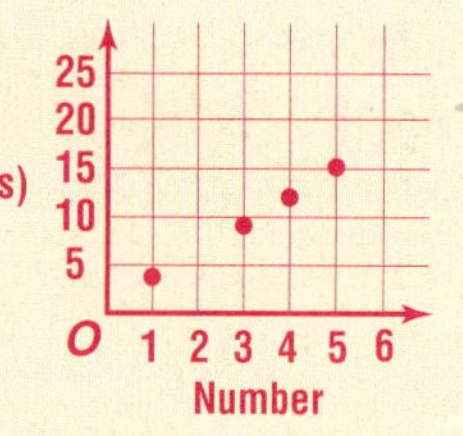

***Skills Practice*, p. 57, and *Practice*, p. 58 (shown)**

2-2 **Practice**

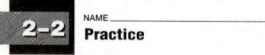

NAME ______ DATE ______ PERIOD ______

Student Edition Pages 58–63

The Coordinate Plane

Write the ordered pair that names each point.

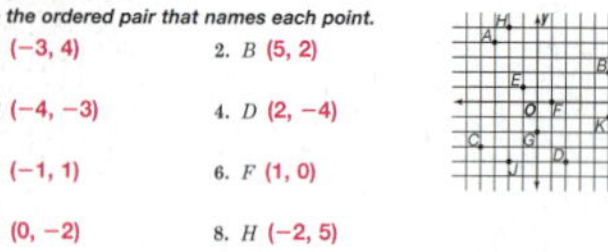

1. *A* (−3, 4)
2. *B* (5, 2)
3. *C* (−4, −3)
4. *D* (2, −4)
5. *E* (−1, 1)
6. *F* (1, 0)
7. *G* (0, −2)
8. *H* (−2, 5)
9. *J* (−2, −4)
10. *K* (5, −1)

Graph each point on the coordinate plane.

11. *K*(0, −3)
12. *L*(−2, 3)
13. *M*(4, 4)
14. *N*(−3, 0)
15. *P*(−4, −1)
16. *Q*(1, −2)
17. *R*(−5, 5)
18. *S*(3, 2)
19. *T*(2, 1)
20. *W*(−1, −4)

Name the quadrant in which each point is located.

21. (1, 9) I
22. (−2, −7) III
23. (0, −1) none
24. (−4, 6) II
25. (5, −3) IV
26. (−3, 0) none
27. (−1, −1) III
28. (6, −5) IV
29. (−8, 4) II
30. (−9, −2) III

Exercises

Practice

Homework Help

For Exercises	See Examples
11–16, 29, 36–37	1–3
17–22, 30, 34	4, 5
23–28, 31–33	6, 7
35	8

Extra Practice
See page 694.

Write the ordered pair that names each point.

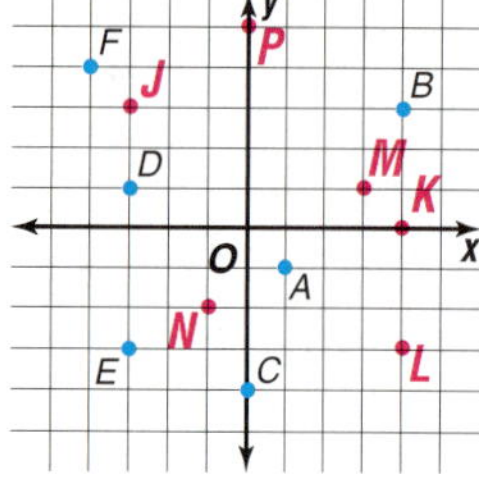

11. A **(1, −1)**
12. B **(4, 3)**
13. C **(0, −4)**
14. D **(−3, 1)**
15. E **(−3, −3)**
16. F **(−4, 4)**

Graph each point on a coordinate plane.

17. $J(-3, 3)$
18. $K(4, 0)$
19. $L(4, -3)$
20. $M(3, 1)$
21. $N(-1, -2)$
22. $P(0, 5)$

Name the quadrant in which each point is located.

23. $(-3, -4)$ **III**
24. $(6, -2)$ **IV**
25. $(0, 4)$ **none**
26. $(11, 15)$ **I**
27. $(-15, 25)$ **II**
28. $(-18, 0)$ **none**

29. What point lies on both the x-axis and the y-axis? **origin**
30. Graph three ordered pairs in which the x- and y-coordinates are equal. Describe the graph. **See margin.**

If the graph of $A(x, y)$ satisfies the given conditions, name the quadrant in which point A is located.

C

31. $x > 0, y > 0$ **I**
32. $x < 0, y < 0$ **III**
33. $x > 0, y < 0$ **IV**

Applications and Problem Solving

34. See margin for graph; rectangle.

34. **Geometry** Graph the points $A(-1, 1)$, $B(4, 1)$, $C(4, 0)$, and $D(-1, 0)$ on the same coordinate plane. Connect the points in alphabetical order and then connect A and D. Describe the figure.
35. **Entertainment** It costs \$3 to rent a video for a day. **a. \$3, \$9, \$15**
    a. Find the total cost of renting 1, 3, and 5 videos for a day.
    b. Graph the ordered pairs (number of videos, cost). **See margin.**
    c. Make a prediction about the location of the graph of (4, 12). Check your prediction by graphing (4, 12). **See margin.**

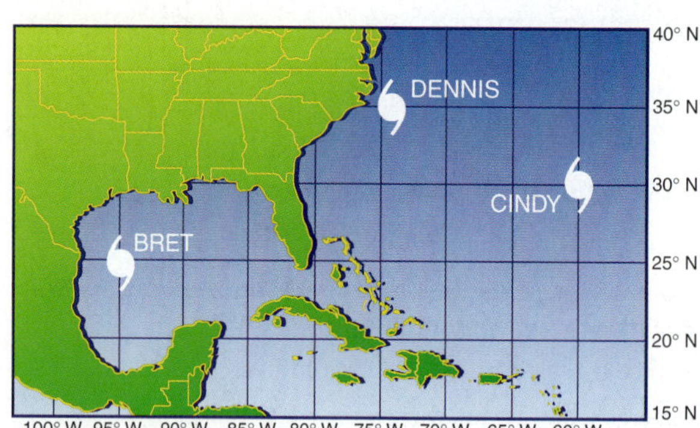

36. **Meteorology** Weather forecasters use a coordinate system composed of latitude (horizontal) and longitude (vertical) lines to locate hurricanes. For example, the position of Hurricane Dennis is 35°N latitude and 74°W longitude, or (35°N, 74°W). Write the position of each hurricane as an ordered pair.
    a. Bret **(25°N, 95°W)**
    b. Cindy **(30°N, 60°W)**

37a. $A(2, 1)$, $B(6, 1)$, $C(6, 5)$

37b. See margin.

37c. The triangles have the same shape, but the second is larger.

37. **Geometry** A *vertex* of a triangle is a point where two sides of the triangle meet.
    a. Identify the coordinates of the vertices in the triangle at the right.
    b. Multiply each x- and y-coordinate of the vertices by 2 and graph the new ordered pairs. Connect the points.
    c. Compare the two figures. Write a sentence that tells how the figures are the same and how they are different.

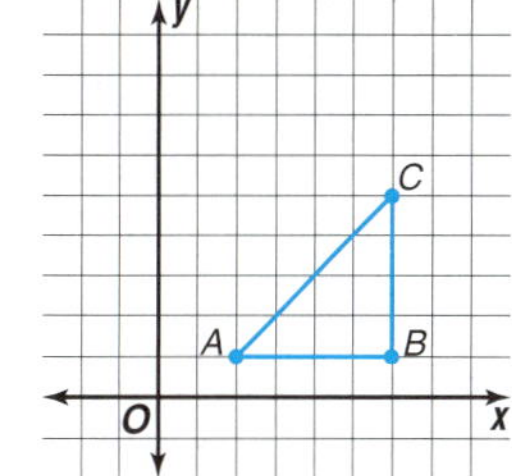

38. **Critical Thinking** Where are all of the possible locations for the graph of (x, y) if $x = y$? **Quadrant I, Quadrant III, or the origin**

Mixed Review

39. **Geography** The Caribbean Sea has an average depth of 8685 feet below sea level. Use an integer to express this depth. *(Lesson 2–1)* **−8685**

40. **Compare and contrast** a histogram and a cumulative frequency histogram. *(Lesson 1–7)* **See margin.**

Name the property shown by each statement. *(Lesson 1–3)*

41. $15 + 4 = 4 + 15$ **Comm. (+)**
42. $a(bc) = (ab)c$ **Assoc. (×)**
43. $3 + 4 = 7$, 7 is a whole number **Closure (+)**
44. $4 + (5 + 6) = 4 + (6 + 5)$ **Comm. (+)**

Standardized Test Practice

45. **Multiple Choice** Evaluate $8x - 3y$ if $x = 2$ and $y = 3$. *(Lesson 1–2)* **A**

Ⓐ 7 Ⓑ 25 Ⓒ 39 Ⓓ 57

Quiz 1 Lessons 2–1 and 2–2

Replace each ● with < or > to make a true sentence. *(Lesson 2–1)*

1. 3 ● −2 **>**
2. 0 ● −5 **>**
3. −6 ● −2 **<**
4. Order −6, −10, 10, 5, −7, and 0 from least to greatest. *(Lesson 2–1)* **−10, −7, −6, 0, 5, 10**
5. Evaluate $|-3| + |-8|$. *(Lesson 2–1)* **11**

Name the quadrant in which each point is located. *(Lesson 2–2)*

6. $A(4, -2)$ **IV**
7. $B(-5, -5)$ **III**
8. $C(0, -4)$ **none**
9. $D(-8, 6)$ **II**
10. **Entertainment** It costs $4 to buy a student ticket to the movies.
    a. Find the cost of 2, 4, and 5 tickets. **$8, $16, $20**
    b. Graph the ordered pairs (number of tickets, cost). *(Lesson 2–2)* **See margin.**

www.algconcepts.com/self_check_quiz

4 ASSESS

Open-Ended Assessment

Speaking Have students explain step-by-step how to graph the point (−3, 4) on a coordinate plane.

Quiz 1

The Quiz provides students with a brief review of the concepts and skills in Lessons 2–1 and 2–2. Lesson numbers are given to the right of the exercises or instruction lines so students can review concepts not yet mastered.

Chapter 2, Quiz A (Lessons 2–1 and 2–2) is available in the *Chapter 2 Resource Masters*, p. 91.

Answers

37b.

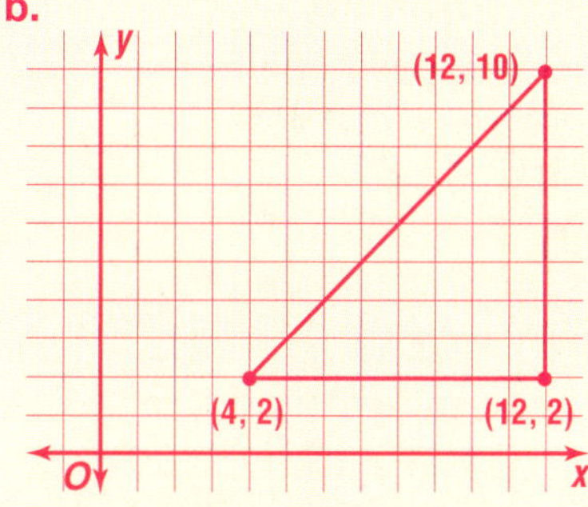

40. Both graphs display data over equal intervals. A histogram uses data from a frequency table and a cumulative frequency histogram uses data from a cumulative frequency table.

***Enrichment*, p. 60**

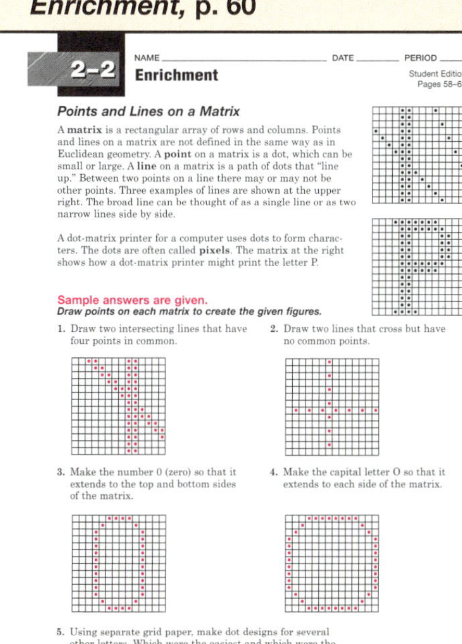

2–2 NAME ______ DATE ______ PERIOD ______

Enrichment

Student Edition Pages 58–63

Points and Lines on a Matrix

A **matrix** is a rectangular array of rows and columns. Points and lines on a matrix are not defined in the same way as in Euclidean geometry. A **point** on a matrix is a dot, which can be small or large. A **line** on a matrix is a path of dots that "line up." Between two points on a line there may or may not be other points. Three examples of lines are shown at the upper right. The broad line can be thought of as a single line or as two narrow lines side by side.

A dot-matrix printer for a computer uses dots to form characters. The dots are often called **pixels**. The matrix at the right shows how a dot-matrix printer might print the letter P.

Sample answers are given.

Draw points on each matrix to create the given figures.

1. Draw two intersecting lines that have four points in common.
2. Draw two lines that cross but have no common points.
3. Make the number 0 (zero) so that it extends to the top and bottom sides of the matrix.
4. Make the capital letter O so that it extends to each side of the matrix.
5. Using separate grid paper, make dot designs for several other letters. Which were the easiest and which were the most difficult? **See students' work.**

Extra Credit

Suppose the x-coordinate of point K is greater than its y-coordinate. Which is the only quadrant where point K could *not* be located? **II**

Answer *Quiz 1*

10b.

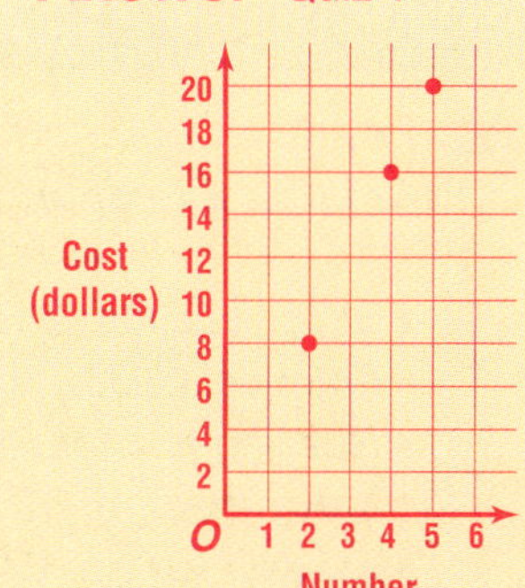

Lesson 2–3

1 FOCUS

5-Minute Check
Lesson 2–2

Write the ordered pair that names each point.

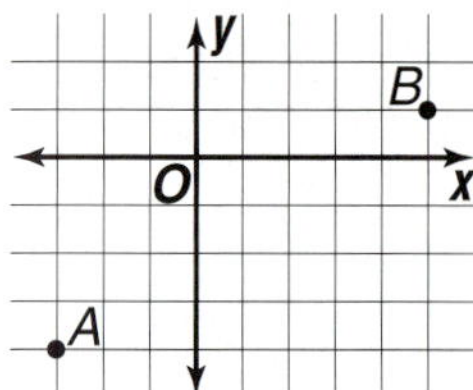

1. *A* **(−3, −4)**
2. *B* **(5, 1)**

Graph each point on a coordinate plane.

3. *C*(−2, 0)
4. *D*(3, −5)

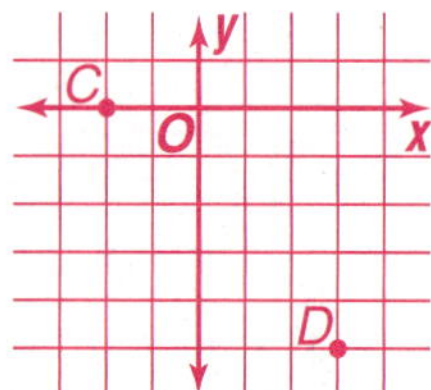

5. Refer to your graph for Exercises 3 and 4. Name the quadrant in which each point is located. ***C*: none; *D*: IV**

Motivating the Lesson

Real-World Connection Present students with this situation. *Your best friend loans you $6 so you can buy an item you need for school. As you leave the store, you find a $5 bill in the parking lot. After you have repaid your friend, how much of the $5 is yours?* Students should conclude that they would still be $1 in debt. Inform the class that the student's situation can be modeled by the mathematical sentence $-6 + 5 = -1$.

2–3 Adding Integers

What You'll Learn
You'll learn to add integers.

Why It's Important
Banking Banks use integers in checking accounts.
See Example 5.

There are several ways to add integers. One way is to use the 1-tiles from a set of algebra tiles.

Find $3 + 2$.
Combine 3 positive tiles with 2 positive tiles on a mat.

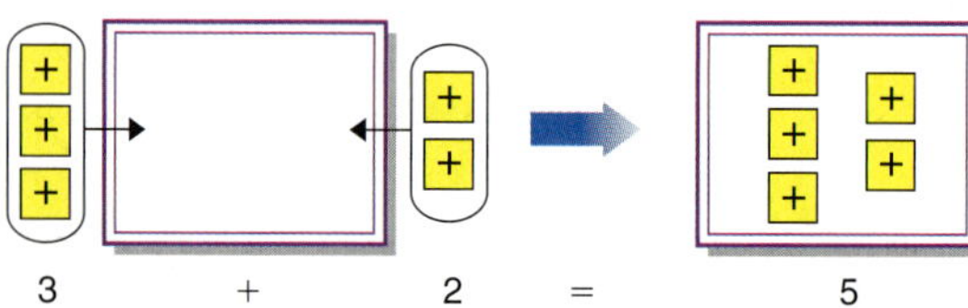

There are 5 positive tiles on the mat. Therefore, $3 + 2 = 5$.

Find $-3 + (-2)$.
Combine 3 negative tiles with 2 negative tiles.

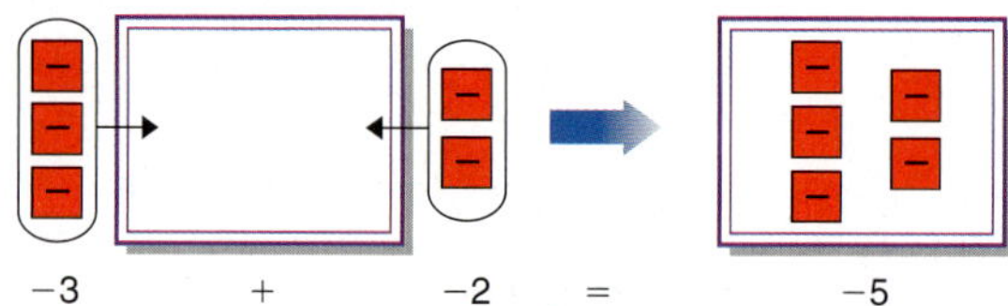

There are 5 negative tiles on the mat. Therefore, $-3 + (-2) = -5$.

You can also add integers on a number line. Start at 0. Positive integers are represented by arrows pointing *right*. Negative integers are represented by arrows pointing *left*. Start at 0. Move 3 units to the right. From there, move another 2 units to the right.

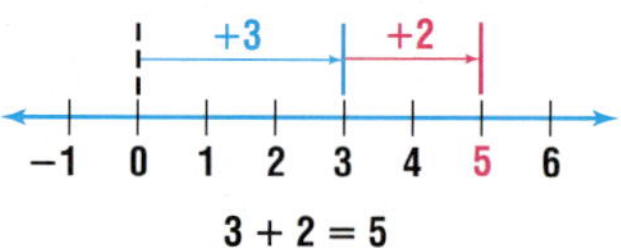

Start at 0. Move 3 units to the left. From there, move another 2 units to the left.

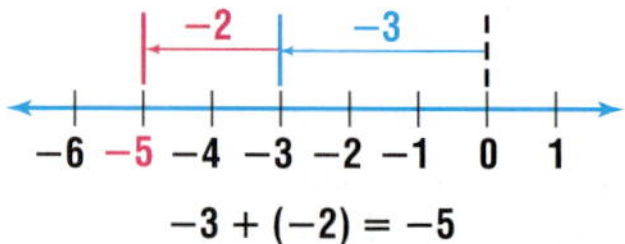

Resource Manager

Reproducible Masters
Chapter 2 Resource Masters
- *Study Guide,* p. 61
- *Skills Practice,* p. 62
- *Practice,* p. 63
- *Reading to Learn Mathematics,* p. 64
- *Enrichment,* p. 65
- *Assessment,* p. 90

Hands-On Algebra, pp. 34–35
School-to-Workplace, p. 2

Transparencies
5-Minute Check, 2–3

Technology/Multimedia
Interactive Chalkboard CD-ROM

These and other similar examples suggest the following rule for adding integers with the same sign.

Adding Integers with the Same Sign	
Words:	To add integers with the same sign, add their absolute values. Give the result the same sign as the integers.
Numbers:	$3 + 2 = 5$, $-3 + (-2) = -5$

Examples

Find each sum.

1 $4 + 5$

$4 + 5 = 9$ *Both numbers are positive, so the sum is positive.*

2 $-6 + (-2)$

$-6 + (-2) = -8$ *Both numbers are negative, so the sum is negative.*

Your Turn

b. -6 c. -15

a. $8 + 9$ **17** b. $-2 + (-4)$ c. $-5 + (-10)$ d. $11 + 6$ **17**

What is the result when you add two numbers that differ only in sign, like 3 and -3?

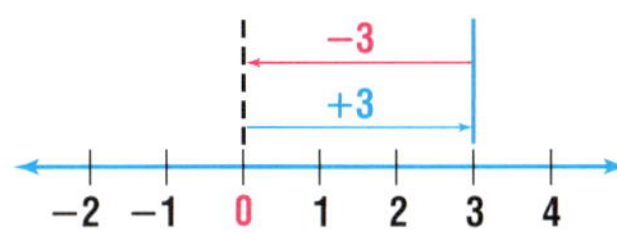

Start at zero. Move 3 units to the right. From there, move 3 units to the left.

$3 + (-3) = 0$

You can also use tiles. When one positive tile is paired with one negative tile, the result is a **zero pair**. You can remove zero pairs from the mat because removing zero does not change the value.

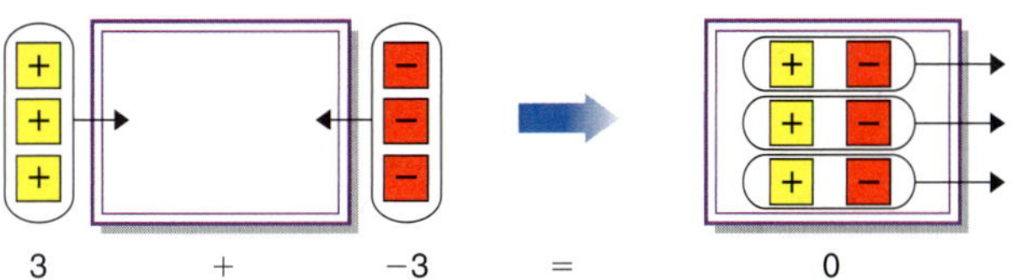

The models above show $3 + (-3) = 0$. If the sum of two numbers is 0, the numbers are called **opposites** or **additive inverses**.

-3 is the additive inverse, or opposite, of 3. $3 + (-3) = 0$
7 is the additive inverse, or opposite, of -7. $-7 + 7 = 0$

www.algconcepts.com/extra_examples

2 TEACH

In-Class Examples

Examples 1–2

Find each sum.

1 $6 + 7$ **13**

2 $-5 + (-8)$ **-13**

Teaching Tip As you discuss the number line shown below Example 2, relate the addition of 3 and -3 to the real-world situation where you begin with no money, you earn \$3, but then you spend \$3. Point out that after spending the money, you again have \$0.

Inclusion Strategies

Students with hearing impairments may benefit from sitting close to the front and audiotaping the lesson. If listening takes all their concentration, they will not be able to take notes. By taping the lesson, they will be able to review the material later.

Teaching Tip Before presenting Example 3, discuss the rule for adding integers with different signs. Relate the two sums in the rule box to the scenario of money. The sum $3 + (-2) = 1$ can be represented by having \$3 and spending \$2. The sum $-3 + 2 = -1$ can be represented by owing a friend \$3 and earning \$2. After giving your earnings to your friend, you still owe the friend \$1.

In-Class Example

Example 3

Find the sum $-9 + 5$. **-4**

Additive Inverse Property	
Words:	The sum of any number and its additive inverse is 0.
Symbols:	$a + (-a) = 0$
Numbers:	$3 + (-3) = 0$, $-7 + 7 = 0$

In the following activity, you'll use tiles to find a rule for adding two integers with different signs.

Materials: algebra tiles, integer mat

Find the sum $3 + (-2)$ using 1-tiles.

Step 1 Place 3 positive tiles and 2 negative tiles on the mat.

Step 2 Make as many zero pairs as you can. Remove them from the mat. The remaining tiles represent the sum.

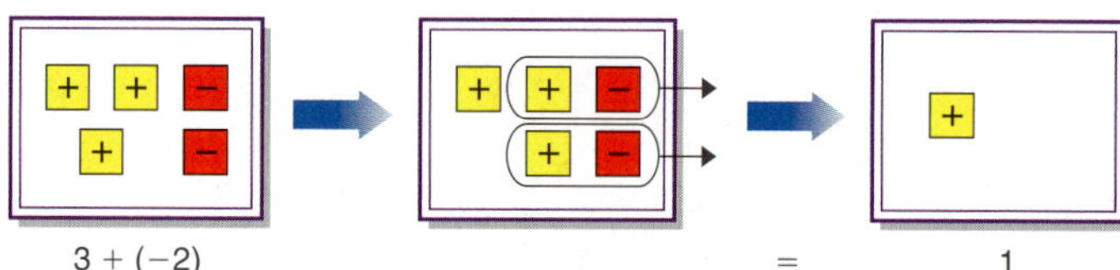

Try These

1. Is the sum $3 + (-2)$ positive or negative? **positive**
2. Which number, 3 or -2, has the greater absolute value? **3**
3. Use tiles to find the sum $-3 + 2$. Compare the sign of the sum with the sign of the number with the greater absolute value. **3. -1; They are the same.**
4. **Make a conjecture** about the sign of each sum. Verify using tiles.
   a. $4 + (-6)$ **−** b. $-7 + 1$ **−** c. $8 + (-2)$ **+** d. $-5 + 9$ **+**

The results of the activity suggest this rule.

Adding Integers with Different Signs	
Words:	To add integers with different signs, find the difference of their absolute values. Give the result the same sign as the integer with the greater absolute value.
Numbers:	$3 + (-2) = 1$, $-3 + 2 = -1$

Examples

Find each sum.

 **$5 + (-3)$**

$|5| - |3| = 5 - 3$ or 2

$|5| > |-3|$, so the sum is positive.

Therefore, $5 + (-3) = 2$.

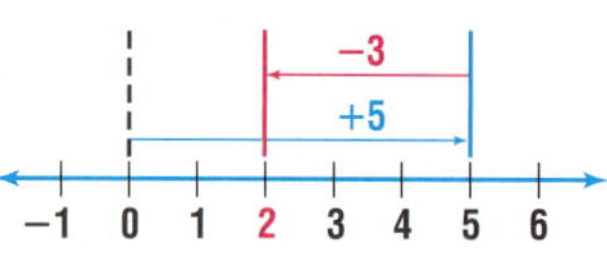

Hands-On Algebra

Cooperative Learning Make sure students understand why they can remove zero pairs. Emphasize the connection between zero pairs and additive inverses.

An additional Hands-On Algebra activity using a paper number line to model integer addition is available in the *Hands-On Algebra Masters*, p. 34.

Hands-On Algebra Masters, p. 35

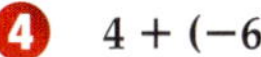

4 $4 + (-6)$

$|-6| - |4| = 6 - 4$ or 2

$|-6| > |4|$, so the sum is negative.

Therefore, $4 + (-6) = -2$.

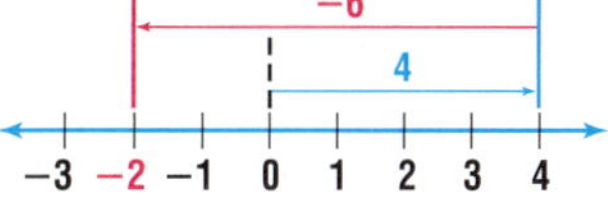

Your Turn

e. $-7 + 5$ **−2** f. $6 + (-8)$ **−2** g. $-4 + 9$ **5** h. $11 + (-8)$ **3**

Example 5 **Banking Link**

Talisa opened a checking account with a deposit of $25. During the next two weeks, she wrote checks for $20 and $15 and made a deposit of $30. Find the balance in her account.

Explore You know that Talisa made deposits of $25 and $30. She wrote checks for $20 and $15. You want to find the balance in her account.

Plan Deposits are represented by positive integers (+25 and +30). Checks are represented by negative integers (−20 and −15). Write an addition sentence and solve.

Solve Let x represent the balance in her account.

$x = 25 + (-20) + (-15) + 30$

$x = 5 + (-15) + 30$ *$25 + (-20) = 5$*

$x = -10 + 30$ *$5 + (-15) = -10$*

$x = 10$ *$-10 + 30 = 20$*

The balance in Talisa's account is $20.

Examine Addition of integers is commutative. So, you can check the solution by adding the integers in a different order. One way is to group all of the positive numbers and all of the negative numbers.

$x = 25 + 30 + (-20) + (-15)$

$x = 55 + (-35)$ *$25 + 30 = 55; -20 + (-15) = -35$*

$x = 20$ ✓

Look Back

Commutative Property: Lesson 1–3

You can use the rules for adding integers to simplify expressions.

Example 

Simplify $5x + (-3x)$.

$5x + (-3x) = [5 + (-3)]x$ *Use the Distributive Property.*

$= 2x$ *$5 + (-3) = 2$*

Your Turn

Simplify each expression. **j. 8m**

i. $-8y + 3y$ **−5y** j. $6m + 4m + (-2m)$ k. $-5x + 4x$ **−x**

In-Class Examples

Example 4

Find the sum $(-9) + 8$. **−1**

Example 5

Traci opened a checking account with a deposit of $100. During the next week, she wrote checks for $45 and $65 and made a deposit of $28. Find the balance in her account. **The balance in Traci's account is $18.**

Example 6

Simplify $-7y + 6y$. **−y**

***Study Guide*, p. 61**

NAME ______ DATE ______ PERIOD ______

Study Guide Student Edition Pages 64–69

Adding Integers

You can use a number line to add integers. Start at 0. Then move to the right for positive integers and move to the left for negative integers.

$2 + 1 = 3$ $-2 + (-1) = -3$

Both integers are positive. First move 2 units right from 0. Then move 1 more unit right.

Both integers are negative. First move 2 units left from 0. Then move 1 more unit left.

When you add one positive integer and one negative integer on the number line, you change directions, which results in one move being subtracted from the other move.

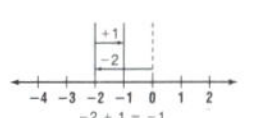

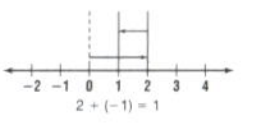

Move 2 units left, then 1 unit right. Move 2 units right, then 1 unit left.

Use the following rules to add two integers and to simplify expressions.

Rule	Examples
To add integers with the same sign, add their absolute values. Give the result the same sign as the integers.	$7 + 4 = 11$ $-8 + (-2) = -10$ $-5x + (-3x) = -8x$
To add integers with different signs, subtract their absolute values. Give the result the same sign as the integer with the greater absolute value.	$9 + (-6) = 3$ $1 + (-5) = -4$ $-2x + 9x = 7x$ $3y + (-4y) = -y$

Find each sum.

1. $5 + 8$ **13** 2. $-8 + (-9)$ **−17** 3. $12 + (-8)$ **4** 4. $-16 + 5$ **−11**
5. $5 + (-8) + (-5)$ **−8** 6. $-8 + (-8) + 20$ **4** 7. $12 + 5 + (-1)$ **16**

Simplify each expression.

8. $3x + (-6x)$ **−3x** 9. $-5y + (-7y)$ **−12y** 10. $2m + (-4m) + (-2m)$ **−4m**

3 PRACTICE/APPLY

Error Analysis

Watch for students who find a sum of 2 in Exercise 17 because they think of the sum as $4 + (-2) + x + (-x)$.

Prevent by having students think of $4x$ as $x + x + x + x$ and $-2x$ as $(-x) + (-x)$, similar to how the expressions are modeled with algebra tiles. Show students that x and $(-x)$ form a "zero pair," which can be removed. Point out that removing two such zero pairs leaves the sum as $x + x$, or $2x$.

Assignment Guide

Basic: 21–61 odd, 62–73
Average: 22–58 even, 60–73

Answers

1.

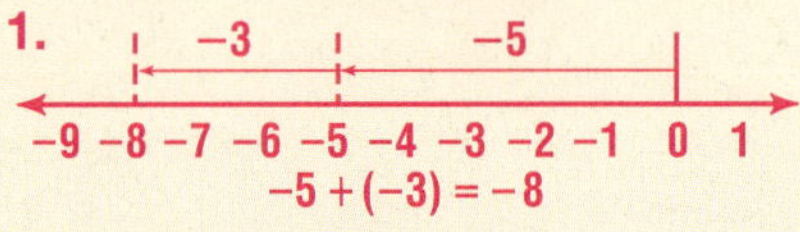

3.

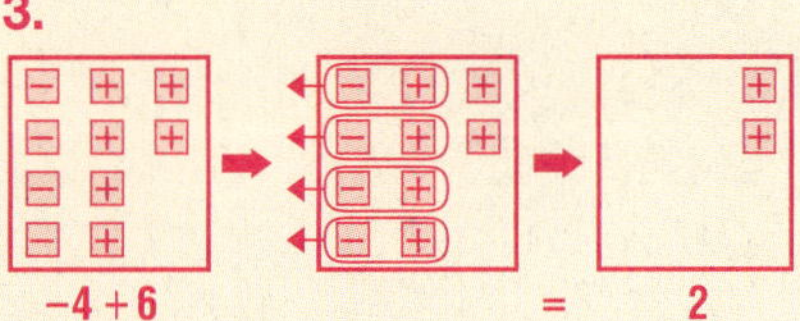

***Skills Practice,* p. 62, and *Practice,* p. 63 (shown)**

NAME ______ DATE ______ PERIOD ______

2-3 Practice — Student Edition Pages 64–69

Adding Integers

Find each sum.

1. 8 + 4 — 12	2. −3 + 5 — 2	3. 9 + (−2) — 7
4. −5 + 11 — 6	5. −7 + (−4) — −11	6. 12 + (−4) — 8
7. −9 + 10 — 1	8. −4 + 4 — 0	9. 2 + (−8) — −6
10. 17 + (−4) — 13	11. −13 + 3 — −10	12. 6 + (−7) — −1
13. −8 + (−9) — −17	14. −2 + 11 — 9	15. −9 + (−2) — −11
16. −1 + 3 — 2	17. 6 + (−5) — 1	18. −11 + 7 — −4
19. −8 + (−8) — −16	20. −6 + 3 — −3	21. 2 + (−2) — 0
22. 7 + (−5) + 2 — 4	23. −4 + 8 + (−3) — 1	24. −5 + (−5) + 5 — −5

Simplify each expression.

25. $5a + (-3a)$ — $2a$	26. $-7y + 2y$ — $-5y$	27. $-9m + (-4m)$ — $-13m$
28. $-2z + (-4z)$ — $-6z$	29. $8x + (-4x)$ — $4x$	30. $-10p + 5p$ — $-5p$
31. $5b + (-2b)$ — $3b$	32. $-4s + 7s$ — $3s$	33. $2n + (-4n)$ — $-2n$
34. $5a + (-6a) + 4a$ — $3a$	35. $-6x + 3x + (-5x)$ — $-8x$	36. $7z + 2z + (-3z)$ — $6z$

Check for Understanding

Communicating Mathematics

2. $-10 + 10 = 0$

Vocabulary
zero pair
opposites
additive inverse

1. **Show** how to find the sum of −5 and −3 on a number line. **See margin.**
2. **Explain** why −10 and 10 are additive inverses.
3. **Draw** a diagram that shows how to find the sum of −4 and 6 using tiles. **See margin.**
4. *Writing Math* Write a paragraph that describes how to add two integers. Be sure to include examples with your description. **See students' work.**

Guided Practice

Getting Ready Tell whether each sum is *positive* or *negative*.

Sample 1: $-4 + (-3)$
Solution: Both integers are negative, so the sum is negative.

Sample 2: $-9 + 11$
Solution: $|11| > |-9|$, so the sum is positive.

5. 5 + 12 **+** 6. 12 + (−15) **−** 7. −3 + (−7) **−**
8. −3 + 9 **+** 9. −5 + (−2) **−** 10. −8 + 12 **+**

Examples 1–4 Find each sum.

11. 7 + 9 **16** 12. −2 + (−8) **−10** 13. 8 + (−9) **−1**
14. −12 + 15 **3** 15. −10 + 5 **−5** 16. 11 + (−2) **9**

Example 6 Simplify each expression. **18. $-11y$ 19. $2a$**

17. $4x + (-2x)$ **$2x$** 18. $-9y + (-2y)$ 19. $3a + (-4a) + 3a$

Example 5

20. **Games** On a famous TV game show, contestants earn money for each correct answer and lose money for each incorrect answer. Suppose a contestant answered questions worth \$100, \$200, and \$400 correctly, but answered questions worth \$300, \$300, and \$400 incorrectly. What was the contestant's final score? **−\$300**

Exercises

Practice 

Homework Help

For Exercises	See Examples
21–26	1, 2
27–38	3, 4
39–44	5
45–47	1–4
48–56	6

Extra Practice See page 695.

Find each sum.

21. 3 + 9 **12** 22. 8 + 6 **14** 23. 5 + 16 **21**
24. −3 + (−10) **−13** 25. −5 + (−6) **−11** 26. −11 + (−7) **−18**
27. −13 + 5 **−8** 28. 12 + (−7) **5** 29. −6 + 15 **9**
30. 6 + (−6) **0** 31. 5 + (−18) **−13** 32. −9 + (−9) **−18**
33. −15 + 7 **−8** 34. 16 + (−11) **5** 35. −10 + (−11) **−21**
36. 30 + (−15) **15** 37. −20 + (−35) **−55** 38. −40 + 26 **−14**
39. 8 + (−5) + 10 **13** 40. 3 + 15 + (−6) **12**
41. −10 + (−4) + (−8) **−22** 42. 15 + 7 + (−7) + (−13) **2**
43. −6 + 12 + (−11) + 1 **−4** 44. 17 + (−21) + 10 + (−17) **−11**

Reteaching Activity

Verbal/Linguistic Learners Have pairs of students take turns explaining or modeling for their partner how to add integers with the same sign and with different signs using algebra tiles.

45. Find the value of y if $y = -3 + 2$. **−1**

46. What is the value of w if $-7 + (-2) = w$? **−9**

47. Find the value of b if $b = 3 + (-6)$. **−3**

Simplify each expression. **49. −15*x***

48. $-9a + 3a$ **−6*a***
49. $-5x + (-10x)$
50. $-16y + 15y$ **−1*y***
51. $-11m + 14m$ **3*m***
52. $4z + (-3z)$ **1*z* or *z***
53. $8c + (-8c)$ **0**
54. $-8b + 4b + (-2b)$ **−6*b***
55. $3y + 8y + (-3y)$ **8*y***
56. $-2n + (-4n) + 3n$ **−3*n***

Evaluate each expression if $x = -4$, $y = -5$, and $z = 4$.

C

57. $x + 4 + (-9)$ **−9**
58. $-7 + y + z$ **−8**
59. $|x| + y$ **−1**

Applications and Problem Solving

60. **Sports** In golf, a score of 0 is called *even par*. One over par is represented by +1, and one under par is represented by −1. In the 1999 U.S. Open, Tiger Woods had scores represented by −2, +1, +2, and 0. What was his final score? **+1 or 1 over par**

61. **Geometry** The points $A(2, 3)$, $B(3, -3)$, and $C(-3, -2)$ are connected with line segments to form a triangle.

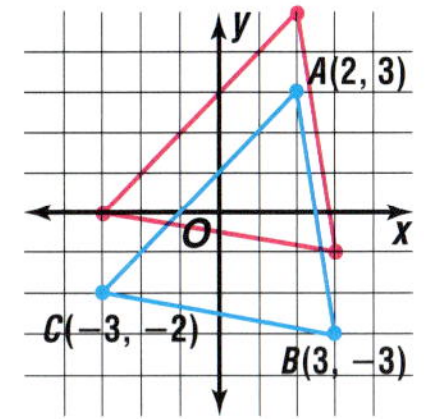

a. Add 2 to each y-coordinate and draw another triangle. **See right.**

b. How did the position of the triangle change? **It moved up 2 units.**

62. **Critical Thinking** Refer to Exercise 61. What change would you make to the ordered pairs so that the triangle would move to the right? **Add a positive integer to the *x*-coordinate.**

Mixed Review

Name the quadrant in which each point is located. *(Lesson 2–2)*

63. $A(6, -5)$ **IV**
64. $B(-2, -2)$ **III**
65. $C(-5, 3)$ **II**
66. $(0, 4)$ **none**

Write an integer for each situation. *(Lesson 2–1)* **68. +2**

67. a debt of \$5 **−5**
68. 2 inches more rain than normal
69. a loss of 10 yards **−10**
70. a deposit of \$17 **+17**
71. maintaining your present weight **0**

Standardized Test Practice

72. **Extended Response** One hundred people were surveyed outside a movie theater to determine the favorite leisure-time activity for a large population. Is this a good sample? Explain your reasoning. *(Lesson 1–6)* **No, the sample might be biased toward going to movies.**

73. **Multiple Choice** Use the pattern in the perimeter P of each rectangle to determine the perimeter of a rectangle made up of ten unit squares. *(Lesson 1–5)* **C**

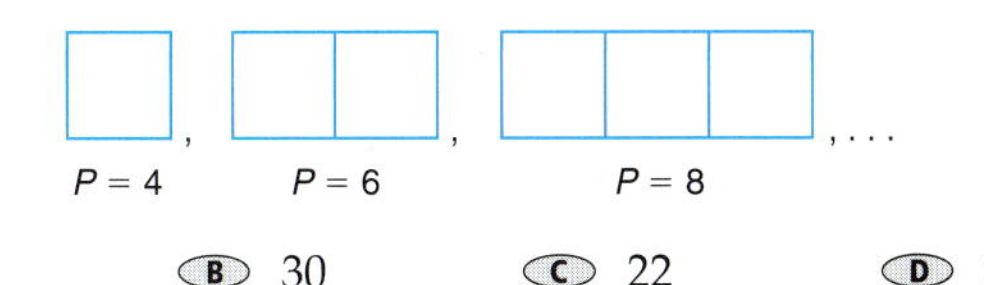

Ⓐ 14 Ⓑ 30 Ⓒ 22 Ⓓ 28

www.algconcepts.com/self_check_quiz

4 ASSESS

Open-Ended Assessment

Modeling Have students demonstrate how to find −6 + 5 using a number line model or algebra tiles.

Mid-Chapter Test (Lessons 2–1 through 2–3) is available in the *Chapter 2 Resource Masters*, p. 90.

***Enrichment*, p. 65**

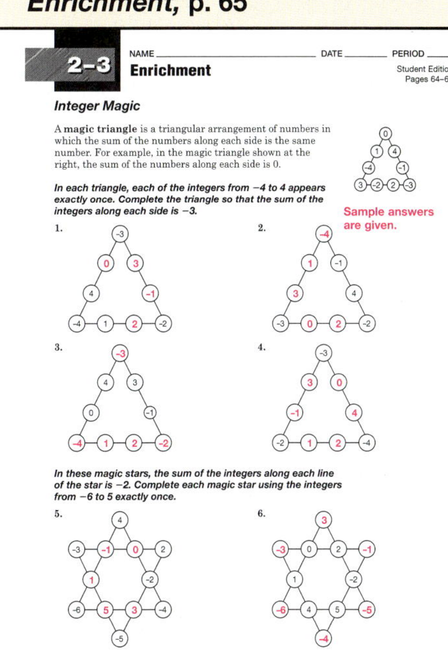

NAME ______ DATE ______ PERIOD ______

2–3 Enrichment Student Edition Pages 64–69

Integer Magic

A **magic triangle** is a triangular arrangement of numbers in which the sum of the numbers along each side is the same number. For example, in the magic triangle shown at the right, the sum of the numbers along each side is 0.

In each triangle, each of the integers from −4 to 4 appears exactly once. Complete the triangle so that the sum of the integers along each side is −3.

Sample answers are given.

1. 2. 3. 4.

In these magic stars, the sum of the integers along each line of the star is −2. Complete each magic star using the integers from −6 to 5 exactly once.

5. 6.

Extra Credit

What is the sum of all the integers from −100 to 100? **0**

2-4 Subtracting Integers

1 FOCUS

5-Minute Check
Lesson 2–3

Find each sum.

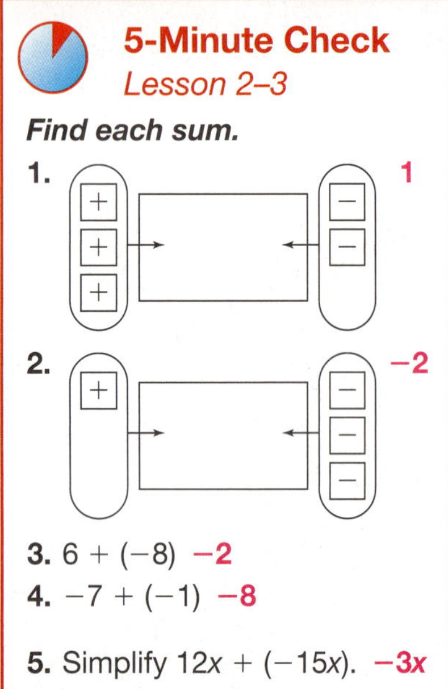

3. 6 + (−8) **−2**

4. −7 + (−1) **−8**

5. Simplify 12*x* + (−15*x*). **−3x**

Motivating the Lesson

Real-World Connection Present this situation to your students. *You owe two friends $5 each. You spend an afternoon helping one friend with a chore so he tells you to forget about paying him the money you owe.* Discuss with students whether they now owe more money or less. **less** Ask students how they could represent this situation using integers.

2 TEACH

Teaching Tip When discussing the algebra tile models, emphasize that subtraction is shown by removing one or more tiles. Stress that this removal is not the same as removing a zero pair.

What You'll Learn
You'll learn to subtract integers.

Why It's Important
Budgeting Families often need to find the difference between the amount of money in a budget and the actual amount spent. *See Exercise 48.*

When you add or subtract two integers, the sum or the difference is also an integer. Algebra tiles can be used as a model for subtraction of integers. In the examples below, you can see how addition and subtraction of integers are related.

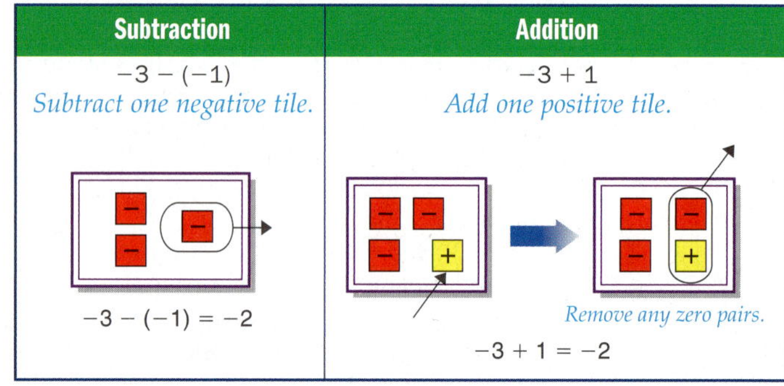

In the examples above, notice that −3 − (−1) = −3 + 1.

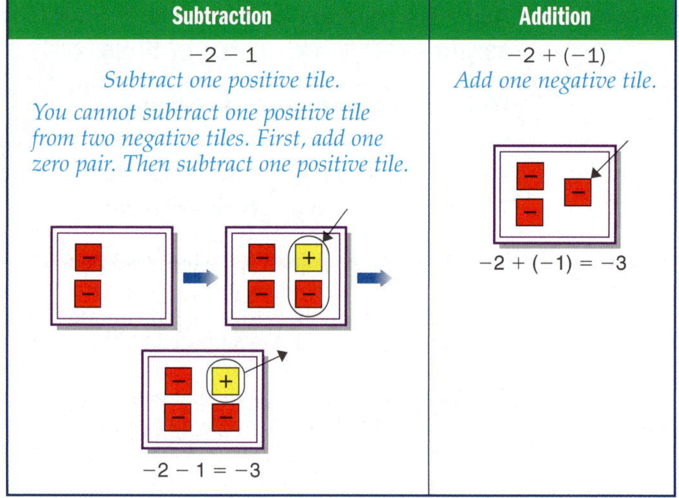

In the examples above, notice that −2 − 1 = −2 + (−1). This example shows that subtracting 1 from −2 is the same as adding −1 to −2.

Resource Manager

Reproducible Masters

Chapter 2 Resource Masters
- *Study Guide,* p. 66
- *Skills Practice,* p. 67
- *Practice,* p. 68
- *Reading to Learn Mathematics,* p. 69
- *Enrichment,* p. 70

Hands-On Algebra, p. 36

Transparencies

5-Minute Check, 2–3

Technology/Multimedia

Interactive Chalkboard CD-ROM

The examples on the previous page suggest that subtracting an integer is the same as adding the additive inverse or opposite of the integer.

Subtraction	Addition	Subtraction	Addition
additive inverses		*additive inverses*	
$-3 - (-1) = -2$	$-3 + 1 = -2$	$-2 - 1 = -3$	$-2 + (-1) = -3$
same result		*same result*	

Subtracting Integers		
	Words:	To subtract an integer, add its additive inverse.
	Model:	$a - b = a + (-b)$
	Numbers:	$9 - 7 = 9 + (-7)$ or 2

Examples

Find each difference.

1 $6 - 4$

$6 - 4 = 6 + (-4)$ *To subtract 4, add −4.*
$= 2$ *Simplify.*

2 $-5 - (-3)$

$-5 - (-3) = -5 + 3$ *To subtract −3, add 3.*
$= -2$ *Simplify.*

3 $-3 - 2$

$-3 - 2 = -3 + (-2)$ *To subtract 2, add −2.*
$= -5$ *Simplify.*

4 $4 - (-1)$

$4 - (-1) = 4 + 1$ *To subtract −1, add 1.*
$= 5$ *Simplify.*

5 $2 - 5$

$2 - 5 = 2 + (-5)$ *To subtract 5, add −5.*
$= -3$ *Simplify.*

6 $-3 - (-6)$

$-3 - (-6) = -3 + 6$ *To subtract −6, add 6.*
$= 3$ *Simplify.*

Your Turn

a. $9 - 3$ **6** b. $-7 - (-2)$ **−5** c. $-8 - 3$ **−11**
d. $5 - (-1)$ **6** e. $-4 - 6$ **−10** f. $-7 - (-11)$ **4**

www.algconcepts.com/extra_examples

In-Class Examples

Examples 1–6

Find each difference.

1. $10 - 3$ **7**
2. $-7 - (-6)$ **−1**
3. $-1 - 8$ **−9**
4. $3 - (-5)$ **8**
5. $4 - 6$ **−2**
6. $-7 - (-10)$ **3**

In-Class Example

Example 7

Evaluate $a - b$ if $a = -8$ and $b = -2$. **−6**

Teaching Tip As you work Example 8, point out to students that they can use the properties of addition to simplify the expression after inserting the values for the variables.

$6 + 2 + (-6)$
$= 6 + (-6) + 2$ *Commutative (+)*
$= 0 + 2$ *Additive Inverse*
$= 2$ *Additive Identity*

In-Class Example

Example 8

Evaluate $p + q - r$ if $p = -7$, $q = -3$, and $r = 2$. **−12**

Teaching Tip As you conclude your discussion of Example 9, show students that they can confirm the answer is reasonable by looking at the numbers and arrows on the map.

In-Class Example

Example 9

During one school year, 23 new students moved into a school district and 52 students moved out of the school district. Find the change in the number of students resulting from these moves. **The number of students in the school district decreased by 29.**

***Study Guide*, p. 66**

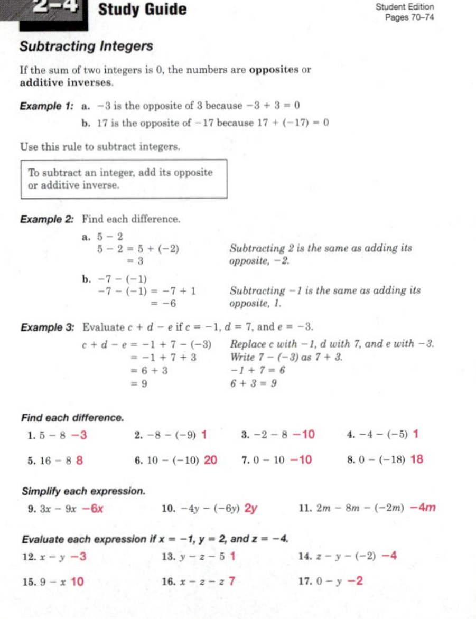

2-4 NAME ______________ DATE ________ PERIOD ______

Study Guide Student Edition Pages 70–74

Subtracting Integers

If the sum of two integers is 0, the numbers are **opposites** or **additive inverses**.

Example 1: a. -3 is the opposite of 3 because $-3 + 3 = 0$
b. 17 is the opposite of -17 because $17 + (-17) = 0$

Use this rule to subtract integers.

> To subtract an integer, add its opposite or additive inverse.

Example 2: Find each difference.

a. $5 - 2$
$5 - 2 = 5 + (-2)$ *Subtracting 2 is the same as adding its opposite, −2.*
$= 3$

b. $-7 - (-1)$
$-7 - (-1) = -7 + 1$ *Subtracting −1 is the same as adding its opposite, 1.*
$= -6$

Example 3: Evaluate $c + d - e$ if $c = -1$, $d = 7$, and $e = -3$.

$c + d - e = -1 + 7 - (-3)$ *Replace c with −1, d with 7, and e with −3.*
$= -1 + 7 + 3$ *Write 7 − (−3) as 7 + 3.*
$= 6 + 3$ *−1 + 7 = 6*
$= 9$ *6 + 3 = 9*

Find each difference.

1. $5 - 8$ **−3**
2. $-8 - (-9)$ **1**
3. $-2 - 8$ **−10**
4. $-4 - (-5)$ **1**
5. $16 - 8$ **8**
6. $10 - (-10)$ **20**
7. $0 - 10$ **−10**
8. $0 - (-18)$ **18**

Simplify each expression.

9. $3x - 9x$ **−6x**
10. $-4y - (-6y)$ **2y**
11. $2m - 8m - (-2m)$ **−4m**

Evaluate each expression if x = −1, y = 2, and z = −4.

12. $x - y$ **−3**
13. $y - z - 5$ **1**
14. $z - y - (-2)$ **−4**
15. $9 - x$ **10**
16. $x - z - z$ **7**
17. $0 - y$ **−2**

When you evaluate expressions, it is helpful to write any subtraction expressions as addition expressions first.

Examples

7 **Evaluate $x - y$ if $x = -2$ and $y = 1$.**

$x - y = -2 - 1$ *Replace x with −2 and y with 1.*
$= -2 + (-1)$ *Write −2 − 1 as −2 + (−1).*
$= -3$ *−2 + (−1) = −3*

8 **Evaluate $a - b + c$ if $a = 6$, $b = -2$, and $c = -6$.**

$a - b + c = 6 - (-2) + (-6)$ *Replace a with 6, b with −2, and c with −6.*
$= 6 + 2 + (-6)$ *Write 6 − (−2) as 6 + 2.*
$= 8 + (-6)$ *6 + 2 = 8*
$= 2$ *8 + (−6) = 2*

Your Turn

g. Evaluate $m - n$ if $m = 5$ and $n = -3$. **8**

h. Evaluate $w - x + y - z$ if $w = -5$, $x = -7$, $y = 10$, and $z = -5$. **17**

Integers are often used to show how data has changed for a given time.

Example
Population Link

9 **The map shows the number of people who moved to Ohio from Indiana in a recent year. It also shows the number of people who left Ohio for Indiana. The change in Ohio's population p can be found by using the formula $p = m - l$, where m is the number of people moving to Ohio and l is the number of people leaving Ohio.**

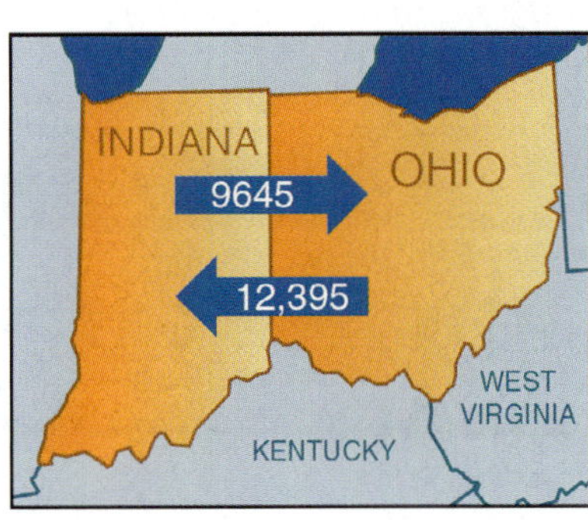

Find the net change in Ohio's population resulting from people moving to and from Indiana.

$p = m - l$

$p = 9645 - 12{,}395$ *Replace m with 9645 and l with 12,395.*

9645 [−] 12395 [ENTER] −2750

$p = -2750$ Ohio's population decreased by 2750 people.

Technology Tip

The [−] key is used for subtraction. The [(−)] key is used to enter negative integers.

Check for Understanding

Communicating Mathematics

1. **Explain** how additive inverses are used in subtraction.
2. **Draw** a diagram using algebra tiles that shows how $2 - 5$ and $2 + (-5)$ have the same result.

1–2. See margin.

Guided Practice

Getting Ready Write each expression as an addition expression.

Sample: $4 - (-3)$ **Solution:** $4 + 3$

3. $10 - 3$ $10 + (-3)$ 4. $-2 - (-5)$ $-2 + 5$ 5. $-4 - 8$ $-4 + (-8)$

Examples 1–6

Find each difference.

6. $8 - 2$ 6
7. $-6 - (-4)$ -2
8. $-5 - 4$ -9
9. $7 - (-4)$ 11
10. $8 - 11$ -3
11. $-4 - (-9)$ 5

Examples 7 & 8

Evaluate each expression if $a = -2$, $b = 6$, $c = -3$, and $d = -1$.

12. $a - b$ -8
13. $b - c + d$ 8

Example 9

14. **Population** In a recent year, 5899 people moved to Ohio from West Virginia, and 5394 people left Ohio for West Virginia. Find the net change in Ohio's population. +505 people

Exercises

Practice

Find each difference.

15. $15 - 2$ 13
16. $11 - 6$ 5
17. $14 - 7$ 7
18. $-9 - (-3)$ -6
19. $-10 - (-2)$ -8
20. $-15 - (-4)$ -11
21. $-10 - 3$ -13
22. $-8 - 4$ -12
23. $-9 - 2$ -11
24. $5 - (-2)$ 7
25. $5 - (-11)$ 16
26. $9 - (-8)$ 17
27. $4 - 10$ -6
28. $9 - 16$ -7
29. $0 - 9$ -9
30. $-4 - (-10)$ 6
31. $0 - (-12)$ 12
32. $-8 - (-14)$ 6
33. Find the value of x if $3 - (-4) = x$. 7
34. What is the value of y if $y = -3 - (-12)$? 9
35. Find the value of v if $v = 2 - 19$. -17

Homework Help

For Exercises	See Examples
15–35, 45–48	1–6
36–44	7, 8

Extra Practice

See page 695.

Evaluate each expression if $x = 10$, $y = -7$, $z = -10$, and $w = 12$.

36. $x - y$ 17
37. $y - z$ 3
38. $15 - w$ 3
39. $7 - x + y$ -10
40. $x - z - w$ 8
41. $x + z - w$ -12

Simplify each expression.

42. $5y - 2y$ $3y$
43. $20n - (-5n)$ $25n$
44. $4a - 9a + 3a$ $-2a$
45. What is the difference of 25 and -25? 50
46. Write $a - (-b)$ as an addition expression. $a + b$

3 PRACTICE/APPLY

Error Analysis

Watch for students who give $-2 + (-5)$ as their answer to Exercise 4.

Prevent by pointing out that two symbols must change when rewriting an integer difference as a sum; namely, the subtraction symbol and the sign of the integer being subtracted. Remind students that an integer is positive if there is no sign preceding it.

Assignment Guide

Basic: 15–47 odd, 49–60
Average: 16–46 even, 47–60

Answers

1. To subtract an integer, add its additive inverse.
2.

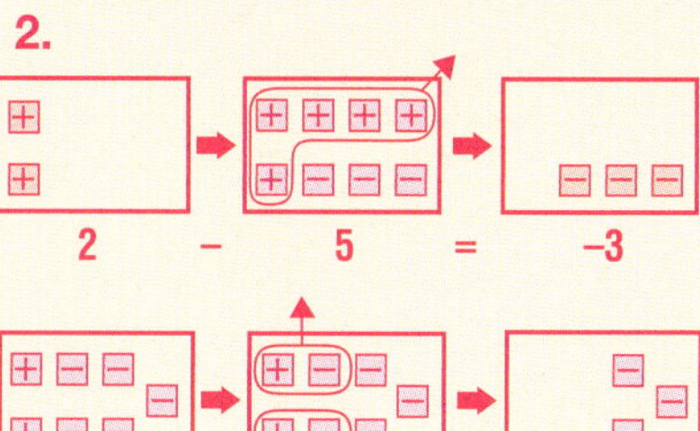

Skills Practice, **p. 67, and** ***Practice,*** **p. 68 (shown)**

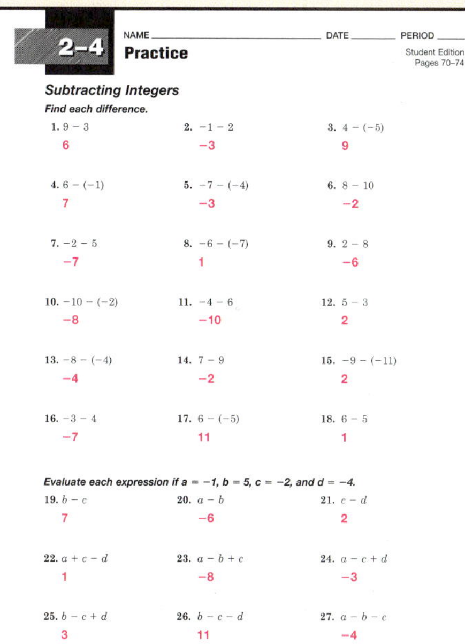

NAME ______ DATE ______ PERIOD ______

2-4 Practice

Student Edition Pages 70–74

Subtracting Integers

Find each difference.

1. $9 - 3$ 6	2. $-1 - 2$ -3	3. $4 - (-5)$ 9
4. $6 - (-1)$ 7	5. $-7 - (-4)$ -3	6. $8 - 10$ -2
7. $-2 - 5$ -7	8. $-6 - (-7)$ 1	9. $2 - 8$ -6
10. $-10 - (-2)$ -8	11. $-4 - 6$ -10	12. $5 - 3$ 2
13. $-8 - (-4)$ -4	14. $7 - 9$ -2	15. $-9 - (-11)$ 2
16. $-3 - 4$ -7	17. $6 - (-5)$ 11	18. $6 - 5$ 1

Evaluate each expression if $a = -1$, $b = 5$, $c = -2$, and $d = -4$.

19. $b - c$ 7	20. $a - b$ -6	21. $c - d$ 2
22. $a + c - d$ 1	23. $a - b + c$ -8	24. $a - c + d$ -3
25. $b - c + d$ 3	26. $b - c - d$ 11	27. $a - b - c$ -4

Reteaching Activity

Interpersonal Learners Have pairs of students work together to model several of Exercises 15–32 using algebra tiles.

4 ASSESS

Open-Ended Assessment

Writing Have students write an explanation of each step of their work as they find the difference $-7 - (-8)$.

Answers

48a. food: $-\$15$; electric: $\$1$; telephone: $-\$6$; heating fuel: $\$35$; water $-\$7$; cable TV: $\$0$

48b. More money was spent than was budgeted.

48c. less, $8

56b.

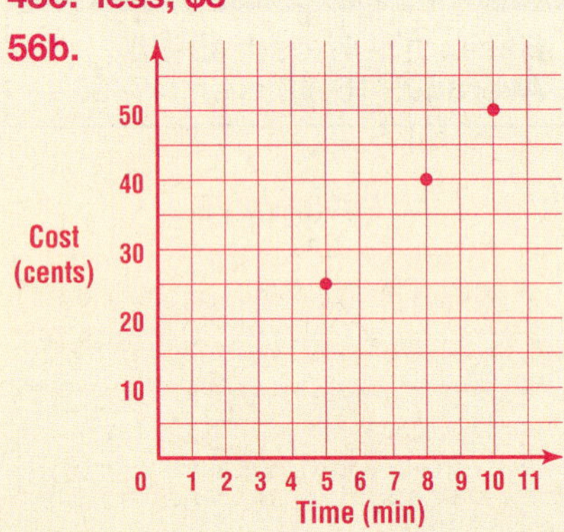

Enrichment, p. 70

2–4 NAME ______ DATE ______ PERIOD ______

Enrichment

Student Edition Pages 70–74

Closure

A **binary operation** matches two numbers in a set to just one number. Addition is a binary operation on the set of whole numbers. It matches two numbers such as 4 and 5 to a single number, their sum.

If the result of a binary operation is always a member of the original set, the set is said to be **closed** under the operation. For example, the set of whole numbers is not closed under subtraction because $3 - 6$ is not a whole number.

Is each operation binary? Write yes or no.

1. the operation ↵, where a ↵ b means to choose the lesser number from a and b **yes**
2. the operation ©, where a © b means to cube the sum of a and b **yes**
3. the operation *sq*, where $sq(a)$ means to square the number a **no**
4. the operation *exp*, where $exp(a, b)$ means to find the value of a^b **yes**
5. the operation $\Uparrow$, where $a \Uparrow b$ means to match a and b to any number greater than either number **no**
6. the operation $\Rightarrow$, where $a \Rightarrow b$ means to round the product of a and b up to the nearest 10 **yes**

Is each set closed under addition? Write yes or no. If your answer is no, give an example.

7. even numbers **yes**
8. odd numbers **no; $3 + 7 = 10$**
9. multiples of 3 **yes**
10. multiples of 5 **yes**
11. prime numbers **no; $3 + 5 = 8$**
12. nonprime numbers **no; $22 + 9 = 31$**

Is the set of whole numbers closed under each operation? Write yes or no. If your answer is no, give an example.

13. multiplication: $a \times b$ **yes**
14. division: $a \div b$ **no; $4 \div 3$ is not a whole number**
15. exponentiation: a^b **yes**
16. squaring the sum: $(a + b)^2$ **yes**

Applications and Problem Solving

47. **Meteorology** The record high temperature in Minneapolis-St. Paul, Minnesota, is 108°F. The record low temperature is 142°F lower. What is the record low temperature? **−34°F**

Expenditures for July

Expenses	Amount Budgeted (dollars)	Amount Spent (dollars)
Food	160	175
Electric	45	44
Telephone	35	41
Heating Fuel	50	15
Water	25	32
Cable TV	25	25

48. **Budgeting** The table shows the Thomas family's budget and expense summary for food and household utilities for July.
    a. For each item, find the difference between the budgeted amount and the amount spent.
    b. What does a negative difference indicate?
    c. Was the total amount spent for these items more or less than the amount budgeted? by how much?
    a–c. See margin.

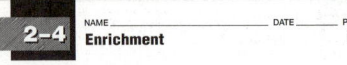

49. **Critical Thinking** Determine whether each statement is *true* or *false*. If *false*, give a counterexample.
    a. Subtraction of integers is commutative. **false; $2 - 5 \neq 5 - 2$**
    b. Subtraction of integers is associative. **false; $(6 - 2) - 3 \neq 6 - (2 - 3)$**
    c. The set of integers is closed under the operation of subtraction. **true**

Mixed Review

Find each sum. *(Lesson 2–3)*

50. $16 + (-5)$ **11**
51. $-12 + (-8)$ **−20**
52. $9 + (-15)$ **−6**
53. $-24 + (-3)$ **−27**
54. $18 + 6$ **24**
55. $-12 + 4$ **−8**

56. **Communications** A new long-distance plan charges a flat rate of 5¢ per minute. *(Lesson 2–2)* **a. 25¢, 40¢, 50¢**
    a. Find the amount spent for calls of 5, 8, and 10 minutes.
    b. Graph the ordered pairs (time, cost). **See margin.**

Replace each ● with < or > to make a true sentence. *(Lesson 2–1)*

57. 2 ● −3 **>**
58. −4 ● −8 **>**
59. −15 ● −14 **<**

Standardized Test Practice

60. **Short Response** The table shows the record high temperatures for each state in the United States. Make a histogram of the data. Use 100–104, 105–109, 110–114, 115–119, 120–124, 125–129, and 130–134 as categories for the histogram. *(Lesson 1–7)* **See margin.**

Record High Temperatures (°F)

112	100	128	120	134	118	106	110	109	112
100	118	117	116	118	121	114	114	105	109
107	112	114	115	118	117	118	125	106	110
122	108	110	121	113	120	119	111	104	111
120	113	120	117	105	110	118	112	114	114

Source: *World Almanac*

Extra Credit

Find $5 + 4 - 3 + 2 - 1 + 0 - (-1) + (-2) - (-3) + (-4) - (-5)$. **10**

Answer

60. Sample answer:

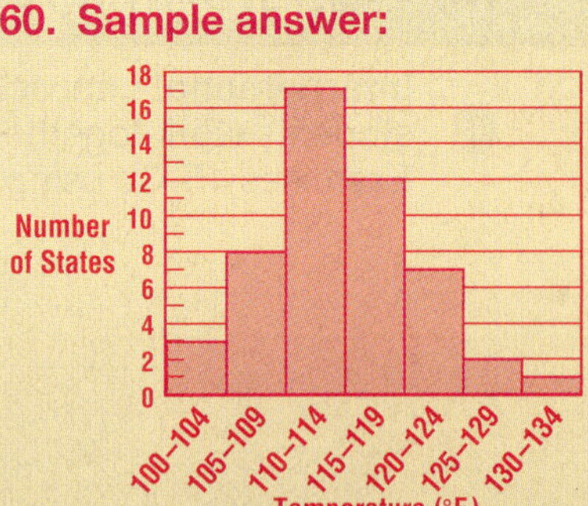

2-5 Multiplying Integers

What You'll Learn
You'll learn to multiply integers.

Why It's Important

Health workers use negative integers to describe declining death rates. *See Exercise 48.*

Multiplying integers can be modeled by repeat addition. The multiplication of integers can be represented on a number line.

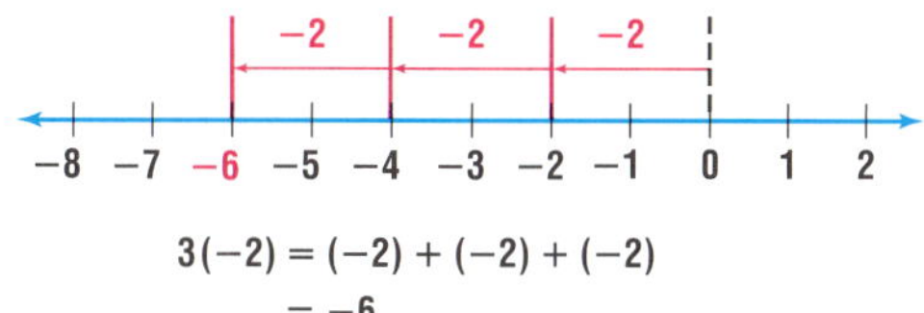

$$3(-2) = (-2) + (-2) + (-2)$$
$$= -6$$

Therefore, $3(-2) = -6$.

The number line below models the product $2(-3)$.

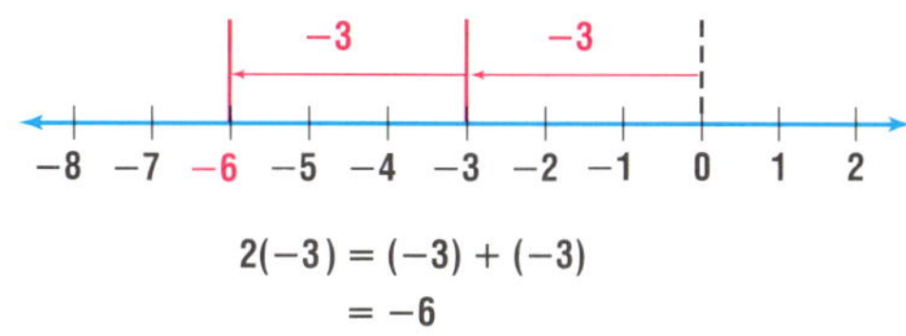

$$2(-3) = (-3) + (-3)$$
$$= -6$$

Look Back
Commutative Property of Multiplication Lesson 1-3

What happens if the order of the factors is changed to $(-2)3$? The Commutative Property of Multiplication guarantees that $3(-2) = (-2)3$. Therefore, $-2(3) = -6$.

In $3(-2) = -6$ and $-2(3) = -6$, one factor is positive, one factor is negative, and the product is negative. These examples suggest the following rule for multiplying two integers with different signs.

Multiplying Two Integers with Different Signs		
	Words:	The product of two integers with different signs is negative.
	Numbers:	$3(-2) = -6$, $-2(3) = -6$

Examples

Find each product.

$6(-8)$

$6(-8) = -48$ *The factors have different signs. The product is negative.*

$-5(9)$

$-5(9) = -45$ *The factors have different signs. The product is negative.*

Your Turn

a. $10(-3)$ **−30** b. $-7(7)$ **−49** c. $15(-3)$ **−45**

1 FOCUS

5-Minute Check
Lesson 2-4

Find each difference.

1. $-9 - (-6)$ **−3**
2. $-11 - 7$ **−18**
3. $12 - 16$ **−4**
4. Evaluate $y - z$ if $y = -3$ and $z = 4$. **−7**
5. Evaluate $r - s + t$ if $r = -1$, $s = -3$, and $t = 7$. **9**

Motivating the Lesson

Hands-On Activity Provide each student with 20 manipulatives of any kind, such as counters. Have students model losing 3 of their manipulatives every day beginning on Monday. Ask students how many counters they will have by the end of Friday. Show students that the daily loss can be represented by the integer −3, and that 5 days' losses can be modeled by the product 5(−3).

2 TEACH

Teaching Tip Before presenting Example 1, show students how to model the multiplication of two integers with different signs using algebra tiles.

In-Class Examples
Examples 1–2

Find each product.

1 $4(-3)$ **−12**
2 $-2(7)$ **−14**

Resource Manager

Reproducible Masters
Chapter 2 Resource Masters
- *Study Guide,* p. 71
- *Skills Practice,* p. 72
- *Practice,* p. 73
- *Reading to Learn Mathematics,* p. 74
- *Enrichment,* p. 75

Hands-On Algebra, pp. 37–38

Transparencies
5-Minute Check, 2-3

Technology/Multimedia
Interactive Chalkboard CD-ROM

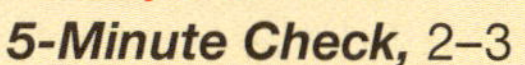

Teaching Tip Another way to introduce the product of two negative numbers at the top of page 76 is to write the following column of products.

$(-2)(3) = -6$
$(-2)(2) = -4$
$(-2)(1) = -2$
$(-2)(0) = 0$
$(-2)(-1) = ???$

Have students look for a pattern in the factors and in their products. By seeing a pattern, students should suggest that the product of −2 and −1 is 2.

In-Class Examples
Examples 3–4
Find each product.
3 10(4) **40**
4 −8(−6) **48**

Teaching Tip As you work Example 5, point out that the order in which you group the factors does not matter. Ask students what property justifies this fact. **Associative Property of Multiplication** Suggest that Example 5 could have been solved by first multiplying −10 and −4, and then multiplying the product, 40, by 8.

In-Class Examples
Examples 5–6
Find each product.
5 7(−3)(−6) **126**
6 4(−5)(−12)(−5) **−1200**

You already know that the product of two positive numbers is positive. What is the sign of the product of two negative numbers? Consider the product −2(−3).

Look Back
Multiplicative Property of Zero: Lesson 1–4

$0 = -2(0)$	*Multiplicative Property of Zero*
$0 = -2[3 + (-3)]$	*Replace 0 with 3 + (−3) or any zero pair.*
$0 = -2(3) + (-2)(-3)$	*Distributive Property*
$0 = -6 + ?$	$-2(3) = -6$

By the Additive Inverse Property, −6 + 6 = 0. Therefore, −2(−3) must be equal to 6. This example suggests the following rule for multiplying two integers with the same sign.

Multiplying Two Integers with the Same Sign	
Words:	The product of two integers with the same sign is positive.
Numbers:	2(3) = 6, −2(−3) = 6

Examples

Find each product.

3 **15(2)**
15(2) = 30 *The factors have the same sign. The product is positive.*

4 **−5(−6)**
−5(−6) = 30 *The factors have the same sign. The product is positive.*

Your Turn

d. 11(9) **99** **e.** −6(−7) **42** **f.** −10(−8) **80**

To find the product of three or more numbers, multiply the first two numbers. Then multiply the result by the next number, until you come to the end of the expression.

Examples

Find each product.

5 **8(−10)(−4)**
8(−10)(−4) = −80(−4) *8(−10) = −80*
= 320 *−80(−4) = 320*

6 **5(−3)(−2)(−2)**
5(−3)(−2)(−2) = −15(−2)(−2) *5(−3) = −15*
= 30(−2) *−15(−2) = 30*
= −60 *30(−2) = −60*

Your Turn

g. −2(−3)(4) **24** **h.** 6(−2)(3) **−36** **i.** (−1)(−5)(−2)(−3) **30**

www.algconcepts.com/extra_examples

You can use the rules for multiplying integers to evaluate algebraic expressions and to simplify expressions.

Examples

7 **Evaluate $2xy$ if $x = -4$ and $y = -2$.**

$2xy = 2(-4)(-2)$ *Replace x with −4 and y with −2.*

$= -8(-2)$ *2(−4) = −8*

$= 16$ *−8(−2) = 16*

8 **Simplify $(2a)(-5b)$.**

$(2a)(-5b) = (2)(a)(-5)(b)$ *2a = (2)(a); −5b = (−5)(b)*

$= (2)(-5)(a)(b)$ *Commutative Property*

$= -10ab$ *(2)(−5) = −10; (a)(b) = ab*

Your Turn

j. Evaluate $-5n$ if $n = -7$. **35** **k.** Simplify $12(-3z)$. **−36z**

Example

Geometry Link

9 **The graphs of $A(3, 5)$, $B(1, 2)$, and $C(5, -1)$ are connected with line segments to form a triangle. Multiply each x-coordinate by -1 and redraw the triangle. Describe how the position of the triangle changed.**

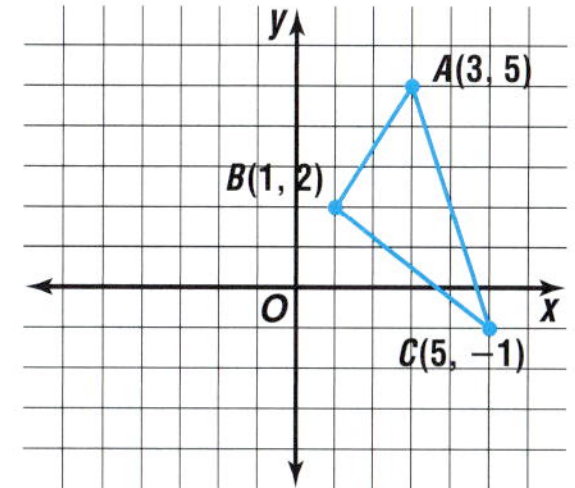

$A(3, 5) \rightarrow (3 \times -1, 5) \rightarrow A'(-3, 5)$

$B(1, 2) \rightarrow (1 \times -1, 2) \rightarrow B'(-1, 2)$

$C(5, -1) \rightarrow (5 \times -1, -1) \rightarrow C'(-5, -1)$

Triangle $A'B'C'$ is shown in green. It is the same size and shape as triangle ABC, but it is reflected, or flipped, over the y-axis.

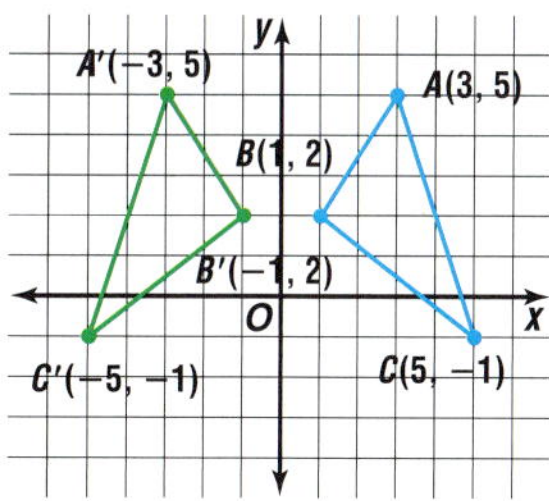

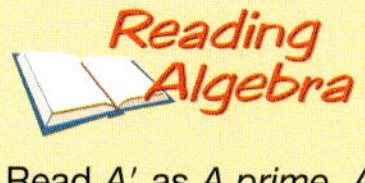

Read *A′* as *A prime. A′* corresponds to point *A* on the original triangle.

Check for Understanding

Communicating Mathematics

1. **Write** the multiplication sentence represented by the model.

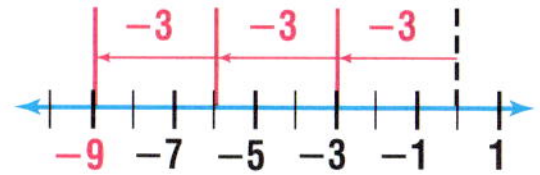

2. **Name** the property that allows you to write $-5(6)$ as $6(-5)$. **Comm. (×)**

1. 3(−3) = −9

In-Class Examples

Example 7

Evaluate $4ab$ if $a = -3$ and $b = -5$. **60**

Example 8

Simplify $(4m)(-7n)$. **−28mn**

Example 9

The graphs of $A(3, 5)$, $B(-2, 4)$, and $C(0, 1)$ are connected with line segments to form a triangle. Multiply each y-coordinate by -1 and redraw the triangle. Describe how the position of the triangle changed. **The triangle was reflected over the x-axis.**

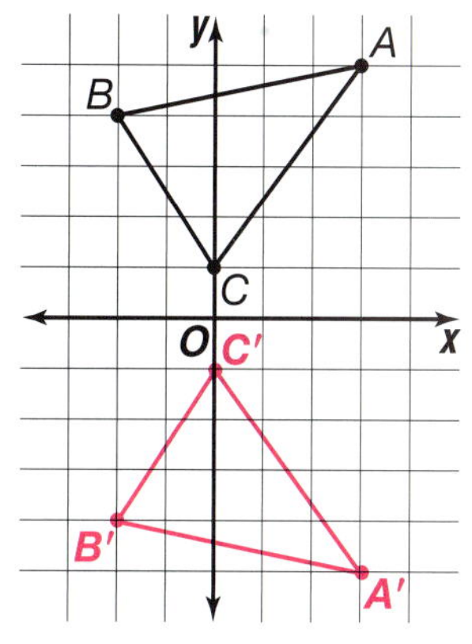

Reteaching Activity

Logical Learners Have volunteers explain in their own words how they would convince a friend that the product of two negative numbers is positive.

***Study Guide*, p. 71**

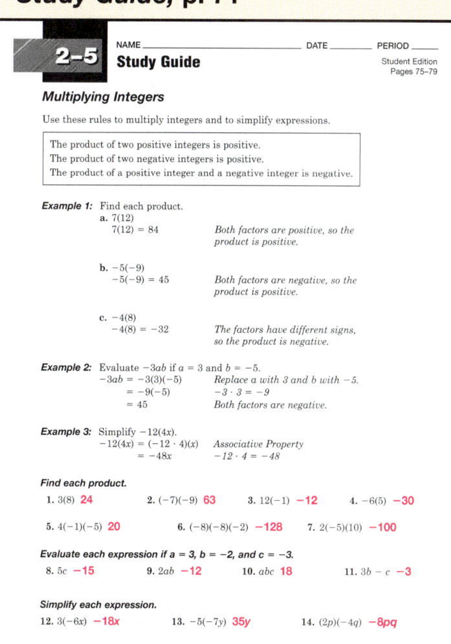

2–5 NAME ______ DATE ______ PERIOD ______

Study Guide

Student Edition Pages 75–79

Multiplying Integers

Use these rules to multiply integers and to simplify expressions.

> The product of two positive integers is positive.
> The product of two negative integers is positive.
> The product of a positive integer and a negative integer is negative.

Example 1: Find each product.

a. $7(12)$

$7(12) = 84$ *Both factors are positive, so the product is positive.*

b. $-5(-9)$

$-5(-9) = 45$ *Both factors are negative, so the product is positive.*

c. $-4(8)$

$-4(8) = -32$ *The factors have different signs, so the product is negative.*

Example 2: Evaluate $-3ab$ if $a = 3$ and $b = -5$.

$-3ab = -3(3)(-5)$ *Replace a with 3 and b with −5.*

$= -9(-5)$ *−3 · 3 = −9*

$= 45$ *Both factors are negative.*

Example 3: Simplify $-12(4x)$.

$-12(4x) = (-12 \cdot 4)(x)$ *Associative Property*

$= -48x$ *−12 · 4 = −48*

Find each product.

1. $3(8)$ **24** **2.** $(-7)(-9)$ **63** **3.** $12(-1)$ **−12** **4.** $-6(5)$ **−30**

5. $4(-1)(-5)$ **20** **6.** $(-8)(-8)(-2)$ **−128** **7.** $2(-5)(10)$ **−100**

Evaluate each expression if a = 3, b = −2, and c = −3.

8. $5c$ **−15** **9.** $2ab$ **−12** **10.** abc **18** **11.** $3b - c$ **−3**

Simplify each expression.

12. $3(-6x)$ **−18x** **13.** $-5(-7y)$ **35y** **14.** $(2p)(-4q)$ **−8pq**

3 PRACTICE/APPLY

Error Analysis

Watch for students who forget that the product of two negative numbers is positive in Exercises 6–8.

Prevent by suggesting students can verify the sign of their answer by counting the number of negative signs and then using this rule: If there is an even number of negative signs among the factors, then the product is positive. When there is odd number of negative signs, the product will be negative.

Assignment Guide

Basic: 15–49 odd, 50–61
Average: 14–44 even, 46–61
All: Quiz 2, 1–10

***Skills Practice,* p. 72, and *Practice,* p. 73 (shown)**

2–5 NAME ______ DATE ______ PERIOD ______

Practice

Student Edition Pages 75–79

Multiplying Integers

Find each product.

1. 3(−7) **−21**
2. −2(8) **−16**
3. 4(5) **20**
4. −7(−7) **49**
5. −9(3) **−27**
6. 8(−6) **−48**
7. 6(2) **12**
8. −5(−7) **35**
9. 2(−8) **−16**
10. −10(−2) **20**
11. 9(−8) **−72**
12. 12(0) **0**
13. −4(−4)(2) **32**
14. 7(−9)(−1) **63**
15. −3(5)(2) **−30**
16. 3(−4)(−2)(2) **48**
17. 6(−1)(2)(1) **−12**
18. −5(−3)(−2)(−1) **30**

Evaluate each expression if a = −3 and b = −5.

19. $-6b$ **30**
20. $8a$ **−24**
21. $4ab$ **60**
22. $-3ab$ **−45**
23. $-9a$ **27**
24. $-2ab$ **−30**

Simplify each expression.

25. $5(-5y)$ **$-25y$**
26. $-7(-3b)$ **$21b$**
27. $-3(6n)$ **$-18n$**
28. $(6a)(-2b)$ **$-12ab$**
29. $(-4m)(-9n)$ **$36mn$**
30. $(-8x)(7y)$ **$-56xy$**

Guided Practice

Examples 1–6

Find each product. 8. −40

3. 2(−6) **−12**
4. −4(9) **−36**
5. 10(8) **80**
6. −7(−11) **77**
7. 2(−6)(−3) **36**
8. 4(−1)(−5)(−2)

Example 7

Evaluate each expression if $a = -4$ and $b = -6$.

9. $-7a$ **28**
10. $-3ab$ **−72**

Example 8

Simplify each expression.

11. $9(-2x)$ **$-18x$**
12. $(-3m)(-2n)$ **$6mn$**

Example 9

13. **Geometry** The graphs of $A(4, 2)$, $B(-3, 4)$, and $C(-1, 1)$ are connected with line segments to form a triangle.
    a. Multiply each y-coordinate by −1 and redraw the triangle.
    b. Describe how the position of the triangle changed. **It was reflected over the x-axis.**

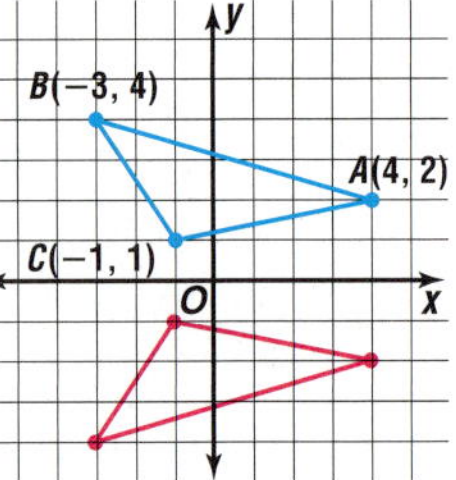

Exercises

Practice

Homework Help	
For Exercises	See Examples
14–25, 32–34	1–4
26–31	5, 6
35–40, 44	7
41–43, 45	8
Extra Practice	
See page 695.	

Find each product.

14. 5(8) **40**
15. 12(−4) **−48**
16. −1(−1) **1**
17. 9(−1) **−9**
18. −6(5) **−30**
19. 3(15) **45**
20. 5(−15) **−75**
21. 13(0) **0**
22. −8(−9) **72**
23. −3(8) **−24**
24. −12(−5) **60**
25. −13(3) **−39**
26. 3(−2)(4) **−24**
27. −1(−3)(9) **27**
28. −2(−2)(−2) **−8**
29. 3(4)(−7) **−84**
30. −2(4)(−5)(2) **80**
31. −1(−1)(1)(−1) **−1**
32. Find the value of a if $a = -3(14)$. **−42**
33. What is the value of n if $n = (-11)(-9)$? **99**
34. Find the value of p if $12(-10) = p$. **−120**

Evaluate each expression if $x = 2$, $y = -3$, and $z = -5$.

35. $-4x$ **−8**
36. $7xy$ **−42**
37. xyz **30**

38. $2y + z$ **−11**
39. $5x - y$ **13**
40. $3y + 4z$ **−29**

Simplify each expression.

41. $4(-2a)$ **$-8a$**
42. $-8(5m)$ **$-40m$**
43. $(-4m)(-8n)$ **$32mn$**
44. What is the product of −3, −4, and −5? **−60**
45. Evaluate $8a - 2b$ if $a = -2$ and $b = 3$. **−22**

Applications and Problem Solving

46. **Patterns** Find the next term in the pattern −1, 2, −4, 8, . . . **−16**
47. **Oceanography** A research submarine descends to the ocean floor at a rate of 100 feet per minute. Write a multiplication equation that tells how far the submarine moves in 5 minutes. **−100(5) = −500**

48. **Health** From 1995 through 1998, deaths from AIDS decreased by an average of about 11,000 per year. If 49,351 people died in 1995, about how many died in 1998? **about 16,351**

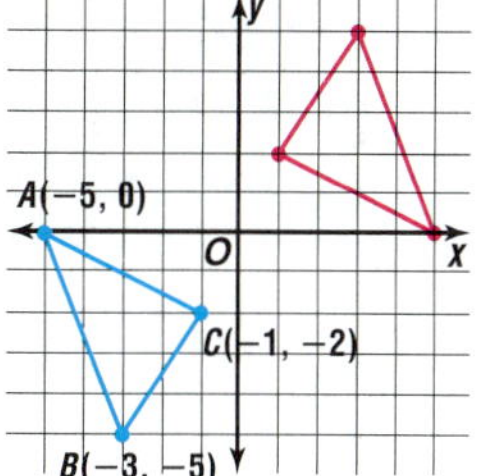

49. **Geometry** $A(-5, 0)$, $B(-3, -5)$, and $C(-1, -2)$ are connected with line segments to form a triangle.
   a. Multiply each x- and y-coordinate by -1 and draw another triangle. **See right.**
   b. Describe how the position of the triangle changed. **It was reflected over both the x- and y-axis.**

50. **Critical Thinking** If the product of three integers is negative, what can you conclude about the signs of the integers? Write a rule for determining the sign of the product of three nonzero integers. **See margin.**

Mixed Review

Evaluate each expression if $a = -3$, $b = 7$, $c = -8$, and $d = -15$. *(Lessons 2–3 & 2–4)*

51. $a + b$ **4**
52. $b - (-1)$ **8**
53. $c - (-3)$ **−5**
54. $c + d$ **−23**
55. $d + b$ **−8**
56. $d - b$ **−22**
57. $5 - b$ **−2**
58. $a - b$ **−10**
59. $c + 8$ **0**

Standardized Test Practice

60. **Short Response** The melting point of several common elements are shown. Which element has the lowest melting point? *(Lesson 2–1)* **helium**

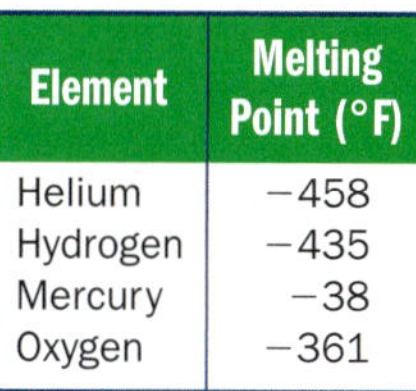

Element	Melting Point (°F)
Helium	−458
Hydrogen	−435
Mercury	−38
Oxygen	−361

Exercise 60

61. **Multiple Choice** Which verbal expression represents the algebraic expression $5x - 3$? *(Lesson 1–1)* **C**
   (A) three minus five times a number x
   (B) a number x decreased by three
   (C) three less than five times a number x
   (D) five more than a number x minus three

Quiz 2 — Lessons 2–3 through 2–5

Find each sum, difference, or product. *(Lessons 2–3, 2–4, & 2–5)*

1. $-5 + (-2)$ **−7**
2. $3 - 8$ **−5**
3. $-4(-8)$ **32**
4. $6(-9)$ **−54**
5. $9 - (-4)$ **13**
6. $-10 + 5$ **−5**
7. $-15(3)$ **−45**
8. $18 + (-2)$ **16**
9. $-11 - (-2)$ **−9**

10. **Meteorology** The temperature between the ground and 11 kilometers above the ground drops about 7°C for each kilometer higher in altitude. Suppose the ground temperature is 0°C. Find the temperature 2 kilometers above the ground. *(Lesson 2–5)* **−14°C**

www.algconcepts.com/self_check_quiz

Extra Credit

The graphs of $A(0, 0)$, $B(1, 1)$, and $C(-1, 2)$ are connected with line segments to form a triangle. Multiply each x- and y-coordinate by -2 and redraw the triangle. Describe how the triangle changed. **Sample answer: The triangle is flipped over the x- and y-axes and it doubled in size.**

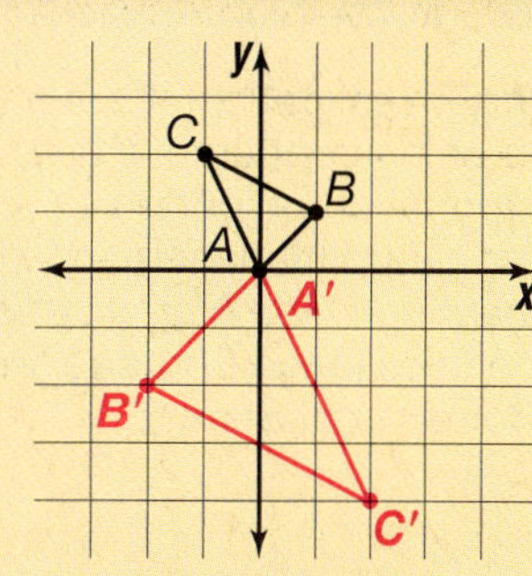

4 ASSESS

Open-Ended Assessment

Writing Have students write an explanation of why $(3)(-4) = (-3)(4)$. Students can sketch algebra tile models or number line models of the products as part of their explanation.

Quiz 2

The Quiz provides students with a brief review of the concepts and skills in Lessons 2–3 through 2–5. Lesson numbers are given to the right of the exercises or instruction lines so students can review concepts not yet mastered.

Answer

50. **If either one factor or all three factors are negative, the product is negative.**

***Enrichment*, p. 75**

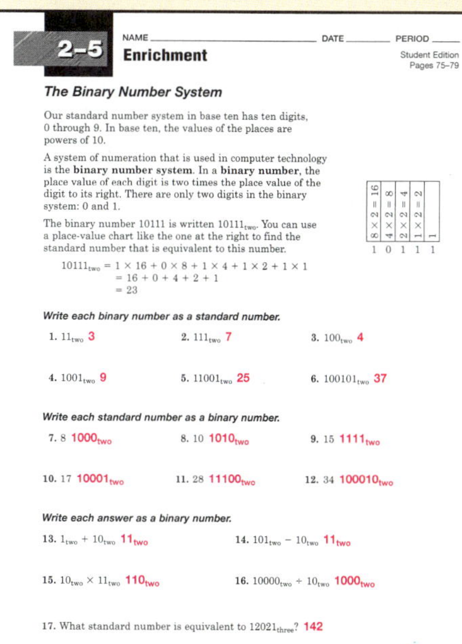

2–5 NAME ________ DATE ______ PERIOD ____

Enrichment Student Edition Pages 75–79

The Binary Number System

Our standard number system in base ten has ten digits, 0 through 9. In base ten, the values of the places are powers of 10.

A system of numeration that is used in computer technology is the **binary number system**. In a **binary number**, the place value of each digit is two times the place value of the digit to its right. There are only two digits in the binary system: 0 and 1.

The binary number 10111 is written 10111_{two}. You can use a place-value chart like the one at the right to find the standard number that is equivalent to this number.

$8 \times 2 = 16$	$4 \times 2 = 8$	$2 \times 2 = 4$	$1 \times 2 = 2$	1
1	0	1	1	1

$$10111_{two} = 1 \times 16 + 0 \times 8 + 1 \times 4 + 1 \times 2 + 1 \times 1 = 16 + 0 + 4 + 2 + 1 = 23$$

Write each binary number as a standard number.

1. 11_{two} **3**
2. 111_{two} **7**
3. 100_{two} **4**
4. 1001_{two} **9**
5. 11001_{two} **25**
6. 100101_{two} **37**

Write each standard number as a binary number.

7. 8 **1000_{two}**
8. 10 **1010_{two}**
9. 15 **1111_{two}**
10. 17 **10001_{two}**
11. 28 **11100_{two}**
12. 34 **100010_{two}**

Write each answer as a binary number.

13. $1_{two} + 10_{two}$ **11_{two}**
14. $101_{two} - 10_{two}$ **11_{two}**
15. $10_{two} \times 11_{two}$ **110_{two}**
16. $10000_{two} \div 10_{two}$ **1000_{two}**

17. What standard number is equivalent to 12021_{three}? **142**

PREPARE

This optional investigation is designed to be completed by pairs of students over 1–2 days.

Objective

Students investigate the use of matrices, learn the terms related to matrices, represent and organize data in matrices, and perform matrix addition, subtraction, and scalar multiplication. Students present their findings by creating a poster or booklet.

Mathematical Overview

This investigation utilizes the following concepts:

- organizing data,
- looking for patterns, and
- adding, subtracting, and multiplying integers.

Suggested Time Management	
Investigation	30–45 min
Extension: Gathering Data	30–45 min
Extension: Summarizing Data	20–30 min

Motivating the Lesson

At the beginning of the investigation, students learn how matrices are organized by rows and columns. You may want to have students identify each element by its row number and column number. Emphasize that when adding or subtracting two matrices, you must pay close attention to the location of the elements.

Answer

1. $\begin{bmatrix} 5{,}035{,}118 \\ 2{,}776{,}144 \\ 2{,}738{,}588 \\ 2{,}930{,}235 \\ 1{,}276{,}249 \\ 2{,}219{,}395 \end{bmatrix}$

Chapter 2 Investigation

Bits, Bytes, and BUGS!

Materials

 calculator

Matrices

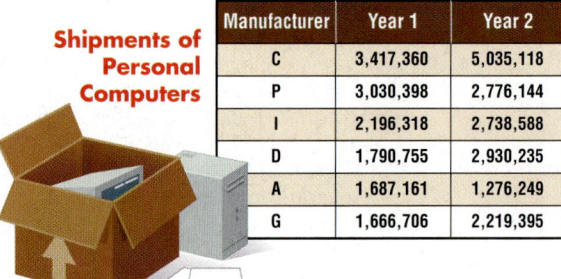

Shipments of Personal Computers

Manufacturer	Year 1	Year 2
C	3,417,360	5,035,118
P	3,030,398	2,776,144
I	2,196,318	2,738,588
D	1,790,755	2,930,235
A	1,687,161	1,276,249
G	1,666,706	2,219,395

Source: *The Wall Street Journal Almanac*

Investigate

1. A **matrix** is a rectangular arrangement of numbers in rows and columns. Each number in a matrix is called an **element**. A matrix is an **ordered array** because the order of the elements matters. The **dimensions** of a matrix tell how many rows and columns it has. The data about the computers shipped in Year 1 could be organized in a 6 × 1 matrix as shown. Write the data for Year 2 as a matrix. **See margin.**

$$\begin{bmatrix} 3{,}417{,}360 \\ 3{,}030{,}398 \\ 2{,}196{,}318 \\ 1{,}790{,}755 \\ 1{,}687{,}161 \\ 1{,}666{,}706 \end{bmatrix}$$

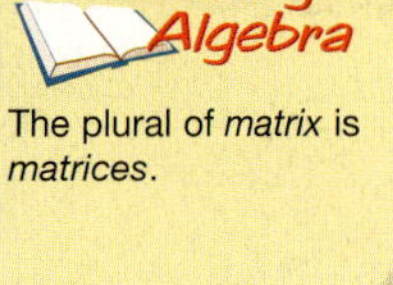

Reading Algebra

The plural of *matrix* is *matrices*.

2. Two matrices can be added as shown below.

$$\begin{bmatrix} 2 & -3 \\ -1 & 8 \\ 0 & 5 \end{bmatrix} + \begin{bmatrix} -5 & 3 \\ 7 & -7 \\ -10 & 3 \end{bmatrix} = \begin{bmatrix} 2+(-5) & -3+3 \\ -1+7 & 8+(-7) \\ 0+(-10) & 5+3 \end{bmatrix} = \begin{bmatrix} -3 & 0 \\ 6 & 1 \\ -10 & 8 \end{bmatrix}$$

a. Write your own rule for adding two matrices. **2a. Add the corresponding elements.**

b. Use matrix addition to find the total number of personal computers shipped by the manufacturers in both years. **See margin.**

Cooperative Learning

This investigation offers an excellent opportunity for using cooperative groups. For more information on cooperative learning strategies and group management, see *Cooperative Learning in the Mathematics Classroom,* one of the titles in the Glencoe Mathematics Professional Series.

Answer

2b. $\begin{bmatrix} 8{,}452{,}478 \\ 5{,}806{,}542 \\ 4{,}934{,}906 \\ 4{,}720{,}990 \\ 2{,}963{,}410 \\ 3{,}886{,}101 \end{bmatrix}$

3a. Subtract the corresponding elements.

3c. The manufacturer shipped fewer computers in 1997 than in 1996.

3. Two matrices can be subtracted as shown below.

$$\begin{bmatrix} -1 & 0 \\ 4 & -2 \end{bmatrix} - \begin{bmatrix} 1 & 5 \\ -3 & 6 \end{bmatrix} = \begin{bmatrix} -1-1 & 0-5 \\ 4-(-3) & -2-6 \end{bmatrix} = \begin{bmatrix} -2 & -5 \\ 7 & -8 \end{bmatrix}$$

a. Write your own rule for subtracting two matrices.

b. Use matrix subtraction to find how many more personal computers were shipped by each manufacturer in Year 2 than in Year 1. **See margin.**

c. What do negative elements indicate?

4. You can multiply any matrix by a number called a **scalar**. When **scalar multiplication** is performed, each element is multiplied by the scalar, and a new matrix is formed.

$$6\begin{bmatrix} 8 & -2 & 10 \\ -5 & 4 & 6 \end{bmatrix} = \begin{bmatrix} 6(8) & 6(-2) & 6(10) \\ 6(-5) & 6(4) & 6(6) \end{bmatrix} = \begin{bmatrix} 48 & -12 & 60 \\ -30 & 24 & 36 \end{bmatrix}$$

Suppose the computer industry predicted a 20% increase in shipments compared to the number of shipments in Year 2. Use scalar multiplication to find the predicted number of computer shipments. (*Hint:* Multiply the matrix by 1.2 to show an increase of 20%.) **See margin.**

Extending the Investigation

In this extension, you will investigate how matrices are used in the real world. Here are some suggestions.

- The matrices below show the sales and expenses for two different companies for 2003 and 2004. Use the information in the matrices to find a matrix that shows each company's profits in 2003 and 2004. (*Hint:* Profits = Sales − Expenses)

	Sales (million dollars) 2003	2004	Expenses (million dollars) 2003	2004	**Profits (million dollars) 2003**	**2004**
Company A →	4761	6471	4362	5917	**399**	**554**
Company B →	5061	3483	4904	4838	**157**	**−1355**

- Find some data that can be organized using matrices. Then write a problem using the data that can be solved by adding, subtracting, or using scalar multiplication.

Presenting Your Conclusions

Here are some ideas to help you present your conclusions to the class.

- Prepare a poster presenting matrices in a creative manner. Show how you solved problems using matrices.
- Make a booklet of your problems, matrices, and solutions.

Investigation For more information on matrix addition, visit: www.algconcepts.com

MANAGE

Teaching Tip If some students need a more hands-on approach to matrices, have them use small sticky notes to "build" each matrix. Have students write one matrix element on each note and then arrange the sticky notes to form the "matrix."

Working in Pairs For Exercises 2b and 3b, suggest that one student recite the matrix addition or subtraction while their partner records the entries.

Working as a Class If students choose to make booklets, you could have them exchange their booklets and then evaluate each other's work.

ASSESS

Students should add, subtract, and multiply the matrices correctly. In the extension, they should find data suited to matrix operations. Also, they should write a problem that can be solved by one of the matrix operations.

Students should add their poster or booklet to their portfolios at this time.

Answers

3b. $\begin{bmatrix} 1,617,758 \\ -254,254 \\ 542,270 \\ 1,139,480 \\ -410,912 \\ 552,689 \end{bmatrix}$

4. $\begin{bmatrix} 6,042,142 \\ 3,331,373 \\ 3,286,306 \\ 3,516,282 \\ 1,531,499 \\ 2,663,274 \end{bmatrix}$

Lesson 2–6

1 FOCUS

5-Minute Check

Lesson 2–5

Find each product.

1. $(-8)(-9)$ **72**
2. $-7(-10)(-5)$ **−350**
3. $-3(2)(-4)$ **24**
4. The graphs of $A(5, -2)$, $B(2, 0)$, and $C(-3, -3)$ are connected with line segments to form a triangle. Multiply each x-coordinate by -1 and redraw the triangle.

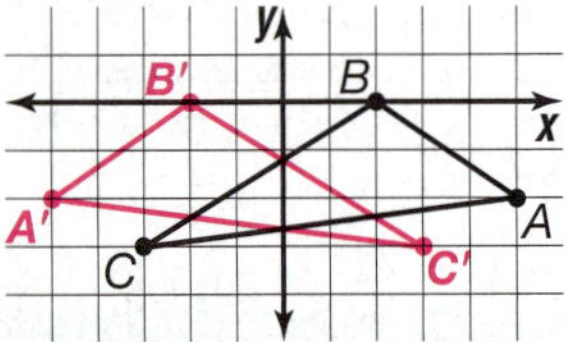

5. Describe how the position of the triangle in Exercise 4 changed. **The triangle was reflected over the y-axis.**

Motivating the Lesson

Hands-On Activity Provide each student with 8 positive algebra tiles. Have students separate the tiles into four groups. Ask them to state the value of each group. Now give each student 8 negative algebra tiles. Again have students separate the tiles into four groups. Ask students the value of each group.

2 TEACH

Teaching Tip Model the related sentences in the middle of page 82 on the board or overhead using algebra tile models. Model all of the multiplication sentences first. Then beside each of them model the corresponding division sentence.

2–6 Dividing Integers

What You'll Learn
You'll learn to divide integers.

Why It's Important
Farming The average change in populations can be found by dividing integers. *See Exercise 42.*

Dividing two integers can be modeled by separating objects into new groups. In the example below, algebra tiles are used to represent the division of integers.

Find $-6 \div 2$. *The expression $-6 \div 2$ means to separate six negative tiles into 2 groups.*

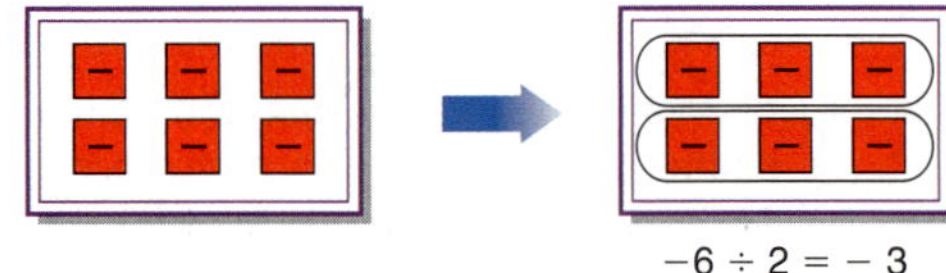

$-6 \div 2 = -3$

Therefore, $-6 \div 2 = -3$. Is a negative integer divided by a positive integer always negative? Recall that division is related to multiplication.

$$-6 \div 2 = -3 \qquad 2 \times (-3) = -6$$

What if a negative integer is divided by a negative integer?

$$-9 \div (-3) = 3 \qquad -3 \times 3 = -9$$

Study the pairs of related sentences in the table below. Look for a pattern in the signs.

Related Sentences	
Multiplication	**Division**
$2 \times (-3) = -6$	$-6 \div 2 = -3$
$-2 \times (-3) = 6$	$6 \div (-2) = -3$
$-2 \times 3 = -6$	$-6 \div (-2) = 3$
$2 \times 3 = 6$	$6 \div 2 = 3$

Signs are different. The quotients are negative.
Signs are the same. The quotients are positive.

The pattern suggests the following rule for dividing integers.

Dividing Integers		
	Words:	The quotient of two integers with the same sign is positive.
	Numbers:	$6 \div 2 = 3$, $-6 \div (-2) = 3$
	Words:	The quotient of two integers with different signs is negative.
	Numbers:	$-6 \div 2 = -3$, $6 \div (-2) = -3$

Resource Manager

Reproducible Masters
Chapter 2 Resource Masters
- *Study Guide,* p. 76
- *Skills Practice,* p. 77
- *Practice,* p. 78
- *Reading to Learn Mathematics,* p. 79
- *Enrichment,* p. 80
- *Assessment,* p. 91

Transparencies
5-Minute Check, 2–6

Technology/Multimedia
AlgePASS, Lesson 8
Interactive Chalkboard CD-ROM

Examples

Find each quotient.

$-10 \div 2$

The signs are different.
The quotient is negative.
$-10 \div 2 = -5$

$-32 \div (-8)$

The signs are the same.
The quotient is positive.
$-32 \div (-8) = 4$

Your Turn

a. $-9 \div 3$ **−3** b. $-20 \div (-4)$ **5** c. $16 \div (-2)$ **−8**

Recall that fractions are another way of showing division.

Example

Evaluate $\frac{6x}{y}$ if $x = -4$ and $y = 8$.

$\frac{6x}{y} = \frac{6(-4)}{8}$ *Replace x with −4 and y with 8.*

$= \frac{-24}{8}$ *$6(-4) = -24$*

$= -3$ *$\frac{-24}{8}$ means $-24 \div 8$.*

Your Turn

Evaluate each expression if $x = -3$ and $y = 6$.

d. $-12 \div x$ **4** e. $\frac{xy}{-2}$ **9** f. $\frac{36}{3x}$ **−4**

Example

Media Link

4 The table shows the number of CDs and cassettes that were shipped in 1990 and 2000. What was the average change in the number of cassettes that were shipped for each of those ten years?

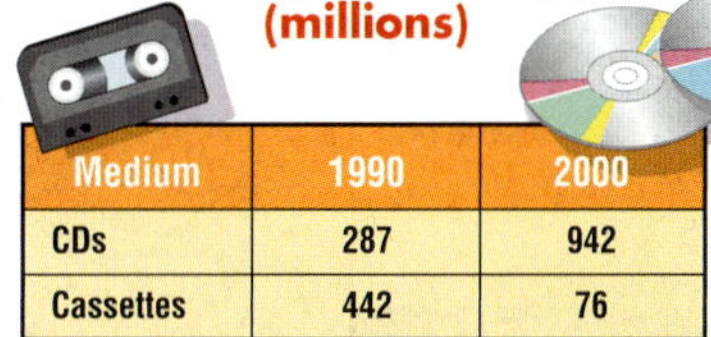

Recording Media Shipped (millions)

Medium	1990	2000
CDs	287	942
Cassettes	442	76

Source: *Statistical Abstract of the United States*

First, find the change in the number of cassettes that were shipped.

$76 - 442 = -366$ *There were 366 million fewer cassettes shipped in 2000 than in 1990.*

To find the average change, divide −366 by 10.

$-366 \div 10 = -36.6$

The average change in the number of cassettes that were shipped was −36.6 per year. This means that each year there were about 36,600,000 fewer cassettes shipped than the year before.

www.algconcepts.com/extra_examples

Teaching Tip While you discuss the rules on page 82 for dividing integers, point out that the rules about the signs of the quotients are the same as the rules about the signs of the products in Lesson 2–5.

In-Class Examples

Examples 1–2

Find each quotient.

1 $-12 \div 3$ **−4**

2 $-50 \div (-10)$ **5**

Example 3

Evaluate $\frac{3a}{b}$ if $a = -6$ and $b = 9$. **−2**

Example 4

In the last 5 years at a high school, the number of students with no tardies during the entire school year dropped from 315 to 95. What was the average change in the number of students without a tardy for each of those 5 years? **The average change in the number of students with no tardies for the entire school year was −44 per year.**

***Study Guide,* p. 76**

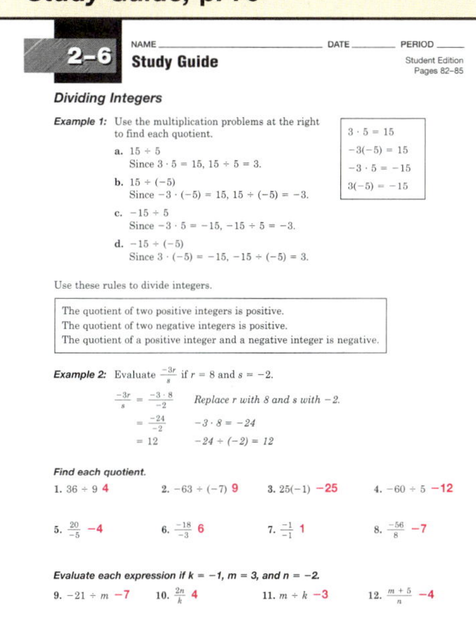

2–6 NAME ______ DATE ______ PERIOD ______

Study Guide

Student Edition Pages 82–85

Dividing Integers

Example 1: Use the multiplication problems at the right to find each quotient.

$3 \cdot 5 = 15$
$-3(-5) = 15$
$-3 \cdot 5 = -15$
$3(-5) = -15$

a. $15 \div 5$
Since $3 \cdot 5 = 15$, $15 \div 5 = 3$.

b. $15 \div (-5)$
Since $-3 \cdot (-5) = 15$, $15 \div (-5) = -3$.

c. $-15 \div 5$
Since $-3 \cdot 5 = -15$, $-15 \div 5 = -3$.

d. $-15 \div (-5)$
Since $3 \cdot (-5) = -15$, $-15 \div (-5) = 3$.

Use these rules to divide integers.

The quotient of two positive integers is positive.
The quotient of two negative integers is positive.
The quotient of a positive integer and a negative integer is negative.

Example 2: Evaluate $\frac{-3r}{s}$ if $r = 8$ and $s = -2$.

$\frac{-3r}{s} = \frac{-3 \cdot 8}{-2}$ *Replace r with 8 and s with −2.*

$= \frac{-24}{-2}$ *$-3 \cdot 8 = -24$*

$= 12$ *$-24 \div (-2) = 12$*

Find each quotient.

1. $36 \div 9$ **4** 2. $-63 \div (-7)$ **9** 3. $25(-1)$ **−25** 4. $-60 \div 5$ **−12**

5. $\frac{20}{-5}$ **−4** 6. $\frac{-18}{-3}$ **6** 7. $\frac{-1}{-1}$ **1** 8. $\frac{-56}{8}$ **−7**

Evaluate each expression if k = −1, m = 3, and n = −2.

9. $-21 \div m$ **−7** 10. $\frac{2n}{k}$ **4** 11. $m \div k$ **−3** 12. $\frac{m+5}{n}$ **−4**

Check for Understanding

Communicating Mathematics

1. **Write** two division sentences related to the multiplication sentence $-5 \times 2 = -10$. $-10 \div 2 = -5$, $-10 \div (-5) = 2$

2. **You Decide** Joel claims that a positive number divided by a negative number is a positive number. Abbey claims that a negative number divided by a negative number is a negative number. Who is correct? Explain. See margin.

Guided Practice

Examples 1 & 2

Find each quotient.

3. $-55 \div 11$ -5
4. $-14 \div (-2)$ 7
5. $15 \div (-3)$ -5
6. $16 \div 4$ 4
7. $-20 \div (-5)$ 4
8. $\frac{-8}{2}$ -4

Example 3

Evaluate each expression if $a = 3$, $b = -12$, and $c = -6$.

9. $-24 \div a$ -8
10. $\frac{ab}{9}$ -4
11. $\frac{6b}{c}$ 12

Example 4

12. **Economy** In July, 1998, about 6,200,000 people were unemployed in the United States. Twelve months later, this figure dropped to 5,900,000. What was the average change in unemployment for each of the last twelve months? $-25,000$ people

Exercises

Practice

Find each quotient.

13. $-12 \div (-12)$ 1
14. $-18 \div 3$ -6
15. $36 \div 6$ 6
16. $-10 \div (-2)$ 5
17. $30 \div (-5)$ -6
18. $15 \div 5$ 3
19. $-25 \div (-5)$ 5
20. $-21 \div 7$ -3
21. $45 \div (-5)$ -9
22. $24 \div (-24)$ -1
23. $-20 \div (-2)$ 10
24. $-72 \div 9$ -8
25. $64 \div (-8)$ -8
26. $-48 \div (-4)$ 12
27. $-40 \div 8$ -5
28. $\frac{-49}{-7}$ 7
29. $\frac{60}{-5}$ -12
30. $\frac{-26}{2}$ -13
31. Find the value of a if $-42 \div 7 = a$. -6
32. What is the value of m if $m = -81 \div (-9)$? 9
33. Find the value of w if $w = 85 \div (-17)$. -5

Homework Help

For Exercises	See Examples
13–33, 40, 41	1, 2
34–39, 45	3
42–44	4

Extra Practice

See page 696.

Evaluate each expression if $x = 5$, $y = -6$, $z = 2$, and $w = -3$.

34. $18 \div y$ -3
35. $y \div z$ -3
36. $\frac{x}{5}$ 1
37. $\frac{-4w}{2}$ 6
38. $\frac{x - z}{w}$ -1
39. $\frac{y - 8}{z}$ -7
40. What is the quotient of -42 and -7? 6
41. Divide 100 by -50. -2

3 PRACTICE/APPLY

Error Analysis

Watch for students who write the incorrect sign for the quotient in Exercises 3–8.
Prevent by reminding them that the rule for the sign of the quotient of two integers is the same as the rule for the sign of the product of two integers.

Assignment Guide

Basic: 13–45 odd, 46–57
Average: 14–40 even, 42–57

Answer

2. neither; positive ÷ negative = negative; negative ÷ negative = positive

Skills Practice, **p. 77, and *Practice,* p. 78 (shown)**

2-6 NAME ______ DATE ______ PERIOD ______

Practice

Student Edition Pages 82–85

Dividing Integers

Find each quotient.

1. $28 \div 7$ 4
2. $-33 \div 3$ -11
3. $42 \div (-6)$ -7
4. $-81 \div (-9)$ 9
5. $12 \div 4$ 3
6. $72 \div (-9)$ -8
7. $15 \div 15$ 1
8. $-30 \div 5$ -6
9. $-40 \div (-8)$ 5
10. $56 \div (-7)$ -8
11. $-21 \div (-3)$ 7
12. $-64 \div 8$ -8
13. $-8 \div 8$ -1
14. $-22 \div (-2)$ 11
15. $32 \div (-8)$ -4
16. $-54 \div (-9)$ 6
17. $60 \div (-6)$ -10
18. $63 \div 9$ 7
19. $-45 \div (-9)$ 5
20. $-60 \div 5$ -12
21. $24 \div (-3)$ -8
22. $\frac{-12}{6}$ -2
23. $\frac{40}{-10}$ -4
24. $\frac{-45}{-9}$ 5

Evaluate each expression if $a = 4$, $b = -9$, and $c = -6$.

25. $-48 \div a$ -12
26. $b \div 3$ -3
27. $9c \div b$ 6
28. $\frac{ab}{c}$ 6
29. $\frac{bc}{-6}$ -9
30. $\frac{3c}{b}$ 2
31. $\frac{12a}{c}$ -8
32. $\frac{-4b}{a}$ 9
33. $\frac{ac}{6}$ -4

Reteaching Activity

Visual/Spatial Learners Have several students become "live" algebra tiles by giving them large pieces of paper to hold indicating their sign (either + or −). Then arrange the students in smaller groups to model dividing integers.

Applications and Problem Solving

42. **Animals** Experts estimate that there were about 100,000 tigers living 100 years ago. Today, there are only about 6000. What was the average change in tiger population for each of the last 100 years? **−940 tigers**

43. **Farming** The table shows the number of thousands of farms in the U.S. according to their size.

U.S. Farms (thousands)

Acres	1997	2002
1–9	205	197
10–49	531	563
50–179	694	659
180–499	428	389
500–999	179	162
1000–1999	103	99
2000 or more	74	78

Source: Census of Agriculture

a. Find the average yearly change in the number of farms that are between 50 and 179 acres in size.

43a. −7000

b. For farms of which size(s) is the average change a positive number? **10 to 49 and 2000 or more**

44. **Media** Refer to the table on page 83. **a. 66 million CDs**

a. What was the average change in the number of CDs that were shipped for each of the seven years from 1990 to 1997?

b. If this trend continues, estimate the number of CDs that will be shipped in 2005. **Sample answer: 1,279,000,000 CDs**

45. **Energy** A measure called *degree days* is used to estimate the energy needed for heating on cold days. The formula $d = 65 - \frac{h + l}{2}$ can be used to find degree days. In the formula, d represents degree days, h represents the high temperature of a given day, and l represents the low temperature of that day. Find the degree days for a day in which the high temperature was −2°F and the low temperature was −16°F. **74**

46. **Critical Thinking** Explain why division by zero is not possible. **See margin.**

Mixed Review

Find each sum, difference, or product. *(Lessons 2–3, 2–4, & 2–5)*

47. 9(−6) **−54**
48. −11 + (−4) **−15**
49. 9 − (−7) **16**
50. 15 + (−25) **−10**
51. −10(−8) **80**
52. 8 − 10 **−2**
53. −7 − (−5) **−2**
54. −8(9) **−72**
55. −16 + 20 **4**

Standardized Test Practice

56. **Short Response** A competition swimming pool is 75 feet long and 72 feet wide. It is filled to a depth of 6 feet. Use the formula $V = \ell wh$, where ℓ is the length, w is the width, and h is the depth, to find the volume V in cubic feet of water in the pool. *(Lesson 1–5)* **32,400**

57. **Multiple Choice** Which property of real numbers allows you to conclude that if $2t + 4t = 36$, then $4t + 2t = 36$? *(Lesson 1–3)* **B**

Ⓐ Distributive Property
Ⓑ Commutative Property
Ⓒ Associative Property
Ⓓ Additive Inverse Property

www.algconcepts.com/self_check_quiz

4 ASSESS

Open-Ended Assessment

Speaking Ask students to describe how to find the quotient −48 ÷ (−6).

Chapter 2, Quiz B (Lessons 2–3 through 2–6) is available in the *Chapter 2 Resource Masters,* p. 91.

Answer

46. Sample answer: Consider 10 ÷ 0 = x. This can be written as x(0) = 10. There is no number for x that makes this a true statement. Therefore, division by zero is undefined.

***Enrichment,* p. 80**

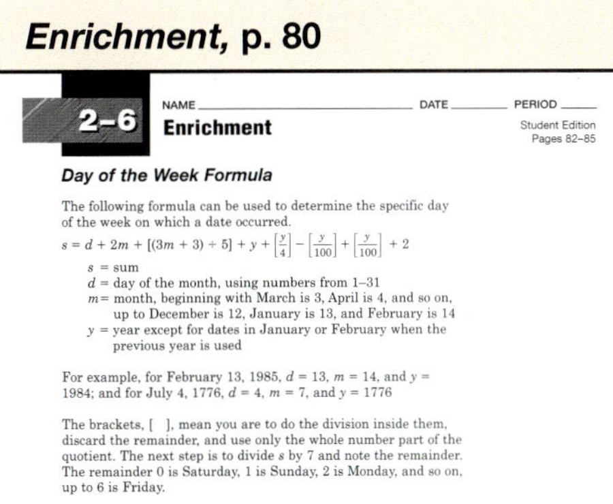

NAME ______ DATE ______ PERIOD ______

2–6 Enrichment

Student Edition Pages 82–85

Day of the Week Formula

The following formula can be used to determine the specific day of the week on which a date occurred.

$s = d + 2m + [(3m + 3) \div 5] + y + \left[\frac{y}{4}\right] - \left[\frac{y}{100}\right] + \left[\frac{y}{100}\right] + 2$

s = sum
d = day of the month, using numbers from 1–31
m = month, beginning with March is 3, April is 4, and so on, up to December is 12, January is 13, and February is 14
y = year except for dates in January or February when the previous year is used

For example, for February 13, 1985, $d = 13$, $m = 14$, and $y = 1984$; and for July 4, 1776, $d = 4$, $m = 7$, and $y = 1776$

The brackets, [ ], mean you are to do the division inside them, discard the remainder, and use only the whole number part of the quotient. The next step is to divide s by 7 and note the remainder. The remainder 0 is Saturday, 1 is Sunday, 2 is Monday, and so on, up to 6 is Friday.

Example: What day of the week was October 3, 1854?

For October 3, 1854, $d = 3$, $m = 10$, and $y = 1854$.

$s = 3 + [\underbrace{2(10)}] + [\underbrace{(3 \times 10 + 3) \div 5}] + 1854 + \left[\frac{1854}{4}\right] - \left[\frac{1854}{100}\right] + \left[\frac{1854}{400}\right] + 2$
$= 3 + 20 + 6 + 1854 + 463 - 18 + 4 + 2$
$= 2334$

$s \div 7 = 2334 \div 7$
$= 333$ R3

Since the remainder is 3, the day of the week was Tuesday.

Solve.

1. See if the formula works for today's date. **Answers will vary.**
2. On what day of the week were you born? **Answers will vary.**
3. What will be the day of the week on April 13, 2006?
$s = 13 + 2(4) + [(3 \times 4 + 3) \div 5] + 2006 + \left[\frac{2006}{4}\right] - \left[\frac{2006}{100}\right] + \left[\frac{2006}{400}\right] + 2$
$= 13 + 8 + 3 + 2006 + 501 - 20 + 4 + 2 = 2518$; $2518 \div 7 = 359$ R5 → Thursday
4. On what day of the week was July 4, 1776?
$s = 4 + 2(7) + [(3 \times 7 + 3) \div 5] + 1776 + \left[\frac{1776}{4}\right] - \left[\frac{1776}{100}\right] + \left[\frac{1776}{400}\right] + 2$
$= 4 + 14 + 4 + 1776 + 444 - 17 + 4 + 2 = 2231$; $2231 \div 7 = 318$ R5 → Thursday

Extra Credit

If the quotient $a \div b$ is negative, what can you say about the values of a and b? **Either a is negative or b is negative, but not both of them.**

Study Guide and Assessment

Understanding and Using the Vocabulary

This section provides a listing of the new terms, properties, and phrases that were introduced in this chapter. The exercises check students' understanding of the terms by using a variety of verbal formats including matching, completion, and true/false.

Glossary A complete glossary of terms appears on pages 762–783.

MindJogger Videoquizzes

MindJogger Videoquizzes provide an alternative review of concepts presented in this chapter. Students work in teams to answer questions, gaining points for correct answers.

CHAPTER 2 Study Guide and Assessment

Understanding and Using the Vocabulary

After completing this chapter, you should be able to define each term, property, or phrase and give an example or two of each.

interNET CONNECTION **Review Activities**
For more review activities, visit: www.algconcepts.com

absolute value *(p. 54)*
additive inverse *(p. 65)*
coordinate *(p. 53)*
coordinate plane *(p. 58)*
coordinate system *(p. 58)*
dimensions *(p. 80)*
element *(p. 80)*
graph *(p. 53)*
integers *(p. 52)*
matrix *(p. 80)*
natural numbers *(p. 52)*
negative numbers *(p. 52)*
number line *(p. 52)*
opposites *(p. 65)*
ordered array *(p. 80)*
ordered pair *(p. 58)*
origin *(p. 58)*
quadrants *(p. 60)*
scalar *(p. 81)*
scalar multiplication *(p. 81)*
Venn diagrams *(p. 52)*
x-axis *(p. 58)*
x-coordinate *(p. 58)*
y-axis *(p. 58)*
y-coordinate *(p. 58)*
zero pair *(p. 65)*

Complete each sentence using a term from the vocabulary list.

1. On a number line, the numbers to the left of zero are __?__. **negative numbers**
2. The __?__ is the plane that contains the x-axis and the y-axis. **coordinate plane**
3. If the sum of two numbers is 0, the numbers are called __?__. **opposites, additive inverses, or zero pairs**
4. The __?__ is the first number in an ordered pair. **x-coordinate**
5. The distance a number is from 0 on the number line is the __?__. **absolute value**
6. The numbers 1, 2, 3, 4, . . . are __?__. **natural numbers**
7. Whole numbers are a subset of __?__. **integers**
8. The __?__ of a point is the number corresponding to that point on a number line. **coordinate**
9. A(n) __?__ is used to locate any point on a coordinate plane. **ordered pair**
10. The x-axis and y-axis separate the coordinate plane into four regions called __?__. **quadrants**

Skills and Concepts

Objectives and Examples

• **Lesson 2–1** Graph integers on a number line and compare and order integers.

Replace the ● with < or > to make a true sentence.

$$7 \bullet -3$$

7 is to the right of −3 on the number line, so $7 > -3$.

Review Exercises

Replace each ● with < or > to make a true sentence.

11. $0 \bullet -5$ **>**
12. $-3 \bullet 3$ **<**
13. $-9 \bullet -7$ **<**
14. $|-12| \bullet -12$ **>**
15. Order −4, 7, 4, −2, −3, and 0 from least to greatest. **−4, −3, −2, 0, 4, 7**
16. Order −15, −23, −18, and −20 from greatest to least. **−15, −18, −20, −23**

www.algconcepts.com/vocabulary_review

Resource Manager

Reproducible Masters
Chapter 2 Resource Masters
• *Assessment*, pp. 81–89, 92–94

Technology/Multimedia
MindJogger Videoquizzes
ExamView® Pro

Objectives and Examples

• **Lesson 2–2** Graph points on a coordinate plane.

Write the ordered pair that names each point and name the quadrant in which each point is located.

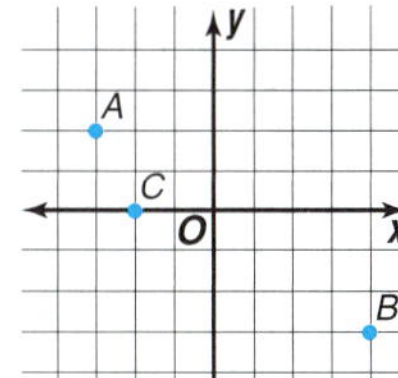

$A(-3, 2)$, II
$B(4, -3)$, IV
$C(-2, 0)$, none

Review Exercises

Write the ordered pair that names each point.

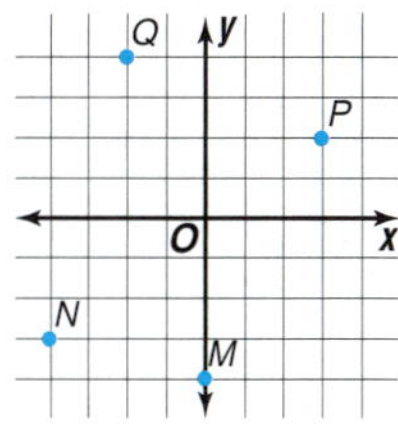

17. P **(3, 2)**
18. Q **(−2, 4)**
19. N **(−4, −3)**
20. M **(0, −4)**

Name the quadrant in which each point is located.

21. $(6, 10)$ **I**
22. $(-4, 8)$ **II**
23. $(0, -12)$ **none**
24. $(13, -7)$ **IV**

• **Lesson 2–3** Add integers.

Find $-2 + (-3)$.
Both numbers are negative, so the sum is negative.
$-2 + (-3) = -5$

Find $4 + (-12)$.
$|-12| > |4|$, so the sum is negative.
$4 + (-12) = -8$

Find each sum.

25. $8 + (-14)$ **−6**
26. $-7 + 5$ **−2**
27. $-8 + (-2)$ **−10**
28. $-8 + 8$ **0**
29. $23 + (-18)$ **5**
30. $-14 + (-12)$ **−26**
31. $-10 + 3 + (-6) + 8$ **−5**
32. $7 + (-5) + (-7) + 15$ **10**

Simplify each expression.

33. $7x + (-5x)$ **2x**
34. $-4y + (-y)$ **−5y**
35. $14m + (-10m)$ **4m**
36. $-31x + 27x$ **−4x**

• **Lesson 2–4** Subtract integers.

Find $7 - (-3)$.
$7 - (-3) = 7 + 3$ *To subtract −3, add 3.*
$= 10$

Find $-4 - 8$.
$-4 - 8 = -4 + (-8)$ *To subtract 8, add −8.*
$= -12$

Find each difference.

37. $6 - 14$ **−8**
38. $-11 - (-5)$ **−6**
39. $4 - (-5)$ **9**
40. $-3 - 5$ **−8**
41. $-6 - (-2)$ **−4**
42. $10 - (-10)$ **20**

Evaluate each expression if $x = 3$, $y = -5$, and $z = -1$.

43. $2 - x$ **−1**
44. $y - z$ **−4**
45. $x + y - z$ **−1**
46. $x - y + z$ **7**

Skills and Concepts

The **Objectives and Examples** section reviews the skills and concepts of the chapter and shows completely worked examples.

The **Review Exercises** provide practice for the corresponding objectives.

ExamView® Pro

Use the networkable ***ExamView® Pro Testmaker CD-ROM*** to:

- Create **multiple versions** of tests.
- Create **modified** tests for **inclusion** students with one mouse click.
- **Edit** existing questions and **add** your own questions.
- Build tests aligned with state standards using built-in **state curriculum correlations**.
- Change **English** tests to **Spanish** with one mouse click and vice versa.

Mixed Problem Solving
See pages 724–731.

Applications and Problem Solving

This section provides additional practice in solving real-world problems that involve the concepts of this chapter.

Objectives and Examples	Review Exercises
• **Lesson 2–5** Multiply integers. $-6(-4) = 24$ *The integers have the same sign, so the product is positive.* $3(-5) = -15$ *The integers have different signs, so the product is negative.*	**Find each product.** **47.** $-7(-5)$ **35** **48.** $8(-4)$ **−32** **49.** $-3(-2)(6)$ **36** **50.** $-1(-4)(-5)$ **−20** **Evaluate each expression if $a = -3$ and $b = -6$.** **51.** $-9a$ **27** **52.** $-7ab$ **−126** **Simplify each expression.** **53.** $-7(6m)$ **−42*m*** **54.** $(-3x)(-15y)$ **45*xy***
• **Lesson 2–6** Divide integers. $-9 \div (-3) = 3$ *The integers have the same sign, so the quotient is positive.* $\frac{8}{-2} = -4$ *The integers have different signs, so the quotient is negative.*	**Find each quotient.** **55.** $42 \div (-6)$ **−7** **56.** $-63 \div -7$ **9** **57.** $-24 \div (4)$ **−6** **58.** $-40 \div (-5)$ **8** **Evaluate each expression if $a = -4$, $b = -2$, and $c = 3$.** **59.** $\frac{6a}{c}$ **−8** **60.** $\frac{a+b}{2c}$ **−1**

Applications and Problem Solving

61. Banking Mikaela opened a checking account by depositing \$250. She later wrote a check for \$25 for the phone bill and \$32 for a magazine subscription. Then Mikaela received \$20 for her birthday and deposited it into her account. What was her balance after her birthday deposit? *(Lesson 2–3)* **\$213**

62. Geometry The graphs of $M(5, 4)$, $N(-4, 3)$, and $P(0, 1)$ are connected with line segments to form a triangle. *(Lesson 2–5)*

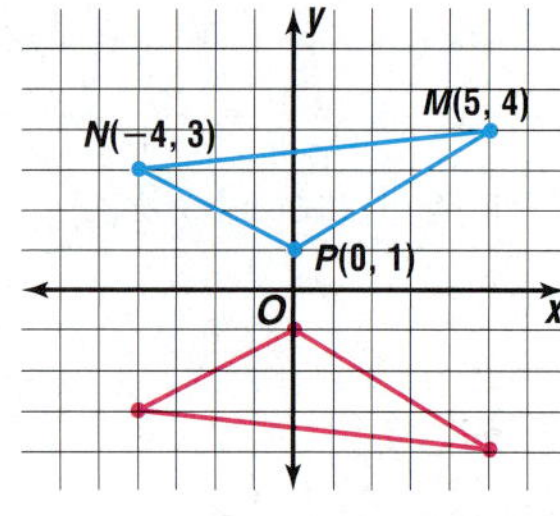

a. Multiply each y-coordinate by -1 and draw another triangle. **See above.**

b. Describe how the position of the triangle changed. **It was reflected over the *x*-axis.**

63. Scuba Diving A scuba diver descends at a rate of 40 feet per minute. Write a multiplication equation that tells how far the scuba diver moves in 2 minutes. *(Lesson 2–5)* **−40(2) = −80**

Assessment

Four forms of Chapter 2 Test are available in the *Chapter 2 Resource Masters.*

	Form of Test		Level
1A	Multiple Choice	pp. 81–82	Average
1B	Multiple Choice	pp. 83–84	Basic
2A	Free Response	pp. 85–86	Average
2B	Free Response	pp. 87–88	Basic

Chapter 2 Test, Form 1B, is shown at the left. Chapter 2 Test, Form 2B, is shown on the next page.

Assessment, pp. 83–84

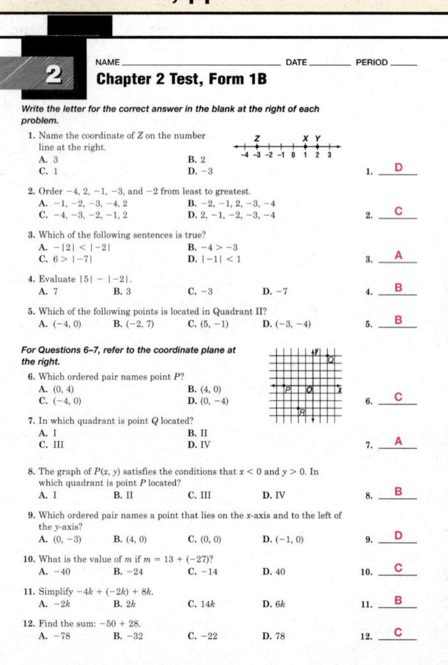

2 NAME ______ DATE ______ PERIOD ______

Chapter 2 Test, Form 1B

Write the letter for the correct answer in the blank at the right of each problem.

1. Name the coordinate of Z on the number line at the right.
A. 3 B. 2
C. 1 D. −3 1. **D**

2. Order −4, 2, −1, −3, and −2 from least to greatest.
A. −1, −2, −3, −4, 2 B. −2, −1, 2, −3, −4
C. −4, −3, −2, −1, 2 D. 2, −1, −2, −3, −4 2. **C**

3. Which of the following sentences is true?
A. $-|2| < |-2|$ B. $-4 > -3$
C. $6 > |-7|$ D. $|-1| < 1$ 3. **A**

4. Evaluate $|5| - |-2|$.
A. 7 B. 3 C. −3 D. −7 4. **B**

5. Which of the following points is located in Quadrant II?
A. (−4, 0) B. (−2, 7) C. (5, −1) D. (−3, −4) 5. **B**

For Questions 6–7, refer to the coordinate plane at the right.

6. Which ordered pair names point P?
A. (0, 4) B. (4, 0)
C. (−4, 0) D. (0, −4) 6. **C**

7. In which quadrant is point Q located?
A. I B. II
C. III D. IV 7. **A**

8. The graph of $P(x, y)$ satisfies the conditions that $x < 0$ and $y > 0$. In which quadrant is point P located?
A. I B. II C. III D. IV 8. **B**

9. Which ordered pair names a point that lies on the x-axis and to the left of the y-axis?
A. (0, −3) B. (4, 0) C. (0, 0) D. (−1, 0) 9. **D**

10. What is the value of m if $m = 13 + (-27)$?
A. −40 B. −24 C. −14 D. 40 10. **C**

11. Simplify $-4k + (-2k) + 8k$.
A. $-2k$ B. $2k$ C. $14k$ D. $6k$ 11. **B**

12. Find the sum: −50 + 28.
A. −78 B. −32 C. −22 D. 78 12. **C**

CHAPTER 2 Test

1. **Write** two division sentences related to the multiplication sentence $6 \times (-7) = -42$.
2. **Graph** $\{4, -2, 1\}$ on a number line. **1–2. See margin.**

Replace each ● with < or > to make a true sentence.

3. -5 ● -8 **>**
4. 9 ● -2 **>**

Graph each point on a coordinate plane and name the quadrant in which each point is located. **5–7. See margin for graphs.**

5. $X(4, 3)$ **I**
6. $M(-3, 2)$ **II**
7. $A(0, -4)$ **none**

Find each sum, difference, product, or quotient.

8. $-16 + 9$ **−7**
9. $5 - (-2)$ **7**
10. $-3 - 5$ **−8**
11. $4(-6)$ **−24**
12. $-14 \div (-7)$ **2**
13. $-8(-7)$ **56**
14. $\frac{-32}{8}$ **−4**
15. $\frac{-25}{-5}$ **5**
16. $-8 - 2$ **−10**
17. $-7 + 5 + (-12)$ **−14**
18. $8 + (-14) + (-6)$ **−12**
19. $5(-2)(3)$ **−30**

Evaluate each expression if $m = 5$, $n = -8$, and $p = -3$.

20. $n - p$ **−5**
21. $m + n$ **−3**
22. $2(n)$ **−16**
23. $p - m + n$ **−16**
24. $\frac{n}{-2}$ **4**
25. $\frac{m + n}{p}$ **1**

Simplify each expression.

26. $3x - 8x$ **$-5x$**
27. $10y - (-3y)$ **$13y$**
28. $9(-4x)$ **$-36x$**
29. $-2(-5y)$ **$10y$**
30. $(-7m)(3n)$ **$-21mn$**
31. $3x + 4y + x - 2y$ **$4x + 2y$**

32a. A: 81, B: 63, C: 75, D: 93, J: 61, K: 38

32. **Weather** The table shows record high and low temperatures for six cities for the month of November.
   a. Find the differences in temperatures for each city.
   b. Which city had the greatest difference in its record temperatures? **Duluth**

33. **Sports** The Tigers football team had a gain of 7 yards on their first run. They lost 3 yards on their second run and gained 12 yards on their third run. What was the total gain or loss of yardage in the three runs? **16 yd gain**

Record Temperatures in November (°F)

City	High	Low
Atlanta, GA	84	3
Boston, MA	78	15
Columbus, OH	80	5
Duluth, MN	70	−23
Juneau, AK	56	−5
Kahului, HI	93	55

Exercise 32

www.algconcepts.com/chapter_test

Chapter Test

Answers

1. $-42 \div (-7) = 6$, $-42 \div 6 = -7$

2.

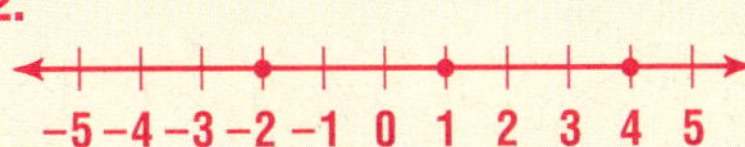

5–7.

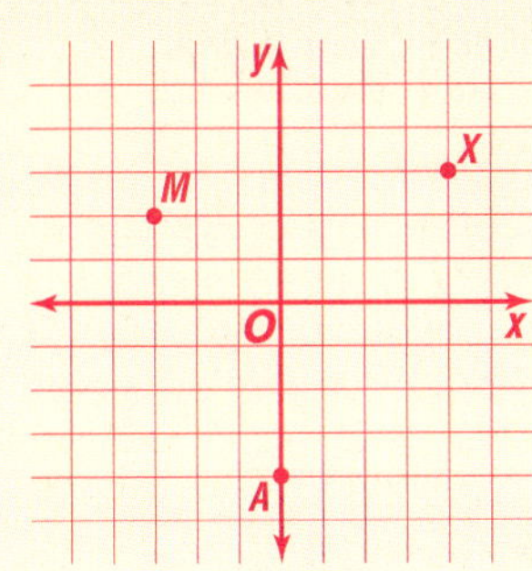

Chapter Test Bonus Question

The points with coordinates $(-3, 4)$, $(-3, -2)$, and $(3, -2)$ are three of the four vertices of a square. What are the coordinates of the point that is the fourth vertex of the square? **(3, 4)**

Assessment, pp. 87–88

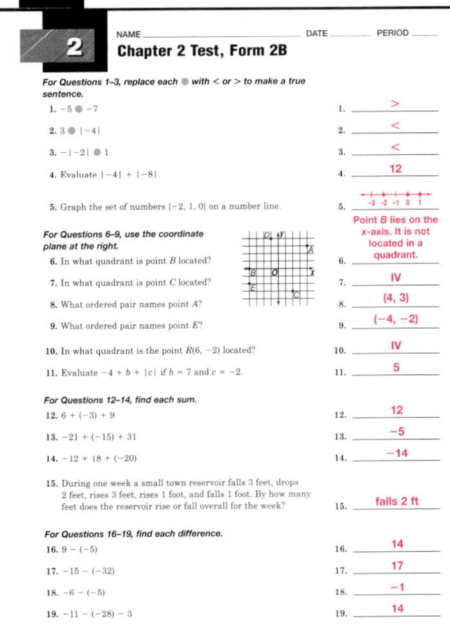

2

NAME ________ DATE ______ PERIOD ____

Chapter 2 Test, Form 2B

For Questions 1–3, replace each ● with < or > to make a true sentence.

1. -5 ● -7 — 1. **>**
2. 3 ● $|-4|$ — 2. **<**
3. $-|-2|$ ● 1 — 3. **<**
4. Evaluate $|-4| + |-8|$. — 4. **12**
5. Graph the set of numbers $\{-2, 1, 0\}$ on a number line. — 5. (number line: −3 −2 −1 0 1)

For Questions 6–9, use the coordinate plane at the right.

6. In what quadrant is point B located? — 6. **Point B lies on the x-axis. It is not located in a quadrant.**
7. In what quadrant is point C located? — 7. **IV**
8. What ordered pair names point A? — 8. **(4, 3)**
9. What ordered pair names point E? — 9. **(−4, −2)**
10. In what quadrant is the point $R(6, -2)$ located? — 10. **IV**
11. Evaluate $-4 + b + |c|$ if $b = 7$ and $c = -2$. — 11. **5**

For Questions 12–14, find each sum.

12. $6 + (-3) + 9$ — 12. **12**
13. $-21 + (-15) + 31$ — 13. **−5**
14. $-12 + 18 + (-20)$ — 14. **−14**
15. During one week a small town reservoir falls 3 feet, drops 2 feet, rises 3 feet, rises 1 foot, and falls 1 foot. By how many feet does the reservoir rise or fall overall for the week? — 15. **falls 2 ft**

For Questions 16–19, find each difference.

16. $9 - (-5)$ — 16. **14**
17. $-15 - (-32)$ — 17. **17**
18. $-6 - (-5)$ — 18. **−1**
19. $-11 - (-28) - 3$ — 19. **14**

Preparing for Standardized Tests

Pages 90–91 are part of a complete test preparation course that is described in detail on page T9 of the Teacher's Handbook. The test items on these pages were written in the same style as those in state proficiency tests and standardized tests like ACT and SAT.

Diagnosis and Prescription

Each of the 10 test questions on page 91 is cross-referenced to the lesson where that SAT or ACT skill is covered. If students miss a particular type of problem, you can have them study that skill.

(See chart at the bottom of page 91. Note that SPT = State Proficiency Test, SAT = Scholastic Assessment Test, and ACT = American College Test.)

Assessment, p. 92

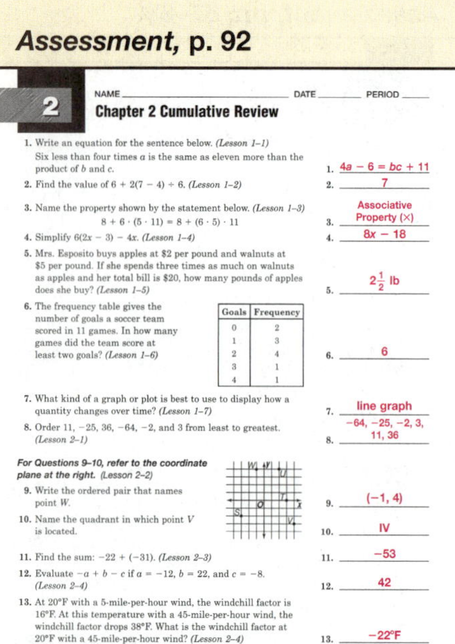

NAME ________ DATE ______ PERIOD ______

2 Chapter 2 Cumulative Review

1. Write an equation for the sentence below. *(Lesson 1–1)* Six less than four times a is the same as eleven more than the product of b and c. — 1. $4a - 6 = bc + 11$
2. Find the value of $6 + 2(7 - 4) \div 6$. *(Lesson 1–2)* — 2. 7
3. Name the property shown by the statement below. *(Lesson 1–3)* $8 + 6 \cdot (5 \cdot 11) = 8 + (6 \cdot 5) \cdot 11$ — 3. Associative Property (×)
4. Simplify $6(2x - 3) - 4x$. *(Lesson 1–4)* — 4. $8x - 18$
5. Mrs. Esposito buys apples at \$2 per pound and walnuts at \$5 per pound. If she spends three times as much on walnuts as apples and her total bill is \$20, how many pounds of apples does she buy? *(Lesson 1–5)* — 5. $2\frac{1}{2}$ lb
6. The frequency table gives the number of goals a soccer team scored in 11 games. In how many games did the team score at least two goals? *(Lesson 1–6)* — 6. 6

Goals	Frequency
0	2
1	3
2	4
3	1
4	1

7. What kind of a graph or plot is best to use to display how a quantity changes over time? *(Lesson 1–7)* — 7. line graph
8. Order 11, −25, 36, −64, −2, and 3 from least to greatest. *(Lesson 2–1)* — 8. −64, −25, −2, 3, 11, 36

For Questions 9–10, refer to the coordinate plane at the right. (Lesson 2–2)

9. Write the ordered pair that names point W. — 9. (−1, 4)
10. Name the quadrant in which point V is located. — 10. IV
11. Find the sum: $-22 + (-31)$. *(Lesson 2–3)* — 11. −53
12. Evaluate $-a + b - c$ if $a = -12$, $b = 22$, and $c = -8$. *(Lesson 2–4)* — 12. 42
13. At 20°F with a 5-mile-per-hour wind, the windchill factor is 16°F. At this temperature with a 45-mile-per-hour wind, the windchill factor drops 38°F. What is the windchill factor at 20°F with a 45-mile-per-hour wind? *(Lesson 2–4)* — 13. −22°F
14. Find the product of −2, 3, −1, and 8. *(Lesson 2–5)* — 14. 48
15. Evaluate $\frac{3xy}{-4}$ if $x = -6$ and $y = 2$. *(Lesson 2–6)* — 15. 9
16. Find the quotient: $(-126) \div (-9)$. *(Lesson 2–6)* — 16. 14

CHAPTER 2 Preparing for Standardized Tests

Data Analysis Problems

You will need to create and interpret frequency tables as well as data graphs. This includes bar graphs, histograms, line graphs, and stem-and-leaf plots.

Test-Taking Tip

Read the graph or table before you read the question.

Example 1

Use the information on movie-making costs in the table. Make two line graphs on one grid, one for average production costs and the other for average marketing costs. Title the graph, label the axes, use appropriate scales, and accurately graph the data.

Movie-Making Costs ($ millions)

Year	Average Production Costs	Average Marketing Costs
1980	9.4	4.3
1985	16.8	6.5
1990	26.8	12.0
1995	36.3	17.7

Hint In open-ended questions, you may need to construct a graph, draw a diagram, or explain your answer.

Solution The x-axis shows the years. Decide on a scale for the y-axis, which represents costs. Since the lowest cost is 4.3 and the highest is 36.6, use a scale of 0 − 40 with intervals of 5. Mark each point (year, cost). Connect the points with line segments.

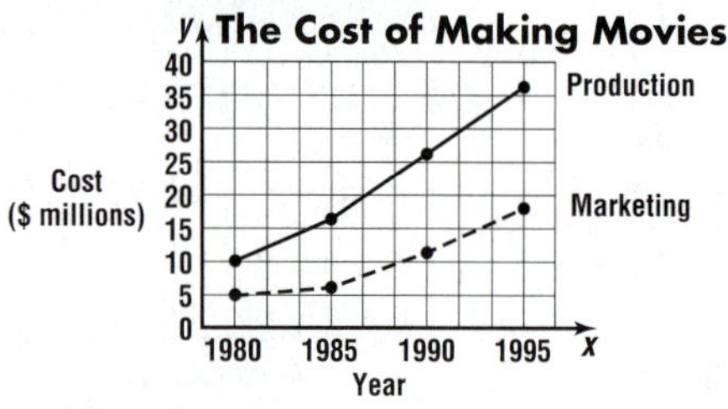

Example 2

The graph below represents the amount of money each person earns per day. How many days must Andy work to earn as much as Jill would earn in four days?

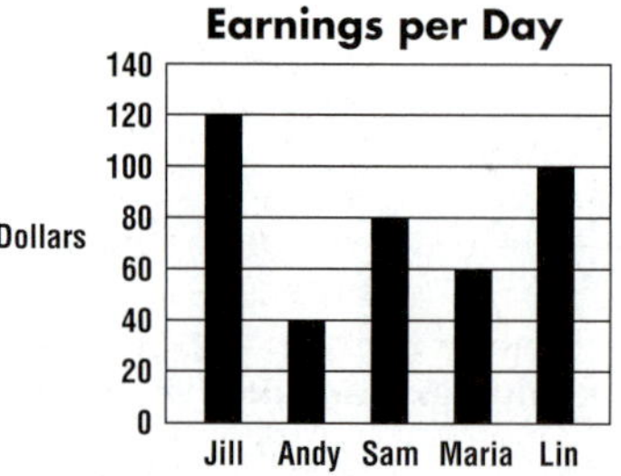

Ⓐ 480 Ⓑ 120 Ⓒ 80 Ⓓ 12 Ⓔ 3

Hint Look carefully at the graph. Read the title and labels. The range of the y-axis scale is 140.

Solution Calculate the amount that Jill earns in four days. The graph shows that she earns \$120 per day. In four days, she will earn $4 \times \$120$ or \$480.

Now calculate how many days it will take Andy to earn the amount of \$480. The graph shows that Andy earns \$40 per day.

Divide 480 by 40.

$$\frac{\$480}{\$40 \text{ per day}} = 12 \text{ days}$$

Andy needs to work 12 days to earn \$480. So, the answer is D.

Resource Manager

Reproducible Masters

Chapter 2 Resource Masters

- *Assessment*, pp. 92–94

Preparing for Standardized Tests

For test-taking strategies and more practice, see pages 732–749.

After you work each problem, record your answer on the answer sheet provided or on a sheet of paper.

Multiple Choice

1. One winter night the temperature dropped 3° every hour. If the temperature was 0° at midnight, what was the temperature at 4:00 A.M.? *(Lesson 2–5)* **A**

   (A) −12° (B) −15° (C) 12° (D) 32°

2. Which of the following numbers, when subtracted from −8, gives a result greater than −8? *(Lesson 2–4)* **A**

   (A) −2 (B) 0 (C) 2 (D) 3

3. How many even integers are there between −4 and 4? *(Lesson 2–1)* **B**

   (A) 2 (B) 3 (C) 4 (D) 6 (E) 8

4. If hot dogs are sold in packs of 10 and buns are sold in packs of 12, what is the smallest number of each you can buy to have no extra hot dogs or buns? *(Basic Skill)* **B**

   (A) 30 (B) 60 (C) 90 (D) 120

5. Find the greatest number of fat grams of any hamburger shown. *(Lesson 1–7)* **C**

Hamburgers	Stem	Chicken
	0	9
9 8 0	1	0 1 2 5 5 7 9
7 2	2	
5 2 0	3	$1 \mid 2 = 12$ g

   (A) 53 (B) 91 (C) 35 (D) 21

6. When was income closest to $20,000? *(Lesson 1–7)* **B**

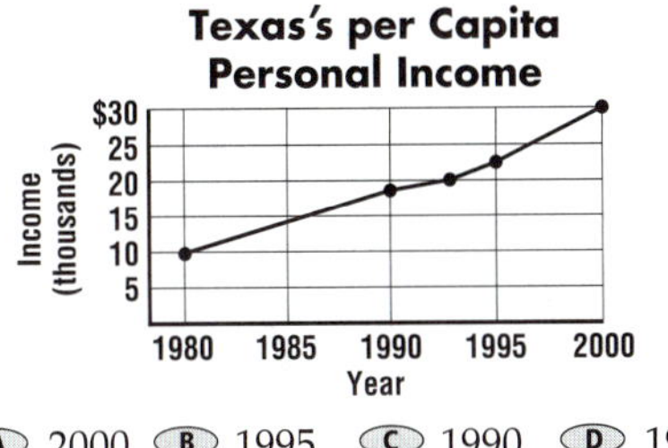

   (A) 2000 (B) 1995 (C) 1990 (D) 1980

7. The frequency table shows how many books each student read over the summer. Which statement is correct? *(Lesson 1–6)* **B**

Number of Books	Tally	Frequency
0	III	3
1	~~IIII~~ II	7
2	~~IIII~~ IIII	9
3	III	3
4	I	1

   (A) The students read 4 different books.
   (B) The most students read only 2 books.
   (C) Two students read 9 books.
   (D) There are a total of 21 students.

8. Which has the greatest value? *(Lesson 2–5)* **A**

   (A) $-38 \times (-10)$ (B) -38×10
   (C) $38 \times (-10)$ (D) 1×38

Short Response

9. Evaluate $\dfrac{-2 \times 4}{36 \div 2 - 5 \times 2}$. *(Lesson 1–2)* **−1**

Extended Response

10. The table below shows the winners of the first 17 World Cup soccer competitions. *(Lesson 1–6)*

Winners of the First 17 World Cup Soccer Competitions

Year	Champion	Year	Champion
1930	Uruguay	1974	West Germany
1934	Italy	1978	Argentina
1938	Italy	1982	Italy
1950	Uruguay	1986	Argentina
1954	West Germany	1990	West Germany
1958	Brazil	1994	Brazil
1962	Brazil	1998	France
1966	England	2002	Brazil
1970	Brazil		

Part A Construct a frequency table of the World Cup soccer champions.

Part B Use the table to make a bar graph. **See margin.**

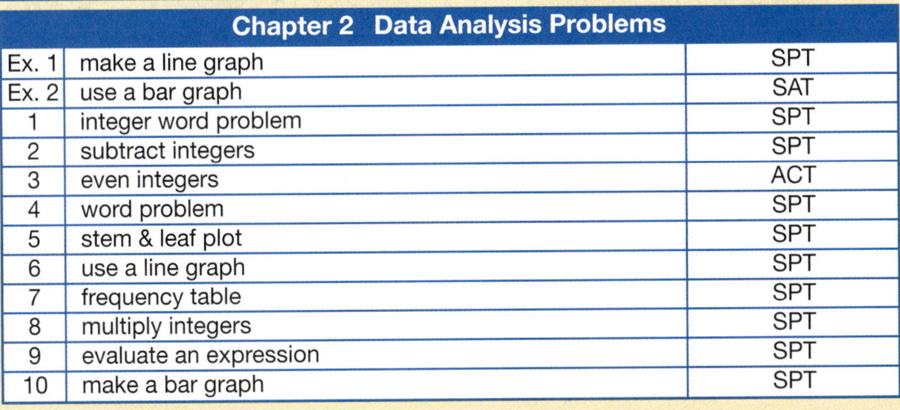

Chapter 2 Data Analysis Problems		
Ex. 1	make a line graph	SPT
Ex. 2	use a bar graph	SAT
1	integer word problem	SPT
2	subtract integers	SPT
3	even integers	ACT
4	word problem	SPT
5	stem & leaf plot	SPT
6	use a line graph	SPT
7	frequency table	SPT
8	multiply integers	SPT
9	evaluate an expression	SPT
10	make a bar graph	SPT

A bubble-in answer sheet for these practice problems is available on page A1 of the *Chapter 2 Resource Masters.*

Additional Practice

Additional test practice questions are available in the *Chapter 2 Resource Masters*, pp. 93–94.

Answers

10A.

World Cup Titles		
Country	Tally	Frequency
Uruguay	II	2
Italy	III	3
West Germany	III	3
Brazil	~~IIII~~	5
England	I	1
Argentina	II	2
France	I	1

10B.

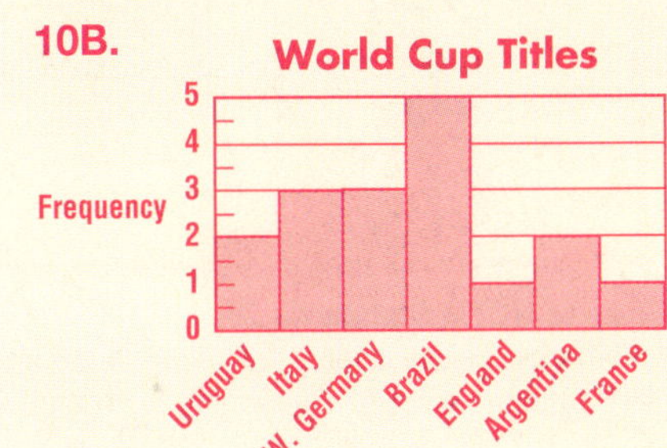

***Assessment*, pp. 93–94**

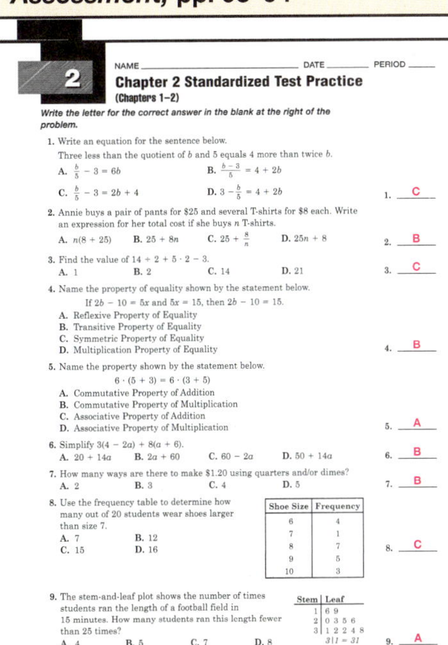

2 NAME ______ DATE ______ PERIOD ______

Chapter 2 Standardized Test Practice
(Chapters 1–2)

Write the letter for the correct answer in the blank at the right of the problem.

1. Write an equation for the sentence below.
   Three less than the quotient of b and 5 equals 4 more than twice b.
   A. $\frac{b}{5} - 3 = 6b$ B. $\frac{b-3}{5} = 4 + 2b$
   C. $\frac{b}{5} - 3 = 2b + 4$ D. $3 - \frac{b}{5} = 4 + 2b$ 1. **C**
2. Annie buys a pair of pants for \$25 and several T-shirts for \$8 each. Write an expression for her total cost if she buys n T-shirts.
   A. $n(8 + 25)$ B. $25 + 8n$ C. $25 + \frac{8}{n}$ D. $25n + 8$ 2. **B**
3. Find the value of $14 \div 2 + 5 \cdot 2 - 3$.
   A. 1 B. 2 C. 14 D. 21 3. **C**
4. Name the property of equality shown by the statement below.
   If $2b - 10 = 5x$ and $5x = 15$, then $2b - 10 = 15$.
   A. Reflexive Property of Equality
   B. Transitive Property of Equality
   C. Symmetric Property of Equality
   D. Multiplication Property of Equality 4. **B**
5. Name the property shown by the statement below.
   $6 \cdot (5 + 3) = 6 \cdot (3 + 5)$
   A. Commutative Property of Addition
   B. Commutative Property of Multiplication
   C. Associative Property of Addition
   D. Associative Property of Multiplication 5. **A**
6. Simplify $3(4 - 2a) + 8(a + 6)$.
   A. $20 + 14a$ B. $2a + 60$ C. $60 - 2a$ D. $50 + 14a$ 6. **B**
7. How many ways are there to make \$1.20 using quarters and/or dimes?
   A. 2 B. 3 C. 4 D. 5 7. **B**
8. Use the frequency table to determine how many out of 20 students wear shoes larger than size 7.
   A. 7 B. 12
   C. 15 D. 16 8. **C**

Shoe Size	Frequency
6	4
7	1
8	7
9	5
10	3

9. The stem-and-leaf plot shows the number of times students ran the length of a football field in 15 minutes. How many students ran this length fewer than 25 times?
   A. 4 B. 5 C. 7 D. 8 9. **A**

Stem	Leaf
1	6 9
2	0 3 5 6
3	1 2 2 4 8
	$3 \mid 1 = 31$

CHAPTER 3

Resource Manager

Addition and Subtraction Equations

Instructional Objectives

Lesson (pages)	Objectives	NCTM Standards 2000	State/Local Objectives
3–1 (94–99)	Compare and order rational numbers.	1, 6, 8, 9	
3–2 (100–103)	Add and subtract rational numbers.	1, 6, 8, 9	
3–3 (104–109)	Find the mean, median, mode, and range of a set of data.	1, 5, 6, 8, 9, 10	
Investigation (110–111)	Explore arithmetic sequences.	1, 2, 6, 7, 8, 9, 10	
3–4 (112–116)	Determine whether a given number is a solution of an equation.	1, 6, 8, 9	
3–5 (117–121)	Solve addition and subtraction equations by using models.	1, 2, 6, 8, 9, 10	
3–6 (122–127)	Solve addition and subtraction equations by using the properties of equality.	1, 6, 8, 9	
3–7 (128–131)	Solve equations involving absolute value.	1, 6, 8, 9	

Key to NCTM Standards 2000

1 Number & Operations; **2** Algebra; **3** Geometry; **4** Measurement; **5** Data Analysis & Probability; **6** Problem Solving; **7** Reasoning & Proof; **8** Communications; **9** Connections; **10** Representation

Suggested Pacing *See page T13 for a complete course-planning calendar.*

Standard refers to schedules that provide 45- to 55-minute periods that meet each day.
Block refers to schedules that provide approximately 90-minute periods which may meet every day for one semester or every other day over two semesters.

<table>
<tr><th>PACING</th><th>DAY 1</th><th>DAY 2</th><th>DAY 3</th><th>DAY 4</th><th>DAY 5</th><th>DAY 6</th></tr>
<tr><td>Standard Core (Chapters 1–13)</td><td colspan="2">Lesson 3–1</td><td>Lesson 3–2</td><td colspan="2">Lesson 3–3</td><td>INV</td></tr>
<tr><td>Standard Enhanced (Chapters 1–15)</td><td>Lesson 3–1</td><td>Lesson 3–2</td><td colspan="2">Lesson 3–3</td><td>INV</td><td>Lesson 3–4</td></tr>
<tr><td>Block Core (Chapters 1–13)</td><td>Chapter 2 Test & Lesson 3–1</td><td>Lessons 3–2 & 3–3</td><td>INV Lesson 3–4</td><td>Lessons 3–5 & 3–6</td><td>Lesson 3–7</td><td>SG+A</td></tr>
<tr><td>Block Enhanced (Chapters 1–15)</td><td>Chapter 2 Test & Lesson 3–1</td><td>Lessons 3–2 & 3–3</td><td>INV Lesson 3–4</td><td>Lessons 3–5 & 3–6</td><td>Lesson 3–7 & SG+A</td><td>Chapter Test & Lesson 4–1</td></tr>
</table>

Instructional Resources

Lesson	Materials and Manipulatives (see below for Glencoe Manipulative Resources)	Blackline Masters (page numbers): Chapter 3 Resource Masters: Study Guide	Practice (Skills & Average)	Reading to Learn Mathematics	Enrichment	Assessment	Hands-On Algebra*	School-to-Workplace*	Graphing Calculator Masters*	5-Minute Check Transparencies
3–1	masking tape	95	96–97	98	99		42			3–1
3–2	inch ruler [1, 2, 4] uncooked rice fraction bars colored pencils or pens	100	101–102	103	104		43	3	9, 10	3–2
3–3	graphing calculator	105	106–107	108	109	141	44		8	3–3
Investigation	sugar cubes or wooden cubes									
3–4		110	111–112	113	114	140				3–4
3–5	algebra tiles [1, 2, 3, 4] equation/product mat [1, 2, 4]	115	116–117	118	119		45, 46			3–5
3–6	algebra tiles [1, 2, 3, 4] equation/product mat [1, 2, 4]	120	121–122	123	124		47			3–6
3–7	algebra tiles [1, 2, 3, 4] equation/product mat [1, 2, 4]	125	126–127	128	129	141	48			3–7
Study Guide & Assessment/ Chapter Test	algebra tiles [1, 2, 3, 4] equation/product mat [1, 2, 4]					131–139, 142–144				

***See page 92c for examples of these instructional materials.**

Key to Glencoe Manipulative Resources

[1]Classroom Manipulative Resources [2]Student Manipulative Resources [3]Overhead Manipulative Resources [4]Hands-On Algebra Masters

INV = Investigation *SG+A = Study Guide and Assessment*

DAY 7	DAY 8	DAY 9	DAY 10	DAY 11	DAY 12	DAY 13
Lesson 3–4	Lesson 3–5	Lesson 3–6		Lesson 3–7	SG+A	Chapter Test
Lesson 3–5	Lesson 3–6		Lesson 3–7	SG+A	Chapter Test	
Chapter Test & Lesson 4–1						

Resource Manager

The pages shown on this page are a small sample of the materials available on *TeacherWorks: All-in-One Lesson Planner and Resource Center.*

This CD-ROM includes all of the blackline masters and transparencies available for this program.

It also includes a lesson planner and interactive Teacher Edition, so you can customize lesson plans and reproduce classroom resources quickly and easily, from just about anywhere.

Applications

School-to-Workplace Masters, p. 3

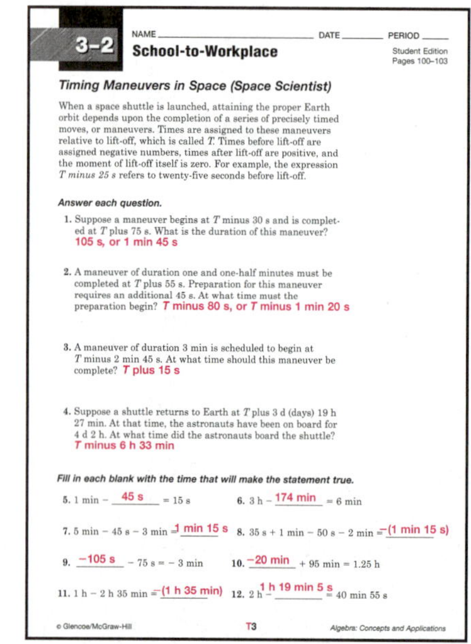

3–2 NAME ________ DATE ______ PERIOD ____

School-to-Workplace

Student Edition Pages 100–103

Timing Maneuvers in Space (Space Scientist)

When a space shuttle is launched, attaining the proper Earth orbit depends upon the completion of a series of precisely timed moves, or maneuvers. Times are assigned to these maneuvers relative to lift-off, which is called *T*. Times before lift-off are assigned negative numbers, times after lift-off are positive, and the moment of lift-off itself is zero. For example, the expression *T minus 25 s* refers to twenty-five seconds before lift-off.

Answer each question.

1. Suppose a maneuver begins at *T* minus 30 s and is completed at *T* plus 75 s. What is the duration of this maneuver? **105 s, or 1 min 45 s**
2. A maneuver of duration one and one-half minutes must be completed at *T* plus 55 s. Preparation for this maneuver requires an additional 45 s. At what time must the preparation begin? ***T* minus 80 s, or *T* minus 1 min 20 s**
3. A maneuver of duration 3 min is scheduled to begin at *T* minus 2 min 45 s. At what time should this maneuver be complete? ***T* plus 15 s**
4. Suppose a shuttle returns to Earth at *T* plus 3 d (days) 19 h 27 min. At that time, the astronauts have been on board for 4 d 2 h. At what time did the astronauts board the shuttle? ***T* minus 6 h 33 min**

Fill in each blank with the time that will make the statement true.

5. 1 min – **45 s** = 15 s
6. 3 h – **174 min** = 6 min
7. 5 min – 45 s – 3 min = **1 min 15 s**
8. 35 s + 1 min – 50 s – 2 min = **–(1 min 15 s)**
9. **–105 s** – 75 s = – 3 min
10. **–20 min** + 95 min = 1.25 h
11. 1 h – 2 h 35 min = **–(1 h 35 min)**
12. 2 h – **1 h 19 min 5 s** = 40 min 55 s

© Glencoe/McGraw-Hill T3 *Algebra: Concepts and Applications*

Manipulatives/Modeling

Hands-On Algebra Masters, pp. 42–48

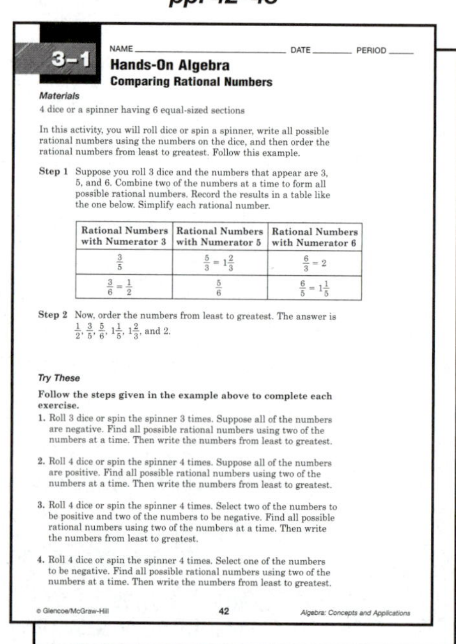

3–1 NAME ________ DATE ______ PERIOD ____

Hands-On Algebra

Comparing Rational Numbers

Materials

4 dice or a spinner having 6 equal-sized sections

In this activity, you will roll dice or spin a spinner, write all possible rational numbers using the numbers on the dice, and then order the rational numbers from least to greatest. Follow this example.

Step 1 Suppose you roll 3 dice and the numbers that appear are 3, 5, and 6. Combine two of the numbers at a time to form all possible rational numbers. Record the results in a table like the one below. Simplify each rational number.

Rational Numbers with Numerator 3	Rational Numbers with Numerator 5	Rational Numbers with Numerator 6
$\frac{3}{5}$	$\frac{5}{3} = 1\frac{2}{3}$	$\frac{6}{3} = 2$
$\frac{3}{6} = \frac{1}{2}$	$\frac{5}{6}$	$\frac{6}{5} = 1\frac{1}{5}$

Step 2 Now, order the numbers from least to greatest. The answer is $\frac{1}{2}, \frac{3}{5}, \frac{5}{6}, 1\frac{1}{5}, 1\frac{2}{3}$, and 2.

Try These

Follow the steps given in the example above to complete each exercise.

1. Roll 3 dice or spin the spinner 3 times. Suppose all of the numbers are negative. Find all possible rational numbers using two of the numbers at a time. Then write the numbers from least to greatest.
2. Roll 4 dice or spin the spinner 4 times. Suppose all of the numbers are positive. Find all possible rational numbers using two of the numbers at a time. Then write the numbers from least to greatest.
3. Roll 4 dice or spin the spinner 4 times. Select two of the numbers to be positive and two of the numbers to be negative. Find all possible rational numbers using two of the numbers at a time. Then write the numbers from least to greatest.
4. Roll 4 dice or spin the spinner 4 times. Select one of the numbers to be negative. Find all possible rational numbers using two of the numbers at a time. Then write the numbers from least to greatest.

© Glencoe/McGraw-Hill 42 *Algebra: Concepts and Applications*

Technology/Multimedia

Graphing Calculator Masters, pp. 8–10

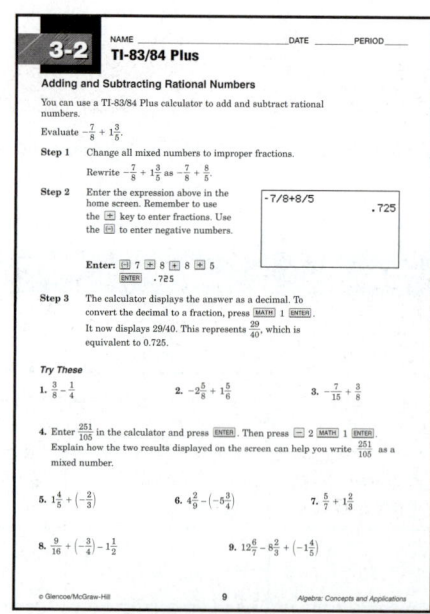

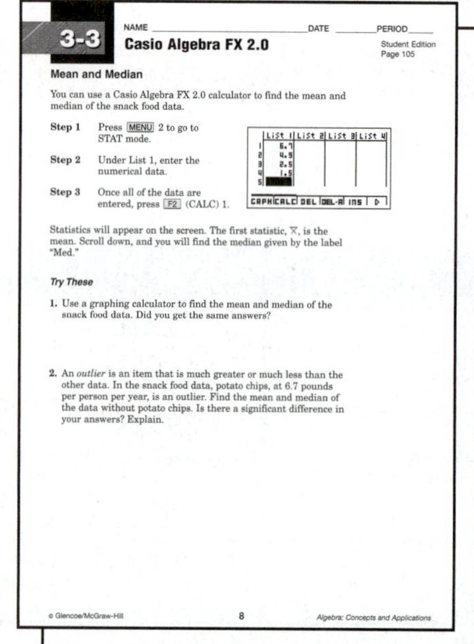

3-3 NAME ________ DATE ______ PERIOD ____

Casio Algebra FX 2.0

Student Edition Page 105

Mean and Median

You can use a Casio Algebra FX 2.0 calculator to find the mean and median of the snack food data.

Step 1 Press [MENU] 2 to go to STAT mode.

Step 2 Under List 1, enter the numerical data.

Step 3 Once all of the data are entered, press [F2] (CALC) 1.

	List 1	List 2	List 3	List 4
1	6.7			
2	4.9			
3	2.5			
4	1.5			
5				

GRPH CALC DEL DEL-A INS ▷

Statistics will appear on the screen. The first statistic, $\overline{x}$, is the mean. Scroll down, and you will find the median given by the label "Med."

Try These

1. Use a graphing calculator to find the mean and median of the snack food data. Did you get the same answers?
2. An *outlier* is an item that is much greater or much less than the other data. In the snack food data, potato chips, at 6.7 pounds per person per year, is an outlier. Find the mean and median of the data without potato chips. Is there a significant difference in your answers? Explain.

© Glencoe/McGraw-Hill 8 *Algebra: Concepts and Applications*

3-2 NAME ________ DATE ______ PERIOD ____

TI-83/84 Plus

Adding and Subtracting Rational Numbers

You can use a TI-83/84 Plus calculator to add and subtract rational numbers.

Evaluate $-\frac{7}{8} + 1\frac{3}{5}$.

Step 1 Change all mixed numbers to improper fractions.

Rewrite $-\frac{7}{8} + 1\frac{3}{5}$ as $-\frac{7}{8} + \frac{8}{5}$.

Step 2 Enter the expression above in the home screen. Remember to use the [÷] key to enter fractions. Use the [(–)] to enter negative numbers.

-7/8+8/5
.725

Enter: [(–)] 7 [÷] 8 [+] 8 [÷] 5 [ENTER] .725

Step 3 The calculator displays the answer as a decimal. To convert the decimal to a fraction, press [MATH] 1 [ENTER]. It now displays 29/40. This represents $\frac{29}{40}$, which is equivalent to 0.725.

Try These

1. $\frac{3}{8} - \frac{1}{4}$
2. $-2\frac{5}{8} + 1\frac{5}{6}$
3. $-\frac{7}{15} + \frac{3}{8}$
4. Enter $\frac{251}{105}$ in the calculator and press [ENTER]. Then press [–] 2 [MATH] 1 [ENTER]. Explain how the two results displayed on the screen can help you write $\frac{251}{105}$ as a mixed number.
5. $1\frac{4}{5} + \left(-\frac{2}{3}\right)$
6. $4\frac{2}{9} - \left(-5\frac{3}{4}\right)$
7. $\frac{5}{7} + 1\frac{2}{3}$
8. $\frac{9}{16} + \left(-\frac{3}{4}\right) - 1\frac{1}{2}$
9. $12\frac{6}{7} - 8\frac{2}{3} + \left(-1\frac{4}{5}\right)$

© Glencoe/McGraw-Hill 9 *Algebra: Concepts and Applications*

Assessment Resources

Type	Student Edition	Teacher's Wraparound Edition	Chapter 3 Resource Masters
Ongoing Assessment	Quizzes 1 and 2, pp. 116, 127	5-Minute Check, pp. 94, 100, 104, 112, 117, 122, 128	Mid-Chapter Test, p. 140 Quizzes A and B, p. 141
Mixed Review	Mixed Review, pp. 99, 103, 109, 116, 121, 127, 131 Standardized Test Practice, Chapters 1–3, pp. 136–137		Cumulative Review, p. 142 Standardized Test Practice, pp. 143–144
Error Analysis	You Decide, pp. 107, 125	Error Analysis, pp. 97, 102, 107, 114, 120, 125, 130	
Standardized Test Prep	Standardized Test Practice, pp. 99, 103, 109, 116, 121, 127, 131 Standardized Test Practice, Chapters 1–3, pp. 136–137		Standardized Test Practice, pp. 143–144
Open-Ended Assessment	Writing Math, pp. 97, 130 Investigation, pp. 110–111 Portfolio, pp. 93, 111	Modeling: pp. 103, 121, 131 Speaking: pp. 116, 127 Writing: pp. 99, 109	Extended Response Assessment, p. 139
Chapter Assessment	Study Guide and Assessment, pp. 132–134 Chapter Test, p. 135		Multiple-Choice Tests (Forms 1A, 1B), pp. 131–134 Free-Response Tests (Forms 2A, 2B), pp. 135–138

Additional Chapter Resources

Student Edition

Graphing Calculator Exploration, p. 105

Teacher's Classroom Resources

Manipulatives/Modeling

Teacher's Guide for Overhead Manipulative Resources

Meeting Individual Needs

Prerequisite Skills Booklet

Spanish Study Guide and Assessment, pp. 22–28, 109–110

Teaching Aids

Solutions Manual

5-Minute Check Transparencies

Glencoe Technology

Instructional

AlgePASS CD-ROM, Lessons 9, 10

Interactive Chalkboard CD-ROM

StudentWorks Plus CD-ROM

Multimedia Applications CD-ROM, Activities 2, 3

Vocabulary PuzzleMaker (online)

Assessment

ExamView® Pro

Teaching Aids

Answer Key Maker CD-ROM

TeacherWorks CD-ROM

*Visit **www.algconcepts.com** for data updates, career information, games, and other interactive activities.*

CHAPTER 3 PREVIEW

Mathematics of the Chapter

This chapter begins with an in-depth study of rational numbers and operations with rational numbers. The study of data interpretation begun in Chapter 1 continues with the introduction of the measures of central tendency: mean, median, and mode, as well as the range, a measure of the variation of a set of data. In the second half of the chapter, students learn to model and solve addition and subtraction equations using algebra tiles. Students then make the transition from using algebra tiles to using the properties of equality to solve these equations. The chapter concludes with a lesson on solving absolute value equations.

What You'll Learn

Have students read over the lists of key ideas and key vocabulary. Have them make a list of any words with which they are not familiar.

Why It's Important

Point out to students that this is only one of many reasons why each objective is important. Others are provided in each lesson.

CHAPTER 3 Addition and Subtraction Equations

What You'll Learn

Key Ideas

- Compare, order, add, and subtract rational numbers. *(Lessons 3–1 and 3–2)*
- Find the mean, median, mode, and range of a set of data. *(Lesson 3–3)*
- Solve addition and subtraction equations. *(Lessons 3–4 to 3–6)*
- Solve absolute value equations. *(Lesson 3–7)*

Key Vocabulary

equivalent equations *(p. 122)*
inequality *(p. 95)*
rational numbers *(p. 94)*
solution *(p. 112)*

Why It's Important

Lighthouses The Outer Banks of North Carolina are home to two of the ten United States National Seashores — Cape Hatteras and Cape Lookout. Tourists enjoy swimming, bird watching, and other activities on their beaches, sand dunes, marshes, and woodlands.

Addition and subtraction equations are used to solve real-life problems like comparing the heights of lighthouses. You will use an equation to compare the heights of the Cape Hatteras and Cape Lookout lighthouses in Lesson 3–6.

CHAPTER 3 LINKS

Lesson	3–1	3–2	3–3	3–4	3–5	3–6	3–7
Math in the Workplace	Shopping	Agriculture	Meteorology	Health	Zoology	History	Science
Applications and Connections	Tools Life Science Geography	Stock Market Personal Finance Sports Shopping Music	Food Sales Sports School Zoos Business	Recreation Life Science Crafts	Sales Sports Weather	Art Hardware Lighthouses Sports Dining	Sports Real Estate Shipping
Math Integration				Geometry		Statistics	

Check Your Readiness

Study these lessons to improve your skills.

Prerequisite Skill, p. 685

Write each fraction in simplest form.

1. $\frac{4}{12}$ $\frac{1}{3}$ 2. $\frac{3}{18}$ $\frac{1}{6}$ 3. $\frac{5}{40}$ $\frac{1}{9}$

4. $\frac{6}{15}$ $\frac{2}{5}$ 5. $\frac{10}{25}$ $\frac{2}{5}$ 6. $\frac{16}{28}$ $\frac{4}{7}$

Prerequisite Skill, p. 684

Find each sum or difference.

7. 3.5 + 2.1 **5.6** 8. 6.6 + 9.8 **16.4** 9. 12.1 + 11.6 **23.7**

10. 11.8 – 3.7 **8.1** 11. 45.8 – 12.7 **33.1** 12. 23.1 – 15.9 **7.2**

13. 1.5 + 2.1 + 3.2 + 1.8 + 2.6 + 3.1 + 2.9 **17.2**

14. 8.9 + 11.1 + 10.7 + 9.8 + 11.0 + 9.3 + 8.7 **69.5**

Prerequisite Skill, p. 684

Find each quotient.

15. 15.6 ÷ 3 **5.2** 16. 18.4 ÷ 4 **4.6** 17. 48.6 ÷ 12 **4.05**

18. 54.9 ÷ 6 **9.15** 19. 188 ÷ 5 **37.6** 20. 225 ÷ 18 **12.5**

Prerequisite Skill, p. 687

Find each sum or difference.

21. $\frac{3}{8} + \frac{1}{8}$ $\frac{1}{2}$ 22. $\frac{1}{6} + \frac{2}{3}$ $\frac{5}{6}$ 23. $\frac{7}{10} + \frac{3}{5}$ $1\frac{3}{10}$

24. $\frac{5}{12} - \frac{1}{12}$ $\frac{1}{3}$ 25. $\frac{5}{6} - \frac{2}{3}$ $\frac{1}{6}$ 26. $\frac{7}{8} - \frac{3}{4}$ $\frac{1}{8}$

FOLDABLES™ Study Organizer

Make this Foldable to help you organize your Chapter 3 notes. Begin with a sheet of 11" by 17" paper.

❶ **Fold** the short sides to meet in the middle.

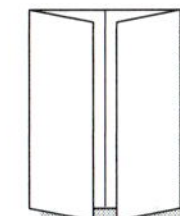

❷ **Fold** the top to the bottom.

❸ **Open.** Cut along the second fold to make four tabs.

❹ **Label** the tabs as shown.

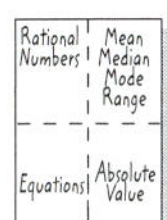

Reading and Writing As you read and study the chapter, write notes and examples under each tab.

www.algconcepts.com/chapter_readiness

Check Your Readiness

This section provides a review of the basic concepts needed before beginning Chapter 3. Lesson references are included for additional student help.

Use Exercises	To Prepare for Lesson(s)
1–6	3–2, 3–4
7–14	3–2, 3–3, 3–6
15–20	3–3
21–26	3–2, 3–6

Quick Review Math Handbook

Glencoe Algebra Lesson	Quick Review Math Handbook Hot Topic
3–1	1–5, 2–2, 2–3, 2–5
3–2	1–5, 2–3, 2–4, 2–6
3–3	4–2, 4–3
Ch. 3 Inv.	6–7
3–4	2–4, 6–4, 6–6
3–5	6–4
3–6	2–4, 2–6, 6–4
3–7	6–4

Noteables™

Interactive Study Notebook with Foldables™

This note-taking guide reinforces excellent note-taking skills. It includes:

- ✓ **Build Your Vocabulary** tool for including mathematical terminology in notes.
- ✓ **Review It** activities with links to prior lessons and items to include in **Foldables™**.
- ✓ **Bringing It All Together** feature to help students review for the Chapter Test.
- ✓ **Are You Ready for the Chapter Test?** feature to allow students to assess their own readiness for the Chapter Test.
- ✓ **Teacher Edition** with transparencies to guide note taking. Corresponds to the Teacher's *Wraparound Edition* In-Class Examples and *Interactive Chalkboard CD-ROM.*

FOLDABLES™ Study Organizer

Use this Foldable for student writing about rational numbers, measures of central tendency, equations, and absolute value. For each tab of the Foldable, ask students to write about a time when they have used the concepts related to that tab to solve a problem.

Lesson 3-1

1 FOCUS

5-Minute Check
Chapter 2

1. Refer to the figure below. Name the coordinate of each point.

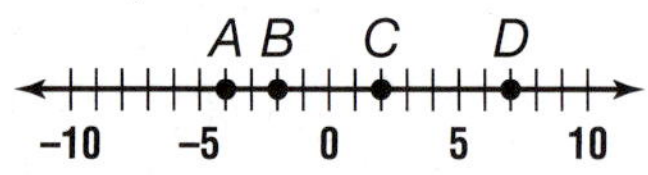

A: −4, B: −2, C: 2, D: 7

Refer to the coordinate plane below.

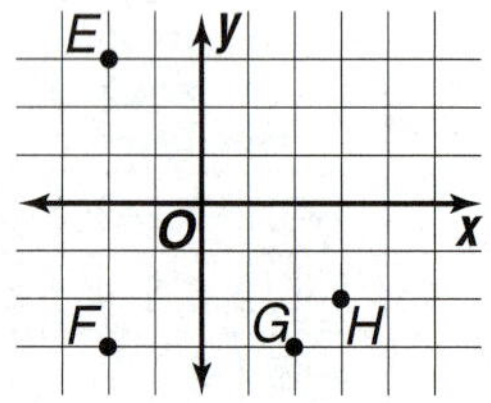

2. Write the ordered pair that names point G. **(2, −3)**
3. Which point on the graph has the coordinates (−2, 3)? ***E***

Find each product or quotient.

4. −3(7) **−21**
5. $\frac{-18}{-3}$ **6**

Motivating the Lesson

Real-World Connection Have students consider the following question. "Suppose you try to loosen a bolt on your bicycle. A wrench marked $\frac{7}{16}$ is too small. Which wrench should you try next, one marked $\frac{5}{8}$ or one marked $\frac{3}{4}$?"

2 TEACH

Teaching Tip When discussing how to write the rational number −2 in the form $\frac{a}{b}$, point out that $-\frac{2}{1}$ can also be written as $\frac{-2}{1}$ and $\frac{2}{-1}$.

3-1 Rational Numbers

What You'll Learn
You'll learn to compare and order rational numbers.

Why It's Important
Shopping Knowing how to compare rational numbers will help you determine better buys.
See Example 6.

Rational numbers are frequently used in daily life. They consist of all positive and negative fractions. All integers are also rational numbers because integers can be expressed as fractions with 1 in the denominator. For example, 2 equals $\frac{2}{1}$ and −3 equals $-\frac{3}{1}$.

Rational Number	
Words:	A rational number is any number that can be expressed as a fraction where the numerator and denominator are integers and the denominator is not zero.
Symbols:	$\frac{a}{b}$, where a and b are integers and $b \neq 0$

Some examples of rational numbers and their form as a fraction are listed in the table below.

Rational Number	−2	$3\frac{1}{4}$	0.625	0	$-0.66\overline{6}$
Form $\frac{a}{b}$	$-\frac{2}{1}$	$\frac{13}{4}$	$\frac{5}{8}$	$\frac{0}{1}$	$-\frac{2}{3}$

Rational numbers can be graphed on a number line in the same manner as integers.

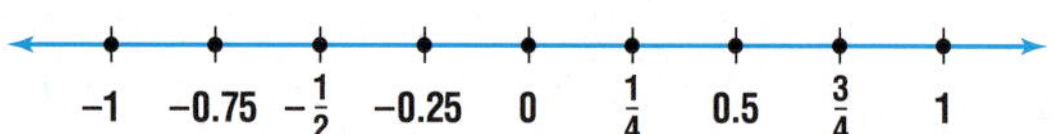

Graphing rational numbers on a number line helps you to compare them. Just as with integers, the numbers on a number line increase in value as you move to the right and decrease in value as you move to the left.

The following statements can be made about the number line above.

Words	Symbols
The graph of −1 is to the left of the graph of $-\frac{1}{2}$. −1 is less than $-\frac{1}{2}$.	$-1 < -\frac{1}{2}$
The graph of 1 is to the right of −0.25. 1 is greater than −0.25.	$1 > -0.25$
The graph of $\frac{1}{4}$ is to the left of 0.5. $\frac{1}{4}$ is less than 0.5.	$\frac{1}{4} < 0.5$

Resource Manager

Reproducible Masters
Chapter 3 Resource Masters
- *Study Guide,* p. 95
- *Skills Practice,* p. 96
- *Practice,* p. 97
- *Reading to Learn Mathematics,* p. 98
- *Enrichment,* p. 99

Hands-On Algebra, p. 42

Transparencies
5-Minute Check, 3-1

Technology/Multimedia
AlgePASS, Lesson 9
Interactive Chalkboard CD-ROM

A mathematical sentence that uses $<$ and $>$ to compare two expressions is called an **inequality**. When you compare two numbers, the following property applies.

Comparison Property		
	Symbols:	For any two numbers a and b, exactly one of the following sentences is true. $a < b$ $a > b$ $a = b$
	Numbers:	If $a = 2$ and $b = 3$, then only $2 < 3$ is true.

Examples

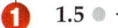

Replace each ● with $<$, $>$, or $=$ to make a true sentence.

1. **1.5 ● -8**

Since any positive number is greater than any negative number, $1.5 > -8$.

2. **$-14 + 8$ ● $2(-3)(5)$**

$-14 + 8$ ● $2(-3)(5)$

-6 ● -30 *Find the value of each side.*

-6 is to the right of -30 on a number line and $-6 > -30$. So, $-14 + 8 > 2(-3)(5)$.

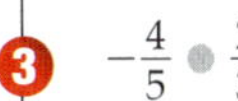

3. $-\frac{4}{5}$ ● $\frac{2}{3}$

Since any negative number is less than any positive number, $-\frac{4}{5} < \frac{2}{3}$.

Your Turn

a. -1 ● -14.2 $>$ b. $-5(0)$ ● $16 + (-16)$ $=$ c. $-\frac{3}{8}$ ● $\frac{5}{8}$ $<$

To compare two fractions with different denominators, you can use **cross products**. Cross products are the products of the diagonal terms of two fractions.

$$\frac{2}{7} \bullet \frac{3}{8}$$

$8(2)$ ● $7(3)$ *Find the cross products.*

$16 < 21$

If $16 < 21$, then $\frac{2}{7} < \frac{3}{8}$.

Teaching Tip After discussing the Comparison Property, ask students to think of characteristics of the symbols that will help them remember their meanings. For example, the inequality symbol $<$ or $>$ always "points" to the lesser number.

In-Class Examples

Examples 1–2

Replace each ● with $<$, $>$, or $=$ to make a true sentence.

1 -1 ● $-\frac{3}{5}$ $<$

2 $-3(2)(0)$ ● $7 + (-8)$ $>$

Teaching Tip Draw students' attention to the arrows that overlay the cross products for the comparison problem shown at the bottom of the page. Point out that following the arrows will help students put the products in the correct order.

Teaching Tip When discussing Example 4, caution students to be sure they assign the negative sign in both rational numbers to *either* the numerator or denominator when forming the cross products, but not to *both* of them.

In-Class Examples

Examples 3–4

Replace each ● with <, >, or = to make a true sentence.

3 $\frac{7}{12}$ ● $\frac{5}{8}$ **<**

4 $-\frac{4}{15}$ ● $-\frac{3}{10}$ **>**

Example 5

Write $\frac{5}{6}$, $\frac{7}{9}$, and $\frac{4}{5}$ in order from least to greatest. **$\frac{7}{9}, \frac{4}{5}, \frac{5}{6}$**

Example 6

Latisha needs to buy snacks for her art club. A package of 12 granola bars costs \$2.69 and a package of 18 granola bars costs \$3.55. Which is the better buy? Explain. **Since \$0.20 < \$0.22, the package of 18 granola bars is the better buy.**

This example illustrates the following property.

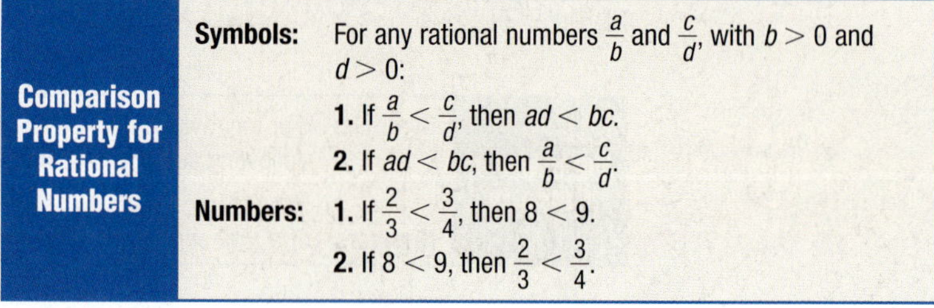

Comparison Property for Rational Numbers

Symbols: For any rational numbers $\frac{a}{b}$ and $\frac{c}{d}$, with $b > 0$ and $d > 0$:

1. If $\frac{a}{b} < \frac{c}{d}$, then $ad < bc$.

2. If $ad < bc$, then $\frac{a}{b} < \frac{c}{d}$.

Numbers: **1.** If $\frac{2}{3} < \frac{3}{4}$, then $8 < 9$.

2. If $8 < 9$, then $\frac{2}{3} < \frac{3}{4}$.

This property also holds true if < is replaced by > or =.

Examples Replace each ● with <, >, or = to make a true sentence.

4 $\frac{3}{8}$ ● $\frac{4}{10}$

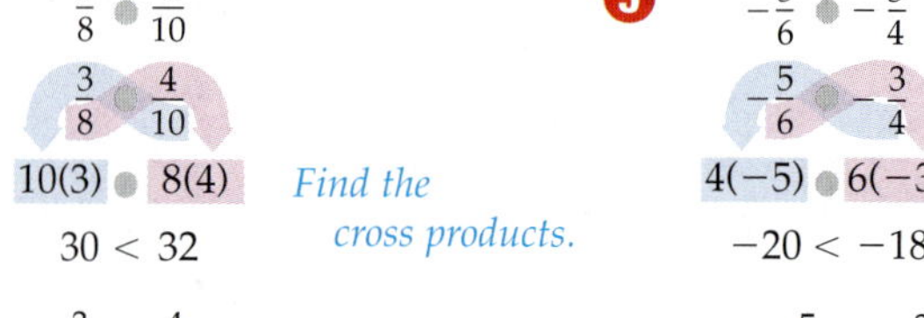

$10(3)$ ● $8(4)$ *Find the cross products.*

$30 < 32$

So, $\frac{3}{8} < \frac{4}{10}$.

5 $-\frac{5}{6}$ ● $-\frac{3}{4}$

$4(-5)$ ● $6(-3)$

$-20 < -18$

So, $-\frac{5}{6} < -\frac{3}{4}$.

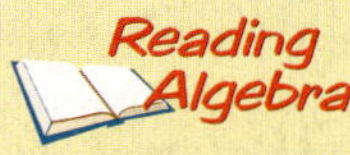

Technology Tip You can use a calculator to express rational numbers as decimals so they can be easily compared.

Your Turn

d. $\frac{4}{5}$ ● $\frac{7}{8}$ **<**

e. $-\frac{5}{10}$ ● $-\frac{3}{5}$ **>**

Another way to compare rational numbers is to express them as terminating or repeating decimals.

Example 6 **Write $\frac{3}{8}$, $\frac{1}{3}$, and $\frac{2}{5}$ in order from least to greatest.**

$\frac{3}{8} = 0.375$ *This is a terminating decimal.*

$\frac{1}{3} = 0.3333\ldots$ or $0.\overline{3}$ *This is a repeating decimal.*

$\frac{2}{5} = 0.4$ *This is a terminating decimal.*

In order from least to greatest, the decimals are $0.\overline{3}$, 0.375, 0.4. So, the fractions in order from least to greatest are $\frac{1}{3}, \frac{3}{8}, \frac{2}{5}$.

Reading Algebra A bar over a number indicates that it repeats indefinitely. The number $0.\overline{3}$ is read as *zero point three repeating.*

Your Turn **Write the numbers in each set from least to greatest.**

f. $\frac{4}{8}, \frac{9}{10}, \frac{2}{3}$ **$\frac{4}{8}, \frac{2}{3}, \frac{9}{10}$**

g. $-\frac{1}{3}, -\frac{5}{8}, -\frac{1}{6}$ **$-\frac{5}{8}, -\frac{1}{3}, -\frac{1}{6}$**

You can use the cost per unit, or **unit cost**, to compare the costs of similar items. Many people use unit cost to comparison shop. The least unit cost is the best buy.

$$\text{unit cost} = \text{total cost} \div \text{number of units}$$

Example
Shopping Link

7 **Rolando needs to buy colored pencils. The cost of a package of 12 pencils is \$6.39. A package of 24 pencils costs \$12.89. Which is the better buy? Explain.**

Find the unit cost of each package. In each case, the unit cost is expressed in cents per pencil.

unit cost of package of 12:
$6.39 \div 12 = 0.5325$ or about \$0.53 per pencil

unit cost of package of 24:
$12.89 \div 24 \approx 0.5371$ or about \$0.54 per pencil

Since \$0.53 < \$0.54, the package of 12 pencils is the better buy.

Check for Understanding

Communicating Mathematics

Vocabulary
rational number
inequality
cross products
unit cost

1. **Write** three examples of rational numbers. Include one in decimal form and one negative number.
2. **Give an example** of two different fractions whose cross products are equal.

1–2. See margin.

3. *Writing Math* Create a list of at least five commonly used fraction-decimal equivalents such as $\frac{1}{2} = 0.5$. Explain why memorizing them is useful. **See students' work.**

Guided Practice

Getting Ready **Express each fraction as a decimal.**

Sample 1: $\frac{1}{4}$
Solution: $1 \div 4 = 0.25$

Sample 2: $\frac{2}{3}$
Solution: $2 \div 3 = 0.\overline{6}$

4. $\frac{8}{10}$ **0.8**
5. $\frac{3}{4}$ **0.75**
6. $\frac{2}{6}$ **$0.\overline{3}$**
7. $\frac{5}{8}$ **0.625**
8. $\frac{7}{9}$ **$0.\overline{7}$**

9. Graph the set of numbers $\left\{-3, 1.5, \frac{1}{4}, 0, -2.\overline{6}\right\}$ on a number line. Use it to explain why $-3 < -2.\overline{6}$. **See margin.**

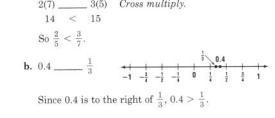

Reteaching Activity

Kinesthetic Learners Create a large number line on the classroom floor using masking tape. Show tick marks for every $\frac{1}{8}$ of a unit from −2 to 2, and label the number line with the integers −2 through 2. Assign each student a rational number between −2 and 2, such as $-\frac{5}{7}$. Have students plot their number on the number line using a piece of masking tape with their rational number written on it. Students who are unsure about the location of their number on the number line should use cross products to compare it to other students' numbers.

3 PRACTICE/APPLY

Error Analysis

Watch for students who are confused by Exercise 1 because they think a rational number must be a fraction.

Prevent by presenting students with fractions and decimals that are equivalent, such as $\frac{3}{10}$ and 0.3. By having students read both numbers aloud, they should realize that if the fraction $\frac{3}{10}$ is rational, then so is the decimal 0.3. Emphasize that a fraction is just one form of a rational number.

Answers

1. Sample answer: $-\frac{1}{2}$, $\frac{4}{3}$, and 0.9
2. Sample answer: $\frac{4}{6}$ and $\frac{2}{3}$
9\.

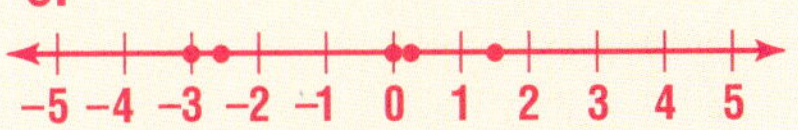

Since −3 is to the left of $-2.\overline{6}$ on the number line, −3 is less than $-2.\overline{6}$.

***Study Guide*, p. 95**

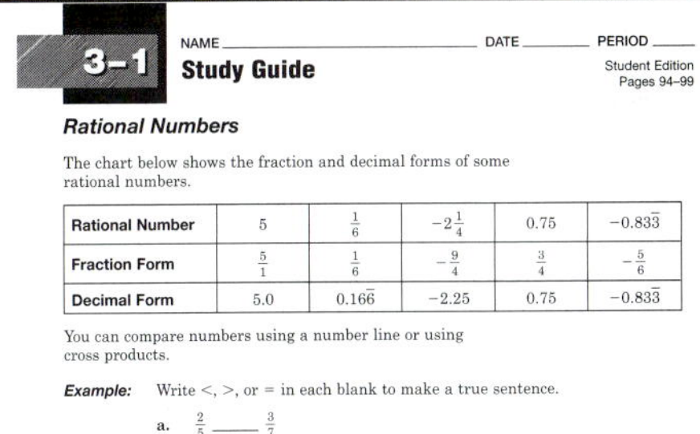

3–1 NAME ______ DATE ______ PERIOD ______

Study Guide

Student Edition Pages 94–99

Rational Numbers

The chart below shows the fraction and decimal forms of some rational numbers.

Rational Number	5	$\frac{1}{6}$	$-2\frac{1}{4}$	0.75	$-0.83\overline{3}$
Fraction Form	$\frac{5}{1}$	$\frac{1}{6}$	$-\frac{9}{4}$	$\frac{3}{4}$	$-\frac{5}{6}$
Decimal Form	5.0	$0.16\overline{6}$	−2.25	0.75	$-0.83\overline{3}$

You can compare numbers using a number line or using cross products.

Example: Write <, >, or = in each blank to make a true sentence.

a. $\frac{2}{5}$ ____ $\frac{3}{7}$

2(7) ____ 3(5) *Cross multiply.*

14 < 15

So $\frac{2}{5} < \frac{3}{7}$.

b. 0.4 ____ $\frac{1}{3}$

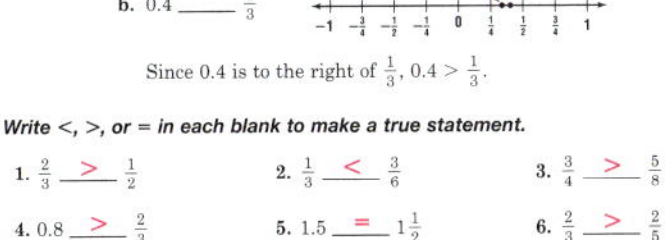

Since 0.4 is to the right of $\frac{1}{3}$, $0.4 > \frac{1}{3}$.

Write <, >, or = in each blank to make a true statement.

1. $\frac{2}{3}$ **>** $\frac{1}{2}$
2. $\frac{1}{3}$ **<** $\frac{3}{6}$
3. $\frac{3}{4}$ **>** $\frac{5}{8}$
4. 0.8 **>** $\frac{2}{3}$
5. 1.5 **=** $1\frac{1}{2}$
6. $\frac{2}{3}$ **>** $\frac{2}{5}$
7. $\frac{3}{5}$ **<** $\frac{5}{7}$
8. 0.63 **<** $\frac{2}{3}$
9. $0.\overline{3}$ **=** $\frac{1}{3}$
10. −4 **=** $-\frac{8}{2}$
11. 1.25 **=** $\frac{5}{4}$
12. 0.445 **>** $0.44\overline{4}$

Write the numbers in each set from least to greatest.

13. $\frac{1}{3}, \frac{1}{8}, \frac{1}{2}$ **$\frac{1}{8}, \frac{1}{3}, \frac{1}{2}$**
14. $\frac{3}{8}, 0.4, -\frac{2}{5}$ **$-\frac{2}{5}, \frac{3}{8}, 0.4$**
15. $\frac{3}{4}, \frac{1}{5}, 0.8$ **$\frac{1}{5}, \frac{3}{4}, 0.8$**
16. $-\frac{1}{2}, -\frac{2}{3}, -2$ **$-2, -\frac{2}{3}, -\frac{1}{2}$**

Assignment Guide

Basic: 17–43 odd, 44–52
Average: 18–40 even, 41–52

Answer

31.

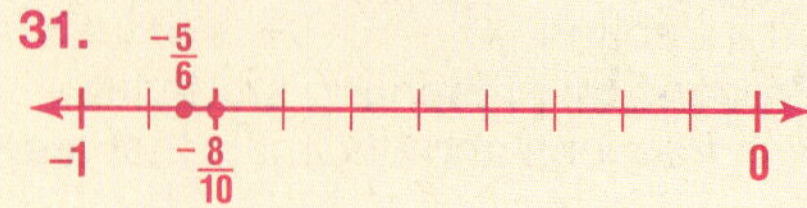

Since $-\frac{5}{6}$ is to the left of $-\frac{8}{10}$ on the number line, $-\frac{8}{10} > -\frac{5}{6}$.

***Skills Practice*, p. 96, and *Practice*, p. 97 (shown)**

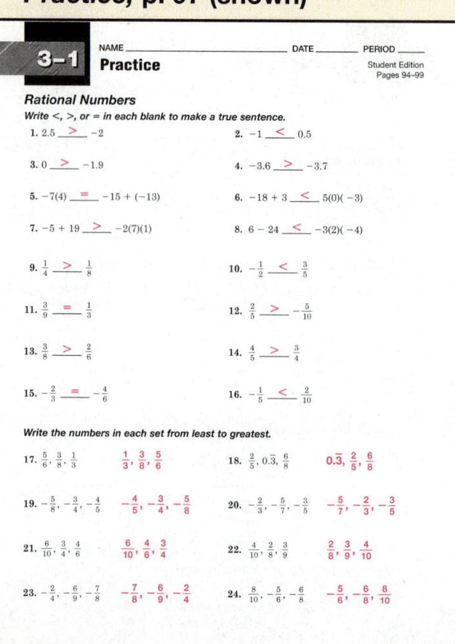

3–1 NAME ________ DATE ______ PERIOD _____

Practice — Student Edition Pages 94–99

Rational Numbers

Write <, >, or = in each blank to make a true sentence.

1. 2.5 __>__ -2
2. -1 __<__ 0.5
3. 0 __>__ -1.9
4. -3.6 __>__ -3.7
5. $-7(4)$ __=__ $-15 + (-13)$
6. $-18 + 3$ __<__ $5(0)(-3)$
7. $-5 + 19$ __>__ $-2(7)(1)$
8. $6 - 24$ __<__ $-3(2)(-4)$
9. $\frac{1}{4}$ __>__ $\frac{1}{8}$
10. $-\frac{1}{2}$ __<__ $\frac{3}{5}$
11. $\frac{3}{9}$ __=__ $\frac{1}{3}$
12. $\frac{2}{5}$ __>__ $-\frac{5}{10}$
13. $\frac{3}{8}$ __>__ $\frac{2}{6}$
14. $\frac{4}{5}$ __>__ $\frac{3}{4}$
15. $-\frac{2}{3}$ __=__ $-\frac{4}{6}$
16. $-\frac{1}{5}$ __<__ $\frac{2}{10}$

Write the numbers in each set from least to greatest.

17. $\frac{5}{6}, \frac{3}{8}, \frac{1}{3}$ **$\frac{1}{3}, \frac{3}{8}, \frac{5}{6}$**
18. $\frac{2}{5}, 0.\overline{3}, \frac{6}{8}$ **$0.\overline{3}, \frac{2}{5}, \frac{6}{8}$**
19. $-\frac{5}{8}, -\frac{3}{4}, -\frac{4}{5}$ **$-\frac{4}{5}, -\frac{3}{4}, -\frac{5}{8}$**
20. $-\frac{2}{3}, -\frac{5}{7}, -\frac{3}{5}$ **$-\frac{5}{7}, -\frac{2}{3}, -\frac{3}{5}$**
21. $\frac{6}{10}, \frac{3}{4}, \frac{4}{6}$ **$\frac{6}{10}, \frac{4}{6}, \frac{3}{4}$**
22. $\frac{4}{10}, \frac{2}{8}, \frac{3}{9}$ **$\frac{2}{8}, \frac{3}{9}, \frac{4}{10}$**
23. $-\frac{2}{4}, -\frac{6}{9}, -\frac{7}{8}$ **$-\frac{7}{8}, -\frac{6}{9}, -\frac{2}{4}$**
24. $\frac{8}{10}, -\frac{5}{6}, -\frac{6}{8}$ **$-\frac{5}{6}, -\frac{6}{8}, \frac{8}{10}$**

Examples 1–4 Replace each ● with <, >, or = to make a true sentence.

10. $0 ● -4.3$ **>**
11. $15 - 24 ● -3(2)(-6)$ **<**
12. $\frac{1}{2} ● \frac{1}{10}$ **>**
13. $\frac{2}{8} ● \frac{1}{4}$ **=**

Example 5 Write the numbers in each set from least to greatest.

14. $-\frac{2}{3}, -\frac{7}{8}, -\frac{3}{5}$ **$-\frac{7}{8}, -\frac{2}{3}, -\frac{3}{5}$**
15. $\frac{4}{5}, \frac{7}{10}, \frac{6}{8}$ **$\frac{7}{10}, \frac{6}{8}, \frac{4}{5}$**

Example 6

16. **Shopping** Thompson's Market sells a 25-pound bag of Happy Chow dog food for $15.49 and a 30-pound bag of Super Chow dog food for $17.39. Which brand is the better buy? Explain.
Super Chow; Happy Chow: $0.62/lb, Super Chow: $0.58/lb

Exercises

Practice

Replace each ● with <, >, or = to make a true sentence.

17. $-12 ● -2$ **<**
18. $8.2 ● -7$ **>**
19. $-0.88 ● -0.86$ **<**
20. $-3 ● 3(0)$ **<**
21. $-5 - 3 ● 9$ **<**
22. $-8 + 3 ● 7(2)(-3)$ **>**
23. $\frac{1}{8} ● 0.124$ **>**
24. $\frac{4}{5} ● 0.8$ **=**
25. $-\frac{3}{8} ● \frac{3}{4}$ **<**
26. $-\frac{1}{5} ● -\frac{2}{10}$ **=**
27. $\frac{4}{6} ● \frac{2}{5}$ **>**
28. $\frac{5}{8} ● \frac{5}{6}$ **<**

Homework Help	
For Exercises	See Examples
17–19, 29	1
20–22	2
23–28, 30	3, 4, 5
32–37, 42, 43	6
38–41	7

Extra Practice
See page 696.

29. Compare the numbers -2.002 and -2.02 using an inequality.

29. $-2.002 > -2.02$

30. Write an inequality that compares $\frac{5}{8}$ and $\frac{2}{3}$. **$\frac{2}{3} > \frac{5}{8}$**

31. Using a number line, explain why $-\frac{8}{10} > -\frac{5}{6}$. **See margin.**

Write the numbers in each set from least to greatest.

32. $0.2, -\frac{5}{6}, \frac{4}{5}$ **$-\frac{5}{6}, 0.2, \frac{4}{5}$**
33. $\frac{1}{2}, \frac{5}{8}, 0.6$ **$\frac{1}{2}, 0.6, \frac{5}{8}$**

34. $-\frac{3}{5}, -\frac{8}{10}, \frac{4}{6}$ **$-\frac{8}{10}, -\frac{3}{5}, \frac{4}{6}$**
35. $-\frac{1}{4}, -\frac{2}{6}, -\frac{1}{8}$ **$-\frac{2}{6}, -\frac{1}{4}, -\frac{1}{8}$**
36. $\frac{3}{10}, -\frac{2}{8}, -\frac{1}{3}$ **$-\frac{1}{3}, -\frac{2}{8}, \frac{3}{10}$**
37. $\frac{3}{5}, 0.\overline{6}, \frac{3}{8}$ **$\frac{3}{8}, \frac{3}{5}, 0.\overline{6}$**

38. \$0.05 < \$0.06
39. \$0.13 < \$0.14
40. \$0.40 < \$0.43
41. \$0.0004 and \$0.0003 are both > \$0.0002

C

Which is the better buy? Explain.

38. a 16-ounce bottle of juice for \$0.89 or a 20-ounce bottle for \$1.09
39. a dozen eggs for \$1.59 or 18 eggs for \$2.49
40. a 25-pack of computer diskettes for \$9.95 or a 30-pack for \$12.97
41. an $8\frac{1}{3}$-yd by 12-in. roll of aluminum foil for \$1.29, a $16\frac{2}{3}$-yd by 12-in. roll for \$2.89, or a 25-yd by 12-in. roll for \$2.99

Applications and Problem Solving

42. **Tools** Julie is using a socket wrench to tighten a bolt on her car. She needs to use the next size smaller than the $\frac{7}{16}$-inch socket. Which socket should she use, $\frac{1}{2}$-inch, $\frac{3}{8}$-inch, or $\frac{9}{16}$-inch? **$\frac{3}{8}$-inch**

43. **Life Science** Scientists have identified over 360,000 species of beetles. The table below lists the average lengths of six beetles.

Species	Firefly	Confused Flour Beetle	Elm Leaf Beetle	Mealworm	Lady Beetle	Japanese Beetle
Length (in.)	$\frac{3}{4}$	$\frac{1}{8}$	$\frac{1}{4}$	$\frac{5}{8}$	$\frac{1}{3}$	$\frac{1}{2}$

a. Firefly, Japanese Beetle, Mealworm

a. Which beetles are larger than the Lady Beetle?
b. Name the beetle whose length is half that of the Japanese Beetle. **Elm Leaf Beetle**

44. **Critical Thinking** Does $\frac{8}{6.4}$ name a rational number? Explain. **See margin.**

Mixed Review

Find each quotient. *(Lesson 2–6)*

45. $-42 \div 6$ **−7**
46. $-16 \div (-4)$ **4**
47. $-\frac{12}{3}$ **−4**

Evaluate each expression if $a = 4$, $b = -5$, and $c = 2$. *(Lesson 2–5)*

48. $-3b$ **15**
49. $6ac$ **48**
50. $a + bc$ **−6**

Standardized Test Practice

51. **Grid In** The deepest lake in the world is Lake Baikal, in Siberia. Its deepest point is 5315 feet deep. If its surface is 1493 feet above sea level, how many feet below sea level is the deepest point? *(Lesson 2–3)* **3822**

52. **Multiple Choice** Use the graph to name the coordinates of point F. *(Lesson 2–2)* **A**

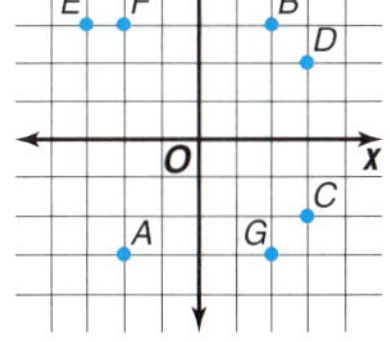

Ⓐ $(-2, 3)$ Ⓑ $(3, 2)$
Ⓒ $(-2, -3)$ Ⓓ $(2, 3)$

www.algconcepts.com/self_check_quiz

Extra Credit

Between any two rational numbers, there is always another rational number. Name a rational number that is between $\frac{9}{10}$ and $0.\overline{9}$.
Sample answer: 0.95

4 ASSESS

Open-Ended Assessment

Writing Ask students to write a paragraph explaining how to determine the greater of two rational numbers.

Answer

44. yes; $\frac{8}{6.4} = \frac{80}{64}$ or $\frac{5}{4}$

***Enrichment*, p. 99**

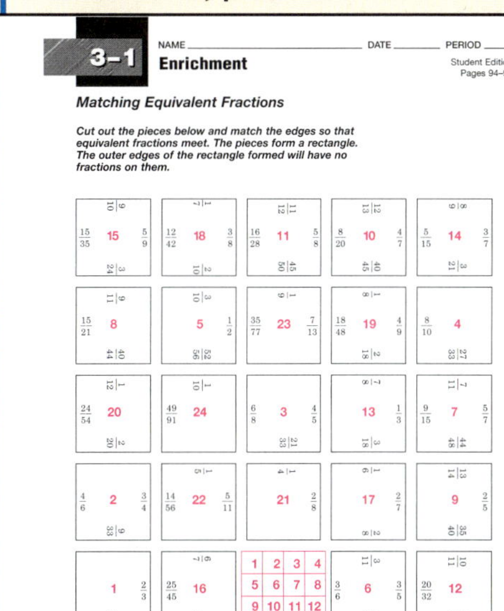
3-1 NAME ______ DATE ______ PERIOD ______
Enrichment Student Edition Pages 94–99

Matching Equivalent Fractions

Cut out the pieces below and match the edges so that equivalent fractions meet. The pieces form a rectangle. The outer edges of the rectangle formed will have no fractions on them.

1 FOCUS

5-Minute Check
Lesson 3–1

Replace each ● with <, >, or = to make a true sentence.

1. $-0.\overline{6}$ ● $-\frac{2}{3}$ =
2. $-6 + (3 + (-2))$ ● $(-3)(-5)$ <
3. $-\frac{7}{9}$ ● $-\frac{5}{7}$ <
4. Write $-\frac{4}{7}$, $-\frac{3}{8}$, and $-\frac{5}{9}$ in order from greatest to least. $-\frac{3}{8}, -\frac{5}{9}, -\frac{4}{7}$

Motivating the Lesson

Hands-On Activity Have students work with a partner. Give each pair a ruler and a container filled with uncooked rice. Have the students record the height of the rice to the nearest quarter of an inch. Then have them remove some of the rice and remeasure the height. Ask students how they could determine the change in height. Use any references to subtraction to introduce the lesson.

2 TEACH

In-Class Examples
Examples 1–2

Find each sum.

1 $0.41 + (-1.3)$ -0.89

2 $-5\frac{3}{10} + 2\frac{3}{5}$ $-2\frac{7}{10}$

TECHNOLOGY

An alternative technology option using a graphing calculator is available for teaching this lesson.

3–2 Adding and Subtracting Rational Numbers

What You'll Learn
You'll learn to add and subtract rational numbers.

Why It's Important
Agriculture Knowing how to add and subtract rational numbers can help you solve agriculture problems.
See Example 4.

To add rational numbers, you can use the same rules you used to add integers.

Examples

1 **Find each sum.**

$-4.8 + (-8.7)$

$-4.8 + (-8.7) = -(|-4.8| + |-8.7|)$ *The numbers have the same sign. Add their absolute values.*
$= -(4.8 + 8.7)$
$= -13.5$ *Simplify.*

2 $3\frac{2}{3} + \left(-4\frac{1}{6}\right)$

$3\frac{2}{3} + \left(-4\frac{1}{6}\right) = 3\frac{4}{6} + \left(-4\frac{1}{6}\right)$ *The LCD is 6. Replace $3\frac{2}{3}$ with $3\frac{4}{6}$.*
$= -\left(\left|-4\frac{1}{6}\right| - \left|3\frac{4}{6}\right|\right)$ *The signs differ. Subtract the lesser absolute value from the greater.*
$= -\left(4\frac{1}{6} - 3\frac{4}{6}\right)$ *The sign is negative. Why?*
$= -\left(3\frac{7}{6} - 3\frac{4}{6}\right)$ *Replace $4\frac{1}{6}$ with $3\frac{7}{6}$. Then subtract.*
$= -\frac{3}{6}$ or $-\frac{1}{2}$ *Simplify.*

Prerequisite Skills Review
Operations with Decimals, p. 684

Your Turn

a. $-0.76 + (-1.34)$ -2.1

b. $-1\frac{3}{4} + \left(-4\frac{1}{8}\right)$ $-5\frac{7}{8}$

When adding three or more rational numbers, you can use the Commutative and Associative Properties to rearrange the addends.

Example

3 **Find 14.8 + (−7.2) + 30.7.**

$14.8 + (-7.2) + 30.7$
$= (14.8 + 30.7) + (-7.2)$ *Comm. & Assoc. Properties (+)*
$= 45.5 + (-7.2)$ *Add.*
$= 38.3$ *Simplify.*

Your Turn **Find each sum.** c. -70.5

c. $28.3 + (-56.1) + 32.4 + (-75.1)$

d. $-\frac{4}{3} + \frac{5}{8} + \left(-\frac{7}{3}\right)$ $-\frac{73}{24}$ or $-3\frac{1}{24}$

Resource Manager

Reproducible Masters

Chapter 3 Resource Masters
- *Study Guide,* p. 100
- *Skills Practice,* p. 101
- *Practice,* p. 102
- *Reading to Learn Mathematics,* p. 103
- *Enrichment,* p. 104

Graphing Calculator, pp. 9–10
Hands-On Algebra, p. 43
School-to-Workplace, p. 3

Transparencies
5-Minute Check, 3–2

Technology/Multimedia
Interactive Chalkboard CD-ROM

Example 4 — Agriculture Link

Jason Wilson has kept track of the level of water in his farm's pond over the past five years. He compared the water level to the overall average. Find the net change in the water level of the pond.

Year	Level Above or Below Average (feet)
2002	2
2003	$1\frac{1}{3}$
2004	$-\frac{3}{4}$
2005	$-2\frac{1}{2}$
2006	$1\frac{1}{4}$

Prerequisite Skills Review
Adding and Subtracting Fractions, p. 687

To find the net change in the water level, add.

$$2 + 1\tfrac{1}{3} + \left(-\tfrac{3}{4}\right) + \left(-2\tfrac{1}{2}\right) + 1\tfrac{1}{4}$$

Group positive numbers together and negative numbers together.

$$2 + 1\tfrac{1}{3} + \left(-\tfrac{3}{4}\right) + \left(-2\tfrac{1}{2}\right) + 1\tfrac{1}{4}$$

$= 2 + 1\frac{1}{3} + 1\frac{1}{4} + \left(-\frac{3}{4}\right) + \left(-2\frac{1}{2}\right)$ *Commutative Property*

$= 2 + 1\frac{4}{12} + 1\frac{3}{12} + \left(-\frac{9}{12}\right) + \left(-2\frac{6}{12}\right)$ *The LCD is 12.*

$= \frac{24}{12} + \frac{16}{12} + \frac{15}{12} + \left(-\frac{9}{12}\right) + \left(-\frac{30}{12}\right)$ *Rewrite using the LCD.*

$= \frac{55}{12} + \left(-\frac{39}{12}\right)$ *Add positives and negatives.*

$= \frac{16}{12}$ *Add.*

$= \frac{16 \div 4}{12 \div 4}$ *Divide 16 and 12 by their GCF, 4. Then simplify.*

$= \frac{4}{3}$ or $1\frac{1}{3}$ *Simplify.*

You subtract rational numbers the same way that you subtract integers.

Example 5

Find −4.5 − 6.8.

$-4.5 - 6.8 = -4.5 + (-6.8)$ *To subtract 6.8, add −6.8.*

$= -11.3$ *The numbers have the same sign. Add their absolute values.*

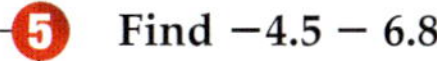

Find each difference.

e. $-72.5 - 81.3$ **−153.8**

f. $7\frac{3}{8} - \left(-4\frac{1}{3}\right)$ **$11\frac{17}{24}$**

www.algconcepts.com/extra_examples

In-Class Examples

Example 3

Find $-3.7 + 12.5 + (-1.3)$. **7.5**

Teaching Tip When discussing Example 4, students may state that the *net change* is the difference between the beginning value and the ending value. Point out that those values are not known.

In-Class Examples

Example 4

Suppose the water level in another pond was measured over a 4-year period. The level above or below average for this pond for each of the 4 years is given in the table. Find the net change in the water level of this pond. **$-1\frac{3}{4}$ ft**

Year	Level Above or Below Average (feet)
1997	$1\frac{1}{2}$
1998	$\frac{3}{4}$
1999	$-3\frac{3}{8}$
2000	$-\frac{5}{8}$

Example 5

Find $-4.7 - (-5.9)$. **1.2**

***Study Guide*, p. 100**

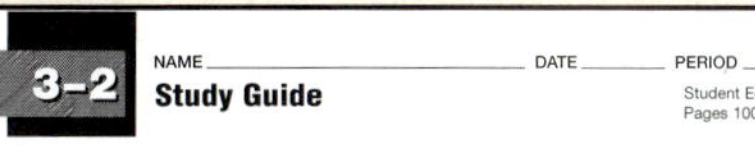

NAME ______ DATE ______ PERIOD ______

Study Guide

Student Edition Pages 100–103

Adding and Subtracting Rational Numbers

During track practice, Sheila recorded the time it took her to run two consecutive miles. She ran the first mile in 7.26 seconds. She ran the second mile in 7.01 seconds. The net change in times from the first mile to the second mile is 7.01 − 7.26 or −0.25 seconds. Sheila ran the second mile 0.25 seconds faster.

To add and subtract rational numbers, you use the same rules that you learned for adding and subtracting integers.

Adding Rational Numbers	To add rational numbers with the *same* sign, add their absolute values. Give the result the same sign as the rational numbers. To add rational numbers with *different* signs, subtract their absolute values. Give the result the same sign as the number with the greater absolute value.
Subtracting Rational Numbers	To subtract a rational number, add its opposite.

Examples: Find each sum or difference.

a. $-7.4 + (-10.3)$
$= -(|-7.4| + |-10.3|)$
$= -(7.4 + 10.3)$
$= -17.7$

b. $-5.2 - 9.1$
$= -5.2 + (-9.1)$
$= -14.3$

c. $-8.2 + 5.2 + (-9.1)$
$= [-8.2 + 5.2] + (-9.1)$
$= -3 + (-9.1)$
$= -|3 + 9.1|$
$= -12.1$

Find each sum or difference.

1. $3.1 + 1.2$ **4.3**
2. $-1.4 + 5.6$ **4.2**
3. $4.2 - 1.7$ **2.5**
4. $8.4 + 36.8$ **45.2**
5. $-6.3 + (-0.12)$ **−6.42**
6. $13.5 + (-10.2)$ **3.3**
7. $-4.3 - 16.8$ **−21.1**
8. $75.25 - 125.55$ **−50.3**
9. $18.12 - (-5.66)$ **23.78**
10. $-11.89 + 25.1$ **13.21**
11. $14.6 + 23.4 + (-3.6)$ **34.4**
12. $\frac{1}{2} + \frac{1}{3}$ **$\frac{5}{6}$**
13. $\frac{2}{3} - \frac{1}{4}$ **$\frac{5}{12}$**
14. $\frac{3}{8} + \frac{2}{7}$ **$\frac{37}{56}$**
15. $-\frac{3}{10} + \frac{3}{4}$ **$\frac{9}{20}$**

Evaluate each expression if $a \equiv \frac{1}{5}$, $b \equiv \mp 2\frac{1}{2}$, $c \equiv \mp 9.5$, and $d \equiv 15.6$.

16. $a + b$ **$-2\frac{3}{10}$**
17. $c - d$ **−25.1**
18. $d - c$ **25.1**

In-Class Example

Example 6

Evaluate $c - d$ if $c = 5\frac{1}{6}$ and $d = -3\frac{5}{8}$. **$8\frac{19}{24}$**

3 PRACTICE/APPLY

Error Analysis

Watch for students who use the wrong sign in their answers to Exercises 3–7.

Prevent by encouraging students to use estimation to check the reasonableness of their answers.

Assignment Guide

Basic: 11–33 odd, 35–42
Average: 12–32 even, 33–42

Answer

1. Add and subtract in the order in which they are given; or group the positive numbers together and the negative numbers together and then add.

Skills Practice, **p. 102, and *Practice,* p. 103 (shown)**

3–2 NAME ______ DATE ______ PERIOD ______

Practice

Student Edition Pages 100–103

Adding and Subtracting Rational Numbers

Find each sum or difference.

1. $6.2 + (-9.4)$ **-3.2**
2. $-7.9 + 8.5$ **0.6**
3. $-2.7 - 3.4$ **-6.1**
4. $5.6 - 7.1$ **-1.5**
5. $-8.3 + (-4.6)$ **-12.9**
6. $4.2 - 1.9$ **2.3**
7. $3.7 + (-5.8)$ **-2.1**
8. $-1.5 - 2.93$ **-4.43**
9. $6.8 + (-4.6) + 5.3$ **7.5**
10. $-4.7 - 8.2 + (-2.5)$ **-15.4**
11. $-\frac{1}{4} - \frac{3}{8}$ **$-\frac{5}{8}$**
12. $\frac{1}{3} + \left(-\frac{5}{9}\right)$ **$-\frac{2}{9}$**
13. $-3\frac{3}{8} + \left(-4\frac{1}{2}\right)$ **$-7\frac{7}{8}$**
14. $-2\frac{2}{3} + 2\frac{1}{2}$ **$-\frac{1}{6}$**
15. $-7\frac{3}{10} - 2\frac{2}{5}$ **$-9\frac{7}{10}$**
16. $5\frac{1}{3} + \left(-3\frac{1}{6}\right)$ **$2\frac{1}{6}$**
17. $2\frac{5}{6} - 6\frac{1}{2}$ **$-3\frac{2}{3}$**
18. $-6\frac{1}{5} + 4\frac{7}{10} + \left(-\frac{3}{5}\right)$ **$-2\frac{1}{10}$**
19. $3\frac{1}{2} + \left(-5\frac{5}{8}\right) + 3\frac{3}{4}$ **$1\frac{5}{8}$**
20. $2\frac{2}{3} - 9\frac{1}{2} - 8\frac{5}{6}$ **$-15\frac{2}{3}$**
21. Evaluate $m + 4\frac{1}{8}$ if $m = -1\frac{3}{4}$. **$2\frac{3}{8}$**
22. Find the value of k if $k = -7\frac{1}{3} - 1\frac{5}{6} + 4\frac{2}{3}$. **$-4\frac{1}{2}$**

Example 6 Evaluate $a - b$ if $a = -\frac{7}{8}$ and $b = -\frac{3}{5}$.

$$a - b = -\frac{7}{8} - \left(-\frac{3}{5}\right) \quad \textit{Replace } a \textit{ with } -\frac{7}{8} \textit{ and } b \textit{ with } -\frac{3}{5}.$$
$$= -\frac{7}{8} + \frac{3}{5} \quad \textit{To subtract } -\frac{3}{5}, \textit{ add } \frac{3}{5}.$$
$$= -\frac{35}{40} + \frac{24}{40} \quad \textit{The LCD is 40.}$$
$$= -\frac{11}{40} \quad \textit{Simplify.}$$

Your Turn

g. Evaluate $x - y$ if $x = 25.8$ and $y = -13.9$. **39.7**

h. If $m = -\frac{3}{4}$ and $n = -\frac{2}{5}$, what is the value of $m - n$? **$-\frac{7}{20}$**

Check for Understanding

Communicating Mathematics

1. **Explain** how the sum $-4.78 + 10.23 + (-7.04) + 0.92$ can be solved in more than one way. **See margin.**
2. **Describe** an example of adding rational numbers from a newspaper. **See students' work.**

Guided Practice

Examples 1–5 **Find each sum or difference.**

3. $4.5 + (-3.6)$ **0.9**
4. $-1.6 - 3.8$ **-5.4**
5. $-5.6 - 9.45$ **-15.05**
6. $-5\frac{7}{8} - 2\frac{3}{4}$ **$-8\frac{5}{8}$**
7. $4\frac{1}{2} + \left(-7\frac{3}{4}\right) + 3\frac{1}{4}$ **0**

Example 6 **Evaluate each expression if $a = -\frac{1}{2}$, $b = -0.85$, $c = 1.36$, and $d = -\frac{5}{6}$.**

8. $b - c$ **-2.21**
9. $a - d$ **$\frac{1}{3}$**

Example 6

10. **Stock Market** On Monday, stock in JAB Corporation rose $3\frac{1}{8}$ points. The next day, the stock dropped $1\frac{3}{4}$ points. What was the net change in the price of the stock? **$1\frac{3}{8}$ points**

Exercises

Practice

Find each sum or difference.

11. $-4.7 + (-8.6)$ **-13.3**
12. $89.3 + (-14.4)$ **74.9**
13. $-2.68 - 3.14$ **-5.82**
14. $-18.7 + 12.0 + (-9.2)$ **-15.9**
15. $-0.23 - 0.13 + (-0.9)$ **-1.26**
16. $-3.12 + 4.33 + (-1.89)$ **-0.68**
17. $-18.9 + (-3.15) - 7.43$ **-29.48**
18. $-0.8 + 3.5 + (-7.6) + 2.8$ **-2.1**
19. $-\frac{1}{2} + \frac{1}{8}$ **$-\frac{3}{8}$**
20. $-\frac{3}{5} - \left(-\frac{5}{8}\right)$ **$\frac{1}{40}$**
21. $3\frac{4}{5} - 9\frac{1}{10}$ **$-5\frac{3}{10}$**

Reteaching Activity

Visual/Spatial Learners Have students practice adding rational numbers using fraction bars to represent the numbers. Positive numbers should be represented using one color (or shading) and negative numbers using another. Point out that pairs of differently-colored, same-size bars cancel each other out and can be removed, leaving a model of the solution. For example, the model above can be used to compute $1\frac{1}{4} + \left(-2\frac{3}{4}\right)$.

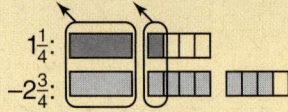

$1\frac{1}{4}$:
$-2\frac{3}{4}$:

22. $-\frac{5}{6}+\frac{5}{8}-\left(-\frac{1}{2}\right)$ $\frac{7}{24}$

23. $-\frac{2}{7}+\frac{2}{5}+\frac{3}{7}$ $\frac{19}{35}$

24. $\frac{1}{4}+2+\left(-\frac{3}{4}\right)$ $1\frac{1}{2}$

25. $-2\frac{3}{4}+5\frac{1}{2}-\left(-\frac{3}{8}\right)$ $3\frac{1}{8}$

Homework Help	
For Exercises	See Examples
11, 12	1
13	5
14–18	3
19–25	2
26–32	6
33, 35	3, 4

Extra Practice
See page 696.

Evaluate each expression if $a=-\frac{2}{5}$, $b=1\frac{3}{4}$, $c=14.6$, and $d=-5.9$.

26. $a+b$ $1\frac{7}{20}$

27. $a-b$ $-2\frac{3}{20}$

28. $c-d$ 20.5

29. $1-d$ 6.9

30. Find the value of $y-0.5$ if $y=-0.8$. −1.3

31. Evaluate $x-3\frac{1}{3}$ if $x=-2\frac{5}{6}$. $-6\frac{1}{6}$

32. Find the value of a if $a=\frac{3}{4}+\left(-\frac{4}{5}\right)-\frac{2}{5}$. $-\frac{9}{20}$

Applications and Problem Solving

33. **Banking** On Wednesday, Lakeesha wrote checks for $289.53, $312.41, and $76.89. On Thursday, she deposited $210.08 and $315.17 into her checking account. What is the net increase or decrease in Lakeesha's account? **−$153.58 or a decrease of $153.58**

34. **Sports** The game of *quoits* is played with metal rings. As in horseshoes, the object of the game is to throw the quoit onto or as close as possible to a vertical metal pole. The dimensions of the quoit are shown at the right. What is the diameter of the hole in the center of the quoit? $2\frac{3}{4}$ **in.**

$1\frac{5}{8}$ in. $1\frac{5}{8}$ in.
6 in.

35. $\frac{1}{4}$ foot above normal

35. **Critical Thinking** The net snow pack level in the Sierra Nevadas over the past five years has been $1\frac{1}{4}$ feet above normal. The following snow pack levels were recorded for the first four years: $1\frac{1}{2}$ feet above normal, $1\frac{7}{8}$ feet below normal, $\frac{3}{4}$ foot below normal, and $2\frac{1}{8}$ feet above normal. What was the snow pack level for the fifth year?

Mixed Review

36. **Shopping** Which is a better buy: a 184-gram can of peanuts for $1.09 or a 340-gram can for $1.99? Explain. *(Lesson 3–1)*
$0.00592 > $0.00585

Evaluate each expression if $g=-4$, $h=5$, and $j=-6$. *(Lesson 2–6)*

37. $-20 \div (h+g)$ −20

38. $\frac{j-9}{h}$ −3

39. Simplify $(4b)(-3c)$. *(Lesson 2–5)* $-12bc$

40. **Music** Monica has 16 CDs that are either country or jazz. She has 8 more jazz CDs than country CDs. How many of each does Monica have? *(Lesson 1–5)* **4 country, 12 jazz**

Standardized Test Practice

41. **Short Response** Simplify $5y+2(7+3y)$. *(Lesson 1–4)* $11y+14$

42. **Grid In** Find the value of $16 \div [2(13-11)]$. *(Lesson 1–2)* 4

4 ASSESS

Open-Ended Assessment

Modeling Ask students to measure two different common objects such as the length of their pencils or the width of a desk. Have them record the measurements to the nearest eighth of an inch. Then ask them to find the difference between the lengths of the two objects.

Extra Credit

The net change in the balance of Theresa's checking account for the week was +$67.39. During the week, she wrote checks for $45.75, $53.05, and $32.21, and made one deposit of $231.95. She also wrote one additional check, but she forgot to write the amount of this check in her check register. For how much was this check written?
$33.55

***Enrichment*, p. 104**

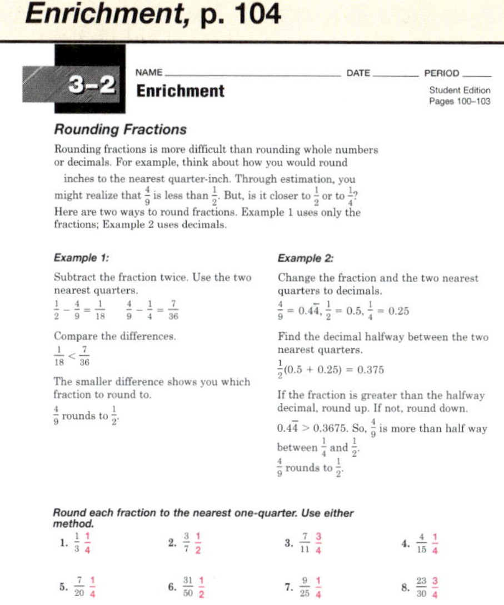

3–2 NAME ________ DATE ______ PERIOD _____

Enrichment

Student Edition
Pages 100–103

Rounding Fractions

Rounding fractions is more difficult than rounding whole numbers or decimals. For example, think about how you would round inches to the nearest quarter-inch. Through estimation, you might realize that $\frac{4}{9}$ is less than $\frac{1}{2}$. But, is it closer to $\frac{1}{2}$ or to $\frac{1}{4}$? Here are two ways to round fractions. Example 1 uses only the fractions; Example 2 uses decimals.

Example 1:

Subtract the fraction twice. Use the two nearest quarters.

$\frac{1}{2}-\frac{4}{9}=\frac{1}{18}$ $\frac{4}{9}-\frac{1}{4}=\frac{7}{36}$

Compare the differences.

$\frac{1}{18}<\frac{7}{36}$

The smaller difference shows you which fraction to round to.

$\frac{4}{9}$ rounds to $\frac{1}{2}$.

Example 2:

Change the fraction and the two nearest quarters to decimals.

$\frac{4}{9}=0.4\overline{4}$, $\frac{1}{2}=0.5$, $\frac{1}{4}=0.25$

Find the decimal halfway between the two nearest quarters.

$\frac{1}{2}(0.5+0.25)=0.375$

If the fraction is greater than the halfway decimal, round up. If not, round down.

$0.4\overline{4}>0.3675$. So, $\frac{4}{9}$ is more than half way between $\frac{1}{4}$ and $\frac{1}{2}$.
$\frac{4}{9}$ rounds to $\frac{1}{2}$.

Round each fraction to the nearest one-quarter. Use either method.

1. $\frac{1}{3}$ $\frac{1}{4}$
2. $\frac{3}{7}$ $\frac{1}{2}$
3. $\frac{7}{11}$ $\frac{3}{4}$
4. $\frac{4}{15}$ $\frac{1}{4}$
5. $\frac{7}{20}$ $\frac{1}{4}$
6. $\frac{31}{50}$ $\frac{1}{2}$
7. $\frac{9}{25}$ $\frac{1}{4}$
8. $\frac{23}{30}$ $\frac{3}{4}$

Round each decimal or fraction to the nearest one-eighth.

9. 0.6 $\frac{5}{8}$
10. 0.1 $\frac{1}{8}$
11. 0.45 $\frac{1}{2}$
12. 0.85 $\frac{7}{8}$
13. $\frac{5}{7}$ $\frac{3}{4}$
14. $\frac{3}{20}$ $\frac{1}{8}$
15. $\frac{23}{25}$ $\frac{7}{8}$
16. $\frac{5}{9}$ $\frac{1}{2}$

3-3 Mean, Median, Mode, and Range

1 FOCUS

5-Minute Check
Lesson 3–2

Find each sum or difference.

1. $-\frac{5}{6} - \left(-\frac{2}{5}\right)$ $-\frac{13}{30}$
2. $-11.2 + 5.4 + (-32.7)$ -38.5
3. $-2\frac{4}{5} - 3\frac{7}{10}$ $-6\frac{1}{2}$
4. A picture is to be put in a frame that measures $15\frac{3}{16}$ inches across. If the frame is $1\frac{7}{8}$ inches wide, what is the width of the artwork that will be visible? $11\frac{7}{16}$ **in.**

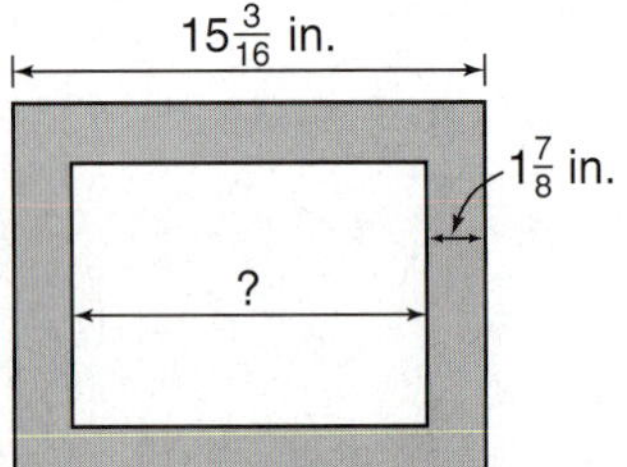

Motivating the Lesson

Real-World Connection Write five headings on the board such as "Number of Siblings," "Number of First Cousins," "Number of Pets," and "Age of One Parent." As students come into class, ask them to record the information about themselves under each heading. Divide the class into 5 groups and have each group find the mean of one of the sets of data. Save the data for use later when discussing median, mode, and range.

What You'll Learn You'll learn to find the mean, median, mode, and range of a set of data.

Why It's Important *Meteorology* A meteorologist can compare climates by comparing the mean, median, mode, and range of a set of temperature data. *See Example 5.*

A set of data can contain many numbers. To help understand the data, you can let one number describe the set of data. This number is called a **measure of central tendency** because it represents the center, or middle, of the data. The most commonly used measures of central tendency are the **mean**, **median**, and **mode**.

Mean The mean, or *average*, of a set of data is the sum of the data divided by the number of pieces of data.

The table below shows how much snack food Americans consume.

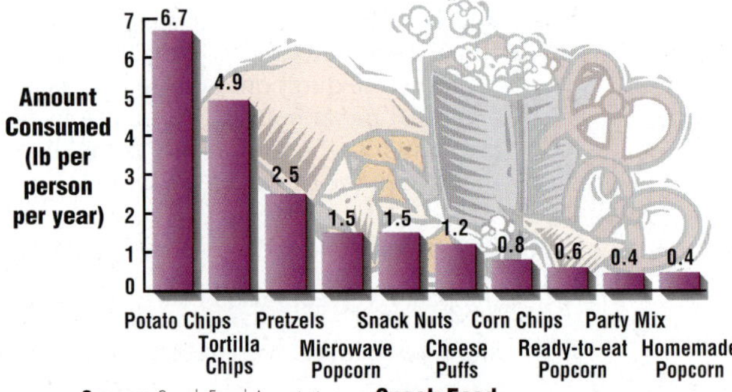

Example 1 Food Link

Find the mean of the snack food data.

First, find the sum of the amounts consumed. Then divide by the number of items of data. In this case there are 10 items of data.

$$\text{mean} = \frac{6.7 + 4.9 + 2.5 + 1.5 + 1.5 + 1.2 + 0.8 + 0.6 + 0.4 + 0.4}{10}$$

$$= \frac{20.5}{10} \text{ or } 2.05$$ The mean of the data is 2.05 pounds.

Prerequisite Skills Review
Operations with Decimals, p. 684

Your Turn **Find the mean of each set of data.**

a. 19, 21, 18, 17, 18, 22, 46 **23**

b. **92.5**

Stem	Leaf
7	3 5 6
8	2 2 4
9	0 4 7 9
10	5 8
11	4 6

$9|4 = 94$

MODELING

An alternative hands-on option using centimeter grid paper, scissors, and tape is available for teaching this lesson.

Resource Manager

Reproducible Masters

Chapter 3 Resource Masters
- *Study Guide,* p. 105
- *Skills Practice,* p. 106
- *Practice,* p. 107
- *Reading to Learn Mathematics,* p. 108
- *Enrichment,* p. 109
- *Assessment,* p. 141

Hands-On Algebra, p. 44
Graphing Calculator, p. 8

Transparencies
5-Minute Check, 3-3

Technology/Multimedia
Interactive Chalkboard CD-ROM

Notice that the amount of potato chips consumed is far greater than the amounts of other snacks consumed. Because the mean is an average of several numbers, a single number that is so much greater or less than the others can affect the mean a great deal. In such cases, the mean becomes less representative of the values in a set of data.

The median is another measure of central tendency.

Median The median of a set of data is the middle number when the data in the set are arranged in numerical order.

Food Link

Find the median of the snack food data.

Arrange the numbers in order from least to greatest.

0.4 0.4 0.6 0.8 1.2 1.5 1.5 2.5 4.9 6.7

median

Since there is an even number of data items, the median is the mean of the two middle values, 1.5 and 1.2. *If there were an odd number of data items, the middle one would be the median.*

$$\text{median} = \frac{1.5 + 1.2}{2} = \frac{2.7}{2} \text{ or } 1.35$$

The median is 1.35 pounds. *The number of values that are greater than the median is the same as the number of values that are less than the median.*

Your Turn **Find the median of each set of data.**

c. 4, 6, 12, 5, 8 **6**

d. 10, 3, 17, 1, 8, 6, 12, 15 **9**

Graphing Calculator Tutorial
See pp. 750–753.

Graphing Calculator Exploration

You can use a graphing calculator to find the mean and median of the snack food data. First press STAT ENTER. Under L1, enter in the numerical data. Once all of the data are entered, press STAT ▶ ENTER ENTER. Statistics will appear on the screen. The first statistic, $\bar{x}$ the mean. Scroll down, and you will find the median given by the label "Med."

Try These

1. Use a graphing calculator to find the mean and median of the snack food data. Did you get the same answers? **yes**
2. An *outlier* is an item that is much greater or much less than the other data. In the snack food data, potato chips, at 6.7 pounds per person per year, is an outlier. Find the mean and median of the data without potato chips. Is there a significant difference in your answers? Explain.

2. Yes; while the median, 1.2, is fairly close to the median of the entire set of data, the mean, 1.53, is much less than that of the entire set. That is because the mean is affected much more by an outlier.

www.algconcepts.com/extra_examples

2 TEACH

Teaching Tip When discussing the definition of *mean*, point out that although the word *average* is most commonly used to indicate the mean, the median and mode are also considered averages.

In-Class Examples

Examples 1–2

The stem-and-leaf plot below shows the number of children enrolled in each of 9 gymnastics classes offered at a local recreation center.

Stem	Leaf
0	4 5 8
1	0 2 5 5 8
2	1 *2|1 = 21*

1 Find the mean of the gymnastics data. **12 students**

2 Find the median of the gymnastics data. **12 students**

Teaching Tip After discussing In-Class Example 2, draw students' attention to the different methods for finding a median if the data set has an even or odd number of items.

Graphing Calculator Exploration

Once students have used the calculator to display the snack food statistics, you may wish to explain the meanings of the other statistics that appear on the calculator screen. The Σx statistic is the sum of all the data items, and n is the number of data items. The greatest and least data values are given by maxX and minX, and these values can be used to calculate the range of the set of data.

In-Class Example

Example 3

Refer to the gymnastics information given in In-Class Examples 1–2 on the previous page. Find the mode of the gymnastics data. **15 students**

Teaching Tip If students have studied functions in a previous mathematics course, they may confuse the statistical range with the range of a function. Stress that the statistical range is a single value.

In-Class Examples

Example 4

Refer to the gymnastics information given in In-Class Examples 1–2 on the previous page. Find the range of the gymnastics data. **17 students**

Example 5

The table below shows the test results of two different classes on the same test. How do the results for Class A compare to the results for Class B?

	Class A	Class B
Mean	79.5	70.5
Median	75	74
Mode	76	77
Range	29	50

The medians of the classes are similar with about half of each class earning a 75 or better. The mean of class B is significantly lower than the mean of Class A, and the range of scores in Class B is much greater. There may have been one or two low scores that reduced the mean score for Class B.

A third measure of central tendency is the mode.

Mode The mode of a set of data is the number that occurs most often in the set.

Sometimes a set of data has only one mode. In other cases, every item in a data set occurs the same number of times. When this happens, the set has no mode.

Example 3 **Food Link**

Find the mode of the snack food data.

Look for the number that occurs most often.

0.4 0.4 0.6 0.8 1.2 1.5 1.5 2.5 4.9 6.7

In this set, 0.4 and 1.5 each appear twice. So, the set of data has two modes, 0.4 pound and 1.5 pounds.

Your Turn **Find the mode of each set of data.**

e. 7, 19, 9, 4, 7, 2 **7**

f. 300, 34, 40, 50, 60 **none**

The snack food data has a mean of 2.05 pounds, a median of 1.35 pounds, and two modes of 0.4 pound and 1.5 pounds. As you can see, the mean, median, and mode may not be the same value.

Measures of central tendency may not give an accurate description of a set of data. Often, **measures of variation** are used to describe the *distribution* of the data. A common measure of variation is the **range**.

Range The range of a set of data is the difference between the greatest and the least values of the set.

To find the range of the snack food data, subtract the least value of the data set from the greatest value.

The greatest value is 6.7.
The least value is 0.4.
So, the range of the data is 6.7 − 0.4 or 6.3.

Example 4 **Find the range of the data set {19, 21, 18, 17, 18, 22, 46}.**

The greatest value is 46. The least value is 17.
So, the range is 46 − 17 or 29.

Your Turn **Find the range of each set of data.**

g. 4, 6, 12, 5, 8 **8**

h.

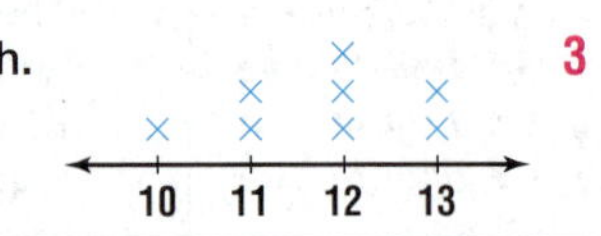

3

From the Classroom of ...

Nicki Hudson
West Linn High School
West Linn, Oregon

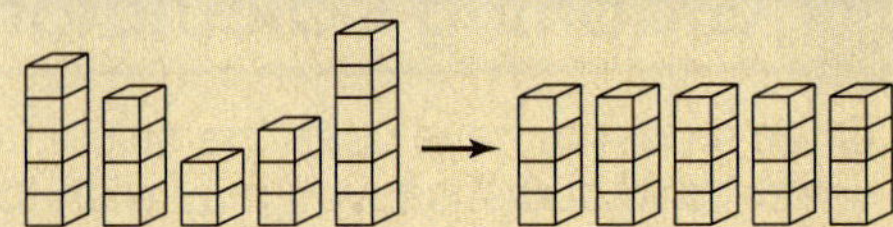

I use centimeter cubes to help my students understand the concept of *mean*, or average. For example, have students model the set of data 5, 4, 2, 3, and 6 by stacking cubes to represent each number. Then tell the students to *level the stacks* to determine the mean. The result is 5 columns of 4 cubes, showing that the mean of the set of data is 4.

Example 5

Meteorology Link

Suppose the temperature in a building is always 10°F cooler than the temperature outside. The average outside temperature for two weeks is given below. Find the mean, median, mode, and range for both the inside and outside temperatures. Compare these values.

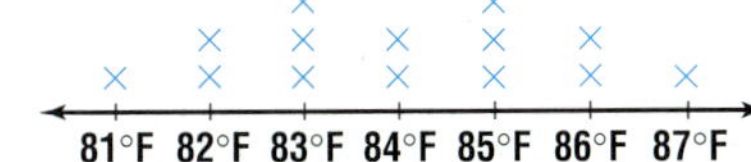

Temperature	Mean	Median	Mode	Range
Outside	84°F	84°F	83°F, 85°F	6°F
Inside	74°F	74°F	73°F, 75°F	6°F

Note that the mean, mode, and median of the outside and inside temperatures differ by 10°F, while the range is the same.

Data can be classified in several ways. Suppose a person tracks the number of men and women who walk into a theater. These data are **univariate** because there is only one variable being measured, gender. The data are also **categorical** because they fit into one of two categories, man or woman. If the person also asks each theatergoer's age, the data would be **bivariate** because there are two quantities being measured, gender and age. The age data are classified as **measurement** because there are several values for age rather than just a few categories.

Check for Understanding

Communicating Mathematics

1–3. See margin.

1. **Illustrate** how the mean of a set of data is affected by an extremely high or low value.
2. **Describe** the steps you would take to find the mean, median, mode, and range of the data below.

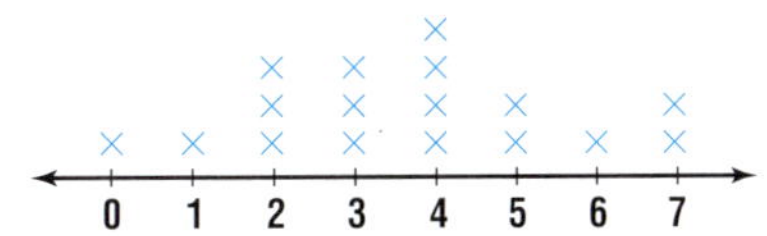

Vocabulary

measure of central tendency
mean, median, mode
measure of variation
range
univariate
catagorical
bivariate

3. **You Decide** Eric says his mean test score is a good measure of how he is doing in class. Sonia disagrees. Who is correct? Explain.

Guided Practice

Getting Ready **Find each quotient without using a calculator.**

Sample 1: 156 ÷ 8

Solution:

```
     19.5
  8)156
   −8
    76
   −72
    40
   −40
     0
```

Sample 2: 98.7 ÷ 3

Solution:

```
     32.9
  3)98.7
   −9
     8
    −6
     2 7
    −2 7
       0
```

(continued on the next page)

3 PRACTICE/APPLY

Error Analysis

Watch for students who have trouble calculating measures of central tendency when the data is given in a graph such as the line plot in Exercise 2.

Prevent by having students write out each value represented in the graph.

Answers

1. When a set of data has one or more extremely high or low values, the mean does not accurately describe the values in the set of data. Consider the set of data 3, 10, 5, 16, 148. The mean is 36.4.
2. mean: Add the values of the ×'s and divide by the number of ×'s.
   median: Arrange the values of the ×'s from least to greatest. Find the middle value.
   mode: Find the value that has the most ×'s above it.
   range: Subtract the lowest value for an × from the highest value for an ×.
3. Sonia; a few really high or really low values could influence the mean so that it is not representative of Eric's scores.

***Study Guide,* p. 105**

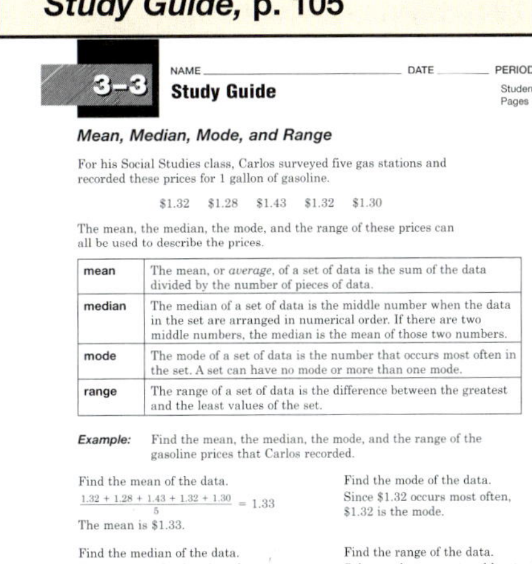
3–3 NAME ______ DATE ______ PERIOD ______

Study Guide

Student Edition Pages 104–109

Mean, Median, Mode, and Range

For his Social Studies class, Carlos surveyed five gas stations and recorded these prices for 1 gallon of gasoline.

\$1.32 \$1.28 \$1.43 \$1.32 \$1.30

The mean, the median, the mode, and the range of these prices can all be used to describe the prices.

mean	The mean, or *average*, of a set of data is the sum of the data divided by the number of pieces of data.
median	The median of a set of data is the middle number when the data in the set are arranged in numerical order. If there are two middle numbers, the median is the mean of those two numbers.
mode	The mode of a set of data is the number that occurs most often in the set. A set can have no mode or more than one mode.
range	The range of a set of data is the difference between the greatest and the least values of the set.

Example: Find the mean, the median, the mode, and the range of the gasoline prices that Carlos recorded.

Find the mean of the data.

$\frac{1.32 + 1.28 + 1.43 + 1.32 + 1.30}{5} = 1.33$

The mean is \$1.33.

Find the median of the data.
First arrange the data in order.
Then identify the middle number.
1.28, 1.30, **1.32**, 1.32, 1.43
The median is \$1.32.

Find the mode of the data.
Since \$1.32 occurs most often, \$1.32 is the mode.

Find the range of the data.
Subtract the greatest and least values.
\$1.43 − \$1.28 = \$0.15
The range of the data is \$0.15.

Find the mean, median, mode, and range of each set of data.

1. 48, 25, 29, 42, 36, 36 mean: 36, median: 36, mode: 36, range: 23
2. 5.1, 2.7, 2.7, 2.7 mean: 3.3, median: 2.7, mode: 2.7, range: 2.4
3. 101, 113, 98 mean: 104, median: 101, mode: none, range: 15
4. 18.2, 20.4, 18.2, 11.6, 20.4
   mean: 17.76, median: 18.2, modes: 18.2 and 20.4, range: 8.8

Reteaching Activity

Auditory/Musical Learners Write two data sets on the board or overhead, one with an odd number of values and the other with an even number of values. Group students in pairs and have them take turns instructing each other on how to find the mean, median, mode, and range of the data. One student should verbalize the steps while their partner performs them.

Assignment Guide

Basic: 17–35 odd, 36–42
Average: 16–30 even, 32–42

4. 87 ÷ 6 **14.5**
5. 126 ÷ 3 **42**
6. 2.04 ÷ 8 **0.255**
7. 0.453 ÷ 3 **0.151**
8. 35.6 ÷ 2 **17.8**
9. 189 ÷ 4 **47.25**

Examples 1–4

10. 46; 45; none; 42
11. 15.5; 12.45; 12.5; 21.1
13. 62.3; 60; 72; 17
14. 75.5; 75; 75; 20

Find the mean, median, mode, and range of each set of data.

10. 26, 30, 45, 61, 68
11. 12.5, 11.2, 12.4, 12.1, 12.5, 32.3
12. 3.8, 6, 4.5, 9, 6.2, 4.9, 7, 7.6, 3.8, 4, 7 **5.8; 6; 3.8, 7; 5.2**
13.

Stem	Leaf
5	5 6 7 8
6	0 0 1
7	2 2 2 $7\|2 = 72$

14.

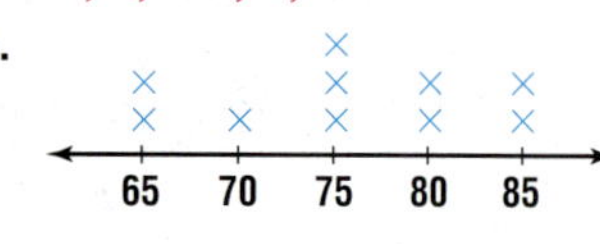

Example 5

15. **Sales** The district manager of a cellular phone company wants to promote an employee to sales manager. The quarterly sales for the two employees for the last two years are listed below.

Name	Quarterly Sales (thousands of dollars)							
Ms. Diaz	200.8	190.9	200.0	210.0	200.1	200.5	200.7	210.0
Mr. Cruz	210.0	180.1	191.2	239.2	210.8	190.8	180.9	210.0

15a. Ms. Diaz: 201.625; 200.6; 210.0; 19.1; Mr. Cruz: 201.625; 200.6; 210.0; 59.1

a. Find the mean, median, mode, and range of the quarterly sales data for each employee.
b. Based on sales, who should be promoted to sales manager? Explain. **Ms. Diaz; her sales are more consistent.**

Exercises

Practice

Homework Help

For Exercises	See Examples
16–27	1–4
28	2
29	3
30–34	1
35, 36	1-3

Extra Practice
See page 697.

Find the mean, median, mode, and range of each set of data.

16. 7, 9, 7, 9, 8, 9, 7 **8; 8; 7, 9; 2**
17. 10, 3, 14, 1, 8, 7, 20 **9; 8; none;**
18. 1.5, 1.2, 1.1, 1.3, 1.6, 1.1
19. 0.3, 0.3, 0.3, 0.1 **0.25; 0.3; 0.3; 0.2**
20. 45, 32, 17, 65, 80, 55
21. 200, 24, 20, 40, 20
22. 95, 76, 88, 82, 73, 65, 76, 76, 84, 90 **80.5, 79, 76, 30**
23. 10.2, 8.8, 10.0, 9.4, 12.6, 10.2, 9.8, 12.0 **10.375, 10.1, 10.2, 3.8**

24.

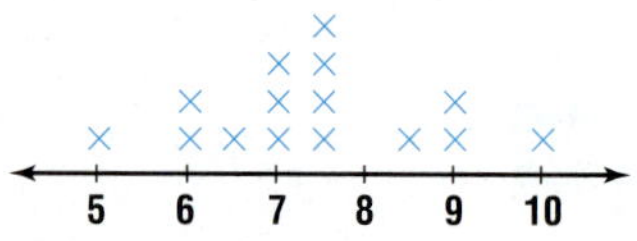

25. **10.8$\overline{3}$; 10.5; 10; 2.5**

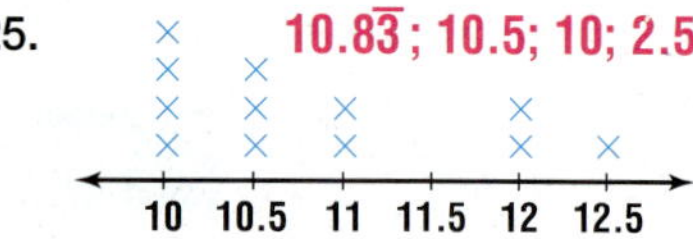

26.

Stem	Leaf
5	6 6 7
6	0 1 1 8
7	0 2 4 4 4
8	0 9 $7\|2 = 72$

27.

Stem	Leaf
18	3 3 4
19	0 0 2 5
20	0 8
21	0 0 $20\|8 = 20.8$

18. 1.3; 1.25; 1.1; 0.5
20. 49; 50; none; 63
21. 60.8; 24; 20; 180
24. 7.4; 7.5; 7.5; 5
26. 68; 69; 74; 33
27. 19.5; 19.2; 18.3, 19.0, 21.0; 2.7

28. Find the median of \$12.99, \$5.89, \$6.75, \$2.45, \$9.25, and \$5.88. **\$6.32**
29. Give an example of a set of data that has no mode. **Sample answer: 3.1, 1.2, 1.0, 1.5**

***Skills Practice,* p. 106, and *Practice,* p. 107 (shown)**

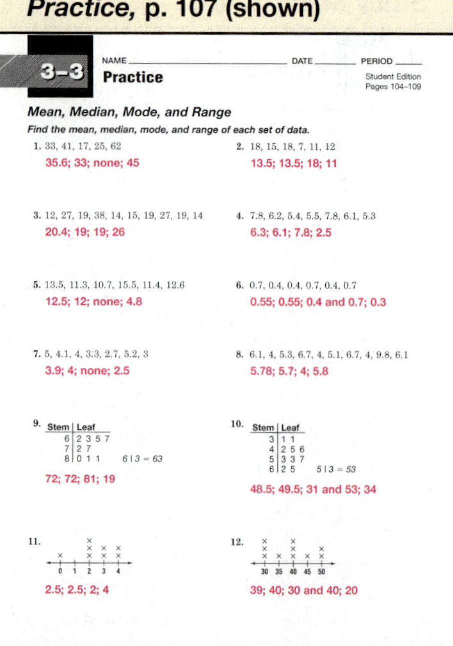

3-3 NAME ______ DATE ______ PERIOD ______
Practice
Student Edition Pages 104–109

Mean, Median, Mode, and Range

Find the mean, median, mode, and range of each set of data.

1. 33, 41, 17, 25, 62 **35.6; 33; none; 45**
2. 18, 15, 18, 7, 11, 12 **13.5; 13.5; 18; 11**
3. 12, 27, 19, 38, 14, 15, 19, 27, 19, 14 **20.4; 19; 19; 26**
4. 7.8, 6.2, 5.4, 5.5, 7.8, 6.1, 5.3 **6.3; 6.1; 7.8; 2.5**
5. 13.5, 11.3, 10.7, 15.5, 11.4, 12.6 **12.5; 12; none; 4.8**
6. 0.7, 0.4, 0.4, 0.7, 0.4, 0.7 **0.55; 0.55; 0.4 and 0.7; 0.3**
7. 5, 4.1, 4, 3.3, 2.7, 5.2, 3 **3.9; 4; none; 2.5**
8. 6.1, 4, 5.3, 6.7, 4, 5.1, 6.7, 4, 9.8, 6.1 **5.78; 5.7; 4; 5.8**
9.

Stem	Leaf
6	2 3 5 7
7	2 7
8	0 1 1 $6\|3 = 63$

72; 72; 81; 19

10.

Stem	Leaf
3	1 1
4	2 5 6
5	3 3 7
6	2 5 $5\|3 = 53$

48.5; 49.5; 31 and 53; 34

11. (line plot: 0, 1, 2, 3, 4) **2.5; 2.5; 2; 4**
12. (line plot: 30, 35, 40, 45, 50) **39; 40; 30 and 40; 20**

Family Activity

With the help of a family member, have students find the amount their family spent each month on telephone bills during the last six months. Have them find the mean, median, mode, and range of the data, and then write a paragraph explaining what these values indicate about the family's telephone usage.

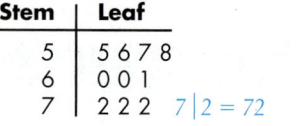

Write a set of data with six numbers that satisfies each set of conditions. **30. Sample answer: 1, 1, 1, 1, 1, 7**

30. The mean is greater than all but one of the numbers.
31. The mean is smaller than all but one of the numbers. **Sample answer: 1, 7, 7, 7, 7, 7**

Applications and Problem Solving

32. **Sports** The batting averages for 10 players on a baseball team are 0.234, 0.253, 0.312, 0.333, 0.281, 0.240, 0.183, 0.222, 0.297, and 0.275. Find the mean batting average for these players. **0.263**
33. **School** Jack's last four test scores in French were 88, 77, 81, and 83. What must he score on the next test so that his average is exactly 85?

33. 96

34. **Zoos** The table shows the number of acres for the five largest zoos in the United States.

Zoo	San Diego Wild Animal Park	Minnesota Zoo	Miami Metrozoo	Bronx Zoo	Albuquerque Biological Park
Acres	2200	500	300	265	240

Source: *World Almanac*

a. What was the mean size of the zoos? **701 acres**

b. Does the mean size accurately describe the values in the set of data? Explain.

34b. No; the mean is higher than all data except one value.

35. **Business** The salaries of the 12 employees at the Hanson Company are \$28,500, \$32,000, \$29,500, \$31,200, \$28,600, \$38,500, \$20,100, \$85,000, \$36,000, \$25,350, \$26,500, and \$19,850.

a. Find the mean, median, and mode of the data.

35a. \$33,425; \$29,050; none

b. Suppose the president is interviewing an applicant. Should the president quote the mean, median, or mode as the "average" salary? Explain. **mean; highest value**

c. If the employees are asking for a pay raise, should they quote the mean, median, or mode as their "average" salary? Explain.

35c. median; lowest value

36. **Critical Thinking** List six numbers whose mean is 50, median is 40, and mode is 20. **Sample answer: 20, 20, 20, 60, 80, 100**

Mixed Review

Find each sum or difference. *(Lesson 3–2)*

37. $-2.5 - 7.4$ **-9.9** 38. $6\frac{1}{3} + \left(-3\frac{3}{4}\right)$ **$2\frac{7}{12}$** 39. $\frac{3}{4} + \left(-\frac{5}{4}\right) - \frac{1}{2}$ **-1**

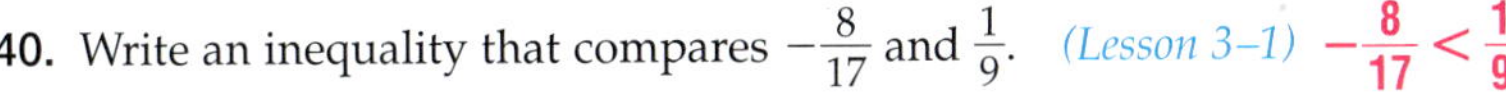

40. Write an inequality that compares $-\frac{8}{17}$ and $\frac{1}{9}$. *(Lesson 3–1)* **$-\frac{8}{17} < \frac{1}{9}$**

Standardized Test Practice

41. **Short Response** Graph the set of points $\{-3, 4, 2, -2, 0, -1\}$ on a number line. *(Lesson 2–1)* **See margin.**
42. **Grid In** You have 40 pieces of fencing that are each 1 meter long. What is the largest rectangular area in square meters you can enclose with these pieces of fencing? (*Hint*: The formula for perimeter is $P = 2(\ell + w)$, and the formula for area is $A = \ell w$, where ℓ is the length and w is the width.) *(Lesson 1–5)* **100**

www.algconcepts.com/self_check_quiz

4 ASSESS

Open-Ended Assessment

Writing Ask students to write a paragraph explaining the difference between the mean, median, mode, and range of a set of data.

Chapter 3, Quiz A (Lessons 3–1 through 3–3) is available in the *Chapter 3 Resource Masters*, p. 141.

Answer

41.

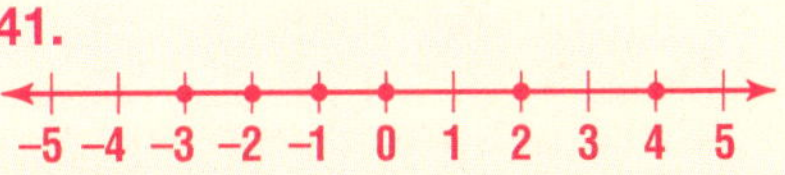

From the Classroom of ...

Nicki Hudson
West Linn High School
West Linn, Oregon

I have my students complete a project. They are instructed to gather a set of data such as *the number of hours spent listening to music* or *the number of pets in the household* from every student in the class. Then I ask them to find the mean, median, mode, and range of the data and display the data in a histogram or bar graph.

Extra Credit

A data set is made up of three integers and has a range of 10.

a. What are the three integers if the mean and median of the data are both 0? **–5, 0, 5**

b. What are the three integers if the mean is 1? **Sample answer: –5, 3, 5**

***Enrichment*, p. 108**

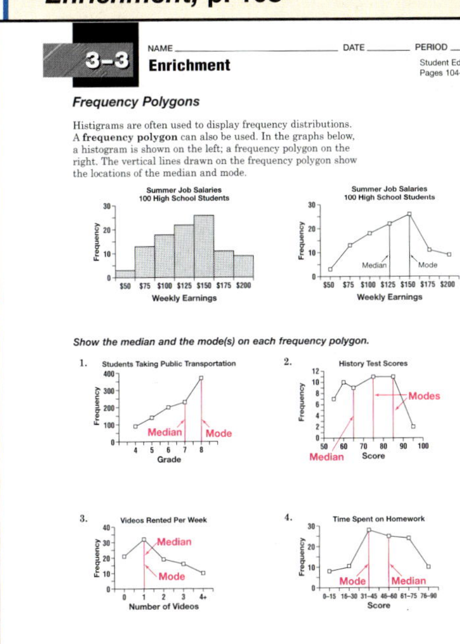

3–3 NAME ______ DATE ______ PERIOD ______

Enrichment Student Edition Pages 104–109

Frequency Polygons

Histigrams are often used to display frequency distributions. A **frequency polygon** can also be used. In the graphs below, a histogram is shown on the left; a frequency polygon on the right. The vertical lines drawn on the frequency polygon show the locations of the median and mode.

Show the median and the mode(s) on each frequency polygon.

PREPARE

This optional investigation is designed to be completed by a group of 3 or 4 students over 1–2 days.

Objective

Explore arithmetic sequences and investigate how to represent them with a recursive formula.

Mathematical Overview

The investigation utilizes the following concepts:

- looking for patterns,
- identifying the difference in an arithmetic sequence, and
- writing a recursive formula for an arithmetic sequence.

Suggested Time Management	
Investigation	30–45 min
Extension: Gathering Data	30–45 min
Extension: Summarizing Data	20–30 min

Motivating the Lesson

Demonstrate how to make the first four figures in Step 1 using cubes. Explain that the figures form a pattern which will be utilized throughout the Investigation. Point out that the numbers of cubes in the figures form an ordered set of numbers called a *sequence*.

Chapter 3 Investigation

Materials

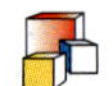
sugar cubes or wooden cubes

Arithmetic Sequences

A **sequence** is a set of numbers in a specific order.

Investigate

1. Use cubes to form the four figures shown. Examine how Figure 2 is different from Figure 1, Figure 3 from Figure 2, and so on.

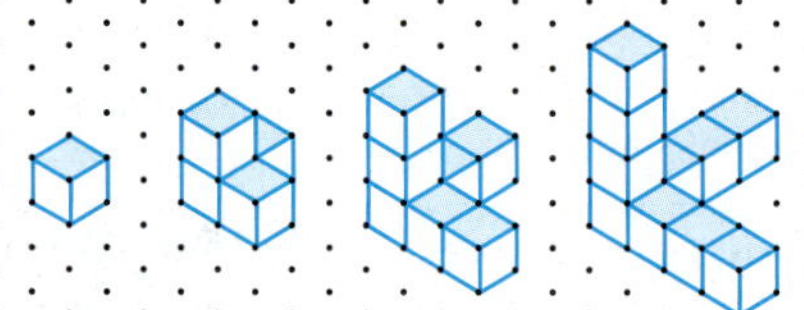

1a–b. See Solutions Manual.

a. Make a table like the one below. Write the numbers 1–10 in columns 1 and 2. Record the number of cubes in each of the four figures. Build a fifth figure and record the number of cubes used. Look for a pattern and complete column 3.

Figure Number	Term Number	Number of Cubes	New Number of Cubes– Previous Number of Cubes
1	1	1	—
2	2	4	4 − 1 = ?

b. Find the difference between the new number of cubes and the number of cubes from the previous row to complete column 4.

c. List the numbers in column 3 from least to greatest, with commas separating the numbers. This list is a sequence. Each number in the sequence is called a **term**. **1, 4, 7, 10, 13, 16, 19, 22, 25, 28**

d. The sequence in part c is called an **arithmetic sequence** because the difference between any two consecutive terms is always the same. What is this difference? **3**

2a. See Solutions Manual.

2. An arithmetic sequence can be written as a formula in several ways.

a. Copy and complete a table like the one on the next page for terms 1–10.

Cooperative Learning

This investigation offers an excellent opportunity for using cooperative groups. For more information on cooperative learning strategies and group management, see *Cooperative Learning in the Mathematics Classroom,* one of the titles in the Glencoe Mathematics Professional Series.

Term Number	Term Value	Form 1	Form 2	Form 3
1	1	1	1	$1 + (0 \cdot 3)$
2	4	$1 + 3$	$1 + 3$	$1 + (1 \cdot 3)$
3	7	$4 + 3$	$1 + 3 + 3$	$1 + (2 \cdot 3)$

2b. term value = 1 + (3 times the term number minus 1).

b. A **recursive formula** for a sequence describes the new term as it relates to the previous term. Look at the column labeled "Form 1." For this arithmetic sequence, you can describe a recursive formula as *term value = previous term value + 3*. Another way to write this formula is $f_{n+1} = f_n + 3$. f_{n+1} is the value you are trying to find. f_n is the value of the previous term. Describe Forms 2 and 3 using a formula like $f_{n+1} = f_n + 3$.

3a. See Solutions Manual.
3b. Sample answer: $7.99, $11.99, $15.99, $19.99, $23.99, $27.99, $31.99, $35.99, $39.99, $43.99; cost of pizza = $4 + previous cost

3. A pizza chain offers the deal *Buy one pizza at regular price and get additional pizzas for only $4 each*. The first pizza costs $7.99.

Pizza Number	Term Number	Cost of Pizzas	New Cost of Pizza – Previous Cost of Pizza
1	1	$7.99	—
2	2	$11.99	$11.99 – $7.99 = ?

a. Copy and complete a table like the one at the left for terms 1–10.

b. Write a sequence representing the cost of the pizzas. Then write a recursive formula representing the sequence.

Extending the Investigation

1. See students' work.

In this extension, you will examine other sequences and their formulas.

1. Use cubes to make the figures shown.
2. Make a table similar to the one in Exercise 1. See Solutions Manual.
3. Write a recursive formula for the sequence. Sample answer: term value = previous term value + 5

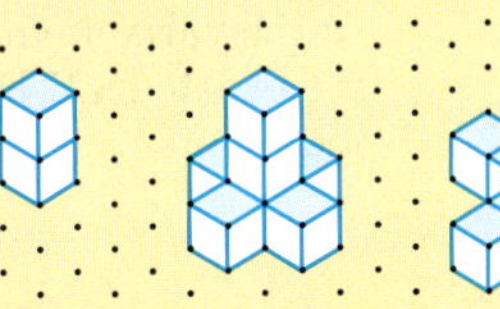

Figure 1 Figure 2 Figure 3

Presenting Your Investigation

Here are some ideas to help you present your conclusions to the class.

- Make a brochure showing the sequences you wrote in this investigation. Include your tables, formulas, and any diagrams.
- Describe a situation that results in a sequence similar to those in this investigation. Make a poster showing tables, formulas, and diagrams for this sequence.

Investigation For more information on sequences, visit: www.algconcepts.com

MANAGE

Teaching Tip You may need to define *consecutive terms* in Exercise 1d. Also, in Exercise 2b, stress that in order to write a recursive formula, the first term must be known.

Working in Groups Students should take turns forming the figures, completing the tables, and writing the formulas. Each group member should record the tables and formulas in their own notebooks.

Working as a Class Have each group designate a representative to present its findings. Have the first representative talk about the first sequence. The next representative can comment on what the first said and then add some new information. Continue in this way until the entire investigation has been reviewed. Hearing how other groups approached the problem will help all students develop a greater insight into arithmetic sequences and their recursive formulas.

ASSESS

Students' work should show they understand that the difference between successive terms in an arithmetic sequence is always the same. Be sure that in students' recursive formulas, each term is defined in terms of the previous term.

Students should add their work to their portfolios at this time.

Lesson 3–4

1 FOCUS

5-Minute Check
Lesson 3–3

The poster below gives the price of 8 items sold at a high school football game.

Refreshments	
Pizza, Hotdogs	*$2.00*
Cookies, Granola Bars, Ice Cream	*$1.00*
Popcorn, Soda	*$1.25*
Bottled Water	*$0.50*

1. Find the mean price of the items. **$1.25**
2. Find the median price of the items. **$1.125**
3. Find the mode of the prices. **$1.00**
4. Find the range of the prices. **$1.50**
5. Suppose a new item is added whose price changes the median price to $1.25 but which does *not* change the mean price. What is the price of this new item? **$1.25**

Motivating the Lesson

Real-World Connection Present students with the following situation. *Suppose you went to the store and bought 2 granola bars and a carton of ice cream. The total cost was $5.40. You know the ice cream was $3.50. How can you find the cost of each granola bar?* After discussing students' suggestions, explain that in this lesson they will learn to write and solve open sentences to answer questions such as this.

3–4 Equations

What You'll Learn
You'll learn to determine whether a given number is a solution of an equation.

Why It's Important
Health Knowing how to find the solution of an equation can help you determine your normal blood pressure. *See Example 3.*

A **statement** is any sentence that is either true or false, but not both. Look at the sentences below. Are they true or false?

Statement	True or False?
The sum of 4 and 5 is 9.	true
A number divided by 0 is a rational number.	false
A square has 4 sides of equal lengths.	true
The sum of an even and odd number is even.	false
The sum of angles measures in a triangle equals 180.	true

You can choose symbols to express mathematical statements.

Words	Symbols
A number m minus 5 is equal to 12.	$m - 5 = 12$
20 equals a number y plus 4.	$20 = y + 4$
Eight increased by a number d is 6.	$8 + d = 6$
Seven minus a number equals 5.	$7 - x = 5$
The sum of a number and 11 equals 20.	$z + 11 = 20$

Mathematical sentences like these are called **open sentences**. An open sentence is neither true nor false until the variable is replaced by a value. A set of numbers from which replacements for a variable may be chosen is called a **replacement set**. Finding the replacements for the variable that results in a true sentence is called **solving** the open sentence. The replacement values are called a **solution** of the open sentence.

Examples

Find the solution of $m - 5 = 12$ if the replacement set is {15, 16, 17, 18}.

Value for m	$m - 5 = 12$	True or False?
15	$15 - 5 \stackrel{?}{=} 12$	false
16	$16 - 5 \stackrel{?}{=} 12$	false
17	$17 - 5 \stackrel{?}{=} 12$	true ✓
18	$18 - 5 \stackrel{?}{=} 12$	false

Since 17 makes the sentence $m - 5 = 12$ true, the solution is 17.

Resource Manager

Reproducible Masters
Chapter 3 Resource Masters
- *Study Guide,* p. 110
- *Skills Practice,* p. 111
- *Practice,* p. 112
- *Reading to Learn Mathematics,* p. 113
- *Enrichment,* p. 114
- *Assessment,* p. 140

Transparencies
5-Minute Check, 3–4

Technology/Multimedia
Interactive Chalkboard CD-ROM

2 Find the solution of $2n + 3 = 9$ if the replacement set is {2, 3, 4, 5}.

Value for n	$2n + 3 = 9$	True or False?
2	$2(2) + 3 \stackrel{?}{=} 9$	false
3	$2(3) + 3 \stackrel{?}{=} 9$	true ✓
4	$2(4) + 3 \stackrel{?}{=} 9$	false
5	$2(5) + 3 \stackrel{?}{=} 9$	false

Therefore, the solution of $2n + 3 = 9$ is 3.

Your Turn a. 5

a. Find the solution of $3n - 2 = 13$ if the replacement set is {4, 5, 6, 7}.
b. Find the solution of $-2x + 1 = 7$ if the replacement set is {−3, −2, 2, 3}. −3

Open sentences and replacement sets are used frequently in real-life situations.

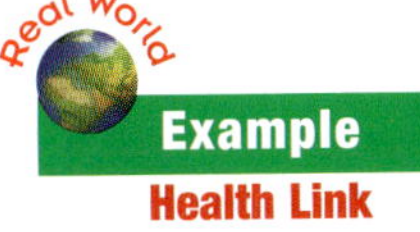

Example
Health Link

3 **The normal systolic blood pressure p of a woman who is A years old is given by the formula $p = \frac{A(A + 5)}{100} + 107$. Ms. Adams recently had her annual physical. Her blood pressure was normal. If Ms. Adams' blood pressure was 121, how old is she: 34, 35, 36, or 37?**

You need to find the replacement for the variable that results in a true sentence. First, replace p with 121.

$121 = \frac{A(A + 5)}{100} + 107$

Then replace A with each of the replacements.

Value for A	$121 = \frac{A(A + 5)}{100} + 107$	True or False?
34	$121 \stackrel{?}{=} \frac{34(34 + 5)}{100} + 107$	false
35	$121 \stackrel{?}{=} \frac{35(35 + 5)}{100} + 107$	true ✓
36	$121 \stackrel{?}{=} \frac{36(36 + 5)}{100} + 107$	false
37	$121 \stackrel{?}{=} \frac{37(37 + 5)}{100} + 107$	false

The solution is 35. So, Ms. Adams is 35 years old.

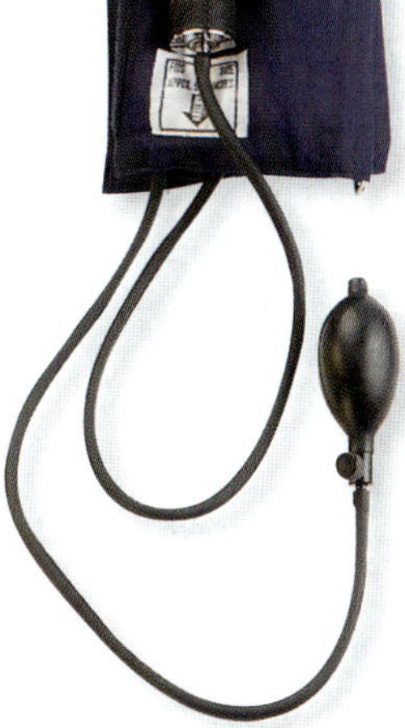
Blood pressure cuff

www.algconcepts.com/extra_examples

2 TEACH

In-Class Examples

Example 1

Find the solution of $13 = 33 + 4d$ if the replacement set is {−6, −5, −4, −3}. −5

Example 2

Find the solution of $\frac{n + 1}{3} = n - 1$ if the replacement set is {0, 1, 2, 3}. 2

Teaching Tip While discussing Example 3, point out to students that if a variable occurs more than once in an open sentence, all occurrences of the variable must be replaced by the same element of the replacement set.

In-Class Example

Example 3

The temperature C, in degrees Celsius, that is equivalent to a temperature of F degrees Fahrenheit is given by $C = \frac{5(F - 32)}{9}$. If the thermometer reads 25°C, what is the temperature in degrees Fahrenheit: 76°F, 77°F, 78°F, or 79°F? 77°F

In-Class Examples

Examples 4–5

Solve each equation.

4 $h = [24 \div (-3)(2)] - 1$ **−5**

5 $\dfrac{4 \cdot 5 - 6}{(3 \cdot 4) - 5} = c$ **2**

3 PRACTICE/APPLY

Error Analysis

Watch for students who confuse an open sentence with a statement in Exercise 1.
Prevent by pointing out that just as an English sentence must have a verb, an open sentence in mathematics has an equals sign or inequality symbol that acts as the "verb."

Answers

1. An open sentence is neither true nor false; a statement is either true or false.
2. A solution is a replacement for the variable in an open sentence that results in a true statement.
3. Replace n with each element of the replacement set to see which replacement results in a true statement.

***Study Guide*, p. 110**

NAME ________ DATE ______ PERIOD _____

Study Guide

Student Edition
Pages 112–116

Equations

Jason has won 3 gold medals in swim meets this year. Next Saturday he will swim in 3 events. How many events will he win if he has 5 gold medals at the end of the meet?

Let m = the number of gold medals Jason wins next Saturday. Then the equation $3 + m = 5$ models the number of gold medals Jason will have at the end of the meet. The replacement set for m is {1, 2, 3}. Since 2 is the only number from the replacement set that makes the equation true, 2 is the solution of the equation.

Example 1: Find the solution of $x + 10 = -21$ if the replacement set is {−30, −31, −32}.

Find the value in the replacement set that makes the equation true.

$x + 10 = -21$	$x + 10 = -21$	$x + 10 = -21$
$-30 + 10 = -21$	$-31 + 10 = -21$	$-32 + 10 = -21$
false	*true*	*false*

Since −31 makes the equation $x + 10 = -21$ true, the solution is −31.

Sometimes you can solve equations by applying the order of operations.

Example 2: Solve each equation.

a. $b = 27 - 2(3)$ *Multiply.*
$b = 27 - 6$ *Subtract.*
$b = 21$
The solution is 21.

b. $\frac{5+1}{8(2) - 14} - 8 = h$ *Evaluate the fraction.*
$\frac{6}{2} - 8 = h$ *Divide.*
$3 - 8 = h$ *Subtract.*
$-5 = h$
The solution is −5.

Find the solution of each equation if the replacement sets are x = {1, 2, 3}, y = {−5, −4, −3}, and z = {−2, 0, 2}.

1. $x + 1 = 4$ **3**
2. $5 + z = 3$ **−2**
3. $-6 = -3z$ **2**
4. $7y - 2 = -37$ **−5**
5. $x + 12x = 26$ **2**
6. $1 = 7z - (-1)$ **0**
7. $\frac{y}{5-7} = 2$ **−4**
8. $\frac{10}{x} + x = 7$ **2**
9. $8z + 5 = z - 9$ **−2**

Solve each equation.

10. $\frac{49}{7} = g$ **7**
11. $-6 + 9 = p$ **3**
12. $n = 5 - 24 \div 3$ **−3**
13. $\frac{2 \cdot 4 - 8}{1 - 2} = d$ **0**
14. $\frac{8 - 9}{4(5)} = w$ **$-\frac{1}{20}$**
15. $s = \frac{4 + 15 \div 3}{-4(1) + 1}$ **−3**

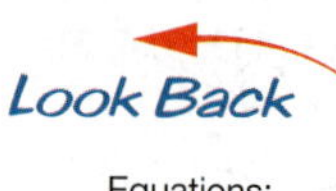

Look Back

Equations: Lesson 1–1

Recall that an open sentence containing an equals sign, =, is an *equation*. Sometimes, you can solve an equation by simply applying the order of operations.

Examples

Solve each equation.

4 $4(8) + 6 = a$

$4(8) + 6 = a$ *Original equation*
$32 + 6 = a$ *Multiply 4 and 8.*
$38 = a$ *Add 32 and 6.*

The solution is 38.

Prerequisite Skills Review
Simplifying Fractions, p. 685

5 $m = \dfrac{3 + 4 - 2}{28 - 3}$

$m = \dfrac{3 + 4 - 2}{28 - 3}$ *Evaluate the numerator and denominator separately.*

$m = \dfrac{5}{25}$ *$3 + 4 - 5 = 7 - 2$ or 5*
$28 - 3 = 25$

$m = \dfrac{1}{5}$ *Simplify.*

The solution is $\frac{1}{5}$.

Your Turn

c. $t = -3(4 \cdot 3 + 6)$ **−54**

d. $\dfrac{5 \cdot 3 + 3}{4 \cdot 2 - 2} = h$ **3**

Check for Understanding

Communicating Mathematics

1. **Describe** the difference between a statement and an open sentence. **1–3. See margin.**
2. **Define** the term *solution*.
3. **Explain** how to find the solution of $2n - 3 = 25$ if the replacement set is {12, 13, 14, 15}.

Vocabulary
statement
open sentences
replacement set
solving
solution

Guided Practice

Examples 1 & 2

Find the solution of each equation if the replacement sets are a = {5, 6, 7}, b = {−1, 0, 1}, and c = {−2, −1, 0, 1}.

4. $6 = c + 8$ **−2**
5. $2a + 5 = 17$ **6**
6. $\dfrac{3 + 15}{6} = 3b$ **1**

Reteaching Activity

Logical Learners Have students create a flow chart or algorithm that leads them to find the solution of an equation like those in Examples 4 and 5.

Examples 4 & 5

Solve each equation. 8. 48

7. $12 - 0.3 = y$ **11.7**
8. $p = 5(10) + 2 - 4$
9. $n = \frac{24 \div 8 + 1}{15 \div 3}$ $\frac{4}{5}$

Example 3

10. **Health** The normal systolic blood pressure p of a man who is A years old is given by the formula $p = \frac{A(3A - 10)}{500} + 120$. How old is Mr. Nichols if his normal blood pressure is 122: 18, 19, 20, or 21? **20**

Exercises

Practice

A

Find the solution of each equation if the replacement sets are x = {2, 3, 4, 5}, m = {−5, −4, −3}, and d = {−1, 0, 1, 2}.

11. $x + 5 = 9$ **4**
12. $2d = -2$ **−1**
13. $8 - m = 11$ **−3**
14. $-11 = -7 + m$ **−4**
15. $4x + (-2) = 10$ **3**
16. $9d - 20 = -20$ **0**
17. $6x + 7 = 31$ **4**
18. $-23 = 3d - 4(5)$ **−1**
19. $3 - m = 8$ **−5**
20. $\frac{3 + 15}{m} - 6 = -12$
21. $\frac{6d}{4} + 3 = 3d$ **2**
22. $\frac{2(x - 2)}{3} = \frac{4}{7 - 5}$ **5**

20. −3

Homework Help	
For Exercises	See Examples
11–22	1, 2
23–34	4, 5
35	4
36	5
37–39	3

Extra Practice

See page 697.

Solve each equation.

23. $6 \cdot 7 = y$ **42**
24. $g = 14.8 - 3.75$
25. $h = 5 + 2 \cdot 3$ **11**
26. $-7.7 - 5.2 = k$
27. $3 + 18 \div 6 = w$ **6**
28. $m = -2(5) + 4 \cdot 5$ **10**
29. $n = (18 \div 6) - 4$
30. $20 - 3 \cdot 3 - 14 = m$
31. $7 \cdot 3 - 15 \div 5 = b$ **18**
32. $\frac{7 \cdot 5 + 5}{(9 \cdot 3) - 7} = p$ **2**
33. $a = \frac{14 + 7}{2 \cdot 2}$ $5\frac{1}{4}$
34. $\frac{4 \cdot 7 - 4}{2 \div 2 + 5} = d$ **4**

24. 11.05
26. −12.9
29. −1
30. −3

B

35. Find the solution of $2(7) - 12 = v$. **2**
36. What is the value of g if $g = \frac{24 \div 2}{6}$? **2**

Applications and Problem Solving

C

37. **Recreation** At Sinclair's Balloon Rental, the cost of renting a hot air balloon is given by the formula $C = 85 + 36h$, where C is the cost in dollars and h is the number of hours. Michael and Keisha paid a total of $301 for a balloon ride. Did they rent the balloon for 5, 6, 7, or 8 hours? **6 h**

38. **Geometry** To find the area of a trapezoid, you can use the formula $A = \frac{1}{2}h(a + b)$, where h is the height of the trapezoid and a and b are the lengths of the bases. If a trapezoid has an area of 35 square feet and bases of 4 feet and 10 feet, is the height 3, 4, 5, or 6 feet? **5 ft**

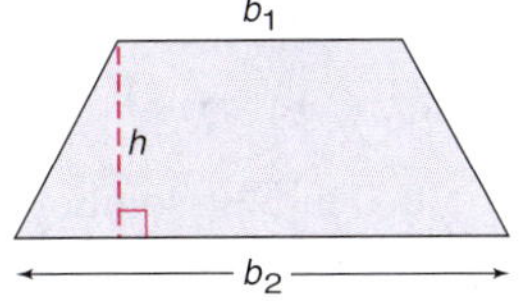

Assignment Guide

Basic: 11–39 odd, 40–47
Average: 12–36 even, 37–47
All: Quiz 1, 1–10

***Skills Practice*, p. 111, and *Practice*, p. 112 (shown)**

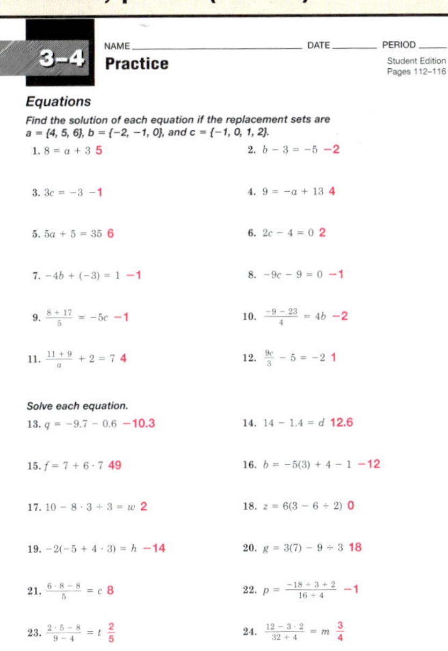

3–4 NAME ______ DATE ______ PERIOD ______

Practice

Student Edition Pages 112–116

Equations

Find the solution of each equation if the replacement sets are a = {4, 5, 6}, b = {−2, −1, 0}, and c = {−1, 0, 1, 2}.

1. $8 = a + 3$ **5**
2. $b - 3 = -5$ **−2**
3. $3c = -3$ **−1**
4. $9 = -a + 13$ **4**
5. $5a + 5 = 35$ **6**
6. $2c - 4 = 0$ **2**
7. $-4b + (-3) = 1$ **−1**
8. $-9c - 9 = 0$ **−1**
9. $\frac{8 + 17}{5} = -5c$ **−1**
10. $\frac{-9 - 23}{4} = 4b$ **−2**
11. $\frac{11 + 9}{a} + 2 = 7$ **4**
12. $\frac{9c}{3} - 5 = -2$ **1**

Solve each equation.

13. $q = -9.7 - 0.6$ **−10.3**
14. $14 - 1.4 = d$ **12.6**
15. $f = 7 + 6 \cdot 7$ **49**
16. $b = -5(3) + 4 - 1$ **−12**
17. $10 - 8 \cdot 3 \div 3 = w$ **2**
18. $z = 6(3 - 6 \div 2)$ **0**
19. $-2(-5 + 4 \cdot 3) = h$ **−14**
20. $g = 3(7) - 9 \div 3$ **18**
21. $\frac{6 \cdot 8 - 8}{5} = c$ **8**
22. $p = \frac{-18 \div 3 + 2}{16 \div 4}$ **−1**
23. $\frac{2 \cdot 5 - 8}{9 - 4} = t$ $\frac{2}{5}$
24. $\frac{12 - 3 \cdot 2}{32 \div 4} = m$ $\frac{3}{4}$

4 ASSESS

Open-Ended Assessment

Speaking Have students work with a partner. Ask one student to give a verbal definition of open sentence while their partner provides verbal feedback. The second student then explains how to solve an open sentence while the first provides verbal feedback.

Quiz 1

The Quiz provides students with a brief review of the concepts and skills in Lessons 3–1 through 3–4. Lesson numbers are given to the right of the exercises or instruction lines so students can review concepts not yet mastered.

Mid-Chapter Test (Lessons 3–1 through 3–4) is available in the *Chapter 3 Resource Masters*, p. 140.

***Enrichment*, p. 114**

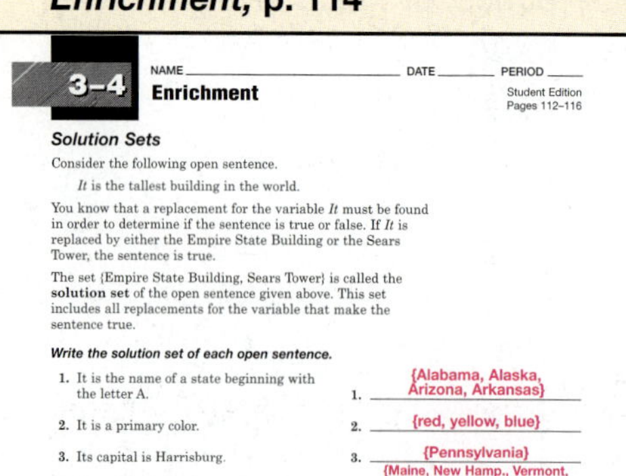

3–4 NAME ______ DATE ______ PERIOD ______

Enrichment

Student Edition Pages 112–116

Solution Sets

Consider the following open sentence.

It is the tallest building in the world.

You know that a replacement for the variable *It* must be found in order to determine if the sentence is true or false. If *It* is replaced by either the Empire State Building or the Sears Tower, the sentence is true.

The set {Empire State Building, Sears Tower} is called the **solution set** of the open sentence given above. This set includes all replacements for the variable that make the sentence true.

Write the solution set of each open sentence.

1. It is the name of a state beginning with the letter A. 1. {Alabama, Alaska, Arizona, Arkansas}
2. It is a primary color. 2. {red, yellow, blue}
3. Its capital is Harrisburg. 3. {Pennsylvania}
4. It is a New England state. 4. {Maine, New Hamp., Vermont, Mass., Rhode Is., Conn.}
5. $x + 4 = 10$ 5. {6}
6. It is the name of a month that contains the letter *r*. 6. {Jan, Feb, Mar, Apr, Sept, Oct, Nov, Dec}
7. During the 1970s, she was the wife of a U.S. President. 7. {Pat Nixon, Betty Ford, Rosalyn Carter}
8. It is an even number between 1 and 13. 8. {2, 4, 6, 8, 10,12}
9. $31 = 72 - k$ 9. {41}
10. It is the square of 2, 3, or 4. 10. {4, 9, 16}

Write an open sentence for each solution set.

11. {A, E, I, O, U} 11. It is a vowel.
12. {1, 3, 5, 7, 9} 12. It is an odd number between 0 and 10.
13. {June, July, August} 13. It is a summer month.
14. {Atlantic, Pacific, Indian, Arctic} 14. It is an ocean.

Blue Whales

39. **Life Science** At birth, a baby blue whale weighs 4000 pounds and gains 200 pounds each day while nursing. The formula $W = 4000 + 200d$ represents this situation, where W is the weight of the baby and d is the number of days it has nursed. How many days has a baby blue whale been nursing if it weighs 17,000 pounds: 63, 64, or 65? **65 days**

40. **Critical Thinking** Write four different open sentences that each have 3 as a solution. **See students' work.**

Mixed Review

Find the mean, median, mode, and range of each set of data. *(Lesson 3–3)*

41. 38, 42, 46, 40, 40 **41.2; 40; 40; 8**

42. 3.1, 3.2, 3.3, 3.4, 3.5, 3.6, 3.7, 3.8 **3.45; 3.45; none; 0.7**

Evaluate each expression if $x = -\frac{1}{3}$, $y = \frac{1}{2}$, and $z = 3\frac{1}{4}$. *(Lesson 3–2)*

43. $2 - x$ $\mathbf{2\frac{1}{3}}$

44. $y + z$ $\mathbf{3\frac{3}{4}}$

45. $x - y - z$ $\mathbf{-4\frac{1}{12}}$

Standardized Test Practice

46. **Short Response** Thomas buys $\frac{2}{3}$ yard of ribbon, $\frac{3}{8}$ yard of plaid fabric, and $\frac{1}{2}$ yard of striped fabric. Write the lengths in order from least to greatest. *(Lesson 3–1)* $\mathbf{\frac{3}{8}, \frac{1}{2}, \frac{2}{3}}$

47. **Multiple Choice** Which of the following is equal to the expression $[(8 - 7) - (6 - 5)] - [(8 - 7 - 6) - 5]$? *(Lesson 2–4)* **D**

Ⓐ -31 Ⓑ -11 Ⓒ 0 Ⓓ 10

Quiz 1 Lessons 3–1 through 3–4

Replace each ● with <, >, or = to make a true sentence. *(Lesson 3–1)*

1. $-16.3 \bullet -16.03$ **<**

2. $3(-4) + 7 \bullet -12 + 6$ **>**

3. $\frac{2}{3} \bullet \frac{4}{6}$ **=**

4. Which is the better buy, a 12-pack of soda for \$2.79 or a 24-pack of soda for \$5.69? *(Lesson 3–1)* **12-pack; \$0.23 < \$0.24**

5. Find $3.54 - (-4.78) + 1.02 + (-7.65)$. *(Lesson 3–2)* **1.69**

Find the mean, median, mode, and range for each set of data. *(Lesson 3–3)*

6. 45, 12, 67, 22, 34
36; 34; none; 55

7.

Stem	Leaf
1	1 3 7
2	1 6 7
3	0 0 4 6
4	4 7 $1\mid 1 = 1.1$

2.8; 2.85; 3.0; 3.6

8.

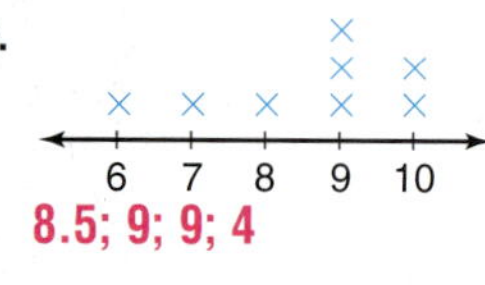

8.5; 9; 9; 4

9. Find the solution of $3d - 7 = -13$ if the replacement set is $\{-4, -3, -2, -1\}$. *(Lesson 3–4)* **−2**

10. Solve $m = \frac{2 + 4 \cdot 6}{3 + 10} - 8$. *(Lesson 3–4)* **−6**

www.algconcepts.com/self_check_quiz

Extra Credit

Jacob's age is three more than his sister Jessica's age. The sum of their ages is 15. How old is Jessica? Let j represent Jessica's age. Write an open sentence to represent this situation. Give a reasonable replacement set and then use it to find the solution of your open sentence.
$j + (j + 3) = 15$; Sample set: {4, 5, 6, 7}; Solution: 6

3–5 Solving Equations by Using Models

What You'll Learn
You'll learn to solve addition and subtraction equations by using models.

Why It's Important
Zoology Knowing how to solve equations can help in finding unknown values. *See Example 3.*

Many real-world situations can be approached by developing a model, writing an equation, and solving it. You can use algebra tiles to solve equations. The green tile represents the variable, x.

Examples

1 **Use algebra tiles to solve each equation.**

$x + (-2) = -7$

Step 1 Model $x + (-2) = -7$ by placing 1 green tile and 2 red square tiles on one side of the mat to represent $x + (-2)$. Place 7 red square tiles on the other side to represent -7.

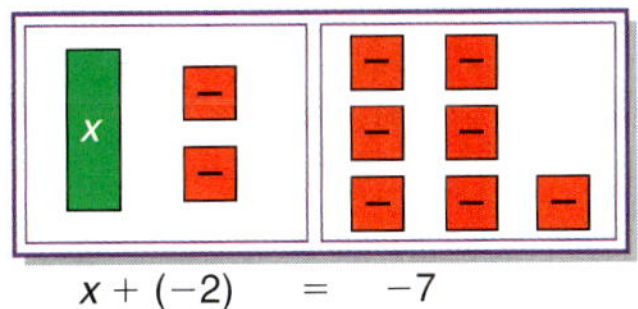

Step 2 To get the green tile by itself, remove 2 red square tiles from each side.

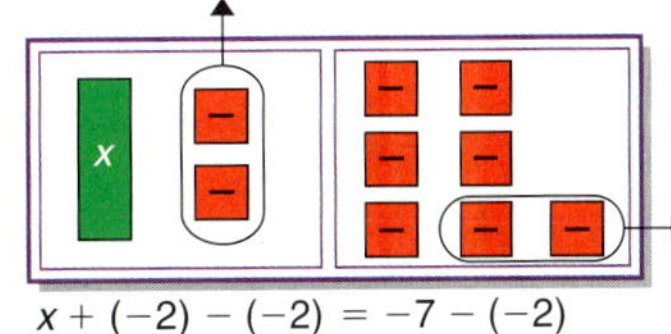

Step 3 The green tile on the left side of the mat is matched with 5 red square tiles. Therefore, $x = -5$.

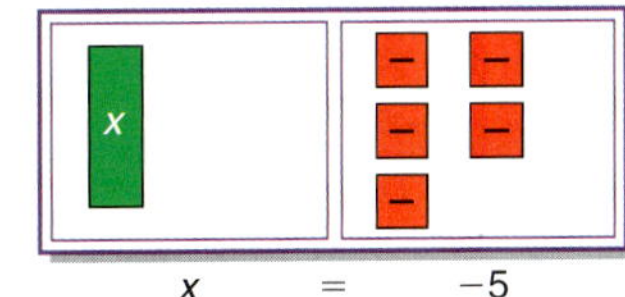

2 $x - 3 = 5$

Step 1 Write the equation in the form $x + (-3) = 5$. Place 1 green tile and 3 red square tiles on one side of the mat to represent $x + (-3)$. Place 5 yellow square tiles on the other side of the mat to represent $+5$.

(continued on the next page)

Resource Manager

Reproducible Masters
Chapter 3 Resource Masters
- *Study Guide*, p. 115
- *Skills Practice*, p. 116
- *Practice*, p. 117
- *Reading to Learn Mathematics*, p. 118
- *Enrichment*, p. 119

Hands-On Algebra, pp. 45–46

Transparencies
5-Minute Check, 3–5

Technology/Multimedia
Interactive Chalkboard CD-ROM

1 FOCUS

5-Minute Check
Lesson 3–4

1. Kenyon found the ad shown below in his local newspaper. He wants to spend exactly $105 renting a canoe. Using the information in the ad, he wrote the formula $C = 30 + 15h$ for the cost C of renting a canoe for h hours including delivery and pick-up. Can he rent a canoe for 3, 4, 5, or 6 hours for $105? **5 h**

Canoe Rentals
$15 per hour
$30 delivery and pick-up charge

2. If Kenyon rents the canoe for 8 hours, the formula becomes $C = 30 + 15(8)$. Find the cost of renting a canoe for 8 hours. **$150**

Find the solution of each equation if the replacement sets are v = {−3, −2, −1, 0} and z = {20, 21, 22, 23}.

3. $7v + 2 = -12$ **−2**
4. $\frac{3(z + 2)}{5} = 15$ **23**
5. Find the solution of $\frac{18 \div 3 + 2}{4} = y$. **2**

Motivating the Lesson

Hands-On Activity Provide each student several positive and negative integer tiles. Have students practice modeling integer addition problems, such as $-4 + 7$ and $-3 + (-9)$, with the tiles. Make sure they understand the concept of a *zero pair*. Also have them model some subtraction problems, such as $-6 - (-2)$.

2 TEACH

Teaching Tip If algebra tiles are not available, they can be made from colored paper.

Teaching Tip Encourage students to think of the algebra tile mat as a balance scale with the center line as the balance point. Emphasize that the initial model of an algebraic equation is considered to be balanced. When modeling the steps for solving the equation, remind students that they must keep the mat in balance by performing the same operations on both sides of the mat. In Example 1, point out that the same number and color of tiles are removed from both sides of the mat to maintain the balance. In Example 2, stress that an equal number of yellow tiles are added to each side to create three zero pairs on the left side. Make sure that students understand that the zero pairs can be removed because they have a value of 0 and thus their removal from the mat does not effect the balance.

In-Class Examples

Examples 1–2

Use algebra tiles to solve each equation.

1 $t + 3 = 1$ **−2**

2 $b - 4 = -3$ **1**

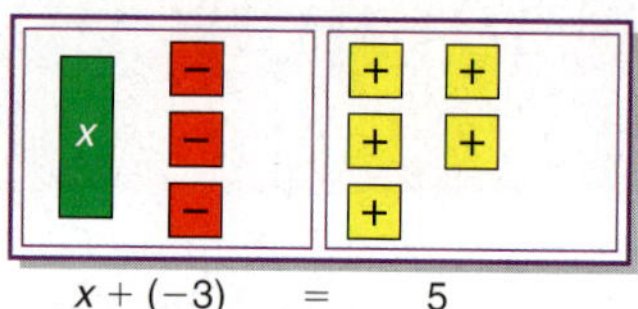

Step 2 To get the green tile by itself, you need to remove 3 red square tiles from each side. Since there are no red square tiles on the right side of the mat, you will need to add 3 yellow square tiles to each side to make 3 zero pairs on the left side of the mat.

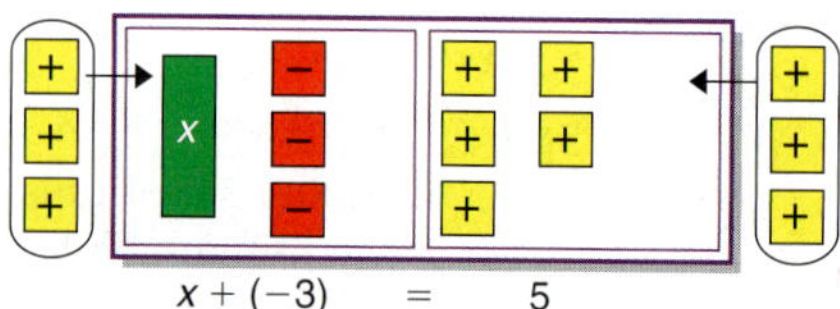

Step 3 Remove the zero pairs.

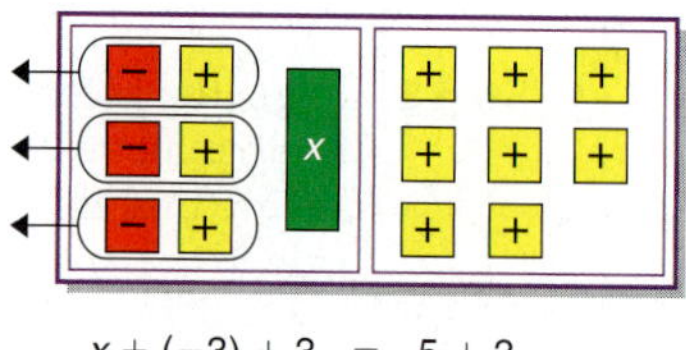

Step 4 The green tile is matched with 8 yellow square tiles. Therefore, $x = 8$.

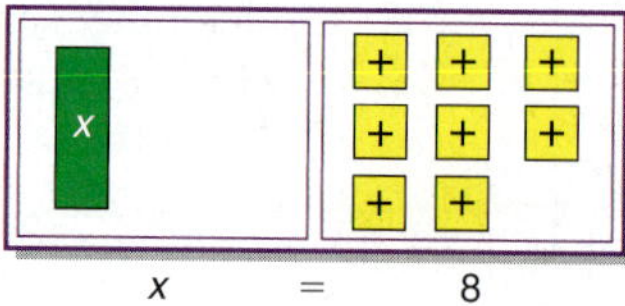

Your Turn **Use algebra tiles to solve each equation.**

a. $x - 4 = -8$ **−4** b. $2 + x = -7$ **−9** c. $x + 6 = 5$ **−1**

In Examples 1 and 2, you solved each equation for the variable by isolating the variable on one side of the equation.

Inclusion Strategies

Students with visual impairments may have difficulty distinguishing the colors that are used to represent the negative and positive integer tiles. For these students, you can adapt a set of algebra tiles by placing a small sticker on each positive tile. The sticker can be felt with the student's fingers. The negative integer tiles can be used without stickers added to them.

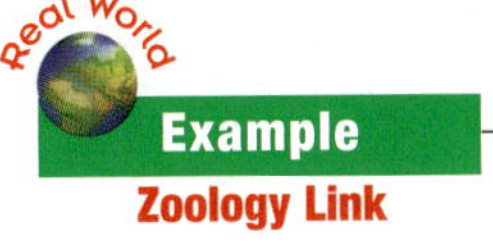

Example 3 — Zoology Link

The largest bear is the Alaskan brown bear. When fully grown, it has a height of 9 feet. The smallest bear is the sun bear. The Alaskan brown bear is 6 feet longer than the sun bear. Find the length of the sun bear.

Explore You know that the Alaskan brown bear is 6 feet longer than the sun bear. You need to find the length of the sun bear.

Plan Let x represent the length of the sun bear. Translate the problem into an equation using the variable x. Model the equation and then solve.

Solve

length of sun bear	*plus*	*6 feet*	*equals*	*length of Alaskan brown bear*
x	$+$	6	$=$	9

$x + 6 = 9$

To solve for x, subtract 6 from each side.

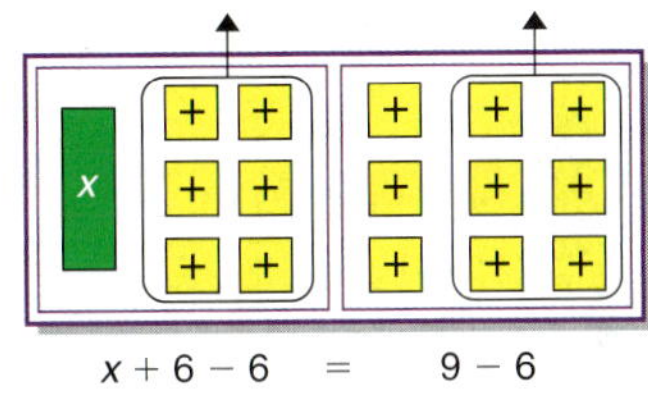

$x + 6 - 6 = 9 - 6$

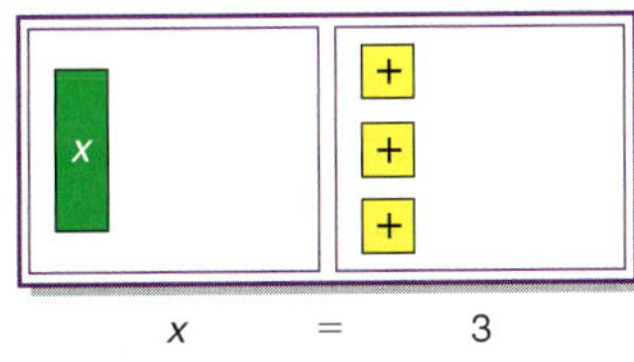

$x = 3$

The solution is 3. So, the length of the sun bear is 3 feet.

Examine Check the solution.

The length of the Alaskan brown bear is equal to the length of the sun bear plus 6 feet. The length of the sun bear is 3 feet, and 3 feet plus 6 feet is 9 feet. ✓

The answer is correct.

Alaskan Brown Bear

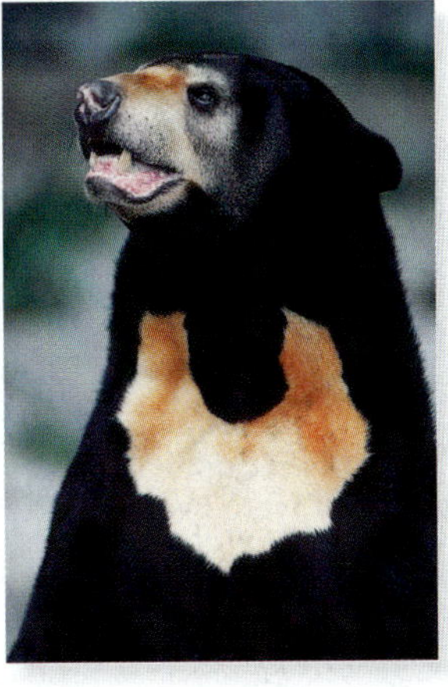

Sun Bear

Teaching Tip Stress to students the importance of checking the solution to a problem such as that in Example 3.

In-Class Example

Example 3

Beth made \$5 less than her brother Seth at their family yard sale. If Beth made \$7, how much did Seth make? **\$12**

Study Guide, p. 115

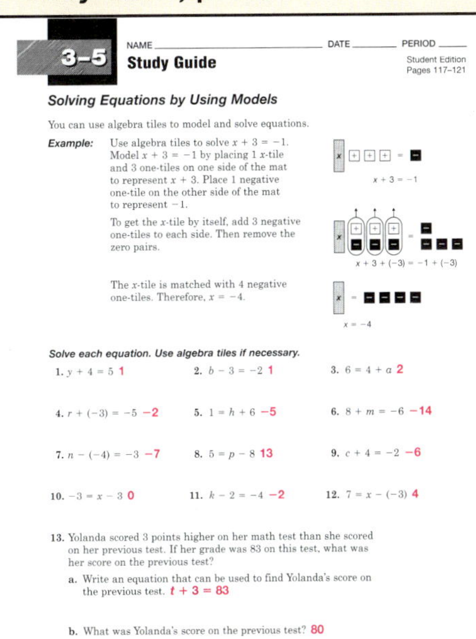

3–5 NAME ______ DATE ______ PERIOD ______

Study Guide

Student Edition Pages 117–121

Solving Equations by Using Models

You can use algebra tiles to model and solve equations.

Example: Use algebra tiles to solve $x + 3 = -1$. Model $x + 3 = -1$ by placing 1 x-tile and 3 one-tiles on one side of the mat to represent $x + 3$. Place 1 negative one-tile on the other side of the mat to represent -1.

$x + 3 = -1$

To get the x-tile by itself, add 3 negative one-tiles to each side. Then remove the zero pairs.

$x + 3 + (-3) = -1 + (-3)$

The x-tile is matched with 4 negative one-tiles. Therefore, $x = -4$.

$x = -4$

Solve each equation. Use algebra tiles if necessary.

1. $y + 4 = 5$ **1**
2. $b - 3 = -2$ **1**
3. $6 = 4 + a$ **2**
4. $r + (-3) = -5$ **−2**
5. $1 = h + 6$ **−5**
6. $8 + m = -6$ **−14**
7. $n - (-4) = -3$ **−7**
8. $5 = p - 8$ **13**
9. $c + 4 = -2$ **−6**
10. $-3 = x - 3$ **0**
11. $k - 2 = -4$ **−2**
12. $7 = x - (-3)$ **4**
13. Yolanda scored 3 points higher on her math test than she scored on her previous test. If her grade was 83 on this test, what was her score on the previous test?
    a. Write an equation that can be used to find Yolanda's score on the previous test. **$t + 3 = 83$**
    b. What was Yolanda's score on the previous test? **80**

3 PRACTICE/APPLY

Error Analysis

Watch for students who have trouble with the model of the equation in Exercise 2.
Prevent by emphasizing to students that equations involving subtraction should be rewritten as addition equations prior to using algebra tiles.

Answers

1.
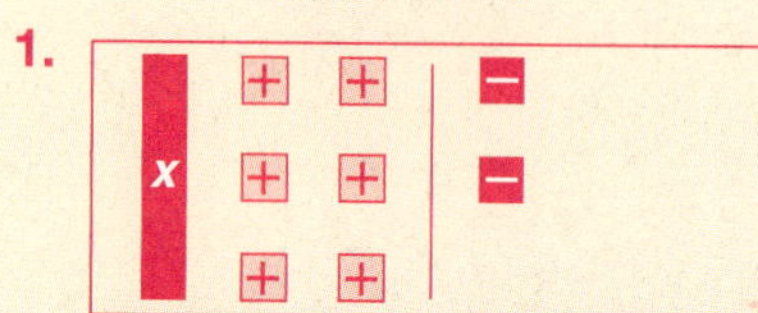

2. Subtracting a negative is the same as adding a positive. So, $y - (-8) = -6$ is the same as $y + 8 = -6$. The model shows a representation of a variable and a positive 8 on one side of an equation and a −6 on the other side of the equation.

Skills Practice, **p. 116, and** ***Practice,*** **p. 117 (shown)**

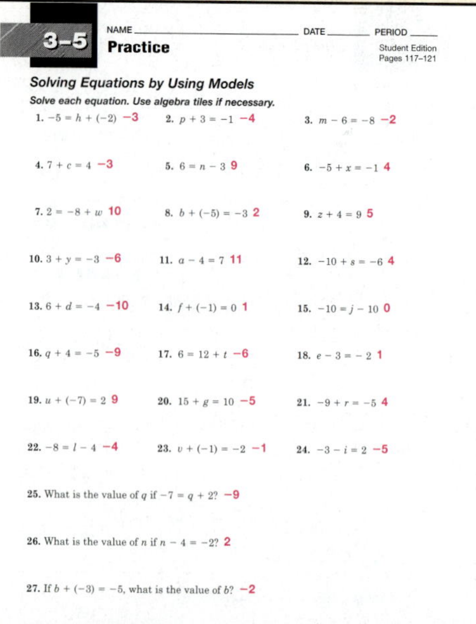

3–5 NAME ______________ DATE ________ PERIOD ______

Practice

Student Edition Pages 117–121

Solving Equations by Using Models

Solve each equation. Use algebra tiles if necessary.

1. $-5 = h + (-2)$ **−3**	2. $p + 3 = -1$ **−4**	3. $m - 6 = -8$ **−2**
4. $7 + c = 4$ **−3**	5. $6 = n - 3$ **9**	6. $-5 + x = -1$ **4**
7. $2 = -8 + w$ **10**	8. $b + (-5) = -3$ **2**	9. $z + 4 = 9$ **5**
10. $3 + y = -3$ **−6**	11. $a - 4 = 7$ **11**	12. $-10 + s = -6$ **4**
13. $6 + d = -4$ **−10**	14. $f + (-1) = 0$ **1**	15. $-10 = j - 10$ **0**
16. $q + 4 = -5$ **−9**	17. $6 = 12 + t$ **−6**	18. $e - 3 = -2$ **1**
19. $u + (-7) = 2$ **9**	20. $15 + g = 10$ **−5**	21. $-9 + r = -5$ **4**
22. $-8 = l - 4$ **−4**	23. $v + (-1) = -2$ **−1**	24. $-3 - i = 2$ **−5**

25. What is the value of q if $-7 = q + 2$? **−9**

26. What is the value of n if $n - 4 = -2$? **2**

27. If $b + (-3) = -5$, what is the value of b? **−2**

Check for Understanding

Communicating Mathematics

1. **Show** how to model $6 + x = -2$ using algebra tiles.
2. **Explain** why the model below represents $x - (-8) = -6$.

1–2. See margin.

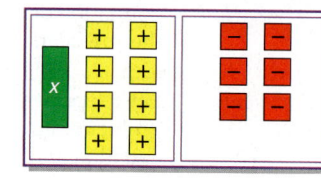

Guided Practice

Getting Ready **Write an equation for each model.**

Sample 1:

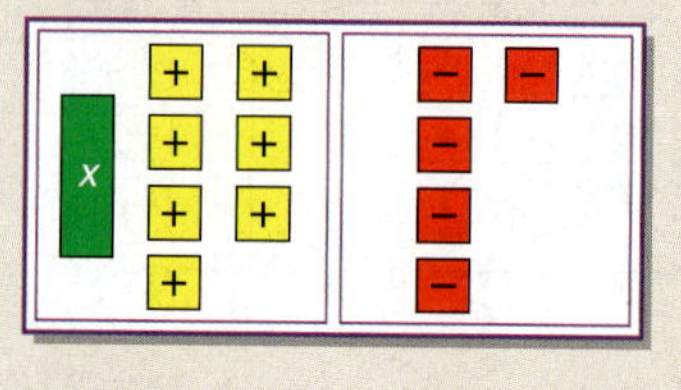

Solution: $x + 7 = -5$

Sample 2:

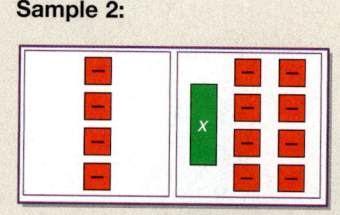

Solution: $-4 = x + (-8)$

3.
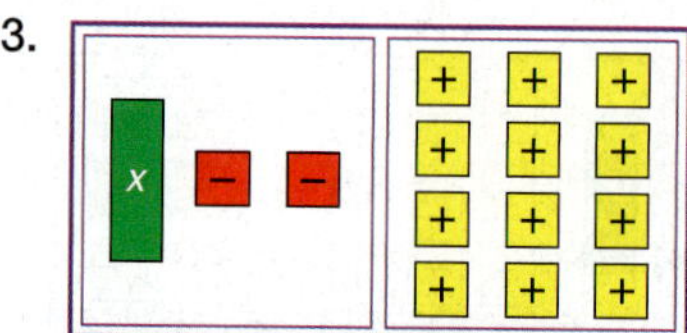

4.
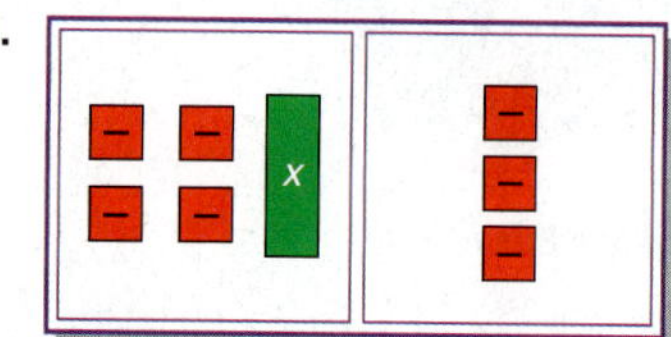

3. $x + (-2) = 12$
4. $-4 + x = -3$

5.
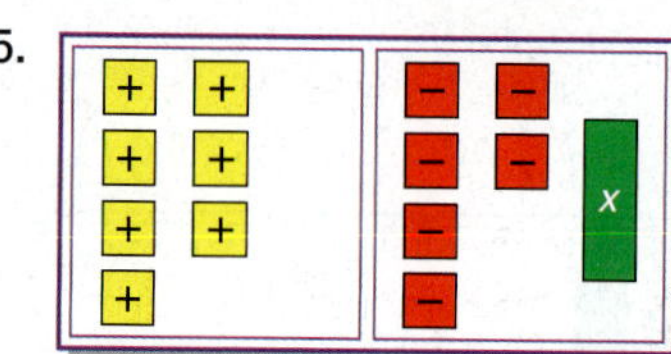

6.
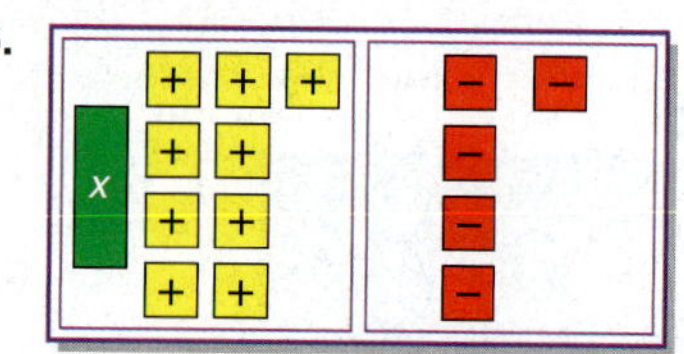

5. $7 = -6 + x$
6. $x + 9 = -5$

Examples 1 & 2

Solve each equation. Use algebra tiles if necessary.

7. $m + 2 = -2$ **−4**
8. $3 = n - 6$ **9**
9. $t - 4 = -8$ **−4**
10. $y + (-3) = -1$ **2**
11. $4 = x - 2$ **6**
12. $-5 + c = -5$ **0**
13. Find the value of h if $-4 = h - 7$. **3**

Example 3

14. **Sales** Jeff Simons sold 27 cars last month. This is 36 fewer cars than he sold during the same time period one year ago. What were his sales one year ago?
    a. Write an equation that can be used to find Mr. Simons' sales one year ago. $27 = c - 36$
    b. What were Mr. Simons' sales one year ago? **63 cars**

Reteaching Activity

Interpersonal Learners Have students work with a partner. One student should write down an equation and the other should solve it using algebra tiles. The partners should then switch roles and repeat the activity.

Exercises

Practice

Homework Help	
For Exercises	See Examples
15–35	1, 2
36–37	3

Extra Practice
See page 697.

Solve each equation. Use algebra tiles if necessary.

15. $-2 + y = -8$ **−6**
16. $w - 1 = 3$ **4**
17. $n + 5 = -2$ **−7**
18. $7 = q - 4$ **11**
19. $p - 6 = 2$ **8**
20. $3 = z + 10$ **−7**
21. $-3 + k = -3$ **0**
22. $4 = g + 6$ **−2**
23. $2 + f = -1$ **−3**
24. $10 = 8 + b$ **2**
25. $-2 = d + (-3)$ **1**
26. $x + 6 = -23$ **−29**
27. $-7 = a + (-9)$ **2**
28. $k - 7 = -8$ **−1**
29. $p - 9 = 7$ **16**
30. $-12 + r = -2$ **10**
31. $v + 9 = 2$ **−7**
32. $-13 = c + (-4)$ **−9**
33. What is the value of m if $8 + m = -13$? **−21**
34. If $a + (-14) = -16$, what is the value of a? **−2**

35. When 6 is subtracted from d, the result is 5. Find the value of d. **11**

Applications and Problem Solving

36. **Sports** The 1998 NASCAR season consisted of 33 races. Out of the 33 races, driver Jeff Gorden finished in one of the top five positions 26 times. This number is 15 more than the number of times driver Bobby Labonte finished in the top five positions. **Source:** NASCAR
    a. Write an equation that represents this situation. **$n + 15 = 26$**

    b. How many times did Bobby Labonte finish in one of the top five positions? **11 times**
37. **Weather** Jaclyn needs to record the outside temperature for her science project. At 6:00 P.M., the temperature was 8°F. The difference between the temperature she recorded in the morning and the evening temperature was −2°F.
    a. Write an equation that Jaclyn can use to determine the morning temperature. **$t - 8 = -2$**
    b. What was the morning temperature? **6°F**
38. **Critical Thinking** Explain how algebra tiles can be used to solve the equation $2n + 1 = 7$. **See margin.**

Mixed Review

Find the solution of each equation if the replacement set is $x = \{-3, -2, -1, 0, 1, 2, 3\}$. *(Lesson 3–4)*

39. $18 + x = 20$ **2**
40. $x \div 9 = 0$ **0**
41. $5x - 7 = -2$ **1**
42. **Sports** The stem-and-leaf plot shows Tanya's bowling scores for the past eight games. Find Tanya's mean bowling score. *(Lesson 3–3)* **106**

Stem	Leaf
9	0 1
10	4 5 8
11	5 6 9

10 | 5 = 105

43. Find the product of −9 and −6. *(Lesson 2–5)* **54**
44. Simplify $(4m - 5n) + (3m + 3n)$. *(Lesson 1–4)* **$7m - 2n$**

Standardized Test Practice

Ⓐ Ⓑ Ⓒ Ⓓ

45. **Short Response** Identify the property shown by $24 + 7 = 7 + 24$. *(Lesson 1–3)* **Commutative (+)**
46. **Multiple Choice** Choose the expression that represents *5 subtracted from the sum of a number and 2.* *(Lesson 1–1)* **B**
    Ⓐ $5 - x + 2$
    Ⓑ $x + 2 - 5$
    Ⓒ $2x - 5$
    Ⓓ $5 - 2x$

www.algconcepts.com/self_check_quiz

Assignment Guide

Basic: 15–37 odd, 38–46
Average: 16–34 even, 36–46

4 ASSESS

Open-Ended Assessment

Modeling Write the equations $b + 5 = 7$ and $b + 7 = 5$ on the board or overhead. Have students model each equation and its solution. Then discuss how the models of the two solutions differ.

Answer

38. Place 2 green tiles and 1 yellow tile on one side of the model to show $2n + 1$. Place 7 yellow tiles on the other side. Remove 1 yellow tile from each side. Then pair up each green tile with 3 yellow tiles. So, $n = 3$.

***Enrichment,* p. 119**

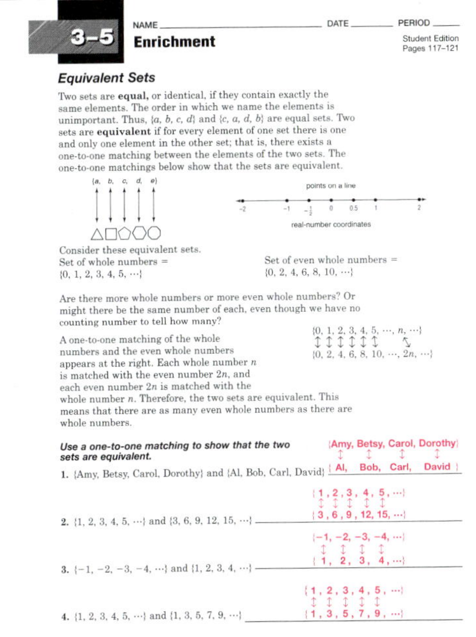

3–5 NAME ______ DATE ______ PERIOD ______

Enrichment

Student Edition Pages 117–121

Equivalent Sets

Two sets are **equal,** or identical, if they contain exactly the same elements. The order in which we name the elements is unimportant. Thus, $\{a, b, c, d\}$ and $\{c, a, d, b\}$ are equal sets. Two sets are **equivalent** if for every element of one set there is one and only one element in the other set; that is, there exists a one-to-one matching between the elements of the two sets. The one-to-one matchings below show that the sets are equivalent.

Consider these equivalent sets.
Set of whole numbers = $\{0, 1, 2, 3, 4, 5, \cdots\}$
Set of even whole numbers = $\{0, 2, 4, 6, 8, 10, \cdots\}$

Are there more whole numbers or more even whole numbers? Or might there be the same number of each, even though we have no counting number to tell how many?

A one-to-one matching of the whole numbers and the even whole numbers appears at the right. Each whole number n is matched with the even number $2n$, and each even number $2n$ is matched with the whole number n. Therefore, the two sets are equivalent. This means that there are as many even whole numbers as there are whole numbers.

$\{0, 1, 2, 3, 4, 5, \cdots, n, \cdots\}$
$\{0, 2, 4, 6, 8, 10, \cdots, 2n, \cdots\}$

Use a one-to-one matching to show that the two sets are equivalent.

1. {Amy, Betsy, Carol, Dorothy} and {Al, Bob, Carl, David} **{Amy, Betsy, Carol, Dorothy} ↕ {Al, Bob, Carl, David}**
2. $\{1, 2, 3, 4, 5, \cdots\}$ and $\{3, 6, 9, 12, 15, \cdots\}$ **$\{1, 2, 3, 4, 5, \cdots\}$ ↕ $\{3, 6, 9, 12, 15, \cdots\}$**
3. $\{-1, -2, -3, -4, \cdots\}$ and $\{1, 2, 3, 4, \cdots\}$ **$\{-1, -2, -3, -4, \cdots\}$ ↕ $\{1, 2, 3, 4, \cdots\}$**
4. $\{1, 2, 3, 4, 5, \cdots\}$ and $\{1, 3, 5, 7, 9, \cdots\}$ **$\{1, 2, 3, 4, 5, \cdots\}$ ↕ $\{1, 3, 5, 7, 9, \cdots\}$**

Extra Credit

Make up a real-world situation that could be represented by the equation $11 = z - 9$. Solve the equation using algebra tiles and explain how the solution relates to the situation. **Sample answer: I have \$11 left after spending \$9 on a game. How much money did I have originally? The solution of the equation, 20, is the amount of money I had originally.**

3-6 Solving Addition and Subtraction Equations

1 FOCUS

5-Minute Check

Lesson 3–5

Write an equation for each model.

1.
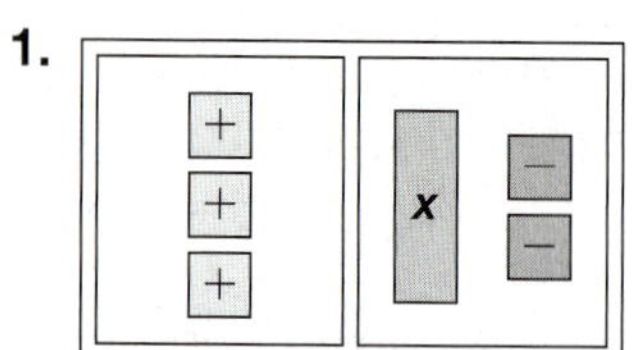

$3 = x + (-2)$

2.
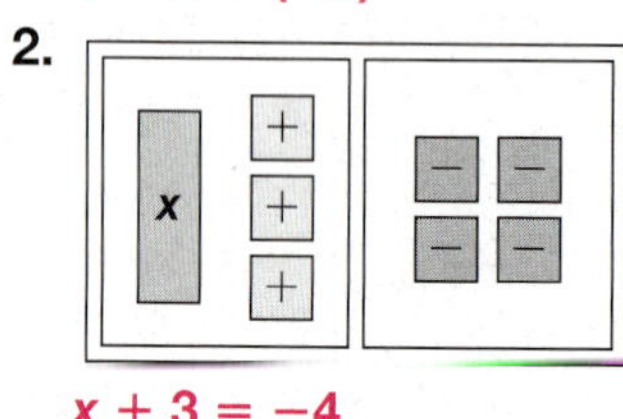

$x + 3 = -4$

Solve each equation. Use algebra tiles if necessary.

3. $6 = m - 8$ **14**
4. $-3 + g = -4$ **−1**
5. The low temperature today is 10 degrees higher than the low temperature yesterday. If the low temperature today is −2°F, write an equation to represent this situation and then solve it using algebra tiles. $x + 10 = -2$; **−12°F**

Motivating the Lesson

Real-World Connection Present students with the following situation. *Suppose you go to a store to buy a bag of potato chips. You give the cashier a $10 bill and receive $8.93 in change.* Tell students that the equation $c + 8.93 = 10.00$ can be used to find the cost c of the chips. Discuss with them why this problem cannot be solved using algebra tiles. **Algebra tiles can only be used to model equations that involve only integers.**

What You'll Learn
You'll learn to solve addition and subtraction equations by using the properties of equality.

Why It's Important
History Knowing how to solve equations can help you find important dates in history. *See Exercise 17.*

The Addition Property of Equality and the Subtraction Property of Equality can be used to solve equations. The example below illustrates the **Addition Property of Equality**.

$$14 = 14$$

$$14 + 3 = 14 + 3 \quad \text{Add 3 to each side.}$$

$$17 = 17$$

Addition Property of Equality	
Words:	If you add the same number to each side of an equation, the two sides remain equal.
Symbols:	For any numbers a, b, and c, if $a = b$, then $a + c = b + c$.
Numbers:	If $x = 2$, then $x + 3 = 2 + 3$.

Note that c can be positive, negative, or 0. Look at the equations below. In the equation on the left, $c = 3$. In the equation on the right, $c = -3$.

$$a + c = b + c \qquad\qquad a + c = b + c$$

$$15 + 3 = 15 + 3 \qquad\qquad 15 + (-3) = 15 + (-3)$$

When the same number is added to each side of an equation, the result is an **equivalent equation**. Equivalent equations have the same solution.

$x + 2 = 6$ — *The solution of the equation is 4.*

$x + 2 + 3 = 6 + 3$ — *Using the Addition Property of Equality, add 3 to each side.*

$x + 5 = 9$ — *The solution of this equation is also 4.*

The Addition Property of Equality can be modeled with algebra tiles. Model the equation $x - 3 = 5$. Each side of the mat represents a side of the equation. When you add 3 tiles to each side, the result is an equivalent equation, $x = 8$.

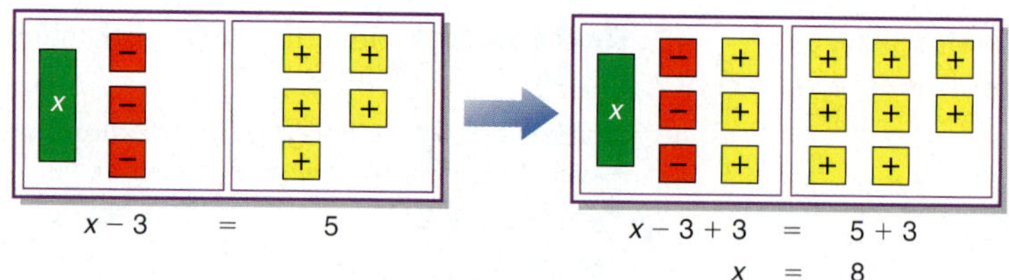

Resource Manager

Reproducible Masters

Chapter 3 Resource Masters
- *Study Guide*, p. 120
- *Skills Practice*, p. 121
- *Practice*, p. 122
- *Reading to Learn Mathematics*, p. 123
- *Enrichment*, p. 124

Hands-On Algebra, p. 47

Transparencies

5-Minute Check, 3–6

Technology/Multimedia

AlgePASS, Lesson 10

Interactive Chalkboard CD-ROM

Examples

1 **Solve each equation. Check your solution.**

$m - 10 = 14$

$$\begin{aligned} m - 10 &= 14 && \textit{Original equation} \\ m - 10 + 10 &= 14 + 10 && \textit{Add 10 to each side.} \\ m + 0 &= 24 && \textit{−10 + 10 = 0} \\ m &= 24 && \textit{Simplify.} \end{aligned}$$

To check that 24 is the solution, replace m in the original equation with 24.

Check:

$$\begin{aligned} m - 10 &= 14 && \textit{Original equation} \\ 24 - 10 &\stackrel{?}{=} 14 && \textit{Replace m with 24.} \\ 14 &= 14 \ \checkmark && \textit{24 − 10 = 14} \end{aligned}$$

The solution is 24.

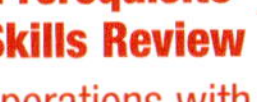

Prerequisite Skills Review
Operations with Decimals, p. 684

2 $-6.2 + x = 15.5$

$$\begin{aligned} -6.2 + x &= 15.5 && \textit{Original equation} \\ -6.2 + x + 6.2 &= 15.5 + 6.2 && \textit{Add 6.2 to each side.} \\ x + 0 &= 21.7 && \textit{Simplify.} \\ x &= 21.7 && \textit{Simplify.} \end{aligned}$$

To check that 21.7 is the solution, replace x with 21.7.

Check:

$$\begin{aligned} -6.2 + x &= 15.5 && \textit{Original equation} \\ -6.2 + 21.7 &\stackrel{?}{=} 15.5 && \textit{Replace x with 21.7.} \\ 15.5 &= 15.5 && \checkmark \ \textit{−6.2 + 21.7 = 15.5} \end{aligned}$$

The solution is 21.7.

Your Turn

a. $a - 5 = -14$ **−9**

b. $-20.4 = k + 7.8$ **−28.2**

In addition to the Addition Property of Equality, there is also a **Subtraction Property of Equality**.

Subtraction Property of Equality		
	Words:	If you subtract the same number from each side of an equation, the two sides remain equal.
	Symbols:	For any numbers a, b, and c, if $a = b$, then $a - c = b - c$.
	Numbers:	If $x = 5$, then $x - 3 = 5 - 3$.

www.algconcepts.com/extra_examples

2 TEACH

Teaching Tip Discuss the algebra tile models at the top of the page in detail, emphasizing that the Addition Property of Equality allows them to do symbolically what they did physically with algebra tiles in the previous lesson.

Teaching Tip In Example 1, make sure that students know why 10 was added to each side. Also, emphasize that the goal is to isolate the variable, which means to get the variable alone on one side of the equation. Point out the similarity of this process to algebra tile modeling where the goal was to get the x-tile alone on one side of the mat.

In-Class Examples

Examples 1–2

Solve each equation. Check your solution.

1 $r + (-13) = 15$ **28**

2 $-4.8 + y = -13.7$ **−8.9**

In-Class Examples

Example 3

Solve $k + 12 = -6$. Check your solution. **−18**

Example 4

Yoko was born in 1982 and her great-grandfather Hideo was born in 1917. Use the equation $1917 + n = 1982$ to find the number of years between their births. **65 years**

Example 3 Solve $9 + a = -3$. Check your solution.

$9 + a = -3$	*Original equation*
$9 + a - 9 = -3 - 9$	*Subtract 9 from each side.*
$a = -12$	*Simplify.*

Check:

$9 + a = -3$	*Original equation*
$9 + (-12) \stackrel{?}{=} -3$	*Replace a with −12.*
$-3 = -3$ ✓	*Add −12 to 9.*

The solution is −12.

Sometimes the Subtraction Property of Equality can be used instead of the Addition Property of Equality. Recall that subtracting a number is the same as adding its inverse.

Example 4 **Art Link**

Romanian Constantin Brancusi and American David Smith are successful twentieth-century sculptors. Brancusi's *Bird in Space* and Smith's *Cubi XVIII* were created in 1918 and 1964, respectively. Use the equation $1918 + y = 1964$ to find the number of years between these two sculptures.

Bird in Space

Method 1: Use the Subtraction Property of Equality.

$1918 + y = 1964$	*Original equation*
$1918 + y - 1918 = 1964 - 1918$	*Subtract 1918 from each side.*
$y = 46$	

Check:

$1918 + y = 1964$	*Original equation*
$1918 + 46 \stackrel{?}{=} 1964$	*Replace y with 46.*
$1964 = 1964$ ✓	

Method 2: Use the Addition Property of Equality.

$1918 + y = 1964$	*Original equation*
$1918 + y + (-1918) = 1964 + (-1918)$	*Add −1918 to each side.*
$y = 46$	*The answers are the same.*

The solution is 46. So, the sculptures were created 46 years apart.

Example 5

Solve $\frac{5}{6} = y - \left(-\frac{2}{3}\right)$. Check your solution.

Prerequisite Skills Review
Adding and Subtracting Fractions, p. 687

$\frac{5}{6} = y - \left(-\frac{2}{3}\right)$	*Original equation*
$\frac{5}{6} = y + \frac{2}{3}$	*Rewrite the equation. Why?*
$\frac{5}{6} - \frac{2}{3} = y + \frac{2}{3} - \frac{2}{3}$	*Subtract $\frac{2}{3}$ from each side.*
$\frac{5}{6} - \frac{4}{6} = y$	*Rewrite $\frac{2}{3}$ as $\frac{4}{6}$.*
$\frac{1}{6} = y$	*Simplify.*

Check:

$\frac{5}{6} = y - \left(-\frac{2}{3}\right)$	*Original equation*
$\frac{5}{6} \stackrel{?}{=} \frac{1}{6} - \left(-\frac{2}{3}\right)$	*Replace y with $\frac{1}{6}$.*
$\frac{5}{6} \stackrel{?}{=} \frac{1}{6} + \frac{2}{3}$	*Rewrite the equation.*
$\frac{5}{6} \stackrel{?}{=} \frac{1}{6} + \frac{4}{6}$	*Rewrite $\frac{2}{3}$ as $\frac{4}{6}$.*
$\frac{5}{6} = \frac{5}{6}$ ✓	*Add $\frac{1}{6}$ and $\frac{4}{6}$.*

The solution is $\frac{1}{6}$.

Your Turn Solve each equation. Check your solution.

c. $h + 8.5 = 3.4$ **−5.1**

d. $-\frac{4}{7} = b - \left(-\frac{3}{7}\right)$ **−1**

Check for Understanding

Communicating Mathematics

Vocabulary
equivalent equations

1. **Explain** how you can show that −4 is a solution of $12 + m = 8$.
2. **Write** two equivalent equations.
3. **Complete** the following statement. Justify your answer.
   If $x - 5 = 12$, then $x - 2 =$ ___?___ .
4. **You Decide** Malcom thinks that $y - (-24) = -36$ is equivalent to $y + (-24) = -36$. Adita disagrees. Who is correct, and why?

1–4. See margin.

Guided Practice

Getting Ready State the number you would add to or subtract from each side of the equation to solve it.

Sample 1: $c + 16 = 14$ **Solution:** Subtract 16 from each side.
Sample 2: $-3 + h = -13$ **Solution:** Add 3 to each side.

5. $y + 21 = -7$ **− 21**
6. $-10 + p = 25$ **+ 10**
7. $34 = d + (-9)$ **+ 9**
8. $14 = 16 + s$ **− 16**
9. $u - (-9.3) = -53.1$ **− 9.3**
10. $8 = a + (-5)$ **+ 5**

5–10. Sample answers given.

In-Class Example

Example 5

Solve $x - \left(-\frac{2}{5}\right) = \frac{7}{10}$. Check your solution. $\frac{3}{10}$

3 PRACTICE/APPLY

Error Analysis

Watch for students who have trouble identifying what number to add to or subtract from each side of an equation in Exercises 7, 9, and 10.
Prevent by stressing that the operation symbol changes but the sign of the number remains the same.

Answers

1. Replace *m* with −4 to see if 12 + *m* is equal to 8.
2. Sample answer: 9 − (−8) = 4 and 9 + 8 = 4
3. 15; In the first equation, add 5 to each side to get *x* = 17. By replacing *x* with 17 in the second equation, the result is 17 − 2 or 15.
4. Adita; *y* − (−24) = −36 is equivalent to *y* + 24 = −36.

***Study Guide*, p. 120**

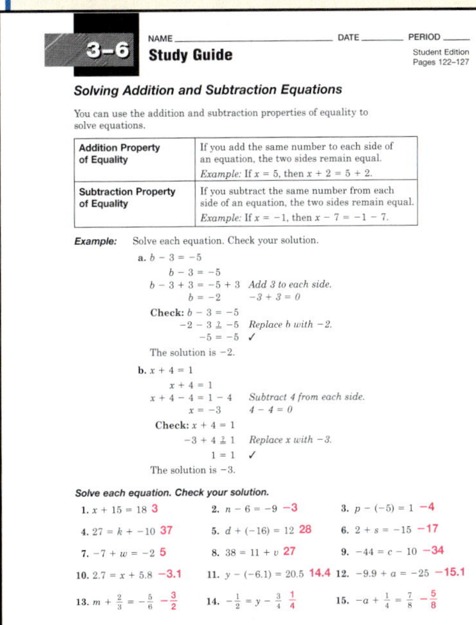

NAME ______________ DATE ______ PERIOD _____

3–6 Study Guide

Student Edition Pages 122–127

Solving Addition and Subtraction Equations

You can use the addition and subtraction properties of equality to solve equations.

Addition Property of Equality	If you add the same number to each side of an equation, the two sides remain equal. *Example:* If $x = 5$, then $x + 2 = 5 + 2$.
Subtraction Property of Equality	If you subtract the same number from each side of an equation, the two sides remain equal. *Example:* If $x = -1$, then $x - 7 = -1 - 7$.

Example: Solve each equation. Check your solution.

a. $b - 3 = -5$

$b - 3 = -5$
$b - 3 + 3 = -5 + 3$ *Add 3 to each side.*
$b = -2$ *$-3 + 3 = 0$*

Check: $b - 3 = -5$
$-2 - 3 \stackrel{?}{=} -5$ *Replace b with −2.*
$-5 = -5$ ✓

The solution is −2.

b. $x + 4 = 1$

$x + 4 = 1$
$x + 4 - 4 = 1 - 4$ *Subtract 4 from each side.*
$x = -3$ *$4 - 4 = 0$*

Check: $x + 4 = 1$
$-3 + 4 \stackrel{?}{=} 1$ *Replace x with −3.*
$1 = 1$ ✓

The solution is −3.

Solve each equation. Check your solution.

1. $x + 15 = 18$ **3**
2. $n - 6 = -9$ **−3**
3. $p - (-5) = 1$ **−4**
4. $27 = k + -10$ **37**
5. $d + (-16) = 12$ **28**
6. $2 + s = -15$ **−17**
7. $-7 + w = -2$ **5**
8. $38 = 11 + v$ **27**
9. $-44 = c - 10$ **−34**
10. $2.7 = x + 5.8$ **−3.1**
11. $y - (-6.1) = 20.5$ **14.4**
12. $-9.9 + a = -25$ **−15.1**
13. $m + \frac{2}{3} = -\frac{5}{6}$ $-\frac{3}{2}$
14. $-\frac{1}{2} = y - \frac{3}{4}$ $\frac{1}{4}$
15. $-a + \frac{1}{4} = \frac{7}{8}$ $-\frac{5}{8}$

Reteaching Activity

Intrapersonal Learners Have students make a two-column table. In the first column, have them list all the ways they might make a mistake when solving addition and subtraction equations. In the second column, directly opposite each possible mistake they listed, have students write a suggestion to themselves about how they could avoid making the mistake.

Assignment Guide

Basic: 19–41 odd, 43–51
Average: 18–40 even, 41–51
All: Quiz 2, 1–5

Examples 1, 2, 4, & 5

Solve each equation. Check your solution.

11. $h + 8 = -7$ **−15**
12. $4 = -5 + d$ **9**
13. $2.75 + x = 3.85$ **1.1**
14. $x + 12 = -32$ **−44**
15. $h - \frac{1}{2} = -\frac{2}{3}$ $\mathbf{-\frac{1}{6}}$
16. $8 = x - (-12)$ **−4**

Example 3

17. **History** The state of Georgia entered the union in 1788. New Mexico entered the union 124 years later.
    a. Write an equation that could be used to find the year New Mexico entered the union. **Sample answer: $x - 1788 = 124$**
    b. When did New Mexico enter the union? **1912**

Exercises

Practice

Solve each equation. Check your solution.

A

18. $k + (-8) = 29$ **37**
19. $y - (-7) = 3$ **−4**
20. $-9 = 8 + q$ **−17**
21. $b - 14 = -11$ **3**
22. $-21 = 5 + c$ **−26**
23. $-3 + m = 10$ **13**
24. $27 + y = -49$ **−76**
25. $7 + w = -6$ **−13**
26. $11 = r + (-9)$ **20**
27. $-15 + y = -63$ **−48**
28. $-39 = n - 24$ **−15**
29. $52 = 18 + j$ **34**
30. $k - 14.6 = 9.2$ **23.8**
31. $3.4 = m + 7.2$ **−3.8**
32. $-3.41 + n = 0.23$
33. $\frac{5}{6} = g - \left(-\frac{1}{6}\right)$ $\mathbf{\frac{2}{3}}$
34. $\frac{3}{4} + y = \frac{7}{8}$ $\mathbf{\frac{1}{8}}$
35. $g + \left(-\frac{4}{9}\right) = \frac{2}{3}$ $\mathbf{\frac{10}{9}}$ **or** $\mathbf{1\frac{1}{9}}$

Homework Help	
For Exercises	See Examples
18–32, 36, 37	1–3
33–35	5
38–41	4

Extra Practice

See page 698.

32. 3.64

36. Solve for s if $7 + s = -10$. **−17**

B

37. What is the solution of the equation $g + (-2) = 11$? **13**

Write an equation and solve. **38. $n + (-22) = 45$; $n = 67$**

38. A number increased by −22 is 45. Find the number.
39. What number decreased by −15 is 20? **$m - (-15) = 20$; $m = 5$**
40. The sum of a number and 8 is −47. What is the number?
    $p + 8 = -47$; $p = -55$

Applications and Problem Solving

41. **Inflation** Today, the cost of a 50-pound box of nails is about \$32.50. This is \$27.16 more than the cost of the same size box of nails sold in 1800.
    Source: *Family Handyman*
    a. Write an equation that could be used to find the cost of a 50-pound box of nails sold in 1800. **$c + 27.16 = 32.50$**
    b. In 1800, how much did a 50-pound box of nails cost? **\$5.34**

***Skills Practice,* p. 121, and *Practice,* p. 122 (shown)**

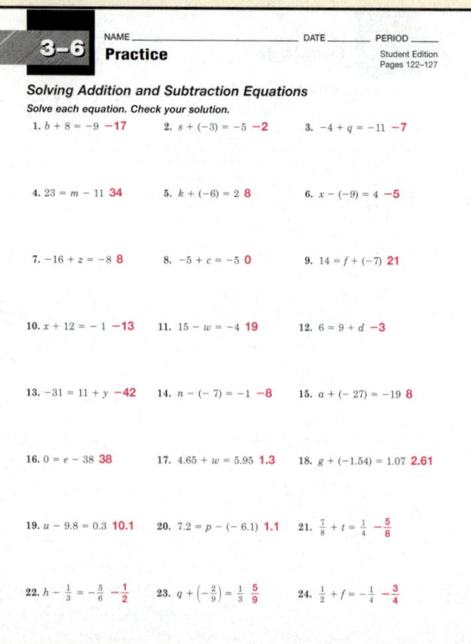

3-6 NAME ______________ DATE ________ PERIOD _____

Practice

Student Edition Pages 122–127

Solving Addition and Subtraction Equations

Solve each equation. Check your solution.

1. $b + 8 = -9$ **−17**
2. $s + (-3) = -5$ **−2**
3. $-4 + q = -11$ **−7**
4. $23 = m - 11$ **34**
5. $k + (-6) = 2$ **8**
6. $x - (-9) = 4$ **−5**
7. $-16 + z = -8$ **8**
8. $-5 + c = -5$ **0**
9. $14 = f + (-7)$ **21**
10. $x + 12 = -1$ **−13**
11. $15 - w = -4$ **19**
12. $6 = 9 + d$ **−3**
13. $-31 = 11 + y$ **−42**
14. $n - (-7) = -1$ **−8**
15. $a + (-27) = -19$ **8**
16. $0 = e - 38$ **38**
17. $4.65 + w = 5.95$ **1.3**
18. $g + (-1.54) = 1.07$ **2.61**
19. $u - 9.8 = 0.3$ **10.1**
20. $7.2 = p - (-6.1)$ **1.1**
21. $\frac{7}{8} + t = \frac{1}{4}$ $\mathbf{-\frac{5}{8}}$
22. $h - \frac{1}{3} = -\frac{5}{6}$ $\mathbf{-\frac{1}{2}}$
23. $q + \left(-\frac{2}{9}\right) = \frac{1}{3}$ $\mathbf{\frac{5}{9}}$
24. $\frac{1}{2} + f = -\frac{1}{4}$ $\mathbf{-\frac{3}{4}}$

interNET CONNECTION For the latest information on lighthouses, visit: www.algconcepts.com

42. **Lighthouses** There are an estimated 850 lighthouses in the United States. The tallest is located in Cape Hatteras, North Carolina. It is 27 feet taller than the one located in Cape Lookout, North Carolina. The height of the Cape Lookout lighthouse is 169 feet.
    a. Write an equation that could be used to find the height of the lighthouse located in Cape Hatteras, North Carolina.
    b. What is the height of the lighthouse? **196 ft**

42a. $t - 27 = 169$

43. **Critical Thinking** What value of x makes $x + x = x$ true? **0**

Mixed Review

44. **Sports** On the first day of an LPGA golf tournament, professional golfer Pearl Sinn's score was +4 or 4 over par. Her score on the second day was added to the first day's score, and the total was +1 or 1 over par. What was Sinn's score on the second day? *(Lesson 3–5)*
    a. Let s represent Sinn's score on the second day. Then translate the problem into an equation using the variable s. $4 + s = 1$
    b. Solve for s to find Sinn's score on the second day. **−3 or 3 under par**

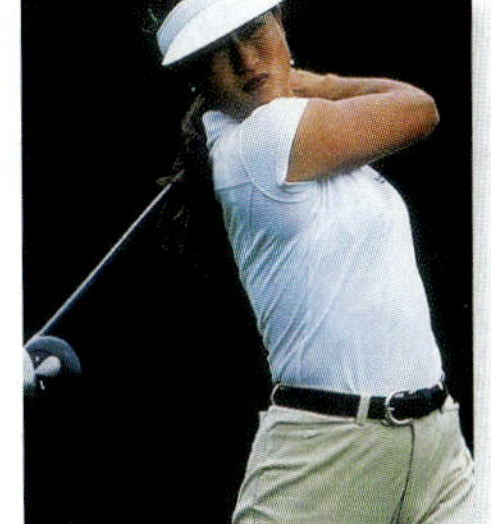

Pearl Sinn

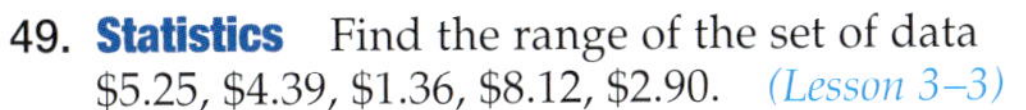

Solve each equation. *(Lesson 3–4)*

45. $3.75 - 0.5 = m$ **3.25**
46. $x = -4(8) + 6$ **−26**
47. $(-15 \div -3) + 1 = y$ **6**
48. $t = [3 - (-5)] \cdot 7$ **56**
49. **Statistics** Find the range of the set of data \$5.25, \$4.39, \$1.36, \$8.12, \$2.90. *(Lesson 3–3)* **\$6.76**

Standardized Test Practice

50. **Short Response** Find the quotient of 36 and −4. *(Lesson 2–6)* **−9**
51. **Multiple Choice** Choose the set of numbers that is correctly ordered from greatest to least. *(Lesson 2–1)* **D**
    (A) {−3, 2, 1, 0}
    (B) {5, −4, 0, −1}
    (C) {−4, −3, −2, 0}
    (D) {7, 4, −1, −3}

Quiz 2 Lessons 3–5 and 3–6

Solve each equation. Check your solution. *(Lessons 3–5 & 3–6)*

1. $y - 3 = -7$ **−4**
2. $6 = x + 9$ **−3**
3. $17.8 = s + (-5.2)$ **23**
4. $-\frac{5}{6} = p - \frac{1}{2}$ $-\frac{1}{3}$

5a. Let c = cost of Chicago meal; $3.89 = c - 0.62$.

5. **Dining** In Sacramento, California, the average cost of a fast-food meal was \$3.89 in a recent year. This was 62 cents less than the average cost of a fast-food meal in Chicago, Illinois. What was the average price of a fast-food meal in Chicago? **Source:** Runzheimer International, 1999
    a. Choose a variable and write an equation that can be used to find the average cost of a fast-food meal in Chicago.
    b. Find the average cost of a fast-food meal in Chicago. **\$4.51**

www.algconcepts.com/self_check_quiz

4 ASSESS

Open-Ended Assessment

Speaking Have students explain how the Addition Property of Equality and the Subtraction Property of Equality are used to solve equations.

Quiz 2

The Quiz provides students with a brief review of the concepts and skills in Lessons 3–5 and 3–6. Lesson numbers are given to the right of the exercises or instruction lines so students can review concepts not yet mastered.

Extra Credit

To solve an *addition* equation, you *subtract* the same number from both sides of an equation. What do you think you should do to solve equations like $2n = 8$ and $-3x = 12$? Write down your idea, try it, and then check your solution. **Divide both sides of the equation by the integer that is multiplied times the variable.**

***Enrichment*, p. 124**

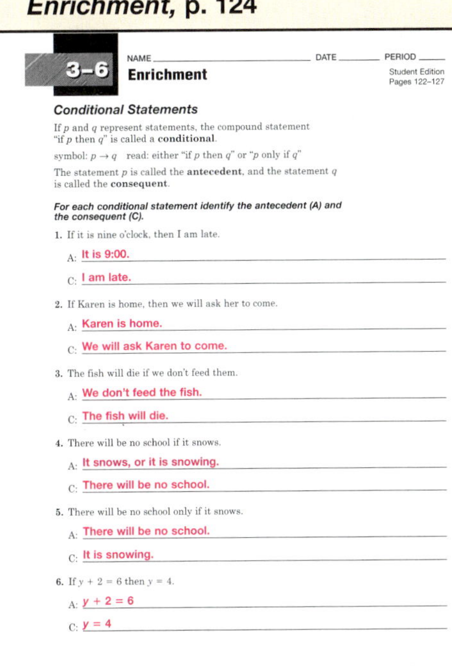

3–6 NAME ____ DATE ____ PERIOD ____

Enrichment

Student Edition Pages 122–127

Conditional Statements

If p and q represent statements, the compound statement "if p then q" is called a **conditional**.

symbol: $p \rightarrow q$ read: either "if p then q" or "p only if q"

The statement p is called the **antecedent**, and the statement q is called the **consequent**.

For each conditional statement identify the antecedent (A) and the consequent (C).

1. If it is nine o'clock, then I am late.
   A: It is 9:00.
   C: I am late.
2. If Karen is home, then we will ask her to come.
   A: Karen is home.
   C: We will ask Karen to come.
3. The fish will die if we don't feed them.
   A: We don't feed the fish.
   C: The fish will die.
4. There will be no school if it snows.
   A: It snows, or it is snowing.
   C: There will be no school.
5. There will be no school only if it snows.
   A: There will be no school.
   C: It is snowing.
6. If $y + 2 = 6$ then $y = 4$.
   A: $y + 2 = 6$
   C: $y = 4$

Lesson 3-7

1 FOCUS

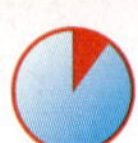

5-Minute Check
Lesson 3–6

Maria made the timeline below, which gives the birth years of three members of her family.

Year of Birth

1932 Alison born; 1956 Emma born; 1984 Maria born

1930 1940 1950 1960 1970 1980 1990

1. Write and solve an equation to represent the number of years between Maria's birth and her mother Emma's birth. **1956 + y = 1984; 28 years**
2. Write and solve an equation to represent the number of years between Maria's birth and her grandmother Alison's birth. **1932 + x = 1984; 52 years**

Solve each equation. Check your solution.

3. $-16 = 6 + t$ **−22**
4. $k - 18.9 = 8.4$ **27.3**
5. $\frac{5}{6} = x - \left(-\frac{1}{2}\right)$ $\frac{1}{3}$

Motivating the Lesson

Hands-On Activity Separate the class into small groups. Have each group draw a number line labeled with the integers from −10 to 10. Have them plot all of the points that are 4 units away from 0. Repeat the activity using 5 units and 10 units in place of 4 units. Have each person in the group write a sentence to summarize their work.

2 TEACH

In-Class Example
Example 1

Solve $|d - 4| = 3$. Check your solution. **{1, 7}**

3–7 Solving Equations Involving Absolute Value

What You'll Learn
You'll learn to solve equations involving absolute value.

Why It's Important
Science Knowing how to solve equations involving absolute value can help you determine greatest and least values.
See Example 4.

Some equations involve absolute value. Consider the graph of the equation $|a| = 4$.

$$|a| = 4$$

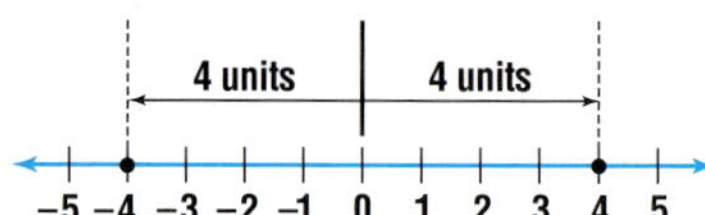

The distance from 0 to a is 4 units. Therefore, $a = -4$ or $a = 4$. When an equation has more than one solution, the solutions are often written as a set $\{a, b\}$. The solution set is $\{-4, 4\}$.

Look Back
Absolute Value: Lesson 2–1

Equations that involve absolute value can be solved by using a number line or by writing them as compound sentences and solving them.

Example 1 **Solve $|x - 3| = 5$. Check your solution.**

Method 1: Use a number line.

$|x - 3| = 5$ means the distance between x and 3 is 5 units. So, to find x on the number line, start at 3 and move 5 units in either direction.

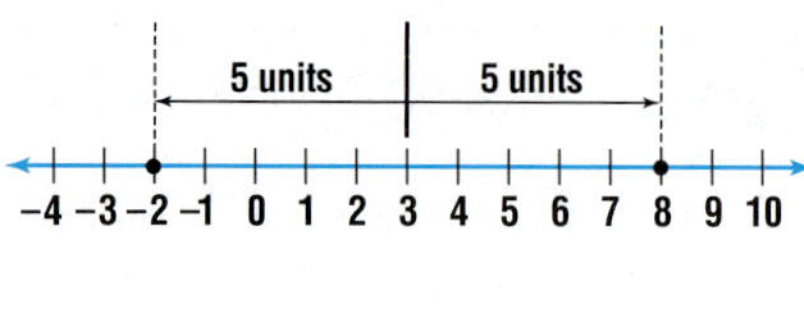

$x = -2$ $\qquad$ $x = 8$

Method 2: Write and solve a compound sentence.

$|x - 3| = 5$ also means $x - 3 = 5$ or $x - 3 = -5$.

$x - 3 = 5$	or	$x - 3 = -5$
$x - 3 + 3 = 5 + 3$	*Add 3 to each side.*	$x - 3 + 3 = -5 + 3$
$x = 8$		$x = -2$

Resource Manager

Reproducible Masters

Chapter 3 Resource Masters
- *Study Guide,* p. 125
- *Skills Practice,* p. 126
- *Practice,* p. 127
- *Reading to Learn Mathematics,* p. 128
- *Enrichment,* p. 129
- *Assessment,* p. 141

Hands-On Algebra, p. 48

Transparencies
5-Minute Check, 3–7

Technology/Multimedia
Interactive Chalkboard CD-ROM

Check: *Replace x with 8.*

$$|x - 3| = 5$$
$$|8 - 3| \stackrel{?}{=} 5$$
$$|5| \stackrel{?}{=} 5$$
$$5 = 5 \ \checkmark$$

Replace x with −2.

$$|x - 3| = 5$$
$$|-2 - 3| \stackrel{?}{=} 5$$
$$|-5| \stackrel{?}{=} 5$$
$$5 = 5 \ \checkmark$$

The solution set is $\{-2, 8\}$.

Your Turn **Solve each equation. Check your solution.**

a. $|c - 4| = 2$ **{2, 6}** b. $6 = |5 + h|$ **{−11, 1}**

When solving an equation involving an absolute value symbol, you can think of the absolute value bars as grouping symbols.

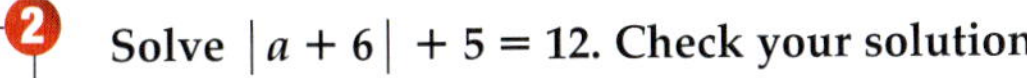

Example 2 Solve $|a + 6| + 5 = 12$. Check your solution.

To solve the equation, first simplify the expression.

$$|a + 6| + 5 = 12 \quad \text{Original equation}$$
$$|a + 6| + 5 - 5 = 12 - 5 \quad \text{Subtract 5 from each side.}$$
$$|a + 6| = 7 \quad \text{Simplify.}$$

Next, write a compound sentence and solve it.

$a + 6 = 7$ or $a + 6 = -7$

$a + 6 - 6 = 7 - 6$ *Subtract 6 from each side.* $a + 6 - 6 = -7 - 6$

$a = 1$ $a = -13$

Check: *Replace a with 1.*

$$|a + 6| + 5 = 12$$
$$|1 + 6| + 5 \stackrel{?}{=} 12$$
$$|7| + 5 \stackrel{?}{=} 12$$
$$7 + 5 \stackrel{?}{=} 12$$
$$12 = 12 \ \checkmark$$

Replace a with −13.

$$|a + 6| + 5 = 12$$
$$|-13 + 6| + 5 \stackrel{?}{=} 12$$
$$|-7| + 5 \stackrel{?}{=} 12$$
$$7 + 5 \stackrel{?}{=} 12$$
$$12 = 12 \ \checkmark$$

The solution set is $\{1, -13\}$.

Your Turn **Solve each equation. Check your solution.**

c. $|m + 5| - 4 = 18$ **{−27, 17}** d. $13 = |-8 + d| + 2$ **{−3, 19}**

Sometimes an equation has no solution. For example, $|x| = -6$ is never true. The absolute value of a number is always positive or zero. So, there is no replacement for x that will make the sentence true. The solution set has no members. It is called the **empty set**, and its symbols are { } or ∅.

www.algconcepts.com/extra_examples

In-Class Example

Example 2

Solve $|g + 3| - 2 = 6$. Check your solution. **{−11, 5}**

Teaching Tip Show students how they can use the number line method shown in Example 1 to complete Example 2 by rewriting the equation $|a + 6| = 7$ as $|a - (-6)| = 7$. This equation can be interpreted verbally as "the distance between a and -6 is 7." Locating -6 on a number line and then moving 7 units in both directions gives the solutions of the equation, 1 and -13.

***Study Guide*, p. 125**

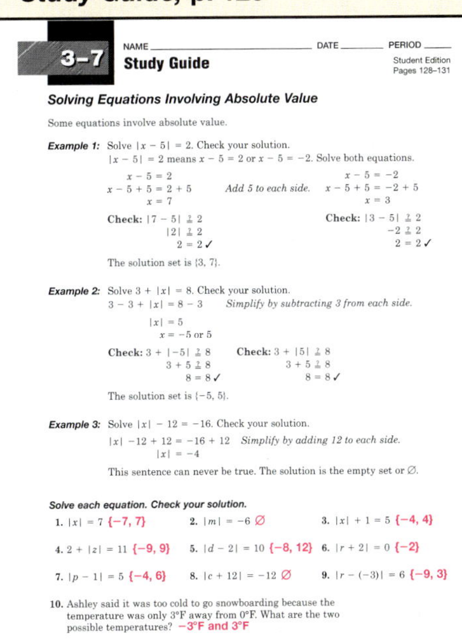

3–7 NAME ______ DATE ______ PERIOD ______

Study Guide Student Edition Pages 128–131

Solving Equations Involving Absolute Value

Some equations involve absolute value.

Example 1: Solve $|x - 5| = 2$. Check your solution.
$|x - 5| = 2$ means $x - 5 = 2$ or $x - 5 = -2$. Solve both equations.

$x - 5 = 2$; $x - 5 + 5 = 2 + 5$ *Add 5 to each side.* $x = 7$ | $x - 5 = -2$; $x - 5 + 5 = -2 + 5$; $x = 3$

Check: $|7 - 5| \stackrel{?}{=} 2$; $|2| \stackrel{?}{=} 2$; $2 = 2 \checkmark$ | **Check:** $|3 - 5| \stackrel{?}{=} 2$; $-2 \stackrel{?}{=} 2$; $2 = 2 \checkmark$

The solution set is {3, 7}.

Example 2: Solve $3 + |x| = 8$. Check your solution.
$3 - 3 + |x| = 8 - 3$ *Simplify by subtracting 3 from each side.*
$|x| = 5$
$x = -5$ or 5

Check: $3 + |-5| \stackrel{?}{=} 8$; $3 + 5 \stackrel{?}{=} 8$; $8 = 8 \checkmark$ | **Check:** $3 + |5| \stackrel{?}{=} 8$; $3 + 5 \stackrel{?}{=} 8$; $8 = 8 \checkmark$

The solution set is {−5, 5}.

Example 3: Solve $|x| - 12 = -16$. Check your solution.
$|x| - 12 + 12 = -16 + 12$ *Simplify by adding 12 to each side.*
$|x| = -4$

This sentence can never be true. The solution is the empty set or ∅.

Solve each equation. Check your solution.

1. $|x| = 7$ **{−7, 7}**
2. $|m| = -6$ **∅**
3. $|x| + 1 = 5$ **{−4, 4}**
4. $2 + |z| = 11$ **{−9, 9}**
5. $|d - 2| = 10$ **{−8, 12}**
6. $|r + 2| = 0$ **{−2}**
7. $|p - 1| = 5$ **{−4, 6}**
8. $|c + 12| = -12$ **∅**
9. $|r - (-3)| = 6$ **{−9, 3}**
10. Ashley said it was too cold to go snowboarding because the temperature was only 3°F away from 0°F. What are the two possible temperatures? **−3°F and 3°F**

In-Class Examples

Example 3

Solve the equation $|y + 5| - 2 = -7$. $\varnothing$

Example 4

In a survey, it was found that 78% of voters in a school district favored building a new high school. It is estimated that the actual number of voters that favor building the school differs from 78% by no more than 5%. Write and then solve an equation that could be used to find the least and greatest percentage of voters that favor building a new high school. **$|x - 78| = 5$; The least percentage is 73% and the greatest percentage is 83%.**

3 PRACTICE/APPLY

Error Analysis

Watch for students who try to apply the Addition Property of Equality to solve equations such as those given in Exercises 6–9. ***Prevent by*** having students practice finding solutions to absolute value equations using number lines. This method will emphasize that there is usually more than one solution to an absolute value equation.

***Skills Practice,* p. 126, and *Practice,* p. 127 (shown)**

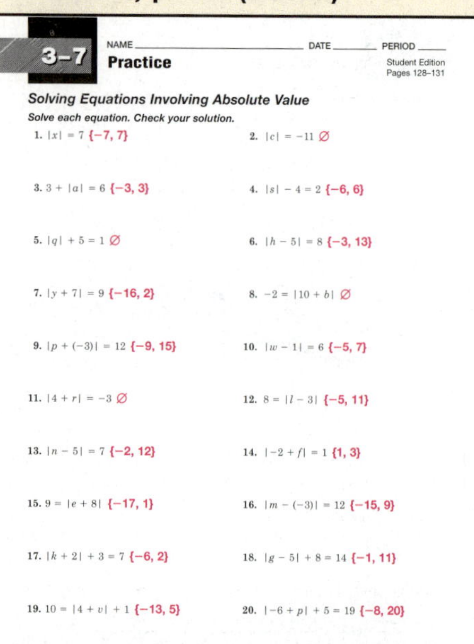

3–7 **Practice**

NAME ______ DATE ______ PERIOD ______

Student Edition Pages 128–131

Solving Equations Involving Absolute Value

Solve each equation. Check your solution.

1. $|x| = 7$ **{−7, 7}**
2. $|c| = -11$ $\varnothing$
3. $3 + |a| = 6$ **{−3, 3}**
4. $|s| - 4 = 2$ **{−6, 6}**
5. $|q| + 5 = 1$ $\varnothing$
6. $|h - 5| = 8$ **{−3, 13}**
7. $|y + 7| = 9$ **{−16, 2}**
8. $-2 = |10 + b|$ $\varnothing$
9. $|p + (-3)| = 12$ **{−9, 15}**
10. $|w - 1| = 6$ **{−5, 7}**
11. $|4 + r| = -3$ $\varnothing$
12. $8 = |l - 3|$ **{−5, 11}**
13. $|n - 5| = 7$ **{−2, 12}**
14. $|-2 + f| = 1$ **{1, 3}**
15. $9 = |e + 8|$ **{−17, 1}**
16. $|m - (-3)| = 12$ **{−15, 9}**
17. $|k + 2| + 3 = 7$ **{−6, 2}**
18. $|g - 5| + 8 = 14$ **{−1, 11}**
19. $10 = |4 + v| + 1$ **{−13, 5}**
20. $|-6 + p| + 5 = 19$ **{−8, 20}**

Example 3 Solve $|d| + 7 = 2$. Check your solution.

First, simplify the expression.

$|d| + 7 = 2$ *Original equation*

$|d| + 7 - 7 = 2 - 7$ *Subtract 7 from each side.*

$|d| = -5$ *Simplify.*

This sentence can never be true. The solution is the empty set or $\varnothing$.

Your Turn

e. $3 = |s - 4| + 9$ $\varnothing$

f. $|w| - 18 = -6$ **{−12, 12}**

Science Link

4 **Hydrogen can exist as a solid, liquid, or gas. For hydrogen to be a liquid, its temperature must be within 2° of −257°C. Write and then solve an equation that can be used to find the least and greatest temperatures at which hydrogen is a liquid.**

Let t represent temperature. t differs from −257 by exactly 2 degrees. Write an equation that represents the least and greatest temperatures.

The difference between t and −257 *is* *2 degrees.*

$$|t - (-257)| = 2$$

$t + 257 = 2$	*Rewrite the equation.*	$t + 257 = -2$
$t + 257 - 257 = 2 - 257$	*Subtract 257.*	$t + 257 - 257 = -2 - 257$
$t = -255$	*Simplify.*	$t = -259$

The solutions are −255 and −259. So, the least temperature for hydrogen to remain a liquid is −259°C, and the greatest temperature is −255°C.

Check for Understanding

Communicating Mathematics

1. **Write** an absolute value equation that has no solution. **1–3. See margin.**
2. **Draw** a number line to show the solution of $|d - 2| = 1$.
3. Writing Math Explain why an absolute value equation can have two solutions.

Vocabulary

empty set

Guided Practice

Examples 1 & 3

Solve each equation. Check your solution.

4. $|a| = 6$ **{−6, 6}**
5. $|w| - 2 = 1$ **{−3, 3}**
6. $|x + 1| = 4$ **6. {−5, 3}**
7. $-5 = |w - 9|$ $\varnothing$
8. $|3 + g| = 7$ **{−10, 4}**
9. $|h - 3| + 6 = 16$ **{−7, 13}**

Reteaching Activity

Visual/Spatial Learners Have students use number lines to solve several of the equations in Exercises 12–29. When necessary, be sure that students simplify the equation using the Addition and Subtraction Properties of Equality so that the absolute value expression is isolated on one side of the equation before using the number line. Stress that they also need to convert any addition expressions within the absolute value symbols to subtraction expressions. For example, $|z + 5|$ should be rewritten as $|z - (-5)|$ and thought of as the distance between z and -5.

Example 3

10. The absolute value of the sum of 6 and a number is 2. What is the number? **−8 or −4**

Example 4

11. **Sports** Ricardo's bowling score was within 6 points of his average score of 145. Write and solve an equation that can be used to find the least and greatest bowling score Ricardo could have received in this game. **$|s - 145| = 6$; {139, 151}**

15. {−6, 6}
16. {−2, 14}
18. {−3, −1}
20. {1, 17}
23. {−21, 1}

Exercises

Practice

Homework Help	
For Exercises	See Examples
12, 13, 16–20, 22–25, 30–33	1
14, 15, 21	3
26–29, 34	2
Extra Practice	
See page 698.	

Solve each equation. Check your solution.

12. $|\ell| = -8$ **∅**
13. $|b| = 2$ **{−2, 2}**
14. $4 + |x| = 7$ **{−3, 3}**
15. $|y| - 10 = -4$
16. $|x - 6| = 8$
17. $|10 + n| = -2$ **∅**
18. $1 = |2 + p|$
19. $|-1 + y| = 0$ **{1}**
20. $8 = |-9 + q|$
21. $|s| + 12 = 6$ **∅**
22. $-9 = |2 + h|$ **∅**
23. $11 = |10 + d|$
24. $17 = |y + (-3)|$
25. $|z - (-6)| = 13$
26. $9 = 4 + |k + 1|$
27. $|b - 2| + 5 = 16$ **{−9, 13}**
28. $6 + |-5 + f| = 18$ **{−7, 17}**
29. $|2 + r| + 4 = 1$ **∅**
30. How many solutions exist for $|a + 3| = 6$? **2**
31. How many solutions exist for $|4 + b| = 0$? **1**
32. Is $-2 = |3 + c|$ sometimes, always, or never true? Explain. **Never, absolute value is always positive.**

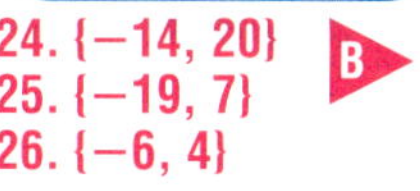

24. {−14, 20}
25. {−19, 7}
26. {−6, 4}

Applications and Problem Solving

33. **$|p - 165{,}000| = 10{,}000$; {155,000, 175,000}**
34. **$|w - 32| = 6$; {26, 38}**

For Exercises 33 and 34, write and solve an equation.

33. **Real Estate** Tamika Samuel is going to place an offer on a home. If she comes within \$10,000 of the asking price of \$165,000, she has a good chance of getting the home. Write and solve an equation to find out what her greatest and least offers should be.

34. **Shipping** The Jones Packaging Company needs to ship a crate of books to a book store. The crate must weigh within 6 kilograms of 40 kilograms. If the crate alone weighs 8 kilograms, what are the greatest and least amounts the books can weigh?

35. **Critical Thinking** Which number line represents the solution set of the equation $1 = -2 + |c + 1|$? **A**

A 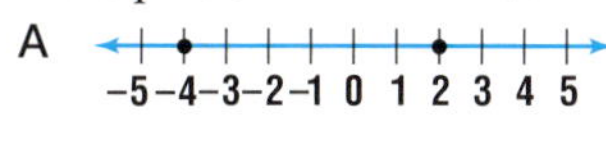

B 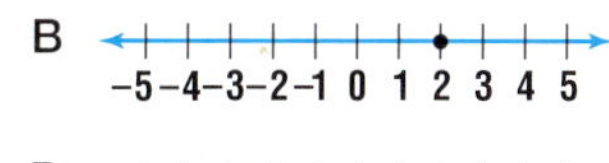

C 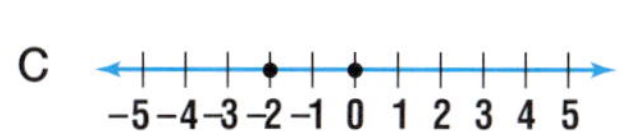

D 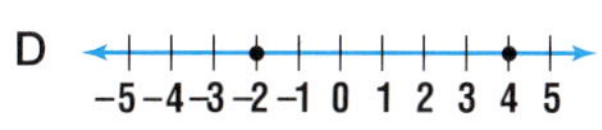

Mixed Review

Solve each equation. Check your solution. *(Lesson 3–6)*

36. $13 + w = -6$ **−19**
37. $t - 3 = 45$ **48**
38. $11.3 = h - 5.7$ **17**
39. What is the value of x if $9 + x = -5$? *(Lesson 3–5)* **−14**

Standardized Test Practice

40. **Grid In** If $y - (-8) = 12$, what is the value of y? *(Lesson 3–5)* **4**

41. **Multiple Choice** A store stocks five different brands of cookies priced at \$2.25, \$2.50, \$2.00, \$2.25, and \$1.85. What is the difference between the mode price and the mean price? *(Lesson 3–3)*

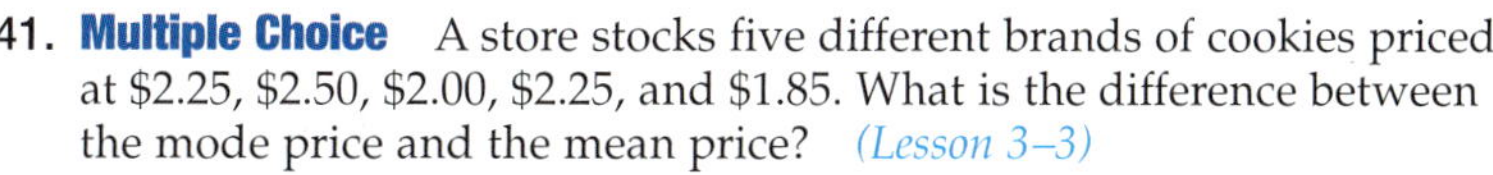

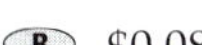 \$0.00 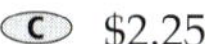 \$0.08  \$2.25 Ⓓ \$4.42 **B**

www.algconcepts.com/self_check_quiz

Assignment Guide

Basic: 13–33 odd, 35–41
Average: 12–32 even, 33–41

4 ASSESS

Open-Ended Assessment

Modeling Have students work with a partner. Write two absolute value equations on the board or overhead. One student should solve the first equation using a number line while the other student uses the compound sentence method. The students should compare their results and then switch roles to solve the second equation.

Chapter 3, Quiz B (Lessons 3–4 through 3–7) is available in the *Chapter 3 Resource Masters*, p. 141.

Answers

Page 130

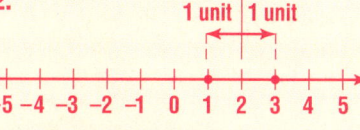

1. Sample answer: $|a - 2| = -5$
2. 1 unit | 1 unit; number line −5 −4 −3 −2 −1 0 1 2 3 4 5 with points at 1 and 3
   $d = 1$ or $d = 3$
3. Sample answer: The value inside the absolute value sign can either be positive or negative. You have to consider both cases, so there may be up to 2 solutions.

***Enrichment*, p. 129**

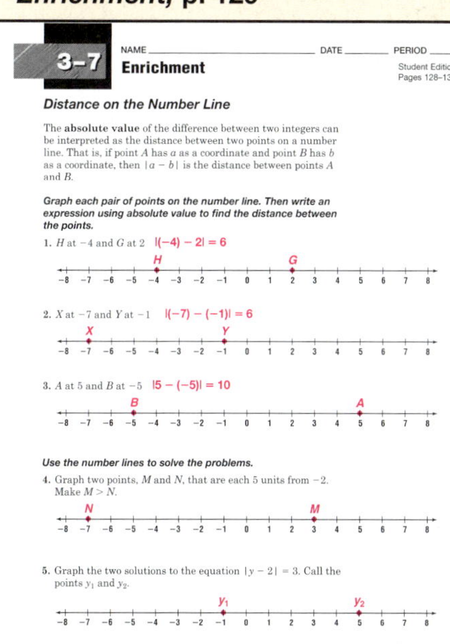

3–7 NAME ______ DATE ______ PERIOD ______

Enrichment

Student Edition Pages 128–131

Distance on the Number Line

The **absolute value** of the difference between two integers can be interpreted as the distance between two points on a number line. That is, if point A has a as a coordinate and point B has b as a coordinate, then $|a - b|$ is the distance between points A and B.

Graph each pair of points on the number line. Then write an expression using absolute value to find the distance between the points.

1. H at −4 and G at 2 $|(-4) - 2| = 6$
2. X at −7 and Y at −1 $|(-7) - (-1)| = 6$
3. A at 5 and B at −5 $|5 - (-5)| = 10$

Use the number lines to solve the problems.

4. Graph two points, M and N, that are each 5 units from −2. Make $M > N$.
5. Graph the two solutions to the equation $|y - 2| = 3$. Call the points y_1 and y_2.

Extra Credit

A polling company has surveyed some voters after a local election. They expect the actual voting results to differ by no more than 8 percentage points from the survey results. For mayor, 56% of those voters polled cast ballots for Reinhardt, while 42% of those polled voted for Chang. If the polling company's expectation about their data is correct, can you tell from the poll who won the race? Explain. **No. Sample explanation: Since the actual results can differ by 8%, Reinhardt may have gotten as little as 48% of the vote while Chang could have gotten as much as 50% of the vote. In this case Chang would win. If the survey results were closer to the actual results, Reinhardt would win.**

Understanding and Using the Vocabulary

This section provides a listing of the new terms, properties, and phrases that were introduced in this chapter. The exercises check students' understanding of the terms by using a variety of verbal formats including matching, completion, and true/false.

Glossary A complete glossary of terms appears on pages 762–783.

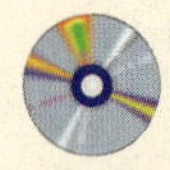

MindJogger Videoquizzes

MindJogger Videoquizzes provide an alternative review of concepts presented in this chapter. Students work in teams to answer questions, gaining points for correct answers.

CHAPTER 3 Study Guide and Assessment

Understanding and Using the Vocabulary

interNET CONNECTION **Review Activities** For more review activities, visit: www.algconcepts.com

After completing this chapter, you should be able to define each term, property, or phrase and give an example of each.

arithmetic sequence *(p. 110)*
bivariate *(p. 107)*
categorical *(p. 107)*
cross products *(p. 95)*
empty set *(p. 129)*
equivalent equations *(p. 122)*
inequality *(p. 95)*
mean *(p. 104)*
measurement *(p. 107)*
measure of central tendency *(p. 104)*
measure of variation *(p. 106)*
median *(p. 104)*
mode *(p. 104)*
open sentences *(p. 112)*
range *(p. 106)*
rational numbers *(p. 94)*
recursive formula *(p. 111)*
replacement set *(p. 112)*
sequence *(p. 110)*
solution *(p. 112)*
solving *(p. 112)*
statement *(p. 112)*
term *(p. 110)*
unit cost *(p. 97)*
univariate *(p. 107)*

Choose the correct term to complete each sentence.

1. The (median, mean) is the middle number when data are arranged in numerical order.
2. The (range, unit cost) is used to compare the price of similar items.
3. The difference between the greatest and least values of a set is the (mode, range).
4. When the same number is subtracted from each side of an equation, the result is a(n) (replacement set, equivalent equation).
5. A (rational number, mean) can be expressed in the form $\frac{a}{b}$, where $b \neq 0$.
6. A solution set with no members is the (replacement set, empty set).
7. An (inequality, equivalent equation) uses the symbols < and > to compare two expressions.
8. (Measures of central tendency, measures of variation) describe the distribution of data.
9. A (solution, rational number) can always be expressed as a fraction.
10. You can use (cross products, open sentences) to compare two fractions with different denominators.

Skills and Concepts

Objectives and Examples

- **Lesson 3–1** Compare and order rational numbers.

Replace the ● with <, >, or = to make a true sentence.

$\frac{2}{3}$ ● $\frac{4}{5}$

$5 \cdot 2$ ● $3 \cdot 4$

If $10 < 12$, then $\frac{2}{3} < \frac{4}{5}$.

Review Exercises

Replace each ● with <, >, or = to make a true sentence.

11. -9 ● -11 **>**
12. -13 ● 13 **<**
13. 0.35 ● -3.5 **>**
14. 2.2 ● 2.20 **=**
15. $\frac{3}{8}$ ● $\frac{3}{4}$ **<**
16. $-\frac{7}{4}$ ● -1.75 **=**
17. Write $-1.1, \frac{1}{8}, 0.25, 0$ in order from least to greatest. **$-1.1, 0, \frac{1}{8}, 0.25$**

www.algconcepts.com/vocabulary_review

Resource Manager

Reproducible Masters
Chapter 3 Resource Masters
- *Assessment*, pp. 131–139, 142–144

Technology/Multimedia
MindJogger Videoquizzes
ExamView® Pro

Objectives and Examples

- **Lesson 3-2** Add and subtract rational numbers.

Find the difference.
$-0.3 - (-1.8) = -0.3 + 1.8$
$= 1.5$

Review Exercises

Find each sum or difference. 18. −12.6

18. $-4.5 + (-8.1)$
19. $3.7 - (-6.8)$ 10.5
20. $\frac{8}{9} - \frac{1}{3}$ $\frac{5}{9}$
21. $-\frac{13}{7} + \frac{6}{7}$ −1
22. $-\frac{4}{3} + \frac{5}{6} + \left(-\frac{7}{3}\right)$ $-\frac{17}{6}$ or $-2\frac{5}{6}$

- **Lesson 3-3** Find the mean, median, mode, and range of a set of data.

Find the mean, median, mode, and range of 2, 4, 5, 6, 6, 13, 14, 14, 14, 16, and 16.

mean: $\frac{110}{11} = 10$

median: The 6th or middle value is 13.
mode: The most frequent number is 14.
range: 16 − 2 or 14

Find the mean, median, mode, and range of each set of data.

23. 6, 8, 6, 5, 5, 7, 9, 2, 7, 5, 3 $5.\overline{72}$; 6; 5; 7
24. 8.6, 7.5, 9.9, 5.1, 7.1 7.64; 7.5; none; 4.8
25.

Stem	Leaf
0	3 6 7 7 7 7 9
1	0 0 0 0 0 0 3 4 4 6 6 7 7 9 9
2	0 1
3	1 3 | 1 = 31

12.92; 10; 10; 28

- **Lesson 3-4** Determine whether a given number is a solution of an equation.

Find the solution of $x + (-3) = 2$ if the replacement set is $x = \{4, 5, 6\}$.

The solution is 5, since $5 + (-3) = 2$.

Find the solution of each equation if the replacement sets are $x = \{-3, -2, -1\}$, $y = \{0, 2, 4\}$, and $z = \{1, 3, 5\}$.

26. $3x = -6$ −2
27. $5y - 4 = -4$ 0
28. $\frac{3}{5}z = 3$ 5
29. $\frac{2+8}{x} + 9 = -1$ −1

Solve each equation. 31. 45

30. $a = 21 - 5$ 16
31. $5(6) - 3(-5) = w$

- **Lesson 3-5** Solve addition and subtraction equations using algebra tiles.

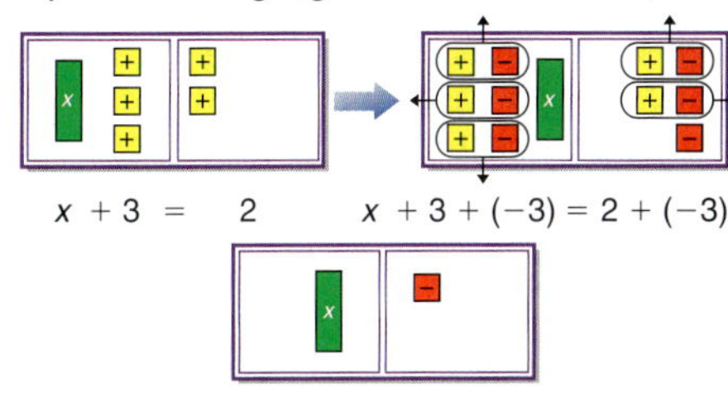

Solve each equation. Use algebra tiles if necessary.

32. $x + (-3) = 5$ 8
33. $4 = 6 + y$ −2
34. $-7 + z = 3$ 10
35. $p + 2 = -2$ −4
36. $s - (-6) = 1$ −5
37. $-1 = g + (-1)$ 0

38. What is the value of h if $h + 4 = 2$? −2

39. A number increased by 4 is 11. Write an equation. Then solve for the number.
$x + 4 = 11$; $x = 7$

Skills and Concepts

The **Objectives and Examples** section reviews the skills and concepts of the chapter and shows completely worked examples.

The **Review Exercises** provide practice for the corresponding objectives.

ExamView® Pro

Use the networkable ***ExamView® Pro Testmaker CD-ROM*** to:

- Create **multiple versions** of tests.
- Create **modified** tests for **inclusion** students with one mouse click.
- **Edit** existing questions and **add** your own questions.
- Build tests aligned with state standards using built-in **state curriculum correlations**.
- Change **English** tests to **Spanish** with one mouse click and vice versa.

Mixed Problem Solving
See pages 724–731.

Applications and Problem Solving

This section provides additional practice in solving real-world problems that involve the concepts of this chapter.

Objectives and Examples

- **Lesson 3–6** Solve addition and subtraction equations by using the properties of equality.

Solve $m + 16 = 8$. Check your solution.

$$m + 16 = 8$$
$$m + 16 - 16 = 8 - 16 \quad \textit{Subtract 16 from each side.}$$
$$m = -8$$

Check:
$$m + 16 = 8$$
$$-8 + 16 \stackrel{?}{=} 8 \quad \textit{Replace m with } -8.$$
$$8 = 8 \quad \checkmark$$

Review Exercises

40. −24 42. 38 44. −8.5

Solve each equation. Check your solution.

40. $z + 15 = -9$
41. $19 = y + 7$ **12**
42. $p + (-7) = 31$
43. $m - (-4) = 21$ **17**
44. $r - (-1.2) = -7.3$
45. $j + (-2.4) = 6.6$ **9**
46. $\frac{1}{2} + t = \frac{1}{2}$ **0**
47. $-\frac{1}{6} = v - \frac{2}{3}$ $\frac{1}{2}$

Write an equation and solve.

48. Some number added to −16 is equal to 39. What is the number? **$x + (-16) = 39$; 55**

49. Twelve more than a number is −108. What is the number? **$x + 12 = -108$; −120**

- **Lesson 3–7** Solve equations involving absolute value.

Solve $|x + 6| = 14$. Check your solution.

$$x + 6 = 14 \qquad \text{or} \qquad x + 6 = -14$$
$$x + 6 - 6 = 14 - 6 \qquad x + 6 - 6 = -14 - 6$$
$$x = 8 \qquad x = -20$$

Check:
$$|8 + 6| = 14 \qquad |-20 + 6| = 14$$
$$|14| \stackrel{?}{=} 14 \qquad |-14| \stackrel{?}{=} 14$$
$$14 = 14 \ \checkmark \qquad 14 = 14 \ \checkmark$$

Solve each equation. Check your solution.

50. $|x| = 22$ **{−22, 22}**
51. $|c| - 13 = -5$ **{−8, 8}**
52. $7 = 10 + |y|$ **∅**
53. $6 = |-2 + f|$ **{−4, 8}**
54. $|h - 1| = 0$ **{1}**
55. $|a - 3| = -8$ **∅**

56. Write an equation to represent the situation *The distance between a number x and −4 is 10 units.* **$|x - (-4)| = 10$**

Applications and Problem Solving

57. Shopping The approximate prices for 16 different types of athletic shoes are shown on the line plot below. *(Lesson 3–3)*

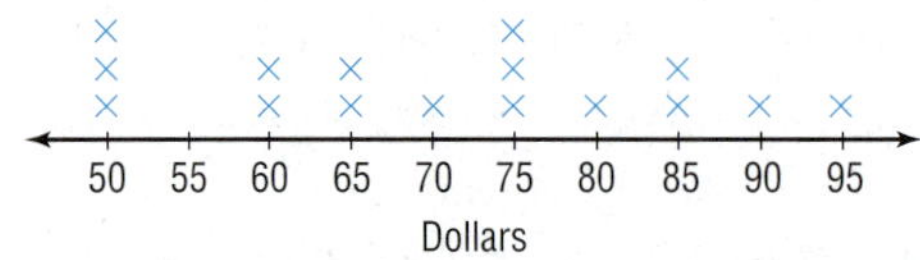

Find the mean, median, mode, and range for the prices. **\$70.625; \$72.5; \$50, \$75; \$45**

58. Manufacturing A certain bolt used in lawn mowers will work properly only if its diameter differs from 2 centimeters by no more than 0.04 centimeter. Write and then solve an equation that could be used to find the least and greatest diameter of the bolt. *(Lesson 3–7)* **$|d - 2| = 0.04$; {1.96, 2.04}**

Assessment and Evaluation

Four forms of Chapter 3 Test are available in the *Chapter 3 Resource Masters.*

Chapter 3 Test, Form 1B, is shown at the left. Chapter 3 Test, Form 2B, is shown on the next page.

	Form of Test		Level
1A	Multiple Choice	pp. 131–132	Average
1B	Multiple Choice	pp. 133–134	Basic
2A	Free Response	pp. 135–136	Average
2B	Free Response	pp. 137–138	Basic

Assessment, pp. 133–134

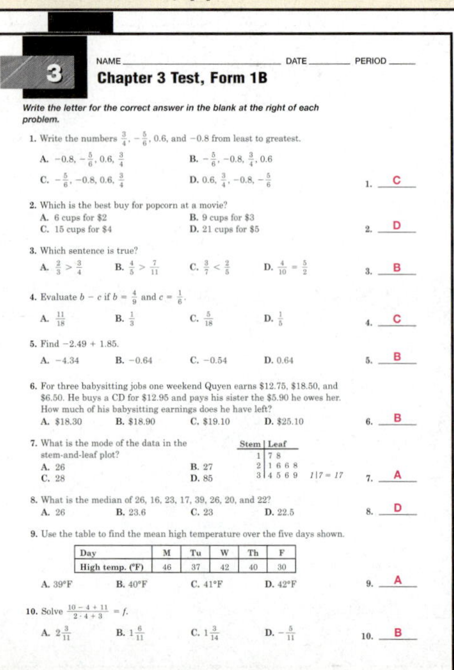

3

NAME __________ DATE _____ PERIOD _____

Chapter 3 Test, Form 1B

Write the letter for the correct answer in the blank at the right of each problem.

1. Write the numbers $\frac{3}{4}$, $-\frac{5}{6}$, 0.6, and −0.8 from least to greatest.
A. $-0.8, -\frac{5}{6}, 0.6, \frac{3}{4}$ B. $-\frac{5}{6}, -0.8, \frac{3}{4}, 0.6$
C. $-\frac{5}{6}, -0.8, 0.6, \frac{3}{4}$ D. $0.6, \frac{3}{4}, -0.8, -\frac{5}{6}$ 1. C

2. Which is the best buy for popcorn at a movie?
A. 6 cups for \$2 B. 9 cups for \$3
C. 15 cups for \$4 D. 21 cups for \$5 2. D

3. Which sentence is true?
A. $\frac{2}{3} > \frac{3}{4}$ B. $\frac{4}{5} > \frac{7}{11}$ C. $\frac{3}{7} < \frac{2}{5}$ D. $\frac{4}{10} = \frac{5}{2}$ 3. B

4. Evaluate $b - c$ if $b = \frac{4}{9}$ and $c = \frac{1}{6}$.
A. $\frac{11}{18}$ B. $\frac{1}{3}$ C. $\frac{5}{18}$ D. $\frac{1}{5}$ 4. C

5. Find −2.49 + 1.85.
A. −4.34 B. −0.64 C. −0.54 D. 0.64 5. B

6. For three babysitting jobs one weekend Quyen earns \$12.75, \$18.50, and \$6.50. He buys a CD for \$12.95 and pays his sister the \$5.90 he owes her. How much of his babysitting earnings does he have left?
A. \$18.30 B. \$18.90 C. \$19.10 D. \$25.10 6. B

7. What is the mode of the data in the stem-and-leaf plot?
A. 26 B. 27
C. 28 D. 85 7. A

Stem	Leaf
1	7 8
2	1 6 6 8
3	4 5 6 9

$1|7 = 17$

8. What is the median of 26, 16, 23, 17, 39, 26, 20, and 22?
A. 26 B. 23.6 C. 23 D. 22.5 8. D

9. Use the table to find the mean high temperature over the five days shown.

Day	M	Tu	W	Th	F
High temp. (°F)	46	37	42	40	30

A. 39°F B. 40°F C. 41°F D. 42°F 9. A

10. Solve $\frac{10 - 4 + 11}{2 \cdot 4 + 3} = f$.
A. $2\frac{3}{11}$ B. $1\frac{6}{11}$ C. $1\frac{3}{14}$ D. $-\frac{5}{11}$ 10. B

CHAPTER 3 Test

1. **List** at least two methods you can use to compare fractions with unlike denominators. **cross products and number lines**
2. **Describe** the process of finding the mean, median, mode, and range of a set of data. **See margin.**

Replace each ● with <, >, or = to make a true sentence.

3. $-4 \bullet -3$ **<**
4. $-1.01 \bullet -1.1$ **>**
5. $\frac{3}{5} \bullet \frac{8}{15}$ **>**

6. Write $-\frac{1}{2}$, $-\frac{2}{9}$, and $-\frac{3}{8}$ in order from least to greatest. $-\frac{1}{2}, -\frac{3}{8}, -\frac{2}{9}$

Find each sum or difference.

7. $-4.0 - 2.8$ **−6.8**
8. $0.32 - (-0.45)$ **0.77**
9. $6.1 + (-3.2)$ **2.9**
10. $\frac{7}{10} + \left(-\frac{3}{10}\right)$ $\frac{2}{5}$
11. $\frac{1}{5} - \left(-\frac{3}{4}\right)$ $\frac{19}{20}$
12. $\frac{3}{8} + 2\frac{1}{2}$ $2\frac{7}{8}$

13. **Finance** Sally recorded the amount of money she spent on lunch for five school days in the table at the right.
    a. Find Sally's mean, median, mode, and range of lunch money. **\$1.95; \$1.50; \$0.99; \$3.00**
    b. Explain which measure of central tendency is the least useful in describing Sally's lunch costs. **The mode is the least useful because it is the smallest amount.**

Day	Cost
Monday	\$0.99
Tuesday	\$1.50
Wednesday	\$3.99
Thursday	\$0.99
Friday	\$2.28

Solve each equation.

14. $10 \div 2 + 6 = x$ **11**
15. $8(0.3) - 0.5 = y$ **1.9**
16. $a = \frac{2 \cdot 6 + 4}{8 \div 2}$ **4**

Solve each equation. Use algebra tiles if necessary.

17. $g + 7 = 2$ **−5**
18. $-6 + h = -1$ **5**

Solve each equation. Check your solution.

19. $a - (-76) = 44$ **−32**
20. $5.2 + f = 16.4$ **11.2**
21. $\frac{1}{2} = b - \frac{7}{8}$ $\frac{11}{8}$ **or** $1\frac{3}{8}$
22. $3 + |c| = 6$ **{−3, 3}**
23. $|m - 4| = 1$ **{3, 5}**
24. $-12 = |t - (-7)|$ $\varnothing$

25. Taryn visited the "Guess Your Age" booth at the state fair. If the person in the booth could not guess Taryn's age within three years, Taryn would win a stuffed animal. Taryn is 16 years old. Write and solve an equation that can be used to find the person's greatest and least guesses, without Taryn's winning a prize. $|a - 16| = 3$; **{13, 19}**

www.algconcepts.com/chapter_test

Chapter Test

Answer

2. mean: Find the sum of the values in the data set and then divide by the number of values.
median: Arrange the data values in order. The median is the middle value or the mean of the two middle values.
mode: Find the value that appears most often.
range: Subtract the least value from the greatest value.

Chapter Test Bonus Question

For what value or values of x is $|x| = x$? For what value or values of x is $|x| = -x$? **all nonnegative numbers; negative numbers**

Assessment, pp. 137–138

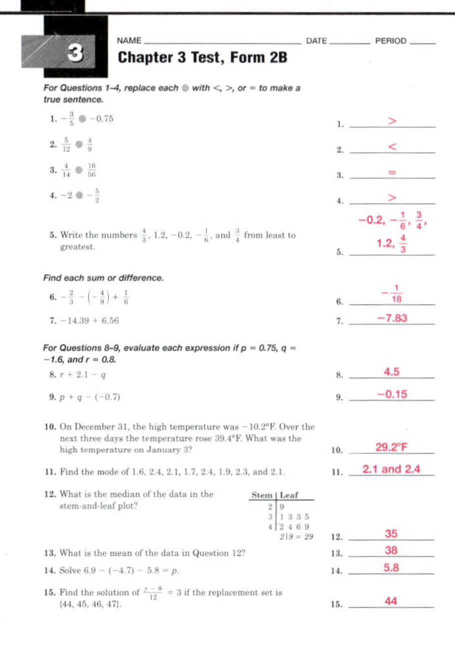

NAME ______ DATE ______ PERIOD ______

Chapter 3 Test, Form 2B

For Questions 1–4, replace each ● with <, >, or = to make a true sentence.

1. $-\frac{3}{5} \bullet -0.75$ 1. >
2. $\frac{5}{12} \bullet \frac{4}{9}$ 2. <
3. $\frac{4}{14} \bullet \frac{16}{56}$ 3. =
4. $-2 \bullet -\frac{5}{2}$ 4. >
5. Write the numbers $\frac{4}{3}$, 1.2, −0.2, $-\frac{1}{6}$, and $\frac{3}{4}$ from least to greatest. 5. $-0.2, -\frac{1}{6}, \frac{3}{4}, 1.2, \frac{4}{3}$

Find each sum or difference.

6. $-\frac{2}{3} - \left(-\frac{4}{9}\right) + \frac{1}{6}$ 6. $-\frac{1}{18}$
7. $-14.39 + 6.56$ 7. −7.83

For Questions 8–9, evaluate each expression if p = 0.75, q = −1.6, and r = 0.8.

8. $r + 2.1 - q$ 8. 4.5
9. $p + q - (-0.7)$ 9. −0.15
10. On December 31, the high temperature was −10.2°F. Over the next three days the temperature rose 39.4°F. What was the high temperature on January 3? 10. 29.2°F
11. Find the mode of 1.6, 2.4, 2.1, 1.7, 2.4, 1.9, 2.3, and 2.1. 11. 2.1 and 2.4
12. What is the median of the data in the stem-and-leaf plot? 12. 35

Stem	Leaf
2	9
3	1 3 3 5
4	2 4 6 9

$2|9 = 29$

13. What is the mean of the data in Question 12? 13. 38
14. Solve $6.9 - (-4.7) - 5.8 = p$. 14. 5.8
15. Find the solution of $\frac{y - 8}{12} = 3$ if the replacement set is {44, 45, 46, 47}. 15. 44

Preparing for Standardized Tests

Pages 136–137 are part of a complete test preparation course that is described in detail on page T9 of the Teacher's Handbook. The test items on these pages were written in the same style as those in state proficiency tests and standardized tests like ACT and SAT.

Diagnosis and Prescription

Each of the 10 test questions on page 137 is cross-referenced to the lesson where that SAT or ACT skill is covered. If students miss a particular type of problem, you can have them study that skill.

(See chart at the bottom of page 137. Note that SPT = State Proficiency Test, SAT = Scholastic Assessment Test, and ACT = American College Test.)

***Assessment*, p. 142**

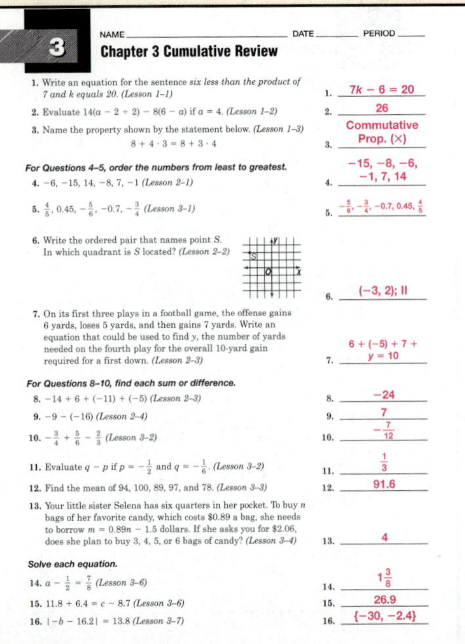

3 NAME ______ DATE ______ PERIOD ______

Chapter 3 Cumulative Review

1. Write an equation for the sentence *six less than the product of 7 and k equals 20. (Lesson 1–1)* — 1. $7k - 6 = 20$
2. Evaluate $14(a - 2 \div 2) - 8(6 - a)$ if $a = 4$. *(Lesson 1–2)* — 2. 26
3. Name the property shown by the statement below. *(Lesson 1–3)* $8 + 4 \cdot 3 = 8 + 3 \cdot 4$ — 3. Commutative Prop. (×)

For Questions 4–5, order the numbers from least to greatest.

4. $-6, -15, 14, -8, 7, -1$ *(Lesson 2–1)* — 4. $-15, -8, -6, -1, 7, 14$
5. $\frac{4}{5}, 0.45, -\frac{5}{6}, -0.7, -\frac{3}{4}$ *(Lesson 3–1)* — 5. $-\frac{5}{6}, -\frac{3}{4}, -0.7, 0.45, \frac{4}{5}$
6. Write the ordered pair that names point *S*. In which quadrant is *S* located? *(Lesson 2–2)* — 6. $(-3, 2)$; II
7. On its first three plays in a football game, the offense gains 6 yards, loses 5 yards, and then gains 7 yards. Write an equation that could be used to find *y*, the number of yards needed on the fourth play for the overall 10-yard gain required for a first down. *(Lesson 2–3)* — 7. $6 + (-5) + 7 + y = 10$

For Questions 8–10, find each sum or difference.

8. $-14 + 6 + (-11) + (-5)$ *(Lesson 2–3)* — 8. -24
9. $-9 - (-16)$ *(Lesson 2–4)* — 9. 7
10. $-\frac{3}{4} + \frac{5}{6} - \frac{2}{3}$ *(Lesson 3–2)* — 10. $-\frac{7}{12}$
11. Evaluate $q - p$ if $p = -\frac{1}{2}$ and $q = -\frac{1}{6}$. *(Lesson 3–2)* — 11. $\frac{1}{3}$
12. Find the mean of 94, 100, 89, 97, and 78. *(Lesson 3–3)* — 12. 91.6
13. Your little sister Selena has six quarters in her pocket. To buy *n* bags of her favorite candy, which costs \$0.89 a bag, she needs to borrow $m = 0.89n - 1.5$ dollars. If she asks you for \$2.06, does she plan to buy 3, 4, 5, or 6 bags of candy? *(Lesson 3–4)* — 13. 4

Solve each equation.

14. $a - \frac{1}{2} = \frac{7}{8}$ *(Lesson 3–6)* — 14. $1\frac{3}{8}$
15. $11.8 + 6.4 = c - 8.7$ *(Lesson 3–6)* — 15. 26.9
16. $|-b - 16.2| = 13.8$ *(Lesson 3–7)* — 16. $\{-30, -2.4\}$

CHAPTER 3 Preparing for Standardized Tests

Number Concept Problems

Standardized tests include many questions that use fractions and decimals. You'll need to convert between fractions and decimals. You'll also need to add, subtract, multiply, and divide fractions and decimals. Here are some terms to remember.

reciprocal   repeating decimal

Test-Taking Tip

Be sure you practice using your calculator before the test. Look at each number you enter to avoid mistakes.

Example 1

Which of the following numbers, when divided by $\frac{1}{2}$, gives a result less than $\frac{1}{2}$?

(A) $\frac{2}{8}$ 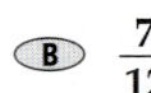 (B) $\frac{7}{12}$  (C) $\frac{2}{3}$  (D) $\frac{5}{24}$

Hint In some multiple-choice questions, you can check each answer choice to see if it is correct.

Solution Recall that dividing by a fraction is equivalent to multiplying by its reciprocal. Dividing by $\frac{1}{2}$ is the same as multiplying by 2. Remember that 2 is also written as $\frac{2}{1}$.

A $\frac{2}{8} \times \frac{2}{1} = \frac{4}{8}$ or $\frac{1}{2}$ **B** $\frac{7}{12} \times \frac{2}{1} = \frac{7}{6}$ or $1\frac{1}{6}$

C $\frac{2}{3} \times \frac{2}{1} = \frac{4}{3}$ or $1\frac{1}{3}$ **D** $\frac{5}{24} \times \frac{2}{1} = \frac{10}{24}$ or $\frac{5}{12}$

Decide which of the four resulting numbers is less than $\frac{1}{2}$. Answer choice A is equal to $\frac{1}{2}$, choices B and C are greater than $\frac{1}{2}$, but choice D is less than $\frac{1}{2}$.

$$\frac{5}{12} < \frac{6}{12} = \frac{1}{2}$$

The answer is D.

Example 2

If $\frac{2}{11}$ is written as a decimal carried out to 100 places, what is the sum of the first 50 digits to the right of the decimal point?

(A) 100 (B) 225 (C) 350 (D) 450 (E) 900

Hint Your calculator can be useful for some types of questions.

Solution First write the fraction as a decimal. Use either long division or your calculator.

$$\frac{2}{11} = 0.1818181818\ldots \text{ or } 0.\overline{18}$$

Since this is a repeating decimal, the first 50 digits are 25 pairs of the digits 1 and 8.

The question asks for the *sum* of the first 50 digits. The sum of each pair of 1 and 8 is 9. The sum of 25 pairs is 25×9 or 225.

The answer is B.

Resource Manager

Reproducible Masters

Chapter 3 Resource Masters

- *Assessment*, pp. 142–144

Preparing for Standardized Tests
For test-taking strategies and more practice, see pages 732–749.

After you work each problem, record your answer on the answer sheet provided or on a sheet of paper.

Multiple Choice

1. In Store X, which color button costs the most per individual unit? *(Lesson 3–1)* **E**

Price of Buttons in Store X	
Color	**Cost**
Black	\$2 per 5 buttons
Blue	\$2 per 6 buttons
Brown	\$3 per 8 buttons
Orange	\$4 per 12 buttons
Red	\$4 per 7 buttons

Ⓐ black Ⓑ blue Ⓒ brown
Ⓓ orange Ⓔ red

2. You burn 15 additional Calories each hour by standing instead of sitting. Suppose you burn x Calories per hour while sitting. How many Calories would you burn in one hour of standing? *(Lesson 1–1)* **B**

Ⓐ $15x$ Ⓑ $x + 15$
Ⓒ $x - 15$ Ⓓ $15 - x$

3. The total number of students in a class is represented by $5x - 6$. If $2x + 6$ represents the number of boys, which expression represents the number of girls? *(Lesson 1–3)* **C**

Ⓐ $3x$ Ⓑ $12 - 3x$
Ⓒ $3x - 12$ Ⓓ $3x + 12$

4. Order $-\frac{3}{4}, \frac{3}{4}$, and $-\frac{4}{3}$ from least to greatest. *(Lesson 3–1)* **D**

Ⓐ $\frac{3}{4}, -\frac{4}{3}, -\frac{3}{4}$ Ⓑ $-\frac{4}{3}, \frac{3}{4}, -\frac{3}{4}$
Ⓒ $-\frac{3}{4}, -\frac{4}{3}, \frac{3}{4}$ Ⓓ $-\frac{4}{3}, -\frac{3}{4}, \frac{3}{4}$

5. What is the GCF of the numbers in the sequence 6, 12, 18, 24, 30, . . . ? *(Basic Skill)* **B**

Ⓐ 3 Ⓑ 6 Ⓒ 12 Ⓓ 18

6. Two years ago, Luna was ℓ years of age. How old will Luna be 2 years from now? *(Lesson 1–1)* **A**

Ⓐ $\ell + 4$ Ⓑ $\ell + 2$ Ⓒ 2ℓ Ⓓ ℓ

7. If you evaluate the following expression according to the order of operations, in what order should the operations be performed? *(Lesson 1–2)* **A**

$$12.5 \times 3 + 5 + 20 - 12$$

Ⓐ $\times, +, +, -$ Ⓑ $+, -, \times, +$
Ⓒ $+, \times, -, +$ Ⓓ $-, +, +, \times$

8. Which expression finds multiples of 5? *(Lesson 1–1)* **C**

Ⓐ $x + 5$ Ⓑ $x - 5$ Ⓒ $5x$ Ⓓ $5 - x$

Grid In

9. Yvette's goal is to run 30 miles a week. So far this week, she has run 6.5, 5.2, 7.8, 3, and 6.9 miles. How many more miles does she need to run to reach her goal? *(Lesson 3–2)* **0.6**

Extended Response

10. Band members Scott and Ben are taking the bus to a football game. The bus leaves at 5:45 P.M. They have to arrive 15 minutes before the bus leaves to load instruments. It takes Scott 10 minutes to drive to Ben's house and 25 minutes to drive from there to the school. What is the latest time that Scott can leave his house? *(Lesson 3–2)*

Part A Explain the steps you use in solving this problem. **See margin.**

Part B Calculate the latest time Scott can leave home. **4:55 P.M.**

www.algconcepts.com/standardized_test

Chapter 3 More Number Concept Problems		
Ex. 1	dividing fractions	SPT
Ex. 2	repeating decimals	SAT
1	fraction and decimal word problem	SAT
2	write an expression	SPT
3	write an expression	ACT
4	order rational numbers	SPT
5	factors	SPT
6	write an expression	SAT
7	order of operations	SPT
8	write an equation	SPT
9	decimal word problem	SPT
10	time word problem	SPT

A bubble-in answer sheet for these practice problems is available on page A1 of the *Chapter 3 Resource Masters.*

Additional Practice

Additional test practice questions are available in the *Chapter 3 Resource Masters,* pp. 143–144.

Answer

10A. Sample answer: Subtract the 15 minutes from 5:45 to find the time they should arrive at school. Subtract the 10 minutes and then the 25 minutes driving time to find the time Scott should leave his house.

***Assessment,* pp. 143–144**

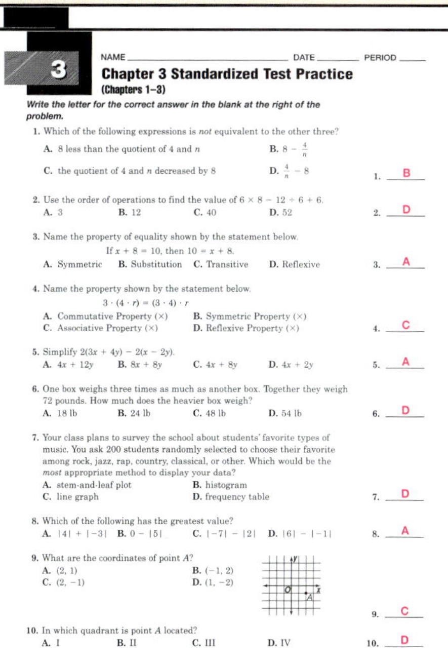
3 NAME ____________ DATE ______ PERIOD ______

Chapter 3 Standardized Test Practice
(Chapters 1–3)

Write the letter for the correct answer in the blank at the right of the problem.

1. Which of the following expressions is *not* equivalent to the other three?
A. 8 less than the quotient of 4 and n B. $8 - \frac{4}{n}$
C. the quotient of 4 and n decreased by 8 D. $\frac{4}{n} - 8$ 1. B

2. Use the order of operations to find the value of $6 \times 8 - 12 \div 6 + 6$.
A. 3 B. 12 C. 40 D. 52 2. D

3. Name the property of equality shown by the statement below.
If $x + 8 = 10$, then $10 = x + 8$.
A. Symmetric B. Substitution C. Transitive D. Reflexive 3. A

4. Name the property shown by the statement below.
$3 \cdot (4 \cdot r) = (3 \cdot 4) \cdot r$
A. Commutative Property (×) B. Symmetric Property (×)
C. Associative Property (×) D. Reflexive Property (×) 4. C

5. Simplify $2(3x + 4y) - 2(x - 2y)$.
A. $4x + 12y$ B. $8x + 8y$ C. $4x + 8y$ D. $4x + 2y$ 5. A

6. One box weighs three times as much as another box. Together they weigh 72 pounds. How much does the heavier box weigh?
A. 18 lb B. 24 lb C. 48 lb D. 54 lb 6. D

7. Your class plans to survey the school about students' favorite types of music. You ask 200 students randomly selected to choose their favorite among rock, jazz, rap, country, classical, or other. Which would be the *most* appropriate method to display your data?
A. stem-and-leaf plot B. histogram
C. line graph D. frequency table 7. D

8. Which of the following has the greatest value?
A. $|4| + |-3|$ B. $0 - |5|$ C. $|-7| - |2|$ D. $|6| - |-1|$ 8. A

9. What are the coordinates of point A?
A. (2, 1) B. (−1, 2)
C. (2, −1) D. (1, −2) 9. C

10. In which quadrant is point A located?
A. I B. II C. III D. IV 10. D

CHAPTER 4

Resource Manager

Multiplication and Division Equations

Instructional Objectives

Lesson (pages)	Objectives	NCTM Standards 2000	State/Local Objectives
4–1 (140–145)	Multiply rational numbers.	1, 6, 8, 9	
4–2 (146–151)	Use tree diagrams or the Fundamental Counting Principle to count outcomes.	1, 2, 5, 6, 7, 8, 9, 10	
Investigation (152–153)	Explore permutations and combinations.	1, 2, 5, 6, 7, 8, 9	
4–3 (154–159)	Divide rational numbers.	1, 6, 8, 9	
4–4 (160–164)	Solve multiplication and division equations by using the properties of equality.	1, 6, 8, 9	
4–5 (165–170)	Solve equations involving more than one operation.	1, 6, 8, 9	
4–6 (171–175)	Solve equations with variables on both sides.	1, 6, 8, 9	
4–7 (176–179)	Solve equations with grouping symbols.	1, 6, 8, 9	

Key to NCTM Standards 2000
1 Number & Operations; **2** Algebra; **3** Geometry; **4** Measurement; **5** Data Analysis & Probability; **6** Problem Solving; **7** Reasoning & Proof; **8** Communications; **9** Connections; **10** Representation

Suggested Pacing

See page T13 for a complete course-planning calendar.

Standard refers to schedules that provide 45- to 55-minute periods that meet each day.
Block refers to schedules that provide approximately 90-minute periods which may meet every day for one semester or every other day over two semesters.

PACING	DAY 1	DAY 2	DAY 3	DAY 4	DAY 5	DAY 6
Standard Core (Chapters 1–13)	Lesson 4–1		Lesson 4–2	INV	Lesson 4–3	
Standard Enhanced (Chapters 1–15)	Lesson 4–1		Lesson 4–2	INV	Lesson 4–3	Lesson 4–4
Block Core (Chapters 1–13)	Chapter 3 Test & Lesson 4–1	Lesson 4–2 & INV	Lessons 4–3 & 4–4	Lesson 4–5	Lessons 4–6 & 4–7	SG+A
Block Enhanced (Chapters 1–15)	Chapter 3 Test & Lesson 4–1	Lesson 4–2 & INV	Lessons 4–3 & 4–4	Lessons 4–5 & 4–6	Lesson 4–7 & SG+A	Chapter Test & Lesson 5–1

Instructional Resources

Lesson	Materials and Manipulatives (see below for Glencoe Manipulative Resources)	Study Guide	Practice (Skills & Average)	Reading to Learn Mathematics	Enrichment	Assessment	Hands-On Algebra*	School-to-Workplace*	Graphing Calculator Masters*	5-Minute Check Transparencies
		Blackline Masters (page numbers): Chapter 4 Resource Masters					Blackline Masters (page numbers)			
4–1	grid paper [1, 4] colored pencils straightedge [1, 2, 4]	145	146–147	148	149		53			4–1
4–2	coins dice	150	151–152	153	154					4–2
Investigation	index cards green, blue, pink, and yellow self-adhesive notes graphing calculator									
4–3		155	156–157	158	159	191	54, 55			4–3
4–4	paper clips or pennies	160	161–162	163	164	190				4–4
4–5	graphing calculator algebra tiles [1, 2, 3, 4]	165	166–167	168	169		56, 57, 58	4	11	4–5
4–6	index cards	170	171–172	173	174		59		12, 13	4–6
4–7		175	176–177	178	179	191				4–7
Study Guide & Assessment/ Chapter Test						181–189, 192–194				

See page 138c for examples of these instructional materials.

Key to Glencoe Manipulative Resources

[1]Classroom Manipulative Resources [2]Student Manipulative Resources [3]Overhead Manipulative Resources [4]Hands-On Algebra Masters

INV = Investigation *SG+A = Study Guide and Assessment*

DAY 7	DAY 8	DAY 9	DAY 10	DAY 11	DAY 12	DAY 13
Lesson 4–4	Lesson 4–5		Lesson 4–6	Lesson 4–7	SG+A	Chapter Test
Lesson 4–5		Lesson 4–6	Lesson 4–7	SG+A	Chapter Test	
Chapter Test & Lesson 5–1						

Resource Manager

The pages shown on this page are a small sample of the materials available on *TeacherWorks: All-in-One Lesson Planner and Resource Center.*

This CD-ROM includes all of the blackline masters and transparencies available for this program.

It also includes a lesson planner and interactive Teacher Edition, so you can customize lesson plans and reproduce classroom resources quickly and easily, from just about anywhere.

Applications

School-to-Workplace Masters, p. 4

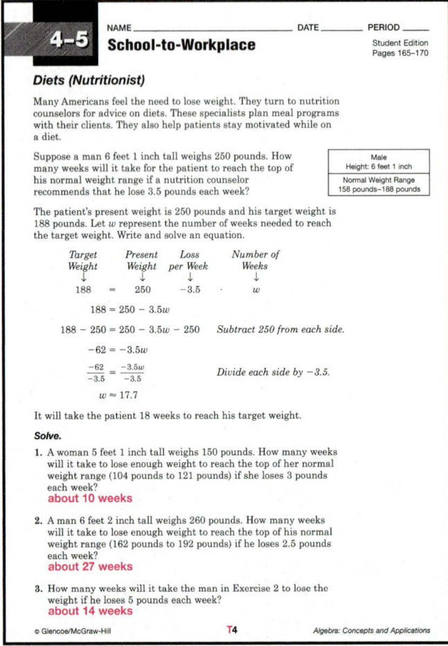

4-5 NAME ______ DATE ______ PERIOD ______

School-to-Workplace Student Edition Pages 165–170

Diets (Nutritionist)

Many Americans feel the need to lose weight. They turn to nutrition counselors for advice on diets. These specialists plan meal programs with their clients. They also help patients stay motivated while on a diet.

Suppose a man 6 feet 1 inch tall weighs 250 pounds. How many weeks will it take for the patient to reach the top of his normal weight range if a nutrition counselor recommends that he lose 3.5 pounds each week?

Male Height: 6 feet 1 inch
Normal Weight Range 158 pounds–188 pounds

The patient's present weight is 250 pounds and his target weight is 188 pounds. Let w represent the number of weeks needed to reach the target weight. Write and solve an equation.

Target Weight		Present Weight	Loss per Week		Number of Weeks
188	=	250	−3.5	·	w

$188 = 250 - 3.5w$

$188 - 250 = 250 - 3.5w - 250$ *Subtract 250 from each side.*

$-62 = -3.5w$

$\frac{-62}{-3.5} = \frac{-3.5w}{-3.5}$ *Divide each side by −3.5.*

$w \approx 17.7$

It will take the patient 18 weeks to reach his target weight.

Solve.

1. A woman 5 feet 1 inch tall weighs 150 pounds. How many weeks will it take to lose enough weight to reach the top of her normal weight range (104 pounds to 121 pounds) if she loses 3 pounds each week? **about 10 weeks**
2. A man 6 feet 2 inch tall weighs 260 pounds. How many weeks will it take to lose enough weight to reach the top of his normal weight range (162 pounds to 192 pounds) if he loses 2.5 pounds each week? **about 27 weeks**
3. How many weeks will it take the man in Exercise 2 to lose the weight if he loses 5 pounds each week? **about 14 weeks**

© Glencoe/McGraw-Hill T4 *Algebra: Concepts and Applications*

Manipulatives/Modeling

Hands-On Algebra Masters, pp. 53–59

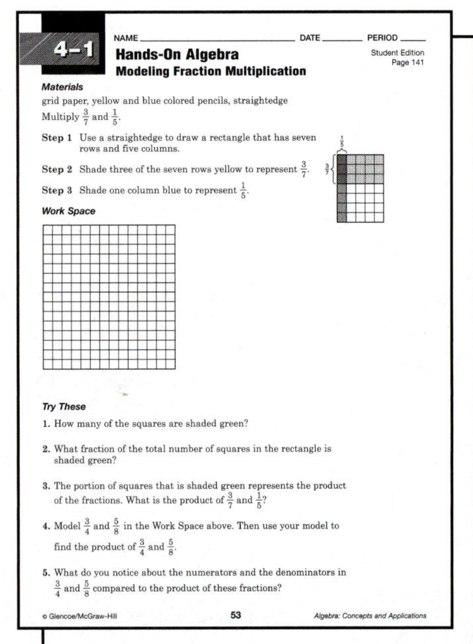

4-1 NAME ______ DATE ______ PERIOD ______

Hands-On Algebra Student Edition Page 141

Modeling Fraction Multiplication

Materials

grid paper, yellow and blue colored pencils, straightedge

Multiply $\frac{3}{7}$ and $\frac{1}{5}$.

Step 1 Use a straightedge to draw a rectangle that has seven rows and five columns.

Step 2 Shade three of the seven rows yellow to represent $\frac{3}{7}$.

Step 3 Shade one column blue to represent $\frac{1}{5}$.

Work Space

Try These

1. How many of the squares are shaded green?
2. What fraction of the total number of squares in the rectangle is shaded green?
3. The portion of squares that is shaded green represents the product of the fractions. What is the product of $\frac{3}{7}$ and $\frac{1}{5}$?
4. Model $\frac{3}{4}$ and $\frac{5}{8}$ in the Work Space above. Then use your model to find the product of $\frac{3}{4}$ and $\frac{5}{8}$.
5. What do you notice about the numerators and the denominators in $\frac{3}{4}$ and $\frac{5}{8}$ compared to the product of these fractions?

© Glencoe/McGraw-Hill 53 *Algebra: Concepts and Applications*

Technology/Multimedia

Graphing Calculator Masters, pp. 11–13

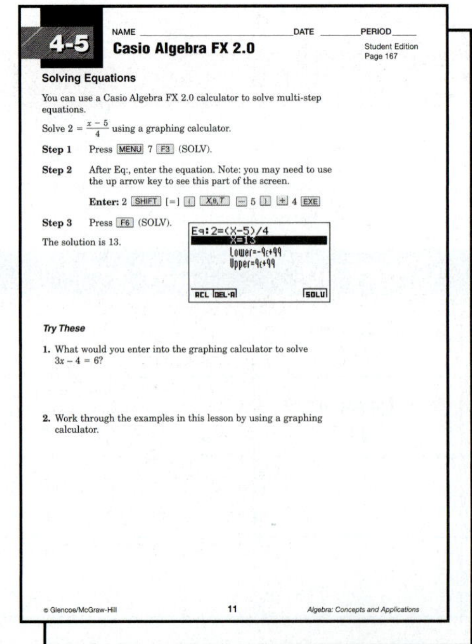

4-5 NAME ______ DATE ______ PERIOD ______

Casio Algebra FX 2.0 Student Edition Page 167

Solving Equations

You can use a Casio Algebra FX 2.0 calculator to solve multi-step equations.

Solve $2 = \frac{x-5}{4}$ using a graphing calculator.

Step 1 Press MENU 7 F3 (SOLV).

Step 2 After Eq:, enter the equation. Note: you may need to use the up arrow key to see this part of the screen.

Enter: 2 SHIFT [=] (X,θ,T − 5) ÷ 4 EXE

Step 3 Press F6 (SOLV).

The solution is 13.

Try These

1. What would you enter into the graphing calculator to solve $3x - 4 = 6$?
2. Work through the examples in this lesson by using a graphing calculator.

© Glencoe/McGraw-Hill 11 *Algebra: Concepts and Applications*

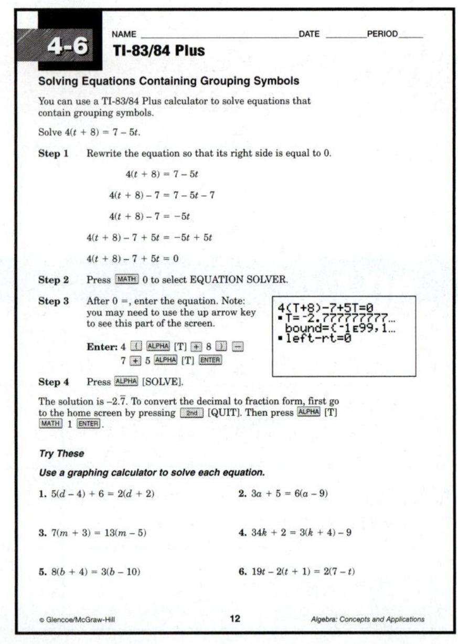

4-6 NAME ______ DATE ______ PERIOD ______

TI-83/84 Plus

Solving Equations Containing Grouping Symbols

You can use a TI-83/84 Plus calculator to solve equations that contain grouping symbols.

Solve $4(t + 8) = 7 - 5t$.

Step 1 Rewrite the equation so that its right side is equal to 0.

$4(t + 8) = 7 - 5t$

$4(t + 8) - 7 = 7 - 5t - 7$

$4(t + 8) - 7 = -5t$

$4(t + 8) - 7 + 5t = -5t + 5t$

$4(t + 8) - 7 + 5t = 0$

Step 2 Press MATH 0 to select EQUATION SOLVER.

Step 3 After 0 =, enter the equation. Note: you may need to use the up arrow key to see this part of the screen.

Enter: 4 (ALPHA [T] + 8) − 7 + 5 ALPHA [T] ENTER

Step 4 Press ALPHA [SOLVE].

The solution is $-2.\overline{7}$. To convert the decimal to fraction form, first go to the home screen by pressing 2nd [QUIT]. Then press ALPHA [T] MATH 1 ENTER.

Try These

Use a graphing calculator to solve each equation.

1. $5(d - 4) + 6 = 2(d + 2)$
2. $3a + 5 = 6(a - 9)$
3. $7(m + 3) = 13(m - 5)$
4. $34k + 2 = 3(k + 4) - 9$
5. $8(b + 4) = 3(b - 10)$
6. $19t - 2(t + 1) = 2(7 - t)$

© Glencoe/McGraw-Hill 12 *Algebra: Concepts and Applications*

Assessment Resources

Type	Student Edition	Teacher's Wraparound Edition	Chapter 4 Resource Masters
Ongoing Assessment	Quizzes 1 and 2, pp. 159, 175	5-Minute Check, pp. 140, 146, 154, 160, 165, 171, 176	Mid-Chapter Test, p. 190 Quizzes A and B, p. 191
Mixed Review	Mixed Review, pp. 145, 151, 158, 164, 170, 174, 179 Standardized Test Practice, Chapters 1–4, pp. 184–185		Cumulative Review, p. 192 Standardized Test Practice, pp. 193–194
Error Analysis	You Decide, pp. 148, 168	Error Analysis, pp. 143, 148, 157, 163, 168, 173, 178	
Standardized Test Prep	Standardized Test Practice, pp. 145, 151, 159, 164, 170, 175, 179 Standardized Test Practice, Chapters 1–4, pp. 184–185		Standardized Test Practice, pp. 193–194
Open-Ended Assessment	Writing Math, pp. 156, 178 Investigation, pp. 152–153 Portfolio, pp. 139, 153	Modeling: p. 151 Speaking: pp. 159, 164, 175 Writing: pp. 145, 170, 179	Extended Response Assessment, p. 189
Chapter Assessment	Study Guide and Assessment, pp. 180–182 Chapter Test, p. 183		Multiple-Choice Tests (Forms 1A, 1B), pp. 181–184 Free-Response Tests (Forms 2A, 2B), pp. 185–188

Additional Chapter Resources

Student Edition

Hands-On Algebra, p. 141
Graphing Calculator Exploration, p. 167

Teacher's Classroom Resources

Manipulatives/Modeling

Teacher's Guide for Overhead Manipulative Resources

Meeting Individual Needs

Prerequisite Skills Booklet
Spanish Study Guide and Assessment, pp. 29–35, 111–112

Teaching Aids

Solutions Manual
5-Minute Check Transparencies

Glencoe Technology

Instructional

AlgePASS CD-ROM, Lessons 10–12
Interactive Chalkboard CD-ROM
StudentWorks Plus CD-ROM
Multimedia Applications CD-ROM, Activities 2, 3
Vocabulary PuzzleMaker (online)

Assessment

ExamView® Pro

Teaching Aids

Answer Key Maker CD-ROM
TeacherWorks CD-ROM

Visit ***www.algconcepts.com*** *for data updates, career information, games, and other interactive activities.*

CHAPTER 4 PREVIEW

Mathematics of the Chapter

This chapter begins with multiplication and division of rational numbers. Students count the number of outcomes of an experiment. In the second half of the chapter, students solve multiplication and division equations. Multi-step equations are then explored. Next, students combine like terms to solve problems with variables on each side of the equation, and they use the Distributive Property to solve equations involving grouping symbols.

What You'll Learn

Have students read over the lists of key ideas and key vocabulary. Have them make a list of any words with which they are not familiar.

Why It's Important

Point out to students that this is only one of many reasons why each objective is important. Others are provided in each lesson.

What's MATH Got To Do With It?

 **Real-Life Videos** engage students, showing them how math is used in everyday situations. Use Video 1 to discuss how equations and formulas are used in buying surfboards and baking cakes.

CHAPTER

Multiplication and Division Equations

What You'll Learn

Key Ideas

- Multiply and divide rational numbers. *(Lessons 4–1 and 4–3)*
- Use tree diagrams or the Fundamental Counting Principle to count outcomes. *(Lesson 4–2)*
- Solve equations involving multiplication and division, more than one operation, variables on both sides, or grouping symbols. *(Lessons 4-4 to 4-7)*

Key Vocabulary

combination *(p. 152)*
event *(p. 147)*
multiplicative inverses *(p. 154)*
permutation *(p. 152)*
tree diagram *(p. 146)*

Why It's Important

The Tower of Pisa is one of the most-visited tourist attractions in Europe. Soon after its construction in the twelfth century, the foundation sank, causing the tower to lean.

Equations involving multiplication and division are used to model real-life situations. You will multiply rational numbers to analyze an experiment Galileo performed at the Tower of Pisa in Lesson 4–1.

138 Chapter 4 Multiplication and Division Equations

CHAPTER 4 LINKS

Lesson	4–1	4–2	4–3	4–4	4–5	4–6	4–7
Math in the Workplace	Health	Manufacturing	Cooking	Plumbing	Shopping	Sports	Painting
Applications and Connections	Physical Science Purchases Space History Aviation Height	Dance Clothing Bicycles Dining History Family	Lunch Track Nutrition Arts Uniforms Sales	Consumer Geography Entertainment Technology Animals Sports	Technology Weather Animals Consumer Spending Gardening	Temperature School Supplies Entertainment Health Basketball	Building Gardening Astronomy
Math Integration				Geometry	Number Theory Geometry	Number Theory Geometry	Number Theory Geometry

Check Your Readiness

Study these lessons to improve your skills.

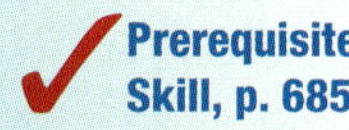
Lesson 2-5, pp. 75-79

Find each product.

1. $6(9)$ **54**
2. $-4(2)$ **−8**
3. $-5(0)$ **0**
4. $-11(-2)$ **22**
5. $-15(-4)$ **60**
6. $8(-11)$ **−88**
7. $4(-1)(-5)$ **20**
8. $4(2)(3)$ **24**
9. $-3(3)(-6)$ **54**

Lesson 2-6, pp. 82-85

Find each quotient.

10. $18 \div 2$ **9**
11. $-14 \div 2$ **−7**
12. $-36 \div (-9)$ **4**
13. $42 \div (-6)$ **-7**
14. $24 \div 6$ **4**
15. $-56 \div 8$ **−7**
16. $\frac{-16}{-4}$ **4**
17. $\frac{56}{-7}$ **−8**
18. $\frac{-81}{3}$ **−27**

Prerequisite Skill, p. 685

Write each fraction in simplest form.

19. $\frac{2}{8}$ $\frac{1}{4}$
20. $\frac{9}{12}$ $\frac{3}{4}$
21. $\frac{16}{24}$ $\frac{2}{3}$
22. $\frac{50}{60}$ $\frac{5}{6}$
23. $\frac{21}{35}$ $\frac{3}{5}$
24. $\frac{60}{96}$ $\frac{5}{8}$

Lesson 1-4, pp. 19-23

Simplify each expression.

25. $8a - a$ $7a$
26. $9x - 3x$ $6x$
27. $7t - 6t$ t
28. $13y - 5y$ $8y$
29. $4(x - 2)$ $4x - 8$
30. $3(n + 6)$ $3n + 18$
31. $2(b - 1.6)$ $2b - 3.2$
32. $5(3b + 6)$ $15b + 30$
33. $4(8m - 3)$ $32m - 12$

FOLDABLES™ Study Organizer

Make this Foldable to help you organize your Chapter 4 notes. Begin with seven sheets of grid paper.

❶ **Fold** each sheet in half along the width.

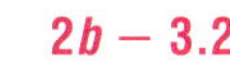

❷ **Unfold** and cut four rows from the left side of each sheet, from the top to the fold.

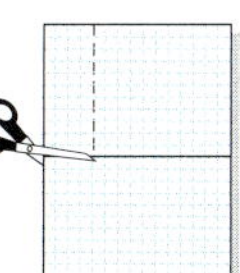

❸ **Stack** the sheets and staple to form a book.

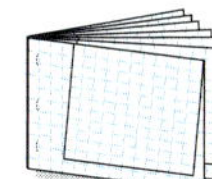

❹ **Label** each page with a lesson number and title.

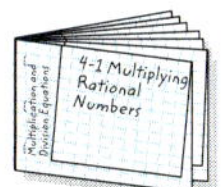

Reading and Writing As you read and study the chapter, unfold each page and fill the journal with notes, graphs, and examples.

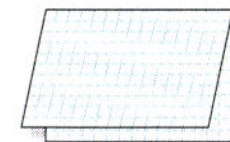

www.algconcepts.com/chapter_readiness

Check Your Readiness

This section provides a review of the basic concepts needed before beginning Chapter 4. Lesson references are included for additional student help.

Use Exercises	To Prepare for Lesson(s)
1–9	4-1
10–18	4-3
19–24	4-1, 4-3
25–33	4-7

Quick Review Math Handbook (hot words hot topics)

Glencoe Algebra Lesson	Quick Review Math Handbook Hot Topic
4-1	1-5, 2-4, 2-6
4-2	4-5, 4-6
Ch. 4 Inv.	2-3, 2-4, 4-5
4-3	1-5, 2-4, 2-6, 4-4
4-4	1-3, 6-4
4-5	6-2, 6-4
4-6, 4-7	2-1, 6-4

Noteables™ Interactive Study Notebook with Foldables™

This note-taking guide reinforces excellent note-taking skills. It includes:

✓ **Build Your Vocabulary** tool for including mathematical terminology in notes.
✓ **Review It** activities with links to prior lessons and items to include in **Foldables™**.
✓ **Bringing It All Together** feature to help students review for the Chapter Test.
✓ **Are You Ready for the Chapter Test?** feature to allow students to assess their own readiness for the Chapter Test.
✓ **Teacher Edition** with transparencies to guide note taking. Corresponds to the *Teacher's Wraparound Edition* In-Class Examples and *Interactive Chalkboard CD-ROM.*

FOLDABLES™ Study Organizer On the pages of this Foldable, students should write what they have learned in each lesson. Then have students create and solve a problem for each tab.

Lesson 4–1

4–1 Multiplying Rational Numbers

1 FOCUS

5-Minute Check
Chapter 3

1. The ages of 14 students in a college biology lab are represented in the line plot below. Find the mean, median, mode, and range for the ages.

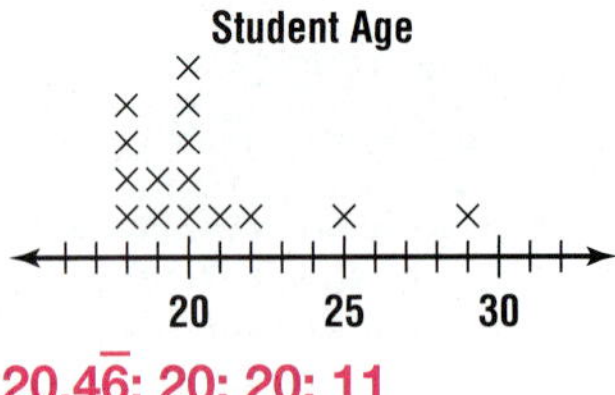

20.4$\overline{6}$; 20; 20; 11

Solve each equation.

2. $k + 11 = -4$ **−15**
3. $-2.7 + f = 59.1$ **61.8**
4. $|b - 8| - 3 = 18$ **−13, 29**
5. Joshua is planning a rectangular garden. The dimensions of the garden are shown in the figure below. How many feet of fencing will be needed to enclose his garden?

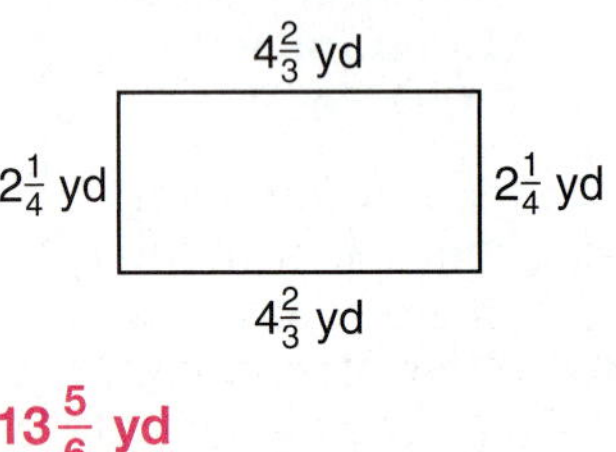

$13\frac{5}{6}$ yd

Motivating the Lesson

Real-World Connection Have students consider the following situation. "Suppose you deposited all your earnings from your summer job in a savings account. You plan to withdraw \$50 in spending money each month for the 9 months of the school year. You can represent each \$50 withdrawal as −50. How can you use multiplication to find the total amount withdrawn during the 9 months?" **9(−50) = −450**

What You'll Learn You'll learn to multiply rational numbers.

Why It's Important *Health* One way to determine safe backpack weights is to multiply rational numbers. *See Exercise 16.*

Knowing how to multiply rational numbers is essential in simplifying algebraic expressions and solving equations. The multiplication of rational numbers can be modeled by repeated addition.

Find $4 \times (-0.2)$.

$$4 \times (-0.2) = -0.2 + (-0.2) + (-0.2) + (-0.2)$$
$$= -0.8$$

This example illustrates the rule for multiplying two rational numbers.

Multiplying Two Rational Numbers	Words	Numbers
Different Signs	The product of two rational numbers with different signs is negative.	$1.2(-4.5) = -5.4$ $-1.2(4.5) = -5.4$
Same Sign	The product of two rational numbers with the same sign is positive.	$3(2.8) = 8.4$ $-3(-2.8) = 8.4$

Examples

Prerequisite Skills Review
Operations with Decimals, p. 684

Find each product.

1. **2.3(−4)**

$2.3(-4) = -9.2$ *The factors have different signs. The product is negative.*

2. **−3.5(−0.8)**

$-3.5(-0.8) = 2.8$ *The factors have the same sign. The product is positive.*

Your Turn

a. $-5(-1.3)$ **6.5** b. $-2.4(7.5)$ **−18** c. $8 \cdot (-3.2)$ **−25.6**

Real World **Example** **Sports Link**

3. **A skydiver jumps from 12,000 feet. Solve the equation $h = 12{,}000 + (0.5)(-32.1)(576)$ to find his height after he free-falls for 24 seconds.**

$h = 12{,}000 + (0.5)(-32.1)(576)$
$= 12{,}000 + (-16.05)(576)$ *Multiply 0.5 and −32.1.*
$= 12{,}000 + (-9244.8)$ *Multiply −16.05 and 576.*
$= 2755.2$ *Add 12,000 and −9244.8.*

After 24 seconds, the skydiver's height is 2755.2 feet.

Resource Manager

Reproducible Masters

Chapter 4 Resource Masters
- *Study Guide,* p. 145
- *Skills Practice,* p. 146
- *Practice,* p. 147
- *Reading to Learn Mathematics,* p. 148
- *Enrichment,* p. 149

Hands-On Algebra, p. 53

Transparencies
5-Minute Check, 4–1

Technology/Multimedia
Interactive Chalkboard CD-ROM

You can use grid paper to model the multiplication of fractions.

Materials:  grid paper  colored pencils  straightedge

Multiply $\frac{3}{7}$ and $\frac{1}{5}$ using a model.

Step 1 Use a straightedge to draw a rectangle that has seven rows and five columns.

Step 2 Shade three of the seven rows yellow to represent $\frac{3}{7}$.

Step 3 Shade one column blue to represent $\frac{1}{5}$.

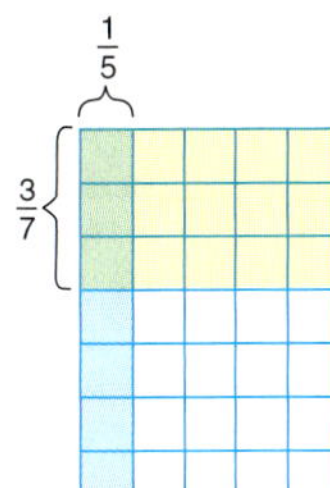

Try These

1. How many of the squares are shaded green? **3 squares**
2. What fraction of the total number of squares in the rectangle is shaded green? $\frac{3}{35}$
3. The portion of squares that is shaded green represents the product of the fractions. What is the product of $\frac{3}{7}$ and $\frac{1}{5}$? $\frac{3}{35}$
4. Use grid paper to model $\frac{3}{4}$ and $\frac{5}{8}$. Then use your model to find the product of $\frac{3}{4}$ and $\frac{5}{8}$. **See margin for model; $\frac{15}{32}$.**
5. What do you notice about the relationship between the numerators and denominators of $\frac{3}{4}$ and $\frac{5}{8}$ and their product?

5. The product of the numerators is the numerator of the product. The product of the denominators is the denominator of the product.

The Hands-On Algebra activity suggests the following rule for multiplying fractions.

Multiplying Fractions		
	Words:	To multiply fractions, multiply the numerators and multiply the denominators.
	Symbols:	$\frac{a}{b} \cdot \frac{c}{d} = \frac{ac}{bd}$, where $b, d \neq 0$
	Numbers:	$\frac{1}{5} \cdot \frac{3}{4} = \frac{1 \cdot 3}{5 \cdot 4}$ or $\frac{3}{20}$

www.algconcepts.com/extra_examples

2 TEACH

In-Class Examples

Examples 1–2

Find each product.

1 $-3.2(5)$ **−16**

2 $-4.7(-0.4)$ **1.88**

Example 3

Refer to Example 3. Solve the equation $h = 12{,}000 + (0.5)(-32.1)(144)$ to find the skydiver's height after he free-falls for 12 seconds. **9688.8 ft**

Answer

4.

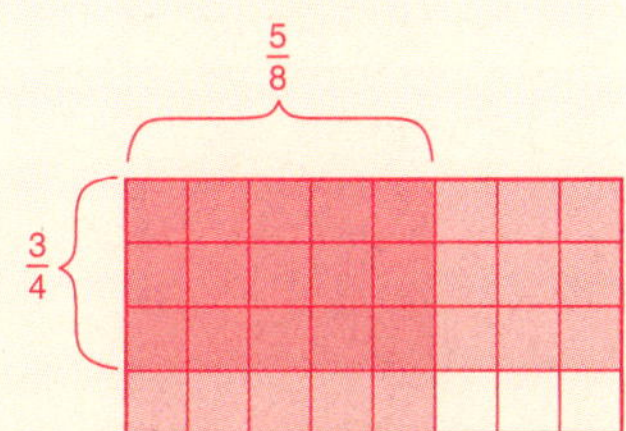

Hands-On Algebra

Cooperative Learning As an extension to the activity, have students work with a partner and ask each student to think of a fraction less than 1. Instruct one partner to use the denominator of his or her fraction to draw one side of a rectangle on grid paper. The other partner then draws an adjacent side of the rectangle using his or her denominator. After completing the rectangle, have each student model their fraction by shading a portion of the rectangle as in the activity. The partners should then discuss how the product of their fractions is shown in the model.

Hands-On Algebra Masters, p. 53

In-Class Example

Example 4

Find $\frac{2}{7}\left(-\frac{3}{5}\right)$. $-\frac{6}{35}$

Teaching Tip When discussing Examples 5 and 6, make sure students understand they must first change a whole number or mixed number to its improper form before multiplying.

In-Class Examples

Examples 5–6

Find each product.

5 $-8 \cdot \left(-\frac{3}{7}\right)$ $3\frac{3}{7}$

6 $-3\frac{1}{3} \cdot \left(\frac{1}{7}\right)$ $-\frac{10}{21}$

Examples

Find each product.

4 $-\frac{4}{5} \cdot \frac{1}{3}$

$-\frac{4}{5} \cdot \frac{1}{3} = -\frac{4 \cdot 1}{5 \cdot 3}$ *Multiply the numerators and multiply the denominators.*

$= -\frac{4}{15}$ *The factors have different signs. The product is negative.*

5 $6 \cdot \left(\frac{2}{5}\right)$

$6 \cdot \left(\frac{2}{5}\right) = \frac{6}{1} \cdot \frac{2}{5}$ *Rewrite 6 as an improper fraction.*

$= \frac{6 \cdot 2}{1 \cdot 5}$ *Multiply the numerators and multiply the denominators.*

$= \frac{12}{5}$ or $2\frac{2}{5}$ *The factors have the same sign. The product is positive.*

Prerequisite Skills Review
Simplifying Fractions, p. 685

6 $-\frac{1}{4} \cdot \left(-4\frac{1}{2}\right)$

$-\frac{1}{4} \cdot \left(-4\frac{1}{2}\right) = -\frac{1}{4} \cdot \left(-\frac{9}{2}\right)$ *Rewrite $-4\frac{1}{2}$ as an improper fraction.*

$= \left(\frac{1 \cdot 9}{4 \cdot 2}\right)$ *Multiply the numerators and multiply the denominators.*

$= \frac{9}{8}$ or $1\frac{1}{8}$ *The factors have the same sign. The product is positive.*

Your Turn

d. $\frac{3}{4}\left(\frac{5}{7}\right)$ $\frac{15}{28}$

e. $\frac{4}{5} \cdot (-3)$ $-\frac{12}{5}$ or $-2\frac{2}{5}$

f. $-\frac{1}{2}\left(-1\frac{3}{8}\right)$ $\frac{11}{16}$

Study these products.

$-1(5) = -5$ *−1 times 5 equals −5.*

$(-1)(-2.4) = 2.4$ *−1 times −2.4 equals 2.4.*

Notice that multiplying a number by −1 results in the opposite of the number. This suggests the **Multiplicative Property of −1**.

Multiplicative Property of −1		
Words:	The product of −1 and any number is the number's additive inverse.	
Symbols:	$-1 \cdot a = -a$	**Numbers:** $-1 \cdot \frac{2}{7} = -\frac{2}{7}$

Reteaching Activity

Interpersonal Learners Have students work with a partner. Each student should use colored pencils and grid paper to represent a multiplication problem such as those in the Hands-On Algebra activity on page 141. They should then switch papers with their partner. After receiving their partner's paper, instruct them to write out the multiplication represented by the model and determine the product. Students should then check their partner's work.

You can use what you know about multiplying rational numbers to simplify algebraic expressions.

Example

7 **Simplify $-7.2t(5u)$.**

$$
\begin{aligned}
-7.2t(5u) &= (-7.2)(t)(5)(u) && -7.2t = (-7.2)(t);\ (5u) = (5)(u)\\
&= (-7.2)(5)(t)(u) && \text{Commutative Property}\\
&= (-7.2 \cdot 5)(t \cdot u) && \text{Associative Property}\\
&= -36tu && \text{Simplify.}
\end{aligned}
$$

8 **Simplify $\frac{2}{5}m\left(\frac{3}{5}n\right)$.**

$$
\begin{aligned}
\frac{2}{5}m\left(\frac{3}{5}n\right) &= \left(\frac{2}{5}\right)(m)\left(\frac{3}{5}\right)(n) && \frac{2}{5}m = \left(\frac{2}{5}\right)(m);\ \left(\frac{3}{5}n\right) = \left(\frac{3}{5}\right)(n)\\
&= \left(\frac{2}{5}\right)\left(\frac{3}{5}\right)(m)(n) && \text{Commutative Property}\\
&= \left(\frac{2}{5} \cdot \frac{3}{5}\right)(m \cdot n) && \text{Associative Property}\\
&= \frac{6}{25}mn && \text{Simplify.}
\end{aligned}
$$

Your Turn

g. $2x(-4.5y)$ **$-9xy$**

h. $\frac{5}{7}g\left(\frac{2}{3}h\right)$ **$\frac{10}{21}gh$**

When you add any two rational numbers, is the sum always a rational number? What happens when you subtract or multiply rational numbers? In all three cases, the result is a rational number.

Closure Property of Rational Numbers	The sum, difference, or product of two rational numbers is always a rational number. Therefore, the set of rational numbers is closed under addition, subtraction, and multiplication.

For example, you can add, subtract, or multiply $\frac{5}{7}$ and $\frac{1}{7}$, and the result will be a rational number. *Check this.*

Check for Understanding

Communicating Mathematics

1. **State** whether $1.2bc$ is *sometimes*, *always*, or *never* a rational number if b and c are rational numbers. Explain.
2. **Determine** whether the product of three negative rational numbers can ever be positive. Explain.
3. **Write** a multiplication equation that can be shown using the model. **$\frac{1}{4} \cdot \frac{5}{6} = \frac{5}{24}$**

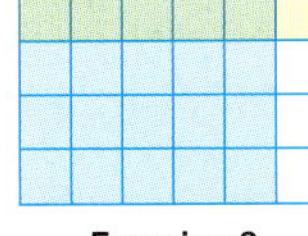

Exercise 3

1. Always, because rational numbers are closed under multiplication.
2. See margin.

Teaching Tip When discussing the Multiplicative Property of -1, point out that the variable a can be replaced by negative numbers and zero as well as positive numbers.

In-Class Example

Example 7

Simplify $5b(-2.2y)$. **$-11by$**

3 PRACTICE/APPLY

Error Analysis

Watch for students who try to apply addition rules, such as finding a common denominator or multiplying only the numerators, when multiplying the fractions in Exercises 10, 12, and 14.

Prevent by discussing the area model shown in the Hands-On Algebra activity again, to help students understand why both the numerators and the denominators are multiplied.

Answer

2. No; the product of any two negative rational numbers is positive. The product of a positive number and a negative rational number is negative.

***Study Guide*, p. 145**

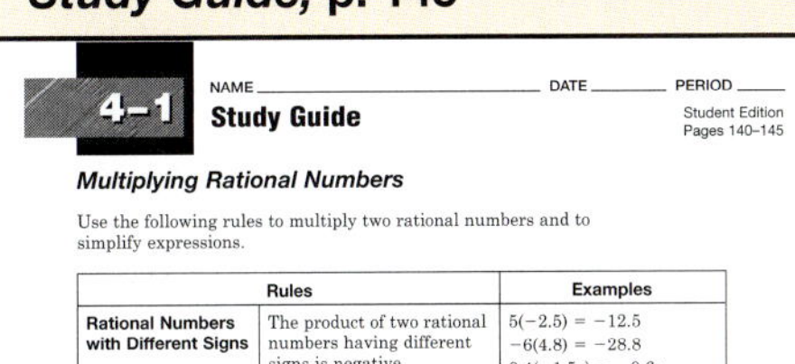

4–1 NAME ______________ DATE ________ PERIOD _____

Study Guide

Student Edition Pages 140–145

Multiplying Rational Numbers

Use the following rules to multiply two rational numbers and to simplify expressions.

Rules		Examples
Rational Numbers with Different Signs	The product of two rational numbers having different signs is negative.	$5(-2.5) = -12.5$ $-6(4.8) = -28.8$ $0.4(-1.5x) = -0.6x$
Rational Numbers with the Same Sign	The product of two rational numbers having the same sign is positive.	$(-4.2)(-8) = 33.6$ $12(3.2) = 38.4$ $(3w)(2.1y) = 6.3wy$
Fractions	To multiply fractions, multiply the numerators and multiply the denominators.	$-\frac{3}{5} \cdot \frac{1}{4} = -\frac{3}{20}$ $-\frac{1}{2} \cdot \left(-2\frac{1}{3}\right) = -\frac{1}{2} \cdot \left(-\frac{7}{3}\right)$ $= \frac{7}{6}$ or $1\frac{1}{6}$

Find each product.

1. $3.8(-4)$ **-15.2**
2. $-6(-1.2)$ **7.2**
3. $6(6.1)$ **36.6**
4. $-0.5(-5.2)$ **2.6**
5. $(-8.4) \cdot 0$ **0**
6. $-2.4 \cdot (10.5)$ **-25.2**
7. $\frac{1}{7} \cdot \frac{2}{3}$ **$\frac{2}{21}$**
8. $-2 \cdot \left(-\frac{2}{5}\right)$ **$\frac{4}{5}$**
9. $-\frac{2}{3} \cdot \frac{2}{9}$ **$-\frac{4}{27}$**
10. $-\frac{5}{3} \cdot \left(-\frac{2}{7}\right)$ **$\frac{10}{21}$**
11. $0 \cdot \left(-\frac{1}{5}\right)$ **0**
12. $\frac{1}{3}\left(-1\frac{1}{4}\right)$ **$-\frac{5}{12}$**

Simplify each expression.

13. $3(1.4a)$ **$4.2a$**
14. $-0.6x(-3y)$ **$1.8xy$**
15. $\frac{1}{4} \cdot \left(-\frac{1}{2}b\right)$ **$-\frac{1}{8}b$**

Assignment Guide

Basic: 17–47 odd, 48–58
Average: 18–44 even, 45–58

Guided Practice

Getting Ready Find each product.

Sample: $(-4)(-6)$ **Solution:** $(-4)(-6) = 24$

4. $-1 \cdot 8$ **−8**
5. $3(-3)$ **−9**
6. $-7 \cdot (-4)$ **28**

Examples 1–6

Find each product.

7. $2.1(-7)$ **−14.7**
8. $(-1.8)(-3.5)$ **6.3**
9. $0.5(-4.6)$ **−2.3**
10. $-\frac{1}{3}\left(-\frac{4}{7}\right)$ $\frac{4}{21}$
11. $-3 \cdot \frac{3}{4}$ $-\frac{9}{4}$ or $-2\frac{1}{4}$
12. $2\frac{3}{5}\left(\frac{1}{4}\right)$ $\frac{13}{20}$

Example 7

Simplify each expression.

13. $3\left(\frac{1}{5}y\right)$ $\frac{3}{5}y$
14. $-\frac{1}{4} \cdot \frac{7}{10}k$ $-\frac{7}{40}k$
15. $-2.5c(-2.4d)$ **6cd**

Example 5

16. **Health** Students who carry backpacks that are too heavy can develop spinal problems. The maximum recommended backpack weight y is given by the equation $y = \frac{3}{20}x$, where x is a student's weight. Would a 12-pound backpack be too heavy for a student who weighs 110 pounds? Explain.
Source: *Zillions* **No; the maximum backpack weight**

for a 110-pound student would be $16\frac{1}{2}$ pounds.

Exercises

Practice

Homework Help	
For Exercises	See Examples
17–25, 48	1, 2
26–34	4, 5, 6
35–42	3, 7, 8
43, 45–47	3
Extra Practice	
See page 698.	

25. 57.6

Find each product.

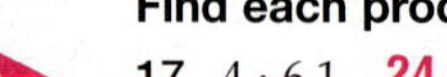

A

17. $4 \cdot 6.1$ **24.4**
18. $3.8(8)$ **30.4**
19. $7.1(-1)$ **−7.1**
20. $(-0.5)(-5.6)$ **2.8**
21. $-6.0(8.5)$ **−51**
22. $-0.4(1.5)$ **−0.6**
23. $-2.9 \cdot 10$ **−29**
24. $-7.3 \cdot 0$ **0**
25. $(-12.8)(-1)(4.5)$
26. $\frac{1}{5}\left(\frac{2}{3}\right)$ $\frac{2}{15}$
27. $-1 \cdot \left(-\frac{5}{8}\right)$ $\frac{5}{8}$
28. $-\frac{4}{7} \cdot \frac{3}{5}$ $-\frac{12}{35}$
29. $(0)\left(-\frac{6}{7}\right)$ **0**
30. $-\frac{3}{5}\left(-\frac{9}{10}\right)$ $\frac{27}{50}$
31. $\frac{7}{3}\left(-\frac{3}{7}\right)$ **−1**
32. $\left(-\frac{3}{5}\right)\left(\frac{-2}{7}\right)(-1)$ $-\frac{6}{35}$
33. $3 \cdot \frac{7}{8}$ $\frac{21}{8}$ or $2\frac{5}{8}$
34. $\frac{6}{7}\left(-2\frac{1}{3}\right)$ **−2**

Simplify each expression.

B

35. $2(-1.5x)$ **−3x**
36. $0.6(-15y)$ **−9y**
37. $(4r)(2.2s)$ **8.8rs**
38. $\frac{5}{6}r(-18s)$ **−15rs**
39. $\left(\frac{3}{5}a\right)\left(\frac{2}{5}b\right)$ $\frac{6}{25}ab$
40. $\frac{1}{3}y\left(-\frac{1}{4}z\right)$ $-\frac{1}{12}yz$

C

41. $-\frac{1}{2}s\left(-\frac{1}{2}\right) + \frac{2}{3}\left(\frac{3}{2}s\right)$ $\frac{5}{4}s$ or $1\frac{1}{4}s$
42. $\frac{5}{6}\left(\frac{3}{4}a\right) - \left(\frac{1}{8}a\right)\left(-\frac{1}{3}\right)$ $\frac{16}{24}a$ or $\frac{2}{3}a$

***Skills Practice,* p. 146, and *Practice,* p. 147 (shown)**

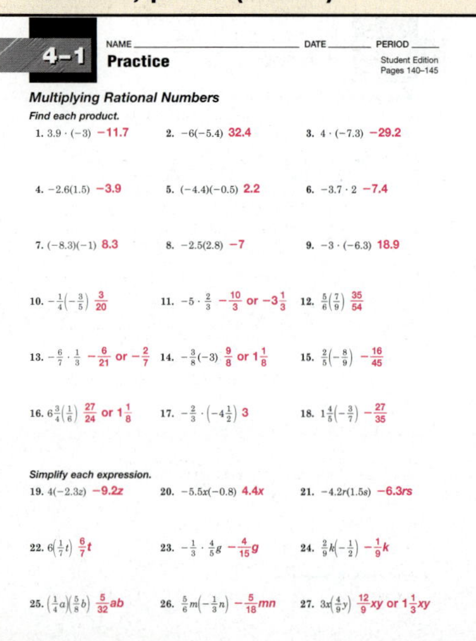

4–1 NAME ______ DATE ______ PERIOD ______
Practice
Student Edition Pages 140–145

Multiplying Rational Numbers

Find each product.

1. $3.9 \cdot (-3)$ **−11.7**
2. $-6(-5.4)$ **32.4**
3. $4 \cdot (-7.3)$ **−29.2**
4. $-2.6(1.5)$ **−3.9**
5. $(-4.4)(-0.5)$ **2.2**
6. $-3.7 \cdot 2$ **−7.4**
7. $(-8.3)(-1)$ **8.3**
8. $-2.5(2.8)$ **−7**
9. $-3 \cdot (-6.3)$ **18.9**
10. $-\frac{1}{4}\left(-\frac{3}{5}\right)$ $\frac{3}{20}$
11. $-5 \cdot \frac{2}{3}$ $-\frac{10}{3}$ or $-3\frac{1}{3}$
12. $\frac{5}{6}\left(\frac{7}{9}\right)$ $\frac{35}{54}$
13. $-\frac{6}{7} \cdot \frac{1}{3}$ $-\frac{6}{21}$ or $-\frac{2}{7}$
14. $-\frac{3}{8}(-3)$ $\frac{9}{8}$ or $1\frac{1}{8}$
15. $\frac{2}{5}\left(-\frac{8}{9}\right)$ $-\frac{16}{45}$
16. $6\frac{3}{4}\left(\frac{1}{6}\right)$ $\frac{27}{24}$ or $1\frac{1}{8}$
17. $-\frac{2}{3} \cdot \left(-4\frac{1}{2}\right)$ **3**
18. $1\frac{4}{5}\left(-\frac{3}{7}\right)$ $-\frac{27}{35}$

Simplify each expression.

19. $4(-2.3z)$ **−9.2z**
20. $-5.5x(-0.8)$ **4.4x**
21. $-4.2r(1.5s)$ **−6.3rs**
22. $6\left(\frac{1}{7}t\right)$ $\frac{6}{7}t$
23. $-\frac{1}{3} \cdot \frac{4}{5}g$ $-\frac{4}{15}g$
24. $\frac{2}{9}k\left(-\frac{1}{2}\right)$ $-\frac{1}{9}k$
25. $\left(\frac{1}{4}a\right)\left(\frac{5}{8}b\right)$ $\frac{5}{32}ab$
26. $\frac{5}{6}m\left(-\frac{1}{3}n\right)$ $-\frac{5}{18}mn$
27. $3x\left(\frac{4}{9}y\right)$ $\frac{12}{9}xy$ or $1\frac{1}{3}xy$

43. Solve $x = 2(-3.2) - 0.5(-4.8)$ to find the value of x. **-4**

44. Evaluate $7a(-4.5b) + (-2ab)$ if $a = 2$ and $b = -1$. **67**

Applications and Problem Solving

45. **Shopping** Teri wants to buy a bicycle with the money she earns from her after-school job. If she can save \$22.50 a week, how much money will she have after six weeks? **\$135**

46. **Space** A day on Jupiter is much shorter than a day on Earth because Jupiter rotates on its axis about $2\frac{2}{5}$ times as fast as Earth rotates on its axis. During our month of June, how many days will Jupiter have?

46. 72 days

47. $-\frac{2401}{40}$ m or $-60\frac{1}{4}$ m

47. **History** Galileo (1564–1642) was a mathematician and scientist who is said to have experimented with falling objects from the Leaning Tower of Pisa. Suppose Galileo dropped a stone from the tower and it took $3\frac{1}{2}$ seconds to reach the ground. Solve $d = -\frac{1}{2}\left(9\frac{4}{5}\right)\left(3\frac{1}{2}\right)\left(3\frac{1}{2}\right)$ to find the distance d in meters that the stone fell.

A negative answer means that the final position is lower than the initial position.

48. **Critical Thinking** Describe the values of a and b for each statement to be true. **See margin.**

a. ab is positive. b. ab is negative. c. $ab = 0$

Mixed Review

50. $\{-4, 4\}$

51. $\{-12, -6\}$

Solve each equation. Check your solution. *(Lesson 3–7)*

49. $|t| = -5$ $\varnothing$ 50. $2 + |x| = 6$ 51. $3 = |-9 - a|$

52. **Aviation** A traffic helicopter descended 160 meters to observe road conditions. It leveled off at 225 meters. *(Lesson 3–6)*

a. Let a represent the original altitude. Write an equation to represent this problem. **$a - 160 = 225$**

b. What was the helicopter's original altitude? **385 m**

Find each sum or difference. *(Lesson 3–2)*

53. $3.5 + (-2.1)$ **1.4** 54. $-1.7 - 4.0$ **-5.7** 55. $-\frac{3}{4} + \frac{1}{8}$ **$-\frac{5}{8}$**

56. Find $24 \div (-6)$. *(Lesson 2–6)* **-4**

Standardized Test Practice

57. **Extended Response** The stem-and-leaf plot shows the heights of 30 students. *(Lesson 1–7)*

a. What is the height of the tallest student? **73 in.**

b. Which height occurs most frequently? **64 in.**

Class Heights

Stem	Leaf
5	7 8 8 9 9 9
6	0 0 1 2 2 2 3 3 4 4 4 4 4 5 5 7 8 9 9
7	0 0 2 3 3 $7\|2 = 72$ in.

58. **Multiple Choice** The expression $(220 - y) \times 0.8 \div 4$ gives the target 15-second heart rate for an athlete y years old during a workout. Find the target 15-second heart rate for a 17-year-old athlete during a workout. *(Lesson 1–2)* **B**

(A) 27 (B) 40.6 (C) 216.6 (D) 219.7

www.algconcepts.com/self_check_quiz

4 ASSESS

Open-Ended Assessment

Writing Ask students to write a paragraph explaining how to multiply rational numbers. They should include an explanation of how to determine the sign of the product.

Answers

48a. a and b are both positive or a and b are both negative.

48b. a is positive and b is negative or a is negative and b is positive.

48c. a is 0 or b is 0 or both a and b are 0.

***Enrichment*, p. 149**

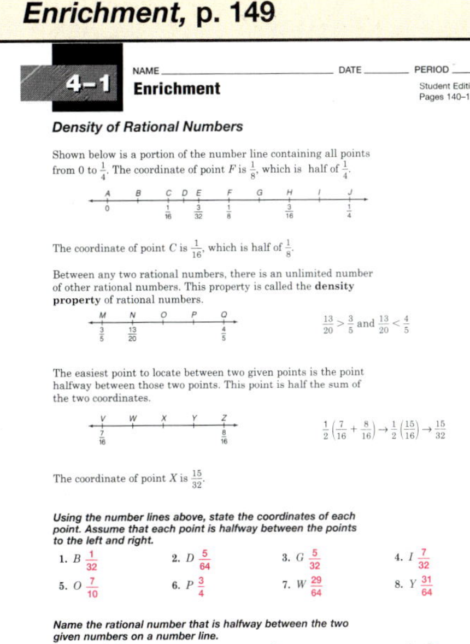

4–1 NAME ______ DATE ______ PERIOD ______

Enrichment

Student Edition Pages 140–145

Density of Rational Numbers

Shown below is a portion of the number line containing all points from 0 to $\frac{1}{4}$. The coordinate of point F is $\frac{1}{8}$, which is half of $\frac{1}{4}$.

The coordinate of point C is $\frac{1}{16}$, which is half of $\frac{1}{8}$.

Between any two rational numbers, there is an unlimited number of other rational numbers. This property is called the **density property** of rational numbers.

$\frac{13}{20} > \frac{3}{5}$ and $\frac{13}{20} < \frac{4}{5}$

The easiest point to locate between two given points is the point halfway between those two points. This point is half the sum of the two coordinates.

$\frac{1}{2}\left(\frac{7}{16} + \frac{8}{16}\right) \rightarrow \frac{1}{2}\left(\frac{15}{16}\right) \rightarrow \frac{15}{32}$

The coordinate of point X is $\frac{15}{32}$.

Using the number lines above, state the coordinates of each point. Assume that each point is halfway between the points to the left and right.

1. B $\frac{1}{32}$ 2. D $\frac{5}{64}$ 3. G $\frac{5}{32}$ 4. I $\frac{7}{32}$

5. O $\frac{7}{10}$ 6. P $\frac{3}{4}$ 7. W $\frac{29}{64}$ 8. Y $\frac{31}{64}$

Name the rational number that is halfway between the two given numbers on a number line.

9. $\frac{1}{2}$ and $\frac{3}{4}$ $\frac{5}{8}$ 10. 0 and 5 $2\frac{1}{2}$ 11. 4 and $5\frac{1}{4}$ $4\frac{5}{8}$

12. $1\frac{1}{2}$ and $\frac{5}{8}$ $1\frac{1}{16}$ 13. $\frac{1}{2}$ and $\frac{2}{3}$ $\frac{7}{12}$ 14. $\frac{7}{8}$ and $\frac{8}{9}$ $\frac{127}{144}$

Extra Credit

Evaluate the following expressions.

$-1 \cdot (-1) = (-1)^2 = ?$

$-1 \cdot (-1) \cdot (-1) = (-1)^3 = ?$

$-1 \cdot (-1) \cdot (-1) \cdot (-1) = (-1)^4 = ?$

$-1 \cdot (-1) \cdot (-1) \cdot (-1) \cdot (-1) = (-1)^5 = ?$

Based on your observations, what can you say about the value of $(-1)^n$, where n is an integer? **If n is even, $(-1)^n = 1$, and if n is odd, $(-1)^n = -1$.**

Lesson 4–2

1 FOCUS

5-Minute Check
Lesson 4–1

Write a multiplication equation that can be shown using each model.

1.

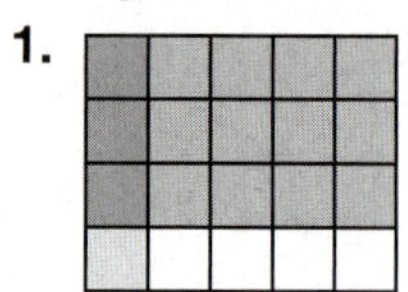

$\frac{3}{4} \cdot \frac{1}{5} = \frac{3}{20}$

2.

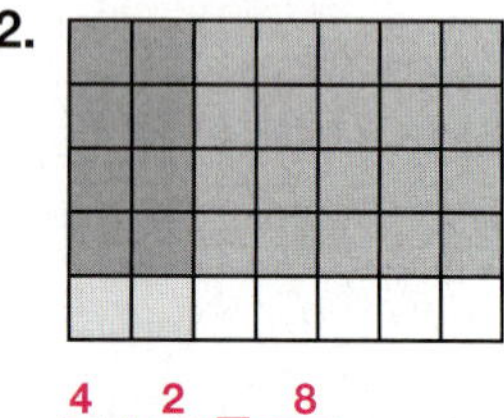

$\frac{4}{5} \cdot \frac{2}{7} = \frac{8}{35}$

Find each product.

3. $-3(1.2)$ $\mathbf{-3.6}$

4. $0.4(-3.5)(-1)$ $\mathbf{1.4}$

5. Simplify $2y\left(-\frac{1}{5}z\right)$. $-\frac{2}{5}yz$

Motivating the Lesson

Hands-On Activity Have students work with a partner. Give each pair of students a coin and a die. Ask them to determine all the different outcomes that can occur when the coin is flipped and the die is rolled. For example, one outcome is heads on the coin and a 2 on the die. Then discuss ways to organize their list of outcomes, leading into a discussion of drawing a tree diagram to model this situation.

4–2 Counting Outcomes

What You'll Learn
You'll learn to use tree diagrams and the Fundamental Counting Principle to count outcomes.

Why It's Important
Manufacturing Car manufacturers count outcomes to determine the number of different key combinations that are possible.
See Exercise 23.

Baroness Martine de Beausoleil was a French scientist in the 17th century who spent 30 years studying geology and mathematics. She determined which rocks were valuable by the minerals they contained. Some other ways to classify rocks are by their texture and by their color.

How many different rocks are possible having the characteristics shown in the table? We can represent this situation by using a **tree diagram**.

Rock Characteristics

Mineral	Texture	Color
gold quartz iron silver	glassy dull	brown gray

Quartz

Mineral	Texture	Color	Outcome
gold	glassy	brown	gold, glassy, brown
		gray	gold, glassy, gray
	dull	brown	gold, dull, brown
		gray	gold, dull, gray
quartz	glassy	brown	quartz, glassy, brown
		gray	quartz, glassy, gray
	dull	brown	quartz, dull, brown
		gray	quartz, dull, gray
iron	glassy	brown	iron, glassy, brown
		gray	iron, glassy, gray
	dull	brown	iron, dull, brown
		gray	iron, dull, gray
silver	glassy	brown	silver, glassy, brown
		gray	silver, glassy, gray
	dull	brown	silver, dull, brown
		gray	silver, dull, gray

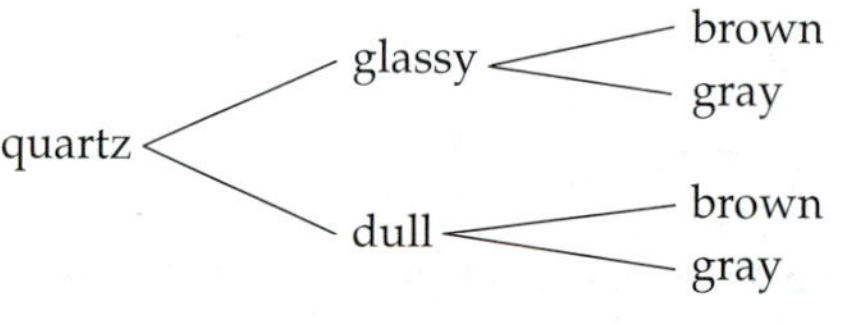

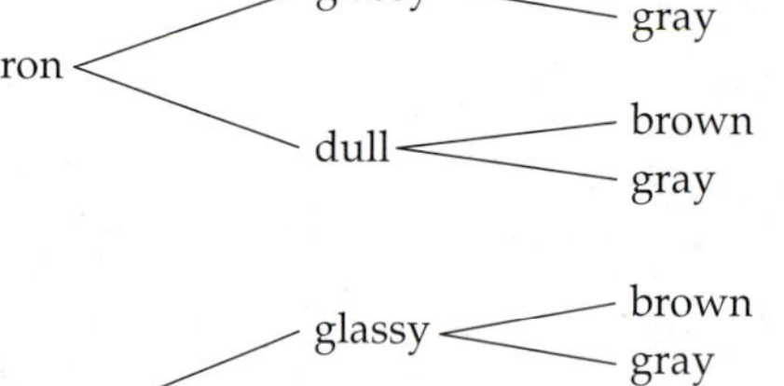

Gold

By using the tree diagram, we find that there are 16 different rocks possible. These results are called **outcomes**. For example, the first outcome is a rock that contains gold, has a glassy texture, and is brown. The list of all the possible outcomes is called the **sample space**.

Resource Manager

Reproducible Masters
Chapter 4 Resource Masters
- *Study Guide*, p. 150
- *Skills Practice*, p. 151
- *Practice*, p. 152
- *Reading to Learn Mathematics*, p. 153
- *Enrichment*, p. 154

Transparencies
5-Minute Check, 4–2

Technology/Multimedia
Interactive Chalkboard CD-ROM

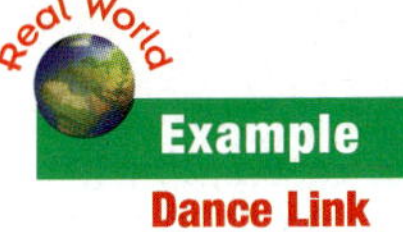

Example 1
Dance Link

The Greenwood Dance Company has three lead female dancers, Teresa, Kendra, and Angelica, and two lead male dancers, Keith and Jamal. How many different ways can the director choose one female dancer and one male dancer to lead the next production?

Make a tree diagram to find the number of combinations.

Female	Male	Outcome
Teresa	Keith	Teresa, Keith
	Jamal	Teresa, Jamal
Kendra	Keith	Kendra, Keith
	Jamal	Kendra, Jamal
Angelica	Keith	Angelica, Keith
	Jamal	Angelica, Jamal

The sample space contains six outcomes. So, there are six different ways that the director can choose the leads for the next production.

Your Turn **a. 12; See margin for tree diagram.**

a. The Ice Cream Parlor offers the choices below for making sundaes. Draw a tree diagram to find the number of different sundaes that can be made.

Ice Cream	Topping	Whipped Cream
chocolate vanilla rocky road	chocolate butterscotch	yes no

An **event** is a subset of the possible outcomes, or sample space. In Example 1, the choice of female dancers is one event, and the choice of male dancers is another event. Notice that the product of the number of choices in each event is $3 \cdot 2$ or 6. This method of finding the number of possible outcomes is called the **Fundamental Counting Principle**.

Fundamental Counting Principle	If event M can occur in m ways and is followed by event N that can occur in n ways, then the event M followed by event N can occur in $m \times n$ ways.

2 TEACH

In-Class Example
Example 1
Brooke is shopping for a new computer system. She has a list of 2 different CPUs, 3 different monitors, and 3 different printers. How many different ways can she choose one CPU, one monitor, and one printer from her list? 18

Teaching Tip You may need to remind students of the meaning of the term *subset* when discussing the definition of an *event*.

Answer
Your Turn

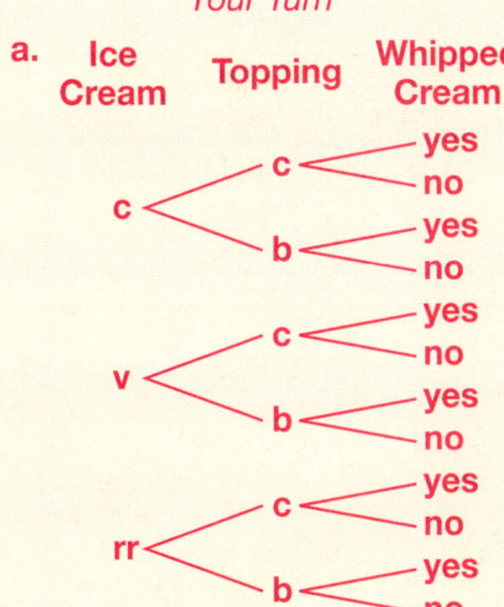

Teaching Tip Students should quickly realize that while a tree diagram is useful for finding the number of possible outcomes when there are a few possibilities, it is impractical when the number of possible outcomes is great. In a situation such as that in Example 2, point out that the Fundamental Counting Principle is a more efficient method for finding the total number of outcomes. However, stress that a tree diagram actually lists the possible outcomes, while the Fundamental Counting Principle only gives the number of possible outcomes.

In-Class Example

Example 2

How many different kinds of photo processing are possible? **12**

Process Time	Paper Type	Photo Size
1 hour 1 day	regular glossy deluxe	3 by 5 4 by 6

3 PRACTICE/APPLY

Error Analysis

Watch for students who have trouble listing the possible outcomes for an experiment such as the one presented in Exercise 3. Students may list the outcomes as being 0 heads, 1 head, 2 heads, and 3 heads.

Prevent by encouraging students to draw tree diagrams and list the entries along each different branch as an outcome. Point out that getting heads on the first coin and then tails on the second and third coins is a different outcome than getting tails on the first two coins and then a head on the third coin.

Example **2** **Clothing Link**

How many different kinds of school sweatshirts are possible?

Style	Color	Size
school name school logo team graphic	red tan	small medium large 1X 2X

There are 3 styles, 2 colors, and 5 sizes, so the number of different sweatshirts is $3 \times 2 \times 5$ or 30.

Your Turn

b. Catina wants to buy a red sweatshirt in size large or 1X in any style. How many choices does she have? **6**

Check for Understanding

Communicating Mathematics

Vocabulary
tree diagram
outcome
sample space
event

1. **Explain** the Fundamental Counting Principle in your own words.
2. **Represent** a real-life situation by drawing a tree diagram. Include a description of each event and a list of all the outcomes in the sample space. **See students' work.**
3. **You Decide** Ling says that if three coins are tossed, then the number of outcomes with 2 heads and 1 tail is the same as the number with 1 head and 2 tails. Lorena thinks he is wrong. Who is correct and why? **Ling; three outcomes have 2 heads: HHT, HTH, THH. Three outcomes have 2 tails: TTH, THT, HTT.**

1. Sample answer: the number of ways two or more events can occur is the product of the number of ways each event can occur.

Guided Practice

Getting Ready **Suppose you spin the spinner twice. Determine whether each is an *outcome* or a *sample space*.**

Sample 1: (red, blue)
Solution: Spinning a red and then a blue (red, blue) is an outcome because it is only one possible result.
Sample 2: (red, red), (red, blue), (blue, red), (blue, blue)
Solution: This is a sample space because it is a list of all possible outcomes.

Determine whether each is an *outcome* or a *sample space* for the given experiment.

4. (5, 2, 2); rolling a number cube three times **outcome**
5. (H, T); tossing a coin once **sample space**
6. (H, H), (H, T), (T, H), (T, T); tossing a coin twice **sample space**
7. (4, 10, J, 3, 7); choosing five cards from a standard deck **outcome**

Reteaching Activity

Verbal/Linguistic Learners Tell students to imagine that a classmate was absent from class during this lesson. Have them write a paragraph explaining how to use the Fundamental Counting Principle to determine the number of possible outcomes. Instruct students to write their paragraph with the expectation that it would be read by the absent student.

Example 1 8. Suppose you can order a burrito or a taco with beef, chicken, or bean filling. Find the number of possible outcomes by drawing a tree diagram. **6; See margin for tree diagram.**

Example 2 9. Suppose you roll a die twice. Find the number of possible outcomes by using the Fundamental Counting Principle. **36**

Example 1 10. **Shopping** Enrique wants to buy a bicycle, but he is having trouble deciding what kind to buy.

Type	Speed	Handlebars
road mountain	21 16 14 10	touring racing

a. Draw a tree diagram to represent his choices. **See margin.**

b. How many different kinds of bicycles are possible? **16**

c. How many different kinds of bicycles have racing handlebars and 16 or greater gear speeds? **4**

Exercises

Practice

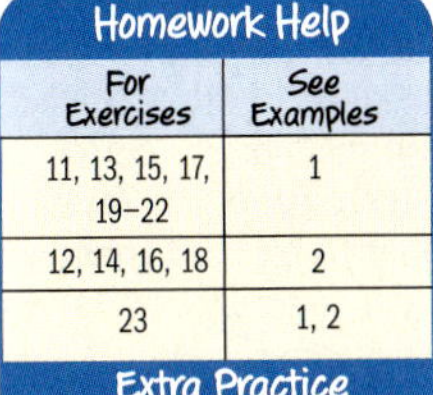

For Exercises	See Examples
11, 13, 15, 17, 19–22	1
12, 14, 16, 18	2
23	1, 2

Extra Practice
See page 699.

Find the number of possible outcomes by drawing a tree diagram.

11. three tosses of a coin **8**

12. choosing one marble from each box shown in the table at the right **8**

11–12. See margin for tree diagrams.

Box A	Box B	Box C
blue green	green red	green white

13. Anita purchased five T-shirts and three pairs of jeans for school, as shown in the table at the right. How many different T-shirt and jeans outfits are possible? **15; See Solutions Manual for tree diagram.**

T-Shirt	Jeans
red orange green white striped	black white blue

Find the number of possible outcomes by using the Fundamental Counting Principle.

14. different cars with options shown at the right **16**

15. possible sequences of answers on a 5-question true-false quiz **32**

Color	Engine	Transmission
red blue white green	4-cylinder 6-cylinder	manual automatic

16. different 1-topping pizzas with the choices shown at the right **72**

Crust	Size	Topping	
thin regular thick	individual small medium large	pepperoni mushroom sausage	onion olive pepper

Answers

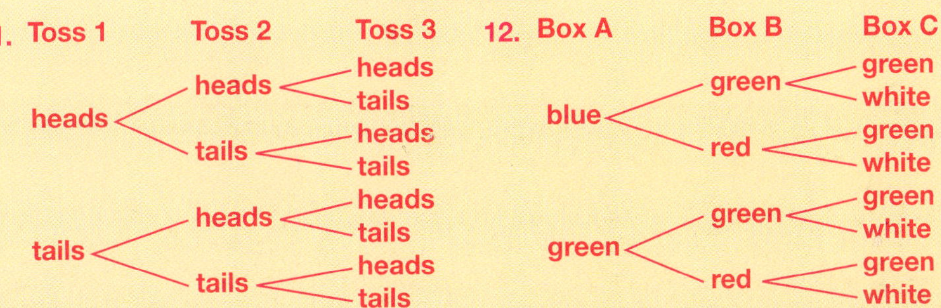

Assignment Guide

Basic: 11–23 odd, 24–31
Average: 12–18 even, 20–31

Answers

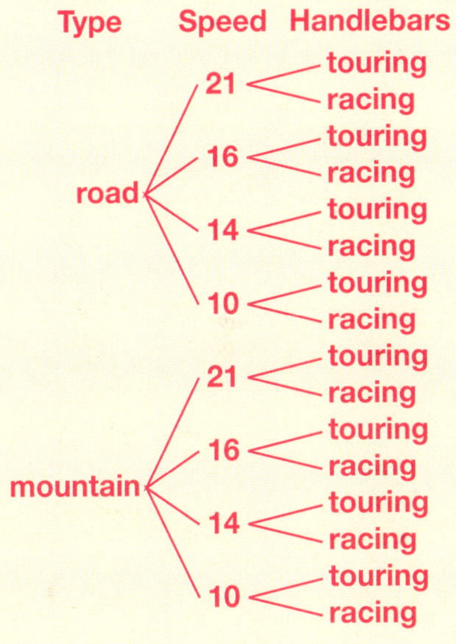

***Study Guide*, p. 150**

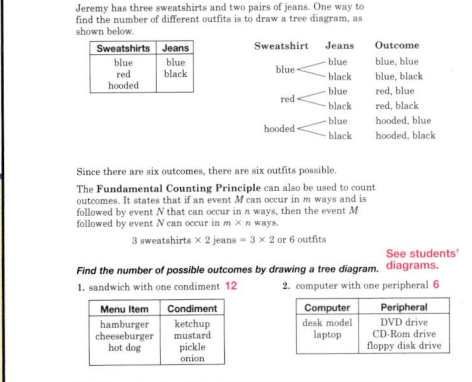

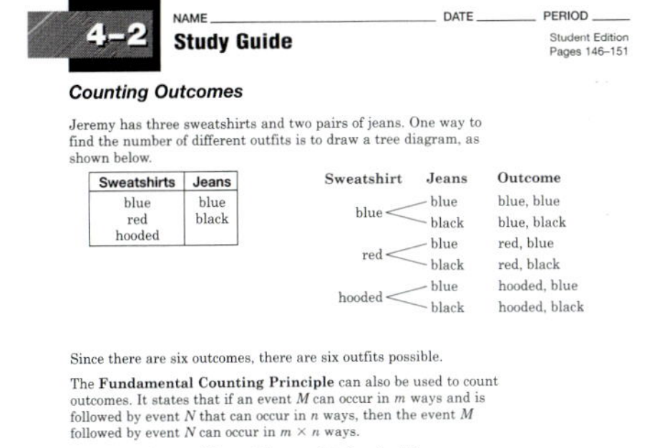

4–2 NAME ____ DATE ____ PERIOD ____

Study Guide Student Edition Pages 146–151

Counting Outcomes

Jeremy has three sweatshirts and two pairs of jeans. One way to find the number of different outfits is to draw a tree diagram, as shown below.

Sweatshirts	Jeans
blue red hooded	blue black

Sweatshirt	Jeans	Outcome
blue	blue	blue, blue
	black	blue, black
red	blue	red, blue
	black	red, black
hooded	blue	hooded, blue
	black	hooded, black

Since there are six outcomes, there are six outfits possible.

The **Fundamental Counting Principle** can also be used to count outcomes. It states that if an event M can occur in m ways and is followed by event N that can occur in n ways, then the event M followed by event N can occur in $m \times n$ ways.

3 sweatshirts × 2 jeans = 3 × 2 or 6 outfits

Find the number of possible outcomes by drawing a tree diagram. **See students' diagrams.**

1. sandwich with one condiment **12**

Menu Item	Condiment
hamburger cheeseburger hot dog	ketchup mustard pickle onion

2. computer with one peripheral **6**

Computer	Peripheral
desk model laptop	DVD drive CD-Rom drive floppy disk drive

Find the number of possible outcomes by using the Fundamental Counting Principle.

3. A certain game includes cards with 3 different pictures, 4 colors, and 6 numbers. Find the number of cards in the game. **72**

4. Find the number of possible ways of answering a true-false test that has twelve questions. **24**

5. Suppose you can order sweaters from a catalog in 8 sizes, 3 colors, and 2 delivery methods. Find the number of possible sweater orders. **48**

17. Refer to Example 1. How many different ways can the director choose one female dancer and one male dancer if there are five female dancers and six male dancers? **30**

18a. {(R, R), (R, B), (R, Y), (B, R), (B, B), (B, Y), (Y, R), (Y, B), (Y, Y)}

18. Suppose the spinner at the right is spun twice. Assume that the spinner will not land on a border.
    a. Let R represent red, B represent blue, and Y represent yellow. Write all possible outcomes in the sample space as ordered pairs.
    b. How many outcomes have at least one blue? **5**

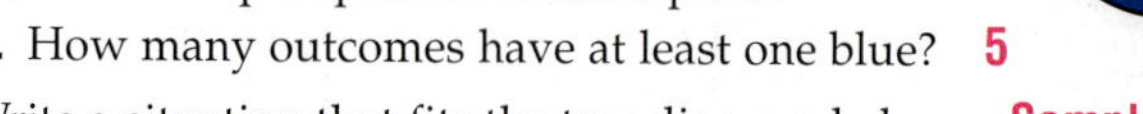

C 19. Write a situation that fits the tree diagram below. **Sample answer: choosing a red, green, or white car with or without a sunroof**

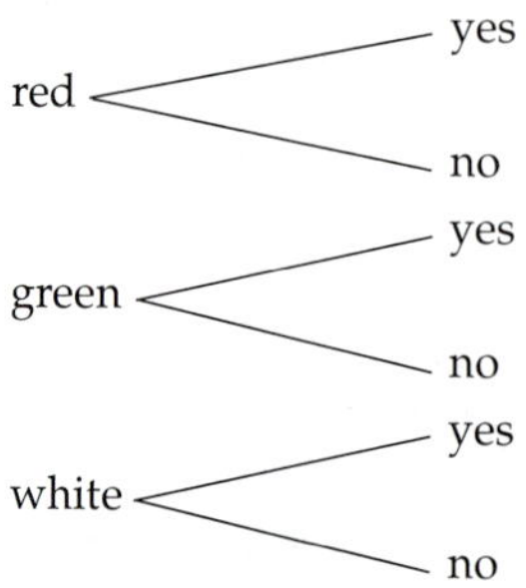

Applications and Problem Solving

20. **Dining** A free continental breakfast offers one type of bread and one beverage.
    a. How many different breakfasts are possible? **24**
    b. How many of the breakfasts include a muffin? **8**

Bread	Beverage
bagel	tea
bran muffin	coffee
English muffin	juice
white toast	milk
wheat toast	
raisin toast	

21. **History** In 1869, Fanny Jackson Coppin became the first African-American school principal. One of her favorite teaching poems began:

> A noun is the name of anything,
> As school, or garden, hoop, or swing.
> Adjectives tell the kind of noun,
> As great, small, pretty, white, or brown.

How many two-word phrases can you make using one of the underlined nouns and one of the underlined adjectives from the poem? **20**

22. **Family** Valerie and Jessie just got married. They hope to eventually have two girls and a boy, in any order.
    a. How many combinations of three children are possible? **8**
    b. How many outcomes will give them two girls and a boy? **3**
    c. In how many of the outcomes from part b will the girls be born in consecutive order? **2**

***Skills Practice,* p. 151, and *Practice,* p. 152 (shown)**

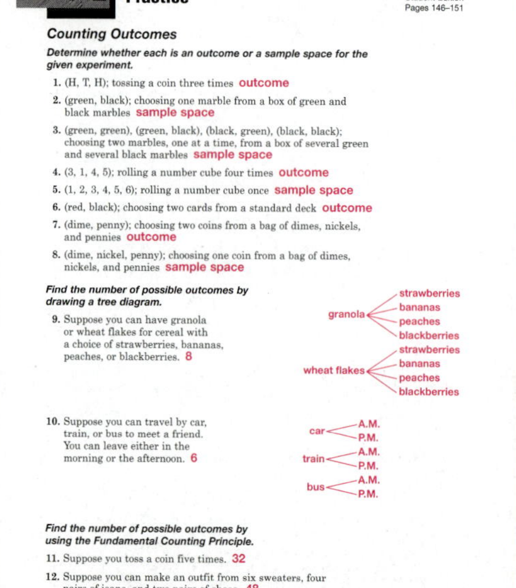

NAME ______ DATE ______ PERIOD ______

4–2 Practice

Student Edition Pages 146–151

Counting Outcomes

Determine whether each is an outcome or a sample space for the given experiment.

1. (H, T, H); tossing a coin three times **outcome**
2. (green, black); choosing one marble from a box of green and black marbles **sample space**
3. (green, green), (green, black), (black, green), (black, black); choosing two marbles, one at a time, from a box of several green and several black marbles **sample space**
4. (3, 1, 4, 5); rolling a number cube four times **outcome**
5. (1, 2, 3, 4, 5, 6); rolling a number cube once **sample space**
6. (red, black); choosing two cards from a standard deck **outcome**
7. (dime, penny); choosing two coins from a bag of dimes, nickels, and pennies **outcome**
8. (dime, nickel, penny); choosing one coin from a bag of dimes, nickels, and pennies **sample space**

Find the number of possible outcomes by drawing a tree diagram.

9. Suppose you can have granola or wheat flakes for cereal with a choice of strawberries, bananas, peaches, or blackberries. **8**

granola — strawberries, bananas, peaches, blackberries
wheat flakes — strawberries, bananas, peaches, blackberries

10. Suppose you can travel by car, train, or bus to meet a friend. You can leave either in the morning or the afternoon. **6**

car — A.M., P.M.
train — A.M., P.M.
bus — A.M., P.M.

Find the number of possible outcomes by using the Fundamental Counting Principle.

11. Suppose you toss a coin five times. **32**
12. Suppose you can make an outfit from six sweaters, four pairs of jeans, and two pairs of shoes. **48**

Data Update For the latest information on the number of key combinations, visit: www.algconcepts.com

23. **Keys** A car manufacturer makes keys with six sections.
    a. Until the 1960s, there were only two patterns for each section. How many different keys were possible? **64**
    b. After the 1960s, the car manufacturer made keys having three different patterns for each section. How many different keys were possible after the 1960s? **729**

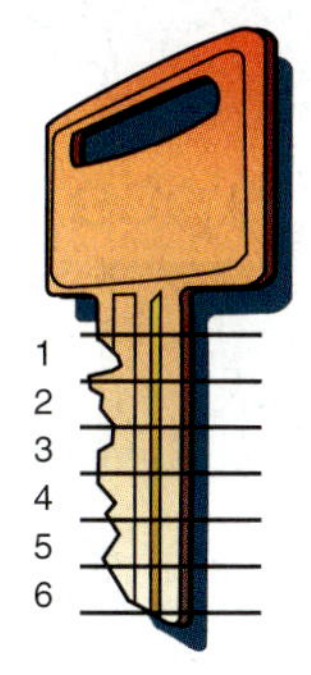

24. **Critical Thinking** The president, vice president, secretary, and treasurer of the Drama Club pose for a yearbook picture. If they sit in four chairs, how many different seating arrangements are possible? **24**

Mixed Review

Find each product. *(Lesson 4–1)*

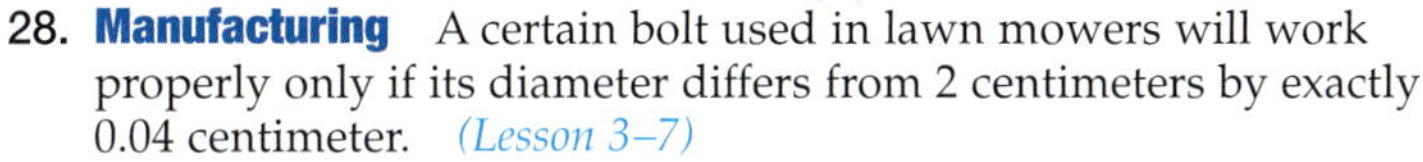

25. $-3 \cdot 5.4$ **−16.2** 26. $-7.2(-1.5)$ **10.8** 27. $\frac{3}{5} \cdot \frac{2}{7}$ $\mathbf{\frac{6}{35}}$

28. **Manufacturing** A certain bolt used in lawn mowers will work properly only if its diameter differs from 2 centimeters by exactly 0.04 centimeter. *(Lesson 3–7)*
    a. Let d represent the diameter. Write an equation to represent this problem. $|d - 2| = 0.04$
    b. What are the least and greatest diameters for this bolt? **1.96 cm, 2.04 cm**

29. Solve $x - \frac{1}{3} = -\frac{3}{4}$. Check your solution. *(Lesson 3–6)* $-\frac{5}{12}$

Standardized Test Practice

30. **Grid In** In the graph below, how many degrees did the temperature rise from 1:00 to 5:00? *(Lesson 2–4)* **16**

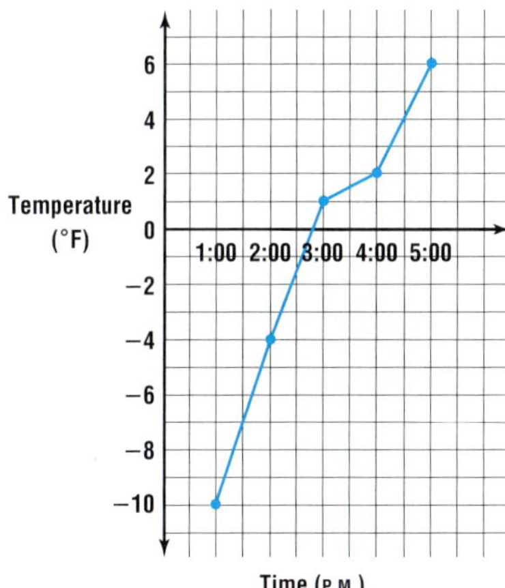

31. **Multiple Choice** Ayani has $600 in the bank at an annual interest rate of 3%. How much money will he have in his account after four years? (*Hint:* Use the formula for simple interest $I = prt$.) *(Lesson 1–5)* **B**

    (A) $72 (B) $672 (C) $607 (D) $720

www.algconcepts.com/self_check_quiz

Extra Credit

An ice cream store offers 6 different flavors of ice cream. You can get a single scoop or double scoop. How many different ice cream cones can you order if the flavors can be used for more than one scoop and if double-scoop ice cream cones with the same two flavors but in different orders are counted separately? **42 cones**

4 ASSESS

Open-Ended Assessment

Modeling Provide students with a list of the elective courses available to students at each grade level in your high school. Have students draw a tree diagram representing the possible ways to select one elective course during each year they are in high school.

***Enrichment*, p. 154**

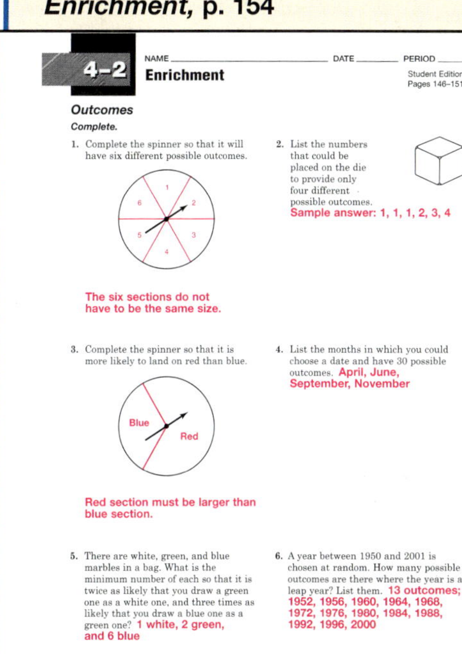

NAME ______ DATE ______ PERIOD ______

4–2 Enrichment

Student Edition Pages 146–151

Outcomes

Complete.

1. Complete the spinner so that it will have six different possible outcomes.
   The six sections do not have to be the same size.

2. List the numbers that could be placed on the die to provide only four different possible outcomes. Sample answer: 1, 1, 1, 2, 3, 4

3. Complete the spinner so that it is more likely to land on red than blue.
   Red section must be larger than blue section.

4. List the months in which you could choose a date and have 30 possible outcomes. April, June, September, November

5. There are white, green, and blue marbles in a bag. What is the minimum number of each so that it is twice as likely that you draw a green one as a white one, and three times as likely that you draw a blue one as a green one? 1 white, 2 green, and 6 blue

6. A year between 1950 and 2001 is chosen at random. How many possible outcomes are there where the year is a leap year? List them. 13 outcomes; 1952, 1956, 1960, 1964, 1968, 1972, 1976, 1980, 1984, 1988, 1992, 1996, 2000

Investigation

PREPARE

This optional investigation is designed to be completed by pairs of students over 1–2 days.

Objective

Students become familiar with permutations and combinations and the formulas used to calculate their values. Students present their results by creating a pamphlet.

Mathematical Overview

The investigation utilizes the following concepts:

- the Fundamental Counting Principle,
- modeling permutations and combinations, and
- using formulas.

Suggested Time Management	
Investigation	25–35 min
Extension: Gathering Data	20–30 min
Extension: Summarizing Data	35–55 min

Motivating the Lesson

Have students think about double scoop ice cream cones. Suggest that there are two ways to count the number of possible cones, one way if you consider the order of the scoops to be important and another if the order is not considered important. Explain to students that in this investigation they will be exploring the distinction between situations where the order of choices is and is not important, and how this distinction affects the number of possible choices.

Chapter 4 Investigation

Materials

 index cards

 green, blue, pink, and yellow self-adhesive notes

 graphing calculator

Permutations and Combinations

Mandy makes and sells stuffed rabbits at craft shows. There are three sets of pieces in the rabbit pattern, and each set is cut from a different color fabric. One set of pieces is for the body, one is for the arms, and the other is for the legs. How many different rabbits can she make?

Investigate

1. Label the index cards Set 1, Set 2, and Set 3.
   a. Suppose Mandy has three different fabrics. Let a green, a blue, and a pink self-adhesive note represent the three fabrics.
   b. Lay the cards on a table and place a different color note on each.

This represents one arrangement of colors.

Set 1 (Body)

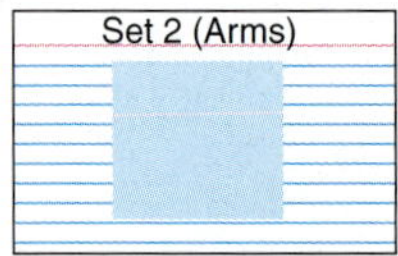

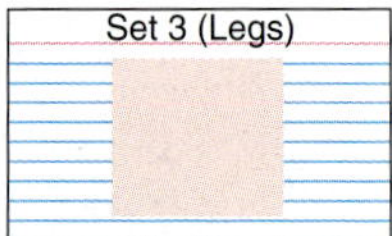

   c. Find and record all of the different arrangements that are possible using each color only once. Two different arrangements are shown.

Arrangement	Set 1 (body)	Set 2 (arms)	Set 3 (legs)
A	blue	green	pink
B	blue	pink	green

   d. An arrangement of objects in which order is important is called a **permutation**. How many permutations of the three colors are there? **6**
   e. Suppose the order of the three colors does not matter. For example, a rabbit with a green body, blue arms, and pink legs is considered the same as a rabbit with a blue body, pink arms, and green legs. An arrangement of objects in which order is not important is called a **combination**. How many combinations are possible using green, blue, and pink fabrics? **1**

Cooperative Learning

This investigation offers an excellent opportunity for using cooperative groups. For more information on cooperative learning strategies and group management, see *Cooperative Learning in the Mathematics Classroom,* one of the titles in the Glencoe Mathematics Professional Series.

2. Suppose Mandy has four different fabrics. Use a yellow self-adhesive note to represent the fourth fabric.
   a. Use the index cards to find the number of different rabbits that are possible when there are four colors from which to choose. Record your results in a new table. Remember that you can use only three colors at a time. Two possible arrangements are shown below.

Arrangement	Set 1 (body)	Set 2 (arms)	Set 3 (legs)
A	green	blue	pink
B	yellow	blue	pink

   How many permutations are possible? **24**

   b. How many different rabbits are possible if the order of the colors is not important? **4**

Extending the Investigation

In this extension, you will use formulas to find permutations and combinations. The mathematical notation 4! means $4 \cdot 3 \cdot 2 \cdot 1$. The symbol 4! is read *four* ***factorial***. $n!$ means the product of all counting numbers beginning with n and counting backward to 1.

- The number of permutations of n objects taken r at a time is defined as ${}_nP_r = \frac{n!}{(n-r)!}$.

  The number of permutations of 4 colors taken 3 at a time can be written ${}_4P_3$. Find ${}_4P_3$ and compare your answer to the answer in Exercise 2a. **24; They are the same.**
- The number of combinations of n objects taken r at a time is defined ${}_nC_r = \frac{n!}{(n-r)!r!}$.

  The number of combinations of 4 colors taken 3 at a time can be written ${}_4C_3$. Find ${}_4C_3$ and compare your answer to the answer in Exercise 2b. **4; They are the same.**
- Use the formulas to find the number of permutations and combinations of 5 colors taken 3 at a time and 6 colors taken 3 at a time. **${}_5P_3 = 60$, ${}_5C_3 = 10$; ${}_6P_3 = 120$, ${}_6C_3 = 20$**
- We define 0! as 1. **Make a conjecture** comparing ${}_nP_n$ and ${}_nC_n$. **Sample answer: ${}_nP_n = n!$ and ${}_nC_n = 1$**

Presenting Your Conclusions

Here are some ideas to help you present your conclusions to the class.

- Make a pamphlet showing the permutations and combinations for the rabbit pieces and fabrics.
- Include diagrams and tables as needed. Show how the formulas can be used to make the calculations.

Investigation For more information on permutations and combinations, visit: www.algconcepts.com

From the Classroom of ...

Don McGurrin
Wake County Public Schools
Raleigh, North Carolina

One of my favorite discussions involving permutations and combinations is why we call the locks on lockers *combination* locks. Certainly, the order of the numbers is important. Shouldn't they be called *permutation* locks? Students seem to remember this better than anything.

MANAGE

Teaching Tip In Step 1c, you may want to stress the importance of making an organized list. Stress that if students do not organize the possible arrangements in some manner, they may have difficulty identifying all of the possible permutations.

Teaching Tip When discussing the Extending the Investigation section, you may need to remind students that when they evaluate expressions such as $(n - r)!$, they must perform the operation inside the parentheses first.

Working in Pairs Students should each make their own lists of the possible arrangements. They can then compare their lists. This will help students discover if they have missed any possible arrangements of colors and sets of pieces. The students should work together to answer the questions in the extension and to create their pamphlet.

Working as a Class You may want to have each pair of students display their pamphlet on a class bulletin board.

ASSESS

Students' work should show that they understand the difference between a permutation and a combination. They should be able to list the permutations and combinations for a given situation and to use the formulas for calculating the number of permutations and combinations.

Students should add their pamphlets to their portfolios at this time.

Lesson 4-3

1 FOCUS

5-Minute Check
Lesson 4–2

On a game show, a contestant is given the choice of spinning either Spinner A or Spinner B. Each of the two spinners has one red, one green, and one yellow section. The tree diagram below shows the possible outcomes.

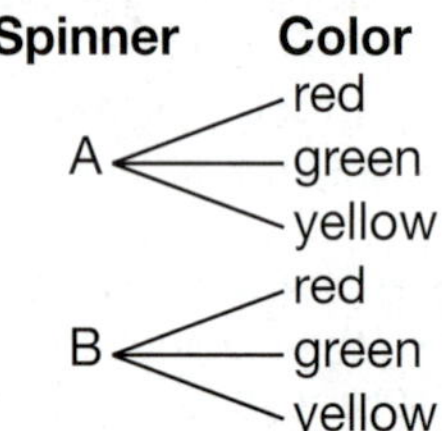

1. List the elements of the sample space. **(A, red), (A, green), (A, yellow), (B, red), (B, green), (B, yellow)**
2. How many different outcomes are possible? **6**
3. If a contestant had a choice of 5 different spinners each with 6 different sections, how many outcomes would be possible? **30**
4. Joelle needs to choose a 2-digit code for her bicycle lock. If both digits can be any number from 0 to 9, how many codes are possible? **100**

4-3 Dividing Rational Numbers

What You'll Learn

You'll learn to divide rational numbers.

Why It's Important

Cooking To adjust recipes for different serving sizes, cooks must know how to divide rational numbers. *See Example 5.*

When you divide two rational numbers, the quotient is also a rational number. You can use the rules below to find the sign of the quotient. The sign rules are the same as the ones used for dividing integers.

Dividing Rational Numbers	Words	Numbers
Different Signs	The quotient of two numbers with different signs is negative.	$-2.1 \div 7 = -0.3$ $2.1 \div (-7) = -0.3$
Same Sign	The quotient of two numbers with the same sign is positive.	$4.5 \div 0.9 = 5$ $-4.5 \div (-0.9) = 5$

Examples

Prerequisite Skills Review
Operations with Decimals, p. 684

Find each quotient.

1 $-6 \div 0.5$

$-6 \div 0.5 = -12$ *Numbers have different signs. The quotient is negative.*

2 $-7.4 \div (-2)$

$-7.4 \div (-2) = 3.7$ *Numbers have the same sign. The quotient is positive.*

Your Turn

a. $16 \div (-2.5)$ **−6.4** b. $-3.9 \div 3$ **−1.3** c. $-8.4 \div (-1.2)$ **7**

Two numbers whose product is 1 are called **multiplicative inverses** or **reciprocals**.

$\frac{7}{8}$ and $\frac{8}{7}$ are reciprocals because $\frac{7}{8} \cdot \frac{8}{7} = 1$.

-3 and $-\frac{1}{3}$ are reciprocals because $-3 \cdot \left(-\frac{1}{3}\right) = 1$.

a and $\frac{1}{a}$, where $a \neq 0$, are reciprocals because $a \cdot \frac{1}{a} = 1$.

These examples demonstrate the **Multiplicative Inverse Property**.

Multiplicative Inverse Property		
	Words:	The product of a number and its multiplicative inverse is 1.
	Symbols:	For every number $\frac{a}{b}$, where $a, b \neq 0$, there is exactly one number $\frac{b}{a}$ such that $\frac{a}{b} \cdot \frac{b}{a} = 1$.
	Numbers:	$-\frac{4}{9}$ and $-\frac{9}{4}$ are multiplicative inverses.

MODELING

Alternative hands-on options using self-adhesive notes, paper, tangram patterns, and a straightedge are available for teaching this lesson.

Resource Manager

Reproducible Masters

Chapter 4 Resource Masters

- *Study Guide,* p. 155
- *Skills Practice,* p. 156
- *Practice,* p. 157
- *Reading to Learn Mathematics,* p. 158
- *Enrichment,* p. 159
- *Assessment,* p. 191

Hands-On Algebra, pp. 54–55

Transparencies

5-Minute Check, 4–3

Technology/Multimedia

Interactive Chalkboard CD-ROM

In Lesson 2–4, you learned that subtracting a number is the same as adding its additive inverse. In the same way, dividing by a number is the same as multiplying by its multiplicative inverse or reciprocal.

Dividing Fractions		
	Words:	To divide a fraction by any nonzero number, multiply by the reciprocal of the number.
	Symbols:	$\frac{a}{b} \div \frac{c}{d} = \frac{a}{b} \cdot \frac{d}{c}$, where b, c, $d \neq 0$
	Numbers:	$\frac{3}{4} \div \frac{2}{5} = \frac{3}{4} \cdot \frac{5}{2}$ or $\frac{15}{8}$

Use the rules for dividing rational numbers to determine the sign of the quotient.

Examples

Find each quotient.

3 $\frac{4}{7} \div (-8)$

$\frac{4}{7} \div (-8) = \frac{4}{7} \cdot \left(-\frac{1}{8}\right)$ *To divide by −8, multiply by its reciprocal, $-\frac{1}{8}$.*

$= -\frac{4}{56}$ *The numbers have different signs. The product is negative.*

$= -\frac{1}{14}$ *Write the fraction in simplest form.*

Prerequisite Skills Review

Simplifying Fractions, p. 685

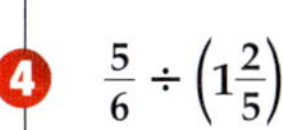

4 $\frac{5}{6} \div \left(1\frac{2}{5}\right)$

$\frac{5}{6} \div \left(1\frac{2}{5}\right) = \frac{5}{6} \div \frac{7}{5}$ *Rewrite $1\frac{2}{5}$ as an improper fraction.*

$= \frac{5}{6} \cdot \frac{5}{7}$ *To divide by $\frac{7}{5}$, multiply by its reciprocal, $\frac{5}{7}$.*

$= \frac{25}{42}$ *The numbers have the same sign. The product is positive.*

Example

Cooking Link

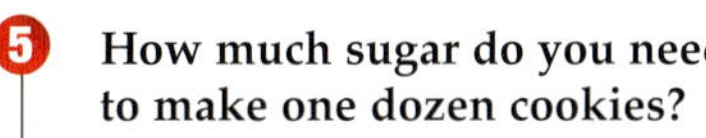

5 **How much sugar do you need to make one dozen cookies?**

Divide the recipe by 2. The recipe calls for $\frac{2}{3}$ cup of sugar.

Raspberry Almond Cookies

1 cup butter
$\frac{2}{3}$ cup sugar
$\frac{1}{2}$ tsp. almond extract
2 cups flour
$\frac{1}{2}$ cup raspberry jam
Makes two dozen cookies.

$\begin{pmatrix}\text{amount}\\ \text{of sugar}\end{pmatrix} \div 2 = \frac{2}{3} \div 2$

$= \frac{2}{3} \cdot \frac{1}{2}$

$= \frac{2}{6}$ or $\frac{1}{3}$ *Write the fraction in simplest form.*

To make one dozen cookies, you need $\frac{1}{3}$ cup of sugar.

www.algconcepts.com/extra_examples

Motivating the Lesson

Real-World Connection Ask students the following question. "Suppose you work in an art gallery and you want to hang two paintings so that the distance between the centers of the paintings is the same as the distance from either center to the end of the wall. The paintings are both 6 feet wide and the wall measures $22\frac{1}{4}$ feet across. Where should the center of each painting be placed?" Discuss how division must be used to find the answer.

2 TEACH

Teaching Tip Point out the similarity between the rules for dividing rational numbers on page 154 and the rules for multiplying rational numbers presented on page 140 of Lesson 4–1.

In-Class Examples

Examples 1–2

Find each quotient.

1 $8 \div (-2.5)$ **−3.2**

2 $-9.3 \div (-0.3)$ **31**

Teaching Tip When discussing Examples 3 and 4, stress that the reciprocal of the divisor is used when changing to multiplication and the dividend remains unchanged. You may need to review the meanings of the terms *dividend* and *divisor*.

In-Class Examples

Examples 3–4

Find each quotient.

3 $-12 \div \frac{2}{5}$ **−30**

4 $2\frac{3}{7} \div \frac{1}{2}$ $4\frac{6}{7}$

In-Class Examples

Example 5

Refer to the real-world problem posed in the *Motivating the Lesson* feature on page 155. How far from the end of the wall closest to it should the center of each painting be located? **$7\frac{5}{12}$ ft**

Example 6

Evaluate $\frac{3}{x}$ if $x = -\frac{3}{4}$. **−4**

Answers

1. No; the product of a negative number and a positive number can never equal 1.
2. $n \div \frac{1}{2}$ is the greatest and $n \cdot \frac{1}{2}$ is the least.

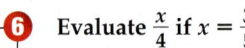

Your Turn Find each quotient.

d. $14 \div \left(-\frac{2}{3}\right)$ **−21** e. $-\frac{2}{7} \div \left(-\frac{5}{9}\right)$ **$\frac{18}{35}$** f. $\frac{1}{6} \div 2\frac{4}{5}$ **$\frac{5}{84}$**

You can use what you know about dividing rational numbers to evaluate algebraic expressions.

Example 6 Evaluate $\frac{x}{4}$ if $x = \frac{3}{5}$.

$\frac{x}{4} = \frac{\frac{3}{5}}{4}$ *Replace x with $\frac{3}{5}$.*

$= \frac{3}{5} \div 4$ *Rewrite the fraction as a division sentence.*

$= \frac{3}{5} \cdot \frac{1}{4}$ *To divide by 4, multiply by its reciprocal, $\frac{1}{4}$.*

$= \frac{3}{20}$ *Multiply the numerators and multiply the denominators.*

Your Turn Evaluate if $x = \frac{3}{5}$.

g. $\frac{6}{x}$ **10** h. $-\frac{x}{2}$ **$-\frac{3}{10}$** i. $\frac{4}{7x}$ **$\frac{20}{21}$**

Check for Understanding

Communicating Mathematics

1–2. See margin.

1. **Determine** whether the reciprocal of a negative rational number can ever be positive. Explain.
2. **Compare** a positive rational number n with the value of $n \div \frac{1}{2}$ and with the value of $n \cdot \frac{1}{2}$. Which is the greatest? the least?
3. Writing Math Describe real-life situations in which you would divide fractions to solve problems. **See students' work.**

Vocabulary
multiplicative inverse
reciprocal

Guided Practice

Getting Ready Name the reciprocal of each number.

Sample: $-\frac{1}{4}$ **Solution:** $-\frac{1}{4}\left(-\frac{4}{1}\right) = 1$, so the reciprocal is $-\frac{4}{1}$ or -4.

4. 3 **$\frac{1}{3}$** 5. $-\frac{2}{3}$ **$-\frac{3}{2}$** 6. $2\frac{3}{4}$ **$\frac{4}{11}$** 7. $-\frac{1}{x}, x \neq 0$ **$-x$**

Reteaching Activity

Visual/Spatial Learners Have students use rectangular models to represent problems involving the division of two rational numbers. For example, to model the problem $2\frac{1}{4} \div \frac{3}{4}$, draw three equal-sized rectangles and divide each of them into quarters. Shade the rectangles to represent $2\frac{1}{4}$. Then ask students how many $\frac{3}{4}$-models there are in the shaded regions.

Examples 1–5

Find each quotient.

8. $-6 \div 1.5$ **-4**
9. $3.8 \div 19$ **0.2**
10. $-4.7 \div (-0.5)$ **9.4**
11. $-\frac{1}{2} \div \left(\frac{7}{3}\right)$ **$-\frac{3}{14}$**
12. $4 \div \left(\frac{7}{8}\right)$ **$\frac{32}{7}$ or $4\frac{4}{7}$**
13. $2\frac{2}{5} \div \frac{1}{3}$ **$\frac{36}{5}$ or $7\frac{1}{5}$**

Example 6

Evaluate each expression if $a = \frac{1}{4}$ and $b = -\frac{2}{3}$.

14. $\frac{a}{3}$ **$\frac{1}{12}$**
15. $\frac{5}{b}$ **$-\frac{15}{2}$ or $-7\frac{1}{2}$**
16. $\frac{a}{b}$ **$-\frac{3}{8}$**

Example 4

17. **Lunch** Students working on decorations for homecoming ordered pizza for lunch. They ordered five 8-slice pizzas, and each person ate $2\frac{1}{2}$ slices. There was no pizza left over. How many students were there? **16 students**

Exercises

Practice

Find each quotient.

A

18. $8 \div (-1.6)$ **-5**
19. $-7.5 \div 3$ **-2.5**
20. $9.6 \div 3.2$ **3**
21. $2.7 \div -27$ **-0.1**
22. $0.4 \div -0.4$ **-1**
23. $15.6 \div 1.3$ **12**
24. $\frac{1}{8} \div \frac{1}{7}$ **$\frac{7}{8}$**
25. $0 \div \frac{3}{8}$ **0**
26. $\frac{3}{4} \div \left(-\frac{4}{5}\right)$ **$-\frac{15}{16}$**
27. $-\frac{5}{6} \div \frac{5}{6}$ **-1**
28. $\frac{1}{-7} \div \frac{4}{9}$ **$-\frac{9}{28}$**
29. $-\frac{2}{3} \div \left(-\frac{5}{7}\right)$ **$\frac{14}{15}$**
30. $-\frac{7}{9} \div \left(-\frac{9}{7}\right)$ **$\frac{49}{81}$**
31. $5 \div \frac{1}{9}$ **45**
32. $-\frac{1}{5} \div 10$ **$-\frac{1}{50}$**

33. $\frac{-3}{4} \div 8$ **$-\frac{3}{32}$**
34. $-14 \div \left(-2\frac{4}{5}\right)$ **5**
35. $-3\frac{1}{2} \div \frac{1}{6}$ **-21**

Homework Help	
For Exercises	See Examples
18–23	1, 2
24–35	3, 4
36–49	5, 6

Extra Practice

See page 699.

Evaluate each expression if $j = -\frac{1}{2}$, $k = \frac{3}{4}$, and $n = \frac{1}{5}$.

36. $\frac{3}{n}$ **15**
37. $\frac{9}{k}$ **12**
38. $\frac{k}{2}$ **$\frac{3}{8}$**
39. $\frac{-j}{5}$ **$\frac{1}{10}$**
40. $\frac{1}{2n}$ **$\frac{5}{2}$ or $2\frac{1}{2}$**
41. $\frac{n}{j}$ **$-\frac{2}{5}$**
42. $\frac{k}{j}$ **$-\frac{6}{4}$ or $-1\frac{1}{2}$**
43. $\frac{2n}{j}$ **$-\frac{4}{5}$**
44. $\frac{jk}{n}$ **$-\frac{15}{8}$ or $-1\frac{7}{8}$**

3 PRACTICE/APPLY

Error Analysis

Watch for students whose answers for Exercises 8, 10, and 11 have the wrong sign.
Prevent by suggesting that students determine the sign of their answer before doing any computation. Remind students that the quotient of two negative numbers is positive and the quotient of a positive number and a negative number is negative.

Assignment Guide

Basic: 19–49 odd, 50–63
Average: 18–46 even, 47–63
All: Quiz 1, 1–10

***Study Guide*, p. 155**

4–3 NAME ________ DATE ______ PERIOD ____

Study Guide

Student Edition Pages 154–159

Dividing Rational Numbers

Use the following rules to divide rational numbers or fractions.

Rule or Property		Examples
Dividing Rational Numbers	The quotient of two numbers having different signs is negative The quotient of two numbers having the same sign is positive.	$-4.8 \div 6 = -0.8$ $4.8 \div (-6) = -0.8$ $2.8 \div (0.4) = 7$ $-2.8 \div (-0.4) = 7$
Multiplicative Inverse Property	For every number $\frac{a}{b}$, where $a, b \neq 0$, there is exactly one number $\frac{b}{a}$ such that $\frac{a}{b} \cdot \frac{b}{a} = 1$.	$-\frac{3}{4}$ and $-\frac{4}{3}$ are multiplicative inverses.
Dividing Fractions	$\frac{a}{b} \div \frac{c}{d} = \frac{a}{b} \cdot \frac{d}{c}$, where $b, c, d \neq 0$	$\frac{2}{5} \div \frac{3}{4} = \frac{2}{5} \cdot \frac{4}{3}$ or $\frac{8}{15}$ $\frac{3}{4} \div (-6) = \frac{3}{4} \cdot \left(-\frac{1}{6}\right)$ $= -\frac{3}{24}$ or $-\frac{1}{8}$

Find each quotient.

1. $-8 \div 2.5$ -3.2
2. $-1.6 \div (-2)$ 0.8
3. $3.6 \div 0.6$ 6
4. $5.5 \div (-5.5)$ -1
5. $0 \div -0.6$ 0
6. $-18.7 \div 5.5$ -3.4
7. $-42 \div (-0.5)$ 84
8. $0 \div \frac{2}{7}$ 0
9. $-\frac{4}{5} \div \frac{4}{5}$ -1
10. $\frac{1}{7} \div \frac{2}{5}$ $\frac{5}{14}$
11. $-\frac{2}{5} \div \frac{1}{2}$ $-\frac{4}{5}$
12. $\frac{1}{9} \div (-4)$ $-\frac{1}{36}$
13. $-\frac{3}{8} \div \frac{11}{8}$ $-\frac{3}{11}$
14. $-\frac{3}{7} \div \left(-\frac{7}{3}\right)$ $\frac{9}{49}$
15. $\frac{3}{4} \div \frac{1}{5}$ $\frac{15}{4} = 3\frac{3}{4}$

45. 18

45. Find the quotient of -12 and $-\frac{2}{3}$.

46. Solve the equation $t = -\frac{7}{18} \div 3\frac{1}{2}$ to find the value of t. $-\frac{1}{9}$

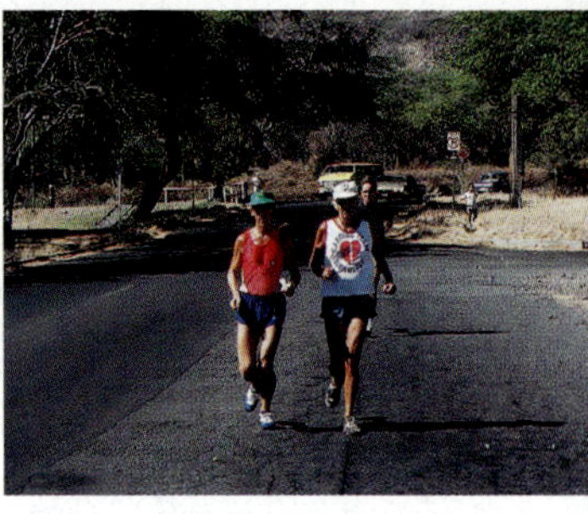

Applications and Problem Solving

47. **Track** Mr. Vance is training for a 10-kilometer race in a senior citizen track meet. Suppose $3\frac{1}{4}$ laps equals 1 kilometer and he wants to finish the race in 65 minutes. In how many minutes will he have to run each lap? **2 min**

48. **Nutrition** A 20-ounce bottle of soda contains $16\frac{2}{3}$ teaspoons of sugar. A 12-ounce can of soda contains $16\frac{2}{3} \div \frac{5}{3}$ teaspoons of sugar. How much sugar is in a 12-ounce can of soda? **10 tsp**

49. **Sales** The performing arts series at Central State University offers a package ticket to students for $62.50. If there are five performances, how much do students pay for each performance? **$12.50**

50. **Critical Thinking**
   a. Is the division of two rational numbers always a rational number? If not, give a counterexample. **No; $\frac{2}{3} \div 0$ is undefined.**
   b. Is the set of rational numbers closed under division? Explain.
   c. What number has no reciprocal? Explain. **0; $\frac{1}{0}$ is undefined.**

50b. No; if 0 is the divisor, the quotient is undefined.

Mixed Review

51. **Uniforms** The school dress for female students is a uniform, a blouse, and socks. A sweater may be added, if necessary. Find the number of different outfits that female students can wear. *(Lesson 4–2)* **48**

Uniform	Blouse	Socks	Sweater
skirt jumper	white yellow blue pink	white blue gray	yes no

52. **Sales** Ms. Ortiz wants to buy a washer and dryer from Bargain Appliances. If she buys on the "90 days same as cash" plan, she will have to make 5 equal payments of $146.12 each. If she writes a check for each payment, what is the net effect on her checking account? *(Lesson 4–1)* **−$730.60**

Find the mean, median, mode, and range for each set of data. *(Lesson 3–3)*

53. 6, 8, 8, 8, 6, 4, 3, 7, 4 **6; 6; 8; 5**
54. 9, 4, 13, 11, 7, 6, 20 **10; 9; none; 16**
55. 20, 11, 20, 23, 20, 29 **20.5; 20; 20; 18**

***Skills Practice,* p. 156, and *Practice,* p. 157 (shown)**

4–3 NAME ______ DATE ______ PERIOD ______

Practice — Student Edition Pages 154–159

Dividing Rational Numbers

Find each quotient.

1. $-8.5 \div 5$ **−1.7**
2. $4.2 \div 14$ **0.3**
3. $2.8 \div (-0.5)$ **−5.6**
4. $3.6 \div (-6)$ **−0.6**
5. $-5.1 \div (-1.7)$ **3**
6. $7.8 \div (-0.3)$ **26**
7. $-4.8 \div 1.2$ **−4**
8. $7.5 \div (-1.5)$ **−5**
9. $-3.7 \div (-0.1)$ **37**
10. $-\frac{3}{4} \div \frac{5}{2}$ $-\frac{3}{10}$
11. $\frac{1}{5} \div \frac{1}{3}$ $\frac{3}{5}$
12. $4 \div \frac{9}{10}$ $\frac{40}{9}$ or $4\frac{4}{9}$
13. $\frac{5}{6} \div \left(-\frac{2}{3}\right)$ $-\frac{5}{4}$ or $-1\frac{1}{4}$
14. $-\frac{3}{8} \div 6$ $-\frac{1}{16}$
15. $-\frac{2}{7} \div (-3)$ $\frac{2}{21}$
16. $-\frac{4}{5} \div 4\frac{1}{2}$ $-\frac{8}{45}$
17. $-2\frac{2}{3} \div \frac{3}{4}$ $-\frac{32}{9}$ or $-3\frac{5}{9}$
18. $-1\frac{1}{8} \div \left(-\frac{5}{7}\right)$ $\frac{63}{40}$ or $1\frac{23}{40}$

Evaluate each expression if $m = \frac{1}{5}$ and $n = -\frac{3}{4}$.

19. $\frac{m}{4}$ $\frac{1}{20}$
20. $\frac{5}{n}$ $-\frac{20}{3}$ or $-6\frac{2}{3}$
21. $-\frac{m}{7}$ $-\frac{1}{35}$
22. $\frac{6}{m}$ **30**
23. $\frac{n}{3}$ $-\frac{1}{4}$
24. $\frac{n}{m}$ $-\frac{15}{4}$ or $-3\frac{3}{4}$
25. $\frac{m}{n}$ $-\frac{4}{15}$
26. $-\frac{2m}{3}$ $-\frac{2}{15}$
27. $-\frac{1}{3n}$ $\frac{4}{9}$

Replace each ● with <, >, or = to make a true sentence. *(Lesson 3–1)*

56. $-3.6 \bullet 0$ **<**

57. $\frac{3}{10} \bullet 0.3$ **=**

58. $-7(-4) \bullet 7 - 15$ **>**

Which is the better buy? Explain. *(Lesson 3–1)* **59. \$6.30 < \$6.92**

59. a $\frac{1}{2}$-pound bag of cashews for \$3.15 or $\frac{3}{4}$-pound bag for \$5.19

60. three liters of soda for \$2.25 or two liters for \$1.98 **\$0.75 < \$0.99**

61. a 48-ounce bottle of dishwashing liquid for \$2.69 or a 22-ounce bottle for \$1.09 **\$0.06 > \$0.05**

62. **Grid In** Evaluate $16 + k$ if $k = -11$. *(Lesson 2–3)* **5**

63. **Multiple Choice** Write the ordered pair that names point Q. *(Lesson 2–2)* **A**

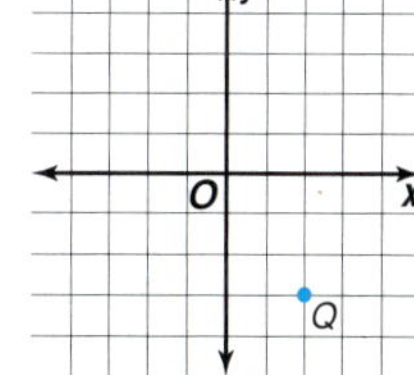

Ⓐ $(2, -3)$ Ⓑ $(-2, -3)$
Ⓒ $(-3, 2)$ Ⓓ $(-3, -2)$

Quiz 1 Lessons 4–1 through 4–3

Find each product. *(Lesson 4–1)*

1. $9.4 \cdot (-2)$ **−18.8**

2. $-\frac{3}{8}\left(-\frac{1}{7}\right)$ $\frac{3}{56}$

3. $4 \cdot \frac{5}{8}$ $\frac{20}{8}$ **or** $2\frac{1}{2}$

Find the number of possible outcomes. *(Lesson 4–2)*

4. choosing a sandwich, a side, and a beverage from the table **36**

5. different three-digit security codes using the numbers 1, 2, and 3 (You can use the numbers more than once.) **27**

6. rolling a number cube three times **216**

Sandwich	Side	Beverage
ham	French fries	milk
turkey	cole slaw	tea
egg salad	fruit	soda
		juice

Exercise 4

Find each quotient. *(Lesson 4–3)*

7. $-8.1 \div (-3)$ **2.7**

8. $\frac{1}{2} \div \left(-\frac{3}{4}\right)$ $-\frac{2}{3}$

9. $2\frac{1}{3} \div \frac{3}{4}$ $\frac{28}{9}$ **or** $3\frac{1}{9}$

10. **World Records** The longest loaf of bread ever baked was 2132 feet $2\frac{1}{2}$ inches. If this loaf were cut into $\frac{1}{2}$-inch wide slices, how many slices of bread would there have been? *(Lesson 4–3)* **51,173 slices**

www.algconcepts.com/self_check_quiz

Extra Credit

Simplify the expression $1 \div x \div x$ for any nonzero number x. $\frac{1}{x \cdot x}$ **or** $\frac{1}{x^2}$

4 ASSESS

Open-Ended Assessment

Speaking On the board or overhead, write a variety of division problems involving rational numbers. Have students choose a problem and explain how to complete it.

Quiz 1

The Quiz provides students with a brief review of the concepts and skills in Lessons 4–1 through 4–3. Lesson numbers are given to the right of the exercises or instruction lines so that students can review concepts not yet mastered.

Chapter 4, Quiz A (Lessons 4–1 through 4–3) is available in the *Chapter 9 Resource Masters*, p. 191.

***Enrichment,* p. 159**

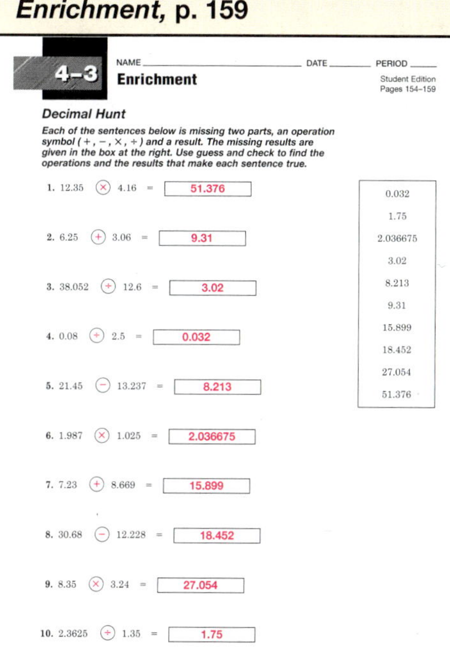
NAME ______ DATE ______ PERIOD ______

4–3 **Enrichment**

Student Edition Pages 154–159

Decimal Hunt

Each of the sentences below is missing two parts, an operation symbol (+ , − , × , ÷) and a result. The missing results are given in the box at the right. Use guess and check to find the operations and the results that make each sentence true.

1. 12.35 (×) 4.16 = **51.376**
2. 6.25 (+) 3.06 = **9.31**
3. 38.052 (÷) 12.6 = **3.02**
4. 0.08 (÷) 2.5 = **0.032**
5. 21.45 (−) 13.237 = **8.213**
6. 1.987 (×) 1.025 = **2.036675**
7. 7.23 (+) 8.669 = **15.899**
8. 30.68 (−) 12.228 = **18.452**
9. 8.35 (×) 3.24 = **27.054**
10. 2.3625 (÷) 1.35 = **1.75**

0.032
1.75
2.036675
3.02
8.213
9.31
15.899
18.452
27.054
51.376

Lesson 4–4

1 FOCUS

5-Minute Check
Lesson 4–3

Find each quotient.

1. $-2.7 \div (-0.9)$ **3**
2. $-9 \div 2\frac{1}{3}$ **$-3\frac{6}{7}$**
3. Evaluate $\frac{x}{5}$ if $x = \frac{3}{7}$. **$\frac{3}{35}$**

A mother buys a Family Craft Kit from a catalog. The contents of the kit are listed below.

Family Craft Kit
32 craft sticks
$15\frac{1}{2}$ $\frac{1}{2}$-oz blocks of clay
64 sheets of colored paper
$8\frac{2}{3}$ yd of yarn
24 pipe cleaners

4. If the contents of the kit are divided equally among her three children, how many $\frac{1}{2}$-oz blocks of clay will each child receive? **$5\frac{1}{6}$ blocks**
5. If two neighbor children join in, how many yards of yarn can each child use? **$1\frac{11}{15}$ yd**

Motivating the Lesson

Hands-On Activity Give each student 10 items, such as paper clips or pennies, to use as counters. Have students use the counters and a guess-and-check strategy to solve equations such as $2x = 6$ and $\frac{1}{2}x = 5$. Encourage students to read these problems as *2 times what number is 6* and *half of what number is 5.*

4–4 Solving Multiplication and Division Equations

What You'll Learn
You'll learn to solve multiplication and division equations by using the properties of equality.

Why It's Important
Plumbing Plumbers solve equations to find correct pipe weights. *See Exercise 36.*

In Lesson 3-6 you learned about the Addition Property of Equality and the Subtraction Property of Equality. There are properties of equality for multiplication and division as well. One way to solve algebraic equations is to use the **Division Property of Equality**.

Division Property of Equality		
	Words:	If you divide each side of an equation by the same nonzero number, the two sides remain equal.
	Symbols:	For any numbers a, b, and c, with $c \neq 0$, if $a = b$, then $\frac{a}{c} = \frac{b}{c}$.
	Numbers:	If $3x = 12$, then $\frac{3x}{3} = \frac{12}{3}$.

Examples

Solve each equation. Check your solution.

$3y = 45$

$3y = 45$ *Original equation*

$\frac{3y}{3} = \frac{45}{3}$ *Divide each side by 3.*

$y = 15$ *Simplify.*

Check: $3y = 45$ *Original equation*

$3(15) \stackrel{?}{=} 45$ *$y = 15$*

$45 = 45$ ✓

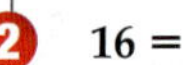

$16 = -2h$

$16 = -2h$ *Original equation*

$\frac{16}{-2} = \frac{-2h}{-2}$ *Divide each side by -2.*

$-8 = h$ *Simplify.*

Check: $16 = -2h$ *Original equation*

$16 \stackrel{?}{=} -2(-8)$ *$h = -8$*

$16 = 16$ ✓

3 **$7.4a = -37$**

$7.4a = -37$ *Original equation*

$\frac{7.4a}{7.4} = \frac{-37}{7.4}$ *Divide each side by 7.4.*

$a = -5$ *Check by substituting into the original equation.*

Prerequisite Skills Review
Operations with Decimals, p. 684

Resource Manager

Reproducible Masters
Chapter 4 Resource Masters
- *Study Guide,* p. 160
- *Skills Practice,* p. 161
- *Practice,* p. 162
- *Reading to Learn Mathematics,* p. 163
- *Enrichment,* p. 164
- *Assessment,* p. 190

Transparencies
5-Minute Check, 4–4

Technology/Multimedia
AlgePASS, Lesson 10
Interactive Chalkboard CD-ROM

Your Turn Solve each equation. Check your solution.

a. $4t = 28$ 7 b. $19.2 = -6d$ −3.2 c. $-0.1m = -7$ 70

Sometimes, you will have to write and solve an equation to solve a problem.

Example 4 Consumer Link

Xue Wu is in charge of buying sandwiches for a hiking trip. She has $24, and sandwiches cost $3.80 each. How many sandwiches can she buy?

Explore You know that the sandwiches cost $3.80 and Xue Wu has $24.

Plan Let s represent the number of sandwiches and write an equation to represent the problem.

price per sandwich	times	number of sandwiches	equals	total cost
3.80	×	s	=	24

Solve Solve the equation for s.

$3.80s = 24$ *Original equation*

$\frac{3.80s}{3.80} = \frac{24}{3.80}$ *Divide each side by 3.80.*

$s \approx 6.3$ *Simplify.*

Since Xue Wu can't buy part of a sandwich, she has enough money to buy 6 sandwiches.

Examine Does the answer make sense? Round 3.80 to 4 and multiply by 6.

$4 \times 6 = 24$

So 6 sandwiches would cost about $24. The answer makes sense.

Another useful property in solving equations is the **Multiplication Property of Equality**.

Multiplication Property of Equality		
	Words:	If you multiply each side of an equation by the same number, the two sides remain equal.
	Symbols:	For any numbers a, b, and c, if $a = b$, then $ac = bc$.
	Numbers:	If $\frac{1}{5}x = 8$, then $5\left(\frac{1}{5}x\right) = 5(8)$.

www.algconcepts.com/extra_examples

2 TEACH

Teaching Tip When discussing Examples 1–3, encourage students to focus on the variable in the expression. Stress that the goal is to isolate the variable and that this is accomplished using inverse operations. Point out in Example 1, for instance, that since the variable is multiplied by 3, you need to divide each side of the equation by 3.

In-Class Examples

Examples 1–3

Solve each equation. Check your solution.

1 $5b = 30$ **6**

2 $-24 = 3g$ **−8**

3 $-5.5z = -22$ **4**

Teaching Tip When presenting Example 4, emphasize to students the need to examine their answer. Stress that for all word problems they should check to be sure their answer makes sense in the situation described in the problem.

In-Class Example

Example 4

Brian received a $25 gift certificate from his grandparents for his birthday. How many $2.35 packages of trading cards can he buy with the gift certificate?
10 packages

In-Class Examples

Examples 5–7

Solve each equation. Check your solution.

5 $\frac{w}{7} = 6$ **42**

6 $-9 = \frac{1}{2}m$ **−18**

7 $-\frac{2}{5}x = -8$ **20**

Example 8

The manager of a movie theater estimates that $\frac{5}{7}$ of the people who attend a matinee are children. How many people attended the 1:00 P.M. matinee today if 250 children's tickets were sold? **350 people**

Examples

Solve each equation. Check your solution.

5 $\frac{g}{4} = 9$

$\frac{g}{4} = 9$	*Original equation*	Check: $\frac{g}{4} = 9$
$4\left(\frac{g}{4}\right) = 4(9)$	*Multiply each side by 4.*	$\frac{36}{4} \stackrel{?}{=} 9$ *g = 36*
$g = 36$	*Simplify.*	$9 = 9$ ✓

6 $-5 = -\frac{1}{6}t$

$-5 = -\frac{1}{6}t$	*Original equation*	Check: $-5 = -\frac{1}{6}t$
$-6(-5) = -6\left(-\frac{1}{6}t\right)$	*Multiply each side by −6.*	$-5 \stackrel{?}{=} -\frac{1}{6}(30)$
$30 = t$	*Simplify.*	$-5 = -5$ ✓

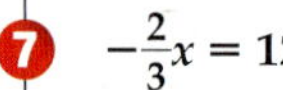

7 $-\frac{2}{3}x = 12$

$-\frac{2}{3}x = 12$	*Original equation*
$-\frac{3}{2}\left(-\frac{2}{3}x\right) = -\frac{3}{2}(12)$	*Multiply each side by $-\frac{3}{2}$.*
$x = -18$	*Check by substituting into the original equation.*

Your Turn

d. $\frac{t}{4} = 8$ **32** **e.** $\frac{5}{6}a = 25$ **30** **f.** $36 = -\frac{3}{4}x$ **−48**

Real World Example

Geography Link

8 **In 1999, Nunavut (NU-na-voot) became Canada's newest territory in fifty years. There are 15,000 people in Nunavut under age 25. This is about $\frac{3}{5}$ of the total population. What is the population of Nunavut?**

Let n represent the total population. Then $\frac{3}{5}n = 15{,}000$ represents the problem.

$\frac{3}{5}n = 15{,}000$	*Original equation*
$\frac{5}{3}\left(\frac{3}{5}n\right) = \frac{5}{3}(15{,}000)$	*Multiply each side by $\frac{5}{3}$. Why?*
$n = 25{,}000$	*Simplify.*

The total population of Nunavut is 25,000.

Study Guide, p. 160

4-4 NAME ______ DATE ______ PERIOD ______

Study Guide Student Edition Pages 160–164

Solving Multiplication and Division Equations

Use the following rules to solve algebraic equations.

Rule or Property		Examples
Division Property of Equality	For any numbers a, b, and c, with $c \neq 0$, if $a = b$, then $\frac{a}{c} = \frac{b}{c}$.	$4x = -24$ $\frac{4x}{4} = \frac{-24}{4}$ $x = 6$
Multiplication Property of Equality	For any numbers a, b, and c, if $a = b$, then $ac = bc$.	$\frac{1}{4}x = 12$ $4 \cdot \frac{1}{4}x = 4 \cdot 12$ $x = 48$

Solve each equation.

1. $8y = 48$ **6**
2. $-6x = 42$ **−7**
3. $3.6 = 12n$ **0.3**
4. $1.5 = -3w$ **−0.5**
5. $0 = -0.6y$ **0**
6. $18 = 0.5x$ **36**
7. $-6 = -8y$ **$\frac{3}{4}$**
8. $\frac{7}{4}n = 1$ **$\frac{4}{7}$**
9. $\frac{1}{5}x = 10$ **50**
10. $\frac{3}{4}y = 12$ **16**
11. $-\frac{3}{5}y = -3$ **5**
12. $4 = -\frac{1}{2}x$ **−8**
13. $8 = \frac{x}{-2}$ **−16**
14. $-\frac{4}{5}n = -\frac{1}{4}$ **$\frac{5}{16}$**
15. $\frac{1}{2} = -\frac{5}{8}x$ **$-\frac{4}{5}$**
16. $1.2y = 9$ **7.5**
17. $-4x = 12.8$ **−3.2**
18. $-\frac{n}{3} = \frac{1}{3}$ **−1**

Reteaching Activity

Intrapersonal Learners Have each student divide a sheet of paper into two columns. In the first column, instruct them to write down an example of the three types of equations found in this lesson: those where the variable was multiplied by an integer, those where the variable was divided by an integer, and those where the variable was multiplied by a fraction. In the second column, tell students to write notes to themselves about how to solve each type of equation.

Check for Understanding

Communicating Mathematics

1. **Describe** how to solve $4x = 36$. **Divide each side by 4.**
2. **List** the steps you take when you check the solution of $4x = 36$.
3. **Write** a multiplication equation whose solution is -7. **Sample answer: $-8a = 56$**

2. Replace x in the equation with the solution. Simplify.

Guided Practice
Examples 1–8

Solve each equation. Check your solution.

4. $8x = -24$ **−3**
5. $-5v = -40$ **8**
6. $6 = 0.3y$ **20**
7. $\frac{r}{15} = 3$ **45**
8. $-7 = \frac{f}{8}$ **−56**
9. $-9 = -\frac{9}{4}n$ **4**

Example 3

10. **Geometry** The area of the rectangle is 51.2 square centimeters. Find the length. **8 cm**

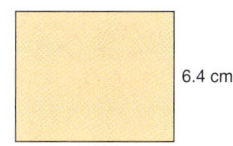

Exercises

Practice

Solve each equation. Check your solution.

11. $4c = 24$ **6**
12. $9p = 63$ **7**
13. $30 = 2x$ **15**
14. $-6a = 60$ **−10**
15. $-8t = 56$ **−7**
16. $-3 = -5n$ **$\frac{3}{5}$**
17. $14 = 0.5x$ **28**
18. $0 = 3.9c$ **0**
19. $2.2r = 11$ **5**
20. $\frac{9}{2}t = 1$ **$\frac{2}{9}$**
21. $\frac{1}{2}y = 32$ **64**
22. $\frac{b}{3} = -5$ **−15**
23. $10 = \frac{a}{-7}$ **−70**
24. $-\frac{2}{7}x = -6$ **21**
25. $-18 = -\frac{2}{3}p$ **27**
26. $-\frac{8}{5}p = 40$ **−25**
27. $-\frac{24}{5}k = 12$ **$-\frac{5}{2}$ or $-2\frac{1}{2}$**
28. $-\frac{63}{10} = -\frac{21}{10}b$ **3**

Homework Help	
For Exercises	See Examples
11–13	1
14–16	2
17–19	3
20–22	5
23–28	6, 7
29–31	5–7
36–39	4

Extra Practice
See page 699.

29. What is the solution of $-18.4 = -9.2n$? **2**
30. Solve $\frac{k}{6} = -6$. Then check your solution. **−36**

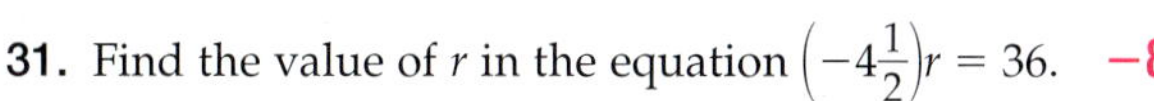

31. Find the value of r in the equation $\left(-4\frac{1}{2}\right)r = 36$. **−8**
32. If $7h - 5 = 4$, then $21h - 15 =$ __?__. **12**

Write an equation and solve.

33. Eight times a number x is 112. What is the number? **$8x = 112$; 14**
34. Negative 154 equals the product of negative 7 and a number p. What is p? **$-154 = -7p$; 22**
35. Three fifths of a number y is negative 9. What is the number?

35. $\frac{3}{5}y = -9$; −15

3 PRACTICE/APPLY

Error Analysis

Watch for students who divide each side of the equation by the constant term rather than the coefficient of the variable in Exercises 4–6.

Prevent by encouraging students to use mental math strategies to think about the problem and also to check their solution. In Exercise 4 for example, students should think *8 times what number is −24?* In problems such as Exercise 6, students can decide whether the solution should be greater than or less than 6.

Assignment Guide

Basic: 11–39 odd, 40–49
Average: 12–34 even, 36–49

***Skills Practice,* p. 161, and *Practice,* p. 162 (shown)**

4-4 NAME ________ DATE ________ PERIOD ______

Practice

Student Edition Pages 160–164

Solving Multiplication and Division Equations

Solve each equation.

1. $7p = -42$ **−6**
2. $-3z = 27$ **−9**
3. $-8q = -56$ **7**
4. $-28 = 2a$ **−14**
5. $5f = 40$ **8**
6. $-9g = 18$ **−2**
7. $-48 = -12r$ **4**
8. $4 = 0.8w$ **5**
9. $-2.4t = 6$ **−2.5**
10. $0 = 5.3k$ **0**
11. $-1.6s = -8$ **5**
12. $2.5d = -11$ **−4.4**
13. $\frac{m}{9} = 2$ **18**
14. $-8 = \frac{y}{4}$ **−32**
15. $\frac{2}{5}s = 18$ **45**
16. $-2 = -\frac{8}{3}b$ **$\frac{3}{4}$**
17. $-\frac{c}{6} = 6$ **−36**
18. $-\frac{v}{12} = -5$ **60**
19. $\frac{1}{8}d = -1$ **−8**
20. $4 = \frac{4}{5}x$ **5**
21. $-\frac{7}{6}r = 28$ **−24**
22. $\frac{9}{10}z = -9$ **−10**
23. $-\frac{b}{18} = 2$ **−36**
24. $-\frac{3}{7}n = -21$ **49**

4 ASSESS

Open-Ended Assessment

Speaking On the board or overhead, write several equations similar to those in the lesson. Ask students to explain how they would solve the equations.

Mid-Chapter Test (Lessons 4–1 through 4–4) is available in the *Chapter 4 Resource Masters*, p. 190.

Answer

40. Dividing rational numbers is the same as multiplying by the reciprocal of the divisor.

***Enrichment*, p. 164**

4–4 NAME ______ DATE ______ PERIOD ______

Enrichment Student Edition Pages 160–164

Puzzle

Solve each equation. The first one is completed.

Equation	Solution	Letter
1. $\frac{m}{12} = 13$	1. 156	U
2. $17v = -578$	2. −34	E
3. $\frac{c}{75} = 18$	3. 1350	R
4. $-252d = -5796$	4. 23	H
5. $64 \cdot w = 5568$	5. 87	O
6. $g \div 29 = 61$	6. 1769	B
7. $p(85) = -7225$	7. −85	T
8. $39x = 663$	8. 17	I
9. $\frac{k}{18} = 30$	9. 540	A
10. $\frac{z}{-94} = -32$	10. 3008	N
11. $-112q = 1456$	11. −13	F
12. $201y = -1608$	12. −8	S

Put the letter that is next to each solution in the box below the corresponding solution shown above.

What is a RINGLEADER?

-85	23	-34		-13	17	1350	-8	-85		87	3008	-34
T	H	E		F	I	R	S	T		O	N	E

17	3008		-85	23	-34		1769	540	-85	23	-85	156	1769
I	N		T	H	E		B	A	T	H	T	U	B

Applications and Problem Solving

36. **Plumbing** Two meters of copper tubing weigh 0.25 kilogram. How much do 50 meters of the same tubing weigh? **6.25 kg**

37. **Entertainment** James is in charge of purchasing student tickets for an Atlanta Braves baseball game. Each student paid \$6 per ticket, and he collected \$288.
    a. Write an equation to represent the number of students n who bought tickets. **$6n = 288$**
    b. How many students are going to the game? **48**

38. **Geometry** Find the length of the base b of the triangle if its area A is 182 square centimeters. Use the formula $A = \frac{1}{2}bh$. **26 cm**

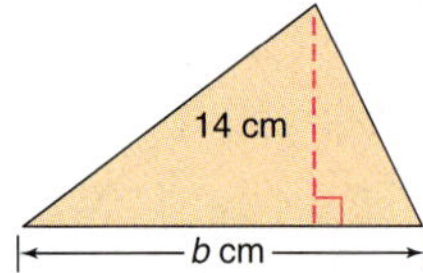

39. **Technology** Engineering students at The Ohio State University built an experimental electric race car that went around a 1.1-mile lap in 36 seconds. How fast did it go? Find the rate r in miles per hour by using the formula $d = rt$, where d is the distance and t is the time. (*Hint:* Convert seconds to hours.) **110 mph**

40. **Critical Thinking** Explain why the Multiplication and Division Properties of Equality for fractions can be thought of as one property. **See margin.**

Mixed Review

Find each quotient. *(Lesson 4–3)*

41. $4 \div 0.2$ **20**
42. $-8.1 \div 9$ **−0.9**
43. $\frac{1}{4} \div \frac{3}{7}$ **$\frac{7}{12}$**

Solve each equation. *(Lesson 3–4)*

44. $n = 5 + 16 \div 4$ **9**
45. $2.6(5) - 13 = x$ **0**
46. $\frac{5 + 3}{3 \cdot 3} = t$ **$\frac{8}{9}$**

47. **Animals** Giant deep-sea squids can be up to 19 yards long. How many feet long can they be? *(Lesson 2–5)* **57 ft**

Standardized Test Practice

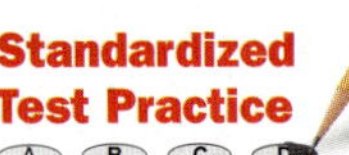

48. **Extended Response** A field hockey playing field is shown at the right. *(Lesson 1–4)*
    a. Write an equation to represent the area A of the field.
    b. Simplify the expression and find the area if a is 60 yards and b is 25 yards. **6000 yd²**

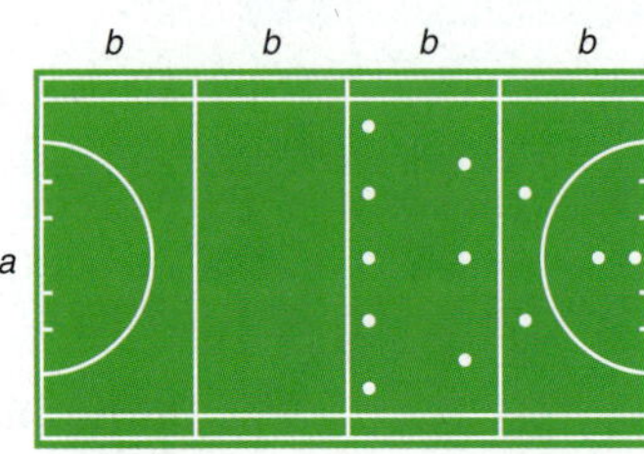

48a. $A = a(b + b + b + b)$ or $A = ab + ab + ab + ab$ or $A = 4ab$

49. **Multiple Choice** Write an equation for the following sentence. Three less than six times y equals 14. *(Lesson 1–1)* **D**
    (A) $3 + 6y = 14$
    (B) $3 - 6y = 14$
    (C) $6y = 14 - 3$
    (D) $6y - 3 = 14$

Extra Credit

Kevin claims that you can solve $3.2x = 8$ by multiplying each side of the equation by $\frac{5}{16}$. Do you think he is right? Explain your response.

Yes; sample answer: 3.2 is the same as $3\frac{1}{5}$ or $\frac{16}{5}$, so multiplying each side by the reciprocal of $\frac{16}{5}$ will isolate the variable.

4–5 Solving Multi-Step Equations

Lesson 4–5

What You'll Learn
You'll learn to solve equations involving more than one operation.

Why It's Important
Shopping Multi-step equations are used to calculate purchase orders.
See Exercise 40.

Equations that have more than one operation require more than one step to solve. To solve this type of problem, the best strategy is to undo each operation in reverse order. In other words, work backward.

Examples

1 **Solve each equation. Check your solution.**

$\frac{x}{3} + 5 = 14$

$\frac{x}{3} + 5 = 14$	*Original equation*
$\frac{x}{3} + 5 - 5 = 14 - 5$	*Subtract 5 from each side.*
$\frac{x}{3} = 9$	*Simplify.*
$3\left(\frac{x}{3}\right) = 3(9)$	*Multiply each side by 3.*
$x = 27$	*Simplify.*

Check:

$\frac{x}{3} + 5 = 14$	*Original equation*
$\frac{27}{3} + 5 \stackrel{?}{=} 14$	*Replace x with 27.*
$9 + 5 \stackrel{?}{=} 14$	*Simplify $\frac{27}{3}$.*
$14 = 14 \ \checkmark$	*Add 9 and 5.*

2 $2k - 7 = 23$

$2k - 7 = 23$	*Original equation*
$2k - 7 + 7 = 23 + 7$	*Add 7 to each side.*
$2k = 30$	*Simplify.*
$\frac{2k}{2} = \frac{30}{2}$	*Divide each side by 2.*
$k = 15$	*Check the solution.*

3 **Solve $\frac{h-7}{2.5} = -10$. Check your solution.**

$\frac{h-7}{2.5} = -10$	*Original equation*
$2.5\left(\frac{h-7}{2.5}\right) = 2.5(-10)$	*Multiply each side by 2.5.*
$h - 7 = -25$	*Simplify.*
$h - 7 + 7 = -25 + 7$	*Add 7 to each side.*
$h = -18$	*Simplify.*

(continued on the next page)

Resource Manager

Reproducible Masters
Chapter 4 Resource Masters
- *Study Guide*, p. 165
- *Skills Practice*, p. 166
- *Practice*, p. 167
- *Reading to Learn Mathematics*, p. 168
- *Enrichment*, p. 169

Graphing Calculator, p. 11
Hands-on Algebra, p. 56–58
School-To Workplace, p. 4

Transparencies
5-Minute Check, 4–5

Technology/Multimedia
AlgePASS, Lesson 11
Interactive Chalkboard CD-ROM

1 FOCUS

5-Minute Check
Lesson 4–4

Solve each equation. Check your solution.

1. $-21 = 9n$ $-2\frac{1}{3}$
2. $\frac{r}{6} = -3$ -18
3. $-\frac{5}{9}p = 10$ -18
4. If the area of the rectangle shown below is 4.2 square centimeters, what is its length x? **3.5 cm**

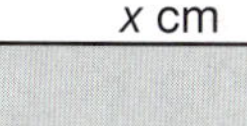

Motivating the Lesson

Real-World Connection Tell students to imagine their parent went shopping for school supplies and purchased a package of pencils priced at $1.50 and several three-ring binders priced at $3.25 each. The total cost of the items before sales tax was added was $14.50. Ask students to describe some ways they could determine the number of binders that were purchased.

2 TEACH

In-Class Examples
Examples 1–2

Solve each equation. Check your solution.

1 $\frac{x}{6} - 4 = 2$ 36
2 $3m + 12 = 27$ 5

In-Class Examples

Example 3

Solve $-6.2 = \frac{3 + n}{15}$. Check your solution. **−96**

Example 4

In a city, the tallest building is 1268 feet tall. This is 35 feet greater than 3 times the height of the fifth tallest building. How tall is the fifth tallest building? **411 ft**

Check:

$\frac{h - 7}{2.5} = -10$ *Original equation*

$\frac{-18 - 7}{2.5} \stackrel{?}{=} -10$ *Replace h with −18.*

$\frac{-25}{2.5} \stackrel{?}{=} -10$ *Simplify.*

$-10 = -10$ ✓

Your Turn **Solve each equation. Check your solution.**

a. $11 + 9v = 119$ **12** **b.** $\frac{a}{8} - 5.2 = 3$ **65.6** **c.** $\frac{b + 4}{7} = 6$ **38**

Technology Link

4 **Trains that float on magnets above rails have been developed in Japan. They are designed to run at a maximum speed of 341 miles per hour. This is 71 miles per hour faster than twice the speed of Japan's bullet trains. Find the speed of the bullet trains in Japan and graph the solution.**

Explore You know that the speed of the floating trains is 341 mph. This is 71 mph faster than twice Japan's bullet trains.

Plan Let x represent the speed of the bullet trains. Then translate the given information into an equation and solve.

Solve

speed of floating train	*equals*	*71 miles per hour*	*plus*	*twice the speed of bullet trains*
341	=	71	+	$2x$

$341 = 71 + 2x$ *Original equation*

$341 - 71 = 71 + 2x - 71$ *Subtract 71 from each side.*

$270 = 2x$ *Simplify.*

$\frac{270}{2} = \frac{2x}{2}$ *Divide each side by 2.*

$135 = x$ *Simplify.*

The bullet trains in Japan travel at 135 miles per hour.

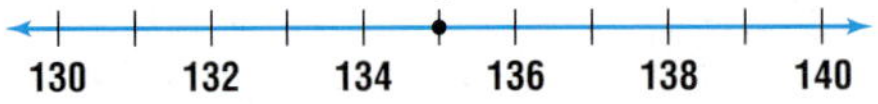

[Number line from 130 to 140 with a point at 135; labels 130, 132, 134, 136, 138, 140]

Examine Check by substituting 135 into the original equation.

$341 = 71 + 2x$ *Original equation*

$341 \stackrel{?}{=} 71 + 2(135)$ *Replace x with 135.*

$341 \stackrel{?}{=} 71 + 270$ *Multiply 2 and 135.*

$341 = 341$ ✓ *Add 71 and 270.*

www.algconcepts.com/extra_examples

Instruct students to play the game *I'm Thinking of a Number* with a family member. The student should begin by thinking of a number and then giving their family member an equation whose solution is the number. For example, if the student thinks of the number 5, then the player's clue might be "12 is 2 more than twice the number." The family member could then determine the number by writing and solving the equation $12 = 2x + 2$. The players should play the game several times, switching roles each time.

You can use a graphing calculator to solve multi-step equations.

Graphing Calculator Tutorial
See pp. 750–753.

To edit or replace an equation, press ▲ until the equation editor is displayed. Then edit or replace the equation.

Graphing Calculator Exploration

Solve $2 = \frac{x-5}{4}$ using a graphing calculator.

Step 1 Rewrite the equation so that one side is equal to 0.

$$2 = \frac{x-5}{4} \quad \textit{Original equation}$$

$$2 - 2 = \frac{x-5}{4} - 2 \quad \textit{Subtract 2 from each side.}$$

$$0 = \frac{x-5}{4} - 2 \quad \textit{Simplify.}$$

Step 2 Press MATH and select 0:Solver.

Step 3 Enter the equation after the 0 =. Press

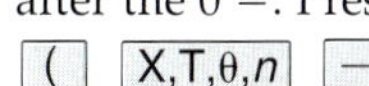

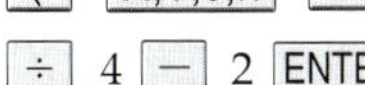

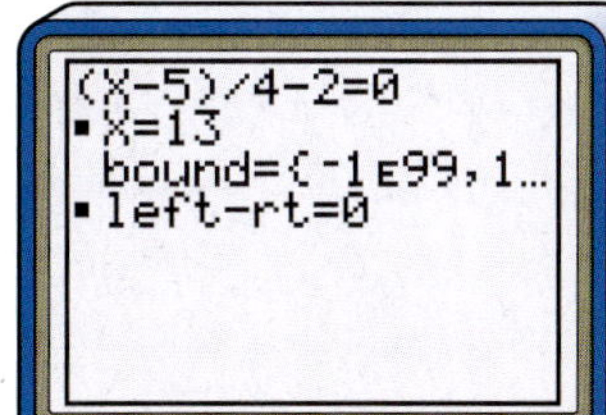

Step 4 Press ALPHA [SOLVE].

The solution is 13.

Try These **1. 3*x* − 4 − 6 or 3*x* − 10**

1. What would you enter into the calculator to solve $3x - 4 = 6$?
2. Work through the examples in this lesson with a calculator. **See students' work.**

Consecutive integers are integers in counting order, such as 3, 4, and 5. Beginning with an even integer and counting by two will result in *consecutive even integers*. For example, −6, −4, −2, 0, and 2 are consecutive even integers. Beginning with an odd integer and counting by two will result in *consecutive odd integers*. For example, −1, 1, 3, and 5 are consecutive odd integers.

Consecutive Integers		
Integers	**Even Integers**	**Odd Integers**
$n = 1$	$n = 2$	$n = 3$
$n + 1 = 2$	$n + 2 = 4$	$n + 2 = 5$
$n + 2 = 3$	$n + 4 = 6$	$n + 4 = 7$
⋮	⋮	⋮

Example 5 **Number Theory Link**

Find three consecutive even integers whose sum is −18.

Explore You know that the sum of three consecutive even integers is −18. You need to find the three numbers.

Plan Let n represent the first even integer. Then $n + 2$ represents the second even integer, and $n + 4$ represents the third. *Why?*

(continued on the next page)

Teaching Tip When discussing the chart above Example 5, be sure students understand that the same expressions can be used to represent consecutive even integers and consecutive odd integers because the difference is 2 in both cases. Stress that it is the value of n (odd or even) that differs between the two representations.

Graphing Calculator Exploration

When the student presses MATH 0 to access the Solver feature, the display may not have the words EQUATION SOLVER at the top of the screen. This means that there is already an equation stored from a previous use of the calculator. The student should press ▲ CLEAR to go to the EQUATION SOLVER screen and clear the expression from the previous equation. The student can then enter the new expression and press ENTER ALPHA [SOLVE] to solve the new equation.

In-Class Example

Example 5

Find four consecutive odd integers whose sum is -8. **$-5, -3, -1, 1$**

3 PRACTICE/APPLY

Error Analysis

Watch for students who perform the steps for solving a multi-step equation in the wrong order in Exercise 8, multiplying 7 by -3 and then adding 4 to the product. ***Prevent by*** encouraging students to think of equation solving as undoing an expression, using the order of operations in reverse order. That is, undoing subtractions and additions first, followed by multiplications and divisions, and finally grouping symbols.

***Study Guide*, p. 165**

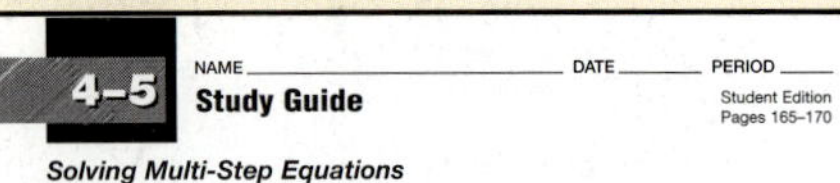

NAME ____________ DATE ______ PERIOD ______

4–5 Study Guide

Student Edition Pages 165–170

Solving Multi-Step Equations

Some equations require more than one step to solve. To solve a problem that has more than one step, the best strategy is to undo the operations in reverse order. Always check your solution.

Example 1: Solve $3x - 8 = 31$.

$$3x - 8 = 31$$
$$3x - 8 + 8 = 31 + 8$$
$$3x = 39$$
$$\frac{3x}{3} = \frac{39}{3}$$
$$x = 13$$

Check:
$$3x - 8 = 31$$
$$3(13) - 8 \stackrel{?}{=} 31$$
$$39 - 8 \stackrel{?}{=} 31$$
$$31 = 31 \checkmark$$

Example 2: Solve $\frac{x}{4} + 6 = 12$.

$$\frac{x}{4} + 6 = 12$$
$$\frac{x}{4} + 6 - 6 = 12 - 6$$
$$\frac{x}{4} = 6$$
$$4 \cdot \frac{x}{4} = 4 \cdot 6$$
$$x = 24$$

Check:
$$\frac{x}{4} + 6 = 12$$
$$\frac{24}{4} + 6 \stackrel{?}{=} 12$$
$$6 + 6 \stackrel{?}{=} 12$$
$$12 = 12 \checkmark$$

Solve each equation. Check your solution.

1. $6y - 4 = 50$ **9**
2. $-8x - 2 = 38$ **-5**
3. $-15 = -12n + 9$ **2**
4. $-2m + 6 = 22$ **-8**
5. $-4y - 8 = 30$ **-9.5**
6. $-1.5x + 6 = -54$ **40**
7. $8.2y + 4 = 65.5$ **7.5**
8. $-1.5 = 3w + 6$ **-2.5**
9. $-8 = -2.5x + 5$ **5.2**
10. $\frac{x}{2} + 7 = 9$ **4**
11. $\frac{x}{3} - 12 = -2$ **30**
12. $\frac{x}{-6} - 12 = 2$ **-84**
13. $3x + 7 = 7$ **0**
14. $-8 = 1.5y + 4$ **-8**
15. $\frac{y}{2} + 2 = 25$ **46**

Solve	$n + (n + 2) + (n + 4) = -18$	*Write an equation.*
	$3n + 6 = -18$	*Add like terms.*
	$3n + 6 - 6 = -18 - 6$	*Subtract 6 from each side.*
	$3n = -24$	*Simplify.*
	$\frac{3n}{3} = \frac{-24}{3}$	*Divide each side by 3.*
	$n = -8$	*Simplify.*

$n = -8$ $\quad$ $n + 2 = -8 + 2 = -6$ $\quad$ $n + 4 = -8 + 4 = -4$

The numbers are -8, -6, and -4.

Examine Check by adding the three numbers.

$$-8 + (-6) + (-4) \stackrel{?}{=} -18$$
$$-18 = -18 \checkmark$$

Your Turn

d. Find three consecutive integers whose sum is 27. **8, 9, 10**

Check for Understanding

Communicating Mathematics

1. **Describe** how you work backward to solve multi-step equations.
2. **Write** an equation that requires more than one operation to solve. **See students' work.**

Vocabulary
consecutive integers

3. **You Decide** Soto solved the equation $\frac{t + 16}{4} = 1$ using the steps at the right. Jean does not agree with his answer. Who is correct? Explain. **Jean; Soto did not multiply each side by 4 in the second step.**

$$\frac{t + 16}{4} = 1$$
$$t + 16 = 1$$
$$t + 16 - 16 = 1 - 16$$
$$t = -15$$

1. Sample answer: If necessary, first undo addition and/or subtraction of constants, then undo multiplication or division of variables.

Guided Practice

4. Subtract 7 from each side.
5. Add 5 to each side.
6. Multiply each side by 3.

Examples 1–4

Getting Ready **State the first step in solving each equation.**

Sample: $3x - 4 = -10$ $\quad$ **Solution:** Add 4 to each side.

4. $7 + 6v = 43$
5. $11 = \frac{g}{8} - 5$
6. $\frac{-y - 2}{3} = 15$

Solve each equation. Check your solution.

7. $6t - 1 = 11$ **2**
8. $7 = 4 - \frac{n}{3}$ **-9**
9. $-5r + 2 = 27$ **-5**
10. $-5.6 + 4d = 2$ **1.9**
11. $2 = \frac{f + 8}{-6}$ **-20**
12. $\frac{7n - 1}{8} = 6$ **7**

Reteaching Activity

Kinesthetic Learners Have students use algebra tiles to model the solution steps for simple multi-step equations, such as $2x + 1 = 7$, $3x - 1 = 5$, and $4x - 2 = -6$.

Example 1

13. **Animals** A teacher in Wisconsin came up with a formula that more accurately gives a cat's or dog's age in people years.

> Take your cat's/dog's age, subtract 1, multiply by 6, add 21, and that will give you the equivalent in people years.

Source: *The Mathematics Teacher*

a. Write an equation to represent the age of a dog or cat in people years y if the dog or cat is x years old. $y = 6(x - 1) + 21$

b. How old is a dog if he is 17 in people years? $\frac{1}{3}$ yr or 4 months

Exercises

Practice

Solve each equation. Check your solution.

14. $6 = 4n + 2$ **1**
15. $8 + 3k = 5$ **−1**
16. $3b - 7 = 2$ **3**
17. $15 = 1 - 2t$ **−7**
18. $12 = -5h + 2$ **−2**
19. $-13 - 9y = -13$ **0**
20. $8 + 1.6a = 0$ **−5**
21. $0.3x + 3 = 4.8$ **6**
22. $6.5 = 2x + 4.1$ **1.2**
23. $27 = 3 + 2.5t$ **9.6**
24. $-4 - 0.7m = 2.3$
25. $-6z + 8 = -40.6$
26. $7 = \frac{x}{2} + 5$ **4**
27. $\frac{y}{3} - 6 = 4$ **30**
28. $8 + \frac{c}{-4} = 12$ **−16**
29. $28 = 7 - \frac{3}{2}t$ **−14**
30. $\frac{3 + n}{7} = -5$ **−38**
31. $-10 = \frac{s - 8}{-6}$ **68**

Homework Help

For Exercises	See Examples
14–25, 33	2
26–29	1
30–32	3
34–42	4, 5

Extra Practice

See page 700.

32. What is the solution of $6 = \frac{s - 8}{-7}$? **−34**

33. Find the value of y in the equation $-2.5y - 4 = 14$. **−7.2**

34. $\frac{3}{4}c + 5 = 26$; **28**

24. −9
25. 8.1
34–36. See margin for graphs.

For Exercises 34–36, write an equation and solve each problem. Graph the solution.

34. Five more than three-fourths of a number c is 26. Find the number.

35. The steps below are applied to a number n to get $\frac{1}{3}$. What is the number? $\frac{n - 2}{-3} = \frac{1}{3}$; 1
    - subtract 2
    - divide by −3

36. Start with the number x. If you multiply the number by 2, divide by 5, and add 20, you get 10. What is the value of x? $\frac{2x}{5} + 20 = 10$; −25

Applications and Problem Solving

37. **Geometry** Find the value of x if the perimeter of the square is 20 inches. **3**

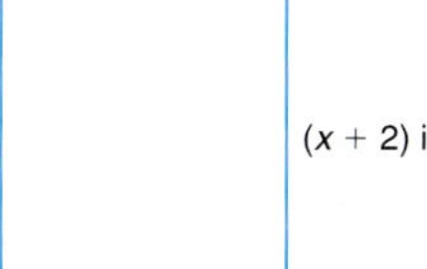

Exercise 37

38. **Number Theory** Four consecutive odd integers have a sum of 56.

a. Define a variable and write an equation to represent the problem.

b. Find the numbers. **11, 13, 15, 17**

38a. x = first odd integer; $x + (x + 2) + (x + 4) + (x + 6) = 56$

Assignment Guide

Basic: 15–41 odd, 42–50
Average: 14–36 even, 37–50

Answers

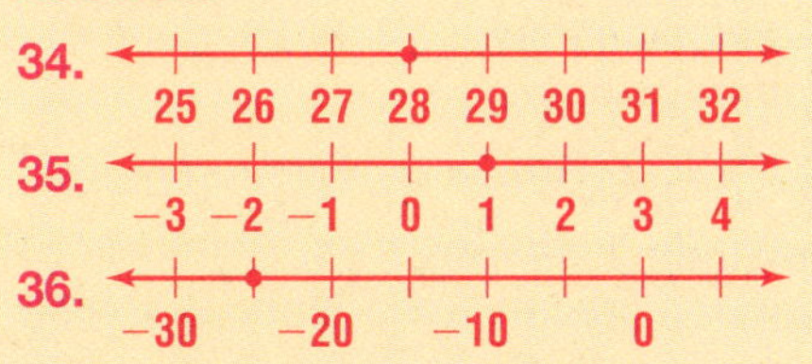

***Skills Practice,* p. 166, and *Practice,* p. 167 (shown)**

4–5 NAME ______ DATE ______ PERIOD ______

Practice

Student Edition Pages 165–170

Solving Multi-Step Equations

Solve each equation. Check your solution.

1. $8z - 6 = 18$ **3**
2. $-4s + 1 = 9$ **−2**
3. $12 = -3k + 3$ **−3**
4. $5 - 2f = 19$ **−7**
5. $-31 = -6w - 7$ **4**
6. $6 + 7r = 13$ **1**
7. $-8 = 8 - 2c$ **8**
8. $0.4u + 1 = 6.6$ **14**
9. $3b - 2.5 = 5$ **2.5**
10. $4.7 + 2g = 7.3$ **1.3**
11. $-2.1q - 1 = -1$ **0**
12. $-2 = \frac{t}{4} - 3$ **4**
13. $\frac{p}{9} + 4 = 7$ **27**
14. $7 - \frac{m}{2} = 0$ **14**
15. $8 = 5 - \frac{c}{6}$ **−18**
16. $\frac{y - 5}{3} = 2$ **11**
17. $1 = \frac{c + 1}{-8}$ **−9**
18. $\frac{-4a + 4}{5} = -4$ **6**
19. $-4 = \frac{x}{7} + 3$ **−49**
20. $\frac{8h - 2}{9} = 6$ **7**
21. $9 - \frac{1}{4}j = 5$ **16**

4 ASSESS

Open-Ended Assessment

Writing Have students write a description of the procedure for solving multi-step equations, such as those in this lesson. Then have them exchange their papers with a classmate and test each other's procedure on one of the equations on page 169.

Answer

42. Sample answer: The son inherited twice as much as the wife, and the wife inherited twice as much as the daughter. If x represents the daughter's inheritance, then $2x$ represents the wife's, and $4x$ represents the son's. Solving $x + 2x + 4x = 1$, daughter $= \frac{1}{7}$, wife $= \frac{2}{7}$, and son $= \frac{4}{7}$.

***Enrichment*, p. 169**

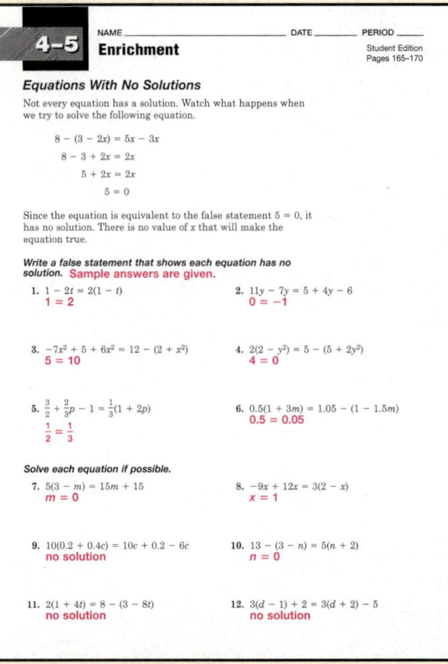

NAME ______ DATE ______ PERIOD ______

4-5 Enrichment

Student Edition Pages 165–170

Equations With No Solutions

Not every equation has a solution. Watch what happens when we try to solve the following equation.

$$8 - (3 - 2x) = 5x - 3x$$
$$8 - 3 + 2x = 2x$$
$$5 + 2x = 2x$$
$$5 = 0$$

Since the equation is equivalent to the false statement $5 = 0$, it has no solution. There is no value of x that will make the equation true.

Write a false statement that shows each equation has no solution. Sample answers are given.

1. $1 - 2t = 2(1 - t)$ $1 = 2$
2. $11y - 7y = 5 + 4y - 6$ $0 = -1$
3. $-7x^2 + 5 + 6x^2 = 12 - (2 + x^2)$ $5 = 10$
4. $2(2 - y^2) = 5 - (5 + 2y^2)$ $4 = 0$
5. $\frac{3}{2} + \frac{2}{3}p - 1 = \frac{1}{3}(1 + 2p)$ $\frac{1}{2} = \frac{1}{3}$
6. $0.5(1 + 3m) = 1.05 - (1 - 1.5m)$ $0.5 = 0.05$

Solve each equation if possible.

7. $5(3 - m) = 15m + 15$ $m = 0$
8. $-9x + 12x = 3(2 - x)$ $x = 1$
9. $10(0.2 + 0.4c) = 10c + 0.2 - 6c$ no solution
10. $13 - (3 - n) = 5(n + 2)$ $n = 0$
11. $2(1 + 4t) = 8 - (3 - 8t)$ no solution
12. $3(d - 1) + 2 = 3(d + 2) - 5$ no solution

39. **Consumer Spending** Dwayne was checking his expenses and found that in eight visits to the barbershop, he had spent $143 for haircuts. Of that amount, $15 was tips. **a. $8c + 15 = 143$**
    a. Write an equation to represent the cost c of the haircuts.
    b. How much does Dwayne pay for each haircut before the tip? **$16**

40. **Shopping** Tanisa and her sister ordered a total of 5 pairs of shoes from a catalog. Each pair cost the same amount. The sisters had to pay $10 for shipping and handling. If their total purchase was $170, how much did each pair of shoes cost? **$32**

41. **Gardening** Mr. Green purchased 39 plants for his garden. His purchase included an equal number of ferns, irises, and gladiolas, as well as 12 impatiens. **a. $3x + 12 = 39$**
    a. Write an equation representing the number of items he purchased.
    b. How many ferns did he buy? **9**

Impatiens

42. **Critical Thinking** According to legend, Islamic mathematician al-Khowarizmi (810 A.D. – ?) stated in his will that if he had a son, the son should receive twice as much of the estate as al-Khowarizmi's wife when he died. If he had a daughter, the daughter should receive half as much as al-Khowarizmi's wife. However, al-Khowarizmi had twins, a boy and a girl. What fraction of his estate should each one get? Explain how you found your answer. **See margin.**

Mixed Review

Geometry **Find each missing measure.** *(Lesson 4–4)*

43. $A = 15 \text{ in}^2$ **3**
44. $A = 49 \text{ cm}^2$ **7**
45. $A = 27 \text{ m}^2$ **6**

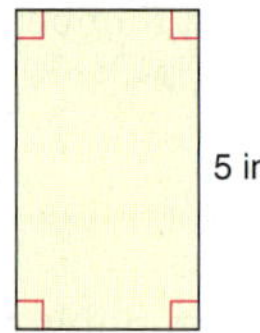

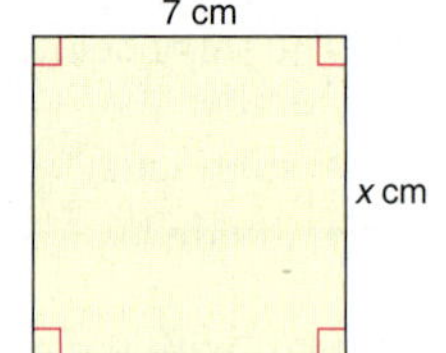

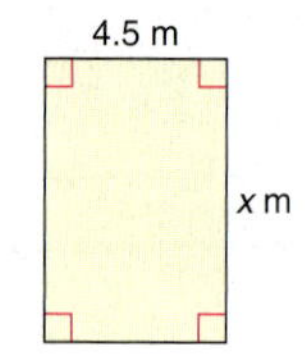

Solve each equation. *(Lesson 3–5)*

46. $-1 + x = -5$ **−4**
47. $n - 3 = 1$ **4**
48. $t + 4 = -2$ **−6**

Standardized Test Practice

49. **Short Response** Find the next term in the pattern −1, 3, −9, 27, ... *(Lesson 2–5)* **−81**

50. **Short Response** Graph two positive integers and two negative integers on a number line. *(Lesson 2–1)* **See students' work.**

Extra Credit

Make up a real-life problem that could be solved using the equation $6y + 15 = 63$. Solve the equation and explain how the solution relates to the real-life situation. **Sample answer: I paid $63 for Internet service this month, which included the $15 monthly fee and a $6 per hour charge. How many hours did I use the Internet? The solution is 8 hours.**

4–6 Variables on Both Sides

What You'll Learn
You'll learn to solve equations with variables on both sides.

Why It's Important
Sports Equations comparing athletes' performances have variables on both sides. *See Example 3.*

Many equations contain variables on both sides. To solve these equations, first use the Addition or Subtraction Property of Equality to write an equivalent equation that has all of the variables on one side. Then solve.

Examples

Solve each equation. Check your solution.

1 $5x = x - 12$

$5x = x - 12$	*Original equation*
$5x - x = x - 12 - x$	*Subtract x from each side.*
$4x = -12$	*Simplify.*
$\frac{4x}{4} = \frac{-12}{4}$	*Divide each side by 4.*
$x = -3$	*Simplify.*

Check:

$5x = x - 12$	*Original equation*
$5(-3) \stackrel{?}{=} -3 - 12$	*Replace x with −3.*
$-15 = -15$ ✓	

2 $16 + \frac{1}{4}n = \frac{3}{4}n$

$16 + \frac{1}{4}n = \frac{3}{4}n$	*Original equation*
$16 + \frac{1}{4}n - \frac{1}{4}n = \frac{3}{4}n - \frac{1}{4}n$	*Subtract $\frac{1}{4}n$ from each side.*
$16 = \frac{1}{2}n$	$\frac{3}{4}n - \frac{1}{4}n = \frac{2}{4}n$ *or* $\frac{1}{2}n$
$2(16) = 2\left(\frac{1}{2}n\right)$	*Multiply each side by 2.*
$32 = n$	*Check the solution.*

Your Turn c. 1.7

a. $a + 9 = 4a$ **3**   b. $\frac{2}{3}n = \frac{1}{3}n - 2$ **−6**   c. $5k - 2.4 = 3k + 1$

You can solve an equation with a variable on both sides to determine when a quantity that is increasing and one that is decreasing will be the same.

Resource Manager

Reproducible Masters

Chapter 4 Resource Masters
- *Study Guide,* p. 170
- *Skills Practice,* p. 171
- *Practice,* p. 172
- *Reading to Learn Mathematics,* p. 173
- *Enrichment,* p. 174

Graphing Calculator, pp. 12–13
Hands-On Algebra, p. 59

Transparencies
5-Minute Check, 4–6

Technology/Multimedia
AlgePASS, Lesson 12
Interactive Chalkboard CD-ROM

1 FOCUS

5-Minute Check
Lesson 4–5

Solve each equation. Check your solution.

1. $5b - 1 = 4$ **1**
2. $3 + \frac{c}{-2} = 5$ **−4**
3. $-1 = \frac{2 + x}{7}$ **−9**
4. Stacy spent $161 on floor tickets. If two of them were for children, how many adult tickets did she buy? Write an equation and solve.

Ticket Prices

	Location	
	Floor	*Balcony*
Adult	$25	$20
Child	$18	$12

161 = 36 + 25x; 5 adult floor tickets

Motivating the Lesson

Real-World Connection Present the following situation to students. Suppose you are signing up for cable television service. There are two service plans. One plan costs $10 per month plus $8 for each premium channel. The second plan costs $20 per month plus $3 for each premium channel. For what number of premium channels will the two plans cost the same? Work with students to model the situation with the equation $10 + 8x = 20 + 3x$, where x is the number of premium channels.

2 TEACH

In-Class Examples
Examples 1–2

Solve each equation. Check your solution.

1 $y + 8 = 9y$ **1**

2 $\frac{2}{5}x = 6 - \frac{1}{5}x$ **10**

Teaching Tip While discussing the equation $52.3 - 0.1x = 59.1 - 0.3x$ in Example 3, point out that x represents the number of years *after* 1996.

In-Class Example

Example 3

In the 1996 Olympics, the winning times for the 100-meter freestyle were about 48.7 seconds for men and 54.5 seconds for women. Suppose the men's times decrease 0.2 second per year and the women's times decrease 0.3 second per year. Solve $48.7 - 0.2x = 54.5 - 0.3x$ to find when men and women would have the same winning times. **58 yr after 1996, or in 2054**

Teaching Tip When discussing the definition of *identity* above Example 4, you may want to tell students that this type of equation is sometimes said to have an *infinite* number of solutions. You may need to define *infinite,* if some students are unfamiliar with the term.

In-Class Examples

Examples 4–5

Solve each equation.

4 $3 - 4y = 6 - 4y$ **no solution**

5 $8m + 2 = 2 - 3m + 11m$ **identity**

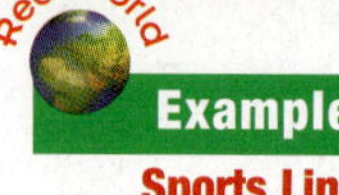

Example 3 Real World Sports Link

The 2003 World Championship 100-meter dash times were 10.07 seconds for men and 10.85 seconds for women. If each year the men's times decrease 0.08 second and the women's times decrease 0.14 second, when would they have the same winning times?

Let y represent the number of years. Write an equation.

Words	*Men's times decrease 0.08 second per year.*	*Women's times decrease 0.14 second per year.*
Variables	$10.07 - 0.08y$	$10.85 - 0.14y$

Equation

$10.07 - 0.08y = 10.85 - 0.14y$ *Write an equation.*

$10.07 - 0.08y + 0.14y = 10.85 - 0.14y + 0.14y$ *Add 0.14y to each side.*

$10.07 + 0.06y = 10.85$ *Simplify.*

$10.07 - 10.07 + 0.06y = 10.85 - 10.07$ *Subtract 10.07 from each side.*

$0.06y = 0.78$ *Simplify.*

$\frac{0.06y}{0.06} = \frac{0.78}{0.06}$ *Divide each side by 0.2.*

$y = 13$ *Simplify.*

At these rates, men and women have the same times 13 years after 2003, or in 2016.

Some equations have no solution. This means that there is no value for the variable that will make the equation true. Other equations may have every number as the solution. An equation that is true for every value of the variable is called an **identity**.

Examples 4 **Solve each equation.**

$5x + 3 = 8 + 5x$

$5x + 3 = 8 + 5x$ *Original equation*

$5x + 3 - 5x = 8 + 5x - 5x$ *Subtract 5x from each side.*

$3 = 8$ *False statement*

The equation has no solution. *3 = 8 is never true.*

5 **$3 + 3t - 5 = 3t - 2$**

$3 + 3t - 5 = 3t - 2$ *Original equation*

$3t - 2 = 3t - 2$ *Simplify.*

The equation is an identity. *3t − 2 = 3t − 2 is true for all values of t.*

Your Turn

d. $2t + 4 - t = 4 + t$ **identity** **e.** $16h + 7 = 16 + 16h$ **no solution**

www.algconcepts.com/extra_examples

Inclusion Strategies

Students with a learning disability may feel overwhelmed with the more complex equations in this lesson and again in Lesson 4–7. Work with these students to write down the steps for solving these equations on an index card. They should then keep the index card readily available and follow the steps as they do their homework and in-class assignments.

Check for Understanding

Communicating Mathematics

1–3. See margin.

1. **List** the steps you would take to solve the equation $3x + 5 = -2x - 16$.
2. **Choose** the correct term and justify your answer. Equations with variables on both sides (*sometimes, always, never*) have a solution.
3. **Explain** the difference between an equation that is an identity and an equation with no solution.

Vocabulary
identity

Guided Practice

Getting Ready State the first step in solving each equation.

Sample: $9r = -10 + 7r$ **Solution:** Subtract $7r$ from each side.

4. $2y = 4 - 3y$ 4. Add $3y$ to each side.
5. $0.5p + 4 = 1.5p$ 5. Subtract $0.5p$ from each side.
6. $\frac{-y - 6}{3} = 7y$ 6. Multiply each side by 3.

Examples 1–5

Solve each equation. Check your solution.

7. $10x = 3x + 14$ **2**
8. $2r - 21 = 3 - 4r$ **4**
9. $8y + 1 = 1 + 8y$ **identity**
10. $7 - 6x = 41 - 6x$ **no solution**
11. $3.4a + 3 = 2a - 4$ **−5**
12. $\frac{1}{3}m - 10 = \frac{2}{3}m + 4$ **−42**

Example 1

13. **Number Theory** Three times a number n is 24 less than five times the number.
   a. Write an equation to represent the problem. **$3n = 5n - 24$**
   b. Find the value of n. **12**

Exercises

Practice

Solve each equation. Check your solution.

14. $10h = 8h + 6$ **3**
15. $7u - 19 = 6u$ **19**
16. $6x + 2 = 3x - 1$ **−1**
17. $5t - 9 = -3t + 7$ **2**
18. $2n + 15 = 2n + 20 - 5$ **identity**
19. $3x + 6 = 7x - 4$ **$\frac{5}{2}$ or $2\frac{1}{2}$**
20. $8y - 5 = -y + 2$ **$\frac{7}{9}$**
21. $18 + 3k = 22 + 3k$ **no solution**
22. $12 + 2.5a = 5a$ **4.8**
23. $a + 3.1 = 6.6 - 4a$ **0.7**
24. $6.2y + 7 = 3y - 1$ **−2.5**
25. $3.2x - 1.8 = 6 + 11x$ **−1**
26. $4.5 - x = 2x - 8.1$ **4.2**
27. $1.6 - 4.3y = -5.2 - 2.3y$ **3.4**

B

28. $\frac{1}{2}d = 2 - \frac{1}{2}d$ **2**
29. $\frac{1}{5}p - 4 = \frac{2}{5}p$ **−20**
30. $\frac{1}{3}z - 4 = -\frac{1}{3}z + 18$ **33**
31. $\frac{1}{9}p - 9 = \frac{4}{9}p - 29$ **60**

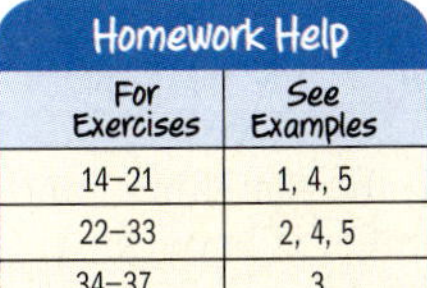

Homework Help

For Exercises	See Examples
14–21	1, 4, 5
22–33	2, 4, 5
34–37	3

Extra Practice
See page 700.

3 PRACTICE/APPLY

Error Analysis

Watch for students who have trouble combining like terms when adding or subtracting a variable expression on each side of an equation such as those in Exercises 4–12.

Prevent by encouraging students to write out all the steps while solving these equations. Some students may also find using parentheses helpful. See the Teaching Tip on page 172.

Assignment Guide

Basic: 15–35 odd, 37–48
Average: 14–34 even, 35–48
All: Quiz 2, 1–10

Answers

1. Add $2x$ to each side, subtract 5 from each side, divide each side by 5.
2. Sometimes; these equations can also have no solutions.
3. An equation that is an identity is true for every value of the variable; an equation with no solution is never true.

Reteaching Activity

Auditory/Musical Learners Have students work with a partner. Give each pair of students a list of equations like those in Exercises 14–31. The partners should take turns describing orally what the first few steps would be to solve an equation. The student who is listening should critique their partner's description.

***Study Guide*, p. 170**

NAME ______ DATE ______ PERIOD ______

Study Guide

Student Edition Pages 171–175

Variables on Both Sides

Some equations contain variables on both sides and require more than one step to solve. To solve these equations, first use the Addition or Subtraction Property of Equality to write an equivalent equation that has all of the variables on one side. Then solve and check.

Example 1: Solve $2x - 6 = x + 4$.

$$2x - 6 = x + 4$$
$$2x - 6 - x = x + 4 - x$$
$$x - 6 = 4$$
$$x - 6 + 6 = 4 + 6$$
$$x = 10$$

Check: $2x - 6 = x + 4$

$$2(10) - 6 \stackrel{?}{=} 10 + 4$$
$$20 - 6 \stackrel{?}{=} 14$$
$$14 = 14 \checkmark$$

Example 2: Solve $\frac{1}{4}x - 12 = \frac{3}{4}x$.

$$\frac{1}{4}x - 12 = \frac{3}{4}x$$
$$\frac{1}{4}x - 12 - \frac{1}{4}x = \frac{3}{4}x - \frac{1}{4}x$$
$$-12 = \frac{1}{2}x$$
$$2 \cdot (-12) = 2 \cdot \frac{1}{2}x$$
$$-24 = x$$

Check: $\frac{1}{4}x - 12 = \frac{3}{4}x$

$$\frac{1}{4}(-24) - 12 \stackrel{?}{=} \frac{3}{4}(-24)$$
$$-6 - 12 \stackrel{?}{=} -18$$
$$-18 = -18 \checkmark$$

Solve each equation. Check your solution.

1. $6m - 40 = m$ **8**
2. $-5y - 2 = y + 10$ **−2**
3. $-15n = -12n + 9$ **−3**
4. $-4y + 6 = -3y + 12$ **−6**
5. $6y - 8 = 6y - 6 - 2$ **identity**
6. $-15x + 8 = -15x - 7$ **no solution**
7. $4.2y + 4.4 = 3.1y$ **−4**
8. $w = 3.8w - 7$ **2.5**
9. $-8 - m = -3.5m + 5$ **5.2**
10. $\frac{1}{5}x + 12 = \frac{2}{5}x$ **60**
11. $\frac{1}{3}x - 8 = -\frac{1}{3}x$ **12**
12. $\frac{3}{7}x - 14 = -\frac{5}{7}x + 2$ **14**

32. Solve $\frac{3}{4}n + 16 = 2 - \frac{1}{8}n$. **−16**

33. Find the solution of $18 - 3.8x = 7.5 - 1.3x$. **4.2**

34. Eight times a number n is 51 more than five times the number.
   a. Write an equation to represent the problem. **$8n = 5n + 51$**
   b. Find the number. **17**

Applications and Problem Solving

35. **Geometry** The length of the rectangle is twice the width. **a. $3w - 18 = 2w$**

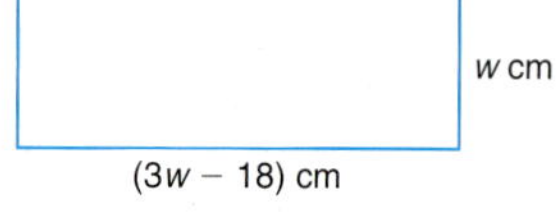

   a. Write an equation to represent the relationship between the length and width.
   b. What are the dimensions of the rectangle? **18 cm by 36 cm**

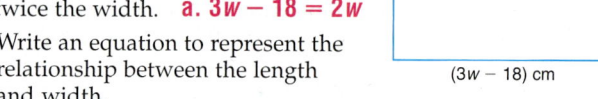

36. **Temperature** On the Fahrenheit scale, water freezes at 32°. On the Celsius scale, water freezes at 0°. The formula $F = \frac{9}{5}C + 32$ gives the temperature in degrees Fahrenheit when the temperature in degrees Celsius is known. Is the temperature ever the same on both scales? To find a temperature when both are the same, let $F = C$ in the equation $F = \frac{9}{5}C + 32$ and solve $C = \frac{9}{5}C + 32$. **yes; −40°**

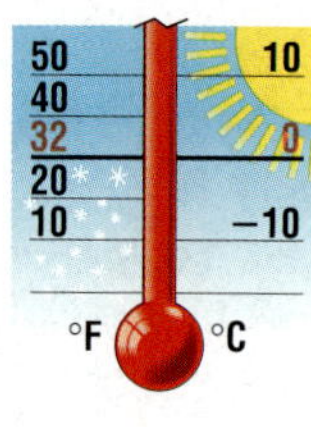

37. **Critical Thinking** Write an equation with variables on both sides that has a solution of 6. **Sample answer: $2x = x + 6$**

Mixed Review

38. Solve $3 + 6k = 45$. Check your solution. *(Lesson 4–5)* **7**

Solve each equation. Check your solution. *(Lesson 4–4)*

39. $-7t = 63$ **−9**
40. $0 = 4.5c$ **0**
41. $-\frac{2}{3}x = -8$ **12**

42. **School Supplies** How many different kinds of transparent tape are possible? *(Lesson 4–2)* **36**

Transparent Tape		
Type	**Size**	**Color**
permanent	$\frac{1}{2}$ in.	clear
removable	$\frac{3}{4}$ in.	lime
double-sided	1 in.	purple
		pink

43. **Entertainment** To raise money for new uniforms, the marching band set up a lottery using three numbers. The first number is between 1 and 4, inclusive. The second number is between 3 and 8, inclusive. The third number is between 5 and 15, inclusive. How many different lottery numbers are possible? *(Lesson 4–2)* **264 numbers**

Solve each equation. Check your solution. *(Lesson 3–6)*

44. $8 + g = 17$ **9**
45. $b + (-5) = 26$ **31**
46. $k - 6 = -11$ **−5**

***Skills Practice*, p. 171, and *Practice*, p. 172 (shown)**

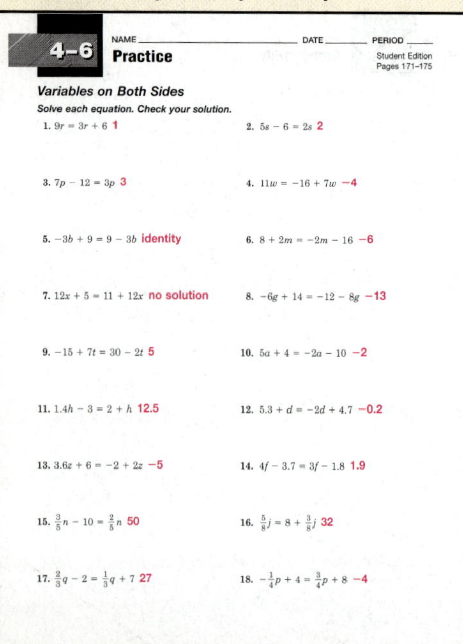

4–6 NAME ______ DATE ______ PERIOD ______

Practice

Student Edition Pages 171–175

Variables on Both Sides

Solve each equation. Check your solution.

1. $9r = 3r + 6$ **1**
2. $5s - 6 = 2s$ **2**
3. $7p - 12 = 3p$ **3**
4. $11w = -16 + 7w$ **−4**
5. $-3b + 9 = 9 - 3b$ **identity**
6. $8 + 2m = -2m - 16$ **−6**
7. $12x + 5 = 11 + 12x$ **no solution**
8. $-6g + 14 = -12 - 8g$ **−13**
9. $-15 + 7t = 30 - 2t$ **5**
10. $5a + 4 = -2a - 10$ **−2**
11. $1.4h - 3 = 2 + h$ **12.5**
12. $5.3 + d = -2d + 4.7$ **−0.2**
13. $3.6z + 6 = -2 + 2z$ **−5**
14. $4f - 3.7 = 3f - 1.8$ **1.9**
15. $\frac{3}{5}n - 10 = \frac{2}{5}n$ **50**
16. $\frac{5}{8}j = 8 + \frac{3}{8}j$ **32**
17. $\frac{2}{3}q - 2 = \frac{1}{3}q + 7$ **27**
18. $-\frac{1}{4}p + 4 = \frac{3}{4}p + 8$ **−4**

47. **Grid In** Most people drink more than the recommended eight glasses of beverage each day. However, some beverages actually make people thirsty. The graph shows the daily average glasses of beverages per person. Find the difference between the glasses of water and the glasses of caffeine soda that a person drinks on average per day. *(Lesson 3–2)* **3.3**

Some Drinks Leave You Thirsty

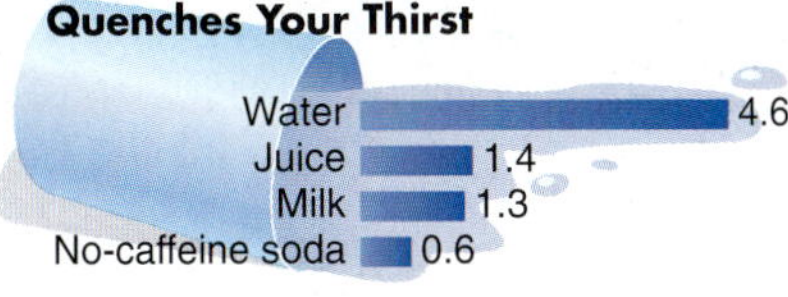

Source: Yankelovich for New York Hospital Nutrition Information Center

48. **Multiple Choice** The frequency table shows the results of a survey that asked students how many times a week they buy lunch at school. Determine which statement is correct. *(Lesson 1–6)* **B**

Ⓐ Most students buy lunch four times a week.

Ⓑ The greatest number of students buys lunch three times a week.

Ⓒ The least number of students buys lunch every day.

Ⓓ Five students buy lunch twice a week.

Buying Lunch

Number of Days	Tally	Frequency
0	III	3
1	~~IIII~~	5
2	~~IIII~~ I	6
3	~~IIII~~ ~~IIII~~	10
4	~~IIII~~ III	8
5	~~IIII~~	5

Quiz 2 Lessons 4–4 through 4–6

Solve each equation. Check your solution.

1. $-3t = 18$ **−6**
2. $\frac{c}{-10} = -5$ **50**
3. $\frac{5}{7}d = 45$ **63** *(Lesson 4–4)*
4. $2g + 11 = -7$ **−9**
5. $\frac{b}{4} - 11 = 7$ **72**
6. $-9 = \frac{d + 5}{3}$ **−32** *(Lesson 4–5)*
7. $5n + 4 = 7n - 20$ **12**
8. $8t + 15 = -9t - 2$ **−1**
9. $6a - 4 = 15 + 6a$ **no solution** *(Lesson 4–6)*
10. **Sports** According to NBA standards, a basketball should bounce back $\frac{2}{3}$ the distance from which it is dropped. How high should a basketball bounce when it is dropped from a height of $3\frac{3}{4}$ feet? *(Lesson 4–4)* **$2\frac{1}{2}$ ft**

www.algconcepts.com/self_check_quiz

4 ASSESS

Open-Ended Assessment

Speaking Ask students to describe the different types of solutions that are possible when solving equations and what results indicate each type of solution.

Quiz 2

The Quiz provides students with a brief review of the concepts and skills in Lessons 4–4 through 4–6. Lesson numbers are given to the right of the exercises or instruction lines so students can review concepts not yet mastered.

Extra Credit

Karla and her sister each start a savings account. Karla begins with $50 and plans to deposit an additional $20 per month. Her sister Beth begins with $200 and plans to deposit an additional $15 per month. After how many *years* will the two sisters have the same amount in their accounts? **2.5 yr**

***Enrichment*, p. 174**

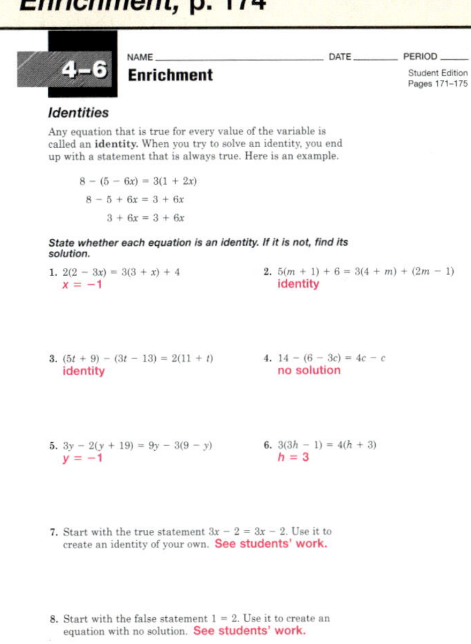

NAME ______ DATE ______ PERIOD ______

4–6 Enrichment

Student Edition Pages 171–175

Identities

Any equation that is true for every value of the variable is called an **identity.** When you try to solve an identity, you end up with a statement that is always true. Here is an example.

$$8 - (5 - 6x) = 3(1 + 2x)$$
$$8 - 5 + 6x = 3 + 6x$$
$$3 + 6x = 3 + 6x$$

State whether each equation is an identity. If it is not, find its solution.

1. $2(2 - 3x) = 3(3 + x) + 4$ **$x = -1$**
2. $5(m + 1) + 6 = 3(4 + m) + (2m - 1)$ **identity**
3. $(5t + 9) - (3t - 13) = 2(11 + t)$ **identity**
4. $14 - (6 - 3c) = 4c - c$ **no solution**
5. $3y - 2(y + 19) = 9y - 3(9 - y)$ **$y = -1$**
6. $3(3h - 1) = 4(h + 3)$ **$h = 3$**
7. Start with the true statement $3x - 2 = 3x - 2$. Use it to create an identity of your own. **See students' work.**
8. Start with the false statement $1 = 2$. Use it to create an equation with no solution. **See students' work.**

4–7 Grouping Symbols

1 FOCUS

5-Minute Check
Lesson 4–6

Solve each equation. Check your solution.

1. $3 + 5m = 8m - 9$ **4**
2. $9w + 6 = 9w + 3 - 5$ **no solution**
3. $x + 14.8 = 1.2 - 7x$ **−1.7**
4. The width of the rectangle below is half of the length. Write and solve an equation to find the dimensions of the rectangle.

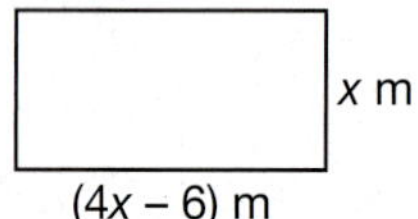

$x = \frac{1}{2}(4x - 6)$; 3 m by 6 m

5. The height of the triangle below is 2.5 times the length of the base. Write and solve an equation to find the dimensions of the triangle.

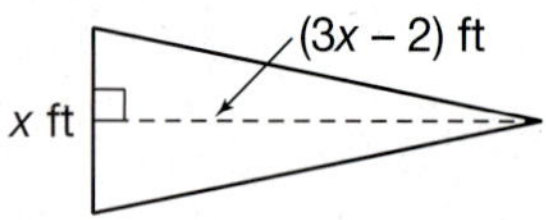

$3x - 2 = 2.5x$; base: 4 ft, height: 10 ft

Motivating the Lesson

Real-World Connection Tell students to imagine they are going to paint the walls of their room at home. Inform them that paint is sold in gallons and pints, with each container indicating how many square feet of wall space the paint in it will cover. Ask students how they would calculate the square footage of the wall space in their room. Use the suggestions to lead into the situation presented at the beginning of the lesson.

What You'll Learn
You'll learn to solve equations with grouping symbols.

Why It's Important
Painting Painters can solve equations by grouping symbols to determine how much paint to buy.

A gallon of paint covers about 350 square feet. Figure the room's square footage by multiplying the combined wall lengths by wall height and subtracting 15 square feet for each window and door.

Source: *The Family Handyman Magazine Presents Handy Hints for Home, Yard, and Workshop*

Suppose a room measures 16 feet long by 12 feet wide by 7 feet high and has two windows and a door. How many gallons of paint are needed to paint the room? To solve this problem, first find the square footage of the room. *The sum of the wall lengths is the perimeter of the room.*

square footage		*perimeter of room*		*wall height*		*window + window + door*
x	$=$	$2(16 + 12)$	$\cdot$	7	$-$	$(15 + 15 + 15)$

Now solve the equation to find x.

$x = 2(16 + 12) \cdot 7 - (15 + 15 + 15)$	*Original equation*
$x = 2(28) \cdot 7 - 45$	*Simplify inside the parentheses.*
$x = 56 \cdot 7 - 45$	*Multiply 2 and 28.*
$x = 392 - 45$	*Multiply 56 and 7.*
$x = 347$	*Subtract 45 from 392.*

The room has an area of 347 square feet. Therefore, only one gallon of paint is needed to paint the room.

In solving the equation above, the first step was to remove the parentheses. This is often the first step in solving any equation that contains grouping symbols.

Examples

Solve each equation. Check your solution.

1 $8 = 4(3x + 5)$

$8 = 4(3x + 5)$	*Original equation*
$8 = 12x + 20$	*Distributive Property*
$8 - 20 = 12x + 20 - 20$	*Subtract 20 from each side.*
$-12 = 12x$	*Simplify.*
$\frac{-12}{12} = \frac{12x}{12}$	*Divide each side by 12.*
$-1 = x$	*Simplify.*

Resource Manager

Reproducible Masters
Chapter 4 Resource Masters
- *Study Guide,* p. 175
- *Skills Practice,* p. 176
- *Practice,* p. 177
- *Reading to Learn Mathematics,* p. 178
- *Enrichment,* p. 179
- *Assessment,* p. 191

Transparencies
5-Minute Check, 4–7

Technology/Multimedia
AlgePASS, Lesson 12
Interactive Chalkboard CD-ROM

$$\begin{aligned}
\textbf{Check: } 8 &= 4(3x + 5) && \textit{Original equation} \\
8 &\stackrel{?}{=} 4[3(-1) + 5] && \textit{Replace } x \textit{ with } -1. \\
8 &\stackrel{?}{=} 4(-3 + 5) && \textit{Multiply.} \\
8 &\stackrel{?}{=} 4(2) && \textit{Add.} \\
8 &= 8 \ \checkmark && \textit{Multiply.}
\end{aligned}$$

2 $2(3n - 5) - 3(1 - n) = 5$

$$\begin{aligned}
2(3n - 5) - 3(1 - n) &= 5 && \textit{Original equation} \\
6n - 10 - 3 + 3n &= 5 && \textit{Distributive Property} \\
9n - 13 &= 5 && \textit{Add like terms.} \\
9n - 13 + 13 &= 5 + 13 && \textit{Add 13 to each side.} \\
9n &= 18 && \textit{Simplify.} \\
\frac{9n}{9} &= \frac{18}{9} && \textit{Divide each side by 9.} \\
n &= 2 && \textit{Simplify.}
\end{aligned}$$

$$\begin{aligned}
\textbf{Check: } 2(3n - 5) - 3(1 - n) &= 5 && \textit{Original equation} \\
2[3(2) - 5) - 3(1 - 2)] &\stackrel{?}{=} 5 && \textit{Replace } n \textit{ with 2.} \\
2(6 - 5) - 3(-1) &\stackrel{?}{=} 5 && \textit{Simplify.} \\
2(1) + 3 &\stackrel{?}{=} 5 && \textit{Simplify.} \\
5 &= 5 \ \checkmark
\end{aligned}$$

Your Turn

a. $7 = 3(x + 1) - 2$ **2**

b. $4(t + 5) + 6(2t - 3) = 12$ $\frac{5}{8}$

Building Link

3 The Rock and Roll Hall of Fame in Cleveland, Ohio, has a trapezoid-shaped cinema with an area of about 4675 square feet. Find x, the approximate width of the movie screen in the cinema.

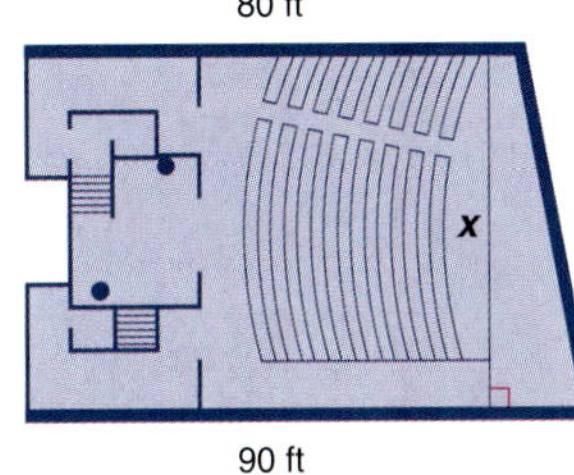

Rock and Roll Hall of Fame

$$\begin{aligned}
A &= \frac{1}{2}(b_1 + b_2)h && \textit{Formula for the area of a trapezoid} \\
4675 &= \frac{1}{2}(90 + 80)x && \textit{Replace } A \textit{ with 4675, } b_1 \textit{ with 90, } b_2 \textit{ with 80, and } h \textit{ with } x. \\
4675 &= \frac{1}{2}(170)x && \textit{Add 90 and 80.} \\
4675 &= 85x && \textit{Multiply } \tfrac{1}{2} \textit{ and 170.} \\
\frac{4675}{85} &= \frac{85x}{85} && \textit{Divide each side by 85.} \\
55 &= x && \textit{Simplify.}
\end{aligned}$$

The width of the movie screen is 55 feet.

www.algconcepts.com/extra_examples

2 TEACH

Teaching Tip When you present Example 1, some students may point out that the value of the expression $3x + 5$ must be 2 since $4 \cdot 2 = 8$. Point out that this is an alternative method for solving this *particular* equation, but that most other equations, like the one given in Example 2, cannot be solved in this way. Ask these students to apply their method again when presenting In-Class Example 1 below.

In-Class Examples

Examples 1–2

Solve each equation. Check your solution.

1 $5(2x - 1) = -25$ **−2**

2 $5(h + 6) - 6 = 3(5h - 2)$ **3**

Example 3

The area of the trapezoid below is 64 square millimeters. Find the value of x. **5**

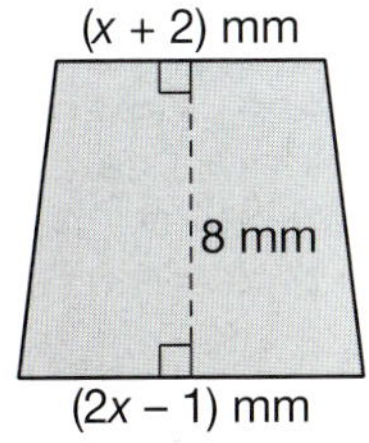

***Study Guide*, p. 175**

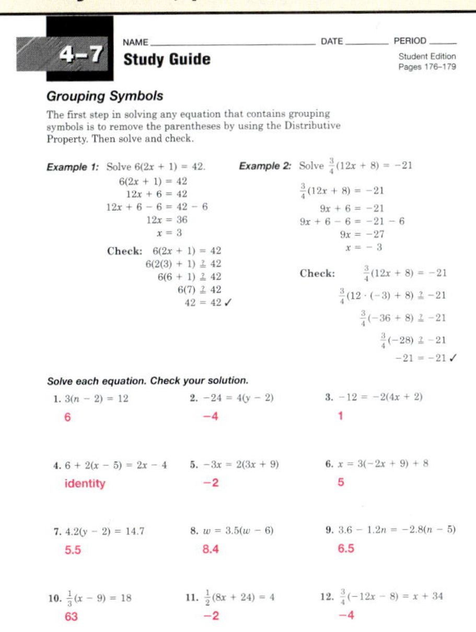

4–7 NAME ________ DATE ______ PERIOD ______

Study Guide

Student Edition Pages 176–179

Grouping Symbols

The first step in solving any equation that contains grouping symbols is to remove the parentheses by using the Distributive Property. Then solve and check.

Example 1: Solve $6(2x + 1) = 42$.

$$\begin{aligned}
6(2x + 1) &= 42 \\
12x + 6 &= 42 \\
12x + 6 - 6 &= 42 - 6 \\
12x &= 36 \\
x &= 3
\end{aligned}$$

$$\begin{aligned}
\textbf{Check: } 6(2x + 1) &= 42 \\
6(2(3) + 1) &\stackrel{?}{=} 42 \\
6(6 + 1) &\stackrel{?}{=} 42 \\
6(7) &\stackrel{?}{=} 42 \\
42 &= 42 \ \checkmark
\end{aligned}$$

Example 2: Solve $\frac{3}{4}(12x + 8) = -21$

$$\begin{aligned}
\frac{3}{4}(12x + 8) &= -21 \\
9x + 6 &= -21 \\
9x + 6 - 6 &= -21 - 6 \\
9x &= -27 \\
x &= -3
\end{aligned}$$

$$\begin{aligned}
\textbf{Check: } \frac{3}{4}(12x + 8) &= -21 \\
\frac{3}{4}(12 \cdot (-3) + 8) &\stackrel{?}{=} -21 \\
\frac{3}{4}(-36 + 8) &\stackrel{?}{=} -21 \\
\frac{3}{4}(-28) &\stackrel{?}{=} -21 \\
-21 &= -21 \ \checkmark
\end{aligned}$$

Solve each equation. Check your solution.

1. $3(n - 2) = 12$ **6**
2. $-24 = 4(y - 2)$ **−4**
3. $-12 = -2(4x + 2)$ **1**
4. $6 + 2(x - 5) = 2x - 4$ **identity**
5. $-3x = 2(3x + 9)$ **−2**
6. $x = 3(-2x + 9) + 8$ **5**
7. $4.2(y - 2) = 14.7$ **5.5**
8. $w = 3.5(w - 6)$ **8.4**
9. $3.6 - 1.2n = -2.8(n - 5)$ **6.5**
10. $\frac{1}{3}(x - 9) = 18$ **63**
11. $\frac{1}{2}(8x + 24) = 4$ **−2**
12. $\frac{3}{4}(-12x - 8) = x + 34$ **−4**

3 PRACTICE/APPLY

Error Analysis

Watch for students who have difficulty with Exercise 1 where a negative sign is being distributed. ***Prevent by*** having students first rewrite the expression as an addition expression: $w + (-8)(w + 15) = -9$. This makes it clear that the number being distributed is -8 and students should be able to see that they must distribute the negative sign to both terms inside the parentheses.

Assignment Guide

Basic:	11–33 odd, 34–40
Average:	12–30 even, 31–40

Answer

1. Use the Distributive Property to remove the parentheses.

Skills Practice, **p. 176, and *Practice,* p. 177 (shown)**

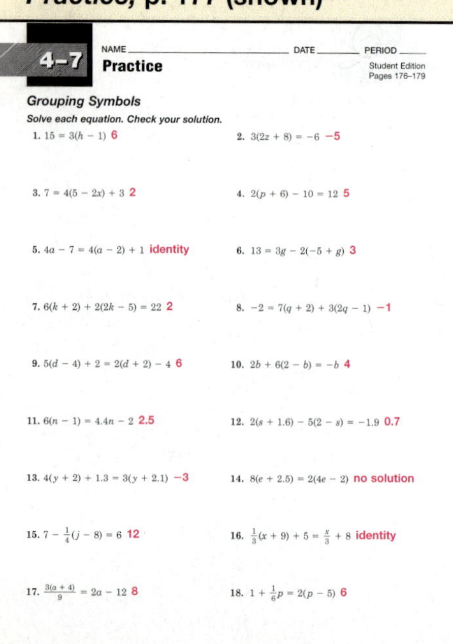

4-7 NAME ____________ DATE ______ PERIOD ____

Practice

Student Edition Pages 176–179

Grouping Symbols

Solve each equation. Check your solution.

1. $15 = 3(h - 1)$ **6**
2. $3(2z + 8) = -6$ **−5**
3. $7 = 4(5 - 2x) + 3$ **2**
4. $2(p + 6) - 10 = 12$ **5**
5. $4a - 7 = 4(a - 2) + 1$ **identity**
6. $13 = 3g - 2(-5 + g)$ **3**
7. $6(k + 2) + 2(2k - 5) = 22$ **2**
8. $-2 = 7(q + 2) + 3(2q - 1)$ **−1**
9. $5(d - 4) + 2 = 2(d + 2) - 4$ **6**
10. $2b + 6(2 - b) = -b$ **4**
11. $6(n - 1) = 4.4n - 2$ **2.5**
12. $2(s + 1.6) - 5(2 - s) = -1.9$ **0.7**
13. $4(y + 2) + 1.3 = 3(y + 2.1)$ **−3**
14. $8(e + 2.5) = 2(4e - 2)$ **no solution**
15. $7 - \frac{1}{4}(j - 8) = 6$ **12**
16. $\frac{1}{3}(x + 9) + 5 = \frac{x}{3} + 8$ **identity**
17. $\frac{3(a + 4)}{9} = 2a - 12$ **8**
18. $1 + \frac{1}{6}p = 2(p - 5)$ **6**

Check for Understanding

Communicating Mathematics

1. **State** the first step in solving $w - 8(w + 15) = -9$. **See margin.**
2. **Describe** a way to remove grouping symbols from an equation.
3. *Writing Math* Write a summary of the techniques you learned in this chapter for solving equations. **See students' work.**

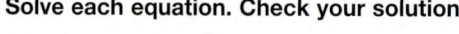

2. Sample answer: Use the Distributive Property.

Guided Practice Examples 1–3

Solve each equation. Check your solution.

4. $12 = 2(k + 1)$ **5**
5. $3n - 2 = 3(n + 1) - 5$ **identity**
6. $7 + 6x = 2(10 + 3x)$ **no solution**
7. $3p - 8(1 + p) = 12$ **−4**
8. $5(2k - 3) + 2(k + 4) = 17$ **2**
9. $5 - \frac{1}{2}(b - 6) = 4$ **8**

Example 3

10. **Geometry** Find the value of x in the trapezoid if its area is 39 square centimeters. **4**

x cm
6 cm
(x + 5) cm

Exercises

Practice

Solve each equation. Check your solution.

11. $3(a + 2) = 24$ **6**
12. $14 = 2(4n - 1)$ **2**
13. $7 = 5(x - 4) + 2$ **5**
14. $6(2 + y) - 4 = -10$ **−3**
15. $-2(6 + p) = 8$ **−10**
16. $5(z - 1) = 5z$ **no solution**
17. $t + 4(1 - t) = 9t$ $\frac{1}{3}$
18. $3(b - 2) - 4(2b + 5) = 9$ **−7**
19. $18 = 2(g + 4) + 3(2g - 6)$
20. $7(y - 2) + 8 = 3(y - 4) - 2$
21. $2.4h + 1 = 7(2h - 4)$ **2.5**
22. $2(f + 3) - 2.7 = 3(f + 1.1)$ **0**

B

23. $6(3.5 + r) - 1 = 7r - r$
24. $4(5 - x) - 2.1(6x) = -4.9$ **1.5**
25. $4(2a - 8) = 7a + 10$ **42**
26. $3 + \frac{5}{8}x = 2(x - 4)$ **8**

C

27. $3(1 - 4k) = \frac{1}{2}(-24k + 6)$ **identity**
28. $\frac{5(n - 1)}{3} = n + 1$ **4**
29. Find the solution of $2[3 + 4(g + 1)] = -2$. Check your solution. **−2**
30. What is the value of y in $6(3 + 5y) + 5(y - 8) = 13$? **1**

19. $\frac{7}{2}$ **or** $3\frac{1}{2}$
20. −2
23. no solution

Homework Help

For Exercises	See Examples
11–17, 29	1
18–28, 30	2
31–34	3

Extra Practice See page 700.

Applications and Problem Solving

31. **Geometry** The area of a trapezoid is 70 square inches. Find the lengths of the bases if base 1 is four more than twice the length of base 2, and the altitude is 5 inches. **base 1 = 20 in.; base 2 = 8 in.**

Reteaching Activity

Logical Learners Have students write an algorithm to use when solving an equation. Classmates can then exchange algorithms and test them on some of the equations in this lesson.

32. **Gardening** Carlos is building a fence around his garden to keep rabbits out. One side of the garden is against the garage, as shown in the diagram at the right. He has 88 feet of fencing to fit around the three other sides.

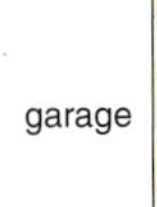

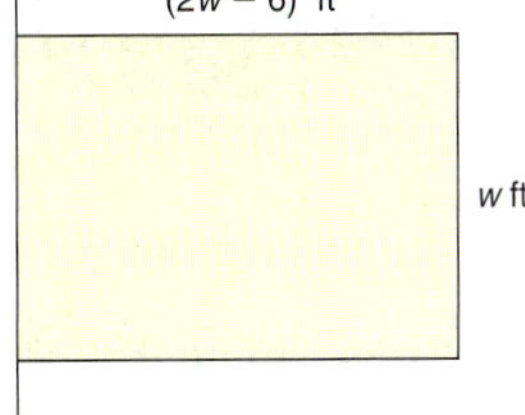

Full Rectangle
P = 2L + 2W

32a. $w + 2(2w - 6) = 88$

P = 2L + W

88 =

a. Write an equation to find the width w of the garden.

b. What are the dimensions of the garden? **20 ft by 34 ft**

33. **Number Theory** Twice the greater of two consecutive odd integers is 13 less than three times the lesser. Find the integers. **17, 19**

34. **Critical Thinking** *Complementary angles* are two angles whose measures have a sum of 90°. Each angle is said to be the *complement* of the other angle. The measure of angle A is one-half the measure of its complement. Find the measure of angle A and the measure of its complement. **30, 60**

Mixed Review

Solve each equation. Check your solution. *(Lesson 4–6)*

35. $2n + 3 = 6 - n$ **1**

36. $5t - 12 = 8t$ **−4**

37. **Number Theory** Four less than one-half of a number x is 7. *(Lesson 4–5)*

a. Write an equation to represent the problem. $\frac{1}{2}x - 4 = 7$

b. What is the number? **22**

38. Evaluate $\frac{ab}{-3}$ if $a = 6$ and $b = -5$. *(Lesson 2–6)* **10**

Standardized Test Practice

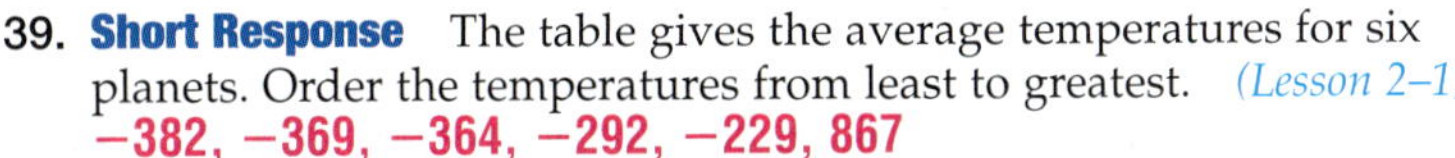

39. **Short Response** The table gives the average temperatures for six planets. Order the temperatures from least to greatest. *(Lesson 2–1)*
−382, −369, −364, −292, −229, 867

Planet	Average Temperature (°F)
Jupiter	−229
Neptune	−364
Pluto	−382
Saturn	−292
Uranus	−369
Venus	867

40. **Multiple Choice** A pitcher's earned-run average (ERA) is given by the formula $ERA = 9\left(\frac{a}{b}\right)$, where a is the number of earned runs the pitcher has allowed and b is the number of innings pitched. Find a pitcher's ERA to the nearest hundredth if he has allowed 23 earned runs in 180 innings. *(Lesson 1–5)* **C**

(A) 0.01 (B) 0.13 (C) 1.15 (D) 7.83

www.algconcepts.com/self_check_quiz

4 ASSESS

Open-Ended Assessment

Writing Write an equation similar to the exercises in this lesson on the board or overhead. Ask students to solve the equation, identifying where in their solution steps they used the Distributive Property.

Chapter 4, Quiz B (Lessons 4–4 through 4–7) is available in the *Chapter 4 Resource Masters*, p. 191.

Extra Credit

The area of a trapezoid is 90 square inches. If the length of the shorter base is twice the height of the trapezoid and the length of the longer base is 3 times the height, what is the height of the trapezoid? **6 in.**

***Enrichment*, p. 179**

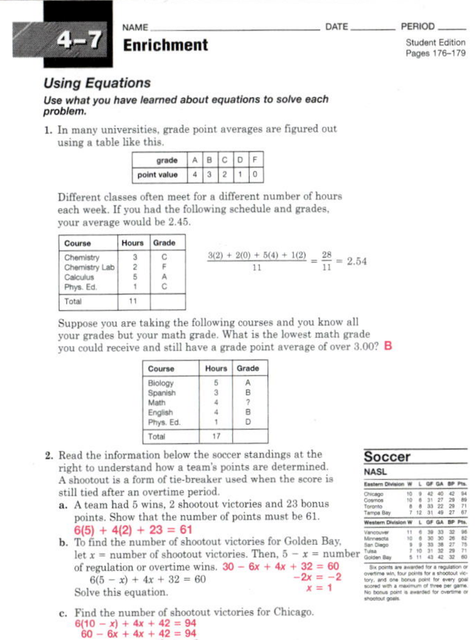

4–7 NAME ______________ DATE ________ PERIOD ______

Enrichment

Student Edition
Pages 176–179

Using Equations

Use what you have learned about equations to solve each problem.

1. In many universities, grade point averages are figured out using a table like this.

grade	A	B	C	D	F
point value	4	3	2	1	0

Different classes often meet for a different number of hours each week. If you had the following schedule and grades, your average would be 2.45.

Course	Hours	Grade
Chemistry	3	C
Chemistry Lab	2	F
Calculus	5	A
Phys. Ed.	1	C
Total	11	

$$\frac{3(2) + 2(0) + 5(4) + 1(2)}{11} = \frac{28}{11} = 2.54$$

Suppose you are taking the following courses and you know all your grades but your math grade. What is the lowest math grade you could receive and still have a grade point average of over 3.00? **B**

Course	Hours	Grade
Biology	5	A
Spanish	3	B
Math	4	?
English	4	B
Phys. Ed.	1	D
Total	17	

2. Read the information below the soccer standings at the right to understand how a team's points are determined. A shootout is a form of tie-breaker used when the score is still tied after an overtime period.

a. A team had 5 wins, 2 shootout victories and 23 bonus points. Show that the number of points must be 61.
$6(5) + 4(2) + 23 = 61$

b. To find the number of shootout victories for Golden Bay, let x = number of shootout victories. Then, $5 - x$ = number of regulation or overtime wins.
$6(5 - x) + 4x + 32 = 60$
Solve this equation.
$30 - 6x + 4x + 32 = 60$
$-2x = -2$
$x = 1$

c. Find the number of shootout victories for Chicago.
$6(10 - x) + 4x + 42 = 94$
$60 - 6x + 4x + 42 = 94$
$-2x = -8$
$x = 4$

Soccer

NASL

Eastern Division	W	L	GF	GA	BP	Pts.
Chicago	10	9	42	40	42	94
Cosmos	10	6	31	27	29	89
Toronto	8	8	33	22	29	71
Tampa Bay	7	12	31	49	27	67
Western Division	**W**	**L**	**GF**	**GA**	**BP**	**Pts.**
Vancouver	11	6	39	33	32	96
Minnesota	10	6	30	30	26	82
San Diego	9	9	33	38	27	75
Tulsa	7	10	31	32	29	71
Golden Bay	5	11	43	42	32	60

Six points are awarded for a regulation or overtime win, four points for a shootout victory, and one bonus point for every goal scored with a maximum of three per game. No bonus point is awarded for overtime or shootout goals.

Study Guide and Assessment

Understanding and Using the Vocabulary

This section provides a listing of the new terms, properties, and phrases that were introduced in this chapter. The exercises check students' understanding of the terms by using a variety of verbal formats including matching, completion, and true/false.

Glossary A complete glossary of terms appears on pages 762–783.

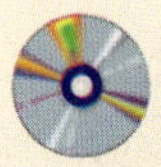

MindJogger Videoquizzes

MindJogger Videoquizzes provide an alternative review of concepts presented in this chapter. Students work in teams to answer questions, gaining points for correct answers.

Understanding and Using the Vocabulary

After completing this chapter, you should be able to define each term, property, or phrase and give an example or two of each.

interNET CONNECTION **Review Activities** For more review activities, visit: www.algconcepts.com

Algebra

consecutive integers *(p. 167)*
identity *(p. 172)*
multiplicative inverses *(p. 154)*
reciprocals *(p. 154)*

Counting

combination *(p. 152)*
factorial *(p. 153)*
event *(p. 147)*
outcomes *(p. 146)*
permutation *(p. 152)*
sample space *(p. 146)*
tree diagram *(p. 146)*

Complete each sentence using a term from the vocabulary list.

1. A subset of the possible outcomes of a sample space is called a(n) __?__. **event**
2. Two numbers whose product is 1 are called __?__. **multiplicative inverses or reciprocals**
3. To find the sample space of a particular situation, a(n) __?__ may be used. **tree diagram**
4. Integers in counting order are called __?__. **consecutive integers**
5. An equation that is true for every value of the variable is called a(n) __?__. **identity**
6. The __?__ is the list of all the possible outcomes. **sample space**
7. The results shown by a tree diagram are called __?__. **outcomes**
8. The numbers $-\frac{4}{3}$ and $-\frac{3}{4}$ are called __?__. **multiplicative inverses or reciprocals**
9. To find the possible number of __?__ by multiplying, the Fundamental Counting Principle is used. **outcomes**
10. The equation $\frac{1}{2}x + 1 = \frac{1}{2}x + 1$ is an example of a(n) __?__. **identity**

Skills and Concepts

Objectives and Examples

- **Lesson 4–1** Multiply rational numbers.

$-3.2(-4) = 12.8$ *The signs are the same. The product is positive.*

$\frac{4}{5}\left(-\frac{2}{9}\right) = -\frac{4 \cdot 2}{5 \cdot 9}$

$= -\frac{8}{45}$ *The signs are different. The product is negative.*

Review Exercises

Find each product.

11. $6(3.8)$ **22.8**
12. $-9.2 \cdot 0$ **0**
13. $-1\left(\frac{3}{4}\right)$ $-\frac{3}{4}$
14. $-\frac{5}{6}\left(-\frac{1}{3}\right)$ $\frac{5}{18}$

Simplify each expression. **18.** $\frac{9}{40}mn$

15. $-2.7(4y)$ **$-10.8y$**
16. $5a(3.4b)$ **$17ab$**
17. $6s\left(-\frac{1}{7}\right)$ $-\frac{6}{7}s$
18. $\left(-\frac{3}{8}m\right)\left(-\frac{3}{5}n\right)$

Resource Manager

Reproducible Masters

Chapter 4 Resource Masters

- *Assessment,* pp. 181–189, 192–194

Technology/Multimedia

MindJogger Videoquizzes

ExamView® Pro

Objectives and Examples

- **Lesson 4–2** Use tree diagrams or the Fundamental Counting Principle to count outcomes.

Shirt	Pants	Socks
red green white	black blue	red green white

There are 3 shirts, 2 pants, and 3 socks, so the number of different combinations is $3 \times 2 \times 3$ or 18.

Review Exercises

19. A sandwich can be made with wheat or white bread. It can have ham, turkey, or beef. Find the number of possible outcomes by drawing a tree diagram. **See margin.**
20. Find the number of different team jerseys. **32**

Shirt	Stripe	Numbers
white blue	green yellow gold red	green black gold red

- **Lesson 4–3** Divide rational numbers.

Evaluate $\frac{t}{3}$ if $t = \frac{1}{5}$.

$\frac{t}{3} = \frac{\frac{1}{5}}{3}$ *Replace t with $\frac{1}{5}$.*

$= \frac{1}{5} \div 3$ *Rewrite as division.*

$= \frac{1}{5} \cdot \frac{1}{3}$ or $\frac{1}{15}$ *Multiply by reciprocal.*

Find each quotient.

21. $5 \div (-2.5)$ **−2**
22. $-7.2 \div (-1.6)$ **4.5**
23. $\frac{4}{5} \div \frac{3}{2}$ $\frac{8}{15}$
24. $3\frac{1}{2} \div \left(-\frac{1}{2}\right)$ **−7**

Evaluate each expression if $x = \frac{2}{3}$ and $y = -\frac{1}{2}$.

25. $\frac{y}{4}$ $-\frac{1}{8}$
26. $\frac{5}{x}$ $\frac{15}{2}$ or $7\frac{1}{2}$
27. $\frac{y}{x}$ $-\frac{3}{4}$

- **Lesson 4–4** Solve multiplication and division equations by using the properties of equality.

$\frac{x}{3} = -12$

$3\left(\frac{x}{3}\right) = 3(-12)$ *Multiply each side by 3.*

$x = -36$ *Simplify.*

Solve each equation. Check your solution.

28. $5n = 30$ **6**
29. $16 = 2a$ **8**
30. $-k = 5$ **−5**
31. $-8y = 32$ **−4**
32. $3t = 4.2$ **1.4**
33. $-2x = -24.6$ **12.3**
34. $\frac{m}{4} = -6$ **−24**
35. $\frac{5}{9}r = 5$ **9**

- **Lesson 4–5** Solve equations involving more than one operation.

$\frac{x}{4} + 6 = 18$

$\frac{x}{4} + 6 - 6 = 18 - 6$ *Subtract 6 from each side.*

$4\left(\frac{x}{4}\right) = 4(12)$ *Multiply each side by 4.*

$x = 48$ *Simplify.*

Solve each equation. Check your solution.

36. $7t + 3 = 24$ **3**
37. $1 + 6y = -11$ **−2**
38. $8 - 2b = 10$ **−1**
39. $-4y + 1 = 1$ **0**
40. $2.1 - 3r = 4.8$
41. $-1.8 = 5.1w - 12$
42. $14 = 6 + \frac{x}{5}$ **40**
43. $\frac{4 + x}{7} = -3$ **−25**

40. −0.9 41. 2

Skills and Concepts

The **Objectives and Examples** section reviews the skills and concepts of the chapter and shows completely worked examples.

The **Review Exercises** provide practice for the corresponding objectives.

Answer

19. 6 outcomes

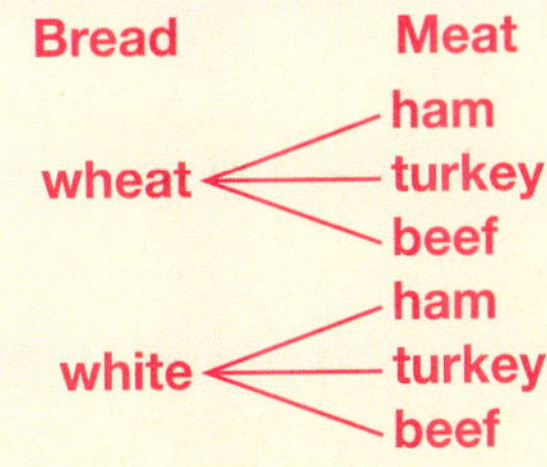

Mixed Problem Solving
See pages 724–731.

Applications and Problem Solving

This section provides additional practice in solving real-world problems that involve the concepts of this chapter.

Objectives and Examples

• **Lesson 4-6** Solve equations with variables on both sides.

$$\begin{aligned} 7x &= 4x - 3 \\ 7x - 4x &= 4x - 3 - 4x && \text{Subtract } 4x \text{ from each side.} \\ 3x &= -3 \\ \frac{3x}{3} &= \frac{-3}{3} && \text{Divide each side by 3.} \\ x &= -1 && \text{Simplify.} \end{aligned}$$

Review Exercises

Solve each equation. Check your solution.

44. $3a = a + 4$ **2**
45. $2t + 4 = 10 + t$ **6**
46. $11 + 4m = 4m + 8$ **no solution**
47. $5x - 5.2 = 9.4 + 3x$ **7.3**
48. $6 + \frac{1}{3}t = -2 + \frac{2}{3}t$ **24**

• **Lesson 4-7** Solve equations with grouping symbols.

$$\begin{aligned} 5(x - 7) &= 10 \\ 5x - 35 &= 10 && \text{Distributive Property} \\ 5x - 35 + 35 &= 10 + 35 && \text{Add 35 to each side.} \\ 5x &= 45 && \text{Simplify.} \\ \frac{5x}{5} &= \frac{45}{5} && \text{Divide each side by 5.} \\ x &= 9 && \text{Simplify.} \end{aligned}$$

Solve each equation. Check your solution.

49. $6a = 2(4 - a)$ **1**
50. $10 = 5(n + 3)$ **−1**
51. $4(3h - 2) = 7h + 2$ **2**
52. $2(1 + 2x) + 3(x - 4) = 11$ **3**
53. $\frac{1}{3}(12 - 6t) = 4 - 2t$ **identity**

Applications and Problem Solving

54. **Dining** With each shrimp, salmon, or perch dinner at the Seafood Palace, you may have soup or salad. You may also have broccoli, baked potato, or rice. How many different dinners are possible? *(Lesson 4–2)* **18**

55. **Space** Halley's Comet flashes through the sky every 76.3 years. How many times will it appear in 381.5 years? *(Lesson 4–3)* **5 times**

56. **Number Theory** Four more than two-thirds of a number n is zero. *(Lesson 4–5)*
    a. Write an equation to represent the problem. $\frac{2}{3}n + 4 = 0$
    b. What is the value of n? **−6**

57. **Geometry** The width of the rectangle is equal to one-half the length. What is the width of the rectangle? *(Lesson 4–6)* **6 in.**

(x + 4) in.

6x in.

Assessment, pp. 183–184

4

NAME ________ DATE ______ PERIOD ____

Chapter 4 Test, Form 1B

Write the letter for the correct answer in the blank at the right of each problem.

1. Simplify $(9x)(4.5y)$.
   A. $13.5xy$ B. $2xy$ C. $40.5xy$ D. $40.5x + y$ 1. C
2. Find $\left(\frac{2}{5}\right)\left(\frac{15}{8}\right)$.
   A. $\frac{16}{75}$ B. $\frac{3}{4}$ C. $\frac{6}{7}$ D. $\frac{17}{13}$ 2. B
3. Evaluate $8.4a - 0.5ab$ if $a = 5$ and $b = 4$.
   A. 22 B. 30 C. 32 D. 40 3. C
4. It takes $\frac{3}{8}$ cup of syrup to make one gallon of punch. How many cups of syrup does it take to make six gallons of punch?
   A. $1\frac{1}{8}$ B. $2\frac{1}{8}$ C. $2\frac{1}{4}$ D. $2\frac{1}{2}$ 4. C
5. At the appetizer table are carrots, celery, broccoli, and red pepper along with two dips, dill and blue cheese. How many different choices of dip and vegetable are available?
   A. 6 B. 8 C. 10 D. 12 5. B
6. Find the number of possible arrangements of two songs if you pick the first song from a CD that has 12 tracks and the second from a CD that has 9 tracks.
   A. 21 B. 81 C. 108 D. 120 6. C
7. Evaluate $-\frac{j}{k}$ if $j = \frac{2}{5}$ and $k = -\frac{3}{10}$.
   A. $\frac{4}{3}$ B. $\frac{3}{25}$ C. $-\frac{3}{25}$ D. $-\frac{4}{3}$ 7. A
8. Find $-6.8 \div 17$.
   A. −4 B. −0.4 C. −0.34 D. 4 8. B
9. It takes $\frac{5}{8}$ yard of canvas to make a tote bag. If Sara has 10 yards of canvas, how many tote bags can she make?
   A. 12 B. 14 C. 16 D. 18 9. C
10. Solve $-1.6r = 4$.
    A. −6.4 B. −2.5 C. −2.4 D. 6.4 10. B
11. How many \$8.75 calendars can you buy with \$100?
    A. 9 B. 10 C. 11 D. 85 11. C

Assessment

Four forms of Chapter 4 Test are available in the *Chapter 4 Resource Masters.*

Chapter 4 Test, Form 1B, is shown at the left. Chapter 4 Test, Form 2B, is shown on the next page.

	Form of Test		Level
1A	Multiple Choice	pp. 181–182	Average
1B	Multiple Choice	pp. 183–184	Basic
2A	Free Response	pp. 185–186	Average
2B	Free Response	pp. 187–188	Basic

CHAPTER 4 Test

1. Write an equation using grouping symbols that results in $x = -5$. **Sample answer: $2(x + 1) = -8$**
2. Draw a tree diagram to show the different calculators that are possible from the choices given in the table.

Brand	Type	Power
A B C	scientific graphing	solar battery

Exercise 2

3. Explain whether the equation $3(2x - 3) - 1 = 6x - 10$ is an identity or has no solution. **Identity; the equation simplifies to $6x - 10 = 6x - 10$, which is true for every value of x.**

2. 12 calculators; See margin for tree diagram.

Simplify each expression.

4. $7.2(-4x)$ **$-28.8x$**
5. $-3.9(-10t)$ **$39t$**
6. $-\frac{4}{5}a\left(\frac{2}{3}b\right)$ **$-\frac{8}{15}ab$**
7. $\frac{2}{3}\left(\frac{1}{2}g\right) + \frac{1}{3}g$ **$\frac{2}{3}g$**

8. Find the number of possible outfits by drawing a tree diagram. **9 outfits; See margin for tree diagram.**

Shirt	Pants
red	blue
yellow	black
green	gray

Find the number of possible outcomes by using the Fundamental Counting Principle.

9. rolling an eight-sided die twice **64**
10. spinning the spinner three times, and it never lands on a border **64**

Evaluate each expression if $f = \frac{1}{2}$, $g = \frac{2}{5}$, and $h = -\frac{1}{3}$.

11. $\frac{f}{5}$ **$\frac{1}{10}$**
12. $\frac{3}{h}$ **-9**
13. $\frac{g}{f}$ **$\frac{4}{5}$**
14. $\frac{h}{g}$ **$-\frac{5}{6}$**

Solve each equation.

15. $5n = 35$ **7**
16. $18.4 = -2m$ **-9.2**
17. $\frac{3}{4}x = 6$ **8**
18. $20 = -\frac{5}{4}a$ **-16**
19. $3r + 1 = 7$ **2**
20. $13 - 4y = 29$ **-4**
21. $10 = 7 + \frac{5}{8}w$ **$\frac{24}{5}$ or $4\frac{4}{5}$**
22. $7b + 8 = 5b - 4$ **-6**
23. $2(h - 3) = 5 + 13h$ **-1**

24. **Number Theory** Find three consecutive even integers whose sum is -12. **$-6, -4, -2$**

25. **Geometry** Find the dimensions of the rectangle if the perimeter is 220 feet. **50 ft by 60 ft**

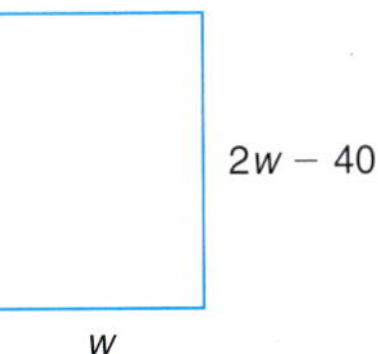

www.algconcepts.com/chapter_test

Chapter Test Bonus Question

Your favorite professional baseball team is involved in a best-of-five playoff series. Whichever team wins three games first wins the series. How many ways are possible for your team to win the series? **10**

Chapter Test

Answers

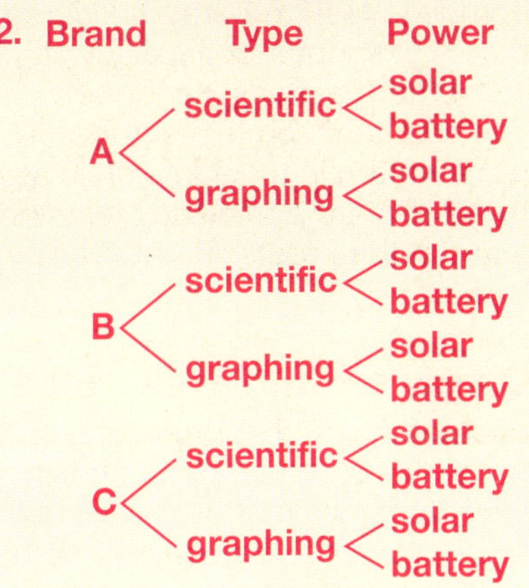

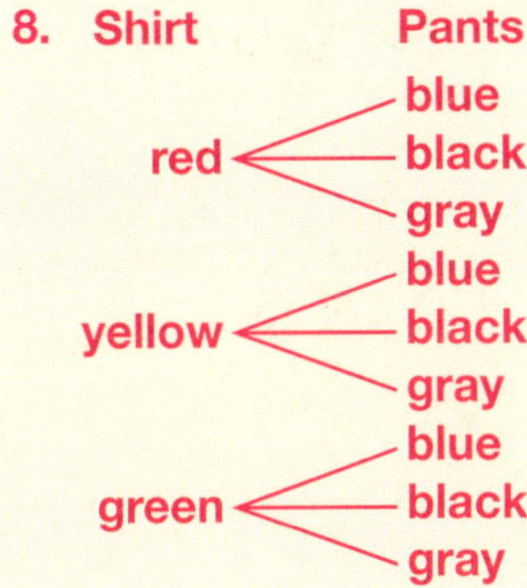

***Assessment*, pp. 187–188**

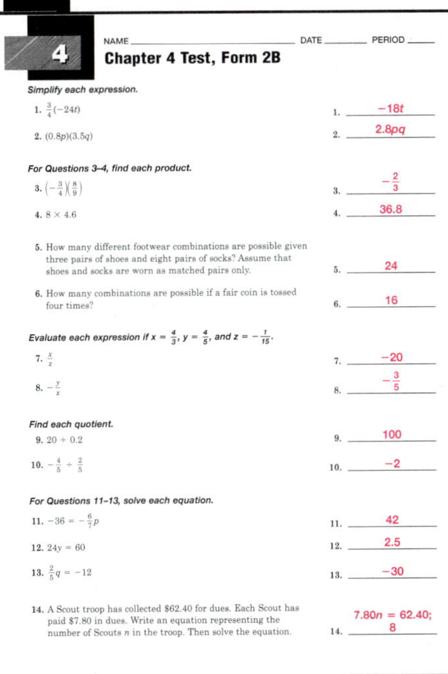

NAME ______ DATE ______ PERIOD ______

4 Chapter 4 Test, Form 2B

Simplify each expression.

1. $\frac{3}{4}(-24t)$ — 1. $-18t$
2. $(0.8p)(3.5q)$ — 2. $2.8pq$

For Questions 3–4, find each product.

3. $\left(-\frac{3}{4}\right)\left(\frac{8}{9}\right)$ — 3. $-\frac{2}{3}$
4. 8×4.6 — 4. 36.8
5. How many different footwear combinations are possible given three pairs of shoes and eight pairs of socks? Assume that shoes and socks are worn as matched pairs only. — 5. 24
6. How many combinations are possible if a fair coin is tossed four times? — 6. 16

Evaluate each expression if $x = \frac{4}{3}$, $y = \frac{4}{5}$, and $z = -\frac{1}{15}$.

7. $\frac{x}{z}$ — 7. -20
8. $-\frac{y}{x}$ — 8. $-\frac{3}{5}$

Find each quotient.

9. $20 \div 0.2$ — 9. 100
10. $-\frac{4}{5} \div \frac{2}{5}$ — 10. -2

For Questions 11–13, solve each equation.

11. $-36 = -\frac{6}{7}p$ — 11. 42
12. $24y = 60$ — 12. 2.5
13. $\frac{2}{5}q = -12$ — 13. -30
14. A Scout troop has collected \$62.40 for dues. Each Scout has paid \$7.80 in dues. Write an equation representing the number of Scouts n in the troop. Then solve the equation. — 14. $7.80n = 62.40$; 8

Preparing for Standardized Tests

Pages 184–185 are part of a complete test preparation course that is described in detail on page T9 of the Teacher's Handbook. The test items on these pages were written in the same style as those in state proficiency tests and standardized tests like ACT and SAT.

Diagnosis and Prescription

Each of the 10 test questions on page 185 is cross-referenced to the lesson where that SAT or ACT skill is covered. If students miss a particular type of problem, you can have them study that skill.

(See chart at the bottom of page 185. Note that SPT = State Proficiency Test, SAT = Scholastic Assessment Test, and ACT = American College Test.)

***Assessment*, p. 192**

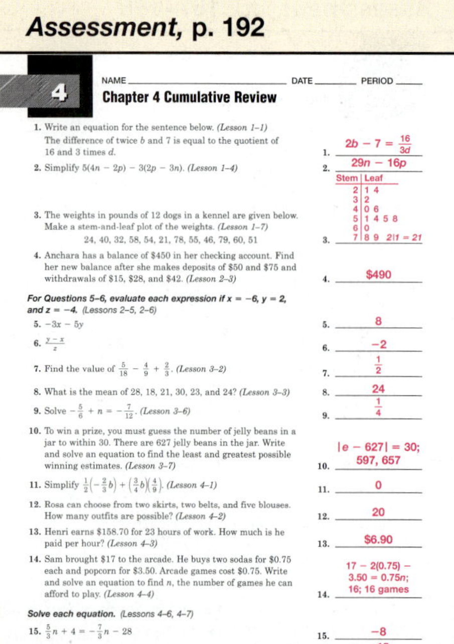

NAME ______ DATE ______ PERIOD ______

4 Chapter 4 Cumulative Review

1. Write an equation for the sentence below. *(Lesson 1–1)* The difference of twice b and 7 is equal to the quotient of 16 and 3 times d. — 1. $2b - 7 = \frac{16}{3d}$
2. Simplify $5(4n - 2p) - 3(2p - 3n)$. *(Lesson 1–4)* — 2. $29n - 16p$
3. The weights in pounds of 12 dogs in a kennel are given below. Make a stem-and-leaf plot of the weights. *(Lesson 1–7)* 24, 40, 32, 58, 54, 21, 78, 55, 46, 79, 60, 51 — 3.

Stem	Leaf
2	1 4
3	2
4	0 6
5	1 4 5 8
6	0
7	8 9

$2|1 = 21$

4. Anchara has a balance of $450 in her checking account. Find her new balance after she makes deposits of $50 and $75 and withdrawals of $15, $28, and $42. *(Lesson 2–3)* — 4. $490

For Questions 5–6, evaluate each expression if x = −6, y = 2, and z = −4. *(Lessons 2–5, 2–6)*

5. $-3x - 5y$ — 5. 8
6. $\frac{y - x}{z}$ — 6. −2
7. Find the value of $\frac{5}{18} - \frac{4}{9} + \frac{2}{3}$. *(Lesson 3–2)* — 7. $\frac{1}{2}$
8. What is the mean of 28, 18, 21, 30, 23, and 24? *(Lesson 3–3)* — 8. 24
9. Solve $-\frac{5}{6} + n = -\frac{7}{12}$. *(Lesson 3–6)* — 9. $\frac{1}{4}$
10. To win a prize, you must guess the number of jelly beans in a jar to within 30. There are 627 jelly beans in the jar. Write and solve an equation to find the least and greatest possible winning estimates. *(Lesson 3–7)* — 10. $|e - 627| = 30$; 597, 657
11. Simplify $\frac{1}{2}\left(-\frac{2}{3}b\right) + \left(\frac{3}{4}b\right)\left(\frac{4}{9}\right)$. *(Lesson 4–1)* — 11. 0
12. Rosa can choose from two skirts, two belts, and five blouses. How many outfits are possible? *(Lesson 4–2)* — 12. 20
13. Henri earns $158.70 for 23 hours of work. How much is he paid per hour? *(Lesson 4–3)* — 13. $6.90
14. Sam brought $17 to the arcade. He buys two sodas for $0.75 each and popcorn for $3.50. Arcade games cost $0.75. Write and solve an equation to find n, the number of games he can afford to play. *(Lesson 4–4)* — 14. $17 - 2(0.75) - 3.50 = 0.75n$; 16; 16 games

Solve each equation. *(Lessons 4–6, 4–7)*

15. $\frac{5}{3}n + 4 = -\frac{7}{3}n - 28$ — 15. −8
16. $180 = 15(2 - t)$ — 16. −10

Statistics Problems

Test-Taking Tip

On standardized tests, *average* is the same as *mean*.

Many standardized test questions provide a table or a list of numerical data and ask you to find the *mean*, *median*, *mode*, and *range*.

$$\text{mean} = \frac{\text{sum of the numbers}}{\text{number of numbers}}$$

Example 1

The average speeds of the winners of the Indianapolis 500 from 1994 to 2003 are listed in the table. What is the median of the speeds?

Indianapolis 500 Winners			
Year	**Average Speed (mph)**	**Year**	**Average Speed (mph)**
1994	160.9	1999	153.2
1995	153.6	2000	167.6
1996	148.0	2001	141.6
1997	145.8	2002	166.5
1998	145.2	2003	156.3

Ⓐ 153.2 Ⓑ 153.4
Ⓒ 153.6 Ⓓ 153.87

Hint The median of an even number of values is the average (mean) of the middle two values when they are arranged in numeric order.

Solution Don't be confused by the phrase *average speed*; think of it as the winning speed. Write the winning speeds in order from least to greatest.

141.6, 145.2, 145.8, 148.0, <u>153.2</u>, <u>153.6</u>, 156.3, 160.9, 165.5, 167.6

There are ten values, so the median is the mean of the two middle values.

$$\frac{153.2 + 153.6}{2} = \frac{306.8}{2} \text{ or } 153.4$$

The median is 153.4 mph. The answer is B.

Example 2

A survey of the town of Niceville found a mean of 3.2 persons per household and a mean of 1.2 televisions per household. If 48,000 people live in Niceville, how many televisions are there in Niceville?

Ⓐ 15,000 Ⓑ 16,000 Ⓒ 18,000
Ⓓ 40,000 Ⓔ 57,600

Hint Use your calculator for large numbers.

Solution Solve this problem one step at a time.

You need to find the number of televisions. The survey information is based on households, so first calculate the number of households.

number of people	*divided by*	*people per household*	*equals*	*number of households*
48,000	÷	3.2	=	15,000

The survey information says there is a mean (average) of 1.2 televisions per household. Multiply the number of households by the televisions per household.

number of households	*times*	*televisions per household*	*equals*	*number of televisions*
15,000	×	1.2	=	18,000

The answer is C.

Resource Manager

Reproducible Masters

Chapter 4 Resource Masters

- *Assessment*, pp. 192–194

Preparing for Standardized Tests
For test-taking strategies and more practice, see pages 732–749.

After you work each problem, record your answer on the answer sheet provided or on a sheet of paper.

Multiple Choice

1. A company plans to open a new fitness center. It conducts a survey of the number of hours people exercise weekly. The results for 12 people chosen at random are 7, 2, 4, 8, 3, 0, 3, 1, 0, 5, 3, and 4 hours. What is the mode of the data? *(Lesson 3–3)* **A**

Ⓐ 3 Ⓑ 3.3 Ⓒ 4
Ⓓ There is no mode.

2. For their science fiction book reports, students must choose one book from a list of eight. The books have the following numbers of pages: 272, 188, 164, 380, 442, 216, 360, and 262. What is the median number of pages? *(Lesson 3–3)* **A**

Ⓐ 267 Ⓑ 285.5 Ⓒ 303 Ⓓ 442

3. $\frac{0.5 + 0.5 + 0.5 + 0.5}{4} =$ *(Lesson 4–3)* **C**

Ⓐ 0.05 Ⓑ 0.125 Ⓒ 0.5
Ⓓ 1 Ⓔ 2.0

4. The graph shows the percent of men and women who held multiple jobs. In what year was the total percent of men and women holding multiple jobs greatest? *(Lesson 3–2)* **C**

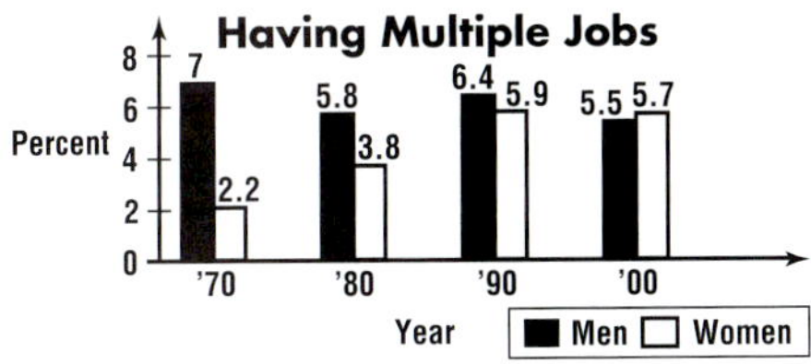

Source: U.S. Bureau of Labor Statistics

Ⓐ 1970 Ⓑ 1980
Ⓒ 1990 Ⓓ 2000

5. If $\frac{12}{4} = x$, then $4x + 2 =$ *(Lesson 4–3)* **E**

Ⓐ 2. Ⓑ 3. Ⓒ 4. Ⓓ 12. Ⓔ 14.

6. Over nine games, a baseball team had an average of 8 runs per game. In the first seven games, they had an average of 6 runs per game. They scored the same number of runs in each of the last two games. How many runs did they score in the last game? *(Lesson 3–3)* **B**

Ⓐ 5 Ⓑ 15 Ⓒ 26 Ⓓ 30 Ⓔ 46

7. The stem-and-leaf plot shows the populations of ten cities in Oklahoma. Which statement is correct? *(Lesson 3–3)* **D**

Stem	Leaf
35	6
36	2 4
37	5 6 7
38	1 2 3 6

35|6 = 35,600

Ⓐ The mode is 38,000.
Ⓑ The range of the data is 35,600.
Ⓒ Most of these cities have fewer than 36,000 people.
Ⓓ The median population is 37,650.

8. If $5 + \frac{n}{4} = 9$, then what is the value of n? *(Lesson 4–5)* **C**

Ⓐ 3.5 Ⓑ 4 Ⓒ 16 Ⓓ 56

Grid In

9. If $3x = 12$, then $8 \div x = ?$ *(Lesson 4–4)* **2**

Extended Response

10. The number of daily visitors to the local history museum during the month of June is shown below. *(Lesson 1–7)*

11 19 61 18 43 22 18 14 21 8
19 41 36 16 16 14 24 31 64 7
29 24 27 33 31 71 89 61 41 34

Part A Construct a frequency table that represents the data. **See margin.**

Part B Construct a histogram from the frequency table. **See margin.**

A bubble-in answer sheet for these practice problems is available on page A1 of the *Chapter 4 Resource Masters.*

Additional Practice

Additional test practice questions are available in the *Chapter 4 Resource Masters*, pp. 193–194.

Answers

10A. Sample answer:

Museum Visitors		
Number	**Tally**	**Frequency**
0–19	卌 卌 I	11
20–39	卌 卌 I	11
40–59	III	3
60–79	IIII	4
80–99	I	1

10B.

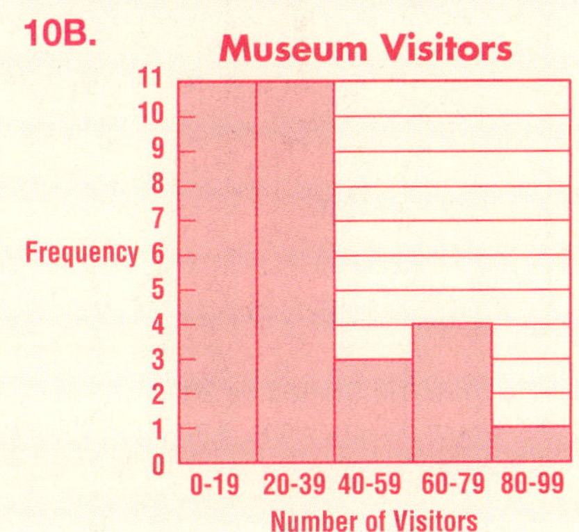

***Assessment*, pp. 193–194**

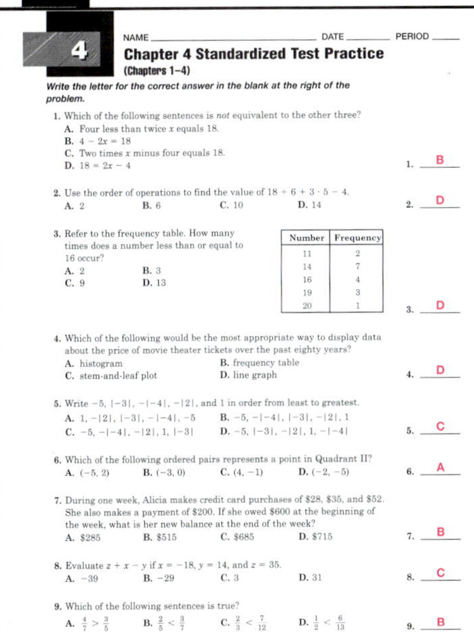

NAME ________ DATE ______ PERIOD ______

4 **Chapter 4 Standardized Test Practice**
(Chapters 1–4)

Write the letter for the correct answer in the blank at the right of the problem.

1. Which of the following sentences is *not* equivalent to the other three?
   A. Four less than twice x equals 18.
   B. $4 - 2x = 18$
   C. Two times x minus four equals 18.
   D. $18 = 2x - 4$ 1. **B**

2. Use the order of operations to find the value of $18 \div 6 + 3 \cdot 5 - 4$.
   A. 2 B. 6 C. 10 D. 14 2. **D**

3. Refer to the frequency table. How many times does a number less than or equal to 16 occur?
   A. 2 B. 3
   C. 9 D. 13

Number	Frequency
11	2
14	7
16	4
19	3
20	1

3. **D**

4. Which of the following would be the most appropriate way to display data about the price of movie theater tickets over the past eighty years?
   A. histogram B. frequency table
   C. stem-and-leaf plot D. line graph 4. **D**

5. Write -5, $|-3|$, $-|-4|$, $-|2|$, and 1 in order from least to greatest.
   A. $1, -|2|, |-3|, -|-4|, -5$ B. $-5, -|-4|, |-3|, -|2|, 1$
   C. $-5, -|-4|, -|2|, 1, |-3|$ D. $-5, |-3|, -|2|, 1, -|-4|$ 5. **C**

6. Which of the following ordered pairs represents a point in Quadrant II?
   A. $(-5, 2)$ B. $(-3, 0)$ C. $(4, -1)$ D. $(-2, -5)$ 6. **A**

7. During one week, Alicia makes credit card purchases of \$28, \$35, and \$52. She also makes a payment of \$200. If she owed \$600 at the beginning of the week, what is her new balance at the end of the week?
   A. \$285 B. \$515 C. \$685 D. \$715 7. **B**

8. Evaluate $z + x - y$ if $x = -18$, $y = 14$, and $z = 35$.
   A. -39 B. -29 C. 3 D. 31 8. **C**

9. Which of the following sentences is true?
   A. $\frac{4}{7} > \frac{3}{5}$ B. $\frac{2}{5} < \frac{3}{7}$ C. $\frac{2}{3} < \frac{7}{12}$ D. $\frac{1}{2} < \frac{6}{13}$ 9. **B**

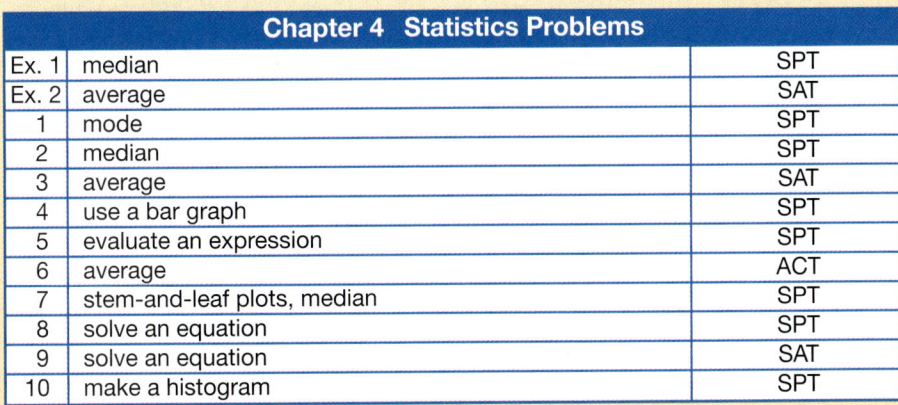

Chapter 4 Statistics Problems		
Ex. 1	median	SPT
Ex. 2	average	SAT
1	mode	SPT
2	median	SPT
3	average	SAT
4	use a bar graph	SPT
5	evaluate an expression	SPT
6	average	ACT
7	stem-and-leaf plots, median	SPT
8	solve an equation	SPT
9	solve an equation	SAT
10	make a histogram	SPT

CHAPTER 5

Resource Manager

Proportional Reasoning and Probability

Instructional Objectives

Lesson (pages)	Objectives	NCTM Standards 2000	State/Local Objectives
5–1 (188–193)	Solve proportions.	1, 6, 8, 9	
5–2 (194–197)	Solve problems involving scale drawings and models.	1, 2, 3, 4, 6, 8, 9, 10	
5–3 (198–203)	Solve problems by using the percent proportion.	1, 6, 8, 9	
5–4 (204–209)	Solve problems by using the percent equation.	1, 6, 8, 9	
Investigation (210–211)	Explore box-and-whisker plots.	1, 2, 6, 7, 8, 9, 10	
5–5 (212–217)	Solve problems involving percent of increase or decrease.	1, 2, 6, 8, 9	
5–6 (218–223)	Find the probability and odds of a simple event.	1, 5, 6, 8, 9, 10	
5–7 (224–229)	Find the probability of mutually exclusive and inclusive events.	1, 5, 6, 8, 9	

Key to NCTM Standards 2000
1 Number & Operations; **2** Algebra; **3** Geometry; **4** Measurement; **5** Data Analysis & Probability; **6** Problem Solving; **7** Reasoning & Proof; **8** Communications; **9** Connections; **10** Representation

Suggested Pacing *See page T13 for a complete course-planning calendar.*

Standard refers to schedules that provide 45- to 55-minute periods that meet each day.
Block refers to schedules that provide approximately 90-minute periods which may meet every day for one semester or every other day over two semesters.

PACING	DAY 1	DAY 2	DAY 3	DAY 4	DAY 5	DAY 6
Standard Core (Chapters 1–13)	Lesson 5–1		Lesson 5–2	Lesson 5–3	Lesson 5–4	INV
Standard Enhanced (Chapters 1–15)	Lesson 5–1		Lesson 5–2	Lesson 5–3	Lesson 5–4	INV
Block Core (Chapters 1–13)	Chapter 4 Test & Lesson 5–1	Lessons 5–2 & 5–3	Lesson 5–4 & INV	Lesson 5–5	Lessons 5–6 & 5–7	SG+A
Block Enhanced (Chapters 1–15)	Chapter 4 Test & Lesson 5–1	Lessons 5–2 & 5–3	Lesson 5–4 & INV	Lessons 5–5 & 5–6	Lesson 5–7 & SG+A	Chapter Test & Lesson 6–1

Instructional Resources

Lesson	Materials and Manipulatives (see below for Glencoe Manipulative Resources)	Blackline Masters (page numbers): Chapter 5 Resource Masters: Study Guide	Practice (Skills & Average)	Reading to Learn Mathematics	Enrichment	Assessment	Hands-On Algebra*	School-to-Workplace*	Graphing Calculator Masters*	5-Minute Check Transparencies
5–1	recipe	195	196–197	198	199					5–1
5–2	grid paper [1, 4] state or local map	200	201–202	203	204		64			5–2
5–3	compass [1, 2] protractor [1, 2, 3, 4] 10-by-10 grids	205	206–207	208	209	241	65, 66		15, 16	5–3
5–4		210	211–212	213	214	240		5		5–4
Investigation	ruler [1, 2, 4] calculator									
5–5	graphing calculator advertisement showing discounts	215	216–217	218	219				14	5–5
5–6	paper bag two-inch slips of paper coins	220	221–222	223	224		67			5–6
5–7	paper bags counters [1, 2, 3, 4] pennies and nickels	225	226–227	228	229	241	68			5–7
Study Guide & Assessment/ Chapter Test						231–239, 242–244				

***See page 186c for examples of these instructional materials.**

Key to Glencoe Manipulative Resources

[1]Classroom Manipulative Resources [2]Student Manipulative Resources [3]Overhead Manipulative Resources [4]Hands-On Algebra Masters

INV = Investigation *SG+A = Study Guide and Assessment*

<table>
<tr><th>DAY 7</th><th>DAY 8</th><th>DAY 9</th><th>DAY 10</th><th>DAY 11</th><th>DAY 12</th><th>DAY 13</th></tr>
<tr><td colspan="2">Lesson 5–5</td><td>Lesson 5–6</td><td colspan="2">Lesson 5–7</td><td>SG+A</td><td>Chapter Test</td></tr>
<tr><td colspan="2">Lesson 5–5</td><td>Lesson 5–6</td><td>Lesson 5–7</td><td>SG+A</td><td>Chapter Test</td><td></td></tr>
<tr><td>Chapter Test & Lesson 6–1</td><td></td><td></td><td></td><td></td><td></td><td></td></tr>
<tr><td></td><td></td><td></td><td></td><td></td><td></td><td></td></tr>
</table>

Resource Manager

The pages shown on this page are a small sample of the materials available on *TeacherWorks: All-in-One Lesson Planner and Resource Center.*

This CD-ROM includes all of the blackline masters and transparencies available for this program.

It also includes a lesson planner and interactive Teacher Edition, so you can customize lesson plans and reproduce classroom resources quickly and easily, from just about anywhere.

Applications

School-to-Workplace Masters, p. 5

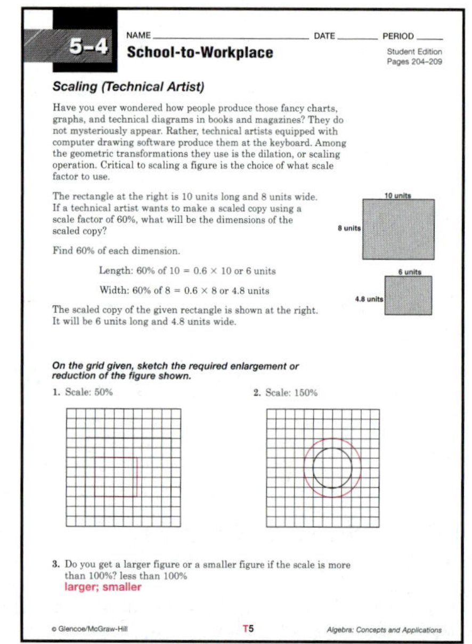

5-4 NAME ______ DATE ______ PERIOD ______

School-to-Workplace

Student Edition Pages 204–209

Scaling (Technical Artist)

Have you ever wondered how people produce those fancy charts, graphs, and technical diagrams in books and magazines? They do not mysteriously appear. Rather, technical artists equipped with computer drawing software produce them at the keyboard. Among the geometric transformations they use is the dilation, or scaling operation. Critical to scaling a figure is the choice of what scale factor to use.

The rectangle at the right is 10 units long and 8 units wide. If a technical artist wants to make a scaled copy using a scale factor of 60%, what will be the dimensions of the scaled copy?

10 units
8 units

Find 60% of each dimension.

Length: 60% of 10 = 0.6×10 or 6 units

Width: 60% of 8 = 0.6×8 or 4.8 units

6 units
4.8 units

The scaled copy of the given rectangle is shown at the right. It will be 6 units long and 4.8 units wide.

On the grid given, sketch the required enlargement or reduction of the figure shown.

1. Scale: 50%
2. Scale: 150%
3. Do you get a larger figure or a smaller figure if the scale is more than 100%? less than 100%
larger; smaller

© Glencoe/McGraw-Hill T5 *Algebra: Concepts and Applications*

Manipulatives/Modeling

Hands-On Algebra Masters, pp. 64–68

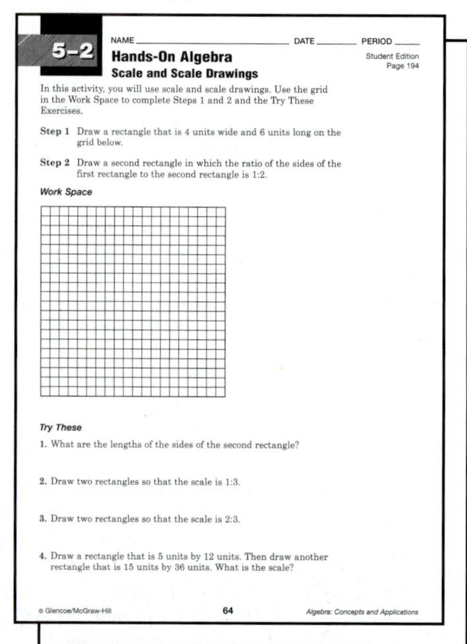

5-2 NAME ______ DATE ______ PERIOD ______

Hands-On Algebra

Scale and Scale Drawings

Student Edition Page 194

In this activity, you will use scale and scale drawings. Use the grid in the Work Space to complete Steps 1 and 2 and the Try These Exercises.

Step 1 Draw a rectangle that is 4 units wide and 6 units long on the grid below.

Step 2 Draw a second rectangle in which the ratio of the sides of the first rectangle to the second rectangle is 1:2.

Work Space

Try These

1. What are the lengths of the sides of the second rectangle?
2. Draw two rectangles so that the scale is 1:3.
3. Draw two rectangles so that the scale is 2:3.
4. Draw a rectangle that is 5 units by 12 units. Then draw another rectangle that is 15 units by 36 units. What is the scale?

© Glencoe/McGraw-Hill 64 *Algebra: Concepts and Applications*

Technology/Multimedia

Graphing Calculator Masters, pp. 14–16

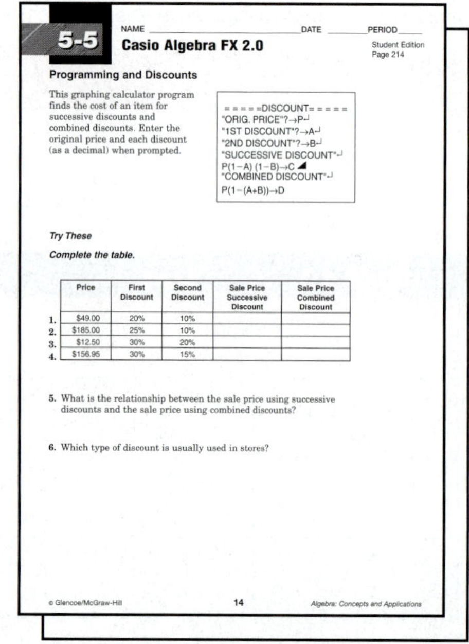

5-5 NAME ______ DATE ______ PERIOD ______

Casio Algebra FX 2.0

Student Edition Page 214

Programming and Discounts

This graphing calculator program finds the cost of an item for successive discounts and combined discounts. Enter the original price and each discount (as a decimal) when prompted.

```
= = = = =DISCOUNT= = = = =
"ORIG. PRICE"?→P↵
"1ST DISCOUNT"?→A↵
"2ND DISCOUNT"?→B↵
"SUCCESSIVE DISCOUNT"↵
P(1−A) (1−B)→C ◢
"COMBINED DISCOUNT"↵
P(1−(A+B))→D
```

Try These

Complete the table.

	Price	First Discount	Second Discount	Sale Price Successive Discount	Sale Price Combined Discount
1.	$49.00	20%	10%		
2.	$185.00	25%	10%		
3.	$12.50	30%	20%		
4.	$156.95	30%	15%		

5. What is the relationship between the sale price using successive discounts and the sale price using combined discounts?
6. Which type of discount is usually used in stores?

© Glencoe/McGraw-Hill 14 *Algebra: Concepts and Applications*

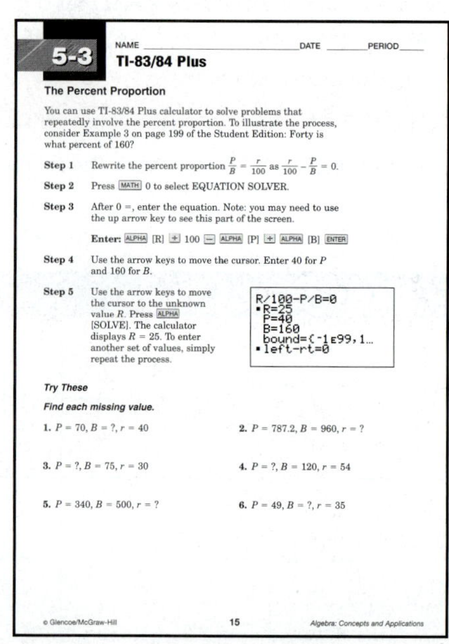

5-3 NAME ______ DATE ______ PERIOD ______

TI-83/84 Plus

The Percent Proportion

You can use TI-83/84 Plus calculator to solve problems that repeatedly involve the percent proportion. To illustrate the process, consider Example 3 on page 199 of the Student Edition: Forty is what percent of 160?

Step 1 Rewrite the percent proportion $\frac{P}{B} = \frac{r}{100}$ as $\frac{r}{100} - \frac{P}{B} = 0$.

Step 2 Press MATH 0 to select EQUATION SOLVER.

Step 3 After 0 =, enter the equation. Note: you may need to use the up arrow key to see this part of the screen.

Enter: ALPHA [R] ÷ 100 − ALPHA [P] ÷ ALPHA [B] ENTER

Step 4 Use the arrow keys to move the cursor. Enter 40 for P and 160 for B.

Step 5 Use the arrow keys to move the cursor to the unknown value R. Press ALPHA [SOLVE]. The calculator displays $R = 25$. To enter another set of values, simply repeat the process.

```
R/100-P/B=0
▪R=25
 P=40
 B=160
 bound={-1E99,1…
▪left-rt=0
```

Try These

Find each missing value.

1. $P = 70, B = ?, r = 40$
2. $P = 787.2, B = 960, r = ?$
3. $P = ?, B = 75, r = 30$
4. $P = ?, B = 120, r = 54$
5. $P = 340, B = 500, r = ?$
6. $P = 49, B = ?, r = 35$

© Glencoe/McGraw-Hill 15 *Algebra: Concepts and Applications*

Assessment Resources

Type	Student Edition	Teacher's Wraparound Edition	Chapter 4 Resource Masters
Ongoing Assessment	Quizzes 1 and 2, pp. 203, 217	5-Minute Check, pp. 188, 194, 198, 204, 212, 219, 224	Mid-Chapter Test, p. 240 Quizzes A and B, p. 241
Mixed Review	Mixed Review, pp. 193, 197, 203, 209, 217, 223, 229 Standardized Test Practice, Chapters 1–5, pp. 234–235		Cumulative Review, p. 242 Standardized Test Practice, pp. 243–244
Error Analysis	You Decide, pp. 201, 207	Error Analysis, pp. 191, 196, 201, 208, 215, 222, 228	
Standardized Test Prep	Standardized Test Practice, pp. 193, 197, 203, 209, 217, 223, 229 Standardized Test Practice, Chapters 1–5, pp. 234–235		Standardized Test Practice, pp. 243–244
Open-Ended Assessment	Writing Math, pp. 215, 227 Investigation, pp. 210–211 Portfolio, pp. 187, 211	Modeling: p. 203 Speaking: pp. 197, 217, 229 Writing: pp. 193, 209, 223	Extended Response Assessment, p. 239
Chapter Assessment	Study Guide and Assessment, pp. 230–232 Chapter Test, p. 233		Multiple-Choice Tests (Forms 1A, 1B) pp. 231–234 Free-Response Tests (Forms 2A, 2B), pp. 235–238

Additional Chapter Resources

Student Edition
Math in the Workplace, p. 218
Hands-On Algebra, pp. 194, 220, 224
Graphing Calculator Exploration, p. 214

Teacher's Classroom Resources
Manipulatives/Modeling
Teacher's Guide for Overhead Manipulative Resources

Meeting Individual Needs
Prerequisite Skills Booklet
Spanish Study Guide and Assessment, pp. 36–42, 113–114

Teaching Aids
Solutions Manual
5-Minute Check Transparencies

Glencoe Technology

Instructional
AlgePASS CD-ROM, Lessons 13, 14
Interactive Chalkboard CD-ROM
StudentWorks Plus CD-ROM
Multimedia Applications CD-ROM, Activity 4
Vocabulary PuzzleMaker (online)

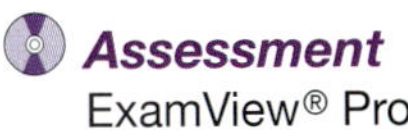

Assessment
ExamView® Pro

Teaching Aids
Answer Key Maker CD-ROM
TeacherWorks CD-ROM

Visit ***www.algconcepts.com*** *for data updates, career information, games, and other interactive activities.*

Mathematics of the Chapter

This chapter provides students with explanations, illustrations, and experience using ratios, percent, proportions, and probability. Students will begin by solving proportions to convert measurements, find rates, and use scales. Then they will solve several types of percent problems, including percent of increase and percent of decrease problems. Finally, they explore concepts involving probability, including odds, mutually exclusive events, inclusive events, and compound events.

What You'll Learn

Have students read over the lists of key ideas and key vocabulary. Have them make a list of any words with which they are not familiar.

Why It's Important

Point out to students that this is only one of many reasons why each objective is important. Others are provided in each lesson.

What's MATH Got To Do With It? Real-Life Videos

What's Math Got to Do With It? Real-Life Videos engage students, showing them how math is used in everyday situations. Use Video 4 to discuss how probability is used in selling frozen yogurt and landing robots on Mars.

CHAPTER 5 Proportional Reasoning and Probability

What You'll Learn

Key Ideas

- Solve problems involving proportions. *(Lessons 5–1 and 5–2)*
- Solve problems involving percents. *(Lessons 5–3 to 5–5)*
- Find the probability of mutually exclusive events and inclusive events. *(Lesson 5–7)*

Key Vocabulary

odds *(p. 221)*
percent *(p. 198)*
probability *(p. 219)*
proportion *(p. 188)*
ratio *(p. 188)*

Why It's Important

Agriculture Corn is America's number one crop and one of the most versatile crops grown today. It is not just used for food. Products made from corn are all around us. The raw materials from processed corn are used to make antibiotics, glue, paint, fuel, and many other products.

Proportional reasoning is used in many branches of mathematics and other fields such as business and navigation. You will use a proportion to predict the amount of corn a farmer can expect from a crop in Lesson 5–1.

CHAPTER 5 LINKS

Lesson	5–1	5–2	5–3	5–4	5–5	5–6	5–7
Math in the Workplace	Cooking	Movies	Health	Business	Retail Sales Accountant	Manufacturing	Marketing
Applications and Connections	Science Travel Agriculture Health City Planning Construction	Maps Transportation Bridges Model Cars Architecture Life Science	Business Government Education Pharmacy Nursing Maps	Banking Sales Chemistry Credit Cards Health	Business Sales Travel Biology Real Estate Economics	Population Basketball Marketing Allowance Banking	Meteorology Savings Sports Games Population
Math Integration	Measurement		Statistics	Statistics	Statistics		

Study these lessons to improve your skills.

✓ Check Your Readiness

Lesson 4–4, pp. 160–164

Solve each equation. Check your solution.

1. $6a = 18$ **3**
2. $5d = 100$ **20**
3. $9x = 81$ **9**
4. $8c = 32$ **4**
5. $4n = 60$ **15**
6. $7k = 98$ **14**

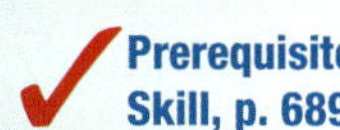
Prerequisite Skill, p. 689

Write each percent as a decimal.

7. 25% **0.25**
8. 60% **0.6**
9. 45% **0.45**
10. 8% **0.08**
11. 125% **1.25**
12. 33.3% **0.333**

Prerequisite Skill, p. 687

Find each sum or difference.

13. $\frac{1}{6} + \frac{1}{6}$ $\mathbf{\frac{1}{3}}$
14. $\frac{1}{10} + \frac{3}{10}$ $\mathbf{\frac{2}{5}}$
15. $\frac{3}{10} + \frac{3}{5}$ $\mathbf{\frac{9}{10}}$
16. $\frac{5}{6} + \frac{1}{12}$ $\mathbf{\frac{11}{12}}$
17. $\frac{3}{4} + \frac{1}{12}$ $\mathbf{\frac{5}{6}}$
18. $1 - \frac{3}{5}$ $\mathbf{\frac{2}{5}}$

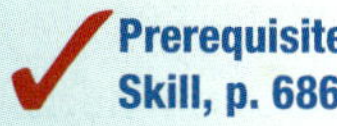
Prerequisite Skill, p. 686

Find each product.

19. $\frac{1}{10} \cdot \frac{1}{10}$ $\mathbf{\frac{1}{100}}$
20. $\frac{3}{8} \cdot \frac{1}{8}$ $\mathbf{\frac{3}{64}}$
21. $\frac{3}{10} \cdot \frac{1}{2}$ $\mathbf{\frac{3}{20}}$
22. $\frac{1}{6} \cdot \frac{2}{3}$ $\mathbf{\frac{1}{9}}$
23. $\frac{7}{10} \cdot \frac{2}{5}$ $\mathbf{\frac{7}{25}}$
24. $\frac{4}{13} \cdot \frac{1}{2}$ $\mathbf{\frac{2}{13}}$

Prerequisite Skill, p. 684

Find each sum or product.

25. $0.65 + 0.12$ **0.77**
26. $0.15 + 0.06$ **0.21**
27. $0.34 + 0.28$ **0.62**
28. $0.5 \cdot 0.5$ **0.25**
29. $0.4 \cdot 0.6$ **0.24**
30. $0.125 \cdot 0.125$ **0.015625**

Make this Foldable to help you organize your Chapter 5 notes. Begin with a sheet of notebook paper.

❶ **Fold** lengthwise to the holes.

❷ **Cut** four tabs.

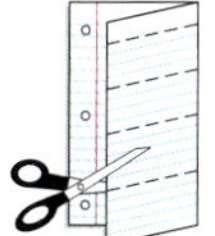

❸ **Label** the tabs as shown.

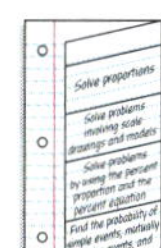

Reading and Writing Store your foldable in a 3-ring binder. As you read and study the chapter, write notes and examples under each tab.

www.algconcepts.com/chapter_readiness

FOLDABLES™ Study Organizer Use this Foldable for student writing about proportion, percent, and probability. On each tab have students write about any time when they have used the methods from the chapter to solve problems. Then have students create real-world problems for which they might use the methods of the chapter.

Check Your Readiness

This section provides a review of the basic concepts needed before beginning Chapter 5. Lesson references are included for additional student help.

Use Exercises	To Prepare for Lesson(s)
1–6	5–1, 5–2, 5–3
7–12	5–3, 5–4
13–18	5–7
19–24	5–7
25–30	5–4

Quick Review Math Handbook

Glencoe Algebra Lesson	Quick Review Math Handbook Hot Topic
5–1	2–8, 6–5
5–2	2–8, 6–5
Ch. 5 Inv.	4–2
5–5	2–8, 6–4, 6–5
5–6	3–4, 4–5, 4–6
5–7	2–9, 4–6

Noteables™

Interactive Study Notebook with Foldables™

This note-taking guide reinforces excellent note-taking skills. It includes:

- ✓ **Build Your Vocabulary** tool for including mathematical terminology in notes.
- ✓ **Review It** activities with links to prior lessons and items to include in **Foldables™**.
- ✓ **Bringing It All Together** feature to help students review for the Chapter Test.
- ✓ **Are You Ready for the Chapter Test?** feature to allow students to assess their own readiness for the Chapter Test.
- ✓ **Teacher Edition** with transparencies to guide note taking. Corresponds to the *Teacher's Wraparound Edition* In-Class Examples and *Interactive Chalkboard CD-ROM.*

Lesson 5-1

1 FOCUS

5-Minute Check
Chapter 4

The figure below shows the dimensions of a small rectangular garden.

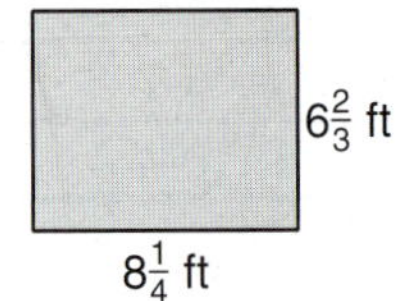

1. What is the area of the garden? **55 ft^2**
2. One of the longer sides of the garden has a fence along it. If the fence was made by joining 3 sections of fencing that were each the same length, how long is each section of the fence? **$2\frac{3}{4}$ ft**

Solve each equation.

3. $8.4x + 2.1 = 3.4x - 0.4$ **-0.5**
4. $18x = -6(4 - x)$ **-2**
5. A stationery store offers the following options for personal stationery.
   Lettering: script, block, formal
   Ink: black, blue, gold
   Envelope: return address, no return address
   Find the number of different types of stationery using the Fundamental Counting Principle. **18 types**

Motivating the Lesson

Real-World Connection Show students a recipe, from a magazine or newspaper, that includes the number of servings the recipe makes. Discuss with students how to determine the amount of each ingredient needed to make 4.5 times that number of servings for a family-reunion picnic.

5-1 Solving Proportions

What You'll Learn
You'll learn to solve proportions.

Why It's Important
Cooking Caterers use proportions to adjust their recipes. *See Exercise 50.*

In mathematics, a **ratio** is a comparison of two numbers by division. For example, the ratio of 15 and 10 can be expressed in the following three ways.

15 to 10   15:10 $\frac{15}{10}$

You can express $\frac{15}{10}$ in simplest form as $\frac{3}{2}$.

$$\frac{15}{10} = \frac{3}{2} \quad (\div 5 \text{ numerator and denominator})$$

An equation stating that two ratios are equal is called a **proportion**. So, $\frac{15}{10} = \frac{3}{2}$ is a proportion. Every proportion has two equal cross products.

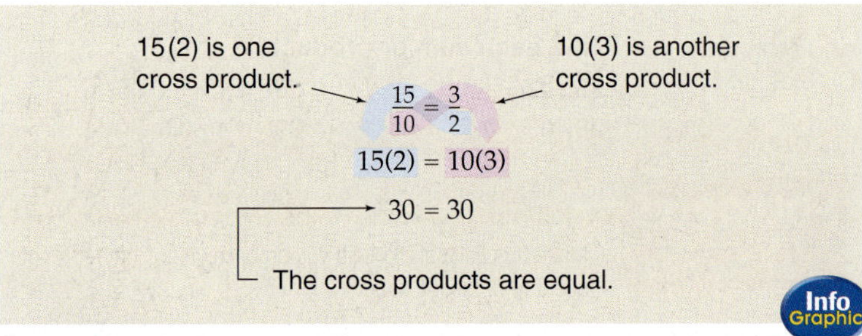

15(2) is one cross product.
10(3) is another cross product.
$\frac{15}{10} = \frac{3}{2}$
15(2) = 10(3)
30 = 30
The cross products are equal.

In the proportion, $b \neq 0$ and $d \neq 0$.

Property of Proportions

Words: The cross products of a proportion are equal.

Symbols: If $\frac{a}{b} = \frac{c}{d}$, then $ad = bc$. If $ad = bc$, then $\frac{a}{b} = \frac{c}{d}$.

Numbers: If $\frac{15}{10} = \frac{3}{2}$, then $15(2) = 10(3)$.
If $15(2) = 10(3)$, then $\frac{15}{10} = \frac{3}{2}$.

You can use cross products to solve proportions. Sometimes you'll use the Distributive Property.

Resource Manager

Reproducible Masters
Chapter 5 Resource Masters
- *Study Guide,* p. 195
- *Skills Practice,* p. 196
- *Practice,* p. 197
- *Reading to Learn Mathematics,* p. 198
- *Enrichment,* p. 199

Transparencies
5-Minute Check, 5–1

Technology/Multimedia
AlgePASS, Lesson 13
Interactive Chalkboard CD-ROM

Examples

Solve each proportion.

1 $\frac{5}{3} = \frac{75}{n}$

$\frac{5}{3} = \frac{75}{n}$	*Original equation*
$5n = 3(75)$	*Find the cross products.*
$5n = 225$	*Multiply.*
$\frac{5n}{5} = \frac{225}{5}$	*Divide each side by 5.*
$n = 45$	*Simplify.*

Check:

$\frac{5}{3} = \frac{75}{n}$	*Original equation*
$\frac{5}{3} \stackrel{?}{=} \frac{75}{45}$	*Replace n with 45.*
$5(45) \stackrel{?}{=} 3(75)$	*Find the cross products.*
$225 = 225$ ✓	

The solution is 45.

2 $\frac{y}{8} = \frac{y+4}{10}$

$\frac{y}{8} = \frac{y+4}{10}$	*Original equation*
$10y = 8(y + 4)$	*Find the cross products.*
$10y = 8y + 32$	*Distributive Property*
$10y - 8y = 8y + 32 - 8y$	*Subtract 8y from each side.*
$2y = 32$	*Simplify.*
$\frac{2y}{2} = \frac{32}{2}$	*Divide each side by 2.*
$y = 16$	*Simplify.*

Check:

$\frac{y}{8} = \frac{y+4}{10}$	*Original equation*
$\frac{16}{8} \stackrel{?}{=} \frac{16+4}{10}$	*Replace y with 16.*
$\frac{16}{8} \stackrel{?}{=} \frac{20}{10}$	*Add 16 and 4.*
$2 = 2$ ✓	*Simplify.*

The solution is 16.

Your Turn

a. $\frac{3}{a} = \frac{18}{24}$ **4**   b. $\frac{r}{100} = \frac{7}{8}$ **87.5**   c. $\frac{5}{c} = \frac{13}{c+8}$ **5**

Technology Tip

A graphing calculator follows the order of operations. So, it is not necessary to press ENTER after multiplying 3 and 75 below.

3 [×] 75 [÷] 5 ENTER 45

2 TEACH

Teaching Tip When discussing the proportion $\frac{a}{b} = \frac{c}{d}$ in the Property of Proportions box, show students that the proportion can also be written using the notation $a:b = c:d$. ("a is to b as c is to d.") Then point out that b and c are sometimes called the *means* of the proportion and a and d are called the *extremes*.

$a:b = c:d$ — means (b, c); extremes (a, d)

Suggest that students can use these two terms to also remember the Property of Proportions as "the product of the means equals the product of the extremes."

In-Class Examples

Examples 1–2

Solve each proportion.

1 $\frac{9}{m} = \frac{63}{35}$ **5**

2 $\frac{z-9}{4} = \frac{z}{5}$ **45**

Teaching Tip When discussing Example 3, point out that the situation can also be solved using the proportion $\frac{1 \text{ pound}}{x \text{ pounds}} = \frac{16 \text{ ounces}}{72 \text{ ounces}}$, because it has the same cross products as the proportion $\frac{72 \text{ ounces}}{x \text{ pounds}} = \frac{16 \text{ ounces}}{1 \text{ pound}}$.

In-Class Examples

Examples 3–4

Convert each measurement as indicated.

3 15 pints to quarts $7\frac{1}{2}$ qt

4 4750 millimeters to meters 4.75 m

Example 5

The density of copper is 8.96 grams per cubic centimeter. Suppose you have a piece of copper whose volume is 25 cubic centimeters. How many grams of copper do you have? 224 g

You can use proportions to find equivalent measurements within the English or metric systems.

Examples

Convert each measurement as indicated.

3 72 ounces to pounds

16 ounces = 1 pound

Let x represent the number of pounds. Write a proportion.

$$\frac{72 \text{ ounces}}{x \text{ pounds}} = \frac{16 \text{ ounces}}{1 \text{ pound}}$$

$$\frac{72}{x} = \frac{16}{1}$$

$$72(1) = 16x$$

$$72 = 16x$$

$$\frac{72}{16} = \frac{16x}{16}$$

$$4.5 = x$$

So, 72 ounces = 4.5 pounds.

4 15 meters to centimeters

1 meter = 100 centimeters

Let x represent the number of centimeters. Write a proportion.

$$\frac{1 \text{ meter}}{100 \text{ centimeters}} = \frac{15 \text{ meters}}{x \text{ centimeters}}$$

$$\frac{1}{100} = \frac{15}{x}$$

$$1x = 100(15)$$

$$x = 1500$$

So, 15 meters = 1500 centimeters.

Your Turn

d. 96 centimeters to meters **0.96 m** **e.** 2 gallons to quarts **8 qt**

A **rate** is a ratio of two measurements having different units of measure. For example, 120 miles in 2 hours is a rate. When a rate is simplified so it has a denominator of 1, it is called a **unit rate**. To find the unit rate of 120 miles in 2 hours, divide by 2. The result is a unit rate of 60 miles in 1 hour, or 60 miles per hour.

To solve problems involving rates, you can use **dimensional analysis**. This is the process of carrying units throughout a computation.

Example

Science Link

5 Density is a measure of the amount of matter that occupies a given volume. The density of wood is 0.71 gram per cubic centimeter. Suppose you have a piece of wood whose volume is 50 cubic centimeters. How many grams of wood does it contain?

0.71 gram per cubic centimeter = $\frac{0.71 \text{ gram}}{1 \text{ cubic centimeter}}$

Multiply the unit rate by the number of cubic centimeters of wood.

$$\frac{0.71 \text{ gram}}{1 \text{ cubic centimeter}} \cdot \frac{50 \text{ cubic centimeters}}{1} = \frac{35.5 \text{ grams}}{1} \text{ or } 35.5 \text{ grams}$$

Note that the unit of measure, cubic centimeters, can be canceled.

So, the piece of wood contains 35.5 grams of wood.

Real World

Example 6

Travel Link

Juanita's family drove 400 miles in 8 hours. The next day, they need to drive another 700 miles. At the same rate, how long will it take them to drive the 700 miles?

Explore You know that the family drove 400 miles in 8 hours. You need to find how long it will take to drive 700 miles farther.

Plan Write the rate 400 miles in 8 hours as a unit rate. Then divide 700 miles by the unit rate.

Solve $\frac{400 \text{ miles}}{8 \text{ hours}} = \frac{50 \text{ miles}}{1 \text{ hour}}$ *The unit rate is 50 miles per hour.*

$700 \text{ miles} \div \frac{50 \text{ miles}}{1 \text{ hour}} = \frac{700 \text{ miles}}{1} \cdot \frac{1 \text{ hour}}{50 \text{ miles}}$

$= \frac{\overset{14}{\cancel{700 \text{ miles}}}}{1} \cdot \frac{1 \text{ hour}}{\underset{1}{\cancel{50 \text{ miles}}}}$ *Note that the units cancel.*

$= 14 \text{ hours}$

Examine It takes 8 hours to travel 400 miles. A trip of 700 miles would take nearly twice as long. So, 14 hours is reasonable.

You can also solve this problem by solving the proportion $\frac{400 \text{ miles}}{8 \text{ hours}} = \frac{700 \text{ miles}}{h \text{ hours}}$.

Check for Understanding

Communicating Mathematics

1. **Explain** how to use the Property of Proportions to solve $\frac{x}{12} = \frac{5}{6}$. **1. Multiply 12 and 5. Then divide by 6.**
2. **Write** two ratios that form a proportion and two other ratios that do *not* form a proportion. **See margin.**

Vocabulary
ratio
proportion
rate
unit rate
dimensional analysis

Guided Practice

Getting Ready Find the cross products for each proportion.

Sample: $\frac{2}{8} = \frac{4}{16}$ **Solution:** $2(16) = 8(4)$

3. $\frac{75}{100} = \frac{3}{4}$ **3. 75(4) = 100(3)**
4. $\frac{16}{12} = \frac{12}{9}$ **4. 16(9) = 12(12)**
5. $\frac{8}{12} = \frac{10}{15}$ **5. 8(15) = 12(10)**

Examples 1 & 2

Solve each proportion.

6. $\frac{6}{2} = \frac{9}{a}$ **3**
7. $\frac{3}{5} = \frac{b}{100}$ **60**
8. $\frac{5}{c} = \frac{25}{35}$ **7**
9. $\frac{2}{1.9} = \frac{4}{x}$ **3.8**
10. $\frac{y}{4} = \frac{y+2}{5}$ **8**
11. $\frac{z}{5} = \frac{z+2}{7}$ **5**

Examples 3 & 4

Convert each measurement as indicated.

12. 3 pounds to ounces **48 oz**
13. 2000 grams to kilograms **2 kg**

In-Class Example

Example 6

A trucker drove 210 miles in 5 hours. At this rate, how far will she travel in 8 hours? **336 mi**

3 PRACTICE/APPLY

Error Analysis

Watch for students who confuse solving a proportion (finding cross products) with multiplying fractions (multiply the numerators and multiply the denominators) in Exercises 3–11.

Prevent by showing two problems such as $\frac{x}{12} = \frac{5}{6}$ and $\frac{x}{12} \times \frac{5}{6}$. Ask students to identify and compare the symbols between each pair of ratios. Stress that only when solving a proportion (an equals sign between two ratios) should they calculate the cross products.

Answer

2. Sample answer: $\frac{3}{4}, \frac{6}{8}; \frac{1}{2}, \frac{2}{3}$

Reteaching Activity

Verbal/Linguistic Learners Have students state the Property of Proportions in words and symbols. Then have them show how the property is used to solve a proportion such as $\frac{8}{15} = \frac{w}{60}$, explaining each step of the procedure.

***Study Guide*, p. 195**

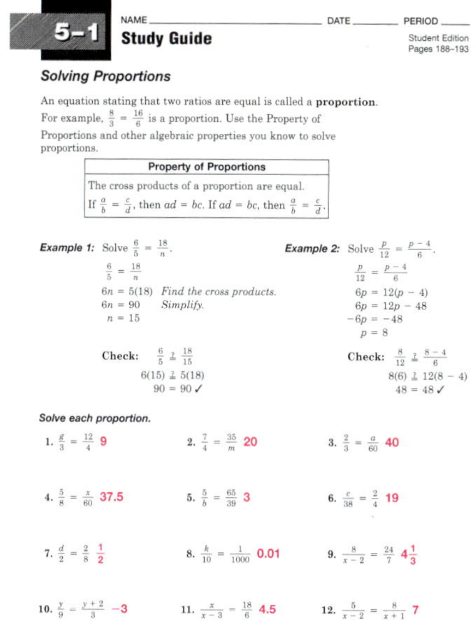

5–1 NAME ______ DATE ______ PERIOD ______

Study Guide Student Edition Pages 188–193

Solving Proportions

An equation stating that two ratios are equal is called a **proportion**. For example, $\frac{8}{3} = \frac{16}{6}$ is a proportion. Use the Property of Proportions and other algebraic properties you know to solve proportions.

Property of Proportions
The cross products of a proportion are equal. If $\frac{a}{b} = \frac{c}{d}$, then $ad = bc$. If $ad = bc$, then $\frac{a}{b} = \frac{c}{d}$.

Example 1: Solve $\frac{6}{5} = \frac{18}{n}$.

$\frac{6}{5} = \frac{18}{n}$
$6n = 5(18)$ *Find the cross products.*
$6n = 90$ *Simplify.*
$n = 15$

Check: $\frac{6}{5} \stackrel{?}{=} \frac{18}{15}$
$6(15) \stackrel{?}{=} 5(18)$
$90 = 90$ ✓

Example 2: Solve $\frac{p}{12} = \frac{p-4}{6}$.

$\frac{p}{12} = \frac{p-4}{6}$
$6p = 12(p - 4)$
$6p = 12p - 48$
$-6p = -48$
$p = 8$

Check: $\frac{8}{12} \stackrel{?}{=} \frac{8-4}{6}$
$8(6) \stackrel{?}{=} 12(8 - 4)$
$48 = 48$ ✓

Solve each proportion.

1. $\frac{g}{3} = \frac{12}{4}$ **9**
2. $\frac{7}{4} = \frac{35}{m}$ **20**
3. $\frac{2}{3} = \frac{a}{60}$ **40**
4. $\frac{5}{8} = \frac{x}{60}$ **37.5**
5. $\frac{5}{b} = \frac{65}{39}$ **3**
6. $\frac{c}{38} = \frac{2}{4}$ **19**
7. $\frac{d}{2} = \frac{2}{8}$ $\frac{1}{2}$
8. $\frac{k}{10} = \frac{1}{1000}$ **0.01**
9. $\frac{8}{x-2} = \frac{24}{7}$ $4\frac{1}{3}$
10. $\frac{y}{9} = \frac{y+2}{3}$ **−3**
11. $\frac{x}{x-3} = \frac{18}{6}$ **4.5**
12. $\frac{5}{x-2} = \frac{8}{x+1}$ **7**

Teaching Tip In Exercise 14, inform students that a bushel is a dry measure equivalent to 32 quarts, and that an acre is a land measure equivalent to 43,560 square feet or 4840 square yards.

Assignment Guide

Basic: 17–51 odd, 53–64
Average: 16–46 even, 47–64

Example 5

14. **Agriculture** A farmer gets 300 bushels of corn for each acre of land that he plants. At this rate, how many bushels will he expect to get if he plants 90 acres of corn? **27,000 bushels**

Example 6

15. **Travel** Katie knows that she can drive her car an average of 40 miles while using one gallon of gasoline. How many gallons of gasoline will she need for a trip of 180 miles? **4.5 gal**

Exercises

Practice

Homework Help

For Exercises	See Examples
16–38	1, 2
39–46, 49	3, 4
47, 48	5, 6

Extra Practice
See page 701.

Solve each proportion.

16. $\frac{14}{6} = \frac{21}{m}$ **9**
17. $\frac{5}{4} = \frac{y}{12}$ **15**
18. $\frac{5}{8} = \frac{x}{100}$ **62.5**
19. $\frac{3}{a} = \frac{18}{24}$ **4**
20. $\frac{18}{27} = \frac{t}{3}$ **2**
21. $\frac{8}{20} = \frac{30}{x}$ **75**
22. $\frac{6}{1} = \frac{1200}{r}$ **200**
23. $\frac{45}{5} = \frac{81}{p}$ **9**
24. $\frac{5}{g} = \frac{6}{3}$ **2.5**
25. $\frac{2}{3} = \frac{7}{y}$ **10.5**
26. $\frac{a}{3.5} = \frac{14}{7}$ **7**
27. $\frac{n}{2} = \frac{0.7}{0.4}$ **3.5**
28. $\frac{10}{y} = \frac{2.5}{4}$ **16**
29. $\frac{0.35}{3} = \frac{c}{18}$ **2.1**
30. $\frac{y+2}{y} = \frac{8}{6}$ **6**

B

31. $\frac{8}{x+3} = \frac{5}{x}$ **5**
32. $\frac{x+6}{5} = \frac{x}{4}$ **24**
33. $\frac{x}{x+4} = \frac{3}{5}$ **6**
34. $\frac{5}{x} = \frac{9}{x+4}$ **5**
35. $\frac{x-7}{x+8} = \frac{2}{3}$ **37**
36. $\frac{4}{x-3} = \frac{5}{x+5}$ **35**

37. Are $\frac{15}{2}$ and $\frac{7.5}{1}$ equivalent ratios? Explain your reasoning.

37. Yes; the cross products are equal.

38. Find the value of x that makes $\frac{18}{9} = \frac{28}{x}$ a proportion. **14**

Convert each measurement as indicated.

39. 5 feet to inches **60 in.**
40. 3 quarts to pints **6 pt**
41. 3 liters to milliliters **3000 mL**
42. 2.4 grams to milligrams
43. 16 quarts to gallons **4 gal**
44. 4200 pounds to tons **2.1 t**
45. 520 meters to kilometers **0.52 km**
46. 1600 milliliters to liters **1.6 L**

42. 2400 mg

Applications and Problem Solving

47. **Health** An average adult's body contains about 5 quarts of blood. If an adult donates 1 pint of blood, how many pints are left? **9 pt**

48. **City Planning** A water treatment plant can filter about 350,000 gallons of water in seven days. At this rate, about how many gallons of water can be filtered in 30 days? **1,500,000 gal**

***Skills Practice*, p. 196, and *Practice*, p. 197 (shown)**

5–1 NAME ______ DATE ______ PERIOD ______

Practice

Student Edition Pages 188–193

Solving Proportions

Solve each proportion.

1. $\frac{1}{7} = \frac{a}{14}$ **2**
2. $\frac{8}{2} = \frac{12}{h}$ **3**
3. $\frac{5}{b} = \frac{3}{9}$ **15**
4. $\frac{9}{6} = \frac{15}{b}$ **10**
5. $\frac{3}{y} = \frac{9}{21}$ **7**
6. $\frac{21}{24} = \frac{c}{8}$ **7**
7. $\frac{8}{3} = \frac{32}{x}$ **12**
8. $\frac{4}{5} = \frac{p}{25}$ **20**
9. $\frac{24}{g} = \frac{3}{8}$ **64**
10. $\frac{5}{2} = \frac{13}{a}$ **5.2**
11. $\frac{1.5}{c} = \frac{15}{5}$ **0.5**
12. $\frac{6}{2.4} = \frac{x}{4}$ **10**
13. $\frac{4}{2} = \frac{x+2}{3}$ **4**
14. $\frac{c+5}{15} = \frac{8}{6}$ **15**
15. $\frac{7}{7} = \frac{10}{b+3}$ **7**
16. $\frac{y}{y+3} = \frac{2}{5}$ **2**
17. $\frac{z+6}{12} = \frac{z}{4}$ **3**
18. $\frac{5}{4} = \frac{a+4}{2a}$ **$2\frac{2}{3}$**

Convert each measurement as indicated.

19. 5 pounds to ounces **80 oz**
20. 3000 grams to kilograms **3 kg**
21. 7 feet to inches **84 in.**
22. 4 meters to centimeters **400 cm**
23. 6 quarts to gallons **1.5 gal**
24. 250 centimeters to meters **2.5 m**

49. **Measurement** Is a 2-quart pitcher large enough to hold 1 batch of fruit punch? **no**

Fruit Punch
Mix together:
$2\frac{1}{2}$ cups cherry juice
2 cups orange juice
$1\frac{1}{2}$ cups pineapple juice
3 cups ginger ale

Exercise 49

50. **Cooking** A recipe for 72 chocolate chip cookies needs $4\frac{1}{2}$ cups of flour. How much flour is needed to make 48 cookies? **3 c**

51. **Construction** A batch of concrete is made by mixing 1 part cement, $2\frac{1}{2}$ parts sand, and $3\frac{1}{2}$ parts gravel. How many pounds of sand and gravel are needed to make 300 pounds of cement? **750 lb, 1050 lb**

52a. TX: 19.4; OSU: 11.6; UCLA: 10.9; GA: 9.7; TN: 17.5

*inter***NET** CONNECTION

Data Update For the latest information on university enrollments, visit: www.algconcepts.com

52. **Education** A student-to-teacher ratio is used to compare the number of students at a university with the number of faculty.

University	Students	Faculty
Texas	48,857	2517
Ohio State	48,278	4151
UCLA	35,558	3276
Georgia	29,693	3075
Tennessee	25,397	1453

Source: *The World Almanac*

a. To the nearest tenth, express the student-to-teacher ratio for each school as a decimal.

b. Rank the universities in the chart in order from greatest student-to-teacher ratio to least. **TX, TN, OSU, UCLA, GA**

53. **Critical Thinking** If $\frac{6}{a} = \frac{b}{8}$, what happens to the value of b as the value of a increases? **It decreases.**

Mixed Review

Solve each equation. *(Lessons 4–6 and 4–7)*

54. $5a - 3 = 8a + 12$ **−5**
55. $10n + 6 = 8n - 6$ **−6**
56. $6x + 7 = 8x - 9$ **8**
57. $3y - 4 = -2y - 14$ **−2**
58. $4(m - 2) = 6m$ **−4**
59. $-4(a - 2) = 10$ $-\frac{1}{2}$
60. $6(z + 2) - 4 = -10$ **−3**
61. $3x - 2(x + 3) = 6$ **12**

62. **Food** The Pasta Palace offers a choice of six types of pasta with five types of sauce. Each dinner comes with or without a meatball. How many different dinners are possible? *(Lesson 4–2)* **60 dinners**

Standardized Test Practice

63. **Short Response** Find the product of $\frac{2}{3}$ and $-\frac{3}{5}$. *(Lesson 4–1)* $-\frac{2}{5}$

64. **Multiple Choice** For what value(s) of x is the equation $|x - 2| = 5$ true? *(Lesson 3–7)* **C**

(A) −3 only (B) 7 only (C) −3 and 7 (D) −7 and 3

www.algconcepts.com/self_check_quiz

Teaching Tip In Exercise 52, be sure students understand that "student-to-teacher ratio" refers to the number of students divided by the number of teachers.

4 ASSESS

Open-Ended Assessment

Writing Ask students to describe how to recognize a proportion and how to use cross products to solve a proportion.

***Enrichment*, p. 199**

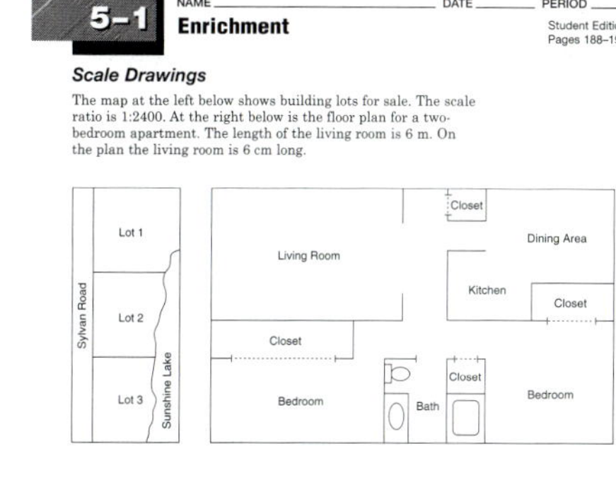

Answer each question.

1. On the map, how many feet are represented by an inch? **200 ft**
2. On the map, measure the frontage of Lot 2 on Sylvan Road in inches. What is the actual frontage in feet? **200 ft**
3. What is the scale ratio represented on the floor plan? **1:100**
4. On the floor plan, measure the width of the living room in centimeters. What is the actual width in meters? **4 m**
5. About how many square meters of carpeting would be needed to carpet the living room? **24 m^2**
6. Make a scale drawing of your classroom using an appropriate scale. **See students' work.**
7. On the scale for a map of Lancaster, Pennsylvania, 2.5 cm equals 3 km. Find the scale ratio. **1:120,000**

Extra Credit

In a proportion, two of the numbers are 8 and one of the other numbers is 4. Find all possible values for the fourth number. **4, 16**

Lesson 5-2

1 FOCUS

5-Minute Check
Lesson 5–1

Solve each proportion.

1. $\frac{n}{7} = \frac{15}{35}$ $n = 3$
2. $\frac{8}{p} = \frac{2}{7}$ $p = 28$

Each scale shows a measurement in pounds. Convert each measurement to ounces.

3. **64 oz**

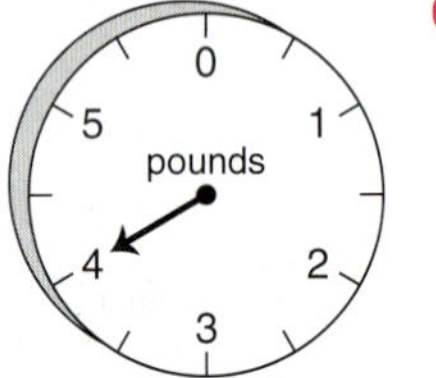

4. **40 oz**

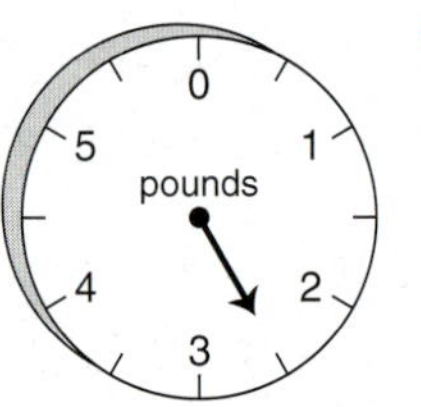

5. A train traveled 390 miles in 6 hours. At the same rate, how long would it take the train to travel 520 miles? **8 h**

Motivating the Lesson

Real-World Connection Ask students if they have built, collected, or played with model trains, planes, dollhouses, or other such objects. Ask those who have to describe some of the relationships between the models and the full-size objects.

5-2 Scale Drawings and Models

What You'll Learn
You'll learn to solve problems involving scale drawings and models.

Why It's Important
Movies Movie directors use models for special effects. *See Example 2.*

Leonardo da Vinci was commissioned to construct a huge bronze horse for the Duke of Milan. Unfortunately, he died before he could complete it.

Five hundred years later, master sculptor Nina Akamu created an 8-foot clay model from da Vinci's sketches. The model was then enlarged by sections to become the 24-foot *Il Cavallo*—the largest bronze horse in the world.

Il Cavallo

A **scale drawing** or **scale model** is used to represent an object that is too large or too small to be drawn or built at actual size. The **scale** is the ratio of a length on the drawing or model to the corresponding length of the real object.

length of clay model → *length of bronze statue* → $\frac{8 \text{ ft}}{24 \text{ ft}} = \frac{1 \text{ ft}}{3 \text{ ft}}$

The scale is 1 foot = 3 feet or 1:3.
If the scale uses the same units, it is not necessary to include them.

In the following activity, you will investigate scales with rectangles.

Hands-On Algebra

Materials:  grid paper

Step 1 Draw a rectangle that is 4 units wide and 6 units long on your grid paper.

Step 2 Draw a second rectangle so that the ratio of the sides of the first rectangle to the second rectangle is 1:2.

Try These

1. What are the lengths of the sides of the second rectangle? **1. 8 units, 12 units**
2. Draw two rectangles so that the scale is 1:3. **See students' work.**
3. Draw two rectangles so that the scale is 2:3. **See students' work.**
4. Draw a rectangle that is 5 units by 12 units. Then draw another rectangle that is 15 units by 36 units. What is the scale? **1:3**

Resource Manager

Reproducible Masters
Chapter 5 Resource Masters
- *Study Guide,* p. 200
- *Skills Practice,* p. 201
- *Practice,* p. 202
- *Reading to Learn Mathematics,* p. 203
- *Enrichment,* p. 204

Hands-On Algebra, p. 64

Transparencies
5-Minute Check, 5–2

Technology/Multimedia
Interactive Chalkboard CD-ROM

One of the most common types of scale drawings is a map.

Example 1 Real World — **Map Link**

The scale on a map of Texas is 1 inch = 50 miles. Find the actual distance between Dallas and Houston if the distance between them on the map is 4.5 inches.

Use the scale and the distance given on the map to write a proportion.

$$\frac{1 \text{ inch}}{50 \text{ miles}} = \frac{4.5 \text{ inches}}{x \text{ miles}} \quad \begin{matrix} \leftarrow \textit{map distance} \\ \leftarrow \textit{actual distance} \end{matrix}$$

$1x = 50(4.5)$ *Find the cross products.*

$x = 225$ *Simplify.*

The distance between Dallas and Houston is about 225 miles.

Your Turn

Find the actual distance between San Antonio and Houston if the distance between them on the map is $3\frac{3}{4}$ inches. **187.5 mi**

You can determine the scale if you know the dimensions of the model and the dimensions of the actual object.

Example 2 Real World — **Movie Link**

The ship *Titanic*, which sank in 1912, was 880 feet long. When a movie was made about it more than 80 years later, a 44-foot-long model of the ship was created for special effects shots. Find the scale that was needed to design other parts of the model.

Write the ratio of the length of the model to the length of the ship. Then solve a proportion in which the length of the model is 1 foot and the length of the ship is x feet.

$$\frac{44 \text{ feet}}{880 \text{ feet}} = \frac{1 \text{ foot}}{x \text{ feet}} \quad \begin{matrix} \leftarrow \textit{model length} \\ \leftarrow \textit{actual length} \end{matrix}$$

$44x = 880(1)$ *Find the cross products.*

$44x = 880$ *Simplify.*

$\frac{44x}{44} = \frac{880}{44}$ *Divide each side by 44.*

$x = 20$ *Simplify.*

The scale is 1 foot = 20 feet or 1:20.

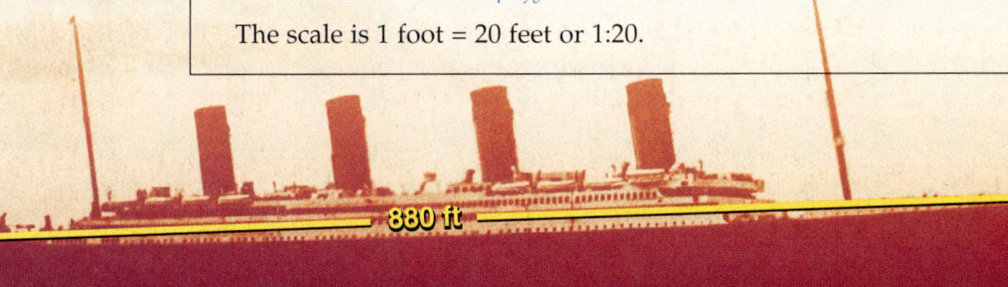

2 TEACH

Teaching Tip Before discussing Example 1, have students look at an actual map of your state or region. Ask them to locate and describe the scale information for the map, which is usually presented both in words and as a ruler or number line.

In-Class Examples

Example 1

The scale on a map of the upper Midwest is 1 inch = 15 miles. Find the distance between Chicago and Milwaukee on the map if the distance between the two cities is 90 miles. **6 in.**

Example 2

A railroad car is 36 feet long and a scale model of the railroad car is 1.5 feet long. What is the scale for the model? **1 ft = 24 ft or 1 : 24**

Study Guide, p. 200

NAME ______ DATE ______ PERIOD ______

Study Guide

Student Edition Pages 194–197

Scale Drawings and Models

A scale drawing or scale model is used to represent an object that is too large or too small to be drawn or built at its actual size.

Example 1: The Statue of Liberty is about 150 feet tall. A model is 15 inches tall. Find the scale used.

The scale used is the ratio of the length of the model to the actual length, or 15 in. = 150 feet. Reduce the ratio to get a scale of 1 in. = 10 feet.

Another method is to convert 150 feet to inches, or 150 ft = 150(12) in. = 1800 in. The scale is 15 in. = 1800 in. Reduce the ratio to get 1:120. The units do not need to be included if they are the same.

Example 2: The scale of a map of Florida is 1 inch = 20 miles. Find the actual distance between Miami and St. Petersburg if the distance between them on the map is 13 inches.

$$\frac{1 \text{ inch}}{20 \text{ miles}} = \frac{13 \text{ inches}}{x \text{ miles}} \quad \textit{Use a proportion.}$$

$1 \cdot x = 20(13)$

$x = 260$

The distance from Miami to St. Petersburg is 260 miles.

Find each scale or distance.

1. On the blueprint of a house, the kitchen is 5 inches long. If the actual kitchen is 20 feet long, find the scale of the blueprint. **1 in. = 4 ft**
2. On a map, the scale is 1 inch = 25 miles. Find the actual distance for each map distance. **130 mi; 250 mi; 210 mi**

From	To	Map Distance
Portsmouth, OH	Springfield, OH	5.2 inches
Chicago, IL	Lawrenceville, IL	10 inches
Santa Fe, NM	Clovis, NM	$8\frac{2}{5}$ inches

3. The Sears Tower is 1450 feet tall. If a model is 25 inches tall, find the scale. **1 in. = 58 ft**
4. In an HO scale model of a train, the length of the engine is 6 inches. If the HO scale is 1 : 87, find the actual length. Write the answer in feet. **43.5 ft**
5. Las Vegas, Nevada, is 445 miles from Reno, Nevada. If the distance on the map is $11\frac{1}{8}$ inches, find the scale used for the map. **1 in. = 40 mi**

Hands-On Algebra

Cooperative Learning Refer to the Hands-On Algebra on page 194. In Step 2, watch for students who draw a rectangle that is 2 units wide and 3 units long. Stress to them that the ratio 1 : 2 is given as the *ratio of the sides of the first rectangle to the second rectangle,* meaning that the sides of the second rectangle are twice as long as those of the first rectangle. You might ask students to repeat Exercises 2 and 3, using different rectangles than they did the first time.

Hands-On Algebra Masters, p. 64

3 PRACTICE/APPLY

Error Analysis

Watch for students who reverse one of the ratios when they attempt to write a proportion in Exercises 3–4.
Prevent by having students start each proportion by writing the ratios in words. In Exercises 3 and 4 for example, students could start by writing the ratio $\frac{\text{distance on map in inches}}{\text{actual distance in miles}}$.

Assignment Guide

Basic: 7–15 odd, 17–22
Average: 6–10 even, 12–22

Answer

1. Sample answer:

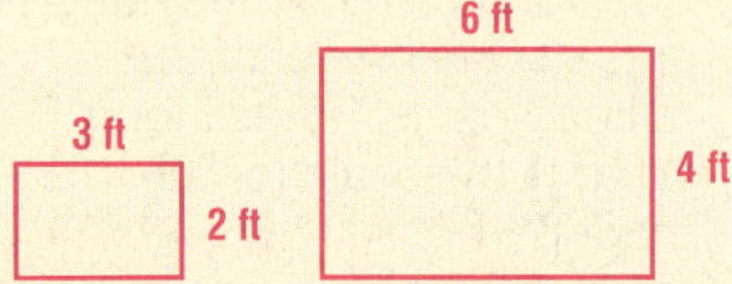

Skills Practice, **p. 201, and** ***Practice,*** **p. 202 (shown)**

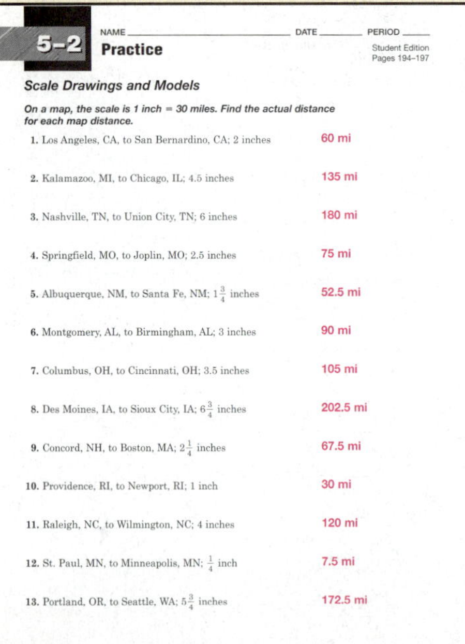

5-2 NAME ________ DATE ______ PERIOD ______

Practice

Student Edition Pages 194–197

Scale Drawings and Models

On a map, the scale is 1 inch = 30 miles. Find the actual distance for each map distance.

1. Los Angeles, CA, to San Bernardino, CA; 2 inches — 60 mi
2. Kalamazoo, MI, to Chicago, IL; 4.5 inches — 135 mi
3. Nashville, TN, to Union City, TN; 6 inches — 180 mi
4. Springfield, MO, to Joplin, MO; 2.5 inches — 75 mi
5. Albuquerque, NM, to Santa Fe, NM; $1\frac{3}{4}$ inches — 52.5 mi
6. Montgomery, AL, to Birmingham, AL; 3 inches — 90 mi
7. Columbus, OH, to Cincinnati, OH; 3.5 inches — 105 mi
8. Des Moines, IA, to Sioux City, IA; $6\frac{3}{4}$ inches — 202.5 mi
9. Concord, NH, to Boston, MA; $2\frac{1}{4}$ inches — 67.5 mi
10. Providence, RI, to Newport, RI; 1 inch — 30 mi
11. Raleigh, NC, to Wilmington, NC; 4 inches — 120 mi
12. St. Paul, MN, to Minneapolis, MN; $\frac{1}{4}$ inch — 7.5 mi
13. Portland, OR, to Seattle, WA; $5\frac{3}{4}$ inches — 172.5 mi

Check for Understanding

Vocabulary
scale drawing
scale model
scale

Communicating Mathematics

1. **Sketch** two rectangles in which the ratio of the sides of the first rectangle to the sides of the second rectangle is 1:2. **See margin.**
2. **Name** two careers in which you would use scale drawings or scale models. **Sample answer: architect, sculptor**

Guided Practice

Example 1

On a map, the scale is 1 inch = 50 miles. Find the actual distance for each map distance.

	From	To	Map Distance	
3.	Sacramento, CA	Fresno, CA	3 inches	**150 mi**
4.	Kansas City, MO	St. Louis, MO	$4\frac{3}{4}$ inches	**237.5 mi**

Example 2

5. **Transportation** A scale model of a Boeing 747 jumbo jet is 2 meters long. If an actual jet is 70.4 meters long, find the scale of the model. **1 m = 35.2 m**

Exercises

Practice 

On a map, the scale is 1 inch = 40 miles. Find the actual distance for each map distance.

	From	To	Map Distance	
6.	Tampa, FL	Orlando, FL	2 inches	**80 mi**
7.	Albany, NY	Syracuse, NY	3 inches	**120 mi**
8.	Jacksonville, FL	Pensacola, FL	8 inches	**320 mi**
9.	Washington, DC	Richmond, VA	2.5 inches	**100 mi**
10.	Cleveland, OH	Pittsburgh, PA	$2\frac{3}{4}$ inches	**110 mi**
11.	Hartford, CT	New Haven, CT	$\frac{3}{4}$ inch	**30 mi**

Homework Help

For Exercises	See Examples
6–11, 16	1
12–15	2

Extra Practice
See page 701.

Applications and Problem Solving 

12. **Bridges** The 1500-foot Natchez Trace Bridge in Natchez, Mississippi, is the longest precast segmental arch bridge in North America. If a model of the bridge is 12 inches long, find the scale of the model.

12. 1 in. = 125 ft

13. **Model Cars** In a scale model of a race car, the wheels have a diameter of 2 inches. If the actual wheels have a diameter of 30 inches, find the scale of the model. **1 in. = 15 in.**
14. **Architecture** On a blueprint of a house, one bedroom is 4 inches long. When the house is built, the room will be 16 feet long. Find the scale of the blueprint. **1 in. = 4 ft**

Reteaching Activity

Kinesthetic Learners Prior to class, draw a figure made from several geometric shapes. Make numerous copies of the figure on a copy machine. Then make several enlarged copies and several reduced copies of the figure. Have students work with a partner. Provide each pair with one full-size copy of your figure and one of the enlarged or reduced copies. Have students use a ruler to measure several pairs of corresponding lengths on their figures, and use the results to determine the scale that relates the two figures.

15. **Life Science** In a photograph in a science textbook, the length of a ladybug is 2.5 centimeters. The scale of the photograph is 1 centimeter = 0.2 centimeter. What is the actual length of the ladybug? **0.5 cm**

16. **Travel** On a map of Colorado, the scale is 1 centimeter = 40 kilometers. The distance on the map between the northern and southern boundaries of Colorado is 11.5 centimeters.

16a. 460 km

a. How many kilometers is the northern boundary from the southern boundary?

b. Suppose you travel at an average speed of 80 kilometers per hour. How long will it take you to travel through the state along Route 25? Round your answer to the nearest hour. **6 h**

17. **Critical Thinking** The state of Pennsylvania is 332 miles long and 179 miles wide. Suppose you want to draw a map of Pennsylvania on an $8\frac{1}{2}$-by-11 inch piece of paper. Find the scale that will allow you to draw the map as large as possible. **1 in. = 30.2 mi**

Mixed Review

18. **Monuments** A monument of Crazy Horse, the famous Oglala chief, has been under construction for more than 50 years. The sculptors use a model to calculate how to carve the statue. The length of the model is 18.8 feet, the height of the model is 16.6 feet, and the height of the monument is 563 feet. To the nearest foot, what is the length of the monument? *(Lesson 5–1)* **638 ft**

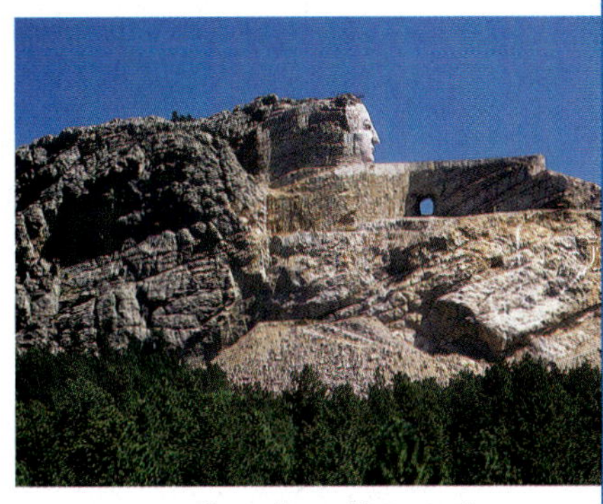

Crazy Horse Monument

Solve each equation. *(Lesson 4–7)*

19. $3(-2x + 5) = 7$ $\frac{4}{3}$ 20. $-2(3y - 4) = 3 - y$ **1** 21. $2(3x + 1) - 8x = 10$ **−4**

22. **Multiple Choice** What relationship exists between the x- and y-coordinates of each of the points shown on the graph? *(Lesson 2–2)* **D**

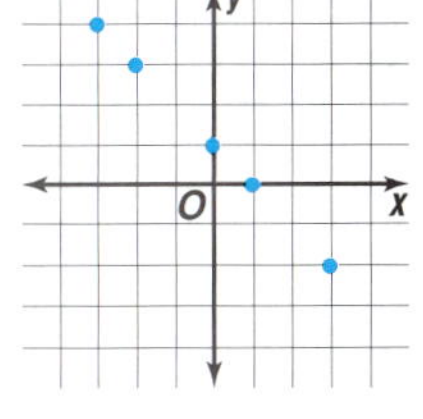

Ⓐ They are opposites.

Ⓑ Their sum is 2.

Ⓒ The y-coordinate is 1 more than the x-coordinate.

Ⓓ Their sum is 1.

www.algconcepts.com/self_check_quiz

4 ASSESS

Open-Ended Assessment

Speaking Ask students to describe what they would look for to determine whether a dollhouse, toy car, or toy train is a scale model of an actual object.

Extra Credit

A rectangular sculpture is 3 feet wide and 5 feet long. In a scale drawing of the sculpture, the perimeter of the rectangle is 8 inches. Find the width and length of the scale drawing.
1.5 in. wide; 2.5 in. long

***Enrichment*, p. 204**

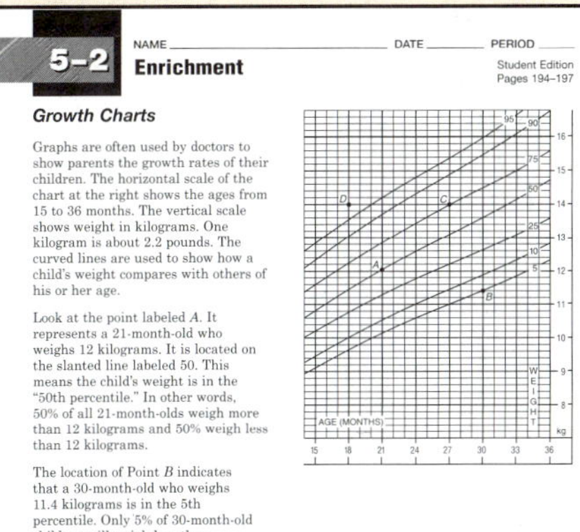

NAME ______ DATE ______ PERIOD ______

5–2 Enrichment

Student Edition Pages 194–197

Growth Charts

Graphs are often used by doctors to show parents the growth rates of their children. The horizontal scale of the chart at the right shows the ages from 15 to 36 months. The vertical scale shows weight in kilograms. One kilogram is about 2.2 pounds. The curved lines are used to show how a child's weight compares with others of his or her age.

Look at the point labeled A. It represents a 21-month-old who weighs 12 kilograms. It is located on the slanted line labeled 50. This means the child's weight is in the "50th percentile." In other words, 50% of all 21-month-olds weigh more than 12 kilograms and 50% weigh less than 12 kilograms.

The location of Point B indicates that a 30-month-old who weighs 11.4 kilograms is in the 5th percentile. Only 5% of 30-month-old children will weigh less than 11.4 kilograms.

Solve.

1. Look at the point labeled C. How much does the child weigh? How old is he? What percent of children his age will weigh more than he does? **14 kg; 27 months; 25%**

2. Look at the point labeled D. What is the child's age and weight? What percent of children her age will weigh more than she does? **18 months; 14 kg; less than 5%**

3. What is the 50th percentile weight for a child 27 months old? **13.1 kg**

4. How much weight would child B have to gain to be in the 50th percentile? **2.2 kg**

5. If child D did not gain any weight for four months, what percentile would he be in? **90th**

6. How much heavier is a $2\frac{1}{2}$-year-old in the 90th percentile than one in the 10th percentile? **3.6 kg**

1 FOCUS

5-Minute Check
Lesson 5–2

1. The distance between two cities is 180 miles. On a map, this distance is 5 inches. Find the scale used for drawing the map. **1 in. = 36 mi**

On a map, the scale is 1 inch = 40 miles.

2. Find the actual distance between two locations if the distance between them on the map is $3\frac{3}{4}$ inches. **150 mi**
3. Find the map distance for an actual distance of 220 miles. **$5\frac{1}{2}$ in.**

On a map, the scale is 1 inch = 20 miles. Find the actual distance for each map distance.

	From	To	Map Distance
4.	Miami, FL	Orlando, FL	$11\frac{1}{2}$ in.
5.	Omaha, NE	Wichita, KS	15 in.

4. 230 mi 5. 300 mi

TECHNOLOGY

An alternative technology option using a graphing calculator is available for teaching this lesson.

5-3 The Percent Proportion

What You'll Learn
You'll learn to solve problems by using the percent proportion.

Why It's Important
Health Nurses use percents to calculate the correct dosage of medicines. *See Exercise 42.*

Quantities are often expressed as percents. A **percent** is a ratio that compares a number to 100. Percent also means *per hundred* or *hundredths*. For example, a 5% glucose solution contains 5 grams of glucose in 100 milliliters of solution.

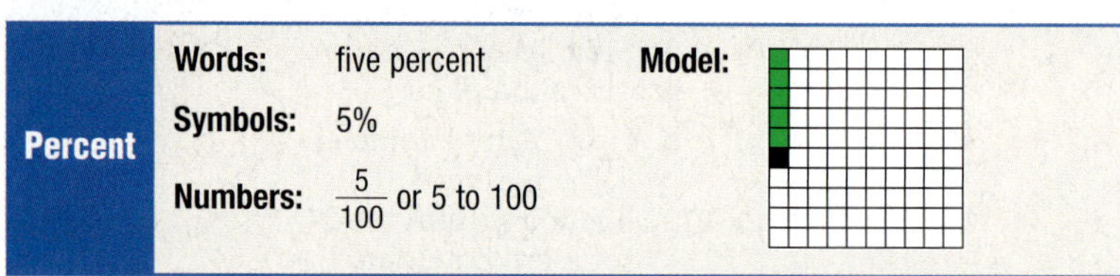

Percent	
Words:	five percent
Symbols:	5%
Numbers:	$\frac{5}{100}$ or 5 to 100
Model:	

You can express a fraction or a ratio as a percent by using a proportion in which one denominator is 100.

Examples

Express each fraction or ratio as a percent.

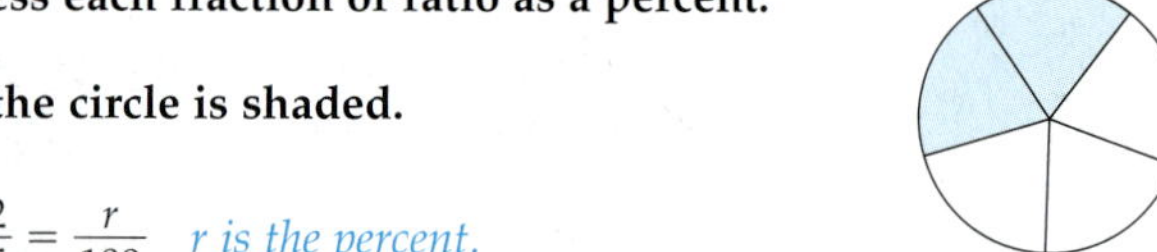

1 **$\frac{2}{5}$ of the circle is shaded.**

$$\frac{2}{5} = \frac{r}{100} \quad \textit{r is the percent.}$$
$$2(100) = 5r \quad \textit{Find the cross products.}$$
$$200 = 5r \quad \textit{Multiply 2 and 100.}$$
$$\frac{200}{5} = \frac{5r}{5} \quad \textit{Divide each side by 5.}$$
$$40 = r \quad \textit{Simplify.}$$

So, $\frac{2}{5} = 40\%$.

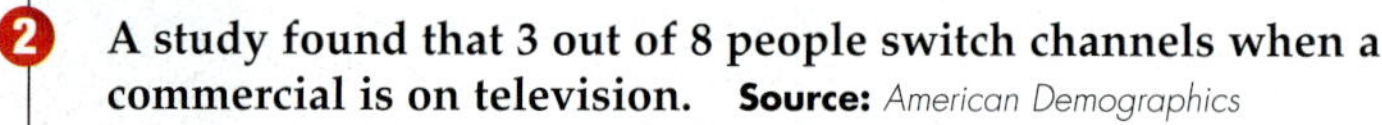

2 **A study found that 3 out of 8 people switch channels when a commercial is on television.** **Source:** *American Demographics*

$$\frac{3}{8} = \frac{r}{100} \quad \textit{Write a proportion.}$$
$$3(100) = 8r \quad \textit{Find the cross products.}$$
$$300 = 8r \quad \textit{Multiply 3 and 100.}$$
$$\frac{300}{8} = \frac{8r}{8} \quad \textit{Divide each side by 8.}$$
$$37.5 = r$$

37.5% of people switch channels.

Your Turn

a. $\frac{73}{50}$ **146%**

b. 2 out of 3 **$66\frac{2}{3}\%$ or $66.\overline{6}\%$**

Resource Manager

Reproducible Masters
Chapter 5 Resource Masters
- *Study Guide,* p. 205
- *Skills Practice,* p. 206
- *Practice,* p. 207
- *Reading to Learn Mathematics,* p. 208
- *Enrichment,* p. 209
- *Assessment,* p. 241

Graphing Calculator, pp. 15–16
Hands-On Algebra, pp. 65–66

Transparencies
5-Minute Check, 5-3

Technology/Multimedia
AlgePASS, Lesson 14
Interactive Chalkboard CD-ROM

In Example 1, the number of parts shaded, 2, is called the **percentage** (P). It is being compared to the total number of parts, 5, which is called the **base** (B).

$$\begin{matrix} \text{percentage} \rightarrow \\ \text{base} \quad\quad \rightarrow \end{matrix} \frac{2}{5} = \frac{40}{100} \leftarrow \text{percent}$$

This is an example of a **percent proportion**.

Percent Proportion	If P is the percentage, B is the base, and r is the percent, the percent proportion is $\frac{P}{B} = \frac{r}{100}$.

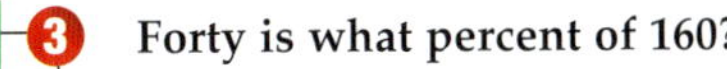

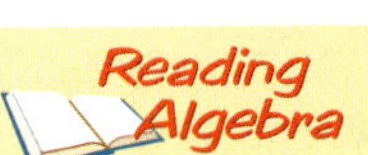

In percent problems, the base follows the word *of*. It is *not* always the larger number.

3 **Forty is what percent of 160?**

$\frac{P}{B} = \frac{r}{100}$ *Use the percent proportion. Write "what percent" as* $\frac{r}{100}$.

$\frac{40}{160} = \frac{r}{100}$ *Replace P with 40 and B with 160.*

$40(100) = 160r$ *Find the cross products.*

$4000 = 160r$ *Multiply 40 and 100.*

$\frac{4000}{160} = \frac{160r}{160}$ *Divide each side by 160.*

$25 = r$ *Simplify.*

So, 40 is 25% of 160.

4 **50% of what number is 42.5?**

$\frac{P}{B} = \frac{r}{100}$ *Use the percent proportion.*

$\frac{42.5}{B} = \frac{50}{100}$ *Replace P with 42.5 and r with 50.*

$42.5(100) = 50B$ *Find the cross products.*

$4250 = 50B$ *Multiply 425 and 100.*

$\frac{4250}{50} = \frac{50B}{50}$ *Divide each side by 50.*

$85 = B$ *Simplify.*

So, 50% of 85 is 42.5.

Your Turn

c. 7 is what percent of 20? **35%**
d. 60 is 15% of what number? **400**
e. Find 25% of 66. **16.5**
f. What number is 10% of 88? **8.8**

Motivating the Lesson

Hands-On Activity Distribute copies of a 10-by-10 grid. Give students the following directions for drawing a closed figure on their grid: choose a starting point, draw only along the grid lines, do not intersect any previously-drawn lines, and complete the figure by ending at the starting point. After everyone has completed their figures, ask them to determine what percent of the area of the grid is inside their figure and what percent is outside of it. When the students have finished, ask them if it was necessary to count the squares inside and outside their figure to find both percents, or if they calculated one percent and then used it to find the other.

2 TEACH

In-Class Examples

Examples 1–2

Express each fraction or ratio as a percent.

1 $\frac{5}{8}$ of the square is shaded.

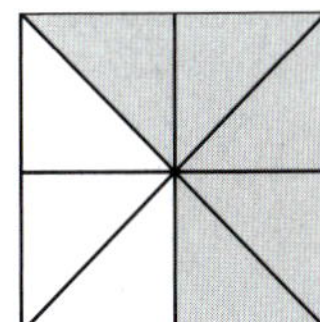

62.5%

2 In a classroom, 18 out of 24 students used a computer at home last week. **75%**

Example 3

What percent of 175 is 35? **20%**

Example 4

20 is 40% of what number? **50**

In-Class Example

Example 5

A family recently moved to a new home. The table below shows the number of days they spent packing boxes, cleaning the two homes, and then unpacking boxes.

Project	Days
Packing	7
Cleaning	5
Unpacking	8

What percent of the total did they spend on each activity? **packing: 35%; cleaning: 25%; unpacking: 40%**

Teaching Tip After presenting Example 5, ask students to redo the percent column, rounding the values to the nearest whole percent. Have them add the percents, and then discuss how rounding can result in the sum of the percents being greater than or less than 100%.

In-Class Example

Example 6

Make a circle graph of the data in In-Class Example 5.

Moving Activities

Businesses can use a time study to determine what percent of an employee's time is spent on various activities.

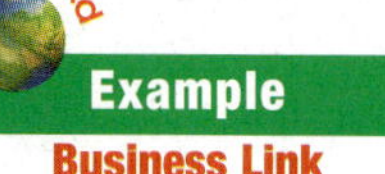

Example 5 Business Link

The manager of The Furniture Store conducted a time study of the employees in its warehouse. The chart at the right shows the average number of hours spent on each activity during the workday. What percent of the time do the employees spend on each activity?

Activity	Time (hours)
Loading furniture	3
Delivering furniture	4
Lunch	1

The employees work 3 + 4 + 1 or 8 hours each day. This is the base. To find each percent, write and solve the percent proportion for each activity.

Loading: $\frac{3}{8} = \frac{r}{100}$ 3 [×] 100 [÷] 8 [ENTER] 37.5

Delivering: $\frac{4}{8} = \frac{r}{100}$ 4 [×] 100 [÷] 8 [ENTER] 50

Lunch: $\frac{1}{8} = \frac{r}{100}$ 1 [×] 100 [÷] 8 [ENTER] 12.5

The results are summarized in the chart.

Note that the total of the percent column is 100%.

Activity	Time (hours)	Percent
Loading furniture	3	37.5
Delivering furniture	4	50.0
Lunch	1	12.5

Data are often displayed using circle graphs. A **circle graph** is a graph that shows the relationship between parts of the data and the whole.

Example 6 Statistics Link

Make a circle graph of the data in Example 5.

Loading furniture accounts for 37.5% of the time. A circle is composed of 360°. To find the number of degrees for this section of the graph, find 37.5% of 360. Repeat this process for the other sections. *Replace B with 360 and r with each percent.*

Loading: $\frac{P}{360} = \frac{37.5}{100}$ 360 [×] 37.5 [÷] 100 [ENTER] 135

Delivering: $\frac{P}{360} = \frac{50}{100}$ 360 [×] 50 [÷] 100 [ENTER] 180

Lunch: $\frac{P}{360} = \frac{12.5}{100}$ 360 [×] 12.5 [÷] 100 [ENTER] 45

Use a compass to draw a circle. Then draw a radius. Use a protractor to draw a 135° angle. *You can start with any of the angles.*

135°
radius

From the new radius, draw a 180° angle. The remaining section should be 45°. Label each section of the graph with the category and percent. Give the graph a title.

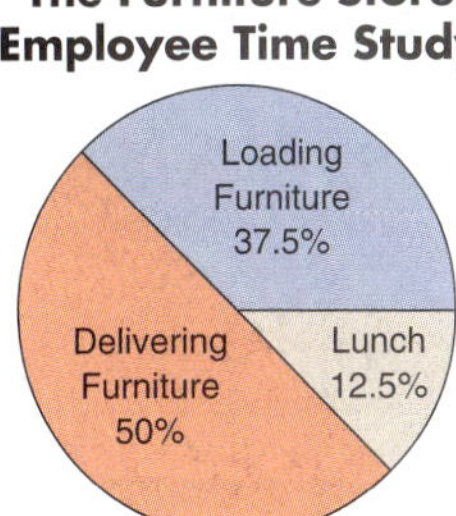

Check for Understanding

Communicating Mathematics

1–2. See margin.

1. **Explain** what P, B, and r represent in the percent proportion.
2. **You Decide?** Brian and Julia are trying to find what percent 30 is of 20. Brian uses the proportion $\frac{30}{20} = \frac{r}{100}$. Julia uses the proportion $\frac{20}{30} = \frac{r}{100}$. Who is correct? Explain your reasoning.

Vocabulary
percent
percentage
base
percent proportion
circle graph

Guided Practice

Write a proportion that can be used to find each number. 3–6. See margin.

Sample: 5 is what percent of 40? **Solution:** $\frac{5}{40} = \frac{r}{100}$

3. What percent of 24 is 12?
4. 80 is 75% of what number?
5. Find 45% of 60.
6. What percent of 50 is 100?

Examples 1 & 2

Express each fraction or ratio as a percent.

7. 3 out of 5 **60%**
8. $\frac{1}{3}$ **$33\frac{1}{3}$%**
9. 7 to 10 **70%**
10. In a recent year, 17 out of 20 students used the Internet as a reference source. **85%**

3 PRACTICE/APPLY

Error Analysis

Watch for students who incorrectly write the percent as a decimal in the percent proportion, such as writing $\frac{r}{100}$ as $\frac{0.25}{100}$ when $r = 25$.

Prevent by stressing that the % symbol means "per hundred," and encouraging students to develop the habit of reading 25% as "25 per hundred" or "25 over 100."

Answers

1. **P is the percentage, B is the base, and r is the percent or the rate per hundred.**
2. **Brian; the problem is comparing 30 to 20. Therefore, 20 is the base.**
3. $\frac{12}{24} = \frac{r}{100}$
4. $\frac{80}{B} = \frac{75}{100}$
5. $\frac{P}{60} = \frac{45}{100}$
6. $\frac{100}{50} = \frac{r}{100}$

***Study Guide,* p. 205**

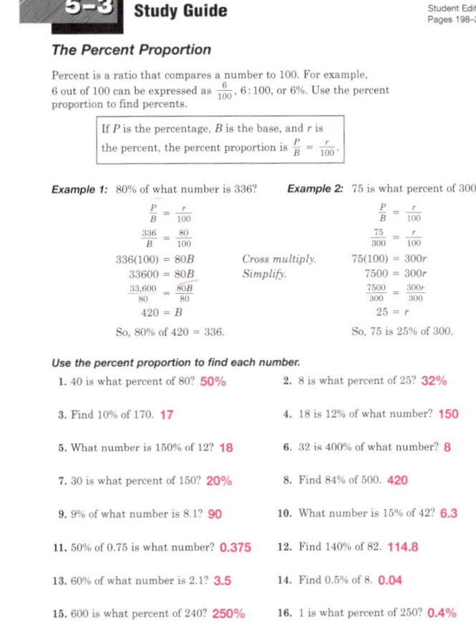

NAME ______ DATE ______ PERIOD ______

5-3 Study Guide

Student Edition Pages 198–203

The Percent Proportion

Percent is a ratio that compares a number to 100. For example, 6 out of 100 can be expressed as $\frac{6}{100}$, 6:100, or 6%. Use the percent proportion to find percents.

If P is the percentage, B is the base, and r is the percent, the percent proportion is $\frac{P}{B} = \frac{r}{100}$.

Example 1: 80% of what number is 336?

$$\frac{P}{B} = \frac{r}{100}$$
$$\frac{336}{B} = \frac{80}{100}$$
$$336(100) = 80B \quad \textit{Cross multiply.}$$
$$33600 = 80B \quad \textit{Simplify.}$$
$$\frac{33{,}600}{80} = \frac{80B}{80}$$
$$420 = B$$

So, 80% of 420 = 336.

Example 2: 75 is what percent of 300?

$$\frac{P}{B} = \frac{r}{100}$$
$$\frac{75}{300} = \frac{r}{100}$$
$$75(100) = 300r$$
$$7500 = 300r$$
$$\frac{7500}{300} = \frac{300r}{300}$$
$$25 = r$$

So, 75 is 25% of 300.

Use the percent proportion to find each number.

1. 40 is what percent of 80? **50%**
2. 8 is what percent of 25? **32%**
3. Find 10% of 170. **17**
4. 18 is 12% of what number? **150**
5. What number is 150% of 12? **18**
6. 32 is 400% of what number? **8**
7. 30 is what percent of 150? **20%**
8. Find 84% of 500. **420**
9. 9% of what number is 8.1? **90**
10. What number is 15% of 42? **6.3**
11. 50% of 0.75 is what number? **0.375**
12. Find 140% of 82. **114.8**
13. 60% of what number is 2.1? **3.5**
14. Find 0.5% of 8. **0.04**
15. 600 is what percent of 240? **250%**
16. 1 is what percent of 250? **0.4%**

Reteaching Activity

Auditory/Musical Learners Have students write out their own description for the percent proportion, such as "the percentage is to the base as the percent is to one hundred" or "the ratio of the percentage to the base is the same as the ratio of the percent to one hundred." Have them put their description to music as a means for memorizing the correct relationships between the parts of a proportion.

Assignment Guide

Basic: 17–43 odd, 44–52
Average: 16–38 even, 40–52
All: Quiz 1, 1–5

Answer

15b.

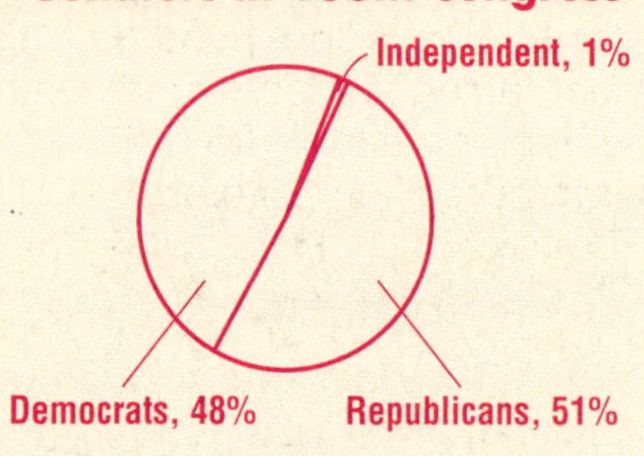

***Skills Practice*, p. 206, and *Practice*, p. 207 (shown)**

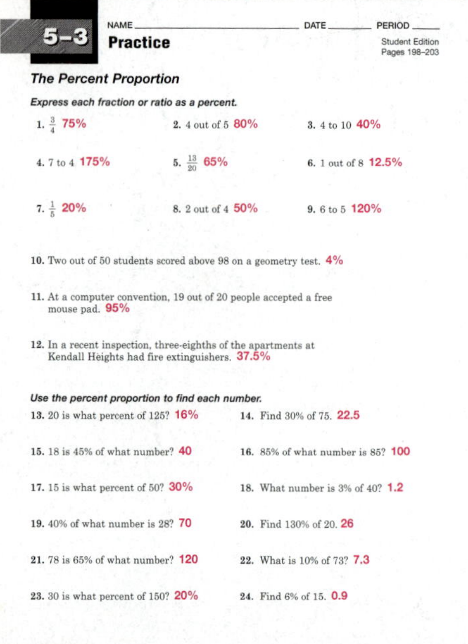

5–3 NAME ______ DATE ______ PERIOD ______

Practice

Student Edition Pages 198–203

The Percent Proportion

Express each fraction or ratio as a percent.

1. $\frac{3}{4}$ 75%
2. 4 out of 5 80%
3. 4 to 10 40%
4. 7 to 4 175%
5. $\frac{13}{20}$ 65%
6. 1 out of 8 12.5%
7. $\frac{1}{5}$ 20%
8. 2 out of 4 50%
9. 6 to 5 120%
10. Two out of 50 students scored above 98 on a geometry test. 4%
11. At a computer convention, 19 out of 20 people accepted a free mouse pad. 95%
12. In a recent inspection, three-eighths of the apartments at Kendall Heights had fire extinguishers. 37.5%

Use the percent proportion to find each number.

13. 20 is what percent of 125? 16%
14. Find 30% of 75. 22.5
15. 18 is 45% of what number? 40
16. 85% of what number is 85? 100
17. 15 is what percent of 50? 30%
18. What number is 3% of 40? 1.2
19. 40% of what number is 28? 70
20. Find 130% of 20. 26
21. 78 is 65% of what number? 120
22. What is 10% of 73? 7.3
23. 30 is what percent of 150? 20%
24. Find 6% of 15. 0.9

Examples 3 & 4

Use the percent proportion to find each number.

11. 30 is what percent of 120? **25%**
12. 16 is 40% of what number? **40**
13. Find 20% of 82. **16.4**
14. Find 250% of 8. **20**

15. Government In the 108th Congress, there were 48 Democratic Senators, 51 Republican Senators, and 1 Independent Senator.

Example 5 a. What percent of the Senators were Democrats? What percent were Republicans? What percent were Independents? **48%, 51%, 1%**

Example 6 b. Make a circle graph of the data. **See margin.**

Exercises

Practice

Express each fraction or ratio as a percent.

A

16. $\frac{1}{4}$ **25%**
17. 5 out of 8 **62.5%**
18. 9 to 10 **90%**
19. 4 to 5 **80%**
20. $\frac{5}{4}$ **125%**
21. $\frac{9}{12}$ **75%**
22. 9 out of 20 **45%**
23. 3 to 2 **150%**
24. 1 out of 6 **$16\frac{2}{3}\%$**

25. Three out of 5 people have eaten in an Italian restaurant during the past six months. **60%**

26. Two-thirds of all households own their homes. **$66\frac{2}{3}\%$**

27. About 3 of every 50 people in the United States are between the ages of 14 and 17. **6%**

Homework Help

For Exercises	See Examples
16–27	1–2
28–39	3–4
40–41	3
42	5
43	6

Extra Practice

See page 701.

Use the percent proportion to find each number.

28. 60 is what percent of 150? **40%**
29. 9 is what percent of 25? **36%**
30. 15 is 20% of what number? **75**
31. 50% of what number is 95? **190**
32. What number is 60% of 5? **3**
33. Find 10% of 125. **12.5**

B

34. What percent of 50 is 75? **150%**
35. What number is 125% of 48? **60**
36. 5% of what number is 6.5? **130**
37. 1 is what percent of 200? **0.5%**
38. Find 0.1% of 450. **0.45**
39. 18.6 is 200% of what number? **9.3**

Applications and Problem Solving

40. Education If you answer 35 problems correctly on a 40-problem test, what percent did you answer correctly? **87.5%**

41. Pharmacy A peroxide solution contains 5 milliliters of peroxide in 10 milliliters of solution. What is the percent of peroxide in the solution? **50%**

42. Nursing A nurse prepares a 5% glucose solution by dissolving 5 grams of glucose in 100 milliliters of solution. How many grams of glucose are in 20 milliliters of the solution? **1 g**

43. **Sales** About 21 million people give floral gifts for Mother's Day. The chart shows the percent of fresh-cut flowers that are typically purchased. Make a circle graph of the data. **See margin.**

Mother's Day Flowers (percent)

Mother's Day Flowers (percent)	
Mixed Bouquets	47
Roses	23
Carnations	16
Single-Stem Flowers	14

Source: California Cut Flower Commission

44. Sample answer: Greta's income is $50 per week, and last week she spent $75. She spent 150% of her weekly income.

44. **Critical Thinking** Describe a real-life situation in which the percentage is greater than the base.

Mixed Review

45. **Maps** The distance from St. Louis, Missouri, to Dallas, Texas, is 630 miles. On a map, the distance is 9 inches. What is the scale of the map? *(Lesson 5–2)* **1 in. = 70 mi**

Convert each measurement as indicated. *(Lesson 5–1)*

46. 54 inches to yards **1.5 yd**

47. 2.5 kilometers to meters **2500 m**

48. **Health** An ulcer medication has 300 milligrams in 2 tablets. How many milligrams are in 3 tablets? *(Lesson 5–1)* **450 mg**

Add or subtract. *(Lesson 3–2)*

49. $-1.5 - 3.7$ **-5.2**

50. $\frac{3}{8} - \left(-\frac{1}{8}\right)$ **$\frac{1}{2}$**

51. $\frac{3}{5} + \left(-\frac{1}{5}\right)$ **$\frac{2}{5}$**

Standardized Test Practice

52. **Multiple Choice** Which set of numbers is *not* closed under multiplication? *(Lesson 1–3)* **B**

Ⓐ real numbers
Ⓑ negative integers
Ⓒ positive integers
Ⓓ whole numbers

Quiz 1 — Lessons 5–1 through 5–3

Solve each proportion. *(Lesson 5–1)*

1. $\frac{9}{25} = \frac{x}{75}$ **27**

2. $\frac{z}{9} = \frac{z+4}{21}$ **3**

3. **Surveying** A civil engineer uses the scale 1 inch = 40 feet to draw a parcel of land. How many linear feet of land are represented by a 2.5-inch line on the drawing? *(Lesson 5–2)* **100 ft**

Use the percent proportion to find each number. *(Lesson 5–3)*

4. Six is what percent of 15? **40%**

5. 95 is 10% of what number? **950**

www.algconcepts.com/self_check_quiz

4 ASSESS

Open-Ended Assessment

Modeling Have students model various percents between 0 and 100 by shading portions of a 10-by-10 grid.

Quiz 1

The Quiz provides students with a brief review of the concepts and skills in Lessons 5–1 through 5–3. Lesson numbers are given to the right of the exercises or instruction lines so students can review concepts not yet mastered.

Chapter 5, Quiz A (Lessons 5–1 through 5–3) is available in the *Chapter 5 Resource Masters*, p. 241.

Answer

43. Mother's Day Flowers

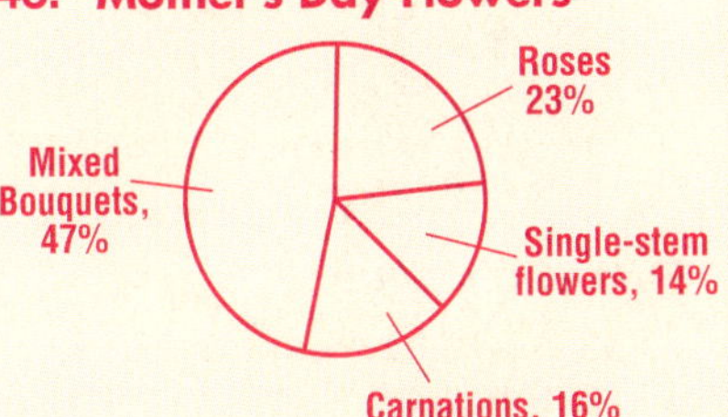

Extra Credit

A circle graph has three sections. If the ratio of the values modeled by the sections is 2:3:4, find the angle measure for each section of the circle graph. **80°, 120°, 160°**

***Enrichment*, p. 209**

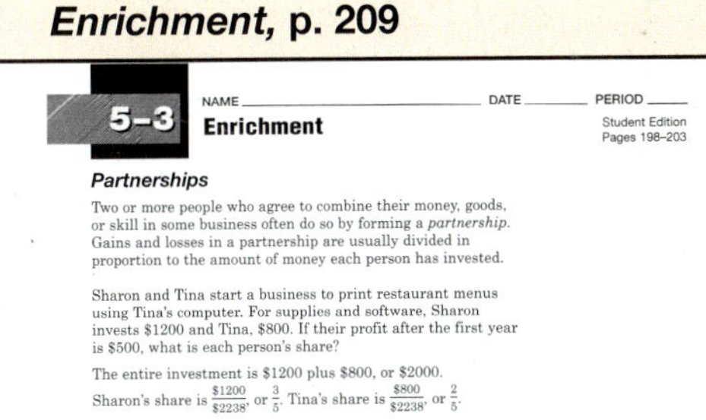

5–3 NAME ______ DATE ______ PERIOD ______

Enrichment

Student Edition Pages 198–203

Partnerships

Two or more people who agree to combine their money, goods, or skill in some business often do so by forming a *partnership*. Gains and losses in a partnership are usually divided in proportion to the amount of money each person has invested.

Sharon and Tina start a business to print restaurant menus using Tina's computer. For supplies and software, Sharon invests $1200 and Tina, $800. If their profit after the first year is $500, what is each person's share?

The entire investment is $1200 plus $800, or $2000. Sharon's share is $\frac{\$1200}{\$2238}$, or $\frac{3}{5}$. Tina's share is $\frac{\$800}{\$2238}$, or $\frac{2}{5}$.

Answer each question.

1. In the problem above, what is Sharon's share of the profit? **$300**

2. What is Tina's share? **$200**

3. Sal and Mario were partners in a construction business. Sal put in $5238 and Mario, $4238. Their profit in 3 years was $4500. What was each partner's share? **Sal: $2500; Mario: $2238**

4. Tim, Bob, and Alice entered into a business partnership for 2 years. Tim put in $3600, Bob put in $2400, and Alice put in $2238. If their profit was $3238, what was each person's share? **Tim: $1350; Bob: $900; Alice: $750**

5. Fred, Vito, and Hal began with a capital of $28,238. Fred furnished $7238; Vito, $6238; and Hal, the remainder. If they gained 14% on their investment, what was each person's share? **Fred: $980; Vito: $840; Hal: $2100**

6. A storeroom belonging to Taylor, Bartinelli, & Wong was entirely destroyed by fire. They received $9675 insurance. What was each person's share, if Taylor owned 15%, Bartinelli, 40%, and Wong the remainder? **Taylor: $1451.25; Bartinelli: $3870; Wong: $4353.75**

7. Divide a profit of $7200 between two partners so that the ratio of the shares is 4 to 5. **$3200, $4238**

8. Divide a profit of $18,238 among three partners so that the ratio of the shares is 1:2:3. **$3238, $6238, $9238**

Lesson 5-4

1 FOCUS

5-Minute Check
Lesson 5–3

Express each fraction as a percent.

1. $\frac{3}{6}$ of the circle is shaded. **50%**

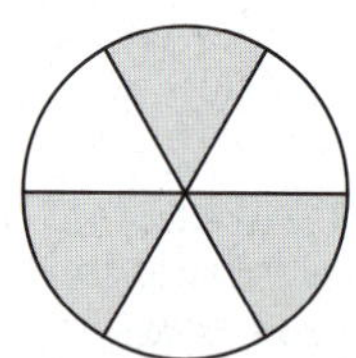

2. $\frac{6}{8}$ of the figure is shaded. **75%**

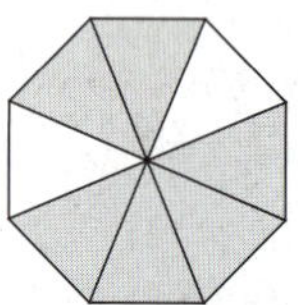

Find each value.

3. 32 is what percent of 200? **16%**
4. 80% of what number is 24? **30**
5. What percent of 90 is 36? **40%**

Motivating the Lesson

Real-World Connection Ask students to describe when the cost of an item or service ends up being more or being less than the marked price. Their answers for "more" may include sales tax, tips, shipping or handling charges, and their answers for "less" may include discounts, coupons, and rebates.

2 TEACH

Teaching Tip As you introduce this lesson, be sure students understand the difference between r in the Percent Proportion and R in the Percent Equation. Emphasize that if r is a percent such as 35%, then R is the decimal number 0.35.

5–4 The Percent Equation

What You'll Learn
You'll learn to solve problems by using the percent equation.

Why It's Important
Business Employers use the percent equation to calculate taxes. *See Example 3.*

Interest and tax payments involve finding percents. You can use the percent proportion to solve these kinds of problems. However, it is usually easier to write the percent proportion as an equation.

$\frac{P}{B} = \frac{r}{100}$	*Start with the percent proportion.*
$\frac{P}{B} = R$	*Replace $\frac{r}{100}$ with R.*
$\frac{P}{B}B = RB$	*Multiply each side of the equation by B.*
$P = RB$	

The equation $P = RB$ is called the **percent equation**. In this equation, R is the **rate**. The rate is the decimal form of the percent.

Percent Equation

Words: The percentage is equal to the rate times the base.

Symbols: $P = RB$, where P is the percentage, B is the base, and R is the rate.

The percent equation is easier to use when the rate and base are known. However, the percent equation can be used to solve any percent problem.

Examples

1 **Find 4% of $160.**

$P = RB$	*Use the percent equation.*
$= 0.04(160)$	*Replace R with 0.04 and B with 160.*
$= 6.4$	0.04 [×] 160 [ENTER] 6.4

So, 4% of $160 is $6.40.

Reading Algebra

To write a percent as a decimal, place or move the decimal point two places to the left and drop the percent sign.

4% = 04% = 0.04
60% = 60% = 0.6
1.8% = 01.8% = 0.018

2 **12 is 60% of what number?**

$P = RB$	*Use the percent equation.*
$12 = 0.6B$	*Replace P with 12 and R with 0.6.*
$\frac{12}{0.6} = \frac{0.6B}{0.6}$	*Divide each side by 0.6.*
$20 = B$	12 [÷] 0.6 [ENTER] 20

So, 12 is 60% of 20.

Your Turn

a. Find 62% of 120. **74.4**

b. 75% of what number is 12? **16**

Resource Manager

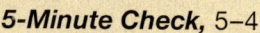

Reproducible Masters
Chapter 5 Resource Masters
- *Study Guide,* p. 210
- *Skills Practice,* p. 211
- *Practice,* p. 212
- *Reading to Learn Mathematics,* p. 213
- *Enrichment,* p. 214
- *Assessment,* p. 240

School-to-Workplace, p. 5

Transparencies
5-Minute Check, 5–4

Technology/Multimedia
AlgePASS, Lesson 14
Interactive Chalkboard CD-ROM

Many real-world problems can be solved by using the percent equation.

Example 3
Tax Link

The Federal Insurance Contributions Act (FICA) requires employers to deduct 6.2% of your income for social security taxes. Suppose your weekly pay is $140. What amount would be deducted from your pay for social security taxes?

To find the amount deducted, find 6.2% of 140.

$P = RB$ *Use the percent equation.*
$= 0.062(140)$ *Replace R with 0.062 and B with 140.*
$= 8.68$ *0.062 [×] 140 [ENTER] 8.68*

So, $8.68 would be deducted from your pay.

Technology Tip

Many scientific calculators have a [%] key.

6.2 [%] [×] 140 [=] *8.68*

Percents are also used in simple interest problems. **Simple interest** is the amount paid or earned for the use of money. If you have a savings account, you earn interest. If you borrow money through a loan or with a credit card, you pay interest.

The formula $I = prt$ is used to solve problems involving interest.

- I represents the interest,
- p represents the amount of money invested or borrowed, which is called the *principal*,
- r represents the annual interest rate, and
- t represents the time in years.

Example 4
Banking Link

Rodney Turner is opening a savings account that earns 4% annual interest. He wants to earn at least $50 in interest after 2 years. How much money should he save in order to earn $50 in interest?

$I = prt$
$50 = p(0.04)(2)$ *Replace I with 50, r with 0.04, and t with 2.*
$50 = 0.08p$ *$0.04 \times 2 = 0.08$*
$\frac{50}{0.08} = \frac{0.08p}{0.08}$ *Divide each side by 0.08.*
$625 = p$ *50 [÷] 0.08 [ENTER] 625*

Rodney should invest at least $625 to earn $50 in interest.

Your Turn

c. Jessica deposited $3000 in a savings account that pays an interest rate of 6%. How long should she leave the money in the account if she wants to earn $90 in interest? **0.5 yr or 6 mo**

www.algconcepts.com/extra_examples

In-Class Examples

Example 1
Find 17% of $250. **$42.50**

Example 2
35% of what number is 105? **300**

Example 3
A store collects 8.5% sales tax on all purchases. How much tax must be paid on a purchase of $188? **$15.98**

Example 4
Riona serves food at a restaurant where she is paid 18% of the diners' bills. She earned $126 last weekend. What was the total of the diners' bills? **$700**

Teaching Tip In Example 5, a difficult step for many students is knowing how to start. Encourage your students to think of a variety of ways to organize the information. For many students, finding a way to organize the information in a problem takes more thought and effort than solving the equation for the problem.

In-Class Example

Example 5

All 208 freshmen at a school went on a field trip. For transportation, buses that each hold 64 students and vans that each hold 8 students were used. Every bus and van was completely filled, and there were 5 vehicles used. How many buses and vans were used? **3 buses, 2 vans**

Mixture problems involve combining two or more parts into a whole. The parts that are combined usually have a different price or a different percent of something.

Example 5 Sales Link

Crystal sold tickets to the Drama Club's spring play. Adult tickets cost \$8.00, and student tickets cost \$5.00. Crystal sold 35 more student tickets than adult tickets. She collected a total of \$1475. How many of each type of ticket did she sell?

Explore Let a be the number of adult tickets that Crystal sold. Since there were 35 more student tickets sold than adult tickets, $a + 35$ is the number of student tickets sold.

Plan Make a chart of the information.

	Number Sold	Price	Sales
Adult Tickets	a	\$8	$8a$
Student Tickets	$a + 35$	\$5	$5(a + 35)$

$$\underbrace{8a}_{\text{Sales of adult tickets}} \underbrace{+}_{\text{plus}} \underbrace{5(a + 35)}_{\text{sales of student tickets}} \underbrace{=}_{\text{equals}} \underbrace{1475}_{\text{total sales.}}$$

Solve

$8a + 5(a + 35) = 1475$	*Original equation*
$8a + 5a + 175 = 1475$	*Distributive Property*
$13a + 175 = 1475$	*$8a + 5a = 13a$*
$13a + 175 - 175 = 1475 - 175$	*Subtract 175 from each side.*
$13a = 1300$	*Simplify.*
$\frac{13a}{13} = \frac{1300}{13}$	*Divide each side by 13.*
$a = 100$	*Simplify.*

Crystal sold 100 adult tickets and 100 + 35 or 135 student tickets.

Examine If 100 adult tickets were sold, the total amount of money collected for them would be 100×8 or \$800.

If 135 student tickets were sold, the total amount of money collected for them would be 135×5 or \$675.

The total sales would be \$800 + \$675 or \$1475. ✓

Look Back

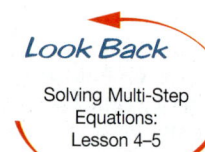

Solving Multi-Step Equations: Lesson 4–5

Mixture problems occur often in chemistry.

Example 6
Chemistry Link

Kelsey is doing a chemistry experiment that calls for a 30% solution of copper sulfate. She has 40 milliliters of a 25% solution. How many milliliters of a 60% solution should she add to make the required 30% solution?

Let x represent the amount of 60% solution to be added. Since she starts with 40 milliliters of solution, the final solution will have $40 + x$ milliliters.

	Amount of Solution (mL)	Amount of Copper Sulfate
25% solution	40	0.25(40)
60% solution	x	$0.60x$
30% solution	$40 + x$	$0.30(40 + x)$

Amount of copper sulfate in 25% solution *plus* *amount of copper sulfate in 60% solution* *equals* *amount of copper sulfate in mixture.*

$$\underbrace{0.25(40)}_{} + \underbrace{0.60x}_{} = \underbrace{0.30(40 + x)}_{}$$

$0.25(40) + 0.60x = 0.30(40 + x)$	*Original equation*
$10 + 0.6x = 12 + 0.3x$	*Distributive Property*
$10 + 0.6x - 0.3x = 12 + 0.3x - 0.3x$	*Subtract 0.3x from each side.*
$10 + 0.3x = 12$	*Simplify.*
$10 + 0.3x - 10 = 12 - 10$	*Subtract 10 from each side.*
$0.3x = 2$	*Simplify.*
$\frac{0.3x}{0.3} = \frac{2}{0.3}$	*Divide each side by 0.3.*
$x \approx 6.7$	2 [÷] 0.3 [ENTER] 6.666666667

Kelsey should add about 6.7 milliliters of the 60% solution.

In-Class Example

Example 6

A chemistry experiment calls for a 20% solution of copper sulfate. You have 150 milliliters of a 30% solution. How many milliliters of an 18% solution should you add to get a 20% solution? **750 mL**

Answers

1. *r* represents the percent rate; *R* is the rate written in decimal form.
3. Nikki; in the percent equation, the *R* is the decimal form of the percent.

Check for Understanding

Communicating Mathematics

Vocabulary
percent equation
rate
simple interest
mixture problem

1. **Explain** how r and R are different in the percent proportion and the percent equation. **1. See margin.**
2. **Write** a percent equation to find the following. 20% of what amount is $500? **$500 = 0.2B$**
3. **You Decide** Akira and Nikki are using the percent equation to find 5% of 17.99. Akira multiplies 17.99 by 5. Nikki multiplies 17.99 by 0.05. Who is correct? Explain. **See margin.**

Reteaching Activity

Interpersonal Learners While working in pairs, have one student select values for two of the three variables, *P*, *R*, and *B*, in the percent equation. Their partner should then use the two values to solve for the third variable. Instruct students to work together to verify the answer. The partners should then reverse roles and repeat the activity.

***Study Guide*, p. 210**

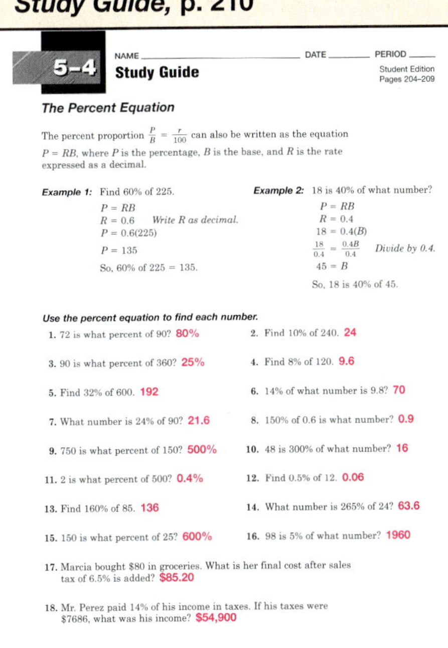
NAME ______ DATE ______ PERIOD ______

5–4 Study Guide

Student Edition Pages 204–209

The Percent Equation

The percent proportion $\frac{P}{B} = \frac{r}{100}$ can also be written as the equation $P = RB$, where P is the percentage, B is the base, and R is the rate expressed as a decimal.

Example 1: Find 60% of 225.
$P = RB$
$R = 0.6$ *Write R as decimal.*
$P = 0.6(225)$
$P = 135$
So, 60% of 225 = 135.

Example 2: 18 is 40% of what number?
$P = RB$
$R = 0.4$
$18 = 0.4(B)$
$\frac{18}{0.4} = \frac{0.4B}{0.4}$ *Divide by 0.4.*
$45 = B$
So, 18 is 40% of 45.

Use the percent equation to find each number.

1. 72 is what percent of 90? **80%**
2. Find 10% of 240. **24**
3. 90 is what percent of 360? **25%**
4. Find 8% of 120. **9.6**
5. Find 32% of 600. **192**
6. 14% of what number is 9.8? **70**
7. What number is 24% of 90? **21.6**
8. 150% of 0.6 is what number? **0.9**
9. 750 is what percent of 150? **500%**
10. 48 is 300% of what number? **16**
11. 2 is what percent of 500? **0.4%**
12. Find 0.5% of 12. **0.06**
13. Find 160% of 85. **136**
14. What number is 265% of 24? **63.6**
15. 150 is what percent of 25? **600%**
16. 98 is 5% of what number? **1960**
17. Marcia bought $80 in groceries. What is her final cost after sales tax of 6.5% is added? **$85.20**
18. Mr. Perez paid 14% of his income in taxes. If his taxes were $7686, what was his income? **$54,900**

3 PRACTICE/APPLY

Error Analysis

Watch for students who have trouble organizing the information in Exercise 10.

Prevent by encouraging the students to think of the values of the coffee as the price per pound times the number of pounds.

Assignment Guide
Basic: 11–29 odd, 31–38
Average: 12–24 even, 25–38

Teaching Tip Exercises 14 and 18–21 involve percents greater than 100%. Be sure students realize that the value of R will be greater than 1 for these exercises.

***Skills Practice*, p. 211, and *Practice*, p. 212 (shown)**

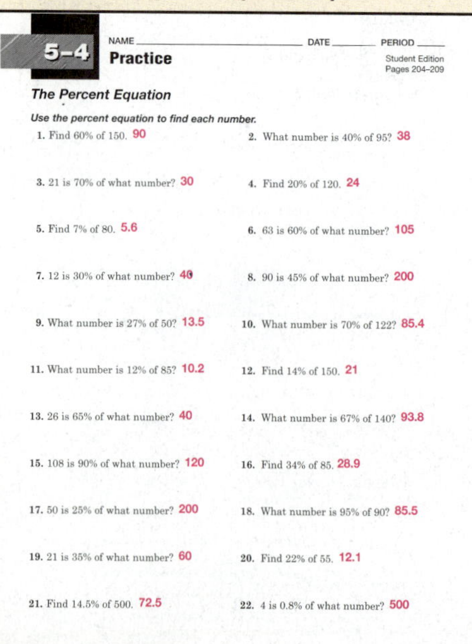

5-4 NAME ____________ DATE ______ PERIOD ______

Practice

Student Edition Pages 204–209

The Percent Equation

Use the percent equation to find each number.

1. Find 60% of 150. 90	2. What number is 40% of 95? 38
3. 21 is 70% of what number? 30	4. Find 20% of 120. 24
5. Find 7% of 80. 5.6	6. 63 is 60% of what number? 105
7. 12 is 30% of what number? 40	8. 90 is 45% of what number? 200
9. What number is 27% of 50? 13.5	10. What number is 70% of 122? 85.4
11. What number is 12% of 85? 10.2	12. Find 14% of 150. 21
13. 26 is 65% of what number? 40	14. What number is 67% of 140? 93.8
15. 108 is 90% of what number? 120	16. Find 34% of 85. 28.9
17. 50 is 25% of what number? 200	18. What number is 95% of 90? 85.5
19. 21 is 35% of what number? 60	20. Find 22% of 55. 12.1
21. Find 14.5% of 500. 72.5	22. 4 is 0.8% of what number? 500

Guided Practice

Use the percent equation to find each number.

Examples 1 & 2

4. Find 90% of 200. **180**
5. What number is 38% of 110? **5. 41.8**
6. 60 is 80% of what number? **75**
7. 220 is 40% of what number? **550**

Example 3

8. **Business** Employers are required to deduct 1.45% of your income for Medicare taxes. If your weekly pay is \$100, what amount would be deducted from your pay? **\$1.45**

Example 4

9. **Banking** Miguel bought a \$500 certificate of deposit that paid an annual interest rate of 5%. If the certificate was issued for three years, how much simple interest will he earn? **\$75**

Examples 5 & 6

10. **Business** Marian works at The Daily Grind coffee shop. She makes a blend of coffee by mixing hazelnut, which sells for \$4 a pound, and chocolate almond, which sells for \$7 a pound. How many pounds of chocolate almond should she mix with 10 pounds of hazelnut if she wants to sell the mixture for \$6 a pound? **20 lb**

Exercises

Practice 

13. 117   16. 550   17. 625

Use the percent equation to find each number.

Homework Help	
For Exercises	See Examples
11–24	1–2
25–26, 28	4
29	5
30	6

Extra Practice
See page 702.

11. Find 40% of 280. **112**
12. What number is 5% of 120? **6**
13. What number is 65% of 180?
14. Find 200% of 90. **180**
15. 18 is 75% of what number? **24**
16. 110 is 20% of what number?
17. 250 is 40% of what number?
18. 60 is 150% of what number? **40**
19. Find 110% of 80. **88**
20. 90 is 200% of what number? **45**
21. What number is 400% of 16? **64**
22. Find 5% of 3200. **160**
23. 25 is 40% of what number? **62.5**
24. Find 0.5% of 240. **1.2**

Applications and Problem Solving

25. **Banking** How much interest will Marissa earn if she invests \$250 at an annual rate of 4% for 5 years? **\$50**

26. **Banking** How long will it take Mr. Albany to earn \$75 if he invests \$3000 at an annual rate of 5%? **6 months**

27a. $P = 0.29(150)$

27. **Statistics** The graph shows ways that teenagers ages 12 to 17 get spending money.
    a. Suppose you survey a group of 150 15-year-olds. Write a percent equation to predict how many students get a regular allowance.
    b. Solve the equation. Round to the nearest whole number. **44**

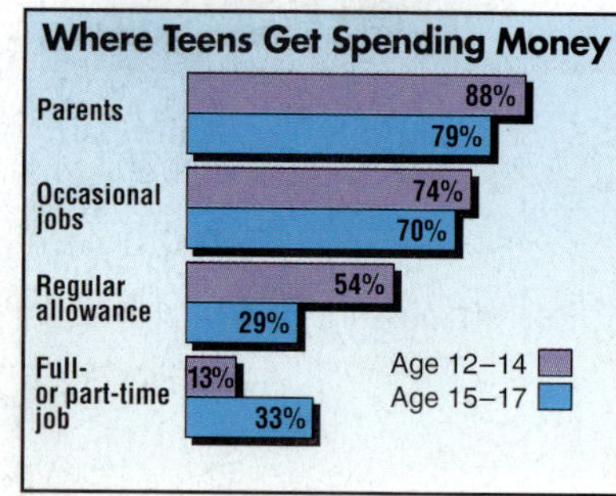

Source: *ICR TeenEXCEL*

28. **Credit Cards** A credit card company charges an annual rate of 18% on the unpaid balance on credit card accounts. Suppose you have an unpaid balance of $800. How much monthly interest will you be charged? (*Hint:* First find the monthly interest rate.) **$12**

29. **Sales** Great Smoky Mountains Tours conducts hiking trips along the Appalachian Trail. The owner charges $250 for adults and $175 for children. He takes groups of up to 15 people on a single trip. If he wants to earn $3150 for his next trip to pay for some new equipment, how many of the 15 people on the trip should be adults? **7 adults**

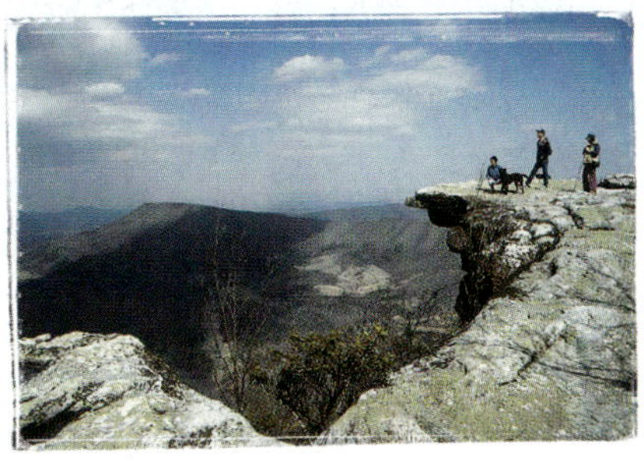

30. **Chemistry** Roberto needs a solution that is 30% silver nitrate. He has 12 ounces of a 25% silver nitrate solution. How many ounces of a 40% silver nitrate solution should he add to make a solution that is 30% silver nitrate? **6 oz**

31. **Critical Thinking** If $a\%$ of $b = x$ and $b\%$ of $a = y$, what is the relationship between x and y? $x = y$

Mixed Review

32. **Health** Nutritionists recommend a diet in which less than 30% of your total Calories come from fat. Which subs listed in the chart have less than 30% of their Calories from fat? *(Lesson 5–3)* **turkey, ham & cheese**

The Sub Shop

Kind of Sub	Total Calories	Calories from Fat
Italian	522	288
Meatball	459	198
Turkey	322	90
Roast Beef	345	108
Ham & Cheese	322	81

33. On a map, the scale is 1 inch = 60 miles. Find the actual distance for each map distance. *(Lesson 5–2)*

	From	To	Map Distance	
a.	Memphis, TN	Nashville, TN	3 inches	**180 mi**
b.	Shreveport, LA	Jackson, MS	$3\frac{1}{2}$ inches	**210 mi**
c.	Little Rock, AR	Memphis, TN	$2\frac{1}{4}$ inches	**135 mi**

Solve each equation. *(Lesson 4–6)*

34. $12x + 15 = 35 + 2x$ **2**
35. $3y + 10 = 2y - 21$ **−31**
36. $6 - 8a = 20a + 20$ **−0.5**
37. $7n - 13 = 3n + 7$ **5**

38. Sample answer: 1, 2, 3, 4, 10

Standardized Test Practice

38. **Extended Response** List a set of data that includes at least five numbers for which the mean is greater than the median. *(Lesson 3–3)*

www.algconcepts.com/self_check_quiz

4 ASSESS

Open-Ended Assessment

Writing Ask students to describe how the percent equation and the percent proportion are related.

Mid-Chapter Test (Lessons 5–1 through 5–4) is available in the *Chapter 5 Resource Masters,* p. 240.

***Enrichment*, p. 214**

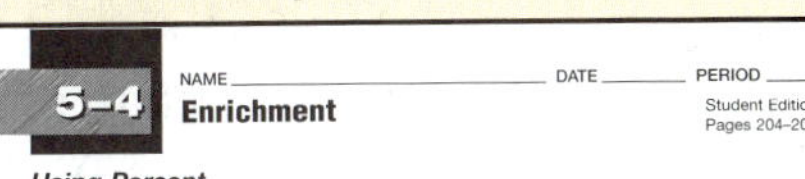

5–4 **Enrichment**

NAME ______ DATE ______ PERIOD ______

Student Edition Pages 204–209

Using Percent

Use what you have learned about percent to solve each problem.

A TV movie had a "rating" of 15 and a 25 "share." The rating of 15 means that 15% of the nation's total TV households were tuned in to this show. The 25 share means that 25% of the homes with TVs turned on were tuned to the movie. How many TV households had their TVs turned off at this time?

To find out, let T = the number of TV households
and x = the number of TV households with the TV off.
Then $T - x$ = the number of TV households with the TV on.

Since $0.15T$ and $0.25(T - x)$ both represent the number of households tuned to the movie,

$$0.15T = 0.25(T - x)$$
$$0.15T = 0.25T - 0.25x.$$

Solve for x.

$$0.25x = 0.10T$$
$$x = \frac{0.10T}{0.25} = 0.40T$$

Forty percent of the TV households had their TVs off when the movie was aired.

Answer each question.

1. During that same week, a sports broadcast had a rating of 22.1 and a 43 share. Show that the percent of TV households with their TVs off was about 48.6%. $0.221T = 0.43T - 0.43x$; $x = \frac{0.221T - 0.43T}{-0.43} = 0.486T$
2. Find the percent of TV households with their TVs turned off during a show with a rating of 18.9 and a 29 share. **34.8%**
3. Show that if T is the number of TV households, r is the rating, and s is the share, then the number of TV households with the TV off is $\frac{(s - r)T}{s}$. **Solve $rT = s(T - x)$ for x.**
4. If the fraction of TV households with no TV on is $\frac{s - r}{s}$ then show that the fraction of TV households with TVs on is $\frac{r}{s}$. $1 - \frac{s - r}{s} = \frac{r}{s}$
5. Find the percent of TV households with TVs on during the most watched serial program in history: the last episode of *M*A*S*H*, which had a 60.3 rating and a 77 share. $\frac{60.3}{77} = 78.3\%$
6. A local station now has a 2 share. Each share is worth $50,000 in advertising revenue per month. The station is thinking of going commercial free for the three months of summer to gain more listeners. What would its new share have to be for the last 4 months of the year to make more money for the year than it would have made had it not gone commercial free? **greater than 3.5**

Extra Credit

If 4% of a number is 17, find 208% of that number. **884**

PREPARE

This optional activity is designed to be completed by groups of 2–3 students over 1–2 days.

Objectives

Students find the lower quartile, median, upper quartile, and extremes for a set of data, and illustrate those five values in a box-and-whisker plot. They compare two data sets by comparing the corresponding quartiles, medians, and extremes, and identify data values at particular percentiles.

Mathematical Overview

This investigation utilizes the following concepts:

- organizing data,
- analyzing data,
- describing data, and
- comparing sets of data.

Suggested Time Management	
Investigation	25–40 min
Extension: Gathering Data	25–40 min
Extension: Summarizing Data	30–40 min

Motivating the Lesson

Have students look through the list of bridges in the left column of the table on page 210, which shows several bridges located inside the United States. Ask students if they have crossed any of the bridges in the list. For those students who have, ask them to give their impressions of the bridge or bridges they have crossed.

Chapter 5 Investigation

BUILDING BRIDGES

Materials

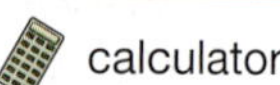 calculator

Box-and-Whisker Plots

The table shows the lengths of the longest suspension bridges.

Name of Bridge (inside U.S.)	Length (ft)	Name of Bridge (outside U.S.)	Length (ft)
Verrazano-Narrows	4260	Akashi Kaikyo	6529
Golden Gate	4200	Izmit Bay	5472
Mackinac Straits	3800	Storebaelt	5328
George Washington	3500	Humber	4626
Tacoma Narrows II	2800	Jiangyin Yangtze	4543
San Francisco-Oakland	2310	Tsing Ma	4518
Bronx-Whitestone	2300	Hardanger Fjord	4347
Delaware Memorial	2150	Hoga Kusten	3970
Seaway Skyway	2150	High Coast	3969
Walt Whitman	2000	Minami Bisan-Seto	3668
Ambassador International	1850	Second Bosporus	3576

Source: *Information Please Almanac*

Let's use a **box-and-whisker plot** to investigate these data.

Investigate

1. To make box-and-whisker plots for these data, you need to find five important values for each set of data: the median (M), the two extremes, the upper quartile (UQ), and the lower quartile (LQ).

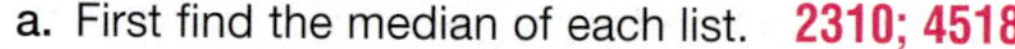

a. First find the median of each list. **2310; 4518**

b. Consider only the upper half of each list. Find the medians. These numbers are the **upper quartiles**. **3800; 5328**

c. Consider only the bottom half of each list. Find the medians. These numbers are the **lower quartiles**. **2150; 3969**

d. The greatest (GV) and least values (LV) in each list are called the **extremes**. Find the extremes. **4260, 1850; 6529, 3576**

Cooperative Learning

This investigation offers an excellent opportunity for using cooperative groups. For more information on cooperative learning strategies and group management, see *Cooperative Learning in the Mathematics Classroom,* one of the titles in the Glencoe Mathematics Professional Series.

2a. See below.

2. Draw a number line.
   a. Make a box-and-whisker plot for the U.S. bridges using the five values from Exercise 1a–1d. Mark dots above their coordinates.
   b. Complete the box-and-whisker plot as shown below.

The lines outside the box are called whiskers.

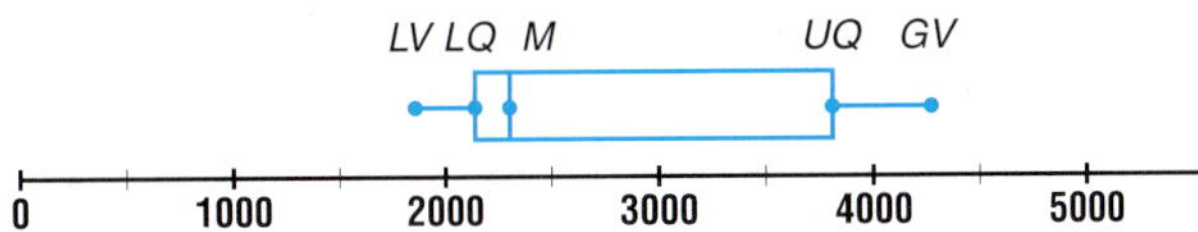

3. Make a box-and-whisker plot for the bridges outside the U.S. **See margin.**

Extending the Investigation

In this extension, you will compare the box-and-whisker plots for the two sets of data. You will also investigate how changes in the data affect a box-and-whisker plot.

1. Use the box-and-whisker plots to answer these questions.
   a. What percent of the data lies below the median? **50%**
   b. What percent of the data lies inside the box? **50%**
   c. What percent of the data lies in one whisker? **25%**
   d. What percent of bridges outside the U.S. are longer than the longest U.S. bridge? Explain your reasoning. **See margin.**
2. When data are arranged in order from least to greatest, you can describe the data with percentiles. A **percentile** is the point below which a given percent of the data lies. For example, 50% of the data falls below the median. So, the median is the 50th percentile for the data.
   a. Which U.S. bridge is at the 50th percentile? **San Francisco-Oakland**
   b. Which bridge outside the U.S. is at the 25th percentile? **High Coast**
   c. Which U.S. bridge is at the 75th percentile? **Mackinac Straits**
3. Suppose a 7000-foot bridge is built in the United States. **a–c. See Solutions Manual.**
   a. Redraw the box-and-whisker plot, including this value.
   b. Describe the change in the box-and-whisker plot.
   c. How do these data change your comparisons of the bridges?

Presenting Your Conclusions

Here are some ideas to help you present your conclusions to the class.

- Prepare a poster displaying your box-and-whisker plots. Write a comparison of the data.
- Find two related sets of data that you would like to compare with box-and-whisker plots. Draw the plots. Then prepare an oral presentation using the plots as visual aids.

interNET CONNECTION **Investigation** For more information on box-and-whisker plots, visit: www.algconcepts.com

Answers

3. 0 1000 2000 3000 4000 5000 6000 7000

Extending the Investigation

1d. 7 out of 11, or 64%, of the bridges outside the U.S. are longer than the longest U.S. bridge.

MANAGE

Teaching Tip For Exercises 2 and 3, consider having students use a single number line for both box-and-whisker plots, drawing one above the number line and the other below the number line. This will make it easier for them to make the comparison in Exercise 1d of the *Extending the Investigation* section.

Working in Groups If students work in groups, they can divide up the tasks of drawing the box-and-whisker plots for the U.S. bridges, non-U.S. bridges, and the altered list of bridges in Exercise 3a of the *Extending the Investigation* section. The group members should work together to present their conclusions.

Working as a Class You can assign half the class to each set of data in Exercise 1. The exercises in the *Extending the Investigation* section should then be discussed as a class.

ASSESS

Students' work should show that they can find a median for a set of data and they can illustrate the quartiles, median, and extremes using a box-and-whisker plot. They should be able to describe each of the four sections of a box-and-whisker plot and explain why each section contains 25% of the data values. Also, students should be able to compare two distinct box-and-whisker plots by comparing corresponding values for the quartiles, median, and extremes.

PORTFOLIO Students should add their poster and written comparison to their portfolios at this time.

Lesson 5-5

1 FOCUS

5-Minute Check
Lesson 5-4

Use the percent equation to find each number.

1. Find 12% of \$45. **\$5.40**
2. 30 is what percent of 75? **40%**
3. 72 is 30% of what number? **240**
4. A credit card company charges an annual rate of 21% interest on the unpaid balance of credit card accounts. If the unpaid balance on an account is \$1240, how much monthly interest will be charged? **\$21.70**
5. If the sales tax rate is 7%, how much sales tax is paid on a purchase of \$35? **\$2.45**

Motivating the Lesson

Real-World Connection Show the class two "coupons" that can be used together to purchase a computer; one coupon should read "\$75 off" and the other should read "10% off." Ask the students if it makes any difference which coupon is applied to the purchase first. Help students realize that the final price is lower if the "10% off" coupon is applied first, followed by the "\$75 off" coupon.

2 TEACH

In-Class Examples
Examples 1–2

Find the percent of increase or decrease. Round to the nearest percent.

1 original: 110
new: 140 **27% increase**

2 original: 150
new: 125 **17% decrease**

5-5 Percent of Change

What You'll Learn
You'll learn to solve problems involving percent of increase or decrease.

Why It's Important
Retail Sales Percent of decrease is used when stores have sales.
See Example 4.

In 1990, the average cost of one dozen large eggs was \$1.00. By 2000, the cost had decreased to \$0.96. The decrease in the cost of goods and services is called *deflation*. **Source:** *Statistical Abstract of the United States*

To find the deflation rate, write a ratio that compares the amount of the decrease to the original price.

amount of decrease: \$1.00 − \$0.96 = \$0.04 or 4¢
original price: \$1.00 or 100¢

$$\frac{\text{amount of decrease} \rightarrow 4}{\text{original price} \rightarrow 100} = 4\% \quad \leftarrow \textit{deflation rate}$$

So, the cost of one dozen eggs decreased 4% from 1990 to 2000.

When an increase or decrease is expressed as a percent, the percent is called the **percent of increase** or the **percent of decrease**. The earlier amount is always used as the base in the percent equation.

Examples

Find the percent of increase or decrease. Round to the nearest percent.

1 **original: 25**
new: 29

Find the amount of increase.
29 − 25 = 4

Use the percent proportion.

$\frac{P}{B} = \frac{r}{100}$	*Percent proportion*
$\frac{4}{25} = \frac{r}{100}$	*P = 4, B = 25*
$4(100) = 25r$	*Cross products*
$400 = 25r$	*Multiply.*
$\frac{400}{25} = \frac{25r}{25}$	*Divide each side by 25.*
$16 = r$	*Simplify.*

The percent of increase is 16%.

2 **original: 18**
new: 12

Find the amount of decrease.
18 − 12 = 6

Use the percent proportion.

$\frac{P}{B} = \frac{r}{100}$	*Percent proportion*
$\frac{6}{18} = \frac{r}{100}$	*P = 6, B = 18*
$6(100) = 18r$	*Cross products*
$600 = 18r$	*Multiply.*
$\frac{600}{18} = \frac{18r}{18}$	*Divide each side by 18.*
$33 \approx r$	*Simplify.*

The percent of decrease is about 33%.

Your Turn

a. original: 12
new: 20 **≈ 67% increase**

b. original: 50
new: 49 **2% decrease**

Resource Manager

Reproducible Masters
Chapter 5 Resource Masters
- *Study Guide,* p. 215
- *Skills Practice,* p. 216
- *Practice,* p. 217
- *Reading to Learn Mathematics,* p. 218
- *Enrichment,* p. 219

Graphing Calculator, p. 14

Transparencies
5-Minute Check, 5–5

Technology/Multimedia
Interactive Chalkboard CD-ROM

Two applications of percent of change are sales tax and discounts. **Sales tax** is a tax that is added to the cost of the item. It is an example of a percent of increase. **Discount** is the amount by which the regular price of an item is reduced. It is an example of a percent of decrease.

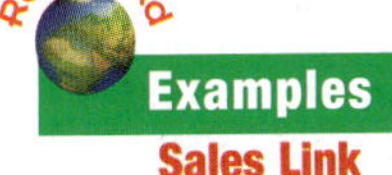

Examples
Sales Link

3 **Ms. Cruz bought a car for $12,500. A state sales tax of 4% is then added to the price of the car. What was the total price?**

Method 1

First, use the percent equation to find the sales tax.

$P = RB$
$= 0.04(12{,}500)$
$= 500$

Then, add the $500 sales tax to $12,500.

$12{,}500 + 500 = 13{,}000$

Method 2

A sales tax of 4% means that Ms. Cruz will pay 100% + 4% or 104% of the price of the car.

Use the percent equation to find the total price.

$P = RB$
$= 1.04(12{,}500)$
$= 13{,}000$

The total price was $13,000.

4 **All shoes at The Runner's Place are on sale at a 25% discount. If a pair of running shoes originally cost $65, what is the sale price?**

Method 1

First, use the percent equation to find the discount.

$P = RB$ *Percent equation*
$= 0.25(65)$ $R = 0.25, B = 65$
$= 16.25$ *Multiply.*

Then, subtract the $16.25 discount from $65.

$65 - 16.25 = 48.75$

Method 2

A discount of 25% means that the buyer will pay 100% − 25% or 75% of the selling price. Use the percent equation to find the sale price.

$P = RB$ *Percent equation*
$= 0.75(65)$ $R = 0.75, B = 65$
$= 48.75$ *Multiply.*

The sale price of the running shoes is $48.75.

Your Turn

c. What is the total cost of a basketball that sells for $45 if the sales tax rate is 7%? **$48.15**

d. All long-sleeve T-shirts are on sale for 40% off. If the original price was $19.95, what is the discount price? **$11.97**

www.algconcepts.com/extra_examples

In-Class Example

Example 3

A family bought a home computer for $890. A sales tax of 5.5% on the purchase was then added. What was the total price? **$938.95**

Teaching Tip After discussing Example 3, point out that there are some states that do not have a sales tax and students in those states need the idea carefully explained. Ask your students how they could use math concepts to explain the idea of a sales tax.

In-Class Example

Example 4

Refer to In-Class Example 3. Suppose the family also bought a printer. The original price of the printer was $140, but they received a 15% discount. What was the sale price of the printer (before the sales tax was added)? **$119**

As a class, make up a list of 5–10 household costs, such as food, utilities, clothing, telephone, entertainment, and so on. Have each student work with a family member to find the amount spent on each category two months ago and last month. Then have them calculate the percent of increase or percent of decrease for several of the categories.

Teaching Tip If a store in your area has recently advertised a sale where successive discounts are being offered, bring a copy of the advertisement to class and have students compute the final prices of several items.

Sometimes discount stores advertise an additional discount. For example, suppose all merchandise in an outlet store is 50% off retail prices every day. If the store has a sale where you can take another 25% off, there are two ways you might interpret this.

- The discounts can be successive. That is, 50% is taken off the original price and then 25% is taken off the resulting price.
- The discounts can be combined. That is, the discount is 50% + 25% or 75% of the original price.

You can use a graphing calculator to examine these two interpretations.

Graphing Calculator Tutorial
See pp. 750–753.

Graphing Calculator Exploration

The graphing calculator program at the right finds the cost of an item for successive discounts and combined discounts.

Copy the program into your calculator. Run the program and enter the original price and each discount (as a decimal) when prompted.

```
PROGRAM:DISCOUNT
:Disp "ORIG. PRICE"
:Input P
:Disp "1ST DISCOUNT?"
:Input A
:Disp "2ND DISCOUNT?"
:Input B
:P(1-A)(1-B)→C
:P(1-(A+B))→D
:Disp "SUCCESSIVE
 DISCOUNT", C
:Disp "COMBINED
DISCOUNT", D
```

Try These

Copy and complete the table.

	Price	First Discount	Second Discount	Sale Price Successive Discount	Sale Price Combined Discount
1.	$ 49.00	20%	10%	$ 35.28	$ 34.30
2.	$185.00	25%	10%	$124.88	$120.25
3.	$ 12.50	30%	20%	$ 7.00	$ 6.25
4.	$156.95	30%	15%	$ 93.39	$ 86.32

5. What is the relationship between the sale price using successive discounts and the sale price using combined discounts?
6. Which type of discount is usually used in stores? **successive**

5. The sale price using combined discounts is less than the sale price using successive discounts.

Graphing Calculator Exploration

If this is the first time students have entered a program on the calculator, they may need some help. Students need to know that pressing PRGM ▶ allows them to access the list containing Disp and Input. Emphasize that they must press ENTER after typing each line of the program in order to go to the next line. Stress that when they run the program, students need to use decimal values for the discount percents when prompted for the values of *A* and *B*.

Percent of change is often used to describe how data change from one year to another. Sometimes the percent of increase is greater than 100%.

Example 5

Statistics Link

The graph shows on-line music sales for 2001 and projected sales for 2006. To the nearest percent, find the percent of increase.

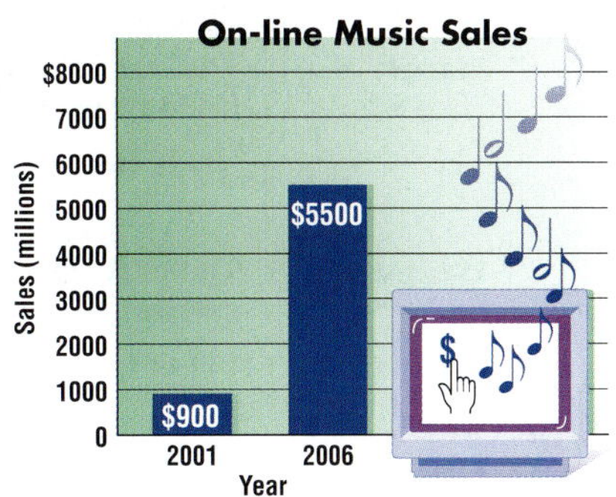

Source: Jupiter Communications

The amount of increase is 5500 − 900 or 4600.

$$\frac{P}{B} = \frac{r}{100}$$

$$\frac{4600}{900} = \frac{r}{100}$$

$$4600(100) = 900r$$

$$460{,}000 = 900r$$

$$\frac{460{,}000}{900} = \frac{900r}{900}$$

460,000 [÷] 900 [ENTER] 511.111111

$$r \approx 511$$

On-line music sales are expected to increase by about 511% from 2001 to 2006.

Check for Understanding

Communicating Mathematics

1–2. See margin.

1. **Describe** two different methods for finding the total price of an item if the sales tax rate and the cost of the item are given.
2. **Explain** how a percent of increase can be greater than 100%.
3.  Find an example of percent of change in a newspaper or magazine. Then write and solve a problem using your example. **See students' work.**

Vocabulary
- percent of increase
- percent of decrease
- sales tax
- discount

Guided Practice

Getting Ready **Write a percent proportion to find the percent of change.**

Sample: original: 15, new: 20 **Solution:** $\frac{5}{15} = \frac{r}{100}$

4–6. See margin.

4. original: 25
new: 23
5. original: 14
new: 18
6. original: 30
new: 70

In-Class Example

Example 5

The price of a share of stock in an Internet company rose from \$32 to \$121 last year. To the nearest percent, find the percent of increase. **278%**

3 PRACTICE/APPLY

Error Analysis

Watch for students who use the new value in the denominator of a percent of change fraction instead of the original in Exercises 4–6. ***Prevent by*** having students begin each exercise by writing "$\frac{\text{change}}{\text{original}} =$."

Answers

1. First method: find the amount of the tax, then add the tax to the cost of the item. Second method: add the tax rate to 100%, then find the total cost.
2. The amount of the increase is greater than the original amount.
4. $\frac{2}{25} = \frac{r}{100}$
5. $\frac{4}{14} = \frac{r}{100}$
6. $\frac{40}{30} = \frac{r}{100}$

***Study Guide*, p. 215**

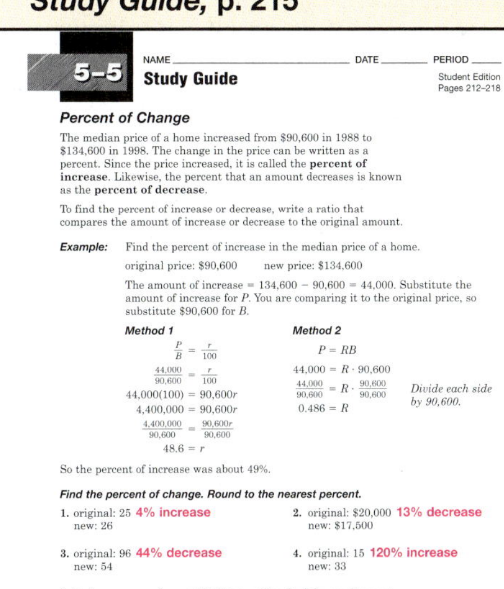

NAME ______ DATE ______ PERIOD ______

5-5 Study Guide

Student Edition Pages 212–218

Percent of Change

The median price of a home increased from \$90,600 in 1988 to \$134,600 in 1998. The change in the price can be written as a percent. Since the price increased, it is called the **percent of increase**. Likewise, the percent that an amount decreases is known as the **percent of decrease**.

To find the percent of increase or decrease, write a ratio that compares the amount of increase or decrease to the original amount.

Example: Find the percent of increase in the median price of a home.

original price: \$90,600 new price: \$134,600

The amount of increase = 134,600 − 90,600 = 44,000. Substitute the amount of increase for P. You are comparing it to the original price, so substitute \$90,600 for B.

Method 1

$$\frac{P}{B} = \frac{r}{100}$$
$$\frac{44{,}000}{90{,}600} = \frac{r}{100}$$
$$44{,}000(100) = 90{,}600r$$
$$4{,}400{,}000 = 90{,}600r$$
$$\frac{4{,}400{,}000}{90{,}600} = \frac{90{,}600r}{90{,}600}$$
$$48.6 = r$$

Method 2

$$P = RB$$
$$44{,}000 = R \cdot 90{,}600$$
$$\frac{44{,}000}{90{,}600} = R \cdot \frac{90{,}600}{90{,}600}$$ *Divide each side by 90,600.*
$$0.486 = R$$

So the percent of increase was about 49%.

Find the percent of change. Round to the nearest percent.

1. original: 25 **4% increase**
new: 26
2. original: \$20,000 **13% decrease**
new: \$17,500
3. original: 96 **44% decrease**
new: 54
4. original: 15 **120% increase**
new: 33
5. In January, gasoline cost \$1.05 per gallon. In July, gasoline cost \$1.20 per gallon. Find the percent of increase. **14%**
6. The original price of a CD player is \$105. A discount of \$21 is given. Find the percent of decrease. **20%**
7. The original cost of a computer is \$900. If the price is decreased by 15%, what is the sale price? **\$765**

Reteaching Activity

Intrapersonal Learners Have students write a journal entry to themselves explaining how to find percent of increase and percent of decrease. Instruct them to include an example of each type of problem. Ask them to include a short list of real-world situations where percent of change occurs.

Assignment Guide

Basic: 15–35 odd, 37–44
Average: 14–32 even, 34–44
All: Quiz 2, 1–5

Examples 1 & 2 Find the percent of increase or decrease. Round to the nearest percent.

7. original: 20 new: 16 **20% dec**
8. original: 8 new: 20 **150% inc**

Example 3 The cost of an item and a sales tax rate are given. Find the total price of each item to the nearest cent.

9. in-line skates: \$90; 5% **\$94.50**
10. make-up: \$14.95; 4% **\$15.55**

Example 4 The original cost of an item and a discount rate are given. Find the sale price of each item to the nearest cent.

11. computer: \$1200; 10% **\$1080**
12. tennis balls: \$4.50; 25% **\$3.38**

Example 5

13. **Sales** In one year, online retail sales increased from \$6 to \$11 billion. Find the percent of increase to the nearest percent. **83%**

Exercises

Practice

Find the percent of increase or decrease. Round to the nearest percent.

A

14. original: 10 new: 12 **20% inc**
15. original: 50 new: 56 **12% inc**
16. original: 30 new: 12 **60% dec**
17. original: 500 new: 420 **16% dec**
18. original: 15 new: 36 **140% inc**
19. original: 48 new: 112 **133% inc**

Homework Help

For Exercises	See Examples
14–19	1, 2
20–25	3
26–34	4

Extra Practice
See page 702.

The cost of an item and a sales tax rate are given. Find the total price of each item to the nearest cent.

20. television: \$500; 6% **\$530**
21. CD player: \$40; 7% **\$42.80**
22. dress: \$55; 6% **\$58.30**
23. jeans: \$38; 4% **\$39.52**
24. CD: \$17.99; 5% **\$18.89**
25. book: \$25.99; 2.5% **\$26.64**

The original cost of an item and a discount rate are given. Find the sale price of each item to the nearest cent.

26. shoes: \$40; 25% **\$30**
27. tires: \$280; 5% **\$266**
28. earrings: \$12; 40% **\$7.20**
29. watch: \$35; 15% **\$29.75**
30. video: \$24.99; 30% **\$17.49**
31. radio: \$18.95; 25% **\$14.21**

B

32. What number is 30% less than 40? **28**
33. Find the percent of increase from \$18 to \$36. **100%**

Applications and Problem Solving

34. **Retail Sales** Georgia bought a new belt that was on sale for 15% off. If the original cost was \$24, what was the sale price? **\$20.40**

***Skills Practice,* p. 216, and *Practice,* p. 217 (shown)**

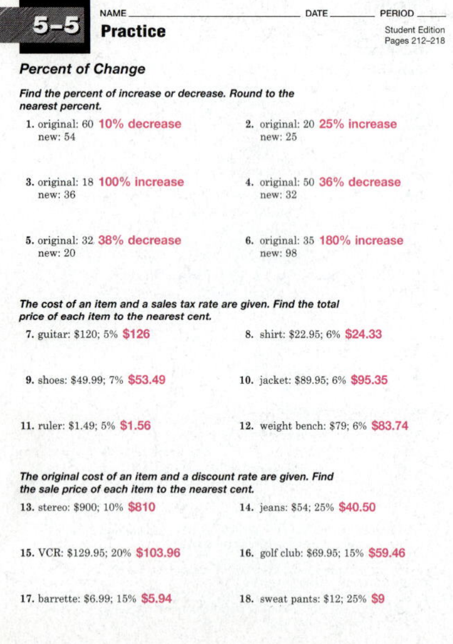

5-5 NAME ______ DATE ______ PERIOD ______

Practice Student Edition Pages 212–218

Percent of Change

Find the percent of increase or decrease. Round to the nearest percent.

1. original: 60 new: 54 **10% decrease**
2. original: 20 new: 25 **25% increase**
3. original: 18 new: 36 **100% increase**
4. original: 50 new: 32 **36% decrease**
5. original: 32 new: 20 **38% decrease**
6. original: 35 new: 98 **180% increase**

The cost of an item and a sales tax rate are given. Find the total price of each item to the nearest cent.

7. guitar: \$120; 5% **\$126**
8. shirt: \$22.95; 6% **\$24.33**
9. shoes: \$49.99; 7% **\$53.49**
10. jacket: \$89.95; 6% **\$95.35**
11. ruler: \$1.49; 5% **\$1.56**
12. weight bench: \$79; 6% **\$83.74**

The original cost of an item and a discount rate are given. Find the sale price of each item to the nearest cent.

13. stereo: \$900; 10% **\$810**
14. jeans: \$54; 25% **\$40.50**
15. VCR: \$129.95; 20% **\$103.96**
16. golf club: \$69.95; 15% **\$59.46**
17. barrette: \$6.99; 15% **\$5.94**
18. sweat pants: \$12; 25% **\$9**

From the Classroom of ...

Peter Pace
Life Center Academy
Burlington, New Jersey

I have students organize the data in a table to find percent of change.

Old	New	Difference	% Change
152	327	175	$\frac{175}{152} \approx 1.15$ or 115%

Students love it! They never miss using this method.

35. **Travel** The table shows the number of new passports issued in the United States. Between which years was the percent of increase greater: 1999 to 2001, or 2001 to 2003? Explain your reasoning. **See margin.**

New Passports (millions)	
1999	6.7
2001	7.1
2003	7.3

Source: U.S. State Department

Data change / earliest data (start) × 100

36. **Biology** Over a period of years, the number of breeding pairs of bald eagles increased from 2.2 thousand pairs to 5.7 thousand pairs. Find the percent of increase to the nearest percent.
Source: Fish and Wildlife Service **159%**

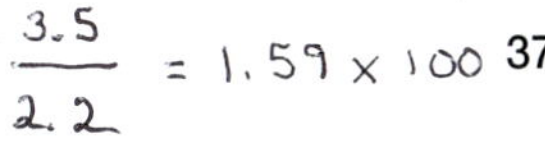

Bald Eagle

37. **Critical Thinking** An amount is increased by 10%. The result is decreased by 10%. Is the final result less than, greater than, or equal to the original amount? **less than**

Mixed Review

38. **Real Estate** A *commission* is a percent of sales that is paid for selling a product. Tom Augustus sells real estate at a 7% commission. Last week, his sales totaled $90,000. What was his commission? *(Lesson 5–4)* **$6300**

Express each fraction or ratio as a percent. *(Lesson 5–3)*

39. 5 out of 8 **62.5%** 40. 2 to 5 **40%** 41. $\frac{7}{20}$ **35%**

Convert each measurement as indicated. *(Lesson 5–1)*

42. 15 inches to feet **1.25 ft** 43. 1500 grams to kilograms **1.5 kg**

Standardized Test Practice

44. **Multiple Choice** Which point corresponds to a number that *cannot* be obtained by subtracting two integers? *(Lesson 2–4)* **C**

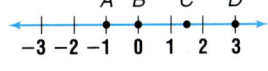

A B C D
−3 −2 −1 0 1 2 3

Ⓐ *A* Ⓑ *B* Ⓒ *C* Ⓓ *D*

Quiz 2 Lessons 5–4 and 5–5

Use the percent equation to find each number. *(Lesson 5–4)*

1. Find 4% of 625. **25**
2. What percent of 50 is 6? **12%**
3. 15 is 75% of what number? **20**
4. Find 105% of 80. **84**
5. **Economics** The inflation rate for groceries is 5%. How much will a cart of groceries cost next year if it costs $134.10 this year? Round to the nearest cent. *(Lesson 5–5)* **$140.81**

www.algconcepts.com/self_check_quiz

4 ASSESS

Open-Ended Assessment

Speaking Ask students to explain how they can tell by inspection whether a percent of increase will be greater than 100%.

Quiz 2

The Quiz provides students with a brief review of the concepts and skills in Lessons 5–4 and 5–5. Lesson numbers are given to the right of the exercises or instruction lines so students can review concepts not yet mastered.

Answer

35. 1999 to 2001; The amount of increase is larger and the base is smaller between these years.

Extra Credit

The original price of a sweater was $20. The price was increased by 6%, but then the new price was decreased by 10%. Find the final price of the sweater. **$19.08**

***Enrichment*, p. 219**

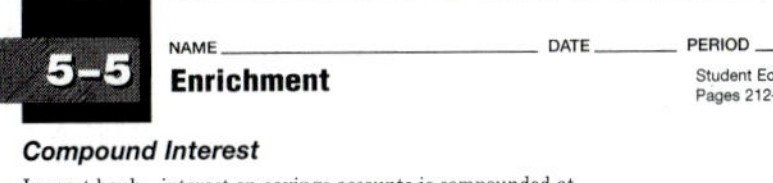

NAME ______ DATE ______ PERIOD ______

5-5 Enrichment Student Edition Pages 212–218

Compound Interest

In most banks, interest on savings accounts is compounded at set time periods such as three or six months. At the end of each period, the bank adds the interest earned to the account. During the next period, the bank pays interest on all the money in the bank, including interest. In this way, the account earns interest on interest.

Suppose Ms. Tanner has $1000 in an account that is compounded quarterly at 5%. Find the balance after the first two quarters.

Use $I = prt$ to find the interest earned in the first quarter if $p = 1000$ and $r = 5\%$. Why is t equal to $\frac{1}{4}$?

First quarter: $I = 1000 \times 0.05 \times \frac{1}{4}$
$I = 12.50$

The interest, $12.50, earned in the first quarter is added to $1000. The principal becomes $1012.50.

Second quarter: $I = 1012.50 \times 0.05 \times \frac{1}{4}$
$I = 12.65625$ The interest in the second quarter is $12.66.

The balance after two quarters is $1012.50 + 12.66 or $1025.16.

Answer each of the following questions.

1. How much interest is earned in the third quarter of Ms. Tanner's account? I = $12.81
2. What is the balance in her account after three quarters? $1037.97
3. What is the balance in her account after one year? $1050.94
4. Suppose Ms. Tanner's account is compounded semiannually. What is the balance at the end of six months? $1025.00
5. What is the balance after one year if her account is compounded semiannually? $1050.63

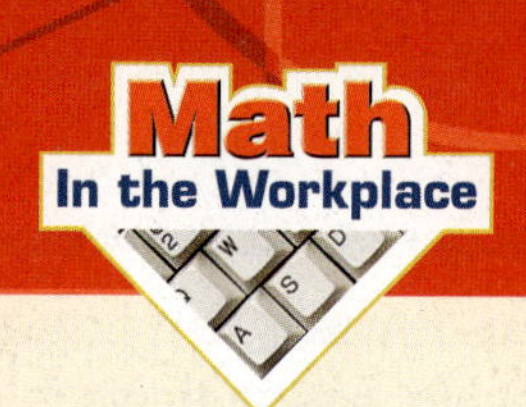

There are three major areas where accountants work: public accounting, management accounting, and government accounting. Public accountants often own their own business, while management accountants work for large companies. Government accountants work for state or federal governmental agencies.

Computers are used widely in accounting now, with special software programs greatly reducing the amount of manual computations that had to be done in years past. So accountants must have good computer skills, as well as good people skills.

Most accountants have at least a bachelor's degree in accounting, and many have a masters degree in accounting or business administration. In addition, there are a variety of licensing agencies for accountants, with the most well-known recognition being the CPA (Certified Public Accountants) designation.

High school courses in mathematics, business education, and computers are beneficial for students considering this field.

Related Careers

- appraisers
- loan officers
- financial analysts
- actuaries
- underwriters

Community Connection

Invite an accountant to visit your class and offer students some insights into the field of accounting. If the accountant has experience working in more than one area of accounting, ask him or her to compare and contrast their experiences. Have students prepared to ask several follow-up questions, such as how accounting has changed during this person's career.

Accountant

Accountants prepare and analyze financial reports. Some accountants track the depreciation of their company's property and inventory. *Depreciation* is the decrease in value of an item due to its age. There are different ways to calculate depreciation.

- *Straight-Line Method:* The depreciation is the same each year.
- *Double-Declining-Balance Method:* The depreciation is determined by a percent of decrease.

Suppose a media production company purchased equipment for \$240,000. The company estimates that the equipment will last for 5 years. At that time, the value of the equipment will be \$20,000. The table and graph show the value each year for a straight-line (SL) depreciation of \$44,000 and a double-declining-balance (DDB) depreciation of 40%.

Value (dollars)		
Year	**SL**	**DDB**
0	240,000	240,000
1	196,000	144,000
2	152,000	86,400
3	108,000	51,840
4	64,000	31,104
5	20,000	20,000

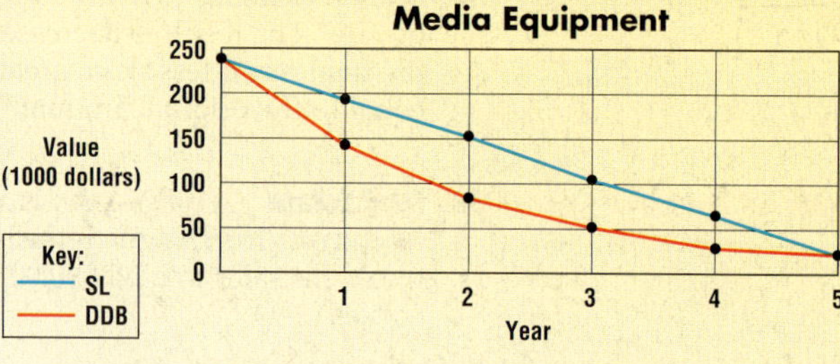

1. Why do you think the first method is called the straight-line method? **Its graph is a straight line.**
2. For which year is the difference between the straight-line value and the double-declining-balance value the greatest? How is this shown on the graph? **2; The distance between the graphs is the greatest.**

FAST FACTS About Accountants

Working Conditions

- usually work in a comfortable environment
- generally work a 40-hour week, but some work 50 hours a week or more

Education

- most have a college degree in accounting
- knowledge of computers a necessity

Earnings

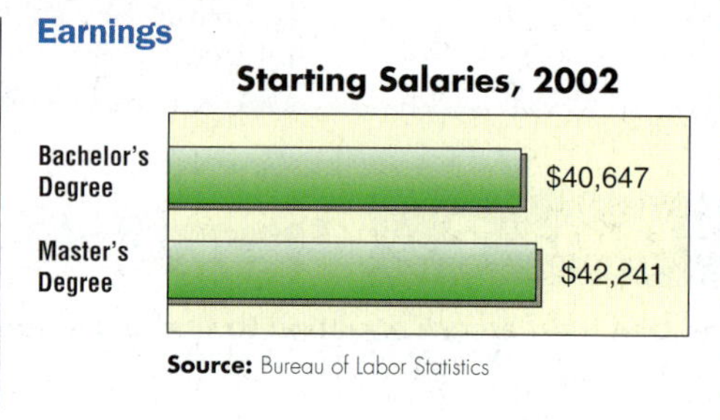

Source: Bureau of Labor Statistics

Career Data For the latest information on accountants, visit: www.algconcepts.com

Not on the Net

If students have limited or no access to the Internet, they can obtain additional information about the field of accounting and the CPA examination by writing to the following organization.

American Institute of Certified Public Accountants
Harborside Financial Center
201 Plaza III
Jersey City, NJ 07311-3881

For information on colleges and universities offering accredited programs in accounting, students can write to the following organization.

The Association to Advance Collegiate Schools of Business
600 Emerson Road, Suite 300
St. Louis, MO 63141-6762

5-6 Probability and Odds

What You'll Learn

You'll learn to find the probability and odds of a simple event.

Why It's Important

Manufacturing Inspectors use experimental probability when they find defective items. *See Examples 2 & 3.*

The table shows all of the possible outcomes when you roll a pair of dice. The highlighted outcomes are doubles.

	1	2	3	4	5	6
1	(1, 1)	(1, 2)	(1, 3)	(1, 4)	(1, 5)	(1, 6)
2	(2, 1)	(2, 2)	(2, 3)	(2, 4)	(2, 5)	(2, 6)
3	(3, 1)	(3, 2)	(3, 3)	(3, 4)	(3, 5)	(3, 6)
4	(4, 1)	(4, 2)	(4, 3)	(4, 4)	(4, 5)	(4, 6)
5	(5, 1)	(5, 2)	(5, 3)	(5, 4)	(5, 5)	(5, 6)
6	(6, 1)	(6, 2)	(6, 3)	(6, 4)	(6, 5)	(6, 6)

There are 36 possible outcomes. If the dice are fair, each outcome is equally likely to occur. Of those 36 outcomes, 6 are doubles. You can measure the chances of an event happening with **probability**.

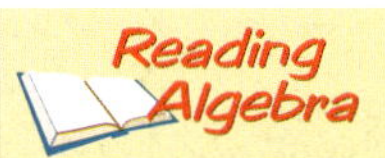

P(doubles) is read as the *probability of rolling doubles.*

Probability

Words: The probability of an event is a ratio that compares the number of favorable outcomes to the number of possible outcomes.

Symbols: $P(\text{event}) = \dfrac{\text{number of favorable outcomes}}{\text{number of possible outcomes}}$

Numbers: $P(\text{doubles}) = \dfrac{6}{36}$ or $\dfrac{1}{6}$

The probability that an event will happen is between 0 and 1 inclusive.

- A probability of 0 means that the event is impossible.
- A probability of 1 means the event is certain to happen.
- The closer a probability is to 1, the more likely it is to happen.

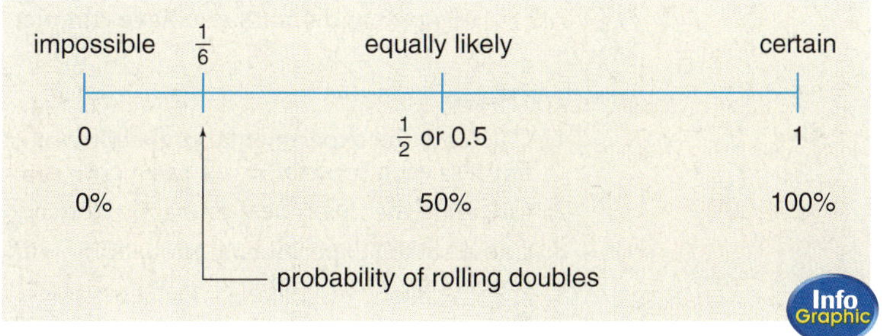

1 FOCUS

5-Minute Check
Lesson 5-5

Find each percent of increase or decrease.

1. original: 25
   new: 21 **16% decrease**
2. original: 120
   new: 174 **45% increase**
3. A brand of running shoes increased in cost by 3%. Find the new price if the old price was $65. **$66.95**

The cost of an item and a sales tax rate are given. Find the total price of each item to the nearest cent.

4. jacket: $78; 6.5% **$83.07**
5. DVD player: $249; 7.5% **$267.68**

Motivating the Lesson

Hands-On Activity Provide each student with a coin. Have the students flip their coin 5 times. Ask whether any students got all heads or all tails. Inform students that the probability of getting all heads (or all tails) when flipping a coin 5 times is 1 out of 32. Use any questions about how this probability is determined to begin the lesson.

2 TEACH

Teaching Tip As part of the discussion of the outcomes table for two dice, ask students if one sum is more likely than any other. Focus students' attention on the unshaded diagonal of entries from the lower left corner of the chart to the upper right. They should see that the probability of rolling a sum of 7 is $\frac{6}{36}$ or $\frac{1}{6}$, and also that this is greater than the probability for any other possible sum.

Resource Manager

Reproducible Masters

Chapter 5 Resource Masters
- *Study Guide,* p. 220
- *Skills Practice,* p. 221
- *Practice,* p. 222
- *Reading to Learn Mathematics,* p. 223
- *Enrichment,* p. 224

Hands-On Algebra, p. 67

Transparencies

5-Minute Check, 5-6

Technology/Multimedia

Interactive Chalkboard CD-ROM

In-Class Example

Example 1

Use the graph shown in Example 1. If a person is chosen at random, what is the probability that the person is age 5–17? $\frac{3}{16}$ **or 18.75%**

When all possible outcomes have an equally likely chance of happening, the outcomes are said to be **random**.

1 **A population distribution for California is shown. If a person is chosen at random, what is the probability that the person is age 65 or older?**

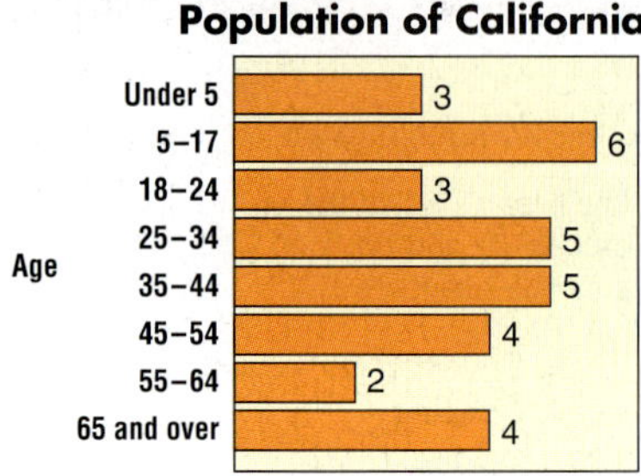

There are 4 million people who are age 65 or older. The total population is $3 + 6 + 3 + 5 + 5 + 4 + 2 + 4$ or 32 million.

$$P(\text{65 or older}) = \frac{\text{number of people age 65 or older}}{\text{total population}}$$

$$= \frac{4}{32} \text{ or } \frac{1}{8}$$

The probability of choosing a person age 65 or older is $\frac{1}{8}$ or 12.5%.

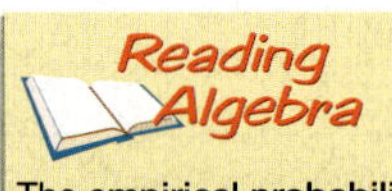

The **empirical probability** is the most accurate probability based upon repeated trials in an experiment.

The probability that was found in Example 1 is called the **theoretical probability**. Theoretical probability is what *should* occur. What *actually* occurs when we conduct an experiment is called the **experimental probability**.

Materials: paper bag  20 two-inch pieces of paper

Work with a partner.

Step 1 Mark 4 slips of paper with an X, 7 slips of paper with a Y, and 9 slips of paper with a Z.

Step 2 Put the slips of paper in the bag and mix well.

Step 3 Draw one slip of paper from the bag and record its letter.

Step 4 Return the slip of paper to the bag and mix well. Repeat Steps 3 and 4 until you have completed 20 trials.

Try These

1. Calculate the experimental probability of choosing each letter. Express each probability as a percent. **See students' work.**
2. Calculate the theoretical probability of choosing each letter.
3. Compare the experimental probability with the theoretical probability. How similar are they? **See students' work.**

2. $P(X) = 20\%$, $P(Y) = 35\%$, $P(Z) = 45\%$

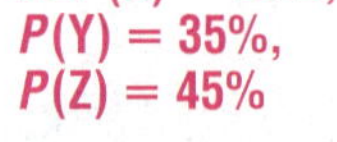

www.algconcepts.com/extra_examples

Hands-On Algebra

Cooperative Learning If time allows after completing the activity, have students repeat Steps 3 and 4 until they have completed 40 additional trials. Have them combine these results with their first 20 trials and compute the experimental probabilities from 60 trials. Have students compare these experimental probabilities to the theoretical probabilities. Ask them if the experimental and theoretical probabilities are more similar for 60 trials than for 20 trials.

Hands-On Algebra Masters, p. 67

Examples
Manufacturing Link

Experimental probability is often used by quality-control inspectors.

2 **A quality-control inspector for Office Suppliers checked a sample of 250 marking pens and found that 14 of them were defective. Find the experimental probability of choosing a defective pen.**

$$P(\text{defective pen}) = \frac{\text{number of defective pens}}{\text{total number of pens}}$$

$$= \frac{14}{250} \text{ or } \frac{7}{125}$$

3 **If the percent of defective pens in Example 2 is greater than 6%, production will be stopped. Should the quality-control inspector stop production?**

First, change $\frac{7}{125}$ to a percent. Then compare it to 6%.

$\frac{P}{B} = \frac{r}{100}$ *Use the percent proportion.*

$\frac{7}{125} = \frac{r}{100}$ *Replace P with 7 and B with 125.*

$7(100) = 125r$ *Find the cross products.*

$700 = 125r$ *Multiply 7 and 100.*

$\frac{700}{125} = \frac{125r}{125}$ *Divide each side by 125.*

$5.6 = r$ *700 [÷] 125 [ENTER] 5.6*

The sample contains 5.6% defective pens. Since this is less than 6%, the quality-control inspector should not stop production.

Another way to measure the chance of an event occurring is with **odds**.

Odds

Words: The odds of an event occurring is a ratio that compares the number of favorable outcomes to the number of unfavorable outcomes.

Symbols: $\text{Odds} = \frac{\text{number of favorable outcomes}}{\text{number of unfavorable outcomes}}$

Example

4 **A bag contains 6 red marbles, 3 blue marbles, and 1 yellow marble. Find the odds of choosing a red marble.**

There are 6 red marbles. So, there are 6 favorable outcomes.

There are 3 + 1 or 4 marbles that are *not* red. So, there are 4 unfavorable outcomes.

odds of choosing a red marble = 6:4 or 3:2

Your Turn

Find the odds of choosing a blue marble. **3:7**

In-Class Examples

Examples 2–3

A systems technician at a large corporation monitored their e-mail system for one hour and found that employees sent 550 e-mail messages, of which 375 were sent to e-mail accounts outside the company.

2 Find the experimental probability that a randomly-chosen e-mail message is being sent outside the company. $\frac{15}{22}$

3 If the percent of e-mails sent outside the company is greater than 65%, the company will increase its number of outside phone lines. Does the company need to increase its number of outside phones lines at this time? **yes**

Example 4

A coin is randomly removed from a change purse that contains 7 pennies, 8 nickels, and 5 quarters. What are the odds that the coin is a nickel? **8:12 or 2:3**

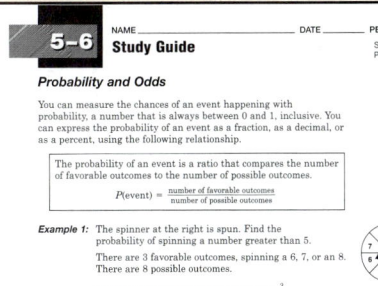

Inclusion Strategies

If there are students with visual impairments in your class, have volunteers develop dice or spinners that can be used and "read" by the visually-impaired students.

Study Guide, p. 220

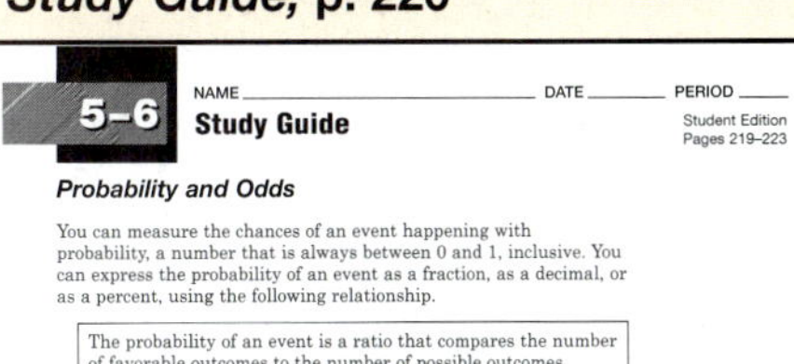

5–6 NAME ______ DATE ______ PERIOD ______

Study Guide

Student Edition Pages 219–223

Probability and Odds

You can measure the chances of an event happening with probability, a number that is always between 0 and 1, inclusive. You can express the probability of an event as a fraction, as a decimal, or as a percent, using the following relationship.

The probability of an event is a ratio that compares the number of favorable outcomes to the number of possible outcomes.

$$P(\text{event}) = \frac{\text{number of favorable outcomes}}{\text{number of possible outcomes}}$$

Example 1: The spinner at the right is spun. Find the probability of spinning a number greater than 5.

There are 3 favorable outcomes, spinning a 6, 7, or an 8. There are 8 possible outcomes.

Therefore $P(\text{a number greater than 5}) = \frac{3}{8}$.

Another way to measure the chance of an event is with odds.

The odds of an event is a ratio that compares the number of favorable outcomes to the number of unfavorable outcomes.

$$\text{Odds} = \frac{\text{number of favorable outcomes}}{\text{number of unfavorable outcomes}}$$

Example 2: Refer to the spinner above. Find the odds of spinning an even number.

There are 4 favorable outcomes: 2, 4, 6, or 8. The number of unfavorable outcomes is 8 − 4 = 4. Therefore the odds of spinning an even number are 4:4 or 1:1.

Refer to the spinner above. Find the probability of each outcome.

1. a 3 $\frac{1}{8}$
2. an even number $\frac{1}{2}$
3. a number less than 3 $\frac{1}{4}$
4. a number greater than 4 $\frac{1}{2}$

Refer to the spinner above. Find the odds of each outcome.

5. an odd number **1:1**
6. a multiple of 3 **1:3**
7. not a 6 **7:1**
8. a number greater than 4 **1:1**

3 PRACTICE/APPLY

Error Analysis

Watch for students who confuse odds and probability in Exercises 2–3.
Prevent by suggesting the following memory aid to students. "There are odd numbers greater than 1, and the odds that an event will occur can also be greater than 1." Remind students that probabilities are values from 0 to 1, inclusive.

Assignment Guide

Basic: 9–25 odd, 26–31
Average: 10–22 even, 23–31

Teaching Tip Exercise 19 previews the content presented in Lesson 5–7 on *Compound Events.*

Answer

1. Both theoretical and experimental probability are ratios that compare the number of favorable outcomes to the number of possible outcomes. Theoretical probability is based on known characteristics. It tells what should happen. Experimental probability is the result of an experiment or simulation.

***Skills Practice,* p. 221, and *Practice,* p. 222 (shown)**

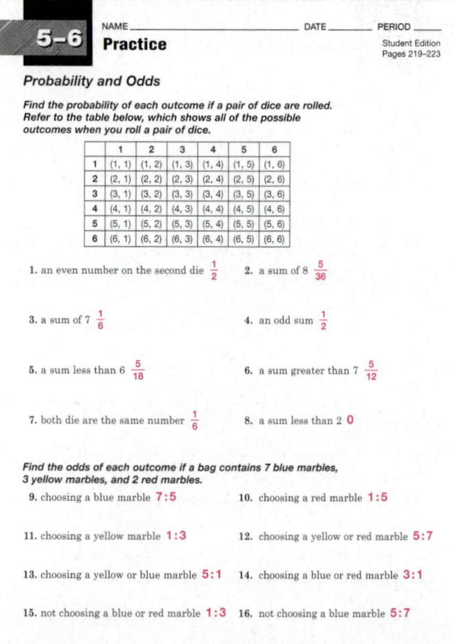
5–6 **Practice**
NAME ______ DATE ______ PERIOD ______
Student Edition Pages 219–223

Probability and Odds

Find the probability of each outcome if a pair of dice are rolled. Refer to the table below, which shows all of the possible outcomes when you roll a pair of dice.

	1	2	3	4	5	6
1	(1, 1)	(1, 2)	(1, 3)	(1, 4)	(1, 5)	(1, 6)
2	(2, 1)	(2, 2)	(2, 3)	(2, 4)	(2, 5)	(2, 6)
3	(3, 1)	(3, 2)	(3, 3)	(3, 4)	(3, 5)	(3, 6)
4	(4, 1)	(4, 2)	(4, 3)	(4, 4)	(4, 5)	(4, 6)
5	(5, 1)	(5, 2)	(5, 3)	(5, 4)	(5, 5)	(5, 6)
6	(6, 1)	(6, 2)	(6, 3)	(6, 4)	(6, 5)	(6, 6)

1. an even number on the second die $\frac{1}{2}$
2. a sum of 8 $\frac{5}{36}$
3. a sum of 7 $\frac{1}{6}$
4. an odd sum $\frac{1}{2}$
5. a sum less than 6 $\frac{5}{18}$
6. a sum greater than 7 $\frac{5}{12}$
7. both die are the same number $\frac{1}{6}$
8. a sum less than 2 **0**

Find the odds of each outcome if a bag contains 7 blue marbles, 3 yellow marbles, and 2 red marbles.

9. choosing a blue marble **7:5**
10. choosing a red marble **1:5**
11. choosing a yellow marble **1:3**
12. choosing a yellow or red marble **5:7**
13. choosing a yellow or blue marble **5:1**
14. choosing a blue or red marble **3:1**
15. not choosing a blue or red marble **1:3**
16. not choosing a blue marble **5:7**

Check for Understanding

Communicating Mathematics

1–3. See margin.

1. **Compare and contrast** theoretical probability and experimental probability.
2. **Describe** how to find the odds of an event happening.
3. Can the odds of an event happening be greater than 1? **Explain** why or why not.

Vocabulary
probability
random
theoretical probability
experimental probability
empirical probability
odds

Guided Practice

Example 1

4. $\frac{1}{2}$ 5. $\frac{1}{36}$

Refer to the application at the beginning of the lesson. Find the probability of each outcome if a pair of dice is rolled.

4. an even number on the first die
5. a sum of 2

Example 4

Find the odds of each outcome if a die is rolled.

6. a number greater than 4 **1:2**
7. an even number **1:1**

Examples 2 & 3

8a. $\frac{7}{8}$

8. **Sports** Michael made 7 free throws out of 8 attempts during the last basketball game.
   a. What was the experimental probability of making a free throw?
   b. Express the probability as a percent. **87.5%**

Exercises

Practice

9. $\frac{1}{2}$ 10. $\frac{1}{36}$

Homework Help

For Exercises	See Examples
9–14, 21	1
15–20	4
23	2, 3

Extra Practice
See page 702.

Refer to the application at the beginning of the lesson. Find the probability of each outcome if a pair of dice are rolled.

9. an odd number on the first die
10. a sum of 12
11. a sum of 5
12. an even sum
13. a sum greater than 12 **0**
14. a sum less than 13 **1**

11. $\frac{1}{9}$ 12. $\frac{1}{2}$

Find the odds of each outcome if the spinner at the right is spun.

15. an even number **1:1**
16. greater than 2 **3:1**
17. *not* a 5 **7:1**
18. red **3:5**
19. red or blue **5:3**
20. *not* red **5:3**

B

21. $\frac{1}{4}$

21. What is the probability that a month picked at random starts with the letter *J*?
22. If the probability that an event occurs is $\frac{1}{2}$, find the odds that it occurs. **1:1**

Reteaching Activity

Intrapersonal Learners Have students write a letter or a diary entry in which they analyze what the words "impossible" and "certain" mean with regard to probability.

Answers

2. Find the number of favorable outcomes and the number of unfavorable outcomes. Then, write a ratio that compares them.
3. Yes; when the number of favorable outcomes is greater than the number of unfavorable outcomes, the odds of the outcome are greater than 1.

Applications and Problem Solving

23. **Manufacturing** A quality-control inspector checked three samples of sweaters during the day. Each sample contained 40 sweaters. The chart lists the number of sweaters that were defective.

Quality Control Results	
Sample (time)	Number Defective
9:00	2
1:00	1
4:00	5

a. For each sample, find the experimental probability of choosing a defective sweater.

23a. 9:00, $\frac{1}{20}$; 1:00, $\frac{1}{40}$; 4:00, $\frac{1}{8}$

b. Production is stopped if the percent of defective sweaters is greater than 5%. For which sample(s) should production be stopped? **4:00**

24. **Marketing** The local video store advertises that 1 out of 4 customers will receive a free box of popcorn when they rent a video.

a. What are the odds of receiving free popcorn? **1:3**

b. What are the odds *against* receiving free popcorn? **3:1**

c. At the end of the first day, 15 customers out of 75 had received free popcorn. Find the experimental probability.

24c. $\frac{1}{5}$

d. Did the store give away as much free popcorn as it had advertised? **no**

25. **Allowance** The graph shows the weekly allowance for students in grades 6 through 12.

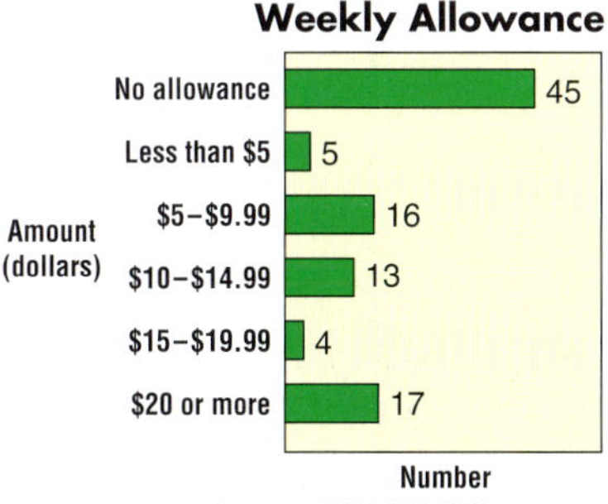

a. If a student is chosen at random, what is the probability the student receives $5 or more as a weekly allowance?

25a. $\frac{1}{2}$

b. What are the odds that a student chosen at random receives no allowance? **9:11**

26. **Critical Thinking** The complement of an event's occurring is the event's *not* occurring. If the probability that an event will occur is $\frac{3}{5}$, what is the probability of its complement? $\frac{2}{5}$

Mixed Review

The original cost of an item and a discount rate are given. Find the sale price of each item to the nearest cent. *(Lesson 5–5)*

27. basketball: $45, 20% **$36**

28. television: $399, 10% **$359.10**

29. **Banking** How much interest will Ben earn if he invests $1000 at a rate of 6% for 3 years? *(Lesson 5–4)* **$180**

Standardized Test Practice

Ⓐ Ⓑ Ⓒ Ⓓ

30. **Grid In** What percent of 25 is 30? *(Lesson 5–3)* **120**

31. **Multiple Choice** Which expression is equivalent to $-2(x + 5) + 6(x + 5)$? *(Lesson 2–5)* **A**

Ⓐ $4(x + 5)$ Ⓑ $4x + 40$ Ⓒ $8x + 40$ Ⓓ $4x - 20$

www.algconcepts.com/self_check_quiz

Extra Credit

The ratio that represents the odds of an event occurring is equal to the ratio representing the probability that the event will occur. Find the probability for the event. **0**

4 ASSESS

Open-Ended Assessment

Writing Ask students to write out a complete outcomes chart (like the one shown on page 219) for spinning two spinners, each divided into four equal sections labeled 10, 20, 30, and 40. Have them use the chart to compute the probability of spinning a sum of 60, and the odds of spinning two multiples of 4.

***Enrichment*, p. 224**

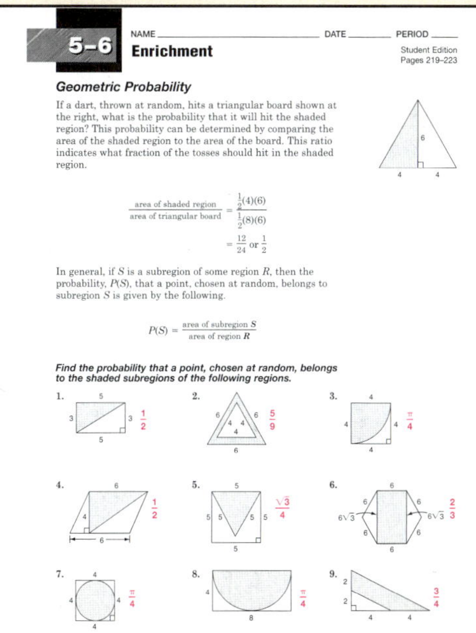

5-6 NAME ______ DATE ______ PERIOD ______

Enrichment

Student Edition Pages 219–223

Geometric Probability

If a dart, thrown at random, hits a triangular board shown at the right, what is the probability that it will hit the shaded region? This probability can be determined by comparing the area of the shaded region to the area of the board. This ratio indicates what fraction of the tosses should hit in the shaded region.

$$\frac{\text{area of shaded region}}{\text{area of triangular board}} = \frac{\frac{1}{2}(4)(6)}{\frac{1}{2}(8)(6)} = \frac{12}{24} \text{ or } \frac{1}{2}$$

In general, if S is a subregion of some region R, then the probability, $P(S)$, that a point, chosen at random, belongs to subregion S is given by the following.

$$P(S) = \frac{\text{area of subregion } S}{\text{area of region } R}$$

Find the probability that a point, chosen at random, belongs to the shaded subregions of the following regions.

1. $\frac{1}{2}$
2. $\frac{5}{9}$
3. $\frac{\pi}{4}$
4. $\frac{1}{2}$
5. $\frac{\sqrt{3}}{4}$
6. $\frac{2}{3}$
7. $\frac{\pi}{4}$
8. $\frac{\pi}{4}$
9. $\frac{3}{4}$

Lesson 5–7

1 FOCUS

5-Minute Check
Lesson 5–6

A bag contains 3 blue marbles, 7 red marbles, and 2 green marbles. One marble is selected at random from the bag.

1. Find the probability that the marble is green. $\frac{2}{12}$ or $\frac{1}{6}$
2. Find the probability that the marble is red. $\frac{7}{12}$
3. Find the odds that the marble is blue. 3:9 or 1:3

During one day, an ATM machine dispensed 500 $20 bills, of which 75 were the older bills with a small picture of President Andrew Jackson on them.

4. Find the experimental probability that a customer who received a $20 bill received one of the older bills. $\frac{75}{500}$ or $\frac{3}{20}$
5. Find the odds that a randomly-chosen user of the ATM who received one $20 bill actually received one of the new $20 bills. 425:75 or 17:3

Motivating the Lesson

Hands-On Activity Provide each student with a penny and a nickel. Instruct them to flip both coins at once. Find out how many students got heads on both coins. Then ask how they might calculate the probability of getting heads for both. $\frac{1}{2} \cdot \frac{1}{2} = \frac{1}{4}$ Stress that this is the probability of getting heads on the penny *and* heads on the nickel.

5–7 Compound Events

What You'll Learn

You'll learn to find the probability of mutually exclusive and inclusive events.

Why It's Important

Marketing

Restaurant owners can use probability to make predictions about their business. *See Exercise 21.*

Compound events consist of two or more simple events that are connected by the words *and* or *or*. Let's investigate a case where simple events are connected by the word *and*.

Hands-On Algebra

Materials:  2 paper bags  red and yellow counters

Work with a partner.

Step 1 Place a red counter and a yellow counter in each bag.

Step 2 Without looking, remove one counter from each bag. Record the color combination in the order that you drew the counters. Return the counters to their respective bags.

Step 3 Repeat 99 times. Count and record the number of red/red, red/yellow, yellow/red, and yellow/yellow combinations.

Try These 1–4. Sample answers: about 25%

Estimate the probability of each outcome.

1. P(red *and* red)
2. P(red *and* yellow)
3. P(yellow *and* red)
4. P(yellow *and* yellow)

Reading Algebra

P(red and yellow) means the probability of choosing red from the first bag and yellow from the second bag.

Choosing a counter from bag 1 did not affect choosing a counter from bag 2. These events are called **independent events** because the outcome of one event does not affect the outcome of the other event.

Look Back

Tree diagrams: Lesson 4–2

You can analyze the experiment with a tree diagram.

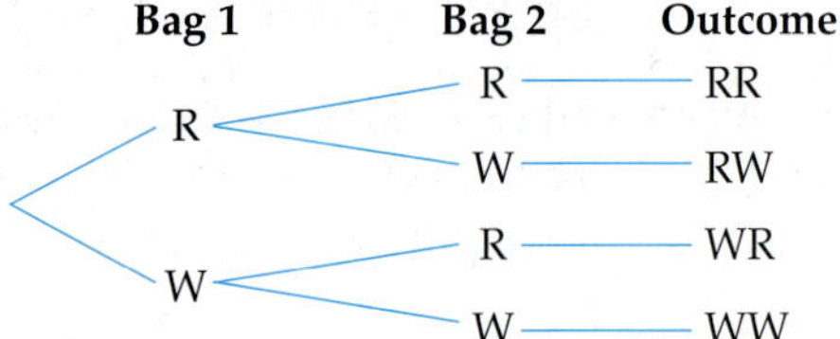

There are four equally-likely outcomes. So, the probability of choosing white on the first draw *and* white on the second draw is $\frac{1}{4}$. You can also multiply to find the probability of two independent events.

$$\underbrace{P(\text{white from bag 1})}_{\frac{1}{2}} \times \underbrace{P(\text{white from bag 2})}_{\frac{1}{2}} = \underbrace{P(\text{white and white})}_{\frac{1}{4}}$$

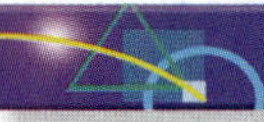

Resource Manager

Reproducible Masters

Chapter 5 Resource Masters

- *Study Guide,* p. 225
- *Skills Practice,* p. 226
- *Practice,* p. 227
- *Reading to Learn Mathematics,* p. 228
- *Enrichment,* p. 229
- *Assessment,* p. 241

Hands-On Algebra, p. 68

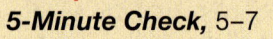

Transparencies

5-Minute Check, 5–7

Technology/Multimedia

Interactive Chalkboard CD-ROM

Probability of Independent Events

Words: The probability of two independent events is found by multiplying the probability of the first event by the probability of the second event.

Symbols: $P(A \text{ and } B) = P(A) \cdot P(B)$

Model:

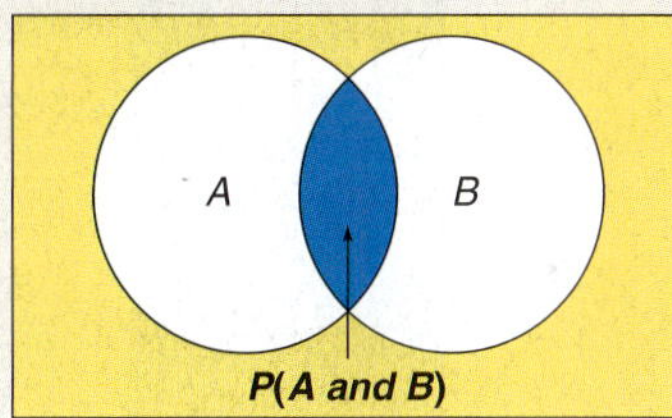

Example 1

Two dice are rolled. Find the probability that an odd number is rolled on the first die and the number 4 is rolled on the second.

$P(\text{odd number}) = \frac{3}{6} \text{ or } \frac{1}{2}$ *There are three ways to roll an odd number.*

$P(4) = \frac{1}{6}$ *There is 1 way to roll a 4.*

$P(\text{odd number } and \text{ } 4) = \frac{1}{2} \cdot \frac{1}{6} \text{ or } \frac{1}{12}$

Your Turn

a. Two dice are rolled. Find the probability that an even number is rolled on the first die and a number greater than 4 on the second die. $\frac{1}{6}$

Two events can also be connected by the word *or*. For example, consider the probability of drawing a jack or a queen from a standard deck of 52 cards. Since a card cannot be both a jack and a queen, the events are **mutually exclusive**. That is, both events cannot occur at the same time.

The probability of two mutually exclusive events is found by adding.

$P(\text{jack or queen}) = P(\text{jack}) + P(\text{queen})$

$= \frac{4}{52} + \frac{4}{52}$

$= \frac{8}{52} \text{ or } \frac{2}{13}$

The probability of drawing a jack or a queen is $\frac{2}{13}$.

www.algconcepts.com/extra_examples

2 TEACH

Teaching Tip After you discuss the definition of *independent events,* ask students to suggest events that could be called *dependent events.* (One example is drawing a second marble from a bag of marbles without replacing the first marble that was drawn.)

Teaching Tip Before discussing the model in the Probability of independent Events box, you may want to review Venn diagrams.

In-Class Example

Example 1

The two spinners shown below are spun at the same time. Find the probability that the left spinner lands on green and the right spinner lands on a number greater than 2. $\frac{2}{9}$

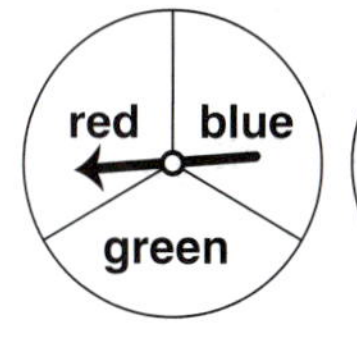

Hands-On Algebra

Cooperative Learning Refer to the Hands-On Algebra on page 224. Have students place all four counters in one bag and repeat the activity by drawing a counter twice from the bag, being sure to place the first counter back in the bag before drawing a second one. Have them compare their results to those from the original activity.

Hands-On Algebra Masters, p. 68

Teaching Tip Compare and contrast the Venn diagram in the Mutually-Exclusive Events box on this page with the Venn diagram for independent events shown on the previous page.

In-Class Example

Example 2

A marble is selected at random from a bag that contains 5 red marbles, 3 blue marbles, and 2 yellow marbles. What is the probability that the marble is either red or yellow? $\frac{7}{10}$

Probability of Mutually Exclusive Events

Words: The probability of two mutually exclusive events is found by adding the probability of the first event and the probability of the second event.

Symbols: $P(A \text{ or } B) = P(A) + P(B)$

Model:

Example 2 **Jamal has 4 quarters, 2 dimes, and 4 nickels in his pocket. He takes one coin from his pocket at random. What is the probability that the coin is either a quarter or a dime?**

A coin cannot be both a quarter and a dime, so the events are mutually exclusive. Find the sum of the individual probabilities.

$P(\text{quarter } or \text{ dime}) = P(\text{quarter}) + P(\text{dime})$

$= \frac{4}{10} + \frac{2}{10}$ *4 of the coins are quarters and 2 are dimes.*

$= \frac{6}{10}$ *Add.*

$= \frac{3}{5}$ *Simplify.*

The probability of choosing a quarter or a dime is $\frac{3}{5}$.

Your Turn

b. Find the probability of Jamal's choosing a quarter or a nickel. $\frac{4}{5}$

Sometimes events are connected by the word *or,* but they are not mutually exclusive. For example, suppose there is a chance of rain on Saturday and there is a chance of rain on Sunday. You want to find the chance of rain over the weekend. Because it could rain on *both* Saturday *and* Sunday, rainfall on Saturday and Sunday are not mutually exclusive events. The two events are called **inclusive events.**

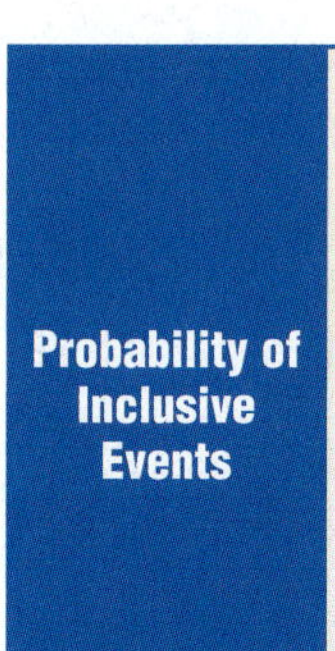

Probability of Inclusive Events

Words: The probability of two inclusive events is found by adding the probabilities of the events, then subtracting the probability of both events.

Symbols: $P(A \text{ or } B) = P(A) + P(B) - P(A \text{ and } B)$

Model:

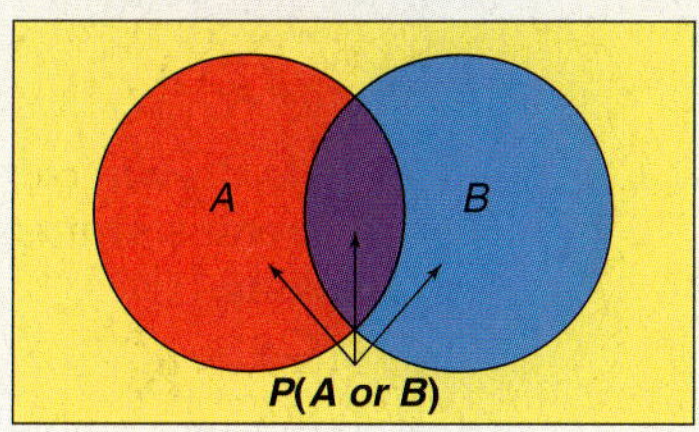

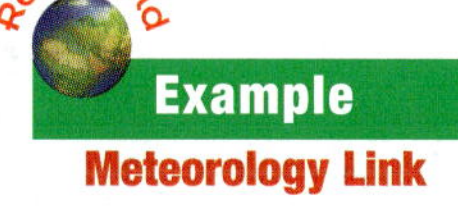

Example 3

Meteorology Link

If there is a 40% chance of rain on Saturday and a 60% chance of rain on Sunday, find the probability that it will rain on either Saturday or Sunday.

Since it is possible to rain on both days, these events are inclusive.

$P(\text{Saturday}) = 0.4$ $\qquad$ $P(\text{Sunday}) = 0.6$

These events are independent since the weather on Saturday does not affect the weather on Sunday.

$P(\text{Saturday } or \text{ Sunday})$

$= P(\text{Saturday}) + P(\text{Sunday}) - P(\text{Saturday } and \text{ Sunday})$

$= 0.4 + 0.6 - (0.4)(0.6)$

$= 1.0 - 0.24$ *Simplify.*

$= 0.76$ or 76% *Write as a percent.*

The probability that it will rain on the weekend is 76%.

Concept Summary				
	Probability	$P(A$ and $B)$	$P(A$ or $B)$	$P(A$ or $B)$
	Type of Events	A and B are independent.	A and B are inclusive (sample spaces overlap).	A and B are mutually exclusive (sample spaces do not overlap).
	Formula	$P(A) \cdot P(B)$	$P(A) + P(B) - P(A \text{ and } B)$	$P(A) + P(B)$

Check for Understanding

Communicating Mathematics

1. **Explain** the difference between $P(A$ and $B)$ and $P(A$ or $B)$. **See margin.**
2. *Writing Math* Compare and contrast independent events, mutually exclusive events, and inclusive events. Give an example of each. **See students' work.**

Vocabulary
compound events
independent events
mutually exclusive
inclusive

Teaching Tip Again, compare and contrast the Venn diagram shown on this page with the two previous diagrams.

In-Class Example

Example 3

If there is a 90% chance of snow in January and a 95% chance of snow in February, find the probability that it will snow sometime in January or February. **0.995**

Answer

1. **$P(A$ and $B)$ is the probability that both A and B occur; $P(A$ or $B)$ is the probability that either A or B occur.**

Study Guide, p. 225

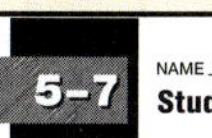

NAME ______ DATE ____ PERIOD ____

Study Guide

Student Edition Pages 224–229

Compound Events

To win a game, you have to roll either an even number or a 3 with one die. What is the probability that you will win? The probability of tossing an even number = $P(\text{even number}) = \frac{1}{2}$, and the probability of tossing a 3 is $P(3) = \frac{1}{6}$. These events cannot occur at the same time and so they are **mutually exclusive**. This is a compound event because there is more than 1 event occurring. You find the probability of mutually exclusive events by adding the probabilities of each event.

$P(\text{even number or a 3}) = P(\text{even number}) + P(3)$

$= \frac{1}{2} + \frac{1}{6}$ or $\frac{2}{3}$

Compound events are called **independent** if the outcome of one event does not affect the outcome of the other event. For example, if two dice are rolled, rolling an odd number with the first die does not affect rolling a 4 on the second die. You find the probability of two independent events by multiplying the probabilities of each event.

$P(\text{odd number and a 4}) = P(\text{odd number}) \cdot P(4)$

$= \frac{1}{2} \cdot \frac{1}{6}$ or $\frac{1}{12}$

Probability of Mutually Exclusive Events *A* and *B*	$P(A \text{ or } B) = P(A) + P(B)$
Probability of Independent Events *A* and *B*	$P(A \text{ and } B) = P(A) \cdot P(B)$

A die is rolled. Find the probability of each compound event.

1. $P(\text{a 4 or a 2})$ $\frac{1}{3}$
2. $P(\text{an odd number or a 1})$ $\frac{2}{3}$
3. $P(\text{a 2 or a 3})$ $\frac{1}{3}$
4. $P(\text{a 1 or a 6})$ $\frac{1}{3}$
5. $P(\text{an even number or a 5})$ $\frac{2}{3}$
6. $P(\text{a 1 or a multiple of 2})$ $\frac{2}{3}$
7. $P(\text{a 1 or a number greater than 4})$ $\frac{1}{2}$
8. $P(\text{an odd number or an even number})$ 1

Two dice are rolled. Find the probability of each compound event.

9. $P(\text{an odd number and a 1})$ $\frac{1}{12}$
10. $P(\text{an odd number and an even number})$ $\frac{1}{4}$
11. $P(\text{two odd numbers})$ $\frac{1}{4}$
12. $P(\text{a 6 and a number greater than 4})$ $\frac{1}{18}$
13. $P(\text{two numbers less than 3})$ $\frac{1}{9}$
14. $P(\text{an even number and a 5})$ $\frac{1}{12}$

Reteaching Activity

Visual/Spatial Learners Have students make a poster showing the results for the Probability of Independent Events, the Probability of Mutually-Exclusive Events, and the Probability of Independent Events. The poster should show the Venn diagrams and symbolic statements for each property, and at least one example illustrating each property.

3 PRACTICE/APPLY

Error Analysis

Watch for students who forget to subtract $P(A \text{ and } B)$ when computing the probability of the inclusive events in Exercises 6–7. ***Prevent by*** referring students back to the Venn diagram on page 227, which shows that circle A and circle B both contain the overlapping region (A and B). Stress that $P(A) + P(B)$ actually "counts" this region twice, so one of the "countings" must be subtracted.

Assignment Guide

Basic: 11–21 odd, 22–27
Average: 10–18 even, 19–27

Teaching Tip For Exercises 10–15, some students may be interested in finding the probability of each outcome if the card is *not* replaced after the first drawing.

***Skills Practice,* p. 226, and *Practice,* p. 227 (shown)**

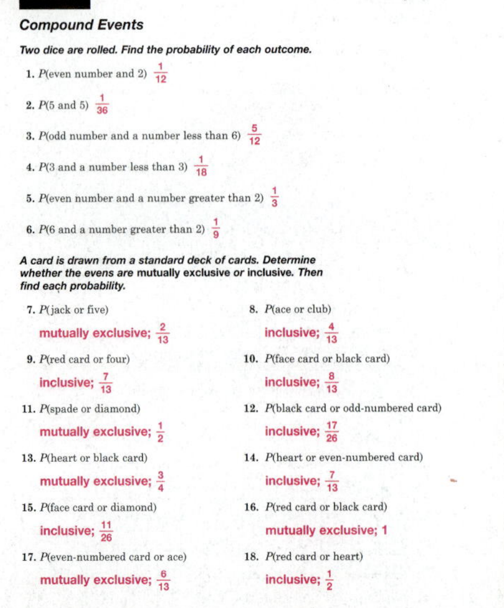

5-7 NAME ______ DATE ______ PERIOD ______

Practice

Student Edition Pages 224–229

Compound Events

Two dice are rolled. Find the probability of each outcome.

1. P(even number and 2) $\frac{1}{12}$
2. P(5 and 5) $\frac{1}{36}$
3. P(odd number and a number less than 6) $\frac{5}{12}$
4. P(3 and a number less than 3) $\frac{1}{18}$
5. P(even number and a number greater than 2) $\frac{1}{3}$
6. P(6 and a number greater than 2) $\frac{1}{9}$

A card is drawn from a standard deck of cards. Determine whether the evens are* mutually exclusive *or* inclusive. *Then find each probability.

7. P(jack or five) **mutually exclusive; $\frac{2}{13}$**
8. P(ace or club) **inclusive; $\frac{4}{13}$**
9. P(red card or four) **inclusive; $\frac{7}{13}$**
10. P(face card or black card) **inclusive; $\frac{8}{13}$**
11. P(spade or diamond) **mutually exclusive; $\frac{1}{2}$**
12. P(black card or odd-numbered card) **inclusive; $\frac{17}{26}$**
13. P(heart or black card) **mutually exclusive; $\frac{3}{4}$**
14. P(heart or even-numbered card) **inclusive; $\frac{7}{13}$**
15. P(face card or diamond) **inclusive; $\frac{11}{26}$**
16. P(red card or black card) **mutually exclusive; 1**
17. P(even-numbered card or ace) **mutually exclusive; $\frac{6}{13}$**
18. P(red card or heart) **inclusive; $\frac{1}{2}$**

Guided Practice

Example 1

A die is rolled and the spinner is spun. Find the probability of each event.

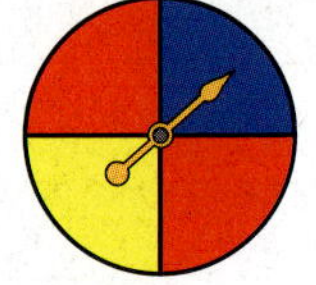

3. P(3 and blue) $\frac{1}{24}$
4. P(even and red) $\frac{1}{4}$

Examples 2 & 3

A card is drawn from a standard deck of cards. Determine whether each event is *mutually exclusive* or *inclusive*. Then find each probability.

5. P(jack or king) **M, $\frac{2}{13}$**
6. P(ace or red card) **I, $\frac{7}{13}$**
7. P(jack or diamond) **I, $\frac{4}{13}$**
8. P(diamond or club) **M, $\frac{1}{2}$**

Example 3

9. **Savings** The chart shows the probability that students in grades 6 through 12 are saving for certain items. What is the probability that a student picked at random is saving for a CD player or a computer? **20.1%**

What Students Save For

Item	Probability
Car	31%
College	27%
Clothing	19%
CDs or CD player	15%
Trip or vacation	6%
Computer or software	6%

Source: *USA WEEKEND*

Exercises

Practice

A card is drawn from a deck of ten cards numbered 1 through 10. The card is replaced in the deck and another card is drawn. Find the probability of each outcome.

10. P(5 and then a 3) $\frac{1}{100}$
11. P(two even numbers) $\frac{1}{4}$
12. P(two numbers greater than 4)
13. P(6 and then an odd number)
14. P(an odd number and then an even number) $\frac{1}{4}$
15. P(a number greater than 7 and then a number less than 6) $\frac{3}{20}$
16. What is the probability of tossing a coin three times and getting heads each time? $\frac{1}{8}$

Homework Help

For Exercises	See Examples
10–16	1
17–21	2, 3

Extra Practice

See page 703.

12. $\frac{9}{25}$

13. $\frac{1}{20}$

17. M, $\frac{2}{3}$

Determine whether each event is *mutually exclusive* or *inclusive*. Then find each probability.

17. There are 3 books about the American Revolution, 2 books about the Civil War, and 4 books about World War II on a shelf. If a book is selected at random, what is the probability of choosing a book about the Civil War or World War II?
18. A card is drawn from a deck of cards. What is the probability that it is a black card or a face card? **I, $\frac{8}{13}$**

Applications and Problem Solving

115/1000

P(0.115) · P(0.115)

19. **Sports** The probability that Barry Bonds hit a home run during an official time at bat during the 2003 season was 0.115. Find the probability that he hit a home run on two consecutive times at bat. Express your answer as a decimal rounded to the nearest thousandth. **0.013**

20. $\frac{1}{22,320}$

20. **Games** A radio station sponsored The Birthday Game, in which it would award $1 million to any caller whose birthday matched a certain month, day, and year. If there were 62 possible years, find the probability that a caller wins $1 million. Assume 30 days in each month.

21. **Marketing** The graph shows the percent of adults who say they've eaten at certain types of restaurants during the last six months. **a. 34.2%** **b. 28%**
   a. What is the probability that a person chosen at random has eaten in both a Chinese and Mexican restaurant?
   b. Suppose no one has eaten at both a French and a Japanese restaurant. What was the probability that a person ate at either a French or a Japanese restaurant?
   c. What is the probability that a person chosen at random has eaten at either an American or Italian restaurant? **88.3%**

Restaurant Choices

Kind	Percent
American	70
Chinese	60
French	10
Italian	61
Japanese	18
Mexican	57
Thai	11

Source: *American Demographics* Inc.

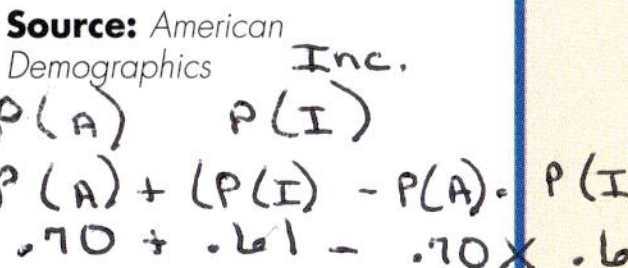

P(A) P(I)

P(A) + (P(I) − P(A) · P(I)

.70 + .61 − .70 × .61

.131 −

22. $\frac{7}{28} \cdot \frac{6}{27}$ or $\frac{1}{18}$

22. **Critical Thinking** A standard domino set has 28 tiles. Seven of these tiles have the same number of dots on each side and are called *doubles*. As the game begins, each player draws one tile. What is the probability that the first and second players each draw doubles?

Mixed Review

Determine the probability of each outcome. *(Lesson 5–6)*

23. A month is chosen at random that ends in *-ber*. $\frac{1}{3}$
24. A die is rolled and shows a number greater than 1. $\frac{5}{6}$
25. A brown-haired student is chosen at random from a class of 25 students. There are 15 brown-haired students in the class. $\frac{3}{5}$

Standardized Test Practice

26. **Short Response** In 2003, the population of North America was about 323 million. It is expected to increase to about 390 million by 2050. By what percent is the population expected to increase? *(Lesson 5–5)* **about 21%**

27. **Multiple Choice** Evaluate the expression $|3| + |-5|$. *(Lesson 2–1)*
   (A) 2 (B) −2 (C) −8 (D) 8 **D**

www.algconcepts.com/self_check_quiz

4 ASSESS

Open-Ended Assessment

Speaking Ask students to describe the difference between mutually exclusive events and inclusive events.

Chapter 5, Quiz B (Lessons 5–4 through 5–7) is available in the *Chapter 5 Resource Masters*, p. 241.

Extra Credit

A coin is tossed three times, and comes up heads all three times. It is going to be tossed two more times. Find the probability that it comes up heads on both of the next two tosses. $\frac{1}{4}$

***Enrichment*, p. 229**

NAME ______ DATE ______ PERIOD ______

5-7 Enrichment

Student Edition Pages 224–229

Conditional Probability

The probability of an event given the occurrence of another event is called **conditional probability.** The conditional probability of event A given event B is denoted $P(A|B)$.

Example: Suppose a pair of number cubes is rolled. It is known that the sum is greater than seven. Find the probability that the number cubes match.

There are 15 sums greater than seven and there are 36 possible pairs altogether.

There are three matching pairs greater than seven, (4, 4), (5, 5), and (6, 6).

$P(B) = \frac{15}{36}$ $P(A \text{ and } B) = \frac{3}{36}$

$P(A|B) = \frac{P(A \text{ and } B)}{P(B)}$

$= \frac{\frac{3}{36}}{\frac{15}{36}}$ or $\frac{1}{5}$

The conditional probability is $\frac{1}{5}$.

A card is drawn from a standard deck of 52 cards and is found to be red. Given that event, find each of the following probabilities.

1. P(heart) $\frac{1}{2}$
2. P(ace) $\frac{1}{13}$
3. P(face card) $\frac{3}{13}$
4. P(jack or ten) $\frac{2}{13}$
5. P(six of spades) 0
6. P(six of hearts) $\frac{1}{26}$

A sports survey taken at Stirers High School shows that 48% of the respondents liked soccer, 66% liked basketball, and 38% liked hockey. Also, 30% liked soccer and basketball, 22% liked basketball and hockey and 28% liked soccer and hockey. Finally, 12% liked all three sports.

7. Find the probability that Meg likes soccer if she likes basketball. $\frac{5}{11}$
8. Find the probability that Juan likes basketball if he likes soccer. $\frac{5}{8}$
9. Find the probability that Mieko likes hockey if she likes basketball. $\frac{1}{3}$
10. Find the probability that Greg likes hockey if he likes soccer. $\frac{7}{12}$

Study Guide and Assessment

Understanding and Using the Vocabulary

This section provides a listing of the new terms, properties, and phrases that were introduced in this chapter. The exercises check students' understanding of the terms by using a variety of verbal formats including matching, completion, and true/false.

Glossary A complete glossary of terms appears on pages 762–783.

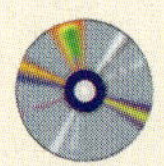

MindJogger Videoquizzes

MindJogger Videoquizzes provide an alternative review of concepts presented in this chapter. Students work in teams to answer questions, gaining points for correct answers.

CHAPTER 5 Study Guide and Assessment

Understanding and Using the Vocabulary

After completing this chapter, you should be able to define each term, property, or phrase and give an example or two of each.

interNET CONNECTION **Review Activities**
For more review activities, visit: www.algconcepts.com

Algebra

base *(p. 199)*
dimensional analysis *(p. 190)*
discount *(p. 213)*
mixture problems *(p. 206)*
percent *(p. 198)*
percent equation *(p. 204)*
percent of decrease *(p. 212)*
percent of increase *(p. 212)*
percent proportion *(p. 199)*
percentage *(p. 199)*
proportion *(p. 188)*
rate *(pp. 190, 204)*
ratio *(p. 188)*
sales tax *(p. 213)*
scale *(p. 194)*
scale drawing *(p. 194)*
scale model *(p. 194)*
simple interest *(p. 205)*
unit rate *(p. 190)*

Statistics

box-and-whisker plot *(p. 210)*
circle graph *(p. 200)*
extremes *(p. 210)*
lower quartile *(p. 210)*
percentile *(p. 211)*
upper quartile *(p. 210)*

Probability

complement *(p. 223)*
compound event *(p. 224)*
empirical probability *(p. 220)*
experimental probability *(p. 220)*
inclusive events *(p. 226)*
independent events *(p. 224)*
mutually exclusive *(p. 225)*
odds *(p. 221)*
probability *(p. 219)*
random *(p. 220)*
theoretical probability *(p. 220)*

Choose the correct term to complete each sentence.

1. The equation $\frac{2}{3} = \frac{10}{15}$ is a (proportion, ratio).
2. A ratio is a comparison of two numbers by (multiplication, division).
3. A probability of 0 means that an event is (certain to happen, impossible).
4. The probability that occurs when you conduct an experiment is called the (empirical probability, theoretical probability).
5. The (odds, probability) that an event will occur is the ratio of the number of favorable outcomes to the number of unfavorable outcomes.
6. In the percent proportion $\frac{3}{8} = \frac{r}{100}$, 8 is the (base, percentage).
7. A (unit rate, ratio) always has a denominator of 1.
8. In a percent equation, the base is (always, not always) the largest number.
9. (Discount, Sales tax) is an example of a percent of decrease.
10. If two events cannot occur at the same time, they are (mutually exclusive, inclusive).

www.algconcepts.com/vocabulary_review

Resource Manager

Reproducible Masters
- *Assessment*, pp. 231–239, 242–244

Technology/Multimedia
MindJogger Videoquizzes
ExamView® Pro

Skills and Concepts

Objectives and Examples

- **Lesson 5-1** Solve proportions.

$\frac{7}{4} = \frac{n}{2}$

$14 = 4n$ *Find the cross products.*
$3.5 = n$ *Divide each side by 4.*

Review Exercises

Solve each proportion.

11. $\frac{5}{8} = \frac{n}{72}$ **45**
12. $\frac{r}{11} = \frac{35}{55}$ **7**
13. $\frac{x}{x-1} = \frac{4}{3}$ **4**
14. $\frac{z-7}{6} = \frac{z+3}{7}$ **67**

- **Lesson 5-2** Solve problems involving scale drawings and models.

In a drawing, 3 cm = 45 m. Find the length of a line representing 75 meters.

$\frac{3 \text{ centimeters}}{45 \text{ meters}} = \frac{x \text{ centimeters}}{75 \text{ meters}}$

$225 = 45x$ *Find the cross products.*
$5 = x$ *Divide each side by 45.*

On a map, 2 inches = 5 miles. Find the actual distance for each map distance.

15. 4 inches **10 mi**
16. 5 inches **12.5 mi**
17. 12 inches **30 mi**
18. 9 inches **22.5 mi**

19. A scale model of the Statue of Liberty is 10 inches high. If the Statue of Liberty is 305 feet tall, find the scale of the model. **1 in. = 30.5 ft**

- **Lesson 5-3** Solve problems by using the percent proportion.

75 is what percent of 250?

$\frac{75}{250} = \frac{r}{100}$ *Use the percent proportion.*
$7500 = 250r$ *Find the cross products.*
$30 = r$ *Divide each side by 250.*

So, 75 is 30% of 250.

Use the percent proportion to find each number.

20. What number is 60% of 80? **48**
21. 21 is 35% of what number? **60**
22. 7 is what percent of 56? **12.5%**
23. 60 is what percent of 40? **150%**
24. Find 12% of 5200. **624**
25. 15 is 30% of what number? **50**

- **Lesson 5-4** Solve problems by using the percent equation.

Find 15% of 82.

$P = RB$ *Use the percent equation.*
$P = 0.15(82)$ *Replace R with 0.15 and B with 82.*
$P = 12.3$

So, 15% of 82 is 12.3.

Use the percent equation to find each number.

26. 54 is 150% of what number? **36**
27. Find 45% of 18. **8.1**
28. What is 8% of 80? **6.4**
29. 21 is 14% of what number? **150**
30. What percent of 34 is 17? **50%**

Skills and Concepts

The **Objectives and Examples** section reviews the skills and concepts of the chapter and shows completely worked examples.

The **Review Exercises** provide practice for the corresponding objectives.

ExamView® Pro

Use the networkable ***ExamView® Pro Testmaker CD-ROM*** to:

- Create **multiple versions** of tests.
- Create **modified** tests for **inclusion** students with one mouse click.
- **Edit** existing questions and **add** your own questions.
- Build tests aligned with state standards using built-in **state curriculum correlations**.
- Change **English** tests to **Spanish** with one mouse click and vice versa.

Mixed Problem Solving
See pages 724–731.

Applications and Problem Solving

This section provides additional practice in solving real-world problems that involve the concepts of this chapter.

Objectives and Examples	Review Exercises

- **Lesson 5-5** Solve problems involving percent of increase or decrease.

original: \$120 new: \$108
amount of decrease: \$120 − \$108 = \$12

$\frac{12}{120} = \frac{r}{100}$ *Use the percent proportion.*
$1200 = 120r$ *Find the cross products.*
$10 = r$ *Divide each side by 120.*

The percent of decrease is 10%.

Find the percent of increase or decrease. Round to the nearest percent.

31. original: 8 new: 10 **25% inc.**
32. original: 10 new: 15 **50% inc.**
33. original: 18 new: 12 **33% dec.**
34. original: 800 new: 300 **63% dec.**
35. Find the percent of increase from \$25 to \$50. **100%**
36. What number is 20% less than 80? **64**

- **Lesson 5-6** Find the probability of a simple event.

Find the probability of randomly choosing the letter I in the word PITTSBURGH.

$\frac{\text{number of favorable outcomes}}{\text{number of possible outcomes}} = \frac{1}{10}$

One letter from the word MISSISSIPPI is chosen at random. Find each probability.

37. $P(M)$
38. $P(I)$
39. $P(\text{consonant})$ $\frac{7}{11}$
40. $P(\text{vowel})$ $\frac{4}{11}$
41. Find the probability that a letter chosen at random from the word OHIO is *not* an O.

37. $\frac{1}{11}$ **38.** $\frac{4}{11}$ **41.** $\frac{1}{2}$

- **Lesson 5-7** Find the probability of mutually exclusive and inclusive events.

mutually exclusive:
$P(A \text{ or } B) = P(A) + P(B)$
inclusive:
$P(A \text{ or } B) = P(A) + P(B) - P(A \text{ and } B)$

A six-sided die is rolled. Determine whether each event is *mutually exclusive* or *inclusive*. Then find each probability.

42. $P(\text{even or less than 5})$ **inclusive,** $\frac{5}{6}$
43. $P(6 \text{ or odd})$ **mutually exclusive,** $\frac{2}{3}$

Applications and Problem Solving

44. Food Of the students surveyed, 40% chose pizza as their favorite lunch. If there are 1250 students in the school, how many would you expect to order pizza? *(Lesson 5–4)* **500 students**

45. Shopping Inez is buying a sound system that costs \$399. Since she is an employee of the store, she receives a 15% discount. How much will Inez pay for the sound system? *(Lesson 5–5)* **\$339.15**

Assessment, pp. 233–234

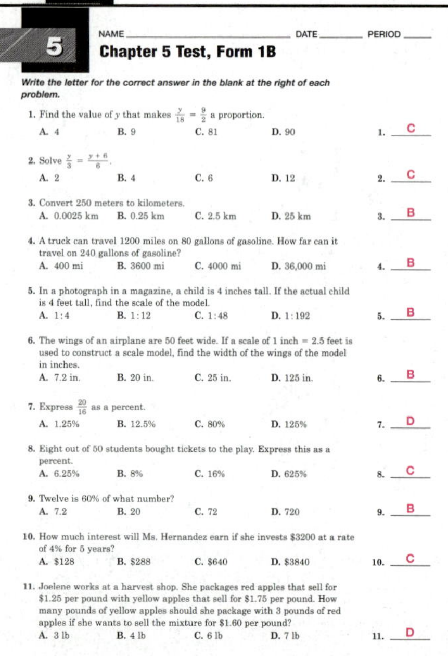

NAME ________ DATE ______ PERIOD _____

5 **Chapter 5 Test, Form 1B**

Write the letter for the correct answer in the blank at the right of each problem.

1. Find the value of y that makes $\frac{y}{18} = \frac{9}{2}$ a proportion.
   A. 4 B. 9 C. 81 D. 90 — 1. C
2. Solve $\frac{y}{3} = \frac{y+6}{6}$.
   A. 2 B. 4 C. 6 D. 12 — 2. C
3. Convert 250 meters to kilometers.
   A. 0.0025 km B. 0.25 km C. 2.5 km D. 25 km — 3. B
4. A truck can travel 1200 miles on 80 gallons of gasoline. How far can it travel on 240 gallons of gasoline?
   A. 400 mi B. 3600 mi C. 4000 mi D. 36,000 mi — 4. B
5. In a photograph in a magazine, a child is 4 inches tall. If the actual child is 4 feet tall, find the scale of the model.
   A. 1:4 B. 1:12 C. 1:48 D. 1:192 — 5. B
6. The wings of an airplane are 50 feet wide. If a scale of 1 inch = 2.5 feet is used to construct a scale model, find the width of the wings of the model in inches.
   A. 7.2 in. B. 20 in. C. 25 in. D. 125 in. — 6. B
7. Express $\frac{20}{16}$ as a percent.
   A. 1.25% B. 12.5% C. 80% D. 125% — 7. D
8. Eight out of 50 students bought tickets to the play. Express this as a percent.
   A. 6.25% B. 8% C. 16% D. 625% — 8. C
9. Twelve is 60% of what number?
   A. 7.2 B. 20 C. 72 D. 720 — 9. B
10. How much interest will Ms. Hernandez earn if she invests \$3200 at a rate of 4% for 5 years?
   A. \$128 B. \$288 C. \$640 D. \$3840 — 10. C
11. Joelene works at a harvest shop. She packages red apples that sell for \$1.25 per pound with yellow apples that sell for \$1.75 per pound. How many pounds of yellow apples should she package with 3 pounds of red apples if she wants to sell the mixture for \$1.60 per pound?
   A. 3 lb B. 4 lb C. 6 lb D. 7 lb — 11. D

Assessment

Four forms of Chapter 5 Test are available in the *Chapter 5 Resource Masters.*

Chapter 5 Test, Form 1B, is shown at the left. Chapter 5 Test, Form 2B, is shown on the next page.

	Form of Test		Level
1A	Multiple Choice	pp. 231–232	Average
1B	Multiple Choice	pp. 233–234	Basic
2A	Free Response	pp. 235–236	Average
2B	Free Response	pp. 237–238	Basic

CHAPTER 5 Test

1. **Write** the percent proportion and the percent equation. Explain the meaning of P, B, r, and R. **1–2. See margin.**
2. **Compare and contrast** experimental probability and theoretical probability.

Solve each proportion.

3. $\frac{3}{x} = \frac{12}{16}$ **4**
4. $\frac{9}{y} = \frac{15}{10}$ **6**
5. $\frac{x}{5} = \frac{x+2}{6}$ **10**
6. $\frac{3}{3a+2} = \frac{3}{8}$ **2**

Find each number.

7. 20% of 40 is what number? **8**
8. 12 is what percent of 60? **20**
9. 20 is what percent of 16? **125**
10. 23 is 25% of what number? **92**
11. Find 120% of 32. **38.4**
12. What is 35% of 60? **21**

13. **Measurement** Convert 42 inches to feet. **3.5 ft**

14. **Recycling** When 2000 pounds of paper are recycled or reused, 17 trees are saved. How many trees would be saved if 8000 pounds of paper are recycled? **68 trees**

15. **Hobbies** Model railroads are scaled-down models of real trains. The scale on an HO model train is 1 inch = 87 inches. An HO model of a modern diesel locomotive is 8 inches long. How long is the real locomotive? **696 in.**

16. **Banking** How long will it take Mr. Roberts to earn \$1500 if he invests \$5000 at a rate of 6%? **5 yr**

17. **Shopping** The Just Skates sporting goods store advertises that all in-line skates are on sale for 20% off the regular price. Find the sale price of a pair of skates that cost \$160. **\$128**

18. **Taxes** What is the cost of a pair of jeans that sells for \$49 if the sales tax rate is 6%? **\$51.94**

19. **Sports** A quarterback threw 18 completed passes out of 30 attempts. Find the experimental probability of making a completed pass. Express the probability as a percent. **60%**

20. **Probability** A die is rolled. What is the probability of rolling a 5 or a number greater than 3? $\frac{1}{2}$

www.algconcepts.com/chapter_test

Chapter Test Bonus Question

Two dice are rolled, and the two numbers on the dice are added. List all the outcomes that have a probability of less than $\frac{1}{9}$. **2, 3, 4, 10, 11, 12**

Chapter Test

Answers

1. $\frac{P}{B} = \frac{r}{100}$ and $P = BR$, where P is the percentage, B is the base, r is the percent, and R is the rate.
2. Both are the ratio of the number of favorable outcomes to the number of possible outcomes. The theoretical probability is what should happen; experimental probability is what actually happens in an experiment.

***Assessment*, pp. 237–238**

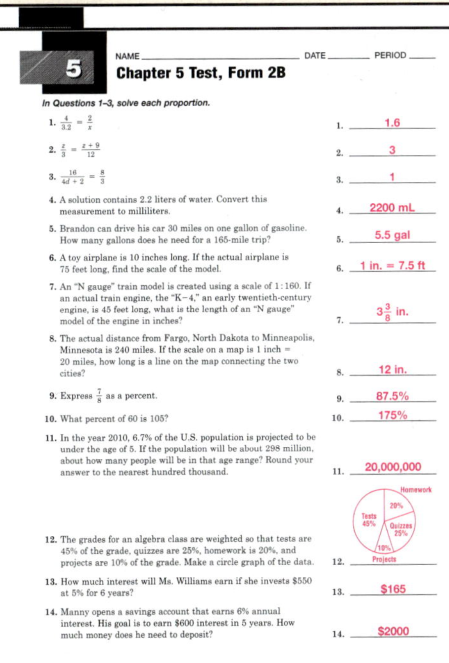
NAME ________ DATE ______ PERIOD ______

5 Chapter 5 Test, Form 2B

In Questions 1–3, solve each proportion.

1. $\frac{4}{3.2} = \frac{2}{x}$ — 1. 1.6
2. $\frac{z}{3} = \frac{z+9}{12}$ — 2. 3
3. $\frac{16}{4d+2} = \frac{8}{3}$ — 3. 1
4. A solution contains 2.2 liters of water. Convert this measurement to milliliters. — 4. 2200 mL
5. Brandon can drive his car 30 miles on one gallon of gasoline. How many gallons does he need for a 165-mile trip? — 5. 5.5 gal
6. A toy airplane is 10 inches long. If the actual airplane is 75 feet long, find the scale of the model. — 6. 1 in. = 7.5 ft
7. An "N gauge" train model is created using a scale of 1:160. If an actual train engine, the "K−4," an early twentieth-century engine, is 45 feet long, what is the length of an "N gauge" model of the engine in inches? — 7. $3\frac{3}{8}$ in.
8. The actual distance from Fargo, North Dakota to Minneapolis, Minnesota is 240 miles. If the scale on a map is 1 inch = 20 miles, how long is a line on the map connecting the two cities? — 8. 12 in.
9. Express $\frac{7}{8}$ as a percent. — 9. 87.5%
10. What percent of 60 is 105? — 10. 175%
11. In the year 2010, 6.7% of the U.S. population is projected to be under the age of 5. If the population will be about 298 million, about how many people will be in that age range? Round your answer to the nearest hundred thousand. — 11. 20,000,000
12. The grades for an algebra class are weighted so that tests are 45% of the grade, quizzes are 25%, homework is 20%, and projects are 10% of the grade. Make a circle graph of the data. — 12.

13. How much interest will Ms. Williams earn if she invests \$550 at 5% for 6 years? — 13. \$165
14. Manny opens a savings account that earns 6% annual interest. His goal is to earn \$600 interest in 5 years. How much money does he need to deposit? — 14. \$2000

Preparing for Standardized Tests

Pages 234–235 are part of a complete test preparation course that is described in detail on page T9 of the Teacher's Handbook. The test items on these pages were written in the same style as those in state proficiency tests and standardized tests like ACT and SAT.

Diagnosis and Prescription

Each of the 10 test questions on page 235 is cross-referenced to the lesson where that SAT or ACT skill is covered. If students miss a particular type of problem, you can have them study that skill.

(See chart at the bottom of page 235. Note that SPT = State Proficiency Test, SAT = Scholastic Assessment Test, and ACT = American College Test.)

***Assessment*, p. 242**

5 NAME ______ DATE ______ PERIOD ______

Chapter 5 Cumulative Review

1. Evaluate $6 \times 9 - \frac{a}{b}$ if $a = 12$ and $b = -2$. *(Lesson 1–2)* — 1. 60

In Questions 2–4, simplify each expression.

2. $3a + 4b + 3(2a + b)$ *(Lesson 1–4)* — 2. $9a + 7b$
3. $-4n + (-2n) + n$ *(Lesson 2–3)* — 3. $-5n$
4. $-3(3x)(-5z)$ *(Lesson 2–5)* — 4. $45xz$
5. Over a 4-day period, the value of a share of stock rises $1\frac{3}{8}$, drops $\frac{3}{4}$, rises $1\frac{1}{2}$, and then drops $2\frac{7}{8}$. What is the net rise or drop of the stock over the period? *(Lesson 3–2)* — 5. drop of $\frac{3}{4}$
6. Find the mean of the data below. *(Lesson 3–3)* 18, 24, 17, 42, 39, 19 — 6. 26.5
7. Solve $|x - 3| + 2 = 10$. *(Lesson 3–7)* — 7. 11; −5
8. Find the product of $\frac{4}{5}$, $-\frac{2}{3}$, and -1. *(Lesson 4–1)* — 8. $\frac{8}{15}$
9. An ice-cream shop offers 8 flavors of ice cream and 5 choices of toppings. You can have the ice cream in a cup, a plain cone, or a sugar cone. How many different combinations of a single flavor, a single topping, and a cup or one of the types of cone are there? *(Lesson 4–2)* — 9. 120

In Questions 10–12, solve each equation or proportion.

10. $-8n + 12 = 4n + 36$ *(Lesson 4–6)* — 10. −2
11. $6x - 3(x + 4) = 9$ *(Lesson 4–7)* — 11. 7
12. $\frac{m}{4} = \frac{m + 12}{24}$ *(Lesson 5–1)* — 12. 2.4
13. A scale model of a house is 2 feet wide. The actual house is 32 feet wide. Find the scale of the model. *(Lesson 5–2)* — 13. 1:16
14. Twelve out of 25 students in a class are girls. Express this ratio as a percent. *(Lesson 5–3)* — 14. 48%
15. Find 24% of 2800. *(Lesson 5–3)* — 15. 672
16. A card is drawn from a standard deck of 52 cards. What is the probability that the card is a 10 or a heart? *(Lesson 5–7)* — 16. $\frac{4}{13}$

CHAPTER 5 Preparing for Standardized Tests

Expression and Equation Problems

All standardized tests include questions that ask you to evaluate expressions and solve equations. You'll need to calculate with positive and negative integers, as well as with fractions and decimals. Be sure that you know and can apply the properties of equality.

Test-Taking Tip

You can use a strategy called "backsolving" on some multiple-choice questions. To use this strategy, substitute each answer choice into the expression or equation to determine which is correct.

Example 1

Solve $y = -14x - 5$ if $x = -1$.

(A) −19 (B) −9 (C) 9  19

Hint Work carefully with negative numbers. Apply the rules for adding and subtracting integers.

Solution Substitute −1 for x and evaluate the right side of the equation.

$y = -14x - 5$	*Original equation*
$y = -14(-1) - 5$	*Substitution*
$y = 14 - 5$	$-14(-1) = 14$
$y = 9$	$14 - 5 = 9$

The answer is C.

You can check your answer by replacing y with 9 and x with −1 in the original equation. If the statement is true, then your answer is correct.

$y = -14x - 5$	*Original equation*
$9 \stackrel{?}{=} -14(-1) - 5$	*Replace x with −1 and y with 9.*
$9 \stackrel{?}{=} 14 - 5$	*Multiply.*
$9 = 9$ ✓	*Subtract.*

Example 2

If $2 + a = 2 - a$, then $a =$

(A) −1 (B) 0 (C) 1 (D) 2 (E) 4

Hint Use the properties of equality to solve an equation.

Solution

$2 + a = 2 - a$	*Original equation*
$2 + a + (-2) = 2 - a + (-2)$	*Add −2 to each side.*
$a = -a$	$2 + (-2) = 0$
$a + a = -a + a$	*Add a to each side.*
$2a = 0$	$-a + a = 0$
$a = 0$	*Divide each side by 2.*

The answer is B.

Alternate Solution Use the strategy known as "backsolving." Substitute each of the answer choices for a and see which value makes the equation true.

A	$2 + (-1) = 2 - (-1)$	$1 = 3$	*not true*
B	$2 + 0 = 2 - 0$	$2 = 2$	*true*
C	$2 + 1 = 2 - 1$	$3 = 1$	*not true*
D	$2 + 2 = 2 - 2$	$4 = 0$	*not true*
E	$2 + 4 = 2 - 4$	$6 = -2$	*not true*

Resource Manager

Reproducible Masters

Chapter 5 Resource Masters

- *Assessment*, pp. 242–244

Preparing for Standardized Tests
For test-taking strategies and more practice, see pages 732–749.

After you work each problem, record your answer on the answer sheet provided or on a sheet of paper.

Multiple Choice

1. The time that a traffic light remains yellow is given by the formula $t = \frac{1}{8}s + 1$, where t is the time in seconds and s is the speed limit (mph). If the speed limit is 40 mph, how long will a light stay yellow? *(Lesson 4–5)* **B**
   Ⓐ 5 seconds
   Ⓑ 6 seconds
   Ⓒ 7 seconds
   Ⓓ 8 seconds

2. The heights of Monica's sunflowers are listed below. What is the median height, in inches, of the sunflowers? *(Lesson 3–3)* **C**

Height of Sunflowers	
2 ft 4 in.	3 ft 11 in.
2 ft 9 in.	4 ft 5 in.
4 ft 3 in.	3 ft 5 in.

   Ⓐ 41 Ⓑ 42.2
   Ⓒ 44 Ⓓ 49

3. If $9b = 81$, then $3 \times 3b =$ *(Lesson 4–4)* **C**
   Ⓐ 9. Ⓑ 27. Ⓒ 81.
   Ⓓ 243. Ⓔ 729.

4. The Tigers baseball team scored four more runs than the Bears. The number of runs the Bears scored is represented by n. Which expression represents the number of runs the Tigers scored? *(Lesson 1–1)* **B**
   Ⓐ $n - 4$ Ⓑ $n + 4$
   Ⓒ $4 - n$ Ⓓ $4n$

5. The formula $F = \frac{9}{5}C + 32$ is used to convert between degrees Celsius and degrees Fahrenheit. If the temperature is 82° F, what is the temperature in degrees Celsius? *(Lesson 1–5)* **A**
   Ⓐ 27.8 Ⓑ 45.6
   Ⓒ 63.3 Ⓓ 90

6. Which expression is equivalent to $\frac{4 + 8x}{12x}$? *(Lesson 4–1)* **A**
   Ⓐ $\frac{1 + 2x}{3x}$ Ⓑ $\frac{1 + 8x}{3x}$ Ⓒ 1
   Ⓓ $\frac{8}{3}$ Ⓔ $\frac{1 + 2x}{3}$

7. The Huang family had weekly grocery bills of \$105, \$115, \$120, and \$98 last month. What was their mean (average) weekly grocery bill last month? *(Lesson 3–3)* **C**
   Ⓐ \$101.50 Ⓑ \$102.00
   Ⓒ \$109.50 Ⓓ \$117.50

8. The equation $y = \frac{1}{3}x - \frac{3}{8}$ relates x and y. When y is $-\frac{1}{2}$, what is the value of x? *(Lesson 4–5)* **D**
   Ⓐ $-\frac{1}{24}$ Ⓑ $-\frac{13}{24}$ Ⓒ $-2\frac{5}{8}$ Ⓓ $-\frac{3}{8}$

Grid In

9. If $\frac{x + 2x + 3x}{2} = 6$, find the value of x. **2**
   (Lesson 4–5) **D**

10B. Sample answer: Distributive, Associative (+), Commutative (+)

Extended Response

10. Consider $3(x - 2) + 4(2x + 1) - 2(1 - x)$. *(Lesson 1–4)*
   Part A Simplify the expression completely. Show your work. **See margin.**
   Part B List the properties you used.

www.algconcepts.com/standardized_test

A bubble-in answer sheet for these practice problems is available on page A1 of the *Chapter 5 Resource Masters.*

Additional Practice

Additional test practice questions are available in the *Chapter 5 Resource Masters*, pp. 243–244.

Answer

10A. Sample answer:
$3(x - 2) + 4(2x + 1) - 2(1 - x)$
$= 3x - 6 + 8x + 4 - 2 + 2x$
$= (3x + 8x + 2x) - 6 + 4 - 2$
$= 13x - 4$

***Assessment*, pp. 243–244**

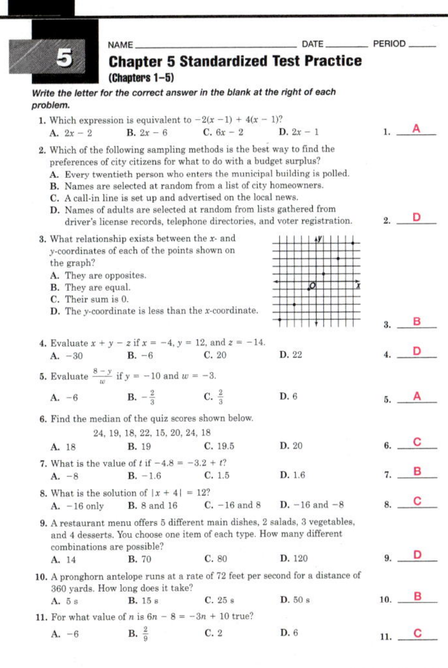

NAME ______ DATE ______ PERIOD ______

5 Chapter 5 Standardized Test Practice
(Chapters 1–5)

Write the letter for the correct answer in the blank at the right of each problem.

1. Which expression is equivalent to $-2(x - 1) + 4(x - 1)$?
   A. $2x - 2$ B. $2x - 6$ C. $6x - 2$ D. $2x - 1$ 1. A
2. Which of the following sampling methods is the best way to find the preferences of city citizens for what to do with a budget surplus?
   A. Every twentieth person who enters the municipal building is polled.
   B. Names are selected at random from a list of city homeowners.
   C. A call-in line is set up and advertised on the local news.
   D. Names of adults are selected at random from lists gathered from driver's license records, telephone directories, and voter registration. 2. D
3. What relationship exists between the x- and y-coordinates of each of the points shown on the graph?
   A. They are opposites.
   B. They are equal.
   C. Their sum is 0.
   D. The y-coordinate is less than the x-coordinate. 3. B
4. Evaluate $x + y - z$ if $x = -4$, $y = 12$, and $z = -14$.
   A. -30 B. -6 C. 20 D. 22 4. D
5. Evaluate $\frac{8 - y}{w}$ if $y = -10$ and $w = -3$.
   A. -6 B. $-\frac{2}{3}$ C. $\frac{2}{3}$ D. 6 5. A
6. Find the median of the quiz scores shown below.
   24, 19, 18, 22, 15, 20, 24, 18
   A. 18 B. 19 C. 19.5 D. 20 6. C
7. What is the value of t if $-4.8 = -3.2 + t$?
   A. -8 B. -1.6 C. 1.5 D. 1.6 7. B
8. What is the solution of $|x + 4| = 12$?
   A. -16 only B. 8 and 16 C. -16 and 8 D. -16 and -8 8. C
9. A restaurant menu offers 5 different main dishes, 2 salads, 3 vegetables, and 4 desserts. You choose one item of each type. How many different combinations are possible?
   A. 14 B. 70 C. 80 D. 120 9. D
10. A pronghorn antelope runs at a rate of 72 feet per second for a distance of 360 yards. How long does it take?
   A. 5 s B. 15 s C. 25 s D. 50 s 10. B
11. For what value of n is $6n - 8 = -3n + 10$ true?
   A. -6 B. $\frac{2}{9}$ C. 2 D. 6 11. C

Chapter 5 Expressions and Equations Problems		
Ex. 1	evaluate an expression	SPT
Ex. 2	solve an equation	SAT
1	evaluate an expression	SPT
2	median	SPT
3	solve an equation	SAT
4	write an expression	SPT
5	solve an equation	SPT
6	simplify an expression	ACT
7	mean	SPT
8	solve an equation	SPT
9	solve an equation	SAT
10	simplify an expression	SPT

CHAPTER 6

Resource Manager

Functions and Graphs

Instructional Objectives

Lesson (pages)	Objectives	NCTM Standards 2000	State/Local Objectives
6–1 (238–243)	Show relations as sets of ordered pairs, as tables, and as graphs.	1, 2, 3, 6, 8, 9, 10	
6–2 (244–249)	Solve linear equations for a given domain.	1, 6, 8, 9	
6–3 (250–255)	Graph linear relations.	1, 2, 3, 6, 8, 9	
6–4 (256–261)	Determine whether a given relation is a function.	1, 6, 8, 9	
Investigation (262–263)	Explore functions.	1, 2, 4, 6, 8, 9, 10	
6–5 (264–269)	Solve problems involving direct variation.	1, 2, 6, 8, 9, 10	
6–6 (270–275)	Solve problems involving inverse variations.	1, 2, 6, 8, 9, 10	

Key to NCTM Standards 2000

1 Number & Operations; **2** Algebra; **3** Geometry; **4** Measurement; **5** Data Analysis & Probability; **6** Problem Solving; **7** Reasoning & Proof; **8** Communications; **9** Connections; **10** Representation

Suggested Pacing

See page T13 for a complete course-planning calendar.

Standard refers to schedules that provide 45- to 55-minute periods that meet each day.

Block refers to schedules that provide approximately 90-minute periods which may meet every day for one semester or every other day over two semesters.

<table>
<tr><th>PACING</th><th>DAY 1</th><th>DAY 2</th><th>DAY 3</th><th>DAY 4</th><th>DAY 5</th><th>DAY 6</th></tr>
<tr><td>Standard Core (Chapters 1–13)</td><td colspan="2">Lesson 6–1</td><td colspan="2">Lesson 6–2</td><td colspan="2">Lesson 6–3</td></tr>
<tr><td>Standard Enhanced (Chapters 1–15)</td><td>Lesson 6–1</td><td>Lesson 6–2</td><td>Lesson 6–3</td><td colspan="2">Lesson 6–4</td><td>INV</td></tr>
<tr><td>Block Core (Chapters 1–13)</td><td>Chapter 5 Test Lesson 6–1</td><td>Lessons 6–2 & 6–3</td><td>Lesson 6–4 & INV</td><td>Lessons 6–5 & 6–6</td><td>SG+A</td><td>Chapter Test & Lesson 7–1</td></tr>
<tr><td>Block Enhanced (Chapters 1–15)</td><td>Chapter 5 Test Lesson 6–1</td><td>Lessons 6–2 & 6–3</td><td>Lesson 6–4 & INV</td><td>Lessons 6–5 & 6–6</td><td>SG+A</td><td>Chapter Test & Lesson 7–1</td></tr>
</table>

Instructional Resources

Lesson	Materials and Manipulatives (see below for Glencoe Manipulative Resources)	Blackline Masters (page numbers): Chapter 6 Resource Masters								
		Study Guide	Practice (Skills & Average)	Reading to Learn Mathematics	Enrichment	Assessment	Hands-On Algebra*	School-to-Workplace*	Graphing Calculator Masters*	5-Minute Check Transparencies
6–1	coordinate planes [4] straightedge [1, 2, 4]	245	246–247	248	249		72	6		6–1
6–2	coordinate planes [4] straightedge [1, 2, 4] dictionary	250	251–252	253	254	285				6–2
6–3	coordinate planes [4] straightedge [1, 2, 4] geoboard	255	256–257	258	259	284	73			6–3
6–4	coordinate planes [4] straightedge [1, 2, 4] metersticks	260	261–262	263	264		74, 75			6–4
Investigation	grid paper [1, 4] meterstick red and blue pens or pencils almanac straightedge [1, 2, 4]									
6–5	coordinate planes [4] straightedge [1, 2, 4]	265	266–267	268	269				18, 19	6–5
6–6	coordinate planes [4] straightedge [1, 2, 4] graphing calculator algebra tiles [1, 2, 3, 4]	270	271–272	273	274	285	76		17	6–6
Study Guide & Assessment/ Chapter Test	coordinate planes [4] straightedge [1, 2, 4]					275–283, 286–288				

***See page 236c for examples of these instructional materials.**

Key to Glencoe Manipulative Resources

[1]Classroom Manipulative Resources [2]Student Manipulative Resources [3]Overhead Manipulative Resources [4]Hands-On Algebra Masters

INV = Investigation *SG+A = Study Guide and Assessment*

DAY 7	DAY 8	DAY 9	DAY 10	DAY 11	DAY 12	DAY 13
Lesson 6–4	INV	Lesson 6–5	Lesson 6–6	SG+A	Chapter Test	
Lesson 6–5	Lesson 6–6	SG+A	Chapter Test			

The pages shown on this page are a small sample of the materials available on *TeacherWorks: All-in-One Lesson Planner and Resource Center.*

This CD-ROM includes all of the blackline masters and transparencies available for this program.

It also includes a lesson planner and interactive Teacher Edition, so you can customize lesson plans and reproduce classroom resources quickly and easily, from just about anywhere.

Applications

School-to-Workplace Masters, p. 6

6–1 NAME ____________ DATE ______ PERIOD ____

School-to-Workplace Student Edition Pages 238–243

Traffic Patterns (Traffic Technician)

Highway planners often find a need to expand an existing highway system. Another bridge over a bay may need to be built. An existing roadway may need to be widened. What was once a country road may need to become a super highway.

One way to estimate how many cars and trucks use a particular route is to count them. Traffic technicians can actually count them and record results on paper. They can also use automatic devices to achieve the same goal. In any case, a statistical record of traffic patterns is necessary to determine whether a highway system needs to be expanded.

The table at the right shows the actual count of cars and trucks that use a particular roadway during rush hour on one weekday selected at random.

Time	Cars	Trucks
6:00 A.M.–6:30 A.M.	133	65
6:30 A.M.–7:00 A.M.	200	60
7:00 A.M.–7:30 A.M.	210	63
7:30 A.M.–8:00 A.M.	210	63
8:00 A.M.–8:30 A.M.	250	55
8:30 A.M.–9:00 A.M.	222	50

What relations does the table suggest? What are the domains and ranges of the relations?

There is a relationship between time intervals and the count of cars.

Domain: set of half-hour time intervals between 6:00 A.M. and 9:00 A.M.
Range: {133, 200, 210, 222, 250}

There is a relationship between time intervals and the count of trucks.

Domain: set of half-hour time intervals between 6:00 A.M. and 9:00 A.M.
Range: {50, 55, 60, 63, 65}

Solve.

1. In what time interval is the count of cars the greatest?
   8:00 A.M.–8:30 A.M.
2. Is there a relationship between the number of cars using the road at one time and the number of trucks using the road at the same time?
   Yes; the relation is the set of ordered pairs (count of cars, count of trucks).
3. Is there a trend in the count of cars as rush hour progresses?
   Yes; the count increases until it reaches a peak of 250 between 8:00 A.M.–8:30 A.M. and then decreases.

© Glencoe/McGraw-Hill T6 *Algebra: Concepts and Applications*

Manipulatives/Modeling

Hands-On Algebra Masters, pp. 72–76

6–1 NAME ____________ DATE ______ PERIOD ____

Hands-On Algebra
Relations Student Edition Page 239

Materials
grid paper, sheet of blank paper

A relation can be shown as a set of ordered pairs, as a table, and as a graph. You can also find the domain and range of a relation. Follow this example to make a table and graph of a relation and to find the domain and range.

Express the relation $\{(-2, 4), (2, 5), (3, 3), (4, 2), (5, -1)\}$ as a table and as a graph. Then determine the domain and range.

Step 1 Fill in the table using the ordered pairs. For the first ordered pair, $(-2, 4)$, the x-coordinate is -2 and the y-coordinate is 4.

x	y
−2	4

Step 2 On the grid below, graph the set of ordered pairs.

Step 3 To find the domain of the relation, place a blank sheet of paper over the graph. Slowly move the paper to the right. For each point that you uncover, write down the x-coordinate of the point.

Step 4 To find the domain of the relation, cover the graph again with the sheet of paper. Slowly move the paper up. For each point that you uncover, write down the y-coordinate of the point.

Work Space

y
O x

Try These

Express each relation as a table and as a graph. Then determine the domain and range.

1. $\{(-4, -1), (-3, -1), (0, 3), (2, -6), (5, 5)\}$
2. $\{(-4, 4), (-2, 2), (0, 0), (2, 2), (4, 4)\}$

© Glencoe/McGraw-Hill 72 *Algebra: Concepts and Applications*

Technology/Multimedia

Graphing Calculator Masters, pp. 17–19

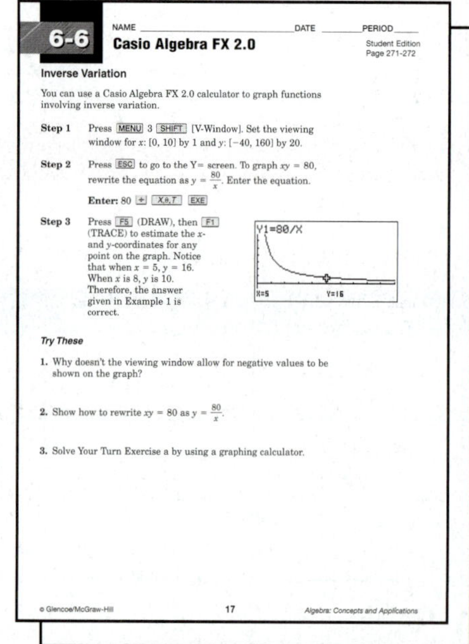

6-6 NAME ____________ DATE ______ PERIOD ____

Casio Algebra FX 2.0 Student Edition Page 271-272

Inverse Variation

You can use a Casio Algebra FX 2.0 calculator to graph functions involving inverse variation.

Step 1 Press [MENU] 3 [SHIFT] [V-Window]. Set the viewing window for x: [0, 10] by 1 and y: [−40, 160] by 20.

Step 2 Press [ESC] to go to the Y= screen. To graph $xy = 80$, rewrite the equation as $y = \frac{80}{x}$. Enter the equation.

Enter: 80 [÷] [X,θ,T] [EXE]

Step 3 Press [F5] (DRAW), then [F1] (TRACE) to estimate the x- and y-coordinates for any point on the graph. Notice that when $x = 5$, $y = 16$. When x is 8, y is 10. Therefore, the answer given in Example 1 is correct.

Try These

1. Why doesn't the viewing window allow for negative values to be shown on the graph?
2. Show how to rewrite $xy = 80$ as $y = \frac{80}{x}$.
3. Solve Your Turn Exercise a by using a graphing calculator.

© Glencoe/McGraw-Hill 17 *Algebra: Concepts and Applications*

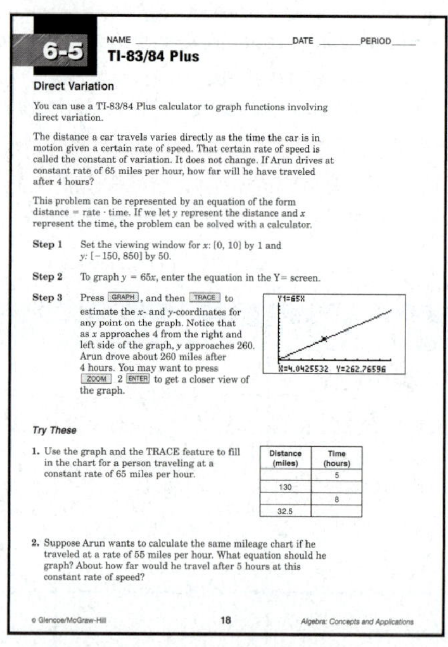

6-5 NAME ____________ DATE ______ PERIOD ____

TI-83/84 Plus

Direct Variation

You can use a TI-83/84 Plus calculator to graph functions involving direct variation.

The distance a car travels varies directly as the time the car is in motion given a certain rate of speed. That certain rate of speed is called the constant of variation. It does not change. If Arun drives at constant rate of 65 miles per hour, how far will he have traveled after 4 hours?

This problem can be represented by an equation of the form distance = rate · time. If we let y represent the distance and x represent the time, the problem can be solved with a calculator.

Step 1 Set the viewing window for x: [0, 10] by 1 and y: [−150, 850] by 50.

Step 2 To graph $y = 65x$, enter the equation in the Y= screen.

Step 3 Press [GRAPH], and then [TRACE] to estimate the x- and y-coordinates for any point on the graph. Notice that as x approaches 4 from the right and left side of the graph, y approaches 260. Arun drove about 260 miles after 4 hours. You may want to press [ZOOM] 2 [ENTER] to get a closer view of the graph.

Try These

1. Use the graph and the TRACE feature to fill in the chart for a person traveling at a constant rate of 65 miles per hour.

Distance (miles)	Time (hours)
	5
130	
	8
32.5	

2. Suppose Arun wants to calculate the same mileage chart if he traveled at a rate of 55 miles per hour. What equation should he graph? About how far would he travel after 5 hours at this constant rate of speed?

© Glencoe/McGraw-Hill 18 *Algebra: Concepts and Applications*

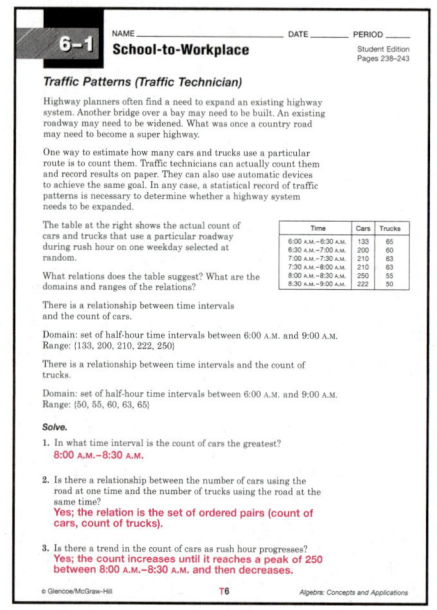

Assessment Resources

Type	Student Edition	Teacher's Wraparound Edition	Chapter 6 Resource Masters
Ongoing Assessment	Quizzes 1 and 2, pp. 255, 269	5-Minute Check, pp. 238, 244, 250, 256, 264, 270	Mid-Chapter Test, p. 284 Quizzes A and B, p. 285
Mixed Review	Mixed Review, pp. 243, 249, 255, 261, 269, 275 Standardized Test Practice, Chapters 1–6, pp. 280–281		Cumulative Review, p. 286 Standardized Test Practice, pp. 287–288
Error Analysis	You Decide, pp. 247, 259	Error Analysis, pp. 241, 247, 254, 259, 268, 273	
Standardized Test Prep	Standardized Test Practice, pp. 243, 249, 255, 261, 269, 275 Standardized Test Practice, Chapters 1–6, pp. 280–281		Standardized Test Practice, pp. 287–288
Open-Ended Assessment	Writing Math, pp. 241, 267, 273 Investigation, pp. 262–263 Portfolio, pp. 237, 263	Modeling: pp. 255, 261 Speaking: pp. 249, 269 Writing: pp. 243, 275	Extended Response Assessment, p. 283
Chapter Assessment	Study Guide and Assessment, pp. 276–278 Chapter Test, p. 279		Multiple-Choice Tests (Forms 1A, 1B), pp. 275–278 Free-Response Tests (Forms 2A, 2B), pp. 279–282

Additional Chapter Resources

Student Edition

Graphing Calculator Exploration, pp. 271–272

Teacher's Classroom Resources

Manipulatives/Modeling

Teacher's Guide for Overhead Manipulative Resources

Meeting Individual Needs

Prerequisite Skills Booklet

Spanish Study Guide and Assessment, pp. 43–48, 115–116

Teaching Aids

Solutions Manual

5-Minute Check Transparencies

Glencoe Technology

AlgePASS CD-ROM, Lessons 15, 16

Interactive Chalkboard CD-ROM

StudentWorks Plus CD-ROM

Multimedia Applications CD-ROM, Activity 5

Vocabulary PuzzleMaker (online)

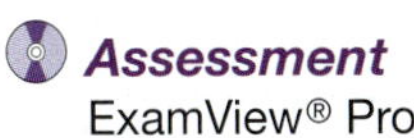

ExamView® Pro

Teaching Aids

Answer Key Maker CD-ROM

TeacherWorks CD-ROM

Visit ***www.algconcepts.com*** *for data updates, career information, games, and other interactive activities.*

CHAPTER PREVIEW 6

Mathematics of the Chapter

This chapter provides students with an in-depth study of functions and graphs. Students will begin by expressing relations as tables, as sets of ordered pairs, and as graphs. Students will also solve linear equations for a given domain and graph linear equations. Then students will learn how to determine if a relation is a function and how to use functional notation. Finally, students solve problems using direct and inverse variation.

What You'll Learn

Have students read over the lists of key ideas and key vocabulary. Have them make a list of any words with which they are not familiar.

Why It's Important

Point out to students that this is only one of many reasons why each objective is important. Others are provided in each lesson.

Answers

1–9.

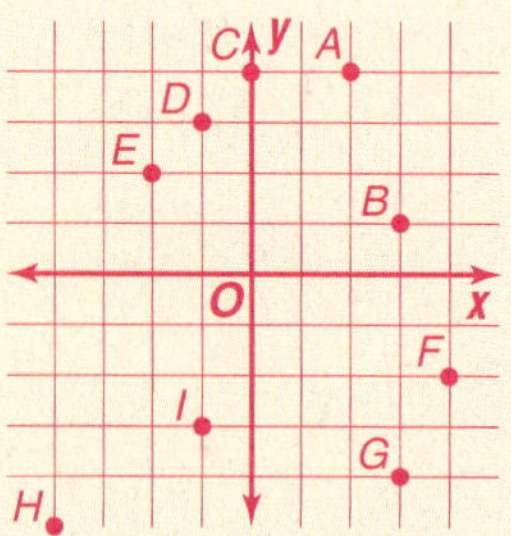

CHAPTER 6 Functions and Graphs

What You'll Learn

Key Ideas

- Show relations as ordered pairs, tables, and graphs. *(Lesson 6–1)*
- Solve and graph linear equations. *(Lessons 6–2 and 6–3)*
- Determine whether a relation is a function. *(Lesson 6–4)*
- Solve problems involving direct variation and inverse variation. *(Lessons 6–5 and 6–6)*

Key Vocabulary

domain *(p. 238)*
function *(p. 256)*
linear equation *(p. 250)*
range *(p. 238)*
relation *(p. 238)*

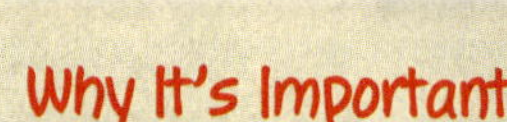

Why It's Important

Plants Even deserts get a little rain each year. When it rains, a cactus plant's leaves, stems, or roots take in as much water as possible to live on during the dry season. A cactus's spines also help it survive. They discourage hungry animal and also slow down wind, making less water evaporate from the cactus.

Functions and relations are used to represent relationships between real-life quantities. You will use a linear relation to predict the number of blooms on a cactus in Lesson 6–3.

CHAPTER 6 LINKS

Lesson	6–1	6–2	6–3	6–4	6–5	6–6
Math in the Workplace	Demographics	Sales	Earth Science	Business	Electronics	Music
Applications and Connections	Economics Counting Food Entertainment Probability Odds	Shopping Life Science Sales Consumerism Banking	Plants Business Life Science	Anatomy Safety Health Consumerism Animals	Astronomy Travel Geography Health Life Science Transportation	Construction Sports Travel
Math Integration		Geometry	Geometry Statistics	Geometry	Measurement	Geometry

Study these lessons to improve your skills.

Check Your Readiness

Lesson 2–2, pp. 58–63

Graph each point on a coordinate plane. **See margin.**

1. $A(2, 4)$
2. $B(3, 1)$
3. $C(0, 4)$
4. $D(-1, 3)$
5. $E(-2, 2)$
6. $F(4, -2)$
7. $G(3, -4)$
8. $H(-4, -5)$
9. $I(-1, -3)$

Lesson 1–2, pp. 8–13

Evaluate each algebraic expression if $x = 2$.

10. $2x$ **4**
11. $-3x$ **−6**
12. $x + 5$ **7**
13. $8 - x$ **6**
14. $2x + 3$ **7**
15. $6x - 2$ **10**
16. $-3x + 4$ **−2**
17. $-4x + 2$ **−6**
18. $7 - 3x$ **1**

Lesson 5–1, pp. 188–193

Solve each proportion.

19. $\frac{5}{6} = \frac{y}{12}$ **10**
20. $\frac{3}{4} = \frac{y}{16}$ **12**
21. $\frac{y}{5} = \frac{12}{15}$ **4**
22. $\frac{4}{24} = \frac{9}{x}$ **54**
23. $\frac{-3}{-9} = \frac{-5}{x}$ **−15**
24. $\frac{0.5}{x} = \frac{2}{20}$ **5**
25. $\frac{12}{1} = \frac{x}{2.5}$ **30**
26. $\frac{150}{2.50} = \frac{225}{x}$ **3.75**
27. $\frac{-2}{3} = \frac{y}{-6}$ **4**

FOLDABLES™ Study Organizer

Make this Foldable to help you organize your Chapter 6 notes. Begin with a sheet of notebook paper.

❶ **Fold** lengthwise to the holes.

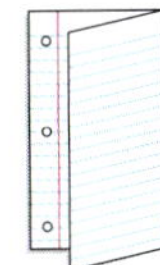

❷ **Cut** along the top line and then cut 10 tabs.

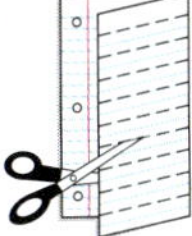

❸ **Label** the tabs using the vocabulary words as shown.

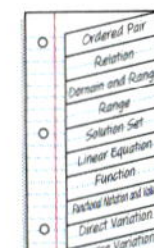

Reading and Writing Store your foldable in a 3-ring binder. As you read and study the chapter, write notes and examples under each tab.

www.algconcepts.com/chapter_readiness

Check Your Readiness

This section provides a review of the basic concepts needed before beginning Chapter 6. Lesson references are included for additional student help.

Use Exercises	To Prepare for Lesson(s)
1–9	6–1, 6–2, 6–3, 6–4
10–18	6–2
19–27	6–5, 6–6

Quick Review Math Handbook

Glencoe Algebra Lesson	Quick Review Math Handbook Hot Topic
6–1	6–7
6–2	6–4, 6–7
6–3	6–2, 6–7
6–4	1–5, 2–3, 6–7
Ch. 6 Inv.	2–8, 5–1, 5–2
6–5	8–2
6–6	1–4

Noteables™

Interactive Study Notebook with Foldables™

This note-taking guide reinforces excellent note-taking skills. It includes:

- ✓ **Build Your Vocabulary** tool for including mathematical terminology in notes.
- ✓ **Review It** activities with links to prior lessons and items to include in **Foldables™**.
- ✓ **Bringing It All Together** feature to help students review for the Chapter Test.
- ✓ **Are You Ready for the Chapter Test?** feature to allow students to assess their own readiness for the Chapter Test.
- ✓ **Teacher Edition** with transparencies to guide note taking. Corresponds to the *Teacher's Wraparound Edition* In-Class Examples and *Interactive Chalkboard CD-ROM.*

FOLDABLES™ Study Organizer

After students make this Foldable, have them record each of the vocabulary words on a separate tab. Then have students write definitions and examples for each of the terms.

Lesson 6-1

1 FOCUS

5-Minute Check
Chapter 5

1. Solve $\frac{15}{45} = \frac{m}{63}$. **21**
2. What percent of the circle is shaded? **$83\frac{1}{3}\%$**

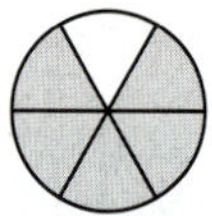

3. Last year, a new digital camera cost $655. This year, the price of that camera has been reduced by 45%. What is this year's price for the camera? **$360.25**

For Exercises 4–5, use the spinner below.

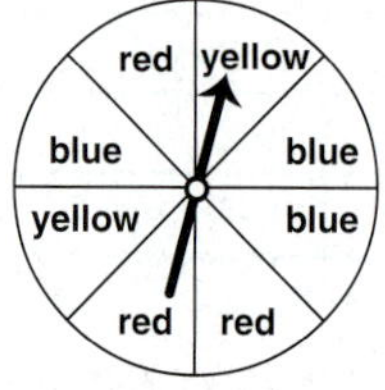

4. Find the probability that the spinner lands on a section that is red. **$\frac{3}{8}$**
5. What are the odds that the spinner lands on a section that is red or blue? **6:2 or 3:1**

Motivating the Lesson

Hands-On Activity Have students graph the points (0, 0), (1, 2), (2, 4), and (3, 6) on a coordinate plane. Suggest that the coordinates of the ordered pairs show the relationship between two values. Ask students what the points could represent. **Sample answer: the number of shoes *y* in *x* pairs**

MODELING

An alternative hands-on option using grid paper and a blank sheet of paper is available for teaching this lesson.

6–1 Relations

What You'll Learn
You'll learn to show relations as sets of ordered pairs, as tables, and as graphs.

Why It's Important
Demographics
Population growth data can be expressed as a relation.
See Example 3.

In Lesson 1-6, you learned that data can be represented in a table. The table at the right shows the estimated number of sea turtles found stranded on the Texas Gulf Coast over several months in a recent year.

Sea Turtles Stranded	
Month	**Number**
January	22
February	29
March	43
April	76
May	41
June	23

Source: HEART

The data can also be represented by a set of ordered pairs, as shown below. Each first coordinate represents the month, and the second coordinate is the number of stranded sea turtles.

$$\{(1, 22), (2, 29), (3, 43), (4, 76), (5, 41), (6, 23)\}$$

The first coordinate in an ordered pair is called the **x-coordinate**. The second coordinate is called the **y-coordinate**.

Each ordered pair can be graphed.

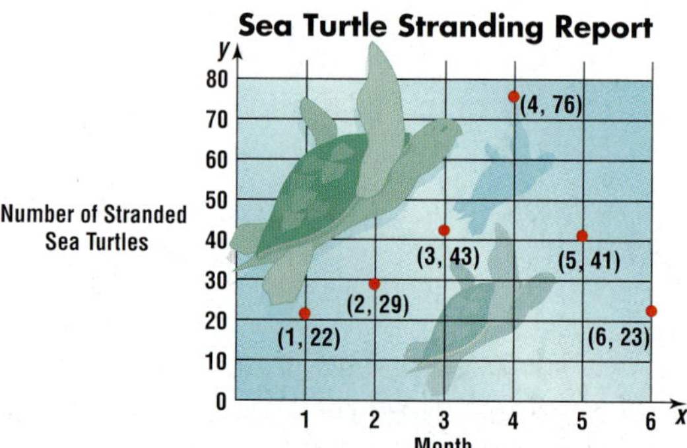

Reading Algebra
The variable x, whose values make up the domain, is called the **independent variable**. The variable y, whose values make up the range, is called the **dependent variable**.

A set of ordered pairs, such as the one for the sea turtle data, is a **relation**. The set of all first coordinates of the ordered pairs, or x-coordinates, is called the **domain** of the relation. The set of all second coordinates, or y-coordinates, is called the **range**.

Domain and Range of a Relation	The domain of a relation is the set of all first coordinates from the ordered pairs of the relation. The range of the relation is the set of all second coordinates from the ordered pairs of the relation.

Resource Manager

Reproducible Masters
Chapter 6 Resource Masters
- *Study Guide,* p. 245
- *Skills Practice,* p. 246
- *Practice,* p. 247
- *Reading to Learn Mathematics,* p. 248
- *Enrichment,* p. 249

Hands-On Algebra, p. 72
School-to-Workplace, p. 6

Transparencies
5-Minute Check, 6–1

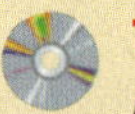

Technology/Multimedia
Interactive Chalkboard CD-ROM

For the relation of the sea turtle data, the domain is {1, 2, 3, 4, 5, 6}, and the range is {22, 29, 43, 76, 41, 23}.

A relation can be shown in several ways. Three ways of representing the relation {(−1, −1), (0, 2), (1, 3)} are shown below.

Concept Summary

Ordered Pairs	Table	Graph
(−1, −1) (0, 2) (1, 3)	x: −1, 0, 1 y: −1, 2, 3	(1, 3), (0, 2), (−1, −1) plotted

x	y
−1	−1
0	2
1	3

Examples

1 **Express the relation {(−2, 4), (2, 5), (3, 3), (4, 2), (5, −1)} as a table and as a graph. Then determine the domain and range.**

x	y
−2	4
2	5
3	3
4	2
5	−1

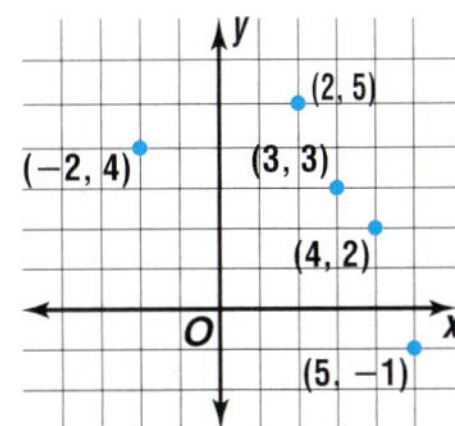

The domain is {−2, 2, 3, 4, 5}, and the range is {4, 5, 3, 2, −1}.

2 **Express the relation shown on the graph as a set of ordered pairs and in a table. Then determine the domain and range.**

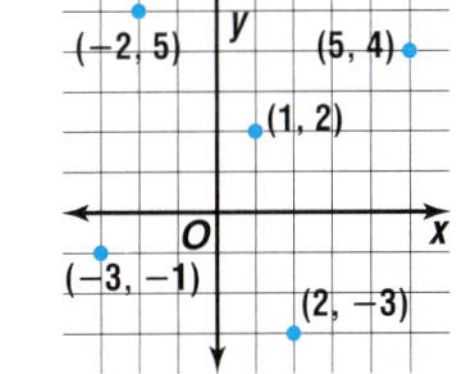

The set of ordered pairs for the relation is {(−3, −1), (−2, 5), (1, 2), (2, −3), (5, 4)}.

x	y
−3	−1
−2	5
1	2
2	−3
5	4

The domain is {−3, −2, 1, 2, 5}, and the range is {−1, 5, 2, −3, 4}.

(continued on the next page)

www.algconcepts.com/extra_examples

2 TEACH

Teaching Tip As you discuss graphing the sea turtle data on page 239, review Lesson 2–2 on coordinate planes if students have difficulty recalling how to graph an ordered pair.

In-Class Examples

Example 1

Express the relation {(−4, 5), (−3, 2), (0, 1), (1, −1), (3, −2)} as a table and as a graph. Then determine the domain and range.

x	y
−4	5
−3	2
0	1
1	−1
3	−2

The domain is {−4, −3, 0, 1, 3}, and the range is {5, 2, 1, −1, −2}.

Example 2

Express the relation shown on the graph as a set of ordered pairs and in a table. Then find the domain and range.

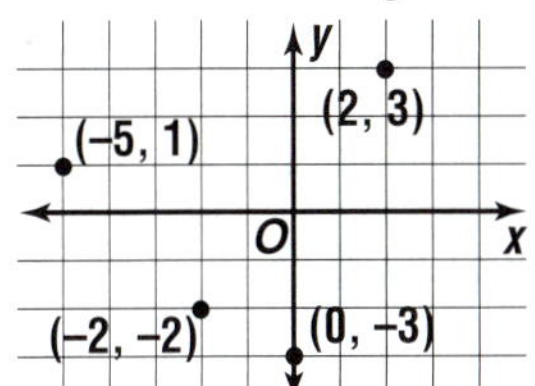

x	y
−5	1
−2	−2
0	−3
2	3

The domain is {−5, −2, 0, 2}, and the range is {1, −2, −3, 3}.

Teaching Tip In Example 3, point out how much easier it is to answer part C using the graph rather than the table. Lead students to recognize that the graph is steeper during periods of greater increase.

In-Class Example

Example 3

The table shows the population of New York City since 1920.

A. Determine the domain and range of the relation.

B. Graph the relation.

C. During which decade was there the greatest increase in population?

Year	Population (millions)
1920	5.6
1930	6.9
1940	7.5
1950	7.9
1960	7.8
1970	7.9
1980	7.1
1990	7.3

A. The domain is {1920, 1930, 1940, 1950, 1960, 1970, 1980, 1990}. The range is {5.6, 6.9, 7.5, 7.9, 7.8, 7.1, 7.3}.

B.

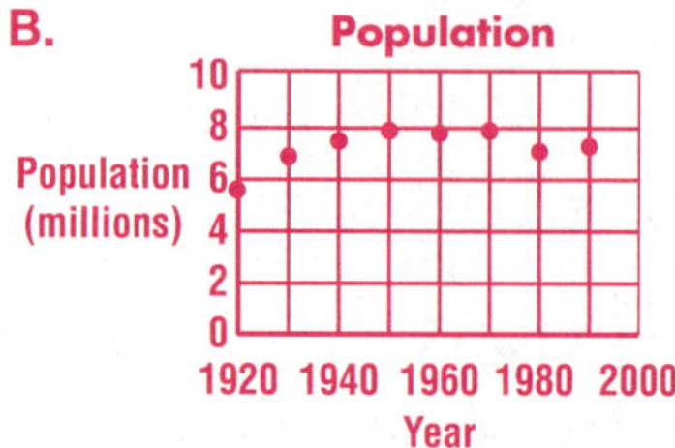

C. The greatest increase in population was during the 1920s.

Your Turn a–b. See margin.

a. Express the relation {(−4, −1), (−3, 1), (0, 3), (2, −6), (5, 5)} as a table and as a graph. Then determine the domain and range.

b. Express the relation shown on the graph as a set of ordered pairs and in a table. Then find the domain and range.

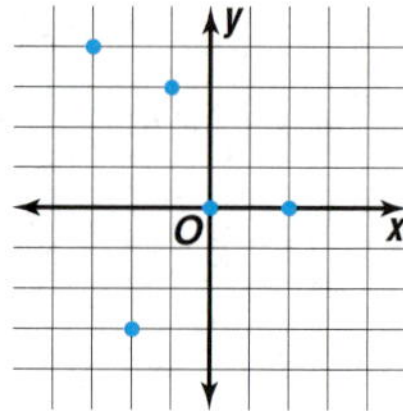

In real-life situations, you may need to select a range of values for the x- or y-axis that does not begin with 0.

Example 3 **Demographics Link**

interNET CONNECTION

Data Update For the latest information on population, visit: www.algconcepts.com

The table shows the population of Kentucky since 1930.

A. Determine the domain and range of the relation.

B. Graph the relation.

C. During which decade was there the greatest increase in population?

Kentucky State Capitol Building

Year	1930	1940	1950	1960	1970	1980	1990	2000
Population (millions)	2.6	2.8	2.9	3.0	3.2	3.7	3.7	4.0

Source: U.S. Census

A. The domain is {1930, 1940, 1950, 1960, 1970, 1980, 1990, 2000}. The range is {2.6, 2.8, 2.9, 3.0, 3.2, 3.7, 4.0}.

B. The x-coordinates go from 1930 to 2000. The scale does not have to begin at 0. The y-coordinates include values from 2.6 to 4.0. You can include 0 and use units of 0.2.

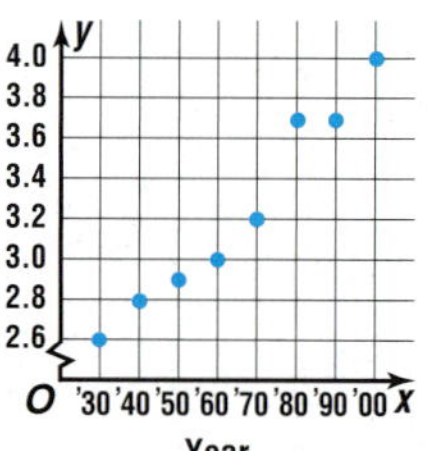

C. The graph shows that the greatest increase in population was during the 1970s.

Answers *Your Turn*

a.

x	y
−4	−1
−3	1
0	3
2	−6
5	5

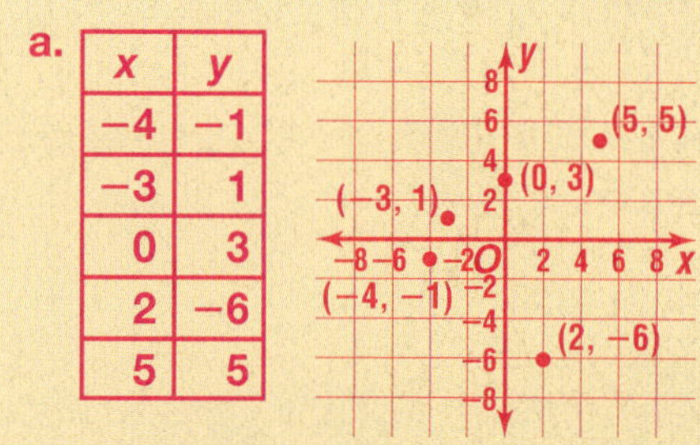

domain: {−4, −3, 0, 2, 5}, range: {−1, 1, 3, −6, 5}

b. {(0, 0), (−1, 3), (−3, 4), (−2, −3), (2, 0)}

x	y
0	0
−1	3
−3	4
−2	−3
2	0

domain: {0, −1, −3, −2, 2}, range: {0, 3, 4, −3}

Check for Understanding

Communicating Mathematics

1–3. See margin.

1. **Describe** three ways to express a relation.
2. **Describe** the following relation.
{(1, Washington), (2, Adams), (3, Jefferson), . . .}
3. *Writing Math* Write a sentence explaining the difference between the domain and the range of a relation.

Vocabulary
relation
domain
range
x-coordinate
y-coordinate

Guided Practice

Example 1

Express each relation as a table and as a graph. Then determine the domain and the range. 4–5. See Solutions Manual.

4. $\{(-4, 2), (-2, 0), (0, 2), (2, 4)\}$
5. $\{(-3, -3.5), (-1, 3.9), (0, 2), (5, 2.5)\}$

Example 2

Express each relation as a set of ordered pairs and in a table. Then determine the domain and the range. 6–7. See margin.

6.

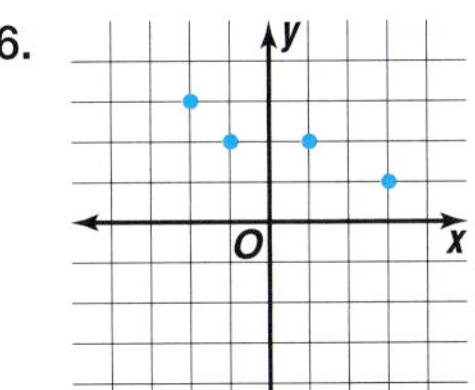

7.

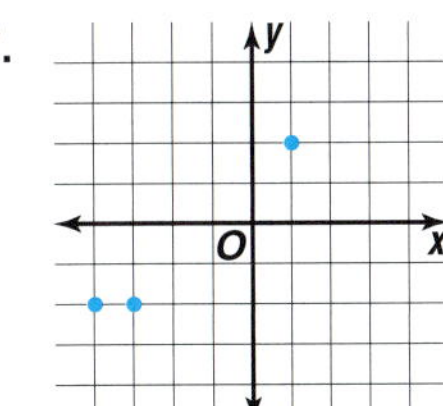

Example 3

8. **Economics** For his economics class, Alonso is researching the growth of the national debt since 1975. He organized his results in a table.
   a. Determine the domain and range of the relation. b. See Solutions Manual.
   b. Graph the relation.
   c. Predict the national debt in 2005. Check your answer on the Internet. about $6800 billion

8a. domain: {1975, 1980, 1985, 1990, 1995, 2000}; range: {577, 930, 1946, 3233, 4974, 5674}

National Debt

Year	Amount (billions)
1975	$577
1980	$930
1985	$1946
1990	$3233
1995	$4974
2000	$5674

Source: Bureau of the Public Debt

Exercises

Practice

Express each relation as a table and as a graph. Then determine the domain and the range. 9–14. See Solutions Manual.

9. $\{(4, 3), (-2, 3), (-2, 4), (4, 4)\}$
10. $\{(2, 3.5), (2.9, 1), (4.5, 7.5)\}$
11. $\{(-2, 0), (3, -7), (2, -5), (-6, 3), (1, 5)\}$
12. $\{(3.1, 0), (4, 1.5), (-1, -1), (-5.5, 1.9)\}$
13. $\left\{\left(-\frac{1}{2}, \frac{1}{2}\right), (1, 0), \left(-\frac{1}{2}, -5\right)\right\}$
14. $\left\{\left(\frac{1}{2}, 3\right), \left(\frac{1}{2}, 1\right), (0, -1), \left(\frac{3}{4}, -2\right)\right\}$

Answers

6. $\{(-2, 3), (-1, 2), (1, 2), (3, 1)\}$

x	y
−2	3
−1	2
1	2
3	1

domain: {−2, −1, 1, 3}, range: {3, 2, 1}

7. $\{(-3, -2), (1, 2), (-4, -2)\}$

x	y
−3	−2
1	2
−4	−2

domain: {−3, 1, −4}, range: {−2, 2}

3 PRACTICE/APPLY

Error Analysis

Watch for students who confuse the domain and range in Exercise 3. ***Prevent by*** encouraging students to create mnemonics they can use to remember which term refers to which variable. For example, since *domain* comes before *range* alphabetically, students could remember that domain comes before range in the same way x comes before y.

Assignment Guide

Basic: 9–25 odd, 27–33
Average: 10–22 even, 24–33

Answers

1. ordered pairs, table, graph
2. The relation matches the order of the president (domain) to the name of the president (range).
3. The domain is the set of all first coordinates from the ordered pairs. The range is the set of all second coordinates from the ordered pairs.

***Study Guide*, p. 245**

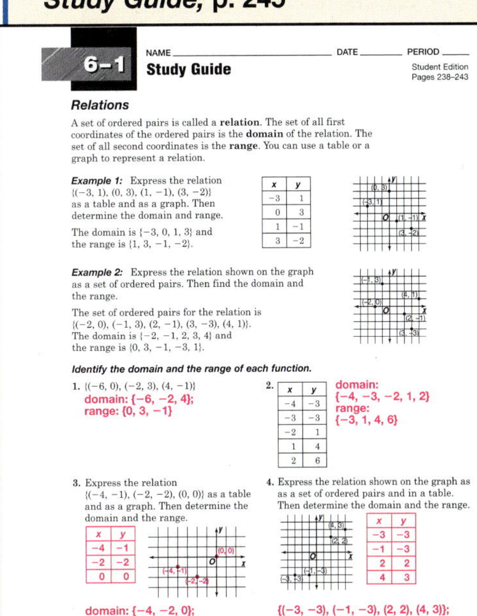

6–1 NAME ________ DATE ______ PERIOD _____

Study Guide

Student Edition Pages 238–243

Relations

A set of ordered pairs is called a **relation**. The set of all first coordinates of the ordered pairs is the **domain** of the relation. The set of all second coordinates is the **range**. You can use a table or a graph to represent a relation.

Example 1: Express the relation $\{(-3, 1), (0, 3), (1, -1), (3, -2)\}$ as a table and as a graph. Then determine the domain and range.

The domain is $\{-3, 0, 1, 3\}$ and the range is $\{1, 3, -1, -2\}$.

x	y
−3	1
0	3
1	−1
3	−2

Example 2: Express the relation shown on the graph as a set of ordered pairs. Then find the domain and the range.

The set of ordered pairs for the relation is $\{(-2, 0), (-1, 3), (2, -1), (3, -3), (4, 1)\}$. The domain is $\{-2, -1, 2, 3, 4\}$ and the range is $\{0, 3, -1, -3, 1\}$.

Identify the domain and the range of each function.

1. $\{(-6, 0), (-2, 3), (4, -1)\}$
domain: {−6, −2, 4}; range: {0, 3, −1}

2.

x	y
−4	−3
−3	−3
−2	1
1	4
2	6

domain: {−4, −3, −2, 1, 2} range: {−3, 1, 4, 6}

3. Express the relation $\{(-4, -1), (-2, -2), (0, 0)\}$ as a table and as a graph. Then determine the domain and the range.

x	y
−4	−1
−2	−2
0	0

domain: {−4, −2, 0}; range: {−2, −1, 0}

4. Express the relation shown on the graph as as a set of ordered pairs and in a table. Then determine the domain and the range.

x	y
−3	−3
−1	−3
2	2
4	3

{(−3, −3), (−1, −3), (2, 2), (4, 3)}; domain: {−3, −1, 2, 4}; range: {−3, 2, 3}

Homework Help

For Exercises	See Examples
9–14, 25	1
15–20, 24, 27	2
26	1, 2

Extra Practice

See page 703.

Express each relation as a set of ordered pairs and in a table. Then determine the domain and the range. **15–20. See Solutions Manual.**

15.

16.

17.

18.

19. 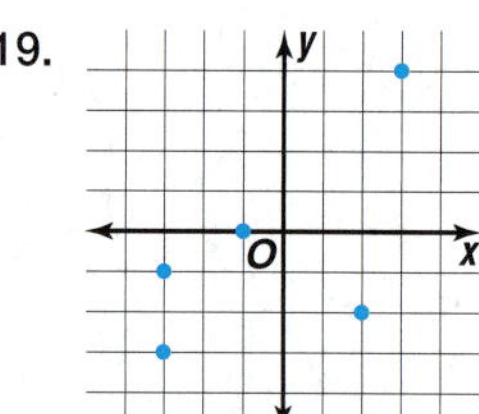

20. 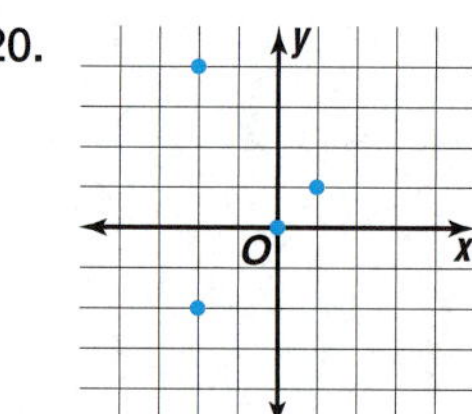

Express each relation as a set of ordered pairs. **21–23. See margin.**

21.

x	y
−1	1
0	2
1	3
2	4

22.

x	y
6	−12
5	−10
4	−8
3	−6

23.

x	y
0	0
−1	−0.5
1	0.5
−2	−1

Applications and Problem Solving

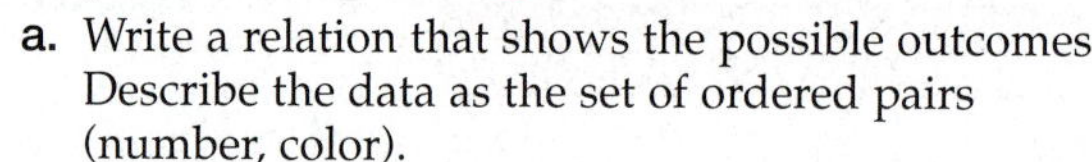

24. **Counting** A die is rolled, and the spinner shown is spun. **a–c. See margin.**

a. Write a relation that shows the possible outcomes. Describe the data as the set of ordered pairs (number, color).

b. Determine the domain and range of the relation.

c. Use this relation to find the number of ways to roll an even number and land on blue.

25. **Food** Kelly buys a dozen bagels to share with her co-workers. She knows that her co-workers prefer blueberry bagels to plain bagels. So, she buys at least twice as many blueberry bagels as plain bagels.

a. Write a relation to show the different possibilities. (*Hint:* Let the domain represent the number of blueberry bagels.)

b. Express the relation in a table. **a–b. See margin.**

Answers

21. {(−1, 1), (0, 2), (1, 3), (2, 4)}

22. {(6, −12), (5, −10), (4, −8), (3, −6)}

23. {(0, 0), (−1, −0.5), (1, 0.5), (−2, −1)}

24a. {(1, blue), (2, blue), (3, blue), (4, blue), (5, blue), (6, blue), (1, green), (2, green), (3, green), (4, green), (5, green), (6, green), (1, blue), (2, blue), (3, blue), (4, blue), (5, blue), (6, blue), (1, yellow), (2, yellow), (3, yellow), (4, yellow), (5, yellow), (6, yellow), (1, green), (2, green), (3, green), (4, green), (5, green), (6, green), (1, red), (2, red), (3, red), (4, red), (5, red), (6, red)}

24b. domain: {1, 2, 3, 4, 5, 6}; range: {blue, green, yellow, red}

24c. 6

25a. {(8, 4), (9, 3), (10, 2), (11, 1), (12, 0)}

25b.

x	y
8	4
9	3
10	2
11	1
12	0

***Skills Practice*, p. 246, and *Practice*, p. 247 (shown)**

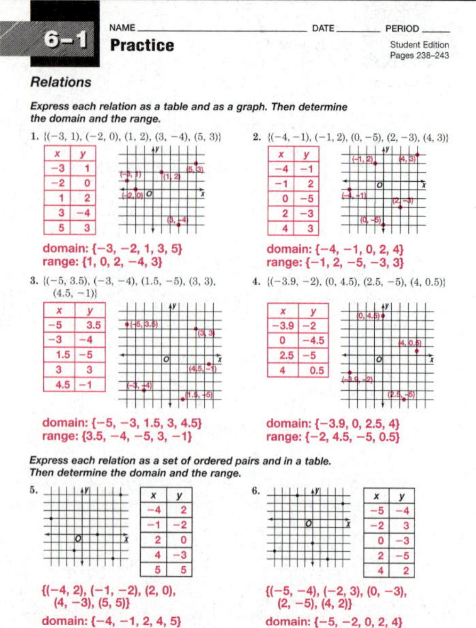

6–1 NAME ______ DATE ______ PERIOD ______

Practice

Student Edition Pages 238–243

Relations

Express each relation as a table and as a graph. Then determine the domain and the range.

1. {(−3, 1), (−2, 0), (1, 2), (3, −4), (5, 3)}

x	y
−3	1
−2	0
1	2
3	−4
5	3

domain: {−3, −2, 1, 3, 5}
range: {1, 0, 2, −4, 3}

2. {(−4, −1), (−1, 2), (0, −5), (2, −3), (4, 3)}

x	y
−4	−1
−1	2
0	−5
2	−3
4	3

domain: {−4, −1, 0, 2, 4}
range: {−1, 2, −5, −3, 3}

3. {(−5, 3.5), (−3, −4), (1.5, −5), (3, 3), (4.5, −1)}

x	y
−5	3.5
−3	−4
1.5	−5
3	3
4.5	−1

domain: {−5, −3, 1.5, 3, 4.5}
range: {3.5, −4, −5, 3, −1}

4. {(−3.9, −2), (0, 4.5), (2.5, −5), (4, 0.5)}

x	y
−3.9	−2
0	−4.5
2.5	−5
4	0.5

domain: {−3.9, 0, 2.5, 4}
range: {−2, 4.5, −5, 0.5}

Express each relation as a set of ordered pairs and in a table. Then determine the domain and the range.

5.

x	y
−4	2
−1	−2
2	0
4	−3
5	5

{(−4, 2), (−1, −2), (2, 0), (4, −3), (5, 5)}
domain: {−4, −1, 2, 4, 5}
range: {2, −2, 0, −3, 5}

6.

x	y
−5	−4
−2	3
0	−3
2	−5
4	2

{(−5, −4), (−2, 3), (0, −3), (2, −5), (4, 2)}
domain: {−5, −2, 0, 2, 4}
range: {−4, 3, −3, −5, 2}

Reteaching Activity

Auditory/Musical Learners Have pairs of students work together to graph points. One student should read the ordered pair quietly aloud while their partner graphs the point on a coordinate plane. After several points have been graphed, the partners should reverse roles and repeat the activity several more times.

26. **Entertainment** The size of a movie on a screen and the distance from the projector to the screen can be expressed as a relation. The diagram below shows that if the distance between the screen and the projector is 3 units, the size of the picture is 9 squares.

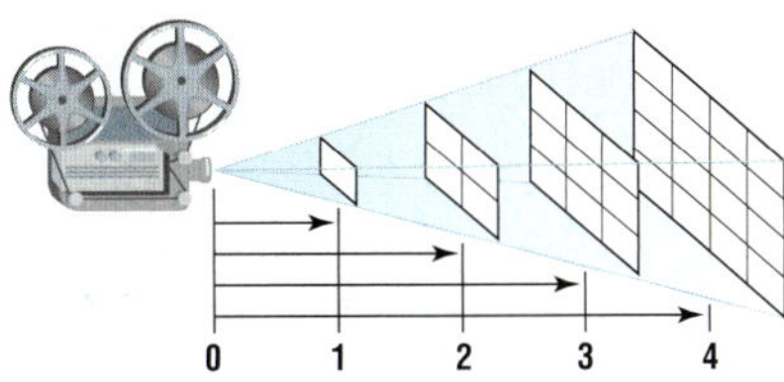

a. Express the relation as a set of ordered pairs. **a–b. See margin.**

b. Describe the pattern between the distance from the projector to the screen and the size of the picture.

c. Find the size of the picture if the projector is 12 units from the screen. **144 squares**

d. Graph the relation. Describe the graph. **See margin.**

27. **Critical Thinking** There are 8 counters numbered 1 through 8 in a bag. Two counters are chosen at random without replacement.

a. What possible sums can you get? **a–b. See margin.**

b. Describe the data as the set of ordered pairs (sum, number of ways to occur).

Mixed Review

28. **Probability** A card is drawn from a standard deck of cards. Find the probability of drawing an ace or a jack. *(Lesson 5–7)* $\frac{2}{13}$

29. **Odds** A bag contains 4 yellow marbles, 3 green marbles, and 3 red marbles. Find the odds of drawing a yellow marble. *(Lesson 5–6)* **4:6 or 2:3**

30. **Food** The Dessert Factory offers the choices shown for ordering a piece of cheesecake. Draw a tree diagram to find the number of different cheesecakes that can be ordered. *(Lesson 4–2)* **See Solutions Manual.**

Cheesecake	Topping
original	blueberry
chocolate chip	strawberry
fudge swirl	raspberry
	cherry

Evaluate each expression. *(Lesson 2–1)*

31. $-|-24|$ **−24**

32. $|8| - |-12|$ **−4**

Standardized Test Practice

33. **Multiple Choice** The stem-and-leaf plot shows the scores for Neshawn's bowling team. What are the highest and lowest scores? *(Lesson 1–7)* **B**

Stem	Leaf
15	3 5 9
16	5 5 8
17	2 3 5 6

$17|3 = 173$

Ⓐ 15, 17 Ⓑ 153, 176
Ⓒ 153, 172 Ⓓ 150, 170

www.algconcepts.com/self_check_quiz

4 ASSESS

Open-Ended Assessment

Writing Ask students to describe how the domain and range of a relation are related to the *x*- and *y*-coordinates of a set of ordered pairs.

Answers

26a. {(1, 1), (2, 4), (3, 9), (4, 16)}

26b. Each domain value is squared to get the corresponding range value.

26d.

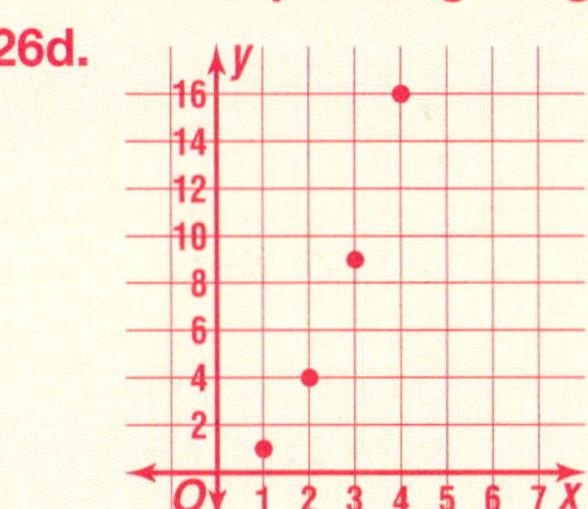

The graph curves upward to the right.

27a. 3, 4, 5, 6, 7, 8, 9, 10, 11, 12, 13, 14, 15

27b. {(3, 1), (4, 1), (5, 2), (6, 2), (7, 3), (8, 3), (9, 4), (10, 3), (11, 3), (12, 2), (13, 2), (14, 1), (15, 1)}

Extra Credit

Both the domain and the range of a relation include only negative values. How will the relation appear when graphed? **Sample answer: All the points on the graph will be in Quadrant III.**

Enrichment, p. 249

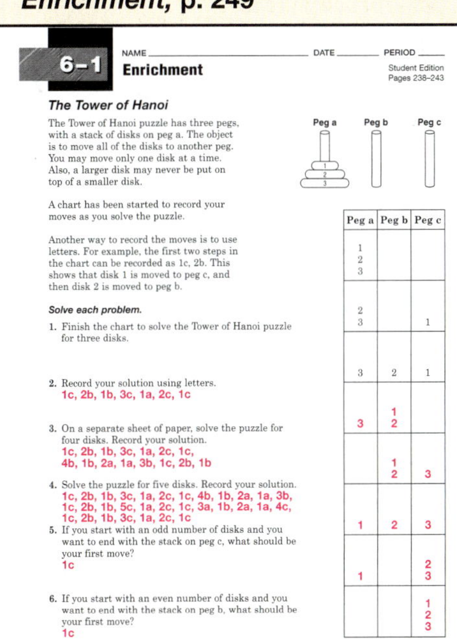

6–1 NAME ______ DATE ______ PERIOD ______

Enrichment

Student Edition Pages 238–243

The Tower of Hanoi

The Tower of Hanoi puzzle has three pegs, with a stack of disks on peg a. The object is to move all of the disks to another peg. You may move only one disk at a time. Also, a larger disk may never be put on top of a smaller disk.

Peg a Peg b Peg c

A chart has been started to record your moves as you solve the puzzle.

Another way to record the moves is to use letters. For example, the first two steps in the chart can be recorded as 1c, 2b. This shows that disk 1 is moved to peg c, and then disk 2 is moved to peg b.

Solve each problem.

1. Finish the chart to solve the Tower of Hanoi puzzle for three disks.

2. Record your solution using letters.
1c, 2b, 1b, 3c, 1a, 2c, 1c

3. On a separate sheet of paper, solve the puzzle for four disks. Record your solution.
1c, 2b, 1b, 3c, 1a, 2c, 1c, 4b, 1b, 2a, 1a, 3b, 1c, 2b, 1b

4. Solve the puzzle for five disks. Record your solution.
1c, 2b, 1b, 3c, 1a, 2c, 1c, 4b, 1b, 2a, 1a, 3b, 1c, 2b, 1b, 5c, 1a, 2c, 1c, 3a, 1b, 2a, 1a, 4c, 1c, 2b, 1b, 3c, 1a, 2c, 1c

5. If you start with an odd number of disks and you want to end with the stack on peg c, what should be your first move?
1c

6. If you start with an even number of disks and you want to end with the stack on peg b, what should be your first move?
1c

Peg a	Peg b	Peg c
1 2 3		
2 3		1
3	2	1
3	**1 2**	
	1 2	**3**
1	**2**	**3**
1		**2 3**
		1 2 3

Lesson 6–2

1 FOCUS

5-Minute Check
Lesson 6–2

Use the relation {(−4, 2), (1, 0), (2, 3), (3, −1), (4, −1)} for Exercises 1–4.

1. Express the relation as a table.

x	−4	1	2	3	4
y	2	0	3	−1	−1

2. Express the relation as a graph.

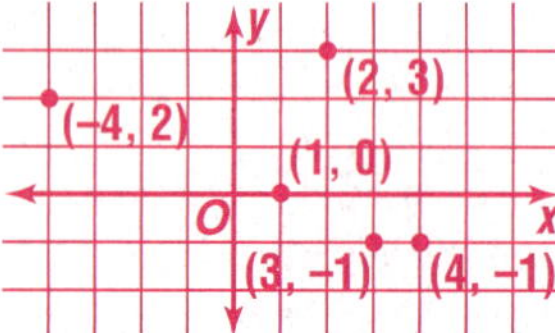

3. What is the domain of the relation? **{−4, 1, 2, 3, 4}**
4. What is the range of the relation? **{−1, 0, 2, 3}**
5. Express the relation as a set of ordered pairs.

x	y
−3	4
−1	3
0	2
1	1
3	0

{(−3, 4), (−1, 3), (0, 2), (1, 1), (3, 0)}

Motivating the Lesson

Hands-On Activity Have students work in pairs. Provide each pair with play money. One member of each pair should give a random amount of money to the other member. The second student computes 10% of the amount as interest and returns the original amount along with the interest to the first student. Ask students to think about the equation involved in each transaction.

6–2 Equations as Relations

What You'll Learn
You'll learn to solve linear equations for a given domain.

Why It's Important
Sales Knowing how to solve linear equations can help salespeople determine their monthly income. *See Exercise 38.*

The table below shows a relationship between two sets of numbers, x and y.

x	$4.2x$	y	(x, y)
0.1	4.2(0.1)	0.42	(0.1, 0.42)
0.25	4.2(0.25)	1.05	(0.25, 1.05)
0.5	4.2(0.5)	2.1	(0.5, 2.1)
0.75	4.2(0.75)	3.15	(0.75, 3.15)
1.0	4.2(1.0)	4.2	(1.0, 4.2)

Each ordered pair (x, y) is a *solution* of the equation $y = 4.2x$.

The set of solutions of the problem is called the **solution set.** In this example, you would write the solution set as {(0.1, 0.42), (0.25, 1.05), (0.5, 2.1), (0.75, 3.15), (1.0, 4.2)}.

The equation $y = 4.2x$ is an example of an **equation in two variables**.

Solution of an Equation in Two Variables
If a true statement results when the numbers in an ordered pair are substituted into an equation in two variables, then the ordered pair is a solution of the equation.

Example

Which of the ordered pairs (0, 1), (2, 3), (−1, 1), or (3, 5) are solutions of $y = 2x - 1$?

Make a table. Substitute the x and y values of each ordered pair into the equation.

x	y	$y = 2x - 1$	True or False?
0	1	$1 = 2(0) - 1$ $1 = -1$	false
2	3	$3 = 2(2) - 1$ $3 = 3$	true ✓
−1	1	$1 = 2(-1) - 1$ $1 = -3$	false
3	5	$5 = 2(3) - 1$ $5 = 5$	true ✓

A true statement results when the ordered pairs (2, 3) and (3, 5) are substituted into the equation. Therefore, ordered pairs (2, 3) and (3, 5) are solutions of the equation $y = 2x - 1$.

Resource Manager

Reproducible Masters
Chapter 6 Resource Masters
- *Study Guide,* p. 250
- *Skills Practice,* p. 251
- *Practice,* p. 252
- *Reading to Learn Mathematics,* p. 253
- *Enrichment,* p. 254
- *Assessment,* p. 285

Transparencies
5-Minute Check, 6–2

Technology/Multimedia
Interactive Chalkboard CD-ROM

Your Turn

a. Which of the ordered pairs (1, 1), (0, 2), (−2, 8), or (−1, −5) are solutions of $y = -3x + 2$? **(0, 2); (−2, 8)**

Since the solutions of an equation in two variables are ordered pairs, this type of equation describes a relation. The set of values of x is the domain of the relation. The set of corresponding values of y is the range of the relation.

Example 2 **Solve $y = 3x$ if the domain is {−2, −1, 0, 1, 3}. Graph the solution set.**

Make a table. Substitute each value of x into the equation to determine the corresponding values of y.

x	3x	y	(x, y)
−2	3(−2)	−6	(−2, −6)
−1	3(−1)	−3	(−1, −3)
0	3(0)	0	(0, 0)
1	3(1)	3	(1, 3)
3	3(3)	9	(3, 9)

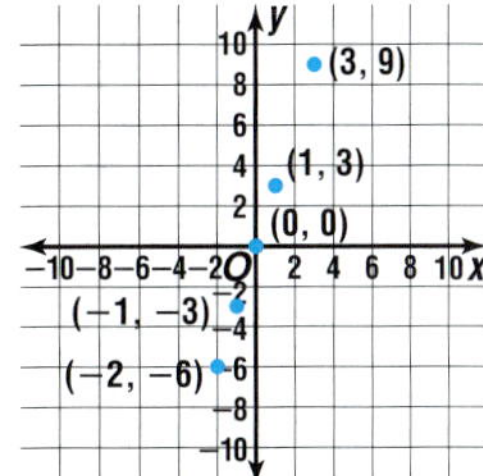

The solution set is {(−2, −6), (−1, −3), (0, 0), (1, 3), (3, 9)}.

Your Turn **b. {(−2, 8), (−1, 4), (0, 0), (1, −4), (2, −8)}; See margin for graph.**

b. Solve $y = -4x$ if the domain is {−2, −1, 0, 1, 2}. Graph the solution set.

Sometimes you can solve an equation for y before substituting each domain value into the equation. This makes creating a table of values easier.

Example 3 **Solve $4x + 2y = 8$ if the domain is {−2, −1, 0, 1, 2}. Graph the solution set.**

First, solve the equation for y in terms of x.

$4x + 2y = 8$ *Original equation*

$2y = 8 - 4x$ *Subtract 4x from each side.*

$\frac{2y}{2} = \frac{8 - 4x}{2}$ *Divide each side by 2.*

$y = \frac{8}{2} - \frac{4x}{2}$ *Write $\frac{8 - 4x}{2}$ as the difference of fractions.*

$y = 4 - 2x$ *Simplify.*

(continued on the next page)

www.algconcepts.com/extra_examples

2 TEACH

Teaching Tip While discussing the table at the top of page 244, make sure students see the connection between the equation $y = 4.2x$ and the values in the table.

In-Class Examples

Example 1

Which of the ordered pairs (0, 0), (1, 4), (2, 1) or (−1, 2) are solutions of $y = -x + 3$? **(2, 1)**

Example 2

Solve $y = 2x + 1$ if the domain is {−2, −1, 0, 1, 2}. Graph the solution set. **{(−2, −3), (−1, −1), (0, 1), (1, 3), (2, 5)}**

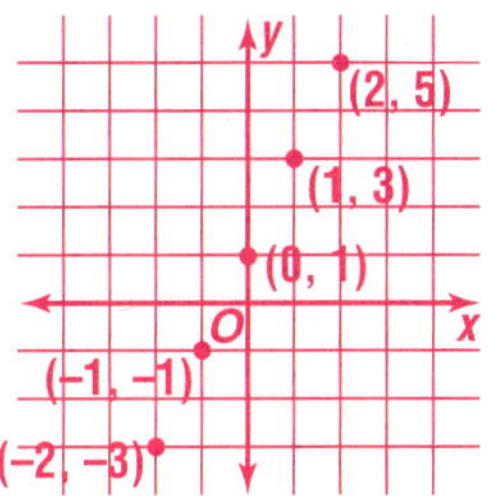

Teaching Tip In Example 3, you might want to show students how to substitute −2 for x in the equation $4x + 2y = 8$ without first solving the equation for y in terms of x. Tell students to imagine doing this for all five values of the domain. Help students recognize how this method involves more work than substituting for x after the equation has been solved for y.

Inclusion Strategies

Pair students with physical handicaps with a partner who can graph points without difficulty. Have the physically-handicapped student read the ordered pairs and direct their partner to the location where the point should be graphed.

Answer

Your Turn

b.

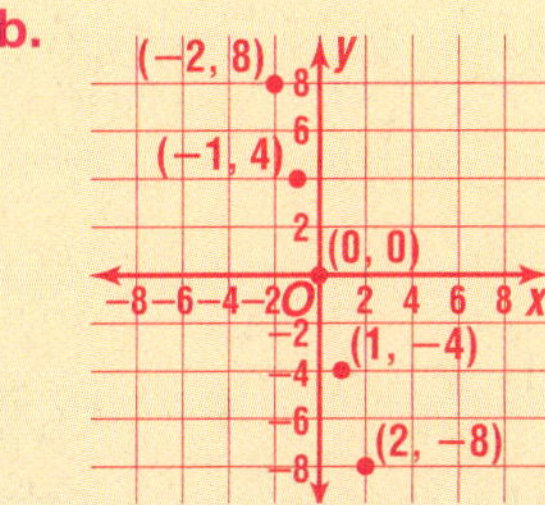

In-Class Example

Example 3

Solve $6x + 3y = 12$ if the domain is $\{-3, -1, 0, 1, 3\}$. Graph the solution set. **$\{(-3, 10), (-1, 6), (0, 4), (1, 2), (3, -2)\}$**

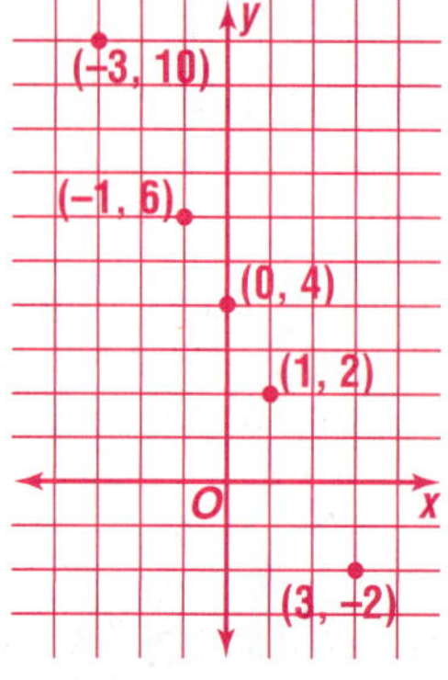

Teaching Tip In Example 4, draw students' attention to the set of values. Stress that these values are the range, not the domain, and therefore are the values of y. Urge students to be careful not to automatically substitute values for x.

In-Class Example

Example 4

Find the domain of $y = 10 - 4x$ if the range is $\{-6, -2, 2, 6, 10\}$. **$\{4, 3, 2, 1, 0\}$**

Answer

Your Turn

c.

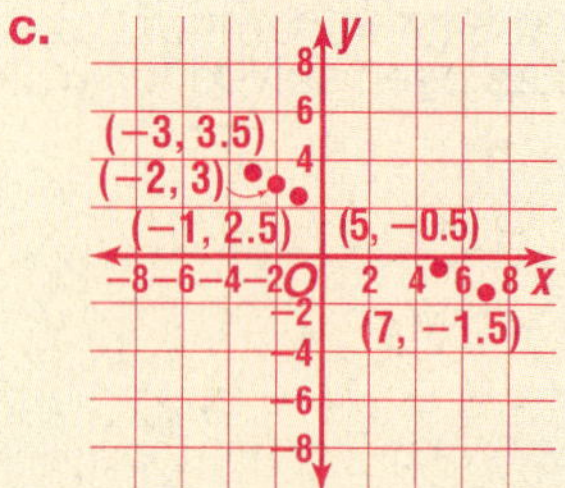

Next, substitute each value of x from the domain to determine the corresponding values of y.

x	$4 - 2x$	y	(x, y)
−2	4 − 2(−2)	8	(−2, 8)
−1	4 − 2(−1)	6	(−1, 6)
0	4 − 2(0)	4	(0, 4)
1	4 − 2(1)	2	(1, 2)
2	4 − 2(2)	0	(2, 0)

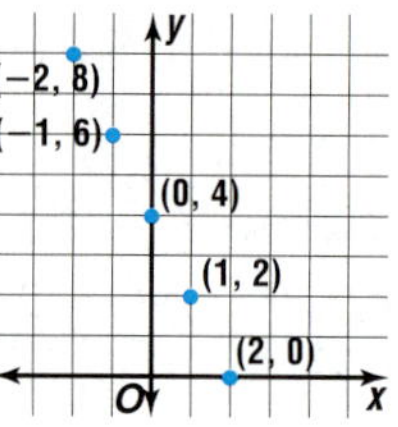

The solution set is $\{(-2, 8), (-1, 6), (0, 4), (1, 2), (2, 0)\}$.

Your Turn **c. $\{(-3, 3.5), (-2, 3), (-1, 2.5), (5, -0.5), (7, -1.5)\}$; See margin for graph.**

c. Solve $3x + 6y = 12$ if the domain is $\{-3, -2, -1, 5, 7\}$.

You can also solve an equation for given range values.

Example  **Find the domain of $y = 2x - 5$ if the range is $\{-3, -1, 1, 3\}$.**

Make a table. Substitute each value of y into the equation. Then solve each equation to determine the corresponding values of x.

y	$y = 2x - 5$	x	(x, y)
−3	−3 = 2x − 5	1	(1, −3)
−1	−1 = 2x − 5	2	(2, −1)
1	1 = 2x − 5	3	(3, 1)
3	3 = 2x − 5	4	(4, 3)

The domain is $\{1, 2, 3, 4\}$.

Your Turn **$\{2, 1, 0, -1, -2\}$**

d. Find the domain of $y = -3x$ if the range is $\{-6, -3, 0, 3, 6\}$.

Sometimes variables other than x and y are used in an equation. In this text, the values of the variable that comes first alphabetically are from the domain.

$m = 2n - 5$ (m: domain, n: range)

$3d = c$ (c: range, d: domain)

$2a + 3b = 6$ (a: domain, b: range)

Example 5 Geometry Link

The equation $2w + 2\ell = P$ can be used to find the perimeter P of a rectangle. Suppose a rectangle has a perimeter of 48 inches. Find the possible dimensions of the rectangle given the domain values {3, 4, 8, 12}.

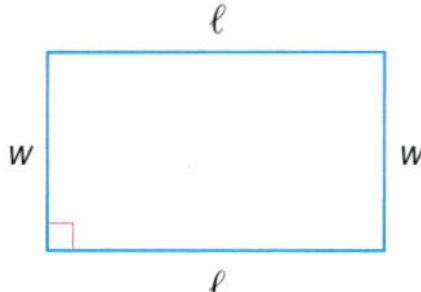

Assume that the values of ℓ come from the domain. Therefore, the equation should be solved for w in terms of ℓ.

$2w + 2\ell = P$	*Perimeter formula*
$2w + 2\ell = 48$	*Replace P with 48.*
$2w = 48 - 2\ell$	*Subtract 2ℓ from each side.*
$\frac{2w}{2} = \frac{48 - 2\ell}{2}$	*Divide each side by 2.*
$w = \frac{48}{2} - \frac{2\ell}{2}$	*Write $\frac{48 - 2\ell}{2}$ as the difference of fractions.*
$w = 24 - \ell$	*Simplify.*

Now, substitute each value of ℓ from the domain into the equation to find the corresponding values of w.

ℓ	$24 - \ell$	w	(ℓ, w)
3	24 − 3	21	(3, 21)
4	24 − 4	20	(4, 20)
8	24 − 8	16	(8, 16)
12	24 − 12	12	(12, 12)

The dimensions of the rectangle could be (3, 21), (4, 20), (8, 16), or (12, 12).

Check for Understanding

Communicating Mathematics

Vocabulary
equation in two variables
solution set

1. **Explain** why (2, 5) is a member of the solution set of $y = 2x + 1$.

   1. When the values are substituted in the equation, the equation is true.

2. **Identify** the set of numbers that make up the range of $y = x + 2$ if the domain is {0, 1, 2, 4}. {2, 3, 4, 6}

3. **You Decide** Lorena and Dan are solving the equation $3n + 2m = 11$ for the domain {−1, 3, 6, 7}. Lorena says to replace n with each of these values. Dan disagrees. He says to replace m with each value. Who is correct? Explain your reasoning. See margin.

Guided Practice

Example 1

Which ordered pairs are solutions of each equation?

4. $3a + b = 8$ a, b   a. (4, −4)   b. (2, 2)   c. (8, 0)   d. (3, 1)
5. $2c + 3d = 11$ c, d   a. (3, 1)   b. (4, −1)   c. (1, 3)   d. (−2, 5)

Reteaching Activity

Interpersonal Learners Have pairs of students discuss Exercise 3 to decide whether Lorena or Dan is correct. Ask students how they would explain to the person who is incorrect about the mistake he or she made and how to correct it.

In-Class Example

Example 5

Suppose a rectangle has a perimeter of 20 inches. Find the possible dimensions of the rectangle given the domain values {1, 2, 4, 5}. **The dimensions of the rectangle could be (1, 9), (2, 8), (4, 6), or (5, 5).**

3 PRACTICE/APPLY

Error Analysis

Watch for students who substitute the given domain values for the wrong variable in Exercises 2, 4, and 5.

Prevent by reminding students that the domain corresponds to x values when the variables x and y are used. When other variables are used in this lesson, the domain corresponds to the variable that comes first alphabetically.

Answer

3. Dan; when using variables other than x and y, it can be assumed that the domain goes with the variable that comes first alphabetically. Thus, m is the domain.

***Study Guide,* p. 250**

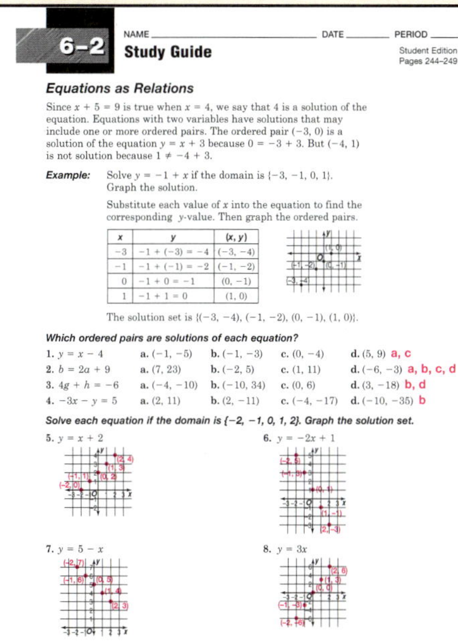

6–2 NAME ______ DATE ______ PERIOD ______

Study Guide Student Edition Pages 244–249

Equations as Relations

Since $x + 5 = 9$ is true when $x = 4$, we say that 4 is a solution of the equation. Equations with two variables have solutions that may include one or more ordered pairs. The ordered pair (−3, 0) is a solution of the equation $y = x + 3$ because $0 = -3 + 3$. But (−4, 1) is not solution because $1 \neq -4 + 3$.

Example: Solve $y = -1 + x$ if the domain is {−3, −1, 0, 1}. Graph the solution.

Substitute each value of x into the equation to find the corresponding y-value. Then graph the ordered pairs.

x	y	(x, y)
−3	−1 + (−3) = −4	(−3, −4)
−1	−1 + (−1) = −2	(−1, −2)
0	−1 + 0 = −1	(0, −1)
1	−1 + 1 = 0	(1, 0)

The solution set is {(−3, −4), (−1, −2), (0, −1), (1, 0)}.

Which ordered pairs are solutions of each equation?

1. $y = x - 4$ a. (−1, −5)   b. (−1, −3)   c. (0, −4)   d. (5, 9) a, c
2. $b = 2a + 9$ a. (7, 23)   b. (−2, 5)   c. (1, 11)   d. (−6, −3) a, b, c, d
3. $4g + h = -6$ a. (−4, −10)   b. (−10, 34)   c. (0, 6)   d. (3, −18) b, d
4. $-3x - y = 5$ a. (2, 11)   b. (2, −11)   c. (−4, −17)   d. (−10, −35) b

Solve each equation if the domain is {−2, −1, 0, 1, 2}. Graph the solution set.

5. $y = x + 2$ (−2, 0), (−1, 1), (0, 2), (1, 3), (2, 4)
6. $y = -2x + 1$ (−2, 5), (−1, 3), (0, 1), (1, −1), (2, −3)
7. $y = 5 - x$ (−2, 7), (−1, 6), (0, 5), (1, 4), (2, 3)
8. $y = 3x$ (−2, −6), (−1, −3), (0, 0), (1, 3), (2, 6)

Assignment Guide

Basic: 13–37 odd, 39–47
Average: 12–34 even, 36–47

Answers

6. {(−2, −8), (−1, −4), (0, 0), (1, 4), (2, 8)}

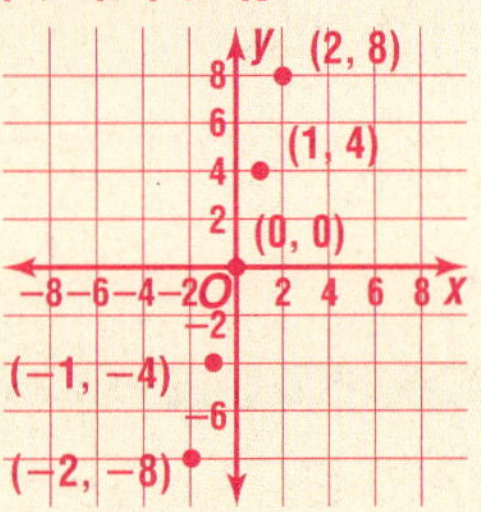

7. {(−2, 7), (−1, 2), (0, −3), (1, −8), (2, −13)}

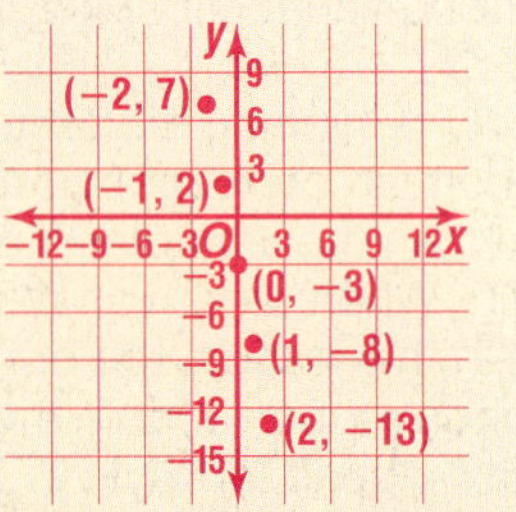

8. {(−2, −2), (−1, 0), (0, 2), (1, 4), (2, 6)}

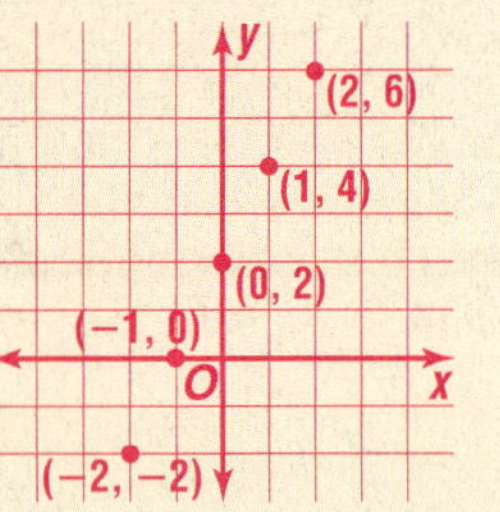

***Skills Practice*, p. 251, and *Practice*, p. 252 (shown)**

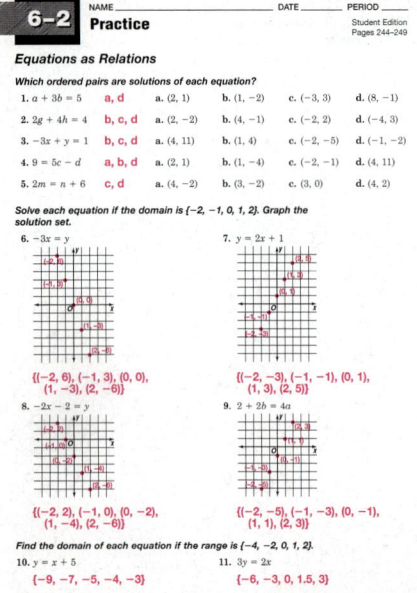

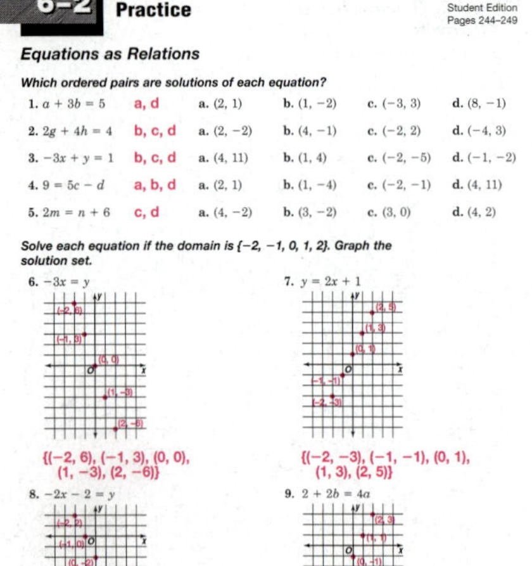

6–2 NAME ______ DATE ______ PERIOD ______

Practice Student Edition Pages 244–249

Equations as Relations

Which ordered pairs are solutions of each equation?

1. $a + 3b = 5$ a, d a. (2, 1) b. (1, −2) c. (−3, 3) d. (8, −1)
2. $2g + 4h = 4$ b, c, d a. (2, −2) b. (4, −1) c. (−2, 2) d. (−4, 3)
3. $-3x + y = 1$ b, c, d a. (4, 11) b. (1, 4) c. (−2, −5) d. (−1, −2)
4. $9 = 5c - d$ a, b, d a. (2, 1) b. (1, −4) c. (−2, −1) d. (4, 11)
5. $2m = n + 6$ c, d a. (4, −2) b. (3, −2) c. (3, 0) d. (4, 2)

Solve each equation if the domain is {−2, −1, 0, 1, 2}. Graph the solution set.

6. $-3x = y$ {(−2, 6), (−1, 3), (0, 0), (1, −3), (2, −6)}
7. $y = 2x + 1$ {(−2, −3), (−1, −1), (0, 1), (1, 3), (2, 5)}
8. $-2x - 2 = y$ {(−2, 2), (−1, 0), (0, −2), (1, −4), (2, −6)}
9. $2 + 2b = 4a$ {(−2, −5), (−1, −3), (0, −1), (1, 1), (2, 3)}

Find the domain of each equation if the range is {−4, −2, 0, 1, 2}.

10. $y = x + 5$ {−9, −7, −5, −4, −3}
11. $3y = 2x$ {−6, −3, 0, 1.5, 3}

Examples 2 & 3 **Solve each equation if the domain is {−2, −1, 0, 1, 2}. Graph the solution set.** 6–9. See margin.

6. $4x = y$
7. $y = -5x - 3$
8. $2x + 2 = y$
9. $5 + 2b = 3a$

Example 4 10. Find the domain of $y = x + 3$ if the range is {−4, −1, 0, 1, 2}. {−7, −4, −3, −2, −1}

Example 5 11. **Geometry** Refer to Example 5. Suppose a rectangle has a perimeter of 56 inches. Find the possible dimensions of the rectangle given the domain values {3, 4, 8, 12}. {(3, 25), (4, 24), (8, 20), (12, 16)}

Exercises

Practice

Which ordered pairs are solutions of each equation?

A

12. $2p - 5q = 1$ c, d a. (7, 3) b. (2, 1) c. (−7, −3) d. (−2, −1)
13. $3g = h + 7$ d a. (2, 4) b. (2, 3) c. (−1, 2) d. (2, −1)
14. $3x + 3y = 0$ a. (2, −2) b. (−2, 2) c. (1, −1) d. (−1, 1)
15. $8 - 2n = 4m$ b a. (0, 2) b. (2, 0) c. (1, −2) d. (0.5, −3)
16. $2c + 4d = 8$ a a. (0, 2) b. (−2, 1) c. (2, 0) d. (−3, 0.5)
17. $8u - 4 = 3v$ c a. (4, 2) b. $\left(0, \frac{1}{2}\right)$ c. (2, 4) d. $\left(\frac{2}{3}, \frac{3}{4}\right)$

Homework Help

For Exercises	See Examples
12–17	1
18–29	2, 3
31–33, 37	4
36–38	3, 5

Extra Practice
See page 703.

14. a, b, c, d

Solve each equation if the domain is {−1, 0, 1, 2, 3}. Graph the solution set. 18–29. See Solutions Manual.

18. $y = 2x$
19. $y = -3x$
20. $y = 7 - x$
21. $x - y = 6$
22. $y = 2x + 1$
23. $y = -5x + 5$
24. $3y = 9x + 3$
25. $2m - 2n = 4$
26. $3 - 5r = 2s$
27. $12 = 3b + 6a$
28. $6w - 3u = 36$
29. $4 = 3a - b$

B

30. Name an ordered pair that is a solution of $y = 6x - 2$. Sample answer: (0, −2)

Find the domain of each equation if the range is {−2, 0, 2, 4}.

31. $y = 2x - 2$ {0, 1, 2, 3}
32. $y = 6 - 2x$ {4, 3, 2, 1}
33. $2y = 4x$ {−1, 0, 1, 2}

C

34. Find the value of d if a solution of $y - dx = 1$ is (2, 5). 2
35. A solution of $4 = ay + 2x$ is (−1, 2). What is the value of a? 3

Applications and Problem Solving

36. **Shopping** Cedric has 90¢ to spend on school supplies. Erasers cost 10¢ each, and pencils cost 15¢ each. Suppose r represents the number of erasers and p represents the number of pencils. The equation $10r + 15p = 90$ represents this situation.
    a. Determine the ordered pairs (p, r) that satisfy the equation if the domain is {0, 2, 4, 6}. {(0, 9), (2, 6), (4, 3), (6, 0)}
    b. Express the relation in a table. b–c. See margin.
    c. Graph the relation.

Answers

9. $\left\{(-2, -5.5), (-1, -4), (0, -2.5), (1, -1), \left(2, \frac{1}{2}\right)\right\}$

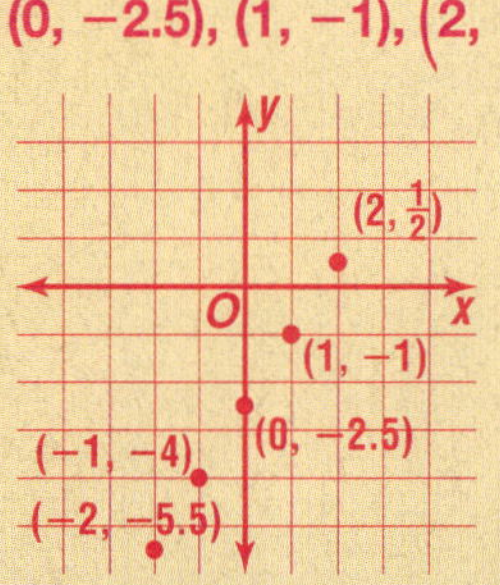

36b.

p	r
0	9
2	6
4	3
6	0

36c. (0, 9), (2, 6), (4, 3), (6, 0)

37. **Life Science** The equation $y = 0.025w$ gives the weight of the brain of an infant at birth, where w is the birth weight in pounds. The table shows the brain weights of five newborns.

	w	y
6.0	?	0.15
6.8	?	0.17
7.2	?	0.18
8.0	?	0.2
8.8	?	0.22

a. Find the birth weight of each infant.

b. Express the relation as a set of ordered pairs.

c. Graph the relation. **See margin.**

b. {(6.0, 0.15), (6.8, 0.17), (7.2, 0.18), (8.0, 0.2), (8.8, 0.22)}

38. **Sales** Mr. Richardson is a sales representative for an electronics store. Each month, he receives a salary of $1800 plus a 6% commission on sales. Suppose s represents his sales. The equation $t = 1800 + 0.06s$ can be used to determine his total monthly income. In July, his sales will either be $800, $1300, or $2000. Find the possible total incomes for July. **$1848, $1878, $1920**

39. **Critical Thinking** The sum of the measures of the two acute angles in a right triangle is 90°. In a given triangle, one acute angle is twice as large as the other acute angle. **a. $x + y = 90$; $x = 2y$**

a. Write two equations that represent these two relations.

b. Determine the ordered pair that is a solution of both equations. **(60, 30)**

Mixed Review

40. Express the relation shown in the table as a set of ordered pairs and as a graph. Then find the domain and range. *(Lesson 6–1)* **See margin.**

x	y
−4	−4
−3	0
2	5
5	−3

Exercise 40

41. **Shopping** At Crafter's Corner, all paint supplies are on sale at a 30% discount. A set of paint brushes normally sells for $15.70. What is the sale price? *(Lesson 5–5)* **$10.99**

42. **Banking** McKenna invests $580 at a rate of 6% for 5 years. How much interest will she earn? *(Lesson 5–4)* **$174**

Solve each equation.

43. $5(4 - 2c) + 3(c + 9) = 12$ *(Lesson 4–7)* **5**

44. $4n - 9 = 5 + 3n$ *(Lesson 4–6)* **14**

45. $3x - 7 = -10$ *(Lesson 4–5)* **−1**

Standardized Test Practice

46. **Short Response** Simplify $5(7ab - 4ab)$. *(Lesson 1–4)* **$15ab$**

47. **Multiple Choice** Name the property of equality shown by the statement *If $6 + 2 = 8$, then $8 = 6 + 2$.* *(Lesson 1–2)* **C**

Ⓐ Reflexive

Ⓑ Transitive

Ⓒ Symmetric

Ⓓ Commutative

www.algconcepts.com/self_check_quiz

? Extra Credit

Find an equation whose solutions include the ordered pairs (2, 1), (3, 2), (4, 3), and (5, 4).

Sample answer: $y = x - 1$

4 ASSESS

Open-Ended Assessment

Speaking Ask students to explain how to tell if the ordered pair (4, 2) is a solution of the equation $2x + 3y = 14$.

Chapter 6, Quiz A (Lessons 6–1 and 6–2) is available in the *Chapter 6 Resource Masters,* p. 285.

Answers

37c.

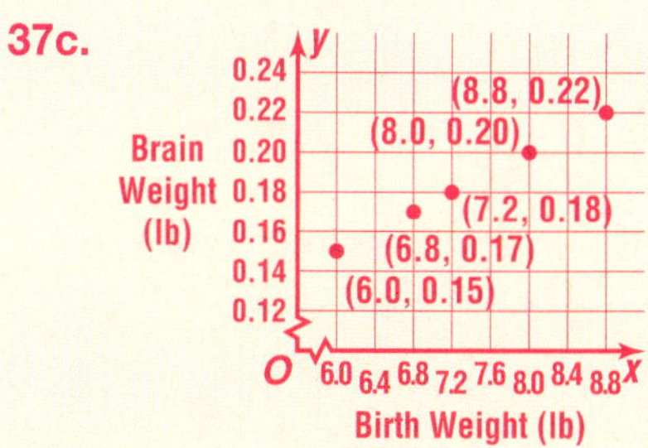

40. **{(−4, −4), (−3, 0), (2, 5), (5, −3)}**

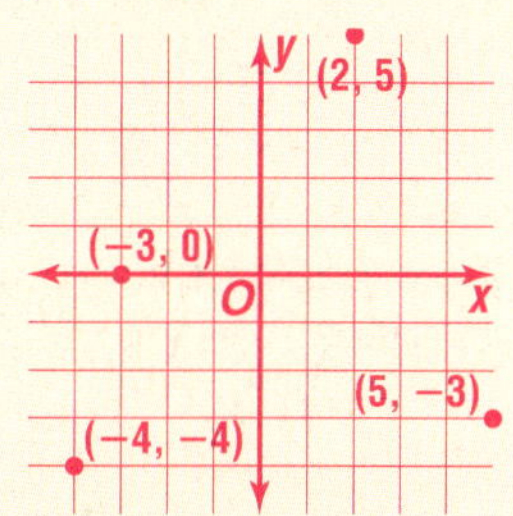

domain: {−4, −3, 2, 5}; range: {−4, 0, 5, −3}

***Enrichment,* p. 254**

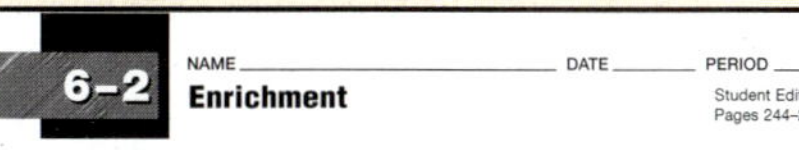

NAME ______ DATE ______ PERIOD ______

6–2 Enrichment

Student Edition Pages 244–249

Diophantine Equations

The first great algebraist, Diophantus of Alexandria (about A.D. 300), devoted much of his work to the solving of indeterminate equations. An indeterminate equation has more than one variable and an unlimited number of solutions. An example is $x + 2y = 4$.

When the coefficients of an indeterminate equation are integers and you are asked to find solutions that must be integers, the equation is called a **diophantine equation**. Such equations can be quite difficult to solve, often involving trial and error—and some luck!

Solve each diophantine equation by finding at least one pair of positive integers that makes the equation true. Some hints are given to help you.

1. $2x + 5y = 32$
   a. First solve the equation for x. $x = 16 - \frac{5y}{2}$
   b. Why must y be an even number? If y is odd, then x won't be an integer.
   c. Find at least one solution. any of these: (11, 2), (6, 4), (1, 6)

2. $5x + 2y = 42$
   a. First solve the equation for x. $x = \frac{42 - 2y}{5}$
   b. Rewrite your answer in the form $x = 8 +$ some expression. $x = 8 + \frac{2 - 2y}{5}$
   c. Why must $(2 - 2y)$ be a multiple of 5? otherwise, x won't be an integer
   d. Find at least one solution. any of these: (8, 1), (6, 6), (4, 11), (2, 16)

3. $2x + 7y = 29$
   (11, 1) or (4, 3)

4. $7x + 5y = 118$
   any of these: (14, 4), (9, 11), (4, 18)

5. $8x - 13y = 100$
   (19, 4), (32, 12) or any pair when $y = 4n$ and n is a positive odd number

6. $3x + 4y = 22$
   (6, 1) or (2, 4)

7. $5x - 14y = 11$
   (5, 1), (19, 6) or any pair when $y = 5m - 4$ and m is a positive number

8. $7x + 3y = 40$
   (4, 4) or (1, 11)

Lesson 6–3

6–3 Graphing Linear Relations

1 FOCUS

5-Minute Check

Lesson 6–2

1. Which of the ordered pairs (0, 0), (1, 2), (2, 3), and (3, 7) are solutions of $y = 2x$? **(0, 0) and (1, 2)**
2. Solve $y = 2x + 3$ if the domain is $\{-3, -1, 0, 1\}$. **$\{(-3, -3), (-1, 1), (0, 3), (1, 5)\}$**
3. Graph the solution set you found in Exercise 2.

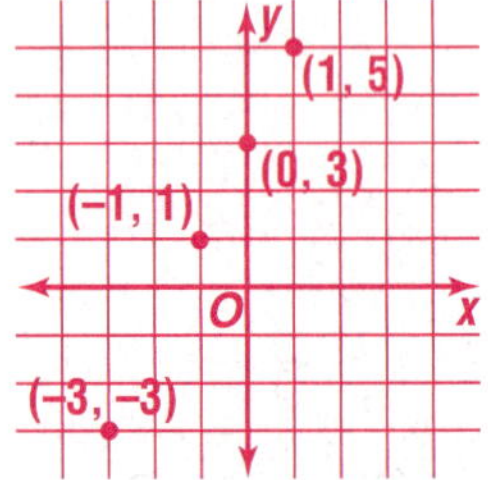

4. Find the domain of $y = \frac{1}{2}x$ if the range is $\{-2, -1, 0, 1, 2\}$. **The domain is $\{-4, -2, 0, 2, 4\}$.**
5. Graph the solution set for the equation in Exercise 4 using the domain values.

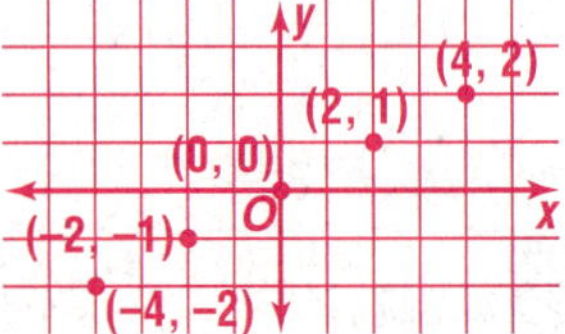

Motivating the Lesson

Hands-On Activity The ordered pairs in the relation {(1.5, 48), (2.0, 55), (2.5, 62), (3.0, 69), (3.5, 76), (4.0, 83), (4.5, 90), (5.0, 97)} represent the length of eruption and time to the next eruption of Old Faithful. Have students graph the points. Have them connect the points using a straightedge. Inform students that the points will all be on one line if they were graphed correctly. Introduce the idea that a relation is called a *linear relation* if the graphs of its solutions form a line.

What You'll Learn

You'll learn to graph linear relations.

Why It's Important

Earth Science Geologists can use the graph of a relation to predict the time at which Old Faithful will erupt. *See Example 8.*

The relation at the right is a solution set for the equation $y = x + 2$. If you graph the relation, the points lie on a straight line, and the graph of the line includes all possible solutions of the equation of the graph.

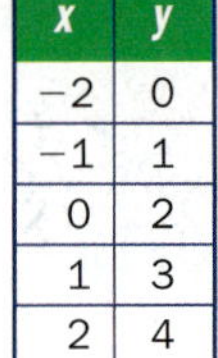

x	y
−2	0
−1	1
0	2
1	3
2	4

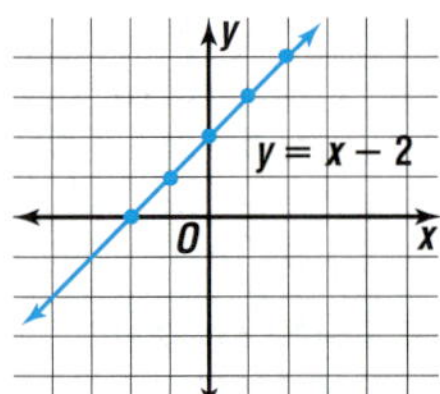

The line contains an infinite number of points. The ordered pair for every point on the line is a solution of the equation $y = x + 2$. An equation whose graph is a straight line is called a **linear equation**.

Linear Equation in Standard Form	A linear equation is an equation that can be written in the form $Ax + By = C$, where A, B, and C are any numbers, and A and B are not both zero. $Ax + By = C$, where $A \geq 0$, is called the *standard form* if A, B, and C are integers whose greatest common factor is 1.

Linear equations can contain one or two variables. However, not all equations in one or two variables are linear equations.

Linear equations	Nonlinear equations
$3x + 4y = 6$	$2x + 4y = 3z^2$
$y = -2$	$2x + 2xy = 3$
$8b = -2a - 5$	$2x^2 = 7$

Examples

Determine whether each equation is a linear equation. Explain. If an equation is linear, identify A, B, and C when written in standard form.

1 $3x = 5 + y$

First, rewrite the equation so that both variables are on the same side of the equation.

$$3x = 5 + y$$

$$3x - y = 5 \quad \textit{Subtract } y \textit{ from each side.}$$

The equation is now in the form $Ax + By = C$, where $A = 3$, $B = -1$, and $C = 5$. Therefore, this is a linear equation.

Resource Manager

Reproducible Masters

Chapter 6 Resource Masters

- *Study Guide*, p. 255
- *Skills Practice*, p. 256
- *Practice*, p. 257
- *Reading to Learn Mathematics*, p. 258
- *Enrichment*, p. 259
- *Assessment*, p. 284

Hands-On Algebra, p. 73

Transparencies

5-Minute Check, 6–3

Technology/Multimedia

Interactive Chalkboard CD-ROM

 $-2x + 3xy = 6$

Since the term $3xy$ has two variables, the equation cannot be written in the form $Ax + By = C$. So, this is not a linear equation.

 $x = -2$

This equation can be written as $x + 0y = -2$. Therefore, it is a linear equation in the form $Ax + By = C$, where $A = 1$, $B = 0$, and $C = -2$.

Your Turn a. yes; $A = -2$, $B = 1$, $C = -8$

a. $8 + y = 2x$ b. $-5z = 4x + 2y$ no c. $y^2 = 4$ no

To graph a linear equation, make a table of ordered pairs that are solutions. Graph the ordered pairs and connect the points with a line. Note that the domain of the equation and its graph includes the values of x and the range includes the values of y.

Examples **Graph each equation.**

 $y = 2x - 3$

Select several values for the domain and make a table. Then graph the ordered pairs and connect them to draw the line.

Usually, values of x such as 0 and integers near 0 are chosen so that calculations are simple.

x	$2x - 3$	y	(x, y)
−2	2(−2) − 3	−7	(−2, −7)
−1	2(−1) − 3	−5	(−1, −5)
0	2(0) − 3	−3	(0, −3)
1	2(1) − 3	−1	(1, −1)
2	2(2) − 3	1	(2, 1)

When graphing an equation, use arrows to show that the graph continues. Also, label the line with the equation.

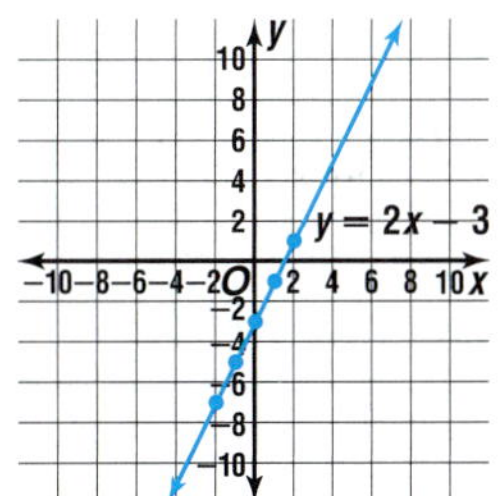

2 TEACH

Teaching Tip After discussing the definition of *linear equation* on page 250, focus students' attention on the examples of nonlinear equations given at the bottom of that page. Point out how each nonlinear equation differs from the linear equation $Ax + By = C$.

In-Class Examples

Examples 1–3

Determine whether each equation is a linear equation. Explain. If an equation is linear, identify A, B, and C when written in standard form.

1 $4xy = 4$ **Since the term 4xy has two variables, the equation cannot be written in the form $Ax + By = C$. So, this is not a linear equation.**

2 $y = x$ **This equation can be written as $x - y = 0$. Therefore, it is a linear equation in the form $Ax + By = C$, where $A = 1$, $B = -1$, and $C = 0$.**

3 $y = 7$ **This equation can be written as $0x + 1y = 7$. Therefore, it is a linear equation in the form $Ax + By = C$, where $A = 0$, $B = 1$, and $C = 7$.**

Teaching Tip While discussing Example 4, point out that finding values for ordered pairs is easier when the equation is solved for y in terms of x than when it is in the standard form $Ax + By = C$.

In-Class Example

Example 4

Graph $y = -x + 2$.

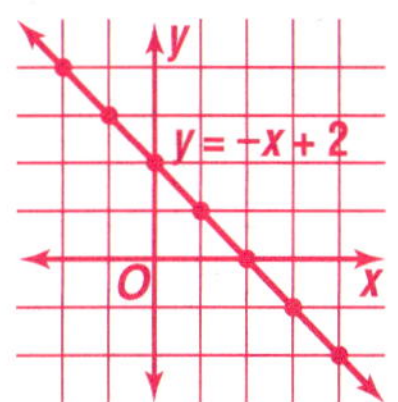

In-Class Examples

Example 5

Graph $3x - y = 2$.

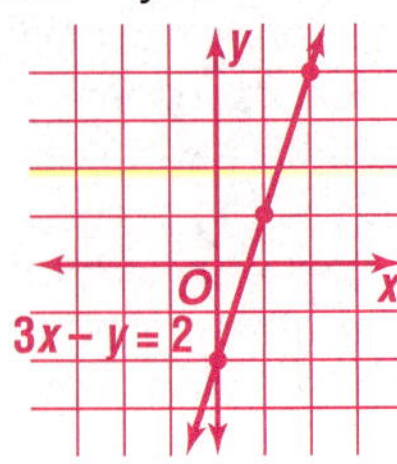

Example 6

Graph $y = -2$. Describe the graph.

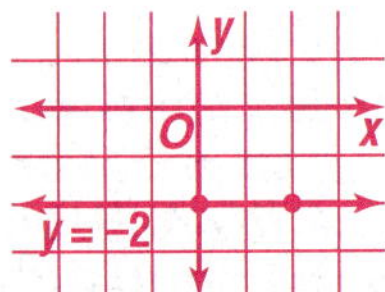

a horizontal line passing through the point at (0, −2)

Answers

Your Turn

d.

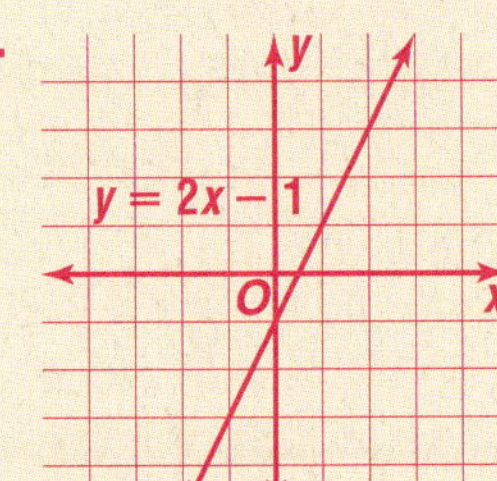

e.

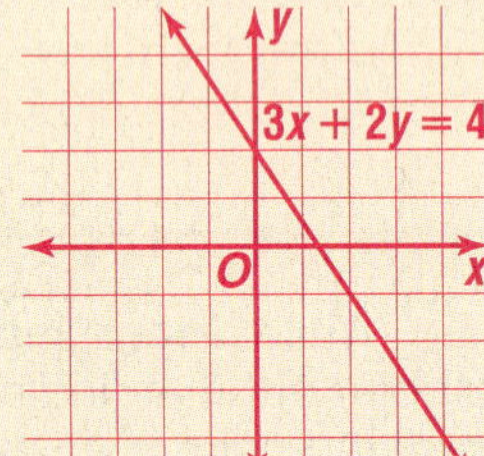

f.

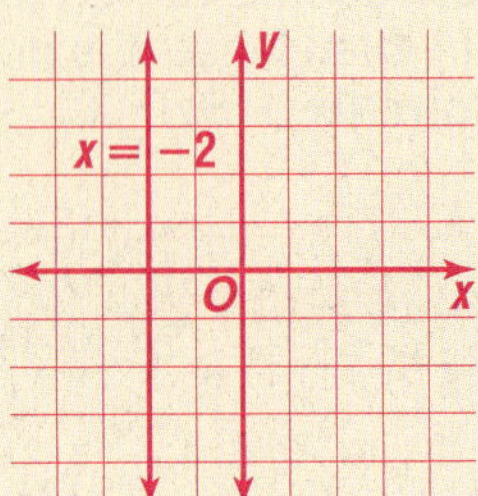

The graph is a vertical line.

5 $2x + 4y = 8$

In order to find values for y more easily, solve the equation for y.

$$2x + 4y = 8$$

$$4y = 8 - 2x \quad \text{Subtract } 2x \text{ from each side.}$$

$$y = \frac{8 - 2x}{4} \quad \text{Divide each side by 4.}$$

Now make a table and draw the graph.

x	$\frac{8-2x}{4}$	y	(x, y)
−4	$\frac{8-2(-4)}{4}$	4	(−4, 4)
−2	$\frac{8-2(-2)}{4}$	3	(−2, 3)
0	$\frac{8-2(0)}{4}$	2	(0, 2)
2	$\frac{8-2(2)}{4}$	1	(2, 1)
4	$\frac{8-2(4)}{4}$	0	(4, 0)

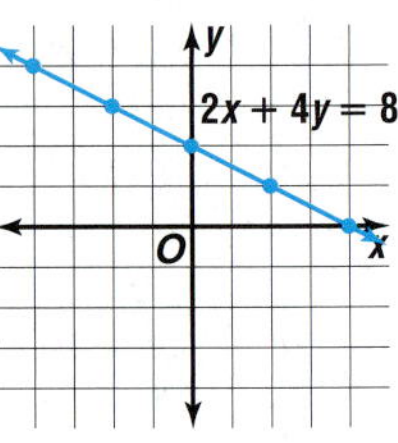

Your Turn **Graph each equation.**

d. $y = 2x - 1$

e. $3x + 2y = 4$ **d–e. See margin.**

Sometimes, the graph of an equation is a vertical or horizontal line.

Example 6 **Graph $y = 3$. Describe the graph.**

In standard form, this equation is written as $0x + y = 3$. So, for any value of x, $y = 3$. For example, if $x = 0$, $y = 3$; if $x = 1$, $y = 3$; if $x = 3$, $y = 3$. By graphing the ordered pairs (0, 3), (1, 3), and (3, 3), you find that the graph of $y = 3$ is a horizontal line.

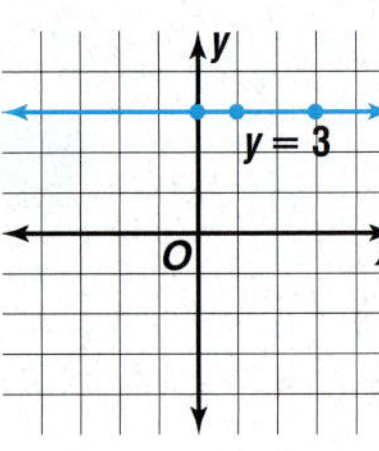

Your Turn

f. Graph the equation $x = -2$. Describe the graph. **See margin.**

Recall that the *domain* of a linear function is all of the x-coordinates of the ordered pairs for points on the line. The *range* of a linear function is all of the y-coordinates of the ordered pairs for points on the line. Since a line extends forever in both directions, if it is not vertical and not horizontal, the domain and the range are both the set of all real numbers.

Example 7 Graph $y = 2x$.

Make a table and draw the graph.

x	$2x$	y	(x, y)
−2	2(−2)	−4	(−2, −4)
−1	2(−1)	−2	(−1, −2)
0	2(0)	0	(0, 0)
1	2(1)	2	(1, 2)
2	2(2)	4	(2, 4)

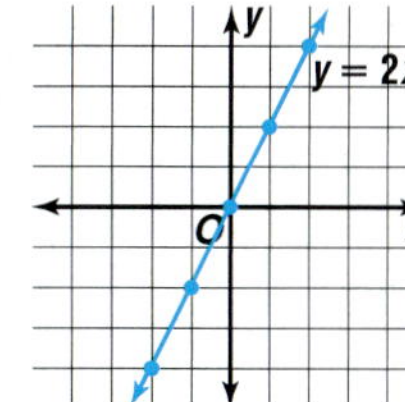

Write $y = 2x$ in standard form.

$y = 2x$ *Original equation*

$0 = 2x - y$ *Subtract y from each side.*

$2x - y = 0$ *Symmetric Property of Equality*

In any equation where $C = 0$, the graph passes through the origin.

Your Turn

g. Graph the equation $y = -3x$. **See margin.**

Example 8 Earth Science

Old Faithful is a famous geyser. The equation $y = 14x + 27$ can be used to predict its eruptions, where y is the time until the next eruption (in minutes) and x is the length of the eruption (in minutes). Suppose Old Faithful erupts at 9:46 A.M. for 3.4 minutes. At about what time will the next eruption occur?

Explore You need to find the time of the next eruption.

Plan Use the graph of $y = 14x + 27$ to solve the problem.

Solve The graph shows that if an eruption lasts 3.4 minutes, the next eruption will occur in about 75 minutes.

Time Until Next Eruption (min): 40, 50, 60, 70, 80, 90, 100
Length of an Eruption (min): O, 1, 2, 3, 4, 5

9 h 46 min → 9 h 46 min
+ 75 min → 1 h 15 min
10 h 61 min → 11 h 1 min

The next eruption should take place at 11:01 A.M.

Examine You can verify the answer by using the equation.

$y = 14x + 27$ *Original equation*

$y \stackrel{?}{=} 14(3.4) + 27$ *Replace x with 3.4.*

$y = 74.6$ *Simplify.*

The answer makes sense because 74.6 minutes added to 9:46 is about 11:01.

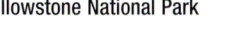

Old Faithful, Yellowstone National Park

In-Class Examples

Example 7

Graph $y = 3x$.

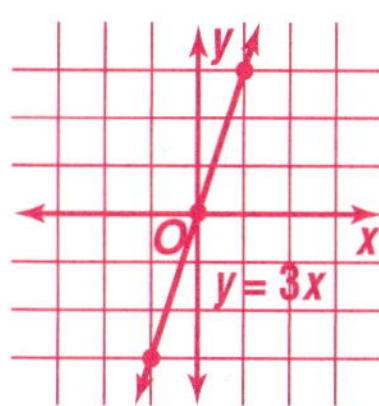

Example 8

Refer to the application at the beginning of the lesson. Suppose Old Faithful has an eruption that lasts 1.4 minutes. About how long will it be until the next eruption? **about 47 min**

Answer

Your Turn

g.

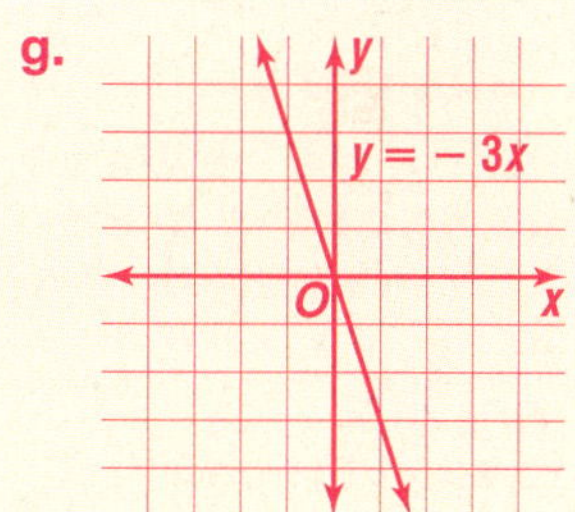

Study Guide, p. 255

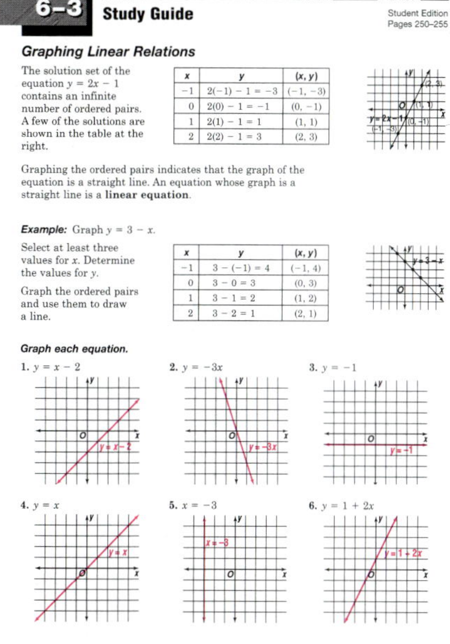

6–3 NAME ______ DATE ______ PERIOD ______

Study Guide

Student Edition Pages 250–255

Graphing Linear Relations

The solution set of the equation $y = 2x - 1$ contains an infinite number of ordered pairs. A few of the solutions are shown in the table at the right.

x	y	(x, y)
−1	2(−1) − 1 = −3	(−1, −3)
0	2(0) − 1 = −1	(0, −1)
1	2(1) − 1 = 1	(1, 1)
2	2(2) − 1 = 3	(2, 3)

Graphing the ordered pairs indicates that the graph of the equation is a straight line. An equation whose graph is a straight line is a **linear equation**.

Example: Graph $y = 3 - x$.

Select at least three values for x. Determine the values for y.

Graph the ordered pairs and use them to draw a line.

x	y	(x, y)
−1	3 − (−1) = 4	(−1, 4)
0	3 − 0 = 3	(0, 3)
1	3 − 1 = 2	(1, 2)
2	3 − 2 = 1	(2, 1)

Graph each equation.

1. $y = x - 2$
2. $y = -3x$
3. $y = -1$
4. $y = x$
5. $x = -3$
6. $y = 1 + 2x$

3 PRACTICE/APPLY

Error Analysis

Watch for students who describe the graph in Exercise 1 as a horizontal line because they associate the variable *x* in the equation with the horizontal *x*-axis. ***Prevent by*** stressing that the graph of $x = -8$ contains all of the points with an *x*-coordinate of 8. Emphasize that there is only one point on the *x*-axis with *x*-coordinate -8, namely the point $(-8, 0)$. Point out that every point on the vertical line through $(-8, 0)$ has *x*-coordinate -8.

Assignment Guide

Basic: 13–41 odd, 42–49
Average: 14–38 even, 40–49
All: Quiz 1, 1–5

Answers

1. The graph of $x = -8$ is a straight line parallel to the *y*-axis. For every value of *y*, $x = -8$.
2. All graphs of the form $y = kx$ pass through the origin.

Skills Practice, **p. 256, and *Practice,* p. 257 (shown)**

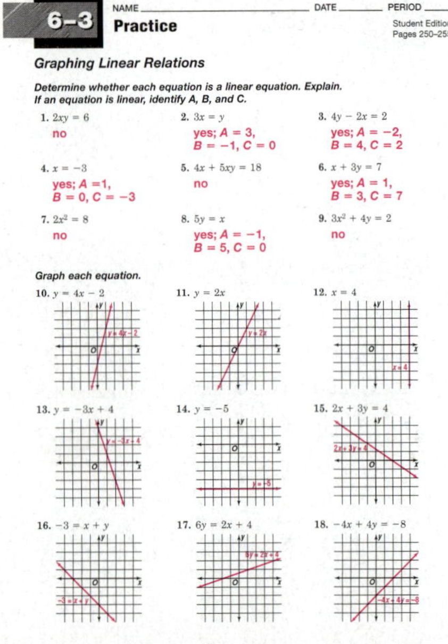

6–3 NAME ______ DATE ______ PERIOD ______
Practice — Student Edition Pages 250–255

Graphing Linear Relations

Determine whether each equation is a linear equation. Explain. If an equation is linear, identify A, B, and C.

1. $2xy = 6$ no
2. $3x = y$ yes; $A = 3$, $B = -1$, $C = 0$
3. $4y - 2x = 2$ yes; $A = -2$, $B = 4$, $C = 2$
4. $x = -3$ yes; $A = 1$, $B = 0$, $C = -3$
5. $4x + 5xy = 18$ no
6. $x + 3y = 7$ yes; $A = 1$, $B = 3$, $C = 7$
7. $2x^2 = 8$ no
8. $5y = x$ yes; $A = -1$, $B = 5$, $C = 0$
9. $3x^2 + 4y = 2$ no

Graph each equation.

10. $y = 4x - 2$
11. $y = 2x$
12. $x = 4$
13. $y = -3x + 4$
14. $y = -5$
15. $2x + 3y = 4$
16. $-3 = x + y$
17. $6y = 2x + 4$
18. $-4x + 4y = -8$

Check for Understanding

Communicating Mathematics

1. **Describe** the graph of $x = -8$. **See margin.**
2. **Explain** how you know that the graph of $y = -2x$ passes through the origin. **See margin.**

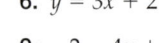

Vocabulary
linear equation

Guided Practice

Examples 1–3
4. yes; $A = 4$, $B = 3$, $C = 9$
Examples 4–7
5. yes; $A = 2$, $B = 0$, $C = 8$

Determine whether each equation is a linear equation. Explain. If an equation is linear, identify *A*, *B*, and *C*.

3. $xy = 3$ **no**
4. $3y + 4x = 9$
5. $2x = 8$

Graph each equation. **6–11. See Solutions Manual.**

6. $y = 3x + 2$
7. $y = -\frac{1}{2}x$
8. $x = -5$
9. $-2 = 4x + y$
10. $3x + 2y = 6$
11. $5y = -2x + 10$

Cactus

12. **Plants** The number of blooms on a cactus is related to the number of days of sun it gets in a month. This relation is given by the equation $y = 7x - 1$, where *x* is the number of days of sun and *y* is the number of blooms. **Example 8**
   a. Graph the equation. **See margin.**
   b. Use the graph to determine the number of blooms on a cactus if it gets two days of sun a month. **13**

Exercises

Practice

Determine whether each equation is a linear equation. Explain. If an equation is linear, identify *A*, *B*, and *C*.

13. $xy = 4$ **no**
14. $6 + 2xy = 4$ **no**
15. $2y - 6x = 0$
16. $4x = 8y$
17. $y = x$
18. $x + 3xy = 5$ **no**
19. $y = x^2 + 8$ **no**
20. $2 - 3x = 8$
21. $4x + x + y = 9$

Homework Help

For Exercises	See Examples
13–21	1–3
22–36	4, 5, 7

Extra Practice
See page 704.

15. yes; $A = -6$, $B = 2$, $C = 0$
16. yes; $A = 4$, $B = -8$, $C = 0$
17. yes; $A = -1$, $B = 1$, $C = 0$
20. yes; $A = -3$, $B = 0$, $C = 6$
21. yes; $A = 5$, $B = 1$, $C = 9$

B

Graph each equation. **22–36. See Solutions Manual.**

22. $x = y$
23. $y = 3x$
24. $-2x = y$
25. $y = -3$
26. $x = -3$
27. $2 = x$
28. $y = 3x + 2$
29. $2x - 5 = y$
30. $y = 4 - 2x$
31. $-1 = y + 4x$
32. $2x + y = -2$
33. $3 = -3x + y$
34. $4x + 2y = 8$
35. $3y - 2x = 6$
36. $-2y = 3x + 1$

C

Each table lists coordinates of points on a linear graph. Copy and complete each table. (*Hint:* Graph the ordered pairs.)

37.

x	y	
0	?	−2
1	−1	
2	0	
3	1	
4	?	2

38.

x	y	
−4	−2	
−2	?	−1
0	0	
2	1	
4	?	2

39.

x	y	
−2	−4	
−1	?	−1
0	2	
1	?	5
2	8	

Reteaching Activity

Verbal/Linguistic Learners Have pairs of students take turns explaining to each other how to make sets of ordered pairs and how to graph equations such as those in Exercises 6–11.

Answer

12a.

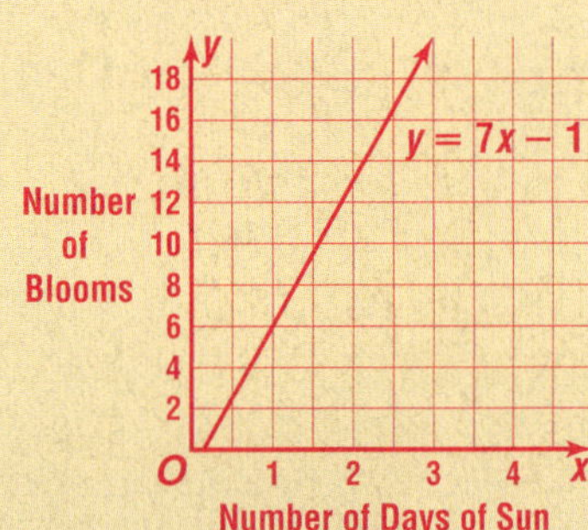

Applications and Problem Solving

40. **Business** Bongo's Clown Service charges a \$50 fee for a clown to perform at a birthday party plus an additional \$2 per guest.
    a. Write an equation that represents this situation. Let x represent the number of guests and let y represent the cost of a clown's services. $y = 50 + 2x$
    b. Graph the equation. See margin.
    c. Serena's parents paid \$68 for a clown to perform at Serena's party. How many people attended Serena's party?

40c. 9 people

41. **Geometry** Complementary angles are two angles whose sum is 90°. Suppose two complementary angles measure $(2x + 2y)°$ and $(x + 4y)°$.
    a. Write an equation for the sum of the measures of these two angles.
    b. Determine five ordered pairs that satisfy this equation.
    c. Graph the equation. See margin.

41a. $3x + 6y = 90$
41b. Sample answer: {(2, 14), (6, 12), (12, 9), (18, 6), (24, 3)}

42. **Critical Thinking** Graph the equations $y = 2x$, $y = 2x + 1$, $y = 2x + 2$, and $y = 2x + 3$ on the same coordinate plane. Then describe the graphs. See margin.

Mixed Review

43. Solve $y = x - 5$ if the domain is $\{-3, 4, 5, 7, 9\}$. *(Lesson 6–2)*

43. $\{-8, -1, 0, 2, 4\}$

44. Express the relation $\{(-1, -1), (0, 5), (1, -2), (3, 4)\}$ as a table and as a graph. *(Lesson 6–1)* See Solutions Manual.

Solve each proportion. *(Lesson 5–1)*

45. $\frac{3}{4} = \frac{x}{8}$ 6
46. $\frac{a}{45} = \frac{3}{15}$ 9
47. $\frac{y}{3} = \frac{y + 4}{6}$ 4

Standardized Test Practice

A B C D

48. **Grid In** Find the range of the data: {17, 24, 16, 25, 38, 57}. *(Lesson 3–3)* 41

49. **Short Response** The temperature outside is 67°F. Suppose it rises an average of 2°F every hour for h hours to reach a high of 79°F. Write an equation to represent this situation. *(Lesson 1–1)* $67 + 2h = 79$

Quiz 1 Lessons 6–1 through 6–3

Express each relation as a table and as a graph. Then determine the domain and the range. *(Lesson 6–1)* 1–2, 4. See Solutions Manual.

1. $\{(1, -1), (3, 5), (0, -2), (3, 3)\}$
2. $\{(-0.5, 3), (1.5, 0), (2.5, 4), (3.5, 2.5)\}$
3. Which ordered pairs are solutions to the equation $2y + 4 = 4x$? *(Lesson 6–2)* b, c
   a. $(-1, 3)$ b. $(0, -2)$ c. $(-2, -6)$ d. $(3, -4)$
4. Solve $y = 4x - 3$ if the domain is $\{-2, -1, 2, 3, 5\}$. Graph the solution set. *(Lesson 6–2)*
5. Anne is saving money to buy a \$2500 used car. She started with \$500 and has been saving \$80 per week since then. The equation $y = 500 + 80x$ represents her total savings y after any number of weeks x. *(Lesson 6–3)*
   a. Graph the equation. See Solutions Manual.
   b. Use the graph to estimate when Anne will have enough money. 25 weeks
   c. Verify your estimate using the equation. See Solutions Manual.

www.algconcepts.com/self_check_quiz

4 ASSESS

Open-Ended Assessment

Modeling Have students use a geoboard to demonstrate how to graph the equation $y = x$.

Quiz 1

The Quiz provides students with a brief review of the concepts and skills in Lessons 6–1 through 6–3. Lesson numbers are given to the right of the exercises or instruction lines so students can review concepts not yet mastered.

Mid-Chapter Test (Lessons 6–1 through 6–3) is available in the *Chapter 6 Resource Masters*, p. 284.

Answers

40b.

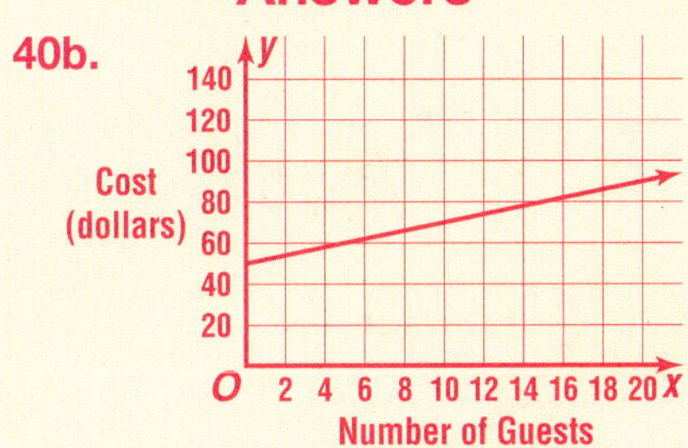

41c.

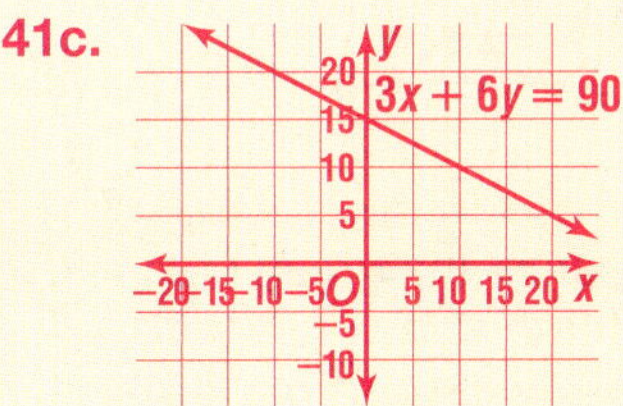

***Enrichment*, p. 259**

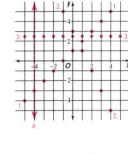

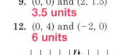

NAME ______ DATE ______ PERIOD ______

Enrichment

Student Edition Pages 250–255

Taxicab Graphs

You have used a rectangular coordinate system to graph equations such as $y = x - 1$ on a coordinate plane. In a coordinate plane, the numbers in an ordered pair (x, y) can be any two real numbers.

A **taxicab plane** is different from the usual coordinate plane. The only points allowed are those that exist along the horizontal and vertical grid lines. You may think of the points as taxicabs that must stay on the streets.

The taxicab graph shows the equations $y = -2$ and $y = x - 1$. Notice that one of the graphs is no longer a straight line. It is now a collection of separate points.

Taxicab Graph of $y = x - 1$

Graph these equations on the taxicab plane at the right.

1. $y = x + 1$
2. $y = -2x + 3$
3. $y = 2.5$
4. $x = -4$

Use your graphs for these problems.

5. Which of the equations would have the same graph in both the usual coordinate plane and the taxicab plane? $x = -4$
6. Describe the form of equations that have the same graph in both the usual coordinate plane and the taxicab plane.
   $x = A$ and $y = B$, where A and B are integers

In the taxicab plane, distances are not measured diagonally, but along the streets. Write the taxi-distance between each pair of points.

7. (0, 0) and (5, 2) 7 units
8. (0, 0) and (−3, 2) 5 units
9. (0, 0) and (2, 1.5) 3.5 units
10. (1, 2) and (4, 3) 4 units
11. (2, 4) and (−1, 3) 4 units
12. (0, 4) and (−2, 0) 6 units

Draw these graphs on the taxicab grid at the right.

13. The set of points whose taxi-distance from (0, 0) is 2 units. indicated by crosses
14. The set of points whose taxi-distance from (2, 1) is 3 units. indicated by dots

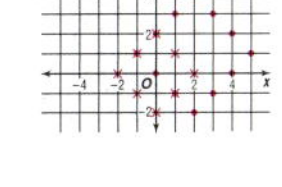

Extra Credit

The equation of the graph at the right has the form $y = ax$. Find the value of a. -2

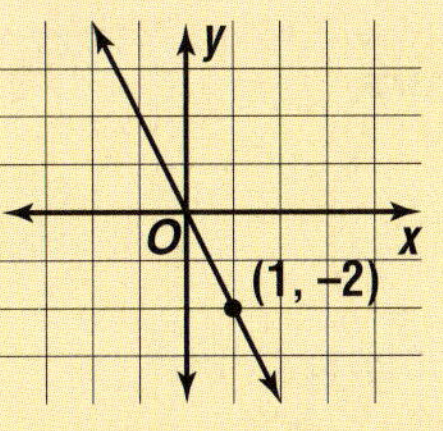

Answer

42.

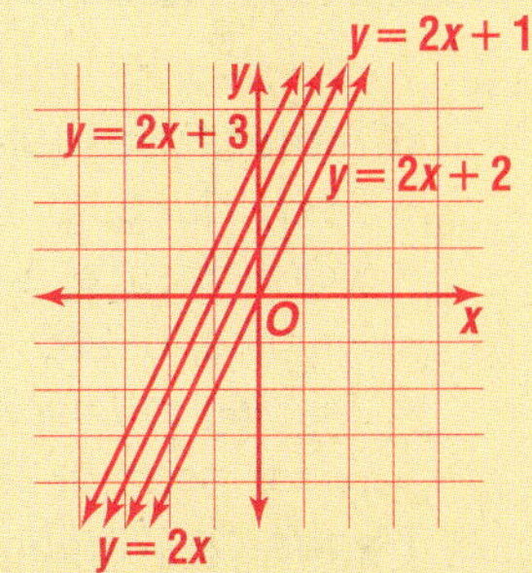

Lesson 6–4

1 FOCUS

5-Minute Check
Lesson 6–3

Determine whether each equation is a linear equation. Explain. If an equation is linear, identify A, B, and C.

1. $x + y = 1$ **yes; $A = 1$, $B = 1$, $C = 1$**
2. $2xy = 4$ **no**

Graph each equation.

3. $y = 2x + 2$

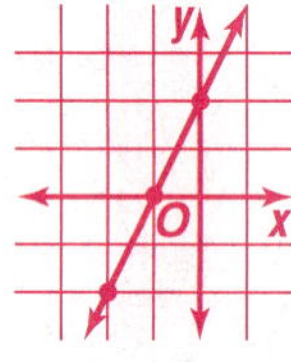

4. $y = -x$

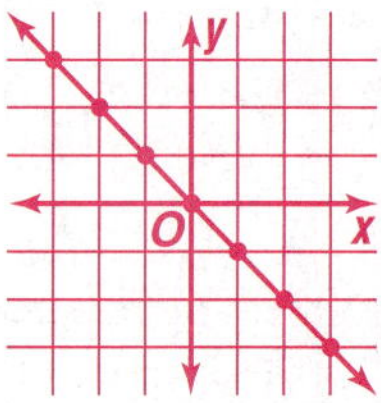

5. Describe the graph of $x = 4$. **Sample answer: a vertical line passing through the point at (4, 0)**

Motivating the Lesson

Real-World Connection Ask students to consider the statement "Your shoe size is a function of the size of your foot." Develop the idea that "is a function of" means "depends on."

2 TEACH

In-Class Example

Example 1

Determine whether the relation {(1, 1), (3, 4), (4, 5), (4, 6)} is a function. Explain your answer. **This is not a function because one member of the domain, 4, is paired with two members of the range, 5 and 6.**

6–4 Functions

What You'll Learn

You'll learn to determine whether a given relation is a function.

Why It's Important

Travel Functions can be used to determine the cost of renting a car. *See Exercise 48.*

Study the graph of the straight line below. Notice that for each value of x, there is exactly one value of y. This type of relation is called a **function**.

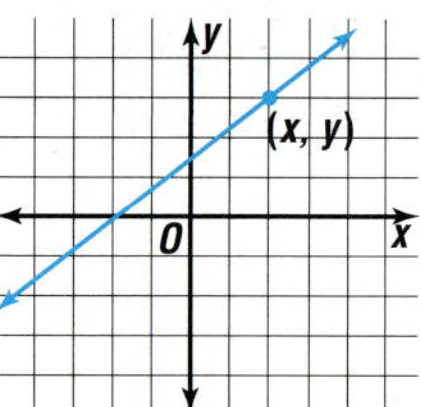

Functions	A function is a relation in which each member of the domain is paired with *exactly* one member of the range.

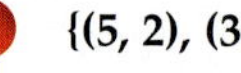

Determine whether each relation is a function. Explain your answer.

1. {(5, 2), (3, 5), (−2, 3), (−5, 1)}

Since each element of the domain is paired with *exactly* one element of the range, this relation is a function.

2.

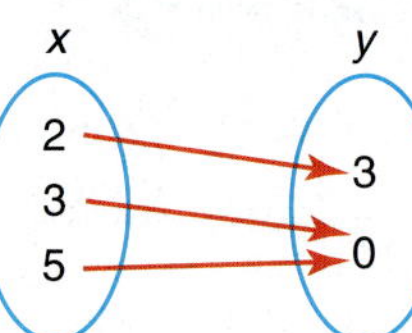

The mapping between x and y represents a function since there is *only one* corresponding element in the range for each element of the domain. *It does not matter if two elements of the domain are paired with the same element in the range.*

3.

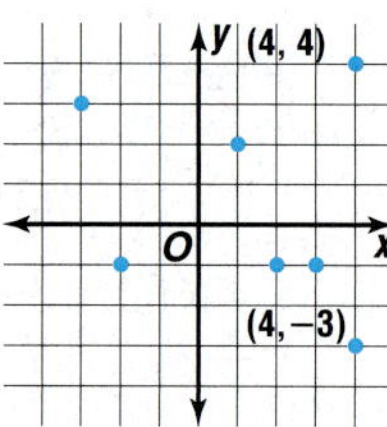

The graph represents a relation that is not a function. Look at the points with the ordered pairs (4, −3) and (4, 4). The member 4 in the domain is paired with both −3 and 4 in the range.

Resource Manager

Reproducible Masters

Chapter 6 Resource Masters
- *Study Guide*, p. 260
- *Skills Practice*, p. 261
- *Practice*, p. 262
- *Reading to Learn Mathematics*, p. 263
- *Enrichment*, p. 264

Hands-On Algebra, pp. 74–75

Transparencies

5-Minute Check, 6–4

Technology/Multimedia

Interactive Chalkboard CD-ROM

a–c. See margin for explanations.

Your Turn

Determine whether each relation is a function. Explain your answer.

a. $\{(5, -2), (3, 2), (4, -1), (-2, 2)\}$ **yes**

b.

x	y
−1	2
0	7
1	4
1	−2
4	5
6	−4

no

c.

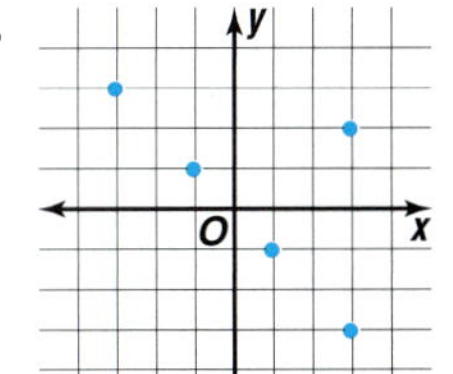

no

To determine whether an equation is a function, you can use the **vertical line test** on the graph of the equation. Consider the graph below.

To perform the test, place a pencil at the left of the graph to represent a vertical line. Move it to the right across the graph.

For each value of x, this vertical line passes through exactly one point on the graph. So, the equation is a function.

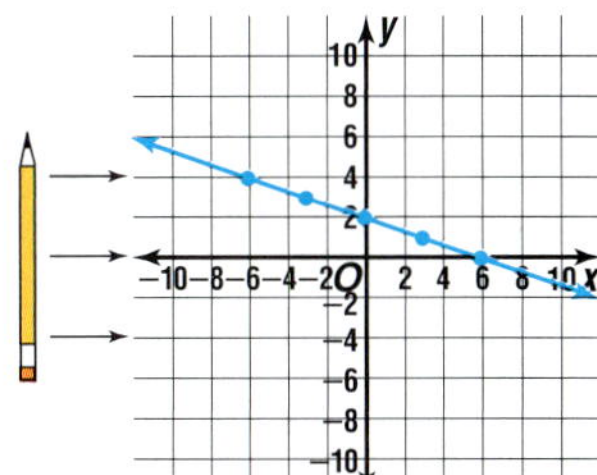

Vertical Line Test for a Function If any vertical line passes through no more than one point of the graph of a relation, then the relation is a function.

Examples

Use the vertical line test to determine whether each relation is a function.

4 $y = |x|$

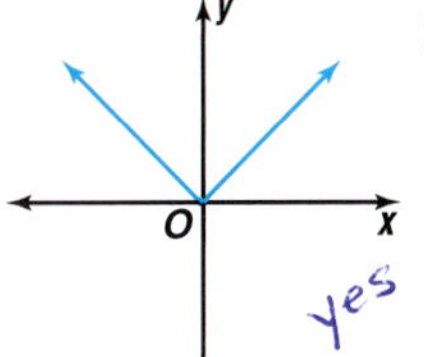

yes

5 $y = \pm\sqrt{x}$

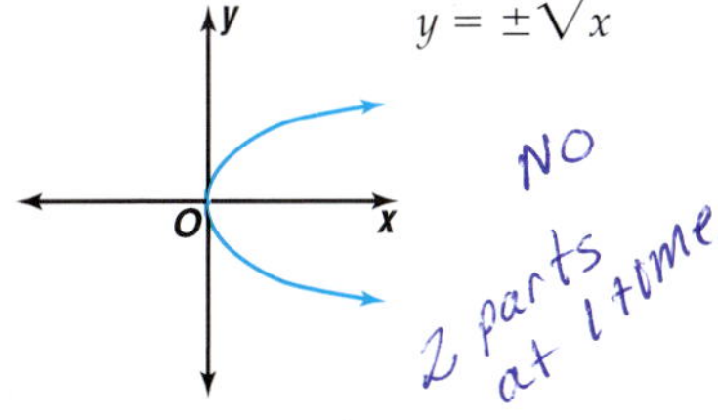

NO 2 parts at 1 time.

Reading Algebra

When $x = 4$ in $y = \pm\sqrt{x}$, y has two values, 2 and −2. So, $y = \pm\sqrt{x}$ does not represent a function.

The relation in Example 4 is a function since any vertical line passes through no more than one point of the graph of the relation. The graph in Example 4 is the graph of the function $y = |x|$. The relation in Example 5 is not a function since a vertical line can pass through more than one point. While the entire graph is not the graph of a function, it can be separated into two parts: the graph of the function $y = \sqrt{x}$ and the graph of the function $y = -\sqrt{x}$.

(continued on the next page)

www.algconcepts.com/extra_examples

In-Class Examples

Examples 2–3

Determine whether each relation is a function. Explain your answer.

2

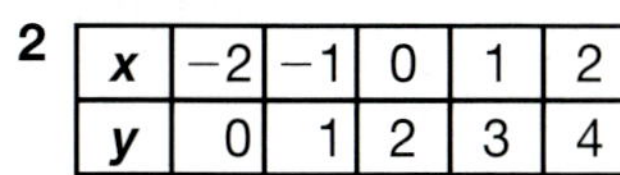

x	−2	−1	0	1	2
y	0	1	2	3	4

The table represents a function since, for each member of the domain, there is only one corresponding member of the range.

3

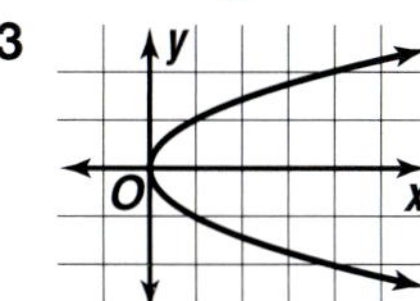

The graph represents a relation that is not a function because there are many members of the domain (the x values) that are paired with two members of the range (the y values); for example, $x = 1$ is paired with both $y = -1$ and $y = 1$.

Teaching Tip As you discuss the vertical line test, point out that any equation with a graph that is a vertical line, such as $x = 2$, is not a function.

In-Class Examples

Examples 4–5

Use the vertical line test to determine whether each relation is a function.

4 **yes**

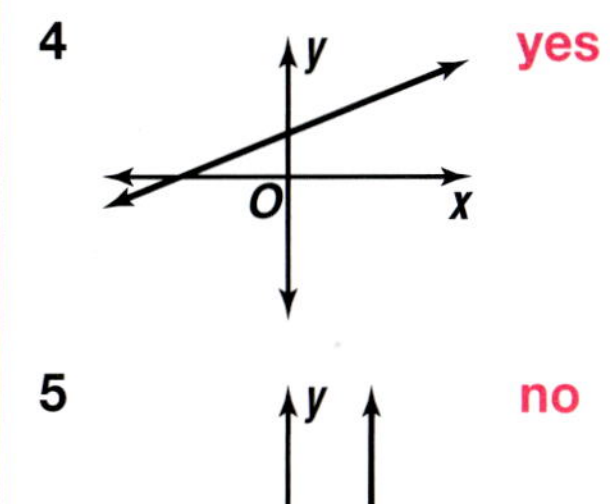

5 **no**

Answers

Your Turn

a. Each element of the domain is paired with exactly one element of the range.

b. The element 1 in the domain is paired with both 4 and −2 in the range.

c. The element 3 in the domain is paired with both 2 and −3 in the range.

Teaching Tip As you discuss *functional notation*, stress that the letter *f* is not used as a variable here. Emphasize the Reading Algebra note midway down page 258. Point out to students that $f(x)$ used in the functional notation takes the place of the variable *y* used in the corresponding equation.

In-Class Examples

Examples 6–8

If f(x) = x − 4, find each value.

6 $f(2)$ -2

7 $f\left(\frac{1}{2}\right)$ $-3\frac{1}{2}$

8 $f(c)$ $c - 4$

Your Turn

Use the vertical line test to determine whether each relation is a function.

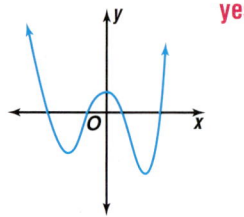

d. no

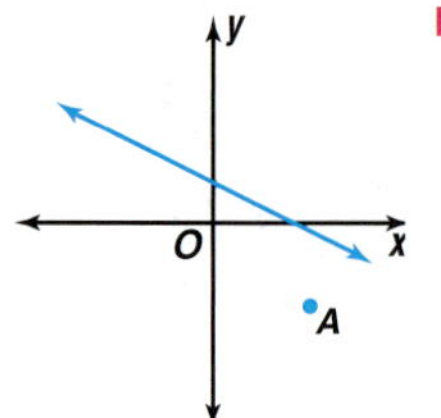

e. yes

Reading Algebra

Read the symbol $f(x)$ as *f at x* or *f of x*. Letters other than *f* are also used for names of functions.

Equations that are functions can be written in a form called **functional notation**. For example, consider the equation $y = 3x + 5$.

equation	functional notation
$y = 3x + 5$	$f(x) = 3x + 5$

In a function, x represents the elements of the domain, and $f(x)$ represents the elements of the range. For example, $f(2)$ is the element in the range that corresponds to the member 2 in the domain. $f(2)$ is called the **functional value** of f for $x = 2$.

You can find $f(2)$ as shown at the right. The ordered pair (2, 11) is a solution of the function of f.

$f(x) = 3x + 5$
$f(2) = 3 \cdot 2 + 5$ *Replace x with 2.*
$f(2) = 6 + 5$ or 11

Examples

If $f(x) = 2x + 3$, find each value.

6 $f(4)$

$f(x) = 2x + 3$
$f(4) = 2(4) + 3$ *Replace x with 4.*
$= 8 + 3$ *Multiply.*
$= 11$ *Add.*

7 $f\left(-\frac{1}{2}\right)$

$f(x) = 2x + 3$
$f\left(-\frac{1}{2}\right) = 2\left(-\frac{1}{2}\right) + 3$ *Replace x with $-\frac{1}{2}$.*
$= -1 + 3$ *Multiply.*
$= 2$ *Add.*

8 $f(6a)$

$f(x) = 2x + 3$
$f(6a) = 2(6a) + 3$ *Replace x with 6a.*
$= 12a + 3$ *Multiply.*

Your Turn

f. If $g(x) = 3x - 7$, find $g(-2)$, $g(1.5)$, and $g(2h)$. $-13, -2.5, 6h - 7$

Refer students to Example 9 on page 259. Have them measure each female family member's height and the length of their femur. Suggest that measuring the length of a person's femur is best done with the person seated, and that the measurement should be done along the outside of the person's thigh from the hip bone to the middle of the knee joint. Point out to students that their measurements must be in centimeters. Ask students to use their measurements to validate the function $h(x) = 61.41 + 2.32x$ given in Example 9 by inserting each female family member's height for $h(x)$ and femur length for x. Have students report their findings.

Functions are often used to solve real-life problems.

Example 9 **Anatomy Link**

Anthropologists use the length of certain bones of a human skeleton to estimate the height of the living person. One of these bones is the femur, which extends from the hip to the knee. To estimate the height in centimeters of a female with a femur of length x, the function $h(x) = 61.41 + 2.32x$ can be used. What was the height of a female whose femur measures 46 centimeters?

$h(x) = 61.41 + 2.32x$ *Original equation*

$h(46) = 61.41 + 2.32(46)$ *Replace x with 46.*

$= 168.13$ *Simplify.*

The woman was about 168 centimeters tall.

Check for Understanding

Communicating Mathematics

1–3. See margin.

1. **Compare and contrast** relations and functions.
2. **Explain** why $f(3) = -1$ if $f(x) = 2x - 7$.
3. **You Decide** Lisa says that the graph of every straight line represents a function. Zina says that there are some graphs of straight lines that do not represent a function. Who is correct? Explain your reasoning.

Vocabulary
function
vertical line test
functional notation
functional value

Guided Practice

Examples 1–3

Determine whether each relation is a function.

4. $\{(2, 3), (3, 5), (4, 6), (2, 5), (1, 4)\}$ no
5. $\{(-1, 6), (1, 4), (-2, 5), (2, 6)\}$ yes
6. yes

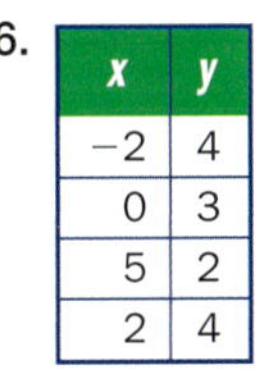
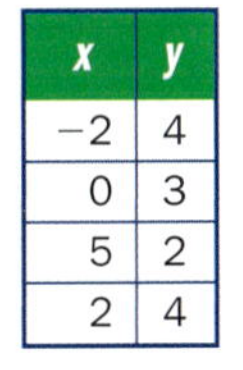

x	y
−2	4
0	3
5	2
2	4

7. no

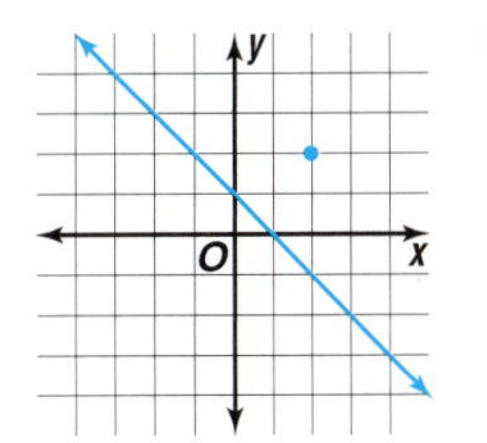

Examples 4 & 5

8. Use the vertical line test to determine whether the relation at the right is a function. yes

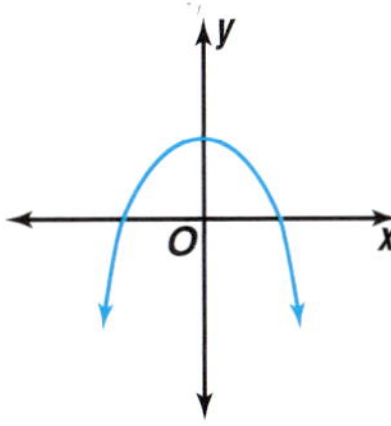

In-Class Example

Example 9

What was the height of a female whose femur measured 20 centimeters?
about 108 cm

3 PRACTICE/APPLY

Error Analysis

Watch for students who substitute −1 for x in Exercise 2. ***Prevent by*** having students point to $f(3) = -1$ in Exercise 2. Read aloud the equality as "*the function when x = 3 has a value of −1.*" Lead students to understand that this is equivalent to saying "*y = −1 when x = 3.*"

Answers

1. Sample answer: A function is a relation. However, in a relation, each member of the domain can be paired with more than one member of the range. In a function, each member of the domain is paired with exactly one member of the range.
2. If $f(x) = 2x - 7$, then $f(3) = -1$ because when you replace x with 3 in the function, the result is −1.

***Study Guide*, p. 260**

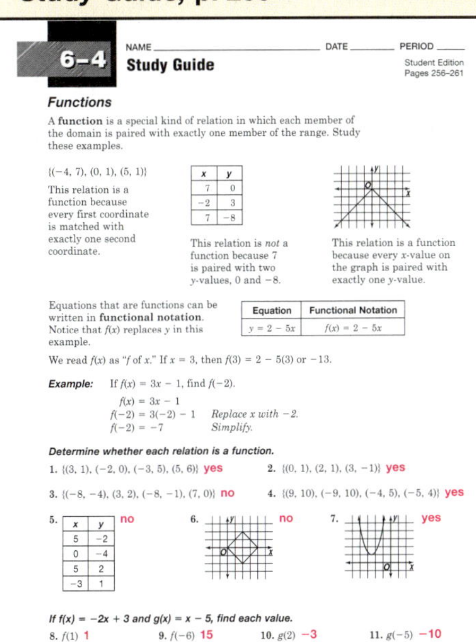

6–4 NAME ______ DATE ______ PERIOD ______

Study Guide

Student Edition Pages 256–261

Functions

A **function** is a special kind of relation in which each member of the domain is paired with exactly one member of the range. Study these examples.

$\{(-4, 7), (0, 1), (5, 1)\}$

This relation is a function because every first coordinate is matched with exactly one second coordinate.

x	y
7	0
−2	3
7	−8

This relation is *not* a function because 7 is paired with two y-values, 0 and −8.

This relation is a function because every x-value on the graph is paired with exactly one y-value.

Equations that are functions can be written in **functional notation**. Notice that $f(x)$ replaces y in this example.

Equation	Functional Notation
$y = 2 - 5x$	$f(x) = 2 - 5x$

We read $f(x)$ as "f of x." If $x = 3$, then $f(3) = 2 - 5(3)$ or -13.

Example: If $f(x) = 3x - 1$, find $f(-2)$.

$f(x) = 3x - 1$

$f(-2) = 3(-2) - 1$ *Replace x with −2.*

$f(-2) = -7$ *Simplify.*

Determine whether each relation is a function.

1. $\{(3, 1), (-2, 0), (-3, 5), (5, 6)\}$ yes
2. $\{(0, 1), (2, 1), (3, -1)\}$ yes
3. $\{(-8, -4), (3, 2), (-8, -1), (7, 0)\}$ no
4. $\{(9, 10), (-9, 10), (-4, 5), (-5, 4)\}$ yes
5. no

x	y
5	−2
0	−4
5	2
−3	1

6. no
7. yes

If $f(x) = -2x + 3$ and $g(x) = x - 5$, find each value.

8. $f(1)$ 1
9. $f(-6)$ 15
10. $g(2)$ −3
11. $g(-5)$ −10

Reteaching Activity

Kinesthetic Learners Have each student measure the length of one of their femurs (in centimeters) by measuring from hip bone to mid-knee while seated. Then have them calculate their heights using the function given in Example 9 on page 259. Have them compare the calculated height to their actual height.

Answer

3. Zina is correct. There are some graphs of straight lines that do not represent a function. The graph of $x = 2$ is a straight line. However, it is not a function since there are many y values for a single x value.

Assignment Guide

Basic: 15–47 odd, 49–58
Average: 14–46 even, 47–58

11. $4b - 3$
12. $-8a - 3$

Examples 6–8 **If $f(x) = 4x - 3$, find each value.**

9. $f(-3)$ -15 10. $f\left(\frac{1}{4}\right)$ -2 11. $f(b)$ 12. $f(-2a)$

Example 9

13. **Safety** The time in seconds that a traffic light remains yellow is given by the function $t(s) = 0.05s + 1$, where s represents the speed limit. How long will a light remain yellow if the speed limit is 45 miles per hour? 3.25 s

Exercises

Practice

Homework Help	
For Exercises	See Examples
14–25	1–3
26–28	4, 5
29–46	6–8
47, 48	6–9

Extra Practice
See page 704.

14. yes
15. yes
16. no
17. no
18. no
19. yes

Determine whether each relation is a function.

A

14. $\{(3, 5), (4, 3), (-2, 1), (-1, 4)\}$
15. $\{(-2, 3), (-3, 2), (5, 2), (7, 2)\}$
16. $\{(2, -3), (3, -2), (5, -2), (2, -5)\}$
17. $\{(0, 2), (-1, 3), (2, 3), (-1, 2)\}$
18. $\{(-3, 5), (-2, 4), (-3, 6), (-2, 3)\}$
19. $\{(1, 2), (2, 1), (-1, 2), (-2, 1)\}$

20. yes

x	y
−2	5
3	−2
4	5
5	5
6	1

21. no

x	y
0	1
0	0
−4	0
2	6
0	−4

22. no

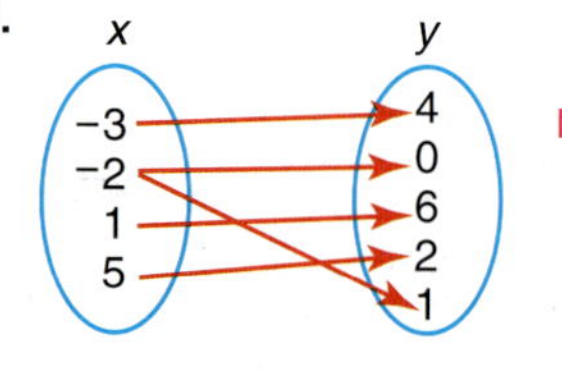

23. no

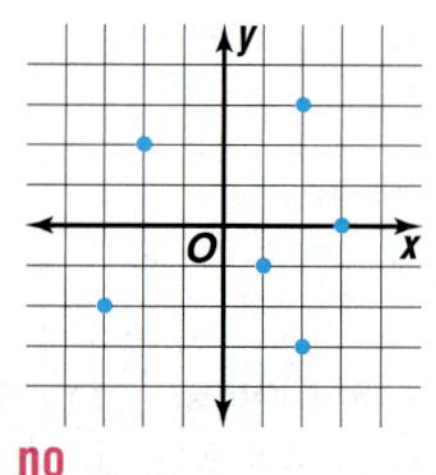

24. no
25. yes

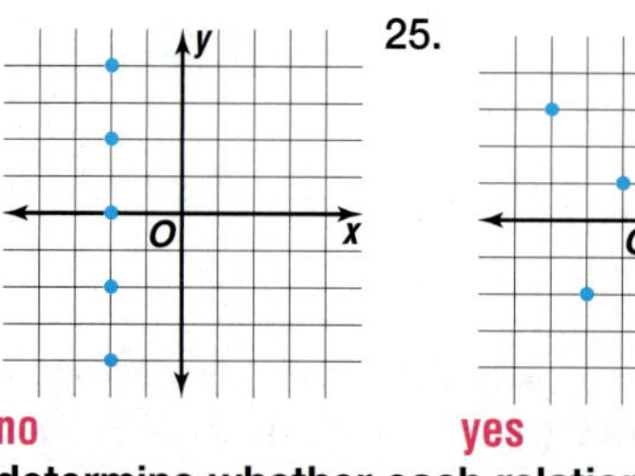

Use the vertical line test to determine whether each relation is a function.

26. no

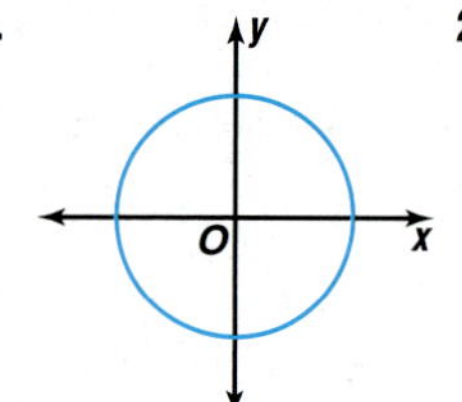

27. yes

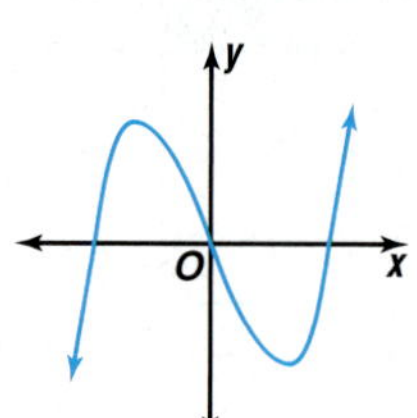

28. yes

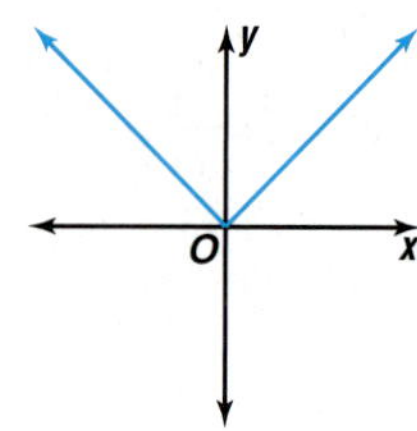

36. -10.2
41. $-3h + 4$
42. $-4d - 5$
43. $6a + 4$
44. $9c - 5$

If $f(x) = 2x - 5$ and $g(x) = -3x + 4$, find each value.

29. $f(2)$ -1 30. $g(4)$ -8 31. $g(-3)$ 13 32. $f(-7)$ -19

33. $g(0.2)$ 3.4 34. $f(0.6)$ -3.8 35. $g(1.9)$ -1.7 36. $f(-2.6)$

B

37. $g\left(\frac{1}{3}\right)$ 3 38. $f\left(-\frac{1}{2}\right)$ -6 39. $f\left(\frac{3}{4}\right)$ $-3\frac{1}{2}$ 40. $g\left(-\frac{5}{6}\right)$ $6\frac{1}{2}$

41. $g(h)$ 42. $f(-2d)$ 43. $g(-2a)$ 44. $f(4.5c)$

***Skills Practice*, p. 261, and *Practice*, p. 262 (shown)**

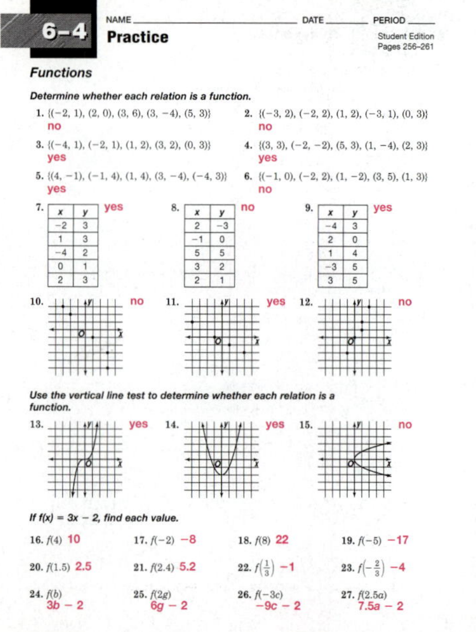

6–4 NAME ______________ DATE ________ PERIOD _____

Practice

Student Edition Pages 256–261

Functions

Determine whether each relation is a function.

1. $\{(-2, 1), (2, 0), (3, 6), (3, -4), (5, 3)\}$ no
2. $\{(-3, 2), (-2, 2), (1, 2), (-3, 1), (0, 3)\}$ no
3. $\{(-4, 1), (-2, 1), (1, 2), (3, 2), (0, 3)\}$ yes
4. $\{(3, 3), (-2, -2), (5, 3), (1, -4), (2, 3)\}$ yes
5. $\{(4, -1), (-1, 4), (1, 4), (3, -4), (-4, 3)\}$ yes
6. $\{(-1, 0), (-2, 2), (1, -2), (3, 5), (1, 3)\}$ no

7. yes

x	y
−2	3
1	3
−4	2
0	1
2	3

8. no

x	y
2	−3
−1	0
5	5
3	2
2	1

9. yes

x	y
−4	3
2	0
1	4
−3	5
3	5

10. no 11. yes 12. no

Use the vertical line test to determine whether each relation is a function.

13. yes 14. yes 15. no

If f(x) = 3x − 2, find each value.

16. $f(4)$ 10 17. $f(-2)$ -8 18. $f(8)$ 22 19. $f(-5)$ -17

20. $f(1.5)$ 2.5 21. $f(2.4)$ 5.2 22. $f\left(\frac{1}{3}\right)$ -1 23. $f\left(-\frac{2}{3}\right)$ -4

24. $f(b)$ $3b - 2$ 25. $f(2g)$ $6g - 2$ 26. $f(-3c)$ $-9c - 2$ 27. $f(2.5a)$ $7.5a - 2$

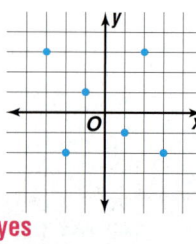

45. Find $f(-4)$ if $f(x) = 5x + 8$. **−12**

46. What is the value of $m(3) + n(-5)$ if $m(x) = 2x$ and $n(x) = 3x - 1$? **−10**

Applications and Problem Solving C

47. **Health** A person's normal systolic blood pressure p depends on the person's age a. To determine the normal systolic blood pressure for an individual, you can use the equation $p = 0.5a + 110$.
   a. Write this relation in functional notation. **$p(a) = 0.5a + 110$**
   b. Find the normal systolic blood pressure for a person who is 36 years old. **128**
   c. Graph the relation. Describe the graph. **See margin.**

48. **Travel** The cost of a one-day car rental from City-Wide Rentals is given by the function $C(x) = 0.18x + 35$, where x is the number of miles that the car is driven. Suppose Ms. Burton drove a distance of 95 miles and back in one day. What is the cost of the car rental? **\$69.20**

49. **Critical Thinking** Consider the functions $h(x) = 2x - 5$ and $g(x) = 3x + 2$.
   a. For what value of x will $h(x)$ and $g(x)$ be equal? **−7**
   b. Graph $h(x)$ and $g(x)$. Describe what you observe. **See margin.**

Mixed Review

50. Graph $y = -2x - 6$. *(Lesson 6–3)* **See margin.**

51. **Geometry** The equation $A = \frac{1}{2}bh$ can be used to find the area of a triangle. Suppose a triangle has an area of 28 square meters. Find the possible heights of the triangle given the domain values {2, 4, 7}. *(Lesson 6–2)* **{(2, 28), (4, 14), (7, 8)}**

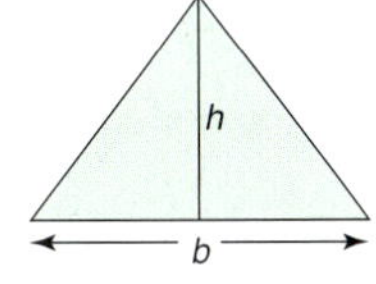

Use the percent proportion to answer each question. *(Lesson 5–3)*

52. What is 145% of 86? **124.7**

53. What percent of 60 is 18? **30%**

54. **Life Science** A scale model of a sequoia tree is 4 feet tall. If the actual sequoia tree is 311 feet tall, what is the scale of the model? *(Lesson 5–2)* **1 ft = $77\frac{3}{4}$ ft**

55. Evaluate $\frac{x}{5}$ if $x = \frac{3}{4}$. *(Lesson 4–3)* **$\frac{3}{20}$**

56. **Animals** The fastest swimming bird is the gentoo penguin of Antarctica. It moves through the water at about 17 miles per hour. Traveling at this rate, how far can the gentoo penguin swim in 45 minutes? *(Lesson 4–1)* **$12\frac{3}{4}$ mi**

Gentoo Penguin

Standardized Test Practice

57. **Grid In** What is the value of w if $w - 13.8 = -12.6$? *(Lesson 3–6)* **1.2**

58. **Multiple Choice** Find the only solution of $3x + (-5) = 10$ if the replacement set is {−5, 3, 5, 8}. *(Lesson 3–4)* **C**
   (A) −5 (B) 3 (C) 5 (D) 8

www.algconcepts.com/self_check_quiz

4 ASSESS

Open-Ended Assessment

Modeling Have students demonstrate how to use a pencil or a straightedge to tell whether the graph of a relation is a function.

Answers

47c.

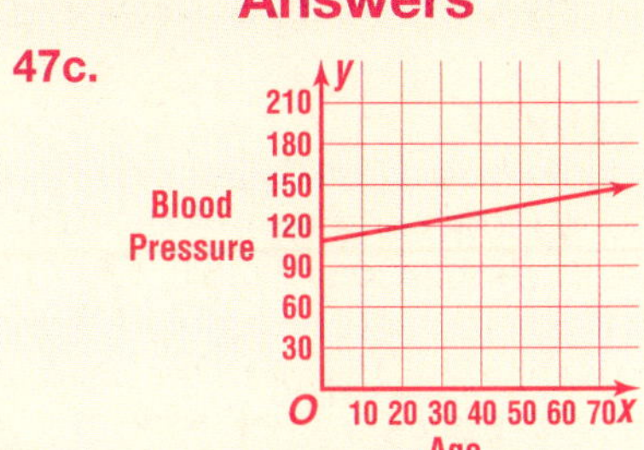

49b.

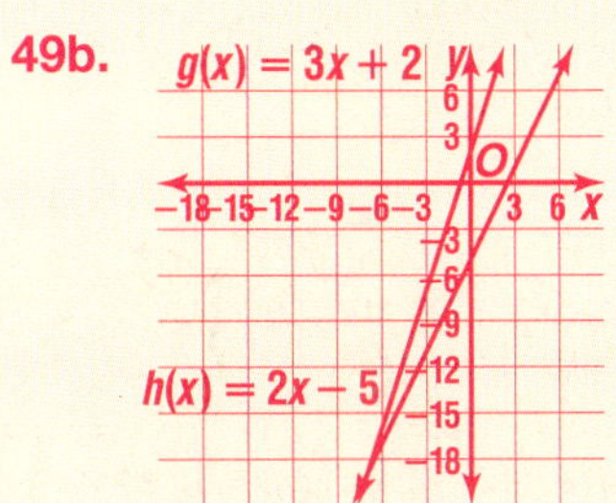

The lines intersect at $x = -7$.

50.

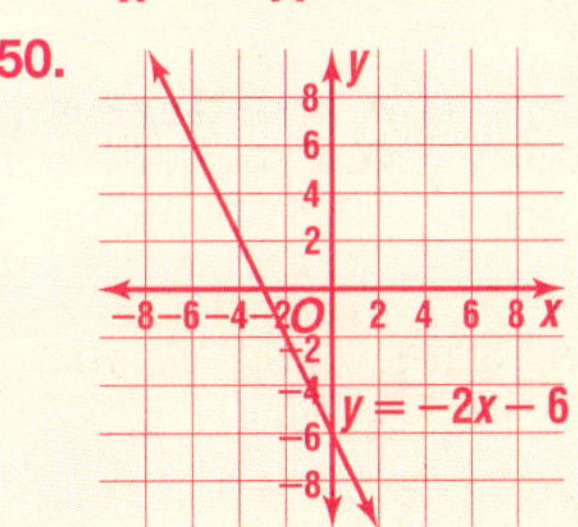

***Enrichment*, p. 264**

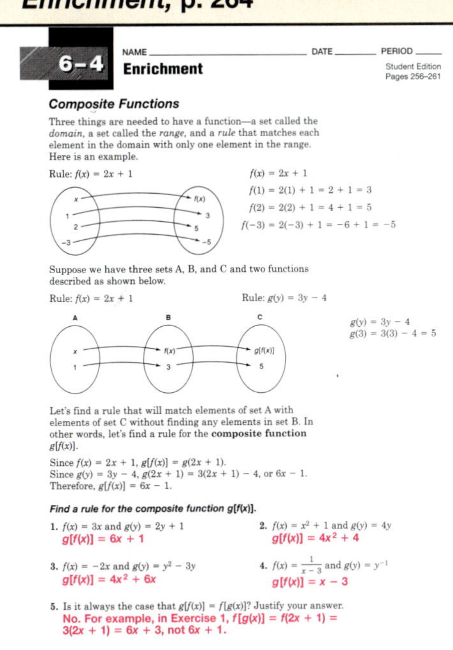

NAME ______ DATE ______ PERIOD ______

6–4 Enrichment

Student Edition Pages 256–261

Composite Functions

Three things are needed to have a function—a set called the *domain*, a set called the *range*, and a *rule* that matches each element in the domain with only one element in the range. Here is an example.

Rule: $f(x) = 2x + 1$

$f(x) = 2x + 1$
$f(1) = 2(1) + 1 = 2 + 1 = 3$
$f(2) = 2(2) + 1 = 4 + 1 = 5$
$f(-3) = 2(-3) + 1 = -6 + 1 = -5$

Suppose we have three sets A, B, and C and two functions described as shown below.

Rule: $f(x) = 2x + 1$ Rule: $g(y) = 3y - 4$

$g(y) = 3y - 4$
$g(3) = 3(3) - 4 = 5$

Let's find a rule that will match elements of set A with elements of set C without finding any elements in set B. In other words, let's find a rule for the **composite function** $g[f(x)]$.

Since $f(x) = 2x + 1$, $g[f(x)] = g(2x + 1)$.
Since $g(y) = 3y - 4$, $g(2x + 1) = 3(2x + 1) - 4$, or $6x - 1$.
Therefore, $g[f(x)] = 6x - 1$.

Find a rule for the composite function g[f(x)].

1. $f(x) = 3x$ and $g(y) = 2y + 1$
   $g[f(x)] = 6x + 1$
2. $f(x) = x^2 + 1$ and $g(y) = 4y$
   $g[f(x)] = 4x^2 + 4$
3. $f(x) = -2x$ and $g(y) = y^2 - 3y$
   $g[f(x)] = 4x^2 + 6x$
4. $f(x) = \frac{1}{x - 3}$ and $g(y) = y^{-1}$
   $g[f(x)] = x - 3$
5. Is it always the case that $g[f(x)] = f[g(x)]$? Justify your answer.
   No. For example, in Exercise 1, $f[g(x)] = f(2x + 1) = 3(2x + 1) = 6x + 3$, not $6x + 1$.

Extra Credit

Graph the functions $g(x) = 2x + 4$ and $h(x) = 2x + 5$ on the same coordinate plane. Are there any values of x for which $g(x)$ and $h(x)$ are the same? Explain. **Check students' graphs. No, there are no such values of x because for every value of x, $h(x) = g(x) + 1$.**

PREPARE

This optional investigation is designed to be completed by a pair of students over 1–2 days.

Objective

Students investigate graphing shoe size data and temperature data. They then study the graphs to help them decide whether there is a relationship between the data pairs. Students describe what relationships they find and present their findings by creating a brochure or poster.

Mathematical Overview

This investigation utilizes the following concepts:

- collecting data,
- writing and graphing ordered pairs, and
- looking for relationships in a set of graphed data.

Suggested Time Management	
Investigation	30–45 min
Extension: Gathering Data	30–45 min
Extension: Summarizing Data	20–30 min

Motivating the Lesson

Suggest to students that linear relationships can be found in many real-world situations, such as between the number of railroad cars on a train and the length of time it takes to clear a railroad crossing. Ask students to think of other real-world linear relationships.

Chapter 6 Investigation

How Many CENTIMETERS in a Foot?

Materials

 grid paper

 meterstick

 red and blue pens or pencils

 almanac

Functions

Many real-world situations represent functions. Is there a relationship between foot length and shoe size? Does a relationship exist between the month of the year and the average temperature? Let's find out!

Investigate

1. Ask eight to ten females and eight to ten males to participate in your data research. **a–d. See students' work.**
   a. Make two tables with the headings shown. Use one table to record data for the females and the other to record data for the males.

Name	Foot Length (cm)	Shoe Size	(Foot Length, Shoe Size)

   b. Measure the foot length in centimeters as each participant stands without shoes on. Find out each person's shoe size. Record the results in your tables.
   c. Write an ordered pair (foot length, shoe size) for each person. Then graph the ordered pairs on a coordinate plane. Use a red pencil to graph the ordered pairs that represent the female data. Use a blue pencil to graph the ordered pairs that represent the male data. Connect each set of ordered pairs with the corresponding color.
   d. Is either relation a function? Why or why not? Describe any relationship that exists between foot length and shoe size.
   e. Save your tables to use later in the investigation.

Cooperative Learning

This investigation offers an excellent opportunity for using cooperative groups. For more information on cooperative learning strategies and group management, see *Cooperative Learning in the Mathematics Classroom,* one of the titles in the Glencoe Mathematics Professional Series.

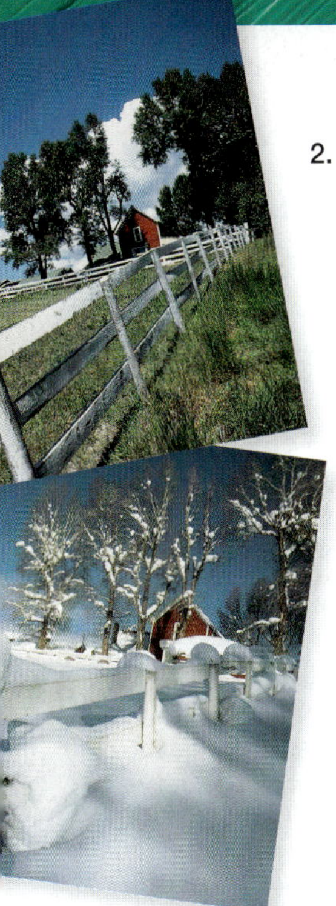

2. Choose a city in the United States. Research temperature data for that city. Find the average high and low temperatures for each month. It may be helpful to use an almanac or the Internet.
   a. Make a table like the one shown. Record the temperature data for the selected city. **a–d. See students' work.**

Month	Average High Temperature (°F)	Average Low Temperature (°F)
January		
February		
⋮		

   b. Write two sets of ordered pairs. The first set should contain the ordered pairs (month, high temperature). Assign January the number 1, February the number 2, and so on. The second set should contain the ordered pairs (month, low temperature).
   c. Graph both sets of data on one coordinate plane. Connect the ordered pairs in the first set using a red pencil. Use a blue pencil to connect the ordered pairs in the second set.
   d. Is either graph a function? Explain. Describe any relationship that exists between the month and the average temperature. Are there any similarities between these two graphs?

Extending the Investigation

In this extension, you will continue to investigate functions.

- Select another city in the United States. Make a table of average high and low temperatures for each month. Then graph the data and determine if the graph is a function. Describe any relationships that exist between the graphs.
- Select a city in Hawaii. Make a table of average high and low temperatures for each month. Then graph the data. How does the graph of the temperature data for Hawaii compare to the graph in Exercise 2?

Presenting Your Investigation

Here are some ideas to help you present your conclusions to the class.

- Make a brochure showing the results of your research. Be sure to include the tables and graphs. List any relationships you have found.
- Discuss any similarities or differences in the graphs.
- Look in an almanac or on the Internet for real-world data whose graph would be a function. Make a poster presenting your data, tables, and graphs.

Investigation For more information on temperatures, visit: www.algconcepts.com

MANAGE

Teaching Tip Students may find that some people are unwilling to participate in the research because they are self-conscious about their shoe size. Encourage students to consider using initials rather than full names in their data tables.

Teaching Tip Students should try to include the largest possible range of ages in their research subjects. If they choose people who are about the same age and have about the same shoe size, they may not be able to draw any conclusions.

Working in Pairs Suggest that one student dictate the ordered pairs while their partner graphs them on a coordinate plane. Remind students to use different colors for the male and female data.

Working as a Class To save time, you can assign half the class to study the shoe data and half the class to study the temperature data.

ASSESS

Students should write ordered pairs and graph them correctly. They should find a relationship in both sets of data and describe the relationship using the appropriate mathematical language.

Students should add their brochure or poster to their portfolios at this time.

From the Classroom of ...

Cathie Oppio
Truckee Meadows Community College High School
Reno, Nevada

I have my students extend part 1 of the Investigation. As a group, we record shoe size data and head circumference data. Then we use a graph to determine whether a relationship exists between these two sets of data. You might also wish to have students determine whether a relationship exists between shoe size and age.

Lesson 6–5

1 FOCUS

5-Minute Check
Lesson 6–4

Use the vertical line test to determine whether each relation is a function.

1. **yes**

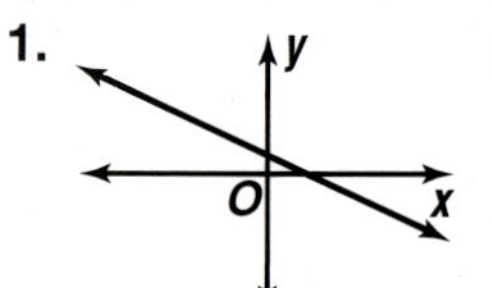

2. **no**

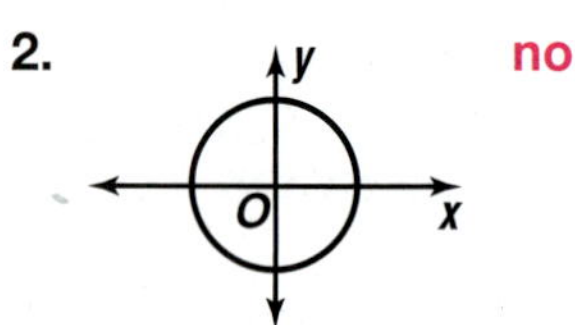

If f(x) = 4x − 1, find each value.

3. $f(4)$ **15**
4. $f\left(\frac{1}{2}\right)$ **1**
5. $f(10k)$ **$40k - 1$**

Motivating the Lesson

Hands-On Activity Put 400 mL of water in a 500 mL beaker. Then add an object such as a spoon or a glass stirring rod to the beaker. Ask students to describe what happens. **The water level rises.** Add additional spoons or rods to the beaker and observe. Challenge students to identify the relationship between the number of objects added and the water level. **The rise of the water level is directly proportional to the number of objects added to the beaker.**

6–5 Direct Variation

What You'll Learn
You'll learn to solve problems involving direct variations.

Why It's Important
Electronics
Electricians can use direct variation to determine the weight of copper wire.
See Exercise 32.

Study the table and graph of the linear function $y = 2x$ shown below.

x	y
−2	−4
−1	−2
0	0
1	2
2	4

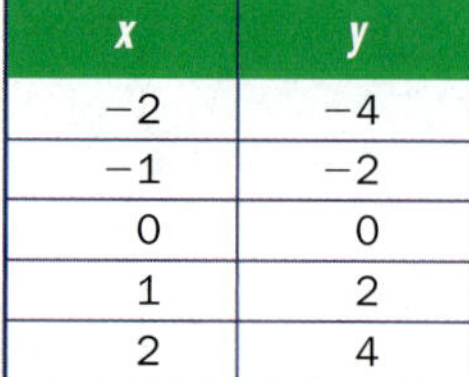

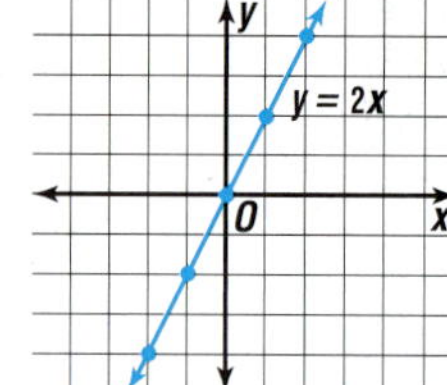

The value of y depends directly on the corresponding value of x. That is, as the value of x increases, the value of y increases. Note that when $x = 0$, $y = 0$. Therefore, the line passes through the origin. The equation $y = 2x$ is called a **direct variation.** We say that *y varies directly as x.*

A direct variation describes a linear relationship because $y = 0.4x$ is a linear equation.

Direct Variation	A direct variation is a linear function that can be written in the form $y = kx$, where $k \neq 0$.

Since the value of y depends on the value of x, y is called the **dependent variable**. x is called the **independent variable**.

As stated, a direct variation can be written in the form $y = x$. Notice that when $x = 0$, $y = 0$. The graph of a direct variation passes through the origin.

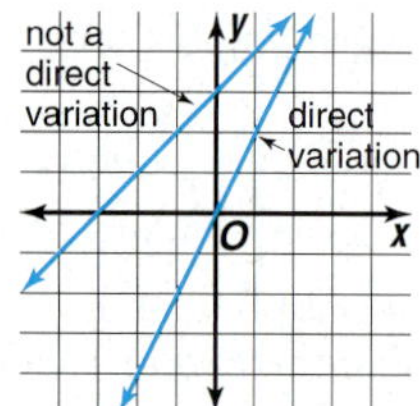

In the equation $y = kx$, k is called the **constant of variation**.

TECHNOLOGY

An alternative technology option using a graphing calculator is available for teaching this lesson.

Resource Manager

Reproducible Masters
Chapter 6 Resource Masters
- *Study Guide,* p. 265
- *Skills Practice,* p. 266
- *Practice,* p. 267
- *Reading to Learn Mathematics,* p. 268
- *Enrichment,* p. 269

Graphing Calculator, pp. 18–19

Transparencies
5-Minute Check, 6–5

Technology/Multimedia
Interactive Chalkboard CD-ROM

Examples

Determine whether the equation is a direct variation.

 $y = 4x$

Graph the equation.

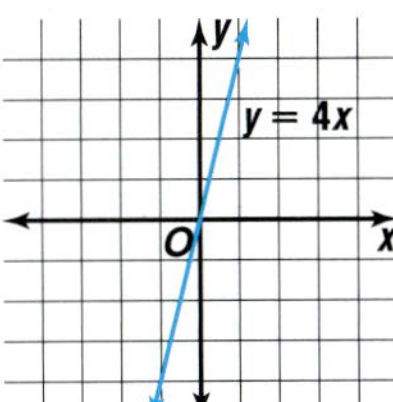

The graph passes through the origin. So, the equation is a direct variation. *The constant of variation is 4.*

 $y = x - 1$

Graph the equation.

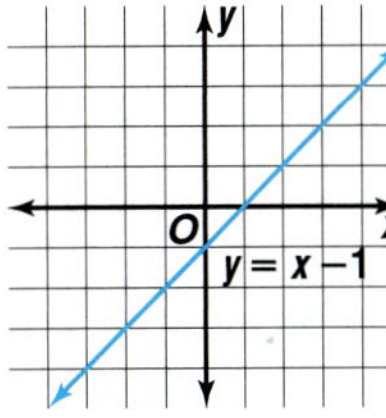

The graph does not pass through the origin. So, the equation is not a direct variation.

Your Turn **Determine whether each equation is a direct variation. Verify your answer with a graph.**

a. $y = 2x + 1$ **no**

b. $y = -3x$ **yes**

a–b. See margin for graphs.

Many real-world problems involve direct variation.

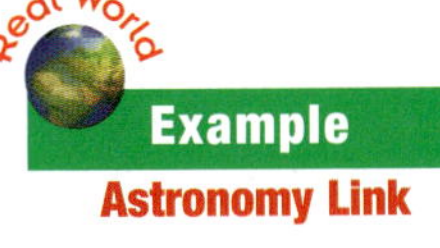

Example 3 — Astronomy Link

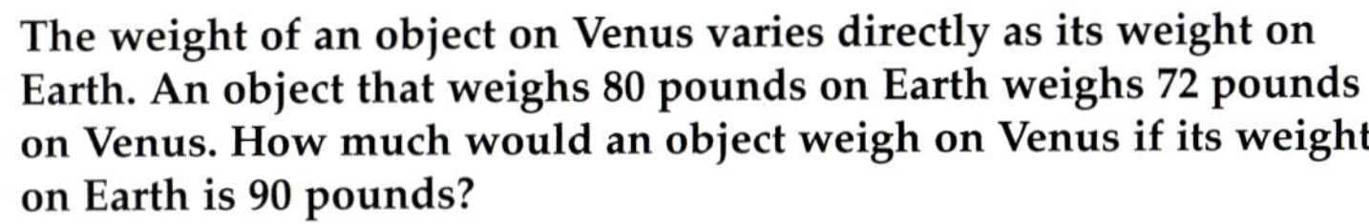

The weight of an object on Venus varies directly as its weight on Earth. An object that weighs 80 pounds on Earth weighs 72 pounds on Venus. How much would an object weigh on Venus if its weight on Earth is 90 pounds?

First, find the constant of variation. Let x = weight on Earth and let y = weight on Venus, since the weight on Venus depends on the weight on Earth.

$y = kx$ *Definition of direct variation*

$72 = k \cdot 80$ *Replace y with 72 and x with 80.*

$\frac{72}{80} = \frac{k \cdot 80}{80}$ *Divide each side by 80.*

$0.9 = k$ *Simplify.*

Next, use the constant of variation to find y when $x = 90$.

$y = kx$ *Definition of direct variation*

$y = 0.9(90)$ *Replace k with 0.9 and x with 90.*

$y = 81$ *Simplify.*

So, the object would weigh 81 pounds on Venus.

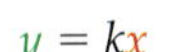

(continued on the next page)

www.algconcepts.com/extra_examples

2 TEACH

Teaching Tip As you introduce the concepts of *direct variation, dependent variable,* and *independent variable,* ask students to think of pairs of items that are dependent on each other and pairs of items that are not dependent. List the students' responses on the board or overhead.

In-Class Examples

Examples 1–2

Determine whether the equation is a direct variation.

1 $y = \frac{1}{2}x$ **yes**

2 $y = x + 2$ **no**

Teaching Tip While discussing Example 3, make sure students understand that the weight of an object on Venus must be represented by the dependent variable, *y*, because it cannot be measured directly.

In-Class Example

Example 3

If an object that weighs 110 pounds on Earth weighs 18.7 pounds on the moon, how much would an object weigh on the moon if its weight on Earth is 275 pounds? **46.75 lb**

Answers

Your Turn

a.

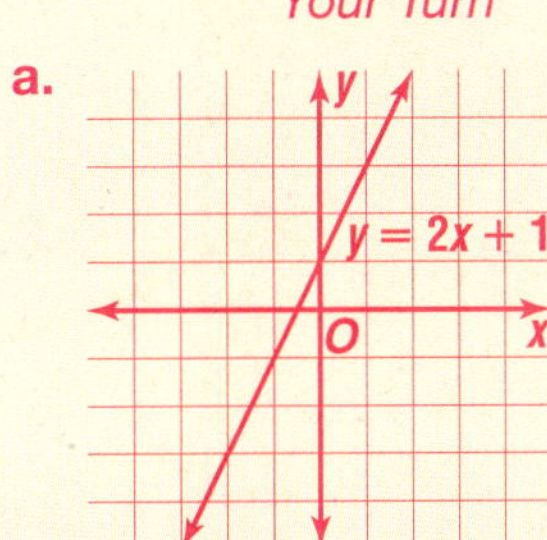

b.

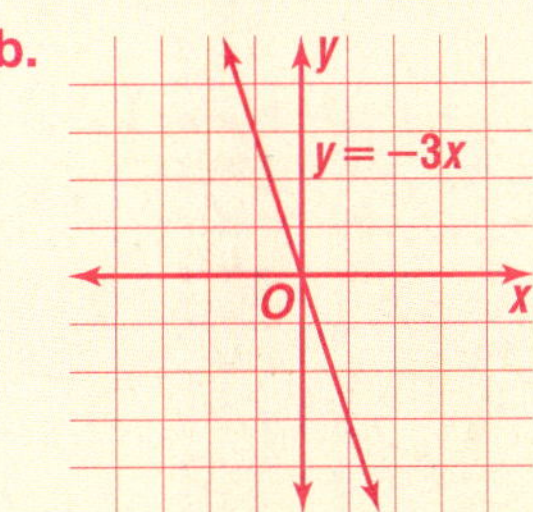

Teaching Tip In Example 4, once you have found the value $k = 30$, lead students to understand that the value of 30 means the car travels 30 miles for each gallon of gas that it uses.

In-Class Example

Example 4

How many gallons of gasoline would be needed for a 550-mile trip if a 66-mile trip used 3 gallons of gasoline? **25 gal**

Your Turn

c. An object that weighs 110 pounds on Earth weighs 18.7 pounds on the moon. How much would an object weigh on the moon if its weight on Earth is 150 pounds? **25.5 lb**

Direct variations can be used to solve **rate problems**.

Example 4 Travel Link

The length of a trip varies directly as the amount of gasoline used. Suppose 4 gallons of gasoline were used for a 120-mile trip. At that rate, how many gallons of gasoline are needed for a 345-mile trip?

Let ℓ represent the length of the trip and let g represent the amount of gasoline used for the trip. The statement *the length varies directly as the amount of gasoline* translates into the equation $\ell = kg$ in the same way as *y varies directly as x* translates into $y = kx$. Find the value of k.

$\ell = kg$ *Direct variation*

$120 = k \cdot 4$ *Replace ℓ with 120 and g with 4.*

$\frac{120}{4} = \frac{4k}{4}$ *Divide each side by 4.*

$30 = k$ *Simplify.*

Next, find the amount of gasoline needed for a 345-mile trip.

$\ell = kg$ *Direct variation*

$345 = 30 \cdot g$ *Replace ℓ with 345 and k with 30.*

$\frac{345}{30} = \frac{30g}{30}$ *Divide each side by 30.*

$11.5 = g$ *Simplify.*

A 345-mile trip would use 11.5 gallons of gasoline.

Your Turn

d. How many gallons of gasoline would be needed for a 363-mile trip if a 72.6-mile trip used 3 gallons of gasoline? **15**

Direct variations are also related to proportions. Using the table on page 264, many proportions can be formed. Two examples are shown.

$$\begin{matrix} x_1 \rightarrow \\ y_1 \rightarrow \end{matrix} \; \frac{5}{2} = \frac{10}{4} \; \begin{matrix} \leftarrow x_2 \\ \leftarrow y_2 \end{matrix}$$

$$\begin{matrix} x_1 \rightarrow \\ x_2 \rightarrow \end{matrix} \; \frac{5}{10} = \frac{2}{4} \; \begin{matrix} \leftarrow y_1 \\ \leftarrow y_2 \end{matrix}$$

Three general forms for proportions like these are $\frac{y_1}{y_2} = \frac{x_1}{x_2}$, $\frac{x_1}{y_1} = \frac{x_2}{y_2}$ and $\frac{y_1}{x_1} = \frac{y_2}{x_2}$. You can use any of these forms to solve direct variation problems.

Example 5 Suppose y varies directly as x and $y = 36$ when $x = 9$. Find x when $y = 48$.

Use $\frac{y_1}{x_1} = \frac{y_2}{x_2}$ to solve the problem.

$\frac{36}{9} = \frac{48}{x_2}$ *Let $y_1 = 36$, $y_2 = 48$, and $x_1 = 9$.*

$36x_2 = 9(48)$ *Find the cross products.*

$36x_2 = 432$ *Multiply.*

$\frac{36x_2}{36} = \frac{432}{36}$ *Divide each side by 36.*

$x_2 = 12$ *Simplify.*

So, $x = 12$ when $y = 48$.

Your Turn

e. Suppose y varies directly as x and $y = 27$ when $x = 6$. Find x when $y = 45$. **10**

You can also use direct variation to convert measurements.

Example 6 **Measurement Link**

If there are 12 inches in 1 foot, how many inches are in 4.5 feet?

Use $\frac{y_1}{y_2} = \frac{x_1}{x_2}$ to solve the problem.

$\frac{12}{1} = \frac{x_1}{4.5}$ *Let $y_1 = 12$, $y_2 = 1$, and $x_2 = 4.5$.*

$12(4.5) = x_1$ *Find the cross products.*

$54 = x_1$ *Simplify.*

There are 54 inches in 4.5 feet.

Your Turn **f. 18.4 pt**

f. How many pints are in 2.3 gallons if there are 8 pints in 1 gallon?

Check for Understanding

Communicating Mathematics

1. **Explain** how you find the constant of variation in a direct variation.
2. **Give an example** of an equation that is a direct variation. **See margin.**
3. *Writing Math* Refer to Example 2. Write a few sentences describing the dependent variable and independent variable. **See margin.**

1. Divide each side of $y = kx$ by x, where $x \neq 0$.

Vocabulary

- direct variation
- independent variable
- dependent variable
- constant of variation
- rate problems

In-Class Examples

Example 5

Suppose y varies directly as x and $y = 27$ when $x = 18$. Find x when $y = 15$. **10**

Example 6

If there are 16 ounces in 1 pound, how many ounces are in 12 pounds? **192 oz**

Answers

2. Sample answer: $y = -4x$
3. Sample answer: The length of the trip depends upon the amount of gasoline used. So, the length of the trip represents the dependent variable and the amount of gasoline used represents the independent variable.

Reteaching Activity

Logical Learners Ask a volunteer to explain to the class why, in a direct variation, if one variable increases, the other variable also increases. Students can use the graph of a direct variation to help explain this fact.

***Study Guide*, p. 265**

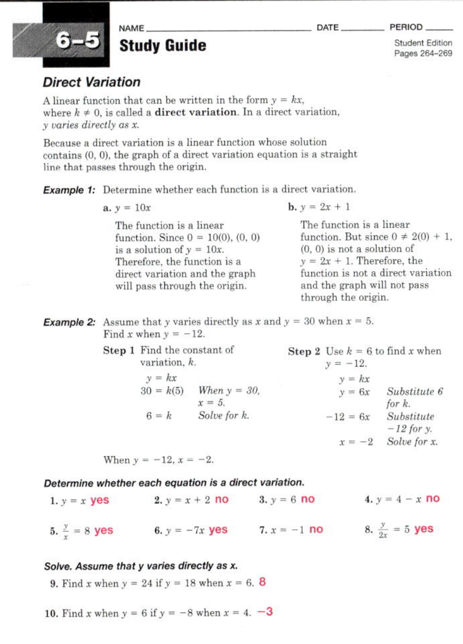
6–5 NAME ______ DATE ______ PERIOD ______

Study Guide Student Edition Pages 264–269

Direct Variation

A linear function that can be written in the form $y = kx$, where $k \neq 0$, is called a **direct variation**. In a direct variation, *y varies directly as x.*

Because a direct variation is a linear function whose solution contains (0, 0), the graph of a direct variation equation is a straight line that passes through the origin.

Example 1: Determine whether each function is a direct variation.

a. $y = 10x$

The function is a linear function. Since $0 = 10(0)$, (0, 0) is a solution of $y = 10x$. Therefore, the function is a direct variation and the graph will pass through the origin.

b. $y = 2x + 1$

The function is a linear function. But since $0 \neq 2(0) + 1$, (0, 0) is not a solution of $y = 2x + 1$. Therefore, the function is not a direct variation and the graph will not pass through the origin.

Example 2: Assume that y varies directly as x and $y = 30$ when $x = 5$. Find x when $y = -12$.

Step 1 Find the constant of variation, k.

$y = kx$

$30 = k(5)$ *When $y = 30$, $x = 5$.*

$6 = k$ *Solve for k.*

Step 2 Use $k = 6$ to find x when $y = -12$.

$y = kx$

$y = 6x$ *Substitute 6 for k.*

$-12 = 6x$ *Substitute -12 for y.*

$x = -2$ *Solve for x.*

When $y = -12$, $x = -2$.

Determine whether each equation is a direct variation.

1. $y = x$ **yes**
2. $y = x + 2$ **no**
3. $y = 6$ **no**
4. $y = 4 - x$ **no**
5. $\frac{y}{x} = 8$ **yes**
6. $y = -7x$ **yes**
7. $x = -1$ **no**
8. $\frac{y}{2x} = 5$ **yes**

Solve. Assume that y varies directly as x.

9. Find x when $y = 24$ if $y = 18$ when $x = 6$. **8**
10. Find x when $y = 6$ if $y = -8$ when $x = 4$. **−3**

3 PRACTICE/APPLY

Error Analysis

Watch for students who set up the proportions incorrectly in Exercises 11–12.

Prevent by demonstrating to students how to write corresponding x and y values on the same side of the proportion. For Exercise 11, suggest students begin by rewriting the phrase "$y = 28$ when $x = 7$" as $\frac{y = 28}{x = 7}$ and the phrase "find x when $y = 52$" as $\frac{y = 52}{x = ?}$. Students should then be able to write the proportion correctly as $\frac{28}{7} = \frac{52}{x}$.

Assignment Guide

Basic: 15–33 odd, 34–41
Average: 16–30 even, 31–41
All: Quiz 2, 1–5

***Skills Practice*, p. 266, and *Practice*, p. 267 (shown)**

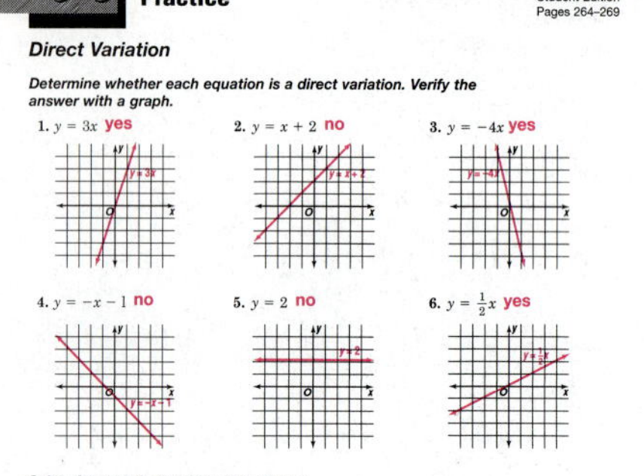
6-5 Practice

NAME ______ DATE ______ PERIOD ______

Student Edition Pages 264–269

Direct Variation

Determine whether each equation is a direct variation. Verify the answer with a graph.

1. $y = 3x$ yes
2. $y = x + 2$ no
3. $y = -4x$ yes
4. $y = -x - 1$ no
5. $y = 2$ no
6. $y = \frac{1}{2}x$ yes

Solve. Assume that y varies directly as x.

7. If $y = 14$ when $x = 5$, find x when $y = 28$. 10
8. Find y when $x = 5$ if $y = -6$ when $x = 2$. −15
9. If $x = 9$ when $y = 18$, find x when $y = 24$. 12
10. If $y = 36$ when $x = -6$, find x when $y = 54$. −9
11. Find y when $x = 3$ if $y = -3$ when $x = 6$. −1.5
12. Find y when $x = 8$ if $y = 4$ when $x = 5$. 6.4

Solve by using direct variation.

13. If there are 4 quarts in a gallon, how many quarts are in 4.5 gallons? 18 qt
14. How many feet are in 62.4 inches if there are 12 inches in a foot? 5.2 ft
15. If there are 2 cups in a pint, how many cups are in 7.2 pints? 14.4 c

Guided Practice

Find the constant of variation for each direct variation.

Sample 1: $y = 3x$
Solution: 3

Sample 2: $n = -0.3m$
Solution: -0.3

6. −4 8. 1.4

4. $y = 2x$ **2** 5. $5c = d$ **5** 6. $h = -4g$ 7. $t = \frac{2}{5}r$ $\frac{2}{5}$ 8. $1.4p = q$

Examples 1 & 2

Determine whether each equation is a direct variation. Verify the answer with a graph. 9–10. See margin for graphs.

9. $y = -x$ yes
10. $y = 5x + 5$ no

Example 5

Solve. Assume that y varies directly as x.

11. If $y = 28$ when $x = 7$, find x when $y = 52$. **13**
12. Find y when $x = 10$ if $y = -8$ when $x = 2$. **−40**

Example 6

13. If there are 3 feet in a yard, how many yards are in 14.1 feet? **4.7 yd**

Examples 3 & 4

14. **Astronomy** The weight of an object on Jupiter varies directly as its weight on Earth. An object that weighs 32.5 pounds on Earth weighs 81.25 pounds on Jupiter. How much would an object weigh on Jupiter if its weight on Earth is 90.2 pounds? **225.5 lb**

Exercises

Practice

For Exercises	See Examples
15–20	1, 2
21–32	5, 6
33, 34	3, 4

Homework Help — Extra Practice: See page 704.

Determine whether each equation is a direct variation. Verify the answer with a graph. 15–20. See Solutions Manual for graphs.

15. $y = x + 4$ no
16. $y = -2x$ yes
17. $y = -3$ no
18. $y = -0.5x$ yes
19. $x = 2$ no
20. $\frac{y}{x} = 6$ yes

Solve. Assume that y varies directly as x.

21. Find x when $y = 5$ if $y = 3$ when $x = 15$. **25**
22. If $x = 24$ when $y = 8$, find y when $x = 12$. **4**
23. If $y = -7$ when $x = -14$, find x when $y = 10$. **20**
24. Find y when $x = 9$ if $y = -4$ when $x = 6$. **−6**

25. If $x = 15$ when $y = 12$, find x when $y = 21$. $26\frac{1}{4}$
26. Find y when $x = 0.5$ if $y = 100$ when $x = 20$. **2.5**
27. If $y = 12$ when $x = 1\frac{1}{8}$, find y when $x = \frac{1}{4}$. $2\frac{2}{3}$

Solve by using direct variation.

C

28. How many inches are in 1.25 yards if there are 36 inches in 1 yard? **28. 45 in.**
29. If 1 bushel is 4 pecks, how many bushels are in 2.4 pecks? **29. 0.6 bu**
30. How many acres are in 0.75 square mile if there are 640 acres in 1 square mile? **480 acres**

Applications and Problem Solving

31. **Shopping** For \$1.89, you can buy 75 square feet of plastic wrap. At this rate, how much does 125 square feet of plastic wrap cost? **\$3.15**

Answers

9.
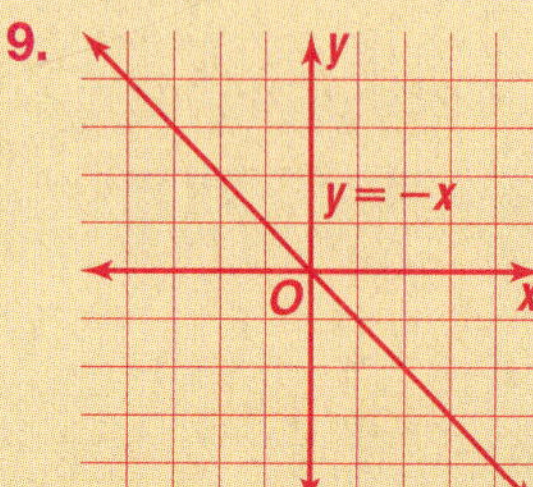

10.
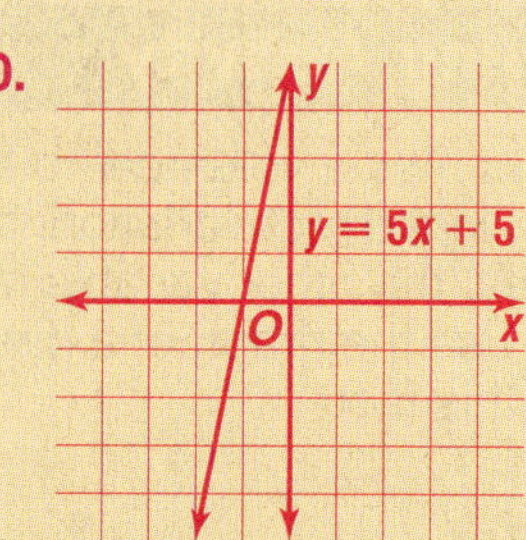

32. **Electronics** How much will 25 meters of copper wire weigh if 3.4 meters of the same copper wire weighs 0.442 kilogram? **3.25 kg**

33. **Health** The height h of an average person varies directly with their foot length f. Suppose the constant of variation is 7.
   a. Write an equation that represents this situation. **$h = 7f$**
   b. Graph the equation. **See margin.**
   c. Use the graph to find the height of a person whose foot length is 9.5 inches. **about 5.5 ft**

34. **Critical Thinking** Assume that y varies directly as x.
   a. Suppose the value of x is doubled. What is the constant of variation? What happens to the value of y? **k; It is doubled.**
   b. Suppose the value of x is tripled. What is the constant of variation? What happens to the value of y? **k; It is tripled.**
   c. Suppose the value of y is doubled. What is the constant of variation? What happens to the value of x? **k; It is doubled.**
   d. Suppose the value of y is tripled. What is the constant of variation? What happens to the value of x? **k; It is tripled.**

Mixed Review

35. Find $d(-3)$ if $d(x) = -5d - 4$. *(Lesson 6–4)* **11**

36. Graph $x = 4$. *(Lesson 6–3)* **See margin.**

37. **Life Science** A full-grown male seal weighs about 550 pounds. This is 50 pounds less than the weight of a full-grown male sea lion. *(Lesson 3–5)* **b. about 600 pounds**
   a. Write an equation that represents the weight of a full-grown male sea lion. **$x - 50 = 550$**
   b. How much does a full-grown male sea lion weigh?

Find each quotient. *(Lesson 2–6)*

38. $-36 \div 4$ **−9**   39. $-64 \div (-8)$ **8**   40. $\frac{-96}{-16}$ **6**

Standardized Test Practice

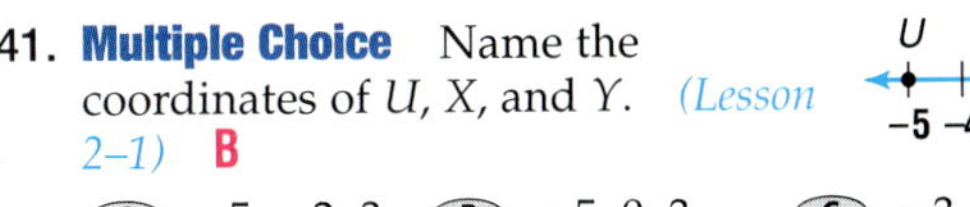

41. **Multiple Choice** Name the coordinates of U, X, and Y. *(Lesson 2–1)* **B**

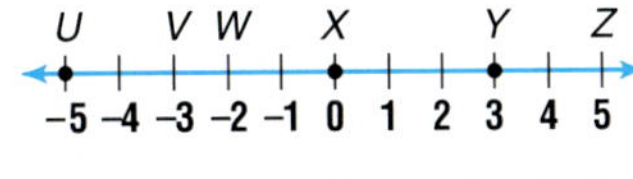

Ⓐ −5, −2, 3   Ⓑ −5, 0, 3   Ⓒ −3, 3, 5   Ⓓ −4, −2, 0

Quiz 2 — Lessons 6–4 and 6–5

Determine whether each relation is a function. *(Lesson 6–4)*

1. $\{(1, 2), (2, -1), (0, 3), (-1, 4)\}$ **yes**   2. $5x = 6 + y$ **yes**
3. If $g(x) = 4x + 3$, find $g(-2.6)$. *(Lesson 6–4)* **−7.4**
4. Find x when $y = -8$ if $y = 4$ when $x = 20$. Assume that y varies directly as x. *(Lesson 6–5)* **−40**
5. **Transportation** The length of a trip varies directly as the amount of gasoline used. Julio's car used 5 gallons of gasoline in the first 145 miles of his trip. How much gasoline should he expect to use in the remaining 232 miles of the trip? *(Lesson 6–5)* **8 gal**

www.algconcepts.com/self_check_quiz

4 ASSESS

Open-Ended Assessment

Speaking Ask students to describe the graph of a direct variation.

Quiz 2

The Quiz provides students with a brief review of the concepts and skills in Lessons 6–4 and 6–5. Lesson numbers are given to the right of the exercises or instruction lines so students can review concepts not yet mastered.

Answers

33b.

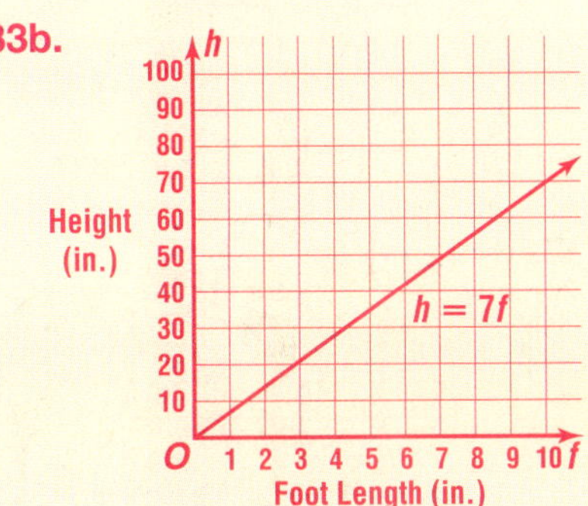

36.

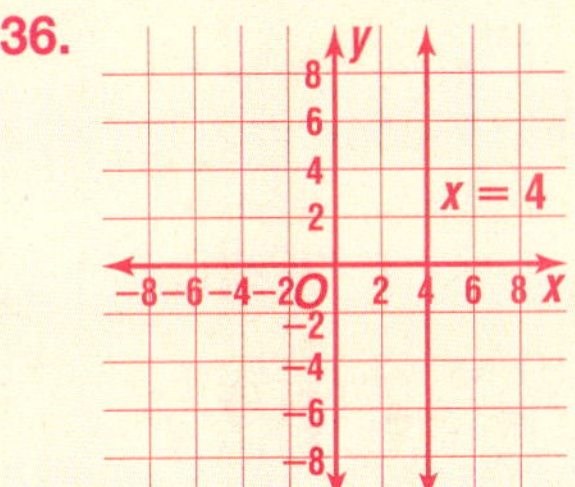

Extra Credit

The graph of a relation is a line that passes through points at (−5, −20), (−2, −8), and (3, 12). Is this the graph of a direct variation? Explain. **he graph is a line passing through the origin, and also the equation of the line can be written in the form $y = 4x$.**

Enrichment, p. 269

NAME ______ DATE ______ PERIOD ______

Enrichment

Student Edition Pages 264–269

***n*th Power Variation**

An equation of the form $y = kx^n$, where $k \neq 0$, describes an nth power variation. The variable n can be replaced by 2 to indicate the second power of x (the square of x) or by 3 to indicate the third power of x (the cube of x).

Assume that the weight of a person of average build varies directly as the cube of that person's height. The equation of variation has the form $w = kh^3$.

The weight that a person's legs will support is proportional to the cross-sectional area of the leg bones. This area varies directly as the square of the person's height. The equation of variation has the form $s = kh^2$.

Answer each question.

1. For a person 6 feet tall who weighs 200 pounds, find a value for k in the equation $w = kh^3$. **$k = 0.93$**
2. Use your answer from Exercise 1 to predict the weight of a person who is 5 feet tall. **about 116 pounds**
3. Find the value for k in the equation $w = kh^3$ for a baby who is 20 inches long and weighs 6 pounds. **$k = 1.296$ for $h = \frac{5}{3}$ ft**
4. How does your answer to Exercise 3 demonstrate that a baby is significantly fatter in proportion to its height than an adult? **k has a greater value.**
5. For a person 6 feet tall who weighs 200 pounds, find a value for k in the equation $s = kh^2$. **$k = 5.55$**
6. For a baby who is 20 inches long and weighs 6 pounds, find an "infant value" for k in the equation $s = kh^2$. **$k = 2.16$ for $h = \frac{5}{3}$ ft**
7. According to the adult equation you found (Exercise 1), how much would an imaginary giant 20 feet tall weigh? **7440 pounds**
8. According to the adult equation for weight supported (Exercise 5), how much weight could a 20-foot tall giant's legs actually support? **only 2222 pounds**
9. What can you conclude from Exercises 7 and 8? **Sample answer: Bone strength limits the size humans can attain.**

Lesson 6–6

1 FOCUS

5-Minute Check
Lesson 6–5

Determine whether each equation is a direct variation. Verify the answer with a graph.

1. $y = 3$ **no**

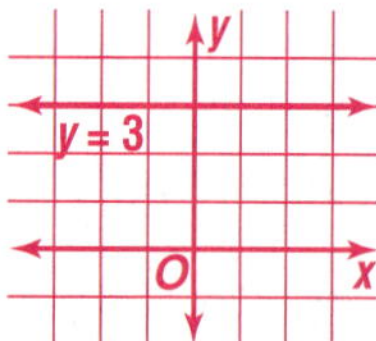

2. $y = \frac{1}{2}x$ **yes**

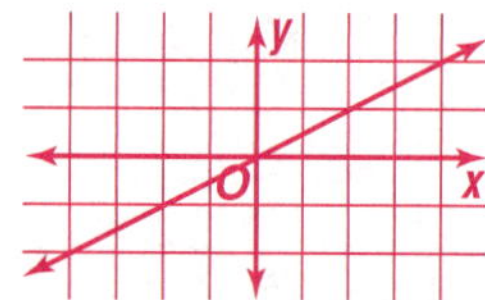

Suppose y varies directly as x and y = 15 when x = 3.

3. Find x when $y = 45$. **9**
4. Find y when $x = 12$. **60**
5. If there are 12 inches in 1 foot, how many inches are in 6 feet? **72 in.**

Motivating the Lesson

Hands-On Activity Have students work with a partner and give each pair of students 12 square algebra tiles. Have students model the 4-by-3 rectangle shown on page 270. Discuss the length, width, and area of this rectangle. Then have students create the 6-by-2 and 12-by-1 rectangles shown.

2 TEACH

Teaching Tip As you discuss the graphs of the direct and inverse variations, point out that an inverse variation is an example of a *nonlinear* function. Also stress that the ordered pair (0, 0) is always a solution of a direct variation but never a solution of an inverse variation.

6–6 Inverse Variation

What You'll Learn
You'll learn to solve problems involving inverse variations.

Why It's Important
Music Musicians can use inverse variations to find the wavelength of a musical tone. *See Exercise 12.*

Each of the rectangles shown has an area of 12 square units. Notice that as the length increases, the width decreases. As the length decreases, the width increases. However, their product stays the same.

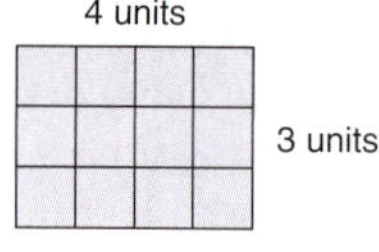

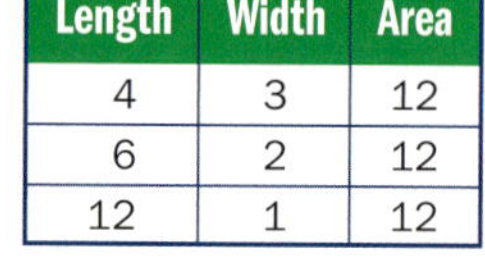

Length	Width	Area
4	3	12
6	2	12
12	1	12

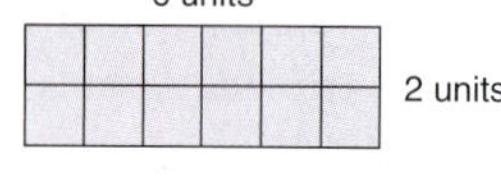

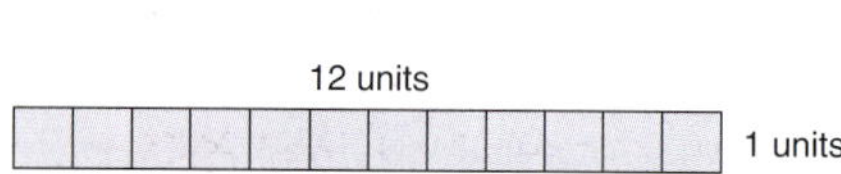

This is an example of an **inverse variation**. We say that y varies inversely as x. This means that as x increases in value, y decreases in value, or as y decreases in value, x increases in value.

Inverse Variation An inverse variation is described by an equation of the form $xy = k$, where $k \neq 0$.

The equation may also be written as $y = \frac{k}{x}$.

The graphs below show how the graph of a direct variation differs from the graph of an inverse variation.

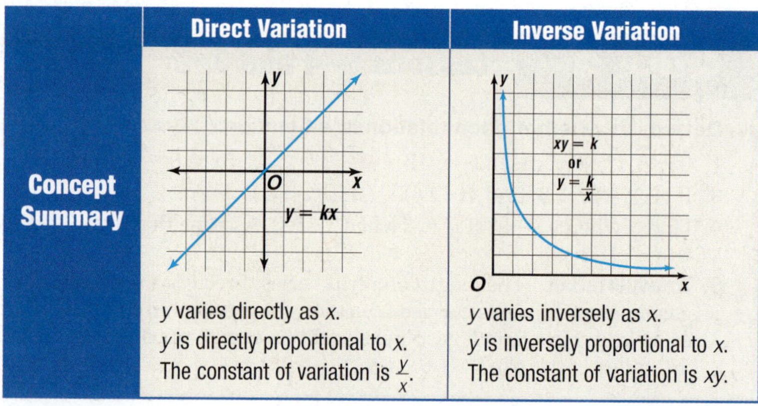

Concept Summary
Direct Variation
y
O
x
y = kx
y varies directly as x.
y is directly proportional to x.
The constant of variation is y/x.
Inverse Variation
y
xy = k
or
y = k/x
O
x
y varies inversely as x.
y is inversely proportional to x.
The constant of variation is xy.

Resource Manager

Reproducible Masters
Chapter 6 Resource Masters
- *Study Guide,* p. 270
- *Skills Practice,* p. 271
- *Practice,* p. 272
- *Reading to Learn Mathematics,* p. 273
- *Enrichment,* p. 274
- *Assessment,* p. 285

Graphing Calculator, p. 17
Hands-On Algebra, p. 76

Transparencies
5-Minute Check, 6–6

Technology/Multimedia
Interactive Chalkboard CD-ROM

Example 1
Construction Link

The number of carpenters needed to frame a house varies inversely as the number of days needed to complete the project. Suppose 5 carpenters can frame a house in 16 days. How many days will 8 carpenters take to frame the house? Assume that they all work at the same rate.

Explore You know that it takes 16 days for 5 carpenters to frame the house. You need to know how many days it will take 8 carpenters to frame the house.

Plan Solve the problem by using inverse variation.

Solve Let x = the number of carpenters. Let y = the number of days. First, find the value of k.

$xy = k$ *Definition of inverse variation*

$(5)(16) = $ *Replace x with 5 and y with 16.*

$80 = $ *The constant of variation is 80.*

Next, find the number of days for 8 carpenters to frame the house.

$y = \frac{k}{x}$ *Divide each side of xy = k by x.*

$= \frac{80}{8}$ *Replace k with 80 and x with 8.*

$= 10$ *Simplify.*

A crew of 8 carpenters can frame the house in 10 days.

Examine It takes 16 days for 5 carpenters to frame a house. It makes sense that 8 carpenters could do the work in less time. 10 days is a reasonable solution.

a. $2\frac{2}{3}$ days

Your Turn

a. The number of masons needed to build a block basement varies inversely as the number of days needed. If 8 masons can build a block basement in 3 days, how long will 9 masons take to do it?

 www.algconcepts.com/extra_examples

Teaching Tip In Example 1, point out that the number of days needed to complete the project depends on the number of carpenters who are working on the job, so the number of days is represented by the dependent variable y.

In-Class Example

Example 1

The number of bricklayers needed to build a brick wall varies inversely as the number of days needed. If 4 bricklayers can build a brick wall in 30 hours, how long would it take 5 bricklayers to do it? **24 h**

Teaching Tip Direct students' attention to the proportion $\frac{x_1}{x_2} = \frac{y_1}{y_2}$ shown in the paragraph above Example 2. Make sure students notice how this proportion differs from the one they used in the previous lesson for solving direct variation problems.

In-Class Example

Example 2

Suppose y varies inversely as x and $y = -2$ when $x = -12$. Find y when $x = 8$. **3**

You can use a graphing calculator to check the answer to Example 1.

Graphing Calculator Tutorial
See pp. 750–753.

Graphing Calculator Exploration

Step 1 Set the viewing window for x: [0, 10] by 1 and y: [0, 200] by 20.

Step 2 To graph $xy = 80$, rewrite the equation as $y = \frac{80}{x}$. Press [Y=] 80 [÷] [X,T,θ,n] to enter the equation.

Step 3 Press [TRACE] to estimate the x- and y-coordinates for any point on the graph. Notice that when $x = 5$, $y = 16$. When x is 8, y is 10. Therefore, the answer given in Example 1 is correct.

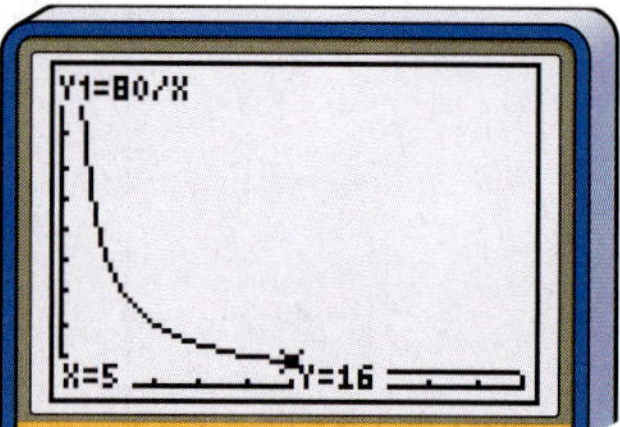

Try These

1. Why doesn't the viewing window allow for negative values to be shown on the graph? **The viewing window is set for positive values.**
2. Show how to rewrite $xy = 80$ as $y = \frac{80}{x}$. **Divide each side by x.**
3. Solve Your Turn Exercise a by using a graphing calculator.

Just as with direct variation, you can use a proportion to solve problems involving inverse variation. The proportion $\frac{x_1}{x_2} = \frac{y_2}{y_1}$ is only one of several that can be formed. *Can you name others?*

Example 

Suppose y varies inversely as x and $y = -6$ when $x = -2$. Find y when $x = 3$.

$\frac{x_1}{x_2} = \frac{y_2}{y_1}$ *Inverse variation proportion*

$\frac{-2}{3} = \frac{y_2}{-6}$ *Let $x_1 = -2$, $y_1 = -6$, and $x_2 = 3$.*

$-2(-6) = 3y_2$ *Find the cross products.*

$12 = 3y_2$ *Multiply.*

$4 = y_2$ *Divide each side by 3.*

Therefore, when $x = 3$, $y = 4$.

Your Turn

b. Suppose y varies inversely as x and $y = 3$ when $x = 12$. Find x when $y = 4$. **9**

Graphing Calculator Exploration

Refer to the Graphing Calculator Exploration on pages 271–272. Another way to display values of the expression while the graph is displayed is to press [2nd] [CALC] and select 1: value. The calculator will display X= in the lower-left corner of the graphing screen. Students can then type a number and press [ENTER] to evaluate the expression for that value of x. Emphasize that the value of x must be within the range of x values specified for the viewing window.

Inverse variations can be used to solve rate problems.

Example 3 **Sports Link**

In the formula $d = rt$, the time t varies inversely as the rate r. A race car traveling 125 miles per hour completed one lap around a race track in 1.2 minutes. How fast was the car traveling if it completed the next lap in 0.8 minute?

First, solve for the distance d, the constant of variation.

$d = rt$ *Formula for distance*

$= (125)(1.2)$ *Replace r with 125 and t with 1.2.*

$= 150$ *The constant of variation is 150.*

Next, find the rate if one lap was completed at 0.8 minute.

$\frac{d}{t} = r$ *Divide each side of d = rt by t.*

$\frac{150}{0.8} =$ *Replace d with 150 and t with 0.8.*

$187.5 =$ *Simplify.*

The car was traveling 187.5 miles per hour.

Check for Understanding

Communicating Mathematics

Vocabulary
inverse variation

1. **Explain** how to find the constant of variation in an inverse variation. **See margin.**
2. **Complete** the table for the inverse variation $xy = 24$.

x	1	2	3	4	6	8	12	24
y	24	?	?	?	?	?	?	?

12 8 6 4 3 2 1

3. **Writing Math** Write a few sentences to explain the difference between inverse variation and direct variation. **See margin.**

Guided Practice

Getting Ready **Determine if each equation is an inverse or a direct variation. Find the constant of variation.**

Sample 1: $cd = -3$

Solution: This equation is of the form $xy = k$. So, it is an inverse variation. The constant of variation is -3.

Sample 2: $h = \frac{1}{2}g$

Solution: This equation is of the form $y = kx$. So, it is a direct variation. The constant of variation is $\frac{1}{2}$.

4. $ab = 4$ **inverse; 4**
5. $s = 5r$ **direct; 5**
6. $mn = -7$ **6. inverse; −7**
7. $y = \frac{1}{3}x$ **direct; $\frac{1}{3}$**
8. $y = \frac{7}{x}$ **inverse; 7**
9. $t = \frac{r}{4}$ **direct; $\frac{1}{4}$**

In-Class Example

Example 3

A student running at 5 miles per hour runs one lap around the school campus in 8 minutes. If a second student takes 10 minutes to run one lap around the school campus, how fast is she running? **4 mph**

3 PRACTICE/APPLY

Error Analysis

Watch for students who confuse inverse and direct variations in Exercises 4–9.

Prevent by having students make themselves a small poster divided into two halves, one labeled *Inverse Variation* and the other labeled *Direct Variation.* In each half, direct students to write the general equation of the relation, a sketch of a sample graph, and several sample equations in different forms.

Answers

1. **Multiply the value of x by the value of y.**
3. **Sample answer: In a direct variation, as x increases, y increases. In an inverse variation, as x increases, y decreases.**

***Study Guide*, p. 270**

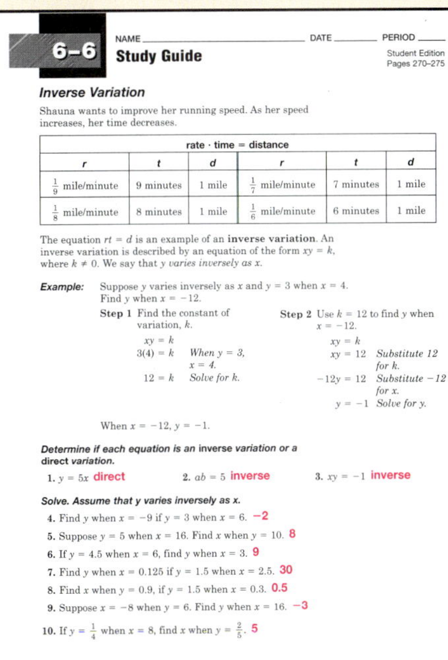

6–6 NAME ______ DATE ______ PERIOD ______

Study Guide

Student Edition Pages 270–275

Inverse Variation

Shauna wants to improve her running speed. As her speed increases, her time decreases.

rate · time = distance					
r	t	d	r	t	d
$\frac{1}{9}$ mile/minute	9 minutes	1 mile	$\frac{1}{7}$ mile/minute	7 minutes	1 mile
$\frac{1}{8}$ mile/minute	8 minutes	1 mile	$\frac{1}{6}$ mile/minute	6 minutes	1 mile

The equation $rt = d$ is an example of an **inverse variation**. An inverse variation is described by an equation of the form $xy = k$, where $k \neq 0$. We say that *y varies inversely as x*.

Example: Suppose y varies inversely as x and $y = 3$ when $x = 4$. Find y when $x = -12$.

Step 1 Find the constant of variation, k.

$xy = k$

$3(4) = k$ *When y = 3, x = 4.*

$12 = k$ *Solve for k.*

Step 2 Use $k = 12$ to find y when $x = -12$.

$xy = k$

$xy = 12$ *Substitute 12 for k.*

$-12y = 12$ *Substitute −12 for x.*

$y = -1$ *Solve for y.*

When $x = -12$, $y = -1$.

Determine if each equation is an inverse variation or a direct variation.

1. $y = 5x$ **direct**
2. $ab = 5$ **inverse**
3. $xy = -1$ **inverse**

Solve. Assume that y varies inversely as x.

4. Find y when $x = -9$ if $y = 3$ when $x = 6$. **−2**
5. Suppose $y = 5$ when $x = 16$. Find x when $y = 10$. **8**
6. If $y = 4.5$ when $x = 6$, find y when $x = 3$. **9**
7. Find y when $x = 0.125$ if $y = 1.5$ when $x = 2.5$. **30**
8. Find x when $y = 0.9$, if $y = 1.5$ when $x = 0.3$. **0.5**
9. Suppose $x = -8$ when $y = 6$. Find y when $x = 16$. **−3**
10. If $y = \frac{1}{4}$ when $x = 8$, find x when $y = \frac{2}{5}$. **5**

Reteaching Activity

Visual/Spatial Learners Have students make a table of ordered pairs that are solutions of the inverse variation $xy = 24$. Then have them graph the ordered pairs. Lead students to recognize that the curved shape of the graph indicates that as the value of one variable increases, the value of the other variable decreases.

Assignment Guide

Basic: 15–29 odd, 30–36
Average: 14–26 even, 28–36

Example 2 **Solve. Assume that *y* varies inversely as *x*.**

10. Suppose $y = 6$ when $x = 2$. Find y when $x = 3$. **4**
11. Find x when $y = 2$ if $y = -4$ when $x = -3$? **6**

Example 1

12. **Music** The frequency of a musical tone varies inversely as its wavelength. If one tone has a frequency of 440 vibrations per second and a wavelength of 2.4 feet, find the wavelength of a tone that has a frequency of 660 vibrations per second. **1.6 ft**

Example 3

13. **Travel** The time it takes to travel a certain distance varies inversely to the speed of travel. Suppose it takes 2.75 hours to drive from one city to another at a rate of 65 miles per hour. How long will the return trip take at a rate of 55 miles per hour? **3.25 hours**

Exercises

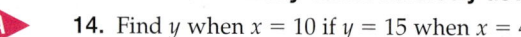

Practice

Homework Help	
For Exercises	See Examples
14–19, 28	2
20–23, 29, 30	1, 3
Extra Practice	
See page 705.	

A

Solve. Assume that *y* varies inversely as *x*.

14. Find y when $x = 10$ if $y = 15$ when $x = 4$. **6**
15. Suppose $y = 20$ when $x = 6$. Find x when $y = 12$. **10**
16. If $y = 1.2$ when $x = 3$, find y when $x = 6$. **0.6**
17. Find y when $x = 2.5$ if $y = 0.15$ when $x = 1.5$. **0.09**
18. If $y = \frac{2}{3}$ when $x = 6$, find x when $y = 1\frac{1}{3}$. **3**
19. Suppose $y = -\frac{1}{2}$ when $x = -\frac{3}{4}$. Find y when $x = \frac{3}{5}$. $\frac{5}{8}$

Find the constant of variation. Then write an equation for each statement.

B

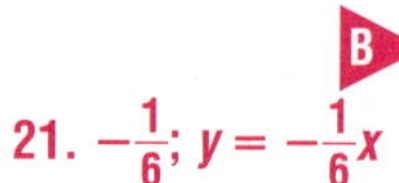

21. $-\frac{1}{6}$; $y = -\frac{1}{6}x$

20. y varies inversely as x, and $y = 5$ when $x = 10$. **50; $xy = 50$**
21. y varies directly as x, and $x = -18$ when $y = 3$.
22. y varies inversely as x, and $y = 4$ when $x = -2$. **−8; $xy = -8$**
23. y varies directly as x, and $x = 7$ when $y = 2.1$. **0.3; $y = 0.3x$**

Match each situation with the equation that represents it. Then tell whether the situation represents a direct variation or an inverse variation.

24. DeShawn purchases a number of pencils for $0.25 each. **d; direct**
25. A winning $125 lottery ticket is shared with a number of people. **a; inverse**
26. Twenty-five pieces of candy are shared with several people. **b; inverse**
27. Each person who has a winning lottery ticket wins $125. **c; direct**

a. $xy = 125$
b. $y = \frac{25}{x}$
c. $y = 125x$
d. $25x = y$

***Skills Practice*, p. 271, and *Practice*, p. 272 (shown)**

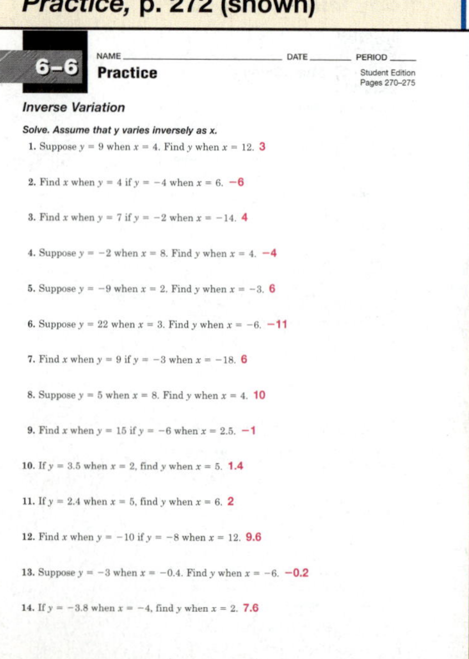

6–6 NAME ______ DATE ______ PERIOD ______

Practice

Student Edition Pages 270–275

Inverse Variation

Solve. Assume that y varies inversely as x.

1. Suppose $y = 9$ when $x = 4$. Find y when $x = 12$. **3**
2. Find x when $y = 4$ if $y = -4$ when $x = 6$. **−6**
3. Find x when $y = 7$ if $y = -2$ when $x = -14$. **4**
4. Suppose $y = -2$ when $x = 8$. Find y when $x = 4$. **−4**
5. Suppose $y = -9$ when $x = 2$. Find y when $x = -3$. **6**
6. Suppose $y = 22$ when $x = 3$. Find y when $x = -6$. **−11**
7. Find x when $y = 9$ if $y = -3$ when $x = -18$. **6**
8. Suppose $y = 5$ when $x = 8$. Find y when $x = 4$. **10**
9. Find x when $y = 15$ if $y = -6$ when $x = 2.5$. **−1**
10. If $y = 3.5$ when $x = 2$, find y when $x = 5$. **1.4**
11. If $y = 2.4$ when $x = 5$, find y when $x = 6$. **2**
12. Find x when $y = -10$ if $y = -8$ when $x = 12$. **9.6**
13. Suppose $y = -3$ when $x = -0.4$. Find y when $x = -6$. **−0.2**
14. If $y = -3.8$ when $x = -4$, find y when $x = 2$. **7.6**

Applications and Problem Solving

28. **Music** The length of a piano string varies inversely as the frequency of its vibrations. A piano string 36 inches long vibrates at a frequency of 480 cycles per second. Find the frequency of a 24-inch string. **720 cycles per second**

29. **Geometry** The length ℓ of the rectangle varies inversely as its width w. **b. 2.4 m**

   a. Find the constant of variation. **36**

   b. What is the width of the rectangle when the length is 15 meters?

   c. Write an equation that represents this inverse variation.

   d. Draw the graph of the equation. **See margin.**

29c. $\ell w = 36$

30. **Critical Thinking** Assume that y varies inversely as x.

   a. Suppose the value of x is doubled. What is the constant of variation? What happens to the value of y? **k; It is halved.**

   b. Suppose the value of y is tripled. What is the constant of variation? What happens to the value of x? **k; It is one third as much.**

Mixed Review

31. Suppose y varies directly as x and $y = 16$ when $x = 12$. Find x when $y = 36$. *(Lesson 6–5)* **27**

32. Is $\{(6, 3), (5, -2), (2, 3), (12, -12)\}$ a function? *(Lesson 6–4)* **yes**

Simplify. *(Lesson 3–2)*

33. $-38.9 + 24.2$ **−14.7**

34. $12.5 + (-15.6) + 22.7 + (-35.8)$ **−16.2**

Standardized Test Practice

35. **Grid In** The surface of Lake Michigan is 176 meters above sea level. Its deepest point is 281 meters below sea level. How many meters below the surface is the deepest part of the lake? *(Lesson 2–4)* **457**

36. **Multiple Choice** Which of the following shows the graph of the relation $\{(-1, 2), (1, 1), (1, 3), (2, 0)\}$? *(Lesson 2–2)* **C**

(A)
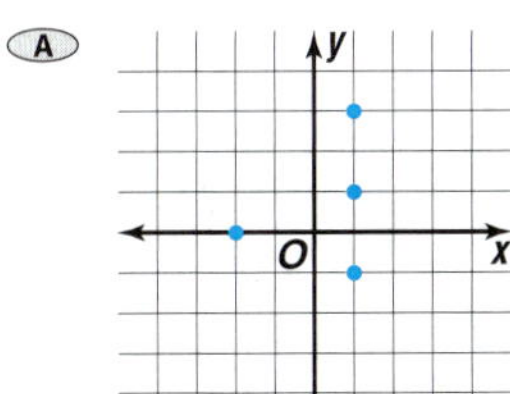

(B)
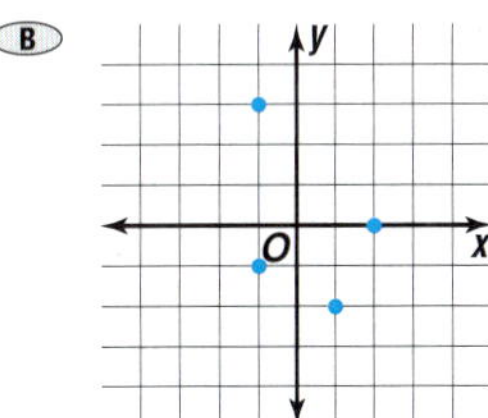

(C)
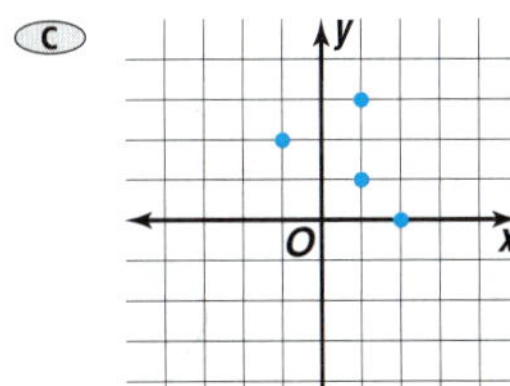

(D)
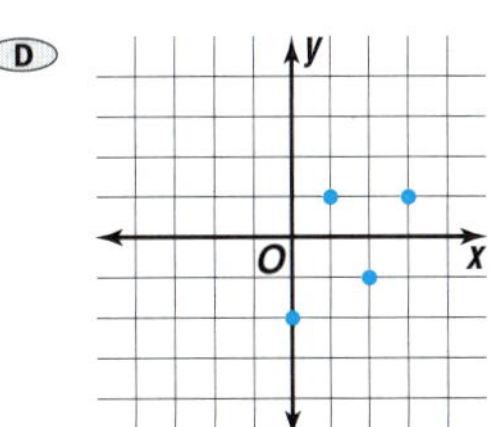

www.algconcepts.com/self_check_quiz

4 ASSESS

Open-Ended Assessment

Writing Explain how to tell if an equation is that of an inverse variation or that of a direct variation.

Chapter 6, Quiz B (Lessons 6–3 through 6–6) is available in the *Chapter 6 Resource Masters*, p. 285.

Answer

29d.

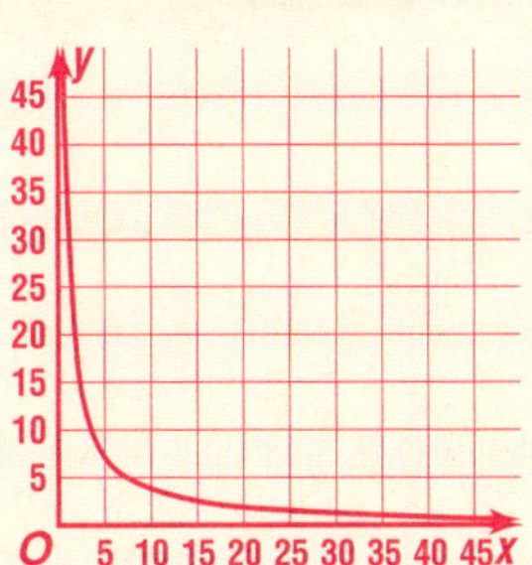

Extra Credit

Assume y varies directly as x. Find y when $x = 0.4$ if $y = 1.125$ when $x = 9$. **0.05**

***Enrichment*, p. 274**

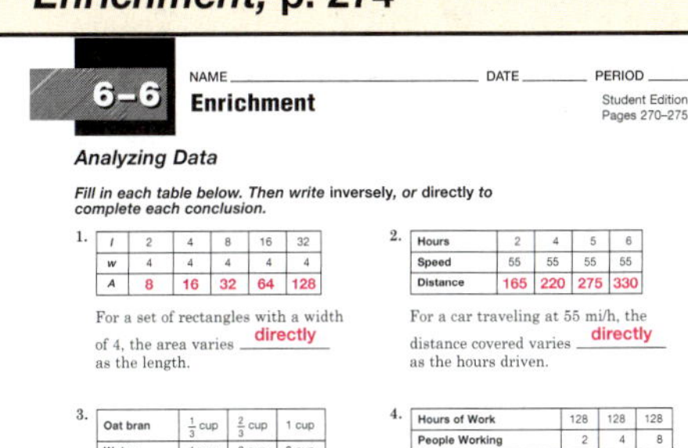

6–6 NAME ______ DATE ______ PERIOD ______

Enrichment

Student Edition Pages 270–275

Analyzing Data

Fill in each table below. Then write inversely, *or* directly *to complete each conclusion.*

1.

l	2	4	8	16	32
w	4	4	4	4	4
A	8	16	32	64	128

For a set of rectangles with a width of 4, the area varies **directly** as the length.

2.

Hours	2	4	5	6
Speed	55	55	55	55
Distance	165	220	275	330

For a car traveling at 55 mi/h, the distance covered varies **directly** as the hours driven.

3.

Oat bran	$\frac{1}{3}$ cup	$\frac{2}{3}$ cup	1 cup
Water	1 cup	2 cup	3 cup
Servings	1	2	3

The number of servings of oat bran varies **directly** as the number of cups of oat bran.

4.

Hours of Work	128	128	128
People Working	2	4	8
Hours per Person	64	32	16

A job requires 128 hours of work. The number of hours each person works varies **inversely** as the number of people working.

5.

Miles	100	100	100	100
Rate	20	25	50	100
Hours	5	4	2	1

For a 100-mile car trip, the time the trip takes varies **inversely** as the average rate of speed the car travels.

6.

b	3	4	5	6
h	10	10	10	10
A	15	20	25	30

For a set of right triangles with a height of 10, the area varies **directly** as the base.

Use the table at the right.

7. x varies **directly** as y.

8. z varies **inversely** as y.

9. x varies **inversely** as z.

x	1	1.5	2	2.5	3
y	2	3	4	5	6
z	60	40	30	24	20

Understanding and Using the Vocabulary

interNET CONNECTION **Review Activities** For more review activities, visit: www.algconcepts.com

After completing this chapter, you should be able to define each term, property, or phrase and give an example or two of each.

constant of variation *(p. 264)*
dependent variable *(pp. 238, 264)*
direct variation *(p. 264)*
domain *(p. 238)*
equation in two variables *(p. 244)*
function *(p. 256)*
functional notation *(p. 258)*
functional value *(p. 258)*
independent variable *(pp. 238, 264)*
inverse variation *(p. 270)*
linear equation *(p. 250)*
range *(p. 238)*
rate problems *(p. 266)*
relation *(p. 238)*
solution set *(p. 244)*
vertical line test *(p. 257)*
x-coordinate *(p. 238)*
y-coordinate *(p. 238)*

Choose the correct term or expression to complete each sentence.

1. A (function, linear equation) is an equation whose graph is a straight line.
2. The set of all second coordinates of a set of ordered pairs is called the (range, domain).
3. Equations that are functions can be written in (functional notation, functional value).
4. The domain of $\{(-2, 3), (0, 4), (1, 5)\}$ is ($\{-2, 4, 1\}$, $\{-2, 0, 1\}$).
5. As x increases in value, y decreases in value in a(n) (direct, inverse) variation.
6. Any set of ordered pairs is a (function, relation).
7. A direct variation is a linear function that can be written in the form ($xy = k$, $y = kx$).
8. In a (relation, function), each member of the domain is paired with one member of the range.
9. The equation ($2 + x = 8$, $y = 2x + 3$) is an example of *an equation in two variables.*
10. The (first, second) coordinate in an ordered pair is called the y-coordinate.

Skills and Concepts

Objectives and Examples

- **Lesson 6–1** Show relations as sets of ordered pairs, as tables, and as graphs.

Express $\{(-1, 0), (0, 3), (1, 2), (2, 1)\}$ as a table and as a graph. Then determine the domain and the range.

x	y
−1	0
0	3
1	2
2	1

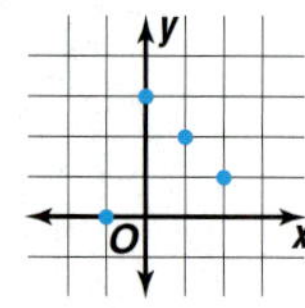

The domain is $\{-1, 0, 1, 2\}$.
The range is $\{0, 3, 2, 1\}$.

Review Exercises

Express each relation as a table and as a graph. Then determine the domain and the range. **11–14. See Solutions Manual.**

11. $\{(-1, 0), (0, 0), (2, 3), (4, 1)\}$
12. $\{(-3, 2), (-2, -1), (0, 4), (3, 3)\}$
13. $\{(-1.5, 2), (1, 3.5), (4.5, -5.5)\}$
14. $\left\{\left(-\frac{1}{2}, 3\right), (1, 2), \left(\frac{1}{2}, 4\right)\right\}$
15. Express the relation as a set of ordered pairs and in a table. Find the domain and range. **See margin.**

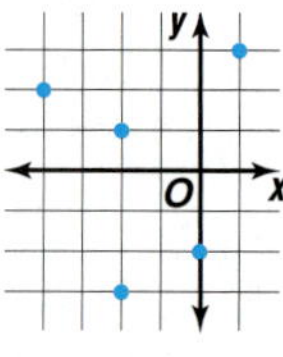

www.algconcepts.com/vocabulary_review

Study Guide and Assessment

Understanding and Using the Vocabulary

This section provides a listing of the new terms, properties, and phrases that were introduced in this chapter. The exercises check students' understanding of the terms by using a variety of verbal formats including matching, completion, and true/false.

Glossary A complete glossary of terms appears on pages 762–783.

MindJogger Videoquizzes

MindJogger Videoquizzes provide an alternative review of concepts presented in this chapter. Students work in teams to answer questions, gaining points for correct answers.

Answer

15. $\{(-4, 2), (-2, 1), (-2, -3), (0, -2), (1, 3)\}$

x	y
−4	2
−2	1
−2	−3
0	−2
1	3

domain: $\{-4, -2, 0, 1\}$, range: $\{2, 1, -3, -2, 3\}$

Resource Manager

Reproducible Masters
Chapter 6 Resource Masters
- *Assessment*, pp. 275–283, 286–288

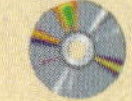

Technology/Multimedia
MindJogger Videoquizzes
ExamView® Pro

Objectives and Examples

- **Lesson 6–2** Solve linear equations for a given domain.

Solve $y = 2x$ if the domain is $\{-1, 0, 2\}$.

x	$2x$	y	(x, y)
−1	2(−1)	−2	(−1, −2)
0	2(0)	0	(0, 0)
2	2(2)	4	(2, 4)

The solution is $\{(-1\ -2), (0, 0), (2, 4)\}$.

Review Exercises

Solve each equation if the domain is {−1, 0, 2, 4}. Graph the solution set.

16. $y = -x$ **17.** $y = \frac{1}{2}x$

18. $y = 3x - 4$ **19.** $3x + 2y = 10$

16–19. See margin.

20. Find the domain of $y = x - 2$ if the range is $\{-3, -1, 4, 6\}$. **{−1, 1, 6, 8}**

21. Find the domain of $y = 2x + 2$ if the range is $\{-2, 0, 2, 4\}$. **{−2, −1, 0, 1}**

- **Lesson 6–3** Graph linear relations.

Graph $y = 2x - 4$.

x	$2x - 4$	y	(x, y)
0	2(0) − 4	−4	(0, −4)
1	2(1) − 4	−2	(1, −2)
2	2(2) − 4	0	(2, 0)
3	2(3) − 4	2	(3, 2)

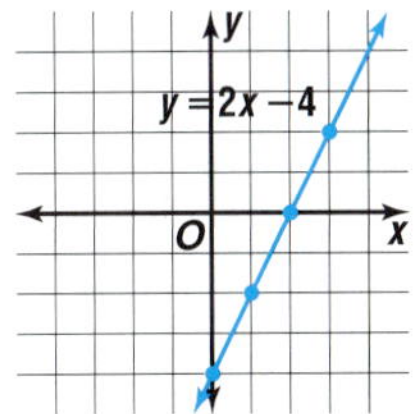

Determine whether each equation is a linear equation. Explain. If an equation is linear, identify *A*, *B*, and *C*. **22–25. See margin.**

22. $yx = -2$ **23.** $-x = y$

24. $7x + y = -2$ **25.** $3x + 4y = 2z$

26–31. See Solutions Manual.

Graph each equation.

26. $y = 2x$ **27.** $y = x - 3$

28. $-1 = y$ **29.** $x = 5$

30. $y = -3x + 4$ **31.** $4 = 8x - 12$

32. The table lists the coordinates of points on a linear graph. Copy and complete the table.

x	y	
?	−2	**−4**
−2	0	
?	2	**0**
1	3	

- **Lesson 6–4** Determine whether a given relation is a function.

Determine whether $\{(3, 2), (5, 3), (4, 3), (5, 2)\}$ is a function.

Because there are two values of y for one value of x, 5, the relation is *not* a function.

Determine whether each relation is a function.

33. $\{(-2, 6), (3, -2), (3, 0), (4, 6)\}$ **no**

34. $\{(2, 8), (9, 3), (-3, 7), (4, 3)\}$ **yes**

35. $\{(-1, 0), (3, 0), (6, 2)\}$ **yes**

If $g(x) = -3x - 6$, find each value.

36. $g(2)$ **−12** **37.** $g(-3.5)$ **4.5** **38.** $g(-4a)$ **$12a - 6$**

ExamView® Pro

Use the networkable ***ExamView® Pro Testmaker CD-ROM*** to:

- Create **multiple versions** of tests.
- Create **modified** tests for **inclusion** students with one mouse click.
- **Edit** existing questions and **add** your own questions.
- Build tests aligned with state standards using built-in **state curriculum correlations.**
- Change **English** tests to **Spanish** with one mouse click and vice versa.

Skills and Concepts

The **Objectives and Examples** section reviews the skills and concepts of the chapter and shows completely worked examples.

The **Review Exercises** provide practice for the corresponding objectives.

Answers

16. $\{(-1, 1), (0, 0), (2, -2), (4, -4)\}$

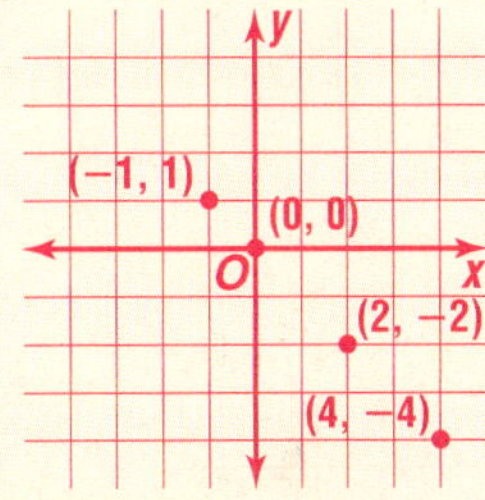

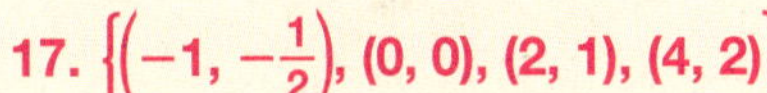

17. $\{(-1, -\frac{1}{2}), (0, 0), (2, 1), (4, 2)\}$

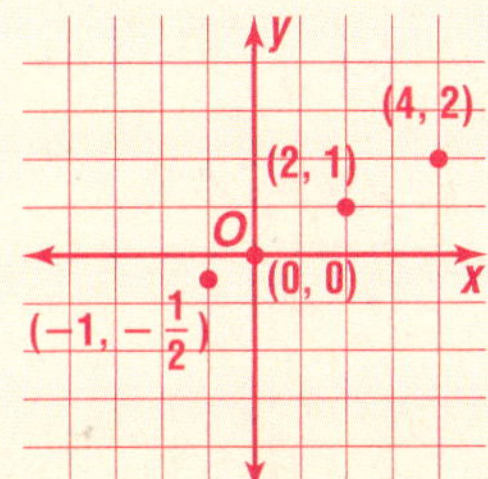

18. $\{(-1, -7), (0, -4), (2, 2), (4, 8)\}$

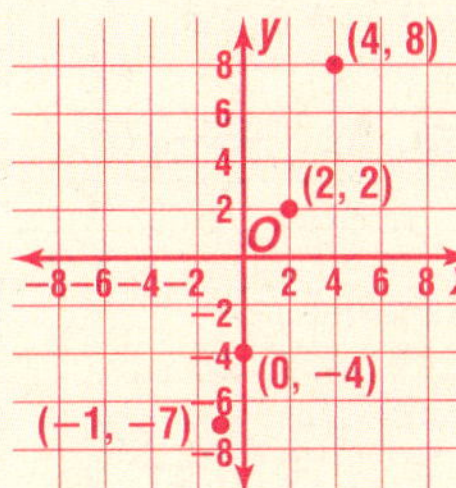

19. $\{(-1, 6.5), (0, 5), (2, 2), (4, -1)\}$

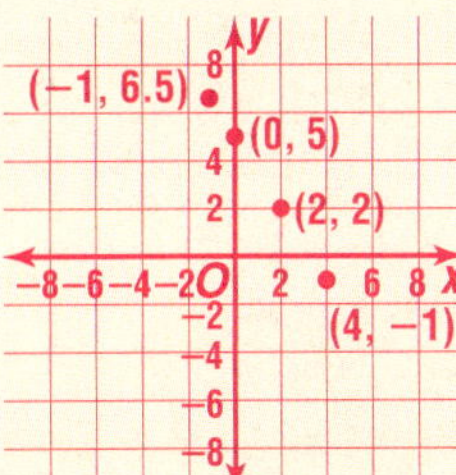

22. no

23. yes; $A = -1, B = -1, C = 0$

24. yes; $A = 7, B = 1, C = -2$

25. no

Applications and Problem Solving

This section provides additional practice in solving real-world problems that involve the concepts of this chapter.

Answers

47. {(5, 1), (10, 2), (15, 3), (20, 4), (25, 5)}; domain: {5, 10, 15, 20, 25}; range: {1, 2, 3, 4, 5}

Page 279

3.

x	−3	−1	0	2	3
y	−2	4	−1	3	3

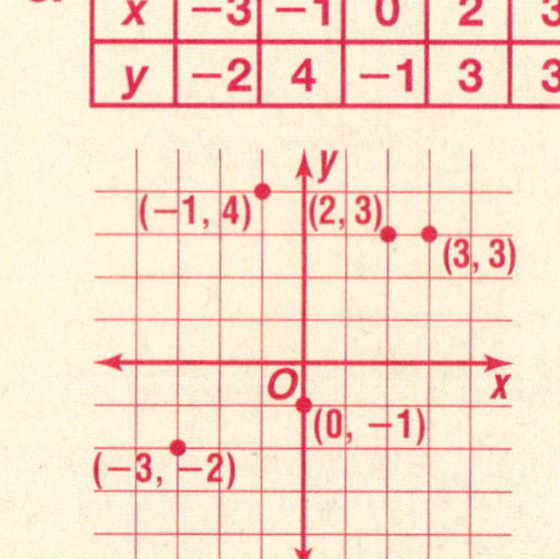

domain: {−3, −1, 0, 2, 3}, range: {−2, 4, −1, 3}

Assessment, pp. 277–278

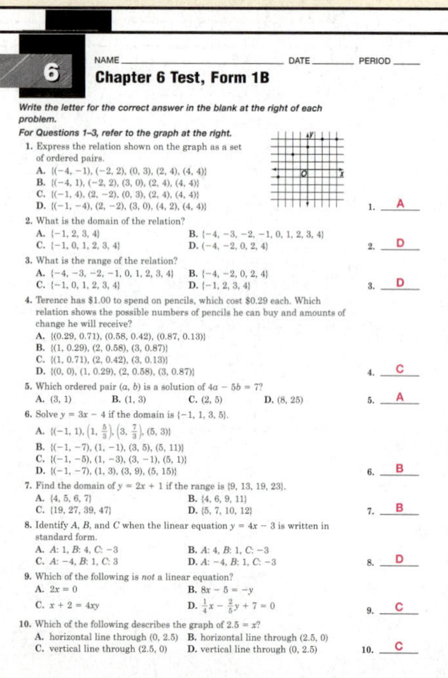

6

NAME ______________ DATE ______ PERIOD ______

Chapter 6 Test, Form 1B

Write the letter for the correct answer in the blank at the right of each problem.

For Questions 1–3, refer to the graph at the right.

1. Express the relation shown on the graph as a set of ordered pairs.
   A. {(−4, −1), (−2, 2), (0, 3), (2, 4), (4, 4)}
   B. {(−4, 1), (−2, 2), (3, 0), (2, 4), (4, 4)}
   C. {(−1, 4), (2, −2), (0, 3), (2, 4), (4, 4)}
   D. {(−1, −4), (2, −2), (3, 0), (4, 2), (4, 4)} 1. A
2. What is the domain of the relation?
   A. {−1, 2, 3, 4} B. {−4, −3, −2, −1, 0, 1, 2, 3, 4}
   C. {−1, 0, 1, 2, 3, 4} D. {−4, −2, 0, 2, 4} 2. D
3. What is the range of the relation?
   A. {−4, −3, −2, −1, 0, 1, 2, 3, 4} B. {−4, −2, 0, 2, 4}
   C. {−1, 0, 1, 2, 3, 4} D. {−1, 2, 3, 4} 3. D
4. Terence has \$1.00 to spend on pencils, which cost \$0.29 each. Which relation shows the possible numbers of pencils he can buy and amounts of change he will receive?
   A. {(0.29, 0.71), (0.58, 0.42), (0.87, 0.13)}
   B. {(1, 0.29), (2, 0.58), (3, 0.87)}
   C. {(1, 0.71), (2, 0.42), (3, 0.13)}
   D. {(0, 0), (1, 0.29), (2, 0.58), (3, 0.87)} 4. C
5. Which ordered pair (a, b) is a solution of $4a - 5b = 7$?
   A. (3, 1) B. (1, 3) C. (2, 5) D. (8, 25) 5. A
6. Solve $y = 3x - 4$ if the domain is {−1, 1, 3, 5}.
   A. $\{(-1, 1), (1, \frac{5}{3}), (3, \frac{7}{3}), (5, 3)\}$
   B. {(−1, −7), (1, −1), (3, 5), (5, 11)}
   C. {(−1, −5), (1, −3), (3, −1), (5, 1)}
   D. {(−1, −7), (1, 3), (3, 9), (5, 15)} 6. B
7. Find the domain of $y = 2x + 1$ if the range is {9, 13, 19, 23}.
   A. {4, 5, 6, 7} B. {4, 6, 9, 11}
   C. {19, 27, 39, 47} D. {5, 7, 10, 12} 7. B
8. Identify A, B, and C when the linear equation $y = 4x - 3$ is written in standard form.
   A. A: 1, B: 4, C: −3 B. A: 4, B: 1, C: −3
   C. A: −4, B: 1, C: 3 D. A: −4, B: 1, C: −3 8. D
9. Which of the following is *not* a linear equation?
   A. $2x = 0$ B. $8x - 5 = -y$
   C. $x + 2 = 4xy$ D. $\frac{1}{4}x - \frac{2}{5}y + 7 = 0$ 9. C
10. Which of the following describes the graph of $2.5 = x$?
    A. horizontal line through (0, 2.5) B. horizontal line through (2.5, 0)
    C. vertical line through (2.5, 0) D. vertical line through (0, 2.5) 10. C

Mixed Problem Solving
See pages 724–731.

Objectives and Examples

- **Lesson 6–5** Solve problems involving direct variation.

Suppose y varies directly as x and $y = 10$ when $x = 15$. Find x when $y = 6$.

Use $\frac{y_1}{x_1} = \frac{y_2}{x_2}$ to solve the proportion.

$\frac{10}{15} = \frac{6}{x_2}$ *Substitute.*

$10x_2 = 15(6)$ *Cross products*

$10x_2 = 90$ *Multiply.*

$x_2 = 9$ So, $x = 9$ when $y = 6$.

Review Exercises

Solve. Assume that *y* varies directly as *x*.

39. Find y when $x = 7$ if $y = 15$ when $x = 5$.
40. If $y = 12$ when $x = -6$, find y when $x = 2$.
41. Find x when $y = -5$ if $x = -37.5$ when $y = 15$. **12.5**

39. 21 40. −4

Solve by using direct variation.

42. If there are 16 ounces in 1 pound, how many ounces are in 2.8 pounds? **44.8 oz**

- **Lesson 6–6** Solve problems involving inverse variation.

Suppose y varies inversely as x and $y = 10$ when $x = 72$. Find x when $y = 24$.

$\frac{x_1}{x_2} = \frac{y_2}{y_1}$ *Inverse variation proportion*

$\frac{72}{x_2} = \frac{24}{10}$ *Substitute.*

$720 = 24x_2$ *Cross products*

$30 = x_2$ So, when $y = 24$, $x = 30$.

Solve. Assume that *y* varies inversely as *x*.

43. If $x = 3$ when $y = 25$, find x when $y = 15$.
44. Find y when $x = 42$ if $y = 21$ when $x = 56$.

43. 5 44. 28

Find the constant of variation. Then write an equation for each statement.

45. y varies inversely as x, and $y = 4$ when $x = 2$. **8; $xy = 8$**
46. y varies directly as x, and $x = -12$ when $y = 3$. **$-\frac{1}{4}$; $y = -\frac{1}{4}x$**

Applications and Problem Solving

47. **Science** The table shows how far away a lightning strike is, given the number of seconds between when the lightning is seen and when thunder is heard. Express the relation as a set of ordered pairs. Then find the domain and range. *(Lesson 6–1)* **See margin.**

Time (s)	5	10	15	20	25
Distance (mi)	1	2	3	4	5

48. **Sales** Jennifer earns a weekly salary of \$255 plus a bonus of \$0.25 for each CD over 150 that she sells each week. The equation $C(r) = 255 + 0.25(r - 150)$ can be used to determine her total weekly income. How much more will she earn by selling 450 CDs than 200 CDs? *(Lesson 6–2)* **\$62.50**
49. **Electronics** In an electrical transformer, voltage varies directly as the number of turns on the coil. What would be the voltage produced by 66 turns, if 110 volts come from 55 turns? *(Lesson 6–5)* **132 volts**

Assessment

Four forms of Chapter 6 Test are available in the *Chapter 6 Resource Masters.*

Chapter 6 Test, Form 1B, is shown at the left. Chapter 6 Test, Form 2B, is shown on the next page.

	Form of Test		Level
1A	Multiple Choice	pp. 275–276	Average
1B	Multiple Choice	pp. 277–278	Basic
2A	Free Response	pp. 279–280	Average
2B	Free Response	pp. 281–282	Basic

CHAPTER 6 Test

1. Write $y = 2x - 3$ in functional notation. $f(x) = 2x - 3$
2. List three ways to show a relation. ordered pairs, table, graph

Express each relation as a table and as a graph. Then determine the domain and range.

3. $\{(-3, -2), (-1, 4), (0, -1), (2, 3), (3, 3)\}$
4. $\{(-1, 1.5), (2, -4), (3, 2.5), (4, -1.5)\}$

3–4. See margin.

Solve each equation if the domain is $\{-2, -1, 0, 4, 6\}$.

5. $y = \frac{1}{2}x$ $\left\{-1, -\frac{1}{2}, 0, 2, 3\right\}$
6. $y = 4 - x$ $\{6, 5, 4, 0, -2\}$
7. $2x - 2y = 6$ $\{-5, -4, -3, 1, 3\}$

Graph each equation. 8–10. See margin.

8. $x = -5$
9. $y = -3x - 1$
10. $6x - 2y = 8$

Determine whether each relation is a function.

11. $\{(4, 1), (2, 4), (-1, 2), (-4, 2)\}$ yes
12. $\{(-3, 2), (0, 4), (1, 5), (3, 0), (-3, -1)\}$ no

13. no

x	y
−1	2
3	−4
−1	0
2	3
1	1

14. yes

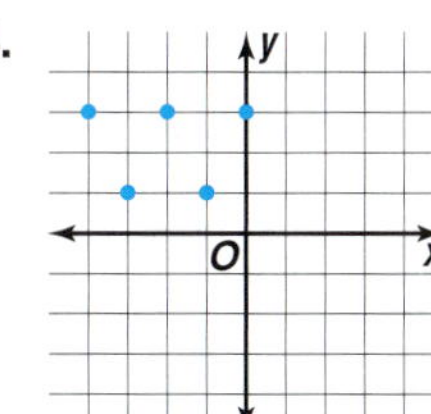

15. yes

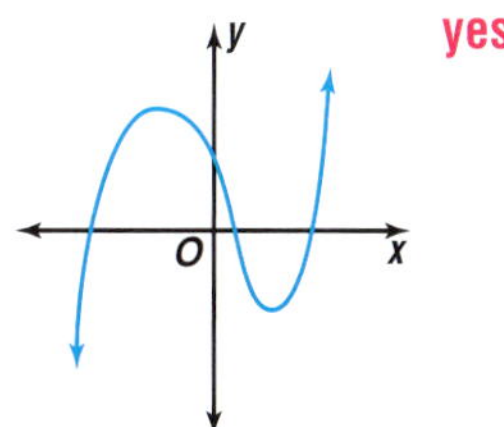

If $f(x) = -3x + 4$ and $g(x) = 2x + 2$, find each value.

16. $f(-1)$ 7
17. $g(3.5)$ 9
18. $g\left(\frac{1}{4}\right)$ $2\frac{1}{2}$
19. $f(2d)$ $-6d + 4$

Solve. Assume that y varies directly as x.

20. Find y when $x = 12$ if $y = 18$ when $x = 27$. 8
21. Suppose $y = -84$ when $x = -12$. Find y when $x = -1$. -7

Solve. Assume that y varies inversely as x.

22. Suppose $x = -16$ when $y = 4$. Find x when $y = 8$. -8
23. Find y when $x = 10$ if $y = 2.5$ when $x = 19.2$. 4.8

24. **Counting** Refer to the spinners. a–b. See margin.
    a. Suppose each spinner is spun once. Write a relation that shows the different possible outcomes.
    b. Use the relation to find the number of ways to land on an odd number and green.

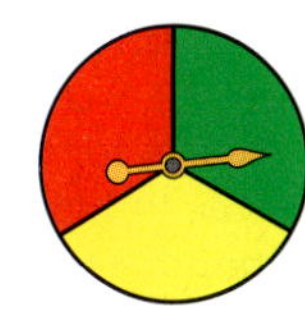

25. **Geometry** To determine the perimeter P of a rectangle, the equation $P = 2w + 2\ell$, where w is the width and ℓ is the length, can be used. Suppose a rectangle has a perimeter of 64 inches. Find the possible dimensions of the rectangle given the domain values $\{3, 6, 7, 10\}$. $\{(3, 29), (6, 26), (7, 25), (10, 22)\}$

www.algconcepts.com/chapter_test

Chapter Test Bonus Question

Find an equation satisfied by the ordered pairs (1, 12), (2, 6), (4, 3), and (6, 2). Is this equation a *direct variation, inverse variation,* or *neither*?

$y = \frac{12}{x}$ or $xy = 12$; inverse variation

Answers

24a. {(1, red), (2, red), (4, red), (5, red), (7, red), (9, red), (1, green), (2, green), (4, green), (5, green), (7, green), (9, green), (1, yellow), (2, yellow), (4, yellow), (5, yellow), (7, yellow), (9, yellow)}

24b. 4

Chapter Test

Answers

4.

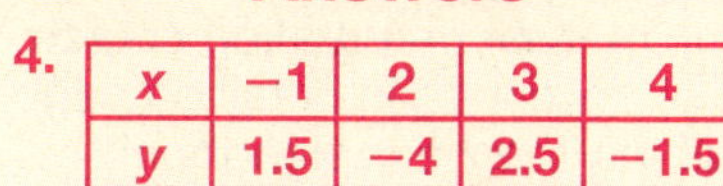

x	−1	2	3	4
y	1.5	−4	2.5	−1.5

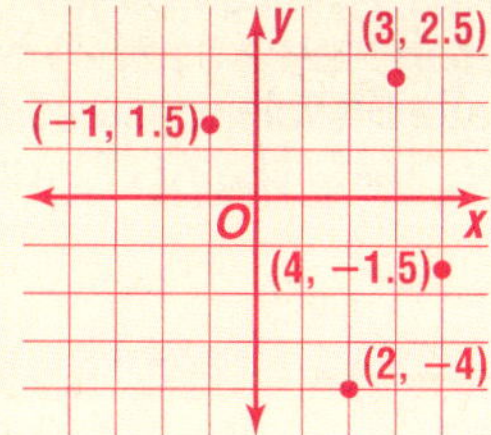

domain: {−1, 2, 3, 4}, range: {1.5, −4, 2.5, −1.5}

8.

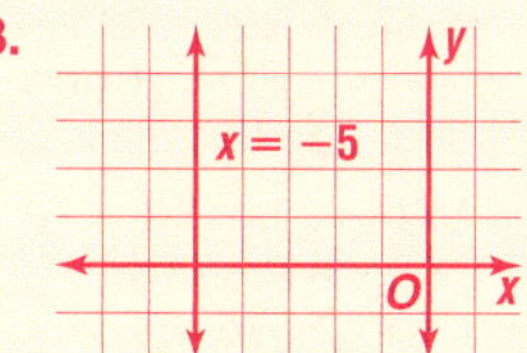

9.

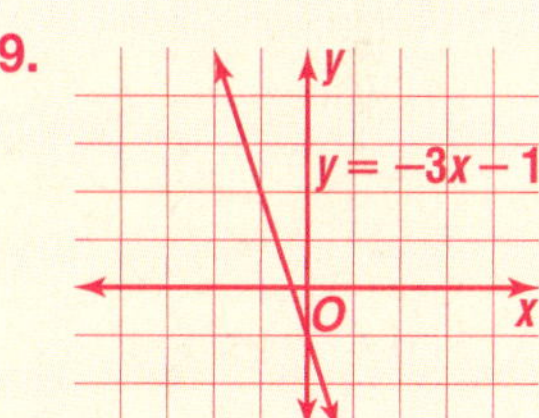

10.

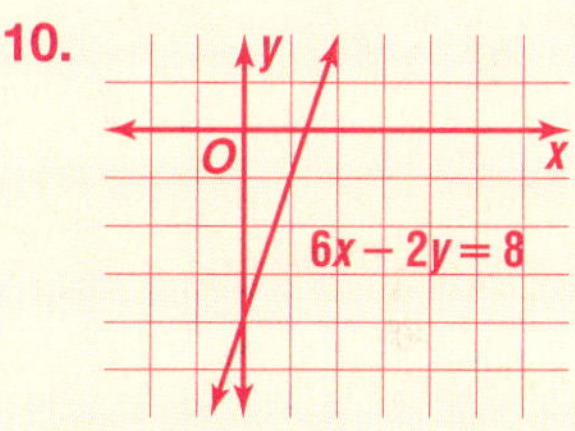

***Assessment*, pp. 281–282**

NAME ______ DATE ______ PERIOD ______

Chapter 6 Test, Form 2B

For Questions 1–3, refer to the relation given by the set of ordered pairs {(1, 4), (0, −2), (2, 3), (−1, 4), (−3, 0)}.

1. Express the relation as a graph.
2. What is the domain of the relation? 2. {−3, −1, 0, 1, 2}
3. What is the range of the relation? 3. {−2, 0, 3, 4}
4. Two dice are rolled. Express a relation in a table for the possible numbers showing on the first die x and the second die y if the outcome is two odd numbers whose sum is at least 8.

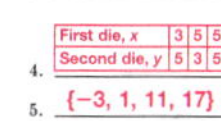

First die, x	3	5	5
Second die, y	5	3	5

5. Find the domain of $x - 2y = 1$ if the range is $\{-2, 0, 5, 8\}$. 5. {−3, 1, 11, 17}
6. Solve $y = -2x + 7$ if the domain is $\{-2, 0, 4, 7\}$. 6. {(−2, 11), (0, 7), (4, −1), (7, −7)}
7. A solution of the equation $4x = 10 - cy$ is $(5, -2)$. What is the value of c? 7. 5
8. A teacher wishes to schedule a quiz or exam during part of a two-hour class. The equation $\ell + e = 120$ can be used to find the available lecture time ℓ given that e minutes are planned for the quiz or exam. Determine the ordered pairs that satisfy the equation if the domain is $\{20, 30, 45, 60\}$. 8. {(20, 100), (30, 90), (45, 75), (60, 60)}

Determine whether each equation is a linear equation. Explain. If the equation is linear, identify A, B, and C.

9. $9y = 4x - 18$ 9. yes; $A = 4, B = -9, C = 18$ or $A = -4, B = 9, C = -18$
10. $xy = 4$ 10. No; the term xy has two variables. The equation cannot be written in the form $Ax + By = C$.
11. $x^2 - 4 = y$ 11. No; the term has a variable to a power other than 1. The term x^2 cannot be written in the form Ax.

Graph each equation.

12. $2 = x - y$
13. $2y + x = -3$

Preparing for Standardized Tests

Pages 280–281 are part of a complete test preparation course that is described in detail on page T9 of the Teacher's Handbook. The test items on these pages were written in the same style as those in state proficiency tests and standardized tests like ACT and SAT.

Diagnosis and Prescription

Each of the 10 test questions on page 281 is cross-referenced to the lesson where that SAT or ACT skill is covered. If students miss a particular type of problem, you can have them study that skill.

(See chart at the bottom of page 281. Note that SPT = State Proficiency Test, SAT = Scholastic Assessment Test, and ACT = American College Test.)

***Assessment*, p. 286**

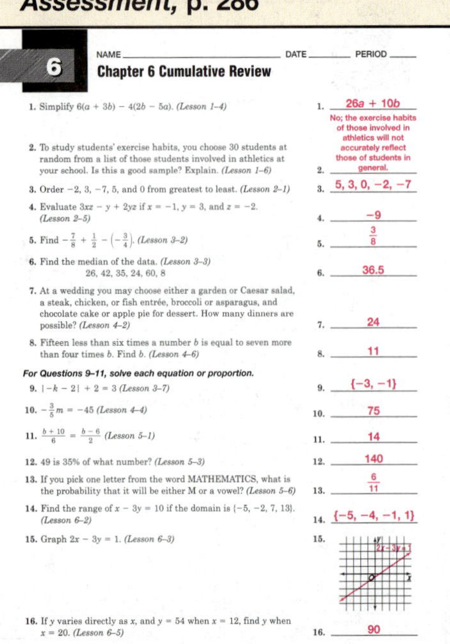

6 NAME ____________ DATE ______ PERIOD ______

Chapter 6 Cumulative Review

1. Simplify $6(a + 3b) - 4(2b - 5a)$. *(Lesson 1–4)* — 1. $26a + 10b$
2. To study students' exercise habits, you choose 30 students at random from a list of those students involved in athletics at your school. Is this a good sample? Explain. *(Lesson 1–6)* — 2. No; the exercise habits of those involved in athletics will not accurately reflect those of students in general.
3. Order −2, 3, −7, 5, and 0 from greatest to least. *(Lesson 2–1)* — 3. 5, 3, 0, −2, −7
4. Evaluate $3xz - y + 2yz$ if $x = -1$, $y = 3$, and $z = -2$. *(Lesson 2–5)* — 4. −9
5. Find $-\frac{7}{8} + \frac{1}{2} - \left(-\frac{3}{4}\right)$. *(Lesson 3–2)* — 5. $\frac{3}{8}$
6. Find the median of the data. *(Lesson 3–3)* 26, 42, 35, 24, 60, 8 — 6. 36.5
7. At a wedding you may choose either a garden or Caesar salad, a steak, chicken, or fish entrée, broccoli or asparagus, and chocolate cake or apple pie for dessert. How many dinners are possible? *(Lesson 4–2)* — 7. 24
8. Fifteen less than six times a number b is equal to seven more than four times b. Find b. *(Lesson 4–6)* — 8. 11

For Questions 9–11, solve each equation or proportion.

9. $|-k - 2| + 2 = 3$ *(Lesson 3–7)* — 9. {−3, −1}
10. $-\frac{3}{5}m = -45$ *(Lesson 4–4)* — 10. 75
11. $\frac{b + 10}{6} = \frac{b - 6}{2}$ *(Lesson 5–1)* — 11. 14
12. 49 is 35% of what number? *(Lesson 5–3)* — 12. 140
13. If you pick one letter from the word MATHEMATICS, what is the probability that it will be either M or a vowel? *(Lesson 5–6)* — 13. $\frac{6}{11}$
14. Find the range of $x - 3y = 10$ if the domain is {−5, −2, 7, 13}. *(Lesson 6–2)* — 14. {−5, −4, −1, 1}
15. Graph $2x - 3y = 1$. *(Lesson 6–3)* — 15. [graph: $2x - 3y = 1$]
16. If y varies directly as x, and $y = 54$ when $x = 12$, find y when $x = 20$. *(Lesson 6–5)* — 16. 90

CHAPTER 6 Preparing for Standardized Tests

Probability and Counting Problems

You will need to solve problems involving combinations, permutations, and probability. Familiarize yourself with the terms *combination*, *permutation*, *tree diagram*, and *outcomes*. You will also need to know the definition of probability.

$$P(\text{event}) = \frac{\text{number of favorable outcomes}}{\text{number of possible outcomes}}$$

Test-Taking Tip

Sketching a picture can help you understand and solve many problems.

Example 1

How many different five-digit ZIP codes can be made if no digit is repeated in a code?

(A) 15,120 (B) 30,240

(C) 60,480 (D) 100,000

Hint Be sure you understand the question before you start to solve it.

Solution A ZIP code consists of five digits. The digits can be 0–9. There are ten possible digits. The problem states that no digit can appear more than once in these ZIP codes. Find the number of possible choices for each of the five positions.

____ ____ ____ ____ ____

There are 10 possible choices for the first position. There are 9 choices for the second position, because one digit was used for the first position. Following the same reasoning, there are 8 choices for the third position, 7 choices for the fourth position, and 6 choices for the fifth position. Multiply these numbers to find the total number of different codes. Use a calculator.

$$10 \times 9 \times 8 \times 7 \times 6 = 30{,}240$$

So, there are 30,240 possible ZIP codes. Thus, the answer is B.

Example 2

Five cards are numbered from 0 to 4. Suppose two cards are selected at random without replacement. What is the probability that the sum of the numbers on the two cards is an even number?

(A) $\frac{1}{5}$ (B) $\frac{3}{10}$ (C) $\frac{2}{5}$ (D) $\frac{1}{2}$ (E) $\frac{3}{5}$

Hint Decide whether or not order matters. In this problem order doesn't matter. So, it is a combination problem.

Solution First, write the possible outcomes of selecting two cards at random.

0 1	0 2	0 3	0 4
1 2	1 3	1 4	
2 3	2 4		
3 4			

There are 10 possible combinations. There are 4 ways to get a sum that is an even number.

$$P(\text{even}) = \frac{4}{10} \text{ or } \frac{2}{5}$$

The answer is C.

Resource Manager

Reproducible Masters

- *Assessment*, pp. 286–288

Preparing for Standardized Tests
For test-taking strategies and more practice, see pages 732–749.

After you work each problem, record your answer on the answer sheet provided or on a sheet of paper.

Multiple Choice

1. Suppose the spinner shown is spun. What is the probability of landing on an even number? *(Lesson 5–6)* **A**

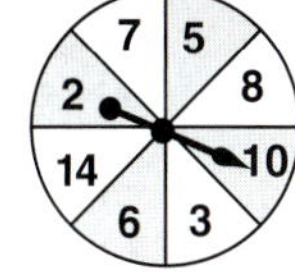

   (A) $\frac{5}{8}$ (B) $\frac{1}{2}$
   (C) $\frac{3}{8}$ (D) $\frac{3}{4}$

2. How many different starting squads of 6 players can be picked from 11 volleyball players? *(Lesson 4–2)* **C**
   (A) 120 (B) 246 (C) 462 (D) 720

3. Diego has a collection of music CDs. Ten are rock, 5 are jazz, and 8 are country. If he chooses a CD at random, what is the probability that he will *not* choose a jazz CD? *(Lesson 5–6)* **E**
   (A) $\frac{5}{23}$ (B) $\frac{5}{18}$ (C) $\frac{15}{23}$
   (D) $\frac{13}{18}$ (E) $\frac{18}{23}$

4. How many different 3-digit security codes can you make using the numbers 1, 2, and 3? You can use the numbers more than once. *(Lesson 4–2)* **D**
   (A) 6 (B) 9 (C) 18 (D) 27

5. A bag contains only white and blue marbles. The probability of selecting a blue marble is $\frac{1}{5}$. The bag contains 200 marbles. If 100 white marbles are added to the bag, what is the probability of selecting a white marble? *(Lesson 5–6)* **E**
   (A) $\frac{2}{15}$ (B) $\frac{7}{15}$ (C) $\frac{8}{15}$
   (D) $\frac{4}{5}$ (E) $\frac{13}{15}$

6. On Tuesday, the temperature fell 6° in 2 hours. Which of these expresses the temperature change per hour? *(Lesson 2–6)*
   (A) −3° (B) −6° (C) 6° (D) 12° **A**

7. Tomorrow 352 students will take a test in the cafeteria. If each table seats 12 students and there are already 15 tables in the cafeteria, how many additional tables will be needed? *(Lesson 1–5)* **B**
   (A) 14 (B) 15 (C) 23 (D) 24

8. Ellen earns $86.80 for working 14 hours. At this rate, how much would she earn if she worked 20 hours? *(Lesson 5–1)* **C**
   (A) $6.20 (B) $60.76
   (C) $124.00 (D) $1,215.20

Grid In

9. If $5x - 4 = x - 1$, what is the value of x? *(Lesson 4–6)* **3/4**

Extended Response

10. The table shows the average monthly low temperatures for Detroit, Michigan. *(Lesson 3–2)*

Month	Temperature (°F)
January	44
February	46
March	54
April	60
May	66
June	72
July	74
August	74
September	70
October	61
November	52
December	45

Source: Excite Travel

Part A Organize the data in a stem-and-leaf plot. **See margin.**
Part B Find the mean, median, and mode. Tell which of these best represents the data. **See margin.**

www.algconcepts.com/standardized_test

A bubble-in answer sheet for these practice problems is available on page A1 of the *Chapter 6 Resource Masters.*

Additional Practice

Additional test practice questions are available in the *Chapter 6 Resource Masters*, pp. 287–288.

Answers

10A.

Stem	Leaf
4	4 5 6
5	2 4
6	0 1 6
7	0 2 4 4 $7|2 = 72°F$

10B. The mean is $59.8\overline{3}$.
The median is the mean of the two middle values, 60 and 61. The median is 60.5.
The mode is 74.
The mean or median best represents the data. The mode is too high.

***Assessment*, pp. 287–288**

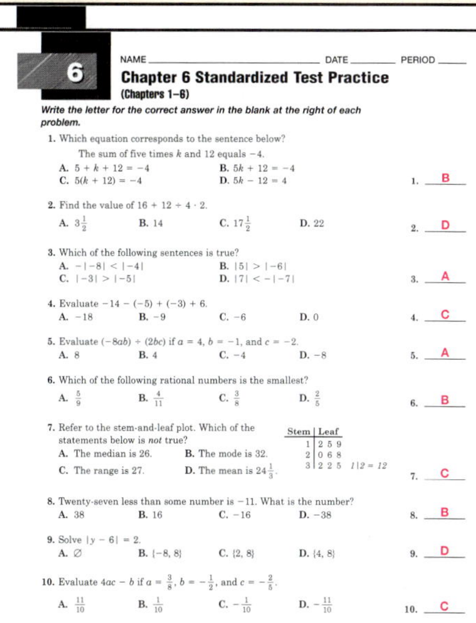
NAME ________ DATE ______ PERIOD _____

6 **Chapter 6 Standardized Test Practice**
(Chapters 1–6)

Write the letter for the correct answer in the blank at the right of each problem.

1. Which equation corresponds to the sentence below?
   The sum of five times k and 12 equals −4.
   A. $5 + k + 12 = -4$ B. $5k + 12 = -4$
   C. $5(k + 12) = -4$ D. $5k - 12 = 4$ 1. **B**

2. Find the value of $16 + 12 \div 4 \cdot 2$.
   A. $3\frac{1}{2}$ B. 14 C. $17\frac{1}{2}$ D. 22 2. **D**

3. Which of the following sentences is true?
   A. $-|-8| < |-4|$ B. $|5| > |-6|$
   C. $|-3| > |-5|$ D. $|7| < -|-7|$ 3. **A**

4. Evaluate $-14 - (-5) + (-3) + 6$.
   A. −18 B. −9 C. −6 D. 0 4. **C**

5. Evaluate $(-8ab) \div (2bc)$ if $a = 4$, $b = -1$, and $c = -2$.
   A. 8 B. 4 C. −4 D. −8 5. **A**

6. Which of the following rational numbers is the smallest?
   A. $\frac{5}{9}$ B. $\frac{4}{11}$ C. $\frac{3}{8}$ D. $\frac{2}{5}$ 6. **B**

7. Refer to the stem-and-leaf plot. Which of the statements below is *not* true?
   A. The median is 26. B. The mode is 32.
   C. The range is 27. D. The mean is $24\frac{1}{3}$. 7. **C**

Stem	Leaf
1	2 5 9
2	0 6 8
3	2 2 5 $1|2 = 12$

8. Twenty-seven less than some number is −11. What is the number?
   A. 38 B. 16 C. −16 D. −38 8. **B**

9. Solve $|y - 6| = 2$.
   A. ∅ B. {−8, 8} C. {2, 8} D. {4, 8} 9. **D**

10. Evaluate $4ac - b$ if $a = \frac{3}{8}$, $b = -\frac{1}{2}$, and $c = -\frac{2}{5}$.
   A. $\frac{11}{10}$ B. $\frac{1}{10}$ C. $-\frac{1}{10}$ D. $-\frac{11}{10}$ 10. **C**

11. Two friends plan an evening of dinner followed by a movie. There are four movies they are interested in seeing and three restaurants near the theater. From how many possibilities of a movie and a restaurant do they have to choose?
   A. 4 B. 7 C. 12 D. 14 11. **C**

Chapter 6 Probability and Counting Problems		
Ex. 1	permutations	SPT
Ex. 2	probability	SAT
1	probability	SPT
2	combinations	SPT
3	probability	SPT
4	permutations	SPT
5	probability	SAT
6	integer division	SPT
7	word problem	ACT
8	rate problem	SPT
9	solve an equation	SAT
10	make a stem & leaf plot	SPT

CHAPTER 7

Resource Manager

Linear Equations

Instructional Objectives

Lesson (pages)	Objectives	NCTM Standards 2000	State/Local Objectives
7–1 (284–289)	Find the slope of a line given the coordinates of two points on the line.	1, 3, 6, 8, 9	
7–2 (290–295)	Write a linear equation in point-slope form given the coordinates of a point on the line and the slope of the line.	1, 6, 8, 9	
7–3 (296–301)	Write a linear equation in slope-intercept form given the slope and y-intercept.	1, 6, 8, 9	
7–4 (302–307)	Graph and interpret points on scatter plots.	1, 2, 3, 5, 6, 8, 9, 10	
Investigation (308–309)	Use best-fit lines to make predictions.	1, 2, 3, 4, 6, 7, 8, 9, 10	
7–5 (310–315)	Graph linear equations by using the x- and y-intercepts or the slope and y-intercept.	1, 3, 6, 8, 9	
7–6 (316–321)	Explore the effects of changing the slopes and y-intercepts of linear functions.	1, 2, 3, 6, 8, 9	
7–7 (322–327)	Write an equation of a line that is parallel or perpendicular to the graph of a given equation and that passes through a given point.	1, 3, 6, 8, 9	

Key to NCTM Standards 2000
1 Number & Operations; **2** Algebra; **3** Geometry; **4** Measurement; **5** Data Analysis & Probability; **6** Problem Solving; **7** Reasoning & Proof; **8** Communications; **9** Connections; **10** Representation

Suggested Pacing

See page T13 for a complete course-planning calendar.

Standard refers to schedules that provide 45- to 55-minute periods that meet each day.
Block refers to schedules that provide approximately 90-minute periods which may meet every day for one semester or every other day over two semesters.

PACING	DAY 1	DAY 2	DAY 3	DAY 4	DAY 5	DAY 6
Standard Core (Chapters 1–13)	Lesson 7–1	Lesson 7–2		Lesson 7–3	Lesson 7–4	
Standard Enhanced (Chapters 1–15)	Lesson 7–1	Lesson 7–2		Lesson 7–3	Lesson 7–4	INV
Block Core (Chapters 1–13)	Chapter 6 Test & Lesson 7–1	Lessons 7–2 & 7–3	Lesson 7–4 & INV	Lessons 7–5 & 7–6	Lesson 7–7	SG+A
Block Enhanced (Chapters 1–15)	Chapter 6 Test & Lesson 7–1	Lessons 7–2 & 7–3	Lesson 7–4 & INV	Lessons 7–5 & 7–6	Lesson 7–7 & SG+A	Chapter Test & Lesson 8–1

Instructional Resources

Lesson	Materials and Manipulatives (see below for Glencoe Manipulative Resources)	Blackline Masters (page numbers)								5-Minute Check Transparencies
		Chapter 7 Resource Masters					Hands-On Algebra*	School-to-Workplace*	Graphing Calculator Masters*	
		Study Guide	Practice (Skills & Average)	Reading to Learn Mathematics	Enrichment	Assessment				
7–1	coordinate planes [4] grid paper [1, 4] straightedge [1, 2, 4] geoboards	289	290–291	292	293		80, 81			7–1
7–2	geoboard or dot paper	294	295–296	297	298					7–2
7–3		299	300–301	302	303	335	82	7		7–3
7–4	straightedge [1, 2, 4] grid paper [1, 4] rulers [1, 2, 4] almanacs	304	305–306	307	308	334	83		21, 22	7–4
Investigation	straightedge [1, 2, 4] grid paper [1, 4] graphing calculator or software									
7–5	straightedge [1, 2, 4] coordinate planes [4]	309	310–311	312	313					7–5
7–6	straightedge [1, 2, 4] coordinate planes [4] graphing calculator	314	315–316	317	318		84		20	7–6
7–7	grid paper [1, 4] protractor straightedge [1, 2, 4] coordinate planes [4]	319	320–321	322	323	335	85, 86			7–7
Study Guide & Assessment/ Chapter Test	straightedge [1, 2, 4] coordinate planes [4]					325–333, 336–338				

*See page 282c for examples of these instructional materials.

Key to Glencoe Manipulative Resources

[1]Classroom Manipulative Resources [2]Student Manipulative Resources [3]Overhead Manipulative Resources [4]Hands-On Algebra Masters

INV = Investigation *SG+A = Study Guide and Assessment*

DAY 7	DAY 8	DAY 9	DAY 10	DAY 11	DAY 12	DAY 13
INV	Lesson 7–5	Lesson 7–6		Lesson 7–7	SG+A	Chapter Test
Lesson 7–5	Lesson 7–6		Lesson 7–7	SG+A	Chapter Test	
Chapter Test & Lesson 8–1						

The pages shown on this page are a small sample of the materials available on *TeacherWorks: All-in-One Lesson Planner and Resource Center.*

This CD-ROM includes all of the blackline masters and transparencies available for this program.

It also includes a lesson planner and interactive Teacher Edition, so you can customize lesson plans and reproduce classroom resources quickly and easily, from just about anywhere.

Applications

School-to-Workplace Masters, p. 7

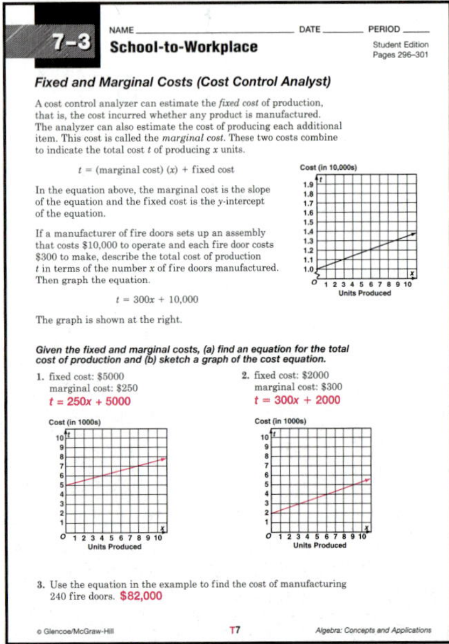

7-3 NAME ________ DATE ______ PERIOD ______

School-to-Workplace Student Edition Pages 296–301

Fixed and Marginal Costs (Cost Control Analyst)

A cost control analyzer can estimate the *fixed cost* of production, that is, the cost incurred whether any product is manufactured. The analyzer can also estimate the cost of producing each additional item. This cost is called the *marginal cost.* These two costs combine to indicate the total cost t of producing x units.

t = (marginal cost) (x) + fixed cost

In the equation above, the marginal cost is the slope of the equation and the fixed cost is the y-intercept of the equation.

If a manufacturer of fire doors sets up an assembly that costs $10,000 to operate and each fire door costs $300 to make, describe the total cost of production t in terms of the number x of fire doors manufactured. Then graph the equation.

$t = 300x + 10{,}000$

The graph is shown at the right.

Given the fixed and marginal costs, (a) find an equation for the total cost of production and (b) sketch a graph of the cost equation.

1. fixed cost: $5000
   marginal cost: $250
   $t = 250x + 5000$
2. fixed cost: $2000
   marginal cost: $300
   $t = 300x + 2000$

3. Use the equation in the example to find the cost of manufacturing 240 fire doors. **$82,000**

© Glencoe/McGraw-Hill T7 *Algebra: Concepts and Applications*

Manipulatives/Modeling

Hands-On Algebra Masters, pp. 80–86

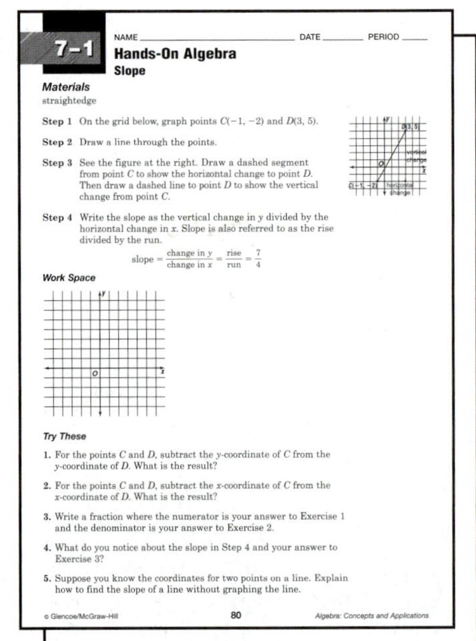

7-1 NAME ________ DATE ______ PERIOD ______

Hands-On Algebra
Slope

Materials
straightedge

Step 1 On the grid below, graph points $C(-1, -2)$ and $D(3, 5)$.

Step 2 Draw a line through the points.

Step 3 See the figure at the right. Draw a dashed segment from point C to show the horizontal change to point D. Then draw a dashed line to point D to show the vertical change from point C.

Step 4 Write the slope as the vertical change in y divided by the horizontal change in x. Slope is also referred to as the rise divided by the run.

$$\text{slope} = \frac{\text{change in } y}{\text{change in } x} = \frac{\text{rise}}{\text{run}} = \frac{7}{4}$$

Work Space

Try These

1. For the points C and D, subtract the y-coordinate of C from the y-coordinate of D. What is the result?
2. For the points C and D, subtract the x-coordinate of C from the x-coordinate of D. What is the result?
3. Write a fraction where the numerator is your answer to Exercise 1 and the denominator is your answer to Exercise 2.
4. What do you notice about the slope in Step 4 and your answer to Exercise 3?
5. Suppose you know the coordinates for two points on a line. Explain how to find the slope of a line without graphing the line.

© Glencoe/McGraw-Hill 80 *Algebra: Concepts and Applications*

Technology/Multimedia

Graphing Calculator Masters, pp. 20–22

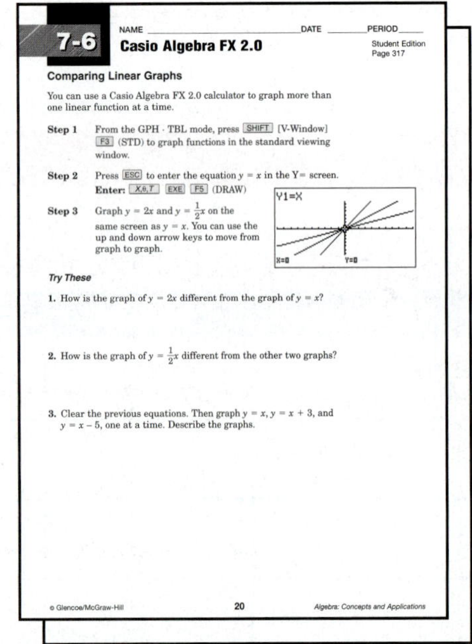

7-6 NAME ________ DATE ______ PERIOD ______

Casio Algebra FX 2.0 Student Edition Page 317

Comparing Linear Graphs

You can use a Casio Algebra FX 2.0 calculator to graph more than one linear function at a time.

Step 1 From the GPH · TBL mode, press SHIFT [V-Window] F3 (STD) to graph functions in the standard viewing window.

Step 2 Press ESC to enter the equation $y = x$ in the Y= screen.
Enter: X,θ,T EXE F5 (DRAW)

Step 3 Graph $y = 2x$ and $y = \frac{1}{2}x$ on the same screen as $y = x$. You can use the up and down arrow keys to move from graph to graph.

Try These

1. How is the graph of $y = 2x$ different from the graph of $y = x$?
2. How is the graph of $y = \frac{1}{2}x$ different from the other two graphs?
3. Clear the previous equations. Then graph $y = x$, $y = x + 3$, and $y = x - 5$, one at a time. Describe the graphs.

© Glencoe/McGraw-Hill 20 *Algebra: Concepts and Applications*

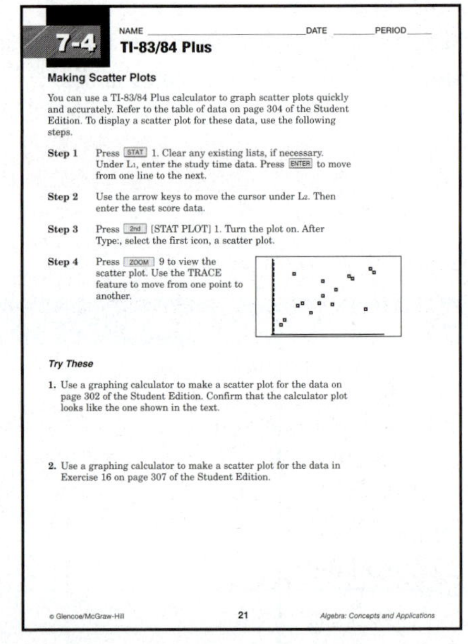

7-4 NAME ________ DATE ______ PERIOD ______

TI-83/84 Plus

Making Scatter Plots

You can use a TI-83/84 Plus calculator to graph scatter plots quickly and accurately. Refer to the table of data on page 304 of the Student Edition. To display a scatter plot for these data, use the following steps.

Step 1 Press STAT 1. Clear any existing lists, if necessary. Under L_1, enter the study time data. Press ENTER to move from one line to the next.

Step 2 Use the arrow keys to move the cursor under L_2. Then enter the test score data.

Step 3 Press 2nd [STAT PLOT] 1. Turn the plot on. After Type:, select the first icon, a scatter plot.

Step 4 Press ZOOM 9 to view the scatter plot. Use the TRACE feature to move from one point to another.

Try These

1. Use a graphing calculator to make a scatter plot for the data on page 302 of the Student Edition. Confirm that the calculator plot looks like the one shown in the text.
2. Use a graphing calculator to make a scatter plot for the data in Exercise 16 on page 307 of the Student Edition.

© Glencoe/McGraw-Hill 21 *Algebra: Concepts and Applications*

Assessment Resources

Type	Student Edition	Teacher's Wraparound Edition	Chapter 7 Resource Masters
Ongoing Assessment	Quizzes 1 and 2, pp. 301, 321	5-Minute Check, pp. 284, 290, 296, 302, 310, 316, 322	Mid-Chapter Test, p. 334 Quizzes A and B, p. 335
Mixed Review	Mixed Review, pp. 289, 295, 301, 307, 315, 321, 327 Standardized Test Practice, Chapters 1–7, pp. 332–333		Cumulative Review, p. 336 Standardized Test Practice, pp. 337–338
Error Analysis	You Decide, pp. 287, 299, 314	Error Analysis, pp. 287, 293, 299, 305, 314, 319, 325	
Standardized Test Prep	Standardized Test Practice, pp. 289, 295, 301, 307, 315, 321, 327 Standardized Test Practice, Chapters 1–7, pp. 332–333		Standardized Test Practice, pp. 337–338
Open-Ended Assessment	Writing Math, pp. 305, 319 Investigation, pp. 308–309 Portfolio, pp. 283, 309	Modeling: pp. 295, 307 Speaking: pp. 301, 315 Writing: pp. 289, 321, 327	Extended Response Assessment, p. 333
Chapter Assessment	Study Guide and Assessment, pp. 328–330 Chapter Test, p. 331		Multiple-Choice Tests (Forms 1A, 1B), pp. 325–328 Free-Response Tests (Forms 2A, 2B), pp. 329–332

Additional Chapter Resources

Student Edition
Hands-On Algebra, p. 324
Graphing Calculator Exploration, p. 317

Teacher's Classroom Resources
Manipulatives/Modeling
Teacher's Guide for Overhead Manipulative Resources

Meeting Individual Needs
Prerequisite Skills Booklet
Spanish Study Guide and Assessment, pp. 49–55, 117–118

Teaching Aids
Solutions Manual
5-Minute Check Transparencies

Glencoe Technology

Instructional
AlgePASS CD-ROM, Lessons 17
Interactive Chalkboard CD-ROM
StudentWorks Plus CD-ROM
Multimedia Applications CD-ROM, Activity 6
Vocabulary PuzzleMaker (online)

Assessment
ExamView® Pro

Teaching Aids
Answer Key Maker CD-ROM
TeacherWorks CD-ROM

*Visit **www.algconcepts.com** for data updates, career information, games, and other interactive activities.*

CHAPTER 7 PREVIEW

Mathematics of the Chapter

This chapter introduces linear equations. The slope of a linear graph is first explored. Then linear functions are represented with equations in point-slope form and slope-intercept form. Scatter plots are created and interpreted. Linear equations are then graphed by using the *x*- and *y*-intercepts, and also by using the slope and *y*-intercept. The graphs of families of linear functions are investigated, and then the equations of parallel and perpendicular lines are explored.

What You'll Learn

Have students read over the lists of key ideas and key vocabulary. Have them make a list of any words with which they are not familiar.

Why It's Important

Point out to students that this is only one of many reasons why each objective is important. Others are provided in each lesson.

CHAPTER 7 Linear Equations

What You'll Learn

Key Ideas

- Find the slope of a line. *(Lesson 7–1)*
- Write linear equations in slope-intercept form and in point-slope form. *(Lessons 7–2 and 7–3)*
- Graph and interpret points on scatter plots. *(Lesson 7–4)*
- Graph and explore linear equations and families of linear equations. *(Lessons 7–5 and 7–6)*
- Write equations of parallel and perpendicular lines. *(Lesson 7–7)*

Key Vocabulary

point-slope form *(p. 290)*
rate of change *(p. 285)*
slope *(p. 284)*
slope-intercept form *(p. 296)*

Why It's Important

Entertainment The Mauch Chunk Switchback Railway, called the first American roller coaster, was built in Pennsylvania in 1827 to carry coal to freight trains. When rail lines were built closer to the mine, the switchback railway became a ride that attracted thousands of tourists every year.

Linear equations represent relationships between real-life quantities. Slope tells you how a quantity changes over time. You will find the slope of the steepest wooden roller coaster in Lesson 7–1.

CHAPTER 7 LINKS

Lesson	**7–1**	**7–2**	**7–3**	**7–4**	**7–5**	**7–6**	**7–7**
Math in the Workplace	Construction	Population	Catering	Insurance	Rates	Business	Surveying
Applications and Connections	Income Entertainment Driving Money Communication	Sports Physical Science Transportation Savings Construction	Business Recycling Work Education	Word Processing School Weather Health Physical Science Architecture	Mail Temperature Physical Science Entertainment	Shipping Population Jewelry Animals Videos	Exercise
Math Integration			Measurement				Geometry

Study these lessons to improve your skills.

Check Your Readiness

Lesson 2–4, pp. 70–74

Find each difference.

1. $6 - 2$ **4**
2. $15 - 7$ **8**
3. $-3 - 2$ **−5**
4. $-5 - (-5)$ **0**
5. $8 - 9$ **−1**
6. $7 - (-2)$ **9**
7. $-2 - 4$ **−6**
8. $-4 - 8$ **−12**
9. $14 - (-4)$ **18**

Lesson 2–6, pp. 82–85

Find the value of each expression.

10. $\frac{5-1}{6-4}$ **2**
11. $\frac{10-1}{7-4}$ **3**
12. $\frac{8-5}{9-5}$ $\frac{3}{4}$
13. $\frac{-2-(-1)}{5-3}$ $-\frac{1}{2}$
14. $\frac{-3-0}{8-6}$ $-\frac{3}{2}$
15. $\frac{-12-(-9)}{2-(-1)}$ **−1**
16. $\frac{-8-(-4)}{-2-(-1)}$ **4**
17. $\frac{3-(-3)}{5-10}$ $-\frac{6}{5}$
18. $\frac{-3-2}{0-3}$ $\frac{5}{3}$

Lesson 2–2, pp. 58–63

Write the ordered pair for each point.

19. A **(3, 4)**
20. B **(−3, −4)**
21. C **(3, 0)**
22. D **(−2, 1)**
23. E **(0, −3)**
24. F **(2, −3)**

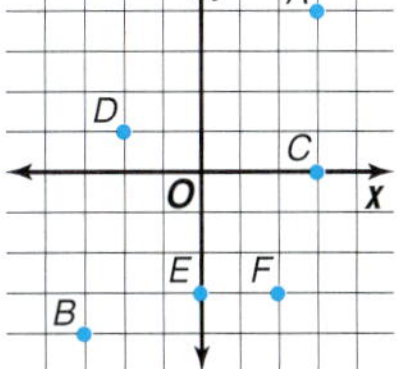

FOLDABLES™ Study Organizer

Make this Foldable to help you organize your Chapter 7 notes. Begin with four sheets of grid paper.

1. **Fold** each sheet in half from top to bottom.

2. **Cut** along each fold. Staple the eight half-sheets together to form a booklet.

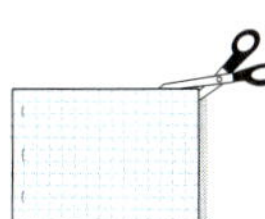

3. **Cut** tabs into the margin. The top tab is 4 lines wide, the next tab is 8 lines wide, and so on.

4. **Label** each page with a lesson number and title. The last tab is for vocabulary.

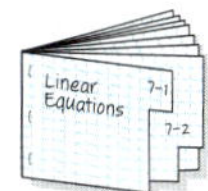

Reading and Writing As you read and study the chapter, fill the journal with notes, diagrams, and examples.

www.algconcepts.com/chapter_readiness

FOLDABLES™ Study Organizer

On each tab of this Foldable students should write important equations and/or sketch graphs using the methods presented in each lesson.

Check Your Readiness

This section provides a review of the basic concepts needed before beginning Chapter 11. Lesson references are included for additional student help.

Use Exercises	To Prepare for Lesson(s)
1–9	7–1, 7–4
10–18	7–1, 7–2, 7–3
19–24	7–2, 7–3

Quick Review Math Handbook

Glencoe Algebra Lesson	Quick Review Math Handbook Hot Topic
7–1	2–4, 6–8
7–2	6–8
7–3	1–5, 6–7, 6–8
7–4	4–3
Ch. 7 Inv.	4–3
7–5	6–2, 6–7
7–6	6–7, 6–8
7–7	6–8

Noteables™

Interactive Study Notebook with Foldables™

This note-taking guide reinforces excellent note-taking skills. It includes:

- ✓ **Build Your Vocabulary** tool for including mathematical terminology in notes.
- ✓ **Review It** activities with links to prior lessons and items to include in **Foldables™**.
- ✓ **Bringing It All Together** feature to help students review for the Chapter Test.
- ✓ **Are You Ready for the Chapter Test?** feature to allow students to assess their own readiness for the Chapter Test.
- ✓ **Teacher Edition** with transparencies to guide note taking. Corresponds to the *Teacher's Wraparound Edition* In-Class Examples and *Interactive Chalkboard CD-ROM.*

Lesson 7-1

1 FOCUS

5-Minute Check
Chapter 6

1. Determine the domain and range of the relation graphed below.

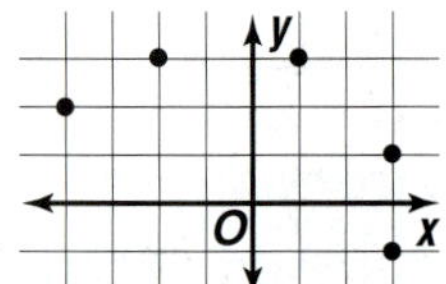

domain: {−4, −2, 1, 3}, range: {−1, 1, 2, 3}

2. Suppose y varies inversely as x and $y = 3$ when $x = 10$. Find y when $x = 5$. 6

Find the domain of each equation if the range is {−2, 0, 4}.

3. $y = 4x$ $\left\{-\frac{1}{2}, 0, 1\right\}$
4. $y = 2x - 1$ $\left\{-\frac{1}{2}, \frac{1}{2}, \frac{5}{2}\right\}$

Motivating the Lesson

Hands-On Activity Draw a coordinate grid on the board or overhead, sketch a line on it, and label two points on the line with their coordinates. Give students a brief introduction to *slope* as the ratio of rise over run. Show them how to draw a right triangle with the two points as the endpoints of the hypotenuse and the two legs of the triangle as the rise and run. Then separate your students into groups and give each group some grid paper. Instruct students to graph two points and connect them with a line segment. Have them draw a triangle similar to your example and count the grid squares to find the rise, run, and slope of the segment.

MODELING

Alternative hands-on options using a straightedge and grid paper are available for teaching this lesson.

7-1 Slope

What You'll Learn

You'll learn to find the slope of a line given the coordinates of two points on the line.

Why It's Important

Construction Building codes require that certain slopes be used for stairways. *See Exercise 28.*

The rows in some movie theaters are so steep that there are no bad seats. We say that the theater has a steep **slope**. Slope is the ratio of the *rise*, or the vertical change, to the *run*, or the horizontal change.

$$\text{slope} = \frac{\text{rise}}{\text{run}} = \frac{9}{10} \quad \frac{\textit{vertical change}}{\textit{horizontal change}}$$

The slope of the floor in the theater at the right is $\frac{9}{10}$.

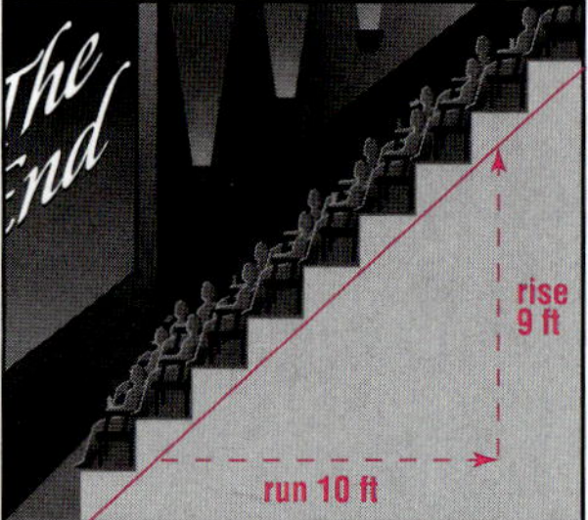

On the graph below, the line passes through points at (4, 0) and (11, 3). The change in y or rise is $3 - 0$ or 3, while the change in x or run is $11 - 4$ or 7. Therefore, the slope of this line is $\frac{3}{7}$.

$$\text{slope} = \frac{\text{rise}}{\text{run}} = \frac{\text{change in } y}{\text{change in } x}$$

Slope

Words: The slope of a line is the ratio of the change in y to the corresponding change in x.

Model:

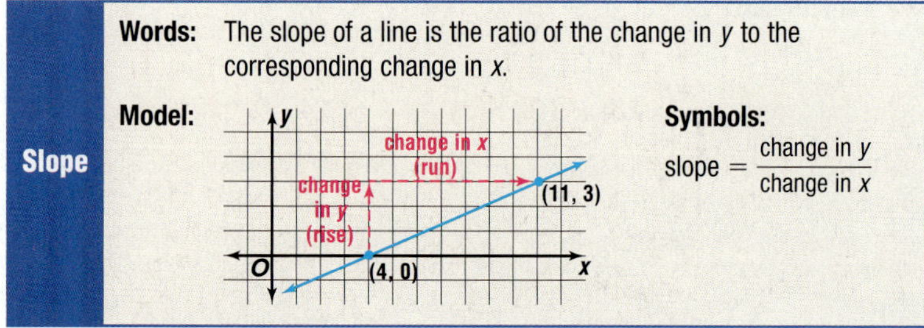

Symbols: $\text{slope} = \frac{\text{change in } y}{\text{change in } x}$

Examples

Determine the slope of each line.

1

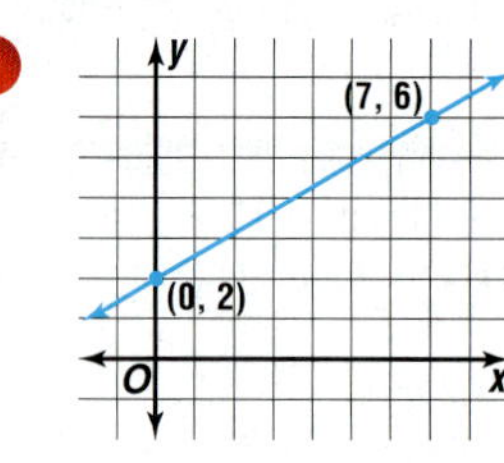

$$\text{slope} = \frac{\text{change in } y}{\text{change in } x} = \frac{6 - 2}{7 - 0} = \frac{4}{7}$$

The slope is $\frac{4}{7}$.

2

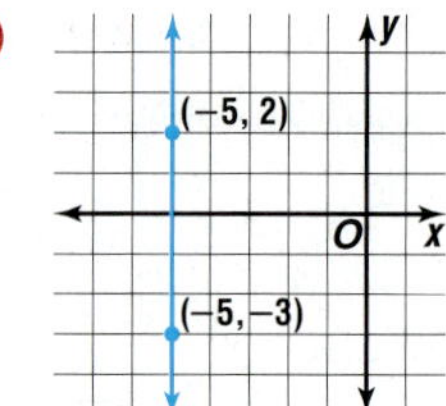

$$\text{slope} = \frac{\text{change in } y}{\text{change in } x} = \frac{-3 - 2}{-5 - (-5)} = \frac{-5}{0}$$

The slope is undefined. *Why?*

Resource Manager

Reproducible Masters

Chapter 7 Resource Masters
- *Study Guide*, p. 289
- *Skills Practice*, p. 290
- *Practice*, p. 291
- *Reading to Learn Mathematics*, p. 292
- *Enrichment*, p. 293

Hands-On Algebra, pp. 80–81

Transparencies

5-Minute Check, 7–1

Technology/Multimedia

AlgePASS, Lesson 17

Interactive Chalkboard CD-ROM

Your Turn

Determine the slope of each line.

a. 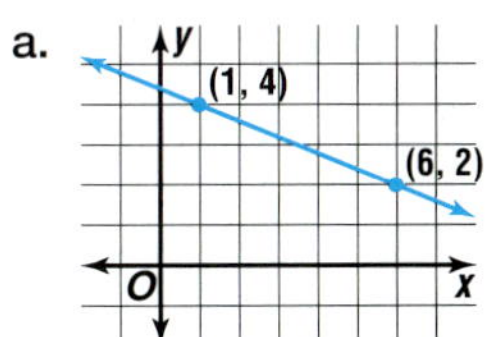

$-\frac{2}{5}$

b.
y
(−2, 3)
(3, 3)
O
x

0

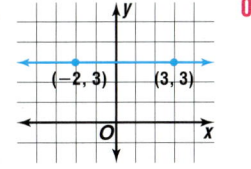

In Example 1, suppose 2 − 6 had been used as the change in y and 0 − 7 had been used as the change in x. Since $\frac{2-6}{0-7}$ is also equal to $\frac{4}{7}$, it does not matter which order is chosen. However, the coordinates of both points must be used in the same order.

In many real-world applications, the slope is the **rate of change**.

Example 3 Income Link

The graph at the right shows the hours worked and income for Helena and Steve. Find the slope of each line. To what does the slope refer?

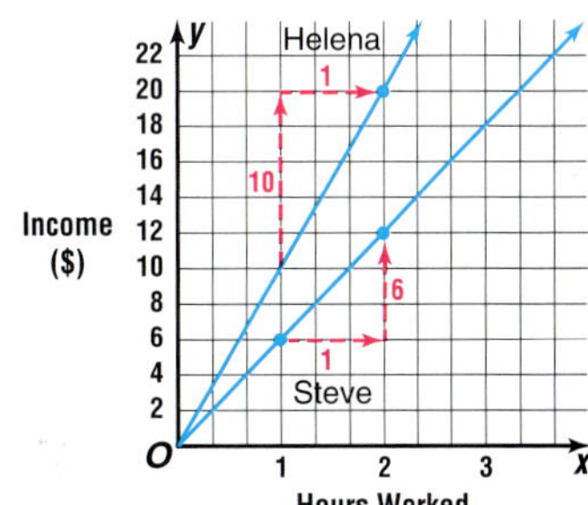

Helena: slope $= \frac{\text{change in } y}{\text{change in } x}$

$= \frac{10}{1}$ or 10

Steve: slope $= \frac{\text{change in } y}{\text{change in } x}$

$= \frac{6}{1}$ or 6

The graph represents Helena's income and Steve's income. The steepness depends on their hourly pay rate. So, the slope refers to each person's pay rate. Helena makes \$10 per hour and Steve makes \$6 per hour. Notice that the domain and range are nonnegative numbers since they cannot work fewer than 0 hours or earn less than \$0.

Look at the pattern in the graph at the right. Each time x increases 3 units, y decreases 1 unit.

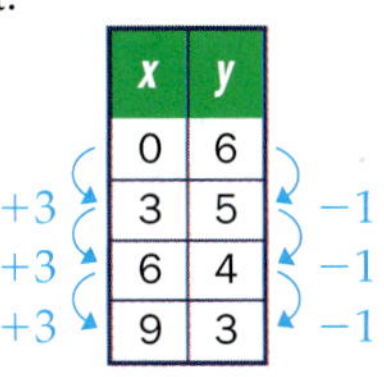

x	y
0	6
3	5
6	4
9	3

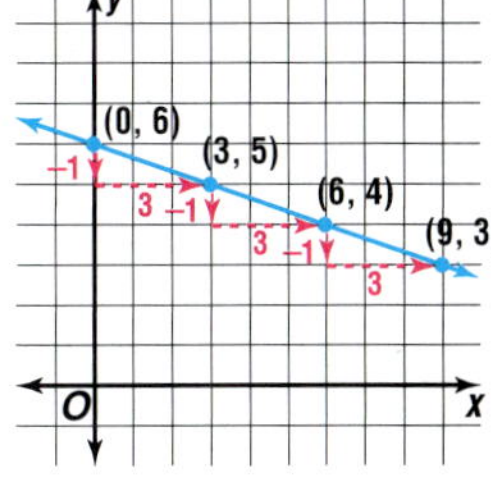

$$\text{slope} = \frac{\text{change in } y}{\text{change in } x} = \frac{-1}{3} \text{ or } -\frac{1}{3}$$

www.algconcepts.com/extra_examples

2 TEACH

In-Class Examples

Examples 1–2

Determine the slope of each line.

1 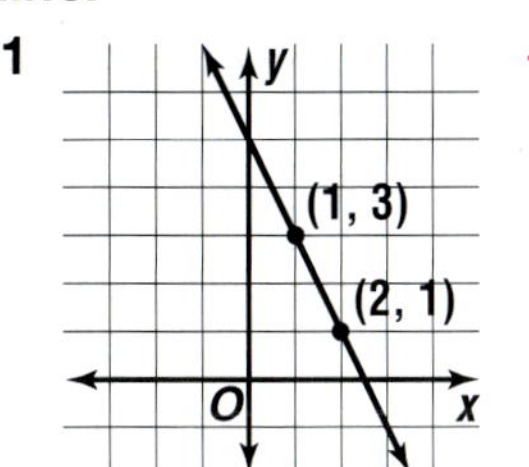

−2

2 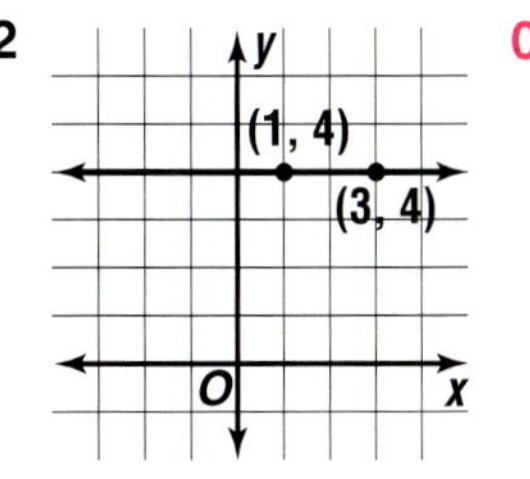

0

Teaching Tip To help students understand the meaning of the slope in an applied problem such as Example 3 have them include the units with the slope. In Example 3, the slope of Helena's graph is $\frac{10 \text{ dollars}}{1 \text{ hour}}$ or \$10 per hour. This will help students to realize that the slope is the rate at which each person is paid.

In-Class Example

Example 3

The graph below shows the distance traveled by Rebecca and Ian during a day-long bicycle ride. Find the slope of each line. To what does the slope refer?

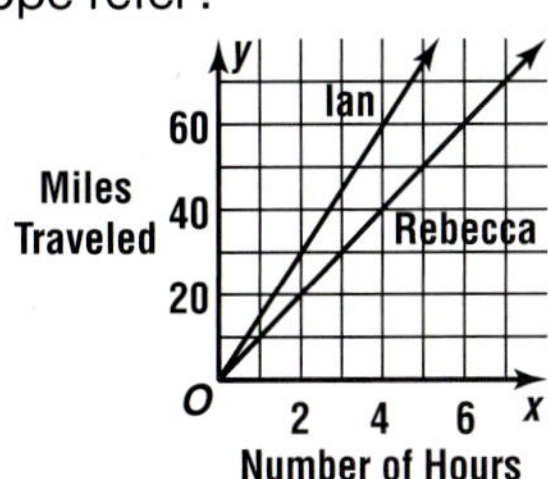

Ian: 15, Rebecca: 10; the slope is the speed in miles per hour for each rider.

From the Classroom of ...

Lisa Thomson
Reed High School
Sparks/Reno, Nevada

I discuss with students other words similar to slope—tilt, steepness, pitch, and so on.

In-Class Example

Example 4

A line contains the points whose coordinates are listed in the table. Determine the slope of the line. **−3**

x	y
−2	7
−1	4
0	1
1	−2

Teaching Tip When reading Examples 5 and 6, students may be confused by the subscripted variables used in the formula for slope. Stress that the symbol y_2 represents a single number, the y-coordinate of the second point. Also, explain that choosing which point is considered the first point and which is considered the second point is arbitrary. Emphasize however that once this choice is made, the two pairs of corresponding coordinates must be subtracted in the same order.

In-Class Examples

Examples 5–6

Determine the slope of each line.

5 the line through points at (3, 8) and (3, 4) **undefined**

6 the line through points at (−4, 1) and (−3, −2) **−3**

Example 4 **A line contains the points whose coordinates are listed in the table. Determine the slope of the line.**

	x	y	
−2	6	−3	+3
−2	4	0	+3
−2	2	3	+3
	0	6	

Each time x decreases 2 units, y increases 3 units.

$$\text{slope} = \frac{\text{change in } y}{\text{change in } x} = \frac{3}{-2}$$

The slope of the line containing these points is $-\frac{3}{2}$.

Your Turn

c. **1**

x	y
0	0
1	1
2	2
3	3

d. $\frac{1}{4}$

x	y
−5	2
−1	3
3	4
7	5

The examples above suggest the following.

Determining Slope Given Two Points

Words: The slope m of a line containing any two points with coordinates (x_1, y_1) and (x_2, y_2) is given by the following formula.

$$\text{slope} = \frac{\text{difference of the } y\text{-coordinates}}{\text{difference of the corresponding } x\text{-coordinates}}$$

Symbols: $m = \frac{y_2 - y_1}{x_2 - x_1}$, where $x_2 \neq x_1$

Examples **Determine the slope of each line.**

5 the line through (2, 9) and (6, 9)

$$m = \frac{y_2 - y_1}{x_2 - x_1} = \frac{9 - 9}{6 - 2} = \frac{0}{4} \text{ or } 0$$

The slope is 0.

6 the line through (3, −2) and (−4, 7)

$$m = \frac{y_2 - y_1}{x_2 - x_1} = \frac{7 - (-2)}{-4 - 3} = \frac{9}{-7} \text{ or } -\frac{9}{7}$$

The slope is $-\frac{9}{7}$.

Your Turn

e. the line through (2, 5) and (3, 9) **4**

f. the line through (−8, 1) and (4, 1) **0**

Reteaching Activity

Kinesthetic Learners Give students geoboards that have a square grid. If geoboards are not available, you can use paper that has a 10-by-10 grid of dots on it. Have students use rubber bands to model a pair of coordinate axes that divide the geoboard into four quadrants. They can then use another rubber band to represent a line segment. Have them use rise over run to find the slope of their segment. Then have them determine the coordinates of the endpoints of their segment and find the slope using the slope formula.

The slopes of lines can be summarized as follows.

Concept Summary

positive slope	negative slope	0 slope	undefined slope
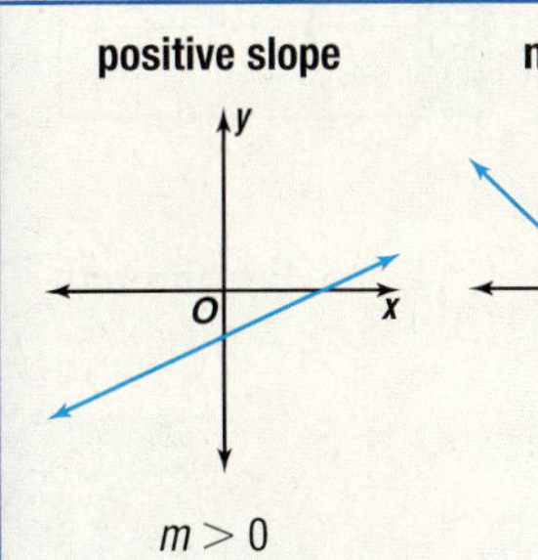	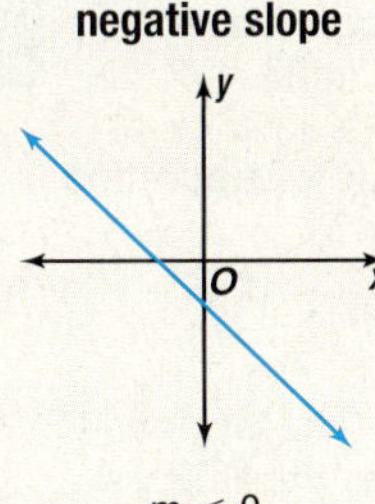	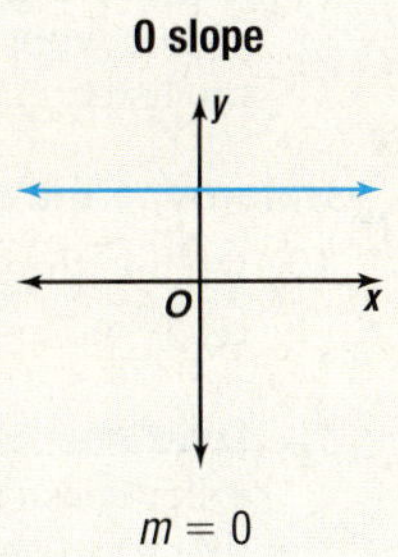	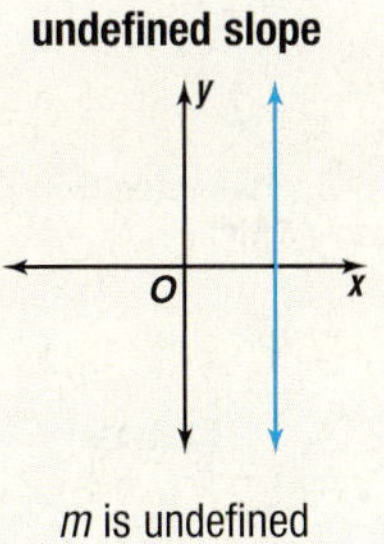
$m > 0$	$m < 0$	$m = 0$	m is undefined

Check for Understanding

Communicating Mathematics

Vocabulary
rate of change
slope

1. **Sketch** the graph of a line having each slope.
   a. negative   b. $\frac{1}{2}$
   c. 0   d. 1

1. See margin.

2. **Choose** the line whose slope is $-\frac{1}{4}$. **c**

a. 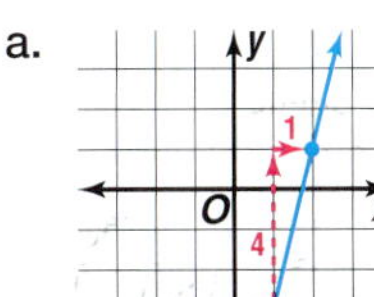

b. 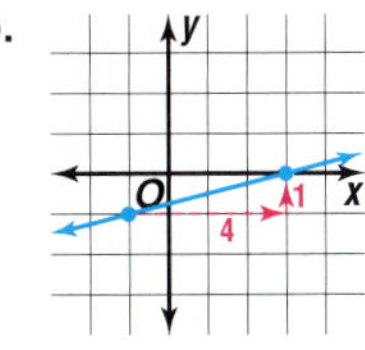

c. 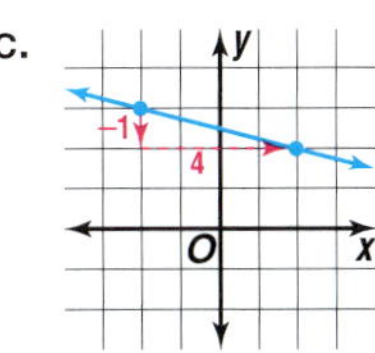

3. **You Decide** Naomi says that only the coordinates of points A and D can be used to find the slope of the line at the right. Percy says he could use the coordinates of B and D to find the slope. Who is correct? Explain.

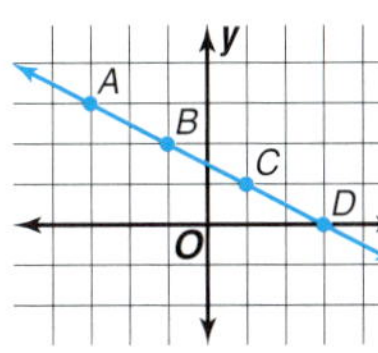

3. Percy; any two points on the line can be used to find the slope.

Guided Practice

Getting Ready Simplify each expression.

Sample: $\frac{-2-1}{4-(-3)}$ **Solution:** $\frac{-2-1}{4-(-3)} = \frac{-3}{7}$ or $-\frac{3}{7}$

4. $\frac{6-1}{-2-5}$ $-\frac{5}{7}$ 5. $\frac{4-(-2)}{7-4}$ 2 6. $\frac{-3-8}{9-(-5)}$ $-\frac{11}{14}$

Examples 1–3

Determine the slope of each line.

7. 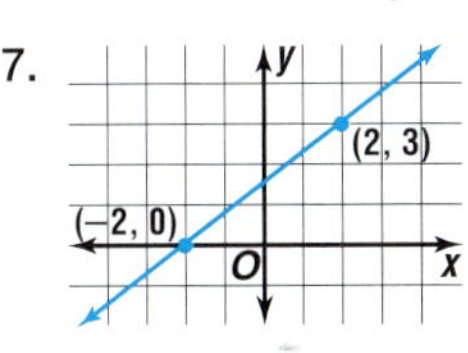

$\frac{3}{4}$

8. 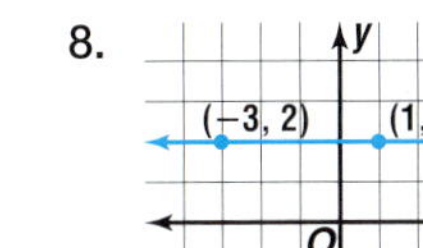

0

Answers

1b. 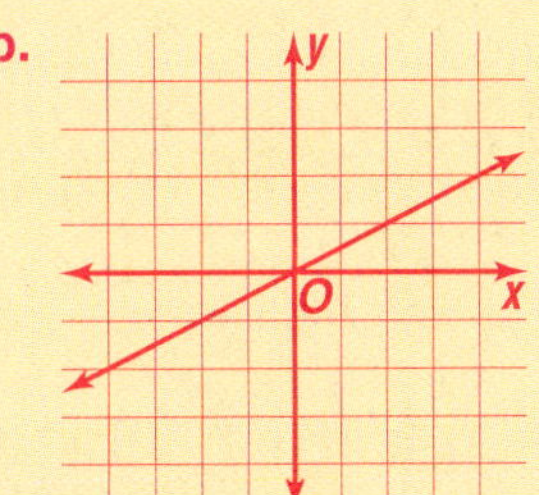

1c. 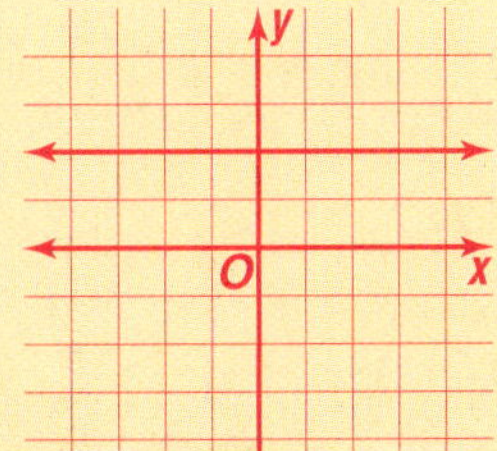

1d. 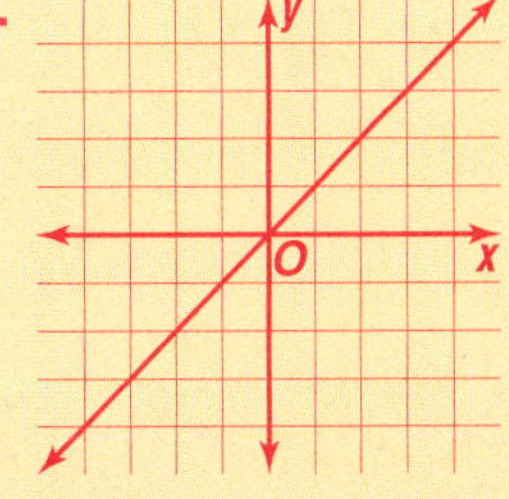

Teaching Tip When discussing the slopes of the lines at the top of the page, use the terms *horizontal* and *vertical* to describe the lines with 0 slope and undefined slope, respectively.

3 PRACTICE/APPLY

Error Analysis

Watch for students who calculate the change in x over the change in y when finding slope in Exercises 7–8.

Prevent by encouraging students to make a quick sketch of the line and look at the rise over run to check their answer.

Answer

1. Sample answers are given.

1a. 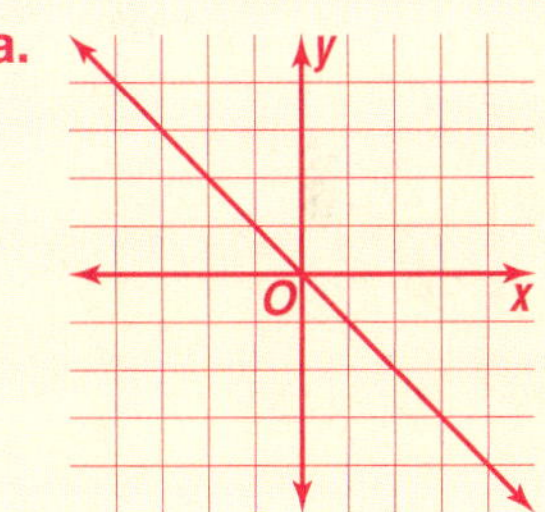

***Study Guide*, p. 289**

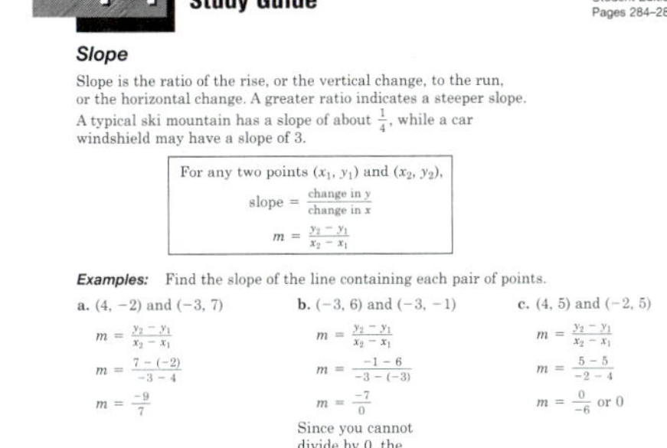

NAME ______ DATE ______ PERIOD ______

7-1 Study Guide

Student Edition
Pages 284–289

Slope

Slope is the ratio of the rise, or the vertical change, to the run, or the horizontal change. A greater ratio indicates a steeper slope. A typical ski mountain has a slope of about $\frac{1}{4}$, while a car windshield may have a slope of 3.

For any two points (x_1, y_1) and (x_2, y_2),
$$\text{slope} = \frac{\text{change in } y}{\text{change in } x}$$
$$m = \frac{y_2 - y_1}{x_2 - x_1}$$

Examples: Find the slope of the line containing each pair of points.

a. (4, −2) and (−3, 7)
$m = \frac{y_2 - y_1}{x_2 - x_1}$
$m = \frac{7-(-2)}{-3-4}$
$m = \frac{-9}{7}$

b. (−3, 6) and (−3, −1)
$m = \frac{y_2 - y_1}{x_2 - x_1}$
$m = \frac{-1-6}{-3-(-3)}$
$m = \frac{-7}{0}$
Since you cannot divide by 0, the slope is *undefined.*

c. (4, 5) and (−2, 5)
$m = \frac{y_2 - y_1}{x_2 - x_1}$
$m = \frac{5-5}{-2-4}$
$m = \frac{0}{-6}$ or 0

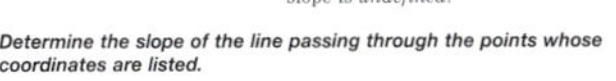

Determine the slope of the line passing through the points whose coordinates are listed.

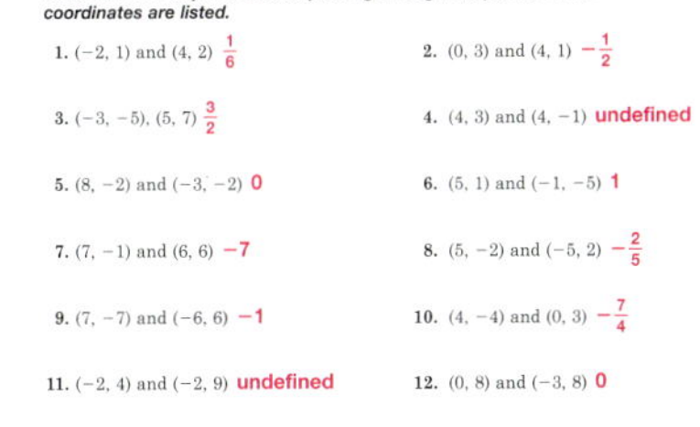

1. (−2, 1) and (4, 2) $\frac{1}{6}$
2. (0, 3) and (4, 1) $-\frac{1}{2}$
3. (−3, −5), (5, 7) $\frac{3}{2}$
4. (4, 3) and (4, −1) undefined
5. (8, −2) and (−3, −2) 0
6. (5, 1) and (−1, −5) 1
7. (7, −1) and (6, 6) −7
8. (5, −2) and (−5, 2) $-\frac{2}{5}$
9. (7, −7) and (−6, 6) −1
10. (4, −4) and (0, 3) $-\frac{7}{4}$
11. (−2, 4) and (−2, 9) undefined
12. (0, 8) and (−3, 8) 0

Assignment Guide

Basic: 13–29 odd, 30–36
Average: 14–26 even, 27–36

Example 4 9. Determine the slope of the line passing through the points whose coordinates are listed in the table. **4**

x	−2	−1	0	1
y	−8	−4	0	4

Examples 5–6 **Determine the slope of each line.**

10. the line through (1, 0) and (−2, 9) **−3**

11. the line through (10, −3) and (5, 3) $-\frac{6}{5}$

Example 1 12. **Entertainment** The Cyclone roller coaster has the steepest first drop of any wooden coaster in the world. It drops about 5 feet for every 3 feet of horizontal change. What is the slope of the first drop? $-\frac{5}{3}$

Cyclone, Coney Island, N.Y.

Exercises

Practice

Determine the slope of each line.

13. $-\frac{5}{3}$

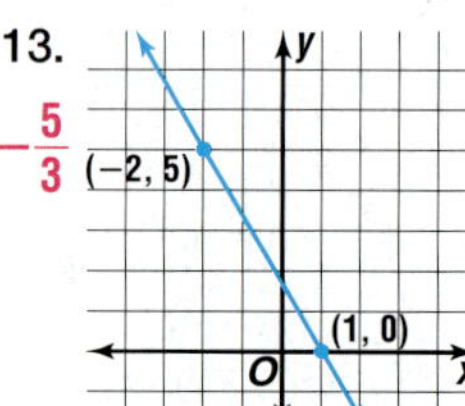

14. $\frac{5}{3}$

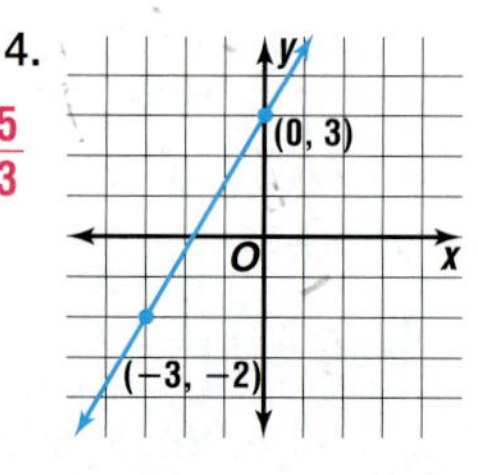

15. $\frac{1}{3}$

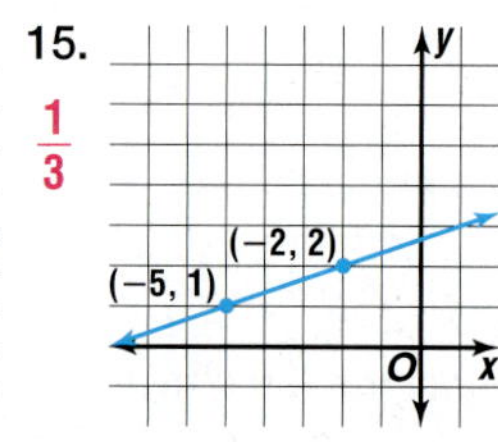

16. undefined

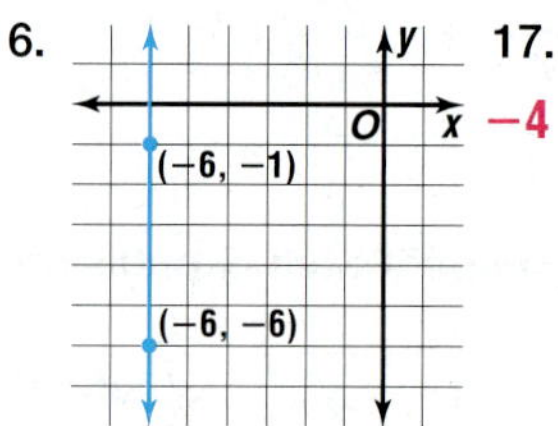

17. **−4**

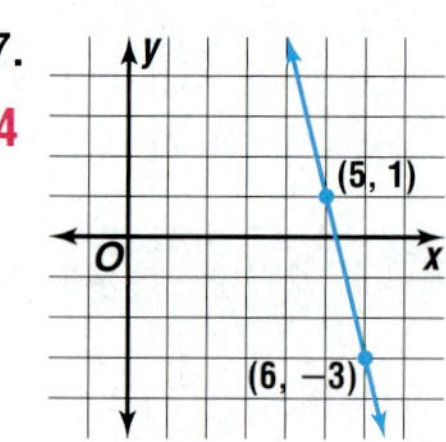

18. $\frac{1}{6}$

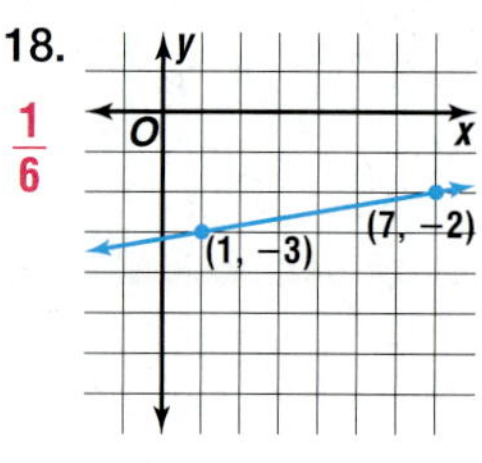

Homework Help	
For Exercises	See Examples
13–18	1, 2
19–21, 27	4
22–26	5, 6
28, 29	3

Extra Practice
See page 705.

Determine the slope of the line passing through the points whose coordinates are listed in each table.

19. **−2**

x	y
0	12
1	10
2	8
3	6

20. $-\frac{1}{2}$

x	y
0	7
2	6
4	5
6	4

21. $\frac{1}{3}$

x	y
−2	−7
1	−6
4	−5
7	−4

***Skills Practice,* p. 290, and *Practice,* p. 291 (shown)**

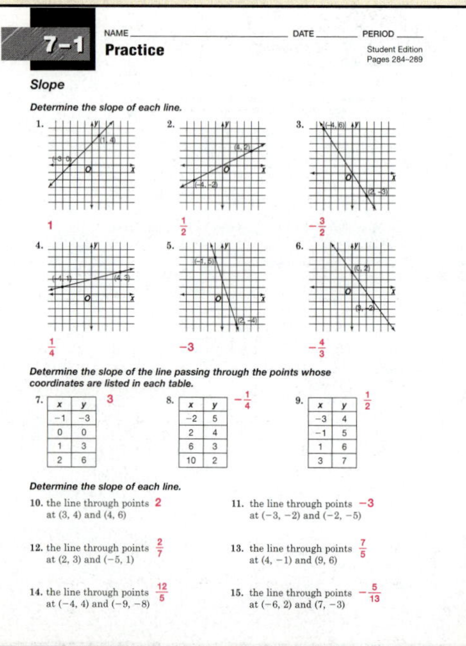

7–1 NAME ______ DATE ______ PERIOD ______

Practice Student Edition Pages 284–289

Slope

Determine the slope of each line.

1. 1
2. $\frac{1}{2}$
3. $-\frac{3}{2}$
4. $\frac{1}{4}$
5. −3
6. $-\frac{4}{3}$

Determine the slope of the line passing through the points whose coordinates are listed in each table.

7. 3

x	y
−1	−3
0	0
1	3
2	6

8. $-\frac{1}{4}$

x	y
−2	5
2	4
6	3
10	2

9. $\frac{1}{2}$

x	y
−3	4
−1	5
1	6
3	7

Determine the slope of each line.

10. the line through points at (3, 4) and (4, 6) **2**
11. the line through points at (−3, −2) and (−2, −5) **−3**
12. the line through points at (2, 3) and (−5, 1) $\frac{2}{7}$
13. the line through points at (4, −1) and (9, 6) $\frac{7}{5}$
14. the line through points at (−4, 4) and (−9, −8) $\frac{12}{5}$
15. the line through points at (−6, 2) and (7, −3) $-\frac{5}{13}$

Determine the slope of each line. 24. undefined

22. the line through (2, 5) and (6, 9) **1**
23. the line through (2, 3) and (−3, 5) $-\frac{2}{5}$
24. the line through (4, −3) and (4, 5)

25. $-\frac{3}{8}$

26. $\frac{6}{13}$

25. Find the slope of a line that passes through (−8, 9) and (0, 6).

26. What is the slope of a line that passes through $A(-8, 2)$ and $B(5, 8)$?

Applications and Problem Solving

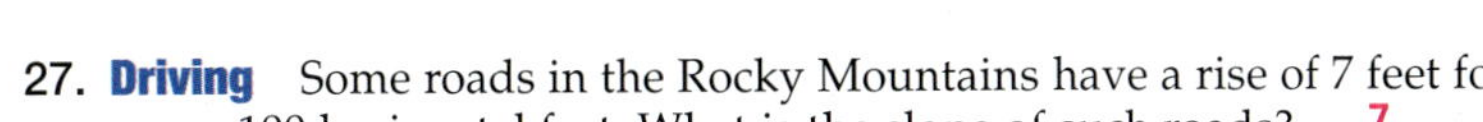

27. **Driving** Some roads in the Rocky Mountains have a rise of 7 feet for every 100 horizontal feet. What is the slope of such roads? $\frac{7}{100}$

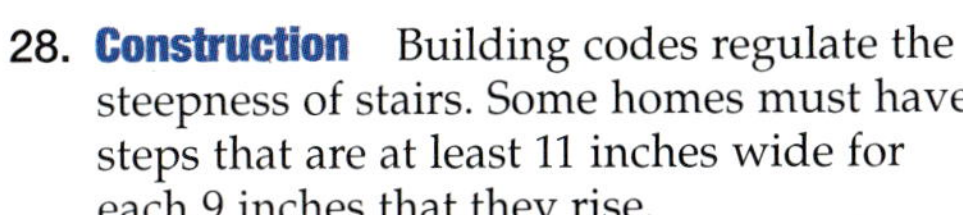

28. **Construction** Building codes regulate the steepness of stairs. Some homes must have steps that are at least 11 inches wide for each 9 inches that they rise.
    a. What is the slope of the stairs? $\frac{9}{11}$
    b. Describe how changing the width or the height affects the steepness of the stairs.

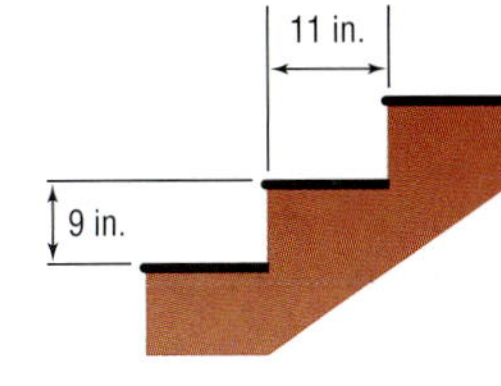

28b. If the width is increased, the stairs become less steep. If the height is increased, the stairs become steeper.

29a. As x increases by 1, $f(x)$ increases by 0.25. The slope is 0.25.

29b. Sample answer: domain: positive integers; range: positive multiples of 0.25

29. **Money** The linear function $f(x) = 0.25x$ represents the value of x quarters.
    a. Explain how changes in x affect $f(x)$. What does this tell you about the slope of the graph of the equation?
    b. Describe a reasonable domain and range for the function.

Number of Quarters x	Value of Quarters $f(x)$
1	0.25
2	0.50
3	0.75
4	1.00

30. **Critical Thinking** Determine whether the table represents a function that is linear or nonlinear. Explain. **See margin.**

x	1	3	5	7
y	2	4	6	16

Mixed Review

31. Assume that y varies inversely as x. Suppose $y = 4$ when $x = 6$. Find y when $x = 2$. *(Lesson 6–6)* **12**

Determine whether each relation is a function. *(Lesson 6–4)*

32. {(4, 8), (5, 5), (−1, 3), (0, 6)} **yes**
33. {(7, 1), (−4, 3), (0, 1), (−4, 0)} **no**

34. **Communication** The coastline of New Zealand is 8161 miles long. This is about 86% of the length of a cable under the ocean that connects New Zealand and Canada. What is the approximate length of the cable? *(Lesson 5–3)* **9490 mi**

35. **Short Response** Solve $9x = 4x - 10$. *(Lesson 4–6)* **−2**

36. **Multiple Choice** Solve $-5 - 0.8y = 7$. *(Lesson 4–5)* **B**
    Ⓐ 15 Ⓑ −15 Ⓒ 9.6 Ⓓ −2.5

www.algconcepts.com/self_check_quiz

4 ASSESS

Open-Ended Assessment

Writing Ask students to write a short summary of the lesson, describing two different methods for finding the slope of a line. Instruct them to include an example of both methods and to identify a situation in which they would want to use each method.

Answer

30. Nonlinear; each time x increases 2 units, y doubles. This is not a linear function.

***Enrichment*, p. 293**

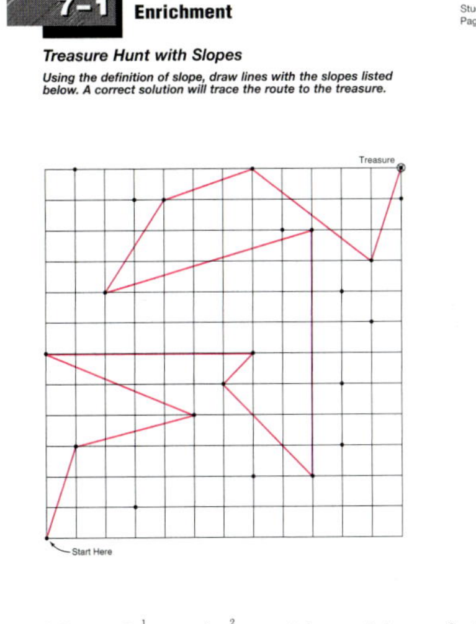

NAME ______ DATE ______ PERIOD ______

7-1 Enrichment Student Edition Pages 284–289

Treasure Hunt with Slopes

Using the definition of slope, draw lines with the slopes listed below. A correct solution will trace the route to the treasure.

1. 3 2. $\frac{1}{4}$ 3. $-\frac{2}{5}$ 4. 0 5. 1 6. −1
7. no slope 8. $\frac{2}{7}$ 9. $\frac{3}{2}$ 10. $\frac{1}{3}$ 11. $-\frac{3}{4}$ 12. 3

Extra Credit

Think of a real-world situation similar to the one described in Example 3 on page 285. Draw a graph to represent your situation. Calculate the slope of each line and describe to what the slope refers. **See students' work.**

7–2 Writing Equations in Point-Slope Form

1 FOCUS

5-Minute Check
Lesson 7–1

Determine the slope of each line.

1. 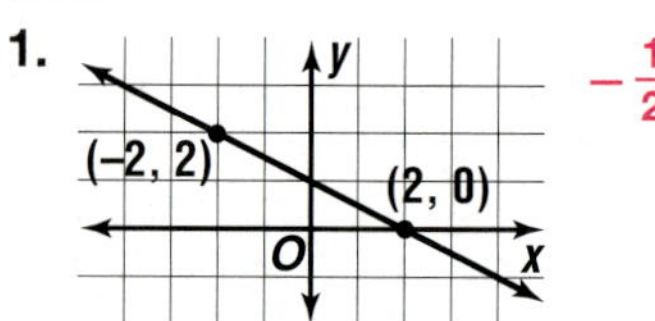

$-\frac{1}{2}$

2. 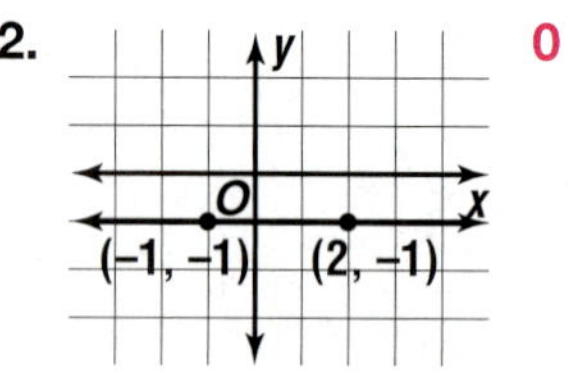

0

3. the line through points at (−9, 1) and (−9, 4) **undefined**
4. the line through points at (−7, 3) and (2, −1) $-\frac{4}{9}$
5. Determine the slope of the line passing through the points whose coordinates are listed in the table. $\frac{5}{2}$

x	y
1	0
3	5
5	10
7	15

Motivating the Lesson

Hands-On Activity Give each student a piece of grid paper and a piece of dry spaghetti. Have them draw a coordinate plane on the grid paper. Then have all the students place the spaghetti on the grid so that it represents a line with slope 2. Discuss the different lines the students graphed. Then have students place the spaghetti so that it represents a line with slope 2 that passes through (3, 1). Point out that they have all now graphed the same line. Explain point-slope form and write the equation of the line.
$y - 1 = 2(x - 3)$

What You'll Learn

You'll learn to write a linear equation in point-slope form given the coordinates of a point on the line and the slope of the line.

Why It's Important

Population Point-slope equations are useful in making predictions about the population. *See Exercise 40.*

You can write an equation of a line if you know its slope and the coordinates of one point on the line.

$$\frac{y_2 - y_1}{x_2 - x_1} = m \quad \textit{Slope of a line}$$

$$\frac{y - y_1}{x - x_1} = m \quad \textit{Replace } (x_2, y_2) \textit{ with } (x, y).$$

$$\left(\frac{y - y_1}{x - x_1}\right)(x - x_1) = m(x - x_1) \quad \textit{Multiply each side by } (x - x_1).$$

$$y - y_1 = m(x - x_1)$$

This equation is said to be in **point-slope form**. Recall that it is called a *linear equation* because its graph is a straight line. *The graph of a nonlinear equation is not a straight line.*

Point-Slope Form

Words: For a nonvertical line through the point at (x_1, y_1) with slope m, the point-slope form of a linear equation is $y - y_1 = m(x - x_1)$.

Numbers: $y - 6 = 4(x - 1)$ *$(x_1, y_1) = (1, 6)$ and $m = 4$*

The equation of a vertical line through the point at (x_1, y_1) is $x = x_1$.

Examples

Write the point-slope form of an equation for each line passing through the given point and having the given slope.

1 $(4, -5)$, $m = \frac{3}{4}$

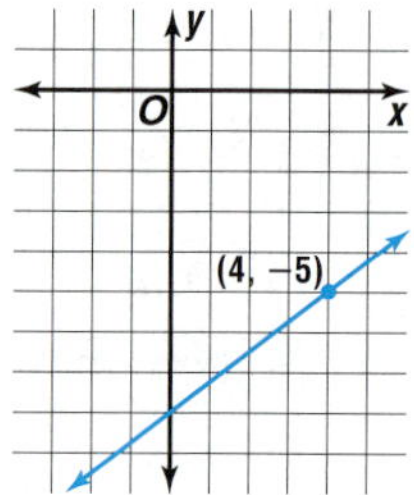

$y - y_1 = m(x - x_1)$ *Point-Slope Form*

$y - (-5) = \frac{3}{4}(x - 4)$ *Replace x_1 with 4, y_1 with −5, and m with $\frac{3}{4}$.*

$y + 5 = \frac{3}{4}(x - 4)$ *Simplify.*

An equation of the line is $y + 5 = \frac{3}{4}(x - 4)$.

Check: Look at the graph. Choose some other point on the line and determine whether it is a solution of $y + 5 = \frac{3}{4}(x - 4)$. Try (0, −8).

$y + 5 = \frac{3}{4}(x - 4)$ *Original equation*

$-8 + 5 \stackrel{?}{=} \frac{3}{4}(0 - 4)$ *Replace x with 0 and y with −8.*

$-3 = -3$ ✓

Resource Manager

Reproducible Masters

Chapter 7 Resource Masters
- *Study Guide,* p. 294
- *Skills Practice,* p. 295
- *Practice,* p. 296
- *Reading to Learn Mathematics,* p. 297
- *Enrichment,* p. 298

Transparencies

5-Minute Check, 7–2

Technology/Multimedia

Interactive Chalkboard CD-ROM

2 $(-7, 2), m = 0$

$y - y_1 = m(x - x_1)$	*Point-Slope Form*
$y - 2 = 0[x - (-7)]$	*Replace x_1 with -7, y_1 with 2, and m with 0.*
$y - 2 = 0$	*Simplify.*
$y - 2 + 2 = 0 + 2$	*Add 2 to each side.*
$y = 2$	*Simplify.*

An equation of the line is $y = 2$.

Your Turn **a.** $y - 2 = 2(x + 3)$ **b.** $y - 4 = -\frac{2}{3}(x - 5)$

a. $(-3, 2), m = 2$ b. $(5, 4), m = -\frac{2}{3}$

You can also write an equation of a line if you know the coordinates of two points on the line.

Example 3 **Write the point-slope form of an equation of the line at the right.**

First, determine the slope of the line.

$m = \frac{y_2 - y_1}{x_2 - x_1} = \frac{5 - 1}{-2 - 3}$ or $-\frac{4}{5}$

The slope is $-\frac{4}{5}$. Now use the slope and either point to write an equation.

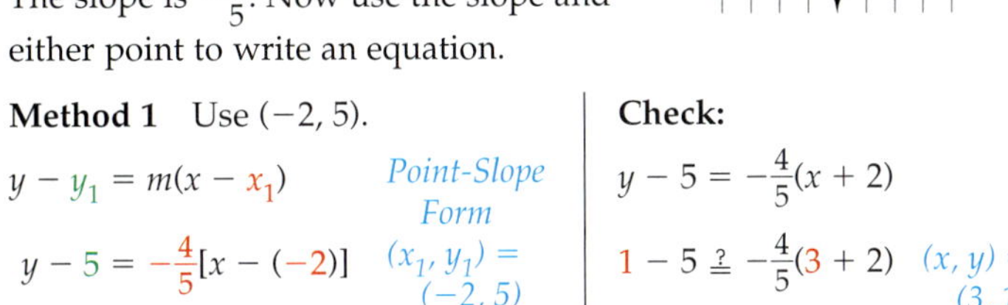

Method 1 Use $(-2, 5)$.

		Check:	
$y - y_1 = m(x - x_1)$	*Point-Slope Form*	$y - 5 = -\frac{4}{5}(x + 2)$	
$y - 5 = -\frac{4}{5}[x - (-2)]$	*$(x_1, y_1) = (-2, 5)$*	$1 - 5 \stackrel{?}{=} -\frac{4}{5}(3 + 2)$	*$(x, y) = (3, 1)$*
$y - 5 = -\frac{4}{5}(x + 2)$	*Simplify.*	$-4 \stackrel{?}{=} -\frac{4}{5}(5)$	
		$-4 = -4$ ✓	

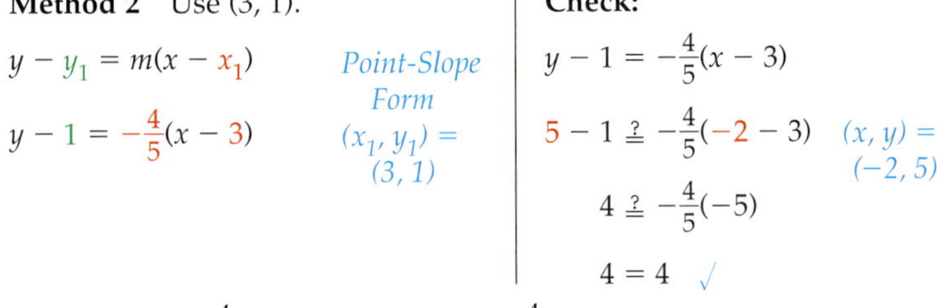

Method 2 Use $(3, 1)$.

		Check:	
$y - y_1 = m(x - x_1)$	*Point-Slope Form*	$y - 1 = -\frac{4}{5}(x - 3)$	
$y - 1 = -\frac{4}{5}(x - 3)$	*$(x_1, y_1) = (3, 1)$*	$5 - 1 \stackrel{?}{=} -\frac{4}{5}(-2 - 3)$	*$(x, y) = (-2, 5)$*
		$4 \stackrel{?}{=} -\frac{4}{5}(-5)$	
		$4 = 4$ ✓	

Both $y - 5 = -\frac{4}{5}(x + 2)$ and $y - 1 = -\frac{4}{5}(x - 3)$ are point-slope forms of an equation for the line passing through (3, 1) and (−2, 5).

2 TEACH

Teaching Tip Go through the development of the point-slope form slowly. Explain that the specific point (x_2, y_2) is replaced with the general point (x, y) so that the equation can be used for any point (x, y) on the line.

In-Class Examples

Examples 1–2

Write the point-slope form of an equation for each line passing through the given point and having the given slope.

1 $(-2, 7), m = -\frac{1}{3}$ **$y - 7 = -\frac{1}{3}(x + 2)$**

2 $(4, 0), m = 4$ **$y = 4(x - 4)$**

Teaching Tip After discussing the two methods shown in Example 3, have students simplify both of the final equations so they can see that the equations are equivalent. Point out also that when one point is used to write the equation, the other point can be used to check that the equation is correct.

In-Class Example

Example 3

Write the point-slope form of an equation of the line below.

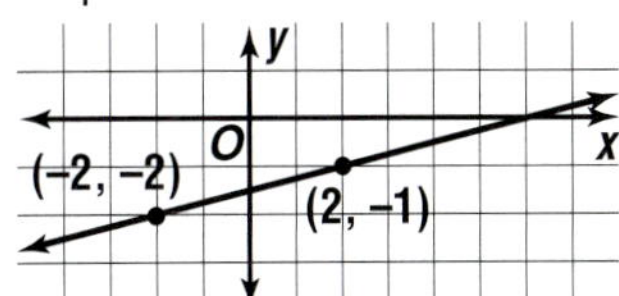

$y + 2 = \frac{1}{4}(x + 2)$ or $y + 1 = \frac{1}{4}(x - 2)$

Teaching Tip When discussing Example 4 it would be a good idea to discuss how the scales of the axes affect the appearance of a graph. The graphs in Your Turn part c and Example 4 appear to have similar slopes. However in truth, the slope of the Your Turn part c graph is more than ten times that of the graph in Example 4. Also, point out that the scales of the axes in Example 4 are not the same. Point out that if both the x- and y-axis were scaled with the same units, the line in the graph would appear almost horizontal.

In-Class Example

Example 4

In 1985, Joe's Sport Shop had sales of 10.6 million dollars. In 1999, the store had sales of 15.2 million dollars. Write the point-slope form of an equation for the line in the graph below.

$y - 10.6 = \frac{23}{70}(x - 1985)$ or

$y - 15.2 = \frac{23}{70}(x - 1999)$

Answers

1. the coordinates of a point on the line and the slope of the line, or the coordinates of two points on the line
2. Sample answer:

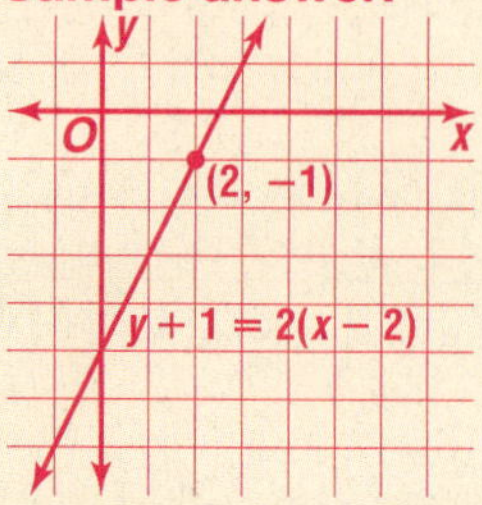

Your Turn

Write the point-slope form of an equation for each line.

c. 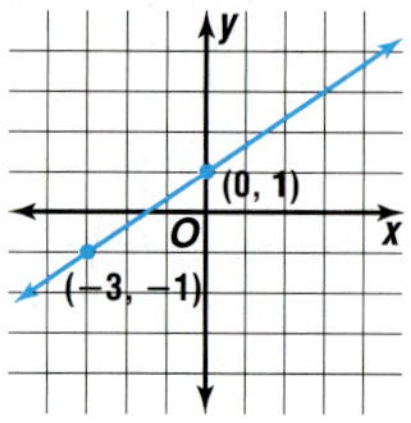

$y + 1 = \frac{2}{3}(x + 3)$ or $y - 1 = \frac{2}{3}x$

d. the line passing through (1, 4) and (3, −5)

$y - 4 = -\frac{9}{2}(x - 1)$ or

$y + 5 = -\frac{9}{2}(x - 3)$

Real World

Example 4 Sports Link

In the 1988 Olympics, the women's winning long jump was about 24.7 feet. It is predicted that by 2020 the record will be about 26.3 feet. Write the point-slope form of an equation for the line in the graph.

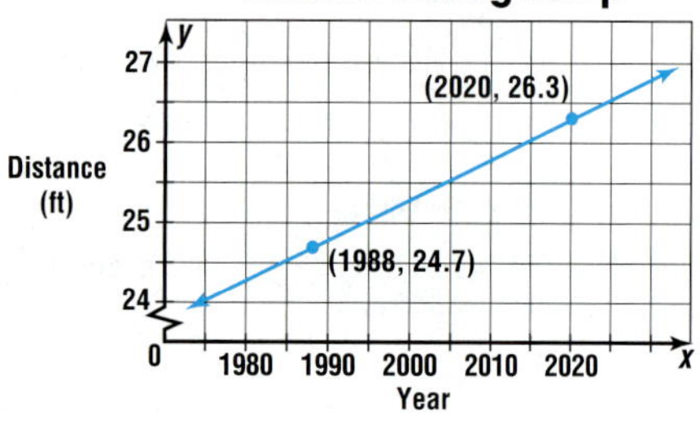

Source: *USA Today*

Prerequisite Skills Review

Operations with Decimals, p. 684

First, determine the slope of the line.

$$m = \frac{y_2 - y_1}{x_2 - x_1} = \frac{26.3 - 24.7}{2020 - 1988} \quad (x_1, y_1) = (1988, 24.7), (x_2, y_2) = (2020, 26.3)$$

$$= \frac{1.6}{32} \text{ or } \frac{1}{20} \qquad \text{The slope is } \frac{1}{20}.$$

Now use the slope and either point to write an equation.

$y - y_1 = m(x - x_1)$ *Point-Slope Form*

$y - 24.7 = \frac{1}{20}(x - 1988)$ *Replace (x_1, y_1) with (1988, 24.7) and m with $\frac{1}{20}$.*

An equation of the line is $y - 24.7 = \frac{1}{20}(x - 1988)$.

Check by substituting (2020, 26.3) into the equation.

Check for Understanding

Communicating Mathematics

1. **Tell** what information is needed to write a linear equation in point-slope form.
2. **Sketch** a line whose slope is 2. Then use a point on the line to write an equation of the line in point-slope form.

1–2. See margin.

Vocabulary

point-slope form

Guided Practice

Getting Ready State the slope and the coordinates of a point on the graph of the equation.

Sample: $y - 2 = 3(x - 5)$

Solution: $y - y_1 = m(x - x_1)$

$y - 2 = 3(x - 5)$

$m = 3, (5, 2)$

3. $y - 6 = 2(x - 1)$ **$m = 2, (1, 6)$**
4. $y + 4 = \frac{2}{5}(x - 8)$ **$m = \frac{2}{5}, (8, -4)$**
5. $y - (-5) = -3(x + 9)$ **$m = -3, (-9, -5)$**

Examples 1 & 2

Write the point-slope form of an equation for each line passing through the given point and having the given slope.

6. $(5, 6), m = 2$ **$y - 6 = 2(x - 5)$**
7. $(-2, 4), m = \frac{3}{4}$ **7. $y - 4 = \frac{3}{4}(x + 2)$**
8. $(3, -4), m = -5$ **$y + 4 = -5(x - 3)$**
9. $(-8, 1), m = 0$ **$y - 1 = 0$ or $y = 1$**

Example 3

Write the point-slope form of an equation for each line.

10.

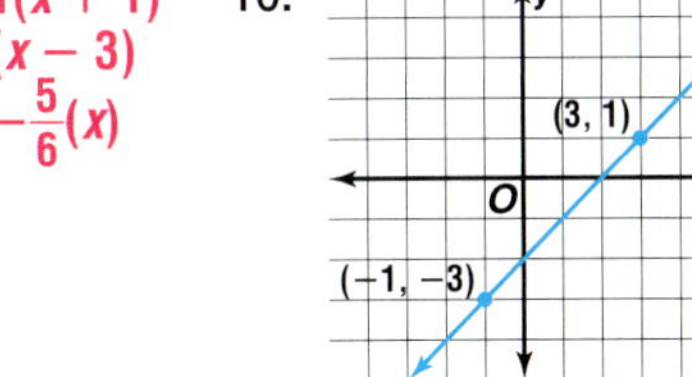

11.

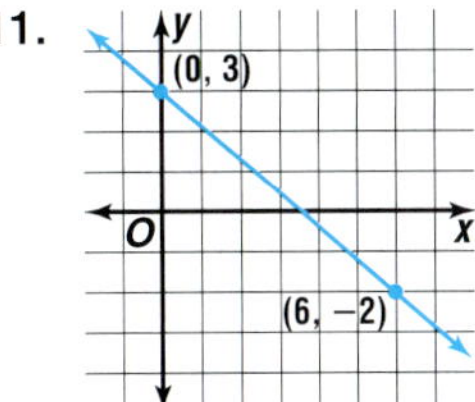

10. $y + 3 = 1(x + 1)$ or $y - 1 = 1(x - 3)$

11. $y - 3 = -\frac{5}{6}(x)$ or $y + 2 = -\frac{5}{6}(x - 6)$

12. the line through (5, 3) and (−2, 1)
13. the line through (−2, 5) and (4, −7)

12. $y - 3 = \frac{2}{7}(x - 5)$ or $y - 1 = \frac{2}{7}(x + 2)$

Example 4

13. $y - 5 = -2(x + 2)$ or $y + 7 = -2(x - 4)$

14. **Physical Science** The table shows the growth of the Pacific giant kelp plant.
    a. Write the point-slope form of an equation that represents the growth.
    b. How much will the plant grow in 7 days?
    c. Would the plant show faster or slower growth if the slope of the equation was 30? **slower**

Time (days)	Growth (cm)
x	y
2	90
3	135

14a. $y - 90 = 45(x - 2)$

14b. 315 cm

Exercises

Practice

Write the point-slope form of an equation for each line passing through the given point and having the given slope. **15–23. See margin.**

15. $(-1, 4), m = -3$
16. $(7, 2), m = 5$
17. $(9, 1), m = 6$
18. $(-3, 6), m = \frac{2}{3}$
19. $(-2, -2), m = \frac{4}{5}$
20. $(-1, -5), m = -7$
21. $(4, -2)$, m is undefined.
22. $(0, 6), m = \frac{3}{5}$
23. $\left(\frac{1}{2}, -4\right), m = -\frac{5}{2}$

3 PRACTICE/APPLY

Error Analysis

Watch for students who write the incorrect signs when finding the point-slope equation of a line in Exercises 7–9.

Prevent by stressing that the formula involves subtraction. Point out that if a coordinate of the point is negative, then the constant will be added to the variable in the equation. Encourage students to always substitute the coordinates of the second point in their equation to check that it is correct.

Answers

15. $y - 4 = -3(x + 1)$
16. $y - 2 = 5(x - 7)$
17. $y - 1 = 6(x - 9)$
18. $y - 6 = \frac{2}{3}(x + 3)$
19. $y + 2 = -\frac{4}{5}(x + 2)$
20. $y + 5 = -7(x + 1)$
21. $x = 4$
22. $y - 6 = \frac{3}{5}(x)$
23. $y + 4 = -\frac{5}{2}\left(x - \frac{1}{2}\right)$

Reteaching Activity

Verbal/Linguistic Learners Tell students to imagine that a classmate was absent during this lesson. Have them write this absent classmate an explanation of how to write an equation for a line when two points on the line are given.

***Study Guide*, p. 294**

7–2 NAME ______ DATE ______ PERIOD ______

Study Guide Student Edition Pages 290–295

Writing Equations in Point-Slope Form

You can write the equation of a line if you know its slope and the coordinates of one point or if you know the coordinates of two points on the line. Use the point-slope form.

Point-Slope Form
For a nonvertical line through the point at (x_1, y_1) with slope m, the point-slope form of a linear equation is $y - y_1 = m(x - x_1)$.

Examples: Write the point-slope form of an equation for each line.

a. the line passing through the point at $(-4, 2)$ and having a slope of $\frac{2}{3}$

$y - y_1 = m(x - x_1)$

$y - 2 = \frac{2}{3}(x - (-4))$

$y - 2 = \frac{2}{3}(x + 4)$

b. the line passing through points at $(4, -4)$ and $(-3, 1)$

$m = \frac{y_2 - y_1}{x_2 - x_1}$ *Find m.*

$m = \frac{1 - (-4)}{-3 - 4}$ or $-\frac{5}{7}$

$y - (-4) = -\frac{5}{7}(x - 4)$ *Substitute m.*

$y + 4 = -\frac{5}{7}(x - 4)$

Write the point-slope form of an equation for each line, given either the coordinates of a point and the slope or the coordinates of two points.

1. $(-2, -1), m = 2$ **$y + 1 = 2(x + 2)$**
2. $(4, -1), m = -\frac{1}{2}$ **$y + 1 = -\frac{1}{2}(x - 4)$**
3. $(-3, -5), m = \frac{3}{2}$ **$y + 5 = \frac{3}{2}(x + 3)$**
4. $(4, 3), m = \frac{1}{5}$ **$y - 3 = \frac{1}{5}(x - 4)$**
5. $(8, -2), m = 0$ **$y + 2 = 0$**
6. $(5, 1), m = -\frac{2}{3}$ **$y - 1 = -\frac{2}{3}(x - 5)$**
7. $(7, -1)$ and $(6, 6)$ **$y + 1 = -7(x - 7)$ or $y - 6 = -7(x - 6)$**
8. $(5, -2)$ and $(-5, 2)$ **$y + 2 = -\frac{2}{5}(x - 5)$ or $y - 2 = -\frac{2}{5}(x + 5)$**
9. $(7, -7)$ and $(-6, 6)$ **$y + 7 = -(x - 7)$ or $y - 6 = -(x + 6)$**
10. $(4, -4)$ and $(0, 3)$ **$y + 4 = -\frac{7}{4}(x - 4)$ or $y - 3 = -\frac{7}{4}x$**
11. $(-2, -4)$ and $(-12, 9)$ **$y + 4 = -\frac{13}{10}(x + 2)$; $y - 9 = -\frac{13}{10}(x + 12)$**
12. $(0, 8)$ and $(-3, 8)$ **$y - 8 = 0$**

Assignment Guide

Basic: 15–39 odd, 41–50
Average: 16–36 even, 38–50

Answers

24. $y - 1 = 5(x - 2)$ or $y - 6 = 5(x - 3)$

25. $y - 4 = \frac{1}{4}(x + 2)$ or $y - 5 = \frac{1}{4}(x - 2)$

26. $y - 6 = 0$ or $y = 6$

27. $y = \frac{3}{4}(x + 2)$ or $y - 3 = \frac{3}{4}(x - 2)$

28. $y + 3 = -\frac{2}{5}(x - 1)$ or $y + 5 = -\frac{2}{5}(x - 6)$

29. $y - 3 = -\frac{5}{2}(x + 3)$ or $y + 2 = -\frac{5}{2}(x + 1)$

30. $y - 1 = 7(x - 2)$ or $y - 8 = 7(x - 3)$

31. $y - 5 = \frac{1}{2}(x - 4)$ or $y - 3 = \frac{1}{2}(x)$

32. $y + 2 = -\frac{1}{4}(x - 4)$ or $y + 1 = -\frac{1}{4}(x)$

33. $y + 4 = 3(x - 1)$ or $y - 5 = 3(x - 4)$

Skills Practice, p. 295, and *Practice*, p. 296 (shown)

7–2 NAME ______ DATE ______ PERIOD ______
Practice
Student Edition Pages 290–295

Writing Equations in Point-Slope Form

Write the point-slope form of an equation for each line passing through the given point and having the given slope.

1. $(4, 7)$, $m = 3$ — $y - 7 = 3(x - 4)$
2. $(-2, 3)$, $m = 5$ — $y - 3 = 5(x + 2)$
3. $(6, -1)$, $m = -2$ — $y + 1 = -2(x - 6)$
4. $(-5, -2)$, $m = 0$ — $y = -2$
5. $(-4, -6)$, $m = \frac{2}{3}$ — $y + 6 = \frac{2}{3}(x + 4)$
6. $(-8, 3)$, $m = -\frac{3}{5}$ — $y - 3 = -\frac{3}{5}(x + 8)$
7. $(7, -9)$, $m = 4$ — $y + 9 = 4(x - 7)$
8. $(-6, 3)$, $m = -\frac{1}{2}$ — $y - 3 = -\frac{1}{2}(x + 6)$
9. $(-2, -5)$, $m = 8$ — $y + 5 = 8(x + 2)$

Write the point-slope form of an equation for each line.

10. 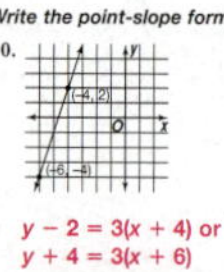

$y - 2 = 3(x + 4)$ or $y + 4 = 3(x + 6)$

11. $y - 3 = -\frac{1}{4}x$ or $y - 4 = -\frac{1}{4}(x + 4)$

12. $y - 1 = \frac{2}{3}(x - 6)$ or $y + 5 = \frac{2}{3}(x + 3)$

13. $y - 1 = -5(x + 3)$ or $y + 4 = -5(x + 2)$

14. the line through points at $(-2, -2)$ and $(-1, -6)$ — $y + 6 = -4(x + 1)$ or $y + 2 = -4(x + 2)$

15. the line through points at $(-7, -3)$ and $(5, -1)$ — $y + 3 = \frac{1}{6}(x + 7)$ or $y + 1 = \frac{1}{6}(x - 5)$

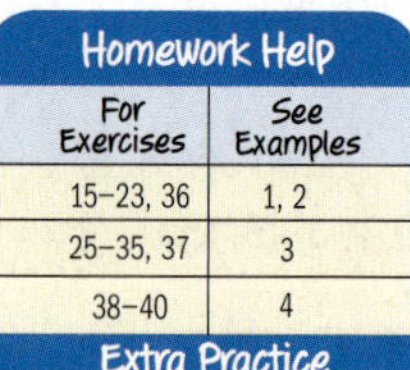

For Exercises	See Examples
15–23, 36	1, 2
25–35, 37	3
38–40	4

Homework Help — **Extra Practice** See page 705.

B Write the point-slope form of an equation for each line.

24. 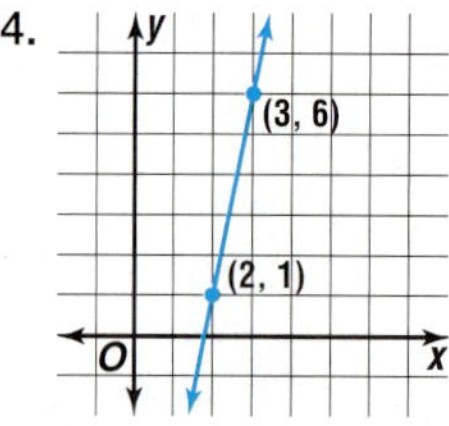

25. 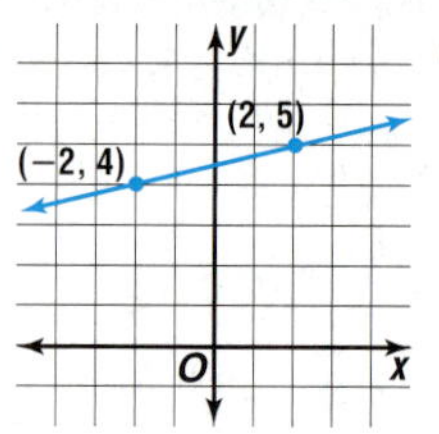

26. 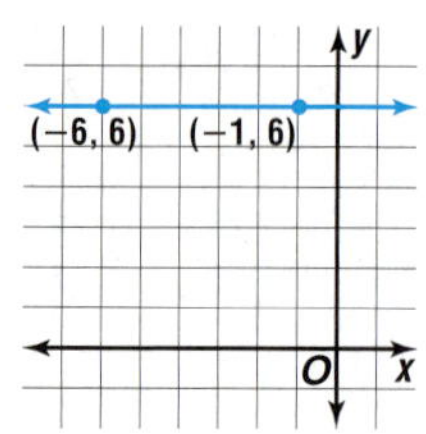

27. 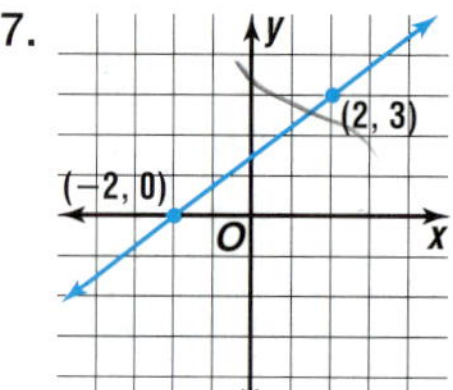

28. (1, −3), (6, −5)

29. 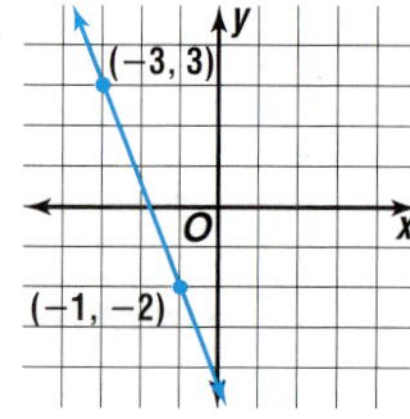

24–35. See margin.

30. the line through (2, 1) and (3, 8)
31. the line through (4, 5) and (0, 3)
32. the line through (4, −2) and (0, −1)
33. the line through (1, −4) and (4, 5)
34. the line through (6, 9) and (−2, 9)
35. the line through (−3, 7) and (5, 2)

36. Write an equation in point-slope form of a line that has a slope of −1.8 and passes through (−5, 11). $y - 11 = -1.8(x + 5)$

37. What is the point-slope form of an equation of the line containing the points $Q(3, 6)$ and $R(-4, 7)$?
$y - 6 = -\frac{1}{7}(x - 3)$ or $y - 7 = -\frac{1}{7}(x + 4)$

Applications and Problem Solving

38. **Transportation** The graph shows the growth in the number of sport-utility vehicles in the United States.

a. Write the point-slope form of an equation for the graph of the line. **a.** $y - 7 = 1.4(x - 1993)$

b. Use the equation to predict the number of sport-utility vehicles in the United States in 2004. **22.4 million**

$\frac{12.6 - 7}{97 - 93} = \frac{5.6}{4}$ million SUV's / Years
$\frac{1.4}{1}$ Per Year
C $m = 1.4$

$y - 7 = 1.4(2004 - 1993)$
$y - 7 = 1.4(11)$

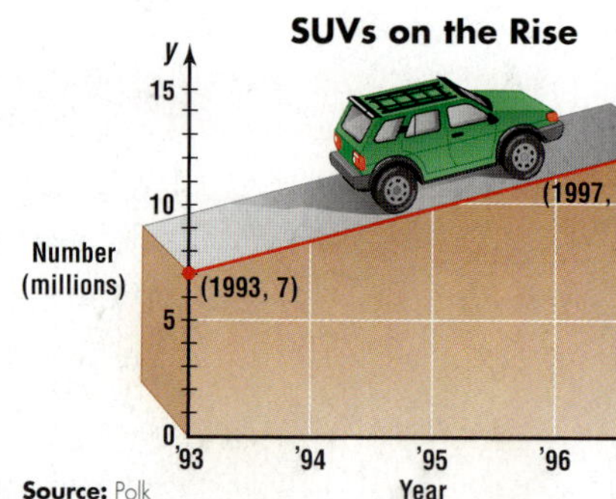

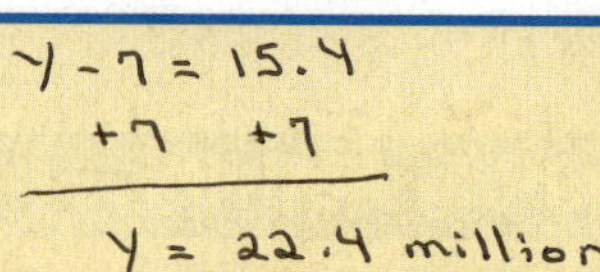

Answers

34. $y - 9 = 0$ or $y = 9$

35. $y - 7 = -\frac{5}{8}(x + 3)$ or $y - 2 = -\frac{5}{8}(x - 5)$

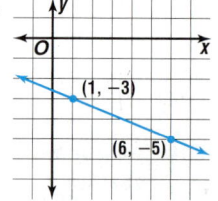

39a. Sample answer: $y - 5 = 5(x - 1)$

39. **Savings** Lynda adds $5 to her savings account each week, as shown in the graph at the right.

a. Write the point-slope form of an equation of the line through the points that represent her savings.

b. Describe a reasonable domain and range for the equation. **Sample answer: domain: positive integers; range: positive multiples of 5**

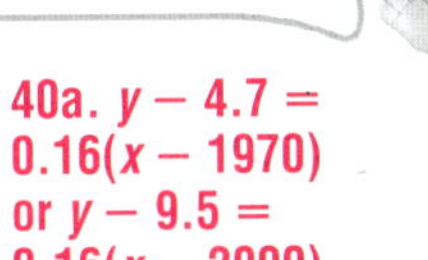

Data Update For the latest information on U.S. population, visit: www.algconcepts.com

40a. $y - 4.7 = 0.16(x - 1970)$ or $y - 9.5 = 0.16(x - 2000)$

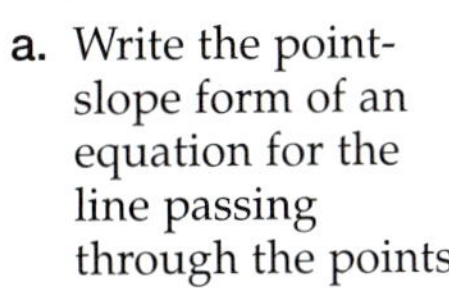

40. **Population** The graph shows the percent of U.S. residents who are born outside the United States.

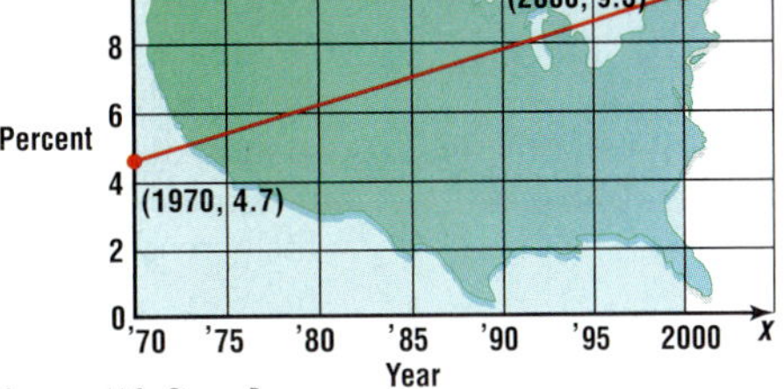

Source: U.S. Census Bureau

a. Write the point-slope form of an equation for the line passing through the points.

b. Use the graph to estimate the percent of residents that were born outside the United States in 1995. Use the equation to check your estimate. **8.7%**

c. Predict the percent of residents that will be born outside the United States in 2005. **10.3%**

41. **Critical Thinking** Find the coordinates of a point through which the graph of $y - 11 = \frac{3}{4}(x - 8)$ passes, if the y-coordinate is twice the x-coordinate. **(4, 8)**

Mixed Review

42. **Construction** A ramp installed to give handicapped people access to a building has a 3-foot rise and a 36-foot run. What is the slope of the ramp? *(Lesson 7–1)* $\frac{1}{12}$

Find the domain of each equation if the range is {0, 2, 4}. *(Lesson 6–2)*

43. $y = 2x$ **{0, 1, 2}**

44. $y = 4 + x$ **{−4, −2, 0}**

45. $y = 2x - 2$ **{1, 2, 3}**

Find the percent of increase or decrease. *(Lesson 5–5)*

46. original: 15
new: 18 **20% increase**

47. original: 6
new: 24 **300% increase**

48. Use the percent equation to find 45% of 160. *(Lesson 5–4)* **72**

Standardized Test Practice

49. **Grid In** Find $(-1.2)(-7)$. *(Lesson 4–1)* **8.4**

50. **Short Response** List six numbers whose median is 12. *(Lesson 3–3)* **Sample answer: 10, 11, 12, 12, 13, 14**

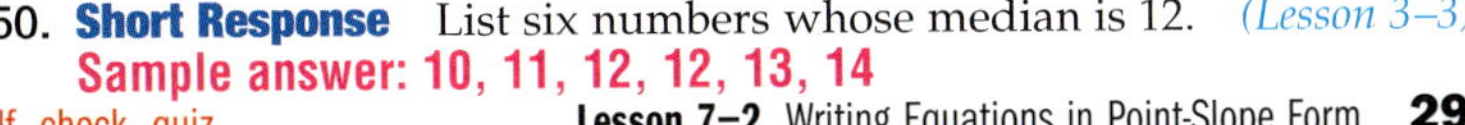

4 ASSESS

Open-Ended Assessment

Modeling Have students work with a partner. Provide each pair of students with a rectangular grid geoboard or rectangular dot paper. After modeling a pair of coordinate axes, one student should model a line on the geoboard or paper while their partner writes its equation. The first student should then check their partner's equation. Students should repeat the exercise several times, switching roles each time.

***Enrichment*, p. 298**

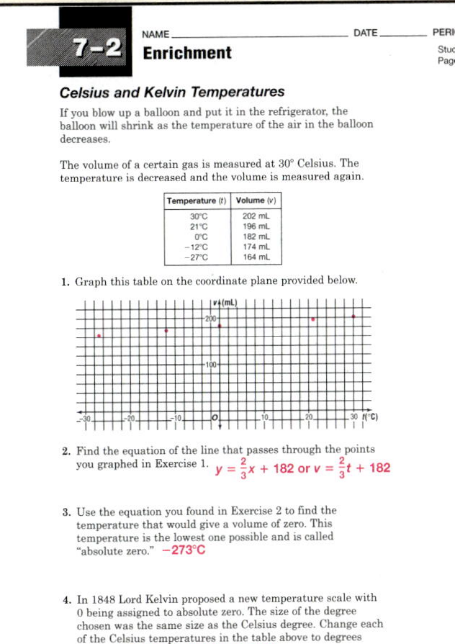

NAME ______ DATE ______ PERIOD ______

7–2 Enrichment

Student Edition Pages 290–295

Celsius and Kelvin Temperatures

If you blow up a balloon and put it in the refrigerator, the balloon will shrink as the temperature of the air in the balloon decreases.

The volume of a certain gas is measured at 30° Celsius. The temperature is decreased and the volume is measured again.

Temperature (t)	Volume (v)
30°C	202 mL
21°C	196 mL
0°C	182 mL
−12°C	174 mL
−27°C	164 mL

1. Graph this table on the coordinate plane provided below.

2. Find the equation of the line that passes through the points you graphed in Exercise 1. $y = \frac{2}{3}x + 182$ or $v = \frac{2}{3}t + 182$

3. Use the equation you found in Exercise 2 to find the temperature that would give a volume of zero. This temperature is the lowest one possible and is called "absolute zero." **−273°C**

4. In 1848 Lord Kelvin proposed a new temperature scale with 0 being assigned to absolute zero. The size of the degree chosen was the same size as the Celsius degree. Change each of the Celsius temperatures in the table above to degrees Kelvin. **303°, 294°, 273°, 261°, 246°**

Extra Credit

On page 290, the point-slope form of a line is defined for a *nonvertical* line. What happens if you use the formula for a vertical line? How does the formula change for a horizontal line? Use examples to support your responses. **Sample answer: For a vertical line, the slope is undefined so you could not substitute a value for m. For a horizontal line, $m = 0$ so the equation becomes $y = y_1$.**

Lesson 7–3

1 FOCUS

5-Minute Check
Lesson 7–2

1. Write the point-slope form of an equation for the line passing through the point at (8, −3) and having slope $-\frac{1}{5}$. $y + 3 = -\frac{1}{5}(x - 8)$

Write the point-slope form of an equation for the line passing through the given points.

2. (3, 2) and (4, 9) $y - 2 = 7(x - 3)$ or $y - 9 = 7(x - 4)$
3. (2, 8) and (−6, 8) $y = 8$
4. Write the point-slope form of an equation for the line.

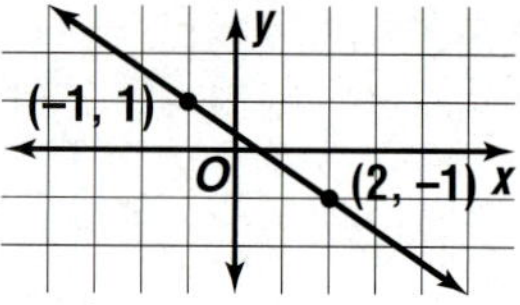

$y - 1 = -\frac{2}{3}(x + 1)$ or

$y + 1 = -\frac{2}{3}(x - 2)$

Motivating the Lesson

Real-World Connection Many companies charge their customers a one-time charge and then a unit charge. For example, a plumber may charge a $25 service charge to come to your house and $50 per hour of service. Situations such as these can be represented using equations in slope-intercept form. Have students brainstorm other businesses that may use a one-time charge and a unit charge. Some examples are phone companies, car rental companies, Internet service providers, and parking lots.

7–3 Writing Equations in Slope-Intercept Form

What You'll Learn
You'll learn to write a linear equation in slope-intercept form given the slope and y-intercept.

Why It's Important
Catering Caterers can use linear equations to represent the total cost. *See Exercise 47.*

The line crosses the y-axis at (0, 5). The number 5 is called the **y-intercept** of the equation of the line. The line crosses the x-axis at (2, 0). Thus, 2 is the **x-intercept** of the equation of the line.

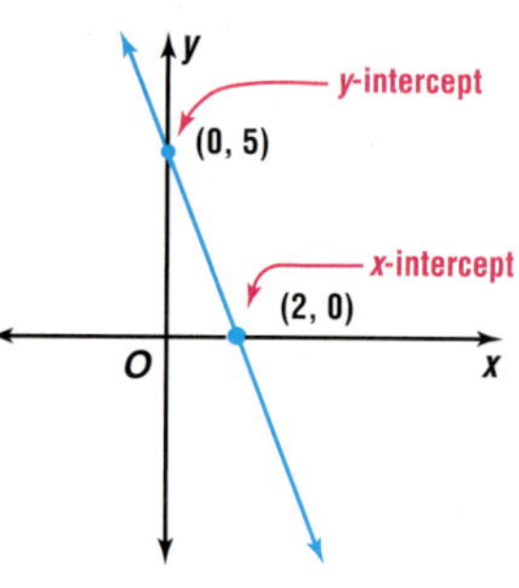

In Lesson 7–2, you learned how to write an equation in point-slope form by using the slope and a point on the line, and two points on the line. You can also write an equation of a line if you know the slope and y-intercept. Consider the graph below, which crosses the y-axis at $(0, b)$.

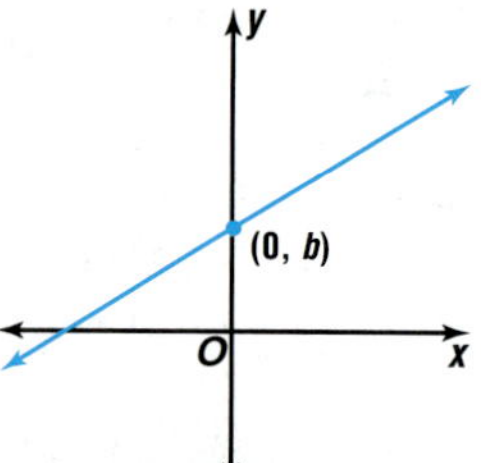

$y - y_1 = m(x - x_1)$	*Point-Slope Form*
$y - b = m(x - 0)$	*Replace (x_1, y_1) with $(0, b)$.*
$y - b = mx$	
$y - b + b = mx + b$	*Add b to each side.*
$y = mx + b$	

(↑ slope: m; ↑ y-intercept: b)

Slope-Intercept Form

Words: Given the slope m and y-intercept b of a line, the slope-intercept form of an equation of the line is $y = mx + b$.

Model:

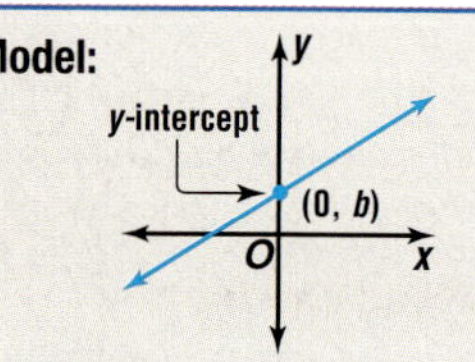

Examples

Write an equation in slope-intercept form of each line with the given slope and y-intercept.

1 $m = -4, b = 5$

$y = mx + b$ *Slope-Intercept Form*

$= -4x + 5$ *Replace m with −4 and b with 5.*

An equation of the line is $y = -4x + 5$.

2 $m = \frac{1}{3}, b = -6$

$y = mx + b$ *Slope-Intercept Form*

$= \frac{1}{3}x + (-6)$ *Replace m with $\frac{1}{3}$ and b with −6.*

$= \frac{1}{3}x - 6$ *Simplify.*

An equation of the line is $y = \frac{1}{3}x - 6$.

Resource Manager

Reproducible Masters
Chapter 7 Resource Masters
- *Study Guide,* p. 299
- *Skills Practice,* p. 300
- *Practice,* p. 301
- *Reading to Learn Mathematics,* p. 302
- *Enrichment,* p. 303
- *Assessment,* p. 335

Hands-On Algebra, p. 82
School-to-Workplace, p. 7

Transparencies
5-Minute Check, 7-3

Technology/Multimedia
Interactive Chalkboard CD-ROM

3 $m = 0, b = 3$

$$\begin{aligned} y &= mx + b && \textit{Slope-Intercept Form} \\ &= 0x + 3 && \textit{Replace m with 0 and b with 3.} \\ &= 3 && \textit{Simplify.} \end{aligned}$$

The equation of the line is $y = 3$. Remember that a line with a slope of 0 is a horizontal line.

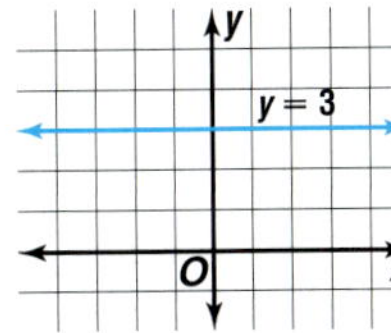

Your Turn **a.** $y = 2x + 1$ **b.** $y = -\frac{5}{3}x$ **c.** $y = -8$

a. $m = 2, b = 1$ b. $m = -\frac{5}{3}, b = 0$ c. $m = 0, b = -8$

Now you can use the methods in Lesson 7–2 to write equations in slope-intercept form.

Examples

Write an equation of the line in slope-intercept form for each situation.

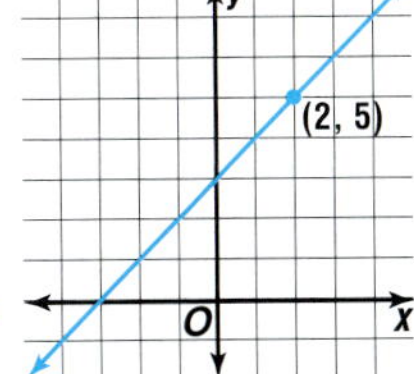

4 **slope 1 and passes through (2, 5)**

$$\begin{aligned} y - y_1 &= m(x - x_1) && \textit{Point-Slope Form} \\ y - 5 &= 1(x - 2) && \textit{Replace } (x_1, y_1) \textit{ with (2, 5) and m with 1.} \\ y - 5 &= x - 2 \end{aligned}$$

$$\begin{aligned} y - 5 + 5 &= x - 2 + 5 && \textit{Add 5 to each side.} \\ y &= x + 3 && \textit{Slope-Intercept Form} \end{aligned}$$

An equation of the line is $y = x + 3$. *What are the x- and y-intercepts?*

Check:
$$\begin{aligned} y &= x + 3 && \textit{Original equation} \\ 5 &\stackrel{?}{=} 2 + 3 && \textit{Replace x with 2 and y with 5.} \\ 5 &= 5 \ \checkmark \end{aligned}$$

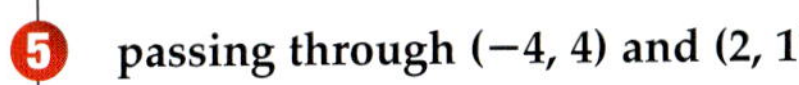

5 **passing through (−4, 4) and (2, 1)**

First, determine the slope of the line.

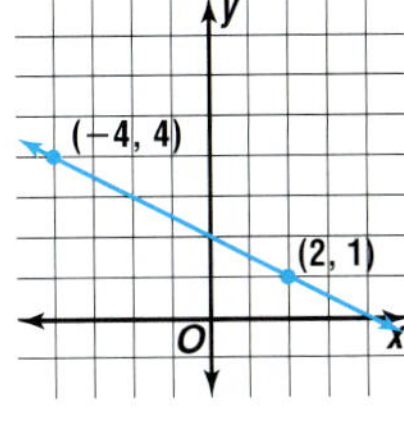

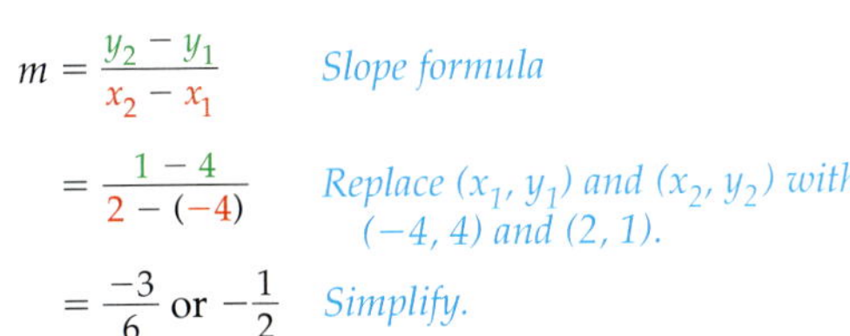

$$\begin{aligned} m &= \frac{y_2 - y_1}{x_2 - x_1} && \textit{Slope formula} \\ &= \frac{1 - 4}{2 - (-4)} && \textit{Replace } (x_1, y_1) \textit{ and } (x_2, y_2) \textit{ with (−4, 4) and (2, 1).} \\ &= \frac{-3}{6} \text{ or } -\frac{1}{2} && \textit{Simplify.} \end{aligned}$$

(continued on next page)

2 TEACH

Teaching Tip When introducing the slope-intercept form of an equation, stress that it is the y-intercept that is used. Stress that it is not necessary to know the x-intercept in order to write the equation in this form.

In-Class Examples

Examples 1–3

Write an equation in slope-intercept form of each line with the given slope and y-intercept.

1 $m = 3, b = -1$ $y = 3x - 1$

2 $m = -\frac{2}{3}, b = 0$ $y = -\frac{2}{3}x$

3 $m = 0, b = -4$ $y = -4$

Teaching Tip When discussing Examples 4 and 5, ask students how these examples differ from the examples in Lesson 7–2. Explain that it is common to leave all linear equations in slope-intercept form. You might also have students choose another point on the graph to verify the equation.

In-Class Example

Example 4

Write an equation of a line in slope-intercept form that has slope −3 and passes through the point at (1, −4).
$y = -3x - 1$

In-Class Examples

Example 5

Write an equation in slope-intercept form of a line passing through points at (6, 2) and (3, −2). $y = \frac{4}{3}x - 6$

Example 6

A plumber charges $25 for a service call plus $50 per hour of service. The graph below represents the plumber's charges. Write an equation of the line in slope-intercept form. $y = 50x + 25$

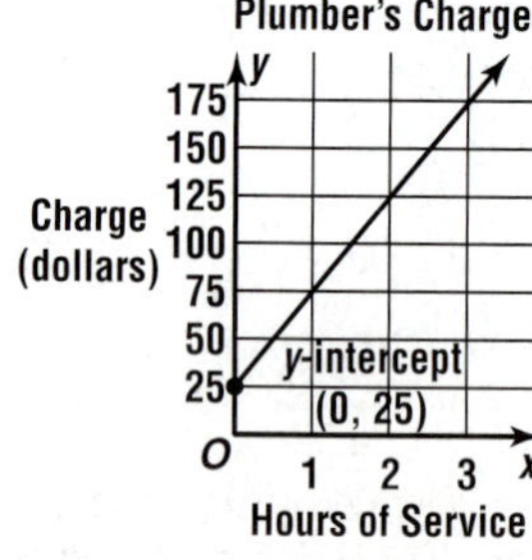

Now substitute the known values into the point-slope form.

$y - y_1 = m(x - x_1)$	*Point-Slope Form*
$y - 4 = -\frac{1}{2}[x - (-4)]$	*Replace (x_1, y_1) with $(-4, 4)$ and m with $-\frac{1}{2}$.*
$y - 4 = -\frac{1}{2}x - 2$	*Distributive Property*

Then write in slope-intercept form.

$y - 4 + 4 = -\frac{1}{2}x - 2 + 4$	*Add 4 to each side.*
$y = -\frac{1}{2}x + 2$	*Slope-Intercept Form*

An equation of the line is $y = -\frac{1}{2}x + 2$. You can see from the graph that the y-intercept is 2. *You can also check by substituting the coordinates of one of the points into the equation.*

Your Turn

Write an equation in slope-intercept form of each line.

d. the line whose slope is $\frac{3}{4}$ and passes through (8, −2) $y = \frac{3}{4}x - 8$

e. the line passing through (2, 4) and (0, 5) $y = -\frac{1}{2}x + 5$

Example 6 Science Link

A California inventor designed the *Skycar*, a car that can fly. The graph represents its landing from 50 meters off the ground. Write an equation of the line in slope-intercept form.

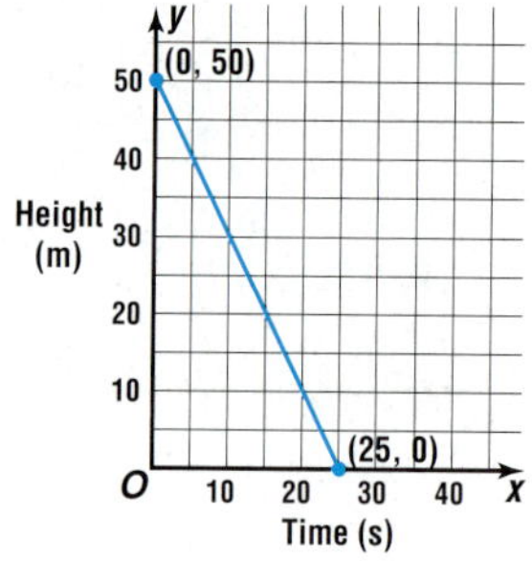

The y-intercept of the line is 50. Determine the slope.

$$m = \frac{y_2 - y_1}{x_2 - x_1} = \frac{0 - 50}{25 - 0} \text{ or } -2$$

Now substitute these values into the slope-intercept form.

$y = mx + b$	*Slope-Intercept Form*
$= -2x + 50$	*Replace m with −2 and b with 50.*

An equation of the line is $y = -2x + 50$.

Check for Understanding

Communicating Mathematics

1. **Sketch** a line that has a y-intercept of 4 and a negative slope. **See margin.**
2. **Determine** whether the graph of $y = -3x + 8$ passes through $(-5, 23)$. Explain how you found your answer.
3. **You Decide?** Andrew says that every line has an x-intercept. Jacquie disagrees. Who is correct? Give an example to support your answer. **Jacquie; the line $y = 2$ does not intersect the x-axis.**

Vocabulary
- y-intercept
- x-intercept
- slope-intercept form

2. Yes; substitute $(-5, 23)$ into the equation and then check.

Guided Practice

Examples 1–3

Write an equation in slope-intercept form of the line with each slope and y-intercept.

4. $m = \frac{3}{2}, b = 4$ **$y = \frac{3}{2}x + 4$**
5. $m = \frac{1}{4}, b = -3$ **$y = \frac{1}{4}x - 3$**
6. $m = 0, b = 12$ **$y = 12$**

Example 4

Write an equation in slope-intercept form of the line having the given slope and passing through the given point.

7. $m = 4, (2, 1)$ **$y = 4x - 7$**
8. $m = \frac{3}{5}, (0, 5)$ **$y = \frac{3}{5}x + 5$**
9. $m = -\frac{1}{2}, (4, -5)$ **$y = -\frac{1}{2}x - 3$**

Example 5

Write an equation in slope-intercept form of the line passing through each pair of points.

10. $(2, 3)$ and $(1, 0)$ **$y = 3x - 3$**
11. $(2, -4)$ and $(5, 2)$ **$y = 2x - 8$**
12. $(-4, 9)$ and $(5, 9)$ **$y = 9$**

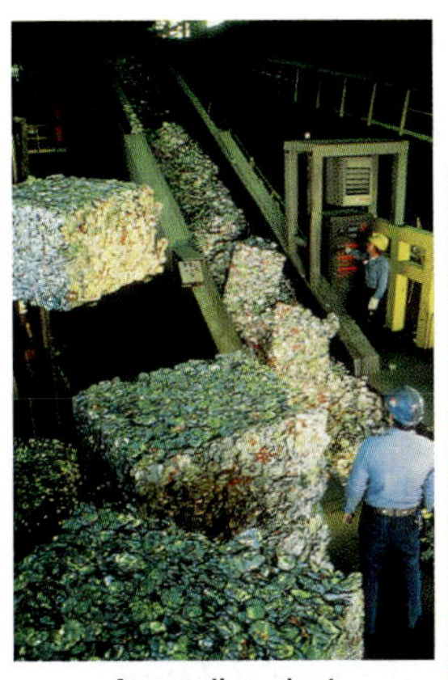

A recycling plant

13. **Recycling** Yoshi collected 100 pounds of cans to recycle. He plans to collect an additional 25 pounds each week. The graph shows the amount of cans he plans to collect. **Example 6**
    a. Write an equation in slope-intercept form of the line. **$y = 25x + 100$**
    b. What does the slope represent?
    c. Use the equation to predict the total amount of cans Yoshi will have collected after 12 weeks. **400 lb**

    13b. the pounds of cans he plans to collect each week

Cans (lb): 50, 100, 150, 200, 250, 300, 350; y; (4, 200); (0, 100); O; 1 2 3 4 5 6 7 8 9 x; Number of Weeks

Exercises

Practice

Write an equation in slope-intercept form of the line with each slope and y-intercept. **14. $y = 3x + 1$ 15. $y = -2x + 5$ 16. $y = 4x - 2$**

14. $m = 3, b = 1$
15. $m = -2, b = 5$
16. $m = 4, b = -2$
17. $m = \frac{3}{2}, b = 7$
18. $m = -\frac{1}{4}, b = -8$
19. $m = -\frac{2}{5}, b = 3$
20. $m = -3, b = 0$ **$y = -3x$**
21. $m = 0, b = 4$ **$y = 4$**
22. $m = -1.6, b = -12$ **$y = -1.6x - 12$**

17. $y = \frac{3}{2}x + 7$
18. $y = -\frac{1}{4}x - 8$
19. $y = -\frac{2}{5}x + 3$

A

3 PRACTICE/APPLY

Error Analysis

Watch for students who make computational errors when writing equations of lines in Exercises 10–12.
Prevent by encouraging students to check their equation using one of the given points.

Assignment Guide

Basic: 15–47 odd, 48–57
Average: 14–44 even, 46–57
All: Quiz 1, 1–5

Answer

1. Sample answer:

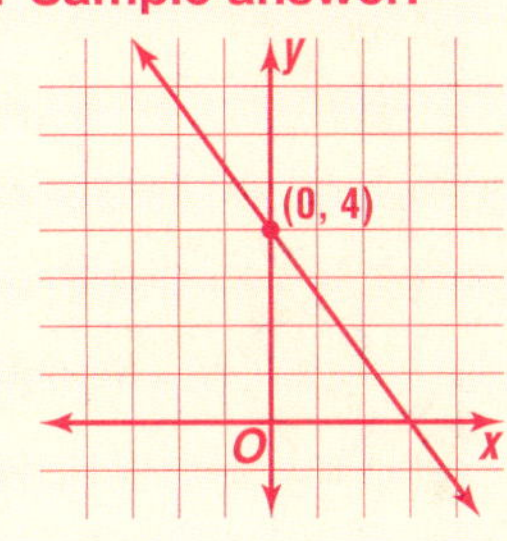

Reteaching Activity

Logical Learners Have students work with a partner. Give each student three index cards and have them write one of these three headings on top of each card: *Slope and y-intercept, Slope and One Point,* and *Two Points.* On the side of each card that has the heading written on it, students should work together to write a series of steps that need to be followed to solve the type of problem identified by the heading. On the other side of each card, students should write an example of that problem type and show how the equation is written using the steps on the card.

***Study Guide*, p. 299**

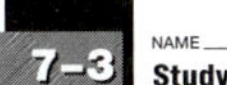

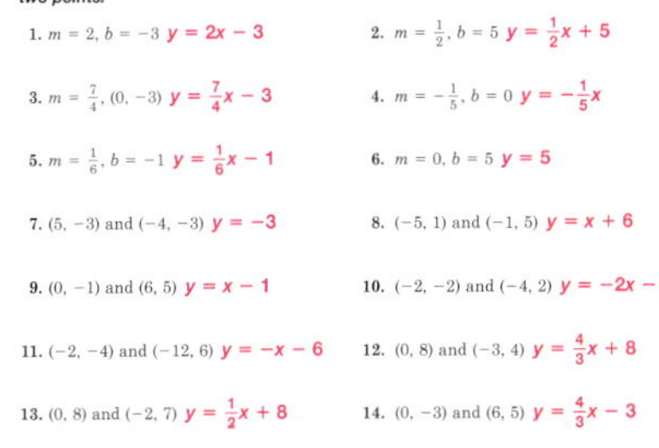

7–3 NAME ______ DATE ______ PERIOD ______

Study Guide Student Edition Pages 296–301

Writing Equations in Slope-Intercept Form

Given the slope m and the y-intercept b of a line, the slope-intercept form of an equation of the line is $y = mx + b$. Sometimes it is more convenient to express the equation of a line in slope-intercept form instead of point-slope form. This form is especially useful when graphing lines.

Examples: Write the point-intercept form of an equation for each line.

a. the line with $m = -\frac{2}{3}$ and $b = 4$

$y = mx + b$

$y = -\frac{2}{3}x + 4$

b. the line passing through points at $(4, -2)$ and $(6, 2)$

$m = \frac{y_2 - y_1}{x_2 - x_1}$ *Find m.*

$m = \frac{2 - (-2)}{6 - 4} = \frac{4}{2} = 2$

$y - 2 = 2(x - 6)$

$y - 2 = 2x - 12$ *Multiply.*

$y - 2 + 2 = 2x - 12 + 2$ *Add 2 to each side.*

$y = 2x - 10$ *Simplify.*

Write the slope-intercept form of an equation for each line, given either the slope and the y-intercept or the coordinates of two points.

1. $m = 2, b = -3$ **$y = 2x - 3$**
2. $m = \frac{1}{2}, b = 5$ **$y = \frac{1}{2}x + 5$**
3. $m = \frac{7}{4}, (0, -3)$ **$y = \frac{7}{4}x - 3$**
4. $m = -\frac{1}{5}, b = 0$ **$y = -\frac{1}{5}x$**
5. $m = \frac{1}{6}, b = -1$ **$y = \frac{1}{6}x - 1$**
6. $m = 0, b = 5$ **$y = 5$**
7. $(5, -3)$ and $(-4, -3)$ **$y = -3$**
8. $(-5, 1)$ and $(-1, 5)$ **$y = x + 6$**
9. $(0, -1)$ and $(6, 5)$ **$y = x - 1$**
10. $(-2, -2)$ and $(-4, 2)$ **$y = -2x - 6$**
11. $(-2, -4)$ and $(-12, 6)$ **$y = -x - 6$**
12. $(0, 8)$ and $(-3, 4)$ **$y = \frac{4}{3}x + 8$**
13. $(0, 8)$ and $(-2, 7)$ **$y = \frac{1}{2}x + 8$**
14. $(0, -3)$ and $(6, 5)$ **$y = \frac{4}{3}x - 3$**

Answers

23. $y = 3x - 17$

24. $y = -\frac{1}{4}x + 8$

25. $y = 5x$

26. $y = 6$

27. $y = -2x + 13$

28. $y = 3x - 15$

29. $y = -5x + 26$

30. $y = -\frac{3}{7}x + \frac{10}{7}$

31. $y = \frac{1}{2}x - \frac{9}{2}$

Page 301

51.

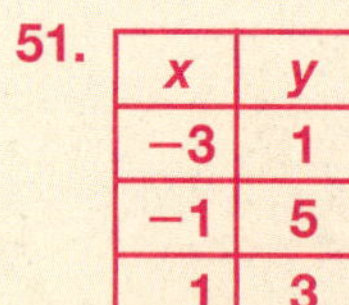

x	y
−3	1
−1	5
1	3
2	6

domain: {−3, −1, 1, 2}, range: {1, 5, 3, 6}

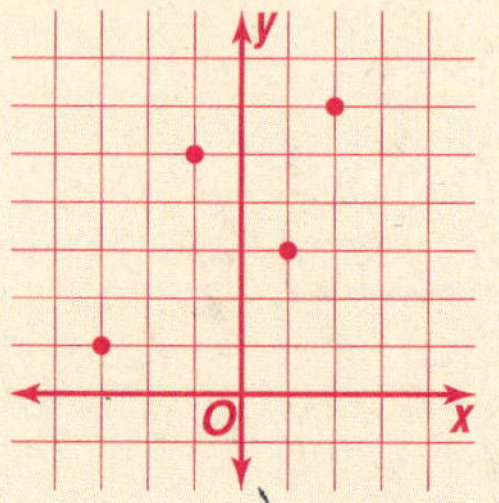

***Skills Practice*, p. 300, and *Practice*, p. 301 (shown)**

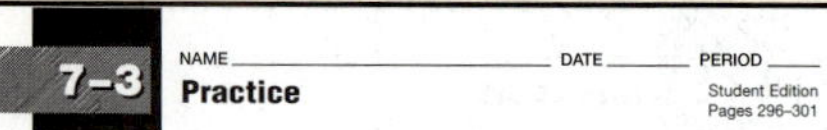

NAME ______ DATE ______ PERIOD ______

7–3 Practice

Student Edition Pages 296–301

Writing Equations in Point-Intercept Form

Write an equation in slope-intercept form of the line with each slope and y-intercept.

1. $m = -3, b = 5$ — $y = -3x + 5$
2. $m = 6, b = 2$ — $y = 6x + 2$
3. $m = 4, b = -1$ — $y = 4x - 1$
4. $m = 0, b = 4$ — $y = 4$
5. $m = \frac{2}{5}, b = -7$ — $y = \frac{2}{5}x - 7$
6. $m = -\frac{3}{4}, b = 8$ — $y = -\frac{3}{4}x + 8$
7. $m = -\frac{4}{3}, b = -2$ — $y = -\frac{4}{3}x - 2$
8. $m = -5, b = 6$ — $y = -5x + 6$
9. $m = \frac{1}{2}, b = -9$ — $y = \frac{1}{2}x - 9$

Write an equation in slope-intercept form of the line having the given slope and passing through the given point.

10. $m = 3, (4, 2)$ — $y = 3x - 10$
11. $m = -2, (-1, 3)$ — $y = -2x + 1$
12. $m = 4, (0, -7)$ — $y = 4x - 7$
13. $m = -\frac{3}{5}, (-5, -3)$ — $y = -\frac{3}{5}x - 6$
14. $m = \frac{1}{4}, (-8, 6)$ — $y = \frac{1}{4}x + 8$
15. $m = -\frac{2}{3}, (9, -4)$ — $y = -\frac{2}{3}x + 2$
16. $m = \frac{5}{6}, (6, -6)$ — $y = \frac{5}{6}x - 11$
17. $m = 0, (-8, -7)$ — $y = -7$
18. $m = -\frac{3}{2}, (-8, 9)$ — $y = -\frac{3}{2}x - 3$

Write an equation in slope-intercept form of the line passing through each pair of points.

19. (1, 3) and (− 3, −5) — $y = 2x + 1$
20. (0, 5) and (3, −4) — $y = -3x + 5$
21. (2, 1) and (3, 6) — $y = 5x - 9$
22. (− 3, 0) and (6, −6) — $y = -\frac{2}{3}x - 2$
23. (4, 5) and (− 5, 5) — $y = 5$
24. (0, 6) and (− 4, 3) — $y = \frac{3}{4}x + 6$
25. (−3, 2) and (3, −6) — $y = -\frac{4}{3}x - 2$
26. (−7, −6) and (−5, −3) — $y = \frac{3}{2}x + \frac{9}{2}$
27. (6, −4) and (0, 2) — $y = -x + 2$

Homework Help

For Exercises	See Examples
14–22, 44–46	1–3
23–31	4
32–40, 47	5
41–43	6

Extra Practice

See page 706.

Write an equation in slope-intercept form of the line having the given slope and passing through the given point. **23–31. See margin.**

23. $m = 3, (5, -2)$
24. $m = -\frac{1}{4}, (0, 8)$
25. $m = 5, (0, 0)$
26. $m = 0, (5, 6)$
27. $m = -2, (4, 5)$
28. $m = 3, (4, -3)$
29. $m = -5, (6, -4)$
30. $m = -\frac{3}{7}, (1, 1)$
31. $m = \frac{1}{2}, (-3, -6)$

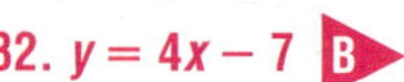

Write an equation in slope-intercept form of the line passing through each pair of points.

32. (3, 5) and (2, 1) — **32.** $y = 4x - 7$
33. (4, 0) and (0, 5) — **33.** $y = -\frac{5}{4}x + 5$
34. (4, −3) and (2, 1) — **34.** $y = -2x + 5$
35. (3, 0) and (2, −4) — **35.** $y = 4x - 12$
36. (1, 6) and (−2, 6) — **36.** $y = 6$
37. (9, −3) and (−1, −3) — **37.** $y = -3$
38. (7, −2) and (14, −4) $y = -\frac{2}{7}x$
39. (1, 0) and (6, −4) $y = -\frac{4}{5}x + \frac{4}{5}$
40. (−1, 8) and (3, 6) $y = -\frac{1}{2}x + \frac{15}{2}$

Write an equation in slope-intercept form for each line.

41.

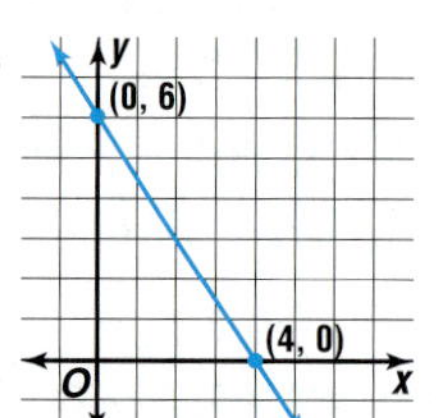

42.

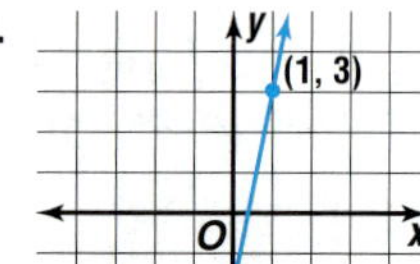

43.

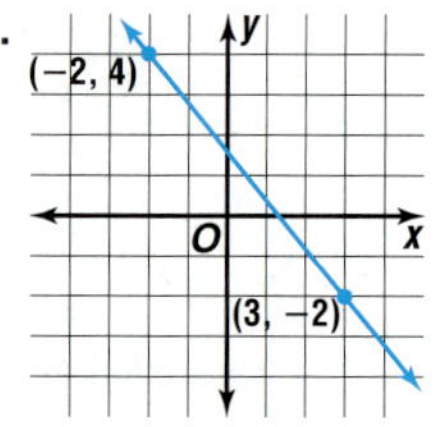

41. $y = -\frac{3}{2}x + 6$

42. $y = 5x - 2$

43. $y = -\frac{6}{5}x + \frac{8}{5}$

44. Write an equation in slope-intercept form of a line with slope $-\frac{2}{3}$ and y-intercept the same as the line whose equation is $y = 4x - 7$. **44.** $y = -\frac{2}{3}x - 7$

45. Write an equation in slope-intercept form of a line with a y-intercept of 11 and slope the same as the line whose equation is $y = -3x + 9$. $y = -3x + 11$

Applications and Problem Solving

46. **Work** For babysitting, Nicole charges a flat fee of \$3, plus \$5 per hour. The graph shows how much she earns.

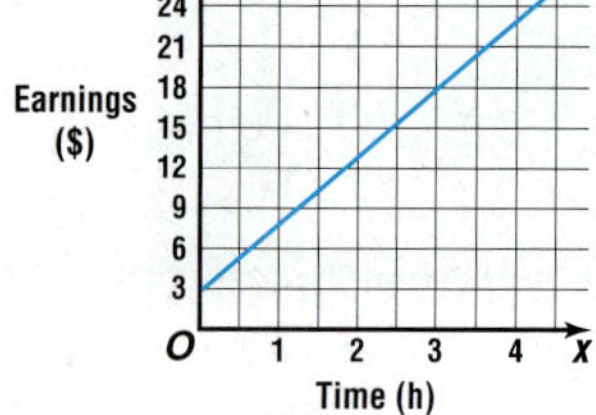

a. Write an equation in slope-intercept form of the line. $y = 5x + 3$

b. What do you think the slope and y-intercept represent? **46b. hourly rate and flat fee**

c. How much money will she make if she babysits 5 hours? **\$28**

d. What are the effects if Nicole changes her rates so that the equation is $y = 6x + 2$? **46d. The flat fee is \$2 and the hourly rate is \$6.**

47. **Catering** A caterer charges \$120 to cater a party for 15 people and \$200 for 25 people. Assume that the cost y is a linear function of the number of people x.

a. Write an equation in slope-intercept form for this function. $y = 8x$

b. Explain what the slope represents. **cost per person**

c. How much would a party for 40 people cost? **\$320**

Family Activity

Talk with a family member about bills that have a fixed charge and a unit charge. Try to think of a charge your family has incurred recently that was structured this way. Write an equation that represents this situation. Some charges that are commonly structured this way are utility bills, repair bills, and rental fees. Some loans can also be structured this way with a down payment and a repeated monthly payment.

48. **Critical Thinking** A line contains points at (5, 5) and (9, 1). Write a convincing argument showing that the x-intercept of the line is 10. Check by sketching the line. **See Solutions Manual.**

Mixed Review

Write the point-slope form of an equation for each line passing through the given point and having the given slope. *(Lesson 7–2)*

49. $(4, 5)$, $m = -2$ $\quad y - 5 = -2(x - 4)$

50. $(6, -2)$, $m = \frac{1}{2}$ $\quad y + 2 = \frac{1}{2}(x - 6)$

Express each relation as a table and as a graph. Then determine the domain and the range. *(Lesson 6–1)* **51–52. See margin.**

51. $\{(-3, 1), (-1, 5), (1, 3), (2, 6)\}$

52. $\{(-3, 0), (-1, 3), (0, -3), (3, 2)\}$

53. **Measurement** Convert 9144 meters to kilometers. *(Lesson 5–1)* **9.144 km**

Find each quotient. *(Lesson 4–3)*

54. $7.2 \div 72$ **0.1**

55. $6 \div (-1.2)$ **−5**

56. $-0.5 \div (-0.2)$ **2.5**

Standardized Test Practice

57. **Multiple Choice** Solve $|c - 3| + 6 = 17$. *(Lesson 3–7)* **A**

Ⓐ $\{-8, 14\}$ Ⓑ $\{8, 14\}$ Ⓒ $\{-14, 8\}$ Ⓓ $\{-14, -8\}$

Quiz 1 — Lessons 7–1 through 7–3

Determine the slope of each line. *(Lesson 7–1)*

1. 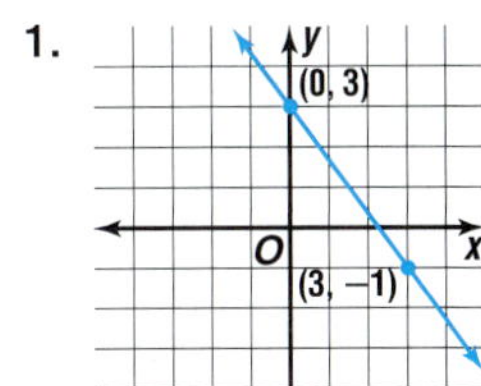

$-\frac{4}{3}$

2. the line through points at $(0, -2)$ and $(2, 2)$ **2**

Write the point-slope form of an equation for each line passing through the given point and having the given slope. *(Lesson 7–2)*

3. $(5, 4)$, $m = -2$ $\quad y - 4 = -2(x - 5)$

4. $(-1, 6)$, $m = \frac{8}{5}$ $\quad y - 6 = \frac{8}{5}(x + 1)$

5. **Education** In order to "curve" a set of test scores, a teacher uses the equation $y = 2.5x + 10$, where y is the curved test score and x is the number of problems answered correctly. *(Lesson 7–3)*
   a. Find the test score of a student who answers 32 problems correctly. **90**
   b. Explain what the slope and y-intercept mean in the equation. **See margin.**

 www.algconcepts.com/self_check_quiz

4 ASSESS

Open-Ended Assessment

Speaking On the board or overhead, write various problems like those in the lesson. Ask students to explain how to complete each of them.

Quiz 1

The Quiz provides students with a brief review of the concepts and skills in Lessons 7–1 through 7–3. Lesson numbers are given to the right of the exercises or instruction lines so that students can review concepts not yet mastered.

Chapter 7, Quiz A (Lessons 7–1 through 7–3) is available in the *Chapter 7 Resource Masters*, p. 335.

Answer

52.

x	y
−3	0
−1	3
0	−3
3	2

domain: $\{-3, -1, 0, 3\}$, range: $\{0, 3, -3, 2\}$

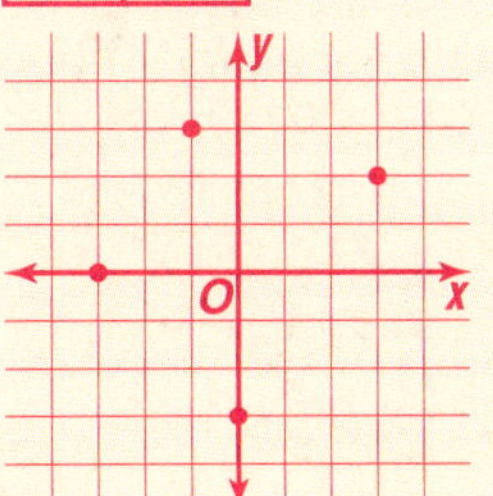

***Enrichment*, p. 303**

7–3 NAME ______ DATE ______ PERIOD ______

Enrichment — Student Edition Pages 296–301

Ideal Weight

You can find your ideal weight as follows.
A woman should weigh 100 pounds for the first 5 feet of height and 5 additional pounds for each inch over 5 feet (5 feet = 60 inches). A man should weigh 106 pounds for the first 5 feet of height and 6 additional pounds for each inch over 5 feet. These formulas apply to people with normal bone structures.

To determine your bone structure, wrap your thumb and index finger around the wrist of your other hand. If the thumb and finger just touch, you have normal bone structure. If they overlap, you are small-boned. If they don't overlap, you are large-boned. Small-boned people should decrease their calculated ideal weight by 10%. Large-boned people should increase the value by 10%.

Calculate the ideal weights of these people.

1. woman, 5 ft 4 in., normal-boned **120 lb**
2. man, 5 ft 11 in., large-boned **189.2 lb**
3. man, 6 ft 5 in., small-boned **187.2 lb**
4. you, if you are at least 5 ft tall **See students' work.**

Suppose a normal-boned man is x inches tall. If he is at least 5 feet tall, then $x - 60$ represents the number of inches this man is over 5 feet tall. For each of these inches, his ideal weight is increased by 6 pounds. Thus, his proper weight (y) is given by the formula $y = 6(x - 60) + 106$ or $y = 6x - 254$. If the man is large-boned, the formula becomes $y = 6x - 254 + 0.10(6x - 254)$.

5. Write the formula for the weight of a large-boned man in slope-intercept form. $y = 6.6x - 279.4$
6. Derive the formula for the ideal weight (y) of a normal-boned female with height x inches. Write the formula in slope-intercept form. $y = 5x - 200$
7. Derive the formula in slope-intercept form for the ideal weight (y) of a large-boned female with height x inches. $y = 5.5x - 220$
8. Derive the formula in slope-intercept form for the ideal weight (y) of a small-boned male with height x inches. $y = 5.4x - 228.6$
9. Find the heights at which normal-boned males and large-boned females would weigh the same. **68 in., or 5 ft 8 in.**

Extra Credit

An organization that had 4100 members in 1988 grew to 11,500 members by 1998. Assume that the number of members y is a linear function of the year x after 1980. (This means that for 1988, $x = 8$.) Write an equation for this function. Explain what the slope represents, and use the equation to predict the number of members in 2002. **$y = 740x - 1820$; The slope is the average number of new members per year; 14,460 members.**

Answer

Quiz 1

5b. The slope represents how much each problem is worth and the y-intercept represents the extra points students get with the curve.

Lesson 7-4

7-4 Scatter Plots

1 FOCUS

5-Minute Check
Lesson 7-3

Write an equation in slope-intercept form of the line described.

1. $m = -2, b = 5$
   $y = -2x + 5$
2. slope $\frac{3}{4}$, passes through the point at $(-8, 1)$
   $y = \frac{3}{4}x + 7$
3. passing through points at $(4, -5)$ and $(-2, -5)$
   $y = -5$
4. A canoe rental service charges a \$20 transportation fee and \$30 per hour to rent a canoe. The situation is represented in the graph below. Write an equation to represent the cost y of renting a canoe for x hours.
   $y = 30x + 20$

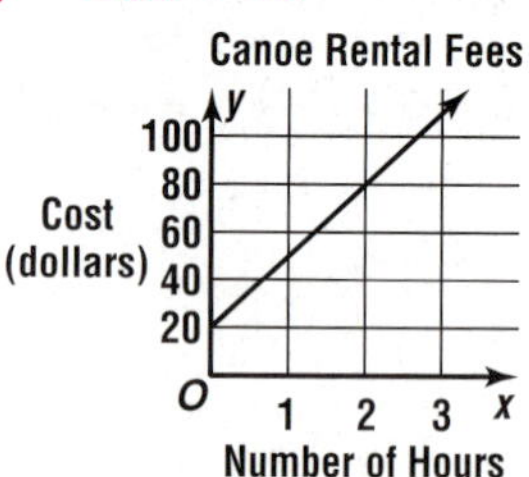

What You'll Learn
You'll learn to graph and interpret points on scatter plots.

Why It's Important
Insurance Insurance companies can use scatter plots to determine policy rates. *See Exercise 17.*

The Caribbean islands have many different species of birds.

To determine if there is a relationship between area and number of bird species, we can graph the data points in a **scatter plot**. In a scatter plot, two sets of data are plotted as ordered pairs in the coordinate plane.

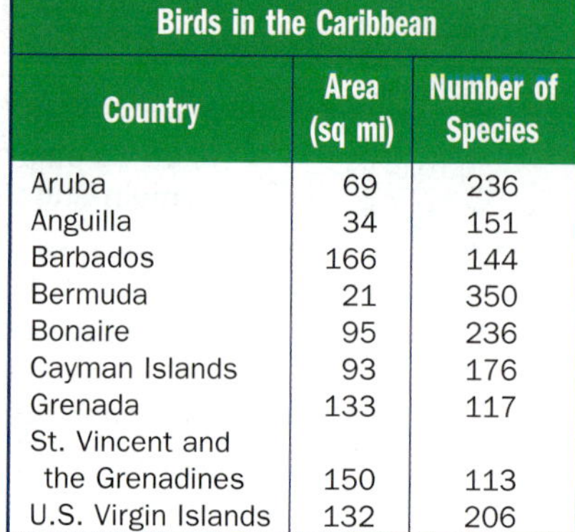

Birds in the Caribbean

Country	Area (sq mi)	Number of Species
Aruba	69	236
Anguilla	34	151
Barbados	166	144
Bermuda	21	350
Bonaire	95	236
Cayman Islands	93	176
Grenada	133	117
St. Vincent and the Grenadines	150	113
U.S. Virgin Islands	132	206

Source: *Thayer Birder's Diary*

(area, number of species)
↑ *x-coordinate* ↑ *y-coordinate*

For example, the point with the box around it is at (95, 236). *The area is the independent quantity, and the number of species is the dependent quantity.*

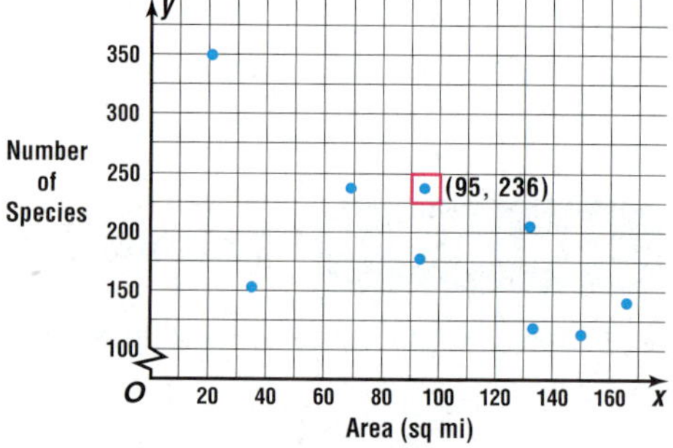

You can use the scatter plot to draw conclusions and make predictions about the data.

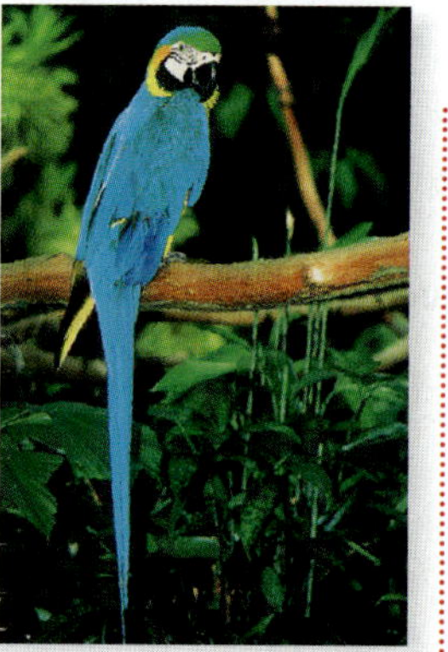

A gold and blue macaw

- The country with the least area has 350 species of birds.
- In general, as area increases, the number of species of birds appears to decrease. This shows a relationship between area and the number of species. That is, the number of species is a function of area. *The country of Anguilla is an exception to this generalization.*
- The domain is the set of all x-coordinates:
  {21, 34, 69, 93, 95, 132, 133, 150, 166}.
- The range is the set of all corresponding y-coordinates:
  {350, 151, 236, 176, 236, 206, 117, 113, 144}.
- Based on the data in the scatter plot, you could predict that a Caribbean country with an area of 50 square miles would have fewer than 350 species of birds.

Resource Manager

Reproducible Masters
Chapter 7 Resource Masters
- *Study Guide,* p. 304
- *Skills Practice,* p. 305
- *Practice,* p. 306
- *Reading to Learn Mathematics,* p. 307
- *Enrichment,* p. 308
- *Assessment,* p. 334

Graphing Calculator, pp. 21–22
Hands-On Algebra, p. 83

Transparencies
5-Minute Check, 7–4

Technology/Multimedia
Interactive Chalkboard CD-ROM

TECHNOLOGY
An alternative technology option using a graphing calculator is available for teaching this lesson.

Often, when data are displayed in a scatter plot, you can determine whether there is a pattern, trend, or relationship between the variables.

- Data points that appear to go uphill show a relationship that is *positive.*
- Data points that appear to go downhill show a relationship that is *negative.*

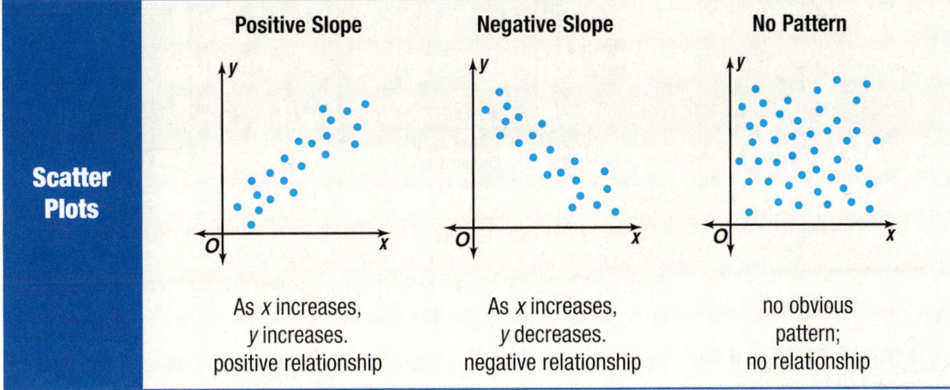

Scatter Plots
Positive Slope
As x increases, y increases. positive relationship
Negative Slope
As x increases, y decreases. negative relationship
No Pattern
no obvious pattern; no relationship

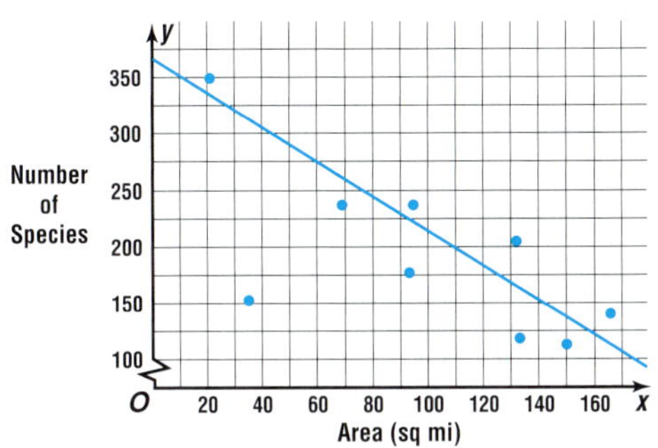

In the opening application, suppose a line were drawn close to most of the data points. The line would have a negative slope. That is, as the area increases, the number of species decreases. Thus, area and number of species would have a negative relationship.

1 **The scatter plot at the right shows the word processing speeds of 12 students and the number of weeks they have studied word processing. Determine whether the scatter plot shows a *positive* relationship, *negative* relationship, or *no* relationship. If there is a relationship, describe it.**

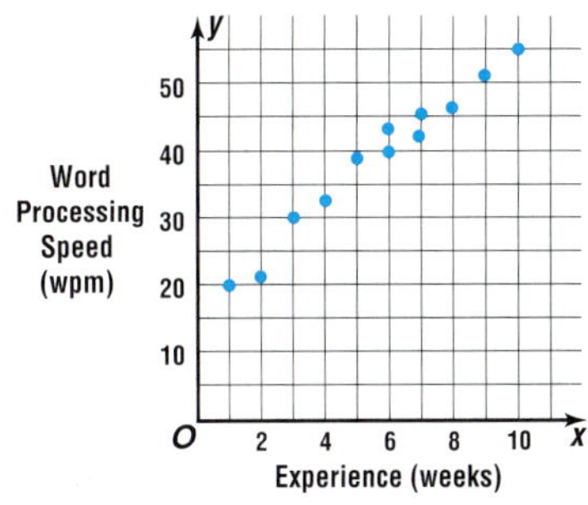

The plot indicates that as the number of weeks of experience increases, the word processing speed increases. Thus, there is a positive relationship between experience and word processing speed.

www.algconcepts.com/extra_examples

Motivating the Lesson

Hands-On Activity As students enter the room at the beginning of class, have several rulers available. Have each student measure the length of their hand (from the heel of their hand to the tip of the longest finger) and the length of their forearm (from the tip of their elbow to the tip of their longest finger). Make a chart on the board or overhead and have students record their measurements as an ordered pair with hand length as x and forearm length as y. Make a scatter plot of the data on the board or overhead. Ask students if they think there is a relationship between these two measurements, and if so, how this relationship is depicted in the graph.

2 TEACH

Teaching Tip When discussing scatter plots, emphasize that students are looking for a general trend in the points. Point out that there may be one or two points that do not fit the trend such as the point for Anguilla in the graph on page 302. However, most of the points should exhibit a clear upwards or downwards trend. Point out also that these are linear relationships.

In-Class Example

Example 1

The scatter plot below shows the number of years of experience and the salary for each employee in a small company. Determine whether the scatter plot shows a *positive* relationship, *negative* relationship, or *no* relationship. If there is a relationship, describe it.

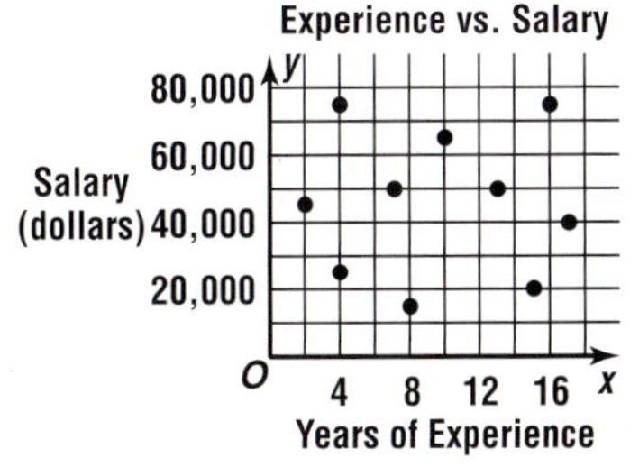

no relationship

In-Class Example

Example 2

The table shows the average number of minutes a pediatric dentist spends during each appointment instructing the patient in proper dental care, and the number of cavities for each patient.

Instruction Time (min)	Number of Cavities
6	1
4	3
7	2
10	1
1	5
1	6
5	3
2	4
2	3

A. Make a scatter plot of the data. Let the horizontal axis represent instruction time and let the vertical axis represent the number of cavities.

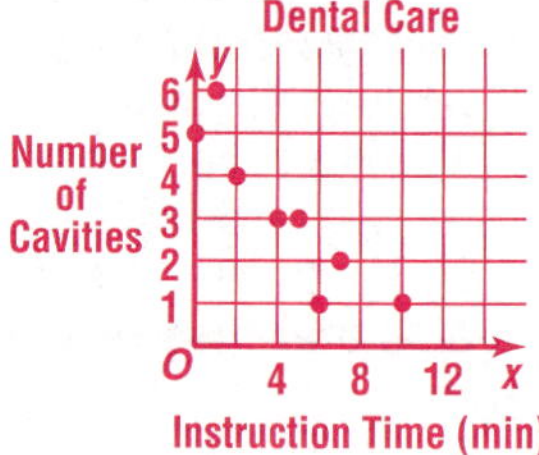

B. Does the scatter plot show a relationship between instruction time and cavities? Explain. **Yes; it appears that a lesser number of cavities is directly related to a greater amount of instruction time. There is a negative relationship.**

C. Describe the independent and dependent variables. Then state the domain and the range. **The number of cavities depends on the amount of instruction time, so the number of cavities is the dependent variable. Instruction time is the independent variable. The domain is the set of all instruction times and the range is the set of all numbers of cavities.**

Your Turn

Determine whether each scatter plot shows a *positive* relationship, *negative* relationship, or *no* relationship. If there is a relationship, describe it.

a.

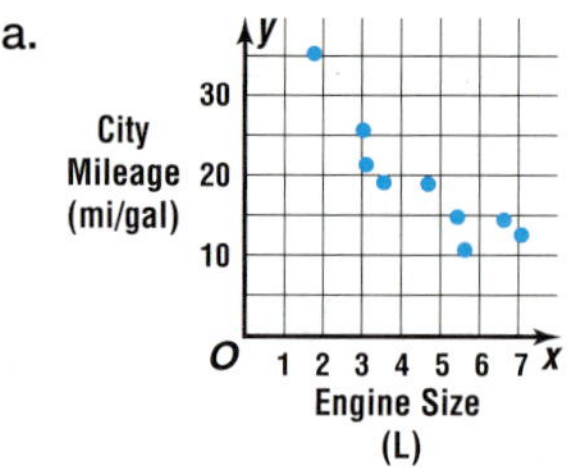

b.

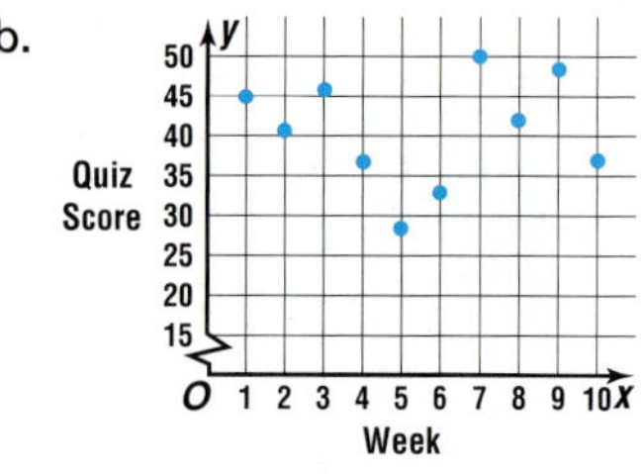

a. Negative relationship; as engine size increases, city mileage decreases. **b. no relationship**

Organizing data in a scatter plot makes it easier to observe patterns.

Example 2
School Link

The table shows the amount of time students in math class spent studying and the scores they received on the test.

Study Time (min)	Test Score	Study Time (min)	Test Score
60	92	30	77
55	79	95	94
10	65	60	83
75	87	15	68
120	98	35	79
90	95	70	78
110	75	25	97
45	73	115	100

A. Make a scatter plot of the data.

Let the horizontal axis represent study time, and let the vertical axis represent test scores. Then plot the data.

B. Does the scatter plot show a relationship between study time and test scores? Explain.

In general, the scatter plot seems to show that a higher grade is directly related to the amount of time spent studying. There is a positive relationship.

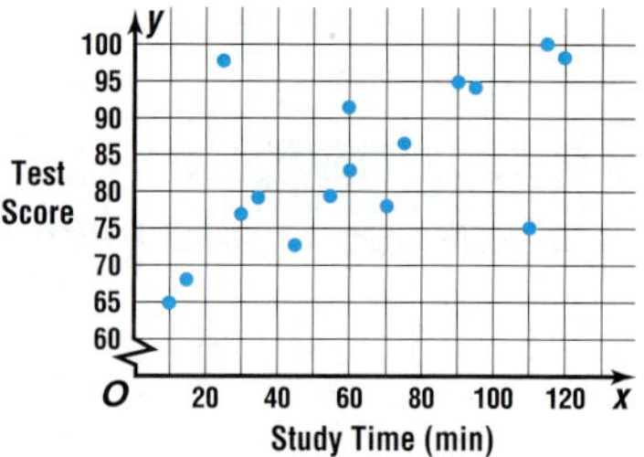

C. Describe the independent and dependent variables. Then state the domain and the range.

Because test scores depend on the amount of study time, test scores is the dependent variable, and study time is the independent variable. The domain is the set of all study times and the range is the set of all test scores.

Reteaching Activity

Interpersonal Learners Have each student work with a partner. Each pair of students should look in an almanac or other research book to find some data presented in a table. They should make a scatter plot of the data and describe the relationship between the variables.

Your Turn c. See margin.

c. Make a scatter plot of the data. Determine whether there is a relationship between annual rainfall and weight of pumpkins. Explain.

Annual Rainfall (in.)	43	35	38	42	30	29	44	40	41
Weight of Pumpkin (lb)	204	195	198	200	190	190	206	202	200

Check for Understanding

Communicating Mathematics

Vocabulary
scatter plot

1. **Describe** how you can use a scatter plot to display two sets of related data. Then explain what the domain, range, independent variable, and dependent variable are. 1–2. See margin.

2. **Explain** how you know whether there appears to be a linear relationship between the two sets of data plotted on a scatter plot. Include the concept of slope in your explanation.

3. Writing Math Describe situations in which data graphed in a scatter plot would have a positive relationship, negative relationship, and no relationship. Tell what the labels on the horizontal and vertical axes for each scatter plot would be. See students' work.

Guided Practice

Example 1

4. Determine whether the scatter plot has a *positive* relationship, *negative* relationship, or *no* relationship. If there is a relationship, describe it. Negative relationship; as TV time increases, physical activity decreases.

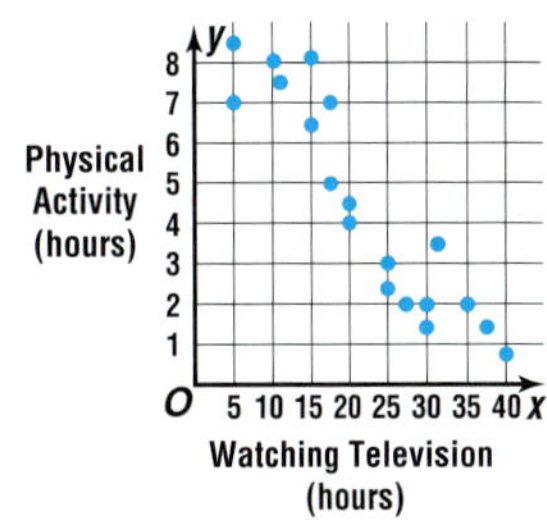

Example 2

5. **Weather** The average maximum and minimum monthly temperatures in July of ten cities are given in the table at the right.

a. Make a scatter plot of the data.

b. Does the scatter plot show a linear relationship between maximum and minimum temperatures? Explain.

5a. See Solutions Manual.
5b. Yes; as maximum temperature increases, so does the minimum temperature. There is a positive relationship.

City	Minimum (°F)	Maximum (°F)
Atlanta, GA	70	88
Lexington, KY	66	86
Galveston, TX	79	87
Columbus, OH	63	84
Jackson, MS	71	92
Albany, NY	60	84
Los Angeles, CA	65	84
Miami, FL	76	89
Portland, OR	57	80
Chicago, IL	63	84

Source: *The World Almanac*

3 PRACTICE/APPLY

Error Analysis

Watch for students who have trouble choosing an appropriate scale for the axes when graphing a scatter plot, such as in Exercise 5a. ***Prevent by*** encouraging students find the least and greatest value of each variable. They should then find the difference of these values to help them determine an appropriate scale.

Answer

Your Turn

c.

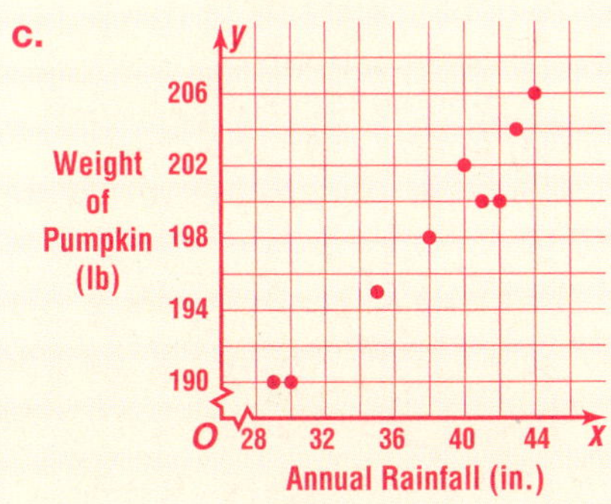

There is a positive relationship; as annual rainfall increases, pumpkin weight increases.

***Study Guide*, p. 304**

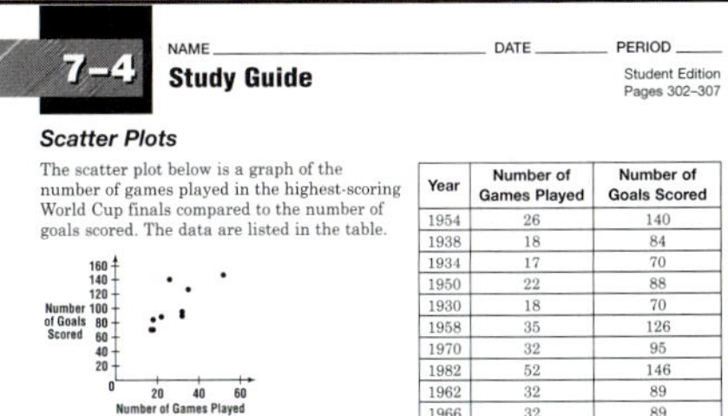
NAME ______ DATE ______ PERIOD ______

7-4 Study Guide

Student Edition Pages 302–307

Scatter Plots

The scatter plot below is a graph of the number of games played in the highest-scoring World Cup finals compared to the number of goals scored. The data are listed in the table.

Year	Number of Games Played	Number of Goals Scored
1954	26	140
1938	18	84
1934	17	70
1950	22	88
1930	18	70
1958	35	126
1970	32	95
1982	52	146
1962	32	89
1966	32	89

You can use the scatter plot to draw conclusions about the data. The data points fall roughly along a line with a positive slope. Therefore, we say there is a positive relationship between the number of games played and the number of goals scored. A line with a negative slope would indicate a negative relationship, and no line would indicate no relationship between the variables. A valid conclusion from this scatter plot is that as the number of games played increases, more goals are scored.

Determine whether each scatter plot has a positive relationship, negative relationship, or no relationship. If there is a relationship, describe it.

1. Negative; as the price of cameras increases, the number of cameras sold decreases.

2. no relationship

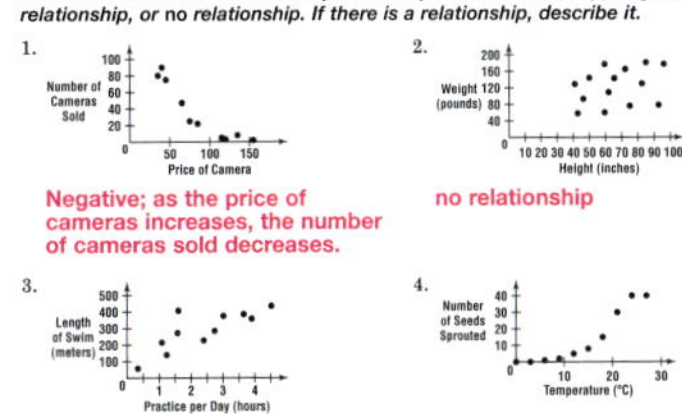

3. Positive; the longer you practice swimming, the further you can swim.

4. Positive; the higher the temperature, the more seeds sprout.

Answers

1. Graph the two sets of data as ordered pairs, where the *x*-coordinate is a member of one data set and the *y*-coordinate is a member of the other data set. The set of *x*-coordinates is the domain and the independent variable. The set of *y*-coordinates is the range and the dependent variable.

2. If a line drawn through the points has a positive slope, then there is a positive relationship between the data sets. If a line drawn through the points has a negative slope, then there is a negative relationship between the data sets.

Assignment Guide

Basic: 7–17 odd, 18–23
Average: 8–14 even, 16–23

Answers

9.

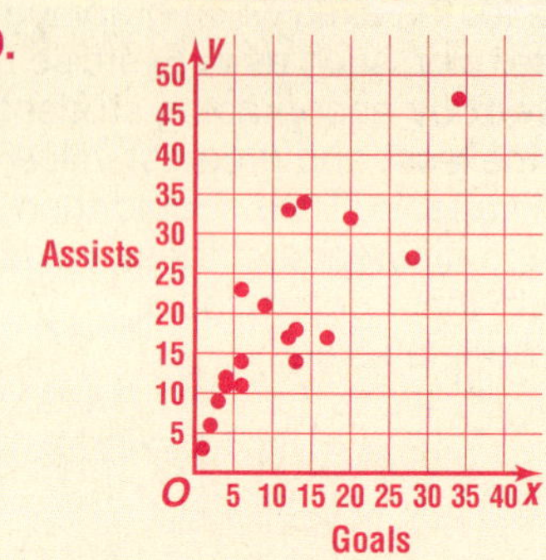

There is a positive relationship; as the number of goals increases, the number of assists increases.

10.

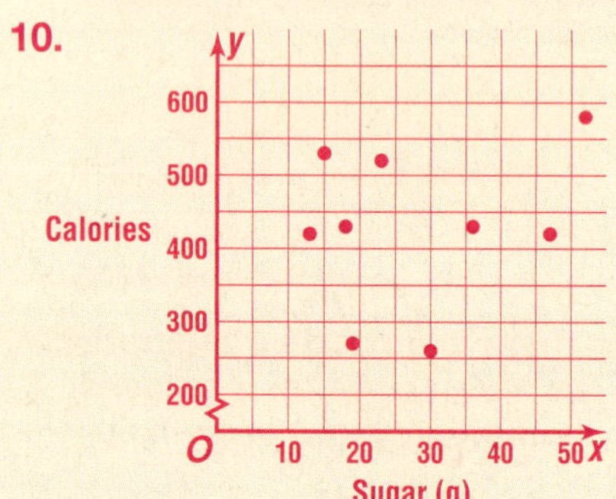

There is no relationship.

***Skills Practice,* p. 305, and *Practice,* p. 306 (shown)**

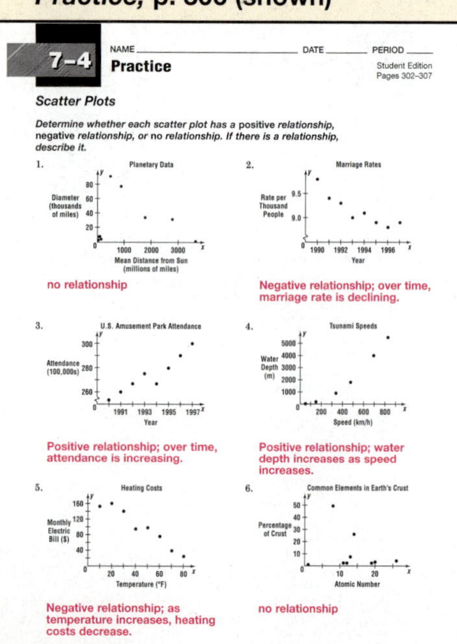

7-4 Practice

NAME ______ DATE ______ PERIOD ______

Student Edition Pages 302–307

Scatter Plots

Determine whether each scatter plot has a positive relationship, negative relationship, or no relationship. If there is a relationship, describe it.

1. Planetary Data — no relationship

2. Marriage Rates — Negative relationship; over time, marriage rate is declining.

3. U.S. Amusement Park Attendance — Positive relationship; over time, attendance is increasing.

4. Tsunami Speeds — Positive relationship; water depth increases as speed increases.

5. Heating Costs — Negative relationship; as temperature increases, heating costs decrease.

6. Common Elements in Earth's Crust — no relationship

Example 2

6. Yes; as weekly exercise increases, resting heart rate decreases. There is a negative relationship.

6. **Health** The scatter plot at the right shows data that compare a person's weekly exercise to that person's resting heart rate. Does there appear to be any relationship between hours of exercise and resting heart rate? If so, describe it.

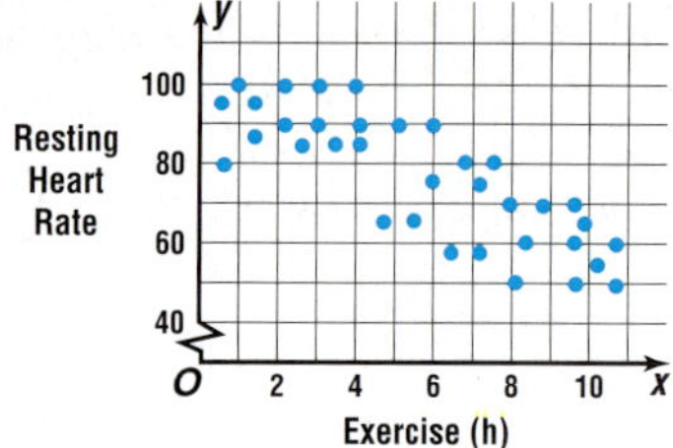

Exercises

Practice 

Homework Help

For Exercises	See Examples
7–8	1
9–17	2

Extra Practice
See page 706.

Determine whether each scatter plot has a *positive* relationship, *negative* relationship, or *no* relationship. If there is a relationship, describe it.

7.

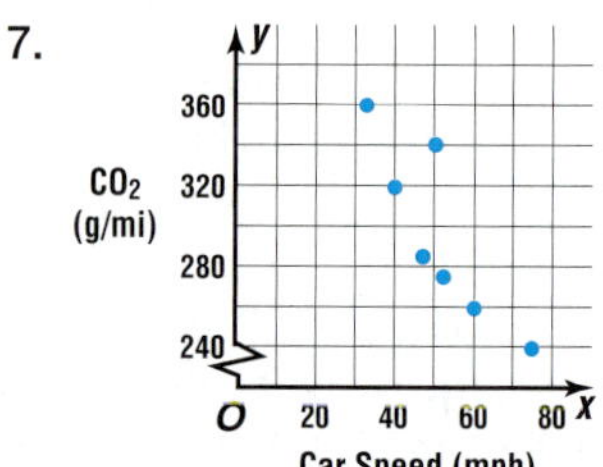

8.

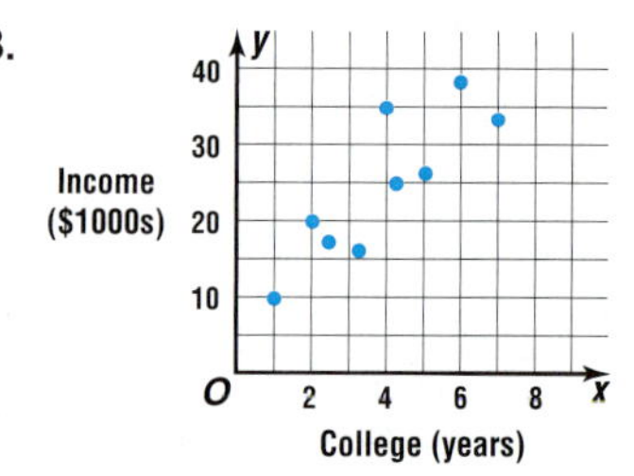

7. Negative relationship; as speed increases, CO_2 decreases.

8. Positive relationship; as the number of years of college education increases, income increases.

B

Make a scatter plot of each set of data. Then describe any trend shown in the scatter plot. 9–10. See margin.

9.

Dallas Stars Hockey Team, Selected Players

Goals	Assists	Goals	Assists
34	47	3	9
28	27	2	6
12	33	20	32
13	18	14	34
12	17	17	17
1	3	9	21
13	14	6	23
4	11	6	14
6	11	4	12

Source: *Infoplease.com*

10.

Breakfast Sweets

Sugar (g)	Calories
30	260
36	430
13	420
19	270
15	530
18	430
47	420
52	580
23	520

Source: *Nutrition Action Healthletter*

11. Sample answer: Have higher maximum speed limits because CO_2 emissions decrease as speed increases.

11. Emission of CO_2 from car exhaust is harmful to the environment. Refer to the scatter plot in Exercise 7. What recommendation would you make about speed limits based on the data? Explain.

Determine whether a scatter plot of the data for the following would show a *positive*, *negative*, or *no* relationship between the variables.

12. playing time and points scored **positive**
13. your height and month of birth **no**
14. size of household and pounds of garbage produced **positive**
15. age of car and its value **negative**

Applications and Problem Solving

16. **Insurance** The table shows data on the age of drivers and the number of traffic accidents they have for each 100,000 drivers of that age.

Age	Number of Accidents	Age	Number of Accidents
16	168	23	120
17	181	24	117
18	190	25	92
19	195	26	80
20	145	27	69
21	126	28	70
22	114	29	75

a. Make a scatter plot of the data. Determine if there is a *positive*, *negative*, or *no* relationship. If there is a relationship, describe it.

b. Based on the scatter plot, what recommendation would you make to insurance companies about how to determine insurance rates for different age drivers?

16a. See margin.

16b. Sample answer: Charge higher rates for younger drivers.

17. **Earth Science** The table below shows the elevation and the average annual precipitation for places around the world.

Elevation (ft)	Average Yearly Precipitation (in.)	Elevation (ft)	Average Yearly Precipitation (in.)
20	49	138	21
66	59	164	47
118	37	79	33
82	45	108	27
79	50	125	59
200	32	10	46
23	44	82	27
118	55	171	21
190	23		

Source: *The World Almanac*

a. Make a scatter plot of the data. **See margin.**

b. Does there seem to be a relationship? If so, describe it.

c. Describe the slope of a line passing through the points. **zero**

17b. No; there does not appear to be a relationship between elevation and precipitation.

18. **Critical Thinking** Refer to the scatter plot in Exercise 6. Would the relationship between the variables change if you relabeled the axes so that resting heart rate was on the horizontal axis? Explain.

Mixed Review

18. No; the relationship would still be negative because as resting heart rate increases, weekly exercise decreases.

19. Write an equation in slope-intercept form of the line passing through $(-2, -8)$ and $(1, 7)$. *(Lesson 7–3)* $y = 5x + 2$

20. Graph $y = x + 4$. *(Lesson 6–3)* **See margin.**

21. **Architecture** The Chrysler Building in New York is approximately 300 meters high. If a model of the building is 60 centimeters high, find the scale of the model. *(Lesson 5–2)* **1 cm = 5 m or 1:500**

22. Solve $-3(5 + x) = 12$. Check your solution. *(Lesson 4–7)* -9

Standardized Test Practice

23. **Multiple Choice** The area of the rectangle is 185.9 square centimeters. Find the length. *(Lesson 4–4)* **D**

(A) 22 cm (B) 163.9 cm

(C) 174.9 cm (D) 16.9 cm

11 cm

x cm

www.algconcepts.com/self_check_quiz

4 ASSESS

Open-Ended Assessment

Modeling Write a set of data on the board or overhead. Ask students to make a scatter plot of the data and to describe the relationship, if any, found in the data.

Mid-Chapter Test (Lessons 7–1 through 7–4) is available in the *Chapter 7 Resource Masters*, p. 334.

Answers

16a.

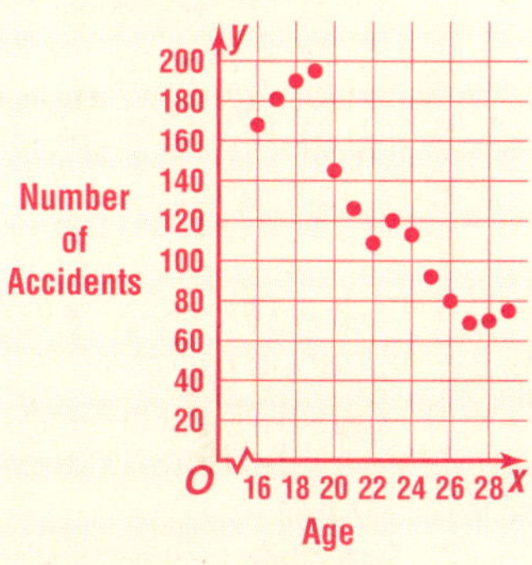

There is a negative relationship. As age increases, the number of accidents decreases.

17a.

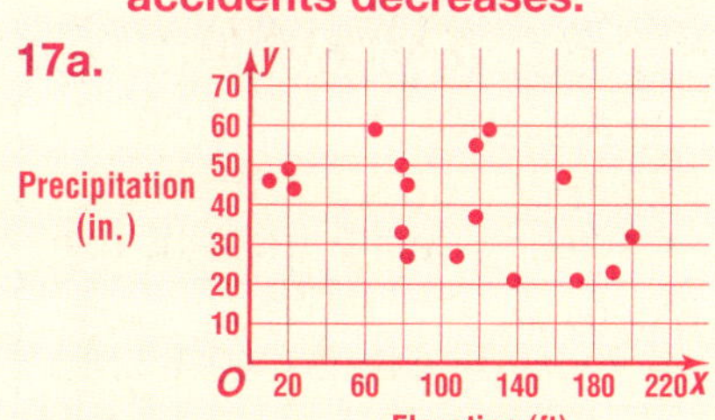

***Enrichment*, p. 308**

7-4 Enrichment

NAME ______ DATE ______ PERIOD ______

Student Edition Pages 302–307

A Scatter Plot

Each point on the graph shows the relation between the number of people attending the Roxy Cinema and the number of cars in the parking lot.

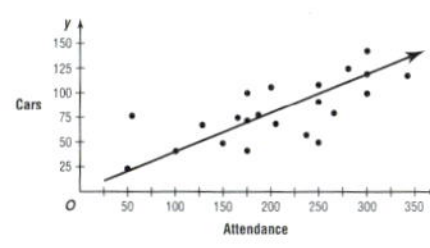

A line is drawn that appears to lie close to most of the points. Here is how to find the equation of this line.

The line passes through (100, 40) and (300, 120). Use the slope-intercept form.

$m = \frac{120 - 40}{300 - 100} = \frac{80}{200} = \frac{2}{5}$

$y = mx + b$
$40 = \frac{2}{5}(100) + b$
$0 = b$

An equation for the line is $y = \frac{2}{5}x$.

Solve each problem.

1. Suppose the owner of the Roxy decides to increase the seating capacity of the theater to 1000. How many cars should the parking lot be prepared to accommodate? **400 cars**

2. The points (240, 60) and (340, 120) lie on the scatter plot. Write an equation for the line through these points. $y = \frac{3}{5}x - 84$

3. Do you think the equation in Exercise 2 is a good representation of the relationship in this problem? **No. The graph of the equation does not appear to go through the center of the data points.**

4. Suppose the equation for the relationship between attendance at the theater and cars in the parking lot is $y = 2x + 20$. What might you suspect about the users of the parking lot? **Many people are parking in the lot who are not going to the Roxy.**

Extra Credit

Describe a real-world situation for each type of relationship: two variables are positively related, two variables are negatively related, and two variables are unrelated. **See students' work.**

Answer

20. $y = x + 4$

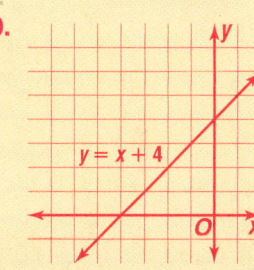

Investigation

PREPARE

This optional investigation is designed to be completed by pairs of students over 1–2 days.

Objective

Students will use data sets to determine functional relationships between quantities and represent these relationships using graphs, verbal descriptions, and equations. They will make scatter plots and use the equation for a best-fit line to make predictions.

Mathematical Overview

The investigation utilizes the following concepts:

- graphing on the coordinate plane,
- slope of a line,
- writing equations given two points on a line, and
- making predictions using the equation for a function.

Suggested Time Management	
Investigation	20–30 min
Extension: Gathering Data	20–30 min
Extension: Summarizing Data	40–55 min

Motivating the Lesson

After reading the opening paragraph ask students if they think humidity and the heat index will have a positive or negative relationship. Explain that you must hold the air temperature constant when examining the relationship between humidity and heat index.

Answer

1a. (0, 69), (5, 69), (10, 70), (15, 71), (20, 72), (25, 72), (30, 73), (35, 73), (40, 74), (45, 74), (50, 75)

Chapter 7 Investigation

Humidity Heats Things Up

Best-Fit Lines and Prediction

Materials

 grid paper

 straightedge

 graphing calculator or graphing software

In the summertime, have you ever noticed that it feels hotter on days that are humid? The *heat index* describes what the temperature feels like to the body when there is humidity.

Investigate

1. The table shows the heat index for various air temperatures and percents of relative humidity. For example, at 100°F and 30% relative humidity, the heat index is 104°F. So, it feels like 104°F.

Air Temperature	Heat Index at Relative Humidities										
	0%	5%	10%	15%	20%	25%	30%	35%	40%	45%	50%
70°F	64	65	65	65	66	66	67	67	68	68	69
75°F	69	69	70	71	72	72	73	73	74	74	75
80°F	73	74	75	76	77	77	78	79	79	80	81
85°F	78	79	80	81	82	83	84	85	86	87	88
90°F	83	84	85	86	87	88	90	91	93	95	96
95°F	87	88	90	91	93	94	96	98	101	104	107
100°F	91	93	95	97	99	101	**104**	107	110	115	120

   a. Look at the row for 75°F. Write ordered pairs of the form (percent humidity, heat index) for the entire row. For example, the first ordered pair is (0, 69). **a–b. See margin.**
   b. Make a scatter plot of the data.
   c. Does the scatter plot show any relationship between humidity and heat index? If so, describe it.
   d. Describe the domain and the range using words and using numbers.

2. When there is a relationship between two sets of data, you can draw a **best-fit line** through the points, as you did in Lesson 7–4. This line is close to most of the data points. Use a straightedge to draw a best-fit line on the scatter plot you made in Exercise 1. **See margin.**

1c. Yes, a positive relationship; as humidity increases, heat index increases.

1d. The domain is the set of relative humidity values, or $0 \leq x \leq 50$. The range is the set of heat index values, or $64 \leq y \leq 120$.

Answers

1b.

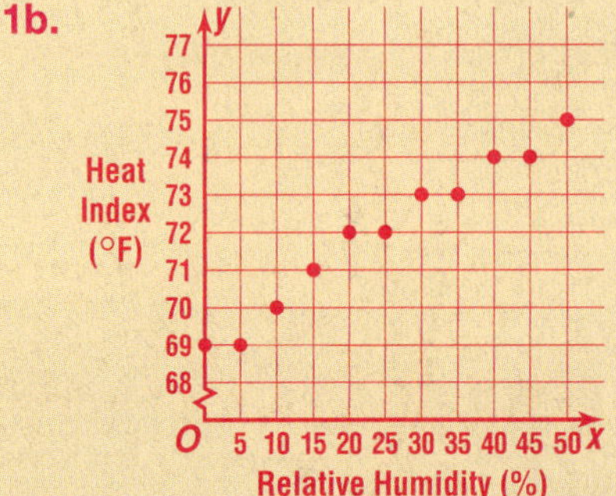

2.

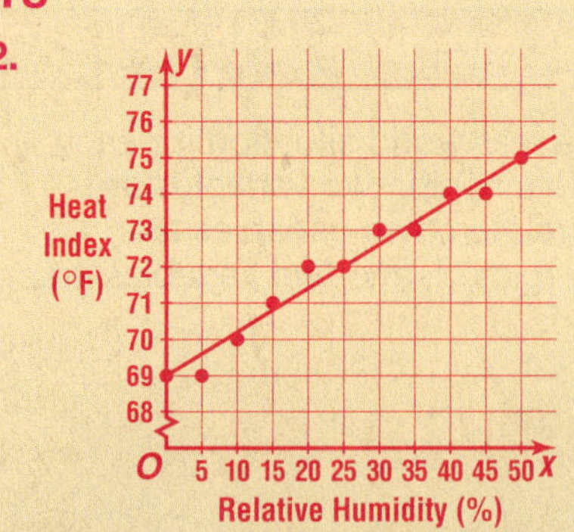

3b. Sample answer: $y = 0.12x + 69$

3c. Heat index depends on humidity, so it is the dependent variable. Humidity is the independent variable.

Another way to fit a line to data is with a **median-median line.** To make a median-median line, you divide the data set into 3 equal parts and find the median for each part. You then use these three points to draw a line to approximate the data.

3. You can use the best-fit line to write a linear equation that describes a relationship between humidity and heat index.
   a. Select two points on the line that you drew in Exercise 2. Use the points to determine the slope of the line. *The points may or may not be points in the scatter plot.* **a. Sample answer: $\frac{3}{25}$ or 0.12**
   b. Write an equation of the line in slope-intercept form.
   c. Describe the independent and dependent variables.
   d. A best-fit line can help you predict values even when you don't have data for every point. **Interpolation** is used to predict values between the greatest and least data points. **Extrapolation** is used to predict values that are below the least value or above the greatest value. Use your equation to predict the heat index at 75°F when the relative humidity is 60%. **Sample answer: 76.2°F**

Extending the Investigation

The process of finding a line that best fits a set of data is called **linear regression.** The slope of the best-fit line is called the **correlation coefficient.** The best-fit line will probably not pass through all of the data points. The distances from data points to the best-fit line are called **residuals.**

In this extension, you will use a graphing calculator or graphing software to explore the line of best fit. On a track or football field, collect data for running times for the following distances: 10 yards, 20 yards, 70 yards, and 100 yards.

- Make a scatter plot and draw best-fit lines for the data you collected. Write an equation in slope-intercept form for the best-fit line. **See students' work.**
- Use your equations to predict the times for running 50, 80, and 150 yards. Are your predictions reasonable? **See students' work.**
- Use a graphing calculator or graphing software to find linear equations for the data. Compare your equation for the best-fit line to the equation given by the calculator or software. **See students' work.**
- Use the best-fit line equation to predict the time to run 1760 yards. **See students' work.**
- Explain whether the time prediction for running 1760 yards is reasonable. **No; the best-fit line is based on data for short distances only.**

Presenting Your Conclusions

- Make a poster displaying the data table, your scatter plots, best-fit line, and equation.

Investigation For more information on windchill factor, visit: www.algconcepts.com

MANAGE

Teaching Tip Students should use an entire sheet of grid paper for the scatter plot. Students' equations for the best-fit line will vary slightly due to the placement of the line.

Teaching Tip Instruct students to draw their line so half of the points lie above the line and half of the points lie below the line. When students find the equation of the best-fit line, make sure they use points *on the line*, which may or may not be plotted data points.

Working in Pairs Lead the groups as a class through the steps to find the equation of their best-fit line in Step 3. Check their work on this step before having them begin the Extension. Then suggest that the pairs of students split the work by each graphing a different humidity and then checking each other's graph.

Working as a Class You may want to conduct this investigation as a whole class activity. Use an overhead projector to display the graph and best-fit line. You may want to have the class work together to make a bulletin board display about heat index.

ASSESS

Students' work should show that they understand how to draw a best-fit line, find its equation, and use the equation to make predictions.

Cooperative Learning

This investigation offers an excellent opportunity for using cooperative groups. For more information on cooperative learning strategies and group management, see *Cooperative Learning in the Mathematics Classroom,* one of the titles in the Glencoe Mathematics Professional Series.

Students should add their posters to their portfolios at this time.

Lesson 7-5

1 FOCUS

5-Minute Check
Lesson 7–4

Determine whether each scatter plot shows a* positive *relationship,* negative *relationship, or* no *relationship. If there is a relationship, describe it.

1.

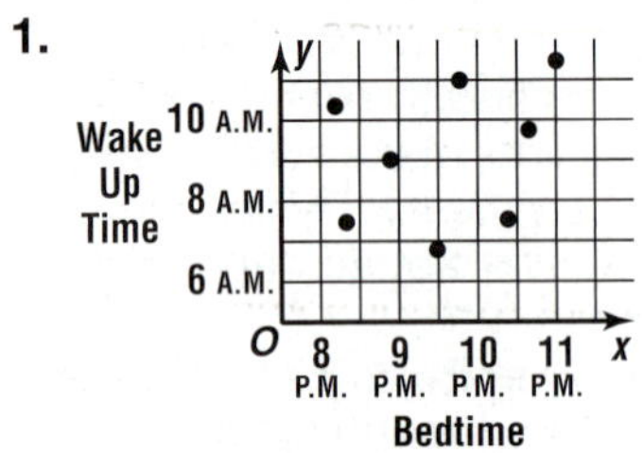

no relationship

2.

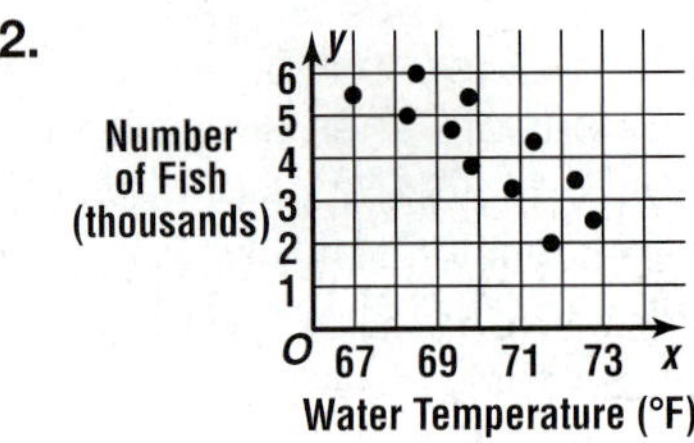

Negative relationship; the number of fish decreases as the temperature of the water increases.

3. The table shows the number of times five tomato plants were fertilized and the number of tomatoes each produced.

Fertilizer Applications	Number of Tomatoes
2	15
1	10
3	18
4	14
2	11

Make a scatter plot of the data. Does the scatter plot show a relationship between the variables? Explain. **See students' graphs; a positive relationship is shown. As the number of fertilizer applications increases, so does the number of tomatoes.**

7-5 Graphing Linear Equations

What You'll Learn
You'll learn to graph linear equations by using the x- and y-intercepts or the slope and y-intercept.

Why It's Important
Rates Linear graphs are helpful in showing phone costs. *See Exercise 38.*

A simple method of graphing a linear equation is using the points where the line crosses the x-axis and the y-axis. Consider the equations $x + y = 8$.

To find the x-intercept, let $y = 0$.

$x + y = 8$	*Original equation*
$x + 0 = 8$	*$y = 0$*
$x = 8$	*Simplify.*

To find the y-intercept, let $x = 0$.

$x + y = 8$	*Original equation*
$0 + y = 8$	*$x = 0$*
$y = 8$	*Simplify.*

The x-intercept is 8, and the y-intercept is 8. This means that the graph intersects the x-axis at (8, 0) and the y-axis at (0, 8). Graph these ordered pairs. Then draw the line that passes through these points. Each point on the graph represents two numbers whose sum is 8. Notice that the table of values confirms these intercepts.

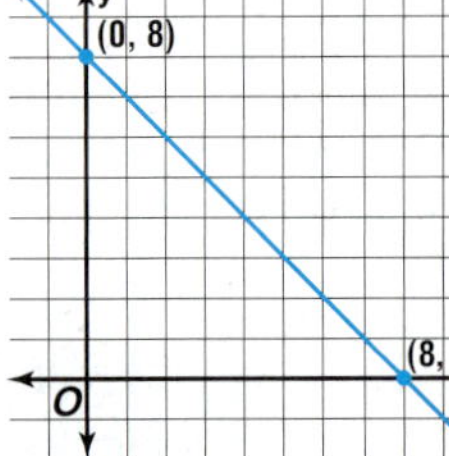

x	y
−1	9
0	8
1	7
2	6
3	5
4	4
5	3
6	2
7	1
8	0
9	−1

Examples

Determine the x-intercept and y-intercept of the graph of each equation. Then graph the equation.

1 $5y - x = 10$

To find the x-intercept, let $y = 0$.

$5y - x = 10$	*Original equation*
$5(0) - x = 10$	*Replace y with 0.*
$-x = 10$	*Simplify.*
$\frac{-x}{-1} = \frac{10}{-1}$	*Divide each side by 5.*
$x = -10$	*Simplify.*

To find the y-intercept, let $x = 0$.

$5y - x = 10$	*Original equation*
$5y - 0 = 10$	*Replace x with 0.*
$5y = 10$	*Simplify.*
$\frac{5y}{5} = \frac{10}{5}$	*Divide each side by 5.*
$y = 2$	*Simplify.*

The x-intercept is −10, and the y-intercept is 2. This means that the graph intersects the x-axis at (−10, 0) and the y-axis at (0, 2). Graph these ordered pairs. Then draw the line that passes through these points.

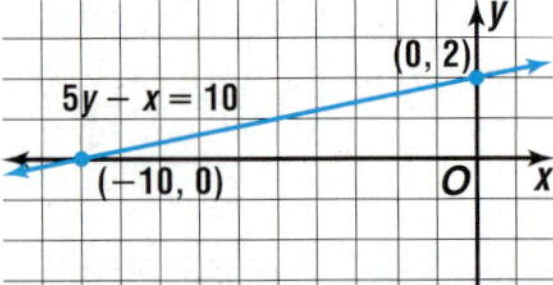

Resource Manager

Reproducible Masters
Chapter 7 Resource Masters
- *Study Guide*, p. 309
- *Skills Practice*, p. 310
- *Practice*, p. 311
- *Reading to Learn Mathematics*, p. 312
- *Enrichment*, p. 313

Transparencies
5-Minute Check, 7–5

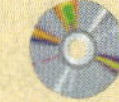

Technology/Multimedia
Interactive Chalkboard CD-ROM

2 $2x - 4y = 8$

To find the x-intercept, let $y = 0$.		To find the y-intercept, let $x = 0$.	
$2x - 4y = 8$	*Original equation*	$2x - 4y = 8$	*Original equation*
$2x - 4(0) = 8$	*Replace y with 0.*	$2(0) - 4y = 8$	*Replace x with 0.*
$2x = 8$	*Simplify.*	$-4y = 8$	*Simplify.*
$\frac{2x}{2} = \frac{8}{2}$	*Divide each side by −4.*	$\frac{-4y}{-4} = \frac{8}{-4}$	*Divide each side by −4.*
$x = 4$	*Simplify.*	$y = -2$	*Simplify.*

The x-intercept is 4, and the y-intercept is -2. This means that the graph intersects the x-axis at (4, 0) and the y-axis at (0, -2). Graph these ordered pairs. Then draw the line that passes through these points.

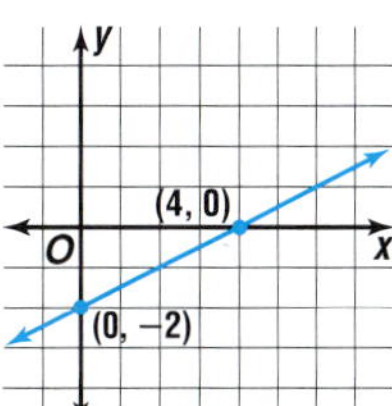

Check: Look at the graph. Choose some other point on the line and determine whether it is a solution of $2x - 4y = 8$. Try (2, -1).

$2x - 4y = 8$ *Original equation*
$2(2) - 4(-1) \stackrel{?}{=} 8$ *Replace x with 2 and y with −1.*
$4 - (-4) \stackrel{?}{=} 8$ *Multiply.*
$8 = 8$ ✓

Your Turn **a–c. See margin for graphs.**

a. $x + y = 2$ **2, 2** b. $3x + y = 3$ **1, 3** c. $4x - 5y = 20$ **5, −4**

You can easily find the slope and y-intercept of the graph of an equation that is written in slope-intercept form.

Example 3
Mail Link

To mail a letter in 2004, it cost \$0.37 for the first ounce and \$0.23 for each additional ounce. This can be represented by $y = 0.37 + 0.23x$. Determine the slope and y-intercept of the graph of the equation.

$y = mx + b$ *Slope-Intercept Form*
$y = 0.23x + 0.37$

The slope is 0.23, and the y-intercept is 0.37. So the slope represents the cost per ounce after the first ounce, and the y-intercept represents the cost of the first ounce of mail.

www.algconcepts.com/extra_examples

Motivating the Lesson

Hands-On Activity Write two simple linear equations on the board or overhead, such as $x + y = 4$ and $y = 2x + 1$. As students come to class have them show the solutions of each equation using a graph. Most students will make a table of values and plot points. Use their work to lead into a discussion about graphing linear equations using the intercepts or the slope and y-intercept.

2 TEACH

Teaching Tip When discussing the application at the beginning of the lesson and Example 1, ask students why the x- and y-intercepts were chosen as the two points to graph. Point out that for some equations, like $2x - 3y = 4$, the intercepts are fractions.

In-Class Examples

Examples 1–2

Determine the x-intercept and y-intercept of the graph of each equation. Then graph the equation.

1 $2y - x = 8$ **−8, 4**

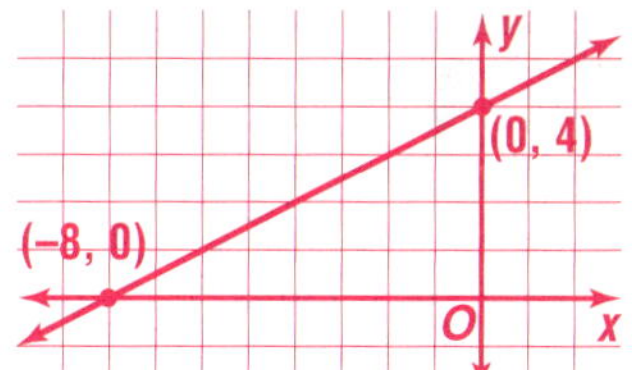

2 $3x - 2y = 12$ **4, −6**

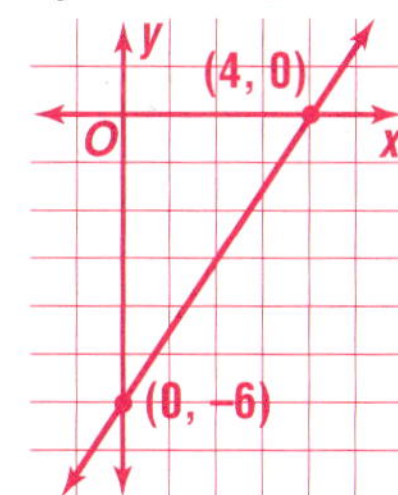

Example 3

Suppose to ship a package it costs \$2.05 for the first pound and \$1.55 for each additional pound. This can be represented by $y = 2.05 + 1.55x$. Determine the slope and y-intercept of the graph of the equation.
$m = 1.55, b = 2.05$

Answers

Your Turn

a.

b.

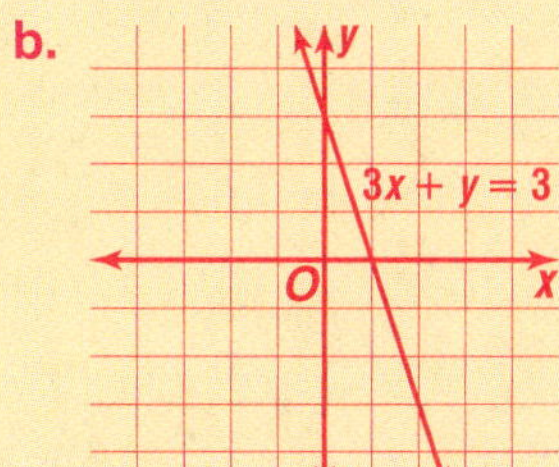

c.

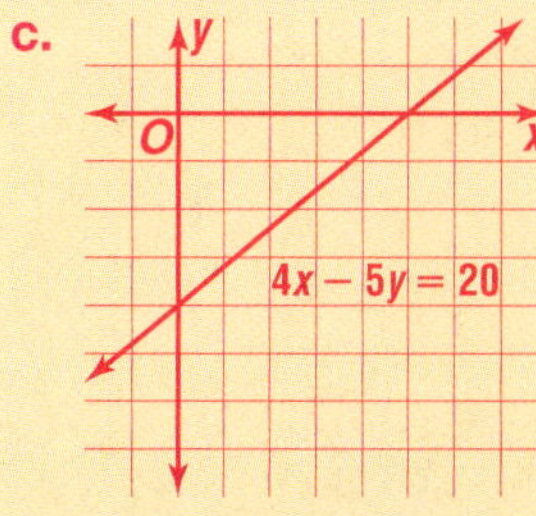

In-Class Examples

Example 4

Determine the slope and y-intercept of the graph of $6x - 9y = 18$.

$m = \frac{2}{3}, b = -2$

Examples 5–6

Graph each equation by using the slope and y-intercept.

5 $y = -\frac{1}{5}x + 2$

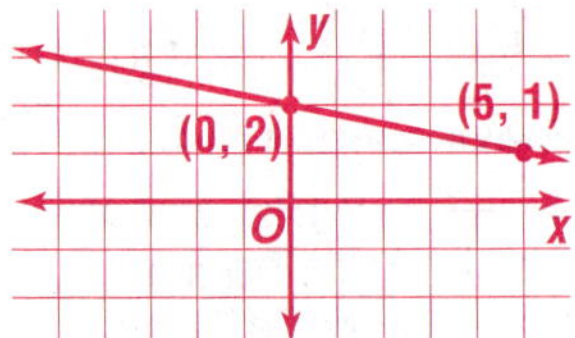

6 $3x - y = 4$

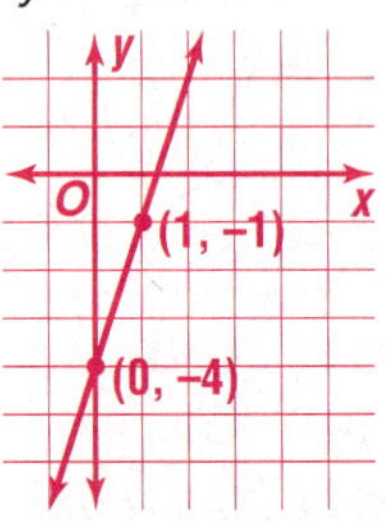

Example 4 Determine the slope and y-intercept of the graph of $10 + 5y = 2x$.

Write the equation in slope-intercept form to find the slope and y-intercept.

$10 + 5y = 2x$	*Original equation*
$10 + 5y - 10 = 2x - 10$	*Subtract 10 from each side.*
$5y = 2x - 10$	*Simplify.*
$\frac{5y}{5} = \frac{2x - 10}{5}$	*Divide each side by 5.*
$y = \frac{2}{5}x - 2$	*Simplify.*

The slope is $\frac{2}{5}$, and the y-intercept is -2.

Your Turn

d. $y = 5x + 9$ $m = 5, b = 9$ **e.** $4x + 3y = 6$ $m = -\frac{4}{3}, b = 2$

You can also graph a linear equation by using the slope and y-intercept.

Examples Graph each equation by using the slope and y-intercept.

5 $y = \frac{2}{3}x - 5$

$y = mx + b$ *Slope-Intercept Form*

$y = \frac{2}{3}x + (-5)$

The slope is $\frac{2}{3}$, and the y-intercept is -5. Graph the point at $(0, -5)$. Then go up 2 units and right 3 units. This will be the point at $(3, -3)$. Then draw the line through points at $(0, -5)$ and $(3, -3)$.

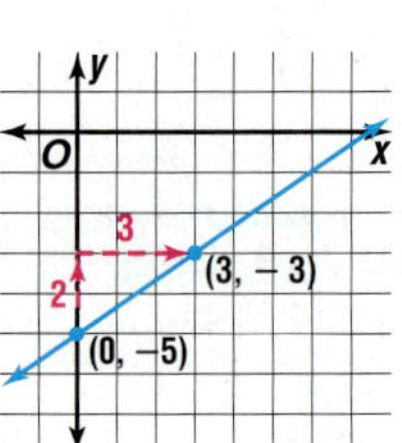

Check: The graph appears to go through the point at $(6, -1)$. Substitute $(6, -1)$ into $y = \frac{2}{3}x + (-5)$.

$y = \frac{2}{3}x + (-5)$	*Original equation*
$-1 \stackrel{?}{=} \frac{2}{3}(6) + (-5)$	*Replace x with 6 and y with −1.*
$-1 \stackrel{?}{=} 4 - 5$	*Simplify.*
$-1 = -1$ ✓	

6 $3x + 2y = 6$

First, write the equation in slope-intercept form.

$$3x + 2y = 6 \quad \textit{Original equation}$$
$$3x + 2y - 3x = 6 - 3x \quad \textit{Subtract 3x from each side.}$$
$$2y = 6 - 3x \quad \textit{Simplify.}$$
$$\frac{2y}{2} = \frac{6 - 3x}{2} \quad \textit{Divide each side by 2.}$$
$$y = -\frac{3}{2}x + 3$$ The slope is $-\frac{3}{2}$, and the y-intercept is 3.

Graph the point at (0, 3). Then go up 3 units and left 2 units. This will be the point at (−2, 6). Then draw the line through (0, 3) and (−2, 6). *Check by substituting the coordinates of another point that appears to lie on the line, such as (2, 0).*

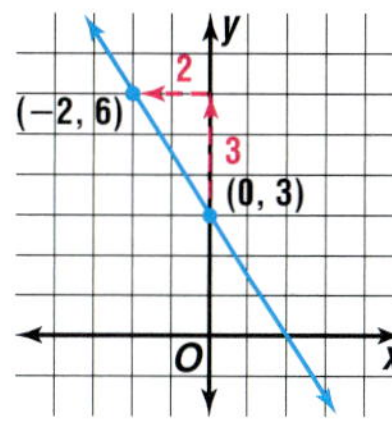

Your Turn f–g. See margin.

f. $y = \frac{1}{2}x + 3$ **g.** $x + 4y = -8$

The graph of a horizontal line has a slope of 0 and no x-intercept. The graph of a vertical line has an undefined slope and no y-intercept.

Examples **Graph each equation.**

7 $y = 4$

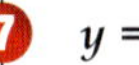

$y = mx + b$

$y = 0x + 4$ *slope = 0, y-intercept = 4*

No matter what the value of x, $y = 4$. So, all ordered pairs are of the form $(x, 4)$. Some examples are (0, 4) and (−3, 4).

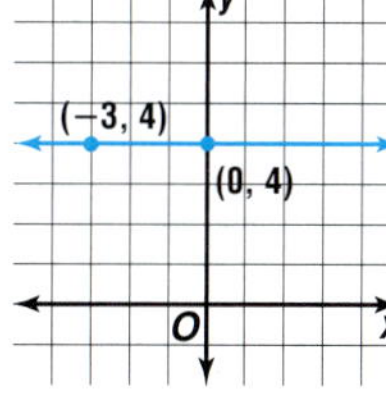

8 $x = -2$

slope: undefined, y-intercept: none

No matter what the value of y, $x = -2$. So, all ordered pairs are of the form $(-2, y)$. Some examples are (−2, −1) and (−2, 3).

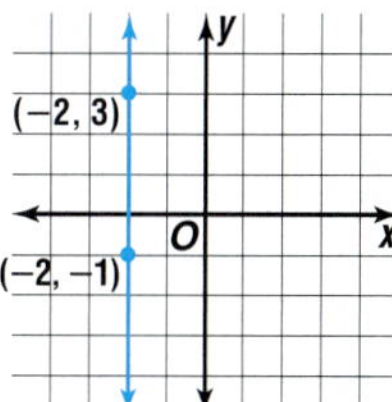

Your Turn h–i. See margin.

h. $y = -1$ **i.** $x = 3$

Answers
Your Turn

g.

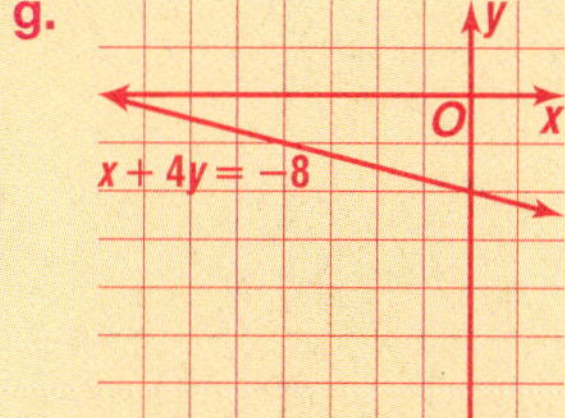

h.

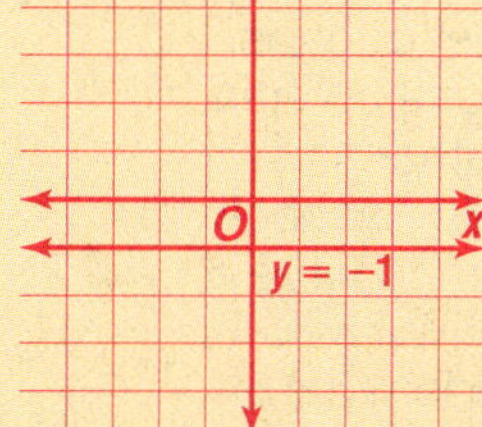

i.

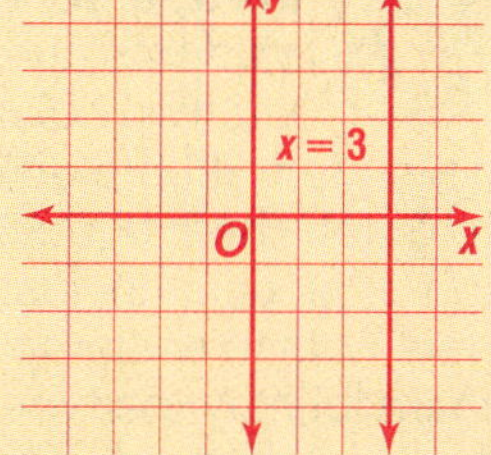

In-Class Examples
Examples 7–8

Graph each equation.

7 $y = -3$

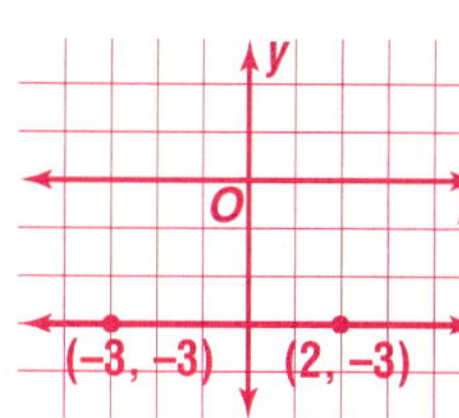

8 $x = 4$

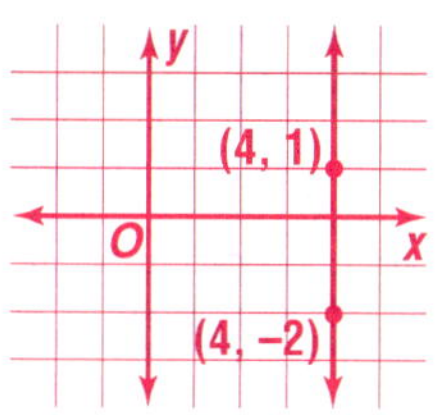

Answer
Your Turn

f.

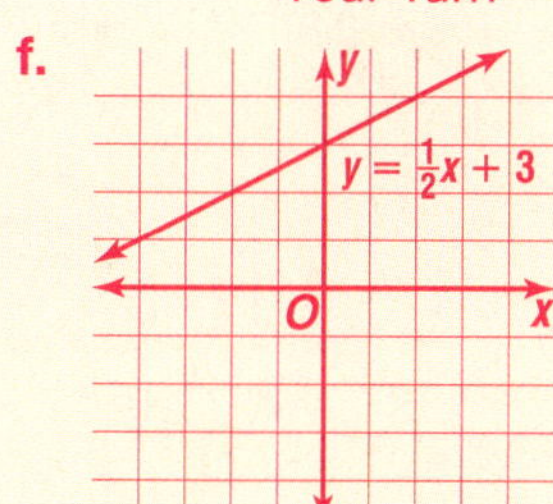

***Study Guide*, p. 309**

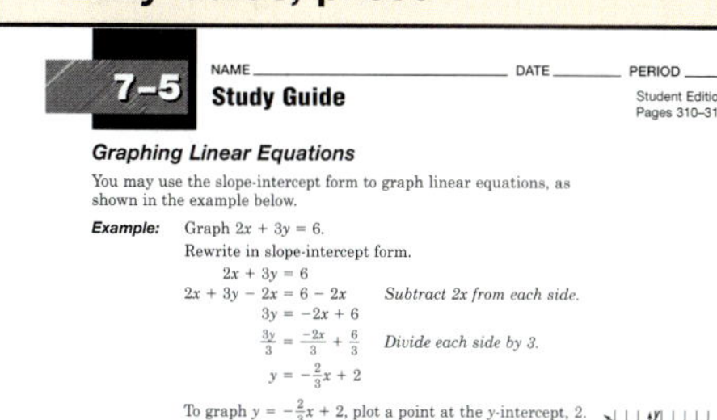

7–5 NAME ______ DATE ______ PERIOD ______

Study Guide Student Edition Pages 310–315

Graphing Linear Equations

You may use the slope-intercept form to graph linear equations, as shown in the example below.

Example: Graph $2x + 3y = 6$.

Rewrite in slope-intercept form.

$$2x + 3y = 6$$
$$2x + 3y - 2x = 6 - 2x \quad \textit{Subtract 2x from each side.}$$
$$3y = -2x + 6$$
$$\frac{3y}{3} = \frac{-2x}{3} + \frac{6}{3} \quad \textit{Divide each side by 3.}$$
$$y = -\frac{2}{3}x + 2$$

To graph $y = -\frac{2}{3}x + 2$, plot a point at the y-intercept, 2.

Then use the slope, $-\frac{2}{3}$. From (0, 2), go down 2 units.

Then go right 3 units. Graph a point at (3, 0). Draw the line through the points.

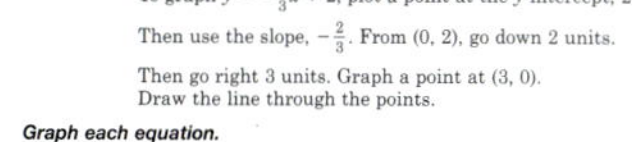

Graph each equation.

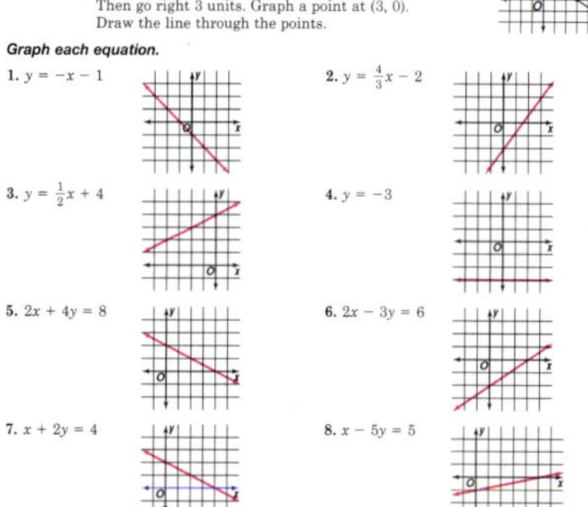

1. $y = -x - 1$
2. $y = \frac{4}{3}x - 2$
3. $y = \frac{1}{2}x + 4$
4. $y = -3$
5. $2x + 4y = 8$
6. $2x - 3y = 6$
7. $x + 2y = 4$
8. $x - 5y = 5$

3 PRACTICE/APPLY

Error Analysis

Watch for students who confuse the x- and y-intercepts in Exercises 3–6.

Prevent by stressing that the x-intercept is the point where the graph crosses the x-axis. Emphasize that at this point the value of y is 0, so students should substitute 0 for y to find the corresponding value of x, the x-intercept. Point out that to find the y-intercept, 0 is substituted for x.

Assignment Guide

Basic: 13–39 odd, 40–49
Average: 12–36 even, 38–49

Answer
Page 315

36a.

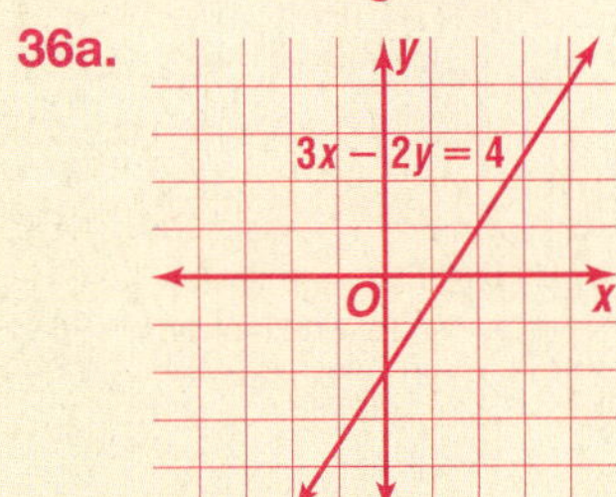

***Skills Practice*, p. 310, and *Practice*, p. 311 (shown)**

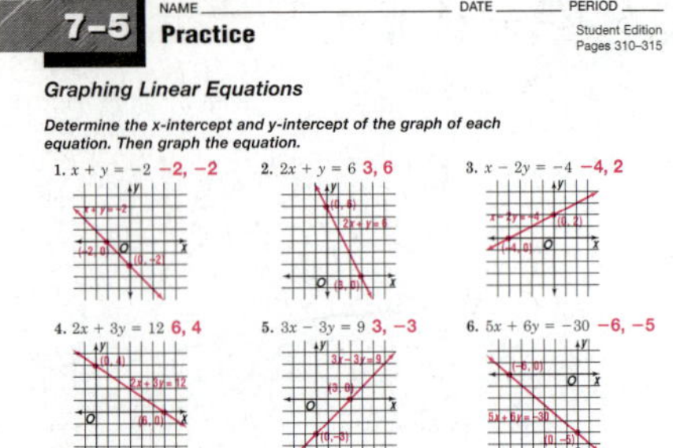
NAME ______ DATE ______ PERIOD ______

7-5 Practice

Student Edition Pages 310–315

Graphing Linear Equations

Determine the x-intercept and y-intercept of the graph of each equation. Then graph the equation.

1. $x + y = -2$ −2, −2
2. $2x + y = 6$ 3, 6
3. $x - 2y = -4$ −4, 2
4. $2x + 3y = 12$ 6, 4
5. $3x - 3y = 9$ 3, −3
6. $5x + 6y = -30$ −6, −5

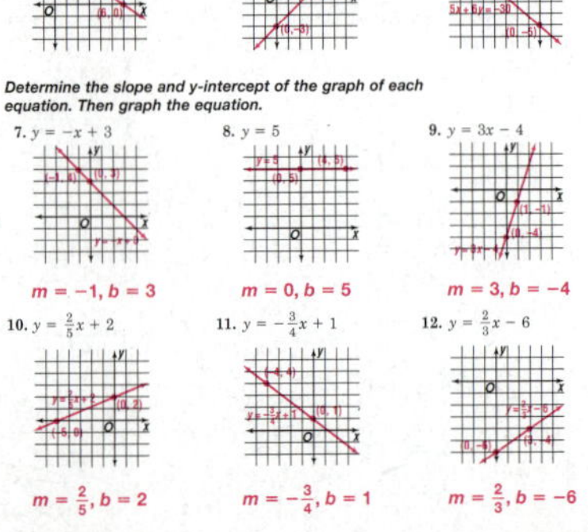
Determine the slope and y-intercept of the graph of each equation. Then graph the equation.

7. $y = -x + 3$ $m = -1, b = 3$
8. $y = 5$ $m = 0, b = 5$
9. $y = 3x - 4$ $m = 3, b = -4$
10. $y = \frac{2}{5}x + 2$ $m = \frac{2}{5}, b = 2$
11. $y = -\frac{3}{4}x + 1$ $m = -\frac{3}{4}, b = 1$
12. $y = \frac{2}{3}x - 6$ $m = \frac{2}{3}, b = -6$

Check for Understanding

Communicating Mathematics

1. **Explain** how many points are needed to draw the graph of a linear equation. **two points, then draw a line through them**

2.  Gwen says that $y = 8$ and $x = 8$ are both functions. Darnell says that only one of the equations is a function. Who is correct? Explain. **Darnell; only $y = 8$ is a function because $x = 8$ does not pass the vertical line test.**

Guided Practice
Examples 1 & 2
3–10. See Solutions Manual.

Determine the x-intercept and y-intercept of the graph of each equation. Then graph the equation.

3. $x + y = -3$ **−3, −3**
4. $x + 4y = 4$ **4, 1**
5. $5x + 2y = 10$ **2, 5**
6. $2x - 6y = 12$ **6, −2**

Examples 3–8

Determine the slope and y-intercept of the graph of each equation. Then graph the equation.

7. $y = x + 1$ $m = 1, b = 1$
8. $y = -4x$ $m = -4, b = 0$
9. $y = \frac{1}{4}x - 3$ $m = \frac{1}{4}, b = -3$
10. $y = -2$ $m = 0, b = -2$

Example 3

11. **Temperature** The equation $F = \frac{9}{5}C + 32$ gives the temperature in degrees Fahrenheit if you know the degrees Celsius. What are the slope and y-intercept of the graph of the equation? $m = \frac{9}{5}, b = 32$

Exercises

12–23. See Solutions Manual for graphs.

Practice **A**

Determine the x-intercept and y-intercept of the graph of each equation. Then graph the equation.

12. $x + y = 4$ **4, 4**
13. $x + y = -5$ **−5, −5**
14. $x + 2y = 4$ **4, 2**
15. $x - 3y = 3$ **3, −1**
16. $5x + y = 5$ **1, 5**
17. $2x + y = -6$ **−3, −6**
18. $4x + 5y = 20$ **5, 4**
19. $-3x + 4y = 12$ **−4, 3**
20. $6x - 3y = 6$ **1, −2**
21. $7x - 2y = 14$ **2, −7**
22. $2x + 5y = -10$ **−5, −2**
23. $x + \frac{1}{2}y = 4$ **4, 8**

Homework Help

For Exercises	See Examples
12–23	1, 2
24–35	4–6
36, 37	7–8
38, 39	3

Extra Practice
See page 706.

Determine the slope and y-intercept of the graph of each equation. Then graph the equation. **24–35. See Solutions Manual for graphs.**

B

24. $y = x + 3$ $m = 1, b = 3$
25. $y = -x + 2$ $m = -1, b = 2$
26. $y = 2x + 1$ $m = 2, b = 1$
27. $y = 3x - 1$ $m = 3, b = -1$
28. $x = 6$ m = undefined, b = none
29. $y = 4$ $m = 0, b = 4$
30. $y = \frac{1}{2}x + 3$
31. $y = \frac{3}{4}x - 4$
32. $y = -\frac{3}{2}x + 5$
33. $-2x + y = 3$ $m = 2, b = 3$
34. $x + 2y = 4$ $m = -\frac{1}{2}, b = 2$
35. $3x + 5y = 10$ $m = -\frac{3}{5}, b = 2$

30. $m = \frac{1}{2}, b = 3$
31. $m = \frac{3}{4}, b = -4$
32. $m = -\frac{3}{2}, b = 5$

Reteaching Activity

Intrapersonal Learners Have students divide a sheet of paper into two columns. In one column, have them list the different methods used to graph linear equations in this lesson. In the other column, instruct them to write a note next to each method that will help them identify the type of equation that is best graphed using this method.

Answer
Page 315

37a.

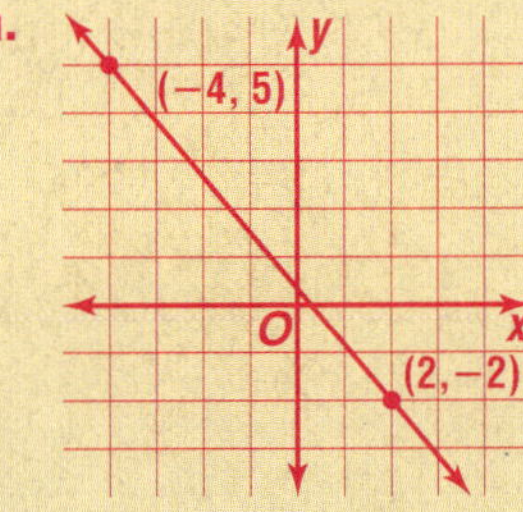

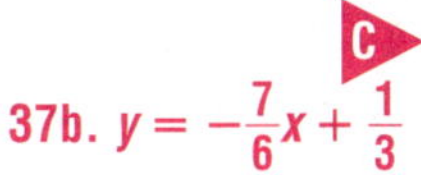

37b. $y = -\frac{7}{6}x + \frac{1}{3}$

38c. Slope is the charge per minute, the y-intercept is the monthly fee.

36. a. Graph $3x - 2y = 4$. **See margin.**
   b. What is the slope of the line? $\frac{3}{2}$
   c. Where does the line intersect the x-axis? $\frac{4}{3}$

37. a. Graph a line that goes through points at $(-4, 5)$ and $(2, -2)$. **a. See margin.**
   b. Write an equation of the line in slope-intercept form.
   c. Name the slope and y-intercept. $m = -\frac{7}{6}$, $b = \frac{1}{3}$

Applications and Problem Solving

38. **Rates** A long-distance phone company charges \$5 per month plus \$0.10 per minute.
   a. Write a linear equation to represent the total monthly cost $f(x)$ as a function of the number of minutes x. $f(x) = 0.10x + 5$
   b. Graph the equation. **See margin.**
   c. Explain what the slope and y-intercept represent.

A long-distance call

39. **Physical Science** The weight of a bucket of water is a linear function of the depth of the water. When there are 4 centimeters of water in the bucket, it weighs 2 pounds. When there are 12 centimeters of water in the bucket, it weighs 4 pounds.
   a. Write an equation for the weight of the bucket $f(x)$ as a function of the depth of the water x. $f(x) = \frac{1}{4}x + 1$
   b. Graph the equation. **See margin.**
   c. Estimate how much a bucket of water will weigh if there are 18 centimeters of water in it. about $5\frac{1}{2}$ lb

40. **Critical Thinking** Explain how the pattern in the arithmetic sequence 2, 2.5, 3, 3.5, 4, 4.5, 5, . . . is related to the graph at the right.
The values in the sequence are the same as the range values of the equation of the line when the domain values are integers.

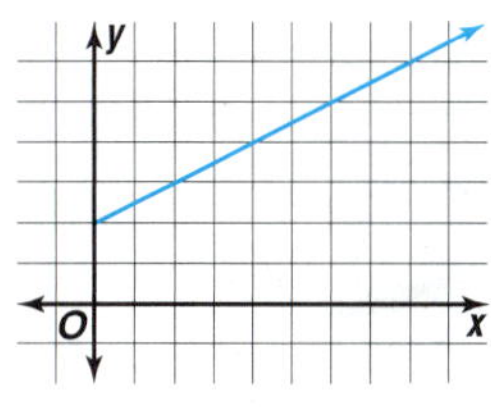

Mixed Review

41. Sketch a scatter plot that shows a positive relationship between two sets of related data. *(Lesson 7–4)* **See margin.**

42. **Entertainment** It costs \$60 to rent jet skis for 2.5 hours. At that rate, how much will it cost to rent jet skis for 6 hours? *(Lesson 6–5)* **\$144**

Find the probability of each outcome if a die is rolled. *(Lesson 5–6)*

43. a 4 $\frac{1}{6}$
44. an even number $\frac{1}{2}$
45. a number greater than 6 **0**

Solve each equation. Check your solution. *(Lesson 3–6)*

46. $-6 + x = 20$ **26**
47. $12 = b - 7$ **19**
48. $0 = k + 1.5$ **−1.5**

Standardized Test Practice

49. **Multiple Choice** Which fractions are ordered from least to greatest? *(Lesson 3–1)* **A**

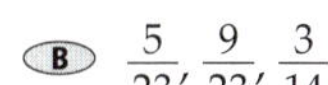

Ⓐ $\frac{3}{14}, \frac{5}{23}, \frac{9}{23}$ Ⓑ $\frac{5}{23}, \frac{9}{23}, \frac{3}{14}$ Ⓒ $\frac{9}{23}, \frac{5}{23}, \frac{3}{14}$ Ⓓ $\frac{3}{14}, \frac{9}{23}, \frac{5}{23}$

www.algconcepts.com/self_check_quiz

4 ASSESS

Open-Ended Assessment

Speaking Write several different types of linear equations on the board or overhead. Ask students to identify which method they would use to graph each equation, and to explain why they chose the method. Then have the student describe the method they chose.

Answers

38b.

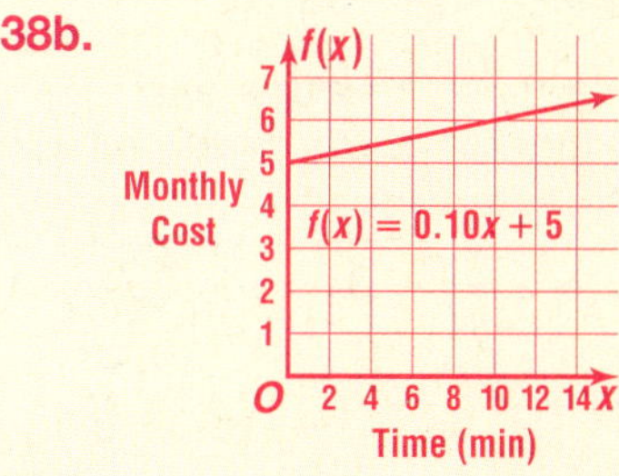

39b.

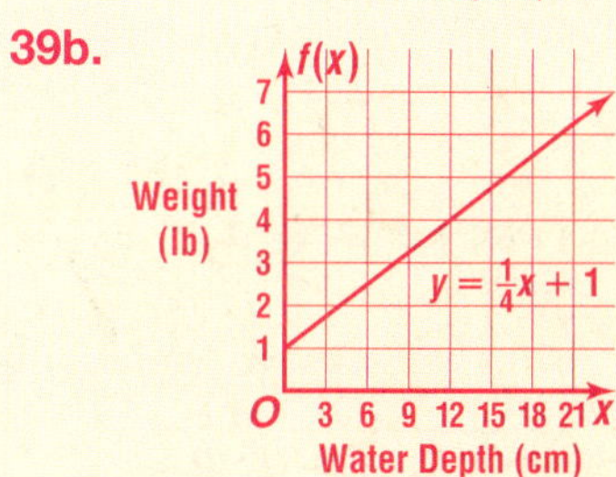

41. Sample answer:

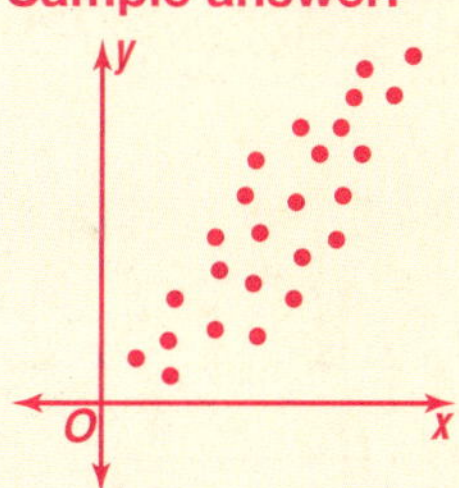

Enrichment, p. 313

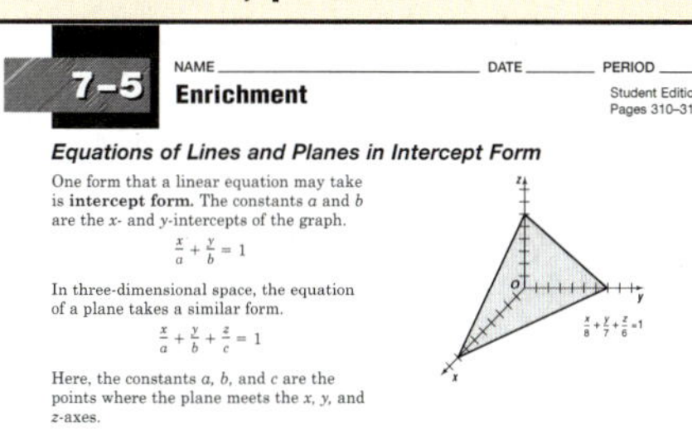

7–5 NAME ______ DATE ______ PERIOD ______

Enrichment Student Edition Pages 310–315

Equations of Lines and Planes in Intercept Form

One form that a linear equation may take is **intercept form.** The constants a and b are the x- and y-intercepts of the graph.

$$\frac{x}{a} + \frac{y}{b} = 1$$

In three-dimensional space, the equation of a plane takes a similar form.

$$\frac{x}{a} + \frac{y}{b} + \frac{z}{c} = 1$$

Here, the constants a, b, and c are the points where the plane meets the x, y, and z-axes.

$\frac{x}{8} + \frac{y}{7} + \frac{z}{6} = 1$

Solve each problem.

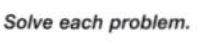

1. Graph the equation $\frac{x}{3} + \frac{y}{2} + \frac{z}{1} = 1$.

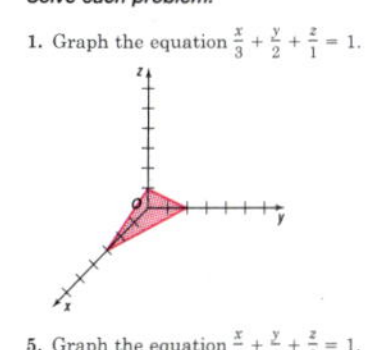

2. For the plane in Exercise 1, write an equation for the line where the plane intersects the xy-plane. Use intercept forms. $\frac{x}{3} + \frac{y}{2} = 1$
3. Write an equation for the line where the plane intersects the xz-plane. $\frac{x}{3} + \frac{z}{1} = 1$
4. Write an equation for the line where the plane intersects the yz-plane. $\frac{y}{2} + \frac{z}{1} = 1$
5. Graph the equation $\frac{x}{1} + \frac{y}{4} + \frac{z}{2} = 1$.
6. Write an equation for the xy-plane. $z = 0$
7. Write an equation for the yz-plane. $x = 0$
8. Write an equation for a plane parallel to the xy-plane with a z-intercept of 2. $z = 2$
9. Write an equation for a plane parallel to the yz-plane with an x-intercept of -3. $x = -3$

Extra Credit

A concession stand sells sodas for \$1 each and hot dogs for \$2 each. You have \$12 to spend on your family at the stand. How many of each item can you buy if you want to spend all of the money? Write an equation to represent this situation. Graph the equation using the x- and y-intercepts. Name one point on the graph that is *not* an intercept and explain what this point represents.
Sample answer: $x + 2y = 12$; See students' graphs; (2, 5) represents buying 2 sodas and 5 hot dogs.

Lesson 7–6

1 FOCUS

5-Minute Check
Lesson 7–5

1. Determine the x-intercept and y-intercept of the graph of $6x - 5y = 30$. **5, −6**

Determine the slope and y-intercept of the graph of each equation.

2. $y = -\frac{4}{5}x - 3$

   $m = -\frac{4}{5}, b = -3$

3. $-5x + 6y = 12$

   $m = \frac{5}{6}, b = 2$

Graph each equation.

4. $y = 2$

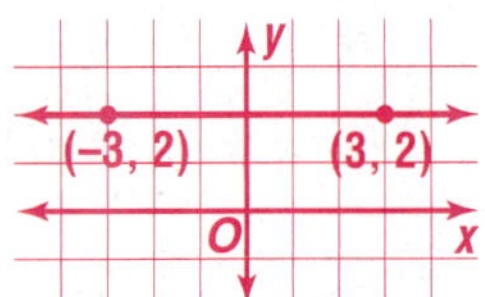

5. $y = -\frac{4}{3}x + 1$

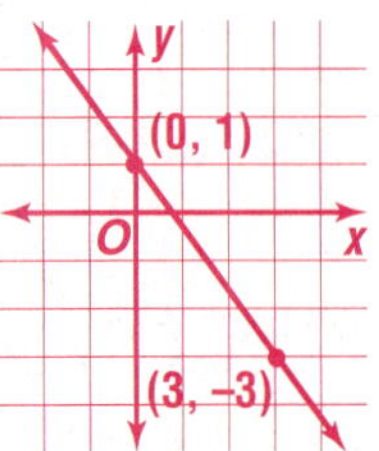

Motivating the Lesson

Real-World Connection
Generate a discussion on families of linear graphs by describing the following situation. Suppose you go to a gourmet coffee shop to buy coffee beans. One type of beans costs \$6.00 per pound and another type costs \$8.00 per pound. If for each type of coffee beans, you were to graph the number of pounds of beans on the x-axis and the cost of the coffee on the y-axis, how would the y-intercepts and slopes of the graphs compare? **The y-intercepts are the same, 0. The slopes are different, 6 and 8.**

7–6 Families of Linear Graphs

What You'll Learn
You'll learn to explore the effects of changing the slopes and y-intercepts of linear functions.

Why It's Important
Business Families of graphs can display different fees. *See Exercise 10.*

The graph representing a cheetah's speed is much steeper than the graph representing a spider's speed or an elephant's speed. This is because the cheetah runs faster and therefore covers greater distance in each unit of time.

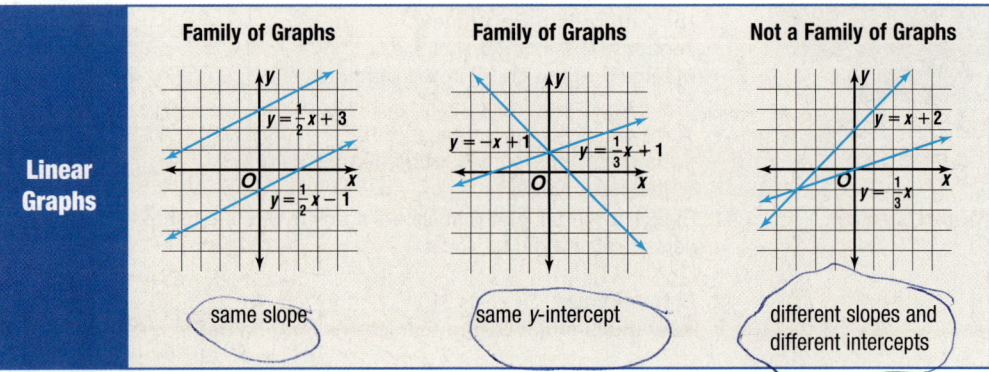

Maximum Animal Speeds

Distance (mi) vs. Time (hr): cheetah 62 mph; elephant 25 mph; spider 1 mph

What do the graphs have in common? They have the same y-intercept, 0. They are called a **family of graphs** because they have at least one characteristic in common that makes them different from other groups of graphs.

Families of linear graphs often fall into two categories—those with the same slope or those with the same x- or y-intercept.

Linear Graphs

Family of Graphs	Family of Graphs	Not a Family of Graphs
$y = \frac{1}{2}x + 3$, $y = \frac{1}{2}x - 1$	$y = -x + 1$, $y = \frac{1}{3}x + 1$	$y = x + 2$, $y = \frac{1}{3}x$
same slope	same y-intercept	different slopes and different intercepts

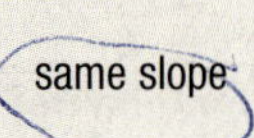

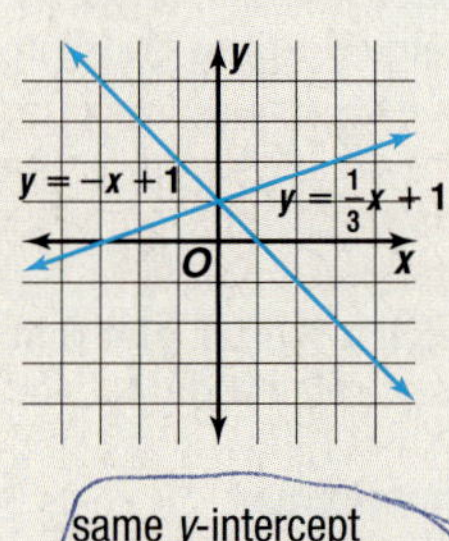

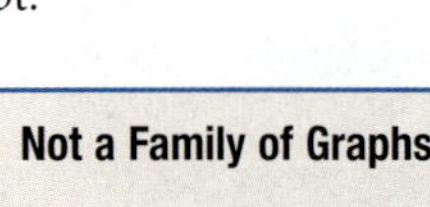

Examples

Graph each pair of equations. Describe any similarities or differences. Explain why they are a family of graphs.

1. $y = 3x + 4$
   $y = 3x - 2$

   The graphs have y-intercepts of 4 and −2, respectively.

   They are a family of graphs because the slope of each line is 3.

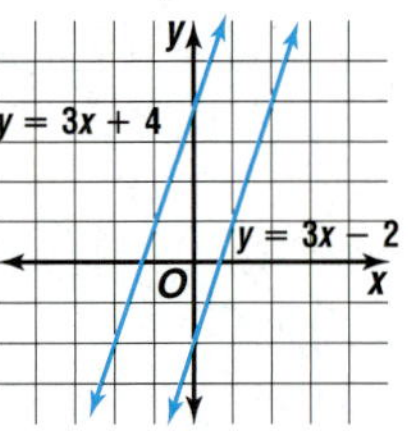

Resource Manager

Reproducible Masters
Chapter 7 Resource Masters
- *Study Guide,* p. 314
- *Skills Practice,* p. 315
- *Practice,* p. 316
- *Reading to Learn Mathematics,* p. 317
- *Enrichment,* p. 318

Graphing Calculator, p. 20
Hands-On Algebra, p. 84

Transparencies
5-Minute Check, 7–6

Technology/Multimedia
Interactive Chalkboard CD-ROM

2 $y = x + 3$

$y = -\frac{1}{2}x + 3$

Each graph has a different slope. Each graph has a y-intercept of 3. Thus, they are a family of graphs.

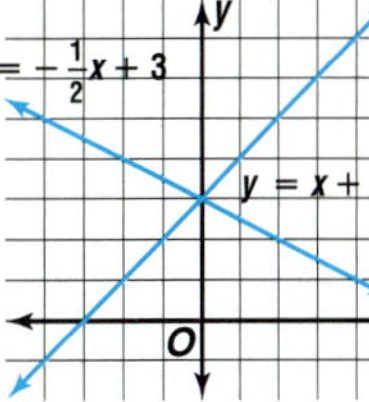

Your Turn a–b. See margin.

a. $y = 2x - 1$
$y = 2x + 5$

b. $y = x + 1$
$y = 3x + 1$

A graphing calculator is a good tool for exploring families of graphs.

Graphing Calculator Exploration

Step 1 Graph $y = x$ in the standard viewing window.

[Y=] [X,T,θ,n] [ZOOM] 6

Step 2 Graph $y = 2x$ and $y = \frac{1}{2}x$.

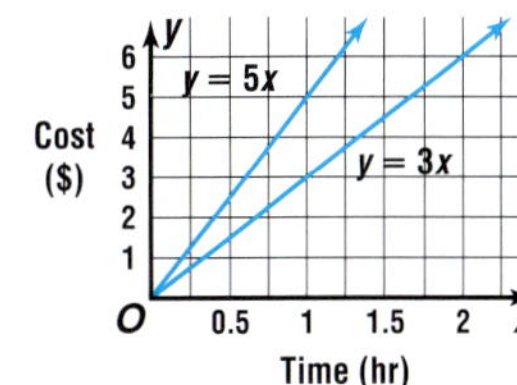

Try These 1. It is steeper.

1. Describe the differences in the graphs of $y = 2x$ and $y = x$ in terms of their slopes and intercepts.
2. Describe the differences in the graph of $y = \frac{1}{2}x$ and the graphs in Exercise 1 in terms of their slopes and intercepts.
3. Clear the Y = list of the previous equations. Then graph $y = |x|$, $y = |x| + 3$, and $y = |x + 2|$. Describe the differences in these graphs in terms of their slopes and intercepts.

Graphing Calculator Tutorial
See pp. 750–753.

2. It is less steep.
3. See margin for graphs; $y = |x|$ rests at the origin, $y = |x| + 3$ intersects the y-axis at (0, 3), and $y = |x + 2|$ intersects the x-axis at (−2, 0).

You can compare graphs of lines by looking at their equations.

Real World **Example 3** Business Link

Matthew and Juan are starting their own pet care business. Juan wants to charge \$5 an hour. Matthew thinks they should charge \$3 an hour. Suppose x represents the number of hours. Then $y = 5x$ and $y = 3x$ represent how much they would charge, respectively. Compare and contrast the graphs of the equations.

The equations have the same y-intercept, but the graph of $y = 5x$ is steeper. This is because its slope, which represents \$5 per hour, is greater than the slope of the graph of $y = 3x$.

(continued on the next page)

Graphing Calculator Exploration

To be sure which graph goes with which equation, students can press [TRACE], use the [▶] key to move the cursor to a good location, and then use the up and down arrow keys to move the cursor from one graph to another. If ExprOn is active, the equation for the graph will appear in the upper-left corner of the screen. To activate ExprOn, the student should press [2nd] [FORMAT], highlight ExprOn, and then press [ENTER].

Answer

3.

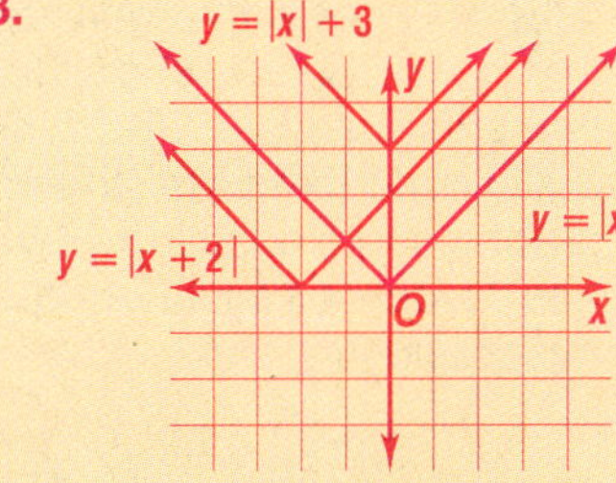

2 TEACH

In-Class Examples

Examples 1–2

Graph each pair of equations. Describe any similarities or differences. Explain why they are a family of graphs.

1 $y = -\frac{1}{2}x + 2$, $y = -\frac{1}{2}x - 1$

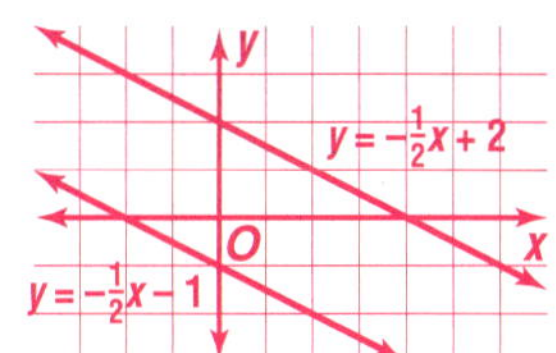

The graphs have y-intercepts of 2 and −1, respectively. They are a family of graphs because the slope of each line is $-\frac{1}{2}$.

2 $y = 5x - 1$, $y = -x - 1$

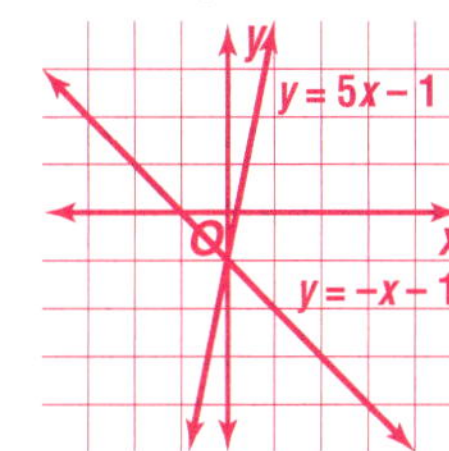

Each graph has a different slope. Each graph has a y-intercept of −1. Thus, they are a family of graphs.

Answers

Your Turn

a.

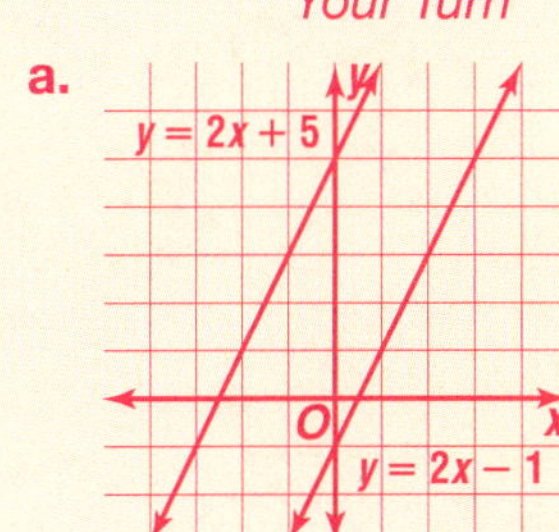

same slope, different y-intercepts

b.

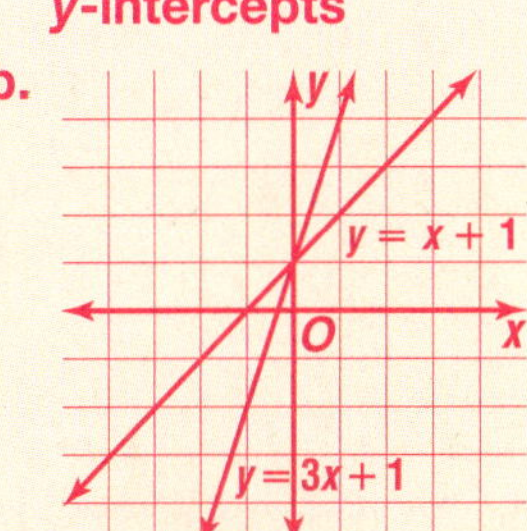

same y-intercept, different slopes

In-Class Examples

Example 3

Gretchen and Max each have a savings account and plan to save \$20 per month. The current balance in Gretchen's account is \$150 and the balance in Max's account is \$100. Then $y = 20x + 150$ and $y = 20x + 100$ represent how much money each has in their account, respectively, after x months. Compare and contrast the graphs of the equations. **The equations have the same slope but the y-intercept of Gretchen's graph is greater than the y-intercept of Max's graph. Gretchen's account will always have more money than Max's.**

Examples 4–5

Change $y = -3x - 1$ so that the graph of the new equation fits each description.

4 same y-intercept, less steep positive slope **Sample answer: $y = x - 1$**

5 same slope, y-intercept is shifted down 2 units **$y = -3x - 3$**

Your Turn Compare and contrast the graphs of the equations. Verify by graphing the equations.

c. $y = -3x + 4$
$y = -x + 4$
c. same y-intercept, different slopes

d. $y = \frac{2}{3}x + 3$
$y = \frac{2}{3}x - 1$
same slope, different y-intercepts

A **parent graph** is the simplest of the graphs in a family. Let's summarize how changing the m or b in $y = mx + b$ affects the graph of the equation.

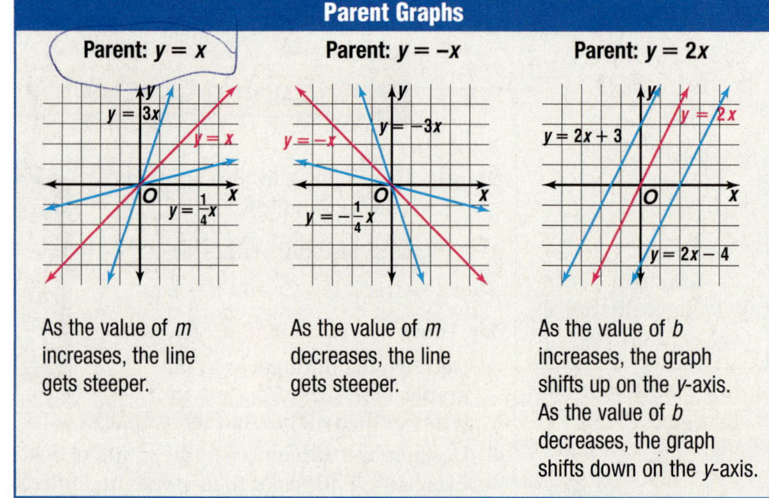

Parent Graphs
Parent: y = x
y = 3x
y = x
y = 1/4 x
O
x
y
As the value of m increases, the line gets steeper.
Parent: y = −x
y = −3x
y = −x
y = −1/4 x
As the value of m decreases, the line gets steeper.
Parent: y = 2x
y = 2x + 3
y = 2x
y = 2x − 4
As the value of b increases, the graph shifts up on the y-axis. As the value of b decreases, the graph shifts down on the y-axis.

You can change a graph by changing the slope or y-intercept.

Examples

Change $y = -\frac{1}{2}x + 3$ so that the graph of the new equation fits each description.

4 same y-intercept, steeper negative slope

The y-intercept is 3, and the slope is $-\frac{1}{2}$. The new equation will also have a y-intercept of 3. In order for the slope to be steeper and still be negative, its value must be less than $-\frac{1}{2}$, such as -2. The new equation is $y = -2x + 3$.

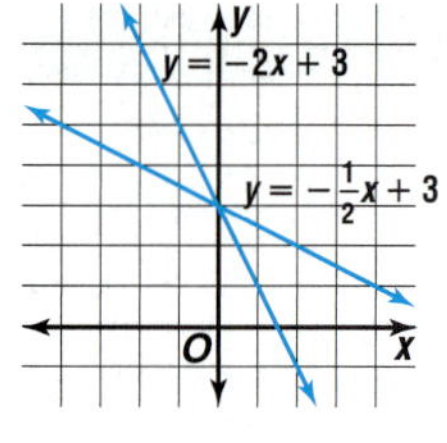

5 same slope, y-intercept is shifted up 4 units

The slope of the new equation will be $-\frac{1}{2}$. Since the current y-intercept is 3, the new y-intercept will be $3 + 4$ or 7. The new equation is $y = -\frac{1}{2}x + 7$. *Check by graphing.*

Reteaching Activity

Auditory/Musical Learners Have students work with a partner. Give each pair of students several families of equations such as those in Exercises 11–22. Students should take turns orally describing the similarities and differences in the graphs and explaining why they are a family of graphs. The students should critique their partners' descriptions.

Your Turn Change $y = 2x + 1$ so that the graph of the new equation fits each description.

e. same slope, shifted down 1 unit **$y = 2x$**

f. same y-intercept, less steep positive slope

Check for Understanding

Communicating Mathematics

1. **Describe** how the graph of each equation is different from the graph of $y = \frac{1}{4}x + 1$.

   a. $y = \frac{1}{4}x - 3$ **shifted down 4 units**
   b. $y = \frac{1}{8}x + 1$ **less steep**
   c. $y = 2x + 1$ **steeper**
   d. $y = \frac{1}{4}x + 4$ **shifted up 3 units**

Vocabulary
family of graphs
parent graph

2. **Sketch** a family of graphs. Identify the parent graph and explain how the graphs are similar. **See students' work.**

3. a. **Explain** the connection between a rate, such as miles per hour or dollars per hour, and slope. **Slope represents rate.**
   b. **Describe** how changing the rate affects the slope of the graph.
   c. **Include** some examples and sketches of graphs. **See students' work.**

3b. A greater rate results in a steeper positive slope.

Guided Practice

Examples 1 & 2

Graph each pair of equations. Describe any similarities or differences and explain why they are a family of graphs. **4–5. See margin.**

4. $y = 3x + 3$
   $y = 3x - 2$

5. $y = \frac{1}{2}x - 4$
   $y = \frac{3}{2}x - 4$

Example 3

Compare and contrast the graphs of each pair of equations. Verify by graphing the equations.

6. $2x + 1 = y$
   $2x = y$
   same slope, different y-intercepts

7. $y = -\frac{1}{4}x + 3$
   $y = -x + 3$
   same y-intercept, different slopes

Examples 4 & 5

Change $y = x - 3$ so that the graph of the new equation fits each description. **8. Sample answer: $y = 2x - 3$**

8. same y-intercept, steeper positive slope

9. same slope, shifted up 3 units **$y = x$**

Example 2

10. **Pet Care** Refer to Example 3. Suppose Juan and Matthew both agree to charge \$3 an hour, but Juan wants to charge an additional \$5 fee per visit. Then $y = 3x + 5$ represents how much Juan would charge. **b. same slope, different y-intercepts**

    a. Graph $y = 3x$ and $y = 3x + 5$. **See margin.**
    b. Explain how the graphs are similar and how they are different.
    c. What does the slope represent? **rate in dollars per hour**

Answer

10a.

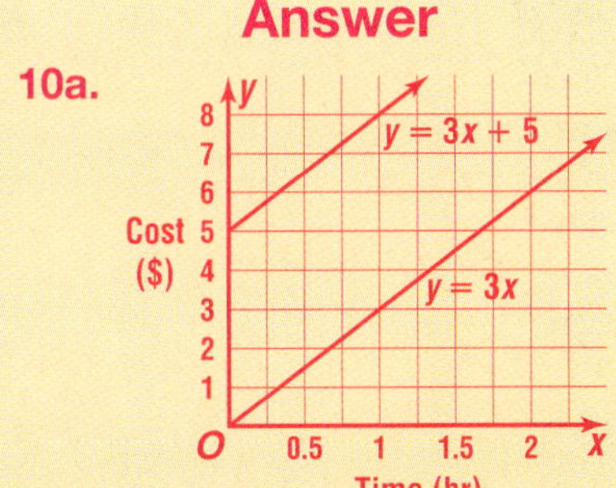

3 PRACTICE/APPLY

Error Analysis

Watch for students who have trouble writing equations for graphs that have varying degrees of steepness such as in Exercise 1. ***Prevent by*** encouraging students to associate the slope with a rate of change. Point out that if the absolute value of the slope is small, the graph changes more gradually and thus is less steep. Stress that the sign of the slope indicates the direction of the graph but does not affect its steepness.

Answers

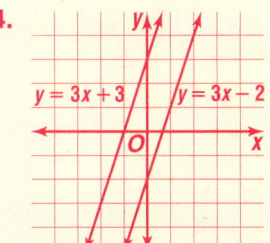

4. $y = 3x + 3$, $y = 3x - 2$
same slope, different y-intercepts

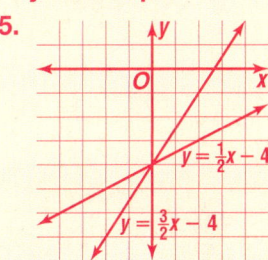

5. $y = \frac{1}{2}x - 4$, $y = \frac{3}{2}x - 4$
same y-intercept, different slopes

Study Guide, **p. 314**

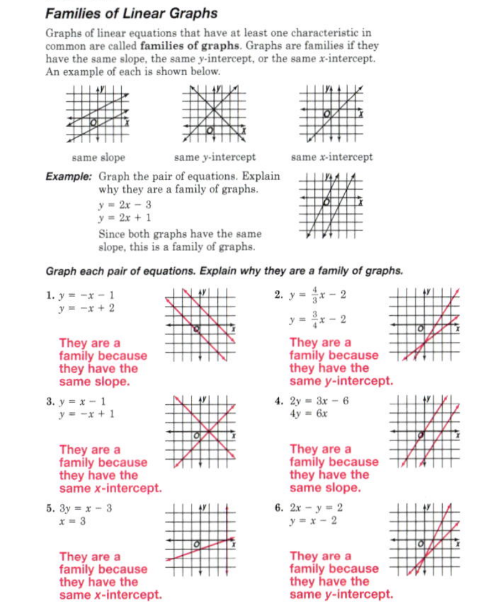

NAME ______ DATE ______ PERIOD ______

7-6 Study Guide Student Edition Pages 316–321

Families of Linear Graphs

Graphs of linear equations that have at least one characteristic in common are called **families of graphs**. Graphs are families if they have the same slope, the same y-intercept, or the same x-intercept. An example of each is shown below.

same slope — same y-intercept — same x-intercept

Example: Graph the pair of equations. Explain why they are a family of graphs.
$y = 2x - 3$
$y = 2x + 1$
Since both graphs have the same slope, this is a family of graphs.

Graph each pair of equations. Explain why they are a family of graphs.

1. $y = -x - 1$
   $y = -x + 2$
   They are a family because they have the same slope.

2. $y = \frac{4}{3}x - 2$
   $y = \frac{3}{4}x - 2$
   They are a family because they have the same y-intercept.

3. $y = x - 1$
   $y = -x + 1$
   They are a family because they have the same x-intercept.

4. $2y = 3x - 6$
   $4y = 6x$
   They are a family because they have the same slope.

5. $3y = x - 3$
   $x = 3$
   They are a family because they have the same x-intercept.

6. $2x - y = 2$
   $y = x - 2$
   They are a family because they have the same y-intercept.

Assignment Guide

Basic: 11–31 odd, 33–39
Average: 12–30 even, 31–39
All: Quiz 2, 1–5

Answers

17. same y-intercept, different slopes
18. same y-intercept, different slopes
19. same slope, different y-intercepts
20. same slope, different y-intercepts
21. same y-intercept, different slopes
22. same slope, different y-intercepts

Exercises

Practice

Graph each pair of equations. Describe any similarities or differences. Explain why they are a family of graphs. 11–16. See Solutions Manual.

Homework Help	
For Exercises	See Examples
11–16	1, 2
17–22, 32	3
23–31	4, 5
Extra Practice	
See page 707.	

11. $y = 5x$
    $y = 2x$
12. $y = 3x + 4$
    $y = x + 4$
13. $y = 4x$
    $y = 4x - 2$
14. $-\frac{2}{3}x + 1 = y$
    $\frac{1}{4}x + 1 = y$
15. $y = -\frac{1}{2}x$
    $y = -3x$
16. $y = -2x + 2$
    $y = -2x - 3$

Compare and contrast the graphs of each pair of equations. Verify by graphing the equations. 17–22. See margin.

17. $y = -x$
    $y = x$
18. $\frac{1}{5}x = y$
    $5x = y$
19. $y = 3x + 1$
    $y = 3x - 8$
20. $y = -\frac{1}{3}x + 4$
    $y = -\frac{1}{3}x$
21. $y = -2x - 2$
    $y = -\frac{3}{4}x - 2$
22. $y = \frac{1}{2}x - 1$
    $x - 3 = 2y$

Change $y = -\frac{4}{5}x + 6$ so that the graph of the new equation fits each description. 23. $y = -\frac{4}{5}x$ 24. Sample answer: $y = \frac{4}{5}x + 6$

23. y-intercept is 0, same slope
24. positive slope, same y-intercept
25. shifted down 4 units, same slope
26. shifted up 2 units, same slope
27. steeper negative slope, same y-intercept
    Sample answer: $y = -x + 6$
28. less steep negative slope, same y-intercept
    Sample answer: $y = -\frac{1}{5}x + 6$

25. $y = -\frac{4}{5}x + 2$

26. $y = -\frac{4}{5}x + 8$

29. Write an equation in slope-intercept form of a line passing through points at $(-3, -5)$ and $(3, 7)$. Then write an equation that has the same slope but a different y-intercept. $y = 2x + 1$; Sample answer: $y = 2x$

30. Write an equation of a line whose graph lies between the graphs of $y = 3x + 4$ and $y = 3x + 2$. Sample answer: $y = 3x + 3$

31a. Yes; they have the same y-intercept, 0.
31b. miles per gallon

Applications and Problem Solving

31. **Shipping** Most cargo is transported by barge, train, or truck. The graph represents the distances traveled for the same amount of fuel.

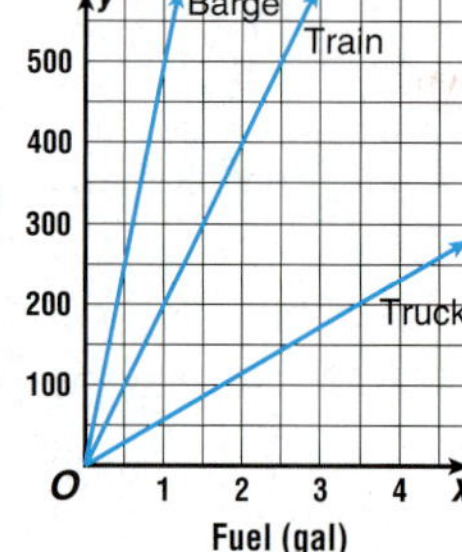

a. Are these graphs a family of graphs? Explain.

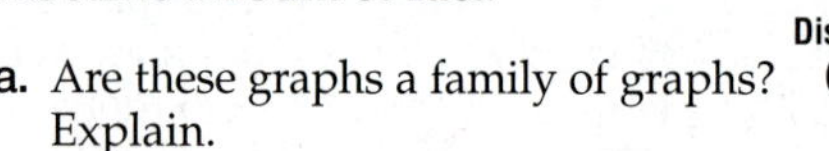

b. What does the slope represent?

c. Which form of transportation gets the most miles per gallon? Explain.
A barge gets 500 miles per gallon of fuel; its graph has the steepest slope.

Inclusion Strategies

Students who are communicably disabled may have trouble describing the changes in a graph when the slope or y-intercept is altered. You might want to allow these students to show you how the graph is affected by drawing a graph. Another way to show the change would be to use a thin plastic straw to represent the line on a grid. The student can manipulate the straw to explain how the graph is affected.

***Skills Practice*, p. 315, and *Practice*, p. 316 (shown)**

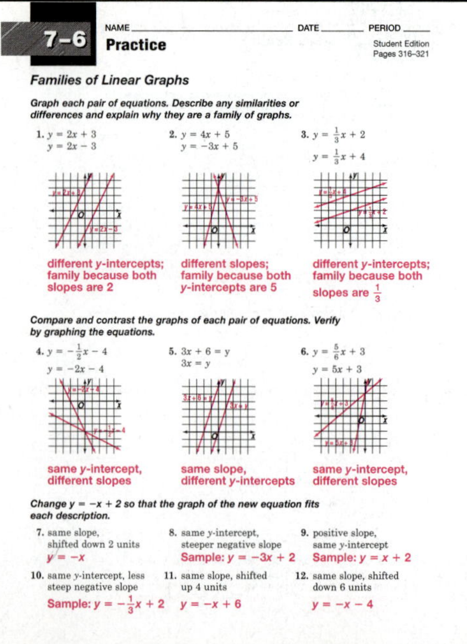

7–6 NAME ______ DATE ______ PERIOD ______
Practice
Student Edition Pages 316–321

Families of Linear Graphs

Graph each pair of equations. Describe any similarities or differences and explain why they are a family of graphs.

1. $y = 2x + 3$
   $y = 2x - 3$
   different y-intercepts; family because both slopes are 2
2. $y = 4x + 5$
   $y = -3x + 5$
   different slopes; family because both y-intercepts are 5
3. $y = \frac{1}{3}x + 2$
   $y = \frac{1}{3}x + 4$
   different y-intercepts; family because both slopes are $\frac{1}{3}$

Compare and contrast the graphs of each pair of equations. Verify by graphing the equations.

4. $y = -\frac{1}{2}x - 4$
   $y = -2x - 4$
   same y-intercept, different slopes
5. $3x + 6 = y$
   $3x = y$
   same slope, different y-intercepts
6. $y = \frac{5}{6}x + 3$
   $y = 5x + 3$
   same y-intercept, different slopes

Change $y = -x + 2$ so that the graph of the new equation fits each description.

7. same slope, shifted down 2 units
   $y = -x$
8. same y-intercept, steeper negative slope
   Sample: $y = -3x + 2$
9. positive slope, same y-intercept
   Sample: $y = x + 2$
10. same y-intercept, less steep negative slope
    Sample: $y = -\frac{1}{3}x + 2$
11. same slope, shifted up 4 units
    $y = -x + 6$
12. same slope, shifted down 6 units
    $y = -x - 4$

32. The graph representing Latino population growth will be steeper than the graph representing growth of the general population.

32. **Population**

> The U.S. Latino population … is growing at a rate five times faster than that of the general population, according to the Census Bureau.

Source: *USA TODAY*

Compare and contrast the graphs that represent the growth of the Latino population and the growth of the general population.

33. **Critical Thinking** Given $A(0, 7)$, $B(2, -3)$, $C(-4, 6)$, and $D(0, 7)$, determine whether $\overline{AB}$ and $\overline{CD}$ are a family of graphs. Explain how you know. **Yes; they have the same y-intercept, 7.**

Mixed Review

Determine the slope and y-intercept of the graph of each equation. *(Lesson 7–5)* **34. $m = 5$, $b = 1$ 35. $m = -2$, $b = 0$**

36. $m = \frac{3}{4}$, $b = -7$

34. $y = 5x + 1$ 35. $y = -2x$ 36. $y = \frac{3}{4}x - 7$

37. **Jewelry** Necklace charms can be gold or silver, boy-shaped or girl-shaped, and contain a different birthstone for each month of the year. How many different charms are possible? *(Lesson 4–2)* **48**

Standardized Test Practice

38. **Short Response** Find $-14 + (-3)$. *(Lesson 3–2)* **−17**

39. **Multiple Choice** What is the property shown by $5(x + 2) = 5x + 10$? *(Lesson 1–4)* **C**

Ⓐ Commutative (×) Ⓑ Identity

Ⓒ Distributive Ⓓ Associative (+)

Quiz 2 Lessons 7–4 through 7–6

1. **Animals** Determine whether a scatter plot showing the temperature of a glass of water at a given air temperature has a *positive* relationship, *negative* relationship, or *no* relationship. If there is a relationship, describe it. *(Lesson 7–4)* **Positive; as air temperature increases, the temperature of a glass of water would increase.**

Determine the x-intercept and y-intercept of the graph of each equation. Then graph the equation. *(Lesson 7–5)* **2–4. See margin for graphs.**

2. $2x + y = 4$ **2, 4** 3. $x + 4y = 4$ **4, 1** 4. $2x - 3y = -12$ **−6, 4**

5. **Videos** The graph shows how much two different video stores pay for movies. *(Lesson 7–6)*

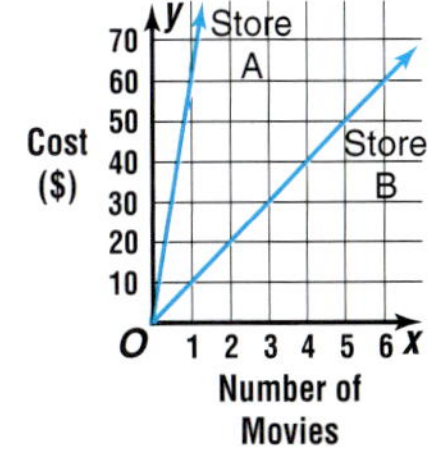

a. Describe any similarities or differences and explain why they are a family of graphs.

b. Explain what the slope represents. **cost per movie**

c. Which store pays less per movie? **Store B**

a. Different slopes; they are a family because they have the same y-intercept, 0.

4 ASSESS

Open-Ended Assessment

Writing Ask students to write a description of a family of linear graphs. They should illustrate their description with an example.

Quiz 2

The Quiz provides students with a brief review of the concepts and skills in Lessons 7–4 through 7–6. Lesson numbers are given to the right of the exercises or instruction lines so students can review concepts not yet mastered.

Answers

Quiz 2

2.

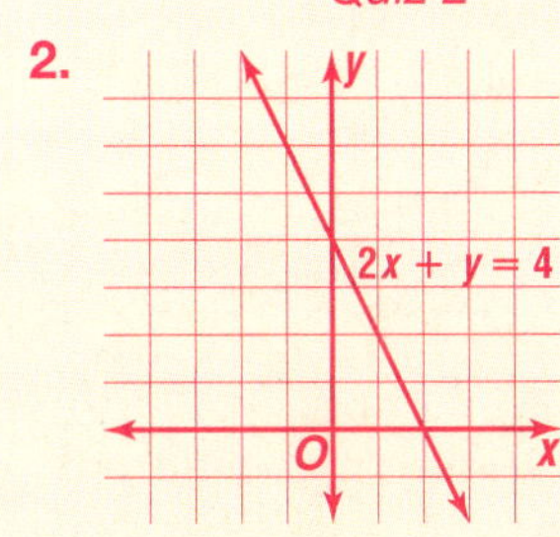

3.

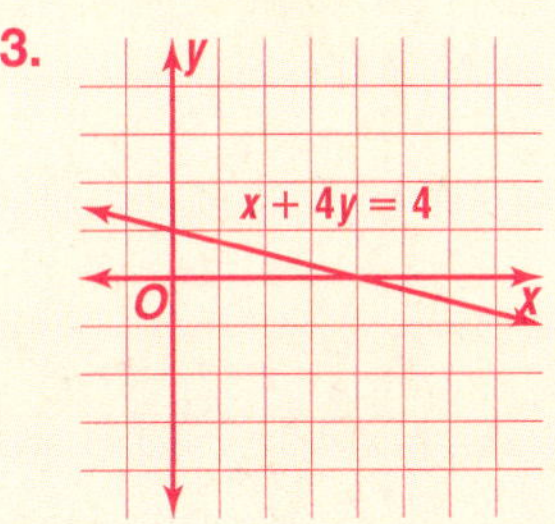

Extra Credit

Think of a real-world situation that could be represented by a family of linear graphs that have different slopes but the same y-intercept. Graph the equations. Repeat this exercise for a family of graphs that have different y-intercepts but the same slope. **See students' work.**

Answer

Quiz 2

4.

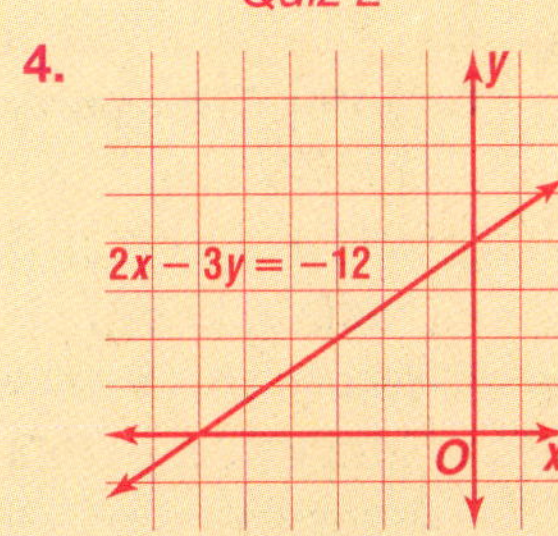

***Enrichment*, p. 318**

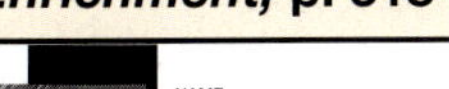

NAME ______ DATE ______ PERIOD ______

7–6 Enrichment

Student Edition Pages 316–321

Inverse Relations

On each grid below, plot the points in Sets A and B. Then connect the points in Set A with the corresponding points in Set B. Then find the inverses of Set A and Set B, plot the two sets, and connect those points.

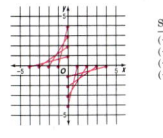

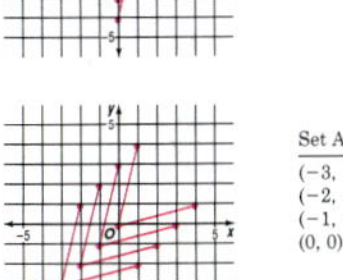

Set A	Set B
(−4, 0)	(0, 1)
(−3, 0)	(0, 2)
(−2, 0)	(0, 3)
(−1, 0)	(0, 4)

Inverse

	Set A	Set B
1.	(0, −4)	(1, 0)
2.	(0, −3)	(2, 0)
3.	(0, −2)	(3, 0)
4.	(0, −1)	(4, 0)

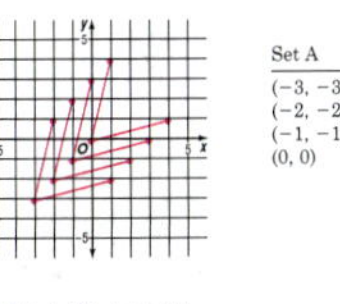

Set A	Set B
(−3, −3)	(−2, 1)
(−2, −2)	(−1, 2)
(−1, −1)	(0, 3)
(0, 0)	(1, 4)

Inverse

	Set A	Set B
5.	(−3, −3)	(1, −2)
6.	(−2, −2)	(2, −1)
7.	(−1, −1)	(3, 0)
8.	(0, 0)	(4, 1)

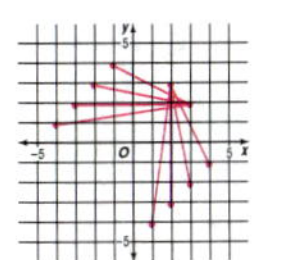

Set A	Set B
(−4, 1)	(3, 2)
(−3, 2)	(3, 2)
(−2, 3)	(3, 2)
(−1, 4)	(3, 2)

Inverse

	Set A	Set B
9.	(1, −4)	(2, 3)
10.	(2, −3)	(2, 3)
11.	(3, −2)	(2, 3)
12.	(4, −1)	(2, 3)

13. What is the graphical relationship between the line segments you drew connecting points in Sets A and B and the line segments connecting points in the inverses of those two sets? **Sample answer: The graphs are reflected across the line $x = y$ by their inverses.**

Lesson 7–7

1 FOCUS

5-Minute Check
Lesson 7–6

Compare and contrast the graphs of each pair of equations. Verify by graphing the equations.

1. $y = -\frac{1}{4}x$ and $y = -\frac{1}{4}x - 2$
   same slope; different y-intercepts

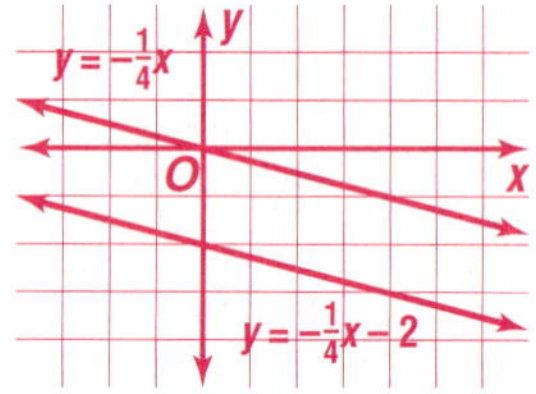

2. $y = -3x + 2$ and $y = 2 - 2x$
   same y-intercept; different slopes

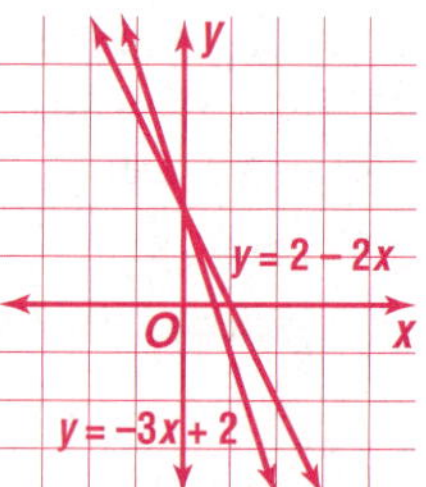

Change $y = 3x - 1$ so that the graph of the new equation fits each description.

3. y-intercept is 0, same slope **$y = 3x$**
4. steeper negative slope, same y-intercept **Sample answer: $y = -4x - 1$**

Motivating the Lesson

Hands-On Activity Write the equations $y = 2x$, $y = 2x - 1$, and $y = 2x + 2$ on the board or overhead. Have students graph all three of these lines on the same coordinate grid. Ask them to describe the lines, and also ask them if the lines will ever intersect. This discussion can lead to the definition of *parallel lines*.

7–7 Parallel and Perpendicular Lines

What You'll Learn
You'll learn to write an equation of a line that is parallel or perpendicular to the graph of a given equation and that passes through a given point.

Why It's Important
Surveying Surveyors use parallel and perpendicular lines to plan construction. *See Exercise 39.*

The graphs of the equations shown at the right are a family of graphs because they have the same slope. Because $4x$ is never equal to $4x + 5$, the value of y will never be the same for any given value of x, and the graphs will never intersect. These lines are **parallel**.

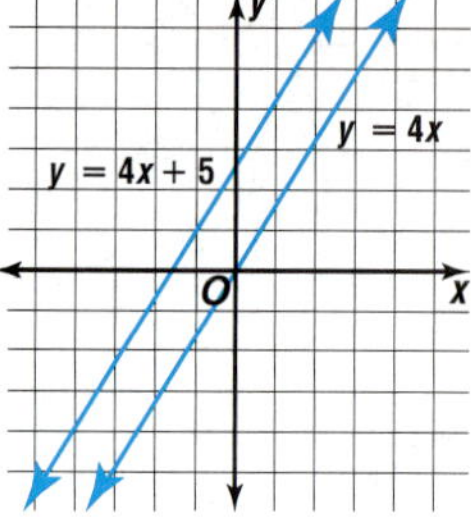

Parallel Lines

Words: If two lines have the same slope, then they are parallel.

Model:

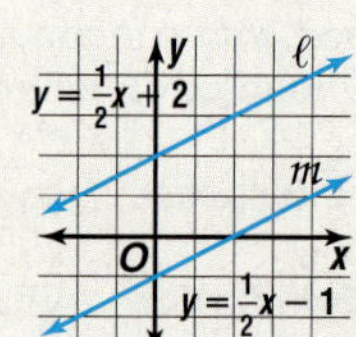

Symbols: $\ell \parallel m$

All vertical lines are parallel.

Example  **1**

Determine whether the graphs of the equations are parallel.

$y = -\frac{3}{4}x - 2$

$4y = -3x + 12$

First, determine the slopes of the lines. Write each equation in slope-intercept form.

$y = -\frac{3}{4}x - 2$ *Slope-Intercept Form*

The slope is $-\frac{3}{4}$.

$4y = -3x + 12$

$\frac{4y}{4} = \frac{-3x + 12}{4}$ *Divide each side by 4.*

$y = -\frac{3}{4}x + 3$ The slope is $-\frac{3}{4}$.

The slopes are the same, so the lines are parallel. *Check by graphing.*

Your Turn

a. $y = 2x$
   $7 = 2x - y$ **yes**

b. $y = -3x + 3$
   $2y = 6x - 5$ **no**

Resource Manager

Reproducible Masters

Chapter 7 Resource Masters
- *Study Guide*, p. 319
- *Skills Practice*, p. 320
- *Practice*, p. 321
- *Reading to Learn Mathematics*, p. 322
- *Enrichment*, p. 323
- *Assessment*, p. 335

Hands-On Algebra, pp. 85–86

Transparencies
5-Minute Check, 7–7

Technology/Multimedia
Interactive Chalkboard CD-ROM

A *parallelogram* is a four-sided figure with two sets of parallel sides.

Example 2 — **Geometry Link**

Determine whether figure *ABCD* is a parallelogram.

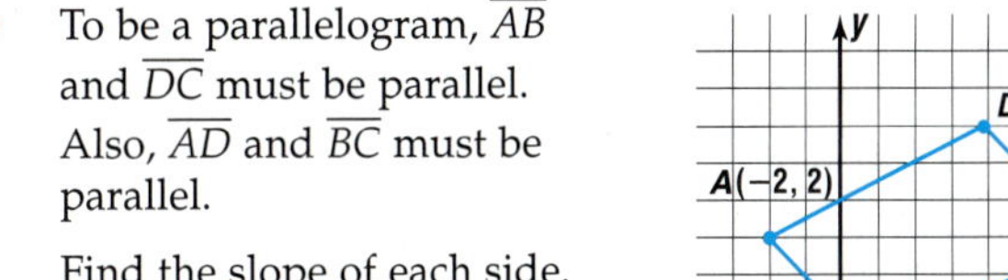

Explore To be a parallelogram, $\overline{AB}$ and $\overline{DC}$ must be parallel. Also, $\overline{AD}$ and $\overline{BC}$ must be parallel.

Plan Find the slope of each side.

Solve

$$\overline{AB}: m = \frac{-1-2}{1-(-2)} = \frac{-3}{3} \text{ or } -1$$

$$\overline{BC}: m = \frac{-1-2}{1-7} = \frac{-3}{-6} \text{ or } \frac{1}{2}$$

$$\overline{DC}: m = \frac{2-5}{7-4} = \frac{-3}{3} \text{ or } -1$$

$$\overline{AD}: m = \frac{5-2}{4-(-2)} = \frac{3}{6} \text{ or } \frac{1}{2}$$

$\overline{AB}$ is parallel to $\overline{DC}$ because their slopes are both −1.
$\overline{BC}$ is parallel to $\overline{AD}$ because their slopes are both $\frac{1}{2}$.
Therefore, *ABCD* is a parallelogram.

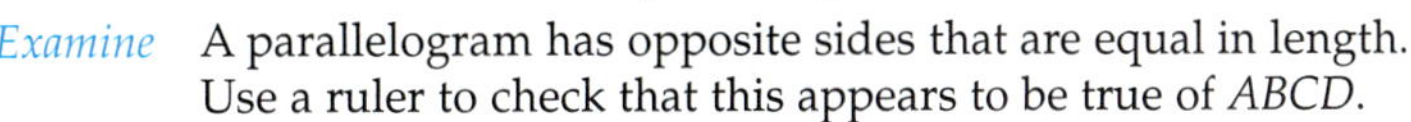

Examine A parallelogram has opposite sides that are equal in length. Use a ruler to check that this appears to be true of *ABCD*.

Prerequisite Skills Review
Simplifying Fractions, p. 685

You can use the slope of a line to write an equation of a line that is parallel to it.

Example 

Write an equation in slope-intercept form of the line that is parallel to the graph of $y = -4x + 8$ and passes through (1, 3).

The slope of the given line is −4. So, the slope of the new line will also be −4. Find the new equation by using the point-slope form.

$y - y_1 = m(x - x_1)$	*Point-Slope Form*
$y - 3 = -4(x - 1)$	*Replace (x_1, y_1) with (1, 3) and m with −4.*
$y - 3 = -4x + 4$	*Distributive Property*
$y - 3 + 3 = -4x + 4 + 3$	*Add 3 to each side.*
$y = -4x + 7$	*Simplify.*

An equation whose graph is parallel to the graph of $4x + y = 8$ and passes through (1, 3) is $y = -4x + 7$. *Check by substituting (1, 3) into $y = -4x + 7$ or by graphing.*

Your Turn

Write an equation in slope-intercept form of the line that is parallel to the graph of each equation and passes through the given point.

c. $y = 6x - 4$; (2, 3) $y = 6x - 9$ d. $3x + 2y = 9$; (2, 0) $y = -\frac{3}{2}x + 3$

www.algconcepts.com/extra_examples

2 TEACH

Teaching Tip Remind students that parallel lines never intersect.

In-Class Examples

Example 1

Determine whether the graphs of the equations are parallel.
$y = 3x + 4$
$9x + 3y = 12$ **not parallel**

Example 2

Determine whether figure *EFGH* is a parallelogram. **no**

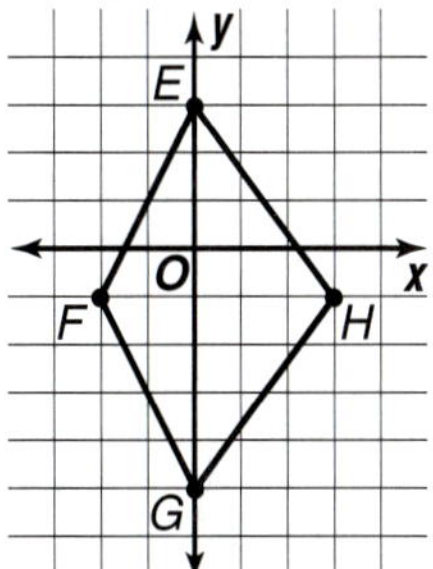

Example 3

Write an equation in slope-intercept form of the line that is parallel to the graph of $y = \frac{2}{3}x - 3$ and passes through the point at (−3, 1).
$y = \frac{2}{3}x + 3$

Teaching Tip It may help students to use the term *negative reciprocals* when referring to the slopes of two perpendicular lines.

Teaching Tip Remind students that perpendicular lines intersect to form a 90° angle.

In-Class Example

Example 4

Determine whether the graphs of the equations are perpendicular.

$y + 2x = -4$

$y = \frac{1}{2}x + 3$ **yes**

Hands-On Algebra

Materials:  grid paper  protractor

Step 1 Draw line ℓ through the origin and $P(5, 3)$.

Step 2 Use a protractor to rotate the line 90°. Label the new line ℓ'.

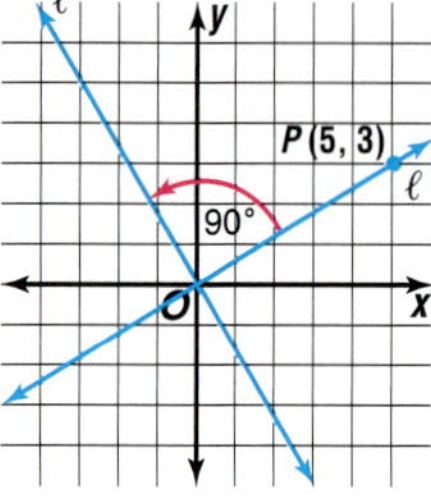

Try These

1. What are the slopes of ℓ and ℓ'? $\frac{3}{5}$; $-\frac{5}{3}$
2. Compare the slopes. What is their product?

2. They are negative reciprocals; −1.

The results of the Hands-On Algebra activity lead to the following definition of **perpendicular lines**.

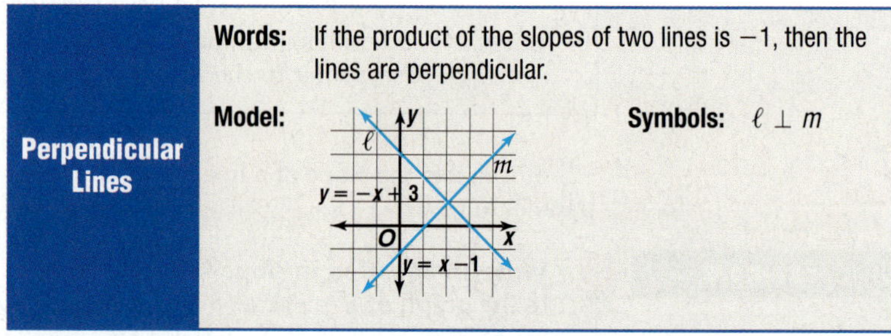

Perpendicular Lines	**Words:** If the product of the slopes of two lines is −1, then the lines are perpendicular.
	Model: (graph) **Symbols:** $\ell \perp m$

In a plane, vertical lines are perpendicular to horizontal lines.

Example 4 **Determine whether the graphs of the equations are perpendicular.**

$y = \frac{2}{3}x + 1$

$y = -\frac{3}{2}x + 2$

The graphs are perpendicular because the product of their slopes is $\frac{2}{3} \cdot \left(-\frac{3}{2}\right)$ or -1.

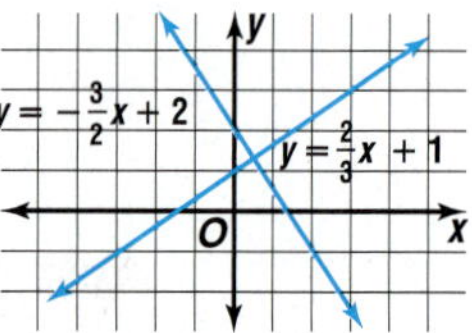

Your Turn

e. $y = \frac{1}{5}x + 2$
   $y = 5x + 1$ **no**

f. $y = -4x + 3$
   $4y = x - 5$ **yes**

Hands-On Algebra

Cooperative Learning Students should use a point with integer coordinates when calculating the slope of the rotated line. You may want to have them use the point $Q(-3, 5)$ and the origin.

An additional Hands-On Algebra activity using tracings of parallel and perpendicular lines is available in the *Hands-On Algebra Masters,* p. 85.

Hands-On Algebra Masters, p. 86

Example 5 Write an equation in slope-intercept form of the line that is perpendicular to the graph of $y = \frac{1}{3}x - 2$ and passes through $(-4, 2)$.

The slope is $\frac{1}{3}$. A line perpendicular to the graph of $y = \frac{1}{3}x - 2$ has slope -3. Find the new equation by using the point-slope form.

$y - y_1 = m(x - x_1)$	*Point-Slope Form*
$y - 2 = -3[x - (-4)]$	*Replace (x_1, y_1) with $(-2, 4)$ and m with -3.*
$y - 2 = -3x - 12$	*Distributive Property*
$y - 2 + 2 = -3x - 12 + 2$	*Add 2 to each side.*
$y = -3x - 10$	

The new equation is $y = -3x - 10$. *Check by substituting $(-4, 2)$ into the equation or by graphing.*

Your Turn Write an equation in slope-intercept form of the line that is perpendicular to the graph of each equation and passes through the given point.

g. $y = 2x + 6$; $(0, 0)$ $y = -\frac{1}{2}x$ h. $2x + 3y = 2$; $(3, 0)$

h. $y = \frac{3}{2}x - \frac{9}{2}$

In-Class Example

Example 5

Write an equation in slope-intercept form of the line that is perpendicular to the graph of $y = -2x + 5$ and passes through the point at $(2, -3)$.
$y = \frac{1}{2}x - 4$

Teaching Tip When discussing the answer to Exercise 1, you might want to point out that any two vertical lines are parallel because their slopes are both undefined. Also, any pair of horizontal and vertical lines are perpendicular even though their slopes are not negative reciprocals (since the slope of the vertical line is undefined).

Check for Understanding

Communicating Mathematics

1. **Compare and contrast** the slopes of lines that are parallel and the slopes of lines that are perpendicular.
2. **Choose** the graph that is parallel to the graph of $3x + 4y = -12$. Explain.

Vocabulary
parallel lines
perpendicular lines

1. Parallel lines have equal slopes. Perpendicular lines have slopes that are negative reciprocals.
2. c; They each have a slope of $-\frac{3}{4}$.

a.
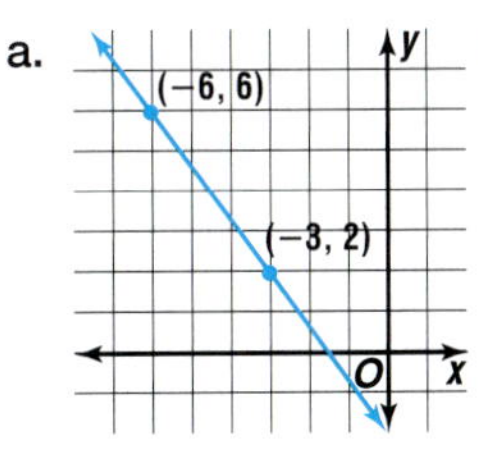

b.
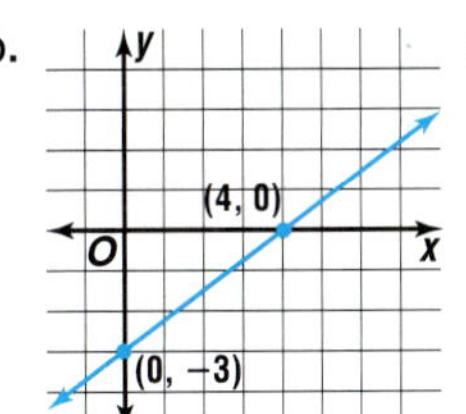

c.
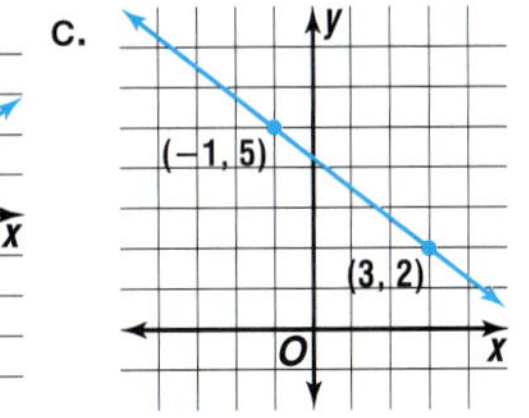

Guided Practice

Getting Ready State the slopes of the lines parallel to and perpendicular to the graph of each equation.

Sample: $y = -3x + 6$
Solution: The slope is -3. Lines parallel to the graph have a slope of -3. Lines perpendicular to the graph have a slope of $\frac{1}{3}$.

3. $y = 4x + 2$ $4, -\frac{1}{4}$
4. $5x + y = 8$ $-5, \frac{1}{5}$
5. $2x - 3y = 7$ $\frac{2}{3}, -\frac{3}{2}$

3 PRACTICE/APPLY

Error Analysis

Watch for students who use the incorrect slope in Exercises 4–5. ***Prevent by*** cautioning students that they must rewrite an equation in slope-intercept form before they can take the coefficient of x as the slope of the graph. Also, remind them that the slope of a perpendicular line is the *negative* reciprocal of the given slope.

***Study Guide*, p. 319**

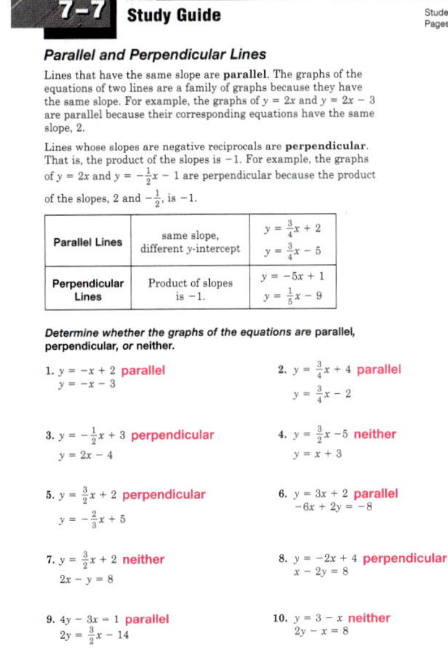

7–7 NAME ______ DATE ______ PERIOD ______
Study Guide Student Edition Pages 322–327

Parallel and Perpendicular Lines

Lines that have the same slope are **parallel**. The graphs of the equations of two lines are a family of graphs because they have the same slope. For example, the graphs of $y = 2x$ and $y = 2x - 3$ are parallel because their corresponding equations have the same slope, 2.

Lines whose slopes are negative reciprocals are **perpendicular**. That is, the product of the slopes is -1. For example, the graphs of $y = 2x$ and $y = -\frac{1}{2}x - 1$ are perpendicular because the product of the slopes, 2 and $-\frac{1}{2}$, is -1.

Parallel Lines	same slope, different y-intercept	$y = \frac{3}{4}x + 2$ $y = \frac{3}{4}x - 5$
Perpendicular Lines	Product of slopes is -1.	$y = -5x + 1$ $y = \frac{1}{5}x - 9$

***Determine whether the graphs of the equations are* parallel, perpendicular, *or* neither.**

1. $y = -x + 2$, $y = -x - 3$ parallel
2. $y = \frac{3}{4}x + 4$, $y = \frac{3}{4}x - 2$ parallel
3. $y = -\frac{1}{2}x + 3$, $y = 2x - 4$ perpendicular
4. $y = \frac{3}{2}x - 5$, $y = x + 3$ neither
5. $y = \frac{3}{2}x + 2$, $y = -\frac{2}{3}x + 5$ perpendicular
6. $y = 3x + 2$, $-6x + 2y = -8$ parallel
7. $y = \frac{3}{2}x + 2$, $2x - y = 8$ neither
8. $y = -2x + 4$, $x - 2y = 8$ perpendicular
9. $4y - 3x = 1$, $2y = \frac{3}{2}x - 14$ parallel
10. $y = 3 - x$, $2y - x = 8$ neither

Reteaching Activity

Visual/Spatial Learners Give students some problems like those in Exercises 30–35. Have students first graph the original equation and the point prior to finding the equation of a parallel or perpendicular line. They should then find the equation of the desired line and graph it to verify that it meets the requirements.

Assignment Guide

Basic: 15–39 odd, 41–50
Average: 16–38 even, 39–50

Examples 1 & 4 **Determine whether the graphs of each pair of equations are *parallel*, *perpendicular*, or *neither*.**

6. $y = 4x - 9$
$y = 4x - 11$ **parallel**

7. $y = -\frac{3}{2}x + 5$
$y = 2x + 1$ **neither**

8. $y = 5x + 3$
$x + 5y = 2$ **perpendicular**

Example 4

9. **Geometry** Determine $m\angle STQ$ if the slope of $\overline{RQ}$ is $\frac{2}{5}$ and the slope of $\overline{SP}$ is $-\frac{5}{2}$. Explain. **90; The lines are perpendicular.**

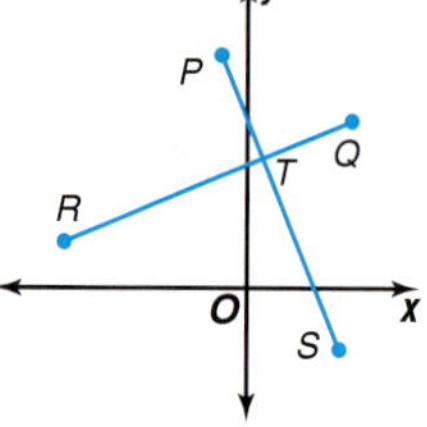

Exercise 9

Example 3 **Write an equation in slope-intercept form of the line that is parallel to the graph of each equation and passes through the given point.**

10. $y = x + 5$; (7, 2) **$y = x - 5$**
11. $y - \frac{3}{4}x = 4$; (2, 0) **$y = \frac{3}{4}x - \frac{3}{2}$**

Example 5 **Write an equation in slope-intercept form of the line that is perpendicular to the graph of each equation and passes through the given point.**

12. $y = 2x + 3$; $(3, -4)$ **$y = -\frac{1}{2}x - \frac{5}{2}$**
13. $3x + 8y = 4$; (0, 4) **$y = \frac{8}{3}x + 4$**

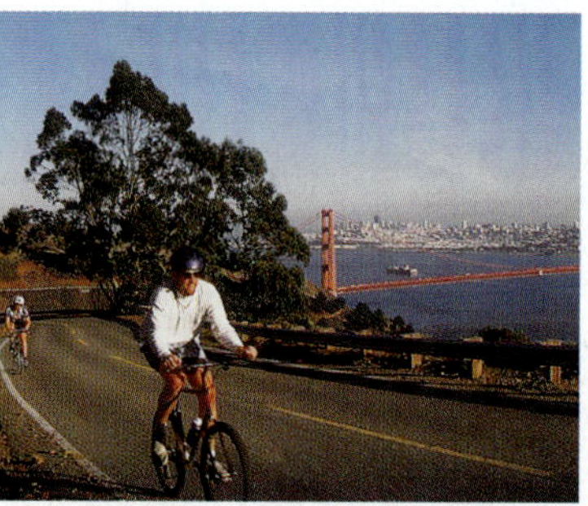
Bicycling near San Francisco

14. **Bicycling** Joey and Cortez are both riding their bikes at 12 miles per hour. However, Cortez started 5 miles ahead of Joey. The equations $y = 12x$ and $y = 12x + 5$ represent the positions of Joey and Cortez, respectively. **Example 1**
    a. If they ride for 3 hours, will Joey catch up to Cortez? Explain your answer in terms of slopes of the graphs of the equations.
    b. Are the graphs a family of graphs? Explain. **yes; same slope**

a. No; the graphs of the equations are parallel lines. They will never intersect.

Exercises

Practice

Determine whether the graphs of each pair of equations are *parallel*, *perpendicular*, or *neither*.

Homework Help	
For Exercises	See Examples
15–23, 39	1, 4
24–29, 36	3
30–35, 37	5
36–37, 40	2

Extra Practice
See page 707.

15. $y = \frac{3}{5}x + 2$
$y = -\frac{5}{3}x - 1$ **perpendicular**

16. $y = 5x + 4$
$y = 5x + 8$ **parallel**

17. $x + 3y = 7$
$y = \frac{1}{3}x - 9$ **neither**

18. $y = -2x + 6$
$y = 2x$ **neither**

19. $y = 2x + 3$
$x + 2y = 4$ **perpendicular**

20. $y = \frac{2}{3}x$
$3x + 2y = 0$ **perpendicular**

21. $3y = 8$
$y = 3$ **parallel**

22. $-\frac{1}{4}x + 5 = 0$
$2y - 9 = 0$ **perpendicular**

23. $7x + 3y = 4$
$3x - 7y = 1$ **perpendicular**

***Skills Practice*, p. 320, and *Practice*, p. 321 (shown)**

7–7 NAME ______ DATE ______ PERIOD ______

Practice Student Edition Pages 322–327

Parallel and Perpendicular Lines

Determine whether the graphs of each pair of equations are parallel, perpendicular, or neither.

1. $y = 3x + 4$, $y = 3x + 7$ **parallel**
2. $y = -4x + 1$, $4y = x + 3$ **perpendicular**
3. $y = 2x - 5$, $y = 5x - 5$ **neither**
4. $y = -\frac{1}{3}x + 2$, $y = 3x - 5$ **perpendicular**
5. $y = \frac{3}{5}x - 3$, $5y = 3x - 10$ **parallel**
6. $y = 4$, $4y = 6$ **parallel**
7. $y = 7x + 2$, $x + 7y = 8$ **perpendicular**
8. $y = \frac{5}{6}x - 6$, $x + 5y = 4$ **neither**
9. $y = -\frac{3}{8}x - 9$, $y = \frac{8}{3}x + 3$ **perpendicular**

Write an equation in slope-intercept form of the line that is parallel to the graph of each equation and passes through the given point.

10. $y = 3x + 6$; (4, 7) **$y = 3x - 5$**
11. $y = x - 4$; $(-2, 3)$ **$y = x + 5$**
12. $y = \frac{1}{2}x + 5$; $(4, -5)$ **$y = \frac{1}{2}x - 7$**
13. $y + \frac{2}{3}x = 3$; $(-6, 1)$ **$y = -\frac{2}{3}x - 3$**
14. $y - \frac{2}{5}x = -5$; (5, 3) **$y = \frac{2}{5}x + 1$**
15. $y + 2x = 4$; $(-1, 2)$ **$y = -2x$**

Write an equation in slope-intercept form of the line that is perpendicular to the graph of each equation and passes through the given point.

16. $y = -5x + 1$; $(2, -1)$ **$y = \frac{1}{5}x - \frac{7}{5}$**
17. $y = 2x - 3$; $(-5, 3)$ **$y = -\frac{1}{2}x + \frac{1}{2}$**
18. $4x + 7y = 3$; $(-4, -7)$ **$y = \frac{7}{4}x$**
19. $3x - 4y = 2$; (6, 0) **$y = -\frac{4}{3}x + 8$**
20. $y = -4x - 2$; $(4, -4)$ **$y = \frac{1}{4}x - 5$**
21. $6x + 5y = -3$; $(-6, 2)$ **$y = \frac{5}{6}x + 7$**

Write an equation in slope-intercept form of the line that is parallel to the graph of each equation and passes through the given point.

24. $y = 4x + 5$; $(2, -3)$ 25. $y = -\frac{1}{2}x - 3$; $(0, 0)$ 26. $y = 4$; $(2, 5)$

27. $4x + y = 1$; $(-2, 1)$ $y = -4x - 7$ 28. $3x + y = 3$; $(3, 5)$ $y = -3x + 14$ 29. $x = 3$; $(2, 6)$ $x = 2$

Write an equation in slope-intercept form of the line that is perpendicular to the graph of each equation and passes through the given point.

30. $y = 2x + 3$; $(3, -4)$ 31. $y = -4x + 5$; $(1, 1)$ 32. $2x - 5y = 3$; $(-2, 7)$

33. $y = 2x$; $(2, 0)$ 34. $x = 6$; $(4, 2)$ $y = 2$ 35. $3x - 5y = 2$; $(-1, 0)$

24. $y = 4x - 11$
25. $y = -\frac{1}{2}x$
26. $y = 5$
30. $y = -\frac{1}{2}x - \frac{5}{2}$
31. $y = \frac{1}{4}x + \frac{3}{4}$
32. $y = -\frac{5}{2}x + 2$
33. $y = -\frac{1}{2}x + 1$
35. $y = -\frac{5}{3}x - \frac{5}{3}$
38a. $y = \frac{5}{2}x - 12$

Determine whether $\overline{AB}$ and $\overline{CD}$ are *parallel*, *perpendicular*, or *neither*.

36. $A(4, 3)$, $B(5, 2)$, $C(8, 3)$, $D(9, 2)$ **parallel**

37. $A(-2, 3)$, $B(4, 0)$, $C(2, 5)$, $D(-1, 11)$ **neither**

38. Write an equation in slope-intercept form of each line.
   a. passes through the point at $(4, -2)$ and is parallel to the graph of $5x - 2y = 6$
   b. perpendicular to the line through points at $(1, 2)$ and $(8, 6)$ and passes through the point at $(0, 5)$ $y = -\frac{7}{4}x + 5$

Applications and Problem Solving

39. **Surveying** Determine whether roads $\overleftrightarrow{AB}$ and $\overleftrightarrow{BC}$ are perpendicular in the survey at the right. Explain.

40. **Geometry** Determine whether quadrilateral $JKLM$ is a rectangle if its vertices are $J(0, 5)$, $K(9, 2)$, $L(7, -4)$, and $M(-2, -1)$. Explain how you know.

41. **Critical Thinking** Suppose the line through points at $(-2, 2)$ and $(x, 5)$ is parallel to the graph of $2x - 4y = 9$. What is x? **4**

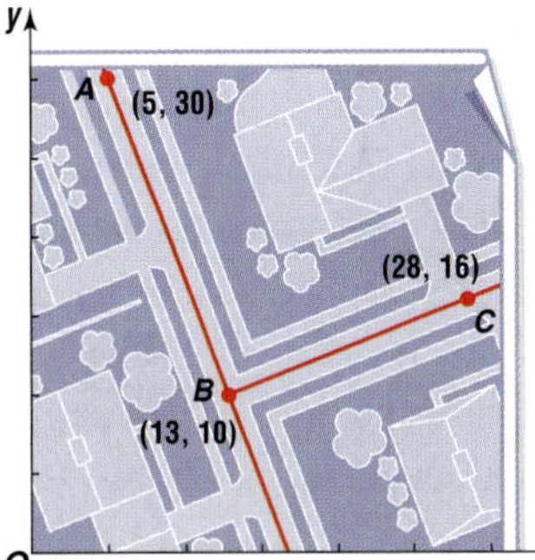

Exercise 39

39. Yes; the product of their slopes is −1.
40. Yes; consecutive sides are perpendicular, so the figure has four right angles.

Mixed Review

Compare and contrast the graphs of each pair of equations. Verify by graphing the equations. *(Lesson 7–6)* **42–44. See margin.**

42. $y = 4x$, $y = 4x - 3$ 43. $y = -3x + 2$, $y = 3x + 2$ 44. $y = x + 2$, $7y = -2x + 14$

A card is drawn from a standard deck of cards. Determine whether each event is *mutually exclusive* or *inclusive*. *(Lesson 5–7)*

45. P(queen or ace) **mutually exclusive** 46. P(8 or red card) **inclusive** 47. P(heart or spade) **mutually exclusive**

48. Solve $6 = q - 5$. *(Lesson 3–5)* **11**

Standardized Test Practice

49. **Grid In** Solve $g = 15.6 - 2.75$. *(Lesson 3–4)* **12.85**

50. **Multiple Choice** In which quadrant does the graph of $G(-15, 4)$ lie? *(Lesson 2–2)* **B**

 I  II  III D IV

www.algconcepts.com/self_check_quiz

4 ASSESS

Open-Ended Assessment

Writing Ask students to write the definitions of parallel lines and perpendicular lines, and to give a written explanation of the process for finding the slope of a line parallel or perpendicular to a given line.

Chapter 7, Quiz B (Lessons 7–4 through 7–7) is available in the *Chapter 7 Resource Masters*, p. 335.

Answers

42. same slope, different y-intercepts
43. same y-intercept, different slopes
44. same y-intercept, different slopes

***Enrichment*, p. 323**

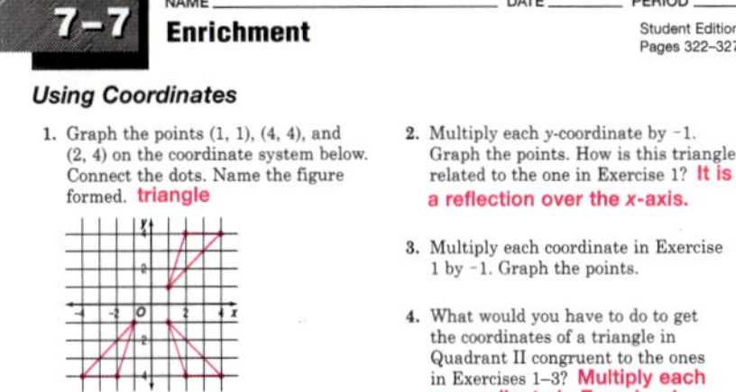

7–7 NAME ______ DATE ____ PERIOD ____
Enrichment Student Edition Pages 322–327

Using Coordinates

1. Graph the points (1, 1), (4, 4), and (2, 4) on the coordinate system below. Connect the dots. Name the figure formed. **triangle**
2. Multiply each y-coordinate by −1. Graph the points. How is this triangle related to the one in Exercise 1? **It is a reflection over the x-axis.**
3. Multiply each coordinate in Exercise 1 by −1. Graph the points.
4. What would you have to do to get the coordinates of a triangle in Quadrant II congruent to the ones in Exercises 1–3? **Multiply each x-coordinate in Exercise 1 by −1.**
5. Graph the points (3, 1), (2, 3), (4, 6), and (5, 4) on the coordinate plane at the right. Connect the dots. The figure formed is called a parallelogram.
6. Add 2 to both coordinates of each point and graph the new coordinates. Name the figure formed. **parallelogram**
7. Add −4 to the x-coordinate of each point in Exercise 5 and graph the new coordinates. Is this figure also a parallelogram? **yes**
8. Graph the points (5, −2), (6, −3), (5, −5), and (3, −3) on the coordinate plane at the right. Name the figure formed. **trapezoid**
9. Multiply both coordinates of each point in Exercise 8 by 2 and graph the new coordinates. This is an enlargement.
10. Multiply both coordinates of each point in Exercise 8 by $\frac{1}{2}$ and graph the new coordinates. This is a reduction.

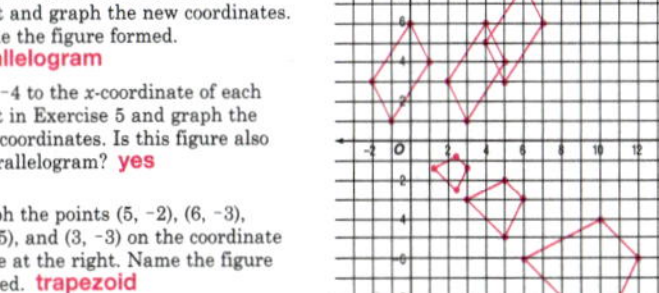

Extra Credit

Write a procedure for testing whether a quadrilateral is a rectangle when the coordinates of its four vertices are given. **Sample answer: Calculate the slopes of all four sides of the quadrilateral. Determine if both pairs of opposite sides have the same slope. Then determine if each pair of adjacent sides have slopes that are negative reciprocals.**

Understanding and Using the Vocabulary

This section provides a listing of the new terms, properties, and phrases that were introduced in this chapter. The exercises check students' understanding of the terms by using a variety of verbal formats including matching, completion, and true/false.

Glossary A complete glossary of terms appears on pages 762–783.

MindJogger Videoquizzes

MindJogger Videoquizzes provide an alternative review of concepts presented in this chapter. Students work in teams to answer questions, gaining points for correct answers.

Answers

Page 329

15. $y - 5 = 3(x - 2)$
16. $y - 6 = \frac{1}{4}(x + 1)$
17. $y - 3 = -5(x)$
18. $y - 8 = 0$ or $y = 8$
19. $y - 4 = -\frac{1}{6}(x - 10)$ or $y - 7 = -\frac{1}{6}(x + 8)$
20. $y = x + 4$
21. $y = 3x - 9$
22. $y = \frac{1}{2}x + 5$
23. $y = 2x$
24. $y = 7$
25. $y = 11x - 6$
26. Negative relationship; as TV time increases, grades decrease.

Understanding and Using the Vocabulary

After completing this chapter, you should be able to define each term, property, or phrase and give an example or two of each.

interNET CONNECTION **Review Activities** For more review activities, visit: www.algconcepts.com

best-fit line *(p. 308)*
correlation coefficient *(p. 309)*
extrapolation *(p. 309)*
family of graphs *(p. 316)*
interpolation *(p. 309)*
linear regression *(p. 309)*
median-median line *(p. 309)*
parallel lines *(p. 322)*
parent graph *(p. 318)*
perpendicular lines *(p. 324)*
point-slope form *(p. 290)*
rate of change *(p. 285)*
residual *(p. 309)*
scatter plot *(p. 302)*
slope *(p. 284)*
slope-intercept form *(p. 296)*
x-intercept *(p. 296)*
y-intercept *(p. 296)*

State whether each sentence is *true* or *false*. If false, replace the underlined word(s) to make a true statement.

1. In the equation $y = mx + b$, b is the <u>x-intercept</u> of the line. **false; y-intercept**
2. A line that appears to go downhill from left to right has a <u>negative</u> slope. **true**
3. <u>Perpendicular</u> lines have the same slopes. **false; parallel**
4. The graph of a <u>linear equation</u> is a straight line. **true**
5. You can write an equation of a line if you know the slope and a <u>point on the line</u>. **true**
6. The <u>point-slope form</u> of a linear equation is $y - y_1 = m(x - x_1)$. **true**
7. The graph of a line having a slope of 0 is a <u>vertical</u> line. **false; horizontal**
8. To determine the y-intercept of the graph of an equation, let <u>y</u> $= 0$ and solve. **false; x**
9. A <u>scatter plot</u> is a graph in which data are plotted as ordered pairs. **true**
10. Data points that appear to go uphill in a scatter plot show a <u>positive</u> relationship. **true**

Skills and Concepts

Objectives and Examples

- **Lesson 7–1** Find the slope of a line given the coordinates of two points on the line.

Determine the slope of the line through points at $(-3, 1)$ and $(7, 4)$.

$$m = \frac{y_2 - y_1}{x_2 - x_1} \quad \textit{Slope formula}$$
$$= \frac{4 - 1}{7 - (-3)} \quad (x_1, y_1) = (-3, 1), (x_2, y_2) = (7, 4)$$
$$= \frac{3}{10} \quad \textit{Simplify.}$$

The slope is $\frac{3}{10}$.

Review Exercises

Determine the slope of each line.

11. 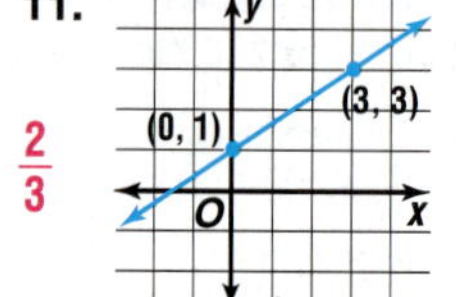

$\frac{2}{3}$

12. 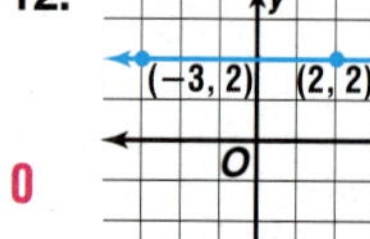

0

13. the line through (6, 0) and (2, 8) **−2**
14. the line through (−4, 9) and (−4, 5) **undefined**

www.algconcepts.com/vocabulary_review

Resource Manager

Reproducible Masters
Chapter 7 Resource Masters
- *Assessment*, pp. 325–333, 336–338

Technology/Multimedia
MindJogger Videoquizzes
ExamView® Pro

Objectives and Examples

- **Lesson 7-2** Write a linear equation in point-slope form given the coordinates of a point on the line and the slope of the line.

$y - y_1 = m(x - x_1)$ *Point-Slope Form*

$y - 8 = 2(x - 1)$ $(x_1, y_1) = (1, 8)$ *and* $m = 2$

Review Exercises

Write the point-slope form of an equation for each line passing through the given point and having the given slope.

15. $(2, 5), m = 3$ **16.** $(-1, 6), m = \frac{1}{4}$

17. $(0, 3), m = -5$ **18.** $(4, 8), m = 0$

19. Write an equation in point-slope form of a line passing through (10, 4) and (−8, 7).

15–19. See margin.

- **Lesson 7-3** Write a linear equation in slope-intercept form given the slope and y-intercept.

$y = mx + b$ *Slope-Intercept Form*

$= 4x + 12$ $m = 4$ *and* $b = 12$

slope *y-intercept*

Write an equation in slope-intercept form of the line with each slope and y-intercept.

20. $m = 1, b = 4$ **21.** $m = 3, b = -9$

22. $m = \frac{1}{2}, b = 5$ **23.** $m = 2, b = 0$

24. $m = 0, b = 7$ **25.** $m = 11, b = -6$

20–25. See margin.

- **Lesson 7-4** Graph and interpret points on scatter plots.

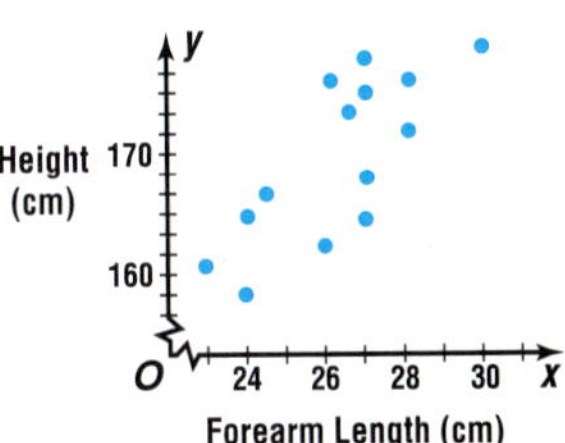

There is a positive relationship between forearm length and height of students. As forearm length increases, height increases.

Determine whether each scatter plot has a *positive* relationship, *negative* relationship, or *no* relationship. If there is a relationship, describe it.

26.

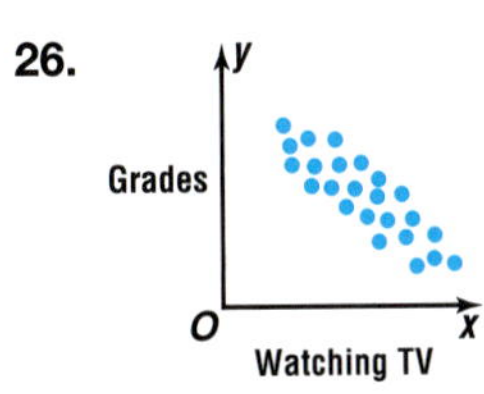

See margin.

27.

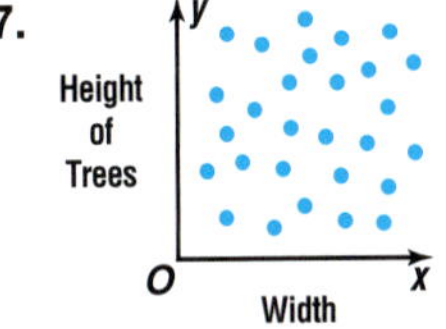

no relationship

- **Lesson 7-5** Graph linear equations by using the x- and y-intercepts or the slope and y-intercept.

slope $= \frac{2}{3}$

y-intercept $= 2$

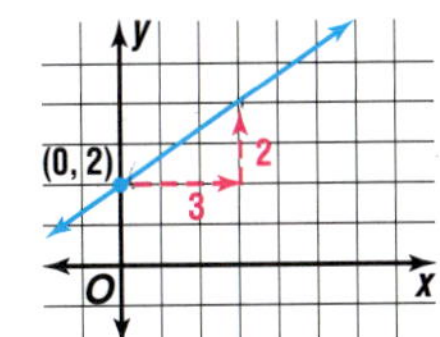

28–31. See margin for graphs.

Determine the x-intercept and y-intercept of the graph of each equation. Then graph.

28. $x + 3y = 6$ **6, 2** **29.** $2x - y = 4$ **2, −4**

Determine the slope and y-intercept of the graph of each equation. Then graph.

30. $y = x + 5$ **$m = 1$, $b = 5$** **31.** $y = -2x + 3$ **$m = -2$, $b = 3$**

Skills and Concepts

The **Objectives and Examples** section reviews the skills and concepts of the chapter and shows completely worked examples.

The **Review Exercises** provide practice for the corresponding objectives.

Answers

28.

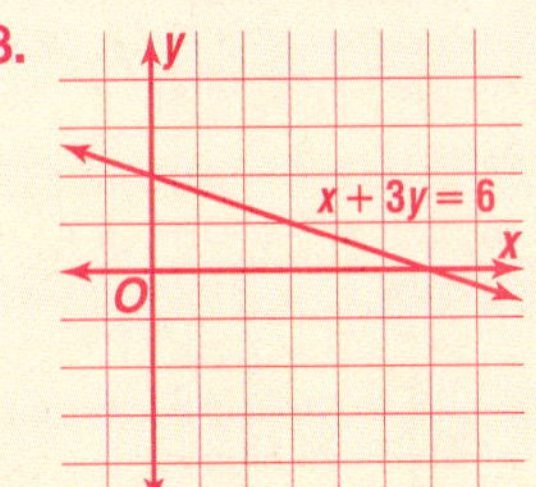

29.

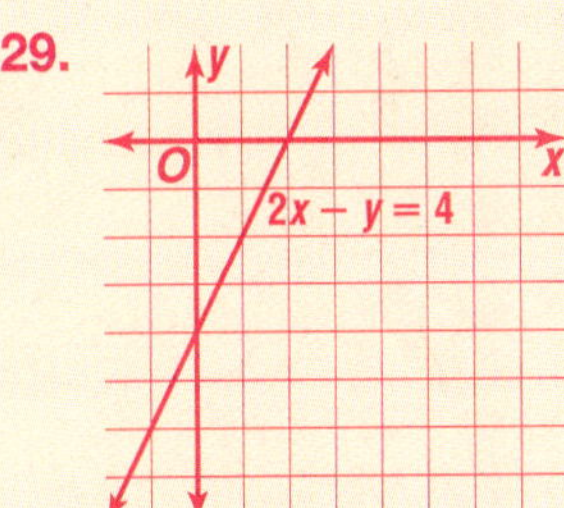

30.

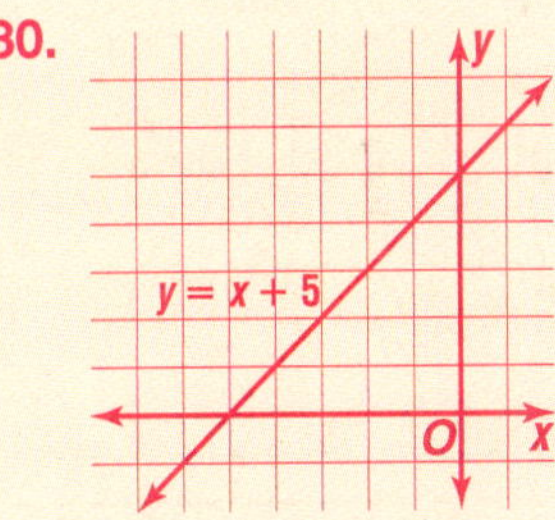

31.

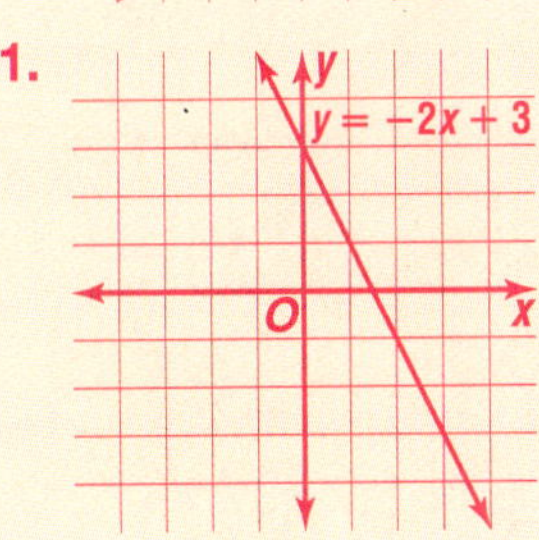

Mixed Problem Solving
See pages 724–731.

Applications and Problem Solving

This section provides additional practice in solving real-world problems that involve the concepts of this chapter.

Answers

32. same y-intercept, different slopes
33. same y-intercept, different slopes
34. same slope, different y-intercepts
35. same slope, different y-intercepts

41a.

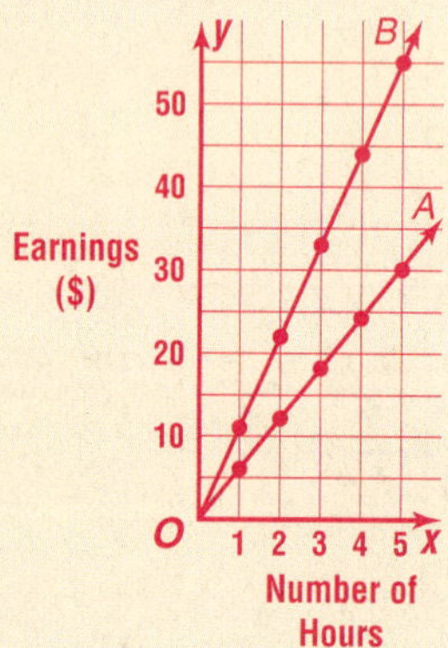

Objectives and Examples

• **Lesson 7–6** Explore the effects of changing the slopes and y-intercepts of linear functions.

The graphs have the same y-intercept and different negative slopes. They are a family of graphs.

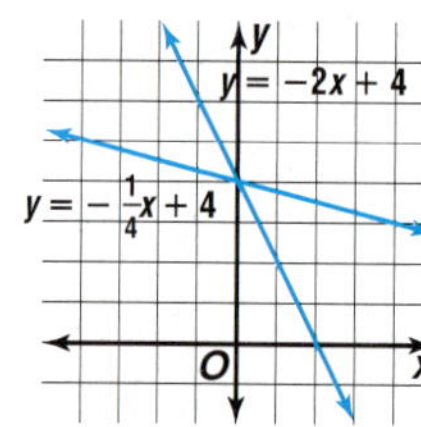

Review Exercises

Compare and contrast the graphs of each pair of equations. Verify by graphing the equations. **32–35. See margin.**

32. $y = x + 4$
$y = 3x + 4$

33. $y = \frac{1}{3}x - 1$
$y = x - 1$

34. $y = 2x$
$y = 2x + 5$

35. $\frac{1}{2}x + 3 = y$
$\frac{1}{2}x - 4 = y$

• **Lesson 7–7** Write an equation of a line that is parallel or perpendicular to the graph of a given equation and that passes through a given point.

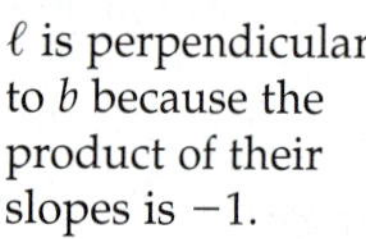

ℓ is parallel to a because their slopes are each $\frac{1}{2}$.

ℓ is perpendicular to b because the product of their slopes is -1.

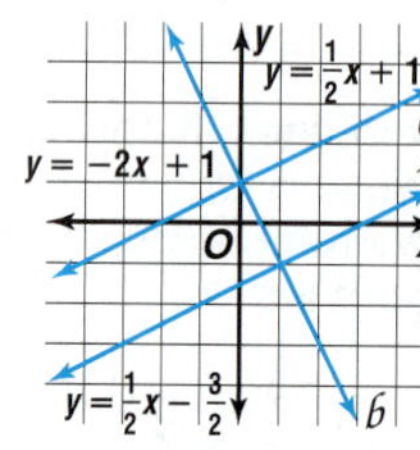

Determine whether the graphs of each pair of equations are *parallel*, *perpendicular*, or *neither*.

36. $y = 2x + 8$
$y = 2x - 4$ **parallel**

37. $y = -3x$ **neither**
$y = 3x + 1$

38. $y = \frac{1}{7}x - 1$
$y = -7x - 1$
perpendicular

39. $\frac{3}{4}x - 5 = y$
$y = \frac{3}{4}x + 6$
parallel

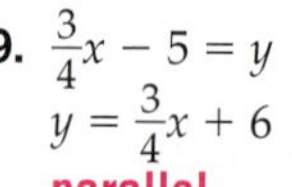

Applications and Problem Solving

40. Building Paved areas are usually slightly inclined so that puddles do not form. The pavement at the right rises 1 inch for every 25 inches of horizontal change. What is the slope? *(Lesson 7–1)* $\frac{1}{25}$

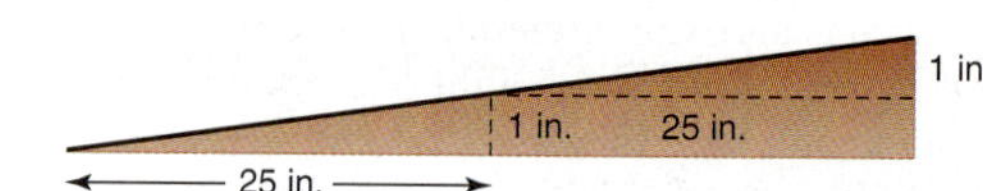

41. Employment The table shows earnings for Employees A and B. *(Lesson 7–6)*

Number of Hours	Earnings ($) A	Earnings ($) B
1	6	11
2	12	22
3	18	33
4	24	44
5	30	55

a. Sketch a graph of each employee's earnings by plotting points with the coordinates (hours, earnings). **See margin.**

b. Suppose Employee C earns more money per hour than Employee A and less money per hour than Employee B. How would a graph of this employee's earnings compare with the graphs in part a? **It would be steeper than graph A and less steep than graph B.**

Assessment

Four forms of Chapter 7 Test are available in the *Chapter 7 Resource Masters.*

	Form of Test		Level
1A	Multiple Choice	pp. 325–326	Average
1B	Multiple Choice	pp. 327–328	Basic
2A	Free Response	pp. 329–330	Average
2B	Free Response	pp. 331–332	Basic

Chapter 7 Test, Form 1B, is shown at the left. Chapter 7 Test, Form 2B, is shown on the next page.

***Assessment*, pp. 327–328**

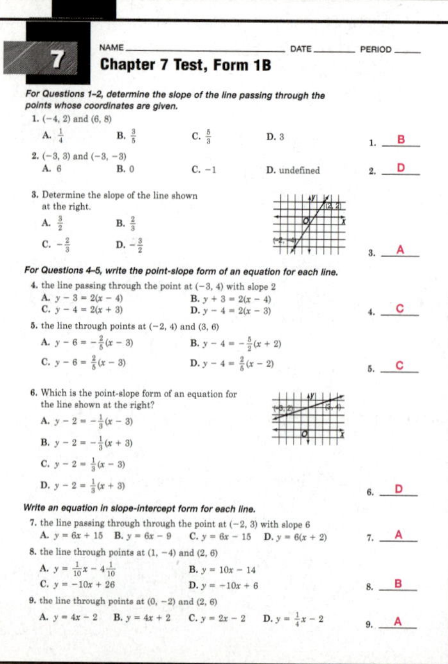

7 NAME ________ DATE ________ PERIOD ________

Chapter 7 Test, Form 1B

For Questions 1–2, determine the slope of the line passing through the points whose coordinates are given.

1. (−4, 2) and (6, 8)
A. $\frac{1}{4}$ B. $\frac{3}{5}$ C. $\frac{5}{3}$ D. 3 — 1. B

2. (−3, 3) and (−3, −3)
A. 6 B. 0 C. −1 D. undefined — 2. D

3. Determine the slope of the line shown at the right.
A. $\frac{3}{2}$ B. $\frac{2}{3}$ C. $-\frac{2}{3}$ D. $-\frac{3}{2}$ — 3. A

For Questions 4–5, write the point-slope form of an equation for each line.

4. the line passing through the point at (−3, 4) with slope 2
A. $y - 3 = 2(x - 4)$ B. $y + 3 = 2(x - 4)$
C. $y - 4 = 2(x + 3)$ D. $y - 4 = 2(x - 3)$ — 4. C

5. the line through points at (−2, 4) and (3, 6)
A. $y - 6 = -\frac{2}{5}(x - 3)$ B. $y - 4 = -\frac{5}{2}(x + 2)$
C. $y - 6 = \frac{2}{5}(x - 3)$ D. $y - 4 = \frac{2}{5}(x - 2)$ — 5. C

6. Which is the point-slope form of an equation for the line shown at the right?
A. $y - 2 = -\frac{1}{3}(x - 3)$
B. $y - 2 = -\frac{1}{3}(x + 3)$
C. $y - 2 = \frac{1}{3}(x - 3)$
D. $y - 2 = \frac{1}{3}(x + 3)$ — 6. D

Write an equation in slope-intercept form for each line.

7. the line passing through through the point at (−2, 3) with slope 6
A. $y = 6x + 15$ B. $y = 6x - 9$ C. $y = 6x - 15$ D. $y = 6(x + 2)$ — 7. A

8. the line through points at (1, −4) and (2, 6)
A. $y = \frac{1}{10}x - 4\frac{1}{10}$ B. $y = 10x - 14$
C. $y = -10x + 26$ D. $y = -10x + 6$ — 8. B

9. the line through points at (0, −2) and (2, 6)
A. $y = 4x - 2$ B. $y = 4x + 2$ C. $y = 2x - 2$ D. $y = \frac{1}{4}x - 2$ — 9. A

CHAPTER 7 Test

1. **Describe** how changing each of the following affects the graph of a linear function. **See margin.**
   a. slope
   b. y-intercept

2. **Compare and contrast** the graphs of $y = 4x - 1$, $y = x - 1$, and $y = 3x - 1$. Verify by graphing the equations. **Each graph has a positive slope and a y-intercept of -1. The graph of $y = 4x - 1$ is the steepest, and the graph of $y = x - 1$ is the least steep.**

Determine the slope of each line.

3. 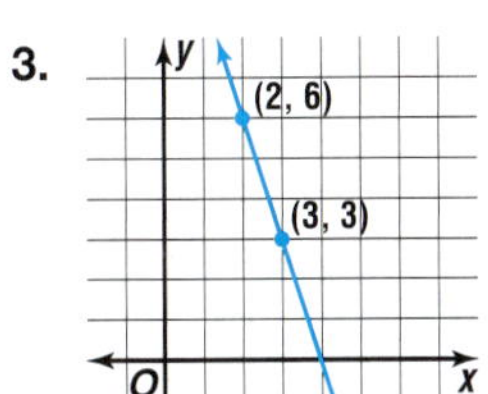

−3

4. 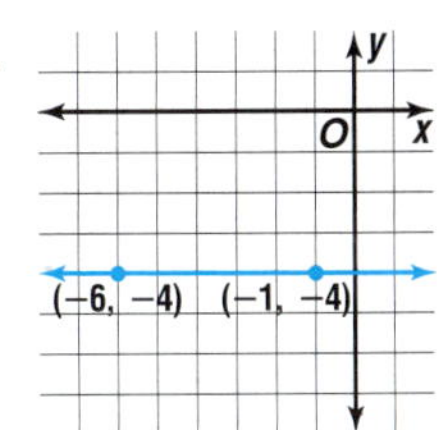

0

5. the line passing through points at $(1, -2)$ and $(6, 0)$ $\frac{2}{5}$

Write the point-slope form of an equation for each line passing through the given point and having the given slope.

6. $(5, 6)$, $m = 3$ $y - 6 = 3(x - 5)$
7. $(-3, 1)$, $m = -2$ $y - 1 = -2(x + 3)$
8. $(-4, 8)$, $m = 0$ $y - 8 = 0$ or $y = 8$

Write an equation in slope-intercept form of the line passing through each pair of points.

9. $(1, 5)$ and $(2, 8)$ $y = 3x + 2$
10. $(3, 1)$ and $(-7, 11)$ $y = -x + 4$
11. $(-4, 0)$ and $(2, 3)$ $y = \frac{1}{2}x + 2$

12. **Music** A disc jockey notices that as the music gets faster, more people start dancing. Would a scatter plot showing speed of music and number of dancers have a *positive* relationship, *negative* relationship, or *no* relationship? **positive relationship**

Graph each equation. **13–18. See Solutions Manual.**

13. $x + y = 2$
14. $x + 3y = 3$
15. $2x - 4y = 12$
16. $y = 2x - 5$
17. $y = \frac{1}{3}x + 2$
18. $y = -1$

19. Are the graphs of the equations *parallel*, *perpendicular*, or *neither*?
$y = -4x + 9$
$y = \frac{1}{4}x - 6$ **perpendicular**

20. **Construction** The *pitch* of a roof describes its steepness. Suppose a roof is 40 feet wide and its pitch is $\frac{1}{2}$. Find x, its height above the rafters at its peak. (*Hint*: Use $\frac{\text{rise}}{\text{run}}$.) **10 ft**

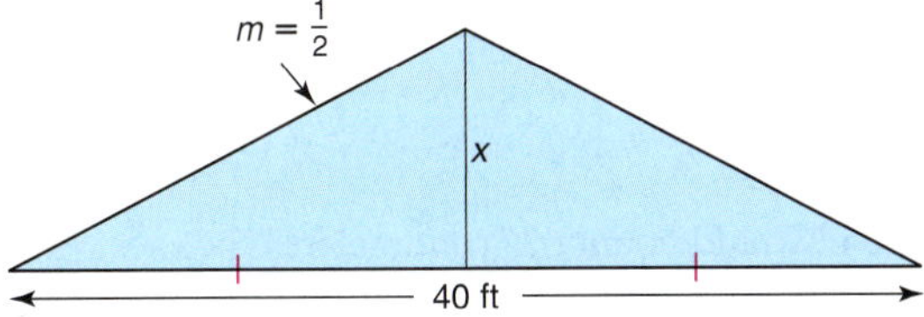

www.algconcepts.com/chapter_test

Chapter Test Bonus Question

On a coordinate plane, the lines containing two of the sides of a square have equations $y = x$ and $y = x + 2$. If the points at $(0, 0)$ and $(0, 2)$ are two of the vertices of the square, write equations for the lines containing the other two sides and find the coordinates of the other two vertices.
$y = -x$, $y = -x + 2$, $(1, 1)$, $(-1, 1)$

Chapter Test

Answer

1. Sample answer:
   a. Changing the slope affects the steepness of the graph. If the value of a positive slope increases or the value of a negative slope decreases, the graph becomes steeper.
   b. Changing the y-intercept shifts the graph up or down. If the value of the y-intercept increases, the graph shifts up.

Assessment, pp. 331–332

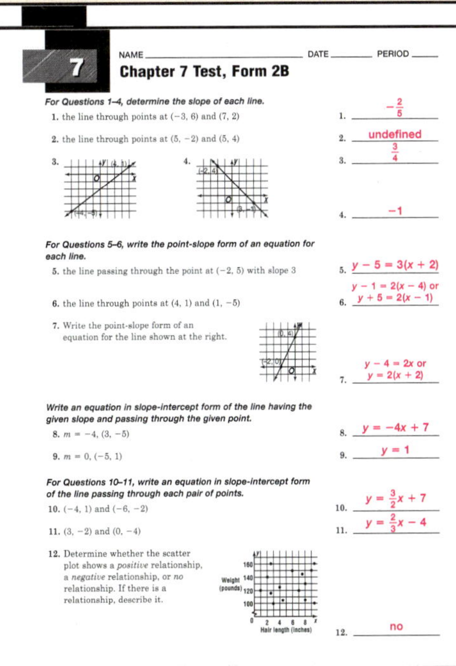

7 NAME ______ DATE ______ PERIOD ______

Chapter 7 Test, Form 2B

For Questions 1–4, determine the slope of each line.

1. the line through points at $(-3, 6)$ and $(7, 2)$ 1. $-\frac{2}{5}$
2. the line through points at $(5, -2)$ and $(5, 4)$ 2. undefined
3. (graph: $(4, 1)$, $(-4, -5)$) 3. $\frac{3}{4}$
4. (graph: $(-2, 4)$, $(3, -1)$) 4. -1

For Questions 5–6, write the point-slope form of an equation for each line.

5. the line passing through the point at $(-2, 5)$ with slope 3 5. $y - 5 = 3(x + 2)$
6. the line through points at $(4, 1)$ and $(1, -5)$ 6. $y - 1 = 2(x - 4)$ or $y + 5 = 2(x - 1)$
7. Write the point-slope form of an equation for the line shown at the right. (graph: $(0, 4)$, $(-2, 0)$) 7. $y - 4 = 2x$ or $y = 2(x + 2)$

Write an equation in slope-intercept form of the line having the given slope and passing through the given point.

8. $m = -4$, $(3, -5)$ 8. $y = -4x + 7$
9. $m = 0$, $(-5, 1)$ 9. $y = 1$

For Questions 10–11, write an equation in slope-intercept form of the line passing through each pair of points.

10. $(-4, 1)$ and $(-6, -2)$ 10. $y = \frac{3}{2}x + 7$
11. $(3, -2)$ and $(0, -4)$ 11. $y = \frac{2}{3}x - 4$
12. Determine whether the scatter plot shows a *positive* relationship, a *negative* relationship, or *no* relationship. If there is a relationship, describe it. (scatter plot: Weight (pounds) vs. Hair length (inches)) 12. no

Preparing for Standardized Tests

Pages 332–333 are part of a complete test preparation course that is described in detail on page T9 of the Teacher's Handbook. The test items on these pages were written in the same style as those in state proficiency tests and standardized tests like ACT and SAT.

Diagnosis and Prescription

Each of the 10 test questions on page 333 is cross-referenced to the lesson where that SAT or ACT skill is covered. If students miss a particular type of problem, you can have them study that skill.

(See chart at the bottom of page 333. Note that SPT = State Proficiency Test, SAT = Scholastic Assessment Test, and ACT = American College Test.)

Assessment, p. 336

7

NAME ________________ DATE ________ PERIOD ______

Chapter 7 Cumulative Review

1. Write an equation for the sentence below. *(Lesson 1–1)*
The difference of twice a number and 16 is equal to the quotient of three times the number and 8.
1. $2n - 16 = \frac{3n}{8}$

2. Yoshi's grade will be attending a brand new school next year. Students who will be attending the school are going to vote on the mascot for the school. To predict which mascot students will choose, Yoshi asks all of the students at her lunch table for their preferences. Is this a representative sample? Why or why not? *(Lesson 1–6)*
2. Sample answer: No; students at the same table may have similar interests, so their opinions may not represent the views of students as a whole.

For Questions 3–5, evaluate each expression.

3. $6 - (-11)$ *(Lesson 2–4)* — 3. 17
4. $-48 \div (-12)$ *(Lesson 2–6)* — 4. 4
5. $-1.8 - 3.2 + 2.4$ *(Lesson 3–2)* — 5. −2.6
6. Find the solution of $x = 16 \div 2 + 2 \cdot 5$. *(Lesson 3–4)* — 6. 18

For Questions 7–8, solve each equation.

7. $-8.2n = -41$ *(Lesson 4–4)* — 7. 5
8. $6 + 4(x + 2) = 18$ *(Lesson 4–7)* — 8. 1
9. A scale model of a house is 18 inches wide. The actual house is 48 feet wide. Find the scale of the model. *(Lesson 5–2)* — 9. 1 in. = $2\frac{2}{3}$ ft or 1:32
10. 80% of what number is 33.6? *(Lesson 5–3)* — 10. 42
11. Determine whether the relation {(3, 2), (4, 5), (−3, 5), (−4, 2)} is a function. *(Lesson 6–4)* — 11. yes
12. Suppose y varies directly as x and $y = 27$ when $x = 3$. Find x when $y = 45$. *(Lesson 6–5)* — 12. 5
13. Suppose y varies inversely as x and $y = 24$ when $x = 3$. Find x when $y = 36$. *(Lesson 6–6)* — 13. 2
14. Write the point-slope form of an equation for the line passing through the point at (3, −6) with slope $-\frac{4}{3}$. *(Lesson 7–2)* — 14. $y + 6 = -\frac{4}{3}(x - 3)$
15. Write an equation in slope-intercept form for the line through points at (2, −4) and (6, −2). *(Lesson 7–3)* — 15. $y = \frac{1}{2}x - 5$
16. Determine the slope and y-intercept of the graph of $12 + 3y = -2x$. *(Lesson 7–5)* — 16. $m = -\frac{2}{3}$; $b = -4$

CHAPTER 7 Preparing for Standardized Tests

Algebra Word Problems

All standardized tests involve writing and solving equations from realistic settings. You'll need to translate words into equations. You'll need to find patterns in number tables and write equations to represent the patterns.

Test-Taking Tip

When a question includes a variable, you can "plug-in" a number for the variable that fits the problem.

Example 1

Ami dropped a rubber ball from several heights and measured how high the ball bounced. Her results are shown in the table. How high would the ball bounce if it were dropped from a height of 36 centimeters?

Height of Drop (cm)	Bounce Height (cm)
40	30
60	45
80	60
100	75

Ⓐ 19 cm Ⓑ 24 cm Ⓒ 27 cm Ⓓ 30 cm

Hint Look for a pattern in the table.

Solution The bounce is 30 when the drop height is 40. The bounce is 75 when the drop height is 100. So, the bounce is $\frac{3}{4}$ of the drop height. Check that this pattern is true for each of the other numbers in the table.

Let b represent the bounce height and let d represent the drop height.

$b = \frac{3}{4}d$ *Write an equation.*

$= \frac{3}{4}(36)$ *Replace d with 36.*

$= \frac{108}{4}$ or 27 *Simplify.*

The ball would bounce 27 centimeters. The answer is C.

Example 2

After N chocolate bars are divided equally among 6 children, 3 bars remain. How many would remain if $(N + 4)$ chocolate bars were divided equally among 6 children?

Ⓐ 0 Ⓑ 1 Ⓒ 2 Ⓓ 3 Ⓔ 4

Hint Sketch a diagram to understand the problem. Use the "plug-in" strategy.

Solution One way to solve this kind of problem is to choose a numeric value for N and "plug it in." Suppose each of the 6 children got 2 chocolate bars, and 3 bars remain. Then $N = 6(2) + 3 = 15$.

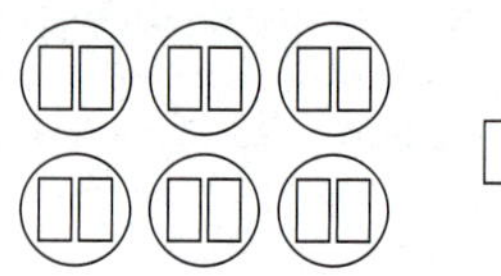 

Now, use the value of 15 for N.

$N + 4 = 15 + 4$ or 19 bars

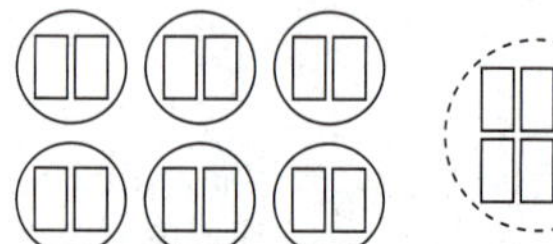

Divide 19 bars among 6 children.

$19 = 18 + 1$ or $3(6) + 1$

The remainder is 1. The answer is B.

Resource Manager

Reproducible Masters

Chapter 7 Resource Masters

- *Assessment*, pp. 336–338

After you work each problem, record your answer on the answer sheet provided or on a sheet of paper.

Multiple Choice

1. Refer to the table and the equation in Example 1 on the opposite page. From what height should Ami drop the ball so that the bounce height is 105 centimeters? *(Lesson 6–1)* **B**
   - (A) 110 cm
   - (B) 140 cm
   - (C) 180 cm
   - (D) 200 cm

2. A cable TV company charges $21.95 per month for basic service. Each premium channel selected costs an additional $5.95 per month. If x represents the number of premium channels selected, which expression can be used to find the monthly cost of cable service? *(Lesson 1–1)* **B**
   - (A) $5.95 + 21.95x$
   - (B) $21.95 + 5.95x$
   - (C) $21.95 + 5.95 + x$
   - (D) $21.95 - 5.95x$

3. Steve ran a 12-mile race at an average speed of 8 miles per hour. If Adam ran the same race at an average speed of 6 miles per hour, how many minutes longer than Steve did Adam take to complete the race? *(Lesson 6–6)*
   - (A) 9
   - (B) 12
   - (C) 16
   - (D) 24
   - (E) 30

   B

4. The bar graph shows the number of people who watch prime time television each day of the week. About how many more people watch on Monday than on Saturday? *(Lesson 1–6)* **B**

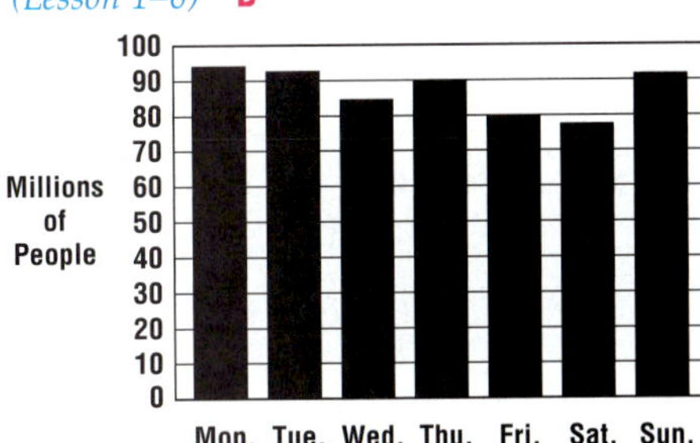

   - (A) 5 million
   - (B) 15 million
   - (C) 5 million
   - (D) 50 million

www.algconcepts.com/standardized_test

5. Kim earns $7 per hour baby-sitting. If x represents the number of hours Kim baby-sits, which expression could be used to find the amount she earns? *(Lesson 1–1)* **C**
   - (A) $x + 7$
   - (B) $x + 5$
   - (C) $7x$
   - (D) $5x$

6. If Nathan is $\frac{1}{4}$ as old as his father and the sum of their ages is 60, then how old is Nathan? *(Lesson 4–5)* **B**
   - (A) 8
   - (B) 12
   - (C) 15
   - (D) 20
   - (E) 48

7. There are 5 different pizza toppings. How many different 3-topping pizzas are possible? *(Lesson 4–2)* **A**
   - (A) 10
   - (B) 15
   - (C) 60
   - (D) 125

8. A group of 5 adults and 3 children sees a play. A child's ticket costs $6.25, and an adult's ticket costs $9.75. Which equation can be used to find c, the amount of change from $100 after paying for the group's tickets? *(Lesson 1–1)* **C**
   - (A) $5(6.25) + 3(9.75) = 100 - c$
   - (B) $5(9.75) + 3(6.25) = c$
   - (C) $5(9.75) + 3(6.25) + c = 100$
   - (D) $5(9.75) + 3(6.25) = 100 + c$

Grid In

9. What number increased by two is equal to two less than twice the number? *(Lesson 4–6)* **4**

Extended Response

10. The fare charged by a taxi driver is a $3 fixed charge plus $0.35 per mile. Beth pays $10 for a ride of m miles. *(Lesson 7–3)*
    $3 + 0.35m = 10$

    Part A Write an equation that can be used to find m. Show your work.

    Part B Use the equation in Part A to find how many miles Beth rode. Show your work. **20 mi**

A bubble-in answer sheet for these practice problems is available on page A1 of the *Chapter 7 Resource Masters.*

Additional Practice

Additional test practice questions are available in the *Chapter 7 Resource Masters*, pp. 337–338.

Chapter 7 Algebra Word Problems		
Ex. 1	write an equation from table	SPT
Ex. 2	algebra word problem	SAT
1	use an equation	SPT
2	write an expression	SPT
3	rate word problem	SAT
4	use a bar graph	SPT
5	write an expression	SPT
6	word problem	ACT
7	combinations	SPT
8	write an expression	SPT
9	algebra word problem	SAT
10	write and use an equation	SPT

***Assessment*, pp. 337–338**

7 NAME ____________ DATE ________ PERIOD ______

Chapter 7 Standardized Test Practice
(Chapters 1–7)

Write the letter for the correct answer in the blank at the right of each problem.

1. Which of the following is an example of the Associative Property of Addition?
   A. $(h + 2) + 3 = h + (2 + 3)$ B. $(h \cdot 2) \cdot 3 = h \cdot (2 \cdot 3)$
   C. $h + 2 = 2 + h$ D. $2(h + 2) = 2h + 4$ 1. A

2. How many ways are there to make a sum of 12 using the numbers 1, 2, and 5?
   A. 5 B. 8 C. 12 D. 13 2. D

3. Jerome's middle school has 280 students. Which is the best way to display data of the heights to the nearest centimeter of all the students in the school?
   A. line graph B. stem-and-leaf plot
   C. histogram D. frequency table 3. C

4. Evaluate $x - z + w$ if $x = -4$, $z = -6$, and $w = -8$.
   A. -18 B. -10 C. -6 D. -2 4. C

5. Evaluate $\frac{xy}{-4}$ if $x = -2$ and $y = 12$.
   A. -6 B. $-\frac{5}{2}$ C. $\frac{5}{2}$ D. 6 5. D

6. Find $-2\frac{1}{3} - 3\frac{1}{6} + 1\frac{5}{6}$.
   A. -4 B. $-3\frac{2}{3}$ C. $-3\frac{1}{2}$ D. -1 6. B

7. Find the mean of the data below, which give the number of days with precipitation recorded in a town over a 10-month period.
   10, 12, 7, 4, 6, 8, 10, 14, 9, 10
   A. 9 B. 9.5 C. 10 D. 12 7. A

8. For what value of x is the equation $\frac{4}{5} = x - \left(-\frac{1}{5}\right)$ true?
   A. $-\frac{3}{5}$ B. $\frac{4}{25}$ C. $\frac{3}{5}$ D. 1 8. C

9. Evaluate $\frac{x}{3}$ if $x = \frac{2}{5}$.
   A. $\frac{2}{15}$ B. $\frac{1}{3}$ C. 1 D. $1\frac{1}{5}$ 9. A

10. Find the value of s if the perimeter of the square is 36 inches. $(s + 4)$ m
    A. 2 B. 4
    C. 5 D. 6 10. C

CHAPTER 8

Resource Manager

Powers and Roots

Instructional Objectives

Lesson (pages)	Objectives	NCTM Standards 2000	State/Local Objectives
8–1 (336–340)	Use powers in expressions.	1, 6, 8, 9	
8–2 (341–345)	Multiply and divide powers.	1, 6, 8, 9	
8–3 (347–351)	Simplify expressions containing negative exponents.	1, 6, 8, 9	
8–4 (352–356)	Express numbers in scientific notation.	1, 2, 6, 8, 9, 10	
8–5 (357–361)	Simplify radicals by using the Product and Quotient Properties of Square Roots.	1, 6, 8, 9	
8–6 (362–365)	Estimate square roots.	1, 6, 8, 9	
8–7 (366–371)	Use the Pythagorean Theorem to solve problems.	1, 3, 4, 6, 8, 9	
Investigation (372–373)	Explore the Pythagorean Theorem.	1, 3, 4, 6, 7, 8, 9, 10	

Key to NCTM Standards 2000
1 Number & Operations; **2** Algebra; **3** Geometry; **4** Measurement; **5** Data Analysis & Probability; **6** Problem Solving; **7** Reasoning & Proof; **8** Communications; **9** Connections; **10** Representation

Suggested Pacing

See page T13 for a complete course-planning calendar.

Standard refers to schedules that provide 45- to 55-minute periods that meet each day.
Block refers to schedules that provide approximately 90-minute periods which may meet every day for one semester or every other day over two semesters.

PACING	DAY 1	DAY 2	DAY 3	DAY 4	DAY 5	DAY 6
Standard Core (Chapters 1–13)	Lesson 8–1	Lesson 8–2		Lesson 8–3	Lesson 8–4	
Standard Enhanced (Chapters 1–15)	Lesson 8–1	Lesson 8–2		Lesson 8–3	Lesson 8–4	Lesson 8–5
Block Core (Chapters 1–13)	Chapter 7 Test & Lesson 8–1	Lesson 8–2	Lessons 8–3 & 8–4	Lessons 8–5 & 8–6	Lesson 8–7 & INV	SG+A
Block Enhanced (Chapters 1–15)	Chapter 7 Test & Lesson 8–1	Lessons 8–2 & 8–3	Lessons 8–4 & 8–5	Lessons 8–6 & 8–7	INV & SG+A	Chapter Test & Lesson 9–1

Instructional Resources

Lesson	Materials and Manipulatives (see below for Glencoe Manipulative Resources)	Blackline Masters (page numbers): Chapter 8 Resource Masters: Study Guide	Practice (Skills & Average)	Reading to Learn Mathematics	Enrichment	Assessment	Hands-On Algebra*	School-to-Workplace*	Graphing Calculator Masters*	5-Minute Check Transparencies
8–1	graphing calculator coordinate planes [4] straightedge [1, 2, 4] index cards	339	340–341	342	343				23	8–1
8–2		344	345–346	347	348		90			8–2
8–3	two-inch squares of paper scissors [1, 2]	349	350–351	352	353	385	91			8–3
8–4	calculators masking tape	354	355–356	357	358	384				8–4
8–5	tape measures	359	360–361	362	363				24, 25	8–5
8–6	base-ten models [1] calculators	364	365–366	367	368		92			8–6
8–7	dot paper scissors [1, 2]	369	370–371	372	373	385	93, 94	8		8–7
Investigation	uncooked spaghetti rulers [1, 2, 4] grid paper [1, 4] scissors [1, 2] calculator									
Study Guide & Assessment/ Chapter Test						375–383, 386–388				

**See page 334c for examples of these instructional materials.*

Key to Glencoe Manipulative Resources
[1]Classroom Manipulative Resources [2]Student Manipulative Resources [3]Overhead Manipulative Resources [4]Hands-On Algebra Masters

INV = Investigation *SG+A = Study Guide and Assessment*

<table>
<tr><th>DAY 7</th><th>DAY 8</th><th>DAY 9</th><th>DAY 10</th><th>DAY 11</th><th>DAY 12</th><th>DAY 13</th></tr>
<tr><td>Lesson 8–5</td><td>Lesson 8–6</td><td colspan="2">Lesson 8–7</td><td>INV</td><td>SG+A</td><td>Chapter Test</td></tr>
<tr><td>Lesson 8–6</td><td colspan="2">Lesson 8–7</td><td>INV</td><td>SG+A</td><td>Chapter Test</td><td></td></tr>
<tr><td>Chapter Test & Lesson 9–1</td><td></td><td></td><td></td><td></td><td></td><td></td></tr>
<tr><td></td><td></td><td></td><td></td><td></td><td></td><td></td></tr>
</table>

Resource Manager

The pages shown on this page are a small sample of the materials available on *TeacherWorks: All-in-One Lesson Planner and Resource Center.*

This CD-ROM includes all of the blackline masters and transparencies available for this program.

It also includes a lesson planner and interactive Teacher Edition, so you can customize lesson plans and reproduce classroom resources quickly and easily, from just about anywhere.

Applications

School-to-Workplace Masters, p. 8

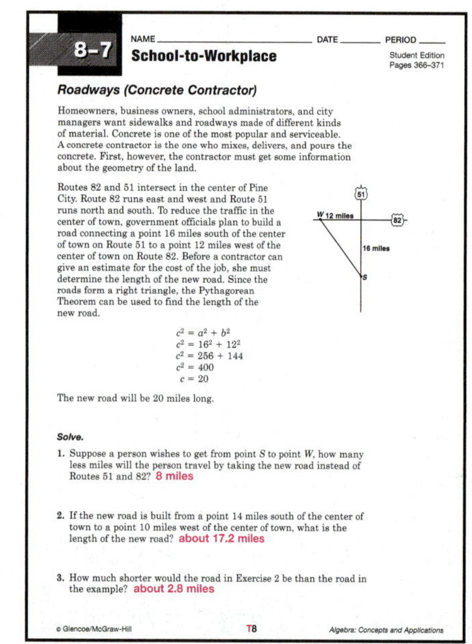

8–7 NAME ________ DATE ______ PERIOD ______

School-to-Workplace

Student Edition Pages 366–371

Roadways (Concrete Contractor)

Homeowners, business owners, school administrators, and city managers want sidewalks and roadways made of different kinds of material. Concrete is one of the most popular and serviceable. A concrete contractor is the one who mixes, delivers, and pours the concrete. First, however, the contractor must get some information about the geometry of the land.

Routes 82 and 51 intersect in the center of Pine City. Route 82 runs east and west and Route 51 runs north and south. To reduce the traffic in the center of town, government officials plan to build a road connecting a point 16 miles south of the center of town on Route 51 to a point 12 miles west of the center of town on Route 82. Before a contractor can give an estimate for the cost of the job, she must determine the length of the new road. Since the roads form a right triangle, the Pythagorean Theorem can be used to find the length of the new road.

$$c^2 = a^2 + b^2$$
$$c^2 = 16^2 + 12^2$$
$$c^2 = 256 + 144$$
$$c^2 = 400$$
$$c = 20$$

The new road will be 20 miles long.

Solve.

1. Suppose a person wishes to get from point S to point W, how many less miles will the person travel by taking the new road instead of Routes 51 and 82? 8 miles
2. If the new road is built from a point 14 miles south of the center of town to a point 10 miles west of the center of town, what is the length of the new road? about 17.2 miles
3. How much shorter would the road in Exercise 2 be than the road in the example? about 2.8 miles

 T8 *Algebra: Concepts and Applications*

Manipulatives/Modeling

Hands-On Algebra Masters, pp. 90–94

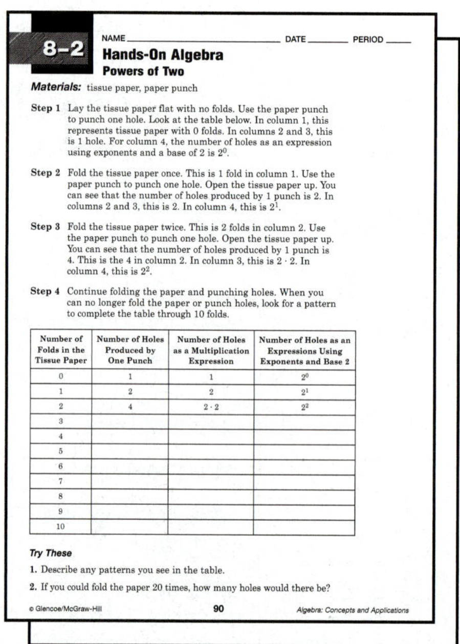

8–2 NAME ________ DATE ______ PERIOD ______

Hands-On Algebra
Powers of Two

Materials: tissue paper, paper punch

Step 1 Lay the tissue paper flat with no folds. Use the paper punch to punch one hole. Look at the table below. In column 1, this represents tissue paper with 0 folds. In columns 2 and 3, this is 1 hole. For column 4, the number of holes as an expression using exponents and a base of 2 is 2^0.

Step 2 Fold the tissue paper once. This is 1 fold in column 1. Use the paper punch to punch one hole. Open the tissue paper up. You can see that the number of holes produced by 1 punch is 2. In columns 2 and 3, this is 2. In column 4, this is 2^1.

Step 3 Fold the tissue paper twice. This is 2 folds in column 2. Use the paper punch to punch one hole. Open the tissue paper up. You can see that the number of holes produced by 1 punch is 4. This is the 4 in column 2. In column 3, this is $2 \cdot 2$. In column 4, this is 2^2.

Step 4 Continue folding the paper and punching holes. When you can no longer fold the paper or punch holes, look for a pattern to complete the table through 10 folds.

Number of Folds in the Tissue Paper	Number of Holes Produced by One Punch	Number of Holes as a Multiplication Expression	Number of Holes as an Expressions Using Exponents and Base 2
0	1	1	2^0
1	2	2	2^1
2	4	$2 \cdot 2$	2^2
3			
4			
5			
6			
7			
8			
9			
10			

Try These

1. Describe any patterns you see in the table.
2. If you could fold the paper 20 times, how many holes would there be?

 90 *Algebra: Concepts and Applications*

Technology/Multimedia

Graphing Calculator Masters, pp. 23–25

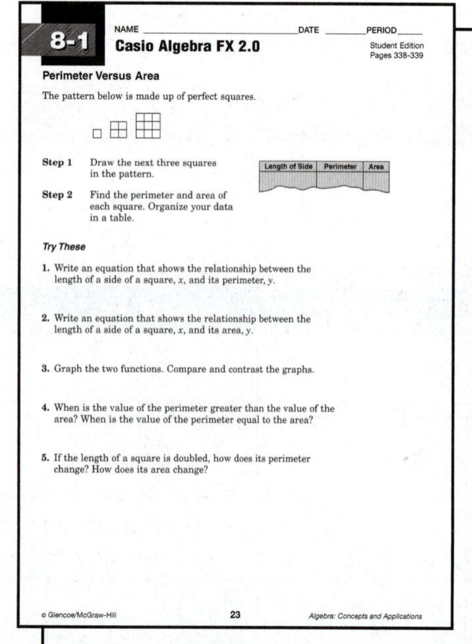

8-1 NAME ________ DATE ______ PERIOD ______

Casio Algebra FX 2.0

Student Edition Pages 338-339

Perimeter Versus Area

The pattern below is made up of perfect squares.

Step 1 Draw the next three squares in the pattern.

Step 2 Find the perimeter and area of each square. Organize your data in a table.

Length of Side	Perimeter	Area

Try These

1. Write an equation that shows the relationship between the length of a side of a square, x, and its perimeter, y.
2. Write an equation that shows the relationship between the length of a side of a square, x, and its area, y.
3. Graph the two functions. Compare and contrast the graphs.
4. When is the value of the perimeter greater than the value of the area? When is the value of the perimeter equal to the area?
5. If the length of a square is doubled, how does its perimeter change? How does its area change?

 23 *Algebra: Concepts and Applications*

8-5 NAME ________ DATE ______ PERIOD ______

TI-83/84 Plus

Graphing Radical Functions

You can use a TI-83/84 Plus calculator to graph functions that involve square roots.

Graph $y = \sqrt{x}$.

Step 1 In the Y= screen, enter [2nd] [√] [X,T,θ,n].

Step 2 Press [ZOOM] 6 to display the graph in the standard viewing window.

Y1=√(X
X=0 Y=0

Try These

1. Trace the graph of $y = \sqrt{x}$ to see what happens for negative values of x. What do you notice about the coordinates displayed at the bottom of the screen?
2. What does your answer for Exercise 1 tell you about the square roots of negative numbers?
3. Press [Y=]. Now enter the equation $y = \sqrt{x^2}$ as Y_1. Press [2nd] [√] [X,T,θ,n] [x^2] [GRAPH]. Describe the graph.
4. On the same screen, graph $y = |x|$ as Y_2. Press [MATH] [▶] 1 [X,T,θ,n] [GRAPH]. Describe the graph.
5. Press [2nd] [QUIT] 1 to go to the home screen. Evaluate $\sqrt{x^2}$ and $|x|$ for $x = 4$. Are the values the same? Choose several other values for x and evaluate both equations. Are the values the same for each value of x that you use? Do you think this will be true for every value of x?

 24 *Algebra: Concepts and Applications*

Assessment Resources

Type	Student Edition	Teacher's Wraparound Edition	Chapter 8 Resource Masters
Ongoing Assessment	Quizzes 1 and 2, pp. 351, 361	5-Minute Check, pp. 336, 341, 347, 352, 357, 362, 366	Mid-Chapter Test, p. 384 Quizzes A and B, p. 385
Mixed Review	Mixed Review, pp. 340, 345, 351, 356, 361, 365, 371 Standardized Test Practice, Chapters 1–8, pp. 378–379		Cumulative Review, p. 386 Standardized Test Practice, pp. 387–388
Error Analysis	You Decide, pp. 339, 344, 350	Error Analysis, pp. 339, 344, 349, 355, 360, 364, 369	
Standardized Test Prep	Standardized Test Practice, pp. 340, 345, 351, 356, 361, 365, 371 Standardized Test Practice, Chapters 1–8, pp. 378–379		Standardized Test Practice, pp. 387–388
Open-Ended Assessment	Writing Math, pp. 355, 360 Investigation, pp. 372–373 Portfolio, pp. 335, 373	Modeling: pp. 361, 371 Speaking: pp. 340, 351, 365 Writing: pp. 345, 356	Extended Response Assessment, p. 383
Chapter Assessment	Study Guide and Assessment, pp. 374–376 Chapter Test, p. 377		Multiple-Choice Tests (Forms 1A, 1B), pp. 375–378 Free-Response Tests (Forms 2A, 2B), pp. 379–383

Additional Chapter Resources

Student Edition
Math in the Workplace, p. 346
Hands-On Algebra, p. 362
Graphing Calculator Exploration, pp. 338–339

Teacher's Classroom Resources
Manipulatives/Modeling
Teacher's Guide for Overhead Manipulative Resources

Meeting Individual Needs
Prerequisite Skills Booklet
Spanish Study Guide and Assessment, pp. 56–62, 119–120

Teaching Aids
Solutions Manual
5-Minute Check Transparencies

Glencoe Technology

Instructional
AlgePASS CD-ROM, Lessons 18–20
Interactive Chalkboard CD-ROM
StudentWorks Plus CD-ROM
Multimedia Applications CD-ROM, Activity 9
Vocabulary PuzzleMaker (online)

Assessment
ExamView® Pro

Teaching Aids
Answer Key Maker CD-ROM
TeacherWorks CD-ROM

*Visit **www.algconcepts.com** for data updates, career information, games, and other interactive activities.*

CHAPTER 8 PREVIEW

Mathematics of the Chapter

This chapter provides students with an exploration of whole-number exponents and square roots, and with an introduction to exponents that are negative integers or zero. After introducing whole-number exponents, students multiply and divide powers, rewrite powers that have negative exponents, and use scientific notation. Then students explore simplifying square roots and estimating the square root of a number. The chapter concludes with the discussion of the Pythagorean Theorem and its converse.

What You'll Learn

Have students read over the lists of key ideas and key vocabulary. Have them make a list of any words with which they are not familiar.

Why It's Important

Point out to students that this is only one of many reasons why each objective is important. Others are provided in each lesson.

CHAPTER

Powers and Roots

What You'll Learn

Key Ideas

- Use, multiply, and divide powers. *(Lessons 8–1 and 8–2)*
- Simplify expressions containing negative exponents. *(Lesson 8–3)*
- Express numbers in scientific notation. *(Lesson 8–4)*
- Simplify radicals. *(Lesson 8–5)*
- Estimate square roots. *(Lesson 8–6)*
- Use the Pythagorean Theorem. *(Lesson 8–7)*

Key Vocabulary

exponent *(p. 336)*
irrational numbers *(p. 362)*
power *(p. 336)*
radical expression *(p. 358)*
square root *(p. 357)*

Why It's Important

Firefighting Every year, fires take thousands of lives and destroy property worth billions. Firefighters help protect us from these dangers. They are frequently the first emergency personnel at a traffic accident or medical emergency to put out a fire, treat injuries, or perform other vital functions.

Functions involving powers and roots are used to describe scientific concepts like motion. You will use a square root function to find the velocity of water in a fire hose in Lesson 8–5.

CHAPTER 8 LINKS

Lesson	8–1	8–2	8–3	8–4	8–5	8–6	8–7
Math in the Workplace	Landscaping	Movie Industry Broadcast Technician	Electronics	Fiber Optics	Aviation	Law	Carpentry Enforcement
Applications and Connections	Science	Earth Science Manufacturing	Biology Physics	Physics Health Biology Astronomy Electronics	Buildings Firefighting Business Architecture	Gardening Meteorology Aviation Travel	Construction Architecture Finance
Math Integration	Number Theory Geometry	Measurement Geometry	Number Theory		Geometry	Geometry	Geometry

Study these lessons to improve your skills.

✓Check Your Readiness

Lesson 2-5, pp. 75-79

Find each product. 6. −32

1. $6 \cdot 6$ 36
2. $10 \cdot 10 \cdot 10$ 1000
3. $(-3)(-3)$ 9
4. $4 \cdot 4 \cdot 4$ 64
5. $5 \cdot 5 \cdot 5 \cdot 5$ 625
6. $(-2)(-2)(-2)(-2)(-2)$
7. $(-3)(-3)(5 \cdot 5)$ 225
8. $-4(6 \cdot 6)$ −144
9. $3(-4)(2)(2)$ −48

Prerequisite Skill, p. 685

Write each fraction in simplest form.

10. $\frac{2}{6}$ $\frac{1}{3}$
11. $\frac{4}{64}$ $\frac{1}{16}$
12. $\frac{24}{32}$ $\frac{3}{4}$
13. $\frac{24}{4}$ 6
14. $\frac{15}{10}$ $\frac{3}{2}$
15. $\frac{9}{72}$ $\frac{1}{8}$
16. $\frac{12}{28}$ $\frac{3}{7}$
17. $\frac{27}{9}$ 3
18. $\frac{12}{60}$ $\frac{1}{5}$

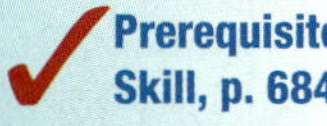
Prerequisite Skill, p. 684

Find each product or quotient.

19. $2.5 \cdot 3$ 7.5
20. $1.6 \cdot 4$ 6.4
21. $5 \cdot 1.2$ 6
22. $7.8 \cdot 1.1$ 8.58
23. $9.4 \cdot 3.2$ 30.08
24. $\frac{4.8}{1.6}$ 3
25. $\frac{10.8}{2.4}$ 4.5
26. $\frac{3.6}{0.6}$ 6
27. $\frac{9.2}{2}$ 4.6

Make this Foldable to help you organize your Chapter 8 notes. Begin with four sheets of grid paper.

❶ **Fold** each sheet in half along the width. Then cut along the crease.

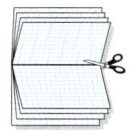

❷ **Staple** the eight half-sheets together to form a booklet.

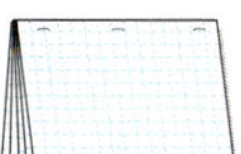

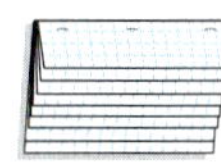

❸ **Cut** seven lines from the bottom of the top sheet, six lines from the second sheet, and so on.

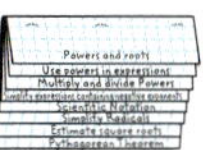

❹ **Label** the tabs with lesson topics as shown.

Reading and Writing As you read and study the chapter, use each page to write notes and graph examples for each lesson.

www.algconcepts.com/chapter_readiness

FOLDABLES™ Study Organizer Students should use the tabs of this Foldable to write rules and properties for working with powers. They should also record an example of each type of problem.

Check Your Readiness

This section provides a review of the basic concepts needed before beginning Chapter 8. Lesson references are included for additional student help.

Use Exercises	To Prepare for Lesson(s)
1–9	8–1, 8–2
10–18	8–2, 8–5
19–27	8–4

Quick Review Math Handbook

hot words hot topics

Glencoe Algebra Lesson	Quick Review Math Handbook Hot Topic
8–1	6–1, 6–3
8–2	2–1, 3–4
8–3	3–4
8–4	3–3, 6–3
8–5	3–2
8–6	3–2
8–7	7–9
Ch. 8. Inv.	7–9

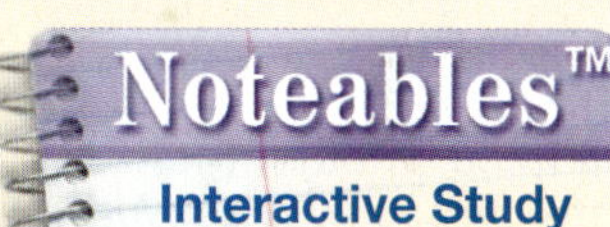

Interactive Study Notebook with Foldables™

This note-taking guide reinforces excellent note-taking skills. It includes:

- ✓ **Build Your Vocabulary** tool for including mathematical terminology in notes.
- ✓ **Review It** activities with links to prior lessons and items to include in **Foldables™**.
- ✓ **Bringing It All Together** feature to help students review for the Chapter Test.
- ✓ **Are You Ready for the Chapter Test?** feature to allow students to assess their own readiness for the Chapter Test.
- ✓ **Teacher Edition** with transparencies to guide note taking. Corresponds to the *Teacher's Wraparound Edition* In-Class Examples and *Interactive Chalkboard CD-ROM.*

Lesson 8–1

1 FOCUS

5-Minute Check
Chapter 7

1. Determine the slope of the line. $\frac{4}{5}$

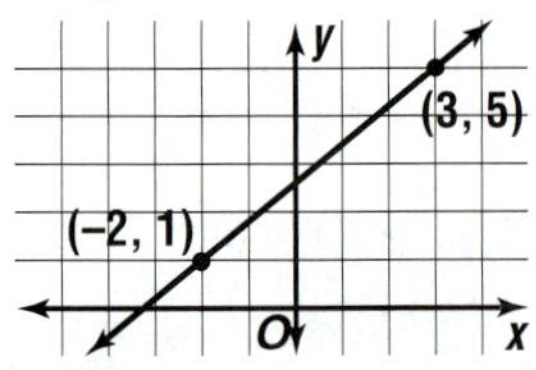

2. Write the point-slope form of an equation for the line passing through the point at (2, −3) and having slope 0. $y = -3$
3. Write an equation in slope-intercept form of the line passing through the points at (−1, 3) and (2, −2). $y = -\frac{5}{3}x + \frac{4}{3}$
4. Determine whether the graphs of the equations are *parallel*, *perpendicular*, or *neither*.
   $y = 2x - 3$
   $4x - 2y = 10$ **parallel**
5. Determine whether the scatter plot shows a *positive* relationship, *negative* relationship, or *no* relationship.

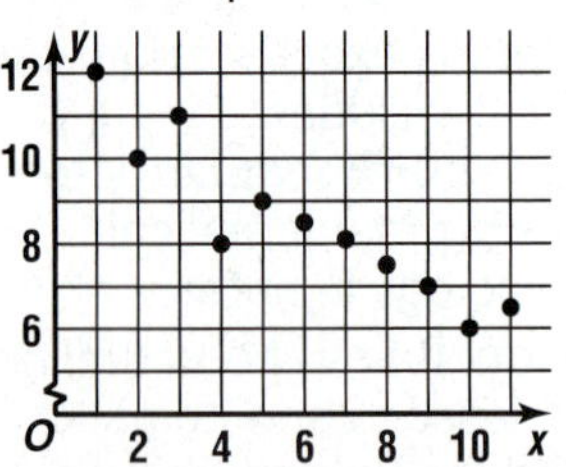

negative relationship

Motivating the Lesson

Real-World Connection Ask students to describe squares they use in games or other activities. As they mention squares that are divided into cells (chess or checker boards, tic-tac-toe forms, and so on), ask them to state the number of cells in each row and column of the square and the total number of cells in the square.

8–1 Powers and Exponents

What You'll Learn
You'll learn to use powers in expressions.

Why It's Important
Landscaping Landscape architects use the formula $A = \pi r^2$ to find the area of a circular region. The 2 in the formula is an exponent. *See Exercise 43.*

Perfect squares like 1, 4, 9, and 16 can be represented by a square array of dots.

1 4 9 16

A perfect square is the product of a number and itself. For example, 16 is a perfect square because $16 = 4 \times 4$. The expression 4×4 can be written using exponents. An **exponent** tells how many times a number, called the **base**, is used as a factor. Numbers that are expressed using exponents are called **powers**. The expression 4×4 can be written as 4^2.

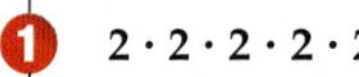

Symbols	Words	Meaning
4^1	4 to the first power	4
4^2	4 to the second power or 4 squared	$4 \cdot 4$
4^3	4 to the third power or 4 cubed	$4 \cdot 4 \cdot 4$
4^4	4 to the fourth power	$4 \cdot 4 \cdot 4 \cdot 4$
4^n	4 to the *n*th power	$\underbrace{4 \cdot 4 \cdot 4 \cdot \ldots \cdot 4}_{n \text{ factors}}$

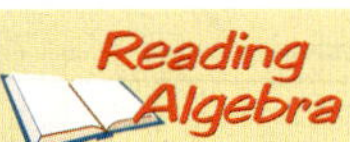

Write each expression using exponents.

1 $2 \cdot 2 \cdot 2 \cdot 2 \cdot 2$

The base is 2. It is a factor 5 times.

$2 \cdot 2 \cdot 2 \cdot 2 \cdot 2 = 2^5$

2 $m \cdot m \cdot m \cdot m$

The base is m. It is a factor 4 times..

$m \cdot m \cdot m \cdot m = m^4$

3 7

The base is 7. It is a factor 1 time.

$7 = 7^1$

Reading Algebra
When no exponent is shown, it is understood to be 1. For example, $10 = 10^1$.

Your Turn

a. $4 \cdot 4 \cdot 4 \cdot 4$ 4^4 b. $x \cdot x \cdot x$ x^3 c. 10 10^1

Resource Manager

Reproducible Masters
Chapter 8 Resource Masters
- *Study Guide,* p. 339
- *Skills Practice,* p. 340
- *Practice,* p. 341
- *Reading to Learn Mathematics,* p. 342
- *Enrichment,* p. 343

Graphing Calculator, p. 23

Transparencies
5-Minute Check, 8–1

Technology/Multimedia
AlgePASS, Lesson 18
Interactive Chalkboard CD-ROM

Example **4** **Write $(2)(2)(2)(-5)(-5)$ using exponents.**

Use the Associative Property to group the factors with like bases.

$$(2)(2)(2)(-5)(-5) = [(2)(2)(2)][(-5)(-5)]$$
$$= (2)^3(-5)^2$$

Your Turn

d. Write $(-1)(-1)(-1)(-1)(3)(3)$ using exponents. $(-1)^4 3^2$

You can use the definition of exponent to write a power as a multiplication expression.

Examples **Write each power as a multiplication expression.**

5 10^2

The base is 10. The exponent 2 means that 10 is a factor 2 times.

$10^2 = 10 \cdot 10$

6 b^3

The base is b. The exponent 3 means that b is a factor 3 times.

$b^3 = b \cdot b \cdot b$

7 $10x^2y^4$

10 is used as a factor once, x is used twice, and y is used 4 times.

$10x^2y^4 = 10 \cdot x \cdot x \cdot y \cdot y \cdot y \cdot y$

Your Turn g. $5 \cdot a \cdot a \cdot a \cdot b \cdot b$

e. 2^3 $2 \cdot 2 \cdot 2$ f. y^4 $y \cdot y \cdot y \cdot y$ g. $5a^3b^2$

You can also use the definition of exponent to evaluate expressions.

Example **Science Link**

8 **The distance between the Earth and the Sun is about 10^8 kilometers. Write this number as a multiplication expression and then evaluate the expression.**

$$10^8 = 10 \cdot 10 \cdot 10 \cdot 10 \cdot 10 \cdot 10 \cdot 10 \cdot 10$$
$$= 100{,}000{,}000$$

The distance between Earth and the sun is about 100 million kilometers.

2 TEACH

In-Class Examples

Examples 1–3

Write each expression using exponents.

1 $5 \cdot 5 \cdot 5$ 5^3

2 $d \cdot d \cdot d \cdot d \cdot d \cdot d$ d^6

3 p p^1

Teaching Tip Before discussing Example 4, students may need to review how the number of negative signs in a product affects the sign of the product.

In-Class Examples

Example 4

Write $(6)(6)(-7)(-7)(-7)(-7)(-7)$ using exponents. $6^2(-7)^5$

Examples 5–7

Write each power as a multiplication expression.

5 6^4 $6 \cdot 6 \cdot 6 \cdot 6$

6 h^5 $h \cdot h \cdot h \cdot h \cdot h$

7 $7a^3b^2$ $7 \cdot a \cdot a \cdot a \cdot b \cdot b$

Example 8

For every 3-digit telephone prefix, the number of possible 4-digit telephone extensions is 10^4. Write this number as a multiplication expression and then evaluate the expression. $10^4 = 10 \cdot 10 \cdot 10 \cdot 10$ or 10,000

From the Classroom of ...

Richard Dube
Taunton High School
Taunton, Massachusetts

When going over order of operations, I like to remind students of the mnemonic device **P**lease **E**xcuse **M**y **D**ear **A**unt **S**ally (**P**arentheses, **E**xponents, **M**ultiplication, **D**ivision, **A**ddition, **S**ubtraction).

Inclusion Strategies

Students with some physical impairments may find it difficult to write an exponent so it appears as a superscript. Explore other ways to indicate exponents, such as the "^" notation used by many graphing calculators or the "E" notation used by some scientific calculators.

Teaching Tip After discussing Example 9, ask students to tell the difference between the values of $4m^3$ and $(4m)^3$ for $m = 2$. They should realize that $4m^3 = 4 \cdot 2^3 = 4 \cdot 8$ or 32 while $(4m)^3 = (4 \cdot 2)^3 = 8^3$ or 512.

In-Class Examples

Examples 9–10

Evaluate each expression.

9 $5a^2$ if $a = 4$ **80**

10 $3p^2 - q^3$ if $p = -4$ and $q = -1$ **49**

Answers

Page 339

Graphing Calculator Exploration

3.

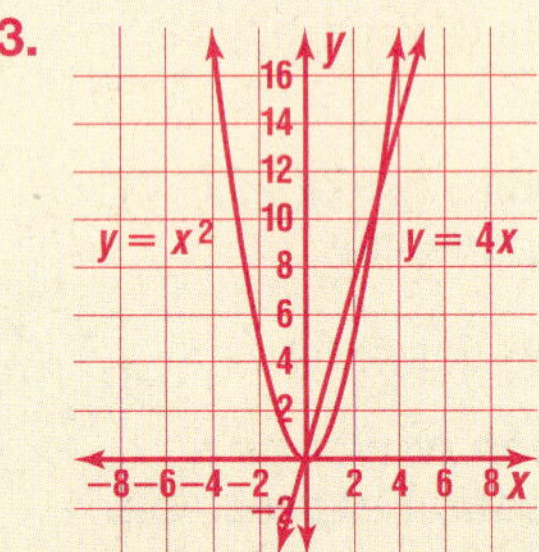

The graph of $y = x^2$ is a curve. The graph of $y = 4x$ is a line. The graphs intersect at (0, 0) and (4, 4).

4. $0 < x < 4$; $x = 4$

***Study Guide*, p. 339**

NAME ______________________ DATE ________ PERIOD ______

Study Guide

Student Edition Pages 336–340

Powers and Exponents

The product of a number with itself is called a **perfect square**. For example, 64 is a perfect square because $64 = 8 \times 8$. The **exponent** 2 can be used to write 8×8 as 8^2. Likewise, the exponent 4 can be used to write $8 \times 8 \times 8 \times 8$ as 8^4. Numbers that are expressed using exponents are called **powers**.

base ⟶ 8^4 ⟵ exponent

Example 1: Write $3 \cdot 3 \cdot 3 \cdot 3 \cdot 3$ using exponents.
The base is 3. There are 5 factors of 3, so the exponent is 5.
$3 \cdot 3 \cdot 3 \cdot 3 \cdot 3 = 3^5$

Example 2: Write $x \cdot x \cdot x \cdot x$ using exponents.
The base is x. There are 4 factors of x, so the exponent is 4.
$x \cdot x \cdot x \cdot x = x^4$

Example 3: Write $5x^2y^3$ as a multiplication expression.
There is 1 factor of 5, 2 factors of x, and 3 factors of y.
$5 \cdot x \cdot x \cdot y \cdot y \cdot y$

Write each expression using exponents.

1. $4 \cdot 4 \cdot 4 \cdot 4 \cdot 4 \cdot 4$ **4^6**

2. $(-3)(-3)(-3)(-3)$ **$(-3)^4$**

3. $x \cdot x \cdot x \cdot x \cdot x$ **x^5**

4. $(-2)\, a \cdot a \cdot a \cdot a \cdot b$ **$-2a^4b$**

5. $10 \cdot 10 \cdot 10 \cdot 10$ **10^4**

6. $5 \cdot x \cdot x \cdot x \cdot x \cdot x \cdot y \cdot y \cdot y$ **$5x^5y^3$**

Write each power as a multiplication expression.

7. 7^3 **$7 \cdot 7 \cdot 7$**

8. $-5y^4$ **$-5 \cdot y \cdot y \cdot y \cdot y$**

9. d^4e^3 **$d \cdot d \cdot d \cdot d \cdot e \cdot e \cdot e$**

10. $9ab^3$ **$9 \cdot a \cdot b \cdot b \cdot b$**

11. wx^2y^3 **$w \cdot x \cdot x \cdot y \cdot y \cdot y$**

12. $(-2)^5$ **$(-2)(-2)(-2)(-2)(-2)$**

Evaluate each expression if $x = -3$, $y = 2$, and $z = -1$.

13. zx^3y^2 **108**

14. $y(x^2 - z)$ **20**

When an expression contains an exponent, simplify the expression using the rules for order of operations.

Order of Operations: Lesson 1–2

Order of Operations

1. Do all operations within grouping symbols first; start with the innermost grouping symbols.
2. Evaluate all powers in order from left to right.
3. Do all multiplications and divisions from left to right.
4. Do all additions and subtractions from left to right.

In any expression, an exponent goes with the number or the quantity in parentheses that immediately precedes it.

$2 \cdot 5^3$ means $2 \cdot 5 \cdot 5 \cdot 5$ *The exponent 3 goes with the 5.*

$(2 \cdot 5)^3$ means $(2 \cdot 5)\,(2 \cdot 5)\,(2 \cdot 5)$ *The exponent goes with $(2 \cdot 5)$.*

Examples

Evaluate each expression.

9 $4m^3$ if $m = 2$

$4m^3 = 4(2)^3$ *Replace m with 2.*
$= 4(8)$ *Evaluate the power: $2 \cdot 2 \cdot 2 = 8$.*
$= 32$ *Multiply.*

10 $3x + y^2$ if $x = -2$ and $y = -3$

$3x + y^2 = 3(-2) + (-3)^2$ *Replace x with −2 and y with −3.*
$= 3(-2) + (9)$ *$(-3)^2 = (-3)(-3)$ or 9*
$= (-6) + (9)$ *Multiply.*
$= 3$ *Add.*

Your Turn

h. $3a^3$ if $a = -2$ **−24**

i. $-5(m + n)^2$ if $m = 4$ and $n = 2$ **−180**

In the following activity, you will use exponents to find perimeters and areas of squares.

Graphing Calculator Tutorial

See pp. 750–753.

Graphing Calculator Exploration

The pattern at the right contains perfect squares.

Step 1 Draw the next three squares in the pattern.

Step 2 Find the perimeter and area of each square. Organize your data in a table.

Length of Side	Perimeter	Area

Graphing Calculator Exploration

To answer Exercise 4, it is necessary to know the coordinates of the point where the graphs intersect. To find these coordinates while a graph is displayed, students can press [2nd] [CALC] and select 5: Intersect. Students can then use the Trace function and the arrow keys to move the cursor close to the point of intersection. Pressing [ENTER] three times will have the calculator display the coordinates of the intersection point.

Try These

3–4. See margin.

1. Write an equation that shows the relationship between the length of a side of a square x and its perimeter y. $y = 4x$
2. Write an equation that shows the relationship between the length of a side of a square x and its area y. $y = x^2$
3. Graph the two equations. Compare and contrast the graphs.
4. When is the value of the perimeter greater than the value of the area? When is the value of the perimeter equal to the area?
5. If the length of each side of a square is doubled, how does its perimeter change? How does its area change? **doubles; quadruples**

Check for Understanding

Communicating Mathematics

1–2. See margin.

1. **Write** a definition of *perfect square*.
2. **Explain** what the 2 represents in 10^2.
3. **You Decide?** Jan thinks that $(6n)^3$ is equal to $6n^3$. Becky thinks they are not equal. Who is correct? Explain your reasoning. **Becky; $(6n)(6n)(6n) = 6 \cdot 6 \cdot 6 \cdot n \cdot n \cdot n$, not $6n^3$.**

> *Vocabulary*
> perfect squares
> exponent
> base
> powers

Guided Practice

Examples 1–3

Write each expression using exponents.

4. $9 \cdot 9 \cdot 9 \cdot 9$ 9^4
5. $a \cdot a \cdot a \cdot a \cdot a$ a^5
6. 3 3^1

Examples 5–7

Write each power as a multiplication expression.

7. 12^4 $12 \cdot 12 \cdot 12 \cdot 12$
8. x^5 $x \cdot x \cdot x \cdot x \cdot x$
9. m^4n^3 $m \cdot m \cdot m \cdot m \cdot n \cdot n \cdot n$

Examples 8–10

Evaluate each expression if $a = 3$, $b = -2$, and $c = 4$.

10. c^3 64
11. $2a^4$ 162
12. $3a^2b$ -54

Example 4

13. **Number Theory** The prime factorization of 360 is $2 \cdot 2 \cdot 2 \cdot 3 \cdot 3 \cdot 5$. Write the prime factorization using exponents. 2^33^25

Exercises

Practice

Write each expression using exponents. 15. $(-2)^4$ 19. 2^23^35

14. $10 \cdot 10 \cdot 10$ 10^3
15. $(-2)(-2)(-2)(-2)$
16. 6 6^1
17. 7 cubed 7^3
18. $4 \cdot 4 \cdot 4 \cdot 6 \cdot 6$ 4^36^2
19. $2 \cdot 3 \cdot 5 \cdot 2 \cdot 3 \cdot 3$
20. $a \cdot a \cdot a \cdot a \cdot b \cdot b$ a^4b^2
21. $3 \cdot x \cdot x \cdot y \cdot y$ $3x^2y^2$
22. $(-5)(m)(m)(m)(n)$

22. $-5m^3n$

3 PRACTICE/APPLY

Error Analysis

Watch for students who, in Exercise 2, describe 10^2 as "ten multiplied by itself two times," because that would mean "start with 10, multiply by 10 once, then multiply by 10 again" which is 10^3. ***Prevent by*** stressing the language "ten appears as a factor two times."

Assignment Guide

Basic: 15–43 odd, 44–53
Average: 14–40 even, 41–53

Answers

1. A perfect square is the product of a number and itself.
2. 2 is an exponent. It tells how many times the base is used as a factor.

Reteaching Activity

Kinesthetic Learners Provide pairs of students with two stacks of 10 index cards. Each card in one stack should have an expression like those in Examples 9 and 10 written on it. The cards in the other stack should each have one of the integers from −5 to 5, excluding zero, written on it. Instruct students to randomly select a card from the "expression" stack and then one card for the value of each variable from the "integer" stack. The students should work together to find the value of the expression. Have the students repeat the activity several times.

***Skills Practice,* p. 340, and *Practice,* p. 341 (shown)**

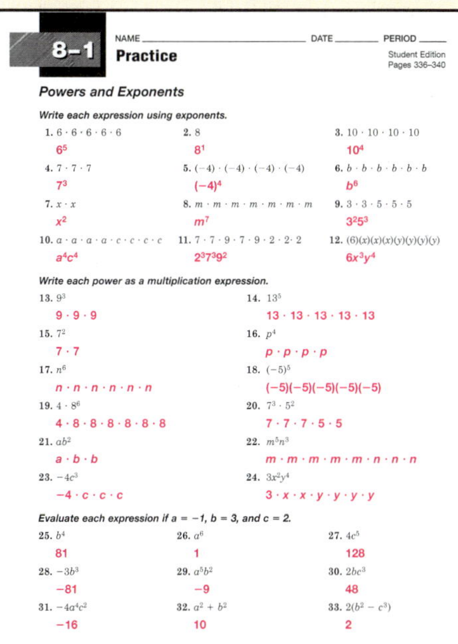

8–1 NAME ______ DATE ______ PERIOD ______

Practice

Student Edition Pages 336–340

Powers and Exponents

Write each expression using exponents.

1. $6 \cdot 6 \cdot 6 \cdot 6 \cdot 6$ 6^5
2. 8 8^1
3. $10 \cdot 10 \cdot 10 \cdot 10$ 10^4
4. $7 \cdot 7 \cdot 7$ 7^3
5. $(-4) \cdot (-4) \cdot (-4) \cdot (-4)$ $(-4)^4$
6. $b \cdot b \cdot b \cdot b \cdot b \cdot b$ b^6
7. $x \cdot x$ x^2
8. $m \cdot m \cdot m \cdot m \cdot m \cdot m \cdot m$ m^7
9. $3 \cdot 3 \cdot 5 \cdot 5 \cdot 5$ 3^25^3
10. $a \cdot a \cdot a \cdot a \cdot c \cdot c \cdot c \cdot c$ a^4c^4
11. $7 \cdot 7 \cdot 9 \cdot 7 \cdot 9 \cdot 2 \cdot 2 \cdot 2$ $2^37^39^2$
12. $(6)(x)(x)(x)(y)(y)(y)(y)$ $6x^3y^4$

Write each power as a multiplication expression.

13. 9^3 $9 \cdot 9 \cdot 9$
14. 13^5 $13 \cdot 13 \cdot 13 \cdot 13 \cdot 13$
15. 7^2 $7 \cdot 7$
16. p^4 $p \cdot p \cdot p \cdot p$
17. n^6 $n \cdot n \cdot n \cdot n \cdot n \cdot n$
18. $(-5)^5$ $(-5)(-5)(-5)(-5)(-5)$
19. $4 \cdot 8^6$ $4 \cdot 8 \cdot 8 \cdot 8 \cdot 8 \cdot 8 \cdot 8$
20. $7^3 \cdot 5^2$ $7 \cdot 7 \cdot 7 \cdot 5 \cdot 5$
21. ab^2 $a \cdot b \cdot b$
22. m^5n^3 $m \cdot m \cdot m \cdot m \cdot m \cdot n \cdot n \cdot n$
23. $-4c^3$ $-4 \cdot c \cdot c \cdot c$
24. $3x^2y^4$ $3 \cdot x \cdot x \cdot y \cdot y \cdot y \cdot y$

Evaluate each expression if $a = -1$, $b = 3$, and $c = 2$.

25. b^4 81
26. a^6 1
27. $4c^5$ 128
28. $-3b^3$ -81
29. a^5b^2 -9
30. $2bc^3$ 48
31. $-4a^4c^2$ -16
32. $a^2 + b^2$ 10
33. $2(b^2 - c^3)$ 2

4 ASSESS

Open-Ended Assessment

Speaking Ask students to explain what it means when a number is written as a base raised to an exponent.

Answers

46.

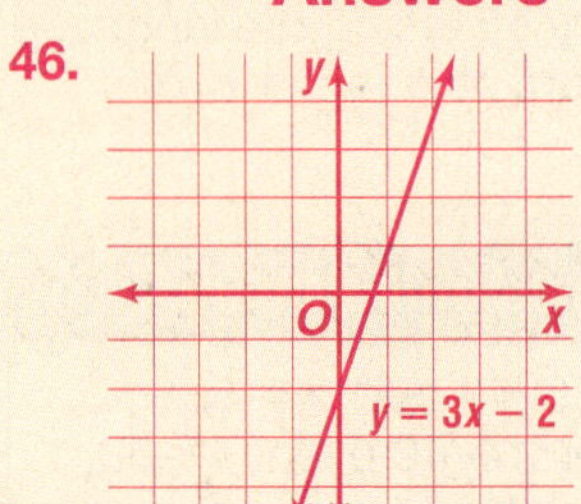

47.

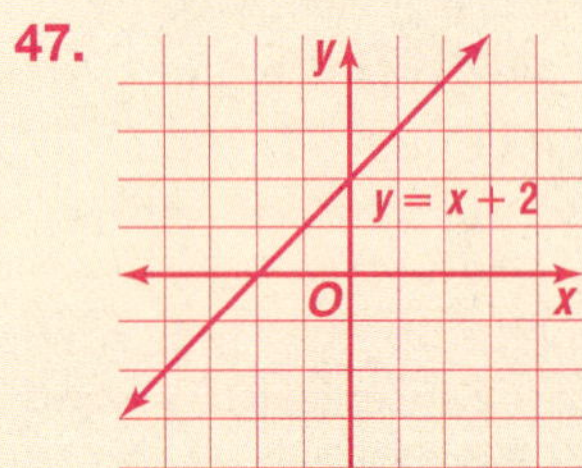

48.

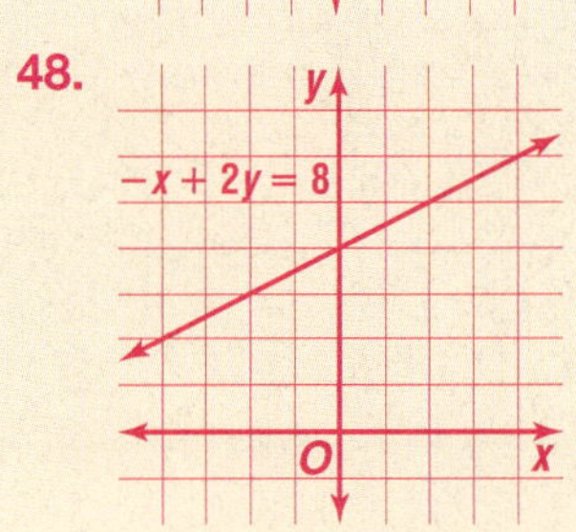

***Enrichment*, p. 343**

8–1 NAME ________________ DATE ________ PERIOD ______

Enrichment

Student Edition Pages 336–340

Patterns with Powers

Use your calculator, if necessary, to complete each pattern.

a.		b.		c.	
2^{10} =	1024	5^{10} =	9,765,625	4^{10} =	1,048,576
2^9 =	512	5^9 =	1,953,125	4^9 =	262,144
2^8 =	256	5^8 =	390,625	4^8 =	65,536
2^7 =	128	5^7 =	78,125	4^7 =	16,384
2^6 =	64	5^6 =	15,625	4^6 =	4096
2^5 =	32	5^5 =	3125	4^5 =	1024
2^4 =	16	5^4 =	625	4^4 =	256
2^3 =	8	5^3 =	125	4^3 =	64
2^2 =	4	5^2 =	25	4^2 =	16
2^1 =	2	5^1 =	5	4^1 =	4

Study the patterns for a, b, and c above. Then answer the questions.

1. Describe the pattern of the exponents from the top of each column to the bottom.
   The exponents decrease by one from each row to the one below.
2. Describe the pattern of the powers from the top of the column to the bottom. To get each power, divide the power on the row above by the base (2, 5, or 4).
3. What would you expect the following powers to be?
   2^0 1 5^0 1 4^0 1
4. Write a rule. Test it on patterns that you obtain using −2, −5, and −4 as bases. Any nonzero number to the zero power equals 1.

Study the pattern below. Then answer the questions.

$0^3 = 0$ $0^2 = 0$ $0^1 = 0$ $0^0 =$ ___?___ 0^{-1} does not exist. 0^{-2} does not exist. 0^{-3} does not exist.

5. Why do 0^{-1}, 0^{-2}, and 0^{-3} not exist?
   Negative exponents are not defined unless the base is nonzero.
6. Based upon the pattern, can you determine whether 0^0 exists?
   No, since the pattern $0^n = 0$ breaks down for $n < 1$.
7. The symbol 0^0 is called an **indeterminate**, which means that it has no unique value. Thus it does not exist as a unique real number. Why do you think that 0^0 cannot equal 1? Sample answer: If $0^0 = 1$, then $1 = \frac{1}{1} = \frac{1^0}{0^0} = \left(\frac{1}{0}\right)^0$, which is a false result, since division by zero is not allowed. Thus, 0^0 cannot equal 1.

Homework Help	
For Exercises	See Examples
14–22	1–4
23–30	5–7
31–41, 43	9, 10
Extra Practice	
See page 707.	

Write each power as a multiplication expression.

23. 3^5 $3 \cdot 3 \cdot 3 \cdot 3 \cdot 3$
24. $(-2)^2$ $(-2)(-2)$
25. $2^4 \cdot 3^2$ $2 \cdot 2 \cdot 2 \cdot 2 \cdot 3 \cdot 3$
26. $2 \cdot 3^5$ $2 \cdot 3 \cdot 3 \cdot 3 \cdot 3 \cdot 3$
27. y^3 $y \cdot y \cdot y$
28. x^2y^2 $x \cdot x \cdot y \cdot y$
29. $6ab^4$ $6 \cdot a \cdot b \cdot b \cdot b \cdot b$
30. $-2y^4$ $-2 \cdot y \cdot y \cdot y \cdot y$

Evaluate each expression if $x = -2$, $y = 3$, $z = -1$, and $w = 0.5$.

31. x^5 **−32**
32. $4y^2$ **36**
33. $-2x^6$ **−128**
34. $x^2 - y^2$ **−5**
35. $3(y^2 + z)$ **24**
36. $-2(x^3 + 1)$ **14**
37. $2w^2$ **0.5**
38. wx^3y **−12**

39. Find the value of $x^2 + 2x + 1$ if $x = -3$. **4**
40. Which is greater, 2^5 or 5^2? 2^5

Applications and Problem Solving

41. **Number Theory** The prime factorization of a number is $2 \cdot 3^5$. Find the number. **486**

42. **Geometry** Use exponents to write an expression that represents the total number of unit cubes in the large cube. Then evaluate the expression. $5^3 = 125$

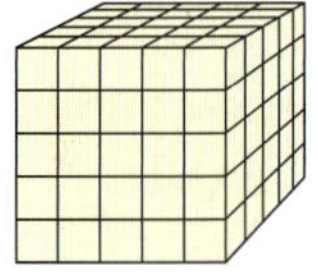

Exercise 42

43. **Landscaping** Landscape architects use the formula $A = \pi r^2$ to find the area of circular flower beds. In the formula, $\pi \approx 3.14$, and r is the radius of the circle.
    a. Estimate the area of a circular flower bed with a radius of 8 feet. **a. 200.96 ft²**
    b. About how many bags of mulch will the landscape architect need to cover the bed if each bag covers about 10 square feet? **21 bags**
44. **Critical Thinking** Suppose you raise a negative integer to a positive power. When is the result negative? When is the result positive? **negative: when the exponent is odd; positive: when the exponent is even**

Mixed Review

45. Write an equation of the line that is parallel to the graph of $y = -2x + 3$ and passes through (1, 4). *(Lesson 7–7)* $y = -2x + 6$

46–48. See margin.

Graph each equation using the slope and y-intercept. *(Lesson 7–6)*

46. $y = 3x - 2$
47. $y = x + 2$
48. $-x + 2y = 8$

Solve each problem. *(Lesson 5–4)*

49. Find 26% of 120. **31.2**
50. 17 is 40% of what number? **42.5**
51. 9 is what percent of 18? **50%**
52. 98% of 40 is what number? **39.2**

Standardized Test Practice

53. **Multiple Choice** A mechanic charges an initial fee of $40 plus $30 for each hour she works. Which equation represents the cost c of a repair job that lasts h hours? *(Lesson 4–5)* **B**
    (A) $c = 30 + 40h$
    (B) $c = 40 + 30h$
    (C) $c = 40 - 30h$
    (D) $c = 30 - 40h$

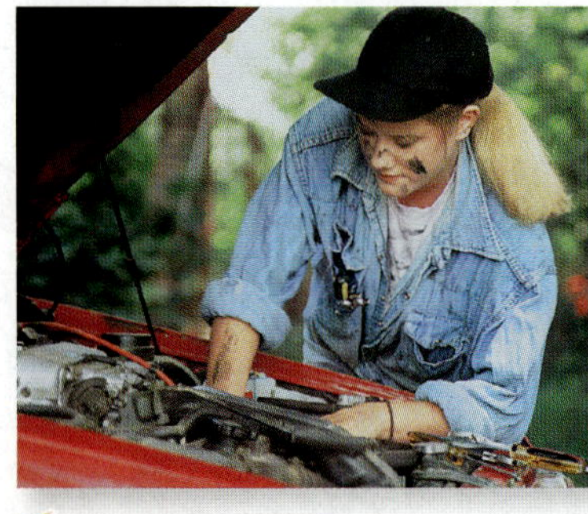

www.algconcepts.com/self_check_quiz

Evaluate $3a$ if $a^3 = 64$. **12**

8-2 Multiplying and Dividing Powers

What You'll Learn
You'll learn to multiply and divide powers.

Why It's Important
Movie Industry The intensity of sound is measured using *decibels*, a unit that is based on powers of ten. *See page 346.*

Powers can be multiplied and divided. In the example below, we will use powers of 2 to form a rule about multiplying exponents. The table shows several powers of 2 and their values.

Power of 2	2^1	2^2	2^3	2^4	2^5	2^6
Value	2	4	8	16	32	64

You can use the table to substitute exponents for the factors of multiplication equations. What do you notice about the exponents in the following products?

Numerical Products	$4 \cdot 2 = 8$	$4 \cdot 8 = 32$	$8 \cdot 8 = 64$
Products of Powers	$2^2 \cdot 2^1 = 2^3$	$2^2 \cdot 2^3 = 2^5$	$2^3 \cdot 2^3 = 2^6$

These examples suggest that you can multiply powers with the same base by adding the exponents. Think about $a^2 \cdot a^3$.

$a^2 \cdot a^3 = (a \cdot a)(a \cdot a \cdot a)$ *a^2 has two factors; a^3 has three factors.*

$= a \cdot a \cdot a \cdot a \cdot a$ *Substitution Property*

$= a^5$ *The product has 2 + 3 or 5 factors.*

Product of Powers

Words: You can multiply powers with the same base by adding the exponents.

Numbers: $3^3 \cdot 3^2 = 3^{3+2}$ or 3^5

Symbols: $a^m \cdot a^n = a^{m+n}$

Examples

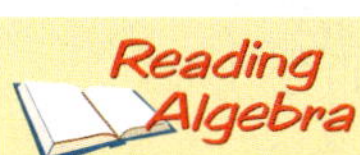

Reading Algebra
An expression is *simplified* when each base appears only once and all of the fractions are in simplest form.

Simplify each expression.

1 $4^3 \cdot 4^5$

$4^3 \cdot 4^5 = 4^{3+5}$ *To multiply powers that have the same base, write the common base, then add the exponents.*

$= 4^8$

2 $x^3 \cdot x^4$

$x^3 \cdot x^4 = x^{3+4}$ *To multiply powers that have the same base, write the common base, then add the exponents.*

$= x^7$

1 FOCUS

5-Minute Check
Lesson 8-1

Write each expression using exponents.

1. $7 \cdot 7 \cdot 7 \cdot 7 \cdot 7 \cdot 7$ 7^6
2. $w \cdot w \cdot w \cdot w$ w^4
3. $(-3)(-3)(a)(a)(b)(b)(b)$ $(-3)^2a^2b^3$

Evaluate each expression if x = 4, y = −3, and z = 5.

4. $5y^3$ -135
5. $-6(2x^2 - z^2)$ -42

Motivating the Lesson

Real-World Connection Ask students if automobile license plates in your state follow a rule such as *4 letters, followed by 3 digits*. If there is such a rule, ask them to use exponents to write the total number of different license plates that are possible. **For the rule above, the number would be $26^4 \cdot 10^3$ or 456,976,000.**

Resource Manager

Reproducible Masters

Chapter 8 Resource Masters
- *Study Guide,* p. 344
- *Skills Practice,* p. 345
- *Practice,* p. 346
- *Reading to Learn Mathematics,* p. 347
- *Enrichment,* p. 348

Hands-On Algebra, p. 90

Transparencies

5-Minute Check, 8-2

Technology/Multimedia

AlgePASS, Lesson 18

Interactive Chalkboard CD-ROM

2 TEACH

Teaching Tip Stress to students that Examples 1 and 2 can be simplified because they have powers to the same base. Have them compare the expressions in Examples 1 and 2 to an expression such as $m^5 \cdot n^2$, which cannot be simplified.

In-Class Examples

Examples 1–4

Simplify each expression.

1 $5^2 \cdot 5^7$ **5^9**

2 $s^{10} \cdot s^5$ **s^{15}**

3 $(10a)(5a^4)$ **$50a^5$**

4 $(m^5n^4)(m^3n^7)$ **m^8n^{11}**

3 $(4y^2)(3y)$

$(4y^2)(3y) = (4 \cdot 3)(y^2 \cdot y)$ *Use the Commutative and Associative Properties.*

$= 12y^{2+1}$ *$y = y^1$*

$= 12y^3$ *Simplify.*

4 $(a^3b^2)(a^2b^4)$

$(a^3b^2)(a^2b^4) = (a^3 \cdot a^2)(b^2 \cdot b^4)$ *Use the Commutative and Associative Properties.*

$= a^{3+2} \cdot b^{2+4}$

$= a^5b^6$ *Simplify.*

Your Turn **c. $-15x^3$ d. x^9y^8**

a. $10^4 \cdot 10^2$ **10^6** b. $y^4 \cdot y^2$ **y^6** c. $(-3x^2)(5x)$ d. $(x^5y^2)(x^4y^6)$

You can use powers of 2 to help find a rule for dividing powers. Study each quotient in the table. What do you notice about the exponents?

Numerical Quotients	$16 \div 8 = 2$	$32 \div 4 = 8$	$64 \div 2 = 32$
Quotient of Powers	$2^4 \div 2^3 = 2^1$	$2^5 \div 2^2 = 2^3$	$2^6 \div 2^1 = 2^5$

These examples suggest that you can divide powers with the same base by subtracting the exponents. Think about $a^5 \div a^2$. Remember that you can write a division expression as a fraction.

$\frac{a^5}{a^2} = \frac{a \cdot a \cdot a \cdot a \cdot a}{a \cdot a}$ *a^5 has five factors; a^2 has two factors.*

$= \frac{\overset{1}{\cancel{a}} \cdot \overset{1}{\cancel{a}} \cdot a \cdot a \cdot a}{\underset{1}{\cancel{a}} \cdot \underset{1}{\cancel{a}}}$ *Notice that $\frac{a \cdot a}{a \cdot a} = 1$.*

$= a \cdot a \cdot a$ *The quotient has $5 - 2$ or 3 factors.*

$= a^3$ *Write using an exponent.*

Quotient of Powers

Words: You can divide powers with the same base by subtracting the exponents.

Numbers: $\frac{5^7}{5^4} = 5^{7-4}$ or 5^3

Symbols: $\frac{a^m}{a^n} = a^{m-n}$ *The value of a cannot be zero.*

www.algconcepts.com/extra_examples

Examples

Simplify each expression.

5 $\frac{4^3}{4^2}$

$\frac{4^3}{4^2} = 4^{3-2}$ *To divide powers that have the same base, write the common base. Then subtract the exponents.*

$= 4^1$ or 4 *Simplify.*

6 $\frac{x^6}{x^4}$

$\frac{x^6}{x^4} = x^{6-4}$ *Write the common base. Then subtract the exponents.*

$= x^2$ *Simplify.*

7 $\frac{8m^4n^5}{2m^3n^2}$

$\frac{8m^4n^5}{2m^3n^2} = \left(\frac{8}{2}\right)\left(\frac{m^4}{m^3}\right)\left(\frac{n^5}{n^2}\right)$ *Group the powers that have the same base.*

$= 4m^{4-3}n^{5-2}$ *Divide.*

$= 4m^1n^3$ *Subtract the exponents.*

$= 4mn^3$ *Simplify.*

Your Turn h. $-3m^2n$

e. $\frac{10^5}{10^2}$ 10^3 f. $\frac{y^5}{y^4}$ y g. $\frac{a^4b^3}{ab^2}$ a^3b h. $\frac{-30m^5n^2}{10m^3n}$

A special case results when you divide a power by itself. Consider the following two ways to simplify $\frac{b^4}{b^4}$, where $b \neq 0$.

Method 1 Definition of Power

$$\frac{b^4}{b^4} = \frac{\overset{1}{\cancel{b}} \cdot \overset{1}{\cancel{b}} \cdot \overset{1}{\cancel{b}} \cdot \overset{1}{\cancel{b}}}{\underset{1}{\cancel{b}} \cdot \underset{1}{\cancel{b}} \cdot \underset{1}{\cancel{b}} \cdot \underset{1}{\cancel{b}}}$$

$$= 1$$

Method 2 Quotient of Powers

$$\frac{b^4}{b^4} = b^{4-4}$$

$$= b^0$$

Since $\frac{b^4}{b^4}$ cannot have two different values, you can conclude that $b^0 = 1$. Therefore, any nonzero number raised to the zero power is equal to 1.

Teaching Tip While discussing the Quotient of Powers rule, ask students why the value of a cannot be zero. Students should recognize that if $a = 0$ then the value of a^n is $0^n = 0$ for any value of n. Remind students that division by zero has no meaning.

In-Class Examples

Examples 5–7

Simplify each expression.

5 $\frac{7^6}{7^2}$ 7^4

6 $\frac{p^{10}}{p}$ p^9

7 $\frac{15a^6b^4}{3a^4b^3}$ $5a^2b$

Teaching Tip After discussing Method 2 at the bottom of the page, have students look back at the first table on page 341. Suggest adding a column with 0 as the entry in the Day row. Have students use the pattern in the Number row to see that the entry corresponding to day 0 is 1. Then have them confirm that the corresponding entry in the Power of 2 row is 2^0. Stress that this shows $2^0 = 1$, or that $b^0 = 1$ when $b = 2$.

***Study Guide*, p. 344**

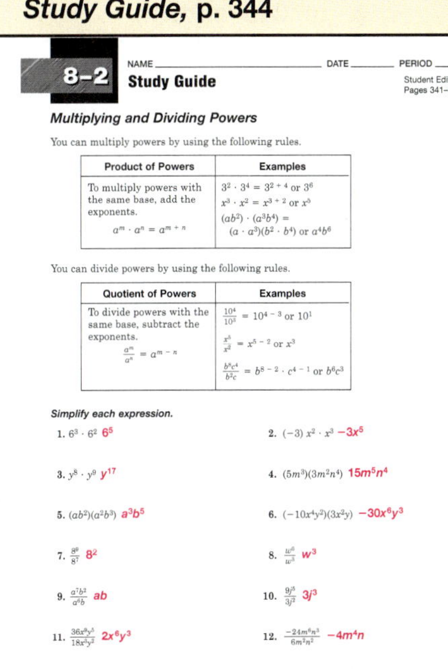

8–2 NAME ______ DATE ______ PERIOD ______

Study Guide Student Edition Pages 341–346

Multiplying and Dividing Powers

You can multiply powers by using the following rules.

Product of Powers	Examples
To multiply powers with the same base, add the exponents. $a^m \cdot a^n = a^{m+n}$	$3^2 \cdot 3^4 = 3^{2+4}$ or 3^6 $x^3 \cdot x^2 = x^{3+2}$ or x^5 $(ab^2) \cdot (a^3b^4) = (a \cdot a^3)(b^2 \cdot b^4)$ or a^4b^6

You can divide powers by using the following rules.

Quotient of Powers	Examples
To divide powers with the same base, subtract the exponents. $\frac{a^m}{a^n} = a^{m-n}$	$\frac{10^4}{10^3} = 10^{4-3}$ or 10^1 $\frac{x^5}{x^2} = x^{5-2}$ or x^3 $\frac{b^8c^4}{b^2c} = b^{8-2} \cdot c^{4-1}$ or b^6c^3

Simplify each expression.

1. $6^3 \cdot 6^2$ 6^5
2. $(-3)\, x^2 \cdot x^3$ $-3x^5$
3. $y^8 \cdot y^9$ y^{17}
4. $(5m^3)(3m^2n^4)$ $15m^5n^4$
5. $(ab^2)(a^2b^3)$ a^3b^5
6. $(-10x^4y^2)(3x^2y)$ $-30x^6y^3$
7. $\frac{8^9}{8^7}$ 8^2
8. $\frac{w^6}{w^3}$ w^3
9. $\frac{a^7b^2}{a^6b}$ ab
10. $\frac{9j^5}{3j^2}$ $3j^3$
11. $\frac{36x^9y^5}{18x^3y^2}$ $2x^6y^3$
12. $\frac{-24m^6n^3}{6m^2n^2}$ $-4m^4n$

In-Class Example

Example 8

Simplify $\frac{12c^5d^7}{4c^2d^4}$. **$3c^3d^3$**

3 PRACTICE/APPLY

Error Analysis

Watch for students who, in Exercise 3, multiply exponents instead of adding them and rewrite $y^2 \cdot y^5$ as y^{10}.

Prevent by having students write an intermediate step, $y^2 \cdot y^5 = y^{2+5} = y^7$, while they say "to multiply two powers with the same base, write the base and add the exponents."

Assignment Guide

Basic: 17–47 odd, 49–59
Average: 16–44 even, 46–59

Answer

1. **In the first case, the bases are the same, but in the second case, the bases are different.**

Skills Practice, **p. 345, and** ***Practice,*** **p. 346 (shown)**

8–2 NAME ________ DATE ______ PERIOD ____

Practice

Student Edition Pages 341–346

Multiplying and Dividing Powers

Simplify each expression.

1. $6^3 \cdot 6^2$ **6^5**
2. $7^6 \cdot 7^4$ **7^{10}**
3. $y^4 \cdot y^8$ **y^{12}**
4. $b \cdot b^4$ **b^5**
5. $(g^2)(g^3)(g)$ **g^6**
6. $m(m^8)$ **m^9**
7. $(a^2b^3)(a^4b)$ **a^6b^4**
8. $(xy^5)(x^3y^3)$ **x^4y^8**
9. $(2c^3)(2c)$ **$4c^4$**
10. $(-3x^2)(6x^2)$ **$-18x^4$**
11. $(-7xy)(-2x)$ **$14x^2y$**
12. $(5m^3n^2)(4m^2n^3)$ **$20m^5n^5$**
13. $(-8ab)(a^2b^5)$ **$-8a^3b^6$**
14. $\frac{9^2}{9}$ **9**
15. $\frac{12^8}{12^3}$ **12^5**
16. $\frac{y^4}{y^2}$ **y^2**
17. $\frac{k^6}{k^6}$ **1**
18. $\frac{x^4y^5}{x^3y^2}$ **xy^3**
19. $\frac{a^9b^6}{a^2b}$ **a^7b^5**
20. $\frac{mn^3}{n^2}$ **mn**
21. $\frac{15a^3}{3a}$ **$5a^2$**
22. $\frac{8x^5y^4}{4x^2y^2}$ **$2x^3y^2$**
23. $\frac{m^2n}{m^2}$ **n**
24. $\frac{6a^5b^7}{-2a^3b^7}$ **$-3a^2$**
25. $\frac{-20x^3y^2}{-5x^3y}$ **$4y$**
26. $\frac{-16ab^4}{4b^3}$ **$-4ab$**
27. $\frac{12x^2y}{2x^2y}$ **6**

Example 8 Simplify $\frac{x^4y^2}{xy^2}$.

$\frac{x^4y^2}{xy^2} = \left(\frac{x^4}{x^1}\right)\left(\frac{y^2}{y^2}\right)$ *Group the powers that have the same base.*

$= x^{4-1}y^{2-2}$ *Divide.*

$= x^3y^0$ *$y^0 = 1$*

$= x^3 \cdot 1$ or x^3 *Simplify.*

Your Turn

i. $\frac{a^3b^4}{a^3b}$ **b^3**
j. $\frac{10x^4y^3}{5x^4y^2}$ **$2y$**
k. $\frac{m^{10}n^5}{m^{10}n^5}$ **1**

Check for Understanding

Communicating Mathematics

1. Explain why $a^4 \cdot a^7$ can be simplified but $a^4 \cdot b^7$ cannot. **See margin.**
2. **You Decide?** Tonia says $10^3 \times 10^2 = 100^5$, but Emilio says $10^3 \times 10^2 = 10^5$. Who is correct? Explain your reasoning. **Emilio; $10^3 \times 10^2 = 10^{3+2}$ or 10^5**

Guided Practice

Simplify each expression.

Examples 1 & 2
3. $y^2 \cdot y^5$ **y^7**
4. $m^5(m)$ **m^6**
5. $(t^2)(t^2)(t)$ **t^5**

Examples 3 & 4
6. $(x^3y)(xy^3)$ **x^4y^4**
7. $(3a^2)(4a^3)$ **$12a^5$**
8. $(-5x^3)(4x^4)$ **$-20x^7$**

Examples 5 & 6
9. $\frac{n^8}{n^5}$ **n^3**
10. $\frac{b^6c^5}{b^3c^2}$ **b^3c^3**
11. $\frac{xy^7}{y^4}$ **xy^3**

Examples 7 & 8
12. $\frac{12x^5}{4x^4}$ **$3x$**
13. $\frac{ab^5c}{ac}$ **b^5**
14. $\frac{22a^4b^5c^7}{-11abc^2}$ **$-2a^3b^4c^5$**

Example 1
15. **Measurement** There are 10^1 millimeters in 1 centimeter and 10^2 centimeters in 1 meter. How many millimeters are in 1 meter? Write your answer as a power and then evaluate the expression. **$10^3 = 1000$**

28. $-16x^7y$

Exercises

Practice 

Simplify each expression. **29. m^7b^2** **30. $a^2b^2c^2$** **31. $m^3n^3a^2$**

16. $2^6 \cdot 2^8$ **2^{14}**
17. $5^3 \cdot 5$ **5^4**
18. $y^7 \cdot y^7$ **y^{14}**
19. $d \cdot d^5$ **d^6**
20. $(b^4)(b^2)$ **b^6**
21. $(a^2b)(ab^4)$ **a^3b^5**
22. $(m^3n)(mn^2)$ **m^4n^3**
23. $(r^3t^4)(r^4t^4)$ **r^7t^8**
24. $(2xy)(3x^2y^2)$ **$6x^3y^3$**
25. $(-5a)(-3a)$ **$15a^2$**
26. $(-10x^3y)(2x^2)$ **$-20x^5y$**
27. $(4x^2y^3)(2xy^2)$ **$8x^3y^5$**
28. $(-8x^3y)(2x^4)$
29. $m^4(m^3b^2)$
30. $(ab)(ac)(bc)$
31. $(m^2n)(am)(an^2)$
32. $\frac{10^9}{10^3}$ **10^6**
33. $\frac{9^6}{9^5}$ **9**
34. $\frac{w^7}{w^3}$ **w^4**
35. $\frac{k^9}{k^4}$ **k^5**

Homework Help

For Exercises	See Examples
16–31, 45, 46, 48	1–4
32–43, 47	5–7
45	8

Extra Practice
See page 708.

Reteaching Activity

Interpersonal Learners Have students work in small groups. Taking turns, one student should state a variable for a base, two other students should each state a whole-number exponent for the base, and then all the students in the group should calculate the product of the two powers.

37. $-12ab^4$
38. $-4y^4z^2$
40. $9ab^2$
41. $15a^3c$

B ▶ 36. $\frac{a^4b^8}{ab^2}$ a^3b^6 37. $\frac{24a^3b^6}{-2a^2b^2}$ 38. $\frac{24x^2y^7z^3}{-6x^2y^3z}$ 39. $\frac{-40mn^2}{-10mn^2}$ 4

40. $\frac{3}{4}a(12b^2)$ 41. $\frac{5}{6}c(18a^3)$ 42. $\left(\frac{1}{2}a^2\right)\left(6ab^2\right)$ $3a^3b^2$ 43. $x^0(2x^3)$ $2x^3$

44. Evaluate 5^0. 1

45. Find the product of $2x$ and $-8x$. $-16x^2$

Applications and Problem Solving

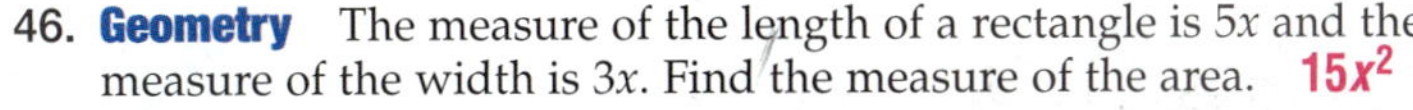

46. **Geometry** The measure of the length of a rectangle is $5x$ and the measure of the width is $3x$. Find the measure of the area. $15x^2$

C ▶ 47. **Earth Science** At a distance of 10^7 meters from Earth, a satellite can see almost all of our planet. At a distance of 10^{13} meters, a satellite can see all of our solar system. How many times as great is a distance of 10^{13} meters as 10^7 meters? 10^6 or 1,000,000 times

48. **Manufacturing** The Pizza Parlor uses square boxes to package their pizzas. The drawing shows that a pizza with radius r just fits inside the box. Write an expression for the area of the bottom of the box. $A = (2r)^2$ or $4r^2$

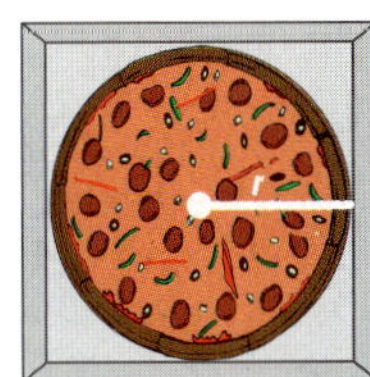

49. **Critical Thinking** Study the following pattern.

$(5^2)^3 = (5^2)(5^2)(5^2)$ or 5^6

Simplify each expression.

a. $(10^3)^4$ 10^{12} b. $(4^5)^3$ 4^{15} c. $(x^2)^4$ x^8

d. Write a rule for finding the *power of a power*. Multiply the exponents; $(x^a)^b = x^{ab}$.

Mixed Review

Evaluate each expression if $x = -1$, $y = 2$, and $z = -3$. *(Lesson 8–1)*

50. z^3 -27 51. $3x^4$ 3 52. $5xy^3z$ 120 53. $3(x^2 + y^2)$ 15

54. Write an equation of the line that is perpendicular to the graph of $y = 3x + 5$ and passes through (0, 0). *(Lesson 7–7)* $y = -\frac{1}{3}x$

Solve each proportion. *(Lesson 5–1)*

55. $\frac{96}{6} = \frac{152}{x}$ 9.5 56. $\frac{9}{m} = \frac{15}{10}$ 6 57. $\frac{8.6}{25.8} = \frac{1}{n}$ 3 58. $\frac{3}{7} = \frac{2.1}{d}$ 4.9

Standardized Test Practice

59. **Multiple Choice** Choose the expression that has a value of 28. *(Lesson 1–2)* D

Ⓐ $4 + 3 \cdot 4$ Ⓑ $75 \div 3 - 2$ Ⓒ $8 \cdot 4 - 8$ Ⓓ $(5 + 3) \cdot 7 \div 2$

www.algconcepts.com/self_check_quiz

Extra Credit

List the powers 4^3, 3^4, 10^0, and 0^{10} in order from least to greatest.
0^{10}, 10^0, 4^3, 3^4

4 ASSESS

Open-Ended Assessment

Writing Ask students to write their own descriptions of how to simplify the quotient of two powers when the bases are the same but the exponents are different.

***Enrichment*, p. 348**

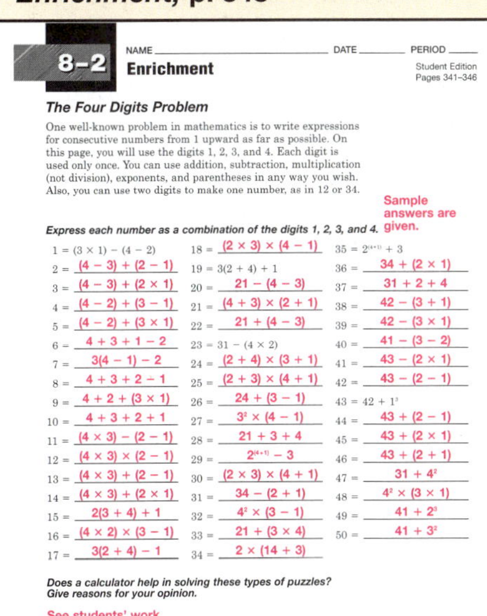

8–2 NAME ________ DATE ______ PERIOD ____

Enrichment

Student Edition Pages 341–346

The Four Digits Problem

One well-known problem in mathematics is to write expressions for consecutive numbers from 1 upward as far as possible. On this page, you will use the digits 1, 2, 3, and 4. Each digit is used only once. You can use addition, subtraction, multiplication (not division), exponents, and parentheses in any way you wish. Also, you can use two digits to make one number, as in 12 or 34.

Sample answers are given.

Express each number as a combination of the digits 1, 2, 3, and 4.

$1 = (3 \times 1) - (4 - 2)$	$18 = (2 \times 3) \times (4 - 1)$	$35 = 2^{(4+1)} + 3$
$2 = (4 - 3) + (2 - 1)$	$19 = 3(2 + 4) + 1$	$36 = 34 + (2 \times 1)$
$3 = (4 - 3) + (2 \times 1)$	$20 = 21 - (4 - 3)$	$37 = 31 + 2 + 4$
$4 = (4 - 2) + (3 - 1)$	$21 = (4 + 3) \times (2 + 1)$	$38 = 42 - (3 + 1)$
$5 = (4 - 2) + (3 \times 1)$	$22 = 21 + (4 - 3)$	$39 = 42 - (3 \times 1)$
$6 = 4 + 3 + 1 - 2$	$23 = 31 - (4 \times 2)$	$40 = 41 - (3 - 2)$
$7 = 3(4 - 1) - 2$	$24 = (2 + 4) \times (3 + 1)$	$41 = 43 - (2 \times 1)$
$8 = 4 + 3 + 2 - 1$	$25 = (2 + 3) \times (4 + 1)$	$42 = 43 - (2 - 1)$
$9 = 4 + 2 + (3 \times 1)$	$26 = 24 + (3 - 1)$	$43 = 42 + 1^3$
$10 = 4 + 3 + 2 + 1$	$27 = 3^2 \times (4 - 1)$	$44 = 43 + (2 - 1)$
$11 = (4 \times 3) - (2 - 1)$	$28 = 21 + 3 + 4$	$45 = 43 + (2 \times 1)$
$12 = (4 \times 3) \times (2 - 1)$	$29 = 2^{(4+1)} - 3$	$46 = 43 + (2 + 1)$
$13 = (4 \times 3) + (2 - 1)$	$30 = (2 \times 3) \times (4 + 1)$	$47 = 31 + 4^2$
$14 = (4 \times 3) + (2 \times 1)$	$31 = 34 - (2 + 1)$	$48 = 4^2 \times (3 \times 1)$
$15 = 2(3 + 4) + 1$	$32 = 4^2 \times (3 - 1)$	$49 = 41 + 2^3$
$16 = (4 \times 2) \times (3 - 1)$	$33 = 21 + (3 \times 4)$	$50 = 41 + 3^2$
$17 = 3(2 + 4) - 1$	$34 = 2 \times (14 + 3)$	

Does a calculator help in solving these types of puzzles? Give reasons for your opinion.

See students' work.

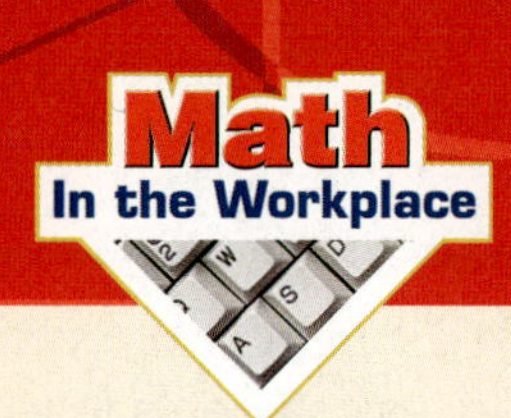

Broadcast technicians install, test, repair, and operate the electronic equipment used to record and transmit radio and television programs. They operate equipment that regulates the signal strength, clarity, and range of sounds and colors of recordings and broadcasts. At small stations, each technician performs a variety of duties, while at large stations technicians are more specialized. New technicians learn skills from experienced technicians and supervisors. They generally begin their careers in small stations and, once experienced, move on to larger ones.

The best way to prepare for a broadcast technician job in radio or television is to obtain training at a technical school, community college, or college. Prospective technicians should take high school courses in math, physics, and electronics.

Related Careers

- air traffic controllers
- radiological technologists
- cardiovascular technologists and technicians

Community Connection

As a class, identify different kinds of local broadcast media and have different groups of students arrange to tour facilities and see a broadcast prepared, packaged, and transmitted. Ask the groups to report on the technical aspects of the work they observed. Then, as a class, discuss the similarities and differences in the broadcasts for various kinds of technicians.

Math In the Workplace

Broadcast Technician

Do you dream of being in the movies? If you don't make it onto the big screen, you may find a career behind the scenes as a sound mixer. Sound mixers are broadcast technicians who develop movie sound tracks. Using a process called *dubbing*, they sit at sound consoles and fade in and fade out each sound by regulating its volume. All of the sounds for each scene are blended on a master sound track.

Sound intensity is measured in *decibels*. The decibel scale is based on powers of ten. The softest audible sound is represented by 10^0. The chart lists several common sounds and their intensity as compared to the softest audible sound.

Sound	Decibels	Intensity
jet airplane	140	10^{14}
rock band	120	10^{12}
motorcycle	110	10^{11}
circular saw	100	10^{10}
busy traffic	80	10^{8}
vacuum cleaner	70	10^{7}
noisy office	60	10^{6}
talking	40	10^{4}
whispering	20	10^{2}
breathing	10	10^{1}
softest sound	0	10^{0}

Find how many times as intense the first sound is as the second.

1. vacuum cleaner, noisy office **10**
2. motorcycle, busy traffic **10^3**
3. rock band, talking **10^8**
4. jet airplane, whispering **10^{12}**
5. Find two sounds, one of which is 10^2 times as intense as the other. **Sample answer: talking and whispering**

FAST FACTS About Broadcast Technicians

Working Conditions

- usually work indoors in pleasant conditions
- work a 40-hour week, but some overtime required to meet deadlines
- evening, weekend, and holiday work

Education

- high school math, physics, and electronics
- postsecondary training in engineering or electronics at a technical school or community college

Earnings

Median Salary by Station Type

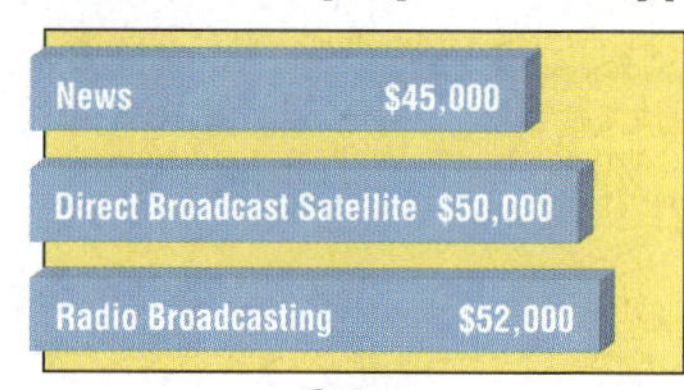

Salary

Source: *PayScale, Inc.*

interNET CONNECTION **Career Data** For up-to-date information about a career as a broadcast technician, visit: www.algconcepts.com

Not on the Net

If students have limited or no access to the Internet, they can find additional information by writing to the following organizations.

National Association of Broadcasters
Career Center
1771 N Street NW
Washington, DC 20036

National Association of Broadcast Employees and Technicians
501 Third Street NW
Washington, DC 20001

8–3 Negative Exponents

What You'll Learn
You'll learn to simplify expressions containing negative exponents.

Why It's Important
Electronics Electric current is measured using the prefixes *micro*, which is 10^{-6}, and *milli*, which is 10^{-3}. *See Exercise 14.*

Not all exponents are positive integers. Some exponents are negative integers. Study the pattern at the right to find the value of 10^{-1} and 10^{-2}. Extending the pattern suggests that $10^{-1} = \frac{1}{10}$ and $10^{-2} = \frac{1}{10^2}$ or $\frac{1}{100}$.

$10^3 = 1000$	$\div 10$
$10^2 = 100$	$\div 10$
$10^1 = 10$	$\div 10$
$10^0 = 1$	$\div 10$
$10^{-1} = ?$	$\div 10$
$10^{-2} = ?$	

When you multiply by the base, the exponent in the result increases by one. For example, $10^3 \times 10 = 10^4$. When you divide by the base, the exponent in the result decreases by one. For example, $10^{-2} \div 10 = 10^{-3}$.

You can use the Quotient of Powers rule and the definition of power to simplify the expression $\frac{x^3}{x^5}$ and write a definition of negative exponents.

Method 1
Quotient of Powers

$$\frac{x^3}{x^5} = x^{3-5} = x^{-2}$$

Method 2
Definition of Power

$$\frac{x^3}{x^5} = \frac{\overset{1}{\cancel{x}} \cdot \overset{1}{\cancel{x}} \cdot \overset{1}{\cancel{x}}}{\underset{1}{\cancel{x}} \cdot \underset{1}{\cancel{x}} \cdot \underset{1}{\cancel{x}} \cdot x \cdot x} = \frac{1}{x \cdot x} \text{ or } \frac{1}{x^2}$$

You can conclude that x^{-2} and $\frac{1}{x^2}$ are equal because $\frac{x^3}{x^5}$ cannot have two different values. This and other examples suggest the following definition.

Negative Exponents	
Numbers:	$5^{-2} = \frac{1}{5^2}; \frac{1}{4^{-3}} = 4^3$
Symbols:	$a^{-n} = \frac{1}{a^n}; \frac{1}{a^{-n}} = a^n$

The value of a cannot be zero.

1 FOCUS

5-Minute Check
Lesson 8–2

Simplify each expression.

1. $12^4 \cdot 12^7$ 12^{11}
2. $\frac{15^{11}}{15^3}$ 15^8
3. $\frac{a^9b^3}{a^2b^2}$ a^7b
4. $(4p^2q)\left(\frac{1}{2}p^7q^6\right)$ $2p^9q^7$
5. $\frac{35m^5n^5}{5n^2}$ $7m^5n^3$

Motivating the Lesson

Hands-On Activity Have each student fold a square piece of paper into 4 smaller squares, and tell them that each smaller square represents one unit. Tell them to cut their paper in half along one of the folds, throw one of the pieces away, and write the number of squares they have left. Have them repeat the process several times. Discuss the sequence of numbers they generate $(4, 2, 1, \frac{1}{2}, \frac{1}{4}, \ldots)$ and how they can use powers of 2 for the first few numbers in the sequence $(2^2, 2^1, 2^0)$. Then ask them to describe the pattern in the exponents.

Resource Manager

Reproducible Masters
Chapter 8 Resource Masters
- *Study Guide,* p. 349
- *Skills Practice,* p. 350
- *Practice,* p. 351
- *Reading to Learn Mathematics,* p. 352
- *Enrichment,* p. 353
- *Assessment,* p. 385

Hands-On Algebra, p. 91

Transparencies
5-Minute Check, 8–3

Technology/Multimedia
AlgePASS, Lesson 18
Interactive Chalkboard CD-ROM

2 TEACH

In-Class Example

Example 1

Write 6^{-2} using positive exponents. Then evaluate the expression. $\frac{1}{6^2} = \frac{1}{36}$

Teaching Tip After discussing Example 1, ask students to write 10^5 as a fraction with 1 as its numerator. Students should realize that they can write 10^5 as $\frac{1}{10^{-5}}$.

In-Class Examples

Examples 2–3

Simplify each expression.

2 q^3r^{-4} $\frac{q^3}{r^4}$

3 $\frac{m^2n^{10}}{m^5n^2}$ $\frac{n^8}{m^3}$

Example 1 Write 10^{-3} using positive exponents. Then evaluate the expression.

$10^{-3} = \frac{1}{10^3}$ *Definition of negative exponent*

$= \frac{1}{1000}$ or 0.001 $10 \cdot 10 \cdot 10 = 1000$

Your Turn

a. 2^{-4} $\frac{1}{2^4} = \frac{1}{16}$ b. 10^{-2} $\frac{1}{10^2} = \frac{1}{100}$ c. 5^{-1} $\frac{1}{5}$

To simplify an expression with a negative exponent, write an equivalent expression that has positive exponents. Each base should appear only once and all fractions should be in simplest form.

Examples Simplify each expression.

2 xy^{-2}

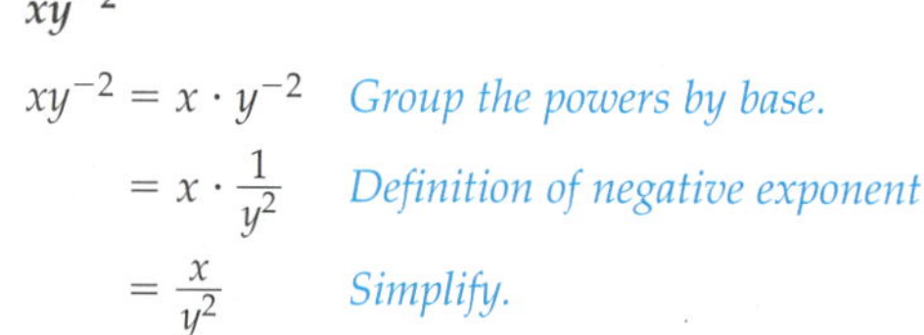

$xy^{-2} = x \cdot y^{-2}$ *Group the powers by base.*

$= x \cdot \frac{1}{y^2}$ *Definition of negative exponent*

$= \frac{x}{y^2}$ *Simplify.*

3 $\frac{a^5b}{a^3b^4}$

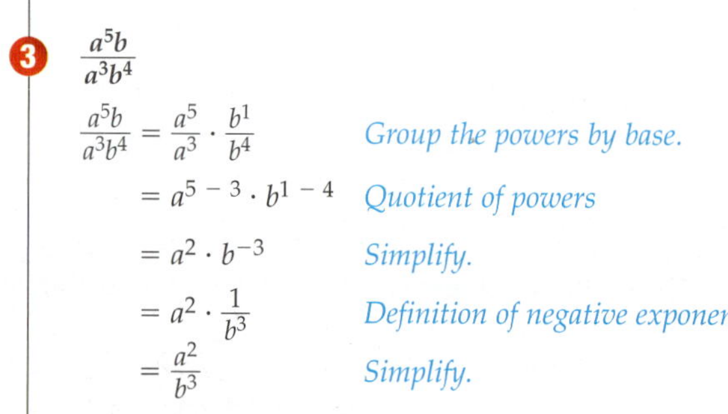

$\frac{a^5b}{a^3b^4} = \frac{a^5}{a^3} \cdot \frac{b^1}{b^4}$ *Group the powers by base.*

$= a^{5-3} \cdot b^{1-4}$ *Quotient of powers*

$= a^2 \cdot b^{-3}$ *Simplify.*

$= a^2 \cdot \frac{1}{b^3}$ *Definition of negative exponent*

$= \frac{a^2}{b^3}$ *Simplify.*

Your Turn

d. mn^{-3} $\frac{m}{n^3}$ e. $\frac{x^4y}{x^2y^5}$ $\frac{x^2}{y^4}$

4 Simplify $\frac{-6r^3s^5}{18r^{-7}s^5t^{-2}}$.

$\frac{-6r^3s^5}{18r^{-7}s^5t^{-2}} = \left(\frac{-6}{18}\right)\left(\frac{r^3}{r^{-7}}\right)\left(\frac{s^5}{s^5}\right)\left(\frac{1}{t^{-2}}\right)$ *Group the powers by base.*

$= \left(\frac{-1}{3}\right)\left(\frac{r^3}{r^{-7}}\right)\left(\frac{s^5}{s^5}\right)\left(\frac{t^0}{t^{-2}}\right)$ $\frac{-6}{18} = -\frac{1}{3}, 1 = t^0$

$= -\frac{1}{3}r^{3-(-7)}s^{5-5}t^{0-(-2)}$ *Quotient of powers*

$= -\frac{1}{3}r^{10}s^0t^2$ $3 - (-7) = 10, 0 - (-2) = 2$

$= -\frac{r^{10}t^2}{3}$ $s^0 = 1$

www.algconcepts.com/extra_examples

Your Turn

Simplify each expression.

f. $\dfrac{5a^4b^6}{-25a^{-2}b^6c^{-3}}$ $-\dfrac{a^6c^3}{5}$

g. $\dfrac{8x^3y^5}{10x^{-3}y^5z}$ $\dfrac{4x^6}{5z}$

Biology Link

5 The *E. coli* bacteria has a width of 10^{-3} millimeter. The head of a pin has a diameter of 1 millimeter. How many *E. coli* bacteria can fit across the head of a pin?

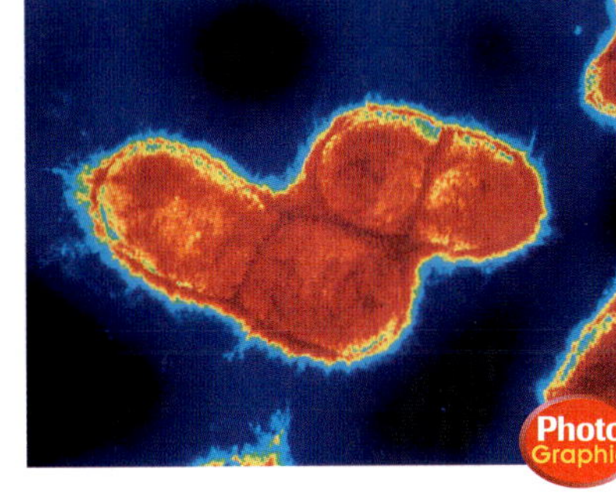

To find the number of bacteria, divide 1 by 10^{-3}.

$$\frac{1}{10^{-3}} = \frac{10^0}{10^{-3}} \quad 1 = 10^0$$
$$= 10^{0-(-3)} \quad \text{Quotient of powers}$$
$$= 10^3 \quad \text{Simplify.}$$

Since $10^3 = 1000$, about 1000 bacteria could fit across the head of a pin.

Your Turn

h. The figure below shows the electromagnetic wave spectrum. An ultraviolet wave has a length of 10^{-5} centimeter. An FM radio wave has a length of 10^2 centimeters. How many times as long is the FM wave as the ultraviolet wave? 10^7

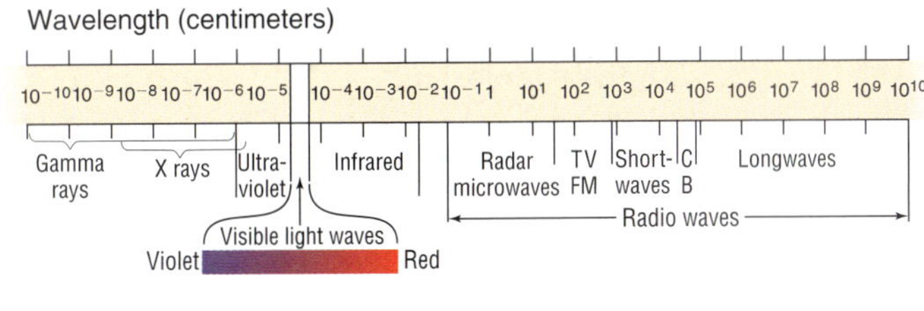

Check for Understanding

Communicating Mathematics

1. Express 5^{-2} using positive integers. Then evaluate the expression. $\frac{1}{5^2} = \frac{1}{25}$
2. Evaluate $6t^{-2}$ if $t = 3$. $\frac{2}{3}$

In-Class Examples

Example 4

Simplify $\dfrac{-10h^{-6}k^4}{25h^3k^7}$. $-\dfrac{2}{5h^9k^3}$

Example 5

A large archery target has a diameter of 1 meter. An arrow tip has a diameter of 10^{-2} meter. How many arrows could fit across the diameter of the target? **100 arrows**

3 PRACTICE/APPLY

Error Analysis

Watch for students who think that a power such as 5^{-2} in Exercise 1 represents a negative number. ***Prevent by*** having students start any problem with negative exponents by rewriting the problem using only positive exponents.

***Study Guide*, p. 349**

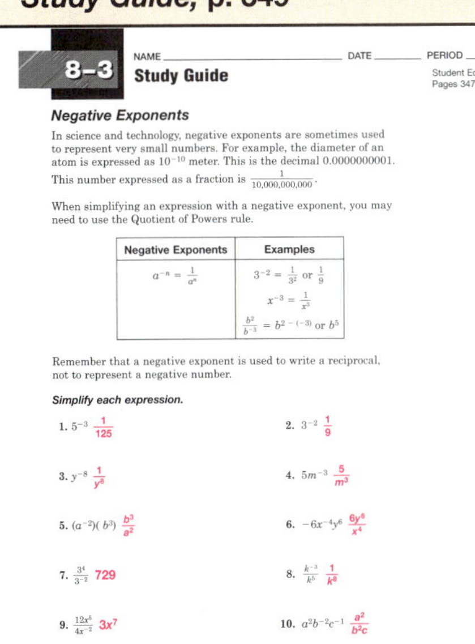

8–3 NAME ______ DATE ______ PERIOD ______

Study Guide

Student Edition Pages 347–351

Negative Exponents

In science and technology, negative exponents are sometimes used to represent very small numbers. For example, the diameter of an atom is expressed as 10^{-10} meter. This is the decimal 0.0000000001. This number expressed as a fraction is $\frac{1}{10,000,000,000}$.

When simplifying an expression with a negative exponent, you may need to use the Quotient of Powers rule.

Negative Exponents	Examples
$a^{-n} = \frac{1}{a^n}$	$3^{-2} = \frac{1}{3^2}$ or $\frac{1}{9}$ $x^{-3} = \frac{1}{x^3}$ $\frac{b^2}{b^{-3}} = b^{2-(-3)}$ or b^5

Remember that a negative exponent is used to write a reciprocal, not to represent a negative number.

Simplify each expression.

1. 5^{-3} $\frac{1}{125}$
2. 3^{-2} $\frac{1}{9}$
3. y^{-8} $\frac{1}{y^8}$
4. $5m^{-3}$ $\frac{5}{m^3}$
5. $(a^{-2})(b^3)$ $\frac{b^3}{a^2}$
6. $-6x^{-4}y^6$ $\frac{6y^6}{x^4}$
7. $\frac{3^4}{3^{-2}}$ 729
8. $\frac{k^{-3}}{k^5}$ $\frac{1}{k^8}$
9. $\frac{12x^5}{4x^{-2}}$ $3x^7$
10. $a^2b^{-2}c^{-1}$ $\frac{a^2}{b^2c}$
11. $\frac{-24m^6n^3}{3m^2n}$ $-8m^4n^2$
12. $\frac{36x^9y^5z^{-2}}{9x^3y^2}$ $\frac{4x^6y^3}{z^2}$

Assignment Guide

Basic: 15–47 odd, 48–57
Average: 16–44 even, 46–57
All: Quiz 1, 1–10

3. **You Decide** Booker and Antonio both correctly simplified $a^{-2} \cdot a^3$.

Booker's Method	**Antonio's Method**
$a^{-2} \cdot a^3 = a^{-2+3}$	$a^{-2} \cdot a^3 = \frac{1}{a^2} \cdot a^3$
$= a^1$ or a	$= \frac{a^3}{a^2}$ or a

a. Which student used the Product of Powers rule? **Booker**

b. Which student used the definition of negative exponents? **Antonio**

c. Whose method do you prefer? Explain your reasoning. **See students' work.**

Guided Practice

Example 1

Write each expression using positive exponents. Then evaluate the expression.

4. 10^{-4} $\frac{1}{10^4} = \frac{1}{10{,}000}$
5. 3^{-3} $\frac{1}{3^3} = \frac{1}{27}$

Simplify each expression.

Example 2

6. z^{-1} $\frac{1}{z}$
7. n^{-3} $\frac{1}{n^3}$
8. $s^{-2}t^3$ $\frac{t^3}{s^2}$
9. $p^{-1}q^{-2}r^2$ $\frac{r^2}{pq^2}$

Examples 3–5

10. $\frac{x^2}{x^3}$ $\frac{1}{x}$
11. $\frac{a^3}{a^{-4}}$ a^7
12. $\frac{m^4n^{-2}}{m^6n^2}$
13. $\frac{3a^3bc^5}{27a^4bc^2}$ $\frac{c^3}{9a}$

12. $\frac{1}{m^2n^4}$

Example 1

14. **Electronics** Electric current can be measured in amperes, milliamperes, or microamperes. The prefixes *milli* and *micro* mean 10^{-3} and 10^{-6}, respectively. Express 10^{-3} and 10^{-6} using positive exponents. $\frac{1}{10^3}$; $\frac{1}{10^6}$

Exercises

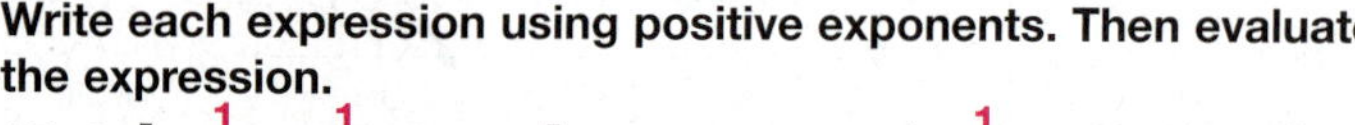

Practice

Write each expression using positive exponents. Then evaluate the expression.

15. 2^{-5} $\frac{1}{2^5} = \frac{1}{32}$
16. 10^{-5} $\frac{1}{10^5} = \frac{1}{100{,}000}$
17. 4^{-1} $\frac{1}{4}$
18. 6^{-2} $\frac{1}{6^2} = \frac{1}{36}$

Homework Help

For Exercises	See Examples
15–18, 45–47	1
19–26	2
27–42	3

Extra Practice
See page 708.

Simplify each expression.

19. r^{-10} $\frac{1}{r^{10}}$
20. p^{-6} $\frac{1}{p^6}$
21. $a^4(a^{-2})$ a^2
22. $x^{-7}(x^4)$ $\frac{1}{x^3}$
23. $s^{-2}t^4$ $\frac{t^4}{s^2}$
24. $a^0b^{-1}c^{-2}$ $\frac{1}{bc^2}$
25. $15rs^{-2}$ $\frac{15r}{s^2}$
26. $10x^{-4}y^{-5}z$
27. $\frac{m^2}{m^{-4}}$ m^6
28. $\frac{x^2}{x^3}$ $\frac{1}{x}$
29. $\frac{k^{-2}}{k^6}$ $\frac{1}{k^8}$
30. $\frac{1}{r^{-3}}$ r^3

26. $\frac{10z}{x^4y^5}$

31. $\frac{an^3}{n^5}$ $\frac{a}{n^2}$
32. $\frac{bm^2}{m^6}$ $\frac{b}{m^4}$
33. $\frac{x^3y^{-3}}{x^3y^6}$ $\frac{1}{y^9}$
34. $\frac{a^5b^{-3}}{a^7b^3}$ $\frac{1}{a^2b^6}$
35. $\frac{12b^5}{4b^{-4}}$ $3b^9$
36. $\frac{24c^6}{4c^{-2}}$ $6c^8$
37. $\frac{7x^4}{28x}$ $\frac{x^3}{4}$
38. $\frac{20y^5}{40y^{-2}}$ $\frac{y^7}{2}$
39. $\frac{4x^3}{28x}$ $\frac{x^2}{7}$
40. $\frac{-15r^5s^8}{5r^5s^2}$ $-3s^6$
41. $\frac{5ac}{8ab^5c^2}$ $\frac{5}{8b^5c}$
42. $\frac{12c^3d^4f^6}{60cd^6f^3}$

42. $\frac{c^2f^3}{5d^2}$

Skills Practice, p. 351, and Practice, p. 352 (shown)

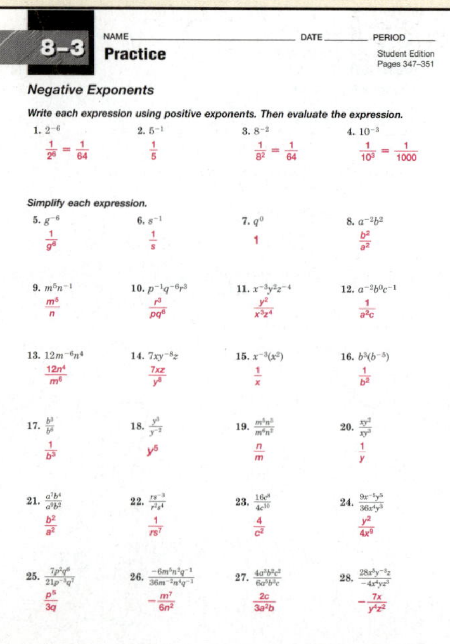

NAME ________ DATE ________ PERIOD ______

8-3 Practice

Student Edition Pages 347–351

Negative Exponents

Write each expression using positive exponents. Then evaluate the expression.

1. 2^{-6} $\frac{1}{2^6} = \frac{1}{64}$
2. 5^{-1} $\frac{1}{5}$
3. 8^{-2} $\frac{1}{8^2} = \frac{1}{64}$
4. 10^{-3} $\frac{1}{10^3} = \frac{1}{1000}$

Simplify each expression.

5. g^{-6} $\frac{1}{g^6}$
6. s^{-1} $\frac{1}{s}$
7. q^0 1
8. $a^{-2}b^2$ $\frac{b^2}{a^2}$
9. m^5n^{-1} $\frac{m^5}{n}$
10. $p^{-1}q^{-6}r^3$ $\frac{r^3}{pq^6}$
11. $x^{-3}y^2z^{-4}$ $\frac{y^2}{x^3z^4}$
12. $a^{-2}b^0c^{-1}$ $\frac{1}{a^2c}$
13. $12m^{-6}n^4$ $\frac{12n^4}{m^6}$
14. $7xy^{-8}z$ $\frac{7xz}{y^8}$
15. $x^{-3}(x^2)$ $\frac{1}{x}$
16. $b^3(b^{-5})$ $\frac{1}{b^2}$
17. $\frac{b^3}{b^6}$ $\frac{1}{b^3}$
18. $\frac{y^3}{y^{-2}}$ y^5
19. $\frac{m^5n^3}{m^6n^2}$ $\frac{n}{m}$
20. $\frac{xy^2}{xy^3}$ $\frac{1}{y}$
21. $\frac{a^7b^4}{a^9b^2}$ $\frac{b^2}{a^2}$
22. $\frac{rs^{-3}}{r^2s^4}$ $\frac{1}{rs^7}$
23. $\frac{16c^8}{4c^{10}}$ $\frac{4}{c^2}$
24. $\frac{9x^{-5}y^5}{36x^4y^3}$ $\frac{y^2}{4x^9}$
25. $\frac{7p^2q^6}{21p^{-3}q^7}$ $\frac{p^5}{3q}$
26. $\frac{-6m^5n^2q^{-1}}{36m^{-2}n^4q^{-1}}$ $-\frac{m^7}{6n^2}$
27. $\frac{4a^3b^2c^2}{6a^5b^3c}$ $\frac{2c}{3a^2b}$
28. $\frac{28x^5y^{-3}z}{-4x^4yz^3}$ $-\frac{7x}{y^4z^2}$

Reteaching Activity

Verbal/Linguistic Learners Group students in pairs or small groups. Have one student read one of the examples in the lesson. The other students should then give a verbal description of how the simplified form was found. Repeat the activity until each group member has read an example from the lesson.

43. 18

44. $-\frac{1}{64}$

45. 2^{-4}

43. Evaluate $4x^{-3}y^2$ if $x = 2$ and $y = 6$.

44. Find the value of $(2b)^{-3}$ if $b = -2$.

45. Which is greater, 2^{-4} or 2^{-6}?

Applications and Problem Solving

46. $10^{-5} = \frac{1}{10^5}$ or $\frac{1}{100{,}000}$; $10^{-4} = \frac{1}{10^4}$ or $\frac{1}{10{,}000}$

46. **Physical Science** Visible light waves have wavelengths between 10^{-5} centimeter and 10^{-4} centimeter. Express 10^{-5} and 10^{-4} using positive exponents. Then evaluate each expression.

47. **Physical Science** Refer to the figure on page 349 in Your Turn, part h. Some infrared waves have lengths of 10^{-3} centimeter. Which kind of wave has a length that is 1000 times as long as an infrared wave? **radar microwaves**

48. **Critical Thinking** Which point on the number line could be the graph of n^{-2} if n is a positive integer? **C**

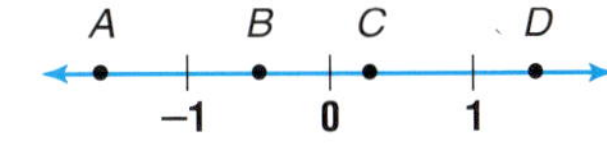

Mixed Review

Simplify each expression. *(Lesson 8–2)*

49. $(5a^3)(-2a^4)$ $-10a^7$
50. $x^4 \cdot x$ x^5
51. $\frac{n^{10}}{n^4}$ n^6
52. $\frac{15a^3b^6c^2}{-5ab^2c^2}$ $-3a^2b^4$

Write each expression using exponents. *(Lesson 8–1)*

53. $z \cdot z \cdot z \cdot z$ z^4
54. $(3)(3)(-2)(-2)(-2)$ $3^2(-2)^3$
55. 9 9^1

Standardized Test Practice

56. **Grid In** Determine the slope of the line passing through $(1, -2)$ and $(6, 2)$. *(Lesson 7–1)* **4/5**

57. **Multiple Choice** The Jaguars softball team played 8 games and scored a total of 96 runs. What was the mean number of runs scored per game? *(Lesson 3–3)* **B**

 8  12 (C) 88 (D) 104

Quiz 1 Lessons 8–1 through 8–3

Write each expression using exponents. *(Lesson 8–1)*

1. $6 \cdot 6 \cdot 6$ 6^3
2. $x \cdot x \cdot x \cdot x$ x^4
3. 10 10^1
4. $(-3)(-3)(-3)$ $(-3)^3$
5. **Number Theory** The prime factorization of 96 is $2 \cdot 2 \cdot 2 \cdot 2 \cdot 2 \cdot 3$. Write the prime factorization using exponents. *(Lesson 8–1)* $2^5 3$

Simplify each expression. *(Lessons 8–2 & 8–3)* 8. $6x^6y^2$

6. $(m^4)(m^6)$ m^{10}
7. $\frac{x^7}{x^2}$ x^5
8. $(-3x^2y^3)(-2x^4y^{-1})$
9. $\frac{3a^2b}{-9a^2b^4}$ $-\frac{1}{3b^3}$

10. **Biology** The length of a *Euglena* protist is about 10^{-2} centimeter. Express 10^{-2} using positive exponents. Then evaluate the expression. $\frac{1}{10^2} = \frac{1}{100}$

www.algconcepts.com/self_check_quiz

4 ASSESS

Open-Ended Assessment

Speaking Ask students to describe the difference between a number multiplied by a negative integer and that number raised to the negative integer.

Quiz 1

The Quiz provides students with a brief review of the concepts and skills in Lessons 8–1 through 8–3. Lesson numbers are given to the right of the exercises or instruction lines so students can review concepts not yet mastered.

Chapter 8, Quiz A (Lessons 8–1 through 8–3) is available in the *Chapter 8 Resource Masters*, p. 385.

Extra Credit

Is the value of the expression $-5a^3b^2$ greater when $a = -3$ and $b = -2$ or when $a = -2$ and $b = -3$? How much greater? **when $a = -3$, $b = -2$; greater by 180**

***Enrichment*, p. 353**

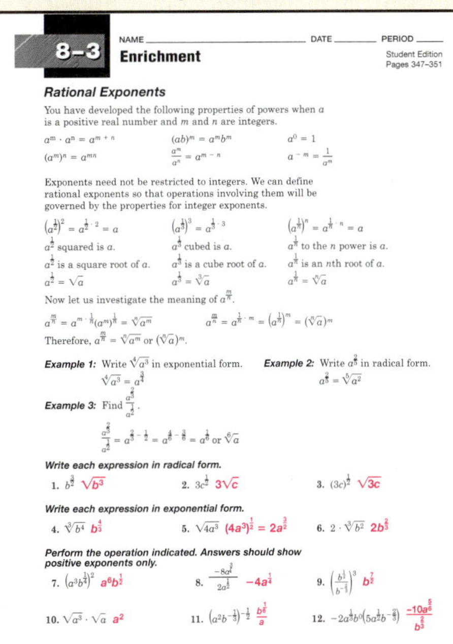

8–3 NAME ______ DATE ______ PERIOD ______

Enrichment Student Edition Pages 347–351

Rational Exponents

You have developed the following properties of powers when a is a positive real number and m and n are integers.

$a^m \cdot a^n = a^{m+n}$ $(ab)^m = a^mb^m$ $a^0 = 1$

$(a^m)^n = a^{mn}$ $\frac{a^m}{a^n} = a^{m-n}$ $a^{-m} = \frac{1}{a^m}$

Exponents need not be restricted to integers. We can define rational exponents so that operations involving them will be governed by the properties for integer exponents.

$\left(a^{\frac{1}{2}}\right)^2 = a^{\frac{1}{2}\cdot 2} = a$	$\left(a^{\frac{1}{3}}\right)^3 = a^{\frac{1}{3}\cdot 3}$	$\left(a^{\frac{1}{n}}\right)^n = a^{\frac{1}{n}\cdot n} = a$
$a^{\frac{1}{2}}$ squared is a.	$a^{\frac{1}{3}}$ cubed is a.	$a^{\frac{1}{n}}$ to the n power is a.
$a^{\frac{1}{2}}$ is a square root of a.	$a^{\frac{1}{3}}$ is a cube root of a.	$a^{\frac{1}{n}}$ is an nth root of a.
$a^{\frac{1}{2}} = \sqrt{a}$	$a^{\frac{1}{3}} = \sqrt[3]{a}$	$a^{\frac{1}{n}} = \sqrt[n]{a}$

Now let us investigate the meaning of $a^{\frac{m}{n}}$.

$a^{\frac{m}{n}} = a^{m\cdot\frac{1}{n}}(a^m)^{\frac{1}{n}} = \sqrt[n]{a^m}$ $a^{\frac{m}{n}} = a^{\frac{1}{n}\cdot m} = \left(a^{\frac{1}{n}}\right)^m = (\sqrt[n]{a})^m$

Therefore, $a^{\frac{m}{n}} = \sqrt[n]{a^m}$ or $(\sqrt[n]{a})^m$.

Example 1: Write $\sqrt[4]{a^3}$ in exponential form.
$\sqrt[4]{a^3} = a^{\frac{3}{4}}$

Example 2: Write $a^{\frac{2}{5}}$ in radical form.
$a^{\frac{2}{5}} = \sqrt[5]{a^2}$

Example 3: Find $\frac{a^{\frac{2}{3}}}{a^{\frac{1}{2}}}$.

$\frac{a^{\frac{2}{3}}}{a^{\frac{1}{2}}} = a^{\frac{2}{3}-\frac{1}{2}} = a^{\frac{4}{6}-\frac{3}{6}} = a^{\frac{1}{6}}$ or $\sqrt[6]{a}$

Write each expression in radical form.

1. $b^{\frac{3}{2}}$ $\sqrt{b^3}$
2. $3c^{\frac{1}{2}}$ $3\sqrt{c}$
3. $(3c)^{\frac{1}{2}}$ $\sqrt{3c}$

Write each expression in exponential form.

4. $\sqrt[3]{b^4}$ $b^{\frac{4}{3}}$
5. $\sqrt{4a^3}$ $(4a^3)^{\frac{1}{2}} = 2a^{\frac{3}{2}}$
6. $2 \cdot \sqrt[3]{b^2}$ $2b^{\frac{2}{3}}$

Perform the operation indicated. Answers should show positive exponents only.

7. $\left(a^3b^{\frac{1}{4}}\right)^2$ $a^6b^{\frac{1}{2}}$
8. $\frac{-8a^{\frac{3}{4}}}{2a^{\frac{1}{2}}}$ $-4a^{\frac{1}{4}}$
9. $\left(\frac{b^{\frac{1}{2}}}{b^{-\frac{2}{3}}}\right)^3$ $b^{\frac{7}{2}}$
10. $\sqrt{a^3} \cdot \sqrt{a}$ a^2
11. $\left(a^2b^{-\frac{1}{3}}\right)^{-\frac{1}{2}}$ $\frac{b^{\frac{1}{6}}}{a}$
12. $-2a^{\frac{1}{3}}b^0\left(5a^{\frac{1}{2}}b^{-\frac{2}{3}}\right)$ $\frac{-10a^{\frac{5}{6}}}{b^{\frac{2}{3}}}$

8–4 Scientific Notation

1 FOCUS

5-Minute Check
Lesson 8–3

Write each expression using positive exponents.

1. 7^{-3} $\frac{1}{7^3} = \frac{1}{343}$
2. 2^{-6} $\frac{1}{2^6} = \frac{1}{64}$

Simplify each expression.

3. $6p^2q^{-2}$ $\frac{6p^2}{q^2}$
4. $\frac{-18w^6z^{-3}}{12w^{-2}z^8}$ $-\frac{3w^8}{2z^{11}}$
5. A space probe has traveled 10^8 kilometers since it was launched. The length of the probe is 10^{-3} kilometers. What is the quotient of the distance traveled divided by the length of the probe? **10^{11} km**

Motivating the Lesson

Hands-On Activity Provide each student with a calculator or use a large calculator that every student can see. Multiply several large numbers together until the calculator display shows the product using exponential notation. Ask the students to continue multiplying and dividing by 10, 100, and 1000, and to describe how the display changes each time.

2 TEACH

Teaching Tip As students read the metric prefixes, they should realize that these prefixes apply to all metric units. Once they know the meaning of the prefix *mega*, they can say that 1 megameter is 10^6 meters, 1 megaliter is 10^6 liters, and 1 megagram is 10^6 grams.

What You'll Learn
You'll learn to express numbers in scientific notation.

Why It's Important
Fiber Optics
Laser technicians use scientific notation when they deal with the speed of light, 3×10^8 meters per second.
See Example 7.

The prefixes *mega*, *giga*, and *kilo* are metric prefixes. They are used with very large measures. Other prefixes are used with very small measures. The chart shows some metric prefixes.

Metric Prefixes

Prefix	Power of 10	Meaning	Prefix	Power of 10	Meaning
tera	10^{12}	1,000,000,000,000	pico	10^{-12}	0.000000000001
giga	10^9	1,000,000,000	nano	10^{-9}	0.000000001
mega	10^6	1,000,000	micro	10^{-6}	0.000001
kilo	10^3	1000	milli	10^{-3}	0.001

Scientific fields and industry use metric units because calculations are easier with powers of ten. This is also why metric units, based on powers of ten, are widely used in science.

When you multiply a number by a power of ten, the nonzero digits in the original number and the product are the same. Only the position of the decimal point is different.

Prerequisite Skills Review
Operations with Decimals, p. 684

$5 \times 10^2 = 5 \times 100$
$= 500$ *2 places right*

$8.23 \times 10^4 = 8.23 \times 10{,}000$
$= 82{,}300$ *4 places right*

$4 \times 10^{-1} = 4 \times 0.1$
$= 0.4$ *1 place left*

$1.23 \times 10^{-3} = 1.23 \times 0.001$
$= 0.00123$ *3 places left*

These and similar examples suggest the following rules for multiplying a number by a power of ten.

Multiplying by Powers of 10	• If the exponent is *positive*, move the decimal point to the *right*. • If the exponent is *negative*, move the decimal point to the *left*.

Resource Manager

Reproducible Masters
Chapter 8 Resource Masters
- *Study Guide*, p. 354
- *Skills Practice*, p. 355
- *Practice*, p. 356
- *Reading to Learn Mathematics*, p. 357
- *Enrichment*, p. 358
- *Assessment*, p. 384

Transparencies
5-Minute Check, 8–4

Technology/Multimedia
AlgePASS, Lesson 19
Interactive Chalkboard CD-ROM

Examples

Express each measurement in standard form.

1 2 megabytes

2 megabytes = 2×10^6 bytes *The prefix mega- means 10^6.*
= 2,000,000 bytes *Move the decimal point 6 places right.*

2 3.6 nanoseconds

3.6 nanoseconds = 3.6×10^{-9} seconds *The prefix nano- means 10^{-9}.*
= 0.0000000036 seconds *Move the decimal point 9 places left.*

Reading Algebra

The absolute value of the exponent tells the number of places to move the decimal point.

Your Turn a. 2,000,000,000 bytes

a. 2 gigabytes b. 3.4 milliseconds **0.0034 s**

When you use very large numbers like 5,800,000 or very small numbers like 0.000076, it is difficult to keep track of the place value. Numbers such as these can be written in **scientific notation**.

Scientific Notation	A number is expressed in scientific notation when it is in the form $a \times 10^n$, where $1 \leq a < 10$ and n is an integer.

Follow these steps to write a number in scientific notation.

- First, move the decimal point after the first nonzero digit.
- Then, find the power of ten by counting the decimal places.
- When the number is greater than one, the exponent of 10 is *positive*.
- When the number is between zero and one, the exponent of 10 is *negative*.

Examples

Express each number in scientific notation.

3 5,800,000

5,800,000 = $5.8 \times 10^?$ *The decimal point moves 6 places.*
= 5.8×10^6 *Since 5,800,000 is greater than one, the exponent is positive.*

4 0.000076 in scientific notation.

0.000076 = $7.6 \times 10^?$ *The decimal point moves 5 places.*
= 7.6×10^{-5} *Since 0.000076 is between zero and one, the exponent is negative.*

Your Turn

c. 3,900,000,000 **3.9×10^9** d. 0.0000035 **3.5×10^{-6}**

www.algconcepts.com/extra_examples

In-Class Examples
Examples 1–2

Express each measurement in standard form.

1 8 kilobytes **8000 bytes**
2 2.5 microseconds **0.0000025 s**

Teaching Tip In the description of scientific notation, students should notice that there are two different inequality symbols, $\leq$ and $<$, used in the range of values for a. Point out that this statement means that a can equal 1 but cannot equal 10.

In-Class Examples
Examples 3–4

Express each number in scientific notation.

3 325,000 **3.25×10^5**
4 0.00028 **2.8×10^{-4}**

You can use scientific notation to simplify computation.

Examples

Evaluate each expression.

5 **$400 \times 2{,}000{,}000{,}000$**

First express each number in scientific notation. Then use the Associative and Commutative Properties to regroup terms.

$400 \times 2{,}000{,}000{,}000 = (4 \times 10^2)(2 \times 10^9)$ *Write in scientific notation.*

$= (4 \times 2)(10^2 \times 10^9)$ *Associative and Commutative Properties*

$= 8 \times 10^{11}$ *Multiply.*

$= 800{,}000{,}000{,}000$ *Write in standard form.*

Technology Tip

On a graphing calculator, 8×10^{11} is shown as 8E11.

6 $\dfrac{4.8 \times 10^3}{1.6 \times 10^1}$

$\dfrac{4.8 \times 10^3}{1.6 \times 10^1} = \left(\dfrac{4.8}{1.6}\right)\left(\dfrac{10^3}{10^1}\right)$ $\quad \dfrac{4.8}{1.6} = 3$

$= 3 \times 10^2$ or 300 *Simplify.*

Your Turn

e. $2000 \times 3{,}000{,}000{,}000$ **6×10^{12} or 6,000,000,000,000**

f. $\dfrac{7.5 \times 10^7}{1.5 \times 10^4}$ **5×10^3 or 5000**

Physics Link

7 **The light from a laser beam travels at a speed of 300,000,000 meters per second. How far does the light travel in 2 nanoseconds? Use the formula $d = rt$, where d is the distance in meters, r is the speed of light, and t is the time in seconds.**

Express 300,000,000 in scientific notation.
Express 2 nanoseconds in seconds.

$300{,}000{,}000 = 3 \times 10^8$ and 2 nanoseconds $= 2 \times 10^{-9}$ seconds

Method 1 Paper and Pencil

$d = rt$ *Formula for distance*

$d = (3 \times 10^8)(2 \times 10^{-9})$ *$r = 3 \times 10^8$ and $t = 2 \times 10^{-9}$*

$d = (3 \times 2)(10^8 \times 10^{-9})$ *Associative Property*

$d = 6 \times 10^{-1}$ *Multiply.*

Method 2 Calculator

3 [2nd] [EE] 8 [×] 2 [2nd] [EE] [(−)] 9 [ENTER] *0.6*

The light travels 6×10^{-1} meter, or 0.6 meter, in 2 nanoseconds.

Teaching Tip In Example 5, be sure students see that the leading digits 4 and 2 are multiplied. Students may sometimes find the correct exponent for 10 but forget to multiply the two other numbers.

In-Class Examples

Examples 5–6

Evaluate each expression.

5 $30 \times 30{,}000{,}000$ **900,000,000**

6 $\dfrac{5.4 \times 10^8}{2 \times 10^5}$ **2.7×10^3 or 2700**

Example 7

A light ray from the sun travels about 186,000 miles per second. How far does it travel in 5×10^2 seconds? Use the formula $d = rt$, where r is the speed of light and t is the time in seconds. **93,000,000 mi**

***Study Guide*, p. 354**

8–4 NAME ______ DATE ______ PERIOD ______

Study Guide

Student Edition Pages 352–356

Scientific Notation

In science and in other applications, **scientific notation** is often used to represent very small or very large numbers. For example, the speed of light is about 3×10^8 meters per second. This represents the number 300,000,000 meters per second.

In scientific notation, a number is expressed in the form $a \times 10^n$, where a is a number greater than 1 and less than 10 and n is an integer.

Scientific Notation	Examples
1. Place the decimal point after the first non-zero digit in the given number. 2. Find the power of 10 by counting decimal places. When the given number is one or greater, the power of 10 is positive. When the given number is between zero and one, the exponent of 10 is negative.	$6{,}200{,}000 = 6.2 \times 10^6$ $0.000056 = 5.6 \times 10^{-5}$

Express each number in standard form.

1. 6.1×10^4 **61,000**
2. 4.8×10^{-2} **0.048**
3. 8.12×10^3 **8120**
4. 5×10^7 **50,000,000**
5. 9×10^{-5} **0.00009**
6. 1.1×10^{-7} **0.00000011**
7. 2.15×10^5 **215,000**
8. 5.1651×10^3 **5165.1**

Express each number in scientific notation.

9. 8400 **8.4×10^3**
10. 3,000,000 **3×10^6**
11. 0.05 **5×10^{-2}**
12. 14.2 **1.42×10**
13. 0.00048 **4.8×10^{-4}**
14. 82,000,000,000 **8.2×10^{10}**
15. 0.0000725 **7.25×10^{-5}**
16. 6 **6×10^0**

Reteaching Activity

Kinesthetic Learners Write each of the powers 10^{-10}, 10^{-9}, 10^{-8}, ..., 10^9, 10^{10}, on a separate sheet of paper. First ask students to display the powers in order, perhaps by taping the sheets along a wall of the classroom. Then have students take turns writing decimals whose values are between 10^{-10} and 10^{10} on the board or overhead. Have volunteers point to the appropriate space between two of the sheets where each value would be placed so it is in numerical order with the powers.

Check for Understanding

Communicating Mathematics

1. **Tell** whether 23.5×10^3 is expressed in scientific notation. Explain your answer. **no, 2.35×10^4**
2. **Explain** an advantage of expressing very large or very small numbers in scientific notation.
3. *Writing Math* Find two very large numbers and two very small numbers in a newspaper. Write each number in standard form and in scientific notation. **See students' work.**

2. Sample answer: to keep track of the place value

Vocabulary
scientific notation

Guided Practice

Getting Ready Find each product.

Sample 1: 2×10^3
Solution: 2000 *three places right*

Sample 2: 3.4×10^{-1}
Solution: 0.34 *one place left*

4. 2.45×10^2 **245**
5. 6.8×10^4 **68,000**
6. 2×10^6 **2,000,000**
7. 6.4×10^{-1} **0.64**
8. 9.23×10^{-2} **0.0923**
9. 3×10^{-4} **0.0003**

Express each measure in standard form.

10. 5 megaohms **5,000,000**
11. 6.5 milliamperes **0.0065**

Express each number in scientific notation.

12. 9500 **9.5×10^3**
13. 56.9 **5.69×10**
14. 0.0087 **8.7×10^{-3}**
15. 0.000023 **2.3×10^{-5}**

Evaluate each expression. Express each result in scientific notation and standard form.

16. $(2 \times 10^5)(3 \times 10^{-8})$ **$6 \times 10^{-3} = 0.006$**
17. $\frac{5.2 \times 10^5}{2 \times 10^2}$ **$2.6 \times 10^3 = 2600$**
18. **Health** The length of the virus that causes AIDS is 0.00011 millimeter. Express 0.00011 in scientific notation. **1.1×10^{-4}**

20. 4,000,000
21. 0.0000000039
Examples 1 & 2
25. 5.28×10^3
26. 2.4×10^5
Examples 3 & 4
27. 2.683×10^2
28. 2.5×10^7
Examples 5–7
29. 3.2×10^{-4}
30. 8×10^{-2}
31. 4.296×10^{-3}
Example 4
32. 1.59×10

Exercises

Practice A

For Exercises	See Examples
19–24, 45–47	1, 2
25–36, 48	3, 4
37–40	5
41–43	6
44, 49	7

Homework Help
Extra Practice
See page 708.

Express each measure in standard form. **19. 5,800,000,000**

19. 5.8 billion dollars
20. 4 megahertz
21. 3.9 nanoseconds
22. 82 kilobytes **82,000**
23. 9 milliamperes **0.009**
24. 2.3 micrograms **0.0000023**

Express each number in scientific notation.

25. 5280
26. 240,000
27. 268.3
28. 25,000,000
29. 0.00032
30. 0.08
31. 0.004296
32. 15.9
33. 0.012 **1.2×10^{-2}**
34. 1,000,000 **1×10^6**
35. 0.000000022 **2.2×10^{-8}**
36. 0.0000946 **9.46×10^{-5}**

Evaluate each expression. Express each result in scientific notation and standard form. **37. $6 \times 10^6 = 6,000,000$** **38. $6 \times 10^8 = 600,000,000$**

B

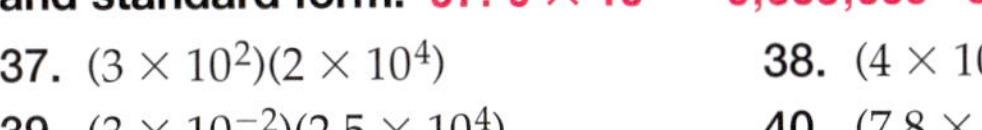

37. $(3 \times 10^2)(2 \times 10^4)$
38. $(4 \times 10^2)(1.5 \times 10^6)$
39. $(3 \times 10^{-2})(2.5 \times 10^4)$ **$7.5 \times 10^2 = 750$**
40. $(7.8 \times 10^{-6})(1 \times 10^{-2})$ **$7.8 \times 10^{-8} = 0.000000078$**

3 PRACTICE/APPLY

Error Analysis

Watch for students who rewrite 9500 as 9.5×10^4 in Exercise 12 because they counted the digits in the number 9500 to determine the exponent.

Prevent by having students draw two arrows under the number 9500, one pointing between the first and second digits and the other pointing to the decimal point. Have students count just the digits between the two arrows to determine the correct exponent.

Assignment Guide

Basic: 19–49 odd, 50–61
Average: 20–44 even, 46–61

***Skills Practice,* p. 355, and *Practice,* p. 356 (shown)**

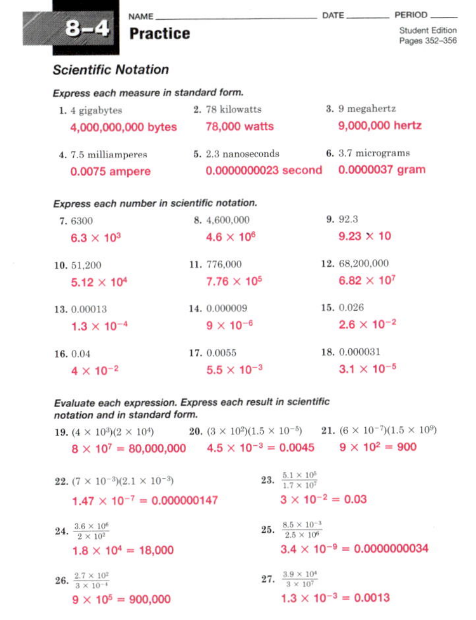

8–4 NAME ______ DATE ______ PERIOD ______

Practice Student Edition Pages 352–356

Scientific Notation

Express each measure in standard form.

1. 4 gigabytes **4,000,000,000 bytes**
2. 78 kilowatts **78,000 watts**
3. 9 megahertz **9,000,000 hertz**
4. 7.5 milliamperes **0.0075 ampere**
5. 2.3 nanoseconds **0.0000000023 second**
6. 3.7 micrograms **0.0000037 gram**

Express each number in scientific notation.

7. 6300 **6.3×10^3**
8. 4,600,000 **4.6×10^6**
9. 92.3 **9.23×10**
10. 51,200 **5.12×10^4**
11. 776,000 **7.76×10^5**
12. 68,200,000 **6.82×10^7**
13. 0.00013 **1.3×10^{-4}**
14. 0.000009 **9×10^{-6}**
15. 0.026 **2.6×10^{-2}**
16. 0.04 **4×10^{-2}**
17. 0.0055 **5.5×10^{-3}**
18. 0.000031 **3.1×10^{-5}**

Evaluate each expression. Express each result in scientific notation and in standard form.

19. $(4 \times 10^3)(2 \times 10^4)$ **$8 \times 10^7 = 80,000,000$**
20. $(3 \times 10^2)(1.5 \times 10^{-5})$ **$4.5 \times 10^{-3} = 0.0045$**
21. $(6 \times 10^{-7})(1.5 \times 10^9)$ **$9 \times 10^2 = 900$**
22. $(7 \times 10^{-3})(2.1 \times 10^{-3})$ **$1.47 \times 10^{-7} = 0.000000147$**
23. $\frac{5.1 \times 10^5}{1.7 \times 10^7}$ **$3 \times 10^{-2} = 0.03$**
24. $\frac{3.6 \times 10^6}{2 \times 10^2}$ **$1.8 \times 10^4 = 18,000$**
25. $\frac{8.5 \times 10^{-3}}{2.5 \times 10^6}$ **$3.4 \times 10^{-9} = 0.0000000034$**
26. $\frac{2.7 \times 10^2}{3 \times 10^{-4}}$ **$9 \times 10^5 = 900,000$**
27. $\frac{3.9 \times 10^4}{3 \times 10^7}$ **$1.3 \times 10^{-3} = 0.0013$**

4 ASSESS

Open-Ended Assessment

Writing Ask students to write a paragraph explaining how they remember when they move the decimal point to the right and when they move the decimal point to the left when they rewrite 6.0×10^3 and 6.0×10^{-3} in standard form.

Mid-Chapter Test (Lessons 8–1 through 8–4) is available in the *Chapter 8 Resource Masters*, p. 384.

Answer

47. Earth, Venus, Mars

Enrichment, p. 358

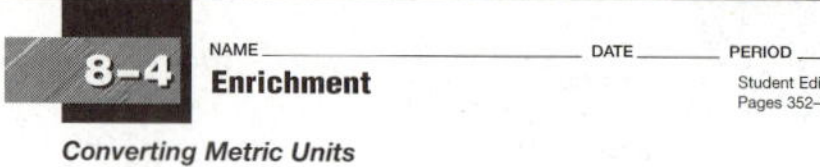

8–4 Enrichment

NAME ______ DATE ______ PERIOD ______

Student Edition Pages 352–356

Converting Metric Units

Scientific notation is convenient to use for unit conversions in the metric system.

Example 1: How many kilometers are there in 4,300,000 meters?

Divide the measure by the number of meters (1000) in one kilometer. Express both numbers in scientific notation.

$\frac{4.3 \times 10^6}{1 \times 10^3} = 4.3 \times 10^3$ The answer is 4.3×10^3 km.

Example 2: Convert 3700 grams into milligrams.

Multiply by the number of milligrams (1000) in 1 gram.

$(3.7 \times 10^3)(1 \times 10^3) = 3.7 \times 10^6$ There are 3.7×10^6 mg in 3700 g.

Complete the following. Express each answer in scientific notation.

1. 250,000 m = 2.5×10^2 km
2. 375 km = 3.75×10^5 m
3. 247 m = 2.47×10^4 cm
4. 5000 m = 5.0×10^6 mm
5. 0.0004 km = 4.0×10^{-1} m
6. 0.01 mm = 1.0×10^{-5} m
7. 6000 m = 6×10^{-6} mm
8. 340 cm = 3.4×10^{-3} km
9. 52,000 mg = 5.2×10^1 g
10. 420 kL = 4.2×10^5 L

Solve.

11. The planet Mars has a diameter of 6.76×10^3 km. What is the diameter of Mars in meters? Express the answer in both scientific and decimal notation.
6,760,000 m; 6.76×10^6 m

12. The distance of the earth from the sun is 149,590,000 km. Light travels 3.0×10^8 meters per second. How long does it take light from the sun to reach the earth in minutes?
8.31 min

13. A light-year is the distance that light travels in one year. (See Exercise 12.) How far is a light year in kilometers? Express your answer in scientific notation.
9.46×10^{12} km

41. $2 \times 10^3 = 2000$
42. $4 \times 10^5 = 400{,}000$
43. $5.5 \times 10^{-5} = 0.000055$

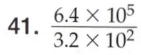

Evaluate each expression. Express each result in scientific notation and standard form.

41. $\frac{6.4 \times 10^5}{3.2 \times 10^2}$
42. $\frac{8 \times 10^2}{2 \times 10^{-3}}$
43. $\frac{13.2 \times 10^{-6}}{2.4 \times 10^{-1}}$
44. Find the product of (1.2×10^5) and (5×10^{-4}) mentally. **60**
45. Express 5×10^6 in standard form. **5,000,000**

Applications and Problem Solving

46. **Biology** The mass of an orchid seed is 3.5×10^{-6} grams. Express 3.5×10^{-6} in standard form. **0.0000035**
47. **Astronomy** The diameter of Venus is 1.218×10^4 km, the diameter of Earth is 1.276×10^4 km, and the diameter of Mars is 6.76×10^3 km. List the planets in order from greatest to least diameter. **See margin.**
48. **Electronics** Engineering notation is similar to scientific notation. However, in engineering notation, the powers of ten are always multiples of 3, such as 10^3, 10^6, 10^{-9}, and 10^{-12}. For example, 240,000 is expressed as 240×10^3. Express 15,000 ohms in scientific and engineering notation. **sci: 1.5×10^4; eng: 15×10^3**

Orchid

49. **Biology** Laboratory technicians look at bacteria through microscopes. A microscope set on 1000× makes an organism appear to be 1000 times larger than its actual size. Most bacteria are between 3×10^{-4} and 2×10^{-3} millimeter in diameter. How large would the bacteria appear under a microscope set on 1000×? **between 0.3 and 2 mm**
50. **Critical Thinking** Express each number in scientific notation.
   a. 32×10^5 **3.2×10^6**
   b. 284×10^3 **2.84×10^5**
   c. 0.76×10^{-2} **7.6×10^{-3}**
   d. 0.09×10^{-3} **9×10^{-5}**

Mixed Review

Simplify each expression. *(Lessons 8–2, 8–3)*

51. $y^5(y^{-2})$ **y^3**
52. $15a^{-1}b^3c^{-2}$ **$\frac{15b^3}{ac^2}$**
53. $\frac{30c^4}{-5c^{-2}}$ **$-6c^6$**
54. $\frac{5r^2s^7}{25r^2s^{10}}$ **$\frac{1}{5s^3}$**
55. $x^2 \cdot x^3 \cdot x$ **x^6**
56. $\frac{2}{3}r(15s^2)$ **$10rs^2$**
57. $\frac{a^4b^5}{ab^2}$ **a^3b^3**
58. $(-15y^2z)(-2y^2z^0)$ **$30y^4z$**
59. $(3x^5y^2)(-2x^{-3}y^4)$ **$-6x^2y^6$**

Standardized Test Practice

60. **Short Response** Julia's wages vary directly as the number of hours she works. If her wages for 5 hours are \$34.75, how much will they be for 30 hours in dollars? *(Lesson 6–5)* **208.50**
61. **Multiple Choice** Choose the graph that represents a function. *(Lesson 6–4)* **C**

(A)

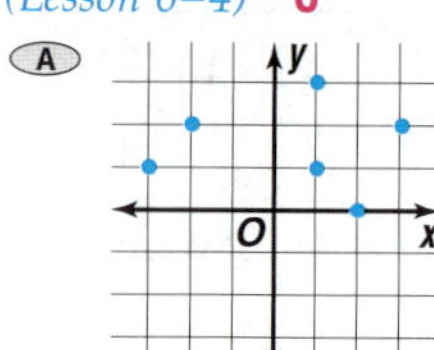

(B)

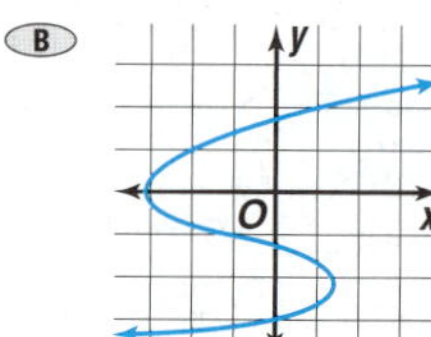

(C)

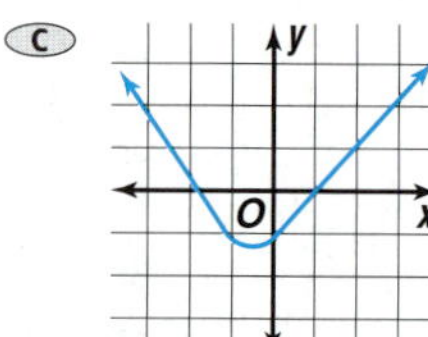

(D)

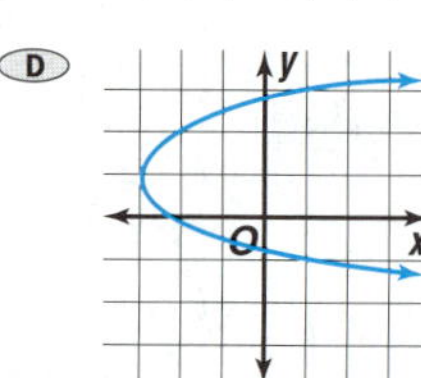

www.algconcepts.com/self_check_quiz

Extra Credit

Which is less, $\frac{1}{10{,}000} \times 10^5$ or $\frac{1}{100{,}000} \times 10^4$? **$\frac{1}{100{,}000} \times 10^4$**

8–5 Square Roots

What You'll Learn
You'll learn to simplify radicals by using the Product and Quotient Properties of Square Roots.

Why It's Important
Aviation Pilots can use the formula $d = 1.5\sqrt{h}$ to determine the distance to the horizon. The formula contains a square root symbol. *See Example 6.*

Can you help this character from *Shoe* take his math test?

SHOE

You will find the square root of 225 in Example 3.

In Lesson 8–1, you learned that *squaring* a number means using that number as a factor twice. The opposite of squaring is finding a **square root**. To find a square root of 36, you must find *two equal factors* whose product is 36.

$6 \times 6 = 36$ → The square root of 36 is 6.

Square Root	**Words:** A square root of a number is one of its two equal factors. **Symbols:** $\sqrt{a} = b$, where $a = b \cdot b$.

The symbol $\sqrt{\ }$, called a **radical sign**, is used to indicate the square root.

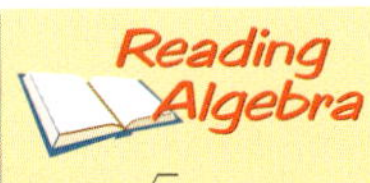

Reading Algebra
Read $\sqrt{a}$ as *the square root of a.*

$\sqrt{36} = 6$ — *$\sqrt{36}$ indicates the positive square root of 36.*
$-\sqrt{36} = -6$ — *$-\sqrt{36}$ indicates the negative square root of 36.*

Exponents can also be used to indicate the square root. $9^{\frac{1}{2}}$ means the same thing as $\sqrt{9}$. $9^{\frac{1}{2}}$ is read *nine to the one half power.* $9^{\frac{1}{2}} = 3$.

Examples

Simplify each expression.

1. $\sqrt{49}$
Since $7^2 = 49$, $\sqrt{49} = 7$.

2. $-\sqrt{64}$
Since $8^2 = 64$, $-\sqrt{64} = -8$.

Your Turn

a. $\sqrt{25}$ **5** b. $\sqrt{121}$ **11** c. $-\sqrt{25}$ **−5** d. $-\sqrt{9}$ **−3**

1 FOCUS

5-Minute Check
Lesson 8–4

Express each number in scientific notation.

1. 1,650,000,000 **1.65×10^9**
2. 0.00141 **1.41×10^{-3}**

Evaluate each expression.

3. $(1.5 \times 10^2)(4 \times 10^7)$ **$6 \times 10^9 = 6{,}000{,}000{,}000$**
4. $\frac{3.6 \times 10^{11}}{1.2 \times 10^9}$ **$3 \times 10^2 = 300$**
5. A signal leaves a cellular phone traveling at a rate of 10^8 feet per second. How far does the signal travel in 4 microseconds? (1 microsecond = 10^{-6} second) **400 ft**

Motivating the Lesson

Real-World Connection Ask students to find the area of any non-square portion of your school, such as a hallway, gym, or classroom floor. Then ask them to find the length of the side of a square room that would have the same area.

2 TEACH

In-Class Examples
Examples 1–2

Simplify each expression.

1 $\sqrt{100}$ **10**
2 $-\sqrt{81}$ **−9**

Resource Manager

Reproducible Masters
Chapter 8 Resource Masters
- *Study Guide,* p. 359
- *Skills Practice,* p. 360
- *Practice,* p. 361
- *Reading to Learn Mathematics,* p. 362
- *Enrichment,* p. 363

Graphing Calculator, pp. 24–25

Transparencies
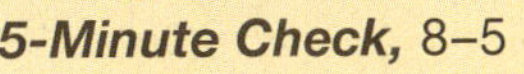
5-Minute Check, 8–5

Technology/Multimedia
Interactive Chalkboard CD-ROM

Teaching Tip When discussing the definitions of *prime number* and *composite number*, stress that every whole number is either prime or composite.

In-Class Examples

Examples 3–4

Simplify each expression.

3 $\sqrt{256}$ **16**

4 $\sqrt{400}$ **20**

A **radical expression** is an expression that contains a square root. You can simplify a radical expression like $\sqrt{225}$ by using prime numbers.

A **prime number** is a whole number that has exactly two factors, the number itself and 1. A **composite number** is a whole number that has more than two factors. Every composite number can be written as the product of prime numbers. The tree diagram shows one way to find the prime factors of 225.

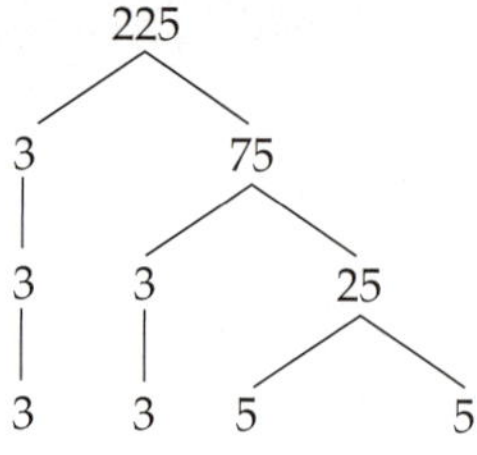

When a number is expressed as a product of prime factors, the expression is called the **prime factorization** of the number. Since 3 and 5 are prime numbers, the prime factorization of 225 is $3 \times 3 \times 5 \times 5$.

To simplify $\sqrt{225}$, use the following property.

Product Property of Square Roots		
	Words:	The square root of a product is equal to the product of each square root.
	Numbers:	$\sqrt{4 \cdot 9} = \sqrt{4} \cdot \sqrt{9}$
	Symbols:	$\sqrt{ab} = \sqrt{a} \cdot \sqrt{b}$ $\quad a \geq 0, b \geq 0$

Examples

Simplify each expression.

3 $\sqrt{225}$

$\sqrt{225} = \sqrt{3 \cdot 3 \cdot 5 \cdot 5}$ *Find the prime factorization of 225.*

$= \sqrt{9 \cdot 25}$ $\quad 3 \times 3 = 9, 5 \times 5 = 25$

$= \sqrt{9} \cdot \sqrt{25}$ *Use the Product Property of Square Roots.*

$= 3 \cdot 5$ or 15 *Simplify each radical.*

4 $\sqrt{576}$

$\sqrt{576} = \sqrt{2 \cdot 2 \cdot 2 \cdot 2 \cdot 2 \cdot 2 \cdot 3 \cdot 3}$ *Find the prime factorization of 576.*

$= \sqrt{64 \cdot 9}$ $\quad 2 \cdot 2 \cdot 2 \cdot 2 \cdot 2 \cdot 2 = 64, 3 \cdot 3 = 9$

$= \sqrt{64} \cdot \sqrt{9}$ *Use the Product Property of Square Roots.*

$= 8 \cdot 3$ or 24 *Simplify each radical.*

Your Turn

e. $\sqrt{144}$ **12**

f. $\sqrt{324}$ **18**

www.algconcepts.com/extra_examples

A similar property for quotients can be used to simplify radicals.

Quotient Property of Square Roots		
	Words:	The square root of a quotient is equal to the quotient of each square root.
	Numbers:	$\sqrt{\frac{4}{9}} = \frac{\sqrt{4}}{\sqrt{9}}$
	Symbols:	$\sqrt{\frac{a}{b}} = \frac{\sqrt{a}}{\sqrt{b}}$ $\quad a \geq 0, b > 0$

Example 5

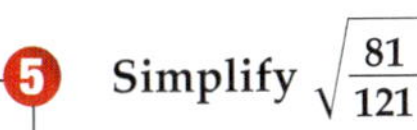

Simplify $\sqrt{\frac{81}{121}}$.

$\sqrt{\frac{81}{121}} = \frac{\sqrt{81}}{\sqrt{121}}$ *Use the Quotient Property of Square Roots.*

$= \frac{9}{11}$ *Simplify.*

Your Turn

g. $\sqrt{\frac{64}{81}}$ $\frac{8}{9}$

h. $\sqrt{\frac{36}{4}}$ 3

You can use a graphing calculator to evaluate square roots. Press [2nd] [$\sqrt{\ }$] and then the number to find its positive square root.

Example 6 Real World

Aviation Link

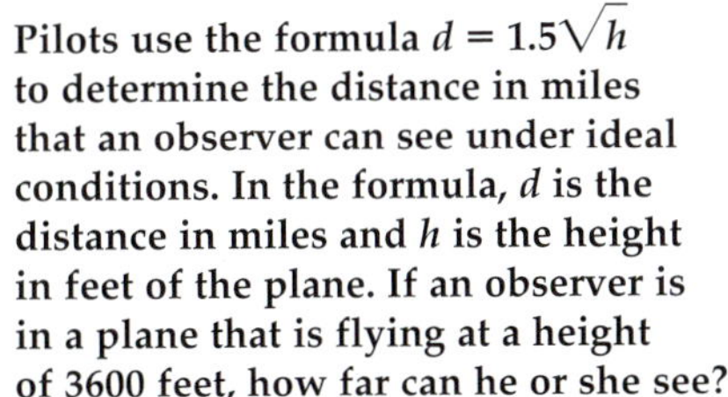

Pilots use the formula $d = 1.5\sqrt{h}$ to determine the distance in miles that an observer can see under ideal conditions. In the formula, d is the distance in miles and h is the height in feet of the plane. If an observer is in a plane that is flying at a height of 3600 feet, how far can he or she see?

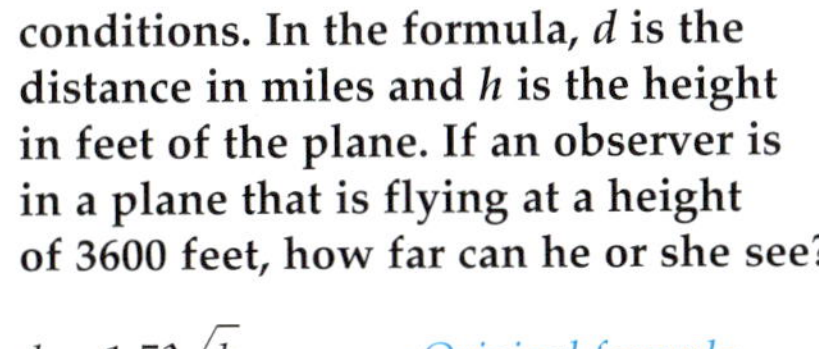

$d = 1.5\sqrt{h}$ *Original formula*

$d = 1.5 \times \sqrt{3600}$ *Replace h with 3600.*

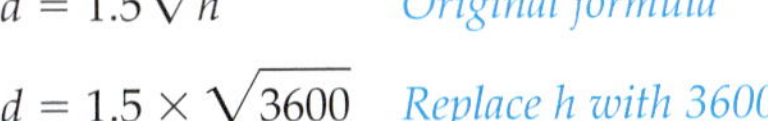

1.5 [2nd] [$\sqrt{\ }$] 3600 [ENTER] 90

The observer can see a distance of 90 miles.

Technology Tip

On a graphing calculator, it is not necessary to use the [×] key when you multiply square roots.

In-Class Examples

Example 5

Simplify $-\sqrt{\frac{100}{9}}$. $-\frac{10}{3}$

Example 6

The formula $r = 0.56\sqrt{A}$ gives the approximate radius r of a circle that has an area of A square units. If a circle has an area of 15 square centimeters, find the radius of the circle to the nearest tenth of a centimeter. **2.2 cm**

***Study Guide*, p. 359**

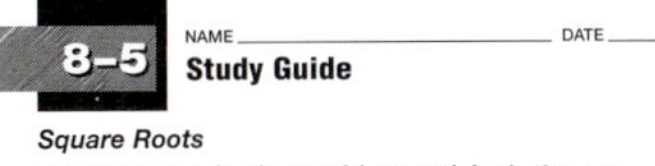

NAME ______ DATE ______ PERIOD ______

8–5 Study Guide

Student Edition Pages 357–361

Square Roots

Suppose you know that the area of the square below is 49 square inches. What is the length of a side? Since 7 × 7 = 49, each side is 7 inches long.

Also, since 7 × 7 = 49, we say that the **square root** of 49 is 7. A shorter way to write this is with the symbol $\sqrt{\ }$, a **radical sign**. Write $\sqrt{49} = 7$. Use the following rules to simplify square roots.

Square Roots	Examples
1. The square root of a number is one of its equal factors.	$\sqrt{49} = 7$
2. The square root of a product is equal to the product of each square root. $\sqrt{ab} = \sqrt{a} \cdot \sqrt{b}$	$\sqrt{225} = \sqrt{9} \cdot \sqrt{25} = 3 \cdot 5 = 15$
3. The square root of a quotient is equal to the quotient of each square root. $\sqrt{\frac{a}{b}} = \frac{\sqrt{a}}{\sqrt{b}}$	$\sqrt{\frac{25}{144}} = \frac{\sqrt{25}}{\sqrt{144}} = \frac{5}{12}$

Simplify.

1. $\sqrt{81}$ 9
2. $-\sqrt{25}$ −5
3. $\sqrt{100}$ 10
4. $-\sqrt{400}$ −20
5. $\sqrt{625}$ 25
6. $-\sqrt{1156}$ −34
7. $\sqrt{\frac{81}{49}}$ $\frac{9}{7}$
8. $\sqrt{\frac{25}{36}}$ $\frac{5}{6}$
9. $-\sqrt{\frac{16}{100}}$ $-\frac{2}{5}$
10. $\sqrt{0.25}$ 0.5
11. $-\sqrt{0.0016}$ −0.04
12. $-\sqrt{\frac{0.16}{0.09}}$ $-\frac{4}{3}$

3 PRACTICE/APPLY

Error Analysis

Watch for students who simplify $\sqrt{4}$ as -2 in Exercise 9, justifying their answer with the true statement that $(-2)^2 = 4$.
Prevent by reminding students that in this lesson the symbol $\sqrt{\ }$ refers to the nonnegative square root. So, $\sqrt{4} = 2$ and $-\sqrt{4} = -2$.

Assignment Guide

Basic: 17–43 odd, 45–55
Average: 16–40 even, 42–55
All: Quiz 2, 1–5

Answer

3. $3^2 = 9$ and $\sqrt{9} = 3$

Skills Practice, **p. 360, and** ***Practice,*** **p. 361 (shown)**

8-5 NAME ______________ DATE ________ PERIOD ______
Practice
Student Edition Pages 357–361

Square Roots

Simplify.

1. $\sqrt{36}$ 6
2. $-\sqrt{16}$ −4
3. $\sqrt{81}$ 9
4. $-\sqrt{144}$ −12
5. $-\sqrt{100}$ −10
6. $-\sqrt{121}$ −11
7. $\sqrt{169}$ 13
8. $-\sqrt{25}$ −5
9. $-\sqrt{529}$ −23
10. $\sqrt{256}$ 16
11. $\sqrt{324}$ 18
12. $-\sqrt{289}$ −17
13. $\sqrt{441}$ 21
14. $-\sqrt{225}$ −15
15. $\sqrt{196}$ 14
16. $\sqrt{400}$ 20
17. $\sqrt{484}$ 22
18. $\sqrt{729}$ 27
19. $-\sqrt{625}$ −25
20. $\sqrt{1225}$ 35
21. $\sqrt{\frac{49}{81}}$ $\frac{7}{9}$
22. $-\sqrt{\frac{16}{25}}$ $-\frac{4}{5}$
23. $\sqrt{\frac{4}{16}}$ $\frac{1}{2}$
24. $-\sqrt{\frac{25}{36}}$ $-\frac{5}{6}$
25. $-\sqrt{\frac{100}{121}}$ $-\frac{10}{11}$
26. $\sqrt{\frac{1}{64}}$ $\frac{1}{8}$
27. $\sqrt{\frac{36}{64}}$ $\frac{3}{4}$
28. $-\sqrt{\frac{144}{36}}$ −2
29. $-\sqrt{\frac{121}{289}}$ $-\frac{11}{17}$
30. $-\sqrt{\frac{225}{625}}$ $-\frac{3}{5}$
31. $\sqrt{\frac{400}{100}}$ 2
32. $\sqrt{\frac{196}{256}}$ $\frac{7}{8}$

Check for Understanding

Communicating Mathematics

1. **Find** the first ten perfect squares. 1. 1, 4, 9, 16, 25, 36, 49, 64, 81, 100
2. **Write** the symbol for the negative square root of 9. $-\sqrt{9}$
3. *Writing Math* Explain why finding a square root and squaring are inverse operations. See margin.

Vocabulary
square root
radical sign
radical expression
prime number
composite number
prime factorization

Guided Practice

Getting Ready Find the prime factorization of each number.

Sample: 81 **Solution:** $81 = 3 \cdot 3 \cdot 3 \cdot 3$ or 3^4

4. 100 $2^2 \cdot 5^2$
5. 121 11^2
6. 169 13^2
7. 196 $2^2 \cdot 7^2$
8. 256 2^8

Simplify.

Examples 1 & 2

9. $\sqrt{4}$ 2
10. $-\sqrt{49}$ −7
11. $-\sqrt{121}$ −11

Examples 2 & 3

12. $\sqrt{256}$ 16
13. $\sqrt{\frac{1}{4}}$ $\frac{1}{2}$
14. $-\sqrt{\frac{49}{121}}$ $-\frac{7}{11}$

Example 5

15. **Buildings** A famous 1933 movie about a gorilla helped make the Empire State Building in New York City a popular tourist attraction. If the gorilla's eyes were at a height of 1225 feet, how far could he see in the distance on a clear day? Use the formula $d = 1.5\sqrt{h}$, where d is the visible distance in miles and h is the height in feet. 52.5 mi

Exercises

Practice

Homework Help

For Exercises	See Examples
16–27, 36–41	1, 2
28–35	5

Extra Practice
See page 709.

Simplify.

A

16. $-\sqrt{81}$ −9
17. $\sqrt{100}$ 10
18. $\sqrt{144}$ 12
19. $-\sqrt{196}$ −14
20. $-\sqrt{169}$ −13
21. $\sqrt{529}$ 23
22. $\sqrt{25}$ 5
23. $-\sqrt{676}$ −26
24. $\sqrt{441}$ 21
25. $-\sqrt{484}$ −22
26. $-\sqrt{1024}$ −32
27. $\sqrt{289}$ 17
28. $\sqrt{\frac{81}{64}}$ $\frac{9}{8}$
29. $-\sqrt{\frac{9}{100}}$ $-\frac{3}{10}$
30. $\sqrt{\frac{36}{196}}$ $\frac{3}{7}$
31. $-\sqrt{\frac{25}{400}}$ $-\frac{1}{4}$
32. $\sqrt{\frac{225}{25}}$ 3
33. $\sqrt{\frac{144}{196}}$ $\frac{6}{7}$
34. $\sqrt{\frac{196}{289}}$ $\frac{14}{17}$
35. $\sqrt{\frac{0.09}{0.16}}$ $\frac{3}{4}$

B

36. $\sqrt{0.16}$ 0.4
37. $-\sqrt{0.0025}$
38. $\sqrt{0.0036}$
39. $\sqrt{0.0009}$
40. Find the negative square root of 49. −7
41. If $x = \sqrt{36}$, what is the value of x? 6

37. −0.05
38. 0.06
39. 0.03

Applications and Problem Solving

C

42. **Geometry** The area of a square is 25 square inches. Find the length of one of its sides. 5 in.

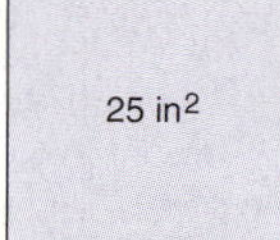

Reteaching Activity

Intrapersonal Learners Have students write a journal entry listing the perfect squares that they can quickly identify, some perfect squares that they do not always identify quickly, and what steps they think they could take to learn to recognize these unfamiliar perfect squares.

43. **Geometry** The formula for the perimeter P of a square is $P = 4s$, where s is the length of a side. The area of a square is 169 square meters. Find its perimeter. **52 m**

44. **Firefighting** The velocity of water sprayed from a nozzle is given by the formula $V = 12.14\sqrt{P}$, where V is the velocity in feet per second and P is the pressure at the nozzle in pounds per square inch. Find the velocity of water if the nozzle pressure is 64 pounds per square inch. **97.12 ft/s**

45. **Critical Thinking** *True* or *false*: $\sqrt{-36} = -6$. Explain. **false; $-6(-6) \neq -36$**

Mixed Review

Express each number in scientific notation. *(Lesson 8–4)*

46. 350 **3.5×10^2**
47. 63,000 **6.3×10^4**
48. 0.023 **2.3×10^{-2}**
49. 0.00076 **7.6×10^{-4}**

Write each expression using positive exponents. *(Lesson 8–3)*

50. 7^{-2} **$\frac{1}{7^2}$**
51. x^{-4} **$\frac{1}{x^4}$**
52. $ab^{-2}c$ **$\frac{ac}{b^2}$**
53. $x^{-2}y^{-3}$ **$\frac{1}{x^2y^3}$**

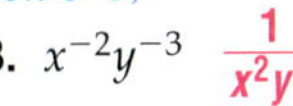

Data Update For the latest information about high school enrollment, visit: www.algconcepts.com

54. **Business** Among high school students ages 15–18, 41% say they are employed. The graph shows the number of hours per week these students work. If you survey 50 high school students who have jobs, predict how many work between 11 and 20 hours per week. *(Lesson 5–3)* **20 students**

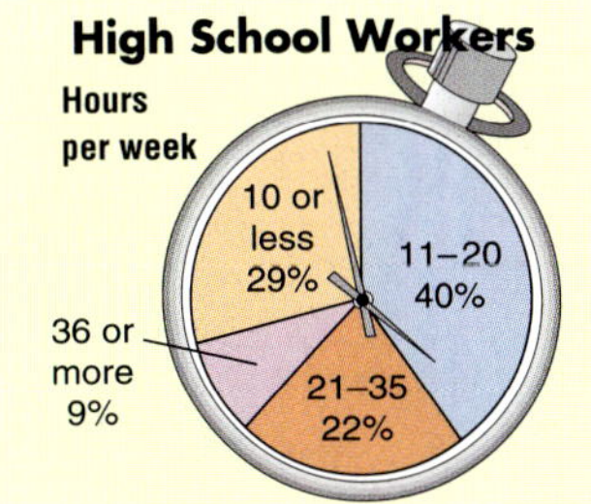

Source: Michaels Opinion Research

Standardized Test Practice

55. **Short Response** Write an equation with variables on both sides in which the solution is −5. *(Lesson 4–6)* **Sample answer: $2x + 6 = x + 1$**

Quiz 2 Lessons 8–4 and 8–5

Express each number in scientific notation. *(Lesson 8–4)*

1. 700,000 **7×10^5**
2. 0.000053 **5.3×10^{-5}**

Simplify each expression. *(Lesson 8–5)*

3. $\sqrt{441}$ **21**
4. $-\sqrt{\frac{9}{36}}$ **$-\frac{1}{2}$**

5. **Architecture** A square house is the most energy-efficient because it has the least outside wall space for its area. What is the length and width of the most energy-efficient house you could build with an area of 900 square feet? *(Lesson 8–5)* **square house, 30 ft × 30 ft**

4 ASSESS

Open-Ended Assessment

Modeling Ask students to model a perfect square and its square root by constructing a square and labeling its dimensions and area.

Quiz 2

The Quiz provides students with a brief review of the concepts and skills in Lessons 8–4 and 8–5. Lesson numbers are given to the right of the exercises or instruction lines so students can review concepts not yet mastered.

Extra Credit

Find the value $\sqrt{1} - \sqrt{4} + \sqrt{9} - \sqrt{16} + \sqrt{25} - \sqrt{36}$. **−3**

Enrichment, p. 363

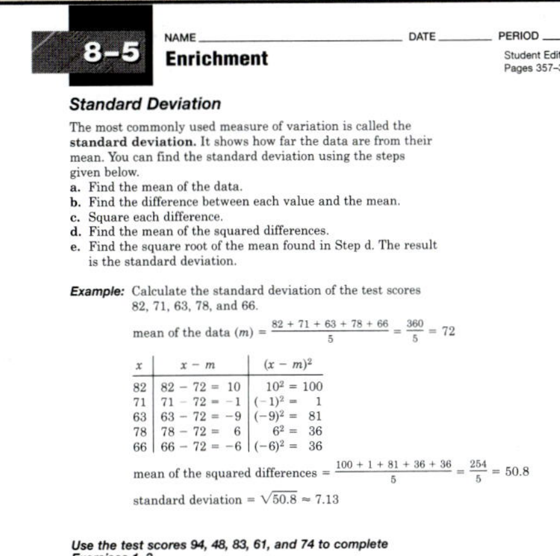

8–5 NAME ______ DATE ______ PERIOD ______

Enrichment

Student Edition Pages 357–361

Standard Deviation

The most commonly used measure of variation is called the **standard deviation.** It shows how far the data are from their mean. You can find the standard deviation using the steps given below.

a. Find the mean of the data.
b. Find the difference between each value and the mean.
c. Square each difference.
d. Find the mean of the squared differences.
e. Find the square root of the mean found in Step d. The result is the standard deviation.

Example: Calculate the standard deviation of the test scores 82, 71, 63, 78, and 66.

mean of the data $(m) = \frac{82 + 71 + 63 + 78 + 66}{5} = \frac{360}{5} = 72$

x	$x - m$	$(x - m)^2$
82	$82 - 72 = 10$	$10^2 = 100$
71	$71 - 72 = -1$	$(-1)^2 = 1$
63	$63 - 72 = -9$	$(-9)^2 = 81$
78	$78 - 72 = 6$	$6^2 = 36$
66	$66 - 72 = -6$	$(-6)^2 = 36$

mean of the squared differences $= \frac{100 + 1 + 81 + 36 + 36}{5} = \frac{254}{5} = 50.8$

standard deviation $= \sqrt{50.8} \approx 7.13$

Use the test scores 94, 48, 83, 61, and 74 to complete Exercises 1–3.

1. Find the mean of the scores. **72**
2. Show that the standard deviation of the scores is about 16.2.

x	$x - m$	$(x - m)^2$
94	$94 - 72 = 22$	484
48	$48 - 72 = -24$	576
83	$83 - 72 = 11$	121
61	$61 - 72 = -11$	121
74	$74 - 72 = 2$	4

mean of $(x - m)^2 = 261.2$
S.D. $= \sqrt{261.2} \approx 16.2$

3. Which had less variations, the test scores listed above or the test scores in the example? **the test scores in the example**

Lesson 8–6

1 FOCUS

5-Minute Check
Lesson 8–5

Simplify.

1. $\sqrt{1600}$ **40**
2. $-\sqrt{225}$ **−15**
3. $-\sqrt{\frac{36}{169}}$ $-\frac{6}{13}$
4. The area of a square is 121 square inches. Find the length of one of its sides. **11 in.**

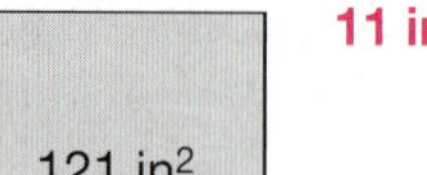

5. The formula $s = 1.5\sqrt{A}$ approximates the length s of each side of an equilateral triangle whose area is A square units. Find the length of each side of an equilateral triangle whose area is 20 square centimeters. Round to the nearest tenth of a centimeter. **6.7 cm**

Motivating the Lesson

Hands-On Activity Ask students to draw two number lines on a sheet of paper, one right above the other. Have them label the lower number line 1 through 15 and label the upper number line with the square of the number that is directly below it on the lower number line. Then, for several non-square numbers between 1 and 225, ask students to locate the number on the upper number line, use a calculator to find its square root, and locate that square root on the lower number line. Discuss the vertical alignment of the numbers and their square roots.

8–6 Estimating Square Roots

What You'll Learn
You'll learn to estimate square roots.

Why It's Important
Law Enforcement
Police officers use the formula $s = \sqrt{30df}$ when they investigate accidents.
See Exercise 38.

Numbers like 25 and 81 are perfect squares because their square roots are whole numbers ($\sqrt{25} = 5$ and $\sqrt{81} = 9$). But there are many other numbers that are *not* perfect squares.

Notice what happens when you find $\sqrt{2}$ and $\sqrt{15}$ with a calculator.

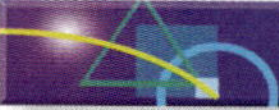

2nd [$\sqrt{\ }$] 2 ENTER 1.414213562...

2nd [$\sqrt{\ }$] 15 ENTER 3.872983346...

Numbers like $\sqrt{2}$ and $\sqrt{15}$ are not integers or rational numbers because their decimal values do not terminate or repeat. They are **irrational numbers**. You can estimate irrational square roots by using perfect squares.

Hands-On Algebra

Materials:  base-ten tiles

You can use base-ten tiles to estimate the square root of 60.

Step 1 Arrange 60 tiles into the largest square possible. The square has 49 tiles, with 11 left over.

Step 2 Add tiles until you have the next larger square. You need to add 4 tiles. This square has 64 tiles.

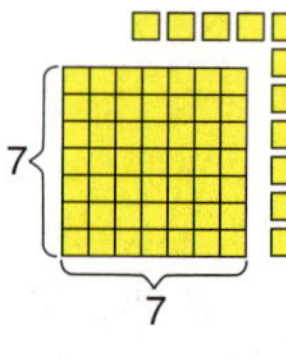

Step 1

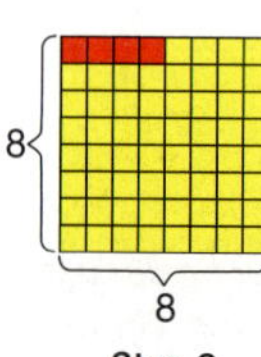

Step 2

Step 3 Now use these models to estimate $\sqrt{60}$.

- 60 is between 49 and 64.
- $\sqrt{60}$ is between 7 and 8.
- Since 60 is closer to 64 than to 49, $\sqrt{60}$ is closer to 8 than to 7.
- To the nearest whole number, $\sqrt{60} \approx 8$.

Try These

For each number, arrange base-ten tiles into the largest square possible. Then add tiles until you have the next larger square. To the nearest whole number, estimate the square root of each number.

1. 20 **4** 2. 76 **9** 3. 150 **12** 4. 3 **2**

Resource Manager

Reproducible Masters
Chapter 8 Resource Masters
- *Study Guide,* p. 364
- *Skills Practice,* p. 365
- *Practice,* p. 366
- *Reading to Learn Mathematics,* p. 367
- *Enrichment,* p. 368

Hands-On Algebra, p. 92

Transparencies
5-Minute Check, 8–6

Technology/Multimedia
Interactive Chalkboard CD-ROM

Examples

Estimate each square root to the nearest whole number.

1 $\sqrt{22}$

List some perfect squares to find the two perfect squares closest to 22.

1, 4, 9, 16, 25, 36, . . .

22 is between 16 and 25.

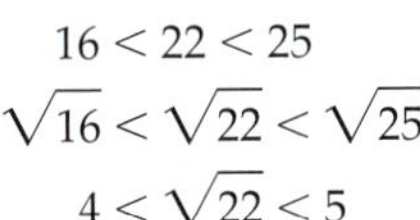

$16 < 22 < 25$

$\sqrt{16} < \sqrt{22} < \sqrt{25}$

$4 < \sqrt{22} < 5$

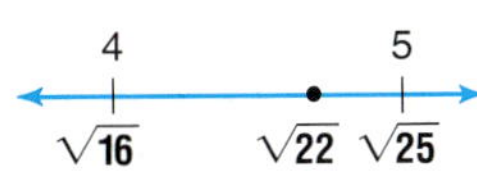

Since 22 is closer to 25 than to 16, the best whole number estimate for $\sqrt{22}$ is 5.

Reading Algebra

The expression $16 < 22 < 25$ means that *16 is less than 22 and 22 is less than 25*. It also means that 22 is *between* 16 and 25.

2 $\sqrt{130}$

. . . 64, 81, 100, 121, 144, . . .

$121 < 130 < 144$

$\sqrt{121} < \sqrt{130} < \sqrt{144}$

$11 < \sqrt{130} < 12$

11 12

√121 √130 √144

Since 130 is closer to 121 than to 144, the best whole number estimate for $\sqrt{130}$ is 11.

Your Turn

a. $\sqrt{45}$ 7

b. $\sqrt{190}$ 14

Gardening Link

3 **A box of fertilizer covers 250 square feet of garden. Find the length in whole feet of the largest square garden that can be fertilized with one box of fertilizer.**

area ≤ 250 ft^2

Find the largest perfect square that is less than 250. Then find its square root.

$225 < 250$ *$15^2 = 225$*
$16^2 = 256$

The square root of 225 is 15. Therefore, the largest square garden that can be fertilized with one box of fertilizer has a length of 15 feet.

www.algconcepts.com/extra_examples

2 TEACH

Teaching Tip In Examples 1 and 2, students will probably realize that they are identifying the first perfect square on each side of the given number. If not, focus students' attention on the two *outer* numbers in the range of values in Example 1, $16 < 22 < 25$.

In-Class Examples

Examples 1–2

Estimate each square root.

1 $\sqrt{48}$ **7**

2 $\sqrt{200}$ **14**

Example 3

A box of ceramic tiles provides 150 square feet of floor covering. Find the length in whole feet of the largest square room that can be covered with one box of the ceramic tiles. **12 ft**

Hands-On Algebra

Cooperative Learning Refer to the Hands-On Algebra on page 362. Encourage students to use 100-square tiles and 10-square tiles whenever possible. For example, in Exercise 3 they could start with one 100-square tile and five 10-square tiles.

Hands-On Algebra Masters, p. 92

***Study Guide,* p. 364**

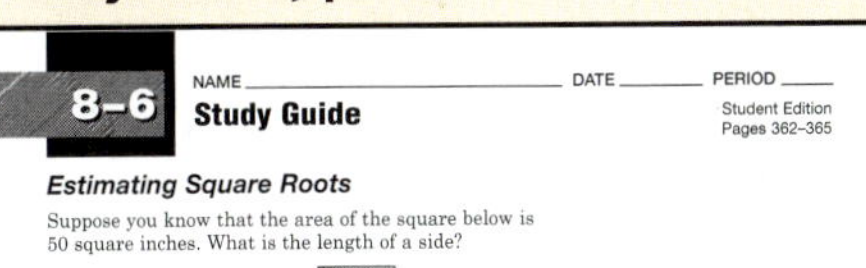

8–6 NAME ______ DATE ______ PERIOD ______

Study Guide

Student Edition Pages 362–365

Estimating Square Roots

Suppose you know that the area of the square below is 50 square inches. What is the length of a side?

x 50 in^2 x

Since there is no rational number whose square is 50, you need to estimate the answer. If you use a calculator to find $\sqrt{50}$, it will return an approximate value of 7.071067812. This represents an **irrational number**, a decimal number that does not repeat or terminate.

You can use perfect squares to estimate irrational square roots. Since 50 is close to 49, $\sqrt{50} \approx 7$, so the length of the side of the square is about 7 inches. Likewise, if the area of a square is 60 square inches, the side length would be $\sqrt{60} \approx 8$ inches, since the actual value is close to $\sqrt{64}$, or 8.

√49 √50 √55 √60 √64

7 8

Estimate each square root to the nearest whole number.

1. $\sqrt{90}$ 9
2. $\sqrt{134}$ 12
3. $\sqrt{17}$ 4
4. $\sqrt{500}$ 22
5. $\sqrt{1000}$ 32
6. $\sqrt{98}$ 10
7. $\sqrt{320}$ 18
8. $\sqrt{5}$ 2
9. $\sqrt{75}$ 9
10. $\sqrt{84.5}$ 9
11. $\sqrt{128.9}$ 11
12. $\sqrt{0.025}$ 0
13. $\sqrt{0.0075}$ 0
14. $\sqrt{10.01}$ 3
15. $\sqrt{0.9988}$ 1

3 PRACTICE/APPLY

Error Analysis

Watch for students who suggest in Exercise 1 that any expression written with a radical sign, such as $\sqrt{100}$ or $\sqrt{\frac{81}{4}}$, is an irrational number.

Prevent by stressing that radicals involving perfect squares, or expressions that can be simplified to perfect squares, can be written as a ratio of integers ($\sqrt{100} = 10 = \frac{10}{1}$ and $\sqrt{\frac{81}{4}} = \frac{9}{2}$) and therefore are rational numbers.

Assignment Guide

Basic: 13–37 odd, 39–51
Average: 14–34 even, 35–51

Answers

1. Its decimal value does not terminate or repeat.
2. 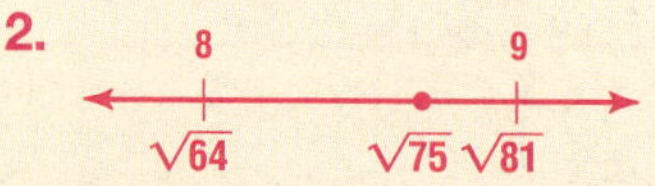

3. Sample answer: The area of a square garden is 200 square feet. Estimate the length of one side of the garden.

***Skills Practice*, p. 365, and *Practice*, p. 366 (shown)**

8-6 NAME ______ DATE ______ PERIOD ______
Practice Student Edition Pages 362–365

Estimating Square Roots

Estimate each square root to the nearest whole number.

1. $\sqrt{10}$ 3	2. $\sqrt{14}$ 4	3. $\sqrt{32}$ 6
4. $\sqrt{19}$ 4	5. $\sqrt{40}$ 6	6. $\sqrt{6}$ 2
7. $\sqrt{53}$ 7	8. $\sqrt{23}$ 5	9. $\sqrt{30}$ 5
10. $\sqrt{21}$ 5	11. $\sqrt{90}$ 9	12. $\sqrt{73}$ 9
13. $\sqrt{72}$ 8	14. $\sqrt{56}$ 7	15. $\sqrt{89}$ 9
16. $\sqrt{135}$ 12	17. $\sqrt{152}$ 12	18. $\sqrt{110}$ 10
19. $\sqrt{162}$ 13	20. $\sqrt{129}$ 11	21. $\sqrt{181}$ 13
22. $\sqrt{174}$ 13	23. $\sqrt{223}$ 15	24. $\sqrt{195}$ 14
25. $\sqrt{240}$ 15	26. $\sqrt{271}$ 16	27. $\sqrt{312}$ 18
28. $\sqrt{380}$ 19	29. $\sqrt{335}$ 18	30. $\sqrt{300}$ 17

Check for Understanding

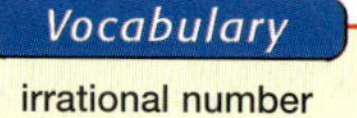

irrational number

Communicating Mathematics

1–3. See margin.

1. **Explain** why $\sqrt{10}$ is an irrational number.
2. **Graph** $\sqrt{75}$ on a number line.
3. **Write a problem** that can be solved by estimating $\sqrt{200}$.

Guided Practice

Getting Ready Find two consecutive perfect squares between which each number lies.

Sample: 60 **Solution:** 60 is between 49 and 64.

4. 56 **49, 64** 5. 85 **81, 100** 6. 175 **169, 196** 7. 500 **484, 529**

Examples 1 & 2 Estimate each square root to the nearest whole number.

8. $\sqrt{85}$ **9** 9. $\sqrt{71}$ **8** 10. $\sqrt{149}$ **12** 11. $\sqrt{255}$ **16**

Example 3

12. **Geometry** The area of a square is 200 square inches. Find the length of each side. Round to the nearest whole number. **14 in.**

Exercises

Practice

Homework Help	
For Exercises	See Examples
13–34	1, 2
35–38	3
Extra Practice	
See page 709.	

Estimate each square root to the nearest whole number.

13. $\sqrt{3}$ **2** 14. $\sqrt{7}$ **3** 15. $\sqrt{13}$ **4** 16. $\sqrt{19}$ **4** 17. $\sqrt{33}$ **6**
18. $\sqrt{56}$ **7** 19. $\sqrt{113}$ **11** 20. $\sqrt{175}$ **13** 21. $\sqrt{410}$ **20** 22. $\sqrt{500}$
23. $\sqrt{575}$ **24** 24. $\sqrt{1000}$ **32** 25. $\sqrt{60.3}$ **8** 26. $\sqrt{94.5}$ **10** 27. $\sqrt{131.4}$
28. $\sqrt{2.314}$ **1** 29. $\sqrt{152.75}$ 30. $\sqrt{189.2}$ 31. $\sqrt{0.08}$ **0** 32. $\sqrt{0.76}$
22. 22 27. 11 29. 12 30. 14 32. 1

33. Tell whether 6 is closer to $\sqrt{34}$ or $\sqrt{44}$. **$\sqrt{34}$**

34. Which is closer to $\sqrt{43}$, 6 or 7? **7**

Applications and Problem Solving

35. **Weather** Meteorologists use the formula $t = \sqrt{\frac{D^3}{216}}$ to describe violent storms such as hurricanes. In the formula, D is the diameter of the storm in miles, and t is the number of hours it will last. A typical hurricane has a diameter of about 40 miles. Estimate how long a typical hurricane lasts. **17 h**

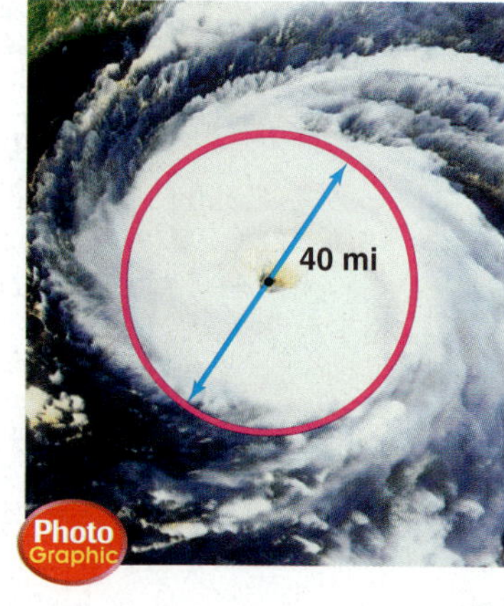

Reteaching Activity

Verbal/Linguistic Learners Have students work in small groups. The members of the group should select a number that is not a perfect square. Then one student in the group should explain how to find the two perfect squares that are closest to the selected number, and then how to approximate the square root of the selected number. The activity should be repeated until every group member has had their turn explaining the process.

36. **Aviation** The British Airways Concorde flew at a height of 60,000 feet. To the nearest mile, about how far could the pilot see when he or she looked out the window on a clear day? Use the formula $d = 1.5\sqrt{h}$, where d is the visible distance in miles and h in the height in feet. **367 mi**

37. **Geometry** You can use the formula $A = \sqrt{s(s-a)(s-b)(s-c)}$ to find the area of a triangle given the measures of its sides. In the formula, a, b, and c, are the measures of the sides, A is the area, and s is one-half the perimeter. The figure shows a triangular plot of land that contains a large lake. Estimate the area of the triangle to the nearest square mile. **31 mi²**

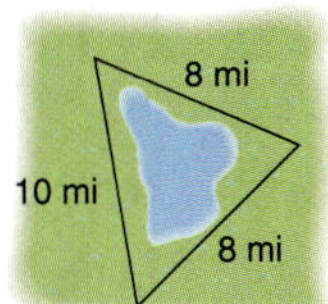

38. **Law Enforcement** When police officers investigate traffic accidents, they need to determine the speed of the vehicles involved in the accident. They can measure the skid marks and then use the formula $s = \sqrt{30df}$. In the formula, s is the speed in miles per hour, d is the length of the skid marks in feet, and f is the friction factor that depends on road conditions. The table gives some values for f.

	Friction Factor (f)	
	concrete	asphalt
wet	0.4	0.5
dry	0.8	1.0

Use the formula to determine the speed in miles per hour for each skid length and road condition. Round to the nearest tenth.

a. 40-foot skid, wet concrete
b. 150-foot skid, dry asphalt
c. 350-foot skid, wet asphalt
d. 45-foot skid, dry concrete

38a. 21.9 mph
38b. 67.1 mph
38c. 72.5 mph
38d. 32.9 mph

39. **Critical Thinking** Find three numbers that have square roots between 3 and 4. **10, 11, 12, 13, 14, 15**

Mixed Review

Find each square root. *(Lesson 8–5)*

40. $\sqrt{0.36}$ **0.6** 41. $\sqrt{256}$ **16** 42. $\sqrt{729}$ **27** 43. $\sqrt{\frac{169}{121}}$ **$\frac{13}{11}$**

Express each measure in standard form. *(Lesson 8–4)*

44. 2 gigabytes **2,000,000,000**
45. 4 nanoseconds **0.000000004**
46. 6.5 megahertz **6,500,000**

Write an equation in slope-intercept form of the line having the given slope that passes through the given point. *(Lesson 7–3)*

47. 5; (3, −2) **$y = 5x - 17$**
48. $\frac{1}{4}$; (0, 8) **$y = \frac{1}{4}x + 8$**
49. −5; (5, 4) **$y = -5x + 29$**

Standardized Test Practice

Ⓐ Ⓑ Ⓒ Ⓓ

50. **Grid In** Tim drove 4 hours at an average speed of 60 miles per hour. How many hours would it take Tim to drive the same distance at an average speed of 50 miles per hour? *(Lesson 6–6)* **4.8**

51. **Multiple Choice** Which statement is *not* correct? **C** *(Lesson 3–1)*

Ⓐ $\frac{1}{2} > \frac{3}{8}$ Ⓑ $\frac{4}{5} < \frac{5}{6}$ Ⓒ $\frac{8}{11} < \frac{7}{13}$ Ⓓ $\frac{1}{3} < \frac{10}{13}$

www.algconcepts.com/self_check_quiz

4 ASSESS

Open-Ended Assessment

Speaking Ask students to explain how to estimate the square root of a number that is not a perfect square.

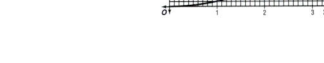

Extra Credit

Which is greater, $\sqrt{10} + \sqrt{50} + \sqrt{150}$ or $\sqrt{10 + 50 + 150}$? How much greater, to the nearest whole number?

The value of $\sqrt{10} + \sqrt{50} + \sqrt{150}$ is greater by about 8.

***Enrichment*, p. 368**

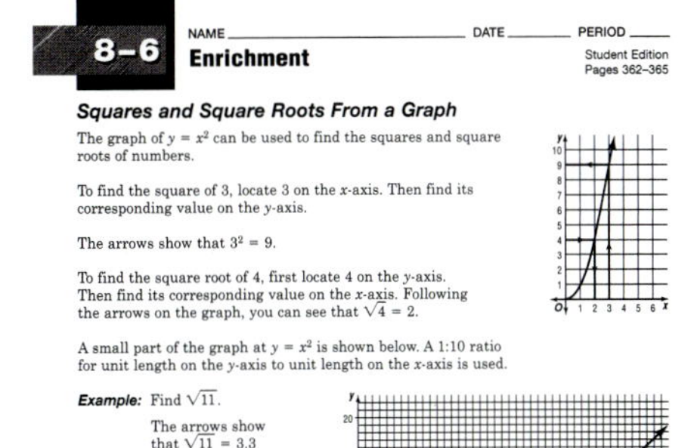
8–6 NAME ______ DATE ______ PERIOD ______

Enrichment Student Edition Pages 362–365

Squares and Square Roots From a Graph

The graph of $y = x^2$ can be used to find the squares and square roots of numbers.

To find the square of 3, locate 3 on the x-axis. Then find its corresponding value on the y-axis.

The arrows show that $3^2 = 9$.

To find the square root of 4, first locate 4 on the y-axis. Then find its corresponding value on the x-axis. Following the arrows on the graph, you can see that $\sqrt{4} = 2$.

A small part of the graph at $y = x^2$ is shown below. A 1:10 ratio for unit length on the y-axis to unit length on the x-axis is used.

Example: Find $\sqrt{11}$.

The arrows show that $\sqrt{11} = 3.3$ to the nearest tenth.

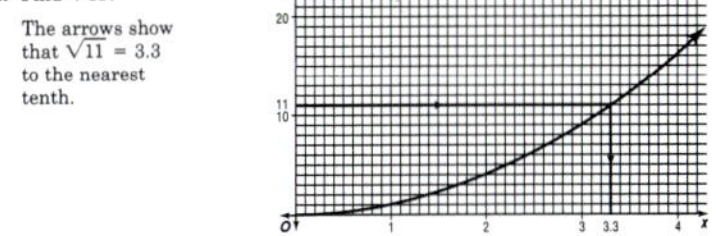

Use the graph above to find each of the following to the nearest whole number.

1. 1.5^2 **2**
2. 2.7^2 **7**
3. 0.9^2 **1**
4. 3.6^2 **13**
5. 4.2^2 **18**
6. 3.9^2 **15**

Use the graph above to find each of the following to the nearest tenth.

7. $\sqrt{15}$ **3.9**
8. $\sqrt{8}$ **2.8**
9. $\sqrt{3}$ **1.7**
10. $\sqrt{5}$ **2.2**
11. $\sqrt{14}$ **3.7**
12. $\sqrt{17}$ **4.1**

Lesson 8–7

1 FOCUS

5-Minute Check

Lesson 8–6

Estimate each square root to the nearest whole number.

1. $\sqrt{75}$ 9
2. $\sqrt{95}$ 10
3. $\sqrt{300}$ 17

A box of ceiling tiles covers 175 square feet of ceiling.

4. Find the length, to the nearest foot, of a side of a square ceiling that can be covered with 1 box of the ceiling tiles. 13 ft
5. Find the length, to the nearest foot, of a side of a square ceiling that can be covered with 4 boxes of the ceiling tiles. 26 ft

Motivating the Lesson

Hands-On Activity Ask students to draw and cut out two squares of different sizes. Then ask them to draw a third square whose area is the sum of the areas of their two squares. When they finish, discuss the kind of triangle that can be formed using the sides of their three squares.

8–7 The Pythagorean Theorem

What You'll Learn

You'll learn to use the Pythagorean Theorem to solve problems.

Why It's Important

Carpentry

Carpenters use the Pythagorean Theorem to determine whether the corners of a deck are right angles. *See Example 4.*

The sides of the right triangle below have lengths of 3, 4, and 5 units. The relationship among these lengths forms the basis for one of the most famous theorems in mathematics.

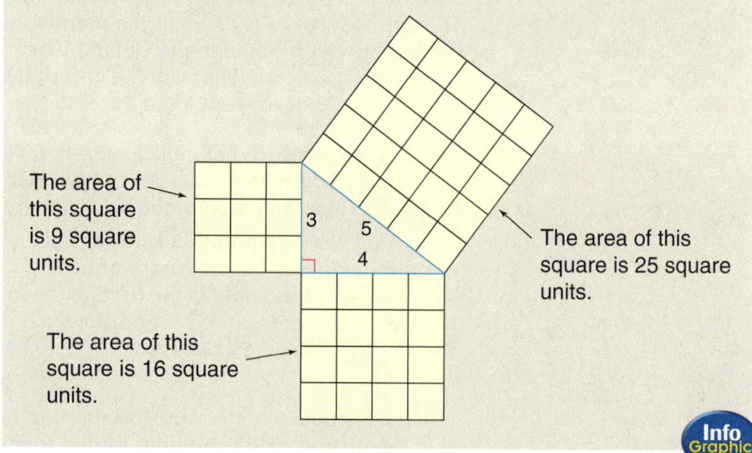

The area of this square is 9 square units.
3
5
4
The area of this square is 25 square units.
The area of this square is 16 square units.
Info Graphic

The two sides that form the right angle are called the **legs**. In the triangle above, the lengths of the legs are 3 units and 4 units. The side opposite the right angle is called the **hypotenuse**. The hypotenuse of this triangle has a length of 5 units.

The squares drawn along each side of the triangle illustrate the Pythagorean Theorem geometrically. Study the areas of the squares. Do you notice a relationship between them? The area of the larger square is equal to the total area of the two smaller squares.

$$25 = 9 + 16$$

$$5^2 = 3^2 + 4^2$$

This relationship is true for *any* right triangle and is called the **Pythagorean Theorem**.

Pythagorean Theorem

Words: In a right triangle, the square of the length of the hypotenuse, c, is equal to the sum of the squares of the lengths of the legs, a and b.

Model:

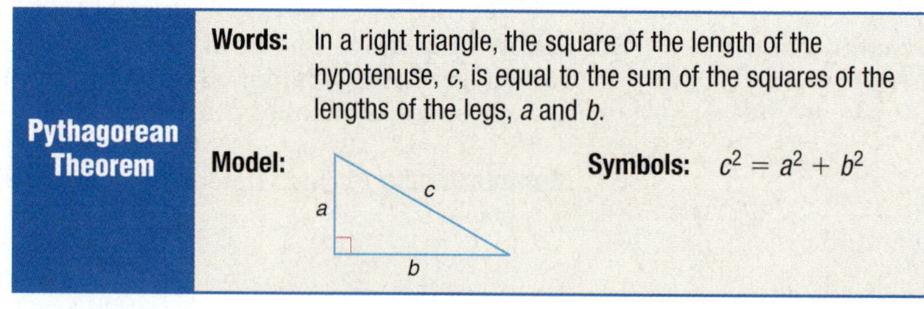

a
c
b

Symbols: $c^2 = a^2 + b^2$

Resource Manager

Reproducible Masters

- *Study Guide,* p. 369
- *Skills Practice,* p. 370
- *Practice,* p. 371
- *Reading to Learn Mathematics,* p. 372
- *Enrichment,* p. 373
- *Assessment,* p. 385

Hands-On Algebra, pp. 93–94
School-to-Workplace, p. 8

Transparencies

5-Minute Check, 8–7

Technology/Multimedia

AlgePASS, Lesson 20
Interactive Chalkboard CD-ROM

MODELING

Alternative hands-on options using grid paper, scissors, tape, markers, rope, and a protractor are available for teaching this lesson.

Example 1 Find the length of the hypotenuse of the right triangle.

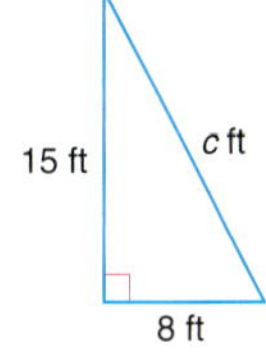

$c^2 = a^2 + b^2$	*Pythagorean Theorem*
$c^2 = 15^2 + 8^2$	*Replace a with 15 and b with 8.*
$c^2 = 225 + 64$	*Multiply.*
$c^2 = 289$	*Add.*
$c = \sqrt{289}$	*Find the square root of each side.*
$c = 17$	*Simplify.*

The length of the hypotenuse is 17 feet.

Your Turn

a. Find the length of the hypotenuse of a right triangle if the lengths of the legs are 6 meters and 8 meters. **10 m**

You can also use the Pythagorean Theorem to find the length of a leg of a right triangle.

Example 2 Find the length of one leg of a right triangle if the length of the hypotenuse is 14 meters and the length of the other leg is 6 meters. Round to the nearest tenth.

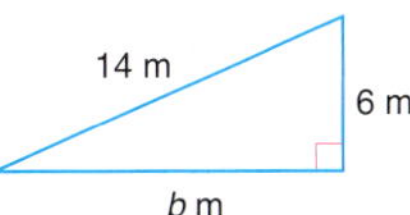

$c^2 = a^2 + b^2$	*Pythagorean Theorem*
$14^2 = 6^2 + b^2$	*Replace c with 14 and a with 6.*
$196 = 36 + b^2$	*Multiply.*
$196 - 36 = 36 - 36 + b^2$	*Subtract 36 from each side.*
$160 = b^2$	***Estimate:** Since $10^2 = 100$ and*
$\sqrt{160} = b$	*$15^2 = 225$, $\sqrt{160}$ is between 10 and 15.*

To the nearest tenth, the length of the leg is 12.6 meters.

Reading Algebra

The hypotenuse is *always* the longest side of a right triangle.

Your Turn

Find each missing measure. Round to the nearest tenth.

b. **24**

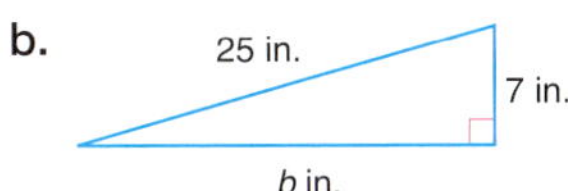

c. **17.3**

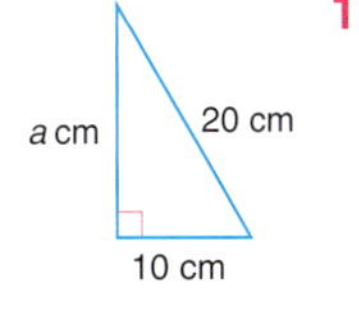

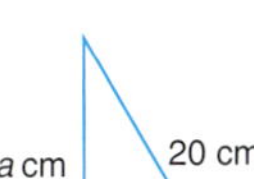

2 TEACH

In-Class Example

Example 1

Find the length of the hypotenuse of the right triangle. **15 in.**

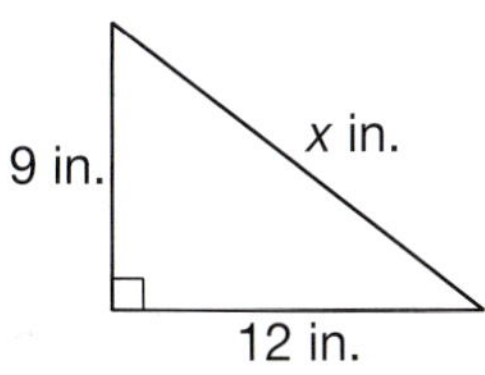

Teaching Tip In Example 2, some students might mistakenly substitute 14 and 6 for the values of *a* and *b*. Stress that if one of the sides whose length is known is the hypotenuse, that value should be substituted for *c*.

In-Class Example

Example 2

Find the length of one leg of a right triangle if the length of the hypotenuse is 22 centimeters and the length of the other leg is 15 centimeters. Round to the nearest tenth. **16.1 cm**

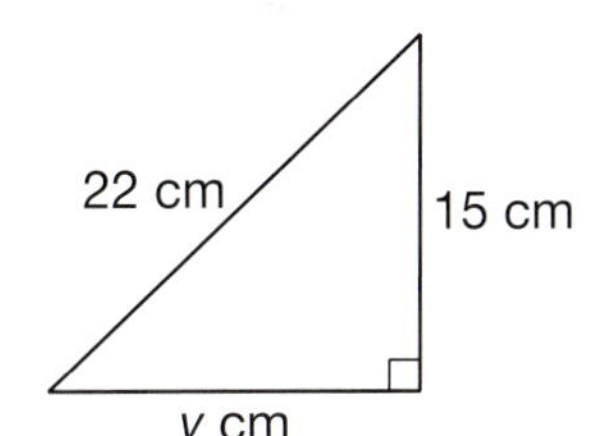

Teaching Tip When discussing the converse of the Pythagorean Theorem, point out how the converse of a theorem is formed by switching the *if* and *then* parts of the original statement.

In-Class Examples

Example 3

The measures of the three sides of a triangle are 6, 11, and 13. Determine whether this triangle is a right triangle. **$13^2 = 169$ and $6^2 + 11^2 = 36 + 121$ or 157; Since $169 \neq 157$, the triangle is *not* a right triangle.**

Example 4

When laying out a rectangular driveway, a concrete worker measures along one side a distance of 16 feet and along the adjacent side a distance of 30 feet. The measure of the hypotenuse is 34 feet. Is the corner of the driveway square? **$16^2 + 30^2 = 256 + 900$ or 1156 and $34^2 = 1156$, so the triangle is a right triangle and the corner is square.**

A **converse** of a theorem is the reverse, or opposite, of the theorem. You can use the converse of the Pythagorean Theorem to test whether a triangle is a right triangle.

Converse of the Pythagorean Theorem	If c is the measure of the longest side of a triangle and $c^2 = a^2 + b^2$, then the triangle is a right triangle.

Examples

3 **The measures of the three sides of a triangle are 5, 7, and 9. Determine whether this triangle is a right triangle.**

$c^2 = a^2 + b^2$ *Pythagorean Theorem*

$9^2 \stackrel{?}{=} 5^2 + 7^2$ *Replace c with 9, a with 5, and b with 7.*

$81 \stackrel{?}{=} 25 + 49$ *Multiply.*

$81 \neq 74$ *Add.*

Since $c^2 \neq a^2 + b^2$, the triangle is *not* a right triangle.

Your Turn

The measures of three sides of a triangle are given. Determine whether each triangle is a right triangle.

d. 20, 21, 28 **no**

e. 37, 12, 34 **no**

Carpentry Link

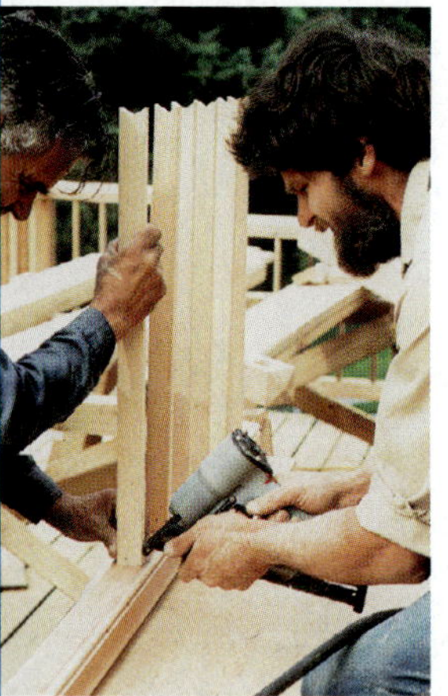

4 **A carpenter checks whether the corners of a deck are square by using a 3-4-5 system. He or she measures along one side in 3-foot units and along the adjacent side in the same number of 4-foot units. If the measure of the hypotenuse is the same number of 5-foot units, the corner of the deck is square.** *If a corner is square, it is formed by a right angle. So, the triangle is a right triangle.*

Suppose a carpenter measures along one side of a deck, a distance of 9 feet, and along the adjacent side, a distance of 12 feet. The measure of the hypotenuse is 15 feet. Is the corner of the deck square?

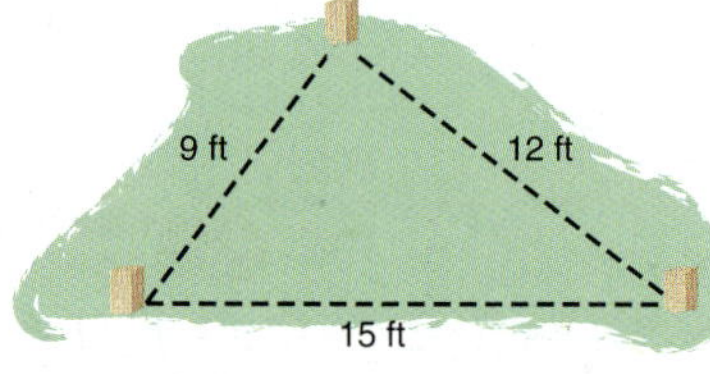

Explore You know that the measures of the sides are 9, 12, and 15. You want to know if the sides form a right triangle.

Plan The measures of the legs are 9 and 12, and the measure of the hypotenuse is 15. Use these numbers in the Pythagorean Theorem.

After students have discussed the converse of the Pythagorean Theorem, tell them to test some of the "square corners" at their home. With the help of family members, they can use triples such as 3, 4, 5 or 5, 12, 13 to check whether a corner of a room, garden, patio, or deck forms a right angle.

Solve	$c^2 = a^2 + b^2$	*Pythagorean Theorem*
	$15^2 \stackrel{?}{=} 9^2 + 12^2$	*Replace c with 15, a with 9, and b with 12.*
	$225 \stackrel{?}{=} 81 + 144$	*Multiply.*
	$225 = 225$	*Add.*

Since $c^2 = a^2 + b^2$, the triangle is a right triangle and the corner of the deck is square.

Examine The measures of the sides are 9, 12, and 15. They are multiples of a 3-4-5 triangle. The answer is reasonable.

Check for Understanding

Communicating Mathematics

1. **Draw** a right triangle and label the right angle, the hypotenuse, and the legs.
2. **Explain** how you know whether a triangle is a right triangle if you know the lengths of the three sides.

1–2. See margin.

Vocabulary
hypotenuse
leg
Pythagorean Theorem
converse

Guided Practice

Getting Ready Determine whether each sentence is *true* or *false*.

Sample: $4^2 + 5^2 = 6^2$ **Solution:** $16 + 25 \neq 36$; false

3. $9^2 + 10^2 = 11^2$ false
4. $7^2 + 24^2 = 25^2$ true
5. $3^2 + 12^2 = 20^2$ false

If *c* is the measure of the hypotenuse and *a* and *b* are the measures of the legs, find each missing measure. Round to the nearest tenth if necessary.

Example 1

6. 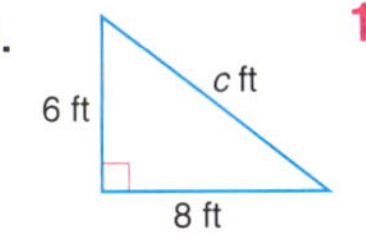

10

7. 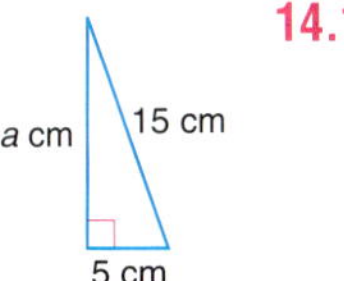

14.1

Example 2

8. $a = 30, c = 34, b = ?$ 16
9. $a = 7, b = 4, c = ?$ 8.1

Example 3

The lengths of three sides of a triangle are given. Determine whether each triangle is a right triangle.

10. 9 ft, 16 ft, 20 ft no
11. 9 mm, 40 mm, 41 mm yes

Example 4

12. **Construction** A builder is laying out the foundation for a house. The measure along one side is 12 feet, the adjacent side is 16 feet, and the hypotenuse is 21 feet. Determine whether the corner is a right angle. no

3 PRACTICE/APPLY

Error Analysis

Watch for students who get a value of 45.3 in Exercise 8 because they found the sum of the squares of the two given numbers and then took the square root of this sum. ***Prevent by*** stressing that it is not always the lengths of the legs that are known. Emphasize that students must always carefully identify which two side lengths are given.

Answers

1. Sample answer:

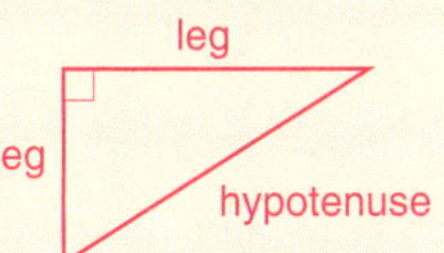

2. If the square of the measure of the hypotenuse is equal to the sum of the squares of the measures of the legs, then it is a right triangle.

***Study Guide*, p. 369**

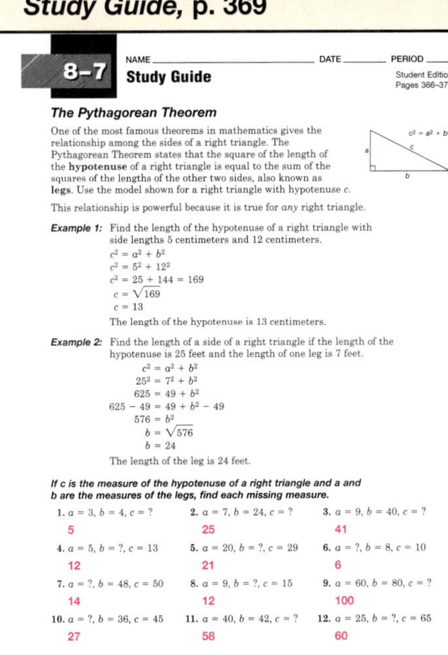

8–7 NAME ______ DATE ______ PERIOD ______

Study Guide

Student Edition Pages 366–371

The Pythagorean Theorem

One of the most famous theorems in mathematics gives the relationship among the sides of a right triangle. The Pythagorean Theorem states that the square of the length of the **hypotenuse** of a right triangle is equal to the sum of the squares of the lengths of the other two sides, also known as **legs**. Use the model shown for a right triangle with hypotenuse *c*.

$c^2 = a^2 + b^2$

This relationship is powerful because it is true for *any* right triangle.

Example 1: Find the length of the hypotenuse of a right triangle with side lengths 5 centimeters and 12 centimeters.

$c^2 = a^2 + b^2$
$c^2 = 5^2 + 12^2$
$c^2 = 25 + 144 = 169$
$c = \sqrt{169}$
$c = 13$

The length of the hypotenuse is 13 centimeters.

Example 2: Find the length of a side of a right triangle if the length of the hypotenuse is 25 feet and the length of one leg is 7 feet.

$c^2 = a^2 + b^2$
$25^2 = 7^2 + b^2$
$625 = 49 + b^2$
$625 - 49 = 49 + b^2 - 49$
$576 = b^2$
$b = \sqrt{576}$
$b = 24$

The length of the leg is 24 feet.

If c is the measure of the hypotenuse of a right triangle and a and b are the measures of the legs, find each missing measure.

1. $a = 3, b = 4, c = ?$ 5
2. $a = 7, b = 24, c = ?$ 25
3. $a = 9, b = 40, c = ?$ 41
4. $a = 5, b = ?, c = 13$ 12
5. $a = 20, b = ?, c = 29$ 21
6. $a = ?, b = 8, c = 10$ 6
7. $a = ?, b = 48, c = 50$ 14
8. $a = 9, b = ?, c = 15$ 12
9. $a = 60, b = 80, c = ?$ 100
10. $a = ?, b = 36, c = 45$ 27
11. $a = 40, b = 42, c = ?$ 58
12. $a = 25, b = ?, c = 65$ 60

Reteaching Activity

Visual/Spatial Learners Have students draw a right triangle of any size. Then have them carefully draw and cut out three squares whose side lengths are those of the three sides of their triangle. Then have them cut the largest square (the square on the hypotenuse) into pieces that will exactly cover the other two squares (the squares on the other two sides).

Assignment Guide

Basic: 13–35 odd, 36–47
Average: 14–32 even, 33–47

Teaching Tip In Exercise 30, you might want to have students try to draw a triangle using the three lengths 1, 1, and 2. They should quickly realize that these lengths cannot be the sides of *any* triangle.

Exercises

Practice

If *c* is the measure of the hypotenuse and *a* and *b* are the measures of the legs, find each missing measure. Round to the nearest tenth if necessary.

13. **13**

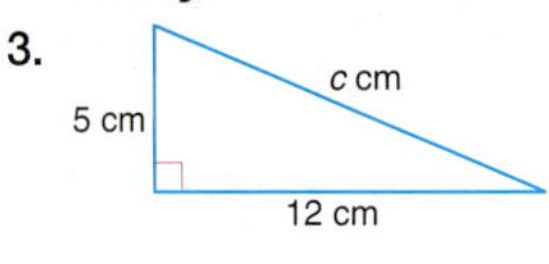

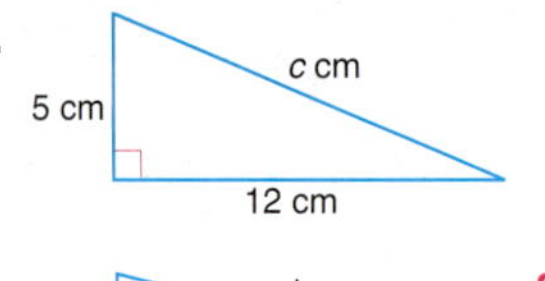

14. **12**

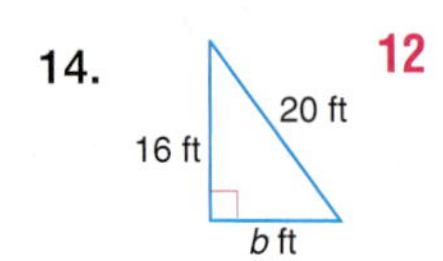

15. **8.2**

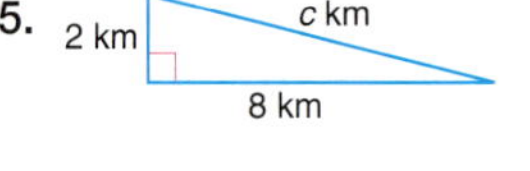

16. **7**

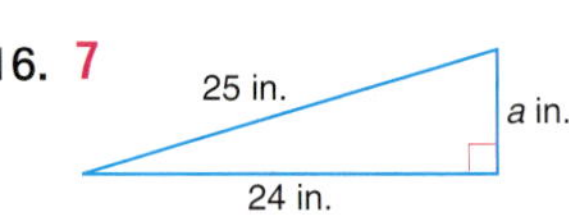

17. **4.2**

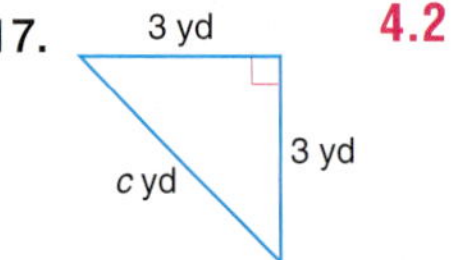

18. **11.2**

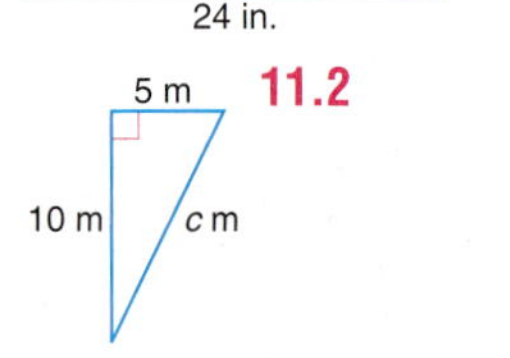

Homework Help	
For Exercises	**See Examples**
13, 15, 17–19, 23, 32, 34, 36	1
14, 16, 20–22, 24, 35	2
25–31	3, 4

Extra Practice
See page 709.

19. $a = 6, b = 3, c = ?$ **6.7**
20. $b = 10, c = 11, a = ?$ **4.6**
21. $c = 29, a = 20, b = ?$ **21**
22. $a = 5, c = 30, b = ?$ **29.6**
23. $a = 7, b = 9, c = ?$ **11.4**
24. $a = 11, c = 20, b = ?$ **16.7**

The lengths of three sides of a triangle are given. Determine whether each triangle is a right triangle.

25. 11 in., 12 in., 16 in. **no**
26. 11 cm, 60 cm, 61 cm **yes**
27. 6 ft, 8 ft, 9 ft **no**
28. 6 mi, 7 mi, 12 mi **no**
29. 45 m, 60 m, 75 m **yes**
30. 1 mm, 1 mm, 2 mm **no**

31. Is a triangle with measures 30, 40, and 50 a right triangle? **yes**

32. Find the length of the hypotenuse of a right triangle if the lengths of the legs are 6 miles and 11 miles. Round to the nearest tenth. **12.5 mi**

Applications and Problem Solving

33. **Geometry** Find the length of the diagonal of a rectangle whose length is 8 meters and whose width is 5 meters. Round to the nearest tenth. **9.4 m**

34. **Carpentry** The rise, the run, and the rafter of a pitched roof form a right triangle. Find the length of the rafter that is needed if the rise of the roof is 6 feet and the run is 12 feet. Round your answer to the nearest tenth. **13.4 ft**

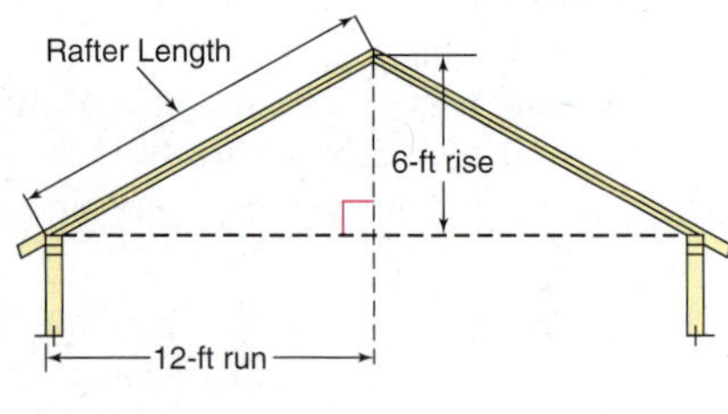

***Skills Practice,* p. 370, and *Practice,* p. 371 (shown)**

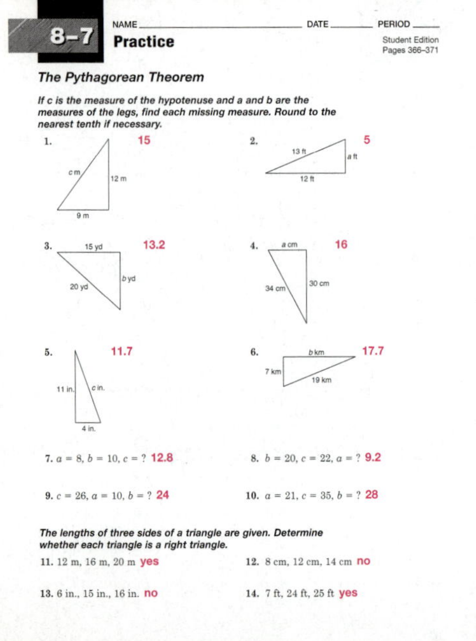

8–7 NAME ______ DATE ______ PERIOD ______

Practice Student Edition Pages 366–371

The Pythagorean Theorem

If c is the measure of the hypotenuse and a and b are the measures of the legs, find each missing measure. Round to the nearest tenth if necessary.

1. **15** (c m, 12 m, 9 m)
2. **5** (13 ft, a ft, 12 ft)
3. **13.2** (15 yd, b yd, 20 yd)
4. **16** (a cm, 34 cm, 30 cm)
5. **11.7** (11 in., c in., 4 in.)
6. **17.7** (b km, 7 km, 19 km)
7. $a = 8, b = 10, c = ?$ **12.8**
8. $b = 20, c = 22, a = ?$ **9.2**
9. $c = 26, a = 10, b = ?$ **24**
10. $a = 21, c = 35, b = ?$ **28**

The lengths of three sides of a triangle are given. Determine whether each triangle is a right triangle.

11. 12 m, 16 m, 20 m **yes**
12. 8 cm, 12 cm, 14 cm **no**
13. 6 in., 15 in., 16 in. **no**
14. 7 ft, 24 ft, 25 ft **yes**

35. **Architecture** Architects often use arches when they design windows and doors. The Early English, or pointed, arch is based on an *equilateral triangle*. An equilateral triangle has three sides of equal measure.

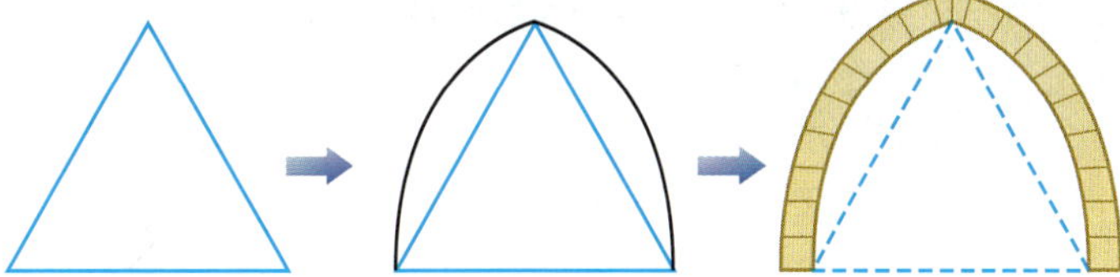

In the figure at the right, the height h separates the equilateral triangle into two right triangles of equal size. Find the height of the arch. Round to the nearest tenth. **6.9 ft**

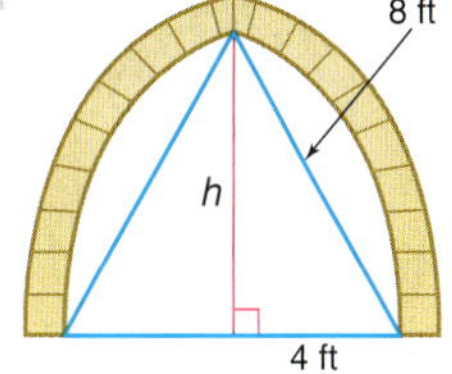

36. **Critical Thinking** The figure shows three squares. The two smaller squares each have an area of 1 square unit.

a. Find the area of the third square.

b. Draw a square on rectangular dot paper with an area of 5 square units.

a–b. See margin.

Mixed Review

Estimate each square root to the nearest whole number. *(Lesson 8–6)*

37. 65 **8** 38. 95 **10** 39. 200 **14** 40. 500 **22**

Simplify. *(Lesson 8–5)*

41. $\sqrt{225}$ **15** 42. $-\sqrt{49}$ **−7** 43. $\sqrt{100}$ **10** 44. $-\sqrt{\frac{9}{25}}$ $-\frac{3}{5}$

45. **Finance** To save for a new bicycle, Katie begins a savings plan. Her plan can be described by the equation $s = 5w + 100$, where s represents the total savings in dollars and w represents the number of weeks since the start of the savings plan. *(Lesson 7–3)* **a. $100**

a. How much money had Katie already saved when the plan started?

b. How much does Katie save each week? **$5**

Standardized Test Practice

46. **Short Response** Write the point-slope form of an equation for the line that passes through the point at (2, −4) and has a slope of −1. *(Lesson 7–2)*

46. $y - (-4) = -1(x - 2)$ or $y + 4 = -1(x - 2)$

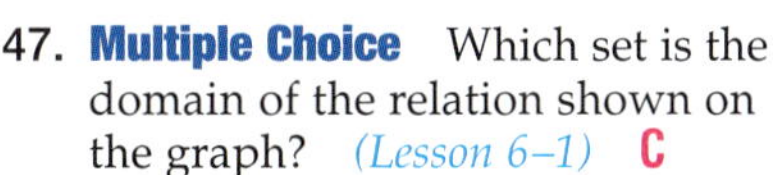

47. **Multiple Choice** Which set is the domain of the relation shown on the graph? *(Lesson 6–1)* **C**

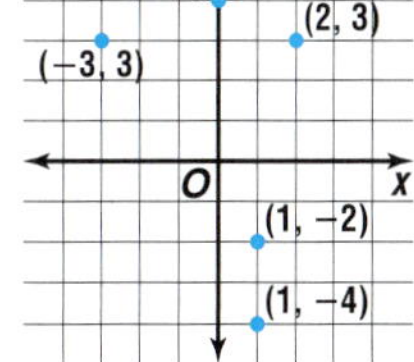

Ⓐ {−3, 0, 2}

Ⓑ {−4, −2, 3, 4}

Ⓒ neither A nor B

Ⓓ both A and B

4 ASSESS

Open-Ended Assessment

Modeling Give students any three numbers that can represent the lengths of the sides of a triangle. Ask them to draw a triangle having three sides of these lengths, tell whether their triangle looks like a right triangle, and then use the converse of the Pythagorean Theorem to verify their response.

Chapter 8, Quiz B (Lessons 8–4 through 8–7) is available in the *Chapter 8 Resource Masters,* p. 385.

Answers

36a. 2 square units

36b.

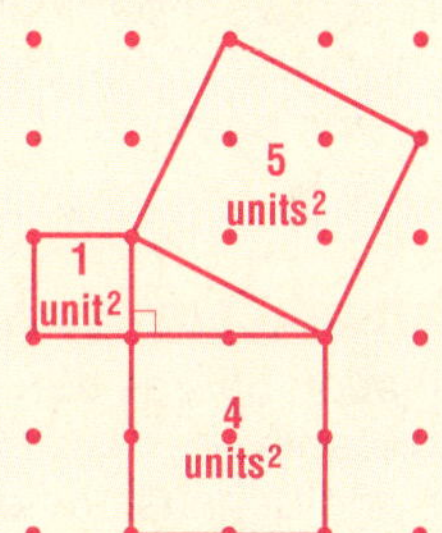

Extra Credit

Two of the three sides of a right triangle have lengths of 15.8 centimeters and 21.3 centimeters. Find two possible values, each to the nearest tenth of a centimeter, for the length of the third side of the right triangle. **26.5 cm and 14.3 cm**

***Enrichment*, p. 373**

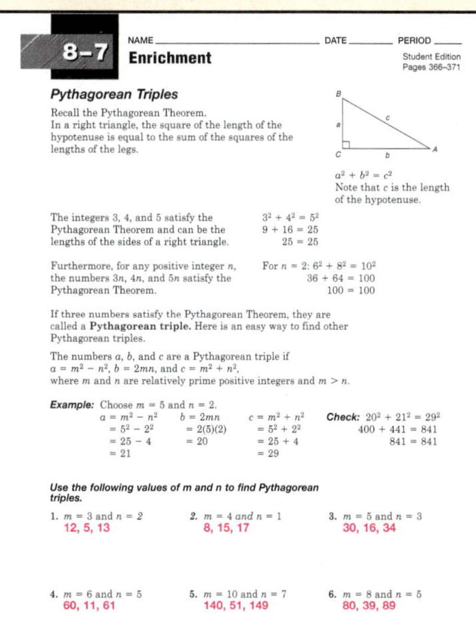

8–7 NAME ________ DATE ______ PERIOD ______

Enrichment

Student Edition Pages 366–371

Pythagorean Triples

Recall the Pythagorean Theorem.
In a right triangle, the square of the length of the hypotenuse is equal to the sum of the squares of the lengths of the legs.

$a^2 + b^2 = c^2$
Note that c is the length of the hypotenuse.

The integers 3, 4, and 5 satisfy the Pythagorean Theorem and can be the lengths of the sides of a right triangle.

$3^2 + 4^2 = 5^2$
$9 + 16 = 25$
$25 = 25$

Furthermore, for any positive integer n, the numbers $3n$, $4n$, and $5n$ satisfy the Pythagorean Theorem.

For $n = 2$: $6^2 + 8^2 = 10^2$
$36 + 64 = 100$
$100 = 100$

If three numbers satisfy the Pythagorean Theorem, they are called a **Pythagorean triple.** Here is an easy way to find other Pythagorean triples.

The numbers a, b, and c are a Pythagorean triple if $a = m^2 - n^2$, $b = 2mn$, and $c = m^2 + n^2$, where m and n are relatively prime positive integers and $m > n$.

Example: Choose $m = 5$ and $n = 2$.

$a = m^2 - n^2 = 5^2 - 2^2 = 25 - 4 = 21$

$b = 2mn = 2(5)(2) = 20$

$c = m^2 + n^2 = 5^2 + 2^2 = 25 + 4 = 29$

Check: $20^2 + 21^2 = 29^2$
$400 + 441 = 841$
$841 = 841$

Use the following values of m and n to find Pythagorean triples.

1. $m = 3$ and $n = 2$ **12, 5, 13**
2. $m = 4$ and $n = 1$ **8, 15, 17**
3. $m = 5$ and $n = 3$ **30, 16, 34**
4. $m = 6$ and $n = 5$ **60, 11, 61**
5. $m = 10$ and $n = 7$ **140, 51, 149**
6. $m = 8$ and $n = 5$ **80, 39, 89**

Investigation

PREPARE

This optional investigation is designed to be completed by groups of 2–3 students over 1–2 days.

Objective

Students use several lengths for the diagonal of a rectangle, and for each diagonal they construct several different-shaped rectangles. For each rectangle, they investigate the ratio of height to length. Finally, they identify rectangles that have a particular height-to-length ratio and present their findings in a poster and report.

Mathematical Overview

This investigation utilizes the following concepts:
- the shape and dimensions of rectangles,
- ratios, and
- the Pythagorean Theorem.

Suggested Time Management	
Investigation	25–40 min
Extension: Gathering Data	25–35 min
Extension: Summarizing Data	30–45 min

Motivating the Lesson

Ask students how the angle between a diagonal and a side of a rectangle changes as the shape of the rectangle changes. Then ask them to reverse the idea, and think about how, as a diagonal changes position, the shape of the rectangle changes. Demonstrate how they will use the rigid spaghetti and grid paper to explore this idea.

Chapter 8 Investigation

Must Be TV

Using the Pythagorean Theorem

Materials

 uncooked spaghetti

 ruler

 grid paper

 scissors

 calculator

Look at an advertisement for an electronics store and you will find television sets that have 10-inch, 32-inch, and 36-inch screens. Did you ever wonder what these numbers represent? They represent the hypotenuse of a right triangle.

A television set is shaped like a rectangle with length ℓ and width w. The diagonal is a line segment from one corner to the opposite corner, which separates the screen into two right triangles of equal size. The diagonal is the hypotenuse of the right triangles. So, a 10-inch television set has a screen in which the diagonal is 10 inches.

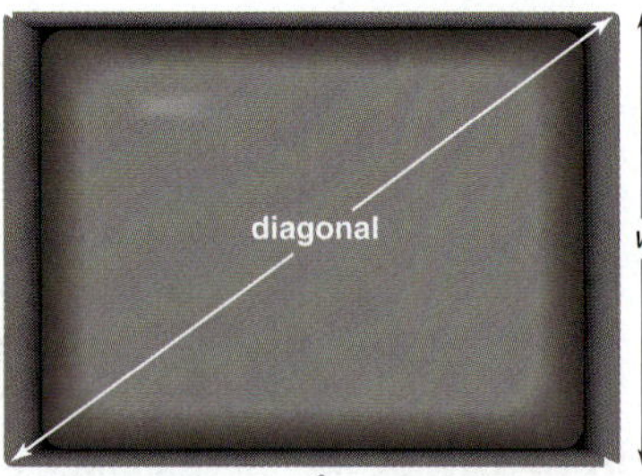

Investigate

1. Use a piece of uncooked spaghetti and grid paper to find several rectangles that have a 10-inch diagonal.
   a. Break the spaghetti so that one piece measures 10 inches.
   b. Place the spaghetti diagonally on the grid paper as shown below.
   c. Using the ends of the spaghetti as endpoints of the diagonal, draw horizontal and vertical lines from the endpoints to complete the rectangle. **See students' work.**

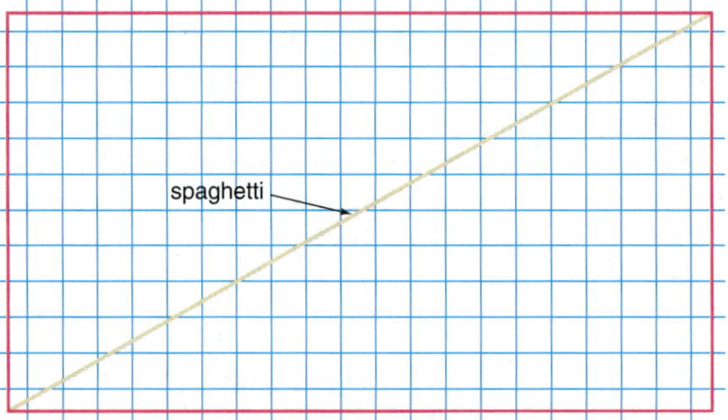

Cooperative Learning

This investigation offers an excellent opportunity for using cooperative groups. For more information on cooperative learning strategies and group management, see *Cooperative Learning in the Mathematics Classroom,* one of the titles in the Glencoe Mathematics Professional Series.

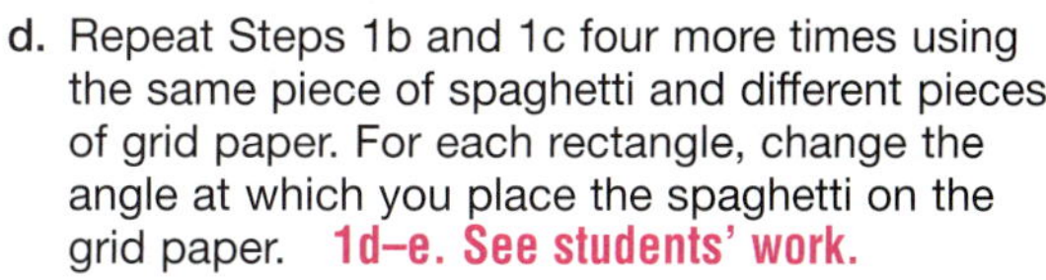

d. Repeat Steps 1b and 1c four more times using the same piece of spaghetti and different pieces of grid paper. For each rectangle, change the angle at which you place the spaghetti on the grid paper. **1d–e. See students' work.**

e. Cut out the rectangles. Choose the rectangle that looks most like a television screen. Describe how this rectangle is different from the others.

2. Television manufacturers have set a standard ratio for the size of a television screen. The ratio of the height to the width is 3:4. For which of your rectangles is the ratio of the height to the width closest to 3:4? **See students' work.**

Extending the Investigation

See margin.

In this extension, you will predict the dimensions of a 19-inch television screen.

- Use a calculator and the Pythagorean Theorem to find the measures of five different rectangles that have a 19-inch diagonal.
- Find the areas of each rectangle and the ratios of each height to width.
- Analyze the data and make a prediction about the dimensions of a 19-inch television screen.

19 in.
?
?

Presenting Your Investigation

Here are some ideas to help you present your conclusions to the class.

- Make a poster that shows scale drawings of your rectangles.
- Write a report in which you explain how you made your prediction. Be sure to include the terms *ratio* and *Pythagorean Theorem* in your report. Then compare your prediction to the dimensions of an actual 19-inch television.

Investigation For more information on ratios, visit: www.algconcepts.com

MANAGE

Teaching Tip Some students may need help preparing a piece of spaghetti so it is close to 10 inches long. Also, you may want to discuss the idea that the size of the grids on the grid paper does not matter because students will be calculating ratios of values.

Working in Groups If students work in groups, they can each create one of the rectangles needed for Exercise 1d. When making a prediction about a 19-inch television screen in the Extension, the group members can work together to suggest the height and width dimensions.

Working as a Class Assign half the class to calculate height-to-width ratios for rectangles cut out of grid paper and the other half to find various rectangles (formed by two right triangles) when the hypotenuse is 19 inches. The students can explain their activities to each other and identify those rectangles and right triangles that were found in both activities.

ASSESS

Students' reports should include information about how the height, length, and diagonal of a rectangle are related and about how the relationship among those three measurements affects the shape of the rectangle. Students should be able to explain how to predict the height and length of a television screen when they know the diagonal measure of the TV screen.

Students should add their posters and reports to their portfolios at this time.

Study Guide and Assessment

Understanding and Using the Vocabulary

This section provides a listing of the new terms, properties, and phrases that were introduced in this chapter. The exercises check students' understanding of the terms by using a variety of verbal formats including matching, completion, and true/false.

Glossary A complete glossary of terms appears on pages 762–783.

MindJogger Videoquizzes

MindJogger Videoquizzes provide an alternative review of concepts presented in this chapter. Students work in teams to answer questions, gaining points for correct answers.

Answers

15. $12 \cdot 12 \cdot 12$
16. $(-2)(-2)(-2)(-2)(-2)$
17. $y \cdot y$
18. $a \cdot a \cdot a \cdot b \cdot b$
19. $5 \cdot x \cdot x \cdot y \cdot y \cdot y \cdot z \cdot z$
20. $-5 \cdot m \cdot m$

CHAPTER 8 Study Guide and Assessment

Understanding and Using the Vocabulary

After completing this chapter, you should be able to define each term, property, or phrase and give an example or two of each.

interNET CONNECTION **Review Activities**
For more review activities, visit: www.algconcepts.com

Algebra

base *(p. 336)*
composite number *(p. 358)*
exponent *(p. 336)*
irrational numbers *(p. 362)*
negative exponent *(p. 348)*
perfect square *(p. 336)*
power *(p. 336)*
prime factorization *(p. 358)*
prime number *(p. 358)*
radical expression *(p. 358)*
radical sign *(p. 357)*
scientific notation *(p. 353)*
square root *(p. 357)*

Geometry

hypotenuse *(p. 366)*
leg *(p. 366)*
Pythagorean Theorem *(p. 366)*

Logic

converse *(p. 268)*

Complete each sentence using a term from the vocabulary list.

1. When a number is expressed as a product of factors that are all prime, the expression is called the ___?___ of the number. **prime factorization**
2. The symbol $\sqrt{\ }$ is called a ___?___. **radical sign**
3. The square roots of numbers such as 2, 3, and 7 are ___?___. **irrational numbers**
4. The ___?___ tells the relationship among the measures of the legs of a right triangle and the hypotenuse. **Pythagorean Theorem**
5. In the expression 2^3, the 3 is called the ___?___. It tells how many times 2 is used as a factor. **exponent**
6. The ___?___ of a number is one of its two equal factors. **square root**
7. The longest side of a right triangle is called the ___?___. **hypotenuse**
8. When you express 5,000,000 as 5×10^6, you are using ___?___. **scientific notation**
9. Numbers like 25, 36, and 100 are called ___?___. **perfect squares**
10. Numbers that are expressed using exponents are called ___?___. **powers**

Skills and Concepts

Objectives and Examples

• **Lesson 8–1** Use powers in expressions.

Write $n \cdot n \cdot n \cdot n$ using exponents.
$n \cdot n \cdot n \cdot n = n^4$

Write $2^3 \cdot 3^2$ as a multiplication expression.
$(2^3)(3^2) = (2)(2)(2)(3)(3)$

Review Exercises

Write each expression using exponents.

11. $9 \cdot 9 \cdot 9 \cdot 9 \cdot 9$ **9^5**
12. 5 squared **5^2**
13. $2 \cdot 2 \cdot 3 \cdot 3 \cdot 3$ **$2^2 3^3$**
14. $(-2)(x)(x)(y)(y)$ **$-2x^2y^2$**

Write each power as a multiplication expression. **15–20. See margin.**

15. 12^3
16. $(-2)^5$
17. y^2
18. a^3b^2
19. $5x^2y^3z^2$
20. $-5m^2$

www.algconcepts.com/vocabulary_review

Resource Manager

Reproducible Masters
Chapter 8 Resource Masters
• *Assessment*, pp. 375–383, 386–388

Technology/Multimedia
MindJogger Videoquizzes
ExamView® Pro

Objectives and Examples	Review Exercises
• **Lesson 8–2** Multiply and divide powers. $(2y^4)(5y^5) = (2 \cdot 5)(y^4 \cdot y^5)$ $= (10)(y^{4+5})$ $= 10y^9$ $\frac{x^7y^4}{x^3y^3} = \left(\frac{x^7}{x^3}\right)\left(\frac{y^4}{y^3}\right)$ $= (x^{7-3})(y^{4-3})$ $= x^4y$	**Simplify each expression.** **24. $-12a^3b^4$** **21.** $b^2 \cdot b^5$ **b^7** **22.** $y^3 \cdot y^3 \cdot y$ **y^7** **23.** $(a^2b)(a^2b^2)$ **a^4b^3** **24.** $(3ab)(-4a^2b^3)$ **25.** $\frac{y^{10}}{y^6}$ **y^4** **26.** $\frac{a^2b^4}{a^3b^2}$ **$\frac{b^2}{a}$** **27.** $\frac{42x^7}{14x^4}$ **$3x^3$** **28.** $\frac{-15x^3y^5z^4}{5x^2y^2z}$ **$-3xy^3z^3$**
• **Lesson 8–3** Simplify expressions containing negative exponents. $a^2b^{-3} = a^2 \cdot \frac{1}{b^3}$ $= \frac{a^2}{b^3}$	**Write each expression using positive exponents. Then evaluate the expression.** **29.** 2^{-3} **$\frac{1}{2^3} = \frac{1}{8}$** **30.** 10^{-2} **$\frac{1}{10^2} = \frac{1}{100}$** **Simplify each expression.** **31.** x^{-4} **$\frac{1}{x^4}$** **32.** $m^{-2}n^5$ **$\frac{n^5}{m^2}$** **33.** $\frac{y^3}{y^{-4}}$ **y^7** **34.** $\frac{rs^2}{s^3}$ **$\frac{r}{s}$** **35.** $\frac{-25a^5b^6}{5a^5b^3}$ **$-5b^3$** **36.** $\frac{27b^{-2}}{14b^{-3}}$ **$\frac{27b}{14}$**
• **Lesson 8–4** Express numbers in scientific notation. $3600 = 3.6 \times 10^3$ $0.023 = 2.3 \times 10^{-2}$ **39. 2.4×10^5** **40. 3.14×10^{-4}** **41. 4.88×10^9** **42. 1.5×10^{-5}**	**Express each measure in standard form.** **37.** 1.5 nanoseconds **0.0000000015** **38.** 5 kilobytes **5000** **Express each number in scientific notation.** **39.** 240,000 **40.** 0.000314 **41.** 4,880,000,000 **42.** 0.000015 **Evaluate each expression.** **43.** $(2 \times 10^5)(3 \times 10^6)$ **6×10^{11}** **44.** $\frac{8 \times 10^{-3}}{2 \times 10^4}$ **4×10^{-7}**
• **Lesson 8–5** Simplify radicals. $\sqrt{400} = \sqrt{2 \cdot 2 \cdot 2 \cdot 2 \cdot 5 \cdot 5}$ $= \sqrt{16 \cdot 25}$ $= \sqrt{16} \cdot \sqrt{25}$ $= 4 \cdot 5$ or 20	**Simplify.** **45.** $\sqrt{121}$ **11** **46.** $-\sqrt{324}$ **-18** **47.** $\sqrt{\frac{4}{81}}$ **$\frac{2}{9}$** **48.** $-\sqrt{\frac{100}{225}}$ **$-\frac{2}{3}$**

Skills and Concepts

The **Objectives and Examples** section reviews the skills and concepts of the chapter and shows completely worked examples.

The **Review Exercises** provide practice for the corresponding objectives.

ExamView® Pro

Use the networkable ***ExamView® Pro Testmaker CD-ROM*** to:

- Create **multiple versions** of tests.
- Create **modified** tests for **inclusion** students with one mouse click.
- **Edit** existing questions and **add** your own questions.
- Build tests aligned with state standards using built-in **state curriculum correlations**.
- Change **English** tests to **Spanish** with one mouse click and vice versa.

Mixed Problem Solving
See pages 724–731.

Applications and Problem Solving

This section provides additional practice in solving real-world problems that involve the concepts of this chapter.

Objectives and Examples

• **Lesson 8–6** Estimate square roots.

Estimate $\sqrt{150}$.

$144 < 150 < 169$ *. . . 100, 121, 144, 169, . . .*

$\sqrt{144} < \sqrt{150} < \sqrt{169}$

$12 < \sqrt{150} < 13$ *150 is closer to 144 than to 169.*

The best whole number estimate for $\sqrt{150}$ is 12.

Review Exercises

Estimate each square root to the nearest whole number.

49. $\sqrt{19}$ **4**
50. $\sqrt{108}$ **10**
51. $\sqrt{200}$ **14**
52. $\sqrt{125.52}$ **11**
53. Which is closer to $\sqrt{50}$, 7 or 8? **7**

• **Lesson 8–7** Use the Pythagorean Theorem to solve problems.

Find the length of the missing side.

$$c^2 = a^2 + b^2$$
$$25^2 = 15^2 + b^2$$
$$625 = 225 + b^2$$
$$625 - 225 = 225 - 225 + b^2$$
$$400 = b^2$$
$$20 = b$$

15 ft
25 ft
b ft

If c is the measure of the hypotenuse and a and b are the measures of the legs, find each missing measure. Round to the nearest tenth if necessary.

54. **15** 55. **11.2**

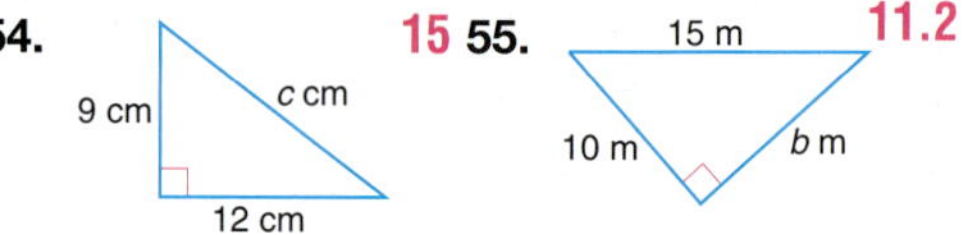

56. $a = 6$ mi, $b = 10$ mi, $c = ?$ **11.7**
57. $b = 6$, $c = 12$, $a = ?$ **10.4**

Applications and Problem Solving

58. **Biology** *E. coli* is a fast-growing bacteria that splits into two identical cells every 15 minutes. If you start with one *E. coli* bacterium, how many bacteria will there be at the end of 2 hours? *(Lesson 8–1)* **256 bacteria**

59. **Geometry** The area of a square is 90 square meters. Estimate the length of its side to the nearest whole number. *(Lesson 8–6)* **9 m**

60. **Construction** Park managers want to construct a road from the park entrance directly to a campsite. Use the figure below to find the length of the proposed road. *(Lesson 8–7)* **17 mi**

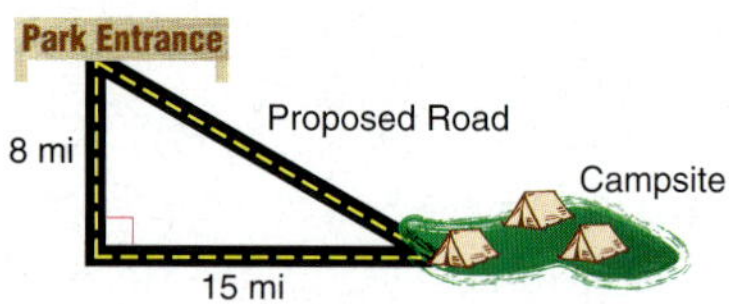

Assessment

Four forms of Chapter 8 Test are available in the *Chapter 8 Resource Masters.*

Chapter 8 Test, Form 1B, is shown at the left. Chapter 8 Test, Form 2B, is shown on the next page.

	Form of Test		Level
1A	Multiple Choice	pp. 375–376	Average
1B	Multiple Choice	pp. 377–378	Basic
2A	Free Response	pp. 379–380	Average
2B	Free Response	pp. 381–382	Basic

***Assessment*, pp. 377–378**

8 NAME ________ DATE ______ PERIOD ____

Chapter 8 Test, Form 1B

Write the letter for the correct answer in the blank at the right of each problem.

1. Write the expression $3 \cdot y \cdot y \cdot y \cdot y$ using exponents.
   A. $3y^2$ B. $3y^4$ C. $81y^4$ D. $(3y)^4$ 1. **B**
2. Which expression is equivalent to $-2h^5$?
   A. $(-2h)(-2h)(-2h)(-2h)(-2h)$ B. $-32h^5$
   C. $-2 \cdot h \cdot h \cdot h \cdot h \cdot h$ D. $32h^5$ 2. **C**
3. Evaluate $-4x^3y^2$ if $x = -2$ and $y = 3$.
   A. −288 B. −96 C. 96 D. 288 3. **D**

For Questions 4–9, simplify each expression.

4. $x^5 \cdot x^3 \cdot x$
   A. x^8 B. x^9 C. x^{15} D. $15x^9$ 4. **B**
5. $(4x^4y)(2x^2)$
   A. $8x^8y$ B. $6x^6y$ C. $8x^6y$ D. $8x^4y^3$ 5. **C**
6. $\frac{-36m^4n^7}{18m^2n^6}$
   A. $-4m^2n$ B. $-2m^6n^{13}$ C. $2m^2n$ D. $-2m^2n$ 6. **D**
7. $b^4(b^{-2})$
   A. b^2 B. $\frac{1}{b^8}$ C. $-b^2$ D. b^8 7. **A**
8. $\frac{h^{-2}}{h^3}$
   A. $\frac{1}{h^6}$ B. $\frac{1}{h^5}$ C. $\frac{1}{h}$ D. $-h^5$ 8. **B**
9. $\frac{-3m^4n^6}{15m^3n^{-2}}$
   A. $-5mn^8$ B. $-\frac{m}{5n^8}$ C. $-\frac{m}{5n^4}$ D. $-\frac{mn^8}{5}$ 9. **D**
10. Express 32,000,000 miles in scientific notation.
    A. 3.2×10^{-7} mi B. 32×10^7 mi
    C. 3.2×10^7 mi D. 3.2×10^8 mi 10. **C**
11. Express 0.000045 grams in scientific notation.
    A. 4.5×10^{-6} g B. 4.5×10^{-5} g
    C. 4.5×10^5 g D. 4.5×10^6 g 11. **B**
12. Evaluate $(4.4 \times 10^3)(2 \times 10^4)$. Express the answer in scientific notation.
    A. 8.8×10^{-7} B. 8.8×10^7
    C. 8.8×10^8 D. 8.8×10^{12} 12. **B**
13. Jupiter's circumference is about 4.5×10^5 kilometers, or about 63 times the circumference of Pluto. What is Pluto's approximate circumference?
    A. 2.8×10^7 km B. 2.8×10^3 km
    C. 7.1×10^3 km D. 1.4×10^4 km 13. **C**

CHAPTER 8 Test

1. **Write** a multiplication problem whose product is $12x^5$. **Sample answer: $(4x^3)(3x^2)$**
2. **Explain** why 3 is the square root of 9. **$3 \times 3 = 9$**
3. **Draw** a figure that can be used to estimate $\sqrt{96}$. **See margin.**

Evaluate each expression if $x = 2$, $y = 3$, $z = -1$, and $w = -2$.

4. $3x^5$ **96**
5. $5w^3y$ **−120**
6. $12y^{-2}$ $\frac{4}{3}$
7. z^0 **1**

Simplify each expression.

8. $(a^3b^2)(a^5b)$ a^8b^3
9. $\frac{xy^3}{y^5}$ $\frac{x}{y^2}$
10. m^{-10} $\frac{1}{m^{10}}$
11. $\frac{-3r^2s^8}{12r^2s^5}$ $-\frac{s^3}{4}$

Express each number in scientific notation.

12. 0.00000125 1.25×10^{-6}
13. 36,000,000,000 3.6×10^{10}

Evaluate each expression. Express each result in scientific notation and standard form.

14. $(8.2 \times 10^{-5})(1 \times 10^{-3})$
$8.2 \times 10^{-8} = 0.000000082$
15. $\frac{6 \times 10^3}{2 \times 10^{-2}}$
$3 \times 10^5 = 300{,}000$

Estimate each square root to the nearest whole number.

16. $\sqrt{5}$ **2**
17. $\sqrt{20}$ **4**
18. $\sqrt{63}$ **8**
19. $\sqrt{395}$ **20**

Find each missing measure. Round to the nearest tenth if necessary.

20. 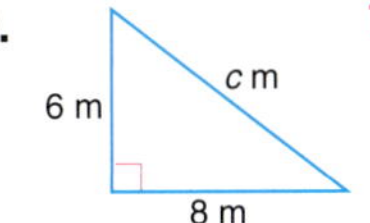

10

21. 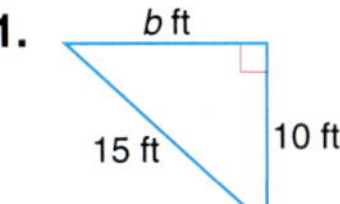

11.2

22. 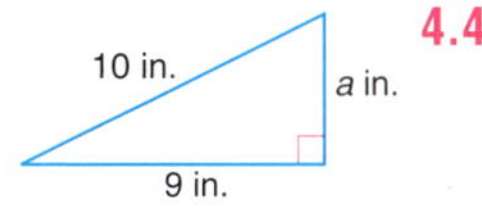

4.4

23. **Chemistry** The diameter of an atom is about 10^{-8} centimeter and the diameter of its nucleus is about 10^{-13} centimeter. How many times as large is the diameter of the atom as the diameter of its nucleus?
10^5 or 100,000 times
24. **Sports** Gymnastic routines are performed on a 40-foot by 40-foot square mat. Gymnasts usually use the diagonal of the mat because it gives them more distance to complete their routine. To the nearest whole number, how much longer is the diagonal of the mat than one of its sides? **17 ft**

25. **Sequences** The first term of the sequence $1, \frac{1}{2}, \frac{1}{4}, \frac{1}{8}, \ldots$ can be written as 2^0. Write the next three terms using negative exponents. $2^{-4}, 2^{-5}, 2^{-6}$

www.algconcepts.com/chapter_test

Chapter Test Bonus Question

Find three whole numbers x, y, and z so that $\sqrt{\frac{1}{x}} + \sqrt{\frac{1}{y}} = \sqrt{\frac{1}{z}}$.

Sample answers: $x = 9, y = 36, z = 4$; $x = 144, y = 36, z = 16$; $x = 225, y = 900, z = 100$

Chapter Test

Answer

3. 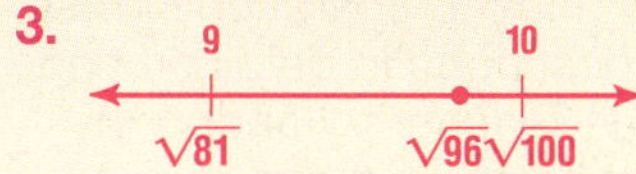

Assessment, pp. 381–382

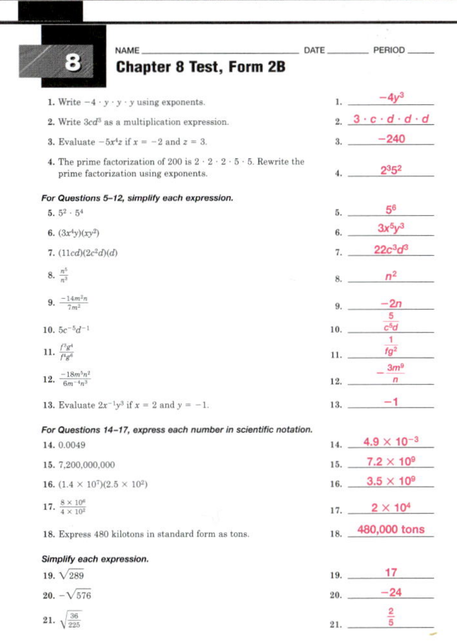

NAME ______ DATE ______ PERIOD ______

Chapter 8 Test, Form 2B

1. Write $-4 \cdot y \cdot y \cdot y$ using exponents. 1. $-4y^3$
2. Write $3cd^3$ as a multiplication expression. 2. $3 \cdot c \cdot d \cdot d \cdot d$
3. Evaluate $-5x^4z$ if $x = -2$ and $z = 3$. 3. -240
4. The prime factorization of 200 is $2 \cdot 2 \cdot 2 \cdot 5 \cdot 5$. Rewrite the prime factorization using exponents. 4. 2^35^2

For Questions 5–12, simplify each expression.

5. $5^2 \cdot 5^4$ 5. 5^6
6. $(3x^4y)(xy^2)$ 6. $3x^5y^3$
7. $(11cd)(2c^2d)(d)$ 7. $22c^3d^3$
8. $\frac{n^5}{n^3}$ 8. n^2
9. $\frac{-14m^2n}{7m^2}$ 9. $-2n$
10. $5c^{-5}d^{-1}$ 10. $\frac{5}{c^5d}$
11. $\frac{f^3g^4}{f^4g^6}$ 11. $\frac{1}{fg^2}$
12. $\frac{-18m^5n^2}{6m^{-4}n^3}$ 12. $-\frac{3m^9}{n}$
13. Evaluate $2x^{-1}y^3$ if $x = 2$ and $y = -1$. 13. -1

For Questions 14–17, express each number in scientific notation.

14. 0.0049 14. 4.9×10^{-3}
15. 7,200,000,000 15. 7.2×10^9
16. $(1.4 \times 10^7)(2.5 \times 10^2)$ 16. 3.5×10^9
17. $\frac{8 \times 10^6}{4 \times 10^2}$ 17. 2×10^4
18. Express 480 kilotons in standard form as tons. 18. 480,000 tons

Simplify each expression.

19. $\sqrt{289}$ 19. 17
20. $-\sqrt{576}$ 20. -24
21. $\sqrt{\frac{36}{225}}$ 21. $\frac{2}{5}$

Preparing for Standardized Tests

Pages 378–379 are part of a complete test preparation course that is described in detail on page T9 of the Teacher's Handbook. The test items on these pages were written in the same style as those in state proficiency tests and standardized tests like ACT and SAT.

Diagnosis and Prescription

Each of the 10 test questions on page 379 is cross-referenced to the lesson where that SAT or ACT skill is covered. If students miss a particular type of problem, you can have them study that skill.

(See chart at the bottom of page 379. Note that SPT = State Proficiency Test, SAT = Scholastic Assessment Test, and ACT = American College Test.)

***Assessment*, p. 386**

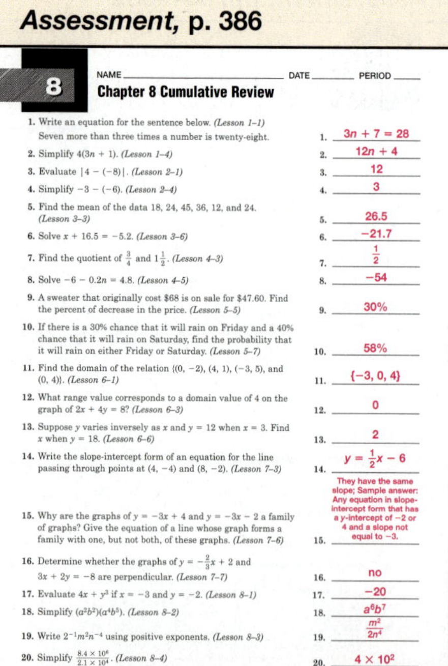

8

NAME ______ DATE ______ PERIOD ______

Chapter 8 Cumulative Review

1. Write an equation for the sentence below. *(Lesson 1–1)* Seven more than three times a number is twenty-eight. — 1. $3n + 7 = 28$
2. Simplify $4(3n + 1)$. *(Lesson 1–4)* — 2. $12n + 4$
3. Evaluate $|4 - (-8)|$. *(Lesson 2–1)* — 3. 12
4. Simplify $-3 - (-6)$. *(Lesson 2–4)* — 4. 3
5. Find the mean of the data 18, 24, 45, 36, 12, and 24. *(Lesson 3–3)* — 5. 26.5
6. Solve $x + 16.5 = -5.2$. *(Lesson 3–6)* — 6. -21.7
7. Find the quotient of $\frac{3}{4}$ and $1\frac{1}{2}$. *(Lesson 4–3)* — 7. $\frac{1}{2}$
8. Solve $-6 - 0.2n = 4.8$. *(Lesson 4–5)* — 8. -54
9. A sweater that originally cost $68 is on sale for $47.60. Find the percent of decrease in the price. *(Lesson 5–5)* — 9. 30%
10. If there is a 30% chance that it will rain on Friday and a 40% chance that it will rain on Saturday, find the probability that it will rain on either Friday or Saturday. *(Lesson 5–7)* — 10. 58%
11. Find the domain of the relation $\{(0, -2), (4, 1), (-3, 5), \text{and } (0, 4)\}$. *(Lesson 6–1)* — 11. $\{-3, 0, 4\}$
12. What range value corresponds to a domain value of 4 on the graph of $2x + 4y = 8$? *(Lesson 6–3)* — 12. 0
13. Suppose y varies inversely as x and $y = 12$ when $x = 3$. Find x when $y = 18$. *(Lesson 6–6)* — 13. 2
14. Write the slope-intercept form of an equation for the line passing through points at $(4, -4)$ and $(8, -2)$. *(Lesson 7–3)* — 14. $y = \frac{1}{2}x - 6$
15. Why are the graphs of $y = -3x + 4$ and $y = -3x - 2$ a family of graphs? Give the equation of a line whose graph forms a family with one, but not both, of these graphs. *(Lesson 7–6)* — 15. They have the same slope; Sample answer: Any equation in slope-intercept form that has a y-intercept of -2 or 4 and a slope not equal to -3.
16. Determine whether the graphs of $y = -\frac{2}{3}x + 2$ and $3x + 2y = -8$ are perpendicular. *(Lesson 7–7)* — 16. no
17. Evaluate $4x + y^3$ if $x = -3$ and $y = -2$. *(Lesson 8–1)* — 17. -20
18. Simplify $(a^2b^2)(a^4b^5)$. *(Lesson 8–2)* — 18. a^6b^7
19. Write $2^{-1}m^2n^{-4}$ using positive exponents. *(Lesson 8–3)* — 19. $\frac{m^2}{2n^4}$
20. Simplify $\frac{8.4 \times 10^6}{2.1 \times 10^4}$. *(Lesson 8–4)* — 20. 4×10^2

Pythagorean Theorem Problems

All standardized tests contain several problems that you can solve using the Pythagorean Theorem. The **Pythagorean Theorem** states that in a right triangle, the sum of the squares of the measures of the legs equals the square of the measure of the hypotenuse.

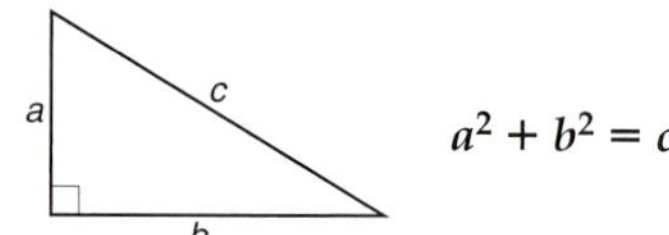

$$a^2 + b^2 = c^2$$

Test-Taking Tip

The 3-4-5 right triangle and its multiples, like 6-8-10 and 9-12-15, occur frequently on standardized tests. Other commonly used Pythagorean triples include 5-12-13 and 7-24-25.

Example 1

Use the information in the figure below. Find *BC* to the nearest centimeter.

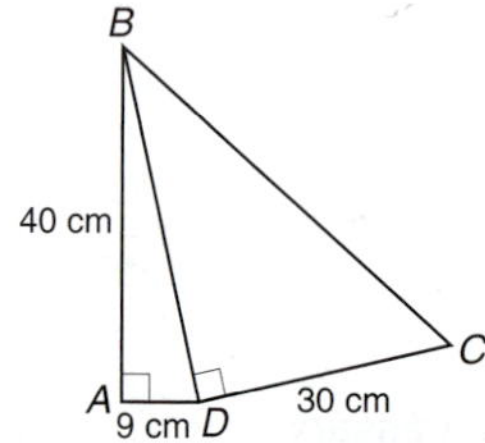

Ⓐ 31 cm Ⓑ 41 cm
Ⓒ 51 cm Ⓓ 80 cm

Hint Use the Pythagorean Theorem twice.

Solution The figure is made up of two right triangles, $\triangle ABD$ and $\triangle BDC$. $\overline{BD}$ is the hypotenuse of $\triangle ABD$. $\overline{BD}$ is also a leg of $\triangle BDC$.

First, find BD. Then use BD to find BC.

$(BD)^2 = 9^2 + 40^2$ $(BC)^2 = 30^2 + 41^2$
$(BD)^2 = 81 + 1600$ $(BC)^2 = 900 + 1681$
$(BD)^2 = 1681$ $(BC)^2 = 2581$
$BD = 41$ $BC \approx 50.80$

The answer is C.

Example 2

A 25-foot ladder is placed against a vertical wall of a building with the bottom of the ladder standing on concrete 7 feet from the base of the building. If the top of the ladder slips down 4 feet, then the bottom of the ladder will slide out how many feet?

Ⓐ 5 ft Ⓑ 6 ft
Ⓒ 7 ft Ⓓ 8 ft

Hint Start by drawing a diagram.

Solution The ladder placed against the wall forms a 7-24-25 right triangle. After the ladder slips down 4 feet, the new right triangle has sides that are multiples of a 3-4-5 right triangle, 15-20-25.

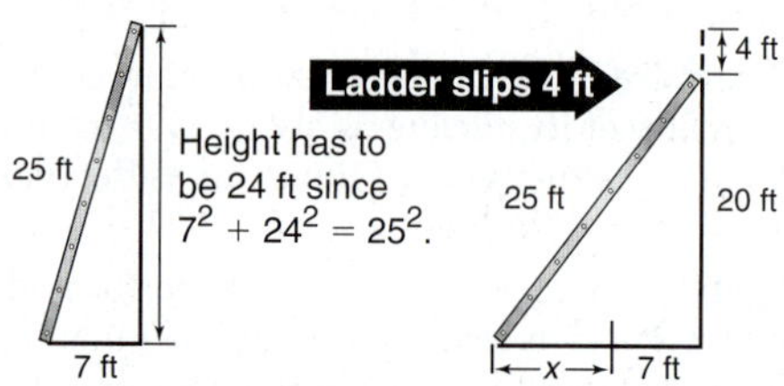

The bottom of the ladder will be 15 feet from the wall. This means the bottom of the ladder will slide out 15 − 7 or 8 feet.

The answer is D.

Resource Manager

Reproducible Masters

Chapter 8 Resource Masters

- *Assessment*, pp. 386–388

Preparing for Standardized Tests
For test-taking strategies and more practice, see pages 732–749.

After you work each problem, record your answer on the answer sheet provided or on a sheet of paper.

Multiple Choice

1. Diego bought 3 fish. Every month, the number of fish doubles. The formula $f = 3 \cdot 2^m$ represents the number of fish he will have after m months. How many fish will he have after 4 months if he keeps all of them and none of them dies? *(Lesson 8–1)* **C**
   - (A) 12
   - (B) 24
   - (C) 48
   - (D) 1296

2. Which number is greater than 2.8? *(Lesson 8–6)* **B**
   - (A) 175%
   - (B) $\sqrt{12}$
   - (C) -3.5
   - (D) 5.6×10^{-1}

3. A wire from the top of a 25-foot flagpole is attached to a point 20 feet from the base of the flagpole. To the nearest whole number, find the length of the wire. *(Lesson 8–7)* **B**

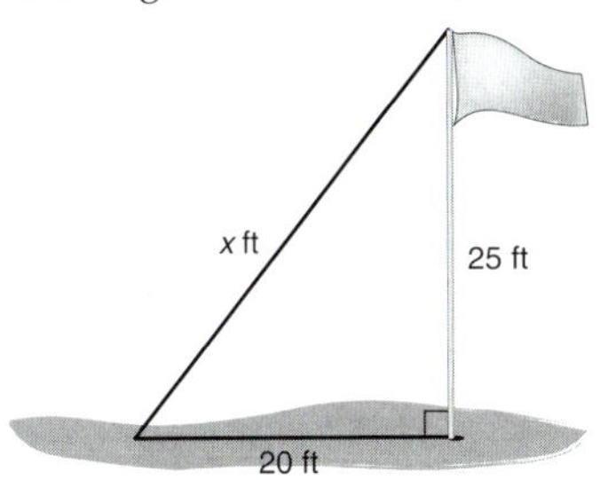

   - (A) 45 ft
   - (B) 32 ft
   - (C) 15 ft
   - (D) 35 ft

4. Which of the following are measures of three sides of a right triangle? *(Lesson 8–7)* **D**
   - (A) 4, 7, 8
   - (B) 10, 15, 20
   - (C) 3, 7, 9
   - (D) 9, 12, 15

5. Which equation does *not* have 6 as a solution? *(Lesson 4–5)* **C**
   - (A) $2(x - 6) = 0$
   - (B) $4x + 3 = 27$
   - (C) $2x + 1 = 14$
   - (D) $2x + x = 18$

6. $\triangle ABC$ and $\triangle ACD$ are right triangles. If $AB = 20$, $BC = 15$, and $AD = 7$, then $CD =$ — *(Lesson 8–7)* **C**

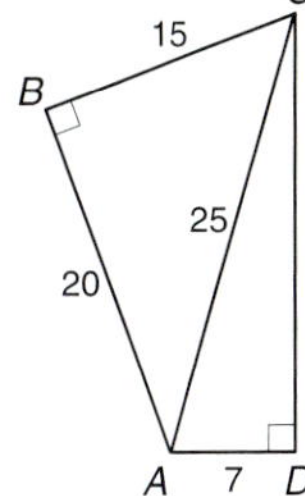

   - (A) 22.
   - (B) 23.
   - (C) 24.
   - (D) 25.

7. A farmer wants to put a fence around a square garden that has a total area of 1600 square feet. If fencing costs $1.50 per foot, how much will it cost to fence the garden? To solve this problem, begin by— *(Lesson 1–5)* **D**
   - (A) dividing 1600 by $1.50.
   - (B) multiplying 1600 by $1.50 and taking the square root of the product.
   - (C) dividing 1600 by 4 and multiplying by $1.50.
   - (D) finding the square root of 1600 and multiplying by 4.

8. If $x = 3$, $y = -2$, $z = -1$, and $w = 5$, then the value of $9xy^2z^{10}w^0$ is— *(Lesson 8–1)* **D**
   - (A) -590.
   - (B) -108.
   - (C) 1.
   - (D) 108.

Grid In

9. On the first test of the semester, Jennifer scored a 60. On the last test of the semester, she scored a 75. By what percent did Jennifer's score improve? *(Lesson 5–5)* **25**

Short Response

10. The table shows the squares of six whole numbers. Describe one way that the digits in the squares are related to the digits in the original number. *(Lesson 1–5)* **See margin.**

$15^2 = 225$	$55^2 = 3025$	$135^2 = 18,225$
$25^2 = 625$	$95^2 = 9025$	$175^2 = 30,625$

www.algconcepts.com/standardized_test

A bubble-in answer sheet for these practice problems is available on page A1 of the *Chapter 8 Resource Masters.*

Additional Practice

Additional test practice questions are available in the *Chapter 8 Resource Masters*, pp. 387–388.

Answer

10. Sample answer: The last two digits in the squares equal the square of the last digit in the original numbers.

Chapter 8 Pythagorean Theorem Problems		
Ex. 1	Pythagorean Theorem	SPT
Ex. 2	ladder word problem	SAT
1	evaluate an equation	SPT
2	comparing rational numbers	SPT
3	Pythagorean Theorem	SPT
4	Pythagorean Theorem	SPT
5	solving an equation	SPT
6	Pythagorean Theorem	ACT
7	word problem	SPT
8	evaluate an expression	SPT
9	percent word problem	SAT
10	algebra word problem	SPT

***Assessment*, pp. 387–388**

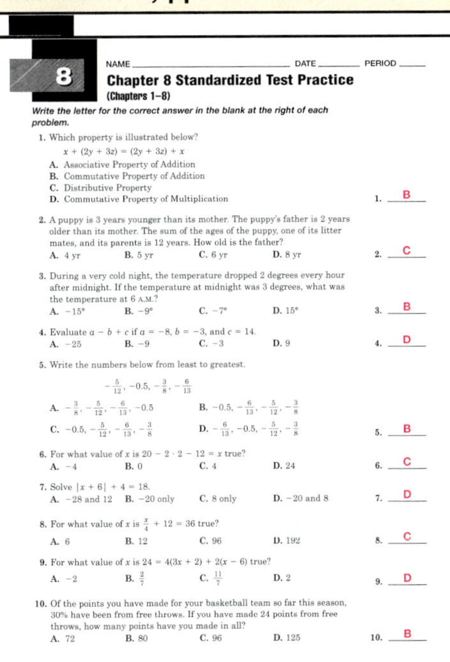
NAME ______ DATE ______ PERIOD ______

8 **Chapter 8 Standardized Test Practice**
(Chapters 1–8)

Write the letter for the correct answer in the blank at the right of each problem.

1. Which property is illustrated below?
   $x + (2y + 3z) = (2y + 3z) + x$
   A. Associative Property of Addition
   B. Commutative Property of Addition
   C. Distributive Property
   D. Commutative Property of Multiplication — 1. B

2. A puppy is 3 years younger than its mother. The puppy's father is 2 years older than its mother. The sum of the ages of the puppy, one of its litter mates, and its parents is 12 years. How old is the father?
   A. 4 yr B. 5 yr C. 6 yr D. 8 yr — 2. C

3. During a very cold night, the temperature dropped 2 degrees every hour after midnight. If the temperature at midnight was 3 degrees, what was the temperature at 6 A.M.?
   A. −15° B. −9° C. −7° D. 15° — 3. B

4. Evaluate $a - b + c$ if $a = -8$, $b = -3$, and $c = 14$.
   A. −25 B. −9 C. −3 D. 9 — 4. D

5. Write the numbers below from least to greatest.
   $-\frac{5}{12}, -0.5, -\frac{3}{8}, -\frac{6}{13}$
   A. $-\frac{3}{8}, -\frac{5}{12}, -\frac{6}{13}, -0.5$ B. $-0.5, -\frac{6}{13}, -\frac{5}{12}, -\frac{3}{8}$
   C. $-0.5, -\frac{5}{12}, -\frac{6}{13}, -\frac{3}{8}$ D. $-\frac{6}{13}, -0.5, -\frac{5}{12}, -\frac{3}{8}$ — 5. B

6. For what value of x is $20 - 2 \cdot 2 - 12 = x$ true?
   A. −4 B. 0 C. 4 D. 24 — 6. C

7. Solve $|x + 6| + 4 = 18$.
   A. −28 and 12 B. −20 only C. 8 only D. −20 and 8 — 7. D

8. For what value of x is $\frac{x}{4} + 12 = 36$ true?
   A. 6 B. 12 C. 96 D. 192 — 8. C

9. For what value of x is $24 = 4(3x + 2) + 2(x - 6)$ true?
   A. −2 B. $\frac{2}{7}$ C. $\frac{11}{7}$ D. 2 — 9. D

10. Of the points you have made for your basketball team so far this season, 30% have been from free throws. If you have made 24 points from free throws, how many points have you made in all?
    A. 72 B. 80 C. 96 D. 125 — 10. B

CHAPTER 9 Resource Manager

Polynomials

Instructional Objectives

Lesson (pages)	Objectives	NCTM Standards 2000	State/Local Objectives
9–1 (382–387)	Identify and classify polynomials and find their degree.	1, 2, 6, 8, 9	
9–2 (388–393)	Add and subtract polynomials.	1, 6, 8, 9	
9–3 (394–398)	Multiply a polynomial by a monomial.	1, 6, 8, 9	
9–4 (399–404)	Multiply two binomials.	1, 6, 8, 9	
9–5 (405–409)	Develop and use the patterns for $(a + b)^2$, $(a - b)^2$, and $(a + b)(a - b)$.	1, 6, 8, 9	
Investigation (410–411)	Investigate the ratios of areas of different-sized squares.	1, 2, 3, 4, 6, 7, 8, 9	

Key to NCTM Standards 2000
1 Number & Operations; **2** Algebra; **3** Geometry; **4** Measurement; **5** Data Analysis & Probability; **6** Problem Solving; **7** Reasoning & Proof; **8** Communications; **9** Connections; **10** Representation

Suggested Pacing

See page T13 for a complete course-planning calendar.

Standard refers to schedules that provide 45- to 55-minute periods that meet each day.
Block refers to schedules that provide approximately 90-minute periods which may meet every day for one semester or every other day over two semesters.

<table>
<tr><th>PACING</th><th>DAY 1</th><th>DAY 2</th><th>DAY 3</th><th>DAY 4</th><th>DAY 5</th><th>DAY 6</th></tr>
<tr><td>Standard Core (Chapters 1–13)</td><td colspan="2">Lesson 9–1</td><td colspan="2">Lesson 9–2</td><td>Lesson 9–3</td><td>Lesson 9–4</td></tr>
<tr><td>Standard Enhanced (Chapters 1–15)</td><td>Lesson 9–1</td><td colspan="2">Lesson 9–2</td><td>Lesson 9–3</td><td>Lesson 9–4</td><td>Lesson 9–5</td></tr>
<tr><td>Block Core (Chapters 1–13)</td><td>Chapter 8 Test & Lesson 9–1</td><td>Lessons 9–2 & 9–3</td><td>Lessons 9–4 & 9–5</td><td>INV</td><td>SG+A</td><td>Chapter Test & Lesson 10–1</td></tr>
<tr><td>Block Enhanced (Chapters 1–15)</td><td>Chapter 8 Test & Lesson 9–1</td><td>Lessons 9–2 & 9–3</td><td>Lessons 9–4 & 9–5</td><td>INV & SG+A</td><td>Chapter Test & Lesson 10–1</td><td></td></tr>
</table>

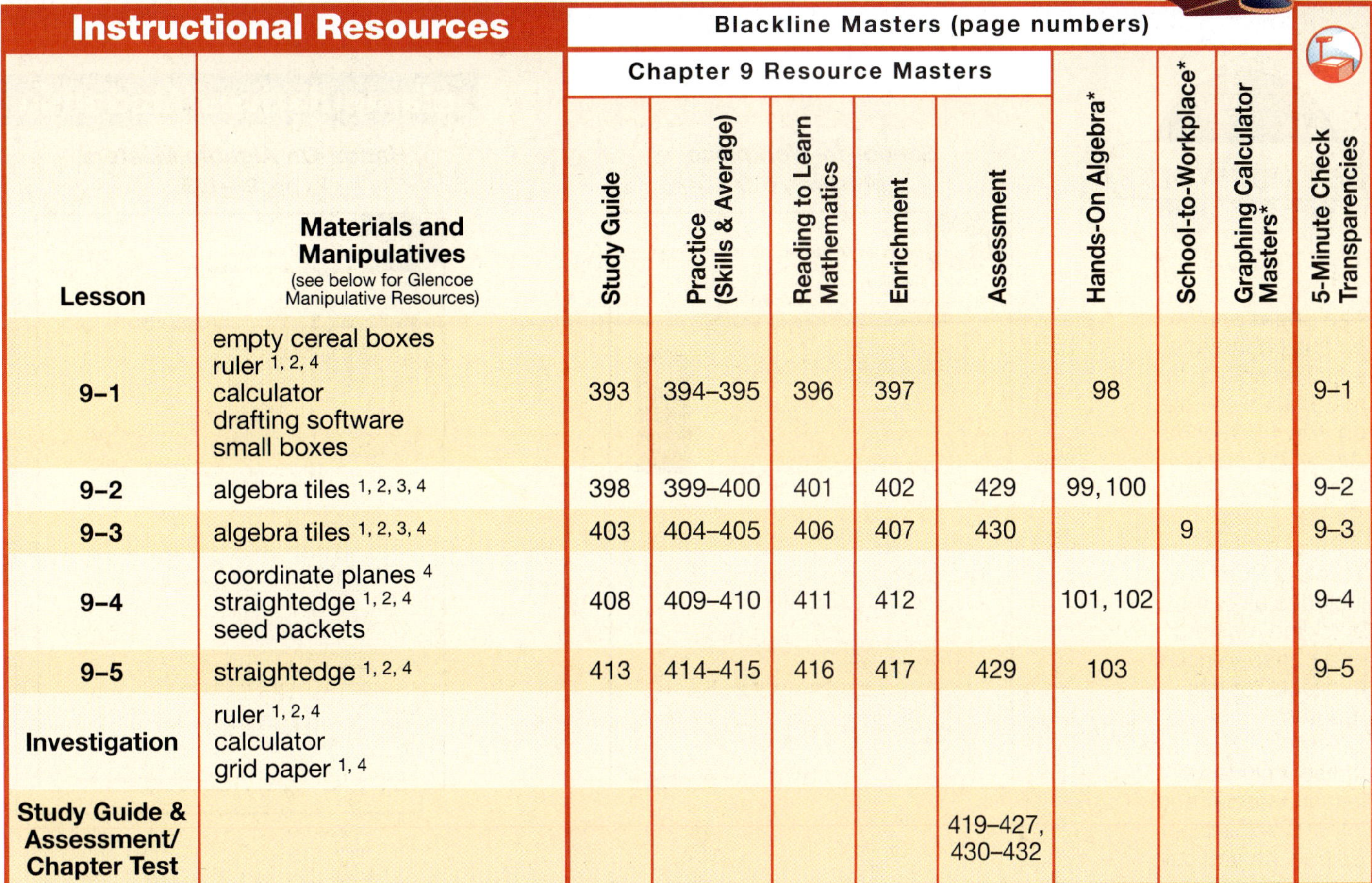

Instructional Resources

Lesson	Materials and Manipulatives (see below for Glencoe Manipulative Resources)	Blackline Masters (page numbers): Chapter 9 Resource Masters: Study Guide	Practice (Skills & Average)	Reading to Learn Mathematics	Enrichment	Assessment	Hands-On Algebra*	School-to-Workplace*	Graphing Calculator Masters*	5-Minute Check Transparencies
9–1	empty cereal boxes ruler [1, 2, 4] calculator drafting software small boxes	393	394–395	396	397		98			9–1
9–2	algebra tiles [1, 2, 3, 4]	398	399–400	401	402	429	99, 100			9–2
9–3	algebra tiles [1, 2, 3, 4]	403	404–405	406	407	430		9		9–3
9–4	coordinate planes [4] straightedge [1, 2, 4] seed packets	408	409–410	411	412		101, 102			9–4
9–5	straightedge [1, 2, 4]	413	414–415	416	417	429	103			9–5
Investigation	ruler [1, 2, 4] calculator grid paper [1, 4]									
Study Guide & Assessment/ Chapter Test						419–427, 430–432				

**See page 380c for examples of these instructional materials.*

Key to Glencoe Manipulative Resources

[1]Classroom Manipulative Resources [2]Student Manipulative Resources [3]Overhead Manipulative Resources [4]Hands-On Algebra Masters

INV = Investigation *SG+A = Study Guide and Assessment*

DAY 7	DAY 8	DAY 9	DAY 10	DAY 11	DAY 12	DAY 13
Lesson 9–4	Lesson 9–5	INV	SG+A	Chapter Test		
INV	SG+A	Chapter Test				

The pages shown on this page are a small sample of the materials available on *TeacherWorks: All-in-One Lesson Planner and Resource Center.*

This CD-ROM includes all of the blackline masters and transparencies available for this program.

It also includes a lesson planner and interactive Teacher Edition, so you can customize lesson plans and reproduce classroom resources quickly and easily, from just about anywhere.

Applications

School-to-Workplace Masters, p. 9

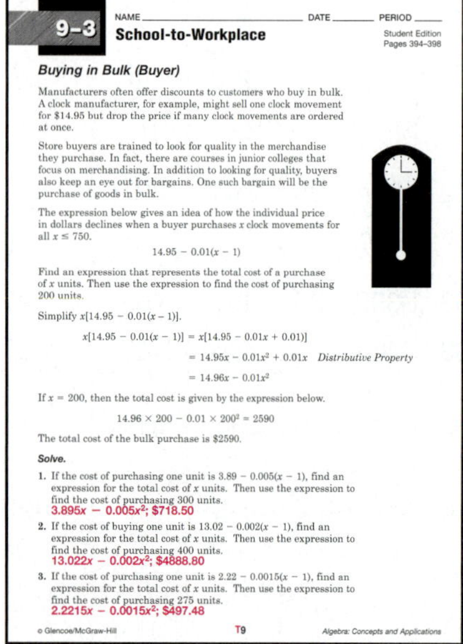

9–3 NAME ______ DATE ______ PERIOD ______

School-to-Workplace

Student Edition Pages 394–398

Buying in Bulk (Buyer)

Manufacturers often offer discounts to customers who buy in bulk. A clock manufacturer, for example, might sell one clock movement for $14.95 but drop the price if many clock movements are ordered at once.

Store buyers are trained to look for quality in the merchandise they purchase. In fact, there are courses in junior colleges that focus on merchandising. In addition to looking for quality, buyers also keep an eye out for bargains. One such bargain will be the purchase of goods in bulk.

The expression below gives an idea of how the individual price in dollars declines when a buyer purchases x clock movements for all $x \leq 750$.

$$14.95 - 0.01(x - 1)$$

Find an expression that represents the total cost of a purchase of x units. Then use the expression to find the cost of purchasing 200 units.

Simplify $x[14.95 - 0.01(x - 1)]$.

$$x[14.95 - 0.01(x - 1)] = x[14.95 - 0.01x + 0.01)]$$
$$= 14.95x - 0.01x^2 + 0.01x \quad \textit{Distributive Property}$$
$$= 14.96x - 0.01x^2$$

If $x = 200$, then the total cost is given by the expression below.

$$14.96 \times 200 - 0.01 \times 200^2 = 2590$$

The total cost of the bulk purchase is $2590.

Solve.

1. If the cost of purchasing one unit is $3.89 - 0.005(x - 1)$, find an expression for the total cost of x units. Then use the expression to find the cost of purchasing 300 units.
   $3.895x - 0.005x^2$; $718.50
2. If the cost of buying one unit is $13.02 - 0.002(x - 1)$, find an expression for the total cost of x units. Then use the expression to find the cost of purchasing 400 units.
   $13.022x - 0.002x^2$; $4888.80
3. If the cost of purchasing one unit is $2.22 - 0.0015(x - 1)$, find an expression for the total cost of x units. Then use the expression to find the cost of purchasing 275 units.
   $2.2215x - 0.0015x^2$; $497.48

© Glencoe/McGraw-Hill T9 *Algebra: Concepts and Applications*

Manipulatives/Modeling

Hands-On Algebra Masters, pp. 98–103

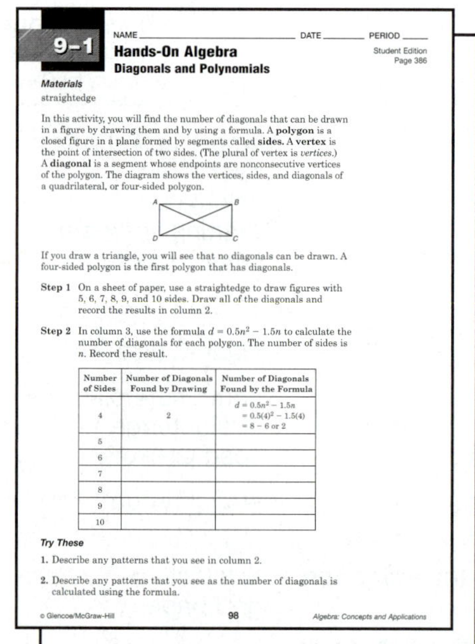

9–1 NAME ______ DATE ______ PERIOD ______

Hands-On Algebra
Diagonals and Polynomials

Student Edition Page 386

Materials
straightedge

In this activity, you will find the number of diagonals that can be drawn in a figure by drawing them and by using a formula. A **polygon** is a closed figure in a plane formed by segments called **sides.** A **vertex** is the point of intersection of two sides. (The plural of vertex is *vertices.*) A **diagonal** is a segment whose endpoints are nonconsecutive vertices of the polygon. The diagram shows the vertices, sides, and diagonals of a quadrilateral, or four-sided polygon.

If you draw a triangle, you will see that no diagonals can be drawn. A four-sided polygon is the first polygon that has diagonals.

Step 1 On a sheet of paper, use a straightedge to draw figures with 5, 6, 7, 8, 9, and 10 sides. Draw all of the diagonals and record the results in column 2.

Step 2 In column 3, use the formula $d = 0.5n^2 - 1.5n$ to calculate the number of diagonals for each polygon. The number of sides is n. Record the result.

Number of Sides	Number of Diagonals Found by Drawing	Number of Diagonals Found by the Formula
4	2	$d = 0.5n^2 - 1.5n$ $= 0.5(4)^2 - 1.5(4)$ $= 8 - 6$ or 2
5		
6		
7		
8		
9		
10		

Try These

1. Describe any patterns that you see in column 2.
2. Describe any patterns that you see as the number of diagonals is calculated using the formula.

© Glencoe/McGraw-Hill 98 *Algebra: Concepts and Applications*

Assessment Resources

GLENCOE'S ASSESSMENT ADVANTAGE

Type	Student Edition	Teacher's Wraparound Edition	Chapter 9 Resource Masters
Ongoing Assessment	Quizzes 1 and 2, pp. 398, 409	5-Minute Check, pp. 382, 388, 388, 394, 399, 405	Mid-Chapter Test, p. 428 Quizzes A and B, p. 429
Mixed Review	Mixed Review, pp. 387, 393, 398, 404, 409 Standardized Test Practice, Chapters 1–9, pp. 416–417		Cumulative Review, p. 430 Standardized Test Practice, pp. 431–432
Error Analysis	You Decide, pp. 396, 408	Error Analysis, pp. 386, 392, 396, 403, 408	
Standardized Test Prep	Standardized Test Practice, pp. 387, 393, 398, 404, 409 Standardized Test Practice, Chapters 1–9, pp. 416–417		Standardized Test Practice, pp. 431–432
Open-Ended Assessment	Writing Math, pp. 385, 403 Investigation, pp. 410–411 Portfolio, pp. 381, 411	Modeling: p. 393 Speaking: pp. 398, 404 Writing: pp. 387, 409	Extended Response Assessment, p. 427
Chapter Assessment	Study Guide and Assessment, pp. 412–414 Chapter Test, p. 415		Multiple-Choice Tests (Forms 1A, 1B), pp. 419–422 Free-Response Tests (Forms 2A, 2B), pp. 423–426

Additional Chapter Resources

Student Edition
Hands-On Algebra, pp. 388–389, 400

Teacher's Classroom Resources
Manipulatives/Modeling
Teacher's Guide for Overhead Manipulative Resources

Meeting Individual Needs
Prerequisite Skills Booklet
Spanish Study Guide and Assessment, pp. 63–67, 121–122

Teaching Aids
Solutions Manual
5-Minute Check Transparencies

Glencoe Technology

Instructional
AlgePASS CD-ROM, Lessons 21, 22
Interactive Chalkboard CD-ROM
StudentWorks Plus CD-ROM
Multimedia Applications CD-ROM, Activity 9
Vocabulary PuzzleMaker (online)

Assessment
ExamView® Pro

Teaching Aids
Answer Key Maker CD-ROM
TeacherWorks CD-ROM

Visit ***www.algconcepts.com***
for data updates, career information, games, and other interactive activities.

Mathematics of the Chapter

This chapter provides students with an in-depth study of polynomials. Students will begin by identifying and classifying polynomials and finding the degree of polynomials and their terms. Students will then add and subtract polynomials, followed by multiplying polynomials by monomials. Finally, they will multiply binomials and explore the special products for $(a + b)^2$, $(a - b)^2$, and $(a + b)(a - b)$.

What You'll Learn

Have students read over the lists of key ideas and key vocabulary. Have them make a list of any words with which they are not familiar.

Why It's Important

Point out to students that this is only one of many reasons why each objective is important. Others are provided in each lesson.

CHAPTER 9

Polynomials

What You'll Learn

Key Ideas

- Identify and classify polynomials. *(Lesson 9–1)*
- Add and subtract polynomials. *(Lesson 9–2)*
- Multiply polynomials. *(Lessons 9–3 and 9–4)*
- Develop and use the patterns for $(a + b)^2$, $(a - b)^2$, and $(a + b)(a - b)$. *(Lesson 9–5)*

Key Vocabulary

binomial *(p. 383)*
FOIL method *(p. 401)*
monomial *(p. 382)*
polynomial *(p. 383)*

Why It's Important

Insects With colors ranging from pale green to bright orange, many moths are as beautiful as butterflies. But the lifespan of an adult moth is short. They live only a few days, just long enough to produce eggs for the next generation.

Polynomials can help you solve nonlinear equations. You will use a polynomial to model the number of eggs a certain type of moth can produce in Lesson 9–1.

380 Chapter 9 Polynomials

CHAPTER 9 LINKS

Lesson	9–1	9–2	9–3	9–4	9–5
Math in the Workplace	Medicine	Framing	Recreation	Packaging	Biology
Applications and Connections	Science Population	Sailing	Sports Ecology	Clocks Food	Photography
Math Integration	Geometry	Geometry	Geometry	Geometry Number Theory	Number Theory Geometry Probability

Check Your Readiness

Study these lessons to improve your skills.

Lesson 1–4, pp. 19–23

Simplify each expression. 6. $15k + 1$

1. $5a + 2a$ $7a$
2. $6x + 8x$ $14x$
3. $12g + 4g$ $16g$
4. $8n - 4n$ $4n$
5. $15t - 7t$ $8t$
6. $20k - 5k + 1$
7. $3(m + 2)$ $3m + 6$
8. $6(d - 1)$ $6d - 6$
9. $3(p + q)$ $3p + 3q$
10. $8(m + 2n)$ $8m + 16n$
11. $a(b + c)$ $ab + ac$
12. $7x(y - 4)$ $7xy - 28x$
13. $5a + 2b + a - b$ $6a + b$
14. $14c - 1 + 2c - 3$ $16c - 4$
15. $3(y - 1) + 6$ $3y + 3$

Lesson 8–2, pp. 341–345

Simplify each expression.

16. $x \cdot x^2$ x^3
17. $n^2 \cdot n^4$ n^6
18. $(w^4)(w^3)$ w^7
19. $(5d)(3d)$ $15d^2$
20. $(-3r)(r)$ $-3r^2$
21. $(2a)(6a^2)$ $12a^3$
22. $(7x)(10x^3)$ $70x^4$
23. $k(12jk)$ $12jk^2$
24. $(ab)(ab^2)$ a^2b^3
25. $(12n)(4m^2)$ $48nm^2$
26. $p^3(p^2q^2)$ p^5q^2
27. $(2g)(3g^2h)(4gh^3)$ $24g^4h^4$

FOLDABLES™ Study Organizer

Make this Foldable to help you organize your Chapter 9 notes. Begin with a sheet of notebook paper.

1. **Fold** lengthwise to the holes.

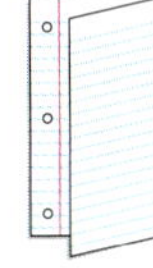

2. **Cut** along the top line and then cut three tabs.

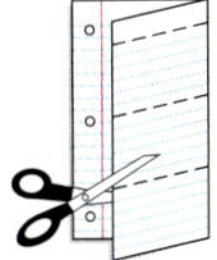

3. **Label** the tabs with the lesson topics as shown.

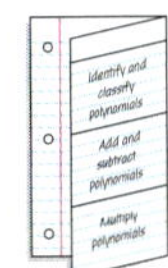

Reading and Writing Store the foldable in a 3-ring binder. As you read and study the chapter, write notes and examples under the tabs.

Check Your Readiness

This section provides a review of the basic concepts needed before beginning Chapter 9. Lesson references are included for additional student help.

Use Exercises	To Prepare for Lesson(s)
1–15	9–2, 9–3, 9–4, 9–5
16–27	9–3, 9–4, 9–5

Quick Review Math Handbook

Glencoe Algebra Lesson	Quick Review Math Handbook Hot Topic
9–1	6–2
9–2	6–2
9–3	6–2
9–4	3–2, 3–4
9–5	
Ch. 9 Inv.	6–5, 7–5

FOLDABLES™ Study Organizer

After students make this Foldable, have them write the rules for classifying polynomials and for performing operations on polynomials. Encourage students to think of and record ways in which they might use polynomials to solve real-world problems.

Noteables™

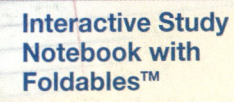

Interactive Study Notebook with Foldables™

This note-taking guide reinforces excellent note-taking skills. It includes:

- **Build Your Vocabulary** tool for including mathematical terminology in notes.
- **Review It** activities with links to prior lessons and items to include in **Foldables™**.
- **Bringing It All Together** feature to help students review for the Chapter Test.
- **Are You Ready for the Chapter Test?** feature to allow students to assess their own readiness for the Chapter Test.
- **Teacher Edition** with transparencies to guide note taking. Corresponds to the *Teacher's Wraparound Edition* In-Class Examples and *Interactive Chalkboard CD-ROM*.

Lesson 9-1

1 FOCUS

5-Minute Check
Chapter 8

Simplify each expression.

1. $(2x^3y^2)(3xy^4)$ $6x^4y^6$
2. $\frac{-18c^4d^{-3}}{12c^2d^2}$ $-\frac{3c^2}{2d^5}$
3. Express the distance from the sun to Jupiter, 778,000,000 kilometers, in scientific notation. 7.78×10^8 km
4. Simplify $-\sqrt{\frac{144}{64}}$. $-\frac{3}{2}$
5. Find the length of the hypotenuse of the right triangle below. Round to the nearest tenth if necessary. 19.2 m

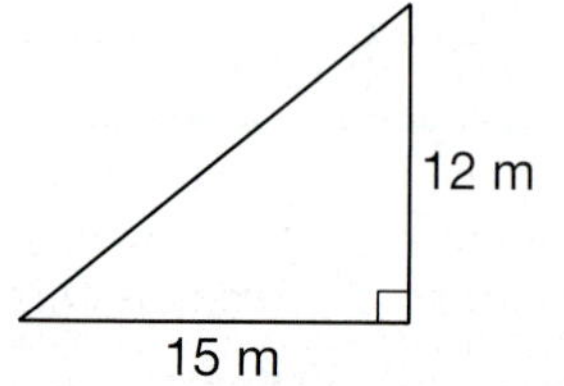

Motivating the Lesson

Hands-On Activity Give small groups of students boxes with all edges labeled x, y, or z. Have them write expressions for the area of each face and then use these results to write an expression for the total surface area of the box.

MODELING

An alternative hands-on option using a straightedge is available for teaching this lesson.

9–1 Polynomials

What You'll Learn
You'll learn to identify and classify polynomials and find their degree.

Why It's Important
Medicine Doctors can use polynomials to study the heart. *See Exercise 57.*

A **monomial** is a number, a variable, or a product of numbers and variables that have only positive exponents. A monomial cannot have a variable as an exponent. The tables below show examples of expressions that are and are not monomials.

Monomials	
-4	a number
y	a variable
a^2	the product of variables
$\frac{1}{2}x^2y$	the product of numbers and variables

Not Monomials	
2^x	has a variable as an exponent
$x^2 + 3$	a sum
$5a^{-2}$	includes a negative exponent
$\frac{3}{x}$	a quotient

Examples

Determine whether each expression is a monomial. Explain why or why not.

1 $-6ab$

$-6ab$ is a monomial because it is the product of a number and variables.

2 $m^2 - 4$

$m^2 - 4$ is *not* a monomial because it includes subtraction.

Your Turn

a. 10 b. $5z^{-3}$ c. $\frac{6}{x}$ d. x^2

a. yes; a number
b. no; has a negative exponent
c. no; includes division
d. yes; product of variables

A monomial or the sum of one or more monomials is called a **polynomial**. For example, $x^3 + x^2 + 3x + 2$ is a polynomial. Each monomial is a term. The terms of the polynomial are x^3, x^2, $3x$, and 2.

Recall that to subtract, you add the opposite.

Resource Manager

Reproducible Masters

Chapter 9 Resource Masters
- *Study Guide*, p. 393
- *Skills Practice*, p. 394
- *Practice*, p. 395
- *Reading to Learn Mathematics*, p. 396
- *Enrichment*, p. 397

Hands-On Algebra, p. 98

Transparencies

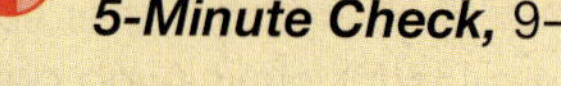

5-Minute Check, 9–1

Technology/Multimedia

Interactive Chalkboard CD-ROM

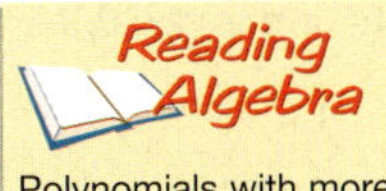

Polynomials with more than three terms have no special name.

Special names are given to polynomials with two or three terms. A polynomial with two terms is a **binomial**. A polynomial with three terms is a **trinomial**. Here are some examples.

Binomial	Trinomial
$x + 2$	$a + b + c$
$5c - 4$	$x^2 + 5x - 7$
$4w^2 - w$	$3a^2 + 5ab + 2b^2$

Examples

State whether each expression is a polynomial. If it is a polynomial, identify it as a *monomial, binomial,* or *trinomial.*

$2m - 7$

The expression $2m - 7$ can be written as $2m + (-7)$. So, it is a polynomial. Since it can be written as the sum of two monomials, $2m$ and -7, it is a binomial.

$x^2 + 3x - 4 - 5$

The expression $x^2 + 3x - 4 - 5$ can be written as $x^2 + 3x + (-9)$. So, it is a polynomial. Since it can be written as the sum of three monomials, it is a trinomial.

$\frac{5}{2x} - 3$

The expression $\frac{5}{2x} - 3$ is not a polynomial since $\frac{5}{2x}$ is not a monomial.

Your Turn

e. $5a - 9 + 3$ f. $4m^{-2} + 2$ **no** g. $3y^2 - 6 + 7y$

e. yes; binomial
g. yes; trinomial

The terms of a polynomial are usually arranged so that the powers of one variable are in descending or ascending order.

Polynomial	Descending Order	Ascending Order	
$2x + x^2 + 1$	$x^2 + 2x + 1$	$1 + 2x + x^2$	
$3y^2 + 5y^3 + y$	$5y^3 + 3y^2 + y$	$y + 3y^2 + 5y^3$	
$x^2 + y^2 + 3xy$	$x^2 + 3xy + y^2$	$y^2 + 3xy + x^2$	*in terms of x*
$2xy + y^2 + x^2$	$y^2 + 2xy + x^2$	$x^2 + 2xy + y^2$	*in terms of y*

www.algconcepts.com/extra_examples

2 TEACH

Teaching Tip As you discuss the definition of monomial, stress that negative values of powers are monomials but that powers with negative exponents are not. For example, $-x^2$ is a monomial but x^{-2} is not.

In-Class Examples

Examples 1–2

Determine whether each expression is a monomial. Explain why or why not.

1 a^2b^3c **yes; product of variables**

2 $\frac{1}{x}$ **no; includes division**

Teaching Tip As you introduce the definition of *polynomial*, write the prefixes *mono-*, *bi-*, *tri-*, and *poly-* on the board or overhead. Ask students to think of words that begin with these prefixes, such as *mono*chrome, *bi*cycle, *tri*ceratops, and *poly*chrome. Using students' contributions, lead them to recognize that *mono* means one, *bi* means two, *tri* means three, and *poly* means many.

Teaching Tip Before Example 3, stress that each term in a polynomial must be a monomial.

In-Class Examples

Examples 3–5

***State whether each expression is a polynomial. If it is a polynomial, identify it as a* monomial, binomial, *or* trinomial.**

3 $-4x + 2$ **yes; binomial**

4 $5 + 3x^2 + x + 2$ **yes; trinomial**

5 $3x^{-2} + 4x^3$ **no**

From the Classroom of ...

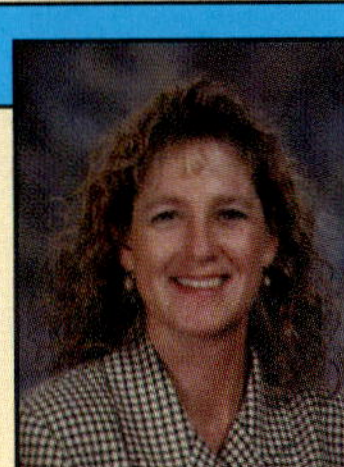

Dr. Deborah Hutchens
Great Bridge Middle School
Chesapeake, Virginia

Give each student a piece of construction paper with one term on it. Have students stand in front of the class and move around until the terms are in ascending order.

In-Class Examples

Examples 6–7

Find the degree of each polynomial.

6 $8b^4 + 9^2$ **4**

7 $2ab + 3a^2b + 5a^4b^2$ **6**

Example 8

About how many eggs would you expect the moth in Example 8 to produce if her abdomen measures 2 millimeters? **46**

Zero Power: Lesson 8–2

The **degree** of a monomial is the sum of the exponents of the variables.

Monomial	Degree
$-3x^2$	2
$5pq^2$	$1 + 2 = 3$
2	0

$p = p^1$
$2 = 2x^0$

To find the degree of a polynomial, find the degree of each term. The greatest of the degrees of its terms is the degree of the polynomial.

Polynomial	Terms	Degree of the Terms	Degree of the Polynomial
$2n + 7$	$2n$, 7	1, 0	1
$3x^2 + 5x$	$3x^2$, $5x$	2, 1	2
$a^6 + 2a^3 + 1$	a^6, $2a^3$, 1	6, 3, 0	6
$5x^4 - 4a^2b^6 + 3x$	$5x^4$, $-4a^2b^6$, $3x$	4, 8, 1	8

Examples

Find the degree of each polynomial.

 $5a^2 + 3$

Term	Degree
$5a^2$	2
3	0

$3 = 3x^0$

The degree of $5a^2 + 3$ is 2.

 $6x^2 - 4x^2y - 3xy$

Term	Degree
$6x^2$	2
$-4x^2y$	2 + 1 or 3
$-3xy$	1 + 1 or 2

$y = y^1$
$x = x^1, y = y^1$

The degree of $6x^2 - 4x^2y - 3xy$ is 3.

Your Turn

h. $3x^2 - 7x$ **2**

i. $8m^3 - 2m^2n^2 + 5$ **4**

Polynomials can be used to represent many real-world situations.

Example 8 **Science Link**

The expression $14x^3 - 17x^2 - 16x + 34$ can be used to estimate the number of eggs that a certain type of female moth can produce. In the expression, x represents the width of the abdomen in millimeters. About how many eggs would you expect this type of moth to produce if her abdomen measures 3 millimeters?

Reteaching Activity

Visual/Spatial Learners Have groups of students make colorful posters that describe and give examples of polynomials, including monomials, binomials, and trinomials. Have them include examples of terms that are and are not monomials.

Explore You know the width of the abdomen. You need to determine the number of eggs that the moth can produce.

Plan In the expression $14x^3 - 17x^2 - 16x + 34$, replace x with 3 to determine the number of eggs that the moth can produce.

Female Cecropia Moth

Solve

$$\begin{aligned} 14x^3 - 17x^2 - 16x + 34 &= 14(3)^3 - 17(3)^2 - 16(3) + 34 \\ &= 14(27) - 17(9) - 16(3) + 34 \\ &= 378 - 153 - 48 + 34 \\ &= 211 \end{aligned}$$

Examine Check your computations by adding in a different order.

$$\begin{aligned} 378 - 153 - 48 + 34 &\stackrel{?}{=} 378 + 34 + (-153) + (-48) \\ &\stackrel{?}{=} 412 + (-201) \\ &= 211 \ \checkmark \end{aligned}$$

The moth can produce about 211 eggs.

Answer

2. To find the degree of a polynomial, first find the degree of each of its terms. The degree of the polynomial is the greatest of the degrees of its terms.

Check for Understanding

Communicating Mathematics

Vocabulary
monomial
polynomial
binomial
trinomial
degree

1. **Write** an expression that is *not* a monomial. Explain why it is *not* a monomial.
2. **Explain** how to find the degree of a polynomial. See margin.
3. **Arrange** the terms of the polynomial $3x^2 + 5xy + 2y^2$ so that the powers of x are in ascending order. $2y^2 + 5xy + 3x^2$
4. Writing Math Write a definition for *monomial, polynomial, binomial,* and *trinomial*. Give three examples of each. See margin.

1. Sample answer: $\frac{2}{x}$; it includes division.

Guided Practice

Examples 1 & 2

Determine whether each expression is a monomial. Explain why or why not.

5. $7xy$ yes; product of numbers and variables
6. $a^2 + 5$ no; includes addition

Examples 3–5

State whether each expression is a polynomial. If it is a polynomial, identify it as a *monomial*, *binomial*, or *trinomial*.

7. $2mn + 3$ yes; binomial
8. 0 yes; monomial
9. $\frac{4}{d} - d^2$ no
10. $h^2 - 3h + 8 + 2$ yes; trinomial

Answer

4. A polynomial is a monomial or a sum of monomials. A monomial is an integer, a variable, or a product. A binomial is a polynomial with two terms. A trinomial is a polynomial with three terms.

Examples			
Polynomial	Monomial	Binomial	Trinomial
$a + b + c$	$-3abc$	$a - b$	$a - b - c$
$2x - 3y + 5z$	$2x^2y^7$	$2x + 3y$	$2z + 2y^5 - 6$
$3m^2 + 6mn - 2n^4$	$6mn^3p$	$4m^3 - 6n^2$	$4mn^2 - 6n + 3m$

***Study Guide*, p. 393**

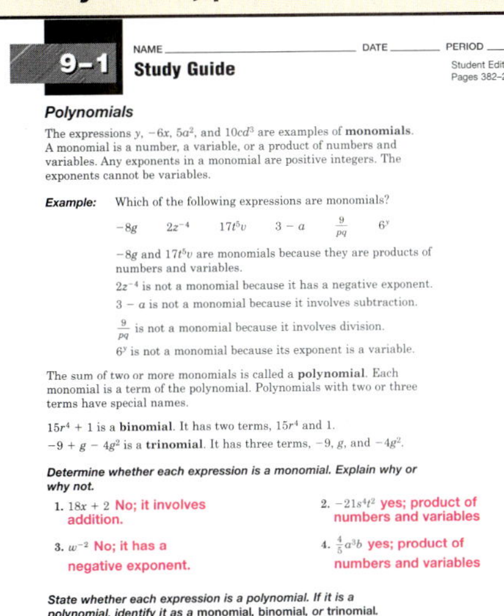
9–1 NAME ______ DATE ______ PERIOD ______

Study Guide

Student Edition Pages 382–287

Polynomials

The expressions y, $-6x$, $5a^2$, and $10cd^3$ are examples of **monomials**. A monomial is a number, a variable, or a product of numbers and variables. Any exponents in a monomial are positive integers. The exponents cannot be variables.

Example: Which of the following expressions are monomials?

$-8g$ $\quad 2z^{-4}$ $\quad 17t^5v$ $\quad 3 - a$ $\quad \frac{9}{pq}$ $\quad 6^y$

$-8g$ and $17t^5v$ are monomials because they are products of numbers and variables.

$2z^{-4}$ is not a monomial because it has a negative exponent.

$3 - a$ is not a monomial because it involves subtraction.

$\frac{9}{pq}$ is not a monomial because it involves division.

6^y is not a monomial because its exponent is a variable.

The sum of two or more monomials is called a **polynomial**. Each monomial is a term of the polynomial. Polynomials with two or three terms have special names.

$15r^4 + 1$ is a **binomial**. It has two terms, $15r^4$ and 1.

$-9 + g - 4g^2$ is a **trinomial**. It has three terms, -9, g, and $-4g^2$.

Determine whether each expression is a monomial. Explain why or why not.

1. $18x + 2$ No; it involves addition.
2. $-21s^4t^2$ yes; product of numbers and variables
3. w^{-2} No; it has a negative exponent.
4. $\frac{4}{5}a^3b$ yes; product of numbers and variables

***State whether each expression is a polynomial. If it is a polynomial, identify it as a* monomial, binomial, *or* trinomial.**

5. $\frac{8}{x}$ no
6. $-7r + 9s - 3$ yes; trinomial
7. $abc^3 - a^3bc$ yes; binomial
8. $35u^5v^6$ yes; monomial
9. $5 + 5^k$ no
10. $8d - 9e \div f$ no
11. $16x - 16y$ yes; binomial
12. $8j^2 + 3j - 7$ yes; trinomial
13. $3m^3 + \frac{1}{3}m$ yes; binomial
14. $-14p + p^{-14}$ no

3 PRACTICE/APPLY

Error Analysis

Watch for students who give the degree of the polynomial in Exercise 14 as 3, the greatest exponent on a variable in a term. ***Prevent by*** stressing that the degree of a term is the *sum* of the exponents of the variables in that term. The degree of the polynomial is then the greatest of the degrees of its terms.

Assignment Guide

Basic: 17–57 odd, 58–69
Average: 16–54 even, 56–69

Answers

16. yes; product of a number and a variable
17. no; includes addition
18. no; includes division
19. yes; product of numbers and variables
20. no; has a negative exponent
21. yes; product of numbers and variables

***Skills Practice*, p. 394, and *Practice*, p. 395 (shown)**

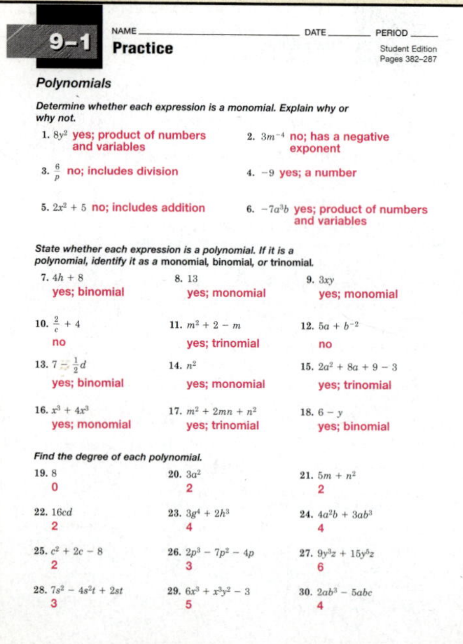

9–1 NAME ______________ DATE ______ PERIOD ______

Practice

Student Edition Pages 382–287

Polynomials

Determine whether each expression is a monomial. Explain why or why not.

1. $8y^2$ yes; product of numbers and variables
2. $3m^{-4}$ no; has a negative exponent
3. $\frac{6}{p}$ no; includes division
4. -9 yes; a number
5. $2x^2 + 5$ no; includes addition
6. $-7a^3b$ yes; product of numbers and variables

State whether each expression is a polynomial. If it is a polynomial, identify it as a monomial, binomial, ***or*** trinomial.

7. $4h + 8$ yes; binomial
8. 13 yes; monomial
9. $3xy$ yes; monomial
10. $\frac{2}{c} + 4$ no
11. $m^2 + 2 - m$ yes; trinomial
12. $5a + b^{-2}$ no
13. $7 - \frac{1}{2}d$ yes; binomial
14. n^2 yes; monomial
15. $2a^2 + 8a + 9 - 3$ yes; trinomial
16. $x^3 + 4x^3$ yes; monomial
17. $m^2 + 2mn + n^2$ yes; trinomial
18. $6 - y$ yes; binomial

Find the degree of each polynomial.

19. 8 0
20. $3a^2$ 2
21. $5m + n^2$ 2
22. $16cd$ 2
23. $3g^4 + 2h^3$ 4
24. $4a^2b + 3ab^3$ 4
25. $c^2 + 2c - 8$ 2
26. $2p^3 - 7p^2 - 4p$ 3
27. $9y^3z + 15y^5z$ 6
28. $7s^2 - 4s^2t + 2st$ 3
29. $6x^3 + x^3y^2 - 3$ 5
30. $2ab^3 - 5abc$ 4

Examples 6 & 7

Find the degree of each polynomial.

11. $25x$ **1**
12. 5 **0**
13. $15m^2 + 4n$ **2**
14. $-6y^3z - 4y^2z$ **4**

Example 8

15. **Geometry** To find the number of diagonals d in an n-sided figure, you can use the formula $d = 0.5n^2 - 1.5n$.
    a. Find the number of diagonals in an octagon. **20**
    b. How many diagonals are in a hexagon? **9**

Exercises

Practice

Homework Help	
For Exercises	See Examples
16–21	1, 2
22–33, 54, 55	3, 4, 5
33–45, 57	6, 7
56	8
Extra Practice	
See page 710.	

Determine whether each expression is a monomial. Explain why or why not. **16–21. See margin.**

16. $5z$
17. $7a + 2$
18. $\frac{5}{x}$
19. $8x^2y$
20. y^{-3}
21. $-10a^2b^2c$

State whether each expression is a polynomial. If it is a polynomial, identify it as a *monomial*, *binomial*, or *trinomial*.

22. y^3 **yes; monomial**
23. $2y^2 + 5y - 7$
24. $a^3 - 3a$
25. $2r + 3s^{-3} - t$ **no**
26. $17 + 12x^3 - 3x^4$
27. $2m^2 + 5 - 1$
28. $-4j + 2k^2 - 3$
29. $x^2 - \frac{1}{2}x$
30. $\frac{x}{y}$ **no**
31. $-12a^2b^3cd^5e^3$
32. $2.5w^5 - v^3w^2$
33. $3x + 4x$

23. yes; trinomial
24. yes; binomial
26. yes; trinomial
27. yes; binomial
28. yes; trinomial
29. yes; binomial
31. yes; monomial
32. yes; binomial
33. yes; monomial

Find the degree of each polynomial.

34. 1 **0**
35. $8x^2$ **2**
36. $-14x^3$ **3**
37. $5p^2 + 3p + 5$ **2**
38. $4s^2 - 6t^8$ **8**
39. $p - 4q + 5r$ **1**
40. $25a + a^2$ **2**
41. $a^3 + a^2 + a$ **3**
42. $15g^3h^4 - 10gh^5$ **7**
43. $10v^4w^2 + vw^5$ **6**
44. $3rs^4 - 2r^2 + 7$ **5**
45. $5cd^4 - 2bcd$ **5**

Arrange the terms of each polynomial so that the powers of *x* are in descending order. **47. $-x^5 + x^2 - x + 25$**

46. $2x^2 + x^4 - 6$ **$x^4 + 2x^2 - 6$**
47. $x^2 - x + 25 - x^5$
48. $3x - 4x^2y + 5 - 3x^5$ **$-3x^5 - 4x^2y + 3x + 5$**
49. $6w^3x + 3wx^2 - 10x^6 + 5x^7$ **$5x^7 - 10x^6 + 3wx^2 + 6w^3x$**

Arrange the terms of each polynomial so that the powers of *x* are in ascending order. **50. $2 + x^2 + x^4 + x^5$** **51. $7 + 2x - x^2 + 5x^3$**

50. $2 + x^4 + x^5 + x^2$
51. $5x^3 - x^2 + 7 + 2x$
52. $a^3x - 6a^3x^3 + 0.5x^5 + 4x^2$ **$a^3x + 4x^2 - 6a^3x^3 + 0.5x^5$**
53. $3x^3y + 5xy^4 - x^2y^3 + y^6$ **$y^6 + 5xy^4 - x^2y^3 + 3x^3y$**
54. *True* or *false*: Every monomial is a polynomial. **true**
55. *True* or *false*: Every polynomial is a monomial. **false**

Applications and Problem Solving

56. **Geometry** The surface area of a right cylinder is given by the polynomial $2\pi rh + 2\pi r^2$, where r is the radius of the base of the cylinder and h is the height of the cylinder.

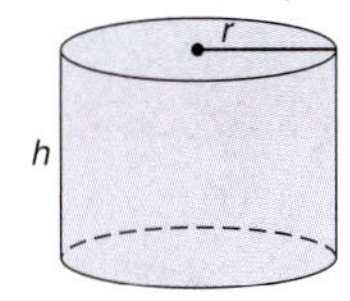

   a. Classify the polynomial. **binomial**
   b. Find the surface area of a right circular cylinder with a height of 5 feet and a radius of 3 feet. Round to the nearest tenth. **150.8 ft^2**
   c. The volume of a cylinder is given by the expression $\pi r^2 h$. Using the same height and radius as in part b, find the volume of the cylinder. Round to the nearest tenth. 141.3

57. **Medicine** Doctors study the heart of a potential heart attack patient by injecting a dye in a vein near the heart. In a healthy heart, the amount of dye in the bloodstream after t seconds is given by the expression $-0.006t^4 + 1.79t - 0.53t^2 + 0.14t^3$.
   a. Arrange the terms of the polynomial so that the powers of t are in descending order. **$-0.006t^4 + 0.14t^3 - 0.53t^2 + 1.79t$**
   b. Find the degree of the polynomial. **4**

58a. See students' work.

58. **Critical Thinking** A number in base 10 can be written in polynomial form. For example, $3247 = 3(10)^3 + 2(10)^2 + 4(10) + 7(10)^0$.
   a. Write the year of your birth in polynomial form.
   b. Suppose 2356 is a number in base a. Write 2356 in polynomial form. **$2a^3 + 3a^2 + 5a + 6$**
   c. What is the degree of the polynomial in part b? **3**

Mixed Review

The lengths of three sides of a triangle are given. Determine whether each triangle is a right triangle. *(Lesson 8–7)*

59. 5 cm, 12 cm, 13 cm **yes**
60. 2 in., 3 in., 4 in. **no**
61. 8 ft, 15 ft, 17 ft **yes**

Estimate each square root to the nearest whole number. *(Lesson 8–6)*

62. $\sqrt{20}$ **4**
63. $\sqrt{50}$ **7**
64. $\sqrt{75}$ **9**
65. $\sqrt{120}$ **11**

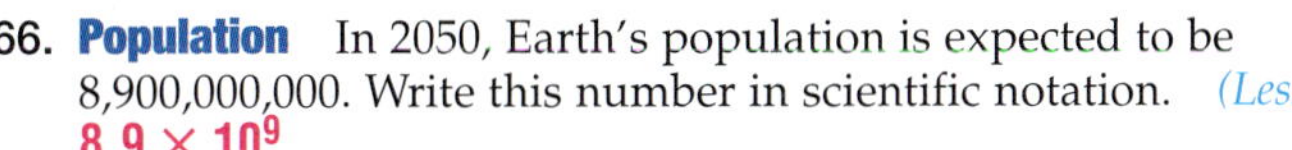

66. **Population** In 2050, Earth's population is expected to be 8,900,000,000. Write this number in scientific notation. *(Lesson 8–4)* **8.9×10^9**

Simplify each expression. *(Lesson 8–3)*

67. xy^{-3} **$\frac{x}{y^3}$**
68. $\frac{k^{-3}}{k^4}$ **$\frac{1}{k^7}$**

Standardized Test Practice

69. **Multiple Choice** The graph of $y = 3x + 6$ is shown at the right. Which is the slope of a line parallel to the graph of $y = 3x + 6$? *(Lesson 7–7)* **B**

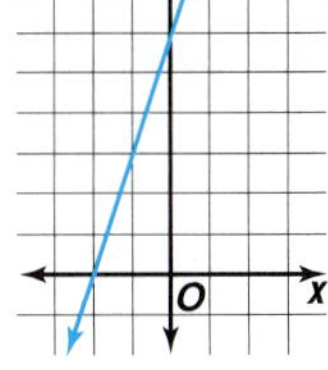

   A -3
   B 3
   C $-\frac{1}{3}$
   D $\frac{1}{6}$

www.algconcepts.com/self_check_quiz

4 ASSESS

Open-Ended Assessment

Writing Have students write a note to a classmate explaining how to find the degree of a polynomial.

***Enrichment*, p. 397**

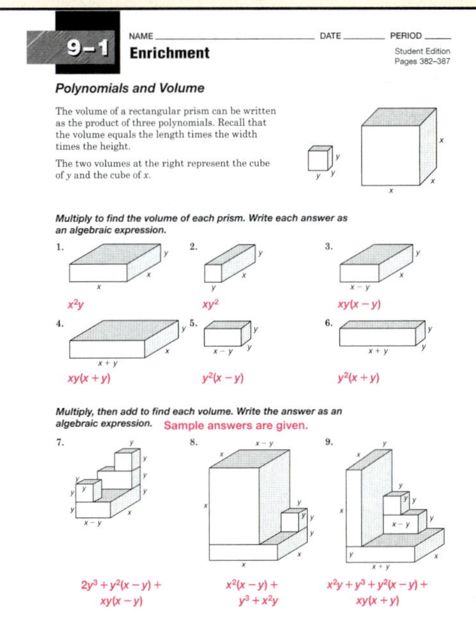

NAME ______ DATE ______ PERIOD ______

9–1 Enrichment

Student Edition Pages 382–387

Polynomials and Volume

The volume of a rectangular prism can be written as the product of three polynomials. Recall that the volume equals the length times the width times the height.

The two volumes at the right represent the cube of y and the cube of x.

Multiply to find the volume of each prism. Write each answer as an algebraic expression.

1. x^2y
2. xy^2
3. $xy(x - y)$
4. $xy(x + y)$
5. $y^2(x - y)$
6. $y^2(x + y)$

Multiply, then add to find each volume. Write the answer as an algebraic expression. Sample answers are given.

7. $2y^3 + y^2(x - y) + xy(x - y)$
8. $x^2(x - y) + y^3 + x^2y$
9. $x^2y + y^3 + y^2(x - y) + xy(x + y)$

Extra Credit

Write a trinomial with degree 3 that has four different variables.

Sample answer: $5 + 3wx + yz^2$

Lesson 9-2

1 FOCUS

5-Minute Check
Lesson 9–1

***State whether each expression is a polynomial. If it is a polynomial, identify it as a* monomial, binomial, *or* trinomial.**

1. $x^{-2} + 4y$ **no**
2. $x^3y + x^2y^2 + y^3$ **yes; trinomial**
3. $2a - 4b$ **yes; binomial**
4. Find the degree of $4g^2h + 3gh^3$. **4**
5. The surface area of a right circular cone is given by the polynomial $\pi r^2 + \pi rs$, where r is the radius of the cone's base, h is the cone's height, and s is the cone's slant height. Classify the polynomial and give its degree. **binomial; 2**

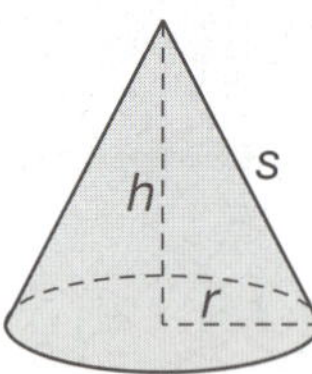

Motivating the Lesson

Hands-On Activity Hand out algebra tiles to small groups of students. Have students form rectangles of various sizes using combinations of tiles. Then have them write expressions in terms of x for the side lengths and perimeters of the rectangles. Point out that when they find the perimeters, they are actually adding polynomials.

9-2 Adding and Subtracting Polynomials

What You'll Learn

You'll learn to add and subtract polynomials.

Why It's Important

Framing Addition of polynomials can be used to find the size of a picture. *See Exercise 43.*

You can use algebra tiles like the ones below to model polynomials.

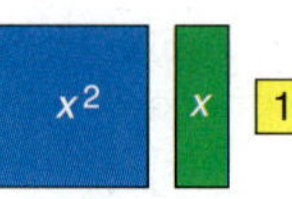

A model of the polynomial $2x^2 + 3x + 1$ is shown below.

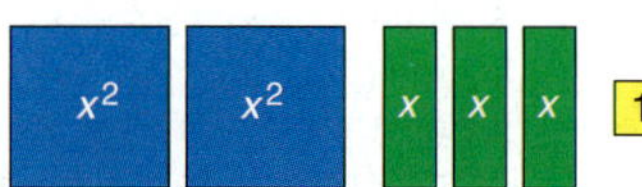

Red tiles are used to represent -1, $-x$, and $-x^2$. The model of the polynomial $-x^2 - 2x - 3$ is shown below.

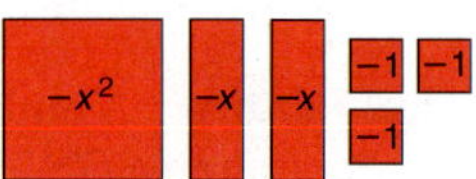

Once you know how to model polynomials using algebra tiles, you can use algebra tiles to add and subtract polynomials.

Materials: algebra tiles

Add $x^2 + 2x - 3$ and $x^2 - x + 4$.

Step 1 Model each polynomial.

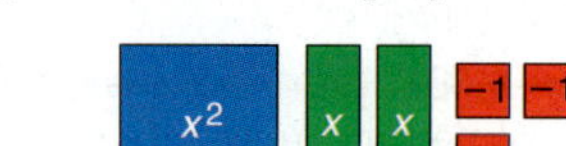
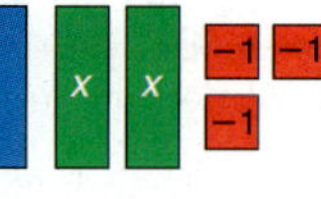

$x^2 + 2x - 3$

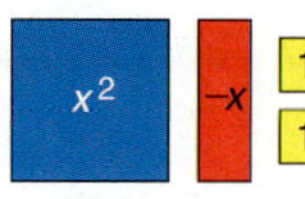

$x^2 - x + 4$

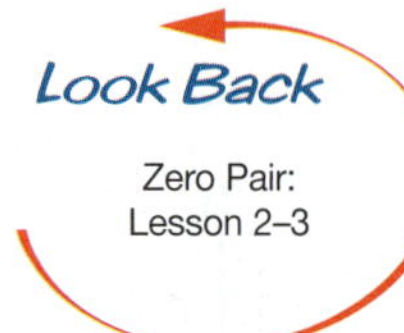

Step 2 Combine like shapes and remove all zero pairs. Recall that a zero pair is formed by pairing one tile with its opposite.

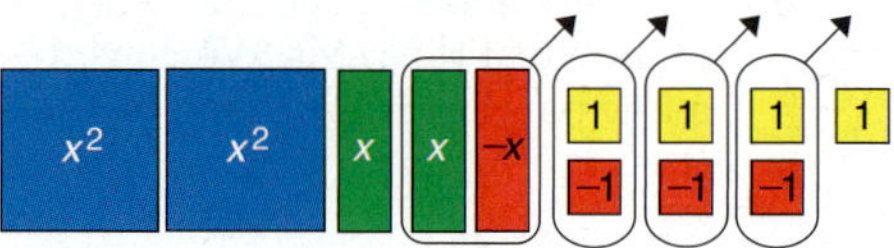

Therefore, $(x^2 + 2x - 3) + (x^2 - x + 4) = 2x^2 + x + 1$.

Resource Manager

Reproducible Masters

Chapter 9 Resource Masters
- *Study Guide,* p. 398
- *Skills Practice,* p. 399
- *Practice,* p. 400
- *Reading to Learn Mathematics,* p. 401
- *Enrichment,* p. 402
- *Assessment,* p. 429

Hands-On Algebra, pp. 99–100

Transparencies

5-Minute Check, 9–2

Technology/Multimedia

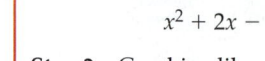

Interactive Chalkboard CD-ROM

3. $3x^2 + 5x + 3$
4. $4x^2 - x - 5$
5. $x^2 + 2x + 2$
6. $3x^2 - 2x + 2$

Try These

Use algebra tiles to find each sum.

1. $(3x + 5) + (2x - 3)$ $5x + 2$
2. $(-2x + 3) + (4x - 3)$ $2x$
3. $(2x^2 + 2x - 4) + (x^2 + 3x + 7)$
4. $(3x^2 + x - 4) + (x^2 - 2x - 1)$
5. $(x^2 - 1) + (2x + 3)$
6. $(2x^2 + 3) + (x^2 - 2x - 1)$
7. **Write** a rule for adding polynomials without models.

7. Sample answer: Combine like terms and add the coefficients.

You can add polynomials by grouping the like terms together and then finding their sum.

Examples

Look Back

Like Terms: Lesson 1–4

Find each sum.

1 $(4x - 3) + (2x + 5)$

Method 1

Group the like terms together.

$$\begin{aligned}(4x - 3) + (2x + 5) &= (4x + 2x) + (-3 + 5) && \textit{Group the like terms.}\\ &= (4 + 2)x + (-3 + 5) && \textit{Distributive Property}\\ &= 6x + 2 && \textit{Simplify.}\end{aligned}$$

Method 2

Add in column form.

$$\begin{array}{rr} & 4x - 3 \\ (+) & 2x + 5 \\ \hline & 6x + 2 \end{array}$$ *Align the like terms.*

2 $(x^2 + 2x - 5) + (3x^2 - x + 4)$

$$\begin{aligned}(x^2 + 2x - 5) + (3x^2 - x + 4) &= (x^2 + 3x^2) + (2x - x) + (-5 + 4)\\ &= (1 + 3)x^2 + (2 - 1)x + (-5 + 4)\\ &= 4x^2 + 1x - 1 \text{ or } 4x^2 + x - 1\end{aligned}$$

Check your answer by adding in column form.

3 $(2x^2 + 5xy + 3y^2) + (8x^2 - 7y^2)$

$$\begin{array}{rr} & 2x^2 + 5xy + 3y^2 \\ (+) & 8x^2 \qquad\quad - 7y^2 \\ \hline & 10x^2 + 5xy - 4y^2 \end{array}$$ *Align the like terms.*

Your Turn

b. $-x^2 - 2x + 7$
c. $8a^2 - 2ab + 2b^2$

a. $(3x + 9) + (5x + 3)$ $8x + 12$

b. $(-2x^2 + x + 5) + (x^2 - 3x + 2)$

c. $$\begin{array}{rr} & a^2 - 2ab + 4b^2 \\ (+) & 7a^2 \qquad\quad - 2b^2 \\ \hline \end{array}$$

d. $(7m^2 - 6) + (5m - 2)$ $7m^2 + 5m - 8$

www.algconcepts.com/extra_examples

2 TEACH

Teaching Tip Before beginning Examples 1–3, review with students how to identify the coefficient of a term and how to identify like terms.

In-Class Examples

Examples 1–3

Find each sum.

1 $(3s + 4t) + (6s - 2t)$ $9s + 2t$

2 $(b^2 + 4b - 6) + (3b^2 - 3b + 1)$ $4b^2 + b - 5$

3 $(2d^2 + 7de - 8e^2) + (-d^2 + 8e^2)$ $d^2 + 7de$

Hands-On Algebra

Cooperative Learning Refer to the Hands-On Algebra on page 388. Draw students' attention to the sign of each term as they model the polynomials. Remind them that subtracting terms is equivalent to adding negative terms. Because 1 and −1 are additive inverses, each combination of a positive and negative tile *of the same dimensions* is equivalent to zero, and can be removed.

An additional Hands-On Algebra activity using algebra tiles to model subtraction of polynomials is available in the *Hands-On Algebra Masters,* p. 100.

Hands-On Algebra Masters, p. 99

Teaching Tip While discussing Example 4, stress that the subtraction applies to both terms of the second binomial. So, the additive inverses both of $3x$ and of 1 must be added to their like terms in the first binomial. Similarly, in Example 5 the subtraction applies to all three terms of the second trinomial.

In-Class Examples

Examples 4–5

Find each difference.

4 $(2g + 7) - (g + 2)$ $g + 5$

5 $(4a^2 - 3a + 4) - (a^2 + 6a + 1)$ $3a^2 - 9a + 3$

Subtracting Integers: Lesson 2–4

Recall that you can subtract an integer by adding its additive inverse or opposite.

$2 - 3 = 2 + (-3)$ *The additive inverse of 3 is −3.*

$5 - (-4) = 5 + 4$ *The additive inverse of −4 is 4.*

Similarly, you can subtract a polynomial by adding its additive inverse. To find the additive inverse of a polynomial, replace each term with its additive inverse.

Polynomial	Additive Inverse	
$a + 2$	$-a - 2$	$-(a + 2) = -a - 2$
$x^2 + 3x - 1$	$-x^2 - 3x + 1$	$-(x^2 + 3x - 1) = -x^2 - 3x + 1$
$2x^2 - 5xy + y^2$	$-2x^2 + 5xy - y^2$	$-(2x^2 - 5xy + y^2) = -2x^2 + 5xy - y^2$

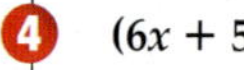

Examples

Find each difference.

4 $(6x + 5) - (3x + 1)$

Method 1

Find the additive inverse of $3x + 1$. Then group the like terms together and add.

The additive inverse of $3x + 1$ is $-(3x + 1)$ or $-3x - 1$.

$$\begin{aligned}(6x + 5) - (3x + 1) &= (6x + 5) + (-3x - 1) && \textit{Add the additive inverse.}\\ &= (6x - 3x) + (5 - 1) && \textit{Group the like terms.}\\ &= (6 - 3)x + (5 - 1) && \textit{Distributive Property}\\ &= 3x + 4\end{aligned}$$

Method 2

Arrange like terms in column form.

$$\begin{array}{r} 6x + 5 \\ (-)\ 3x + 1 \\ \hline \end{array} \quad \textit{Add the additive inverse.} \quad \begin{array}{r} 6x + 5 \\ (+)\ -3x - 1 \\ \hline 3x + 4 \end{array}$$

5 $(2y^2 - 3y + 5) - (y^2 + 2y + 8)$

The additive inverse of $y^2 + 2y + 8$ is $-(y^2 + 2y + 8)$ or $-y^2 - 2y - 8$.

$$\begin{aligned}(2y^2 - 3y + 5) - (y^2 + 2y + 8) &= (2y^2 - 3y + 5) + (-y^2 - 2y - 8)\\ &= (2y^2 - 1y^2) + (-3y - 2y) + (5 - 8)\\ &= (2 - 1)y^2 + (-3 - 2)y + (5 - 8)\\ &= 1y^2 + (-5y) + (-3) \text{ or } y^2 - 5y - 3\end{aligned}$$

To check the answer, add $y^2 - 5y - 3$ and $y^2 + 2y + 8$. The sum should be $2y^2 - 3y + 5$.

Inclusion Strategies

Students with learning disabilities may take longer until they are comfortable adding and subtracting polynomials without models. Keep algebra tiles readily available to students, and encourage them to use algebra tiles to model polynomial sums and differences until they feel confident enough to find answers without models.

6 $(3x^2 + 5) - (-4x + 2x^2 + 3)$

Reorder the terms of the second polynomial so that the powers of x are in descending order.

$-4x + 2x^2 + 3 = 2x^2 - 4x + 3$

Then arrange like terms in column form.

$$\begin{array}{rr} & 3x^2 + 5 \\ (-) & 2x^2 - 4x + 3 \\ \hline \end{array} \quad \textit{Add the additive inverse.} \quad \begin{array}{rr} & 3x^2 + 5 \\ (+) & -2x^2 + 4x - 3 \\ \hline & x^2 + 4x + 2 \end{array}$$

Your Turn

e. $(3x - 2) - (5x - 4)$ $-2x + 2$

f. $(10x^2 + 8x - 6) - (3x^2 + 2x - 9)$

g. $\begin{array}{rr} & 6m^2 + 7 \\ (-) & -2m^2 + 2m - 3 \\ \hline \end{array}$

h. $(5x^2 - 4x) - (2 - 3x)$ $5x^2 - x - 2$

f. $7x^2 + 6x + 3$

g. $8m^2 - 2m + 10$

Polynomials can be used to represent measures of geometric figures.

Example **Geometry Link**

7 **The perimeter of triangle ABC is $7x + 2y$. Find the measure of the third side of the triangle.**

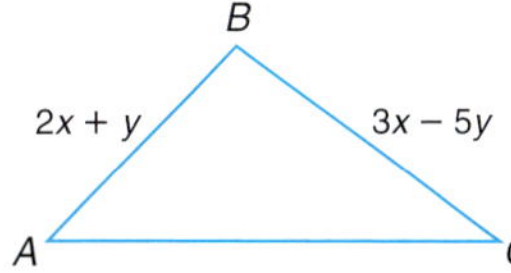

Explore You know the perimeter of the triangle and the measures of two sides. You need to find AC, the measure of the third side.

Plan The perimeter of a triangle is the sum of the measures of the three sides. To find AC, subtract the two given measures from the perimeter.

Solve $AC = \underbrace{\textit{Perimeter}} - \underbrace{AB} - \underbrace{BC}$

$$\begin{aligned} AC &= (7x + 2y) - (2x + y) - (3x - 5y) \\ &= (7x + 2y) + (-2x - y) + (-3x + 5y) \\ &= [7x + (-2x) + (-3x)] + [2y + (-y) + 5y] \\ &= [7 + (-2) + (-3)]x + [2 + (-1) + 5]y \\ &= 2x + 6y \end{aligned}$$

The measure of the third side of the triangle, AC, is $2x + 6y$.

Examine Check by adding the measures of the three sides.

$$\begin{array}{rr} & 2x + y \\ & 3x - 5y \\ (+) & 2x + 6y \\ \hline & 7x + 2y \end{array}$$

The perimeter is $7x + 2y$. The answer checks.

In-Class Example

Example 6

Find $(4p^2 - p) - (8 + 3p - p^2)$. $5p^2 - 4p - 8$

Teaching Tip When discussing Example 7, point out that students can also find the measure of the third side by adding the measures of the two known sides and subtracting the sum from the perimeter.

In-Class Example

Example 7

The measure of the perimeter of a triangle is $9a + 2b$. Two of the sides have lengths of $3a + b$ and $5a$. Find the measure of the third side of the triangle. $a + b$

***Study Guide*, p. 398**

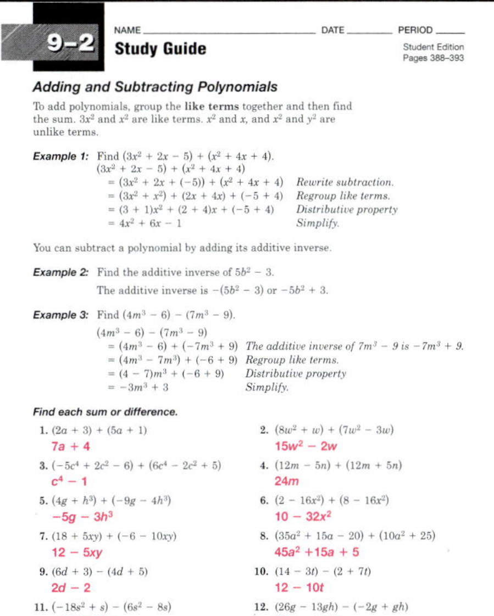

9–2 NAME ______ DATE ______ PERIOD ______

Study Guide

Student Edition Pages 388–393

Adding and Subtracting Polynomials

To add polynomials, group the **like terms** together and then find the sum. $3x^2$ and x^2 are like terms. x^2 and x, and x^2 and y^2 are unlike terms.

Example 1: Find $(3x^2 + 2x - 5) + (x^2 + 4x + 4)$.

$(3x^2 + 2x - 5) + (x^2 + 4x + 4)$

$= (3x^2 + 2x + (-5)) + (x^2 + 4x + 4)$ *Rewrite subtraction.*

$= (3x^2 + x^2) + (2x + 4x) + (-5 + 4)$ *Regroup like terms.*

$= (3 + 1)x^2 + (2 + 4)x + (-5 + 4)$ *Distributive property*

$= 4x^2 + 6x - 1$ *Simplify.*

You can subtract a polynomial by adding its additive inverse.

Example 2: Find the additive inverse of $5b^2 - 3$.

The additive inverse is $-(5b^2 - 3)$ or $-5b^2 + 3$.

Example 3: Find $(4m^3 - 6) - (7m^3 - 9)$.

$(4m^3 - 6) - (7m^3 - 9)$

$= (4m^3 - 6) + (-7m^3 + 9)$ *The additive inverse of $7m^3 - 9$ is $-7m^3 + 9$.*

$= (4m^3 - 7m^3) + (-6 + 9)$ *Regroup like terms.*

$= (4 - 7)m^3 + (-6 + 9)$ *Distributive property*

$= -3m^3 + 3$ *Simplify.*

Find each sum or difference.

1. $(2a + 3) + (5a + 1)$ $7a + 4$
2. $(8w^2 + w) + (7w^2 - 3w)$ $15w^2 - 2w$
3. $(-5c^4 + 2c^2 - 6) + (6c^4 - 2c^2 + 5)$ $c^4 - 1$
4. $(12m - 5n) + (12m + 5n)$ $24m$
5. $(4g + h^3) + (-9g - 4h^3)$ $-5g - 3h^3$
6. $(2 - 16x^2) + (8 - 16x^2)$ $10 - 32x^2$
7. $(18 + 5xy) + (-6 - 10xy)$ $12 - 5xy$
8. $(35a^2 + 15a - 20) + (10a^2 + 25)$ $45a^2 + 15a + 5$
9. $(6d + 3) - (4d + 5)$ $2d - 2$
10. $(14 - 3t) - (2 + 7t)$ $12 - 10t$
11. $(-18s^2 + s) - (6s^2 - 8s)$ $-24s^2 + 9s$
12. $(26g - 13gh) - (-2g + gh)$ $28g - 14gh$
13. $(7y^2 + 2y + 21) - (9y^2 + 6y + 11)$ $-2y^2 - 4y + 10$
14. $(-5m^2 + 2n - 1) - (7m^2 + 16n - 8)$ $-12m^2 - 14n + 7$

3 PRACTICE/APPLY

Error Analysis

Watch for students who subtract some, but not all of the terms in the second polynomial in Exercises 13–16.
Prevent by reminding students that when subtracting a polynomial, they must subtract each term.

Assignment Guide

Basic: 19–43 odd, 44–52
Average: 18–40 even, 42–52

Answer

2. In subtraction, you find the additive inverse of the polynomial being subtracted. Then add.

Skills Practice, **p. 399, and** ***Practice,*** **p. 400 (shown)**

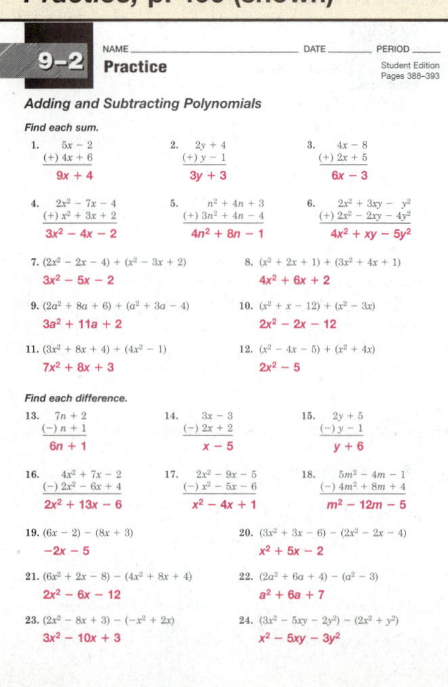

9–2 NAME ________ DATE ______ PERIOD _____

Practice

Student Edition Pages 388–393

Adding and Subtracting Polynomials

Find each sum.

1. $\begin{array}{r} 5x - 2 \\ (+)\ 4x + 6 \\ \hline 9x + 4 \end{array}$
2. $\begin{array}{r} 2y + 4 \\ (+)\ y - 1 \\ \hline 3y + 3 \end{array}$
3. $\begin{array}{r} 4x - 8 \\ (+)\ 2x + 5 \\ \hline 6x - 3 \end{array}$
4. $\begin{array}{r} 2x^2 - 7x - 4 \\ (+)\ x^2 + 3x + 2 \\ \hline 3x^2 - 4x - 2 \end{array}$
5. $\begin{array}{r} n^2 + 4n + 3 \\ (+)\ 3n^2 + 4n - 4 \\ \hline 4n^2 + 8n - 1 \end{array}$
6. $\begin{array}{r} 2x^2 + 3xy - y^2 \\ (+)\ 2x^2 - 2xy - 4y^2 \\ \hline 4x^2 + xy - 5y^2 \end{array}$
7. $(2x^2 - 2x - 4) + (x^2 - 3x + 2)$ $3x^2 - 5x - 2$
8. $(x^2 + 2x + 1) + (3x^2 + 4x + 1)$ $4x^2 + 6x + 2$
9. $(2a^2 + 8a + 6) + (a^2 + 3a - 4)$ $3a^2 + 11a + 2$
10. $(x^2 + x - 12) + (x^2 - 3x)$ $2x^2 - 2x - 12$
11. $(3x^2 + 8x + 4) + (4x^2 - 1)$ $7x^2 + 8x + 3$
12. $(x^2 - 4x - 5) + (x^2 + 4x)$ $2x^2 - 5$

Find each difference.

13. $\begin{array}{r} 7n + 2 \\ (-)\ n + 1 \\ \hline 6n + 1 \end{array}$
14. $\begin{array}{r} 3x - 3 \\ (-)\ 2x + 2 \\ \hline x - 5 \end{array}$
15. $\begin{array}{r} 2y + 5 \\ (-)\ y - 1 \\ \hline y + 6 \end{array}$
16. $\begin{array}{r} 4x^2 + 7x - 2 \\ (-)\ 2x^2 - 6x + 4 \\ \hline 2x^2 + 13x - 6 \end{array}$
17. $\begin{array}{r} 2x^2 - 9x - 5 \\ (-)\ x^2 - 5x - 6 \\ \hline x^2 - 4x + 1 \end{array}$
18. $\begin{array}{r} 5m^2 - 4m - 1 \\ (-)\ 4m^2 + 8m + 4 \\ \hline m^2 - 12m - 5 \end{array}$
19. $(6x - 2) - (8x + 3)$ $-2x - 5$
20. $(3x^2 + 3x - 6) - (2x^2 - 2x - 4)$ $x^2 + 5x - 2$
21. $(6x^2 + 2x - 8) - (4x^2 + 8x + 4)$ $2x^2 - 6x - 12$
22. $(2a^2 + 6a + 4) - (a^2 - 3)$ $a^2 + 6a + 7$
23. $(2x^2 - 8x + 3) - (-x^2 + 2x)$ $3x^2 - 10x + 3$
24. $(3x^2 - 5xy - 2y^2) - (2x^2 + y^2)$ $x^2 - 5xy - 3y^2$

Check for Understanding

Communicating Mathematics

1. **Describe** the first step you take when you add or subtract polynomials in column form. **Arrange like terms in column form.**
2. **Explain** how addition and subtraction of polynomials are related. **See margin.**

Guided Practice

Getting Ready Find the additive inverse of each polynomial.

Sample 1: $6y - 3z$
Solution: $-6y + 3z$

Sample 2: $-a^2 + 2ab - 3b^2$
Solution: $a^2 - 2ab + 3b^2$

3. $2a + 9b$ $\mathbf{-2a - 9b}$
4. $-4m + 6n$ $\mathbf{4m - 6n}$
5. $x^2 + 8x + 5$ **5.** $\mathbf{-x^2 - 8x - 5}$
6. $-3h^2 - 2h - 3$ $\mathbf{3h^2 + 2h + 3}$
7. $4xy^2 + 6x^2y - y^3$ $\mathbf{-4xy^2 - 6x^2y + y^3}$
8. $-2c^3 + c^2 - 3c$ $\mathbf{2c^3 - c^2 + 3c}$

Examples 1–3 Find each sum.

9. $\begin{array}{r} 6y - 5 \\ (+)\ 2y + 7 \\ \hline 8y + 2 \end{array}$
10. $\begin{array}{r} x^2 - 6x + 5 \\ (+)\ x^2 + 4x - 7 \\ \hline 2x^2 - 2x - 2 \end{array}$
11. $(2x^2 + 4x + 5) + (x^2 - 5x - 3)$ $\mathbf{3x^2 - x + 2}$
12. $(2x^2 - 5x + 4) + (3x^2 - 1)$ $\mathbf{5x^2 - 5x + 3}$

Examples 4–6 Find each difference.

13. $\begin{array}{r} 3x + 4 \\ (-)\ x + 2 \\ \hline 2x + 2 \end{array}$
14. $\begin{array}{r} 6m^2 - 5m + 3 \\ (-)\ 5m^2 + 2m - 7 \\ \hline m^2 - 7m + 10 \end{array}$
15. $(5x^2 + 4x - 1) - (4x^2 + x + 2)$ $\mathbf{x^2 + 3x - 3}$
16. $(5x^2 - 4x) - (-3x + 2)$ $\mathbf{5x^2 - x - 2}$

Example 7

17. **Geometry** The perimeter of triangle *DEF* is $12x^2 - 7x + 9$. Find the measure of the third side of the triangle. $\mathbf{x^2 - x + 5}$

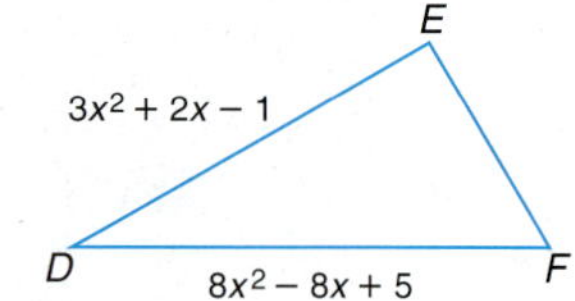

Exercises

Practice

Homework Help	
For Exercises	See Examples
18–26, 37, 38, 40	1, 2, 3
27–36, 39, 41, 44	4, 5, 6
42	7

Extra Practice
See page 710.

Find each sum. **23.** $\mathbf{3n^2 + 13n + 11}$ **24.** $\mathbf{8x^2 - 3x + 3}$

18. $\begin{array}{r} 2x + 3 \\ (+)\ x - 1 \\ \hline 3x + 2 \end{array}$
19. $\begin{array}{r} 2x^2 - 5x + 4 \\ (+)\ 2x^2 + 8x - 1 \\ \hline 4x^2 + 3x + 3 \end{array}$
20. $\begin{array}{r} 2x^2 + 3xy - 4y^2 \\ (+)\ 4x^2 \qquad + 4y^2 \\ \hline 6x^2 + 3xy \end{array}$
21. $(12x + 8y) + (2x - 7y)$ $\mathbf{14x + y}$
22. $(-7y + 3x) + (4x + 3y)$ $\mathbf{7x - 4y}$
23. $(n^2 + 5n + 3) + (2n^2 + 8n + 8)$
24. $(5x^2 - 7x + 9) + (3x^2 + 4x - 6)$
25. $(5n^2 - 4) + (2n^2 - 8n + 9)$ $\mathbf{7n^2 - 8n + 5}$
26. $(-2x^2 + 3xy - 3y^2) + (2x^2 - 5xy)$ $\mathbf{-2xy - 3y^2}$

Find each difference.

27. $\begin{array}{r} 9x + 5 \\ (-)\ 4x - 3 \\ \hline 5x + 8 \end{array}$
28. $\begin{array}{r} 5a^2 + 7a + 9 \\ (-)\ 3a^2 + 4a + 1 \\ \hline 2a^2 + 3a + 8 \end{array}$
29. $\begin{array}{r} 3x^2 + 4x - 1 \\ (-)\ 4x^2 + \ x + 2 \\ \hline -x^2 + 3x - 3 \end{array}$

Reteaching Activity

Auditory/Musical Learners It may help some students to replace variables with nouns and then speak the modified polynomials aloud to themselves. For example, using *ax* for *x* and *yak* for *y*, students would read $3x + 2y$ as *three axes and two yaks*. This may help emphasize that these terms are unlike terms, and so cannot be combined by adding or subtracting.

32. $x^2 + 7x$
33. $a^2 + 3a - 1$
34. $2x^2 + 5x + 5$
35. $5x^2 - xy - 2y^2$
38. $10a + 8$

B

39. $-2x^2y - 7x^2y^2$

40. $6x^2 - x + 2$

30. $(3x - 4) - (5x + 2)$ $-2x - 6$
31. $(9x - 4y) - (12x - 5y)$ $-3x + y$
32. $(2x^2 + 5x + 3) - (x^2 - 2x + 3)$
33. $(3a^2 + 2a - 5) - (2a^2 - a - 4)$
34. $(3x^2 + 5x + 4) - (-1 + x^2)$
35. $(5x^2 - 4xy) - (2y^2 - 3xy)$

Find each sum or difference. 36. $4x^3 - 5x - 17$ 37. $2pq + pr$

36. $(6x^3 - 10) - (2x^3 + 5x + 7)$
37. $(2pq + 3qr - 4pr) + (5pr - 3qr)$
38. $(3 + 2a - a^2) + (a^2 + 8a + 5)$
39. $(-x^2y - 4x^2y^2) - (x^2y + 3x^2y^2)$

40. Find the sum of $(x^2 + x + 5)$, $(3x^2 - 4x - 2)$, and $(2x^2 + 2x - 1)$.
41. What is $x^2 + 6x$ minus $3x^2 + 7$? $-2x^2 + 6x - 7$

Applications and Problem Solving

42. **Geometry** The measure of each side of a triangle is $3x + 1$. Find its perimeter. $9x + 3$

43. **Framing** The standard measurement for a custom-made picture frame is the *united inch*. It refers to the sum of the length and the width of a picture to be framed. Suppose a picture's length is 4 inches longer than its width. Then the width is w and the length would be $w + 4$.

a. Write a polynomial that represents the size of the picture in united inches. $2w + 4$ in.

b. If the width is 15 inches, use the polynomial from part a to find the size of the picture in united inches. 34 in.

44. $-3m^2 - m + 5$

44. **Critical Thinking** One polynomial is subtracted from a second polynomial and the difference is $3m^2 + m - 5$. What is the difference when the second polynomial is subtracted from the first?

Mixed Review

Find the degree of each polynomial. *(Lesson 9–1)*

45. $y^2 + 3y + 2$ 2
46. $n^6 + 3n^3 + 2$ 6
47. $3x^2 + 2x^2y$ 3

48. **Sailing** A rope from the top of a mast on a sailboat extends to a point 6 feet from the base of the mast. Suppose the rope is 24 feet long. To the nearest foot, how high is the mast? *(Lesson 8–7)* 23 ft

Simplify each expression. *(Lesson 8–2)*

49. $(x^2y)(x^3y^2)$ x^5y^3
50. $(4a^2)(2a^4)$ $8a^6$
51. $(-2y)(-5xy)$ $10xy^2$

Standardized Test Practice

52. **Multiple Choice** Which relation is a function? *(Lesson 6–4)* A

A $\{(3, 1), (4, 2), (2, 3), (1, 2)\}$
B $\{(1, 1), (-1, 5), (3, 3), (-1, 2)\}$
C $\{(-1, 3), (1, 4), (3, 2), (1, 2)\}$
D $\{(-2, 1), (4, 2), (3, 3), (3, 2)\}$

www.algconcepts.com/self_check_quiz

4 ASSESS

Open-Ended Assessment

Modeling Have students use algebra tiles to model $(2x^2 + 3x - 4) + (-x^2 - 2x + 2)$.

Chapter 9, Quiz A (Lessons 9–1 and 9–2) is available in the *Chapter 9 Resource Masters*, p. 429.

***Enrichment*, p. 402**

NAME ______ DATE ______ PERIOD ______

9–2 Enrichment

Student Edition Pages 388–393

Geometric Series

The terms of this polynomial form a geometric series.

$$a + ar + ar^2 + ar^3 + ar^4$$

The first term is the constant a. Then each term after that is found by multiplying by a constant multiplier r.

Use the equation $S = a + ar + ar^2 + ar^3 + ar^4$ for Exercises 1–3.

1. Multiply each side of the equation by r. $rS = ar + ar^2 + ar^3 + ar^4 + ar^5$
2. Subtract the original equation from your result in Exercise 1. $rS - S = ar^5 - a$
3. Solve the result from Exercise 2 for the variable S. $S = \frac{a(r^5 - 1)}{r - 1}$

Use the polynomial $a + ar + ar^2 + ar^3 + ar^4 + \cdots + ar^{n-1}$ for Exercises 4–8.

4. Write the 10th term of the polynomial. ar^9
5. If $a = 5$ and $r = 2$, what is the 8th term? $ar^7 = 640$
6. Follow the steps in Exercises 1–3 to write a formula for the sum of this polynomial. $S = \frac{a(r^n - 1)}{r - 1}$
7. If the 3rd term is 20 and the 6th term is 160, solve for r^3 and then find r. Then solve $ar^2 = 20$ for a and find the value of the first six terms of the polynomial. $\frac{ar^5}{ar^2} = \frac{160}{20}$, $r^3 = 8$, $r = 2$; $ar^2 = 20$; $a = 5$; 5, 10, 20, 40, 80, 160
8. Find the sum of the first six terms of the geometric series that begins 3, 6, 12, 24, ... First write the values for a and r. $a = 3, r = 2$; $S = \frac{3(2^6 - 1)}{2 - 1} = 3 \times 63 = 189$

Extra Credit

Find two trinomials that have a sum of $x^3 + 4x^2y - 3y^3$.
Sample answer: $(2x^3 + 3x^2y - 2y^3) + (-x^3 + x^2y - y^3)$

1 FOCUS

5-Minute Check
Lesson 9–2

Find each sum or difference.

1. $(5x + 2) + (x - 3)$ $6x - 1$
2. $(a^2 + 3b - 5) + (3 - 4b)$ $a^2 - b - 2$
3. $(2x^2 - 3x + 4) - (3x^2 - x)$ $-x^2 - 2x + 4$

The measure of the perimeter of triangle XYZ is 3g + h.

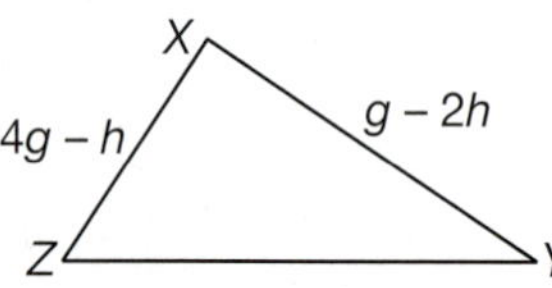

4. Find the sum of the measures of the two known sides. $5g - 3h$
5. Find the measure of the third side of triangle XYZ. $-2g + 4h$

Motivating the Lesson

Hands-On Activity Give pairs of students one set of algebra tiles. Have them represent $x^2 + 3x - 1$. Ask students how they would represent multiplying the polynomial by 2. Then have student pairs model the result of doubling the polynomial. Make sure students realize that all three terms of the polynomial must be doubled.

2 TEACH

Teaching Tip Model the problem at the beginning of the lesson using algebra tiles on an overhead projector.

In-Class Examples
Examples 1–2

Find each product.

1 $x(x + 1)$ $x^2 + x$
2 $g(3g^2 + 4)$ $3g^3 + 4g$

9–3 Multiplying a Polynomial by a Monomial

What You'll Learn
You'll learn to multiply a polynomial by a monomial.

Why It's Important
Recreation You can use monomials and polynomials to solve problems involving recreation. *See Exercise 65.*

Suppose you have a square whose length and width are x units. If you increase the length by 3 units, what is the area of the new figure?

You can model this problem by using algebra tiles. The figures show how to make a rectangle whose length is $x + 3$ units and whose width is x units.

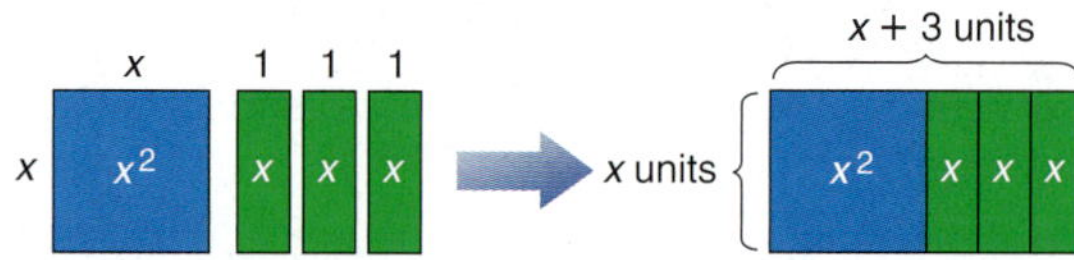

The area of any rectangle is the product of its length and its width. The area can also be found by adding the areas of the tiles.

Formula
$A = \ell w$
$= (x + 3)x$ or $x(x + 3)$
$= x^2 + 3x$

Algebra Tiles
$A = x^2 + x + x + x$
$= x^2 + 3x$

Since the areas are equal, $x(x + 3) = x^2 + 3x$. Another expression for the same area is $x^2 + 3x$ square units.

The example above shows how the Distributive Property can be used to multiply a polynomial by a monomial.

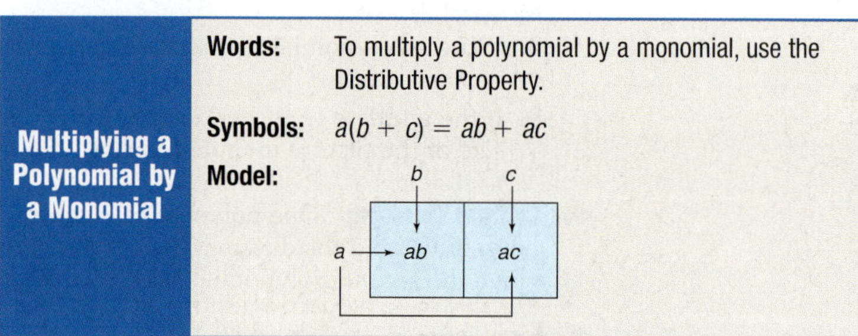

Multiplying a Polynomial by a Monomial

Words: To multiply a polynomial by a monomial, use the Distributive Property.

Symbols: $a(b + c) = ab + ac$

Model:

Examples

Find each product.

1 $y(y + 5)$

$y(y + 5) = y(y) + y(5)$
$= y^2 + 5y$

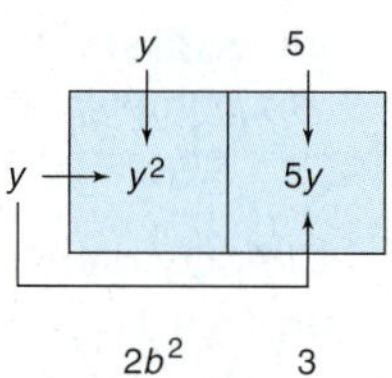

Look Back
Multiplying Powers: Lesson 8–2

2 $b(2b^2 + 3)$

$b(2b^2 + 3) = b(2b^2) + b(3)$
$= 2b^3 + 3b$

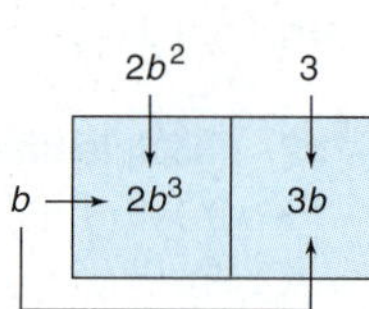

Resource Manager

Reproducible Masters
Chapter 9 Resource Masters
- *Study Guide,* p. 403
- *Skills Practice,* p. 404
- *Practice,* p. 405
- *Reading to Learn Mathematics,* p. 406
- *Enrichment,* p. 407
- *Assessment,* p. 428

Hands-On Algebra, p. 101
School-to-Workplace, p. 9

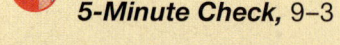

Transparencies
5-Minute Check, 9–3

Technology/Multimedia
Interactive Chalkboard CD-ROM

3 $-2n(7 - 5n^2)$

$-2n(7 - 5n^2) = -2n(7) + (-2n)(-5n^2)$
$= -14n + 10n^3$

	7	$-5n^2$
$-2n$	$-14n$	$10n^3$

4 $3x^3(2x^2 - 5x + 8)$

$3x^3(2x^2 - 5x + 8) = 3x^3(2x^2) + 3x^3(-5x) + 3x^3(8)$
$= 6x^5 - 15x^4 + 24x^3$

Your Turn

a. $7(2x + 5)$ $14x + 35$
b. $4x(3x^2 - 7)$ $12x^3 - 28x$
c. $-5a(6 - 3a^2)$ $-30a + 15a^3$
d. $2m^2(5m^2 - 7m + 8)$

d. $10m^4 - 14m^3 + 16m^2$

Many equations contain polynomials that must be multiplied.

Examples

Look Back Solving Equations with Grouping Symbols: Lesson 4–7

Solve each equation.

5 $11(y - 3) + 5 = 2(y + 22)$

$11(y - 3) + 5 = 2(y + 22)$	*Original equation*
$11y - 33 + 5 = 2y + 44$	*Distributive Property*
$11y - 28 = 2y + 44$	*Combine like terms.*
$11y - 28 - 2y = 2y + 44 - 2y$	*Subtract 2y from each side.*
$9y - 28 = 44$	*Simplify.*
$9y - 28 + 28 = 44 + 28$	*Add 28 to each side.*
$9y = 72$	*Simplify.*
$\frac{9y}{9} = \frac{72}{9}$	*Divide each side by 9.*
$y = 8$	The solution is 8.

6 $w(w + 12) = w(w + 14) + 12$

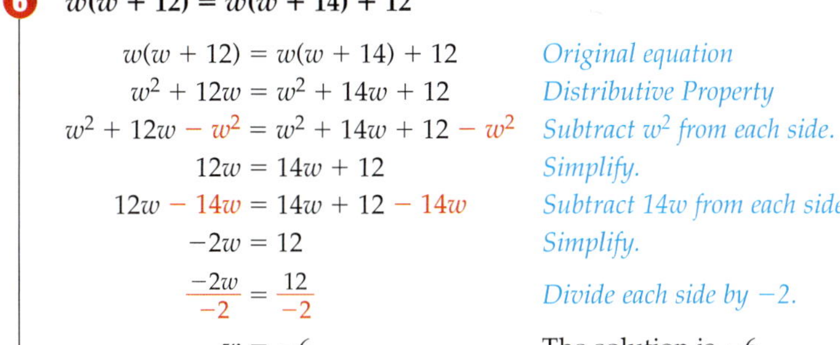

$w(w + 12) = w(w + 14) + 12$	*Original equation*
$w^2 + 12w = w^2 + 14w + 12$	*Distributive Property*
$w^2 + 12w - w^2 = w^2 + 14w + 12 - w^2$	*Subtract w^2 from each side.*
$12w = 14w + 12$	*Simplify.*
$12w - 14w = 14w + 12 - 14w$	*Subtract 14w from each side.*
$-2w = 12$	*Simplify.*
$\frac{-2w}{-2} = \frac{12}{-2}$	*Divide each side by −2.*
$w = -6$	The solution is −6.

Your Turn

e. $2(5x - 12) = 6(-2x + 3) + 2$ 2
f. $a(a + 2) + 3a = a(a - 3) + 8$ 1

In-Class Examples

Examples 3–4

Find each product.

3 $-y(2y - 6)$ $-2y^2 + 6y$

4 $b^2(2b^2 - 4b - 9)$ $2b^4 - 4b^3 - 9b^2$

Examples 5–6

Solve each equation.

5 $3(d - 4) - 8 = 5(5d + 1) - 3$ -1

6 $a(3 + a) - 2 = a(a - 1) + 6$ 2

Teaching Tip Begin Example 7 by replacing the variable expressions with integers. Work the problem with these integers so that students can see the problem-solving method and geometry involved. Then repeat the Example using the variable expressions given.

In-Class Example

Example 7

Find the area of the shaded region in simplest form.

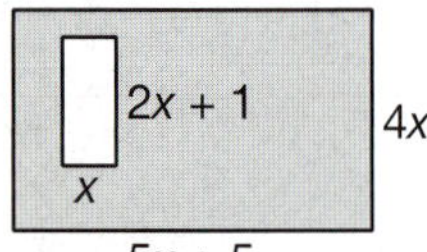

5x + 5

$18x^2 + 19x$

3 PRACTICE/APPLY

Error Analysis

Watch for students who multiply only the first term of the binomial or trinomial by the monomial in Exercises 4–12.

Prevent by stressing that this is an application of the Distributive Property. So, the monomial term outside the parentheses must be multiplied by every term of the polynomial within the parentheses. If a polynomial containing three terms is multiplied by a monomial, there will be three terms in the product.

***Study Guide*, p. 403**

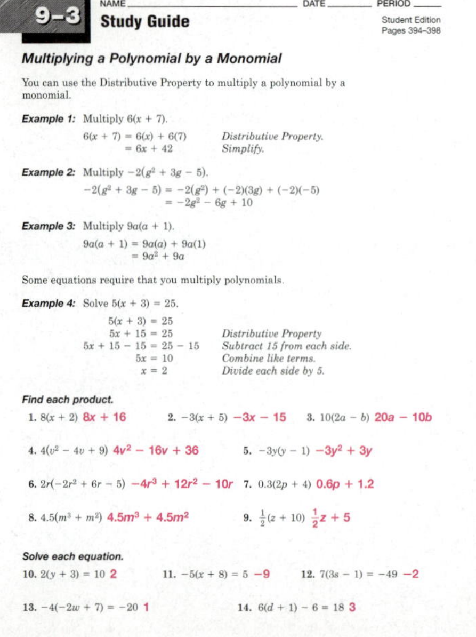

9–3 NAME ____________ DATE ______ PERIOD ____

Study Guide Student Edition Pages 394–398

Multiplying a Polynomial by a Monomial

You can use the Distributive Property to multiply a polynomial by a monomial.

Example 1: Multiply $6(x + 7)$.

$6(x + 7) = 6(x) + 6(7)$ *Distributive Property.*
$= 6x + 42$ *Simplify.*

Example 2: Multiply $-2(g^2 + 3g - 5)$.

$-2(g^2 + 3g - 5) = -2(g^2) + (-2)(3g) + (-2)(-5)$
$= -2g^2 - 6g + 10$

Example 3: Multiply $9a(a + 1)$.

$9a(a + 1) = 9a(a) + 9a(1)$
$= 9a^2 + 9a$

Some equations require that you multiply polynomials.

Example 4: Solve $5(x + 3) = 25$.

$5(x + 3) = 25$
$5x + 15 = 25$ *Distributive Property*
$5x + 15 - 15 = 25 - 15$ *Subtract 15 from each side.*
$5x = 10$ *Combine like terms.*
$x = 2$ *Divide each side by 5.*

Find each product.

1. $8(x + 2)$ $8x + 16$
2. $-3(x + 5)$ $-3x - 15$
3. $10(2a - b)$ $20a - 10b$
4. $4(v^2 - 4v + 9)$ $4v^2 - 16v + 36$
5. $-3y(y - 1)$ $-3y^2 + 3y$
6. $2r(-2r^2 + 6r - 5)$ $-4r^3 + 12r^2 - 10r$
7. $0.3(2p + 4)$ $0.6p + 1.2$
8. $4.5(m^3 + m^2)$ $4.5m^3 + 4.5m^2$
9. $\frac{1}{2}(z + 10)$ $\frac{1}{2}z + 5$

Solve each equation.

10. $2(y + 3) = 10$ 2
11. $-5(x + 8) = 5$ -9
12. $7(3s - 1) = -49$ -2
13. $-4(-2w + 7) = -20$ 1
14. $6(d + 1) - 6 = 18$ 3

You can apply multiplication of a polynomial by a monomial to problems involving area.

Example 7 — Geometry Link

Find the area of the shaded region in simplest form.

3x, 2x, x + 2, x

Subtract the area of the smaller rectangle from the area of the larger rectangle.

area of larger rectangle:	$3x(x + 2)$	*$A = \ell w$*
area of smaller rectangle:	$2x(x)$	
area of shaded region:	$3x(x + 2) - 2x(x)$	

$A = 3x(x + 2) - 2x(x)$ *Area of shaded region*
$= 3x(x) + 3x(2) - 2x(x)$ *Distributive Property*
$= 3x^2 + 6x - 2x^2$ *Multiply.*
$= 1x^2 + 6x$ or $x^2 + 6x$ *Combine like terms.*

The area of the shaded region is $x^2 + 6x$.

Check for Understanding

Communicating Mathematics

1. **Name** the property used to express $3n(2n - 6)$ as $6n^2 - 18n$. **1. Distributive Property**
2. **Write** an expression for the area of the rectangle at the right in the following two ways.
   a. a product of a monomial and a polynomial $x(2x + 3)$
   b. a sum of monomials $2x^2 + 3x$

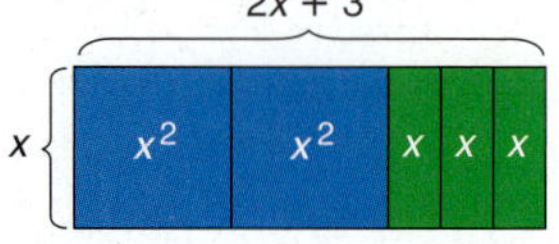

3. 

Consuelo says that $2x(3x + 4) = 6x^2 + 8x$ is a true statement. Shawn says that $2x(3x + 4) = 6x^2 + 4$ is a true statement. Who is correct? Explain. **Consuelo; Shawn forgot to multiply $2x$ and 4.**

Guided Practice

Examples 1–4 Find each product.

4. $2(y + 5)$ $2y + 10$
5. $x(x + 2)$ $x^2 + 2x$
6. $3a(a - 1)$ $3a^2 - 3a$
7. $-4(x - 2)$ $-4x + 8$
8. $-3n(5 - 2n)$
9. $4z(z^2 - 2z)$
10. $2(2d^2 + 3d + 8)$ $4d^2 + 6d + 16$
11. $3(8y^2 + 3y - 5)$ $24y^2 + 9y - 15$
12. $-2a(4a^2 - 3a - 1)$ $-8a^3 + 6a^2 + 2a$

8. $-15n + 6n^2$
9. $4z^3 - 8z^2$

Examples 5 & 6 Solve each equation.

13. $7(x + 2) = 42$ 4
14. $4(y - 8) + 10 = 2y + 12$ 17
15. $-3(2a - 4) + 9 = 3(a + 1)$ 2
16. $x(x + 3) + 5x = x(x + 5) + 9$ 3

Example 7

17. **Geometry** Find the area of the shaded region in simplest form. $3x^2$

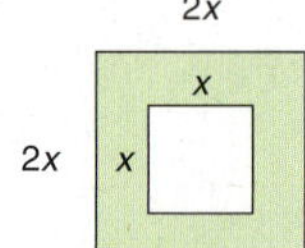

Reteaching Activity

Verbal/Linguistic Learners Encourage students to read each product aloud quietly to help them understand what they are doing. For example, Exercise 18 can be read as *2 multiplied by the sum of x and 6.* Multiplying 2 by a sum indicates that 2 must be multiplied by both terms of the sum.

Exercises

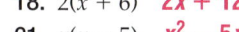

Practice

Find each product.

18. $2(x + 6)$ **$2x + 12$**
19. $-3(y + 3)$ **$-3y - 9$**
20. $7(2a + 3)$ **$14a + 21$**
21. $x(x - 5)$ **$x^2 - 5x$**
22. $n(n + 4)$ **$n^2 + 4n$**
23. $z(3z - 2)$ **$3z^2 - 2z$**
24. $2m(m + 4)$
25. $4x(2x - 3)$
26. $5y(y + 1)$ **$5y^2 + 5y$**
27. $-2a(a - 2)$
28. $-3x(x - 5)$
29. $-5y(6 - 2y)$
30. $3s(4s^2 - 7)$
31. $5d(d^2 + 3)$
32. $-7p(-3p - 6)$
33. $-2a(5a^2 - 7a + 2)$
34. $4x(-2x + 7x^2)$
35. $5n(8n^3 + 7n^2 - 3n)$
36. $-3y(6 - 9y + 4y^2)$
37. $7(-2a^2 + 5a - 11)$
38. $-5x^2(3x^2 - 8x - 12)$
39. $1.2(c^2 - 10)$
40. $0.1(4x^2 - 7x)$
41. $0.25x(4x - 8)$
42. $\frac{1}{2}(2x^2 - 6x)$ **$x^2 - 3x$**
43. $\frac{2}{3}(6y^2 - 9y + 3)$ **$4y^2 - 6y + 2$**
44. $\frac{1}{4}x(12x^2 + 8x)$ **$3x^3 + 2x^2$**

Homework Help	
For Exercises	See Examples
18–26, 62, 65–66	1, 2
27–44, 57–61	3, 4
45–58	5, 6
63, 64	7

Extra Practice
See page 710.

24. $2m^2 + 8m$
25. $8x^2 - 12x$
27. $-2a^2 + 4a$
28. $-3x^2 + 15x$
29. $-30y + 10y^2$
30. $12s^3 - 21s$
31. $5d^3 + 15d$
32. $21p^2 + 42p$
33–41. See margin.

Solve each equation.

45. $8(x - 3) = 16$ **5**
46. $4(y - 3) = -8$ **1**
47. $40 = 5(a + 10)$ **−2**
48. $2(x + 4) + 9 = 5x - 1$ **6**
49. $6(a + 2) - 5 = 2a + 3$ **−1**
50. $8x + 14 = 3(x - 2) + 10$ **−2**
51. $2(5a - 13) = 6(-2a + 3)$ **2**
52. $3(2y - 4) = -2(2y - 9)$ **3**
53. $5(n + 7) = 3(-2n + 1) - 1$ **−3**
54. $b(b + 12) = b(b + 14) + 12$ **−6**
55. $c(c + 8) - c(c + 3) - 23 = 3c + 11$ **17**
56. $m(m - 8) + 3m = -2 + m(m - 9)$ **$-\frac{1}{2}$**

57. What is the product of $4x$ and $x^2 - 2x + 1$? **$4x^3 - 8x^2 + 4x$**
58. Multiply $-2n$ and $3 - 5n$. **$-6n + 10n^2$**

Simplify.

59. $2a(a^2 - 5a + 4) - 6(a^3 + 3a - 11)$ **$-4a^3 - 10a^2 - 10a + 66$**
60. $5t^2(t + 2) - 5t(4t^2 - 3t + 6) + 3(t^2 - 6)$ **$-15t^3 + 28t^2 - 30t - 18$**
61. $3n^2(n - 4) + 6n(3n^2 + n - 7) - 4(n - 7)$ **$21n^3 - 6n^2 - 46n + 28$**

Applications and Problem Solving

62. **Sports** The length of a pool table is twice as long as its width. Suppose the width of a pool table is $4x - 1$. What is the length of the table? **$8x - 2$**

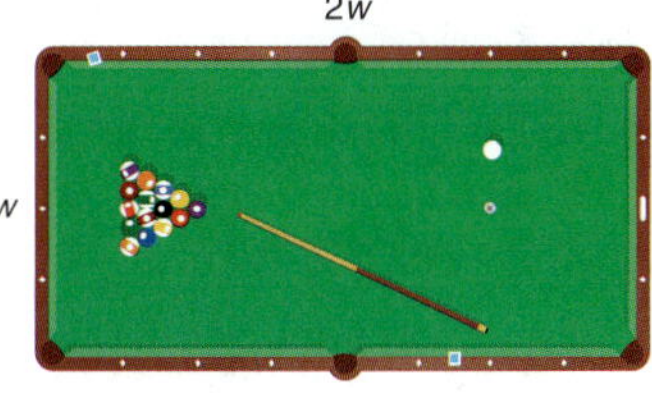

Geometry **Find the area of the shaded region for each figure.**

63.

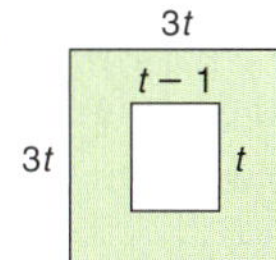

$8t^2 + t$

64.

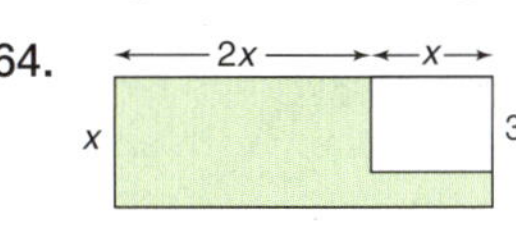

$3x^2 - 3x$

Assignment Guide

Basic: 19–65 odd, 66–78
Average: 18–60 even, 62–78
All: Quiz 1, 1–5

Answers

33. $-10a^3 + 14a^2 - 4a$
34. $-8x^2 + 28x^3$
35. $40n^4 + 35n^3 - 15n^2$
36. $-18y + 27y^2 - 12y^3$
37. $-14a^2 + 35a - 77$
38. $-15x^4 + 40x^3 + 60x^2$
39. $1.2c^2 - 12$
40. $0.4x^2 - 0.7x$
41. $x^2 - 2x$

***Skills Practice,* p. 404, and *Practice,* p. 405 (shown)**

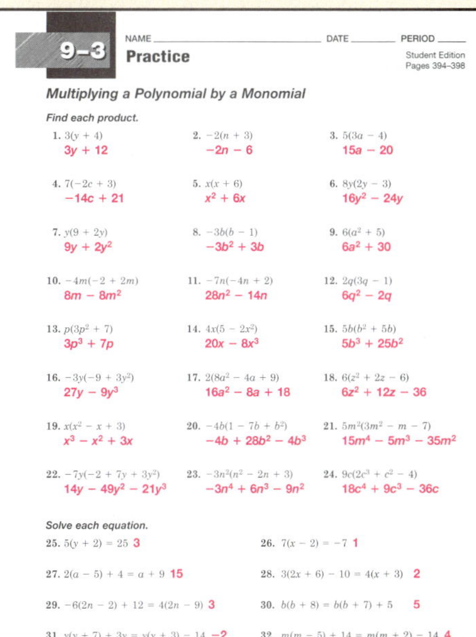

9–3 NAME ______ DATE ______ PERIOD ______

Practice — Student Edition Pages 394–398

Multiplying a Polynomial by a Monomial

Find each product.

1. $3(y + 4)$ $3y + 12$
2. $-2(n + 3)$ $-2n - 6$
3. $5(3a - 4)$ $15a - 20$
4. $7(-2c + 3)$ $-14c + 21$
5. $x(x + 6)$ $x^2 + 6x$
6. $8y(2y - 3)$ $16y^2 - 24y$
7. $y(9 + 2y)$ $9y + 2y^2$
8. $-3b(b - 1)$ $-3b^2 + 3b$
9. $6(a^2 + 5)$ $6a^2 + 30$
10. $-4m(-2 + 2m)$ $8m - 8m^2$
11. $-7n(-4n + 2)$ $28n^2 - 14n$
12. $2q(3q - 1)$ $6q^2 - 2q$
13. $p(3p^2 + 7)$ $3p^3 + 7p$
14. $4x(5 - 2x^2)$ $20x - 8x^3$
15. $5b(b^2 + 5b)$ $5b^3 + 25b^2$
16. $-3y(-9 + 3y^2)$ $27y - 9y^3$
17. $2(8a^2 - 4a + 9)$ $16a^2 - 8a + 18$
18. $6(z^2 + 2z - 6)$ $6z^2 + 12z - 36$
19. $x(x^2 - x + 3)$ $x^3 - x^2 + 3x$
20. $-4b(1 - 7b + b^2)$ $-4b + 28b^2 - 4b^3$
21. $5m^2(3m^2 - m - 7)$ $15m^4 - 5m^3 - 35m^2$
22. $-7y(-2 + 7y + 3y^2)$ $14y - 49y^2 - 21y^3$
23. $-3n^2(n^2 - 2n + 3)$ $-3n^4 + 6n^3 - 9n^2$
24. $9c(2c^3 + c^2 - 4)$ $18c^4 + 9c^3 - 36c$

Solve each equation.

25. $5(y + 2) = 25$ 3
26. $7(x - 2) = -7$ 1
27. $2(a - 5) + 4 = a + 9$ 15
28. $3(2x + 6) - 10 = 4(x + 3)$ 2
29. $-6(2n - 2) + 12 = 4(2n - 9)$ 3
30. $b(b + 8) = b(b + 7) + 5$ 5
31. $y(y + 7) + 3y = y(y + 3) - 14$ −2
32. $m(m - 5) + 14 = m(m + 2) - 14$ 4

From the Classroom of …

Victoria Fortenberry
Lincoln High School
Tallahassee, Florida

When teaching my students how to find the area of a shaded region, I like to place circles inside of squares or rectangles so that students can practice using the area formula of a circle as well as the area formulas of rectangles and squares.

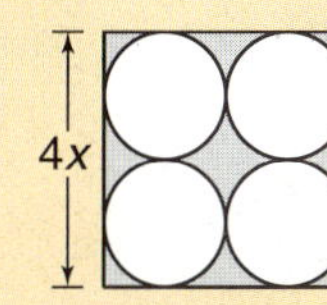

4 ASSESS

Open-Ended Assessment

Speaking Have students explain how to find the product $3(x + 4)$.

Quiz 1

The Quiz provides students with a brief review of the concepts and skills in Lessons 9–1 through 9–3. Lesson numbers are given to the right of the exercises or instruction lines so students can review concepts not yet mastered.

Mid-Chapter Test (Lessons 9–1 through 9–3) is available in the *Chapter 9 Resource Masters*, p. 428.

***Enrichment*, p. 407**

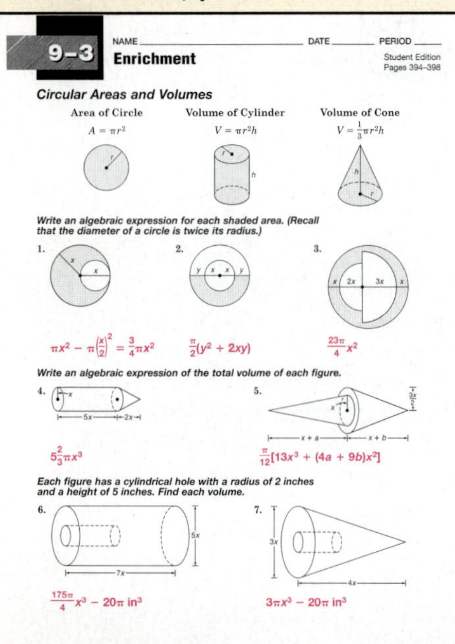

9–3 NAME ______ DATE ______ PERIOD ______

Enrichment — Student Edition Pages 394–398

Circular Areas and Volumes

Area of Circle: $A = \pi r^2$ | Volume of Cylinder: $V = \pi r^2 h$ | Volume of Cone: $V = \frac{1}{3}\pi r^2 h$

Write an algebraic expression for each shaded area. (Recall that the diameter of a circle is twice its radius.)

1. $\pi x^2 - \pi\left(\frac{x}{2}\right)^2 = \frac{3}{4}\pi x^2$
2. $\frac{\pi}{2}(y^2 + 2xy)$
3. $\frac{23\pi}{4}x^2$

Write an algebraic expression of the total volume of each figure.

4. $5\frac{2}{3}\pi x^3$
5. $\frac{\pi}{12}[13x^3 + (4a + 9b)x^2]$

Each figure has a cylindrical hole with a radius of 2 inches and a height of 5 inches. Find each volume.

6. $\frac{175\pi}{4}x^3 - 20\pi \text{ in}^3$
7. $3\pi x^3 - 20\pi \text{ in}^3$

65. **Recreation** On the Caribbean island of Trinidad, children play a form of hopscotch called Jumby. The pattern for this game is shown at the right. Suppose the length of each rectangle is $y + 5$ units long and y units wide.
    a. Write an expression in simplest form for the area of the pattern. $7y^2 + 35y$ units^2
    b. If y represents 10 inches, find the area of the pattern. 1050 in^2

66. **Critical Thinking** Write three different multiplication expressions whose product is $6a^2 + 8a$. $2(3a^2 + 4a)$, $a(6a + 8)$, $2a(3a + 4)$

Mixed Review

Find each sum or difference. *(Lesson 9–2)*

67. $(2x + 1) + (5x - 3)$ $7x - 2$
68. $(3a - 2) - (a + 4)$ $2a - 6$
69. $(x^2 + 2x - 3) + (2x^2 - 4x)$ $3x^2 - 2x - 3$
70. $(y^2 + 8y) - (2y^2 + 5)$ $-y^2 + 8y - 5$
71. **Ecology** The deer population of the Kaibab Plateau in Arizona from 1905 to 1930 can be estimated by the polynomial $-0.13x^5 + 3.13x^4 + 4000$, where x is the number of years after 1900. Find the degree of the polynomial. *(Lesson 9–1)* 5

Simplify. *(Lesson 8–5)*

72. $\sqrt{81}$ 9
73. $-\sqrt{121}$ -11
74. $\sqrt{256}$ 16
75. $\sqrt{\frac{9}{100}}$ $\frac{3}{10}$

Find the odds of each outcome if a die is rolled. *(Lesson 5–6)*

76. a number less than 5 2:1
77. an odd number 1:1

Standardized Test Practice

Ⓐ Ⓑ Ⓒ Ⓓ

78. **Multiple Choice** Which equation is equivalent to $12 + (x - 7) = 21$? *(Lesson 4–7)* C
    Ⓐ $12x - 7 = 21$
    Ⓑ $12x - 84 = 21$
    Ⓒ $x + 5 = 21$
    Ⓓ $19 + x = 21$

Quiz 1 — Lessons 9–1 through 9–3

1. Find the degree of $3y^4 + 2y^3 - y^2 + 5y - 3$. *(Lesson 9–1)* 4
2. Arrange the terms of $2xy + x^2 - y^2$ so that the powers of x are in descending order. *(Lesson 9–1)* $x^2 + 2xy - y^2$

Find each sum or difference. *(Lesson 9–2)*

3. $(4x^2 + 3x) - (6x^2 - 5x + 2)$ $-2x^2 + 8x - 2$
4. $(2w^2 - 4w - 12) + (15 - 3w^2 + 2w)$ $-w^2 - 2w + 3$
5. **Sports** In the National Football League, the length of the playing field is 40 feet longer than twice its width. *(Lesson 9–3)*
    a. Express the width and length of the playing field as polynomials. x ft, $2x + 40$ ft
    b. Find the polynomial that represents the area of the playing field. $2x^2 + 40x$ ft^2

www.algconcepts.com/self_check_quiz

Extra Credit

When Travis is three times as old as he will be in two years, he will be three quarters of a century old. Write an equation using T to represent Travis' current age. Then solve your equation to find T.

$3(T + 2) = \frac{3}{4}(100)$; $T = 23$ yr

9–4 Multiplying Binomials

What You'll Learn
You'll learn to multiply two binomials.

Why It's Important
Packaging The FOIL method can be used to find the dimensions of a cereal box. See Exercise 55.

Katie has a square herb garden in which she grows parsley. The measure of each side is x feet. She wants to increase the length by 2 feet and the width by 1 foot so she can grow sage, rosemary, and thyme. A plan for the garden is shown at the right. Find two different expressions for the area of the new garden.

One way to find the area of the garden is to use the formula for area. Find the product of the length and width of the new garden.

Formula

$A = \ell w$
$\quad = (x + 2)(x + 1)$

You can also find the area by adding the areas of the smaller regions.

Sum of Regions

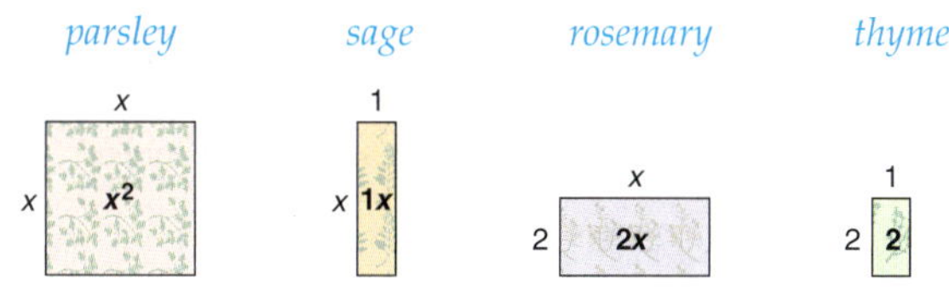

$A = x^2 + 1x + 2x + 2$
$\quad = x^2 + 3x + 2$ *Combine like terms.*

Since the areas are equal, both expressions are equal. Therefore, $(x + 2)(x + 1) = x^2 + 3x + 2$.

The multiplication expression can also be shown in the model below. Notice that the Distributive Property is used twice.

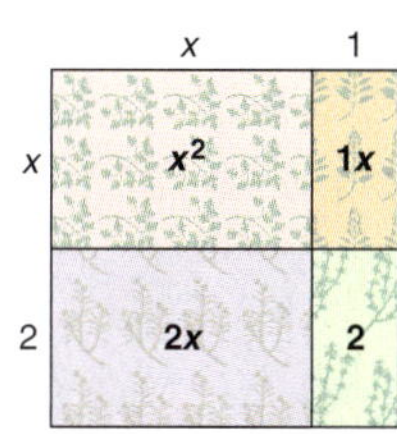

$x(x + 1) = x(x) + x(1)$ *Area of top row*
$\quad = x^2 + 1x$

$2(x + 1) = 2(x) + 2(1)$ *Area of bottom row*
$\quad = 2x + 2$

So, $(x + 2)(x + 1) = x^2 + 1x + 2x + 2$ or $x^2 + 3x + 2$.

1 FOCUS

5-Minute Check
Lesson 9–3

Find each product.

1. $x(2x + 3)$ $2x^2 + 3x$
2. $y(-y + 7)$ $-y^2 + 7y$

Solve each equation.

3. $3(a + 5) + 2 = 4(a + 4) + 3$ -2
4. $4(2y - 3) = 3(3y - 7)$ 9
5. Find the area of the shaded region in simplest form.

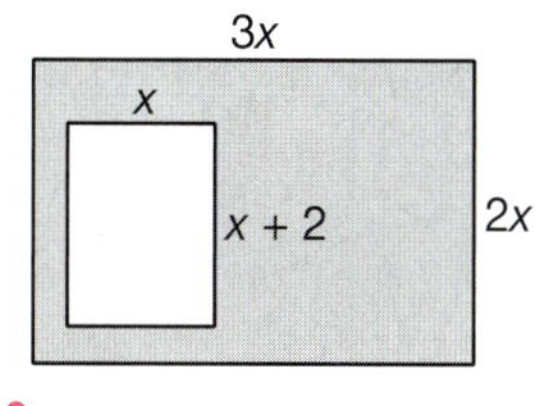

$5x^2 - 2x$

Motivating the Lesson

Real-World Connection Ask students who have gardened or who have family gardens what they have grown and how much space the plants needed. If possible, bring in seed packets or planting information that gives the area of ground different types of plants need. Tie this information to the gardening application at the beginning of the lesson to give students realistic images of gardening.

2 TEACH

Teaching Tip As you discuss the application at the beginning of the lesson, choose a different color to represent each of the four regions for clarity. Write the area of each region in its corresponding color.

Resource Manager

Reproducible Masters

Chapter 9 Resource Masters
- *Study Guide,* p. 408
- *Skills Practice,* p. 409
- *Practice,* p. 410
- *Reading to Learn Mathematics,* p. 411
- *Enrichment,* p. 412

Hands-On Algebra, p. 102

Transparencies

5-Minute Check, 9–4

Technology/Multimedia

AlgePASS, Lesson 21
Interactive Chalkboard CD-ROM

In-Class Examples

Examples 1–2

Find each product.

1 $(x - 1)(x + 5)$ $x^2 + 4x - 5$

2 $(a + 2)(2a - 3)$ $2a^2 + a - 6$

Answers

1.

	x	3
x	x^2	$3x$
2	$2x$	6

2.

	x	4
x	x^2	$4x$
3	$3x$	12

3.

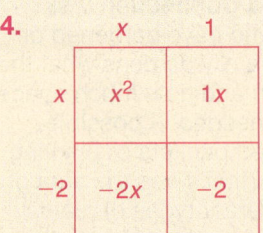

	x	1
$2x$	$2x^2$	$2x$
1	$1x$	1

4.

	x	1
x	x^2	$1x$
-2	$-2x$	-2

5.

	x	-2
x	x^2	$-2x$
-3	$-3x$	6

6.

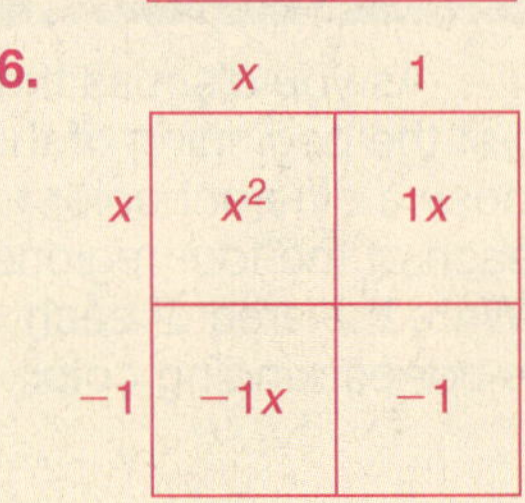

	x	1
x	x^2	$1x$
-1	$-1x$	-1

Hands-On Algebra

Materials: straightedge

Use a model to find the product of $(x + 2)$ and $(x - 1)$.

Step 1 Put the $x + 2$ and $x - 1$ outside the box as shown. *Note that $x - 1 = x + (-1)$.*

Step 2 Use the Distributive Property to multiply x by $x - 1$ and place the products inside the boxes.

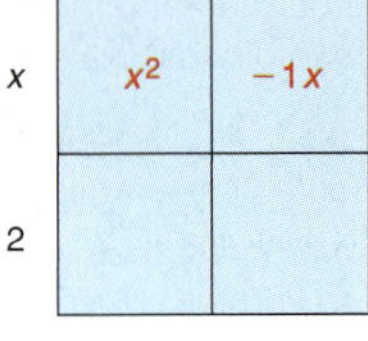

	x	-1
x	x^2	$-1x$
2		

Step 3 Use the Distributive Property to multiply 2 by $x - 1$ and place the products inside the boxes.

	x	-1
x	x^2	$-1x$
2	$2x$	-2

Step 4 Find the sum of the terms inside the boxes: $x^2 - 1x + 2x - 2 = x^2 + 1x - 2$ or $x^2 + x - 2$.

Therefore, $(x + 2)(x - 1) = x^2 + x - 2$.

1–6. See margin for models.

Try These **Use a model to find each product.**

1. $(x + 2)(x + 3)$ **2.** $(x + 3)(x + 4)$ **3.** $(2x + 1)(x + 1)$

4. $(x - 2)(x + 1)$ **5.** $(x - 3)(x - 2)$ **6.** $(x - 1)(x + 1)$

1. $x^2 + 5x + 6$
2. $x^2 + 7x + 12$
3. $2x^2 + 3x + 1$
4. $x^2 - x - 2$
5. $x^2 - 5x + 6$
6. $x^2 - 1$

You can also use the Distributive Property to multiply binomials.

Examples

Find each product.

1 $(x + 3)(x - 4)$

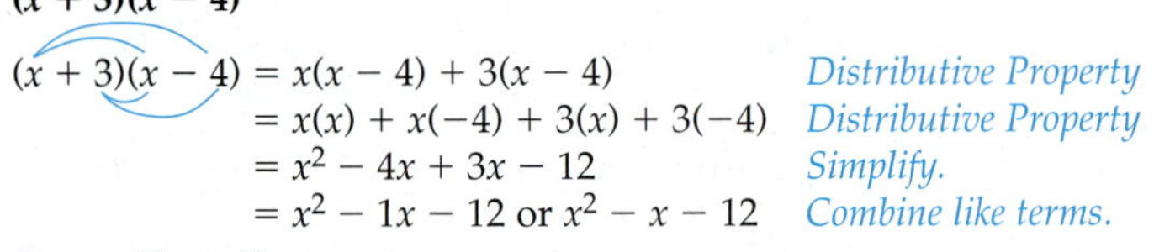

$(x + 3)(x - 4) = x(x - 4) + 3(x - 4)$ *Distributive Property*

$= x(x) + x(-4) + 3(x) + 3(-4)$ *Distributive Property*

$= x^2 - 4x + 3x - 12$ *Simplify.*

$= x^2 - 1x - 12$ or $x^2 - x - 12$ *Combine like terms.*

2 $(2y - 1)(y - 3)$

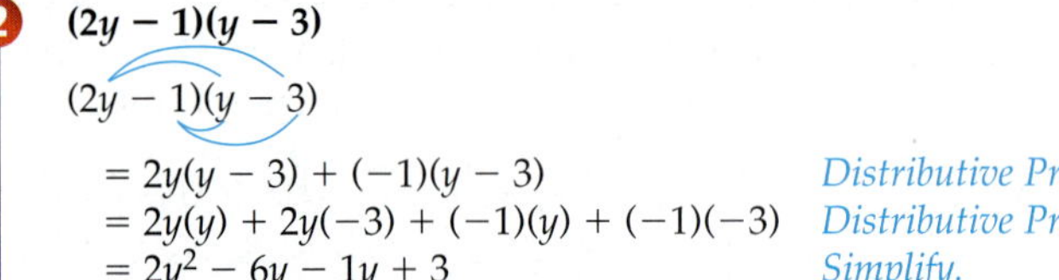

$(2y - 1)(y - 3)$

$= 2y(y - 3) + (-1)(y - 3)$ *Distributive Property*

$= 2y(y) + 2y(-3) + (-1)(y) + (-1)(-3)$ *Distributive Property*

$= 2y^2 - 6y - 1y + 3$ *Simplify.*

$= 2y^2 - 7y + 3$ *Combine like terms.*

Your Turn

a. $(y + 4)(y - 2)$ $y^2 + 2y - 8$ **b.** $(m - 3)(m + 1)$ $m^2 - 2m - 3$

c. $(x - 5)(x - 2)$ $x^2 - 7x + 10$ **d.** $(2a - 3)(a - 2)$ $2a^2 - 7a + 6$

www.algconcepts.com/extra_examples

Hands-On Algebra

Cooperative Learning Make sure that students do not confuse this model with algebra tiles, or they may mistake each square for an x^2-tile. The purpose of this model is to provide a grid to make sure that students include all terms of the product. The values inside the squares change as different binomials are used. Also, in Step 3, watch for any students who accidentally multiply 2 by x^2 and $-1x$ instead of by x and -1.

Hands-On Algebra Masters, p. 102

Two binomials can always be multiplied using the Distributive Property. However, the following shortcut can also be used. It is called the **FOIL method**.

$$(3x + 1)(x + 2) = (3x)(x) + (3x)(2) + (1)(1x) + (1)(2)$$

F	O	I	L
product of FIRST terms	*product of OUTER terms*	*product of INNER terms*	*product of LAST terms*

$= 3x^2 + 6x + 1x + 2$
$= 3x^2 + 7x + 2$

FOIL Method for Multiplying Two Binomials

To multiply two binomials, find the sum of the products of
F the First terms,
O the Outer terms,
I the Inner terms, and
L the Last terms.

Examples

Find each product.

3 $(y + 4)(y + 6)$

$(y + 4)(y + 6) = (2x)(2x) + (2x)(2) + (-3)(2x) + (-3)(2)$ (F, O, I, L)
$= y^2 + 6y + 4y + 24$ *Multiply.*
$= y^2 + 10y + 24$ *Combine like terms.*

4 $(2x - 3)(2x + 2)$

$(2x - 3)(2x + 2) = (2x)(2x) + (2x)(2) + (-3)(2x) + (-3)(2)$ (F, O, I, L)
$= 4x^2 + 4x - 6x - 6$ *Multiply.*
$= 4x^2 - 2x - 6$ *Combine like terms.*

5 $(3a - b)(2a + 4b)$

$(3a - b)(2a + 4b) = (3a)(2a) + (3a)(4b) + (-b)(2a) + (-b)(4b)$
$= 6a^2 + 12ab - 2ab - 4b^2$
$= 6a^2 + 10ab - 4b^2$

Your Turn

e. $(n + 3)(n + 5)$ **f.** $(x - 4)(2x + 3)$ **g.** $(2x + y)(3x - 2y)$

e. $n^2 + 8n + 15$
f. $2x^2 - 5x - 12$
g. $6x^2 - xy - 2y^2$

Teaching Tip As you discuss the FOIL method, point out that students are multiplying two terms by two terms—both terms of one binomial are multiplied by both terms of the other. Because $2 \times 2 = 4$, students should expect to find four products every time they multiply two binomials.

In-Class Examples

Examples 3–5

Find each product.

3 $(d + 2)(d + 8)$
$d^2 + 10d + 16$

4 $(e + 4)(2e - 4)$
$2e^2 + 4e - 16$

5 $(5x + y)(4x - 2y)$
$20x^2 - 6xy - 2y^2$

Family Activity

With a family member, find a box at home. Define the shortest dimension as x. Write expressions for the other two dimensions as the sum of x and the nearest whole number of inches or centimeters. For example, if one side is 2 inches longer than the side of length x, then its measure is $x + 2$. Use your expressions for the dimensions to find the area of each side of the box, the total surface area of the box, and the volume of the box.

Teaching Tip As you discuss Example 6, in which the products of the outer terms and inner terms are not like terms, contrast it with Examples 3–5, in which the products of the outer terms and inner terms are like terms. This may help students avoid mistakes they might make by finding a binomial product hurriedly.

In-Class Examples

Example 6

Find the product of $a - 2$ and $a^2 - 4$. $a^3 - 2a^2 - 4a + 8$

Example 7

Find the volume of a rectangular prism with base dimensions x and $x + 4$ units and height $x + 2$ units. $x^3 + 6x^2 + 8x$ cubic units

Answer

1.

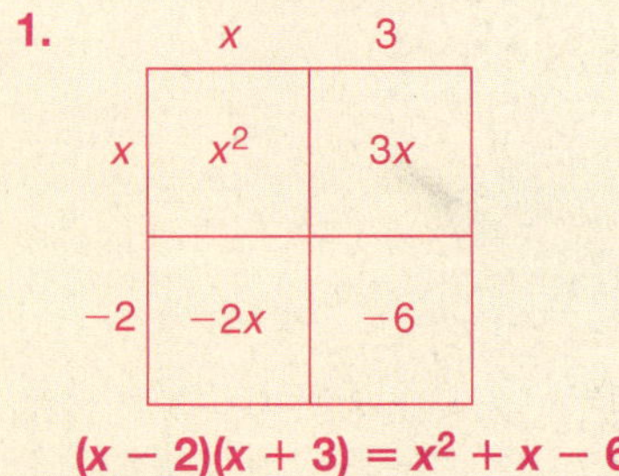

$(x - 2)(x + 3) = x^2 + x - 6$

Study Guide, p. 408

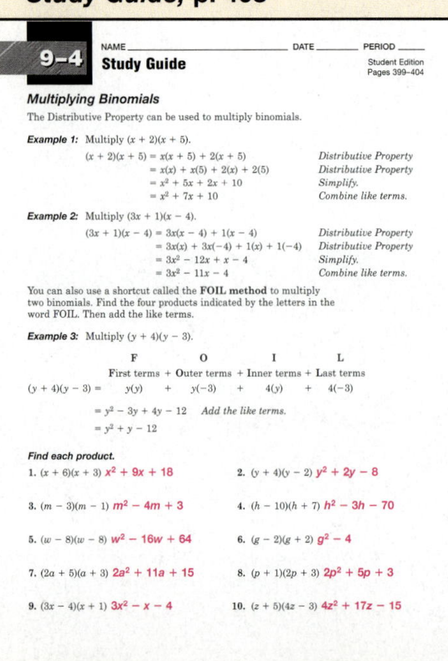

9–4 NAME ______ DATE ______ PERIOD ______

Study Guide

Student Edition Pages 399–404

Multiplying Binomials

The Distributive Property can be used to multiply binomials.

Example 1: Multiply $(x + 2)(x + 5)$.

$(x + 2)(x + 5) = x(x + 5) + 2(x + 5)$	*Distributive Property*
$= x(x) + x(5) + 2(x) + 2(5)$	*Distributive Property*
$= x^2 + 5x + 2x + 10$	*Simplify.*
$= x^2 + 7x + 10$	*Combine like terms.*

Example 2: Multiply $(3x + 1)(x - 4)$.

$(3x + 1)(x - 4) = 3x(x - 4) + 1(x - 4)$	*Distributive Property*
$= 3x(x) + 3x(-4) + 1(x) + 1(-4)$	*Distributive Property*
$= 3x^2 - 12x + x - 4$	*Simplify.*
$= 3x^2 - 11x - 4$	*Combine like terms.*

You can also use a shortcut called the **FOIL method** to multiply two binomials. Find the four products indicated by the letters in the word FOIL. Then add the like terms.

Example 3: Multiply $(y + 4)(y - 3)$.

	F		O		I		L
	First terms	+	Outer terms	+	Inner terms	+	Last terms
$(y + 4)(y - 3) =$	$y(y)$	+	$y(-3)$	+	$4(y)$	+	$4(-3)$

$= y^2 - 3y + 4y - 12$ *Add the like terms.*

$= y^2 + y - 12$

Find each product.

1. $(x + 6)(x + 3)$ $x^2 + 9x + 18$
2. $(y + 4)(y - 2)$ $y^2 + 2y - 8$
3. $(m - 3)(m - 1)$ $m^2 - 4m + 3$
4. $(h - 10)(h + 7)$ $h^2 - 3h - 70$
5. $(w - 8)(w - 8)$ $w^2 - 16w + 64$
6. $(g - 2)(g + 2)$ $g^2 - 4$
7. $(2a + 5)(a + 3)$ $2a^2 + 11a + 15$
8. $(p + 1)(2p + 3)$ $2p^2 + 5p + 3$
9. $(3x - 4)(x + 1)$ $3x^2 - x - 4$
10. $(z + 5)(4z - 3)$ $4z^2 + 17z - 15$

Sometimes it is not possible to simplify the product of two binomials.

Example 6 **Find the product of $x^2 - 4$ and $x + 3$.**

$(x^2 - 4)(x + 3) = (x^2)(x) + (x^2)(3) + (-4)(x) + (-4)(3)$

$= x^3 + 3x^2 - 4x - 12$ *There are no like terms.*

Your Turn

h. Find the product of $x - 2$ and $x^2 + 3$. $x^3 - 2x^2 + 3x - 6$

You can use the FOIL method to solve problems involving volume.

Example 7 **Geometry Link**

The volume V of a rectangular prism is equal to the area of the base B times the height h. Express the volume of the prism as a polynomial. Use $V = Bh$.

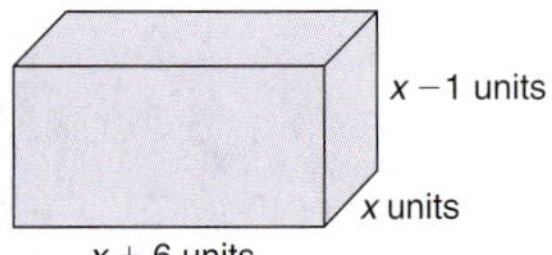

First, find the area of the base. The base is a rectangle.

$B = \ell w$

$= (x + 6)(x)$ *Replace ℓ with $x + 6$ and w with x.*

$= x^2 + 6x$ *Use the Distributive Property.*

To find the volume, multiply the area of the base by the height.

$V = Bh$

$= (x^2 + 6x)(x - 1)$ *Replace B with $x^2 + 6x$ and h with $x - 1$.*

$= (x^2)(x) + (x^2)(-1) + (6x)(x) + (6x)(-1)$ *Use the FOIL method.*

$= x^3 - x^2 + 6x^2 - 6x$ *Multiply.*

$= x^3 + 5x^2 - 6x$ *Combine like terms.*

The volume of the prism is $x^3 + 5x^2 - 6x$ cubic units.

Check for Understanding

Communicating Mathematics

Vocabulary
FOIL method

1. **Draw** a model that represents the product of $x - 2$ and $x + 3$. Then find the product.
2. **Write** the multiplication expression represented by each model.

a.

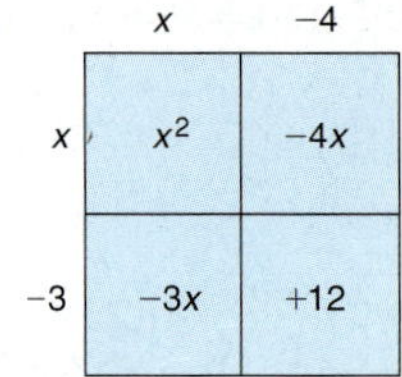

b.

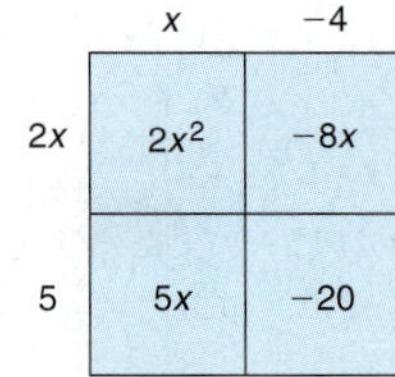

1. See margin.
2a. $(x - 3)(x - 4) = x^2 - 7x + 12$
2b. $(2x + 5)(x - 4) = 2x^2 - 3x - 20$

Reteaching Activity

Kinesthetic Learners Have students stand together in pairs. Then have two pairs stand facing each other. Direct the members of each pair to shake hands with each member of the other pair. Ask students how many handshakes there are in all. Point out that these four handshakes correspond to the four terms in a binomial product found by the FOIL method. The handshakes straight across represent the first and last product terms, and the handshakes diagonally across represent the outer and inner product terms.

3. Writing Math Compare and contrast the procedure for multiplying two binomials and the procedure for multiplying a binomial and a monomial. **See margin.**

Guided Practice

Getting Ready **Find the sum of the products of the inner terms and the outer terms.**

Sample: $(x - 6)(x + 2)$ **Solution:** $(x - 6)(x + 2)$

I, O

$-6x + 2x = -4x$

4. $(a - 5)(a - 3)$ $-8a$ 5. $(x + 1)(x + 5)$ $6x$ 6. $(g + 1)(g - 9)$ $-8g$
7. $(2m + 3)(3m + 2)$ $13m$ 8. $(2j + 1)(j - 3)$ $-5j$ 9. $(y + 5)(2y + 4)$ $14y$

Examples 1–5

Find each product. Use the Distributive Property or the FOIL method.

10. $(x + 2)(x + 4)$ $x^2 + 6x + 8$ 11. $(w - 7)(w + 5)$ $w^2 - 2w - 35$
12. $(a - 3)(a - 6)$ $a^2 - 9a + 18$ 13. $(2y + 3)(y - 1)$ $2y^2 + y - 3$
14. $(3x - 4)(2x - 3)$ 15. $(2y + 7)(3y - 5)$ $6y^2 + 11y - 35$
16. $(x + 2y)(x + 3y)$ $x^2 + 5xy + 6y^2$ 17. $(2a - 5b)(a + 2b)$
18. $(y^2 + 3)(y - 2)$ $y^3 - 2y^2 + 3y - 6$ 19. $(m^3 - 2m)(m + 3)$ $m^4 + 3m^3 - 2m^2 - 6m$

14. $6x^2 - 17x + 12$
17. $2a^2 - ab - 10b^2$

Example 6

20. **Geometry** Express the volume of the rectangular prism as a polynomial. Use the formula $V = Bh$, where B is the area of the base and h is the height of the prism. $2x^3 + 2x^2 - 24x$ in³

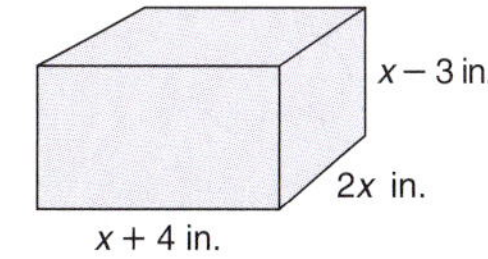

Exercises

21–50. See margin.

A **Find each product. Use the Distributive Property or the FOIL method.**

21. $(x + 4)(x + 8)$ 22. $(r + 7)(r + 2)$ 23. $(a - 3)(a + 7)$
24. $(x - 3)(x - 7)$ 25. $(n - 11)(n - 5)$ 26. $(y - 4)(y + 15)$
27. $(z + 6)(z - 4)$ 28. $(s - 11)(s + 5)$ 29. $(x - 4)(3x + 2)$
30. $(2a + 5)(a - 7)$ 31. $(z - 4)(3z - 5)$ 32. $(y - 3)(2y - 5)$
33. $(5n + 2)(n - 3)$ 34. $(x + 7)(2x - 3)$ 35. $(3a + 1)(3a + 1)$
36. $(2x + 5)(5x + 3)$ 37. $(4h - 3)(3h + 2)$ 38. $(7a - 1)(2a - 3)$
39. $(2x + 7)(x - 3)$

B

40. $(8m + 2n)(6m + 5n)$ 41. $(2y - 4z)(3y - 6z)$
42. $(3a - b)(2a + b)$ 43. $(2x + 5y)(3x - y)$ 44. $(5n - 2p)(5n + 2p)$
45. $(x^2 + 1)(x - 2)$ 46. $(y + 3)(y^2 - 4)$ 47. $(x^2 + 3)(3x^2 - 1)$
48. $(2y^2 + 1)(y + 1)$ 49. $(x^2 + 2x)(x - 3)$ 50. $(x + a)(x + b)$

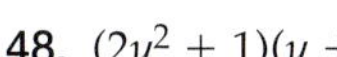

Homework Help

For Exercises	See Examples
21–39	1, 2, 3
40–52	4, 5, 6
53–56	7

Extra Practice
See page 711.

C

51. Find the product of $(x + 3)$ and $(x - 3)$. $x^2 - 9$
52. What is the product of x, $(2x + 1)$, and $(x - 3)$? $2x^3 - 5x^2 - 3x$

3 PRACTICE/APPLY

Error Analysis

Watch for students who make errors of sign when finding the terms of the binomial products in Exercises 10–19.

Prevent by reminding students that subtraction means adding the opposite, so multiplying by any binomial term that is subtracted is the same as multiplying by its opposite.

Assignment Guide

Basic: 21–55 odd, 56–66
Average: 22–52 even, 53–66

Answers

3. The Distributive Property is used to multiply both two binomials and a binomial and a monomial. But when you multiply two binomials, there are four multiplications to perform. With a binomial and a monomial there are only two multiplications to perform.

21. $x^2 + 12x + 32$
22. $r^2 + 9r + 14$
23. $a^2 + 4a - 21$
24. $x^2 - 10x + 21$
25. $n^2 - 16n + 55$
26. $y^2 + 11y - 60$
27. $z^2 + 2z - 24$

Skills Practice, **p. 409, and** ***Practice,*** **p. 410 (shown)**

9-4 NAME ______ DATE ______ PERIOD ______

Practice

Student Edition Pages 399–404

Multiplying Binomials

Find each product. Use the Distributive Property or the FOIL method.

1. $(y + 4)(y + 3)$ $y^2 + 7y + 12$ 2. $(x + 2)(x + 1)$ $x^2 + 3x + 2$ 3. $(b + 5)(b - 2)$ $b^2 + 3b - 10$
4. $(a - 6)(a - 4)$ $a^2 - 10a + 24$ 5. $(z - 5)(z + 3)$ $z^2 - 2z - 15$ 6. $(n - 1)(n - 8)$ $n^2 - 9n + 8$
7. $(x + 7)(x - 4)$ $x^2 + 3x - 28$ 8. $(y - 3)(y + 9)$ $y^2 + 6y - 27$ 9. $(b + 2)(b + 3)$ $b^2 + 5b + 6$
10. $(2c + 5)(c - 4)$ $2c^2 - 3c - 20$ 11. $(4x - 7)(x + 3)$ $4x^2 + 5x - 21$ 12. $(x - 1)(5x - 4)$ $5x^2 - 9x + 4$
13. $(3y + 1)(3y + 2)$ $9y^2 + 9y + 2$ 14. $(2n + 4)(5n - 3)$ $10n^2 + 14n - 12$ 15. $(7h - 3)(4h - 1)$ $28h^2 - 19h + 3$
16. $(2m - 6)(3m + 2)$ $6m^2 - 14m - 12$ 17. $(6a + 2)(2a + 3)$ $12a^2 + 22a + 6$ 18. $(4c + 5)(2c - 2)$ $8c^2 + 2c - 10$
19. $(x + y)(2x + y)$ $2x^2 + 3xy + y^2$ 20. $(3a + 4b)(a - 3b)$ $3a^2 - 5ab - 12b^2$ 21. $(3m - 3n)(3m - 2n)$ $9m^2 - 15mn + 6n^2$
22. $(7p - 4q)(2p + 3q)$ $14p^2 + 13pq - 12q^2$ 23. $(2r + 2s)(2r + 3s)$ $4r^2 + 10rs + 6s^2$ 24. $(3y - 5z)(3y + 3z)$ $9y^2 - 6yz - 15z^2$
25. $(x^2 + 1)(x - 3)$ $x^3 - 3x^2 + x - 3$ 26. $(y - 4)(y^2 + 2)$ $y^3 - 4y^2 + 2y - 8$ 27. $(2c^2 - 5)(c - 4)$ $2c^3 - 8c^2 - 5c + 20$
28. $(a^3 - 3a)(a + 4)$ $a^4 + 4a^3 - 3a^2 - 12a$ 29. $(b^2 + 2)(b^2 + 3)$ $b^4 + 5b^2 + 6$ 30. $(x^3 - 3)(4x + 1)$ $4x^4 + x^3 - 12x - 3$

Answers

28. $s^2 - 6s - 55$
29. $3x^2 - 10x - 8$
30. $2a^2 - 9a - 35$
31. $3z^2 - 17z + 20$
32. $2y^2 - 11y + 15$
33. $5n^2 - 13n - 6$
34. $2x^2 + 11x - 21$
35. $9a^2 + 6a + 1$
36. $10x^2 + 31x + 15$
37. $12h^2 - h - 6$
38. $14a^2 - 23a + 3$
39. $2x^2 + x - 21$
40. $48m^2 + 52mn + 10n^2$
41. $6y^2 - 24yz + 24z^2$
42. $6a^2 + ab - b^2$
43. $6x^2 + 13xy - 5y^2$
44. $25n^2 - 4p^2$
45. $x^3 - 2x^2 + x - 2$
46. $y^3 + 3y^2 - 4y - 12$
47. $3x^4 + 8x^2 - 3$
48. $2y^3 + 2y^2 + y + 1$
49. $x^3 - x^2 - 6x$
50. $x^2 + ax + bx + ab$

4 ASSESS

Open-Ended Assessment

Speaking Ask students to explain how to use the FOIL method to multiply two binomials.

Answers

62.

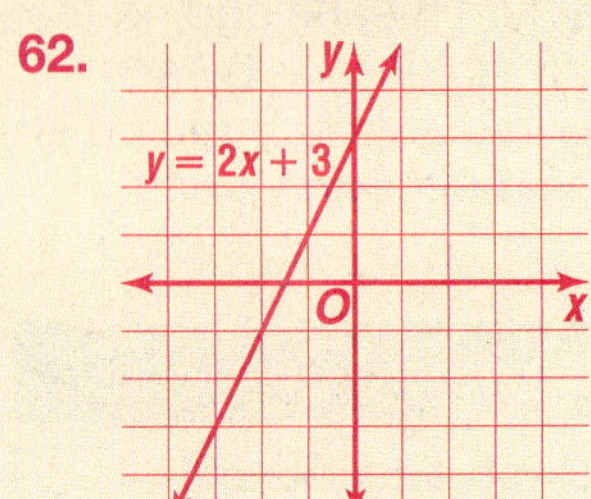

63.

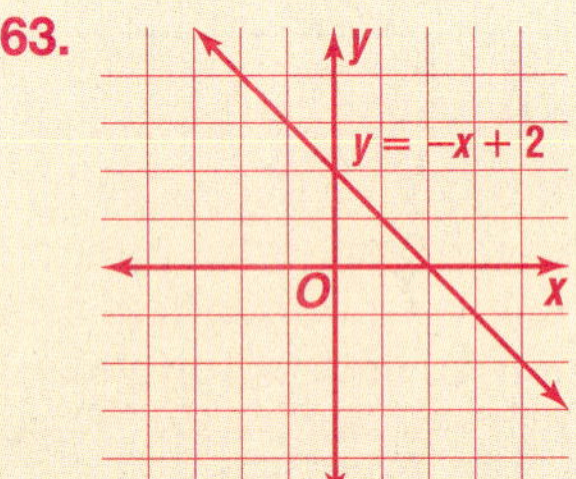

64.

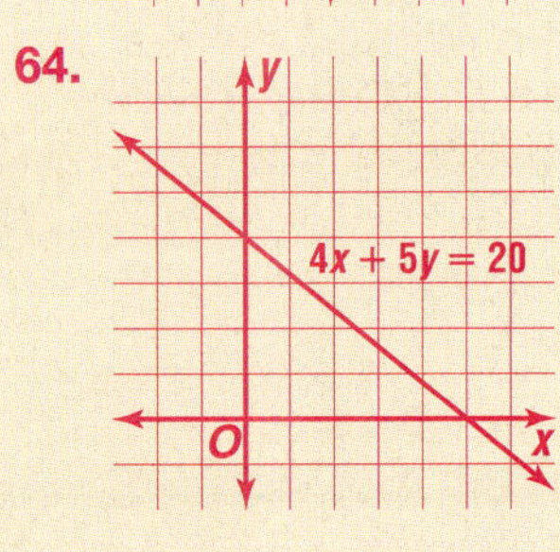

x

Enrichment, p. 412

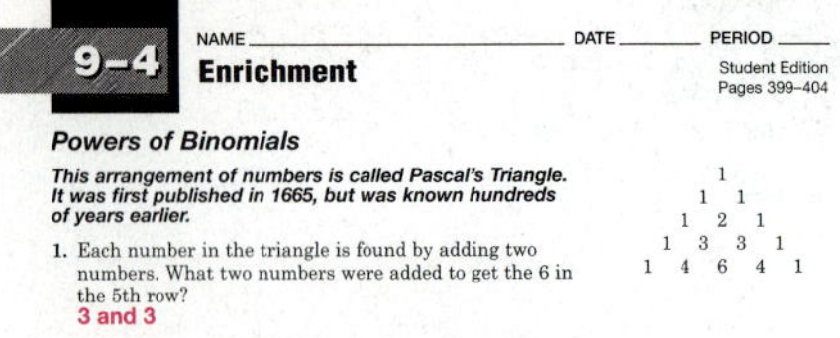

9–4 NAME ______ DATE ______ PERIOD ______

Enrichment

Student Edition Pages 399–404

Powers of Binomials

This arrangement of numbers is called Pascal's Triangle. It was first published in 1665, but was known hundreds of years earlier.

1
1 1
1 2 1
1 3 3 1
1 4 6 4 1

1. Each number in the triangle is found by adding two numbers. What two numbers were added to get the 6 in the 5th row?
3 and 3

2. Describe how to create the 6th row of Pascal's Triangle.
The first and last numbers are 1. Evaluate 1 + 4, 4 + 6, 6 + 4, and 4 + 1 to find the other numbers.

3. Write the numbers for rows 6 through 10 of the triangle.

Row 6: 1 5 10 10 5 1
Row 7: 1 6 15 20 15 6 1
Row 8: 1 7 21 35 35 21 7 1
Row 9: 1 8 28 56 70 56 28 8 1
Row 10: 1 9 37 84 126 126 84 37 9 1

Multiply to find the expanded form of each product.

4. $(a + b)^2$ $a^2 + 2ab + b^2$

5. $(a + b)^3$ $a^3 + 3a^2b + 3ab^2 + b^3$

6. $(a + b)^4$ $a^4 + 4a^3b + 6a^2b^2 + 4ab^3 + b^4$

Now compare the coefficients of the three products in Exercises 4–6 with Pascal's Triangle.

7. Describe the relationship between the expanded form of $(a + b)^n$ and Pascal's Triangle. The coefficients of the expanded form are found in row $n + 1$ of Pascal's Triangle.

8. Use Pascal's Triangle to write the expanded form of $(a + b)^6$.
$a^6 + 6a^5b + 15a^4b^2 + 20a^3b^3 + 15a^2b^4 + 6ab^5 + b^6$

Applications and Problem Solving

53. **Geometry** Find the area of each shaded region in simplest form.

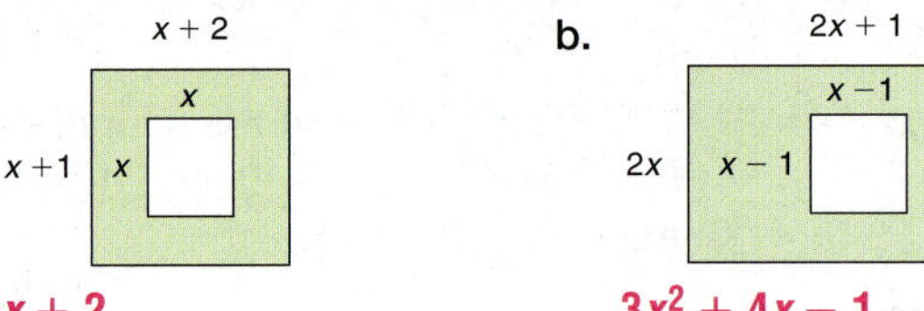

a. $3x + 2$ b. $3x^2 + 4x - 1$

54. **Number Theory** Use the FOIL method to find the product of 16 and 38. (*Hint*: Write 16 as 10 + 6 and 38 as 30 + 8.)
$(10 + 6)(30 + 8) = 300 + 80 + 180 + 48$ or 608

55. **Packaging** A cereal box has a length of $2x$ inches, a width of $x - 2$ inches, and a height of $2x + 5$ inches. **a.** $4x^3 + 2x^2 - 20x$ in.
  a. Express the volume of the package as a polynomial.
  b. Find the volume of the box if $x = 5$. 450 in^3

56. **Critical Thinking** Use the Distributive Property to find the product of $x + 2$ and $x^2 - 3x + 2$. See Solutions Manual.

Mixed Review

57. **Clocks** Before mechanical clocks were invented, candles were used to keep track of time. One formula that was used was $c = 2(5 - t)$, where c is the height of the candle in inches and t is the time in hours that the candle burns. *(Lesson 9–3)*
  a. Use the Distributive Property to multiply 2 and $5 - t$. $10 - 2t$
  b. Find the height of the candle when $t = 2$. 6 in.

58. Find the sum of $x^2 + 5x$ and $-3x - 7$. *(Lesson 9–2)* $x^2 + 2x - 7$

Write each expression using exponents. *(Lesson 8–1)*

59. $2 \cdot 2 \cdot 2 \cdot x \cdot x \cdot x$ 2^3x^3
60. 10 10^1
61. $(3)(3)(-2)(-2)(-2)$ $3^2(-2)^3$

Graph each equation. *(Lesson 7–5)* 62–64. See margin.

62. $y = 2x + 3$
63. $y = -x + 2$
64. $4x + 5y = 20$

interNET CONNECTION

Data Update For the latest information on breakfast foods, visit: www.algconcepts.com

65. **Food** The graph shows the favorite breakfast food for students ages 9–12. Suppose you interview 200 students who are in this age range. How many would you expect to like cereal for breakfast? *(Lesson 5–4)*
84 students

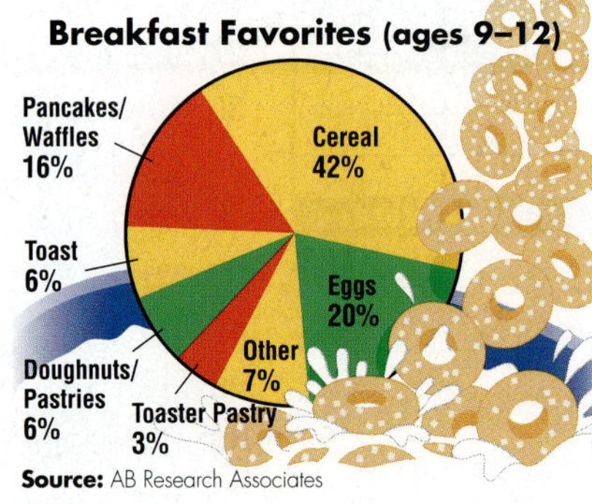

Standardized Test Practice

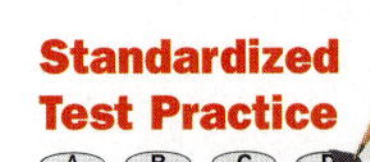

66. **Extended Response** Draw and label a rectangle whose perimeter is 20 centimeters. *(Lesson 1–5)* See margin.

Extra Credit

When the binomial $x + c$ is multiplied by itself, the product is $x^2 + 18x + 81$. What is the value of c? 9

Answer

66.

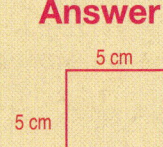

5 cm

5 cm

9–5 Special Products

What You'll Learn
You'll learn to develop and use the patterns for $(a + b)^2$, $(a - b)^2$, and $(a + b)(a - b)$.

Why It's Important
Biology Geneticists use a technique similar to finding $(a + b)^2$ to predict the characteristics of a population. *See Example 5.*

Some products of polynomials appear frequently in real-life problems. Expressions like $(a + b)^2$, $(a - b)^2$, and $(a + b)(a - b)$ occur so often that it is helpful to develop patterns for their products.

To develop these patterns, we will model a simpler expression, $(x + 1)^2$, geometrically.

The area of a square with a side length of $x + 1$ is $(x + 1)(x + 1)$ or $(x + 1)^2$. The total area can also be found by adding the areas of the inner regions together. The right side of the equation below represents the area of the square as the sum of the areas of the four small squares.

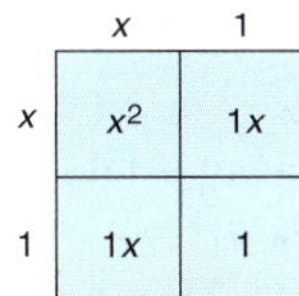

$$(x + 1)^2 = x^2 + 1x + 1x + 1$$
$$= x^2 + 2x + 1$$

twice the product of 1 and x

You can use a similar model to find $(x - 1)^2$. $x - 1 = x + (-1)$

	x	-1
x	x^2	$-1x$
-1	$-1x$	1

$$(x - 1)^2 = x^2 - 1x - 1x + 1$$
$$= x^2 - 2x + 1$$

twice the product of −1 and x

The square of a sum and the square of a difference can be found by using the following rules.

Square of a Sum and Square of a Difference

Symbols: $(a + b)^2 = a^2 + 2ab + b^2$
$(a - b)^2 = a^2 - 2ab + b^2$

Models:

	a	b
a	a^2	ab
b	ab	b^2

	a	$-b$
a	a^2	$-ab$
$-b$	$-ab$	b^2

Resource Manager

Reproducible Masters
Chapter 9 Resource Masters
- *Study Guide,* p. 413
- *Skills Practice,* p. 414
- *Practice,* p. 415
- *Reading to Learn Mathematics,* p. 416
- *Enrichment,* p. 417
- *Assessment,* p. 429

Hands-On Algebra, p. 103

Transparencies
5-Minute Check, 9–5

Technology/Multimedia
AlgePASS, Lesson 22
Interactive Chalkboard CD-ROM

Lesson 9–5

1 FOCUS

5-Minute Check
Lesson 9–4

Find each product. Use the Distributive Property or the FOIL method.

1. $(b - 3)(2b - 1)$ **$2b^2 - 7b + 3$**
2. $(3a + 2)(a - 1)$ **$3a^2 - a - 2$**
3. $(2m + 1)(3m - 1)$ **$6m^2 + m - 1$**
4. What is the product of x^2, $(3x + 1)$, and $(x - 4)$? **$3x^4 - 11x^3 - 4x^2$**
5. Express the volume of the rectangular prism as a polynomial.

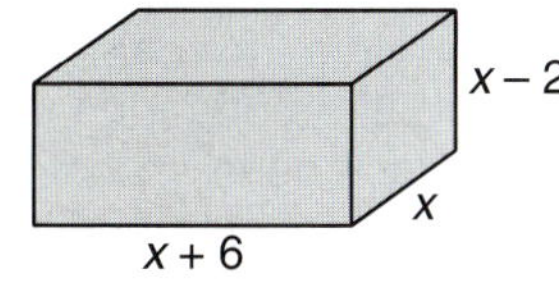

$x^3 + 4x^2 - 12x$ cubic units

Motivating the Lesson

Real-World Connection Have students investigate whether $(a + b)^2 = a^2 + b^2$ by substituting whole numbers for a and b. For example, if $a = 1$ and $b = 2$, $(a + b)^2 = (1 + 2)^2 = 3^2$ or 9 and $a^2 + b^2 = 1^2 + 2^2 = 1 + 4$ or 5. Thus, $(a + b)^2 \neq a^2 + b^2$.

2 TEACH

Teaching Tip As you discuss the Square of a Sum and Square of a Difference formulas, have students describe the products out loud. For example, the square of a sum is each term squared plus twice the product of the terms.

Teaching Tip As you discuss Example 3, emphasize that the square of a difference contains a minus sign and a plus sign, respectively. Some students may expect both signs to be minus signs.

In-Class Examples

Examples 1–4

Find each product.

1 $(b + 5)^2$ $b^2 + 10b + 25$
2 $(c - 3)^2$ $c^2 - 6c + 9$
3 $(2d + 1)^2$ $4d^2 + 4d + 1$
4 $(3e - 3f)^2$ $9e^2 - 18ef + 9f^2$

Teaching Tip As you work Example 5, review how to find the probability of independent events from Lesson 5–7.

In-Class Example

Example 5

In a certain population, a parent has a 20% chance of passing on a certain gene to its offspring. If an offspring receives one gene each from its mother and father, what is the probability that an offspring will receive at least one of these genes? **0.36 or 36%**

Examples

Find each product.

 $(x + 4)^2$

$(a + b)^2 = a^2 + 2ab + b^2$ *Square of a Sum*
$(x + 4)^2 = x^2 + 2(x)(4) + 4^2$ *Replace a with x and b with 4.*
$= x^2 + 8x + 16$ *Simplify.*

 $(4m + n)^2$

$(a + b)^2 = a^2 + 2ab + b^2$ *Square of a Sum*
$(4m + n)^2 = (4m)^2 + 2(4m)(n) + n^2$ *Replace a with 4m and b with n.*
$= 16m^2 + 8mn + n^2$ *Simplify.*

 $(w - 5)^2$

$(a - b)^2 = a^2 - 2ab + b^2$ *Square of a Difference*
$(w - 5)^2 = w^2 - 2(w)(5) + 5^2$ *Replace a with w and b with 5.*
$= w^2 - 10w + 25$ *Simplify.*

 $(3p - 2q)^2$

$(a - b)^2 = a^2 - 2ab + b^2$ *Square of a Difference*
$(3p - 2q)^2 = (3p)^2 - 2(3p)(2q) + (2q)^2$ *Replace a with 3p and b with 2q.*
$= 9p^2 - 12pq + 4q^2$ *Simplify.*

Your Turn

a. $(y + 3)^2$ $y^2 + 6y + 9$
b. $(3g + h)^2$ $9g^2 + 6gh + h^2$
c. $(d - 2)^2$ $d^2 - 4d + 4$
d. $(4x - 3y)^2$ $16x^2 - 24xy + 9y^2$

Biologists use a method that is similar to squaring a sum to find the characteristics of offspring based on genetic information.

Example
Biology Link

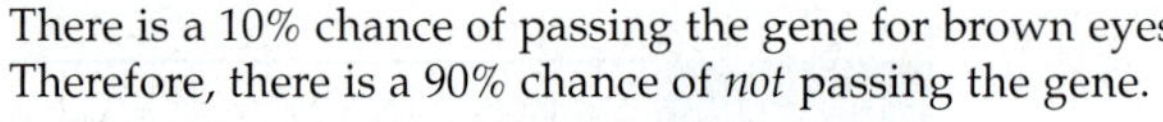

5 **In a certain population, a parent has a 10% chance of passing the gene for brown eyes to its offspring. If an offspring receives one eye-color gene from its mother and one from its father, what is the probability that an offspring will receive at least one gene for brown eyes?**

There is a 10% chance of passing the gene for brown eyes. Therefore, there is a 90% chance of *not* passing the gene.

Look Back
Probability of Independent Events: Lesson 5–7

Use the model at the right to show all possible combinations. In the model, B represents the gene for brown eyes and b represents the gene for *not* brown eyes. *Note that the percents are written as decimals.*

		Father	
		B = 0.1	b = 0.9
Mother	B = 0.1	BB (0.1)(0.1) = 0.01	Bb (0.1)(0.9) = 0.09
	b = 0.9	bB (0.9)(0.1) = 0.09	bb (0.9)(0.9) = 0.81

 www.algconcepts.com/extra_examples

Three of the four small squares in the model contain a B. Add their probabilities.

$0.01 + 0.09 + 0.09 = 0.19$

So, the probability of an offspring receiving at least one gene for brown eyes is 0.19 or 19%.

You can use the FOIL method to find product of the sum and difference of the same two terms. Consider $(x + 3)(x - 3)$.

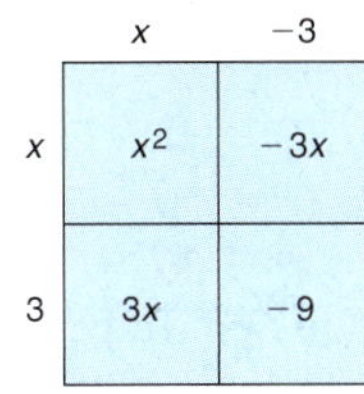

$$(x + 3)(x - 3) = x^2 - 3x + 3x - 9$$
$$= x^2 - 9$$

square of the second term (9)
square of the first term (x^2)

Note that the product is the difference of the squares of the terms. The product of a sum and a difference can be found by using the following rule.

Product of a Sum and a Difference

Symbols: $(a + b)(a - b) = a^2 - b^2$

Model:

	a	$-b$
a	a^2	$-ab$
b	ab	$-b^2$

Examples

Find each product.

6 $(y + 2)(y - 2)$

$(a + b)(a - b) = a^2 - b^2$ *Product of a Sum and a Difference*

$(y + 2)(y - 2) = y^2 - (2)^2$ *Replace a with y and b with 2.*

$= y^2 - 4$ *Simplify.*

7 $(2r + s)(2r - s)$

$(a + b)(a - b) = a^2 - b^2$ *Product of a Sum and a Difference*

$(2r + s)(2r - s) = (2r)^2 - s^2$ *Replace a with 2r and b with s.*

$= 4r^2 - s^2$ *Simplify.*

Your Turn **f. $25m^2 - 36n^2$**

e. $(x + y)(x - y)$ **$x^2 - y^2$**

f. $(5m - 6n)(5m + 6n)$

In-Class Examples

Examples 6–7

Find each product.

6 $(3 + a)(3 - a)$ **$9 - a^2$**

7 $(5b - 2)(5b + 2)$ **$25b^2 - 4$**

***Study Guide*, p. 413**

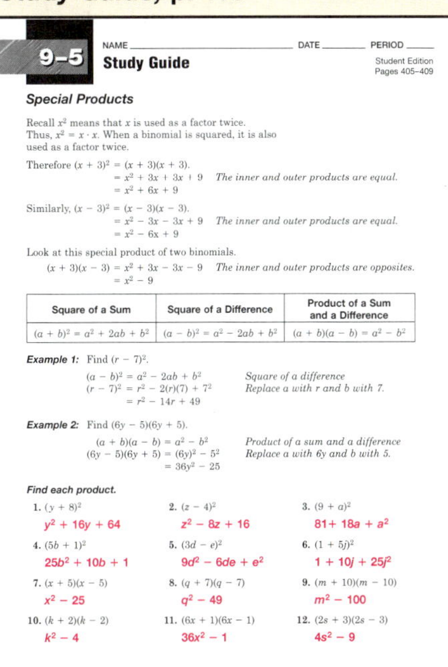

9–5 NAME ______ DATE ______ PERIOD ______

Study Guide

Student Edition Pages 405–409

Special Products

Recall x^2 means that x is used as a factor twice. Thus, $x^2 = x \cdot x$. When a binomial is squared, it is also used as a factor twice.

Therefore $(x + 3)^2 = (x + 3)(x + 3)$.
$= x^2 + 3x + 3x + 9$ *The inner and outer products are equal.*
$= x^2 + 6x + 9$

Similarly, $(x - 3)^2 = (x - 3)(x - 3)$.
$= x^2 - 3x - 3x + 9$ *The inner and outer products are equal.*
$= x^2 - 6x + 9$

Look at this special product of two binomials.
$(x + 3)(x - 3) = x^2 + 3x - 3x - 9$ *The inner and outer products are opposites.*
$= x^2 - 9$

Square of a Sum	Square of a Difference	Product of a Sum and a Difference
$(a + b)^2 = a^2 + 2ab + b^2$	$(a - b)^2 = a^2 - 2ab + b^2$	$(a + b)(a - b) = a^2 - b^2$

Example 1: Find $(r - 7)^2$.
$(a - b)^2 = a^2 - 2ab + b^2$ *Square of a difference*
$(r - 7)^2 = r^2 - 2(r)(7) + 7^2$ *Replace a with r and b with 7.*
$= r^2 - 14r + 49$

Example 2: Find $(6y - 5)(6y + 5)$.
$(a + b)(a - b) = a^2 - b^2$ *Product of a sum and a difference*
$(6y - 5)(6y + 5) = (6y)^2 - 5^2$ *Replace a with 6y and b with 5.*
$= 36y^2 - 25$

Find each product.

1. $(y + 8)^2$ $y^2 + 16y + 64$
2. $(z - 4)^2$ $z^2 - 8z + 16$
3. $(9 + a)^2$ $81 + 18a + a^2$
4. $(5b + 1)^2$ $25b^2 + 10b + 1$
5. $(3d - e)^2$ $9d^2 - 6de + e^2$
6. $(1 + 5j)^2$ $1 + 10j + 25j^2$
7. $(x + 5)(x - 5)$ $x^2 - 25$
8. $(q + 7)(q - 7)$ $q^2 - 49$
9. $(m + 10)(m - 10)$ $m^2 - 100$
10. $(k + 2)(k - 2)$ $k^2 - 4$
11. $(6x + 1)(6x - 1)$ $36x^2 - 1$
12. $(2s + 3)(2s - 3)$ $4s^2 - 9$

3 PRACTICE/APPLY

Error Analysis

Watch for students who have three terms in their products in Exercises 7 and 8.

Prevent by having students review the formula for the Product of a Sum and a Difference. Remind them that the products of the outer terms and inner terms cancel each other, so the final product has only two terms, both of which are squares.

Assignment Guide

Basic: 11–35 odd, 36–43
Average: 10–32 even, 33–43
All: Quiz 2, 1–5

Answers

10. $r^2 + 4r + 4$
11. $x^2 + 10x + 25$
12. $a^2 + 4ab + 4b^2$
13. $w^2 - 16w + 64$
14. $4x^2 - 4xy + y^2$
15. $m^2 - 6mn + 9n^2$
16. $p^2 - 4$
17. $x^2 - 16y^2$
18. $4x^2 - 9$
19. $25 + 10k + k^2$
20. $9x^2 - 12xy + 4y^2$
21. $y^2 - 9z^2$
22. $36 - 24m + 4m^2$

***Skills Practice,* p. 414, and *Practice,* p. 415 (shown)**

9-5 NAME ______ DATE ______ PERIOD ______

Practice — Student Edition Pages 405–409

Special Products

Find each product.

1. $(y + 4)^2$ $y^2 + 8y + 16$	2. $(x + 3)^2$ $x^2 + 6x + 9$	3. $(m + 6)^2$ $m^2 + 12m + 36$
4. $(2b + c)^2$ $4b^2 + 4bc + c^2$	5. $(x + 3y)^2$ $x^2 + 6xy + 9y^2$	6. $(4r + s)^2$ $16r^2 + 8rs + s^2$
7. $(2m + 2n)^2$ $4m^2 + 8mn + 4n^2$	8. $(4a + 2b)^2$ $16a^2 + 16ab + 4b^2$	9. $(3g + 3h)^2$ $9g^2 + 18gh + 9h^2$
10. $(b - 3)^2$ $b^2 - 6b + 9$	11. $(p - 4)^2$ $p^2 - 8p + 16$	12. $(s - 5)^2$ $s^2 - 10s + 25$
13. $(3x - 3)^2$ $9x^2 - 18x + 9$	14. $(2y - 3)^2$ $4y^2 - 12y + 9$	15. $(c - 6d)^2$ $c^2 - 12cd + 36d^2$
16. $(m - 2n)^2$ $m^2 - 4mn + 4n^2$	17. $(5x - y)^2$ $25x^2 - 10xy + y^2$	18. $(a - 4b)^2$ $a^2 - 8ab + 16b^2$
19. $(3p - 5q)^2$ $9p^2 - 30pq + 25q^2$	20. $(2j - 4k)^2$ $4j^2 - 16jk + 16k^2$	21. $(2r - 2s)^2$ $4r^2 - 8rs + 4s^2$
22. $(y + 3)(y - 3)$ $y^2 - 9$	23. $(x + 6)(x - 6)$ $x^2 - 36$	24. $(a + 9)(a - 9)$ $a^2 - 81$
25. $(3a + b)(3a - b)$ $9a^2 - b^2$	26. $(4r + s)(4r - s)$ $16r^2 - s^2$	27. $(2y + 6)(2y - 6)$ $4y^2 - 36$
28. $(5x + 4)(5x - 4)$ $25x^2 - 16$	29. $(2c + 4d)(2c - 4d)$ $4c^2 - 16d^2$	30. $(3m + 6n)(3m - 6n)$ $9m^2 - 36n^2$

Check for Understanding

Communicating Mathematics

1. **Match** each description with an expression.
   a. square of a sum **ii** — i. $(r + t)(r - t)$
   b. square of a difference **iii** — ii. $(z + 1)^2$
   c. product of a sum and difference **i** — iii. $(6 - a)^2$

2. 

Jessica says that the product of two binomials is always a trinomial. Hector disagrees. Who is correct? Explain your reasoning. **Hector; the product of a sum and a difference is a binomial.**

Guided Practice
Examples 1–4, 6, & 7
4. $4a^2 + 16ab + 16b^2$
Example 5
5. $m^2 - 18m + 81$

Find each product.

3. $(y + 2)^2$ $y^2 + 4y + 4$
4. $(2a + 4b)^2$
5. $(m - 9)^2$
6. $(j - 7k)^2$ $j^2 - 14jk + 49k^2$
7. $(x + 7)(x - 7)$ $x^2 - 49$
8. $(5r + 9s)(5r - 9s)$ $25r^2 - 81s^2$
9. **Biology** In a certain population, a parent has a 5% chance of passing the gene for cystic fibrosis to its offspring. Use a model to find the probability that an offspring will *not* receive the gene for cystic fibrosis from either of its parents. **90.25%**

Exercises

Practice

Find each product. 10–30. See margin.

10. $(r + 2)^2$	11. $(x + 5)^2$	12. $(a + 2b)^2$
13. $(w - 8)^2$	14. $(2x - y)^2$	15. $(m - 3n)^2$
16. $(p + 2)(p - 2)$	17. $(x + 4y)(x - 4y)$	18. $(2x - 3)(2x + 3)$
19. $(5 + k)^2$	20. $(3x - 2y)^2$	21. $(y + 3z)(y - 3z)$
22. $(6 - 2m)^2$	23. $(3x + 5)(3x - 5)$	24. $(5 + 2p)(5 + 2p)$
25. $(4 + 2x)^2$		26. $(5a - 2b)(5a + 2b)$
27. $(a - 3b)(a - 3b)$		28. $y(y + 1)(y - 1)$
29. $x(x + 1)^2$		30. $(x + 2)(x - 2)(x - 3)$

Homework Help

For Exercises	See Examples
10–12, 19, 24, 25, 29, 32, 34	1, 2
13–15, 20, 22, 27	3, 4
16–18, 21, 23, 26, 28, 30, 31, 33, 35	6, 7

Extra Practice
See page 711.

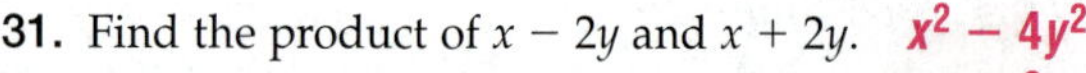

31. Find the product of $x - 2y$ and $x + 2y$. $x^2 - 4y^2$

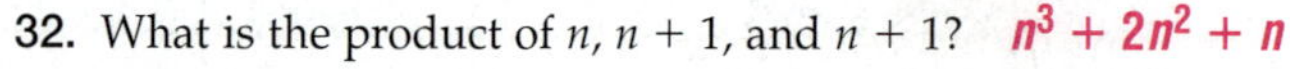

32. What is the product of n, $n + 1$, and $n + 1$? $n^3 + 2n^2 + n$

Applications and Problem Solving

33. **Number Theory** Explain how to find the product of 39 and 41 mentally. (*Hint*: Write 39 as 40 − 1 and 41 as 40 + 1.) **$(40 - 1)(40 + 1) = 1600 - 1$ or 1599**

34. **Photography** Kareem cut off a 1-inch strip from each side of a square photograph so that it would fit into a square picture frame. He removed a total of 20 square inches. **a. See margin.**
    a. Draw and label a diagram that represents this situation. Let x represent the side length of the original photograph.
    b. Write an equation that could be used to find the dimensions of the original photograph. $x^2 - (x - 2)^2 = 20$
    c. Find the original dimensions of the photograph. **6 in. by 6 in.**

Reteaching Activity

Interpersonal Learners Have pairs of students work together to review the formulas in this lesson. Ask them to share methods they can use to help them remember and understand the formulas.

Answers

23. $9x^2 - 25$
24. $25 + 20p + 4p^2$
25. $16 + 16x + 4x^2$
26. $25a^2 - 4b^2$
27. $a^2 - 6ab + 9b^2$
28. $y^3 - y$
29. $x^3 + 2x^2 + x$
30. $x^3 - 3x^2 - 4x + 12$

35. **Geometry** The area of a triangle is given by the expression $\frac{1}{2}bh$, where b represents the length of the base and h represents the height. Suppose a right triangle has a base that measures $x - 3$ units and a height of $x + 3$ units.

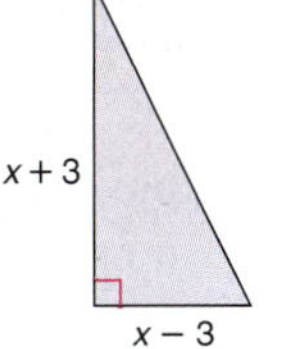

35a. $\frac{1}{2}x^2 + \left(-\frac{9}{2}\right)$

a. Express the area of the triangle as a sum of two monomials.

b. Find the area of the triangle if $x = 5$. **8 square units**

c. What is the length of the hypotenuse if $x = 6$? **about 9.5 units**

36. **Critical Thinking** Use a model to find $(a + b + c)^2$. **See margin.**

Mixed Review

Geometry **Find the area of each shaded region in simplest form.** *(Lessons 9–3 & 9–4)*

37. 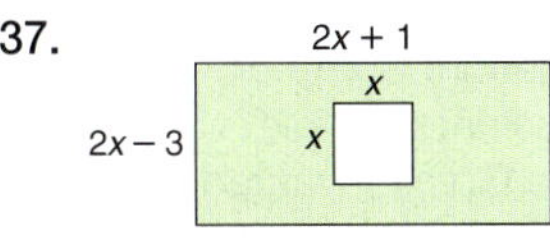

$3x^2 - 4x - 3$

38. 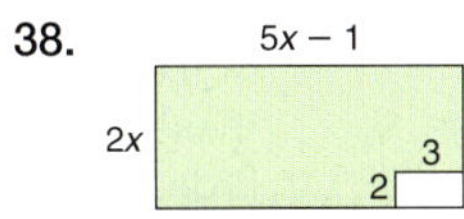

$10x^2 - 2x - 6$

Write each equation in slope-intercept form. *(Lesson 7–3)*

39. $2y = 6x + 8$ **$y = 3x + 4$**

40. $x + y = -4$ **$y = -x - 4$**

41. $2x - 5y = 15$ **$y = \frac{2}{5}x - 3$**

Standardized Test Practice

42. **Grid In** Enrique has 3 dimes, 2 quarters, and 5 nickels in his pocket. He takes one coin from his pocket at random. What is the probability that the coin is either a dime or a nickel? *(Lesson 5–7)* **4/5**

43. **Multiple Choice** For what value(s) of b is the equation $3 + 2|b| = 7$ true? *(Lesson 3–7)* **B**

Ⓐ −5 and 5 Ⓑ −2 and 2 Ⓒ −2 only Ⓓ 2 only

Quiz 2 Lessons 9–4 and 9–5

Find each product. *(Lessons 9–4 & 9–5)*

1. $(c + 2)(c + 8)$ **$c^2 + 10c + 16$**
2. $(5y - 3)(y + 2)$ **$5y^2 + 7y - 6$**
3. $(c - 1)^2$ **$c^2 - 2c + 1$**
4. $(3x - 8)(3x + 8)$ **$9x^2 - 64$**
5. **Geometry** Find the area of the shaded region if the length of a side of the large square is $x + 3$ centimeters and the length of a side of the smaller square is $x - 3$ centimeters. *(Lesson 9–5)* **$12x$ cm²**

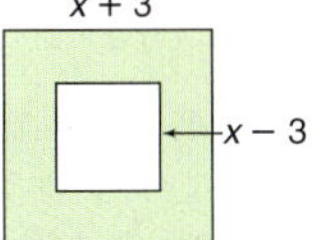

www.algconcepts.com/self_check_quiz

4 ASSESS

Open-Ended Assessment

Writing Have students write a journal entry explaining when to use the Square of a Sum Formula, the Square of a Difference Formula, and the Product of a Sum and a Difference Formula.

Quiz 2

The Quiz provides students with a brief review of the concepts and skills in Lessons 9–4 and 9–5. Lesson numbers are given to the right of the exercises or instruction lines so students can review concepts not yet mastered.

Chapter 9, Quiz B (Lessons 9–3 through 9–5) is available in the *Chapter 9 Resource Masters*, p. 429.

Answer

Page 408

34a. 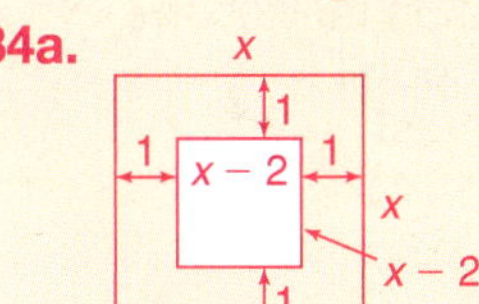

***Enrichment*, p. 417**

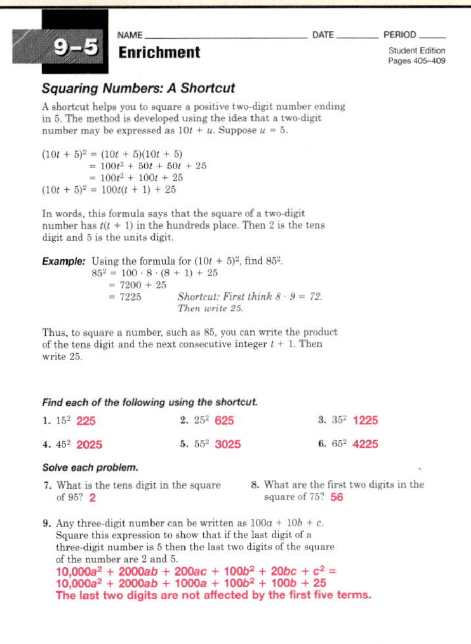

NAME ________ DATE ______ PERIOD ____

9–5 Enrichment

Student Edition Pages 405–409

Squaring Numbers: A Shortcut

A shortcut helps you to square a positive two-digit number ending in 5. The method is developed using the idea that a two-digit number may be expressed as $10t + u$. Suppose $u = 5$.

$$(10t + 5)^2 = (10t + 5)(10t + 5)$$
$$= 100t^2 + 50t + 50t + 25$$
$$= 100t^2 + 100t + 25$$
$$(10t + 5)^2 = 100t(t + 1) + 25$$

In words, this formula says that the square of a two-digit number has $t(t + 1)$ in the hundreds place. Then 2 is the tens digit and 5 is the units digit.

Example: Using the formula for $(10t + 5)^2$, find 85^2.

$$85^2 = 100 \cdot 8 \cdot (8 + 1) + 25$$
$$= 7200 + 25$$
$$= 7225$$

Shortcut: First think $8 \cdot 9 = 72$. Then write 25.

Thus, to square a number, such as 85, you can write the product of the tens digit and the next consecutive integer $t + 1$. Then write 25.

Find each of the following using the shortcut.

1. 15^2 **225**
2. 25^2 **625**
3. 35^2 **1225**
4. 45^2 **2025**
5. 55^2 **3025**
6. 65^2 **4225**

Solve each problem.

7. What is the tens digit in the square of 95? **2**
8. What are the first two digits in the square of 75? **56**
9. Any three-digit number can be written as $100a + 10b + c$. Square this expression to show that if the last digit of a three-digit number is 5 then the last two digits of the square of the number are 2 and 5.
**$10{,}000a^2 + 2000ab + 200ac + 100b^2 + 20bc + c^2 =$
$10{,}000a^2 + 2000ab + 1000a + 100b^2 + 100b + 25$
The last two digits are not affected by the first five terms.**

Extra Credit

Write the product $450 \cdot 550$ as the product of a sum and a difference. Then use your result to evaluate the product.

**$450 \cdot 550 = (500 - 50)(500 + 50) =$
$500^2 - 50^2 = 250{,}000 - 2500 = 247{,}500$**

Answer

36. 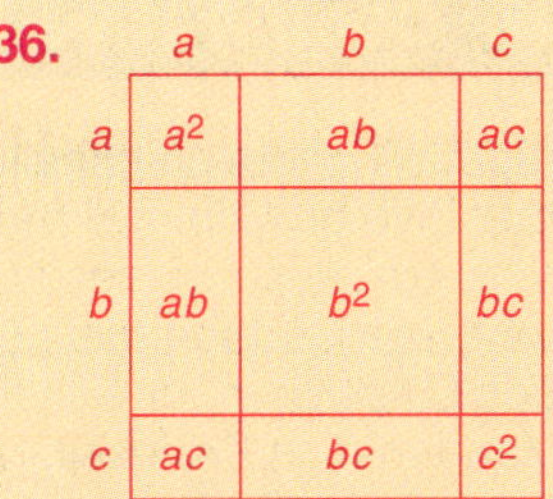

$(a + b + c)^2 = a^2 + 2ab + 2ac + b^2 + 2bc + c^2$

Investigation

PREPARE

This optional investigation is designed to be completed by pairs of students over 1–2 days.

Objectives

Students investigate the effects of doubling or tripling the side length of a square on its area. They make conjectures about the relationship between the ratio of side lengths of squares and the ratio of their areas. They also investigate drawing a square whose area is twice that of a given square. Students present their findings by creating a bulletin board, writing a report, or making a video.

Mathematical Overview

This investigation utilizes the following concepts:

- area of squares,
- ratio, and
- inductive reasoning.

Suggested Time Management	
Investigation	30–45 min
Extension: Gathering Data	30–45 min
Extension: Summarizing Data	20–30 min

Motivating the Lesson

Ask students to give the side lengths of two squares so that the area of one square is twice that of the other. They should quickly realize that this is not a simple problem. You can use squares of side lengths 5 and 7 or 10 and 14 as examples of cases where the area of the second square is *close* to twice that of the first. Stress that this points out the need for a systematic exploration of the areas of squares as you begin this investigation.

Chapter 9 Investigation

It's Greek to Me!

Areas and Ratio

Materials

 ruler

 calculator

 grid paper

Early Greek mathematicians enjoyed solving problems that required a lot of thought and investigation. Some of these problems still fascinate people today. For example, the Greeks wanted to construct a cube whose volume was twice the volume of a given cube. First, let's look at similar problems. What happens to the area of a square when you double the length of the sides of that square? Can you construct a square whose area is twice as great as a given square?

Temple of Poseidon

Investigate

Draw squares with different lengths on grid paper and find the ratios of their areas. Use a table or spreadsheet to record the lengths, areas, and ratios.

1. Draw a 3-centimeter by 3-centimeter square. Label it A. Find the area.
2. Draw a second square that has a side length twice that of square A. Label it B. Find the area.
3. Record your information in a table like the one below.

Square A		Square B		Ratios	
Length of Side	Area	Length of Side	Area	$\frac{\text{Length B}}{\text{Length A}}$	$\frac{\text{Area B}}{\text{Area A}}$
3	9	6	36	2:1	4:1

4. Draw three more pairs of squares and label them A and B. Find the areas and ratios for each pair of squares. **See students' work.**
5. Make a conjecture about the ratio of areas of squares if the ratio of the lengths of their sides is 3:1. **area B: area A = 9:1**

Cooperative Learning

This investigation offers an excellent opportunity for using cooperative groups. For more information on cooperative learning strategies and group management, see *Cooperative Learning in the Mathematics Classroom,* one of the titles in the Glencoe Mathematics Professional Series.

Extending the Investigation

In this extension, you will investigate the ratios of the areas of other squares. You will also try to find the length of the sides of two squares so that the area of one square is twice the other.

first bullet: 9:1

- Draw pairs of squares so that the ratio of the side length of one square to the side length of the second square is 3:1. **Make a conjecture** about the ratio of their areas.
- Draw pairs of squares so that the ratio of the side length of one square to the side length of the second square is *not* 2:1 or 3:1. Find the ratio of the lengths of their sides. **Make a conjecture** about the ratio of their areas. ratio of the side length squared
- Draw pairs of squares so that the lengths of their sides differ by 1. For example, one square might have a side length of 4 units and the second a side length of 5 units. Find the ratio of the lengths of their sides. **Make a conjecture** about the ratios of their areas. ratio of the side length squared
- Draw pairs of squares so that the lengths of their sides differ by a number greater than 1. Find the ratio of the lengths of their sides. **Make a conjecture** about the ratios of their areas. ratio of the side length squared
- Draw a 2-centimeter by 2-centimeter square. Then sketch a square whose area is twice as great as the area of the first square. Find the side length of the second square. $\sqrt{8}$ or about 2.52 cm

Presenting Your Conclusions

Here are some ideas to help you present your conclusions to the class.

- Make a bulletin board showing the results of this investigation.
- Write a report about your conjectures. Include diagrams or computations that help to explain your findings.
- Make a video showing the ratios that you discovered. Present your conjectures in a creative way.

Investigation For more information on three classical problems in Greek mathematics, visit: www.algconcepts.com

MANAGE

Teaching Tip For students who want to know why the ratio of areas is the ratio of side lengths squared, use the following example on the board or overhead. Draw a square with side x. Then the side of a square of twice this length will be $2x$. The area of the first square is x^2; the area of the second square is $(2x)^2 = 4x^2$. The ratio of areas is then $x^2 : 4x^2$ or $1 : 4$. Students can also try this method with squares whose side length ratios are values other than 2.

Working in Pairs If students work in pairs it may be easier for them to make conjectures. If you decide to have students work individually, they are more likely to need help with conjectures.

Working as a Class If students make a bulletin board display of their results, have several pairs work together so they can share and refine their ideas.

ASSESS

Students should be able to calculate the areas correctly and organize their results in tables. They should be able to use the data to write appropriate ratios. They should be able to generalize their results with appropriate conjectures that use mathematical vocabulary correctly. Also, they should be able to use guess-and-check or other problem-solving strategies as they explore drawing a square with twice the area of a given square.

Students should add their bulletin boards, reports, or videos to their portfolios at this time.

Study Guide and Assessment

Understanding and Using the Vocabulary

This section provides a listing of the new terms, properties, and phrases that were introduced in this chapter. The exercises check students' understanding of the terms by using a variety of verbal formats including matching, completion, and true/false.

Glossary A complete glossary of terms appears on pages 762–783.

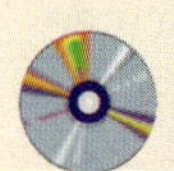

MindJogger Videoquizzes

MindJogger Videoquizzes provide an alternative review of concepts presented in this chapter. Students work in teams to answer questions, gaining points for correct answers.

CHAPTER 9 Study Guide and Assessment

Understanding and Using the Vocabulary

interNET CONNECTION Review Activities For more review activities, visit: www.algconcepts.com

After completing this chapter, you should be able to define each term, property, or phrase and give an example or two of each.

binomial *(p. 383)*
degree *(p. 384)*
FOIL method *(p. 401)*
monomial *(p. 382)*
polynomial *(p. 382)*
trinomial *(p. 383)*

Choose the letter of the term that best matches each statement or phrase. Each letter is used once.

1. $(x - y)^2$ **i**
2. a polynomial with two terms **b**
3. a polynomial with three terms **j**
4. a monomial or a sum of monomials **h**
5. the sum of the exponents of the variables **c**
6. a number, a variable, or a product of numbers and variables **g**
7. Subtract polynomials by adding this. **a**
8. Add polynomials by grouping these together. **f**
9. Use this to multiply any two binomials. **e**
10. Use this to multiply a polynomial by a monomial. **d**

a. additive inverse
b. binomial
c. degree of a polynomial
d. Distributive Property
e. FOIL method
f. like terms
g. monomial
h. polynomial
i. square of a difference
j. trinomial

Skills and Concepts

Objectives and Examples

• **Lesson 9–1** Identify and classify polynomials and find their degree.

Identify $3x^3 - 2x^2$ as a *monomial, binomial,* or *trinomial.*

$3x^3 - 2x^2$ can be written as a sum of two monomials, $3x^3$ and $-2x^2$. It is a binomial.

Review Exercises

State whether each expression is a polynomial. If it is a polynomial, identify it as either a *monomial, binomial,* or *trinomial.*

11. $7m$ **yes; monomial**
12. $5g^{-2}$ **no**
13. $2x^2 + 3x - 4$ **yes; trinomial**
14. $3y + 5y + 8$ **yes; binomial**

Find the degree of each polynomial.

15. 7 **0**
16. $-4m^5$ **5**
17. $10x^2 - x^3y$ **4**
18. $2abc + 9a^5b - 4bc^3$ **6**
19. Arrange $3d^3 - d^2 + 7cd$ so that the powers of d are in ascending order. $7cd - d^2 + 3d^3$

www.algconcepts.com/vocabulary_review

Resource Manager

Reproducible Masters
Chapter 9 Resource Masters
• *Assessment,* pp. 419–427, 430–432

Technology/Multimedia
MindJogger Videoquizzes
ExamView® Pro

Objectives and Examples

• **Lesson 9-2** Add and subtract polynomials.

$(2x + 4) + (5x - 7)$
$= (2x + 5x) + (4 - 7)$
$= (2 + 5)x + (4 - 7)$
$= 7x - 3$

$(7s^2 + 3s - 4) - (3s^2 - 2s + 5)$

The additive inverse of $(3s^2 - 2s + 5)$ is $(-3s^2 + 2s - 5)$.

$(7s^2 + 3s - 4) - (3s^2 - 2s + 5)$
$= (7s^2 + 3s - 4) + (-3s^2 + 2s - 5)$
$= (7s^2 - 3s^2) + (3s + 2s) + (-4 - 5)$
$= (7 - 3)s^2 + (3 + 2)s + (-4 - 5)$
$= 4s^2 + 5s - 9$

Review Exercises

Find each sum or difference.

20. $8x - 7$ $(+)\ 3x + 5$ — **$11x - 2$**

21. $6x^2 \quad + 4$ $(+)\ 3x^2 - 3x + 2$ — **$9x^2 - 3x + 6$**

22. $9m^2 + 6m - 6$ $(-)\ 7m^2 - 3m + 3$ — **$2m^2 + 9m - 9$**

23. $15s^2 \quad - 6$ $(-)\ 8s^2 + 6s - 4$ — **$7s^2 - 6s - 2$**

24. $(17x - 3y) + (-2x + 5y)$ **$15x + 2y$**

25. $(10g^2 + 5g) - (6g^2 - 3g)$ **$4g^2 + 8g$**

26. $(14x^2 + 3x - 6) - (8 - 2x)$ **$14x^2 + 5x - 14$**

27. $(-3s^2t^2 - 4st^2) + (-7s^2t^2 + 6st^2)$

28. What is $5n^2 + 3$ minus $2n^2 - 4$? **$3n^2 + 7$**

27. $-10s^2t^2 + 2st^2$

• **Lesson 9-3** Multiply a polynomial by a monomial.

$2x^2(3x^3 + 2x^2 - x)$
$= 2x^2(3x^3) + 2x^2(2x^2) + 2x^2(-x)$
$= 6x^5 + 4x^4 - 2x^3$

Find each product.

29. $3(x - 5)$ **$3x - 15$**

30. $n(n + 3)$ **$n^2 + 3n$**

31. $2m(m^2 + 4m)$ **$2m^3 + 8m^2$**

32. $x^4(3x^2 - 2x)$ **$3x^6 - 2x^5$**

33. $5(2x^2 + 3x - 2)$ **$10x^2 + 15x - 10$**

34. $-3m^2(m^2 - 2m + 1)$ **$-3m^4 + 6m^3 - 3m^2$**

Solve each equation.

35. $x(x + 8) = x(x + 11) + 9$ **-3**

36. $9(w - 4) + 10 = 2w + 16$ **6**

• **Lesson 9-4** Multiply two binomials.

Find the product of $x + 4$ and $x - 3$.

F L I O

$(x + 4)(x - 3)$
$= (x)(x) + (x)(-3) + (4)(x) + (4)(-3)$
$= x^2 - 3x + 4x - 12$
$= x^2 + x - 12$

Find each product. **45. $10x^2 - 7x - 12$**

37. $(y - 2)(y + 6)$ **$y^2 + 4y - 12$**

38. $(m + 5)(m + 7)$ **$m^2 + 12m + 35$**

39. $(x - 1)(x - 3)$ **$x^2 - 4x + 3$**

40. $(a + 11)(a - 4)$ **$a^2 + 7a - 44$**

41. $(2m - 1)(m - 4)$ **$2m^2 - 9m + 4$**

42. $(x + 4)(3x - 2)$ **$3x^2 + 10x - 8$**

43. $(4y + 2)(3y + 1)$ **$12y^2 + 10y + 2$**

44. $(6x - 3)(2x + 5)$ **$12x^2 + 24x - 15$**

45. Find the product of $5x + 4$ and $2x - 3$.

Skills and Concepts

The **Objectives and Examples** section reviews the skills and concepts of the chapter and shows completely worked examples.

The **Review Exercises** provide practice for the corresponding objectives.

ExamView® Pro

Use the networkable ***ExamView® Pro Testmaker CD-ROM*** to:

- Create **multiple versions** of tests.
- Create **modified** tests for **inclusion** students with one mouse click.
- **Edit** existing questions and **add** your own questions.
- Build tests aligned with state standards using built-in **state curriculum correlations**.
- Change **English** tests to **Spanish** with one mouse click and vice versa.

Mixed Problem Solving
See pages 724–731.

Applications and Problem Solving

This section provides additional practice in solving real-world problems that involve the concepts of this chapter.

Objectives and Examples

- **Lesson 9–5** Develop and use the patterns for $(a + b)^2$, $(a - b)^2$, and $(a + b)(a - b)$.

$$(x + 3)^2 = x^2 + 2(x)(3) + 3^2 = x^2 + 6x + 9 \quad \textit{Square of a sum}$$

$$(y - 5)^2 = y^2 - 2(y)(5) + 5^2 = y^2 - 10y + 25 \quad \textit{Square of a difference}$$

$$(s + 4)(s - 4) = s^2 - 4^2 = s^2 - 16 \quad \textit{Product of a sum and difference}$$

Review Exercises

Find each product.

46. $(w + 6)^2$ **$w^2 + 12w + 36$**

47. $(2x + 3)^2$ **$4x^2 + 12x + 9$**

48. $(8 + g)^2$ **$64 + 16g + g^2$**

49. $(x - 2)^2$ **$x^2 - 4x + 4$**

50. $(3y - 1)^2$ **$9y^2 - 6y + 1$**

51. $(5m - 3n)^2$ **$25m^2 - 30mn + 9n^2$**

52. $(y + 4)(y - 4)$ **$y^2 - 16$**

53. $(2a + 3)(2a - 3)$ **$4a^2 - 9$**

54. $(2p - 3q)(2p + 3q)$ **$4p^2 - 9q^2$**

55. $(5a + 1)(5a + 1)$ **$25a^2 + 10a + 1$**

Applications and Problem Solving

56. Recreation Bocce is a game similar to lawn bowling, but it is played on a rectangular dirt court. If the length of the rectangle is $12x$ and the width is $2x + 2$, find the area of the court in terms of x. Write your answer as a polynomial in simplest form. *(Lesson 9–3)* **$24x^2 + 24x$**

57. Geometry The area of a triangle is given by $\frac{1}{2}bh$, where b is the length of the base of the triangle and h is the measure of the height of the triangle. In triangle ABC, $b = 3x + 8$ and $h = 7x - 4$. Write the polynomial representing the measure of the area of triangle ABC. *(Lesson 9–4)* **$10\frac{1}{2}x^2 + 22x - 16$**

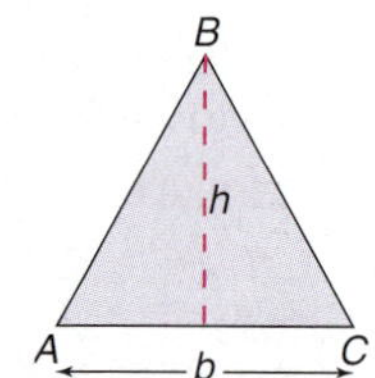

58. Gardening A rectangular garden is 10 feet longer than it is wide. It is surrounded by a brick walkway 3 feet wide, as shown at the right. Suppose the total area of the walkway is 396 square feet. *(Lesson 9–4)*

a. Find the area of the garden. **$x^2 + 10x$**

b. What are the dimensions of the garden? **25 ft by 35 ft**

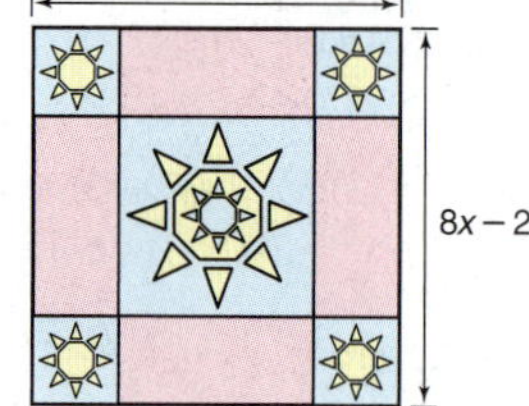

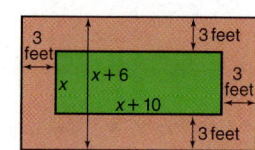

59. Quilting Kirsten is making a quilt to enter in the arts festival. The diagram gives the dimensions of each block of the quilt. *(Lesson 9–5)*

a. Find the area of each block. **$64x^2 - 32x + 4$**

b. The completed quilt is a square of 36 blocks. Write a product of binomials that could be used to determine the area of the entire quilt. **$(48x - 12)(48x - 12)$**

$8x - 2$

$8x - 2$

Assessment, pp. 421–422

NAME ______ DATE ______ PERIOD ______

9 **Chapter 9 Test, Form 1B**

Write the letter for the correct answer in the blank at the right of each problem.

1. Which of the following is a monomial?
   A. $\frac{1}{y}$ B. $2xy$ C. $x + 1$ D. $\frac{a}{b}$ 1. **B**

Find the degree of each polynomial.

2. $x^3 + 4x$
   A. 1 B. 3 C. 4 D. 5 2. **B**
3. $a^2b + 5b^2$
   A. 3 B. 4 C. 5 D. 7 3. **A**

Arrange the terms of each polynomial so that the powers of x are in descending order.

4. $x + 4x^2 + 2x^4$
   A. $x + 4x^2 + 2x^4$ B. $x + 2x^4 + 4x^2$
   C. $2x^4 + 4x^2 + x$ D. $4x^2 + 2x^4 + x$ 4. **C**
5. $3x^3y + 2x + 5x^2y$
   A. $3x^3y + 5x^2y + 2x$ B. $2x + 5x^2y + 3x^3y$
   C. $2x + 3x^3y + 5x^2y$ D. $3x^3y + 2x + 5x^2y$ 5. **A**

Find each sum or difference.

6. $(2x - 5) + (x + 3)$
   A. $x + 8$ B. $x - 2$ C. $3x - 2$ D. $3x + 2$ 6. **C**
7. $(4y + 3x) + (y - 2x)$
   A. $4y + x$ B. $5y + 5x$ C. $5y - x$ D. $5y + x$ 7. **D**
8. $(3x^2 + 4x) + (2x^2 + 9x - 1)$
   A. $5x^2 + 13x - 1$ B. $7x^2 + 11x + 3$
   C. $5x^2 + 12x$ D. $5x^2 + 14x + 1$ 8. **A**
9. $(8a + 5) - (3a + 4)$
   A. $11a + 9$ B. $5a + 9$
   C. $5a - 1$ D. $5a + 1$ 9. **D**
10. $(7x^2 - 2x) - (3x^2 - x)$
    A. $10x^2 - 3x$ B. $4x^2 - x$
    C. $10x^2 - 1$ D. $4x^2 - 3x$ 10. **B**

Find each product.

11. $2(m + 10)$
    A. $2m + 10$ B. $2m + 20$ C. $2m + 12$ D. $m + 20$ 11. **B**
12. $b(b + 12)$
    A. $b^2 + 12$ B. $b^2 + 12b$ C. $2b + 12$ D. $2b + 12b$ 12. **B**
13. $3x(x^2 + 4)$
    A. $3x^3 + 12x$ B. $3x^2 + 12x$ C. $3x + 7$ D. $3x^3 + 12$ 13. **A**

Assessment

Four forms of Chapter 9 Test are available in the *Chapter 9 Resource Masters.*

Chapter 9 Test, Form 1B, is shown at the left. Chapter 9 Test, Form 2B, is shown on the next page.

Form of Test			Level
1A	Multiple Choice	pp. 419–420	Average
1B	Multiple Choice	pp. 421–422	Basic
2A	Free Response	pp. 423–424	Average
2B	Free Response	pp. 425–426	Basic

CHAPTER 9 Test

1. **Explain** how to use the FOIL method to multiply two binomials. **1–2. See margin.**

2. **Explain** how to subtract polynomials.

3. **You Decide** Mark says that $-4wx^{-2}y^3$ is a monomial of degree 2. Linda disagrees. Who is correct? Explain your reasoning. **Linda; $-4wx^{-2}y^3$ is not a monomial.**

State whether each expression is a polynomial. If it is a polynomial, identify it as either a *monomial*, *binomial*, or *trinomial*, and state its degree.

4. $4x^3 + 3x^2$ **yes, binomial, 3**
5. $12r^{-2} - 3r + 6$ **no**
6. m^0 **yes, monomial, 0**
7. $5w^3yz^2$ **yes, monomial, 6**

Find each sum or difference.

8. $\begin{array}{r} 3x^2 - 5x + 4 \\ (+)\ 5x^2 + 7x + 8 \\ \hline \end{array}$ **$8x^2 + 2x + 12$**

9. $\begin{array}{r} 8z^2 - 5z \\ (-)\ 6z^2 - 3z + 4 \\ \hline \end{array}$ **$2z^2 - 2z - 4$**

10. $(y^3 + 6y^2 + 4y) - (2y^3 - 7y)$ **$-y^3 + 6y^2 + 11y$**

11. $(-9h^2 - 5g) + (2g + 7h^2)$ **$-2h^2 - 3g$**

Find each product.

12. $3x(2x^2 + 4x - 5)$ **$6x^3 + 12x^2 - 15x$**
13. $5t^2(-2t + 3t^3 + 4)$ **$-10t^3 + 15t^5 + 20t^2$**
14. $(x + 2)(x + 3)$ **$x^2 + 5x + 6$**
15. $(y - 4)(y - 5)$ **$y^2 - 9y + 20$**
16. $(x - 2)(x + 5)$ **$x^2 + 3x - 10$**
17. $(3n + 4)(2n + 3)$ **$6n^2 + 17n + 12$**
18. $(x + 5)^2$ **$x^2 + 10x + 25$**
19. $(2w - 3)^2$ **$4w^2 - 12w + 9$**
20. $(7r - 4)(7r + 4)$ **$49r^2 - 16$**
21. $(m - 3n)(m + 3n)$ **$m^2 - 9n^2$**

Solve.

22. $2(y - 3) + 9 = 5y - 6$ **3**
23. $x(x + 4) + 5x = x(x - 1) + 20$ **2**

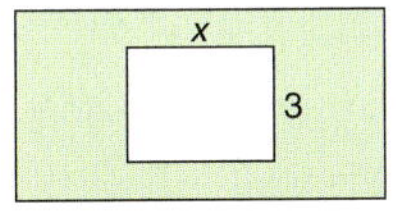

Exercise 24

24. **Geometry** Find the area of the shaded region. **$3x^2 - 2x - 2$ units^2**

25. **Family Life** Mrs. Douglas wants her five children, Aaron, Briana, Casey, Danielle, and Eddie, to spend a total of 35 hours on chores this week. Aaron, the oldest, works twice as many hours as the others. Danielle has earned an hour off from chores by getting an A on her algebra test. If x is the number of hours Danielle will work, then Briana, Casey, and Eddie will each work $x + 1$ hours, and Aaron will work twice as long, $2x + 2$ hours. **a. $x + 3(x + 1) + 2x + 2 = 35$**
    a. Write an equation to represent the number of hours the children will work.
    b. Use the equation to determine the number of hours each child will work on chores this week. **Danielle: 5 hr; Briana, Casey, Eddie: 6 hr each; Aaron: 12 hr**

www.algconcepts.com/chapter_test

Chapter Test Bonus Question

A binomial of degree 3 is squared. What is the degree of the product? **6**

Chapter Test

Answers

1. Find the sum of the products of the First terms, the Outer terms, the Inner terms, and the Last terms.
2. First, find the additive inverse of the polynomial being subtracted. Then, group the like terms together. Finally, combine like terms.

***Assessment*, pp. 425–426**

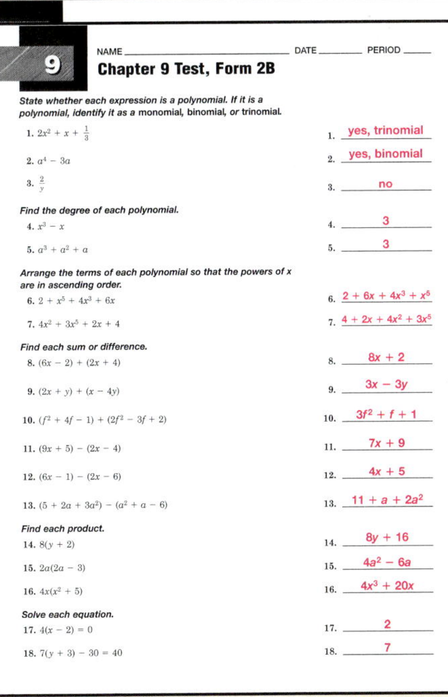

9 NAME ________ DATE ________ PERIOD ________

Chapter 9 Test, Form 2B

State whether each expression is a polynomial. If it is a polynomial, identify it as a monomial, binomial, or trinomial.

1. $2x^2 + x + \frac{1}{3}$ — 1. yes, trinomial
2. $a^4 - 3a$ — 2. yes, binomial
3. $\frac{2}{y}$ — 3. no

Find the degree of each polynomial.

4. $x^3 - x$ — 4. 3
5. $a^3 + a^2 + a$ — 5. 3

Arrange the terms of each polynomial so that the powers of x are in ascending order.

6. $2 + x^5 + 4x^3 + 6x$ — 6. $2 + 6x + 4x^3 + x^5$
7. $4x^2 + 3x^5 + 2x + 4$ — 7. $4 + 2x + 4x^2 + 3x^5$

Find each sum or difference.

8. $(6x - 2) + (2x + 4)$ — 8. $8x + 2$
9. $(2x + y) + (x - 4y)$ — 9. $3x - 3y$
10. $(f^2 + 4f - 1) + (2f^2 - 3f + 2)$ — 10. $3f^2 + f + 1$
11. $(9x + 5) - (2x - 4)$ — 11. $7x + 9$
12. $(6x - 1) - (2x - 6)$ — 12. $4x + 5$
13. $(5 + 2a + 3a^2) - (a^2 + a - 6)$ — 13. $11 + a + 2a^2$

Find each product.

14. $8(y + 2)$ — 14. $8y + 16$
15. $2a(2a - 3)$ — 15. $4a^2 - 6a$
16. $4x(x^2 + 5)$ — 16. $4x^3 + 20x$

Solve each equation.

17. $4(x - 2) = 0$ — 17. 2
18. $7(y + 3) - 30 = 40$ — 18. 7

Preparing for Standardized Tests

Pages 416–417 are part of a complete test preparation course that is described in detail on page T9 of the Teacher's Handbook. The test items on these pages were written in the same style as those in state proficiency tests and standardized tests like ACT and SAT.

Diagnosis and Prescription

Each of the 10 test questions on page 417 is cross-referenced to the lesson where that SAT or ACT skill is covered. If students miss a particular type of problem, you can have them study that skill.

(See chart at the bottom of page 417. Note that SPT = State Proficiency Test, SAT = Scholastic Assessment Test, and ACT = American College Test.)

Assessment, p. 430

9 — NAME ________ DATE ______ PERIOD ______

Chapter 9 Cumulative Review

1. Write an equation for the sentence below. *(Lesson 1–1)*
   Fifteen is equal to a number b multiplied by 7. — 1. $15 = 7b$

Simplify each expression. (Lesson 2–3)

2. $8x + (-7x)$ — 2. x
3. $-4y + (-3y)$ — 3. $-7y$

Solve each equation. (Lesson 3–5)

4. $p + (-8) = 10$ — 4. 18
5. $12 = 20 - z$ — 5. 8

Solve each equation. Check your solution. (Lesson 4–6)

6. $3r + 8 = 4r + 3$ — 6. 5
7. $9q + 2 = 13 + 9q$ — 7. no solution

On a map, the scale is 2 inches = 5 miles. Find the actual distance for each map distance. (Lesson 5–2)

8. 6 inches — 8. 15 miles
9. 15 inches — 9. 37.5 miles

Solve. Assume that y varies directly as x. (Lesson 6–5)

10. Find y when $x = 18$ if $y = 21$ when $x = 14$. — 10. 27
11. If $y = 45$ when $x = 9$, find x when $y = 10$. — 11. 2

Determine whether the graphs of each pair of equations are parallel, perpendicular, or neither. (Lesson 7–7)

12. $y = 8x + 4$, $y = -8x - 4$ — 12. neither
13. $2y = 3x + 4$, $y = -\frac{2}{3}x$ — 13. perpendicular

For Questions 14–15, simplify. (Lesson 8–5)

14. $\sqrt{\frac{1}{4}}$ — 14. $\frac{1}{2}$
15. $-\sqrt{\frac{100}{81}}$ — 15. $-\frac{10}{9}$
16. Find the product of $(4x + 2y)$ and $(x + y)$. *(Lesson 9–4)* — 16. $4x^2 + 6xy + 2y^2$

CHAPTER 9 Preparing for Standardized Tests

Percent Problems

Standardized tests contain several types of percent problems. They may be numeric problems, word problems, or data analysis problems.

Test-Taking Tip

A percent is a fraction whose denominator is 100.

Familiarize yourself with common fractions, decimals, and percents. Below are some examples of common equivalents.

$0.01 = \frac{1}{100} = 1\%$ $\quad$ $0.1 = \frac{1}{10} = 10\%$ $\quad$ $0.333\ldots = \frac{1}{3} \approx 33.3\%$

$0.25 = \frac{1}{4} = 25\%$ $\quad$ $0.5 = \frac{1}{2} = 50\%$ $\quad$ $0.75 = \frac{3}{4} = 75\%$

Example 1

The table shows millions of U.S. travelers to visit foreign countries. What percent of the travelers in 2001 visited Canada? Round to the nearest percent.

U.S. Travelers to Foreign Countries (millions)			
Year	Area		
	Canada	Mexico	Overseas
1999	15.3	17.7	24.6
2000	15.2	18.8	26.9
2001	15.6	17.2	25.2

Source: U.S. Dept. of Commerce

Hint Write a ratio, and then write the fraction as a percent.

Solution First, find the total travelers in 2001.

$$15.6 + 17.2 + 25.2 = 58.0$$

Next, write the ratio of travelers to Canada to total travelers.

$$\frac{15.6}{58.0}$$

Then write this fraction as a decimal and as a percent. Use your calculator.

$$\frac{15.6}{58.0} \approx 0.2690 \text{ or } 27\% \text{ to the nearest percent}$$

Travelers to Canada were about 27% of foreign travel in 2001.

Example 2

A bus company charges \$5 each way to shuttle passengers between the hotel and a shopping mall. On a given day, the bus company has a total capacity of 250 people on the way to the mall and back. If the bus company runs at 90% of capacity, how much money would it collect that day?

Ⓐ \$1147.50 Ⓑ \$1250 Ⓒ \$2250 Ⓓ \$2500 Ⓔ \$2625

Hint Avoid partial answers. Be sure you answer the question that is asked.

Solution First determine how much money the bus company makes when it runs at total capacity. The 250 passengers would pay \$10 each, because the charge is \$5 each way.

$$250(\$10) = \$2500$$

Notice that this total amount is answer choice D, but it is a wrong answer. The question asks for the amount when the bus runs at 90% of capacity.

The word *of* is a clue to multiply. Find 90% of \$2500.

$$(90\%)(\$2500) = (0.90)(2500) \text{ or } 2250$$

The bus company would collect \$2250. Therefore, the answer is C.

Resource Manager

Reproducible Masters

Chapter 9 Resource Masters

- *Assessment*, pp. 430–432

Preparing for Standardized Tests
For test-taking strategies and more practice, see pages 732–749.

After you work each problem, record your answer on the answer sheet provided or on a sheet of paper.

Multiple Choice

1. The areas of two rooms are 150 square feet and 135 square feet. If the total area of the home is 2000 square feet, what percent of the total area is the area of the two rooms? *(Lesson 5–3)* **C**

   (A) $6\frac{3}{4}\%$ (B) $7\frac{1}{2}\%$ (C) $14\frac{1}{4}\%$ (D) $85\frac{3}{4}\%$

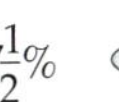
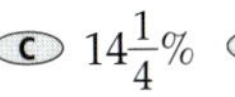

2. A \$9.95 calendar is marked down 40%. Before tax, how much is saved on the purchase of one calendar? *(Lesson 5–4)* **B**

   (A) \$1.99 (B) \$3.98
   (C) \$4.00 (D) \$5.97

3. On a 16-question quiz, Tom answered 2 questions incorrectly. If each question is worth the same number of points, what percent of the total points is his point total? *(Lesson 5–4)* **D**

   (A) 12.5% (B) 16% (C) 85%
   (D) 87.5% (E) 94%

4. An increase in prices has made the cost of remodeling an office 12% more than the original cost. The original cost was \$7145. What is the best estimate of the new cost? *(Lesson 5–5)* **B**

   (A) \$700 (B) \$7700
   (C) \$10,000 (D) \$70,000

5. The graph shows the attendance at a park. Predict the attendance in the year 2013. *(Lesson 6–1)* **B**

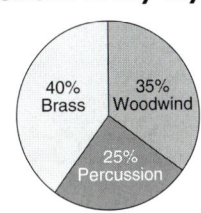

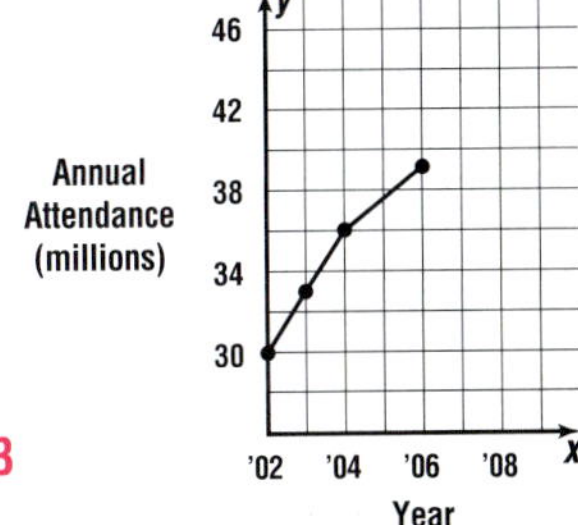
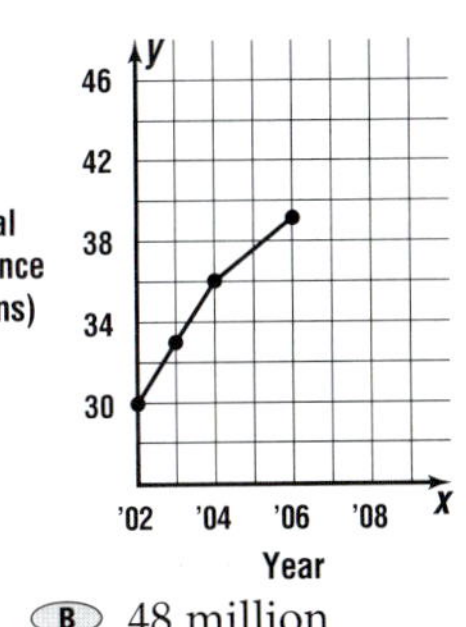

   (A) 38 million (B) 48 million
   (C) 60 million (D) 100 billion

6. A pair of shoes that regularly sold for \$44 is now on sale for \$33. What is the discount rate? *(Lesson 5–5)* **B**

   (A) 11% (B) 25% (C) 75%
   (D) $33\frac{1}{3}\%$ (E) $66\frac{2}{3}\%$

7. There are 60 students in the band. How many play a percussion instrument? *(Lesson 5–4)* **A**

Instruments in Bay City Band

40% Brass
35% Woodwind
25% Percussion

   (A) 15 (B) 21 (C) 24 (D) 25

8. Which expression can be used to find the value of y? *(Lesson 7–3)* **C**

x	y
1	2
2	5
3	8
4	11

   (A) $2x + 1$ (B) $1 - 3x$
   (C) $3x - 1$ (D) $3x + 1$

Grid In

9. A CD player is on sale for \$250. If there is a 6% sales tax, what is the total cost? *(Lesson 5–4)* **\$265**

Extended Response

10. The total land area of a state is 23,159,000 acres. Of that, 13,513,000 acres are cropland, 1,866,000 acres are pastureland, and 3,626,000 acres are forest. *(Lesson 5–4)*

    Part A To the nearest percent, what percent of the state's area is cropland? **58%**

    Part B What percent of the state's area is *not* cropland, pastureland, or forest? **18%**

www.algconcepts.com/standardized_test

A bubble-in answer sheet for these practice problems is available on page A1 of the *Chapter 9 Resource Masters.*

Additional Practice

Additional test practice questions are available in the *Chapter 9 Resource Masters*, pp. 431–432.

Chapter 9 Percent Problems

Ex. 1	using a table and percents	SPT
Ex. 2	percent word problem	ACT
1	percent	SPT
2	percent discount	SPT
3	percent	SAT
4	percent increase	SPT
5	use a line graph	SPT
6	percent discount	SAT
7	a circle graph with percents	SPT
8	table and expression	SPT
9	percent	SPT
10	percent word problem	SPT

***Assessment*, pp. 431–432**

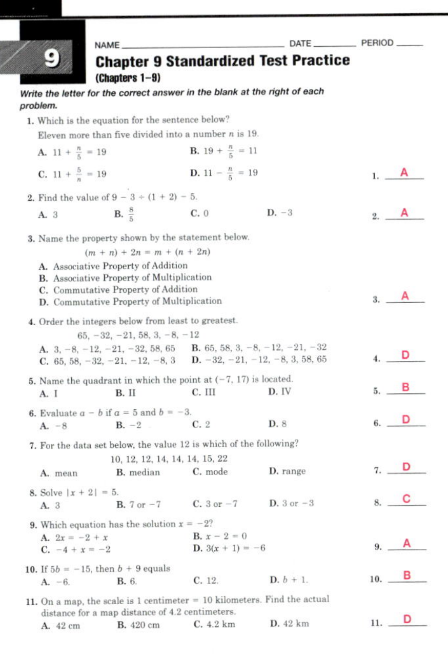

9

NAME ________ DATE ______ PERIOD ____

Chapter 9 Standardized Test Practice
(Chapters 1–9)

Write the letter for the correct answer in the blank at the right of each problem.

1. Which is the equation for the sentence below?
   Eleven more than five divided into a number n is 19.
   A. $11 + \frac{n}{5} = 19$ B. $19 + \frac{n}{5} = 11$
   C. $11 + \frac{5}{n} = 19$ D. $11 - \frac{n}{5} = 19$ 1. A

2. Find the value of $9 - 3 \div (1 + 2) - 5$.
   A. 3 B. $\frac{8}{5}$ C. 0 D. −3 2. A

3. Name the property shown by the statement below.
   $(m + n) + 2n = m + (n + 2n)$
   A. Associative Property of Addition
   B. Associative Property of Multiplication
   C. Commutative Property of Addition
   D. Commutative Property of Multiplication 3. A

4. Order the integers below from least to greatest.
   65, −32, −21, 58, 3, −8, −12
   A. 3, −8, −12, −21, −32, 58, 65 B. 65, 58, 3, −8, −12, −21, −32
   C. 65, 58, −32, −21, −12, −8, 3 D. −32, −21, −12, −8, 3, 58, 65 4. D

5. Name the quadrant in which the point at (−7, 17) is located.
   A. I B. II C. III D. IV 5. B

6. Evaluate $a - b$ if $a = 5$ and $b = -3$.
   A. −8 B. −2 C. 2 D. 8 6. D

7. For the data set below, the value 12 is which of the following?
   10, 12, 12, 14, 14, 14, 15, 22
   A. mean B. median C. mode D. range 7. D

8. Solve $|x + 2| = 5$.
   A. 3 B. 7 or −7 C. 3 or −7 D. 3 or −3 8. C

9. Which equation has the solution $x = -2$?
   A. $2x = -2 + x$ B. $x - 2 = 0$
   C. $-4 + x = -2$ D. $3(x + 1) = -6$ 9. A

10. If $5b = -15$, then $b + 9$ equals
    A. −6. B. 6. C. 12. D. $b + 1$. 10. B

11. On a map, the scale is 1 centimeter = 10 kilometers. Find the actual distance for a map distance of 4.2 centimeters.
    A. 42 cm B. 420 cm C. 4.2 km D. 42 km 11. D

CHAPTER 10

Resource Manager

Factoring

Instructional Objectives

Lesson (pages)	Objectives	NCTM Standards 2000	State/Local Objectives
10–1 (420–425)	Find the greatest common factor of a set of numbers or monomials.	1, 6, 8, 9	
Investigation (426–427)	Explore perimeter and area.	1, 2, 3, 6, 7, 8, 9, 10	
10–2 (428–433)	Use the GCF and the Distributive Property to factor polynomials.	1, 6, 8, 9	
10–3 (434–439)	Factor trinomials of the form $x^2 + bx + c$.	1, 6, 8, 9	
10–4 (440–444)	Factor trinomials of the form $ax^2 + bx + c$.	1, 6, 8, 9	
10–5 (445–449)	Recognize and factor the differences of squares and perfect square trinomials.	1, 6, 8, 9	

Key to NCTM Standards 2000

1 Number & Operations; **2** Algebra; **3** Geometry; **4** Measurement; **5** Data Analysis & Probability; **6** Problem Solving; **7** Reasoning & Proof; **8** Communications; **9** Connections; **10** Representation

Suggested Pacing

See page T13 for a complete course-planning calendar.

Standard refers to schedules that provide 45- to 55-minute periods that meet each day.
Block refers to schedules that provide approximately 90-minute periods which may meet every day for one semester or every other day over two semesters.

<table>
<tr><th>PACING</th><th>DAY 1</th><th>DAY 2</th><th>DAY 3</th><th>DAY 4</th><th>DAY 5</th><th>DAY 6</th></tr>
<tr><td>Standard Core (Chapters 1–13)</td><td colspan="2">Lesson 10–1</td><td>INV</td><td colspan="2">Lesson 10–2</td><td>Lesson 10–3</td></tr>
<tr><td>Standard Enhanced (Chapters 1–15)</td><td>Lesson 10–1</td><td>INV</td><td colspan="2">Lesson 10–2</td><td>Lesson 10–3</td><td>Lesson 10–4</td></tr>
<tr><td>Block Core (Chapters 1–13)</td><td>Chapter 9 Test & Lesson 10–1</td><td>INV</td><td>Lessons 10–2 & 10–3</td><td>Lessons 10–4 & 10–5</td><td>SG+A</td><td>Chapter Test & Lesson 11–1</td></tr>
<tr><td>Block Enhanced (Chapters 1–15)</td><td>Chapter 9 Test & Lesson 10–1</td><td>INV & Lesson 10–2</td><td>Lessons 10–3 & 10–4</td><td>Lesson 10–5 & SG+A</td><td>Chapter Test & Lesson 11–1</td><td></td></tr>
</table>

Instructional Resources		Blackline Masters (page numbers)								
		Chapter 10 Resource Masters								
Lesson	Materials and Manipulatives (see below for Glencoe Manipulative Resources)	Study Guide	Practice (Skills & Average)	Reading to Learn Mathematics	Enrichment	Assessment	Hands-On Algebra*	School-to-Workplace*	Graphing Calculator Masters*	5-Minute Check Transparencies
10–1	graphing calculator colored pencils or pens	433	434–435	436	437		107		26	10–1
Investigation	algebra tiles [1, 2, 3, 4]									
10–2	algebra tiles [1, 2, 3, 4] equation/product mat [1, 2, 4] grid paper [1, 4] tape measure	438	439–440	441	442	469	108	10		10–2
10–3	straightedge [1, 2, 4]	443	444–445	446	447	468	109		27, 28	10–3
10–4	algebra tiles [1, 2, 3, 4] straightedge [1, 2, 4]	448	449–450	451	452		110, 111			10–4
10–5	10 slips of paper	453	454–455	456	457	469	112			10–5
Study Guide & Assessment/ Chapter Test						459–467, 470–472				

***See page 418c for examples of these instructional materials.**

Key to Glencoe Manipulative Resources

[1]Classroom Manipulative Resources [2]Student Manipulative Resources [3]Overhead Manipulative Resources [4]Hands-On Algebra Masters

INV = Investigation *SG+A = Study Guide and Assessment*

DAY 7	DAY 8	DAY 9	DAY 10	DAY 11	DAY 12	DAY 13
Lesson 10–3	Lesson 10–4	Lesson 10–5	SG+A	Chapter Test		
Lesson 10–5	SG+A	Chapter Test				

Resource Manager

The pages shown on this page are a small sample of the materials available on *TeacherWorks: All-in-One Lesson Planner and Resource Center.*

This CD-ROM includes all of the blackline masters and transparencies available for this program.

It also includes a lesson planner and interactive Teacher Edition, so you can customize lesson plans and reproduce classroom resources quickly and easily, from just about anywhere.

Applications

School-to-Workplace Masters, p. 10

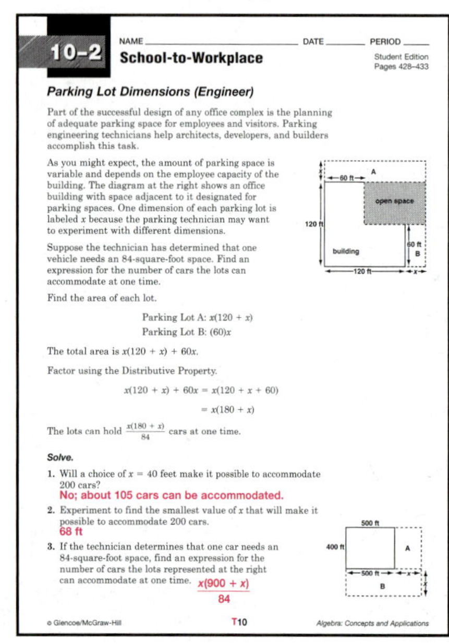

10-2 NAME ______ DATE ______ PERIOD ______

School-to-Workplace

Student Edition Pages 428–433

Parking Lot Dimensions (Engineer)

Part of the successful design of any office complex is the planning of adequate parking space for employees and visitors. Parking engineering technicians help architects, developers, and builders accomplish this task.

As you might expect, the amount of parking space is variable and depends on the employee capacity of the building. The diagram at the right shows an office building with space adjacent to it designated for parking spaces. One dimension of each parking lot is labeled x because the parking technician may want to experiment with different dimensions.

Suppose the technician has determined that one vehicle needs an 84-square-foot space. Find an expression for the number of cars the lots can accommodate at one time.

Find the area of each lot.

Parking Lot A: $x(120 + x)$
Parking Lot B: $(60)x$

The total area is $x(120 + x) + 60x$.

Factor using the Distributive Property.

$$x(120 + x) + 60x = x(120 + x + 60)$$
$$= x(180 + x)$$

The lots can hold $\frac{x(180 + x)}{84}$ cars at one time.

Solve.

1. Will a choice of $x = 40$ feet make it possible to accommodate 200 cars?
   No; about 105 cars can be accommodated.
2. Experiment to find the smallest value of x that will make it possible to accommodate 200 cars.
   68 ft
3. If the technician determines that one car needs an 84-square-foot space, find an expression for the number of cars the lots represented at the right can accommodate at one time. $\frac{x(900 + x)}{84}$

© Glencoe/McGraw-Hill T10 *Algebra: Concepts and Applications*

Manipulatives/Modeling

Hands-On Algebra Masters, pp. 107–112

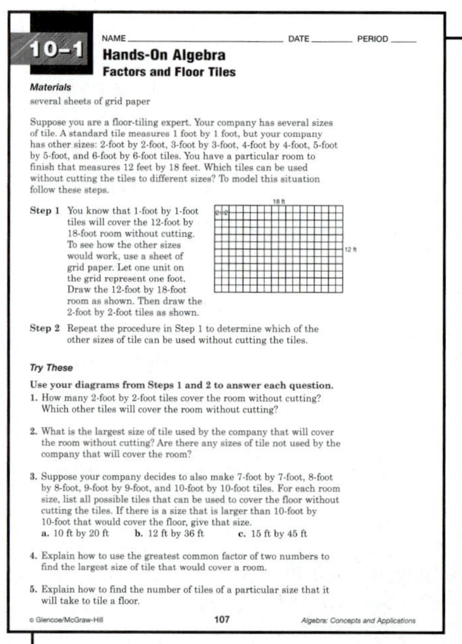

10-1 NAME ______ DATE ______ PERIOD ______

Hands-On Algebra
Factors and Floor Tiles

Materials

several sheets of grid paper

Suppose you are a floor-tiling expert. Your company has several sizes of tile. A standard tile measures 1 foot by 1 foot, but your company has other sizes: 2-foot by 2-foot, 3-foot by 3-foot, 4-foot by 4-foot, 5-foot by 5-foot, and 6-foot by 6-foot tiles. You have a particular room to finish that measures 12 feet by 18 feet. Which tiles can be used without cutting the tiles to different sizes? To model this situation follow these steps.

Step 1 You know that 1-foot by 1-foot tiles will cover the 12-foot by 18-foot room without cutting. To see how the other sizes would work, use a sheet of grid paper. Let one unit on the grid represent one foot. Draw the 12-foot by 18-foot room as shown. Then draw the 2-foot by 2-foot tiles as shown.

Step 2 Repeat the procedure in Step 1 to determine which of the other sizes of tile can be used without cutting the tiles.

Try These

Use your diagrams from Steps 1 and 2 to answer each question.

1. How many 2-foot by 2-foot tiles cover the room without cutting? Which other tiles will cover the room without cutting?
2. What is the largest size of tile used by the company that will cover the room without cutting? Are there any sizes of tile not used by the company that will cover the room?
3. Suppose your company decides to also make 7-foot by 7-foot, 8-foot by 8-foot, 9-foot by 9-foot, and 10-foot by 10-foot tiles. For each room size, list all possible tiles that can be used to cover the floor without cutting the tiles. If there is a size that is larger than 10-foot by 10-foot that would cover the floor, give that size.
   a. 10 ft by 20 ft **b.** 12 ft by 36 ft **c.** 15 ft by 45 ft
4. Explain how to use the greatest common factor of two numbers to find the largest size of tile that would cover a room.
5. Explain how to find the number of tiles of a particular size that it will take to tile a floor.

© Glencoe/McGraw-Hill 107 *Algebra: Concepts and Applications*

Technology/Multimedia

Graphing Calculator Masters, pp. 26–28

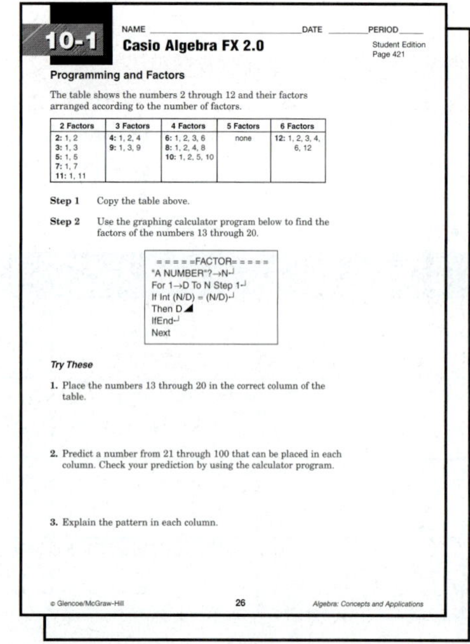

10-1 NAME ______ DATE ______ PERIOD ______

Casio Algebra FX 2.0

Student Edition Page 421

Programming and Factors

The table shows the numbers 2 through 12 and their factors arranged according to the number of factors.

2 Factors	3 Factors	4 Factors	5 Factors	6 Factors
2: 1, 2 **3:** 1, 3 **5:** 1, 5 **7:** 1, 7 **11:** 1, 11	**4:** 1, 2, 4 **9:** 1, 3, 9	**6:** 1, 2, 3, 6 **8:** 1, 2, 4, 8 **10:** 1, 2, 5, 10	none	**12:** 1, 2, 3, 4, 6, 12

Step 1 Copy the table above.

Step 2 Use the graphing calculator program below to find the factors of the numbers 13 through 20.

```
= = = = =FACTOR= = = = =
"A NUMBER"?→N↵
For 1→D To N Step 1↵
If Int (N/D) = (N/D)↵
Then D◢
IfEnd↵
Next
```

Try These

1. Place the numbers 13 through 20 in the correct column of the table.
2. Predict a number from 21 through 100 that can be placed in each column. Check your prediction by using the calculator program.
3. Explain the pattern in each column.

© Glencoe/McGraw-Hill 26 *Algebra: Concepts and Applications*

10-3 NAME ______ DATE ______ PERIOD ______

TI-83/84 Plus

Programming and Factoring

The calculator program, TRINOM, can help you determine whether a polynomial of the form $x^2 + bx + c$, where b and c are integers and $c \neq 0$, can be factored. If it can be factored, the program provides the numbers that allow you to write the factors.

Use the graphing calculator program to find the factors of $x^2 - 6x + 8$. Enter -6 for B and 8 for C.

```
PROGRAM:TRINOM
: Disp "X²+BX+C"
: Disp "ENTER B"
: Input B
: Disp "ENTER C"
: Input C
: For (N, -abs(C), abs (C))
: Disp -N
: End
```

The numbers –2 and –4 should be displayed on the screen. This means that $x^2 - 6x + 8$ in factored form is $(x - 2)(x - 4)$. To run the program again, press ENTER.

Try These

1. Run the program to factor $x^2 + 9x + 14$. Enter 9 for B and 14 for C. What two numbers does the calculator display? What is the factored form of $x^2 + 9x + 14$?
2. If you enter values for B and C and the calculator displays only one number, then both factors are the same. What is the factored form of $x^2 - 6x + 9$?
3. If you enter values for B and C and the calculator displays no numbers, then the trinomial cannot be factored. Run the program for $x^2 - 5x + 2$. Can this trinomial be factored? Explain.
4. Use the program to factor each trinomial.
   a. $x^2 + 3x - 108$ **b.** $x^2 - 26x + 169$ **c.** $x^2 + 27x + 140$
   d. $x^2 - 40x + 400$ **e.** $x^2 - 17x + 72$ **f.** $x^2 + 7x - 78$

© Glencoe/McGraw-Hill 27 *Algebra: Concepts and Applications*

Assessment Resources

Type	Student Edition	Teacher's Wraparound Edition	Chapter 10 Resource Masters
Ongoing Assessment	Quizzes 1 and 2, pp. 433, 444	5-Minute Check, pp. 420, 428, 434, 440, 445	Mid-Chapter Test, p. 468 Quizzes A and B, p. 469
Mixed Review	Mixed Review, pp. 425, 433, 439, 444, 449 Standardized Test Practice, Chapters 1–10, pp. 454–455		Cumulative Review, p. 470 Standardized Test Practice, pp. 471–472
Error Analysis	You Decide, pp. 424, 443	Error Analysis, pp. 424, 432, 438, 443, 448	
Standardized Test Prep	Standardized Test Practice, pp. 425, 433, 439, 444, 449 Standardized Test Practice, Chapters 1–10, pp. 454–455		Standardized Test Practice, pp. 471–472
Open-Ended Assessment	Writing Math, pp. 431, 448 Investigation, pp. 426–427 Portfolio, pp. 419, 427	Modeling: pp. 433, 439 Speaking: pp. 444, 449 Writing: pp. 425	Extended Response Assessment, p. 467
Chapter Assessment	Study Guide and Assessment, pp. 450–452 Chapter Test, p. 453		Multiple-Choice Tests (Forms 1A, 1B), pp. 459–462 Free-Response Tests (Forms 2A, 2B), pp. 463–466

Additional Chapter Resources

Student Edition
Hands-On Algebra, pp. 428, 434–435, 440
Graphing Calculator Exploration, p. 421

Teacher's Classroom Resources
Manipulatives/Modeling
Teacher's Guide for Overhead Manipulative Resources

Meeting Individual Needs
Prerequisite Skills Booklet
Spanish Study Guide and Assessment, pp. 68–72, 123–124

Teaching Aids
Solutions Manual
5-Minute Check Transparencies

Glencoe Technology

Instructional
AlgePASS CD-ROM, Lessons 23, 24
Interactive Chalkboard CD-ROM
StudentWorks Plus CD-ROM
Multimedia Applications CD-ROM, Activity 10
Vocabulary PuzzleMaker (online)

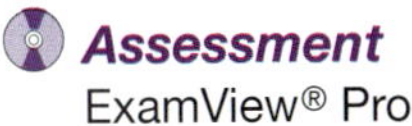

Assessment
ExamView® Pro

Teaching Aids
Answer Key Maker CD-ROM
TeacherWorks CD-ROM

Visit ***www.algconcepts.com*** *for data updates, career information, games, and other interactive activities.*

CHAPTER PREVIEW 10

Mathematics of the Chapter

This chapter provides students with an in-depth study of factoring. Students will begin by finding the greatest common factor of a set of numbers or monomials. Students will then factor polynomials using greatest common factors and the Distributive Property. Next, they will factor trinomials of the form $x^2 + bx + c$ and $ax^2 + bx + c$. Finally, students will learn to recognize and factor the differences of squares and perfect square trinomials.

What You'll Learn

Have students read over the lists of key ideas and key vocabulary. Have them make a list of any words with which they are not familiar.

Why It's Important

Point out to students that this is only one of many reasons why each objective is important. Others are provided in each lesson.

CHAPTER 10 Factoring

What You'll Learn

Key Ideas

- Find greatest common factor (GCF). *(Lesson 10–1)*
- Use the GCF and the Distributive Property to factor polynomials. *(Lesson 10–2)*
- Factor trinomials of the forms $x^2 + bx + c$ and $ax^2 + bx + c$. *(Lessons 10–3 and 10–4)*
- Recognize and factor the differences of squares and perfect square trinomials. *(Lesson 10–5)*

Key Vocabulary

difference of squares *(p. 447)*
factoring *(p. 428)*
greatest common factor (GCF) *(p. 422)*
perfect square trinomial *(p. 430)*

Why It's Important

Landscaping Landscape architects design parks, homes, and businesses to be both functional and beautiful. They plan the locations of buildings, roads, and walkways and the arrangement of plants. They are also often involved with protecting the environment and historical sites.

Factoring polynomials allows you to solve real-world problems involving polynomial equations. You will use factoring to plan a walkway as a landscape architect would in Lesson 10–2.

CHAPTER 10 LINKS					
Lesson	**10–1**	**10–2**	**10–3**	**10–4**	**10–5**
Math in the Workplace	Crafts	Marine Biology	Biology	Manufacturing	Manufacturing
Applications and Connections	Transportation	Landscaping Construction	Gardening Genetics Sales		
Math Integration	Geometry Math History	Geometry	Geometry	Geometry Measurement	Geometry Number Theory

Study these lessons to improve your skills.

✓Check Your Readiness

✓ Lesson 8–1, pp. 336–340

Write each expression using exponents.

1. $3 \cdot 3 \cdot 3 \cdot 3$ **3^4**
2. $9 \cdot 9 \cdot 9$ **9^3**
3. $10 \cdot 10 \cdot 10 \cdot 10 \cdot 10$ **10^5**
4. $2 \cdot 2 \cdot 2 \cdot 3 \cdot 3$ **$2^3 3^2$**
5. $5 \cdot 2 \cdot 2 \cdot 5 \cdot 5 \cdot 5$ **$2^2 5^4$**
6. $n \cdot n \cdot n \cdot n \cdot n \cdot n$ **n^6**
7. $p \cdot p \cdot p \cdot q \cdot q \cdot q$ **$p^3 q^3$**
8. $12 \cdot a \cdot a \cdot a \cdot a \cdot b$ **$12a^4 b$**

✓ Lesson 8–2, pp. 341–345

Simplify each expression.

9. $\frac{5^3}{5}$ **5^2**
10. $\frac{x^6}{x^2}$ **x^4**
11. $\frac{k^{12}}{k^3}$ **k^9**
12. $\frac{a^3b^6}{ab}$ **a^2b^5**
13. $\frac{12m^4n}{2m^2}$ **$6m^2n$**
14. $\frac{45cd^5}{9cd^2}$ **$5d^3$**

✓ Lesson 1–4, pp. 19–23

Simplify each expression.

15. $2(x + y)$ **$2x + 2y$**
16. $3(4d + 1)$ **$12d + 3$**
17. $8(6b + 2)$ **$48b + 16$**
18. $15a - 3a$ **$12a$**
19. $6f + 9f - 8$ **$15f - 8$**
20. $(9m + 3n) + (2m - 6n)$ **$11m - 3n$**

FOLDABLES™ Study Organizer

Make this Foldable to help you organize your Chapter 10 notes. Begin with a sheet of plain $8\frac{1}{2}$" by 11" paper.

❶ **Fold** in half lengthwise.

❷ **Fold** again in thirds.

❸ **Open.** Cut along the second fold to make three tabs.

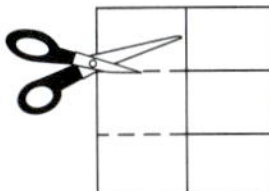

❹ **Label** each tab as shown.

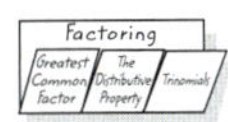

Reading and Writing As you read and study the chapter, unfold each tab and fill the journal with notes and examples.

www.algconcepts.com/chapter_readiness

Check Your Readiness

This section provides a review of the basic concepts needed before beginning Chapter 10. Lesson references are included for additional student help.

Use Exercises	To Prepare for Lesson(s)
1–8	10–1
9–14	10–2, 10–3
15–20	10–2

Quick Review Math Handbook

hot words hot topics

Glencoe Algebra Lesson	Quick Review Math Handbook Hot Topic
10–1	6–2
Ch. 10 Inv.	7–5
10–2	1–2
10–3	
10–4	
10–5	3–2

Noteables™

Interactive Study Notebook with Foldables™

This note-taking guide reinforces excellent note-taking skills. It includes:

- ✓ **Build Your Vocabulary** tool for including mathematical terminology in notes.
- ✓ **Review It** activities with links to prior lessons and items to include in **Foldables™**.
- ✓ **Bringing It All Together** feature to help students review for the Chapter Test.
- ✓ **Are You Ready for the Chapter Test?** feature to allow students assess their own readiness for the Chapter Test.
- ✓ **Teacher Edition** with transparencies to guide note taking. Corresponds to the *Teacher's Wraparound Edition* In-Class Examples and *Interactive Chalkboard CD-ROM.*

FOLDABLES™ Study Organizer Students will use this Foldable journal to record key information about factoring. Have students use each tab to write an explanation of how to perform the objectives on the tabs. Have them include sample problems and solutions to illustrate their explanations.

Lesson 10-1

1 FOCUS

5-Minute Check
Chapter 9

1. Find the degree of the polynomial $2a^3 + 3a^2b^2$. **4**
2. Find $(a^2 - b + 6) - (2a^2 - b)$. $-a^2 + 6$

Find each product.

3. $2b^2(3b^2 - 5b + 6)$ $6b^4 - 10b^3 + 12b^2$
4. $(2x - 5)^2$ $4x^2 - 20x + 25$
5. Find the area of the shaded region in simplest form.

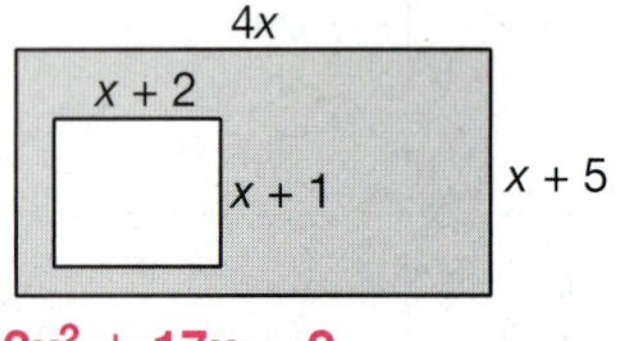

$3x^2 + 17x - 2$

Motivating the Lesson

Real-World Connection Generate a discussion of greatest common factor by describing the following situation. Suppose you are making square pillows for your bedroom. One fabric you choose comes in a width of 48 inches and the other fabric comes in a width of 60 inches. What is the length of the largest pillow you can make if all of the pillows are the same size and there is no fabric wasted? **12 in.**

MODELING

An alternative hands-on option using several sheets of grid paper is available for teaching this lesson.

10–1 Factors

What You'll Learn

You'll learn to find the greatest common factor of a set of numbers or monomials.

Why It's Important

Crafts Quilters use greatest common factors when they cut fabric. *See Exercise 57.*

Recall that when two or more numbers are multiplied, each number is a *factor* of the product. For example, 12 can be expressed as the product of different pairs of whole numbers. Factors can be shown geometrically.

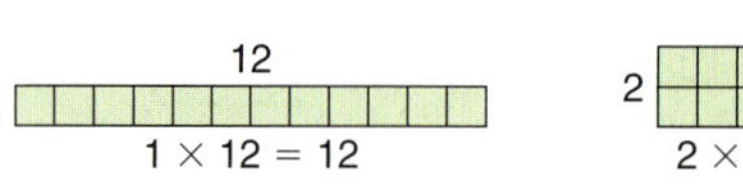

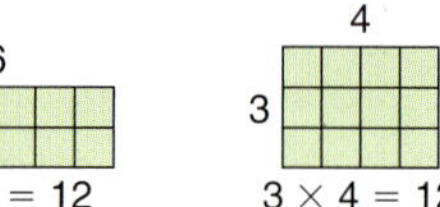

The whole numbers 1, 12, 2, 6, 3, and 4 are the factors of 12.

Some whole numbers have exactly two factors, the number itself and 1. Recall that these numbers are called *prime numbers*. Whole numbers that have more than two factors, such as 12, are called *composite numbers*.

Prime Numbers Less Than 20	2, 3, 5, 7, 11, 13, 17, 19
Composite Numbers Less Than 20	4, 6, 8, 9, 10, 12, 14, 15, 16, 18
Neither Prime nor Composite	0, 1

Examples

Find the factors of each number. Then classify each number as *prime* or *composite*.

1 **72**

To find the factors of 72, list all pairs of whole numbers whose product is 72.

1×72 2×36 3×24 4×18 6×12 8×9

The factors of 72 are 1, 2, 3, 4, 6, 8, 9, 12, 18, 24, 36, and 72. Since 72 has more than two factors, it is a composite number.

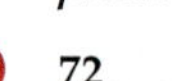

interNET CONNECTION

Data Update For the latest information on prime numbers, visit: www.algconcepts.com

2 **37**

There is only one pair of whole numbers whose product is 37.

1×37

The factors of 37 are 1 and 37. Therefore, 37 is a prime number.

Your Turn **d. 1, 3, 17. 51; C**

a. 25 **1, 5, 25; C** **b.** 23 **1, 23; P** **c.** 79 **1, 79; P** **d.** 51

Resource Manager

Reproducible Masters

Chapter 10 Resource Masters
- *Study Guide,* p. 433
- *Skills Practice,* p. 434
- *Practice,* p. 435
- *Reading to Learn Mathematics,* p. 436
- *Enrichment,* p. 437

Hands-On Algebra, p. 107
Graphing Calculator, p. 26

Transparencies
5-Minute Check, 10–1

Technology/Multimedia
Interactive Chalkboard CD-ROM

You can use a graphing calculator to investigate factor patterns.

Graphing Calculator Exploration

Graphing Calculator Tutorial
See pp. 750–753.

The table below shows the numbers 2 through 12 and their factors arranged by the number of factors.

2 Factors	3 Factors	4 Factors	5 Factors	6 Factors
2: 1, 2 **3:** 1, 3 **5:** 1, 5 **7:** 1, 7 **11:** 1, 11	**4:** 1, 2, 4 **9:** 1, 3, 9	**6:** 1, 2, 3, 6 **8:** 1, 2, 4, 8 **10:** 1, 2, 5, 10	none	**12:** 1, 2, 3, 4, 6, 12

Step 1: Copy the table above.

Step 2: Use the graphing calculator program below to find the factors of the numbers 13 through 20.

```
PROGRAM:FACTOR
:Input "ENTER NUMBER", N
:For (D, 1, N)
:If iPart (N/D) = (N/D)
:Disp D
:END
```

Try These **1–3. See margin.**

1. Place the numbers 13 through 20 in the correct column of the table.
2. Predict a number from 21 through 100 for each column. Check your prediction by using the calculator program.
3. Explain the pattern in each column.

Since $4 \cdot 3 = 12$, 4 is a factor of 12. However, it is not a prime factor of 12 because 4 is not a prime number. Recall that when a number is expressed as a product of prime factors, the expression is called the *prime factorization* of the number.

You can use a *factor tree* to find the prime factorization of a number. Two different factor trees are shown for the prime factorization of 12.

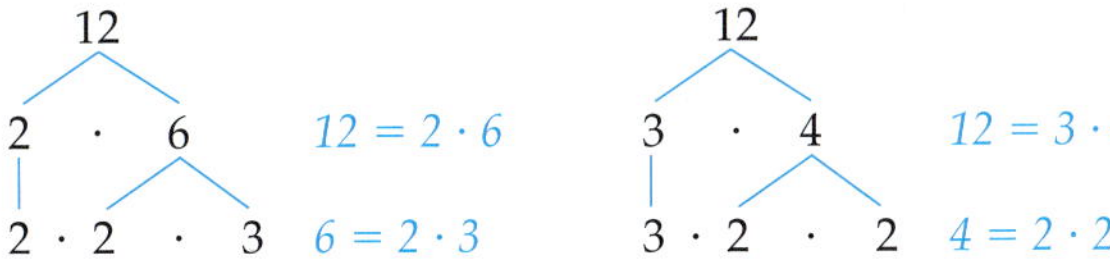

2 TEACH

In-Class Examples
Examples 1–2

Find the factors of each number. Then classify each number as prime or composite.

1 47 **1, 47; prime**

2 35 **1, 5, 7, 35; composite**

Teaching Tip When finding the prime factorization of 12, emphasize that no matter how the factor tree you construct branches, the result is the same. Because you keep branching until all the factors are prime numbers, the ends of the branches will be all the prime factors of a number, no matter what the tree looks like.

Answers

1. 2 factors: 13, 17, 19; 3 factors: none; 4 factors: 14, 15; 5 factors: 16; 6 factors: 18, 20
2. Sample answers: 2 factors: 23; 3 factors: 25; 4 factors: 22; 5 factors: 81; 6 factors: 45
3. 2 factors: prime numbers; 3 factors: squares of prime numbers; 4 factors: numbers with 2 prime factors or the cube of a prime number; 5 factors: a prime number raised to the fourth power; 6 factors: numbers with 3 prime factors, two of which are equal.

Graphing Calculator Exploration

The second and third lines of the program instruct the calculator to test each integer D from 1 to N to see if the integer part of $\frac{N}{D}$ is equal to the value of $\frac{N}{D}$, that is, to see if D divides evenly into N. The fourth line of the program tells the calculator to display each number D that passes this test.

In-Class Examples

Examples 3–5

Factor each monomial.

3 $16b^2c^2$ $2 \cdot 2 \cdot 2 \cdot 2 \cdot b \cdot b \cdot c \cdot c$

4 $9c^3d$ $3 \cdot 3 \cdot c \cdot c \cdot c \cdot d$

5 $-15xy^2$ $-1 \cdot 3 \cdot 5 \cdot x \cdot y \cdot y$

Teaching Tip As you discuss the definition of the *greatest common factor*, some students may find the term intimidating. Point out that the definition means just what it sounds like when the term is broken into its words. The GCF of two or more numbers is the *greatest* number that is a *factor* shared by, or *common* to, both or all of the numbers. Help students practice finding GCFs of easy numbers such as 4 and 6, 12 and 15, or 10 and 15 to help reinforce the definition.

All of the factors in the last row are prime numbers. The factors are in a different order, but the result is the same. Except for the order of the factors, there is only one prime factorization of a number. Thus, the prime factorization of 12 is $2 \cdot 2 \cdot 3$ or $2^2 \cdot 3$.

You can use prime factorization to factor monomials. A monomial is in *factored form* when it is expressed as the product of prime numbers and variables and no variable has an exponent greater than 1.

Examples

Factor each monomial.

Look Back

Monomials: Lesson 9–1

3 $12a^2b$

$12a^2b = 2 \cdot 2 \cdot 3 \cdot a \cdot a \cdot b$ *$12 = 2 \cdot 2 \cdot 3,\ a^2 = a \cdot a$*

4 $100mn^3$

$100mn^3 = 2 \cdot 2 \cdot 5 \cdot 5 \cdot m \cdot n \cdot n \cdot n$ *$100 = 2 \cdot 2 \cdot 5 \cdot 5,\ n^3 = n \cdot n \cdot n$*

5 $-25x^2$

To factor a negative integer, first express it as the product of a whole number and −1. Then find the prime factorization.

$-25x^2 = -1 \cdot 25x^2$ *$-25 = -1 \cdot 25$*

$= -1 \cdot 5 \cdot 5 \cdot x \cdot x$ *$25 = 5 \cdot 5$*

Your Turn

e. $15ab^2$ **f.** $84yz^2$ **g.** $-36b^3$

e. $3 \cdot 5 \cdot a \cdot b \cdot b$
f. $2 \cdot 2 \cdot 3 \cdot 7 \cdot y \cdot z \cdot z$
g. $-1 \cdot 2 \cdot 2 \cdot 3 \cdot 3 \cdot b \cdot b \cdot b$

Two or more numbers may have some common prime factors. Consider the prime factorization of 36 and 42.

$$36 = 2 \cdot 2 \cdot 3 \cdot 3$$
$$42 = 2 \cdot 3 \cdot 7$$

Line up the common factors.

The integers 36 and 42 have 2 and 3 as common prime factors. The product of these prime factors, $2 \cdot 3$ or 6, is called the **greatest common factor (GCF)** of 36 and 42. The GCF is the greatest number that is a factor of both original numbers.

Greatest Common Factor The greatest common factor of two or more integers is the product of the prime factors common to the integers.

The GCF of two or more monomials is the product of their common factors when each monomial is expressed in factored form.

Examples

Find the GCF of each set of numbers or monomials.

6 **24, 60, and 72**

$24 = \boxed{2} \cdot \boxed{2} \cdot 2 \cdot \boxed{3}$ — *Find the prime factorization of each number.*

$60 = \boxed{2} \cdot \boxed{2} \cdot \quad \boxed{3} \cdot 5$ — *Line up as many factors as possible.*

$72 = \boxed{2} \cdot \boxed{2} \cdot 2 \cdot \boxed{3} \cdot 3$ — *Circle the common factors.*

The GCF of 24, 60, and 72 is $2 \cdot 2 \cdot 3$ or 12.

7 **15 and 8**

$15 = 3 \cdot 5$ — *Find the prime factorization of each number.*

$8 = \quad 2 \cdot 2 \cdot 2$ — *Look for common factors.*

There are no common prime factors. The only common factor is 1. So, the GCF of 15 and 8 is 1.

Two numbers or monomials whose greatest common factor is 1 are called *relatively prime*. So, 15 and 8 are relatively prime.

8 **$15a^2b$ and $18ab$**

$15a^2b = \boxed{3} \cdot 5 \cdot \boxed{a} \cdot a \cdot \boxed{b}$ — *Find the prime factorization of each monomial.*

$18ab = 2 \cdot 3 \cdot \boxed{3} \cdot \boxed{a} \cdot \boxed{b}$ — *Circle the common factors.*

The GCF of $15a^2b$ and $18ab$ is $3 \cdot a \cdot b$ or $3ab$.

Your Turn **j. $12abc$**

h. 75, 100, and 150 **25** **i.** $5a$ and $8b$ **1** **j.** $24ab^2c$ and $60a^2bc$

In-Class Examples

Examples 6–8

Find the GCF of each set of numbers or monomials.

6 12, 20, and 24 **4**

7 8 and 9 **1**

8 $21ab^2$ and $9a^2b$ **$3ab$**

Example 9

The area of a rectangle is 24 square inches. Find the length and width so that the rectangle has the least perimeter. Assume that the length and width are both whole numbers. **The least perimeter is 20 inches. The rectangle has a length of 6 inches and a width of 4 inches.**

Knowing the factors of a number can help you with geometry.

Example
Geometry Link

9 **The area of a rectangle is 18 square inches. Find the length and width so that the rectangle has the least perimeter. Assume that the length and width are both whole numbers.**

Explore You know that the area of the rectangle is 18 square inches. You want to find the length and width so that the rectangle has the least perimeter.

Plan Find the factors of 18 and draw rectangles with each length and width. Then find each perimeter.

Solve

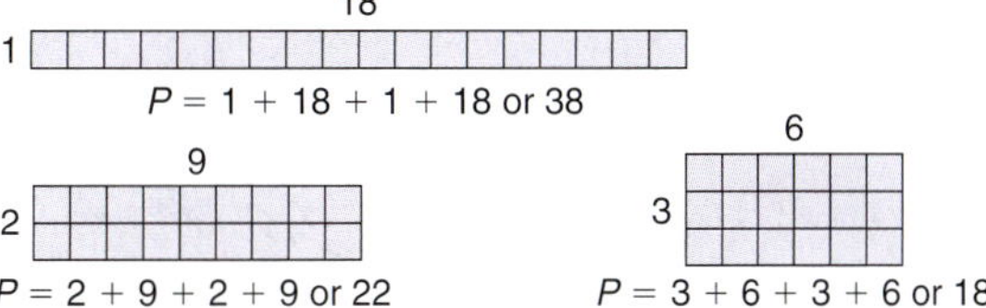

$P = 2 + 9 + 2 + 9$ or 22

$P = 3 + 6 + 3 + 6$ or 18

The least perimeter is 18 inches. The rectangle has a length of 6 inches and a width of 3 inches. *Examine this solution.*

***Study Guide*, p. 433**

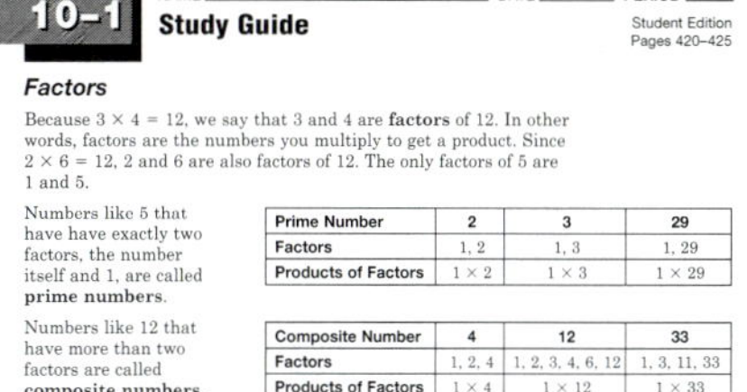

10-1 NAME ______ DATE ______ PERIOD ______

Study Guide

Student Edition Pages 420–425

Factors

Because $3 \times 4 = 12$, we say that 3 and 4 are **factors** of 12. In other words, factors are the numbers you multiply to get a product. Since $2 \times 6 = 12$, 2 and 6 are also factors of 12. The only factors of 5 are 1 and 5.

Numbers like 5 that have have exactly two factors, the number itself and 1, are called **prime numbers**.

Prime Number	2	3	29
Factors	1, 2	1, 3	1, 29
Products of Factors	1×2	1×3	1×29

Numbers like 12 that have more than two factors are called **composite numbers.**

Composite Number	4	12	33
Factors	1, 2, 4	1, 2, 3, 4, 6, 12	1, 3, 11, 33
Products of Factors	1×4 2×2	1×12 2×6 3×4	1×33 3×11

When two numbers are written as the product of their prime factors, they are in factored form.

Example 1: Write 45 in factored form.

$45 = 9 \cdot 5$
$= 3 \cdot 3 \cdot 5$ *Keep factoring until all factors are prime numbers.*

The factored form of 45 is $3 \cdot 3 \cdot 5$.

Example 2: Write $12x^2y$ in factored form.

$12x^2y = 3 \cdot 4 \cdot x \cdot x \cdot y$
$= 3 \cdot 2 \cdot 2 \cdot x \cdot x \cdot y$

The factored form of $12x^2y$ is $3 \cdot 2 \cdot 2 \cdot x \cdot x \cdot y$.

Find the factors of each number. Then classify each number as prime or composite.

1. 10 **1, 2, 5, 10; composite**
2. 7 **1, 7; prime**
3. 15 **1, 3, 5, 15; composite**
4. 21 **1, 3, 7, 21; composite**
5. 31 **1, 31; prime**
6. 49 **1, 7, 49; composite**
7. 47 **1, 47; prime**
8. 39 **1, 3, 13, 39; composite**

Factor each monomial.

9. $18a$ **$2 \cdot 3 \cdot 3 \cdot a$**
10. $35xy$ **$5 \cdot 7 \cdot x \cdot y$**
11. c^3 **$c \cdot c \cdot c$**
12. $20r^2$ **$2 \cdot 2 \cdot 5 \cdot r \cdot r$**
13. $6y^2z$ **$2 \cdot 3 \cdot y \cdot y \cdot z$**

Check for Understanding

Communicating Mathematics

1, 3. See margin.
2. Sample answer: 8, 12

1. **List** the prime numbers between 20 and 50.
2. **Name** two numbers whose GCF is 4.
3. **Explain** how to find the GCF of $8x^2$ and $16x$.
4. **You Decide** Jennifer believes that $2 \cdot 3 \cdot 4 \cdot 5$ is the prime factorization of 120, but Arturo disagrees. Who is correct? Explain. **Arturo; 4 is *not* a prime number.**

Vocabulary
greatest common factor (GCF)

Guided Practice

Getting Ready Find the prime factorization of each number.

Sample 1: 28
Solution: $28 = 2 \cdot 2 \cdot 7$

Sample 2: 60
Solution: $60 = 2 \cdot 2 \cdot 3 \cdot 5$

5. 21 $3 \cdot 7$ 6. 72 $2^3 \cdot 3^2$ 7. 51 $3 \cdot 17$
8. 150 $2 \cdot 3 \cdot 5^2$ 9. 108 $2^2 \cdot 3^3$ 10. 110 $2 \cdot 5 \cdot 11$

Examples 1 & 2
Find the factors of each number. Then classify each number as *prime* or *composite*.
11. 42 **1, 2, 3, 6, 7, 14, 21, 42; C** 12. 47 **1, 47; P**

Examples 3–5
Factor each monomial.
13. $24x^2y$ $2 \cdot 2 \cdot 2 \cdot 3 \cdot x \cdot x \cdot y$ 14. $-16ab^2c$ $-1 \cdot 2 \cdot 2 \cdot 2 \cdot 2 \cdot a \cdot b \cdot b \cdot c$

Examples 6–8
Find the GCF of each set of numbers or monomials.
15. 15, 70 **5** 16. 16, 24, 28 **4** 17. 20, 21 **1**
18. $2x$, $5y$ **1** 19. $7y^2$, $14y^3$ $7y^2$ 20. $-12ab$, $4a^2b^3$ $4ab$

Example 9
21. **Geometry** The area of a rectangle is 72 square centimeters. Find the length and width so that the rectangle has the greatest perimeter. Assume that the length and width are both whole numbers. **72 cm, 1 cm**

Exercises

Practice

Find the factors of each number. Then classify each number as *prime* or *composite*. **23. 1, 2, 4, 5, 10, 20; C**

A
22. 19 **1, 19; P** 23. 20 24. 61 **1, 61; P**
25. 45 **1, 3, 5, 9, 15, 45; C** 26. 49 **1, 7, 49; C** 27. 91 **1, 7, 13, 91; C**

Factor each monomial. **28.** $2 \cdot 2 \cdot 5 \cdot x \cdot x$ **29.** $-1 \cdot 3 \cdot 5 \cdot a \cdot a \cdot b$ **30.** $-1 \cdot 2 \cdot 2 \cdot 2 \cdot 3 \cdot c \cdot c \cdot c$

B
28. $20x^2$ 29. $-15a^2b$ 30. $-24c^3$
31. $50m^2n^2$ $2 \cdot 5 \cdot 5 \cdot m \cdot m \cdot n \cdot n$ 32. $44r^2s$ $2 \cdot 2 \cdot 11 \cdot r \cdot r \cdot s$ 33. $90yz^2$ $2 \cdot 3 \cdot 3 \cdot 5 \cdot y \cdot z \cdot z$

3 PRACTICE/APPLY

Error Analysis

Watch for students who circle the third factor of 2 common to the numbers 16 and 24 in Exercise 16, but not common to the number 28. ***Prevent by*** reminding students that the greatest common factor of a set of numbers is found using only factors that are shared by *all* the numbers. You may want to show as an extra example on the board or overhead finding the GCF of 6, 15, and 20. Though the factors 2, 3, and 5 are each shared by two of the three numbers, their GCF is 1.

Assignment Guide

Basic: 23–55 odd, 57–68
Average: 22–54 even, 55–68

Answers

1. 23, 29, 31, 37, 41, 43, 47
3. Write $8x^2$ and $16x$ as the product of prime numbers and variables where no variable has an exponent. Then multiply the common prime factors and variables. The GCF is $8x$.

Skills Practice, **p. 434, and *Practice,* p. 435 (shown)**

10–1 NAME ________ DATE ______ PERIOD ____
Practice
Student Edition Pages 420–425

Factors

Find the factors of each number. Then classify each number as prime or composite.

1. 36 1, 2, 3, 4, 6, 9, 12, 18, 36; C
2. 31 1, 31; P
3. 28 1, 2, 4, 7, 14, 28; C
4. 70 1, 2, 5, 7, 10, 14, 35, 70; C
5. 43 1, 43; P
6. 27 1, 3, 9, 27; C
7. 28 1, 2, 4, 7, 14; C
8. 97 1, 97; P

Factor each monomial.

9. $30m^2n$ $2 \cdot 3 \cdot 5 \cdot m \cdot m \cdot n$
10. $-12x^2y^3$ $-1 \cdot 2 \cdot 2 \cdot 3 \cdot x \cdot x \cdot y \cdot y \cdot y$
11. $-21ab^2$ $-1 \cdot 3 \cdot 7 \cdot a \cdot b \cdot b$
12. $36r^3s$ $2 \cdot 2 \cdot 3 \cdot 3 \cdot r \cdot r \cdot r \cdot s$
13. $63x^3yz^2$ $3 \cdot 3 \cdot 7 \cdot x \cdot x \cdot x \cdot y \cdot z \cdot z$
14. $-40pq^2r^2$ $-1 \cdot 2 \cdot 2 \cdot 2 \cdot 5 \cdot p \cdot q \cdot q \cdot r \cdot r$

Find the GCF of each set of numbers or monomials.

15. 27, 18 9
16. 9, 12 3
17. 45, 56 1
18. 4, 8, 16 4
19. 32, 36, 38 2
20. 24, 36, 48 12
21. $6x$, $9x$ $3x$
22. $5y^2$, $15y$ $5y$
23. $14c^2$, $-13d$ 1
24. $25mn^2$, $20m$ $5m$
25. $12ab^2$, $18ab$ $6ab$
26. $-28x^2y^3$, $21xy^2$ $7xy^2$
27. $6xy$, $18y^2$ $6y$
28. $18c^2d$, $27cd^2$ $9cd$
29. $7m$, mn m

Reteaching Activity

Visual/Spatial Learners Have students working in groups or as a class write the numbers from 1 to 100 on a piece of poster board. Then have them complete the method of the Sieve of Eratosthenes by striking through the number 1, then striking through the multiples of 2 in a different color, then through the multiples of 3 in a third color, and so on, until only prime numbers remain, which students can then circle. This will help students learn to recognize prime numbers.

Find the GCF of each set of numbers or monomials.

For Exercises	See Examples
22–27, 52, 54, 57	1, 2
28–33	3, 4, 5
34–51, 53	6, 7, 8
56	1, 2, 6, 7, 8

Homework Help

Extra Practice

See page 711.

34. 24, 40 **8**
35. 12, 8 **4**
36. 17, 21 **1**
37. 18, 36 **18**
38. 20, 30 **10**
39. 45, 72 **9**
40. $3x^2, 3x$ **$3x$**
41. $18y^2, 3y$ **$3y$**
42. $-5ab, 6b^2$ **b**
43. $-18, 45mn$ **9**
44. $24a^2, 60ab$ **$12a$**
45. $9x^2y, 10m^2n$ **1**
46. 6, 8, 12 **2**
47. 20, 21, 25 **1**
48. 18, 30, 54 **6**
49. $5m^2, 15n^2, 25mn$ **5**
50. $6ax^2, 18ay^2, 9az^3$ **$3a$**
51. $15r^2, 35s^2, 70rs$ **5**

52. What is the greatest prime number less than 90? **89**

53. Find the greatest common factor of $5x^2$, $5y^2$, and $10xy$. **5**

54. *Twin primes* are prime numbers that differ by 2, such as 5 and 7. Find two other sets of twin primes that are between 25 and 45. **29 and 31, 41 and 43**

Applications and Problem Solving

55. **Crafts** Ashley wants to make a quilt from two different kinds of fabric. One is 60 inches wide, and the other is 48 inches wide. What are the dimensions of the largest squares she can cut from both fabrics so that no fabric is wasted? **12 in. by 12 in.**

56. **Math History** In 1880, English mathematician John Venn (1834–1923) developed a way to show how sets of numbers are related. The *Venn diagram* shows the prime factors of 12 and 28. The common factors are in the overlapping circles, and the GCF of 12 and 28 is $2 \cdot 2$ or 4.

Prime Factors of 12 and 18

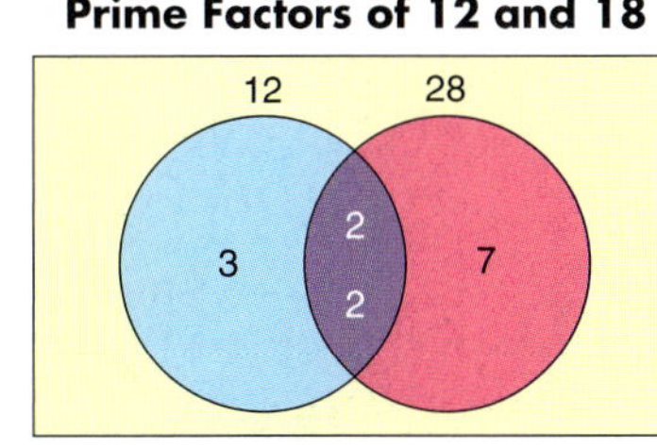

a. Draw a Venn diagram showing the prime factors of 36 and 45. **56a. See margin.**

b. Find the GCF of 36 and 45. **9**

57. **Critical Thinking** Explain why 2 is the only even prime number. **See margin.**

Mixed Review

Find each product. *(Lessons 9–3, 9–4, and 9–5)*

58. $(x + 3)(x - 3)$
59. $(2y + 1)(2y + 1)$
60. $(3a - 2)^2$
61. $(z + 4)(z + 3)$
62. $(x - 5)(x - 4)$
63. $(2n + 1)(n + 4)$
64. $3(x - 5)$ **$3x - 15$**
65. $2a(3 - a^2)$ **$6a - 2a^3$**
66. $4x^2y(3x - 2y)$ **$12x^3y - 8x^2y^2$**

Standardized Test Practice

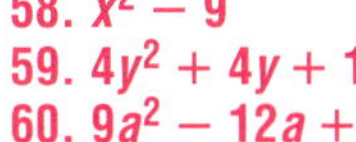

67. **Short Response** In 2025, there are expected to be 817,000,000 cars in use worldwide. Write 817,000,000 in scientific notation. *(Lesson 8–4)* **8.17×10^8**

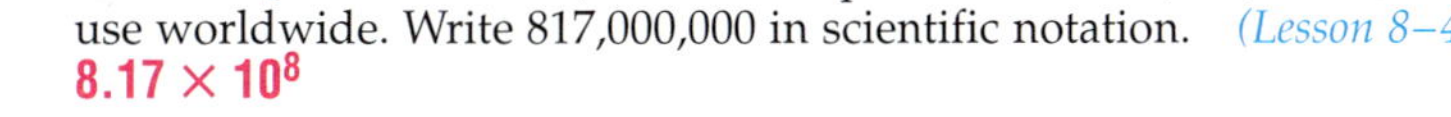

68. **Multiple Choice** Which graph below is *not* the graph of a function? *(Lesson 6–4)* **B**

Ⓐ
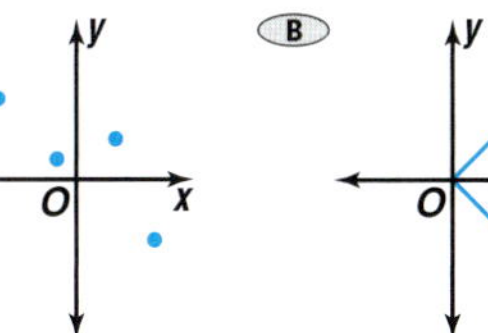

Ⓑ
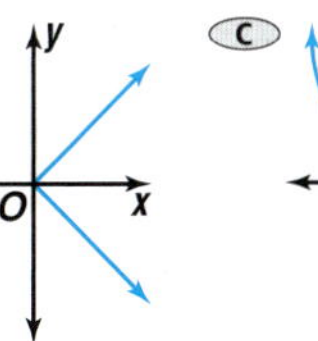

Ⓒ
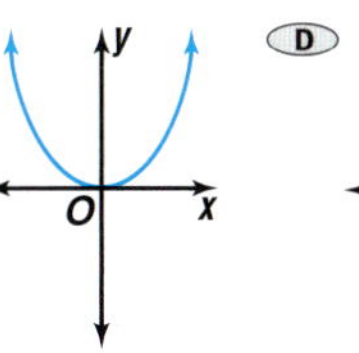

Ⓓ
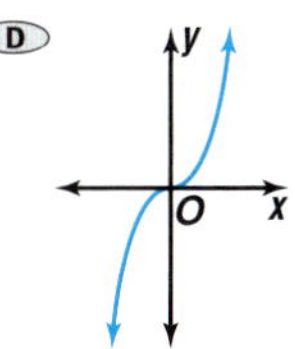

58. $x^2 - 9$
59. $4y^2 + 4y + 1$
60. $9a^2 - 12a + 4$
61. $z^2 + 7z + 12$
62. $x^2 - 9x + 20$
63. $2n^2 + 9n + 4$

www.algconcepts.com/self_check_quiz

4 ASSESS

Open-Ended Assessment

Writing Have students write brief paragraphs describing how to find the greatest common factor of two integers and of two monomials.

Answers

56a.

Prime Factors of 36 and 45

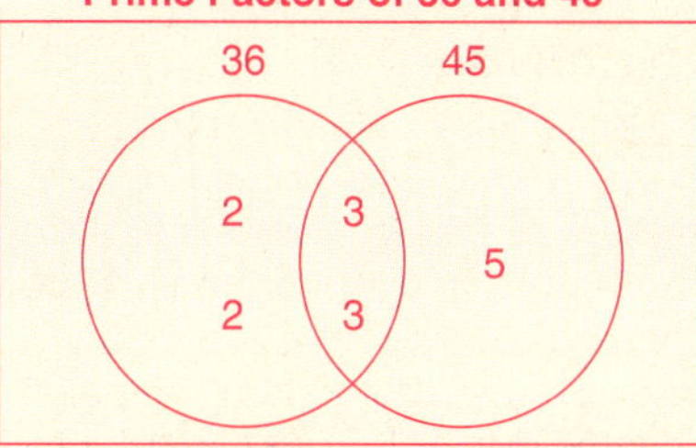

57. Every even number has a factor of 2. So, any even number greater than 2 has at least 3 factors—the number itself, 1, and 2—and is therefore composite.

Extra Credit

Which numbers cannot be expressed as a product only of prime factors? **0, 1, negative integers, and numbers that are not integers**

Enrichment, p. 437

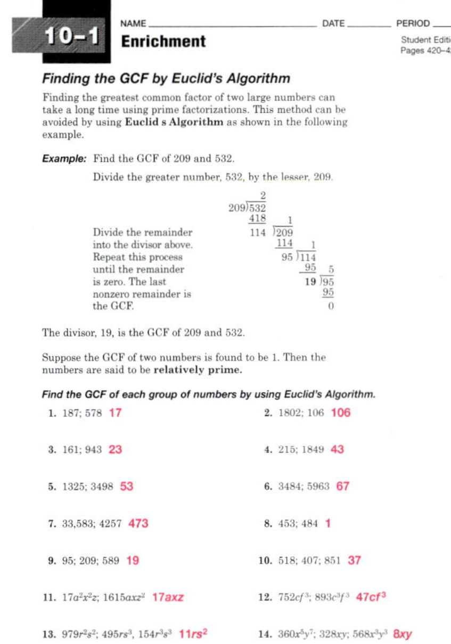

10–1 NAME ______ DATE ______ PERIOD ______

Enrichment

Student Edition Pages 420–425

Finding the GCF by Euclid's Algorithm

Finding the greatest common factor of two large numbers can take a long time using prime factorizations. This method can be avoided by using **Euclid s Algorithm** as shown in the following example.

Example: Find the GCF of 209 and 532.

Divide the greater number, 532, by the lesser, 209.

Divide the remainder into the divisor above. Repeat this process until the remainder is zero. The last nonzero remainder is the GCF.

$$209\overline{)532} = 2,\ 532 - 418 = 114;\quad 114\overline{)209} = 1,\ 209 - 114 = 95;\quad 95\overline{)114} = 1,\ 114 - 95 = 19;\quad 19\overline{)95} = 5,\ 95 - 95 = 0$$

The divisor, 19, is the GCF of 209 and 532.

Suppose the GCF of two numbers is found to be 1. Then the numbers are said to be **relatively prime.**

Find the GCF of each group of numbers by using Euclid's Algorithm.

1. 187; 578 **17**
2. 1802; 106 **106**
3. 161; 943 **23**
4. 215; 1849 **43**
5. 1325; 3498 **53**
6. 3484; 5963 **67**
7. 33,583; 4257 **473**
8. 453; 484 **1**
9. 95; 209; 589 **19**
10. 518; 407; 851 **37**
11. $17a^2x^2z$; $1615axz^2$ **$17axz$**
12. $752cf^3$; $893c^3f^3$ **$47cf^3$**
13. $979r^2s^2$; $495rs^3$, $154r^3s^3$ **$11rs^2$**
14. $360x^5y^7$; $328xy$; $568x^3y^3$ **$8xy$**

Investigation

PREPARE

This optional investigation is designed to be completed by groups of 3–4 students over 1–2 days.

Objective

Students use 1-tiles to form rectangles with counting number areas. They factor these counting numbers to find the dimensions of rectangles with these areas, and then compare the perimeters and areas of the rectangles. Students also investigate patterns in the perimeters of rectangles of various dimensions. They present their findings by making a brochure or a video.

Mathematical Overview

This investigation utilizes the following concepts:

- area of rectangles,
- perimeter of rectangles,
- number patterns, and
- factors of counting numbers.

Suggested Time Management	
Investigation	30–45 min
Extension: Gathering Data	30–45 min
Extension: Summarizing Data	20–30 min

Motivating the Lesson

The task of finding counting number dimensions of a square with equal area and perimeter should make students curious. It may help some students to realize that they are looking for values of ℓ and w so that $\ell w = 2(\ell + w)$. Remind students, however, that though they are looking for equal *numerical values*, the perimeter and area are never actually the same because one is measured in linear units and the other in square units.

Chapter 10 Investigation

Materials

algebra tiles

Perimeter and Area

You know that the perimeter of a rectangle is the distance around the outside of the figure. It is measured in units like inches, centimeters, or feet. The area of a rectangle is the number of square units needed to cover the surface. It is measured in units like square inches or square centimeters. Are there any rectangles in which the measure of the perimeter is equal to the measure of the area? Let's investigate.

Investigate

1. Use the 1-tiles from a set of algebra tiles.

   a. Make a table like the one below with 21 rows.

Area of Rectangle	Possible Dimensions of Rectangle	Perimeter of Rectangle
1		
2		

A 1-by-2 rectangle and a 2-by-1 rectangle have the same perimeter. So, they are listed only once in the table.

   b. Select one tile. The measure of its area is 1. The dimension of this rectangle is 1 by 1. Find the measure of the perimeter of the rectangle and enter it in column 3. **4**

   c. Select two tiles. You can form a 1-by-2 rectangle with the tiles. This rectangle has an area of 2. Write 1 by 2 in column 2. What is the perimeter of this rectangle? Enter it in column 3. **6**

   d. Repeat this process using three tiles. **8**

   e. When you select four tiles, you have two options: a 1-by-4 rectangle or a 2-by-2 rectangle.

   Write 1 by 4 and 2 by 2 in column 2 as the dimensions of the rectangles. Find the perimeters of the two rectangles and enter the results in column 3. **10, 8**

Cooperative Learning

This investigation offers an excellent opportunity for using cooperative groups. For more information on cooperative learning strategies and group management, see *Cooperative Learning in the Mathematics Classroom,* one of the titles in the Glencoe Mathematics Professional Series.

Your table should look like this.

Area of Rectangle	Possible Dimensions of Rectangle	Perimeters of Rectangle
1	1 by 1	4
2	1 by 2	6
3	1 by 3	8
4	1 by 4, 2 by 2	10, 8

2. See students' work.

2. Use tiles to form all possible rectangles with areas from 5 through 20. Write the dimensions in column 2 and the perimeters in column 3.
3. Analyze the data. Which rectangle(s) has the same numerical values for perimeter and area? **4 by 4 or 3 by 6**
4. Make a conjecture about whether there are other rectangles that have the same numerical values for perimeter and area. **There are no other rectangles that satisfy this condition.**

Extending the Investigation

In this extension, you will continue to investigate the area and perimeter of rectangles. Extend your table through at least 30 squares.

- Study the perimeters for only the rectangles whose dimensions are 1 by n. Make an ordered list of the perimeters. Describe the pattern of the perimeters.
- Study the perimeters for only the rectangles whose dimensions are 2 by n, but not 2 by 2. Make an ordered list of the perimeters. Describe the pattern of the perimeters.
- Study the perimeters for only the rectangles whose dimensions are 3 by n, but not 3 by 3. Make an ordered list of the perimeters. Describe the pattern of the perimeters.
- Study the perimeters for rectangles that are also squares. Make an ordered list of these perimeters. Describe the pattern of the perimeters.

Presenting Your Conclusions

Here are some ideas to help you present your conclusions to the class.

- Make a brochure describing your findings. Include figures and tables to illustrate your results.
- Make a video showing the patterns you discovered. You may want to have your classmates take on the roles of rectangles with various dimensions.

Investigation For more information on perimeter and area, visit: www.algconcepts.com

MANAGE

Teaching Tip Remind students that rectangles such as a 1-by-2 rectangle and a 2-by-1 rectangle are really the same because one can be rotated to coincide with the other. Encourage students to list possible dimensions in each row by listing the first dimension in each pair in increasing order. This should help them identify patterns in the Extending the Investigation.

Teaching Tip Ask students what kind of number the perimeter can be. They should come to realize that because the perimeter is twice the sum of the length and width, it must be even. So, any rectangle with an odd value for an area cannot have a perimeter with the same value.

Working in Groups If students work in groups, they can save time by having members split up the task of finding possible dimensions and corresponding perimeters for different areas. Working in groups also brings together more possible observations as students look for patterns and form conjectures.

Working as a Class You can have groups collaborate if they wish to make a video. Students could work on this video with the technology education teacher or comparable person at your school.

Inclusion Strategies

Some students with behavioral difficulties benefit by being isolated from distractions. Instead of having such students work in an unstructured group, make them a part of groups where the tasks are clearly broken down and defined. Then these students can work in an appropriate area away from distractions while completing their tasks. Alternately, such students can be placed in a group that is given close teacher encouragement and guidance through the investigation.

ASSESS

Students should correctly and completely record the rectangle dimensions and perimeters. They should identify the dimensions for which the perimeter and area have the same value, and should offer reasoning for how they made their conjectures. Presentations should describe any patterns observed in the perimeters of rectangles fully.

Students should add their findings and brochures or videos to their portfolios at this time.

Lesson 10–2

1 FOCUS

5-Minute Check
Lesson 10–1

1. Find the factors of 18. Then classify the number as *prime* or *composite*. **1, 2, 3, 6, 9, 18; composite**
2. Find the prime factorization of 126. **$2 \cdot 3 \cdot 3 \cdot 7$**
3. Factor $-36xy^2$. **$-1 \cdot 2 \cdot 2 \cdot 3 \cdot 3 \cdot x \cdot y \cdot y$**
4. Find the GCF of $16ab$ and $20a^2b^2$. **$4ab$**
5. You have two large squares of poster board, one 45 inches wide and the other 54 inches wide. You want to cut both into squares of the same size to use in making a display sign for the science fair. What are the dimensions of the largest squares you can cut from both pieces so that no poster board is left over? **9 in. by 9 in.**

Answers

1.

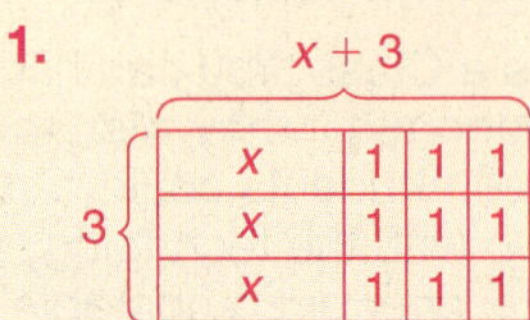

2.

3.

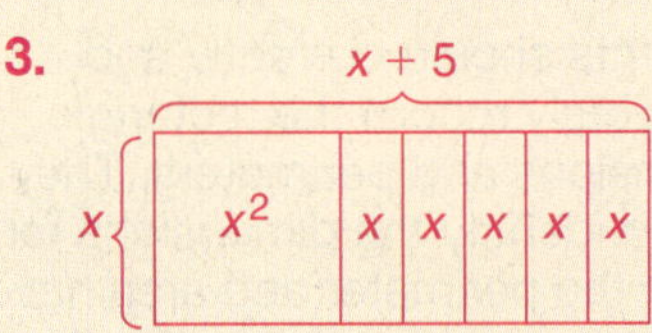

4.

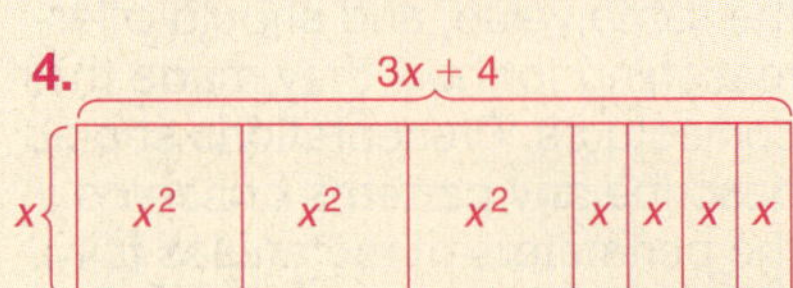

10–2 Factoring Using the Distributive Property

What You'll Learn
You'll learn to use the GCF and the Distributive Property to factor polynomials.

Why It's Important
Marine Biology You can find the height a dolphin jumps out of the water by evaluating an expression that is written in factored form.
See Exercise 46.

Sometimes, you know the product and are asked to find the factors. This process is called **factoring**.

For example, suppose you want to paint a rectangle on a wall and you only have enough paint to cover 20 square feet. If the length of each side must be an integer, what are the dimensions of all the possible rectangles you could paint?

Recall that the formula for the area of a rectangle is $A = \ell w$. If $A = 20$ square feet, then the measures of the length and width of the painted rectangle must be factor pairs of 20. The factor pairs of 20 are 1 and 20, 2 and 10, and 4 and 5. The figures below show rectangles with these factors as measures of length and width.

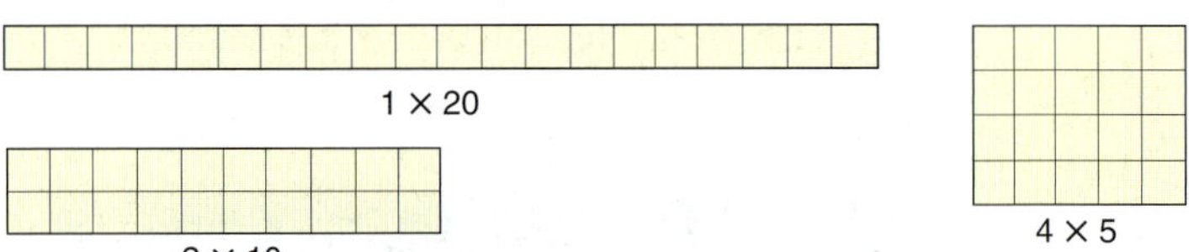

Hands-On Algebra
Algebra Tiles

Materials:  algebra tiles 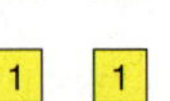 product mat

Use algebra tiles to factor $2x + 8$.

Step 1 Model the polynomial $2x + 8$.

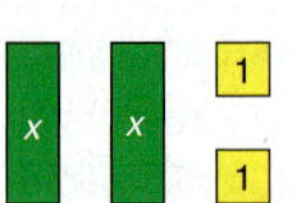

Step 2 Arrange the tiles into a rectangle. The total area of the tiles represents the product. Its length and width represent the factors. The rectangle has a width of 2 and a length of $x + 4$. So, $2x + 8 = 2(x + 4)$.

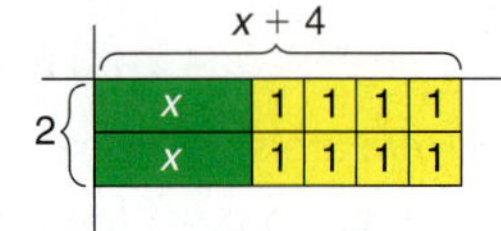

Try These **1–4. See margin.**

Use algebra tiles to factor each binomial.

1. $3x + 9$
2. $4x + 10$
3. $x^2 + 5x$
4. $3x^2 + 4x$

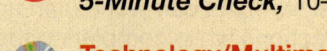

Resource Manager

Reproducible Masters
Chapter 10 Resource Masters
- *Study Guide*, p. 438
- *Skills Practice*, p. 439
- *Practice*, p. 440
- *Reading to Learn Mathematics*, p. 441
- *Enrichment*, p. 442
- *Assessment*, p. 469

Hands-On Algebra, p. 108
School-to-Workplace, p. 10

Transparencies
5-Minute Check, 10–2

Technology/Multimedia
AlgePASS, Lesson 23
Interactive Chalkboard CD-ROM

Look Back

Distributive Property: Lesson 1–4

In Chapter 9, you used the Distributive Property to multiply a polynomial by a monomial.

$$2y(4y + 5) = 2y(4y) + 2y(5)$$
$$= 8y^2 + 10y$$

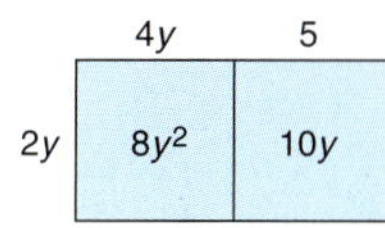

You can reverse this process to express a polynomial in *factored form*. A polynomial is in factored form when it is expressed as the product of polynomials. For example, to factor $8y^2 + 10y$, find the greatest common factor of $8y^2$ and $10y$.

$$8y^2 = 2 \cdot 2 \cdot \textcircled{2} \cdot \textcircled{y} \cdot y$$
$$10y = \textcircled{2} \cdot 5 \cdot \textcircled{y}$$

The GCF of $8y^2$ and $10y$ is $2y$. Write each term as a product of the GCF and its remaining factors. Then use the Distributive Property.

$$8y^2 + 10y = 2y(4y) + 2y(5)$$
$$= 2y(4y + 5) \quad \textit{Distributive Property}$$

$8y^2 + 10y$ written in factored form is $2y(4y + 5)$.

Examples

Factor each polynomial.

1 $30x^2 + 12x$

First, find the GCF of $30x^2$ and $12x$.

$$30x^2 = \textcircled{2} \cdot \textcircled{3} \cdot 5 \cdot \textcircled{x} \cdot x$$
$$12x = 2 \cdot \textcircled{2} \cdot \textcircled{3} \cdot \textcircled{x}$$

	5x	2
6x	30x²	12x

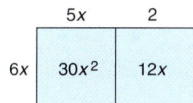

The GCF of $30x^2$ and $12x$ is $6x$. Write each term as a product of the GCF and its remaining factors.

$$30x^2 + 12x = 6x(5x) + 6x(2)$$
$$= 6x(5x + 2) \quad \textit{Distributive Property}$$

2 $15ab^2 - 25abc$

$$15ab^2 = 3 \cdot \textcircled{5} \cdot \textcircled{a} \cdot \textcircled{b} \cdot b$$
$$25abc = \textcircled{5} \cdot 5 \cdot \textcircled{a} \cdot \textcircled{b} \cdot c$$

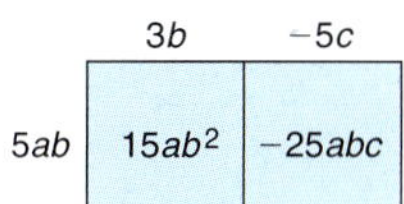

The GCF is $5ab$.

$$15ab^2 - 25abc = 5ab(3b) - 5ab(5c)$$
$$= 5ab(3b - 5c) \quad \textit{Distributive Property}$$

www.algconcepts.com/extra_examples

Motivating the Lesson

Real-World Connection Have students investigate the Distributive Property using whole numbers. Evaluate each pair of expressions. Are the expressions equivalent?

$6 + 8 \stackrel{?}{=} 2(3 + 4)$ **14; yes**
$14 + 28 \stackrel{?}{=} 14(1 + 2)$ **42; yes**
$12 + 8 + 4 \stackrel{?}{=} 4(3 + 2 + 1)$ **24; yes**
$18 + 24 + 12 \stackrel{?}{=} 6(3 + 4 + 2)$ **54; yes**

Discuss how each expression is factored using the greatest common factor and the Distributive Property.

2 TEACH

Teaching Tip Before discussing Examples 1 and 2, emphasize that factoring polynomials is not a completely new concept. Students are simply reversing the steps they followed when multiplying polynomials in Chapter 9.

In-Class Examples

Examples 1–2

Factor each polynomial.

1 $24y + 18y^2$ **$6y(4 + 3y)$**

2 $18fg - 21gh^2$ **$3g(6f - 7h^2)$**

Hands-On Algebra

Cooperative Learning Refer to the Hands-On Algebra on page 428. Remind students that it does not matter exactly how they orient the tiles when forming a rectangle. As long as they form a rectangle, its dimensions are factors of the original polynomial. Also, students may need help modeling Exercise 3 because it is the first exercise to use the x^2-tile.

Hands-On Algebra Masters, p. 108

In-Class Examples

Examples 3–4

Factor each polynomial.

3 $5a + 20ab + 10a^2$ **$5a(1 + 4b + 2a)$**

4 $17de - 15f$ **cannot be factored**

Teaching Tip Before discussing Example 5, review dividing powers. Also, it may help some students to see the division in Example 5 rewritten explicitly as multiplication by the reciprocal to see how the Distributive Property necessitates that both terms be divided by the divisor.

In-Class Example

Example 5

Divide $(24a^2 - 20a)$ by $4a$. **$6a - 5$**

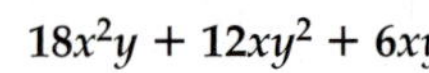

Factor each polynomial.

3 $18x^2y + 12xy^2 + 6xy$

$18x^2y = 2 \cdot 3 \cdot 3 \cdot x \cdot x \cdot y$
$12xy^2 = 2 \cdot 2 \cdot 3 \cdot x \cdot y \cdot y$
$6xy = 2 \cdot 3 \cdot x \cdot y$

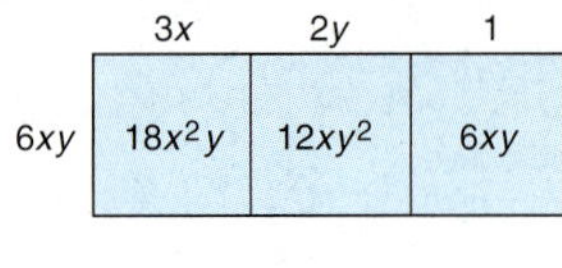

The GCF is $6xy$. When $6xy$ is factored from $6xy$ the remaining factor is 1.

$$18x^2y + 12xy^2 + 6xy = 6xy(3x) + 6xy(2y) + 6xy(1)$$
$$= 6xy(3x + 2y + 1) \quad \textit{Distributive Property}$$

4 $7x^2 + 9yz$

$7x^2 = 7 \cdot x \cdot x$
$9yz = 3 \cdot 3 \cdot y \cdot z$

There are no common factors of $7x^2$ and $9yz$ other than 1. Therefore, $7x^2 + 9yz$ cannot be factored using the GCF. It is a prime polynomial.

Reading Algebra

A polynomial that cannot be written as a product of two polynomials with integral coefficients is called a *prime polynomial*.

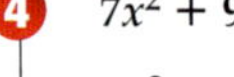

a. $12n^2 - 8n$ **$4n(3n - 2)$**

b. $16a^2b + 10ab^2$ **$2ab(8a + 5b)$**

c. $20rs^2 - 15r^2s + 5rs$ **c. $5rs(4s - 3r + 1)$**

d. $21x + 5y + 16z$ **d. cannot be factored**

If you know a product and one of its factors, you can use division to find the other factor. To divide a polynomial by a monomial, divide each term of the polynomial by the monomial.

5 **Divide $15x^3 + 12x^2$ by $3x$.**

Look Back

Dividing Powers: Lesson 8–2

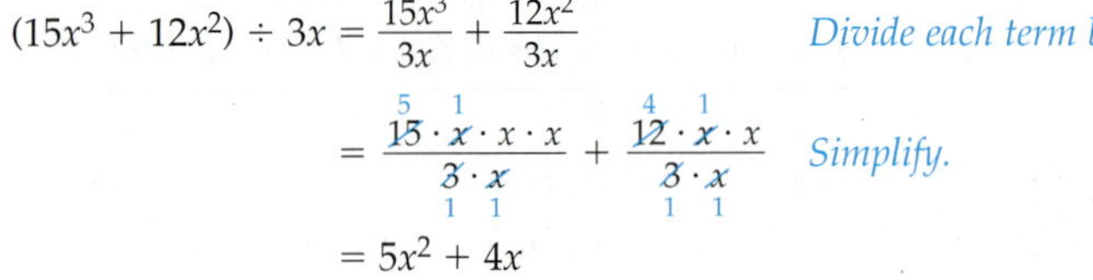

$$(15x^3 + 12x^2) \div 3x = \frac{15x^3}{3x} + \frac{12x^2}{3x} \quad \textit{Divide each term by } 3x.$$
$$= \frac{\overset{5}{\cancel{15}} \cdot \overset{1}{\cancel{x}} \cdot x \cdot x}{\underset{1}{\cancel{3}} \cdot \underset{1}{\cancel{x}}} + \frac{\overset{4}{\cancel{12}} \cdot \overset{1}{\cancel{x}} \cdot x}{\underset{1}{\cancel{3}} \cdot \underset{1}{\cancel{x}}} \quad \textit{Simplify.}$$
$$= 5x^2 + 4x$$

Therefore, $(15x^3 + 12x^2) \div 3x = 5x^2 + 4x$.

Your Turn

Find each quotient.

e. $(9b^2 - 15) \div 3$ **$3b^2 - 5$**

f. $(10x^2y^2 + 5xy) \div 5xy$ **$2xy + 1$**

430 Chapter 10

Factoring a polynomial can help simplify computations.

Example 6 **Landscaping Link**

A stone walkway is to be built around a square planter that contains a shade tree.

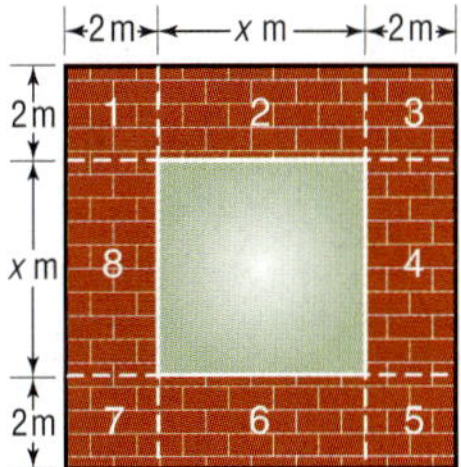

A. If the walkway is 2 meters wide, write an expression in factored form that represents the area of the walkway.

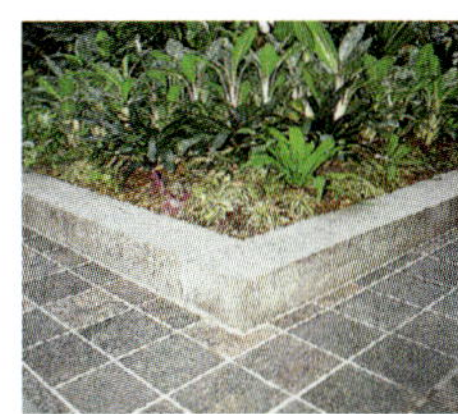

Let x represent the length and width of the planter. You can find the area of the walkway by finding the sum of the areas of the 8 rectangular sections shown in the figure.

The resulting expression can be simplified by using the Distributive Property to combine like terms and then factoring.

Regions 1, 2, 3, 4, 5, 6, 7, 8

$$\begin{aligned} A &= \underbrace{2\cdot 2}_{1} + \underbrace{2\cdot x}_{2} + \underbrace{2\cdot 2}_{3} + \underbrace{2\cdot x}_{4} + \underbrace{2\cdot 2}_{5} + \underbrace{2\cdot x}_{6} + \underbrace{2\cdot 2}_{7} + \underbrace{2\cdot x}_{8} \\ &= 4 + 2x + 4 + 2x + 4 + 2x + 4 + 2x \\ &= 16 + 8x \quad \textit{4 + 4 + 4 + 4 = 16 and 2x + 2x + 2x + 2x = 8x} \\ &= 8(2) + 8(x) \quad \textit{The GCF of 16 and 8x is 8.} \\ &= 8(2 + x) \quad \textit{Distributive Property} \end{aligned}$$

B. If the dimensions of the square planter are 1.5 meters by 1.5 meters, find the area of the walkway.

$$\begin{aligned} A &= 8(2 + x) \quad \textit{Area of a rectangle} \\ &= 8(2 + 1.5) \quad \textit{Replace x with 1.5.} \\ &= 8(3.5) \text{ or } 28 \quad \textit{Simplify.} \end{aligned}$$

The area of the walkway is 28 square meters.

Check for Understanding

Communicating Mathematics

1. **Illustrate** with algebra tiles or a drawing how to factor $x^2 + 2x$. **1–3. See margin.**
2. **Explain** what it means to factor a polynomial.
3. *Writing Math* Write a few sentences explaining how the Distributive Property is used to factor polynomials. Include at least two examples.

Vocabulary
factoring

Teaching Tip When discussing Example 6, you may want to point out to students that the area of the walkway may also be found by subtracting the area of the planter from the total area of the planter and walkway: $(x + 4)^2 - x^2 = x^2 + 8x + 16 - x^2$ or $8x + 16$.

In-Class Example

Example 6

The diagram below shows a walkway that is 2 meters wide surrounding a rectangular planter. Write an expression in factored form that represents the area of the walkway.

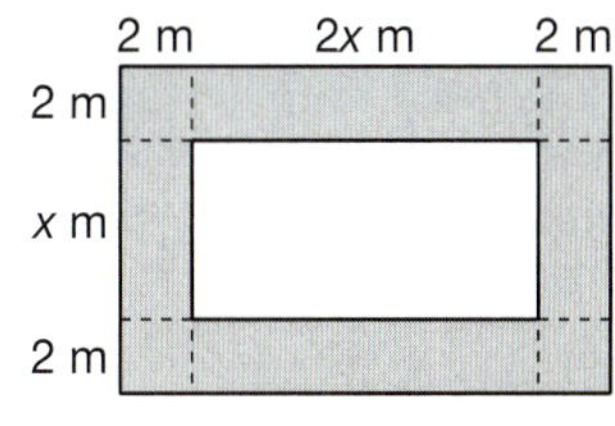

4(4 + 3x) square meters

Answers

1.

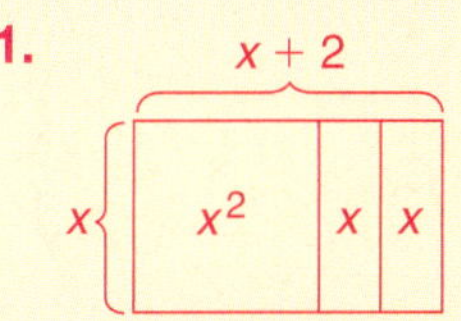

2. to find the factors of the polynomial

Study Guide, p. 438

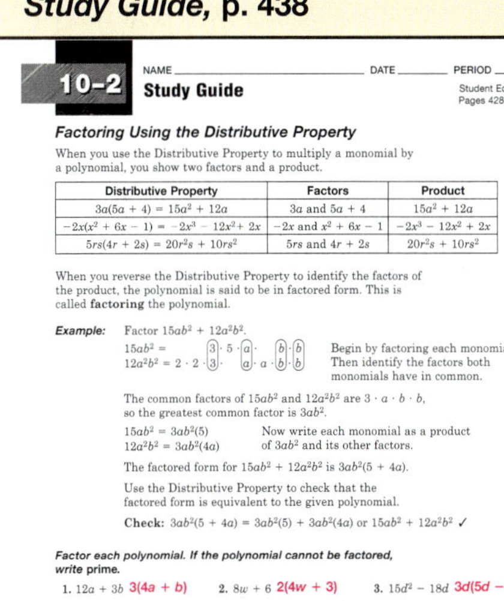

10–2 NAME ____________ DATE ______ PERIOD ______

Study Guide

Student Edition Pages 428–433

Factoring Using the Distributive Property

When you use the Distributive Property to multiply a monomial by a polynomial, you show two factors and a product.

Distributive Property	Factors	Product
$3a(5a + 4) = 15a^2 + 12a$	$3a$ and $5a + 4$	$15a^2 + 12a$
$-2x(x^2 + 6x - 1) = -2x^3 - 12x^2 + 2x$	$-2x$ and $x^2 + 6x - 1$	$-2x^3 - 12x^2 + 2x$
$5rs(4r + 2s) = 20r^2s + 10rs^2$	$5rs$ and $4r + 2s$	$20r^2s + 10rs^2$

When you reverse the Distributive Property to identify the factors of the product, the polynomial is said to be in factored form. This is called **factoring** the polynomial.

Example: Factor $15ab^2 + 12a^2b^2$.

$15ab^2 = 3 \cdot 5 \cdot a \cdot b \cdot b$
$12a^2b^2 = 2 \cdot 2 \cdot 3 \cdot a \cdot a \cdot b \cdot b$

Begin by factoring each monomial. Then identify the factors both monomials have in common.

The common factors of $15ab^2$ and $12a^2b^2$ are $3 \cdot a \cdot b \cdot b$, so the greatest common factor is $3ab^2$.

$15ab^2 = 3ab^2(5)$
$12a^2b^2 = 3ab^2(4a)$

Now write each monomial as a product of $3ab^2$ and its other factors.

The factored form for $15ab^2 + 12a^2b^2$ is $3ab^2(5 + 4a)$.

Use the Distributive Property to check that the factored form is equivalent to the given polynomial.

Check: $3ab^2(5 + 4a) = 3ab^2(5) + 3ab^2(4a)$ or $15ab^2 + 12a^2b^2$ ✓

Factor each polynomial. If the polynomial cannot be factored, write prime.

1. $12a + 3b$ **$3(4a + b)$**
2. $8w + 6$ **$2(4w + 3)$**
3. $15d^2 - 18d$ **$3d(5d - 6)$**
4. $5c^4 + 2c^2$ **$c^2(5c^2 + 2)$**
5. $12mn^3 - 5n$ **$n(12mn^2 - 5)$**
6. $4g + 13h^3$ **prime**
7. $2 - 16x^2$ **$2(1 - 8x^2)$**
8. $35xy^3 + 7x^2y$ **$7xy(5y^2 + x)$**
9. $27pw - 25q$ **prime**
10. $48c^2d^2 + 36c^2d$ **$12c^2d(4d + 3)$**
11. $ad + 3x^2 + 9$ **prime**
12. $7gh^2 + 7g + 14gh$ **$7g(h^2 + 1 + 2h)$**
13. $35a^2 + 15a - 20ab^2$ **$5a(7a + 3 - 4b^2)$**

Reteaching Activity

Naturalist Learners Have students look on the school grounds for a walkway built around a rectangle, a border around a rectangular garden, or some other situation similar to that in Example 6. Then have students write their own example in the style of Example 6 using measurements they have made themselves.

Answer

3. The Distributive Property is used to factor out a common factor from the terms of an expression. Examples: $5xy + 10x^2y^2 = 5xy(1 + 2xy)$, $2a^2 + 8a + 10 = 2(a^2 + 4a + 5)$

3 PRACTICE/APPLY

Error Analysis

Watch for students who don't factor out all of the common factors in Exercise 15.
Prevent by reminding students first to find the greatest common numerical factor shared by all the terms. Then have them identify the highest power of each and every variable common to all the terms. After factoring a monomial from a polynomial, urge students to check that there are no more common factors in the terms of the polynomial in the factored form.

Assignment Guide

Basic: 19–47 odd, 48–57
Average: 20–44 even, 45–57
All: Quiz 1, 1–5

***Skills Practice,* p. 439, and *Practice Masters,* p. 440 (shown)**

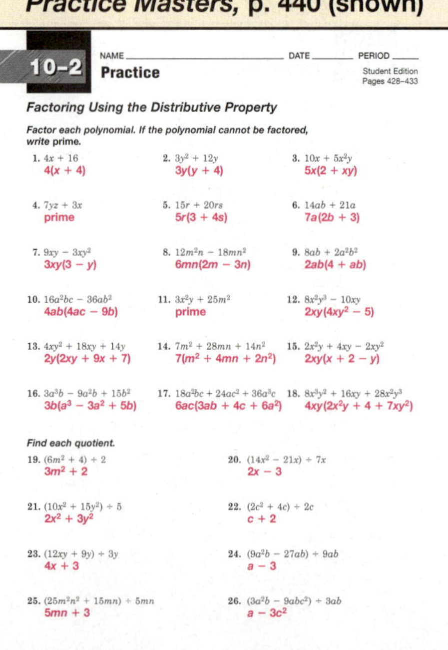
10-2 NAME ____________ DATE ______ PERIOD ____

Practice — Student Edition Pages 428–433

Factoring Using the Distributive Property

***Factor each polynomial. If the polynomial cannot be factored, write* prime.**

1. $4x + 16$ — $4(x + 4)$
2. $3y^2 + 12y$ — $3y(y + 4)$
3. $10x + 5x^2y$ — $5x(2 + xy)$
4. $7yz + 3x$ — prime
5. $15r + 20rs$ — $5r(3 + 4s)$
6. $14ab + 21a$ — $7a(2b + 3)$
7. $9xy - 3xy^2$ — $3xy(3 - y)$
8. $12m^2n - 18mn^2$ — $6mn(2m - 3n)$
9. $8ab + 2a^2b^2$ — $2ab(4 + ab)$
10. $16a^2bc - 36ab^2$ — $4ab(4ac - 9b)$
11. $3x^2y + 25m^2$ — prime
12. $8x^2y^3 - 10xy$ — $2xy(4xy^2 - 5)$
13. $4xy^2 + 18xy + 14y$ — $2y(2xy + 9x + 7)$
14. $7m^2 + 28mn + 14n^2$ — $7(m^2 + 4mn + 2n^2)$
15. $2x^2y + 4xy - 2xy^2$ — $2xy(x + 2 - y)$
16. $3a^3b - 9a^2b + 15b^2$ — $3b(a^3 - 3a^2 + 5b)$
17. $18a^2bc + 24ac^2 + 36a^3c$ — $6ac(3ab + 4c + 6a^2)$
18. $8x^3y^2 + 16xy + 28x^2y^3$ — $4xy(2x^2y + 4 + 7xy^2)$

Find each quotient.

19. $(6m^2 + 4) \div 2$ — $3m^2 + 2$
20. $(14x^2 - 21x) \div 7x$ — $2x - 3$
21. $(10x^2 + 15y^2) \div 5$ — $2x^2 + 3y^2$
22. $(2c^2 + 4c) \div 2c$ — $c + 2$
23. $(12xy + 9y) \div 3y$ — $4x + 3$
24. $(9a^2b - 27ab) \div 9ab$ — $a - 3$
25. $(25m^2n^2 + 15mn) \div 5mn$ — $5mn + 3$
26. $(3a^2b - 9abc^2) \div 3ab$ — $a - 3c^2$

Guided Practice

Getting Ready Find the GCF of the terms in each expression.

Sample: $9a^2 + 3a$ **Solution:** $9a^2 = 3 \cdot 3 \cdot a \cdot a$
$3a = 3 \cdot a$ The GCF is $3a$.

4. $8xy + 12mn$ **4**
5. $x^2 + 2x$ **x**
6. $5xy + y^2$ **y**
7. $18y^2 + 30y$ **$6y$**
8. $3ab - 2a^2b$ **ab**
9. $15m^2n - 20m^2n$ **$5m^2n$**

Examples 1–4

Factor each polynomial. If the polynomial cannot be factored, write *prime*. **12. $6a(2ab + 1)$**

10. $3x + 6$ **$3(x + 2)$**
11. $2x^2 + 4x$ **$2x(x + 2)$**
12. $12a^2b + 6a$
13. $7mn - 13yz$ ***prime***
14. $3x^2y + 6xy + 9y^2$ **$3y(x^2 + 2x + 3y)$**
15. $2a^3b^2 + 8ab + 16a^2b^3$ **$2ab(a^2b + 4 + 8ab^2)$**

Example 5

Find each quotient.

16. $(12m^2 - 15m) \div 3m$ **$4m - 5$**
17. $(3c^2d + 9cd) \div 3cd$ **$c + 3$**

Examples 6, 7

18. **Landscaping** Kiyoshi is planning to build a walkway around her square koi pond. The walkway is 6 feet wide.
    a. If x represents the measure of one side of the pond, write an expression in factored form that represents the area of the walkway. **$24(x + 6)$**
    b. If the dimensions of the pond are 8 feet by 8 feet, find the area of the walkway. **336 ft²**

Exercises

Practice

Homework Help

For Exercises	See Examples
19–30, 46	1, 2
31–36	3, 4
37–43, 45	5
47	6

Extra Practice
See page 712.

Factor each polynomial. If the polynomial cannot be factored, write *prime*. **19–36. See margin.**

19. $9x + 15$
20. $6x + 3x^2$
21. $8x + 2x^2y$
22. $7a^2b^2 + 3ab^3$
23. $3c^2d - 6c^2d^2$
24. $7x - 3y$
25. $36mn - 11mn^2$
26. $18xy^2 + 24x^2y$
27. $19ab + 21xy$
28. $14mn^2 - 2mn$
29. $12xy^3 + y^4$
30. $3a^2b - 6a^2b^2$
31. $24xy + 18xy^2 - 3y$
32. $3x^3y + 9xy + 36xy^2$
33. $x + x^2y^3 + x^3y^2$
34. $6x^2 + 9xy + 24x^2y^2$
35. $12axy - 14ay + 20ax$
36. $42xyz - 12x^2y^2 + 3x^3y^3$

Find each quotient.

37. $(27x^2 - 21y^2) \div 3$ **$9x^2 - 7y^2$**
38. $(5abc + c) \div c$ **$5ab + 1$**

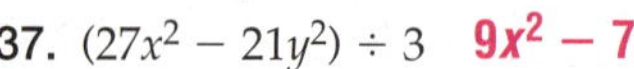

39. $(14ab + 28b) \div 14b$ **$a + 2$**
40. $(16x + 24xy) \div 8x$ **$2 + 3y$**
41. $(4x^2y^2z + 6xz^2) \div 2xz$ **$2xy^2 + 3z$**
42. $(3x^2y + 12xyz^2) \div 3xy$ **$x + 4z^2$**
43. Divide $6x^2 + 9$ by 3. **$2x^2 + 3$**
44. What is the GCF of $14abc^2$ and $18c$? **$2c$**

Applications and Problem Solving

45. **Geometry** The area of a rectangle is $(16x + 4y)$ square feet. If the width is 4 feet, find the length. **$(4x + y)$ ft**

A = 16x + 4y
↓
16x + 4y = L · W

16x/4 + 4y/4 = 4L/4 = 4x + y = L

Answers

19. $3(3x + 5)$
20. $3x(2 + x)$
21. $2x(4 + xy)$
22. $ab^2(7a + 3b)$
23. $3c^2d(1 - 2d)$
24. prime
25. $mn(36 - 11n)$
26. $6xy(3y + 4x)$
27. prime
28. $2mn(7n - 1)$
29. $y^3(12x + y)$
30. $3a^2b(1 - 2b)$

46. **Marine Biology** In a pool at a water park, a dolphin jumps out of the water traveling at 24 feet per second. Its height h, in feet, above the water after t seconds is given by the formula $h = 24t - 16t^2$.
   a. Factor the expression $24t - 16t^2$. **$8t(3 - 2t)$**
   b. Find the height of a dolphin when $t = 0.75$ second. **9 ft**

47. **Geometry** Write an expression in factored form that represents the area of the shaded region.

a. 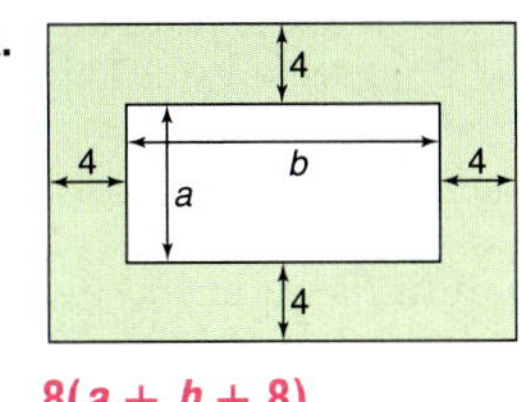

b. 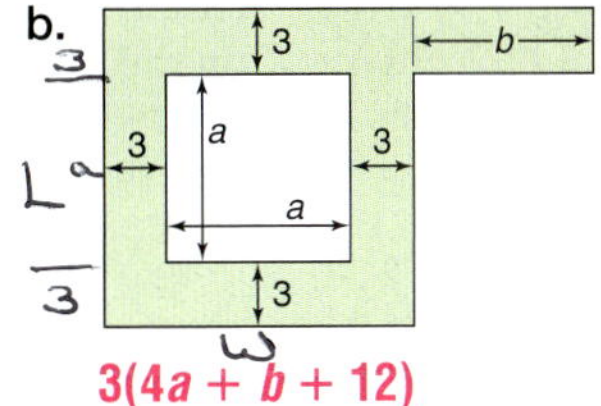

$8(a + b + 8)$ **$3(4a + b + 12)$**

48. **Critical Thinking** The length and width of a rectangle are represented by $2x$ and $9 - 4x$. If x must be an integer, what are the possible measures for the area of this rectangle? **10, 4**

Mixed Review

Classify each number as *prime* or *composite*. *(Lesson 10–1)*

49. 2 **P** 50. 21 **C** 51. 49 **C** 52. 53 **P** 53. 90 **C**

54. **Geometry** The length of a side of a square is $3x + 5$ units. What is the area of the square? *(Lesson 9–5)* **$(9x^2 + 30x + 25)$ square units**

Add or subtract. *(Lesson 9–2)*

55. $(x^2 + 4x - 3) + (2x^2 - 6x - 9)$ **$3x^2 - 2x - 12$**
56. $(2y^2 - 5y + 3) - (5y^2 - 4)$ **$-3y^2 - 5y + 7$**

Standardized Test Practice

57. **Short Response** Write a second degree polynomial. *(Lesson 9–1)* **Sample answer: $x^2 + 3x - 4$**

Quiz 1 Lessons 10–1 and 10–2

1. Find the prime factorization of 24. *(Lesson 10–1)* **$2^3 \cdot 3$**

Find the GCF of the terms in each expression. Then factor the expression. *(Lessons 10–1, 10–2)*

2. $20s + 40s^2$ **$20s$; $20s(1 + 2s)$**
3. $ax^3 + 7bx^3 + 11cx^3$ **x^3; $x^3(a + 7b + 11c)$**
4. $6x^3 + 12x^2 + 6x$ **$6x$; $6x(x^2 + 2x + 1)$**

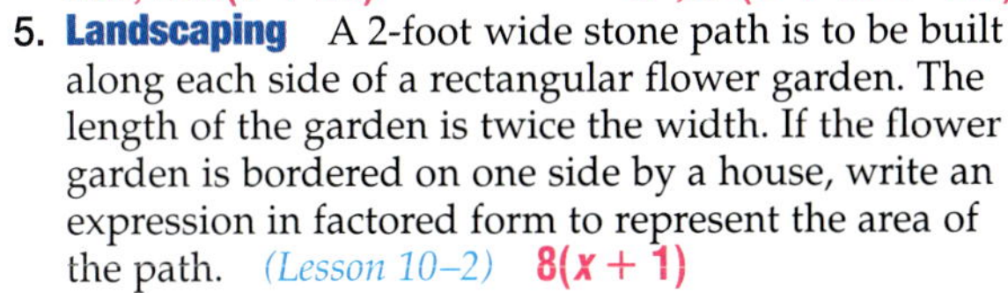

5. **Landscaping** A 2-foot wide stone path is to be built along each side of a rectangular flower garden. The length of the garden is twice the width. If the flower garden is bordered on one side by a house, write an expression in factored form to represent the area of the path. *(Lesson 10–2)* **$8(x + 1)$**

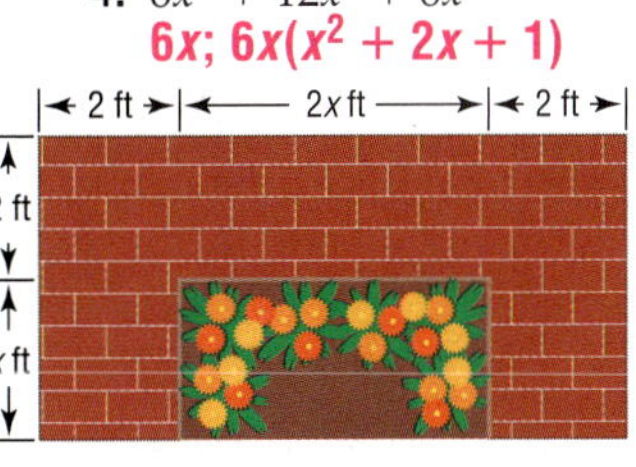

www.algconcepts.com/self_check_quiz

Extra Credit

The area of a circle is found using the formula $A = \pi r^2$, where r is the radius of the circle. Write an expression using π in factored form that represents the area of the shaded region at the right. **$9x^2(25\pi - 9)$**

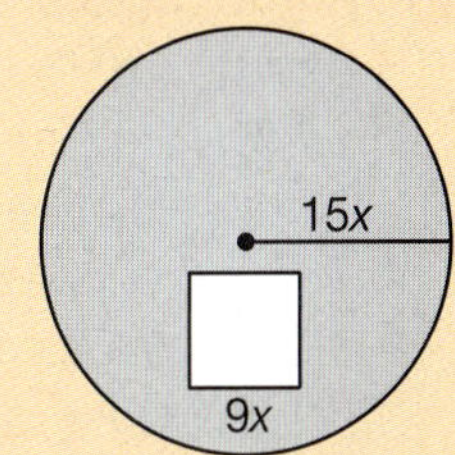

4 ASSESS

Open-Ended Assessment

Modeling Have students use algebra tiles to factor $4x + 8$.

Quiz 1

The Quiz provides students with a brief review of the concepts and skills in Lessons 10–1 and 10–2. Lesson numbers are given to the right of the exercises or instruction lines so students can review concepts not yet mastered.

Chapter 10, Quiz A (Lessons 10–1 and 10–2) is available in the *Chapter 10 Resource Masters*, p. 469.

Answers

31. $3y(8x + 6xy - 1)$
32. $3xy(x^2 + 3 + 12y)$
33. $x(1 + xy^3 + x^2y^2)$
34. $3x(2x + 3y + 8xy^2)$
35. $2a(6xy - 7y + 10x)$
36. $3xy(14z - 4xy + x^2y^2)$

***Enrichment*, p. 442**

10-2 NAME ____ DATE ____ PERIOD ____
Enrichment Student Edition Pages 428–433

Puzzling Primes

A **prime number** has only two factors, itself and 1. The number 6 is not prime because it has 2 and 3 as factors; 5 and 7 are prime. The number 1 is not considered to be prime.

1. Use a calculator to help you find the 25 prime numbers less than 100.
2, 3, 5, 7, 11, 13, 17, 19, 23, 29, 31, 37 41, 43, 47, 53, 59, 61, 67, 71, 73, 79, 83, 89, 97

Prime numbers have interested mathematicians for centuries. They have tried to find expressions that will give all the prime numbers, or only prime numbers. In the 1700s, Euler discovered that the expression $x^2 + x + 41$ will yield prime numbers for values of x from 0 through 39.

2. Find the prime numbers generated by Euler's formula for x from 0 through 7.
41, 43, 47, 53, 61, 71, 83, 97

3. Show that the expression $x^2 + x + 31$ will not give prime numbers for very many values of x.
It works for x = 0, 2, 3, 5, and 6 but not for x = 1, 4, and 7.

4. Find the largest prime number generated by Euler's formula. 1601

Goldbach s Conjecture is that every nonzero even number greater than 2 can be written as the sum of two primes. No one has ever proved that this is always true. No one has disproved it, either.

5. Show that Goldbach's Conjecture is true for the first 5 even numbers greater than 2.
4 = 2 + 2, 6 = 3 + 3, 8 = 3 + 5, 10 = 3 + 7, 12 = 5 + 7

6. Give a way that someone could disprove Goldbach's Conjecture.
Find an even number that cannot be written as the sum of two primes.

Lesson 10–3

1 FOCUS

5-Minute Check
Lesson 10–2

Factor each polynomial.

1. $16y + 12y^2$ **$4y(4 + 3y)$**
2. $3d^2ef - d^3ef^2$ **$d^2ef(3 - df)$**
3. $9ab + 15ab^2 + 3a^2b$ **$3ab(3 + 5b + a)$**

Refer to the diagram below, which shows a walkway built around a statue with a square base.

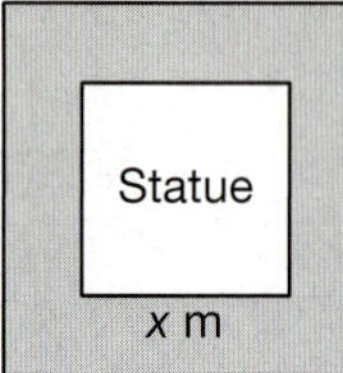

4. If the walkway is 3 meters wide, write an expression in factored form that represents the area of the walkway. **$12(x + 3)$ m^2**
5. If the base of the statue is 3 meters wide, find the area of the walkway using the expression in Question 4. **72 m^2**

Motivating the Lesson

Real-World Connection Ask a biology teacher to make a brief presentation about dominant and recessive genes to the class. Have the teacher bring in real-world examples of gene combinations that can be modeled using Punnett squares.

10–3 Factoring Trinomials: $x^2 + bx + c$

What You'll Learn
You'll learn to factor trinomials of the form $x^2 + bx + c$.

Why It's Important
Biology Geneticists use Punnett squares, which are similar to the models for factoring trinomials.
See Exercise 51.

In biology, *Punnett squares* are used to show possible ways that traits can be passed from parents to their offspring.

Each parent has two genes for each trait. The letters representing the parent's genes are placed on the outside of the Punnett square. The letters inside the boxes show the possible gene combinations for their offspring.

	G	*g*
G	*GG*	*Gg*
g	*Gg*	*gg*

The Punnett square at the right shows the gene combinations for fur color in rabbits.

- *G* represents the dominant gene for gray fur.
- *g* represents the recessive gene for white fur.

Notice that the Punnett square is similar to the model for multiplying binomials. The model below shows the product of $(x + 1)$ and $(x + 3)$.

$$(x + 1)(x + 3) = x^2 + 3x + 1x + 3$$
$$= x^2 + 4x + 3$$

	x	3
x	x^2	$3x$
1	$1x$	3

In this lesson, you will factor a trinomial into the product of two binomials.

Hands-On Algebra

Materials:  straightedge

Use a model to factor $x^2 + 5x + 4$.

Step 1 Draw a square with four sections. Put the first and last terms into the boxes as shown.

x^2	
	4

Step 2 Factor x^2 as $x \cdot x$ and place the factors outside the box.

Now, think of factors of 4 to place outside the box.

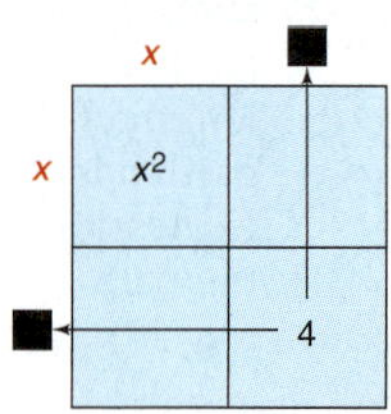

TECHNOLOGY

TECHNOLOGY

An alternative technology option using a graphing calculator is available for teaching this lesson.

Resource Manager

Reproducible Masters
Chapter 10 Resource Masters
- *Study Guide,* p. 443
- *Skills Practice,* p. 444
- *Practice,* p. 445
- *Reading to Learn Mathematics,* p. 446
- *Enrichment,* p. 447
- *Assessment,* p. 468

Graphing Calculator, pp. 27–28
Hands-On Algebra, p. 109

Transparencies
5-Minute Check, 10–3

Technology/Multimedia
AlgePASS, Lesson 23
Interactive Chalkboard CD-ROM

Step 3 The number 4 has two different factor pairs, 2 and 2, and 4 and 1. Try the factor pairs until you find the one that results in a middle term of $5x$.

Try 2 and 2.

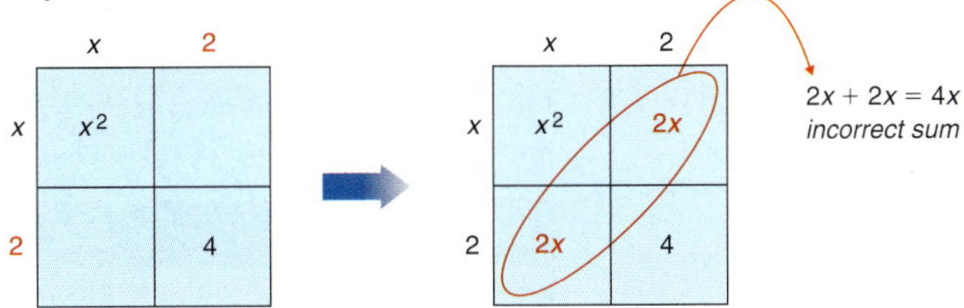

Try 4 and 1.

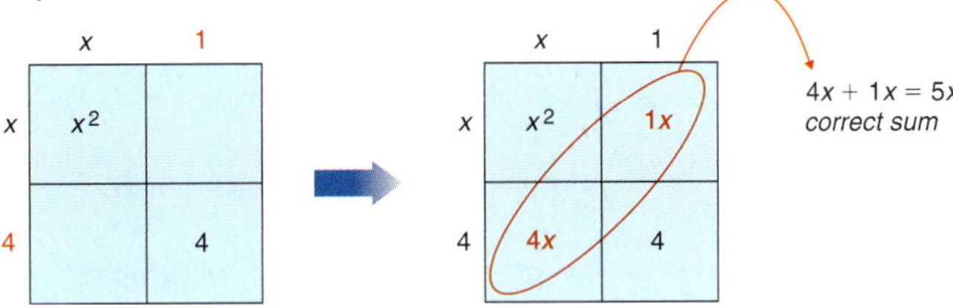

Step 4 The integers 4 and 1 result in the correct middle term, $5x$. Therefore, $x^2 + 5x + 4 = (x + 4)(x + 1)$.

1. $(x + 1)(x + 5)$
2. $(x + 6)(x + 1)$
3. $(x + 6)(x + 2)$
4. $(x - 1)(x - 2)$
5. $(x - 2)(x - 4)$
6. $(x - 3)(x - 3)$

Try These 1–6. See margin for models.

Use a model to factor each trinomial.

1. $x^2 + 6x + 5$
2. $x^2 + 7x + 6$
3. $x^2 + 8x + 12$
4. $x^2 - 3x + 2$
5. $x^2 - 6x + 8$
6. $x^2 - 6x + 9$

FOIL Method: Lesson 9–4

The FOIL method will help you factor trinomials without models. Use the following method to factor $x^2 + 6x + 8$.

Step 1 x^2 is the product of the **F**irst terms, and 8 is the product of the **L**ast terms.

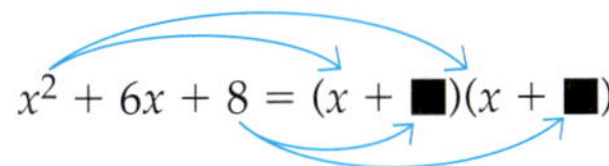

Step 2 Try several factor pairs of 8 until the sum of the products of the **O**uter and **I**nner terms is $6x$. Check by using FOIL.

Try 1 and 8. $(x + 1)(x + 8) = x^2 + 8x + 1x + 8$
$= x^2 + 9x + 8$ *9x is not the correct term.*

Try 2 and 4. $(x + 2)(x + 4) = x^2 + 4x + 2x + 8$
$= x^2 + 6x + 8$ ✓

Therefore, $x^2 + 6x + 8 = (x + 2)(x + 4)$.

Hands-On Algebra

Cooperative Learning Refer to the Hands-On Algebra on page 434. It is vital that students understand how the model relates to the math. Stress that the two numerical terms outside the box can be any two factors whose product is the numerical term of the trinomial. Consider working Step 3 on the board or overhead as students work along with you to help clarify how the model is used.

Hands-On Algebra Masters, p. 109

2 TEACH

Teaching Tip Before beginning the Hands-On Algebra activity, review how the square models are used to find the product of two binomials. Remind students not to confuse these models with algebra tiles, but that the boxes are used to make sure that all binomial product terms are found, not to represent fixed areas.

Answers

1.

	x	5
x	x^2	$5x$
1	$1x$	5

2.

	x	3
x	x^2	$3x$
2	$2x$	6

3.

	x	4
x	x^2	$4x$
3	$3x$	12

4.

	x	-2
x	x^2	$-2x$
-1	$-1x$	2

5.

	x	-4
x	x^2	$-4x$
-2	$-2x$	8

6.

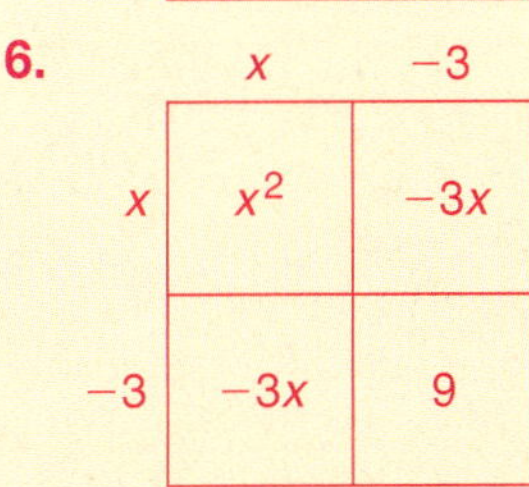

In-Class Examples

Examples 1–3

Factor each trinomial.

1 $x^2 + x - 12$ **$(x - 3)(x + 4)$**

2 $x^2 - 9x + 8$ **$(x - 8)(x - 1)$**

3 $x^2 + x + 1$ **prime**

Examples

Factor each trinomial.

1 $x^2 - 7x + 10$

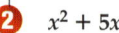

$x^2 - 7x + 10 = (x + ■)(x + ■)$

Find integers whose product is 10 and whose sum is -7. *Recall that the product of two negative integers is positive.*

Product	Integers	Sum
10	$-1, -10$	$-1 + (-10) = -11$
10	$-2, -5$	$-2 + (-5) = -7$ ✓

Therefore, $x^2 - 7x + 10 = (x - 2)(x - 5)$.

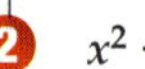

2 $x^2 + 5x - 6$

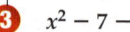

$x^2 + 5x - 6 = (x + ■)(x + ■)$

Find integers whose product is -6 and whose sum is 5. *Recall that the product of a positive integer and a negative integer is negative.*

Product	Integers	Sum
-6	$-2, 3$	$-2 + 3 = 1$
-6	$2, -3$	$2 + (-3) = -1$
-6	$-1, 6$	$-1 + 6 = 5$ ✓

You can stop listing factors when you find a pair that works.

Therefore, $x^2 + 5x - 6 = (x - 1)(x + 6)$.

3 $x^2 - 7 - 3x$

First, write the trinomial as $x^2 - 3x - 7$.

$x^2 - 3x - 7 = (x + ■)(x + ■)$

Find two integers whose product is -7 and whose sum is -3.

Product	Integers	Sum
-7	$-1, 7$	$-1 + 7 = 6$
-7	$1, -7$	$1 + (-7) = -6$

There are no factors of -7 whose sum is -3. Therefore, $x^2 - 3x - 7$ is a prime polynomial.

a. $(x + 2)(x + 1)$ b. $(a + 3)(a + 1)$ c. $(b + 2)(b + 2)$

d. $(y - 3)(y - 4)$
e. $(n - 7)(n + 2)$
f. prime

Your Turn

a. $x^2 + 3x + 2$ b. $a^2 + 4a + 3$ c. $b^2 + 4b + 4$

d. $y^2 - 7y + 12$ e. $n^2 - 5n - 14$ f. $m^2 - m + 1$

From the Classroom of ...

Nicki Hudson
West Linn High School
West Linn, Oregon

I like to introduce this concept with algebra tiles. Students select the tiles that represent the polynomial to be factored. Then they try to arrange them to form a rectangle. The side lengths of the rectangles are the factors of the polynomial. Point out that sometimes the tiles cannot be arranged in a rectangle.

In the previous lesson, you learned that the terms of a polynomial might have a GCF that can be factored using the Distributive Property. When you factor trinomials, always check for a GCF first.

Example 4

Factor $2x^2 - 20x - 22$.

First, check for a GCF.

$2x^2 - 20x - 22 = 2(x^2 - 10x - 11)$ *The GCF is 2.*

Now, factor $x^2 - 10x - 11$. *Find two integers whose product is −11 and whose sum is −10.*

Product	Integers	Sum
−11	−1, 11	−1 + 11 = 10
−11	1, −11	1 + (−11) = −10 ✓

So, $x^2 - 10x - 11 = (x + 1)(x - 11)$.

Therefore, $2x^2 - 20x - 22 = 2(x + 1)(x - 11)$. *Check by using FOIL.*

Your Turn

Factor each polynomial. **h. $5(m + 5)(m + 4)$**

g. $3y^2 - 9y - 54$ **$3(y - 6)(y + 3)$** **h.** $5m^2 + 45m + 100$

The area of a figure can often be expressed as a trinomial.

Example 5
Gardening Link

Tammy is planning a rectangular garden in which the width will be 4 feet less than its length. She has decided to put a birdbath within the garden, occupying a space 3 feet by 4 feet. How many square feet are now left for planting? Express the answer in factored form.

Let ℓ = the length of the original rectangle.

Let $\ell - 4$ = the width of the original rectangle.

(Diagram: outer rectangle labeled ℓ and $\ell - 4$; inner rectangle labeled 3 and 4.)

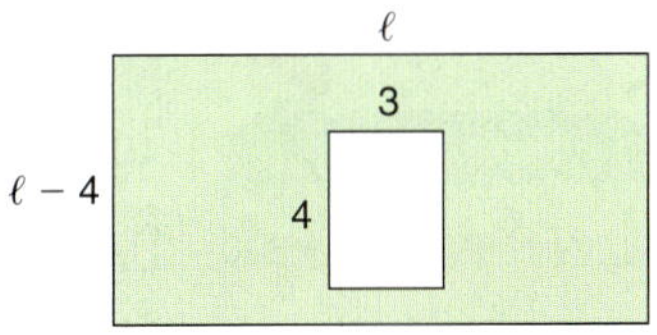

Find the area of the original rectangle.

$A = \ell w$ *Area = length × width*

$A = \ell(\ell - 4)$ *Replace w with $\ell - 4$.*

$A = \ell^2 - 4\ell$ *Distributive Property*

Find the area of the small rectangle.

$A = 4(3)$ or 12

Remaining area = area of original rectangle − area of small rectangle

$= \ell^2 - 4\ell - 12$

The remaining area is $\ell^2 - 4\ell - 12$ or $(\ell - 6)(\ell + 2)$.

Teaching Tip While discussing Example 4, you may want to show students that the trinomial could be factored directly as $(2x + 2)(x - 11)$ or $(x + 1)(2x - 22)$ by using models, but that then the common factor 2 would have to be factored out. Encourage students that it is much easier first to factor out the greatest common factor of the terms.

In-Class Examples

Example 4

Factor $4x^2 - 8x - 60$.
$4(x + 3)(x - 5)$

Example 5

Sahej is planning a rectangular garden in which the width will be 2 feet less than the length. He will put a composting box inside the garden that measures 2 feet by 4 feet. How many square feet are now left for planting? Express the answer in factored form.
$(\ell - 4)(\ell + 2)$

***Study Guide*, p. 443**

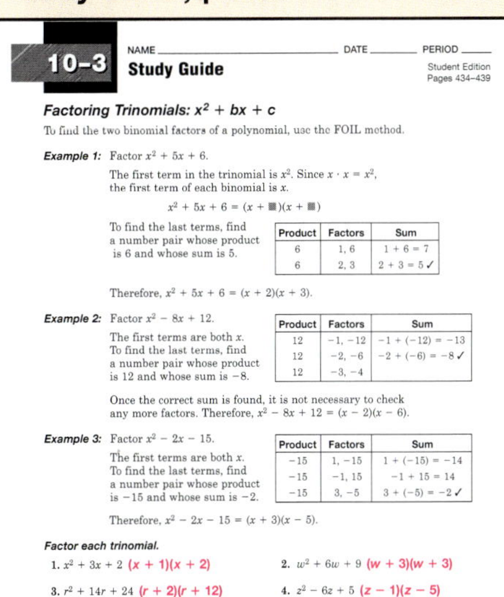

10-3 NAME ______ DATE ______ PERIOD ______

Study Guide Student Edition Pages 434–439

Factoring Trinomials: $x^2 + bx + c$

To find the two binomial factors of a polynomial, use the FOIL method.

Example 1: Factor $x^2 + 5x + 6$.

The first term in the trinomial is x^2. Since $x \cdot x = x^2$, the first term of each binomial is x.

$$x^2 + 5x + 6 = (x + ■)(x + ■)$$

To find the last terms, find a number pair whose product is 6 and whose sum is 5.

Product	Factors	Sum
6	1, 6	1 + 6 = 7
6	2, 3	2 + 3 = 5 ✓

Therefore, $x^2 + 5x + 6 = (x + 2)(x + 3)$.

Example 2: Factor $x^2 - 8x + 12$.

The first terms are both x. To find the last terms, find a number pair whose product is 12 and whose sum is −8.

Product	Factors	Sum
12	−1, −12	−1 + (−12) = −13
12	−2, −6	−2 + (−6) = −8 ✓
12	−3, −4	

Once the correct sum is found, it is not necessary to check any more factors. Therefore, $x^2 - 8x + 12 = (x - 2)(x - 6)$.

Example 3: Factor $x^2 - 2x - 15$.

The first terms are both x. To find the last terms, find a number pair whose product is −15 and whose sum is −2.

Product	Factors	Sum
−15	1, −15	1 + (−15) = −14
−15	−1, 15	−1 + 15 = 14
−15	3, −5	3 + (−5) = −2 ✓

Therefore, $x^2 - 2x - 15 = (x + 3)(x - 5)$.

Factor each trinomial.

1. $x^2 + 3x + 2$ **$(x + 1)(x + 2)$**
2. $w^2 + 6w + 9$ **$(w + 3)(w + 3)$**
3. $r^2 + 14r + 24$ **$(r + 2)(r + 12)$**
4. $z^2 - 6z + 5$ **$(z - 1)(z - 5)$**
5. $f^2 - 6f + 8$ **$(f - 2)(f - 4)$**
6. $x^2 - 15x + 56$ **$(x - 7)(x - 8)$**
7. $v^2 + 15v + 36$ **$(v + 3)(v + 12)$**
8. $k^2 - 23k + 42$ **$(k - 2)(k - 21)$**
9. $y^2 - 20y + 100$ **$(y - 10)(y - 10)$**
10. $a^2 + 4a - 45$ **$(a + 9)(a - 5)$**
11. $x^2 + 7x - 18$ **$(x + 9)(x - 2)$**
12. $m^2 - 21m - 22$ **$(m + 1)(m - 22)$**

Have students explain Punnett squares and how they are used in genetics to a family member. Have students and family members research examples, such as their eye color, the hair color of a pet, the colors of flowers in their yard, and so on, that might be modeled by Punnett squares. Have students explain also how the method used in Punnett squares can be used to factor trinomials.

3 PRACTICE/APPLY

Error Analysis

Watch for students who make errors such as factoring Exercise 9 as $(x + 5)(x + 1)$ or Exercise 13 as $(x - 5)(x + 2)$.

Prevent by having students always multiply their factors to make sure the product is the original trinomial. Review Examples 1 and 2, in which sums of possible factors are listed in a table. Encourage students that they will improve at factoring trinomials mentally, but that it is best to use an organized approach while gaining skill.

Assignment Guide

Basic: 19–51 odd, 52–62
Average: 20–48 even, 49–62

Answers

1.

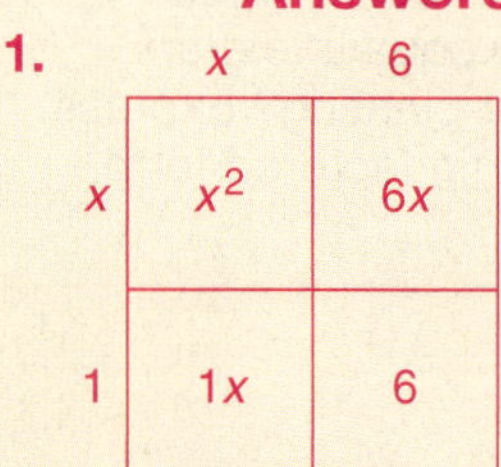

$(x + 1)(x + 6)$

19. $(b + 4)(b + 1)$
20. $(x + 5)(x + 5)$
21. $(a + 3)(a + 4)$
22. prime

***Skills Practice,* p. 444, and *Practice,* p. 445 (shown)**

10-3 NAME ______ DATE ______ PERIOD ______

Practice — Student Edition Pages 434–439

Factoring Trinomials: $x^2 + bx + c$

Factor each trinomial. If the trinomial cannot be factored, write prime.

1. $x^2 + 5x + 6$ $(x + 2)(x + 3)$	2. $y^2 + 5y + 4$ $(y + 4)(y + 1)$	3. $m^2 + 12m + 35$ $(m + 5)(m + 7)$
4. $p^2 + 8p + 15$ $(p + 5)(p + 3)$	5. $a^2 + 8a + 12$ $(a + 6)(a + 2)$	6. $n^2 + 4n + 4$ $(n + 2)(n + 2)$
7. $x^2 + 9x + 18$ $(x + 3)(x + 6)$	8. $x^2 + x + 3$ prime	9. $y^2 - 6y + 8$ $(y - 2)(y - 4)$
10. $c^2 - 8c + 15$ $(c - 5)(c - 3)$	11. $m^2 - 2m + 1$ $(m - 1)(m - 1)$	12. $b^2 - 9b + 20$ $(b - 4)(b - 5)$
13. $x^2 - 8x + 7$ $(x - 1)(x - 7)$	14. $n^2 - 5n + 6$ $(n - 3)(n - 2)$	15. $y^2 - 8y + 12$ $(y - 6)(y - 2)$
16. $c^2 - 4c + 5$ prime	17. $x^2 - x - 12$ $(x + 3)(x - 4)$	18. $m^2 + 5m - 6$ $(m - 1)(m + 6)$
19. $a^2 + 4a - 12$ $(a - 2)(a + 6)$	20. $y^2 - y - 6$ $(y + 2)(y - 3)$	21. $b^2 - 3b - 10$ $(b - 5)(b + 2)$
22. $x^2 + 3x - 4$ $(x + 4)(x - 1)$	23. $c^2 + 2c - 15$ $(c + 5)(c - 3)$	24. $2x^2 + 10x + 8$ $2(x + 4)(x + 1)$
25. $3y^2 - 15y + 18$ $3(y - 2)(y - 3)$	26. $5m^2 - 10m - 40$ $5(m + 2)(m - 4)$	27. $3b^2 + 6b - 9$ $3(b - 1)(b + 3)$
28. $4n^2 + 12n + 8$ $4(n + 2)(n + 1)$	29. $2x^2 + 8x - 24$ $2(x - 2)(x + 6)$	30. $3y^2 - 15y + 12$ $3(y - 4)(y - 1)$

Check for Understanding

Communicating Mathematics

1. **Illustrate** how to factor $x^2 + 7x + 6$ using a model. See margin.
2. **Explain** why the trinomial $x^2 + x + 5$ cannot be factored.
3. **Complete** the following sentence.
When you factor $m^2 - 3m - 10$, you want to find two integers whose product is ___?___ and whose sum is ___?___. $-10, -3$

2. There are no factors of 5 whose sum is 1.

Guided Practice

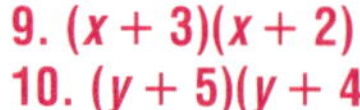

Getting Ready Find two integers whose product is the first number and whose sum is the second number.

Sample: 10, 7 **Solution:** $2 \times 5 = 10$, $2 + 5 = 7$

4. 30, 11 — 6, 5
5. 12, −7 — −4, −3
6. −10, 3 — 5, −2
7. −6, −5 — −6, 1
8. −30, −7 — −10, 3

Factor each trinomial. If the trinomial cannot be factored, write *prime*.

Examples 1–4

9. $x^2 + 5x + 6$
10. $y^2 + 9y + 20$
11. $a^2 - 5a + 4$
12. $z^2 - 8z + 16$
13. $x^2 + 3x - 10$
14. $m^2 - 4m - 21$
15. $w^2 + w + 2$ prime
16. $3a^2 + 15a + 12$ $3(a + 4)(a + 1)$
17. $2c^2 - 12c - 14$

9. $(x + 3)(x + 2)$
10. $(y + 5)(y + 4)$
11. $(a - 1)(a - 4)$
12. $(z - 4)(z - 4)$
13. $(x + 5)(x - 2)$
14. $(m - 7)(m + 3)$
17. $2(c - 7)(c + 1)$

Example 5

18. **Geometry** Find the area of the shaded region. Express the area in factored form. $(x + 3)(x - 1)$

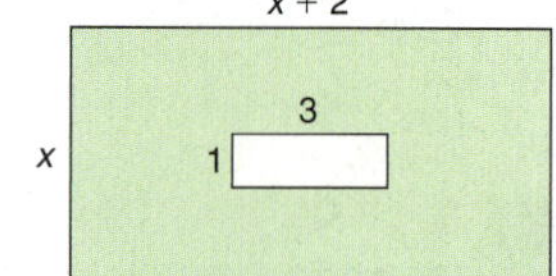

Exercises

Practice

Factor each trinomial. If the trinomial cannot be factored, write *prime*. 19–45. See margin.

Homework Help	
For Exercises	**See Examples**
19–39, 46, 48, 50, 52	1–3
40–45	4
49	5
Extra Practice	
See page 712.	

19. $b^2 + 5b + 4$
20. $x^2 + 10x + 25$
21. $a^2 + 7a + 12$
22. $a^2 + 3a + 5$
23. $y^2 + 12y + 27$
24. $z^2 + 13z + 40$
25. $x^2 - 8x + 15$
26. $a^2 - 4a + 4$
27. $c^2 - 13c + 36$
28. $d^2 - 11d + 28$
29. $m^2 - 5m + 1$
30. $y^2 - 12y + 32$
31. $c^2 + 2c - 3$
32. $x^2 - 5x - 24$
33. $r^2 - 3r - 18$
34. $m^2 - 2m - 24$
35. $n^2 + 13n - 30$
36. $m^2 + 11m - 12$
37. $x^2 - 17x + 72$
38. $a^2 - a - 90$
39. $r^2 + 22r - 48$

40. $4x^2 + 28x + 40$
41. $3y^2 - 21y + 36$
42. $z^3 + z^2 - 12z$
43. $m^3 + 3m^2 + 2m$
44. $3y^3 - 24y^2 + 36y$
45. $2a^3 + 14a^2 - 16a$

46. Express $x^2 + 24x + 95$ as the product of two binomials. $(x + 19)(x + 5)$
47. Write a trinomial that cannot be factored.
48. Complete the trinomial $x^2 + 6x +$ ___?___ with a positive integer so that the resulting trinomial can be factored. 5, 8, or 9

47. Sample answer: $x^2 + 7x - 9$

Reteaching Activity

Logical Learners Have students write in a column all the factor pairs *a* and *b* whose product is 24 or −24. Next to these factors, have students write out $(x + a)(x + b)$ and the terms of each product, labeling each term of the product F, O, I, or L as appropriate. Then have students simplify each product. By comparing the resulting trinomials with their binomial factors in this organized way, students should begin to gain insight into possible factorizations of a trinomial.

Applications and Problem Solving

49. **Geometry** Refer to the figure at the right. **a. $x^2 - 5x - 24$**

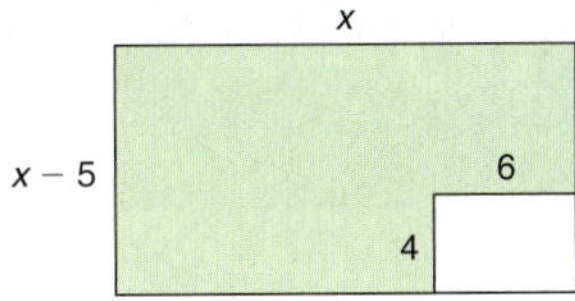

a. Express the area of the shaded region as a polynomial.

b. Express the area in factored form.

49b. $(x - 8)(x + 3)$

50. **Geometry** The volume of a rectangular prism is $x^3 + 4x^2 + 3x$. Find the length, width, and height of the prism if each dimension can be written as a monomial or binomial with integral coefficients. (*Hint:* Use the formula $V = \ell wh$.) **x, $x + 1$, $x + 3$**

51. **Genetics** In guinea pigs, a black coat is a dominant trait over a white coat. Let C represent a black coat and c represent a white coat in the Punnett squares below. Find the missing genes or gene pair.

a.

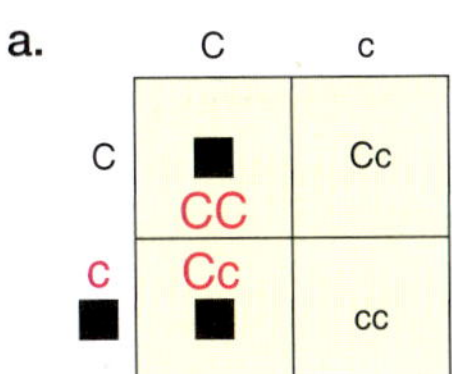

b.

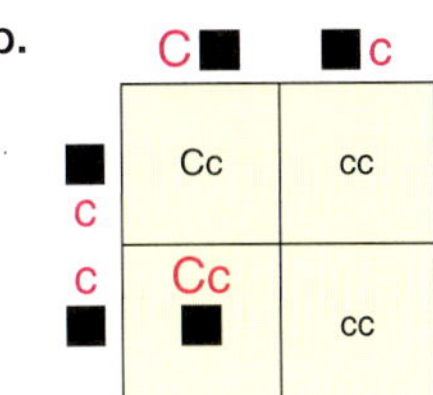

52. **Critical Thinking** Find all values of k so that the trinomial $x^2 + kx + 10$ can be factored. **11, −11, 7, −7**

Mixed Review

Find each quotient. *(Lesson 10–2)*

53. $(10x^2 + 25y^2) \div 5$ **$2x^2 + 5y^2$**

54. $(2y^2 + 4y) \div y$ **$2y + 4$**

55. $(6a^2 + 8ab - 6b^2) \div 2$ **$3a^2 + 4ab - 3b^2$**

56. $(3x^2y^2 + 9x^3y^2z) \div 3x^2y^2$ **$1 + 3xz$**

Find the GCF of each set of numbers or monomials. *(Lesson 10–1)*

57. $12a$, $16b$ **4**

58. $6a^2b$, $9ab^2$ **$3ab$**

59. $15x$, $7y$ **1**

60. 15, 60, 75 **15**

Standardized Test Practice

61. **Extended Response** The graph shows the value of riding lawn mower shipments in 1997 and 2000. *(Lesson 7–4)*

a. Write an equation of the line in slope-intercept form.

b. What does the slope represent?

c. Use the equation to predict the value of riding lawn mower shipments in 2005. **$3 billion**

61a. $y = 0.2x - 398$
61b. the average increase each year

Riding Lawn Mower Shipments (billion dollars)

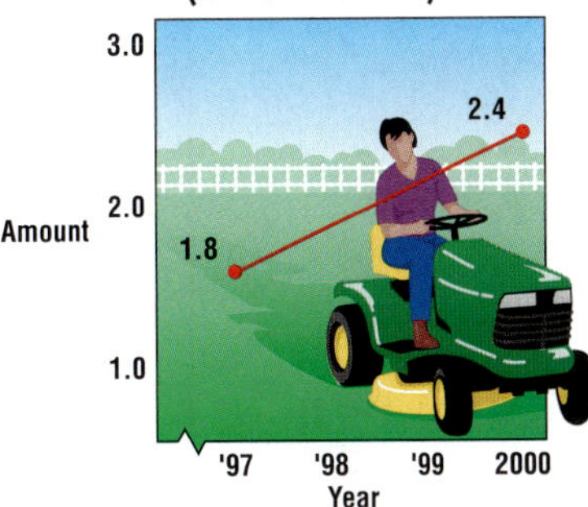

Source: Freedonia Group

62. **Multiple Choice** Evaluate $8x + 3y$ if $x = 9$ and $y = -2$. *(Lesson 2–5)* **A**

Ⓐ 66   Ⓑ 57   Ⓒ 78   Ⓓ 11

www.algconcepts.com/self_check_quiz

4 ASSESS

Open-Ended Assessment

Modeling Have students demonstrate how to use a model to factor $x^2 + x - 20$.

Mid-Chapter Test (Lessons 10–1 through 10–3) is available in the *Chapter 10 Resource Masters*, p. 468.

Answers

23. $(y + 9)(y + 3)$
24. $(z + 8)(z + 5)$
25. $(x - 5)(x - 3)$
26. $(a - 2)(a - 2)$
27. $(c - 9)(c - 4)$
28. $(d - 7)(d - 4)$
29. prime
30. $(y - 8)(y - 4)$
31. $(c + 3)(c - 1)$
32. $(x - 8)(x + 3)$
33. $(r - 6)(r + 3)$
34. $(m - 6)(m + 4)$
35. $(n + 15)(m - 2)$
36. $(m + 12)(m - 1)$
37. $(x - 9)(x - 8)$
38. $(a - 10)(a + 9)$
39. $(r + 24)(r - 2)$
40. $4(x + 2)(x + 5)$
41. $3(y - 4)(y - 3)$
42. $z(z - 3)(z + 4)$
43. $m(m + 1)(m + 2)$
44. $3y(y - 6)(y - 2)$
45. $2a(a + 8)(a - 1)$

***Enrichment*, p. 447**

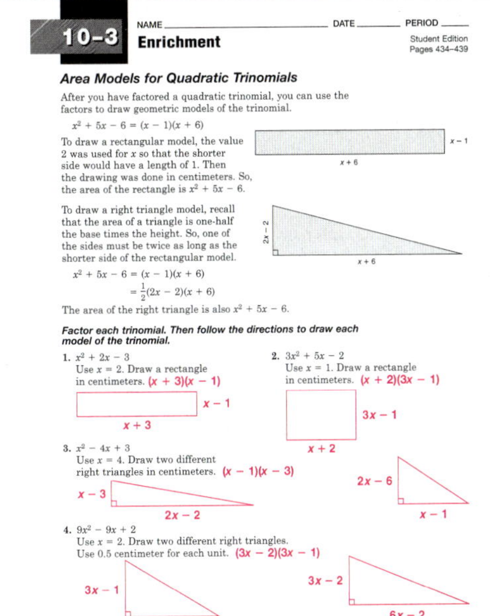

10–3 NAME ______ DATE ______ PERIOD ______

Enrichment

Student Edition Pages 434–439

Area Models for Quadratic Trinomials

After you have factored a quadratic trinomial, you can use the factors to draw geometric models of the trinomial.

$x^2 + 5x - 6 = (x - 1)(x + 6)$

To draw a rectangular model, the value 2 was used for x so that the shorter side would have a length of 1. Then the drawing was done in centimeters. So, the area of the rectangle is $x^2 + 5x - 6$.

To draw a right triangle model, recall that the area of a triangle is one-half the base times the height. So, one of the sides must be twice as long as the shorter side of the rectangular model.

$$x^2 + 5x - 6 = (x - 1)(x + 6)$$
$$= \frac{1}{2}(2x - 2)(x + 6)$$

The area of the right triangle is also $x^2 + 5x - 6$.

Factor each trinomial. Then follow the directions to draw each model of the trinomial.

1. $x^2 + 2x - 3$
Use $x = 2$. Draw a rectangle in centimeters. $(x + 3)(x - 1)$

2. $3x^2 + 5x - 2$
Use $x = 1$. Draw a rectangle in centimeters. $(x + 2)(3x - 1)$

3. $x^2 - 4x + 3$
Use $x = 4$. Draw two different right triangles in centimeters. $(x - 1)(x - 3)$

4. $9x^2 - 9x + 2$
Use $x = 2$. Draw two different right triangles. Use 0.5 centimeter for each unit. $(3x - 2)(3x - 1)$

Extra Credit

A cube has a volume of $2a^4b + 2a^3b - 112a^2b$ cubic units. One dimension is the GCF of the terms of the volume. The other two dimensions are binomials. What are the dimensions of the cube?
$2a^2b$ units, $a - 7$ units, and $a + 8$ units

Lesson 10-4

1 FOCUS

5-Minute Check

Lesson 10–3

***Factor each trinomial. If the trinomial cannot be factored, write* prime.**

1. $x^2 - 9x + 18$ $(x - 3)(x - 6)$
2. $x^2 + 12x + 35$ $(x + 5)(x + 7)$
3. $x^2 - x - 1$ prime
4. $2x^2 - 6x - 8$ $2(x + 1)(x - 4)$
5. Find the area of the shaded region. Express the area in factored form.

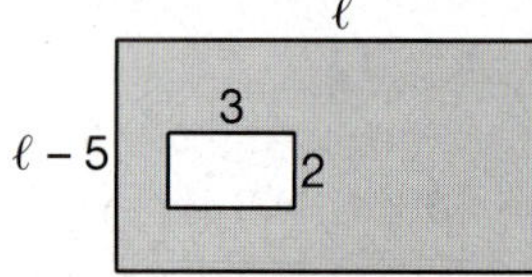

$A = (\ell - 6)(\ell + 1)$

Motivating the Lesson

Hands-On Activity Have students use algebra tiles to find the product of $2x + 1$ and $x + 2$. Then ask them how their result indicates the factored form of the trinomial $2x^2 + 5x + 2$.

Answers

1.

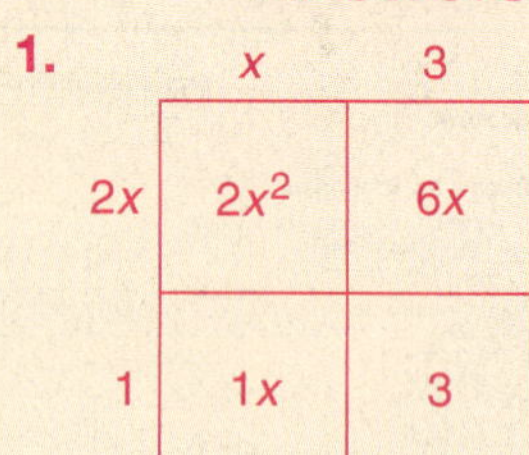

2.

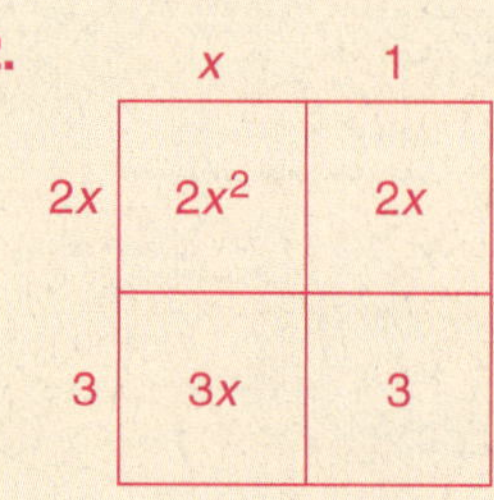

10–4 Factoring Trinomials: $ax^2 + bx + c$

What You'll Learn

You'll learn to factor trinomials of the form $ax^2 + bx + c$.

Why It's Important

Manufacturing The volume of a rectangular crate can be expressed in factored form. *See Example 4.*

In this lesson, you will learn to factor trinomials in which the coefficient of x^2 is a number other than 1.

Hands-On Algebra

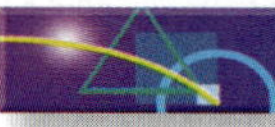

Materials:  straightedge

Use a model to factor $2x^2 + 7x + 6$.

Step 1 Draw a square with four sections. Put the first and last terms into the boxes as shown.

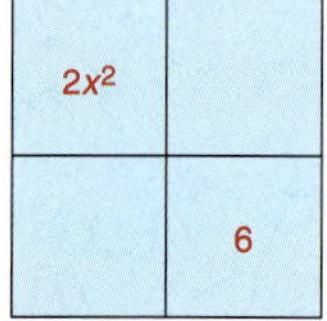

Step 2 Factor $2x^2$ as $2x \cdot x$ and place the factors outside the box.

Think of factors of 6 to place outside the box.

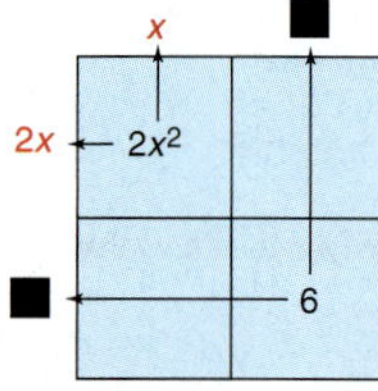

Step 3 The number 6 has two different factor pairs, 2 and 3, and 1 and 6. Try the factor pairs until you find the one that results in a middle term of $7x$. First, try 2 and 3. Note that there are two different ways of placing the 2 and 3 outside of the box.

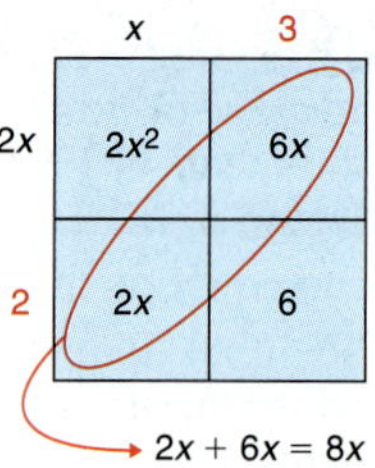

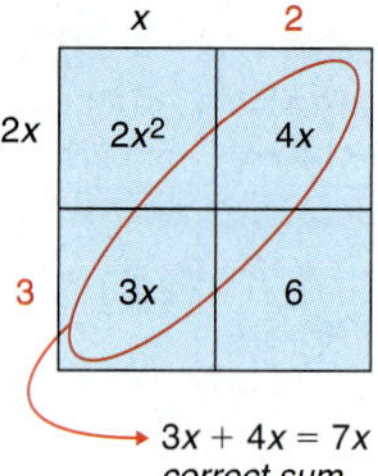

Step 4 The second model results in the correct middle term, $7x$. Therefore, $2x^2 + 7x + 6 = (2x + 3)(x + 2)$.

1–6. See margin for models.

Try These **Use a model to factor each trinomial.**

1. $2x^2 + 7x + 3$
2. $2x^2 + 5x + 3$
3. $3x^2 + 7x + 2$
4. $3x^2 + 8x + 5$
5. $4x^2 + 8x + 3$
6. $4x^2 + 13x + 3$

1. $(2x + 1)(x + 3)$
2. $(2x + 3)(x + 1)$
3. $(3x + 1)(x + 2)$
4. $(3x + 5)(x + 1)$
5. $(2x + 1)(2x + 3)$
6. $(4x + 1)(x + 3)$

Resource Manager

Reproducible Masters

Chapter 10 Resource Masters
- *Study Guide*, p. 448
- *Skills Practice*, p. 449
- *Practice*, p. 450
- *Reading to Learn Mathematics*, p. 451
- *Enrichment*, p. 452

Hands-On Algebra, pp. 110–111

Transparencies

5-Minute Check, 10–4

Technology/Multimedia

AlgePASS, Lesson 24

Interactive Chalkboard CD-ROM

The FOIL method will help you factor trinomials without models.

Examples

Factor each trinomial.

$2x^2 - 7x + 3$

$2x^2$ is the product of the **F**irst terms, and 3 is the product of the **L**ast terms.

$2x^2 - 7x + 3 = (2x + ■)(x + ■)$

The last term, 3, is positive. The sum of the inside and outside terms, -7, is negative. So, both factors of 3 must be negative. Try factor pairs of 3 until the sum of the products of the **O**uter and **I**nner terms is $-7x$.

Try -3 and -1. $(2x - 3)(x - 1) = 2x^2 - 2x - 3x + 3$
$= 2x^2 - 5x + 3$ *$-5x$ is not the correct middle term.*

$(2x - 1)(x - 3) = 2x^2 - 6x - 1x + 3$
$= 2x^2 - 7x + 3$ ✓

Therefore, $2x^2 - 7x + 3 = (2x - 1)(x - 3)$.

$3y^2 + 2y - 5$

$3y^2$ is the product of the **F**irst terms, and -5 is the product of the **L**ast terms.

$3y^2 + 2y - 5 = (3y + ■)(y + ■)$

Find integers whose product is -5. Try factor pairs of -5 until the sum of the products of the **O**uter and **I**nner terms is $2y$.

Try -5 and 1. $(3y - 5)(y + 1) = 3y^2 + 3y - 5y - 5$
$= 3y^2 - 2y - 5$ *$-2y$ is not the correct middle term.*

$(3y + 1)(y - 5) = 3y^2 - 15y + 1y - 5$
$= 3y^2 - 14y - 5$ *$-14y$ is not the correct middle term.*

Try 5 and -1. $(3y + 5)(y - 1) = 3y^2 - 3y + 5y - 5$
$= 3y^2 + 2y - 5$ ✓

Therefore, $3y^2 + 2y - 5 = (3y + 5)(y - 1)$.

Your Turn

a. $2x^2 + 3x + 1$ b. $5y^2 + 2y - 3$ c. $3z^2 - 8z + 4$

a. $(2x + 1)(x + 1)$
b. $(5y - 3)(y + 1)$
c. $(3x - 2)(x - 2)$

2 TEACH

Teaching Tip When discussing Examples 1 and 2, point out to students that there are two possible arrangements for each factor pair, only one of which will provide the correct factorization. This is because the coefficients of the variable terms of the binomial factors are not the same.

In-Class Examples

Examples 1–2

Factor each trinomial.

1 $2x^2 - 9x + 4$
$(2x - 1)(x - 4)$

2 $3y^2 + 7y - 6$
$(3y - 2)(y + 3)$

Answers

Page 440

3.

	x	2
$3x$	$3x^2$	$6x$
1	$1x$	2

4.

	x	1
$3x$	$3x^2$	$3x$
5	$5x$	5

5.

	$2x$	3
$2x$	$4x^2$	$6x$
1	$2x$	3

6.

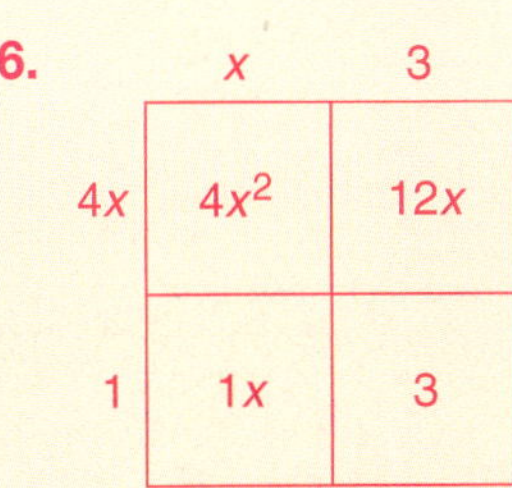

Hands-On Algebra

Cooperative Learning Refer to the Hands-On Algebra on page 440. After Step 3, point out that students do not need to try the factor pair 1 and 6 because the factor pair 2 and 3 works for the given middle term. Some students might have expected 1 and 6 to be the correct factors because their sum is 7. Because the binomial factors have variable terms of x and $2x$ instead of x and x, however, the sum of the coefficients of the outer and inner products using 1 and 6 is either 8 or 13.

An additional Hands-On Algebra activity using algebra tiles to factor trinomials is available in the *Hands-On Algebra Masters,* p. 111.

Hands-On Algebra Masters, p. 110

Sometimes the coefficient of x^2 can be factored into more than one pair of integers.

Example  **Factor $4x^2 + 12x + 5$.**

Number	Factor Pairs
4	4 and 1, 2 and 2
5	5 and 1

Reading Algebra

When factoring this kind of trinomial, it is important to keep an organized list of the factors.

Try 4 and 1. $(4x + 5)(1x + 1) = 4x^2 + 4x + 5x + 5$

$= 4x^2 + 9x + 5$ *9x is not the correct middle term.*

$(4x + 1)(1x + 5) = 4x^2 + 20x + 1x + 5$

$= 4x^2 + 21x + 5$ *21x is not the correct middle term.*

Try 2 and 2. $(2x + 5)(2x + 1) = 4x^2 + 2x + 10x + 5$

$= 4x^2 + 12x + 5$ ✓

Therefore, $4x^2 + 12x + 5 = (2x + 5)(2x + 1)$.

Your Turn

d. $6x^2 + 17x + 5$ **(2x + 5)(3x + 1)** **e.** $4x^2 - 8x - 5$ **(2x + 1)(2x − 5)**

Recall that the first step in factoring any polynomial is to factor out any GCF other than 1.

Example 4 **Manufacturing Link**

The volume of a rectangular shipping crate is $6x^3 - 15x^2 - 36x$. Find possible dimensions for the crate.

The formula for the volume of a rectangular prism is $V = \ell wh$. Find three factors of $6x^3 - 15x^2 - 36x$. First, look for a GCF.

$6x^3 - 15x^2 - 36x = 3x(2x^2 - 5x - 12)$ *The GCF is 3x.*

$3x$ is one factor of $6x^3 - 15x^2 - 36x$. Factor $2x^2 - 5x - 12$ to find the other two factors.

$2x^2 - 5x - 12 = (2x + ■)(x + ■)$

The factors of −12 are −3 and 4, 3 and −4, −2 and 6, 2 and −6, −1 and 12, and 1 and −12. Check several combinations; the correct factors are 3 and −4.

$2x^2 - 5x - 12 = (2x + 3)(x - 4)$

So, $6x^3 - 15x^2 - 36x = 3x(2x + 3)(x - 4)$. Therefore, the dimensions can be $3x$, $2x + 3$, and $x - 4$.

Teaching Tip When discussing Example 3, stress to students that they try factor pairs in an organized, methodical way so that they do not miss any possible pairs or arrangements of factors as they try different combinations.

In-Class Examples

Examples 3–4

3 Factor $4x^2 - 4x - 15$. **(2x + 3)(2x − 5)**

4 The volume of a rectangular shipping crate is $2x^3 - 4x^2 - 30x$. Find possible dimensions for the crate. **Sample answer: 2x, x − 5, and x + 3**

Answers

1a. $2x^2 - 5x - 12 = (x - 4)(2x + 3)$
1b. $6x^2 - 11x + 3 = (2x - 3)(3x - 1)$
1c. $5x^2 + 28x + 15 = (5x + 3)(x + 5)$
13. $(2y + 1)(y + 3)$
14. $(2x + 1)(x + 5)$
15. $(2a + 1)(2a + 3)$
16. $(2x + 1)(x - 5)$
17. $(2q + 3)(q - 6)$
18. $(5x + 2)(x - 3)$
19. $(7a + 1)(a + 3)$
20. prime
21. $(3x + 2)(x + 4)$

***Study Guide*, p. 448**

10-4 NAME ______ DATE ______ PERIOD ______

Study Guide

Student Edition Pages 440–444

Factoring Trinomials: $ax^2 + bx + c$

To find the two binomial factors of a polynomial, use the FOIL method.

Example 1: Factor $5x^2 + 37x + 14$.

The first term in the trinomial is $5x^2$. The only factors of 5 are 5 and 1, so the first terms of the binomials are $5x$ and x.

$$5x^2 + 37x + 14 = (5x + ■)(x + ■)$$

The last term in the trinomial is 14, which has two pairs of factors, 1 and 14, and 2 and 7. Try the factor pairs until you find the one that gives a middle term of $37x$.

First Terms	Last Terms	Binomial Pair	Middle Term	Trinomial
$5x, x$	1, 14	$(5x + 1)(x + 14)$	$x + 70x = 71x$	$5x^2 + 71x + 14$
$5x, x$	14, 1	$(5x + 14)(x + 1)$	$14x + 5x = 19x$	$5x^2 + 19x + 14$
$5x, x$	2, 7	$(5x + 2)(x + 7)$	$2x + 35x = 37x$	$5x^2 + 37x + 14$ ✓

Therefore, $5x^2 + 37x + 14 = (5x + 2)(x + 7)$.

Example 2: Factor $6x^2 - 23x + 7$.

There are two possible factor pairs of the first term, $2x$ and $3x$, and $6x$ and x. The last term is positive. The sum of the inside and outside terms is negative. So, the factors of 7 are −1 and −7. Try the factor pairs until you find the one that gives a middle term of $-23x$.

First Terms	Last Terms	Binomial Pair	Middle Term	Trinomial
$2x, 3x$	−1, −7	$(2x - 1)(3x - 7)$	$-3x - 14x = -17x$	$6x^2 - 17x + 7$
$3x, 2x$	−1, −7	$(3x - 1)(2x - 7)$	$-2x - 21x = -23x$	$6x^2 - 23x + 7$ ✓

Therefore, $6x^2 - 23x + 7 = (3x - 1)(2x - 7)$.

Factor each trinomial.

1. $3x^2 + 4x + 1$ **(3x + 1)(x + 1)**
2. $2w^2 + 3w + 1$ **(2w + 1)(w + 1)**
3. $2r^2 + 5r + 3$ **(2r + 3)(r + 1)**
4. $8z^2 + 14z + 5$ **(2z + 1)(4z + 5)**
5. $5f^2 + 27f + 10$ **(5f + 2)(f + 5)**
6. $2x^2 - 3x + 1$ **(2x − 1)(x − 1)**
7. $7v^2 - 10v + 3$ **(7v − 3)(v − 1)**
8. $9k^2 - 9k + 2$ **(3k − 2)(3k − 1)**
9. $4y^2 + 3y - 1$ **(4y − 1)(y + 1)**
10. $5a^2 + 6a - 8$ **(a + 2)(5a − 4)**

Answers

22. $(2z - 5)(z - 3)$
23. $(3x + 5)(x + 3)$
24. $(3m + 4)(m + 2)$
25. prime
26. $(2x - 1)(2x - 3)$
27. $(7x - 1)(2x + 5)$

Check for Understanding

Communicating Mathematics

1. See margin.

1. **Write** the trinomial and its binomial factors shown by each model.

a.

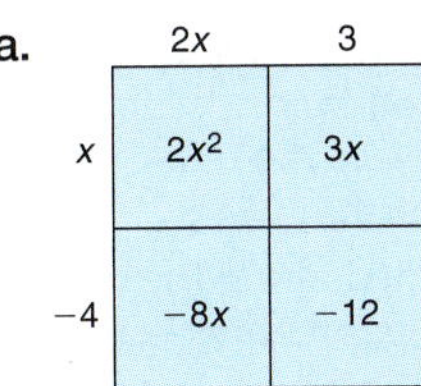

b.

	$3x$	-1
$2x$	$6x^2$	$-2x$
-3	$-9x$	3

c.

	x	5
$5x$	$5x^2$	$25x$
3	$3x$	15

2. **You Decide** Jamal factored the trinomial $18k^2 - 24k + 8$ as $(3k - 2)(6k - 4)$. Jacqui disagrees with Jamal's answer. She says that he did not factor the trinomial completely. Who is correct? Explain your reasoning.

2. Jacqui; Jamal should have factored out the GCF first. The correct answer is $2(3k - 2)(3k - 2)$.

Guided Practice

Examples 1–4

3. $(2a + 3)(a + 1)$
4. $(3y + 1)(y + 2)$
5. $(5x + 3)(x + 2)$
6. $(2x + 3)(x - 1)$

Example 4

7. $(2x + 7)(x - 3)$

Factor each trinomial. If the trinomial cannot be factored, write *prime*. 8. prime

3. $2a^2 + 5a + 3$
4. $3y^2 + 7y + 2$
5. $5x^2 + 13x + 6$
6. $2x^2 + x - 3$
7. $2x^2 + x - 21$
8. $2n^2 - 11n + 7$
9. $10a^2 - 9a + 2$ $(5a - 2)(2a - 1)$
10. $6y^2 - 11y + 4$ $(2y - 1)(3y - 4)$
11. $6x^2 + 16x + 10$ $2(3x + 5)(x + 1)$
12. **Geometry** The measure of the volume of a rectangular prism is $2x^3 + x^2 - 15x$. Find possible dimensions for the prism. Sample answer: x, $2x - 5$, $x + 3$

Exercises

Practice

Homework Help

For Exercises	See Examples
13–21, 23–25, 36, 39, 44	3, 4
22, 26–35, 37–38, 40–41	1, 2
42	4

Extra Practice

See page 712.

B

Factor each trinomial. If the trinomial cannot be factored, write *prime*. 13–39. See margin.

13. $2y^2 + 7y + 3$
14. $2x^2 + 11x + 5$
15. $4a^2 + 8a + 3$
16. $2x^2 - 9x - 5$
17. $2q^2 - 9q - 18$
18. $5x^2 - 13x - 6$
19. $7a^2 + 22a + 3$
20. $3y^2 + 7y + 15$
21. $3x^2 + 14x + 8$
22. $2z^2 - 11z + 15$
23. $3x^2 + 14x + 15$
24. $3m^2 + 10m + 8$
25. $3x^2 + 5x + 1$
26. $4x^2 - 8x + 3$
27. $14x^2 + 33x - 5$
28. $6y^2 - 11y + 4$
29. $8m^2 - 10m + 3$
30. $6r^2 + 9r - 42$
31. $6x^2 + 3x - 30$
32. $4x^2 + 10x - 6$
33. $2x^3 + 5x^2 - 12x$
34. $7x - 5 + 6x^2$
35. $11y + 6y^2 - 2$
36. $15x^3 - 11x^2 - 12x$
37. $2a^2 + 5ab - 3b^2$
38. $15x^2 - 13xy + 2y^2$
39. $9k^2 + 30km + 25m^2$
40. Factor $2x^2 + 5x - 25$. $(2x - 5)(x + 5)$
41. What are the factors of the trinomial $6x^3 + 15x^2 - 9x$? $3x$, $2x - 1$, $x + 3$

Applications and Problem Solving

42. **Measurement** The volume of a rectangular prism is 60 cubic feet. If the measure of the length, width, and height are consecutive integers, find the dimensions. 3 ft, 4 ft, 5 ft

3 PRACTICE/APPLY

Error Analysis

Watch for students who try to factor Exercise 11 directly by trying binomial factors of the form $(6x + a)(x + b)$ or $(3x + a)(2x + b)$, which will lead to an incomplete factorization if left as the product of two binomials.

Prevent by reminding students always to look first for common factors of the terms of a trinomial, so that they can factor out the GCF of the terms before looking for binomial factors. Not only does this ensure complete factorization, it actually makes the process easier.

Assignment Guide

Basic: 13–43 odd, 44–52
Average: 14–40 even, 42–52
All: Quiz 2, 1–5

Answers

28. $(2y - 1)(3y - 4)$
29. $(2m - 1)(4m - 3)$
30. $3(r - 2)(2r + 7)$
31. $3(2x + 5)(x - 2)$
32. $2(2x - 1)(x + 3)$
33. $x(2x - 3)(x + 4)$
34. $(2x - 1)(3x + 5)$
35. $(6y - 1)(y + 2)$
36. $x(5x + 3)(3x - 4)$

***Skills Practice*, p. 449, and *Practice*, p. 450**

10-4 NAME ______ DATE ______ PERIOD ______

Practice

Student Edition Pages 440–444

Factoring Trinomials: $ax^2 + bx + c$

Factor each trinomial. If the trinomial cannot be factored, write prime.

1. $2y^2 + 8y + 6$ $2(y + 1)(y + 3)$
2. $2x^2 + 5x + 2$ $(2x + 1)(x + 2)$
3. $3a^2 - 4a - 4$ $(3a + 2)(a - 2)$
4. $5m^2 - 4m - 1$ $(5m + 1)(m - 1)$
5. $2c^2 + 6c - 8$ $2(c - 1)(c + 4)$
6. $4q^2 + 2q + 3$ prime
7. $3x^2 - 13x + 4$ $(3x - 1)(x - 4)$
8. $4y^2 - 14y + 6$ $2(2y - 1)(y - 3)$
9. $2b^2 - b - 10$ $(2b - 5)(b + 2)$
10. $6a^2 + 8a + 2$ $2(3a + 1)(a + 1)$
11. $3n^2 + 7n - 6$ $(3n - 2)(n + 3)$
12. $3x^2 - 3x - 6$ $3(x + 1)(x - 2)$
13. $2c^2 + 3c - 7$ prime
14. $5y^2 - 17y + 6$ $(5y - 2)(y - 3)$
15. $2b^2 + 2b - 12$ $2(b + 3)(b - 2)$
16. $2x^2 + 10x + 8$ $2(x + 1)(x + 4)$
17. $3m^2 - 19m + 6$ $(3m - 1)(m - 6)$
18. $4a^2 + 10a - 6$ $2(2a - 1)(a + 3)$
19. $7b^2 - 16b + 4$ $(7b - 2)(b - 2)$
20. $3y^2 - y - 10$ $(3y + 5)(y - 2)$
21. $6c^2 + 11c + 4$ $(2c + 1)(3c + 4)$
22. $10x^2 - x - 2$ $(5x + 2)(2x - 1)$
23. $12m^2 - 11m + 2$ $(4m - 1)(3m - 2)$
24. $9y^2 - 3y - 6$ $3(y - 1)(3y + 2)$
25. $8b^2 + 12b + 4$ $4(b + 1)(2b + 1)$
26. $6x^2 + 8x - 8$ $2(3x - 2)(x + 2)$
27. $4n^2 - 14n + 12$ $2(2n - 3)(n - 2)$
28. $6x^2 + 18x + 12$ $6(x + 2)(x + 1)$
29. $4a^2 + 18a - 10$ $2(2a - 1)(a + 5)$
30. $9y^2 - 15y + 6$ $3(3y - 2)(y - 1)$

Reteaching Activity

Verbal/Linguistic Learners Group students in pairs. Have one student describe out loud to the other how to factor trinomials using the model shown in the Hands-On Algebra on page 440. The second student can then describe out loud to the first student how to factor trinomials using the method shown in Example 3.

Answers

37. $(2a - b)(a + 3b)$
38. $(5x - y)(3x - 2y)$
39. $(3k + 5m)(3k + 5m)$

43. Manufacturing The dimensions of a rectangular piece of metal are shown at the right.

2*y* in.
y − 7 in.

a. If a 1-inch by 1-inch square is removed from each corner, write an expression that represents the area of the remaining piece of metal. Express the area in factored form. **$2(y^2 - 7y - 2)$** **b. $2(y - 9)(y - 1)$**

b. If the metal is folded along the dashed lines, an open box is formed. Write an expression that represents the volume of the box.

c. If $y = 10$ inches, find the area of the metal and the volume of the box. **56 in^2, 18 in^3**

44. Critical Thinking Find all values of k so that the trinomial $4y^2 + ky + 5$ can be factored. **21, −21, 12, −12, 9, −9**

Mixed Review

Factor each polynomial. *(Lessons 10–2 & 10–3)*

45. $x^2 + 14x - 32$ **$(x + 16)(x - 2)$**
46. $y^2 - 7x + 12$ **$(y - 4)(y - 3)$**
47. $3a^3 - 15a^2 + 6a$ **$3a(a^2 - 5a + 2)$**
48. $2n^2 + 2n - 24$ **$2(n + 4)(n - 3)$**

49. **Geometry** Find the length of the diagonal of a rectangle whose length is 24 feet and whose height is 7 feet. *(Lesson 8–7)* **25 ft**

Solve. Assume that y varies directly as x. *(Lesson 6–5)*

50. If $y = 28$ when $x = 7$, find x when $y = 52$. **13**
51. Find x when $y = 45$, if $y = 27$ when $x = 6$. **10**

Standardized Test Practice

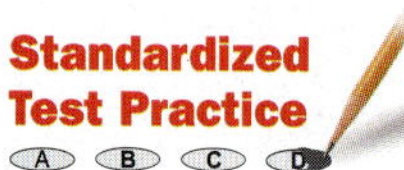

52. **Multiple Choice** What is the solution of $10 - 3(x + 4) = 16$? *(Lesson 4–7)* **A**

Ⓐ -6 Ⓑ $-\frac{12}{7}$ Ⓒ 2 Ⓓ 6

Quiz 2 Lessons 10–3 and 10–4

Factor each trinomial.

1. $x^2 + 3x - 10$ **$(x + 5)(x - 2)$**
2. $x^2 - 5x - 24$ *(Lesson 10–3)* **$(x - 8)(x + 3)$**
3. $2x^2 + 9x + 7$ **$(2x + 7)(x + 1)$**
4. $8x^2 - 16x - 10$ *(Lesson 10–4)* **$2(2x + 1)(2x - 5)$**

5. **Geometry** Find the area of the shaded region. Express the area in factored form. *(Lesson 10–3)* **$(x + 3)(x - 2)$**

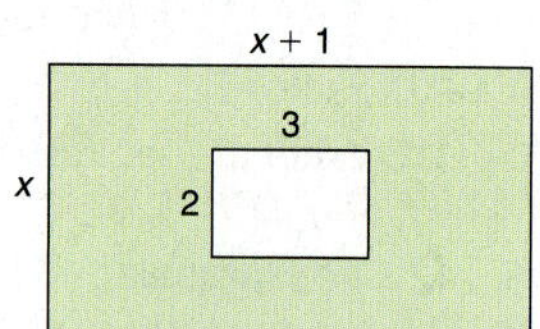

www.algconcepts.com/self_check_quiz

4 ASSESS

Open-Ended Assessment

Speaking Have students describe the kinds of errors they should watch for when factoring polynomials of the form $ax^2 + bx + c$, such as $2x^2 + 5x + 3$ or $3x^2 - 11x + 6$.

Quiz 2

The Quiz provides students with a brief review of the concepts and skills in Lessons 10–3 and 10–4. Lesson numbers are given to the right of the exercises or instruction lines so students can review concepts not yet mastered.

Enrichment, p. 452

10–4 NAME ______ DATE ______ PERIOD ______

Enrichment Student Edition Pages 440–444

Factoring Trinomials of Fourth Degree

Some trinomials of the form $a^4 + a^2b^2 + b^4$ can be written as the difference of two squares and then factored.

Example: Factor $4x^4 - 37x^2y^2 + 9y^4$.

Step 1 Find the square roots of the first and last terms.
$\sqrt{4x^4} = 2x^2$ $\sqrt{9y^4} = 3y^2$

Step 2 Find twice the product of the square roots.
$2(2x^2)(3y^2) = 12x^2y^2$

Step 3 Separate the middle term into two parts. One part is either your answer to Step 2 or its opposite. The other part should be the opposite of a perfect square.
$-37x^2y^2 = -12x^2y^2 - 25x^2y^2$

Step 4 Rewrite the trinomial as the difference of two squares and then factor.

$$\begin{aligned}4x^4 - 37x^2y^2 + 9y^4 &= (4x^4 - 12x^2y^2 + 9y^4) - 25x^2y^2\\ &= (2x^2 - 3y^2)^2 - 25x^2y^2\\ &= [(2x^2 - 3y^2) + 5xy][(2x^2 - 3y^2) - 5xy]\\ &= (2x^2 + 5xy - 3y^2)(2x^2 - 5xy - 3y^2)\end{aligned}$$

Factor each trinomial.

1. $x^4 + x^2y^2 + y^4$ **$(x^2 + xy + y^2)(x^2 - xy + y^2)$**
2. $x^4 + x^2 + 1$ **$(x^2 + x + 1)(x^2 - x + 1)$**
3. $9a^4 - 15a^2 + 1$ **$(3a^2 + 3a - 1)(3a^2 - 3a - 1)$**
4. $16a^4 - 17a^2 + 1$ **$(4a - 1)(a + 1)(4a + 1)(a - 1)$**
5. $4a^4 - 13a^2 + 1$ **$(2a^2 + 3a - 1)(2a^2 - 3a - 1)$**
6. $9a^4 + 26a^2b^2 + 25b^4$ **$(3a^2 + 2ab + 5b^2)(3a^2 - 2ab + 5b^2)$**
7. $4x^4 - 21x^2y^2 + 9y^4$ **$(2x^2 + 3xy - 3y^2)(2x^2 - 3xy - 3y^2)$**
8. $4a^4 - 29a^2c^2 + 25c^4$ **$(2a + 5c)(a - c)(2a - 5c)(a + c)$**

Extra Credit

Is it easier to factor trinomials whose square term coefficient and numerical term are prime or composite? Explain. **Sample answer: Prime; prime numbers have only themselves and one as factors, so there are fewer combinations of factor pairs to try when looking for the correct factorization of the trinomial.**

10–5 Special Factors

What You'll Learn

You'll learn to recognize and factor the differences of squares and perfect square trinomials.

Why It's Important

Manufacturing You can find the area of a washer by using the difference of squares. *See Exercise 52.*

In this lesson, you will learn to recognize and factor polynomials that are **perfect square trinomials**. They have two equal binomial factors.

The model below shows the product $(x + 3)^2$.

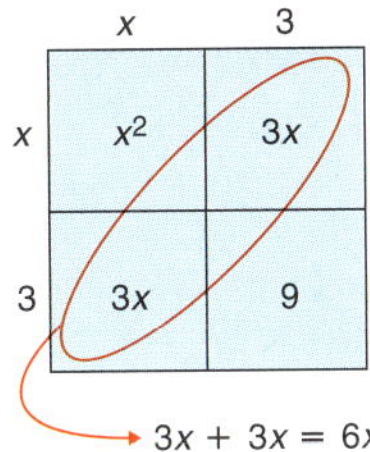

You can also use the FOIL method to find the product.

$$(x + 3)(x + 3) = x^2 + 3x + 3x + 9$$
$$= x^2 + 6x + 9$$

perfect squares

twice the product of x and 3

Square of a Sum: Lesson 9–5

The square of $(x + 3)$ is the sum of

- the square of the first term of the binomial,
- the square of the last term of the binomial, and
- twice the product of the terms of the binomial.

These observations will help you recognize when a trinomial is a perfect square trinomial. They can be factored as shown.

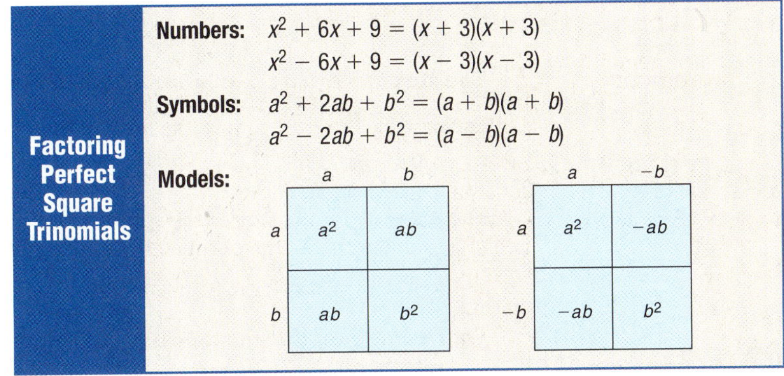

Factoring Perfect Square Trinomials

Numbers: $x^2 + 6x + 9 = (x + 3)(x + 3)$
$x^2 - 6x + 9 = (x - 3)(x - 3)$

Symbols: $a^2 + 2ab + b^2 = (a + b)(a + b)$
$a^2 - 2ab + b^2 = (a - b)(a - b)$

Models:

	a	b
a	a^2	ab
b	ab	b^2

	a	−b
a	a^2	−ab
−b	−ab	b^2

Lesson 10–5

1 FOCUS

5-Minute Check
Lesson 10–4

Factor each trinomial.

1. $2x^2 - 11x - 6$ **(2x + 1)(x − 6)**
2. $3y^2 - 7y + 2$ **(3y − 1)(y − 2)**
3. $5z^2 + 19z + 12$ **(5z + 4)(z + 3)**
4. $4c^2 - 16c + 15$ **(2c − 5)(2c − 3)**
5. Find the dimensions of the shipping crate with height and volume shown below.

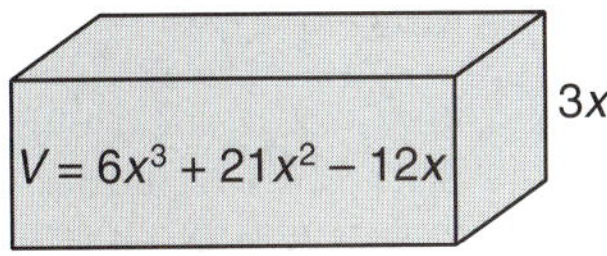

3x, 2x − 1, and x + 4

Motivating the Lesson

Hands-On Activity Have 10 slips of paper on hand, each with a number from 6 to 15 written on it. Mix up the slips and hand them out to 10 students. Have these students keep their numbers hidden and write the square of the number in large print on a piece of paper. These students can then hold up the perfect square and have other students identify the number that was squared. Urge students to memorize the squares so that they recognize them when factoring.

2 TEACH

Teaching Tip As you discuss the square of $x + 3$, ask a volunteer to go to the board or overhead and explain using the FOIL model why the middle term of the product trinomial is twice the product of the terms of each binomial.

Resource Manager

Reproducible Masters

Chapter 10 Resource Masters
- *Study Guide*, p. 453
- *Skills Practice*, p. 454
- *Practice*, p. 455
- *Reading to Learn Mathematics*, p. 456
- *Enrichment*, p. 457
- *Assessment*, p. 469

Hands-On Algebra, p. 112

Transparencies
5-Minute Check, 10–5

Technology/Multimedia
AlgePASS, Lesson 24
Interactive Chalkboard CD-ROM

Teaching Tip As you discuss the chart on factoring perfect square trinomials, make sure students realize that the constant term is always positive, but that the coefficient of the variable term can be positive or negative depending on whether the trinomial is the square of a sum or a difference.

In-Class Examples

Examples 1–3

Determine whether each trinomial is a perfect square trinomial. If so, factor it.

1 $x^2 + 14x + 49$ **yes; $(x + 7)^2$**

2 $9a^2 + 16a + 4$ **no**

3 $16b^2 + 24b + 9$ **yes; $(4b + 3)^2$**

Teaching Tip After Example 3, ask students to recount the three tests they must make of a trinomial when deciding if it is a perfect square. If the trinomial fails any of these tests, it is not a perfect square. If you wish, you can show students a flow diagram of this decision process.

In-Class Example

Example 4

The area of a square is $d^2 - 16d + 64$. Find the perimeter. **$4d - 32$**

Examples

Determine whether each trinomial is a perfect square trinomial. If so, factor it.

1 $x^2 + 10x + 25$

To determine whether $x^2 + 10x + 25$ is a perfect square trinomial, answer each question.

- Is the first term a perfect square? Yes, x^2 is the square of x.
- Is the last term a perfect square? Yes, 25 is the square of 5.
- Is the middle term twice the product of x and 5? Yes, $10x = 2(5x)$.

Therefore, $x^2 + 10x + 25$ is a perfect square trinomial.
$x^2 + 10x + 25 = (x + 5)^2$

2 $4n^2 - 4n + 1$

- Is the first term a perfect square? Yes, $4n^2$ is the square of $2n$.
- Is the last term a perfect square? Yes, 1 is the square of 1 and -1.
- Is the middle term twice the product of $2n$ and -1? Yes, $2(-2n) = -4n$.

Therefore, $4n^2 - 4n + 1$ is a perfect square trinomial.
$4n^2 - 4n + 1 = (2n - 1)^2$

3 $4p^2 - 12p + 36$

- Is the first term a perfect square? Yes, $4p^2$ is the square of $2p$.
- Is the last term a perfect square? Yes, 36 is the square of 6 and -6.
- Is the middle term twice the product of $2p$ and -6? No, $2(-12p) \neq -12p$.

Therefore, $4p^2 - 12p + 36$ is *not* a perfect square trinomial.

Your Turn

a. $a^2 + 2a + 1$ **yes, $(a + 1)^2$**

b. $16x^2 + 20x + 25$ **no**

c. $49x^2 - 14x + 1$ **yes, $(7x - 1)^2$**

Geometry Link

4 **The area of a square is $x^2 + 18x + 81$. Find the perimeter.**

Factor $x^2 + 18x + 81$ to find the measure of one side of the square.

$x^2 + 18x + 81 = (x + 9)^2$

The measure of one side of the square is $x + 9$. A square has four sides of equal length. So, the perimeter is four times the length of a side.

$4(x + 9) = 4x + 36$ *Distributive Property*

The perimeter of the square is $4x + 36$.

www.algconcepts.com/extra_examples

A polynomial like $x^2 - 9$ is called the **difference of squares**. Although this is not a trinomial, it *can* be factored into two binomials. The model shows how to factor $x^2 - 9$.

Product of a Sum and a Difference: Lesson 9–5

perfect squares

$x^2 - 9 = (x - 3)(x + 3)$

difference

product of a sum and a difference

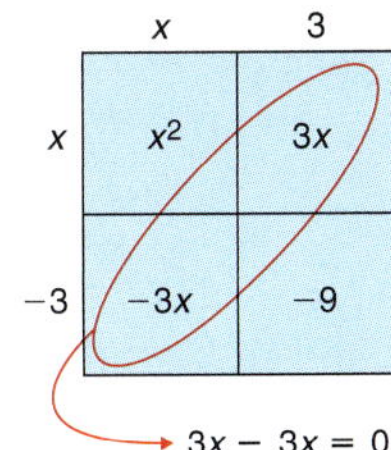

A difference of squares can be factored as shown.

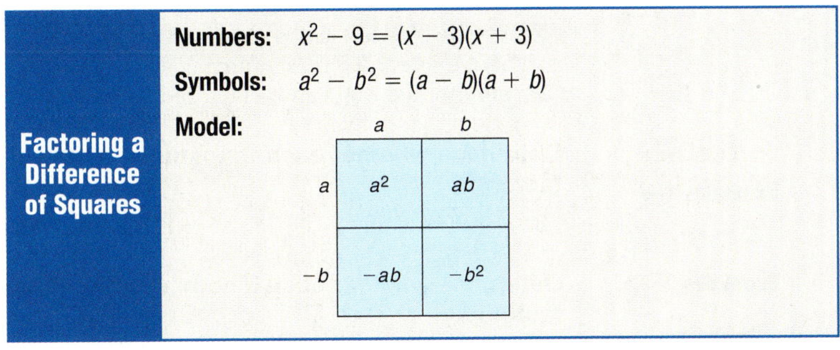

Factoring a Difference of Squares

Numbers: $x^2 - 9 = (x - 3)(x + 3)$

Symbols: $a^2 - b^2 = (a - b)(a + b)$

Model:

Examples

Determine whether each binomial is the difference of squares. If so, factor it.

5 $a^2 - 25$

a^2 and 25 are both perfect squares, and $a^2 - 25$ is a difference.

$a^2 - 25 = (a)^2 - (5)^2$ *$a \cdot a = a^2$, $5 \cdot 5 = 25$*

$= (a - 5)(a + 5)$ *Difference of Squares*

6 $y^2 + 100$

y^2 and 100 are both perfect squares. But $y^2 + 100$ is a sum, not a difference. Therefore $y^2 + 100$ is *not* a difference of squares. It is a prime polynomial.

7 $3n^2 - 48$

First, look for a GCF. Then, determine whether the remaining factor is a difference of squares.

$3n^2 - 48 = 3(n^2 - 16)$ *The GCF of $3n^2$ and 48 is 3.*

$= 3[(n)^2 - (4)^2]$ *$n \cdot n = n^2$, $4 \cdot 4 = 16$*

$= 3(n - 4)(n + 4)$ *Difference of Squares*

Your Turn **d. yes, $(11 - p)(11 + p)$** **e. yes, $25x(x - 2)(x + 2)$**

d. $121 - p^2$ e. $25x^3 - 100x$ f. $4a^2 + 49$ **no**

In-Class Examples

Examples 5–7

Determine whether each binomial is the difference of squares. If so, factor it.

5 $d^2 - 81$ **yes; $(d - 9)(d + 9)$**

6 $f^2 + 64$ **no**

7 $4m^2 - 144$ **yes; $4(m - 6)(m + 6)$**

***Study Guide*, p. 453**

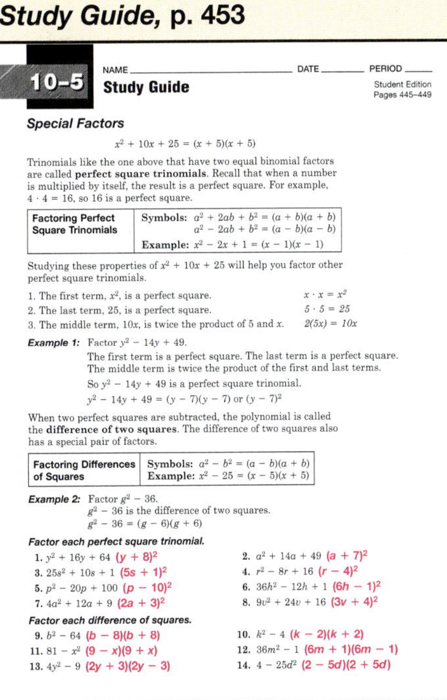

NAME ______ DATE ______ PERIOD ______

10-5 Study Guide

Student Edition Pages 445–449

Special Factors

$x^2 + 10x + 25 = (x + 5)(x + 5)$

Trinomials like the one above that have two equal binomial factors are called **perfect square trinomials**. Recall that when a number is multiplied by itself, the result is a perfect square. For example, $4 \cdot 4 = 16$, so 16 is a perfect square.

Factoring Perfect Square Trinomials	**Symbols:** $a^2 + 2ab + b^2 = (a + b)(a + b)$ $a^2 - 2ab + b^2 = (a - b)(a - b)$ **Example:** $x^2 - 2x + 1 = (x - 1)(x - 1)$

Studying these properties of $x^2 + 10x + 25$ will help you factor other perfect square trinomials.

1. The first term, x^2, is a perfect square. $x \cdot x = x^2$
2. The last term, 25, is a perfect square. $5 \cdot 5 = 25$
3. The middle term, $10x$, is twice the product of 5 and x. $2(5x) = 10x$

Example 1: Factor $y^2 - 14y + 49$.

The first term is a perfect square. The last term is a perfect square. The middle term is twice the product of the first and last terms.

So $y^2 - 14y + 49$ is a perfect square trinomial.

$y^2 - 14y + 49 = (y - 7)(y - 7)$ or $(y - 7)^2$

When two perfect squares are subtracted, the polynomial is called the **difference of two squares**. The difference of two squares also has a special pair of factors.

Factoring Differences of Squares	**Symbols:** $a^2 - b^2 = (a - b)(a + b)$ **Example:** $x^2 - 25 = (x - 5)(x + 5)$

Example 2: Factor $g^2 - 36$.

$g^2 - 36$ is the difference of two squares.

$g^2 - 36 = (g - 6)(g + 6)$

Factor each perfect square trinomial.

1. $y^2 + 16y + 64$ $(y + 8)^2$
2. $a^2 + 14a + 49$ $(a + 7)^2$
3. $25s^2 + 10s + 1$ $(5s + 1)^2$
4. $r^2 - 8r + 16$ $(r - 4)^2$
5. $p^2 - 20p + 100$ $(p - 10)^2$
6. $36h^2 - 12h + 1$ $(6h - 1)^2$
7. $4a^2 + 12a + 9$ $(2a + 3)^2$
8. $9v^2 + 24v + 16$ $(3v + 4)^2$

Factor each difference of squares.

9. $b^2 - 64$ $(b - 8)(b + 8)$
10. $k^2 - 4$ $(k - 2)(k + 2)$
11. $81 - x^2$ $(9 - x)(9 + x)$
12. $36m^2 - 1$ $(6m + 1)(6m - 1)$
13. $4y^2 - 9$ $(2y + 3)(2y - 3)$
14. $4 - 25d^2$ $(2 - 5d)(2 + 5d)$

3 PRACTICE/APPLY

Error Analysis

Watch for students who decide the binomial in Exercise 7 is not a difference of squares because 8 and 50 are not perfect squares. **Prevent by** referring students back to Example 7. Point out that the terms in Exercise 7 have a common factor. Once the factor is divided out, it becomes clear that both terms are perfect squares.

Assignment Guide

Basic: 13–51 odd, 53–63
Average: 14–50 even, 51–63

Answers

1. No; 7 is not a perfect square.
13. $(r + 4)^2$
14. $(x - 8)^2$
15. $(a + 1)^2$
16. $(2a + 1)^2$
17. $(2z - 5)^2$
18. no
19. $(3a + 4)^2$
20. $(d - 11)^2$
21. $(7 + z)^2$
22. $(x - 4)(x + 4)$
23. $(a - 6)(a + 6)$
24. no
25. $(1 - 3m)(1 + 3m)$

Skills Practice, p. 454, and *Practice,* p. 455 (shown)

10-5 NAME ______ DATE ______ PERIOD ______

Practice — Student Edition Pages 445–449

Special Factors

Determine whether each trinomial is a perfect square trinomial. If so, factor it.

1. $y^2 + 6y + 9$ — $(y + 3)^2$
2. $x^2 - 4x + 4$ — $(x - 2)^2$
3. $n^2 + 6n + 3$ — no
4. $m^2 - 12m + 36$ — $(m - 6)^2$
5. $y^2 - 10y + 20$ — no
6. $4a^2 + 16a + 16$ — $(2a + 4)^2$
7. $9x^2 + 6x + 1$ — $(3x + 1)^2$
8. $4n^2 - 20n + 25$ — $(2n - 5)^2$
9. $4y^2 + 9y + 9$ — no

Determine whether each binomial is the difference of squares. If so, factor it.

10. $x^2 - 49$ — $(x - 7)(x + 7)$
11. $b^2 + 16$ — no
12. $y^2 - 81$ — $(y - 9)(y + 9)$
13. $4m^2 - 9$ — $(2m - 3)(2m + 3)$
14. $9a^2 - 16$ — $(3a - 4)(3a + 4)$
15. $25r^2 + 9$ — no
16. $18n^2 - 18$ — $18(n - 1)(n + 1)$
17. $3x^2 - 12y^2$ — $3(x - 2y)(x + 2y)$
18. $8m^2 - 18n^2$ — $2(2m - 3n)(2m + 3n)$

Factor each polynomial. If the polynomial cannot be factored, write prime.

19. $4a - 24$ — $4(a - 6)$
20. $6x + 9$ — $3(2x + 3)$
21. $x^2 + 5x - 10$ — prime
22. $2y^2 + 6y - 20$ — $2(y - 2)(y + 5)$
23. $m^2 - 9n^2$ — $(m - 3n)(m + 3n)$
24. $a^2 - 8a + 16$ — $(a - 4)^2$
25. $5b^2 + 10b$ — $5b(b + 2)$
26. $9y^2 + 12y + 4$ — $(3y + 2)^2$
27. $3x^2 - 3x - 18$ — $(3x + 6)(x - 3)$

The following chart summarizes factoring methods.

Concept Summary	Factoring Method	Number of Terms		
		Two	Three	Four or more
	greatest common factor	✓	✓	✓
	difference of squares	✓		
	perfect square trinomials		✓	
	trinomial with two binomial factors		✓	

Check for Understanding

Communicating Mathematics

1. **State** whether $4c^2 - 7$ can be factored as a difference of squares. Explain. **See margin.**
2. *Writing Math* Copy the chart shown above into your math journal. Then write a polynomial that can be factored by each method. **See students' work.**

Vocabulary
perfect square trinomials
difference of squares

Guided Practice

Examples 1–3 — **Determine whether each trinomial is a perfect square trinomial. If so, factor it.**

3. $y^2 + 14y + 49$ $(y + 7)^2$
4. $x^2 - 10x + 100$ no
5. $a^2 - 10a + 25$ $(a - 5)^2$

Examples 5–7 — **Determine whether each binomial is the difference of squares. If so, factor it.**

6. $16x^2 - 25$ $(4x - 5)(4x + 5)$
7. $8x^2 - 50y^2$ $2(2x - 5y)(2x + 5y)$
8. $49m^2 + 16$ no

Factor each polynomial. If the polynomial cannot be factored, write *prime*.

9. $3x^2 + 15$ $3(x^2 + 5)$
10. $y^2 + 6y - 9$ prime
11. $3y^2 + 21y - 24$ $3(y - 1)(y + 8)$

Example 4

12. **Geometry** The area of a square is $4x^2 + 20xy + 25y^2$. Find the perimeter. $8x + 20y$

Exercises

Practice

Determine whether each trinomial is a perfect square trinomial. If so, factor it.

13. $r^2 + 8r + 16$
14. $x^2 - 16x + 64$
15. $a^2 + 2a + 1$
16. $4a^2 + 4a + 1$
17. $4z^2 - 20z + 25$
18. $9m^2 + 15m + 25$
19. $9a^2 + 24a + 16$
20. $d^2 - 22d + 121$
21. $49 + 14z + z^2$

Determine whether each binomial is the difference of squares. If so, factor it.

22. $x^2 - 16$
23. $a^2 - 36$
24. $y^2 - 20$
25. $1 - 9m^2$
26. $16m^2 - 25n^2$
27. $y^2 + z^2$
28. $8a^2 - 18$
29. $2z^2 - 98$
30. $49 - a^2b^2$

Homework Help

For Exercises	See Examples
13–21, 32	1, 2, 3
22–31	5, 6, 7
33–53	3, 4, 5

Extra Practice
See page 713.

13–30. See margin.

Reteaching Activity

Interpersonal Learners Have small groups of students work together to make posters describing how to recognize and factor perfect square trinomials and differences of squares. Have students include on the posters common mistakes and how to avoid them.

Answers

26. $(4m - 5n)(4m + 5n)$
27. no
28. $2(2a - 3)(2a + 3)$
29. $2(z - 7)(z + 7)$
30. $(7 - ab)(7 + ab)$

31. Sample answer: $25x^2 - 4$; $(5x - 4)(5x + 4)$

31. Write a polynomial that is the difference of two squares. Then factor it.

32. Is $x^2 + x - 1$ a perfect square trinomial? Explain. No; −1 is not a perfect square.

Factor each polynomial. If the polynomial cannot be factored, write *prime*. 33–50. See margin.

33. $5x^2 + 25$	34. $a^2 - 16b^2$	35. $y^2 - 5y + 6$
36. $m^2 + 8m + 16$	37. $2x^2 - 72$	38. $3a^2b + 6ab + 9ab^2$
39. $8xy^2 - 13x^2y$	40. $2r^2 + 3r + 1$	41. $x^2 - 6x - 9$
42. $z^3 + 6z^2 + 9z$	43. $20n^2 + 34n + 6$	44. $b^2 + 6 - 7b$
45. $8w^2 + 14w - 15$	46. $a^3 - 17a^2 + 72a$	47. $5x^2 + 15x + 10$
48. $7a^2 - 21a$	49. $2x^3 - 32x$	50. $2x^2 - 11x - 21$

Applications and Problem Solving

51. **Number Theory** The difference of two numbers is 2. The difference of their squares is 12. Find the numbers. 2, 4

52. **Manufacturing** A metal washer is manufactured by stamping out a circular hole from a metal disk. In the figure, r represents the radius of the metal disk. The radius of the hole is 1 centimeter.

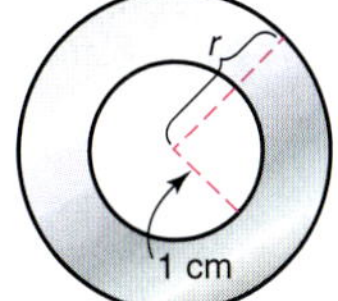

a. Write an expression in factored form for the area of the washer. (*Hint*: Use $A = \pi r^2$.)

b. If $r = 2$ centimeters, find the area of the washer to the nearest hundredth. 9.42 cm²

a. $\pi(r - 1)(r + 1)$

53. **Critical Thinking** The area of a square is $81 - 90x + 25x^2$. If x is a positive integer, what is the least possible measure for the square's perimeter? 4

Mixed Review

Factor each polynomial. *(Lessons 10–3 & 10–4)*

54. $4y^2 + 16y + 15$ $(2y + 3)(2y + 5)$

55. $4x^2 + 11x - 3$ $(4x - 1)(x + 3)$

56. $a^3 - 7a^2 + 12a$ $a(a - 3)(a - 4)$

57. $m^2 - 5m - 14$ $(m - 7)(m + 2)$

Simplify each expression. *(Lesson 8–3)*

58. n^{-2} $\frac{1}{n^2}$

59. $a^5(a^{-3})$ a^2

60. $\frac{1}{r^{-3}}$ r^3

61. $\frac{3c^2d^3f^4}{9c^4d^2f^4}$ $\frac{d}{3c^2}$

Standardized Test Practice

62. **Extended Response** Describe the difference between the graphs of $y = 4x$ and $y = 4x - 5$. *(Lesson 7–6)* The graph moves down 5 units.

63. **Multiple Choice** Which graph is the best example of data that exhibit a linear relationship between the variables x and y? *(Lesson 6–3)* C

Ⓐ
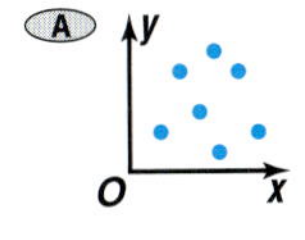

Ⓑ
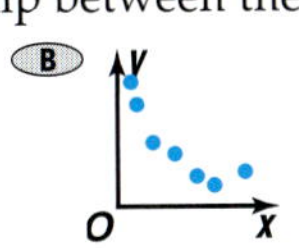

Ⓒ
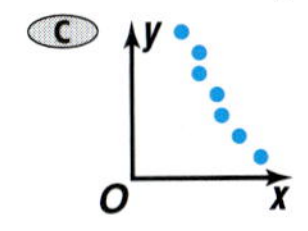

Ⓓ
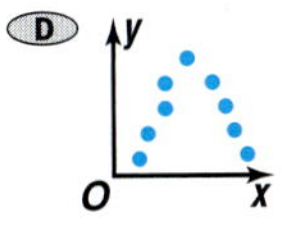

www.algconcepts.com/self_check_quiz

Extra Credit

Factor $x^4 - 16$. $(x - 2)(x + 2)(x^2 + 4)$

4 ASSESS

Open-Ended Assessment

Speaking Have students explain how to determine whether a trinomial is a perfect square trinomial or the difference of squares.

Chapter 10, Quiz B (Lessons 10–3 through 10–5) is available in the *Chapter 10 Resource Masters*, p. 469.

Answers

33. $5(x^2 + 5)$
34. $(a - 4b)(a + 4b)$
35. $(y - 3)(y - 2)$
36. $(m + 4)^2$
37. $2(x - 6)(x + 6)$
38. $3ab(a + 2 + 3b)$
39. $xy(8y - 13x)$
40. $(2r + 1)(r + 1)$
41. prime
42. $z(z + 3)^2$
43. $2(5n + 1)(2n + 3)$
44. $(b - 1)(b - 6)$
45. $(4w - 3)(2w + 5)$
46. $a(a - 8)(a - 9)$
47. $5(x + 1)(x + 2)$
48. $7a(a - 3)$
49. $2x(x - 4)(x + 4)$
50. $(2x + 3)(x - 7)$

***Enrichment*, p. 457**

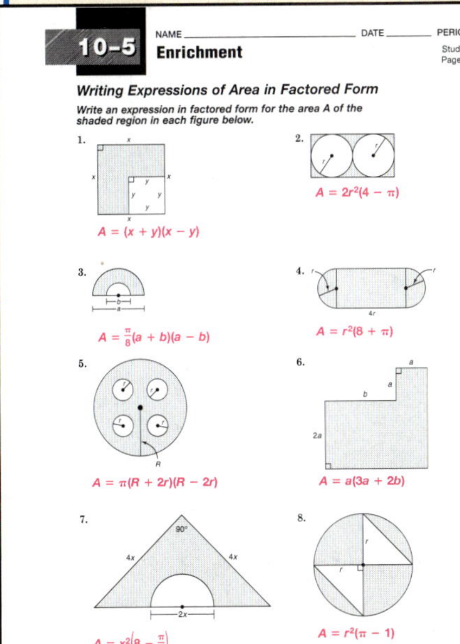
10-5 Enrichment
NAME ______ DATE ______ PERIOD ______
Student Edition Pages 445–449

Writing Expressions of Area in Factored Form

Write an expression in factored form for the area A of the shaded region in each figure below.

1. $A = (x + y)(x - y)$
2. $A = 2r^2(4 - \pi)$
3. $A = \frac{\pi}{8}(a + b)(a - b)$
4. $A = r^2(8 + \pi)$
5. $A = \pi(R + 2r)(R - 2r)$
6. $A = a(3a + 2b)$
7. $A = x^2\left(8 - \frac{\pi}{2}\right)$
8. $A = r^2(\pi - 1)$

Understanding and Using the Vocabulary

This section provides a listing of the new terms, properties, and phrases that were introduced in this chapter. The exercises check students' understanding of the terms by using a variety of verbal formats including matching, completion, and true/false.

Glossary A complete glossary of terms appears on pages 762–783.

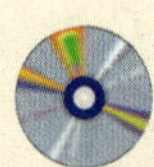

MindJogger Videoquizzes

MindJogger Videoquizzes provide an alternative review of concepts presented in this chapter. Students work in teams to answer questions, gaining points for correct answers.

CHAPTER 10 Study Guide and Assessment

Understanding and Using the Vocabulary

After completing this chapter, you should be able to define each term, property, or phrase and give an example or two of each.

Review Activities
For more review activities, visit: www.algconcepts.com

difference of squares *(p. 447)*
factoring *(p. 428)*
greatest common factor (GCF) *(p. 422)*
perfect square trinomials *(p. 445)*
prime polynomial *(p. 430)*

State whether each sentence is *true* or *false*. If false, replace the underlined word or number to make a true sentence.

1. The prime factorization of 12 is $\underline{3 \cdot 4}$. **false, $2^2 \cdot 3$**
2. $\underline{3x}$ is the greatest common factor of $6x^2$ and $9x$. **true**
3. When two or more numbers are multiplied, each number is a <u>factor</u> of the product. **true**
4. $2y$ and $(y + 3)$ are factors of $\underline{2y^2 + 3}$. **false, $2y^2 + 6y$**
5. When you factor trinomials, always check for a <u>GCF</u> first. **true**
6. The number <u>51</u> is an example of a prime number. **false, sample answer: 5**
7. $(x - 3)(x + 3)$ is the factored form of $\underline{x^2 + 9}$. **false, $x^2 - 9$**
8. Whole numbers that have more than two factors are called <u>composite numbers</u>. **true**
9. A polynomial is in <u>factored form</u> when it is expressed as the product of polynomials. **true**
10. $4a^2 - b^2$ is an example of a <u>perfect square trinomial</u>. **false, difference of squares**

Skills and Concepts

Objectives and Examples

- **Lesson 10–1** Find the greatest common factor of a set of numbers or monomials.

Find the GCF of $12x^2y$ and $30xy^2$.

$12x^2y = 2 \cdot (2) \cdot (3) \cdot (x) \cdot x \cdot (y)$
$30xy^2 = (2) \cdot (3) \cdot 5 \cdot (x) \cdot (y) \cdot y$

The GCF of $12x^2y$ and $30xy^2$ is $2 \cdot 3 \cdot x \cdot y$ or $6xy$.

Review Exercises

Find the GCF of each set of numbers or monomials.

11. 20, 25 **5**
12. 12, 18, 42 **6**
13. 20, 25, 28 **1**
14. $5xy$, $10x$ **$5x$**
15. $9x^2$, $9x$ **$9x$**
16. $6a^2b$, $18a^2b^2$, $9ab^2$ **$3ab$**

www.algconcepts.com/vocabulary_review

Resource Manager

Reproducible Masters
Chapter 10 Resource Masters
- *Assessment*, pp. 459–467, 470–472

Technology/Multimedia
MindJogger Videoquizzes
ExamView® Pro

Objectives and Examples

- **Lesson 10–2** Use the GCF and the Distributive Property to factor polynomials.

Factor $12x^2 - 8xy$.

The GCF of $12x^2$ and $8xy$ is $4x$. Write each term as a product of the GCF and its remaining factors.

$$12x^2 - 8xy = 4x(3x) - 4x(2y)$$
$$= 4x(3x - 2y) \quad \text{Distributive Property}$$

Review Exercises

Factor each polynomial. If the polynomial cannot be factored, write *prime*.

17. $5x + 30y$ **$5(x + 6y)$**
18. $16a^2 + 32b^2$ **$16(a^2 + 2b^2)$**
19. $12ab - 18a^2$ **$6a(2b - 3a)$**
20. $5mn^2 + 10mn$ **$5mn(n + 2)$**
21. $3xy + 12x^2y^2$ **$3xy(1 + 4xy)$**

Find each quotient.

22. $(20x^3 + 15x^2) \div 5x$ **$4x^2 + 3x$**
23. $(40a^2b^2 - 8ab) \div 8ab$ **$5ab - 1$**

- **Lesson 10–3** Factor trinomials of the form $x^2 + bx + c$.

Factor $x^2 + 3x - 10$.

Find integers whose product is -10 and whose sum is 3.

Product	Integers	Sum
−10	2, −5	2 + (−5) = −3
−10	−2, 5	−2 + 5 = 3 ✓

Therefore, $x^2 + 3x - 10 = (x - 2)(x + 5)$.

Factor each trinomial. If the trinomial cannot be factored, write *prime*.

24. $y^2 + 9y + 14$ **$(y + 7)(y + 2)$**
25. $x^2 - 8x + 15$ **$(x - 5)(x - 3)$**
26. $a^2 + 5a - 7$ **prime**
27. $x^2 - 2x - 8$ **$(x - 4)(x + 2)$**
28. $y^2 + 7y + 12$ **$(y + 4)(y + 3)$**
29. $x^2 + 2x - 35$ **$(x + 7)(x - 5)$**
30. $a^2 - a - 1$ **prime**
31. $2n^2 - 8n - 24$ **$2(n - 6)(n + 2)$**

- **Lesson 10–4** Factor trinomials of the form $ax^2 + bx + c$.

Factor $2x^2 + 5x + 3$.

$$2x^2 + 5x + 3 = (2x + \blacksquare)(x + \blacksquare)$$

$$(2x + 3)(x + 1) = 2x^2 + 2x + 3x + 3$$
$$= 2x^2 + 5x + 3$$

Therefore, $2x^2 + 5x + 3 = (2x + 3)(x + 1)$.

Factor each trinomial. If the trinomial cannot be factored, write *prime*.

32. $2z^2 + 7z + 5$ **$(2z + 5)(z + 1)$**
33. $3x^2 + 8x + 5$ **$(3x + 5)(x + 1)$**
34. $3a^2 + 8a + 4$ **$(3a + 2)(a + 2)$**
35. $6a^2 - a - 2$ **$(3a - 2)(2a + 1)$**
36. $3x^2 - 7x - 6$ **$(3x + 2)(x - 3)$**
37. $2y^2 - 9y - 18$ **$(2y + 3)(y - 6)$**
38. $2x^2 + 5x + 6$ **prime**
39. $15a^2 - 20a + 5$ **$5(3a - 1)(a - 1)$**

Skills and Concepts

The **Objectives and Examples** section reviews the skills and concepts of the chapter and shows completely worked examples.

The **Review Exercises** provide practice for the corresponding objectives.

ExamView® Pro

Use the networkable ***ExamView® Pro Testmaker CD-ROM*** to:

- Create **multiple versions** of tests.
- Create **modified** tests for **inclusion** students with one mouse click.
- **Edit** existing questions and **add** your own questions.
- Build tests aligned with state standards using built-in **state curriculum correlations**.
- Change **English** tests to **Spanish** with one mouse click and vice versa.

Mixed Problem Solving
See pages 724–731.

Applications and Problem Solving

This section provides additional practice in solving real-world problems that involve the concepts of this chapter.

Objectives and Examples

- **Lesson 10–5** Recognize and factor the differences of squares and perfect square trinomials.

Factor $a^2 + 6a + 9$.

$$a^2 + 6a + 9 = a^2 + 2(3a) + 3^2$$
$$= (a + 3)^2$$

Factor $x^2 - 25$.

$$x^2 - 25 = x^2 - 5^2$$
$$= (x + 5)(x - 5)$$

Review Exercises

Factor each polynomial. If the polynomial cannot be factored, write *prime*.

40. $y^2 + 8y + 16$ **$(y + 4)^2$**

41. $a^2 - 12a + 36$ **$(a - 6)^2$**

42. $n^2 + 2n - 1$ **prime**

43. $25x^2 + 20x + 4$ **$(5x + 2)^2$**

44. $y^2 - 81$ **$(y - 9)(y + 9)$**

45. $4x^2 - 9$ **$(2x + 3)(2x - 3)$**

46. $a^2 + 49$ **prime**

47. $12c^2 - 12$ **$12(c + 1)(c - 1)$**

Applications and Problem Solving

48. Physics If a flare is launched into the air, its height h feet above the ground after t seconds is given by the formula $h = vt - 16t^2$. In the formula, v represents the initial velocity in feet per second. *(Lesson 10–2)*

a. Factor the expression $vt - 16t^2$. **$t(v - 16t)$**

b. If the flare is launched with an initial velocity of 144 feet per second, find the height after 2 seconds. **224 ft**

49. Genetics The Punnett square below represents the possible gene combinations for hair length in dogs. H represents long hair, and h represents short hair. Find the missing genes for the parents. *(Lesson 10–3)*

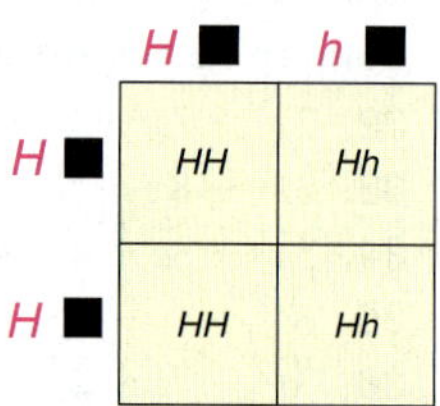

50. Geometry The area of a square is $(25x^2 + 30x + 9)$ square units. Find the perimeter. *(Lesson 10–5)* **$20x + 12$ units**

Assessment, pp. 461–462

10 NAME ________ DATE ______ PERIOD ______

Chapter 10 Test, Form 1B

Write the letter for the correct answer in the blank at the right of each problem.

For Questions 1–2, find the factors of each number and classify the number as prime or composite.

1. 17
   A. 1, 3, 17; composite B. 1, 17; prime
   C. 1, 17; composite D. 1, 7; prime — 1. **B**
2. 32
   A. 1, 2, 4, 8, 16, 32; composite
   B. 1, 2, 4, 6, 12, 32; composite
   C. 1, 2, 3, 4, 16, 32; composite
   D. 1, 2, 32; prime — 2. **A**
3. The area of a rectangle is 54 square centimeters. Find the length and width so that the rectangle has the greatest perimeter. Assume that the length and width are both whole numbers.
   A. 6 cm, 9 cm B. 3 cm, 18 cm C. 2 cm, 27 cm D. 1 cm, 54 cm — 3. **D**

Find the greatest common factor of each set of numbers or monomials.

4. 21, 24
   A. 2 B. 3 C. 4 D. 7 — 4. **B**
5. $25a^2$, $15ab$
   A. $3a$ B. $5a$ C. $15ab$ D. $5a^2$ — 5. **B**

Factor each polynomial. If the polynomial cannot be factored, write prime.

6. $4x + 20$
   A. $4x(1 + 5)$ B. $4x(x + 5)$ C. $4(x + 5)$ D. $2(2x + 5)$ — 6. **C**
7. $7a - 9b$
   A. prime B. $7(a - 2b)$ C. $a(7 - 9b)$ D. $3(4a - 6b)$ — 7. **A**
8. $3c^2 + 12c + 15d$
   A. $3(c + 4)(c + 1)$ B. $3c(c + 4 + 5d)$
   C. $3c(c + 4c + 5d)$ D. $3(c^2 + 4c + 5d)$ — 8. **D**

Find each quotient.

9. $(18xy + 24x) \div 6x$
   A. $3y + 4x$ B. $3xy + 4$ C. $3y + 4$ D. $4y + 4$ — 9. **C**
10. $(20a^2 - 16bc) \div 4$
   A. $6a^2 - 4bc$ B. $5a^2 - 4bc$ C. $5a - 4bc$ D. $5a^2 - 4b$ — 10. **B**

Factor each trinomial. If the trinomial cannot be factored, write prime.

11. $x^2 + 4x + 3$
   A. $(x + 1)(x + 3)$ B. $(x + 2)(x + 2)$ C. prime D. $(x + 4)(x - 1)$ — 11. **A**
12. $y^2 + y - 12$
   A. $(y + 6)(y - 2)$ B. $(y + 3)(y - 4)$ C. $(y + 4)(y - 3)$ D. prime — 12. **C**

Assessment

Four forms of Chapter 10 Test are available in the *Chapter 10 Resource Masters.*

Chapter 10 Test, Form 1B, is shown at the left. Chapter 10 Test, Form 2B, is shown on the next page.

Form of Test			Level
1A	Multiple Choice	pp. 459–460	Average
1B	Multiple Choice	pp. 461–462	Basic
2A	Free Response	pp. 463–464	Average
2B	Free Response	pp. 465–466	Basic

CHAPTER 10 Test

1. **Explain** what it means to *factor* a polynomial. **See margin.**

2. **Write** two monomials whose GCF is 1. **Sample answer: $5x$, $7y$**

3. **Write** the trinomial and its binomial factors shown by the model. **$9x^2 - 12x + 4$, $(3x - 2)(3x - 2)$**

4. **List** two different methods of factoring polynomials. **See margin.**

5. **Classify** the number 15 as *prime* or *composite*. Explain your reasoning. **Composite; it has more than two factors.**

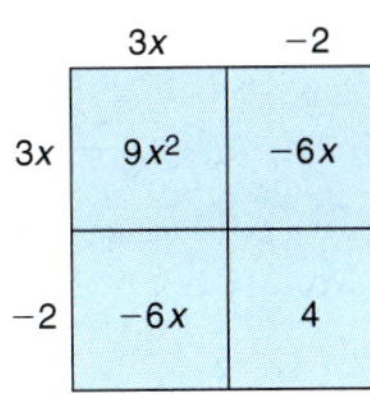

Exercise 3

Factor each monomial.

6. $25x^2y^2$ **$5 \cdot 5 \cdot x \cdot x \cdot y \cdot y$**
7. $-15b^3$ **$-1 \cdot 3 \cdot 5 \cdot b \cdot b \cdot b$**
8. $24a^2b$ **$2 \cdot 2 \cdot 2 \cdot 3 \cdot a \cdot a \cdot b$**

Find the GCF of each set of numbers or monomials.

9. 24, 60 **12**
10. $16a^2$, $30a^3$ **$2a^2$**
11. $20a^2b$, $25a^2b^2$ **$5a^2b$**

Factor each polynomial. If the polynomial cannot be factored, write *prime*.

12. $12x^2 + 18x$ **$6x(2x + 3)$**
13. $3x^2y - 12xy^2$ **$3xy(x - 4y)$**
14. $6a^3 + 8a^2 + 2a$ **$2a(3a^2 + 4a + 1)$**
15. $x^2 + 9x + 8$ **$(x + 1)(x + 8)$**
16. $m^2 - 10m + 24$ **$(m - 6)(m - 4)$**
17. $y^2 - 3y - 18$ **$(y - 6)(y + 3)$**
18. $3x^2 + x - 14$ **$(3x + 7)(x - 2)$**
19. $3m^2 + 17m + 10$ **$(3m + 2)(m + 5)$**
20. $2x^2 - 18$ **$2(x - 3)(x + 3)$**
21. $n^2 - 8n - 16$ **prime**
22. $y^2 + 10y + 25$ **$(y + 5)(y + 5)$**
23. $25m^2 - 16$ **$(5m + 4)(5m - 4)$**
24. $3r^2 + r + 1$ **prime**
25. $6x^3 + 15x^2 - 9x$ **$3x(2x - 1)(x + 3)$**

26. **Geometry** Find the area of the shaded region. Express the area in factored form. **$(x + 4)(x - 3)$**

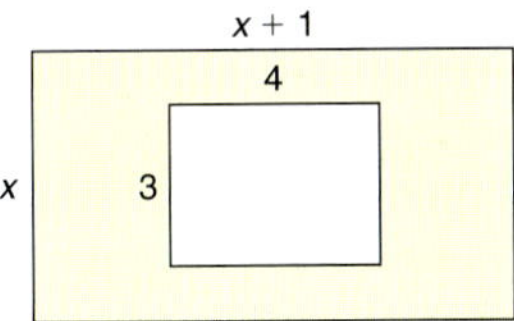

www.algconcepts.com/chapter_test

Chapter Test Bonus Question

Factor the trinomial $x^2y^2 - 12xy + 36$. **$(xy - 6)(xy - 6)$**

Chapter Test

Answers

1. To factor a polynomial means to write it as the product of polynomials.
4. GCF, difference of squares, perfect square trinomial, or trinomial with two binomial factors

***Assessment*, pp. 465–466**

10 NAME ______ DATE ______ PERIOD ______

Chapter 10 Test, Form 2B

Find the factors of each number and classify the number as prime or composite.

1. 11 — 1. 1, 11; prime
2. 50 — 2. 1, 2, 5, 10, 25, 50; composite

Factor each monomial.

3. $9ab$ — 3. $3 \cdot 3 \cdot a \cdot b$
4. $-16xy^2$ — 4. $-1 \cdot 2 \cdot 2 \cdot 2 \cdot 2 \cdot x \cdot y \cdot y$

Find the greatest common factor of each set of numbers or monomials.

5. 20, 24 — 5. 4
6. $10x$, $15x^2$ — 6. $5x$
7. $7x$, $14ax^2$, $28ax$ — 7. $7x$

Factor each polynomial. If the polynomial cannot be factored, write prime.

8. $3 + 12x$ — 8. $3(1 + 4x)$
9. $7ab + 13c$ — 9. prime
10. $6y^2 - 2yz$ — 10. $2y(3y - z)$

For Questions 11–13, find each quotient.

11. $(15x - 10y^2) \div 5$ — 11. $3x - 2y^2$
12. $(6ac + 2bc) \div 2c$ — 12. $3a + b$
13. $(8g^2j - 12g^3hj) \div 4g$ — 13. $2gj - 3g^2hj$

14. The area of a rectangle is $(36xy + 24y)$ square inches. If the width is $12y$ inches, find the length. — 14. $(3x + 2)$ in.

Factor each trinomial. If the trinomial cannot be factored, write prime.

15. $s^2 + 7s + 10$ — 15. $(s + 2)(s + 5)$
16. $x^2 - x - 6$ — 16. $(x + 2)(x - 3)$

CHAPTER 10 Preparing for Standardized Tests

Pages 454–455 are part of a complete test preparation course that is described in detail on page T9 of the Teacher's Handbook. The test items on these pages were written in the same style as those in state proficiency tests and standardized tests like ACT and SAT.

Diagnosis and Prescription

Each of the 10 test questions on page 455 is cross-referenced to the lesson where that SAT or ACT skill is covered. If students miss a particular type of problem, you can have them study that skill.

(See chart at the bottom of page 455. Note that SPT = State Proficiency Test, SAT = Scholastic Assessment Test, and ACT = American College Test.)

Assessment, p. 470

10 NAME ______ DATE ______ PERIOD ______

Chapter 10 Cumulative Review

1. Write an equation for the sentence below. *(Lesson 1–1)* Twenty-four is equal to twice the product of numbers x and z. — 1. $24 = 2xz$
2. Simplify $-9y + (-4y)$. *(Lesson 2–3)* — 2. $-13y$
3. Solve $-4 = g + (-2)$. *(Lesson 3–5)* — 3. -2
4. Find 8(4.1). *(Lesson 4–1)* — 4. 32.8

Use the percent equation to find each number. *(Lesson 5–4)*

5. Find 45% of 36. — 5. 16.2
6. 40 is 5% of what number? — 6. 800

Solve. Assume that y varies directly as x. *(Lesson 6–5)*

7. Find y when $x = 18$ if $y = 33$ when $x = 11$. — 7. 54
8. If $y = 40$ when $x = 32$, find x when $y = 30$. — 8. 24

Determine whether the graphs of each pair of equations are parallel, perpendicular, or neither. *(Lesson 7–7)*

9. $y = -7x + 5$, $y = -7x - 5$ — 9. parallel
10. $3y = 4x$, $y = -\frac{3}{4}x$ — 10. perpendicular

Write each expression using exponents. *(Lesson 8–1)*

11. 7 squared — 11. 7^2
12. $(-3)(x)(x)(x)(y)(y)$ — 12. $-3x^3y^2$

Find each product. *(Lesson 9–3)*

13. $4(x - 2)$ — 13. $4x - 8$
14. $7y(y - y^2)$ — 14. $7y^2 - 7y^3$

Find the greatest common factor of each set of numbers. *(Lesson 10–1)*

15. 17, 18, 19 — 15. 1
16. 32, 36, 44 — 16. 4

Function and Graph Problems

All standardized tests include problems with functions and graphs.

Familiarize yourself with the concepts below.

function	equation of a line
table of values	y-intercept, x-intercept
graph of a line	slope

Test-Taking Tip

Slope is the ratio of the change in y to the change in x. A line that slopes upward from left to right has a positive slope.

Example 1

Martin is paid by the hour for babysitting. His hourly wage is a fixed amount plus an additional amount for each child. The graph shows his hourly wage for up to 5 children. If x represents the number of children, which expression can be used to find Martin's hourly wage, y?

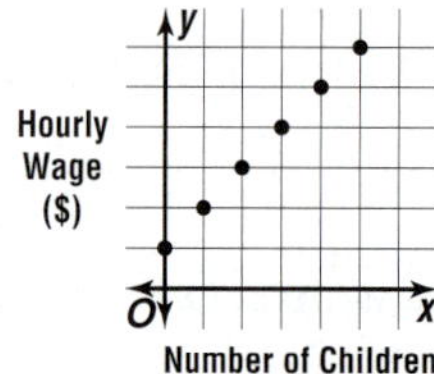

Ⓐ x Ⓑ $x - 1$

Ⓒ $2x + 1$ Ⓓ $x + 1$

Hint Study the graph. The points lie on a line. Find the y-intercept and the slope.

Solution The fixed amount is represented by the y-intercept. It is the point that represents 0 children. The y-intercept is 1.

The slope of the line shows the amount for each child. Moving left to right, each point is one unit higher than the previous point. So the slope is 1.

Martin's hourly wage is the fixed amount, \$1, plus the amount per child, \$1, times the number of children, x. The expression is $1x + 1$ or $x + 1$. The answer is D.

Example 2

What is the equation of a line that is parallel to the line whose equation is $y = \frac{2}{3}x + 5$ and passes through the point at $(-6, 2)$?

Ⓐ $y = \frac{2}{3}x + 5z$ Ⓑ $y = \frac{2}{3}x - 2$

Ⓒ $y = \frac{2}{3}x + 6$ Ⓓ $y = \frac{2}{3}x - \frac{22}{3}$

Ⓔ $y = -\frac{3}{2}x - 7$

Hint Memorize the slope-intercept and point-slope forms of linear equations.

Solution The equation of the given line is in slope-intercept form. The slope is $\frac{2}{3}$. Parallel lines have the same slope. So, the slope of the parallel line must also be $\frac{2}{3}$. This eliminates answer choice E.

The parallel line must pass through $(-6, 2)$. Write the equation of the line in point-slope form.

$y - 2 = \frac{2}{3}[x - (-6)]$	*Point-slope form*
$y - 2 = \frac{2}{3}(x + 6)$	*Simplify.*
$y - 2 = \frac{2}{3}x + \frac{2}{3}(6)$	*Distributive Property*
$y - 2 = \frac{2}{3}x + 4$	*Multiply.*
$y = \frac{2}{3}x + 6$	*Add 2 to each side.*

The answer is C.

Resource Manager

Reproducible Masters

Chapter 10 Resource Masters

- *Assessment*, pp. 470–472

Preparing for Standardized Tests
For test-taking strategies and more practice, see pages 732–749.

After you work each problem, record your answer on the answer sheet provided or on a sheet of paper.

Multiple Choice

1. Use the function table to find the value of y when $x = 5$. *(Lesson 6–1)* **C**

x	y
0	3
1	17
2	31
3	45

(A) 59 (B) 60
(C) 73 (D) 75

2. What is the y-intercept of the line determined by the equation $5x + 2 = 7y - 3$? *(Lesson 7–5)* **D**

(A) -1 (B) $-\frac{1}{7}$ (C) $\frac{1}{7}$ (D) $\frac{5}{7}$ (E) 5

3. Which expression can be used to find the value of y in the graph? *(Lesson 7–3)* **D**

(A) $1 - 2x$ (B) $1 - 3x$
(C) $2x + 1$ (D) $3x - 1$

4. The charge to enter a nature reserve is a fixed amount per vehicle plus a fee for each person in it. The table shows some charges. What would the charge be for a vehicle with 8 people? *(Lesson 6–1)* **C**

People	Charge
1	\$1.50
2	\$2.00
3	\$2.50
4	\$3.00

(A) \$3.50 (B) \$4.00 (C) \$5.00 (D) \$6.00

5. At what point does the line MN cross the y-axis? *(Lesson 7–5)* **D**

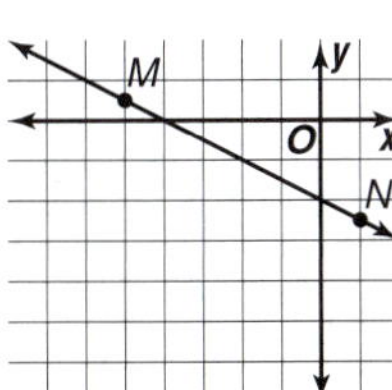

(A) $(-4, 0)$
(B) $(0, -4)$
(C) $(-2, 0)$
(D) $(0, -2)$

www.algconcepts.com/standardized_test

6. The average of two numbers x and y is A. Which of the following is an expression for y? *(Lesson 3–3)* **C**

(A) $\frac{A + x}{2}$ (B) $\frac{A}{2} - x$ (C) $2A - x$
(D) $A - x$ (E) $x - A$

7. Write 4^{-4} without using an exponent. *(Lesson 8–3)* **B**

(A) 0.00039 (B) 0.0039
(C) 0.016 (D) 256

8. Which expression should come next in the pattern $2x, 4x^2, 8x^3, 16x^4, \ldots$? *(Lesson 8–1)* **B**

(A) $24x^5$ (B) $32x^5$
(C) $24x^6$ (D) $32x^6$

Grid In

9. The graph of $y = 4x - 2$ is shown. What is the x-intercept? *(Lesson 7–5)* **0.5**

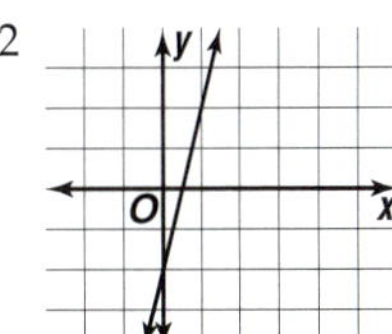

Extended Response

10. The graph shows the distance traveled by an African elephant. *(Lesson 7–1)*

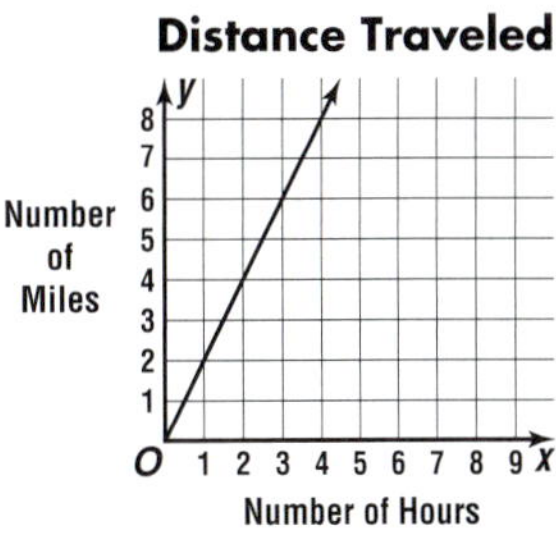

Part A What is the slope of the line? **2**

Part B Explain what the slope represents.
the speed of the elephant

Chapter 10 Function and Graph Problems		
Ex. 1	slope-intercept	SPT
Ex. 2	equation of a parallel line	SAT
1	function table	SPT
2	y-intercept	ACT
3	graph an expression	SPT
4	function table	SPT
5	function table	SPT
6	average	SAT
7	negative exponents	SPT
8	sequence of monomials	SPT
9	x-intercept	SPT
10	slope	SPT

A bubble-in answer sheet for these practice problems is available on page A1 of the *Chapter 10 Resource Masters.*

Additional Practice

Additional test practice questions are available in the *Chapter 10 Resource Masters*, pp. 471–472.

***Assessment*, pp. 471–472**

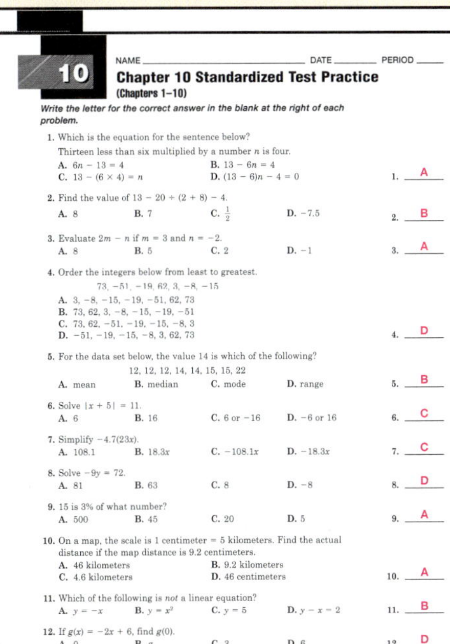
10

NAME ______ DATE ______ PERIOD ______

Chapter 10 Standardized Test Practice
(Chapters 1–10)

Write the letter for the correct answer in the blank at the right of each problem.

1. Which is the equation for the sentence below?
Thirteen less than six multiplied by a number n is four.
A. $6n - 13 = 4$ B. $13 - 6n = 4$
C. $13 - (6 \times 4) = n$ D. $(13 - 6)n - 4 = 0$ 1. A

2. Find the value of $13 - 20 \div (2 + 8) - 4$.
A. 8 B. 7 C. $\frac{1}{2}$ D. -7.5 2. B

3. Evaluate $2m - n$ if $m = 3$ and $n = -2$.
A. 8 B. 5 C. 2 D. -1 3. A

4. Order the integers below from least to greatest.
73, −51, −19, 62, 3, −8, −15
A. 3, −8, −15, −19, −51, 62, 73
B. 73, 62, 3, −8, −15, −19, −51
C. 73, 62, −51, −19, −15, −8, 3
D. −51, −19, −15, −8, 3, 62, 73 4. D

5. For the data set below, the value 14 is which of the following?
12, 12, 12, 14, 14, 15, 15, 22
A. mean B. median C. mode D. range 5. B

6. Solve $|x + 5| = 11$.
A. 6 B. 16 C. 6 or −16 D. −6 or 16 6. C

7. Simplify $-4.7(23x)$.
A. 108.1 B. $18.3x$ C. $-108.1x$ D. $-18.3x$ 7. C

8. Solve $-9y = 72$.
A. 81 B. 63 C. 8 D. −8 8. D

9. 15 is 3% of what number?
A. 500 B. 45 C. 20 D. 5 9. A

10. On a map, the scale is 1 centimeter = 5 kilometers. Find the actual distance if the map distance is 9.2 centimeters.
A. 46 kilometers B. 9.2 kilometers
C. 4.6 kilometers D. 46 centimeters 10. A

11. Which of the following is *not* a linear equation?
A. $y = -x$ B. $y = x^2$ C. $y = 5$ D. $y - x = 2$ 11. B

12. If $g(x) = -2x + 6$, find $g(0)$.
A. 0 B. g C. 3 D. 6 12. D

CHAPTER 11 Resource Manager

Quadratic and Exponential Functions

Instructional Objectives

Lesson (pages)	Objectives	NCTM Standards 2000	State/Local Objectives
11–1 (458–463)	Graph quadratic functions.	1, 2, 3, 6, 8, 9	
11–2 (464–467)	Learn the characteristics of families of parabolas.	1, 2, 3, 6, 8, 9	
11–3 (468–473)	Locate the roots of quadratic equations by graphing the related functions.	1, 3, 6, 8, 9	
11–4 (474–477)	Solve quadratic equations by factoring and by using the Zero Product Property.	1, 6, 8, 9	
11–5 (478–482)	Solve quadratic equations by completing the square.	1, 6, 8, 9	
11–6 (483–487)	Solve quadratic equations by using the Quadratic Formula.	1, 6, 8, 9	
11–7 (488–493)	Graph exponential functions.	1, 2, 3, 6, 8, 9	
Investigation (494–495)	Explore geometric sequences.	1, 2, 3, 6, 7, 8, 9, 10	

Key to NCTM Standards 2000
1 Number & Operations; **2** Algebra; **3** Geometry; **4** Measurement; **5** Data Analysis & Probability; **6** Problem Solving; **7** Reasoning & Proof; **8** Communications; **9** Connections; **10** Representation

Suggested Pacing

See page T13 for a complete course-planning calendar.

Standard refers to schedules that provide 45- to 55-minute periods that meet each day.
Block refers to schedules that provide approximately 90-minute periods which may meet every day for one semester or every other day over two semesters.

PACING	DAY 1	DAY 2	DAY 3	DAY 4	DAY 5	DAY 6
Standard Core (Chapters 1–13)	Lesson 11–1		Lesson 11–2	Lesson 11–3		Lesson 11–4
Standard Enhanced (Chapters 1–15)	Lesson 11–1		Lesson 11–2	Lesson 11–3	Lesson 11–4	Lesson 11–5
Block Core (Chapters 1–13)	Chapter 10 Test & Lesson 11–1	Lessons 11–2 & 11–3	Lessons 11–4 & 11–5	Lessons 11–6 & 11–7	INV	SG+A
Block Enhanced (Chapters 1–15)	Chapter 10 Test & Lesson 11–1	Lessons 11–2 & 11–3	Lessons 11–4 & 11–5	Lessons 11–6 & 11–7	INV & SG+A	Chapter Test & Lesson 12–1

Instructional Resources		Blackline Masters (page numbers)								
		Chapter 11 Resource Masters								
Lesson	**Materials and Manipulatives** (see below for Glencoe Manipulative Resources)	Study Guide	Practice (Skills & Average)	Reading to Learn Mathematics	Enrichment	Assessment	Hands-On Algebra*	School-to-Workplace*	Graphing Calculator Masters*	5-Minute Check Transparencies
11–1	coordinate planes [4]	473	474–475	476	477		118			11–1
11–2	graphing calculator overhead transparency sheets	478	479–480	481	482					11–2
11–3	coordinate planes [4] graphing calculator	483	484–485	486	487	519	119, 120	11	29	11–3
11–4	coordinate planes [4] graphing calculator	488	489–490	491	492	518	121, 122			11–4
11–5	algebra tiles [1, 2, 3, 4] equation/product mat [1, 2, 4]	493	494–495	496	497		123, 124			11–5
11–6	graphing calculator	498	499–500	501	502				31, 32	11–6
11–7	large sheet of paper coordinate planes [4] graphing calculator almanac	503	504–505	506	507	519	125, 126		30	11–7
Investigation	3-in. by 3-in. self-adhesive notes scissors [1, 2] spreadsheet software coordinate planes [4]									
Study Guide & Assessment/ Chapter Test	coordinate planes [4]					509–517, 520–522				

See page 456c for examples of these instructional materials.

Key to Glencoe Manipulative Resources

[1]Classroom Manipulative Resources [2]Student Manipulative Resources [3]Overhead Manipulative Resources [4]Hands-On Algebra Masters

INV = Investigation *SG+A = Study Guide and Assessment*

DAY 7	DAY 8	DAY 9	DAY 10	DAY 11	DAY 12	DAY 13
Lesson 11–5	Lesson 11–6		Lesson 11–7	INV	SG+A	Chapter Test
Lesson 11–6		Lesson 11–7	INV	SG+A	Chapter Test	
Chapter Test & Lesson 12–1						

Resource Manager

The pages shown on this page are a small sample of the materials available on *TeacherWorks: All-in-One Lesson Planner and Resource Center.*

This CD-ROM includes all of the blackline masters and transparencies available for this program.

It also includes a lesson planner and interactive Teacher Edition, so you can customize lesson plans and reproduce classroom resources quickly and easily, from just about anywhere.

Applications

School-to-Workplace Masters, p. 11

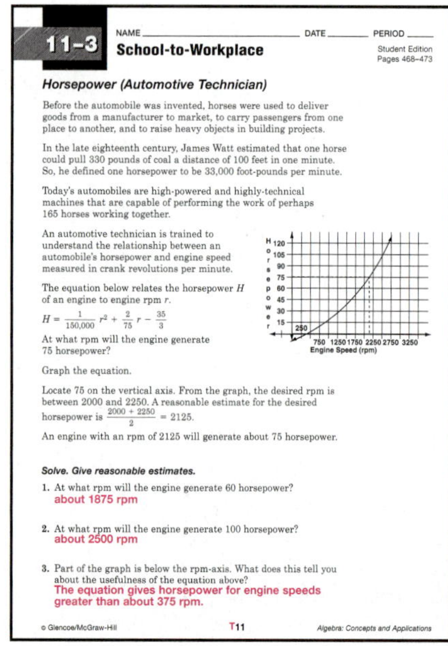

11-3 NAME ______ DATE ______ PERIOD ______

School-to-Workplace

Student Edition Pages 468–473

Horsepower (Automotive Technician)

Before the automobile was invented, horses were used to deliver goods from a manufacturer to market, to carry passengers from one place to another, and to raise heavy objects in building projects.

In the late eighteenth century, James Watt estimated that one horse could pull 330 pounds of coal a distance of 100 feet in one minute. So, he defined one horsepower to be 33,000 foot-pounds per minute.

Today's automobiles are high-powered and highly-technical machines that are capable of performing the work of perhaps 165 horses working together.

An automotive technician is trained to understand the relationship between an automobile's horsepower and engine speed measured in crank revolutions per minute.

The equation below relates the horsepower H of an engine to engine rpm r.

$H = \frac{1}{150{,}000} r^2 + \frac{2}{75} r - \frac{35}{3}$

At what rpm will the engine generate 75 horsepower?

Graph the equation.

Locate 75 on the vertical axis. From the graph, the desired rpm is between 2000 and 2250. A reasonable estimate for the desired horsepower is $\frac{2000 + 2250}{2} = 2125$.

An engine with an rpm of 2125 will generate about 75 horsepower.

Solve. Give reasonable estimates.

1. At what rpm will the engine generate 60 horsepower? **about 1875 rpm**
2. At what rpm will the engine generate 100 horsepower? **about 2500 rpm**
3. Part of the graph is below the rpm-axis. What does this tell you about the usefulness of the equation above? **The equation gives horsepower for engine speeds greater than about 375 rpm.**

© Glencoe/McGraw-Hill T11 *Algebra: Concepts and Applications*

Manipulatives/Modeling

Hands-On Algebra Masters, pp. 118–126

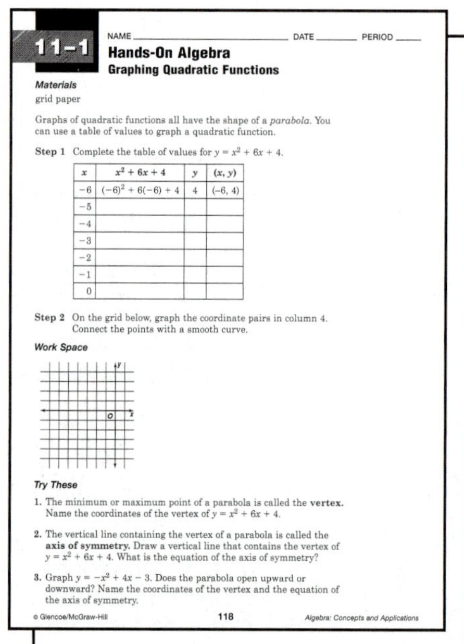

11-1 NAME ______ DATE ______ PERIOD ______

Hands-On Algebra

Graphing Quadratic Functions

Materials

grid paper

Graphs of quadratic functions all have the shape of a *parabola*. You can use a table of values to graph a quadratic function.

Step 1 Complete the table of values for $y = x^2 + 6x + 4$.

x	$x^2 + 6x + 4$	y	(x, y)
−6	$(-6)^2 + 6(-6) + 4$	4	(−6, 4)
−5			
−4			
−3			
−2			
−1			
0			

Step 2 On the grid below, graph the coordinate pairs in column 4. Connect the points with a smooth curve.

Work Space

Try These

1. The minimum or maximum point of a parabola is called the **vertex.** Name the coordinates of the vertex of $y = x^2 + 6x + 4$.
2. The vertical line containing the vertex of a parabola is called the **axis of symmetry.** Draw a vertical line that contains the vertex of $y = x^2 + 6x + 4$. What is the equation of the axis of symmetry?
3. Graph $y = -x^2 + 4x - 3$. Does the parabola open upward or downward? Name the coordinates of the vertex and the equation of the axis of symmetry.

© Glencoe/McGraw-Hill 118 *Algebra: Concepts and Applications*

Technology/Multimedia

Graphing Calculator Masters, pp. 29–32

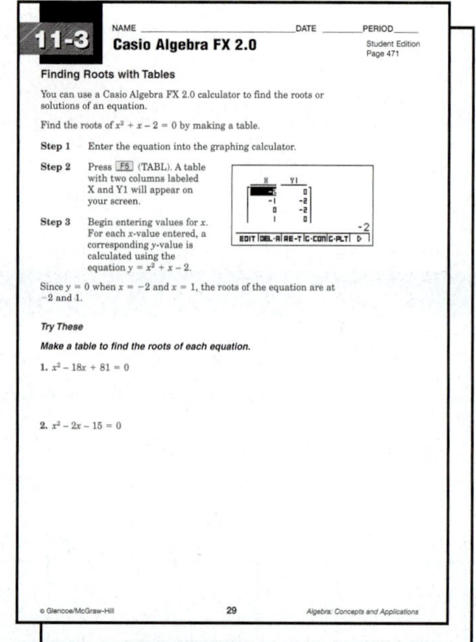

11-3 NAME ______ DATE ______ PERIOD ______

Casio Algebra FX 2.0

Student Edition Page 471

Finding Roots with Tables

You can use a Casio Algebra FX 2.0 calculator to find the roots or solutions of an equation.

Find the roots of $x^2 + x - 2 = 0$ by making a table.

Step 1 Enter the equation into the graphing calculator.

Step 2 Press F5 (TABL). A table with two columns labeled X and Y1 will appear on your screen.

Step 3 Begin entering values for x. For each x-value entered, a corresponding y-value is calculated using the equation $y = x^2 + x - 2$.

Since $y = 0$ when $x = -2$ and $x = 1$, the roots of the equation are at −2 and 1.

Try These

Make a table to find the roots of each equation.

1. $x^2 - 18x + 81 = 0$
2. $x^2 - 2x - 15 = 0$

© Glencoe/McGraw-Hill 29 *Algebra: Concepts and Applications*

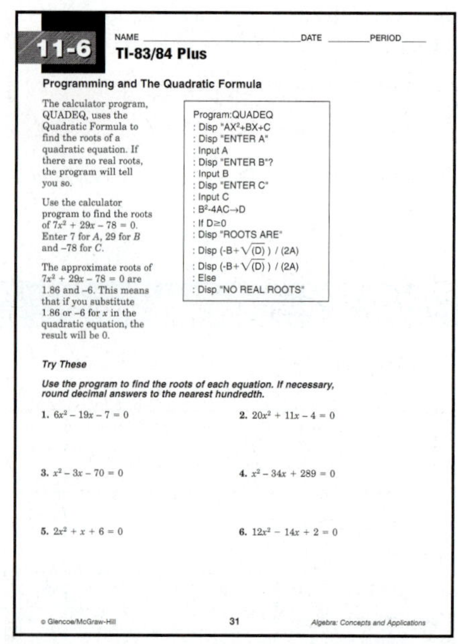

11-6 NAME ______ DATE ______ PERIOD ______

TI-83/84 Plus

Programming and The Quadratic Formula

The calculator program, QUADEQ, uses the Quadratic Formula to find the roots of a quadratic equation. If there are no real roots, the program will tell you so.

Use the calculator program to find the roots of $7x^2 + 29x - 78 = 0$. Enter 7 for A, 29 for B and −78 for C.

The approximate roots of $7x^2 + 29x - 78 = 0$ are 1.86 and −6. This means that if you substitute 1.86 or −6 for x in the quadratic equation, the result will be 0.

```
Program:QUADEQ
: Disp "AX²+BX+C
: Disp "ENTER A"
: Input A
: Disp "ENTER B"?
: Input B
: Disp "ENTER C"
: Input C
: B²-4AC→D
: If D≥0
: Disp "ROOTS ARE"
: Disp (-B+√(D) ) / (2A)
: Disp (-B+√(D) ) / (2A)
: Else
: Disp "NO REAL ROOTS"
```

Try These

Use the program to find the roots of each equation. If necessary, round decimal answers to the nearest hundredth.

1. $6x^2 - 19x - 7 = 0$
2. $20x^2 + 11x - 4 = 0$
3. $x^2 - 3x - 70 = 0$
4. $x^2 - 34x + 289 = 0$
5. $2x^2 + x + 6 = 0$
6. $12x^2 - 14x + 2 = 0$

© Glencoe/McGraw-Hill 31 *Algebra: Concepts and Applications*

GLENCOE'S ASSESSMENT ADVANTAGE

Assessment Resources

Type	Student Edition	Teacher's Wraparound Edition	Chapter 11 Resource Masters
Ongoing Assessment	Quizzes 1 and 2, pp. 473, 487	5-Minute Check, pp. 458, 464, 468, 474, 478, 483, 489	Mid-Chapter Test, p. 518 Quizzes A and B, p. 519
Mixed Review	Mixed Review, pp. 463, 467, 473, 477, 482, 487, 493 Standardized Test Practice, Chapter 1–11, pp. 500–501		Cumulative Review, p. 520 Standardized Test Practice, pp. 521–522
Error Analysis	You Decide, pp. 466, 476	Error Analysis, pp. 461, 466, 471, 476, 481, 486, 492	
Standardized Test Prep	Standardized Test Practice, pp. 463, 467, 473, 477, 482, 487, 493 Standardized Test Practice, Chapters 1–11, pp. 500–501		Standardized Test Practice, pp. 521–522
Open-Ended Assessment	Writing Math, pp. 481, 486 Investigation, pp. 494–495 Portfolio, pp. 457, 495	Modeling: pp. 473, 493 Speaking: pp. 463, 477 Writing: pp. 467, 482, 487	Extended Response Assessment, p. 517
Chapter Assessment	Study Guide and Assessment, pp. 496–498 Chapter Test, p. 499		Multiple-Choice Tests (Forms 1A, 1B), pp. 509–512 Free-Response Tests (Forms 2A, 2B), pp. 513–516

Additional Chapter Resources

Student Edition
Math in the Workplace, p. 488
Hands-On Algebra, pp. 478–479, 489
Graphing Calculator Exploration, pp. 471, 491

Teacher's Classroom Resources
Manipulatives/Modeling
Teacher's Guide for Overhead Manipulative Resources

Meeting Individual Needs
Prerequisite Skills Booklet
Spanish Study Guide and Assessment, pp. 73–79, 125–126

Teaching Aids
Solutions Manual
5-Minute Check Transparencies

Glencoe Technology

Instructional
AlgePASS CD-ROM, Lessons 25–28
Interactive Chalkboard CD-ROM
StudentWorks Plus CD-ROM
Multimedia Applications CD-ROM, Activity 11
Vocabulary PuzzleMaker (online)

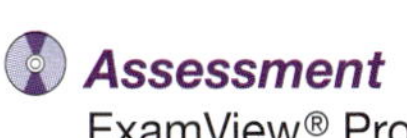
Assessment
ExamView® Pro

Teaching Aids
Answer Key Maker CD-ROM
TeacherWorks CD-ROM

Visit ***www.algconcepts.com*** *for data updates, career information, games, and other interactive activities.*

CHAPTER PREVIEW 11

Mathematics of the Chapter

This chapter provides students with a thorough exploration of quadratic functions and equations. Quadratic functions are graphed and families of quadratic functions are defined. Quadratic equations are solved by graphing, factoring, completing the square, and using the Quadratic Formula. Finally, exponential functions are introduced and graphed.

The study of quadratic equations is central to the Algebra 1 curriculum. Students have studied linear equations, powers and roots, and polynomials. In the previous chapter, they factored polynomials, with special attention to trinomials. This foundation prepares them to solve quadratic equations using various methods. Their understanding of more complex expressions and equations, including radicals and rational expressions and equations, will build upon this study as they advance to later chapters in the text.

What's MATH Got To Do With It?

Real-Life Videos engage students, showing them how math is used in everyday situations. Use Video 3 to discuss how quadratic and exponential functions are used in juggling and glass blowing.

CHAPTER 11 Quadratic and Exponential Functions

What You'll Learn

Key Ideas

- Graph quadratic functions. *(Lesson 11–1)*
- Investigate families of parabolas. *(Lesson 11–2)*
- Solve quadratic equations by graphing, by factoring, by completing the square, and by using the Quadratic Formula. *(Lessons 11–3 to 11–6)*
- Graph exponential functions. *(Lesson 11–7)*

Key Vocabulary

Why It's Important

Space From the Apollo Moon landings to robotic surveys of the Sun, space exploration has changed our view of the universe. It has also led to inventions for everyday life. For example, better athletic shoes, X-ray machines, and scratch-resistant sunglasses were developed from discoveries in the space program.

Quadratic functions are used to solve real-world problems involving gravity. You will use a quadratic function to analyze the effect of gravity on an astronaut in Lesson 11–6.

CHAPTER 11 LINKS

Lesson	11–1	11–2	11–3	11–4	11–5	11–6	11–7
Math in the Workplace	Architecture	Design	Manufacturing	Photography	Construction	Science Civil Engineer	Finance
Applications and Connections	Football Business	Computer Animation Fireworks Baseball Finance	Landmarks Business Skydiving Diving	Recreation Physics	Physics Photography Rockets	Sports Physics Skiing Space Construction	Taxes Population Flooring
Math Integration	Geometry		Number Theory Geometry	Geometry	Geometry	Geometry	

Study these lessons to improve your skills.

✓ Check Your Readiness

Lesson 6–3, pp. 250–255

Graph each equation. See Solutions Manual.

1. $y = x + 1$
2. $y = 3x$
3. $y = -2x$
4. $y = 2x + 4$
5. $y = 6 - x$
6. $y = \frac{1}{2}x - 1$

Lesson 8–2, pp. 341–345

Evaluate each expression.

7. $2x^2$ if $x = 3$ **18**
8. $x^2 + 2x$ if $x = 2$ **8**
9. $-x^2 - x + 4$ if $x = -1$ **4**
10. $\frac{1}{2}x^2 + x - 1$ if $x = -4$ **3**
11. $-\frac{b}{2a}$ if $a = 1$ and $b = -4$ **2**
12. $-\frac{b}{2a}$ if $a = -2$ and $b = 6$ $\frac{3}{2}$

Lessons 10–3 and 10–4 pp. 434–444

Factor each trinomial. If the trinomial cannot be factored, write *prime*. 13. $(x + 2)(x + 8)$ 14. $(n - 2)(n + 6)$ 15. $(b - 5)(b + 3)$

13. $x^2 + 10x + 16$
14. $n^2 + 4n - 12$
15. $b^2 - 2b - 15$
16. $p^2 - 11p + 30$ $(p - 5)(p - 6)$
17. $2c^2 + 5c - 3$ $(2c - 1)(c + 3)$
18. $3m^2 - 5m - 8$ $(3m - 8)(m + 1)$

Make this Foldable to help you organize your Chapter 11 notes. Begin with four sheets of plain $8\frac{1}{2}$" by 11" paper.

❶ **Fold** each sheet in half along the width.

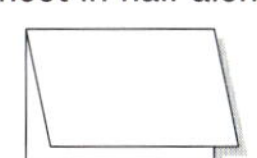

❷ **Unfold** each sheet and tape to form one long sheet.

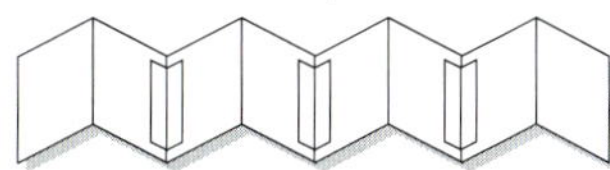

❸ **Label** each page with a lesson number as shown.

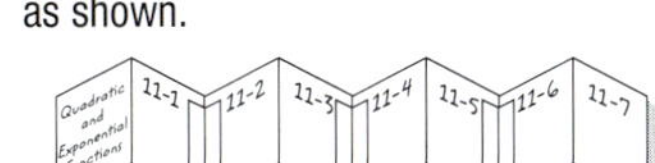

❹ **Refold** to form a booklet. Label the front cover "Quadratic and Exponential Functions."

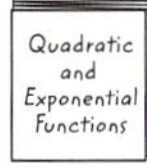

Reading and Writing As you read and study the chapter, write notes and examples for each lesson on each page of the journal.

www.algconcepts.com/chapter_readiness

FOLDABLES™ Study Organizer Have each student make a Foldable accordion book with a page for each of the lessons of Chapter 11. Have students compare and contrast quadratic functions, their graphs, and their solutions to those of linear functions. Have students write about situations where exponential functions are used.

Check Your Readiness

This section provides a review of the basic concepts needed before beginning Chapter 11. Lesson references are included for additional student help.

Use Exercises	To Prepare for Lesson(s)
1–6	11–1, 11–7
7–12	11–1, 11–3
13–18	11–4

Quick Review Math Handbook

Glencoe Algebra Lesson	Quick Review Math Handbook Hot Topic
11–1	6–8
11–2	
11–3	
11–4	
11–5	6–3
11–6	6–3
11–7	6–3

Interactive Study Notebook with Foldables™

This note-taking guide reinforces excellent note-taking skills. It includes:

✓ **Build Your Vocabulary** tool for including mathematical terminology in notes.

✓ **Review It** activities with links to prior lessons and items to include in **Foldables™**.

✓ **Bringing It All Together** feature to help students review for the Chapter Test.

✓ **Are You Ready for the Chapter Test?** feature to allow students to assess their own readiness for the Chapter Test.

✓ **Teacher Edition** with transparencies to guide note taking. Corresponds to the *Teacher's Wraparound Edition* In-Class Examples and *Interactive Chalkboard CD-ROM.*

Lesson 11-1

1 FOCUS

5-Minute Check
Chapter 10

1. Find the GCF of $12ab^3$ and $20a^2b^2$. $4ab^2$
2. Factor $7y + 14y^2$. $7y(1 + 2y)$
3. The width of the rectangle is 8 feet less than the length. If a 3-foot by 3-foot square is removed from the original rectangle, how much of the original rectangle remains?

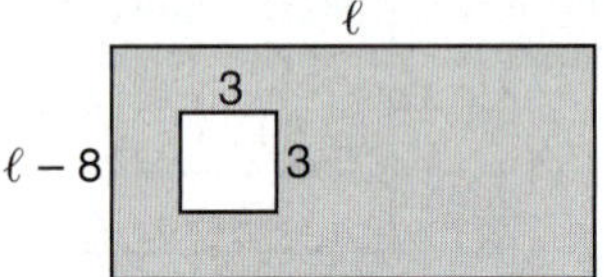

$A = (\ell + 1)(\ell - 9)$

4. Factor the volume to find the dimensions of the crate.

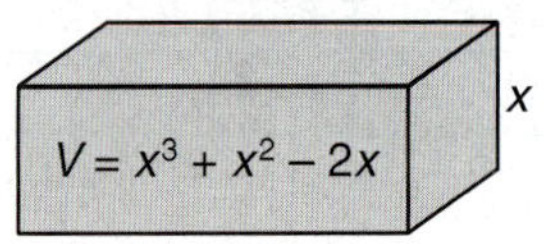

$x(x - 1)(x + 2)$

5. Factor $36x^2 - 81y^2$. $9(2x - 3y)(2x + 3y)$

Motivating the Lesson

Real-World Connection
Parabolas occur frequently in the real world. Tell students about some of the many instances to introduce and motivate this lesson. Some examples are the cross section of objects such as a satellite dish or the reflector in a flashlight. The paths of thrown objects are also parabolas, as are the cables on a suspension bridge such as the Golden Gate Bridge in San Francisco.

MODELING
An alternative hands-on option using grid paper is available for teaching this lesson.

11-1 Graphing Quadratic Functions

What You'll Learn
You'll learn to graph quadratic functions.

Why It's Important
Architecture Architects graph quadratic functions when designing arches. *See Exercise 15.*

The shape of a parabola is like an arch. **Parabolas** are modeled by **quadratic functions** of the form $y = ax^2 + bx + c$. The first term cannot equal zero because it will make the function linear, or a straight line. Therefore, the coefficient a cannot be zero.

Quadratic Function

Words: A quadratic function is a function that can be described by an equation of the form $y = ax^2 + bx + c$, where $a \neq 0$.

Models:

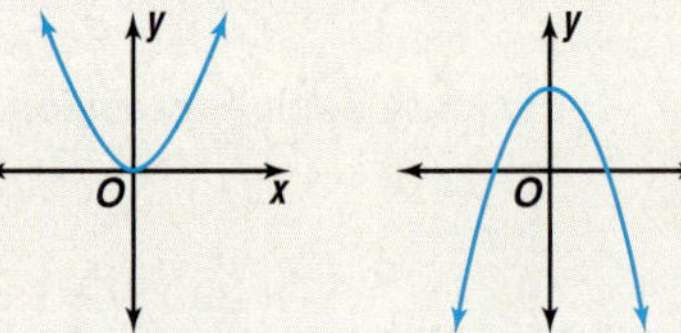

Graphs of all quadratic functions have the shape of a parabola.

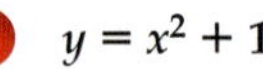

Examples

Graph each quadratic function by making a table of values.

1 $y = x^2 + 1$

First, choose integer values for x. Evaluate the function for each x-value. Graph the points and connect them with a smooth curve.

Look Back
Graphing Relations: Lesson 6–3

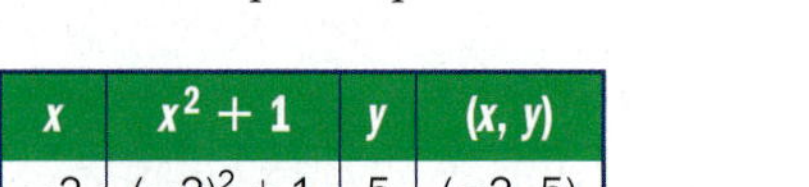

x	$x^2 + 1$	y	(x, y)
−2	$(-2)^2 + 1$	5	(−2, 5)
−1	$(-1)^2 + 1$	2	(−1, 2)
0	$0^2 + 1$	1	(0, 1)
1	$1^2 + 1$	2	(1, 2)
2	$2^2 + 1$	5	(2, 5)

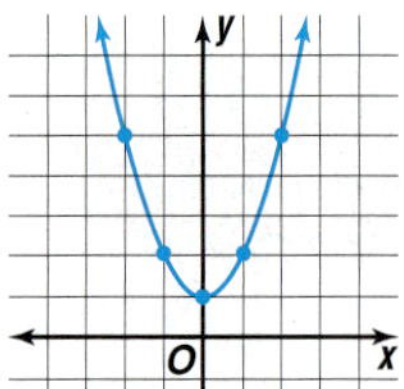

2 $y = -x^2$

x	$-x^2$	y	(x, y)
−2	$-(-2)^2$	−4	(−2, −4)
−1	$-(-1)^2$	−1	(−1, −1)
0	$-(0)^2$	0	(0, 0)
1	$-(1)^2$	−1	(1, −1)
2	$-(2)^2$	−4	(2, −4)

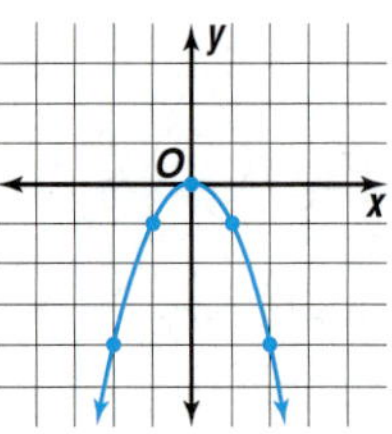

Your Turn a–b. See margin.

a. $y = x^2 + 6$

b. $y = -2x^2 - 1$

Resource Manager

Reproducible Masters
Chapter 11 Resource Masters
- *Study Guide,* p. 473
- *Skills Practice,* p. 474
- *Practice,* p. 475
- *Reading to Learn Mathematics,* p. 476
- *Enrichment,* p. 477

Hands-On Algebra, p. 118

Transparencies
5-Minute Check, 11–1

Technology/Multimedia
AlgePASS, Lesson 25
Interactive Chalkboard CD-ROM

In Example 1, the *lowest point*, or **minimum**, of the graph of $y = x^2 + 1$ is at (0, 1). Since the coefficient of x^2 is positive, the graph opens *upward*. In Example 2, the *highest point*, or **maximum**, of the graph of $y = -x^2$ is at (0, 0). Since the coefficient of x^2 is negative, the graph opens *downward*.

As with linear functions, in a quadratic function x is the independent variable and y is the dependent variable. Since the graph of a quadratic function extends forever to the left and to the right, the domain (x values) of a quadratic function is the set of all real numbers. For a quadratic function with a graph that opens upward, the range (y values) is all real numbers greater than or equal to the minimum value. For a quadratic function with a graph that opens downward, the range is all real numbers less than or equal to the maximum value.

The maximum or minimum point of a parabola is called the **vertex**. The vertical line containing the vertex of a parabola is called the **axis of symmetry**. If you fold the graph of $y = x^2 + 1$ or $y = -x^2$ along the axis of symmetry, the two halves of each graph will coincide. In both examples, the axis of symmetry is the line $x = 0$.

You can use the rule below to find the equation of the axis of symmetry.

Equation of the Axis of Symmetry

Words: The equation of the axis of symmetry for the graph of $y = ax^2 + bx + c$, where $a \neq 0$, is $x = -\frac{b}{2a}$.

Model:

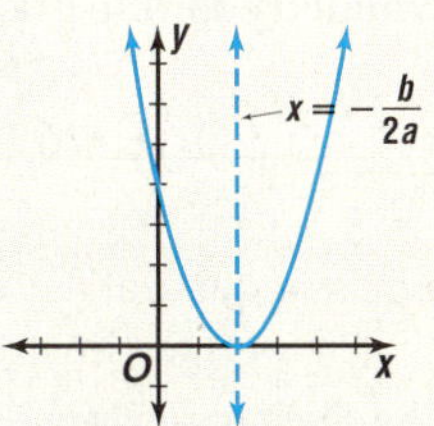

Example 3 Use characteristics of quadratic functions to graph $y = x^2 - 4x - 1$.

A. Find the equation of the axis of symmetry.

B. Find the coordinates of the vertex of the parabola.

C. Graph the function.

A. First identify a, b, and c.

$y = ax^2 + bx + c$

$y = 1x^2 - 4x - 1$ *$x^2 = 1x^2$*

So, $a = 1$, $b = -4$, and $c = -1$.

Now, find the equation of the axis of symmetry.

$x = -\frac{b}{2a}$ *Equation of axis of symmetry*

$x = -\frac{-4}{2(1)}$ *$a = 1, b = -4$*

$x = 2$ *Simplify.*

(continued on the next page)

www.algconcepts.com/extra_examples

2 TEACH

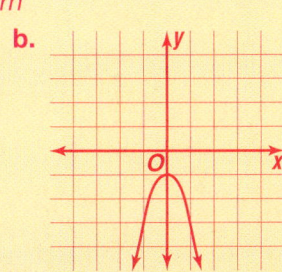

In-Class Examples

Examples 1–2

Graph each quadratic equation by making a table of values.

1 $y = x^2 - 2$

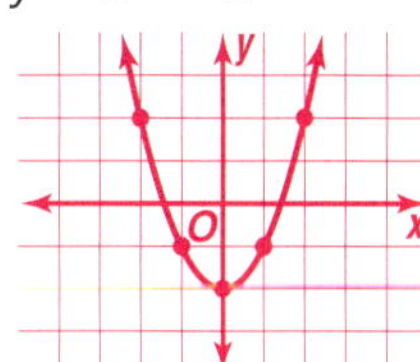

2 $y = -\frac{1}{2}x^2 + 4$

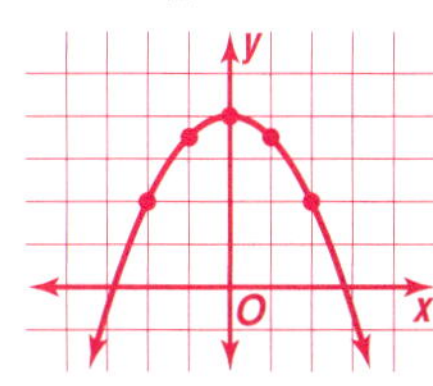

Teaching Tip In Example 3, after students find the coordinates of the vertex, they should use x-values on either side of the vertex when plotting points. Remind students to check their completed graphs to see that the axis of symmetry, vertex, and orientation of the graph are correct.

In-Class Example

Example 3

Use characteristics of quadratic functions to graph $y = -x^2 + 2x + 1$.

A. Find the equation of the axis of symmetry. $x = 1$

B. Find the coordinates of the vertex of the parabola. **(1, 2)**

C. Graph the function.

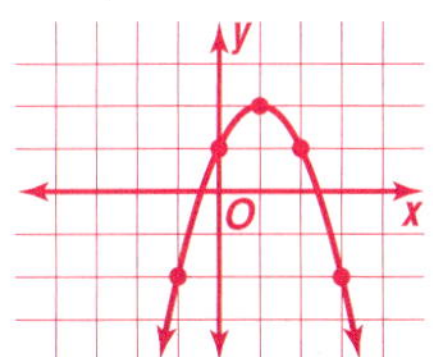

Answers

Your Turn

a.

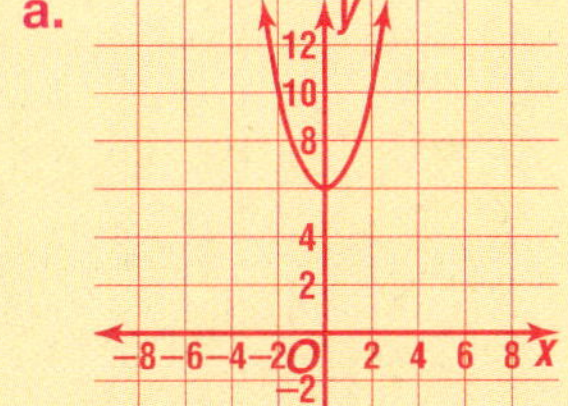

b.

In Class Example

Example 4

A football player throws a short pass. The height y of the ball is given by the equation $y = -16x^2 + 8x + 5$, where x is the number of seconds after the ball was thrown. What is the maximum height reached by the ball? **6 ft**

Answers

Your Turn

c.

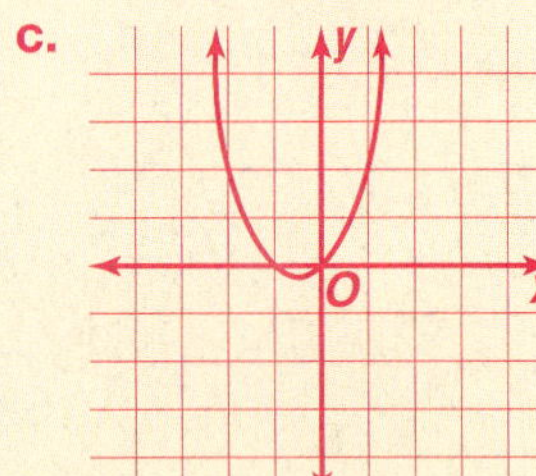

d.

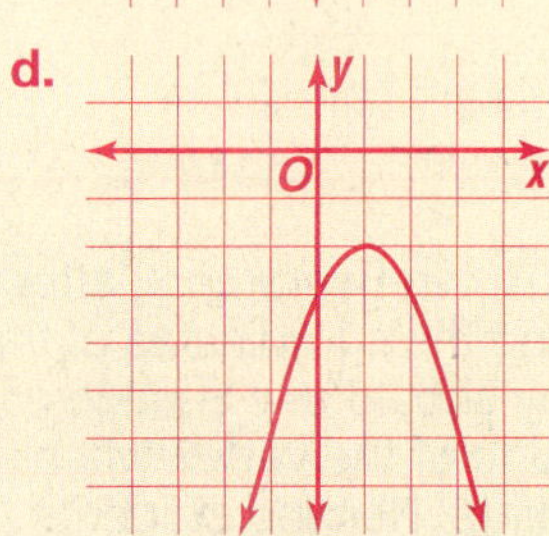

Page 461

1. Sample answer: Quadratic functions have a degree of 2, but linear functions have a degree of 1. The graphs of quadratic functions are not straight lines as with linear functions.
2. The parabola opens downward.

8.

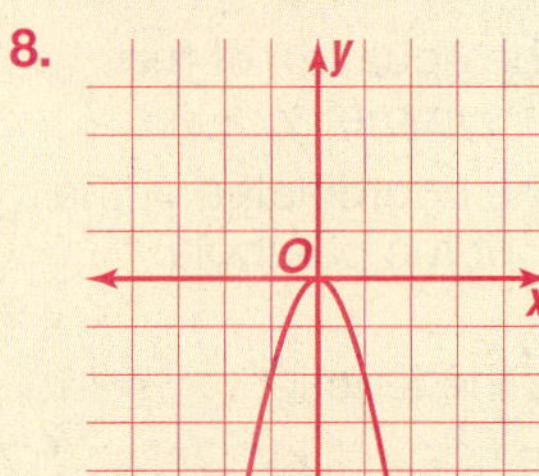

9.

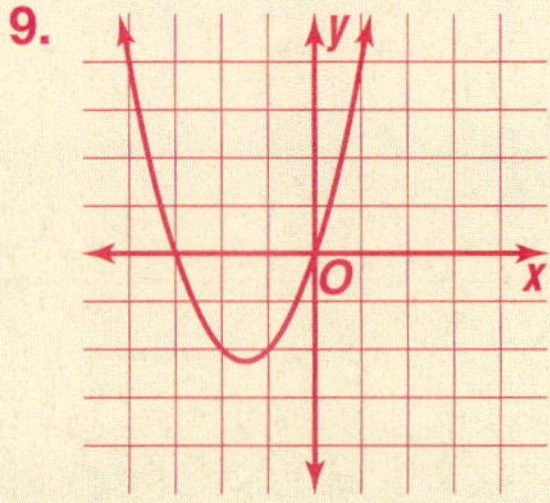

B. Next, find the vertex. Since the equation of the axis of symmetry is $x = 2$, the x-coordinate of the vertex must be 2. Substitute 2 for x in the equation $y = x^2 - 4x - 1$ to solve for y.

$$y = x^2 - 4x - 1$$
$$= (2)^2 - 4(2) - 1$$
$$= 4 - 8 - 1 \text{ or } -5$$

The point at (2, −5) is the vertex. *This point is a minimum*

C. Construct a table. Choose some values for x that are less than 2 and some that are greater than 2. This ensures that points on each side of the axis of symmetry are graphed.

x	$x^2 - 4x - 1$	y	(x, y)
0	$0^2 - 4(0) - 1$	−1	(0, −1)
1	$1^2 - 4(1) - 1$	−4	(1, −4)
2	$2^2 - 4(2) - 1$	−5	(2, −5)
3	$3^2 - 4(3) - 1$	−4	(3, −4)
4	$4^2 - 4(4) - 1$	−1	(4, −1)

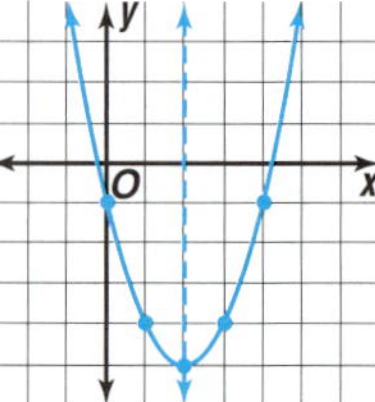

Your Turn **d. $x = 1$; (1, −2)**

Find the coordinates of the vertex and the equation of the axis of symmetry for the graph of each equation. Then graph the function.

c. $y = x^2 + x$ **$x = -\frac{1}{2}$; $\left(-\frac{1}{2}, -\frac{1}{4}\right)$**

d. $y = -x^2 + 2x - 3$

c–d. See margin for graphs.

Models of quadratic functions can be found in the real world.

Real World Example 4

Architecture Link

The Exchange House in London, England, is supported by a steel arch shaped like a parabola. This parabola can be modeled by the quadratic function $y = -0.025x^2 + 2x$, where y represents the height of the arch and x represents the horizontal distance from one end of the base in meters. What is the highest point of the parabolic arch?

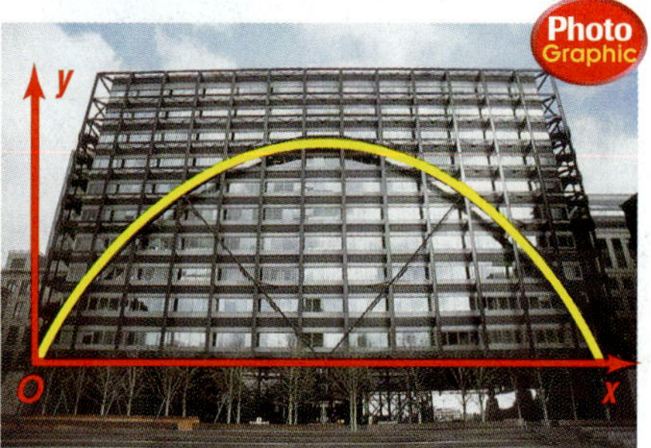

The highest point of the arch is the y-coordinate of the vertex. Find the equation of the axis of symmetry for $h(x) = -0.025x^2 + 2x$.

$$x = -\frac{b}{2a} \quad \text{Equation of axis of symmetry}$$
$$= -\frac{2}{2(-0.025)} \quad a = -0.025, b = 2$$
$$= -\frac{2}{-0.05} \text{ or } 40 \quad \text{Simplify.}$$

Prerequisite Skills Review

Operations with Decimals, p. 684

Next, find the vertex. Since the equation of the axis of symmetry is $x = 40$, the x-coordinate of the vertex must be 40. Substitute 40 for x in the function $h(x) = -0.025x^2 + 2x$. Then solve for h.

$$h(x) = -0.025x^2 + 2x \quad \textit{Original function}$$
$$h(40) = -0.025(40)^2 + 2(40) \quad \textit{Replace x with 40.}$$
$$= -40 + 80 \text{ or } 40 \quad \textit{Simplify.}$$

The point (40, 40) is the vertex. So, the maximum height is 40 meters.

Check for Understanding

Communicating Mathematics

1. **Compare and contrast** quadratic and linear functions. **1–2. See margin.**
2. **Explain** what effect a negative coefficient of x^2 has on the orientation of a parabola.
3. **Name** the point on the graph of a quadratic function that has a unique y-coordinate. **the vertex**

Vocabulary
parabola
quadratic function
minimum
maximum
vertex
axis of symmetry

Guided Practice

Getting Ready Identify the values for a, b, and c for each quadratic function in the form $y = ax^2 + bx + c$.

Sample 1: $y = x^2 - 9$
Solution: $a = 1, b = 0, c = -9$

Sample 2: $y = -4x^2 - 9x + 13$
Solution: $a = -4, b = -9, c = 13$

4. $y = 6x^2 - 3x + 1$ **6, −3, 1**
5. $y = x^2 + 4$ **1, 0, 4**
6. $y = -x^2 - x - 12$ **−1, −1, −12**
7. $y = 3x^2 + 4x$ **3, 4, 0**

Examples 1 & 2 Graph each quadratic function by making a table of values. **8–10. See margin.**

8. $y = -2x^2$
9. $y = x^2 + 3x$
10. $y = -x^2 - 2x + 5$

Example 3 Write the equation of the axis of symmetry and the coordinates of the vertex of the graph of each quadratic function. Then graph the function. **11–14. See Solutions Manual for graphs.**

11. $y = x^2 + 2$ **$x = 0$; (0, 2)**
12. $y = x^2 + 8x + 12$ **12. $x = -4$; (−4, −4)**
13. $y = -x^2 + 4x + 3$ **$x = 2$; (2, 7)**
14. $y = -x^2 + 10x$ **$x = 5$; (5, 25)**

Example 4

15. **Architecture** Mr. Kwan is drafting the windows for a building. Their shape is a parabola modeled by the equation $h = -w^2 + 9$, where h is the height of the window and w is the width in feet.
   a. Graph the function. **See margin.**
   b. Find the maximum height of each window. **9 ft**
   c. Find the width of each window at its base. **6 ft**

3 PRACTICE/APPLY

Error Analysis

Watch for students who give the x-coordinate of the vertex as the maximum or minimum value of the function in problems such as Exercise 15.
Prevent by stressing that the maximum or minimum value is the y-coordinate of the function. Also encourage students to check their responses with the graph.

Answers

10.

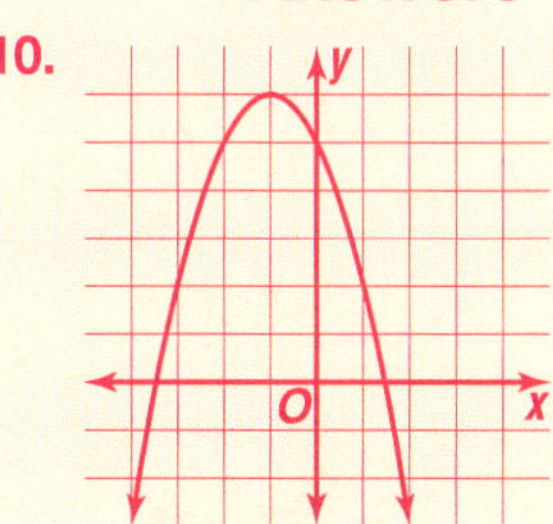

15a.

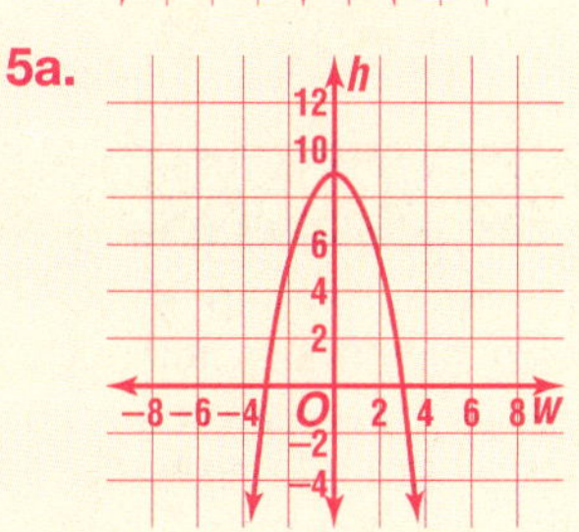

Reteaching Activity

Verbal/Linguistic Learners Have each student work with a partner. Write two quadratic functions on the board or overhead and have each student in the pair graph one of the equations using the method of Example 3. After completing the graph, students should take turns explaining to their partners how they found the equation of the axis of symmetry and the vertex, and how they graphed the equation.

***Study Guide*, p. 473**

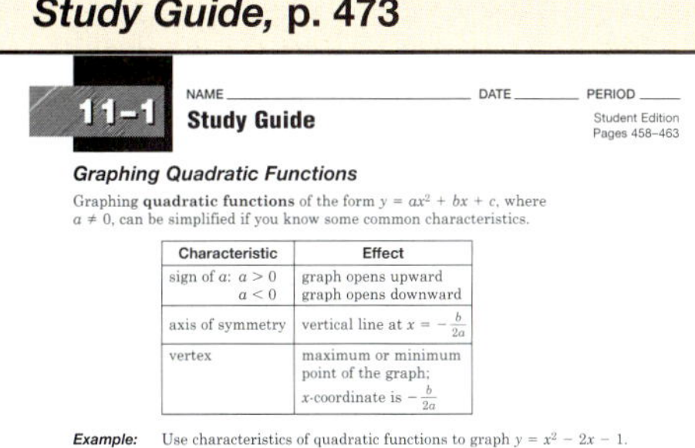

11-1 NAME ______ DATE ______ PERIOD ______

Study Guide Student Edition Pages 458–463

Graphing Quadratic Functions

Graphing **quadratic functions** of the form $y = ax^2 + bx + c$, where $a \neq 0$, can be simplified if you know some common characteristics.

Characteristic	Effect
sign of a: $a > 0$ $a < 0$	graph opens upward graph opens downward
axis of symmetry	vertical line at $x = -\frac{b}{2a}$
vertex	maximum or minimum point of the graph; x-coordinate is $-\frac{b}{2a}$

Example: Use characteristics of quadratic functions to graph $y = x^2 - 2x - 1$.

Step 1 First identify a, b, and c in $y = ax^2 + bx + c$: $a = 1$, $b = -2$, and $c = -1$. Since a > 0, the graph opens upwards.

Step 2 Find the axis of symmetry.

$x = -\frac{b}{2a}$ *Equation of the axis of symmetry*

$x = -\frac{-2}{2(1)}$ or 1 $\quad$ *a = 1 and b = −2*

Step 3 Find the vertex. Since the equation of the axis of symmetry is $x = 1$, the x-coordinate of the vertex is 1. Substitute 1 into the equation $y = x^2 - 2x - 1$ to get $y = (1)^2 - 2(1) - 1$ or -2. The vertex is at $(1, -2)$.

Step 4 Construct a table using values for x that will be on both sides of the axis of symmetry. Choose x-values less than 1 and x-values greater than 1. Graph the points.

x	−1	0	1	2	3
y	2	−1	−2	−1	2

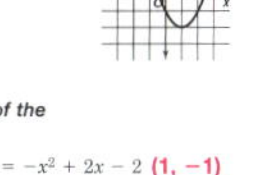

Graph each quadratic equation. Then give the coordinates of the vertex.

1. $y = -3x^2$ **(0, 0)**

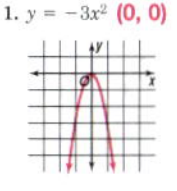

2. $y = x^2 + 2x$ **(−1, −1)**

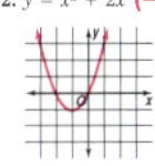

3. $y = -x^2 + 2x - 2$ **(1, −1)**

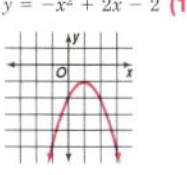

Assignment Guide

Basic: 17–43 odd, 44–52
Average: 16–40 even, 42–52

Answer

42a.

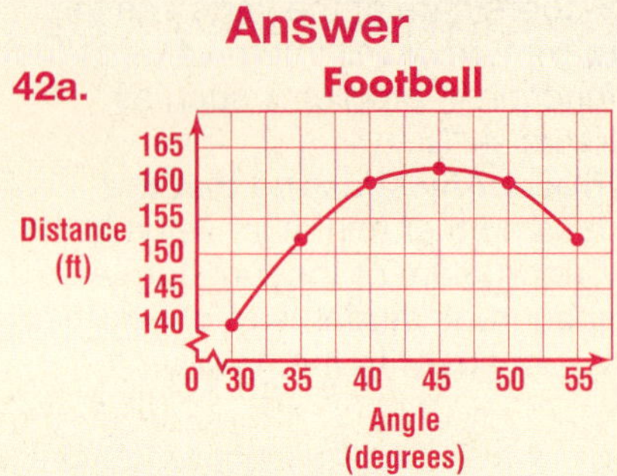

Exercises

Practice

Graph each quadratic function by making a table of values.

16. $y = 2x^2$ 17. $y = 3x^2$ 18. $y = -5x^2$
19. $y = x^2 + 4$ 20. $y = 2x^2 - 1$ 21. $y = -x^2 + 8x$
22. $y = -x^2 + 4x + 1$ 23. $y = 3x^2 - 6x + 1$ 24. $y = \frac{1}{2}x^2 - 6x + 5$

16–36. See Solutions Manual for graphs.

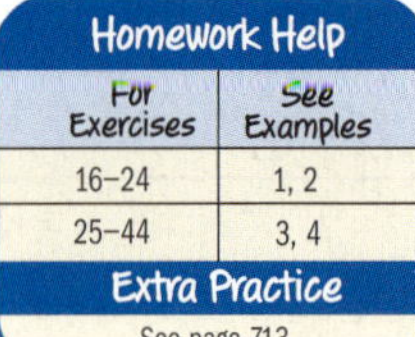

Homework Help	
For Exercises	See Examples
16–24	1, 2
25–44	3, 4
Extra Practice	
See page 713.	

See page 713.

Write the equation of the axis of symmetry and the coordinates of the vertex of the graph of each quadratic function. Then graph the function. **25. $x = 0$; (0, 0) 26. $x = 0$; (0, 0) 27. $x = 0$; (0, 0)**

25. $y = x^2$ 26. $y = -2x^2$ 27. $y = 4x^2$
28. $y = 2x^2 + 3$ 29. $y = -x^2 + 1$ 30. $y = x^2 - 6x$
31. $y = x^2 - 4x + 2$ 32. $y = x^2 - 4x + 10$ 33. $y = -3x^2 - 6x + 4$
34. $y = -\frac{1}{4}x^2 + x - 2$ 35. $y = x^2 + 5x$ 36. $y = 3x^2 - 3x - 2$

34. $x = 2$; (2, −1) 35. $x = -\frac{5}{2}$; $\left(-\frac{5}{2}, -\frac{25}{4}\right)$ 36. $x = \frac{1}{2}$; $\left(\frac{1}{2}, -2\frac{3}{4}\right)$

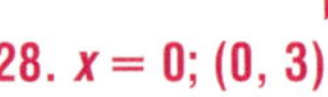

28. $x = 0$; (0, 3)
29. $x = 0$; (0, 1)
30. $x = 3$; (3, −9)
31. $x = 2$; (2, −2)
32. $x = 2$; (2, 6)
33. $x = -1$; (−1, 7)

Match each function with its graph.

37. $y = x^2 + 2x + 1$ **C** 38. $y = x^2 - 2x - 1$ **A** 39. $y = -x^2 + 4x - 3$ **B**

A
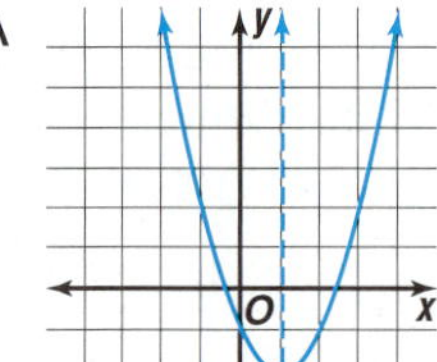

B
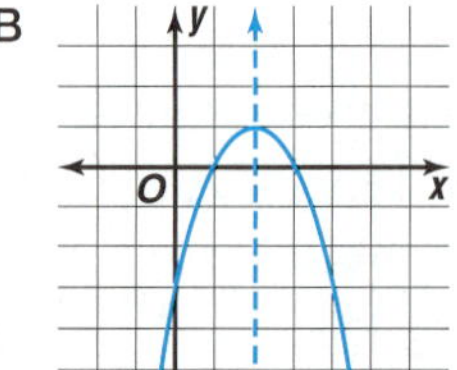

C
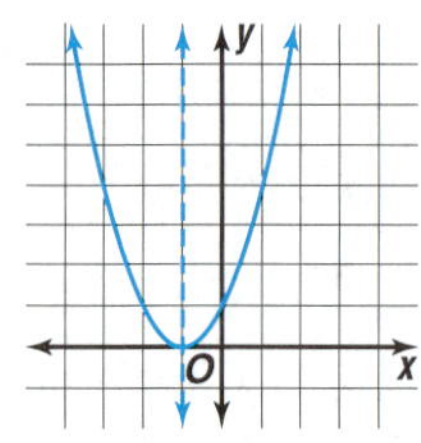

40. Suppose the equation of the axis of symmetry for a quadratic function is $x = -3$ and one of the x-intercepts is −8. What is the other x-intercept? **2**

41. Suppose the points at (−8, 5) and (6, 5) are on the graph of a parabola. What is the equation of the axis of symmetry? **$x = -1$**

Applications and Problem Solving

42. **Sports** Mark wanted to know the angle at which he should kick a football for maximum distance. He used a device that kicked a football at a constant velocity at varying angles. He recorded his results in the table. **a. See margin.**

Angle	Distance (ft)
30°	140
35°	152
40°	160
45°	162
50°	160
55°	152

a. Graph the data in the table.

b. Which angle gives the maximum distance? **45°**

c. Predict how far the football will go if it is kicked at a 60° angle. **140 ft**

Inclusion Strategies

Students with a hearing impairment may have difficulty understanding what the different parts of the parabola are such as the maximum, minimum, vertex, and axis of symmetry. Make sure to illustrate these definitions with a sketch and clearly label each part.

***Skills Practice,* p. 474, and *Practice,* p. 475 (shown)**

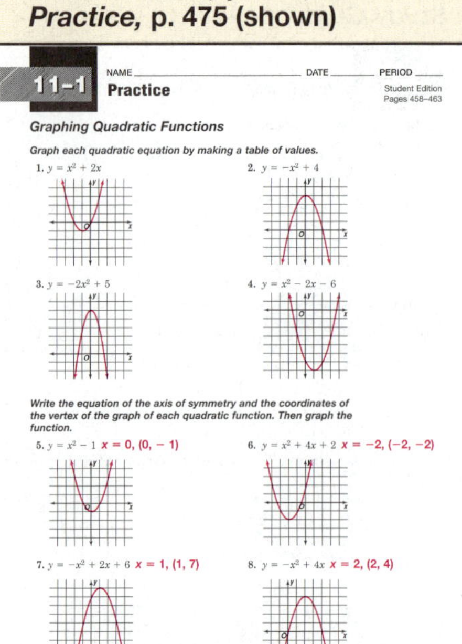

11-1 Practice NAME ______ DATE ______ PERIOD ______
Student Edition Pages 458–463

Graphing Quadratic Functions

Graph each quadratic equation by making a table of values.

1. $y = x^2 + 2x$ 2. $y = -x^2 + 4$
3. $y = -2x^2 + 5$ 4. $y = x^2 - 2x - 6$

Write the equation of the axis of symmetry and the coordinates of the vertex of the graph of each quadratic function. Then graph the function.

5. $y = x^2 - 1$ **$x = 0$, (0, − 1)** 6. $y = x^2 + 4x + 2$ **$x = -2$, (−2, −2)**
7. $y = -x^2 + 2x + 6$ **$x = 1$, (1, 7)** 8. $y = -x^2 + 4x$ **$x = 2$, (2, 4)**

From the Classroom of ...

Daniel Marks
Auburn University at Montgomery
Montgomery, Alabama

I like to discuss why the graph of profit versus the number of employees in Exercise 43 is parabolic. There is an ideal number of employees that leads to maximum profit. With too few employees, not enough work is completed. With too many employees, they get in one another's way and time is wasted.

43. **Business** The profit function of a small business can be expressed as $P(x) = -x^2 + 300x$, where x represents the number of employees.
   a. How many employees will yield the maximum profit? **150**
   b. What is the maximum profit? **$22,500**

44. **Critical Thinking** Graph $y = x^2$. Then graph $y = x^2 - 4$ on the same axes. Describe the difference between the graphs.
$y = x^2 - 4$ is 4 units lower.

Mixed Review

Factor each polynomial. *(Lesson 10–5)*

45. $x^2 - 49$ **$(x - 7)(x + 7)$**
46. $a^2 + 10a + 25$ **$(a + 5)^2$**
47. $2g^2 - 16g + 32$ **$2(g - 4)^2$**

48. **Geometry** The volume of a rectangular prism is $3x^3 - 6x^2 - 24x$. Find dimensions for the prism in terms of x. *(Lesson 10–4)*
Sample answer: $3x$, $x - 4$, $x + 2$

49. Arrange the terms of the polynomial $-4x^2 + 5 - x^3 - 8x$ so that the powers of x are in descending order. Then state the degree of the polynomial. *(Lesson 9–1)* **$-x^3 - 4x^2 - 8x + 5$; 3**

Use the map for Exercises 50–51.

50. You can use the letters and numbers on the map to form ordered pairs and name square areas. State the locations of Fancyburg and Northam Parks as ordered pairs. *(Lesson 2–2)* **(A, 3), (E, 1)**

51. Find the straight-line distance in miles between Fancyburg and Northam Parks. *(Lesson 5–2)* **about 1.4 mi**

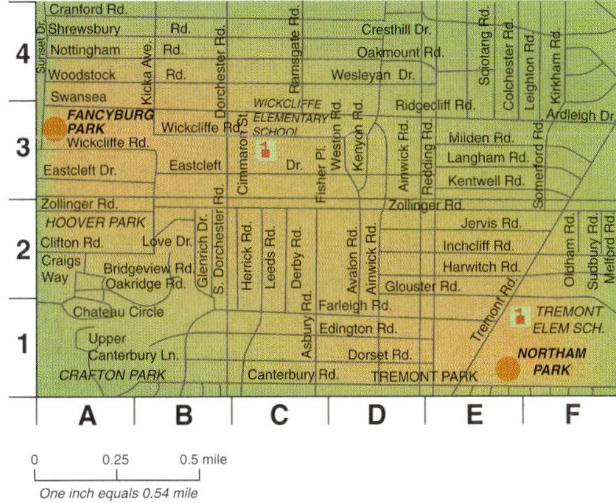

Standardized Test Practice

52. **Multiple Choice** The line graph shows the amount of money spent on magazine advertising of soft drinks over a five-year period. During which of the following periods was the greatest change in spending? *(Lesson 1–7)* **B**

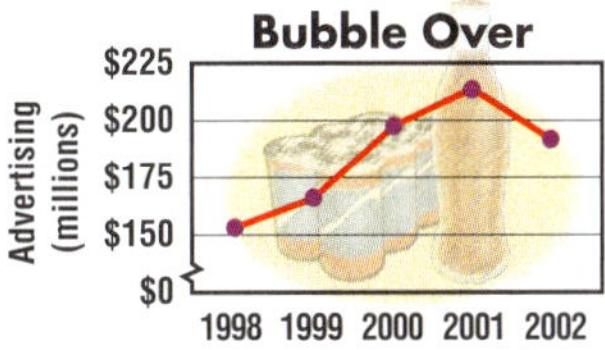

Source: Publishers Information Bureau, Inc.

   Ⓐ 1998–1999 Ⓑ 1999–2000
   Ⓒ 2000–2001 Ⓓ 2001–2002

4 ASSESS

Open-Ended Assessment

Speaking Write several quadratic functions on the board or overhead. Ask students to identify whether they open up or down and have a maximum or minimum value. Have them find *a*, *b*, and *c*, the equation of the axis of symmetry, and the coordinates of the vertex. Then ask which other *x*-values they would use in a table of values when graphing the function.

***Enrichment*, p. 477**

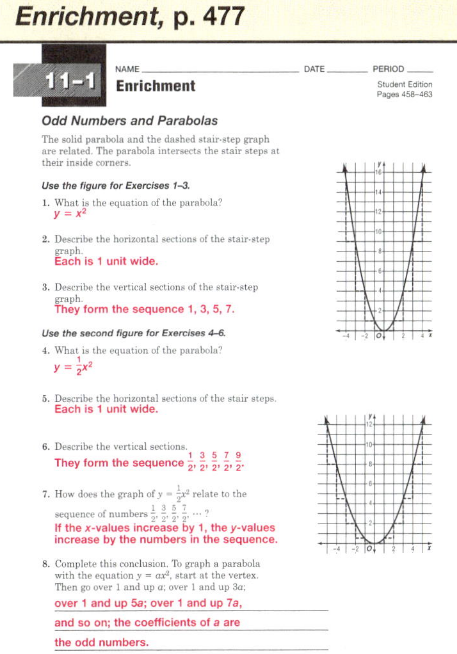
11-1 NAME ______ DATE ______ PERIOD ______

Enrichment Student Edition Pages 458–463

Odd Numbers and Parabolas

The solid parabola and the dashed stair-step graph are related. The parabola intersects the stair steps at their inside corners.

Use the figure for Exercises 1–3.

1. What is the equation of the parabola? $y = x^2$
2. Describe the horizontal sections of the stair-step graph. Each is 1 unit wide.
3. Describe the vertical sections of the stair-step graph. They form the sequence 1, 3, 5, 7.

Use the second figure for Exercises 4–6.

4. What is the equation of the parabola? $y = \frac{1}{2}x^2$
5. Describe the horizontal sections of the stair steps. Each is 1 unit wide.
6. Describe the vertical sections. They form the sequence $\frac{1}{2}, \frac{3}{2}, \frac{5}{2}, \frac{7}{2}, \frac{9}{2}$.
7. How does the graph of $y = \frac{1}{2}x^2$ relate to the sequence of numbers $\frac{1}{2}, \frac{3}{2}, \frac{5}{2}, \frac{7}{2}, \ldots$? If the *x*-values increase by 1, the *y*-values increase by the numbers in the sequence.
8. Complete this conclusion. To graph a parabola with the equation $y = ax^2$, start at the vertex. Then go over 1 and up a; over 1 and up $3a$; over 1 and up 5a; over 1 and up 7a, and so on; the coefficients of a are the odd numbers.

Extra Credit

A child is playing with a toy rocket. The height of the rocket y is given by the equation $y = -16x^2 + 64x + 6$, where x is the number of seconds that have elapsed since the rocket was shot. From what height was the rocket shot? After how many seconds will the rocket fall back to the same height? What is the maximum height of the rocket? **6 ft; 4 s; 70 ft**

Lesson 11-2

1 FOCUS

5-Minute Check

Lesson 11–1

1. Graph $y = -x^2 + 1$.

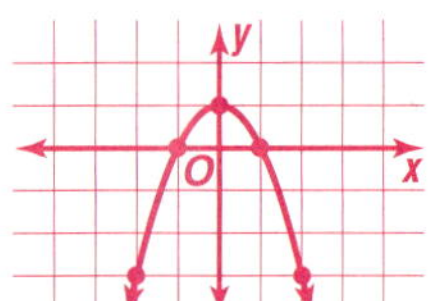

Write the equation of the axis of symmetry and the coordinates of the vertex of the graph of each equation.

2. $y = 2x^2 - 2x - 3$
   $x = \frac{1}{2}, \left(\frac{1}{2}, -3\frac{1}{2}\right)$
3. $y = x^2 - 5x$
   $x = \frac{5}{2}, \left(\frac{5}{2}, -\frac{25}{4}\right)$
4. $y = -x^2 + 7$
   $x = 0, (0, 7)$

Motivating the Lesson

Hands-On Activity On the board or overhead, write a group of related equations like those given in Examples 1 and 2. Have students use graphing calculators to graph the equations on the same screen and then compare the graphs. Ask volunteers to discuss their comparisons.

2 TEACH

In-Class Example

Example 1

Graph the equations $y = -x^2$, $y = -0.5x^2$, and $y = -2x^2$ on the same screen. Compare and contrast the graphs. What conclusions can be drawn? **Each graph opens downward and has its vertex at the origin. The graph of $y = -0.5x^2$ is wider than the graph of $y = -x^2$. The graph of $y = -2x^2$ is narrower than the graph of $y = -x^2$.**

11-2 Families of Quadratic Functions

What You'll Learn

You'll learn the characteristics of families of parabolas.

Why It's Important

Animation Digital artists use families of parabolas to create the illusion of motion. *See Example 5.*

In families of linear graphs, lines either have the same slope or the same y-intercept. However, in families of parabolas, graphs either share a vertex or an axis of symmetry, or both. Also, a family can consist of parabolas of the same shape.

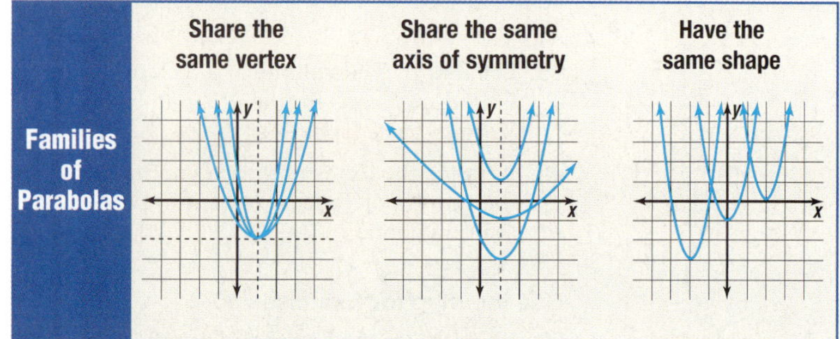

Examples

Graph each group of equations on the same screen. Compare and contrast the graphs. What conclusions can be drawn?

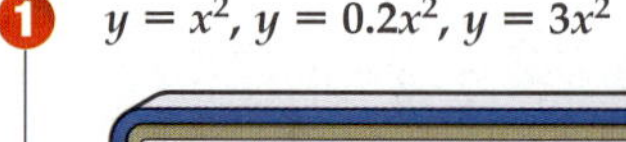

Technology Tip

Use the [Y=] screen to enter functions into the graphing calculator.

The parent graph for each family is $y = x^2$.

1 $y = x^2$, $y = 0.2x^2$, $y = 3x^2$

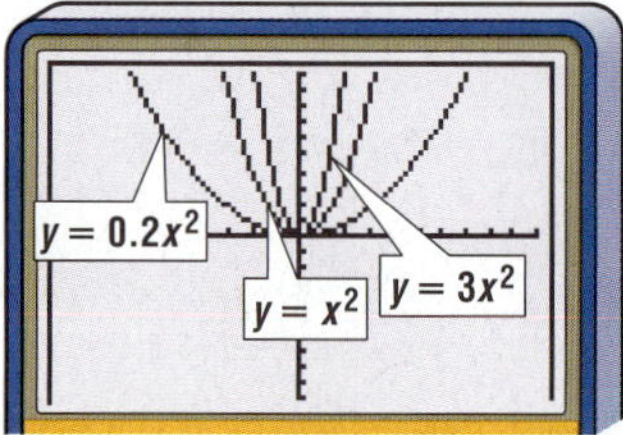

Each graph opens upward and has its vertex at the origin. Therefore, these equations are a family of parabolas. The graph of $y = 0.2x^2$ is wider than the graph of $y = x^2$. The graph of $y = 3x^2$ is more narrow than the graph of $y = x^2$.

The shape of the parabola narrows as the coefficient of x^2 becomes greater. The shape widens as the coefficient of x^2 becomes smaller.

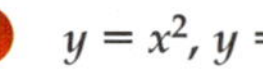

2 $y = x^2$, $y = x^2 - 6$, $y = x^2 + 3$

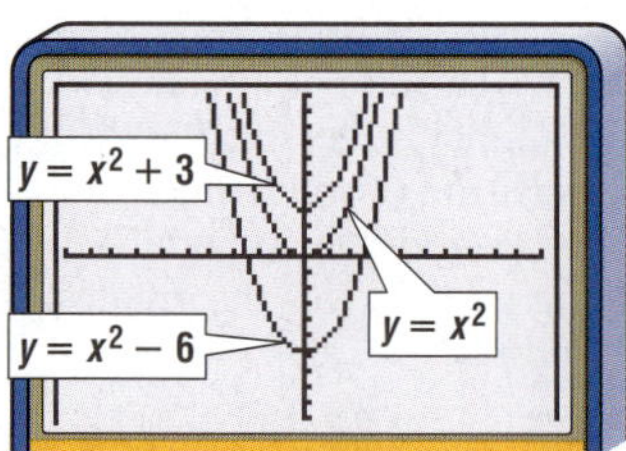

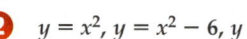

Each graph opens upward and has the same shape as $y = x^2$ so they form a family. Yet, each parabola has a different vertex located along the y-axis.

A constant greater than 0 shifts the graph upward, and a constant less than 0 shifts the graph downward along the axis of symmetry.

Resource Manager

Reproducible Masters

Chapter 11 Resource Masters
- *Study Guide*, p. 478
- *Skills Practice*, p. 479
- *Practice*, p. 480
- *Reading to Learn Mathematics*, p. 481
- *Enrichment*, p. 482

Transparencies

5-Minute Check, 11–2

Technology/Multimedia

AlgePASS, Lesson 26

Interactive Chalkboard CD-ROM

3 $y = x^2, y = (x + 2)^2, y = (x - 4)^2$

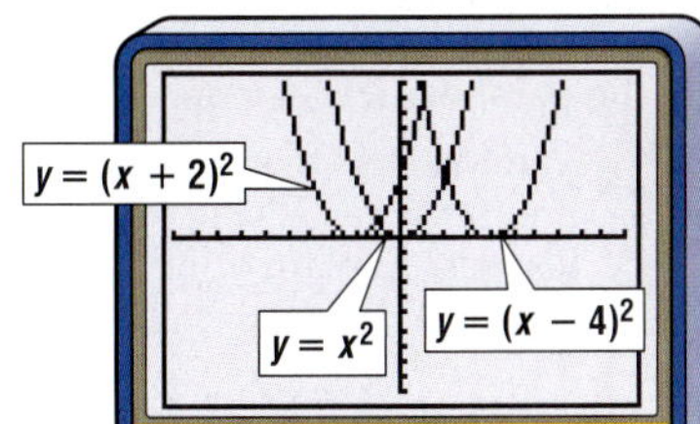

Each graph opens upward and has the same shape as $y = x^2$. However, each parabola has a different vertex located along the x-axis.

Find the number for x that results in 0 inside the parentheses. The graph shifts this number of units to the left or right.

4 $y = x^2, y = (x - 7)^2 + 2$

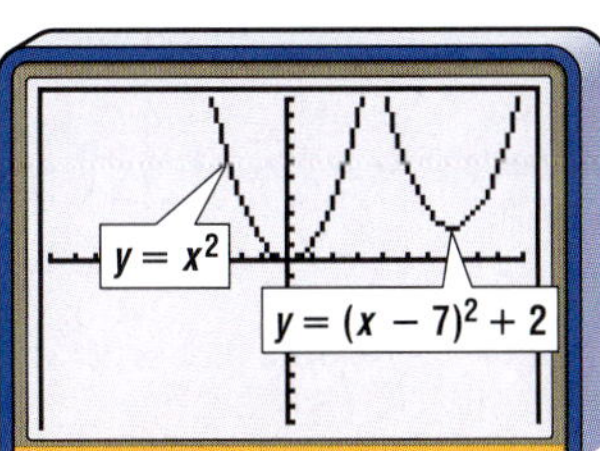

The graph of $y = (x - 7)^2 + 2$ has the same shape as the graph of $y = x^2$. However, it shifts to the right 7 units because a positive 7 will result in zero inside the parentheses. It also shifts upward 2 units because of the constant 2 outside the parentheses.

Your Turn

a. $y = x^2, 2x^2, 4x^2$

b. $y = x^2, x^2 - 1, x^2 - 8$

c. $y = x^2, -x^2$

d. $y = -x^2, y = -(x + 2)^2, y = -(x + 4)^2$

a. As the coefficient of x increases, the graphs become narrower.

b. The graphs shift along the y-axis.

c. The negative sign causes the graph of $y = -x^2$ to open downward.

d. All graphs open downward and shift along the y-axis.

Sometimes computers are used to generate families of graphs.

Example 5

Computer Animation Link

In a computer game, a player dodges space shuttles that are shaped like parabolas. Suppose the vertex of one shuttle is at the origin. The shuttle's initial shape and position are given by the equation $y = 0.5x^2$. It leaves the screen with its vertex at (6, 5). Find an equation to model the final shape and position of the shuttle.

The shape of the shuttle remains the same. However, the vertex shifts from (0, 0) to the right 6 units and up 5 units.

Begin with the original equation.

$y = 0.5x^2$

$y = 0.5(x - 6)^2$

If $x = 6$, then $x - 6 = 0$. Shift the vertex to the right 6 units.

$y = 0.5(x - 6)^2 + 5$

The 5 outside the parentheses shifts the entire parabola up 5 units.

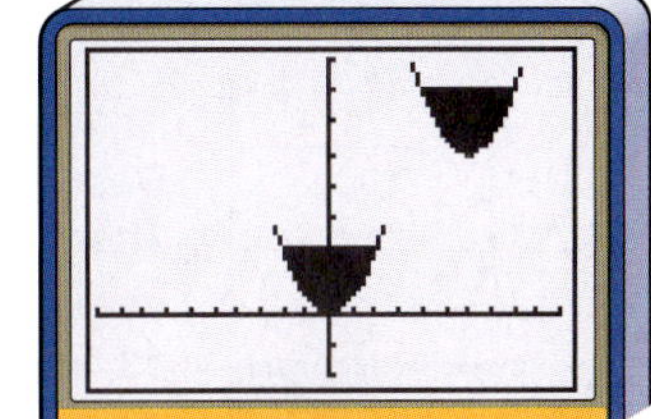

So, the final shape and position of the shuttle can be described by the equation $y = 0.5(x - 6)^2 + 5$.

interNET CONNECTION

Data Update For the latest information on computer animation, visit: www.algconcepts.com

www.algconcepts.com/extra_examples

In-Class Examples

Examples 2–4

Graph each group of equations on the same screen. Compare and contrast the graphs. What conclusions can be drawn?

2 $y = -x^2, y = -x^2 + 1, y = -x^2 + 4$ **Each graph opens downward and has the same shape as $y = -x^2$. Each parabola has a different vertex located along the y-axis. $y = -x^2$ has vertex (0, 0), $y = -x^2 + 1$ has vertex (0, 1) and $y = -x^2 + 4$ has vertex (0, 4).**

3 $y = x^2, y = (x - 3)^2, y = (x + 1)^2$ **Each graph opens upward and has the same shape as $y = x^2$. However, each parabola has a different vertex located along the x-axis. The vertex of $y = x^2$ is (0, 0), the vertex of $y = (x - 3)^2$ is (3, 0) and the vertex of $y = (x + 1)^2$ is (–1, 0).**

4 $y = x^2, y = (x + 2)^2 - 1$ **The graph of $y = (x + 2)^2 - 1$ has the same shape as the graph of $y = x^2$. However, it shifts to the left two units and down 1 unit.**

See p. 466 for In-Class Example 5.

Study Guide, p. 478

NAME ______ DATE ______ PERIOD ______

Study Guide

Student Edition Pages 464–467

Families of Quadratic Functions

Families of parabolas are parabolas that share the same shape, vertex, or axis of symmetry. Graphing calculators make it easy to study families of graphs. Graph each of the following groups of equations on the same screen to compare and contrast the graphs.

$y = x^2, y = 2x^2, y = 0.5x^2$	$y = x^2, y = x^2 - 2, y = x^2 + 2$	$y = (x - 1)^2, y = (x - 2)^2, y = (x - 4)^2$
same vertex as $y = x^2$; open upward; different shapes	same shape as $y = x^2$; open upward; different vertices	same shape; open upward; shift right

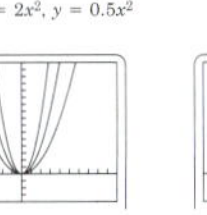

Example: Describe how the graph of $y = (x + 2)^2$ changes from the parent graph of $y = x^2$. Then name the vertex of each graph.

The constant –2 will make the value of the term $(x + 2)^2$ equal to 0. Therefore this graph will shift 2 units to the left. The vertex of $y = x^2$ is at (0, 0), while the vertex of $y = (x + 2)^2$ is at (–2, 0).

Graph each group of equations on the same screen. Compare and contrast the graphs. 1–3. See students' graphs.

1. $y = -x^2$, $y = -4x^2$, $y = -5x^2$ opens down; narrows
2. $y = (x + 2)^2$, $y = (x + 4)^2$, $y = (x + 6)^2$ opens up; shifts left
3. $y = x^2 - 1$, $y = x^2 - 2$, $y = x^2 - 3$ opens up; shifts down

Describe how each graph changes from the parent graph of $y = x^2$. Then name the vertex of each graph.

4. $y = 4x^2$ narrows; (0, 0)
5. $y = x^2 - 2$ down 2; (0, –2)
6. $y = -x^2 - 1$ opens down; down 1; (0, –1)
7. $y = (x + 1)^2$ left 1; (–1, 0)
8. $y = (x - 4)^2$ right 4; (4, 0)
9. $y = 0.4x^2$ widens; (0, 0)
10. $y = x^2 + 3$ up 3; (0, 3)
11. $y = (x + 3)^2$ left 3; (–3, 0)
12. $y = -x^2 - 5$ opens down; down 5; (0, –5)

Reteaching Activity

Kinesthetic Learners Use overhead transparency sheets to practice translating graphs and writing equations for the translated graphs. The parent graph should be sketched on one transparency. A second transparency is placed on top and the parent graph is traced. This second sheet can then be moved to get the translated graph. After writing an equation for the translated graph, students should test points from the graph in the equation. Note that this activity does not work for graphs whose shape is different from the parent graph such as those in Example 1 on page 464.

In-Class Example

Example 5

Refer to Example 5. Another space shuttle begins with original equation $y = -2x^2$. This shuttle moves until its vertex is at $(-2, 3)$. Find an equation to model the shape and position of this shuttle at its final location.
$y = -2(x + 2)^2 + 3$

3 PRACTICE/APPLY

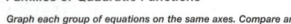

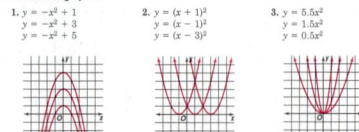

Error Analysis

Watch for students who incorrectly identify the shift of a graph, or use incorrect signs when writing the equation for a shifted image in Exercise 9.
Prevent by encouraging students to test the vertex and one other point on the graph of the transformed image in their equation.

Assignment Guide

Basic: 11–29 odd, 30–35
Average: 10–26 even, 27–35

***Skills Practice,* p. 479, and *Practice,* p. 480 (shown)**

NAME ______ DATE ______ PERIOD ______

11-2 Practice — Student Edition Pages 464–467

Families of Quadratic Functions

Graph each group of equations on the same axes. Compare and contrast the graphs.

1. $y = -x^2 + 1$, $y = -x^2 + 3$, $y = -x^2 + 5$
2. $y = (x + 1)^2$, $y = (x - 1)^2$, $y = (x - 3)^2$
3. $y = 5.5x^2$, $y = 1.5x^2$, $y = 0.5x^2$

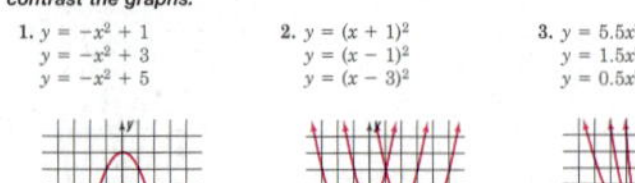

1. All open down; graphs shift up 2 units.
2. All open up; graphs shift right 2 units.
3. All open up; graphs get wider the smaller the coefficient of x^2.

Describe how each graph changes from the parent graph of $y = x^2$. Then name the vertex of each graph.

4. $y = 2x^2$ narrows; (0, 0)
5. $y = x^2 + 3$ up 3 units; (0, 3)
6. $y = -x^2 + 5$ opens down, up 5 units; (0, 5)
7. $y = -0.2x^2$ open down; widens (0, 0)
8. $y = (x + 1)^2$ left 1 unit; (−1, 0)
9. $y = (x - 9)^2$ right 9 units; (9, 0)
10. $y = -4x^2 - 1$ opens down, narrows, down 1 unit; (0, −1)
11. $y = (x - 6)^2 + 5$ right 6 units, up 5 units; (6, 5)
12. $y = -0.5x^2 + 4$ opens down, widens, up 4 units; (0, 4)
13. $y = 5x^2 + 8$ narrows, up 8 units; (0, 8)
14. $y = (x - 2)^2 - 3$ right 2 units, down 3 units; (2, −3)
15. $y = -(x + 1)^2 + 8$ opens down, left 1 unit, up 8 units; (−1, 8)
16. $y = -(x + 3)^2 - 7$ opens down, left 3 units, down 7 units; (−3, −7)
17. $y = -(x - 4)^2 + 5$ opens down, right 4 units, up 5 units; (4, 5)
18. $y = (x + 6)^2 + 2$ left 6 units, up 2 units; (−6, 2)

Check for Understanding

Communicating Mathematics

1. **Describe** the parabola whose equation is $y = 100x^2$. **1. extremely narrow with vertex at (0, 0), opening up**
2. **You Decide** Vanessa says that the graphs of $y = (x - 2)^2$ and $y = x^2 - 2$ are the same. Vickie says that they are different. Who is correct? Explain your answer and sketch the graphs. **2. See margin.**
3. **Match** each equation with its corresponding graph.
   $y = -3x^2$ **B** $y = x^2 + 3$ **A** $y = (x - 3)^2$ **C**

A
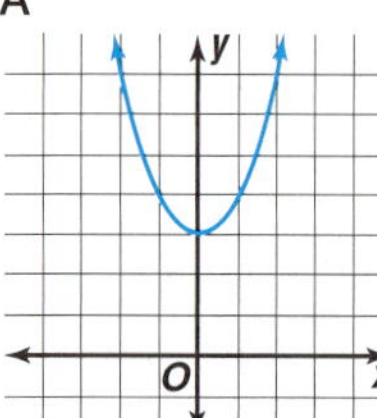

B
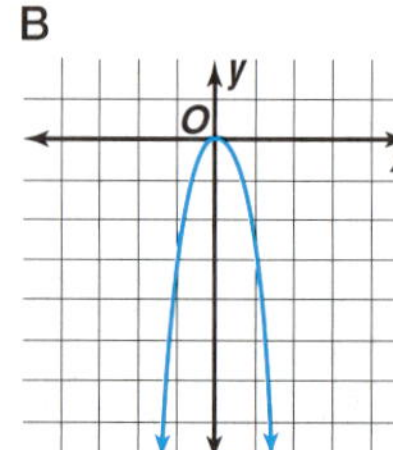

C
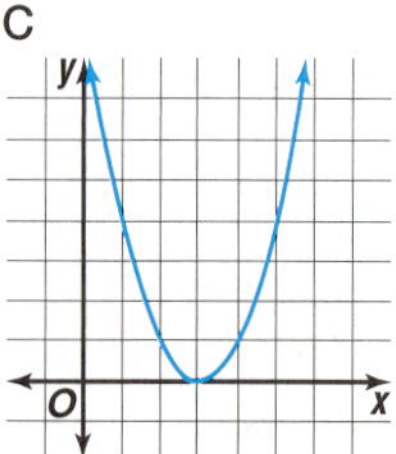

Guided Practice

Example 1

4. Graph $y = x^2$, $y = 0.5x^2$, and $y = 0.1x^2$ on the same axes. Compare and contrast the graphs. **See margin.**

Examples 1–4

Describe how each graph changes from the parent graph of $y = x^2$. Then name the vertex of each graph.

5. $y = 0.7x^2$ **(0, 0)** — **5. widens**
6. $y = x^2 + 10$ **(0, 10)** — **6. up 10 units**
7. $y = (x + 4)^2$ **(−4, 0)** — **7. left 4 units**
8. $y = (x + 2)^2 - 9$ **(−2, −9)** — **8. left 2 units, down 9 units**

Example 5

9. **Computer Animation** Refer to Example 5. Suppose the shuttle is programmed to move to point (4, 3) before leaving the screen. Write the equation that describes its location. $y = 0.5(x - 4)^2 + 3$

Exercises

Practice

Graph each group of equations on the same screen. Compare and contrast the graphs. **10–12. See margin.**

10. $y = -x^2$, $y = -4x^2$, $y = -6x^2$
11. $y = (x + 1)^2$, $y = (x + 2)^2$, $y = (x + 3)^2$
12. $y = -x^2 - 1$, $y = -x^2 - 3$, $y = -x^2 - 5$

Homework Help	
For Exercises	See Examples
10, 13, 17, 18	1
11, 15, 16, 25, 27	3
12, 14, 26	2
23, 28, 29	4, 5

Extra Practice

See page 713.

Describe how each graph changes from the parent graph of $y = x^2$. Then name the vertex of each graph. **13–24. See margin for descriptions. 20. (−4, 0) 21. (0, −8) 23. (−1, −5) 24. (−2, 3)**

13. $y = 5x^2$ **(0, 0)**
14. $y = x^2 - 8$ **(0, −8)**
15. $y = (x - 7)^2$ **(7, 0)**
16. $y = (x - 3)^2$ **(3, 0)**
17. $y = -2x^2$ **(0, 0)**
18. $y = -0.9x^2$ **(0, 0)**

B

19. $y = 2x^2 + 1$ **(0, 1)**
20. $y = -(x + 4)^2$
21. $y = 0.4x^2 - 8$
22. $y = -x^2 + 6$ **(0, 6)**
23. $y = (x + 1)^2 - 5$
24. $y = [x - (-2)]^2 + 3$

Answers

2. Vicki; the graph of $y = (x - 2)^2$ shifts the graph of $y = x^2$ two units to the right. The graph of $y = x^2 - 2$ shifts the graph of $y = x^2$ two units down.

4. As the coefficient of x^2 decreases, the graphs become wider.

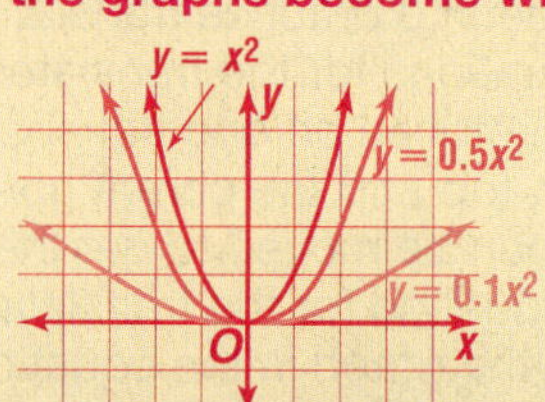

25. What is the equation of the parabola that moves the parent graph $y = x^2$ 8 units to the left? $y = (x + 8)^2$

26. Suppose the parent graph is $y = x^2 + 3$. Write the equation of the parabola that would move it down 5 units. $y = x^2 - 2$

Applications and Problem Solving

27. **Fireworks** In Cincinnati, fireworks are launched from a large barge on the Ohio River for a Labor Day celebration. Their flight can be modeled by the function $h(t) = -4.9(t - 4)^2 + 80$, where h is the height in meters and t is the time in seconds. Suppose that buildings obstruct the fireworks. So, the barge is relocated 30 meters to the east. Write a function to model the path of the fireworks in the new location. $h(t) = -4.9(t - 34)^2 + 80$

Fireworks display

28. **Sports** A baseball player hits a pop-up. The height of the ball can be modeled by the function $h(t) = -16(t - 3.5)^2 + 200$, where h is the height in feet and t is the time in seconds. Another pop-up is hit with the same velocity, but from a half-foot higher.
    a. Write equations to model the height of the ball for each hit.
    b. Graph both equations and find each vertex. **a–b. See margin.**
    c. Does the higher height affect the maximum height? Does the higher height affect the time the ball takes to reach its maximum height? **yes; no**

29. **Critical Thinking** Graphs of two quadratic functions are shown. The graph of $y = x^2$ is the parent graph. Write the equation for the other graph. $y = (x - 2)^2 - 4$

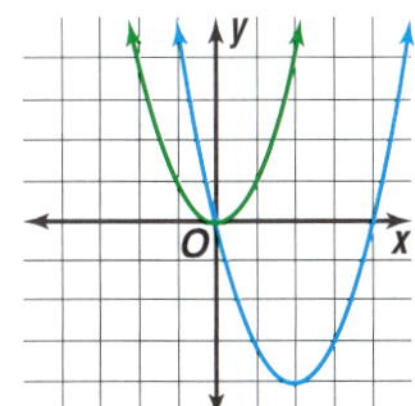

Mixed Review

Find the coordinates of the vertex for each quadratic function. *(Lesson 11–1)*

30. $y = -4x^2 + 8x + 13$ **(1, 17)**

31. $y = 2x^2 + 6x + 3$ **(−1.5, −1.5)**

32. **Finance** Tracy invested \$500, earning an annual interest rate r. The value of her investment at the end of two years can be represented by the polynomial $500r^2 + 1000r + 500$. *(Lesson 10–5)*
    a. Write an expression in factored form for the value of her investment at the end of two years. $500(r + 1)^2$
    b. If $r = 6\%$, find the value of her investment at the end of two years. **\$561.80**

Find each product. *(Lesson 9–4)*

33. $(a + 7)(a - 4)$ $a^2 + 3a - 28$

34. $(3n + 6)(n - 4)$ $3n^2 - 6n - 24$

Standardized Test Practice

35. **Multiple Choice** Nine tickets, numbered 3 through 11, are placed in an empty hat. If one ticket is drawn at random from the hat, what is the probability that a prime number will be on the ticket? *(Lesson 5–6)* **B**

    A $\frac{1}{9}$  B $\frac{4}{9}$ C $\frac{5}{9}$ D $\frac{1}{3}$

www.algconcepts.com/self_check_quiz

4 ASSESS

Open-Ended Assessment

Writing Have students write a description of at least two different types of quadratic function families. Have students explain how changes in the equations are reflected in the graphs of the functions.

Answers

Pages 466–467

10. same vertex, opens down, narrows
11. shifts left, opens up
12. same axis of symmetry, shifts down, opens down
13. narrows
14. down 8 units
15. right 7 units
16. right 3 units
17. opens down, narrows
18. opens down, widens
19. narrows, up 1 unit
20. opens down, left 4 units
21. widens, down 8 units
22. opens down, up 6 units
23. left 1 unit, down 5 units
24. left 2 units, up 3 units

28a. $h(t) = -16(t - 3.5)^2 + 200$, $h(t) = -16(t - 3.5)^2 + 200.5$

Enrichment, p. 482

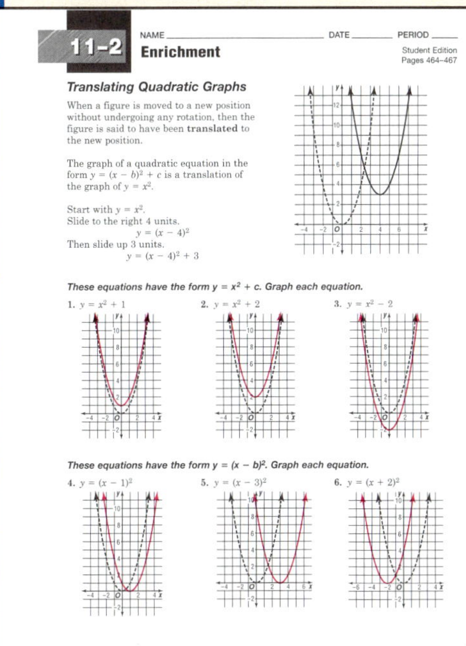

11–2 NAME ______ DATE ______ PERIOD ______

Enrichment

Student Edition Pages 464–467

Translating Quadratic Graphs

When a figure is moved to a new position without undergoing any rotation, then the figure is said to have been **translated** to the new position.

The graph of a quadratic equation in the form $y = (x - b)^2 + c$ is a translation of the graph of $y = x^2$.

Start with $y = x^2$.
Slide to the right 4 units.
$y = (x - 4)^2$
Then slide up 3 units.
$y = (x - 4)^2 + 3$

These equations have the form $y = x^2 + c$. Graph each equation.

1. $y = x^2 + 1$
2. $y = x^2 + 2$
3. $y = x^2 - 2$

These equations have the form $y = (x - b)^2$. Graph each equation.

4. $y = (x - 1)^2$
5. $y = (x - 3)^2$
6. $y = (x + 2)^2$

Extra Credit

The vertex form of a quadratic equation is $y = a(x - h)^2 + k$. Explain what each of the variables a, h, and k tell you about the graph of the equation. Use an example to illustrate your work. **a tells you how wide the parabola is and whether it opens up or down. h is the x-coordinate of the vertex of the parabola and k is the y-coordinate of the vertex. See students' examples.**

Answer

28b.

h(t)
201
200
199
198
0 1 2 3 4 t

Lesson 11-3

1 FOCUS

5-Minute Check
Lesson 11-2

Describe how each graph changes from a parent graph of $y = x^2$. Then name the vertex of each graph.

1. $y = 3x^2$ **narrows; (0, 0)**

2. $y = (x - 3)^2$ **right 3 units; (3, 0)**

Match each equation with its corresponding graph.

3. $y = (x - 2)^2$ **B**

4. $y = x^2 + 2$ **C**

5. $y = (x + 2)^2$ **A**

A.

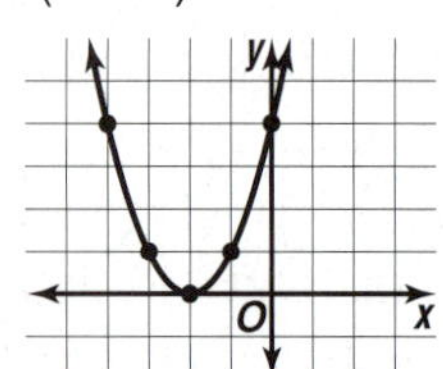

B.

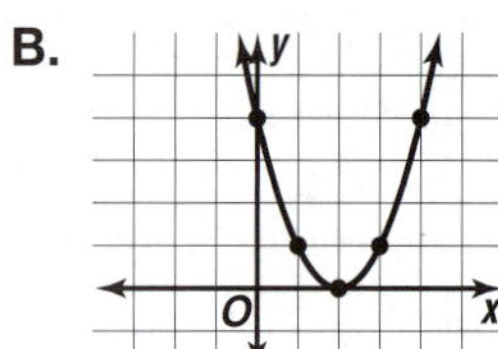

C.

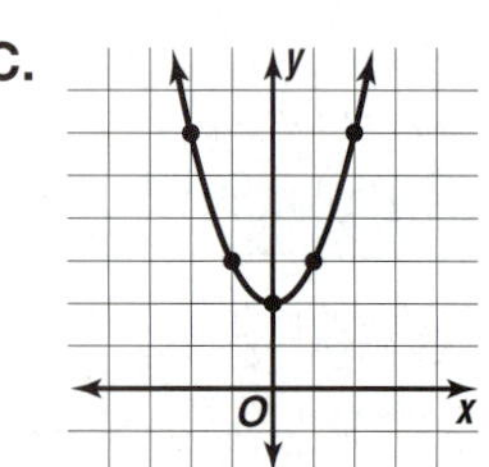

Motivating the Lesson

Real-World Connection Many real-world phenomena such as the path of the water in the fountain in the opening example can be modeled using a quadratic function. The roots of this function represent the distance the water is from its jet when it hits the ground. Another related quadratic function would be the height of the water as a function of time. Ask students what the roots of this function represent. **The time it takes for the water to hit the ground.**

11-3 Solving Quadratic Equations by Graphing

What You'll Learn

You'll learn to locate the roots of quadratic equations by graphing the related functions.

Why It's Important

Manufacturing You can use quadratic equations to solve problems with production and profit. *See Exercise 24.*

In a **quadratic equation**, the value of the related quadratic function is 0. For example, if you substitute 0 for y in the quadratic function, the result is the quadratic equation $0 = ax^2 + bx + c$. The solutions of a quadratic equation are called the **roots** of the equation. The roots of a quadratic equation can be found by finding the x-intercepts or **zeros** of the related quadratic function.

Examples

Landmark Link

1 **The Buckingham Fountain in Chicago has 133 jets through which water flows. The path of water streaming from a jet is in the shape of a parabola. Find the distance from the jet where the water hits the ground by graphing. Use the function $h(d) = -2d^2 + 4d + 6$, where $h(d)$ represents the height of a stream of water at any distance d from its jet in feet.**

Explore Graph the parabola to determine where the stream of water will hit the ground.

Plan Find the solution of the equation by looking at the values of d where $h(d)$ is 0.

Solve Make a table of values to graph the related function $h(d) = -2d^2 + 4d + 6$.

d	$-2d^2 + 4d + 6$	$h(d)$
−1	$-2(-1)^2 + 4(-1) + 6$	0
0	$-2(0)^2 + 4(0) + 6$	6
1	$-2(1)^2 + 4(1) + 6$	8
2	$-2(2)^2 + 4(2) + 6$	6
3	$-2(3)^2 + 4(3) + 6$	0

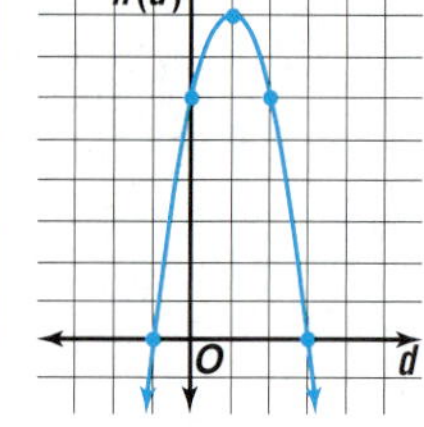

Buckingham Fountain, Chicago

The roots of $0 = -2d^2 + 4d + 6$ are −1 and 3.

Examine Since d represents distance, it cannot be negative. Therefore, $d = -1$ is not a solution, and the only reasonable solution is 3. The water will hit the ground 3 feet from the jet.

Resource Manager

Reproducible Masters

Chapter 11 Resource Masters
- *Study Guide*, p. 483
- *Skills Practice*, p. 484
- *Practice*, p. 485
- *Reading to Learn Mathematics*, p. 486
- *Enrichment*, p. 487
- *Assessment*, p. 519

Graphing Calculator, p. 29
Hands-On Algebra, pp. 119–120
School-to-Workplace, p. 11

Transparencies
5-Minute Check, 11–3

Technology/Multimedia
Interactive Chalkboard CD-ROM

2 **Find the roots of $x^2 - 10x + 16 = 0$ by graphing the related function.**

Press [2nd] [TABLE] 2 to find the roots of an equation graphed on the calculator screen.

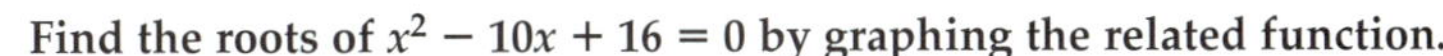

Graph the related function $f(x) = x^2 - 10x + 16$. Before making a table of values, find the equation of the axis of symmetry. This will make selecting x-values for your table easier.

$x = -\frac{b}{2a}$ *Equation of the axis of symmetry*

$x = -\frac{-10}{2(1)}$ or 5 *$a = 1$ and $b = -10$*

The equation of the axis of symmetry is $x = 5$. Now, make a table using x-values around 5. Graph each point on a coordinate plane.

x	$x^2 - 10x + 16$	$f(x)$
2	$2^2 - 10(2) + 16$	0
4	$4^2 - 10(4) + 16$	−8
5	$5^2 - 10(5) + 16$	−9
6	$6^2 - 10(6) + 16$	−8
8	$8^2 - 10(8) + 16$	0

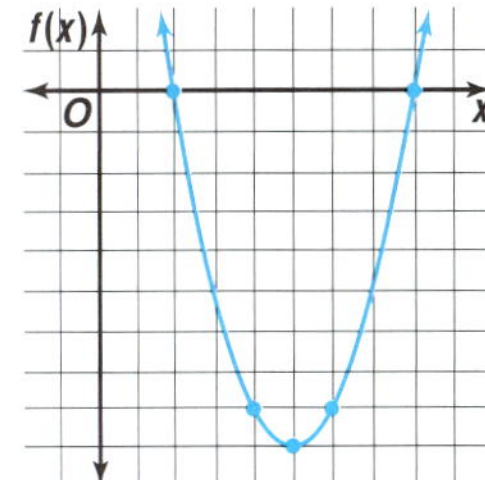

The zeros of the function appear to be 2 and 8. So, the roots are 2 and 8.

Check: Substitute 2 and 8 for x in the equation $x^2 - 10x + 16 = 0$.

$$x^2 - 10x + 16 = 0 \qquad x^2 - 10x + 16 = 0$$
$$2^2 - 10(2) + 16 \stackrel{?}{=} 0 \qquad 8^2 - 10(8) + 16 \stackrel{?}{=} 0$$
$$4 - 20 + 16 \stackrel{?}{=} 0 \qquad 64 - 80 + 16 \stackrel{?}{=} 0$$
$$0 = 0 \ \checkmark \qquad 0 = 0 \ \checkmark$$

Your Turn **a. 4, 1**

a. Find the roots of $0 = x^2 - 5x + 4$ by graphing the related function.

Sometimes exact roots cannot be found by graphing. In this case, estimate solutions by stating the consecutive integers between which the roots are located.

Example **3** **Estimate the roots of $-x^2 + 2x + 1 = 0$.**

Find the equation of the axis of symmetry.

$x = -\frac{b}{2a}$ *Equation of the axis of symmetry*

$x = -\frac{2}{2(-1)}$ or 1 *$a = -1$ and $b = 2$*

The equation of the axis of symmetry is $x = 1$. Now, make a table using x-values around 1. Graph each point on a coordinate plane.

(continued on the next page)

2 TEACH

Teaching Tip Point out to students that although the entire parabola is graphed in Example 1, only the portion of the parabola for which d is between 0 and 3 represents the path of the water.

In-Class Examples

Example 1

Suppose the function $h(t) = -16t^2 + 29t + 6$ represents the height of the water at any time t seconds after it has left its jet. Find the number of seconds it takes for the water to hit the ground by graphing. **2 s**

Example 2

Find the roots of $x^2 + 2x - 15 = 0$ by graphing the related function. **−5 and 3**

Example 3

Estimate the roots of $-x^2 + 4x - 1 = 0$. **One root is between 0 and 1 and the other root is between 3 and 4.**

Teaching Tip In Example 4, have students guess and check pairs of numbers with the sum and product to confirm that there is no solution.

In-Class Example

Example 4

Find two numbers whose sum is 10 and whose product is −24. **−2 and 12**

x	$-x^2 + 2x + 1 = 0$	$f(x)$
−1	$-(-1)^2 + 2(-1) + 1$	−2
0	$-0^2 + 2(0) + 1$	1
1	$-(1)^2 + 2(1) + 1$	2
2	$-(2)^2 + 2(2) + 1$	1
3	$-(3)^2 + 2(3) + 1$	−2

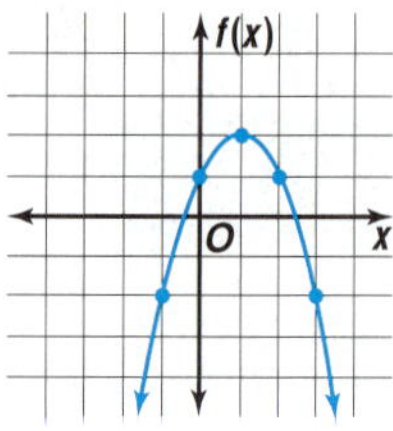

The x-intercepts of the graph are between −1 and 0 and between 2 and 3. So, one root of the equation is between −1 and 0, and the other root is between 2 and 3.

Your Turn **b. between −3 and −2; between 4 and 5**

b. Estimate the roots of $y = x^2 - 2x - 9$.

Example 4

Number Theory Link

Find two numbers whose sum is 4 and whose product is 5.

Explore Let x = one of the numbers.
Then $4 - x$ = the other number.

Plan Since the product of the two numbers is 5, you know that $5 = x(4 - x)$.

$5 = x(4 - x)$

$5 = 4x - x^2$ *Distributive Property*

$0 = -x^2 + 4x - 5$ *Subtract 5 from each side.*

Solve You can solve $0 = -x^2 + 4x - 5$ by graphing the related function $f(x) = -x^2 + 4x - 5$. The equation of the axis of symmetry is $x = -\frac{4}{2(-1)}$ or 2. So, choose x-values around 2 for your table.

x	$-x^2 + 4x - 5$	$f(x)$
0	$-0^2 + 4(0) - 5$	−5
1	$-(1)^2 + 4(1) - 5$	−2
2	$-(2)^2 + 4(2) - 5$	−1
3	$-(3)^2 + 4(3) - 5$	−2
4	$-(4)^2 + 4(4) - 5$	−5

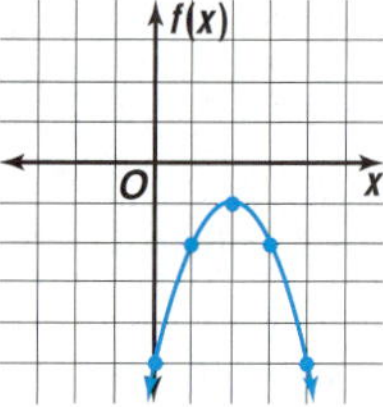

The graph has no x-intercepts since it does not cross the x-axis. This means the equation $x^2 - 4x + 5 = 0$ has no real roots. Thus, it is *not* possible for two numbers to have a sum of 4 and a product of 5. *Examine this solution by testing several pairs of numbers.*

The graphing calculator is a useful tool when making tables in order to solve functions.

Graphing Calculator Exploration

Refer to the Graphing Calculator Exploration on page 471. You may need to explain to students how the TABLE SETUP screen works. To access this screen, press [2nd] [TBLSET]. Move the cursor to the fourth line on the screen and highlight Ask. Then press [ENTER]. On the fifth line, highlight Auto and press [ENTER] again.

Graphing Calculator Exploration

Graphing Calculator Tutorial
See pp. 750–753.

Solve $x^2 + x - 2 = 0$ by making a table.

Enter the equation in the [Y=] screen. Then press [2nd] [TABLE]. A table with two columns labeled X and Y1 will appear on your screen.

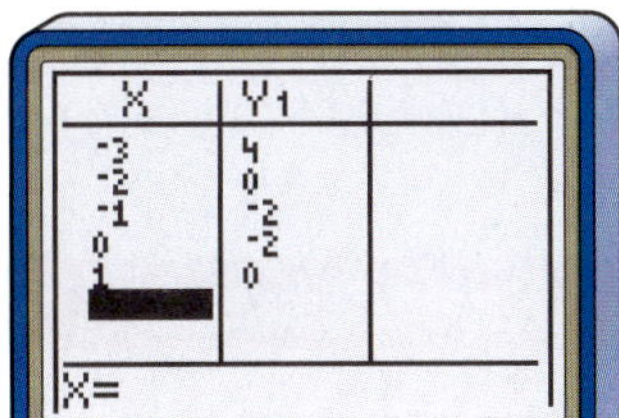

Begin entering values for x. For each x-value entered, a corresponding y-value is calculated using the equation $y = x^2 + x - 2$.

Since $y = 0$ when $x = -2$ and $x = 1$, the roots of the equation are -2 and 1.

Try These

Make a table to find the roots of each equation.

1. $x^2 - 18x + 81 = 0$ **9**
2. $x^2 - 2x - 15 = 0$ **5, −3**

Check for Understanding

Communicating Mathematics

1. **Explain** why finding the x-intercepts of the graph of a quadratic function can be used to solve the related quadratic equation.
2. **Write** the related function for $4 = x^2 - 3x$.
3. **Sketch** a parabola that has one root.

1–3. See margin.

Vocabulary
quadratic equation
roots
zeros

Guided Practice

Getting Ready **State the roots of each quadratic equation.**

Sample:

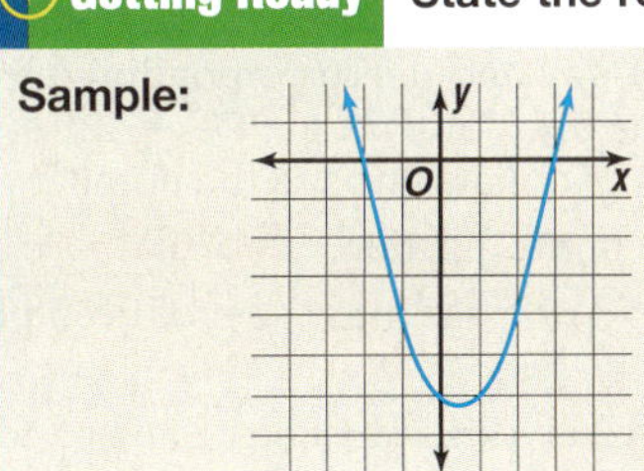

Solution: The x-intercepts of the graph are -2 and 3. So, the roots are -2 and 3.

4.
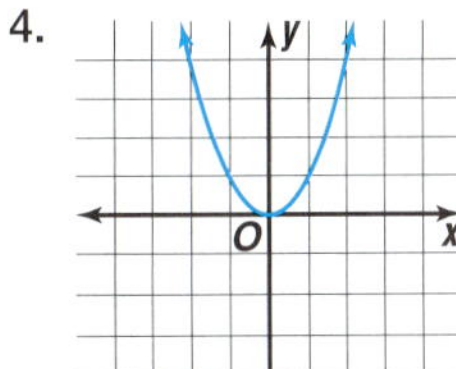

5.
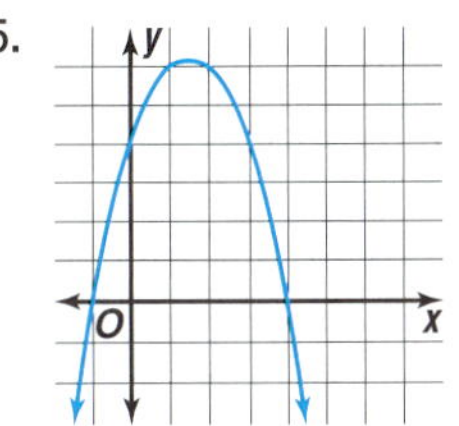

6.
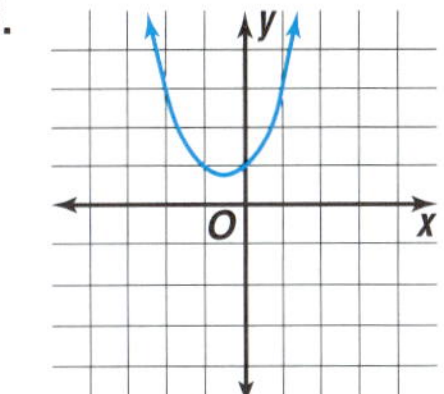

4. 0
5. −1, 4
6. no real roots

3 PRACTICE/APPLY

Error Analysis

Watch for students who do not find the correct related function in Exercise 2.
Prevent by reviewing Exercise 2 in class. Stress that you must rewrite the equation so that is has a zero on one side prior to writing the related function.

Answers

1. Sample answer: The x-intercepts are the values for x where $f(x)$ equals 0. Before solving a quadratic function, you always set it equal to zero.
2. $f(x) = x^2 - 3x - 4$
3. Sample answer:

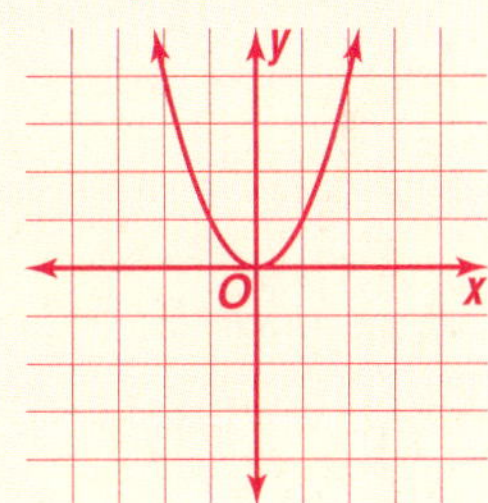

Reteaching Activity

Logical Learners Have students work with a partner. Each student should write out a series of steps they would use to solve a quadratic equation by graphing. Give each pair two quadratic equations to solve. The students should take turns using their partner's steps to solve one of the equations while the partner checks the work.

***Study Guide*, p. 483**

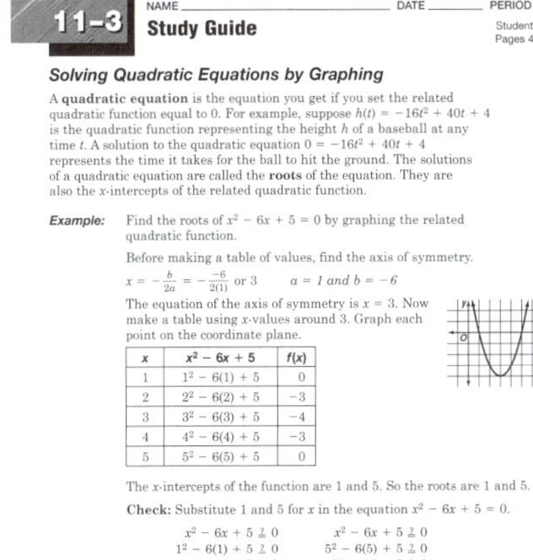

11-3 NAME ______ DATE ______ PERIOD ______

Study Guide

Student Edition
Pages 468–473

Solving Quadratic Equations by Graphing

A **quadratic equation** is the equation you get if you set the related quadratic function equal to 0. For example, suppose $h(t) = -16t^2 + 40t + 4$ is the quadratic function representing the height h of a baseball at any time t. A solution to the quadratic equation $0 = -16t^2 + 40t + 4$ represents the time it takes for the ball to hit the ground. The solutions of a quadratic equation are called the **roots** of the equation. They are also the x-intercepts of the related quadratic function.

Example: Find the roots of $x^2 - 6x + 5 = 0$ by graphing the related quadratic function.

Before making a table of values, find the axis of symmetry.

$x = -\frac{b}{2a} = -\frac{-6}{2(1)}$ or 3   *a = 1 and b = −6*

The equation of the axis of symmetry is $x = 3$. Now make a table using x-values around 3. Graph each point on the coordinate plane.

x	$x^2 - 6x + 5$	$f(x)$
1	$1^2 - 6(1) + 5$	0
2	$2^2 - 6(2) + 5$	−3
3	$3^2 - 6(3) + 5$	−4
4	$4^2 - 6(4) + 5$	−3
5	$5^2 - 6(5) + 5$	0

The x-intercepts of the function are 1 and 5. So the roots are 1 and 5.

Check: Substitute 1 and 5 for x in the equation $x^2 - 6x + 5 = 0$.

$x^2 - 6x + 5 \stackrel{?}{=} 0$ $x^2 - 6x + 5 \stackrel{?}{=} 0$
$1^2 - 6(1) + 5 \stackrel{?}{=} 0$ $5^2 - 6(5) + 5 \stackrel{?}{=} 0$
$1 - 6 + 5 \stackrel{?}{=} 0$ $25 - 30 + 5 \stackrel{?}{=} 0$
$0 = 0$ ✓ $0 = 0$ ✓

Solve each equation by graphing the related function. **1–9. See students' graphs.**

1. $x^2 - 4x + 3 = 0$ **1, 3**
2. $x^2 - 2x + 1 = 0$ **1**
3. $x^2 + 7x + 6 = 0$ **−1, −6**
4. $x^2 + 5x - 14 = 0$ **−7, 2**
5. $x^2 + 10x + 25 = 0$ **−5**
6. $x^2 - 8x - 9 = 0$ **−1, 9**
7. $x^2 + 6x = 0$ **−6, 0**
8. $x^2 + x + 1 = 0$ **no real solutions**
9. $x^2 + 3x + 2 = 0$ **−1, −2**

Assignment Guide

Basic: 11–27 odd, 28–32
Average: 12–22 even, 24–32
All: Quiz 1, 1–10

Answers

22.

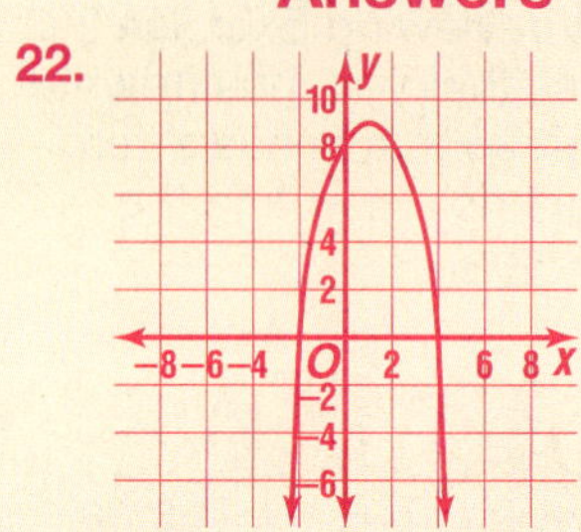

23.

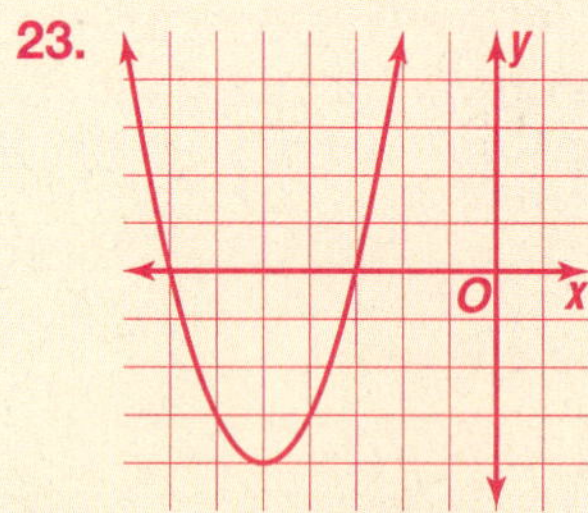

24a.

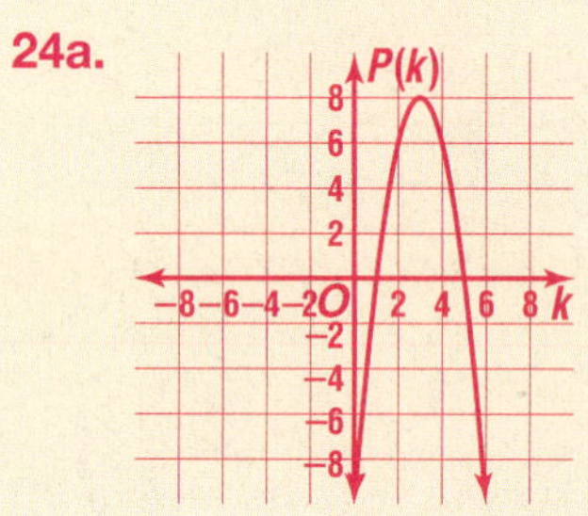

Skills Practice, p. 484, and *Practice,* p. 485 (shown)

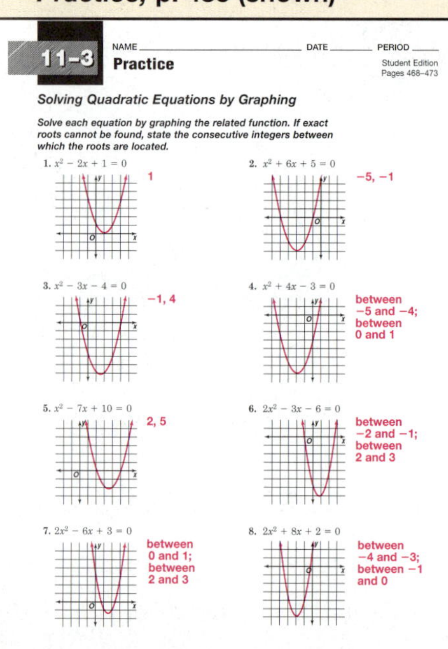

11-3 NAME ________ DATE ________ PERIOD ________

Practice

Student Edition Pages 468–473

Solving Quadratic Equations by Graphing

Solve each equation by graphing the related function. If exact roots cannot be found, state the consecutive integers between which the roots are located.

1. $x^2 - 2x + 1 = 0$ **1**
2. $x^2 + 6x + 5 = 0$ **−5, −1**
3. $x^2 - 3x - 4 = 0$ **−1, 4**
4. $x^2 + 4x - 3 = 0$ **between −5 and −4; between 0 and 1**
5. $x^2 - 7x + 10 = 0$ **2, 5**
6. $2x^2 - 3x - 6 = 0$ **between −2 and −1; between 2 and 3**
7. $2x^2 - 6x + 3 = 0$ **between 0 and 1; between 2 and 3**
8. $2x^2 + 8x + 2 = 0$ **between −4 and −3; between −1 and 0**

Examples 1–3

8. −5, 3

9. between −2 and −1; between 2 and 3

Solve each equation by graphing the related function. If exact roots cannot be found, state the consecutive integers between which the roots are located.

7. $x^2 - 5x + 6 = 0$ **2, 3**
8. $x^2 + 2x - 15 = 0$
9. $2x^2 - 3x - 7 = 0$

Example 4

10. **Number Theory** Use a quadratic equation to find two numbers whose sum is 4 and whose product is −12. **−2, 6**

Exercises

Practice

For Exercises	See Examples
11–19	1, 2
20, 21, 24, 25	3, 4
27	2, 3

Homework Help

Extra Practice

See page 714.

Solve each equation by graphing the related function. If exact roots cannot be found, state the consecutive integers between which the roots are located. **13. −2, 7 14. −1, 7**

11. $x^2 + 2x + 1 = 0$ **−1**
12. $x^2 - 4x + 3 = 0$ **1, 3**
13. $x^2 - 5x - 14 = 0$
14. $-x^2 + 6x + 7 = 0$
15. $x^2 - 10x + 25 = 0$ **5**
16. $x^2 + 3x - 2 = 0$
17. $x^2 + 4x - 2 = 0$
18. $x^2 - 2x + 2 = 0$ **no real solutions**
19. $-2x^2 + 3x + 4 = 0$

16. between −4 and −3; between 0 and 1

17. between −5 and −4; between 0 and 1

19. between −1 and 0; between 2 and 3

Use a quadratic equation to determine the two numbers that satisfy each situation.

20. Their sum is 18 and their product is 81. **9, 9**
21. Their difference is 4 and their product is 32. **−4, −8 or 4, 8**

Use the given roots and vertex of a quadratic equation to graph the related quadratic function. **22–23. See margin.**

22. roots: −2, 4
vertex: (1, 9)

23. roots: −7, −3
vertex: (−5, −4)

Applications and Problem Solving

24. **Business** Mr. Jamison owns a manufacturing company that produces key rings. Last year, he collected data about the number of key rings produced per day and the corresponding profit. He then modeled the data using the function $P(k) = -2k^2 + 12k - 10$, where P is the profit in thousands of dollars and k is the number of key rings in thousands.
   a. Graph the profit function. **See margin.**
   b. How many key rings must be produced per day so that there is no profit and no loss? **1000 or 5000** **c. 3000**
   c. How many key rings must be produced for the maximum profit?
   d. What is the maximum profit? **$8000**

25. **Skydiving** In a recent year, Adrian Nicholas broke two world records by flying 10 miles in 4 minutes 55 seconds without a plane. He jumped from an airplane at 35,000 feet and did not activate his parachute until he was 500 feet above the ground. If air resistance is ignored, how long did Nicholas free-fall? Use the formula $h(t) = -16t^2 + h_0$, where t is the time in seconds and h_0 is the initial height in feet. **between 46 s and 47 s**

Answer

Page 473

26c.

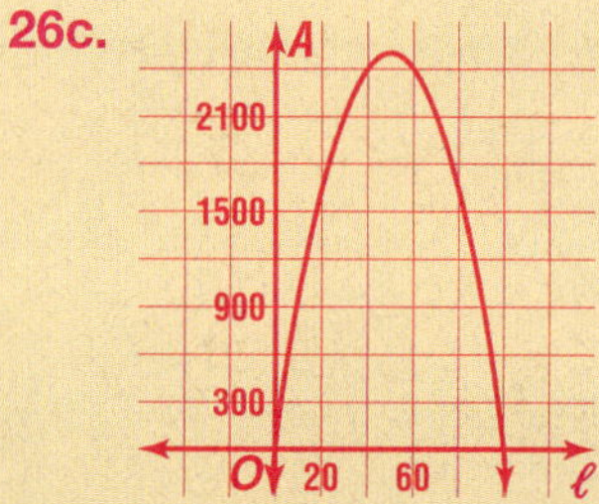

No, if the length is 100 feet, there is no fencing for the width and the area is 0.

26. **Geometry** Mrs. Parker wants to enclose a rectangular running yard at her dog kennel with 200 feet of fencing.

a. Make a drawing of the enclosed yard. Then write a formula for the perimeter and solve for the width. **$P = 2\ell + 2w$, $w = 100 - \ell$**

b. Write a quadratic equation for the area A of the yard. Use the expression for the width from part a. **$A = -\ell^2 + 100\ell$**

c. Show graphically how the area of the yard changes when the length changes. Can the yard have a length of 100 feet? Explain. **26c. See margin.**

27. **Critical Thinking** Suppose the value of a quadratic function is negative when $x = 1$ and positive when $x = 2$. Explain why it is reasonable to assume that the related equation has a root between 1 and 2. **See margin.**

Mixed Review

Describe how each graph changes from its parent graph of $y = x^2$. Then name the vertex. (Lesson 11–2)

28. $y = -x^2$ **(0, 0) opens down**
29. $y = x^2 + 4$ **(0, 4) up 4 units**
30. $y = (x + 6)^2$ **(−6, 0) left 6 units**

Standardized Test Practice

31. **Grid In** If the equation of the axis of symmetry of a quadratic function is $x = 0$, and one of the x-intercepts is -4, what is the other x-intercept? (Lesson 11–1) **4**

32. **Multiple Choice** Evaluate $4^x + (x^2)^3$ if $x = 2$. (Lesson 8–1)

Ⓐ 6 Ⓑ 20 Ⓒ 32 Ⓓ 80 **D**

Quiz 1 Lessons 11–1 through 11–3

Write the equation of the axis of symmetry and the coordinates of the vertex for each quadratic function. Then graph the function. (Lesson 11–1) **1–3. See Solutions Manual for graphs.**

1. $y = x^2 + 6x - 5$ **$x = -3$; $(-3, -14)$**
2. $y = -2x^2 - 9$ **$x = 0$; $(0, -9)$**
3. $y = -3x^2 - 6x + 4$ **$x = -1$; $(-1, 7)$**

Describe how each graph changes from its parent graph of $y = x^2$. Then name the vertex of each graph. (Lesson 11–2)

4. $y = x^2 + 7$ **up 7 units; (0, 7)**
5. $y = (x - 6)^2$ **right 6 units; (6, 0)**
6. $y = -0.8x^2$ **opens down, widens; (0, 0)**

Solve each equation by graphing the related function. If exact roots cannot be found, state the consecutive integers between which the roots are located. (Lesson 11–3) **9. between −1 and 0; between 2 and 3**

7. $x^2 - 5x + 6 = 0$ **2, 3**
8. $x^2 + 6x + 10 = 0$ **no real solutions**
9. $x^2 - 2x - 1 = 0$

10. **Sports** Anna is doing a back dive from a 10-meter platform. Her path of descent is given by the graph of $h(d) = -d^2 + 2d + 10$, where the height h and distance from the platform d are both in meters. (Lessons 11–1 & 11–3)

a. Graph the function. **See Solutions Manual.**

b. At her maximum height, how far away from the platform is she? **1 m**

c. About how far away from the platform will she enter the water? **between 4 m and 5 m**

www.algconcepts.com/self_check_quiz

Extra Credit

Use the table feature of a graphing calculator to solve $x^2 = 3x + 5$ to the nearest tenth. Explain how you solved this problem. **−1.2 and 4.2; Sample answer: I found two integers between which a root is located. Then I changed the increment of the table to 0.1 and looked for the y-value that was closest to zero.**

4 ASSESS

Open-Ended Assessment

Modeling Have students sketch the graphs of related functions for quadratic equations with the following characteristics: two solutions, one solution, no solutions, solution −1, solutions −3 and 2, solutions 0 and 5.

Quiz 1

The Quiz provides students with a brief review of the concepts and skills in Lessons 11–1 through 11–3. Lesson numbers are given to the right of the exercises or instruction lines so that students can review concepts not yet mastered.

Chapter 11, Quiz A (Lessons 11–1 through 11–3) is available in the *Chapter 11 Resource Masters*, p. 519.

Answer

27. The value of the function changes from negative when $x = 1$ to positive when $x = 2$. To do this, the value of the function would have to be 0 for some value of x between 1 and 2. Thus, the value of x represents the x-intercept of the function. This is the root of the related equation.

***Enrichment*, p. 487**

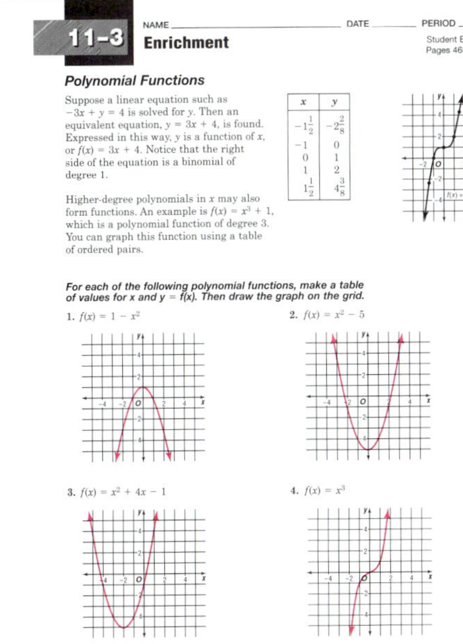

NAME ______ DATE ______ PERIOD ______

11-3 Enrichment

Student Edition Pages 468–473

Polynomial Functions

Suppose a linear equation such as $-3x + y = 4$ is solved for y. Then an equivalent equation, $y = 3x + 4$, is found. Expressed in this way, y is a function of x, or $f(x) = 3x + 4$. Notice that the right side of the equation is a binomial of degree 1.

Higher-degree polynomials in x may also form functions. An example is $f(x) = x^3 + 1$, which is a polynomial function of degree 3. You can graph this function using a table of ordered pairs.

x	y
$-1\frac{1}{2}$	$-2\frac{2}{8}$
-1	0
0	1
1	2
$1\frac{1}{2}$	$4\frac{3}{8}$

For each of the following polynomial functions, make a table of values for x and y = f(x). Then draw the graph on the grid.

1. $f(x) = 1 - x^2$
2. $f(x) = x^2 - 5$
3. $f(x) = x^2 + 4x - 1$
4. $f(x) = x^3$

1 FOCUS

5-Minute Check
Lesson 11–3

State the roots of each quadratic equation.

1. 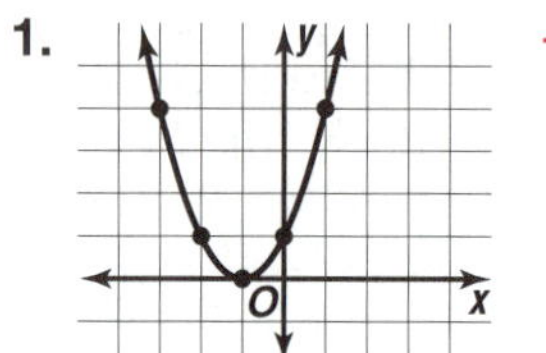

−1

2. 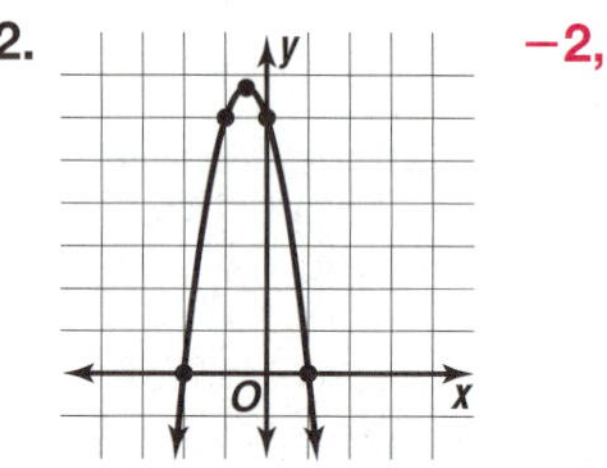

−2, 1

Solve each equation by graphing the related function.

3. $-x^2 + 7x - 10 = 0$ **2 and 5**
4. $x^2 + 3x + 4 = 0$ **no solution**
5. Suppose you drop an object off an 840-foot building. Ignoring air resistance, the height of the object h in feet is given by $h = -16t^2 + 840$, where t is the time in seconds after the object was dropped. Estimate how long it will take for the object to hit the ground. **between 7 and 8 s**

Motivating the Lesson

Hands-On Activity Write the following equations on the board or overhead: $2x = 0$, $5x = 0$, $-4x = 0$. Have students solve these equations. Then ask them: if $ab = 0$, what must be true about the values of a and b? This activity will lead into a discussion of the Zero Product Property.

11–4 Solving Quadratic Equations by Factoring

What You'll Learn

You'll learn to solve quadratic equations by factoring and by using the Zero Product Property.

Why It's Important

Photography You can use quadratic equations to solve problems involving photography. *See Exercise 34.*

In the previous lesson you learned to solve quadratic equations by graphing. Quadratic equations can also be solved by factoring. For example, the equation $8x^2 - 4x = 0$ can be solved by finding factors.

$$8x^2 - 4x = 0$$

$$4x(2x - 1) = 0 \qquad \textit{Factor } 8x^2 - 4x.$$

To solve this equation, find values of x that make the product $4x(2x - 1)$ equal to 0. Since the product of 0 and any number is 0, *at least one* of the factors in the expression must be zero.

$$4x = 0 \quad \text{or} \quad 2x - 1 = 0$$

$$x = 0 \qquad\qquad x = \frac{1}{2}$$

The solutions of $8x^2 - 4x = 0$ are 0 and $\frac{1}{2}$.

This method of solving quadratic equations uses the **Zero Product Property**.

Zero Product Property	For all numbers a and b, if $ab = 0$, then $a = 0$, $b = 0$ or both a and b equal 0.

We can use this property to solve any equation that is written in the form $ab = 0$.

Example 

Solve $3x(x - 1) = 0$. Check your solution.

If $3x(x - 1) = 0$, then $3x = 0$ or $x - 1 = 0$. *Zero Product Property*

$3x = 0$ or $x - 1 = 0$ *Solve each equation.*

$x = 0$ $\qquad x = 1$

Check: Substitute 0 and 1 for x in the original equation.

$3x(x - 1) = 0$	or	$3x(x - 1) = 0$
$3(0)(0 - 1) \stackrel{?}{=} 0$		$3(1)(1 - 1) \stackrel{?}{=} 0$
$0(-1) \stackrel{?}{=} 0$		$3(0) \stackrel{?}{=} 0$
$0 = 0$ ✓		$0 = 0$ ✓

The solutions are 0 and 1.

Your Turn **Solve each equation. Check your solution.**

a. $z(z - 8) = 0$ **0, 8**

b. $(a - 4)(4a + 3) = 0$ **4, −0.75**

Resource Manager

Reproducible Masters

Chapter 11 Resource Masters
- *Study Guide*, p. 488
- *Skills Practice*, p. 489
- *Practice*, p. 490
- *Reading to Learn Mathematics*, p. 491
- *Enrichment*, p. 492
- *Assessment*, p. 518

Hands-On Algebra, pp. 121–122

Transparencies

5-Minute Check, 11–4

Technology/Multimedia

AlgePASS, Lesson 27

Interactive Chalkboard CD-ROM

Example 2 — Recreation Link

At an adventure park, you can jump off a 26-foot cliff into a pool of water below. The equation $h = -16t^2 + 4t + 26$ describes your height h in feet t seconds after your jump. If you jump up and off the cliff, what time will you pass the height of 26 feet again?

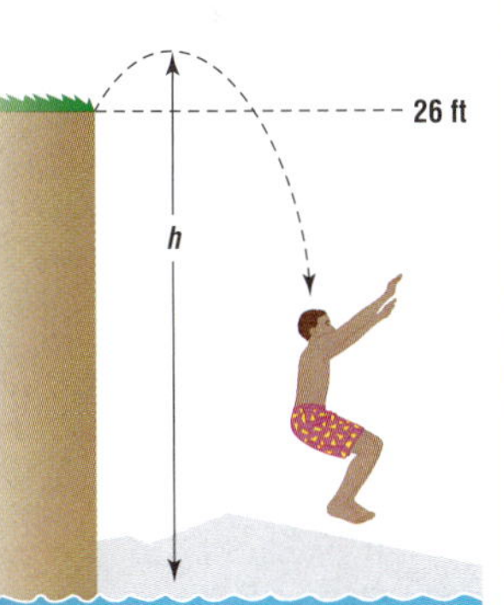

To find the time at which this occurs, let $h = 26$ and solve for t.

$26 = -16t^2 + 4t + 26$

$0 = -16t^2 + 4t$ *Subtract 26 from each side.*

$0 = 4t(-4t + 1)$ *Factor $-16t^2 + 4t$.*

If $4t(-4t + 1) = 0$, then $4t = 0$ or $-4t + 1 = 0$. *Zero Product Property*

$4t = 0$ or $-4t + 1 = 0$ *Solve each equation.*

$t = 0$ $\quad -4t = -1$

$t = \frac{1}{4}$

Check: Graph the equation and find the x intercepts. We will graph the equation $h = 4t(-4t + 1)$ on a graphing calculator.

Enter the equation in the [Y=] screen. Press [GRAPH]. To find the zeros, press [2nd] [CALC] [▼] 2. Use the arrow keys to move the cursor to the left of one of the x-intercepts. Press [ENTER]. Move the cursor to the right of the same x-intercept. Press [ENTER] twice. Coordinates of the root will appear. Repeat for the other root.

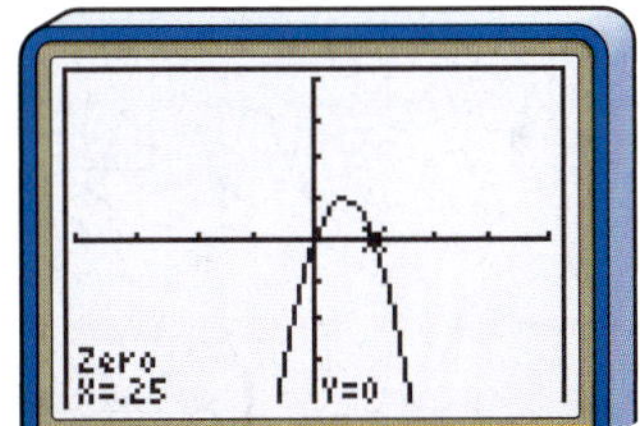

Set viewing window for x: [−1, 1] by 0.25 and y: [−1, 1] by 0.25.

The solutions are 0 and $\frac{1}{4}$. The solution 0 represents the beginning of the jump. So, you would return to your original location after a fourth of a second.

You will often need to factor an equation before using the Zero Product Property to solve the equation.

Example 

Solve $x^2 + 4x - 12 = 0$. Check your solution.

$x^2 + 4x - 12 = 0$

$(x - 2)(x + 6) = 0$ *Factor.*

$x - 2 = 0$ or $x + 6 = 0$ *Zero Product Property*

$x = 2$ $\quad x = -6$ *Check this solution.*

The solutions are −6 and 2.

Look Back

Factoring Trinomials: Lessons 10-3 & 10-4

Your Turn

c. Solve $x^2 - 2x = 3$. Check your solution. **−1, 3**

www.algconcepts.com/extra_examples

2 TEACH

Teaching Tip When discussing the application at the beginning of the lesson, students may question how the formula for the height of the diver was derived. Explain that the equation is derived from the general equation $h = -\frac{1}{2}gt^2 + v_0t + h_0$, where g is the acceleration due to the Earth's gravity and is equal to 32 feet per second squared. The other variables v_0 and h_0 represent the initial velocity and initial height of the object.

In-Class Examples

Example 1

Solve $-2x(x + 5) = 0$. Check your solution. **0, −5**

Example 2

A child throws a ball up in the air. The height h of the ball t seconds after it has been thrown is given by the equation $h = -16t^2 + 8t + 4$. Solve $4 = -16t^2 + 8t + 4$ or $8t(-2t + 1) = 0$ to find how long it would take the ball to reach the height from which it was thrown. **$\frac{1}{2}$ s**

Example 3

Solve $x^2 - 4x - 21 = 0$. Check your solution. **−3, 7**

Reteaching Activity

Interpersonal Learners Have each student work with a partner. On the board or overhead, write several quadratic equations that are easily factored. For each equation, one student should solve it using the Zero Product Property while the other one solves it by graphing on a graphing calculator. The partners should compare their work and switch roles for the next equation.

***Study Guide*, p. 488**

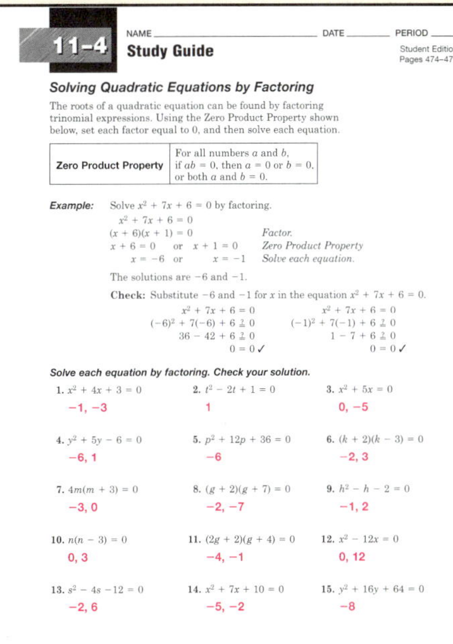

11-4 NAME ______ DATE ______ PERIOD ______

Study Guide — Student Edition Pages 474–477

Solving Quadratic Equations by Factoring

The roots of a quadratic equation can be found by factoring trinomial expressions. Using the Zero Product Property shown below, set each factor equal to 0, and then solve each equation.

Zero Product Property	For all numbers a and b, if $ab = 0$, then $a = 0$ or $b = 0$, or both a and $b = 0$.

Example: Solve $x^2 + 7x + 6 = 0$ by factoring.

$x^2 + 7x + 6 = 0$

$(x + 6)(x + 1) = 0$ *Factor.*

$x + 6 = 0$ or $x + 1 = 0$ *Zero Product Property*

$x = -6$ or $x = -1$ *Solve each equation.*

The solutions are −6 and −1.

Check: Substitute −6 and −1 for x in the equation $x^2 + 7x + 6 = 0$.

$x^2 + 7x + 6 = 0$ $\qquad x^2 + 7x + 6 = 0$

$(-6)^2 + 7(-6) + 6 \stackrel{?}{=} 0$ $\qquad (-1)^2 + 7(-1) + 6 \stackrel{?}{=} 0$

$36 - 42 + 6 \stackrel{?}{=} 0$ $\qquad 1 - 7 + 6 \stackrel{?}{=} 0$

$0 = 0$ ✓ $\qquad 0 = 0$ ✓

Solve each equation by factoring. Check your solution.

1. $x^2 + 4x + 3 = 0$ **−1, −3**
2. $t^2 - 2t + 1 = 0$ **1**
3. $x^2 + 5x = 0$ **0, −5**
4. $y^2 + 5y - 6 = 0$ **−6, 1**
5. $p^2 + 12p + 36 = 0$ **−6**
6. $(k + 2)(k - 3) = 0$ **−2, 3**
7. $4m(m + 3) = 0$ **−3, 0**
8. $(g + 2)(g + 7) = 0$ **−2, −7**
9. $h^2 - h - 2 = 0$ **−1, 2**
10. $n(n - 3) = 0$ **0, 3**
11. $(2g + 2)(g + 4) = 0$ **−4, −1**
12. $x^2 - 12x = 0$ **0, 12**
13. $s^2 - 4s - 12 = 0$ **−2, 6**
14. $x^2 + 7x + 10 = 0$ **−5, −2**
15. $y^2 + 16y + 64 = 0$ **−8**

Teaching Tip Point out to students that in applied situations such as Example 2 and Example 4, they must check to see that solutions to the quadratic equation are reasonable in the context of the application. In many cases, one of the solutions may not answer the question asked.

In-Class Example

Example 4

The length of a rectangle is 4 feet less than three times its width. The area of the rectangle is 55 square feet. Find the measures of the sides.
5 ft by 11 ft

3 PRACTICE/APPLY

Error Analysis

Watch for students who try to apply the Zero Product Property to equations such as those in Exercises 7–9 that are not factored completely or do not equal zero. ***Prevent by*** stressing that to apply the Zero Product Property, you must have a product of factors on one side and zero on the other side of the equation. Also encourage students to check all their answers.

***Skills Practice,* p. 489, and *Practice,* p. 490 (shown)**

11-4 NAME ______ DATE ______ PERIOD ______

Practice — Student Edition Pages 474–477

Solving Quadratic Equations by Factoring

Solve each equation. Check your solution.

1. $s(s + 3) = 0$ **0, −3**	2. $4a(a - 6) = 0$ **0, 6**	3. $3m(m + 5) = 0$ **0, −5**
4. $6t(t - 2) = 0$ **0, 2**	5. $(y + 4)(y - 5) = 0$ **−4, 5**	6. $(p - 2)(p + 3) = 0$ **2, −3**
7. $(x + 5)(x - 6) = 0$ **−5, 6**	8. $(3r + 2)(r - 1) = 0$ **$-\frac{2}{3}$, 1**	9. $(2n - 2)(n + 1) = 0$ **1, −1**
10. $(x - 3)(3x + 6) = 0$ **3, −2**	11. $(y + 4)(2y - 8) = 0$ **−4, 4**	12. $(4c + 3)(c - 7) = 0$ **$-\frac{3}{4}$, 7**
13. $x^2 + 3x - 10 = 0$ **−5, 2**	14. $x^2 - 6x + 8 = 0$ **2, 4**	15. $x^2 + 11x + 30 = 0$ **−6, −5**
16. $x^2 + 4x = 21$ **−7, 3**	17. $x^2 - 5x = 36$ **−4, 9**	18. $x^2 - 5x = 0$ **0, 5**
19. $2a^2 = 6a$ **0, 3**	20. $2x^2 - 10x + 8 = 0$ **1, 4**	21. $3x^2 - 7x - 6 = 0$ **$-\frac{2}{3}$, 3**
22. $5x^2 - x = 4$ **$-\frac{4}{5}$, 1**	23. $3x^2 + 13x = -4$ **$-\frac{1}{3}$, −4**	24. $4x^2 + 7x = 2$ **$\frac{1}{4}$, −2**

Example  **Geometry Link**

The length ℓ of a rectangle is 2 feet more than twice its width w. The area of the rectangle is 144 square feet. Find the measures of the sides.

Explore The formula for the area of a rectangle is $A = \ell w$ and $\ell = 2w + 2$.

Plan *Area* *equals* *length* *times* *width.*

$$144 = (2w + 2) \cdot w$$

Solve

$144 = 2w^2 + 2w$ *Distributive Property*
$0 = 2w^2 + 2w - 144$ *Subtract 144 from each side.*
$0 = 2(w^2 + w - 72)$ *Factor out the GCF, 2.*
$0 = w^2 + w - 72$ *Divide each side by 2.*
$0 = (w + 9)(w - 8)$ *Factor.*

$w + 9 = 0$ or $w - 8 = 0$ *Zero Product Property*
$w = -9$ $\quad w = 8$ *Solve each equation.*

Examine Since width cannot be negative, the width must be equal to 8 feet. Substituting 8 for w, the length of the rectangle is $2(8) + 2$ or 18 feet.

Check for Understanding

Communicating Mathematics

1–3. See margin.

1. **Define** the Zero Product Property in your own words.
2. **Explain** why you should check your answers by substituting values into the original equation rather than into a simplified version.
3.  Lashonda said that she should solve $2x^2 - 3x - 35 = 0$ by graphing. Jerome disagreed. He said that she should solve the equation by factoring. Who is correct, and why?

Guided Practice

Solve each equation. Check your solution.

Examples 1 & 3

4. $(x + 1)(x - 7) = 0$ **4. −1, 7**
5. $3b(b - 5) = 0$ **0, 5**
6. $(2m - 1)(m + 4) = 0$ **6. $\frac{1}{2}$, −4**
7. $x^2 - 6x + 9 = 0$ **3**
8. $x^2 = x$ **0, 1**
9. $7x^2 - 14x = 56$ **9. −2, 4**

Example 2

10. **Geometry** The height of a triangle measures 5 centimeters more than its base. The area of the triangle is 18 square centimeters. Find the measures of the base and the height of the triangle.
$h = 9$ cm, $b = 4$ cm

Exercises

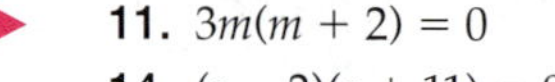

Practice

Solve each equation. Check your solution. 11–28. See margin.

A

11. $3m(m + 2) = 0$
12. $5z(z + 8) = 0$
13. $(p + 7)(p - 6) = 0$
14. $(s - 2)(s + 11) = 0$
15. $(r + 1)(2r - 8) = 0$
16. $(x - 2)(3x + 4) = 0$
17. $x^2 + 10x + 16 = 0$
18. $z^2 + 5z - 6 = 0$
19. $p^2 - 11p + 24 = 0$

B

20. $n^2 + 9n + 18 = 0$
21. $m^2 - m - 12 = 0$
22. $r^2 + 14r = 0$

Answers

1. Any number times 0 is 0. Therefore, if two or more numbers are multiplied and the result is 0, at least one of those numbers has to equal 0.
2. Sample answer: A mathematical error could have been made when simplifying, so the check would not be valid.
3. They are both correct, but factoring can only be used when the equation is not prime.

Homework Help	
For Exercises	See Examples
11–16	1, 2
17–34	3, 4
Extra Practice	
See page 714.	

23. $3w^2 - 9w = 0$
24. $15 = x^2 - 2x$
25. $2x^2 - 8x + 8 = 0$
26. $2a^2 - 70 = 4a$
27. $3b^2 - 12b - 15 = 0$
28. $2v^2 - 17v = 9$

For each problem, define a variable. Then use an equation to solve the problem. **30. 10, 12; −12, −10 31. 8, 11; −11, −8**

29. The length of Luis' house is ten feet longer than it is wide. The area in square feet is 875. Find the dimensions of the house. **25 ft by 35 ft**
30. Find two consecutive even integers whose product is 120.
31. Find two integers whose difference is 3 and whose product is 88.

Applications and Problem Solving

32. **Physics** A flare is launched from a life raft with an initial upward velocity of 192 feet per second. How many seconds will it take for the flare to return to the sea? Use the formula $h = 192t - 16t^2$, where h is the height of the flare in feet and t is the time in seconds. **12 s**

33. **Geometry** Four corners are cut from a rectangular piece of cardboard that measures 3 feet by 5 feet. The cuts are x feet from the corners as shown in the figure at the right. After the cuts are made, the sides are folded up to form an open box. The area of the bottom of the box is 3 square feet. Find the dimensions of the box. **3 ft by 1 ft by 1 ft**

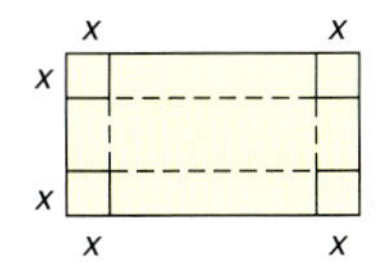

34. 6 in. by 8 in.

34. **Photography** A rectangular photograph is 4 inches wide and 6 inches long. The photograph is enlarged by increasing the length and width by an equal amount. If the area of the new photograph is twice as large as the original area, what are the dimensions of the new photograph?

35. **Critical Thinking** The graph of a quadratic function is shown.
    a. What are the zeros of the function? **−5, 2**
    b. Write an equation for the graph.
    Sample answer: $y = (x + 5)(x - 2)$

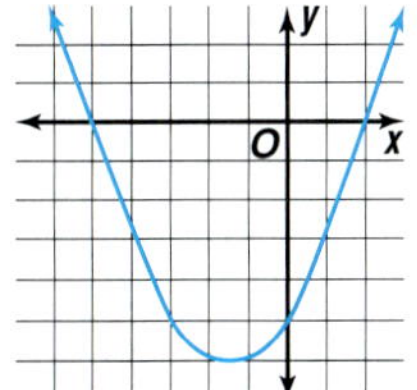

Mixed Review

Solve each equation by graphing the related function. *(Lesson 11–3)*

36. $x^2 - 10x + 21 = 0$ **3, 7**
37. $x^2 - 2x + 5 = 0$ **no real solutions**
38. $x^2 + 4x = 12$ **−6, 2**
39. Graph $y = 0.25x^2$, $y = x^2$, and $y = 4x^2$ on the same set of axes. Compare and contrast the graphs. *(Lesson 11–2)*
    See Solutions Manual.

Factor. *(Lesson 10–3)*

40. $y^2 + 7y + 6$ **$(y + 6)(y + 1)$**
41. $b^2 + b - 42$ **$(b + 7)(b - 6)$**

Standardized Test Practice

42. **Multiple Choice** Which of the following expressions is equivalent to $2^3 \times 4$? *(Lesson 8–2)* **E**

Ⓐ -8^4 Ⓑ 8^3 Ⓒ 6^4 Ⓓ 2^6 Ⓔ 2^5

www.algconcepts.com/self_check_quiz

Assignment Guide

Basic: 11–35 odd, 36–42
Average: 12–30 even, 32–42

4 ASSESS

Open-Ended Assessment

Speaking Write some factored quadratic equations on the board or overhead and ask students to solve them orally. Then write some other quadratic equations and have students explain what they would have to do to them prior to using the Zero Product Property.

Mid-Chapter Test (Lessons 11–1 through 11–4) is available in the *Chapter 11 Resource Masters*, p. 518.

Answers

Pages 476–477

11. 0, −2	12. 0, −8
13. −7, 6	14. 2, −11
15. −1, 4	16. 2, $-\frac{4}{3}$
17. −8, −2	18. −6, 1
19. 3, 8	20. −3, −6
21. −3, 4	22. 0, −14
23. 0, 3	24. −3, 5
25. 2	26. −5, 7
27. −1, 5	28. $-\frac{1}{2}$, 9

***Enrichment*, p. 492**

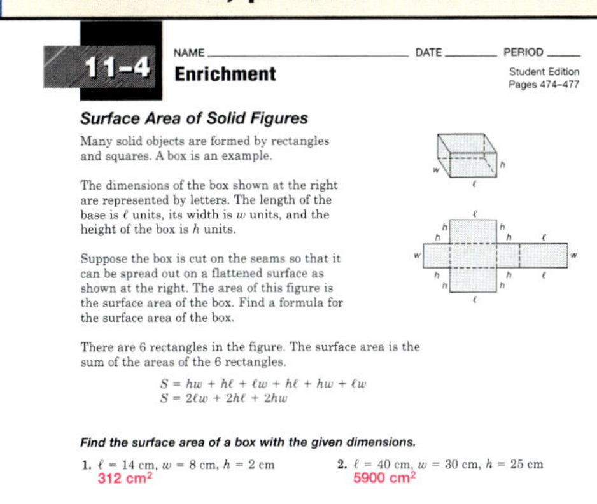

NAME ________ DATE ______ PERIOD _____

11-4 Enrichment

Student Edition Pages 474–477

Surface Area of Solid Figures

Many solid objects are formed by rectangles and squares. A box is an example.

The dimensions of the box shown at the right are represented by letters. The length of the base is ℓ units, its width is w units, and the height of the box is h units.

Suppose the box is cut on the seams so that it can be spread out on a flattened surface as shown at the right. The area of this figure is the surface area of the box. Find a formula for the surface area of the box.

There are 6 rectangles in the figure. The surface area is the sum of the areas of the 6 rectangles.

$$S = hw + h\ell + \ell w + h\ell + hw + \ell w$$
$$S = 2\ell w + 2h\ell + 2hw$$

Find the surface area of a box with the given dimensions.

1. $\ell = 14$ cm, $w = 8$ cm, $h = 2$ cm **312 cm²**
2. $\ell = 40$ cm, $w = 30$ cm, $h = 25$ cm **5900 cm²**
3. $\ell = x$ cm, $w = (x - 3)$ cm, $h = (x + 3)$ cm **$6(x^2 - 3)$ cm²**
4. $\ell = (s + 9)$ cm, $w = (s - 9)$ cm, $h = (s + 9)$ cm **$(6s^2 + 36s - 162)$ cm²**
5. The surface area of a box is 142 cm². The length of the base is 2 cm longer than its width. The height of the box is 2 cm less than the width of the base. Find the dimensions of the box. **$w = 5$ cm; $\ell = w + 2$ or 7 cm; $h = w - 2$ or 3 cm**
6. Write an expression that represents the surface area of the figure shown at the right. Include the surface area of the base. **$22x^2$**

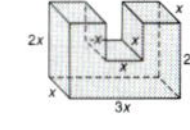

Extra Credit

Make up a problem similar to Example 4 that involves the area of a triangle. Show how to solve your problem. (*Hint:* You may want to start with the answer and work backward to write the problem.)
See students' work.

Lesson 11-5

1 FOCUS

5-Minute Check
Lesson 11-4

Solve each equation. Check your solution.

1. $0 = -4x(x + 6)$ **−6, 0**
2. $(y + 1)(2y - 3) = 0$ **−1, 1.5**
3. $x^2 - 11x + 28 = 0$ **4, 7**
4. The area of the rectangle below is 48 square centimeters. What are the dimensions of the rectangle? **12 cm by 4 cm**

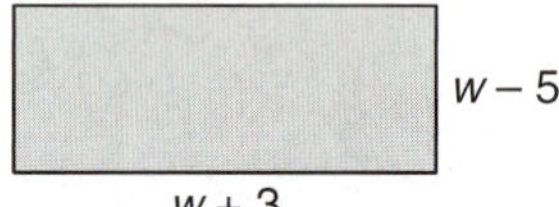

5. At Joyce's block party, some of the neighbors use a large rubber band to shoot water balloons into the air. The height h in feet of a balloon t seconds after launch is given by $h = -16t^2 + 30t + 4$. How many seconds after launch will the water balloon hit the ground? **2 s**

Motivating the Lesson

Hands-On Activity Give each student a set of algebra tiles. Have them represent $x^2 + 2x + 1$. Show them how the tiles can be rearranged to form a square and discuss how this shows that $x^2 + 2x + 1 = (x + 1)^2$. Give them several other perfect square trinomials to practice modeling such as $x^2 + 4x + 4$ and $x^2 + 6x + 9$.

11-5 Solving Quadratic Equations by Completing the Square

What You'll Learn
You'll learn to solve quadratic equations by completing the square.

Why It's Important
Construction You can use quadratic equations to determine the dimensions of a room. *See Exercise 13.*

Carmen planted daylilies on a square piece of land with an area of 49 square feet. She wants to plant petunias around the daylilies to form a border whose area is 72 square feet. What length should she make the sides of the outer square?

Let x represent the length of a side of the outer square. Find a relationship between the variable and given numerical information to write and solve an equation.

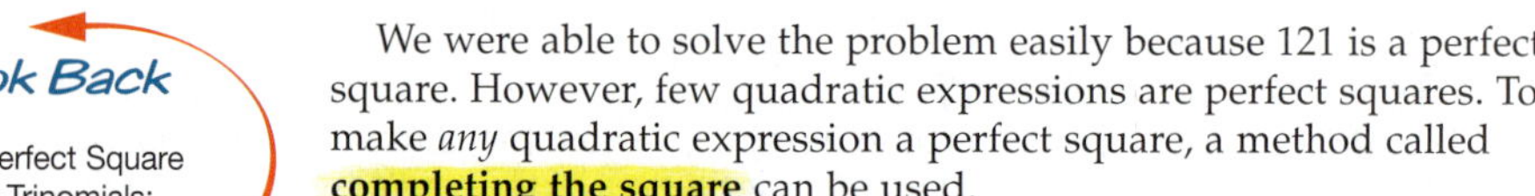

$$x^2 - 49 = 72$$

$$x^2 = 121 \quad \text{Add 49 to each side.}$$

$$\sqrt{x^2} = \pm\sqrt{121} \quad \text{Find the square root of each side.}$$

$$x = \pm 11$$

Length cannot be negative, so the sides of the outer square measure 11 feet.

Look Back
Perfect Square Trinomials: Lesson 10–5

We were able to solve the problem easily because 121 is a perfect square. However, few quadratic expressions are perfect squares. To make *any* quadratic expression a perfect square, a method called **completing the square** can be used.

When given an equation, you can complete the square by writing one side of the equation as a perfect square. This is modeled below.

Materials:  algebra tiles  equation mat

Use algebra tiles to complete the square for $x^2 + 2x - 4 = 0$.

Step 1 Model $x^2 + 2x - 4 = 0$ on the mat.

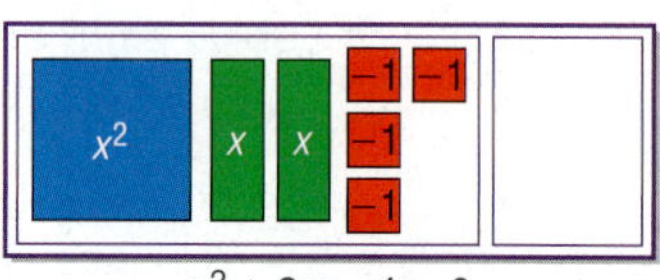

$x^2 + 2x - 4 = 0$

Step 2 Add 4 to each side of the mat. Remove the zero pairs.

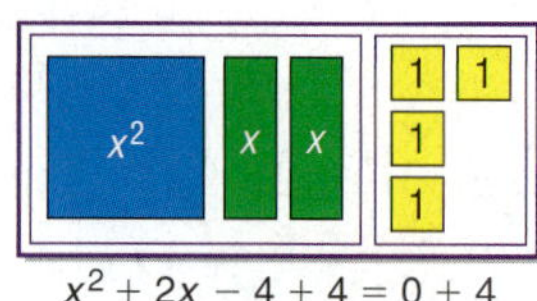

$x^2 + 2x - 4 + 4 = 0 + 4$

Resource Manager

Reproducible Masters
Chapter 11 Resource Masters
- *Study Guide,* p. 493
- *Skills Practice,* p. 494
- *Practice,* p. 495
- *Reading to Learn Mathematics,* p. 496
- *Enrichment,* p. 497

Hands-On Algebra, pp. 123–124

Transparencies
5-Minute Check, 11–5

Technology/Multimedia
Interactive Chalkboard CD-ROM

Step 3 Begin to arrange the x^2 and x-tiles into a square.

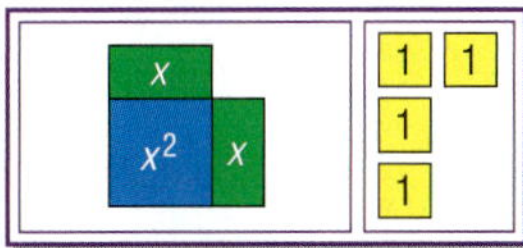

$x^2 + 2x = 4$

Step 4 To complete the square, add 1 yellow tile to each side of the mat. So, the equation is $x^2 + 2x + 1 = 5$ or $(x + 1)^2 = 5$.

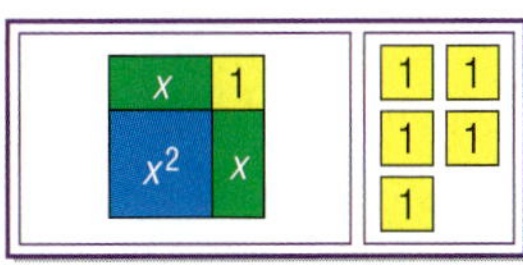

$x^2 + 2x + 1 = 5$

Try These

Use algebra tiles to complete the square of each equation.

1. $x^2 + 4x + 1 = 0$ **$(x + 2)^2 = 3$**
2. $x^2 - 6x = -5$ **$(x - 3)^2 = 4$**

To complete the square for any quadratic expression of the form $x^2 + bx$, you can follow the steps below.

Step 1 Take $\frac{1}{2}$ of b, the coefficient of x.

Step 2 Square the result of Step 1.

Step 3 Add the result of Step 2, $\left(\frac{b}{2}\right)^2$, to $x^2 + bx$.

Example 1

Find the value of c that makes $x^2 + 14x + c$ a perfect square.

Step 1	Find one half of 14.	$\frac{14}{2} = 7$
Step 2	Square the result of Step 1.	$7^2 = 49$
Step 3	Add the result of Step 2 to $x^2 + 14x$.	$x^2 + 14x + 49$

Thus, $c = 49$. Notice that $x^2 + 14x + 49 = (x + 7)^2$.

Your Turn

a. Find the value of c that makes $x^2 - 6x + c$ a perfect square. **9**

Once a perfect square is found, we can solve the equation by taking the square root of each side.

Example 2

Reading Algebra

Read $\pm\sqrt{49}$ as *plus or minus the square root of 49.*

Solve $x^2 + 12x - 13 = 0$ by completing the square.

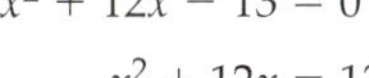

$x^2 + 12x - 13 = 0$	*$x^2 + 12x - 13$ is not a perfect square.*
$x^2 + 12x = 13$	*Add 13 to each side.*
$x^2 + 12x + 36 = 13 + 36$	*Since $\left(\frac{12}{2}\right)^2 = 36$, add 36 to each side.*
$(x + 6)^2 = 49$	*Factor $x^2 + 12x + 36$.*
$\sqrt{(x + 6)^2} = \pm\sqrt{49}$	*Take the square root of each side.*

(continued on the next page)

www.algconcepts.com/extra_examples

2 TEACH

Teaching Tip Focus students' attention on the symbol $\pm$ used in the computation near the top of page 478. Point out that when taking the square root of each side of an equation such as $x^2 = 121$, both the positive and negative values of the result are solutions. Have students consider $(-11)^2 = 121$ to verify this fact.

Teaching Tip Completing the square can be a difficult skill for students. Encourage them to write a summary of the steps used in Examples 1 and 2 that they can follow when doing any completing the square problem.

In-Class Examples

Example 1

Find the value of c that makes $x^2 - 8x + c$ a perfect square. **16**

Example 2

Solve $x^2 + 4x - 5 = 0$ by completing the square. **-5, 1**

Hands-On Algebra

Cooperative Learning Refer to the Hands-On Algebra on pages 478–479. After students complete the first model, ask them to describe in their own words the process used to complete the square. Why is it called completing the square? Show how to solve the equation once the square is completed. Then have students model the exercises. After they complete the square for these problems have students solve the equations and check their answers in the original equation.

An additional Hands-On Algebra activity using a step-by-step method for completing the square for a quadratic expression of the form $x^2 + bx$ is available in the *Hands-On Algebra Masters*, p. 124.

Hands-On Algebra Masters, p. 123

Teaching Tip You may want to caution students not to assume that the negative square root should be eliminated in a real-world problem such as that in Example 3. There are real-world applications for which negative solutions make sense.

Teaching Tip Point out in Example 3 that the equation is divided by the coefficient of w^2 prior to completing the square. This step is necessary whenever the coefficient of the squared term is not 1.

In-Class Example

Example 3

When constructing a room, the width is to be 10 feet more than half the length. Find the dimensions of the room to the nearest tenth of a foot, if its area is to be 135 square feet. **9.2 ft by 14.6 ft**

Study Guide, p. 493

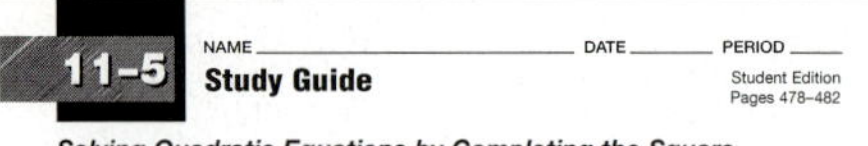

NAME ________ DATE ______ PERIOD ____

11-5 Study Guide

Student Edition Pages 478–482

Solving Quadratic Equations by Completing the Square

Quadratic equations can be solved by **completing the square**. You complete the square for a quadratic expression of the form $x^2 + bx$, according to the steps shown below.

Completing the Square for a Quadratic $x^2 + bx$	Step 1 Take $\frac{1}{2}$ of b and square it. Step 2 Add the result to $x^2 + bx$. The perfect square is $x^2 + bx + \left(\frac{b}{2}\right)^2$.	Example: For $x^2 + 6x$, $b = 6$. $\frac{1}{2} \times 6 = 3$ $3^2 = 9$ The perfect square is $x^2 + 6x + 9$.

Once you determine $\left(\frac{b}{2}\right)^2$, you must add this number to each side of the equation. Finally, you find the square root of each side to solve the resulting equation.

Example: Solve $x^2 + 8x - 9 = 0$ by completing the square.

$x^2 + 8x - 9 = 0$

$x^2 + 8x = 9$ *Add 9 to each side.*

$x^2 + 8x + \left(\frac{8}{2}\right)^2 = 9 + \left(\frac{8}{2}\right)^2$ *Add the square of $\frac{1}{2}b$ to each side.*

$x^2 + 8x + 16 = 25$

$(x + 4)^2 = 25$ *Factor $x^2 + 8x + 16$.*

$x + 4 = \pm\sqrt{25}$ *Take the square root of each side.*

$x = \pm 5 - 4$ *Subtract 4 from each side.*

$x = 5 - 4$ or $x = -5 - 4$

$x = 1$ or $x = -9$ The solutions are 1 and −9.

Solve each equation by completing the square.

1. $x^2 + 6x + 9 = 0$ **−3**
2. $m^2 - 7m - 8 = 0$ **−1, 8**
3. $y^2 + 4y = 6$ **$-2 \pm \sqrt{10}$**
4. $y^2 + 14y = 0$ **−14, 0**
5. $t^2 - 12t - 7 = 0$ **$6 \pm \sqrt{43}$**
6. $z^2 - 8z - 9 = 0$ **−1, 9**
7. $n^2 + 5n = 0$ **−5, 0**
8. $x^2 + 4x = 1$ **$-2 \pm \sqrt{5}$**
9. $k^2 + k - 2 = 0$ **1, −2**

$x + 6 = \pm 7$ *Simplify.*

$x + 6 - 6 = \pm 7 - 6$ *Subtract 6 from each side.*

$x = 7 - 6$ or $x = -7 - 6$ *Simplify each equation.*

$x = 1$ $\quad x = -13$ *Simplify.*

The solutions are −13 and 1. *Check the solution by using the graphing calculator or by substituting −13 and 1 for x in the original equation.*

Your Turn

b. Solve $x^2 + 6x - 16 = 0$ by completing the square. **2, −8**

You can complete the square only if the coefficient of the first term is 1. If the coefficient is *not* 1, first divide each term by the coefficient.

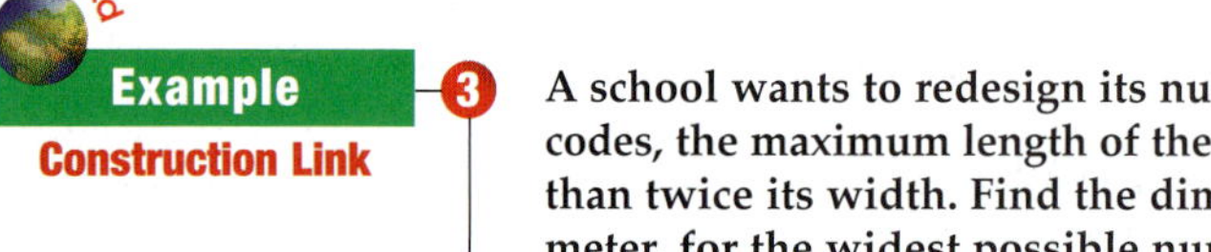

3 **A school wants to redesign its nurse's station. Because of building codes, the maximum length of the rectangular room is 20 meters less than twice its width. Find the dimensions, to the nearest tenth of a meter, for the widest possible nurse's station if its area is to be 60 square meters.**

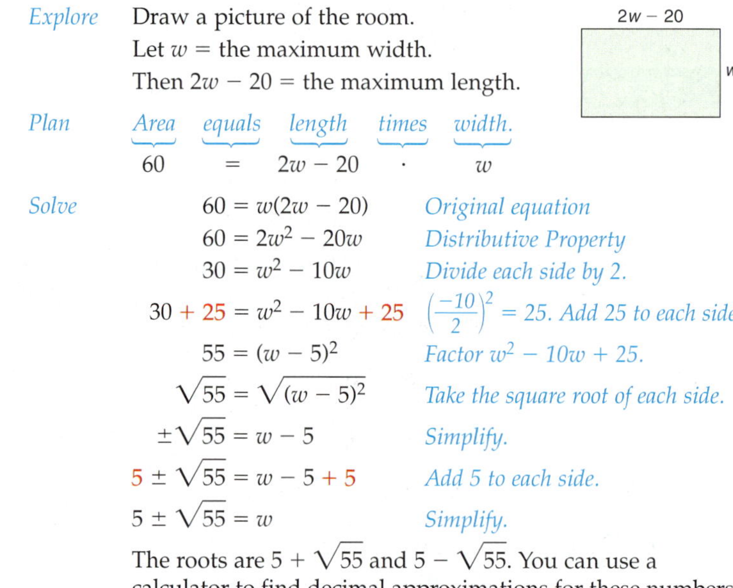

Explore Draw a picture of the room.

Let w = the maximum width.

Then $2w - 20$ = the maximum length.

Plan *Area equals length times width.*

$$60 = 2w - 20 \cdot w$$

Solve

$60 = w(2w - 20)$ *Original equation*

$60 = 2w^2 - 20w$ *Distributive Property*

$30 = w^2 - 10w$ *Divide each side by 2.*

$30 + 25 = w^2 - 10w + 25$ $\left(\frac{-10}{2}\right)^2 = 25$. *Add 25 to each side.*

$55 = (w - 5)^2$ *Factor $w^2 - 10w + 25$.*

$\sqrt{55} = \sqrt{(w - 5)^2}$ *Take the square root of each side.*

$\pm\sqrt{55} = w - 5$ *Simplify.*

$5 \pm \sqrt{55} = w - 5 + 5$ *Add 5 to each side.*

$5 \pm \sqrt{55} = w$ *Simplify.*

The roots are $5 + \sqrt{55}$ and $5 - \sqrt{55}$. You can use a calculator to find decimal approximations for these numbers.

$5 + \sqrt{55} \approx 12.4$ and $5 - \sqrt{55} \approx -2.4$

The solution of −2.4 is not reasonable since there cannot be negative width. The dimensions of the nurse's station should be about 12.4 meters by 2(12.4) − 20 or 4.8 meters.

Examine Since 12.4(4.8) = 59.52, the answer seems reasonable.

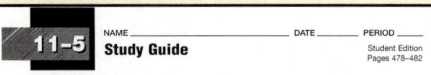

Check for Understanding

Communicating Mathematics

1. **Explain** why completing the square is a good strategy to use in solving the equation $x^2 - 5x - 7 = 0$. **1–2. See margin.**

2. Writing Math List the steps necessary to complete the square of the expression $x^2 - 10x$.

Vocabulary
completing the square

Guided Practice

State whether each trinomial is a perfect square.

Sample 1: $x^2 - 4x - 4$
Solution: No, it is not factorable.

Sample 2: $x^2 + 6x + 9$
Solution: Yes, $x^2 + 6x + 9 = (x + 3)^2$.

3. $x^2 - 2x + 1$ **yes**
4. $x^2 - 18x - 9$ **no**
5. $x^2 + 7x - 14$ **no**
6. $x^2 + 8x + 16$ **yes**

Example 1

Find the value of *c* that makes each trinomial a perfect square.

7. $x^2 + 2x + c$ **1**
8. $z^2 - 18z + c$ **81**
9. $m^2 + 3m + c$ $\frac{9}{4}$

Example 2

11. $-3, -4$
12. $2 \pm \sqrt{13}$

Solve each equation by completing the square. *(Example 2)*

10. $p^2 - 6p = 0$ **0, 6**
11. $r^2 + 7r + 12 = 0$
12. $x^2 - 4x - 9 = 0$

Example 3

13. **Construction** The Thompsons' rectangular bathroom measures 6 feet by 9 feet. They want to double the area of the bathroom by increasing the length and width by the same amount. Find the dimensions, to the nearest foot, of the new bathroom. *(Example 3)* **9 ft by 12 ft**

Exercises

Practice

Find the value of *c* that makes each trinomial a perfect square.

14. $z^2 - 10z + c$ **25**
15. $f^2 + 12f + c$ **36**
16. $h^2 - 20h + c$ **100**
17. $s^2 - 2s + c$ **1**
18. $q^2 + 14q + c$ **49**
19. $x^2 - 24x + c$ **144**
20. $z^2 + 16z + c$ **64**
21. $k^2 + 5k + c$ $\frac{25}{4}$
22. $r^2 + 9r + c$ $\frac{81}{4}$

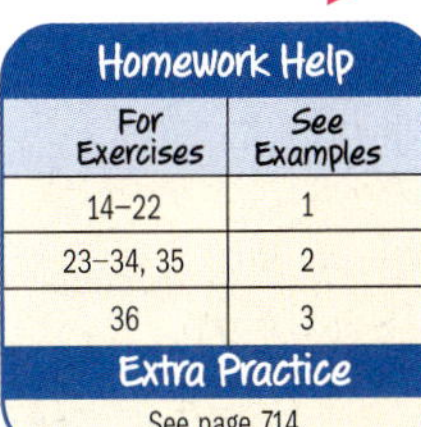

Homework Help
For Exercises | See Examples
14–22 | 1
23–34, 35 | 2
36 | 3
Extra Practice
See page 714.

23. $-3, 1$ 24. $1, 7$ 25. $-5, -1$ 28. $-5 \pm \sqrt{26}$ 29. $1 \pm \sqrt{31}$

Solve each equation by completing the square.

23. $x^2 + 2x - 3 = 0$
24. $x^2 - 8x + 7 = 0$
25. $p^2 + 6p = -5$
26. $r^2 - 16r = 0$ **0, 16**
27. $s^2 - 14s = 0$ **0, 14**
28. $x(x + 10) - 1 = 0$
29. $q^2 - 2q = 30$
30. $4x^2 + 16x + 24 = 0$
31. $3z^2 - 18z = 30$
32. $m^2 - 2m = 6$
33. $d^2 + 3d - 10 = 0$
34. $a^2 - \frac{7}{2}a + \frac{3}{2} = 0$

34. $\frac{1}{2}, 3$

32. $1 \pm \sqrt{7}$ 33. $-5, 2$ 30. no real solutions 31. $3 \pm \sqrt{19}$

Applications and Problem Solving

35. **Physical Science** Omar set off a rocket in a field. The height h of the rocket in feet can be roughly estimated using the formula $h = -16t^2 + 128t$, where t is the time in seconds. How long will it take the rocket to reach a height of 240 feet? **3 s or 5 s**

3 PRACTICE/APPLY

Error Analysis

Watch for students who forget to add the constant to each side of the equation when completing the square in Exercises 10–12. ***Prevent by*** stressing that if you are completing the square in an equation, you must keep the equation in balance. Also encourage students to check their solutions in the original equation.

Assignment Guide

Basic: 15–37 odd, 38–44
Average: 14–34 even, 35–44

Answers

1. The equation is not factorable with integers and the coefficient of x^2 is 1.
2. Divide 10 by 2. Square the result, 5, to give 25. Add 25 to the expression.

Skills Practice, **p. 494, and *Practice,* p. 495 (shown)**

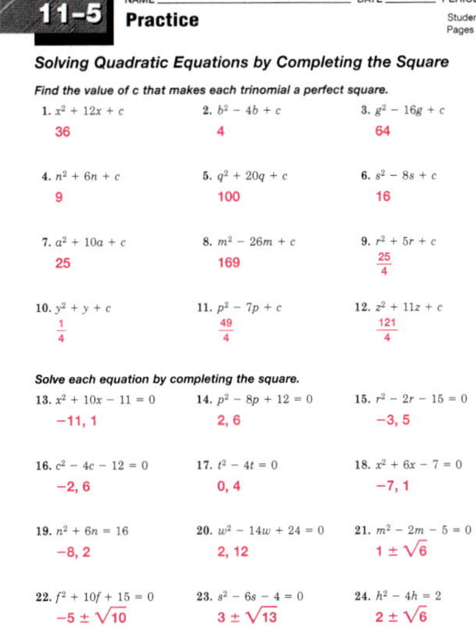

11-5 NAME ______ DATE ______ PERIOD ______
Practice Student Edition Pages 478–482

Solving Quadratic Equations by Completing the Square

Find the value of c that makes each trinomial a perfect square.

1. $x^2 + 12x + c$ 36
2. $b^2 - 4b + c$ 4
3. $g^2 - 16g + c$ 64
4. $n^2 + 6n + c$ 9
5. $q^2 + 20q + c$ 100
6. $s^2 - 8s + c$ 16
7. $a^2 + 10a + c$ 25
8. $m^2 - 26m + c$ 169
9. $r^2 + 5r + c$ $\frac{25}{4}$
10. $y^2 + y + c$ $\frac{1}{4}$
11. $p^2 - 7p + c$ $\frac{49}{4}$
12. $z^2 + 11z + c$ $\frac{121}{4}$

Solve each equation by completing the square.

13. $x^2 + 10x - 11 = 0$ $-11, 1$
14. $p^2 - 8p + 12 = 0$ $2, 6$
15. $r^2 - 2r - 15 = 0$ $-3, 5$
16. $c^2 - 4c - 12 = 0$ $-2, 6$
17. $t^2 - 4t = 0$ $0, 4$
18. $x^2 + 6x - 7 = 0$ $-7, 1$
19. $n^2 + 6n = 16$ $-8, 2$
20. $w^2 - 14w + 24 = 0$ $2, 12$
21. $m^2 - 2m - 5 = 0$ $1 \pm \sqrt{6}$
22. $f^2 + 10f + 15 = 0$ $-5 \pm \sqrt{10}$
23. $s^2 - 6s - 4 = 0$ $3 \pm \sqrt{13}$
24. $h^2 - 4h = 2$ $2 \pm \sqrt{6}$
25. $y^2 - 12y + 7 = 0$ $6 \pm \sqrt{29}$
26. $k^2 - 8k + 13 = 0$ $4 \pm \sqrt{3}$
27. $d^2 + 8d + 9 = 0$ $-4 \pm \sqrt{7}$

Reteaching Activity

Auditory/Musical Learners Have students work with a partner. Give them a list of problems such as Exercises 14–22. Have students take turns explaining the method used to find the value of the constant. The other partner should check the student's response. After gaining confidence in finding the constant *c*, have students move on to solving equations such as those in Exercises 23–34. Again students should explain each step as they complete it.

4 ASSESS

Open-Ended Assessment

Writing Write the equation $x^2 - 16x + 11 = 0$ on the board or overhead. Ask students to write a solution to this equation explaining each step in the process.

Answer

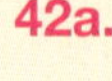

42a.

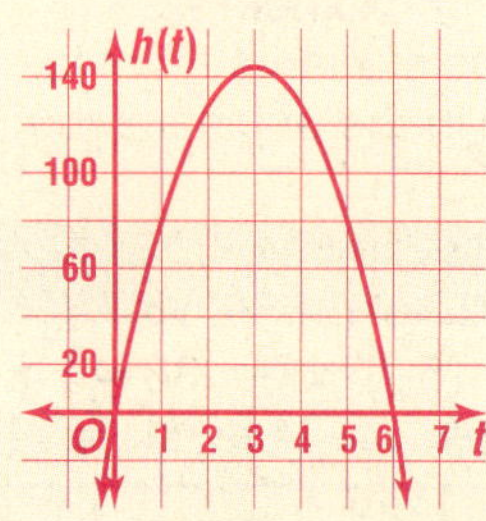

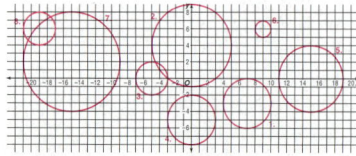

36. **Photography** Sheila places a 54 square inch photo behind a 12-inch-by-12-inch piece of matting. The photograph is positioned so that the matting is twice as wide at the top and bottom as the sides.

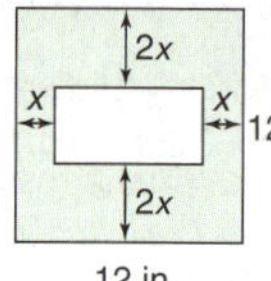

a. Write an equation for the area of the photo in terms of x. $(12 - 4x)(12 - 2x) = 54$

b. Find the dimensions of the photo. **9 in. by 6 in.**

37. **Critical Thinking** Find the values of b that make $x^2 + bx + 81$ a perfect square. -18 **or** 18

Mixed Review

Solve each quadratic equation by factoring. *(Lesson 11–4)*

38. $a^2 + 3a - 28 = 0$ $-7, 4$

39. $x^2 - 7x + 10 = 0$ $2, 5$

40. $z^2 + 5z = -4$ $-4, -1$

41. $2b^2 + 12b + 18 = 0$ -3

42. **Rockets** Jon is launching rockets in an open field. The path of the rocket can be modeled by the quadratic function $h(t) = -16t^2 + 96t$, where $h(t)$ is the height in feet any time t in seconds. *(Lesson 11–3)*

a. Graph the quadratic function. **See margin.**

b. After how many seconds will the rocket reach a height of 80 feet for the *second* time? **5 s**

c. After how many seconds will the rocket hit the ground? **6 s**

d. What is the rocket's maximum height? **144 ft**

Standardized Test Practice

43. **Extended Response** The area of a rectangle is represented by $(2ab + 8b^2)$.

a. Factor the expression to find the dimensions of the rectangle. *(Lesson 10–2)* $2b(a + 4b)$

b. Using the dimensions from part a, write an expression for the perimeter of the rectangle in simplest form. *(Lessons 9–2 & 9–3)* **b.** $2a + 12b$

c. Find the dimensions, area, and perimeter of the rectangle if $a = 5$ inches and $b = 1$ inch. *(Lesson 1–2)*
9 in. by 2 in., 18 in^2**, 22 in.**

44. **Multiple Choice** The graph shows how each Ohio tax dollar was divided among programs and services in a recent year. If Maria paid $32 in state taxes on her last paycheck, what amount of her money went towards education? *(Lessons 5–3 & 5–4)* **C**

Ⓐ $5.12 Ⓑ $10.24
Ⓒ $15.36 Ⓓ $24.96

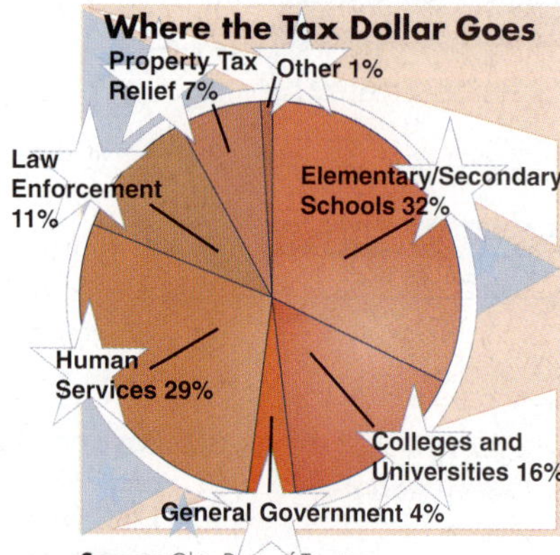

Source: Ohio Dept. of Taxation

www.algconcepts.com/self_check_quiz

Extra Credit

A circus owner believes that by lowering the ticket price to his show, he can increase his total sales from each show by $1000. Currently an average of 500 people pay $15 each to attend a performance. It has been shown that for each $1 decrease in ticket price up to $5, 100 more people will attend the show. Let x be the number of dollar decreases in the ticket price.

a. Write a quadratic equation that represents the sales from one show after the ticket prices are lowered. $(15 - x)(500 + 100x) = 8500$

b. Solve this equation to find the number of dollars (rounded to the nearest dollar) the price must be lowered to achieve the owner's goal. **$1**

Enrichment, **p. 497**

11-5 NAME ______ DATE ______ PERIOD ______

Enrichment

Student Edition Pages 478–482

Graphing Circles by Completing Squares

One use for completing the square is to graph circles. The general equation for a circle with center at the origin and radius r is $x^2 + y^2 = r^2$. An equation represents a circle if it can be transformed into the sum of two squares.

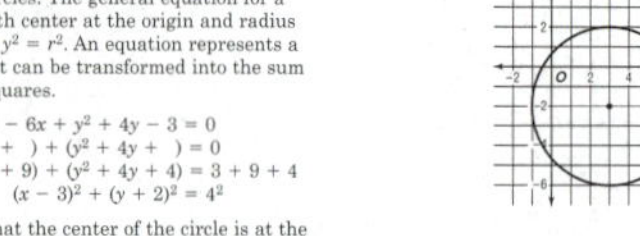

$$x^2 - 6x + y^2 + 4y - 3 = 0$$
$$(x^2 - 6x + \quad) + (y^2 + 4y + \quad) = 0$$
$$(x^2 - 6x + 9) + (y^2 + 4y + 4) = 3 + 9 + 4$$
$$(x - 3)^2 + (y + 2)^2 = 4^2$$

Notice that the center of the circle is at the point $(3, -2)$.

Transform each equation into the sum of two squares. Then graph the circle represented by the equation. Use the coordinate plane provided at the bottom of the page.

1. $x^2 - 14x + y^2 + 6y + 49 = 0$
$(x - 7)^2 + (y + 3)^2 = 3^2$
2. $x^2 + y^2 - 8y - 9 = 0$
$x^2 + (y - 4)^2 = 5^2$
3. $x^2 + 10x + y^2 + 21 = 0$
$(x + 5)^2 + y^2 = 2^2$
4. $x^2 + y^2 + 10y + 16 = 0$
$x^2 + (y + 5)^2 = 3^2$
5. $x^2 - 30x + y^2 + 209 = 0$
$(x - 15)^2 + y^2 = 4^2$
6. $x^2 - 18x + y^2 - 12y + 116 = 0$
$(x^2 - 9)^2 + (y - 6)^2 = 1^2$
7. $x^2 + 30x + y^2 - 4y + 193 = 0$
$(x + 15)^2 + (y - 2)^2 = 6^2$
8. $x^2 + 38x + y^2 - 12y + 393 = 0$
$(x + 19)^2 + (y - 6)^2 = 2^2$

11-6 The Quadratic Formula

What You'll Learn
You'll learn to solve quadratic equations by using the Quadratic Formula.

Why It's Important
Science You can use the Quadratic Formula to solve problems dealing with physics. *See Exercise 10.*

The table below summarizes the ways you have learned to solve quadratic equations. Although these methods can be used to solve all quadratic equations, each method is most useful in particular situations.

Method	When Is the Method Useful?
Graphing	Use only to estimate solutions.
Factoring	Use when the quadratic expression is easy to factor.
Completing the Square	Use when the coefficient of x^2 is 1 and all other coefficients are fairly small.

An alternative method is to develop a general formula for solving *any* quadratic equation. Begin with the general form of a quadratic equation, $ax^2 + bx + c = 0$, where $a \neq 0$, and complete the square.

$$ax^2 + bx + c = 0 \quad \textit{General quadratic equation}$$

$$x^2 + \frac{b}{a}x + \frac{c}{a} = 0 \quad \textit{Divide by a so the coefficient of } x^2 \textit{ is 1.}$$

$$x^2 + \frac{b}{a}x = -\frac{c}{a} \quad \textit{Subtract } \frac{c}{a} \textit{ from each side.}$$

Now complete the square.

$$x^2 + \frac{b}{a}x + \left(\frac{b}{2a}\right)^2 = -\frac{c}{a} + \left(\frac{b}{2a}\right)^2 \quad \frac{1}{2} \cdot \frac{b}{a} = \frac{b}{2a}$$

$$\left(x + \frac{b}{2a}\right)^2 = -\frac{c}{a} + \frac{b^2}{4a^2} \quad \textit{Factor the left side of the equation.}$$

$$\left(x + \frac{b}{2a}\right)^2 = \frac{4a}{4a}\left(-\frac{c}{a}\right) + \frac{b^2}{4a^2} \quad \textit{The common denominator is } 4a^2.$$

$$\left(x + \frac{b}{2a}\right)^2 = \frac{b^2 - 4ac}{4a^2} \quad \textit{Simplify the right side of the equation.}$$

Finally, take the square root of each side and solve for x.

$$\sqrt{\left(x + \frac{b}{2a}\right)^2} = \pm\sqrt{\frac{b^2 - 4ac}{4a^2}} \quad \textit{Take the square root of each side.}$$

$$x + \frac{b}{2a} = \frac{\pm\sqrt{b^2 - 4ac}}{2a} \quad \textit{Simplify: } \sqrt{4a^2} = 2a.$$

$$x = \frac{\pm\sqrt{b^2 - 4ac}}{2a} - \frac{b}{2a} \quad \textit{Subtract } \frac{b}{2a} \textit{ from each side.}$$

$$x = \frac{-b \pm \sqrt{b^2 - 4ac}}{2a} \quad \textit{Combine the terms.}$$

Resource Manager

Reproducible Masters

Chapter 11 Resource Masters
- *Study Guide,* p. 498
- *Skills Practice,* p. 499
- *Practice,* p. 500
- *Reading to Learn Mathematics,* p. 501
- *Enrichment,* p. 502

Graphing Calculator, pp. 31–32

Transparencies

5-Minute Check, 11–6

Technology/Multimedia

AlgePASS, Lesson 28

Interactive Chalkboard CD-ROM

Lesson 11-6

1 FOCUS

5-Minute Check
Lesson 11–5

Find the value of c that makes each trinomial a perfect square.

1. $x^2 + 20x + c$ **100**
2. $x^2 + 3x + c$ $\frac{9}{4}$

Solve each equation by completing the square.

3. $x^2 - 6x - 16 = 0$ **−2, 8**
4. $3m^2 - 6m = 12$ **$1 \pm \sqrt{5}$**
5. If the area of the rectangle below is 136 square meters, what is the value of x? **15**

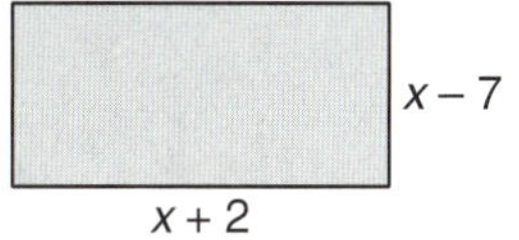

Motivating the Lesson

Real-World Connection
An object is tossed into the air with a velocity of 36 feet per second. If the object is released 3 feet above the ground, the formula $h(t) = -16t^2 + 36t + 3$ can be used to find the height $h(t)$ of the object after t seconds. Discuss solving the equation to find the time that the object reaches a height of 20 feet. Which method is best? Explain that the Quadratic Formula will give an accurate solution for any quadratic equation.

2 TEACH

Teaching Tip Spend some time discussing the chart on this page. It is important for students to understand when each method is best used. Stress that, unlike the other methods, the Quadratic Formula will always give the exact solutions to a quadratic equation or indicate that there are no solutions.

Teaching Tip As you discuss Example 1, encourage students to use parentheses when substituting values for the variables in the Quadratic Formula. This will help students avoid computational and sign errors. Students may need to review the order of operations. You might want to point out to students that the roots of the equation are integers. Emphasize that this means the quadratic equation can be solved by factoring. You might want to have your students check the solution using this method.

In-Class Example

Example 1

Use the Quadratic Formula to solve $2x^2 - 5x + 3 = 0$. **1, 1.5**

The result is called the **Quadratic Formula** and can be used to solve *any* quadratic equation.

The Quadratic Formula	$x = \frac{-b \pm \sqrt{b^2 - 4ac}}{2a},\ a \neq 0$

Examples

Use the Quadratic Formula to solve each equation.

$x^2 - 4x + 3 = 0$

$x = \frac{-b \pm \sqrt{b^2 - 4ac}}{2a}$	*Quadratic Formula*
$= \frac{-(-4) \pm \sqrt{(-4)^2 - 4(1)(3)}}{2(1)}$	*$a = 1$, $b = -4$, and $c = 3$*
$= \frac{4 \pm \sqrt{16 - 12}}{2}$	*$(-4)^2 = 16$ and $4(1)(3) = 12$*
$= \frac{4 \pm \sqrt{4}}{2}$	*Subtract.*
$= \frac{4 \pm 2}{2}$	*Simplify.*
$= \frac{4 + 2}{2}$ or $x = \frac{4 - 2}{2}$	*Find two solutions.*
$= \frac{6}{2}$ or 3 $\quad x = \frac{2}{2}$ or 1	*Simplify.*

Check: Substitute values into the original equation.

$$x^2 - 4x + 3 = 0 \qquad \text{or} \qquad x^2 - 4x + 3 = 0$$
$$3^2 - 4(3) + 3 \stackrel{?}{=} 0 \qquad 1^2 - 4(1) + 3 \stackrel{?}{=} 0$$
$$9 - 12 + 3 \stackrel{?}{=} 0 \qquad 1 - 4 + 3 \stackrel{?}{=} 0$$
$$0 = 0 \ \checkmark \qquad 0 = 0 \ \checkmark$$

The roots are 3 and 1.

$-2x^2 + 3x - 5 = 0$

$x = \frac{-b \pm \sqrt{b^2 - 4ac}}{2a}$	*Quadratic Formula*
$= \frac{-3 \pm \sqrt{3^2 - 4(-2)(-5)}}{2(-2)}$	*$a = -2$, $b = 3$, and $c = -5$*
$= \frac{-3 \pm \sqrt{9 - 40}}{-4}$	*$3^2 = 9$ and $4(-2)(-5) = 40$*
$= \frac{-3 \pm \sqrt{-31}}{-4}$	*Subtract.*

The square root of a negative number, such as $\sqrt{-31}$, is not a real number. So, there are no real solutions for x.

www.algconcepts.com/extra_examples

Have students work with a family member to research football punting and current or past well-known football punters. Have students research typical values for the hang time and upward velocity of a punt. Students can adjust the formula in the application at the beginning of the lesson using the values they found. Invite students to share their findings with the class.

Check: Use a graphing calculator. It is clear that the graph of $f(x) = -2x^2 + 3x - 5$ never crosses the x-axis. Therefore, there are no real roots for the equation $-2x^2 + 3x - 5 = 0$.

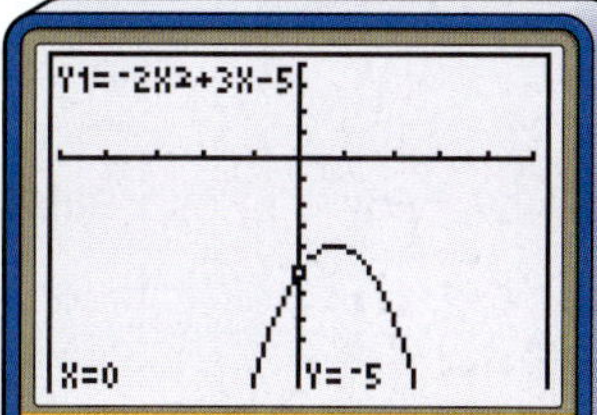

Your Turn

a. $-3x^2 + 6x + 9 = 0$ **−1, 3** b. $x^2 + 4x + 2 = 0$ **$-2 \pm \sqrt{2}$**

When the solutions are irrational numbers, use a calculator to estimate.

Example 3 **Sports Link**

"Hang time" is the total amount of time a ball stays in the air. Manuel can kick a football with an upward velocity of 64 ft/s. His foot meets the ball 2 feet off the ground. What is Manuel's hang time if the ball is not caught? Use the formula $h(t) = -16t^2 + 64t + 2$, where $h(t)$ is the ball's height in feet for any time t, in seconds, after the ball is kicked.

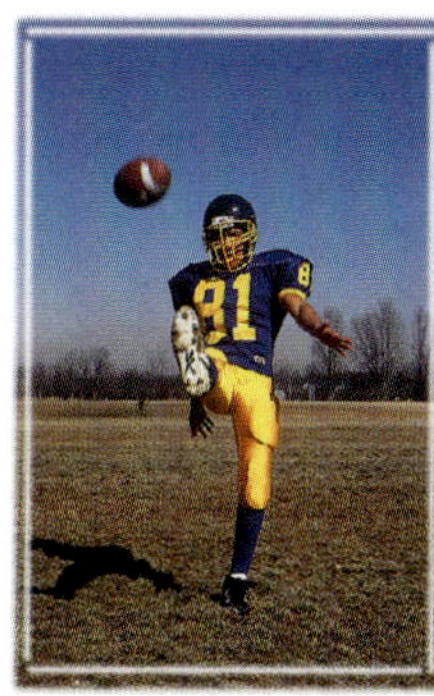

$-16t^2 + 64t + 2 = 0$ *Final height is 0.*

$t = \frac{-b \pm \sqrt{b^2 - 4ac}}{2a}$ *Use the Quadratic Formula.*

$= \frac{-64 \pm \sqrt{64^2 - 4(-16)(2)}}{2(-16)}$ *$a = -16$, $b = 64$, and $c = 2$*

$= \frac{-64 \pm \sqrt{4096 + 128}}{-32}$ *$64^2 = 4096$ and $4(-16)(2) = -128$*

$= \frac{-64 \pm \sqrt{4224}}{-32}$ *$4096 - (-128) = 4096 + 128$*

$t = \frac{-64 + \sqrt{4224}}{-32}$ or $t = \frac{-64 - \sqrt{4224}}{-32}$

≈ -0.031 ≈ 4.031

Since time cannot be negative, the only approximate solution is 4.031. The football has a hang time of about 4 seconds.

Check for Understanding

Communicating Mathematics

1. **Choose** the correct term and justify your answer. Quadratic equations (*sometimes, always, never*) have at least one real solution. **1–2. See margin.**

Vocabulary
Quadratic Formula

2. **Writing Math** List the steps you take when you use the Quadratic Formula to solve a quadratic equation.

In-Class Examples

Example 2

Use the Quadratic Formula to solve $-x^2 + 6x - 9 = 0$. **3**

Example 3

Another punter kicks the football with an upward velocity of 58 ft/s and his foot meets the ball 1 foot off the ground. His formula is $h(t) = -16t^2 + 58t + 1$, where $h(t)$ is the ball's height for any time t after the ball was kicked. What is the hang time? **about 3.64 s**

Answers

1. **Sometimes; quadratic equations can have no real solutions.**
2. **Identify the values for a, b, and c. Substitute those values into the Quadratic Formula. Follow the order of operations to solve for the roots.**

***Study Guide*, p. 498**

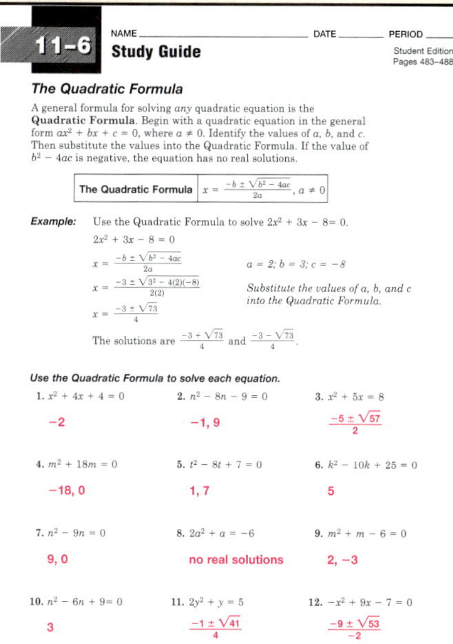

11-6 NAME ______ DATE ______ PERIOD ______

Study Guide Student Edition Pages 483–488

The Quadratic Formula

A general formula for solving *any* quadratic equation is the **Quadratic Formula**. Begin with a quadratic equation in the general form $ax^2 + bx + c = 0$, where $a \neq 0$. Identify the values of a, b, and c. Then substitute the values into the Quadratic Formula. If the value of $b^2 - 4ac$ is negative, the equation has no real solutions.

The Quadratic Formula	$x = \frac{-b \pm \sqrt{b^2 - 4ac}}{2a}, a \neq 0$

Example: Use the Quadratic Formula to solve $2x^2 + 3x - 8 = 0$.

$2x^2 + 3x - 8 = 0$

$x = \frac{-b \pm \sqrt{b^2 - 4ac}}{2a}$ *$a = 2$; $b = 3$; $c = -8$*

$x = \frac{-3 \pm \sqrt{3^2 - 4(2)(-8)}}{2(2)}$ *Substitute the values of a, b, and c into the Quadratic Formula.*

$x = \frac{-3 \pm \sqrt{73}}{4}$

The solutions are $\frac{-3 + \sqrt{73}}{4}$ and $\frac{-3 - \sqrt{73}}{4}$.

Use the Quadratic Formula to solve each equation.

1. $x^2 + 4x + 4 = 0$ **−2**
2. $n^2 - 8n - 9 = 0$ **−1, 9**
3. $x^2 + 5x = 8$ **$\frac{-5 \pm \sqrt{57}}{2}$**
4. $m^2 + 18m = 0$ **−18, 0**
5. $t^2 - 8t + 7 = 0$ **1, 7**
6. $k^2 - 10k + 25 = 0$ **5**
7. $n^2 - 9n = 0$ **9, 0**
8. $2a^2 + a = -6$ **no real solutions**
9. $m^2 + m - 6 = 0$ **2, −3**
10. $n^2 - 6n + 9 = 0$ **3**
11. $2y^2 + y = 5$ **$\frac{-1 \pm \sqrt{41}}{4}$**
12. $-x^2 + 9x - 7 = 0$ **$\frac{-9 \pm \sqrt{53}}{-2}$**

3 PRACTICE/APPLY

Error Analysis

Watch for students who solve Exercises 7–9 incorrectly because they have not memorized the Quadratic Formula.
Prevent by encouraging students to create a poem or rap song describing the formula.

Assignment Guide
Basic: 11–27 odd, 28–33
Average: 12–24 even, 26–33
All: Quiz 2, 1–10

11. −6, −1
12. −9, −1
13. 2, 3
14. −1, 7
15. no real solutions
16. $-\frac{8}{3}, 1$
17. 0, 8
18. 1, 5
19. $\frac{-1 \pm \sqrt{29}}{2}$
20. $\frac{-9 \pm \sqrt{89}}{2}$
21. $\frac{-2 \pm \sqrt{7}}{2}$
22. no real solutions
23. $\frac{-7 \pm \sqrt{69}}{2}$
24. no real solutions
25. $\frac{5 \pm \sqrt{249}}{16}$

Skills Practice, p. 499, and ***Practice,*** p. 500 (shown)

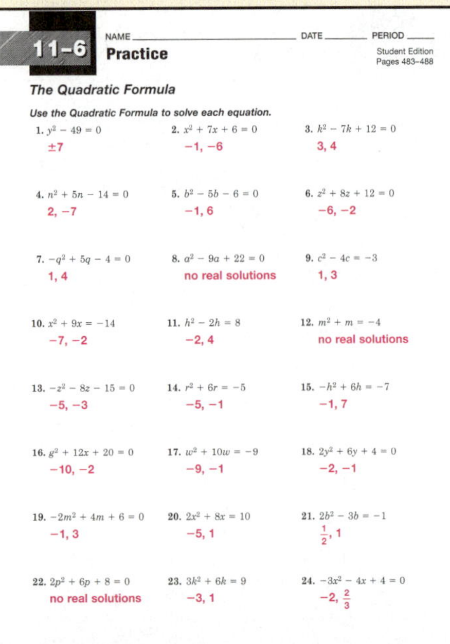

11-6 NAME ______ DATE ______ PERIOD ______
Practice
Student Edition Pages 483–488

The Quadratic Formula

Use the Quadratic Formula to solve each equation.

1. $y^2 - 49 = 0$ ±7
2. $x^2 + 7x + 6 = 0$ −1, −6
3. $k^2 - 7k + 12 = 0$ 3, 4
4. $n^2 + 5n - 14 = 0$ 2, −7
5. $b^2 - 5b - 6 = 0$ −1, 6
6. $z^2 + 8z + 12 = 0$ −6, −2
7. $-q^2 + 5q - 4 = 0$ 1, 4
8. $a^2 - 9a + 22 = 0$ no real solutions
9. $c^2 - 4c = -3$ 1, 3
10. $x^2 + 9x = -14$ −7, −2
11. $h^2 - 2h = 8$ −2, 4
12. $m^2 + m = -4$ no real solutions
13. $-z^2 - 8z - 15 = 0$ −5, −3
14. $r^2 + 6r = -5$ −5, −1
15. $-h^2 + 6h = -7$ −1, 7
16. $g^2 + 12x + 20 = 0$ −10, −2
17. $w^2 + 10w = -9$ −9, −1
18. $2y^2 + 6y + 4 = 0$ −2, −1
19. $-2m^2 + 4m + 6 = 0$ −1, 3
20. $2x^2 + 8x = 10$ −5, 1
21. $2b^2 - 3b = -1$ $\frac{1}{2}$, 1
22. $2p^2 + 6p + 8 = 0$ no real solutions
23. $3k^2 + 6k = 9$ −3, 1
24. $-3x^2 - 4x + 4 = 0$ −2, $\frac{2}{3}$

Guided Practice

Getting Ready Find the value of $b^2 - 4ac$ for each equation.

Sample 1: $-3z^2 + 8z + 5 = 0$
Solution: $8^2 - 4(-3)(5)$
$= 64 + 60$
$= 124$

Sample 2: $6a^2 = 3$
Solution: Rewrite $6a^2 = 3$ as
$6a^2 - 3 = 0$.
$0^2 - 4(6)(-3) = 72$

3. $x^2 + 10x + 13 = 0$ **48**
4. $3x^2 - 9x + 2 = 0$ **57**
5. $4x^2 = 12$ **192**
6. $-2x^2 - 7x = -5$ **89**

Examples 1 & 2

8. no real solutions
9. −3, −1

Use the Quadratic Formula to solve each equation.

7. $z^2 - 25 = 0$ **±5**
8. $r^2 - 5r + 7 = 0$
9. $d^2 + 4d = -3$

Example 3

10. **Physical Science** Josefina tosses a ball upward with an initial velocity of 76 feet per second. She tosses it from an initial height of 2 feet. Find the approximate time that the ball hits the ground. Use the function $h(t) = -16t^2 + 76t + 2$. **about 4.8 s**

Exercises

Practice

A

Use the Quadratic Formula to solve each equation. **11–25. See margin.**

Homework Help	
For Exercises	See Examples
11–25	1, 2
26–27	3
Extra Practice	
See page 715.	

11. $x^2 + 7x + 6 = 0$
12. $r^2 + 10r + 9 = 0$
13. $-a^2 + 5a - 6 = 0$
14. $-b^2 + 6b + 7 = 0$
15. $9d^2 - 4d + 3 = 0$
16. $3v^2 + 5v - 8 = 0$
17. $x^2 - 8x = 0$
18. $-p^2 + 6p - 5 = 0$
19. $w^2 + w - 7 = 0$
20. $z^2 + 9z - 2 = 0$
21. $4g^2 + 8g - 3 = 0$
22. $-2t^2 + 4t = 3$
23. $2c^2 + 14c = 10$
24. $5z^2 + 12 = -6z$

B

25. $8k^2 - 5k = 7$

Applications and Problem Solving

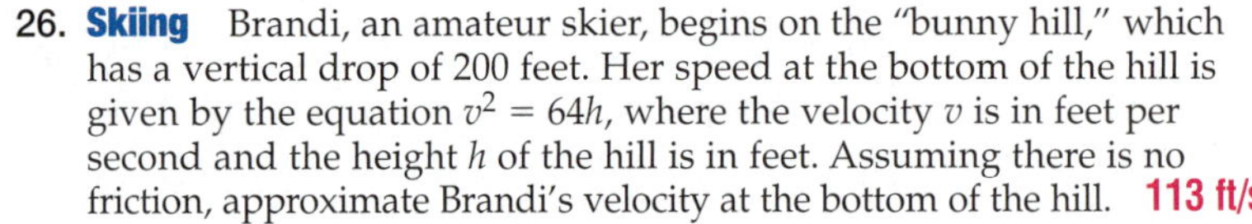

26. **Skiing** Brandi, an amateur skier, begins on the "bunny hill," which has a vertical drop of 200 feet. Her speed at the bottom of the hill is given by the equation $v^2 = 64h$, where the velocity v is in feet per second and the height h of the hill is in feet. Assuming there is no friction, approximate Brandi's velocity at the bottom of the hill. **113 ft/s**

C

27. **Space** Suppose an astronaut can jump vertically with an initial velocity of 5 m/s. The time that it takes him to touch the ground is given by the equation $0 = 5t - 0.5at^2$. The time t is in seconds and the acceleration due to gravity a is in m/s^2.
    a. If the astronaut jumps on the moon where $a = 1.6$ m/s^2, how long will it take him to reach the ground? **6.25 s**
    b. How long will it take him to reach the ground if he jumps on Earth where $a = 9.8$ m/s^2? **about 1.02 s**

Reteaching Activity

Intrapersonal Learners Have students make a two-column table. In the first column, have them list all the ways they might make a mistake when using the Quadratic Formula. In the second column, opposite each mistake, have students write a suggestion to themselves about how they could avoid making the mistake.

28. **Critical Thinking** The expression $b^2 - 4ac$ is called the **discriminant** of a quadratic equation. It determines the number of real solutions you can expect when solving a quadratic equation.

a. Copy and complete the table below.

Equation	$x^2 - 4x + 1 = 0$	$x^2 + 6x + 11 = 0$	$x^2 - 4x + 4 = 0$
Value of the Discriminant	12	−8	0
Number of x-intercepts	2	0	1
Number of Real Solutions	2	0	1 solution

b. Use the table above to describe the discriminant of a quadratic equation for each type of real solution.

i. two solutions **positive** ii. one solution **0** iii. no solutions **negative**

Mixed Review

Find the value of c to make a perfect square. *(Lesson 11–5)*

29. $a^2 + 4a + c$ **4** 30. $x^2 + 8x + c$ **16** 31. $v^2 - 18v + c$ **81**

32. **Geometry** The area of a circle varies directly as the square of the radius r. If the radius of a circle is doubled, how many times as large is the area of the circle? *(Lesson 6–5)* **4 times as large**

Standardized Test Practice

33. **Multiple Choice** If four times a number is equal to that number decreased by 5, what is the number? *(Lesson 4–6)* **B**

Ⓐ -5 Ⓑ $-\frac{5}{3}$ Ⓒ $-\frac{1}{4}$ Ⓓ $\frac{1}{3}$ Ⓔ 4

Quiz 2 Lessons 11–4 through 11–6

Solve each quadratic equation by factoring. *(Lesson 11–4)*

1. $x^2 + 4x - 12 = 0$ **−6, 2** 2. $x^2 - 8x + 16 = 0$ **4** 3. $x^2 - 10x = -21$ **3, 7**

Solve each quadratic equation by completing the square. *(Lesson 11–5)*

4. $x^2 + 4x + 3 = 0$ **−1, −3** 5. $x^2 - 6x + 7 = 0$ **$3 \pm \sqrt{2}$** 6. $2x^2 - 8x = 4$ **$2 \pm \sqrt{6}$**

Solve each equation by using the Quadratic Formula. *(Lesson 11–6)*

7. $x^2 + 3x + 3 = 0$ **no real solutions** 8. $x^2 + 8x + 15 = 0$ **−5, −3** 9. $x^2 + 5x + 3 = 0$ **$\frac{-5 \pm \sqrt{13}}{2}$**

10. **Construction** A company is designing a parabolic arch bridge to span a creek. The height of the bridge is given by the equation $h = -0.02d^2 + 0.8d$, where the height h and the distance d from one end of the base are in feet. Find the width of the bridge. *(Lesson 11–6)* **40 ft**

www.algconcepts.com/self_check_quiz

Teaching Tip In Exercise 28, you might want to inform students that negative discriminants is a topic covered in an Algebra 2 course.

4 ASSESS

Open-Ended Assessment

Writing Ask students to write the Quadratic Formula and explain how to use it to solve a quadratic equation. They should then explain some advantages and disadvantages of the Quadratic Formula when compared to the other methods they have learned for solving a quadratic equation.

Quiz 2

The Quiz provides students with a brief review of the concepts and skills in Lessons 11–4 through 11–6. Lesson numbers are given to the right of the exercises or instruction lines so students can review concepts not yet mastered.

Extra Credit

Recall that the formula for the equation of the axis of symmetry of a parabola is $x = -\frac{b}{2a}$. You saw in Exercise 28 that when the discriminant is zero, the quadratic equation has one solution. Show what happens to the Quadratic Formula when the discriminant is zero. Explain how this result relates to the equation of the axis of symmetry of the parabola. **The Quadratic Formula reduces to $x = -\frac{b}{2a}$ when the discriminant is zero. When the discriminant is zero, there is only one solution. This solution corresponds to the x-intercept of the graph which is also the x-coordinate of the vertex and hence must lie on the axis of symmetry of the parabola.**

***Enrichment*, p. 502**

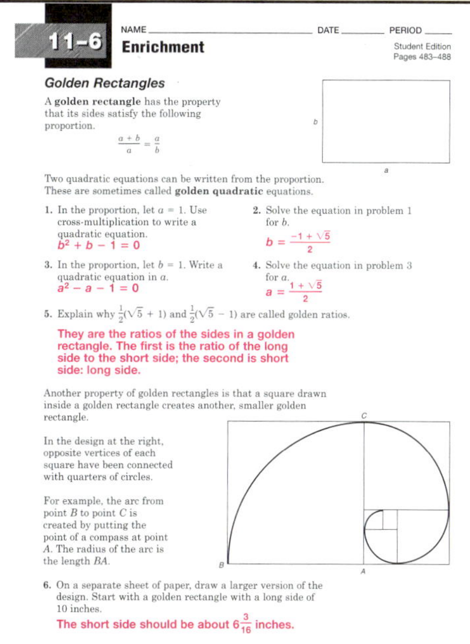

11–6 NAME ________ DATE ______ PERIOD ____

Enrichment Student Edition Pages 483–488

Golden Rectangles

A **golden rectangle** has the property that its sides satisfy the following proportion.

$$\frac{a+b}{a} = \frac{a}{b}$$

Two quadratic equations can be written from the proportion. These are sometimes called **golden quadratic** equations.

1. In the proportion, let $a = 1$. Use cross-multiplication to write a quadratic equation. **$b^2 + b - 1 = 0$**
2. Solve the equation in problem 1 for b. **$b = \frac{-1 + \sqrt{5}}{2}$**
3. In the proportion, let $b = 1$. Write a quadratic equation in a. **$a^2 - a - 1 = 0$**
4. Solve the equation in problem 3 for a. **$a = \frac{1 + \sqrt{5}}{2}$**
5. Explain why $\frac{1}{2}(\sqrt{5} + 1)$ and $\frac{1}{2}(\sqrt{5} - 1)$ are called golden ratios. **They are the ratios of the sides in a golden rectangle. The first is the ratio of the long side to the short side; the second is short side: long side.**

Another property of golden rectangles is that a square drawn inside a golden rectangle creates another, smaller golden rectangle.

In the design at the right, opposite vertices of each square have been connected with quarters of circles.

For example, the arc from point B to point C is created by putting the point of a compass at point A. The radius of the arc is the length BA.

6. On a separate sheet of paper, draw a larger version of the design. Start with a golden rectangle with a long side of 10 inches. **The short side should be about $6\frac{3}{16}$ inches.**

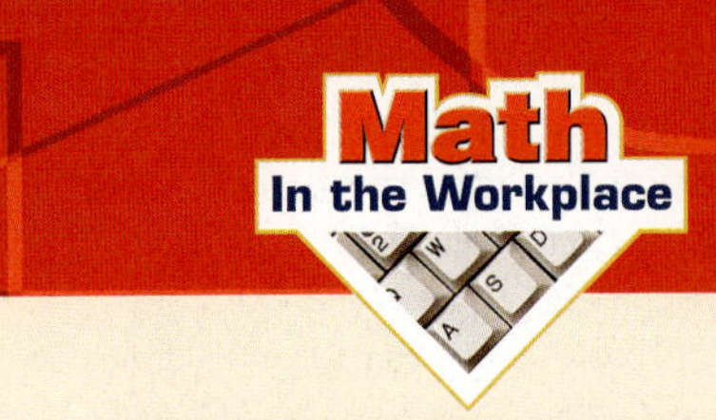

Civil engineers use mathematics and physics to design, plan, and implement the construction of structures such as roads, buildings, bridges, airports, tunnels, and water supply and sewage systems. Some civil engineers work in office buildings while others work at construction sites where they monitor or direct operations or solve on-site problems.

Prospective civil engineers should be interested in mathematics and science and enjoy problem solving. They should also have good interpersonal skills as most large projects require a team approach of engineers, architects and other professionals.

Starting jobs in civil engineering nearly always require a bachelor's degree in engineering or civil engineering.

Related Careers

- computer scientist
- architect
- physical scientist
- science technician

Community Connection

Civil engineers work all over the country in engineering and architectural firms that perform contract work for other businesses and government. Consider assigning one student to interview a civil engineer. Help the student choose an engineering firm from the telephone directory. Have the student write down a few questions to ask the civil engineer about working conditions, how math is used, and the personal traits that are helpful. Have the student share the answers with the class.

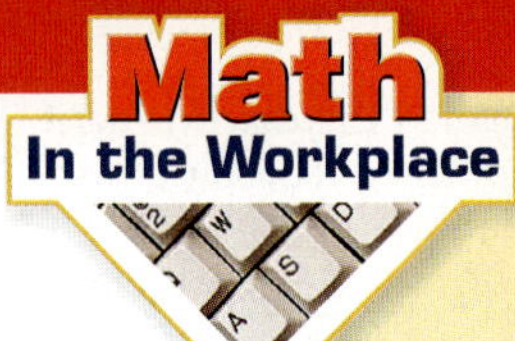

Civil Engineer

Civil engineers work in the oldest branch of engineering. They design and supervise the construction of buildings, bridges, roads, and sewer systems. One of the most innovative structures in bridges, the parabolic arch, was first used extensively by the Romans. Civil engineers continue to use the parabolic arch today for its simple beauty.

A parabolic arch supports the Hulme Arch Bridge in England. The Hulme Arch can be modeled by the quadratic function $h(x) = -0.037x^2 + 25$, where $h(x)$ represents the height of the arch at any point x to the left or right of the center. All measures are in meters.

1. Use the Quadratic Formula to approximate the zeros of the quadratic function $h(x) = -0.037x^2 + 25$ to the nearest whole number. **−26, 26**
2. What do the zeros represent on the Hulme Arch?
3. Find the width of the Hulme Arch to the nearest meter. **52 m**
4. What is the height of the Hulme Arch? **25 m**
5. Graph the Hulme Arch on a coordinate plane. **See Solutions Manual.**

2. where the two ends of the arch meet the ground

FAST FACTS About Civil Engineers

Working Conditions

- usually work in office buildings, industrial plants, or outside on construction sites
- 40-hour work weeks, except around deadlines when more hours are expected

Education

- college degree in engineering, physical science, or mathematics
- creative and detail-oriented personalities are well-suited for the profession

Earnings

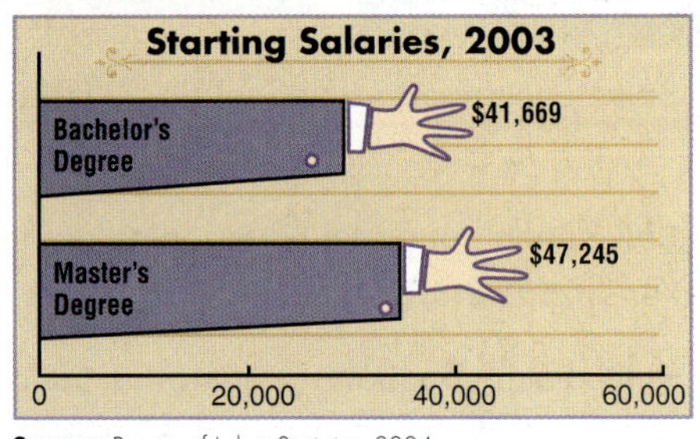

Source: Bureau of Labor Statistics, 2004

interNET CONNECTION **Career Data** For the latest information on civil engineers, visit: www.algconcepts.com

Not on the Net

If students have limited or no access to the Internet, they can find additional information in the following book.

Hagerty, D. Joseph, et al. *Opportunities in Civil Engineering Careers.* Lincolnwood, IL: Vgm Career Horizons, 1997.

For information on ABET accredited engineering programs, students should contact the following organization.

The Accreditation Board for Engineering and Technology, Inc.
111 Market Place, Suite 1050
Baltimore, MD 21202-4012

11–7 Exponential Functions

What You'll Learn
You'll learn to graph exponential functions.

Why It's Important
Finance You can use exponential functions to study the growth of bank accounts and populations. *See Example 4.*

Often, patterns can be used to describe and classify types of functions.

Hands-On Algebra

Materials: ☐ large sheet of paper

Step 1 Fold a large sheet of paper in half. Unfold it and record how many sections are formed by the creases. Refold the paper.

Step 2 Fold the paper in half again. Record how many sections are formed by the creases. Refold the paper.

Step 3 Continue folding in half and recording the number of sections until you can no longer fold the paper.

Try These

1. How many folds could you make? **1–2. See students' work.**
2. How many sections were formed?

The data collected in the Hands-On Algebra activity can be represented in a table to see a pattern more easily. Each fold increases the number of sections by a factor of 2.

Folds	Sections	Pattern
1	2	$2 = 2^1$
2	4	$2(2) = 2^2$
3	8	$2(2)(2) = 2^3$
4	16	$2(2)(2)(2) = 2^4$
5	32	$2(2)(2)(2)(2) = 2^5$

Let x equal the number of folds and let y equal the number of sections. Then the function $y = 2^x$ represents the number of sections for any number of folds. This is an example of an **exponential function**. An exponential function is of the form $y = a^x$, where $a > 0$ and $a \neq 1$. Some people also refer to equations of the form $y = ab^x + c$ as exponential functions. For this form of the exponential function, the value a is called the *coefficient*.

You can make a table to help graph exponential functions.

Example  **1**

Graph $y = 2^x$.

x	2^x	y
−1	2^{-1}	0.5
0	2^0	1
1	2^1	2
2	2^2	4
3	2^3	8

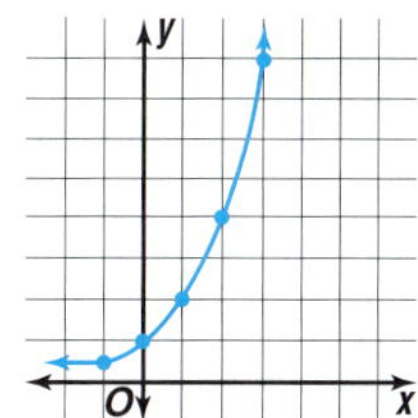

Resource Manager

Reproducible Masters

Chapter 11 Resource Masters
- *Study Guide,* p. 503
- *Skills Practice,* p. 504
- *Practice,* p. 505
- *Reading to Learn Mathematics,* p. 506
- *Enrichment,* p. 507
- *Assessment,* p. 519

Graphing Calculator, p. 30
Hands-On Algebra, pp. 125–126

Transparencies
5-Minute Check, 11–7

Technology/Multimedia
Interactive Chalkboard CD-ROM

Lesson 11–7

1 FOCUS

5-Minute Check
Lesson 11–6

Use the Quadratic Formula to solve each equation.

1. $3x^2 - 4x + 1 = 0$ **$\frac{1}{3}$, 1**
2. $-5x^2 - 10x + 6 = 0$ **$\frac{5 \pm \sqrt{55}}{-5}$**
3. $-x^2 + 2x - 3 = 0$ **no solution**
4. A baseball player hits a baseball with an upward velocity of 70 feet per second from a height of 4 feet. The equation $y = -16t^2 + 70t + 4$ gives the height y of the ball after t seconds. How long will it take the ball to be 7 feet above the ground on the way down? **about 4.3 s**

Motivating the Lesson

Real-World Connection Many real-world phenomena grow at a rate that is faster than a quadratic function. Some examples include population growth and the growth of money in an account paying compound interest. In this lesson, students will learn about exponential functions that will allow them to model these types of phenomena.

2 TEACH

Teaching Tip As you discuss the definition of an *exponential function* before Example 1, stress the difference between an exponential function and a quadratic function. In a quadratic function, the base of the expression is a variable and the exponent is a constant value of 2. In an exponential function, the base is a constant and the exponent is the variable.

In-Class Examples

Example 1

Graph $y = 1.5^x$.

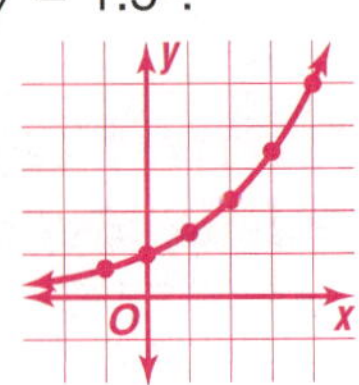

Examples 2–3

Graph each exponential function. Then state the y-intercept.

2 $y = 3^x + 1$ **$y = 2$**

3 $y = 5^x - 4$ **$y = -3$**

Answers

Your Turn

a.

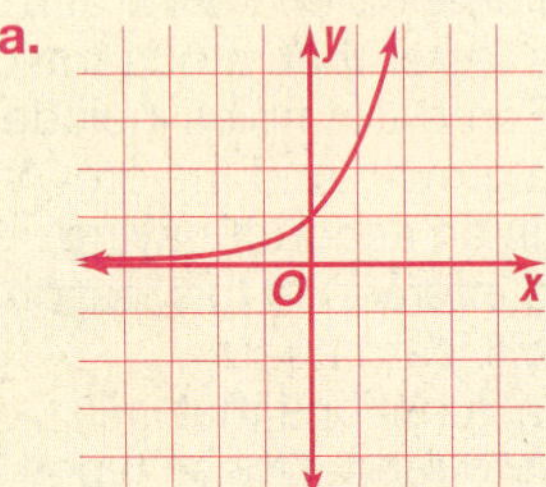

b.

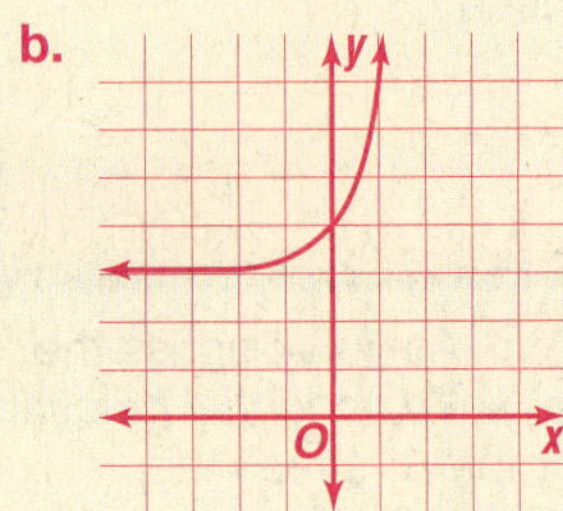

The graph of an exponential function changes little for small x values. However, as the values of x increase, the y values increase quickly.

The **initial value** of an exponential function is the value of the function when $x = 0$. This is the same as the y-intercept. Exponential functions of the form $y = a^x$ have an initial value of 1. Exponential functions of the form $y = ab^x + c$ have an initial value of c.

Examples

Graph each exponential function. Then state the y-intercept.

2 $y = 5^x$

x	5^x	y
-2	5^{-2}	0.04
-1	5^{-1}	0.2
0	5^0	1
1	5^1	5
2	5^2	25

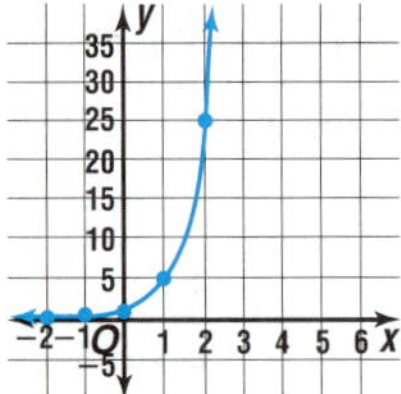

To find the y-intercept, let $x = 0$ and solve for y.

$$y = 5^0 \text{ or } 1$$

In this case, the y-intercept is 1.

3 $y = 2^x + 3$

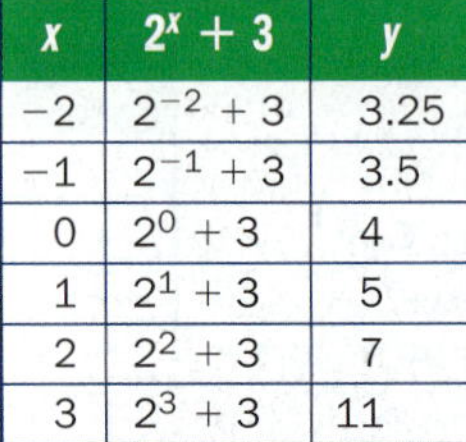

x	$2^x + 3$	y
-2	$2^{-2} + 3$	3.25
-1	$2^{-1} + 3$	3.5
0	$2^0 + 3$	4
1	$2^1 + 3$	5
2	$2^2 + 3$	7
3	$2^3 + 3$	11

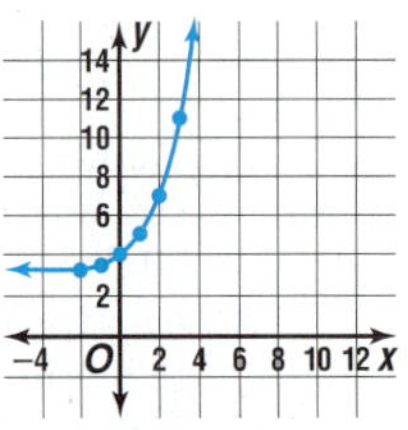

The y-intercept is 4. *The constant is 3: $1 + 3 = 4$.*

Reading Algebra

Read $2^x + 3$ as two to the x plus three.

Your Turn **Graph each function. Then state the y-intercept.**

a. $y = 2.5^x$ **$y = 1$** **b.** $y = 4^x + 3$ **$y = 4$**

a–b. See margin for graphs.

Quantities that increase rapidly have *exponential growth*. For instance, money in the bank may grow at an exponential rate.

Example 4 — Finance Link

When Taina was 10 years old, she received a certificate of deposit (CD) for \$2000 with an annual interest rate of 5%. After eight years, how much money will she have in the account?

After the first year, the CD is worth the initial deposit plus interest.

$2000 + 2000(0.05) = 2000(1 + 0.05)$ *Distributive Property*

$= 2000(1.05)$

www.algconcepts.com/extra_examples

Hands-On Algebra

Cooperative Learning Refer to the Hands-On Algebra on page 489. Have students complete this activity with their books closed so that they do not see the table on page 489. You might draw the headings of the table on the board or overhead to help students organize their work. Ask students to describe the pattern in their own words.

An additional Hands-On Algebra activity using reflection to show the relationship between the graphs of the exponential functions $y = a^x$ and $y = \left(\frac{1}{a}\right)^x$ is available in the *Hands-On Algebra Masters*, p. 126.

Hands-On Algebra Masters, p. 125

Taina can make a table to look for patterns in the growth of her CD.

Year	Balance	Pattern
0	2000	$2000(1.05)^0$
1	2000(**1.05**) = 2100	$2000(1.05)^1$
2	(2000 · **1.05**)**1.05** = 2205.00	$2000(1.05)^2$
3	(2000 · **1.05** · **1.05**)**1.05** = 2315.25	$2000(1.05)^3$

Let x represent the number of years. Then the function that represents the balance in Taina's CD is $B(x) = 2000(1.05)^x$. After eight years, it will have a balance of $2954.91.

For exponential data, you can create a best-fit curve that passes through most of the data points. You can then use this curve to make predictions.

Graphing Calculator Tutorial
See pp. 750–753.

Graphing Calculator Exploration

A city Web site tracks the number of "hits". The table shows the average number of hits per week since the launch.

Years After Launch	"Hits" per Week
4	10,000
5	20,000
6	37,500
7	62,500
8	190,000

Step 1 Use the Edit option in the STAT menu to enter the year data in L1 and the hits data in L2.

Step 2 Find an equation for a best-fit curve.

Enter: [STAT] [▶] 0 [2nd] [L1] [,] [2nd] [L2] [,] [VARS] [▶] [ENTER] [ENTER] [ENTER]

The screen displays the equation $y = a \cdot b^x$ and gives values for a and b.

Step 3 The exponential equation has been stored in the [Y=] menu. To see a graph of the function in the proper window, press [ZOOM] 9.

Make sure that the STATPLOT is turned on.

Try This

Use the graph of the best-fit curve to predict the number of weekly hits during the 15th year after the launch. **around 21,700,000 hits**

Check for Understanding

Communicating Mathematics

Vocabulary
exponential function
initial value

1. **Describe** the general shape of an exponential function. **See margin.**
2. **Explain** why the y-intercept of an exponential function in the form $y = a^x$ is always 1. $y = a^0$ **and** $a^0 = 1$, **so** $y = 1$.

Teaching Tip After completing Example 4, you might want to work with the class to develop the general formula for interest compounded yearly.

In-Class Example

Example 4

When Marcus was 2 years old, his parents invested $1000 in a money market account with an annual average interest rate of 9%. After 15 years, how much money will he have in the account? **$3642.48**

Answer

1. **Sample answer: The graph begins almost flat, but then for increasing x-values, it becomes more and more steep.**

Graphing Calculator Exploration

Students may need to be reminded that the scatter plot icon is the graphing icon that should be highlighted for Plot 1.

Study Guide, p. 503

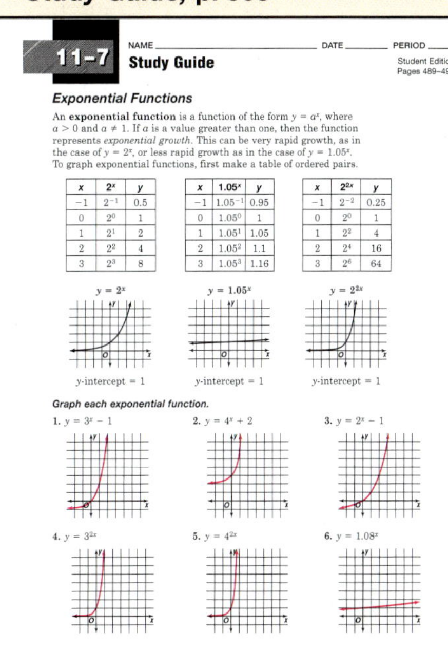

11-7 NAME ______ DATE ______ PERIOD ______

Study Guide

Student Edition Pages 489–493

Exponential Functions

An **exponential function** is a function of the form $y = a^x$, where $a > 0$ and $a \neq 1$. If a is a value greater than one, then the function represents *exponential growth*. This can be very rapid growth, as in the case of $y = 2^x$, or less rapid growth as in the case of $y = 1.05^x$. To graph exponential functions, first make a table of ordered pairs.

x	2^x	y
−1	2^{-1}	0.5
0	2^0	1
1	2^1	2
2	2^2	4
3	2^3	8

x	1.05^x	y
−1	1.05^{-1}	0.95
0	1.05^0	1
1	1.05^1	1.05
2	1.05^2	1.1
3	1.05^3	1.16

x	2^{2x}	y
−1	2^{-2}	0.25
0	2^0	1
1	2^2	4
2	2^4	16
3	2^6	64

$y = 2^x$ — y-intercept = 1

$y = 1.05^x$ — y-intercept = 1

$y = 2^{2x}$ — y-intercept = 1

Graph each exponential function.

1. $y = 3^x - 1$
2. $y = 4^x + 2$
3. $y = 2^x - 1$
4. $y = 3^{2x}$
5. $y = 4^{2x}$
6. $y = 1.08^x$

3 PRACTICE/APPLY

Error Analysis

Watch for students who have trouble scaling the y-axis when graphing exponential functions in Exercises 6–8.

Prevent by encouraging students to make a table of values prior to scaling their axes. The y-axis often needs to be scaled by 2's or 5's.

Assignment Guide

Basic: 11–25 odd, 26–32
Average: 10–20 even, 22–32

Answers

6.

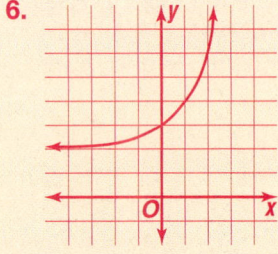

7.

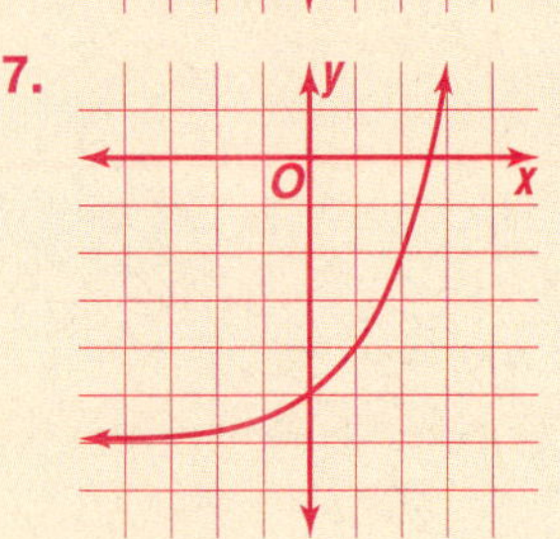

***Skills Practice*, p. 504, and *Practice*, p. 505 (shown)**

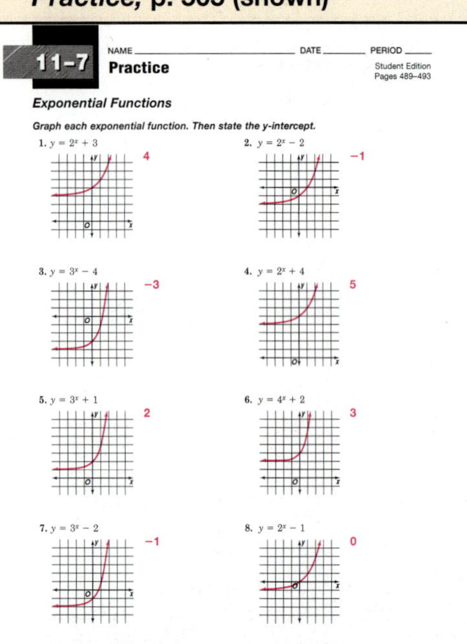

11-7 **Practice**

NAME ______ DATE ______ PERIOD ______

Student Edition Pages 489–493

Exponential Functions

Graph each exponential function. Then state the y-intercept.

1. $y = 2^x + 3$ 4
2. $y = 2^x - 2$ −1
3. $y = 3^x - 4$ −3
4. $y = 2^x + 4$ 5
5. $y = 3^x + 1$ 2
6. $y = 4^x + 2$ 3
7. $y = 3^x - 2$ −1
8. $y = 2^x - 1$ 0

Guided Practice

Getting Ready Use a calculator to find each value to the nearest hundredth.

Sample 1: 3^{-4}
Solution:
3 [^] [(−)] 4 [ENTER] *0.01*

Sample 2: 1.25^5
Solution:
1.25 [^] 5 [ENTER] *3.05*

3. 2^8 **256**   4. 4^{-1} **0.25**   5. 1.08^3 **1.26**

Examples 1–3   **6–8. See margin for graphs.**

Graph each exponential function. Then state the y-intercept.

6. $y = 2^x + 2$ **3**   7. $y = 2^x - 6$ **−5**   8. $y = 3^x + 5$ **6**

Example 4

9. **Finance** When Lindsay was born, her parents opened a savings account for her with \$500. The account pays an annual interest of 2%.
   a. Make a table showing the account's growth for 3 years. **a. See Solutions Manual.**
   b. Write a function that represents the amount of money in Lindsay's account after x number of years. **$T(x) = 500(1.02)^x$** **c. \$714.12**
   c. How much money will be in Lindsay's account after 18 years?
   d. If the initial deposit had been \$1000, how much would Lindsay's account be worth after 18 years? **\$1428.25**

Exercises

Practice

Graph each exponential function. Then state the y-intercept.

10. $y = 3^x$ **1**   11. $y = 4^x$ **1**   12. $y = 2^x - 5$ **−4**
13. $y = 2^x + 1$ **2**   14. $y = 3^x + 2$ **3**   15. $y = 4^x - 3$ **−2**
16. $y = 2^{2x}$ **1**   17. $y = 3^{2x} - 1$ **0**   18. $y = 2^{0.5x}$ **1**

10–18. See Solutions Manual for graphs.

Homework Help

For Exercises	See Examples
10, 11, 16, 18, 25, 26	1, 2
12–15, 17	3
19–21, 24	4

Extra Practice
See page 715.

Find the amount of money in a bank account given the following conditions.

19. initial deposit = \$5000, annual rate = 3%, time = 2 years **\$5304.50**
20. initial deposit = \$1500, annual rate = 10%, time = 10 years **\$3890.61**
21. initial deposit = \$3000, annual rate = 5.5%, time = 5 years **\$3920.88**

Applications and Problem Solving

22. **Taxes** In a recent year, the average tax refund had risen to about \$1700. The table shows the nest egg you would make for yourself by investing that refund at a 10% interest rate. Find your savings after 30 years. **\$29,663.98**

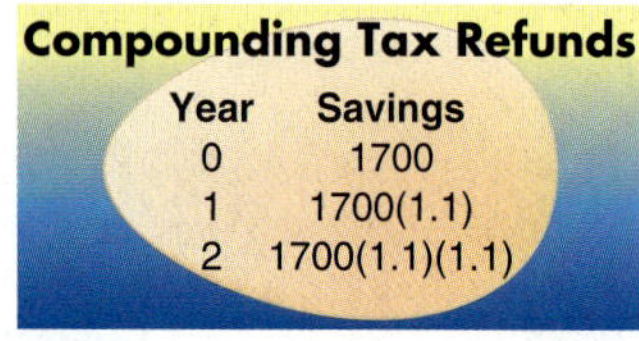

Compounding Tax Refunds

Year	Savings
0	1700
1	1700(1.1)
2	1700(1.1)(1.1)

Source: National Tax Services

23. **Finance** Nuno bought a new car for \$18,000. Once the car is driven off the lot, the car depreciates in value 15% each year. How much will the car be worth in 5 years? **\$7986.70**

Reteaching Activity

Naturalist Learners Have students work in small groups or with a partner. They should find some real-world population data and use the method of the Graphing Calculator Exploration to find an exponential equation to model the data. They can then use the data to make predictions about the population. Students may want to make a poster to display their findings.

Answer

8.

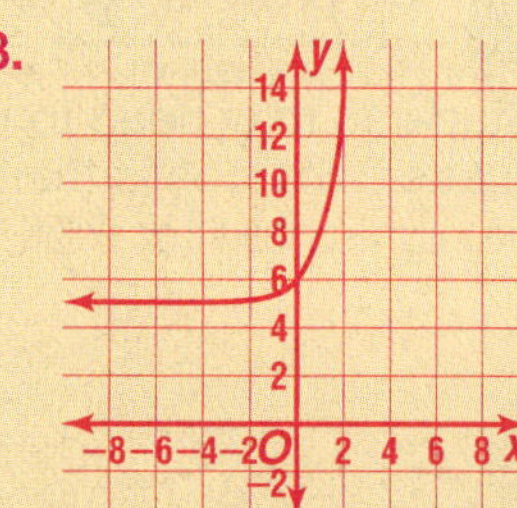

Downtown Atlanta

24. **Population** The Atlanta, Georgia, metropolitan area experienced population growth throughout the 1990s. If the population in 1990 was 2,960,000 and the annual rate of increase was 3%, predict the population in the year 2002. Use the function $P(t) = 2{,}960{,}000(1.03)^t$, where t is the number of years since 1990. **about 4,220,252**

25. **Money** Suppose on the first day of the year, your parents offered you \$500 for the month of January. If you did not take the \$500, they would give you a penny on the first day and then continue to double the amount each following day. Which is the better deal?
    a. Copy and complete the table.
    b. Write a function that represents the amount of money you will receive after x days. $y = 2^{x-1}$
    c. Use the equation in part b to find the day when you receive a payment of more than \$500. **January 17**

Day	Cents	Pattern
1	1	2^0
2	2 · 1	2^1
3	2 · 2 · 1	2^2
4	2 · 2 · 2 · 1	2^3
5	2 · 2 · 2 · 2 · 1	2^4

26. **Critical Thinking** Make a graph of each exponential function listed below. Then describe the shape of an exponential function when a number less than 1 is raised to the x power compared to a number greater than 1. **As x values increase, y values decrease rapidly.**
    a. $y = 2^x$ b. $y = 1.5^x$ c. $y = 0.5^x$ d. $y = 0.25^x$

Mixed Review

Use the Quadratic Formula to solve each equation. *(Lesson 11–6)*

27. $3p^2 - 12 = 0$ **−2, 2**
28. $x^2 + 4x = 10$ $-2 \pm \sqrt{14}$

29. **Flooring** There are 30 square yards in a roll of carpet. If the carpet is unrolled, it forms a rectangle. The length is four yards more than twice its width. *(Lesson 11–5)* **a.** $(2x + 4)x = 30$
    a. Write a quadratic equation to represent the carpet's area.
    b. Find the dimensions. **3 yd by 10 yd**

Write the point-slope form of an equation for each line passing through the given point and having the given slope. *(Lesson 7–2)*

30. (4, −3), 2 $y + 3 = 2(x - 4)$
31. (−2, 6), 0 $y - 6 = 0$

Standardized Test Practice

32. **Extended Response** During a road-trip, Javonte recorded the number of miles he had driven by the end of each hour. He graphed the number of miles driven on the number line shown. *(Lesson 7–1)*

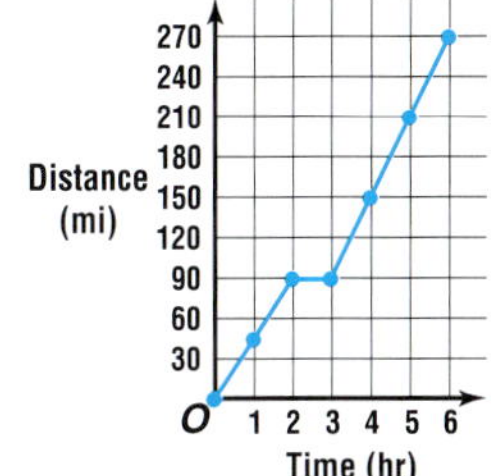

    a. Find the slope of the line between hours 0 and 2. **45**
    b. Find the slope of the line between hours 2 and 3. **0**
    c. Find the slope of the line between hours 3 and 6. **60**
    d. What does the slope represent? (*Hint:* Look at the units.)

32d. the car's velocity or speed

www.algconcepts.com/self_check_quiz

4 ASSESS

Open-Ended Assessment

Modeling Ask students to make three sketches for the exponential function $y = b^x + c$, one for when c is negative, one for when c is positive, and one for when c is zero. In each case, students should label the y-intercept.

Chapter 11, Quiz B (Lessons 11-4 through 11-7) is available in the *Chapter 11 Resource Masters*, p. 519.

Enrichment, p. 507

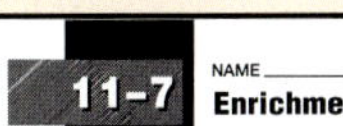

NAME ________ DATE ______ PERIOD ____

Enrichment

Student Edition Pages 489–493

Applications of Geometric Sequence

Populations often grow according to a geometric sequence. If the population of a country grows at the rate of 2% per year, then the common ratio, r, is 1.02. To find the population of a city of 100,000 after 5 years of 2% growth, use the formula ar^{n-1}, where r is the common ratio and n is the number of years.

$$ar^{n-1} = 100{,}000 \times 1.02^{5-1} \quad a = 100{,}000;\ r = 1.02;\ n = 5$$
$$= 100{,}000 \times 1.02^4$$
$$\approx 108{,}243$$

After a few years, a small change in the annual growth rate can cause enormous differences in the population.

Assume that the nation of Grogro had a population of one million in 1990. Using a calculator, find the population of the country in the years 2000, 2050, and 2100 at growth rates of 1%, 3%, and 5% per year. Record your results in the table below.

	Population of Grogro			
	Growth Rate	Year		
		2000	2050	2100
1.	1%	1,093,685	1,798,709	2,958,215
2.	3%	1,304,773	5,720,003	25,075,956
3.	5%	1,551,328	17,789,701	204,001,612

Suppose we want to find the *total* distance a bouncing ball has moved. We need a formula for the *sum* of the terms of a geometric sequence. This is called a **geometric series**.

$$S_n = \frac{a - ar^n}{1 - r}$$
a = *first term*
r = *common ratio* ($r \neq 1$)
n = *number of terms*

Use the formula above to find each sum. Then check your answer by adding.

4. 5 + 10 + 20 + 40 + 80 **155**
5. 80 + 240 + 720 + 2160 + 6480 **9680**

Extra Credit

Graph the exponential function $y = 0.5^x$. Discuss the similarities and differences between this graph and the graph of $y = 2^x$. **The graphs are mirror images of each other. Both graphs have the same y-intercept, (0, 1).**

Investigation

PREPARE

This optional investigation is designed to be completed by pairs of students over 1–2 days.

Objective

Students will create a geometric sequence and use it to define geometric sequences and to investigate their characteristics and graphs. Students also write formulas for geometric sequences and present their findings in a brochure.

Mathematical Overview

The investigation utilizes the following concepts:
- division of fractions,
- writing equations for sequences, and
- graphing exponential functions.

Suggested Time Management	
Investigation	25–35 min
Extension: Gathering Data	20–30 min
Extension: Summarizing Data	35–55 min

Motivating the Lesson

Write the following sequences on the board or overhead.

2, 4, 6, 8, 10, ...
2, 4, 8, 16, 32, ...

Ask students to describe the differences between the two sequences. Remind students that they learned about the first type of sequence, an arithmetic sequence, in the Chapter 3 Investigation. Explain that in this Investigation, they will be looking at the second type of sequence.

Cooperative Learning

This investigation offers an excellent opportunity for using cooperative groups. For more information on cooperative learning strategies and group management, see *Cooperative Learning in the Mathematics Classroom,* one of the titles in the Glencoe Mathematics Professional Series.

Chapter 11 Investigation

Materials:
- five 3-in. by 3-in. self-adhesive notes
- scissors

Geometric Sequences

In the Chapter 3 Investigation, you examined arithmetic sequences. In this investigation, you will look at a different type of sequence.

Investigate

1. Use a self-adhesive note to represent 1 whole unit. Label it "1." Cut a second self-adhesive note in half to represent one half of a unit. Label one of the resulting pieces "$\frac{1}{2}$." Cut the other piece in half again to represent one-fourth of a unit. Label one of the resulting pieces "$\frac{1}{4}$." Continue this process. Arrange the self-adhesive notes as shown below.

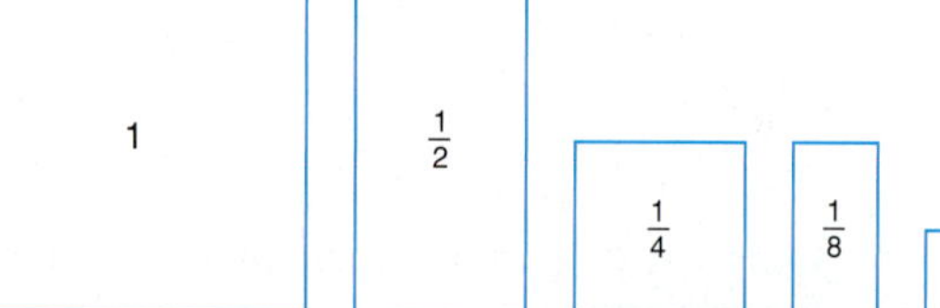

1a–b. See Solutions Manual.

a. Make a table like the one shown. Write the numbers 1–10 in column 1. Record the fractions from each note in column 2.

b. Find the ratio of the new portion and the previous portion to complete column 3.

Term Number	Portion of Self-Adhesive Note	$\frac{\text{New Portion}}{\text{Previous Portion}}$
1	1	—
2	$\frac{1}{2}$	$\frac{1}{2} \div 1 = ?$
3	$\frac{1}{4}$	$\frac{1}{4} \div \frac{1}{2} = ?$

c. List the numbers in column 2 from greatest to least.

1c. $1, \frac{1}{2}, \frac{1}{4}, \frac{1}{8}, \frac{1}{16}, \frac{1}{32}, \frac{1}{64}, \frac{1}{128}, \frac{1}{256}, \frac{1}{512}$

d. The sequence in part c is called a **geometric sequence** because the quotient between any two consecutive terms, called the **common ratio**, is always the same. What is the common ratio? $\frac{1}{2}$

2. A geometric sequence can be written as a formula in several ways.
   a. Copy and complete a table like the one below for terms 1–10.

Term	Portion of Self-Adhesive Note	Form 1	Form 2	Form 3
1	1	1	1	$1 \cdot \left(\frac{1}{2}\right)^0$
2	$\frac{1}{2}$	$1 \cdot \frac{1}{2}$	$1 \cdot \frac{1}{2}$	$1 \cdot \left(\frac{1}{2}\right)^1$
3	$\frac{1}{4}$	$\frac{1}{2} \cdot \frac{1}{2}$	$1 \cdot \frac{1}{2} \cdot \frac{1}{2}$	$1 \cdot \left(\frac{1}{2}\right)^2$

2a. See Solutions Manual.

2b. portion = 1 times one half to the (term number minus one) power

   b. You can describe Form 1 by using the formula *portion = previous portion times one half*. Another way to write this formula is $f_{n+1} = f_n \times \frac{1}{2}$. f_{n+1} is the value you are trying to find. f_n is the value of the previous term. How could you describe Forms 2 and 3 using a formula like $f_{n+1} = f_n \times \frac{1}{2}$?

3. Some bacteria reproduce exponentially by *fission*. One cell splits to become two cells. There are now two bacteria instead of one.

Term Number	Number of Bacteria	New Number of Bacteria / Previous Number of Bacteria
1	3	—
2	6	6 ÷ 3 = ?

3b. Sample answer: 3, 6, 12, 24, 48, 96, 192, 384, 768, 1536; number of bacteria = previous term number times 2

   a. Copy and complete a table like the one at the left for terms 1–10. Use 3 self-adhesive notes to represent 3 bacteria. Cut each in half to produce 6 bacteria. Continue this process. **See Solutions Manual.**
   b. Write a sequence that models the bacteria's growth. Then write a recursive formula representing the sequence.

Extending the Investigation

In this extension, you will examine other sequences and their formulas.

1. Graph the sequence in Exercise 1. Label the *x*-axis "term number" and the *y*-axis "portion of self-adhesive note." What type of function would describe the graph?
2. Use the sequence in Exercise 1 to find the sum of the first 10 terms. Suppose the sequence continued for 100 terms. What would you expect the sum to be?
3. Repeat 1 and 2 for Exercise 3. **1–3. See Solutions Manual.**

Presenting Your Investigation

Here are some ideas to help you present your conclusions to the class.

- Make a brochure showing the sequences you wrote in this investigation.
- Describe a situation that results in a sequence similar to those in this investigation.

Investigation For more information on geometric sequences, visit: www.algconcepts.com

MANAGE

Teaching Tip You might recommend that students fold the notes in half prior to cutting so that the pieces are equal. Encourage students to write their formulas in words rather than using symbols. If you prefer, you can have students use a spreadsheet program for recording the data and doing the calculations in Exercises 1 and 3.

Working in Pairs If students work in pairs, it will give them a chance to discuss the investigation as they complete it and work through any questions together. To make this investigation go more smoothly, have one student make the sticky note sequence while the other records their findings. The students should discuss the questions before writing a response. Check their work on step 3 before having them go to the Extension. They may want to divide up the different parts of the brochure to complete individually and then put it together prior to presenting it to the class.

Working as a Class You may want to conduct this investigation as a whole class activity. Use an overhead projector to display the tables and graphs.

ASSESS

Students' work should show that they understand the basic characteristics of a geometric sequence. They should also see the connection between a geometric sequence and an exponential function.

Students should add their brochures to their portfolios at this time.

Study Guide and Assessment

Understanding and Using the Vocabulary

This section provides a listing of the new terms, properties, and phrases that were introduced in this chapter. The exercises check students' understanding of the terms by using a variety of verbal formats including matching, completion, and true/false.

Glossary A complete glossary of terms appears on pages 762–783.

MindJogger Videoquizzes

MindJogger Videoquizzes provide an alternative review of concepts presented in this chapter. Students work in teams to answer questions, gaining points for correct answers.

CHAPTER 11 Study Guide and Assessment

Understanding and Using the Vocabulary

After completing this chapter, you should be able to define each term, property, or phrase and give an example or two of each.

interNET CONNECTION Review Activities For more review activities, visit: www.algconcepts.com

axis of symmetry *(p. 459)*
completing the square *(p. 478)*
discriminant *(p. 487)*
exponential function *(p. 489)*
geometric sequence *(p. 494)*
initial value *(p. 490)*
maximum *(p. 459)*
minimum *(p. 459)*
parabola *(p. 458)*
quadratic equation *(p. 468)*
Quadratic Formula *(p. 484)*
quadratic function *(p. 458)*
roots *(p. 468)*
vertex *(p. 459)*
zeros *(p. 468)*

Choose the letter of the term that best matches each statement or phrase. Each letter is used once.

1. the maximum or minimum point of a parabola **f**
2. Use this to find the exact roots of any quadratic equation. **b**
3. the method of making a quadratic expression into a perfect square **c**
4. the solutions of a quadratic equation **g**
5. the shape of the graph of any quadratic function **j**
6. a vertical line containing the vertex of a parabola **h**
7. $f(x) = a^x$, $a > 0$ and $a \neq 1$ **a**
8. $f(x) = ax^2 + bx + c$, $a \neq 0$ **i**
9. the x-intercepts of a quadratic function **d**
10. the highest point on the graph of a quadratic function **e**

a. exponential function
b. Quadratic Formula
c. completing the square
d. zeros
e. maximum
f. vertex
g. roots
h. axis of symmetry
i. quadratic function
j. parabola

Skills and Concepts

Objectives and Examples

• **Lesson 11–1** Graph quadratic functions.

Write the equation of the axis of symmetry and the coordinates of the vertex of $f(x) = x^2 - 8x + 12$.

$$x = -\frac{-8}{2(1)} \text{ or } 4$$

Since $4^2 - 8(4) + 12 = -4$, the graph has a vertex at $(4, -4)$.

Review Exercises

Write the equation of the axis of symmetry and the coordinates of the vertex of each quadratic function.

11. $y = x^2 - 3$ **$x = 0$; $(0, -3)$**
12. $y = -x^2 + 4x + 5$ **$x = 2$; $(2, 9)$**
13. $y = 3x^2 + 6x - 17$ **$x = -1$; $(-1, -20)$**
14. $y = x^2 - 3x - 4$ **$x = \frac{3}{2}$; $\left(\frac{3}{2}, -\frac{25}{4}\right)$**

www.algconcepts.com/vocabulary_review

Resource Manager

Reproducible Masters
Chapter 11 Resource Masters
• *Assessment*, pp. 509–517, 520–522

Technology/Multimedia
MindJogger Videoquizzes
ExamView® Pro

Objectives and Examples

- **Lesson 11–1** Graph quadratic functions.

Graph
$f(x) = x^2 - 8x + 12$.

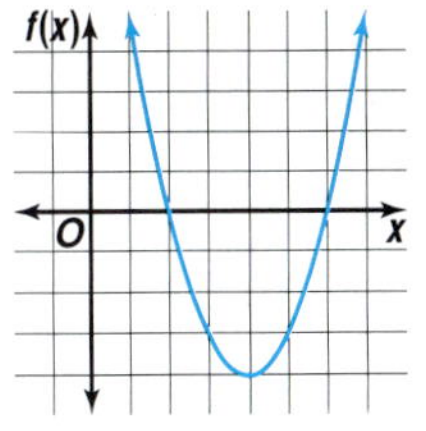

Review Exercises

Use the results from Exercises 11–14 and a table to graph each quadratic function.

15. $y = x^2 - 3$ **15–18. See margin.**
16. $y = -x^2 + 4x + 5$
17. $y = 3x^2 + 6x - 17$
18. $y = x^2 - 3x - 4$

- **Lesson 11–2** Recognize characteristics of families of parabolas.

Describe how the graph of $y = (x + 12)^2 - 5$ changes from the graph of $y = x^2$.

If $x = -12$, then $x + 12 = 0$. The graph shifts 12 units to the left. The -5 outside the parentheses shifts the graph 5 units down. The vertex is at $(-12, -5)$.

Describe how each graph changes from its parent graph of $y = x^2$. Then name the vertex.

19. $y = (x - 8)^2$ **shifts right; (8, 0)**
20. $y = x^2 - 10$ **shifts down; (0, −10)**
21. $y = 0.4x^2$ **widens; (0, 0)**
22. $y = (x + 3)^2$ **shifts left; (−3, 0)**
23. $y = (x - 1)^2 + 7$ **shifts right and up; (1, 7)**
24. $y = -1.5x^2$ **opens down, narrows; (0, 0)**

- **Lesson 11–3** Locate the roots of quadratic equations by graphing.

Find the roots of $0 = x^2 - 8x + 12$.

Based on the graph of the related function shown above, the roots are 2 and 6.

Solve each equation by graphing the related function.

25. $x^2 - x - 12 = 0$ **−3, 4**
26. $x^2 + 10x + 25 = 0$ **−5**
27. $2x^2 - 5x + 4 = 0$ **no real solutions**
28. $x^2 + x - 4 = 0$ **$-3 < x < -2$, $1 < x < 2$**
29. $6x^2 - 13x = 15$ **$-1 < x < 0$, 3**

- **Lesson 11–4** Solve quadratic equations by factoring.

Solve $x^2 - 8x - 20 = 0$.

$$x^2 - 8x - 20 = 0$$
$$(x - 10)(x + 2) = 0$$
$$x - 10 = 0 \quad \text{or} \quad x + 2 = 0$$
$$x = 10 \qquad\qquad x = -2$$

Solve each equation. Check your solution.

30. $y(y + 11) = 0$ **0, −11**
31. $(t + 4)(2t - 10) = 0$ **−4, 5**
32. $z^2 - 36 = 0$ **−6, 6**
33. $x^2 + 16x + 64 = 0$ **−8**
34. $b^2 - 5b - 14 = 0$ **−2, 7**
35. $4g^2 - 4 = 0$ **−1, 1**
36. $2x^2 + 10x + 12 = 0$ **−3, −2**

Skills and Concepts

The **Objectives and Examples** section reviews the skills and concepts of the chapter and shows completely worked examples.

The **Review Exercises** provide practice for the corresponding objectives.

Answers

15. 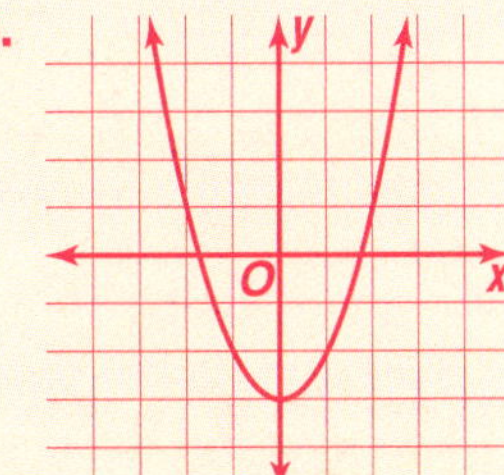

16. 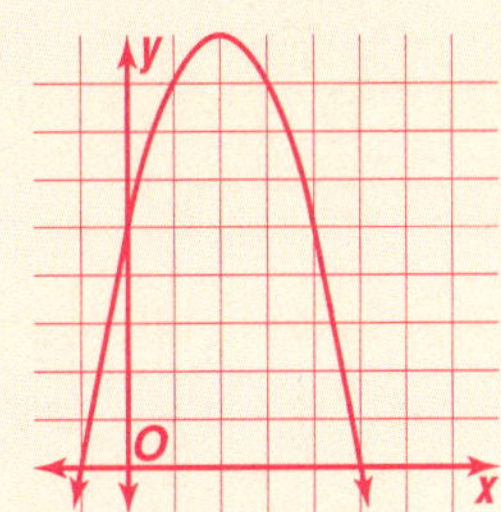

17. 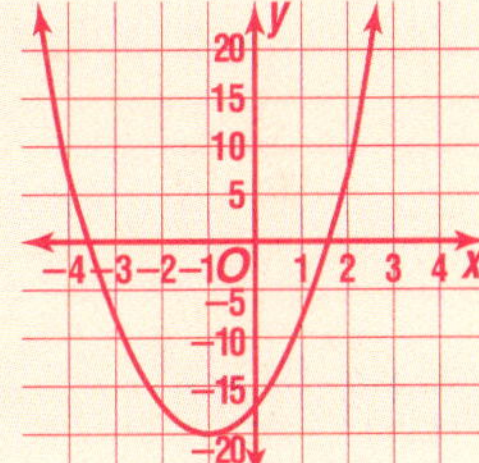

18. 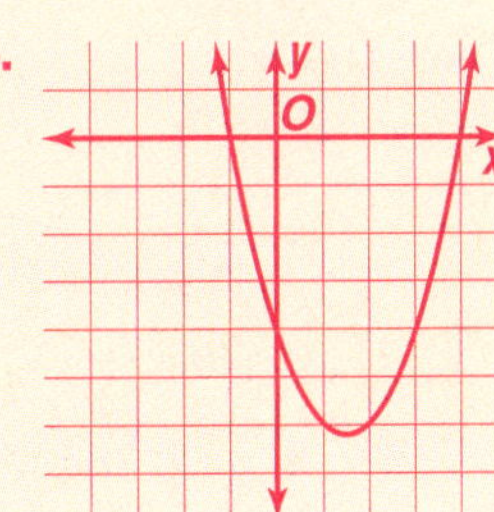

ExamView® Pro

Use the networkable ***ExamView® Pro Testmaker CD-ROM*** to:

- Create **multiple versions** of tests.
- Create **modified** tests for **inclusion** students with one mouse click.
- **Edit** existing questions and **add** your own questions.
- Build tests aligned with state standards using built-in **state curriculum correlations**.
- Change **English** tests to **Spanish** with one mouse click and vice versa.

Mixed Problem Solving
See pages 724–731.

Applications and Problem Solving

This section provides additional practice in solving real-world problems that involve the concepts of this chapter.

Answers

47.

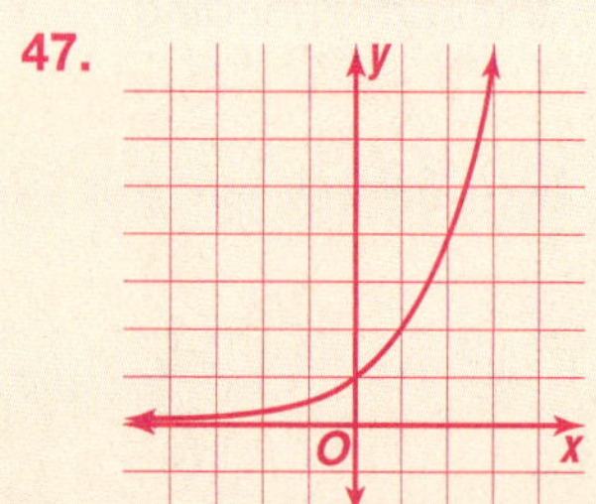

48.

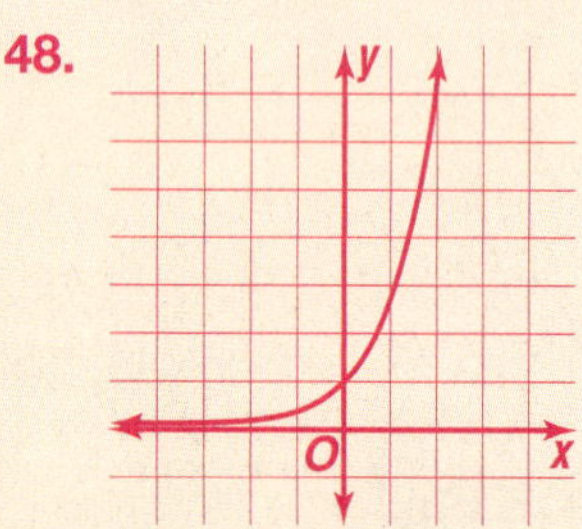

49.

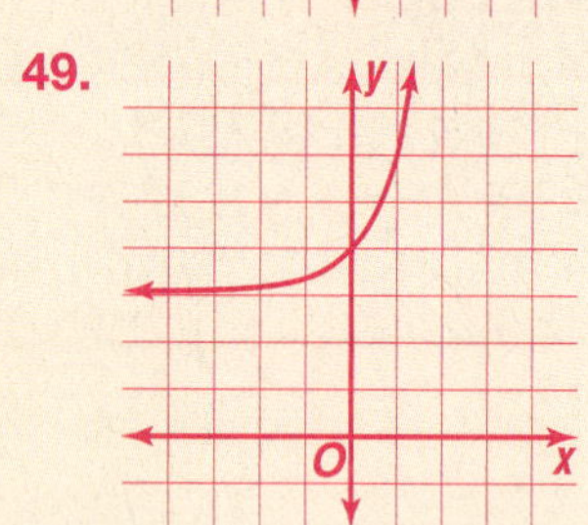

Objectives and Examples | **Review Exercises**

- **Lesson 11–5** Solve quadratic equations by completing the square.

$$x^2 + 6x + 2 = 0$$
$$x^2 + 6x = -2 \quad \textit{Subtract 2.}$$
$$x^2 + 6x + 9 = -2 + 9 \quad \textit{Complete the square.}$$
$$(x + 3)^2 = 7$$
$$x + 3 = \pm\sqrt{7} \quad \textit{Take the square root.}$$
$$x = -3 \pm \sqrt{7} \quad \textit{Subtract 3.}$$

Find the value of *c* that makes each trinomial a perfect square.

37. $r^2 + 4r + c$ **4**
38. $t^2 - 12t + c$ **36**

Solve each quadratic equation by completing the square.

39. $t^2 - 4t - 21 = 0$ **−3, 7**
40. $x^2 - 16x + 23 = 0$ **$8 \pm \sqrt{41}$**

- **Lesson 11–6** Solve equations by using the Quadratic Formula.

Solve $-x^2 + 3x + 40 = 0$.

$$x = \frac{-b \pm \sqrt{b^2 - 4ac}}{2a}$$
$$= \frac{-3 \pm \sqrt{3^2 - 4(-1)(40)}}{2(-1)}$$
$$= \frac{-3 \pm \sqrt{169}}{-2}$$
$$= \frac{-3 + 13}{-2} \quad \text{or} \quad x = \frac{-3 - 13}{-2}$$
$$= -5 \qquad\qquad x = 8$$

Solve each equation by using the Quadratic Formula. Check your solution.

41. $r^2 + 10r + 9 = 0$ **−1, −9**
42. $-2d^2 + 8d = 0$ **0, 4**
43. $3n^2 - 2n - 1 = 0$ **$-\frac{1}{3}$, 1**
44. $s^2 + 2s + 2 = 0$ **no real solutions**
45. $m^2 + 5m = -3$ **$\frac{-5 \pm \sqrt{13}}{2}$**
46. $-4x^2 + 8x + 3 = 0$ **$\frac{2 \pm \sqrt{7}}{2}$**

- **Lesson 11–7** Graph exponential functions.

Graph $y = 2^x - 1$.

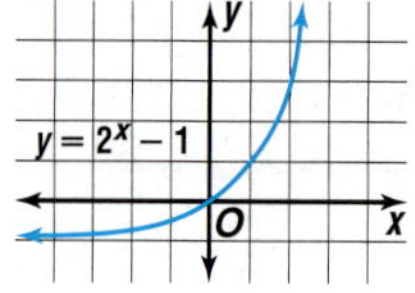

Graph each exponential function. Then state the *y*-intercept.

47. $y = 2^x$ **1**
48. $y = 2.7^x$ **1**
49. $y = 3^x + 3$ **4**
50. $y = 4^x - 4$ **−3**

47–50. See margin for graphs.

Applications and Problem Solving

51. **Finance** Game room tickets cost $5. The function $I(x) = -50x^2 + 100x + 6000$ represents the weekly income when x is the number of 50¢ price increases. Graph the function to find the ticket price that results in the maximum income. *(Lesson 11–1)* **$5.50**

52. **Tuition** A college board voted to increase tuition prices a maximum of 4% annually. If tuition is $6678, in how many years will it exceed $10,000? Use the function $P(t) = 6678(1.04)^t$, where $P(t)$ is tuition and t is the number of years. *(Lesson 11–7)* **11 yr**

***Assessment*, pp. 511–512**

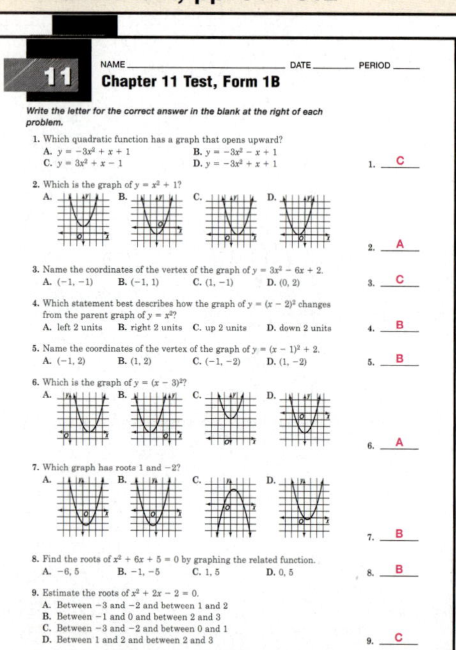

11 NAME ______ DATE ______ PERIOD ______

Chapter 11 Test, Form 1B

Write the letter for the correct answer in the blank at the right of each problem.

1. Which quadratic function has a graph that opens upward?
   A. $y = -3x^2 + x + 1$ B. $y = -3x^2 - x + 1$
   C. $y = 3x^2 + x - 1$ D. $y = -3x^2 + x + 1$ 1. **C**
2. Which is the graph of $y = x^2 + 1$?
   A. B. C. D. 2. **A**
3. Name the coordinates of the vertex of the graph of $y = 3x^2 - 6x + 2$.
   A. (−1, −1) B. (−1, 1) C. (1, −1) D. (0, 2) 3. **C**
4. Which statement best describes how the graph of $y = (x - 2)^2$ changes from the parent graph of $y = x^2$?
   A. left 2 units B. right 2 units C. up 2 units D. down 2 units 4. **B**
5. Name the coordinates of the vertex of the graph of $y = (x - 1)^2 + 2$.
   A. (−1, 2) B. (1, 2) C. (−1, −2) D. (1, −2) 5. **B**
6. Which is the graph of $y = (x - 3)^2$?
   A. B. C. D. 6. **A**
7. Which graph has roots 1 and −2?
   A. B. C. D. 7. **B**
8. Find the roots of $x^2 + 6x + 5 = 0$ by graphing the related function.
   A. −6, 5 B. −1, −5 C. 1, 5 D. 0, 5 8. **B**
9. Estimate the roots of $x^2 + 2x - 2 = 0$.
   A. Between −3 and −2 and between 1 and 2
   B. Between −1 and 0 and between 2 and 3
   C. Between −3 and −2 and between 0 and 1
   D. Between 1 and 2 and between 2 and 3 9. **C**

Assessment

Four forms of Chapter 11 Test are available in the *Chapter 11 Resource Masters.*

Chapter 11 Test, Form 1B, is shown at the left. Chapter 11 Test, Form 2B, is shown on the next page.

Form of Test			Level
1A	Multiple Choice	pp. 509–510	Average
1B	Multiple Choice	pp. 511–512	Basic
2A	Free Response	pp. 513–514	Average
2B	Free Response	pp. 515–516	Basic

CHAPTER 11 Test

For each situation, state whether it is *always* true, *sometimes* true, or *never* true.

1. The graph of a quadratic function opens upward in a "u-shape." **sometimes**
2. You can use the Quadratic Formula to solve quadratic equations. **always**
3. There are two real solutions to any quadratic equation. **sometimes**

Write the equation of axis of symmetry and the coordinates of the vertex of the graph of each quadratic function. Then graph the function. **4–5. See margin for graphs.**

4. $y = x^2 + 6x + 8$ **$x = -3$; $(-3, -1)$**
5. $y = -x^2 + 3x$ **$x = \frac{3}{2}$; $\left(\frac{3}{2}, \frac{9}{4}\right)$**

Describe how each graph changes from its parent graph of $y = x^2$. Then name the vertex of caoh graph.

6. $y = x^2 - 10$ **shifts down; $(0, -10)$**
7. $y = (x - 9)^2$ **shifts right; $(9, 0)$**
8. $y = (x + 7)^2 + 3$ **shifts left and up; $(-7, 3)$**

Solve each equation by graphing the related function. If exact roots cannot be found, state the consecutive integers between which the roots are located.

9. $y = x^2 - 8x + 16$ **4**
10. $y = x^2 - 5x - 1$ **$-1 < x < 0$; $5 < x < 6$**

Solve each equation by factoring. Check your solution.

11. $(r - 3)(r - 8) = 0$ **3, 8**
12. $s^2 - 5s + 4 = 0$ **1, 4**
13. $x^2 + 7x + 10 = 0$ **−5, −2**

Find the value of *c* that makes each trinomial a perfect square.

14. $b^2 + 10b + c$ **25**
15. $x^2 + 8x + c$ **16**
16. $g^2 - 2g + c$ **1**

Solve each equation by completing the square.

17. $a^2 + 14a = -45$ **−9, −5**
18. $v^2 - 6v + 3 = 0$ **$3 \pm \sqrt{6}$**

Solve each equation by using the Quadratic Formula.

19. $2n^2 - n - 3 = 0$ **$-1, \frac{3}{2}$**
20. $3x^2 + x + 5 = 0$ **no real solutions**

Graph each exponential function. Then state the *y*-intercept. **21–22. See margin for graphs.**

21. $y = 1.5^x$ **1**
22. $y = 2^x - 3$ **−2**

23. **Physics** A plane flying 400 feet over Antarctica drops equipment needed by research scientists below. How long will it take the equipment to touch ground? Use the formula $h = -16t^2 + 400$, where the height h is in feet and the time t is in seconds. **5 s**

24. **Geometry** The length of a rectangle is 4 inches more than its width. The area of the rectangle is 60 square inches. Find the dimensions of the rectangle. **6 in. by 10 in.**

25. **Finance** A school district has a \$64 million budget. The school board has voted to increase the budget by only 1% each year. After how many years will the budget be over \$70 million? Use the function $B(t) = 64(1.01)^t$, where $B(t)$ is the budget and t is the time in years. **10 yr**

www.algconcepts.com/chapter_test

Chapter Test Bonus Question

The solutions to the equation $2x^2 + bx - 12 = 0$ are −2 and 3. What is the value of *b*? **−2**

Answer

Page 498

50.

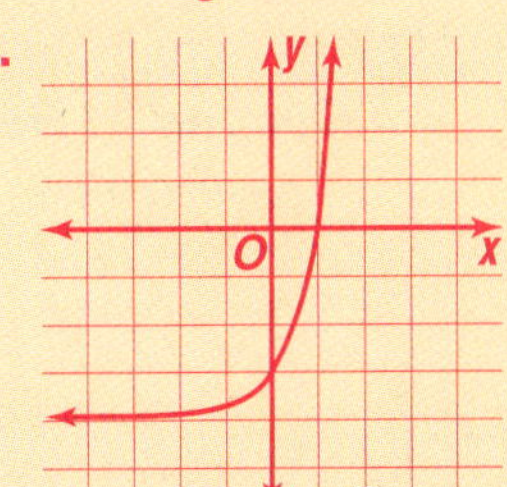

Chapter Test

Answers

4.

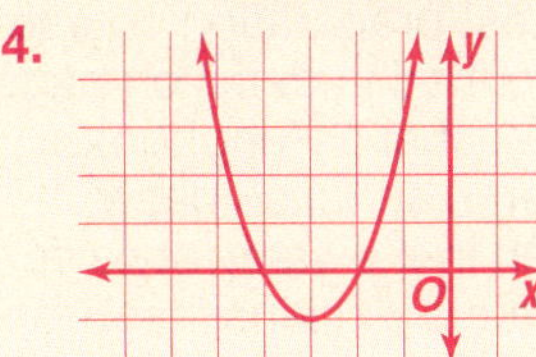

5.

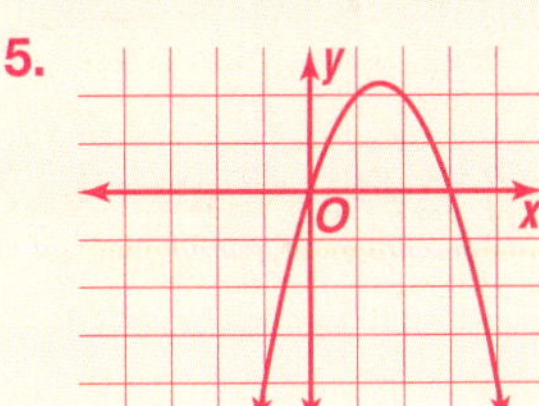

21.

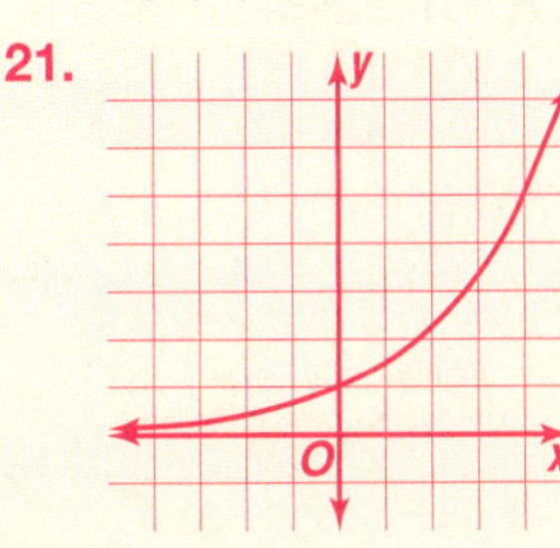

22.

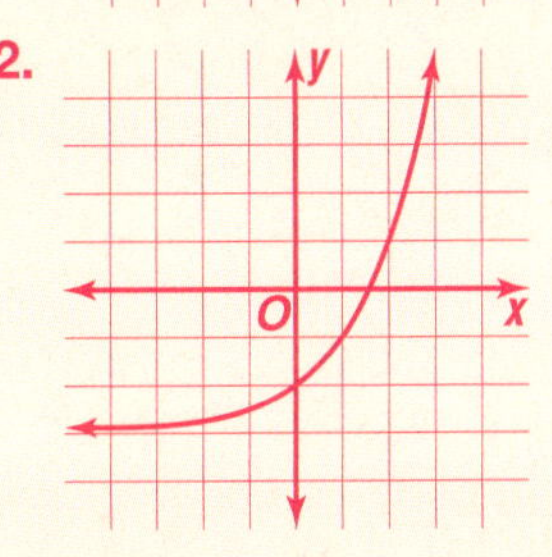

***Assessment*, pp. 515–516**

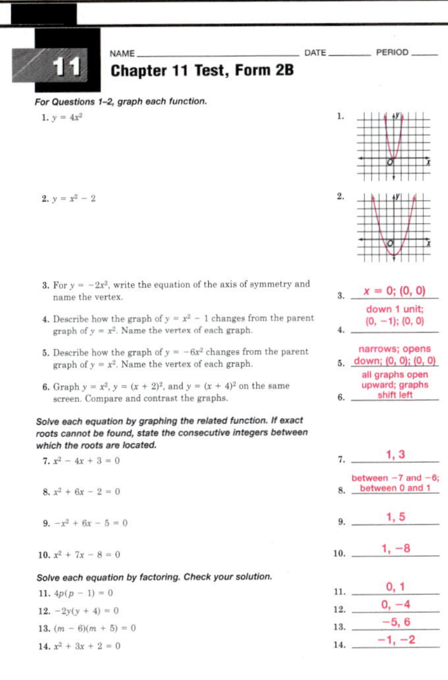

11

NAME ______ DATE ______ PERIOD ______

Chapter 11 Test, Form 2B

For Questions 1–2, graph each function.

1. $y = 4x^2$
2. $y = x^2 - 2$
3. For $y = -2x^2$, write the equation of the axis of symmetry and name the vertex. 3. $x = 0$; $(0, 0)$
4. Describe how the graph of $y = x^2 - 1$ changes from the parent graph of $y = x^2$. Name the vertex of each graph. 4. down 1 unit; $(0, -1)$; $(0, 0)$
5. Describe how the graph of $y = -6x^2$ changes from the parent graph of $y = x^2$. Name the vertex of each graph. 5. narrows; opens down; $(0, 0)$; $(0, 0)$
6. Graph $y = x^2$, $y = (x + 2)^2$, and $y = (x + 4)^2$ on the same screen. Compare and contrast the graphs. 6. all graphs open upward; graphs shift left

Solve each equation by graphing the related function. If exact roots cannot be found, state the consecutive integers between which the roots are located.

7. $x^2 - 4x + 3 = 0$ 7. 1, 3
8. $x^2 + 6x - 2 = 0$ 8. between −7 and −6; between 0 and 1
9. $-x^2 + 6x - 5 = 0$ 9. 1, 5
10. $x^2 + 7x - 8 = 0$ 10. 1, −8

Solve each equation by factoring. Check your solution.

11. $4p(p - 1) = 0$ 11. 0, 1
12. $-2y(y + 4) = 0$ 12. 0, −4
13. $(m - 6)(m + 5) = 0$ 13. −5, 6
14. $x^2 + 3x + 2 = 0$ 14. −1, −2

Preparing for Standardized Tests

Pages 500–501 are part of a complete test preparation course that is described in detail on page T9 of the Teacher's Handbook. The test items on these pages were written in the same style as those in state proficiency tests and standardized tests like ACT and SAT.

Diagnosis and Prescription

Each of the 10 test questions on page 501 is cross-referenced to the lesson where that SAT or ACT skill is covered. If students miss a particular type of problem, you can have them study that skill.

(See chart at the bottom of page 501. Note that SPT = State Proficiency Test, SAT = Scholastic Assessment Test, and ACT = American College Test.)

CHAPTER 11 Preparing for Standardized Tests

Polynomial and Factoring Problems

State proficiency tests may include a few polynomial problems. In addition, many questions on the SAT and ACT ask you to factor or evaluate polynomials.

Memorize these polynomial relationships.

1. difference of squares $a^2 - b^2 = (a + b)(a - b)$
2. perfect square trinomials $a^2 + 2ab + b^2 = (a + b)^2$; $a^2 - 2ab + b^2 = (a - b)^2$

Test-Taking Tip

Look for factors in all problems that include polynomials. Factoring is often the quick way to find an answer.

Example 1

Evaluate $2x^2y - 4y + x^4$ if $x = 2$ and $y = -1$.

(A) 12 (B) 16 (C) 20 (D) 28

Hint Apply the rules for operations carefully when negative numbers are involved.

Solution Substitute 2 for x and -1 for y. Then simplify the expression.

$$\begin{aligned} 2x^2y - 4y + x^4 &= 2(2)^2(-1) - 4(-1) + (2)^4 \\ &= 2(4)(-1) - 4(-1) + 16 \\ &= 8(-1) + 4 + 16 \\ &= -8 + 20 \\ &= 12 \end{aligned}$$

The answer is A.

Example 2

If $y < 0$, which of the polynomials must be less than $y^2 + 150y + 75^2$?

(A) $y^2 + 160y + 75^2$

(B) $(y + 75)^2$

(C) $y^2 - 10y + 75^2$

(D) $(y - 75)^2$

Hint Consider the sign of y and the coefficients of each polynomial.

Solution First expand the binomials in choices B and D. Since $(y + 75)^2 = y^2 + 150y + 75^2$, the polynomial in choice B is equal to the original polynomial, not less than it. This eliminates answer choice B. The binomial in choice D expands to $y^2 - 150y + 75^2$.

The original polynomial and each of the three remaining choices, A, C, and D, have terms y^2 and 75^2. The only differences are in the y terms. The y terms for the original polynomial and the remaining answer choices are $150y$, $160y$, $-10y$, and $-150y$. Since y is negative, the least term is $160y$.

Therefore, the polynomial that must be less than $y^2 + 150y + 75^2$ is $y^2 + 160y + 75^2$.

The answer is A.

Assessment, p. 520

11 NAME ______ DATE ______ PERIOD ______

Chapter 11 Cumulative Review

1. Name the property of equality used to state that $36 + 0 = 36$. *(Lesson 1–2)* — 1. Additive Identity Property
2. Evaluate $8 + (-3) + (-9)$. *(Lesson 2–3)* — 2. -4
3. Simplify $(-3m)(-4n)$. *(Lesson 2–5)* — 3. $12mn$
4. Solve $\frac{2}{3} = z - \left(-\frac{5}{6}\right)$. *(Lesson 3–6)* — 4. $-\frac{1}{6}$
5. Evaluate the expression $\frac{-j}{4}$ if $j = \frac{3}{4}$. *(Lesson 4–3)* — 5. $-\frac{3}{16}$
6. Solve $-8 - 0.6n = -9.5$. *(Lesson 4–5)* — 6. 2.5
7. Solve $\frac{6}{y} = \frac{18}{24}$. *(Lesson 5–1)* — 7. 8
8. Suppose you answered 34 questions correctly on a 40-question test. What percent did you answer correctly? *(Lesson 5–3)* — 8. 85%
9. For the relation $\{(4, -3), (4, 2), (-2, 8), (-3, 6)\}$, find the domain. *(Lesson 6–1)* — 9. $\{-3, -2, 4\}$
10. Write the slope-intercept form of an equation for the line passing through the point at $(2, -4)$ and having a slope of $-\frac{1}{2}$. *(Lesson 7–3)* — 10. $y = -\frac{1}{2}x - 3$
11. For what value of k are the graphs of the lines $y = kx + 5$ and $y = \frac{3}{2}x - 2$ parallel? *(Lesson 7–7)* — 11. $\frac{3}{2}$
12. Evaluate $6x + y^3$ if $x = -2$ and $y = -3$. *(Lesson 8–1)* — 12. -39
13. Simplify $(m^3n^2)(m^5n^4)$. *(Lesson 8–2)* — 13. m^8n^6
14. Simplify $(2x^2 + 4x - 2) + (-2 - x^2)$. *(Lesson 9–2)* — 14. $x^2 + 4x - 4$
15. Find $(x - 2)(x + 4)$. *(Lesson 9–4)* — 15. $x^2 + 2x - 8$
16. Find $(12x^3 + 8x^2) \div 4x$. *(Lesson 10–2)* — 16. $3x^2 + 2x$
17. Solve $-b^2 - b + 9 = 0$. *(Lesson 11–6)* — 17. $\frac{1 \pm \sqrt{37}}{-2}$

Resource Manager

Reproducible Masters

Chapter 11 Resource Masters

- *Assessment*, pp. 520–522

After you work each problem, record your answer on the answer sheet provided or on a sheet of paper.

Multiple Choice

1. Evaluate $h^2 - 5h$ if $h = -3$. *(Lesson 8–2)* **D**
   (A) −24 (B) −6
   (C) 6 (D) 24

2. A swimming pool is shaped like a rectangular prism. Its volume is 1200 cubic feet. The dimensions of a smaller pool are half the size of the larger pool's dimensions. What is the volume, in cubic feet, of the smaller pool? *(Lesson 8–1)* **A**
   (A) 150 (B) 300 (C) 600 (D) 1200

3. For all $x \neq -3$, which of the following is equivalent to the expression $\frac{x^2 - x - 12}{x + 3}$? *(Lesson 10–3)*
   (A) $x - 4$ **A**
   (B) $x - 2$
   (C) $x + 2$
   (D) $x + 4$
   (E) $x + 6$

4. If $x = -2$, then find $3x^2 - 5x - 6$. *(Lesson 8–1)* **D**
   (A) −8 (B) −4 (C) 10 (D) 16

5. The figures below represent a sequence of numbers. What is an expression for the nth term of this sequence? *(Lesson 8–1)* **C**

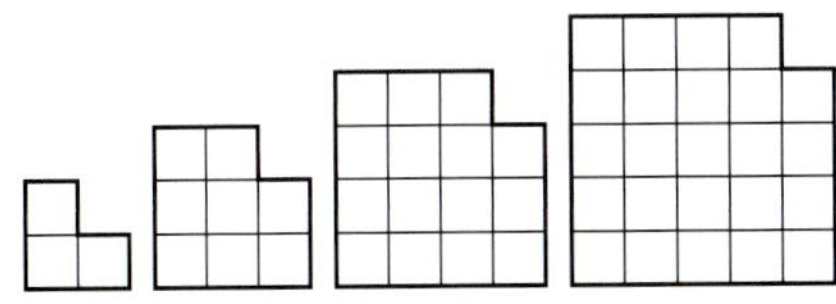

   (A) $n^2 + 2$ (B) $n^2 - 1$
   (C) $n(n + 2)$ (D) $2n^3 - 1$

www.algconcepts.com/standardized_test

6. Which quadratic equation has roots $\frac{1}{2}$ and $\frac{1}{3}$? *(Lesson 11–4)* **E**
   (A) $5x^2 - 5x - 2 = 0$
   (B) $5x^2 - 5x + 1 = 0$
   (C) $6x^2 + 5x + 1 = 0$
   (D) $6x^2 - 6x + 1 = 0$
   (E) $6x^2 - 5x + 1 = 0$

7. Use the function table to find the value of y when $x = 8$. *(Lesson 7–3)* **C**
   (A) 9 (B) 10.5
   (C) 12 (D) 13.5

x	y
1	1.5
2	3.0
3	4.5
4	6.0
5	7.5

8. If $a > 0$ and $a \neq b$, which statement must be true? *(Lesson 10–5)* **A**
   (A) $\frac{a^2 - b^2}{a - b} - b$ is positive.
   (B) $\frac{a^2 - b^2}{a - b} + b$ is positive.
   (C) $\frac{a^2 - b^2}{a - b}$ is positive.
   (D) $\frac{a^2 - b^2}{a - b}$ is negative.

Grid In

9. Solve for x if $\frac{x^2 + 7x + 12}{x + 4} = 5$. *(Lesson 10–3)* **2**

Extended Response

10. Consider $2x + 3y = 15$. *(Lesson 6–3)*
    Part A Make a table of at least six pairs of values that satisfy the equation above.
    Part B Use your table from Part A to graph the equation. **10A–B. See margin.**

A bubble-in answer sheet for these practice problems is available on page A1 of the *Chapter 11 Resource Masters.*

Additional Practice

Additional test practice questions are available in the *Chapter 11 Resource Masters*, pp. 521–522.

Answers

10A. Sample answer:

x	y
−6	9
−3	7
0	5
3	3
6	1
9	−1

10B.

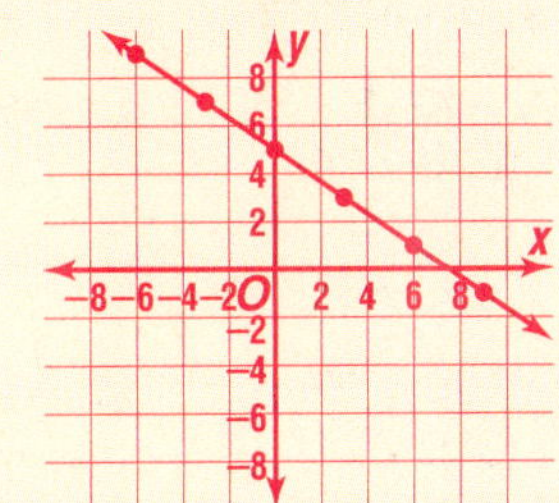

***Assessment*, pp. 521–522**

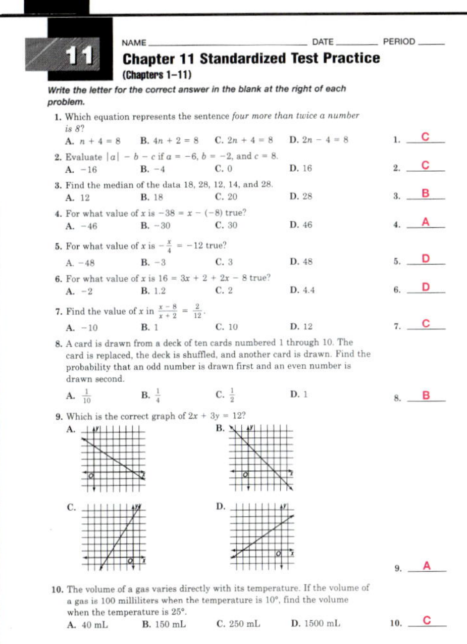

NAME ______ DATE ______ PERIOD ______

11 Chapter 11 Standardized Test Practice
(Chapters 1–11)

Write the letter for the correct answer in the blank at the right of each problem.

1. Which equation represents the sentence *four more than twice a number is 8*?
   A. $n + 4 = 8$ B. $4n + 2 = 8$ C. $2n + 4 = 8$ D. $2n - 4 = 8$ 1. C
2. Evaluate $|a| - b - c$ if $a = -6$, $b = -2$, and $c = 8$.
   A. −16 B. −4 C. 0 D. 16 2. C
3. Find the median of the data 18, 28, 12, 14, and 28.
   A. 12 B. 18 C. 20 D. 28 3. B
4. For what value of x is $-38 = x - (-8)$ true?
   A. −46 B. −30 C. 30 D. 46 4. A
5. For what value of x is $-\frac{x}{4} = -12$ true?
   A. −48 B. −3 C. 3 D. 48 5. D
6. For what value of x is $16 = 3x + 2 + 2x - 8$ true?
   A. −2 B. 1.2 C. 2 D. 4.4 6. D
7. Find the value of x in $\frac{x - 8}{x + 2} = \frac{2}{12}$.
   A. −10 B. 1 C. 10 D. 12 7. C
8. A card is drawn from a deck of ten cards numbered 1 through 10. The card is replaced, the deck is shuffled, and another card is drawn. Find the probability that an odd number is drawn first and an even number is drawn second.
   A. $\frac{1}{10}$ B. $\frac{1}{4}$ C. $\frac{1}{2}$ D. 1 8. B
9. Which is the correct graph of $2x + 3y = 12$?
   A. B. C. D. 9. A
10. The volume of a gas varies directly with its temperature. If the volume of a gas is 100 milliliters when the temperature is 10°, find the volume when the temperature is 25°.
    A. 40 mL B. 150 mL C. 250 mL D. 1500 mL 10. C

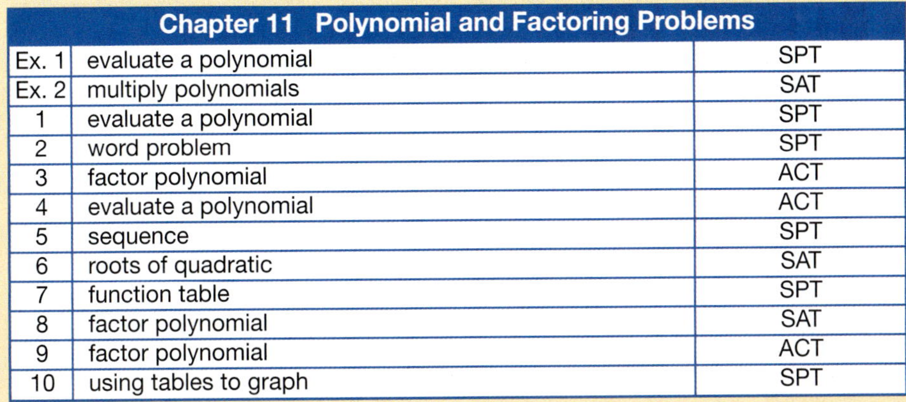

Chapter 11 Polynomial and Factoring Problems		
Ex. 1	evaluate a polynomial	SPT
Ex. 2	multiply polynomials	SAT
1	evaluate a polynomial	SPT
2	word problem	SPT
3	factor polynomial	ACT
4	evaluate a polynomial	ACT
5	sequence	SPT
6	roots of quadratic	SAT
7	function table	SPT
8	factor polynomial	SAT
9	factor polynomial	ACT
10	using tables to graph	SPT

CHAPTER 12

Resource Manager

Inequalities

Instructional Objectives

Lesson (pages)	Objectives	NCTM Standards 2000	State/Local Objectives
12–1 (504–508)	Graph inequalities on a number line.	1, 6, 8, 9	
12–2 (509–513)	Solve inequalities involving addition and subtraction.	1, 6, 8, 9	
12–3 (514–518)	Solve inequalities involving multiplication and division.	1, 6, 8, 9	
12–4 (519–523)	Solve inequalities involving more than one operation.	1, 6, 8, 9	
12–5 (524–529)	Solve compound inequalities.	1, 6, 8, 9	
12–6 (530–534)	Solve inequalities involving absolute value.	1, 6, 8, 9	
12–7 (535–539)	Graph inequalities on the coordinate plane.	1, 3, 6, 8, 9	
Investigation (540–541)	Solve and graph quadratic inequalities.	1, 3, 6, 8, 9, 10	

Key to NCTM Standards 2000
1 Number & Operations; **2** Algebra; **3** Geometry; **4** Measurement; **5** Data Analysis & Probability; **6** Problem Solving; **7** Reasoning & Proof; **8** Communications; **9** Connections; **10** Representation

Suggested Pacing

See page T13 for a complete course-planning calendar.

Standard refers to schedules that provide 45- to 55-minute periods that meet each day.
Block refers to schedules that provide approximately 90-minute periods which may meet every day for one semester or every other day over two semesters.

PACING	DAY 1	DAY 2	DAY 3	DAY 4	DAY 5	DAY 6
Standard Core (Chapters 1–13)	Lesson 12–1	Lesson 12–2		Lesson 12–3	Lesson 12–4	Lesson 12–5
Standard Enhanced (Chapters 1–15)	Lesson 12–1	Lesson 12–2		Lesson 12–3	Lesson 12–4	Lesson 12–5
Block Core (Chapters 1–13)	Chapter 11 Test & Lesson 12–1	Lesson 12–2	Lessons 12–3 & 12–4	Lessons 12–5 & 12–6	Lesson 12–7 & INV	SG+A
Block Enhanced (Chapters 1–15)	Chapter 11 Test & Lesson 12–1	Lessons 12–2 & 12–3	Lessons 12–4 & 12–5	Lessons 12–6 & 12–7	INV & SG+A	Chapter Test & Lesson 13–1

Instructional Resources

Lesson	Materials and Manipulatives (see below for Glencoe Manipulative Resources)	Blackline Masters (page numbers): Chapter 12 Resource Masters: Study Guide	Practice (Skills & Average)	Reading to Learn Mathematics	Enrichment	Assessment	Hands-On Algebra*	School-to-Workplace*	Graphing Calculator Masters*	5-Minute Check Transparencies
12–1	ruler 1, 2, 4 coordinate planes 4 masking tape	523	524–525	526	527		131			12–1
12–2	ruler 1, 2, 4 straws scissors 1, 2 pipe cleaners	528	529–530	531	532		132			12–2
12–3	ruler 1, 2, 4	533	534–535	536	537	569	133, 134	12		12–3
12–4	graphing calculator	538	539–540	541	542	568	135		33	12–4
12–5	ruler 1, 2, 4 coordinate planes 4 overhead transparency sheets permanent marker	543	544–545	546	547		136, 137			12–5
12–6	ruler 1, 2, 4 flowers or leaves	548	549–550	551	552		138			12–6
12–7	coordinate planes 4 ruler 1, 2, 4	553	554–555	556	557	569	139		34, 35	12–7
Investigation	grid paper 1, 4 ruler 1, 2, 4 yellow and blue colored paper									
Study Guide & Assessment/ Chapter Test	ruler 1, 2, 4 coordinate planes 4					559–567, 570–572				

See page 502c for examples of these instructional materials.

Key to Glencoe Manipulative Resources

1Classroom Manipulative Resources 2Student Manipulative Resources 3Overhead Manipulative Resources 4Hands-On Algebra Masters

INV = Investigation *SG+A = Study Guide and Assessment*

DAY 7	DAY 8	DAY 9	DAY 10	DAY 11	DAY 12	DAY 13
Lesson 12–5	Lesson 12–6	Lesson 12–7		INV	SG+A	Chapter Test
Lesson 12–5	Lesson 12–6	Lesson 12–7	INV	SG+A	Chapter Test	
Chapter Test & Lesson 13–1						

Resource Manager

The pages shown on this page are a small sample of the materials available on *TeacherWorks: All-in-One Lesson Planner and Resource Center.*

This CD-ROM includes all of the blackline masters and transparencies available for this program.

It also includes a lesson planner and interactive Teacher Edition, so you can customize lesson plans and reproduce classroom resources quickly and easily, from just about anywhere.

Applications

School-to-Workplace Masters, p. 12

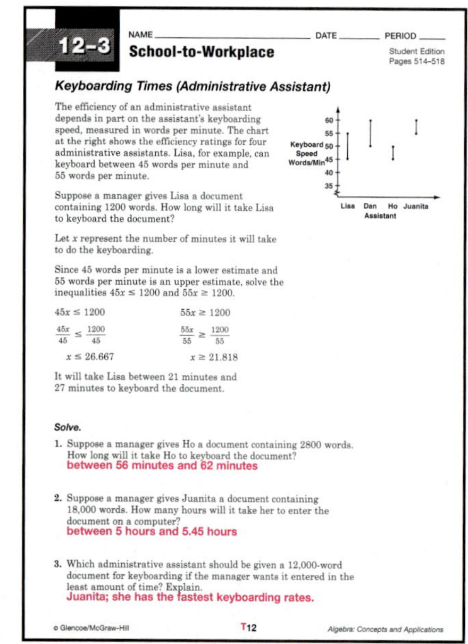

12-3 NAME ______ DATE ______ PERIOD ______

School-to-Workplace

Student Edition Pages 514–518

Keyboarding Times (Administrative Assistant)

The efficiency of an administrative assistant depends in part on the assistant's keyboarding speed, measured in words per minute. The chart at the right shows the efficiency ratings for four administrative assistants. Lisa, for example, can keyboard between 45 words per minute and 55 words per minute.

Suppose a manager gives Lisa a document containing 1200 words. How long will it take Lisa to keyboard the document?

Let x represent the number of minutes it will take to do the keyboarding.

Since 45 words per minute is a lower estimate and 55 words per minute is an upper estimate, solve the inequalities $45x \le 1200$ and $55x \ge 1200$.

$45x \le 1200$ | $55x \ge 1200$

$\frac{45x}{45} \le \frac{1200}{45}$ | $\frac{55x}{55} \ge \frac{1200}{55}$

$x \le 26.667$ | $x \ge 21.818$

It will take Lisa between 21 minutes and 27 minutes to keyboard the document.

Solve.

1. Suppose a manager gives Ho a document containing 2800 words. How long will it take Ho to keyboard the document?
   between 56 minutes and 62 minutes
2. Suppose a manager gives Juanita a document containing 18,000 words. How many hours will it take her to enter the document on a computer?
   between 5 hours and 5.45 hours
3. Which administrative assistant should be given a 12,000-word document for keyboarding if the manager wants it entered in the least amount of time? Explain.
   Juanita; she has the fastest keyboarding rates.

© Glencoe/McGraw-Hill T12 *Algebra: Concepts and Applications*

Manipulatives/Modeling

Hands-On Algebra Masters, pp. 131–139

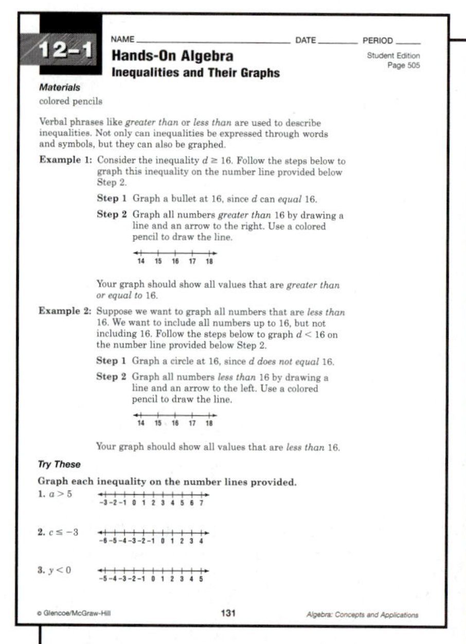

12-1 NAME ______ DATE ______ PERIOD ______

Hands-On Algebra
Inequalities and Their Graphs

Student Edition Page 505

Materials

colored pencils

Verbal phrases like *greater than* or *less than* are used to describe inequalities. Not only can inequalities be expressed through words and symbols, but they can also be graphed.

Example 1: Consider the inequality $d \ge 16$. Follow the steps below to graph this inequality on the number line provided below Step 2.

Step 1 Graph a bullet at 16, since d can *equal* 16.

Step 2 Graph all numbers *greater than* 16 by drawing a line and an arrow to the right. Use a colored pencil to draw the line.

14 15 16 17 18

Your graph should show all values that are *greater than or equal to* 16.

Example 2: Suppose we want to graph all numbers that are *less than* 16. We want to include all numbers up to 16, but not including 16. Follow the steps below to graph $d < 16$ on the number line provided below Step 2.

Step 1 Graph a circle at 16, since *d does not equal* 16.

Step 2 Graph all numbers *less than* 16 by drawing a line and an arrow to the left. Use a colored pencil to draw the line.

14 15 16 17 18

Your graph should show all values that are *less than* 16.

Try These

Graph each inequality on the number lines provided.

1. $a > 5$ −3 −2 −1 0 1 2 3 4 5 6 7
2. $c \le -3$ −6 −5 −4 −3 −2 −1 0 1 2 3 4
3. $y < 0$ −5 −4 −3 −2 −1 0 1 2 3 4 5

© Glencoe/McGraw-Hill 131 *Algebra: Concepts and Applications*

Technology/Multimedia

Graphing Calculator Masters, pp. 33–35

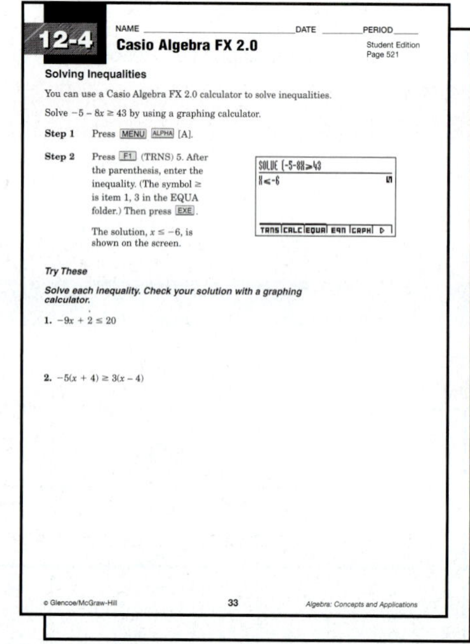

12-4 NAME ______ DATE ______ PERIOD ______

Casio Algebra FX 2.0

Student Edition Page 521

Solving Inequalities

You can use a Casio Algebra FX 2.0 calculator to solve inequalities.

Solve $-5 - 8x \ge 43$ by using a graphing calculator.

Step 1 Press MENU ALPHA [A].

Step 2 Press F1 (TRNS) 5. After the parenthesis, enter the inequality. (The symbol ≥ is item 1, 3 in the EQUA folder.) Then press EXE.

The solution, $x \le -6$, is shown on the screen.

Try These

Solve each inequality. Check your solution with a graphing calculator.

1. $-9x + 2 \le 20$
2. $-5(x + 4) \ge 3(x - 4)$

© Glencoe/McGraw-Hill 33 *Algebra: Concepts and Applications*

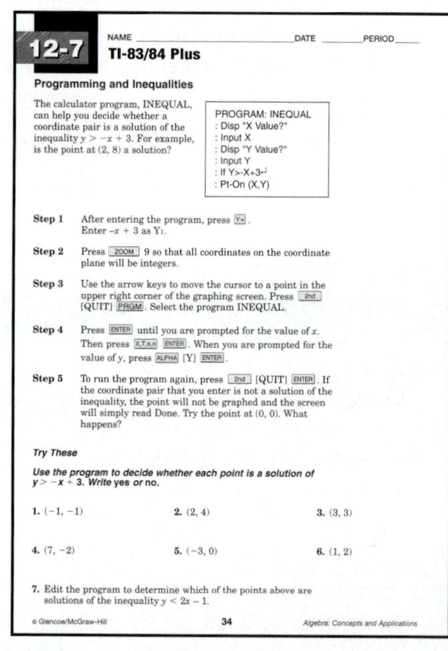

12-7 NAME ______ DATE ______ PERIOD ______

TI-83/84 Plus

Programming and Inequalities

The calculator program, INEQUAL, can help you decide whether a coordinate pair is a solution of the inequality $y > -x + 3$. For example, is the point at (2, 8) a solution?

```
PROGRAM: INEQUAL
: Disp "X Value?"
: Input X
: Disp "Y Value?"
: Input Y
: If Y>-X+3↵
: Pt-On (X,Y)
```

Step 1 After entering the program, press Y=. Enter $-x + 3$ as Y_1.

Step 2 Press ZOOM 9 so that all coordinates on the coordinate plane will be integers.

Step 3 Use the arrow keys to move the cursor to a point in the upper right corner of the graphing screen. Press 2nd [QUIT] PRGM. Select the program INEQUAL.

Step 4 Press ENTER until you are prompted for the value of x. Then press X,T,θ,n ENTER. When you are prompted for the value of y, press ALPHA [Y] ENTER.

Step 5 To run the program again, press 2nd [QUIT] ENTER. If the coordinate pair that you enter is not a solution of the inequality, the point will not be graphed and the screen will simply read Done. Try the point at (0, 0). What happens?

Try These

Use the program to decide whether each point is a solution of $y > -x + 3$. Write yes or no.

1. (−1, −1) 2. (2, 4) 3. (3, 3)

4. (7, −2) 5. (−3, 0) 6. (1, 2)

7. Edit the program to determine which of the points above are solutions of the inequality $y < 2x - 1$.

© Glencoe/McGraw-Hill 34 *Algebra: Concepts and Applications*

GLENCOE'S ASSESSMENT ADVANTAGE

Assessment Resources

Type	Student Edition	Teacher's Wraparound Edition	Chapter 12 Resource Masters
Ongoing Assessment	Quizzes 1 and 2, pp. 518, 534	5-Minute Check, pp. 504, 509, 514, 519, 524, 530, 535	Mid-Chapter Test, p. 568 Quizzes A and B, p. 569
Mixed Review	Mixed Review, pp. 508, 513, 517, 523, 529, 534, 539 Standardized Test Practice, Chapters 1–12, pp. 546–547		Cumulative Review, p. 570 Standardized Test Practice, pp. 571–572
Error Analysis	You Decide, pp. 507, 532	Error Analysis, pp. 507, 512, 516, 522, 527, 533, 538	
Standardized Test Prep	Standardized Test Practice, pp. 508, 513, 518, 523, 529, 534, 539 Standardized Test Practice, Chapters 1–12, pp. 546–547		Standardized Test Practice, pp. 571–572
Open-Ended Assessment	Writing Math, pp. 511, 538 Investigation, pp. 540–541 Portfolio, pp. 503, 541	Act It Out: p. 529 Speaking: pp. 518, 539 Writing: pp. 513, 523, 534 Modeling: p. 508	Extended Response Assessment, p. 567
Chapter Assessment	Study Guide and Assessment, pp. 542–544 Chapter Test, p. 545		Multiple-Choice Tests (Forms 1A, 1B), pp. 559–562 Free-Response Tests (Forms 2A, 2B), pp. 563–566

Additional Chapter Resources

Student Edition
Hands-On Algebra, p. 511
Graphing Calculator Exploration, p. 521

Teacher's Classroom Resources
Manipulatives/Modeling
Teacher's Guide for Overhead Manipulative Resources

Meeting Individual Needs
Prerequisite Skills Booklet
Spanish Study Guide and Assessment, pp. 80–86, 127–128

Teaching Aids
Solutions Manual
5-Minute Check Transparencies

Glencoe Technology

Instructional
AlgePASS CD-ROM, Lessons 29, 30
Interactive Chalkboard CD-ROM
StudentWorks Plus CD-ROM
Multimedia Applications CD-ROM, Activity 7
Vocabulary PuzzleMaker (online)

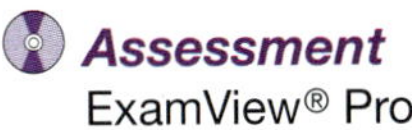

Assessment
ExamView® Pro

Teaching Aids
Answer Key Maker CD-ROM
TeacherWorks CD-ROM

Visit ***www.algconcepts.com*** *for data updates, career information, games, and other interactive activities.*

CHAPTER PREVIEW 12

Mathematics of the Chapter

This chapter provides students with a thorough exploration of inequalities. Students will begin by graphing inequalities on a number line. Then they will learn to solve one-step and multi-step linear inequalities. Next, students will solve compound inequalities and graph their solutions, and will extend these skills to solving inequalities involving absolute value. Finally, students will investigate inequalities in two variables and graph their solutions in the coordinate plane.

What You'll Learn

Have students read over the lists of key ideas and key vocabulary. Have them make a list of any words with which they are not familiar.

Why It's Important

Point out to students that this is only one of many reasons why each objective is important. Others are provided in each lesson.

Real-Life Videos engage students, showing them how math is used in everyday situations. Use Video 2 to discuss how inequalities and systems of equations are used in fighting forest fires and managing an aquarium.

CHAPTER 12 Inequalities

What You'll Learn

Key Ideas

- Graph inequalities on a number line. *(Lesson 12–1)*
- Solve inequalities involving one or two operations. *(Lessons 12–2 to 12–4)*
- Solve compound inequalities. *(Lesson 12–5)*
- Solve inequalities involving absolute value. *(Lesson 12–6)*
- Graph inequalities on the coordinate plane. *(Lesson 12–7)*

Key Vocabulary

boundary *(p. 535)*
compound inequality *(p. 524)*
quadratic inequality *(p. 540)*

Why It's Important

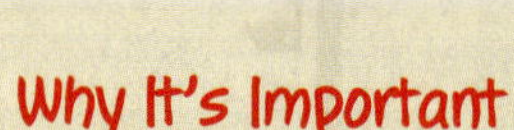

Cycling The Tour de France is the world's most famous and difficult bicycle race. Approximately 20 professional teams of nine riders each participate in the 3-week-long race every July. They race approximately 3600 kilometers (about 2235 miles) traveling through France, Belgium, Italy, Germany, and Spain.

Inequalities represent real-world situations in which a quantity falls within a range of values. You will use an inequality to represent the size of parts for a bicycle in Lesson 12–6.

CHAPTER 12 LINKS

Lesson	12–1	12–2	12–3	12–4	12–5	12–6	12–7
Math in the Workplace	Postal Service	Sports	Savings	School	Nutrition	Manufacturing	Budgeting
Applications and Connections	Sports Conservation Entertainment Safety Finance	Budgeting Fundraising Wrestling Academics Volunteerism	Safety Time Production Budgeting Contests Earthquakes	Weather Employment Recreation Finance Budgeting	Pet Care Construction Taxes Cooking Chemistry Welding	Finance Chemistry Sales Travel	Sales Animals Airlines Travel
Math Integration			Geometry Probability		Geometry	Measurement	

Study these lessons to improve your skills.

✓Check Your Readiness

✓ Lesson 3–1, pp. 94–99

Replace each • with <, >, or = to make a true sentence.

1. $2 \bullet 2.5$ <
2. $-7 \bullet 5$ <
3. $0.56 \bullet 0.05$ >
4. $\frac{4}{12} \bullet \frac{1}{3}$ =
5. $-6 \bullet 0(6)$ <
6. $14 - 7 \bullet (-4)(3)$ >

✓ Lesson 2–1, pp. 52–57

Graph each set of numbers on a number line. See margin.

7. $\{1, 3, 6\}$
8. $\{-2, 3, 4\}$
9. $\{4, 0, -1\}$
10. $\{-5, -3, 2\}$

✓ Lesson 3–6, pp. 122–127

✓ Lesson 3–7, pp. 128–131

✓ Lesson 4–4, pp. 160–164

✓ Lesson 4–5, pp. 165–170

Solve each equation. Check your solution. 21. −10, 10

11. $n + 7 = 12$ 5
12. $d - 12 = 7$ 19
13. $-2 + r = 8$ 10
14. $m + 14 = 20$ 6
15. $p - 18 = 21$ 39
16. $-8 + w = 11$ 19
17. $|x| = 6$ −6, 6
18. $5 + |h| = 9$ −4, 4
19. $14 = |j - 2|$ −12, 16
20. $-|y| = -9$ −9, 9
21. $|g| - 2 = 8$
22. $|k + 4| = 10$ −14, 6
23. $4b = 36$ 9
24. $-16 = 4t$ −4
25. $\frac{w}{5} = 3$ 15
26. $96 = 12d$ 8
27. $-5v = -35$ 7
28. $-12 = \frac{b}{3}$ −36
29. $3z + 11 = 23$ 4
30. $1 + 6c = -5$ −1
31. $\frac{1}{3}r - 5 = 2$ 21

FOLDABLES™ Study Organizer

Make this Foldable to help you organize your Chapter 12 notes. Begin with four sheets of grid paper.

❶ **Fold** each sheet in half from top to bottom.

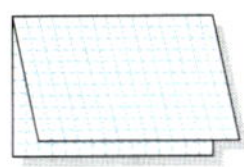

❷ **Cut** each sheet along the fold. Staple the eight half-sheets together to form a booklet.

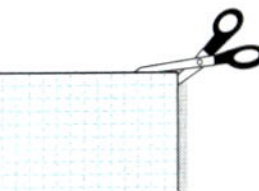

❸ **Label** each page with a lesson number and title.

Reading and Writing As you read and study the chapter, write notes and examples for each lesson in the journal.

FOLDABLES™ Study Organizer On each tab of this Foldable students should write rules for working with inequalities. Then students should solve an inequality of each type listed and graph the solution of the inequality.

Answers

Check Your Readiness

This section provides a review of the basic concepts needed before beginning Chapter 12. Lesson references are included for additional student help.

Use Exercises	To Prepare for Lesson(s)
1–6	12–1, 12–2, 12–3, 12–4, 12–7
7–10	12–1, 12–2, 12–5, 12–6
11–16	12–2
17–22	12–6
23–28	12–3
29–31	12–4

Quick Review Math Handbook hot words hot topics

Glencoe Algebra Lesson	Quick Review Math Handbook Hot Topic
12–1, 12–2, 12–3	6–4, 6–6
12–4, 12–5	6–6
12–6	6–4
12–7	6–7

Interactive Study Notebook with Foldables™

This note-taking guide reinforces excellent note-taking skills. It includes:

✓ **Build Your Vocabulary** tool for including mathematical terminology in notes.

✓ **Review It** activities with links to prior lessons and items to include in **Foldables™**.

✓ **Bringing It All Together** feature to help students review for the Chapter Test.

✓ **Are You Ready for the Chapter Test?** feature to allow students to assess their own readiness for the Chapter Test.

✓ **Teacher Edition** with transparencies to guide note taking. Corresponds to the *Teacher's Wraparound Edition* In-Class Examples and *Interactive Chalkboard CD-ROM*.

12-1 Inequalities and Their Graphs

1 FOCUS

5-Minute Check
Chapter 11

1. Write the equation of the axis of symmetry and the coordinates of the vertex of the graph of the quadratic function $y = -x^2 + 2x + 2$. Then graph the function.

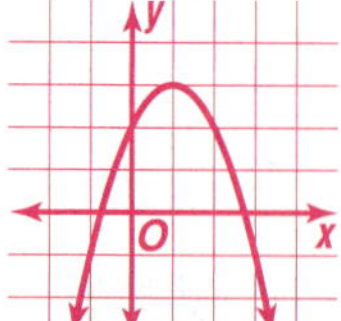

$x = 1$; (1, 3)

2. Describe how the graph of $y = -(x + 3)^2$ changes from the parent graph of $y = x^2$. Then name the vertex of the graph. **same shape, but opens downward and shifted to the left 3 units; (−3, 0)**
3. Solve $s^2 - s - 12 = 0$ by factoring. **−3, 4**
4. Use the Quadratic Formula to solve $z^2 - 4z + 2 = 0$. **$2 \pm \sqrt{2}$**
5. Find the amount of money in a bank account given an initial deposit of \$2500, an annual rate of 6%, and a time of 12 years. **\$5030.49**

Motivating the Lesson

Real-World Connection Point out to students that short-term parking garages often charge in hours, with a portion of an hour charged as a full hour. For example, parking for 3 hours or 3 hours 59 minutes costs the same amount, but at 4 hours, the cost goes up. Have students describe this situation, $3 \leq h < 4$, with as many verbal phrases as they can think of. Point out that these phrases are the language of inequalities, which they will begin to explore in this lesson.

What You'll Learn
You'll learn to graph inequalities on a number line.

Why It's Important
Postal Service Inequalities can be used to describe postal regulations. *See Exercise 12.*

You already know that equations are mathematical statements that describe two expressions with equal values. When the values of the two expressions are *not* equal, their relationship can be described in an inequality.

Verbal phrases like *greater than* or *less than* describe inequalities. For example, 6 is *greater than* 2. This is the same as saying 2 is *less than* 6.

The chart below lists other phrases that indicate inequalities and their corresponding symbols.

Look Back
Inequalities: Lesson 3–1

Inequalities			
<	≤	>	≥
• less than • fewer than	• less than or equal to • at most • no more than • a maximum of	• greater than • more than	• greater than or equal to • at least • no less than • a minimum of

Example  1

Suppose the minimum driving age in your state is 16. Write an inequality to describe people who are *not* of legal driving age in your state.

Let d represent the ages of people who are *not* of legal driving age.

The ages of all drivers — d; *are greater than or equal to* — $\geq$; *16 years.* — 16

$d \geq 16$ is the same as $16 \leq d$.

Then d is *less than* 16, or $d < 16$.

Your Turn **a. Let n = number of papers; $n \leq 50$.**

a. Lisa can carry no more than 50 newspapers on her paper route. Express the number of papers that Lisa can carry as an inequality.

Not only can inequalities be expressed through words and symbols, but they can also be graphed.

Resource Manager

Reproducible Masters
Chapter 12 Resource Masters
- *Study Guide,* p. 523
- *Skills Practice,* p. 524
- *Practice,* p. 525
- *Reading to Learn Mathematics,* p. 526
- *Enrichment,* p. 527

Hands-On Algebra, p. 131

Transparencies
5-Minute Check, 12–1

Technology/Multimedia
Interactive Chalkboard CD-ROM

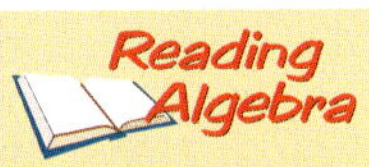

Reading Algebra

On a graph, a circle means that the number is *not* included. A bullet means that the number *is* included.

Consider the inequality $d \geq 16$.

Step 1 Graph a bullet at 16 to show that d can *equal* 16.

Step 2 Graph all numbers *greater than* 16 by drawing a line and an arrow to the right.

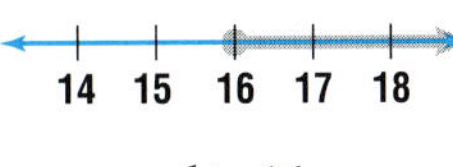

$d \geq 16$

The graph shows all values that are *greater than or equal to* 16.

Now suppose we want to graph all numbers that are *less than* 16. We want to include all numbers up to 16, but not including 16.

Step 1 Graph a circle at 16 to show that d *does not equal* 16.

Step 2 Graph all numbers *less than* 16 by drawing a line and an arrow to the left.

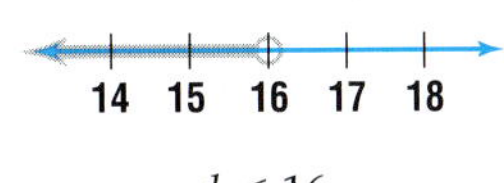

$d < 16$

The graph shows all values that are *less than* 16.

Examples

Graph each inequality on a number line.

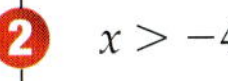

2 $x > -4$

Since x cannot equal -4, graph a circle at -4 and shade to the right.

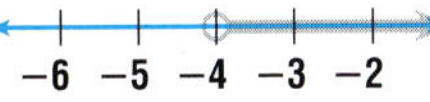

3 $n \geq 1.5$

Since n can equal 1.5, graph a bullet at 1.5 and shade to the right.

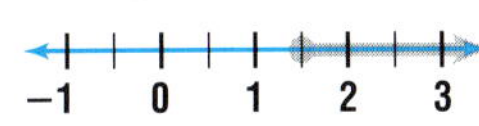

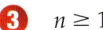

Your Turn

b. $y < 3$ c. $p \leq -\frac{1}{2}$

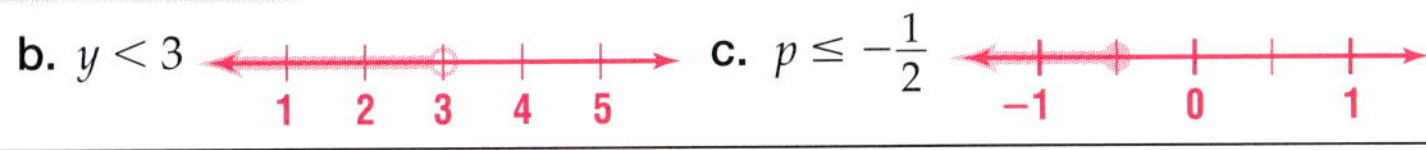

You can also write inequalities given their graphs.

Examples

Write an inequality for each graph.

4

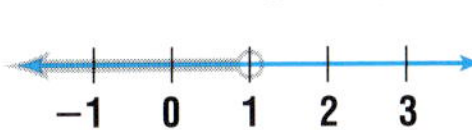

Locate where the graph begins. This graph begins at 1, but 1 is not included. Also note that the arrow is to the left. The graph describes values that are less than 1.

So, $x < 1$.

2 TEACH

Teaching Tip Before discussing Example 1, point out to students that the $\leq$ and $\geq$ symbols each contain the bottom half of an equal sign. This might help students distinguish the meanings of these symbols from those of $<$ or $>$ symbols.

In-Class Example

Example 1

Many movie theaters give a senior-citizen discount to people who are 65 or over. Write an inequality that describes those who are eligible to receive the discount. $a \geq 65$

Teaching Tip While discussing Examples 2–4, point out to students that to check their graphs, they can pick a point in the shaded region. This point should satisfy the inequality.

In-Class Examples

Examples 2–3

Graph each inequality on a number line.

2 $x \leq -1$

−3 −2 −1 0 1

3 $k > \frac{3}{4}$

Example 4

Write an inequality for the graph. $x \geq 2$

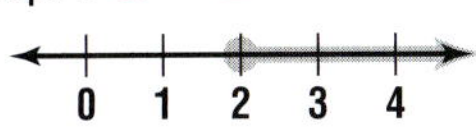

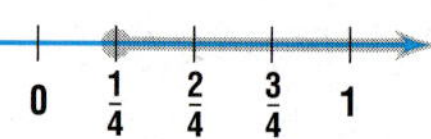

Locate where the graph begins. This graph begins at $\frac{1}{4}$ and includes $\frac{1}{4}$. Note that the arrow is to the right. The graph describes values greater than or equal to $\frac{1}{4}$.

So, $x \geq \frac{1}{4}$.

Your Turn d. $x \leq -2$ e. $x \geq \frac{2}{3}$

d. 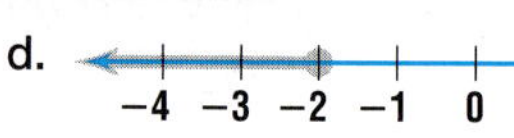

e. 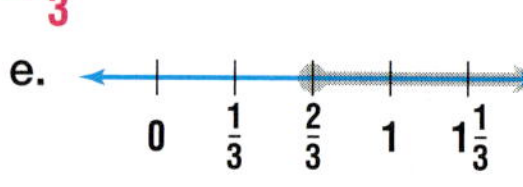

Inequalities are commonly used in the real world.

Real World Example 6 Sports Link

To play junior league soccer, you must be at least 14 years of age.

A. Write an inequality to represent this situation.

Let a represent the ages of people who can play junior league soccer. Then write an inequality using ≥ since the soccer players have to be greater than or equal to 14 years of age.

$a \geq 14$

B. Graph the inequality on a number line.

To graph the inequality, first graph a bullet at 14. Then include all ages *greater than* 14 by drawing a line and an arrow to the right.

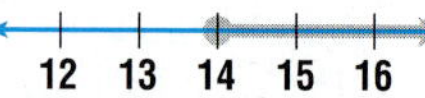

C. The Valdez children are 10, 13, 14, and 16 years old. Which of the children can play junior league soccer?

The set of the children's ages, {10, 13, 14, and 16}, can be called a *replacement set*. It includes possible values of the variable a. In this case, only 14 and 16 satisfy the inequality $a \geq 14$ and are members of the *solution set*. So, the two Valdez children that are 14 and 16 years old can play junior league soccer.

Check for Understanding

Communicating Mathematics

1. **Match** each inequality symbol with its description.
   - i. less than or equal to **b** — a. ≥
   - ii. at least **a** — b. ≤
   - iii. fewer than **d** — c. >
   - iv. greater than **c** — d. <

In Class Examples

Example 5

Write an inequality for the graph.

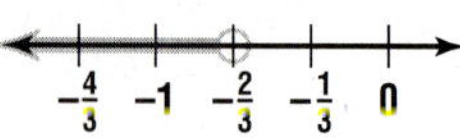

$y < -\frac{2}{3}$

Example 6

No more than 12 students can fit in a large school van. Write an inequality to express this information. Then graph the inequality on a number line.

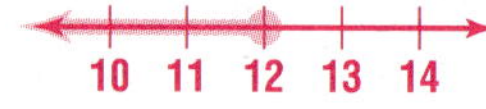

$s \leq 12$

Answers

Page 507

2. There is a bullet at 12 and an arrow pointing to the right.
3. The graph of $x \leq 4$ has a bullet at 4. The graph of $x < 4$ has a circle at 4.

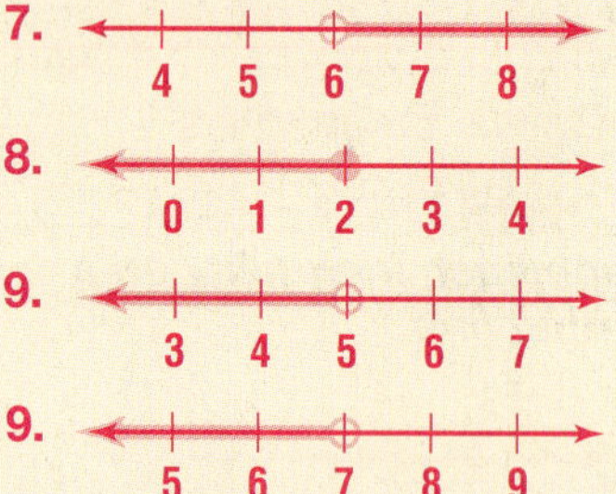

Study Guide, p. 523

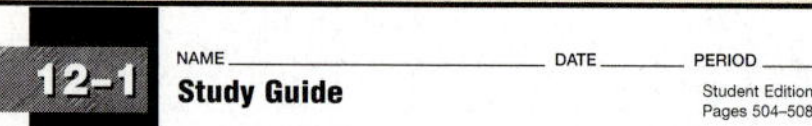

12-1 NAME ______ DATE ______ PERIOD ______

Study Guide — Student Edition Pages 504–508

Inequalities and Their Graphs

Many mathematical relationships can be expressed with inequalities. For example, the President of the United States must be at least 35 years old. If a represents his or her age, this can be expressed with the inequality $a \geq 35$. Some verbal phrases that can be used for inequalities are listed in the chart below.

<	≤	>	≥
• less than • fewer than	• less than or equal to • at most • no more than • a maximum of	• greater than • more than	• greater than or equal to • at least • no less than • a minimum of

Example 1: Write and graph an inequality to describe the age of people who *cannot* be President of the United States.

Let a represent the age of a person who is less than 35 years old. Then $a < 35$. Since a cannot equal 35, graph a *circle* at 35. Then graph all numbers less than 35 by drawing a line and an arrow to the left.

32 33 34 35 36 37

Example 2: Graph $n \geq 2.5$ on a number line.

Since n can equal 2.5, graph a *bullet* at 2.5. Then graph all numbers greater than 2.5 by drawing a line and an arrow to the right.

1 2 3 4 5 6

Write an inequality to describe each number. Then graph the inequality on a number line.

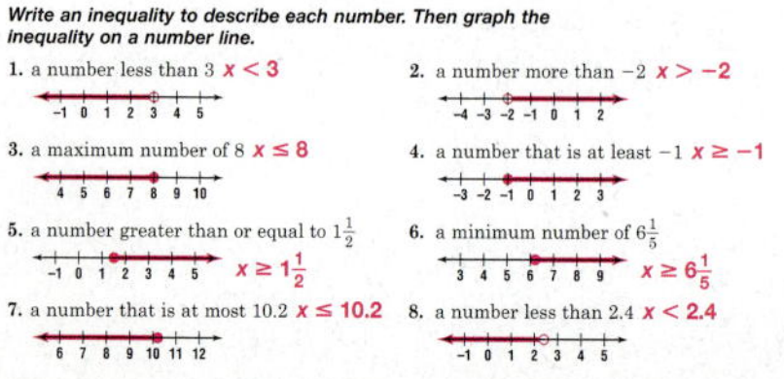

1. a number less than 3 $x < 3$ (−1 0 1 2 3 4 5)
2. a number more than −2 $x > -2$ (−4 −3 −2 −1 0 1 2)
3. a maximum number of 8 $x \leq 8$ (4 5 6 7 8 9 10)
4. a number that is at least −1 $x \geq -1$ (−3 −2 −1 0 1 2 3)
5. a number greater than or equal to $1\frac{1}{2}$ $x \geq 1\frac{1}{2}$ (−1 0 1 2 3 4 5)
6. a minimum number of $6\frac{1}{5}$ $x \geq 6\frac{1}{5}$ (3 4 5 6 7 8 9)
7. a number that is at most 10.2 $x \leq 10.2$ (6 7 8 9 10 11 12)
8. a number less than 2.4 $x < 2.4$ (−1 0 1 2 3 4 5)

Reteaching Activity

Kinesthetic Learners Make a large number line on the floor using masking tape. Mark off the integers from −5 to 5 and ask 11 students to stand on the integers. Write both verbal and mathematical inequalities on the board. For each inequality, have the 11 students raise their hands if their point satisfies the inequality.

2. **Describe** the graph of $x \geq 12$. **2–3. See margin.**

3. **Explain** the difference between the graphs of $x \leq 4$ and $x < 4$.

4. **You Decide** Soto says that $x \geq 7$ is the same as $7 \leq x$. Darrell says that it is not. Who is correct? Explain. **Soto; in both, the inequality symbol is opening toward the x.**

Guided Practice

Example 1

Write an inequality to describe each number.

5. a number less than 14 $x < 14$
6. a number no less than 0 $x \geq 0$

Examples 2 & 3

7–9. See margin.

Graph each inequality on a number line.

7. $x > 6$
8. $x \leq 2$
9. $5 > x$

Examples 4 & 5

Write an inequality for each graph.

10. 6 7 8 9 10 $x \geq 8$
11. −3 −2 −1 0 1 $x < -1$

Example 6

12. **Mail** All letters mailed with one first-class stamp can weigh no more than 1 ounce. Write an inequality to represent this situation. Then graph the inequality. $\ell \leq 1$ 1 2 3 4

Exercises

Practice

Homework Help	
For Exercises	See Examples
13–18	1
19–30	2, 3
31–38	4, 5
39–42	6
Extra Practice	
See page 715.	

Write an inequality to describe each number. **14. $x \leq 13$**

13. a number greater than 4 $x > 4$
14. a number less than or equal to 13
15. a number less than 9 $x < 9$
16. a number no more than 7 $x \leq 7$
17. a number that is at least −8 $x \geq -8$
18. a number more than −6 $x > -6$

Graph each inequality on a number line. **19–30. See margin.**

19. $x < 7$
20. $y > 3$
21. $b \leq 0$
22. $a \leq 5$
23. $x > -4$
24. $1 \leq z$
25. $x \geq 2.5$
26. $d < 1.9$
27. $x < -3.4$
28. $g > \frac{3}{4}$
29. $1\frac{1}{2} \geq x$
30. $h \leq -\frac{1}{3}$

Write an inequality for each graph.

31. 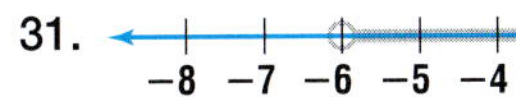

$x > -6$

32. −2 −1 0 1 2
$x \geq 0$

33. 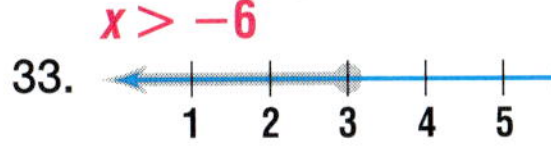

$x \leq 3$

34. −11 −10 −9 −8 −7
$x < -9$

35. 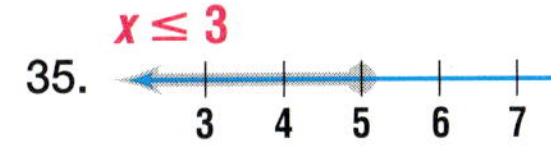

$x \leq 5$

36. 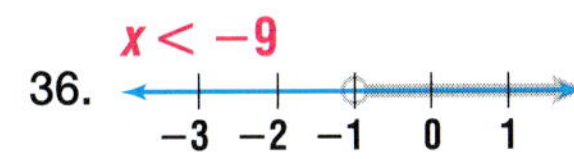

$x > -1$

37. 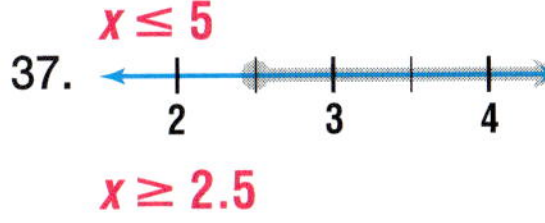

$x \geq 2.5$

38. 1.5 2 2.5
$x > 1.75$

3 PRACTICE/APPLY

Error Analysis

Watch for students who reverse the direction of the inequality sign when translating verbal inequalities that involve a negation, such as *no less than* in Exercise 6 or *not greater than*.

Prevent by encouraging students to translate the original verbal statement to a statement they understand better before writing the mathematical inequality. For example, students may recognize that *no less than* means the same thing as *at least*, which they may feel more comfortable translating into mathematical symbols.

Teaching Tip Discuss the domain of the variable in Exercise 12. Can the weight be a negative number or zero? This real-world example actually requires the compound inequality *greater than 0 and less than or equal to 1*. Students will explore compound inequalities in Lesson 12-5.

Assignment Guide

Basic: 13–41 odd, 42–50
Average: 14–38 even, 39–50

Answer

20. 1 2 3 4 5

***Skills Practice,* p. 524, and *Practice,* p. 525 (shown)**

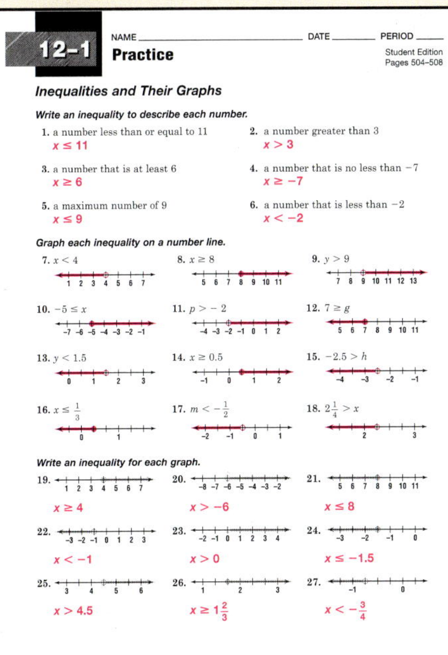

12-1 NAME ______ DATE ______ PERIOD ______

Practice Student Edition Pages 504–508

Inequalities and Their Graphs

Write an inequality to describe each number.

1. a number less than or equal to 11 $x \leq 11$
2. a number greater than 3 $x > 3$
3. a number that is at least 6 $x \geq 6$
4. a number that is no less than −7 $x \geq -7$
5. a maximum number of 9 $x \leq 9$
6. a number that is less than −2 $x < -2$

Graph each inequality on a number line.

7. $x < 4$
8. $x \geq 8$
9. $y > 9$
10. $-5 \leq x$
11. $p > -2$
12. $7 \geq g$
13. $y < 1.5$
14. $x \geq 0.5$
15. $-2.5 > h$
16. $x \leq \frac{1}{3}$
17. $m < -\frac{1}{2}$
18. $2\frac{1}{4} > x$

Write an inequality for each graph.

19. $x \geq 4$
20. $x > -6$
21. $x \leq 8$
22. $x < -1$
23. $x > 0$
24. $x \leq -1.5$
25. $x > 4.5$
26. $x \geq 1\frac{2}{3}$
27. $x < -\frac{3}{4}$

Answers

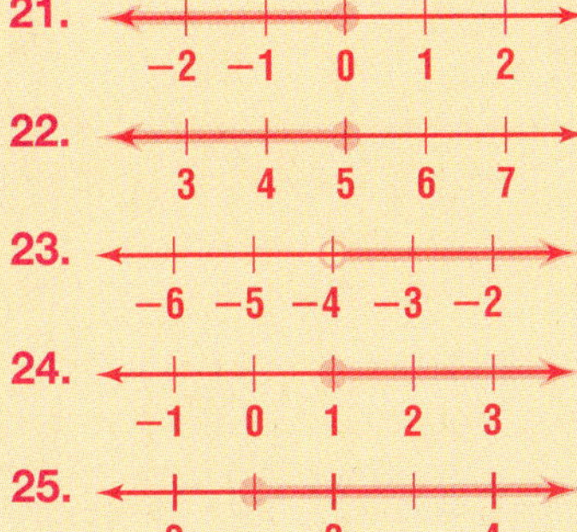

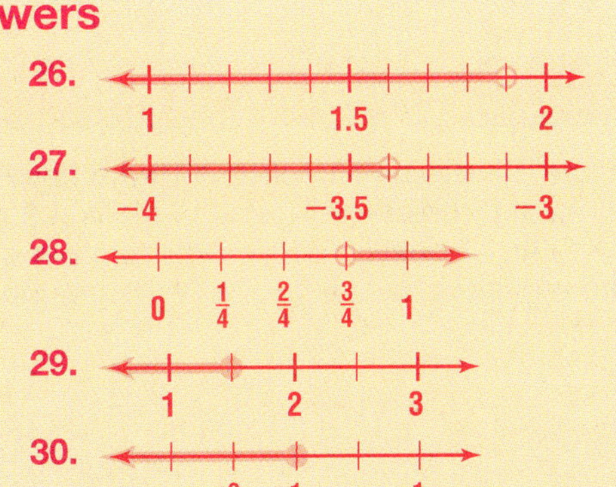

4 ASSESS

Open-Ended Assessment

Modeling Draw a large number line on a poster board or on the chalkboard. You may want to include decimal or fractional values. Make two large arrows out of construction paper, one with an open circle at one end and the other with a closed circle at one end. Have students take turns writing mathematical inequalities on the board or overhead or announcing verbal inequalities. Have other students place the appropriate arrow in the appropriate position on the number line to graph each inequality.

Applications and Problem Solving

For Exercises 39–41, write an inequality to represent each situation. Then graph the inequality on a number line.

39. **Conservation** At a pond, you must return any fish that you catch if it weighs less than 3 pounds. $f < 3$

1 2 3 4 5

40. **Entertainment** A roller coaster requires a person to be at least 42 inches tall to ride it. $h \geq 42$

40 41 42 43 44

41. **Safety** The elevators in an office building have been approved to hold a maximum of 3600 pounds. $w \leq 3600$

3400 3600 3800

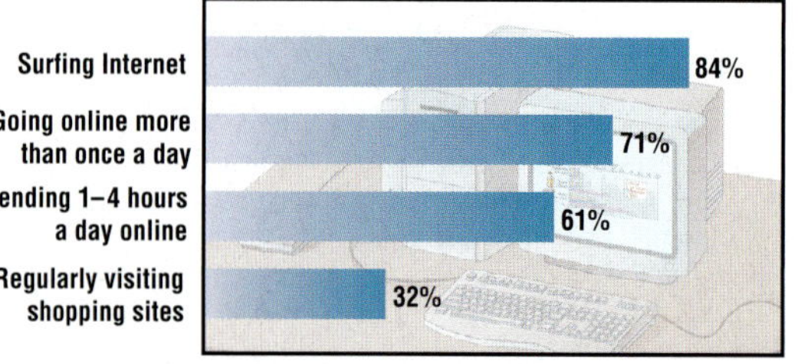

Source: *USA TODAY*

42. **Critical Thinking** The graph shows survey results on the Internet use of 100 college students. Write an inequality to represent the number of students that may spend at least 5 hours online each day. $n \leq 39$

Mixed Review

43. **Finance** Mr. Johnson invested $10,000 at an annual interest rate of 6%. Graph the function $B(t) = 10{,}000(1.06)^t$, where t is the time in years. How many years will it take before the balance is over $15,000? *(Lesson 11–7)* **7 yr**

Solve each equation by using the Quadratic Formula. *(Lesson 11–6)*

44. $t^2 - 3t - 40 = 0$ **−5, 8**
45. $3a^2 + a + 1 = 0$ **no real solutions**
46. $x^2 - 6x + 7 = 0$ $3 \pm \sqrt{2}$

47. Find the prime factorization of 24. How many distinct prime factors are there? *(Lesson 10–1)* $2^3 \cdot 3$; **two**

Simplify each expression.

48. $-\sqrt{16}$ *(Lesson 8–5)* **−4**
49. $\dfrac{-18x^3y}{3x}$ *(Lesson 8–2)* $-6x^2y$

Standardized Test Practice

50. **Multiple Choice** If two lines are perpendicular to each other, which of the following statements could be true? *(Lesson 7–7)* **B**

I. The slopes of both lines are equal.
II. One slope is positive, and the other slope is negative.
III. The slope of one line is the reciprocal of the slope of the other line.

(A) I only (B) II only (C) III only
(D) I and II (E) II and III

www.algconcepts.com/self_check_quiz

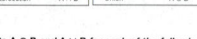

Extra Credit

Three players on a basketball team have each scored at least 16 points per game in the last three games. Write an inequality for the total number of points the three players have scored in the three games.
$p \geq 144$

Enrichment, p. 527

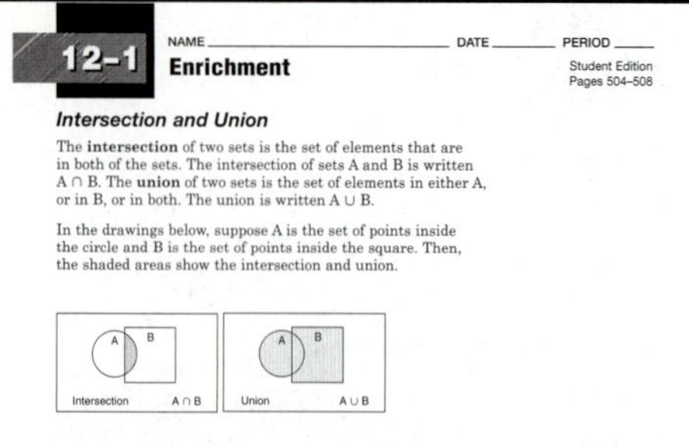

12-1 NAME ______ DATE ______ PERIOD ______

Enrichment Student Edition Pages 504–508

Intersection and Union

The **intersection** of two sets is the set of elements that are in both of the sets. The intersection of sets A and B is written A ∩ B. The **union** of two sets is the set of elements in either A, or in B, or in both. The union is written A ∪ B.

In the drawings below, suppose A is the set of points inside the circle and B is the set of points inside the square. Then, the shaded areas show the intersection and union.

Intersection A ∩ B   Union A ∪ B

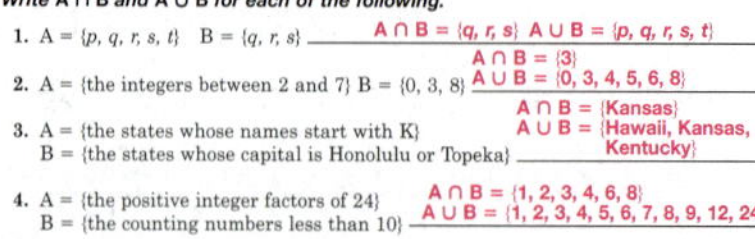

Write A ∩ B and A ∪ B for each of the following.

1. A = {p, q, r, s, t} B = {q, r, s} A ∩ B = {q, r, s} A ∪ B = {p, q, r, s, t}
2. A = {the integers between 2 and 7} B = {0, 3, 8} A ∩ B = {3} A ∪ B = {0, 3, 4, 5, 6, 8}
3. A = {the states whose names start with K} B = {the states whose capital is Honolulu or Topeka} A ∩ B = {Kansas} A ∪ B = {Hawaii, Kansas, Kentucky}
4. A = {the positive integer factors of 24} B = {the counting numbers less than 10} A ∩ B = {1, 2, 3, 4, 6, 8} A ∪ B = {1, 2, 3, 4, 5, 6, 7, 8, 9, 12, 24}

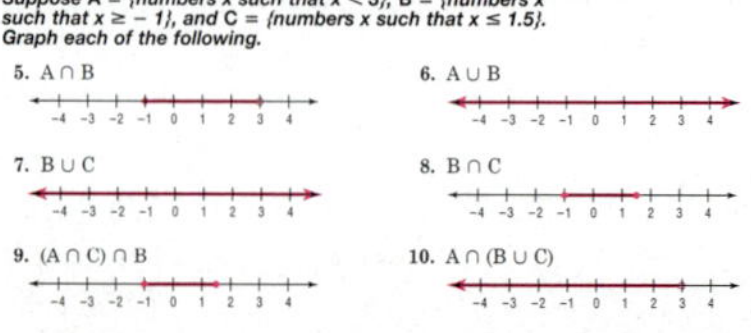

Suppose A = {numbers x such that x < 3}, B = {numbers x such that x ≥ − 1}, and C = {numbers x such that x ≤ 1.5}. Graph each of the following.

5. A ∩ B
6. A ∪ B
7. B ∪ C
8. B ∩ C
9. (A ∩ C) ∩ B
10. A ∩ (B ∪ C)

12–2 Solving Addition and Subtraction Inequalities

What You'll Learn
You'll learn to solve inequalities involving addition and subtraction.

Why It's Important
Sports You can use inequalities to set goals in competitions. *See Example 2.*

The Iditarod, nicknamed "The Last Great Race on Earth," is a dogsled race in Alaska. The 12- to 16-dog teams cover over 1150 miles in subzero weather. The record time for finishing the race is 218 hours (9 days, 2 hours).

A dogsled race

Suppose a team takes 73 hours to reach the first checkpoint of the Iditarod and 98 hours to reach the second. How much time can be spent on the last leg of the race to beat the record time?

Let t represent the time for the last leg of the race. Write an inequality.

The sum of the times	*must be less than*	*the record time.*
$73 + 98 + t$	$<$	218

$$171 + t < 218$$

If this were an equation, we would subtract 171 from each side to solve for t. The same procedure can be used with inequalities, as explained by the properties below. *This problem will be solved in Example 2.*

These properties are also true for $\leq$ and $\geq$.

Addition and Subtraction Properties for Inequalities

Words:	For any inequality, if the same quantity is added or subtracted to each side, the resulting inequality is true.
Symbols:	For all numbers a, b, and c, **1.** if $a > b$, then $a + c > b + c$ and $a - c > b - c$. **2.** if $a < b$, then $a + c < b + c$ and $a - c < b - c$.
Numbers:	$-5 < 1$ $-5 + 2 < 1 + 2$ $-3 < 3$; $2 > -4$ $2 - 3 > -4 - 3$ $-1 > -7$

Example 1 Solve $x + 14 \geq 5$. Check your solution.

$x + 14 \geq 5$	*Original inequality*
$x + 14 - 14 \geq 5 - 14$	*Subtract 14 from each side.*
$x \geq -9$	*Simplify.*

(continued on the next page)

1 FOCUS

5-Minute Check
Lesson 12–1

Write an inequality for each graph.

1. 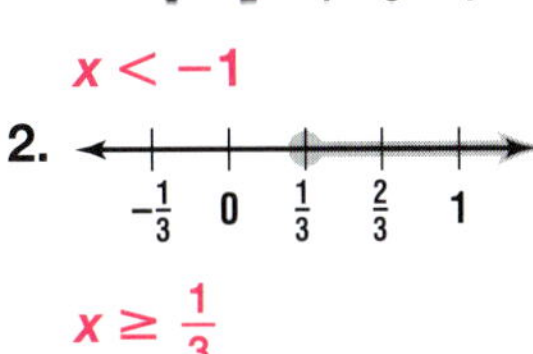
$x < -1$

2. $x \geq \frac{1}{3}$

Graph each inequality on a number line.

3. $b > -1.2$

4. $w \leq 11$

5. A small theater can seat no more than 230 people. Write an inequality to represent this situation. $p \leq 230$

Motivating the Lesson

Real-World Connection Pose the following application to students. Suppose that you plan to run at least 7.5 miles per week during the summer to train for the cross-country season in the fall. During one week, you run 2 miles on Sunday and 2.5 miles on Wednesday. How many more miles do you need to run before the week is over? Have students explain why an inequality is involved, and have them write an inequality to describe the situation.

Resource Manager

Reproducible Masters
Chapter 12 Resource Masters
- *Study Guide,* p. 528
- *Skills Practice,* p. 529
- *Practice,* p. 530
- *Reading to Learn Mathematics,* p. 531
- *Enrichment,* p. 532

Hands-On Algebra, p. 132

Transparencies
5-Minute Check, 12–2

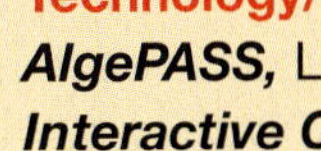

Technology/Multimedia
AlgePASS, Lesson 29
Interactive Chalkboard CD-ROM

2 TEACH

Teaching Tip Students may better understand the check for Example 1 if they first sketch the solution on a number line. Point out that they can then pick the endpoint of the ray and one number on either side of the endpoint to test in the original inequality. Those numbers that make the inequality true should be in the shaded area of the graph.

In-Class Examples

Example 1

Solve $y - 5 < -2$. Check your solution. $y < 3$

Example 2

Refer to the application in the Motivating the Lesson note in the side column of page 509. Solve $2 + 2.5 + m \geq 7.5$ to find how many more miles must be run. $m \geq 3$

Teaching Tip Draw students' attention to the equivalent statements $16 > y$ and $y < 16$ in Example 3. Explain that any inequality can be reversed in this manner. Stress that the symbol is still in the same relationship with the variable. In this case, the small end of the inequality points toward the variable y.

In-Class Example

Example 3

Solve $4w + 8 \leq 3w + 10$. Graph the solution. $\{w|w \leq 2\}$

Check: Substitute a number less than -9, the number -9, and a number greater than -9 into the inequality.

Let $x = -10$.	Let $x = -9$.	Let $x = 0$.
$x + 4 \geq 5$	$x + 14 \geq 5$	$x + 14 \geq 5$
$-10 + 14 \stackrel{?}{\geq} 5$	$-9 + 14 \stackrel{?}{\geq} 5$	$0 + 14 \stackrel{?}{\geq} 5$
$4 \geq 5$ false	$5 \geq 5$ true	$14 \geq 5$ true

The solution is {all numbers greater than or equal to -9}.

Your Turn **Solve each inequality. Check your solution.**

a. $x + 2 < 7$

b. $x - 6 \geq 12$

a. {all numbers less than 5}
b. {all numbers greater than or equal to 18}

A more concise way to express the solution to an inequality is to use **set-builder notation**. The solution in Example 1 in set-builder notation is $\{x \mid x \geq -9\}$.

$\{x$	$\mid$	$x \geq -9\}$
The set of all numbers x	*such that*	*x is greater than or equal to –9.*

In Lesson 12–1, you learned that you can show the solution to an inequality on a line graph. The solution, $\{x \mid x \geq -9\}$, is shown below.

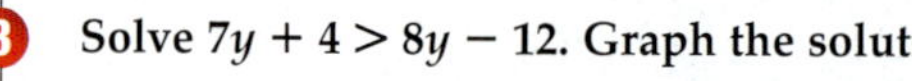

Examples
Sports Link

2 **Refer to the application at the beginning of the lesson. Solve $171 + t < 218$ to find the time needed to finish the last leg of the Iditarod and beat the record.**

$$171 + t < 218 \quad \textit{Original inequality}$$
$$171 + t - 171 < 218 - 171 \quad \textit{Subtract 171 from each side.}$$
$$t < 47 \quad \textit{This means all numbers less than 47.}$$

The solution can be written as $\{t \mid t < 47\}$. So any time less than 47 hours will beat the record.

3 **Solve $7y + 4 > 8y - 12$. Graph the solution.**

$$7y + 4 > 8y - 12 \quad \textit{Original inequality}$$
$$7y + 4 - 7y > 8y - 12 - 7y \quad \textit{Subtract 7y from each side.}$$
$$4 > y - 12 \quad \textit{Simplify.}$$
$$4 + 12 > y - 12 + 12 \quad \textit{Add 12 to each side.}$$
$$16 > y \quad \textit{Simplify.}$$

Since $16 > y$ is the same as $y < 16$, the solution is $\{y \mid y < 16\}$.

The graph of the solution has a circle at 16, since 16 is not included. The arrow points to the left.

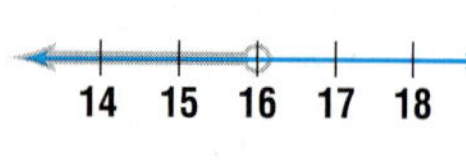

Reteaching Activity

Intrapersonal Learners Have students make a two-column table. In the first column, have them list examples of mistakes to avoid when solving and graphing inequalities. In the second column, opposite each possible mistake, have students write a suggestion to themselves about how to avoid making the mistake.

Your Turn Solve each inequality. Graph the solution.

c. $5y - 3 > 6y - 9$ d. $3r + 7 \leq 2r + 4$

c. $\{y \mid y < 6\}$

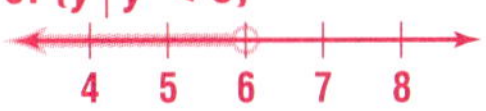

(number line: 4 5 6 7 8)

d. $\{r \mid r \leq -3\}$

(number line: −5 −4 −3 −2 −1)

You can also use inequalities to describe some geometry concepts.

Hands-On Algebra

Materials: 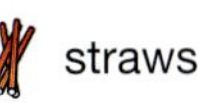 straws  scissors pipe cleaners  ruler

Step 1 Cut one straw so that it is 3 inches long. Cut a second straw 4 inches long. Cut a third straw 5 inches long.

Step 2 Insert a pipe cleaner into each straw. Then form a triangle. Label each side like the one shown at the right.

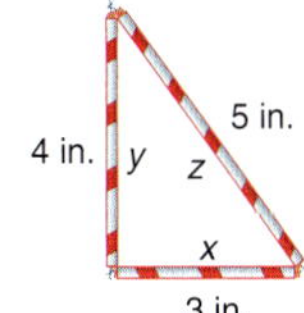

Try These

1. Repeat Steps 1 and 2 for each of the side lengths listed below. In each case, can a triangle be formed? Explain.
   a. $x = 4$ in., $y = 8$ in., $z = 6$ in.
   b. $x = 3$ in., $y = 5$ in., $z = 2$ in. **No, $2 + 3 \not> 5$.**
   c. $x = 1$ in., $y = 6$ in., $z = 3$ in. **No, $1 + 3 \not> 6$.**
2. Study the triangles in Exercise 1. Use an inequality to describe what must be true about the side lengths to form a triangle.
3. Can a triangle be formed with side lengths 9 cm, 12 cm, and 15 cm? Explain. **Yes, the sum of any two sides is greater than the third.**

1a. Yes, the sum of any two sides is greater than the third.

2. side 1 + side 2 > side 3

Check for Understanding

1–2. See margin.

Communicating Mathematics

1. **Compare and contrast** solving inequalities by using addition and subtraction with solving equations by using addition and subtraction.
2. **Write** two inequalities that each have $\{y \mid y > 4\}$ as their solution.
3. *Writing Math* List three situations in which solving an inequality may be helpful. **See students' work.**

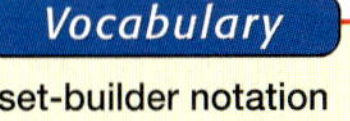

Vocabulary
set-builder notation

Guided Practice

Getting Ready Write an inequality for each statement.

Sample: Five more than a number is greater than sixteen.
Solution: $n + 5 > 16$

4. A number minus three is greater than or equal to ten. **$n - 3 \geq 10$**
5. The sum of 8 and a number is at most 12. **$8 + n \leq 12$**

Answers

1. They are the same.
2. Sample answer: $y > 4$ and $y - 4 > 0$

Hands-On Algebra

Cooperative Learning In this activity students discover the Triangle Inequality, which states that the sum of the lengths of any two sides of a triangle must be greater than the length of the third side. Point out that this requirement must be true of any two sides when compared to the third side. In practice, however, students need test only that the sum of the lengths of the two shortest sides is greater than the length of the third side.

Hands-On Algebra Masters, p. 132

***Study Guide,* p. 528**

12-2 NAME ______ DATE ______ PERIOD ______

Study Guide

Student Edition Pages 509–513

Solving Addition and Subtraction Inequalities

Suppose you already have \$50 and want to earn at least enough money to buy a DVD player for \$325. Let m = the amount of money you earn. You can represent this situation with the inequality $m + 50 \geq 325$. Then the solution to $m + 50 \geq 325$ is the amount of money you must earn. You can use the Addition and Subtraction Properties for Inequalities to solve inequalities involving addition or subtraction. The properties are summarized below.

Addition and Subtraction Properties for Inequalities
For all numbers a, b, and c,
1. if $a > b$, then $a + c > b + c$, and $a - c > b - c$.
2. if $a \geq b$, then $a + c \geq b + c$, and $a - c \geq b - c$.
3. if $a < b$, then $a + c < b + c$, and $a - c < b - c$.
4. if $a \leq b$, then $a + c \leq b + c$, and $a - c \leq b - c$.

Example: Solve $m + 50 \geq 325$. Check your solution.

$m + 50 \geq 325$
$m + 50 - 50 \geq 325 - 50$ *Subtract 50 from each side.*
$m \geq 275$

Check: Substitute a number less than 275, the number 275, and a number greater than 275 into the inequality.

Let $m = 200$.	Let $m = 275$.	Let $m = 300$.
$m + 50 \geq 325$	$m + 50 \geq 325$	$m + 50 \geq 325$
$200 + 50 \overset{?}{\geq} 325$	$275 + 50 \overset{?}{\geq} 325$	$300 + 50 \overset{?}{\geq} 325$
$250 \geq 325$; *false*	$325 \geq 325$; *true*	$350 \geq 325$; *true*

In *set-builder notation* the solution is {all numbers greater than or equal to 275}, or $\{m \mid m \geq 275\}$.

Solve each inequality. Check your solution.

1. $n + 3 > 6$ $\{n \mid n > 3\}$
2. $x - 6 > -2$ $\{x \mid x > 4\}$
3. $-2 + y \leq 8$ $\{y \mid y \leq 10\}$
4. $x - 4 \leq 12$ $\{x \mid x \leq 16\}$
5. $-3 \leq t + 2$ $\{t \mid t \geq -5\}$
6. $1 + p > -1$ $\{p \mid p > -2\}$
7. $y + 1.2 < 3.4$ $\{y \mid y < 2.2\}$
8. $-2.6 + x > 1.9$ $\{x \mid x > 4.5\}$
9. $-1.8 + y \leq 0$ $\{y \mid y \leq 1.8\}$
10. $x - \frac{1}{2} > \frac{3}{4}$ $\{x \mid x > 1\frac{1}{4}\}$
11. $1 \leq y - \frac{2}{3}$ $\{y \mid y \geq 1\frac{2}{3}\}$
12. $p - \frac{1}{8} \geq 1\frac{1}{2}$ $\{p \mid p \geq 1\frac{5}{8}\}$

3 PRACTICE/APPLY

Error Analysis

Watch for students who come up with coefficients of −1 while solving Exercises 12 and 13. ***Prevent by*** encouraging students always to subtract the smaller variable term from each side. Remind students that they may have to rewrite the resulting solution to have the variable on the left in set-builder notation.

Assignment Guide
Basic: 15–43 odd, 45–54
Average: 16–40 even, 41–54

Answers

***Skills Practice,* p. 529, and *Practice,* p. 530 (shown)**

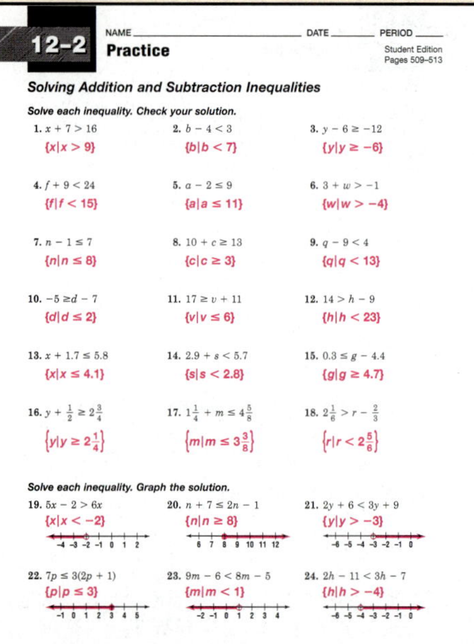

12-2 NAME ______ DATE ______ PERIOD ______

Practice

Student Edition Pages 509–513

Solving Addition and Subtraction Inequalities

Solve each inequality. Check your solution.

1. $x + 7 > 16$ $\{x \mid x > 9\}$
2. $b - 4 < 3$ $\{b \mid b < 7\}$
3. $y - 6 \geq -12$ $\{y \mid y \geq -6\}$
4. $f + 9 < 24$ $\{f \mid f < 15\}$
5. $a - 2 \leq 9$ $\{a \mid a \leq 11\}$
6. $3 + w > -1$ $\{w \mid w > -4\}$
7. $n - 1 \leq 7$ $\{n \mid n \leq 8\}$
8. $10 + c \geq 13$ $\{c \mid c \geq 3\}$
9. $q - 9 < 4$ $\{q \mid q < 13\}$
10. $-5 \geq d - 7$ $\{d \mid d \leq 2\}$
11. $17 \geq v + 11$ $\{v \mid v \leq 6\}$
12. $14 > h - 9$ $\{h \mid h < 23\}$
13. $x + 1.7 \leq 5.8$ $\{x \mid x \leq 4.1\}$
14. $2.9 + s < 5.7$ $\{s \mid s < 2.8\}$
15. $0.3 \leq g - 4.4$ $\{g \mid g \geq 4.7\}$
16. $y + \frac{1}{2} \geq 2\frac{3}{4}$ $\left\{y \mid y \geq 2\frac{1}{4}\right\}$
17. $1\frac{1}{4} + m \leq 4\frac{5}{8}$ $\left\{m \mid m \leq 3\frac{3}{8}\right\}$
18. $2\frac{1}{6} > r - \frac{2}{3}$ $\left\{r \mid r < 2\frac{5}{6}\right\}$

Solve each inequality. Graph the solution.

19. $5x - 2 > 6x$ $\{x \mid x < -2\}$ (−4 −3 −2 −1 0 1 2)
20. $n + 7 \leq 2n - 1$ $\{n \mid n \geq 8\}$ (6 7 8 9 10 11 12)
21. $2y + 6 < 3y + 9$ $\{y \mid y > -3\}$ (−6 −5 −4 −3 −2 −1 0)
22. $7p \leq 3(2p + 1)$ $\{p \mid p \leq 3\}$ (−1 0 1 2 3 4 5)
23. $9m - 6 < 8m - 5$ $\{m \mid m < 1\}$ (−2 −1 0 1 2 3 4)
24. $2h - 11 < 3h - 7$ $\{h \mid h > -4\}$ (−6 −5 −4 −3 −2 −1 0)

Examples 1 & 2 **Solve each inequality. Check your solution.**

6. $x + 9 < 12$ $\{x \mid x < 3\}$
7. $v - 12 \geq 5$ $\{v \mid v \geq 17\}$
8. $a - 6 \leq -14$ $\{a \mid a \leq -8\}$
9. $h - 7 \leq 14$ $\{h \mid h \leq 21\}$
10. $4.6 + x > 2.1$ $\{x \mid x > -2.5\}$
11. $y - \frac{2}{3} < 1\frac{1}{6}$ $\left\{y \mid y < 1\frac{5}{6}\right\}$

Example 3 **Solve each inequality. Graph the solution.**

12. $z - 12 \geq 2z + 4$ $\{z \mid z \leq -16\}$
13. $4x - 1 > 5x$ $\{x \mid x < -1\}$

12–13. See margin for graphs.

interNET CONNECTION

Data Update For the latest prices on computer hardware, visit: www.algconcepts.com

14. **Budgeting** Antonio can spend no more than \$1000 on a new computer system. The hard drive he wants costs \$220. The monitor costs \$300. How much money does Antonio have to spend on other components? **Example 2** $m \leq \$480$

Exercises

Practice

Homework Help	
For Exercises	**See Examples**
15–32, 39,	1
33–38, 40,	3
41–44	2
Extra Practice	
See page 716.	

Solve each inequality. Check your solution.

15. $n + 6 > 9$ $\{n \mid n > 3\}$
16. $x - 7 > -3$ $\{x \mid x > 4\}$
17. $-3 + b < -8$ $\{b \mid b < -5\}$
18. $g + 12 \leq 5$ $\{g \mid g \leq -7\}$
19. $r - 8 < 11$ $\{r \mid r < 19\}$
20. $x + 6 \geq 14$ $\{x \mid x \geq 8\}$
21. $w - 9 > 13$ $\{w \mid w > 22\}$
22. $4 + p \leq 1$ $\{p \mid p \leq -3\}$
23. $t - 5 \leq -5$ $\{t \mid t \leq 0\}$
24. $x + 3 \geq 19$ $\{x \mid x \geq 16\}$
25. $-2 < c + 2$ $\{c \mid c > -4\}$
26. $11 \leq -4 + m$ $\{m \mid m \geq 15\}$
27. $d + 1.4 < 6.8$ $\{d \mid d < 5.4\}$
28. $-3 + x > 11.9$ $\{x \mid x > 14.9\}$
29. $-0.2 \geq 0.3 + z$ $\{z \mid z \leq -0.5\}$
30. $s - (-2) > \frac{3}{4}$ $\left\{s \mid s > -1\frac{1}{4}\right\}$
31. $3\frac{3}{8} + v \leq 5\frac{7}{8}$ $\left\{v \mid v \leq 2\frac{1}{2}\right\}$
32. $\frac{1}{2} > x - \frac{2}{3}$ $\left\{x \mid x < \frac{7}{6}\right\}$

Solve each inequality. Graph the solution.

33–38. See margin for graphs.

35. $\{x \mid x \geq -16\}$

33. $7 > 8g - 7g - 6$ $\{g \mid g < 13\}$
34. $5n < 4n + 10$ $\{n \mid n < 10\}$
35. $3x + 12 \geq 2x - 4$
36. $6a \leq 5(a - 1)$ $\{a \mid a \leq -5\}$
37. $-(-t + 9) \geq 0$ $\{t \mid t \geq 9\}$
38. $3(v - 5) > 2(v - 2)$ $\{v \mid v > 11\}$

Write an inequality for each statement. Then solve.

39. Eight less than a number is not greater than 12. $x - 8 \leq 12, x \leq 20$
40. The sum of 5, 10, and a number is more than 25. $5 + 10 + x > 25, x > 10$

Applications and Problem Solving

41. **Fund-raising** Westfield High School is having a raffle to raise money for Habitat for Humanity. Any homeroom that sells at least 150 tickets will get to help build a home. Ms. Martinez' homeroom is keeping a table of the number of tickets sold each day. How many more tickets do they need to sell to help build a home? $t \geq 99$

Day	Tickets
1	32
2	19
3	?
4	?
5	?

Answers

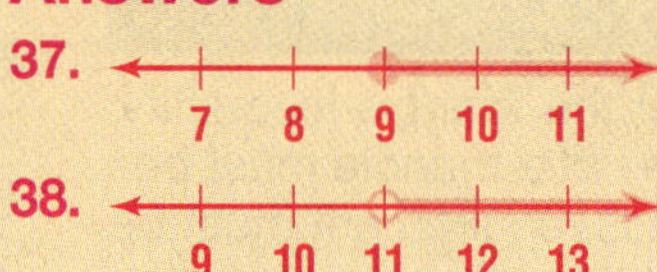

42. **Sports** Carlos weighs 175 pounds. At least how much weight must he lose to wrestle in the 171-pound weight class? $w \geq 4$

43. **Academics** Alissa must earn 475 out of 550 points to receive a grade of B. So far, she has earned 244 test points, 82 quiz points, and 50 homework points. How many points must she score on her final exam to earn at least a B in the class? $p \geq 99$

44. **Volunteerism** Each summer, 70 men bicycle in the Journey of Hope to raise money for people with disabilities. They begin their journey in San Francisco, California, and end in Washington, D.C. The map shows the miles they cycle in California. It takes them at most 285 miles to cycle to Nevada. How many miles is it from Jackson, California, to their first stop in Nevada? $m \leq 115$

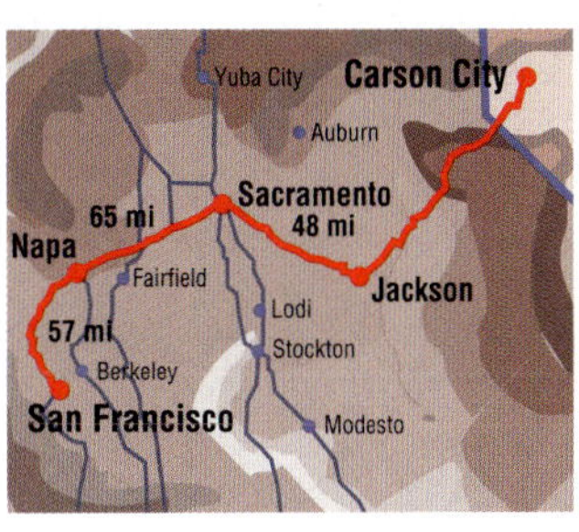

45. **Critical Thinking** Choose the correct term and justify your answer. The solution set of an inequality is (*sometimes, always, never*) the empty set (∅). **Sometimes; examples are $x > x$ and $2x + 1 \leq 2x$.**

Mixed Review

Write an inequality to describe each number. *(Lesson 12–1)*

46. a number no more than 1 $x \leq 1$

47. a number less than −8 $x < -8$

48. a number greater than 3 $x > 3$

49. a minimum number of 5 $x \geq 5$

50. The graph of which of the following equations is shown at the right: $y = 2^x$, $y = 2^x + 1$, $y = 2^x - 1$, or $y = -2^x$? *(Lesson 11–7)* $y = 2^x + 1$

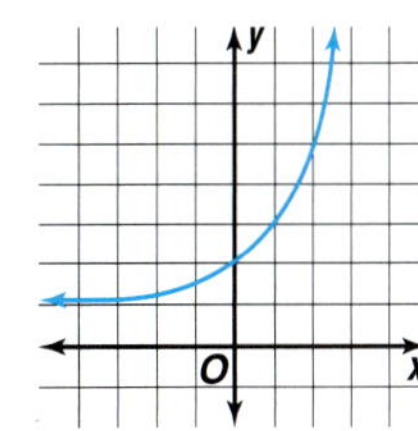

51. Factor the trinomial $3x^2 - 13x - 10$. *(Lesson 10–4)* $(3x + 2)(x - 5)$

52. Write $3x + y = -8$ in slope-intercept form. *(Lesson 7–3)* $y = -3x - 8$

53. **Grid In** If y varies inversely as x and $y = 8$ when $x = 24$, find y when $x = 6$. *(Lesson 6–6)* **32**

54. **Multiple Choice** Suppose **x** $= 4x - 2$. If **x** $= 10$, then what is the value of x? *(Lesson 4–4)* **B**

Ⓐ 2 Ⓑ 3 Ⓒ 38 Ⓓ 40

4 ASSESS

Open-Ended Assessment

Writing Have students write a brief summary describing the properties they can use to solve an addition or subtraction inequality. Have them include examples to illustrate the properties.

Extra Credit

Write a real-world problem that involves an addition or subtraction inequality. Define a variable and write the inequality. Then solve the inequality and graph its solution. **See students' work.**

***Enrichment*, p. 532**

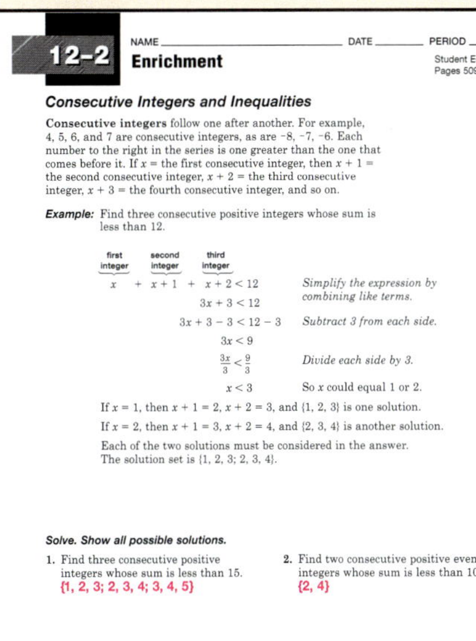

12–2 NAME ______ DATE ______ PERIOD ______

Enrichment

Student Edition Pages 509–513

Consecutive Integers and Inequalities

Consecutive integers follow one after another. For example, 4, 5, 6, and 7 are consecutive integers, as are −8, −7, −6. Each number to the right in the series is one greater than the one that comes before it. If x = the first consecutive integer, then $x + 1$ = the second consecutive integer, $x + 2$ = the third consecutive integer, $x + 3$ = the fourth consecutive integer, and so on.

Example: Find three consecutive positive integers whose sum is less than 12.

$$\underbrace{x}_{\text{first integer}} + \underbrace{x + 1}_{\text{second integer}} + \underbrace{x + 2}_{\text{third integer}} < 12$$

$3x + 3 < 12$ *Simplify the expression by combining like terms.*

$3x + 3 - 3 < 12 - 3$ *Subtract 3 from each side.*

$3x < 9$

$\frac{3x}{3} < \frac{9}{3}$ *Divide each side by 3.*

$x < 3$ So x could equal 1 or 2.

If $x = 1$, then $x + 1 = 2$, $x + 2 = 3$, and {1, 2, 3} is one solution.

If $x = 2$, then $x + 1 = 3$, $x + 2 = 4$, and {2, 3, 4} is another solution.

Each of the two solutions must be considered in the answer. The solution set is {1, 2, 3; 2, 3, 4}.

Solve. Show all possible solutions.

1. Find three consecutive positive integers whose sum is less than 15. **{1, 2, 3; 2, 3, 4; 3, 4, 5}**

2. Find two consecutive positive even integers whose sum is less than 10. **{2, 4}**

3. Find three consecutive positive integers such that the second plus four times the first is less than 21. **{1, 2, 3; 2, 3, 4; 3, 4, 5}**

4. Find three consecutive positive even integers such that the third plus twice the second is less than 26. **{2, 4, 6; 4, 6, 8}**

Lesson 12-3

1 FOCUS

5-Minute Check
Lesson 12–2

Solve each inequality. Check your solution.

1. $r - 10 \leq -2$ **$\{r | r \leq 8\}$**
2. $3 > x + 2.1$ **$\{x | x < 0.9\}$**
3. $4 + 3t < 2t + 7$ **$\{t | t < 3\}$**
4. The chart shows the time Rachel has jogged so far this week. How many more minutes m must she jog this week for her weekly total to be at least 2 hours?

Day	Mon	Tue	Wed
Minutes Jogged	25	30	28

$m \geq 37$

5. The sum of the lengths of any two sides of a triangle is greater than the length of the third side. What length must x be greater than for the figure below to be a triangle? What length must x be less than? **4 in.; 14 in.**

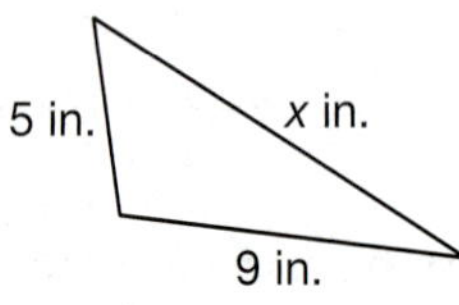

12-3 Solving Multiplication and Division Inequalities

What You'll Learn
You'll learn to solve inequalities involving multiplication and division.

Why It's Important
Savings You can use inequalities when you are trying to budget your money. *See Exercise 37.*

In the previous lesson, you learned that addition and subtraction inequalities are solved with the same procedures as addition and subtraction equations. Can you use the procedures for solving multiplication and division equations to solve inequalities as well?

For example, to solve $4x = 36$, we would divide each side by 4. Will this work when solving inequalities? Consider the inequality $4 < 36$.

$4 < 36$

$\frac{4}{4} \stackrel{?}{<} \frac{36}{4}$ *Divide each side by 4.*

$1 < 9$ true

So, it is possible to solve an inequality when dividing by a positive number. What happens when you divide by a negative number? Consider the inequality $-4 < 36$.

$-4 < 36$

$\frac{-4}{-4} \stackrel{?}{<} \frac{36}{-4}$ *Divide each side by −4.*

$1 < -9$ false

When dividing by a negative number, you must reverse the symbol for the inequality to remain true. This example leads us to the Division Property for Inequalities.

This property is also true for ≤ and ≥ .

Division Property for Inequalities

Words: If you divide each side of an inequality by a positive number, the inequality remains true. If you divide each side of an inequality by a negative number, the inequality symbol must be reversed for the inequality to remain true.

Symbols: For all numbers a, b, and c,
1. if c is positive and $a < b$, then $\frac{a}{c} < \frac{b}{c}$, and
   if c is positive and $a > b$, then $\frac{a}{c} > \frac{b}{c}$.
2. if c is negative and $a < b$, then $\frac{a}{c} > \frac{b}{c}$, and
   if c is negative and $a > b$, then $\frac{a}{c} < \frac{b}{c}$.

Numbers: If $9 > 6$, then $\frac{9}{3} > \frac{6}{3}$ or $3 > 2$.
If $9 > 6$, then $\frac{9}{-3} < \frac{6}{-3}$ or $-3 < -2$.

MODELING
Alternative hands-on options are available for teaching this lesson.

Resource Manager

Reproducible Masters
Chapter 12 Resource Masters
- *Study Guide*, p. 533
- *Skills Practice*, p. 534
- *Practice*, p. 535
- *Reading to Learn Mathematics*, p. 536
- *Enrichment*, p. 537
- *Assessment*, p. 569

Hands-On Algebra, pp. 133–134
School-to-Workplace, p. 12

Transparencies
5-Minute Check, 12-3

Technology/Multimedia
AlgePASS, Lesson 29
Interactive Chalkboard CD-ROM

Examples
Sports Link

1 **Carmen runs at least 15 miles in the park every day. If she runs 5 miles per hour, how long does she run every day? Recall that rate times time equals distance, or $rt = d$.**

$5t \geq 15$ *In $rt = d$, $r = 5$ and $d = 15$.*

$\frac{5t}{5} \geq \frac{15}{5}$ *Divide each side by 5.*

$t \geq 3$ *Keep the symbol facing the same direction.*

Carmen runs at least 3 hours every day.

Prerequisite Skills Review
Operations with Decimals, p. 684

2 **Solve $-10x \leq 25.6$. Check your solution.**

$-10x \leq 25.6$ *Original inequality*

$\frac{-10x}{-10} \geq \frac{25.6}{-10}$ *Divide each side by –10 and reverse the symbol.*

$x \geq -2.56$ *Simplify*

Check: Substitute −2.56 and a number greater than −2.56, such as 0, into the inequality.

Let $x = -2.56$.	Let $x = 0$.
$-10x \leq 25.6$	$-10x \leq 25.6$
$-10(-2.56) \stackrel{?}{\leq} 25.6$	$-10(0) \stackrel{?}{\leq} 25.6$
$25.6 \leq 25.6$ true	$0 \leq 25.6$ true

The solution set is $\{x \mid x \geq -2.56\}$.

Your Turn

Solve each inequality. Check your solution.

a. $8x > 40$ $\{x \mid x > 5\}$ b. $-x \leq 4.7$ $\{x \mid x \geq -4.7\}$

We can also solve inequalities by multiplying. Use the Multiplication Property for Inequalities.

This property is also true for ≤ and ≥.

Multiplication Property for Inequalities

Words: If you multiply each side of an inequality by a positive number, the inequality remains true. If you multiply each side of an inequality by a negative number, the inequality symbol must be reversed for the inequality to remain true.

Symbols: For all numbers a, b, and c,
1. if c is positive and $a < b$, then $ac < bc$, and if c is positive and $a > b$, then $ac > bc$.
2. if c is negative and $a < b$, then $ac > bc$, and if c is negative and $a > b$, then $ac < bc$.

Numbers: If $3 > -7$, then $3(2) > -7(2)$ or $6 > -14$.
If $3 > -7$, then $3(-2) < -7(-2)$ or $-6 < 14$.

www.algconcepts.com/extra_examples

Motivating the Lesson

Hands-On Activity Have students write the integers from −5 to 5 on small squares of paper and order the integers from least to greatest. Have them multiply each integer by 2, write the product in pencil on the back of each square, and order the products. Point out that the order is the same. Next have students erase the original products, write the product of the original integers and −1, and order the products. Seeing that the order reverses should help students understand why an inequality symbol reverses when an inequality is multiplied or divided by a negative number.

2 TEACH

In-Class Example

Example 1

An electric car that needs to be recharged every 260 miles should travel no more than 130 miles from the charger. If you drive at an average speed of 50 miles per hour, what are the lengths of time t you can drive away from the charger and then still make it back without running out of energy? **$t \leq 2.6$ h, or 2 h 36 min**

Teaching Tip As you discuss Example 2, remind students to check their answers in the *original* inequality. Point out that the endpoint and one different number in the solution set should be checked. Also point out that if the two sides of the original inequality do not have the same value for the endpoint students have tested, then they have not found the correct endpoint.

In-Class Example

Example 2

Solve $-2.5x > 10$. Check your solution. **$\{x \mid x < -4\}$**

In-Class Example

Example 3

Solve $-\frac{1}{2}x \geq -5$. Check your solution. $\{x | x \leq 10\}$

3 PRACTICE/APPLY

Error Analysis

Watch for students who make mistakes of sign or inequality direction when dividing each side of an inequality by a negative number as in Exercises 3 5, 6, and 9.

Prevent by suggesting that students who are having trouble solve these inequalities in smaller steps. Have these students take a separate step to multiply each side by -1 and reverse the inequality *before* dividing each side by the coefficient of the variable. This way, they may be less likely to make a mistake.

Answers

1. For both, reverse the symbol when dividing or multiplying by a negative number.
2. Five times four equals twenty. It is not greater than twenty.
3. 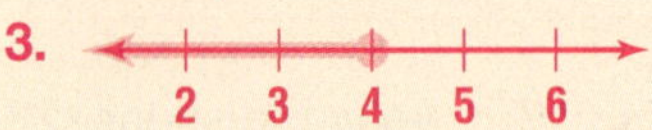

***Study Guide*, p. 533**

12–3 NAME ________ DATE ______ PERIOD _____

Study Guide

Student Edition Pages 514–518

Solving Multiplication and Division Inequalities

Suppose the family car gets 25 miles to a gallon of gasoline and you want to calculate how many gallons of gasoline you will need for a trip that is more than 300 miles long. Let g = the number of gallons of gasoline you will need. You can represent this situation with the inequality $25g > 300$. You can use the Multiplication and Division Properties for Inequalities to solve inequalities involving multiplication or division. The properties are summarized below. They are also true for $\geq$ and $\leq$.

Multiplication and Division Properties for Inequalities
For all numbers a, b, and c,
1. if c is positive and $a > b$, then $ac > bc$ and $\frac{a}{c} > \frac{b}{c}$.
2. if c is positive and $a < b$, then $ac < bc$ and $\frac{a}{c} < \frac{b}{c}$.
3. if c is negative and $a > b$, then $ac < bc$ and $\frac{a}{c} < \frac{b}{c}$.
4. if c is negative and $a < b$, then $ac > bc$ and $\frac{a}{c} > \frac{b}{c}$.

Example 1: Solve $25g \geq 300$.

$25g \geq 300$

$\frac{25g}{25} \geq \frac{300}{25}$ *Divide by 25.*

$g \geq 12$

$\{g | g \geq 12\}$

Example 2: Solve $\frac{y}{-2} \leq 8$.

$\frac{y}{-2} \leq 8$

$-2\left(\frac{y}{-2}\right) \geq -2(8)$ *Multiply by -2 and reverse the symbol.*

$y \geq -16$

$\{y | y \geq -16\}$

Solve each inequality. Check your solution.

1. $3n > 6$ $\{n | n > 2\}$
2. $\frac{x}{4} < 18$ $\{x | x < 72\}$
3. $-2y \leq 8$ $\{y | y \geq -4\}$
4. $3x > -9$ $\{x | x > -3\}$
5. $-8 \leq \frac{t}{-2}$ $\{t | t \leq 16\}$
6. $-2p > -1$ $\left\{p | p < \frac{1}{2}\right\}$
7. $2.4y < -4.8$ $\{y | y < -2\}$
8. $-1.5x > 7.2$ $\{x | x < -4.8\}$
9. $6.2y \leq 3.1$ $\left\{y | y \leq \frac{1}{2}\right\}$
10. $\frac{x}{12} > -3$ $\{x | x > -36\}$
11. $\frac{n}{-3} \geq 1.4$ $\{n | n \leq -4.2\}$
12. $7p > -7$ $\{p | p > -1\}$

Example 3 Solve $-\frac{x}{3} < -6$. Check your solution.

$-\frac{x}{3} < -6$ *Original inequality*

$-3\left(-\frac{x}{3}\right) > -3(-6)$ *Multiply each side by –3 and reverse the symbol.*

$x > 18$ *Simplify.*

Check: Substitute 18 and a number greater than 18, such as 21, into the inequality.

Let $x = 18$.	Let $x = 21$.
$-\frac{x}{3} < -6$	$-\frac{x}{3} < -6$
$-\frac{18}{3} \overset{?}{<} -6$	$-\frac{21}{3} \overset{?}{<} -6$
$-6 < -6$ false	$-7 < -6$ true

The solution is $\{x | x > 18\}$.

Your Turn

Solve each inequality. Check your solution.

c. $\frac{x}{-4} \leq 8$ $\{x | x \geq -32\}$

d. $\frac{2}{3}x > 22$ $\{x | x > 33\}$

Check for Understanding

1–3. See margin.

Communicating Mathematics

1. **Compare and contrast** the Division Property for Inequalities and the Multiplication Property for Inequalities.
2. **Explain** why 5 is *not* a solution of the inequality $4x > 20$.
3. **Graph** the solution of $-2x \geq -8$ on a number line.

Guided Practice

Solve each inequality. Check your solution.

Examples 1 & 2

4. $2z > 12$ $\{z | z > 6\}$
5. $-3x \leq 27$ $\{x | x \geq -9\}$
6. $-7b < 14$ $\{b | b > -2\}$

Example 3

7. $\frac{x}{4} \geq 5$ $\{x | x \geq 20\}$
8. $-\frac{r}{6} < 6$ $\{r | r > -36\}$
9. $\frac{2}{5}a \leq -12$ $\{a | a \leq -30\}$

Example 1

10. **Time** Cherise drives to the restaurant where she works by different routes, but the trip is at most 10 miles. Considering the stops at traffic lights, she thinks her average driving speed is about 40 miles per hour. How much time does it take Cherise to get to work? $t \leq 0.25$ h or 15 min

Reteaching Activity

Logical Learners Have students work with a partner. One student writes a multiplication or division inequality for the other to solve. As the second student solves the inequality, he or she explains in detail to the first student the steps taken to solve the inequality, including why the direction of the inequality symbol is or is not reversed. After completion, the students can reverse roles.

Exercises

Practice

Homework Help

For Exercises	See Examples
11–13, 17–19, 23–25, 29–31, 35–38	1, 2
14–16, 20–22, 26–29, 32–34	3

Extra Practice

See page 716.

Solve each inequality. Check your solution. 11–34. See margin.

11. $-6h \leq 12$
12. $-9n < 18$
13. $8d \leq -24$
14. $\frac{a}{6} > 5$
15. $\frac{g}{3} > -12$
16. $-\frac{b}{8} \geq 5$
17. $7x \geq 49$
18. $-4y \leq -40$
19. $3z > -9$
20. $-\frac{p}{2} > -1$
21. $\frac{c}{9} > -6$
22. $-\frac{a}{8} \leq -4$
23. $3k > 5$
24. $-2t \leq 11$
25. $-3 > -6w$
26. $\frac{3}{4}x < 3$
27. $5 \leq \frac{5}{6}y$
28. $-\frac{2}{5}v \leq 20$
29. $5.3v \geq 10.6$
30. $4.1x < -6.15$
31. $-28 \leq -0.1s$
32. $-\frac{h}{3.8} \geq 2$
33. $\frac{n}{10.5} < 10$
34. $-\frac{x}{0.5} < -7$

Applications and Problem Solving

35. **Geometry** An *acute angle* has a measure less than 90°. If the measure of an acute angle is $2x$, what is the value of x? $x < 45$

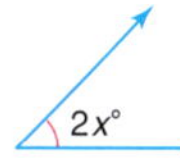

36. **Production** The ink cartridge that Bill just bought for his printer can print up to 900 pages of text. Bill is printing handbooks that are 32 pages each. How many complete handbooks can he print with this cartridge? at most 28 handbooks

37. **Budgeting** Jenny mows lawns to earn money. She wants to earn at least \$200 to buy a new stereo system. If she charges \$12 a lawn, at least how many lawns does she need to mow? at least 17 lawns

38. **Critical Thinking** Use a counterexample to show that if $x < y$, then $x^2 < y^2$ is not always true. Sample answer: $x = -3$, $y = 2$

Mixed Review

Solve each inequality. Check your solution. *(Lesson 12–2)*

39. $z + 1 \leq 5$ $\{z \mid z \leq 4\}$
40. $-3 \geq b + 11$ $\{b \mid b \leq -14\}$
41. **Contests** At a beach museum in San Pedro, California, more than 600 people built a life-size sand sculpture of a whale. Use an inequality to represent the number of people who built the sculpture. *(Lesson 12–1)* $p > 600$

Whale sculpture in San Pedro, California

Assignment Guide

Basic: 11–37 odd, 38–46
Average: 12–34 even, 35–46
All: Quiz 1, 1–10

Answers

11. $\{h \mid h \geq -2\}$
12. $\{n \mid n > -2\}$
13. $\{d \mid d \leq -3\}$
14. $\{a \mid a > 30\}$
15. $\{g \mid g > -36\}$
16. $\{b \mid b \leq -40\}$
17. $\{x \mid x \geq 7\}$
18. $\{y \mid y \geq 10\}$
19. $\{z \mid z > -3\}$
20. $\{p \mid p < 2\}$
21. $\{c \mid c > -54\}$
22. $\{a \mid a \geq 32\}$
23. $\left\{k \mid k > \frac{5}{3}\right\}$
24. $\left\{t \mid t \geq -\frac{11}{2}\right\}$
25. $\left\{w \mid w > \frac{1}{2}\right\}$
26. $\{x \mid x < 4\}$
27. $\{y \mid y \geq 6\}$
28. $\{v \mid v \geq -50\}$
29. $\{v \mid v \geq 2\}$
30. $\{x \mid x < -1.5\}$
31. $\{s \mid s \leq 280\}$
32. $\{h \mid h \leq -7.6\}$
33. $\{n \mid n < 105\}$
34. $\{x \mid x > 3.5\}$

***Skills Practice*, p. 534, and *Practice*, p. 535 (shown)**

12–3 NAME ______ DATE ______ PERIOD ______

Practice

Student Edition Pages 514–518

Solving Multiplication and Division Inequalities

Solve each inequality. Check your solution.

1. $4y < 16$ $\{y \mid y < 4\}$
2. $-3q \leq 18$ $\{q \mid q \geq -6\}$
3. $9g \leq -27$ $\{g \mid g \leq -3\}$
4. $\frac{p}{5} > 5$ $\{p \mid p > 25\}$
5. $\frac{a}{2} < -4$ $\{a \mid a < -8\}$
6. $-\frac{m}{7} \geq 7$ $\{m \mid m \leq -49\}$
7. $-6x \leq 30$ $\{x \mid x \geq -5\}$
8. $-4z > -28$ $\{z \mid z < 7\}$
9. $16 \geq 2e$ $\{e \mid e \leq 8\}$
10. $-\frac{n}{3} \geq -3$ $\{n \mid n \leq 9\}$
11. $4 \leq \frac{f}{6}$ $\{f \mid f \geq 24\}$
12. $-\frac{w}{5} > 8$ $\{w \mid w < -40\}$
13. $-81 < 9v$ $\{v \mid v > -9\}$
14. $6r \leq -42$ $\{r \mid r \leq -7\}$
15. $-12a \leq -60$ $\{a \mid a \leq 5\}$
16. $-4 > \frac{u}{9}$ $\{u \mid u < -36\}$
17. $-\frac{d}{6} < -8.1$ $\{d \mid d > 48.6\}$
18. $\frac{l}{8} > -8$ $\{l \mid l > -64\}$
19. $4k \leq 6$ $\{k \mid k \leq 1.5\}$
20. $-0.9b \geq -2.7$ $\{b \mid b \leq 3\}$
21. $-1.6 < 4t$ $\{t \mid t > -0.4\}$
22. $\frac{2}{3}y > 6$ $\{y \mid y > 9\}$
23. $-\frac{3}{5}c < 15$ $\{c \mid c > -25\}$
24. $-\frac{5}{8}j \geq -10$ $\{j \mid j \leq 16\}$

4 ASSESS

Open-Ended Assessment

Speaking Write several one-step inequalities on the board or overhead. Ask students to describe how they would solve the inequality, including whether or not they would reverse the inequality symbol. Include both addition and subtraction inequalities and multiplication and division inequalities to test students' understanding.

Quiz 1

The Quiz provides students with a brief review of the concepts and skills in Lessons 12–1 through 12–3. Lesson numbers are given to the right of the exercises or instruction lines so that students can review concepts not yet mastered.

Chapter 12, Quiz A (Lessons 12–1 through 12–3) is available in the *Chapter 12 Resource Masters*, p. 569.

Enrichment, p. 537

12-3 NAME ______ DATE ______ PERIOD ______

Enrichment Student Edition Pages 514–518

Some Properties of Inequalities

The two expressions on either side of an inequality symbol are sometimes called the *first* and *second* members of the inequality.

If the inequality symbols of two inequalities point in the same direction, the inequalities have the same sense. For example, $a < b$ and $c < d$ have the same sense; $a < b$ and $c > d$ have opposite senses.

In the problems on this page, you will explore some properties of inequalities.

Three of the four statements below are true for all numbers a and b (or a, b, c, and d). Write each statement in algebraic form. If the statement is true for all numbers, prove it. If it is not true, give an example to show that it is false.

1. Given an inequality, a new and equivalent inequality can be created by interchanging the members and reversing the sense. **If $a > b$, then $b < a$. $a > b$, $a - b > 0$, $-b > -a$, $(-1)(-b) < (-1)(-a)$, $b < a$**

2. Given an inequality, a new and equivalent inequality can be created by changing the signs of both terms and reversing the sense. **If $a > b$, then $-a < -b$. $a > b$, $a - b > 0$, $-b > -a$, $-a < -b$**

3. Given two inequalities with the same sense, the sum of the corresponding members are members of an equivalent inequality with the same sense. **If $a > b$ and $c > d$, then $a + c > b + d$. $a > b$ and $c > d$, so $(a - b)$ and $(c - d)$ are positive numbers, so the sum $(a - b) + (c - d)$ is also positive. $a - b + c - d > 0$, so $a + c > b + d$.**

4. Given two inequalities with the same sense, the difference of the corresponding members are members of an equivalent inequality with the same sense. **If $a > b$ and $c > d$, then $a - c > b - d$. The statement is false. $5 > 4$ and $3 > 2$, but $5 - 3 \ngtr 4 - 2$.**

42. Solve $x^2 + 5x + 4 = 0$ by factoring. *(Lesson 11–3)* **−4, −1**

43. Find the product of $2v - 1$ and $2v + 1$. *(Lesson 9–5)* **$4v^2 - 1$**

44. **Earthquakes** In a recent year, the state of Illinois experienced an earthquake tremor. It measured 3.5 on the Richter scale, releasing about 1.6×10^7 Joules of energy. The largest earthquake ever recorded in Illinois measured 5.5 on the Richter scale, releasing about 5.7×10^{11} Joules of energy. How many times as strong was the largest earthquake as the tremor? *(Lesson 8–4)* **3.5625×10^4 times stronger**

Standardized Test Practice Ⓐ Ⓑ Ⓒ Ⓓ

45. **Short Response** There are 20 students in a class. Each student's name is written on a separate slip of paper and placed in a box. A name is randomly drawn to determine who will read the daily announcements. The slip of paper is then returned to the box. What is the probability that the same name is drawn two days in a row? *(Lesson 5–7)* **$\frac{1}{400}$**

46. **Multiple Choice** If the figure at the right is a square, then what is the value of x? *(Lesson 4–4)* **B**

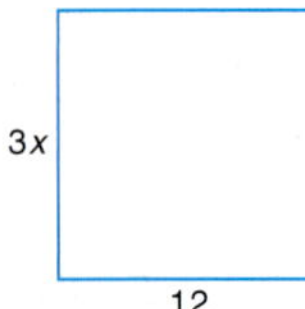

Ⓐ 3 Ⓑ 4
Ⓒ 12 Ⓓ 36

Quiz 1 Lessons 12–1 through 12–3

1. Write an inequality for the graph. *(Lesson 12–1)* **$x < -1$**

   −3 −2 −1 0 1

Solve each inequality. Graph the solution. *(Lesson 12–2)*

2. $a + 2 < 10$ **$\{a \mid a < 8\}$** 6 7 8 9 10
3. $-4 + x > 1$ **$\{x \mid x > 5\}$** 3 4 5 6 7
4. $-5 \leq s - (-3)$ **$\{s \mid s \geq -8\}$** −10 −9 −8 −7 −6
5. $-3 + 13z > 14z$ **$\{z \mid z < -3\}$** −5 −4 −3 −2 −1

Solve each inequality. Check your solution. *(Lesson 12–3)*

6. $3b \geq 30$ **$\{b \mid b \geq 10\}$**
7. $-5x < 25$ **$\{x \mid x > -5\}$**
8. $-\frac{k}{2} > -8$ **$\{k \mid k < 16\}$**
9. $\frac{4}{5}v \leq 6$ **$\left\{v \mid v \leq \frac{15}{2}\right\}$**

10. **Weather** The record high temperature in Jackson, Mississippi, is 100° F. Suppose a recent temperature in Jackson is 84° F. How many degrees must the temperature rise to break the record? *(Lesson 12–2)* **$t > 16°$ F**

www.algconcepts.com/self_check_quiz

Extra Credit

Write a real-world problem that involves a multiplication or division inequality. Define a variable and write the inequality. Then solve the inequality and graph its solution. **See students' work.**

12–4 Solving Multi-Step Inequalities

Lesson 12-4

What You'll Learn
You'll learn to solve inequalities involving more than one operation.

Why It's Important
School You can determine what score is needed to receive a certain class grade. *See Example 5.*

Some inequalities involve more than one operation. The best strategy to solve multi-step inequalities is to undo the operations in reverse order. In other words, work backward just as you did to solve multi-step equations. For example, $2x + 5 > 11$ is a multi-step inequality. You can solve this inequality by following these steps.

Step 1: Undo addition. *Subtract 5 from each side.*

$$2x + 5 - 5 > 11 - 5$$
$$2x > 6$$

Step 2: Undo multiplication. *Divide each side by 2.*

$$\frac{2x}{2} > \frac{6}{2}$$
$$x > 3$$

1 **Solve $9 + 3x < 27$. Check your solution.**

$9 + 3x < 27$	*Original inequality*
$9 + 3x - 9 < 27 - 9$	*Subtract 9 from each side.*
$3x < 18$	*Simplify.*
$\frac{3x}{3} < \frac{18}{3}$	*Divide each side by 3.*
$x < 6$	*Simplify.*

Check: Substitute 0 and 6 into the inequality.

Let $x = 0$.

$9 + 3x < 27$	*Original inequality*
$9 + 3(0) \overset{?}{<} 27$	*Substitute.*
$9 + 0 \overset{?}{<} 27$	*Simplify.*
$9 < 27$	true

Let $x = 6$.

$9 + 3x < 27$	*Original inequality*
$9 + 3(6) \overset{?}{<} 27$	*Substitute.*
$9 + 18 \overset{?}{<} 27$	*Simplify.*
$27 < 27$	false

The solution is $\{x \mid x < 6\}$.

Look Back
Solving Multi-Step Equations: Lesson 4–5

Your Turn

Solve each inequality. Check your solution.

a. $4 + 2x \leq 12$ **$\{x \mid x \leq 4\}$**

b. $8x - 5 \geq 11$ **$\{x \mid x \geq 2\}$**

1 FOCUS

5-Minute Check
Lesson 12–3

Solve each inequality. Check your solution.

1. $-3r < 15$ **$r > -5$**
2. $\frac{b}{4} \geq -5$ **$b \geq -20$**
3. $-\frac{3}{4}y > 9$ **$y < -12$**
4. A toner cartridge for a copy machine can print up to 10,000 copies. A small business makes an average of 250 copies per day. How many days can the company make copies using a single cartridge? **at most 40 days**
5. You want to build a patio with an area of at least 300 square feet along the back of your house as shown. What must be the width w of the patio?

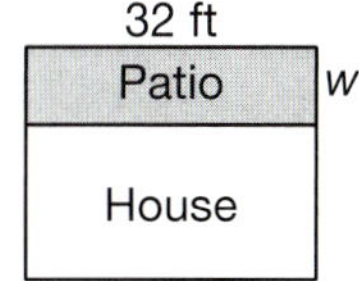

$w \geq 9.375$ ft

Motivating the Lesson

Real-World Connection Have students write and solve an equation for the following situation. The cell phone company charges \$9.99 a month plus \$0.25 per minute. If Janelle's bill is \$29.99, how many minutes did she talk? **$9.99 + 0.25m = 29.99$; 80** Challenge students to compare the steps they followed in writing and solving the equation to the steps they would use to write and solve an inequality to find the number of minutes Janelle could talk if she wants her bill to be less than \$24.99.

Resource Manager

Reproducible Masters
Chapter 12 Resource Masters
- *Study Guide,* p. 538
- *Skills Practice,* p. 539
- *Practice,* p. 540
- *Reading to Learn Mathematics,* p. 541
- *Enrichment,* p. 542
- *Assessment,* p. 568

Graphing Calculator, p. 33
Hands-On Algebra, p. 135

Transparencies
5-Minute Check, 12–4

Technology/Multimedia
AlgePASS, Lesson 29
Interactive Chalkboard CD-ROM

2 TEACH

In-Class Examples

Example 1

Solve $5x - 9 > 21$. Check your solution. **$\{x|x > 6\}$**

Example 2

Kyle is tiling his shower walls with 4-inch square tiles. The expression $9x + 20$ estimates the number of tiles he needs to tile x square feet of shower walls. Kyle has 560 tiles. Solve the inequality $9x + 20 \leq 560$ to find how many square feet Kyle can tile.
$x \leq 60$ square feet

Teaching Tip Stress the paragraph between Examples 2 and 3. Also, as you discuss Example 3, ask students how they could solve the inequality without having to divide each side by a negative number.

In-Class Examples

Example 3

Solve $16 - 2x \leq 3x + 1$. Check your solution. **$\{x|x \geq 3\}$**

Example 4

Solve $3(x + 2) > -75$. Check your solution. **$\{x|x > -27\}$**

Example 2 Weather Link

During Rafael's trip to Mexico, the temperature was always warmer than 30º Celsius. Use the formula $\frac{5}{9}(F - 32) = C$, where F is degrees Fahrenheit and C is degrees Celsius, to write and solve an inequality for the temperature in degrees Farenheit.

$\frac{5}{9}(F - 32) > 30$ — *The temperature is greater than 30.*

$\frac{9}{5} \cdot \frac{5}{9}(F - 32) > \frac{9}{5} \cdot 30$ — *Multiply each side by $\frac{9}{5}$, the reciprocal of $\frac{5}{9}$.*

$F - 32 > 54$ — *Simplify.*

$F - 32 + 32 > 54 + 32$ — *Add 32 to each side.*

$F > 86$ — *Check your solution.*

Therefore, Rafael can expect it to be warmer than 86°F in Mexico.

Remember to *reverse the inequality symbol* if you multiply or divide by a negative number.

Example 3 **Solve $-4x + 3 \geq 23 + 6x$. Check your solution.**

$-4x + 3 \geq 23 + 6x$ — *Original inequality*

$-4x + 3 - 6x \geq 23 + 6x - 6x$ — *Subtract 6x from each side.*

$-10x + 3 \geq 23$ — *Simplify.*

$-10x + 3 - 3 \geq 23 - 3$ — *Subtract 3 from each side.*

$-10x \geq 20$ — *Simplify.*

$\frac{-10x}{-10} \leq \frac{20}{-10}$ — *Divide each side by –10 and reverse the symbol.*

$x \leq -2$ — *Simplify.*

The solution is $\{x \mid x \leq -2\}$. *Check your solution.*

Your Turn **Solve each inequality. Check your solution.**

c. $10 - 5x < 25$ **$\{x \mid x > -3\}$**   d. $-3x + 1 > -17$ **$\{x \mid x < 6\}$**

To solve inequalities that contain grouping symbols, you may use the Distributive Property first.

Example 4 **Solve $8 \leq -2(x - 5)$. Check your solution.**

$8 \leq -2(x - 5)$ — *Original inequality*

$8 \leq -2x + 10$ — *Distributive Property*

$8 - 10 \leq -2x + 10 - 10$ — *Subtract 10 from each side.*

$-2 \leq -2x$ — *Simplify.*

$\frac{-2}{-2} \geq \frac{-2x}{-2}$ — *Divide each side by –2 and reverse the symbol.*

$1 \geq x$ — *Simplify.*

The solution is $\{x \mid x \leq 1\}$. *Check your solution.*

www.algconcepts.com/extra_examples

Talk with family members about relatives, ancestors, or friends who lived or who live in different countries. Do some research to find out the current range of typical temperatures in the areas where they are from. Find out what temperature scale people of that country usually use. Give the temperature ranges both in Fahrenheit and Celsius temperatures. A formula that converts degrees Celsius to degrees Fahrenheit is $F = \frac{9}{5}C + 32$.

Your Turn Solve each inequality. Check your solution.

e. $2 > -(x + 7)$ $\{x \mid x > -9\}$ f. $3(x - 4) \leq x - 5$ $\{x \mid x \leq 3.5\}$

You can use a graphing calculator to solve multi-step inequalities.

Graphing Calculator Tutorial
See pp. 750–753.

Graphing Calculator Exploration

Solve $-5 - 8x \geq 43$ by using a graphing calculator.

Step 1 Clear the [Y=] list to enter the inequality $-5 - 8x \geq 43$. (The $\geq$ symbol is item 4 on the TEST menu.)

Step 2 Press [ZOOM] 6. Use the [TRACE] and arrow keys to to move the cursor along the graph. You should see a line above the x-axis for values of x that are less than or equal to -6.

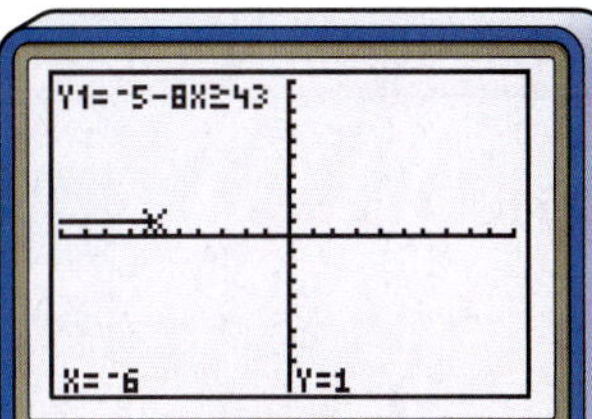

This represents the solution $\{x \mid x \leq -6\}$.

Try These

Solve each inequality. Check your solution with a graphing calculator.

1. $-9x + 2 \leq 20$ $\{x \mid x \geq -2\}$
2. $-5(x + 4) \geq 3(x - 4)$ $\{x \mid x \leq -1\}$

Example 5 School Link

Hannah's scores on the first three of four 100-point tests were 85, 92, and 90. What score must she receive on the fourth test to have a mean score of more than 92 for all tests?

Explore Let s = Hannah's score on the fourth test.

Plan The sum of Hannah's four test scores, divided by 4, will give the mean score. The mean must be *more than* 92.

Solve

$\frac{85 + 90 + 92 + s}{4} > 92$	*The mean is greater than 92.*
$4\left(\frac{85 + 90 + 92 + s}{4}\right) > 4(92)$	*Multiply each side by 4.*
$85 + 90 + 92 + s > 368$	*Simplify.*
$267 + s > 368$	*Add.*
$267 + s - 267 > 368 - 267$	*Subtract 267 from each side.*
$s > 101$	*Simplify.*

(continued on the next page)

Teaching Tip You might want to have students rework the problem in Example 5 with different target mean scores so that students can see situations in which there is an applicable solution. You may also want to have students substitute 100 for s in the mean computation to see what the highest possible mean is that Hannah can have for the four tests.

In-Class Example

Example 5

Karl's point totals in the first four of five basketball games were 15, 12, 19, and 18. How many points t must he score in the fifth game to have a mean point total of more than 16? $t > 16$

Study Guide, p. 538

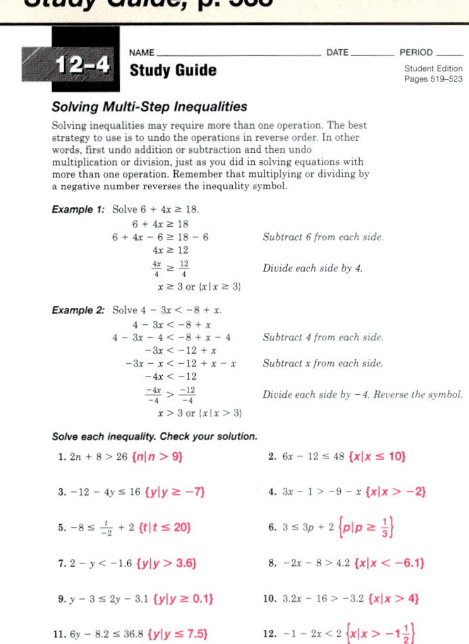

12-4 NAME ______ DATE ______ PERIOD ______

Study Guide

Student Edition Pages 519–523

Solving Multi-Step Inequalities

Solving inequalities may require more than one operation. The best strategy to use is to undo the operations in reverse order. In other words, first undo addition or subtraction and then undo multiplication or division, just as you did in solving equations with more than one operation. Remember that multiplying or dividing by a negative number reverses the inequality symbol.

Example 1: Solve $6 + 4x \geq 18$.

$6 + 4x \geq 18$	
$6 + 4x - 6 \geq 18 - 6$	*Subtract 6 from each side.*
$4x \geq 12$	
$\frac{4x}{4} \geq \frac{12}{4}$	*Divide each side by 4.*
$x \geq 3$ or $\{x \mid x \geq 3\}$	

Example 2: Solve $4 - 3x < -8 + x$.

$4 - 3x < -8 + x$	
$4 - 3x - 4 < -8 + x - 4$	*Subtract 4 from each side.*
$-3x < -12 + x$	
$-3x - x < -12 + x - x$	*Subtract x from each side.*
$-4x < -12$	
$\frac{-4x}{-4} > \frac{-12}{-4}$	*Divide each side by −4. Reverse the symbol.*
$x > 3$ or $\{x \mid x > 3\}$	

Solve each inequality. Check your solution.

1. $2n + 8 > 26$ $\{n \mid n > 9\}$
2. $6x - 12 \leq 48$ $\{x \mid x \leq 10\}$
3. $-12 - 4y \leq 16$ $\{y \mid y \geq -7\}$
4. $3x - 1 > -9 - x$ $\{x \mid x > -2\}$
5. $-8 \leq \frac{t}{-2} + 2$ $\{t \mid t \leq 20\}$
6. $3 \leq 3p + 2$ $\left\{p \mid p \geq \frac{1}{3}\right\}$
7. $2 - y < -1.6$ $\{y \mid y > 3.6\}$
8. $-2x - 8 > 4.2$ $\{x \mid x < -6.1\}$
9. $y - 3 \leq 2y - 3.1$ $\{y \mid y \geq 0.1\}$
10. $3.2x - 16 > -3.2$ $\{x \mid x > 4\}$
11. $6y - 8.2 \leq 36.8$ $\{y \mid y \leq 7.5\}$
12. $-1 - 2x < 2$ $\left\{x \mid x > -1\frac{1}{2}\right\}$

Graphing Calculator Exploration

It is often difficult to use a calculator graph to trace to the exact spot where the solution set of an inequality begins. Zooming in close to this spot, however, can help you make an educated guess. In this case, it appears that the first value of x that makes $-5 - 8x \geq 43$ true is -6. Substitution confirms that the value of $-5 - 8x$ is exactly 43 when $x = -6$. Students can also confirm this using the TABLE feature. At $x = -6$, the value for Y_1 changes from 0 to 1. Zero indicates a false inequality, whereas one indicates a true inequality.

3 PRACTICE/APPLY

Error Analysis

Watch for students who try to begin solving Exercise 12 by dividing each side by 3.
Prevent by stressing that although solving the inequality this way is mathematically correct, it is not efficient because it will introduce fractional values. Stress that it is often best to first use the Distributive Property to eliminate grouping symbols, though there may be exceptions to this. For example, in Exercise 30, the most efficient solution method is to first multiply each side by −6.

Assignment Guide
Basic: 15–37 odd, 38–45
Average: 14–34 even, 35–45

Answers

1. subtraction, then division
2. Sample answer: $2x - 6 > 4$

Skills Practice, **p. 539, and** ***Practice,*** **p. 540 (shown)**

12-4 NAME ______ DATE ______ PERIOD ______

Practice

Student Edition Pages 519–523

Solving Multi-Step Inequalities

Solve each inequality. Check your solution.

1. $3x + 5 < 14$ $\{x|x < 3\}$
2. $3t - 6 > 15$ $\{t|t > 7\}$
3. $-5y + 2 \geq 32$ $\{y|y \leq -6\}$
4. $-2n - 3 \geq -11$ $\{n|n \leq 4\}$
5. $6 \leq 4a + 10$ $\{a|a \geq -1\}$
6. $-28 < 7 + 7w$ $\{w|w > -5\}$
7. $5 - 1.3z \leq 31$ $\{z|z \geq -20\}$
8. $1.7b - 1.1 < 2.3$ $\{b|b < 2\}$
9. $6.4 \geq 8 + 2g$ $\{g|g \leq -0.8\}$
10. $-6 < \frac{k}{2} - 1$ $\{k|k > -10\}$
11. $-\frac{c}{6} + 9 \leq 3$ $\{c|c \geq 36\}$
12. $\frac{5m - 5}{3} \geq -15$ $\{m|m \geq -8\}$
13. $\frac{-2n + 6}{4} > 8$ $\{n|n < -13\}$
14. $\frac{6 - 3n}{6} \leq -5$ $\{n|n \geq 12\}$
15. $9 - 5j < j - 3$ $\{j|j > 2\}$
16. $7p - 4 \geq 3p + 12$ $\{p|p \geq 4\}$
17. $2f - 5 \leq 4f + 13$ $\{f|f \geq -9\}$
18. $5(7 - 2a) \geq -15$ $\{a|a \leq 5\}$
19. $2(q + 2) > 3(q - 6)$ $\{q|q < 22\}$
20. $3(h + 5) < -6(h - 4)$ $\{h|h < 1\}$
21. $-2(b - 3) \leq 4(b - 9)$ $\{b|b \geq 7\}$

Examine Substitute a number greater than 101, such as 102, into the original problem. Hannah's average would be $\frac{85 + 90 + 92 + 102}{4}$ or 92.25. Since $92.25 > 92$ is a true statement, the solution is correct. Hannah must score more than 101 points out of a 100-point test. Without extra credit, this is not possible. So, Hannah cannot have a mean score over 92.

Check for Understanding

Communicating Mathematics

1. **Name** the operations used to solve $5 - 2x < -9$. **1–2. See margin.**
2. **Write** an inequality requiring more than one operation to solve.

Guided Practice

Getting Ready **State which operation you would perform first to solve the inequality.**

Sample: $12x + 4 > 20$ **Solution:** Subtract 4 from each side.

3. $-15z + 7 \geq -10$ **− 7**
4. $24 < 8b - 3$ **+ 3**
5. $4.5a - (-3.1) > 8.2$ **− 3.1**
6. $\frac{n + 3}{9} \leq -11$ **× 9**

Examples 1–4 **Solve each inequality. Check your solution.**

7. $2y + 4 > 12$ $\{y \mid y > 4\}$
8. $8 - 3h \leq 20$ $\{h \mid h \geq -4\}$
9. $5 - 2x > 7$ $\{x \mid x < -1\}$
10. $7z - 4 \geq 10$ $\{z \mid z \geq 2\}$
11. $10 - 5w < w + 22$ $\{w \mid w > -2\}$
12. $3(n - 4) > 2(n + 6)$ $\{n \mid n > 24\}$

Example 5

13. **School** Kira wants her average math grade to be at least 90. Her test scores are 88, 93, and 87. What score does she need on her fourth test to earn an average score of at least 90? $s \geq 92$

Exercises

Practice

Solve each inequality. Check your solution.

A

14. $4t - 8 > 16$ $\{t \mid t > 6\}$
15. $3b + 9 < 45$ $\{b \mid b < 12\}$
16. $2 - 3n \leq 17$ $\{n \mid n \geq -5\}$
17. $-20 \geq 8 - 7x$ $\{x \mid x \geq 4\}$
18. $5 \leq 7c - 2$ $\{c \mid c \geq 1\}$
19. $-6g - 1 < -13$ $\{g \mid g > 2\}$
20. $-12 + 11r \leq 54$ $\{r \mid r \leq 6\}$
21. $8 - 4v \geq 6$ $\{v \mid v \leq 0.5\}$
22. $7 + 0.1a > 9$ $\{a \mid a > 20\}$
23. $0.3m - 2.1 \leq -3.0$ $\{m \mid m \leq -3\}$
24. $\frac{s}{4} - 6 < -11$ $\{s \mid s < -20\}$
25. $\frac{11 - 6d}{5} > 1$ $\{d \mid d < 1\}$
26. $3h - 5 \geq 2h + 4$ $\{h \mid h \geq 9\}$
27. $5x + 3 \leq 2x + 9$ $\{x \mid x \leq 2\}$
28. $6j - 9 > j + 6$ $\{j \mid j > 3\}$
29. $2(7 - 2y) > 10$ $\{y \mid y < 1\}$

B

30. $-\frac{1}{6}(z + 2) \leq -1$ $\{z \mid z \geq 4\}$
31. $\frac{2}{3}(b + 1) < \frac{1}{2}(b + 5)$ $\{b \mid b < 11\}$

Homework Help	
For Exercises	**See Examples**
14–25, 29, 30, 32–36	1, 2, 4
26–28, 31, 37	3
Extra Practice	
See page 716.	

Reteaching Activity

Interpersonal Learners Have each student work with a partner. On the board or overhead, write several inequalities. For each inequality, one student should solve the inequality algebraically while the other solves it using a graphing calculator. The partners should compare their work and then switch roles for the next inequality.

Write and solve an inequality for each situation.

32. The sum of twice a number and 17 is no greater than 41. $\{x \mid x \le 12\}$
33. Five times the sum of a number and 6 is less than 35. $\{x \mid x < 1\}$
34. Two thirds of a number decreased by 7 is at least 9. $\{x \mid x \ge 24\}$

Applications and Problem Solving

35. **Employment** Jeremy is a receptionist at a hair salon. He earns \$7.00 an hour, plus 10% of any hair products he sells. Suppose he works 22 hours a week. How much money in hair products must he sell to earn at least \$180? **at least \$260**

36. **Recreation** The admission fee to a state fair is \$5.00. Each ride costs an additional \$1.50.

 a. Suppose Pilar does not want to spend more than \$20. How many rides can she go on? **36a. 10 rides or less**
 b. The fair has a special admission price for \$14, which includes unlimited rides. For how many rides is this a better deal than paying for each ride separately? **36b. more than 6 rides**

37. **Finance** At a bank, an advertisement reads, "In one year, your earnings will be greater than your original investment plus 6% of the investment." Suppose Diego invests \$1400. How much money can he expect to have at the end of the year? $x > \$1484$

38. **Critical Thinking** Would the solution of $x^2 > 4$ be $x > 2$? Justify your answer. **No; the solution also includes numbers less than −2.**

Mixed Review

Solve each inequality. Check your solution. *(Lesson 12–3)*

39. $5p > 35$ $\{p \mid p > 7\}$
40. $-24 \le -8v$ $\{v \mid v \le 3\}$
41. $-\frac{x}{4} \ge 10$ $\{x \mid x \le -40\}$
42. $\frac{2}{5}z < -6$ $\{z \mid z < -15\}$

43. **Budgeting** Haley earned \$36 babysitting. She plans to buy two books that cost \$8.25 each. With the rest of the money, she plans to go to dinner and see a movie with her friends. At most, how much money can she spend on a movie and dinner? *(Lesson 12–2)* $m \le \$19.50$

Standardized Test Practice

44. **Grid In** Find the value of c that makes $x^2 - 6x + c$ a perfect trinomial square. *(Lesson 11–5)* **9**

45. **Multiple Choice** If $x \ne 5$, then $\frac{x^2 - 3x - 10}{x - 5}$ is equivalent to which of the following? *(Lesson 10–3)* **A**

 (A) $x + 2$ (B) $x - 3$
 (C) $x - 5$ (D) $x + 10$

www.algconcepts.com/self_check_quiz

4 ASSESS

Open-Ended Assessment

Writing Have students give an example of an inequality that relates two binomials of degree one. Then have them write a brief summary of the methods and properties they can use to solve the inequality.

Mid-Chapter Test (Lessons 12-1 through 12-4) is available in the *Chapter 12 Resource Masters*, p. 568.

Extra Credit

Amaar has an "A" average, or an average at or above 90 points out of 100, on his first three history quizzes. Amaar scored 3 points higher on the second quiz than on the first one, and 6 points higher on the third quiz than on the second one. What are Amaar's possible scores s on the first quiz? **$s \ge 86$ and $s \le 91$ (or $86 \le s \le 91$)**

***Enrichment*, p. 542**

NAME ______ DATE ______ PERIOD ______

12-4 Enrichment Student Edition Pages 519–523

Consecutive Integer Problems

Many types of problems and puzzles involve the idea of consecutive integers. Here is an example.

Find four consecutive odd integers whose sum is −80.

An odd integer can be written as $2n + 1$, where n is any of the numbers 0, 1, 2, 3, and so on. Then, the equation for the problem is as follows.

$$(2n + 1) + (2n + 3) + (2n + 5) + (2n + 7) = -80$$

Solve these problems. Write an equation or inequality for each.

1. Complete the solution to the problem in the example. $n = -12$; The integers are −23, −21, −19, −17.
2. Find three consecutive even integers whose sum is 132. $2n + (2n + 2) + (2n + 4) = 132$; $n = 21$; Integers are 42, 44, 46.
3. Find the two least consecutive integers whose sum is greater than 20. $n + (n + 1) > 20$; $n > 9.5$; Integers are 10 and 11.
4. Find the two greatest consecutive integers whose sum is less than 100. $n + (n + 1) < 100$; $n < 49.5$; Integers are 49 and 50.
5. The lesser of two consecutive even integers is 10 more than one-half the greater. Find the integers. $2n = 10 + \frac{1}{2}(2n + 2$; $n = 11$; Integers are 22 and 24.
6. The greater of two consecutive even integers is 6 less than three times the lesser. Find the integers. $2n + 2 = 3(2n) - 6$; $n = 2$; Integers are 4 and 6.
7. Find four consecutive integers such that twice the sum of the two greater integers exceeds three times the first by 91. $2[(n + 2) + (n + 3)] = 3n + 91$; $n = 81$; Integers are 81, 82, 83, 84.
8. Find all sets of four consecutive positive integers such that the greatest integer in the set is greater than twice the least integer in the set. $n + 3 > 2n$; $n < 3$; The two sets are {1, 2, 3, 4} and {2, 3, 4, 5}.

12-5 Solving Compound Inequalities

1 FOCUS

5-Minute Check
Lesson 12–4

Solve each inequality. Check your solution.

1. $8 > -4(x + 6)$ $\{x \mid x > -8\}$
2. $0.5w - 3.7 < 2.0$ $\{w \mid w < 11.4\}$
3. $-\frac{2}{3}x + 4 \geq 12$ $\{x \mid x \leq -12\}$
4. The perimeter of the rectangle below is at most 32 meters. What is the value of w? $w \leq 9$ m

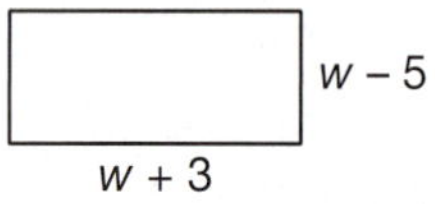

   (rectangle with sides $w - 5$ and $w + 3$)

5. Leticia's scores on the first four of five 100-point tests are shown below. What score s on the fifth test will give her a mean score of at least 90 for all five tests?

Test	1	2	3	4
Score	78	95	88	92

$s \geq 97$

Motivating the Lesson

Hands-On Activity Pick some readily observable distinctions among students in your class, such as eye color, gender, height, shirt color, and so on. By having students who match one or more criteria raise their hands, you can model the use of *and* and *or* as in compound inequalities. For example, have students with brown eyes raise their left hands and boys raise their right hands. Students with both hands raised satisfy the *and* condition, while any student with a hand raised satisfies the *or* condition.

What You'll Learn
You'll learn to solve compound inequalities.

Why It's Important
Nutrition Pharmacists use inequalities to write prescriptions. *See Example 2.*

Lamar is buying vitamins for his dog. The daily dose for the vitamins is based on the dog's weight. Lamar's dog weighs 32 pounds. Since 32 is greater than 25, but less than or equal to 50, he will give his dog 2 tablets.

Daily Dose	Dog's Weight (pounds)
1 tablet	25 or less
2 tablets	greater than 25 and less than or equal to 50
3 tablets	more than 50

Another way to write this information is to use an inequality. Let w represent the weight that requires 2 tablets.

Weight is greater than 25. → $w > 25$ *and* *Weight is less than or equal to 50.* → $w \leq 50$

These two inequalities form a **compound inequality**. The compound inequality $w > 25$ and $w \leq 50$ can be written without using the word *and*.

Method 1 $25 < w \leq 50$
This can be read as 25 is less than w, which is less than or equal to 50.

Method 2 $50 \geq w > 25$
This can be read as 50 is greater than or equal to w, which is greater than 25.

Note that in each, both inequality symbols are facing the same direction.

Example 

Write $x \geq 2$ and $x < 7$ as a compound inequality without using *and*.

$x \geq 2$ and $x < 7$ can be written as $2 \leq x < 7$ or as $7 > x \geq 2$.

Your Turn

a. $x < 10$ and $x \geq -4$ **a. $-4 \leq x < 10$ or $10 > x \geq -4$**

b. $x \leq 6$ and $x \geq 2$ **b. $2 \leq x \leq 6$ or $6 \geq x \geq 2$**

A compound inequality using *and* is true if and only if *both* inequalities are true. Thus, the graph of a compound inequality using *and* is the **intersection** of the graphs of the two inequalities.

Consider the inequality $-2 < x < 3$. It can be written using *and*: $x > -2$ and $x < 3$. To graph, follow the steps below.

Step 1 Graph $x > -2$.

Step 2 Graph $x < 3$.

Step 3 Find the intersection of the graphs.

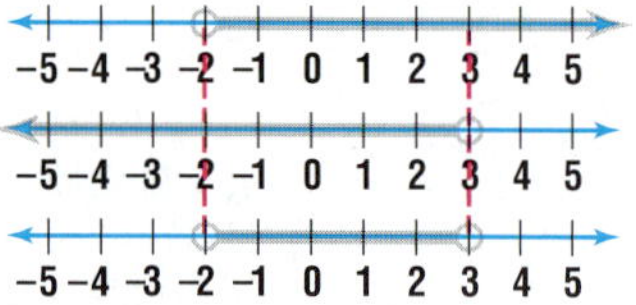

The solution, shown by the graph of the intersection, is $\{x \mid -2 < x < 3\}$.

Resource Manager

Reproducible Masters
Chapter 12 Resource Masters
- *Study Guide*, p. 543
- *Skills Practice*, p. 544
- *Practice*, p. 545
- *Reading to Learn Mathematics*, p. 546
- *Enrichment*, p. 547

Hands-On Algebra, pp. 136–137

Transparencies
5-Minute Check, 12–5

Technology/Multimedia
Interactive Chalkboard CD-ROM

Example 2 Pet Care Link

Refer to the application at the beginning of the lesson. Graph the solution of $25 < w \leq 50$.

Rewrite the compound inequality using *and*.
$25 < w \leq 50$ is the same as $w > 25$ and $w \leq 50$.

Step 1 Graph $w > 25$.

Step 2 Graph $w \leq 50$.

Step 3 Find their intersection.

The solution is $\{w \mid 25 < w \leq 50\}$.

> **Reading Algebra**
> Most of the time, $<$ or $\leq$ symbols are used with compound inequalities.

Your Turn

c. Graph the solution of $3 \leq x \leq 5$.

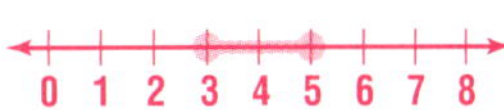

Often, you must solve a compound inequality before graphing it.

Example 3

Solve $4 < x + 3 \leq 12$. Graph the solution.

Step 1 Rewrite the compound inequality using *and*.

$$4 < x + 3 \leq 12$$

$$x + 3 > 4 \quad \text{and} \quad x + 3 \leq 12$$

Step 2 Solve each inequality.

$$x + 3 > 4 \quad \text{and} \quad x + 3 \leq 12$$

$$x + 3 - 3 > 4 - 3 \qquad x + 3 - 3 \leq 12 - 3$$

$$x > 1 \qquad x \leq 9$$

Step 3 Rewrite the inequality as $1 < x \leq 9$.

The solution is $\{x \mid 1 < x \leq 9\}$. The graph of the solution is shown at the right.

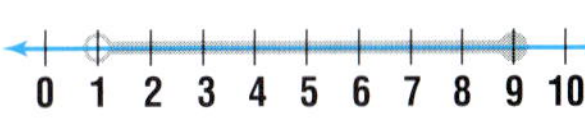

Your Turn $2 < x < 6$

d. Solve $-2 < x - 4 < 2$. Graph the solution.

Another type of compound inequality uses the word *or*. This type of inequality is true if one or more of the inequalities is true. The graph of a compound inequality using *or* is the **union** of the graphs of the two inequalities.

www.algconcepts.com/extra_examples

2 TEACH

Teaching Tip When discussing the two ways to write the compound inequality for the weight requirement in the opening scenario, emphasize that the direction of the inequality symbols in relation to the numbers and variable stays the same. In both inequalities, the point of the $\leq$ or $\geq$ symbol is directed towards w, and the open part of the $<$ or $>$ symbol is directed towards w.

In-Class Examples

Example 1

Write $x < 0$ and $x \geq -3$ as a compound inequality without using *and*. $-3 \leq x < 0$ or $0 > x \geq -3$

Example 2

A veterinarian has a scale for weighing dogs and cats that weigh more than 10 pounds but no more than 65 pounds. The weights w that can be measured on this scale can be written as $10 < w \leq 65$. Graph the solution of this inequality.

Teaching Tip In Example 3, the graphing is completed in a single step. Encourage students to use the three graphing steps shown in Example 2 until they are comfortable graphing a compound inequality in one step.

In-Class Example

Example 3

Solve $-6 \leq x - 3 \leq -1$. Graph the solution.
$\{x \mid -3 \leq x \leq 2\}$

Inclusion Strategies

The problems in this lesson have more steps than in many previous lessons. Students with behavioral difficulties may have trouble staying focused through these problems. You might try pairing these students with stronger students, having each partner take portions or steps of a problem to solve. For example, partners could each solve and graph one of the simple inequalities in a compound inequality and then compare solutions to arrive at the solution to the compound inequality.

In-Class Example

Example 4

Graph the solution of $x \geq 7$ or $x < 3$.

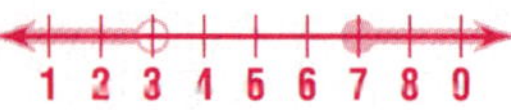

Teaching Tip In Example 5, students may be confused by the fact that the solution of the compound inequality is the same as the solution of one of its simple inequality parts. Explain that this is because every point in the solution of $x \geq 5$ is also a solution of $x > -2$, that is, the solutions of $x \geq 5$ are a *subset* of the solutions of $x > -2$.

In-Class Example

Example 5

Solve $\frac{2}{3}x > 4$ or $-5x < 20$. Graph the solution. **$\{x | x > -4\}$**

−5 −4 −3 −2 −1

Answers

1. **an inequality made of two inequalities**

Page 527

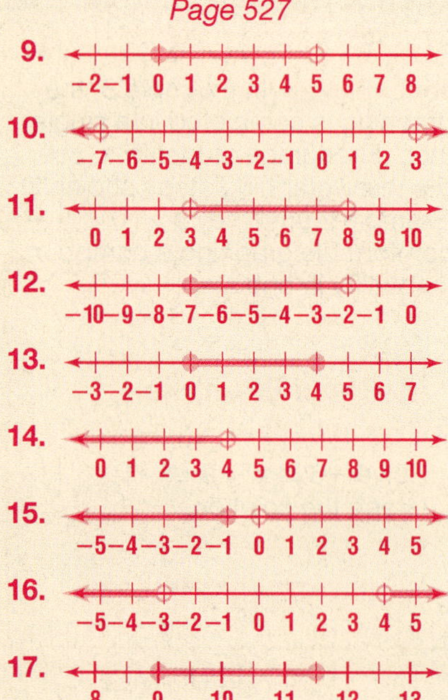

Example 4 Graph the solution of $x > 0$ or $x \leq -1$.

Step 1 Graph $x > 0$.

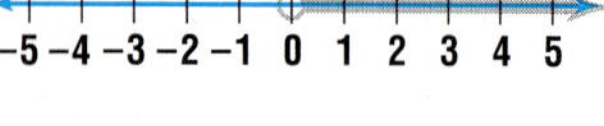

Step 2 Graph $x \leq -1$.

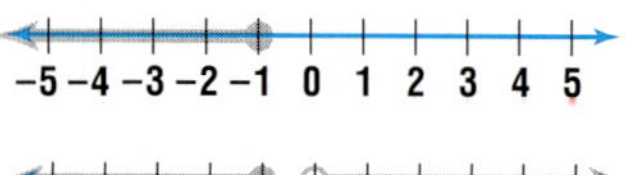

Step 3 Find the union of the graphs.

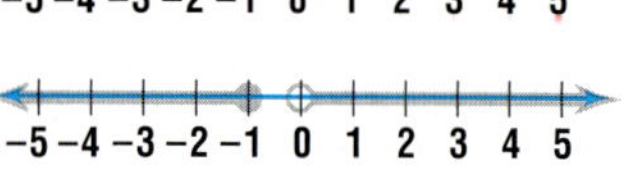

Your Turn

d. Graph the solution of $x > -3$ or $x < -5$.

Sometimes you must solve compound inequalities containing the word *or* before you are able to graph the solution.

Example 5 Solve $3x \geq 15$ or $-2x < 4$. Graph the solution.

$3x \geq 15$	or	$-2x < 4$
$\frac{3x}{3} \geq \frac{15}{3}$		$\frac{-2x}{-2} > \frac{4}{-2}$
$x \geq 5$		$x > -2$

Now graph the solution.

Step 1 Graph $x \geq 5$.

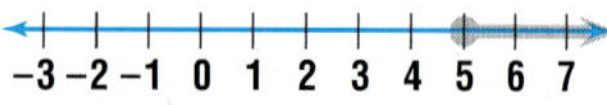

Step 2 Graph $x > -2$

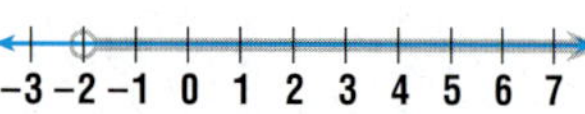

Step 3 Find the union of the graphs.

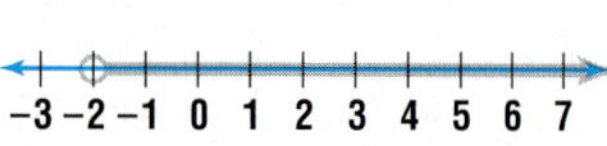

The last graph shows the solution $\{x | x > -2\}$.

Your Turn **$\{x | x < 3\}$**

e. Solve $-6x > 18$ or $x - 2 < 1$. Graph the solution.

Check for Understanding

Communicating Mathematics

1. **Define** *compound inequality* in your own words. **See margin.**
2. **Write** a compound inequality for *x is greater than 3 and less than or equal to 5.* **$3 < x \leq 5$ or $5 \geq x > 3$**

Vocabulary
compound inequality
intersection
union

Reteaching Activity

Visual/Spatial Learners Distribute overhead transparency sheets that have permanent number lines drawn on them to students. Students can graph solutions to each part of an assigned compound inequality on the same number line using different colors of washable markers. The solution to an *and* compound inequality is the portion of the number line where the two colors are blended; the solution to an *or* compound inequality is the portion of the number line marked with either or both colors.

Guided Practice

Getting Ready State whether to find the intersection or union of the two inequalities to graph the solution.

Sample 1: $-8 \leq x + 4 < 12$ **Solution:** Find the intersection since the inequalities can be joined by the word *and*.

Sample 2: $x > -9$ or $x < 5$ **Solution:** Find the union since the inequalities are joined by the word *or*.

3. $x > -10$ and $x < 3$ **intersection**
4. $x < 7$ or $x < 4$ **union**
5. $x - 5 < 2$ or $x \leq 15$ **union**
6. $-13 < 9x \leq -11$ **intersection**

Example 1

Write each compound inequality without using *and*.

7. $\ell > 5$ and $\ell < 10$ **$5 < \ell < 10$**
8. $b < 3$ and $b \geq -2$ **$-2 \leq b < 3$**

Examples 2 & 4

9–16. See margin for graphs.

Graph the solution of each compound inequality.

9. $y < 5$ and $y \geq 0$
10. $x > 3$ or $x < -7$

Examples 3 & 5

12. $\{s \mid -7 \leq s < -2\}$

Solve each compound inequality. Graph the solution.

11. $10 > c + 2 > 5$ **$\{c \mid 3 < c < 8\}$**
12. $2 > s + 4 \geq -3$
13. $0 \leq 2v \leq 8$ **$\{v \mid 0 \leq v \leq 4\}$**
14. $-5x > 10$ or $7x < 28$ **$\{x \mid x < 4\}$**
15. $j + 6 > 6$ or $-4j \geq 4$ **$\{j \mid j \leq -1$ or $j > 0\}$**
16. $-1 + r < -4$ or $-1 + r > 3$ **$\{r \mid r < -3$ or $r > 4\}$**

Example 2

17. **Construction** Odyssey of the Mind competitions encourage students to use creativity in solving difficult problems. One year, students had to construct a balsa-wood structure. The structure needed to be at least 9 inches tall and no more than 11.5 inches tall. Write a compound inequality describing the height of the structure. Graph the solution. **$9 \leq t \leq 11.5$; See margin for graph.**

Exercises

Practice

For Exercises	See Examples
18–23	1
24–29	2, 4
30–35	3
36–55	3, 4, 5
56–59	2, 3

Homework Help — Extra Practice: See page 717.

Write each compound inequality without using *and*.

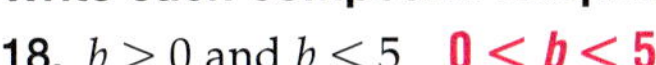

18. $b > 0$ and $b < 5$ **$0 < b < 5$**
19. $h > -8$ and $h < 8$ **$-8 < h < 8$**
20. $y \leq 4$ and $y \geq -1$ **$-1 \leq y \leq 4$**
21. $g \geq 2$ and $g \leq 5$ **$2 \leq g \leq 5$**
22. $-2 < r$ and $1 > r$ **$-2 < r < 1$**
23. $6 < x$ and $x \leq 8$ **$6 < x \leq 8$**

Graph the solution of each compound inequality.

24. $x > 0$ or $x \leq -4$
25. $z \leq 2$ and $z > -2$
26. $k < 7$ and $k > 5$
27. $y \geq 16$ and $y \leq 21$
28. $b > 4$ or $b < 0$
29. $m \leq 10$ or $m < 6$

24–35. See margin for graphs.

Solve each compound inequality. Graph the solution.

30. $2 \leq a + 3 < 7$ **$\{a \mid -1 \leq a < 4\}$**
31. $9 \geq x + 1 \geq 5$ **$\{x \mid 4 \leq x \leq 8\}$**
32. $-16 < 8s < 16$ **$\{s \mid -2 < s < 2\}$**
33. $9 \geq 3w > 0$ **$\{w \mid 0 < w \leq 3\}$**
34. $-6 > r - 2 > -10$
35. $2 \leq h + 5 \leq 8$ **$\{h \mid -3 \leq h \leq 3\}$**

34. $\{r \mid -8 < r < -4\}$

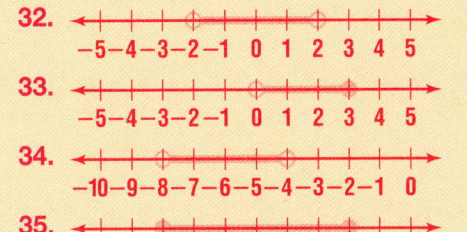

3 PRACTICE/APPLY

Error Analysis

Watch for students who reverse the use of union and intersection when graphing compound inequalities such as those in Exercises 9 and 10.

Prevent by having students pick points in different portions of their graphs to test in the original inequalities. For an *and* compound inequality, any point must satisfy both simple inequalities. For an *or* compound inequality, any point has to satisfy only one of the simple inequalities. Unless the two simple inequalities in the original compound inequality are the same, there will always be points that satisfy one but not both of the simple inequalities.

Assignment Guide

Basic: 19–59 odd, 60–71
Average: 18–54 even, 56–71

Answers

24.–27.

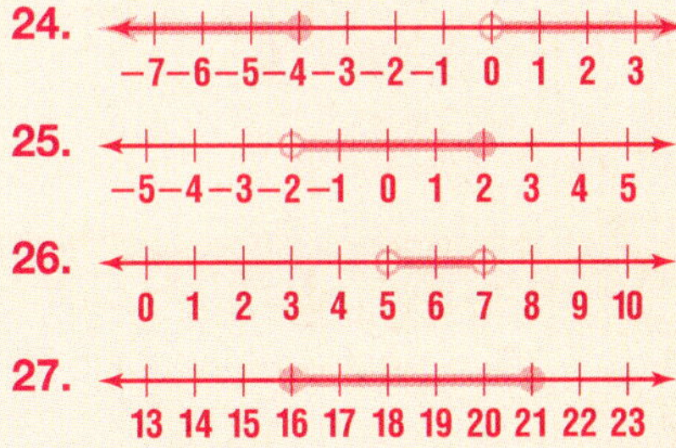

***Study Guide*, p. 543**

12-5 NAME ______ DATE ______ PERIOD ______

Study Guide Student Edition Pages 524–529

Solving Compound Inequalities

In a doctor's office, you may see a sign that displays the normal weight range for a person of your age and height. For example, if you are a 14-year-old girl who is 5 foot 2 inches tall, it may say that your normal weight is between 100 and 120 pounds, inclusive. Another way to write this information is to use an inequality. If w represents weight, then $100 \leq w \leq 120$ is a **compound inequality** that represents this situation. Another way to write the inequality is to write two inequalities using the word *and*: $100 \leq w$ and $w \leq 120$. A compound inequality using *and* is true if and only if *both* inequalities are true. The graph of a compound inequality using *and* is the **intersection** of the graphs of the inequalities, as shown below.

Example 1: Graph $100 \leq w$ and $w \leq 120$.

Step 1 Graph $100 \leq$ w.

Step 2 Graph $w \leq 120$

Step 3 Find the intersection of the graphs.

Another type of compound inequality uses the word *or*. A compound inequality using *or* is true if and only if either or both inequalities are true. Its graph is the **union** of the graphs of the inequalities, as shown below.

Example 2: Graph the solution of $x > 2$ or $x \leq -3$.

Step 1 Graph $x > 2$.

Step 2 Graph $x \leq -3$.

Step 3 Find the union of the graphs.

Graph the solution of each compound inequality.

1. $n > 2$ and $n < 6$
2. $x \leq -2$ or $x > 1$
3. $y \leq -2$ and $y \geq -6$
4. $1 \geq p$ and $p > 0$
5. $2 \leq y$ or $y < -1$
6. $h > 8$ and $h \leq 10$

Answers

28.–31.

32.–35.

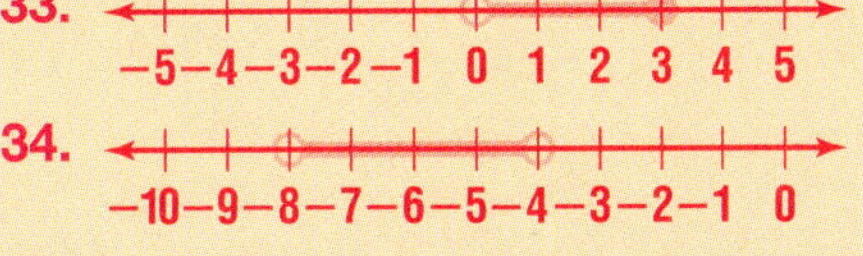

Answers

36.
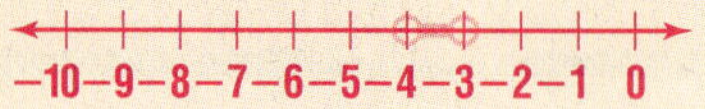

37.
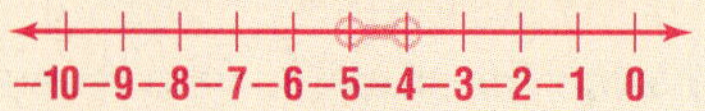

38.
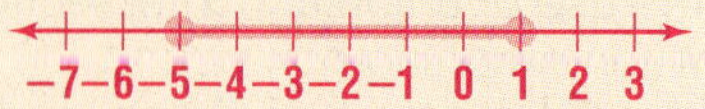

39.
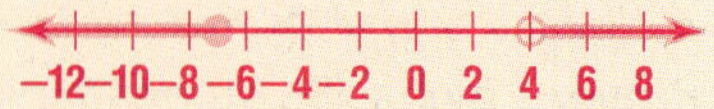

40.
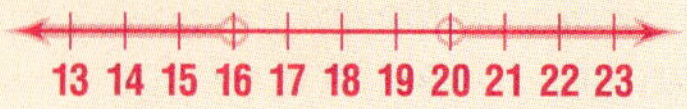

41.

42.

43.

44.
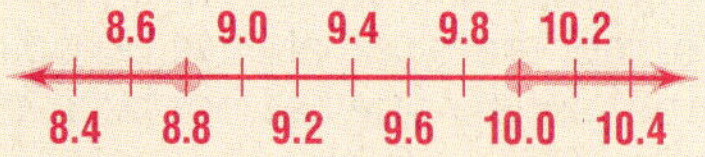

45.
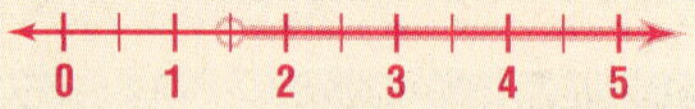

46.

47.
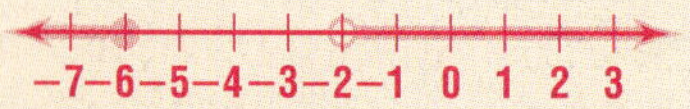

52. $\{y \mid -5 < y < 0\}$

53. $\{x \mid x \leq 3 \text{ or } x > 9\}$

54. $\{x \mid x > -2 \text{ or } x < -3\}$

55. $\{k \mid 11 \leq k \leq 13\}$

57b.

***Skills Practice*, p. 544, and *Practice*, p. 545 (shown)**

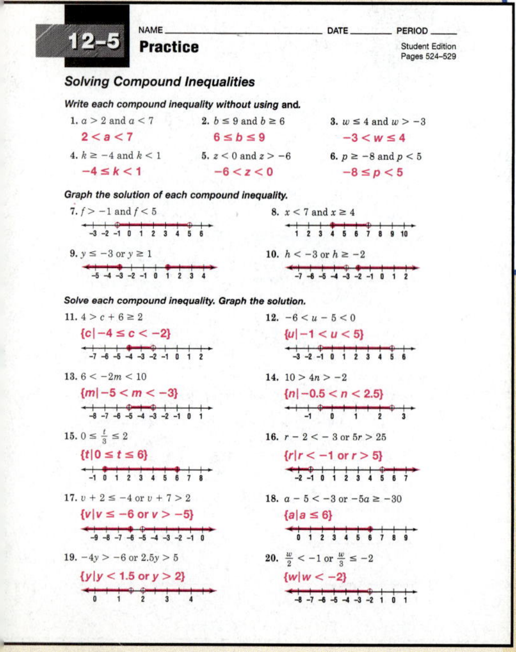

12-5 **Practice**

NAME ______ DATE ______ PERIOD ______

Student Edition Pages 524–529

Solving Compound Inequalities

Write each compound inequality without using and.

1. $a > 2$ and $a < 7$ — $2 < a < 7$
2. $b \leq 9$ and $b \geq 6$ — $6 \leq b \leq 9$
3. $w \leq 4$ and $w > -3$ — $-3 < w \leq 4$
4. $k \geq -4$ and $k < 1$ — $-4 \leq k < 1$
5. $z < 0$ and $z > -6$ — $-6 < z < 0$
6. $p \geq -8$ and $p < 5$ — $-8 \leq p < 5$

Graph the solution of each compound inequality.

7. $f > -1$ and $f < 5$
8. $x < 7$ and $x \geq 4$
9. $y \leq -3$ or $y \geq 1$
10. $h < -3$ or $h \geq -2$

Solve each compound inequality. Graph the solution.

11. $4 > c + 6 \geq 2$ — $\{c \mid -4 \leq c < -2\}$
12. $-6 < u - 5 < 0$ — $\{u \mid -1 < u < 5\}$
13. $6 < -2m < 10$ — $\{m \mid -5 < m < -3\}$
14. $10 > 4n > -2$ — $\{n \mid -0.5 < n < 2.5\}$
15. $0 \leq \frac{t}{3} \leq 2$ — $\{t \mid 0 \leq t \leq 6\}$
16. $r - 2 < -3$ or $5r > 25$ — $\{r \mid r < -1 \text{ or } r > 5\}$
17. $v + 2 \leq -4$ or $v + 7 > 2$ — $\{v \mid v \leq -6 \text{ or } v > -5\}$
18. $a - 5 < -3$ or $-5a \geq -30$ — $\{a \mid a \leq 6\}$
19. $-4y > -6$ or $2.5y > 5$ — $\{y \mid y < 1.5 \text{ or } y > 2\}$
20. $\frac{w}{2} < -1$ or $\frac{w}{3} \leq -2$ — $\{w \mid w < -2\}$

36–47. See margin for graphs.

36. $\{y \mid -4 < y < -3\}$
37. $\{c \mid -5 < c < -4\}$
38. $\{x \mid -5 \leq x \leq 1\}$
39. $\{z \mid z \leq -7 \text{ or } z > 4\}$
40. $\{v \mid v < 16 \text{ or } v > 20\}$
41. $\{r \mid r \leq -1 \text{ or } r > 0\}$
42. $\{j \mid j > 6\}$
43. $\{p \mid p \geq -1\}$
44. $\{c \mid c \leq 8.8 \text{ or } c \geq 10\}$
45. $\{d \mid d > 1.5\}$

C

Solve each compound inequality. Graph the solution.

36. $-5 < y - 1 < -4$
37. $16 < -4c < 20$
38. $6 \geq x - (-5) \geq 0$
39. $z + 3 > 7$ or $z - 5 \leq -12$
40. $v - 8 > 12$ or $v + 4 < 20$
41. $r \leq -1$ or $-8r < 0$
42. $3j > 18$ or $j - 3 \geq 5$
43. $p - 5 > -3$ or $-2p \leq 2$
44. $c - 2.4 \geq 7.6$ or $c - 8.8 \leq 0$
45. $d + 2.1 > 3.6$ or $d - 4 > 0.5$
46. $8x \geq 4$ or $x + \frac{3}{4} < 1$ $\quad \left\{x \mid x < \frac{1}{4} \text{ or } x \geq \frac{1}{2}\right\}$
47. $\frac{w}{3} \leq -2$ or $-w < 2$ $\quad \{w \mid w \leq -6 \text{ or } w > -2\}$

Write a compound inequality for each solution shown below.

48.
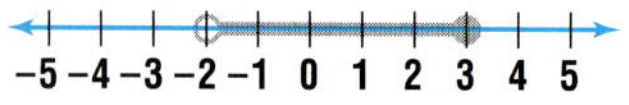

$-2 < x \leq 3$

49.
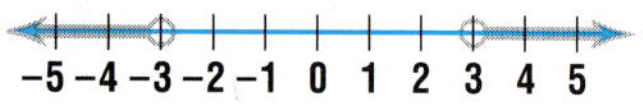

$x < -3$ or $x > 3$

50.
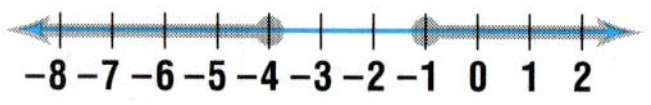

$x \leq -4$ or $x \geq -1$

51.
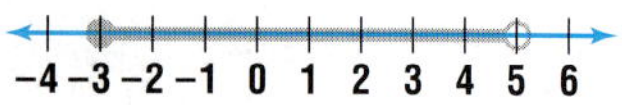

$-3 \leq x < 5$

Solve each compound inequality. **52–55. See margin.**

52. $2 > 3y + 2 > -13$
53. $4x - 7 \leq 5$ or $-(x - 8) < -1$
54. $2x - 1 > -5$ or $-3(x + 1) > 6$
55. $10 \leq 2(k - 6) \leq 14$

Applications and Problem Solving

56. **Taxes** Matthew Brooks is single and has a part-time job while attending college. Last year, he paid \$649 in federal income tax. Write an inequality for his taxable income. Use the table that describes the different tax brackets. **$\$4300 \leq I < \$4350$**

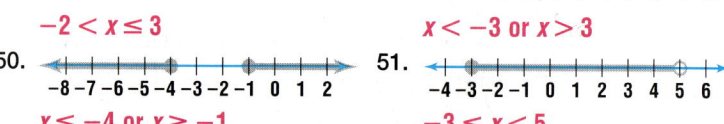

If Form 1040A, line 24, is —		And you are —			
At least	But less than	Single	Married filing jointly	Married filing sepa-rately	Head of a house-hold
		Your tax is —			
4,200	4,250	634	634	634	634
4,250	4,300	641	641	641	641
4,300	4,350	649	649	649	649
4,350	4,400	656	656	656	656
4,400	4,450	664	664	664	664
4,450	4,500	671	671	671	671
4,500	4,550	679	679	679	679
4,550	4,600	686	686	686	686

Source: Ohio Department of Taxation

57. **Cooking** A box of macaroni and cheese lists two sets of directions for cooking. It says to heat the macaroni for 11 to 13 minutes on the stove or 12 to 14 minutes in the microwave.
   a. Write an inequality that represents possible heating times.
   b. Graph the solution. **See margin.**

57a. Sample answer: $11 \leq h \leq 14$

58. **Geometry** To construct any triangle, the sum of the lengths of two sides must be greater than the length of the third. Suppose that two sides of a triangle have lengths of 4 inches and 12 inches. What are the possible values for the length of the third side? Express your answer as a compound inequality. **$8 < s < 16$**

59. **Chemistry** Soil pH is measured on a scale from 0 to 14. The pH level describes the soil as shown below.

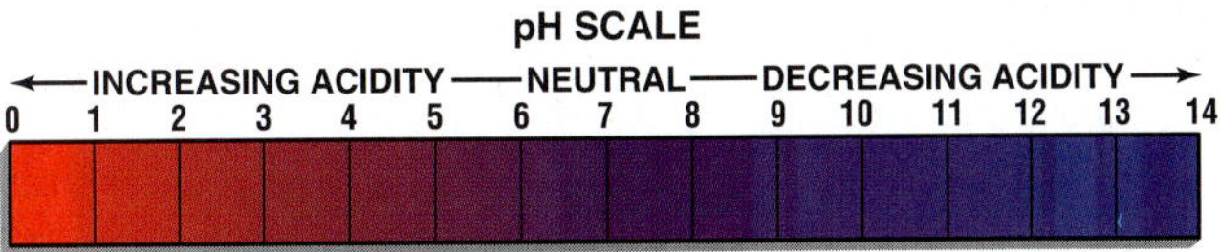

Most plants grow best if the soil pH is between 6.5 and 7.2. Mr. Cohen took samples of soil from his garden and found pH values of 6.3, 6.4, and 6.7. What range of values must the fourth sample have if the soil pH is the best for growing plants? (*Hint:* Find the mean soil pH.) **$6.6 < p < 9.4$**

60. **Critical Thinking** Graph each compound inequality. Then state the solution. **a–b. See margin for graphs.**

a. $x > 4$ or $x \leq 4$ **$\{x \mid x \text{ is a real number}\}$**

b. $2 > 3z + 2 > 14$ **$\varnothing$**

Mixed Review

Solve each inequality. Check your solution. *(Lessons 12–3 & 12–4)*

61. $2y + 4 < 4$ **$\{y \mid y < 0\}$**

62. $-3n - 8 > 22$ **$\{n \mid n < -10\}$**

63. $-2 \leq 0.6x - 5$ **$\{x \mid x \geq 5\}$**

64. $\frac{2}{3}p - 3 \leq 7$ **$\{p \mid p \leq 15\}$**

65. $-10b < -60$ **$\{b \mid b > 6\}$**

66. $3r \geq -12$ **$\{r \mid r \geq -4\}$**

67. **Welding** Maxwell is welding two pieces of iron together. During this process, the iron melts and begins to boil. Small droplets erupt and follow the paths of parabolic arcs. The paths can be modeled by the quadratic function $h(d) = -d^2 + 4d + 30$, where $h(d)$ represents the height of the arc above the ground at any horizontal distance d from the two pieces of iron. All measures are in inches. *(Lesson 11–1)*

a. Graph the function. **See margin.**

b. How high above the pieces of iron do the iron droplets jump? **34 in.**

68. Find the GCF of $8x^2$, $2x$, and $4xy$. *(Lesson 10–1)* **$2x$**

69. Simplify $3(b - 6)$. *(Lesson 9–3)* **$3b - 18$**

Standardized Test Practice

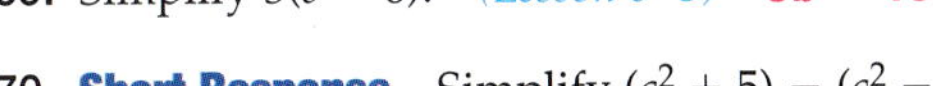

70. **Short Response** Simplify $(c^2 + 5) - (c^2 - 8c + 1)$. *(Lesson 9–2)* **$8c + 4$**

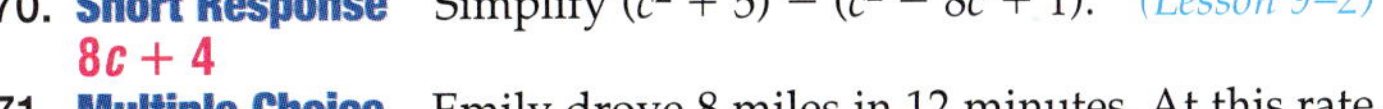

71. **Multiple Choice** Emily drove 8 miles in 12 minutes. At this rate, how many miles will she drive in 1 hour? *(Lesson 5–1)* **C**

Ⓐ 4 mi Ⓑ 20 mi Ⓒ 40 mi
Ⓓ 56 mi Ⓔ 96 mi

www.algconcepts.com/self_check_quiz

4 ASSESS

Open-Ended Assessment

Act It Out Following the order of class seating, assign each student a consecutive integer from a set containing both positive and negative integers. Have students write their numbers in large print on a piece of paper. Then write a pair of simple inequalities for each student to solve. Tell students whether you want to form an *and* or an *or* compound inequality with the simple inequalities. Have each student decide whether her or his number belongs to the solution set. If it does, have those students hold up their numbers. Then have the class decide whether the appropriate students have held up their numbers.

Answers

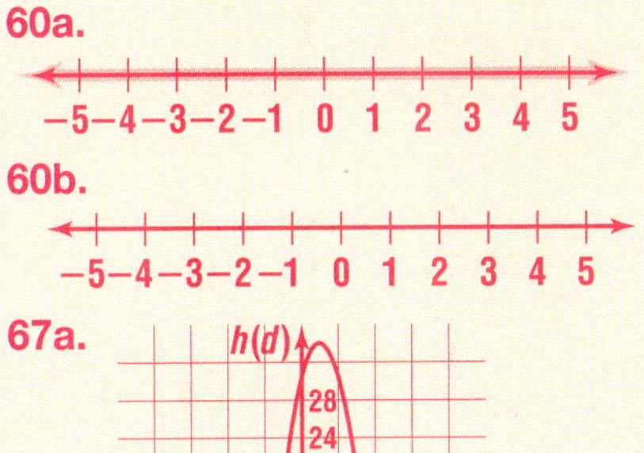

***Enrichment*, p. 547**

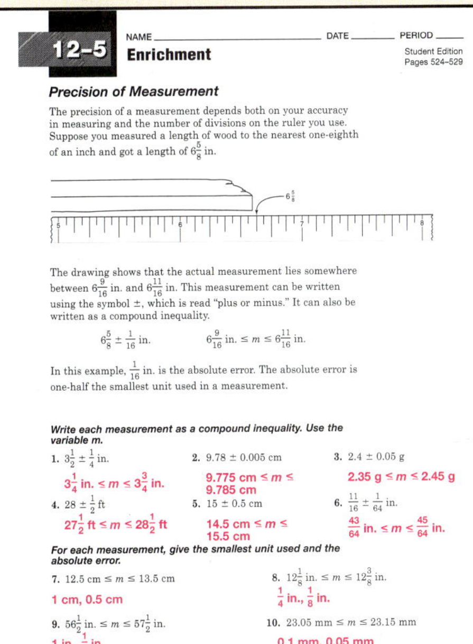

12–5 NAME ______ DATE ______ PERIOD ______

Enrichment Student Edition Pages 524–529

Precision of Measurement

The precision of a measurement depends both on your accuracy in measuring and the number of divisions on the ruler you use. Suppose you measured a length of wood to the nearest one-eighth of an inch and got a length of $6\frac{5}{8}$ in.

The drawing shows that the actual measurement lies somewhere between $6\frac{9}{16}$ in. and $6\frac{11}{16}$ in. This measurement can be written using the symbol ±, which is read "plus or minus." It can also be written as a compound inequality.

$6\frac{5}{8} \pm \frac{1}{16}$ in. $\quad 6\frac{9}{16}$ in. $\leq m \leq 6\frac{11}{16}$ in.

In this example, $\frac{1}{16}$ in. is the absolute error. The absolute error is one-half the smallest unit used in a measurement.

Write each measurement as a compound inequality. Use the variable m.

1. $3\frac{1}{2} \pm \frac{1}{4}$ in. **$3\frac{1}{4}$ in. $\leq m \leq 3\frac{3}{4}$ in.**

2. 9.78 ± 0.005 cm **9.775 cm $\leq m \leq$ 9.785 cm**

3. 2.4 ± 0.05 g **2.35 g $\leq m \leq$ 2.45 g**

4. $28 \pm \frac{1}{2}$ ft **$27\frac{1}{2}$ ft $\leq m \leq 28\frac{1}{2}$ ft**

5. 15 ± 0.5 cm **14.5 cm $\leq m \leq$ 15.5 cm**

6. $\frac{11}{16} \pm \frac{1}{64}$ in. **$\frac{43}{64}$ in. $\leq m \leq \frac{45}{64}$ in.**

For each measurement, give the smallest unit used and the absolute error.

7. 12.5 cm $\leq m \leq$ 13.5 cm **1 cm, 0.5 cm**

8. $12\frac{1}{8}$ in. $\leq m \leq 12\frac{3}{8}$ in. **$\frac{1}{4}$ in., $\frac{1}{8}$ in.**

9. $56\frac{1}{2}$ in. $\leq m \leq 57\frac{1}{2}$ in. **1 in., $\frac{1}{2}$ in.**

10. 23.05 mm $\leq m \leq$ 23.15 mm **0.1 mm, 0.05 mm**

Extra Credit

Using only two different integers, write a compound inequality that has no solution, one that has exactly one solution, one whose graph is two separate rays, and one whose graph is the entire real number line.
Sample answers: $x > 2$ and $x < -2$; $x \geq 2$ and $x \leq 2$; $x < -2$ or $x > 2$; $x > -2$ or $x < 2$

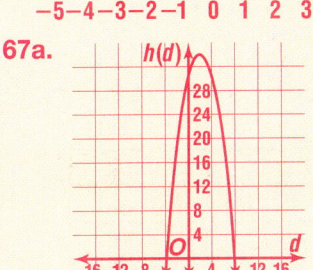

Lesson 12-6

1 FOCUS

5-Minute Check
Lesson 12–5

1. Write the compound inequality $x > 0$ and $x \leq 7$ without using the word *and*. **$0 < x \leq 7$, or $7 \geq x > 0$**

Solve each compound inequality. Graph the solution.

2. $8 < -2x < 18$
$\{x \mid -9 < x < -4\}$

3. $\frac{x}{6} \geq 1$ or $x + 6 < 8$
$\{x \mid x \geq 6$ or $x < 2\}$
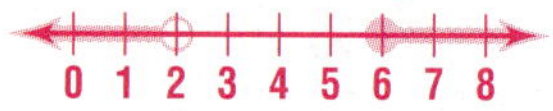

4. The table below gives admission prices for a museum. Write an inequality to represent the age *a* of people charged \$4.00 for admission. **$3 \leq a \leq 12$**

Age	Admission
Under 3	Free
3 to 12	\$4.00
over 12	\$6.00

5. Write a compound inequality for the solution shown below.
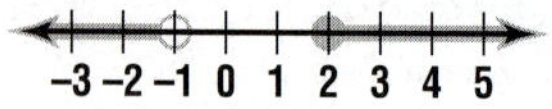

$x < -1$ or $x \geq 2$

Motivating the Lesson

Real-World Connection Most political polls are reported with a margin of error. For example, a poll may predict that a candidate will receive 58% of the vote in an election with a margin of error of ±3%. Ask students if any of them knows what this means. Explain that it means that in an actual election the candidate is very likely to receive between 55% and 61% of the vote. Explain to students how the absolute value inequality $|x - 58| \leq 3$ represents this situation.

12–6 Solving Inequalities Involving Absolute Value

What You'll Learn
You'll learn to solve inequalities involving absolute value.

Why It's Important
Manufacturing Employees use inequalities involving absolute value to determine tolerances. *See Example 3.*

There are three types of open sentences that can involve absolute value. They are listed below. Note that in each case, n is nonnegative since the absolute value of a number can only equal 0 or a positive number.

$|x| = n$ $\qquad |x| < n$ $\qquad |x| > n$

You have already studied equations involving absolute value. Inequalities involving absolute value are similar. Consider the graphs and solutions of the three open sentences below.

Look Back

Absolute Value: Lesson 3–7

Types of Open Sentences with Absolute Value

$|x| = 3$

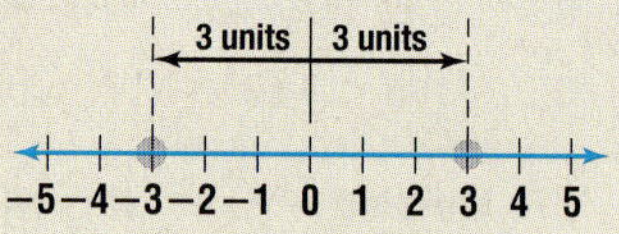

The distance from 0 is 3.
So, $x = 3$ or $x = -3$.

$|x| < 3$

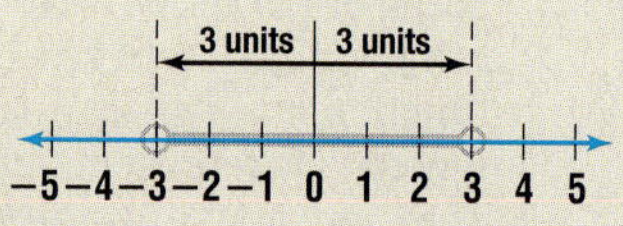

The distance from 0 is less than 3.
So, $x > -3$ and $x < 3$, or $-3 < x < 3$.

$|x| > 3$

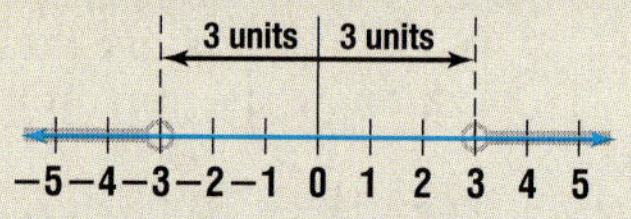

The distance from 0 is greater than 3.
So, $x > 3$ or $x < -3$.

For both equations and inequalities involving absolute value, there are two cases to consider.

Case 1 The value within the absolute value symbols is positive.

Case 2 The value within the absolute value symbols is negative.

Resource Manager

Reproducible Masters

Chapter 12 Resource Masters
- *Study Guide,* p. 548
- *Skills Practice,* p. 549
- *Practice,* p. 550
- *Reading to Learn Mathematics,* p. 551
- *Enrichment,* p. 552

Hands-On Algebra, p. 138

Transparencies

5-Minute Check, 12–6

Technology/Multimedia

Interactive Chalkboard CD-ROM

Example 

Solve $|x - 5| \le 2$. Graph the solution.

Case 1 $x - 5$ is positive.

$x - 5 \le 2$

$x - 5 + 5 \le 2 + 5$ *Add 5.*

$x \le 7$

Case 2 $x - 5$ is negative.

$-(x - 5) \le 2$

$-(x - 5)(-1) \ge 2(-1)$ *Multiply by −1 and reverse the symbol.*

$x - 5 \ge -2$

$5 + 5 \ge -2 + 5$ *Add 5.*

$x \ge 3$

So, the solution is $\{x \mid 3 \le x \le 7\}$.

0 1 2 3 4 5 6 7 8 9 10

The solution makes sense since 3 and 7 are at most 2 units from 5.

Your Turn

2 3 4 5 6 7 8 9 10 11 12

a. Solve $|x - 7| < 4$. Graph the solution. **$\{x \mid 3 < x < 11\}$**

As in Example 1, when solving an inequality involving absolute value and the symbols < or ≤, the solution can be written as an inequality using *and*. However, when solving an inequality involving absolute value and the symbols > or ≥, the solution can be written as an inequality using *or*.

Example 

Solve $|6x| > 18$. Graph the solution.

Case 1 $6x$ is positive.

$6x > 18$

$\frac{6x}{6} > \frac{18}{6}$ *Divide by 6.*

$x > 3$

Case 2 $6x$ is negative.

$-6x > 18$

$\frac{-6x}{-6} < \frac{18}{-6}$ *Divide by −6 and reverse the symbol.*

$x < -3$

So, the solution is $\{x \mid x < -3 \text{ or } x > 3\}$.

Your Turn

b. Solve $|x - 2| \ge 3$. Graph the solution. **$\{x \mid x \le -1 \text{ or } x \ge 5\}$**

Inequalities involving absolute value are often used to indicate *tolerance*. Tolerance is the amount of error or uncertainty that is allowed when taking measurements.

www.algconcepts.com/extra_examples

2 TEACH

Teaching Tip When discussing Example 1, some students may be puzzled that the expression inside the absolute value sign can be negative. Remind them that this expression can be either positive or negative. It is only after the absolute value is taken that the result must be positive. Have students substitute numbers from the solution set into the original inequality to see how the expression inside the absolute value sign can be either positive or negative.

Teaching Tip Interpreting an absolute value inequality in terms of distance, as in the last sentence of Example 1, will bring clarity to some students. Point out that $|x - b|$ is the distance of a point x from b. When students encounter an expression of the form $|x + b|$, as in In-Class Example 1 below, they can see by rewriting it as $|x - (-b)|$ that it expresses the distance of a point x from $-b$.

In-Class Examples

Example 1

Solve $|x + 3| < 4$. Graph the solution. **$\{x \mid -7 < x < 1\}$**

−7 −6 −5 −4 −3 −2 −1 0 1

Example 2

Solve $|4x| \ge 16$. Graph the solution. **$\{x \mid x \le -4 \text{ or } x \ge 4\}$**

−4 −3 −2 −1 0 1 2 3 4

Teaching Tip When discussing the Examine portion of Example 3, point out to students that the phrase *from 0.495 inch to 0.505 inch, inclusive* includes not only all measurements between 0.495 inch and 0.505 inch, but the measurements 0.495 inch and 0.505 inch as well.

In-Class Example

Example 3

A lumber company makes $3\frac{1}{4}$-foot railing posts to use in making decks. The tolerance for the posts is 0.025 foot. What is the range of acceptable post lengths? **$\{\ell \mid 3.225 \text{ ft} \leq \ell \leq 3.275 \text{ ft}\}$**

Answers

1. $|x| < 7$ includes all numbers between -7 and 7. $|x| > 7$ includes all numbers to the left of -7 and to the right of 7.
3. Mia; $|x| \leq 0$ includes only 0. $|x| \geq 0$ includes all real numbers.

Study Guide, **p. 548**

12-6 NAME ______ DATE ______ PERIOD ______

Study Guide Student Edition Pages 530–534

Solving Inequalities Involving Absolute Value

You have already studied equations of the form $|x| = n$ involving absolute value, where n is a nonnegative number. Inequalities involving absolute value are similar. They are of the form $|x| > n$ or $|x| < n$, where n is a nonnegative number.

To solve both equations and inequalities involving absolute value, there are two cases to consider.

Case 1 The value of the expression within the absolute value symbol is positive.

Case 2 The value of the expression within the absolute value symbol is negative.

Example 1: Solve $|x - 4| < 2$. Graph the solution.

Case 1 $x - 4$ is positive.

$x - 4 < 2$
$x - 4 + 4 < 2 + 4$
$x < 6$

Case 2 $x - 4$ is negative.

$-(x - 4) < 2$
$-(x - 4)(-1) > 2(-1)$ *Reverse the symbol.*
$x - 4 > -2$
$x - 4 + 4 > -2 + 4$
$x > 2$

The solution is $\{x \mid 2 < x < 6\}$.

Example 2: Solve $|x + 1| \geq 4$. Graph the solution.

Case 1 $x + 1$ is positive.

$x + 1 \geq 4$
$x + 1 - 1 \geq 4 - 1$
$x \geq 3$

Case 2 $x + 1$ is negative.

$-(x + 1) \geq 4$
$-(x + 1)(-1) \leq 4(-1)$ *Reverse the symbol.*
$x + 1 \leq -4$
$x + 1 - 1 \leq -4 - 1$
$x \leq -5$

The solution is $\{x \mid x \geq 3 \text{ or } x \leq -5\}$.

Solve each inequality. Graph the solution.

1. $|n| \leq 5$ **$\{n \mid -5 \leq n \leq 5\}$**
2. $|4x| \leq 12$ **$\{x \mid -3 \leq x \leq 3\}$**
3. $|y + 1| > 2$ **$\{y \mid y > 1 \text{ or } y < -3\}$**
4. $|6p| > 2.4$ **$\{p \mid p > 0.4 \text{ or } p < -0.4\}$**
5. $|t + 3| \geq 2$ **$\{t \mid t > -1 \text{ or } t < -5\}$**
6. $|h + 2| < 6$ **$\{h \mid -8 < h < 4\}$**

Real World **Example** Measurement Link

3 **When producing $\frac{1}{2}$-inch bolts for bicycle parts, the tolerance is 0.005 inch. What is the range of acceptable bolt measures?**

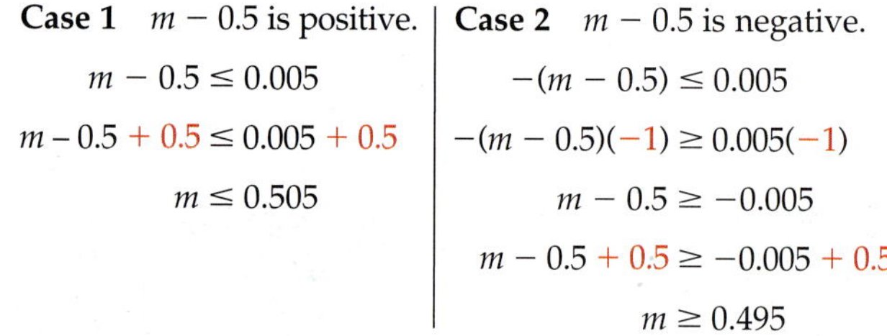

Explore The difference in the actual size of the bolt and its expected size has to be less than or equal to 0.005 inch.

Plan Let m = the actual measure of the bolt.

Then, $\left|m - \frac{1}{2}\right| \leq 0.005$.

Solve $|m - 0.5| \leq 0.005$ *Write $\frac{1}{2}$ as 0.5.*

Case 1 $m - 0.5$ is positive.

$m - 0.5 \leq 0.005$
$m - 0.5 + 0.5 \leq 0.005 + 0.5$
$m \leq 0.505$

Case 2 $m - 0.5$ is negative.

$-(m - 0.5) \leq 0.005$
$-(m - 0.5)(-1) \geq 0.005(-1)$
$m - 0.5 \geq -0.005$
$m - 0.5 + 0.5 \geq -0.005 + 0.5$
$m \geq 0.495$

The solution is $\{m \mid 0.495 \leq m \leq 0.505\}$.

Examine An acceptable bolt must measure from 0.495 inch to 0.505 inch, inclusive. To check the solution, choose a value for m within this range and one outside of this range. Substitute them into the original problem. Which value results in a true inequality?

Check for Understanding

Communicating Mathematics

1. **Compare and contrast** the graphs of the solutions for $|x| < 7$ and $|x| > 7$. **See margin.**
2. **Graph** the solutions for $|x - 2| > 1$ and $|x - 2| < 1$. Which inequality is the *intersection* of the graphs of two inequalities? Which inequality is the *union* of the graphs of two inequalities?
3. **You Decide?** Madison says that the solution for $|x| \leq 0$ is the same as the solution for $|x| \geq 0$. Mia says it is not. Who is correct? Explain. **See margin.**

2. intersection, $|x - 2| < 1$; union, $|x - 2| > 1$

Guided Practice

Getting Ready **Write two inequalities to describe the solution.**

Sample 1: $|x| \leq 5$
Solution: $x \leq 5$ and $x \geq -5$

Sample 2: $|x| > 1$
Solution: $x > 1$ or $x < -1$

4. $|x| < 10$ **$x < 10$ and $x > -10$**
5. $|x| \leq 3$ **$x \leq 3$ and $x \geq -3$**
6. $|x| \geq 2$ **$x \geq 2$ or $x \leq -2$**
7. $|x| > 8$ **$x > 8$ or $x < -8$**

Reteaching Activity

Naturalist Learners Have students bring in natural items such as flowers or leaves that are hard to measure exactly. Point out that there is always some degree of uncertainty in measurement. Have students measure their items, also giving a value to how precisely they think they can find a reliable measure. For example, a student may think her measurement of a leaf is reliable to the nearest eighth of an inch. Then have students write and solve absolute value inequalities to find the expected range of actual measurements.

Examples 1 & 2

8–13. See margin for graphs.

13. $\{s \mid s \geq -1$ or $s \leq -7\}$

Solve each inequality. Graph the solution.

8. $|n - 4| < 5$ $\{n \mid -1 < n < 9\}$
9. $|x - 2| < 6$ $\{x \mid -4 < x < 8\}$
10. $|3j| \leq 12$ $\{j \mid -4 \leq j \leq 4\}$
11. $|t - 5| \geq 3$ $\{t \mid t \geq 8$ or $t \leq 2\}$
12. $|2y| > 2$ $\{y \mid y > 1$ or $y < -1\}$
13. $|s + 4| \geq 3$

Example 3

14. **Measurement** Refer to Example 3. What are the possible measures for the bolt if the tolerance is 0.05 inch? Does a lesser or greater tolerance ensure more accurate measurements? Explain.
$0.45 \leq m \leq 0.55$; lesser tolerance

Exercises

Practice

Solve each inequality. Graph the solution. 15–32. See Solutions Manual.

15. $\lvert m + 1\rvert < 5$	16. $\lvert 3v\rvert < 15$	17. $\lvert z + 7\rvert \leq 2$
18. $\lvert x + 3\rvert \leq 8$	19. $\lvert p - 1\rvert < 2$	20. $\lvert r + 4\rvert < 4$
21. $\lvert 7t\rvert \leq 14$	22. $\lvert a - 3\rvert \leq 4$	23. $\lvert k + 2\rvert < 3.5$
24. $\lvert 5n\rvert \geq 30$	25. $\lvert y + 3\rvert \geq 6$	26. $\lvert z - 1\rvert \geq 1$
27. $\lvert 9x\rvert > 18$	28. $\lvert w + 2\rvert > 5$	29. $\lvert r - 4\rvert \geq 1$
30. $\lvert a - 8\rvert \geq 3$	31. $\lvert h - 3\rvert > 9$	32. $\lvert d + 9\rvert > 0.2$

Homework Help

For Exercises	See Examples
15, 17, 18–20 22, 23, 25, 26, 28–35, 38–43	1, 3
18, 21, 24, 27	2

Extra Practice
See page 717.

Write an inequality involving absolute value for each statement. Do not solve.

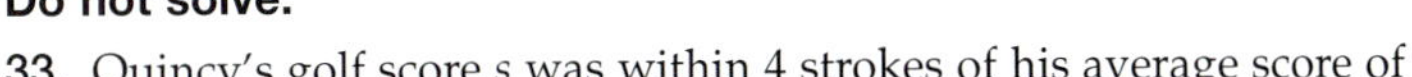

33. Quincy's golf score s was within 4 strokes of his average score of 90.
34. The measure m of a board used to build a cabinet must be within $\frac{1}{4}$ inch of 46 inches, inclusive, to fit properly. $|m - 46| \leq \frac{1}{4}$
35. The cruise control of a car set at 65 mph should keep the speed s within 3 mph, inclusive, of 65 mph. $|s - 65| \leq 3$

33. $|s - 90| < 4$

For each graph, write an inequality involving absolute value.

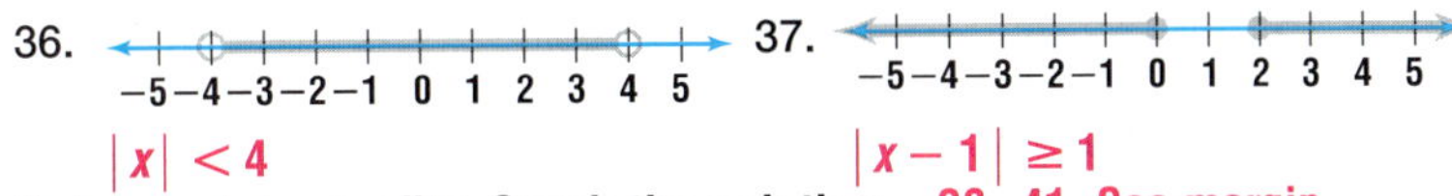

36. $|x| < 4$
37. $|x - 1| \geq 1$

Solve each inequality. Graph the solution. 38–41. See margin.

38. $|2x - 11| \geq 7$
39. $|3x - 12| < 12$
40. $4|x + 3| \leq 8$
41. $10|x + 1| > 90$

Applications and Problem Solving

42. $|b - 8758.20| < 2$; $8756.20 < b < 8760.20$

For Exercises 42–43, write and solve an absolute value inequality.

42. **Finance** Ms. Gibson is a bank teller. She must balance her drawer and be within $2 of the expected balance. If Ms. Gibson's expected balance is $8758.20, what are acceptable balances for her drawer?

43. **Chemistry** For a chemistry project, Marvin must pour 3.25 milliliters of solution into a beaker. If he does not pour within 0.05, inclusive, of 3.25 milliters, the results will be inaccurate. How many milliliters of solution can Marvin use?

43. $|m - 3.25| \leq 0.05$; $3.20 \leq m \leq 3.30$

3 PRACTICE/APPLY

Error Analysis

Watch for students who assign positive and negative values to the side of the inequality opposite the variable expression, such as writing the inequalities $n - 4 < 5$ and $n - 4 < -5$ for Exercise 6, thus forgetting to change the direction of the second inequality. ***Prevent by*** encouraging students to set up their work explicitly as shown in Examples 1–3 to avoid any misunderstanding or errors.

Assignment Guide

Basic: 15–43 odd, 44–52
Average: 16–40 even, 42–52
All: Quiz 2, 1–10

Answers

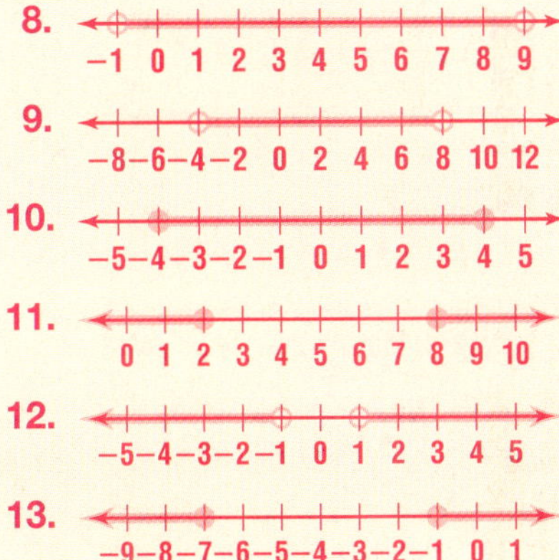

Skills Practice, p. 549, and *Practice,* p. 550 (shown)

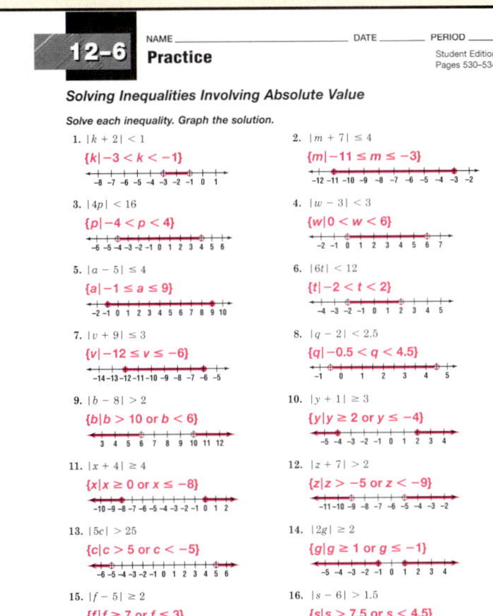

12–6 NAME ______ DATE ______ PERIOD ______

Practice Student Edition Pages 530–534

Solving Inequalities Involving Absolute Value

Solve each inequality. Graph the solution.

1. $|k + 2| < 1$ $\{k \mid -3 < k < -1\}$
2. $|m + 7| \leq 4$ $\{m \mid -11 \leq m \leq -3\}$
3. $|4p| < 16$ $\{p \mid -4 < p < 4\}$
4. $|w - 3| < 3$ $\{w \mid 0 < w < 6\}$
5. $|a - 5| \leq 4$ $\{a \mid -1 \leq a \leq 9\}$
6. $|6t| < 12$ $\{t \mid -2 < t < 2\}$
7. $|v + 9| \leq 3$ $\{v \mid -12 \leq v \leq -6\}$
8. $|q - 2| < 2.5$ $\{q \mid -0.5 < q < 4.5\}$
9. $|b - 8| > 2$ $\{b \mid b > 10$ or $b < 6\}$
10. $|y + 1| \geq 3$ $\{y \mid y \geq 2$ or $y \leq -4\}$
11. $|x + 4| \geq 4$ $\{x \mid x \geq 0$ or $x \leq -8\}$
12. $|z + 7| > 2$ $\{z \mid z > -5$ or $z < -9\}$
13. $|5c| > 25$ $\{c \mid c > 5$ or $c < -5\}$
14. $|2g| \geq 2$ $\{g \mid g \geq 1$ or $g \leq -1\}$
15. $|f - 5| \geq 2$ $\{f \mid f \geq 7$ or $f \leq 3\}$
16. $|s - 6| > 1.5$ $\{s \mid s > 7.5$ or $s < 4.5\}$

Answers

38. $\{x \mid x \leq 2$ or $x \geq 9\}$
39. $\{x \mid 0 < x < 8\}$
40. $\{x \mid -5 \leq x \leq -1\}$
41. $\{x \mid x < -10$ or $x > 8\}$

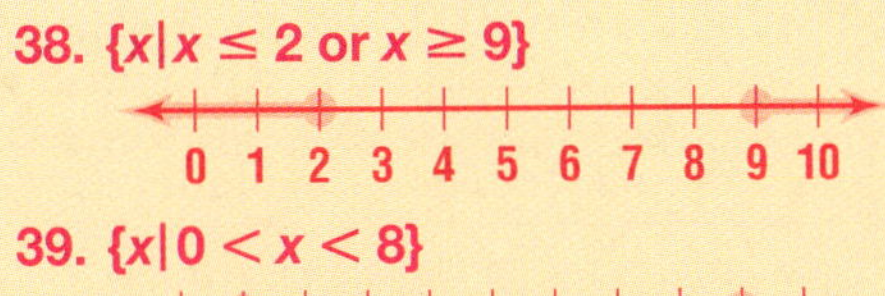

4 ASSESS

Open-Ended Assessment

Writing Have students write a summary explaining the similarities and differences between solving inequalities that do and don't involve absolute value.

Quiz 2

The Quiz provides students with a brief review of the concepts and skills in Lessons 12–4 through 12–6. Lesson numbers are given to the right of exercises or instruction lines so students can review concepts not yet mastered.

Answers

Quiz 2

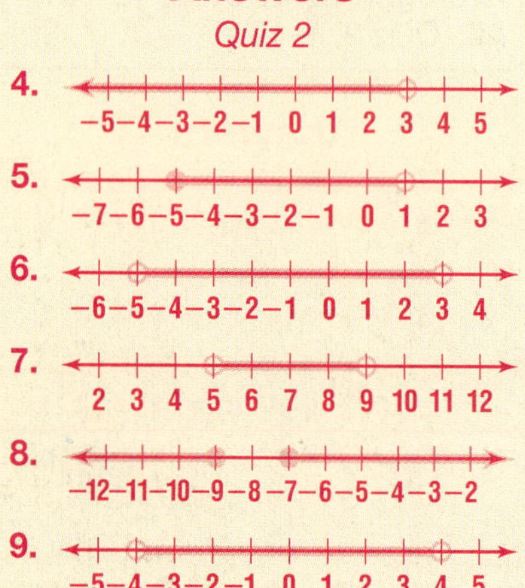

Enrichment, p. 552

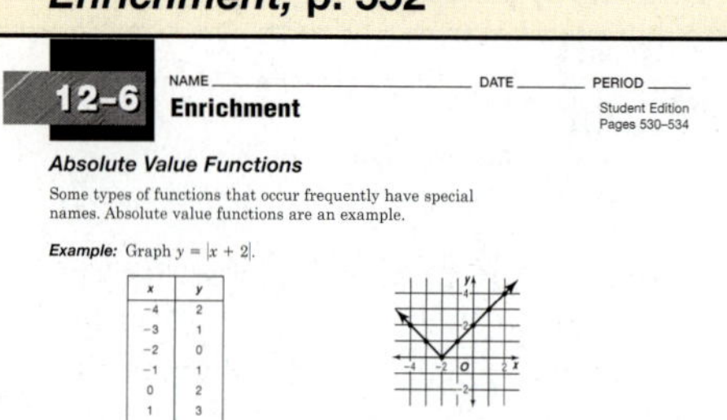

12-6 NAME ______ DATE ______ PERIOD ______

Enrichment Student Edition Pages 530–534

Absolute Value Functions

Some types of functions that occur frequently have special names. Absolute value functions are an example.

Example: Graph $y = |x + 2|$.

x	y
−4	2
−3	1
−2	0
−1	1
0	2
1	3
2	4

Complete the table for each equation. Then, draw the graph.

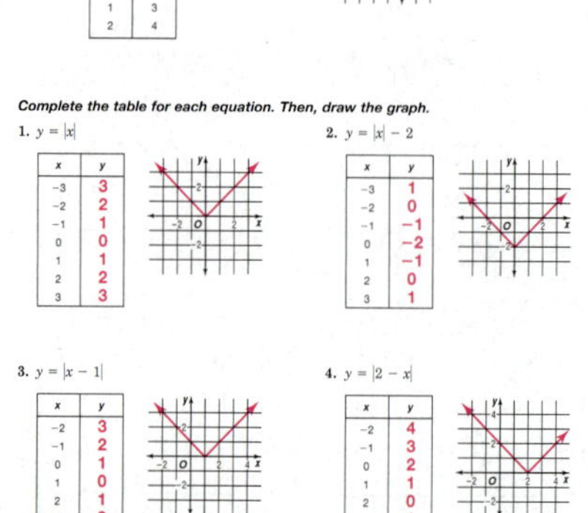

1. $y = |x|$

x	y
−3	3
−2	2
−1	1
0	0
1	1
2	2
3	3

2. $y = |x| - 2$

x	y
−3	1
−2	0
−1	−1
0	−2
1	−1
2	0
3	1

3. $y = |x - 1|$

x	y
−2	3
−1	2
0	1
1	0
2	1
3	2
4	3

4. $y = |2 - x|$

x	y
−2	4
−1	3
0	2
1	1
2	0
3	1
4	2

44. **Critical Thinking** The *percent of error* is the ratio of the greatest possible error (tolerance) to a measurement. You can find the percent of error using this formula.

$$\text{percent of error} = \left|\frac{\text{greatest possible error}}{\text{measurement}}\right| \cdot 100$$

One rating system for in-line skate bearings is based on tolerances. The table shows tolerances for the outside diameter of bearings measuring 22 millimeters.

Rating	Tolerance (mm)
1	0.010
3	0.008
5	0.005

a. Find the percent of error for each rating to the nearest hundredth.

b. What can you conclude about the rating system? **The higher the rating, the better the bearings.**

44a. rating 1, 0.05%; rating 3, 0.04%; rating 5, 0.02%

Mixed Review

Graph the solution of each compound inequality. *(Lesson 12–5)*

45. $m < -7$ or $m \geq 0$
46. $x \geq -2$ and $x \leq 5$
47. $y > -5$ and $y < 0$
48. $r > 2$ or $r < -2$

45–48. See margin.

49. Solve $1 \geq 4y + 5$. *(Lesson 12–4)* $\{y \mid y \leq -1\}$

50. **Sales** Grant bought a sweater for \$44.52. This cost included 6% sales tax. What was the cost of the sweater before tax? *(Lesson 5–5)* **\$42**

51. Solve $|x - 2| = 4$. *(Lesson 3–7)* **−2, 6**

Standardized Test Practice

52. **Short Response** Write three fractions whose sum is $1\frac{5}{8}$. *(Lesson 3–2)* **See students' work.**

Quiz 2 Lessons 12–4 through 12–6

Solve each inequality. *(Lesson 12–4)*

1. $3x + 8 < 11$ $\{x \mid x < 1\}$
2. $5 - 6n > -19$ $\{n \mid n < 4\}$
3. $9d - 4 \geq 8 - d$ $\{d \mid d \geq 1.2\}$

4–9. See margin for graphs.

Solve each compound inequality. Graph the solution. *(Lesson 12–5)*

4. $1 + x < -4$ or $1 + x < 4$ $\{x \mid x < 3\}$
5. $-2 \leq n + 3 < 4$ $\{n \mid -5 \leq n < 1\}$
6. $-6 < -2f < 10$ $\{f \mid -5 < f < 3\}$

Solve each inequality. Graph the solution. *(Lesson 12–6)*

7. $|y - 7| < 2$ $\{y \mid 5 < y < 9\}$
8. $|a + 8| \geq 1$ $\{a \mid a \leq -9$ or $a \geq -7\}$
9. $|5x| < 20$ $\{x \mid -4 < x < 4\}$

10. **Travel** Before the meter begins ticking, the charge for a taxi is \$1.60. For each mile driven, there is an additional charge of 80 cents. What is the greatest distance you can travel in this taxi if you do not want to pay more than \$10.00? *(Lesson 12–4)* **10.5 mi**

www.algconcepts.com/self_check_quiz

Extra Credit

A bakery guarantees that a pan of their cinnamon rolls will weigh from 15.85 ounces to 16.15 ounces, inclusive. What is the bakery's tolerance for the weight? Write an absolute value inequality for the acceptable weight. **0.15 oz; $|w - 16| \leq 0.15$**

Answers

12–7 Graphing Inequalities in Two Variables

What You'll Learn

You'll learn to graph inequalities on the coordinate plane.

Why It's Important

Budgeting By graphing inequalities, you can solve problems where there are many solutions.
See Example 3.

Mr. Wheat is planning to take the Jazz Club to a music festival. Lawn tickets cost \$20, and pavilion tickets cost \$30. If he plans to spend at most \$300, how many of each ticket can Mr. Wheat purchase?

Let x represent the number of lawn tickets. Let y represent the number of pavilion tickets. Then the inequality below represents the solution.

The cost of lawn tickets	*plus*	*the cost of pavilion tickets*	*is at most*	*\$300.*
$20x$	$+$	$30y$	$\leq$	300

The inequality is written in two variables. It is similar to an equation written in two variables. An easy way to show the solution of an inequality is to graph it in the coordinate plane. *This problem will be solved in Example 3.*

The solution set of an inequality in two variables contains many ordered pairs. The graph of these ordered pairs fills an area on the coordinate plane called a **half-plane**. The graph of an equation defines the **boundary** or edge for each half-plane. Use these steps to graph $y > 3$.

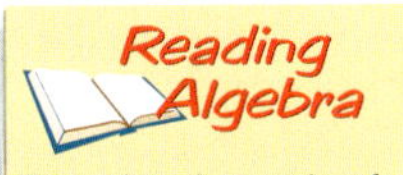

Reading Algebra

The related equation for $y > 3$ is $y = 3$.

Step 1 Determine the boundary by graphing the related equation, $y = 3$.

Step 2 Draw a *dashed* line since the boundary is *not* part of the graph.

Step 3 Determine which half-plane is the solution. To do this, substitute a point from each half-plane into the inequality. Find which point results in a true statement.

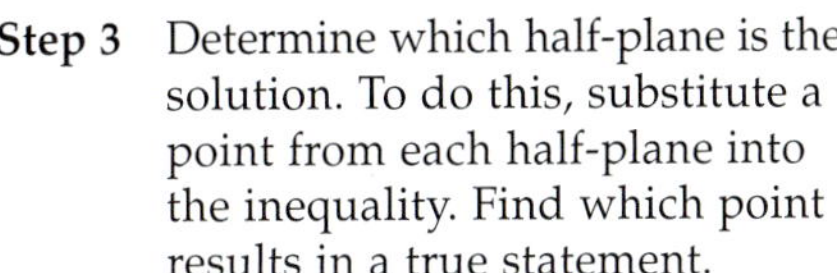

Test the point at (5, 8).	Test the point at (−3, 1).
$y > 3$	$y > 3$
$8 > 3$ *Replace y with 8.*	$1 > 3$ *Replace y with 1.*
true	false

The half-plane that contains (5, 8) is the solution. Shade that half-plane. Any point in the shaded region is a solution of the inequality $y > 3$.

Resource Manager

Reproducible Masters

Chapter 12 Resource Masters

- *Study Guide*, p. 553
- *Skills Practice*, p. 554
- *Practice*, p. 555
- *Reading to Learn Mathematics*, p. 556
- *Enrichment*, p. 557
- *Assessment*, p. 569

Graphing Calculator, pp. 34–35

Hands-On Algebra, p. 139

Transparencies

5-Minute Check, 12–7

Technology/Multimedia

AlgePASS, Lesson 30

Interactive Chalkboard CD-ROM

1 FOCUS

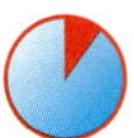

5-Minute Check
Lesson 12–6

Solve each inequality.

1. $|3x| \geq 15$
   $\{x \mid x \leq -5 \text{ or } x \geq 5\}$
2. $|t - 6| < 11$
   $\{t \mid -5 < t < 17\}$
3. Write an inequality involving absolute value for the graph.

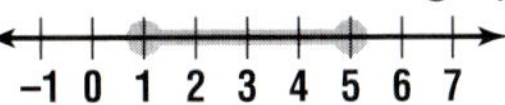

   Sample answer: $|x - 3| \leq 2$
4. Write an inequality involving absolute value for the statement *A box of cookies is acceptable to sell if its weight is within $\frac{1}{4}$ ounce of the stated weight of 18 ounces.* $|w - 18| \leq \frac{1}{4}$
5. A company's tolerance for the diameter of the washer below is 0.003 inch. What is the range of acceptable washer diameters d?

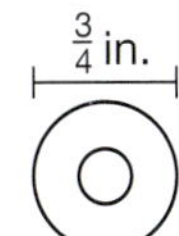

   $\{d \mid 0.747 \text{ in.} \leq d \leq 0.753 \text{ in.}\}$

Motivating the Lesson

Hands-On Activity Have groups of students cut out ten squares of paper, labeling three Q for quarter and seven D for dime. Have groups form and record all combinations (Q, D) they can with the squares so that the indicated value is not more than 75¢. Point out that these are solutions of an *inequality in two variables*, $25Q + 10D \leq 75$. Have students identify the ordered pairs that are not also solutions of $25Q + 10D < 75$. Point out that these points are analogous to endpoints of the graph of an inequality in one variable.

2 TEACH

In-Class Example

Example 1

Graph $y \leq x - 2$.

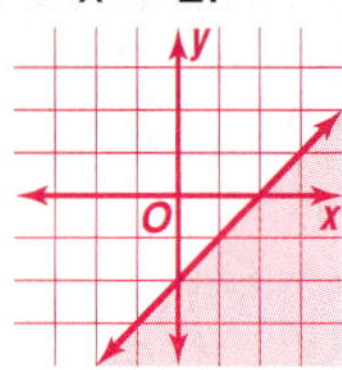

Teaching Tip Prior to discussing Examples 2 and 3, you may want to review graphing linear equations using slope and y-intercept, and using x- and y-intercepts. In Example 2, explain that any point in one of the half-planes can be chosen in Step 3; (0, 0) is chosen when possible because it is the easiest to substitute in the inequality.

In-Class Example

Example 2

Graph $-3x + y > 1$.

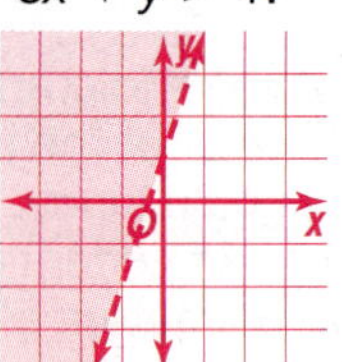

Answer

a.

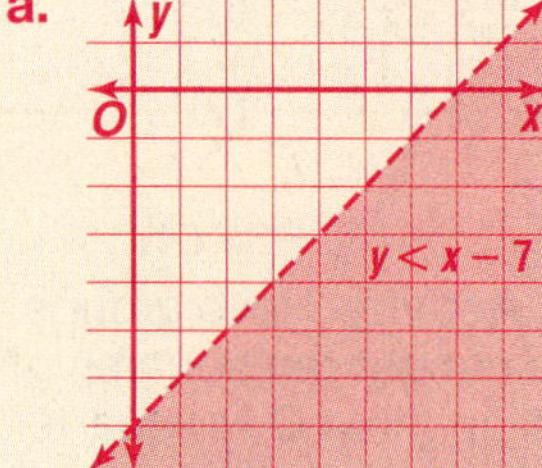

Example 1 Graph $y > -x + 3$.

Step 1 Determine the boundary by graphing the related equation, $y = -x + 3$.

x	$-x + 3$	y
−2	−(−2) + 3	5
−1	−(−1) + 3	4
0	−(0) + 3	3
1	−1 + 3	2
2	−2 + 3	1

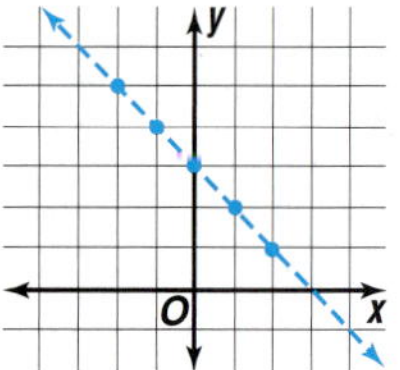

Step 2 Draw a *dashed* line since the boundary is not included.

Step 3 Test any point to find which half-plane is the solution. Use (0, 0) since it is the easiest point to use in calculations.

$y > -x + 3$ *Original inequality*

$0 > -(0) + 3$ *$x = 0, y = 0$*

$0 > 3$ false

Since (0, 0) does *not* result in a true inequality, the half-plane containing (0, 0) is *not* the solution. Thus, shade the other half-plane.

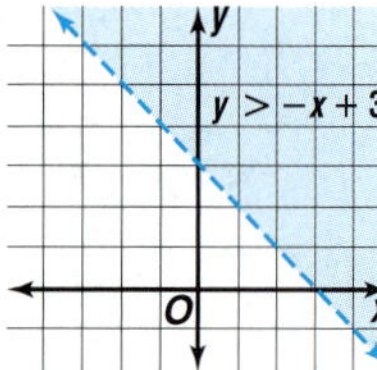

Your Turn

a. Graph $y < x - 7$. **See margin.**

When graphing inequalities, the boundary line is not always dashed. Consider the graph of $y \geq 3$. Since the inequality means $y > 3$ or $y = 3$, the boundary is part of the solution. This is indicated by graphing a solid line.

Example 2 Graph $4x + y \leq 12$.

To make a table or graph for the boundary line, solve the inequality for y in terms of x.

$4x + y \leq 12$ *Original inequality*

$4x + y - 4x \leq 12 - 4x$ *Subtract 4x from each side.*

$y \leq -4x + 12$ *Rewrite 12 − 4x as −4x + 12.*

www.algconcepts.com/extra_examples

From the Classroom of ...

Nicki Hudson
West Linn High School
West Linn, Oregon

When teaching students to graph $y < x + 2$, I like to make sure students understand that for every ordered pair (x, y) whose point is in the shaded region, the y-value will be less than the x-value plus 2.

Step 1 Determine the boundary by graphing $y = -4x + 12$.

Step 2 Draw a *solid* line since the boundary is included.

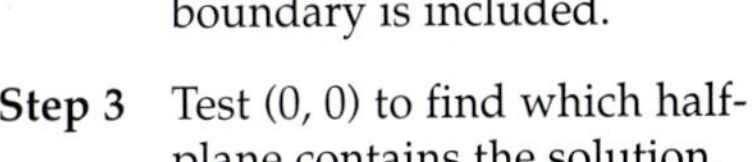

Step 3 Test (0, 0) to find which half-plane contains the solution.

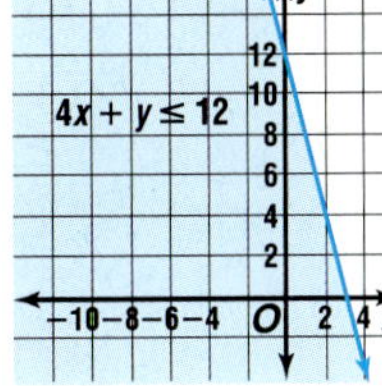

$$4x + y \le 12 \quad \textit{Original inequality}$$
$$4(0) + 0 \le 12 \quad x = 0, y = 0$$
$$0 \le 12 \quad \text{true}$$

The half-plane that contains (0, 0) should be in the solution, which is indicated by the shaded region.

Your Turn

b. Graph $-2x + y \le 10$. **See margin.**

When solving real-life inequalities, the domain and range of the inequality are often restricted to nonnegative numbers or whole numbers.

Example 3 **Budgeting Link**

Refer to the application at the beginning of the lesson. How many lawn and pavilion tickets can Mr. Wheat purchase?

First, solve for y in terms of x.

$$20x + 30y \le 300 \quad \textit{Original inequality}$$
$$20x + 30y - 20x \le 300 - 20x \quad \textit{Subtract 20x from each side.}$$
$$30y \le -20x + 300 \quad \textit{Simplify.}$$
$$\frac{30y}{30} \le \frac{-20x + 300}{30} \quad \textit{Divide each side by 30.}$$
$$y \le \frac{-20x}{30} + \frac{300}{30} \quad \textit{Simplify.}$$
$$y \le -\frac{2}{3}x + 10 \quad \textit{Simplify.}$$

Step 1 Determine the boundary by graphing $y = -\frac{2}{3}x + 10$.

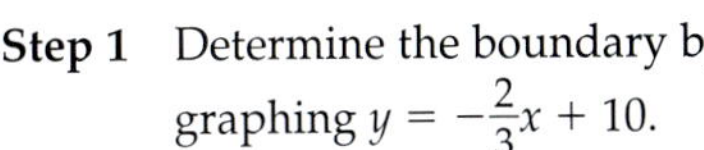

Step 2 Draw a *solid* line since the boundary is included.

Step 3 Test (0, 0) to find which half-plane contains the solution.

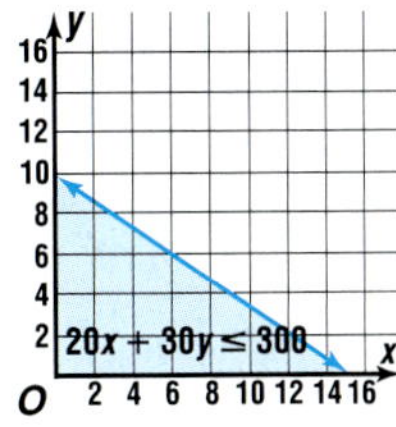

(continued on the next page)

Answer

b.

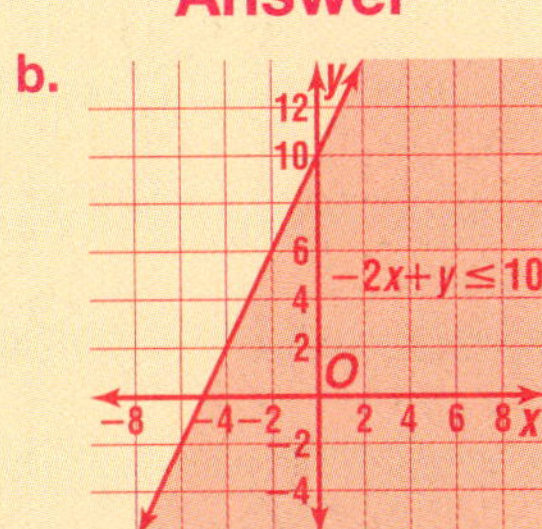

Teaching Tip In Example 3, emphasize that the real-world solution contains only points in the shaded half-plane with whole number coordinates. A number of tickets cannot be negative, so these points are bounded by the x- and y-axes. So, the graph of the real-world solution shows distinct points in the coordinate plane bounded by the x-and y-axes and the line $y = -\frac{2}{3}x + 10$.

In-Class Example

Example 3

Ms. Kwan sells computer systems. She makes a commission of \$50 per home system sold and \$150 per business system sold. The inequality $50x + 150y \ge 450$ represents how many of each system she must sell to make at least \$450 a week in commissions. Graph the inequality. Describe the numbers of systems Ms. Kwan can sell to meet her goal.

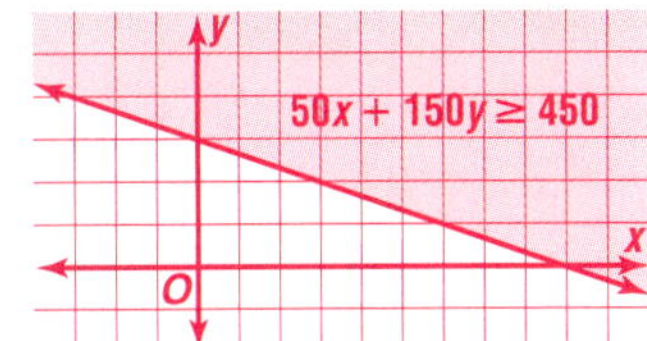

The solutions are ordered pairs in the shaded half-plane that have nonnegative whole-number coordinates, such as (3, 2).

***Study Guide*, p. 553**

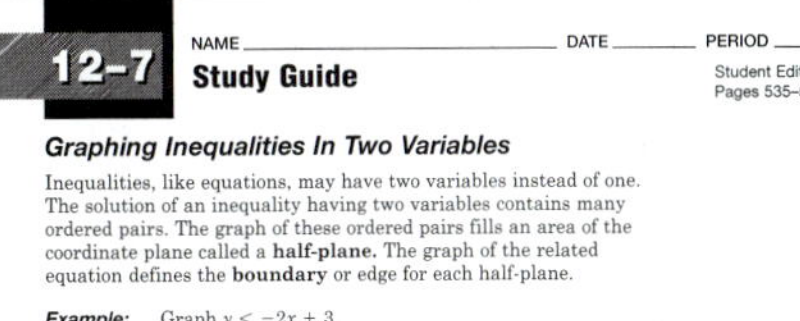

12–7 NAME ________ DATE ______ PERIOD ______

Study Guide Student Edition Pages 535–539

Graphing Inequalities In Two Variables

Inequalities, like equations, may have two variables instead of one. The solution of an inequality having two variables contains many ordered pairs. The graph of these ordered pairs fills an area of the coordinate plane called a **half-plane.** The graph of the related equation defines the **boundary** or edge for each half-plane.

Example: Graph $y < -2x + 3$.

Step 1 Determine the boundary by graphing the related equation, $y = -2x + 3$.

Make a table of values.

x	$-2x + 3$	y
−2	−2(−2) + 3	7
−1	−2(−1) + 3	5
0	−2(0) + 3	3
1	−2(1) + 3	1
2	−2(2) + 3	−1

Step 2 Draw a dashed line because the boundary is not included. *Note:* If the inequality involved ≤ or ≥, the boundary would be included, and you would make the boundary a solid line.

Step 3 Use a point not on the boundary to find which half-plane is the solution. Use (0, 0).

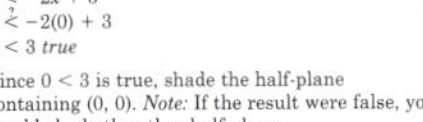

$y < -2x + 3$
$0 \overset{?}{<} -2(0) + 3$
$0 < 3$ *true*

Since $0 < 3$ is true, shade the half-plane containing (0, 0). *Note:* If the result were false, you would shade the other half-plane.

Graph each inequality.

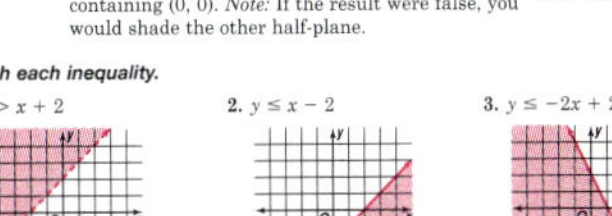

1. $y > x + 2$
2. $y \le x - 2$
3. $y \le -2x + 2$

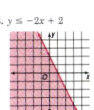

Mr. Wheat cannot buy a negative number of tickets, nor can he buy portions of tickets. The solution is positive ordered pairs that are whole numbers beneath or on the graph of the line $y = -\frac{2}{3}x + 10$.

One solution is (12, 2). This represents 12 lawn tickets and 2 pavilion tickets costing $300.

Check for Understanding

Communicating Mathematics

Vocabulary
half-plane
boundary

1. **Explain** how to determine whether the boundary is a solid line or dashed line when graphing inequalities in two variables.
2. 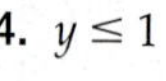

Describe how you could check whether a point is part of the solution of an inequality.

1–2. See margin.

Guided Practice

Getting Ready **Test (0, 0) to find which half-plane is the solution of each inequality.**

Sample: $3x + y > 4$

Solution: $3x + y > 4$

$3(0) + 0 > 4$

$0 > 4$ false

Shade the half-plane *not* containing the point at (0, 0).

3. $x < -2$
4. $y \le 1$
5. $5x - y > 5$

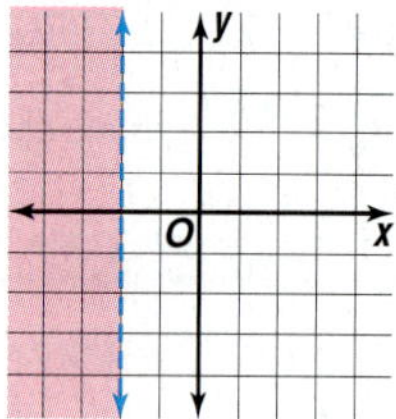

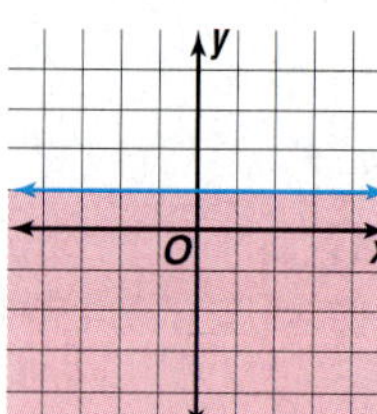

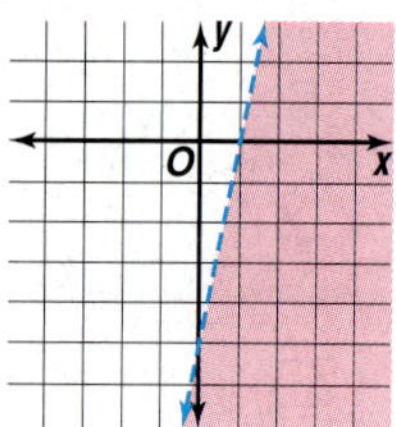

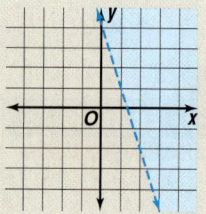

Examples 1 & 2 Graph each inequality. **6–11. See Solutions Manual.**

6. $y \le -1$
7. $y \ge x - 7$
8. $y < 3x + 1$
9. $2x + y < 0$
10. $-4x + y > -8$
11. $x + 3y \ge 9$

Example 3

12. **Sales** Tickets for the winter dance are $5 for singles and $8 for couples. To cover the deejay, photographer, and decoration expenses, a minimum of $1000 must be made from ticket sales. Write and graph an inequality that describes the number of singles' and couples' tickets that must be sold. Name at least one solution. **See margin.**

3 PRACTICE/APPLY

Error Analysis

Watch for students who believe that you always shade the half-plane above the boundary line for > and ≥ inequalities or that you always shade the half-plane below the boundary line for < and ≤ inequalities, which will lead to errors when the coefficient of *y* is negative as in Exercise 5.
Prevent by encouraging students always to test (0, 0) or another point in one of the half-planes before deciding which to shade.

Answers

1. Graph a solid line for ≤ and ≥. Graph a dashed line for < and >.
2. Substitute the coordinates into the original inequality. If the statement is true, then the half-plane containing that point is the solution.
12. 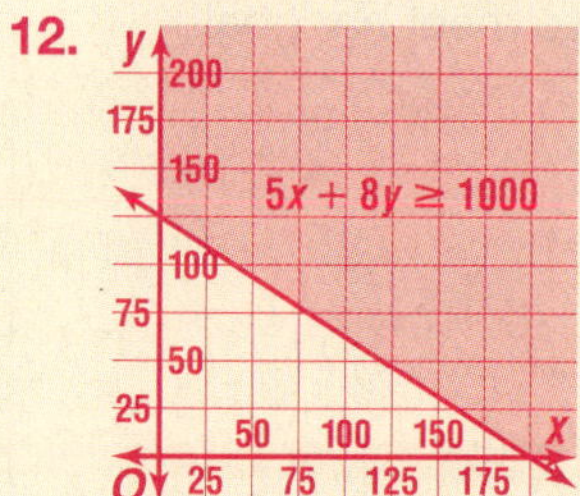

Sample answer: 56 singles, 90 couples

Skills Practice, **p. 554, and *Practice,* p. 555 (shown)**

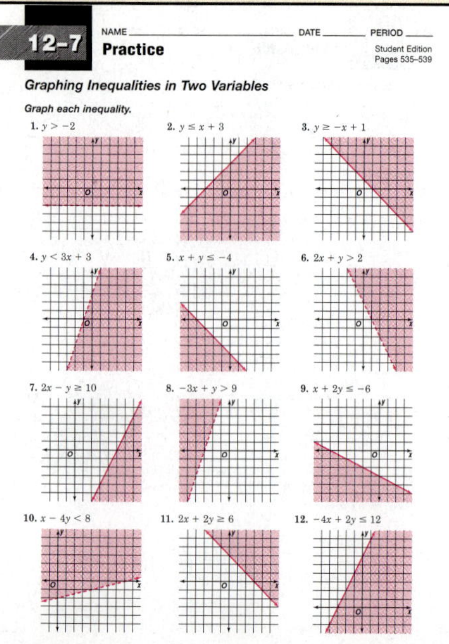

12-7 NAME ________ DATE ______ PERIOD ____

Practice

Student Edition Pages 535–539

Graphing Inequalities in Two Variables

Graph each inequality.

1. $y > -2$
2. $y \le x + 3$
3. $y \ge -x + 1$
4. $y < 3x + 3$
5. $x + y \le -4$
6. $2x + y > 2$
7. $2x - y \ge 10$
8. $-3x + y > 9$
9. $x + 2y \le -6$
10. $x - 4y < 8$
11. $2x + 2y \ge 6$
12. $-4x + 2y \le 12$

Reteaching Activity

Verbal/Linguistic Learners Have students work in pairs. Write several inequalities on the board or overhead. Have one student in each pair verbally guide the second student through the steps in solving and graphing the first inequality. After an inequality is successfully completed, have the students reverse roles.

Exercises

Practice

Graph each inequality. 13–32. See Solutions Manual.

13. $y \leq 7$	14. $x \geq 5$	15. $y < x - 6$
16. $y < -x + 3$	17. $x + y > 8$	18. $y > x$
19. $-y \leq x$	20. $y \leq 2x$	21. $y < -3x + 4$
22. $-x + y < -5$	23. $x + 2y < 10$	24. $2x + y \leq 6$
25. $-3x + y \geq -1$	26. $3x - 2y > -12$	27. $y \leq 2(2x + 1)$
28. $3y > 3(4x - 3)$	29. $2(x + y) \leq 14$	30. $-3(5x - y) > 0$

Homework Help

For Exercises	See Examples
13–21, 27, 28, 32	1
22–26, 29–31, 33–34	2, 3

Extra Practice
See page 717.

For Exercises 31–32, write an inequality and graph the solution.

31. The sum of two numbers is greater than four.
32. Twice a number is less than or equal to another number.

Applications and Problem Solving

33. **Animals** Amara and Toshi have set a goal to find homes for more than twelve pets through the Humane Society.
    a. Write and graph an inequality to determine how many homes each girl must find to reach the goal.
    b. List three of the solutions.
    c. Describe the limitations on the solution set.

33–34. See Solutions Manual.

34. **Airlines** For each mile that Mr. Burnett flies, he earns one point toward a free flight. His credit card company is associated with the airline, and he earns one-half point for each dollar charged. Suppose he needs at least 20,000 points for a free flight. Show with a graph the miles that Mr. Burnett must fly and the money he must charge to get a free flight.

35. **Critical Thinking** Graph the intersection of the solutions of $4x + 2y \geq 8$ and $y < x$. See Solutions Manual.

Mixed Review

36–39. See margin.

Solve each inequality. Graph the solution. *(Lessons 12–5 & 12–6)*

36. $\lvert t + 4 \rvert \geq 3$	37. $\lvert h - 7 \rvert < 2$
38. $-1 < p + 1 \leq 6$	39. $-5b > 10$ or $b + 4 > 5$

40. **Travel** Ben and Pam left the park at the same time. Ben traveled north at 45 miles per hour. Pam traveled east at 60 miles per hour. After 1 hour, how far apart are Ben and Pam? *(Lesson 8–7)* 75 mi

Standardized Test Practice

41. **Short Response** 848.3 in scientific notation. *(Lesson 8–4)*
41. 8.483×10^2

42. **Multiple Choice** Suppose you toss 2 coins at the same time. What is the probability that both land heads up? *(Lesson 5–6)* B

 Ⓐ 0 Ⓑ 0.25 Ⓒ 0.50 Ⓓ 0.75 Ⓔ 1

4 ASSESS

Open-Ended Assessment

Speaking Write several inequalities on the board or overhead. Have students take turns describing how they would graph the boundary line (using slope and y-intercept or x- and y-intercept) and whether the boundary line is solid or dashed. Then have them describe how to determine which half-plane to shade.

Chapter 12, Quiz B (Lessons 12–4 through 12–7) is available in the *Chapter 12 Resource Masters*, p. 569.

Assignment Guide

Basic: 13–33 odd, 35–42
Average: 14–32 even, 33–42

Answers

36. $\{t \mid t \leq -7 \text{ or } t \geq -1\}$

−9 −8 −7 −6 −5 −4 −3 −2 −1 0 1

37. $\{h \mid 5 < h < 9\}$

2 3 4 5 6 7 8 9 10 11 12

38. $\{p \mid -2 < p \leq 5\}$

−3 −2 −1 0 1 2 3 4 5 6 7

39. $\{b \mid b < -2 \text{ or } b > 1\}$

−5 −4 −3 −2 −1 0 1 2 3 4 5

***Enrichment*, p. 557**

12-7 NAME ______ DATE ______ PERIOD ______

Enrichment Student Edition Pages 535–539

Inequalities with Triangles

Recall that a line segment can be named by the letters of its endpoints. Line segment AB (written as "$\overline{AB}$") has points A and B for endpoints. The *length* of AB is written without the bar as AB.

$AB < BC$ $\angle A < \angle B$

The statement on the left above shows that $\overline{AB}$ is shorter than $\overline{BC}$. The statement on the right above shows that the measure of angle A is less than that of angle B.

These three inequalities are true for any triangle ABC, no matter how long the sides are.

a. $AB + BC > AC$
b. If $AB > AC$, then $\angle C > \angle B$.
c. If $\angle C > \angle B$, then $AB > AC$.

Use the three triangle inequalities for these problems.

1. List the sides of triangle DEF in order of increasing length. $\overline{DF}, \overline{DE}, \overline{EF}$
2. In the figure below, which line segment is the shortest? $\overline{LM}$
3. Explain why the lengths 5 cm, 10 cm, and 20 cm could not be used to make a triangle. 5 + 10 is not greater than 20.
4. Two sides of a triangle measure 3 in. and 7 in. Between which two values must the third side be? 4 in. and 10 in.
5. In triangle XYZ, $XY = 15$, $YZ = 12$, and $XZ = 9$. Which is the greatest angle? Which is the least? $\angle Z; \angle Y$
6. List the angles $\angle A$, $\angle C$, $\angle ABC$, and $\angle ABD$, in order of increasing size. $\angle ABD, \angle A, \angle ABC, \angle C$

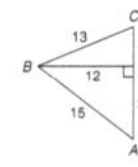

Extra Credit

Guillermo has a package that will cost \$3.20 to mail. He finds some old sheets of 21¢ and 9¢ stamps in a file cabinet. Write an inequality for the numbers of 21¢ stamps x and 9¢ stamps y he can use to mail the package. Then give an example of a solution that uses at least fifteen 9¢ stamps that does not exceed the required postage by more than 1¢.
$21x + 9y \geq 320$; (2, 31), (5, 24), or (8, 17)

Investigation

PREPARE

This optional investigation is designed to be completed by pairs of students over 1–2 days.

Objective

Students graph quadratic equations. By evaluating the quadratic equations for points on, inside, and outside of the resulting parabolas, they learn to identify solutions of quadratic inequalities. Students apply a quadratic inequality to a real-world situation and use the graph of its related function to identify solutions of the inequality. Students summarize what they have learned in a poster and make a brochure of how they can apply this knowledge to develop and solve an original real-world application.

Mathematical Overview

The investigation utilizes the following concepts:

- graphing quadratic equations,
- evaluating quadratic equations,
- identifying solution sets of inequalities, and
- writing quadratic inequalities to represent real-world situations.

Suggested Time Management	
Investigation	25–40 min
Extension: Gathering Data	20–30 min
Extension: Summarizing Data	35–50 min

Motivating the Lesson

Ask students to think of real-world situations they have represented using *linear* inequalities. Explain that many real-world situations such as those involving area are represented by *quadratic* inequalities. Have students recall what they know about quadratic functions and their graphs. Ask them how they think a quadratic inequality will be represented.

Chapter 12 Investigation

Parabolas and Pavilions

Materials

 grid paper

 ruler

yellow and blue colored pencils

1a–d. See Solutions Manual.

Quadratic Inequalities

In this investigation, you will learn how to solve and graph **quadratic inequalities**.

Investigate

1. Graph the quadratic equation $y = x^2 + 2x - 3$ on a piece of graph paper.
   a. With a yellow colored pencil, shade the region inside the parabola.
   b. Make a table like the one below. Fill in column 1 with five points that appear in the yellow region, such as (0, 0). Compare the value of the y-coordinate with the value of $y = x^2 + 2x - 3$ when it is evaluated at the x-coordinate. When the quadratic equation is evaluated at 0, the result is -3. The y-coordinate, 0, is greater than -3. Place the correct inequality symbol in column 3 for each of the other four points that fall in the yellow region.

Point	y-coordinate	$<$ or $>$	$y = x^2 + 2x - 3$
(0, 0)	0	$>$	$y = (0)^2 + 2(0) - 3$ or -3

   c. Write a quadratic inequality that compares the points of the yellow region with the quadratic equation $y = x^2 + 2x - 3$.
   d. With a blue colored pencil, shade the region outside the parabola. Repeat parts b–c for the blue region.
2. Graph the quadratic equation $y = -x^2 - 2x + 3$ on a separate piece of graph paper. **a. See Solutions Manual.**
   a. Shade the region inside the parabola yellow. Then shade the region outside the parabola blue. Make tables similar to those in Step 1. Write inequalities describing the yellow and blue regions.
   b. Suppose you wanted to include the values on the boundary line of a quadratic inequality. Explain how you would write the inequality to include the boundary line. **Use the $\leq$ or $\geq$ symbols.**
   c. Now suppose you did *not* want to include the boundary line of a quadratic inequality. Explain how you would draw the graph to show that the boundary line was *not* included in the solution. **Draw the graph of the parabola with a dashed line.**

Cooperative Learning

This investigation offers an excellent opportunity for using cooperative groups. For more information on cooperative learning strategies and group management, see *Cooperative Learning in the Mathematics Classroom,* one of the titles in the Glencoe Mathematics Professional Series.

3. Spring Town is building a community center, the Spring Town Pavilion. The building is to be 20 feet longer than it is wide and to have an area greater than or equal to 1500 square feet. **a. See margin.**

a. Let w represent the width of the building. Then $w + 20$ represents the length. The area of the building, $w(w + 20)$, must be greater than or equal to 1500 square feet. That is, $w^2 + 20w - 1500 \geq 0$. Graph $f(w) = w^2 + 20w - 1500$.

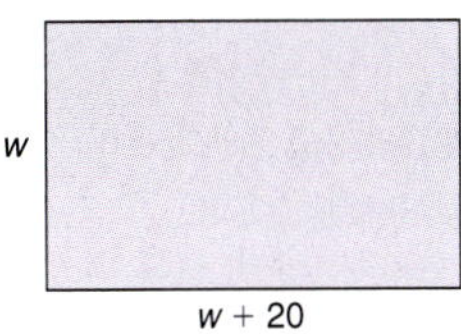

b. Find the values for w that satisfy the inequality in part a. To do this, choose values for w both inside and outside of your parabola. $\{w \mid w \leq -50 \text{ or } w \geq 30\}$

c. What restrictions are placed on your solution for w since it represents width? **It must be positive.**

d. Use your graph to find two possible dimensions for the Spring Town Pavilion. **Sample answer: 30 ft by 50 ft; 35 ft by 55 ft**

Extending the Investigation

In this extension, you will continue to investigate quadratic inequalities.

Graph each quadratic inequality. **1–4. See Solutions Manual.**

1. $y < x^2 + 4x - 8$
2. $y \geq -x^2 + 2x + 15$
3. $y \leq -3x^2 + 3$

4. The Spring Town Pavilion will have a deck for small outdoor gatherings. The length of the deck is to be 6 feet more than the width. The area of the deck is at least one-fifth of the area of the community center building. Write and graph a quadratic inequality that describes this situation. Then give three possible dimensions for the new deck.

Presenting Your Investigation

Here are some ideas to help you present your conclusions to the class.

- Make a poster that describes how to solve quadratic inequalities. Include at least two different inequalities. Show how the graphs are shaded to represent the solutions.
- Write and solve a problem similar to the one in Step 3. Make a brochure that describes the problem, the graph of its solution, and a list of possible solutions.

Investigation For more information on graphing quadratic inequalities, visit: www.algconcepts.com

Answer

3a. $w^2 + 20w \geq 1500$

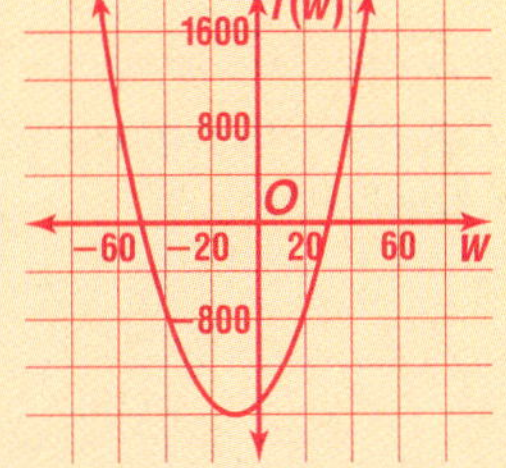

MANAGE

Teaching Tip Review graphing quadratic functions with students. The graphs of the functions in the body of the investigation can be sketched by finding the vertex and factoring to find the x-intercepts. Point out the difference between the quadratic inequalities in two variables in Exercises 1 and 2, where the solution is a region of the coordinate plane, and the quadratic inequality in one variable in Exercise 3 solved by graphing the related function, in which the solution is a portion of a parabola.

Working in Pairs By working in pairs, one student can sketch a graph while the other tests points and records findings. The students can work together on exercise questions. Students can then switch roles for Exercise 2. Partners may want to divide up work on the brochure and poster to complete individually and then check and revise their work together.

Working as a Class You may want to complete the real-world application in Exercise 3 as a class, since the concepts are not as straightforward as in the first two exercises. Make sure students understand the solutions of Exercise 3 before they go on to the Extending the Investigation.

ASSESS

Students' work should show that they understand how the solution set of a quadratic inequality in two variables relates to the graph of the corresponding quadratic equation. They should understand how to solve an application of a quadratic inequality in one variable using the graph of the related quadratic function.

Students should add their posters and brochures to their portfolios at this time.

Study Guide and Assessment

Understanding and Using the Vocabulary

This section provides a listing of the new terms, properties, and phrases that were introduced in this chapter. The exercises check students' understanding of the terms by using a variety of verbal formats including matching, completion, and true/false.

Glossary A complete glossary of terms appears on pages 762–783.

MindJogger Videoquizzes

MindJogger Videoquizzes provide an alternative review of concepts presented in this chapter. Students work in teams to answer questions, gaining points for correct answers.

Answers

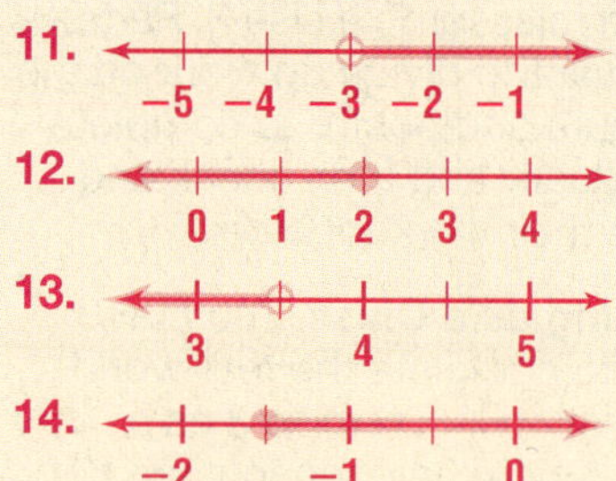

Understanding and Using the Vocabulary

After completing this chapter, you should be able to define each term, property, or phrase and give an example or two of each.

interNET CONNECTION **Review Activities** For more review activities, visit: www.algconcepts.com

boundary *(p. 535)*
compound inequality *(p. 524)*
half-plane *(p. 535)*
intersection *(p. 524)*
quadratic inequalities *(p. 540)*
set-builder notation *(p. 510)*
union *(p. 525)*

Complete each sentence using a term from the vocabulary list.

1. The solution of a compound inequality using *or* can be found by the __?__ of the graphs of the two inequalities. **union**
2. Graph the related equation of an inequality to find the __?__ of the half-plane. **boundary**
3. Use a test point from each __?__ to find the solution of an inequality in two variables. **half-plane**
4. An inequality of the form $x < y < z$ is called a(n) __?__. **compound inequality**
5. The __?__ for the graph of an inequality in two variables will either be a dashed or solid line. **boundary**
6. The __?__ of two graphs is the area where they overlap. **intersection**
7. A(n) __?__ is an area on the coordinate plane representing the solution for an inequality in two variables. **half-plane**
8. $2x + 3 < 5$ or $x \geq -1$ is an example of a(n) __?__. **compound inequality**
9. A solution written in the form $\{x \mid x \leq 3\}$ is written in __?__. **set-builder notation**
10. The solution of a compound inequality using *and* can be found by the __?__ of the graphs of the two inequalities. **intersection**

Skills and Concepts

Objectives and Examples

- **Lesson 12–1** Graph inequalities on a number line.

Graph $x < 5$ on a number line.

The graph begins at 5, but 5 is not included. The arrow is to the left. The graph describes values that are less than 5.

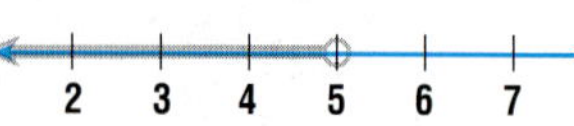

Review Exercises

Graph each inequality on a number line.

11. $x > -3$
12. $z \leq 2$
13. $3.5 > x$
14. $a \geq -1\frac{1}{2}$

11–14. See margin.

Write an inequality for each graph.

15. −7 −6 −5 −4 −3 **$x \geq -5$**

16. 5 6 7 8 9 **$x < 7$**

www.algconcepts.com/vocabulary_review

Resource Manager

Reproducible Masters
Chapter 12 Resource Masters
- *Assessment*, pp. 559–567, 570–572

Technology/Multimedia
MindJogger Videoquizzes
ExamView® Pro

Objectives and Examples

Review Exercises

• **Lesson 12–2** Solve inequalities involving addition and subtraction.

Solve $x - 6 > 2$.

$x - 6 > 2$	*Original inequality*
$x - 6 + 6 > 2 + 6$	*Add 6 to each side.*
$x > 8$	*Simplify.*

The solution is $\{x \mid x > 8\}$.

Solve each inequality. Check your solution.

17. $x + 3 > 7$ **$\{x \mid x > 4\}$**

18. $a - 4 \geq 2$ **$\{a \mid a \geq 6\}$**

19. $\frac{2}{3} \leq y - \frac{1}{2}$ **$\left\{y \mid y \geq 1\frac{1}{6}\right\}$**

20. $12x - 11x + 5 < 3$ **$\{x \mid x < -2\}$**

21. $3(x + 1) \geq 4x$ **$\{x \mid x \leq 3\}$**

• **Lesson 12–3** Solve inequalities involving multiplication and division.

Solve $-4x < -8$.

$-4x < -8$	*Original inequality*
$\frac{-4x}{-4} > \frac{-8}{-4}$	*Divide each side by −4 and reverse the symbol.*
$x > 2$	*Simplify.*

The solution is $\{x \mid x > 2\}$.

Solve each inequality. Check your solution.

22. $3y \geq 12$ **$\{y \mid y \geq 4\}$**

23. $-6n > 30$ **$\{n \mid n < -5\}$**

24. $0.2w \leq -1.8$ **$\{w \mid w \leq -9\}$** **26. $\{t \mid t \leq 28\}$**

25. $\frac{x}{5} > 3$ **$\{x \mid x > 15\}$**

26. $-\frac{t}{2} \geq -14$

27. $\frac{3}{4}w \leq 6$ **$\{w \mid w \leq 8\}$**

28. $\frac{h}{2.3} < -7$ **$\{h \mid h < -16.1\}$**

• **Lesson 12–4** Solve inequalities involving more than one operation.

Solve $2(x + 1) < 4 - 3x$.

$2(x + 1) < 4 - 3x$	*Original inequality*
$2x + 2 < 4 - 3x$	*Distributive Property*
$5x + 2 < 4$	*Add 3x to each side.*
$5x < 2$	*Subtract 2 from each side.*
$x < \frac{2}{5}$	*Divide each side by 5.*

The solution is $\left\{x \mid x < \frac{2}{5}\right\}$.

Solve each inequality. Check your solution.

29. $2x + 6 \geq 14$ **$\{x \mid x \geq 4\}$**

30. $9 < -0.2y - 1$ **$\{y \mid y < -50\}$**

31. $\frac{2}{3}t - 4 \leq 2$ **$\{t \mid t \leq 9\}$**

32. $3(4 + n) < 21$ **$\{n \mid n < 3\}$**

Write and solve an inequality.

33. Four times a number decreased by 3 is greater than 25. **$4x - 3 > 25$; $\{x \mid x > 7\}$**

34. Seven minus two times a number is no less than nine. **$7 - 2x \geq 9$; $\{x \mid x \leq -1\}$**

• **Lesson 12–5** Solve compound inequalities and graph the solution.

Solve $3 < x + 2 \leq 8$.

$x + 2 > 3$	and	$x + 2 \leq 8$
$x + 2 - 2 > 3 - 2$		$x + 2 - 2 \leq 8 - 2$
$x > 1$		$x \leq 6$

The solution is $\{x \mid 1 < x \leq 6\}$.

Solve each compound inequality. Graph the solution. **35–38. See margin for graphs.**

35. $-4 \leq y + 3 < 2$ **$\{y \mid -7 \leq y \leq -1\}$**

36. $2 \leq 3 - t$ or $3 - t > 5$ **$\{t \mid t \leq 1\}$**

37. $3a \geq 6$ or $-5 - a \geq -6$ **$\{a \mid a \leq 1$ or $a \geq 2\}$**

38. $9 < 2x + 1 < 13$ **$\{x \mid 4 < x < 6\}$**

Skills and Concepts

The **Objectives and Examples** section reviews the skills and concepts of the chapter and shows completely worked examples.

The **Review Exercises** provide practice for the corresponding objectives.

Answers

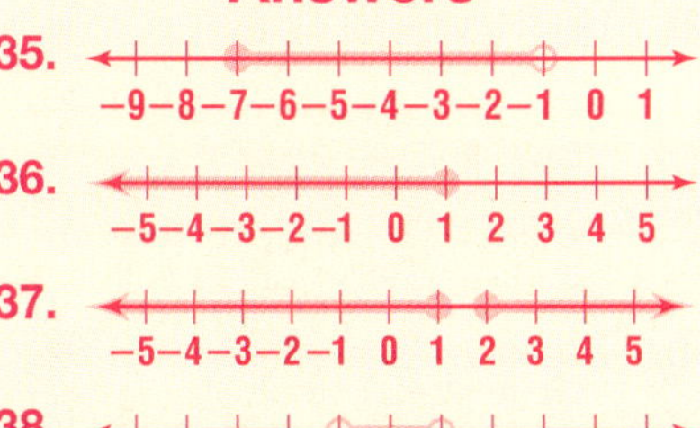

ExamView® Pro

Use the networkable ***ExamView® Pro Testmaker CD-ROM*** to:

- Create **multiple versions** of tests.
- Create **modified** tests for **inclusion** students with one mouse click.
- **Edit** existing questions and **add** your own questions.
- Build tests aligned with state standards using built-in **state curriculum correlations**.
- Change **English** tests to **Spanish** with one mouse click and vice versa.

Mixed Problem Solving
See pages 724–731.

Applications and Problem Solving

This section provides additional practice in solving real-world problems that involve the concepts of this chapter.

Answers

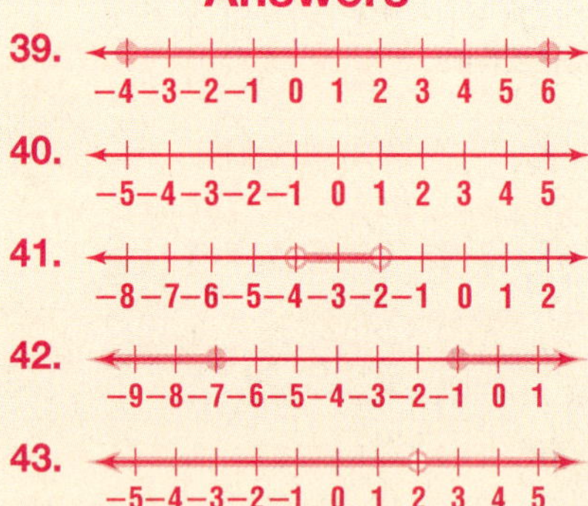

Objectives and Examples

- **Lesson 12–6** Solve inequalities involving absolute value and graph the solution.

Solve $|x + 1| < 1$.

Case 1 $x + 1$ is positive.
$x + 1 < 1$
$x < 0$ *Subtract 1 from each side.*

Case 2 $x + 1$ is negative.
$-(x + 1) < 1$
$-(x + 1)(-1) > 1(-1)$ *Multiply by −1 and reverse the symbol.*
$x + 1 > -1$
$x > -2$ *Subtract 1 from each side.*

The solution is $\{x \mid -2 < x < 0\}$.

Review Exercises

Solve each inequality. Graph the solution.

39. $|t - 1| \le 5$ $\{t \mid -4 \le t \le 6\}$
40. $|a + 7| < -2$ $\varnothing$
41. $|x + 3| < 1$ $\{x \mid -4 < x < -2\}$
42. $|s + 4| \ge 3$ $\{s \mid s \le -7 \text{ or } s \ge -1\}$
43. $|y - 2| > 0$ $\{y \mid y \ne 2\}$
39–43. See margin for graphs.

Write an inequality involving absolute value for each statement. Do not solve.

44. Bianca's guess g was within \$6 of the actual value of \$25. $|g - 25| < 6$
45. The difference between Greg's score s on his final exam and his mean grade of 80 is more than 5 points. $|s - 80| > 5$

- **Lesson 12–7** Graph inequalities in the coordinate plane.

Graph $2x + y < 5$.

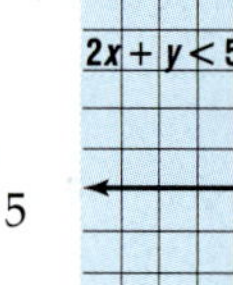

First, solve for y.
$2x + y < 5$
$2x + y - 2x < 5 - 2x$
$y < -2x + 5$

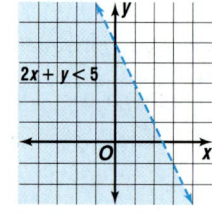

Graph the related function $y = -2x + 5$ as a dashed line and shade.

Graph each inequality.

46. $y \ge x - 1$
47. $y < -2x + 4$
48. $\frac{y}{2} > x - 2$
49. $x - y \ge 3$
46–49. See Solutions Manual.

Applications and Problem Solving

50. Shipping An empty book crate weighs 30 pounds. The weight of a book is 1.5 pounds. For shipping, the crate can weigh no more than 60 pounds. What is the acceptable number of books that can be packed in a crate? *(Lesson 12–4)* **less than or equal to 20 books**

51. Geometry An obtuse angle measures more than 90° but less than 180°. If the measure of an obtuse angle is $3x$, what are the possible values for x? *(Lesson 12–5)* $\{x \mid 30 < x < 60\}$

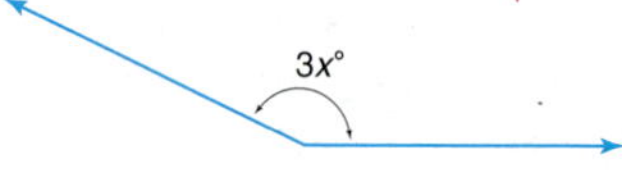

Assessment, pp. 561–562

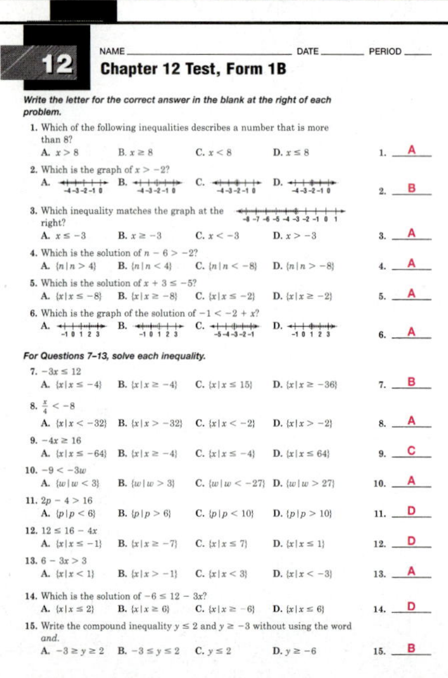

NAME ______ DATE ______ PERIOD ______

12 Chapter 12 Test, Form 1B

Write the letter for the correct answer in the blank at the right of each problem.

1. Which of the following inequalities describes a number that is more than 8?
A. $x > 8$ B. $x \ge 8$ C. $x < 8$ D. $x \le 8$ 1. A
2. Which is the graph of $x > -2$?
A. [graph] B. [graph] C. [graph] D. [graph] 2. B
3. Which inequality matches the graph at the right?
A. $x \le -3$ B. $x \ge -3$ C. $x < -3$ D. $x > -3$ 3. A
4. Which is the solution of $n - 6 > -2$?
A. $\{n \mid n > 4\}$ B. $\{n \mid n < 4\}$ C. $\{n \mid n < -8\}$ D. $\{n \mid n > -8\}$ 4. A
5. Which is the solution of $x + 3 \le -5$?
A. $\{x \mid x \le -8\}$ B. $\{x \mid x \ge -8\}$ C. $\{x \mid x \le -2\}$ D. $\{x \mid x \ge -2\}$ 5. A
6. Which is the graph of the solution of $-1 < -2 + x$?
A. [graph] B. [graph] C. [graph] D. [graph] 6. A

For Questions 7–13, solve each inequality.

7. $-3x \le 12$
A. $\{x \mid x \le -4\}$ B. $\{x \mid x \ge -4\}$ C. $\{x \mid x \le 15\}$ D. $\{x \mid x \ge -36\}$ 7. B
8. $\frac{x}{4} < -8$
A. $\{x \mid x < -32\}$ B. $\{x \mid x > -32\}$ C. $\{x \mid x < -2\}$ D. $\{x \mid x > -2\}$ 8. A
9. $-4x \ge 16$
A. $\{x \mid x \le -64\}$ B. $\{x \mid x \ge -4\}$ C. $\{x \mid x \le -4\}$ D. $\{x \mid x \le 64\}$ 9. C
10. $-9 < -3w$
A. $\{w \mid w < 3\}$ B. $\{w \mid w > 3\}$ C. $\{w \mid w < -27\}$ D. $\{w \mid w > 27\}$ 10. A
11. $2p - 4 > 16$
A. $\{p \mid p < 6\}$ B. $\{p \mid p > 6\}$ C. $\{p \mid p < 10\}$ D. $\{p \mid p > 10\}$ 11. D
12. $12 \le 16 - 4x$
A. $\{x \mid x \le -1\}$ B. $\{x \mid x \ge -7\}$ C. $\{x \mid x \le 7\}$ D. $\{x \mid x \le 1\}$ 12. D
13. $6 - 3x > 3$
A. $\{x \mid x < 1\}$ B. $\{x \mid x > -1\}$ C. $\{x \mid x < 3\}$ D. $\{x \mid x < -3\}$ 13. A
14. Which is the solution of $-6 \le 12 - 3x$?
A. $\{x \mid x \le 2\}$ B. $\{x \mid x \ge 6\}$ C. $\{x \mid x \ge -6\}$ D. $\{x \mid x \le 6\}$ 14. D
15. Write the compound inequality $y \le 2$ and $y \ge -3$ without using the word *and*.
A. $-3 \ge y \ge 2$ B. $-3 \le y \le 2$ C. $y \le 2$ D. $y \ge -6$ 15. B

Assessment

Four forms of Chapter 12 Test are available in the *Chapter 12 Resource Masters.*

Chapter 12 Test, Form 1B, is shown at the left. Chapter 12 Test, Form 2B, is shown on the next page.

	Form of Test		Level
1A	Multiple Choice	pp. 559–560	Average
1B	Multiple Choice	pp. 561–562	Basic
2A	Free Response	pp. 563–564	Average
2B	Free Response	pp. 565–566	Basic

CHAPTER 12 Test

1. **List** at least three verbal phrases that are used to describe inequalities.
2. If you multiply or divide each side of an inequality by a negative number, what must happen to the symbol for the inequality to remain true?
3. Before solving an inequality involving absolute value, which two cases must you consider? **1–3. See margin.**

Write an inequality for each graph.

4. 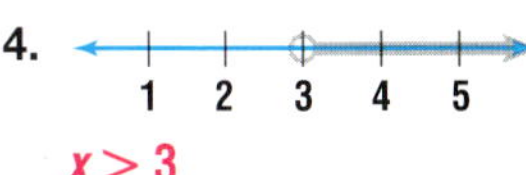

$x > 3$

5. −4 −3 −2 −1 0

$x \geq -2$

Solve each inequality. Check your solution.

6. $2 + x \geq 12$ $\{x \mid x \geq 10\}$
7. $5t + 6 \leq 4t - 3$ $\{t \mid t \leq -9\}$
8. $8 < -4t$ $\{t \mid t < -2\}$
9. $-0.2x > -6$ $\{x \mid x < 30\}$
10. $\frac{t}{4} > 1$ $\{t \mid t > 4\}$
11. $-\frac{2}{5}m \leq 10$ $\{m \mid m \geq -25\}$
12. $-3r - 1 \geq -16$ $\{r \mid r \leq 5\}$
13. $7x - 12 < 30$ $\{x \mid x < 6\}$
14. $2(h - 3) > 6$ $\{h \mid h > 6\}$
15. $8(1 - 2z) \leq 25 + z$ $\{z \mid z \geq -1)$

Solve each inequality. Graph the solution. **16–21. See margin for graphs.**

16. $x + 1 > -2$ and $x + 1 < 6$
17. $4 < 3j - 2 \leq 7$ $\{j \mid 2 < j \leq 3\}$
18. $2n + 5 \geq 15$ or $2n + 5 \leq 3$
19. $-6c > -24$ or $c + 0.25 < 1.3$ $\{c \mid c < 4\}$
20. $|x + 3| \geq 4$ $\{x \mid x \geq 1 \text{ or } x \leq -7\}$
21. $|4b| \leq 16$ $\{b \mid -4 \leq b \leq 4\}$

16. $\{x \mid -3 < x < 5\}$ **18.** $\{n \mid n \geq 5 \text{ or } n \leq -1\}$

Graph each inequality. **22–23. See margin.**

22. $y \geq 5x - 6$
23. $4x - 2y > -6$

24. **Car Rental** Justine is renting a car that costs $32 a day with free unlimited mileage. Since she is under the age of 25, it costs her an additional $10 per day. Justine does not want to pay any more than $200 on car rental costs. For what number of days can she rent a car? **4 or fewer days**

25. **Manufacturing** Ball bearings are used to connect moving parts and minimize friction. Ball bearings for an automobile will work properly only if their diameter is within 0.01 inch, inclusive, of 5 inches. Write and solve an inequality to represent the range of acceptable diameters for these ball bearings. $|m - 5| \leq 0.01; \{m \mid 4.99 \leq m \leq 5.01\}$

www.algconcepts.com/chapter_test

Chapter Test Bonus Question

Write an inequality in slope-intercept form that describes the shaded region. $y > -\frac{1}{2}x + 2$

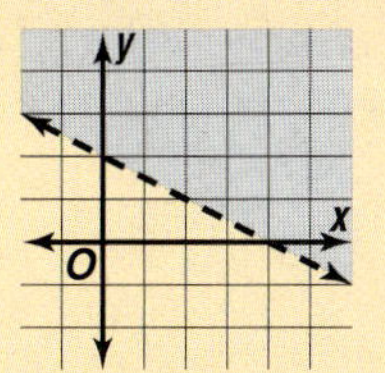

Answer

23. 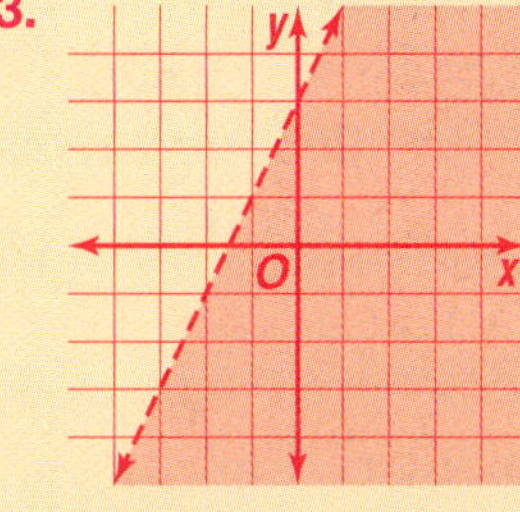

Chapter Test

Answers

1. Sample answer: at least, no more than, at most
2. Reverse the symbol.
3. whether or not the expression in the absolute value is positive or negative

16. 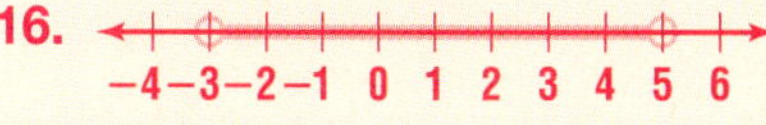

17. −5 −4 −3 −2 −1 0 1 2 3 4 5

18. −3 −2 −1 0 1 2 3 4 5 6 7

19. −2 −1 0 1 2 3 4 5 6 7 8

20. −8 −7 −6 −5 −4 −3 −2 −1 0 1 2

21. −5 −4 −3 −2 −1 0 1 2 3 4 5

22. 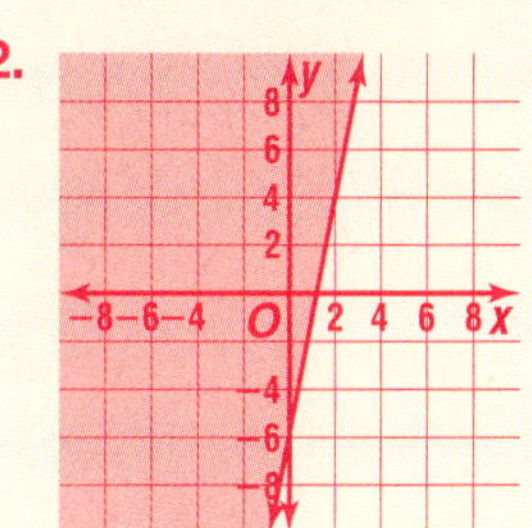

Assessment, pp. 563–564

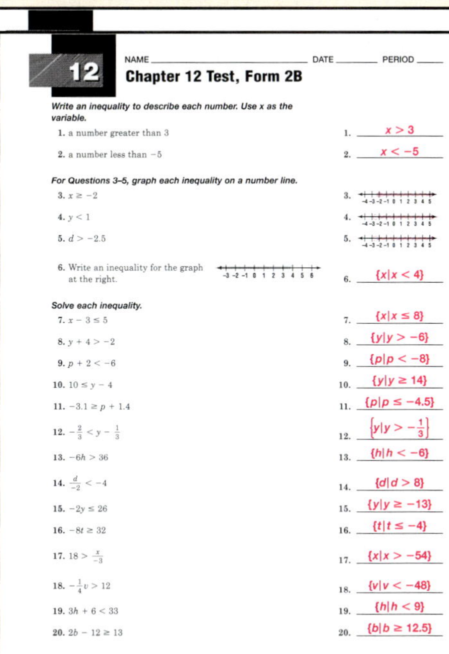

NAME ______ DATE ______ PERIOD ______

12 Chapter 12 Test, Form 2B

Write an inequality to describe each number. Use x as the variable.

1. a number greater than 3 — 1. $x > 3$
2. a number less than −5 — 2. $x < -5$

For Questions 3–5, graph each inequality on a number line.

3. $x \geq -2$ — 3. −4 −3 −2 −1 0 1 2 3 4 5
4. $y < 1$ — 4. −4 −3 −2 −1 0 1 2 3 4 5
5. $d > -2.5$ — 5. −4 −3 −2 −1 0 1 2 3 4 5
6. Write an inequality for the graph at the right. −3 −2 −1 0 1 2 3 4 5 6 — 6. $\{x|x < 4\}$

Solve each inequality.

7. $x - 3 \leq 5$ — 7. $\{x|x \leq 8\}$
8. $y + 4 > -2$ — 8. $\{y|y > -6\}$
9. $p + 2 < -6$ — 9. $\{p|p < -8\}$
10. $10 \leq y - 4$ — 10. $\{y|y \geq 14\}$
11. $-3.1 \geq p + 1.4$ — 11. $\{p|p \leq -4.5\}$
12. $-\frac{2}{3} < y - \frac{1}{3}$ — 12. $\{y|y > -\frac{1}{3}\}$
13. $-6h > 36$ — 13. $\{h|h < -6\}$
14. $\frac{d}{-2} < -4$ — 14. $\{d|d > 8\}$
15. $-2y \leq 26$ — 15. $\{y|y \geq -13\}$
16. $-8t \geq 32$ — 16. $\{t|t \leq -4\}$
17. $18 > \frac{x}{-3}$ — 17. $\{x|x > -54\}$
18. $-\frac{1}{4}v > 12$ — 18. $\{v|v < -48\}$
19. $3h + 6 < 33$ — 19. $\{h|h < 9\}$
20. $2b - 12 \geq 13$ — 20. $\{b|b \geq 12.5\}$

Preparing for Standardized Tests

Pages 546–547 are part of a complete test preparation course that is described in detail on page T9 of the Teacher's Handbook. The test items on these pages were written in the same style as those in state proficiency tests and standardized tests like ACT and SAT.

Diagnosis and Prescription

Each of the 10 test questions on page 547 is cross-referenced to the lesson where that SAT or ACT skill is covered. If students miss a particular type of problem, you can have them study that skill.

(See chart at the bottom of page 547. Note that SPT = State Proficiency Test, SAT = Scholastic Assessment Test, and ACT = American College Test.)

***Assessment,* p. 570**

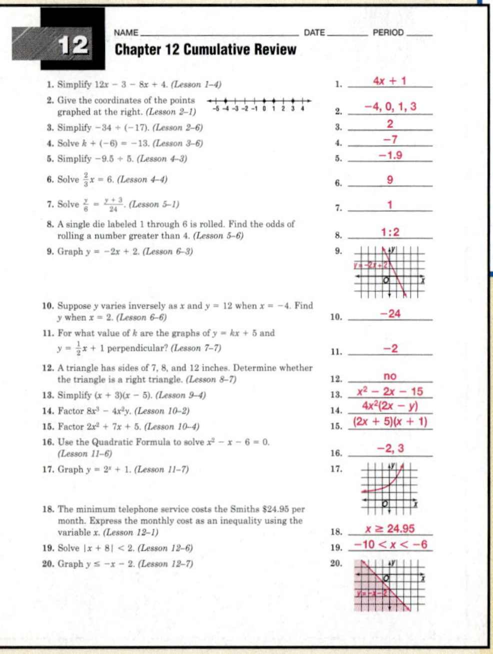

12 NAME ____________ DATE ______ PERIOD ______

Chapter 12 Cumulative Review

1. Simplify $12x - 3 - 8x + 4$. *(Lesson 1–4)* — 1. $4x + 1$
2. Give the coordinates of the points graphed at the right. *(Lesson 2–1)* [number line −5 to 4] — 2. $-4, 0, 1, 3$
3. Simplify $-34 \div (-17)$. *(Lesson 2–6)* — 3. 2
4. Solve $k + (-6) = -13$. *(Lesson 3–6)* — 4. -7
5. Simplify $-9.5 \div 5$. *(Lesson 4–3)* — 5. -1.9
6. Solve $\frac{2}{3}x = 6$. *(Lesson 4–4)* — 6. 9
7. Solve $\frac{y}{6} = \frac{y+3}{24}$. *(Lesson 5–1)* — 7. 1
8. A single die labeled 1 through 6 is rolled. Find the odds of rolling a number greater than 4. *(Lesson 5–6)* — 8. $1:2$
9. Graph $y = -2x + 2$. *(Lesson 6–3)* — 9. [graph: $y = -2x + 2$]
10. Suppose y varies inversely as x and $y = 12$ when $x = -4$. Find y when $x = 2$. *(Lesson 6–6)* — 10. -24
11. For what value of k are the graphs of $y = kx + 5$ and $y = \frac{1}{2}x + 1$ perpendicular? *(Lesson 7–7)* — 11. -2
12. A triangle has sides of 7, 8, and 12 inches. Determine whether the triangle is a right triangle. *(Lesson 8–7)* — 12. no
13. Simplify $(x + 3)(x - 5)$. *(Lesson 9–4)* — 13. $x^2 - 2x - 15$
14. Factor $8x^3 - 4x^2y$. *(Lesson 10–2)* — 14. $4x^2(2x - y)$
15. Factor $2x^2 + 7x + 5$. *(Lesson 10–4)* — 15. $(2x + 5)(x + 1)$
16. Use the Quadratic Formula to solve $x^2 - x - 6 = 0$. *(Lesson 11–6)* — 16. $-2, 3$
17. Graph $y = 2^x + 1$. *(Lesson 11–7)* — 17. [graph]
18. The minimum telephone service costs the Smiths $24.95 per month. Express the monthly cost as an inequality using the variable x. *(Lesson 12–1)* — 18. $x \geq 24.95$
19. Solve $|x + 8| < 2$. *(Lesson 12–6)* — 19. $-10 < x < -6$
20. Graph $y \leq -x - 2$. *(Lesson 12–7)* — 20. [graph: $y = -x - 2$]

CHAPTER 12 Preparing for Standardized Tests

Angle, Triangle, and Quadrilateral Problems

Test-Taking Tip

If there is no figure or diagram, draw one yourself.

You are already aware of some geometry concepts that you need to know for standardized tests. Be sure that you also know the meanings and definitions of the following terms.

right	obtuse	acute	triangle	quadrilateral
equilateral	isosceles	similar (≈)	congruent (≅)	collinear
reflection	translation	rotation	dilation	

Example 1

The triangles below are similar. Find the length of side $\overline{KL}$.

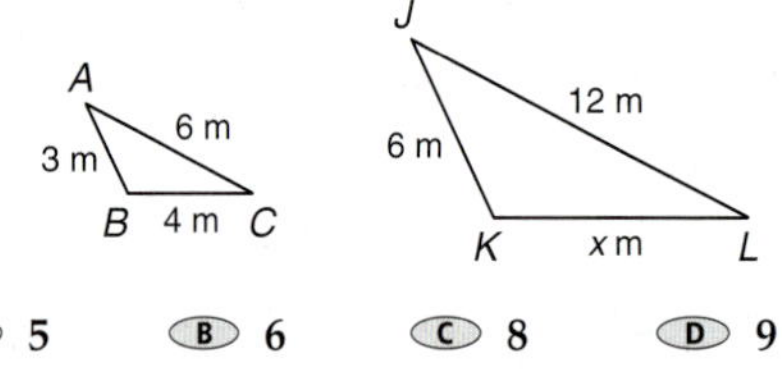

Ⓐ 5 Ⓑ 6 Ⓒ 8 Ⓓ 9

Hint The measures of corresponding sides of similar polygons are proportional.

Solution Find the corresponding sides of the two triangles. Side $\overline{AB}$ of $\triangle ABC$ corresponds to side $\overline{JK}$ of $\triangle JKL$. Side $\overline{BC}$ of $\triangle ABC$ corresponds to side $\overline{KL}$ of $\triangle JKL$. Using these two pairs of corresponding sides, write a proportion.

$\frac{AB}{JK} = \frac{BC}{KL}$ *Sides are proportional.*

$\frac{3}{6} = \frac{4}{x}$ *Substitute side measures.*

$3x = 4(6)$ *Cross multiply.*

$3x = 24$ *Multiply.*

$x = 8$ *Divide each side by 3.*

Side $\overline{KL}$ measures 8 meters. The answer is C.

Example 2

In the figure below, $\overline{ON}$ is congruent to $\overline{LN}$, $m\angle LON = 30$, and $m\angle LMN = 40$. What is $m\angle NLM$?

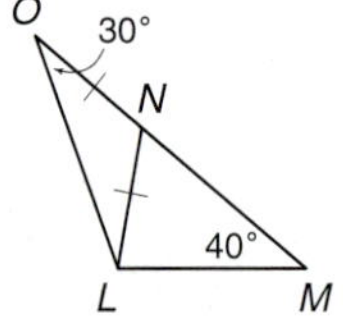

Ⓐ 40 Ⓑ 80 Ⓒ 90 Ⓓ 120

Hint Examine *all* of the triangles in the figure.

Solution Label the unknown angles as 1 and 2. Find $m\angle 2$.

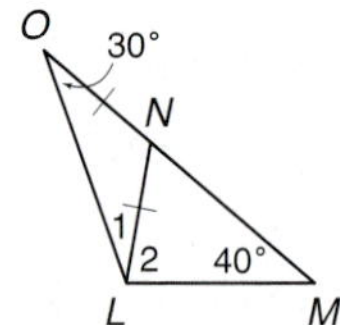

Since $\overline{ON} \cong \overline{LN}$, $\triangle ONL$ is isosceles and the base angles are equal. So, $m\angle 1 = 30$. Since the sum of the angle measures in any triangle is 180, find the sum of the angle measures in $\triangle OML$.

$180 = 30 + 40 + (30 + m\angle 2)$

$180 = 100 + m\angle 2$

$80 = m\angle 2$ *Subtract 100 from each side.*

The answer is B.

Resource Manager

Reproducible Masters

Chapter 12 Resource Masters

- *Assessment*, pp. 570–572

Preparing for Standardized Tests
For test-taking strategies and more practice, see pages 732–749.

After you work each problem, record your answer on the answer sheet provided or on a sheet of paper.

Multiple Choice

1. The rectangles are similar. Find the value of x. *(Lesson 5–1)* **A**

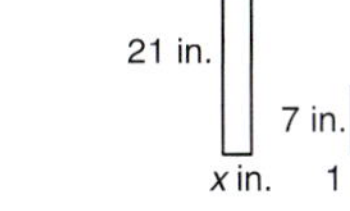

Ⓐ 3 Ⓑ 7
Ⓒ 14 Ⓓ 15

2. $\triangle ABC$ is isosceles. What is $m\angle C$? *(Lesson 4–5)* **B**

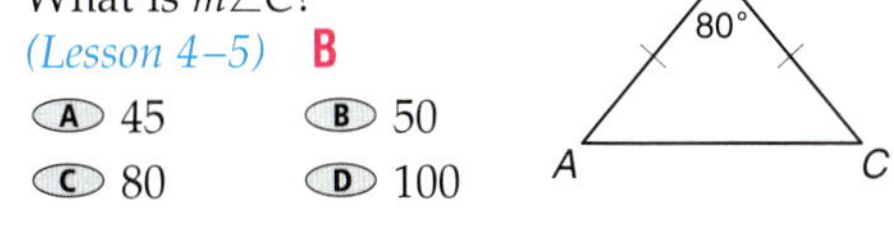

Ⓐ 45 Ⓑ 50
Ⓒ 80 Ⓓ 100

3. What is the sum of a, b, and c? *(Basic Skill)* **C**

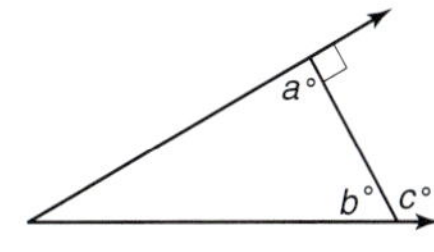

Ⓐ 180 Ⓑ 240
Ⓒ 270 Ⓓ 360
Ⓔ It cannot be determined from the information given.

4. The formula for the area of a trapezoid is $A = \frac{1}{2}h(b_1 + b_2)$, where b_1 and b_2 are the lengths of the bases and h is the height. If the area of trapezoid $QRST$ is 72 square centimeters, then what is the height? *(Lesson 1–5)* **D**

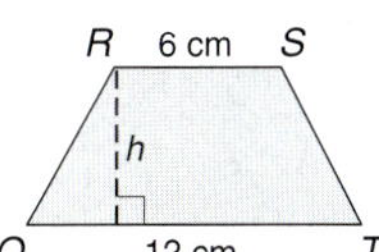

Ⓐ 1 cm Ⓑ 2 cm Ⓒ 4 cm Ⓓ 8 cm

5. Find the x-intercept of $y = -\frac{2}{3}x + 4$. *(Lesson 7–5)* **A**

Ⓐ 6 Ⓑ −6 Ⓒ 4 Ⓓ 0

6. Simplify the expression $\frac{(-5)(4)|-6|}{-3}$. *(Lesson 2–6)* **C**

Ⓐ −120 Ⓑ −40
Ⓒ 40 Ⓓ 120

7. The square of a number is 255 greater than twice the number. What is the number? *(Lesson 11–4)* **B**

Ⓐ 15 or −17 Ⓑ 17 or −15
Ⓒ 31 or −33 Ⓓ 33 or −31

8. What is the value of $b + c$? *(Basic Skill)* **A**

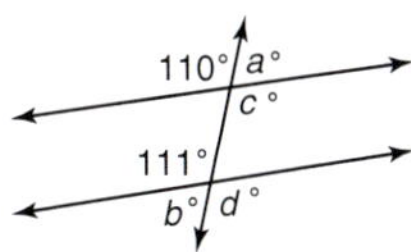

Ⓐ 179 Ⓑ 180
Ⓒ 181 Ⓓ 221

Grid In

9. In the figure, what is the sum of a, b, and c? *(Basic Skill)* **270**

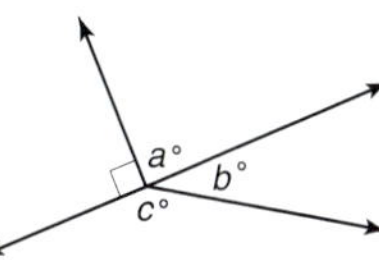

Extended Response

10. Draw a triangle that fits each description. If a drawing is not possible, explain why. *(Basic Skill)*

Part A **See margin.**

i. an obtuse equilateral triangle

ii. an acute equilateral triangle

Part B **See margin.**

i. an obtuse isosceles triangle

ii. an acute isosceles triangle

www.algconcepts.com/standardized_test

A bubble-in answer sheet for these practice problems is available on page A1 of the *Chapter 12 Resource Masters.*

Additional Practice

Additional test practice questions are available in the *Chapter 12 Resource Masters*, pp. 571–572.

Answers

10A. Sample answer:
i) Not possible; the angles of an equilateral triangle all measure 60°.
ii)

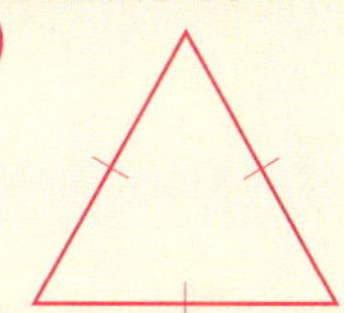

10B. Sample answer:
i)

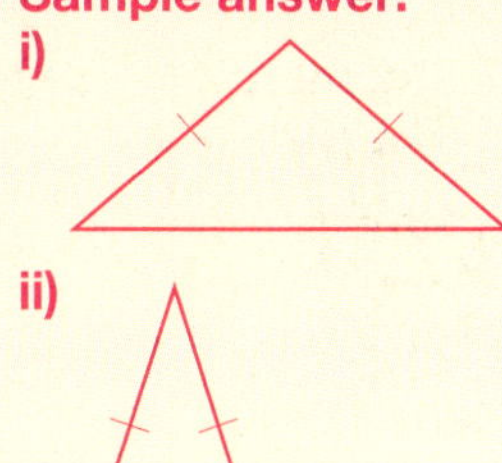

ii)

Chapter 12 Angle, Triangle, and Quadrilateral Problems		
Ex. 1	similar triangles	SPT
Ex. 2	isosceles triangle	ACT
1	similar rectangles	SPT
2	isosceles triangle	SPT
3	straight angles	SAT
4	trapezoid area formula	SPT
5	x-intercept	SPT
6	absolute value	SPT
7	quadratic word problem	SPT
8	angles	SAT
9	angles	SAT
10	triangles	SPT

***Assessment*, pp. 571–572**

12 NAME ________ DATE ________ PERIOD ________

Chapter 12 Standardized Test Practice
(Chapters 1–12)

Write the letter for the correct answer in the blank at the right of each problem.

1. Tell which sentence represents the equation $n + 8 = 12$.
A. Eight more than a number is twelve.
B. Twelve more than a number is eight.
C. Eight is twelve less than a number.
D. The difference of a number and eight is twelve. 1. A

2. Suppose you wanted to buy a music CD and DVD movie. The movie costs $10 more than the CD. The CD costs $15. Which of the following statements is true?
A. The movie costs $25. B. The movie costs $10.
C. Together, they cost $25. D. None are true. 2. A

3. Evaluate $|-5| - |3|$.
A. −8 B. −2 C. 2 D. 8 3. C

4. In a football game, the team first lost 18 yards, then gained 16 yards. Which integer represents the net gain in yards?
A. −34 B. −2 C. 2 D. 34 4. B

5. For what value in the replacement set $\{-1, 0, 1, 2\}$ is the equation $x + 4 = 3x$ true?
A. −1 B. 0 C. 1 D. 2 5. D

6. For what value of x is the equation $-24 = x - (-6)$ true?
A. −30 B. −18 C. 18 D. 30 6. A

7. A sandwich can be made of turkey, tuna, or ham and be put on white or wheat bread. Find the number of possible sandwiches.
A. 2 B. 3 C. 5 D. 6 7. D

8. For what value of x is $-23 = 3x + 4$ true?
A. −27 B. −19 C. −9 D. 9 8. C

9. A coat was on sale for $75. It originally cost $90. Which of the following represents the percent of change?
A. 15% B. 16.7% C. 20% D. 120% 9. B

10. Ten cards in a deck are numbered from 1 to 10. A card is drawn and then replaced. Another card is drawn. Find the probability of drawing a 4 and then a 3.
A. $\frac{1}{100}$ B. $\frac{1}{10}$ C. $\frac{1}{5}$ D. $\frac{1}{2}$ 10. A

11. Which ordered pair is *not* a solution to $2x - 3y = 6$?
A. (4, 0) B. (0, −2) C. (3, 0) D. (−3, −4) 11. A

12. The number of gallons of gas needed to drive on a trip varies directly as the number of miles traveled. If 20 gallons of gas are used to drive 250 miles, how many gallons are used to drive 400 miles?
A. 30 gal B. 32 gal C. 40 gal D. 64 gal 12. B

CHAPTER 13

Resource Manager

Systems of Equations and Inequalities

Instructional Objectives

Lesson (pages)	Objectives	NCTM Standards 2000	State/Local Objectives
13–1 (550–553)	Solve systems of equations by graphing.	1, 3, 6, 8, 9	
13–2 (554–559)	Determine whether a system of equations has one solution, no solution, or infinitely many solutions by graphing.	1, 3, 6, 8, 9, 10	
13–3 (560–565)	Solve systems of equations by the substitution method.	1, 6, 8, 9	
13–4 (566–571)	Solve systems of equations by the elimination method using addition and subtraction.	1, 6, 8, 9	
13–5 (572–577)	Solve systems of equations by the elimination method using multiplication and addition.	1, 6, 8, 9	
Investigation (578–579)	Use matrices to solve systems of equations.	1, 6, 7, 8, 9, 10	
13–6 (580–585)	Solve systems of quadratic and linear equations.	1, 6, 8, 9	
13–7 (586–590)	Solve systems of inequalities by graphing.	1, 3, 6, 8, 9	

Key to NCTM Standards 2000
1 Number & Operations; **2** Algebra; **3** Geometry; **4** Measurement; **5** Data Analysis & Probability; **6** Problem Solving; **7** Reasoning & Proof; **8** Communications; **9** Connections; **10** Representation

Suggested Pacing *See page T13 for a complete course-planning calendar.*

Standard refers to schedules that provide 45- to 55-minute periods that meet each day.
Block refers to schedules that provide approximately 90-minute periods which may meet every day for one semester or every other day over two semesters.

PACING	DAY 1	DAY 2	DAY 3	DAY 4	DAY 5	DAY 6
Standard Core (Chapters 1–13)	Lesson 13–1	Lesson 13–2		Lesson 13–3	Lesson 13–4	
Standard Enhanced (Chapters 1–15)	Lesson 13–1	Lesson 13–2		Lesson 13–3	Lesson 13–4	Lesson 13–5
Block Core (Chapters 1–13)	Chapter 12 Test & Lesson 13–1	Lessons 13–2 & 13–3	Lessons 13–4 & 13–5	INV	Lessons 13–6 & 13–7	SG+A
Block Enhanced (Chapters 1–15)	Chapter 12 Test & Lesson 13–1	Lessons 13–2 & 13–3	Lessons 13–4 & 13–5	INV & Lesson 13–6	Lesson 13–7 & SG+A	Chapter Test & Lesson 14–1

Instructional Resources

Lesson	Materials and Manipulatives (see below for Glencoe Manipulative Resources)	Blackline Masters (page numbers): Chapter 13 Resource Masters: Study Guide	Practice (Skills & Average)	Reading to Learn Mathematics	Enrichment	Assessment	Hands-On Algebra*	School-to-Workplace*	Graphing Calculator Masters*	5-Minute Check Transparencies
13–1	coordinate planes [4] straightedge [1, 2, 4] graphing calculator	573	574–575	576	577		144	13	36	13–1
13–2	coordinate planes [4] straightedge [1, 2, 4] large U.S. map	578	579–580	581	582		145			13–2
13–3	algebra tiles [1, 2, 3, 4] equation/product mat [1, 2, 4] coordinate planes [4] straightedge [1, 2, 4]	583	584–585	586	587	619	146, 147			13–3
13–4		588	589–590	591	592	618				13–4
13–5	compass [1, 2, 3]	593	594–595	596	597					13–5
Investigation										
13–6	coordinate planes [4] straightedge [1, 2, 4] large U.S. map	598	599–600	601	602		148		38, 39	13–6
13–7	coordinate planes [4] straightedge [1, 2, 4] graphing calculator large U.S. map colored transparency sheets	603	604–605	606	607	619			37	13–7
Study Guide & Assessment/ Chapter Test	coordinate planes [4] straightedge [1, 2, 4]					609–617, 620–622				

**See page 548c for examples of these instructional materials.*

Key to Glencoe Manipulative Resources

[1]Classroom Manipulative Resources [2]Student Manipulative Resources [3]Overhead Manipulative Resources [4]Hands-On Algebra Masters

INV = Investigation *SG+A = Study Guide and Assessment*

DAY 7	DAY 8	DAY 9	DAY 10	DAY 11	DAY 12	DAY 13
Lesson 13–5	INV	Lesson 13–6		Lesson 13–7	SG+A	Chapter Test
INV	Lesson 13–6		Lesson 13–7	SG+A	Chapter Test	
Chapter Test						

Resource Manager

The pages shown on this page are a small sample of the materials available on *TeacherWorks: All-in-One Lesson Planner and Resource Center.*

This CD-ROM includes all of the blackline masters and transparencies available for this program.

It also includes a lesson planner and interactive Teacher Edition, so you can customize lesson plans and reproduce classroom resources quickly and easily, from just about anywhere.

Applications

School-to-Workplace Masters, p. 13

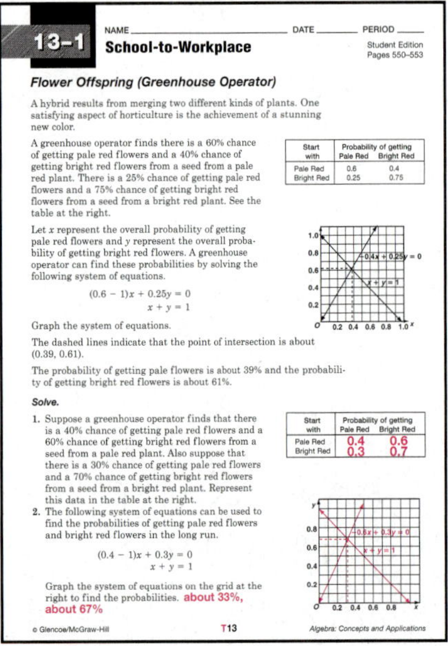

13-1 NAME ______ DATE ______ PERIOD ______

School-to-Workplace

Student Edition Pages 550–553

Flower Offspring (Greenhouse Operator)

A hybrid results from merging two different kinds of plants. One satisfying aspect of horticulture is the achievement of a stunning new color.

A greenhouse operator finds there is a 60% chance of getting pale red flowers and a 40% chance of getting bright red flowers from a seed from a pale red plant. There is a 25% chance of getting pale red flowers and a 75% chance of getting bright red flowers from a seed from a bright red plant. See the table at the right.

Start with	Probability of getting Pale Red	Bright Red
Pale Red	0.6	0.4
Bright Red	0.25	0.75

Let x represent the overall probability of getting pale red flowers and y represent the overall probability of getting bright red flowers. A greenhouse operator can find these probabilities by solving the following system of equations.

$$(0.6 - 1)x + 0.25y = 0$$
$$x + y = 1$$

Graph the system of equations.

The dashed lines indicate that the point of intersection is about (0.39, 0.61).

The probability of getting pale flowers is about 39% and the probability of getting bright red flowers is about 61%.

Solve.

1. Suppose a greenhouse operator finds that there is a 40% chance of getting pale red flowers and a 60% chance of getting bright red flowers from a seed from a pale red plant. Also suppose that there is a 30% chance of getting pale red flowers and a 70% chance of getting bright red flowers from a seed from a bright red plant. Represent this data in the table at the right.

Start with	Probability of getting Pale Red	Bright Red
Pale Red	0.4	0.6
Bright Red	0.3	0.7

2. The following system of equations can be used to find the probabilities of getting pale red flowers and bright red flowers in the long run.

$$(0.4 - 1)x + 0.3y = 0$$
$$x + y = 1$$

Graph the system of equations on the grid at the right to find the probabilities. about 33%, about 67%

© Glencoe/McGraw-Hill T13 *Algebra: Concepts and Applications*

Manipulatives/Modeling

Hands-On Algebra Masters, pp. 144–149

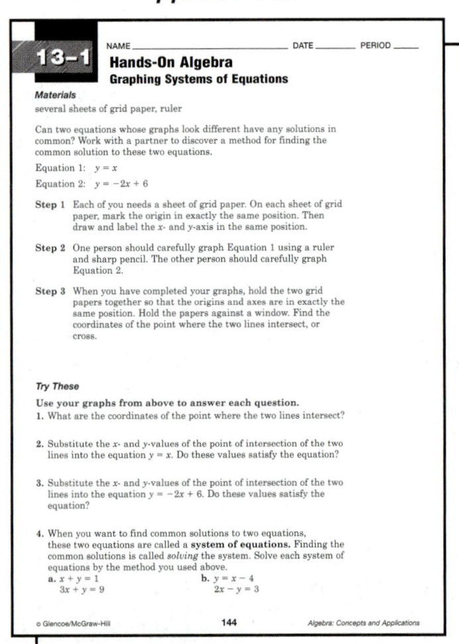

13-1 NAME ______ DATE ______ PERIOD ______

Hands-On Algebra

Graphing Systems of Equations

Materials

several sheets of grid paper, ruler

Can two equations whose graphs look different have any solutions in common? Work with a partner to discover a method for finding the common solution to these two equations.

Equation 1: $y = x$

Equation 2: $y = -2x + 6$

Step 1 Each of you needs a sheet of grid paper. On each sheet of grid paper, mark the origin in exactly the same position. Then draw and label the x- and y-axis in the same position.

Step 2 One person should carefully graph Equation 1 using a ruler and sharp pencil. The other person should carefully graph Equation 2.

Step 3 When you have completed your graphs, hold the two grid papers together so that the origins and axes are in exactly the same position. Hold the papers against a window. Find the coordinates of the point where the two lines intersect, or cross.

Try These

Use your graphs from above to answer each question.

1. What are the coordinates of the point where the two lines intersect?
2. Substitute the x- and y-values of the point of intersection of the two lines into the equation $y = x$. Do these values satisfy the equation?
3. Substitute the x- and y-values of the point of intersection of the two lines into the equation $y = -2x + 6$. Do these values satisfy the equation?
4. When you want to find common solutions to two equations, these two equations are called a **system of equations.** Finding the common solutions is called *solving* the system. Solve each system of equations by the method you used above.
   a. $x + y = 1$, $3x + y = 9$
   b. $y = x - 4$, $2x - y = 3$

© Glencoe/McGraw-Hill 144 *Algebra: Concepts and Applications*

Technology/Multimedia

Graphing Calculator Masters, pp. 36–39

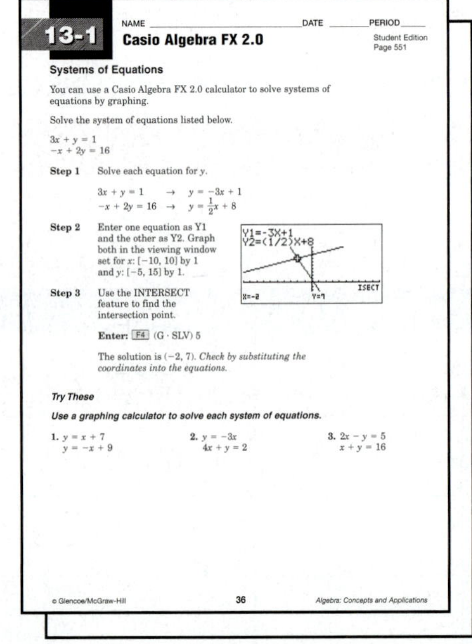

13-1 NAME ______ DATE ______ PERIOD ______

Casio Algebra FX 2.0

Student Edition Page 551

Systems of Equations

You can use a Casio Algebra FX 2.0 calculator to solve systems of equations by graphing.

Solve the system of equations listed below.

$3x + y = 1$
$-x + 2y = 16$

Step 1 Solve each equation for y.

$3x + y = 1 \rightarrow y = -3x + 1$
$-x + 2y = 16 \rightarrow y = \frac{1}{2}x + 8$

Step 2 Enter one equation as Y1 and the other as Y2. Graph both in the viewing window set for x: [−10, 10] by 1 and y: [−5, 15] by 1.

Step 3 Use the INTERSECT feature to find the intersection point.

Enter: F4 (G · SLV) 5

The solution is (−2, 7). *Check by substituting the coordinates into the equations.*

Try These

Use a graphing calculator to solve each system of equations.

1. $y = x + 7$, $y = -x + 9$
2. $y = -3x$, $4x + y = 2$
3. $2x - y = 5$, $x + y = 16$

© Glencoe/McGraw-Hill 36 *Algebra: Concepts and Applications*

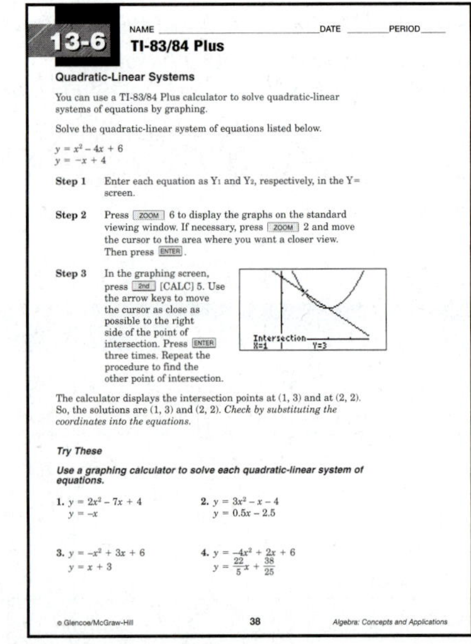

13-6 NAME ______ DATE ______ PERIOD ______

TI-83/84 Plus

Quadratic-Linear Systems

You can use a TI-83/84 Plus calculator to solve quadratic-linear systems of equations by graphing.

Solve the quadratic-linear system of equations listed below.

$y = x^2 - 4x + 6$
$y = -x + 4$

Step 1 Enter each equation as Y_1 and Y_2, respectively, in the Y= screen.

Step 2 Press ZOOM 6 to display the graphs on the standard viewing window. If necessary, press ZOOM 2 and move the cursor to the area where you want a closer view. Then press ENTER.

Step 3 In the graphing screen, press 2nd [CALC] 5. Use the arrow keys to move the cursor as close as possible to the right side of the point of intersection. Press ENTER three times. Repeat the procedure to find the other point of intersection.

The calculator displays the intersection points at (1, 3) and at (2, 2). So, the solutions are (1, 3) and (2, 2). *Check by substituting the coordinates into the equations.*

Try These

Use a graphing calculator to solve each quadratic-linear system of equations.

1. $y = 2x^2 - 7x + 4$, $y = -x$
2. $y = 3x^2 - x - 4$, $y = 0.5x - 2.5$
3. $y = -x^2 + 3x + 6$, $y = x + 3$
4. $y = -4x^2 + 2x + 6$, $y = \frac{22}{5}x + \frac{38}{25}$

© Glencoe/McGraw-Hill 38 *Algebra: Concepts and Applications*

Assessment Resources

Type	Student Edition	Teacher's Wraparound Edition	Chapter 13 Resource Masters
Ongoing Assessment	Quizzes 1 and 2, pp. 565, 585	5-Minute Check, pp. 550, 554, 560, 566, 572, 580, 586	Mid-Chapter Test, p. 618 Quizzes A and B, p. 619
Mixed Review	Mixed Review, pp. 553, 559, 565, 571, 577, 585, 590 Standardized Test Practice, Chapters 1–13, pp. 596–597		Cumulative Review, p. 620 Standardized Test Practice, pp. 621–622
Error Analysis	You Decide, pp. 564, 589	Error Analysis, pp. 552, 557, 564, 569, 576, 584, 589	
Standardized Test Prep	Standardized Test Practice, pp. 553, 559, 565, 571, 577, 585, 590 Standardized Test Practice, Chapters 1–13, pp. 596–597		Standardized Test Practice, pp. 621–622
Open-Ended Assessment	Writing Math, pp. 557, 584 Investigation, pp. 578–579 Portfolio, pp. 549, 579	Modeling: pp. 585, 590 Speaking: pp. 553, 565 Writing: pp. 559, 571, 577	Extended Response Assessment, p. 617
Chapter Assessment	Study Guide and Assessment, pp. 592–594 Chapter Test, p. 595		Multiple-Choice Tests (Forms 1A, 1B), pp. 609–612 Free-Response Tests (Forms 2A, 2B), pp. 613–616

Additional Chapter Resources

Student Edition
Math in the Workplace, p. 591
Hands-On Algebra, p. 560
Graphing Calculator Exploration, pp. 551, 588

Teacher's Classroom Resources
Manipulatives/Modeling
Teacher's Guide for Overhead Manipulative Resources

Meeting Individual Needs
Prerequisite Skills Booklet
Spanish Study Guide and Assessment, pp. 87–93, 129–130

Teaching Aids
Solutions Manual
5-Minute Check Transparencies

Glencoe Technology

Instructional
AlgePASS CD-ROM, Lessons 31
Interactive Chalkboard CD-ROM
StudentWorks Plus CD-ROM
Multimedia Applications CD-ROM, Activity 8
Vocabulary PuzzleMaker (online)

Assessment
ExamView® Pro

Teaching Aids
Answer Key Maker CD-ROM
TeacherWorks CD-ROM

Visit ***www.algconcepts.com*** *for data updates, career information, games, and other interactive activities.*

CHAPTER PREVIEW 13

Mathematics of the Chapter

Students begin graphing pairs of linear equations and seeing whether the lines intersect, coincide, or are parallel. They explore several methods for solving systems of equations: graphing, substitution, and elimination. Then they graph quadratic-linear systems of equations and systems of linear inequalities.

What You'll Learn

Have students read over the lists of key ideas and key vocabulary. Have them make a list of any words with which they are not familiar.

Why It's Important

Point out to students that this is only one of many reasons why each objective is important. Others are provided in each lesson.

Real-Life Videos engage students, showing them how math is used in everyday situations. Use Video 2 to discuss how inequalities and systems of equations are used in fighting forest fires and managing an aquarium.

CHAPTER 13 Systems of Equations and Inequalities

What You'll Learn

Key Ideas

- Solve systems of equations by graphing. *(Lesson 13–1)*
- Determine how many solutions a system of equations has by graphing. *(Lesson 13–2)*
- Solve systems of equations by substitution and by elimination. *(Lessons 13–3 to 13–5)*
- Solve systems of quadratic and linear equations. *(Lesson 13–6)*
- Solve systems of inequalities by graphing. *(Lesson 13–7)*

Key Vocabulary

Why It's Important

Travel After the Louisiana Purchase in 1803, a flood of settlers that included riverboat captains, loggers, trappers, and millers came to live on the Mississippi River. Today the river still provides transportation for industry.

Systems of equations and inequalities are used to represent real-world situations in which there are many conditions that must be met. You will use a system of equations to find the rate of the current of the Mississippi River in Lesson 13–5.

CHAPTER 13 LINKS

Lesson	13–1	13–2	13–3	13–4	13–5	13–6	13–7
Math in the Workplace	Fund-raising	Trains	Metallurgy	Entertainment	Transportation	Meteorology	Money Car Dealer
Applications and Connections	Sales Shopping Submarines School	Transportation Animals Ballooning	Metals Mixtures Driving Exercise Sales	Spas	Entertainment Traveling Sports	Business Transportation	Spending Shopping Basketball Communication Mixtures
Math Integration	Geometry	Number Theory		Number Theory		Geometry Number Theory	

Study these lessons to improve your skills.

Check Your Readiness

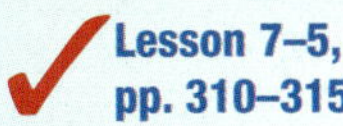
Lesson 7–5, pp. 310–315

Graph each equation. 1–6. See Solutions Manual.

1. $y = x + 2$
2. $y = 2x - 1$
3. $y = \frac{1}{4}x + 2$
4. $x + y = 4$
5. $x - 2y = 6$
6. $2x + y = 4$

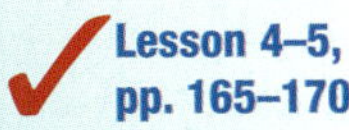
Lesson 4–5, pp. 165–170

Solve each equation. Check your solution.

7. $3n + 2 = 11$ **3**
8. $2c - 6 = -8$ **−1**
9. $2 + 5r = 22$ **4**
10. $6x - 8 = 22$ **5**
11. $5d - 1 = 9$ **2**
12. $-14 = 4k + 2$ **−4**

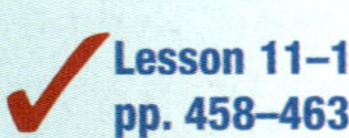
Lesson 11–1, pp. 458–463

13–18. See Solutions Manual.

Graph each quadratic function by making a table of values.

13. $y = x^2$
14. $y = x^2 + 2$
15. $y = 3x^2$
16. $y = -x^2 + 3$
17. $y = x^2 - 3x$
18. $y = \frac{1}{2}x^2$

Lesson 12–7, pp. 535–539

Graph each inequality. 19–24. See Solutions Manual.

19. $y \leq x + 2$
20. $y > 4x + 5$
21. $x \geq 3$
22. $y < \frac{1}{3}x + 2$
23. $x + y \leq 3$
24. $3x - y > 5$

Make this Foldable to help you organize your Chapter 13 notes. Begin with four sheets of grid paper.

1. **Stack** sheets of paper with edges 4 grids apart to create tabs.

2. **Fold** up the bottom edges. All tabs should be the same size.

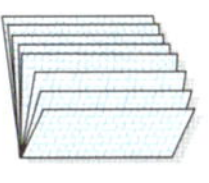

3. **Staple** along the fold.

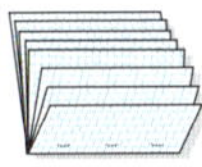

4. **Label** the tabs using lesson numbers and titles.

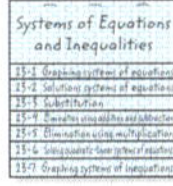

Reading and Writing As you read and study the chapter, use each page to write notes and examples.

www.algconcepts.com/chapter_readiness

Check Your Readiness

This section provides a review of the basic concepts needed before beginning Chapter 1. Lesson references are included for additional student help.

Use Exercises	To Prepare for Lesson(s)
1–6	13–1, 13–2
7–12	13–3, 13–4, 13–5
13–18	13–6
19–24	13–7

Quick Review Math Handbook

Glencoe Algebra Lesson	Quick Review Math Handbook Hot Topic
13–1	6–4
13–2	6–4
13–3	6–2
13–4	6–2
13–5	6–6

Have each student make a Foldable with a tab for each lesson in Chapter 13. Have students explain the relationship between solutions of systems of equations and graphs of the systems. Then have students explain how to solve a system using each method presented in the chapter.

Noteables™

Interactive Study Notebook with Foldables™

This note-taking guide reinforces excellent note-taking skills. It includes:

- ✓ **Build Your Vocabulary** tool for including mathematical terminology in notes.
- ✓ **Review It** activities with links to prior lessons and items to include in **Foldables™**.
- ✓ **Bringing It All Together** feature to help students review for the Chapter Test.
- ✓ **Are You Ready for the Chapter Test?** feature to allow students to assess their own readiness for the Chapter Test.
- ✓ **Teacher Edition** with transparencies to guide note taking. Corresponds to the *Teacher's Wraparound Edition* In-Class Examples and *Interactive Chalkboard CD-ROM.*

Lesson 13-1

1 FOCUS

5-Minute Check
Chapter 12

1. Write an inequality for the graph. $x < -1$

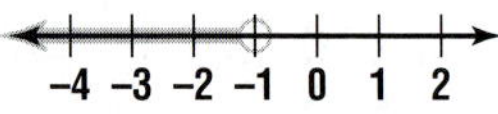

Solve each inequality. Check your solution.

2. $6 < -3d$ $d < -2$
3. $5(x - 1) \geq 20$ $x \geq 5$
4. Solve $s + 7 > 10$ or $2s + 3 < 5$. Graph the solution. $s < 1$ or $s > 3$

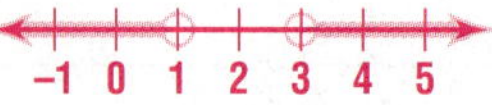

Motivating the Lesson

Hands-On Activity Select four students, and ask them to stand next to different walls of the classroom. Ask a fifth student to imagine that there are lines connecting opposite pairs of students, and to move to the intersection of the two lines. Then ask the four students to move, one at a time, and have the fifth student adjust her or his position each time.

2 TEACH

In-Class Example
Example 1

Solve the system of equations by graphing.
$y = x + 6$
$y = 3x$ (3, 9)

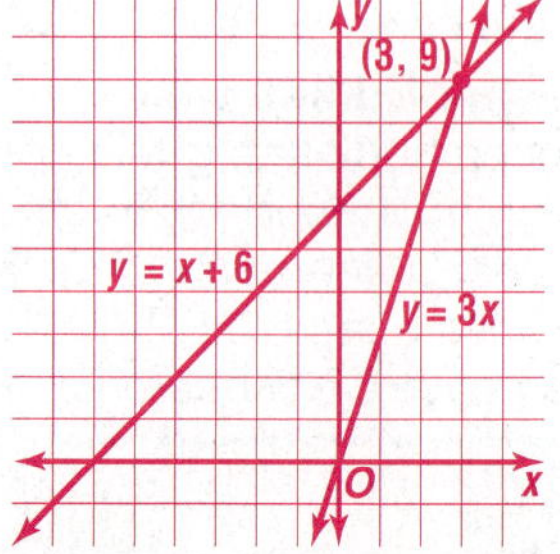

13–1 Graphing Systems of Equations

What You'll Learn
You'll learn to solve systems of equations by graphing.

Why It's Important
Sales Systems of equations can be used to determine how many items need to be sold in order to make a profit.
See Example 3.

A **system of equations** is a set of two or more equations with the same variables. Below are some examples of systems.

$$x + y = 5 \qquad \frac{a}{b} = 2$$
$$3x - 4y = -13 \qquad a + 2b = 8$$

The solution of a system of equations in two variables is an ordered pair that satisfies both equations.

System of Equations

Words: A system of equations is a set of two or more equations with the same variables. The solution is the ordered pair that satisfies all of the equations.

Model:

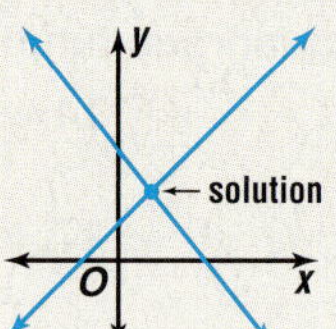

One method for solving a system of equations is to graph the equations on the same coordinate plane. The coordinates of the point of intersection are the solution.

Examples

Solve each system of equations by graphing.

$y = 2x$
$y = -x + 3$

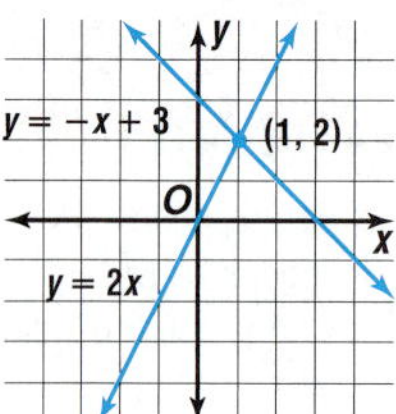

The graphs appear to intersect at (1, 2). Check this estimate by substituting the coordinates into each equation.

Look Back
Graphing Linear Equations: Lesson 7–5

Check:

$y = 2x$	*Original equation*	$y = -x + 3$	*Original equation*
$2 \stackrel{?}{=} 2(1)$	*Replace x with 1 and y with 2.*	$2 \stackrel{?}{=} -1 + 3$	*Replace x with 1 and y with 2.*
$2 = 2$ ✓		$2 = 2$ ✓	

The solution of the system of equations is (1, 2).

Resource Manager

Reproducible Masters

Chapter 13 Resource Masters
- *Study Guide,* p. 573
- *Skills Practice,* p. 574
- *Practice,* p. 575
- *Reading to Learn Mathematics,* p. 576
- *Enrichment,* p. 577

Graphing Calculator, p. 36

Hands-On Algebra, p. 144
School-to-Workplace, p. 13

Transparencies
5-Minute Check, 13–1

Technology/Multimedia
Interactive Chalkboard CD-ROM

2 $x - y = -2$
$x + y = 4$

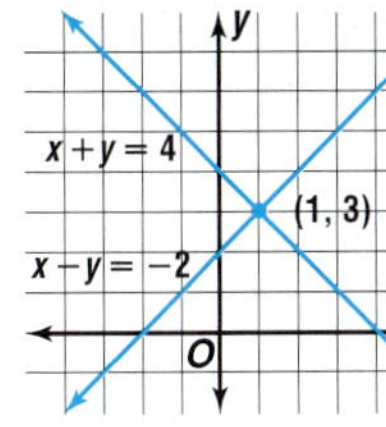

The graphs appear to intersect at (1, 3). *Check this estimate.*

The solution of the system of equations is (1, 3).

Your Turn **a–b. See margin for graphs.**

a. $y = 2x$
$y = -x + 6$ **(2, 4)**

b. $x + y = 1$
$2x + y = 4$ **(3, −2)**

A graphing calculator can be used to solve systems of equations or to check solutions.

Graphing Calculator Tutorial
See pp. 750–753.

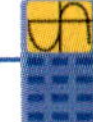

Graphing Calculator Exploration

You can use a graphing calculator to solve systems of equations.

$3x + y = 1$
$-x + 2y = 16$

Step 1 Solve each equation for y.

$3x + y = 1 \rightarrow y = -3x + 1$

$-x + 2y = 16 \rightarrow y = \frac{1}{2}x + 8$

Step 2 Make a table of values for the two equations.

X	Y1	Y2
-5	16	5.5
-4	13	6
-3	10	6.5
-2	7	7
-1	4	7.5
0	1	8
1	-2	8.5

X= -5

Press [Y=] to enter $-3x + 1$ as Y_1. Scroll down to enter $\frac{1}{2}x + 8$ as Y_2. Then press [2nd] [TBLSET] and enter −5 as TblStart and 1 as ΔTbl.

Press [2nd] [TABLE] to display the table.

Step 3 Look in the table for an x value that has the same y value in both equations. The solution is (−2, 7). *Check by graphing on the calculator and finding the intersection point.*

Try These

Use a graphing calculator to solve each system of equations.

1. $y = x + 7$
   $y = -x + 9$ **(1, 8)**
2. $y = -3x$
   $4x + y = 2$ **(2, −6)**
3. $2x - y = 5$
   $x + y = 16$ **(7, 9)**

www.algconcepts.com/extra_examples

Teaching Tip In Example 2, students may want to write the equations in slope-intercept form to check the graphs.

In-Class Example

Example 2

Solve the system of equations by graphing.

$x - y = 3$
$x + y = 1$ **(2, −1)**

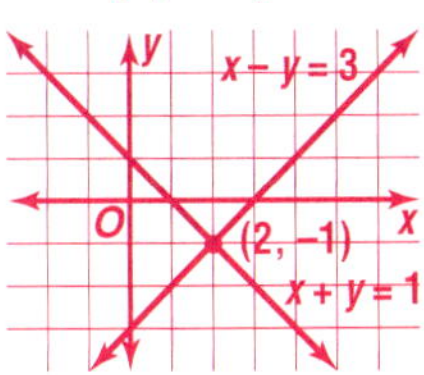

Answers

Your Turn

a.

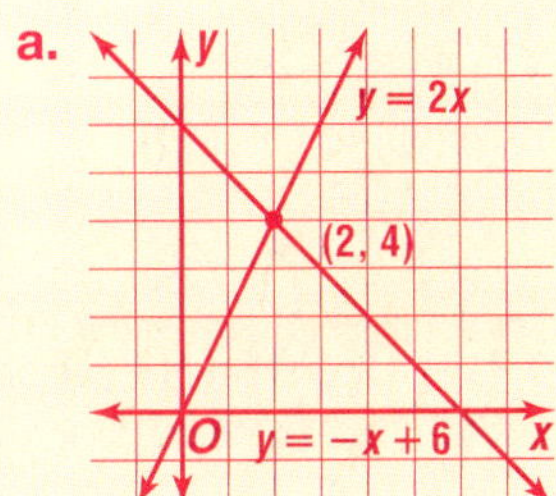

b.

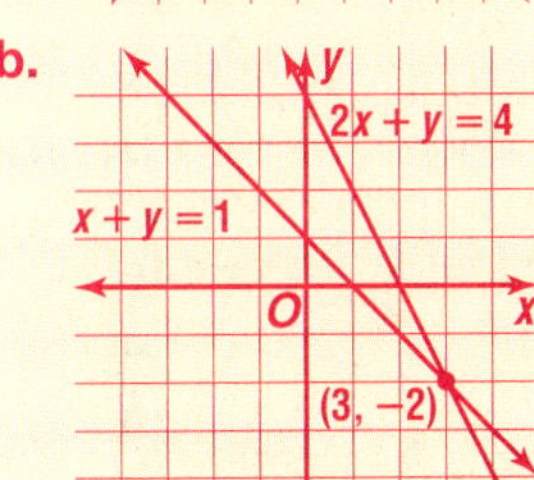

***Study Guide*, p. 573**

13–1 NAME ______ DATE ______ PERIOD ______

Study Guide Student Edition Pages 550–553

Graphing Systems of Equations

The ordered pair (−1, −3) is the solution of the system of equations

$y = x - 2$
$y = 3x$

because when −1 is substituted for x and −3 is substituted for y, both equations are true.

$y = x - 2$	$y = 3x$
$-3 \stackrel{?}{=} -1 - 2$	$-3 \stackrel{?}{=} 3(-1)$
$-3 = -3$ ✓	$-3 = -3$ ✓

You can also graph both equations to show that (−1, −3) is the solution of the system.

The graphs appear to intersect at (−1, −3). Since (−1, −3) is the solution of each equation, it is the solution of the system of equations.

You can also use a graphing calculator to solve the system of equations.

Step 1 Enter these keystrokes in the Y= screen:
[X, T, θ, n] [−] 2 [ENTER]
3 [X, T, θ, n] [ENTER] [GRAPH]

Step 2 Use the INTERSECT feature to find the intersection point.
[2nd] [CALC] 5 [ENTER] [ENTER] [ENTER]

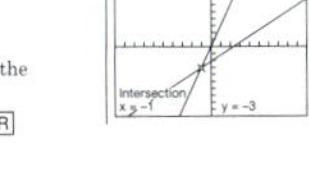

The solution is (−1, −3).

Solve each system of equations by graphing.

1. $x = -1$
   $y = 3$
2. $y = 2$
   $y = x$
3. $y = -3$
   $x = 2$
4. $y = -3x$
   $y = x + 4$
5. $y = 1 - x$
   $y = 2x - 5$
6. $y = -x - 2$
   $y = 3x + 2$

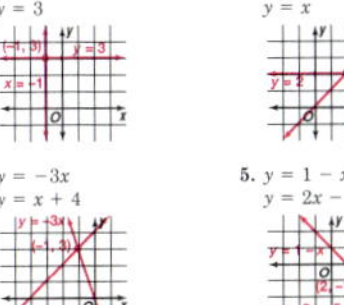
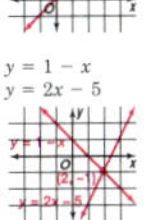
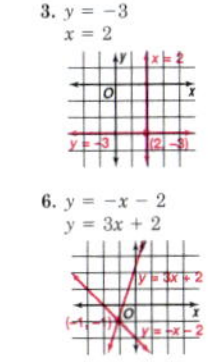

Graphing Calculator Exploration

Encourage students to graph the equations in a viewing window that allows them to see where the graphs intersect. This is a good check for finding the solution using a table.

Teaching Tip In the coordinate plane for Example 3, make sure students understand that the two graphs are restricted to Quadrant I and the positive axes because the values of x and y cannot be negative.

In-Class Example

Example 3

Refer to the situation in Example 3. How many mascot beanbags and caps should the math club try to sell if they sell a total of 50 items and want to make a $240 profit?

10 beanbags, 40 caps

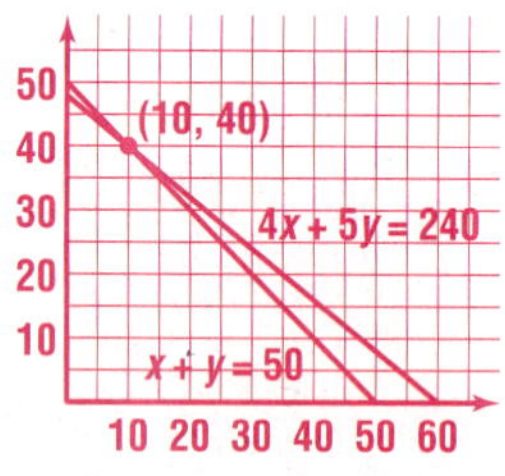

3 PRACTICE/APPLY

Error Analysis

Watch for students who, in Exercise 3, confuse the intersection of two lines with the intersection of a line and either the x-axis or y-axis.

Prevent by stressing that the solution must lie on both lines that represent the two equations.

Skills Practice, **p. 574, and** ***Practice,*** **p. 575 (shown)**

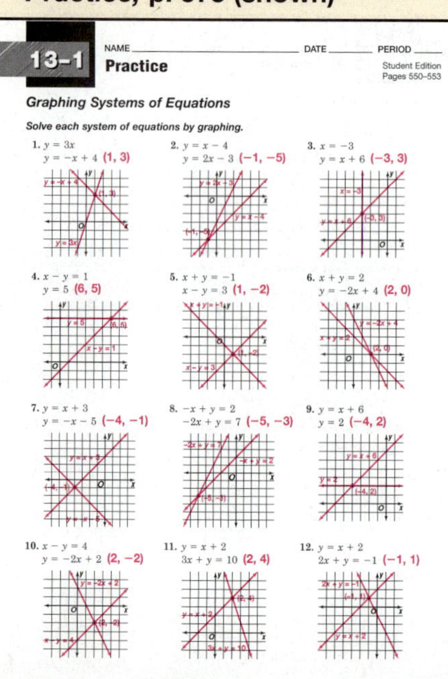

13-1 Practice — NAME ____ DATE ____ PERIOD ____ — Student Edition Pages 550–553

Graphing Systems of Equations

Solve each system of equations by graphing.

1. $y = 3x$, $y = -x + 4$ (1, 3)
2. $y = x - 4$, $y = 2x - 3$ (−1, −5)
3. $x = -3$, $y = x + 6$ (−3, 3)
4. $x - y = 1$, $y = 5$ (6, 5)
5. $x + y = -1$, $x - y = 3$ (1, −2)
6. $x + y = 2$, $y = -2x + 4$ (2, 0)
7. $y = x + 3$, $y = -x - 5$ (−4, −1)
8. $-x + y = 2$, $-2x + y = 7$ (−5, −3)
9. $y = x + 6$, $y = 2$ (−4, 2)
10. $x - y = 4$, $y = -2x + 2$ (2, −2)
11. $y = x + 2$, $3x + y = 10$ (2, 4)
12. $y = x + 2$, $2x + y = -1$ (−1, 1)

Example 3 Sales Link

The Math Club is selling T-shirts for a profit of $4 each and caps for a profit of $5 each. The club wants to sell 50 items and make a profit of $230. How many of each item should the club try to sell?

Let x = the number of T-shirts and y = the number of caps. You can write two equations to represent this situation.

$x + y = 50$ ← *the number of items*

$4x + 5y = 230$ ← *the total profit*

Graph $x + y = 50$ and $4x + 5y = 230$. The graphs appear to intersect at (20, 30). Check this estimate.

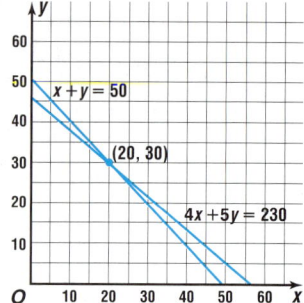

Check:

$x + y = 50$ *Original equation*
$20 + 30 \stackrel{?}{=} 50$ *(x, y) = (20, 30)*
$50 = 50$ ✓

$4x + 5y = 230$ *Original equation*
$4(20) + 5(30) \stackrel{?}{=} 230$ *(x, y) = (20, 30)*
$80 + 150 \stackrel{?}{=} 230$
$230 = 230$ ✓

They should try to sell 20 T-shirts and 30 caps.

Check for Understanding

Vocabulary: system of equations

Communicating Mathematics

1. **Describe** the solution of a system of equations. **the ordered pair for the point at which the graphs of the equations intersect**
2. **Sketch** a system of linear equations that has (2, 3) as its solution. **See margin.**
3. **Determine** the solution of the system of equations represented by each pair of lines.
   a. ℓ and m **(−4, 2)**
   b. m and n **(2, −2)**
   c. n and ℓ **(2, 4)**

Guided Practice Examples 1–3

4–7. See Solutions Manual for graphs.

Solve each system of equations by graphing.

4. $y = x - 3$, $y = -x + 3$ **(3, 0)**
5. $x = -4$, $y = \frac{1}{2}x + 3$ **(−4, 1)**
6. $x + y = 6$, $y = 2$ **(4, 2)**
7. $x + y = -2$, $x - y = 4$ **(1, −3)**

Example 3

8. **Shopping** Randall wants to buy 10 bouquets. The standard bouquet costs $7, and the deluxe one costs $12. He can afford to spend $100.
   a. Write a system of equations for the number of standard bouquets x and the number of deluxe bouquets y that he can buy. **a. $x + y = 10$ and $7x + 12y = 100$**
   b. Use a graphing calculator to find the number of each type of bouquet he can buy. **4 standard, 6 deluxe**

Reteaching Activity

Naturalist Learners Have students make a sketch of some of the main streets around the school. Then have them impose a coordinate grid on their sketches, placing the school at the origin. Select any two intersecting streets, and have them write an equation to represent each street.

Answer

2. Sample answer:

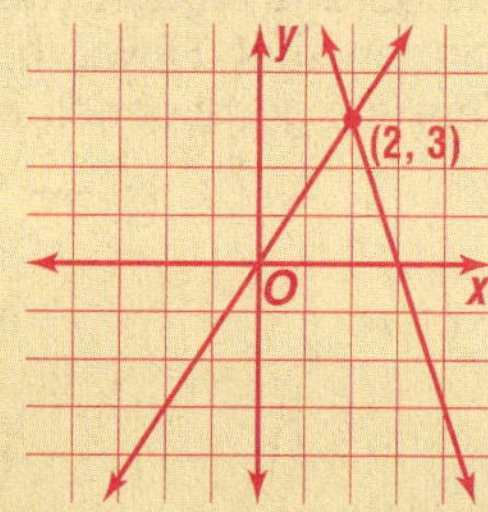

Exercises 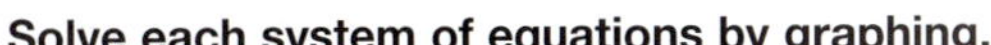

Practice

Homework Help

For Exercises	See Examples
9–14	1
15–24	2, 3

Extra Practice

See page 718.

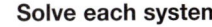

9–20. See Solutions Manual for graphs.

Solve each system of equations by graphing.

9. $x = 5$
 $y = -4$ **(5, −4)**

10. $x = 4$
 $y = x - 5$ **(4, −1)**

11. $y = x + 2$
 $y = -x + 2$ **(0, 2)**

12. $y = -2x + 3$
 $y = -\frac{1}{2}x$ **(2, −1)**

13. $y = 6$
 $y = \frac{4}{3}x + 2$ **(3, 6)**

14. $y = 2x - 7$
 $y = \frac{3}{2}x - 6$ **(2, −3)**

15. $y = 2$
 $x + y = 7$ **(5, 2)**

16. $x + y = 3$
 $x - y = 1$ **(2, 1)**

17. $x - y = 1$ **(−2, −3)**
 $y = -2x - 7$

18. $2x - y = 4$
 $x + y = -4$ **(0, −4)**

19. $x + y = -3$
 $\frac{1}{2}x + y = 0$ **(−6, 3)**

20. $5x + 4y = 10$
 $2x + y = 1$ **(−2, 5)**

21. Find the solution of the system $y = x + 4$ and $3x + 2y = 18$. **(2, 6)**
22. What is the solution of the system $2x + 10y = 0$ and $x + y = 4$?

22. (5, −1)

Applications and Problem Solving

23. **Submarines** Two submarines began dives in the same vertical position to meet at a designated point. If one submarine was on a course approximated by the equation $x + 4y = -14$ and the other was on a course approximated by the equation $x + 3y = -8$, at what location would they meet? Write the coordinates of the point. **(10, −6)**

24. **Geometry** The graphs of the equations $y = x + 2$, $3x + y = 6$, and $y = 5x + 6$ contain the sides of a triangle. **b. (0, 6), (1, 3), (−1, 1)**
 a. Graph the equations. **See margin.**
 b. Find the coordinates of the vertices of the triangle.

25. Sample answer: $y = x + 4$ and $y = -x + 4$

25. **Critical Thinking** Use your knowledge of slope and y-intercepts to write a system of equations with a solution of (0, 4).

Mixed Review

26. Which ordered pair, (−2, 0), (−1, 2), (0, −1), or (2, −2), is a solution of $3x + 4y \geq 2$? *(Lesson 12–7)* **(−1, 2)**

27–29. See margin.

Solve each inequality. Then graph the solution set. *(Lesson 12–6)*

27. $|x - 3| < 2$
28. $|a + 5| \geq 3$
29. $|2m + 2| > 6$

30. **School** Everyone in Mr. McClain's algebra class is at least 15 years old. Write an inequality to represent this information and then make a graph of the inequality. *(Lesson 12–1)* **$x \geq 15$; See margin for graph.**

Standardized Test Practice

31. **Short Response** Solve $2x^2 - 7x + 6 = 0$ by factoring. *(Lesson 11–4)* **$\frac{3}{2}$, 2**

32. **Multiple Choice** Which graph represents the function $y = x^2 + 2x - 3$? *(Lesson 11–1)* **B**

Ⓐ 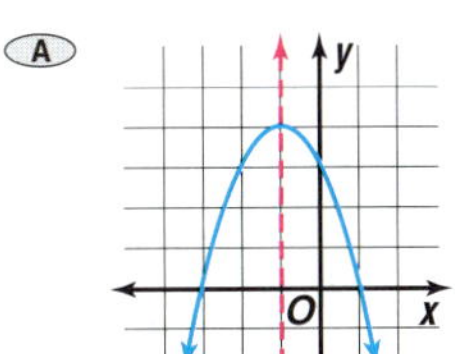

Ⓑ 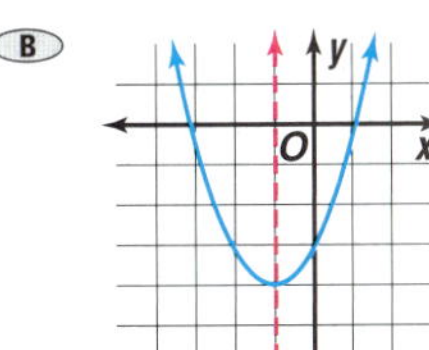

Ⓒ 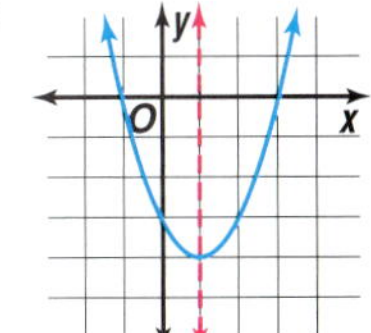

www.algconcepts.com/self_check_quiz

Assignment Guide

Basic: 9–23 odd, 25–32
Average: 10–22 even, 23–32

4 ASSESS

Open-Ended Assessment

Speaking Ask students to explain why a point that is on only one of the two lines (and not on the other) cannot be the solution to a system of equations.

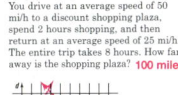

Answers

24a.

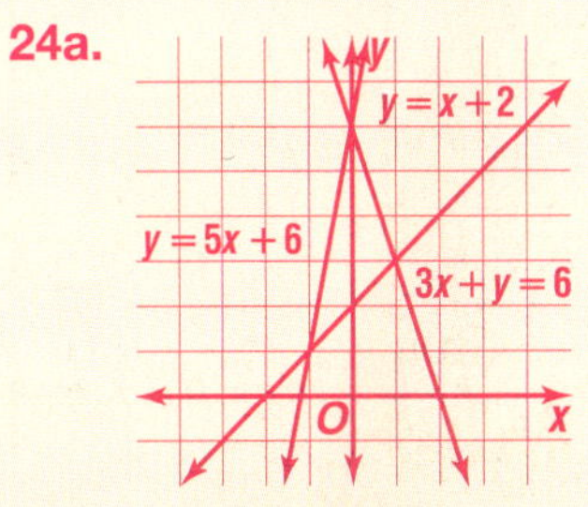

27. $\{x \mid 1 < x < 5\}$

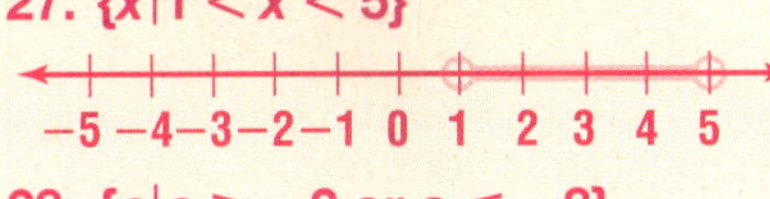

28. $\{a \mid a \geq -2 \text{ or } a \leq -8\}$

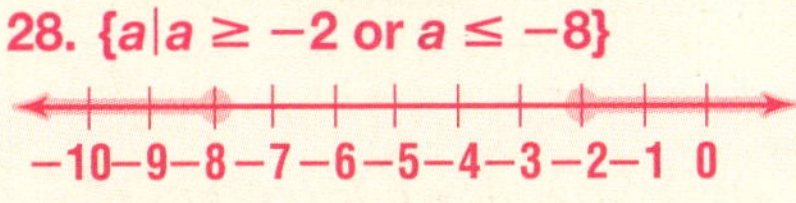

29. $\{m \mid m > 2 \text{ or } m < -4\}$

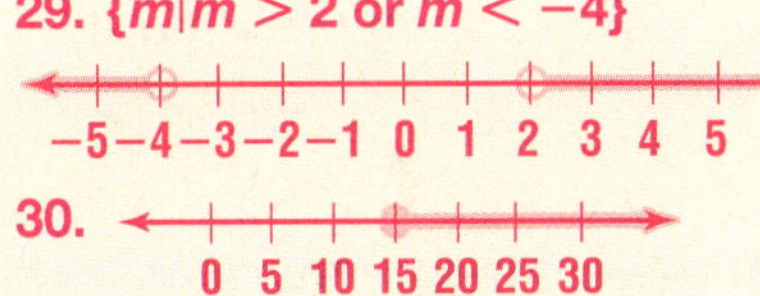

30. 0 5 10 15 20 25 30

Enrichment, p. 577

NAME ______ DATE ______ PERIOD ______

Enrichment

Student Edition Pages 550–553

Graphing a Trip

The formula $d = rt$ is used to solve many types of problems. If you graph an equation such as $d = 50t$, the graph is a model for a car going at 50 mi/h. The time the car travels is t; the distance in miles the car covers is d. The slope of the line is the speed.

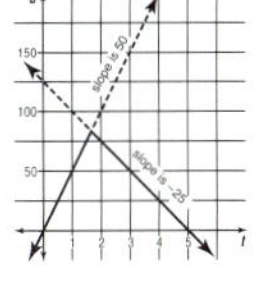

Suppose you drive to a nearby town and return. You average 50 mi/h on the trip out but only 25 mi/h on the trip home. The round trip takes 5 hours. How far away is the town?

The graph at the right represents your trip. Notice that the return trip is shown with a negative slope because you are driving in the opposite direction.

Solve each problem.

1. Estimate the answer to the problem in the above example. About how far away is the town? **about 80 miles**
2. Graph this trip and solve the problem. An airplane has enough fuel for 3 hours of safe flying. On the trip out the pilot averages 200 mi/h flying against a headwind. On the trip back, the pilot averages 250 mi/h. How long a trip out can the pilot make? **about $1\frac{2}{3}$ hours and 330 miles**
3. Graph this trip and solve the problem. You drive to a town 100 miles away. On the trip out you average 25 mi/h. On the trip back you average 50 mi/h. How many hours do you spend driving? **6 hours**

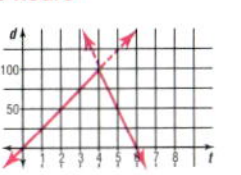

4. Graph this trip and solve the problem. You drive at an average speed of 50 mi/h to a discount shopping plaza, spend 2 hours shopping, and then return at an average speed of 25 mi/h. The entire trip takes 8 hours. How far away is the shopping plaza? **100 miles**

Extra Credit

Find the equations of a vertical line and a horizontal line that intersect at the solution of the system of equations $y = \frac{1}{2}x + 5$ and $y = -x + 2$. **$x = -2$, $y = 4$**

Lesson 13-2

1 FOCUS

5-Minute Check
Lesson 13–1

Use the two equations $y = -2x + 1$ and $y = -3x - 1$ and their graphs to complete the following.

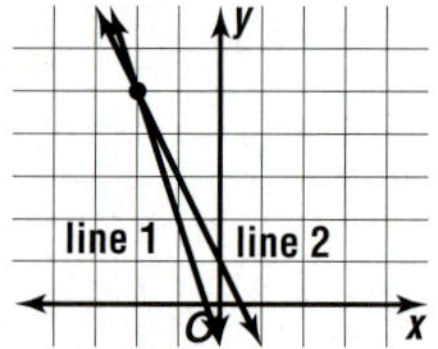

1. Which of the two graphs has the equation $y = -2x + 1$? **line 2**
2. Which of the two graphs has the equation $y = -3x - 1$? **line 1**
3. Find the solution to the system of equations below.
   $y = -2x + 1$
   $y = -3x - 1$ **(−2, 5)**

Use the following system of equations to complete the following.
$x - y = -1$
$2x + y = 7$

4. Graph the two equations.

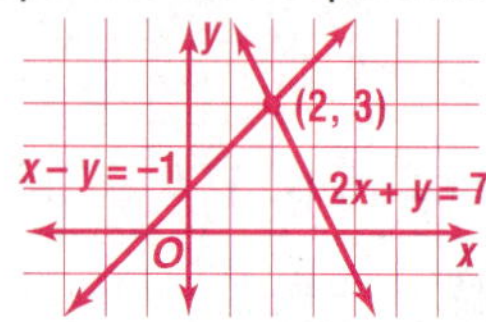

5. Find the point of intersection of the two graphs. **(2, 3)**

13-2 Solutions of Systems of Equations

What You'll Learn
You'll learn to determine whether a system of equations has one solution, no solution, or infinitely many solutions by graphing.

Why It's Important
Trains Train engineers must know if the tracks they are running on intersect another track. *See Example 6.*

Graphs of systems of linear equations may be intersecting lines, parallel lines, or the same line. Systems of equations can be described by the number of solutions they have.

Systems of Equations
- **consistent** at least one solution
  - **independent** exactly one solution
  - **dependent** infinitely many solutions
- **inconsistent** no solution

The different possibilities for the graphs of two linear equations are summarized in the following table.

Graph	Description of Graph	Slopes and Intercepts of Lines	Number of Solutions	Type of System
$y = -x + 3$, $y = x + 1$, (1, 2)	intersecting lines	different slopes	1	consistent and independent
$x + 2y = 2$, $2x + 4y = 4$	same line	same slope, same intercepts	infinitely many	consistent and dependent
$y = -2x + 4$, $y = -2x$	parallel lines	same slope, different intercepts	0	inconsistent

Resource Manager

Reproducible Masters
Chapter 13 Resource Masters
- *Study Guide,* p. 578
- *Skills Practice,* p. 579
- *Practice,* p. 580
- *Reading to Learn Mathematics,* p. 581
- *Enrichment,* p. 582

Hands-On Algebra, p. 145

Transparencies
5-Minute Check, 13–2

Technology/Multimedia
Interactive Chalkboard CD-ROM

MODELING
An alternative hands-on option using grid paper and a ruler is available for teaching this lesson.

Examples

State whether each system is *consistent and independent, consistent and dependent,* or *inconsistent.*

1

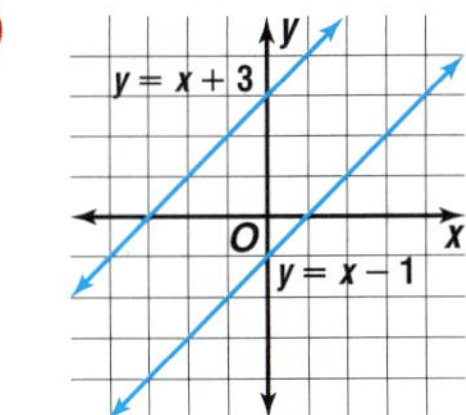

The graphs appear to be parallel lines. Since they do not intersect, there is no solution. This system is inconsistent.

2

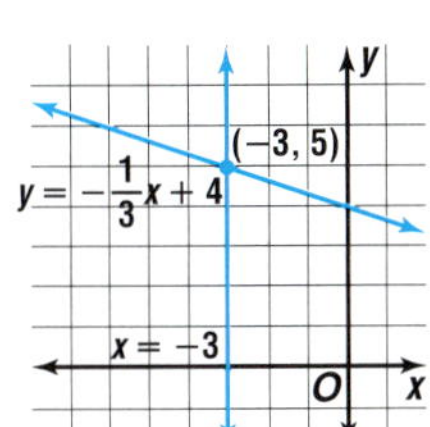

The graphs appear to intersect at the point at $(-3, 5)$. Because there is one solution, this system of equations is consistent and independent.

3

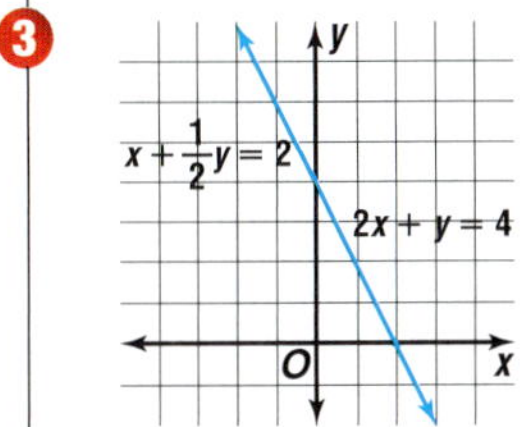

Both equations have the same graph. Because any ordered pair on the graph will satisfy both equations, there are infinitely many solutions. The system is consistent and dependent.

Your Turn

a. 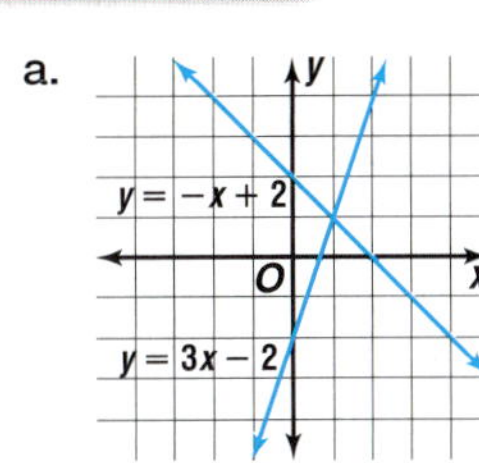

b. 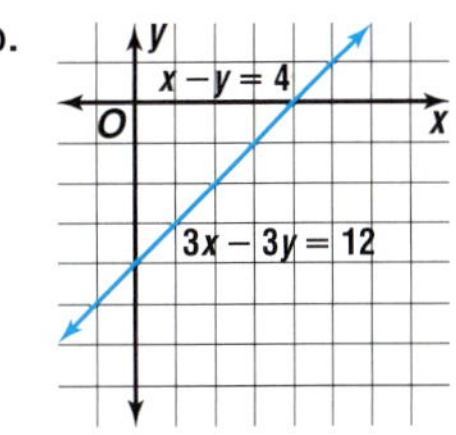

a. consistent and independent
b. consistent and dependent

You can determine the number of solutions to a system of equations by graphing.

Motivating the Lesson

Real-World Connection Have students look at a large map of the United States. Have them select a pair of cities and ask them to indicate a straight-line path that an airplane would follow if it started outside the U.S., flew over both cities, and then left U.S. airspace. Then ask students to select other pairs of U.S. cities so the second airplane would have a path that was parallel to, coincided with, or intersected with the first plane's path.

2 TEACH

In-Class Examples

Examples 1–2

State whether each system is **consistent and independent, consistent and dependent,** ***or*** **inconsistent.**

1

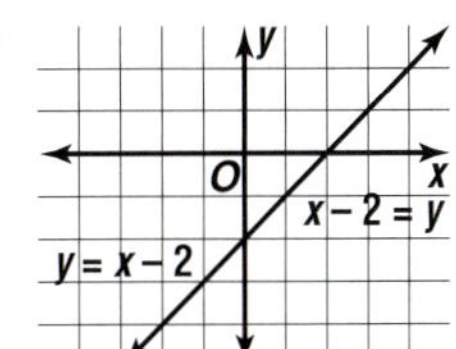

consistent and dependent

2

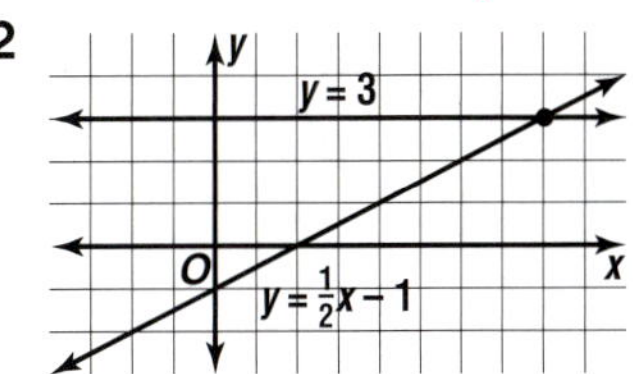

consistent and independent

Example 3

State whether the system is *consistent and independent, consistent and dependent,* or *inconsistent.* inconsistent

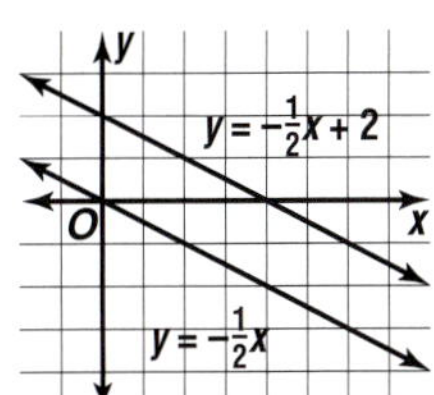

In-Class Example

Example 4

Determine whether the system of equations has *one* solution, *no* solution, or *infinitely many* solutions by graphing. If the system has one solution, name it.

$3x - y = 1$
$6x - 2y = 2$

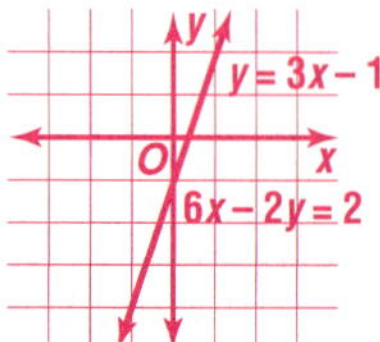

infinitely many

Teaching Tip In Example 5, students may need to review the fact that two lines that have the same slope are parallel.

In-Class Example

Example 5

Determine whether the system of equations has *one* solution, *no* solution, or *infinitely many* solutions by graphing. If the system has one solution, name it.

$y = -x$
$y = x - 2$

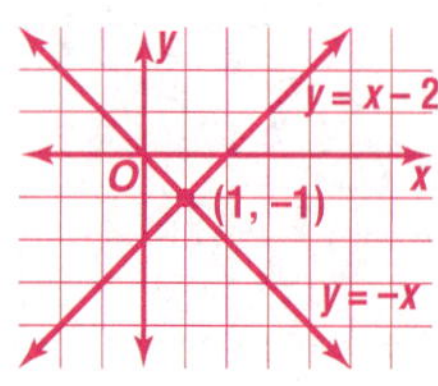

(1, −1)

Look Back

Graphing Linear Equations: Lesson 7–5

Determine whether each system of equations has *one* solution, *no* solution, or *infinitely many* solutions by graphing. If the system has one solution, name it.

4 $y = x + 2$
$y = -3x - 6$

The graphs appear to intersect at $(-2, 0)$. Therefore, this system of equations has one solution, $(-2, 0)$. Check that $(-2, 0)$ is a solution to each equation.

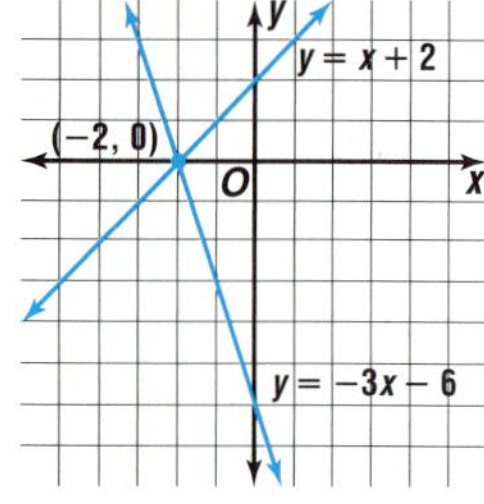

Check:

$y = x + 2$	*Original equation*	$y = -3x - 6$	*Original equation*
$0 \stackrel{?}{=} -2 + 2$	*Replace x with −2 and y with 0.*	$0 \stackrel{?}{=} -3(-2) - 6$	*Replace x with −2 and y with 0.*
$0 = 0$ ✓		$0 = 0$ ✓	

The solution of the system of equations is $(-2, 0)$.

5 $2x + y = 4$
$2x + y = 6$

Write each equation in slope-intercept form.

$2x + y = 4 \rightarrow y = -2x + 4$
$2x + y = 6 \rightarrow y = -2x + 6$

The graphs have the same slope and different y-intercepts. The system of equations has no solution.

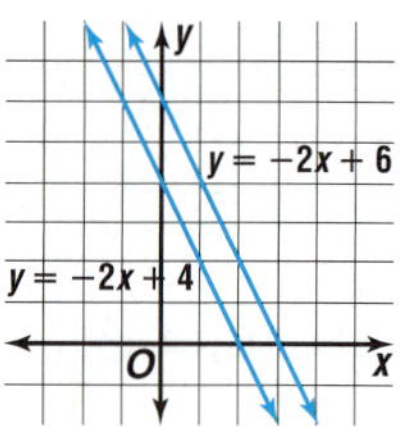

Your Turn **c–f. See margin for graphs.**

c. $y = x + 3$
$y = -2x + 3$ **(0, 3)**

d. $2x + y = 6$
$4x + 2y = 12$ **infinitely many**

e. $3x - y = 3$
$3x - y = 0$ **no solution**

f. $y - 2x = 0$
$y + x = 6$ **(2, 4)**

Answers

Your Turn

c.

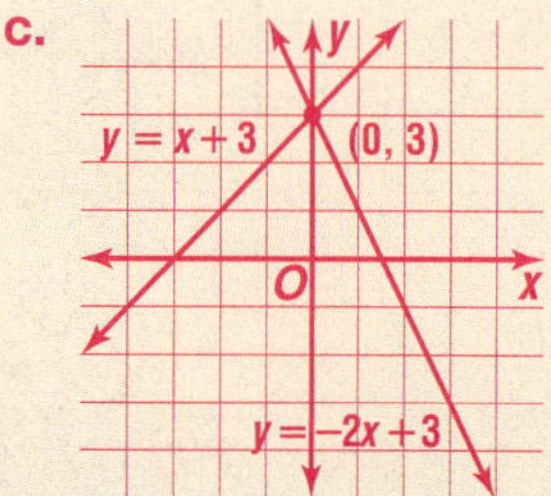

d.

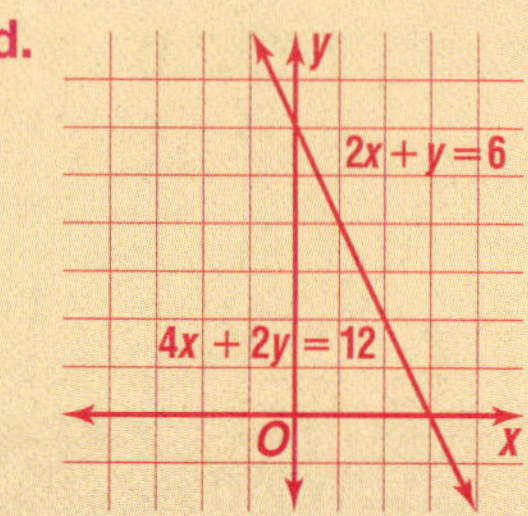

e.

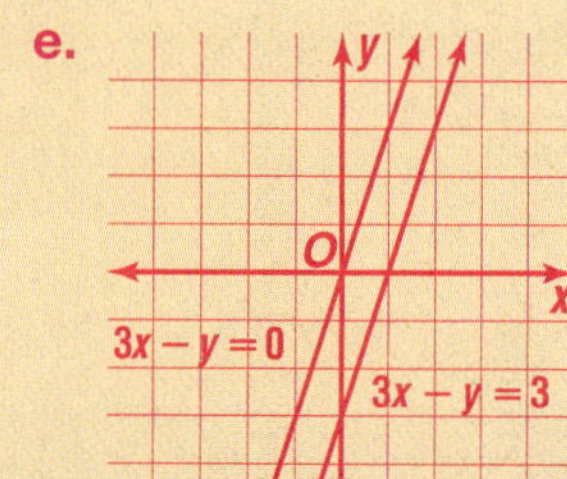

f.

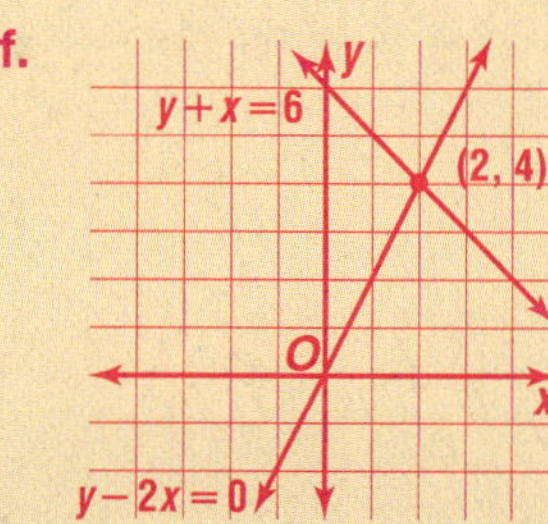

Example 6 — Transportation Link

The system of equations below represents the tracks of two trains. Do the tracks intersect, run parallel, or are the trains running on the same track? Explain.

$x + 2y = 4$
$3x + 6y = 12$

$x + 2y = 4 \qquad 3x + 6y = 12$
$3(x + 2y) = 3(4)$
$x + 2y = 4$ *Divide each side by 3.*

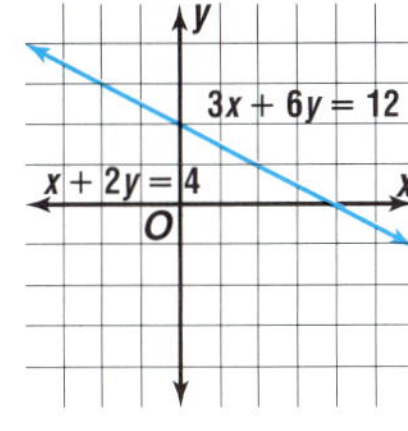

One equation is a multiple of the other. Each equation has the same graph and there are infinitely many solutions. Therefore, the trains are running on the same track.

Check for Understanding

Vocabulary
consistent
independent
dependent
inconsistent

Communicating Mathematics

1. **Describe** the possible graphs of a system of two linear equations. **two parallel lines, two intersecting lines, the same line**
2. **State** the number of solutions for each system of equations described below.
   a. One equation is a multiple of the other. **infinitely many**
   b. The equations have the same y-intercept and different slopes. **1**
   c. The equations have the same slope and different y-intercepts. **0**
3. Writing Math Create memory devices or other ways to remember the definitions of the terms in the Vocabulary box. Describe them in your journal. **See students' work.**

Guided Practice — Examples 1–3

State whether each system is *consistent and independent*, *consistent and dependent*, or *inconsistent*.

4. 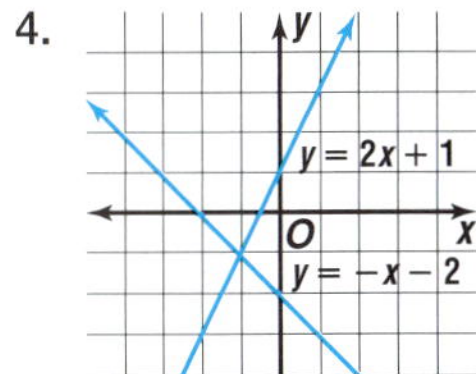

5. 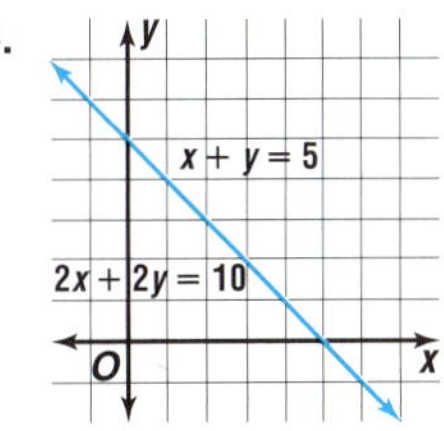

4. consistent and independent
5. consistent and dependent

In-Class Example

Example 6

The system of equations below represents the tracks of two trains. Do the tracks intersect, run parallel, or are the trains running on the same track? Explain.

$4x + 2y = 6$
$8x + 4y = 20$

In slope-intercept form, the two equations are $y = -2x + 3$ and $y = -2x + 5$. The graphs have the same slope and different y-intercepts, so the lines are parallel and the tracks do not intersect.

3 PRACTICE/APPLY

Error Analysis

Watch for students who, in Exercise 5, think that there are no solutions because there is no apparent intersection.

Prevent by using two colors, dashed and dotted lines, or other ways to stress that the graph shows two lines, but the lines coincide.

***Study Guide*, p. 578**

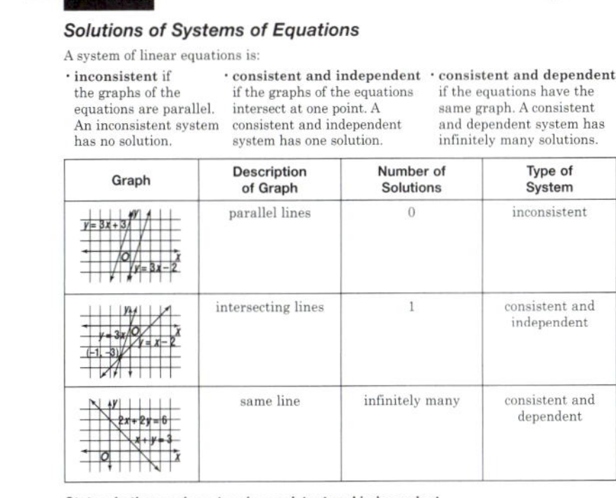

13–2 NAME ______ DATE ______ PERIOD ______

Study Guide — Student Edition Pages 554–559

Solutions of Systems of Equations

A system of linear equations is:

- **inconsistent** if the graphs of the equations are parallel. An inconsistent system has no solution.
- **consistent and independent** if the graphs of the equations intersect at one point. A consistent and independent system has one solution.
- **consistent and dependent** if the equations have the same graph. A consistent and dependent system has infinitely many solutions.

Graph	Description of Graph	Number of Solutions	Type of System
$y = 3x + 3$, $y = 3x - 2$	parallel lines	0	inconsistent
$y = 3x$, $y = x - 2$, (−1, −3)	intersecting lines	1	consistent and independent
$2x + 2y = 6$, $x + y = 3$	same line	infinitely many	consistent and dependent

***State whether each system is* consistent and independent, consistent and dependent, *or* inconsistent.**

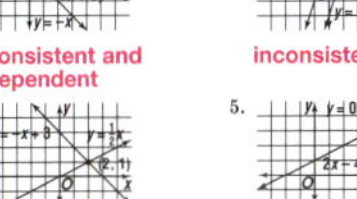

1. $3x + 3y = 0$, $y = -x$ **consistent and dependent**
2. $y = 3x + 2$, $y = 3x - 2$ **inconsistent**
3. $x = 1$, $y = -2$, (1, −2) **consistent and independent**
4. $y = -x + 3$, $y = \frac{1}{2}x$, (2, 1) **consistent and independent**
5. $y = 0.5x + 1$, $2x - 4y = -4$ **consistent and dependent**
6. $x = -2$, $x = 3$ **inconsistent**

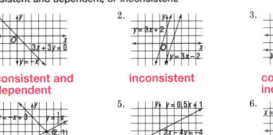

Reteaching Activity

Kinesthetic Learners Ask pairs of students to model part of a line by saying *endpoint*, walking about 10 paces, then saying *endpoint.* Then ask them to model 3 other lines so the pairs of lines (if extended) represent a system that has one solution, many solutions, or no solution.

Assignment Guide

Basic: 11–27 odd, 29–36
Average: 10–26 even, 27–36

Answers

6.

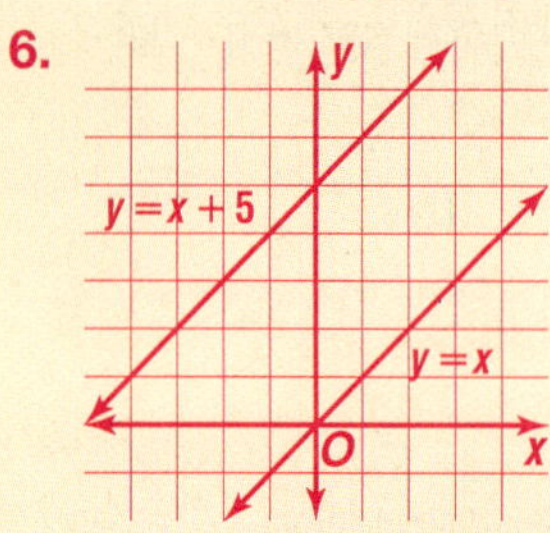

7.

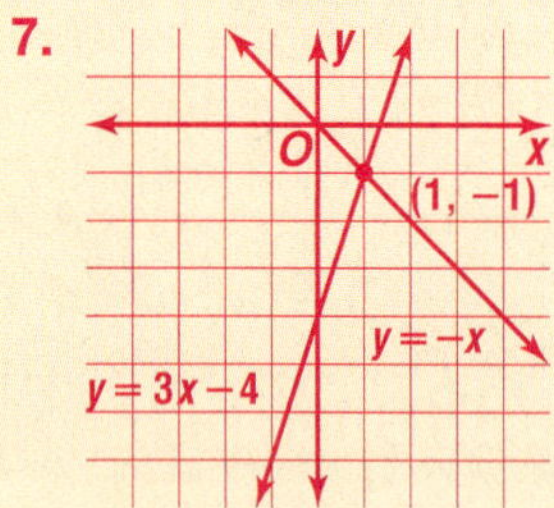

8.

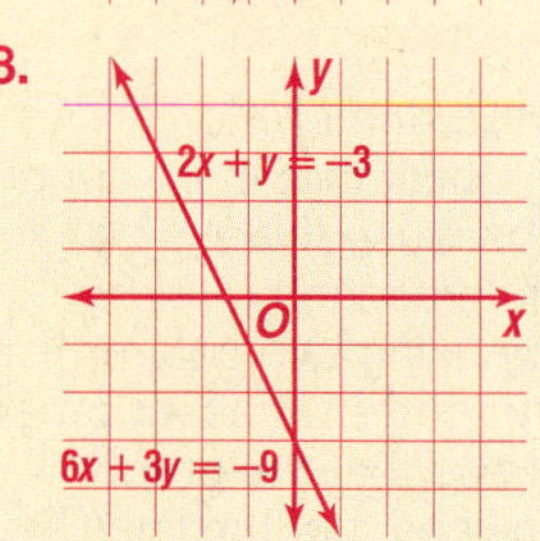

***Skills Practice*, p. 579, and *Practice*, p. 580 (shown)**

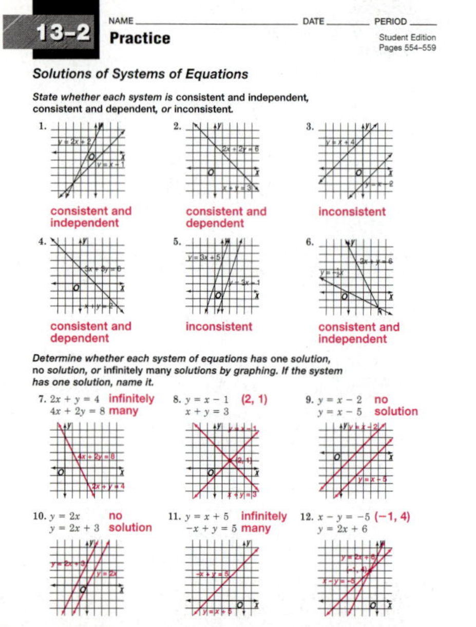

13-2 NAME ______ DATE ______ PERIOD ______

Practice Student Edition Pages 554–559

Solutions of Systems of Equations

State whether each system is consistent and independent, consistent and dependent, or inconsistent.

1. consistent and independent
2. consistent and dependent
3. inconsistent
4. consistent and dependent
5. inconsistent
6. consistent and independent

Determine whether each system of equations has one solution, no solution, or infinitely many solutions by graphing. If the system has one solution, name it.

7. $2x + y = 4$, $4x + 2y = 8$ infinitely many
8. $y = x - 1$, $x + y = 3$ (2, 1)
9. $y = x - 2$, $y = x - 5$ no solution
10. $y = 2x$, $y = 2x + 3$ no solution
11. $y = x + 5$, $-x + y = 5$ infinitely many
12. $x - y = -5$, $y = 2x + 6$ (−1, 4)

Examples 4 & 5 **Determine whether each system of equations has *one* solution, *no* solution, or *infinitely many* solutions by graphing. If the system has one solution, name it.** **6–8. See margin for graphs.**

6. $y = x$
$y = x + 5$
no solution

7. $y = -x$
$y = 3x - 4$ **(1, −1)**

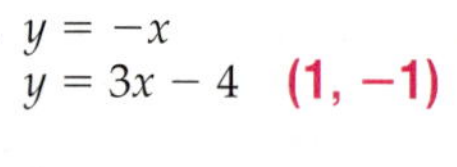

8. $2x + y = -3$
$6x + 3y = -9$
infinitely many

Example 6 9. **Animals** A dog sees a cat 60 feet away and starts running after it at 50 feet per second. At the same time, the cat runs away at 30 feet per second. This situation can be represented by the following system of equations and the graph at the right.

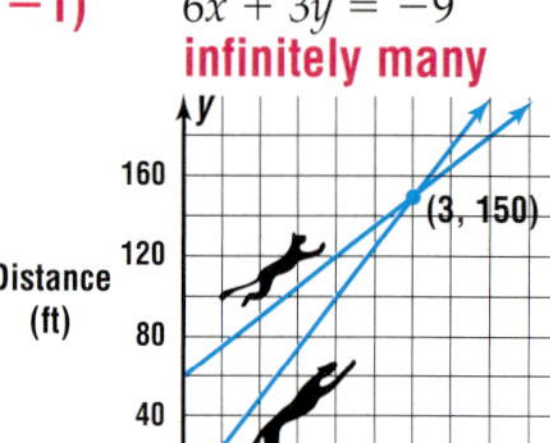

$y = 50x$ ← *dog* $y = 30x + 60$ ← *cat*

a. Is this system of equations *consistent and independent*, *consistent and dependent*, or *inconsistent*? Explain.
a. Consistent and independent; there is one solution, (3, 150).

b. Explain what the point at (3, 150) represents.
After 3 seconds, both the cat and dog will be 150 feet from the dog's original starting position.

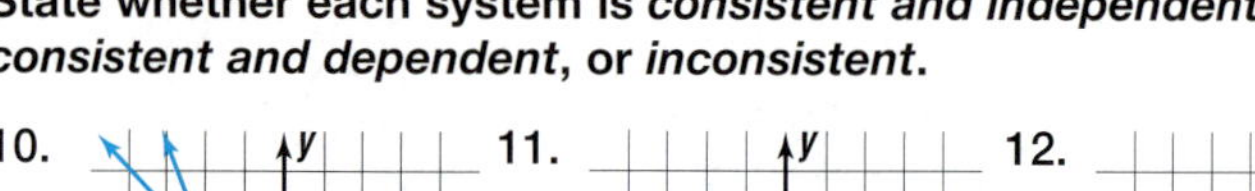

Practice

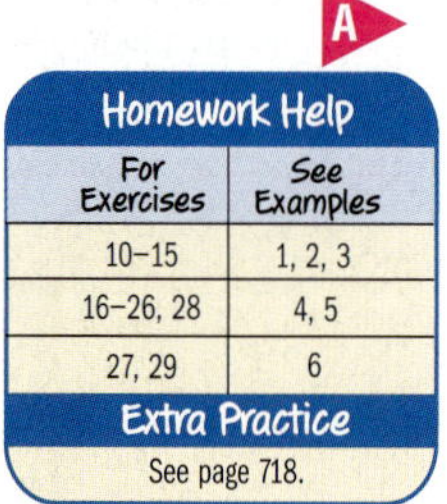

Homework Help	
For Exercises	See Examples
10–15	1, 2, 3
16–26, 28	4, 5
27, 29	6
Extra Practice	
See page 718.	

State whether each system is *consistent and independent*, *consistent and dependent*, or *inconsistent*.

A

10.

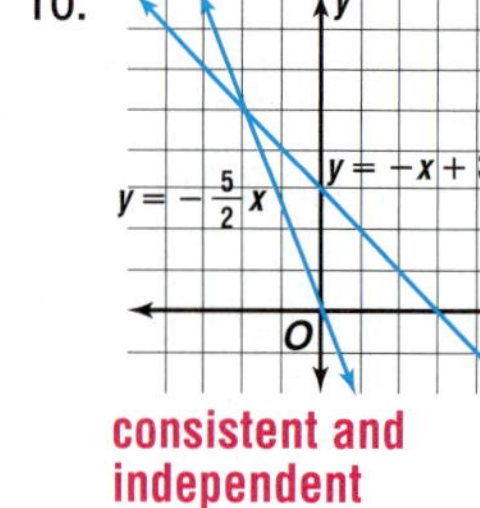

consistent and independent

11.

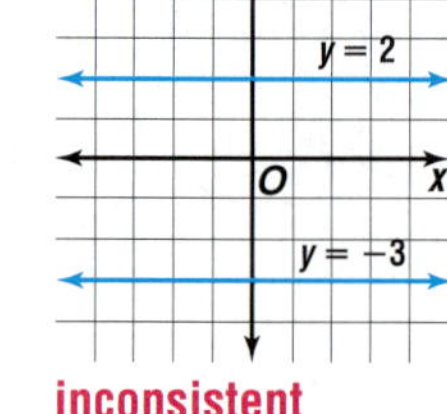

inconsistent

12.

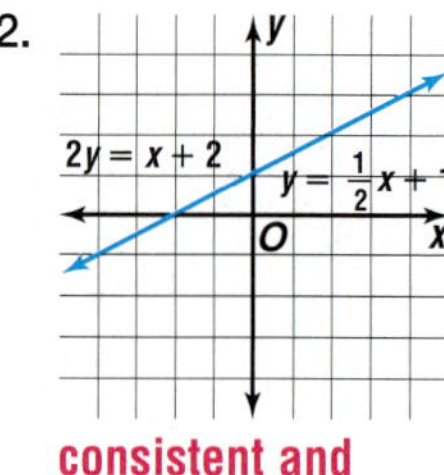

consistent and dependent

13. (graph: $x + y = -2$, $4x + 4y = -8$)

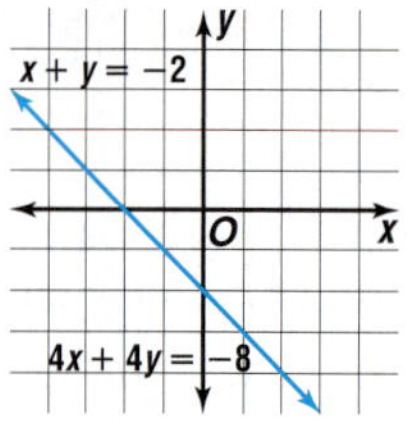

14.

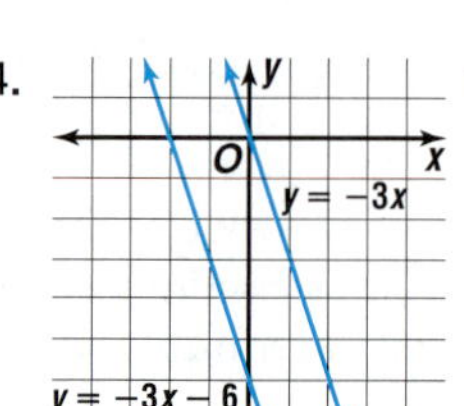

15.

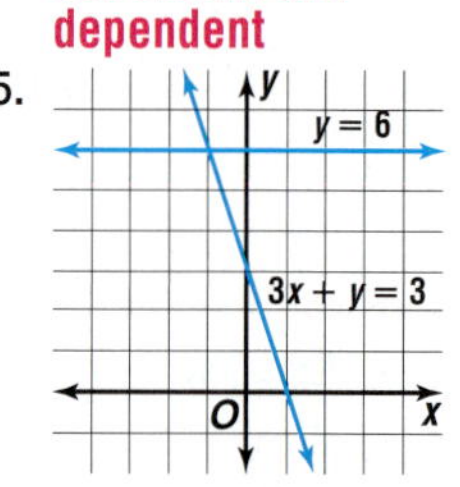

13. consistent and dependent
14. inconsistent
15. consistent and independent

Determine whether each system of equations has *one* solution, *no* solution, or *infinitely many* solutions by graphing. If the system has one solution, name it. **16–24. See Solutions Manual for graphs.**

16. $y = 4x - 6$
$y = 4x - 1$
no solution

17. $y = 3x - 2$
$4y = 12x - 8$
infinitely many

18. $y = \frac{1}{2}x$
$y = -2x + 5$ **(2, 1)**

19. $y = -x + 3$
$y = \frac{1}{5}x - 3$ **(5, −2)**

20. $x = 3$
$2x - 3y = 0$ **(3, 2)**

21. $3x - 2y = -6$
$3x - 2y = 6$
no solution

22. $x + y = 5$
$2x + 2y = 8$
no solution

23. $6x + y = -3$
$-x + y = 4$ **(−1, 3)**

24. $x - 4y = -4$
$\frac{1}{2}x - 2y = -2$
infinitely many

25. Does the system $x - y = 4$ and $x - 3y = 2$ have *one* solution, *no* solution, or *infinitely many* solutions? If the system has one solution, name it. **one; (5, 1)**

26. Without graphing, determine whether the system $x - 3y = 11$ and $2x - 6y = -5$ has *one* solution, *no* solution, or *infinitely many* solutions. Explain how you know. **See margin.**

Applications and Problem Solving

27. **Animals** Refer to Exercise 9. Suppose the dog is chasing another dog whose distance y can be represented by the equation $y = 50x + 20$. What is the solution? Explain what it represents in terms of the dogs. **See margin.**

28. **Ballooning** A hot air balloon is 10 meters above the ground and rising at a rate of 15 meters per minute. Another balloon is 150 meters above the ground and descending at a rate of 20 meters per minute.
    a. Write a system of equations to represent the balloons.
    b. What is the solution of the system of equations? **(4, 70)**
    c. Explain what the solution means. **See margin.**
    a. $y = 15x + 10$ **and** $y = -20x + 150$

29. **Critical Thinking** Write an equation of a line in slope-intercept form that, together with the equation $x + 3y = 9$, forms a system that is inconsistent. **Sample answer:** $y = -\frac{1}{3}x$

Mixed Review

30. What is the solution of the system $y = x + 4$ and $y = -3x - 4$? *(Lesson 13–1)* **(−2, 2)**

31. Graph $y \geq 2x + 1$. *(Lesson 12–7)* **See margin.**

32. **Number Theory** Find two numbers if one number is 6 less than the other and whose product is 7. *(Lesson 11–5)* **1, 7; −7, −1**

Factor each polynomial. If the polynomial cannot be factored, write *prime*. *(Lesson 10–2)*

33. $4x - 8$ **$4(x - 2)$**

34. $13x + 2m$ **prime**

35. $3a^2b^2 + 6ab - 9a$ **$3a(ab^2 + 2b - 3)$**

Standardized Test Practice

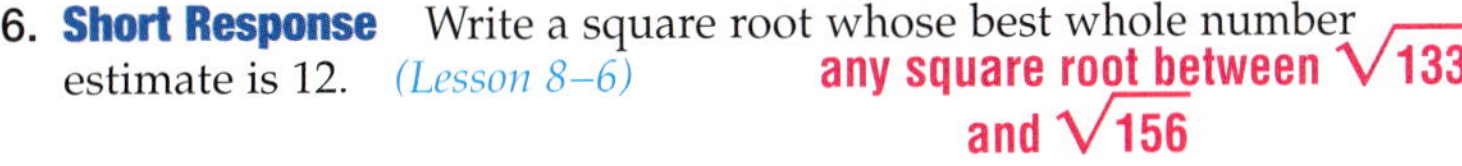

36. **Short Response** Write a square root whose best whole number estimate is 12. *(Lesson 8–6)* **any square root between $\sqrt{133}$ and $\sqrt{156}$**

www.algconcepts.com/self_check_quiz

4 ASSESS

Open-Ended Assessment

Writing Ask students to write, in their own words, explanations of systems of equations that are *consistent and independent*, *consistent and dependent*, and *inconsistent.*

Answers

26. No solution; the graphs have the same slope and different y-intercepts, so they are parallel lines.

27. There is no solution because the graphs of the equations are parallel lines. The second dog never catches up with the first.

28c. After 4 minutes, the balloons will be at the same height, 70 meters.

31.

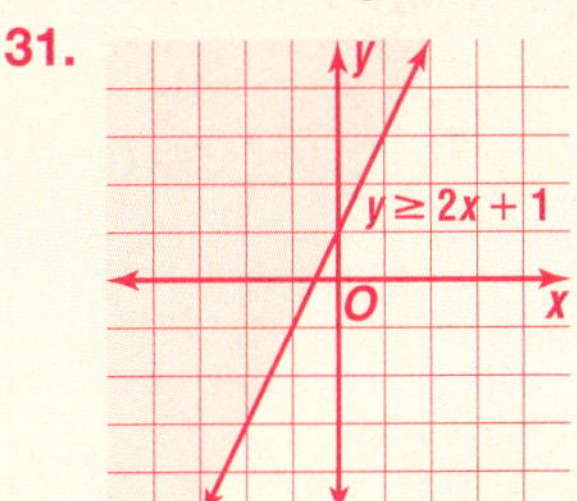

Extra Credit

Here are three linear equations.

$x - 3y = -3$
$3x - 2y = -10$
$4y = 6x - 4$

Which two of the equations would result in a system of equations that is inconsistent? **$3x - 2y = -10$ and $4y = 6x - 4$**

***Enrichment*, p. 582**

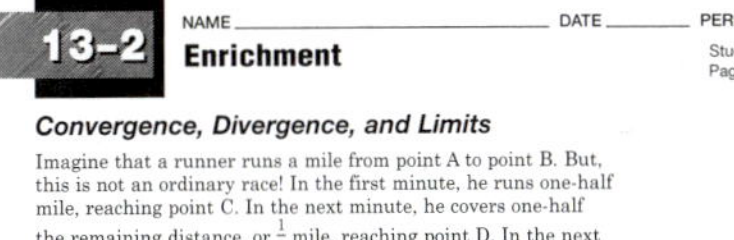

NAME ______ DATE ______ PERIOD ______

13-2 Enrichment

Student Edition Pages 554–559

Convergence, Divergence, and Limits

Imagine that a runner runs a mile from point A to point B. But, this is not an ordinary race! In the first minute, he runs one-half mile, reaching point C. In the next minute, he covers one-half the remaining distance, or $\frac{1}{4}$ mile, reaching point D. In the next minute he covers one-half the remaining distance, or $\frac{1}{8}$ mile, reaching point E.

0 mile, $\frac{1}{2}$ mile, $\frac{3}{4}$ mile, $\frac{7}{8}$ mile, 1 mile
A (3:00 P.M.), C (3:01 P.M.), D (3:02 P.M.), E (3:03 P.M.), B (?)

In this strange race, the runner approaches closer and closer to point B, but never gets there. However close he is to B, there is still some distance remaining, and in the next minute he can cover only half of that distance.

This race can be modeled by the infinite sequence
$\frac{1}{2}, \frac{3}{4}, \frac{7}{8}, \frac{15}{16}, \cdots$.

The terms of the sequence get closer and closer to 1. An infinite sequence that gets arbitrarily close to some number is said to **converge** to that number. The number is the limit of the sequence.

Not all infinite sequences converge. Those that do not are called **divergent.**

Write C if the sequence converges and D if it diverges. If the sequence converges, make a reasonable guess for its limit.

1. $2, 4, 6, 8, 10, \cdots$ **D**
2. $0, 3, 0, 3, 0, 3, \cdots$ **D**
3. $1, \frac{1}{2}, \frac{1}{3}, \frac{1}{4}, \frac{1}{5}, \cdots$ **C, 0**
4. $0.9, 0.99, 0.999, 0.9999, \cdots$ **C, 1**
5. $-5, 5, -5, 5, -5, 5, \cdots$ **D**
6. $0.1, 0.2, 0.3, 0.4, \cdots$ **D**

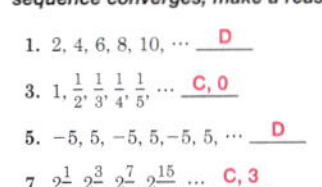

7. $2\frac{1}{4}, 2\frac{3}{4}, 2\frac{7}{8}, 2\frac{15}{16}, \cdots$ **C, 3**
8. $6, 5\frac{1}{2}, 5\frac{1}{3}, 5\frac{1}{4}, 5\frac{1}{5}, \cdots$ **C, 5**
9. $1, 4, 9, 16, 25, \cdots$ **D**
10. $1, -\frac{1}{2}, \frac{1}{3}, -\frac{1}{4}, \frac{1}{5}, -\frac{1}{6}, \cdots$ **C, 0**
11. Create one convergent sequence and one divergent sequence. Give the limit for your convergent sequence. **See students' work.**

Lesson 13–3

1 FOCUS

5-Minute Check
Lesson 13–2

State whether each system is consistent and independent, consistent and dependent, or inconsistent.

1.
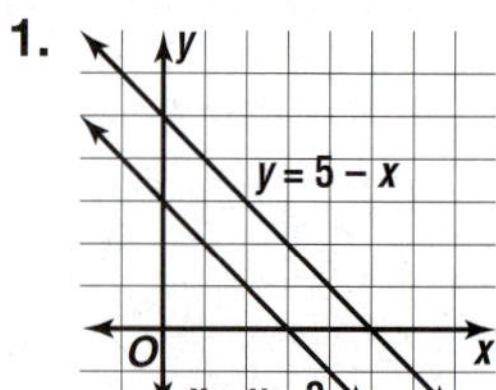

inconsistent

2.
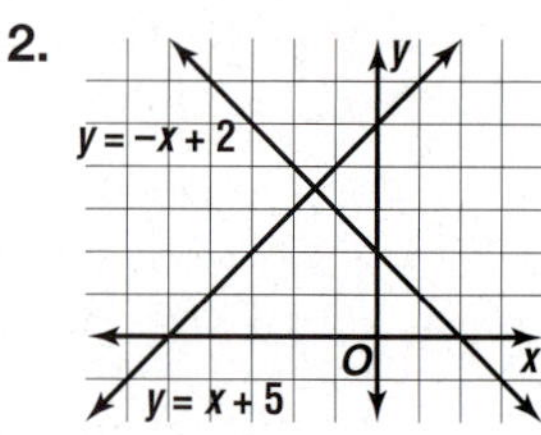

consistent and independent

Determine whether each system of equations has one solution, no solution, or infinitely many solutions by graphing. If the system has one solution, name it.

3. $3x - 4y = 2$ **infinitely many**
$y = \frac{3}{4}x - \frac{1}{2}$

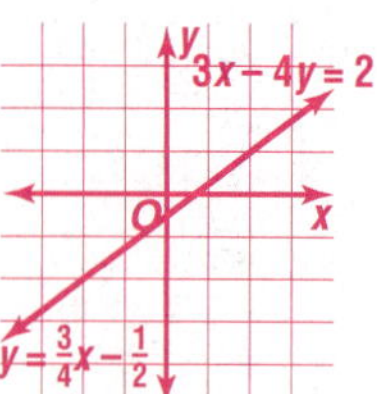

4. $x + y = 3$
$y = 6$ **one solution; (−3, 6)**

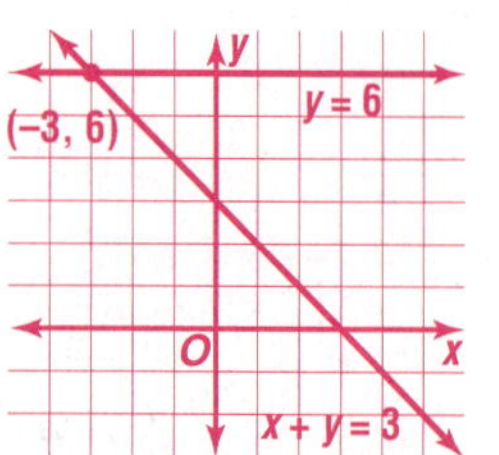

13–3 Substitution

What You'll Learn
You'll learn to solve systems of equations by the substitution method.

Why It's Important
Metallurgy Systems of equations can be used to make metal alloys.
See Example 6.

Look Back
Solving Equations with Algebra Tiles: Lesson 3–5

In Lesson 13–2, you learned to solve systems of equations by graphing. But sometimes the exact coordinates of the point where lines intersect cannot be easily determined from a graph. The solution of a system can also be found by using an algebraic method called **substitution**.

Materials:  algebra tiles  equation mat

Use a model to solve the system $y = x + 1$ and $3x + y = 9$.

Step 1 Let a green tile represent x. Then, since $y = x + 1$, 1 green tile and 1 yellow tile represent y.

$x =$ [x] $y =$ [x] [1]

Step 2 Represent $3x + y = 9$ on the equation mat. On one side of the mat, place three green tiles for $3x$ and 1 green tile and 1 yellow tile for y. On the other side, place 9 yellow tiles.

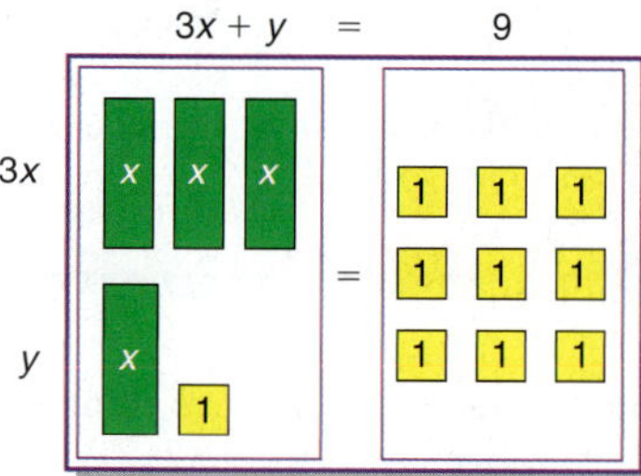

Try These

1. Use what you know about equation mats and zero pairs to solve the equation. What is the value of x? **2**
2. Use the value of x from Exercise 1 and the equation $y = x + 1$ to find the value of y. **3**
3. What is the solution of the system of equations? **(2, 3)**

Resource Manager

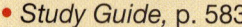

Reproducible Masters
Chapter 13 Resource Masters
- *Study Guide,* p. 583
- *Skills Practice,* p. 584
- *Practice,* p. 585
- *Reading to Learn Mathematics,* p. 586
- *Enrichment,* p. 587
- *Assessment,* p. 619

Hands-On Algebra, pp. 146–147

Transparencies
5-Minute Check, 13–3

Technology/Multimedia
AlgePASS, Lesson 31
Interactive Chalkboard CD-ROM

Examples

Use substitution to solve each system of equations.

$y = 2x$
$3x + y = 5$

The first equation tells you that y is equal to $2x$. So, substitute $2x$ for y in the second equation. Then solve for x.

$3x + y = 5$	*Original equation*
$3x + 2x = 5$	*Replace y with 2x.*
$5x = 5$	*Simplify.*
$\frac{5x}{5} = \frac{5}{5}$	*Divide each side by 5.*
$x = 1$	*Simplify.*

Now substitute 1 for x in either equation and solve for y. *Choose the equation that is easier to solve.*

$y = 2x$	*Original equation*
$y = 2(1)$ or 2	*Replace x with 1.*

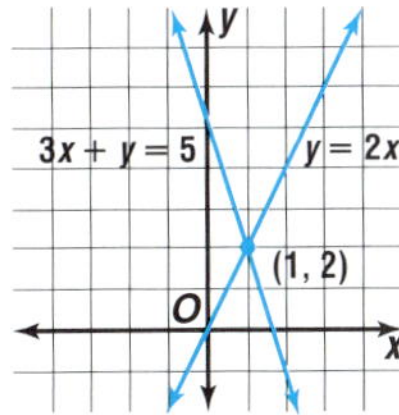

The solution of this system of equations is (1, 2). You can see from the graph that the solution is correct. *You can also check by substituting (1, 2) into each of the original equations.*

$x + y = 1$
$x = y + 6$

Substitute $y + 6$ for x in the first equation. Then solve for y.

$x + y = 1$	*Original equation*
$y + 6 + y = 1$	*Replace x with y + 6.*
$2y + 6 = 1$	*Simplify.*
$2y + 6 - 6 = 1 - 6$	*Subtract 6 from each side.*
$2y = -5$	*Simplify.*
$\frac{2y}{2} = \frac{-5}{2}$	*Divide each side by 2.*
$y = -\frac{5}{2}$	*Simplify.*

Now substitute $-\frac{5}{2}$ for y in either equation and solve for x.

$x = y + 6$	*Original equation*
$x = -\frac{5}{2} + 6$	*Replace y with $-\frac{5}{2}$.*
$x = \frac{7}{2}$	*Simplify.*

The solution of this system of equations is $\left(\frac{7}{2}, -\frac{5}{2}\right)$. *Check by substituting $\left(\frac{7}{2}, -\frac{5}{2}\right)$ into each of the original equations.*

Motivating the Lesson

Hands-On Activity Have a volunteer write a linear equation in x and y on the board or overhead. Ask the class to state any linear equation in x and y in the form "$x = \ldots$" or "$y = \ldots$". Then have the volunteer rewrite the equation on the board by substituting the new value for x or y. Help students see that the written equation, now in one variable, can easily be solved.

2 TEACH

Teaching Tip To reinforce the statement in the paragraph above the Hands-On Algebra activity on page 560, have students graph the equations for Anita and Tionna on the same screen of a graphing calculator.

In-Class Examples

Examples 1–2

Use substitution to solve each system of equations.

1 $x = 2y$
$2x - 3y = 5$ **(10, 5)**

2 $5x + y = 1$
$y = 3x - 3$ $\left(\frac{1}{2}, -\frac{3}{2}\right)$

Hands-On Algebra

Cooperative Learning Refer to the Hands-On Algebra on page 560. Be prepared with some additional systems that students can solve using algebra tiles. If the rectangular tile is used to represent the variable x, then one equation should be in the form "$y = \ldots$" so the variable y can be represented by a combination of x tiles and unit tiles. Some other systems are $y = 2x + 1$ and $2x + 5y = 29$ [solution: (2, 5)] and $y = x + 2$ and $4x + 2y = 10$ [solution: (1, 3)].

An additional Hands-On Algebra activity using a chart to organize the relationships in a mixture problem is available in the *Hands-On Algebra Masters,* p. 147.

Hands-On Algebra Masters, p. 146

In-Class Examples
Examples 3–4

Use substitution to solve each system of equations.

3 $-2x + y = -7$
$3x - 2y = 12$ **(2, −3)**

4 $3x + 1 = y$
$9x - 3y = -3$ **infinitely many**

Example 3

$x - 3y = 3$
$2x - y = 11$

Solve the first equation for x since the coefficient of x is 1.

$x - 3y = 3 \rightarrow x = 3 + 3y$

Next, find the value of y by substituting $3 + 3y$ for x in the second equation.

$$2x - y = 11$$
$$2(3 + 3y) - y = 11$$
$$6 + 6y - y = 11$$
$$6 + 5y = 11$$
$$6 + 5y - 6 = 11 - 6$$
$$5y = 5$$
$$\frac{5y}{5} = \frac{5}{5}$$
$$y = 1$$

Now substitute 1 for y in either equation and solve for x.

$$x - 3y = 3$$
$$x - 3(1) = 3 \quad \textit{Replace y with 1.}$$
$$x - 3 = 3$$
$$x - 3 + 3 = 3 + 3$$
$$x = 6$$

The solution is (6, 1).

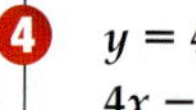

Your Turn

a. $x = 1 + 6y$
$x + 2y = 9$ **(7, 1)**

b. $x + y = 3$
$-2x - 7y = 4$ **(5, −2)**

In Lesson 13–2, you learned how to tell whether a system has one solution, no solution, or infinitely many solutions by looking at the graph. You can also determine this information algebraically.

Examples

Use substitution to solve each system of equations.

4

$y = 4x + 1$
$4x - y = 7$

Find the value of x by substituting $4x + 1$ for y in the second equation.

$$4x - y = 7 \quad \textit{Original equation}$$
$$4x - (4x + 1) = 7 \quad \textit{Replace y with 4x + 1.}$$
$$4x - 4x - 1 = 7 \quad \textit{Distributive Property}$$
$$-1 = 7 \quad \textit{Simplify.}$$

The statement $-1 = 7$ is false. This means that there are no ordered pairs that are solutions to both equations. Compare the slope-intercept forms of the equations, $y = 4x + 1$ and $y = 4x - 7$. Notice that the graphs of these equations have the same slope but different y-intercepts. Thus, the lines are parallel, and the system has no solution.

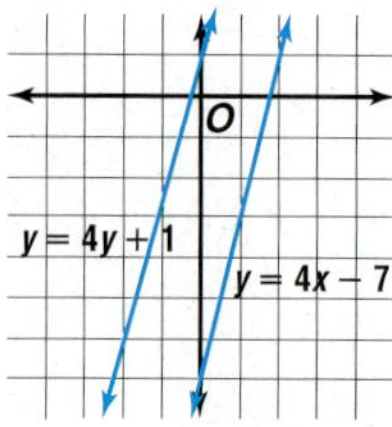

5 $x = 3 - 2y$
$2x + 4y = 6$

$2x + 4y = 6$	*Original equation*
$2(3 - 2y) + 4y = 6$	*Replace x with 3 − 2y.*
$6 - 4y + 4y = 6$	*Distributive Property*
$6 = 6$	*Simplify.*

The statement $6 = 6$ is true. This means that an ordered pair for any point on either line is a solution to both equations. The system has infinitely many solutions.

Your Turn

Use substitution to solve each system of inequalities.

c. $y = 2x + 1$
$4x - 2y = -9$ **no solution**

d. $x - 6y = 5$
$2x = 12y + 10$ **infinitely many**

Example **6** **Metals Link**

Systems of equations can be used to solve mixture problems.

A certain metal alloy is 25% copper. Another metal alloy is 50% copper. How much of each alloy should be used to make 1000 grams of a metal alloy that is 45% copper?

Explore Let a = the number of grams of the 25% copper alloy.
Let b = the number of grams of the 50% copper alloy.

	25% Copper	50% Copper	45% Copper
Total Grams	a	b	1000
Grams of Copper	$0.25a$	$0.50b$	0.45(1000)

Prerequisite Skills Review
Decimals and Percents, p. 689

Plan Write two equations to represent the information.

$a + b = 1000$ ← *total grams of alloy*
$0.25a + 0.50b = 0.45(1000)$ ← *grams of copper*

Solve Use substitution to solve this system. Since $a + b = 1000$, $a = 1000 - b$.

$0.25a + 0.50b = 0.45(1000)$	*Original equation*
$0.25(1000 - b) + 0.50b = 0.45(1000)$	*Replace a with 1000 − b.*
$250 - 0.25b + 0.50b = 450$	*Distributive Property*
$250 + 0.25b = 450$	*Simplify.*
$250 + 0.25b - 250 = 450 - 250$	*Subtract 250.*
$0.25b = 200$	*Simplify.*
$\frac{0.25b}{0.25} = \frac{200}{0.25}$	*Divide each side by 0.25.*
$b = 800$	*Simplify.*

(continued on the next page)

In-Class Examples

Example 5

Use substitution to solve the system of equations.
$y = 3x - 5$
$6x - 2y = 4$ **no solution**

Example 6

One kind of cat litter is 20% gravel. Another litter is 10% gravel. How much of each kind of litter should be used to make 30 tons of litter that is 18% gravel? **24 tons of 20% litter, 6 tons of 10% litter**

Study Guide, p. 583

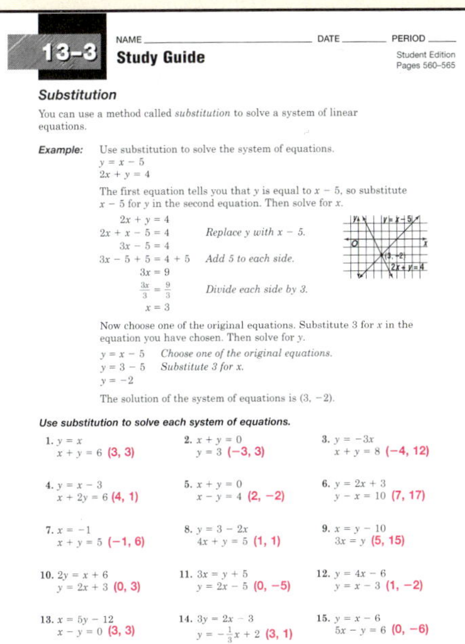

NAME ________ DATE ______ PERIOD ______

13-3 Study Guide

Student Edition Pages 560–565

Substitution

You can use a method called *substitution* to solve a system of linear equations.

Example: Use substitution to solve the system of equations.
$y = x - 5$
$2x + y = 4$

The first equation tells you that y is equal to $x - 5$, so substitute $x - 5$ for y in the second equation. Then solve for x.

$2x + y = 4$	
$2x + x - 5 = 4$	*Replace y with x − 5.*
$3x - 5 = 4$	
$3x - 5 + 5 = 4 + 5$	*Add 5 to each side.*
$3x = 9$	
$\frac{3x}{3} = \frac{9}{3}$	*Divide each side by 3.*
$x = 3$	

Now choose one of the original equations. Substitute 3 for x in the equation you have chosen. Then solve for y.

$y = x - 5$ *Choose one of the original equations.*
$y = 3 - 5$ *Substitute 3 for x.*
$y = -2$

The solution of the system of equations is $(3, -2)$.

Use substitution to solve each system of equations.

1. $y = x$; $x + y = 6$ **(3, 3)**
2. $x + y = 0$; $y = 3$ **(−3, 3)**
3. $y = -3x$; $x + y = 8$ **(−4, 12)**
4. $y = x - 3$; $x + 2y = 6$ **(4, 1)**
5. $x + y = 0$; $x - y = 4$ **(2, −2)**
6. $y = 2x + 3$; $y - x = 10$ **(7, 17)**
7. $x = -1$; $x + y = 5$ **(−1, 6)**
8. $y = 3 - 2x$; $4x + y = 5$ **(1, 1)**
9. $x = y - 10$; $3x = y$ **(5, 15)**
10. $2y = x + 6$; $y = 2x + 3$ **(0, 3)**
11. $3x = y + 5$; $y = 2x - 5$ **(0, −5)**
12. $y = 4x - 6$; $y = x - 3$ **(1, −2)**
13. $x = 5y - 12$; $x - y = 0$ **(3, 3)**
14. $3y = 2x - 3$; $y = -\frac{1}{3}x + 2$ **(3, 1)**
15. $y = x - 6$; $5x - y = 6$ **(0, −6)**

3 PRACTICE/APPLY

Error Analysis

Watch for students who, in Exercise 4, replace y with $x - 4$ in the first equation, get the result "$0 = 0$," and stop.
Prevent by stressing that when they solve for a variable in one equation, the substitution must take place in the other equation. Also, point out that the true result "$0 = 0$" usually suggests that they made this kind of mistake.

Assignment Guide

Basic: 11–31 odd, 33–39
Average: 12–30 even, 31–39
All: Quiz 1, 1–5

Answers

1. when the exact coordinates are not easily determined from the graph
2. The graphs are the same line, and there are infinitely many solutions.
3. Both are correct because you can substitute for either variable.

Skills Practice, **p. 584, and** ***Practice,*** **p. 585 (shown)**

13-3 NAME ______ DATE ______ PERIOD ______
Practice Student Edition Pages 560–565

Substitution

Use substitution to solve each system of equations.

1. $y = x + 8$ $x + y = 2$ **(−3, 5)**	2. $y = 2x$ $5x - y = 9$ **(3, 6)**	3. $y = x + 2$ $3x + 3y = 6$ **(0, 2)**
4. $x = 3y$ $2x + 4y = 10$ **(3, 1)**	5. $x = y + 9$ $x + y = -7$ **(1, −8)**	6. $y = 2x + 1$ $2x - y = 3$ **no solution**
7. $x = 3y$ $2x + 3y = 15$ $\left(5, \frac{5}{3}\right)$	8. $x - 2y = 4$ $3x = 6y + 12$ **infinitely many**	9. $x = 5y - 2$ $2x + 2y = 4$ $\left(\frac{4}{3}, \frac{2}{3}\right)$
10. $4y + 2x = 24$ $x = 3y + 2$ **(8, 2)**	11. $y = 3x + 8$ $4x + 2y = 6$ **(−1, 5)**	12. $x = 3y + 10$ $2x + 2y = -12$ **(−2, −4)**
13. $x + 2y = -4$ $-2x - 3y = 9$ **(−6, 1)**	14. $5x + 2y = 7$ $4x + y = 8$ **(3, −4)**	15. $x = 2y + 11$ $3x + 2y = 9$ **(5, −3)**
16. $x - 2y = -7$ $5x - 7y = -8$ **(11, 9)**	17. $6x - 4y = -5$ $2x + y = 3$ $\left(\frac{1}{2}, 2\right)$	18. $x + 3y = 10$ $4x - 5y = 6$ **(4, 2)**

Now substitute 800 for b in either equation and solve for a.

$$a + b = 1000 \quad \textit{Original equation}$$
$$a + 800 = 1000 \quad \textit{Replace b with 800.}$$
$$a + 800 - 800 = 1000 - 800 \quad \textit{Subtract 800 from each side.}$$
$$a = 200 \quad \textit{Simplify.}$$

So, 200 grams of the 25% copper alloy and 800 grams of the 50% copper alloy should be used.

Examine There should be more of the 50% alloy than the 25% alloy to make a 45% alloy. So the solution (200, 800) is reasonable.

Check for Understanding

Communicating Mathematics

Vocabulary
substitution

1. **Explain** when you might use substitution instead of graphing to solve a system of equations. **1–3. See margin.**
2. **State** what you would conclude if the solution of a system of equations yields the equation $4 = 4$.
3. **You Decide** Faith and Todd are using substitution to solve the system $x + 3y = 8$ and $4x - y = 9$. Faith says that the first step is to solve for x in $x + 3y = 8$. Todd disagrees. He says to solve for y in $4x - y = 9$. Who is right and why?

Guided Practice

Examples 1–5

Use substitution to solve each system of equations.

4. $y = x - 4$
$3x + 2y = 2$ **(2, −2)**
5. $y = 3x$
$7x - y = 16$ **(4, 12)**
6. $x = 2y$
$4x + 2y = 15$ $\left(3, \frac{3}{2}\right)$
7. $3x - 7y = 12$
$x - 2y = 4$ **(4, 0)**
8. $x = 2y + 5$
$3x - 6y = 15$ **infinitely many**
9. $4y = -3x + 8$
$3x + 4y = 6$ **no solution**

Example 6

10. **Mixtures** MX Labs needs to make 500 gallons of a 34% acid solution. The only solutions available are 25% acid and 50% acid. How many gallons of each should be mixed to make the 34% solution? **320 gal of 25% solution, 180 gal of 50% solution**

Exercises

Practice

Homework Help

For Exercises	See Examples
11, 13, 15–17, 31	1, 4
12, 14, 18	2, 5
19–30	3

Extra Practice
See page 718.

15. no solution **20. infinitely many**

Use substitution to solve each system of equations.

11. $y = x$
$3x + y = 4$ **(1, 1)**
12. $x = 2y$
$x + y = 3$ **(2, 1)**
13. $y = 3x$ **(−3, −9)**
$x + 2y = -21$
14. $2y = x$
$x - y = 10$ **(20, 10)**
15. $y = 3x - 8$
$3x - y = 12$
16. $y = x + 7$
$x + y = 1$ **(−3, 4)**
17. $y = \frac{1}{2}x + 3$
$y - 5 = x$ **(−4, 1)**
18. $x = 7 - y$
$2x - y = 8$ **(5, 2)**
19. $x - 3y = -9$
$5x - 2y = 7$ **(3, 4)**
20. $x + y = 0$
$4x + 4y = 0$
21. $x - 2y = -2$
$5x - 4y = 2$ **(2, 2)**
22. $2x - y = 6$
$3x - 5y = 9$ **(3, 0)**

Reteaching Activity

Intrapersonal Learners Have students write a journal entry explaining why *substitution* is a good name for the method of solution described in Lesson 13-3.

26. $\left(-\frac{5}{3}, -\frac{11}{3}\right)$

28. $\left(\frac{23}{8}, -\frac{13}{8}\right)$

30. $(-9, -7)$

23. $x - y = 12$, $3x - y = 16$ **(2, −10)**
24. $4x - 3y = -6$, $x + 5y = 10$ **(0, 2)**
25. $x - 3y = 3$, $2x + 9y = 11$ $\left(4, \frac{1}{3}\right)$

B

26. $4x - y = -3$, $y + 2 = x$
27. $x - 6y = 5$, $2x - 12y = 10$ **infinitely many**
28. $x - 5y = 11$, $3x + y = 7$

29. Use substitution to solve $2x - y = -4$ and $-3x + y = -9$. **(13, 30)**

C

30. What is the solution of the system $x - 2y = 5$ and $3x - 5y = 8$?

Applications and Problem Solving

31. **Driving** Anita and Tionna were both driving the same route from college to their hometown. Anita left an hour before Tionna. Anita drove at an average speed of 55 miles per hour, and Tionna averaged 65 miles per hour.
    a. After how many hours did Tionna catch up with Anita? **5.5 hr**
    b. How many miles did she drive? **357.5 mi**

32. $\frac{2}{3}$ hr or 40 min

32. **Exercise** Laura and Ji-Yong were jogging on a 10-mile path. Laura had a 2-mile head start on Ji-Yong. If Laura ran at an average rate of 5 miles per hour and Ji-Yong ran at an average rate of 8 miles per hour, how long would it take for Ji-Yong to catch up with Laura?

33. See margin.

33. **Critical Thinking** Suppose you have three equations *A*, *B*, and *C*. Suppose a system containing any two of the equations is consistent and independent. Must the three equations together be consistent? Give an example or a counterexample to support your answer.

Mixed Review

34. Infinitely many; see margin for graph.
35. $(-1, 4)$; See margin for graph.
36. $x \leq -2$ or $x > 1$
37. $-3 < x \leq 2$

34. Determine whether the system $y = -2x$ and $2y + 4x = 0$ has *one* solution, *no* solution, or *infinitely many* solutions by graphing. If the system has one solution, name it. *(Lesson 13–2)*

35. Solve $x - y = -5$ and $x + y = 3$ by graphing. *(Lesson 13–1)*

Write a compound inequality for each solution set shown. *(Lesson 12–5)*

36. [number line from −5 to 5]
37. 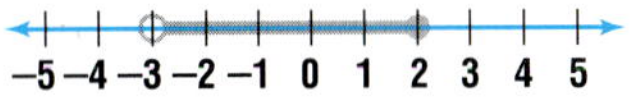 [number line from −5 to 5]

Standardized Test Practice

38. **Short Response** Solve $-5 + x > 12.8$. *(Lesson 12–2)* $x > 17.8$

39. **Grid In** Mr. Drew is a salesperson. The formula for his daily income is $t(s) = 0.15s + 25$, where s is his total sales for the day. Suppose Mr. Drew earns \$94 on Monday. What were his total sales in dollars for that day? *(Lesson 6–4)* **460**

Quiz 1 — Lessons 13–1 through 13–3

Determine whether each system of equations has *one* solution, *no* solution, or *infinitely many* solutions by graphing. If the system has one solution, name it. *(Lessons 13–1 & 13–2)* **1–4. See Solutions Manual for graphs. 4. infinitely many**

1. $y = x - 5$, $y = -\frac{1}{4}x$ **(4, −1)**
2. $y = 2x + 5$, $y = x + 3$ **(−2, 1)**
3. $x + y = 3$, $x + y = 1$ **no solution**
4. $y = -2x + 3$, $2y = -4x + 6$

5. **Mixtures** Ms. Williams mixed nuts that cost \$3.90 per pound with nuts that cost \$4.30 per pound to make a 50-pound mixture of nuts worth \$4.20 per pound. How many pounds of each type of nut did she use? *(Lesson 13–3)* **12.5 lb of the \$3.90, 37.5 lb of the \$4.30**

www.algconcepts.com/self_check_quiz

4 ASSESS

Open-Ended Assessment

Speaking Ask students to explain the steps they take to solve a system of equations by substitution.

Quiz 1

The Quiz provides students with a brief review of the concepts and skills in Lessons 13–1 through 13–3. Lesson numbers are given to the right of the exercises or instruction lines so students can review concepts not yet mastered.

Chapter 13, Quiz A (Lessons 13–1 through 13–3) is available in the *Chapter 13 Resource Masters*, p. 619.

Answers

33. No; the graphs of equations A, B, and C could form a triangle. Any two of the lines would be intersecting, but there is no point where all three lines intersect.

34.

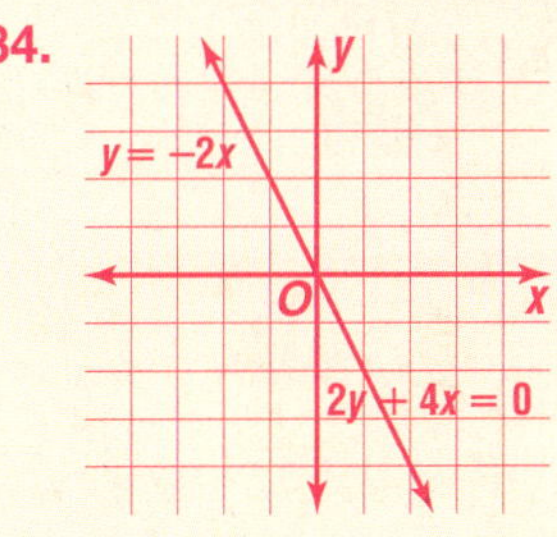

Enrichment, p. 587

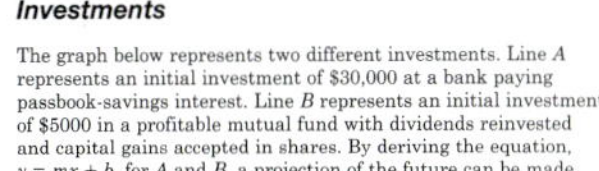

13-3 NAME ______ DATE ______ PERIOD ______

Enrichment

Student Edition Pages 560–565

Investments

The graph below represents two different investments. Line *A* represents an initial investment of \$30,000 at a bank paying passbook-savings interest. Line *B* represents an initial investment of \$5000 in a profitable mutual fund with dividends reinvested and capital gains accepted in shares. By deriving the equation, $y = mx + b$, for *A* and *B*, a projection of the future can be made.

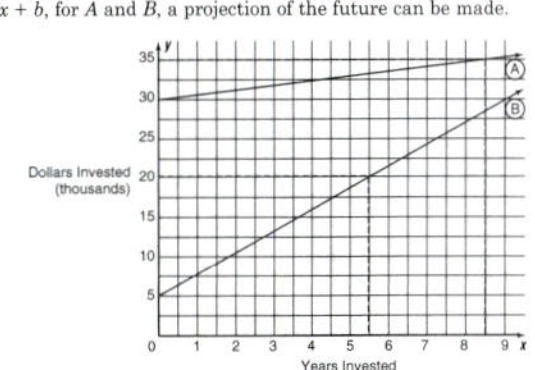

Solve.

1. The y-intercept, b, is the initial investment. Find b for each of the following.
   a. line *A* 30,000 b. line *B* 5000
2. The slope of the line, m, is the rate of return. Find m for each of the following.
   a. line *A* $\frac{35{,}000 - 30{,}000}{8.5 - 0} \approx 588$ b. line *B* $\frac{20{,}000 - 5{,}000}{5.5 - 0} \approx 2727$
3. What are the equations of each of the following lines?
   a. line *A* $y = 588x + 30{,}000$ b. line *B* $y = 2727x + 5000$

Answer each of the following, assuming that the growth of each investment continues in the same pattern.

4. What will be value of the mutual fund after the 11th year? $y = 2727(11) + 5000 = \$34{,}997$
5. What will be the value of the bank account after the 11th year? $y = 588(11) + 30{,}000 = \$36{,}468$
6. When will the mutual fund and the bank account be of equal value? $588x + 30{,}000 = 2727x + 5000 \rightarrow x \approx 11.7$ years
7. In the long term, which investment has the greater payoff? mutual fund

Extra Credit

The ordered pair $(5, -7)$ is the solution to the following system of equations.

$2x + ay = 3$
$bx - 3y = 1$

Find the values of a and b. $a = 1, b = -4$

Answer

35.

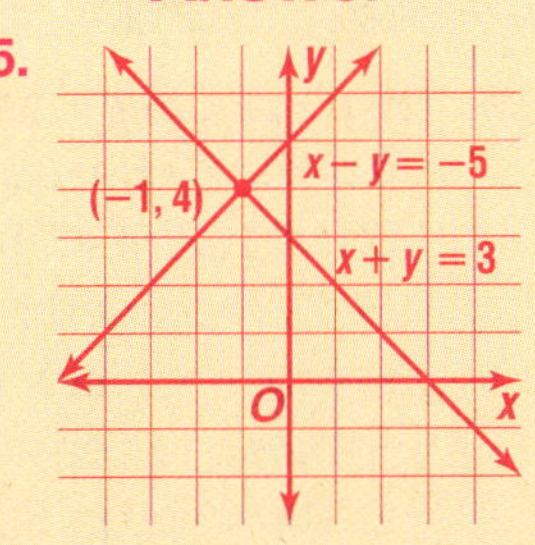

Lesson 13-4

1 FOCUS

5-Minute Check
Lesson 13–3

Use the following system of equations.
$7x - y = 1$
$2x + 2y = 6$

1. Rewrite the first equation so it is in the form "$y = \ldots$". $y = 7x - 1$
2. Substitute your value of y into the second equation and solve the resulting equation for x. $x = \frac{1}{2}$
3. Find the solution to the original system of equations. $\left(\frac{1}{2}, 2\frac{1}{2}\right)$
4. Graph the equations to check your solution.

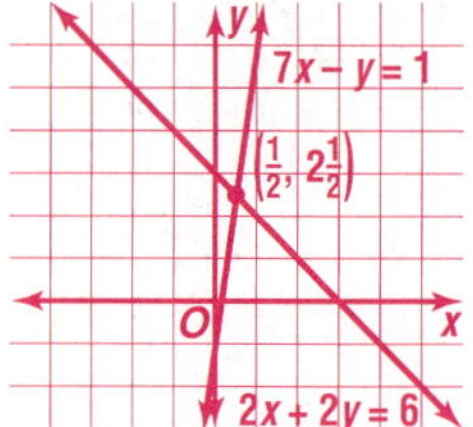

5. Use substitution to solve the following system of equations. Graph the equations to check your solution.
$5x + 3y = -3$
$5x + y = 4$ $\left(1\frac{1}{2}, -3\frac{1}{2}\right)$

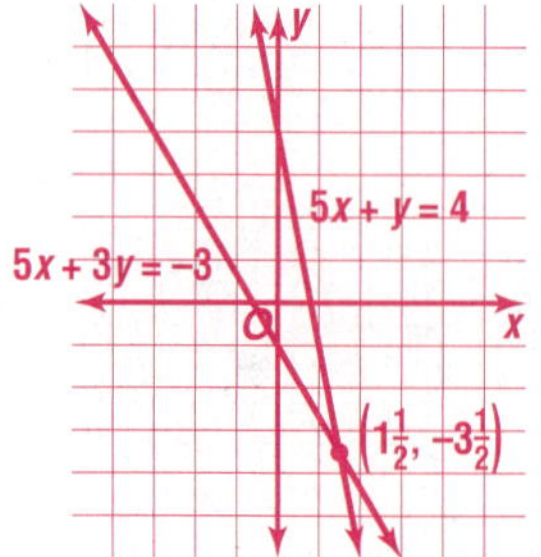

13-4 Elimination Using Addition and Subtraction

What You'll Learn
You'll learn to solve systems of equations by the elimination method using addition and subtraction.

Why It's Important
Entertainment
Solving systems of equations using elimination could be used to determine entertainment costs. *See Exercise 35.*

Another algebraic method for solving systems of equations is called **elimination**. You can eliminate one of the variables by adding or subtracting the equations.

Use elimination when the coefficients of one of the variables are equal or additive inverses. For example, consider the system below.

$$5x + 11y = 12 \qquad 2x + 11y = 36$$

Notice that the coefficient of y in both equations is the same. You can solve this system of equations in three steps.

Step 1: Subtract the two equations, so that y is eliminated. The result is an equation only in x: $3x = -24$.

Step 2: The next step is to solve for x: $x = -8$.

Step 3: Substitute the value of x in any one of the two original equations and solve for y: $y = \frac{52}{11}$.

Example 1 **Use elimination to solve the system of equations.**

$2x + y = 3$
$x + y = 1$

$2x + y = 3$	*Write the equations in column form.*
$(-)\ x + y = 1$	*Subtract the equations to eliminate the y terms.*
$x + 0 = 2$	
$x = 2$	The value of x is 2.

Now substitute in either equation to find the value of y. *Choose the equation that is easier for you to solve.*

$x + y = 1$	*Original equation*
$2 + y = 1$	*Replace x with 2.*
$2 + y - 2 = 1 - 2$	*Subtract 2 from each side.*
$y = -1$	The value of y is -1.

The solution of the system of equations is $(2, -1)$.

Check:

$2x + y = 3$	*Original equation*
$2(2) + (-1) \stackrel{?}{=} 3$	*Replace (x, y) with (2, −1).*
$4 + (-1) \stackrel{?}{=} 3$	*Multiply.*
$3 = 3$	✓
$x + y = 1$	*Original equation*
$2 + (-1) \stackrel{?}{=} 1$	*Replace (x, y) with (2, −1).*
$1 = 1$	✓

Resource Manager

Reproducible Masters
Chapter 13 Resource Masters
- *Study Guide,* p. 588
- *Skills Practice,* p. 589
- *Practice,* p. 590
- *Reading to Learn Mathematics,* p. 591
- *Enrichment,* p. 592
- *Assessment,* p. 618

Transparencies
5-Minute Check, 13–4

Technology/Multimedia
Interactive Chalkboard CD-ROM

Your Turn Use elimination to solve each system of equations.

a. $x + y = 5$
$2x + y = 4$ **(−1, 6)**

b. $3x + 5y = -2$
$3x - 2y = -16$ **(−4, 2)**

2 **A group of 3 adults and 10 students paid $102 for a cavern tour. Another group of 3 adults and 7 students paid $84 for the tour. Find the admission price for an adult and for a student.**

Let a = the price for an adult and s = the price for a student.

Write two equations to represent this situation.

$3a + 10s = 102$ ← *total cost for the first van*
$3a + 7s = 84$ ← *total cost for the second van*

$$\begin{aligned} 3a + 10s &= 102 && \text{Write the equations in column form.} \\ (-)\ 3a + 7s &= 84 && \text{Subtract the equations to eliminate the } a \text{ terms.} \\ \hline 0 + 3s &= 18 \\ 3s &= 18 && \text{Simplify.} \\ \frac{3s}{3} &= \frac{18}{3} && \text{Divide each side by 3.} \\ s &= 6 && \text{The value of } s \text{ is 6.} \end{aligned}$$

Now substitute in either equation to find the value of a.

$$\begin{aligned} 3a + 7s &= 84 && \text{Original equation} \\ 3a + 7(6) &= 84 && \text{Replace } s \text{ with 6.} \\ 3a + 42 &= 84 && \text{Multiply.} \\ 3a + 42 - 42 &= 84 - 42 && \text{Subtract 42 from each side.} \\ 3a &= 42 && \text{Simplify.} \\ \frac{3a}{3} &= \frac{42}{3} && \text{Divide each side by 3.} \\ a &= 14 && \text{The value of } a \text{ is 14.} \end{aligned}$$

The solution of the system of equations is (14, 6). This means that the cost for adults was $14 and the cost for students was $6.

Check:

$$\begin{aligned} 3a + 10s &= 102 && \text{Original equation} \\ 3(14) + 10(6) &\stackrel{?}{=} 102 && \text{Replace } (a, s) \text{ with } (14, 6). \\ 42 + 60 &\stackrel{?}{=} 102 && \text{Multiply.} \\ 102 &= 102 && \checkmark \end{aligned}$$

$$\begin{aligned} 3a + 7s &= 84 && \text{Original equation} \\ 3(14) + 7(6) &\stackrel{?}{=} 84 && \text{Replace } (a, s) \text{ with } (14, 6). \\ 42 + 42 &\stackrel{?}{=} 84 && \text{Multiply.} \\ 84 &= 84 && \checkmark \end{aligned}$$

www.algconcepts.com/extra_examples

Motivating the Lesson

Hands-On Activity Ask a volunteer to write any linear equation in standard form $Ax + By = C$, with $A \neq 0$ and $B \neq 0$. Ask the class to think of several equations, also in standard form, so that subtracting the two equations eliminates the x terms or eliminates the y terms. Discuss how the students could solve the equation that results from subtracting the two equations.

2 TEACH

Teaching Tip In Example 1, focus students' attention on the two equations written in *column form*. Stress that the equations must be in standard form ($Ax + By = C$) when written in column form so that like terms align vertically. Have students look back at Lesson 6–3 if they need to review writing linear equations in standard form.

In-Class Examples

Example 1

Use elimination to solve the system of equations.
$3x - 2y = 21$
$3x + 4y = 3$ **(5, −3)**

Example 2

For a special event at the Kartchner Caverns, the cost of 4 adults and 3 students was $84 and the cost of 2 adults and 4 students was $62. Find the admission price for an adult and for a student at the special event. **$15 for an adult, $8 for a student**

In-Class Example

Example 3

Use elimination to solve the system of equations.

$3x + 4y = 6$

$5x - 4y = -22$ **(−2, 3)**

Teaching Tip In Your Turn part d, be sure students rewrite the second equation so the like variables align vertically.

In some systems of equations, the coefficients of one of the variables are additive inverses. For these systems, apply the elimination method by adding the equations.

Example 3 **Use elimination to solve the system of equations.**

$4x - 6y = 10$

$3x + 6y = 4$

$4x - 6y = 10$	*−6y and +6y are additive inverses.*
$(+)\ 3x + 6y = 4$	*Add the equations to eliminate the y terms.*
$7x + 0 = 14$	
$7x = 14$	*Simplify.*
$\frac{7x}{7} = \frac{14}{7}$	*Divide each side by 7.*
$x = 2$	The value of x is 2.

Now substitute x into either equation to find the value of y.

$4x - 6y = 10$	*Original equation*
$4(2) - 6y = 10$	*Replace x with 2.*
$8 - 6y = 10$	*Multiply.*
$8 - 6y - 8 = 10 - 8$	*Subtract 8 from each side.*
$-6y = 2$	*Simplify.*
$\frac{-6y}{-6} = \frac{2}{-6}$	*Divide each side by −6.*
$y = -\frac{2}{6}$	*Simplify.*
$y = -\frac{1}{3}$	The value of y is $-\frac{1}{3}$.

Prerequisite Skills Review

Simplifying Fractions, p. 685

The solution of the system is $\left(2, -\frac{1}{3}\right)$. *Check this result.*

Your Turn

Use elimination to solve each system of equations.

c. $x + y = 8$
$x - y = -2$ **(3, 5)**

d. $-4x + y = 15$
$3y = 5 - 4x$ $\left(-\frac{5}{2}, 5\right)$

Systems of equations can be used to solve **digit problems**. Digit problems explore the relationships between digits of a number.

Example 4 **Number Theory Link**

The sum of the digits of a two-digit number is 8. If the tens digit is 4 more than the units digit, what is the number?

Let t represent the tens digit and let u represent the units digit.

$t + u = 8$ ← *the sum of the digits*

$t = u + 4$ ← *the relationship between the digits*

Rewrite the second equation so that the t and u are on the same side of the equation.

$$t = u + 4 \rightarrow t - u = 4$$

Then use elimination to solve.

$$\begin{aligned} t + u &= 8 && \textit{Write the equations in column form.} \\ (+)\ t - u &= 4 && \textit{Add the equations to eliminate the u terms.} \\ \hline 2t + 0 &= 12 \\ 2t &= 12 && \textit{Simplify.} \\ \frac{2t}{2} &= \frac{12}{2} && \textit{Divide each side by 2.} \\ t &= 6 && \text{The tens digit is 6.} \end{aligned}$$

Now substitute to find the units digit.

$$\begin{aligned} t + u &= 8 && \textit{Original equation} \\ 6 + u &= 8 && \textit{Replace t with 6.} \\ 6 + u - 6 &= 8 - 6 && \textit{Subtract 6 from each side.} \\ u &= 2 && \text{The units digit is 2.} \end{aligned}$$

Since t is 6 and u is 2, the number is 62. *Check this solution.*

In-Class Example

Example 4

The sum of the digits of a two-digit number is 14. If the units digit is 2 more than the tens digit, what is the number? **68**

Check for Understanding

Communicating Mathematics

1–2. See margin.

1. **Explain** when you would use elimination to solve a system of equations.
2. **Describe** the result when you add each pair of equations. What does the result tell you about the system of equations?
   a. $2x - y = 12$
      $-2x + y = 14$
   b. $x + 5y = 3$
      $-x - 5y = -3$

Vocabulary
elimination
digit problems

Guided Practice

Getting Ready State whether ***addition***, ***subtraction***, ***either***, or ***neither*** should be used to solve each system.

Sample: $3x + y = 6$
$4x + y = 7$

Solution:
$$\begin{aligned} 3x + y &= 6 \\ (-)\ 4x + y &= 7 \\ \hline -x \quad &= -1 \end{aligned}$$
Subtraction should be used because it eliminates the y terms.

3. $x - 2y = 3$
   $3x + 2y = 8$ **addition**
4. $x + y = 5$
   $x - y = 9$ **either**
5. $4x - 5y = 18$
   $2x - 5y = 1$ **subtraction**
6. $x + 7y = 3$
   $2x + y = 4$ **neither**

3 PRACTICE/APPLY

Error Analysis

Watch for students who, in Exercise 5, subtract one set of terms but add, rather than subtract, the other terms or the constants. ***Prevent by*** suggesting to students that, to subtract an equation such as $2x - 5y = 1$, they use the definition of subtraction as "adding the opposite". They can write "subtract $2x - 5y = 1$" as "add $-(2x - 5y = 1)$" or "add $-2x + 5y = -1$."

Answers

1. when the coefficients of terms with the same variable are the same or additive inverses
2a. $0 = 26$; The system is parallel lines with no solution.
2b. $0 = 0$; The system is the same line with infinitely many solutions.

Reteaching Activity

Interpersonal Learners Have pairs of students review some of the systems of equations in this lesson. One student reads aloud the two equations and then identifies the two variables whose coefficients are equal or opposite. The partner then tells whether the equations should be added or subtracted, and which variable that operation would eliminate.

***Study Guide*, p. 588**

13–4 NAME ______ DATE ______ PERIOD ______

Study Guide Student Edition Pages 566–571

Elimination Using Addition and Subtraction

In Lesson 13–3 you used substitution to solve systems of linear equations. You can also use the *elimination method* to solve systems of linear equations. When you use the elimination method, you eliminate one of the variables by adding or subtracting the equations. Add the equations to eliminate the variable whose coefficients are additive inverses. Subtract the equations to eliminate the variable whose coefficients are the same.

Example: Use elimination to solve the system of equations.
$2x - y = -3$
$2x + y = -9$

Step 1 The coefficients of y are -1 and 1, so add the equations to eliminate the y terms. Then solve for x.

$$\begin{aligned} 2x - y &= -3 \\ (+)2x + y &= -9 && \textit{Add the equations.} \\ \hline 4x + 0 &= -12 && \textit{The y terms are eliminated.} \\ 4x &= -12 && \textit{Divide each side by 4.} \\ x &= -3 && \textit{The value of x is } -3. \end{aligned}$$

Step 2 Replace x in one of the original equations with -3. Then solve for y.

$$\begin{aligned} 2x + y &= -9 && \textit{Choose an equation.} \\ 2(-3) + y &= -9 && \textit{Replace x with } -3. \\ -6 + y &= -9 \\ -6 + y + 6 &= -9 + 6 && \textit{Add 6 to each side.} \\ y &= -3 && \textit{The value of y is } -3. \end{aligned}$$

The solution of the system of equations is $(-3, -3)$.

You could also use subtraction to eliminate the x terms in the example.

Step 1 The coefficients of x are both 2, so subtract to eliminate the x terms.

$$\begin{aligned} 2x - y &= -3 \\ (-)2x + y &= -9 && \textit{Subtract the equations.} \\ \hline 0 - 2y &= 6 && \textit{The x terms are eliminated.} \\ -2y &= 6 \\ y &= -3 \end{aligned}$$

Step 2 Solve for x.

$$\begin{aligned} 2x + y &= -9 \\ 2x + (-3) &= -9 && \textit{Replace y with } -3. \\ 2x &= -6 \\ x &= -3 \end{aligned}$$

The solution is $(-3, -3)$.

Use elimination to solve the system of equations.

1. $x - y = 2$
   $x + y = 0$ **(1, −1)**
2. $3x + 2y = 6$
   $-3x + y = 0$ **$\left(\frac{2}{3}, 2\right)$**
3. $2x - y = -4$
   $-3x - y = 6$ **(−2, 0)**
4. $2x + y = 6$
   $3x + y = 5$ **(−1, 8)**
5. $3x - 4y = 11$
   $3x + 5y = -7$ **(1, −2)**
6. $x + y = 6$
   $-2x + y = -3$ **(3, 3)**

Assignment Guide

Basic: 15–37 odd, 38–50
Average: 14–32 even, 34–50

Examples 1–3

13a. $x + y = 42$ and $x = 2y + 3$

Use elimination to solve each system of equations.

7. $x + y = 7$
   $x - y = 9$ **(8, −1)**
8. $x + y = 5$
   $2x + y = 4$ **(−1, 6)**
9. $x + 2y = 6$
   $3x - 2y = 2$ **(2, 2)**
10. $-7x + 4y = 6$
    $7x + y = 19$ **(2, 5)**
11. $3x - 2y = 10$
    $3x - 5y = 1$ $\left(\frac{16}{3}, 3\right)$
12. $4x = 5y - 9$
    $4x + 3y = -1$ **(−1, 1)**

Example 4

13. **Number Theory** The sum of two numbers is 42. The greater number is three more than twice the lesser number.
    a. Write a system of equations to represent the problem.
    b. What are the numbers? **29, 13**

Exercises

Practice

Use elimination to solve each system of equations.

Homework Help	
For Exercises	See Examples
14, 15, 18–25, 32	3
16, 17, 34, 35	1, 2
26–31, 33, 36, 37	4

Extra Practice
See page 719.

14. $x + y = 3$
    $x - y = 3$ **(3, 0)**
15. $x - y = 9$
    $x + y = 19$ **(14, 5)**
16. $x + 4y = 6$
    $x + 3y = 5$ **(2, 1)**
17. $x + y = 10$
    $3x + y = 0$ **(−5, 15)**
18. $3x + y = 13$
    $2x - y = 2$ **(3, 4)**
19. $9x + 2y = 12$
    $7x - 2y = -12$ **(0, 6)**
20. $-3x + 4y = 5$
    $3x - 2y = -7$ **(−3, −1)**
21. $2x - 3y = 19$
    $2x + 3y = 13$ **(8, −1)**
22. $11x - 3y = 10$
    $-2x + 3y = 8$ **(2, 4)**
23. $2x - y = -8$
    $x - y = 7$ **(−15, −22)**
24. $2x - 5y = 9$
    $2x - 3y = 11$ **(7, 1)**
25. $3x - 2y = 10$
    $2x - 2y = 5$ $\left(5, \frac{5}{2}\right)$

B

26. $x + 2y = 8$
    $3x = 6 - 2y$ $\left(-1, \frac{9}{2}\right)$
27. $x + y = 7$
    $21 - 3x = 3y$ **infinitely many**
28. $y = 3x - 5$
    $x - y = 13$ **(−4, −17)**
29. $x = y - 7$
    $2x - 5y = -2$ **(2, 8)**
30. $5x - y = -3$
    $2y = 10x - 7$ **no solution**
31. $x = 10 - y$
    $2x - y = -4$ **(−11, −4)**

32. What is the solution of the system $3x + 5y = -16$ and $3x - 2y = -2$? **32. (−2, −2)**
33. Find the solution of the system $5s + 4t = 12$ and $3s = 4 + 4t$. $\left(2, \frac{1}{2}\right)$

Applications and Problem Solving

34. **Spas** The Feel Better Spa has two specials for new members. They can receive 3 facials and 5 manicures for \$114 or 3 facials and 2 manicures for \$78. What are the prices for facials and manicures? **facial, \$18; manicure, \$12**
35. **Entertainment** The cost of admission to Water World was \$137.50 for 13 children and 2 adults in one party. The admission was \$103.50 for 9 children and 2 adults in another party.
    a. Write a system of equations to represent this problem.
    b. How much is admission for children and adults? **\$8.50 child, \$13.50 adult**

35a. $13x + 2y = 137.50$ and $9x + 2y = 103.50$

***Skills Practice*, p. 589, and *Practice*, p. 590 (shown)**

NAME ______ DATE ______ PERIOD ______

13-4 Practice

Student Edition Pages 566–571

Elimination Using Addition and Subtraction

Use elimination to solve each system of equations.

1. $x + y = 4$; $x - y = -6$ (−1, 5)
2. $x - y = 7$; $x + y = 1$ (4, −3)
3. $3x + y = 12$; $x + y = 8$ (2, 6)
4. $x + 5y = -12$; $x + 2y = -9$ (−7, −1)
5. $x + 2y = 9$; $3x - 2y = 3$ (3, 3)
6. $4x + 2y = 2$; $-4x - 3y = 3$ (3, −5)
7. $4x - 3y = 10$; $2x - 3y = 2$ (4, 2)
8. $2x + 5y = 1$; $2x + 10y = 10$ $\left(-4, \frac{9}{5}\right)$
9. $3y = x + 4$; $2x + 3y = 19$ (5, 3)
10. $2x = y - 4$; $2x + 6y = 3$ $\left(-\frac{3}{2}, 1\right)$
11. $4y = 2x + 8$; $5x - 4y = 22$ (10, 7)
12. $2x + y = 6$; $2x - 2y = -12$ (0, 6)
13. $-3x - y = 24$; $3x - 2y = 3$ (−5, −9)
14. $2x + 3y = 8$; $y = 2x + 8$ (−2, 4)
15. $-7x = y - 4$; $5x - y = 8$ (1, −3)
16. $3x + 5y = 7$; $4x + 5y = 1$ (−6, 5)
17. $6x - 3y = 3$; $6x - 5y = -3$ (2, 3)
18. $y = 2x + 4$; $2x - 4y = 8$ (−4, −4)

36a. $x - y = 15$ and $x = 2y - 2$

36. **Number Theory** The difference between two numbers is 15. The greater number is two less than twice the lesser number.
   a. Write a system of equations to represent this situation.
   b. Find the numbers. **32 and 17**

37a. $x + 2y = 29$ and $y = 3x - 10$

37. **Number Theory** The sum of a number and twice a second number is 29. The second number is ten less than three times the first number.
   a. Write a system of equations to represent this situation.
   b. Find both numbers. **7 and 11**

38. **Critical Thinking** The solution of a system of equations is $(-2, 5)$, and the first equation is $3x + 4y = 14$. **a. Sample answer: $x + y = 3$**
   a. Write a second equation for this system.
   b. Is this the only equation that could be in the system? Explain. **See margin.**

Mixed Review

39. Use substitution to solve the system of equations. *(Lesson 13–3)*
   $y = -3x - 5$
   $4x + y = 6$ **(11, −38)**

State whether each system is *consistent and independent*, *consistent and dependent*, or *inconsistent*. *(Lesson 13–2)*

40. 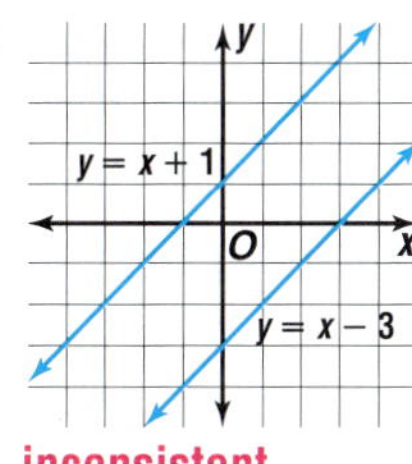

inconsistent

41. 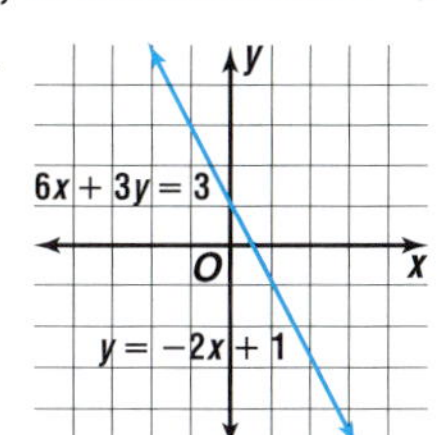

consistent and dependent

42. 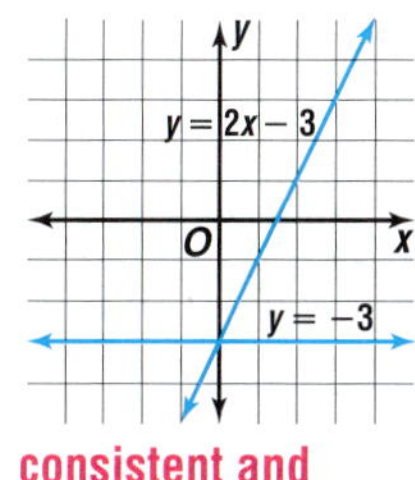

consistent and independent

Solve each inequality. *(Lesson 12–4)*

43. $3y < 12$ **$y < 4$**
44. $6 - 4m \leq 22$ **$m \geq -4$**
45. $9 - 3x > 15$ **$x < -2$**

Find each product. *(Lesson 9–4)*

46. $(m + 4)(m + 7)$ **$m^2 + 11m + 28$**
47. $(a - 3)(a - 3)$ **$a^2 - 6a + 9$**
48. $(2x - 2)(3x + 6)$ **$6x^2 + 6x - 12$**

Standardized Test Practice

49. **Short Response** Write the point-slope form of an equation for a line passing through points (6, 1) and (7, −4). *(Lesson 7–2)*
$y - 1 = -5(x - 6)$ or $y + 4 = -5(x - 7)$

50. **Multiple Choice**
In a scale model of a classic car, the front bumper measures 1.5 inches. If the actual bumper measures 3 feet, what is the scale of the model? *(Lesson 5–2)* **C**

 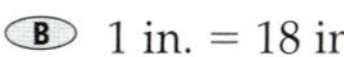

(A) 1 in. = 16 in.
(B) 1 in. = 18 in.
(C) 1 in. = 24 in.
(D) 1 in. = 43 in.

www.algconcepts.com/self_check_quiz

4 ASSESS

Open-Ended Assessment

Writing Ask students to describe how they know that a system of equations can be solved by elimination using addition or subtraction.

Mid-Chapter Test (Lessons 13-1 through 13-4) is available in the *Chapter 13 Resource Masters*, p. 618.

Answer

38b. No; there are infinitely many lines that can contain a point at (−2, 5).

Extra Credit

Assume that the following system of equations can be solved using elimination by addition.

$4x - 6y = -25$
$2x + ay = 10$

Find the value of a and find the solution.
***a* = 6; (−2.5, 2.5)**

***Enrichment*, p. 592**

13-4 NAME ______ DATE ______ PERIOD ______
Enrichment Student Edition Pages 566–571

Arithmetic Series

An **arithmetic series** is a series in which each term after the first may be found by adding the same number to the preceding term. Let S stand for the following series in which each term is 3 more than the preceding one.

$S = 2 + 5 + 8 + 11 + 14 + 17 + 20$

The series remains the same if we reverse the order of all the terms. So let us reverse the order of the terms and add one series to the other, term by term. This is shown at the right.

$S = 2 + 5 + 8 + 11 + 14 + 17 + 20$
$S = 20 + 17 + 14 + 11 + 8 + 5 + 2$
$2S = 22 + 22 + 22 + 22 + 22 + 22 + 22$
$2S = 7(22)$
$S = \frac{7(22)}{2} = 7(11) = 77$

Let a represent the first term of the series.
Let l represent the last term of the series.
Let n represent the number of terms in the series.
In the preceding example, $a = 2$, $l = 20$, and $n = 7$. Notice that when you add the two series, term by term, the sum of each pair of terms is 22. That sum can be found by adding the first and last terms, $2 + 20$ or $a + l$. Notice also that there are 7, or n, such sums. Therefore, the value of $2S$ is 7(22), or $n(a + l)$ in the general case. Since this is twice the sum of the series, you can use the following formula to find the sum of any arithmetic series.

$$S = \frac{n(a + l)}{2}$$

Example 1: Find the sum: $1 + 2 + 3 + 4 + 5 + 6 + 7 + 8 + 9$
$a = 1, l = 9, n = 9$, so $S = \frac{9(1 + 9)}{2} = \frac{9 \cdot 10}{2} = 45$

Example 2: Find the sum: $-9 + (-5) + (-1) + 3 + 7 + 11 + 15$
$a = -9, l = 15, n = 7$, so $S = \frac{7(-9 + 15)}{2} = \frac{7 \cdot 6}{2} = 21$

Find the sum of each arithmetic series.

1. $3 + 6 + 9 + 12 + 15 + 18 + 21 + 24$ **108**
2. $10 + 15 + 20 + 25 + 30 + 35 + 40 + 45 + 50$ **270**
3. $-21 + (-16) + (-11) + (-6) + (-1) + 4 + 9 + 14$ **−28**
4. even whole numbers from 2 through 100 **2550**
5. odd whole numbers between 0 and 100 **2500**

Lesson 13-5

1 FOCUS

5-Minute Check
Lesson 13-4

Use the system of equations to complete the following.
$8x + 3y = -15$
$3y - 4x = -6$

1. Rewrite the two equations in column form, aligning the like variables.
   $8x + 3y = -15$
   $-4x + 3y = -6$
2. Subtract the two equations from Exercise 1 and solve the resulting equation.
   $x = -\frac{3}{4}$
3. Find the value of the other variable. $y = -3$
4. Write the solution to the original system of equations.
   $\left(-\frac{3}{4}, -3\right)$
5. Use elimination to solve the system of equations.
   $3x + y = 6$
   $3x + 5y = 0$ $\left(2\frac{1}{2}, -1\frac{1}{2}\right)$

Motivating the Lesson

Real-World Connection Ask students how they would use algebra to represent the sentence "The total cost for 2 adults and 5 children to go to a movie was \$32." After they suggest $2x + 5y = 32$, ask them how they could represent the cost for that same group going to 3 movies. Discuss how to multiply both sides of $2x + 5y = 32$ by 3 to get $6x + 15y = 96$.

2 TEACH

In-Class Example
Example 1

Use elimination to solve the system of equations.
$5x + 2y = -2$
$3x - 6y = -30$ $(-2, 4)$

13-5 Elimination Using Multiplication

What You'll Learn

You'll learn to solve systems of equations by the elimination method using multiplication and addition.

Why It's Important

Transportation Systems of equations can be used to determine the rate of a river's current. *See Example 5.*

You have learned when and how to solve systems of equations by graphing, substitution, and elimination using addition or subtraction. The best times to use these methods are summarized in the table below.

Method	The Best Time to Use	Example
graphing	if the equations are easy to graph, or if you want an estimate because graphing usually does not give an exact solution	$y = x + 4$ $y = -x + 1$
substitution	if one of the variables in either equation has a coefficient of 1 or −1	$y = 2x + 3$ $3x - 5y = 8$
elimination using addition	if the coefficients of one variable are additive inverses	$4x - y = 9$ $-4x + 3y = 6$
elimination using subtraction	if the coefficients of one variable are the same	$2x + y = 4$ $3x + y = 1$

Sometimes neither of the variables in a system of equations can be eliminated by simply adding or subtracting the equations. In this case, another method is to multiply one or both of the equations by some number so that adding or subtracting eliminates one of the variables.

Examples

1 **Use elimination to solve the system of equations.**
$2x + 3y = 9$
$8x - 5y = 19$

Multiply the first equation by −4 so that the x terms are additive inverses.

$$2x + 3y = 9 \quad \xrightarrow{\text{Multiply by } -4.} \quad -8x - 12y = -36$$
$$8x - 5y = 19 \qquad (+)\ 8x - 5y = 19$$
$$0 - 17y = -17$$
$$\frac{-17y}{-17} = \frac{-17}{-17}$$
$$y = 1$$

Now find the value of x by replacing y with 1 in either equation.

$2x + 3y = 9$	*Original equation*
$2x + 3(1) = 9$	*Replace y with 1.*
$2x + 3 = 9$	*Multiply.*
$2x + 3 - 3 = 9 - 3$	*Subtract 3 from each side.*
$2x = 6$	*Simplify.*
$\frac{2x}{2} = \frac{6}{2}$	*Divide each side by 2.*
$x = 3$	*Simplify.*

The solution of this system of equations is (3, 1). *Check this solution.*

Resource Manager

Reproducible Masters
Chapter 13 Resource Masters
- *Study Guide*, p. 593
- *Skills Practice*, p. 594
- *Practice*, p. 595
- *Reading to Learn Mathematics*, p. 596
- *Enrichment*, p. 597

Transparencies
5-Minute Check, 13–5

Technology/Multimedia
Interactive Chalkboard CD-ROM

 **Use elimination to solve the system of equations.**

$5x - 6y = 25$
$4x + 2y = 3$

Multiply the second equation by 3 so that the y terms are additive inverses.

$$\begin{array}{ll} 5x - 6y = 25 & \\ 4x + 2y = 3 & \end{array} \quad \text{Multiply by 3.} \quad \begin{array}{r} 5x - 6y = 25 \\ (+)\ 12x + 6y = 9 \\ \hline 17x + 0 = 34 \\ \frac{17x}{17} = \frac{34}{17} \\ x = 2 \end{array}$$

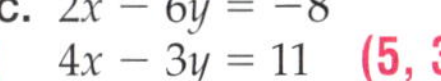

Now find the value of y by replacing x with 2 in either equation.

$4x + 2y = 3$	*Original equation*
$4(2) + 2y = 3$	*Replace x with 2.*
$8 + 2y = 3$	*Multiply.*
$8 + 2y - 8 = 3 - 8$	*Subtract 8 from each side.*
$2y = -5$	*Simplify.*
$\frac{2y}{2} = \frac{-5}{2}$	*Divide each side by 2.*
$y = -\frac{5}{2}$	*Simplify.*

The solution of the system of equations is $\left(2, -\frac{5}{2}\right)$. *Check this solution.*

Your Turn **a. (4, −8)**

a. $2x - y = 16$
$5x + 3y = -4$

b. $4x - 3y = -2$
$2x + 7y = 16$ **(1, 2)**

c. $2x - 6y = -8$
$4x - 3y = 11$ **(5, 3)**

Example 3

Transportation Link

Chris and Alana both take the Metro train to work. In May, Chris took the train 15 times during rush hour and 29 times during non-rush hour for $64.80. Alana took the train 30 times during rush hour and 14 times during non-rush hour for $76.80. What are the rush hour and non-rush hour fares?

r = the rush hour fare n = the non-rush hour fare

$15r + 29n = 64.80$ ← *Chris' expenses*
$30r + 14n = 76.80$ ← *Alana's expenses*

Multiply the first equation by −2 to eliminate the r terms.

$$\begin{array}{l} 15r + 29n = 64.80 \\ 30r + 14n = 76.80 \end{array} \quad \text{Multiply by } -2. \quad \begin{array}{r} -30r - 58n = -129.60 \\ (+)\ 30r + 14n = 76.80 \\ \hline 0 - 44n = -52.80 \\ \frac{-44n}{-44} = \frac{-52.80}{-44} \\ n = 1.20 \end{array}$$

(continued on the next page)

Prerequisite Skills Review
Operations with Decimals, p. 684

In-Class Examples

Example 2

Use elimination to solve the system of equations.
$3x + 2y = 4$
$9x - 4y = 7$ $\left(1, \frac{1}{2}\right)$

Example 3

A music store has one price for all CDs except for the CDs in the Sale section. One customer bought 8 regular CDs and 2 sale CDs and paid $79.50. Another customer bought 4 regular CDs and 5 sale CDs and paid $62.75. What are the costs of regular and sale CDs? **$8.50 regular, $5.75 sale**

 www.algconcepts.com/extra_examples

Teaching Tip In Example 4, the solution that is illustrated eliminates the x terms. Point out to students that if they wanted to eliminate the y terms, they could multiply the first equation by 3 and the second equation by 4.

In-Class Example

Example 4

Use elimination to solve the system of equations.

$4x - 3y = -8$

$7x + 5y = 27$ **(1, 4)**

Now find the value of r by replacing n with 1.20 in either equation.

$15r + 29n = 64.80$	*Original equation*
$15r + 29(1.20) = 64.80$	*Replace n with 1.20.*
$15r + 34.80 = 64.80$	*Multiply.*
$15r + 34.80 - 34.80 = 64.80 - 34.80$	*Subtract 34.80 from each side.*
$15r = 30$	*Simplify.*
$\frac{15r}{15} = \frac{30}{15}$	*Divide each side by 15.*
$r = 2$	*Simplify.*

The solution is (2, 1.20). This means that the rush hour fare is \$2.00 and the non-rush hour fare is \$1.20. Do these fares seem reasonable? *Check this solution by substituting (2, 120) for (r, n) in the original equations.*

Sometimes it is necessary to multiply each equation by a different number and then add in order to eliminate one of the variables.

Example  **4** **Use elimination to solve the system of equations.**

$\mathbf{3x + 4y = -25}$

$\mathbf{2x - 3y = 6}$

Multiply the first equation by 2 and the second equation by -3 so that the x terms are additive inverses.

$3x + 4y = -25$ **Multiply by 2.** $\rightarrow$ $6x + 8y = -50$

$2x - 3y = 6$ **Multiply by −3.** $\rightarrow$ $(+)-6x + 9y = -18$

$$0 + 17y = -68$$

$$\frac{17y}{17} = \frac{-68}{17}$$

$$y = -4$$

Now find the value of x by replacing y with -4 in either equation.

$3x + 4y = -25$	*Original equation*
$3x + 4(-4) = -25$	*Replace y with −4.*
$3x - 16 = -25$	*Multiply.*
$3x - 16 + 16 = -25 + 16$	*Add 16 to each side.*
$3x = -9$	*Simplify.*
$\frac{3x}{3} = \frac{-9}{3}$	*Divide each side by 3.*
$x = -3$	*Simplify.*

The solution of the system of equations is $(-3, -4)$. You can also solve this system of equations by multiplying the first equation by 3 and the second equation by 4. *Why?*

From the Classroom of ...

Don McGurrin
Wake County Public Schools
Raleigh, North Carolina

The elimination method is similar to the addition and subtraction of fractions. With fractions, you must have a common denominator. With systems, you must have a common (or opposite) coefficient. Finding the common coefficient involves the same thinking process as common denominators.

Your Turn

Use elimination to solve each system of equations.

d. $5x + 3y = 12$
$4x - 5y = 17$ **(3, −1)**

e. $4x + 3y = 19$
$3x - 4y = 8$ **(4, 1)**

Systems of equations can be used to solve rate problems.

Example 5

Transportation Link

A barge on the Mississippi River travels 36 miles upstream in 6 hours. The return trip takes the barge only 4 hours. Find the rate of the current.

Explore Let r represent the rate of the barge in still water. Let c represent the rate of the current.

The rate of the barge traveling downstream *with* the current is $r + c$, and the rate of the barge traveling upstream *against* the current is $r - c$.

Plan Use the formula distance = rate × time, or $d = rt$, to write a system of equations. Then solve the system to find c.

	d	r	t	$d = rt$	
Downstream	36	$r + c$	4	$36 = 4r + 4c$	$4(r + c) = 4r + 4c$
Upstream	36	$r - c$	6	$36 = 6r - 6c$	$6(r - c) = 6r - 6c$

Solve

$4r + 4c = 36$ **Multiply by 3.** → $12r + 12c = 108$

$6r - 6c = 36$ **Multiply by −2.** → $(+)\ -12r + 12c = -72$

$$0 + 24c = 36$$
$$24c = 36$$
$$\frac{24c}{24} = \frac{36}{24}$$
$$c = 1.5$$

The rate of the current is 1.5 miles per hour.

Examine Find the value of r for this system and then check the solution.

Barge on the Mississippi

Check for Understanding

Communicating Mathematics

1–2. See margin.

1. **Explain** when you would use elimination with multiplication to solve a system of equations.
2. **Write** a system of equations in which you can eliminate the variable x by multiplying one equation by 3 and then adding the equations.

Reteaching Activity

Logical Learners Write the system of equations $ax + by = c$ and $dx + ey = f$, explaining that the letters a through f represent constants. Give them the solutions $x = \frac{ce - df}{ae - bd}$ and $y = \frac{cd - af}{bd - ae}$, and ask them to verify that those solutions work for any of the linear-linear systems of equations in the chapter.

In-Class Example

Example 5

A barge on the Chicago River travels 24 miles upstream in 4 hours. The return trip takes the barge only 3 hours. Find the rate of the current. **1 mph**

Answers

1. **when neither of the variables in a system of equations can be eliminated by simply adding or subtracting the equations**
2. **Sample answer: $-x - 4y = 1$ and $3x + 5y = 2$**

Study Guide, p. 593

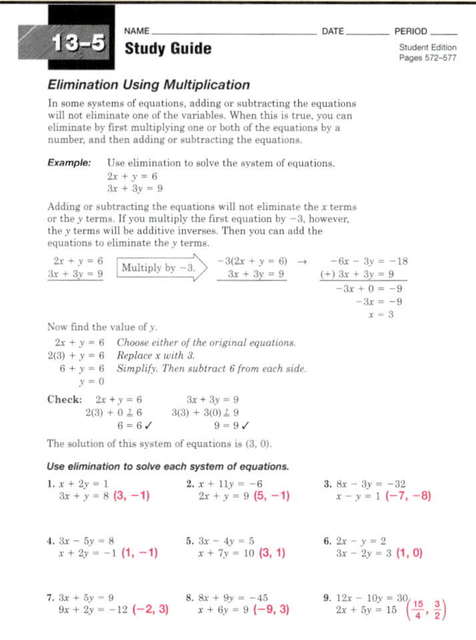

13–5 NAME ________ DATE ______ PERIOD _____

Study Guide Student Edition Pages 572–577

Elimination Using Multiplication

In some systems of equations, adding or subtracting the equations will not eliminate one of the variables. When this is true, you can eliminate by first multiplying one or both of the equations by a number, and then adding or subtracting the equations.

Example: Use elimination to solve the system of equations.
$2x + y = 6$
$3x + 3y = 9$

Adding or subtracting the equations will not eliminate the x terms or the y terms. If you multiply the first equation by −3, however, the y terms will be additive inverses. Then you can add the equations to eliminate the y terms.

$2x + y = 6$, $3x + 3y = 9$ — Multiply by −3. → $-3(2x + y = 6)$, $3x + 3y = 9$ → $-6x - 3y = -18$, $(+)\ 3x + 3y = 9$

$$-3x + 0 = -9$$
$$-3x = -9$$
$$x = 3$$

Now find the value of y.

$2x + y = 6$ *Choose either of the original equations.*
$2(3) + y = 6$ *Replace x with 3.*
$6 + y = 6$ *Simplify. Then subtract 6 from each side.*
$y = 0$

Check: $2x + y = 6$; $2(3) + 0 \stackrel{?}{=} 6$; $6 = 6$ ✓ — $3x + 3y = 9$; $3(3) + 3(0) \stackrel{?}{=} 9$; $9 = 9$ ✓

The solution of this system of equations is (3, 0).

Use elimination to solve each system of equations.

1. $x + 2y = 1$, $3x + y = 8$ **(3, −1)**
2. $x + 11y = -6$, $2x + y = 9$ **(5, −1)**
3. $8x - 3y = -32$, $x - y = 1$ **(−7, −8)**
4. $3x - 5y = 8$, $x + 2y = -1$ **(1, −1)**
5. $3x - 4y = 5$, $x + 7y = 10$ **(3, 1)**
6. $2x - y = 2$, $3x - 2y = 3$ **(1, 0)**
7. $3x + 5y = 9$, $9x + 2y = -12$ **(−2, 3)**
8. $8x + 9y = -45$, $x + 6y = 9$ **(−9, 3)**
9. $12x - 10y = 30$, $2x + 5y = 15$ **$\left(\frac{15}{4}, \frac{3}{2}\right)$**

3 PRACTICE/APPLY

Error Analysis

Watch for students who, in Exercise 4, multiply $5x + 3y$ by -2 but fail to multiply the right-hand side of the equation by -2. ***Prevent by*** stressing that when you multiply an equation by a number, you have to multiply both sides of the equation by that number.

Assignment Guide

Basic: 15–35 odd, 36–46
Average: 14–32 even, 34–46

***Skills Practice,* p. 594, and *Practice,* p. 595 (shown)**

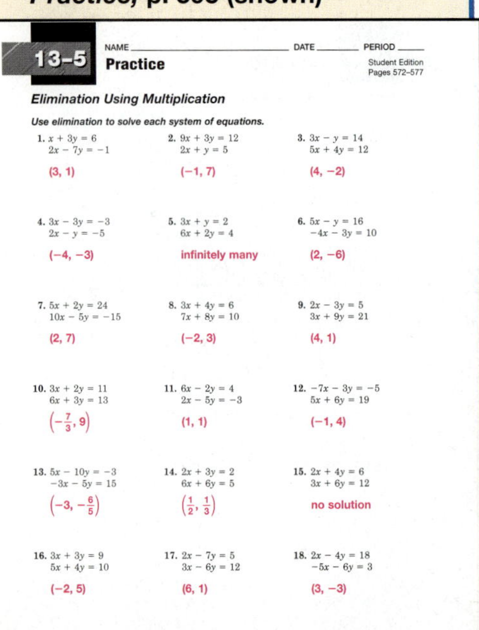

13-5 NAME ______ DATE ______ PERIOD ______

Practice

Student Edition Pages 572–577

Elimination Using Multiplication

Use elimination to solve each system of equations.

1. $x + 3y = 6$, $2x - 7y = -1$ (3, 1)
2. $9x + 3y = 12$, $2x + y = 5$ (−1, 7)
3. $3x - y = 14$, $5x + 4y = 12$ (4, −2)
4. $3x - 3y = -3$, $2x - y = -5$ (−4, −3)
5. $3x + y = 2$, $6x + 2y = 4$ infinitely many
6. $5x - y = 16$, $-4x - 3y = 10$ (2, −6)
7. $5x + 2y = 24$, $10x - 5y = -15$ (2, 7)
8. $3x + 4y = 6$, $7x + 8y = 10$ (−2, 3)
9. $2x - 3y = 5$, $3x + 9y = 21$ (4, 1)
10. $3x + 2y = 11$, $6x + 3y = 13$ $\left(-\frac{7}{3}, 9\right)$
11. $6x - 2y = 4$, $2x - 5y = -3$ (1, 1)
12. $-7x - 3y = -5$, $5x + 6y = 19$ (−1, 4)
13. $5x - 10y = -3$, $-3x - 5y = 15$ $\left(-3, -\frac{6}{5}\right)$
14. $2x + 3y = 2$, $6x + 6y = 5$ $\left(\frac{1}{2}, \frac{1}{3}\right)$
15. $2x + 4y = 6$, $3x + 6y = 12$ no solution
16. $3x + 3y = 9$, $5x + 4y = 10$ (−2, 5)
17. $2x - 7y = 5$, $3x - 6y = 12$ (6, 1)
18. $2x - 4y = 18$, $-5x - 6y = 3$ (3, −3)

3a. substitution
3b. elim. (×)

3. **Match** each system of equations with the method listed below that would best solve the system.

a. $y = 3x$
$6x - 2y = 9$

b. $2x + 4y = 5$
$3x - 2y = -3$

c. $10x - 2y = 12$
$-10x + 3y = -13$
elim. (+)

d. $x + 7y = 5$
$3x + 7y = 1$
elim. (−)

substitution
elimination (+)
elimination (−)
elimination (×)

Guided Practice

Getting Ready **Explain the steps you would take to eliminate the variable *y* in each system of equations.**

Sample: $4x + 3y = -7$
$3x + y = 1$
Solution: Multiply second equation by −3. Then add.

4. $2x + 6y = 10$
$5x + 3y = 1$
5. $3x - 2y = 3$
$8x + y = 27$
6. $2x + y = 4$
$7x - 4y = 29$

4. Multiply second equation by −2. Then add.

Examples 1–4

5. Multiply second equation by 2. Then add.
6. Multiply first equation by 4. Then add.

Use elimination to solve each system of equations.

7. $x + 2y = 6$
$-4x + 5y = 2$ (2, 2)

8. $x - 5y = 0$
$2x - 3y = 7$ (5, 1)

9. $5x + 8y = 1$
$2x - 4y = -14$ (−3, 2)

10. $x + 2y = 5$
$3x + y = 7$ $\left(\frac{9}{5}, \frac{8}{5}\right)$

11. $3x - y = -5$
$6x - 2y = 8$ no solution

12. $2x - 5y = -10$
$7x - 3y = -6$ (0, 2)

Example 5

13. **Recreation** A fishing boat traveled 48 miles upstream in 4 hours. Returning at the same rate, it took 3 hours.
a. Find the rate of the current. 13a. 2 mph
b. Find the rate of the boat in still water. 14 mph

Exercises

Practice

A

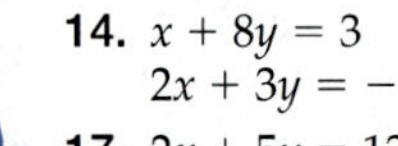

Homework Help

For Exercises	See Examples
14, 15, 17, 20, 29	1
16, 18, 30	2
23, 24, 26–28	4
34	5
35	3

Extra Practice
See page 719.

14. (−5, 1) 17. (−1, 3) 20. $\left(\frac{3}{5}, 3\right)$

Use elimination to solve each system of equations.

14. $x + 8y = 3$
$2x + 3y = -7$

15. $4x + y = 8$
$x - 7y = 2$ (2, 0)

16. $2x + y = 6$
$3x - 7y = 9$ (3, 0)

17. $2x + 5y = 13$
$4x - 3y = -13$

18. $-3x + 2y = 10$
$-2x - y = -5$ (0, 5)

19. $3x - 2y = 0$
$x + 6y = 5$ $\left(\frac{1}{2}, \frac{3}{4}\right)$

20. $-5x + 8y = 21$
$10x + 3y = 15$

21. $6x - 4y = 11$
$2x + 2y = 7$ $\left(\frac{5}{2}, 1\right)$

22. $6x - 5y = -6$
$3x + 10y = 72$ (4, 6)

23. $2x - 7y = 9$
$-3x + 4y = 6$

24. $5x + 3y = 4$
$-4x + 5y = -18$

25. $6x - 3y = 7$
$18x - 9y = 21$

26. $7x - 4y = 16$
$2x + 3y = 17$ (4, 3)

27. $-5x - 2y = 12$
$11x - 5y = -17$

28. $7x - 3y = -9$
$5x - 4y = -25$

23. (−6, −3) 24. (2, −2) 25. infinitely many

27. (−2, −1)
28. (3, 10)
29. $\left(\frac{7}{9}, 0\right)$

B

29. What is the solution of the system $9x + 8y = 7$ and $18x - 15y = 14$?

C

30. Use elimination to find the solution of the system $\frac{1}{3}x - y = -1$ and $\frac{1}{5}x - \frac{2}{5}y = -1$. (−9, −2)

Determine the best method to solve each system of equations. Then solve. 31–33. Sample solution methods are given.

31. $x + 2y = 6$
$2x - 4y = -20$
sub. or mult.; $(-2, 4)$

32. $y = 3x - 2$
$x - 5y = -4$
substitution; $(1, 1)$

33. $y = \frac{1}{2}x + 6$
$2x - 4y = 8$
graphing; no solution

Applications and Problem Solving

34. **Travel** A riverboat on the Mississippi River travels 30 miles upstream in 2 hours and 30 minutes. The return trip downstream takes only 2 hours.
   a. Find the rate of the current. 1.5 mph
   b. Find the rate of the riverboat in still water. 13.5 mph

35. **Entertainment** The science club purchased tickets for a magic show. They paid \$108 for 6 tickets in section A and 10 tickets in section B. The following week, they paid \$104 for 4 tickets in section A and 12 tickets in section B.
   a. Write a system of equations to represent the problem.
   b. What are the prices of the tickets in section A and B? \$8, \$6

35a. $6a + 10b = 108$ and $4a + 12b = 104$

36. **Critical Thinking** The solution of the system $5x + 6y = -9$ and $10x + 8y = c$ is $(9, b)$. Find values for c and b. $c = 18$, $b = -9$

Mixed Review

Use elimination to solve each system of equations. *(Lesson 13–4)*

37. $x - y = 8$
$x + y = 6$
$(7, -1)$

38. $x + y = -4$
$4x + y = 2$
$(2, -6)$

39. $2x + 4y = 7$
$2x - y = 2$
$\left(\frac{3}{2}, 1\right)$

40. Use substitution to solve $x = 5 + 2y$ and $3x - 4y = 3$. *(Lesson 13–3)* $(-7, -6)$

Solve each inequality. *(Lesson 12–3)*

41. $-4m > 16$ $m < -4$

42. $\frac{x}{2} < 6$ $x < 12$

43. $\frac{3a}{4} \geq -3$ $a \geq -4$

44. Use the Quadratic Formula to solve $n^2 - 5n + 12 = 0$. *(Lesson 11–6)* no real solutions

Standardized Test Practice

45. **Grid In** Suppose a baseball player hits a fly ball into right field above first base. The player hits the ball with his bat at a height of 2 feet above the ground and sends the ball upward at a velocity of 25 meters per second. The height of the ball t seconds after the hit can be approximated by the formula $h = -5t^2 + 25t + 2$. If the ball is not caught, how many seconds will it take to hit the ground? Round to the nearest second. *(Lesson 11–3)* 5

46. **Multiple Choice** The graph shows the top turkey producing states. What percent of turkeys does the state of North Carolina produce? Round to the nearest percent. *(Lesson 5–3)* B
   Ⓐ 5% Ⓑ 19%
   Ⓒ 33% Ⓓ 45%

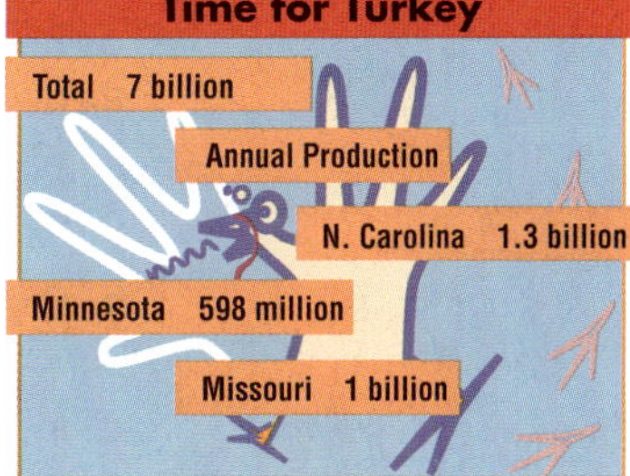

Source: U.S. Census Bureau

www.algconcepts.com/self_check_quiz

4 ASSESS

Open-Ended Assessment

Writing Ask students to write two different systems of equations that can be solved by elimination if they multiply one equation by 4 and the other equation by 3.

Enrichment, p. 597

13-5 NAME ________ DATE ______ PERIOD _____

Enrichment

Student Edition Pages 572–577

Parabolas Through Three Given Points

If you know two points on a straight line, you can find the equation of the line. To find the equation of a parabola, you need three points on the curve.

For example, here is how to approximate an equation of the parabola through the points $(0, -2)$, $(3, 0)$, and $(5, 2)$.

Use the general equation $y = ax^2 + bx + c$. By substituting the given values for x and y, you get three equations.

$(0, -2)$: $-2 = c$
$(3, 0)$: $0 = 9a + 3b + c$
$(5, 2)$: $2 = 25a + 5b + c$

First, substitute -2 for c in the second and third equations. Then solve those two equations as you would any system of two equations. Multiply the second equation by 5 and the third equation by -3.

$$\begin{aligned} 0 &= 45a + 15b - 10 \\ -6 &= -75a - 15b + 6 \\ \hline -6 &= -30a - 4 \\ a &= \frac{1}{15} \end{aligned}$$

To find b, substitute $\frac{1}{15}$ for a in either the second or third equation.

$$0 = 9\left(\frac{1}{15}\right) + 3b - 2$$
$$b = \frac{7}{15}$$

The equation of a parabola through the three points is $y = \frac{1}{15}x^2 + \frac{7}{15}x - 2$.

Find the equation of a parabola through each set of three points.

1. $(1, 5), (0, 6), (2, 3)$ $y = -\frac{1}{2}x^2 - \frac{1}{2}x + 6$
2. $(-5, 0), (0, 0), (8, 100)$ $y = \frac{25}{26}x^2 + \frac{125}{26}x$
3. $(4, -4), (0, 1), (3, -2)$ $y = -\frac{1}{4}x^2 - \frac{1}{4}x + 1$
4. $(1, 3), (6, 0), (0, 0)$ $y = -\frac{3}{5}x^2 + \frac{18}{5}x$
5. $(2, 2), (5, -3), (0, -1)$ $y = -\frac{19}{30}x^2 + \frac{83}{30}x - 1$
6. $(0, 4), (4, 0), (-4, 4)$ $y = \frac{1}{8}x^2 - \frac{1}{2}x + 4$

Extra Credit

Find the y value of the solution of the following system of equations.

$3x + 5y = k$
$3x - 5y = k$ $y = 0$

PREPARE

This optional investigation is designed to be completed by groups of 2–3 students over 1–2 days.

Objective

Students use matrices to summarize types of computers, the numbers of computer sold, and total daily sales figures. They use row operations on augmented matrices to calculate the cost of each type of computer. Then they apply the same procedures to other types, numbers, and prices of computers.

Mathematical Overview

This investigation utilizes the following concepts:

- how matrices and augmented matrices represent equations,
- row operations on matrices, and
- identity matrices.

Suggested Time Management	
Investigation	30–40 min
Extension: Gathering Data	25–40 min
Extension: Summarizing Data	25–40 min

Motivating the Lesson

Ask students how they might organize sales data that consists of several products that have different prices and different numbers of items sold. Discuss how the linear equations they have been studying can be used to represent the sales data.

Chapter 13 Investigation

Crazy Computers

Matrices

Look Back

Matrices: Pages 80–81

In this investigation, you will use matrices to solve systems of equations. Consider the following problem.

Kristin and Scott are sales associates at Crazy Computers. On Monday, Kristin sold 4 Model A computers and 1 Model B computer for a total of \$6395. Scott sold 2 Model A computers and 3 Model B computers for a total of \$7195. What was the price of each computer?

Investigate

You can write a system of equations to solve this problem. You may want to use matrices to solve this system of equations.

1. Write each equation in standard form. Then write the coefficients and constants of the system in a special matrix called an **augmented matrix.** Write the coefficients in the matrix to the left of the dashed line. Write the constants to the right of the dashed line.

$$\begin{aligned} 3x + 5y &= 7 \\ 6x - 1y &= -8 \end{aligned} \quad \Rightarrow \quad \left[\begin{array}{cc|c} 3 & 5 & 7 \\ 6 & -1 & -8 \end{array}\right]$$

   a. Write $2x - 3y = 5$ and $x + 4y = -7$ as an augmented matrix.
   b. Write $x + 6y = -1$ and $y = 5$ as an augmented matrix.

1a. $\left[\begin{array}{cc|c} 2 & -3 & 5 \\ 1 & 4 & -7 \end{array}\right]$

1b. $\left[\begin{array}{cc|c} 1 & 6 & -1 \\ 0 & 1 & 5 \end{array}\right]$

2. You can use **row operations** to simplify an augmented matrix.
   - Switch any two rows.
   - Replace any row with a nonzero multiple of that row.
   - Replace any row with the sum or difference of that row and a multiple of another row.

   a. Switch rows 1 and 2 of $\left[\begin{array}{cc|c} 0 & -5 & 2 \\ 1 & 3 & 4 \end{array}\right]$.
   b. Multiply row 2 of $\left[\begin{array}{cc|c} 1 & -3 & 5 \\ 0 & 2 & -8 \end{array}\right]$ by 0.5.
   c. Replace row 2 of $\left[\begin{array}{cc|c} 1 & -2 & -5 \\ -1 & 3 & 4 \end{array}\right]$ with the sum of rows 1 and 2.
   d. In $\left[\begin{array}{cc|c} 1 & -3 & 2 \\ 0 & 3 & -5 \end{array}\right]$, what row operation would you use to get a 0 in the second column of row 1?

2a. $\left[\begin{array}{cc|c} 1 & 3 & 4 \\ 0 & -5 & 2 \end{array}\right]$

2b. $\left[\begin{array}{cc|c} 1 & -3 & 5 \\ 0 & 1 & -4 \end{array}\right]$

2c. $\left[\begin{array}{cc|c} 1 & -2 & -5 \\ 0 & 1 & -1 \end{array}\right]$

2d. Replace row 1 with the sum of rows 1 and 2.

Cooperative Learning

This investigation offers an excellent opportunity for using cooperative groups. For more information on cooperative learning strategies and group management, see *Cooperative Learning in the Mathematics Classroom,* one of the titles in the Glencoe Mathematics Professional Series.

3. The goal of using row operations to solve a system of equations is to get to the **identity matrix**, which is $\begin{bmatrix} 1 & 0 \\ 0 & 1 \end{bmatrix}$. Suppose the solution of a system of equations is $\left[\begin{array}{cc|c} 1 & 0 & -4 \\ 0 & 1 & 2 \end{array}\right]$. What does this matrix represent?
$x = -4$, $y = 2$

4. Solve the problem. Let x = the cost of the Model A computer and let y = the cost of the Model B computer.

$$4x + 1y = 6395 \quad \leftarrow \textit{Kristin's sales}$$
$$2x + 3y = 7195 \quad \leftarrow \textit{Scott's sales}$$

Step 1 Write the system as an augmented matrix. $\left[\begin{array}{cc|c} 4 & 1 & 6395 \\ 2 & 3 & 7195 \end{array}\right]$

Step 2 Multiply row 2 by 2.
$2 \times 2 = 4$; $2 \times 3 = 6$; $2 \times 7195 = 14{,}390$ $\left[\begin{array}{cc|c} 4 & 1 & 6395 \\ 4 & 6 & 14{,}390 \end{array}\right]$

Step 3 Replace row 2 with the difference of row 1 and row 2. *$4 - 4 = 0$; $1 - 6 = -5$; $6395 - 14{,}390 = -7995$* $\left[\begin{array}{cc|c} 4 & 1 & 6395 \\ 0 & -5 & -7995 \end{array}\right]$

Step 4 Divide row 2 by -5. *$0 \div (-5) = 0$; $-5 \div (-5) = 1$; $-7995 \div (-5) = 1599$* $\left[\begin{array}{cc|c} 4 & 1 & 6395 \\ 0 & 1 & 1599 \end{array}\right]$

Step 5 Replace row 1 with the difference of row 1 and row 2. *$4 - 0 = 4$; $1 - 1 = 0$; $6395 - 1599 = 4796$* $\left[\begin{array}{cc|c} 4 & 0 & 4796 \\ 0 & 1 & 1599 \end{array}\right]$

Step 6 Divide row 1 by 4.
$4 \div 4 = 1$; $0 \div 4 = 0$; $4796 \div 4 = 1199$ $\left[\begin{array}{cc|c} 1 & 0 & 1199 \\ 0 & 1 & 1599 \end{array}\right]$

This means that Model A computer is $1199 and Model B computer is $1599.

Extending the Investigation

In this extension, you will use matrices as a problem-solving tool.

1. Suppose Scott sells 5 Model TX laser printers and 6 Model DM ink jet printers for a total of $4185. Kristin sells 6 Model TX laser printers and 5 Model DM ink jet printers for a total of $4549. What is the cost of each printer? **$579, $215**
2. At Crazy Computers, there are two prices for computer software, Price A and Price B. Suppose Aislyn purchases 3 Price A software and 1 Price B software for a total of $250 and Devin purchases 5 Price A software and 2 Price B software for a total of $450. What are the two prices for the software? **$50, $100**

Presenting Your Investigation

Write a real-world problem that can be solved by using matrices and a system of equations. Present your problem and solution on a poster board.

Investigation For more information on matrices, visit: www.algconcepts.com

Students can work with their families to organize some aspects of their belongings using matrices. The example shown below uses matrices to organize the kinds of clothes that belong to family members.

Clothes				
Family Member	**Shoes**	**Shirts**	**Pants**	**Sweaters**
A				
B				
C				

MANAGE

Teaching Tip Make sure students understand that row operations can only be performed on rows. Stress that similar operations cannot be performed on matrix columns. If students do not understand why this is so, show them how matrix row operations are related to operations on equations that result in equivalent equations.

Working in Groups If students work in groups, each group can work on the exercises in Extending the Investigation. Then, as a class, the groups can describe and discuss any problems that they found, how they dealt with the problems, and what they learned from the Investigation.

Working as a Class Assign half the class to work on each exercise in Extending the Investigation. Then, as a class, students can explain their activities to each other. After the discussion, students can work individually on the Presenting Your Investigation section.

ASSESS

Students should be able to explain their steps each time they perform a row operation. They also should be able to explain why the final augmented matrix, which contains an identity matrix, lets them identify the value of each variable.

Students should add their posters to their portfolios at this time.

13-6 Solving Quadratic-Linear Systems of Equations

1 FOCUS

5-Minute Check
Lesson 13-5

Use the system of equations to complete the following.

10x + 3y = 5

5x + 4y = −5

1. Multiply the second equation by −2. **−10x − 8y = 10**
2. Add your result in Exercise 1 and the first original equation, and solve the resulting equation. **y = −3**
3. Find the value of the other variable. **x = 1.4**
4. Write the solution to the original system of equations. **(1.4, −3)**
5. Use elimination to solve the system of equations.
   $2x - 7y = 3$
   $6x + 3y = 1$ $\left(\frac{1}{3}, -\frac{1}{3}\right)$

Motivating the Lesson

Real-World Connection Have students look at a large map of the United States. Have them place a string on the map so the string traces a parabola. Then ask students to select three pairs of cities on the map such that a line through each pair of cities would miss the parabola, just touch the parabola, or intersect the parabola at two points.

2 TEACH

Teaching Tip Remind students that the curve that represents a quadratic equation is called a *parabola*.

TECHNOLOGY

An alternative technology option using a graphing calculator is available for teaching this lesson.

What You'll Learn

You'll learn to solve systems of quadratic and linear equations.

Why It's Important

Meteorology Quadratic-linear systems of equations can be used to determine when airplanes will cross jet streams and experience turbulence. *See Exercise 24.*

In late November, the jet stream moving across North America could be described by the quadratic equation $y = \frac{1}{4}x^2 - 12$, where Chicago was at the origin. Suppose a plane's route is described by the linear equation $y = -\frac{1}{2}x$. What are the coordinates of the point at which turbulence will occur? *This problem will be solved in Example 7.*

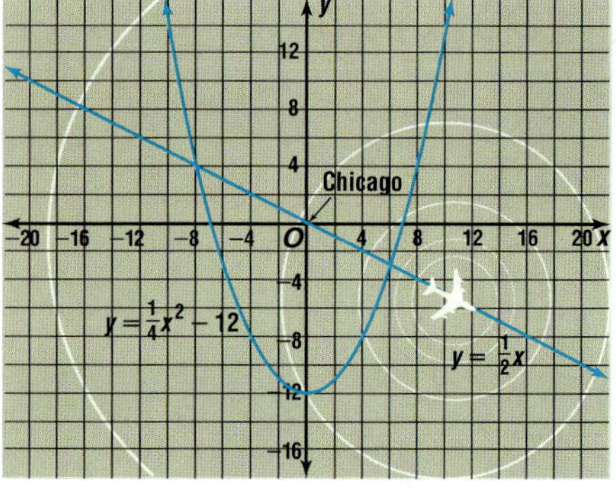

Like a linear system of equations, the solution of a **quadratic-linear** system of equations is the ordered pair that satisfies both equations. A quadratic-linear system can have 0, 1, or 2 solutions, as shown below.

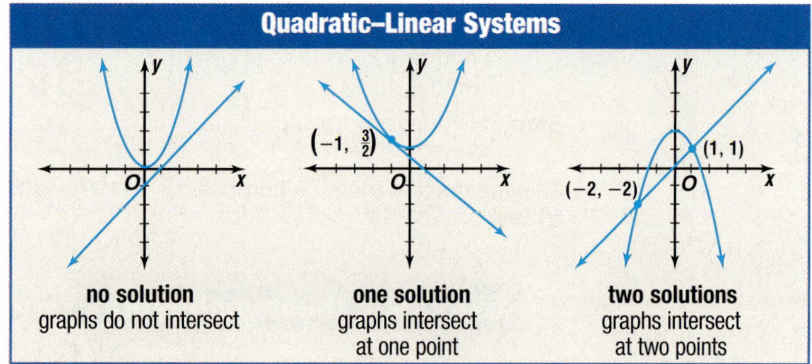

You can solve quadratic-linear systems of equations by using some of the methods you used for solving systems of linear equations. One method is graphing.

Look Back

Graphing Quadratic Equations: Lesson 11–1

Determine whether each system of equations has *one* solution, *two* solutions, or *no* solution by graphing. If the system has one solution or two solutions, name them.

$y = x^2$

$y = x + 2$

The graphs appear to intersect at (−1, 1) and (2, 4). Check this estimate by substituting the coordinates into each equation.

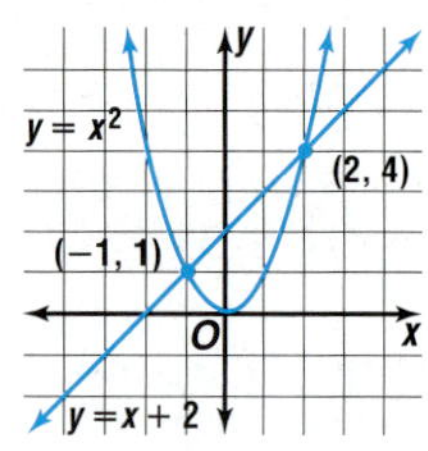

Resource Manager

Reproducible Masters

Chapter 13 Resource Masters
- *Study Guide,* p. 598
- *Skills Practice,* p. 599
- *Practice,* p. 600
- *Reading to Learn Mathematics,* p. 601
- *Enrichment,* p. 602

Graphing Calculator, pp. 38–39

Hands-On Algebra, p. 148

Transparencies

5-Minute Check, 13–6

Technology/Multimedia

Interactive Chalkboard CD-ROM

Check: $y = x^2$ | $y = x + 2$

$1 \stackrel{?}{=} (-1)^2$ $(x, y) = (-1, 1)$ | $1 \stackrel{?}{=} -1 + 2$ $(x, y) = (-1, 1)$

$1 = 1$ ✓ | $1 = 1$ ✓

Check that the ordered pair (2, 4) satisfies both equations.

The solutions of the system of equations are $(-1, 1)$ and $(2, 4)$.

2 $y = 2x^2 + 5$
$y = -x + 3$

Because the graphs do not intersect, there is no solution to this system of equations.

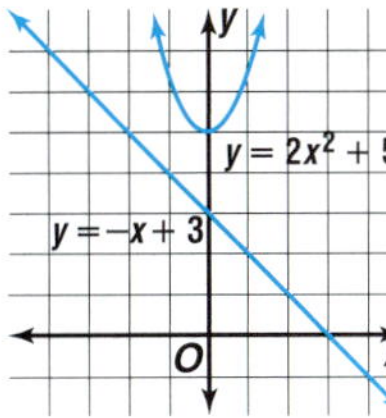

3 $y = -x^2 + 3$
$y = 2x + 4$

The graphs appear to intersect at $(-1, 2)$.

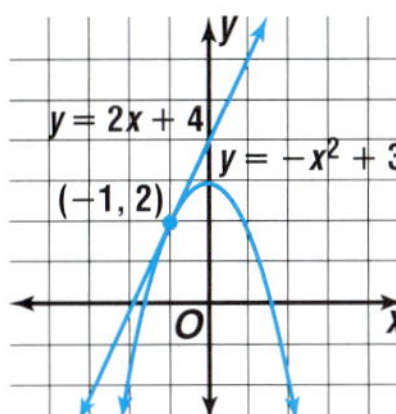

Check: $y = -x^2 + 3$ | $y = 2x + 4$

$2 \stackrel{?}{=} -(-1)^2 + 3$ $(x, y) = (-1, 2)$ | $2 \stackrel{?}{=} 2(-1) + 4$ $(x, y) = (-1, 2)$

$2 \stackrel{?}{=} -(1) + 3$ *Simplify.* | $2 \stackrel{?}{=} -2 + 4$ *Simplify.*

$2 = 2$ ✓ | $2 = 2$ ✓

The solution of the system of equations is $(-1, 2)$.

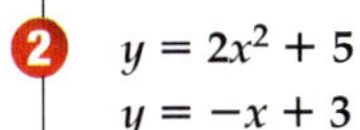

a–c. See margin for graphs. c. (−1, 2), (3, −6)

a. $y = x^2$ **no solution**
$y = x - 2$

b. $y = -2x^2 + 1$
$y = 1$ **(0, 1)**

c. $y = -x^2 + 3$
$y = -2x$

You can also solve quadratic-linear systems of equations by using the substitution method.

Examples

4 **Use substitution to solve each system of equations.**

$y = -4$
$y = x^2 - 4$

Substitute -4 for y in the second equation. Then solve for x.

$y = x^2 - 4$	*Original equation*
$-4 = x^2 - 4$	*Replace y with −4.*
$-4 + 4 = x^2 - 4 + 4$	*Add 4 to each side.*
$0 = x^2$	*Simplify.*
$0 = x$	*Take the square root of each side.*

The solution of the system of equations is $(0, -4)$.

(continued on the next page)

www.algconcepts.com/extra_examples

In-Class Examples

Examples 1–3

Determine whether each system of equations has one solution, two solutions, or no solution by graphing. If the system has one solution or two solutions, name them.

1 $y = x^2$ **two solutions:**
$y = -x$ **(0, 0), (−1, 1)**

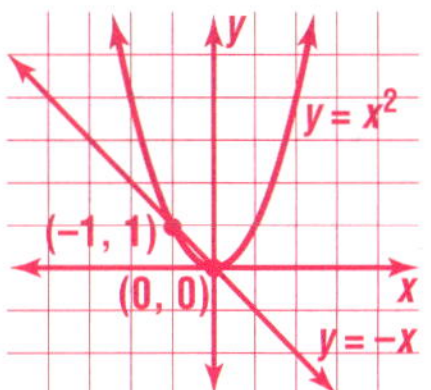

2 $y = -x^2 + 1$
$y = 1$ **one solution; (0, 1)**

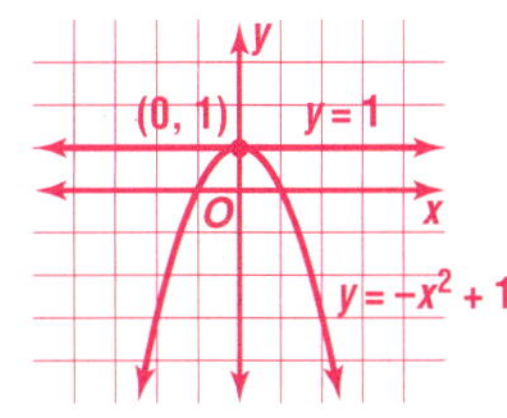

3 $y = -2x^2 + 1$
$x + y = 4$ **no solution**

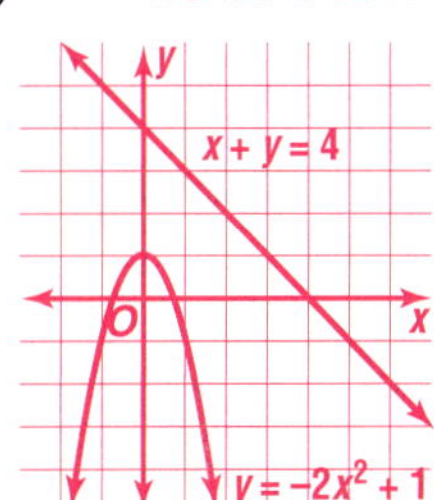

Example 4

Use substitution to solve the system of equations.
$y = 2x^2 - 3$
$y = -1$ **(−1, −1), (1, −1)**

Answers

Your Turn

a.

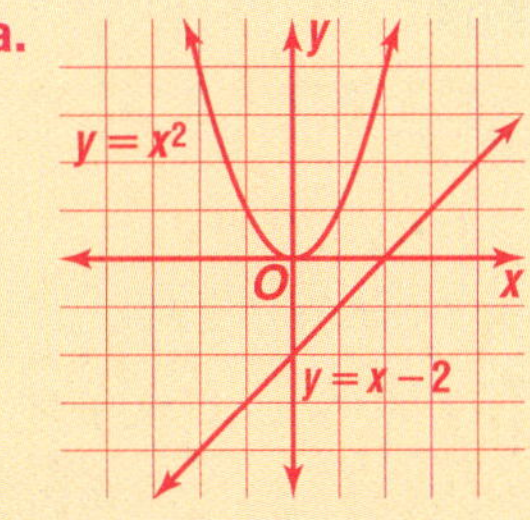

b.

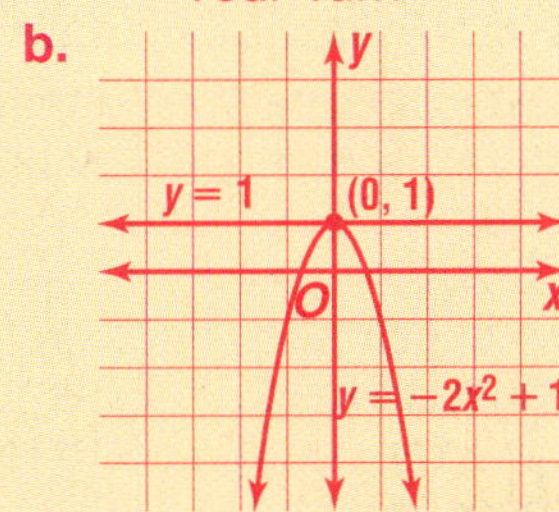

c.

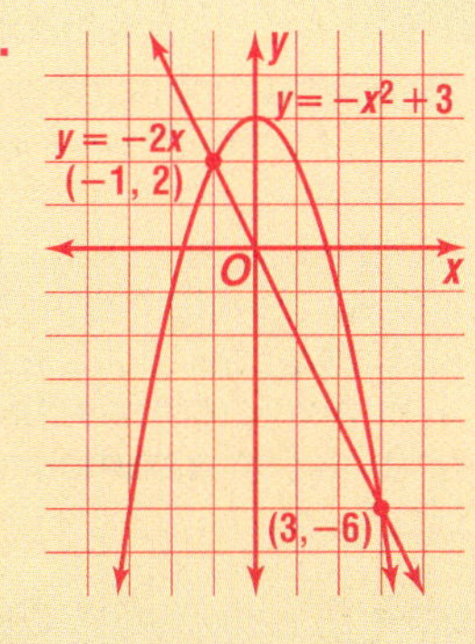

In-Class Example

Example 5

Use substitution to solve the system of equations.
$y = 2x^2 + 3$
$y = 1$ **no real solution**

Teaching Tip In Example 6, be sure students understand that when a quadratic-linear system has two solutions, each solution is an ordered pair. The two y-values are found by substituting *both* x-values, one at a time, into *either* equation.

In-Class Example

Example 6

Use substitution to solve the system of equations.
$y = x^2 - 4x + 4$
$y = x$ **(1, 1), (4, 4)**

Check:
Sketch the graphs of the equations. The parabola and line appear to intersect at (0, −4). The solution is correct.

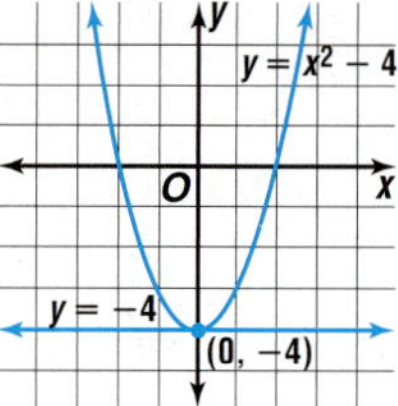

5 $y = x^2$
$y = -3$

Substitute −3 for y in the first equation.

$y = x^2$ *Original equation*
$-3 = x^2$ *Replace y with −3.*
$\sqrt{-3} = x$ *Take the square root of each side.*

There is no real solution because the square root of a negative number is not a real number. *Check by graphing.*

6 $y = x^2 - 4x + 6$
$y = -x + 4$

Substitute $-x + 4$ for y in the first equation.

$y = x^2 - 4x + 6$ *Original equation*
$-x + 4 = x^2 - 4x + 6$ *Replace y with −x + 4.*
$-x + 4 - 4 = x^2 - 4x + 6 - 4$ *Subtract 4 from each side*
$-x = x^2 - 4x + 2$ *Simplify.*
$-x + x = x^2 - 4x + 2 + x$ *Add x to each side.*
$0 = x^2 - 3x + 2$ *Simplify.*
$0 = (x - 2)(x - 1)$ *Factor.*

$0 = x - 2$ or $0 = x - 1$ *Zero Product Property*
$x = 2$ $x = 1$ *Simplify.*

Substitute the values of x in either equation to find the corresponding values of y. *Choose the equation that is easier for you to solve.*

$y = -x + 4$ *Original equation*
$y = -2 + 4$ *Replace x with 2.*
$y = 2$ *Add.*

$y = -x + 4$ *Original equation*
$y = -1 + 4$ *Replace x with 1.*
$y = 3$ *Add.*

The solutions of the system of equations are (2, 2) and (1, 3). The graph shows that the solutions are probably correct. *You can also check by substituting the ordered pairs into the original equations.*

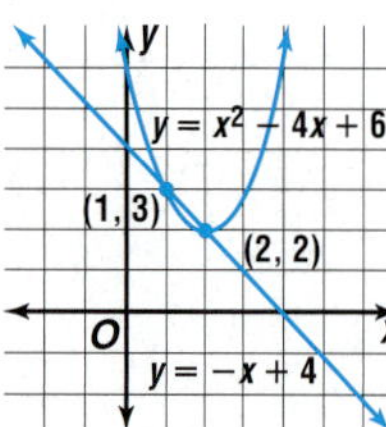

Inclusion Strategies

Exercises in this lesson, such as Example 6, are lengthy and have several sequential steps. The concentration and perseverance required may be especially difficult for students with behavioral difficulties and students with limited attention spans. It may be helpful to write a step-by-step flowchart for these problems, to help students follow their own progress through a problem or for you to refer to or focus on a particular part of a problem.

Your Turn e. no solution f. (−1, −2), (6, 12)

Use substitution to solve each system of equations.

d. $y = x^2 - 12$
$x = 3$ (3, −3)

e. $y = -x^2$
$y = x + 2$

f. $y = x^2 - 3x$
$y = 2x$

Example 7 **Meteorology Link**

Refer to the application at the beginning of the lesson. What are the coordinates of the point at which turbulence will occur?

Use substitution to solve the system $y = \frac{1}{4}x^2 - 12$ and $y = -\frac{1}{2}x$.

Substitute $-\frac{1}{2}x$ for y in the first equation.

$y = \frac{1}{4}x^2 - 12$ *Original equation*

$-\frac{1}{2}x = \frac{1}{4}x^2 - 12$ *Replace y with $-\frac{1}{2}x$.*

$4\left(-\frac{1}{2}x\right) = 4\left(\frac{1}{4}x^2 - 12\right)$ *Multiply each side by 4.*

$-2x = x^2 - 48$ *Simplify.*

$-2x + 2x = x^2 - 48 + 2x$ *Add 2x to each side.*

$0 = x^2 + 2x - 48$ *Simplify.*

$0 = (x + 8)(x - 6)$ *Factor.*

$0 = x + 8$ or $0 = x - 6$ *Zero Product Property*

$x = -8$ $\quad$ $x = 6$ *Simplify.*

Substitute the values of x to find the corresponding values of y.

$y = -\frac{1}{2}x$ *Original equation* $\quad$ $y = -\frac{1}{2}x$ *Original equation*

$y = -\frac{1}{2}(-8)$ or 4 *Replace x with −8.* $\quad$ $y = -\frac{1}{2}(6)$ or −3 *Replace x with 6.*

The solutions of the system are (−8, 4) and (6, −3). This means that turbulence will occur as the plane passes through points having these coordinates. *Check by substituting the coordinates into each equation and by looking at the graph at the beginning of the lesson.*

interNET CONNECTION

Data Update For the latest information about the jet stream, visit: www.algconcepts.com

In-Class Example

Example 7

Refer to the application on page 580, where the jet stream is described by the quadratic equation $y = \frac{1}{4}x^2 - 12$. The route of a U.S. Air Force jet is described by the linear equation $y = \frac{1}{2}x$. What are the coordinates at which turbulence will occur? (8, 4), (−6, −3)

Answer

2. Sample answer:

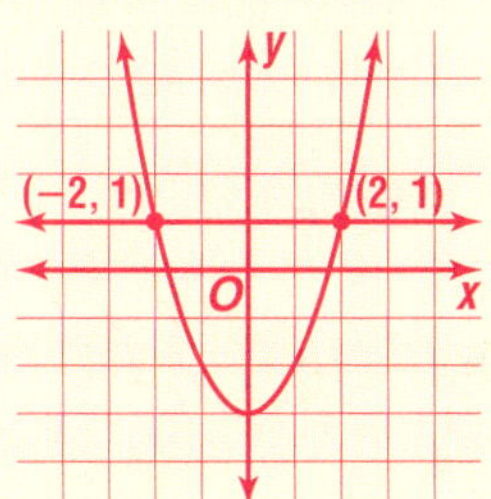

Check for Understanding

Communicating Mathematics

1. (−1, −7), (2, 0)

1. **State** the solution of the system of equations shown at the right.
2. **Sketch** a system of quadratic-linear equations that has solutions (−2, 1) and (2, 1). See margin.
3. *Writing Math* List the different methods of solving linear and quadratic-linear systems of equations. Describe the situation in which each method is most useful. See margin.

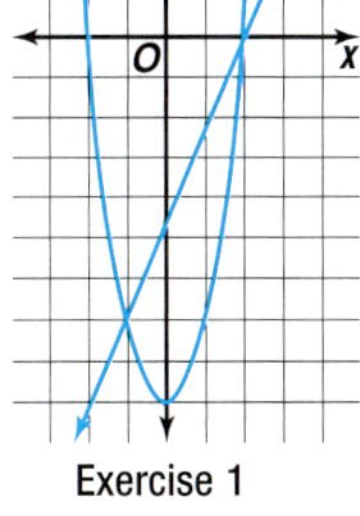

Exercise 1

Reteaching Activity

Visual/Spatial Learners Ask students to demonstrate the ways the graphs of a quadratic equation and linear equation can be related. Students should include diagrams and specific examples in their demonstrations.

Answer

3. Sample answer: The following methods can be used to solve linear systems of equations: graphing, substitution, elimination using addition or subtraction, and elimination using multiplication. Graphing and substitution can be used to solve quadratic-linear systems of equations. Graphing is useful when you want to estimate the solution. The other methods are useful when you want to find the exact solution.

***Study Guide*, p. 598**

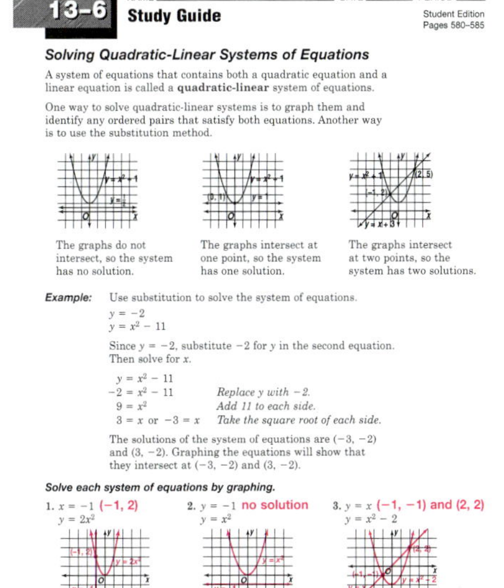
NAME ________ DATE ______ PERIOD ______

13–6 Study Guide Student Edition Pages 580–585

Solving Quadratic-Linear Systems of Equations

A system of equations that contains both a quadratic equation and a linear equation is called a **quadratic-linear** system of equations.

One way to solve quadratic-linear systems is to graph them and identify any ordered pairs that satisfy both equations. Another way is to use the substitution method.

The graphs do not intersect, so the system has no solution. The graphs intersect at one point, so the system has one solution. The graphs intersect at two points, so the system has two solutions.

Example: Use substitution to solve the system of equations.
$y = -2$
$y = x^2 - 11$

Since $y = -2$, substitute −2 for y in the second equation. Then solve for x.

$y = x^2 - 11$
$-2 = x^2 - 11$ *Replace y with −2.*
$9 = x^2$ *Add 11 to each side.*
$3 = x$ or $-3 = x$ *Take the square root of each side.*

The solutions of the system of equations are (−3, −2) and (3, −2). Graphing the equations will show that they intersect at (−3, −2) and (3, −2).

Solve each system of equations by graphing.

1. $x = -1$, $y = 2x^2$ (−1, 2)
2. $y = -1$, $y = x^2$ no solution
3. $y = x$, $y = x^2 - 2$ (−1, −1) and (2, 2)

Use substitution to solve each system of equations.

4. $y = 4$, $y = x^2$ (−2, 4) and (2, 4)
5. $x = -1$, $y = x^2$ (−1, 1)
6. $y = 5$, $y = x^2 - 4$ (−3, 5) and (3, 5)

3 PRACTICE/APPLY

Error Analysis

Watch for students who, in Exercise 6, find the two x values, 1 and -6, and write their answer as the ordered pair $(1, -6)$ or $(-6, 1)$. ***Prevent by*** stressing that every solution is always an (x, y) pair; when they find two values for x, then the next step is to find a y value for each of the two x values.

Assignment Guide

Basic: 9–23 odd, 25–32
Average: 10–22 even, 23–32
All: Quiz 2, 1–5

Answers

4.

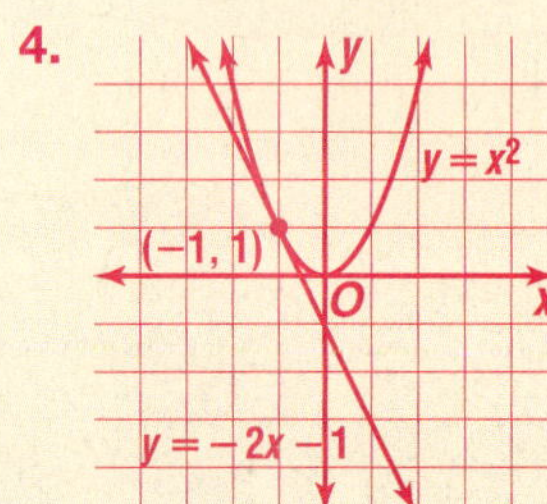

5.

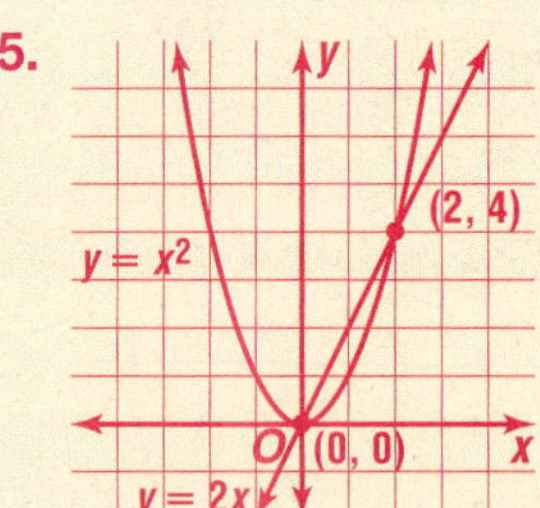

Skills Practice, **p. 599, and** ***Practice,*** **p. 600 (shown)**

13-6 NAME ______ DATE ______ PERIOD ______
Practice Student Edition Pages 580–585

Solving Quadratic-Linear Systems of Equations

Solve each system of equations by graphing.

1. $y = x^2 + 2$, $y = x + 4$ — $(-1, 3), (2, 6)$
2. $y = x^2 - 1$, $y = x - 2$ — no solution
3. $y = -x^2 + 3$, $y = 3$ — $(0, 3)$
4. $y = x^2 + 1$, $y = -x - 1$ — no solution
5. $y = -x^2$, $y = -2x + 1$ — $(1, -1)$
6. $y = x^2 - 2$, $y = x + 4$ — $(-2, 2), (3, 7)$

Use substitution to solve each system of equations.

7. $y = -x^2 + 1$, $y = x - 1$ — $(-2, -3), (1, 0)$
8. $y = x^2 + 2$, $y = -4$ — no solution
9. $y = x^2 - 5$, $x = -3$ — $(-3, 4)$
10. $y = -6x^2 + 1$, $y = x + 1$ — $(0, 1), \left(-\frac{1}{6}, \frac{5}{6}\right)$
11. $y = 2x^2 + 3$, $y = x + 2$ — no solution
12. $y = x^2 + x - 4$, $y = x - 3$ — $(-1, -4), (1, -2)$

Guided Practice

Solve each system of equations by graphing.

Examples 1–3

4. $y = x^2$
$y = -2x - 1$ **(−1, 1)**

5. $y = x^2$
$y = 2x$ **(0, 0), (2, 4)**

4–5. See margin for graphs.

Examples 4–6

Use substitution to solve each system of equations.

6. $y = x^2 - 6$
$y = -5x$ **(−6, 30), (1, −5)**

7. $y = x^2 + 5$
$y = -3$ **no solution**

Example 7

8. **Business** Students in an Algebra I class at Banneker High School are simulating the start-up of a company. The income y can be described by the equation $y = \frac{1}{8}x^2$, where x represents the time in months. The expenses have been growing at a constant rate and can be defined by the equation $y = x$. When will income equal expenses?
at 0 and 8 months

Exercises

Practice

9–14. See Solutions Manual for graphs.

Solve each system of equations by graphing.

A

9. $y = x^2 - 4$
$y = 2x - 4$

10. $x = 2$
$y = x^2 + 1$ **(2, 5)**

11. $y = 2x + 1$
$y = x^2 + 4$

12. $y = x^2 - 6$
$y = x - 4$
(−1, −5), (2, −2)

13. $y = -x^2 + 5$
$y = -x + 3$
(−1, 4), (2, 1)

14. $y = -x^2 + 4$
$y = \frac{1}{2}x + 5$
no solution

Homework Help	
For Exercises	See Examples
9–14	1, 2, 3
15–22, 24	4, 5, 6
23–25	7

Extra Practice
See page 719.

Use substitution to solve each system of equations.

15. $y = \frac{1}{8}x^2 + 5$
$x = -4$ **(−4, 7)**

16. $y = -3x^2$
$y = 2$ **no solution**

17. $y = \frac{1}{2}x^2$
$y = 3x + 8$

B

18. $y = \frac{1}{2}x^2 - 4$
$y = 3x + 4$
(−2, −2), (8, 28)

19. $y = x^2 + 3$
$y = -\frac{1}{2}x - 5$
no solution

20. $y = x^2 - x - 3$
$y = -1$
(−1, −1), (2, −1)

21. What is the solution of the system $y = x^2 - 9$ and $y = 3x - 9$?

22. Find the solution of the system $y = -x^2 + 2x - 3$ and $y = -2x + 1$.
(2, −3)

9. (2, 0), (0, −4)
11. no solution
17. (8, 32), (−2, 2)
21. (3, 0), (0, −9)

Applications and Problem Solving

C

23. **Geometry** Four corners are cut from a rectangular piece of cardboard that is 10 feet by 4 feet. The cuts are x feet from the corners, as shown in the figure at the right. After the cuts are made, the sides of the rectangle are folded to form an open box. The area of the bottom of the box is 16 square feet.

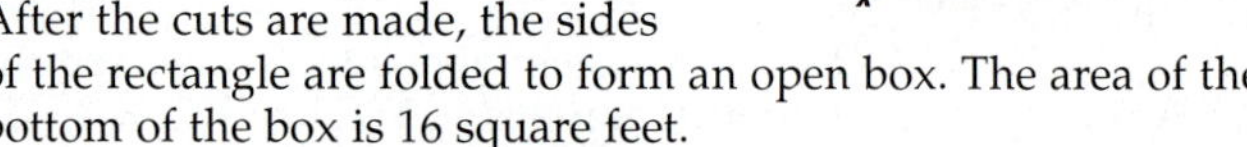

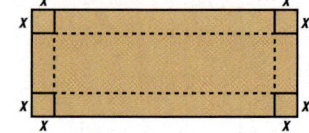

a. Write two equations to represent the area A of the bottom of the box. $A = 16$ **and** $A = 4x^2 - 28x + 40$
b. What are the dimensions of the box? $\ell = 8$ **ft,** $w = 2$ **ft,** $h = 1$ **ft**
c. What is the volume of the box? **16 ft³**

24. **Meteorology** Refer to the application at the beginning of the lesson. If a plane's route is described by the linear equation $y = x - 14$, will it experience turbulence by the jet stream that is described by the quadratic equation $y = \frac{1}{4}x^2 - 12$? Explain. **No, the graphs of the equations do not intersect.**

25. **Critical Thinking** Suppose the perimeter and area of a square have the same measure. Find the length of the sides of the square. Explain how you can use a system of quadratic-linear equations to find the answer. **See margin.**

Mixed Review

26. **Sports** Diego kayaks 16 miles downstream in 2 hours. It takes him 8 hours to make the return trip. What is the rate of the current? *(Lesson 13–5)* **3 mph**

27. Solve the system $4x + 3y = 5$ and $-3x - 2y = -4$ by using elimination. *(Lesson 13–4)* **(2, −1)**

28. Describe how the graph of $y = -(x - 3)^2$ changes from its parent graph of $y = x^2$. *(Lesson 11–2)* **right 3 units, opens down**

Solve each equation. *(Lesson 3–7)*

29. $2 + |x| = 9$ **{−7, 7}**

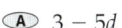

30. $|y| + 4 = 3$ **∅**

31. $12 = 6 + |w + 2|$ **{4, −8}**

Standardized Test Practice

32. **Multiple Choice** Which expression shows *3 less than the product of 5 and d*? *(Lesson 1–1)* **C**

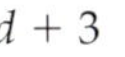

Ⓐ $3 - 5d$ Ⓑ $5d + 3$ Ⓒ $5d - 3$ Ⓓ $5 + d - 3$

Quiz 2 Lessons 13–4 through 13–6

Use elimination to solve each system of equations. *(Lessons 13–4 & 13–5)*

1. $x + y = 6$
   $2x - y = 6$ **(4, 2)**

2. $2x - 3y = -9$
   $x - 2y = -5$ **(−3, 1)**

3. $3x - 5y = 8$
   $4x - 7y = 10$ **(6, 2)**

4. Use substitution to find the solution of the system $y = x^2 + 5$ and $y = 6x - 3$. *(Lesson 13–6)* **(4, 21), (2, 9)**

5. **Number Theory** The difference between two numbers is 38. The greater number is three times the lesser number minus two. *(Lesson 13–4)*
   a. Write a system of equations to represent the situation. **$x - y = 38$, $x = 3y - 2$**
   b. What are the numbers? **58, 20**

4 ASSESS

Open-Ended Assessment

Modeling Have students graph a quadratic equation. Then ask them to use a pencil to model the graphs of lines that intersect the parabola in two points, one point, or do not intersect the parabola.

Quiz 2

The Quiz provides students with a brief review of the concepts and skills in Lessons 13–4 through 13–6. Lesson numbers are given to the right of the exercises or instruction lines so students can review concepts not yet mastered.

Answer

25. Side length 4 units; solve the quadratic-linear system of equations $p = s^2$ and $p = 4s$, where p represents perimeter and s represents side length. The solutions are (0, 0) and (4, 16).

***Enrichment*, p. 602**

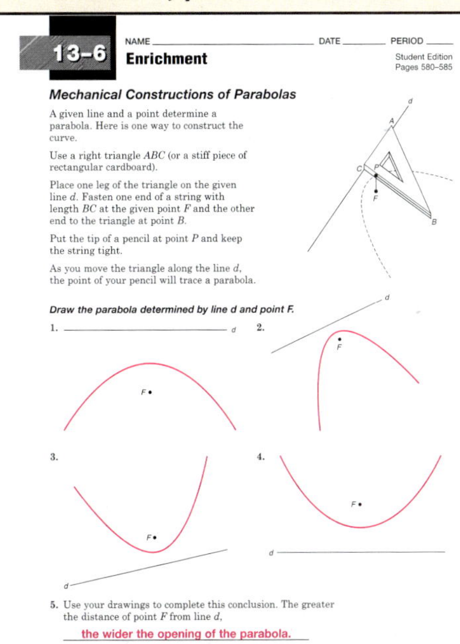

13–6 NAME ______ DATE ______ PERIOD ______

Enrichment

Student Edition Pages 580–585

Mechanical Constructions of Parabolas

A given line and a point determine a parabola. Here is one way to construct the curve.

Use a right triangle *ABC* (or a stiff piece of rectangular cardboard).

Place one leg of the triangle on the given line *d*. Fasten one end of a string with length *BC* at the given point *F* and the other end to the triangle at point *B*.

Put the tip of a pencil at point *P* and keep the string tight.

As you move the triangle along the line *d*, the point of your pencil will trace a parabola.

Draw the parabola determined by line d and point F.

1.

2.

3.

4.

5. Use your drawings to complete this conclusion. The greater the distance of point *F* from line *d*,
the wider the opening of the parabola.

Extra Credit

The solution of the system whose equations are $y = x^2 - 1$ and a line are the two points whose x values are -2 and 1. Write an equation for the line.
$y = -x + 1$

13–7 Graphing Systems of Inequalities

What You'll Learn

You'll learn to solve systems of inequalities by graphing.

Why It's Important

Money Systems of inequalities can be useful in helping you get the most for your money. *See Example 3.*

Thomas Edison was a newsboy during the Civil War. One day, he persuaded the editor to give him 300 copies of the paper instead of the usual 100. He went to the train station where people were eager for news of the war. He was able to sell the papers for 10¢ and 25¢ instead of the usual 5¢. If he wanted to earn at least \$52, how many papers could he have sold for 10¢ and 25¢?

x = the number of 10¢ papers $\qquad$ y = the number of 25¢ papers

The following **system of inequalities** can be used to represent the conditions of this problem.

$x + y \leq 300$ *He can sell as many as 300 papers.*

$0.10x + 0.25y \geq 52$ *He wants to earn at least \$52.*

Since both x and y represent the number of papers, neither can be a negative number. Thus, $x \geq 0$ and $y \geq 0$. The solution of the system is the set of all ordered pairs that satisfy both inequalities and lie in the first quadrant. The solution can be determined by graphing each inequality in the same coordinate plane as shown below.

Look Back

Graphing Linear Inequalities in Two Variables: Lesson 12–7

Recall that the graph of each inequality is called a *half-plane.*

- Any point in the yellow region satisfies $x + y \leq 300$.
- Any point in the blue region satisfies $0.10x + 0.25y \geq 52$.
- Any point in the green region satisfies both inequalities. The intersection of the two half-planes represents the solution to the system of inequalities.
- The graphs of $x + y = 300$ and $0.10x + 0.25y = 52$ are the boundaries of the region and are included in the graph of the system.

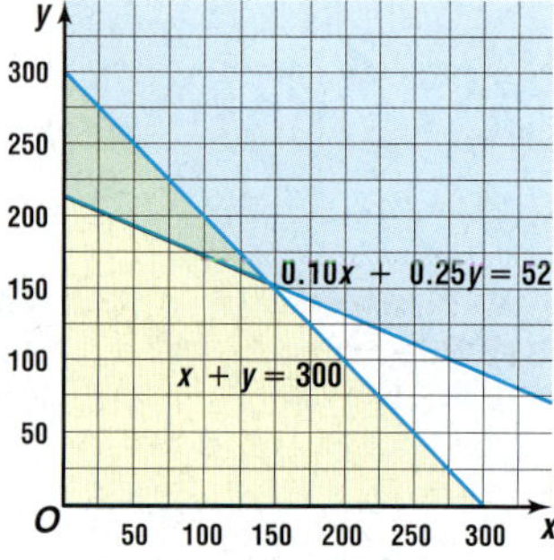

This solution is a region that contains the graphs of an infinite number of ordered pairs. An example is (50, 225). This means that Edison could have sold 50 papers at 10¢ and 225 papers at 25¢ to earn at least \$52.

Check:

$x + y \leq 300$ *Original inequality*

$50 + 225 \overset{?}{\leq} 300$ *(x, y) = (50, 225)*

$275 \leq 300$ ✓

$0.10x + 0.25y \geq 52$ *Original inequality*

$0.10(50) + 0.25(225) \overset{?}{\geq} 52$ *(x, y) = (50, 225)*

$61.25 \geq 52$ ✓

1 FOCUS

5-Minute Check

Lesson 13–6

Determine whether each system of equations has one ***solution, two solutions, or*** no ***solution by graphing. If the system has one solution or two solutions, name them.***

1. $y = -x^2 + 3$, $y = -x - 3$ **two; (−2, −1), (3, −6)**

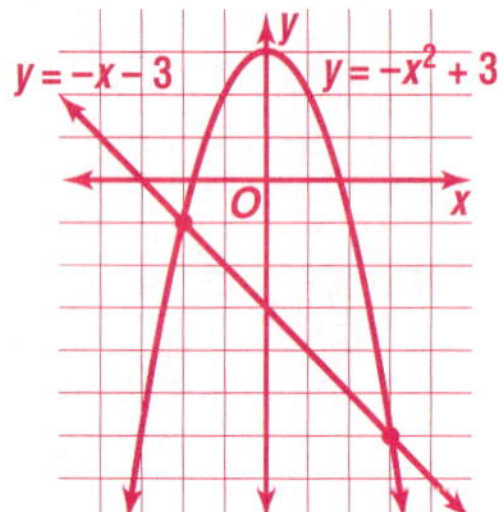

2. $y = \frac{1}{2}x^2 + 3$, $y = 2x + 1$ **one; (2, 5)**

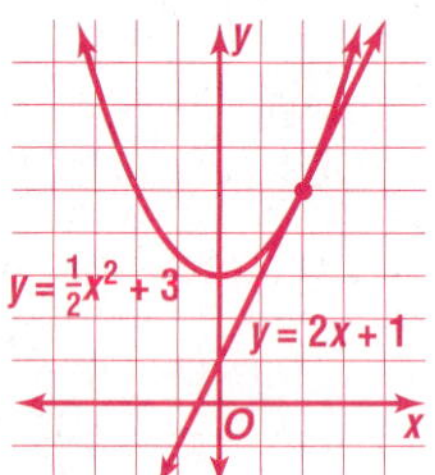

Use the system of equations to complete the following.

$y = x^2 + 2$

$y = -x + 4$

3. Substitute $-x + 4$ for y in the first equation, and then solve for x. **−2 or 1**
4. Find the solution(s) to the system of equations. **(−2, 6), (1, 3)**

Motivating the Lesson

Real-World Connection Use a map of the United States that shows lines of latitude and longitude. Ask students to state one condition for latitude (such as *south of 85° latitude*) and one condition for longitude (such as *east of 35° longitude*). Then ask students to find all the states that satisfy both conditions.

Resource Manager

Reproducible Masters

Chapter 13 Resource Masters

- *Study Guide,* p. 603
- *Skills Practice,* p. 604
- *Practice,* p. 605
- *Reading to Learn Mathematics,* p. 606
- *Enrichment,* p. 607
- *Assessment,* p. 619

Graphing Calculator, p. 37

Hands-On Algebra, p. 149

Transparencies

5-Minute Check, 13–7

Technology/Multimedia

Interactive Chalkboard CD-ROM

Remember that the boundary lines of inequalities are only included in the solution if the inequality symbol is greater than or equal to, $\geq$, or less than or equal to, $\leq$.

Examples

Solve each system of inequalities by graphing. If the system does not have a solution, write *no solution*.

1 $y \leq -1$
$y > x - 3$

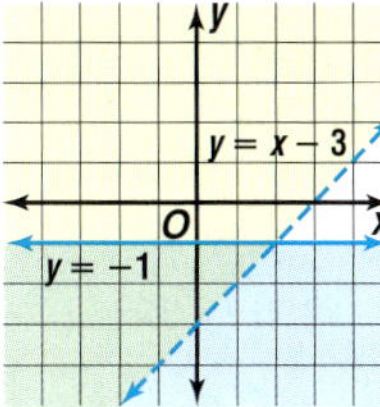

The solution is the ordered pairs in the intersection of the graphs of $y \leq -1$ and $y > x - 3$. The region is shaded in green at the right. The graphs of $y = -1$ and $y = x - 3$ are the boundaries of this region. The graph of $y = x - 3$ is a dashed line and is not included in the solution of the system. *Choose a point and check the solution.*

2 $2y < x - 2$
$-3x + 6y \geq 12$

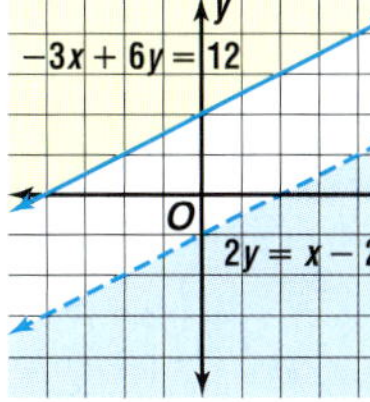

The graphs of $2y = x - 2$ and $-3x + 6y = 12$ are parallel lines. *Check this by graphing or by comparing the slopes.*

Because the regions in the solution of $2y < x - 2$ and $-3x + 6y \geq 12$ have no points in common, the system of inequalities has no solution.

Your Turn **a–b. See margin.**

a. $x \geq -1$
$y < x - 2$

b. $y \geq 4x + 5$
$x + y \geq 3$

Example

Spending Link

3 **Luisa has \$96 to spend on gifts for the holidays. She must buy at least 9 gifts. She plans to buy puzzles that cost \$8 or \$12. How many of each puzzle can she buy?**

x = the number of \$8 puzzles $\quad$ y = the number of \$12 puzzles

The following system of inequalities can be used to represent the conditions of this problem.

$x + y \geq 9$ *She wants to buy at least 9 puzzles.*
$8x + 12y \leq 96$ *The total cost must be no more than \$96.*

Because the number of puzzles she can buy cannot be negative, both $x \geq 0$ and $y \geq 0$.

(continued on the next page)

www.algconcepts.com/extra_examples

2 TEACH

In-Class Examples

Examples 1–2

***Solve each system of inequalities by graphing. If the system does not have a solution, write* no solution.**

1 $x \geq 3$
$y < x + 4$

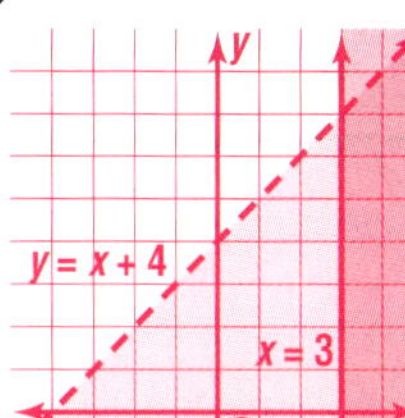

2 $x + y > 2$
$2y \leq -2x - 2$ **no solution**

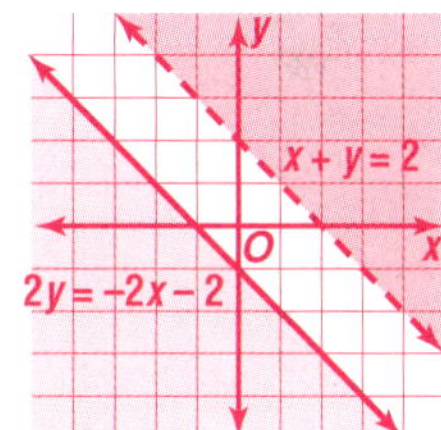

Example 3

A radio station is giving away tickets to a play. They plan to give away tickets to seats that cost \$10 or \$20. They want to give away at least 20 tickets, and the total cost of all the tickets can be no more than \$300. How many tickets at each price can they give away?

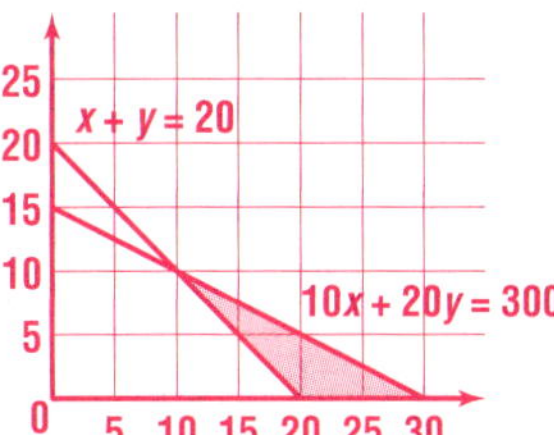

Any point in the shaded region is a possible solution. For example, they can give away 20 of the \$10 tickets and 5 of the \$20 tickets.

Answers

Your Turn

a.

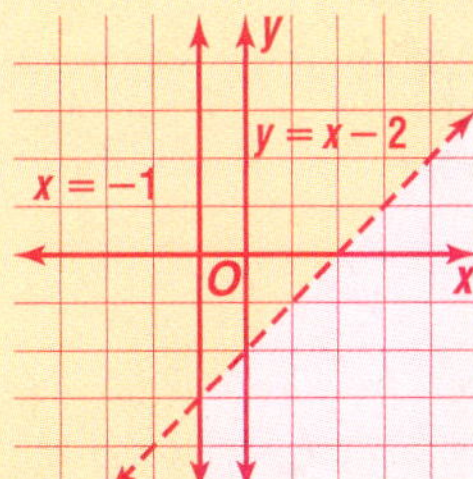

b.

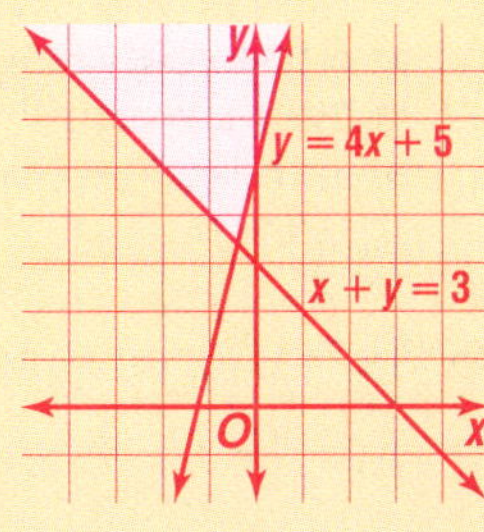

Answers

Graphing Calculator Exploration

1.

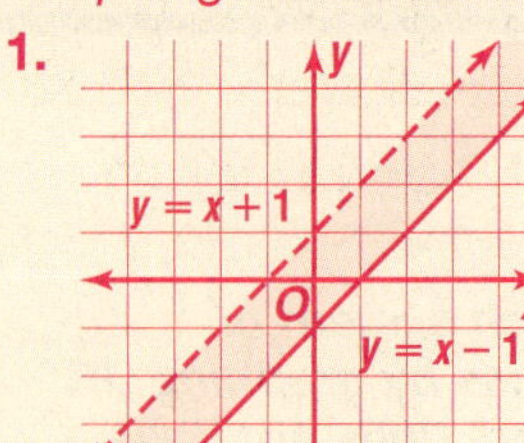

Possible solution: (0, 0)

2.

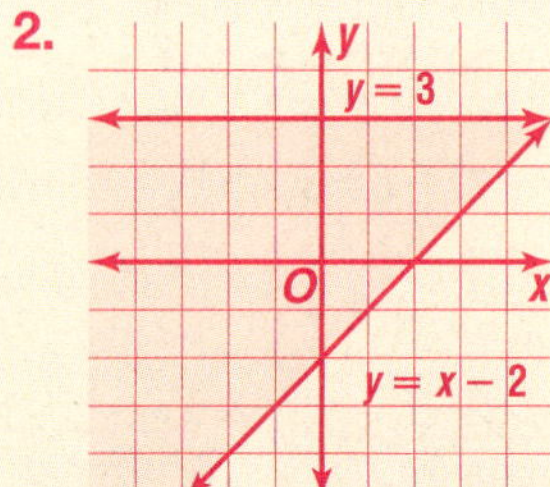

Possible solution: (−1, 3)

3.

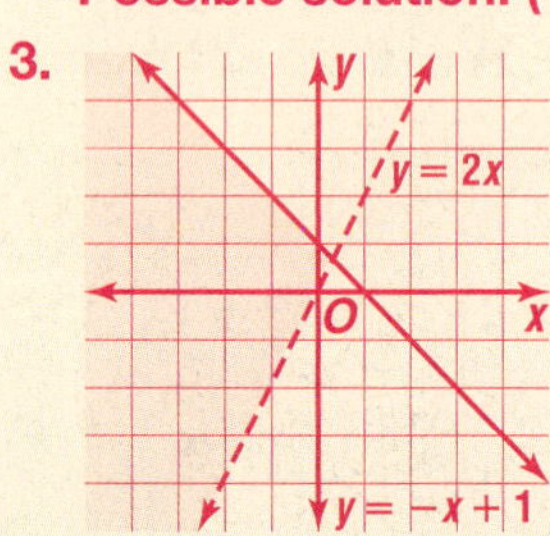

Possible solution: (−2, −2)

Study Guide, p. 603

13–7 NAME ________ DATE ______ PERIOD ______

Study Guide

Student Edition Pages 586–591

Graphing Systems of Inequalities

To graph a system of linear inequalities, first graph a boundary line for each inequality. Then shade on one side of each boundary line. The region where the shaded areas overlap contains the solutions of the system of inequalities.

Example: Solve the system of inequalities by graphing.

$y > -2$
$y \leq x + 1$

Step 1 Graph the boundary lines $y = -2$ and $y = x + 1$. Since y is *greater than* -2, make the line $y = -2$ dashed. The line $y = x + 1$ is solid because y is *less than or equal to* $x + 1$.

Step 2 Because $y > -2$ has a *greater than* symbol, shade above the boundary line. Shade below the boundary line for $y \leq x + 1$ because this inequality contains a *less than* symbol.

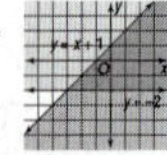

The region where the shaded areas intersect is the solution of the system of inequalities.

Use the following rules to help you graph inequalities.

Inequality	Example	Boundary Line	Where to Shade
$y < ax + b$	$y < 2x$	dashed	below the line
$y \leq ax + b$	$y \leq -x + 1$	solid	below the line
$y > ax + b$	$y > 5x - 2$	dashed	above the line
$y \geq ax + b$	$y \geq -4x$	solid	above the line

***Solve each system of inequalities by graphing. If the system does not have a solution, write* no solution.**

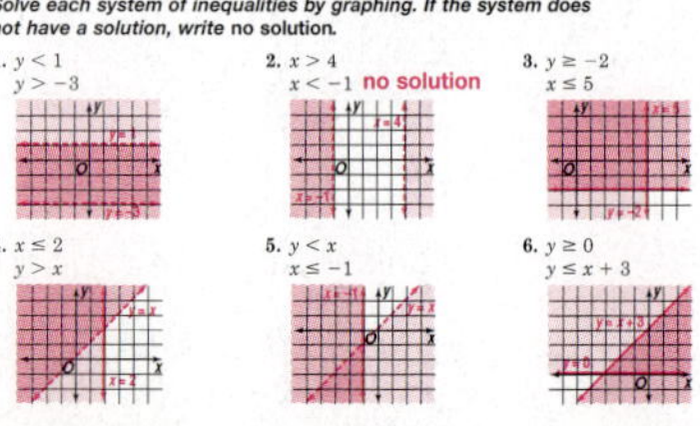

1. $y < 1$, $y > -3$
2. $x > 4$, $x < -1$ **no solution**
3. $y \geq -2$, $x \leq 5$
4. $x \leq 2$, $y > x$
5. $y < x$, $x \leq -1$
6. $y \geq 0$, $y \leq x + 3$

The solutions are all of the ordered pairs in the intersection of the graphs of these inequalities. Only the first quadrant is used because $x \geq 0$ and $y \geq 0$.

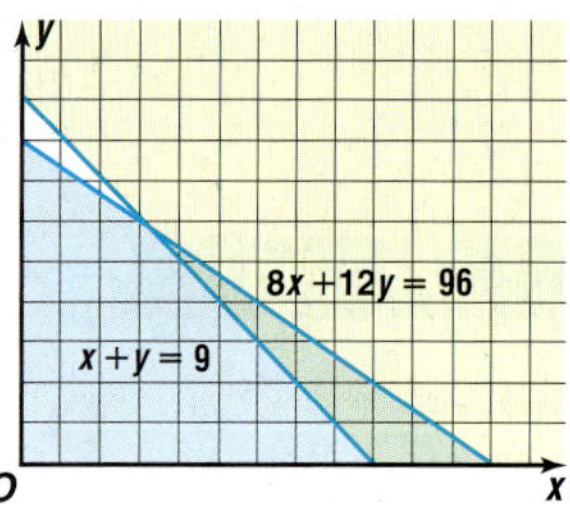

Any point with whole-number coordinates in the region is a possible solution. For example, since the point at (9, 1) is in the region, Luisa could buy 9 puzzles for \$8 and 1 for \$12. The greatest number of gifts that she could buy would be 12 puzzles for \$8 and no puzzles for \$12.

A graphing calculator can be helpful in solving systems of inequalities or in checking solutions.

Graphing Calculator Tutorial
See pp. 750–753.

Graphing Calculator Exploration

Use a graphing calculator to solve the system of inequalities.

$y \geq 4x - 3$
$y \leq -2x + 1$

The graphing calculator graphs equations and shades above the first equation entered and below the second equation entered. (Note that inequalities that have $>$ or $\geq$ are lower boundaries and inequalities that have $<$ or $\leq$ are upper boundaries.)

Step 1 Press [2nd] [DRAW] 7 to choose the SHADE feature.

Step 2 Enter the equation that is the lower boundary of the region to be shaded, 4 [X,T,θ,n] [−] 3.

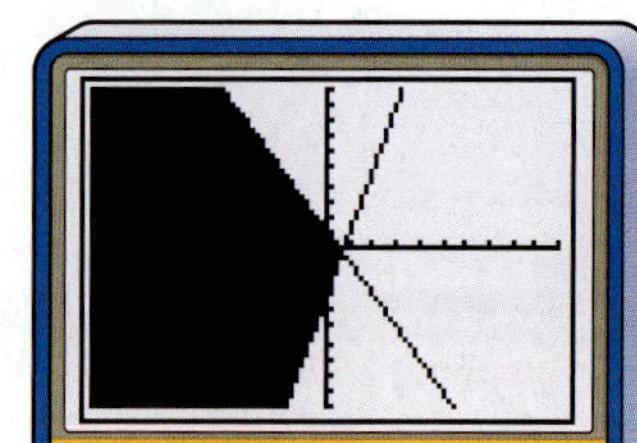

Step 3 Press [,] and enter the equation that is the upper boundary of the region to be shaded, [(−)] 2

Try These **1–3. See margin.**

Use a graphing calculator to graph each system of inequalities. State one possible solution.

1. $y \geq x - 1$
   $y < x + 1$
2. $y \geq x - 2$
   $y \leq 3$
3. $y > 2x$
   $y \leq -x + 1$

Graphing Calculator Exploration

You may wish to demonstrate how to use the shading options on the Y= screen. Move the cursor to the equation for which shading is desired. Press the left arrow key to go to the symbol to the left of the letter Y. Press [ENTER] repeatedly to view the options. The meaning of each option symbol is explained in the TI-83 Plus Guidebook.

Check for Understanding

Communicating Mathematics

1–2. See margin.

1. **Compare and contrast** the solution of a system of equations and the solution of a system of inequalities.
2. **Sketch** a system of linear equations that has no solution.
3. **You Decide** Kyle says that the solution of $y \geq 2x + 2$ and $y \leq -x - 1$ is all of the points in region B. Tarika says that the solution is all of the points in region D. Who is correct? Explain. **Kyle; all of the points in region B satisfy both inequalities.**

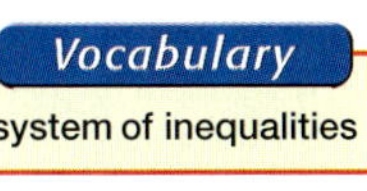

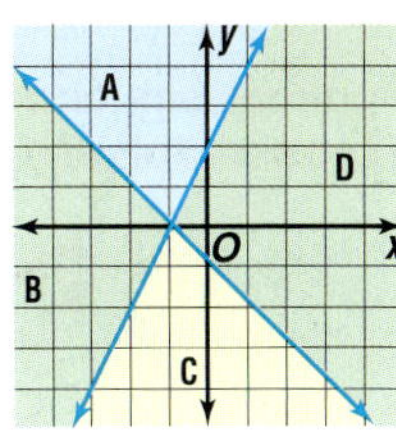

Exercise 3

Guided Practice

Getting Ready State whether each ordered pair is a solution of the system of inequalities $x \geq -2$ and $y \leq 5$.

Sample: $(-3, 5)$ **Solution:** $x \geq -2$
$-3 \overset{?}{\geq} -2$ *Replace x with −3.*
$(-3, 5)$ is not a solution.

4. $(6, -1)$ **yes**
5. $(8, 7)$ **no**
6. $(0, 0)$ **yes**

Examples 1 & 2

Solve each system of inequalities by graphing. If the system does not have a solution, write *no solution*.

7–10. See Solutions Manual.

7. $x \geq -1$
   $y \geq -4$
8. $y \leq 5$
   $y \geq -x + 1$
9. $x + y > 2$
   $y < x + 6$
10. $y \geq \frac{1}{2}x - 2$
    $x - y < 3$

Example 3

11. **Shopping** Nathaniel must buy two types of cookies for a banquet dinner. He only has \$25 to spend and needs at least 6 dozen cookies. The grocery store has small sugar cookies for \$3 a dozen and chocolate chip cookies for \$4 a dozen. How many cookies of each type can he buy? List three possible solutions. **Sample answer: 2 dozen sugar, 4 dozen choc. chip; 4 dozen sugar, 3 dozen choc. chip; 5 dozen sugar, 2 dozen choc. chip**

Exercises

Practice

Homework Help	
For Exercises	See Examples
12–25	1, 2
26	3
28	2

Extra Practice
See page 720.

Solve each system of inequalities by graphing. If the system does not have a solution, write *no solution*. **12–23. See Solutions Manual.**

12. $x < 2$
    $y > -1$
13. $y > 0$
    $x \leq 0$
14. $y > 2$
    $y > -x + 2$
15. $y \geq x$
    $y \leq -x$
16. $y \leq 2x$
    $y \geq -3x$
17. $y \geq -3$
    $y < -3x + 1$
18. $x > 2$
    $y > -2x + 3$
19. $y \leq 2x - 1$
    $y \geq 2x + 2$
20. $x + y \geq 2$
    $x + y \leq 6$
21. $y \leq 2x + 2$
    $2x + y \leq 4$
22. $y + 4 < x$
    $2y + 4 > -3x$
23. $2y + x < 4$
    $3x - y > 1$

Reteaching Activity

Verbal/Linguistic Learners Have students give an oral presentation on how to graph two inequalities, including how to tell whether each boundary line is included in its half-plane, and how to tell which part of the plane represents the solution to both inequalities.

3 PRACTICE/APPLY

Error Analysis

Watch for students who, in Exercises 7–10, graph the equations correctly but make mistakes choosing which region to shade.
Prevent by having students choose a point in the region and substituting the coordinates into the two inequalities to check that the point fulfills both conditions.

Assignment Guide

Basic: 13–27 odd, 28–37
Average: 12–24 even, 26–37

Answers

1. A system of linear equations may have at most one solution if the equations are distinct. A system of inequalities may have an infinite number of solutions.
2. Sample answer:

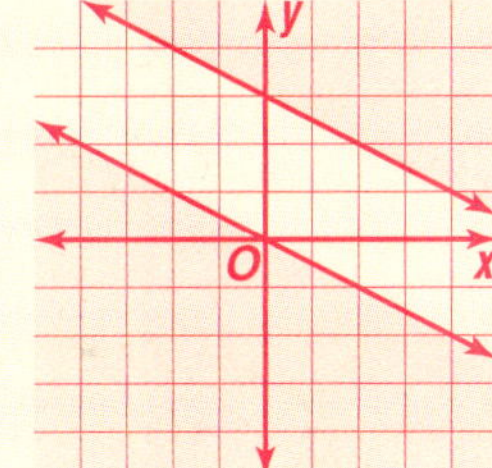

Skills Practice, **p. 604, and *Practice,* p. 605 (shown)**

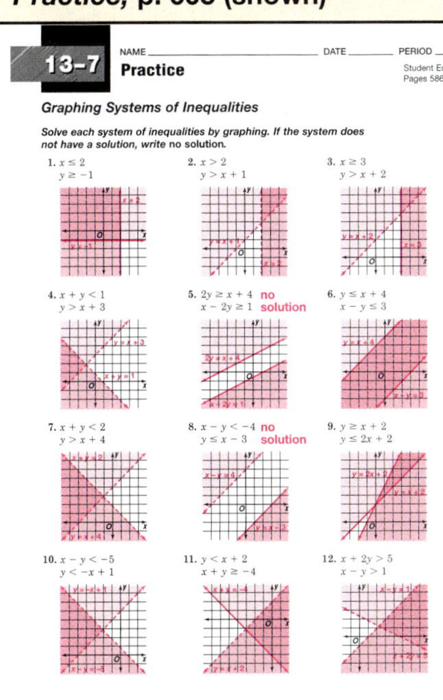

13-7 NAME ______ DATE ______ PERIOD ______

Practice

Student Edition Pages 586–591

Graphing Systems of Inequalities

Solve each system of inequalities by graphing. If the system does not have a solution, write no solution.

1. $x \leq 2$
   $y \geq -1$
2. $x > 2$
   $y > x + 1$
3. $x \geq 3$
   $y > x + 2$
4. $x + y < 1$
   $y > x + 3$
5. $2y \geq x + 4$
   $x - 2y \geq 1$ no solution
6. $y \leq x + 4$
   $x - y \leq 3$
7. $x + y < 2$
   $y > x + 4$
8. $x - y < -4$
   $y \leq x - 3$ no solution
9. $y \geq x + 2$
   $y \leq 2x + 2$
10. $x - y < -5$
    $y < -x + 1$
11. $y < x + 2$
    $x + y \geq -4$
12. $x + 2y > 5$
    $x - y > 1$

4 ASSESS

Open-Ended Assessment

Modeling Have students use lightly colored transparency sheets or cellophane to model inequalities on an overhead. Name simple inequalities such as $y > x$ and $y < 2$ for students to model and have students point out the region of the graph that fulfills both inequalities.

Chapter 13, Quiz B (Lessons 13–4 through 13–7) is available in the *Chapter 13 Resource Masters*, p. 619.

Answer

27. People who use the phone less than 75 minutes a month should select the \$0.15 per minute plan. People who use the phone more than 75 minutes a month should select the \$0.05 per minute plan. The middle plan is never the cheapest.

***Enrichment*, p. 607**

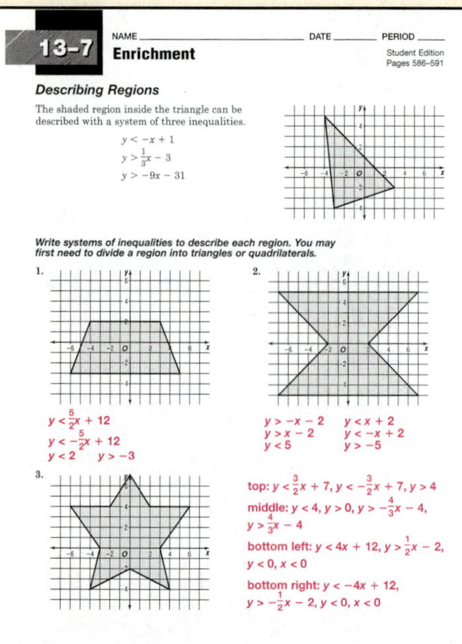

13-7 NAME ______ DATE ______ PERIOD ______

Enrichment

Student Edition Pages 586–591

Describing Regions

The shaded region inside the triangle can be described with a system of three inequalities.

$y < -x + 1$
$y > \frac{1}{3}x - 3$
$y > -9x - 31$

Write systems of inequalities to describe each region. You may first need to divide a region into triangles or quadrilaterals.

1. $y < \frac{5}{2}x + 12$, $y < -\frac{5}{2}x + 12$, $y < 2$, $y > -3$

2. $y > -x - 2$, $y > x - 2$, $y < 5$, $y < x + 2$, $y < -x + 2$, $y > -5$

3. top: $y < \frac{3}{2}x + 7, y < -\frac{3}{2}x + 7, y > 4$
middle: $y < 4, y > 0, y > -\frac{4}{3}x - 4, y > \frac{4}{3}x - 4$
bottom left: $y < 4x + 12, y > \frac{1}{2}x - 2, y < 0, x < 0$
bottom right: $y < -4x + 12, y > -\frac{1}{2}x - 2, y < 0, x < 0$

24–25. See Solutions Manual.

24. Use the graphs of $y \geq x - 3$ and $y \geq -x - 1$ to determine the solution of the system.

25. Find the solution of the system $3x + 2y \leq -6$ and $y > x + 2$ by graphing.

Applications and Problem Solving

26. **Sports** Nykia must score at least 20 points in the last basketball game of the season to tie the school record for total points in a season. Usually she has no more than 40 opportunities to shoot the ball (both field goals and free throws combined) and makes half of her shots. What combination of shots can Nykia make to score the 20 points? Field goals are worth 2 points and free throws are worth 1 point. **Sample answer: 8 field goals, 4 free throws**

27. **Communication** A long-distance carrier offers three fee plans to their customers.
- 15 cents a minute with no monthly service charge
- 10 cents a minute and a monthly service charge of \$5.25
- 5 cents a minute and a monthly service charge of \$7.50

Depending on how much people use the phone each month, which plan should they select? **See margin.**

28. **Critical Thinking** Write an inequality involving absolute value that has no solution. **Sample answer: $|x| < -3$**

Mixed Review

Determine whether each system of equations has *one* solution, *two* solutions, or *no* solution by graphing. If the system has one or two solutions, name them. *(Lesson 13–6)*

29–31. See Solutions Manual for graphs.

29. $y = x^2$
$y = 4$
(2, 4), (−2, 4)

30. $y = x^2 + 3$
$y = 2x - 1$
no solution

31. $y = -x^2 + 2$
$y = -x + 5$
no solution

32. Use elimination to find the solution of the system $5x + 4y = -2$ and $3x - 2y = 1$. *(Lesson 13–5)* $\left(0, -\frac{1}{2}\right)$

33. **Mixtures** The Coffee Hut mixes coffee beans that cost \$5.25 per pound with coffee beans that cost \$6.50 per pound. They need a mixture of 5 pounds of coffee beans that costs \$5.50 per pound. How many pounds of each type of coffee bean should they use? *(Lesson 13–3)* **4 lb of the \$5.25, 1 lb of the \$6.50**

Solve each equation. *(Lesson 4–4)*

34. $-4w = 32$ **−8**

35. $13 = 2.6a$ **5**

36. $\frac{2}{5}y = -8$ **−20**

Standardized Test Practice

37. **Multiple Choice** Find $-15 + (-3) + (-7) + 6$. *(Lesson 2–3)* **A**

(A) −19 (B) 1
(C) 11 (D) 16

www.algconcepts.com/self_check_quiz

Extra Credit

Write the system of inequalities whose solution is the graph at the right. $y \leq x + 3, y > \frac{3}{4}x$

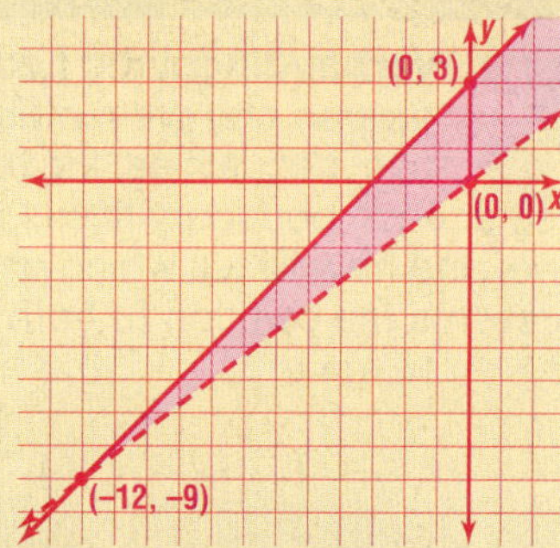

Car Dealer

Do you work well with people? Do cars interest you? If so, you may want to consider a career as a car dealer. Car dealers must keep track of the number of cars they need to sell each week to ensure maximum profit.

Suppose a car dealer receives a profit of \$500 for each mid-sized car m sold and \$750 for each sport-utility vehicle s sold. The dealer must sell at least two mid-sized cars for each sport-utility vehicle and must earn at least \$3500 per week. The table and graph shown represent this situation.

Situation	Inequality
At least 2 mid-sized cars must be sold for each sport-utility vehicle.	$m \geq 2s$
Earn a profit of at least \$3500.	$500m + 750s \geq 3500$

Weekly Sales

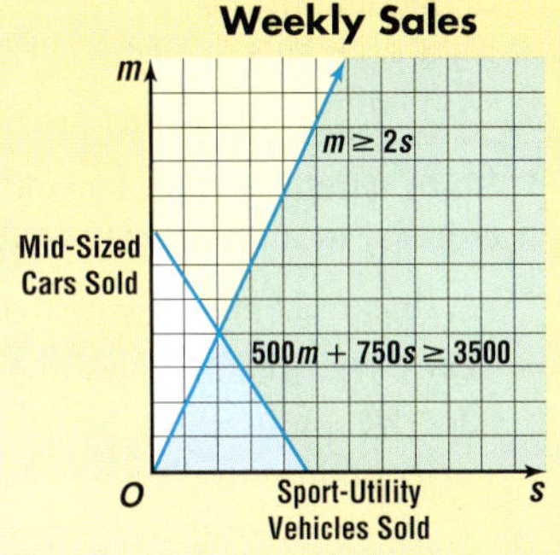

1. Suppose a car dealer sells 2 sport-utility vehicles. How many mid-sized cars must be sold to earn at least \$3500? **4 cars**
2. If a car dealer sells only one sport-utility vehicle, how many mid-sized cars must be sold to meet the goal? **6 cars**

FAST FACTS About Car Dealers

Working Conditions

- usually work in a comfortable environment
- evening, weekend, and holiday work
- work a 40-hour week, with some overtime

Education

- good communication skills and computer skills are helpful
- high school math and business classes
- college degree in business is helpful

Job Outlook

Expected Growth in Jobs

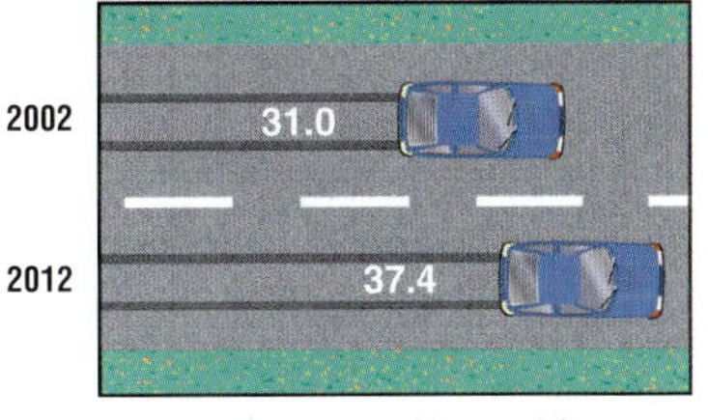

Employment (thousands)

Source: Bureau of Labor Statistics

interNET CONNECTION **Career Data** For the latest information on car dealers, visit: www.algconcepts.com

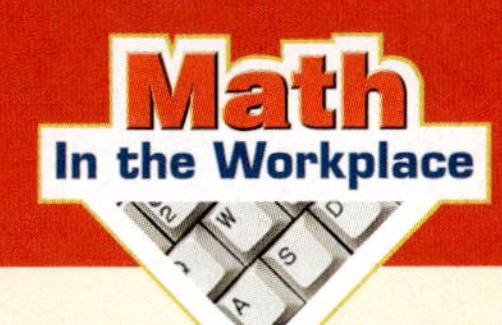

Car dealers work together as a team to make the dealership successful. Car dealers must know about not only the products they sell but also finance, insurance, state and federal laws, warranties, and the automobile industry in general. To keep sales staff up to date on the latest product developments and sales techniques, dealers and manufacturers conduct regular training sessions and encourage salespeople to take advantage of a wide variety of outside sales and business courses.

Prospective car dealers should take high school classes in finance and business, as well as classes that make them familiar with automotive design and maintenance.

Related Careers

- car finance and insurance managers
- service center managers
- fleet managers

Community Connection

Almost all cities, large and small, have car dealerships. If possible, invite a car dealer to visit the class to discuss the nature of the work and the kinds of math skills used on the job.

Not on the Net

If students have limited or no access to the Internet, they can find additional information by contacting the following organziations.

National Automobile Dealers Association
8400 Westpark Drive
McLean, VA 22102

Vocational Industrial Clubs of America
P.O. Box 3000
Leesburg, VA 20177

Understanding and Using the Vocabulary

This section provides a listing of the new terms, properties, and phrases that were introduced in this chapter. The exercises check students' understanding of the terms by using a variety of verbal formats including matching, completion, and true/false.

Glossary A complete glossary of terms appears on pages 762–783.

MindJogger Videoquizzes

MindJogger Videoquizzes provide an alternative review of concepts presented in this chapter. Students work in teams to answer questions, gaining points for correct answers.

Answers

Page 593

17. (3, −2)

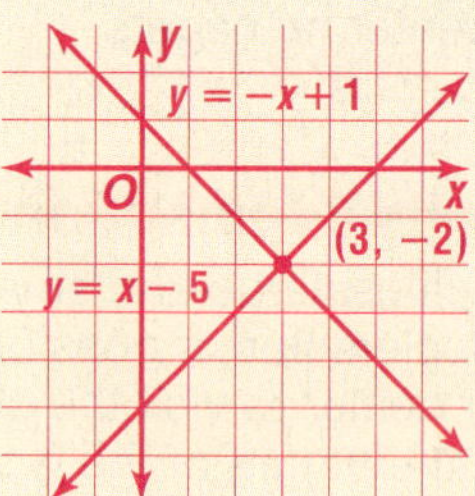

18. no solution

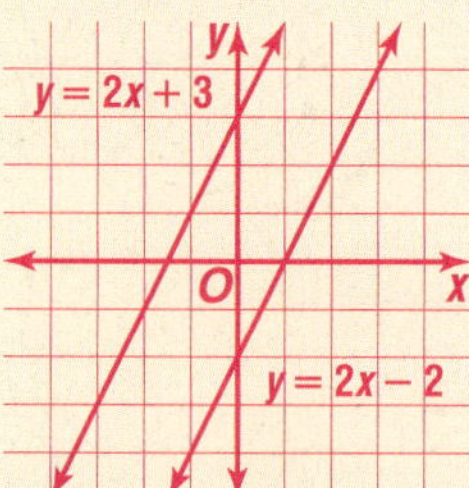

19. infinitely many

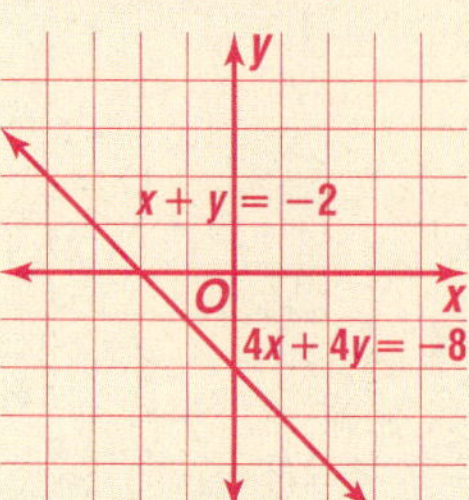

Understanding and Using the Vocabulary

After completing this chapter, you should be able to define each term, property, or phrase and give an example or two of each.

interNET CONNECTION Review Activities For more review activities, visit: www.algconcepts.com

augmented matrix *(p. 578)*
consistent *(p. 554)*
dependent *(p. 554)*
digit problems *(p. 568)*
elimination *(p. 566)*
identity matrix *(p. 579)*
inconsistent *(p. 554)*
independent *(p. 554)*
quadratic-linear *(p. 580)*
row operations *(p. 578)*
substitution *(p. 560)*
system of equations *(p. 550)*
system of inequalities *(p. 586)*

State whether each sentence is *true* or *false*. If false, replace the underlined word(s) to make a true statement.

1. Inconsistent is the description used for a system of equations that has <u>no solution</u>. **true**
2. A system of inequalities can be solved by <u>graphing</u> the inequalities. **true**
3. The solution to a system of equations is the ordered pair that satisfies <u>two</u> of the equations in the system. **false; all**
4. A system of linear equations that has <u>at least</u> one solution is called independent. **false; exactly**
5. The exact coordinates of the point where two or more lines intersect can be determined by using <u>substitution</u>. **true**
6. A set of two or more equations is called a <u>system of equations</u>. **true**
7. <u>Dependent</u> systems of equations have infinitely many solutions. **true**
8. The description of a system of linear equations depends on the number of <u>solutions</u>. **true**
9. To solve a system of equations by <u>substitution</u>, one of the variables is eliminated by either adding or subtracting the equations. **false; elimination**
10. A system of equations that has at least one solution can be described as <u>inconsistent</u>. **false; consistent**

Skills and Concepts

Objectives and Examples

- **Lesson 13–1** Solve systems of equations by graphing.

$y = x + 1$
$x + y = 3$

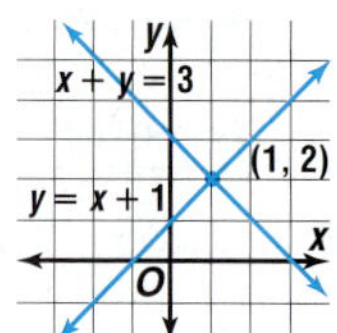

The solution is (1, 2).

Review Exercises

Solve each system of equations by graphing. 11–16. See Solutions Manual for graphs.

11. $y = 4x$; $y = -x + 5$ **(1, 4)**
12. $x = y$; $y = 2x - 3$ **(3, 3)**
13. $y = \frac{1}{2}x + 3$; $y = -x$ **(−2, 2)**
14. $y = 5$; $2y = x + 3$ **(7, 5)**
15. $x - 3y = -6$; $y = -\frac{1}{2}x - 3$ **(−6, 0)**
16. $y = 3x + 1$; $y = -3x + 1$ **(0, 1)**

www.algconcepts.com/vocabulary_review

Resource Manager

Reproducible Masters
Chapter 13 Resource Masters
- *Assessment*, pp. 609–617, 620–622

Technology/Multimedia
MindJogger Videoquizzes
ExamView® Pro

Objectives and Examples

• **Lesson 13–2** Determine whether a system of equations has one solution, no solution, or infinitely many solutions by graphing.

$y = x + 5$
$y = -4x$

This system has one solution, $(-1, 4)$.

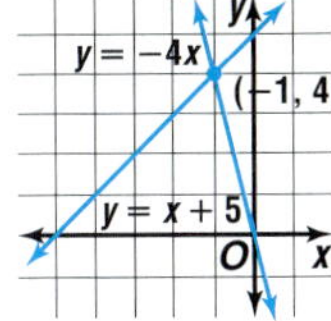

• **Lesson 13–3** Solve systems of equations by the substitution method.

To solve $y = 3x$ and $2x + y = 10$, substitute $3x$ for y in the second equation.

$2x + 3x = 10$
$5x = 10$
$x = 2$

$y = 3x$
$y = 3(2)$ or 6
The solution is $(2, 6)$.

• **Lesson 13–4** Solve systems of equations by the elimination method using addition and subtraction.

$$\begin{array}{r} 3x - 4y = 5 \\ (-)\ 3x + 2y = -7 \\ \hline -6y = 12 \\ y = -2 \end{array}$$

$$\begin{array}{r} 3x - 4(-2) = 5 \\ 3x + 8 = 5 \\ 3x = -3 \\ x = -1 \end{array}$$

The solution is $(-1, -2)$.

• **Lesson 13–5** Solve systems of equations by the elimination method using multiplication and addition.

$3x + 2y = 5$
$x + y = 1$ **Multiply by –2.**

$$\begin{array}{r} 3x + 2y = 5 \\ (+)\ -2x - 2y = -2 \\ \hline x = 3 \end{array}$$

Substitute to find the value of y. The solution is $(3, -2)$.

Review Exercises

Determine whether each system of equations has *one* solution, *no* solution, or *infinitely many* solutions by graphing. If the system has one solution, name it.

17. $y = -x + 1$, $y = x - 5$
18. $y = 2x + 3$, $y = 2x - 2$
19. $x + y = -2$, $4x + 4y = -8$
20. $y = \frac{1}{2}x - 1$, $y = x$

17–20. See margin.

Use substitution to solve each system of equations. 21–26. See margin.

21. $y = 2x$, $4x + y = 3$
22. $3y = x$, $4x + 2y = 14$
23. $y = 3x + 1$, $3x - y = 4$
24. $y = x + 1$, $3x - 2y = -12$
25. $2x - 3y = 4$, $4x - 6y = 8$
26. $y - 3 = x$, $2x + y = 4$

Use elimination to solve each system of equations. 27–32. See margin.

27. $2x + y = 2$, $x + y = -2$
28. $x - 2y = 4$, $2x + 2y = 2$
29. $x - y = 5$, $-x + y = 3$
30. $5x + 3y = 11$, $2x + 3y = 2$
31. $4x + 3y = 3$, $-4x - 3y = -3$
32. $y = 3 - 4x$, $x + y = 2$

Use elimination to solve each system of equations. 33–38. See margin.

33. $2x + 3y = 4$, $x + 2y = 2$
34. $4x + 3y = 3$, $-2x + 4y = 4$
35. $3x + 4y = 2$, $2x - y = 1$
36. $x - 3y = 2$, $-3x + 5y = 2$
37. $3x - 2y = 4$, $4x - 3y = -5$
38. $5x + 3y = 4$, $6x + 4y = 3$

Skills and Concepts

The **Objectives and Examples** section reviews the skills and concepts of the chapter and shows completely worked examples.

The **Review Exercises** provide practice for the corresponding objectives.

Answers

20. $(-2, -2)$

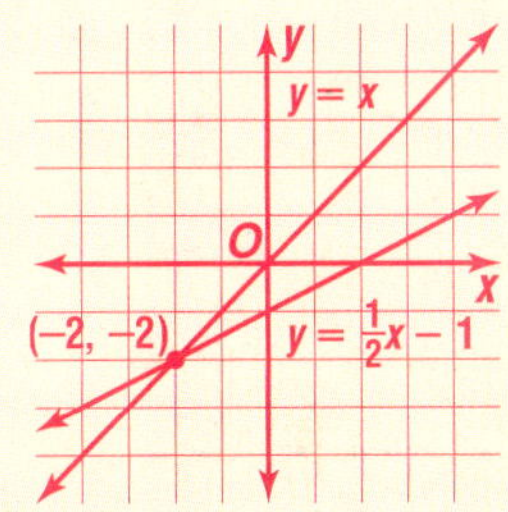

21. $\left(\frac{1}{2}, 1\right)$
22. $(3, 1)$
23. no solution
24. $(-10, -9)$
25. infinitely many
26. $\left(\frac{1}{3}, \frac{10}{3}\right)$
27. $(4, -6)$
28. $(2, -1)$
29. no solution
30. $\left(3, -\frac{4}{3}\right)$
31. infinitely many
32. $\left(\frac{1}{3}, \frac{5}{3}\right)$
33. $(2, 0)$
34. $(0, 1)$
35. $\left(\frac{6}{11}, \frac{1}{11}\right)$
36. $(-4, -2)$
37. $(22, 31)$
38. $\left(\frac{7}{2}, -\frac{9}{2}\right)$

ExamView® Pro

Use the networkable ***ExamView® Pro Testmaker CD-ROM*** to:

- Create **multiple versions** of tests.
- Create **modified** tests for **inclusion** students with one mouse click.
- **Edit** existing questions and **add** your own questions.
- Build tests aligned with state standards using built-in **state curriculum correlations**.
- Change **English** tests to **Spanish** with one mouse click and vice versa.

Mixed Problem Solving
See pages 724–731.

Applications and Problem Solving

This section provides additional practice in solving real-world problems that involve the concepts of this chapter.

Answers

Page 595

1. Sample answer:

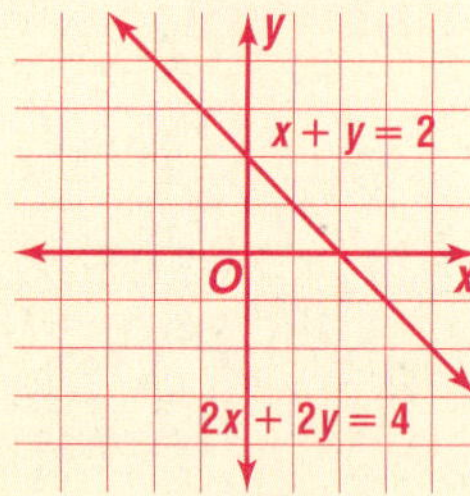

3.

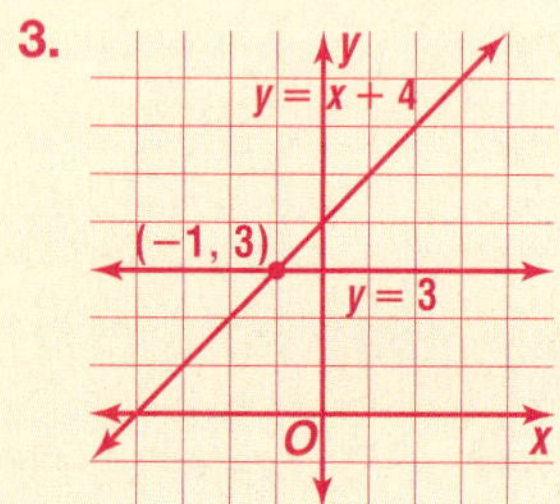

Objectives and Examples

- **Lesson 13–6** Solve systems of quadratic and linear equations.

$y = 2x + 3$
$y = x^2 - 5$

Substitute $x^2 - 5$ for y in the first equation.

$$y = 2x + 3$$
$$x^2 - 5 = 2x + 3$$
$$x^2 - 5 - 2x - 3 = 2x + 3 - 2x - 3$$
$$x^2 - 2x - 8 = 0$$
$$(x + 2)(x - 4) = 0$$

$x + 2 = 0$ or $x - 4 = 0$
$x = -2$ $\quad$ $x = 4$

Solve by finding the corresponding values of y. The solutions are $(-2, -1)$ and $(4, 11)$.

Review Exercises

39–42. See Solutions Manual for graphs.

Determine whether each system of equations has *one* solution, *two* solutions, or *no* solution by graphing. If the system has one or two solutions, name them.

39. $y = -x^2$
$y = x - 2$
$(-2, -4), (1, -1)$

40. $y = x^2 + 2$
$y = 6$
$(-2, 6), (2, 6)$

41. $y = 2x^2$
$y = -x - 4$
no solution

42. $y = x^2 - 1$
$y = -4x - 5$
$(-2, 3)$

Use substitution to solve each system of equations.

43. $y = x^2 + 2$
$y = 3x + 6$
$(4, 18), (-1, 3)$

44. $y = -x^2 + 4$
$y = 4x + 10$
no solution

45. Solve the system $y = x^2 - 2x$ and $y = 4x - 9$ by using substitution. $(3, 3)$

- **Lesson 13–7** Solve systems of inequalities by graphing.

Solve the system of inequalities $x \geq -2$ and $y < x - 3$ by graphing.

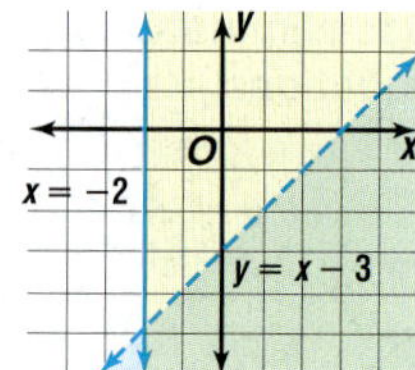

Solve each system of inequalities by graphing. 46–50. See Solutions Manual.

46. $x < 4$
$y \geq -2$

47. $x \geq -3$
$y \leq x + 2$

48. $x + y > -4$
$x + y \leq 2$

49. $y \geq 2x + 1$
$y > -x - 1$

50. Use the graphs of $-6x + 3y > 9$ and $y < 2x - 4$ to determine the solution of the system. If the system does not have a solution, write *no solution*.

Applications and Problem Solving

51. Mixtures Mr. Collins mixed almonds that cost \$4.25 per pound with cashews that cost \$6.50 per pound. He now has a mixture of 20 pounds of nuts that costs \$5.60 per pound. How many pounds of each type of nut did he use? *(Lesson 13–3)* 8 lb of almonds, 12 lb of cashews

52. Number Theory The difference between the tens digit and the units digit of a two-digit number is 3. Suppose the tens digit is one less than twice the units digit. What is the number? *(Lesson 13–4)* 74

53. Geometry The difference between the length and width of a rectangle is 7 feet. Find the dimensions of the rectangle if its perimeter is 50 feet. *(Lesson 13–5)* 16 ft by 9 ft

***Assessment*, pp. 611–612**

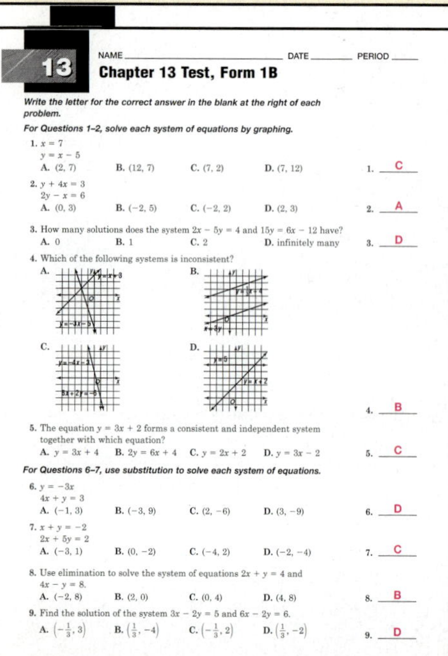

13 NAME ________ DATE ________ PERIOD ________

Chapter 13 Test, Form 1B

Write the letter for the correct answer in the blank at the right of each problem.

For Questions 1–2, solve each system of equations by graphing.

1. $x = 7$
$y = x - 5$
A. (2, 7) B. (12, 7) C. (7, 2) D. (7, 12) — 1. C

2. $y + 4x = 3$
$2y - x = 6$
A. (0, 3) B. (−2, 5) C. (−2, 2) D. (2, 3) — 2. A

3. How many solutions does the system $2x - 5y = 4$ and $15y = 6x - 12$ have?
A. 0 B. 1 C. 2 D. infinitely many — 3. D

4. Which of the following systems is inconsistent?
A. [graph: $y = x + 3$, $y = -3x - 5$] B. [graph: $y = \frac{1}{3}x + 4$, $x + 3y$] C. [graph: $y = -4x - 3$, $8x + 2y = -6$] D. [graph: $y = 5$, $y = x + 2$] — 4. B

5. The equation $y = 3x + 2$ forms a consistent and independent system together with which equation?
A. $y = 3x + 4$ B. $2y = 6x + 4$ C. $y = 2x + 2$ D. $y = 3x - 2$ — 5. C

For Questions 6–7, use substitution to solve each system of equations.

6. $y = -3x$
$4x + y = 3$
A. (−1, 3) B. (−3, 9) C. (2, −6) D. (3, −9) — 6. D

7. $x + y = -2$
$2x + 5y = 2$
A. (−3, 1) B. (0, −2) C. (−4, 2) D. (−2, −4) — 7. C

8. Use elimination to solve the system of equations $2x + y = 4$ and $4x - y = 8$.
A. (−2, 8) B. (2, 0) C. (0, 4) D. (4, 8) — 8. B

9. Find the solution of the system $3x - 2y = 5$ and $6x - 2y = 6$.
A. $\left(-\frac{1}{3}, 3\right)$ B. $\left(\frac{1}{3}, -4\right)$ C. $\left(-\frac{1}{3}, 2\right)$ D. $\left(\frac{1}{3}, -2\right)$ — 9. D

Assessment

Four forms of Chapter 13 Test are available in the *Chapter 13 Resource Masters.*

	Form of Test		Level
1A	Multiple Choice	pp. 609–610	Average
1B	Multiple Choice	pp. 611–612	Basic
2A	Free Response	pp. 613–614	Average
2B	Free Response	pp. 615–616	Basic

Chapter 13 Test, Form 1B, is shown at the left. Chapter 13 Test, Form 2B, is shown on the next page.

CHAPTER 13 Test

1. **Graph** a system of equations that has infinitely many solutions. **See margin.**
2. **Explain** when you would use elimination with subtraction to solve a system of equations. **when the coefficients of one variable are the same**

Solve each system of equations by graphing. **3–5. See margin for graphs.**

3. $y = 3$
$y = x + 4$ **(−1, 3)**
4. $x + y = -2$
$2x - y = -4$ **(−2, 0)**
5. $y = -x^2 - 1$
$y = -5$ **(2, −5), (−2, −5)**

State whether each system is *consistent and independent*, *consistent and dependent*, or *inconsistent*.

6.

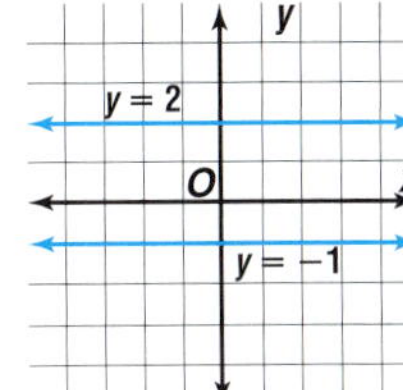

inconsistent

7.

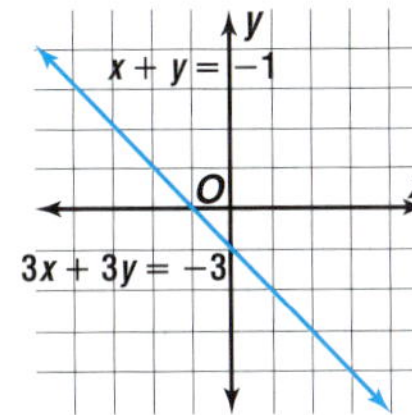

consistent and dependent

8. Use graphing to determine whether the system $y = -2x$ and $2x + y = 4$ has *one* solution, *no* solution, or *infinitely many* solutions. If the solution has one solution, name it. **No solution; see margin for graph.**

Use substitution to solve each system of equations.

9. $y = 3x$
$x + y = 4$ **(1, 3)**
10. $x + y = -2$
$x = y + 10$ **(4, −6)**
11. $y = 5x - 3$
$10x - 2y = -2$ **no solution**
12. $y = x^2 - 15$
$y = 2x$ **(5, 10), (−3, −6)**
13. $y = 5x + 4$
$y = x^2 + 5x$ **(−2, −6), (2, 14)**
14. $y = 3x + 2$
$y = x^2 + 6$ **no solution**

Use elimination to solve each system of equations.

15. $x + y = 5$
$x - y = -9$ **(−2, 7)**
16. $4x - 5y = 7$
$x + 5y = 8$ **(3, 1)**
17. $2x - y = 32$
$y = 60 - 2x$ **(23, 14)**
18. $x + 3y = -1$
$2x + 4y = -2$ **(−1, 0)**
19. $5x - 2y = 3$
$15x - 6y = 9$ **infinitely many**
20. $-5x + 8y = 21$
$10x + 3y = 15$ $\left(\frac{3}{5}, 3\right)$

Solve each system of inequalities by graphing. **21–23. See Solutions Manual.**

21. $y \leq -3$
$y > -x - 2$
22. $y < x + 4$
$y > x - 2$
23. $x \leq 2y$
$2x + 3y \leq 6$

24. Find two numbers whose sum is 64 and whose difference is 42. **11, 53**

25. **Transportation** Two trains travel toward each other on parallel tracks at the same time from towns 450 miles apart. Suppose one train travels 6 miles per hour faster than the other train. What is the rate of each train if they meet in 5 hours? **42 mph, 48 mph**

www.algconcepts.com/chapter_test

Chapter Test Bonus Question

The three lines whose equations are $x + 2y = 1$, $x - y = 4$, and $x + y = -2$ form a triangle. Find the coordinates of the vertices of the triangle, and write a system of three inequalities that describes the triangle and its interior.

(−5, 3), (3, −1), (1, −3); $y \leq -\frac{1}{2}x + \frac{1}{2}$, $y \geq -x - 2$, $y \geq x - 4$

Chapter Test

Answers

4.

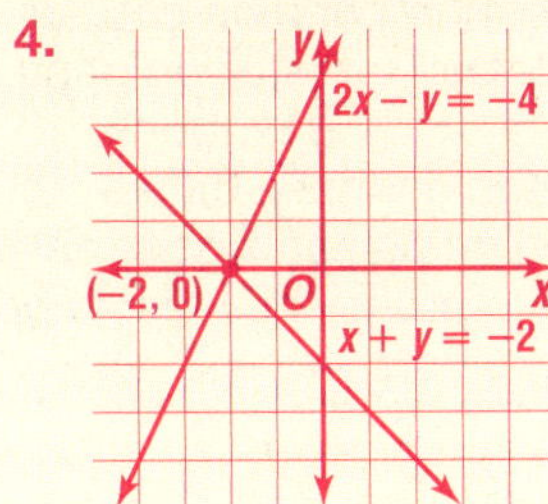

5.

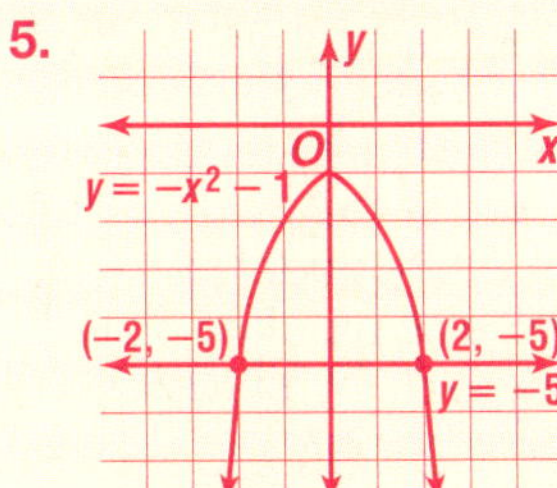

8.

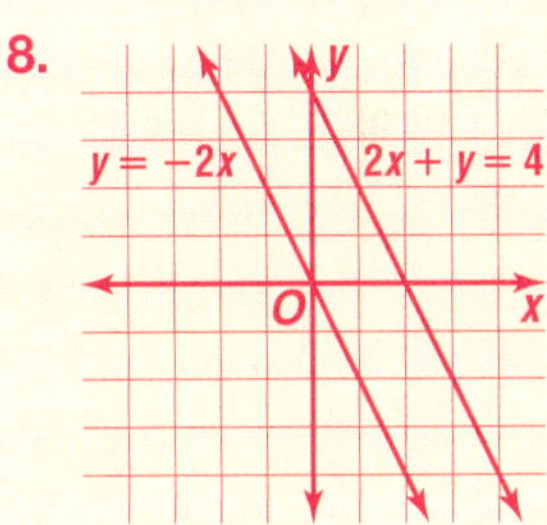

***Assessment*, pp. 615–616**

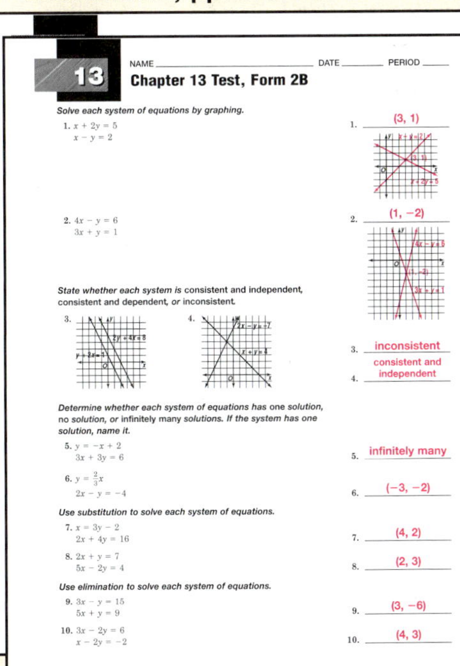

13 NAME ______ DATE ______ PERIOD ______

Chapter 13 Test, Form 2B

Solve each system of equations by graphing.

1. $x + 2y = 5$
$x - y = 2$ — 1. (3, 1)
2. $4x - y = 6$
$3x + y = 1$ — 2. (1, −2)

State whether each system is consistent and independent, consistent and dependent, or inconsistent.

3. — 3. inconsistent
4. — 4. consistent and independent

Determine whether each system of equations has one solution, no solution, or infinitely many solutions. If the system has one solution, name it.

5. $y = -x + 2$
$3x + 3y = 6$ — 5. infinitely many
6. $y = \frac{2}{3}x$
$2x - y = -4$ — 6. (−3, −2)

Use substitution to solve each system of equations.

7. $x = 3y - 2$
$2x + 4y = 16$ — 7. (4, 2)
8. $2x + y = 7$
$5x - 2y = 4$ — 8. (2, 3)

Use elimination to solve each system of equations.

9. $3x - y = 15$
$5x + y = 9$ — 9. (3, −6)
10. $3x - 2y = 6$
$x - 2y = -2$ — 10. (4, 3)

Pages 596–597 are part of a complete test preparation course that is described in detail on page T9 of the Teacher's Handbook. The test items on these pages were written in the same style as those in state proficiency tests and standardized tests like ACT and SAT.

Diagnosis and Prescription

Each of the 10 test questions on page 597 is cross-referenced to the lesson where that SAT or ACT skill is covered. If students miss a particular type of problem, you can have them study that skill.

(See chart at the bottom of page 597. Note that SPT = State Proficiency Test, SAT = Scholastic Assessment Test, and ACT = American College Test.)

Assessment, p. 620

13 NAME ______ DATE ______ PERIOD ______

Chapter 13 Cumulative Review

1. Write an equation for the sentence *Eight less than three times a number m equals 10. (Lesson 1–1)* — 1. $3m - 8 = 10$
2. Evaluate $\frac{2a - 5b}{c} + 4d$ if $a = -2$, $b = 4$, $c = -8$, and $d = 5$. *(Lessons 2–3 through 2–6)* — 2. 23
3. Find the mean of the following set of data: 24, 48, 38, 50, 45. *(Lesson 3–3)* — 3. 41
4. Find $6 \div (-0.02)$. *(Lesson 4–3)* — 4. −300
5. Find the percent of decrease from \$140 to \$91. *(Lesson 5–5)* — 5. 35%
6. Find the domain of $y = 2x - 3$ if the range is $\{-7, -3, 5, 11\}$. *(Lesson 6–2)* — 6. $\{-2, 0, 4, 7\}$
7. Determine the slope of the line passing through points at $(-2, 4)$ and $(6, -2)$. *(Lesson 7–1)* — 7. $-\frac{3}{4}$
8. Simplify $\frac{8x^2y^{-3}}{2x^{-1}y^4}$. *(Lesson 8–3)* — 8. $\frac{4x^3}{y^7}$
9. Find $(8x^2 - 3x + 1) - (4x^2 - 5x + 3)$. *(Lesson 9–2)* — 9. $4x^2 + 2x - 2$
10. Factor $x^2 - 2x - 15$. *(Lesson 10–3)* — 10. $(x - 5)(x + 3)$
11. Write an equation of the axis of symmetry of the graph of $y = 3x^2 - 12x + 7$. *(Lesson 11–1)* — 11. $x = 2$
12. Solve $x^2 + 8x - 20 = 0$ by completing the square. *(Lesson 11–5)* — 12. −10, 2
13. Solve $-3 < 2x + 1 < 9$. *(Lesson 12–5)* — 13. $\{x \mid -2 < x < 4\}$
14. Mr. Wilson's bakery donated 240 cookies to a school bake sale. The cookies were packaged in bags of three, which sold for \$0.50, and bags of 12, which sold for \$1.50. When all the cookies were sold, the total collected was \$36. How many bags of three cookies were sold? *(Lesson 13–3)* — 14. 48
15. Use elimination to solve the system of equations $3x + 2y = 10$ and $5x - 2y = 14$. *(Lesson 13–4)* — 15. $\left(3, \frac{1}{2}\right)$
16. Use substitution to solve the system of equations $y = x^2 - 3x + 11$ and $y = 3x + 2$. *(Lesson 13–6)* — 16. (3, 11)

CHAPTER 13 Preparing for Standardized Tests

Perimeter, Area, and Volume Problems

Standardized tests often include questions on perimeter, area, circumference, and volume. You'll need to know and apply formulas for each of these measurements. Be sure you know these terms.

area	circumference	height	perimeter	width
base	diameter	length	radius	

Test-Taking Tip

Use the information given in the figure.

Example 1

Richard plans to increase the floor area of his health club's weight room. The figure below shows the existing floor area with a solid line and the additional floor area with a dotted line. Find the length, in feet, of the new weight room.

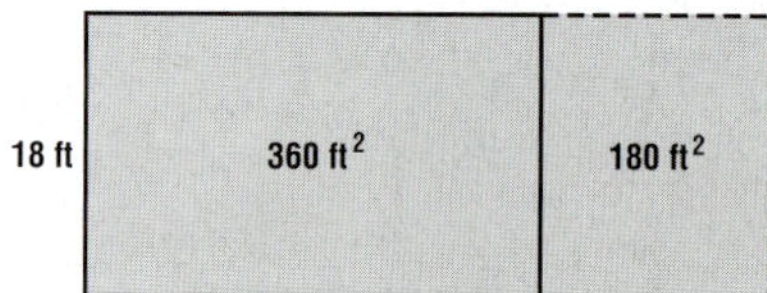

Hint Familiarize yourself with the formulas for the area and perimeter of quadrilaterals.

Solution The existing room has an area of 360 square feet and a width of 18 feet. The addition has an area of 180 square feet and a width of 18 feet.

To find the length of the new weight room, first find the area of the new room.

$$A = 360 + 180 \text{ or } 540 \text{ square feet}$$

Next, use the formula for the area of a rectangle to find the length of the room.

$A = \ell w$	*Area formula*
$540 = \ell \cdot 18$	*Replace A with 540 and w with 18.*
$\frac{540}{18} = \frac{18\ell}{18}$	*Divide each side by 18.*
$30 = \ell$	*Simplify.*

The length of the new room is 30 feet.

Example 2

What is the diameter of a circle with a circumference of 5 inches?

A $\frac{5}{\pi}$ in. **B** $\frac{10}{\pi}$ in. **C** 5 in. **D** 5π in. **E** 10π in.

Hint If a geometry problem has no figure, sketch one.

Solution

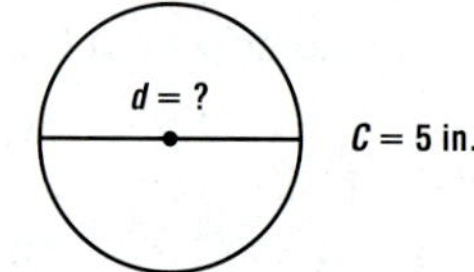

You know the circumference of the circle. You need to find the diameter of the circle.

Use the formula $C = \pi d$, where C is the circumference and d is the diameter to find the circumference of the circle.

First, replace C with 5. Then solve the equation for d.

$C = \pi d$	*Circumference formula*
$5 = \pi d$	*Replace C with 5.*
$\frac{5}{\pi} = \frac{\pi d}{\pi}$	*Divide each side by π.*
$\frac{5}{\pi} = d$	*Simplify.*

The answer is A.

Resource Manager

Reproducible Masters

Chapter 13 Resource Masters

- *Assessment* pp. 620–622

Preparing for Standardized Tests
For test-taking strategies and more practice, see pages 732–749.

After you work each problem, record your answer on the answer sheet provided or on a sheet of paper.

Multiple Choice

1. Quinn's neighborhood is a rectangle 2 miles by 1.5 miles. How many miles does Quinn jog if he jogs around the boundary of his neighborhood? *(Basic Skill)* **D**

   (A) 3 (B) 3.5 (C) 6 (D) 7

2. If $AC = 4$, what is the area of $\triangle ABC$? *(Lesson 1–5)* **B**

   (A) $\frac{1}{2}$ (B) 2 (C) 4 (D) 8

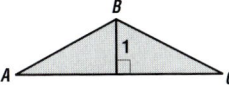

3. Micela is making a poster that has a length of 36 inches. If the maximum perimeter is 96 inches, which inequality can be used to determine the width w of the poster? *(Lesson 12–4)* **A**

   (A) $96 \geq 2(36) + 2w$ (B) $96 \leq 2(36) + 2w$
   (C) $96 \geq 36 + 2w$ (D) $96 \geq 36 + w$

4. If the area of a circle is 16 square meters, what is its radius in meters? *(Lesson 8–1)* **C**

   (A) $\frac{8}{\pi}$ (B) $\frac{16}{\pi}$ (C) $\frac{4\sqrt{\pi}}{\pi}$
   (D) 12π (E) $144\pi^2$

5. If you double the length and the width of a rectangle, how does its perimeter change? *(Lesson 8–1)* **B**

   (A) It increases by $1\frac{1}{2}$.
   (B) It doubles.
   (C) It quadruples.
   (D) It does not change.

6. What is the area of the figure? *(Lesson 1–5)* **B**

   (A) 192 ft^2
   (B) 360 ft^2
   (C) 456 ft^2
   (D) 720 ft^2

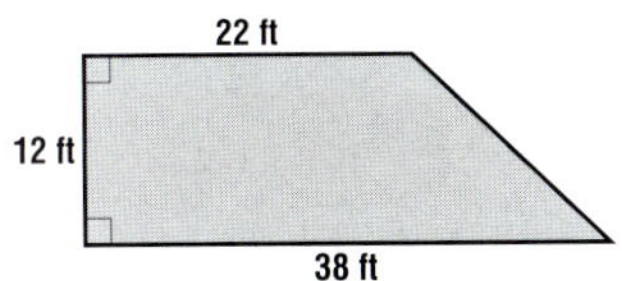

7. The table shows the speed of a car and the distance needed to safely stop the car. Which equation represents the data? *(Lesson 11–2)* **D**

   (A) $y = x + 25$
   (B) $y = x^2 + 20$
   (C) $y = x^2 \times 20$
   (D) $y = x^2 \div 20$
   (E) $y = x^2 \div 25$

Speed (x)	Distance (y)
20	20
30	45
40	80
50	125

8. Which number could be the diameter of the circle? *(Lesson 1–5)* **A**

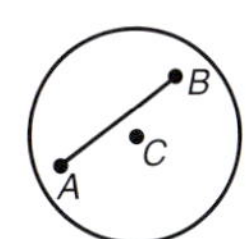

$AB = 8$

   (A) 12 (B) 8 (C) 6 (D) 4

Grid In

9. Allie sketched her walking route. The curved part is a semicircle with a radius of 10.5 meters. If Allie walks this route 10 times, how many meters does she walk? Round to the nearest meter. *(Lesson 1–5)*
   934

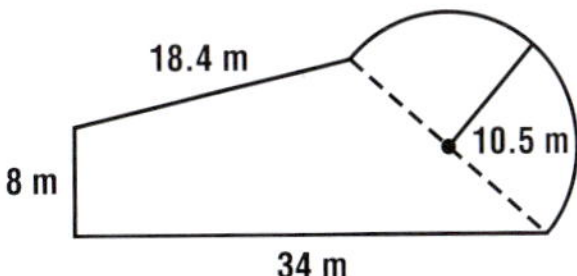

Extended Response

10. Suppose you have 120 feet of fence.
    Part A What is the greatest possible rectangular area you can enclose? **900 sq ft**
    Part B What are the dimensions of the rectangle that has this area? *(Lesson 1–5)*
    30 ft by 30 ft

www.algconcepts.com/standardized_test

Chapter 13 Perimeter, Area, and Volume Problems		
Ex. 1	area of rectangle	SPT
Ex. 2	circumference	SAT
1	perimeter of rectangle	SPT
2	area of triangle	SPT
3	perimeter and inequality	SPT
4	area of circle	ACT
5	perimeter of rectangle	SPT
6	area of polygon	SPT
7	function table	SPT
8	radius	SAT
9	perimeter	SPT
10	area of rectangle	SPT

A bubble-in answer sheet for these practice problems is available on page A1 of the *Chapter 13 Resource Masters.*

Additional Practice

Additional test practice questions are available in the *Chapter 13 Resource Masters*, pp. 621–622.

***Assessment*, pp. 621–622**

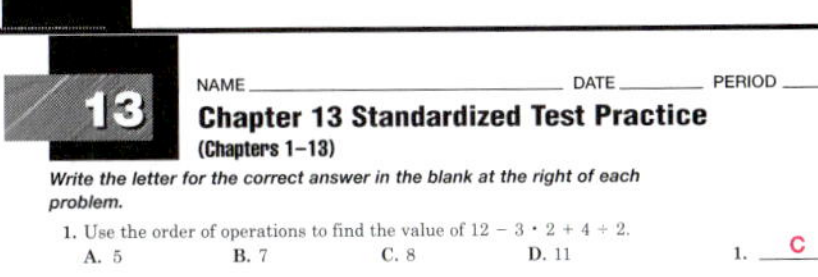
13 NAME ______ DATE ______ PERIOD ______

Chapter 13 Standardized Test Practice
(Chapters 1–13)

Write the letter for the correct answer in the blank at the right of each problem.

1. Use the order of operations to find the value of $12 - 3 \cdot 2 + 4 \div 2$.
   A. 5 B. 7 C. 8 D. 11 — 1. **C**
2. Simplify $4x(-2) - 5(-3x)$.
   A. $-7x$ B. $7x$ C. $-23x$ D. $23x$ — 2. **B**
3. Evaluate $\frac{7}{12} - \left(-\frac{1}{3}\right) - \frac{5}{6}$.
   A. $-\frac{7}{6}$ B. $\frac{1}{12}$ C. $\frac{5}{12}$ D. 1 — 3. **B**
4. Evaluate $ab + \frac{c}{d}$ when $a = -\frac{1}{2}$, $b = \frac{2}{3}$, $c = \frac{-1}{8}$, and $d = \frac{3}{4}$.
   A. $-\frac{1}{2}$ B. $-\frac{1}{3}$ C. $\frac{1}{3}$ D. $\frac{3}{4}$ — 4. **A**
5. For a book review, students must pick one novel from a list of fifteen. They may choose to do a traditional written book report, an oral report, or a book advertisement. Use the Fundamental Counting Principle to determine the number of possible choices for the book review.
   A. 18 B. 30 C. 45 D. 60 — 5. **C**
6. What number is 70% of 130?
   A. 81 B. 91 C. 101 D. 185 — 6. **B**
7. Two dice are rolled. Find the probability that the sum of the two numbers is either a multiple of 3 or an even number.
   A. $\frac{1}{3}$ B. $\frac{1}{2}$ C. $\frac{2}{3}$ D. $\frac{5}{6}$ — 7. **C**
8. Suppose y varies inversely as x and $y = 16$ when $x = 3$. Find y when $x = 10$.
   A. 4.8 B. $5\frac{1}{3}$ C. 48 D. $53\frac{1}{3}$ — 8. **A**
9. Write the point-slope form of an equation for the line passing through points at (3, 8) and (−1, −4).
   A. $y - 3 = 3(x - 8)$ B. $y + 4 = \frac{1}{3}(x + 1)$
   C. $y - 8 = -3(x - 3)$ D. $y - 8 = 3(x - 3)$ — 9. **D**
10. Write an equation in slope-intercept form of the line perpendicular to the graph of $2x + y = 7$ and passing through the point at (6, 0).
    A. $y = -\frac{1}{2}x + 6$ B. $y = \frac{1}{2}x - 3$
    C. $y = 2x - 9$ D. $y = -2x + 15$ — 10. **B**

CHAPTER 14

Resource Manager

Radical Expressions

Instructional Objectives

Lesson (pages)	Objectives	NCTM Standards 2000	State/Local Objectives
14–1 (600–605)	Describe the relationships among sets of numbers.	1, 2, 6, 8, 9, 10	
14–2 (606–611)	Find the distance between two points in the coordinate plane.	1, 3, 6, 8, 9	
Investigation (612–613)	Explore the Midpoint Formula.	1, 3, 6, 7, 8, 9	
14–3 (614–619)	Simplify radical expressions.	1, 6, 8, 9	
14–4 (620–623)	Add and subtract radical expressions.	1, 6, 8, 9	
14–5 (624–629)	Solve simple radical equations in which only one radical contains a variable.	1, 6, 8, 9	

Key to NCTM Standards 2000
1 Number & Operations; **2** Algebra; **3** Geometry; **4** Measurement; **5** Data Analysis & Probability; **6** Problem Solving; **7** Reasoning & Proof; **8** Communications; **9** Connections; **10** Representation

Suggested Pacing

See page T13 for a complete course-planning calendar.

Standard refers to schedules that provide 45- to 55-minute periods that meet each day.
Block refers to schedules that provide approximately 90-minute periods which may meet every day for one semester or every other day over two semesters.

<table>
<tr><th>PACING</th><th>DAY 1</th><th>DAY 2</th><th>DAY 3</th><th>DAY 4</th><th>DAY 5</th><th>DAY 6</th></tr>
<tr><td>Standard Core
(Chapters 1–13)</td><td></td><td></td><td></td><td></td><td></td><td></td></tr>
<tr><td>Standard Enhanced
(Chapters 1–15)</td><td>Lesson 14–1</td><td>Lesson 14–2</td><td>INV</td><td colspan="2">Lesson 14–3</td><td>Lesson 14–4</td></tr>
<tr><td>Block Core
(Chapters 1–13)</td><td></td><td></td><td></td><td></td><td></td><td></td></tr>
<tr><td>Block Enhanced
(Chapters 1–15)</td><td>Chapter 13 Test & Lesson 14–1</td><td>Lesson 14–2 & INV</td><td>Lessons 14–3 & 14–4</td><td>Lesson 14–5 & SG+A</td><td>Chapter Test & Lesson 15–1</td><td></td></tr>
</table>

Instructional Resources

Lesson	Materials and Manipulatives (see below for Glencoe Manipulative Resources)	Blackline Masters (page numbers): Chapter 14 Resource Masters: Study Guide	Practice (Skills & Average)	Reading to Learn Mathematics	Enrichment	Assessment	Hands-On Algebra*	School-to-Workplace*	Graphing Calculator Masters*	5-Minute Check Transparencies
14–1	ruler [1, 2, 4] calculator coordinate planes [4]	623	624–625	626	627					14–1
14–2	grid paper [1, 4] straightedge [1, 2, 4] coordinate planes [4]	628	629–630	631	632	659	153, 154	14		14–2
Investigation	grid paper [1, 4] ruler [1, 2, 4] scissors [1, 2] coordinate planes [4]									
14–3	metric ruler [1, 2, 4]	633	634–635	636	637	658	155		41, 42	14–3
14–4	coordinate planes [4]	638	639–640	641	642		156, 157			14–4
14–5	graphing calculator ruler [1, 2, 4] coordinate planes [4]	643	644–645	646	647	659			40	14–5
Study Guide & Assessment/ Chapter Test	ruler [1, 2, 4] coordinate planes [4]					649–657, 660–662				

See page 598c for examples of these instructional materials.

Key to Glencoe Manipulative Resources

[1]Classroom Manipulative Resources [2]Student Manipulative Resources [3]Overhead Manipulative Resources [4]Hands-On Algebra Masters

INV = Investigation *SG+A = Study Guide and Assessment*

DAY 7	DAY 8	DAY 9	DAY 10	DAY 11	DAY 12	DAY 13
Lesson 14–5	SG+A	Chapter Test				

Resource Manager

The pages shown on this page are a small sample of the materials available on *TeacherWorks: All-in-One Lesson Planner and Resource Center.*

This CD-ROM includes all of the blackline masters and transparencies available for this program.

It also includes a lesson planner and interactive Teacher Edition, so you can customize lesson plans and reproduce classroom resources quickly and easily, from just about anywhere.

Applications

School-to-Workplace Masters, p. 14

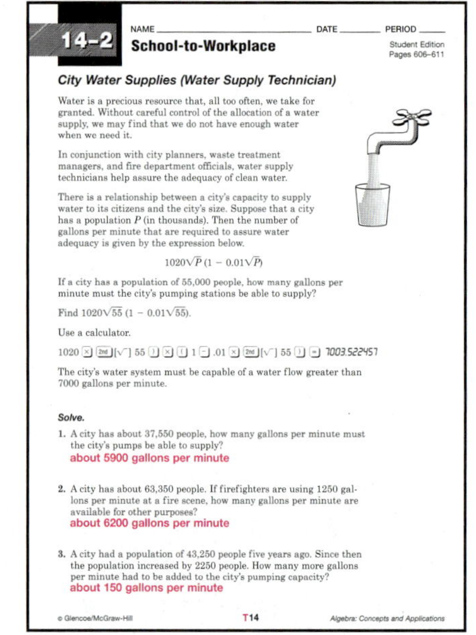

14-2 NAME ______ DATE ______ PERIOD ______

School-to-Workplace

Student Edition Pages 606–611

City Water Supplies (Water Supply Technician)

Water is a precious resource that, all too often, we take for granted. Without careful control of the allocation of a water supply, we may find that we do not have enough water when we need it.

In conjunction with city planners, waste treatment managers, and fire department officials, water supply technicians help assure the adequacy of clean water.

There is a relationship between a city's capacity to supply water to its citizens and the city's size. Suppose that a city has a population P (in thousands). Then the number of gallons per minute that are required to assure water adequacy is given by the expression below.

$$1020\sqrt{P}\,(1 - 0.01\sqrt{P})$$

If a city has a population of 55,000 people, how many gallons per minute must the city's pumping stations be able to supply?

Find $1020\sqrt{55}\,(1 - 0.01\sqrt{55})$.

Use a calculator.

1020 [×] [2nd] [√‾] 55 [)] [×] [(] 1 [−] .01 [×] [2nd] [√‾] 55 [)] [=] 7003.522457

The city's water system must be capable of a water flow greater than 7000 gallons per minute.

Solve.

1. A city has about 37,550 people, how many gallons per minute must the city's pumps be able to supply?
   about 5900 gallons per minute
2. A city has about 63,350 people. If firefighters are using 1250 gallons per minute at a fire scene, how many gallons per minute are available for other purposes?
   about 6200 gallons per minute
3. A city had a population of 43,250 people five years ago. Since then the population increased by 2250 people. How many more gallons per minute had to be added to the city's pumping capacity?
   about 150 gallons per minute

© Glencoe/McGraw-Hill T14 *Algebra: Concepts and Applications*

Manipulatives/Modeling

Hands-On Algebra Masters, pp. 153–157

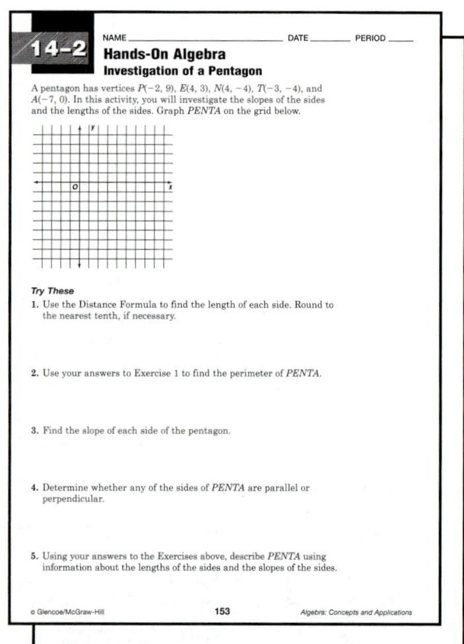

14-2 NAME ______ DATE ______ PERIOD ______

Hands-On Algebra
Investigation of a Pentagon

A pentagon has vertices $P(-2, 9)$, $E(4, 3)$, $N(4, -4)$, $T(-3, -4)$, and $A(-7, 0)$. In this activity, you will investigate the slopes of the sides and the lengths of the sides. Graph *PENTA* on the grid below.

Try These

1. Use the Distance Formula to find the length of each side. Round to the nearest tenth, if necessary.
2. Use your answers to Exercise 1 to find the perimeter of *PENTA*.
3. Find the slope of each side of the pentagon.
4. Determine whether any of the sides of *PENTA* are parallel or perpendicular.
5. Using your answers to the Exercises above, describe *PENTA* using information about the lengths of the sides and the slopes of the sides.

© Glencoe/McGraw-Hill 153 *Algebra: Concepts and Applications*

Technology/Multimedia

Graphing Calculator Masters, pp. 40–42

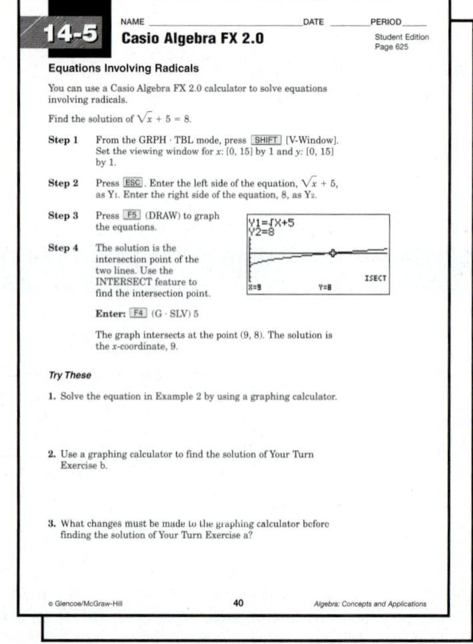

14-5 NAME ______ DATE ______ PERIOD ______

Casio Algebra FX 2.0

Student Edition Page 625

Equations Involving Radicals

You can use a Casio Algebra FX 2.0 calculator to solve equations involving radicals.

Find the solution of $\sqrt{x} + 5 = 8$.

Step 1 From the GRPH · TBL mode, press [SHIFT] [V-Window]. Set the viewing window for x: [0, 15] by 1 and y: [0, 15] by 1.

Step 2 Press [ESC]. Enter the left side of the equation, $\sqrt{x} + 5$, as Y_1. Enter the right side of the equation, 8, as Y_2.

Step 3 Press [F5] (DRAW) to graph the equations.

Step 4 The solution is the intersection point of the two lines. Use the INTERSECT feature to find the intersection point.

Enter: [F4] (G · SLV) 5

The graph intersects at the point (9, 8). The solution is the x-coordinate, 9.

Try These

1. Solve the equation in Example 2 by using a graphing calculator.
2. Use a graphing calculator to find the solution of Your Turn Exercise b.
3. What changes must be made to the graphing calculator before finding the solution of Your Turn Exercise a?

© Glencoe/McGraw-Hill 40 *Algebra: Concepts and Applications*

14-3 NAME ______ DATE ______ PERIOD ______

TI-83/84 Plus

Programming and Simplifying Radical Expressions

You can use a TI-83/84 Plus calculator and a program to simplify radical expressions.

Use the graphing calculator program to express $\sqrt{192}$ in simplest form.

Enter the number 192.

The last line displayed is {8 3}. This means that $\sqrt{192}$ in simplest form is $8\sqrt{3}$. In other words, $\sqrt{192} = 8\sqrt{3}$.

```
PROGRAM: RADICAL
: Disp "ENTER N"
: Input N
: For (D, 1, N)
: If iPart (N/D²) = (N/D²)
: Disp {D, N/D²}
: END
```

Try These

Use the program to simplify each radical expression.

1. $\sqrt{48}$
2. $\sqrt{150}$
3. $\sqrt{80}$
4. $\sqrt{147}$
5. $\sqrt{72}$
6. $\sqrt{125}$
7. $\sqrt{128}$
8. $\sqrt{192}$
9. $\sqrt{175}$
10. $\sqrt{245}$
11. $\sqrt{50}$
12. $\sqrt{320}$
13. In step 2 of the activity, explain what the first three pairs of whole numbers and radicals signify.

© Glencoe/McGraw-Hill 41 *Algebra: Concepts and Applications*

Glencoe's Assessment Advantage

Assessment Resources

Type	Student Edition	Teacher's Wraparound Edition	Chapter 14 Resource Masters
Ongoing Assessment	Quizzes 1 and 2, pp. 611, 629	5-Minute Check, pp. 600, 606, 614, 620, 624	Mid-Chapter Test, p. 658 Quizzes A and B, p. 659
Mixed Review	Mixed Review, pp. 605, 611, 619, 623, 629 Standardized Test Practice, Chapters 1–14, pp. 634–635		Cumulative Review, p. 660 Standardized Test Practice, pp. 661–662
Error Analysis	You Decide, pp. 609, 618	Error Analysis, pp. 603, 609, 618, 622, 628	
Standardized Test Prep	Standardized Test Practice, pp. 605, 611, 619, 623, 629 Standardized Test Practice, Chapters 1–14, pp. 634–635		Standardized Test Practice, pp. 661–662
Open-Ended Assessment	Writing Math, pp. 621, 627 Investigation, pp. 612–613 Portfolio, pp. 599, 613	Speaking: pp. 605, 629 Writing: pp. 611, 619, 623	Extended Response Assessment, p. 657
Chapter Assessment	Study Guide and Assessment, pp. 630–632 Chapter Test, p. 633		Multiple-Choice Tests (Forms 1A, 1B), pp. 649–652 Free-Response Tests (Forms 2A, 2B), pp. 653–656

Additional Chapter Resources

Student Edition
Hands-On Algebra, p. 606
Graphing Calculator Exploration, p. 625

Teacher's Classroom Resources
Manipulatives/Modeling
Teacher's Guide for Overhead Manipulative Resources

Meeting Individual Needs
Prerequisite Skills Booklet
Spanish Study Guide and Assessment, pp. 94–98, 131–132

Teaching Aids
Solutions Manual
5-Minute Check Transparencies

Glencoe Technology

Instructional
AlgePASS CD-ROM, Lessons 32
Interactive Chalkboard CD-ROM
StudentWorks Plus CD-ROM
Multimedia Applications CD-ROM, Activity 13
Vocabulary PuzzleMaker (online)

Assessment
ExamView® Pro

Teaching Aids
Answer Key Maker CD-ROM
TeacherWorks CD-ROM

Visit ***www.algconcepts.com*** *for data updates, career information, games, and other interactive activities.*

CHAPTER 14 PREVIEW

Mathematics of the Chapter

In this chapter students investigate irrational numbers and radical expressions. At the beginning of the chapter, students classify real numbers as natural numbers, whole numbers, integers, rational numbers, or irrational numbers. They then calculate the distance between pairs of points in the plane. In the remainder of the chapter, students explore irrational numbers by simplifying, adding, and subtracting radical expressions and by solving simple radical equations.

What You'll Learn

Have students read over the lists of key ideas and key vocabulary. Have them make a list of any words with which they are not familiar.

Why It's Important

Point out to students that this is only one of many reasons why each objective is important. Others are provided in each lesson.

Answers

19. $y^2 - 9$
20. $q^2 - 16$
21. $z^2 - 25$
22. $r^2 - t^2$
23. $9c^2 - 25$
24. $4a^2 - 49b^2$

CHAPTER 14 Radical Expressions

What You'll Learn

Key Ideas

- Describe the relationships among sets of numbers. *(Lesson 14–1)*
- Find the distance between two points in the coordinate plane. *(Lesson 14–2)*
- Simplify, add, and subtract radical expressions. *(Lessons 14–3 and 14–4)*
- Solve simple radical equations. *(Lesson 14–5)*

Key Vocabulary

Distance Formula *(p. 616)*
radical equations *(p. 624)*
real numbers *(p. 600)*
rationalizing the denominator *(p. 615)*

Why It's Important

Meteorology At any moment about 2000 thunderstorms are in progress on Earth. Some storms are a few miles in diameter, and some span hundreds of miles. Scientists use sophisticated equipment and mathematical models to monitor and predict these storms.

Radical expressions are used to represent some real-world situations that cannot be modeled with linear equations. You will use a radical equation to find how long a thunderstorm is expected to last in Lesson 14–1.

CHAPTER 14 LINKS					
Lesson	**14–1**	**14–2**	**14–3**	**14–4**	**14–5**
Math in the Workplace	Meteorology	Engineering	Science	Hobbies	Engineering
Applications and Connections	Science Electricity Money	Maps School	Sports Quilting	Media	Science Animals
Math Integration	Geometry	Geometry	Geometry	Geometry	Geometry

Study these lessons to improve your skills.

Check Your Readiness

Lesson 8–5, pp. 357–361

Simplify.

1. $\sqrt{\frac{1}{4}}$ $\frac{1}{2}$
2. $\sqrt{\frac{4}{9}}$ $\frac{2}{3}$
3. $\sqrt{\frac{16}{4}}$ 2
4. $\sqrt{\frac{36}{49}}$ $\frac{6}{7}$
5. $-\sqrt{\frac{1}{121}}$ $-\frac{1}{11}$
6. $-\sqrt{\frac{25}{144}}$ $-\frac{5}{12}$

Lesson 8–6, pp. 362–365

Estimate each square root to the nearest whole number.

7. $\sqrt{2}$ **1**
8. $\sqrt{6}$ **2**
9. $\sqrt{14}$ **4**
10. $\sqrt{26}$ **5**
11. $\sqrt{52}$ **7**
12. $\sqrt{104}$ **10**

Lesson 1–4, pp. 19–23

Simplify each expression.

13. $6p + 7p$ $13p$
14. $14c - 3c$ $11c$
15. $5v - v$ $4v$
16. $7d - 3d + 5d$ $9d$
17. $12r - 6r + 3$ $6r + 3$
18. $5m - 4n + m + 6n$ $6m + 2n$

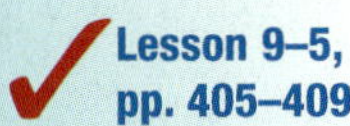
Lesson 9–5, pp. 405–409

Find each product. **See margin.**

19. $(y - 3)(y + 3)$
20. $(q + 4)(q - 4)$
21. $(z - 5)(z + 5)$
22. $(r - t)(r + t)$
23. $(3c - 5)(3c + 5)$
24. $(2a + 7b)(2a - 7b)$

Make this Foldable to help you organize your Chapter 14 notes. Begin with a sheet of 11" by 17" paper.

❶ **Fold** the short sides to meet in the middle.

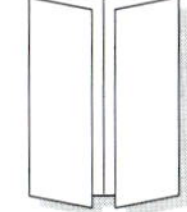

❷ **Fold** the top to the bottom.

❸ **Open.** Cut along the second fold to make four tabs.

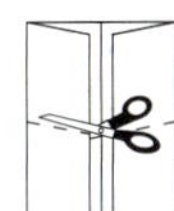

❹ **Label** each tab as shown.

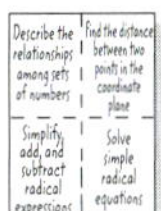

Reading and Writing As you read and study the chapter, write notes and examples under each tab.

FOLDABLES™ Study Organizer On each tab of this Foldable, have students record the ideas or problems from the chapter that are most difficult. The students can refer to their journals later when reviewing for chapter, unit, or semester tests. Also have students record times when they have seen radical expressions and how they can use radical expressions to solve problems.

Check Your Readiness

This section provides a review of the basic concepts needed before beginning Chapter 14. Lesson references are included for additional student help.

Use Exercises	To Prepare for Lesson(s)
1–6	14–1, 14–3
7–12	14–1, 14–2
13–18	14–4
19–24	14–3

Quick Review Math Handbook

Glencoe Algebra Lesson	Quick Review Math Handbook Hot Topic
14–1	3–2
14–2	
Ch. 14 Inv.	6–7
14–3	3–2
14–4	3–2
14–5	

Noteables™

Interactive Study Notebook with Foldables™

This note-taking guide reinforces excellent note-taking skills. It includes:

- ✓ **Build Your Vocabulary** tool for including mathematical terminology in notes.
- ✓ **Review It** activities with links to prior lessons and items to include in **Foldables™**.
- ✓ **Bringing It All Together** feature to help students review for the Chapter Test.
- ✓ **Are You Ready for the Chapter Test?** feature to allow students to assess their own readiness for the Chapter Test.
- ✓ **Teacher Edition** with transparencies to guide note taking. Corresponds to the *Teacher's Wraparound Edition* In-Class Examples and *Interactive Chalkboard CD-ROM.*

Lesson 14–1

1 FOCUS

5-Minute Check
Chapter 13

1. Solve the system of equations by graphing.
   $y = x + 3$
   $x + y = 5$ **(1, 4)**

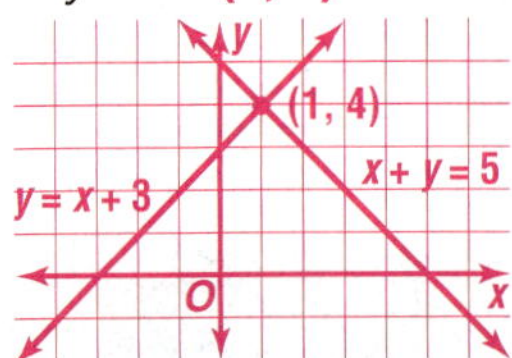

2. Use substitution to solve the system of equations.
   $y = x^2 - 4$
   $y - x + 2$ **(−2, 0), (3, 5)**

Use elimination to solve each system of equations.

3. $2x - 3y = 16$
   $-x + 3y = -11$ **(5, −2)**
4. $4x + 3y = 17$
   $5x - y = -12$ **(−1, 7)**
5. Solve the system of inequalities by graphing.
   $y < 3$
   $x + y \geq 1$

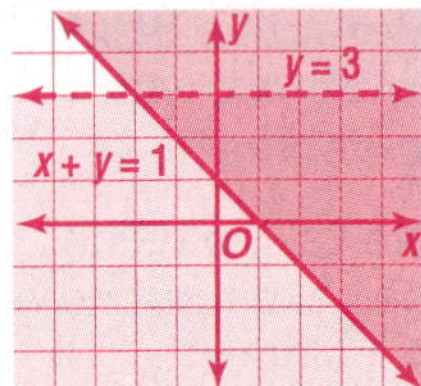

Motivating the Lesson

Real-World Connection Ask the students how they could categorize the vehicles owned or operated by their family. Offer this sample categorization plan.

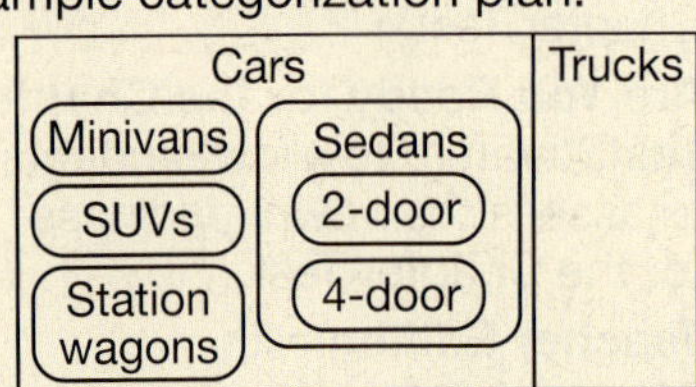

Stress the fact that a particular vehicle can fit into several categories, such as car, sedan, and 4-door.

14–1 The Real Numbers

What You'll Learn
You'll learn to describe the relationships among sets of numbers.

Why It's Important
Meteorology Meteorologists use real numbers when determining the duration of a thunderstorm. *See Exercise 16.*

In Lesson 3–1, you learned about *rational numbers*. Natural numbers, whole numbers, and integers are all rational numbers. These sets are listed below.

Natural Numbers:	$\{1, 2, 3, 4, \ldots\}$
Whole Numbers:	$\{0, 1, 2, 3, \ldots\}$
Integers:	$\{\ldots, -2, -1, 0, 1, 2, \ldots\}$
Rational Numbers:	{all numbers that can be expressed in the form $\frac{a}{b}$, where a and b are integers and $b \neq 0$}

Recall that repeating or terminating decimals are also rational numbers because they can be expressed as $\frac{a}{b}$, where a and b are integers and $b \neq 0$. The square roots of perfect squares are also rational numbers. For example, $\sqrt{0.16}$ is a rational number since $\sqrt{0.16} = 0.4$. However, $\sqrt{21}$ is irrational because 21 is not a perfect square.

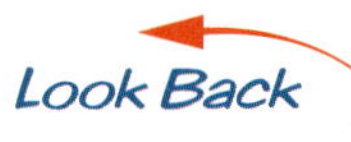

Look Back
Venn Diagram: Lesson 2–1

The Venn diagram shows the relationship among the different types of rational numbers. For example, the set of whole numbers is a subset of the integers. This means that all whole numbers are integers. Similarly, all rational numbers are real numbers.

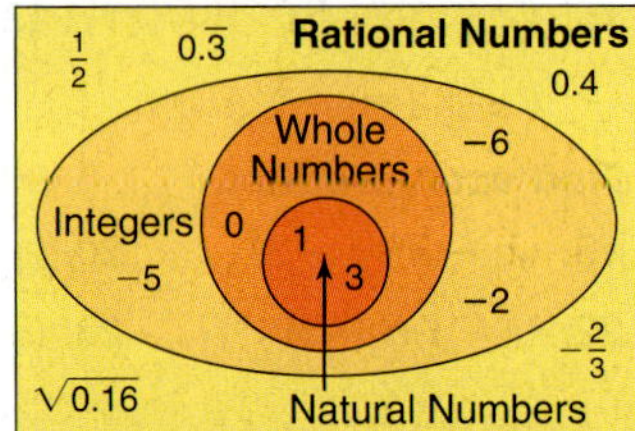

In Lesson 8–6, you learned about *irrational numbers*. A few examples of irrational numbers are shown below.

$0.1010010001\ldots$ π $0.153768\ldots$ $-\sqrt{23}$

The set of rational numbers and the set of irrational numbers together form the set of **real numbers**. *Numbers such as $\sqrt{-1}$ and $4 + \sqrt{-9}$ are called **complex numbers**. The set of complex numbers includes all of the real numbers as well as numbers involving square roots of negative numbers. We will not deal with complex numbers in this text.*

Natural numbers are also called *counting numbers*.

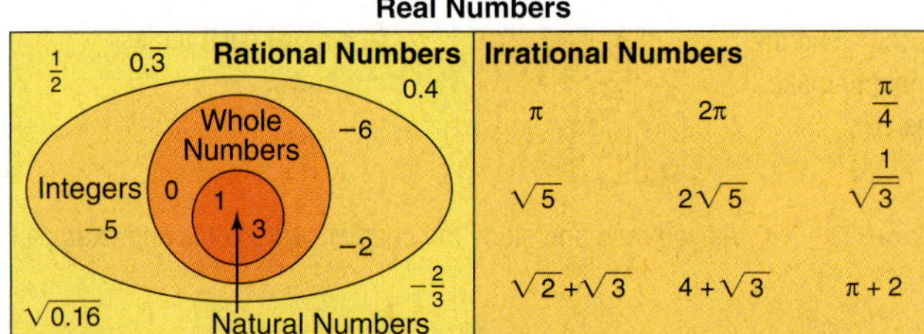

Resource Manager

Reproducible Masters
Chapter 14 Resource Masters
- *Study Guide*, p. 623
- *Skills Practice*, p. 624
- *Practice*, p. 625
- *Reading to Learn Mathematics*, p. 626
- *Enrichment*, p. 627

Transparencies
5-Minute Check, 14–1

Technology/Multimedia
AlgePASS, Lesson 32
Interactive Chalkboard CD-ROM

Examples

Name the set or sets of numbers to which each real number belongs.

1. -5 This number is an integer and a rational number.

2. $\frac{12}{3}$ Since $\frac{12}{3} = 4$, this number is a natural number, a whole number, an integer, and a rational number.

3. $\sqrt{95}$ $\sqrt{95} = 9.746794345\ldots$ It is not the square root of a perfect square. So, it is irrational.

4. $0.\overline{7}$ This repeating decimal is a rational number since it is equivalent to $\frac{7}{9}$. $7 \div 9 = 0.7777777\ldots$

5. $-\sqrt{4}$ Since $-\sqrt{4} = -2$, this number is an integer and a rational number.

Your Turn **Name the set or sets of numbers to which each real number belongs. Let N = natural numbers, W = whole numbers, Z = integers, Q = rational numbers, and I = irrational numbers.**

a. $3.141592\ldots$ b. $-\sqrt{9}$ c. $0.1666666\ldots$ d. $\frac{20}{4}$ e. 0

a. I
b. Z, Q
c. Q
d. N, W, Z, Q
e. W, Z, Q

If you graph all of the rational numbers, you will still have some "holes" in the number line. The irrational numbers "fill in" the number line. The graph of all real numbers is the entire number line without any "holes."

This property of real numbers is called the Completeness Property.

Completeness Property for Points on the Number Line	Each real number corresponds to exactly one point on the number line. Each point on the number line corresponds to exactly one real number.

You have learned how to graph rational numbers. Irrational numbers can also be graphed. Therefore, every real number can be graphed. Use a calculator or a table of squares and square roots to find approximate values of square roots that are irrational. These values can be used to approximate the graphs of square roots.

www.algconcepts.com/extra_examples

2 TEACH

Teaching Tip When discussing rational numbers on page 600, stress to students that a terminating decimal such as 2.1578 is a rational number because $2.1578 = \frac{21{,}578}{10{,}000}$. Students may enjoy using calculators to find fractions that represent repeating decimals such as 0.222..., 0.454545..., and 0.505050.... **$\frac{2}{9}$, $\frac{5}{11}$, and $\frac{50}{99}$, respectively**

Teaching Tip Students should realize that the sets of rational and irrational numbers are *mutually exclusive*, meaning that a number cannot be in both sets.

In-Class Examples

Examples 1–5

Name the set or sets of numbers to which each real number belongs.

1. 10 **natural number, whole number, integer, rational number**
2. $-\sqrt{100}$ **integer, rational number**
3. 5^2 **natural number, whole number, integer, rational number**
4. $-0.666\ldots$ **rational number**
5. $\sqrt{17}$ **irrational number**

Family Activity

Have students work with a family member to find twenty numbers on the labels of food, clothing, or other items in their home. They should classify each number using the diagram on page 600.

In-Class Examples

Examples 6–7

Find an approximation, to the nearest tenth, for each square root. Then graph the square root on a number line.

6 $\sqrt{10}$ **3.2**

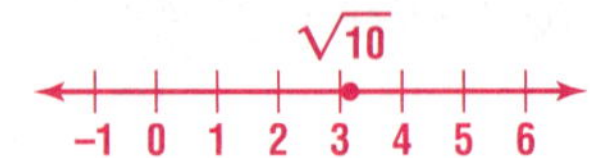

7 $-\sqrt{8}$ **−2.8**

−√8
−5 −4 −3 −2 −1 0 1 2

Examples 8–9

Determine whether each number is rational or irrational. If it is irrational, find two consecutive integers between which its graph lies on the number line.

8 $-\sqrt{59}$ **irrational, between −7 and −8**

9 $\sqrt{121}$ **rational**

Answers

Your Turn

f.

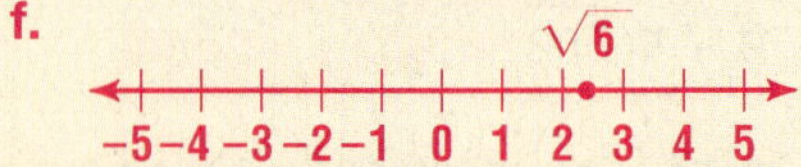

g. −√23
−5 −4 −3 −2 −1 0 1 2 3 4 5

Examples

Find an approximation, to the nearest tenth, for each square root. Then graph the square root on a number line.

6 $\sqrt{3}$

Enter: [2nd] [$\sqrt{\ }$] 3 [ENTER] 1.732050808

An approximate value for $\sqrt{3}$ is 1.7.

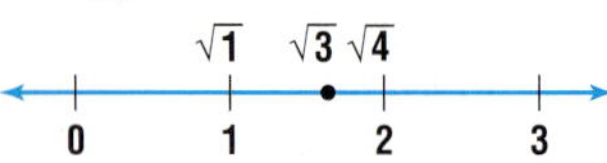

7 $-\sqrt{12}$

Enter: [(−)] [2nd] [$\sqrt{\ }$] 12 [ENTER] −3.464101615

An approximate value for $-\sqrt{12}$ is -3.5.

−√16 −√12 −√9
−5 −4 −3 −2

Your Turn **f–g. See margin for graphs.**

f. $\sqrt{6}$ **2.4**

g. $-\sqrt{23}$ **−4.8**

Determine whether each number is rational or irrational. If it is irrational, find two consecutive integers between which its graph lies on the number line.

8 $\sqrt{47}$

$\sqrt{36} < \sqrt{47} < \sqrt{49}$ — *$\sqrt{47}$ is between $\sqrt{36}$ and $\sqrt{49}$.*

$6 < \sqrt{47} < 7$ — *$\sqrt{36} = 6$ and $\sqrt{49} = 7$.*

47 is not a perfect square. So, its square root is irrational. The graph of $\sqrt{47}$ lies between 6 and 7.

9 $-\sqrt{26}$

$-\sqrt{36} < -\sqrt{26} < -\sqrt{25}$ — *$-\sqrt{26}$ is between $-\sqrt{36}$ and $-\sqrt{25}$.*

$-6 < -\sqrt{26} < -5$ — *$-\sqrt{36} = -6$ and $-\sqrt{25} = -5$.*

26 is not a perfect square. So, its square root is irrational. The graph of $-\sqrt{26}$ lies between -6 and -5.

Your Turn

h. $\sqrt{64}$ **rational**

i. $-\sqrt{28}$ **irrational; −5 and −6**

Reteaching Activity

Visual/Spatial Learners Have students make their own version of the diagram of number sets on page 601. Instruct them that their diagrams should have different examples than the ones shown in the textbook, and should be accompanied by an explanation of how to determine where to place any particular number.

You have solved equations with rational number solutions. Some equations have solutions that are irrational numbers.

Example 10 — Science Link

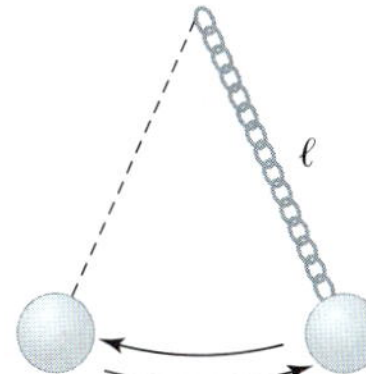

The time t in seconds it takes for a pendulum to complete one full swing (back and forth) is given by the equation $t = 2\pi\sqrt{\frac{\ell}{9.8}}$, where ℓ is the length of the pendulum in meters. Suppose a pendulum has a length of 4.9 meters. How long does it take the pendulum to complete one full swing?

$t = 2\pi\sqrt{\frac{\ell}{9.8}}$ *Original formula*

$= 2\pi\sqrt{\frac{4.9}{9.8}}$ *Replace ℓ with 4.9.*

$= 2\pi\sqrt{0.5}$ *Simplify.*

Enter: 2 [2nd] [π] [2nd] [$\sqrt{\ }$] 0.5 [ENTER] 4.442882938

The pendulum will complete one full swing in about 4.4 seconds.

Check for Understanding

Communicating Mathematics

> **Vocabulary**
> real numbers

1. **Give an example** of a number that is an integer and a rational number.
2. **Give** two counterexamples for the statement *all square roots are irrational numbers.* **2–3. See margin.**
3. **Write** a square root that is an irrational number.

1. Sample answer: −4

Guided Practice

Examples 1–5

Name the set or sets of numbers to which each real number belongs. Let N = natural numbers, W = whole numbers, Z = integers, Q = rational numbers, and I = irrational numbers.

4. $-\sqrt{36}$ **Z, Q**
5. 0.3131131113 . . . **I**
6. 4 **N, W, Z, Q**
7. $0.\overline{2}$ **Q**

Examples 6 & 7

Find an approximation, to the nearest tenth, for each square root. Then graph the square root on a number line.

8–11. See margin for graphs.

8. $\sqrt{7}$ **2.6**
9. $-\sqrt{20}$ **−4.5**
10. $\sqrt{32}$ **5.7**
11. $\sqrt{54}$ **7.3**

Examples 8 & 9

Determine whether each number is *rational* or *irrational*. If it is irrational, find two consecutive integers between which its graph lies on the number line. **14. irrational; −12 and −13**

12. $\sqrt{63}$ **irrational; 7 and 8**
13. $-\sqrt{25}$ **rational**
14. $-\sqrt{147}$
15. $\sqrt{49}$ **rational**

Answers

In-Class Example

Example 10

The radius r of a circle with area A is given by the equation $r = \sqrt{\frac{A}{\pi}}$. Suppose a circle has an area of 15 square inches. What is the measure of the radius? **about 2.2 in.**

3 PRACTICE/APPLY

Error Analysis

Watch for students who approximate $-\sqrt{20}$ as −4.5 but then graph $-\sqrt{20}$ between −4 and −3 on the number line in Exercise 9.

Prevent by stressing how to locate negative values on a number line.

Answers

2. Sample answer: $\sqrt{4} = 2$ and $\sqrt{1.21} = 1.1$
3. Sample answer: $\sqrt{13}$
8.

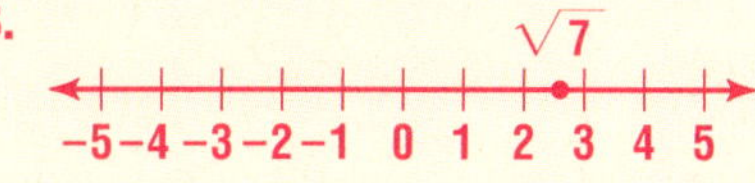

9. $-\sqrt{20}$
−5 −4 −3 −2 −1 0 1 2 3 4 5

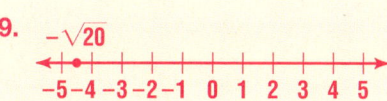

***Study Guide*, p. 623**

14-1 NAME ______ DATE ______ PERIOD ______

Study Guide

Student Edition Pages 600–605

The Real Numbers

The numbers that we use every day belong to the set of real numbers. The real numbers can be divided into rational numbers and irrational numbers. Rational numbers can be expressed as fractions. Irrational numbers cannot be expressed as fractions. The set of rational numbers contains the natural numbers, the whole numbers, and the integers.

Natural Numbers: {1, 2, 3, 4, ...}
Whole Numbers: {0, 1, 2, 3, ...}
Integers: {..., −2, −1, 0, 1, 2, ...}

The Venn diagram shows how subsets of the real numbers are related.

Rational Numbers: $0.\overline{3}$, $\sqrt{0.16}$, 0.4, $\frac{1}{2}$, $-\frac{2}{3}$; Integers: −5, −6; Whole Numbers: 0, $\sqrt{9}$; Natural Numbers: 1, 3; Irrational Numbers: π, $\sqrt{15}$

- 3 is a natural number, a whole number, an integer, and a rational number.
- 0 is a whole number, an integer, and a rational number.
- $\sqrt{9}$ or 3 is an integer and a rational number.
- $\sqrt{15}$ or 3.872983346... is an irrational number.
- $0.\overline{3}$ or $\frac{1}{3}$ is a rational number.

Example: Find an approximation to the nearest tenth of $\sqrt{12}$.

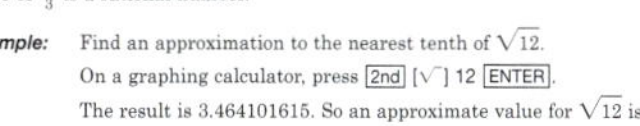

On a graphing calculator, press [2nd] [$\sqrt{\ }$] 12 [ENTER].

The result is 3.464101615. So an approximate value for $\sqrt{12}$ is 3.5.

Name the set or sets of numbers to which each real number belongs. Let N = natural numbers, W = whole numbers, Z = integers, Q = rational numbers, and I = irrational numbers.

1. −4 *Z, Q*
2. $\frac{2}{5}$ *Q*
3. $-\sqrt{25}$ *Z, Q*
4. 10 *N, W, Z, Q*
5. 2.3 *Q*
6. $\sqrt{3}$ *I*
7. −4.324781... *I*
8. $-\frac{24}{8}$ *Z, Q*
9. $\sqrt{100}$ *N, W, Z, Q*
10. $\frac{1}{9}$ *Q*
11. −0.25 *Q*
12. $\sqrt{15}$ *I*

Find an approximation to the nearest tenth for each square root.

13. $\sqrt{2}$ 1.4
14. $\sqrt{14}$ 3.7
15. $-\sqrt{20}$ −4.5
16. $\sqrt{55}$ 7.4

Assignment Guide

Basic: 17–57 odd, 59–73
Average: 18–54 even, 55–73

Answers

41. rational
42. irrational; -3 and -4
43. irrational; 6 and 7
44. irrational; 4 and 5
45. rational
46. irrational; 8 and 9
47. irrational; -2 and -3
48. irrational; 9 and 10
49. irrational; 11 and 12
50. rational
51. rational
52. irrational; 14 and 15
53.

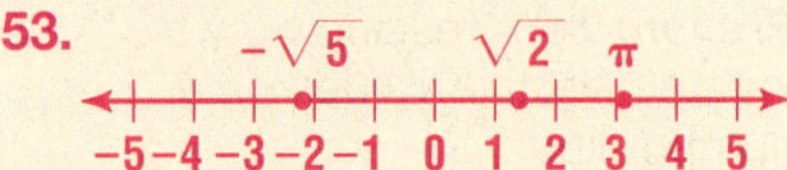

Example 10

16. **Meteorology** To estimate the amount of time t in hours that a thunderstorm will last, meteorologists use the formula $t = \sqrt{\frac{d^3}{216}}$, where d is the diameter of the storm in miles.
   a. Estimate how long a thunderstorm will last if its diameter is 8.4 miles. **about 1.7 h**
   b. Find the diameter of a storm that will last for 1 hour. **6 mi**

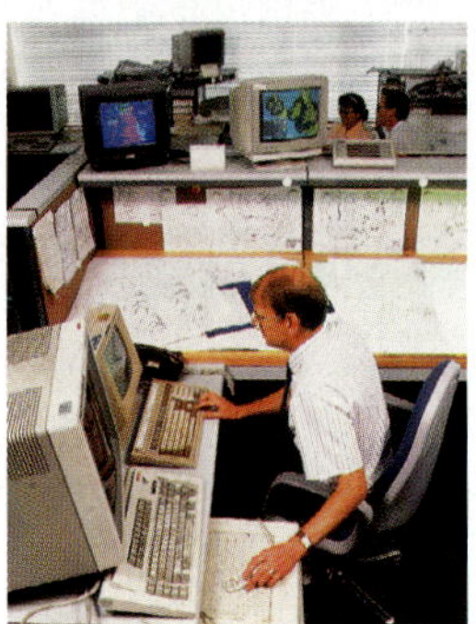

Exercises

Practice 

Name the set or sets of numbers to which each real number belongs. Let N = natural numbers, W = whole numbers, Z = integers, Q = rational numbers, and I = irrational numbers.

17. $-\frac{1}{2}$ **Q**
18. $\sqrt{17}$ **I**
19. 0 **W, Z, Q**
20. $\frac{15}{3}$
21. -5 **Z, Q**
22. 1.202002 . . .
23. $\sqrt{81}$
24. 0.125 **Q**
25. $-\sqrt{121}$ **Z, Q**
26. $\frac{6}{18}$ **Q**
27. 0.834834 . . . **Q**
28. $-\frac{3}{1}$ **Z, Q**

Homework Help	
For Exercises	See Examples
17–28	1–5
29–40, 53, 54, 57, 58	6, 7
41–52, 56	8, 9
55	10

Extra Practice
See page 720.

Find an approximation, to the nearest tenth, for each square root. Then graph the square root on a number line.

29. $\sqrt{3}$ **1.7**
30. $\sqrt{8}$ **2.8**
31. $-\sqrt{10}$ **−3.2**
32. $\sqrt{17}$ **4.1**
33. $-\sqrt{40}$ **−6.3**
34. $\sqrt{37}$ **6.1**
35. $\sqrt{52}$ **7.2**
36. $-\sqrt{99}$ **−9.9**
37. $\sqrt{108}$ **10.4**
38. $-\sqrt{112}$ **−10.6**
39. $\sqrt{250}$ **15.8**
40. $\sqrt{300}$ **17.3**

20. N, W, Z, Q
22. I
23. N, W, Z, Q

29–40. See Solutions Manual for graphs.

Determine whether each number is *rational* or *irrational*. If it is irrational, find two consecutive integers between which its graph lies on the number line. **41–52. See margin.**

41. $\sqrt{16}$
42. $-\sqrt{12}$
43. $\sqrt{41}$
44. $\sqrt{24}$
45. $-\sqrt{81}$
46. $\sqrt{66}$
47. $-\sqrt{7}$
48. $\sqrt{98}$
49. $\sqrt{125}$
50. $\sqrt{144}$
51. $-\sqrt{169}$
52. $\sqrt{220}$

B

53. Graph $\sqrt{2}$, π, and $-\sqrt{5}$ on a number line. **See margin.**
54. Write $2.\bar{4}$, 2.41, and $\sqrt{6}$ in order from least to greatest. **2.41, $2.\bar{4}$, $\sqrt{6}$**

Applications and Problem Solving

55. **Science** Refer to Example 10. Suppose a pendulum has a length of 18 meters. How long does it take the pendulum to complete one full swing? **about 8.5 s**

56. **Geometry** The radius of a circle is given by $r = \sqrt{\frac{A}{\pi}}$, where A is the area of the circle. Suppose a circle has an area of 25 square centimeters. What is the measure of the radius? **about 2.8 cm**

***Skills Practice*, p. 624, and *Practice*, p. 625 (shown)**

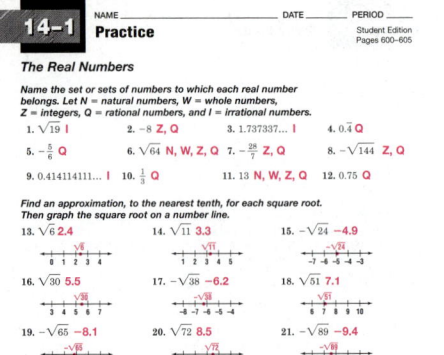

14–1 NAME ______________ DATE ______ PERIOD ______
Practice
Student Edition Pages 600–605

The Real Numbers

Name the set or sets of numbers to which each real number belongs. Let N = natural numbers, W = whole numbers, Z = integers, Q = rational numbers, and I = irrational numbers.

1. $\sqrt{19}$ **I**
2. -8 **Z, Q**
3. 1.737337... **I**
4. $0.\bar{4}$ **Q**
5. $-\frac{5}{6}$ **Q**
6. $\sqrt{64}$ **N, W, Z, Q**
7. $-\frac{28}{7}$ **Z, Q**
8. $-\sqrt{144}$ **Z, Q**
9. 0.414114111... **I**
10. $\frac{1}{3}$ **Q**
11. 13 **N, W, Z, Q**
12. 0.75 **Q**

Find an approximation, to the nearest tenth, for each square root. Then graph the square root on a number line.

13. $\sqrt{6}$ **2.4**
14. $\sqrt{11}$ **3.3**
15. $-\sqrt{24}$ **−4.9**
16. $\sqrt{30}$ **5.5**
17. $-\sqrt{38}$ **−6.2**
18. $\sqrt{51}$ **7.1**
19. $-\sqrt{65}$ **−8.1**
20. $\sqrt{72}$ **8.5**
21. $-\sqrt{89}$ **−9.4**
22. $\sqrt{118}$ **10.9**
23. $-\sqrt{131}$ **−11.4**
24. $\sqrt{104}$ **10.2**

Determine whether each number is rational or irrational. If it is irrational, find two consecutive integers between which its graph lies on the number line.

25. $\sqrt{28}$ **irrational; 5 and 6**
26. $-\sqrt{9}$ **rational**
27. $\sqrt{56}$ **irrational; 7 and 8**
28. $-\sqrt{14}$ **irrational; −4 and −3**
29. $\sqrt{36}$ **rational**
30. $\sqrt{99}$ **irrational; 9 and 10**
31. $-\sqrt{73}$ **irrational; −9 and −8**
32. $\sqrt{196}$ **rational**
33. $\sqrt{77}$ **irrational; 8 and 9**
34. $-\sqrt{100}$ **rational**
35. $\sqrt{88}$ **irrational; 9 and 10**
36. $-\sqrt{46}$ **irrational; −7 and −6**

57. 5.2 ohms and 1200 watts or 4.6 ohms and 1500 watts

57. **Electricity** The voltage V in a circuit is given by $V = \sqrt{PR}$, where P is the power in watts and R is the resistance in ohms. A circuit is designed with two resistance settings, 4.6 ohms and 5.2 ohms, and two power settings, 1200 watts and 1500 watts. Which settings are required so the voltage is between 75 and 85 volts?

58. **Geometry** Find the area of each rectangle. Round answers to the nearest tenth. (*Hint:* Use the Pythagorean Theorem.)

a.

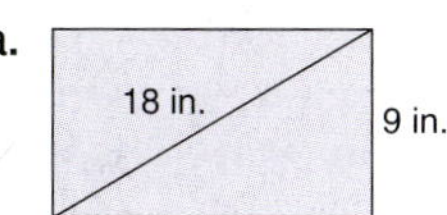

140.3 in^2

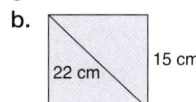

b. 22 cm, 15 cm

241.4 cm^2

59. **Critical Thinking** Roots can also be indicated with *rational exponents*. For any real number b, $\sqrt[n]{b} = b^{\frac{1}{n}}$. So, a square root is given by $\sqrt{b}$ or $b^{\frac{1}{2}}$. A *cube root* is given by $\sqrt[3]{b}$ or $b^{\frac{1}{3}}$. For example, $\sqrt[3]{216} = 216^{\frac{1}{3}} = 6$ because $6 \times 6 \times 6 = 216$. To what power do you raise

a. 9 to get 3? $\frac{1}{2}$ b. 64 to get 4? $\frac{1}{3}$ c. x^4 to get x? $\frac{1}{4}$

Mixed Review

60. Find the solution of the system of inequalities $y \leq x + 3$ and $y > -x - 4$. *(Lesson 13–7)* **See margin.**

61. What is the solution of the system of equations $y = x^2 + 2$ and $y = 4x - 1$? *(Lesson 13–6)* **(3, 11), (1, 3)**

62. Graph the solution of $|4n| > 16$ on a number line. *(Lesson 12–6)* **See margin.**

Solve each inequality. *(Lesson 12–2)*

63. $x + 3 < -4$ $\{x \mid x < -7\}$

64. $w - 14 \geq -8$ $\{w \mid w \geq 6\}$

65. $3.4 + m \leq 1.6$ $\{m \mid m \leq -1.8\}$

66. Write an inequality for the graph shown at the right. *(Lesson 12–1)* $x < 4.5$

−3 −2 −1 0 1 2 3 4 5 6 7

67. Graph $y = 2^x - 1$. Then state the y-intercept. *(Lesson 11–7)* **See margin for graph; 0.**

Factor each polynomial. If the polynomial cannot be factored, write *prime*. *(Lesson 10–5)*

68. $x^2 - 25y^2$ $(x + 5y)(x - 5y)$

69. $4a^2 - 36$ $4(a + 3)(a - 3)$

70. $m^2 + 2m + 4$ prime

71. Find the degree of $5x^3 - 3x^3y - 6xy$. *(Lesson 9–4)* **4**

Standardized Test Practice

72. **Short Response** Cleavon has $35 to spend on CDs. Suppose CDs cost $13.85 each, including tax. Write an equation that can be used to determine how many CDs Cleavon can buy. *(Lesson 4–4)* $13.85x \leq 35$

73. **Multiple Choice** What is the value of a if $6 - (-8) = a$? *(Lesson 2–3)* **A**

Ⓐ 14 Ⓑ 6 Ⓒ −14 Ⓓ −6

Extra Credit

Graph π, $3\frac{1}{7}$, and $3\frac{10}{71}$ on a number line.

Sample answer:

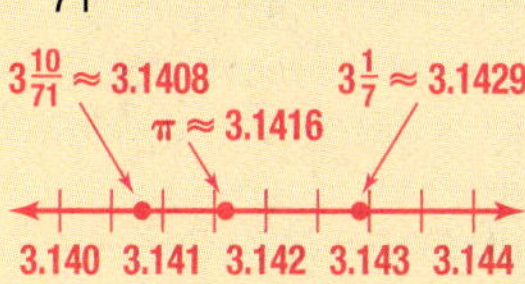

4 ASSESS

Open-Ended Assessment

Speaking Ask students to explain how the Venn diagram on page 601 shows that a number can be a member of more than one set of numbers.

Answers

60.

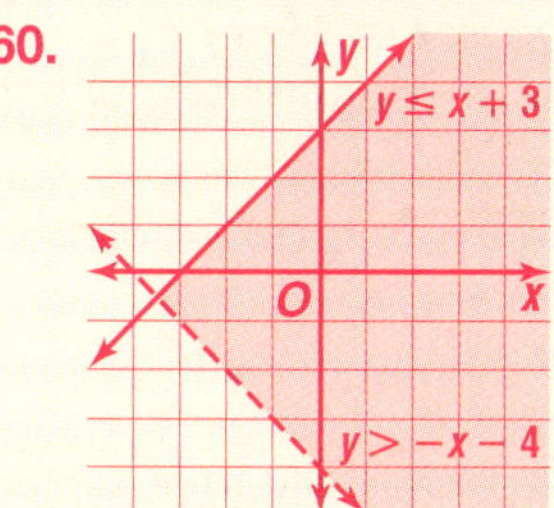

62. $\{x \mid x < -4 \text{ or } x > 4\}$

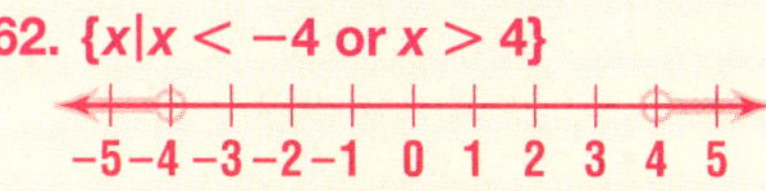

67.

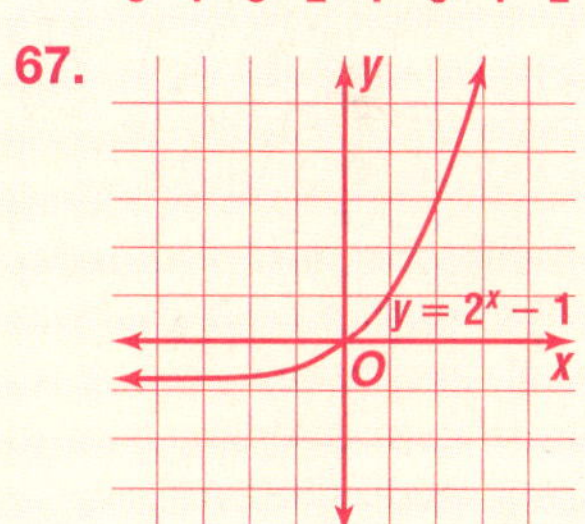

***Enrichment*, p. 627**

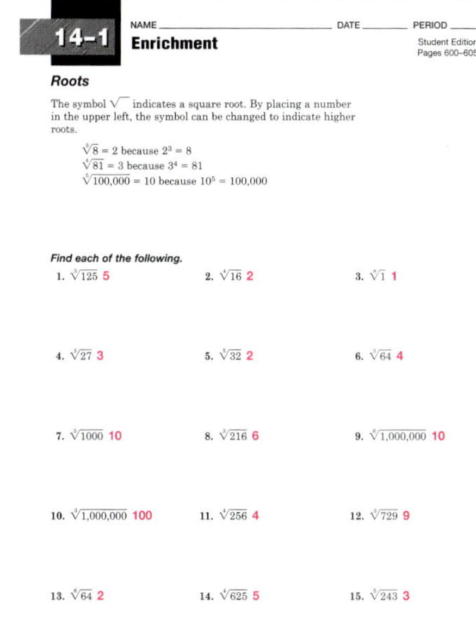

14-1 NAME ______ DATE ______ PERIOD ______

Enrichment

Student Edition Pages 600–605

Roots

The symbol $\sqrt{\ }$ indicates a square root. By placing a number in the upper left, the symbol can be changed to indicate higher roots.

$\sqrt[3]{8} = 2$ because $2^3 = 8$

$\sqrt[4]{81} = 3$ because $3^4 = 81$

$\sqrt[5]{100{,}000} = 10$ because $10^5 = 100{,}000$

Find each of the following.

1. $\sqrt[3]{125}$ 5	2. $\sqrt[4]{16}$ 2	3. $\sqrt[7]{1}$ 1
4. $\sqrt[3]{27}$ 3	5. $\sqrt[5]{32}$ 2	6. $\sqrt[3]{64}$ 4
7. $\sqrt[3]{1000}$ 10	8. $\sqrt[3]{216}$ 6	9. $\sqrt[6]{1{,}000{,}000}$ 10
10. $\sqrt[3]{1{,}000{,}000}$ 100	11. $\sqrt[4]{256}$ 4	12. $\sqrt[3]{729}$ 9
13. $\sqrt[6]{64}$ 2	14. $\sqrt[4]{625}$ 5	15. $\sqrt[5]{243}$ 3

Lesson 14–2

1 FOCUS

5-Minute Check
Lesson 14–1

Name the set or sets of numbers to which each real number belongs.

1. $-\sqrt{144}$ **integers, rational numbers**
2. $-\sqrt{5}$ **irrational numbers**
3. Find an approximation, to the nearest tenth, for $\sqrt{80}$. Then graph $\sqrt{80}$ on a number line. **8.9**

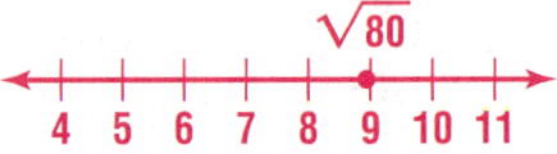

4. Determine whether $\sqrt{38}$ is *rational* or *irrational*. If it is irrational, find two consecutive integers between which its graph lies on the number line. **irrational, between 6 and 7**
5. The equation $d = \sqrt{2A}$ gives the length d of a diagonal of a square that has an area of A square units. Find the length of the diagonal of a square, to the nearest tenth of a centimeter, if the area of the square is 210 square centimeters. **20.5 cm**

Motivating the Lesson

Real-World Connection Ask two students who live near your school to describe where they live in relation to the school. Have them give the location as the number of blocks (or miles) east or west of the school and the number of blocks (or miles) north or south of the school. Then ask the class how they can use that information to find the distance, in blocks (or miles), between the two students' homes.

14–2 The Distance Formula

What You'll Learn
You'll learn to find the distance between two points in the coordinate plane.

Why It's Important
Engineering Engineers can use the distance formula to determine the amount of cable needed to install a cable system. *See Exercise 9.*

A coordinate system is superimposed over a map of Washington, D.C. Jessica and Omar walk from the Metro Center to 15th Street and then up 15th Street to McPherson Square. How far is the Metro center from McPherson Square? *This problem will be solved in Example 2.*

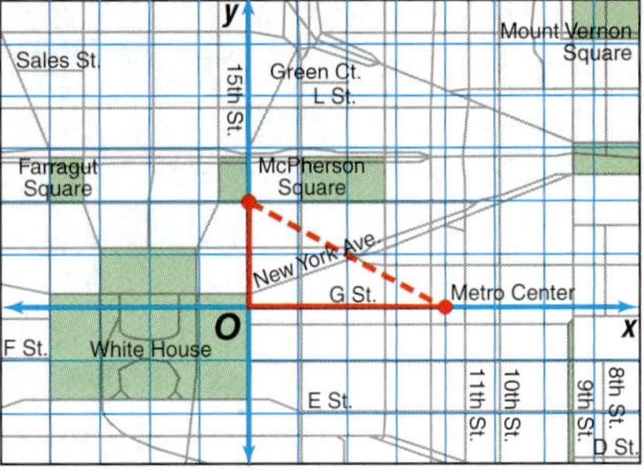

Recall that subtraction is used to find the distance between two points that lie on a vertical or horizontal line.

In the following activity, you will find the distance between two points that do not lie on a horizontal or vertical line.

Hands-On Algebra

Materials:  grid paper  straightedge

Step 1 Graph $M(-4, 3)$ and $N(2, -5)$ on a coordinate plane.

Step 2 Draw a vertical segment from M and a horizontal segment from N. Label the point of intersection P.

Step 3 Find the coordinates of P. **(−4, −5)**

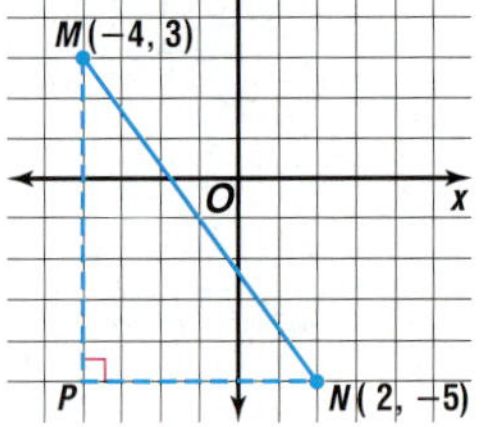

Try These **4. Pythagorean Theorem**

1. Find the distance between M and P. **8 units**
2. Find the distance between N and P. **6 units**
3. What kind of triangle is $\triangle MNP$? **right**
4. What theorem can be used to find MN if MP and NP are known?
5. Find the distance between M and N. **10 units**

Resource Manager

 **Reproducible Masters**

Chapter 14 Resource Masters
- *Study Guide*, p. 628
- *Skills Practice*, p. 629
- *Practice*, p. 630
- *Reading to Learn Mathematics*, p. 631
- *Enrichment*, p. 632
- *Assessment*, p. 659

Hands-On Algebra, pp. 153–154
School-to-Workplace, p. 14

 **Transparencies**
5-Minute Check, 14–2

 **Technology/Multimedia**
Interactive Chalkboard CD-ROM

In the activity, you discovered that $(MN)^2 = (MP)^2 + (NP)^2$. You can solve this equation for MN to find the distance between points M and N.

$$(MN)^2 = (MP)^2 + (NP)^2$$

$$\sqrt{(MN)^2} = \sqrt{(MP)^2 + (NP)^2} \quad \textit{Take the positive square root of each side.}$$

$$MN = \sqrt{(MP)^2 + (NP)^2}$$

The **Distance Formula** can be used to find the distance between any two points (x_1, y_1) and (x_2, y_2).

The Distance Formula

Words: The distance d between any two points with coordinates (x_1, y_1) and (x_2, y_2) is given by $d = \sqrt{(x_2 - x_1)^2 + (y_2 - y_1)^2}$.

Model: $A(x_1, y_1)$, $B(x_2, y_2)$, O, x, y

Example 1 Find the distance between points $A(1, 2)$ and $B(-3, 7)$.

$$d = \sqrt{(x_2 - x_1)^2 + (y_2 - y_1)^2} \quad \textit{Distance Formula}$$

$$= \sqrt{(-3 - 1)^2 + (7 - 2)^2} \quad (x_1, y_1) = (1, 2), \text{ and } (x_2, y_2) = (-3, 7)$$

$$= \sqrt{(-4)^2 + 5^2} \quad \textit{Subtract.}$$

$$= \sqrt{16 + 25} \quad (-4)^2 = 16 \text{ and } 5^2 = 25.$$

$$= \sqrt{41} \text{ or about 6.4 units}$$

Your Turn a. $\sqrt{10}$ or about 3.2 units

a. Find the distance between points $C(3, 5)$ and $D(6, 4)$.

Example 2 **Map Link**

Refer to the application at the beginning of the lesson. The Metro Center is located 4 blocks east of the intersection of 15th Street and G Street. McPherson Square is located 2 blocks north of the intersection. How far is the Metro Center from McPherson Square?

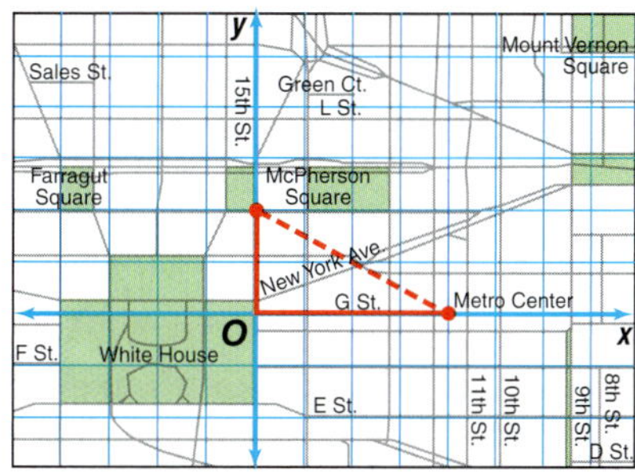

(continued on the next page)

www.algconcepts.com/extra_examples

2 TEACH

Teaching Tip Focus students' attention on the second and third lines of the solution at the top of the page. Inform students that $\sqrt{(MN)^2}$ can be replaced by just the positive value of MN because MN represents a length, which cannot be negative. (Later, in Lesson 14.3, students will write $\sqrt{x^2} = |x|$ because x could be negative.)

In-Class Examples

Example 1

Find the distance between points $E(-3, 8)$ and $F(5, 1)$.
$\sqrt{113}$ or about 10.6 units

Example 2

Refer to the application at the beginning of the lesson. Two taxis leave the intersection of 15th Street and G Street. One taxi travels 7 blocks north and the other taxi travels 11 blocks east. How far apart are the taxis when they stop?
$\sqrt{170}$ or about 13.0 blocks

Hands-On Algebra

Cooperative Learning Refer to the Hands-On Algebra on page 606. Ask students to repeat the activity, including the exercises, using a horizontal line from M and vertical line from N. Instruct them to label this point of intersection Q. Have them compare these results with those from the original activity.

An additional Hands-On Algebra activity using a pentagon in a coordinate grid to explore slope and the Distance Formula is available in the *Hands-On Algebra Masters,* p. 153.

Hands-On Algebra Masters, p. 154

Teaching Tip After discussing Example 3, ask students to graph the three points (4, 7), (7, 3), and (1, 3). The graph can help them see that both of the points (7, 3) and (1, 3) are 5 units from (4, 7).

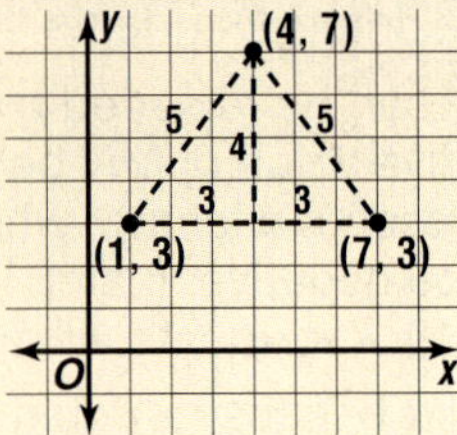

In-Class Example

Example 3

Find the value of a if $M(a, 6)$ and $N(2, 3)$ are $\sqrt{18}$ units apart. **−1 or 5**

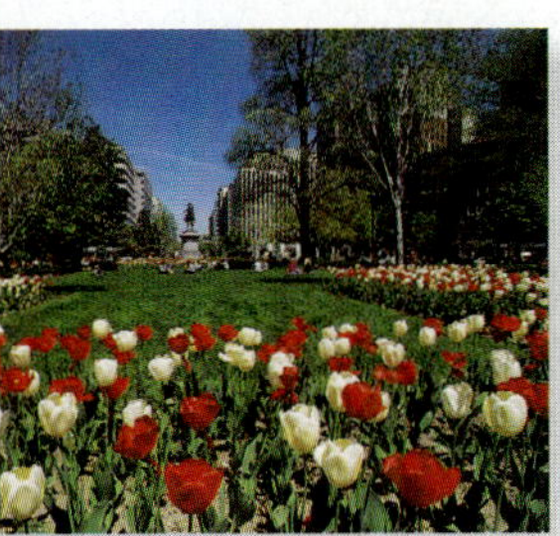

McPherson Square, Washington, D.C.

Let the Metro Center be represented by (4, 0) and McPherson Square by (0, 2). So, $x_1 = 4$, $y_1 = 0$, $x_2 = 0$, and $y_2 = 2$.

$d = \sqrt{(x_2 - x_1)^2 + (y_2 - y_1)^2}$	*Distance Formula*
$= \sqrt{(0 - 4)^2 + (2 - 0)^2}$	$(x_1, y_1) = (4, 0)$, *and* $(x_2, y_2) = (0, 2)$
$= \sqrt{(-4)^2 + (2)^2}$	*Subtract.*
$= \sqrt{16 + 4}$	$(-4)^2 = 16$ *and* $2^2 = 4$.
$= \sqrt{20}$ or about 4.5 blocks	

The Metro Center is about 4.5 blocks from McPherson Square.

Suppose you know the coordinates of a point, one coordinate of another point, and the distance between the two points. You can use the Distance Formula to find the missing coordinate.

Example 3 **Find the value of a if $G(4, 7)$ and $H(a, 3)$ are 5 units apart.**

$d = \sqrt{(x_2 - x_1)^2 + (y_2 - y_1)^2}$	*Distance Formula*
$5 = \sqrt{(a - 4)^2 + (3 - 7)^2}$	$(x_1, y_1) = (4, 7)$, *and* $(x_2, y_2) = (a, 3)$
$5 = \sqrt{(a - 4)^2 + (-4)^2}$	*Subtract.*
$5 = \sqrt{a^2 - 8a + 16 + 16}$	$(a - 4)^2 = a^2 - 8a + 16$ *and* $(-4)^2 = 16$
$5 = \sqrt{a^2 - 8a + 32}$	*Simplify.*
$5^2 = (\sqrt{a^2 - 8a + 32})^2$	*Square each side.*
$25 = a^2 - 8a + 32$	*Simplify.*
$0 = a^2 - 8a + 7$	*Subtract 25 from each side.*
$0 = (a - 7)(a - 1)$	*Factor.*
$a - 7 = 0$ or $a - 1 = 0$	*Zero Product Property*
$a = 7$ $a = 1$	*Solve.*

The value of a is 7 or 1.

Your Turn **b. −2 or −12**

b. Find the value of a if $J(a, 5)$ and $K(-7, 3)$ are $\sqrt{29}$ units apart.

Check for Understanding

Communicating Mathematics

1. Name the theorem that is used to derive the Distance Formula. **Pythagorean Theorem**

Vocabulary
Distance Formula

From the Classroom of ...

Nicki Hudson
West Linn High School
West Linn, Oregon

I have often done an inductive approach to the Distance Formula that is very similar to the Hands-On Algebra activity on page 606. I give my students the coordinates of the endpoints of many diagonal segments and ask them to determine the process for finding the length of each segment. They always get it on their own discovery path if they are given enough time.

2–3. See margin.

2. **Tell** how to find the distance between the points $A(10, 3)$ and $B(2, 3)$ without using the Distance Formula.

3. **You Decide** Anna says that to find the distance between $A(5, -8)$ and $B(7, -6)$, you should take the square root of $(7 - 5)^2 + [-6 - (-8)]^2$. Nick disagrees. He says to take the square root of $(5 - 7)^2 + [-8 - (-6)]^2$. Who is correct? Explain your reasoning.

Guided Practice

Find the distance between each pair of points. Round to the nearest tenth, if necessary.

Example 1

4. $C(5, -1), D(2, 2)$ $\sqrt{18}$ or 4.2
5. $M(6, 8), N(3, 4)$ 5
6. $R(-3, 8), S(5, 4)$ $\sqrt{80}$ or 8.9

Example 3

Find the value of *a* if the points are the indicated distance apart.

7. $A(3, -1), B(a, 7); d = 10$ 9 or −3
8. $J(5, a), K(6, 1); d = \sqrt{10}$ −2 or 4

Example 2

9. **Engineering** The MAXTEC Company is installing a fiber optic cable system between two of their office buildings. One of the buildings, MAXTEC West, is located 5 miles west and 2 miles north of the corporate office. The other building, MAXTEC East, is located 4 miles east and 5 miles north of the corporate office. How much cable will be needed to connect MAXTEC West and MAXTEC East? Round to the nearest tenth. 9.5 mi

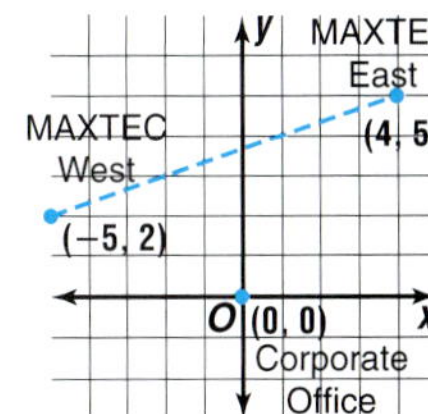

Exercises

Practice

Find the distance between each pair of points. Round to the nearest tenth, if necessary.

10. $X(5, 0), Y(12, 0)$ 7
11. $A(-6, -4), B(-6, 8)$ 12
12. $M(-2, -5), N(3, 7)$ 13
13. $P(-4, 0), Q(3, -3)$ $\sqrt{58}$ or 7.6
14. $V(-3, -4), W(-1, -2)$
15. $C(7, 2), D(-4, 10)$
16. $E(3, -6), F(9, -2)$ $\sqrt{52}$ or 7.2
17. $G(-4, -6), H(-7, -3)$

14. $\sqrt{8}$ or 2.8 15. $\sqrt{185}$ or 13.6 17. $\sqrt{18}$ or 4.2

Homework Help	
For Exercises	See Examples
10–17, 24, 25	1, 2
18–23, 26, 27	3
Extra Practice	
See page 720.	

Find the value of *a* if the points are the indicated distance apart.

B

18. $A(a, -5), B(-3, -2); d = 5$
19. $D(-3, a), E(5, 2); d = 17$
20. $Q(7, 2), R(-1, a); d = 10$
21. $G(7, -3), H(5, a); d = \sqrt{85}$
22. $T(6, -3), U(-3, a); d = \sqrt{130}$ −10 or 4
23. $U(1, -6), V(10, a); d = \sqrt{145}$ 2 or −14

18. −7 or 1
19. 17 or −13
20. 8 or −4
21. −12 or 6

3 PRACTICE/APPLY

Error Analysis

Watch for students who omit the negative sign that is part of the *y*-coordinate of point *C* in Exercise 4 and write $(y_2 - y_1)^2$ as $(2 - 1)^2 = 1$.
Prevent by having students use nested parentheses to indicate substitution, so they would write $(y_2 - y_1)^2$ as $(2 - (-1))^2$ or $(2 + 1)^2 = 3^2$ or 9.

Assignment Guide

Basic: 11–31 odd, 32–42
Average: 10–26 even, 28–42
All: Quiz 1, 1–10

Answers

2. Because the two points are on a horizontal line, the distance between them is the absolute value of the difference of their *x*-coordinates $|10 - 2|$ or 8 units.
3. Both are correct since it does not matter which ordered pair is first when using the Distance Formula as long as the coordinates are used in the same order.

***Study Guide*, p. 628**

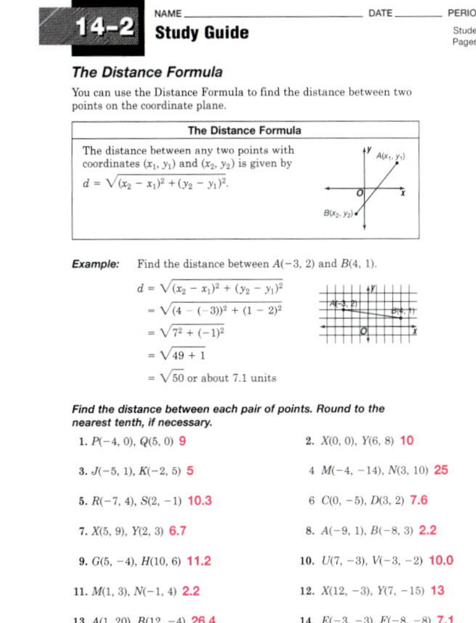
14-2 NAME ______ DATE ______ PERIOD ______

Study Guide Student Edition Pages 606–611

The Distance Formula

You can use the Distance Formula to find the distance between two points on the coordinate plane.

The Distance Formula

The distance between any two points with coordinates (x_1, y_1) and (x_2, y_2) is given by $d = \sqrt{(x_2 - x_1)^2 + (y_2 - y_1)^2}$.

Example: Find the distance between $A(-3, 2)$ and $B(4, 1)$.

$$d = \sqrt{(x_2 - x_1)^2 + (y_2 - y_1)^2}$$
$$= \sqrt{(4 - (-3))^2 + (1 - 2)^2}$$
$$= \sqrt{7^2 + (-1)^2}$$
$$= \sqrt{49 + 1}$$
$$= \sqrt{50} \text{ or about 7.1 units}$$

Find the distance between each pair of points. Round to the nearest tenth, if necessary.

1. $P(-4, 0), Q(5, 0)$ 9
2. $X(0, 0), Y(6, 8)$ 10
3. $J(-5, 1), K(-2, 5)$ 5
4. $M(-4, -14), N(3, 10)$ 25
5. $R(-7, 4), S(2, -1)$ 10.3
6. $C(0, -5), D(3, 2)$ 7.6
7. $X(5, 9), Y(2, 3)$ 6.7
8. $A(-9, 1), B(-8, 3)$ 2.2
9. $G(5, -4), H(10, 6)$ 11.2
10. $U(7, -3), V(-3, -2)$ 10.0
11. $M(1, 3), N(-1, 4)$ 2.2
12. $X(12, -3), Y(7, -15)$ 13
13. $A(1, 20), B(12, -4)$ 26.4
14. $E(-3, -3), F(-8, -8)$ 7.1
15. $A(5, 4), B(0, 6)$ 5.4
16. $S(-6, -6), T(-5, -1)$ 5.1

Reteaching Activity

Interpersonal Learners Have students work in small groups. One student should name the coordinates of two points that are in different quadrants of the coordinate plane. Another student then describes how to find the distance between the two points. All the students calculate that distance and compare their results. The activity should be repeated several times, with different students selecting points and describing how to find the distance.

Answers

30a.

(−35, −55) Webber Hall

(42, −30) Packard Hall

32. No; $\overline{AC}$ and $\overline{BD}$ are the diagonals of the trapezoid. $AC = \sqrt{157}$ and $BD = \sqrt{101}$. Since $\sqrt{157} \neq \sqrt{101}$, trapezoid $ABCD$ is not isosceles.

***Skills Practice*, p. 629, and *Practice*, p. 630 (shown)**

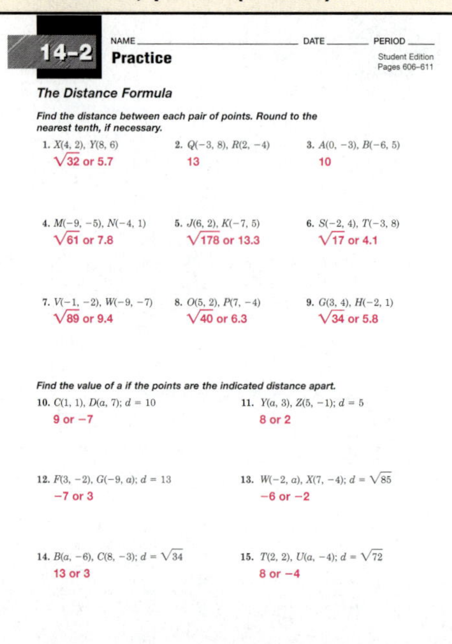

14-2 NAME ______ DATE ______ PERIOD ______

Practice

Student Edition Pages 606–611

The Distance Formula

Find the distance between each pair of points. Round to the nearest tenth, if necessary.

1. $X(4, 2), Y(8, 6)$ — $\sqrt{32}$ or 5.7
2. $Q(-3, 8), R(2, -4)$ — 13
3. $A(0, -3), B(-6, 5)$ — 10
4. $M(-9, -5), N(-4, 1)$ — $\sqrt{61}$ or 7.8
5. $J(6, 2), K(-7, 5)$ — $\sqrt{178}$ or 13.3
6. $S(-2, 4), T(-3, 8)$ — $\sqrt{17}$ or 4.1
7. $V(-1, -2), W(-9, -7)$ — $\sqrt{89}$ or 9.4
8. $O(5, 2), P(7, -4)$ — $\sqrt{40}$ or 6.3
9. $G(3, 4), H(-2, 1)$ — $\sqrt{34}$ or 5.8

Find the value of a if the points are the indicated distance apart.

10. $C(1, 1), D(a, 7); d = 10$ — 9 or −7
11. $Y(a, 3), Z(5, -1); d = 5$ — 8 or 2
12. $F(3, -2), G(-9, a); d = 13$ — −7 or 3
13. $W(-2, a), X(7, -4); d = \sqrt{85}$ — −6 or −2
14. $B(a, -6), C(8, -3); d = \sqrt{34}$ — 13 or 3
15. $T(2, 2), U(a, -4); d = \sqrt{72}$ — 8 or −4

24. Find the distance between $J(-9, 5)$ and $K(-4, -2)$. $\sqrt{74}$ **or 8.6**
25. What is the distance between $C(-8, 1)$ and $D(5, 6)$? $\sqrt{194}$ **or 13.9**
26. What is the value of c if $W(1, c)$ and $V(-4, 9)$ are 13 units apart? **26. −3 or 21**
27. Suppose $M(b, 9)$ and $N(20, -5)$ are $\sqrt{340}$ units apart. What is the value of b? **8 or 32**

Applications and Problem Solving C

28. **Geometry** Triangle MNP has vertices $M(-6, 14)$, $N(2, -1)$, and $P(-2, 2)$. Find the perimeter of $\triangle MNP$. Round to the nearest tenth. **34.6**

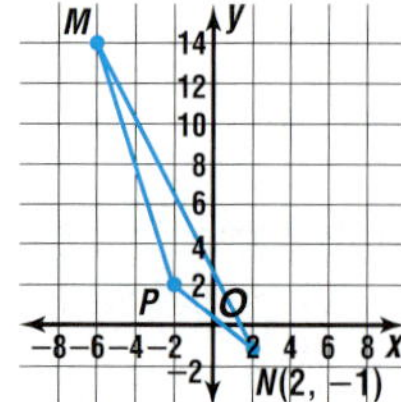

29. **Geometry** An isosceles triangle is a triangle with at least two congruent sides. Determine whether $\triangle CDE$ with vertices $C(1, 6)$, $D(2, 1)$, and $E(4, 1)$ is isosceles. Explain your reasoning. **No; no two sides have the same measure.**

30. **School** At Dannis State University, Webber Hall is located 35 meters west and 55 meters south of the student union. Packard Hall is located 42 meters east and 30 meters south of the student union.

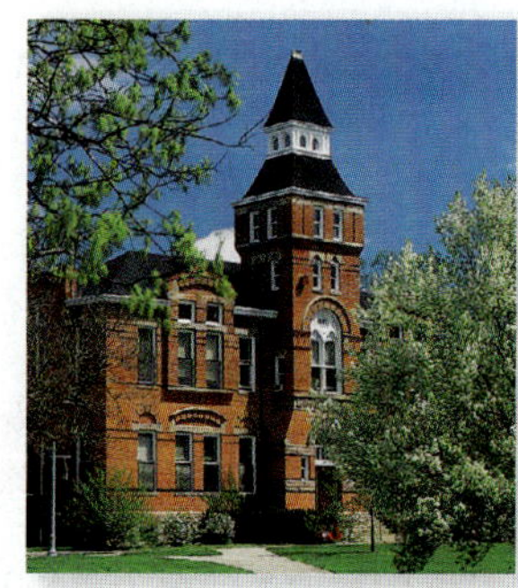

   a. Draw a diagram on a coordinate grid to represent this situation.
   b. How far is Webber Hall from Packard Hall? Round to the nearest tenth. **81.0 m**

   a. See margin.

31. **Engineering** Refer to Exercise 9. Suppose the MAXTEC Company builds another office building, MAXTEC North, 4 miles east and 2 miles north of the corporate office. How much cable will be needed to connect each of the buildings to the corporate office? Round to the nearest tenth. **16.3 mi**

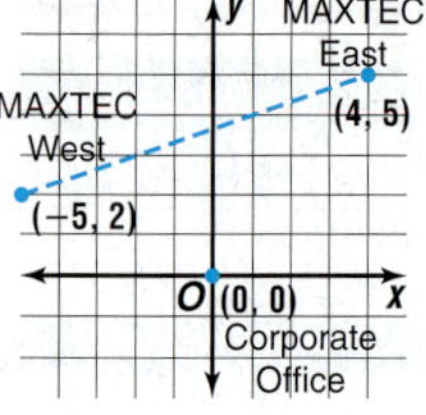

32. **Critical Thinking** If the diagonals of a trapezoid have the same length, then the trapezoid is isosceles. The vertices of trapezoid $ABCD$ are $A(-2, 2)$, $B(10, 6)$, $C(9, 8)$, and $D(0, 5)$. Is the trapezoid isosceles? Explain. **See margin.**

Inclusion Strategies

Students who are communicably disabled may have special difficulty describing the general form of the Distance Formula and some of the substitution steps. One way to help include them could be to write out the general form, and ask them to indicate each number that is substituted for x_2, x_1, y_2, and y_1.

Mixed Review

33–35. See margin for graphs.

Find an approximation, to the nearest tenth, for each square root. Then graph the square root on a number line. *(Lesson 14–1)*

33. $\sqrt{12}$ **3.5**
34. $-\sqrt{27}$ **−5.2**
35. $\sqrt{135}$ **11.6**

36. Solve $y \geq -4$ and $y < -2x + 1$ by graphing. *(Lesson 13–7)* **See margin.**

Solve each inequality. Check your solution. *(Lesson 12–3)*

37. $\frac{k}{3} \geq 2$ **$\{k \mid k \geq 6\}$**
38. $\frac{-n}{4} > 3$ **$\{n \mid n < -12\}$**
39. $\frac{3}{4}b \leq -6$ **$\{b \mid b \leq -8\}$**

40. Factor $12m^2 + 7m - 10$. *(Lesson 10–4)* **$(3m - 2)(4m + 5)$**

Standardized Test Practice

41. **Short Response** Find the solution of $|3 + w| + 4 = 2$. *(Lesson 3–7)* **∅**

42. **Multiple Choice** The table shows the number of pets that have lived at the White House. Which two pets account for exactly half of the pets that have lived at the White House? *(Lesson 1–7)* **D**

   Ⓐ dog and cat Ⓑ horse and bird
   Ⓒ bird and dog Ⓓ dog and horse

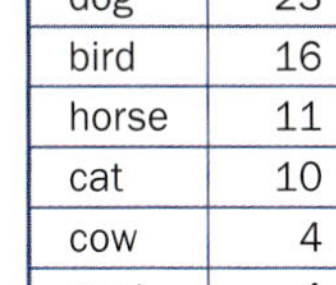
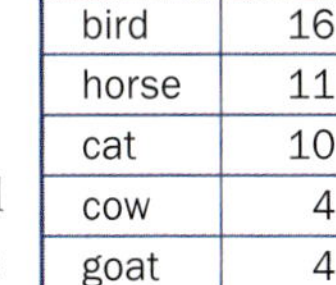

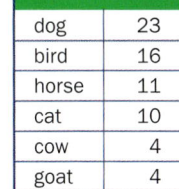

White House Pets	
dog	23
bird	16
horse	11
cat	10
cow	4
goat	4

Source: *USA TODAY*

Quiz 1 Lessons 14–1 and 14–2

Name the set or sets of numbers to which each real number belongs. Let N = natural numbers, W = whole numbers, Z = integers, Q = rational numbers, and I = irrational numbers. *(Lesson 14–1)*

1. 0 **W, Z, Q**
2. $\sqrt{36}$ **N, W, Z, Q**
3. 0.121231234 . . . **I**
4. Find an approximation, to the nearest tenth, for $-\sqrt{35}$. *(Lesson 14–1)* **−5.9**
5. Write $\sqrt{2}$, 1.22, and $1.\overline{2}$ in order from least to greatest. *(Lesson 14–1)* **1.22, $1.\overline{2}$, $\sqrt{2}$**

Find the distance between each pair of points. Round to the nearest tenth, if necessary. *(Lesson 14–2)*

6. $A(2, 0)$, $B(-1, 3)$ **$\sqrt{18}$ or 4.2**
7. $C(4, 5)$, $D(-3, -2)$ **$\sqrt{98}$ or 9.9**
8. $E(0, -4)$, $F(8, 7)$ **$\sqrt{185}$ or 13.6**
9. What is the value of m if $A(m, 8)$ and $B(3, 4)$ are 5 units apart? *(Lesson 14–2)* **0 or 6**
10. **Geometry** A scalene triangle is a triangle with no congruent sides. Determine whether $\triangle XYZ$ with vertices $X(5, 4)$, $Y(1, 5)$, and $Z(-1, 1)$ is scalene. Explain. *(Lesson 14–2)* **Yes; all three sides have different measures.**

www.algconcepts.com/self_check_quiz

4 ASSESS

Open-Ended Assessment

Writing Ask students to write expressions for the distance between $A(0, 0)$ and $B(x, y)$, and between $M(a, b)$ and $N(x, y)$. **$\sqrt{x^2 + y^2}$ and $\sqrt{(a - x)^2 + (b - y)^2}$, respectively**

Quiz 1

The Quiz provides students with a brief review of the concepts and skills in Lessons 14–1 and 14–2. Lesson numbers are given to the right of the exercises or instruction lines so students can review concepts not yet mastered.

Chapter 14, Quiz A (Lessons 14–1 and 14–2) is available in the *Chapter 14 Resource Masters*, p. 659.

Answers

33.

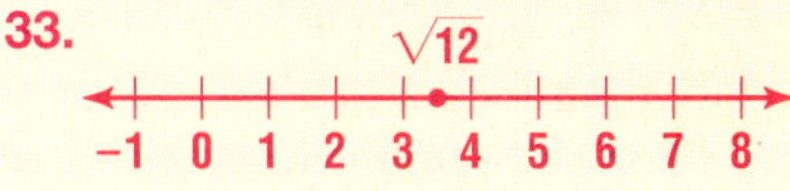

34.

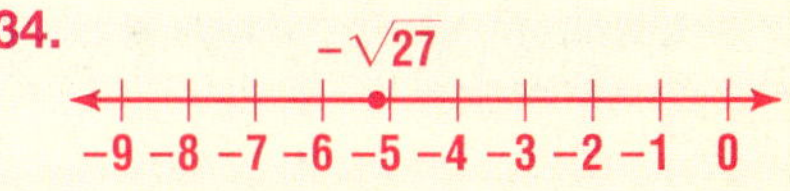

35.

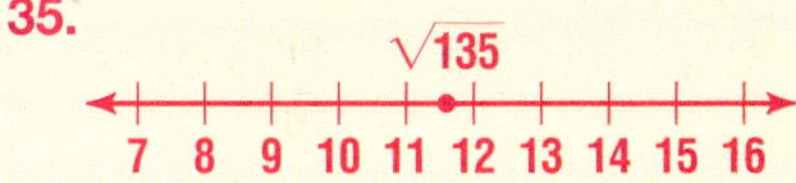

***Enrichment*, p. 632**

14–2 NAME ______ DATE ______ PERIOD ______

Enrichment Student Edition Pages 606–611

Lengths on a Grid

You can easily find segment lengths on a grid if the endpoints are grid-line intersections. For horizontal or vertical segments, simply count squares. For diagonal segments, use the Pythagorean Theorem. This theorem states that in any right triangle, if the length of the longest side (the side opposite the right angle) is c and the two shorter sides have lengths a and b, then $c^2 = a^2 + b^2$.

Example: Find the measure of $\overline{EF}$ on the grid at the right. Locate a right triangle with $\overline{EF}$ as its longest side.

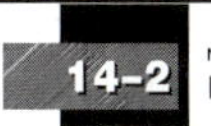

$$EF = \sqrt{2^2 + 5^2} = \sqrt{29} \approx 5.4 \text{ units}$$

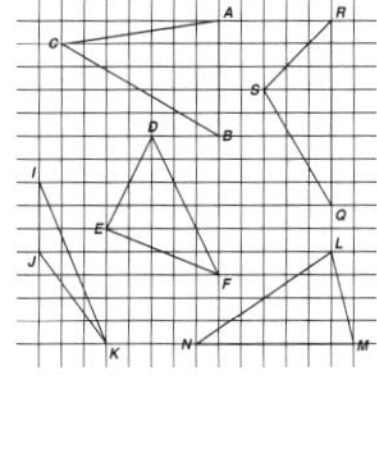

Find each measure to the nearest tenth of a unit.

1. IJ **3**
2. MN **7**
3. RS **4.2**
4. QS **5.8**
5. IK **7.6**
6. JK **5**
7. LM **4.1**
8. LN **7.2**

Use the grid above. Find the perimeter of each triangle to the nearest tenth of a unit.

9. $\triangle ABC$ **20.2**
10. $\triangle QRS$ **18**
11. $\triangle DEF$ **16.6**
12. $\triangle LMN$ **18.3**

Answers shown are found by rounding segment lengths before adding.

13. Of all the segments shown on the grid, which is longest? What is its length? **$BC = 8.1$**
14. On the grid, 1 unit = 0.5 cm. How can the answers above be used to find the measures in centimeters? **Divide by 2 or multiply by 0.5.**
15. Use your answer from Exercise 8 to calculate the length of segment LN in centimeters. Check by measuring with a centimeter ruler. **3.6 cm**
16. Use a centimeter ruler to find the perimeter of triangle IJK to the nearest tenth of a centimeter. **7.8 cm**

Extra Credit

Which point is farther from $A(1, 3)$: $X(-5, 1)$ or $Y(5, -2)$? **$Y(5, -2)$**

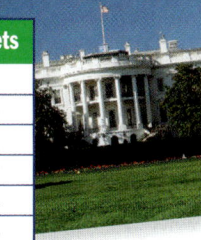

Answer

36.

y $y < -2x + 1$ O x $y \geq -4$

Investigation

PREPARE

This optional investigation is designed to be competed by groups of 3–4 students over 1–2 days.

Objectives

Students plot points and use paper folding to develop a formula for the midpoint of a segment. Then students use their midpoint formula and the Distance Formula to explore relationships between parts of triangles.

Mathematical Overview

This investigation utilizes the following concepts:

- finding the mean of two values,
- the diagonals of a parallelogram bisect each other, and
- the midpoints of the sides of a right triangle are the vertices of another right triangle.

Suggested Time Management	
Investigation	25–35 min
Extension: Gathering Data	25–35 min
Extension: Summarizing Data	30–50 min

Motivating the Lesson

Ask students how many points on a segment are exactly halfway between the endpoints of the segment. Then ask students to come up with ways to find that point, which is called the *midpoint*. Tell students that this investigation uses paper folding to find the midpoint of a segment.

Chapter 14 Investigation

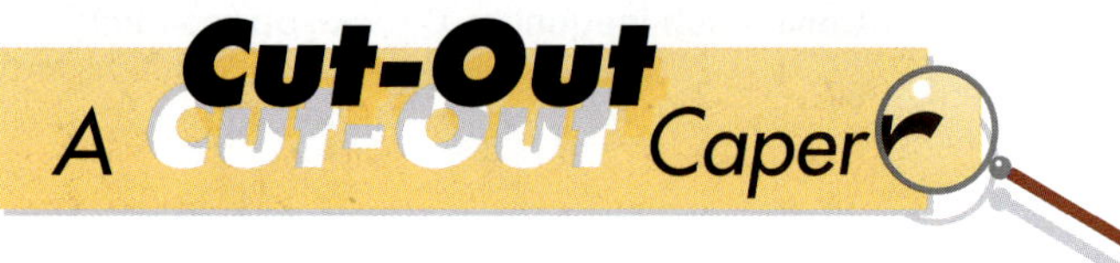

Materials

 grid paper

 ruler

 scissors

Midpoints

In Lesson 14–2, you learned how to find the distance between two points in a coordinate plane. In this Investigation, you will learn how to find the **midpoint** of a segment. The midpoint of a line segment is the point on the segment that separates it into two segments of equal length.

Investigate

1. Graph $A(-2, -1)$, $B(6, -1)$, $C(10, 3)$, and $D(2, 3)$ on a coordinate plane.
   a. Connect the points. What type of figure is formed? **parallelogram**
   b. Make a table and record the coordinates as shown.

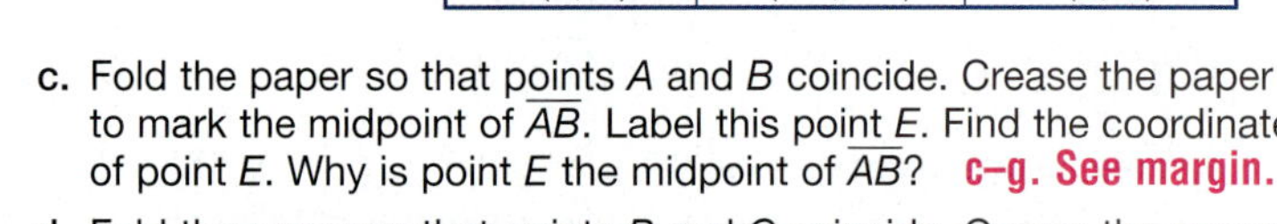

Coordinates of First Point	Coordinates of Second Point	Coordinates of Midpoint
$A(-2, -1)$	$B(6, -1)$	$E(\mathbf{2}, \mathbf{-1})$
$B(6, -1)$	$C(10, 3)$	$F(\mathbf{8}, \mathbf{1})$
$C(10, 3)$	$D(2, 3)$	$G(\mathbf{6}, \mathbf{3})$
$D(2, 3)$	$A(-2, -1)$	$H(\mathbf{0}, \mathbf{1})$

   c. Fold the paper so that points A and B coincide. Crease the paper to mark the midpoint of $\overline{AB}$. Label this point E. Find the coordinates of point E. Why is point E the midpoint of $\overline{AB}$? **c–g. See margin.**
   d. Fold the paper so that points B and C coincide. Crease the paper to mark the midpoint of $\overline{BC}$. Label this point F. Find the coordinates of point F. Continue this process with points C and D and points D and A. Label the midpoints G and H, respectively. Record the coordinates of these midpoints in your table.
   e. Draw $\overline{AC}$ and $\overline{BD}$. These are the diagonals of the figure. Find the midpoints of $\overline{AC}$ and $\overline{DB}$. Label the midpoints M and N, respectively. What are the coordinates of M and N? What appears to be true about the midpoints of the diagonals?
   f. Draw $\overline{EG}$ and $\overline{FH}$. What are the coordinates of their midpoints?
   g. Notice the triangles formed on the inside of the figure. Explain any relationships that exist between the triangles. You may cut out the triangles to help see the relationships.

Cooperative Learning

This investigation offers an excellent opportunity for using cooperative groups. For more information on cooperative learning strategies and group management, see *Cooperative Learning in the Mathematics Classroom,* one of the titles in the Glencoe Mathematics Professional Series.

2a. right triangle

2. Graph $X(-3, -3)$, $Y(3, -3)$, and $Z(3, 5)$ on a coordinate plane.
   a. Connect the points. What type of figure is formed?
   b. Make a table and record the coordinates as shown.

Coordinates of First Point	Coordinates of Second Point	Coordinates of Midpoint
$X(-3, -3)$	$Y(3, -3)$	$U(0, -3)$
$Y(3, -3)$	$Z(3, 5)$	$V(3, 1)$
$Z(3, 5)$	$X(-3, -3)$	$W(0, 1)$

2c. $U(0, -3)$; $V(3, 1)$; $W(0, 1)$

2d. They have the same length; see students' work.

2e. They have the same size and shape.

   c. Use paper folding to find the midpoints of the segments. Label the midpoints U, V, and W as shown in the table. What are the coordinates of the midpoints? Record the results.
   d. Draw $\overline{YW}$. Compare the measures of $\overline{YW}$, $\overline{XW}$, and $\overline{ZW}$. What appears to be true? Use the Distance Formula to verify your results.
   e. Draw $\overline{WU}$ and $\overline{WV}$. Four triangles are formed. Explain any relationships that exist between the triangles. You may cut out the triangles to help see the relationships.
3. Refer to the tables. Study each pair of first point coordinates, second point coordinates, and the corresponding midpoint coordinates.
   a. Explain how the concept of *mean*, or average, can be used to find the midpoint of a segment. **a–b. See margin.**
   b. Write a formula to find the midpoint of two points $X(a, b)$ and $Y(c, d)$.

Extending the Investigation

In this extension, you will continue to investigate midpoints.

- Graph points $A(3, 2)$, $B(-1, -6)$, and $C(-5, -4)$ on a coordinate plane. Connect the points to form a figure. Using the Midpoint Formula you discovered in Exercise 3b, determine the coordinates of the midpoints of each side of the figure. Label the points D, E, and F. Use the Distance Formula to verify that the midpoints separate each segment into two equal lengths. $D(1, -2)$, $E(-3, -5)$, $F(-1, -1)$
- Use the Midpoint Formula to determine the midpoint of each pair of points.
  a. $M(4, 3)$, $N(2, 5)$ **(3, 4)** b. $R(1, 0)$, $S(-3, 2)$ **(−1, 1)** c. $C(3, -6)$, $D(1, -4)$ **(2, −5)**

Presenting Your Investigation

Here are some ideas to help you present your conclusions to the class.

- Make a poster showing the results of your research. Include the tables, graphs, and formula for finding the midpoint of a segment. List any relationships you discovered.
- Discuss any similarities in the triangles formed inside of the figures.

Investigation For more information on midpoints, visit: www.algconcepts.com

MANAGE

Teaching Tip Students may need to hold their graphs up to a strong light to help them match up the endpoints when they fold to mark the midpoint. You might want to have students make a copy of their work before allowing them to cut out the triangles in Exercise 1g.

Working in Groups If students work in groups they can share the tasks of finding points E through H and points U through W. After sharing the coordinates of the midpoints with the rest of the group, they can explore and discuss the relationships involving the triangles formed by connecting the midpoints.

Working as a Class Assign half the class to work on Exercise 1 and the other half to work on Exercise 2. Each half of the class can present its findings in a class discussion. Then students can work individually or in small groups on Exercise 3 and the Extending the Investigation section.

ASSESS

Students' posters and reports should show a clear explanation of their formula for the midpoint of a segment. Students should be able to find the coordinates of the midpoints of the sides of a parallelogram or right triangle. They should also be able to explain how to use the Distance Formula to find the length of a segment.

Students should add their posters and support materials to their portfolios at this time.

Answers

1c. $E(2, -1)$; It separates the segment into two segments of equal length.

1d. $F(8, 1)$; $G(6, 3)$; $H(0, 1)$

1e. $M(4, 1)$, $N(4, 1)$; The midpoints of the diagonals are the same point.

1f. The midpoints are located at (4, 1).

1g. Triangles *AEM*, *AHM*, *CGM*, and *CFM* have the same size and shape. Triangles *BEM*, *BFM*, *DHM*, and *DGM* also have the same size and shape.

3a. To find the midpoint of a segment, find the mean, or average, of the x-coordinates and the mean, or average, of the y-coordinates.

3b. midpoint $= \left(\frac{a+c}{2}, \frac{b+d}{2}\right)$

14-3 Simplifying Radical Expressions

1 FOCUS

5-Minute Check
Lesson 14–2

1. Find the value of a if $X(5, a)$ and $Y(8, 4)$ are $\sqrt{34}$ units apart. **−1 or 9**

Find the distance between each pair of points. Round to the nearest tenth, if necessary.

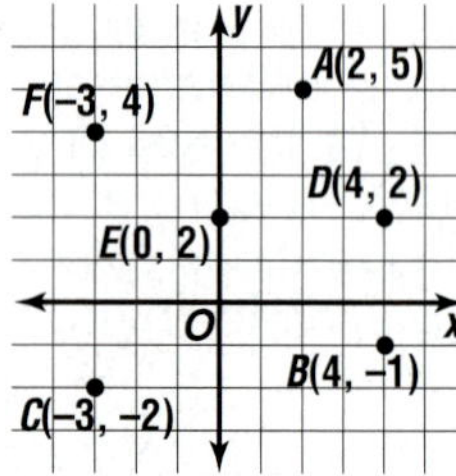

2. A and B **$\sqrt{40}$ or 6.3**
3. A and C **$\sqrt{74}$ or 8.6**
4. D and F **$\sqrt{53}$ or 7.3**
5. E and F **$\sqrt{13}$ or 3.6**

Motivating the Lesson

Hands-On Activity Ask students to draw a rectangle and measure the length and width to the nearest millimeter. Then ask them to find the side length of a square having the same area as their rectangle. Ask how they calculated the side length of the square, and discuss how a radical expression could be given for the length.

What You'll Learn
You'll learn to simplify radical expressions.

Why It's Important
Science Scientists can use radical expressions to find the speed of a river. *See Exercise 41.*

Why does the character from the comic yell *"the square root of sixteen"* when hitting the golf ball?

FoxTrot

The expression $\sqrt{16}$ is a **radical expression**. Since the radicand, 16, is a perfect square, $\sqrt{16} = 4$. In Lesson 8–5, you learned to simplify radical expressions using the Product Property of Square Roots and prime factorization. You can simplify radical expressions in which the radicand is not a perfect square in a similar manner. *Recall that the radicand is the number or expression under the square root symbol.*

Examples

Simplify $\sqrt{75}$.

$\sqrt{75} = \sqrt{3 \cdot 5 \cdot 5}$	*Prime factorization*
$= \sqrt{3 \cdot 25}$	*$5 \times 5 = 25$*
$= \sqrt{3} \cdot \sqrt{25}$	*Product Property of Square Roots*
$= \sqrt{3} \cdot 5$ or $5\sqrt{3}$	*Simplify $\sqrt{25}$.*

Your Turn **Simplify each square root. Leave in radical form.**

a. $\sqrt{68}$ **$2\sqrt{17}$** b. $\sqrt{375}$ **$5\sqrt{15}$**

Sports Link

On a softball field, the distance from second base to home plate is $\sqrt{800}$ meters. Express $\sqrt{800}$ in simplest radical form.

$\sqrt{800} = \sqrt{2 \cdot 2 \cdot 2 \cdot 2 \cdot 2 \cdot 5 \cdot 5}$	*Prime factorization*
$= \sqrt{16 \cdot 2 \cdot 25}$	*$4 \times 4 = 16$, $5 \times 5 = 25$*
$= \sqrt{16} \cdot \sqrt{2} \cdot \sqrt{25}$	*Product Property of Square Roots*
$= 4 \cdot \sqrt{2} \cdot 5$	*Simplify $\sqrt{16}$ and $\sqrt{25}$.*
$= 20\sqrt{2}$	*Multiply.*

The distance is $20\sqrt{2}$ meters.

interNET CONNECTION
Data Update For the latest information on softball, visit: www.algconcepts.com

TECHNOLOGY

An alternative technology option using a graphing calculator is available for teaching this lesson.

Resource Manager

Reproducible Masters
Chapter 14 Resource Masters
- *Study Guide*, p. 633
- *Skills Practice*, p. 634
- *Practice*, p. 635
- *Reading to Learn Mathematics*, p. 636
- *Enrichment*, p. 637
- *Assessment*, p. 658

Graphing Calculator, pp. 41–42
Hands-On Algebra, pp. 155–156

Transparencies
5-Minute Check, 14–3

Technology/Multimedia
Interactive Chalkboard CD-ROM

The Product Property can also be used to multiply square roots.

Example 3 Simplify $\sqrt{5} \cdot \sqrt{35}$.

$\sqrt{5} \cdot \sqrt{35} = \sqrt{5} \cdot \sqrt{5 \cdot 7}$ *Prime factorization*

$= \sqrt{5} \cdot \sqrt{5} \cdot \sqrt{7}$ *Product Property of Square Roots*

$= \sqrt{5^2} \cdot \sqrt{7}$ *$\sqrt{5} \times \sqrt{5} = \sqrt{5^2}$*

$= 5 \cdot \sqrt{7}$ or $5\sqrt{7}$ *Simplify.*

Your Turn **Simplify each expression. Leave in radical form.**

c. $\sqrt{3} \cdot \sqrt{15}$ $3\sqrt{5}$

d. $\sqrt{10} \cdot \sqrt{30}$ $10\sqrt{3}$

Look Back

Quotient Property of Square Roots: Lesson 8–5

To divide square roots and simplify radical expressions that involve division, use the Quotient Property of Square Roots. A fraction containing radicals is in simplest form if no radicals are left in the denominator.

Example 4 Simplify $\frac{\sqrt{32}}{\sqrt{4}}$.

$\frac{\sqrt{32}}{\sqrt{4}} = \sqrt{\frac{32}{4}}$ *Quotient Property of Square Roots*

$= \sqrt{8}$ *$32 \div 4 = 8$*

$= \sqrt{2 \cdot 2 \cdot 2}$ *Prime factorization*

$= \sqrt{2^3}$ *$2 \cdot 2 \cdot 2 = 2^3$*

$= \sqrt{2^2} \cdot \sqrt{2}$ *Product Property of Square Roots*

$= 2\sqrt{2}$ *Simplify.*

Your Turn **Simplify each expression. Leave in radical form.**

e. $\frac{\sqrt{72}}{\sqrt{6}}$ $2\sqrt{3}$

f. $\frac{\sqrt{160}}{\sqrt{2}}$ $4\sqrt{5}$

To eliminate radicals from the denominator of a fraction, you can use a method for simplifying radical expressions called **rationalizing the denominator**.

Example 5 Simplify $\frac{\sqrt{5}}{\sqrt{10}}$.

$\frac{\sqrt{5}}{\sqrt{10}} = \frac{\sqrt{5}}{\sqrt{10}} \cdot \frac{\sqrt{10}}{\sqrt{10}}$ *$\frac{\sqrt{10}}{\sqrt{10}} = 1$*

$= \frac{\sqrt{5 \cdot 10}}{\sqrt{10 \cdot 10}}$ *Product Property of Square Roots*

(continued on the next page)

2 TEACH

In-Class Examples

Example 1

Simplify $\sqrt{48}$. $4\sqrt{3}$

Example 2

The diagonal of a square game board is $\sqrt{700}$ centimeters long. Express $\sqrt{700}$ in simplest radical form. $10\sqrt{7}$

Example 3

Simplify $\sqrt{15} \cdot \sqrt{75}$. $15\sqrt{5}$

Teaching Tip In Example 4, students may immediately recognize $\sqrt{4}$ as being equal to 2 and wonder why this was not the first step taken. Show them the following alternate steps, emphasizing that more than one sequence of steps is possible.

$\frac{\sqrt{32}}{\sqrt{4}} = \frac{\sqrt{32}}{2} = \frac{\sqrt{16 \cdot 2}}{2} = \frac{\sqrt{16} \cdot \sqrt{2}}{2} = \frac{4 \cdot \sqrt{2}}{2} = 2\sqrt{2}$

In-Class Example

Example 4

Simplify $\frac{\sqrt{300}}{\sqrt{15}}$. $2\sqrt{5}$

Teaching Tip In Example 5, an alternate method is to rewrite the expression as a fraction inside a single radical, reduce the fraction, rewrite it as the quotient of two radicals, and then multiply by the appropriate form of 1 to simplify.

$\frac{\sqrt{5}}{\sqrt{10}} = \sqrt{\frac{5}{10}} = \sqrt{\frac{1}{2}} = \frac{\sqrt{1}}{\sqrt{2}} = \frac{1}{\sqrt{2}} = \frac{1 \cdot \sqrt{2}}{\sqrt{2} \cdot \sqrt{2}} = \frac{\sqrt{2}}{2}$

In-Class Example

Example 5

Simplify $\frac{\sqrt{6}}{\sqrt{30}}$. $\frac{\sqrt{5}}{5}$

Teaching Tip When discussing *conjugates* and Example 6, point out that multiplying by a form of 1 created using a conjugate works because the product of conjugates in the denominator is the product of a sum and a difference. Stress that such a product yields the difference of two squares, one of which is the square of a square root thus eliminating the radical in the denominator.

In-Class Example

Example 6

Simplify $\frac{5}{4 + \sqrt{3}}$. $\frac{20 - 5\sqrt{3}}{13}$

$= \frac{\sqrt{50}}{\sqrt{100}}$ *Simplify.*

$= \frac{\sqrt{25 \cdot 2}}{10}$ *$50 = 25 \cdot 2$*

$= \frac{\sqrt{25} \cdot \sqrt{2}}{10}$ *Product Property of Square Roots*

$= \frac{5 \cdot \sqrt{2}}{10}$ or $\frac{\sqrt{2}}{2}$ *Simplify.*

Your Turn **Simplify each expression. Leave in radical form.**

g. $\frac{\sqrt{6}}{\sqrt{8}}$ $\frac{\sqrt{3}}{2}$

h. $\frac{\sqrt{32}}{\sqrt{3}}$ $\frac{4\sqrt{6}}{3}$

Binomials of the form $a\sqrt{b} + c\sqrt{d}$ and $a\sqrt{b} - c\sqrt{d}$ are **conjugates** of each other because their product is a rational number.

Product of a Sum and a Difference: Lesson 9–5

$(6 + \sqrt{3})(6 - \sqrt{3}) = 6^2 - (\sqrt{3})^2$ *Use the pattern $(a + b)(a - b) = a^2 - b^2$ to simplify the product.*

$= 36 - 3$

$= 33$

Conjugates are useful for simplifying radical expressions because their product is always a rational number.

Example 6 **Simplify $\frac{6}{3 - \sqrt{2}}$.**

To rationalize the denominator, multiply both the numerator and denominator by $3 + \sqrt{2}$, which is the conjugate of $3 - \sqrt{2}$.

$\frac{6}{3 - \sqrt{2}} = \frac{6}{3 - \sqrt{2}} \cdot \frac{3 + \sqrt{2}}{3 + \sqrt{2}}$ *Notice that $\frac{3 + \sqrt{2}}{3 + \sqrt{2}} = 1$.*

$= \frac{6(3) + 6\sqrt{2}}{3^2 - (\sqrt{2})^2}$ *Distributive Property $(a - b)(a + b) = a^2 - b^2$*

$= \frac{18 + 6\sqrt{2}}{9 - 2}$ *Simplify.*

$= \frac{18 + 6\sqrt{2}}{7}$ *Subtract.*

Your Turn **Simplify each expression. Leave in radical form.**

i. $\frac{3}{3 - \sqrt{5}}$ $\frac{9 + 3\sqrt{5}}{4}$

j. $\frac{4}{5 + \sqrt{6}}$ $\frac{20 - 4\sqrt{6}}{19}$

Radical expressions are in simplest form if the following conditions are met.

Simplified Form for Radicals

A radical expression is in simplest form when the following three conditions have been met.

1. No radicands have perfect square factors other than 1.
2. No radicands contain fractions.
3. No radicals appear in the denominator of a fraction.

Consider the expression $\sqrt{x^2}$. It appears that $\sqrt{x^2} = x$. However, if $x = -3$, then $\sqrt{(-3)^2}$ is 3, not -3. For radical expressions like $\sqrt{x^2}$, use absolute value to ensure nonnegative results. The results of simplifying a few radical expressions are listed below.

$\sqrt{x^2} = |x|$ $\quad \sqrt{x^3} = x\sqrt{x}$ $\quad \sqrt{x^4} = x^2$ $\quad \sqrt{x^5} = x^2\sqrt{x}$ $\quad \sqrt{x^6} = |x^3|$

For $\sqrt{x^3}$, absolute value is not necessary. If x were negative, then x^3 would be negative, and $\sqrt{x^3}$ is not a real number. *Why is absolute value not used for $\sqrt{x^4}$?*

Example 7 Simplify $\sqrt{98ab^2c^4}$. Use absolute value symbols if necessary.

$\sqrt{98ab^2c^4}$	
$= \sqrt{2 \cdot 7 \cdot 7 \cdot a \cdot b^2 \cdot c^4}$	*Prime factorization*
$= \sqrt{2 \cdot 49 \cdot a \cdot b^2 \cdot c^4}$	*$7 \times 7 = 49$*
$= \sqrt{2} \cdot \sqrt{49} \cdot \sqrt{a} \cdot \sqrt{b^2} \cdot \sqrt{c^4}$	*Product Property of Square Roots*
$= \sqrt{2} \cdot 7 \cdot \sqrt{a} \cdot |b| \cdot c^2$	*Simplify.*
$= 7|b|c^2\sqrt{2a}$	*The absolute value of b ensures a nonnegative result.*

Your Turn

Simplify each expression. Use absolute value symbols if necessary.

k. $\sqrt{63ab^2}$ **$3|b|\sqrt{7a}$** $\qquad$ l. $\sqrt{200x^2y^3}$ **$10|x|y\sqrt{2y}$**

Check for Understanding

Communicating Mathematics

1. **Explain** why absolute values are sometimes needed when simplifying radical expressions containing variables. **1. to ensure nonnegative results**
2. **Explain** how you can show whether $8 + \sqrt{2}$ and $8 - \sqrt{2}$ are conjugates. **See margin.**

Vocabulary
- radical expression
- rationalizing the denominator
- conjugates

Teaching Tip In Example 7, students should realize that $\sqrt{c^4}$ is c^2, and not $|c^2|$, because c^2 is always positive regardless of the value of c.

In-Class Example

Example 7

Simplify $\sqrt{48y^3z^6}$. Use absolute value symbols if necessary. **$4y|z^3|\sqrt{3y}$**

Answer

2. Find the product of $8 + \sqrt{2}$ and $8 - \sqrt{2}$. Since the product (62) is rational, they are conjugates.

Reteaching Activity

Intrapersonal Learners Ask students to write a journal entry explaining what a *conjugate* is and how it is used to convert the denominator of some rational expressions into an integer.

Study Guide, p. 633

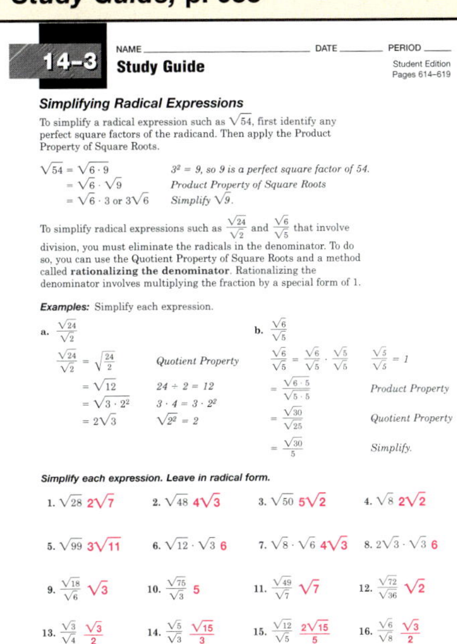

14–3 NAME ______ DATE ______ PERIOD ______

Study Guide Student Edition Pages 614–619

Simplifying Radical Expressions

To simplify a radical expression such as $\sqrt{54}$, first identify any perfect square factors of the radicand. Then apply the Product Property of Square Roots.

$\sqrt{54} = \sqrt{6 \cdot 9}$	*$3^2 = 9$, so 9 is a perfect square factor of 54.*
$= \sqrt{6} \cdot \sqrt{9}$	*Product Property of Square Roots*
$= \sqrt{6} \cdot 3$ or $3\sqrt{6}$	*Simplify $\sqrt{9}$.*

To simplify radical expressions such as $\frac{\sqrt{24}}{\sqrt{2}}$ and $\frac{\sqrt{6}}{\sqrt{5}}$ that involve division, you must eliminate the radicals in the denominator. To do so, you can use the Quotient Property of Square Roots and a method called **rationalizing the denominator**. Rationalizing the denominator involves multiplying the fraction by a special form of 1.

Examples: Simplify each expression.

a. $\frac{\sqrt{24}}{\sqrt{2}}$

$\frac{\sqrt{24}}{\sqrt{2}} = \sqrt{\frac{24}{2}}$	*Quotient Property*
$= \sqrt{12}$	*$24 \div 2 = 12$*
$= \sqrt{3 \cdot 2^2}$	*$3 \cdot 4 = 3 \cdot 2^2$*
$= 2\sqrt{3}$	*$\sqrt{2^2} = 2$*

b. $\frac{\sqrt{6}}{\sqrt{5}}$

$\frac{\sqrt{6}}{\sqrt{5}} = \frac{\sqrt{6}}{\sqrt{5}} \cdot \frac{\sqrt{5}}{\sqrt{5}}$	*$\frac{\sqrt{5}}{\sqrt{5}} = 1$*
$= \frac{\sqrt{6 \cdot 5}}{\sqrt{5 \cdot 5}}$	*Product Property*
$= \frac{\sqrt{30}}{\sqrt{25}}$	*Quotient Property*
$= \frac{\sqrt{30}}{5}$	*Simplify.*

Simplify each expression. Leave in radical form.

1. $\sqrt{28}$ **$2\sqrt{7}$**
2. $\sqrt{48}$ **$4\sqrt{3}$**
3. $\sqrt{50}$ **$5\sqrt{2}$**
4. $\sqrt{8}$ **$2\sqrt{2}$**
5. $\sqrt{99}$ **$3\sqrt{11}$**
6. $\sqrt{12} \cdot \sqrt{3}$ **6**
7. $\sqrt{8} \cdot \sqrt{6}$ **$4\sqrt{3}$**
8. $2\sqrt{3} \cdot \sqrt{3}$ **6**
9. $\frac{\sqrt{18}}{\sqrt{6}}$ **$\sqrt{3}$**
10. $\frac{\sqrt{75}}{\sqrt{3}}$ **5**
11. $\frac{\sqrt{49}}{\sqrt{7}}$ **$\sqrt{7}$**
12. $\frac{\sqrt{72}}{\sqrt{36}}$ **$\sqrt{2}$**
13. $\frac{\sqrt{3}}{\sqrt{4}}$ **$\frac{\sqrt{3}}{2}$**
14. $\frac{\sqrt{5}}{\sqrt{3}}$ **$\frac{\sqrt{15}}{3}$**
15. $\frac{\sqrt{12}}{\sqrt{5}}$ **$\frac{2\sqrt{15}}{5}$**
16. $\frac{\sqrt{6}}{\sqrt{8}}$ **$\frac{\sqrt{3}}{2}$**

3 PRACTICE/APPLY

Error Analysis

Watch for students who make a mistake in Exercise 4 when they multiply two conjugates and write the product $(5 - \sqrt{6})(5 + \sqrt{6})$ as $25 - 36 = -11$ or as $5 - 6 = -1$. ***Prevent by*** having students show all the steps in the FOIL method when multiplying conjugates. In this case, they should write $(5 - \sqrt{6})(5 + \sqrt{6})$ as $(5)(5) + 5(\sqrt{6}) + (-\sqrt{6})(5) + (-\sqrt{6})(\sqrt{6}) = 25 + 5\sqrt{6} - 5\sqrt{6} + (-\sqrt{36}) = 25 + (-6) = 19$.

Assignment Guide

Basic: 17–41 odd, 42–50
Average: 16–38 even, 40–50

Answer

3. Greg is correct. In its simplest form, the expression is written as $\frac{3 - \sqrt{5}}{2}$ since the numbers 6, 2, and 4 in $\frac{6 - 2\sqrt{5}}{4}$ are all divisible by 2.

Skills Practice, **p. 634, and** ***Practice,*** **p. 635 (shown)**

14-3 NAME ______________ DATE ________ PERIOD ______

Practice

Student Edition Pages 614–619

Simplifying Radical Expressions

Simplify each expression. Leave in radical form.

1. $\sqrt{28}$ $2\sqrt{7}$
2. $\sqrt{48}$ $4\sqrt{3}$
3. $\sqrt{72}$ $6\sqrt{2}$
4. $\sqrt{90}$ $3\sqrt{10}$
5. $\sqrt{175}$ $5\sqrt{7}$
6. $\sqrt{245}$ $7\sqrt{5}$
7. $\sqrt{7} \cdot \sqrt{14}$ $7\sqrt{2}$
8. $\sqrt{2} \cdot \sqrt{10}$ $2\sqrt{5}$
9. $\sqrt{10} \cdot \sqrt{60}$ $10\sqrt{6}$
10. $\frac{\sqrt{48}}{\sqrt{2}}$ $2\sqrt{6}$
11. $\frac{\sqrt{54}}{\sqrt{3}}$ $3\sqrt{2}$
12. $\frac{\sqrt{96}}{\sqrt{8}}$ $2\sqrt{3}$
13. $\frac{\sqrt{20}}{\sqrt{3}}$ $\frac{2\sqrt{15}}{3}$
14. $\frac{\sqrt{2}}{\sqrt{10}}$ $\frac{\sqrt{5}}{5}$
15. $\frac{\sqrt{8}}{\sqrt{6}}$ $\frac{2\sqrt{3}}{3}$
16. $\frac{5}{4 - \sqrt{7}}$ $\frac{20 + 5\sqrt{7}}{9}$
17. $\frac{4}{3 + \sqrt{2}}$ $\frac{12 - 4\sqrt{2}}{7}$
18. $\frac{3}{3 - \sqrt{3}}$ $\frac{3 + \sqrt{3}}{2}$

Simplify each expression. Use absolute value symbols if necessary.

19. $\sqrt{50x^2}$ $5|x|\sqrt{2}$
20. $\sqrt{27ab^3}$ $3|b|\sqrt{3ab}$
21. $\sqrt{49c^6d^4}$ $7|c^3|d^2$
22. $\sqrt{63x^2y^5z^2}$ $3|xz|y^2\sqrt{7y}$
23. $\sqrt{56m^2n^4p^3}$ $2|mp|n^2\sqrt{14p}$
24. $\sqrt{108r^2s^3t^6}$ $6|rt^3|s\sqrt{3s}$

3. See margin.

3.  LaToya says that, in simplest form, the expression $\frac{2}{3 + \sqrt{5}}$ is written as $\frac{6 - 2\sqrt{5}}{4}$. Greg disagrees. He says it should be written as $\frac{3 - \sqrt{5}}{2}$. Who is correct? Explain your reasoning.

Guided Practice

Getting Ready **State the conjugate of each expression. Then multiply the expression by its conjugate.**

Sample: $1 + \sqrt{7}$ **Solution:** The conjugate is $1 - \sqrt{7}$.
$(1 + \sqrt{7})(1 - \sqrt{7}) = 1^2 - (\sqrt{7})^2$
$= 1 - 7$ or -6

4. $5 - \sqrt{6}$ $5 + \sqrt{6}$; 19
5. $2 + \sqrt{8}$ $2 - \sqrt{8}$; -4
6. $2\sqrt{7} - 3\sqrt{2}$ $2\sqrt{7} + 3\sqrt{2}$; 10

Examples 1–6 **Simplify each expression. Leave in radical form.**

7. $\sqrt{75}$ $5\sqrt{3}$
8. $\sqrt{96}$ $4\sqrt{6}$
9. $\sqrt{6} \cdot \sqrt{15}$ $3\sqrt{10}$
10. $\frac{\sqrt{36}}{\sqrt{3}}$ $2\sqrt{3}$
11. $\frac{\sqrt{3}}{\sqrt{7}}$ $\frac{\sqrt{21}}{7}$
12. $\frac{1}{6 - \sqrt{3}}$ $\frac{6 + \sqrt{3}}{33}$

Example 7 **Simplify each expression. Use absolute value symbols if necessary.**

13. $\sqrt{36x^2y}$ $6|x|\sqrt{y}$
14. $\sqrt{50m^4n^5}$ $5m^2n^2\sqrt{2n}$

Example 2

15. **Geometry** The radius of the circle is $\sqrt{32}$ units. Express the radius in simplest form. $4\sqrt{2}$ units

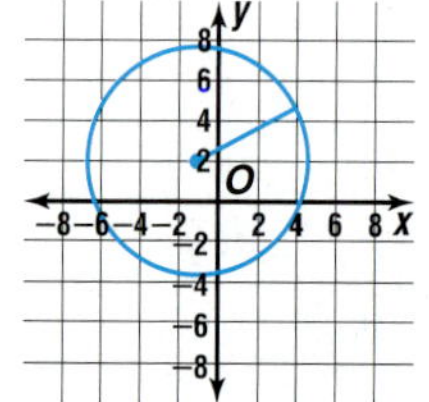

Exercises

Practice 

Simplify each expression. Leave in radical form.

Homework Help	
For Exercises	See Examples
16–20, 40, 41	1, 2
21–24	3
25–30	4, 5
31–33	6
34–39	7

Extra Practice
See page 721.

16. $\sqrt{45}$ $3\sqrt{5}$
17. $\sqrt{98}$ $7\sqrt{2}$
18. $\sqrt{280}$ $2\sqrt{70}$
19. $\sqrt{500}$ $10\sqrt{5}$
20. $\sqrt{1000}$ $10\sqrt{10}$
21. $\sqrt{2} \cdot \sqrt{8}$ 4
22. $\sqrt{6} \cdot \sqrt{18}$ $6\sqrt{3}$
23. $3\sqrt{5} \cdot \sqrt{5}$ 15
24. $\sqrt{8} \cdot \sqrt{12}$ $4\sqrt{6}$
25. $\frac{\sqrt{48}}{\sqrt{3}}$ 4
26. $\frac{\sqrt{52}}{\sqrt{4}}$ $\sqrt{13}$
27. $\frac{\sqrt{80}}{\sqrt{2}}$ $2\sqrt{10}$
28. $\frac{\sqrt{3}}{\sqrt{8}}$ $\frac{\sqrt{6}}{4}$
29. $\frac{\sqrt{4}}{\sqrt{5}}$ $\frac{2\sqrt{5}}{5}$
30. $\frac{\sqrt{5}}{\sqrt{12}}$ $\frac{\sqrt{15}}{6}$
31. $\frac{2}{4 - \sqrt{3}}$ $\frac{8 + 2\sqrt{3}}{13}$
32. $\frac{5}{3 + \sqrt{2}}$ $\frac{15 - 5\sqrt{2}}{7}$
33. $\frac{4}{6 - \sqrt{7}}$ $\frac{24 + 4\sqrt{7}}{29}$

35. $2|m|\sqrt{10}$ **B**
36. $|c^3|\sqrt{47d}$

Simplify each expression. Use absolute value symbols if necessary.

34. $\sqrt{16gh^4}$ $4h^2\sqrt{g}$
35. $\sqrt{40m^2}$
36. $\sqrt{47c^6d}$
37. $\sqrt{54a^2b^3}$ $3|a|b\sqrt{6b}$
38. $\sqrt{125rst}$ $5\sqrt{5rst}$
39. $\sqrt{36x^3y^4z^5}$ $6|x|y^2z^2\sqrt{xz}$

Applications and Problem Solving

C

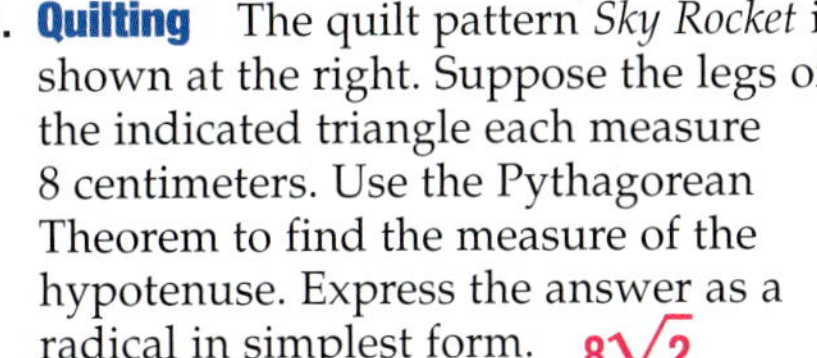

40. **Quilting** The quilt pattern *Sky Rocket* is shown at the right. Suppose the legs of the indicated triangle each measure 8 centimeters. Use the Pythagorean Theorem to find the measure of the hypotenuse. Express the answer as a radical in simplest form. $8\sqrt{2}$

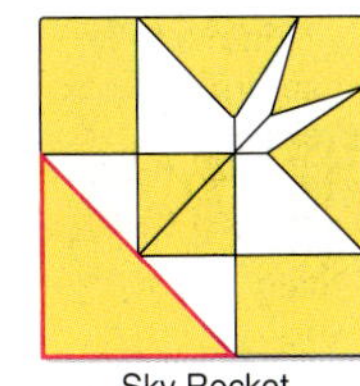

Sky Rocket

41. **Science** An L-shaped tube like the one shown can be used to measure the speed V in miles per hour of water in a river. By using the formula $V = \sqrt{2.5h}$, where h is the height in inches of the column of water above the surface, the speed of the water can be found.

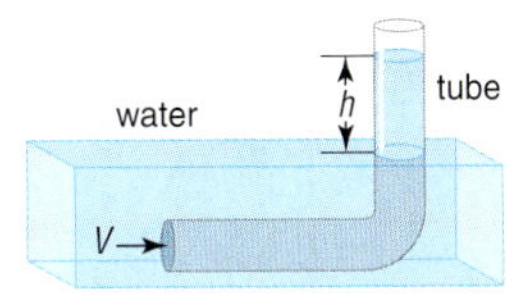

a. Suppose the tube is placed in a river and the height of the column of water is 4.8 inches. What is the speed of the water? $2\sqrt{3}$ mph

b. What would h be if the speed is exactly 5 miles per hour? 10 in.

42. **Critical Thinking** Is the sentence $\sqrt{a \cdot b} = \sqrt{a} \cdot \sqrt{b}$ true for negative values of a and b? Explain your reasoning. **See margin.**

Mixed Review

43. Point J is located at $(-1, 5)$, and point K is located at $(m, 2)$. What is the value of m if the points are $\sqrt{18}$ units apart? *(Lesson 14–2)* 2, −4

Name the set or sets of numbers to which each real number belongs. Let N = natural numbers, W = whole numbers, Z = integers, Q = rational numbers and I = irrational numbers. *(Lesson 14–1)*

44. 0.75 Q
45. $-\sqrt{144}$ Z, Q
46. $\frac{20}{4}$ N, W, Z, Q

47. no solution

47. Use substitution to solve $y = 2x + 3$ and $y - 2x = -5$. *(Lesson 13–3)*

48. Solve $15 > a - 3 > 10$. Graph the solution. *(Lesson 12–5)* $\{a \mid 13 < a < 18\}$; See margin for graph.

Standardized Test Practice

49. **Short Response** Tell whether 9 is closer to $\sqrt{79}$ or $\sqrt{89}$. *(Lesson 8–6)* $\sqrt{79}$

50. **Multiple Choice** Two dice are rolled. Find the probability that a number less than 5 is rolled on one die and an odd number is rolled on the other die. *(Lesson 5–7)* C

Ⓐ $\frac{1}{2}$ Ⓑ $\frac{2}{3}$ Ⓒ $\frac{1}{3}$ Ⓓ $\frac{3}{4}$

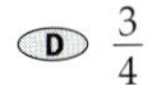

www.algconcepts.com/self_check_quiz

4 ASSESS

Open-Ended Assessment

Writing Ask students to write a radical expression similar to the one shown in Example 6 and then show how to simplify their expression.

Mid-Chapter Test (Lessons 14–1 through 14–3) is available in the *Chapter 14 Resource Masters*, p. 658.

Answers

42. No, because square roots of negative numbers are not defined in the set of real numbers; $\sqrt{(-2)\cdot(-3)} \neq \sqrt{-2}\cdot\sqrt{-3}$.

48. (number line from 10 to 20 with open circles at 13 and 18, shaded between)

10 11 12 13 14 15 16 17 18 19 20

***Enrichment*, p. 637**

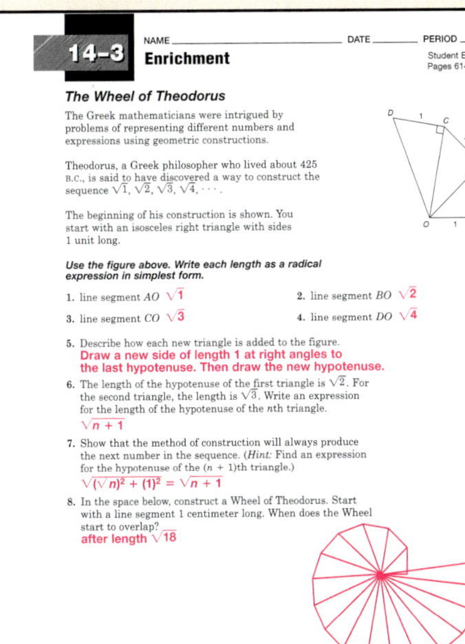

14-3 **Enrichment** NAME ____ DATE ____ PERIOD ____ Student Edition Pages 614–619

The Wheel of Theodorus

The Greek mathematicians were intrigued by problems of representing different numbers and expressions using geometric constructions.

Theodorus, a Greek philosopher who lived about 425 B.C., is said to have discovered a way to construct the sequence $\sqrt{1}, \sqrt{2}, \sqrt{3}, \sqrt{4}, \cdots$.

The beginning of his construction is shown. You start with an isosceles right triangle with sides 1 unit long.

Use the figure above. Write each length as a radical expression in simplest form.

1. line segment AO $\sqrt{1}$
2. line segment BO $\sqrt{2}$
3. line segment CO $\sqrt{3}$
4. line segment DO $\sqrt{4}$
5. Describe how each new triangle is added to the figure. Draw a new side of length 1 at right angles to the last hypotenuse. Then draw the new hypotenuse.
6. The length of the hypotenuse of the first triangle is $\sqrt{2}$. For the second triangle, the length is $\sqrt{3}$. Write an expression for the length of the hypotenuse of the nth triangle. $\sqrt{n+1}$
7. Show that the method of construction will always produce the next number in the sequence. (*Hint:* Find an expression for the hypotenuse of the $(n + 1)$th triangle.) $\sqrt{(\sqrt{n})^2 + (1)^2} = \sqrt{n+1}$
8. In the space below, construct a Wheel of Theodorus. Start with a line segment 1 centimeter long. When does the Wheel start to overlap? after length $\sqrt{18}$

Extra Credit

Simplify $\frac{1}{2+\sqrt{3}} \div \frac{1}{4-\sqrt{3}}$. $11 - 6\sqrt{3}$

1 FOCUS

5-Minute Check
Lesson 14–3

Simplify each expression.

1. $\sqrt{6} \cdot \sqrt{18}$ **$6\sqrt{3}$**
2. $\frac{\sqrt{48}}{\sqrt{4}}$ **$2\sqrt{3}$**
3. $\frac{\sqrt{5}}{\sqrt{50}}$ **$\frac{\sqrt{10}}{10}$**
4. $\frac{5}{10 - \sqrt{2}}$ **$\frac{50 + 5\sqrt{2}}{98}$**
5. Simplify $\sqrt{72w^2x^3y^4}$. Use absolute value symbols if necessary. **$6|w|xy^2\sqrt{2x}$**

Motivating the Lesson

Hands-On Activity Have students draw a vertical and horizontal segment on a coordinate plane so that one segment intersects the other at its midpoint. Then have them connect the four endpoints to form a quadrilateral (the figure is called a *kite*). Their diagrams should look similar to one of the kites in the figure below.

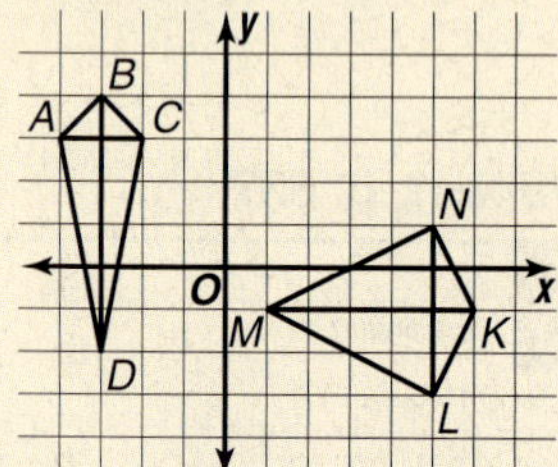

Have them find the exact length (as radical expressions) of each side of their quadrilateral. Then ask them to write a radical expression for the perimeter of their quadrilateral.

MODELING
Alternative hands-on options using scissors are available for teaching this lesson.

14–4 Adding and Subtracting Radical Expressions

What You'll Learn
You'll learn to add and subtract radical expressions.

Why It's Important
Hobbies Knowing how to add radical expressions can help you find the perimeter of a sail on a sailboat. *See Exercise 37.*

To find the exact perimeter of quadrilateral *ABCD*, you will need to add radical expressions.

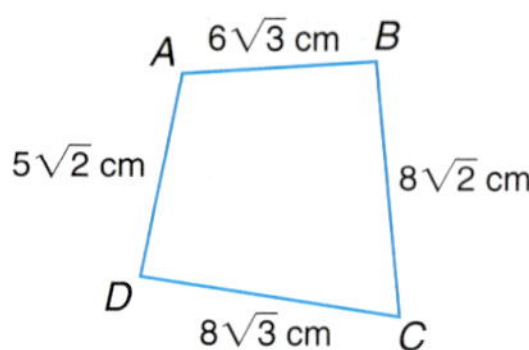

Radical expressions with the same radicands can be added or subtracted in the same way that monomials are added or subtracted.

Look Back
Adding and Subtracting Monomials: Lesson 9–2

Monomials	**Radical Expressions**
$5x + 3x = (5 + 3)x$ $= 8x$	$5\sqrt{2} + 3\sqrt{2} = (5 + 3)\sqrt{2}$ $= 8\sqrt{2}$
$8y - 2y = (8 - 2)y$ $= 6y$	$8\sqrt{3} - 2\sqrt{3} = (8 - 2)\sqrt{3}$ $= 6\sqrt{3}$

Notice that the Distributive Property was used to simplify each radical expression.

Examples

Simplify each expression.

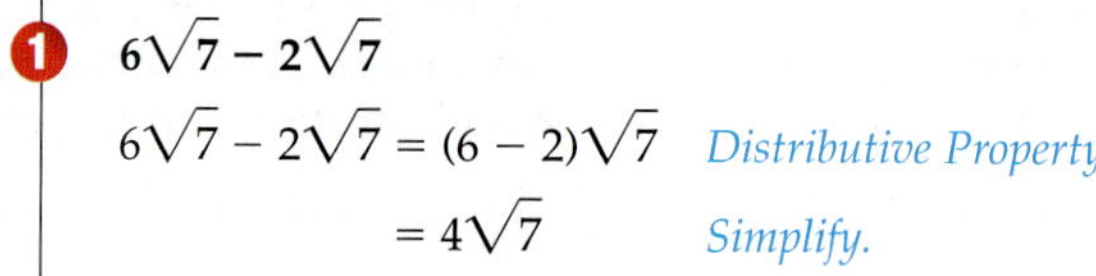

1 $6\sqrt{7} - 2\sqrt{7}$

$6\sqrt{7} - 2\sqrt{7} = (6 - 2)\sqrt{7}$ *Distributive Property*
$= 4\sqrt{7}$ *Simplify.*

2 $5\sqrt{5} + 3\sqrt{5} - 18\sqrt{5}$

$5\sqrt{5} + 3\sqrt{5} - 18\sqrt{5} = (5 + 3 - 18)\sqrt{5}$ *Distributive Property*
$= -10\sqrt{5}$ *Simplify.*

Your Turn **d. $-5\sqrt{13}$**

a. $8\sqrt{6} + 3\sqrt{6}$ **$11\sqrt{6}$**
b. $5\sqrt{2} - 12\sqrt{2}$ **$-7\sqrt{2}$**
c. $4\sqrt{3} + 7\sqrt{3} - 2\sqrt{3}$ **$9\sqrt{3}$**
d. $3\sqrt{13} - 2\sqrt{13} - 6\sqrt{13}$

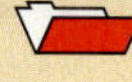

Resource Manager

Reproducible Masters
Chapter 14 Resource Masters
- *Study Guide*, p. 638
- *Skills Practice*, p. 639
- *Practice*, p. 640
- *Reading to Learn Mathematics*, p. 641
- *Enrichment*, p. 642

Hands-On Algebra, pp. 156–157

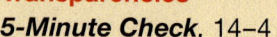

Transparencies
5-Minute Check, 14–4

Technology/Multimedia
Interactive Chalkboard CD-ROM

Example 3 **Geometry Link**

Refer to the beginning of the lesson. Find the exact perimeter of quadrilateral *ABCD*.

$P = 6\sqrt{3} + 8\sqrt{2} + 8\sqrt{3} + 5\sqrt{2}$ *Like terms: $6\sqrt{3}$ and $8\sqrt{3}$; $8\sqrt{2}$ and $5\sqrt{2}$*

$= 6\sqrt{3} + 8\sqrt{3} + 8\sqrt{2} + 5\sqrt{2}$ *Commutative Property*

$= (6 + 8)\sqrt{3} + (8 + 5)\sqrt{2}$ *Distributive Property*

$= 14\sqrt{3} + 13\sqrt{2}$ *Simplify.*

The exact perimeter of quadrilateral *ABCD* is $14\sqrt{3} + 13\sqrt{2}$ centimeters.

In Example 3, the expression $14\sqrt{3} + 13\sqrt{2}$ cannot be simplified further for the following reasons.

- The radicands are different.
- There are no common factors.
- Each radicand is in simplest form.

If the radicals in a radical expression are not in simplest form, simplify them first. Then use the Distributive Property wherever possible to further simplify the expression.

Example 4

Simplify $2\sqrt{20} - \sqrt{45}$.

$2\sqrt{20} - \sqrt{45} = 2\sqrt{2^2 \cdot 5} - \sqrt{3^2 \cdot 5}$ *Prime factorization*

$= 2(2\sqrt{5}) - (3\sqrt{5})$ *Simplify.*

$= 4\sqrt{5} - 3\sqrt{5}$ *Multiply.*

$= (4 - 3)\sqrt{5}$ *Distributive Property*

$= 1\sqrt{5}$ or $\sqrt{5}$ *Simplify.*

Your Turn **Simplify each expression.**

e. $4\sqrt{27} - 5\sqrt{12}$ **$2\sqrt{3}$**

f. $7\sqrt{18} + 3\sqrt{50}$ **$36\sqrt{2}$**

Check for Understanding

1–3. See margin.

1. **Describe** in your own words how to add radical expressions.
2. **Explain** why you should simplify each radical in a radical expression before adding or subtracting.
3.  Writing Math Explain how you use the Distributive Property to simplify the sum or difference of like radicals.

www.algconcepts.com/extra_examples

2 TEACH

In-Class Examples

Examples 1–2

Simplify each expression.

1 $12\sqrt{10} + 15\sqrt{10}$ **$27\sqrt{10}$**

2 $3\sqrt{2} - 9\sqrt{2} + \sqrt{2}$ **$-5\sqrt{2}$**

Example 3

The lengths of the three sides of a triangle are $10\sqrt{3}$ centimeters, $5\sqrt{5}$ centimeters, and $9\sqrt{3}$ centimeters. Find the exact perimeter of the triangle. **$(19\sqrt{3} + 5\sqrt{5})$ cm**

Example 4

Simplify $2\sqrt{48} - 3\sqrt{75}$. **$-7\sqrt{3}$**

Answers

1. **Sample answer: Add like terms.**
2. **to determine whether there are any like radicands**
3. **The Distributive Property allows you to add like terms. Radicals with like radicands can be added or subtracted.**

***Study Guide,* p. 638**

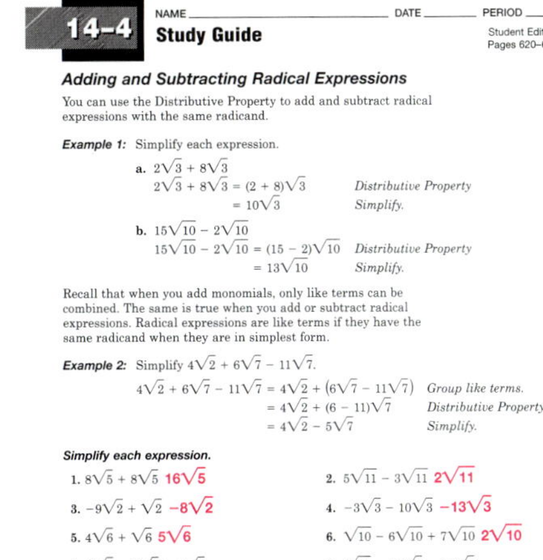

NAME ________ DATE ______ PERIOD ______

14-4 Study Guide

Student Edition Pages 620–623

Adding and Subtracting Radical Expressions

You can use the Distributive Property to add and subtract radical expressions with the same radicand.

Example 1: Simplify each expression.

a. $2\sqrt{3} + 8\sqrt{3}$

$2\sqrt{3} + 8\sqrt{3} = (2 + 8)\sqrt{3}$ *Distributive Property*

$= 10\sqrt{3}$ *Simplify.*

b. $15\sqrt{10} - 2\sqrt{10}$

$15\sqrt{10} - 2\sqrt{10} = (15 - 2)\sqrt{10}$ *Distributive Property*

$= 13\sqrt{10}$ *Simplify.*

Recall that when you add monomials, only like terms can be combined. The same is true when you add or subtract radical expressions. Radical expressions are like terms if they have the same radicand when they are in simplest form.

Example 2: Simplify $4\sqrt{2} + 6\sqrt{7} - 11\sqrt{7}$.

$4\sqrt{2} + 6\sqrt{7} - 11\sqrt{7} = 4\sqrt{2} + (6\sqrt{7} - 11\sqrt{7})$ *Group like terms.*

$= 4\sqrt{2} + (6 - 11)\sqrt{7}$ *Distributive Property*

$= 4\sqrt{2} - 5\sqrt{7}$ *Simplify.*

Simplify each expression.

1. $8\sqrt{5} + 8\sqrt{5}$ **$16\sqrt{5}$**
2. $5\sqrt{11} - 3\sqrt{11}$ **$2\sqrt{11}$**
3. $-9\sqrt{2} + \sqrt{2}$ **$-8\sqrt{2}$**
4. $-3\sqrt{3} - 10\sqrt{3}$ **$-13\sqrt{3}$**
5. $4\sqrt{6} + \sqrt{6}$ **$5\sqrt{6}$**
6. $\sqrt{10} - 6\sqrt{10} + 7\sqrt{10}$ **$2\sqrt{10}$**
7. $2\sqrt{2} - 5\sqrt{2} + 4\sqrt{5}$ **$-3\sqrt{2} + 4\sqrt{5}$**
8. $\sqrt{11} - 15\sqrt{3} - 10\sqrt{3}$ **$\sqrt{11} - 25\sqrt{3}$**
9. $8\sqrt{13} + 3\sqrt{13} - 4\sqrt{7} - 3\sqrt{7}$ **$11\sqrt{13} - 7\sqrt{7}$**
10. $-4\sqrt{6} + 2\sqrt{6} + \sqrt{3} + \sqrt{3}$ **$2\sqrt{6} + 2\sqrt{3}$**
11. $-3\sqrt{5} + 9\sqrt{2} + 5\sqrt{2} + 5\sqrt{5}$ **$2\sqrt{5} + 14\sqrt{2}$**
12. $7\sqrt{2} + \sqrt{8}$ **$9\sqrt{2}$**
13. $3\sqrt{3} + \sqrt{27}$ **$6\sqrt{3}$**
14. $5\sqrt{32} - 6\sqrt{2}$ **$14\sqrt{2}$**

Reteaching Activity

Verbal/Linguistic Learners Have students give an oral explanation of how to identify like radicals and how to add or subtract like radicals.

3 PRACTICE/APPLY

Error Analysis

Watch for students who confuse adding like radicals with multiplying radicals, writing the sum $7\sqrt{6} + 4\sqrt{6}$ in Exercise 8 as $28\sqrt{36}$ or 168.

Prevent by showing students how to use the Distributive Property to rewrite $7\sqrt{6} + 4\sqrt{6}$ as $7 \cdot \sqrt{6} + 4 \cdot \sqrt{6} = (7 + 4)\sqrt{6} = 11\sqrt{6}$.

Assignment Guide

Basic: 15–37 odd, 38–46
Average: 16–34 even, 35–46

***Skills Practice,* p. 639, and *Practice,* p. 640 (shown)**

14-4 NAME ______ DATE ______ PERIOD ______
Practice Student Edition Pages 620–623

Adding and Subtracting Radical Expressions

Simplify each expression.

1. $3\sqrt{7} + 4\sqrt{7}$ **$7\sqrt{7}$**
2. $9\sqrt{2} - 4\sqrt{2}$ **$5\sqrt{2}$**
3. $-5\sqrt{17} + 12\sqrt{17}$ **$7\sqrt{17}$**
4. $7\sqrt{3} - 3\sqrt{3}$ **$4\sqrt{3}$**
5. $-8\sqrt{5} + 2\sqrt{5}$ **$-6\sqrt{5}$**
6. $-7\sqrt{11} - 2\sqrt{11}$ **$-9\sqrt{11}$**
7. $13\sqrt{10} - 5\sqrt{10}$ **$8\sqrt{10}$**
8. $-6\sqrt{7} + 4\sqrt{7}$ **$-2\sqrt{7}$**
9. $3\sqrt{7} + \sqrt{3}$ **in simplest form**
10. $2\sqrt{6} + 4\sqrt{6} + 5\sqrt{6}$ **$11\sqrt{6}$**
11. $5\sqrt{3} + 4\sqrt{3} - 7\sqrt{3}$ **$2\sqrt{3}$**
12. $3\sqrt{2} - 2\sqrt{2} + 5\sqrt{2}$ **$6\sqrt{2}$**
13. $11\sqrt{5} - 3\sqrt{5} - 2\sqrt{5}$ **$6\sqrt{5}$**
14. $6\sqrt{13} + 3\sqrt{13} - 12\sqrt{13}$ **$-3\sqrt{13}$**
15. $4\sqrt{10} - 3\sqrt{10} - 5\sqrt{10}$ **$-4\sqrt{10}$**
16. $4\sqrt{6} - 2\sqrt{6} + 3\sqrt{6}$ **$5\sqrt{6}$**
17. $7\sqrt{7} + 4\sqrt{3} - 5\sqrt{7}$ **$2\sqrt{7} + 4\sqrt{3}$**
18. $-9\sqrt{2} + 4\sqrt{6} + 2\sqrt{2}$ **$-7\sqrt{2} + 4\sqrt{6}$**
19. $\sqrt{12} + 2\sqrt{27}$ **$8\sqrt{3}$**
20. $5\sqrt{63} - \sqrt{28}$ **$13\sqrt{7}$**
21. $-4\sqrt{96} + 6\sqrt{24}$ **$-4\sqrt{6}$**
22. $-3\sqrt{45} + 3\sqrt{180}$ **$9\sqrt{5}$**
23. $-4\sqrt{56} + 3\sqrt{126}$ **$\sqrt{14}$**
24. $2\sqrt{72} - 3\sqrt{50}$ **$-3\sqrt{2}$**
25. $7\sqrt{32} + 3\sqrt{75}$ **$28\sqrt{2} + 15\sqrt{3}$**
26. $\sqrt{32} + \sqrt{8} + \sqrt{18}$ **$9\sqrt{2}$**
27. $2\sqrt{20} - \sqrt{80} + \sqrt{45}$ **$3\sqrt{5}$**

Guided Practice

Getting Ready Express each radical in simplest form. Then determine the like radicals.

Sample: $5\sqrt{6}, 3\sqrt{5}, 3\sqrt{20}$

Solution: $5\sqrt{6} = 5\sqrt{6}$
$3\sqrt{5} = 3\sqrt{5}$ ✓
$3\sqrt{20} = 6\sqrt{5}$ ✓

4. $4\sqrt{2}, -9\sqrt{2}$

4. $4\sqrt{2}, -3\sqrt{18}, 2\sqrt{6}$
5. $-3\sqrt{10}, 4\sqrt{5}, -5\sqrt{3}$ **none**
6. $-2\sqrt{7}, 6\sqrt{14}, 5\sqrt{28}$ **$-2\sqrt{7}, 10\sqrt{7}$**
7. $4\sqrt{12}, -7\sqrt{6}, -6\sqrt{27}$ **$8\sqrt{3}, -18\sqrt{3}$**

Examples 1, 2, & 4

Simplify each expression.

8. $7\sqrt{6} + 4\sqrt{6}$ **$11\sqrt{6}$**
9. $4\sqrt{5} - 2\sqrt{5}$ **$2\sqrt{5}$**
10. $8\sqrt{3} - 3\sqrt{3} + 7\sqrt{3}$ **$12\sqrt{3}$**
11. $3\sqrt{7} - 6\sqrt{7} - 2\sqrt{7}$ **$-5\sqrt{7}$**
12. $5\sqrt{27} + 2\sqrt{48}$ **$23\sqrt{3}$**
13. $2\sqrt{50} - 4\sqrt{32}$ **$-6\sqrt{2}$**

Example 3

14. **Geometry** Find the exact perimeter of triangle ABC. **$6\sqrt{5} + 6$ m**

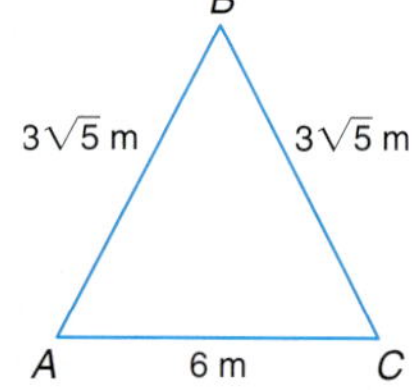

Exercises

Practice

Simplify each expression.

15. $4\sqrt{3} + 2\sqrt{3}$ **$6\sqrt{3}$**
16. $12\sqrt{2} + 3\sqrt{2}$ **$15\sqrt{2}$**
17. $15\sqrt{5} - \sqrt{5}$ **$14\sqrt{5}$**
18. $5\sqrt{2} - 8\sqrt{3}$ **in simplest form**
19. $-6\sqrt{7} + 10\sqrt{7}$ **$4\sqrt{7}$**
20. $-7\sqrt{3} - 2\sqrt{3}$ **$-9\sqrt{3}$**
21. $5\sqrt{5} + 3\sqrt{5} - 10\sqrt{5}$ **$-2\sqrt{5}$**
22. $6\sqrt{7} - 5\sqrt{7} + 8\sqrt{7}$ **$9\sqrt{7}$**
23. $7\sqrt{2} - 16\sqrt{2} + 8\sqrt{5}$
24. $5\sqrt{11} + 2\sqrt{11} - 4\sqrt{11}$ **$3\sqrt{11}$**
25. $-\sqrt{6} - 4\sqrt{6} + 12\sqrt{6}$ **$7\sqrt{6}$**
26. $-9\sqrt{14} + 2\sqrt{14} - 6\sqrt{14}$
27. $4\sqrt{3} + 5\sqrt{12}$ **$14\sqrt{3}$**
28. $-2\sqrt{28} + 3\sqrt{7}$ **$-\sqrt{7}$**

B

29. $2\sqrt{18} - 5\sqrt{8}$ **$-4\sqrt{2}$**
30. $3\sqrt{63} - \sqrt{112}$ **$5\sqrt{7}$**
31. $\sqrt{50} + \sqrt{108} + \sqrt{18}$ **$8\sqrt{2} + 6\sqrt{3}$**
32. $-4\sqrt{27} - 5\sqrt{32} + 2\sqrt{75}$ **$-2\sqrt{3} - 20\sqrt{2}$**
33. Simplify $2\sqrt{20} - \sqrt{180} - 3\sqrt{24}$. **$-2\sqrt{5} - 6\sqrt{6}$**
34. Write $-2\sqrt{50} + 5\sqrt{48} + 7\sqrt{98}$ in simplest form. **$39\sqrt{2} + 20\sqrt{3}$**

23. $-9\sqrt{2} + 8\sqrt{5}$
26. $-13\sqrt{14}$

Homework Help	
For Exercises	**See Examples**
15–17, 19–22, 24–26, 37	1, 2
18, 23, 27–34	4
35	3
Extra Practice	
See page 721.	

Applications and Problem Solving

35. **Geometry** The perimeter of quadrilateral $DEFG$ is $14\sqrt{7}$ feet. What is the missing measure? **$5\sqrt{7}$**

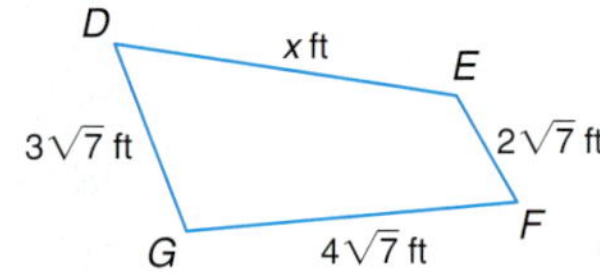

36a. $\sqrt{7} + \sqrt{7} + \sqrt{7} + \sqrt{7}$

Close to Home

COUNT OFF BY THE SQUARE ROOT OF 7!
UHH....

36. **Media** Refer to the comic. Suppose the first team member says "$\sqrt{7}$," the second says "$2\sqrt{7}$," and so on.
    a. Write an expression that could be used to determine the response of the fourth student in line.
    b. Determine the response of the fourth person in line. $4\sqrt{7}$
    c. Suppose the coach asks the students to count off by $\sqrt{5}$. Find the response of the last person in line. $13\sqrt{5}$

37. **Hobbies** Ling is making a model of a sailboat. The dimensions of one of the sails are shown in the diagram.
    a. Find the missing measure. (*Hint:* Use the Pythagorean Theorem.) $4\sqrt{3}$
    b. Determine the exact perimeter of the sail. $6\sqrt{3} + 6$

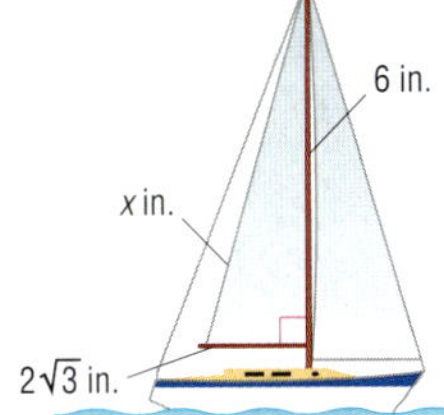

38. No; for example, $-\sqrt{7} + \sqrt{7} = 0$.

38. **Critical Thinking** Is the set of irrational numbers closed under addition? Explain your reasoning.

Mixed Review

39. Simplify $\sqrt{45mn^3p^2}$. *(Lesson 14–3)* $3n|p|\sqrt{5mn}$

Find the distance between each pair of points. Round to the nearest tenth, if necessary. *(Lesson 14–2)* 41. $\sqrt{101}$ or 10.0

40. $C(-3, 5)$, $D(0, 8)$ $\sqrt{18}$ or 4.2
41. $J(4, -2)$, $K(-6, -1)$

42. **Geometry** The base of a triangle measures 3 feet more than its height. The area of the triangle is 20 square feet. Write and solve a quadratic equation to find the height and base of the triangle. *(Lesson 11–4)* $h^2 + 3h - 40 = 0$; $h = 5$ ft, $b = 8$ ft

Factor each monomial. *(Lesson 10–1)*

43. $-26a^3b^2$ $-1 \cdot 2 \cdot 13 \cdot a \cdot a \cdot a \cdot b \cdot b$
44. $36xy^4z$ $2 \cdot 2 \cdot 3 \cdot 3 \cdot x \cdot y \cdot y \cdot y \cdot y \cdot z$

Standardized Test Practice

45. **Short Response** Find the product of $a + 2c$ and $a - 2c$. *(Lesson 9–5)* $a^2 - 4c^2$

46. **Multiple Choice** The scatter plot shows the relationship between the cost of a stereo system and the rating it received by a consumer group. Which of the following best describes the relationship? *(Lesson 7–4)* A

    Ⓐ positive Ⓑ dependent
    Ⓒ no pattern Ⓓ negative

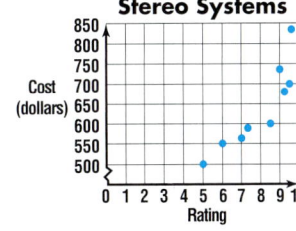

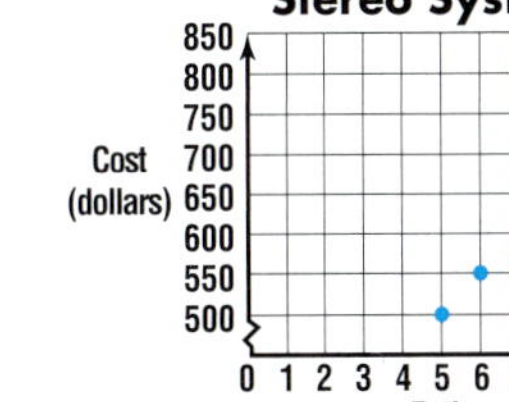

www.algconcepts.com/self_check_quiz

4 ASSESS

Open-Ended Assessment

Writing Ask students to write three expressions that have different radicands but that can be simplified to become like radicals.

Extra Credit

Simplify $a\sqrt{1} + b\sqrt{2} + c\sqrt{4} + d\sqrt{8} + e\sqrt{16} + f\sqrt{32}$.
$(a + 2c + 4e) + (b + 2d + 4f)\sqrt{2}$

***Enrichment*, p. 642**

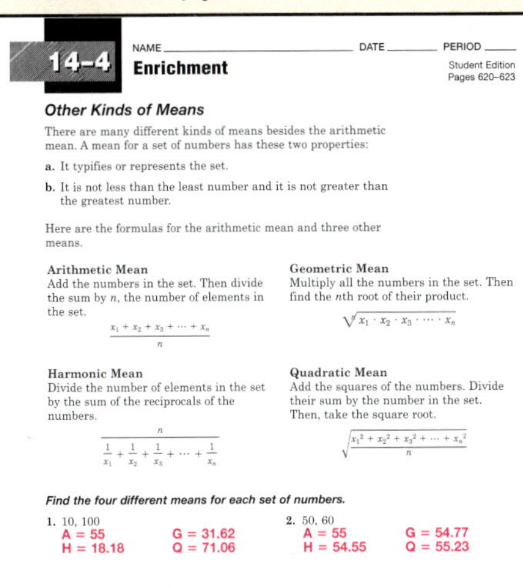

14–4 NAME ______ DATE ______ PERIOD ______

Enrichment Student Edition Pages 620–623

Other Kinds of Means

There are many different kinds of means besides the arithmetic mean. A mean for a set of numbers has these two properties:

a. It typifies or represents the set.

b. It is not less than the least number and it is not greater than the greatest number.

Here are the formulas for the arithmetic mean and three other means.

Arithmetic Mean
Add the numbers in the set. Then divide the sum by n, the number of elements in the set.

$$\frac{x_1 + x_2 + x_3 + \cdots + x_n}{n}$$

Geometric Mean
Multiply all the numbers in the set. Then find the nth root of their product.

$$\sqrt[n]{x_1 \cdot x_2 \cdot x_3 \cdot \cdots \cdot x_n}$$

Harmonic Mean
Divide the number of elements in the set by the sum of the reciprocals of the numbers.

$$\frac{n}{\frac{1}{x_1} + \frac{1}{x_2} + \frac{1}{x_3} + \cdots + \frac{1}{x_n}}$$

Quadratic Mean
Add the squares of the numbers. Divide their sum by the number in the set. Then, take the square root.

$$\sqrt{\frac{x_1^2 + x_2^2 + x_3^2 + \cdots + x_n^2}{n}}$$

Find the four different means for each set of numbers.

1. 10, 100
   A = 55 G = 31.62
   H = 18.18 Q = 71.06
2. 50, 60
   A = 55 G = 54.77
   H = 54.55 Q = 55.23
3. 1, 2, 3, 4, 5,
   A = 3 G = 2.61
   H = 2.19 Q = 3.32
4. 2, 2, 4, 4
   A = 3 G = 2.83
   H = 2.67 Q = 3.16
5. Use the results from Exercises 1 to 4 to compare the relative sizes of the four types of means.
   From least to greatest, the means are the harmonic, geometric, arithmetic, and quadratic means.

Lesson 14-5

1 FOCUS

5-Minute Check
Lesson 14–4

Simplify each expression.

1. $6\sqrt{15} - 4\sqrt{15}$ $2\sqrt{15}$

2. $11\sqrt{11} - 25\sqrt{11} + 7\sqrt{11}$ $-7\sqrt{11}$

3. $3\sqrt{80} + 2\sqrt{45}$ $18\sqrt{5}$

Find the exact perimeter of each figure.

4.

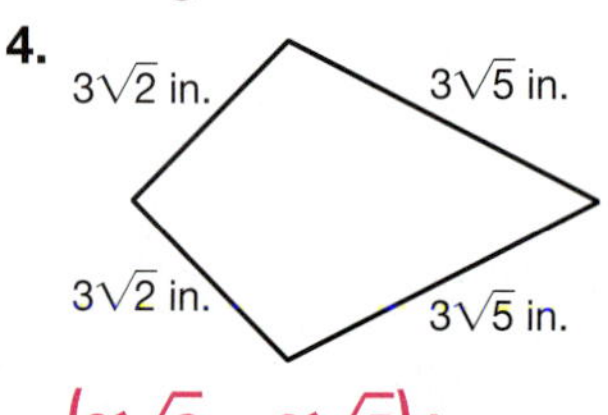

$(6\sqrt{2} + 6\sqrt{5})$ in.

5.

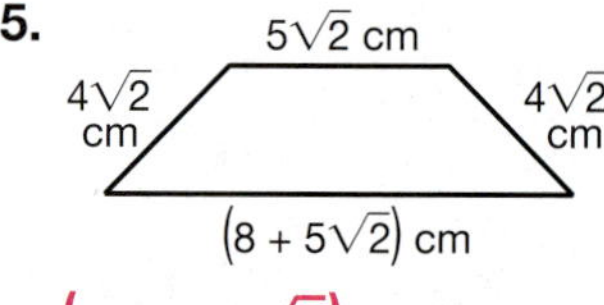

$(8 + 18\sqrt{2})$ cm

Motivating the Lesson

Hands-On Activity Ask students to graph this system of equations on the same coordinate plane.

$$y = x^2 + 1$$
$$y = 0$$

Students should discover that there are no points of intersection for the two graphs. Discuss what this means. **There is no point whose coordinates satisfy both equations.** Inform students that they will be exploring a situation in this lesson where a possible solution value may not check in an equation.

14–5 Solving Radical Equations

What You'll Learn

You'll learn to solve simple radical equations in which only one radical contains a variable.

Why It's Important

Engineering

Engineers use radical equations to determine the velocity of roller coasters.

See Example 5.

The speed of a roller coaster as it travels through a loop depends on the height of the hill from which the coaster has just descended. The equation $s = 8\sqrt{h - 2r}$ gives the speed s in feet per second where h is the height of the hill and r is the radius of the loop.

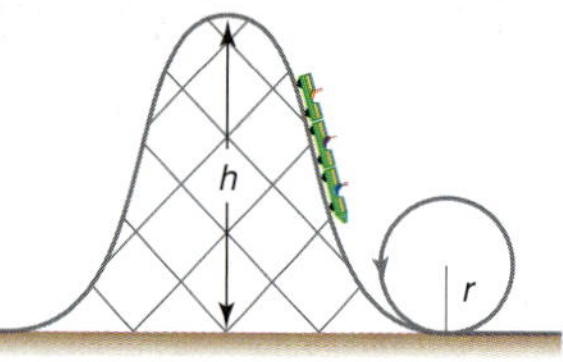

Suppose the owner of an amusement park wants to design a roller coaster that will travel at a speed of 40 feet per second as it goes through a loop with a radius of 30 feet. How high should the hill be? *This problem will be solved in Example 5.*

Equations like $s = 8\sqrt{h - 2r}$ that contain radicals with variables in the radicand are called **radical equations**. To solve these equations, first isolate the radical on one side of the equation. Then square each side of the equation to eliminate the radical.

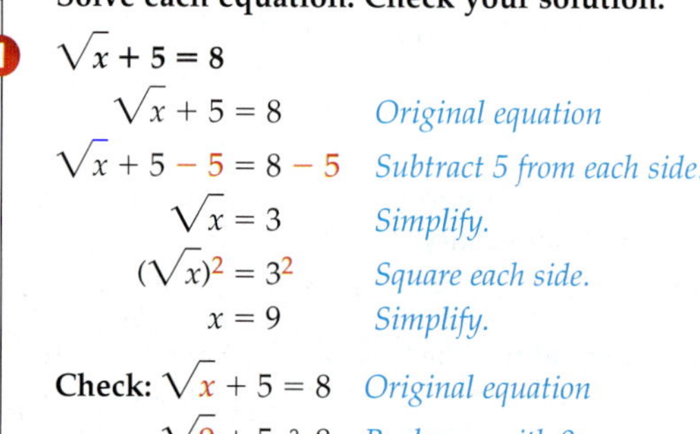

Examples

Solve each equation. Check your solution.

1 $\sqrt{x} + 5 = 8$

$\sqrt{x} + 5 = 8$	*Original equation*
$\sqrt{x} + 5 - 5 = 8 - 5$	*Subtract 5 from each side.*
$\sqrt{x} = 3$	*Simplify.*
$(\sqrt{x})^2 = 3^2$	*Square each side.*
$x = 9$	*Simplify.*

Check:

$\sqrt{x} + 5 = 8$	*Original equation*
$\sqrt{9} + 5 \stackrel{?}{=} 8$	*Replace x with 9.*
$3 + 5 \stackrel{?}{=} 8$	$\sqrt{9} = 3$
$8 = 8$	✓

The solution is 9.

2 $\sqrt{m + 3} + 2 = 6$

$\sqrt{m + 3} + 2 = 6$	*Original equation*
$\sqrt{m + 3} + 2 - 2 = 6 - 2$	*Subtract 2 from each side.*
$\sqrt{m + 3} = 4$	*Simplify.*
$(\sqrt{m + 3})^2 = 4^2$	*Square each side.*
$m + 3 = 16$	*Simplify.*
$m + 3 - 3 = 16 - 3$	*Subtract 3 from each side.*
$m = 13$	*Check this result.*

The solution is 13.

Resource Manager

Reproducible Masters

Chapter 14 Resource Masters
- *Study Guide,* p. 643
- *Skills Practice,* p. 644
- *Practice,* p. 645
- *Reading to Learn Mathematics,* p. 646
- *Enrichment,* p. 647
- *Assessment,* p. 659

Graphing Calculator, p. 40

Transparencies

5-Minute Check, 14–5

Technology/Multimedia

Interactive Chalkboard CD-ROM

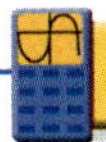

Your Turn Solve each equation. Check your solution.

a. $\sqrt{y} - 6 = 4$ **100**   b. $\sqrt{a-1} + 5 = 7$ **5**

You can use a graphing calculator to solve radical equations.

Graphing Calculator Tutorial
See pp. 750–753.

Graphing Calculator Exploration

Find the solution of $\sqrt{x} + 5 = 8$.

Step 1 Set the viewing window for x: [0, 15] by 1 and y: [0, 15] by 1.

Step 2 Press Y=. Enter the left side of the equation, $\sqrt{x} + 5$, as Y_1. Enter the right side of the equation, 8, as Y_2.

Step 3 Press GRAPH to graph the equations.

Step 4 The solution of the equation is the intersection point of the two lines. Use the INTERSECT feature to find the intersection point.

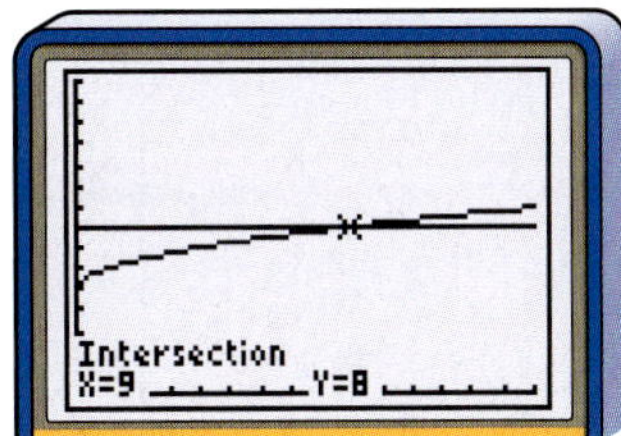

Enter: 2nd [CALC] 5

The graph intersects at the point (9, 8). The solution is the x-coordinate, 9.

Try These **1–2. See students' work.**

1. Solve the equation in Example 2 by using a graphing calculator.
2. Use a graphing calculator to find the solution of Your Turn Exercise b.
3. What changes must be made to the graphing calculator before finding the solution of Your Turn Exercise a?

3. Sample answer: Set the viewing window for x:[0, 150] by 15 and y:[0, 10] by 1.

Squaring each side of an equation may produce results that do not satisfy the *original* equation. So, you must check all solutions when you solve radical equations.

Examples

Solve each equation. Check your solution.

3 $\sqrt{n+2} = n - 4$

$\sqrt{n+2} = n - 4$	*Original equation*
$(\sqrt{n+2})^2 = (n-4)^2$	*Square each side.*
$n + 2 = n^2 - 8n + 16$	*Simplify.*
$n - n + 2 - 2 = n^2 - 8n + 16 - n - 2$	*Subtract n and 2 from each side.*
$0 = n^2 - 9n + 14$	*Simplify.*
$0 = (n-7)(n-2)$	*Factor.*

(continued on the next page)

www.algconcepts.com/extra_examples

2 TEACH

In-Class Examples

Examples 1–2

Solve each equation. Check your solution.

1 $3 + \sqrt{x} = 12$ **81**

2 $-6 + \sqrt{y-2} = 1$ **51**

Graphing Calculator Exploration

Inform students that the graphing calculator automatically supplies a left parenthesis in Step 2 for the expression that appears under the radical symbol. Warn students that they must supply the right parenthesis to close the parentheses after the expression under the radical symbol has been entered. Failure to do so will often result in an incorrect graph.

In-Class Examples

Examples 3–4

Solve each equation. Check your solution.

3 $\sqrt{3m + 4} = m$ **4**

4 $\sqrt{n + 5} = n + 5$ **−4, −5**

$$n - 7 = 0 \quad \text{or} \quad n - 2 = 0 \qquad \textit{Use the Zero Product Property.}$$
$$n = 7 \qquad\qquad n = 2 \qquad \textit{Solve.}$$

Check:

$$\sqrt{n + 2} = n - 4 \qquad\qquad \sqrt{n + 2} = n - 4$$
$$\sqrt{7 + 2} \stackrel{?}{=} 7 - 4 \qquad\qquad \sqrt{2 + 2} \stackrel{?}{=} 2 - 4$$
$$\sqrt{9} \stackrel{?}{=} 3 \qquad\qquad \sqrt{4} \stackrel{?}{=} -2$$
$$3 = 3 \ \checkmark \qquad\qquad 2 \neq -2$$

Since 2 does not satisfy the original equation, 7 is the only solution.

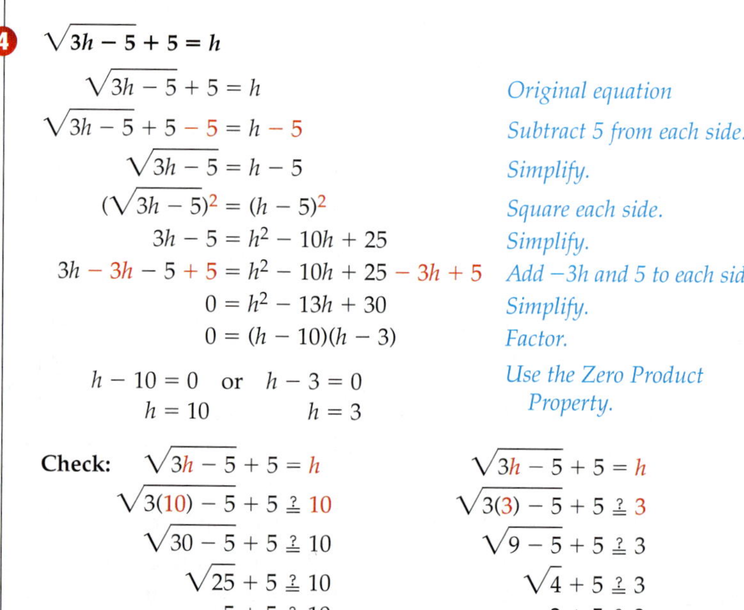

4 $\sqrt{3h - 5} + 5 = h$

$\sqrt{3h - 5} + 5 = h$	*Original equation*
$\sqrt{3h - 5} + 5 - 5 = h - 5$	*Subtract 5 from each side.*
$\sqrt{3h - 5} = h - 5$	*Simplify.*
$(\sqrt{3h - 5})^2 = (h - 5)^2$	*Square each side.*
$3h - 5 = h^2 - 10h + 25$	*Simplify.*
$3h - 3h - 5 + 5 = h^2 - 10h + 25 - 3h + 5$	*Add −3h and 5 to each side.*
$0 = h^2 - 13h + 30$	*Simplify.*
$0 = (h - 10)(h - 3)$	*Factor.*
$h - 10 = 0$ or $h - 3 = 0$	*Use the Zero Product Property.*
$h = 10 \qquad h = 3$	

Check:

$$\sqrt{3h - 5} + 5 = h \qquad\qquad \sqrt{3h - 5} + 5 = h$$
$$\sqrt{3(10) - 5} + 5 \stackrel{?}{=} 10 \qquad\qquad \sqrt{3(3) - 5} + 5 \stackrel{?}{=} 3$$
$$\sqrt{30 - 5} + 5 \stackrel{?}{=} 10 \qquad\qquad \sqrt{9 - 5} + 5 \stackrel{?}{=} 3$$
$$\sqrt{25} + 5 \stackrel{?}{=} 10 \qquad\qquad \sqrt{4} + 5 \stackrel{?}{=} 3$$
$$5 + 5 \stackrel{?}{=} 10 \qquad\qquad 2 + 5 \stackrel{?}{=} 3$$
$$10 = 10 \ \checkmark \qquad\qquad 7 \neq 3$$

Since 3 does not satisfy the original equation, 10 is the only solution.

Your Turn

c. $\sqrt{d + 1} = d - 1$ **3** 　　 **d.** $\sqrt{3x - 14} + x = 6$ **5**

Radical equations are used in many real-life situations.

Example 5 **Engineering Link**

Refer to the application at the beginning of the lesson. How high should the hill be?

Explore You know the speed of the coaster and the radius of the loop. You need to know the height of the hill.

interNET CONNECTION

Data Update For the latest information on roller coasters, visit: www.algconcepts.com

Plan Use the equation $s = 8\sqrt{h - 2r}$ to find the height of the hill.

Solve

$$s = 8\sqrt{h - 2r} \quad \text{Original equation}$$
$$40 = 8\sqrt{h - 2(30)} \quad \text{Replace } s \text{ with 40 and } r \text{ with 30.}$$
$$40 = 8\sqrt{h - 60} \quad \text{Multiply.}$$
$$\frac{40}{8} = \frac{8\sqrt{h - 60}}{8} \quad \text{Divide each side by 8.}$$
$$5 = \sqrt{h - 60} \quad \text{Simplify.}$$
$$5^2 = (\sqrt{h - 60})^2 \quad \text{Square each side.}$$
$$25 = h - 60 \quad \text{Simplify.}$$
$$25 + 60 = h - 60 + 60 \quad \text{Add 60 to each side.}$$
$$85 = h \quad \text{Simplify.}$$

The height of the hill is 85 feet.

Examine Check the result.

$$s = 8\sqrt{h - 2r} \quad \text{Original equation}$$
$$40 \stackrel{?}{=} 8\sqrt{85 - 2(30)} \quad \text{Replace } s \text{ with 40, } h \text{ with 85, and } r \text{ with 30.}$$
$$40 \stackrel{?}{=} 8\sqrt{85 - 60} \quad \text{Multiply.}$$
$$40 \stackrel{?}{=} 8\sqrt{25} \quad \text{Subtract.}$$
$$40 \stackrel{?}{=} 8 \cdot 5 \quad \sqrt{25} = 5$$
$$40 = 40 \ \checkmark$$

The answer is correct.

In-Class Example

Example 5

The equation $s = 8\sqrt{h - 2r}$ gives the speed s in feet per second of a roller coaster as it travels through a loop of radius r feet after coming down a hill of height h feet. Find h if s is 48 feet per second when r is 32 feet. **100 ft**

Answers

1. Isolate the radical.
2. Because squaring each side of an equation does not necessarily produce results that satisfy the original equation.
3. See students' work.

Check for Understanding

Communicating Mathematics

1–3. See margin.

Vocabulary
radical equation

1. **Tell** the first step you should take when solving a radical equation.
2. **Explain** why it is important to check radical equations after finding possible solutions.
3. *Writing Math* Write an example of a radical equation. Then write the steps for solving the equation.

Guided Practice

Getting Ready Square each side of the following equations.

Sample: $\sqrt{y - 1} = 4$

Solution: $\sqrt{y - 1} = 4$
$(\sqrt{y - 1})^2 = 4^2$
$y - 1 = 16$

4. $\sqrt{x} = 7$ **$x = 49$**
5. $\sqrt{a + 5} = 2$ **$a + 5 = 4$**
6. $9 = \sqrt{2c - 3}$ **$81 = 2c - 3$**

Study Guide, p. 643

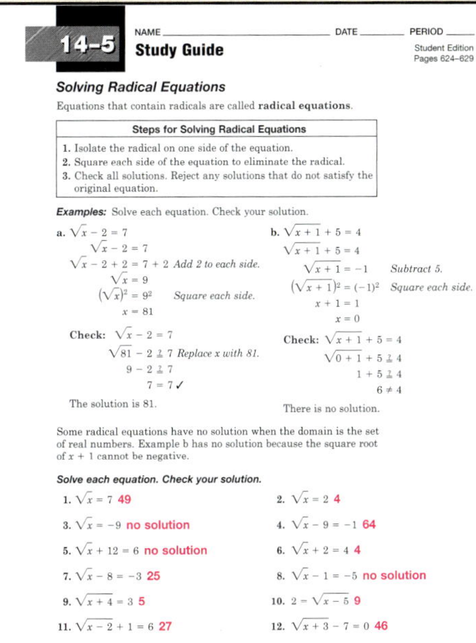

14-5 NAME ________ DATE ______ PERIOD _____

Study Guide Student Edition Pages 624–629

Solving Radical Equations

Equations that contain radicals are called **radical equations**.

Steps for Solving Radical Equations
1. Isolate the radical on one side of the equation. 2. Square each side of the equation to eliminate the radical. 3. Check all solutions. Reject any solutions that do not satisfy the original equation.

Examples: Solve each equation. Check your solution.

a. $\sqrt{x} - 2 = 7$
$\sqrt{x} - 2 = 7$
$\sqrt{x} - 2 + 2 = 7 + 2$ *Add 2 to each side.*
$\sqrt{x} = 9$
$(\sqrt{x})^2 = 9^2$ *Square each side.*
$x = 81$

Check: $\sqrt{x} - 2 = 7$
$\sqrt{81} - 2 \stackrel{?}{=} 7$ *Replace x with 81.*
$9 - 2 \stackrel{?}{=} 7$
$7 = 7 \checkmark$

The solution is 81.

b. $\sqrt{x + 1} + 5 = 4$
$\sqrt{x + 1} + 5 = 4$
$\sqrt{x + 1} = -1$ *Subtract 5.*
$(\sqrt{x + 1})^2 = (-1)^2$ *Square each side.*
$x + 1 = 1$
$x = 0$

Check: $\sqrt{x + 1} + 5 = 4$
$\sqrt{0 + 1} + 5 \stackrel{?}{=} 4$
$1 + 5 \stackrel{?}{=} 4$
$6 \neq 4$

There is no solution.

Some radical equations have no solution when the domain is the set of real numbers. Example b has no solution because the square root of $x + 1$ cannot be negative.

Solve each equation. Check your solution.

1. $\sqrt{x} = 7$ **49**
2. $\sqrt{x} = 2$ **4**
3. $\sqrt{x} = -9$ **no solution**
4. $\sqrt{x} - 9 = -1$ **64**
5. $\sqrt{x} + 12 = 6$ **no solution**
6. $\sqrt{x} + 2 = 4$ **4**
7. $\sqrt{x} - 8 = -3$ **25**
8. $\sqrt{x} - 1 = -5$ **no solution**
9. $\sqrt{x + 4} = 3$ **5**
10. $2 = \sqrt{x - 5}$ **9**
11. $\sqrt{x - 2} + 1 = 6$ **27**
12. $\sqrt{x + 3} - 7 = 0$ **46**

3 PRACTICE/APPLY

Error Analysis

Watch for students who simply begin to solve a radical equation by trying to square both sides of the equations in Exercises 7–10. ***Prevent by*** stressing that the first step is to determine if the radical expression is isolated, and if not, to isolate it on one side of the equation by squaring both sides.

Assignment Guide

Basic: 13–33 odd, 34–45
Average: 12–30 even, 32–45
All: Quiz 2, 1–10

***Skills Practice*, p. 644, and *Practice*, p. 645 (shown)**

14-5 NAME ______________ DATE ________ PERIOD ______

Practice

Student Edition Pages 624–629

Solving Radical Equations

Solve each equation. Check your solution.

1. $\sqrt{x} - 6 = 3$ **81**
2. $\sqrt{k} + 7 = 20$ **169**
3. $\sqrt{p + 3} + 4 = 7$ **6**
4. $\sqrt{n + 11} - 8 = -3$ **14**
5. $\sqrt{w - 2} - 1 = 6$ **51**
6. $\sqrt{y - 5} + 9 = 14$ **30**
7. $\sqrt{2r + 1} - 10 = -1$ **40**
8. $\sqrt{3h - 11} + 2 = 9$ **20**
9. $\sqrt{a + 4} = a - 8$ **12**
10. $\sqrt{z - 3} + 5 = z$ **7**
11. $\sqrt{3b + 9} + 3 = b$ **9**
12. $\sqrt{5f - 5} + 1 = f$ **6 and 1**
13. $\sqrt{8 + 2c} = c - 8$ **14**
14. $\sqrt{3s - 6} = s - 2$ **5 and 2**
15. $\sqrt{4h + 4} + h = 7$ **3**
16. $\sqrt{5m + 4} = m + 2$ **1 and 0**
17. $\sqrt{2y - 7} - y = -5$ **8**
18. $\sqrt{3k + 4} + k = 8$ **4**

Examples 1–4

Solve each equation. Check your solution.

7. $\sqrt{x} - 4 = 7$ **121**
8. $\sqrt{a - 2} - 5 = 3$ **66**
9. $\sqrt{m + 5} + 1 = m$ **4**
10. $\sqrt{1 + 2x} = x + 1$ **0**

Example 5

11. **Engineering** Refer to the application at the beginning of the lesson. Suppose the owner of an amusement park wants to design a roller coaster that will travel through a loop at a speed of 56 feet per second after descending a 120-foot hill. What should the radius of the loop be? **35.5 ft**

Exercises

Practice

Solve each equation. Check your solution.

Homework Help	
For Exercises	See Examples
12–23, 30–33	1, 2
24–29	3, 4
Extra Practice	
See page 721.	

12. $\sqrt{x} = 3$ **9**
13. $\sqrt{a} = -4$ **no solution**
14. $7 = \sqrt{7m}$ **7**
15. $\sqrt{-3d} = 6$ **−12**
16. $\sqrt{y} + 5 = 0$ **no solution**
17. $0 = \sqrt{2c} - 2$ **2**
18. $\sqrt{m - 4} = 6$ **40**
19. $\sqrt{w + 6} = 9$ **75**
20. $3 = \sqrt{4n + 1}$ **2**
21. $11 = \sqrt{2z - 5}$ **63**
22. $\sqrt{8h + 1} - 5 = 0$ **3**
23. $2 = \sqrt{3b - 5} + 6$ **no solution**
24. $\sqrt{x + 6} = x$ **3**
25. $p = \sqrt{5p - 6}$ **2 and 3**

B

26. $\sqrt{8 - b} = b - 2$ **4**
27. $\sqrt{k - 2} + 4 = k$ **6**
28. $t - 1 = \sqrt{2t + 6}$ **5**
29. $5 + \sqrt{3j - 5} = j$ **10**

30. Solve $\sqrt{\frac{a}{4}} = 6$. **144**

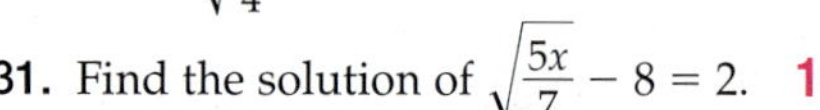

31. Find the solution of $\sqrt{\frac{5x}{7}} - 8 = 2$. **140**

Applications and Problem Solving

32. **Science** The speed of sound S, in meters per second, near Earth's surface can be determined using the formula $S = 20\sqrt{t + 273}$, where t is the surface temperature in degrees Celsius. Suppose a racing team has designed a car that can travel 340 meters per second, in hopes of breaking the sound barrier. At what temperature will the speed of sound be 340 meters per second? **16° C**

33. **Science** The formula $t = \sqrt{\frac{2s}{g}}$ can be used to determine the time t, in seconds, it takes an object initially at rest to fall s meters. In this formula, g is the acceleration due to gravity in meters per second squared.
   a. Suppose a rock falls 7.2 meters in 3 seconds on the moon. What is the acceleration due to gravity on the moon? **1.6 m/s²**
   b. Suppose a rock falls 78.4 meters in 4 seconds on Earth. What is the acceleration due to gravity on Earth? **9.8 m/s²**

Reteaching Activity

Auditory/Musical Learners Have students write a poem or song about why they should always check each solution to a radical equation.

34. **Critical Thinking** Find two numbers such that the square root of their sum is 5 and the square root of their product is 12. **9 and 16**

Mixed Review

Simplify each expression. *(Lesson 14–4)*

35. $-7\sqrt{5} + 2\sqrt{5}$ **$-5\sqrt{5}$**
36. $4\sqrt{7} - 3\sqrt{28}$ **$-2\sqrt{7}$**
37. $-\sqrt{54} - \sqrt{18} + \sqrt{24}$ **$-\sqrt{6} - 3\sqrt{2}$**

Simplify each expression. Leave in radical form. *(Lesson 14–3)*

38. $\sqrt{5} \cdot \sqrt{12}$ **$2\sqrt{15}$**
39. $\dfrac{\sqrt{15}}{\sqrt{3}}$ **$\sqrt{5}$**
40. $\dfrac{5}{4 - \sqrt{7}}$ **$\dfrac{20 + 5\sqrt{7}}{9}$**

41. Find the solution of $y = \frac{1}{2}x - 4$ and $y = -3$ by graphing. *(Lesson 13–1)* **(2, −3); See margin for graph.**

Write an inequality for each graph. *(Lesson 12–1)*

42. 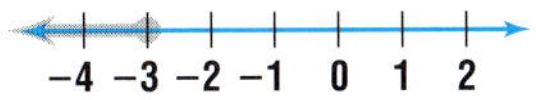

$x \leq -3$

43. 

$x > 1$

Standardized Test Practice

44. **Short Response** There are two field mice living in a barn. Suppose the number of mice triples every 4 months. How many mice will there be after 3 years if none of them die? *(Lesson 11–7)* **39,366**

45. **Grid In** Suppose y varies inversely as x and $y = 18$ when $x = 15$. Find y when $x = 12$. *(Lesson 6–6)* **22.5**

Quiz 2 Lessons 14–3 through 14–5

Simplify each expression. Leave in radical form. *(Lesson 14–3)*

1. $\sqrt{90}$ **$3\sqrt{10}$**
2. $\sqrt{5} \cdot \sqrt{25}$ **$5\sqrt{5}$**
3. $\dfrac{1}{7 + \sqrt{5}}$ **$\dfrac{7 - \sqrt{5}}{44}$**
4. Write $\sqrt{16x^2y^3}$ in simplest form. Use absolute value symbols if necessary. *(Lesson 14–3)* **$4|x|y\sqrt{y}$**

Simplify each expression. *(Lesson 14–4)*

5. $3\sqrt{5} + 9\sqrt{5}$ **$12\sqrt{5}$**
6. $2\sqrt{40} - 7\sqrt{2} + 6\sqrt{10}$ **$10\sqrt{10} - 7\sqrt{2}$**
7. **Geometry** The perimeter of $\triangle ABC$ is $16\sqrt{2}$ inches. Find the value of x. *(Lesson 14–4)* **$7\sqrt{2}$**

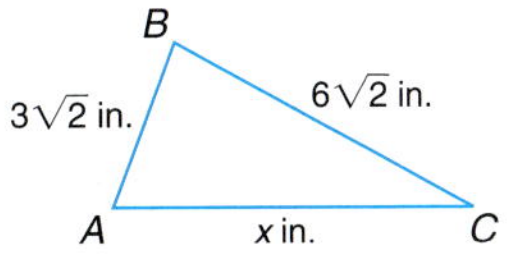

Exercise 7

Solve each equation. Check your solution. *(Lesson 14–5)*

8. $\sqrt{x} = 4$ **16**
9. $\sqrt{a} + 8 = 0$ **no solution**
10. $\sqrt{m + 3} = 5$ **22**

www.algconcepts.com/self_check_quiz

Extra Credit

Solve $\sqrt{x} + \sqrt{x + 1} = 5$. (*Hint:* You have to isolate a radical, and square both sides in two different steps.)
5.76

4 ASSESS

Open-Ended Assessment

Speaking Ask students to explain how they can tell that a value is not a solution to an equation.

Quiz 2

The Quiz provides students with a brief review of the concepts and skills in Lessons 14–3 through 14–5. Lesson numbers are given to the right of the exercises or instruction lines so students can review concepts not yet mastered.

Chapter 14, Quiz B (Lessons 14–3 through 14–5) is available in the *Chapter 14 Resource Masters*, p. 659.

Answer

41. 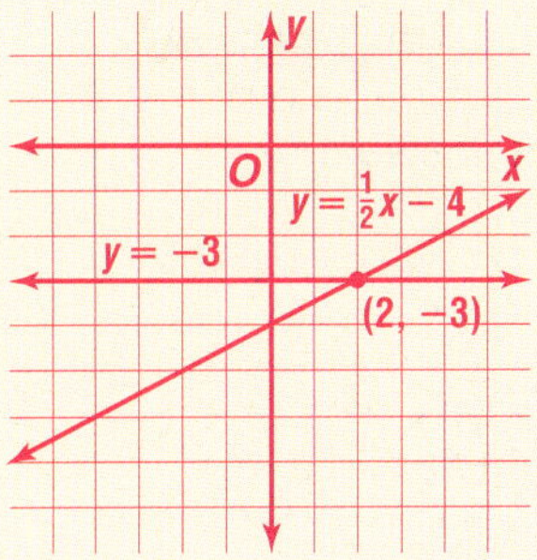

***Enrichment*, p. 647**

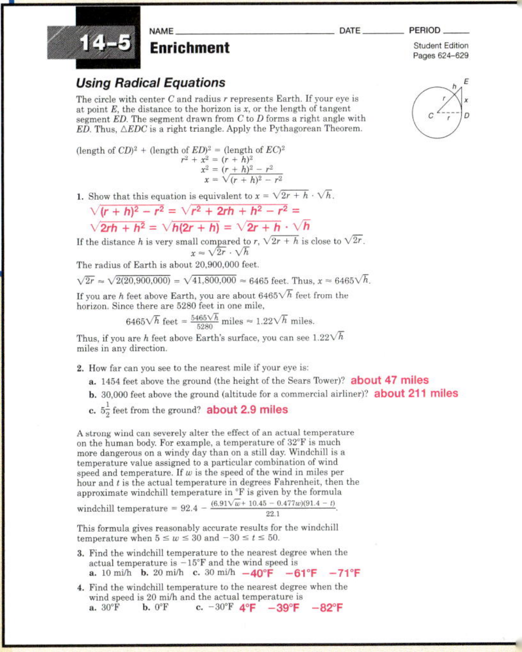

NAME ______ DATE ______ PERIOD ______

14–5 Enrichment

Student Edition Pages 624–629

Using Radical Equations

The circle with center C and radius r represents Earth. If your eye is at point E, the distance to the horizon is x, or the length of tangent segment ED. The segment drawn from C to D forms a right angle with ED. Thus, $\triangle EDC$ is a right triangle. Apply the Pythagorean Theorem.

(length of CD)^2 + (length of ED)^2 = (length of EC)^2

$$r^2 + x^2 = (r + h)^2$$
$$x^2 = (r + h)^2 - r^2$$
$$x = \sqrt{(r + h)^2 - r^2}$$

1. Show that this equation is equivalent to $x = \sqrt{2r + h} \cdot \sqrt{h}$.
$\sqrt{(r + h)^2 - r^2} = \sqrt{r^2 + 2rh + h^2 - r^2} = \sqrt{2rh + h^2} = \sqrt{h(2r + h)} = \sqrt{2r + h} \cdot \sqrt{h}$

If the distance h is very small compared to r, $\sqrt{2r + h}$ is close to $\sqrt{2r}$.

$$x \approx \sqrt{2r} \cdot \sqrt{h}$$

The radius of Earth is about 20,900,000 feet.

$\sqrt{2r} \approx \sqrt{2(20{,}900{,}000)} = \sqrt{41{,}800{,}000} \approx 6465$ feet. Thus, $x \approx 6465\sqrt{h}$.

If you are h feet above Earth, you are about $6465\sqrt{h}$ feet from the horizon. Since there are 5280 feet in one mile,

$$6465\sqrt{h} \text{ feet} = \frac{5465\sqrt{h}}{5280} \text{ miles} \approx 1.22\sqrt{h} \text{ miles.}$$

Thus, if you are h feet above Earth's surface, you can see $1.22\sqrt{h}$ miles in any direction.

2. How far can you see to the nearest mile if your eye is:
   a. 1454 feet above the ground (the height of the Sears Tower)? **about 47 miles**
   b. 30,000 feet above the ground (altitude for a commercial airliner)? **about 211 miles**
   c. $5\frac{1}{2}$ feet from the ground? **about 2.9 miles**

A strong wind can severely alter the effect of an actual temperature on the human body. For example, a temperature of 32°F is much more dangerous on a windy day than on a still day. Windchill is a temperature value assigned to a particular combination of wind speed and temperature. If w is the speed of the wind in miles per hour and t is the actual temperature in degrees Fahrenheit, then the approximate windchill temperature in °F is given by the formula

$$\text{windchill temperature} = 92.4 - \frac{(6.91\sqrt{w} + 10.45 - 0.477w)(91.4 - t)}{22.1}.$$

This formula gives reasonably accurate results for the windchill temperature when $5 \leq w \leq 30$ and $-30 \leq t \leq 50$.

3. Find the windchill temperature to the nearest degree when the actual temperature is −15°F and the wind speed is
   a. 10 mi/h b. 20 mi/h c. 30 mi/h **−40°F −61°F −71°F**
4. Find the windchill temperature to the nearest degree when the wind speed is 20 mi/h and the actual temperature is
   a. 30°F b. 0°F c. −30°F **4°F −39°F −82°F**

Study Guide and Assessment

Understanding and Using the Vocabulary

This section provides a listing of the new terms, properties, and phrases that were introduced in this chapter. The exercises check students' understanding of the terms by using a variety of verbal formats including matching, completion, and true/false.

Glossary A complete glossary of terms appears on pages 762–783.

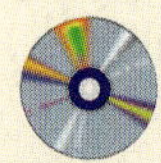

MindJogger Videoquizzes

MindJogger Videoquizzes provide an alternative review of concepts presented in this chapter. Students work in teams to answer questions, gaining points for correct answers.

Answers

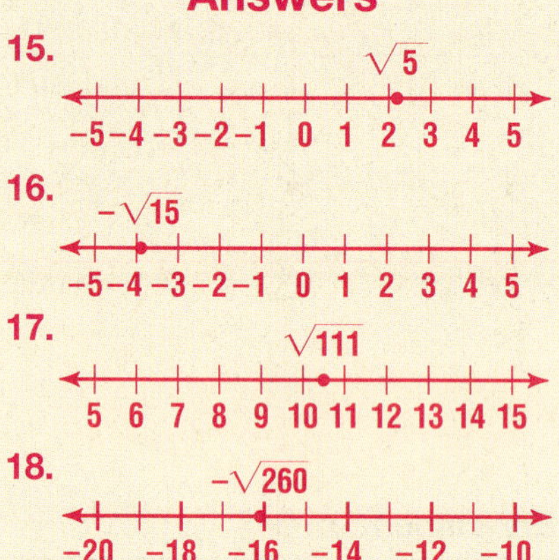

CHAPTER 14 Study Guide and Assessment

Understanding and Using the Vocabulary

After completing this chapter, you should be able to define each term, property, or phrase and give an example or two of each.

interNET CONNECTION Review Activities For more review activities, visit: www.algconcepts.com

conjugates *(p. 616)*
Distance Formula *(p. 607)*
radical equations *(p. 624)*
real numbers *(p. 600)*
rationalizing the denominator *(p. 615)*

Choose the correct term to complete each sentence.

1. To find the distance d between any two points (x_1, y_1) and (x_2, y_2), use the formula $\left(\sqrt{(x_2 - x_1)^2 - (y_2 - y_1)^2}, \underline{\sqrt{(x_2 - x_1)^2 + (y_2 - y_1)^2}}\right)$.
2. The product of conjugates is (sometimes, <u>always</u>) a rational number.
3. Rationalizing the denominator is a way of eliminating (<u>radicals</u>, perfect squares) from the denominator of a fraction.
4. Natural numbers, whole numbers, and integers are all (irrational, <u>rational</u>) numbers.
5. The expression ($\underline{\sqrt{7ab}}$, $\sqrt{8mn}$) is in simplest form.
6. Radical equations (<u>sometimes</u>, always) have more than one solution.
7. The number 0 is a (<u>whole</u>, natural) number.
8. Binomials of the form $a\sqrt{b} + c\sqrt{d}$ and $a\sqrt{b} - c\sqrt{d}$ are (<u>conjugates</u>, radical equations).
9. The equation $\sqrt{x} + 4 = 0$ has (one, <u>no</u>) real solution.
10. The set of (integers, <u>real numbers</u>) contains the sets of rational and irrational numbers.

Skills and Concepts

Objectives

- **Lesson 14–1** Describe the relationships among sets of numbers.

$\sqrt{16} = 4$; so $\sqrt{16}$ is a natural number, a whole number, an integer, and a rational number.

The number −3 is an integer and a rational number.

2.151151115 . . . is not the square root of a perfect square. So, it is an irrational number.

Examples

Name the set or sets of numbers to which each real number belongs. Let N = natural numbers, W = whole numbers, Z = integers, Q = rational numbers, and I = irrational numbers.

11. $-\frac{1}{4}$ **Q**
12. π **I**
13. $-\sqrt{25}$ **Z, Q**
14. $\frac{24}{6}$ **N, W, Z, Q**

15–18. See margin for graphs.

Find an approximation, to the nearest tenth, for each square root. Then graph the square root on a number line.

15. $\sqrt{5}$ **2.2**
16. $-\sqrt{15}$ **−3.9**
17. $\sqrt{111}$ **10.5**
18. $-\sqrt{260}$ **−16.1**

www.algconcepts.com/vocabulary_review

Resource Manager

Reproducible Masters
Chapter 14 Resource Masters
- *Assessment*, pp. 649–657, 660–662

Technology/Multimedia
MindJogger Videoquizzes
ExamView® Pro

Objectives and Examples

• **Lesson 14–2** Find the distance between two points in the coordinate plane.

Find the distance between points $A(3, -2)$ and $B(5, 8)$.

$$\begin{aligned} d &= \sqrt{(x_2 - x_1)^2 + (y_2 - y_1)^2} \\ &= \sqrt{(5 - 3)^2 + (8 - (-2))^2} \\ &= \sqrt{(2)^2 + (10)^2} \\ &= \sqrt{4 + 100} \\ &= \sqrt{104} \\ &\approx 10.2 \end{aligned}$$

Review Exercises

Find the distance between each pair of points. Round to the nearest tenth, if necessary.

19. $P(1, 2), Q(4, 6)$ **5**
20. $J(-6, -3), K(1, 0)$ **$\sqrt{58}$ or 7.6**
21. $X(4, -1), Y(-2, 5)$ **$\sqrt{72}$ or 8.5**
22. $R(-9, -5), S(-2, 19)$ **25**

Find the value of *a* if the points are the indicated distance apart.

23. $G(a, 2), H(-3, 5); d = \sqrt{13}$ **−1 or −5**
24. $C(4, -7), D(-6, a); d = \sqrt{116}$ **−11 or −3**

• **Lesson 14–3** Simplify radical expressions.

$$\begin{aligned} \sqrt{288} &= \sqrt{12^2 \cdot 2} \\ &= 12\sqrt{2} \end{aligned}$$

$$\begin{aligned} \sqrt{32x^3y^2} &= \sqrt{32x^3y^2} \\ &= \sqrt{2 \cdot 2 \cdot 2 \cdot 2 \cdot 2 \cdot x^3 \cdot y^2} \\ &= \sqrt{2 \cdot 16 \cdot x^3 \cdot y^2} \\ &= \sqrt{2} \cdot \sqrt{16} \cdot \sqrt{x^3} \cdot \sqrt{y^2} \\ &= \sqrt{2} \cdot 4 \cdot x \cdot \sqrt{x} \cdot |y| \\ &= 4x|y|\sqrt{2x} \end{aligned}$$

Simplify each expression. Leave in radical form. **29–30. See margin.**

25. $\sqrt{50}$ **$5\sqrt{2}$**
26. $\sqrt{3} \cdot \sqrt{6}$ **$3\sqrt{2}$**
27. $\dfrac{\sqrt{54}}{\sqrt{2}}$ **$3\sqrt{3}$**
28. $\dfrac{\sqrt{7}}{\sqrt{13}}$ **$\dfrac{\sqrt{91}}{13}$**
29. $\dfrac{1}{4 - \sqrt{5}}$
30. $\dfrac{3}{13 - \sqrt{2}}$

Simplify each expression. Use absolute value symbols if necessary.

31. $\sqrt{300a^2bc^4}$ **$10|a|c^2\sqrt{3b}$**
32. $\sqrt{121m^5n^4p^3}$ **$11m^2n^2|p|\sqrt{mp}$**

• **Lesson 14–4** Add and subtract radical expressions.

$$\begin{aligned} 8\sqrt{7} - 6\sqrt{7} &= (8 - 6)\sqrt{7} \\ &= 2\sqrt{7} \end{aligned}$$

$$\begin{aligned} 13\sqrt{2} + 4\sqrt{2} - 7\sqrt{2} &= (13 + 4 - 7)\sqrt{2} \\ &= 10\sqrt{2} \end{aligned}$$

$$\begin{aligned} \sqrt{54} + \sqrt{96} &= \sqrt{3^2 \cdot 6} + \sqrt{4^2 \cdot 6} \\ &= 3\sqrt{6} + 4\sqrt{6} \\ &= (3 + 4)\sqrt{6} \text{ or } 7\sqrt{6} \end{aligned}$$

Simplify each expression.

33. $2\sqrt{5} + 5\sqrt{5}$ **$7\sqrt{5}$**
34. $3\sqrt{8} - 6\sqrt{8}$ **$-6\sqrt{2}$**
35. $-5\sqrt{6} + 14\sqrt{6} - 9\sqrt{6}$ **0**
36. $12\sqrt{3} - 14\sqrt{3} + 6\sqrt{3}$ **$4\sqrt{3}$**
37. $7\sqrt{2} + 5\sqrt{18}$ **$22\sqrt{2}$**
38. $-5\sqrt{40} + 10\sqrt{10}$ **0**
39. Simplify $2\sqrt{45} - 5\sqrt{80}$. **$-14\sqrt{5}$**
40. Write $2\sqrt{45} + 3\sqrt{75} - \sqrt{50}$ in simplest form. **$6\sqrt{5} + 15\sqrt{3} - 5\sqrt{2}$**

Skills and Concepts

The **Objectives and Examples** section reviews the skills and concepts of the chapter and shows completely worked examples.

The **Review Exercises** provide practice for the corresponding objectives.

Answers

29. $\dfrac{4 + \sqrt{5}}{11}$

30. $\dfrac{39 + 3\sqrt{2}}{167}$

ExamView® Pro

Use the networkable ***ExamView® Pro Testmaker CD-ROM*** to:

- Create **multiple versions** of tests.
- Create **modified** tests for **inclusion** students with one mouse click.
- **Edit** existing questions and **add** your own questions.
- Build tests aligned with state standards using built-in **state curriculum correlations**.
- Change **English** tests to **Spanish** with one mouse click and vice versa.

Mixed Problem Solving
See pages 724–731.

Applications and Problem Solving

This section provides additional practice in solving real-world problems that involve the concepts of this chapter.

Answer

52a.

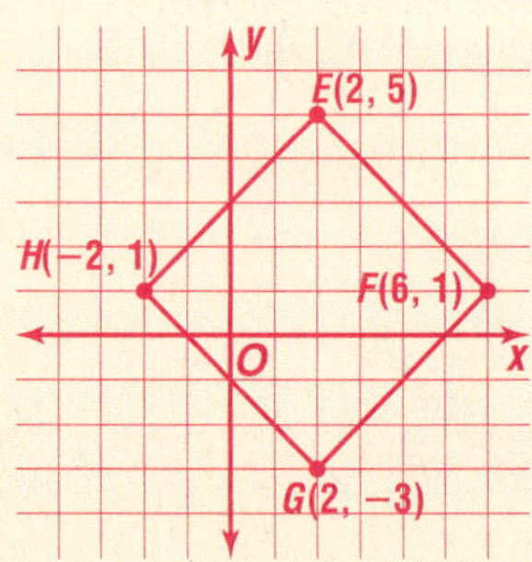

Assessment, pp. 651–652

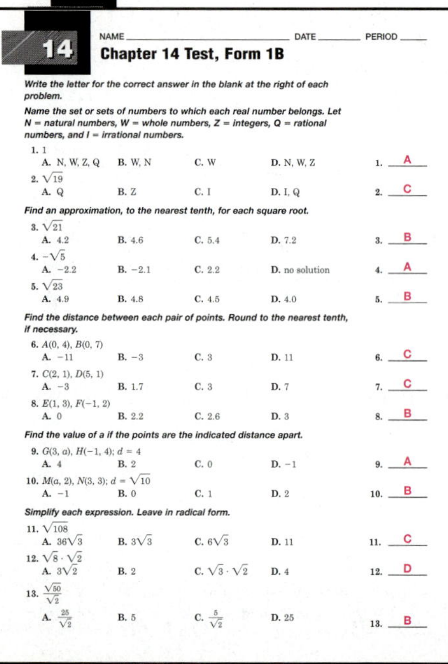

14

NAME ______ DATE ______ PERIOD ______

Chapter 14 Test, Form 1B

Write the letter for the correct answer in the blank at the right of each problem.

Name the set or sets of numbers to which each real number belongs. Let N = natural numbers, W = whole numbers, Z = integers, Q = rational numbers, and I = irrational numbers.

1. 1
   A. N, W, Z, Q B. W, N C. W D. N, W, Z — 1. A
2. $\sqrt{19}$
   A. Q B. Z C. I D. I, Q — 2. C

Find an approximation, to the nearest tenth, for each square root.

3. $\sqrt{21}$
   A. 4.2 B. 4.6 C. 5.4 D. 7.2 — 3. B
4. $-\sqrt{5}$
   A. −2.2 B. −2.1 C. 2.2 D. no solution — 4. A
5. $\sqrt{23}$
   A. 4.9 B. 4.8 C. 4.5 D. 4.0 — 5. B

Find the distance between each pair of points. Round to the nearest tenth, if necessary.

6. $A(0, 4)$, $B(0, 7)$
   A. −11 B. −3 C. 3 D. 11 — 6. C
7. $C(2, 1)$, $D(5, 1)$
   A. −3 B. 1.7 C. 3 D. 7 — 7. C
8. $E(1, 3)$, $F(-1, 2)$
   A. 0 B. 2.2 C. 2.6 D. 3 — 8. B

Find the value of a if the points are the indicated distance apart.

9. $G(3, a)$, $H(-1, 4)$; $d = 4$
   A. 4 B. 2 C. 0 D. −1 — 9. A
10. $M(a, 2)$, $N(3, 3)$; $d = \sqrt{10}$
   A. −1 B. 0 C. 1 D. 2 — 10. B

Simplify each expression. Leave in radical form.

11. $\sqrt{108}$
   A. $36\sqrt{3}$ B. $3\sqrt{3}$ C. $6\sqrt{3}$ D. 11 — 11. C
12. $\sqrt{8} \cdot \sqrt{2}$
   A. $3\sqrt{2}$ B. 2 C. $\sqrt{3} \cdot \sqrt{2}$ D. 4 — 12. D
13. $\frac{\sqrt{50}}{\sqrt{2}}$
   A. $\frac{25}{\sqrt{2}}$ B. 5 C. $\frac{5}{\sqrt{2}}$ D. 25 — 13. B

Objectives and Examples

- **Lesson 14–5** Solve simple radical equations in which only one radical contains a variable.

Solve $\sqrt{x + 4} - 8 = -5$.

$$\begin{aligned} \sqrt{x + 4} - 8 &= -5 \\ \sqrt{x + 4} - 8 + 8 &= -5 + 8 \\ \sqrt{x + 4} &= 3 \\ (\sqrt{x + 4})^2 &= 3^2 \\ x + 4 &= 9 \\ x + 4 - 4 &= 9 - 4 \\ x &= 5 \end{aligned}$$

The solution is 5.

Review Exercises

Solve each equation. Check your solution.

41. $\sqrt{y} = 5$ **25**

42. $\sqrt{m} = -3$

43. $\sqrt{9h} = 9$ **9**

44. $\sqrt{-4a} = 8$ **−16**

45. $7 = \sqrt{a} + 10$

46. $\sqrt{n - 12} = 15$ **237**

47. $t - 1 = \sqrt{3t + 7}$ **6**

48. $\sqrt{4x - 3} = x$ **3 and 1**

49. Solve $\sqrt{\frac{g}{6}} = 5$. **150**

50. Find the solution of $\sqrt{\frac{3x}{4}} + 2 = 5$. **12**

42. no solution **45. no solution**

Applications and Problem Solving

51. Weather The graph shows record low temperatures. Name the set or sets of numbers to which the temperatures belong. *(Lesson 14–1)* **integers, rational numbers**

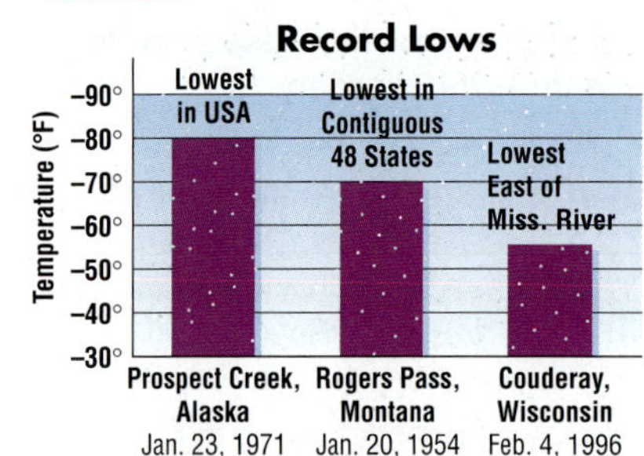

Source: National Oceanic and Atmospheric Administration

52. Geometry Quadrilateral *EFGH* is a square with vertices $E(2, 5)$, $F(6, 1)$, $G(2, -3)$, and $H(-2, 1)$. *(Lesson 14–2)*

a. Draw quadrilateral *EFGH* on a coordinate plane. **See margin.**

b. Determine the perimeter of the figure. Round to the nearest tenth. **22.6 units**

53. Gems The weight of the largest pearl ever found in a giant clam is about $\sqrt{200}$ pounds. Express the weight in simplest radical form. *(Lesson 14–3)* **53. $10\sqrt{2}$ lb**

54. Geometry Triangle *XYZ* has a perimeter of $28\sqrt{5}$ inches. What is the measure of side *YZ*? *(Lesson 14–4)* **$12\sqrt{5}$ in.**

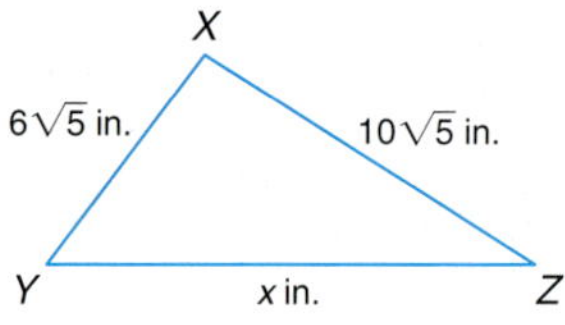

55. Nature The equation $S = 3\sqrt{d}$ gives the speed *S* of a tidal wave in meters per second if *d* is the depth of the water in meters. *(Lesson 14–5)*

a. Suppose a tidal wave is traveling 186 meters per second. What is the depth of the water? **3844 m**

b. What is the depth of the water if a tidal wave is traveling 153 meters per second? **2601 m**

Assessment and Evaluation

Four forms of Chapter 14 Test are available in the *Chapter 14 Resource Masters.*

Chapter 14 Test, Form 1B, is shown at the left. Chapter 14 Test, Form 2B, is shown on the next page.

	Form of Test		Level
1A	Multiple Choice	pp. 649–650	Average
1B	Multiple Choice	pp. 651–652	Basic
2A	Free Response	pp. 653–654	Average
2B	Free Response	pp. 655-656	Basic

CHAPTER 14 Test

1. **Name** two examples of each of the following. 1a–c. See margin.
   a. integer b. irrational number c. rational number
2. **Write** a pair of conjugates. Sample answer: $3 + \sqrt{5}$ and $3 - \sqrt{5}$

Name the set or sets of numbers to which each real number belongs. Let N = natural numbers, W = whole numbers, Z = integers, Q = rational numbers, and I = irrational numbers.

3. $\frac{5}{3}$ Q
4. 0 W, Z, Q
5. $-0.45275563\ldots$ I
6. $-\sqrt{81}$ Z, Q

7. Graph $\sqrt{5}$, $-\sqrt{3}$, and 4.5 on a number line. See margin.

8. Write 3.22, -3.22, $\sqrt{11}$, and $3.\overline{2}$ in order from least to greatest. $-3.22, 3.22, 3.\overline{2}, \sqrt{11}$

Find the distance between each pair of points. Round to the nearest tenth, when necessary. 11. $\sqrt{40}$ or 6.3

9. $G(7, 4), H(-2, 4)$ 9
10. $X(3, -3), Y(-9, 2)$ 13
11. $M(1, -1), N(-5, 1)$

12. Suppose $A(5, m)$ and $B(8, 1)$ are 5 units apart. What is the value of m? 5 or -3

Simplify each expression. Leave in radical form.

13. $\sqrt{80}$ $4\sqrt{5}$
14. $\sqrt{3} \cdot \sqrt{27}$ 9
15. $4\sqrt{3} \cdot \sqrt{6}$ $12\sqrt{2}$
16. $\frac{\sqrt{96}}{\sqrt{8}}$ $2\sqrt{3}$
17. $\frac{\sqrt{7}}{\sqrt{11}}$ $\frac{\sqrt{77}}{11}$
18. $\frac{7}{7 + \sqrt{5}}$ $\frac{49 - 7\sqrt{5}}{44}$
19. $-4\sqrt{7} + 6\sqrt{7} - 12\sqrt{7}$ $-10\sqrt{7}$
20. $5\sqrt{18} - 2\sqrt{50}$ $5\sqrt{2}$

Simplify each expression. Use absolute value symbols if necessary.

21. $\sqrt{25a^3b^2}$ $5a|b|\sqrt{a}$
22. $\sqrt{56m^4np}$ $2m^2\sqrt{14np}$

Solve each equation. Check your solution.

23. $\sqrt{b} + 8 = 5$ no solution
24. $\sqrt{4x - 3} = 6 - x$ 3

25. **Science** The distance d in miles a person can see on any planet is given by the formula $d = \sqrt{\frac{rh}{2640}}$, where r is the radius of the planet in miles and h is the height of the person in feet. Suppose a person 6 feet tall is standing on Mars. If the radius of the planet is 2109 miles, how far can the person see? Round to the nearest tenth. 2.2 mi

www.algconcepts.com/chapter_test

Chapter Test Bonus Question

Point $C(4, n)$ is the same distance from point $A(2, 4)$ as it is from point $B(8, 6)$. Find the value of n and the length of each side of triangle ABC.

$n = 8$; $AB = 2\sqrt{10}$ or about 6.3 units, $BC = AC = 2\sqrt{5}$ or about 4.5 units

Chapter Test

Answers

1a. Sample answer: -3 and 6

1b. Sample answer: 5.2323323332... and π

1c. Sample answer: 0.25 and $\frac{1}{2}$

7. Number line from -3 to 7 with points graphed at $-\sqrt{3}$, $\sqrt{5}$, and 4.5.

Assessment, pp. 655–656

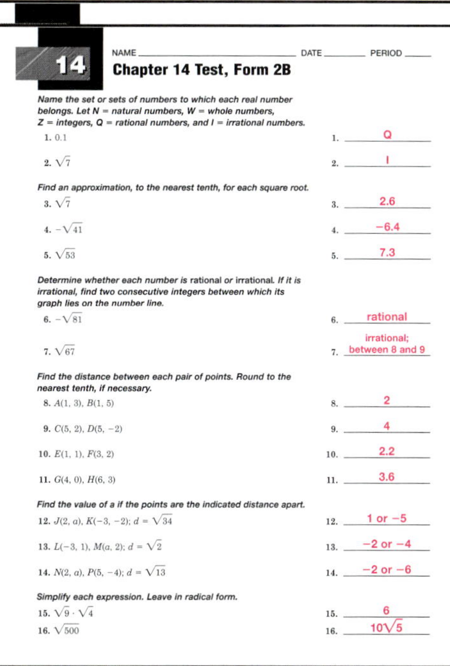

NAME ______ DATE ______ PERIOD ______

14 Chapter 14 Test, Form 2B

Name the set or sets of numbers to which each real number belongs. Let N = natural numbers, W = whole numbers, Z = integers, Q = rational numbers, and I = irrational numbers.

1. 0.1 — 1. Q
2. $\sqrt{7}$ — 2. I

Find an approximation, to the nearest tenth, for each square root.

3. $\sqrt{7}$ — 3. 2.6
4. $-\sqrt{41}$ — 4. -6.4
5. $\sqrt{53}$ — 5. 7.3

Determine whether each number is rational or irrational. If it is irrational, find two consecutive integers between which its graph lies on the number line.

6. $-\sqrt{81}$ — 6. rational
7. $\sqrt{67}$ — 7. irrational; between 8 and 9

Find the distance between each pair of points. Round to the nearest tenth, if necessary.

8. $A(1, 3), B(1, 5)$ — 8. 2
9. $C(5, 2), D(5, -2)$ — 9. 4
10. $E(1, 1), F(3, 2)$ — 10. 2.2
11. $G(4, 0), H(6, 3)$ — 11. 3.6

Find the value of a if the points are the indicated distance apart.

12. $J(2, a), K(-3, -2); d = \sqrt{34}$ — 12. 1 or -5
13. $L(-3, 1), M(a, 2); d = \sqrt{2}$ — 13. -2 or -4
14. $N(2, a), P(5, -4); d = \sqrt{13}$ — 14. -2 or -6

Simplify each expression. Leave in radical form.

15. $\sqrt{9} \cdot \sqrt{4}$ — 15. 6
16. $\sqrt{500}$ — 16. $10\sqrt{5}$

Pages 634–635 are part of a complete test preparation course that is described in detail on page T9 of the Teacher's Handbook. The test items on these pages were written in the same style as those in state proficiency tests and standardized tests like ACT and SAT.

Diagnosis and Prescription

Each of the 10 test questions on page 635 is cross-referenced to the lesson where that SAT or ACT skill is covered. If students miss a particular type of problem, you can have them study that skill.

(See chart at the bottom of page 635. Note that SPT = State Proficiency Test, SAT = Scholastic Assessment Test, and ACT = American College Test.)

Assessment, p. 660

14 NAME ____________ DATE ______ PERIOD ______

Chapter 14 Cumulative Review

1. Name the property shown by the statement below. *(Lesson 1–3)*
   $7 + x + 13 = 13 + 7 + x$
   1. Commutative Property of Addition
2. Name the quadrant in which point $(-2, 5)$ is located. *(Lesson 2–2)*
   2. quadrant II
3. Find the solution of $x - \frac{1}{3} = \frac{1}{3}$ if the replacement set is $x = \left\{0, \frac{1}{3}, \frac{2}{3}, 1\right\}$. *(Lesson 3–4)*
   3. $\frac{2}{3}$
4. Find $5\frac{1}{2} \div \frac{1}{2}$. *(Lesson 4–3)*
   4. 11
5. Solve $\frac{3}{8} = \frac{y}{56}$. *(Lesson 5–1)*
   5. 21
6. If there are 16 ounces in one pound, how many ounces are in $3\frac{1}{4}$ pounds? *(Lesson 6–5)*
   6. 52
7. Write an equation in point-slope form of the line with slope -2 and y-intercept 3. *(Lesson 7–2)*
   7. $y - 3 = -2x$
8. Simplify $\frac{16x^6y^2}{8x^2y^4}$. *(Lesson 8–3)*
   8. $\frac{2x^4}{y^2}$
9. Find $(8c + 18d) + (4c - 8d)$. *(Lesson 9–2)*
   9. $12c + 10d$
10. Find the greatest common factor of $8x^2y^3$, $12x^3y^6$, and $36x^4y^9$. *(Lesson 10–1)*
    10. $4x^2y^3$
11. Solve $7g^2 - 7 = 0$. Check your solution. *(Lesson 11–4)*
    11. $-1, 1$
12. Solve $\frac{x}{2} \leq 12$. Check your solution. *(Lesson 12–3)*
    12. $\{x \mid x \leq 24\}$
13. What is the solution of the system $y = \frac{1}{2}x + 3$ and $2x = y$? *(Lesson 13–3)*
    13. $(2, 4)$
14. Solve $\sqrt{\frac{500}{y}} = 10$. *(Lesson 14–5)*
    14. 5

CHAPTER 14 Preparing for Standardized Tests

Systems of Equations Problems

On standardized tests, you will often need to translate problems into systems of equations and inequalities. You will also need to solve systems of equations using substitution or addition methods.

Test-Taking Tip

To solve systems of equations on standardized tests, try adding them first. This technique may work because the systems often are of the form $x + y = 6$ and $x - y = 4$.

Example 1

Hector has 20 coins. They are all quarters and nickels. The total value is \$2.20. Which system of equations can be used to determine the number of quarters q and the number of nickels n?

Ⓐ $q + n = 20$
$0.30qn = 2.20$

Ⓑ $q + n = 20$
$0.25q + 0.05n = 2.20$

Ⓒ $q + n = 20$
$0.05q + 0.25n = 2.20$

Ⓓ $q + n = 20$
$0.25q + 0.05n = 20$

Hint First write the equations in words and then translate them into symbols.

Solution The number of quarters plus the number of nickels is 20.

$$q + n = 20$$

The value of the quarters plus the value of the nickels equals the total value. The value of the quarters is 25 cents times the number of quarters or $0.25q$. The value of the nickels is 5 cents times the number of nickels or $0.05n$.

$$0.25q + 0.05n = 2.20$$

The answer is B.

Example 2

If $2x + 3y = 20$ and $3x + 2y = 40$, what is the value of $x + y$?

Hint Look for ways to solve a problem without lengthy calculations.

Solution Read carefully. You must find the value of $x + y$, not the individual values of x and y.

Add the two equations.

$$\begin{aligned} 2x + 3y &= 20 \\ +\ 3x + 2y &= 40 \\ \hline 5x + 5y &= 60 \end{aligned}$$

Notice that the coefficients of x and y are both 5. Divide each side of the equation by 5. Simplify each term.

$$\frac{5x + 5y}{5} = \frac{60}{5}$$

$$\frac{\overset{1}{\cancel{5}}(x + y)}{\underset{1}{\cancel{5}}} = \frac{\overset{12}{\cancel{60}}}{\underset{1}{\cancel{5}}}$$

$$x + y = 12$$

The value of $x + y$ is 12. The answer is 12.

Resource Manager

Reproducible Masters

Chapter 14 Resource Masters

- *Assessment*, pp. 660–662

Preparing for Standardized Tests
For test-taking strategies and more practice, see pages 732–749.

After you work each problem, record your answer on the answer sheet provided or on a sheet of paper.

Multiple Choice

1. Jared's total SAT score was 1340. His math score m was 400 points less than twice his verbal score v. Which system of equations can determine his scores? *(Lesson 13–1)* **A**

   (A) $m + v = 1340$, $m = 2v - 400$ (B) $m + v = 1340$, $m = 400 - 2v$

   (C) $m + v = 1340$, $400m = 2v$ (D) $m - v = 1340$, $m = 2v - 400$

2. The flag for Monroe High School has an area of 120 square feet. A new flag will have a width $1\frac{1}{2}$ times the width of the old flag. The length is the same. What is the area, in square feet, of the new flag? *(Lesson 13–2)* **B**

   (A) 160 (B) 180 (C) 240 (D) 360

3. Where do the lines with equations $y = 2x - 2$ and $7x - 3y = 11$ intersect? *(Lesson 13–3)* **A**

   (A) (5, 8) (B) (8, 5) (C) $\left(\frac{5}{8}, -1\right)$

   (D) $\left(\frac{5}{8}, 1\right)$ (E) $\left(\frac{25}{16}, \frac{9}{8}\right)$

4. $\triangle LMN \sim \triangle PQR$. What is PQ? *(Lesson 5–1)* **D**

   L, 7 in., M, 5 in., N; P, x in., Q, 8 in., R

   (A) 4.4 in.
   (B) 7 in.
   (C) 10 in.
   (D) 11.2 in.

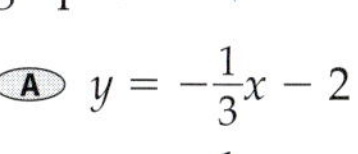

5. Which equation represents the line graphed? *(Lesson 7–5)* **A**

   (A) $y = -\frac{1}{3}x - 2$

   (B) $y = -\frac{1}{3}x + 2$

   (C) $y = \frac{1}{3}x - 2$ (D) $y = \frac{1}{3}x + 2$

6. Gina has x marbles and Dawn has y marbles. Together they have q marbles. If Gina gives Dawn 3 of her marbles, then they will have an equal number. How many marbles does Dawn have? *(Lesson 1–1)* **A**

   (A) $\frac{q-6}{2}$ (B) $\frac{q-3}{2}$ (C) $\frac{q-6}{2}$

   (D) $q - 3$ (E) $q + 3$

7. What is the slope of a line parallel to the line graphed? *(Lesson 7–7)* **C**

   (A) -3 (B) $-\frac{1}{3}$

   (C) $\frac{1}{3}$ (D) 3

8. What value of n makes the equation true? *(Lesson 8–1)* **C**

   $$9^n - 8^n = 1^n$$

   (A) -1 (B) 0 (C) 1 (D) 2

Grid In

9. $\angle ABC$ is a right angle, $\overrightarrow{BC}$ bisects $\angle EBD$. If $m\angle ABD$ is 130, then what is x? *(Lesson 3–6)* **50**

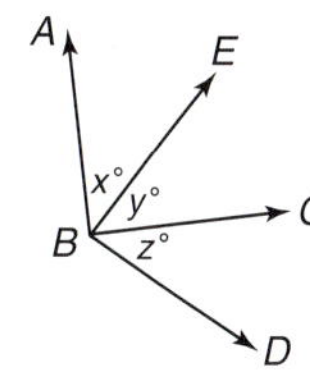

Extended Response

10. Talent show tickets are \$3 for adults and \$2 for students. Ticket sales of at least \$600 are needed to cover costs. At least 5 times as many students as adults attend. *(Lesson 13–7)* **See margin.**

    Part A Write a system of inequalities to find the number of adult and student tickets that need to be sold. Let x represent the number of adult tickets and let y represent the number of student tickets.

    Part B Graph the system of inequalities. Give one example of adult and student ticket sales that would solve the system.

www.algconcepts.com/standardized_test

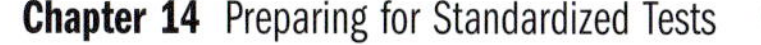

A bubble-in answer sheet for these practice problems is available on page A1 of the *Chapter 14 Resource Masters.*

Additional Practice

Additional test practice questions are available in the *Chapter 14 Resource Masters*, pp. 661–662.

Answers

10A. $3x + 2y \geq 600$
$y \geq 5x$

10B.

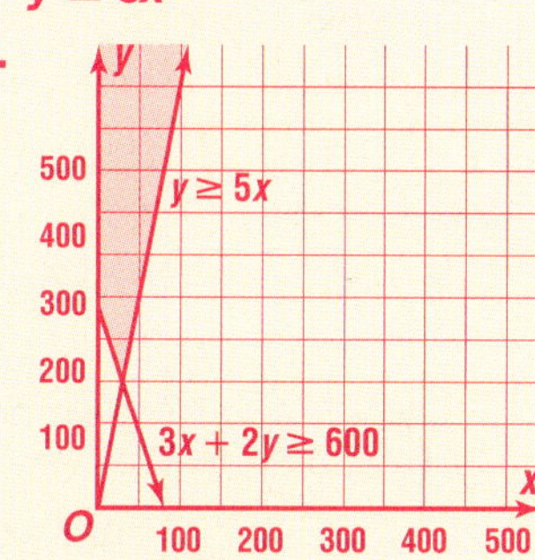

One solution is 50 adults and 300 students.

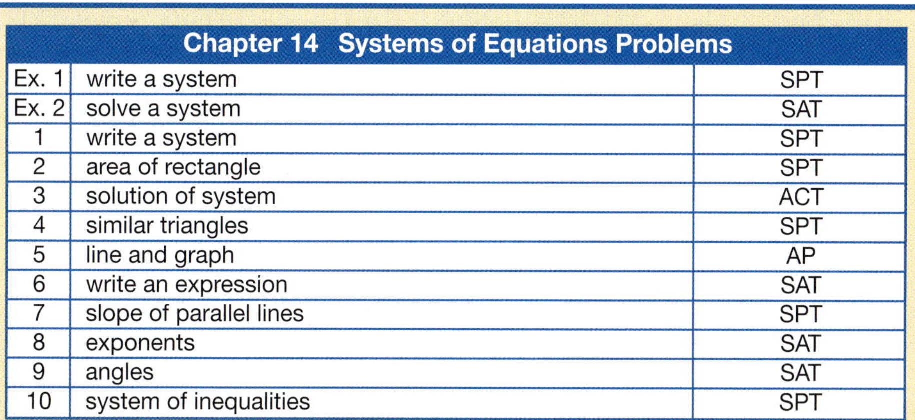

Chapter 14 Systems of Equations Problems		
Ex. 1	write a system	SPT
Ex. 2	solve a system	SAT
1	write a system	SPT
2	area of rectangle	SPT
3	solution of system	ACT
4	similar triangles	SPT
5	line and graph	AP
6	write an expression	SAT
7	slope of parallel lines	SPT
8	exponents	SAT
9	angles	SAT
10	system of inequalities	SPT

***Assessment*, pp. 661–662**

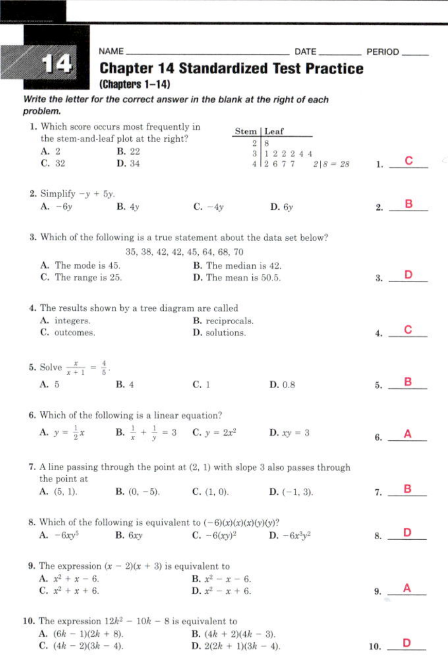

14 NAME ______ DATE ______ PERIOD ______

Chapter 14 Standardized Test Practice
(Chapters 1–14)

Write the letter for the correct answer in the blank at the right of each problem.

1. Which score occurs most frequently in the stem-and-leaf plot at the right?
   A. 2 B. 22 C. 32 D. 34 — 1. C

Stem	Leaf
2	8
3	1 2 2 2 4 4
4	2 6 7 7

$2|8 = 28$

2. Simplify $-y + 5y$.
   A. $-6y$ B. $4y$ C. $-4y$ D. $6y$ — 2. B

3. Which of the following is a true statement about the data set below?
   35, 38, 42, 42, 45, 64, 68, 70
   A. The mode is 45. B. The median is 42. C. The range is 25. D. The mean is 50.5. — 3. D

4. The results shown by a tree diagram are called
   A. integers. B. reciprocals. C. outcomes. D. solutions. — 4. C

5. Solve $\frac{x}{x+1} = \frac{4}{5}$.
   A. 5 B. 4 C. 1 D. 0.8 — 5. B

6. Which of the following is a linear equation?
   A. $y = \frac{1}{2}x$ B. $\frac{1}{x} + \frac{1}{y} = 3$ C. $y = 2x^2$ D. $xy = 3$ — 6. A

7. A line passing through the point at (2, 1) with slope 3 also passes through the point at
   A. (5, 1). B. (0, −5). C. (1, 0). D. (−1, 3). — 7. B

8. Which of the following is equivalent to $(-6)(x)(x)(x)(y)(y)$?
   A. $-6xy^5$ B. $6xy$ C. $-6(xy)^2$ D. $-6x^3y^2$ — 8. D

9. The expression $(x - 2)(x + 3)$ is equivalent to
   A. $x^2 + x - 6$. B. $x^2 - x - 6$. C. $x^2 + x + 6$. D. $x^2 - x + 6$. — 9. A

10. The expression $12k^2 - 10k - 8$ is equivalent to
    A. $(6k - 1)(2k + 8)$. B. $(4k + 2)(4k - 3)$. C. $(4k - 2)(3k - 4)$. D. $2(2k + 1)(3k - 4)$. — 10. D

CHAPTER 15

Resource Manager

Rational Expressions and Equations

Instructional Objectives

Lesson (pages)	Objectives	NCTM Standards 2000	State/Local Objectives
15–1 (638–643)	Simplify rational expressions.	1, 6, 8, 9	
15–2 (644–649)	Multiply and divide rational expressions.	1, 6, 8, 9	
15–3 (650–655)	Divide polynomials by binomials.	1, 6, 8, 9	
15–4 (656–661)	Add and subtract rational expressions with like denominators.	1, 6, 8, 9	
15–5 (662–667)	Add and subtract rational expressions with unlike denominators.	1, 6, 8, 9	
15–6 (668–673)	Solve rational equations.	1, 6, 8, 9, 10	
Investigation (674–675)	Explore work problems.	1, 2, 4, 6, 8, 9 10	

Key to NCTM Standards 2000
1 Number & Operations; **2** Algebra; **3** Geometry; **4** Measurement; **5** Data Analysis & Probability; **6** Problem Solving; **7** Reasoning & Proof; **8** Communications; **9** Connections; **10** Representation

Suggested Pacing *See page T13 for a complete course-planning calendar.*

Standard refers to schedules that provide 45- to 55-minute periods that meet each day.
Block refers to schedules that provide approximately 90-minute periods which may meet every day for one semester or every other day over two semesters.

PACING	DAY 1	DAY 2	DAY 3	DAY 4	DAY 5	DAY 6
Standard Core (Chapters 1–13)						
Standard Enhanced (Chapters 1–15)	Lesson 15–1	Lesson 15–2	Lesson 15–3	Lesson 15–4	Lesson 15–5	
Block Core (Chapters 1–13)						
Block Enhanced (Chapters 1–15)	Chapter 14 Test & Lesson 15–1	Lessons 15–2 & 15–3	Lessons 15–4 & 15–5	Lesson 5–6 & INV	SG+A & Chapter Test	

Instructional Resources

Lesson	Materials and Manipulatives (see below for Glencoe Manipulative Resources)	Blackline Masters (page numbers): Chapter 15 Resource Masters: Study Guide	Practice (Skills & Average)	Reading to Learn Mathematics	Enrichment	Assessment	Hands-On Algebra*	School-to-Workplace*	Graphing Calculator Masters*	5-Minute Check Transparencies
15–1	graphing calculator grid paper [1, 4]	663	664–665	666	667		161		43	15–1
15–2	coordinate planes [4] straightedge [1, 2, 4]	668	669–670	671	672	703				15–2
15–3	algebra tiles [1, 2, 3, 4] equation/product mat [1, 2, 4] grid paper [1, 4]	673	674–675	676	677	702	162		44, 45	15–3
15–4	coordinate planes [4] straightedge [1, 2, 4]	678	679–680	681	682					15–4
15–5	grid paper [1, 4]	683	684–685	686	687		163, 164	15		15–5
15–6	coordinate planes [4] straightedge [1, 2, 4]	688	689–690	691	692	702	165			15–6
Investigation	calculator spreadsheet software									
Study Guide & Assessment/ Chapter Test						693–701, 704–706				

**See page 636c for examples of these instructional materials.*

Key to Glencoe Manipulative Resources

[1]Classroom Manipulative Resources [2]Student Manipulative Resources [3]Overhead Manipulative Resources [4]Hands-On Algebra Masters

INV = Investigation *SG+A = Study Guide and Assessment*

DAY 7	DAY 8	DAY 9	DAY 10	DAY 11	DAY 12	DAY 13
Lesson 15–6	INV	SG+A	Chapter Test			

Resource Manager

The pages shown on this page are a small sample of the materials available on *TeacherWorks: All-in-One Lesson Planner and Resource Center.*

This CD-ROM includes all of the blackline masters and transparencies available for this program.

It also includes a lesson planner and interactive Teacher Edition, so you can customize lesson plans and reproduce classroom resources quickly and easily, from just about anywhere.

Applications

School-to-Workplace Masters, p. 15

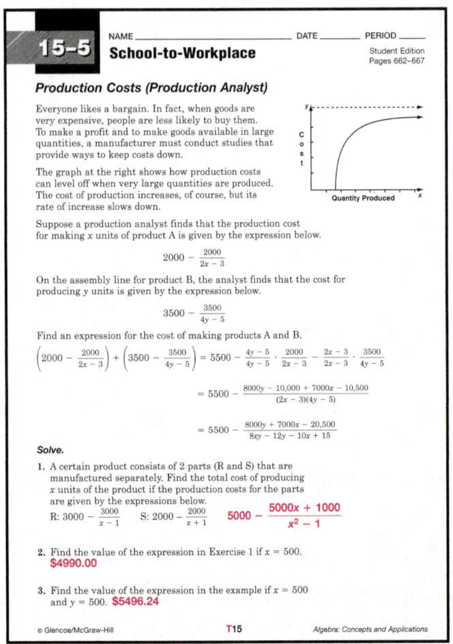

15–5 NAME ______ DATE ______ PERIOD ______

School-to-Workplace

Student Edition Pages 662–667

Production Costs (Production Analyst)

Everyone likes a bargain. In fact, when goods are very expensive, people are less likely to buy them. To make a profit and to make goods available in large quantities, a manufacturer must conduct studies that provide ways to keep costs down.

The graph at the right shows how production costs can level off when very large quantities are produced. The cost of production increases, of course, but its rate of increase slows down.

Suppose a production analyst finds that the production cost for making x units of product A is given by the expression below.

$$2000 - \frac{2000}{2x-3}$$

On the assembly line for product B, the analyst finds that the cost for producing y units is given by the expression below.

$$3500 - \frac{3500}{4y-5}$$

Find an expression for the cost of making products A and B.

$$\left(2000 - \frac{2000}{2x-3}\right) + \left(3500 - \frac{3500}{4y-5}\right) = 5500 - \frac{4y-5}{4y-5}\cdot\frac{2000}{2x-3} - \frac{2x-3}{2x-3}\cdot\frac{3500}{4y-5}$$

$$= 5500 - \frac{8000y - 10{,}000 + 7000x - 10{,}500}{(2x-3)(4y-5)}$$

$$= 5500 - \frac{8000y + 7000x - 20{,}500}{8xy - 12y - 10x + 15}$$

Solve.

1. A certain product consists of 2 parts (R and S) that are manufactured separately. Find the total cost of producing x units of the product if the production costs for the parts are given by the expressions below.
   R: $3000 - \frac{3000}{x-1}$ S: $2000 - \frac{2000}{x+1}$ $5000 - \frac{5000x + 1000}{x^2 - 1}$
2. Find the value of the expression in Exercise 1 if $x = 500$. $4990.00
3. Find the value of the expression in the example if $x = 500$ and $y = 500$. $5496.24

© Glencoe/McGraw-Hill T15 *Algebra: Concepts and Applications*

Manipulatives/Modeling

Hands-On Algebra Masters, pp. 161–165

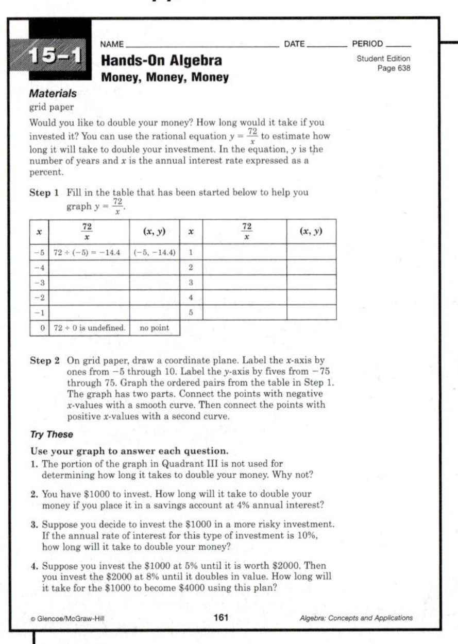

15–1 NAME ______ DATE ______ PERIOD ______

Hands-On Algebra
Money, Money, Money

Student Edition Page 638

Materials
grid paper

Would you like to double your money? How long would it take if you invested it? You can use the rational equation $y = \frac{72}{x}$ to estimate how long it will take to double your investment. In the equation, y is the number of years and x is the annual interest rate expressed as a percent.

Step 1 Fill in the table that has been started below to help you graph $y = \frac{72}{x}$.

x	$\frac{72}{x}$	(x, y)	x	$\frac{72}{x}$	(x, y)
−5	72 ÷ (−5) = −14.4	(−5, −14.4)	1		
−4			2		
−3			3		
−2			4		
−1			5		
0	72 ÷ 0 is undefined.	no point			

Step 2 On grid paper, draw a coordinate plane. Label the x-axis by ones from −5 through 10. Label the y-axis by fives from −75 through 75. Graph the ordered pairs from the table in Step 1. The graph has two parts. Connect the points with negative x-values with a smooth curve. Then connect the points with positive x-values with a second curve.

Try These

Use your graph to answer each question.

1. The portion of the graph in Quadrant III is not used for determining how long it takes to double your money. Why not?
2. You have $1000 to invest. How long will it take to double your money if you place it in a savings account at 4% annual interest?
3. Suppose you decide to invest the $1000 in a more risky investment. If the annual rate of interest for this type of investment is 10%, how long will it take to double your money?
4. Suppose you invest the $1000 at 5% until it is worth $2000. Then you invest the $2000 at 8% until it doubles in value. How long will it take for the $1000 to become $4000 using this plan?

© Glencoe/McGraw-Hill 161 *Algebra: Concepts and Applications*

Technology/Multimedia

Graphing Calculator Masters, pp. 43–45

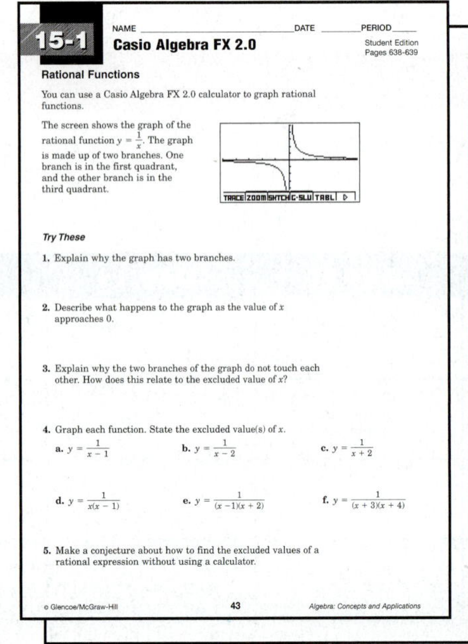

15-1 NAME ______ DATE ______ PERIOD ______

Casio Algebra FX 2.0

Student Edition Pages 638-639

Rational Functions

You can use a Casio Algebra FX 2.0 calculator to graph rational functions.

The screen shows the graph of the rational function $y = \frac{1}{x}$. The graph is made up of two branches. One branch is in the first quadrant, and the other branch is in the third quadrant.

Try These

1. Explain why the graph has two branches.
2. Describe what happens to the graph as the value of x approaches 0.
3. Explain why the two branches of the graph do not touch each other. How does this relate to the excluded value of x?
4. Graph each function. State the excluded value(s) of x.
   a. $y = \frac{1}{x-1}$ **b.** $y = \frac{1}{x-2}$ **c.** $y = \frac{1}{x+2}$
   d. $y = \frac{1}{x(x-1)}$ **e.** $y = \frac{1}{(x-1)(x+2)}$ **f.** $y = \frac{1}{(x+3)(x+4)}$
5. Make a conjecture about how to find the excluded values of a rational expression without using a calculator.

© Glencoe/McGraw-Hill 43 *Algebra: Concepts and Applications*

15-3 NAME ______ DATE ______ PERIOD ______

TI-83/84 Plus

Dividing Polynomials

You can use a TI-83/84 Plus calculator to check whether you divided two polynomials correctly. To illustrate this procedure, refer to Example 4 on page 652 of the Student Edition.

Step 1 In the Y= screen, enter the original problem, $(x^2 - 4) \div (x + 1)$, as Y_1. Then enter the answer, $x - 1 + \frac{-3}{x+1}$, as Y_2.

Step 2 Press [2nd] [TABLE]. Enter any values for x if they have not already been entered. Compare the numbers in the Y_1 and Y_2 columns. These are the y-values for each equation evaluated at the given values of x. All calculations has been done for you.

X	Y1	Y2
-1	ERROR	ERROR
0	-4	-4
1	-1.5	-1.5
2	0	0
3	1.25	1.25
4	2.4	2.4
5	3.5	3.5

X=5

For each value of x, the values of Y_1 and Y_2 are the same. So, it appears that $x - 1 + \frac{-3}{x+1}$ is equivalent to $(x^2 - 4) \div (x + 1)$. The answer is correct.

Try These

Use the method above to complete Exercises 1-4.

1. A student divided $x^2 + 1$ by $x + 5$ and got $x - 5 + \frac{24}{x+5}$. Did the student divide correctly? How do you know?
2. Check whether $(2x^2 - 5x) \div (x + 3)$ is equal to $2x - 11 + \frac{33}{x+3}$.
3. Check whether $(5x^2 + 7x + 2) \div (x - 2)$ is equal to $5x + 17 + \frac{35}{x-2}$.
4. Check whether $(-3x^2 + 8x + 7) \div (x + 4)$ is equal to $-3x + 20 + \frac{-73}{x+4}$.

© Glencoe/McGraw-Hill 44 *Algebra: Concepts and Applications*

Assessment Resources

Type	Student Edition	Teacher's Wraparound Edition	Chapter 15 Resource Masters
Ongoing Assessment	Quizzes 1 and 2, pp. 655, 667	5-Minute Check, pp. 638, 644, 650, 656, 662, 668	Mid-Chapter Test, p. 702 Quizzes A and B, p. 703
Mixed Review	Mixed Review, pp. 643, 649, 654, 661, 667, 673 Standardized Test Practice, Chapters 1–15, pp. 680–681		Cumulative Review, p. 704 Standardized Test Practice, pp. 705–706
Error Analysis	You Decide, pp. 641, 665	Error Analysis, pp. 642, 647, 653, 660, 665, 672	
Standardized Test Prep	Standardized Test Practice, pp. 643, 649, 655, 661, 667, 673 Standardized Test Practice, Chapters 1–15, pp. 680–681		Standardized Test Practice, pp. 705–706
Open-Ended Assessment	Writing Math, pp. 647, 671 Investigation, pp. 674–675 Portfolio, pp. 637, 675	Modeling: p. 655 Speaking: pp. 661, 667, 673 Writing: pp. 643, 649	Extended Response Assessment, p. 701
Chapter Assessment	Study Guide and Assessment, pp. 676–678 Chapter Test, p. 679		Multiple-Choice Tests (Forms 1A, 1B), pp. 693–694 Free-Response Tests (Forms 2A, 2B), pp. 697–700

Additional Chapter Resources

Student Edition
Hands-On Algebra, p. 650
Graphing Calculator Exploration, pp. 638–639

Teacher's Classroom Resources
Manipulatives/Modeling
Teacher's Guide for Overhead Manipulative Resources

Meeting Individual Needs
Prerequisite Skills Booklet
Spanish Study Guide and Assessment, pp. 99–104, 133–134

Teaching Aids
Solutions Manual
5-Minute Check Transparencies

Glencoe Technology

Instructional
AlgePASS CD-ROM, Lessons 33
Interactive Chalkboard CD-ROM
StudentWorks Plus CD-ROM
Multimedia Applications CD-ROM, Activity 12
Vocabulary PuzzleMaker (online)

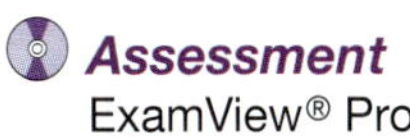

Assessment
ExamView® Pro

Teaching Aids
Answer Key Maker CD-ROM
TeacherWorks CD-ROM

Visit ***www.algconcepts.com*** *for data updates, career information, games, and other interactive activities.*

CHAPTER PREVIEW 15

Mathematics of the Chapter

Students will begin this chapter by simplifying rational expressions. Then they will learn to multiply and divide rational expressions, followed by dividing polynomials by binomials using long division. Students will learn to add and subtract rational expressions, both with like denominators and with unlike denominators. Finally, students will explore solving rational equations, including uniform motion problems.

What You'll Learn

Have students read over the lists of key ideas and key vocabulary. Have them make a list of any words with which they are not familiar.

Why It's Important

Point out to students that this is only one of many reasons why each objective is important. Others are provided in each lesson.

Answers

13. $w^2 + 4w$
14. $3k^2 + 6k$
15. $15c - 20$
16. $5mn + 10m$
17. $r^3 + 3r^2 - r$
18. $20t^4 + 5t^2 - 30t$
19. $y^2 + 2y - 3$
20. $w^2 - 6w + 8$
21. $p^2 - p - 30$
22. $2r^2 - 7r - 30$
23. $2s^2 + s - 10$
24. $2g^3 - 2g^2 + g - 1$

CHAPTER 15 Rational Expressions and Equations

What You'll Learn

Key Ideas

- Simplify, multiply, and divide rational expressions. *(Lessons 15–1 and 15–2)*
- Divide polynomials by binomials. *(Lesson 15–3)*
- Add and subtract rational expressions. *(Lessons 15–4 and 15–5)*
- Solve rational equations. *(Lessons 15–6)*

Key Vocabulary

rational equation *(p. 668)*
rational expression *(p. 638)*
rational function *(p. 638)*

Why It's Important

Transportation As the United States spread from east to west in the 1800s, systems were built to connect the nation. Rivers and roads and then canals and railroads moved people and goods. By 1930, more than half of American families owned a car. Now airplanes provide even faster transportation.

Rational expressions are fractions in which the numerator and denominator are polynomials. You will use rational expressions to find travel times in Lesson 15–3.

CHAPTER 15 LINKS						
Lesson	**15–1**	**15–2**	**15–3**	**15–4**	**15–5**	**15–6**
Math in the Workplace	Photography	Carpentry	Aviation	Entertainment	Teaching	Aviation
Applications and Connections	History Sports Investing Entertainment	Sports Sewing Cycling Business	Transportation Sports Fast Food		Astronomy Entertainment Fitness Parades	Travel Transportation Quizzes Cycling Space Farming
Math Integration		Probability	Geometry	Geometry	Geometry	

Study these lessons to improve your skills.

✓ Check Your Readiness

✓ **Lesson 11–4, pp. 474–477**

Solve each equation. Check your solution.

1. $4n(n - 1) = 0$ **0, 1**
2. $2b(b + 4) = 0$ **0, −4**
3. $(h - 1)(h + 2) = 0$ **1, −2**
4. $(2x - 1)(x + 6) = 0$ **$\frac{1}{2}$, −6**
5. $a^2 - 8a + 12 = 0$ **6, 2**
6. $x^2 + 4x - 45 = 0$ **−9, 5**

Lesson 8–2, 341–345

Simplify each expression.

7. $k \cdot k^6$ **k^7**
8. $b \cdot 6b^3$ **$6b^4$**
9. $3d^2 \cdot 4d^3$ **$12d^5$**
10. $(p^3q)(p^2q^2)$ **p^5q^3**
11. $5m^2 \cdot 6n^2$ **$30m^2n^2$**
12. $6a^2 \cdot 10ab$ **$60a^3b$**

Lesson 9–3, 394–398

Find each product. See margin.

13. $w(w + 4)$
14. $3k(k + 2)$
15. $5(3c - 4)$
16. $5m(n + 2)$
17. $r(r^2 + 3r - 1)$
18. $5t(4t^3 + t - 6)$

✓ **Lesson 9–4, pp. 399–404**

19. $(y - 1)(y + 3)$
20. $(w - 2)(w - 4)$
21. $(p - 6)(p + 5)$
22. $(2r + 5)(r - 6)$
23. $(s - 2)(2s + 5)$
24. $(2g^2 + 1)(g - 1)$

FOLDABLES™ Study Organizer

Make this Foldable to help you organize your Chapter 15 notes. Begin with a sheet of notebook paper.

❶ **Fold** lengthwise to the holes.

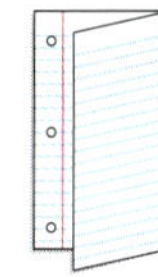

❷ **Cut** along the top line and then cut ten tabs.

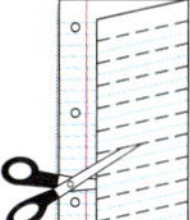

❸ **Label** the tabs using vocabulary words from the chapter as shown.

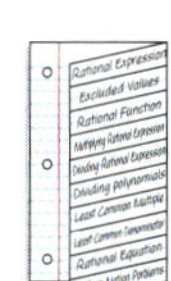

Reading and Writing Store the Foldable in a 3-ring binder. As you read and study the chapter, write notes and examples under each tab.

www.algconcepts.com/chapter_readiness

Check Your Readiness

This section provides a review of the basic concepts needed before beginning Chapter 15. Lesson references are included for additional student help.

Use Exercises	To Prepare for Lesson(s)
1–6	15–1
7–12	15–2
13–24	15–2, 15–5

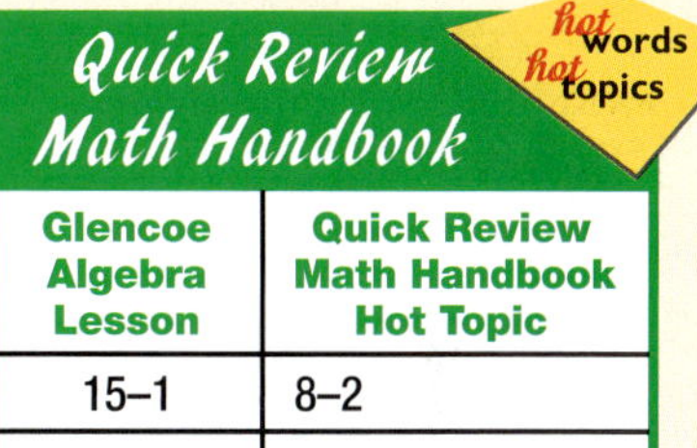

Glencoe Algebra Lesson	Quick Review Math Handbook Hot Topic
15–1	8–2
15–4	1–4

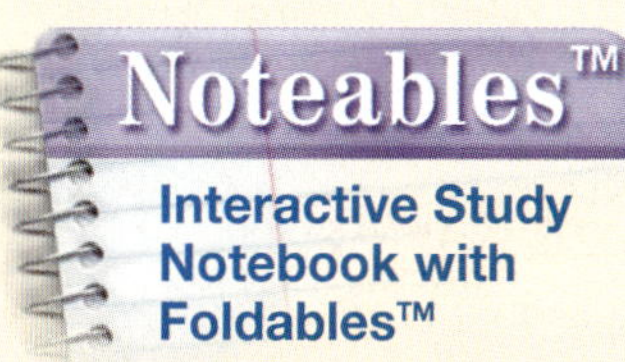

This note-taking guide reinforces excellent note-taking skills. It includes:

- ✓ **Build Your Vocabulary** tool for including mathematical terminology in notes.
- ✓ **Review It** activities with links to prior lessons and items to include in **Foldables™**.
- ✓ **Bringing It All Together** feature to help students review for the Chapter Test.
- ✓ **Are You Ready for the Chapter Test?** feature to allow students to assess their own readiness for the Chapter Test.
- ✓ **Teacher Edition** with transparencies to guide note taking. Corresponds to the *Teacher's Wraparound Edition* In-Class Examples and *Interactive Chalkboard CD-ROM.*

FOLDABLES™ Study Organizer

After students make this Foldable, have them record each of the vocabulary words on a separate tab. Then have students write definitions and examples of each of the terms. They can use this book to study the important terms from the chapter.

Lesson 15-1

1 FOCUS

5-Minute Check
Chapter 14

1. Determine whether the number $-\sqrt{186}$ is *rational* or *irrational*. If it is irrational, find two consecutive integers between which its graph lies on the number line. **irrational; −13 and −14**
2. Find the distance between $P(-1, 3)$ and $Q(-6, 8)$. **$\sqrt{50}$, or about 7.1 units**
3. Simplify $\frac{\sqrt{8}}{\sqrt{12}}$. **$\frac{\sqrt{6}}{3}$**
4. Find the exact perimeter of quadrilateral *WXYZ*.

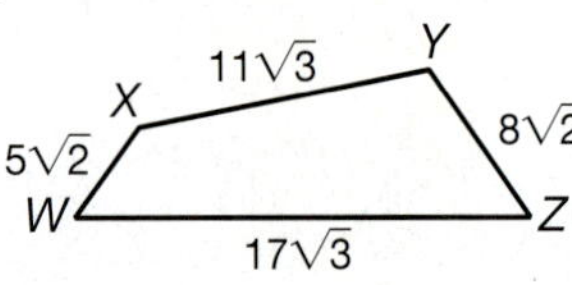

$28\sqrt{3} + 13\sqrt{2}$

5. Solve $\sqrt{x - 5} + 3 = 10$. Check your solution. **54**

Motivating the Lesson

Hands-On Activity Have students use grid paper to draw or cut out rectangles with whole number dimensions and an area of 24 square units. See who recognizes that the dimensions are in an inverse variation relationship. Ask students to express the length ℓ as a function of the width w. Point out that $\frac{24}{w}$ is a *rational expression*, and ask students for what values of x this expression of length makes sense. Tell students that graphs of rational functions such as $\ell = \frac{24}{w}$ have their own distinctive shapes that they will explore in this lesson.

15-1 Simplifying Rational Expressions

What You'll Learn
You'll learn to simplify rational expressions.

Why It's Important
Photography Photographers use a rational expression to find the amount of light that falls on an object. *See Exercise 63.*

A fraction denotes a quotient. In algebra, the fraction $\frac{2x}{x + 3}$ is called a **rational expression.** In a rational expression, both the numerator and denominator are polynomials.

Rational Expression	A rational expression is an algebraic fraction whose numerator and denominator are polynomials.

Every polynomial is a rational expression because it can be written as a quotient with 1 in the denominator.

Zero cannot be the denominator of a fraction because division by zero is undefined. In the expression $\frac{2x}{x + 3}$, if $x = -3$, the denominator equals zero. So, any value assigned to a variable that results in a denominator of zero must be excluded from the domain of the variable. These values are called **excluded values**.

A function that contains a rational expression is called a **rational function.** You can use the graph of the rational function of a rational expression to investigate excluded values of the variable.

Graphing Calculator Tutorial
See pp. 750–753.

Graphing Calculator Exploration

The screen shows the graph of the rational function $y = \frac{1}{x}$. The graph is made up of two branches. One branch is in the first quadrant, and the other branch is in the third quadrant.

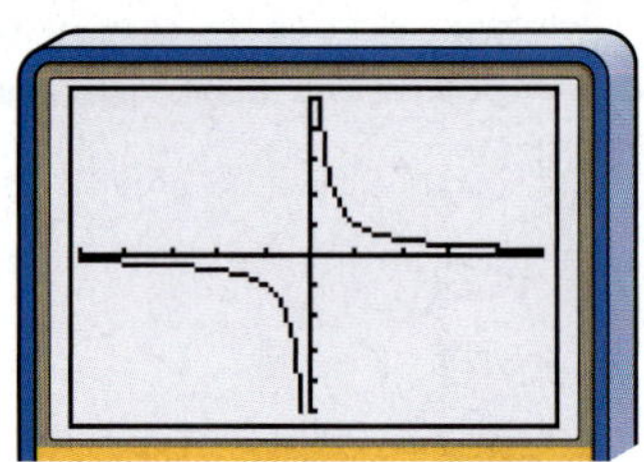

Try These
1. Explain why the graph has two branches.
2. Describe what happens to the graph as the value of x approaches 0.
3. Explain why the two branches of the graph do not touch each other. How does this relate to the excluded value of x? **See margin.**

1. One branch shows positive values for x and y; the other branch shows negative values for x and y.
2. See margin.

Resource Manager

Reproducible Masters
Chapter 15 Resource Masters
- *Study Guide*, p. 663
- *Skills Practice*, p. 664
- *Practice*, p. 665
- *Reading to Learn Mathematics*, p. 666
- *Enrichment*, p. 667

Graphing Calculator, p. 43
Hands-On Algebra, p. 161

Transparencies
5-Minute Check, 15–1

Technology/Multimedia
AlgePASS, Lesson 33
Interactive Chalkboard CD-ROM

4. Graph each function. State the excluded value(s) of x.

a. $y = \frac{1}{x-1}$ **1**

b. $y = \frac{1}{x-2}$ **2**

c. $y = \frac{1}{x+2}$ **−2**

d. $y = \frac{1}{x(x-1)}$ **0, 1**

e. $y = \frac{1}{(x-1)(x+2)}$ **1, −2**

f. $y = \frac{1}{(x+3)(x+4)}$ **−3, −4**

5. **Make a conjecture** about how to find the excluded values of a rational expression without using a calculator.

5. Find the values of the variable when the denominator is 0.

Examples

Find the excluded value(s) for each rational expression.

1 $\frac{5m}{2+m}$

Exclude the values for which $2 + m = 0$.

$2 + m = 0$

$m = -2$

So, m cannot equal -2.

Look Back

Zero Product Property: Lesson 11–4

$\frac{6}{a(a-4)}$

Exclude the values for which $a(a - 4) = 0$.

$a(a - 4) = 0$

$a = 0$ or $a - 4 = 0$ *Zero Product Property*

So, a cannot equal 0 or 4.

$\frac{5n}{n^2-25}$

Exclude the values for which $n^2 - 25 = 0$.

$n^2 - 25 = 0$

$(n + 5)(n - 5) = 0$ *Factor $n^2 - 25$.*

$n = -5$ or $n = 5$ *Zero Product Property*

So, n cannot equal -5 or 5.

Your Turn **b. $x \neq 0$ or -2 c. $a \neq 3$ or -2**

a. $\frac{2s}{s-3}$ **$s \neq 3$**

b. $\frac{3x}{x(x+2)}$

c. $\frac{2a-3}{a^2-a-6}$

Recall that you can simplify a fraction by using the following steps.

- First, factor the numerator and denominator.
- Then, divide the numerator and denominator by the greatest common factor.

www.algconcepts.com/extra_examples

2 TEACH

Teaching Tip Inform students that the relationship $y = \frac{72}{x}$ described in the lesson's opening paragraph is called the "Rule of 72" by economists. You may want to have students find out interest rates on savings accounts, bonds, or certificates of deposit so that they can evaluate the rule for actual interest rates.

Teaching Tip While discussing Example 1, make sure that students understand that it is okay for the variable m to be 0 as long as this doesn't make the value of the denominator 0.

In-Class Examples

Examples 1–3

Find the excluded value(s) for each rational expression.

1 $\frac{7a}{5-a}$ **$a \neq 5$**

2 $\frac{20}{(x+3)x}$ **$x \neq -3$ or 0**

3 $\frac{10y}{y^2-4}$ **$y \neq -2$ or 2**

Answers

Graphing Calculator Exploration

2. It approaches the y-axis.
3. Since $x = 0$ is excluded as a value, the graphs cannot cross either axis, $y = 0$ or $x = 0$. Thus, they cannot touch each other.

Graphing Calculator Exploration

Refer to the Graphing Calculator Exploration on pages 638–639. Be sure students put parentheses around the *entire* denominator when entering each equation in Exercise 4 in the calculator's function editor. If students use the calculator's [ZStandard] window, the calculator will connect the branches of each graph unless students set the calculator to Dot mode. You may want to have students use the [ZDecimal] window, which will not connect the branches, since the asymptotes determined by the excluded values have integer values. It may take some trial and error for students to find optimal viewing windows in Exercises 4d–4f, for which each graph has two asymptotes.

Teaching Tip In Examples 4 and 5, another approach is to rewrite each rational expression as a product of 1 and another fraction by dividing out the GCF of the numerator and denominator:

$$\frac{60}{100} = \frac{20}{20} \cdot \frac{3}{5} = 1 \cdot \frac{3}{5} = \frac{3}{5}$$

$$\frac{8a^2b}{12ab^3} = \frac{4ab}{4ab} \cdot \frac{2a}{3b^2} = 1 \cdot \frac{2a}{3b^2} = \frac{2a}{3b^2}$$

In-Class Examples

Example 4

Express 35% as a fraction in simplest form. $\frac{7}{20}$

Examples 5–7

Simplify each rational expression.

5 $\frac{20x^5y}{25x^2y^3}$ $\frac{4x^3}{5y^2}$

6 $\frac{4x + 12}{7x + 21}$ $\frac{4}{7}$

7 $\frac{m(m-2)}{m^2 - 5m + 6}$ $\frac{m}{m-3}$

Example 4 History Link

In the 1960 presidential election, more than 60% of the registered voters cast ballots. No presidential election since 1960 has had a greater voter turnout. Express 60% as a fraction in simplest form.

$60\% = \frac{60}{100}$

$= \frac{2 \cdot 2 \cdot 3 \cdot 5}{2 \cdot 2 \cdot 5 \cdot 5}$ *Factor 60 and 100.*

$= \frac{\overset{1}{\cancel{2}} \cdot \overset{1}{\cancel{2}} \cdot 3 \cdot \overset{1}{\cancel{5}}}{\underset{1}{\cancel{2}} \cdot \underset{1}{\cancel{2}} \cdot 5 \cdot \underset{1}{\cancel{5}}}$ or $\frac{3}{5}$ *The GCF of 60 and 100 is 2 · 2 · 5.*

Your Turn Simplify each fraction.

d. $\frac{9}{24}$ $\frac{3}{8}$ e. $\frac{15}{20}$ $\frac{3}{4}$ f. $\frac{12}{30}$ $\frac{2}{5}$

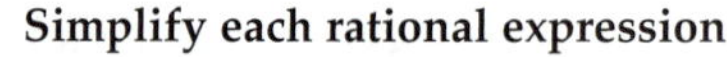

You can use the same procedure to simplify rational expressions that have polynomials in the numerator and denominator. To *simplify* means that the numerator and denominator have no factors in common, except 1.

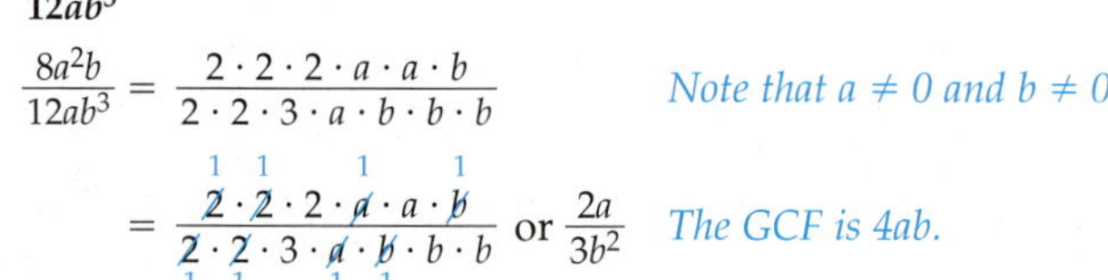

Examples

From this point on, you can assume that all values that give a denominator of zero are excluded.

Simplify each rational expression.

5 $\frac{8a^2b}{12ab^3}$

$\frac{8a^2b}{12ab^3} = \frac{2 \cdot 2 \cdot 2 \cdot a \cdot a \cdot b}{2 \cdot 2 \cdot 3 \cdot a \cdot b \cdot b \cdot b}$ *Note that $a \neq 0$ and $b \neq 0$.*

$= \frac{\overset{1}{\cancel{2}} \cdot \overset{1}{\cancel{2}} \cdot 2 \cdot \overset{1}{\cancel{a}} \cdot a \cdot \overset{1}{\cancel{b}}}{\underset{1}{\cancel{2}} \cdot \underset{1}{\cancel{2}} \cdot 3 \cdot \underset{1}{\cancel{a}} \cdot \underset{1}{\cancel{b}} \cdot b \cdot b}$ or $\frac{2a}{3b^2}$ *The GCF is 4ab.*

6 $\frac{2x-2}{5x-5}$

$\frac{2x-2}{5x-5} = \frac{2(x-1)}{5(x-1)}$ *Factor 2x − 2 and 5x − 5.*

$= \frac{2\overset{1}{\cancel{(x-1)}}}{5\underset{1}{\cancel{(x-1)}}}$ or $\frac{2}{5}$ *The GCF is (x − 1).*

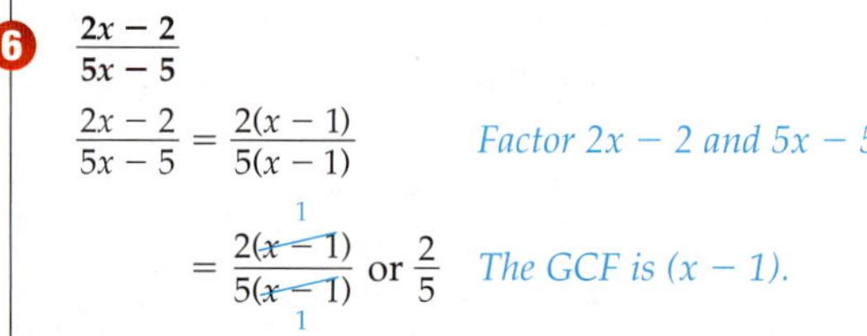

Look Back

Factoring Trinomials: Lesson 10–3

7 $\frac{a(a+1)}{a^2 + 3a + 2}$

$\frac{a(a+1)}{a^2 + 3a + 2} = \frac{a(a+1)}{(a+2)(a+1)}$ *Factor $a^2 + 3a + 2$.*

$= \frac{a\overset{1}{\cancel{(a+1)}}}{(a+2)\underset{1}{\cancel{(a+1)}}}$ or $\frac{a}{a+2}$ *The GCF is (a + 1).*

8 $\dfrac{x^2 - 9}{x^2 - x - 6}$

$$\frac{x^2 - 9}{x^2 - x - 6} = \frac{(x - 3)(x + 3)}{(x + 2)(x - 3)} \quad \textit{Factor } x^2 - 9 \textit{ and } x^2 - x - 6.$$

$$= \frac{\overset{1}{\cancel{(x - 3)}}(x + 3)}{(x + 2)\underset{1}{\cancel{(x - 3)}}} \text{ or } \frac{x + 3}{x + 2} \quad \textit{The GCF is } (x - 3).$$

9 $\dfrac{4 - 2x}{x^2 - 4x + 4}$

$$\frac{4 - 2x}{x^2 - 4x + 4} = \frac{2(2 - x)}{(x - 2)(x - 2)} \quad \textit{Factor } 4 - 2x \textit{ and } x^2 - 4x + 4.$$

$$= \frac{2(-1)(x - 2)}{(x - 2)(x - 2)} \quad \textit{Factor } -1 \textit{ from } (2 - x).$$

$$= \frac{2(-1)\overset{1}{\cancel{(x - 2)}}}{(x - 2)\underset{1}{\cancel{(x - 2)}}} \text{ or } \frac{-2}{x - 2} \quad \textit{The GCF is } (x - 2).$$

Your Turn Simplify each rational expression.

g. $\dfrac{14y^3z}{8y^2z^4}$ $\dfrac{7y}{4z^3}$ h. $\dfrac{-12a^2b^2}{18a^5}$ $\dfrac{-2b^2}{3a^3}$ i. $\dfrac{3y - 9}{4y - 12}$ $\dfrac{3}{4}$

j. $\dfrac{x(x + 3)}{x^2 + 6x + 9}$ $\dfrac{x}{x + 3}$ k. $\dfrac{a^2 - 2a - 8}{a^2 + 3a + 2}$ $\dfrac{a - 4}{a + 1}$ l. $\dfrac{x^2 - 4x}{16 - x^2}$ $\dfrac{-x}{4 + x}$

Check for Understanding

Communicating Mathematics

1. **Explain** why 2 is an excluded value for $\dfrac{x}{x - 2}$.
The denominator is 0 when $x = 2$.
2. **List** the steps you would use to simplify a rational expression.

Vocabulary
rational expression
excluded values
rational function

2–3. See margin.

3.  Sam and Darnell are trying to simplify $\dfrac{x + 3}{x + 4}$. Sam says the answer is $\dfrac{3}{4}$. Darnell says it is already in simplest form. Who is correct? Explain your reasoning.

Guided Practice

Getting Ready Factor each expression.

Sample 1: $5y + 15$
Solution: $5(y + 3)$

Sample 2: $x^2 + 5x + 6$
Solution: $(x + 3)(x + 2)$

4. $4x - 20$ $4(x - 5)$
5. $16x^2 - 20x$
6. $y^2 + 6y + 8$
7. $a^2 - a - 20$
8. $n^2 - 3n - 18$
9. $x^2 - 10x + 24$

5. $4x(4x - 5)$
6. $(y + 2)(y + 4)$
7. $(a - 5)(a + 4)$
8. $(n - 6)(n + 3)$
9. $(x - 6)(x - 4)$

Examples 1–3 Find the excluded value(s) for each rational expression.

10. $\dfrac{7}{x - 2}$ 2
11. $\dfrac{4a}{a(a + 6)}$ 0, −6
12. $\dfrac{2x + 3}{x^2 - 2x - 15}$ 5, −3

Teaching Tip While discussing Example 8, make sure that students understand that $\dfrac{x + 3}{x + 2}$ cannot be simplified as $\dfrac{3}{2}$. Only factors can be divided out; terms of a factor cannot be divided out independently.

In-Class Examples
Examples 8–9

Simplify each rational expression.

8 $\dfrac{a^2 - 25}{a^2 - 2a - 15}$ $\dfrac{a + 5}{a + 3}$

9 $\dfrac{6 - 2x}{x^2 + x - 12}$ $\dfrac{-2}{x + 4}$

Answers

2. First, factor the numerator and denominator. Then, divide the numerator and denominator by the greatest common factor.
3. Darnell; x is not a factor of $x + 3$ and $x + 4$. Therefore, it cannot be divided out.

***Study Guide*, p. 663**

15–1 NAME ______ DATE ______ PERIOD ______
Study Guide
Student Edition
Pages 638–643

Simplifying Rational Expressions

A **rational expression** is an algebraic fraction whose numerator and denominator are polynomials. For example, $\frac{3}{x}$, $\frac{3}{x - 2}$, $\frac{x - 2}{3}$, and $\frac{x}{x - 2}$ are all rational expressions because 3, x, and $x - 2$ are all polynomials.

To simplify a rational expression, use the same steps that you use to simplify any fraction.

- First factor the numerator and denominator. The factors may be polynomials.
- Then divide the numerator and denominator by the greatest common factor. The greatest common factor may be a polynomial.

Example 1: Simplify $\frac{16a^4b^2}{20a^2b^5}$.

$$\frac{16a^4b^2}{20a^2b^5} = \frac{2 \cdot 2 \cdot 2 \cdot 2 \cdot a \cdot a \cdot a \cdot a \cdot b \cdot b}{2 \cdot 2 \cdot 5 \cdot a \cdot a \cdot b \cdot b \cdot b \cdot b \cdot b}$$

$$= \frac{\overset{1}{\cancel{2}} \cdot \overset{1}{\cancel{2}} \cdot 2 \cdot 2 \cdot \overset{1}{\cancel{a}} \cdot \overset{1}{\cancel{a}} \cdot a \cdot a \cdot \overset{1}{\cancel{b}} \cdot \overset{1}{\cancel{b}}}{\underset{1}{\cancel{2}} \cdot \underset{1}{\cancel{2}} \cdot 5 \cdot \underset{1}{\cancel{a}} \cdot \underset{1}{\cancel{a}} \cdot \underset{1}{\cancel{b}} \cdot \underset{1}{\cancel{b}} \cdot b \cdot b \cdot b} \quad \textit{The GCF is } 4a^2b^2.$$

$$= \frac{4a^2}{5b^3}$$

Example 2: Simplify $\frac{m(m + 1)}{m^2 - 5m - 6}$.

$$\frac{m(m + 1)}{m^2 - 5m - 6} = \frac{m(m + 1)}{(m - 6)(m + 1)} \quad \textit{Factor } m^2 - 5m - 6.$$

$$= \frac{m\overset{1}{\cancel{(m + 1)}}}{(m - 6)\underset{1}{\cancel{(m + 1)}}} \quad \textit{The GCF is } (m + 1).$$

$$= \frac{m}{m - 6}$$

Simplify each rational expression.

1. $\frac{10}{25}$ $\frac{2}{5}$
2. $\frac{3m}{6n}$ $\frac{m}{2n}$
3. $\frac{12xy}{6x}$ $2y$
4. $\frac{-8abc}{16ac}$ $\frac{-b}{2}$
5. $\frac{6a^2b^3}{-2a^4b^3}$ $\frac{-3}{a^2}$
6. $\frac{3(x + 2)}{6(x + 2)}$ $\frac{1}{2}$
7. $\frac{m - 4}{m(m - 4)}$ $\frac{1}{m}$
8. $\frac{d(d - 4)}{c(d - 4)}$ $\frac{d}{c}$
9. $\frac{(c - 2)(d + 3)}{(c - 2)(d + 6)}$ $\frac{d + 3}{d + 6}$
10. $\frac{x^2 - 2x}{x - 2}$ x
11. $\frac{y^2 + 3y}{y(y - 3)}$ $\frac{y + 3}{y - 3}$
12. $\frac{x + 3}{x^2 + x - 6}$ $\frac{1}{x - 2}$

Reteaching Activity

Visual/Spatial Learners Have one student in a group of four or more write a rational expression such as $\dfrac{(x + 5)(x - 2)}{(x - 2)(x + 3)}$, in which the numerator and denominator have one pair of shared binomial factors and one pair of different binomial factors. Have a second student rewrite the expression by multiplying the binomials, a third simplify the expression, and a fourth find the excluded values. Have students alternate roles until each student has performed each role. Have each group make a display of its results on a large piece of paper or poster board, making columns from left to right for the expanded expression, the factored expression, the simplified expression, and the excluded values.

Examples 4–9

Simplify each rational expression.

13. $\frac{8}{20}$ $\frac{2}{5}$
14. $\frac{5x}{25x^2}$ $\frac{1}{5x}$
15. $\frac{12y^2z}{18yz^4}$ $\frac{2y}{3z^3}$
16. $\frac{x(x+3)}{7(x+3)}$ $\frac{x}{7}$
17. $\frac{3y+9}{4y+12}$ $\frac{3}{4}$
18. $\frac{x^2-3x}{x^2+x-12}$ $\frac{x}{x+4}$
19. $\frac{x^2+2x-8}{x^2+5x+4}$ $\frac{x-2}{x+1}$
20. $\frac{x^2+6x+5}{x^2+3x-10}$ $\frac{x+1}{x-2}$
21. $\frac{25-x^2}{x^2+x-30}$ $\frac{-(x+5)}{x+6}$

Example 4

22. **Sports** The National Hockey League team that wins the first game of the best-of-seven series for the Stanley Cup has about an 80% chance of winning the series. Express 80% as a fraction in simplest form. **Source:** NHL $\frac{4}{5}$

Exercises

Practice

Homework Help	
For Exercises	See Examples
23–31, 60, 64	1–3
32–44, 54	5, 6
45–58	7–9
61–63	4

Extra Practice
See page 722.

Find the excluded value(s) for each rational expression.

23. $\frac{x}{x+5}$ -5
24. $\frac{3a}{2a+4}$ -2
25. $\frac{n+6}{n-10}$ 10
26. $\frac{5}{a(a+8)}$ $-8, 0$
27. $\frac{x+2}{(x+2)(x-3)}$ $-2, 3$
28. $\frac{5m}{(m-2)(m-3)}$ $2, 3$
29. $\frac{9x}{x^2+5x}$ $0, -5$
30. $\frac{x^2+6x+5}{x^2+3x-10}$ $-5, 2$
31. $\frac{y+3}{y^2-16}$ $4, -4$

Simplify each rational expression.

32. $\frac{10}{16}$ $\frac{5}{8}$
33. $\frac{12}{36}$ $\frac{1}{3}$
34. $\frac{15}{20}$ $\frac{3}{4}$
35. $\frac{4c}{6d}$ $\frac{2c}{3d}$
36. $\frac{8xy}{24x^2}$ $\frac{y}{3x}$
37. $\frac{-36abc^2}{9ac}$ $-4bc$
38. $\frac{2(x+1)}{8(x+1)}$ $\frac{1}{4}$
39. $\frac{n-3}{5(n-3)}$ $\frac{1}{5}$
40. $\frac{x(x+5)}{y(x+5)}$ $\frac{x}{y}$

41. $\frac{(x+3)(x-2)}{(x-2)(x+1)}$ $\frac{x+3}{x+1}$
42. $\frac{(y+6)(y-6)}{(y+6)(y+6)}$ $\frac{y-6}{y+6}$
43. $\frac{x^2-3x}{2(x-3)}$ $\frac{x}{2}$
44. $\frac{a^2-a}{a-1}$ a
45. $\frac{r^2-r-6}{3r-9}$ $\frac{r+2}{3}$
46. $\frac{x^2+3x+2}{x^2+2x+1}$ $\frac{x+2}{x+1}$
47. $\frac{y^2+7y+12}{y^2-16}$ $\frac{y+3}{y-4}$
48. $\frac{x-3}{x^2+x-12}$ $\frac{1}{x+4}$
49. $\frac{y^2+4y+4}{y^2+y-2}$ $\frac{y+2}{y-1}$
50. $\frac{r^2-4r-5}{r^2-2r-15}$ $\frac{r+1}{r+3}$
51. $\frac{z^2-z-20}{z^2+7z+12}$ $\frac{z-5}{z+3}$
52. $\frac{6x^2+24x}{x^2+8x+16}$ $\frac{6x}{x+4}$

C

53. $\frac{8m^2-16m}{m^2-4m+4}$ $\frac{8m}{m-2}$
54. $\frac{r^2+6r+5}{2r^2-2}$ $\frac{r+5}{2(r-1)}$
55. $\frac{c^2-c-20}{c^3+10c^2+24c}$ 55. $\frac{c-5}{c(c+6)}$
56. $\frac{9-3x}{x^2-6x+9}$ $\frac{-3}{x-3}$
57. $\frac{x^2-4x}{16-x^2}$ $\frac{-x}{4+x}$
58. $\frac{12-4y}{y^2+y-12}$ $\frac{-4}{y+4}$
59. Write $\frac{4a}{3a+a^2}$ in simplest form. $\frac{4}{3+a}$
60. What are the excluded values of x for $\frac{x^2-9}{x^2+5x+6}$? $-2, -3$

3 PRACTICE/APPLY

Error Analysis

Watch for students who simplify a rational expression before evaluating the denominator for excluded values in problems such as Exercise 11.

Prevent by stressing that students use the denominator of the original rational expression to determine excluded values. In Exercise 11, the original rational expression $\frac{4a}{a(a+6)}$ is not defined when $a = 0$. The simplified expression $\frac{4}{a+6}$ is not equivalent to the original expression when $a = 0$.

Assignment Guide

Basic: 23–63 odd, 64–75
Average: 24–60 even, 61–75

***Skills Practice,* p. 664, and *Practice,* p. 665 (shown)**

NAME ______ DATE ______ PERIOD ______

15-1 Practice

Student Edition Pages 638–643

Simplifying Rational Expressions

Find the excluded value(s) for each rational expression.

1. $\frac{2n}{n-4}$ 4
2. $\frac{6}{x+3}$ -3
3. $\frac{3b}{b(b+9)}$ $0, -9$
4. $\frac{y+2}{y^2-4}$ $2, -2$
5. $\frac{4x+6}{(x+6)(x-5)}$ $-6, 5$
6. $\frac{2a-2}{a^2-3a-28}$ $7, -4$

Simplify each rational expression.

7. $\frac{6}{15}$ $\frac{2}{5}$
8. $\frac{12m}{18m^3}$ $\frac{2}{3m^2}$
9. $\frac{16x^2y}{36xy^3}$ $\frac{4x}{9y^2}$
10. $\frac{25ab}{30b^2}$ $\frac{5a}{6b}$
11. $\frac{-8y^4z}{20y^6z^2}$ $\frac{-2}{5y^2z}$
12. $\frac{5(x-1)}{8(x-1)}$ $\frac{5}{8}$
13. $\frac{y(y+7)}{9(y+7)}$ $\frac{y}{9}$
14. $\frac{x^2-4x}{3(x-4)}$ $\frac{x}{3}$
15. $\frac{x^2+2x}{5x+10}$ $\frac{x}{5}$
16. $\frac{x^2+5x}{(x+5)(x-7)}$ $\frac{x}{x-7}$
17. $\frac{x^2-6x}{x^2-4x-12}$ $\frac{x}{x+2}$
18. $\frac{(x+4)(x+4)}{(x+4)(x-2)}$ $\frac{x+4}{x-2}$
19. $\frac{b^2+6b+9}{b^2-2b-15}$ $\frac{b+3}{b-5}$
20. $\frac{y^2-36}{y^2+9y+18}$ $\frac{y-6}{y+3}$
21. $\frac{x^2-16}{x^2+x-12}$ $\frac{x-4}{x-3}$
22. $\frac{y^2+4y+4}{y^2-4}$ $\frac{y+2}{y-2}$
23. $\frac{a^2+3a}{a^2-3a-18}$ $\frac{a}{a-6}$
24. $\frac{y^2+7y+10}{y^2+5y}$ $\frac{y+2}{y}$
25. $\frac{x^2+4x+3}{x^2+3x+2}$ $\frac{x+3}{x+2}$
26. $\frac{x^2-6x+8}{x^2+x-6}$ $\frac{x-4}{x+3}$
27. $\frac{9-x^2}{x^2+6x-27}$ $\frac{-(x+3)}{x+9}$

Applications and Problem Solving

61. **Investing** The equation $y = \frac{72}{x}$ can be used to estimate how long it will take to double an investment. In the equation, y is the number of years, and x is the annual interest rate expressed as a percent. How long would it take to double your money if it is invested at an annual rate of 8%? **9 yr**

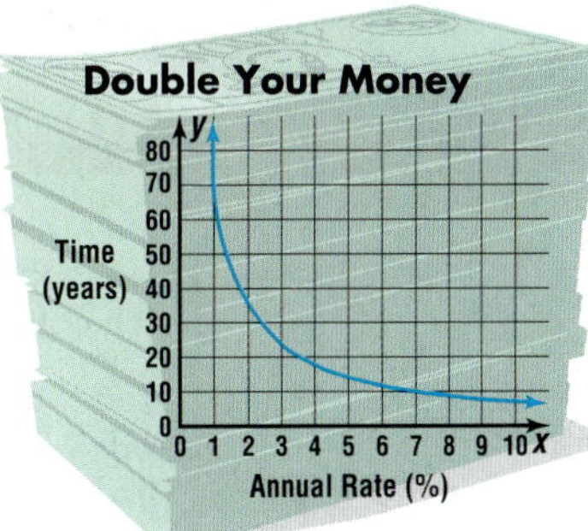

62a. $\frac{6}{25}$
62b. $\frac{1}{10}$

62. **Entertainment** The graph shows the number of films that adults see in a theater in a typical month.
    a. What fraction of adults see one movie?
    b. What fraction of adults see three or more movies?

63. **Photography** The intensity of light that falls on an object is given by the formula $I = \frac{P}{d^2}$. In the formula, d is the distance from the light source, P is the power in lumens, and I is the intensity of light in lumens per square meter. An object being photographed is 3 meters from a 72-lumen light source. Find the intensity. **8 lumens/m²**

64. **Critical Thinking** Write rational expressions that have the following as excluded values. **See margin.**
    a. 0 b. -3 c. -2 and 7

Mixed Review

Solve each equation. *(Lesson 14–6)*

65. $\sqrt{x} = 4$ **16**
66. $\sqrt{y} = 2\sqrt{5}$ **20**
67. $\sqrt{n - 3} = 4$ **19**
68. $\sqrt{a} + 4 = 29$ **625**

Simplify each expression. *(Lesson 14–5)*

69. $2\sqrt{3} + \sqrt{3} + 4\sqrt{3}$ $\mathbf{7\sqrt{3}}$
70. $6\sqrt{5} - 3\sqrt{5} + 10\sqrt{5}$ $\mathbf{13\sqrt{5}}$
71. $4\sqrt{3} + \sqrt{12}$ $\mathbf{6\sqrt{3}}$
72. $2\sqrt{50} + 3\sqrt{5}$ $\mathbf{10\sqrt{2} + 3\sqrt{5}}$

73. Find the distance between $J(2, 1)$ and $K(5, 5)$. *(Lesson 14–2)* **5**

Standardized Test Practice

74. **Short Response** The sum of two numbers is 9. Their difference is 3. Find the numbers. *(Lesson 13–4)* **3, 6**

75. **Multiple Choice** Which property or properties of inequalities allow(s) you to prove that if $2y + 8 > 18$, then $y > 5$? *(Lesson 12–4)* **D**
    (A) Distributive Property
    (B) Addition Property
    (C) Division Property
    (D) Addition and Division Properties

www.algconcepts.com/self_check_quiz

4 ASSESS

Open-Ended Assessment

Writing Have students write a brief paragraph describing what an *excluded value* means for a rational expression and its graph, and how excluded values can be found.

Answers

64a. Sample answer: $\frac{2}{x}$

64b. Sample answer: $\frac{x}{x + 3}$

64c. Sample answer: $\frac{x}{(x + 2)(x - 7)}$

***Enrichment*, p. 667**

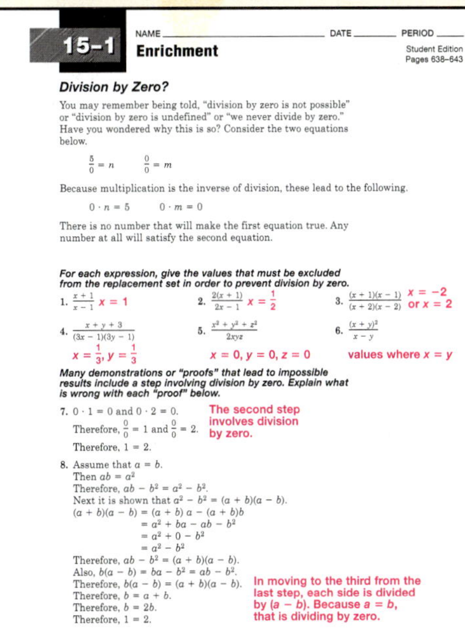

15-1 NAME ________ DATE ______ PERIOD _____

Enrichment

Student Edition Pages 638–643

Division by Zero?

You may remember being told, "division by zero is not possible" or "division by zero is undefined" or "we never divide by zero." Have you wondered why this is so? Consider the two equations below.

$\frac{5}{0} = n$ $\frac{0}{0} = m$

Because multiplication is the inverse of division, these lead to the following.

$0 \cdot n = 5$ $0 \cdot m = 0$

There is no number that will make the first equation true. Any number at all will satisfy the second equation.

For each expression, give the values that must be excluded from the replacement set in order to prevent division by zero.

1. $\frac{x + 1}{x - 1}$ $x = 1$
2. $\frac{2(x + 1)}{2x - 1}$ $x = \frac{1}{2}$
3. $\frac{(x + 1)(x - 1)}{(x + 2)(x - 2)}$ $x = -2$ or $x = 2$
4. $\frac{x + y + 3}{(3x - 1)(3y - 1)}$ $x = \frac{1}{3}, y = \frac{1}{3}$
5. $\frac{x^2 + y^2 + z^2}{2xyz}$ $x = 0, y = 0, z = 0$
6. $\frac{(x + y)^2}{x - y}$ values where $x = y$

Many demonstrations or "proofs" that lead to impossible results include a step involving division by zero. Explain what is wrong with each "proof" below.

7. $0 \cdot 1 = 0$ and $0 \cdot 2 = 0$.
   Therefore, $\frac{0}{0} = 1$ and $\frac{0}{0} = 2$.
   Therefore, $1 = 2$.
   The second step involves division by zero.

8. Assume that $a = b$.
   Then $ab = a^2$
   Therefore, $ab - b^2 = a^2 - b^2$.
   Next it is shown that $a^2 - b^2 = (a + b)(a - b)$.
   $(a + b)(a - b) = (a + b)a - (a + b)b$
   $= a^2 + ba - ab - b^2$
   $= a^2 + 0 - b^2$
   $= a^2 - b^2$
   Therefore, $ab - b^2 = (a + b)(a - b)$.
   Also, $b(a - b) = ba - b^2 = ab - b^2$.
   Therefore, $b(a - b) = (a + b)(a - b)$.
   Therefore, $b = a + b$.
   Therefore, $b = 2b$.
   Therefore, $1 = 2$.
   In moving to the third from the last step, each side is divided by $(a - b)$. Because $a = b$, that is dividing by zero.

Extra Credit

Find the excluded value(s) for the rational expression $\frac{5}{x^2 + 3}$.
There are no excluded values because the denominator cannot equal zero.

1 FOCUS

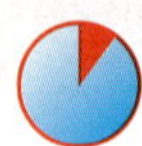

5-Minute Check
Lesson 15–1

1. Find the excluded value(s) for $\frac{10x^2}{(x - 0.5)(2x + 5)}$. $x \neq -2.5$ or 0.5

Simplify each rational expression.

2. $\frac{10x^2}{5x^5y}$ $\frac{2}{x^3y}$
3. $\frac{2x + 6}{8x + 24}$ $\frac{1}{4}$
4. $\frac{x^2 + x - 12}{x^2 + 7x + 12}$ $\frac{x - 3}{x + 3}$
5. The interest rate r on a principal of p dollars that earns I dollars of simple interest in t years can be found using the rational function $r = \frac{I}{pt}$. If $10,000 earns $1400 in 2 years, what is r? $r = 0.07$, or 7%

Motivating the Lesson

Real-World Connection Present the following scenario to students. They expect to average $1\frac{1}{2}$ miles per hour on a hiking trip. The distance they can travel in a given time t hours is $1\frac{1}{2}t$ miles. For $t = 2\frac{1}{2}$ hours, this is $\frac{3}{2} \cdot \frac{5}{2} = \frac{15}{4}$ miles. The time it takes to cover a given distance d is $d \div 1\frac{1}{2}$. For $d = 6$ miles, this is $6 \div \frac{3}{2} = 6 \cdot \frac{2}{3} = 4$ hours. Point out to students that the methods they use to multiply and divide fractions in this scenario apply whether they are working with simple fractions or with rational expressions with several variables.

15-2 Multiplying and Dividing Rational Expressions

What You'll Learn

You'll learn to multiply and divide rational expressions.

Why It's Important

Carpentry Carpenters divide rational expressions to find how many boards of a certain size can be cut from a piece of wood. *See Exercise 42.*

Two rational expressions can be multiplied or divided just like two rational numbers.

You can use two methods to multiply rational numbers.

- **Method 1** Multiply numerators and multiply denominators. Then divide each numerator and denominator by the greatest common factor.
- **Method 2** Divide numerators and denominators by any common factors. Then multiply numerators and denominators.

Method 1

Multiply, then simplify.

$$\frac{2}{5} \cdot \frac{1}{2} = \frac{2}{10}$$
$$= \frac{\overset{1}{\cancel{2}}}{\underset{5}{\cancel{10}}} \text{ or } \frac{1}{5} \quad \textit{The GCF is 2.}$$

Method 2

Simplify, then multiply.

$$\frac{2}{5} \cdot \frac{1}{2} = \frac{\overset{1}{\cancel{2}}}{5} \cdot \frac{1}{\underset{1}{\cancel{2}}} \quad \textit{2 is a common factor.}$$
$$= \frac{1}{5}$$

Both methods have the same result.

You can use the same methods to multiply rational expressions. Multiply $\frac{2x}{(x + 1)^2}$ and $\frac{(x + 1)}{x}$.

Method 1

Multiply, then simplify.

$$\frac{2x}{(x + 1)^2} \cdot \frac{x + 1}{x} = \frac{2x(x + 1)}{x(x + 1)^2}$$
$$= \frac{2\cancel{x}\cancel{(x + 1)}}{\cancel{x}\cancel{(x + 1)}(x + 1)}$$
$$= \frac{2}{(x + 1)}$$

Method 2

Simplify, then multiply.

$$\frac{2x}{(x + 1)^2} \cdot \frac{x + 1}{x} = \frac{2\cancel{x}}{(x + 1)\cancel{(x + 1)}} \cdot \frac{\cancel{(x + 1)}}{\cancel{x}}$$
$$= \frac{2}{(x + 1)}$$

Examples

Find each product.

1 $\frac{6r^2}{5s^2} \cdot \frac{10rs}{6r^3}$

Method 1 $\frac{6r^2}{5s^2} \cdot \frac{10rs}{6r^3} = \frac{60r^3s}{30r^3s^2}$ *Multiply.*

$$= \frac{\overset{1}{\cancel{2}} \cdot 2 \cdot \overset{1}{\cancel{3}} \cdot \overset{1}{\cancel{5}} \cdot \overset{1}{\cancel{r}} \cdot \overset{1}{\cancel{r}} \cdot \overset{1}{\cancel{r}} \cdot \overset{1}{\cancel{s}}}{\underset{1}{\cancel{2}} \cdot \underset{1}{\cancel{3}} \cdot \underset{1}{\cancel{5}} \cdot \underset{1}{\cancel{r}} \cdot \underset{1}{\cancel{r}} \cdot \underset{1}{\cancel{r}} \cdot \underset{1}{\cancel{s}} \cdot s} \text{ or } \frac{2}{s} \quad \textit{Simplify.}$$

Method 2 $\frac{6r^2}{5s^2} \cdot \frac{10rs}{6r^3} = \frac{\overset{1\,1}{\cancel{6}\cancel{r^2}}}{\underset{1\,s}{\cancel{5}\cancel{s^2}}} \cdot \frac{\overset{2\,1\,1}{\cancel{10}\cancel{r}\cancel{s}}}{\underset{1\,1}{\cancel{6}\cancel{r^3}}}$ *Simplify.*

$$= \frac{2}{s} \quad \textit{Multiply.}$$

Resource Manager

Reproducible Masters

Chapter 15 Resource Masters
- *Study Guide*, p. 668
- *Skills Practice*, p. 669
- *Practice*, p. 670
- *Reading to Learn Mathematics*, p. 671
- *Enrichment*, p. 672
- *Assessment*, p. 703

Transparencies

5-Minute Check, 15–2

Technology/Multimedia

Interactive Chalkboard CD-ROM

2 $\dfrac{3m}{n+2} \cdot \dfrac{n+2}{9m^2}$

$$\frac{3m}{n+2} \cdot \frac{n+2}{9m^2} = \frac{\overset{1\,1}{\cancel{3m}}}{\underset{1}{\cancel{n+2}}} \cdot \frac{\overset{1}{\cancel{n+2}}}{\underset{3\,m}{\cancel{9m^2}}}$$ *3, m, and n + 2 are common factors.*

$$= \frac{1}{3m}$$ *Simplify.*

3 $\dfrac{y-3}{y+5} \cdot \dfrac{2y^2+10y}{2y-6}$

$$\frac{y-3}{y+5} \cdot \frac{2y^2+10y}{2y-6} = \frac{y-3}{y+5} \cdot \frac{2y(y+5)}{2(y-3)}$$ *Factor $2y^2 + 10y$ and $2y - 6$.*

$$= \frac{\overset{1}{\cancel{y-3}}}{\underset{1}{\cancel{y+5}}} \cdot \frac{\overset{1}{\cancel{2}}y\overset{1}{\cancel{(y+5)}}}{\underset{1}{\cancel{2}}\underset{1}{\cancel{(y-3)}}}$$ *2, y − 3, and y + 5 are common factors.*

$$= \frac{y}{1} \text{ or } y$$ *Simplify.*

4 $\dfrac{x^2-25}{x^2-3x-10} \cdot \dfrac{x+2}{x}$

$$\frac{x^2-25}{x^2-3x-10} \cdot \frac{x+2}{x} = \frac{(x+5)(x-5)}{(x-5)(x+2)} \cdot \frac{x+2}{x}$$ *Factor $x^2 - 25$ and $x^2 - 3x - 10$.*

$$= \frac{(x+5)\overset{1}{\cancel{(x-5)}}}{\underset{1}{\cancel{(x-5)}}\underset{1}{\cancel{(x+2)}}} \cdot \frac{\overset{1}{\cancel{x+2}}}{x}$$ *x − 5 and x + 2 are common factors.*

$$= \frac{x+5}{x}$$ *Simplify.*

Your Turn Find each product.

a. $\dfrac{12x}{5y} \cdot \dfrac{20y^2}{36x^2}$ $\dfrac{4y}{3x}$

b. $\dfrac{x+2}{8} \cdot \dfrac{12x}{(x+2)(x-2)}$ $\dfrac{3x}{2(x-2)}$

c. $\dfrac{n-2}{6} \cdot \dfrac{3}{n^2-2n}$ $\dfrac{1}{2n}$

d. $\dfrac{a^2+6a+9}{a^2-4} \cdot \dfrac{a-2}{a+3}$ $\dfrac{a+3}{a+2}$

Look Back

Reciprocal: Lesson 4–3

To divide a rational number by any nonzero number, multiply by its reciprocal. You can use the same method to multiply rational expressions.

$$\frac{2}{3} \div \frac{1}{2} = \frac{2}{3} \cdot \frac{2}{1}$$ *The reciprocal of $\frac{1}{2}$ is $\frac{2}{1}$.*

$$= \frac{4}{3}$$

$$\frac{5}{x} \div \frac{y}{z} = \frac{5}{x} \cdot \frac{z}{y}$$ *The reciprocal of $\frac{y}{z}$ is $\frac{z}{y}$.*

$$= \frac{5z}{xy}$$

2 TEACH

In-Class Examples

Examples 1–4

Find each product.

1. $\dfrac{10a^2}{8ab} \cdot \dfrac{4ab}{10a^5}$ $\dfrac{1}{2a^3}$
2. $\dfrac{4a^2}{b-5} \cdot \dfrac{b-5}{12a}$ $\dfrac{a}{3}$
3. $\dfrac{z+3}{z-5} \cdot \dfrac{3z^2-15z}{4z+12}$ $\dfrac{3z}{4}$
4. $\dfrac{m-4}{m^2} \cdot \dfrac{m^2-1}{m^2-3m-4}$ $\dfrac{m-1}{m^2}$

Teaching Tip To show students why the reciprocal of $\frac{y}{z}$ is $\frac{z}{y}$, you may want to guide them through the steps below.

$$\frac{1}{\frac{y}{z}} = \frac{1 \cdot \frac{z}{y}}{\frac{y}{z} \cdot \frac{z}{y}} = \frac{\frac{z}{y}}{\frac{yz}{zy}} = \frac{\frac{z}{y}}{1} = \frac{z}{y}$$

In-Class Examples

Example 5

One lap around a school campus is $1\frac{1}{4}$ miles. How many laps will you have to complete to run 10 miles? **8 laps**

Examples 6–7

Find each quotient.

6 $\frac{15a^3}{b^3} \div \frac{5a^2}{2b}$ $\frac{6a}{b^2}$

7 $\frac{6x-6}{x-3} \div (x-1)$ $\frac{6}{x-3}$

Real World

Example 5 Sports Link

The Indianapolis 500 is a 500-mile automobile race. Each lap is $2\frac{1}{2}$ miles long. How many laps does a driver complete to race the entire 500 miles?

To find the number of laps, divide 500 by $2\frac{1}{2}$.

$500 \div 2\frac{1}{2} = \frac{500}{1} \div \frac{5}{2}$ *Express 500 and $2\frac{1}{2}$ as improper fractions.*

$= \frac{500}{1} \cdot \frac{2}{5}$ *The reciprocal of $\frac{5}{2}$ is $\frac{2}{5}$.*

$= \frac{\cancelto{100}{500}}{1} \cdot \frac{2}{\cancelto{1}{5}}$ *5 is a common factor.*

$= \frac{200}{1}$ or 200 *Simplify.*

A driver completes 200 laps.

Examples

Find each quotient.

6 $\frac{6x^3}{y} \div \frac{2x}{y^2}$

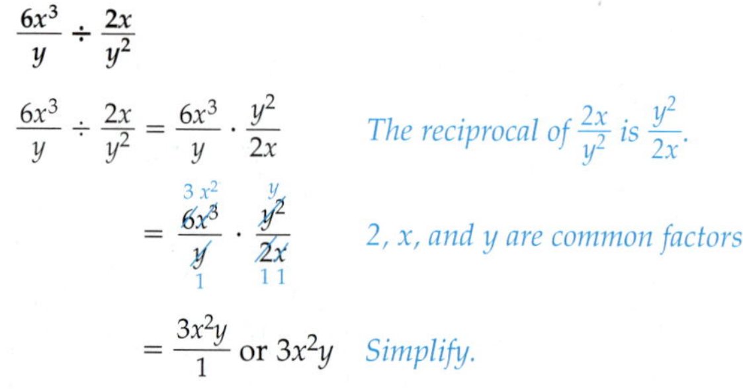

$\frac{6x^3}{y} \div \frac{2x}{y^2} = \frac{6x^3}{y} \cdot \frac{y^2}{2x}$ *The reciprocal of $\frac{2x}{y^2}$ is $\frac{y^2}{2x}$.*

$= \frac{\cancelto{3x^2}{6x^3}}{\cancelto{1}{y}} \cdot \frac{\cancelto{y}{y^2}}{\cancelto{1\,1}{2x}}$ *2, x, and y are common factors.*

$= \frac{3x^2y}{1}$ or $3x^2y$ *Simplify.*

7 $\frac{2m+8}{m+5} \div (m+4)$

$\frac{2m+8}{m+5} \div (m+4) = \frac{2m+8}{m+5} \cdot \frac{1}{m+4}$ *The reciprocal of $(m+4)$ is $\frac{1}{m+4}$.*

$= \frac{2(m+4)}{m+5} \cdot \frac{1}{m+4}$ *Factor $2m+8$.*

$= \frac{2\cancelto{1}{(m+4)}}{m+5} \cdot \frac{1}{\cancelto{1}{m+4}}$ *$(m+4)$ is a common factor.*

$= \frac{2}{m+5}$ *Simplify.*

Your Turn

e. $\frac{12y}{7z^2} \div \frac{3xy}{2z}$ $\frac{8}{7xz}$

f. $\frac{x^2+4x+3}{x^2} \div (x+3)$ $\frac{x+1}{x^2}$

Inclusion Strategies

The process of calling for the reciprocal of a rational expression is a part of each division exercise. If there are hearing impaired students in your class, work with them to come up with a symbol that you or they can show to represent calling for a reciprocal.

Sometimes it is necessary to factor -1 from one of the terms.

Example  Find $\frac{x^2-4}{2y} \div \frac{2-x}{6xy}$.

$\frac{x^2-4}{2y} \div \frac{2-x}{6xy} = \frac{x^2-4}{2y} \cdot \frac{6xy}{2-x}$ *The reciprocal of $\frac{2-x}{6xy}$ is $\frac{6xy}{2-x}$.*

$= \frac{(x+2)(x-2)}{2y} \cdot \frac{6xy}{2-x}$ *Factor x^2-4.*

$= \frac{(x+2)(x-2)}{2y} \cdot \frac{6xy}{-1(x-2)}$ *Factor -1 from $2-x$.*

$= \frac{(x+2)\cancel{(x-2)}^{1}}{\cancel{2y}_{1\,1}} \cdot \frac{\cancel{6xy}^{3\,1}}{-1\cancel{(x-2)}_{1}}$ *$x-2$, y, and 2 are common factors.*

$= \frac{3x(x+2)}{-1}$ or $-3x(x+2)$ *Simplify.*

Reading Algebra

Instead of factoring -1 from $2-x$, you could have factored -1 from $x-2$ and obtained the same results.

Your Turn

g. Find $\frac{x^2-9}{8x^3} \div \frac{6-2x}{4x}$ $\frac{x+3}{-4x^2}$

Check for Understanding

Communicating Mathematics

1. **Identify** and correct the error that was made while finding the quotient. **See margin.**

$$\frac{a^2-25}{3a} \div \frac{a+5}{15a^2} = \frac{3a}{(a+5)(a-5)} \cdot \frac{a+5}{15a^2}$$

$$= \frac{\cancel{3a}^{1\,1}}{\cancel{(a+5)}_{1}(a-5)} \cdot \frac{\cancel{a+5}^{1}}{\cancel{15a^2}_{5\,a}} \text{ or } \frac{1}{5a(a-5)}$$

2. *Writing Math* Write a short paragraph explaining which method you prefer when finding the product of rational expressions: to simplify first and then multiply or to multiply first and then simplify. Include examples to support your point of view. **See students' work.**

Guided Practice

Getting Ready **Find the reciprocal of each expression.**

Sample 1: $\frac{m}{2}$	**Sample 2:** $(a+b)$
Solution: The reciprocal is $\frac{2}{m}$.	**Solution:** The reciprocal is $\frac{1}{a+b}$.

3. $\frac{x^2}{4}$ $\frac{4}{x^2}$
4. y $\frac{1}{y}$
5. $\frac{x-4}{x+5}$ $\frac{x+5}{x-4}$
6. $y+6$ $\frac{1}{y+6}$

In-Class Example

Example 8

Find $\frac{x^2-36}{4x^2y} \div \frac{6-x}{4x^2y^5}$. $-y^4(x+6)$

3 PRACTICE/APPLY

Error Analysis

Watch for students who write the reciprocal of $y+6$ in Exercise 6 as $\frac{1}{y} + \frac{1}{6}$.

Prevent by stressing that the reciprocal of an *expression* is the number 1 divided by the entire expression.

Answer

1. The reciprocal of $\frac{a^2-25}{3a}$ was used instead of the reciprocal of $\frac{a+5}{15a^2}$.

***Study Guide*, p. 668**

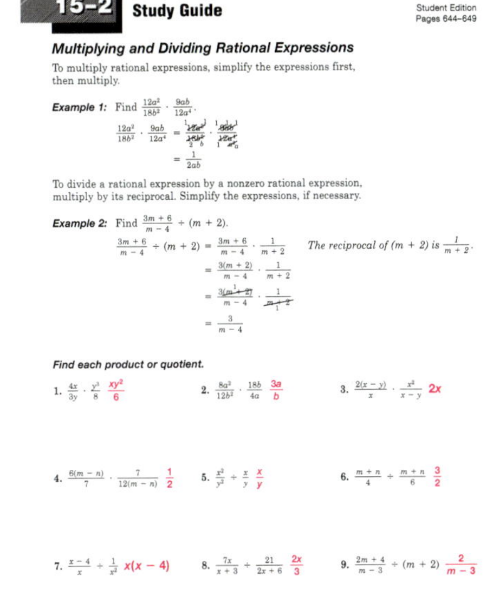

15–2 NAME ______ DATE ______ PERIOD ______

Study Guide Student Edition Pages 644–649

Multiplying and Dividing Rational Expressions

To multiply rational expressions, simplify the expressions first, then multiply.

Example 1: Find $\frac{12a^2}{18b^2} \cdot \frac{9ab}{12a^4}$.

$\frac{12a^2}{18b^2} \cdot \frac{9ab}{12a^4} = \frac{\cancel{12a^2}}{\cancel{18b^2}} \cdot \frac{\cancel{9ab}}{\cancel{12a^4}}$

$= \frac{1}{2ab}$

To divide a rational expression by a nonzero rational expression, multiply by its reciprocal. Simplify the expressions, if necessary.

Example 2: Find $\frac{3m+6}{m-4} \div (m+2)$.

$\frac{3m+6}{m-4} \div (m+2) = \frac{3m+6}{m-4} \cdot \frac{1}{m+2}$ *The reciprocal of $(m+2)$ is $\frac{1}{m+2}$.*

$= \frac{3(m+2)}{m-4} \cdot \frac{1}{m+2}$

$= \frac{3\cancel{(m+2)}^{1}}{m-4} \cdot \frac{1}{\cancel{m+2}_{1}}$

$= \frac{3}{m-4}$

Find each product or quotient.

1. $\frac{4x}{3y} \cdot \frac{y^3}{8}$ $\frac{xy^2}{6}$
2. $\frac{8a^2}{12b^2} \cdot \frac{18b}{4a}$ $\frac{3a}{b}$
3. $\frac{2(x-y)}{x} \cdot \frac{x^2}{x-y}$ $2x$
4. $\frac{6(m-n)}{7} \cdot \frac{7}{12(m-n)}$ $\frac{1}{2}$
5. $\frac{x^2}{y^2} \div \frac{x}{y}$ $\frac{x}{y}$
6. $\frac{m+n}{4} \div \frac{m+n}{6}$ $\frac{3}{2}$
7. $\frac{x-4}{x} \div \frac{1}{x^2}$ $x(x-4)$
8. $\frac{7x}{x+3} \div \frac{21}{2x+6}$ $\frac{2x}{3}$
9. $\frac{2m+4}{m-3} \div (m+2)$ $\frac{2}{m-3}$

Reteaching Activity

Verbal/Linguistic Learners Have students working in pairs or small groups select one of the lesson exercises. Have one student explain in detail the steps of the solution to the rest of the group. Have groups repeat the activity with different problems until all students have a chance to explain the solution to a problem.

Assignment Guide

Basic: 17–43 odd, 44–54
Average: 16–40 even, 42–54

Examples 1–4 Find each product.

7. $\frac{a^2b}{b^2c} \cdot \frac{c}{d}$ $\frac{a^2}{bd}$

8. $\frac{5(n-1)}{-3} \cdot \frac{9}{n-1}$ -15

9. $\frac{2x-10}{x^2+x-12} \cdot \frac{x-3}{x-5}$ $\frac{2}{x+4}$

10. $\frac{y^2+3y-10}{2y} \cdot \frac{y^2-3y}{y^2-5y+6}$ $\frac{y+5}{2}$

Examples 6–8 Find each quotient.

11. $\frac{12a^3}{bc} \div \frac{3a^2}{bc}$ $4a$

12. $\frac{9xy}{5} \div 3x^2y^2$ $\frac{3}{5xy}$

13. $\frac{3x^2+6x}{x} \div \frac{2x+4}{x^2}$ $\frac{3x^2}{2}$

14. $\frac{x^2-5x+6}{2x^2} \div \frac{3-x}{4}$ $\frac{-2(x-2)}{x^2}$

Example 5

15. **Sewing** It takes $1\frac{1}{2}$ yards of fabric to make one flag for the school color guard. How many flags can be made from 18 yards of fabric? **12 flags**

Exercises

Practice

For Exercises	See Examples
16–18	1
19–27, 40	2–4
28–39, 41	6–8
42–43	5

Homework Help

Extra Practice
See page 722.

Find each product.

16. $\frac{ab}{ac} \cdot \frac{c}{d}$ $\frac{b}{d}$

17. $\frac{3x}{2y} \cdot \frac{y^2}{6}$ $\frac{xy}{4}$

18. $\frac{6a^2}{8n^2} \cdot \frac{12n}{9a}$ $\frac{a}{n}$

19. $\frac{3(a-b)}{a} \cdot \frac{a^2}{a-b}$ $3a$

20. $\frac{2(a+2b)}{5} \cdot \frac{5}{3(a+2b)}$

21. $\frac{7s}{s+2} \cdot \frac{2s+4}{21}$ $\frac{2s}{3}$

22. $\frac{3x+30}{2x} \cdot \frac{4x}{4x+40}$ $\frac{3}{2}$

23. $\frac{x+3}{x+4} \cdot \frac{x}{x^2+7x+12}$ $\frac{x}{(x+4)^2}$

24. $\frac{3a-6}{a^2-9} \cdot \frac{a+3}{a^2-2a}$ $\frac{3}{a(a-3)}$

25. $\frac{x^2-9}{x+7} \cdot \frac{2x+14}{x^2+6x+9}$ $\frac{2(x-3)}{x+3}$

26. $\frac{x}{x^2+8x+15} \cdot \frac{2x+10}{x^2}$ $\frac{2}{x(x+3)}$

27. $\frac{n^2}{n^2-4} \cdot \frac{n^2-5n+6}{n^2-3n}$ $\frac{n}{n+2}$

Find each quotient.

28. $\frac{a^2}{b} \div \frac{a^2}{b^2}$ b

29. $\frac{7a^2b}{xy} \div \frac{7}{6xy}$ $6a^2b$

30. $2xz \div \frac{4xy}{z}$ $\frac{z^2}{2y}$

31. $\frac{2a^3}{a+1} \div \frac{a^2}{a+1}$ $2a$

32. $\frac{b^2-9}{4b} \div (b-3)$

33. $\frac{y^2+8y+16}{y^2} \div (y+4)$

34. $\frac{y^2}{y+2} \div \frac{y}{y+2}$ y

35. $\frac{m^2+2m+1}{2} \div \frac{m+1}{m-1}$

36. $\frac{x^2-4x+4}{3x} \div \frac{x^2-4}{6}$ $\frac{2(x-2)}{x(x+2)}$

37. $\frac{x^2+7x+10}{x-1} \div \frac{x^2+2x-15}{1-x}$

38. $\frac{x^2-16}{16-x^2} \div \frac{7}{x}$ $\frac{-x}{7}$

39. $\frac{a^2-9}{9} \div \frac{6-2a}{27a^2}$ $\frac{-3a^2(a+3)}{2}$

40. Find the product of $\frac{y-3}{8}$ and $\frac{12}{y-3}$. $\frac{3}{2}$

41. What is the quotient when $\frac{x^2}{x^2-y^2}$ is divided by $\frac{x^2}{x+y}$? $\frac{1}{x-y}$

20. $\frac{2}{3}$

32. $\frac{b+3}{4b}$

33. $\frac{y+4}{y^2}$

35. $\frac{(m+1)(m-1)}{2}$

37. $\frac{-(x+2)}{x-3}$

Family Activity

Many grocery stores' pricing displays include a cost-per-unit price. Either by using these displays or a calculator, students can find the cost-per-unit price for various items while shopping with a family member. For each cost-per-unit price, have students find and interpret the reciprocal, which will give the number of units that can be bought per dollar or per cent. Students can share their family's findings with the class.

***Skills Practice*, p. 669, and *Practice*, p. 670 (shown)**

15-2 NAME ______ DATE ______ PERIOD ______

Practice

Student Edition Pages 644–649

Multiplying and Dividing Rational Expressions

Find each product.

1. $\frac{3x^2}{2y} \cdot \frac{y^2}{9}$ $\frac{x^2y}{6}$

2. $\frac{4a^2b}{6b^2c} \cdot \frac{3ab}{2c}$ $\frac{a^3}{c^2}$

3. $\frac{7n}{n-2} \cdot \frac{3(n-2)}{28}$ $\frac{3n}{4}$

4. $\frac{2}{m(m+3)} \cdot \frac{3m+9}{6}$ $\frac{1}{m}$

5. $\frac{4y+8}{y^2-2y} \cdot \frac{y-2}{y+2}$ $\frac{4}{y}$

6. $\frac{x^2-49}{x^2+5x} \cdot \frac{x+5}{x+7}$ $\frac{x-7}{x}$

7. $\frac{5x+25}{x^2-5x+6} \cdot \frac{x-3}{x+5}$ $\frac{5}{x-2}$

8. $\frac{a+5}{3a+6} \cdot \frac{3a^2+6a}{a^2+2a-15}$ $\frac{a}{a-3}$

9. $\frac{x^2+8x+12}{4x-12} \cdot \frac{2x-6}{x^2+4x-12}$ $\frac{x+2}{2(x-2)}$

10. $\frac{2n^2-10n}{n^2-9n+20} \cdot \frac{n^2-8n+16}{4n^2}$ $\frac{n-4}{2n}$

Find each quotient.

11. $\frac{4a^3}{b^2c} \div \frac{2a}{bc}$ $\frac{2a^2}{b}$

12. $\frac{15x^2y^2}{3} \div 3xy$ $\frac{5xy}{3}$

13. $\frac{3y+9}{y+2} \div (y+3)$ $\frac{3}{y+2}$

14. $\frac{8n^3}{n-4} \div \frac{4n}{n-4}$ $2n^2$

15. $\frac{6x^2y}{3y} \div 2xy$ $\frac{x}{y}$

16. $\frac{b^2-81}{b} \div (b+9)$ $\frac{b-9}{b}$

17. $\frac{6x^2+36x}{4x} \div \frac{4x+24}{2x^2}$ $\frac{3x^2}{4}$

18. $\frac{y^2+5y-14}{9y} \div \frac{y^2-8y+12}{3y^2}$ $\frac{y(y+7)}{3(y-6)}$

19. $\frac{x^2-2x-15}{x-2} \div \frac{x^2-10x+25}{2-x}$ $\frac{-(x+3)}{x-5}$

20. $\frac{y^2-8y+7}{5y^2} \div \frac{7-y}{10y}$ $\frac{-2(y-1)}{y}$

Applications and Problem Solving

42. **Carpentry** How many boards, each 2 feet 8 inches long, can be cut from a board 16 feet 6 inches long? **6 boards**

43. **Probability** Two darts are randomly thrown one at a time. Assume that both hit the target.
   a. What is the probability that both will hit the shaded region?
   b. Find the probability if $x = 3$. $\frac{81}{625}$ **or about 13%**

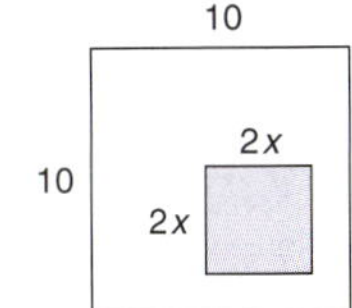

43a. $\frac{x^4}{625}$

44. **Critical Thinking** Find two different pairs of rational expressions whose product is $\frac{5x^2}{8y^3}$. **Sample answers:** $\frac{5x}{2y} \cdot \frac{x}{4y^2}$, $\frac{5x^2}{8} \cdot \frac{1}{y^3}$

Mixed Review

Simplify. *(Lesson 15–1)*

45. $\frac{14a^2b^4c}{2a^3bc}$ $\frac{7b^3}{a}$

46. $\frac{m^2 - 16}{m^2 - 8m + 16}$ $\frac{m + 4}{m - 4}$

47. $\frac{8 - 4x}{x^2 - 4x + 4}$ $\frac{-4}{x - 2}$

48. **Cycling** You can use the formula $s = 4\sqrt{r}$ to find the fastest speed at which a cyclist can turn a corner safely and not tip over. In the formula, s is the speed in miles per hour, and r is the radius of the corner in feet. If $s = 8$ miles per hour, find r. *(Lesson 14–5)* **4 ft**

Use elimination to solve each system of equations. *(Lesson 13–5)*

49. $x + y = 4$
$2x - 3y = -7$ **(1, 3)**

50. $x + 3y = -4$
$x - 2y = 6$ **(2, −2)**

51. $3a + 4b = -25$
$2a - 3b = 6$ **(−3, −4)**

52. Graph $2x + 3y \leq 12$. *(Lesson 12–7)* **See margin.**

Standardized Test Practice

53. **Extended Response** Paul wants to start a small lawn-care company to earn money in the summer. He estimates that his expenses will be \$200 for gasoline and equipment. He plans to charge \$25 per lawn. The equation $P = 25n - 200$ can be used to find his profits. In the equation, n represents the number of lawns he cuts. *(Lesson 6–2)*
   a. Determine the ordered pairs that satisfy the equation if the domain is {5, 8, 10}. **(5, −75), (8, 0), (10, 50)**
   b. Graph the relation. **See margin.**

54. **Multiple Choice** A 60-kilogram mass is 140 centimeters from the fulcrum of a lever. How far from the fulcrum must an 80-kilogram mass be to balance the lever? *(Lesson 4–4)* **D**

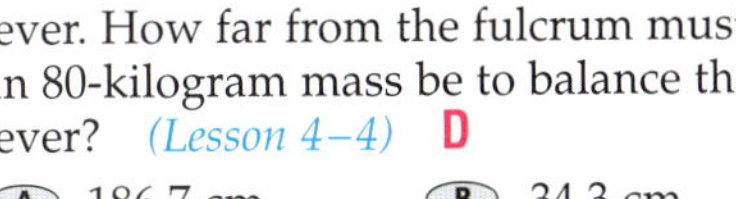

A 186.7 cm
B 34.3 cm
C 120 cm
D 105 cm

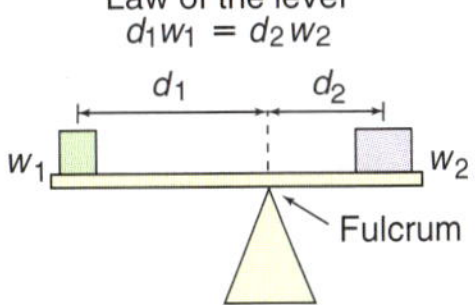

www.algconcepts.com/self_check_quiz

4 ASSESS

Open-Ended Assessment

Writing Have students write a brief summary detailing how to find the quotient of two rational expressions.

Chapter 15, Quiz A (Lessons 15–1 and 15–2) is available in the *Chapter 15 Resource Masters*, p. 703.

Answers

52.

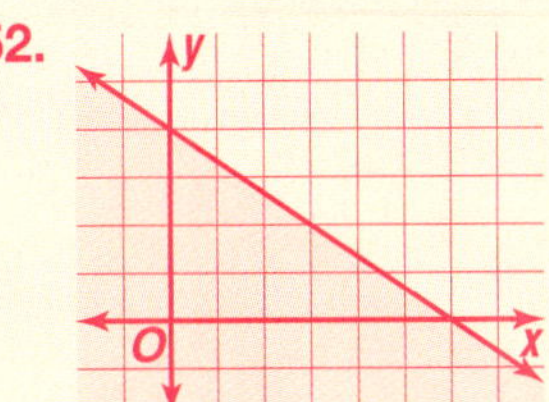

53b.

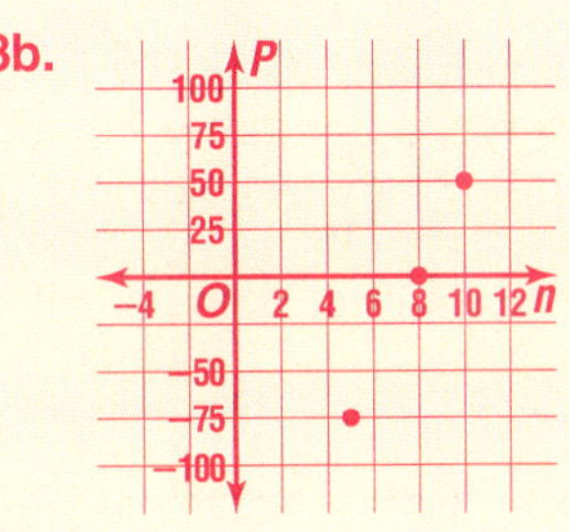

***Enrichment*, p. 672**

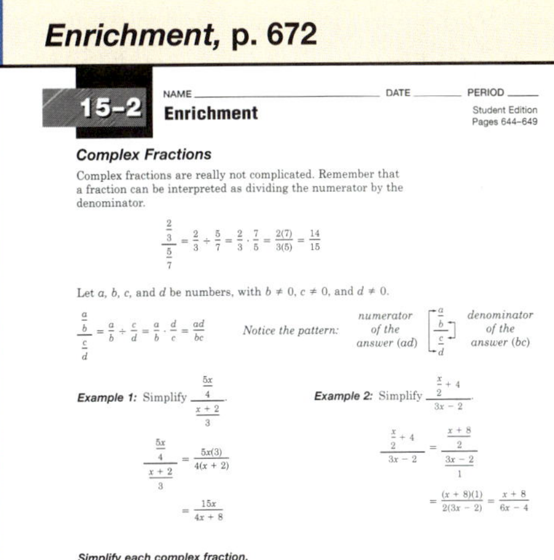

NAME ______ DATE ______ PERIOD ______

15–2 Enrichment

Student Edition Pages 644–649

Complex Fractions

Complex fractions are really not complicated. Remember that a fraction can be interpreted as dividing the numerator by the denominator.

$$\frac{\frac{2}{3}}{\frac{5}{7}} = \frac{2}{3} \div \frac{5}{7} = \frac{2}{3} \cdot \frac{7}{5} = \frac{2(7)}{3(5)} = \frac{14}{15}$$

Let a, b, c, and d be numbers, with $b \neq 0$, $c \neq 0$, and $d \neq 0$.

$$\frac{\frac{a}{b}}{\frac{c}{d}} = \frac{a}{b} \div \frac{c}{d} = \frac{a}{b} \cdot \frac{d}{c} = \frac{ad}{bc}$$

Notice the pattern: *numerator of the answer (ad)* $\left[\frac{\frac{a}{b}}{\frac{c}{d}}\right]$ *denominator of the answer (bc)*

Example 1: Simplify $\frac{\frac{5x}{4}}{\frac{x+2}{3}}$.

$$\frac{\frac{5x}{4}}{\frac{x+2}{3}} = \frac{5x(3)}{4(x+2)} = \frac{15x}{4x+8}$$

Example 2: Simplify $\frac{\frac{x}{2}+4}{3x-2}$.

$$\frac{\frac{x}{2}+4}{3x-2} = \frac{\frac{x+8}{2}}{\frac{3x-2}{1}} = \frac{(x+8)(1)}{2(3x-2)} = \frac{x+8}{6x-4}$$

Simplify each complex fraction.

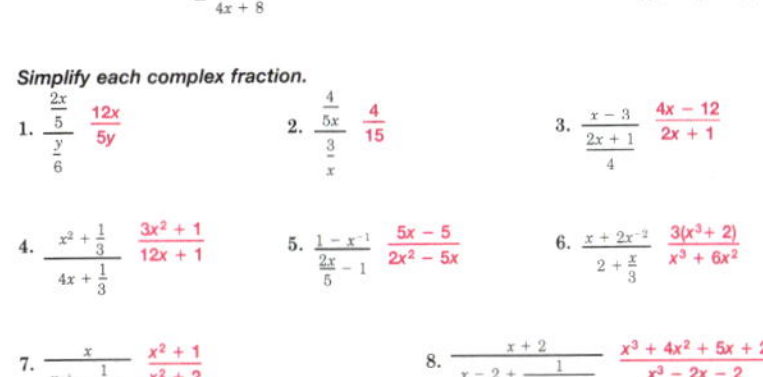

1. $\frac{\frac{2x}{5}}{\frac{y}{6}}$ $\frac{12x}{5y}$

2. $\frac{\frac{4}{5x}}{\frac{3}{x}}$ $\frac{4}{15}$

3. $\frac{x-3}{\frac{2x+1}{4}}$ $\frac{4x-12}{2x+1}$

4. $\frac{x^2+\frac{1}{3}}{4x+\frac{1}{3}}$ $\frac{3x^2+1}{12x+1}$

5. $\frac{1-x^{-1}}{\frac{2x}{5}-1}$ $\frac{5x-5}{2x^2-5x}$

6. $\frac{x+2x^{-2}}{2+\frac{x}{3}}$ $\frac{3(x^3+2)}{x^3+6x^2}$

7. $\frac{x}{x+\frac{1}{x+\frac{1}{x}}}$ $\frac{x^2+1}{x^2+2}$

8. $\frac{x+2}{x-2+\frac{1}{x+2+\frac{1}{x}}}$ $\frac{x^3+4x^2+5x+2}{x^3-2x-2}$

Extra Credit

What is the quotient of the expressions $\frac{a^2 - 2a - 3}{2a^2 + 4a}$ and $\frac{a^2 - 5a + 6}{8a^2 + 16a}$ divided by $\frac{4a + 4}{a - 2}$? **1**

Lesson 15–3

15–3 Dividing Polynomials

1 FOCUS

5-Minute Check
Lesson 15–2

Find each product.

1. $\frac{8t}{2t+4} \cdot \frac{5t+10}{2t^2}$ $\frac{10}{t}$
2. $\frac{n^2+n-12}{n+2} \cdot \frac{3n+6}{n^2-16}$ $\frac{3(n-3)}{n-4}$

Find each quotient.

3. $\frac{a^2b^2}{ac} \div \frac{b^2c^2}{bc}$ $\frac{ab}{c^2}$
4. $\frac{12x^2y}{x^2-81} \div \frac{24xy^2}{9-x}$ $\frac{-x}{2y(x+9)}$
5. Two darts are randomly thrown one at a time. Assume that both hit the target. What is the probability that both will hit the shaded region? $\frac{x^4}{144}$

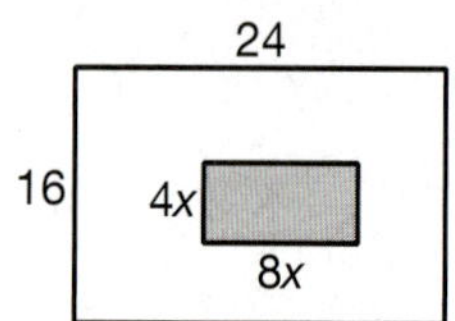

Motivating the Lesson

Hands-On Activity Have students use grid paper to divide 36 by 3, 4, and 6. To do this, have them draw a width for a rectangle equal to the divisor. Then have them extend the length until the rectangle formed has an area of 36 square units. Have students try dividing 36 by 5 in this way. Ask them what the extra tile represents if they form a 5-by-7 rectangle. This activity will lead directly to the Hands-On Algebra activity, in which students will use algebra tiles to divide polynomials.

TECHNOLOGY

An alternative technology option using a graphing calculator is available for teaching this lesson.

What You'll Learn
You'll learn to divide polynomials by binomials.

Why It's Important
Aviation You can use polynomials to determine distance, rate, and time. *See Exercise 32.*

In the previous lesson, you learned that some divisions can be performed using factoring.

$$(x^2 + 4x + 3) \div (x + 3) = \frac{x^2 + 4x + 3}{x + 3}$$

$$= \frac{\overset{1}{\cancel{(x+3)}}(x+1)}{\underset{1}{\cancel{(x+3)}}} \quad \textit{Factor the dividend.}$$

$$= x + 1$$

Therefore, $(x^2 + 4x + 3) \div (x + 3) = x + 1$. Since the remainder is 0, the divisor is a factor of the dividend.

You can also use algebra tiles to model the division shown above.

Materials:  algebra tiles  product mat

Step 1 Model the polynomial $x^2 + 4x + 3$. This represents the dividend.

Step 2 Place the x^2-tile at the corner of the mat. Arrange the three 1-tiles to make a length of $x + 3$. This represents the divisor.

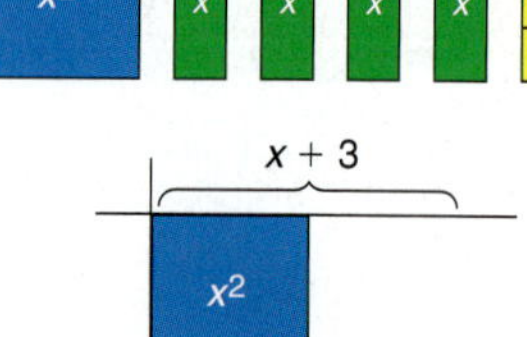

Step 3 Use the remaining tiles to make a rectangle. The width of the rectangle, $x + 1$, is the quotient.

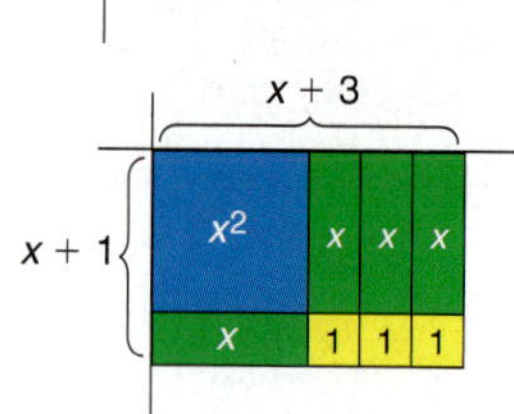

Try These

Use algebra tiles to find each quotient. Recall that you can add zero-pairs without changing the value of the polynomial.

1–4. See margin for models.

1. $(x^2 + 6x + 9) \div (x + 3)$ $x + 3$
2. $(x^2 - 4x + 4) \div (x - 2)$ $x - 2$
3. $(x^2 + x - 6) \div (x + 3)$ $x - 2$
4. $(x^2 - 9) \div (x + 3)$ $x - 3$
5. What happens when you try to model $(x^2 + 5x + 9) \div (x + 3)$? What do you think your result means? **See margin.**

Resource Manager

Reproducible Masters
Chapter 15 Resource Masters
- *Study Guide,* p. 673
- *Skills Practice,* p. 674
- *Practice,* p. 675
- *Reading to Learn Mathematics,* p. 676
- *Enrichment,* p. 677
- *Assessment,* p. 702

Graphing Calculator, pp. 44–45
Hands-On Algebra, p. 162

Transparencies
5-Minute Check, 15–3

Technology/Multimedia
Interactive Chalkboard CD-ROM

You can also divide polynomials using long division. Follow these steps to divide $2x^2 + 7x + 3$ by $2x + 1$.

Step 1 To find the first term of the quotient, divide the first term of the dividend, $2x^2$, by the first term of the divisor, $2x$.

$$\begin{array}{r l} x & \\ 2x + 1\overline{)2x^2 + 7x + 3} & 2x^2 \div 2x = x \\ \underline{(-)\ 2x^2 + 1x} & \textit{Multiply x and 2x + 1.} \\ 6x & \textit{Subtract.} \end{array}$$

Step 2 To find the next term of the quotient, divide the first term of the partial dividend, $6x$, by the first term of the divisor, $2x$.

$$\begin{array}{r l} x + 3 & \\ 2x + 1\overline{)2x^2 + 7x + 3} & \\ \underline{(-)\ 2x^2 + 1x} & \\ 6x + 3 & \textit{Bring down 3; } 6x \div 2x = 3. \\ \underline{(-)\ 6x + 3} & \textit{Multiply 3 and 2x + 1.} \\ 0 & \textit{Subtract.} \end{array}$$

Therefore, $(2x^2 + 7x + 3) \div (2x + 1) = x + 3$.

Examples

Find each quotient.

1 $(6y - 3) \div (2y - 1)$

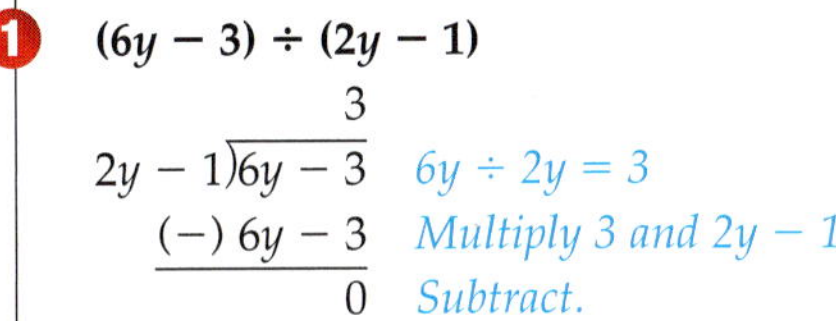

$$\begin{array}{r l} 3 & \\ 2y - 1\overline{)6y - 3} & 6y \div 2y = 3 \\ \underline{(-)\ 6y - 3} & \textit{Multiply 3 and 2y - 1.} \\ 0 & \textit{Subtract.} \end{array}$$

Therefore, $(6y - 3) \div (2y - 1) = 3$.

2 $(x^2 - 2x - 8) \div (x + 2)$

$$\begin{array}{r l} x - 4 & \\ x + 2\overline{)x^2 - 2x - 8} & \\ \underline{(-)\ x^2 + 2x} & \textit{Multiply x and x + 2.} \\ -4x - 8 & \textit{Subtract: } -2x - 2x = -4x\textit{; bring down } -8. \\ \underline{(-)\ -4x - 8} & \textit{Multiply } -4 \textit{ and x + 2.} \\ 0 & \textit{Subtract.} \end{array}$$

Therefore, $(x^2 - 2x - 8) \div (x + 2) = x - 4$.

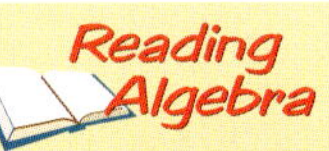

Reading Algebra

Since $x - 4$ is a factor of $x^2 - 2x - 8$, you can check the quotient by simplifying $\frac{x^2 - 2x - 8}{x + 2}$ or by multiplying $(x + 2)$ by $(x - 4)$.

Your Turn

a. $(15a - 6) \div (5a - 2)$ **3**

b. $(x^2 - 7x + 10) \div (x - 2)$ **$x - 5$**

2 TEACH

Teaching Tip Encourage students to use multiplication to check the division problems in Examples 1 and 2 using the fact that *quotient* × *divisor* = *dividend*.

In-Class Examples

Examples 1–2

Find each quotient.

1 $(15x + 10) \div (3x + 2)$ **5**

2 $(x^2 - x - 12) \div (x + 3)$ **$x - 4$**

Answers

Hands-On Algebra

1.

x^2	x	x	x
x	1	1	1
x	1	1	1
x	1	1	1

2.

x^2	$-x$	$-x$
$-x$	1	1
$-x$	1	1

3.

x^2	x	x	x
$-x$	-1	-1	-1
$-x$	-1	-1	-1

4.

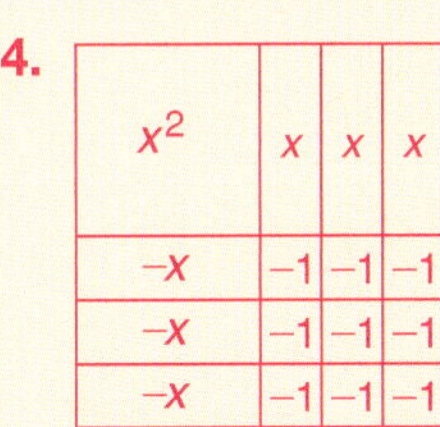

x^2	x	x	x
$-x$	-1	-1	-1
$-x$	-1	-1	-1
$-x$	-1	-1	-1

5. Sample answer: There are tiles left over. The quotient has a remainder.

Hands-On Algebra

Cooperative Learning Refer to the Hands-On Algebra on page 650. After students complete the exercises, you may want to have them try to model the polynomial division problem $(2x^2 + 7x + 3) \div (x + 3)$, in which the coefficient of the square term is not 1. Two models are shown at the right.

Hands-On Algebra Masters, p. 162

x^2	x^2	x
1	x	x
1	x	x
1	x	x

x^2	x	x	x
x^2	x	x	x
x	1	1	1

Teaching Tip Remind students that they can check division problems by multiplication even when there is a remainder, as in Examples 3 and 4, using the fact that *quotient* × *divisor* + *remainder* = *dividend*.

In-Class Examples

Examples 3–4

Find each quotient.

3 $(8a^2 + 14a + 9) \div (2a + 3)$ $4a + 1 + \frac{6}{2a+3}$

4 $(x^2 - 20) \div (x - 5)$ $x + 5 + \frac{5}{x-5}$

If the divisor is *not* a factor of the dividend, the remainder will not be 0. The quotient can be expressed as follows.

$$\text{quotient} = \text{partial quotient} + \frac{\text{remainder}}{\text{divisor}}$$

Example 3 **Find $(8y^2 - 2y + 1) \div (2y - 1)$.**

$$\begin{array}{r l}
4y + 1 & \\
2y - 1\overline{)8y^2 - 2y + 1} & \\
(-)\ 8y^2 - 4y & \textit{Multiply } 4y \textit{ and } 2y - 1. \\
2y + 1 & \textit{Subtract. Then bring down } 1. \\
(-)\ 2y - 1 & \textit{Multiply } 1 \textit{ and } 2y - 1. \\
2 & \textit{Subtract. The remainder is } 2.
\end{array}$$

The quotient is $4y + 1$ with remainder 2.

So, $(8y^2 - 2y + 1) \div (2y - 1) = 4y + 1 + \frac{2}{2y-1}$.

Your Turn

c. Find $(2a^2 + 7a + 3) \div (a + 2)$. $2a + 3 + \frac{-3}{a+2}$

In an expression like $x^2 - 4$ there is no x term. In such situations, rename the dividend using zero as the coefficient of the missing term.

Example 4 **Find $(x^2 - 4) \div (x + 1)$.**

$$\begin{array}{r l}
x - 1 & \\
x + 1\overline{)x^2 + 0x - 4} & \textit{Rename } x^2 - 4 \textit{ as } x^2 + 0x - 4. \\
(-)\ x^2 + 1x & \textit{Multiply } x \textit{ and } x + 1. \\
-1x - 4 & \textit{Subtract. Then bring down } -4. \\
-1x - 1 & \textit{Multiply } -1 \textit{ and } x + 1. \\
-3 & \textit{Subtract. The remainder is } -3.
\end{array}$$

Therefore, $(x^2 - 4) \div (x + 1) = x - 1 + \frac{-3}{x+1}$.

Your Turn

d. Find $(a^3 + 8a - 20) \div (a - 2)$. $a^2 + 2a + 12 + \frac{4}{a-2}$

If you know the area of a rectangle and the length of one side, you can find the width by dividing polynomials.

From the Classroom of ...

Jerome D. Hayden
Normal Community High School
Normal, Illinois

I introduce this lesson by reviewing arithmetic long division. Students can use similar steps when they divide polynomials.

Example 5 **Geometry Link**

Find the width of a rectangle if its area is $10x^2 + 29x + 21$ square units and its length is $2x + 3$ units.

$2x + 3$
$10x^2 + 29x + 21$

To find the width, divide the area $10x^2 + 29x + 21$ by the length $2x + 3$.

$$\begin{array}{r l}
5x + 7 & \\
2x + 3\overline{)10x^2 + 29x + 21} & \\
(-)\ 10x^2 + 15x & \textit{Multiply 5x and 2x + 3.} \\
14x + 21 & \textit{Subtract. Then bring down 21.} \\
(-)\ 14x + 21 & \textit{Multiply 7 and 2x + 3.} \\
0 & \textit{The remainder is 0.}
\end{array}$$

Therefore, the width of the rectangle is $5x + 7$ units.
You can check your answer by multiplying $(2x + 3)$ and $(5x + 7)$.

Check for Understanding

Communicating Mathematics

1. **Identify** the dividend, divisor, quotient, and remainder. **See margin.**
   $(3k^2 - 7k - 5) \div (3k + 2) = k - 3 + \frac{1}{3k + 2}$
2. **Write** a division problem represented by the model.
3. **Explain** how you know whether the quotient is a factor of the dividend. **See margin.**

2. $(x^2 - 3x - 4) \div (x - 4) = (x + 1)$

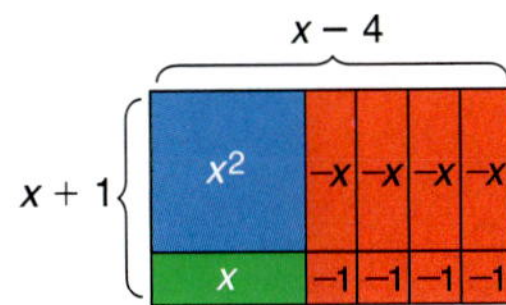

Guided Practice

Getting Ready Find each quotient.

Sample 1: $x^2 \div x$
Solution: x

Sample 2: $4x^2 \div 2x$
Solution: $2x$

4. $3y^2 \div y$ **$3y$**
5. $6a^2 \div 2a$ **$3a$**
6. $10x^3 \div 5x$ **$2x^2$**

Examples 1–4

Find each quotient.

7. $(x^2 + 4x) \div (x + 4)$ **x**
8. $(a^2 + 3a + 2) \div (a + 1)$ **$a + 2$**
9. $(2x^2 + 3x - 3) \div (2x - 1)$
   $x + 2 - \frac{1}{2x - 1}$
10. $(s^3 + 9) \div (s - 3)$
   $s^2 + 3s + 9 + \frac{36}{s - 3}$

Example 5

11. **Geometry** Find the length of a rectangle if its area is $2x^2 - 5x - 12$ square inches and its width is $x - 4$ inches. **$2x + 3$ in.**

$x - 4$	$2x^2 - 5x - 12$

In-Class Example

Example 5

Find the length of a rectangle if its area is $12x^2 + 13x + 3$ square units and its width is $3x + 1$ units. **$4x + 3$ units**

$3x + 1$	$12x^2 + 13x + 3$

3 PRACTICE/APPLY

Error Analysis

Watch for students who use the quotient as the denominator of a remainder, as in incorrectly writing the remainder in Exercise 1 as $\frac{1}{k - 3}$.

Prevent by having students write the general relationship $\frac{\text{dividend}}{\text{divisor}} = \text{quotient} + \frac{\text{remainder}}{\text{divisor}}$, and then rewriting this relationship using the specific dividend, divisor, quotient, and remainder.

Answers

1. **dividend: $3k^2 - 7k - 5$; divisor: $3k + 2$; quotient: $k - 3$; remainder: $\frac{1}{3k + 2}$**
3. **The quotient is a factor of the dividend if the remainder is 0.**

***Study Guide*, p. 673**

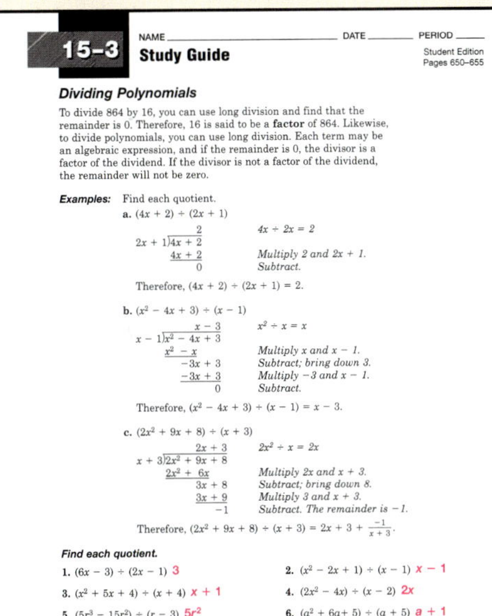

15–3 NAME ______ DATE ______ PERIOD ______

Study Guide Student Edition Pages 650–655

Dividing Polynomials

To divide 864 by 16, you can use long division and find that the remainder is 0. Therefore, 16 is said to be a **factor** of 864. Likewise, to divide polynomials, you can use long division. Each term may be an algebraic expression, and if the remainder is 0, the divisor is a factor of the dividend. If the divisor is not a factor of the dividend, the remainder will not be zero.

Examples: Find each quotient.

a. $(4x + 2) \div (2x + 1)$

$$\begin{array}{r l}
2 & 4x \div 2x = 2 \\
2x + 1\overline{)4x + 2} & \\
4x + 2 & \textit{Multiply 2 and 2x + 1.} \\
0 & \textit{Subtract.}
\end{array}$$

Therefore, $(4x + 2) \div (2x + 1) = 2$.

b. $(x^2 - 4x + 3) \div (x - 1)$

$$\begin{array}{r l}
x - 3 & x^2 \div x = x \\
x - 1\overline{)x^2 - 4x + 3} & \\
x^2 - x & \textit{Multiply x and x − 1.} \\
-3x + 3 & \textit{Subtract; bring down 3.} \\
-3x + 3 & \textit{Multiply −3 and x − 1.} \\
0 & \textit{Subtract.}
\end{array}$$

Therefore, $(x^2 - 4x + 3) \div (x - 1) = x - 3$.

c. $(2x^2 + 9x + 8) \div (x + 3)$

$$\begin{array}{r l}
2x + 3 & 2x^2 \div x = 2x \\
x + 3\overline{)2x^2 + 9x + 8} & \\
2x^2 + 6x & \textit{Multiply 2x and x + 3.} \\
3x + 8 & \textit{Subtract; bring down 8.} \\
3x + 9 & \textit{Multiply 3 and x + 3.} \\
-1 & \textit{Subtract. The remainder is −1.}
\end{array}$$

Therefore, $(2x^2 + 9x + 8) \div (x + 3) = 2x + 3 + \frac{-1}{x + 3}$.

Find each quotient.

1. $(6x - 3) \div (2x - 1)$ 3
2. $(x^2 - 2x + 1) \div (x - 1)$ $x - 1$
3. $(x^2 + 5x + 4) \div (x + 4)$ $x + 1$
4. $(2x^2 - 4x) \div (x - 2)$ $2x$
5. $(5r^3 - 15r^2) \div (r - 3)$ $5r^2$
6. $(a^2 + 6a + 5) \div (a + 5)$ $a + 1$
7. $(2a^2 - 5a - 3) \div (2a + 1)$ $a - 3$
8. $(6x^2 - 2x + 5) \div (x + 1)$ $6x - 8 + \frac{13}{x + 1}$

Reteaching Activity

Intrapersonal Learners Have students write a journal entry describing the similarities and differences between dividing whole numbers and dividing polynomials.

Assignment Guide

Basic: 13–31 odd, 33–42
Average: 12–28 even, 30–42
All: Quiz 1, 1–10

Answers

20. $c + 3 + \frac{9}{c+9}$

21. $t - 2 - \frac{32}{3t-4}$

22. $2m + 3 - \frac{3}{m+2}$

23. $b + 2 - \frac{4}{2b-1}$

24. $a^2 + 2a + 12 + \frac{3}{a-2}$

Exercises

Practice

Find each quotient.

12. $(12y - 4) \div (3y - 1)$ **4**
13. $(x^2 - 3x) \div (x - 3)$ **x**
14. $(8a^2 + 6a) \div (4a + 3)$ **$2a$**
15. $(6r^3 - 15r^2) \div (2r - 5)$ **$3r^2$**
16. $(a^2 + 6a + 5) \div (a + 5)$ **$a + 1$**
17. $(x^2 + x - 12) \div (x - 3)$ **$x + 4$**
18. $(s^2 + 11s + 18) \div (s + 2)$ **$s + 9$**
19. $(a^2 - 2a - 35) \div (a - 7)$ **$a + 5$**
20. $(c^2 + 12c + 36) \div (c + 9)$
21. $(3t^2 - 10t - 24) \div (3t - 4)$
22. $(2m^2 + 7m + 3) \div (m + 2)$
23. $(2b^2 + 3b - 6) \div (2b - 1)$
24. $(a^3 + 8a - 21) \div (a - 2)$
25. $(x^3 + 27) \div (x + 3)$ **$x^2 - 3x + 9$**
26. $(x^3 - 8) \div (x - 2)$ **$x^2 + 2x + 4$**
27. $(4x^4 - 2x^2 + x + 1) \div (x - 1)$ **27. $4x^3 + 4x^2 + 2x + 3 + \frac{4}{x-1}$**

20–24. See margin.

28. Find the quotient when $x^2 + 9x + 20$ is divided by $x + 4$. **$x + 5$**
29. What is the quotient when $2x^2 - 9x + 9$ is divided by $2x - 3$? **$x - 3$**

Homework Help	
For Exercises	See Examples
12–15	1
16–29, 32, 33	2–4
30	5
Extra Practice	
See page 722.	

Applications and Problem Solving

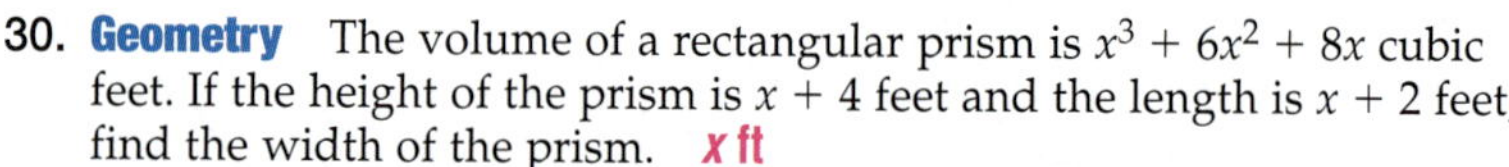

30. **Geometry** The volume of a rectangular prism is $x^3 + 6x^2 + 8x$ cubic feet. If the height of the prism is $x + 4$ feet and the length is $x + 2$ feet, find the width of the prism. **x ft**

31. **Transportation** The distance from San Francisco, CA, to New York, NY, is 2807 miles. To the nearest hour, find the number of hours it would take to travel this distance using each method of transportation at the given average speed. Use the formula $d = rt$.

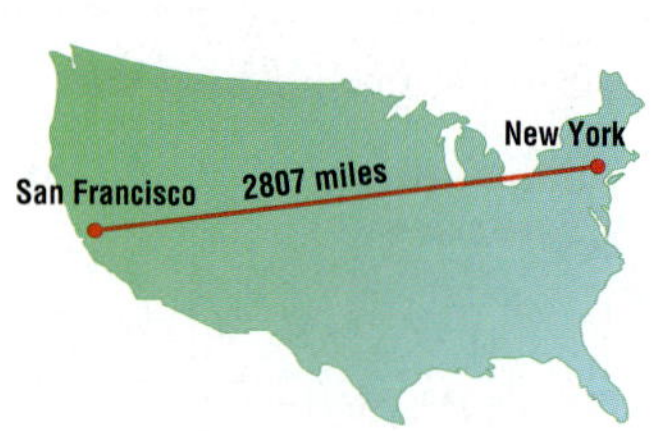

a. walking, 3 mph **936 h**
b. stagecoach, 20 mph **140 h**
c. automobile, 55 mph **51 h**
d. jumbo jet, 608 mph **5 h**

32. **Aviation** The distance d in miles flown by an airplane is given by the polynomial $x^2 + 501x + 500$. Suppose that $x + 500$ represents the speed r in miles per hour of the airplane. Use the formula $d = rt$ to find the following.
    a. Find the polynomial that represents the time t in hours. **$x + 1$**
    b. If $x = 5$, find the distance, rate, and time for the airplane. **3030 mi; 505 mph; 6 h**

33. **Critical Thinking** Find the value of k if the remainder is 15 when $x^3 - 7x^2 + 4x + k$ is divided by $x - 2$. **27**

Mixed Review

Find each product or quotient. *(Lesson 15–2)*

34. $\frac{3x+9}{x} \cdot \frac{x^2}{x^2-9}$ **$\frac{3x}{x-3}$**
35. $\frac{a^2}{b^2} \div \frac{a^2}{b^2}$ **1**

36. **Sports** When a professional football team from the west coast plays a professional football team from the east coast on television on Monday nights, the west coast team wins 64% of the time. Express 64% as a fraction in simplest form. **Source:** Stanford University Sleep Disorders Clinic *(Lesson 15–1)* **$\frac{16}{25}$**

***Skills Practice*, p. 674, and *Practice*, p. 675 (shown)**

15-3 NAME ______ DATE ______ PERIOD ______
Practice
Student Edition Pages 650–655

Dividing Polynomials

Find each quotient.

1. $(4x - 2) \div (2x - 1)$ **2**
2. $(y^2 + 5y) \div (y + 5)$ **y**
3. $(9a^2 + 6a) \div (3a + 2)$ **$3a$**
4. $(8n^3 - 4n^2) \div (4n - 2)$ **$2n^2$**
5. $(x^2 - 9x + 18) \div (x - 6)$ **$x - 3$**
6. $(b^2 - b - 20) \div (b - 5)$ **$b + 4$**
7. $(y^2 + 4y + 4) \div (y + 2)$ **$y + 2$**
8. $(m^2 - 5m - 6) \div (m + 1)$ **$m - 6$**
9. $(b^2 + 11b + 30) \div (b + 4)$ **$b + 7 + \frac{2}{b+4}$**
10. $(x^2 - 6x + 9) \div (x - 2)$ **$x - 4 + \frac{1}{x-2}$**
11. $(r^2 - 4) \div (r + 3)$ **$r - 3 + \frac{5}{r+3}$**
12. $(4x^2 + 6x + 5) \div (2x - 2)$ **$2x + 5 + \frac{15}{2x-2}$**
13. $(3n^2 - 11n + 8) \div (n - 3)$ **$3n - 2 + \frac{2}{n-3}$**
14. $(6y^2 + 5y - 3) \div (3y + 1)$ **$2y + 1 + \frac{-4}{3y+1}$**
15. $(s^3 - 1) \div (s - 1)$ **$s^2 + s + 1$**
16. $(a^3 + 4a + 16) \div (a + 2)$ **$a^2 - 2a + 8$**
17. $(m^3 - 9) \div (m - 2)$ **$m^2 + 2m + 4 + \frac{-1}{m-2}$**
18. $(x^3 - 7x - 8) \div (x + 1)$ **$x^2 - x - 6 + \frac{-2}{x+1}$**

Determine whether each system of equations has *one* solution, *no* solution, or *infinitely many* solutions. *(Lesson 13–2)*

37. 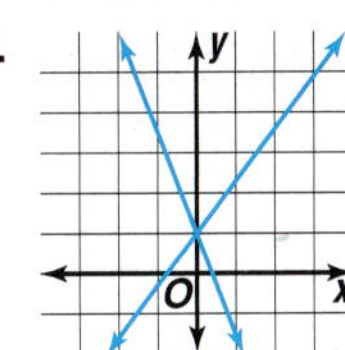

one

38. 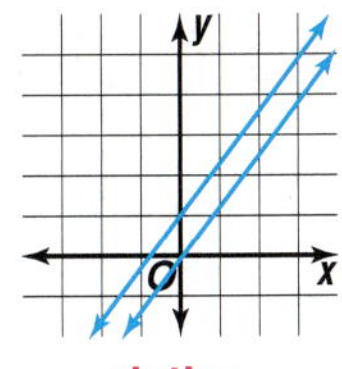

no solution

39. 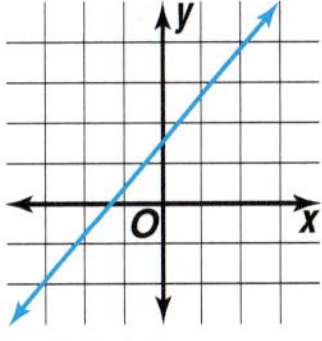

infinitely many

Solve each inequality. *(Lesson 12–4)*

40. $3x - 1 > 14$ $x > 5$

41. $-7y + 6 \leq 48$ $y \geq -6$

Standardized Test Practice

A B C D

42. **Multiple Choice** Which is the graph of $y = x^2 + 2$? *(Lesson 11–1)* D

A 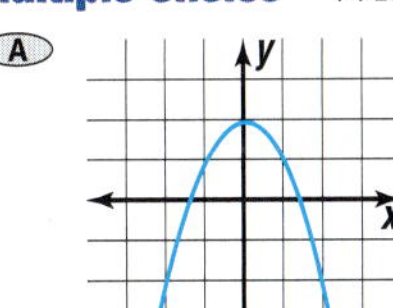

B 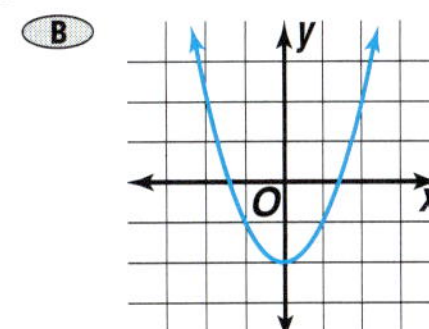

C 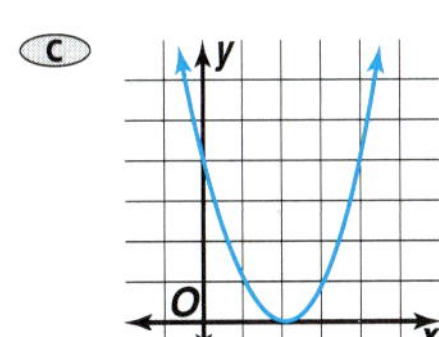

D 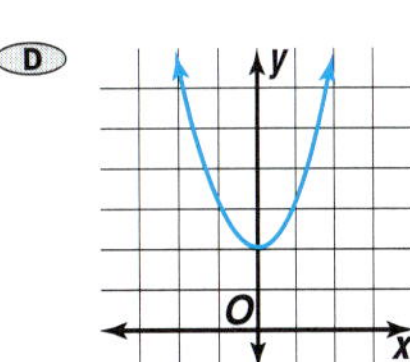

Quiz 1 Lessons 15–1 through 15–3

Find the excluded value(s) for each rational expression. Then simplify. *(Lesson 15–1)*

1. $\frac{m^2 - 2m}{m - 2}$ 2; m

2. $\frac{x^2 - 4x + 4}{x^2 + 4x - 12}$ 2, −6; $\frac{x - 2}{x + 6}$

Find each product or quotient. *(Lesson 15–2)*

3. $\frac{y + 3}{2y + 8} \cdot \frac{y^2 + 5y + 4}{y^2 + 4y + 3}$ $\frac{1}{2}$

4. $\frac{a^2b^2}{6a + 18} \div \frac{ab}{a^2 - 9}$ $\frac{ab(a - 3)}{6}$

5. $\frac{6 - 2a}{a^2 + a - 2} \cdot \frac{a^2 + 3a + 2}{a - 3}$ $\frac{-2(a + 1)}{a - 1}$

6. $\frac{x^2}{x^2 - 16} \div \frac{x}{x^2 - 16}$ x

Find each quotient. *(Lesson 15–3)*

7. $(y^2 - 5y) \div (y - 5)$ y

8. $(n^2 + 2n - 10) \div (n - 2)$ $n + 4 + \frac{-2}{n - 2}$

9. $(x^3 - 1) \div (x - 1)$ $x^2 + x + 1$

10. **Fast Food** About 90% of Americans eat fast food in a given month. Of these, 60% use the drive-through window at least once. What percent of Americans use a fast-food drive-through at least once a month?
Source: Maritz Marketing Research Inc. *(Lesson 15–2)* 54%

www.algconcepts.com/self_check_quiz

4 ASSESS

Open-Ended Assessment

Modeling Have students write a trinomial of the form $x^2 + bx + c$, where b is 0, ±1, or ±3 and c is ±1, ±2, or ±4. Then have students use algebra tiles to model dividing the trinomial by a binomial of the form $x + d$, where $d = \pm 1$ or ± 2. Alternately, you may assign specific trinomials and binomials. When students find a factor, have them write the trinomial in factored form. When the trinomial is not evenly divisible by a binomial, have students state this.

Quiz 1

The Quiz provides students with a brief review of the concepts and skills in Lessons 15–1 through 15–3. Lesson numbers are given to the right of the exercises or instruction lines so students can review concepts not yet mastered.

Mid-Chapter Test (Lessons 15–1 through 15–3) is available in the *Chapter 15 Resource Masters*, p. 702.

Extra Credit

Which of the expressions $x - 1$, $x - 2$, or $x - 3$ is *not* the result of dividing $x^3 - 7x + 6$ by some polynomial? $x - 3$

***Enrichment*, p. 677**

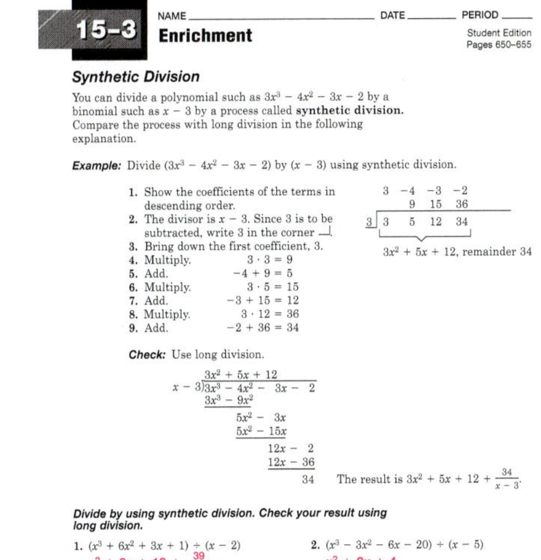

15-3 NAME ________ DATE ______ PERIOD ______

Enrichment

Student Edition Pages 650–655

Synthetic Division

You can divide a polynomial such as $3x^3 - 4x^2 - 3x - 2$ by a binomial such as $x - 3$ by a process called **synthetic division.** Compare the process with long division in the following explanation.

Example: Divide $(3x^3 - 4x^2 - 3x - 2)$ by $(x - 3)$ using synthetic division.

1. Show the coefficients of the terms in descending order.
2. The divisor is $x - 3$. Since 3 is to be subtracted, write 3 in the corner ⌋.
3. Bring down the first coefficient, 3.
4. Multiply. $3 \cdot 3 = 9$
5. Add. $-4 + 9 = 5$
6. Multiply. $3 \cdot 5 = 15$
7. Add. $-3 + 15 = 12$
8. Multiply. $3 \cdot 12 = 36$
9. Add. $-2 + 36 = 34$

	3	−4	−3	−2
		9	15	36
3	3	5	12	34

$3x^2 + 5x + 12$, remainder 34

Check: Use long division.

$$\begin{array}{r} 3x^2 + 5x + 12 \\ x - 3\overline{)3x^3 - 4x^2 - 3x - 2} \\ \underline{3x^3 - 9x^2} \\ 5x^2 - 3x \\ \underline{5x^2 - 15x} \\ 12x - 2 \\ \underline{12x - 36} \\ 34 \end{array}$$

The result is $3x^2 + 5x + 12 + \frac{34}{x - 3}$.

Divide by using synthetic division. Check your result using long division.

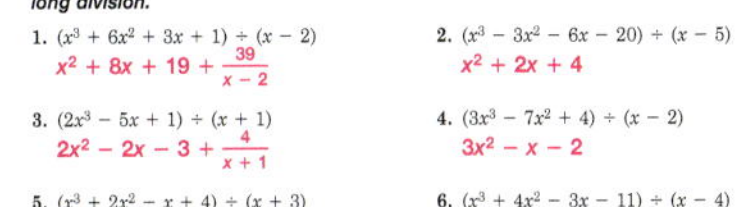

1. $(x^3 + 6x^2 + 3x + 1) \div (x - 2)$
$x^2 + 8x + 19 + \frac{39}{x - 2}$

2. $(x^3 - 3x^2 - 6x - 20) \div (x - 5)$
$x^2 + 2x + 4$

3. $(2x^3 - 5x + 1) \div (x + 1)$
$2x^2 - 2x - 3 + \frac{4}{x + 1}$

4. $(3x^3 - 7x^2 + 4) \div (x - 2)$
$3x^2 - x - 2$

5. $(x^3 + 2x^2 - x + 4) \div (x + 3)$
$x^2 - x + 2 - \frac{2}{x + 3}$

6. $(x^3 + 4x^2 - 3x - 11) \div (x - 4)$
$x^2 + 8x + 29 + \frac{105}{x - 4}$

Lesson 15-4

1 FOCUS

5-Minute Check
Lesson 15–3

Find each quotient.

1. $(30y + 12) \div (5y + 2)$ **6**
2. $(x^2 - 11x + 30) \div (x - 5)$ **$x - 6$**
3. $(2x^2 - 5x - 9) \div (2x + 3)$ **$x - 4 + \frac{3}{2x + 3}$**
4. What is the quotient of $(2x^2 - 5)$ divided by $(x - 2)$? **$2x + 4 + \frac{3}{x - 2}$**
5. Find the width of the rectangle if its area is $9x^2 + 9x - 10$ square meters and its length is $3x + 5$ meters. **$3x - 2$ m**

$9x^2 + 9x - 10$

$3x + 5$

Motivating the Lesson

Real-World Connection One apple pie recipe for a 10-inch pie dish calls for $1\frac{1}{3}$ cups of flour for the shell and $\frac{1}{3}$ cup of flour for the filling. To find the total amount of flour in one pie students can add to find that $\frac{4}{3} + \frac{1}{3} = \frac{5}{3}$ cups, or $1\frac{2}{3}$ cups. Have students note that in x apple pies, there are $\frac{4x}{3} + \frac{x}{3}$ cups of flour. Tell students that they can add these rational expressions just as they would any two fractions with like denominators.

15–4 Combining Rational Expressions with Like Denominators

What You'll Learn
You'll learn to add and subtract rational expressions with like denominators.

Why It's Important
Entertainment
Rational expressions can be combined to determine concert profits.
See Exercise 35.

Rational expressions with like denominators are added or subtracted in the same way as number fractions with like denominators. In the following example, two steps are used to subtract $\frac{1}{10}$ from $\frac{4}{10}$.

Step 1 Add or subtract the numerators. $\frac{4}{10} - \frac{1}{10} = \frac{4 - 1}{10}$

Step 2 Write the sum or difference over the common denominator. $= \frac{3}{10}$

You can use this same method to add or subtract rational expressions with like denominators. Recall that in rational expressions, both the numerator and denominator can have variables.

Examples

Find each sum or difference.

1 $\frac{2}{a} + \frac{4}{a}$

$\frac{2}{a} + \frac{4}{a} = \frac{2 + 4}{a}$ *The common denominator is a. Add the numerators.*

$= \frac{6}{a}$ *Simplify.*

2 $\frac{5x}{11} - \frac{2x}{11}$

$\frac{5x}{11} - \frac{2x}{11} = \frac{5x - 2x}{11}$ *The common denominator is 11. Subtract the numerators.*

$= \frac{3x}{11}$ *Simplify.*

3 $\frac{b}{b + 3} - \frac{3}{b + 3}$

$\frac{b}{b + 3} - \frac{3}{b + 3} = \frac{b - 3}{b + 3}$ *The common denominator is b + 3.*

Your Turn

a. $\frac{4m}{7} + \frac{m}{7}$ **$\frac{5m}{7}$**

b. $\frac{7}{x} - \frac{3}{x}$ **$\frac{4}{x}$**

c. $\frac{n}{n - 2} - \frac{1}{n - 2}$ **$\frac{n - 1}{n - 2}$**

Resource Manager

Reproducible Masters
Chapter 15 Resource Masters
- *Study Guide,* p. 678
- *Skills Practice,* p. 679
- *Practice,* p. 680
- *Reading to Learn Mathematics,* p. 681
- *Enrichment,* p. 682

Transparencies
5-Minute Check, 15–4

Technology/Multimedia
Interactive Chalkboard CD-ROM

When adding or subtracting fractions, the result is not always in simplest form. To find the simplest form, divide the numerator and the denominator by the greatest common factor (GCF).

$$\frac{3}{8} + \frac{1}{8} = \frac{4}{8}$$
$$= \frac{\overset{1}{\cancel{4}}}{\underset{2}{\cancel{8}}} \quad \textit{The GCF of 4 and 8 is 4.}$$
$$= \frac{1}{2}$$

$$\frac{9}{16} - \frac{3}{16} = \frac{6}{16}$$
$$= \frac{\overset{3}{\cancel{6}}}{\underset{8}{\cancel{16}}} \quad \textit{The GCF of 6 and 16 is 2.}$$
$$= \frac{3}{8}$$

You can use similar steps to combine rational expressions.

Examples

Find each sum or difference. Write in simplest form.

4 $\frac{7}{10m} + \frac{1}{10m}$

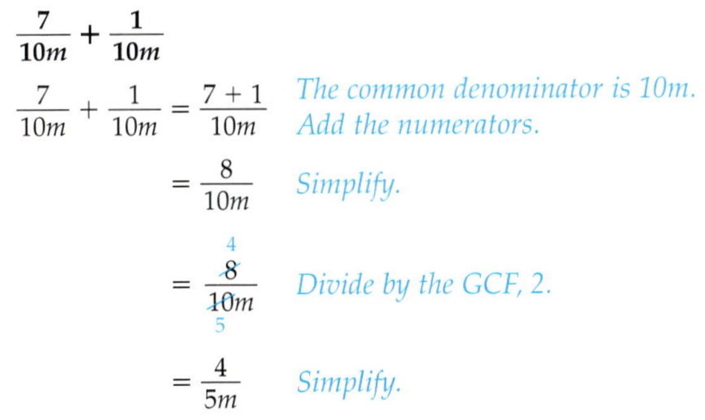

$\frac{7}{10m} + \frac{1}{10m} = \frac{7+1}{10m}$ *The common denominator is 10m. Add the numerators.*

$= \frac{8}{10m}$ *Simplify.*

$= \frac{\overset{4}{\cancel{8}}}{\underset{5}{\cancel{10}m}}$ *Divide by the GCF, 2.*

$= \frac{4}{5m}$ *Simplify.*

5 $\frac{3}{3t} - \frac{9}{3t}$

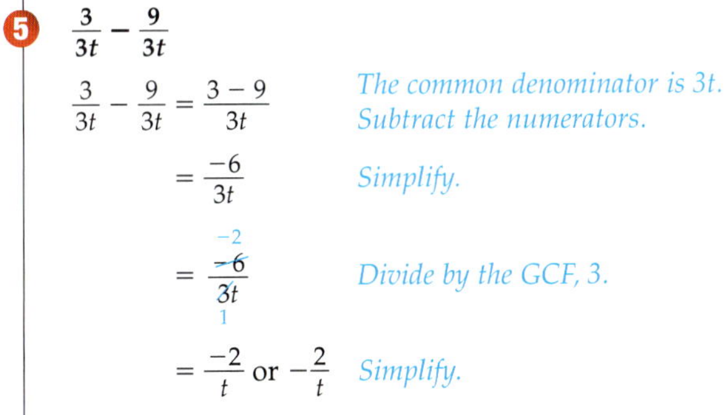

$\frac{3}{3t} - \frac{9}{3t} = \frac{3-9}{3t}$ *The common denominator is 3t. Subtract the numerators.*

$= \frac{-6}{3t}$ *Simplify.*

$= \frac{\overset{-2}{\cancel{-6}}}{\underset{1}{\cancel{3}t}}$ *Divide by the GCF, 3.*

$= \frac{-2}{t}$ or $-\frac{2}{t}$ *Simplify.*

Your Turn

d. $\frac{a}{12} + \frac{2a}{12}$ $\frac{a}{4}$

e. $\frac{8}{9y} - \frac{5}{9y}$ $\frac{1}{3y}$

Sometimes, the denominators of rational expressions are binomials.

2 TEACH

In-Class Examples

Examples 1–2

Find each sum or difference.

1 $\frac{7}{y} - \frac{10}{y}$ $-\frac{3}{y}$

2 $\frac{13a}{17} + \frac{2a}{17}$ $\frac{15a}{17}$

Teaching Tip Another way students can reduce the sum or difference in Examples 3 and 4 is to rewrite the expression as the product of 1 and another fraction by dividing out the GCF of the numerator and denominator.

$$\frac{8}{10m} = \frac{2}{2} \cdot \frac{4}{5m} = 1 \cdot \frac{4}{5m} = \frac{4}{5m}$$

$$\frac{-6}{3t} = \frac{3}{3} \cdot \frac{-2}{t} = 1 \cdot \frac{-2}{t} = -\frac{2}{t}$$

In-Class Examples

Examples 3–4

Find each sum or difference. Write in simplest form.

3 $\frac{8}{15n} + \frac{4}{15n}$ $\frac{4}{5n}$

4 $\frac{5}{2p} - \frac{13}{2p}$ $\frac{-4}{p}$, or $-\frac{4}{p}$

In-Class Examples

Examples 5–7

Find each sum or difference. Write in simplest form.

5 $\frac{10}{3x+1} - \frac{3}{3x+1}$ $\frac{7}{3x+1}$

6 $\frac{a-3}{a+3} + \frac{4}{a+3}$ $\frac{a+1}{a+3}$

7 $\frac{2x-5}{x-3} + \frac{x-4}{x-3}$ 3

Teaching Tip In Example 8, students may need reminding that the entire numerator of the second expression must be subtracted. It may help some students to think of the subtraction as adding the second expression's opposite.

In-Class Example

Example 8

Find $\frac{12a}{2a-1} - \frac{2a-5}{2a-1}$. Write in simplest form. $\frac{10a+5}{2a-1}$

Examples

Find each sum or difference. Write in simplest form.

6 $\frac{3}{x+2} + \frac{1}{x+2}$

$\frac{3}{x+2} + \frac{1}{x+2} = \frac{3+1}{x+2}$ *The common denominator is x + 2. Add the numerators.*

$= \frac{4}{x+2}$ *Simplify.*

7 $\frac{y+2}{y-1} - \frac{8}{y-1}$

$\frac{y+2}{y-1} - \frac{8}{y-1} = \frac{y+2-8}{y-1}$ *The common denominator is y − 1. Subtract the numerators.*

$= \frac{y-6}{y-1}$ *Simplify.*

Your Turn

f. $\frac{4m}{2m+3} + \frac{5}{2m+3}$ $\frac{4m+5}{2m+3}$ g. $\frac{5x}{x+2} - \frac{2x}{x+2}$ $\frac{3x}{x+2}$

Sometimes you must factor in order to simplify the sum or difference of rational expressions.

Examples

Find each sum or difference. Write in simplest form.

8 $\frac{b}{b+4} + \frac{b+8}{b+4}$

$\frac{b}{b+4} + \frac{b+8}{b+4} = \frac{b+(b+8)}{b+4}$ *The common denominator is b + 4. Add the numerators.*

$= \frac{2b+8}{b+4}$ *Simplify.*

$= \frac{2(b+4)}{b+4}$ *Factor the numerator.*

$= \frac{2\overset{1}{\cancel{(b+4)}}}{\underset{1}{\cancel{b+4}}}$ *Divide by the GCF, b + 4.*

$= 2$ *Simplify.*

9 $\frac{15x}{4x-1} - \frac{3x+3}{4x-1}$

$\frac{15x}{4x-1} - \frac{3x+3}{4x-1} = \frac{15x-(3x+3)}{4x-1}$ *The common denominator is 4x − 1. Subtract the numerators.*

$= \frac{15x-3x-3}{4x-1}$ *Distributive Property*

$= \frac{12x-3}{4x-1}$ *Simplify.*

$$= \frac{3(4x - 1)}{4x - 1} \quad \textit{Factor the numerator.}$$

$$= \frac{3\overset{1}{\cancel{(4x - 1)}}}{\underset{1}{\cancel{4x - 1}}} \quad \textit{Divide by the GCF, } 4x - 1.$$

$$= 3 \quad \textit{Simplify.}$$

Your Turn **Find each sum or difference. Write in simplest form.**

h. $\frac{3r}{r + 5} + \frac{15}{r + 5}$ **3**

i. $\frac{m}{m + 3} - \frac{3m + 6}{m + 3}$ **−2**

You can solve some geometry problems by adding rational expressions.

Example 
Geometry Link

Find the perimeter of rectangle *ABCD*.

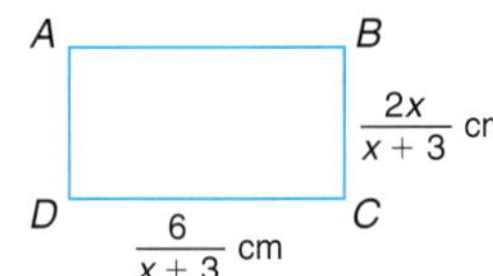

$$P = 2\ell + 2w$$

$$= 2\left(\frac{6}{x + 3}\right) + 2\left(\frac{2x}{x + 3}\right) \quad \textit{Replace } \ell \textit{ with } \frac{6}{x + 3} \textit{ and } w \textit{ with } \frac{2x}{x + 3}.$$

$$= \frac{2(6)}{x + 3} + \frac{2(2x)}{x + 3} \quad \textit{Multiply.}$$

$$= \frac{12}{x + 3} + \frac{4x}{x + 3} \quad \textit{Simplify.}$$

$$= \frac{12 + 4x}{x + 3} \quad \textit{Add the numerators.}$$

$$= \frac{4(3 + x)}{x + 3} \quad \textit{Factor the numerator.}$$

$$= \frac{4\overset{1}{\cancel{(x + 3)}}}{\underset{1}{\cancel{x + 3}}} \quad \textit{Divide by the GCF, } x + 3.$$

$$= 4 \quad \textit{Simplify.}$$

The perimeter of rectangle *ABCD* is 4 centimeters.

Check for Understanding

Communicating Mathematics

1. **Explain** how the sum of two fractions with the same denominator can have a sum of zero. **They are additive inverses.**
2. **Identify** the mistake in the solution to the problem below. Then write the correct sum. **The denominators should not have been added; 2.**

$$\frac{4x}{2x + 1} + \frac{2}{2x + 1} = \frac{4x + 2}{4x + 2} \text{ or } 1$$

Reteaching Activity

Logical Learners You can reinforce logical learners' understanding of adding and subtracting rational expressions by showing them a form of the distributive property: $\frac{m}{d} \pm \frac{n}{d} = \frac{1}{d}(m \pm n) = \frac{m \pm n}{d}$. Have students apply this property to any of the examples in the lesson to confirm that they get the same results.

In-Class Example

Example 9

Find the perimeter of rectangle *PQRS*. **8 in.**

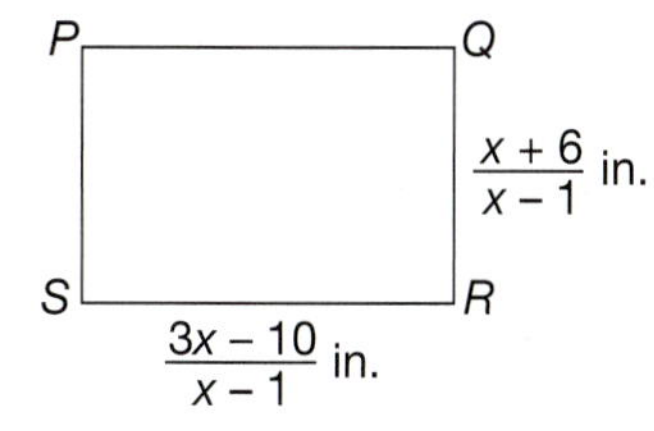

***Study Guide*, p. 678**

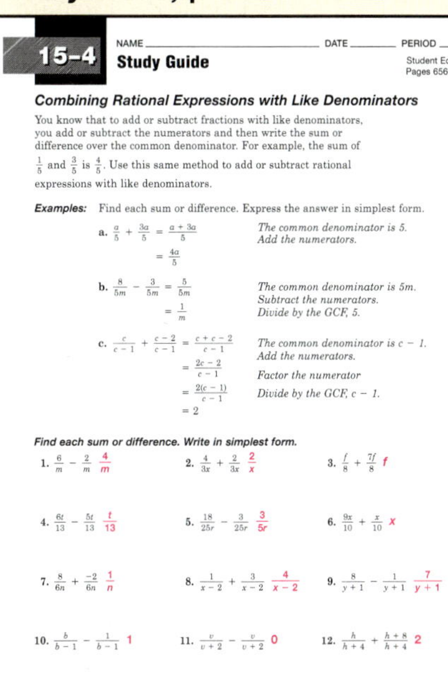

15–4 NAME ______ DATE ______ PERIOD ______

Study Guide Student Edition Pages 656–661

Combining Rational Expressions with Like Denominators

You know that to add or subtract fractions with like denominators, you add or subtract the numerators and then write the sum or difference over the common denominator. For example, the sum of $\frac{1}{5}$ and $\frac{3}{5}$ is $\frac{4}{5}$. Use this same method to add or subtract rational expressions with like denominators.

Examples: Find each sum or difference. Express the answer in simplest form.

a. $\frac{a}{5} + \frac{3a}{5} = \frac{a + 3a}{5}$ *The common denominator is 5. Add the numerators.*
$= \frac{4a}{5}$

b. $\frac{8}{5m} - \frac{3}{5m} = \frac{5}{5m}$ *The common denominator is 5m. Subtract the numerators.*
$= \frac{1}{m}$ *Divide by the GCF, 5.*

c. $\frac{c}{c - 1} + \frac{c - 2}{c - 1} = \frac{c + c - 2}{c - 1}$ *The common denominator is c − 1. Add the numerators.*
$= \frac{2c - 2}{c - 1}$ *Factor the numerator*
$= \frac{2(c - 1)}{c - 1}$ *Divide by the GCF, c − 1.*
$= 2$

Find each sum or difference. Write in simplest form.

1. $\frac{6}{m} - \frac{2}{m}$ $\frac{4}{m}$
2. $\frac{4}{3x} + \frac{2}{3x}$ $\frac{2}{x}$
3. $\frac{f}{8} + \frac{7f}{8}$ f
4. $\frac{6t}{13} - \frac{5t}{13}$ $\frac{t}{13}$
5. $\frac{18}{25r} - \frac{3}{25r}$ $\frac{3}{5r}$
6. $\frac{9x}{10} + \frac{x}{10}$ x
7. $\frac{8}{6n} + \frac{-2}{6n}$ $\frac{1}{n}$
8. $\frac{1}{x - 2} + \frac{3}{x - 2}$ $\frac{4}{x - 2}$
9. $\frac{8}{y + 1} - \frac{1}{y + 1}$ $\frac{7}{y + 1}$
10. $\frac{b}{b - 1} - \frac{1}{b - 1}$ 1
11. $\frac{v}{v + 2} - \frac{v}{v + 2}$ 0
12. $\frac{h}{h + 4} + \frac{h + 8}{h + 4}$ 2

3 PRACTICE/APPLY

Error Analysis

Watch for students who add or multiply denominators, such as in Exercise 7, where they could incorrectly add $\frac{5}{x} + \frac{2}{x}$ as $\frac{7}{2x}$ or $\frac{7}{x^2}$.

Prevent by reminding students that adding rational expressions with like denominators is just like adding fractions with like denominators—you add the numerators, and write the result over the shared denominator.

Assignment Guide

Basic: 15–35 odd, 36–44
Average: 14–32 even, 34–44

Skills Practice, **p. 679, and *Practice,* p. 680 (shown)**

15-4 NAME ______ DATE ______ PERIOD ______

Practice

Student Edition Pages 656–661

Combining Rational Expressions with Like Denominators

Find each sum or difference. Write in simplest form.

1. $\frac{8}{n} + \frac{4}{n}$ $\frac{12}{n}$
2. $\frac{3x}{9} + \frac{4x}{9}$ $\frac{7x}{9}$
3. $\frac{7}{2k} - \frac{5}{2k}$ $\frac{1}{k}$
4. $\frac{6n}{n} - \frac{3n}{n}$ 3
5. $\frac{-5a}{2} + \frac{4a}{2}$ $-\frac{a}{2}$
6. $\frac{2y}{3} + \frac{y}{3}$ y
7. $\frac{9x}{11} - \frac{8x}{11}$ $\frac{x}{11}$
8. $\frac{6p}{5} - \frac{p}{5}$ p
9. $\frac{9}{16q} + \frac{3}{16q}$ $\frac{3}{4q}$
10. $\frac{4t}{9} - \frac{t}{9}$ $\frac{t}{3}$
11. $\frac{1}{4m} - \frac{3}{4m}$ $-\frac{1}{2m}$
12. $\frac{-2}{10x} + \frac{6}{10x}$ $\frac{2}{5x}$
13. $\frac{6s}{7} + \frac{8s}{7}$ $2s$
14. $\frac{8}{3y} - \frac{2}{3y}$ $\frac{2}{y}$
15. $\frac{4}{x-7} - \frac{2}{x-7}$ $\frac{2}{x-7}$
16. $\frac{-2}{x+3} + \frac{3}{x+3}$ $\frac{1}{x+3}$
17. $\frac{5}{y-4} - \frac{8}{y-4}$ $-\frac{3}{y-4}$
18. $\frac{3m}{m+2} - \frac{m}{m+2}$ $\frac{2m}{m+2}$
19. $\frac{3n}{n-1} + \frac{2}{n-1}$ $\frac{3n+2}{n-1}$
20. $\frac{5a}{a+4} - \frac{7}{a+4}$ $\frac{5a-7}{a+4}$
21. $\frac{4g}{g+3} + \frac{12}{g+3}$ 4
22. $\frac{2r+2}{r-5} - \frac{r-4}{r-5}$ $\frac{r+6}{r-5}$
23. $\frac{s-3}{s+1} + \frac{4s+8}{s+1}$ 5
24. $\frac{-11b}{5b+3} + \frac{12b-2}{5b+3}$ $\frac{b-2}{5b+3}$
25. $\frac{15y}{4y-2} - \frac{3y+6}{4y-2}$ 3
26. $\frac{5c+3}{2c+1} + \frac{9c+4}{2c+1}$ 7
27. $\frac{2x+3}{3x+4} - \frac{8x+11}{3x+4}$ -2

3. **Complete** the table.

a	b	$a+b$	$a-b$	$a \cdot b$	$a \div b$
$\frac{2}{t}$	$\frac{1}{t}$	$\frac{3}{t}$	$\frac{1}{t}$	$\frac{2}{t^2}$	2
$\frac{12}{y}$	$\frac{4}{y}$	$\frac{16}{y}$	$\frac{8}{y}$	$\frac{48}{y^2}$	3
$\frac{x}{x-1}$	$\frac{1}{x-1}$	$\frac{x+1}{x-1}$	1	$\frac{x}{x^2-2x+1}$	x

Guided Practice

Getting Ready Find each sum or difference. Write in simplest form.

Sample 1: $\frac{4}{15} + \frac{3}{15}$

Solution: $\frac{4}{15} + \frac{3}{15} = \frac{4+3}{15}$
$= \frac{7}{15}$

Sample 2: $\frac{8}{9} - \frac{2}{9}$

Solution: $\frac{8}{9} - \frac{2}{9} = \frac{8-2}{9}$
$= \frac{6}{9}$ or $\frac{2}{3}$

4. $\frac{5}{7} + \frac{1}{7}$ $\frac{6}{7}$
5. $\frac{4}{12} + \frac{5}{12}$ $\frac{3}{4}$
6. $\frac{6}{10} - \frac{1}{10}$ $\frac{1}{2}$

Examples 1–9

Find each sum or difference. Write in simplest form.

7. $\frac{5}{x} + \frac{2}{x}$ $\frac{7}{x}$
8. $\frac{2t}{3} - \frac{t}{3}$ $\frac{t}{3}$
9. $\frac{7y}{y} - \frac{8y}{y}$ -1
10. $\frac{a}{12} + \frac{2a}{12}$ $\frac{a}{4}$
11. $\frac{2x}{x-3} - \frac{6}{x-3}$ 2
12. $\frac{2c+3}{c-4} - \frac{c-2}{c-4}$

12. $\frac{c+5}{c-4}$

Example 10

13. **Geometry** Find the perimeter of the rectangle. $\frac{10y}{7x-2y}$ in.

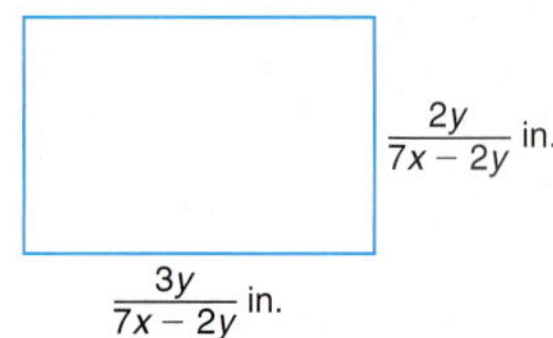

Exercises

Practice 

Homework Help	
For Exercises	See Examples
14–25, 35	1–5
26–33	6–9
34	10
Extra Practice	
See page 723.	

Find each sum or difference. Write in simplest form.

14. $\frac{7}{m} + \frac{4}{m}$ $\frac{11}{m}$
15. $\frac{x}{8} + \frac{6x}{8}$ $\frac{7x}{8}$
16. $\frac{7}{15z} - \frac{3}{15z}$ $\frac{4}{15z}$
17. $\frac{5t}{2} - \frac{4t}{2}$ $\frac{t}{2}$
18. $\frac{8p}{13} - \frac{3p}{13}$ $\frac{5p}{13}$
19. $\frac{1}{6r} + \frac{6}{6r}$ $\frac{7}{6r}$
20. $\frac{a}{2} + \frac{a}{2}$ a
21. $\frac{n}{3} + \frac{2n}{3}$ n
22. $\frac{5}{3y} - \frac{2}{3y}$ $\frac{1}{y}$
23. $\frac{5x}{24} - \frac{3x}{24}$ $\frac{x}{12}$
24. $\frac{8}{27m} + \frac{4}{27m}$ $\frac{4}{9m}$
25. $\frac{5}{2z} + \frac{-7}{2z}$ $-\frac{1}{z}$

28. $\frac{1}{x+2}$

26. $\frac{8}{y-2} - \frac{6}{y-2}$ $\frac{2}{y-2}$
27. $\frac{y}{a+1} - \frac{y}{a+1}$ 0
28. $\frac{3}{x+2} - \frac{2}{x+2}$
29. $\frac{2x}{x+3} + \frac{6}{x+3}$ 2
30. $\frac{8n+3}{3n+4} - \frac{2n-5}{3n+4}$ 2
31. $\frac{10m-1}{4m-3} - \frac{8-2m}{4m-3}$ 3

C

32. What is $\frac{3c-7}{2c-3}$ minus $\frac{c-4}{2c-3}$? 1
33. Find the sum of $\frac{x+y}{y-2}$ and $\frac{y-x}{y-2}$. $\frac{2y}{y-2}$

Applications and Problem Solving

34. **Geometry** Find the perimeter of the isosceles triangle. 5 m

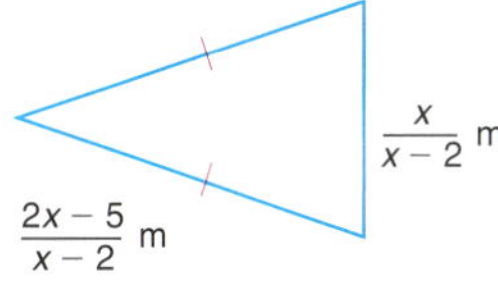

35. **Entertainment** A musical group receives \$15,000 for each concert. The five members receive equal shares of the money. They must pay 0.3 of their earnings for other salaries and expenses.
    a. Write a rational expression that represents how much money each member of the music group receives for a concert after expenses.
    b. Simplify the expression you wrote in part a to find how much each member of the group receives for a concert. \$2100

    a. Sample answer: $\frac{15{,}000}{5} - \frac{0.3(15{,}000)}{5}$

36. **Critical Thinking** Which rational expression is not equivalent to the others? c
    a. $\frac{4}{x-5}$ b. $-\frac{4}{5-x}$ c. $\frac{-4}{x-5}$ d. $\frac{-4}{5-x}$

Mixed Review

40. $t + 3 + \frac{9}{t+9}$

42. (−2, 2), (1, −1); See margin for graph.

Find each quotient. *(Lesson 15–3)*

37. $(a^2 + 9a + 20) \div (a + 5)$ $a + 4$
38. $(x^2 - 2x - 35) \div (x - 7)$ $x + 5$
39. $(2y^2 - 3y - 35) \div (2y + 7)$ $y - 5$
40. $(t^2 + 12t + 36) \div (t + 9)$
41. Find the product of $\frac{5x^2y}{8ab}$ and $\frac{12a^2b}{25x}$. *(Lesson 15–2)* $\frac{3axy}{10}$
42. Solve the system $y = x^2 - 2$ and $y = -x$ by graphing. *(Lesson 13–6)*

43. **Short Response** Find $(3x + y)^2$. *(Lesson 9–5)* $9x^2 + 6xy + y^2$
44. **Multiple Choice** About 25% of all identical twins are "mirror" twins—a reflection of each other. In a group of identical twins, 35 pairs were found to be "mirror" twins. About how many pairs of identical twins would you expect to be in the group? *(Lesson 5–4)* B

Ⓐ 9 Ⓑ 140
Ⓒ 70 Ⓓ 100

4 ASSESS

Open-Ended Assessment

Speaking Ask students to explain how they can tell whether or not a sum or difference of rational expressions is in simplest form or can be reduced.

Answer

42.

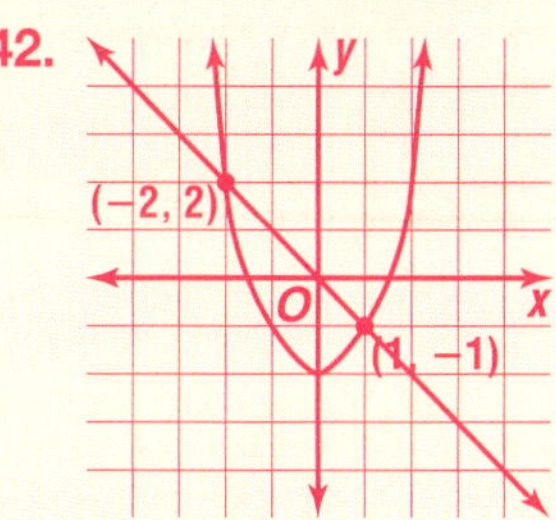

***Enrichment*, p. 682**

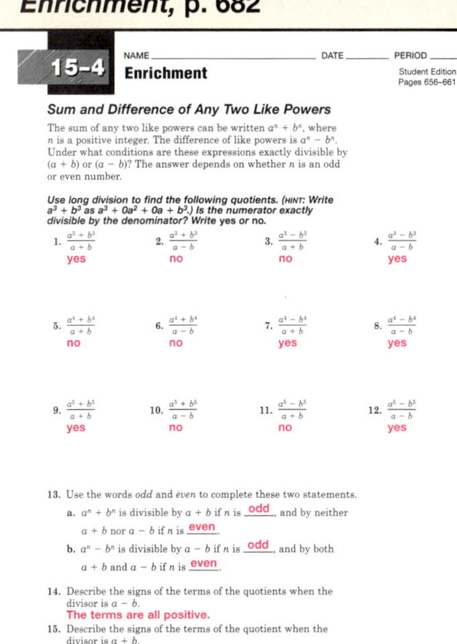

NAME ______ DATE ______ PERIOD ______

15-4 Enrichment

Student Edition Pages 656–661

Sum and Difference of Any Two Like Powers

The sum of any two like powers can be written $a^n + b^n$, where n is a positive integer. The difference of like powers is $a^n - b^n$. Under what conditions are these expressions exactly divisible by $(a + b)$ or $(a - b)$? The answer depends on whether n is an odd or even number.

Use long division to find the following quotients. (Hint: Write $a^3 + b^3$ as $a^3 + 0a^2 + 0a + b^3$.) Is the numerator exactly divisible by the denominator? Write yes or no.

1. $\frac{a^3+b^3}{a+b}$ yes
2. $\frac{a^3+b^3}{a-b}$ no
3. $\frac{a^3-b^3}{a+b}$ no
4. $\frac{a^3-b^3}{a-b}$ yes
5. $\frac{a^4+b^4}{a+b}$ no
6. $\frac{a^4+b^4}{a-b}$ no
7. $\frac{a^4-b^4}{a+b}$ yes
8. $\frac{a^4-b^4}{a-b}$ yes
9. $\frac{a^5+b^5}{a+b}$ yes
10. $\frac{a^5+b^5}{a-b}$ no
11. $\frac{a^5-b^5}{a+b}$ no
12. $\frac{a^5-b^5}{a-b}$ yes

13. Use the words *odd* and *even* to complete these two statements.
    a. $a^n + b^n$ is divisible by $a + b$ if n is odd, and by neither $a + b$ nor $a - b$ if n is even.
    b. $a^n - b^n$ is divisible by $a - b$ if n is odd, and by both $a + b$ and $a - b$ if n is even.
14. Describe the signs of the terms of the quotients when the divisor is $a - b$.
    The terms are all positive.
15. Describe the signs of the terms of the quotient when the divisor is $a + b$.
    The terms are alternately positive and negative.

Extra Credit

Find the mean of the expressions $\frac{4p^3 + p^2 + 5p + 3}{3p - 1}$, $\frac{-p^3 - 3p^2 - 2p - 2}{3p - 1}$, and $\frac{p^2 - 3p - 1}{3p - 1}$. $\frac{p^2}{3}$

Lesson 15–5

1 FOCUS

5-Minute Check
Lesson 15–4

Find each sum or difference. Write in simplest form.

1. $\frac{15}{w} + \frac{3}{w}$ $\frac{18}{w}$
2. $\frac{17}{5v} - \frac{2}{5v}$ $\frac{3}{v}$
3. $\frac{4r}{r-3} - \frac{12}{r-3}$ 4
4. $\frac{2m-7}{2m+1} + \frac{8m+12}{2m+1}$ 5
5. Find the perimeter of the isosceles triangle. 6 cm

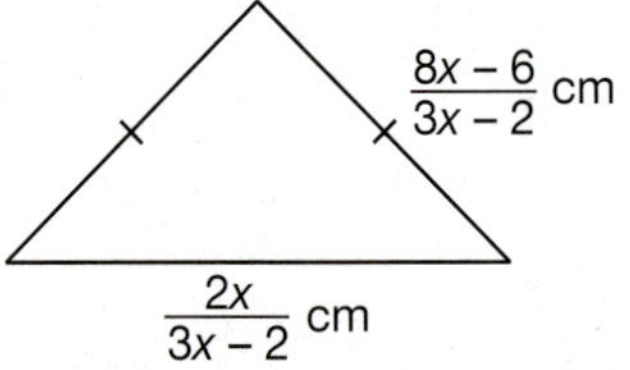

Motivating the Lesson

Hands-On Activity Have students use grid paper to mark or cut out two different-sized squares whose sides are not multiples. Then have them draw or construct a "train" of each size square using copies of the original squares so that the two trains are the same length, as shown below.

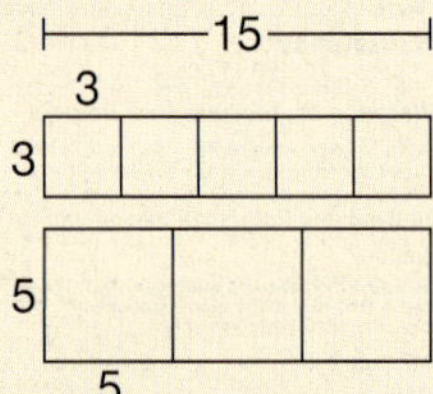

Students should realize that the shortest length for a train is the LCM of the lengths of the sides of the two squares.

15–5 Combining Rational Expressions with Unlike Denominators

What You'll Learn
You'll learn to add and subtract rational expressions with unlike denominators.

Why It's Important
Teaching Teachers can use LCM when dividing their classes into groups. *See Example 3.*

The **least common multiple (LCM)** is the least number that is a common multiple of two or more numbers. Suppose you want to find the LCM of 4, 6, and 9. Write multiples of each number until you find a common multiple.

multiples of 4: 0, 4, 8, 12, 16, 20, 24, 28, 32, **36**, . . .
multiples of 6: 0, 6, 12, 18, 24, 30, **36**, . . .
multiples of 9: 0, 9, 18, 27, **36**, . . .

Zero is a multiple of every number, but it cannot be the LCM. The LCM of 4, 6, and 9 is 36.

You can also use prime factorization to find the LCM.

$$4 = 2 \cdot 2 \qquad 6 = 2 \cdot 3 \qquad 9 = 3 \cdot 3$$

The prime factors are 2 and 3. The greatest number of times 2 appears is twice (in 4). The greatest number of times 3 appears is twice (in 9). So, the LCM of 4, 6, and 9 is $2 \cdot 2 \cdot 3 \cdot 3$ or 36. This is the same answer you got by listing the multiples.

Examples

Find the LCM for each pair of expressions.

1 $15a^2b$, $18a^3$

$15a^2b = 3 \cdot 5 \cdot a \cdot a \cdot b$ *Factor each expression.*
$18a^3 = 2 \cdot 3 \cdot 3 \cdot a \cdot a \cdot a$

$\text{LCM} = 2 \cdot 3 \cdot 3 \cdot 5 \cdot a \cdot a \cdot a \cdot b$ *Use each factor the greatest number of times it appears in either factorization.*
$= 90a^3b$

2 $x^2 + x - 6$, $2x^2 - 3x - 2$

$x^2 + x - 6 = (x + 3)(x - 2)$ *Factor each expression.*
$2x^2 - 3x - 2 = (2x + 1)(x - 2)$

$\text{LCM} = (x + 3)(x - 2)(2x + 1)$ *Use each factor the greatest number of times it appears in either factorization.*

Your Turn **b. $(x - 3)(x + 3)(x + 1)$**

a. $12a$, $8ab$ **$24ab$**

b. $x^2 - 9$, $x^2 - 2x - 3$

Resource Manager

Reproducible Masters
Chapter 15 Resource Masters
- *Study Guide*, p. 683
- *Skills Practice*, p. 684
- *Practice*, p. 685
- *Reading to Learn Mathematics*, p. 686
- *Enrichment*, p. 687

Hands-On Algebra, pp. 163–164
School-to-Workplace, p. 15

Transparencies
5-Minute Check, 15–5

Technology/Multimedia
Interactive Chalkboard CD-ROM

Example **Teaching Link**

3 **Mrs. Reimer wants student groups of 4, 6, or 8. What is the minimum number of desks needed?**

Find the LCM of 4, 6, and 8.

$4 = 2 \cdot 2$

$6 = 2 \cdot 3$

$8 = 2 \cdot 2 \cdot 2$

LCM $= 2 \cdot 2 \cdot 2 \cdot 3$ or 24 *Use each factor the greatest number of times it appears in any of the factorizations.*

Mrs. Reimer must have at least 24 desks in her classroom.

To add or subtract fractions with unlike denominators, first rename the fractions so the denominators are alike. Any common denominator could be used. However, the computation is usually easier if you use the **least common denominator (LCD)**. Recall that the least common denominator is the LCM of the denominators.

Examples

Write each pair of rational expressions with the same LCD.

4 $\frac{9}{2m^2}, \frac{5}{6m}$

First find the LCD.

$2m^2 = 2 \cdot m \cdot m$ *Factor each expression.*

$6m = 2 \cdot 3 \cdot m$

LCD $= 2 \cdot 3 \cdot m \cdot m$ or $6m^2$

Then write each fraction with the same LCD.

$\frac{9}{2m^2} \cdot \frac{3}{3} = \frac{27}{6m^2}$ $\qquad$ $\frac{5}{6m} \cdot \frac{m}{m} = \frac{5m}{6m^2}$

5 $\frac{1}{x+2}, \frac{3x}{4x+8}$

First find the LCD.

$x + 2 = x + 2$ *Factor each expression.*

$4x + 8 = 4(x + 2)$

LCD $= 4(x + 2)$

Then write each fraction with the same LCD.

$\frac{1}{x+2} \cdot \frac{4}{4} = \frac{4}{4(x+2)}$ $\qquad$ $\frac{3x}{4x+8} = \frac{3x}{4(x+2)}$

Your Turn d. $\frac{x(x-2)}{(x-6)(x-2)}, \frac{(x-6)(x-3)}{(x-6)(x-2)}$

c. $\frac{6}{b^3}, \frac{7}{ab}$ $\frac{6a}{ab^3}, \frac{7b^2}{ab^3}$ $\qquad$ d. $\frac{x}{x-6}, \frac{x-3}{x-2}$

2 TEACH

In-Class Examples

Examples 1–2

Find the LCM for each pair of expressions.

1 $12m^4n^5$, $14m^2n$ $\mathbf{84m^4n^5}$

2 $x^2 + 3x - 10$, $3x^2 - 7x + 2$ $(x-2)(x+5)(3x-1)$

Teaching Tip Have students check the results of Example 3 by sketching possible desk groupings. Their sketches should show 6 groups of 4 desks, 4 groups of 6 desks, and 3 groups of 8 desks.

In-Class Examples

Example 3

Suppose a teacher wants to be able to arrange the desks in a classroom either in groups of 2, 5, or 6. What is the minimum number of desks needed? **30**

Examples 4–5

Write each pair of rational expressions with the same LCD.

4 $\frac{3}{5m}, \frac{4}{2m^2}$ $\frac{6m}{10m^2}, \frac{20}{10m^2}$

5 $\frac{1}{x-3}, \frac{5x}{2x-6}$ $\frac{2}{2x-6}, \frac{5x}{2x-6}$

In-Class Examples

Examples 6–7

Find each sum or difference. Write in simplest form.

6 $\frac{3}{4m^3} + \frac{5}{8m^2}$ $\frac{5m+6}{8m^3}$

7 $\frac{3}{x+6} - \frac{4x}{x^2-36}$ $\frac{-x-18}{x^2-36}$ or $-\frac{x+18}{x^2-36}$

Use the following steps to add or subtract rational expressions with unlike denominators.

Step 1 Find the LCD.

Step 2 Change each rational expression into an equivalent expression using the LCD.

Step 3 Add or subtract as with rational expressions with like denominators.

Step 4 Simplify if necessary.

Examples

Find each sum or difference. Write in simplest form.

6 $\frac{5}{6x} + \frac{7}{12x^2}$

Step 1 Find the LCD.

$6x = 2 \cdot 3 \cdot x$

$12x^2 = 2 \cdot 2 \cdot 3 \cdot x \cdot x$

$\text{LCM} = 2 \cdot 2 \cdot 3 \cdot x \cdot x$ or $12x^2$

$\text{LCD} = 12x^2$

Step 2 Rename each expression with the LCD as the denominator. The denominator of $\frac{7}{12x^2}$ is already $12x^2$, so only $\frac{5}{6x}$ needs to be renamed.

$\frac{5}{6x} \cdot \frac{2x}{2x} = \frac{10x}{12x^2}$ *Why do you multiply $\frac{5}{6x}$ by $\frac{2x}{2x}$?*

Step 3 Add.

$\frac{5}{6x} + \frac{7}{12x^2} = \frac{10x}{12x^2} + \frac{7}{12x^2}$ *Use the LCD.*

$= \frac{10x+7}{12x^2}$ *This expression is in simplest form.*

7 $\frac{m}{m^2-9} - \frac{3}{m-3}$

$m^2 - 9 = (m-3)(m+3)$ *Factor each expression.*

$m - 3 = m - 3$

$\text{LCM} = (m-3)(m+3)$

$\text{LCD} = (m-3)(m+3)$

$\frac{m}{m^2-9} - \frac{3}{m-3}$

$= \frac{m}{(m-3)(m+3)} - \frac{3}{m-3} \cdot \frac{m+3}{m+3}$ *Multiply by $\frac{3}{m-3}$ by $\frac{m+3}{m+3}$.*

$= \frac{m}{(m-3)(m+3)} - \frac{3m+9}{(m-3)(m+3)}$ *The LCD is $(m-3)(m+3)$.*

$= \frac{m-(3m+9)}{(m-3)(m+3)}$ *Subtract the numerators.*

$= \frac{m-3m-9}{(m-3)(m+3)}$ *Distributive Property*

$= \frac{-2m-9}{(m-3)(m+3)}$ *Simplify.*

8 $\frac{5}{x-3} + \frac{7}{2x-6}$

$$\frac{5}{x-3} + \frac{7}{2x-6} = \frac{5}{x-3} + \frac{7}{2(x-3)}$$ *The LCD is $2(x-3)$.*

$$= \frac{5}{x-3} \cdot \frac{2}{2} + \frac{7}{2(x-3)}$$ *Multiply $\frac{5}{x-3}$ by $\frac{2}{2}$.*

$$= \frac{10}{2(x-3)} + \frac{7}{2(x-3)}$$ *Simplify.*

$$= \frac{10+7}{2(x-3)}$$ *Add the numerators.*

$$= \frac{17}{2(x-3)}$$ *Simplify.*

Your Turn Find each sum or difference. Write in simplest form.

e. $\frac{3}{xy} + \frac{2}{y^2}$ $\frac{3y+2x}{xy^2}$

f. $\frac{5a}{2a+6} - \frac{a}{a+3}$ $\frac{3a}{2(a+3)}$

Check for Understanding

Communicating Mathematics

1. **Write** two rational expressions with unlike denominators in which one of the denominators is the LCD of the rational expressions.
2. **Explain** the steps you would take to find the LCD for the expression $\frac{3}{8x^2} - \frac{5}{12xy} + \frac{7}{6y^2}$. Then simplify the expression.
3. **You Decide** Ashley found the sum of $\frac{4}{x-1}$ and $\frac{5}{x}$ to be $\frac{9}{2x-1}$. Malik found the sum to be $\frac{9x-5}{x^2-x}$. Who is correct? Explain why.

Vocabulary
least common multiple
least common denominator

1. Sample answer: $\frac{1}{2x}$, $\frac{1}{4x}$
2–3. See Solutions Manual.

Guided Practice

Getting Ready Find the LCM for each pair of numbers.

Sample: 10, 12 **Solution:** $10 = 2 \cdot 5$ $12 = 2 \cdot 2 \cdot 3$
LCM $= 2 \cdot 2 \cdot 3 \cdot 5$ or 60

4. 6, 10 **30**
5. 8, 18 **72**
6. 12, 15 **60**

Examples 1 & 2 Find the LCM for each pair of expressions.

7. $6xy$, $15y^2$ $\mathbf{30xy^2}$
8. $x + 1$, $x^2 + 5x + 4$ $\mathbf{(x+1)(x+4)}$

Examples 4 & 5 Write each pair of rational expressions with the same LCD.

9. $\frac{4}{a^2}, \frac{5}{a}$ $\frac{4}{a^2}, \frac{5a}{a^2}$
10. $\frac{7}{t+3}, \frac{13}{2t+6}$ $\frac{14}{2(t+3)}, \frac{13}{2(t+3)}$

In-Class Example

Example 8

Find $\frac{8}{5x+15} - \frac{3}{x+3}$. Write in simplest form.

$\frac{-7}{x+3}$ or $-\frac{7}{x+3}$

3 PRACTICE/APPLY

Error Analysis

Watch for students who multiply by incorrect or unnecessary factors when writing rational expressions with the same LCD, as in Exercise 10, where some students may be tempted to multiply the numerator of the second expression by $t + 3$. ***Prevent by*** reminding students that to rewrite an expression means to multiply it by some form of the number 1. For each expression, the numerator and denominator of that number 1 are the product of any factors that you must multiply the denominator of that expression by so that it equals the LCD of the two expressions. In Exercise 10, the second expression is already written with the LCD, so it can be left alone.

Reteaching Activity

Interpersonal Learners Have students work in pairs or small groups. Have one student write two rational expressions, and another student explain to the first student or the rest of the group how to find the LCM for the two expressions. Have students repeat this activity until each student gets a chance to present an explanation.

***Study Guide*, p. 683**

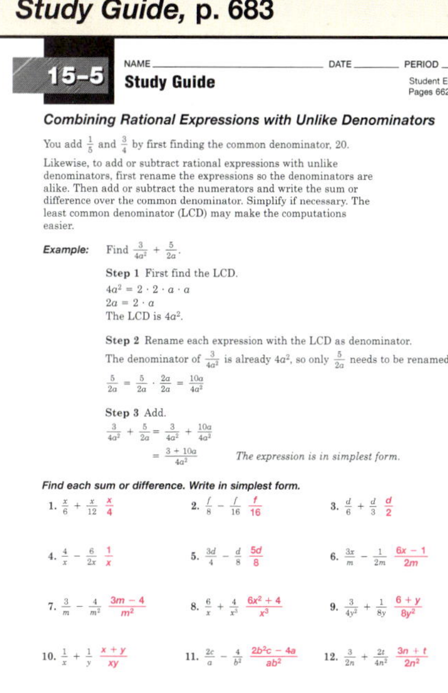

15–5 NAME ______ DATE ______ PERIOD ______

Study Guide Student Edition Pages 662–667

Combining Rational Expressions with Unlike Denominators

You add $\frac{1}{5}$ and $\frac{3}{4}$ by first finding the common denominator, 20.

Likewise, to add or subtract rational expressions with unlike denominators, first rename the expressions so the denominators are alike. Then add or subtract the numerators and write the sum or difference over the common denominator. Simplify if necessary. The least common denominator (LCD) may make the computations easier.

Example: Find $\frac{3}{4a^2} + \frac{5}{2a}$.

Step 1 First find the LCD.
$4a^2 = 2 \cdot 2 \cdot a \cdot a$
$2a = 2 \cdot a$
The LCD is $4a^2$.

Step 2 Rename each expression with the LCD as denominator.
The denominator of $\frac{3}{4a^2}$ is already $4a^2$, so only $\frac{5}{2a}$ needs to be renamed.
$\frac{5}{2a} = \frac{5}{2a} \cdot \frac{2a}{2a} = \frac{10a}{4a^2}$

Step 3 Add.
$\frac{3}{4a^2} + \frac{5}{2a} = \frac{3}{4a^2} + \frac{10a}{4a^2}$
$= \frac{3+10a}{4a^2}$ *The expression is in simplest form.*

Find each sum or difference. Write in simplest form.

1. $\frac{x}{6} + \frac{x}{12}$ $\frac{x}{4}$
2. $\frac{f}{8} - \frac{f}{16}$ $\frac{f}{16}$
3. $\frac{d}{6} + \frac{d}{3}$ $\frac{d}{2}$
4. $\frac{4}{x} - \frac{6}{2x}$ $\frac{1}{x}$
5. $\frac{3d}{4} - \frac{d}{8}$ $\frac{5d}{8}$
6. $\frac{3x}{m} - \frac{1}{2m}$ $\frac{6x-1}{2m}$
7. $\frac{3}{m} - \frac{4}{m^2}$ $\frac{3m-4}{m^2}$
8. $\frac{6}{x} + \frac{4}{x^3}$ $\frac{6x^2+4}{x^3}$
9. $\frac{3}{4y^2} + \frac{1}{8y}$ $\frac{6+y}{8y^2}$
10. $\frac{1}{x} + \frac{1}{y}$ $\frac{x+y}{xy}$
11. $\frac{2c}{a} - \frac{4}{b^2}$ $\frac{2b^2c-4a}{ab^2}$
12. $\frac{3}{2n} + \frac{2t}{4n^2}$ $\frac{3n+t}{2n^2}$

Assignment Guide

Basic: 19–53 odd, 54–64
Average: 18–50 even, 52–64
All: Quiz 2, 1–5

Examples 6–8 Find each sum or difference. Write in simplest form.

11. $\frac{t}{6} + \frac{3t}{12}$ $\frac{5t}{12}$
12. $\frac{2}{3a} - \frac{4}{9a}$ $\frac{2}{9a}$
13. $\frac{2}{ab^2} + \frac{3}{ab}$ $\frac{2 + 3b}{ab^2}$
14. $\frac{6}{m} + \frac{5}{n}$ $\frac{6n + 5m}{mn}$
15. $\frac{x}{x^2 - 4} - \frac{4}{x + 2}$ $\frac{-3x + 8}{x^2 - 4}$
16. $\frac{2}{3a - 1} - \frac{7}{15a - 5}$ $\frac{3}{5(3a - 1)}$

17. **Astronomy** Earth, Jupiter, and Saturn revolve around the Sun about once every 1, 12, and 30 years, respectively. The last time Earth, Jupiter, and Saturn were lined up with the sun was in 1982, and Jupiter and Saturn could be seen close together, high in the sky at midnight. In what year will this happen again? **Example 3** 2042

Exercises

Practice 

Homework Help	
For Exercises	See Examples
18–23	1, 2
24–29	4, 5
30–51	6–8
52–54	3

Extra Practice
See page 723.

Find the LCM for each pair of expressions.

18. $3t, 9t^2$ $9t^2$
19. $12a^2, 2ab^2$ $12a^2b^2$
20. $16m, 6mn$ $48mn$
21. $y + 2, y^2 - 4$ $(y + 2)(y - 2)$
22. $x^2 + 9x + 14, x^2 + 3x + 2$ $(x + 1)(x + 2)(x + 7)$
23. $x^2 - 9, 3x^2 - 8x - 3$ $(x + 3)(x - 3)(3x + 1)$

Write each pair of rational expressions with the same LCD.

24. $\frac{4}{m^3}, \frac{1}{m}$ $\frac{4}{m^3}, \frac{m^2}{m^3}$
25. $\frac{7}{2t}, \frac{8}{10t}$ $\frac{35}{10t}, \frac{8}{10t}$
26. $\frac{3}{6ab}, \frac{14}{4a^2}$ $\frac{6a}{12a^2b}, \frac{42b}{12a^2b}$
27. $\frac{5}{xy}, \frac{6}{yz}$ $\frac{5z}{xyz}, \frac{6x}{xyz}$
28. $\frac{x}{x^2 - 1}, \frac{-2}{x - 1}$ $\frac{x}{x^2 - 1}, \frac{-2(x + 1)}{x^2 - 1}$
29. $\frac{10k}{3k + 1}, \frac{k}{3 + 9k}$ $\frac{30k}{3(3k + 1)}, \frac{k}{3(3k + 1)}$

Find each sum or difference. Write in simplest form.

30. $\frac{a}{5} - \frac{a}{15}$ $\frac{2a}{15}$
31. $\frac{d}{2} - \frac{d}{5}$ $\frac{3d}{10}$
32. $\frac{t}{3} + \frac{2t}{7}$ $\frac{13t}{21}$
33. $\frac{9}{4x} + \frac{3}{2x}$ $\frac{15}{4x}$
34. $\frac{5}{2a} - \frac{3}{6a}$ $\frac{2}{a}$
35. $\frac{m}{3t} + \frac{1}{t}$ $\frac{m + 3}{3t}$

B

36. $\frac{4}{a^3} - \frac{2}{a}$ $\frac{4 - 2a^2}{a^3}$
37. $\frac{7}{3x} - \frac{1}{6x^2}$ $\frac{14x - 1}{6x^2}$
38. $\frac{2t}{5n^2} - \frac{1}{3n}$ $\frac{6t - 5n}{15n^2}$
39. $\frac{2}{x} + \frac{x + 3}{y}$
40. $\frac{6z}{7w} + \frac{2z}{w^3}$
41. $\frac{1}{4xy} + \frac{y}{10x^2}$
42. $\frac{9}{y + 1} - \frac{3}{4y + 4}$
43. $\frac{4a}{2a + 6} + \frac{3}{a + 3}$
44. $\frac{7}{x - 2} + \frac{3}{x}$ $\frac{10x - 6}{x(x - 2)}$

C

45. $\frac{y}{y^2 - 2y + 1} - \frac{1}{y - 1}$
46. $\frac{3a}{a^2 + 4a + 4} - \frac{2}{a + 2}$
47. $\frac{5x + 2}{2x - 1} + \frac{2x - 3}{8x - 4}$
48. $\frac{7}{a^2 + 2a - 3} - \frac{5}{a + 5a + 6}$
49. $\frac{x^2 + 4x - 5}{x^2 - 2x - 3} + \frac{2}{x + 1}$
50. Add $\frac{2x + 3}{x^2 - 4}$ and $\frac{6}{x + 2}$. $\frac{8x - 9}{x^2 - 4}$
51. Simplify $\frac{m}{m - n} - \frac{5}{m}$. $\frac{m^2 - 5m + 5n}{m(m - n)}$

39. $\frac{2y + x^2 + 3x}{xy}$
40. $\frac{6w^2z + 14z}{7w^3}$
41. $\frac{5x + 2y^2}{20x^2y}$
42. $\frac{33}{4(y + 1)}$
43. $\frac{2a + 3}{a + 3}$
45. $\frac{1}{(y - 1)^2}$
46. $\frac{a - 4}{(a + 2)^2}$
47. $\frac{22x + 5}{4(2x - 1)}$
48. $\frac{2a + 19}{(a - 1)(a + 2)(a + 3)}$
49. $\frac{x^2 + 6x - 11}{(x + 1)(x - 3)}$

***Skills Practice,* p. 684, and *Practice,* p. 685 (shown)**

15-5 NAME __________ DATE _____ PERIOD _____

Practice

Student Edition Pages 662–667

Combining Rational Expressions with Unlike Denominators

Find the LCM for each pair of expressions.

1. $4ab, 18b$ $36ab$
2. $6x^2y, 9xy$ $18x^2y$
3. $10a^2, 12ab^2$ $60a^2b^2$
4. $y + 2, y^2 - 4$ $(y + 2)(y - 2)$
5. $x^2 - 9, x^2 + 5x + 6$ $(x + 2)(x + 3)(x - 3)$
6. $x^2 - 3x - 4, 2x^2 - 2x - 8$ $(x + 1)(x - 4)(2x + 2)$

Write each pair of expressions with the same LCD.

7. $\frac{4}{b}, \frac{5}{ab}$ $\frac{4a}{ab}, \frac{5}{ab}$
8. $\frac{5}{6c^2}, \frac{3}{8c}$ $\frac{20}{24c^2}, \frac{9c}{24c^2}$
9. $\frac{6}{7x^2y}, \frac{4}{5xy}$ $\frac{30}{35x^2y}, \frac{28x}{35x^2y}$
10. $\frac{3}{r + 4}, \frac{7}{2r + 8}$ $\frac{6}{2(r + 4)}, \frac{7}{2(r + 4)}$
11. $\frac{x}{x - 2}, \frac{x + 1}{x - 5}$ $\frac{x(x - 5)}{(x - 2)(x - 5)}, \frac{(x + 1)(x - 2)}{(x - 2)(x - 5)}$
12. $\frac{3}{y - 4}, \frac{2y}{y^2 - 16}$ $\frac{3(y + 4)}{y^2 - 16}, \frac{2y}{y^2 - 16}$

Find each sum or difference. Write in simplest form.

13. $\frac{2k}{8} + \frac{3k}{16}$ $\frac{7k}{16}$
14. $\frac{n}{2} - \frac{n}{7}$ $\frac{5n}{14}$
15. $\frac{7}{3b} - \frac{3}{b}$ $-\frac{2}{3b}$
16. $\frac{5}{x} + \frac{3}{y}$ $\frac{5y + 3x}{xy}$
17. $\frac{2}{m^2n} - \frac{6}{mn}$ $\frac{2 - 6m}{m^2n}$
18. $\frac{c}{4c} + \frac{8}{c}$ $\frac{c + 32}{4c}$
19. $\frac{1}{6a} - \frac{2}{9a^2}$ $\frac{3a - 4}{18a^2}$
20. $\frac{2}{3ab} + \frac{3b}{4ab}$ $\frac{8 + 9b}{12ab}$
21. $\frac{p}{4p^2q} + \frac{3}{5pq}$ $\frac{17p}{20p^2q}$
22. $\frac{2x}{3xy^2} - \frac{2}{5xy}$ $\frac{10x - 6y}{15xy^2}$
23. $\frac{2s}{s^2 - 4} + \frac{4}{s + 2}$ $\frac{6s - 8}{s^2 - 4}$
24. $\frac{b}{b^2 - 9} - \frac{5}{b - 3}$ $\frac{-4b - 15}{b^2 - 9}$
25. $\frac{-5}{2r + 3} + \frac{7}{6r + 9}$ $\frac{-8}{3(2r + 3)}$
26. $\frac{6}{y + 2} + \frac{3}{y}$ $\frac{9y + 6}{y(y + 2)}$
27. $\frac{x}{x - 3} - \frac{2}{x - 4}$ $\frac{x^2 - 6x + 6}{(x - 3)(x - 4)}$

Applications and Problem Solving

52. **Entertainment** The choreographer of a musical needs enough dancers so that they can be arranged in groups of 6, 9, and 12, with no one sitting out. What is the least number of dancers needed? **36**

53. **Fitness** Cynthia jogs around an oval track in 150 seconds. Minya walks around the track in 210 seconds. Suppose they start at the same time. After how many minutes will they meet back at their starting point? **17.5 min**

54. **Critical Thinking** Consider the numbers 18 and 20.
   a. Find the GCF and LCM of the numbers. **2, 180**
   b. What is the value of GCF · LCM? **360**
   c. Describe the relationship between 18 · 20 and GCF · LCM.
   d. Describe how you could find the GCF of two numbers if you already know the LCM. **Divide the product of the two numbers by the LCM.**

54c. The product of the numbers is equal to the product of the GCF and LCM.

Mixed Review

Find each sum or difference. Write in simplest form. *(Lesson 15–4)*

55. $\frac{9}{a} + \frac{7}{a}$ $\frac{16}{a}$
56. $\frac{2}{5t} + \frac{3}{5t}$ $\frac{1}{t}$
57. $\frac{4n}{n+1} - \frac{2n-2}{n+1}$ **2**

58. Find $(m^2 + 6m - 7) \div (m + 7)$. *(Lesson 15–3)* $m - 1$

Simplify each expression. *(Lesson 14–4)*

59. $7\sqrt{19} + 2\sqrt{19}$ $9\sqrt{19}$
60. $6\sqrt{12} + 4\sqrt{3}$ $16\sqrt{3}$
61. $8\sqrt{5} - 3\sqrt{2}$

61. in simplest form

62. **Geometry** The length of the hypotenuse of a right triangle is $\sqrt{\frac{56}{25}}$ centimeters. Express the length as a radical in simplest form. *(Lesson 14–3)*

62. $\frac{2\sqrt{14}}{5}$ **cm**

Standardized Test Practice

63. **Short Response** What is the solution of the system $y = 7 - x$ and $x - y = -3$? *(Lesson 13–3)* **(2, 5)**

64. **Multiple Choice** Determine which polynomial can be the measure of the area of a square. *(Lesson 10–5)* **C**
   Ⓐ $2x^2 - 13x + 36$
   Ⓑ $2n^2 + 20n - 100$
   Ⓒ $4a^2 + 12a + 9$
   Ⓓ $4a^2 - 6a + 9$

Quiz 2 Lessons 15–3 through 15–5

Find each quotient. *(Lesson 15–3)*

1. $(x^2 + 6x - 16) \div (x - 2)$ $x + 8$
2. $(2x^2 - 11x - 20) \div (2x + 3)$ $x - 7 + \frac{1}{2x+3}$

Find each sum or difference. Write in simplest form. *(Lesson 15–4)*

3. $\frac{a}{9} + \frac{7a}{9}$ $\frac{8a}{9}$
4. $\frac{5}{21t} + \frac{4}{21t}$ $\frac{3}{7t}$

5. **Parades** At the Veteran's Day parade, members of the Veterans of Foreign Wars (VFW) found that they could arrange themselves in rows of 6, 7, or 8, with no one left over. What is the least number of VFW members in the parade? *(Lesson 15–5)* **168**

Extra Credit

Find the mean length of the sides of triangle *ABC*. $\frac{22xy}{9}$

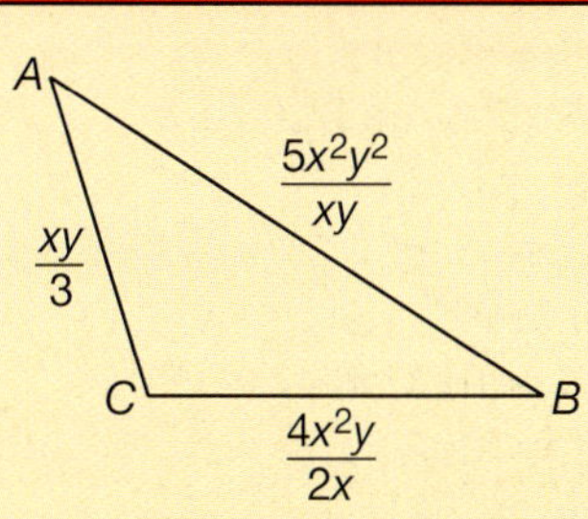

4 ASSESS

Open-Ended Assessment

Speaking Have students present two rational expressions with different denominators. Then ask them to explain how they would find the sum or difference of the rational expressions, tracing through all steps including finding the LCM of the denominators, rewriting the expressions using the LCD, performing the indicated operation, and simplifying.

Quiz 2

The Quiz provides students with a brief review of the concepts and skills in Lessons 15–3 through 15–5. Lesson numbers are given to the right of the exercises or instruction lines so students can review concepts not yet mastered.

***Enrichment*, p. 687**

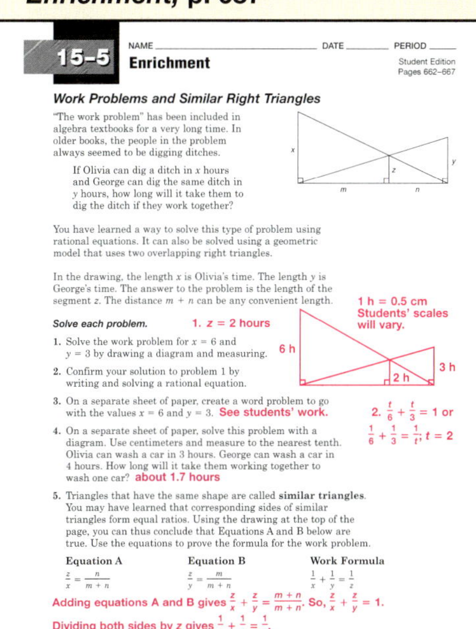

15–5 NAME ______ DATE ______ PERIOD ______

Enrichment Student Edition Pages 662–667

Work Problems and Similar Right Triangles

"The work problem" has been included in algebra textbooks for a very long time. In older books, the people in the problem always seemed to be digging ditches.

If Olivia can dig a ditch in x hours and George can dig the same ditch in y hours, how long will it take them to dig the ditch if they work together?

You have learned a way to solve this type of problem using rational equations. It can also be solved using a geometric model that uses two overlapping right triangles.

In the drawing, the length x is Olivia's time. The length y is George's time. The answer to the problem is the length of the segment z. The distance $m + n$ can be any convenient length.

Solve each problem. **1. z = 2 hours**

1. Solve the work problem for $x = 6$ and $y = 3$ by drawing a diagram and measuring.
2. Confirm your solution to problem 1 by writing and solving a rational equation.
3. On a separate sheet of paper, create a word problem to go with the values $x = 6$ and $y = 3$. **See students' work.**
4. On a separate sheet of paper, solve this problem with a diagram. Use centimeters and measure to the nearest tenth. Olivia can wash a car in 3 hours. George can wash a car in 4 hours. How long will it take them working together to wash one car? **about 1.7 hours**
5. Triangles that have the same shape are called **similar triangles**. You may have learned that corresponding sides of similar triangles form equal ratios. Using the drawing at the top of the page, you can thus conclude that Equations A and B below are true. Use the equations to prove the formula for the work problem.

1 h = 0.5 cm Students' scales will vary. (6 h, 3 h, 2 h)

2. $\frac{t}{6} + \frac{t}{3} = 1$ **or** $\frac{1}{6} + \frac{1}{3} = \frac{1}{t}; t = 2$

Equation A	Equation B	Work Formula
$\frac{z}{x} = \frac{n}{m+n}$	$\frac{z}{y} = \frac{m}{m+n}$	$\frac{1}{x} + \frac{1}{y} = \frac{1}{z}$

Adding equations A and B gives $\frac{z}{x} + \frac{z}{y} = \frac{m+n}{m+n}$. So, $\frac{z}{x} + \frac{z}{y} = 1$.

Dividing both sides by z gives $\frac{1}{x} + \frac{1}{y} = \frac{1}{z}$.

Lesson 15-6

1 FOCUS

5-Minute Check
Lesson 15–5

1. Find the LCM for the expressions $x^2 - 1$ and $x^2 + 2x + 1$. $(x - 1)(x + 1)(x + 1)$

Write each pair of rational expressions with the same LCD.

2. $\frac{2}{5b^3}, \frac{-7}{15b}$ $\frac{6}{15b^3}, \frac{-7b^2}{15b^3}$
3. $\frac{y}{y-3}, \frac{5y}{4y-12}$ $\frac{4y}{4(y-3)}, \frac{5y}{4(y-3)}$

Each rectangle below is shown with its area.

$\frac{2y+9}{2y+5}$	$\frac{4y-6}{8y+20}$

4. What is the sum of the areas of the rectangles? $\frac{3}{2}$
5. What is the difference of the areas of the rectangles? $\frac{2y+21}{4y+10}$

MODELING

An alternative hands-on option is available for teaching this lesson.

15-6 Solving Rational Equations

What You'll Learn
You'll learn to solve rational equations.

Why It's Important
Aviation Pilots can use rational equations to find wind speed. *See Exercise 9.*

Every week there is a 10-point quiz in math class. For the first five quizzes, Julia scored a total of 36 points, which gave her an average of 7.2. She is determined to get 10 points on each of the next quizzes until she brings her average up to an 8. On how many quizzes must she score 10 points in order to have an overall quiz average of 8 points?

Let x represent the number of quizzes on which she must score 10 points.

total number of quizzes $= 5 + x$

sum of scores $= 36 + 10x$

average $= \frac{36 + 10x}{5 + x}$ ← *sum of scores* ← *number of quizzes*

$8 = \frac{36 + 10x}{5 + x}$ *Julia wants to have an average of 8.*

The equation above is a **rational equation** because it contains at least one rational expression. *You will solve this problem in Exercise 39.*

There are three steps in solving rational equations.

Step 1 Find the LCD of each term.
Step 2 Multiply each side of the equation by the LCD.
Step 3 Use the Distributive Property to simplify.

Examples

Solve each equation. Check your solution.

1 $\frac{3a}{5} + \frac{3}{2} = \frac{7a}{10}$

$\frac{3a}{5} + \frac{3}{2} = \frac{7a}{10}$ *The LCD is 10.*

$10\left(\frac{3a}{5} + \frac{3}{2}\right) = 10\left(\frac{7a}{10}\right)$ *Multiply each side by the LCD.*

$10\left(\frac{3a}{5}\right) + 10\left(\frac{3}{2}\right) = 10\left(\frac{7a}{10}\right)$ *Distributive Property*

$\overset{2}{\cancel{10}}\left(\frac{3a}{\underset{1}{\cancel{5}}}\right) + \overset{5}{\cancel{10}}\left(\frac{3}{\underset{1}{\cancel{2}}}\right) = \overset{1}{\cancel{10}}\left(\frac{7a}{\underset{1}{\cancel{10}}}\right)$ *Divide each by the GCF.*

$6a + 15 = 7a$ *Simplify.*

$6a + 15 - 6a = 7a - 6a$ *Subtract 6a from each side.*

$15 = a$ *Simplify.*

Resource Manager

Reproducible Masters

Chapter 15 Resource Masters
- *Study Guide,* p. 688
- *Skills Practice,* p. 689
- *Practice,* p. 690
- *Reading to Learn Mathematics,* p. 691
- *Enrichment,* p. 692
- *Assessment,* p. 703

Hands-On Algebra, p. 165

Transparencies
5-Minute Check, 15–6

Technology/Multimedia
Interactive Chalkboard CD-ROM

Check: $\frac{3a}{5} + \frac{3}{2} = \frac{7a}{10}$ *Original equation*

$\frac{3(15)}{5} + \frac{3}{2} \stackrel{?}{=} \frac{7(15)}{10}$ *Replace a with 15.*

$9 + \frac{3}{2} \stackrel{?}{=} \frac{21}{2}$ *Simplify.*

$\frac{21}{2} = \frac{21}{2}$ ✓

2 $\frac{2}{3x} - \frac{1}{2x} = \frac{1}{6}$

$\frac{2}{3x} - \frac{1}{2x} = \frac{1}{6}$ *The LCD is 6x.*

$6x\left(\frac{2}{3x} - \frac{1}{2x}\right) = 6x\left(\frac{1}{6}\right)$ *Multiply each side by the LCD.*

$6x\left(\frac{2}{3x}\right) - 6x\left(\frac{1}{2x}\right) = 6x\left(\frac{1}{6}\right)$ *Distributive Property*

$\overset{2}{\cancel{6x}}\left(\frac{2}{\underset{1}{\cancel{3x}}}\right) - \overset{3}{\cancel{6x}}\left(\frac{1}{\underset{1}{\cancel{2x}}}\right) = \overset{1}{\cancel{6}}x\left(\frac{1}{\underset{1}{\cancel{6}}}\right)$ *Divide each by the GCF.*

$4 - 3 = x$ *Simplify.*

$1 = x$ *Subtract.*

Check: $\frac{2}{3x} - \frac{1}{2x} = \frac{1}{6}$ *Original equation*

$\frac{2}{3(1)} - \frac{1}{2(1)} \stackrel{?}{=} \frac{1}{6}$ *Replace x with 1.*

$\frac{2}{3} - \frac{1}{2} \stackrel{?}{=} \frac{1}{6}$ *The LCD is 6.*

$\frac{4}{6} - \frac{3}{6} \stackrel{?}{=} \frac{1}{6}$ $\frac{2}{3} = \frac{4}{6}$

$\frac{1}{6} = \frac{1}{6}$ ✓

3 $\frac{3x}{x+2} + \frac{1}{x+2} = 4$

$\frac{3x}{x+2} + \frac{1}{x+2} = 4$ *The LCD is x + 2.*

$(x+2)\left(\frac{3x}{x+2} + \frac{1}{x+2}\right) = (x+2)4$ *Multiply each side by the LCD.*

$(x+2)\left(\frac{3x}{x+2}\right) + (x+2)\left(\frac{1}{x+2}\right) = 4x + 8$ *Distributive Property*

$\overset{1}{\cancel{(x+2)}}\left(\frac{3x}{\underset{1}{\cancel{x+2}}}\right) + \overset{1}{\cancel{(x+2)}}\left(\frac{1}{\underset{1}{\cancel{x+2}}}\right) = 4x + 8$ *Divide each by the GCF.*

$3x + 1 = 4x + 8$ *Simplify.*

(continued on the next page)

Motivating the Lesson

Real-World Connection Students who have flown frequently in commercial airlines may have observed that flying times are often shorter flying from west-to-east across the United States than flying east-to-west. This is because strong jet stream winds often blow from west to east at altitudes where the planes fly. Flying with the wind, the speed of the wind is added to the plane's *airspeed*, increasing its *ground speed*. The opposite is true flying against the wind. Tell students that solving problems involving distance, rate, and time in *uniform motion* situations such as described above often involves solving equations that contain rational expressions, or *rational equations*.

2 TEACH

In-Class Example

Example 1

Solve $\frac{5x}{4} + \frac{2}{3} = \frac{7x}{12}$. Check your solution. **−1**

Teaching Tip Sometimes, as in Example 2, it may be easy to add or subtract first before solving. In this Example, the result will be the simple proportion $\frac{1}{6x} = \frac{1}{6}$. Until students hone their observational skills, however, multiplying both sides by the LCD provides a sure way for students to eliminate fractions and avoid errors.

In-Class Examples

Examples 2–3

Solve each equation. Check your solution.

2 $\frac{5}{3x} - \frac{4}{5x} = \frac{26}{15}$ $\frac{1}{2}$

3 $\frac{7}{x-1} + \frac{2x}{x-1} = 5$ **4**

In-Class Example

Example 4

Solve $\frac{7}{p} + \frac{2}{p+5} = \frac{4}{p+5}$.

Check your solution. -7

Teaching Tip Students often have a fear of uniform motion problems such as Example 5. Point out that the key is often identifying the quantity that remains the same, which in Example 5 is the time. Once this is done, organizing the information in a table will give clarity. Also, strongly emphasize the examination of the problem at the end. Not only will checking a potential solution prevent errors, it will help students understand the problem more fully as they substitute values they have found.

In-Class Example

Example 5

A rowing crew practices at a steady rate of 9 miles per hour. When they practice in a steadily flowing river, a 10-mile trip downstream takes the same amount of time as a 5-mile trip upstream. What is the rate of the river current? **3 mph**

$3x + 1 - 3x = 4x + 8 - 3x$ *Subtract 3x from each side.*

$1 = x + 8$ *Simplify.*

$1 - 8 = x + 8 - 8$ *Subtract 8 from each side.*

$-7 = x$ *Check the solution.*

4 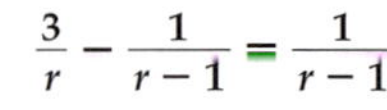 $\frac{3}{r} - \frac{1}{r-1} = \frac{1}{r-1}$

$\frac{3}{r} - \frac{1}{r-1} = \frac{1}{r-1}$ *The LCD is r(r − 1).*

$r(r-1)\left(\frac{3}{r} - \frac{1}{r-1}\right) = r(r-1)\left(\frac{1}{r-1}\right)$ *Multiply each side by the LCD.*

$r(r-1)\left(\frac{3}{r}\right) - r(r-1)\left(\frac{1}{r-1}\right) = r(r-1)\left(\frac{1}{r-1}\right)$ *Distributive Property*

$\overset{1}{\cancel{r}}(r-1)\left(\frac{3}{\underset{1}{\cancel{r}}}\right) - r\overset{1}{\cancel{(r-1)}}\left(\frac{1}{\underset{1}{\cancel{r-1}}}\right) = r\overset{1}{\cancel{(r-1)}}\left(\frac{1}{\underset{1}{\cancel{r-1}}}\right)$ *Divide by the GCF.*

$(r-1)3 - r = r$ *Simplify.*

$3r - 3 - r = r$ *Distributive Property*

$2r - 3 = r$ *Subtract.*

$2r - 3 - r = r - r$ *Subtract r from each side.*

$r - 3 = 0$ *Simplify.*

$r - 3 + 3 = 0 + 3$ *Add 3 to each side.*

$r = 3$ *Check the solution.*

Your Turn **Solve each equation. Check your solution.**

a. $\frac{4x}{3} + \frac{7}{2} = \frac{9x}{12}$ **−6** b. $\frac{18}{b} = \frac{3}{b} + 3$ **5** c. $\frac{4}{n+1} - \frac{2}{n} = \frac{5}{n}$ $-\frac{7}{3}$

Recall that **uniform motion problems** can be solved by using the formula below.

$$\underbrace{\text{distance}}_{d} = \underbrace{\text{rate}}_{r} \cdot \underbrace{\text{time}}_{t}$$

Example 5 Travel Link

The Milbys rented a houseboat on the Sacramento River. The maximum speed of the boat in still water is 8 miles per hour. At this rate, a 30-mile trip downstream took the same amount of time as an 18-mile trip against the current. What was the rate of the current?

Explore Let c = the rate of the current.

Let $8 + c$ = the rate of the boat traveling downstream with the current. *When the boat is traveling with the current, you add the speed of the current to the speed of the boat.*

Let $8 - c$ = the rate of the boat traveling upstream against the current. *When the boat is traveling against the current, you subtract the speed of the current from the speed of the boat.*

Data Update For the latest information on various rivers, visit: www.algconcepts.com

Plan Since $d = rt$, then $t = \frac{d}{r}$.

	d	r	$t = \frac{d}{r}$
Downstream	30	$8 + c$	$\frac{30}{8 + c}$
Upstream	18	$8 - c$	$\frac{18}{8 - c}$

Solve

$$\frac{30}{8 + c} = \frac{18}{8 - c} \quad \textit{The time downstream equals the time upstream.}$$

$$(8 + c)(8 - c)\left(\frac{30}{8 + c}\right) = (8 + c)(8 - c)\left(\frac{18}{8 - c}\right) \quad \textit{The LCD is } (8 + c)(8 - c).$$

$$\overset{1}{\cancel{(8 + c)}}(8 - c)\left(\frac{30}{\underset{1}{\cancel{8 + c}}}\right) = (8 + c)\overset{1}{\cancel{(8 - c)}}\left(\frac{18}{\underset{1}{\cancel{8 - c}}}\right) \quad \textit{Divide by GCF.}$$

$$(8 - c)(30) = (8 + c)(18) \quad \textit{Simplify.}$$

$$240 - 30c = 144 + 18c \quad \textit{Distributive Property}$$

$$240 - 30c + 30c = 144 + 18c + 30c \quad \textit{Add 30c to each side.}$$

$$240 = 144 + 48c \quad \textit{Simplify.}$$

$$240 - 144 = 144 + 48c - 144 \quad \textit{Subtract 144 from each side.}$$

$$96 = 48c \quad \textit{Simplify.}$$

$$\frac{96}{48} = \frac{48c}{48} \quad \textit{Divide each side by 48.}$$

$$2 = c \quad \textit{Simplify.}$$

The rate of the current was 2 miles per hour.

A houseboat on a river

Examine Check the solution to see if it makes sense. The houseboat goes downstream at 8 + 2 or 10 miles per hour. A 30-mile trip would take 30 ÷ 10 or 3 hours. The houseboat goes upstream at 8 − 2 or 6 miles per hour. An 18-mile trip would take 18 ÷ 6 or 3 hours. Both trips take the same amount of time, so the solution is correct.

Check for Understanding

Communicating Mathematics

1. **List** two differences between linear equations and rational equations.
2. Writing Math Describe two different ways in which $\frac{4}{x + 1} = \frac{8}{x - 1}$ can be solved. Then determine whether there are other steps that you could use to solve the examples in this lesson. **1–2. See margin.**

Vocabulary
rational equation
uniform motion problems

Answers

1. Sample answer: In a linear equation, the variable cannot appear in the denominator and it cannot be raised to a power higher than 1.
2. Sample answer: $\frac{4}{x + 1} = \frac{8}{x - 1}$ could be solved by multiplying each side by the LCD $(x + 1)(x - 1)$ and then simplifying. It could also be solved by cross multiplying. See students' work.

Study Guide, p. 688

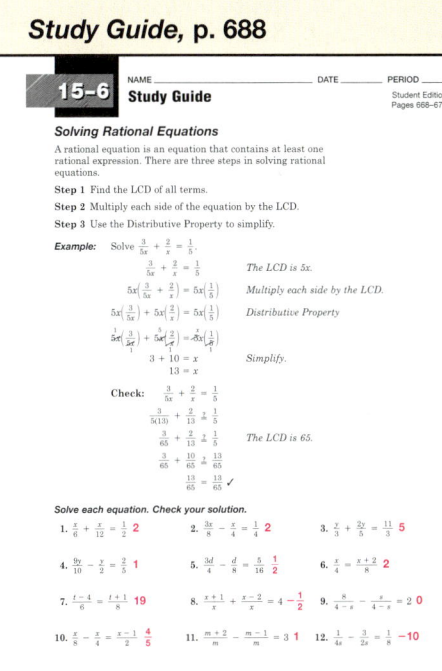

NAME ________ DATE ____ PERIOD ____

15-6 Study Guide

Student Edition Pages 668–673

Solving Rational Equations

A rational equation is an equation that contains at least one rational expression. There are three steps in solving rational equations.

Step 1 Find the LCD of all terms.

Step 2 Multiply each side of the equation by the LCD.

Step 3 Use the Distributive Property to simplify.

Example: Solve $\frac{3}{5x} + \frac{2}{x} = \frac{1}{5}$.

$$\frac{3}{5x} + \frac{2}{x} = \frac{1}{5} \quad \textit{The LCD is 5x.}$$

$$5x\left(\frac{3}{5x} + \frac{2}{x}\right) = 5x\left(\frac{1}{5}\right) \quad \textit{Multiply each side by the LCD.}$$

$$5x\left(\frac{3}{5x}\right) + 5x\left(\frac{2}{x}\right) = 5x\left(\frac{1}{5}\right) \quad \textit{Distributive Property}$$

$$\overset{1}{\cancel{5x}}\left(\frac{3}{\underset{1}{\cancel{5x}}}\right) + \overset{5}{\cancel{5x}}\left(\frac{2}{\underset{1}{\cancel{x}}}\right) = \overset{x}{\cancel{5x}}\left(\frac{1}{\underset{1}{\cancel{5}}}\right)$$

$$3 + 10 = x \quad \textit{Simplify.}$$

$$13 = x$$

Check:

$$\frac{3}{5x} + \frac{2}{x} = \frac{1}{5}$$

$$\frac{3}{5(13)} + \frac{2}{13} \stackrel{?}{=} \frac{1}{5}$$

$$\frac{3}{65} + \frac{2}{13} \stackrel{?}{=} \frac{1}{5} \quad \textit{The LCD is 65.}$$

$$\frac{3}{65} + \frac{10}{65} \stackrel{?}{=} \frac{13}{65}$$

$$\frac{13}{65} = \frac{13}{65} \checkmark$$

Solve each equation. Check your solution.

1. $\frac{x}{6} + \frac{x}{12} = \frac{1}{2}$ **2**
2. $\frac{3x}{8} - \frac{x}{4} = \frac{1}{4}$ **2**
3. $\frac{y}{3} + \frac{2y}{5} = \frac{11}{3}$ **5**
4. $\frac{9y}{10} - \frac{y}{2} = \frac{2}{5}$ **1**
5. $\frac{3d}{4} - \frac{d}{8} = \frac{5}{16}$ $\mathbf{\frac{1}{2}}$
6. $\frac{x}{4} = \frac{x + 2}{8}$ **2**
7. $\frac{t - 4}{6} = \frac{t + 1}{8}$ **19**
8. $\frac{x + 1}{x} + \frac{x - 2}{x} = 4$ $\mathbf{-\frac{1}{2}}$
9. $\frac{8}{4 - s} - \frac{s}{4 - s} = 2$ **0**
10. $\frac{x}{8} - \frac{x}{4} = \frac{x - 1}{2}$ $\mathbf{\frac{4}{5}}$
11. $\frac{m + 2}{m} - \frac{m - 1}{m} = 3$ **1**
12. $\frac{1}{4s} - \frac{3}{2s} = \frac{1}{8}$ **−10**

Reteaching Activity

Auditory/Musical Learners Have students write a catchy jingle or acronym incorporating the three steps on page 668. An example is given below.

To solve a rational equation,
LCD, then multiplication,
to both sides, you see,
then Distributive Property.

3 PRACTICE/APPLY

Error Analysis

Watch for students who forget to multiply the right-hand side of the equations in Exercises 5–8 by the LCD when they multiply the left-hand side by the LCD.

Prevent by encouraging students always to write out the step that shows both sides of the equation being multiplied by the same expression. Even though there are no fractions on the right-hand side of these equations, whatever is done to the left side must be done to the right.

Assignment Guide

Basic: 11–39 odd, 41–50

Average: 10–38 even, 39–50

Skills Practice, **p. 689, and *Practice,* p. 690 (shown)**

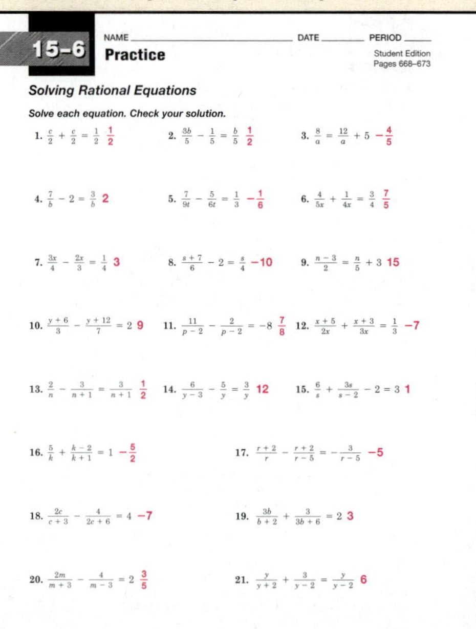

15-6 NAME ________ DATE ______ PERIOD ____

Practice

Student Edition Pages 668–673

Solving Rational Equations

Solve each equation. Check your solution.

1. $\frac{c}{2} + \frac{c}{2} = \frac{1}{2}$ $\frac{1}{2}$
2. $\frac{3b}{5} - \frac{1}{5} = \frac{b}{5}$ $\frac{1}{2}$
3. $\frac{8}{a} = \frac{12}{a} + 5$ $-\frac{4}{5}$
4. $\frac{7}{b} - 2 = \frac{3}{b}$ 2
5. $\frac{7}{9t} - \frac{5}{6t} = \frac{1}{3}$ $-\frac{1}{6}$
6. $\frac{4}{5x} + \frac{1}{4x} = \frac{3}{4}$ $\frac{7}{5}$
7. $\frac{3x}{4} - \frac{2x}{3} = \frac{1}{4}$ 3
8. $\frac{s+7}{6} - 2 = \frac{s}{4}$ −10
9. $\frac{n-3}{2} = \frac{n}{5} + 3$ 15
10. $\frac{y+6}{3} - \frac{y+12}{7} = 2$ 9
11. $\frac{11}{p-2} - \frac{2}{p-2} = -8$ $\frac{7}{8}$
12. $\frac{x+5}{2x} + \frac{x+3}{3x} = \frac{1}{3}$ −7
13. $\frac{2}{n} - \frac{3}{n+1} = \frac{3}{n+1}$ $\frac{1}{2}$
14. $\frac{6}{y-3} - \frac{5}{y} = \frac{3}{y}$ 12
15. $\frac{6}{s} + \frac{3s}{s-2} - 2 = 3$ 1
16. $\frac{5}{k} + \frac{k-2}{k+1} = 1$ $-\frac{5}{2}$
17. $\frac{r+2}{r} - \frac{r+2}{r-5} = -\frac{3}{r-5}$ −5
18. $\frac{2c}{c+3} - \frac{4}{2c+6} = 4$ −7
19. $\frac{3b}{b+2} + \frac{3}{3b+6} = 2$ 3
20. $\frac{2m}{m+3} - \frac{4}{m-3} = 2$ $\frac{3}{5}$
21. $\frac{y}{y+2} + \frac{3}{y-2} = \frac{y}{y-2}$ 6

Guided Practice

Solve each equation. Check your solution.

Examples 1–4

3. $\frac{2x}{7} + \frac{1}{2} = \frac{x}{14}$ $-\frac{7}{3}$
4. $\frac{6}{c} = \frac{10}{c} + 4$ −1
5. $\frac{5}{m+3} - \frac{2}{m+3} = -9$ 5. $-\frac{10}{3}$
6. $\frac{a+1}{a} + \frac{a+4}{a} = 6$ $\frac{5}{4}$
7. $\frac{2x}{x+3} + \frac{3}{x} = 2$ 3
8. $\frac{r-1}{r+1} - \frac{2r}{r-1} = -1$ 0

Example 5

9. **Transportation** An airplane can fly at a rate of 600 miles per hour in calm air. It can fly 2520 miles with the wind in the same time it can fly 2280 miles against the wind. Find the speed of the wind. **30 mph**

Exercises

Practice

Solve each equation. Check your solution.

Homework Help

For Exercises	See Examples
10–20, 22, 23	1, 2
21, 24–39, 41	3, 4
40	5

Extra Practice

See page 723.

10. $\frac{a}{6} + \frac{2a}{3} = -\frac{5}{2}$ −3
11. $\frac{1}{4} + \frac{5r}{8} = \frac{r}{4}$ $-\frac{2}{3}$
12. $\frac{b}{5} - \frac{2b}{15} = 1$ 15
13. $\frac{x}{7} = \frac{x+3}{10}$ 7
14. $\frac{t-1}{4} = \frac{t}{3}$ −3
15. $\frac{2a-3}{6} = \frac{2a}{3} + \frac{1}{2}$ −3
16. $\frac{4}{3y} + \frac{1}{y} = 7$ $\frac{1}{3}$
17. $\frac{6}{x} - \frac{3}{2x} = \frac{1}{2}$ 9
18. $\frac{1}{2a} + \frac{3}{4a} = \frac{1}{4}$ 5
19. $\frac{m+1}{m} + \frac{m+3}{m} = 5$ 19. $\frac{4}{3}$
20. $\frac{t+1}{t} + \frac{t+4}{t} = 6$ $\frac{5}{4}$
21. $\frac{5}{5-p} - \frac{1}{5-p} = -2$ 21. 7
22. $\frac{5}{2x} - \frac{1}{6x} = 2$ $\frac{7}{6}$
23. $\frac{1}{4x} + \frac{1}{6x} = 5$ $\frac{1}{12}$
24. $\frac{3}{x} + \frac{4x}{x-3} = 4$ $\frac{3}{5}$
25. $\frac{n-3}{n} = \frac{n-3}{n-6}$ 3
26. $\frac{3}{r+4} - \frac{1}{r} = \frac{1}{r}$ 8
27. $\frac{5}{x+1} + \frac{1}{x} = \frac{2}{x^2+x}$ 27. $\frac{1}{6}$

B

28. $\frac{5}{n-2} + \frac{2}{n} = \frac{1}{n}$ $\frac{1}{3}$
29. $\frac{6}{t+1} - \frac{3}{4t+4} = \frac{3}{4}$ 29. 6
30. $\frac{1}{2a+6} + \frac{3a}{a+3} = \frac{1}{2}$ 30. $\frac{2}{5}$
31. $\frac{7}{a-1} = \frac{5}{a+3}$ −13
32. $\frac{1}{m+1} + \frac{5}{m-1} = \frac{1}{m^2-1}$ $-\frac{1}{2}$

33. $\frac{j}{j+1} + \frac{5}{j-1} = 1$ $-\frac{3}{2}$
34. $\frac{6}{z+2} + \frac{3}{z^2-4} = \frac{7}{z+2}$ 5
35. $\frac{1}{4m} - \frac{2}{m-3} = \frac{2}{m}$ $\frac{7}{5}$
36. $\frac{x+2}{2x-1} + \frac{3x-6}{8x-4} = \frac{9}{4}$ 1
37. What is the value of w in the equation $\frac{2w-3}{w-3} - 2 = \frac{12}{w+3}$? **5**
38. Solve $\frac{3n}{n^2-5n+4} = \frac{2}{n-4} + \frac{3}{n-1}$. **7**

Applications and Problem Solving

39. **Quizzes** Refer to the application at the beginning of the lesson. Solve $8 = \frac{36+10x}{5+x}$ to find the number of quizzes on which Julia must score 10 points in order to have an overall quiz average of 8 points. **2**

40. **Cycling** A long-distance cyclist pedaling at a steady rate travels 30 miles with the wind. She can travel only 18 miles against the wind in the same amount of time. If the rate of the wind is 3 miles per hour, what is the cyclist's rate without the wind? (*Hint:* Use the formula $d = rt$.) **12 mph**

41. **Critical Thinking** What number would you add to the numerator and denominator of $\frac{2}{11}$ to make a fraction equivalent to $\frac{1}{2}$? **7**

Mixed Review

42. **Space** Scientists can predict when an asteroid might hit Earth or come very close to Earth by studying their orbits. Toro is an asteroid that orbits the sun every 584 days. Earth orbits the sun every 365 days. Suppose Earth and Toro are lined up with the sun today. Find the LCM of 584 and 365 to determine how long before Earth and Toro arrive together back at that same point in their orbits. *(Lesson 15–5)* **2920 days or 8 years**

An asteroid

43. Find the sum of $\frac{11}{4n^2}$ and $\frac{7}{4n^2}$. Write your answer in simplest form. *(Lesson 15–4)* $\frac{9}{2n^2}$

Name the set or sets of numbers to which each real number belongs. Let N = natural numbers, W = whole numbers, Z = integers, Q = rational numbers, and I = irrational numbers. *(Lesson 14–1)*

44. $\frac{16}{2}$ **N, W, Z, Q**

45. -7 **Z, Q**

46. $\sqrt{15}$ **I**

47. **Farming** In order to have enough time in the growing season, a farmer has at most 16 days left to plant his corn and soybean crops. He can plant corn at a rate of 10 acres per day and soybeans at a rate of 15 acres per day. Suppose he has at most 200 acres available and he wants to plant both crops. *(Lesson 13–7)*

a. Let c represent the number of days that corn will be planted and let s represent the number of days that soybeans will be planted. Write a system of inequalities to represent this situation.

b. Graph the system of inequalities. **See margin.**

c. List two ordered pairs that are possible solutions and explain what each ordered pair represents.

47a. $c + s \leq 16$, $10c + 15s \leq 200$

47c. Sample answer: (7, 8), he could plant corn for 7 days and soybeans for 8 days; (2, 12), he could plant corn for 2 days and soybeans for 12 days

Solve each system of equations by graphing. *(Lesson 13–1)*

48. $y = 2x - 1$
$y = -x + 5$ **(2, 3)**

49. $x = 4$
$x + y = 3$ **(4, −1)**

48–49. See margin for graphs.

Standardized Test Practice

50. **Short Response** Write an irrational square root whose graph is between points A and B on the number line below. *(Lesson 8–6)* **Sample answer:** $\sqrt{75}$

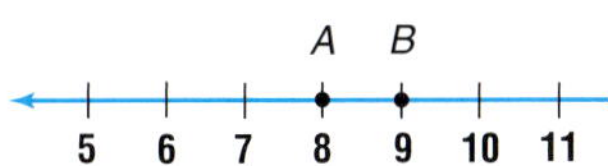

www.algconcepts.com/self_check_quiz

Extra Credit

Solve the system of equations.

$$\frac{5}{x} - \frac{9}{y} = \frac{24}{xy}$$

$$\frac{-6}{x} + \frac{12}{y} = \frac{-30}{xy}$$ **(−1, 3)**

Answer

49.

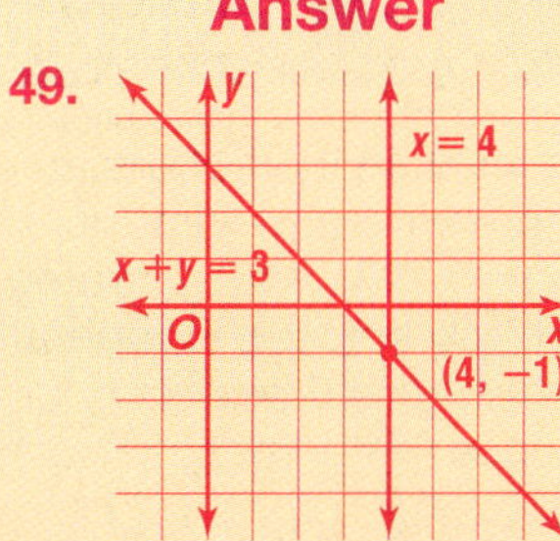

4 ASSESS

Open-Ended Assessment

Speaking Ask students to explain why multiplying both sides of a rational equation by the LCD of all the denominators results in an equivalent equation with no fractions.

Chapter 15, Quiz B (Lessons 15–3 through 15–6) is available in the *Chapter 15 Resource Masters*, p. 703.

Answers

47b.

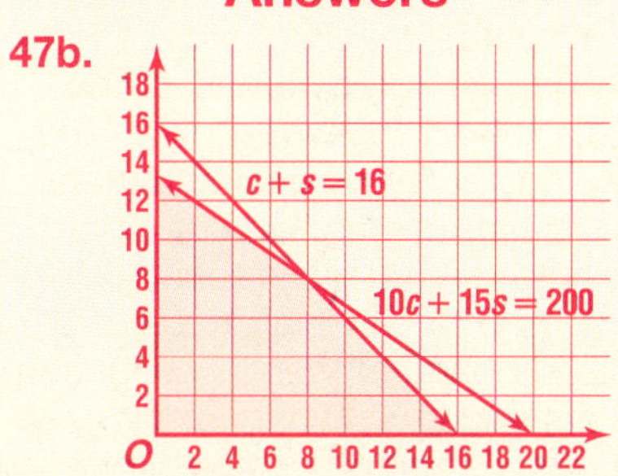

48.

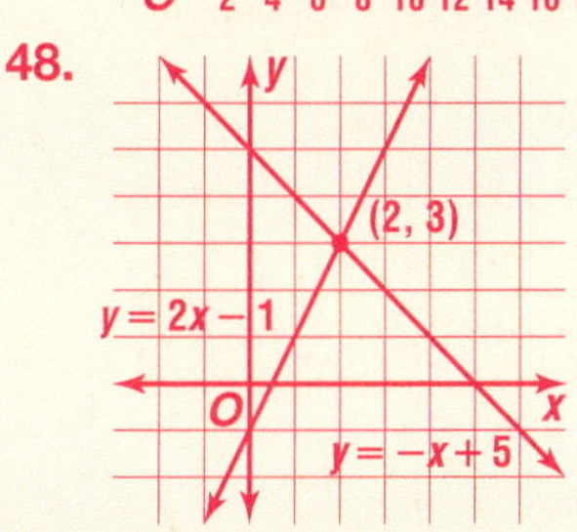

***Enrichment*, p. 692**

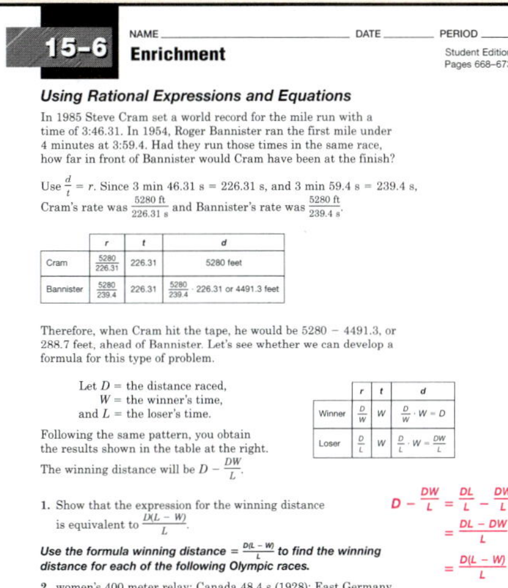

15–6 NAME ______ DATE ______ PERIOD ______

Enrichment

Student Edition Pages 668–673

Using Rational Expressions and Equations

In 1985 Steve Cram set a world record for the mile run with a time of 3:46.31. In 1954, Roger Bannister ran the first mile under 4 minutes at 3:59.4. Had they run those times in the same race, how far in front of Bannister would Cram have been at the finish?

Use $\frac{d}{t} = r$. Since 3 min 46.31 s = 226.31 s, and 3 min 59.4 s = 239.4 s, Cram's rate was $\frac{5280 \text{ ft}}{226.31 \text{ s}}$ and Bannister's rate was $\frac{5280 \text{ ft}}{239.4 \text{ s}}$.

	r	t	d
Cram	$\frac{5280}{226.31}$	226.31	5280 feet
Bannister	$\frac{5280}{239.4}$	226.31	$\frac{5280}{239.4} \cdot 226.31$ or 4491.3 feet

Therefore, when Cram hit the tape, he would be 5280 − 4491.3, or 288.7 feet, ahead of Bannister. Let's see whether we can develop a formula for this type of problem.

Let D = the distance raced,
W = the winner's time,
and L = the loser's time.

Following the same pattern, you obtain the results shown in the table at the right.

	r	t	d
Winner	$\frac{D}{W}$	W	$\frac{D}{W} \cdot W = D$
Loser	$\frac{D}{L}$	W	$\frac{D}{L} \cdot W = \frac{DW}{L}$

The winning distance will be $D - \frac{DW}{L}$.

1. Show that the expression for the winning distance is equivalent to $\frac{D(L - W)}{L}$.

$$D - \frac{DW}{L} = \frac{DL}{L} - \frac{DW}{L} = \frac{DL - DW}{L} = \frac{D(L - W)}{L}$$

Use the formula winning distance = $\frac{D(L - W)}{L}$ ***to find the winning distance for each of the following Olympic races.***

2. women's 400 meter relay: Canada 48.4 s (1928); East Germany 41.6 s (1980) **56.2 meters**

3. men's 200 meter freestyle swimming: Mark Spitz 1 min 52.78 s (1972); Michael Gross 1 min 47.44 s (1984) **9.5 meters**

4. men's 50,000 meter walk: Thomas Green 4 h 50 min 10 s (1932); Hartwig Gauter 3 h 49 min 24 s (1980) **10,471 meters**

5. women's 400 meter freestyle relay: Great Britain 5 min 52.8 s (1912); East Germany 3 min 42.71 s (1980) **147.5 meters**

Investigation

PREPARE

This optional investigation is designed to be completed by groups of 2–3 students over 1–2 days.

Objective

Students make a table to solve a work problem by combining the effects of two rates they have determined. They find rates of work done by using reciprocals of the time a job or process requires. They write and solve rational equations for work problems that combine two rates. Students present their findings with a display, such as a bulletin board, of their investigation results and by creating, solving, and discussing an original work problem.

Mathematical Overview

This investigation utilizes the following concepts:

- rates,
- reciprocals,
- the work formula *rate of work* · *time* = *work done*,
- combined rates, and
- solving rational equations.

Suggested Time Management	
Investigation	30–45 min
Extension: Gathering Data	30–45 min
Extension: Summarizing Data	20–30 min

Motivating the Lesson

Ask students to think about the time it takes to finish a project if two people work together. Would it take a longer or shorter time than if the faster worker does it alone? Would it take more than half the time or less than half the time of the faster worker? Is the combined work rate the same as the sum of the individual work rates? Inform students that in this investigation they will see how problem-solving skills and rational equations will help them find precise answers for this type of problem.

Chapter 15 Investigation

Down the Drain

Materials

calculator

Work Problems

Suppose you have a big job to do, like painting a house. How much faster could you get the job done if you had help? In this investigation, you will use algebraic techniques to solve **work problems**.

Investigate

1. A public swimming pool holds 100,000 gallons of water and has two drain outlets. If only outlet A is open, the pool will drain in 20 hours. If only outlet B is open, the pool will drain in 16 hours.
   a. The drainage rate of outlet A is 100,000 gallons ÷ 20 hours or 5000 gallons per hour. What is the drainage rate of outlet B? **6250 gal/h**
   b. Make a table like the one below and continue filling in columns 2, 3, and 4 until the total water drained in column 4 is more than 100,000 gallons. **See Solutions Manual.**

You could also use a spreadsheet for your calculations.

Number of Hours	Water Drained by Outlet A (gal)	Water Drained by Outlet B (gal)	Total Water Drained (gal)
1	5000	6250	11,250
2	10,000	12,500	22,500

   c. Approximately how long will it take to drain the pool with both outlets open? **Sample answer: 8–9 h**

2. The problem above also can be solved using an algebraic equation.
   a. Since outlet A can drain the pool in 20 hours, it can drain $\frac{1}{20}$ of the pool in 1 hour. How much of the pool can outlet B drain in 1 hour?
   b. Use the table at the right and the following formula to write rational expressions for the work done w by outlets 1 and 2 in t hours.

2a. $\frac{1}{16}$

2b. $\frac{t}{20}, \frac{t}{16}$

$$\underbrace{\text{rate of work}}_{r} \cdot \underbrace{\text{time}}_{t} = \underbrace{\text{work done}}_{w}$$

	r	t	w
Outlet A	$\frac{1}{20}$	t	?
Outlet B	$\frac{1}{16}$	t	?

Cooperative Learning

This investigation offers an excellent opportunity for using cooperative groups. For more information on cooperative learning strategies and group management, see *Cooperative Learning in the Mathematics Classroom,* one of the titles in the Glencoe Mathematics Professional Series.

c. If both outlets are open, then the following equation is true.

work of outlet A + work of outlet B = the whole job

Write an equation to represent the time it takes the pool to drain if both outlets are open. Assume that 1 represents the whole job. *Why?*

d. Solve the equation in part b to find how long it takes to drain the pool if both outlets are open. Compare this answer to the estimate you made in Exercise 1d.

2c. $\frac{t}{20} + \frac{t}{16} = 1$

2d. $8\frac{8}{9}$ h; This is close to the estimate.

3. Micheal and Gus work for a ski resort. Their job is to prepare Bear Tooth Run for skiers. It takes Micheal 4 hours to prepare the run and it takes Gus 5 hours. How long would it take them to prepare the run if they work together? **a.** $\frac{1}{5}$ **per hour**

a. In the swimming pool problem, you knew the total amount of gallons to be drained. In this problem since you do not know the size of the trail, write the work as a fraction done per hour. Micheal's rate is $\frac{1}{4}$ of the job per hour. What is Gus' rate?

b. Write an algebraic equation to represent the work done if Micheal and Gus work together. $\frac{t}{4} + \frac{t}{5} = 1$

c. How long would it take them both to prepare the run? $2\frac{2}{9}$ h

Extending the Investigation

$1\frac{1}{3}$ **h or 1 h and 20 min**

Use a spreadsheet or write an algebraic equation to solve each problem.

- Ian and Mandy own a lawn care business. Ian can mow, trim, and fertilize one particular yard in 2 hours. Mandy can complete the same tasks for that yard in 4 hours. How long would it take them to complete this yard working together?
- In the summer, Sheila and her son, Cory, paint house exteriors. To paint a typical house, it takes Sheila about 45 hours. To paint the same size house, it takes Cory 55 hours. How long would it take them to paint one house working together?
$24\frac{3}{4}$ **h or 24 h and 45 min**

Presenting Your Conclusions

Here are some ideas to help you present your conclusions to the class.

- Design a creative bulletin board displaying a problem from this investigation. Show two ways in which to solve the problem.
- Write a problem similar to those in this investigation. Solve the problem by using a spreadsheet and by using an equation. Write a one-page paper discussing the advantages and disadvantages of using each method to solve work problems.

Investigation For more information on work problems visit: www.algconcepts.com

MANAGE

Teaching Tip Pick a simple example in which the sum of the rates is 1 to help students' understanding. For example, using the pool draining situation, if one outlet can empty $\frac{2}{3}$ of the pool in an hour and the other can empty $\frac{1}{3}$ of the pool in an hour, the outlets together will empty the pool in one hour. Relate this to the equation $\frac{2t}{3} + \frac{t}{3} = 1$.

Working in Groups If students work in groups, they can split up the calculations for the table. Also, each student can solve the rational equation once a group has derived it. Then they can compare results.

Working as a Class You may want to work Exercises 2a–2c as a class to ensure that students understand the work formula, the table, and how to structure the equation. You may want to draw an analogy between the rate-of-work formula and the distance formula.

ASSESS

Students should be able to fill out a table of times and rates and use it to solve a work problem. They should be able to use reciprocals to find parts of a job completed in an hour, and then use the rate-of-work formula to write and solve a rational equation to find the time. Students should be able to present an original work problem clearly and solve it correctly using either an equation or a spreadsheet.

Students should add their displays, including tables, equations, spreadsheets, and discussions, to their portfolios at this time.

Study Guide and Assessment

Understanding and Using the Vocabulary

This section provides a listing of the new terms, properties, and phrases that were introduced in this chapter. The exercises check students' understanding of the terms by using a variety of verbal formats including matching, completion, and true/false.

Glossary A complete glossary of terms appears on pages 762–783.

MindJogger Videoquizzes

MindJogger Videoquizzes provide an alternative review of concepts presented in this chapter. Students work in teams to answer questions, gaining points for correct answers.

CHAPTER 15 Study Guide and Assessment

Understanding and Using the Vocabulary

After completing this chapter, you should be able to define each term, property, or phrase and give an example or two of each.

interNET CONNECTION **Review Activities**
For more review activities, visit: www.algconcepts.com

excluded value *(p. 638)*
least common denominator *(p. 663)*
least common multiple *(p. 662)*
rational equation *(p. 668)*
rational expression *(p. 638)*
rational function *(p. 638)*
uniform motion problems *(p. 670)*
work problems *(p. 674)*

Choose the correct term or expression to complete each sentence.

1. An excluded value is any value assigned to a variable that results in a (denominator, numerator) of zero.
2. A rational expression is an algebraic fraction whose numerator and denominator are (rational numbers, polynomials).
3. To find $\frac{9}{x+y}$ divided by $\frac{3x+6y}{x}$, multiply $\frac{9}{x+y}$ and $\left(\frac{3x+6y}{x}, \frac{x}{3x+6y}\right)$.
4. The excluded values of $\frac{4x}{x^2-36}$ are (0 and 36, 6 and -6).
5. To simplify a rational expression, divide the numerator and denominator by their (greatest common factor, least common multiple).
6. The rational expression $\frac{x+2}{x+4}$ (is, is not) in simplest form.
7. The reciprocal of $a+b$ is $\left(-a-b, \frac{1}{a+b}\right)$.
8. To add or subtract rational expressions, first rename the expressions using the (greatest common factor, least common multiple) as the denominator.
9. The least common multiple of x^2 and $2x$ is ($2x^2$, $2x^3$).
10. The divisor in $(x^2+8x+15) \div (x+5) = (x+3)$ is ($x^2+8x+15$, $x+5$).

Skills and Concepts

Objectives and Examples

• **Lesson 15–1** Simplify rational expressions.

$$\frac{z^2-3z+2}{z^2-2z} = \frac{(z-1)\overset{1}{\cancel{(z-2)}}}{z\underset{1}{\cancel{(z-2)}}} = \frac{z-1}{z}$$

Review Exercises

Simplify each expression.

11. $\frac{4x^2yz}{16xy^3}$ $\frac{xz}{4y^2}$
12. $\frac{a^2-5a}{a-5}$ a
13. $\frac{n^2-16}{n^2+2n-8}$ $\frac{n-4}{n-2}$
14. $\frac{x^3-2x^2-3x}{x^2+x-12}$ $\frac{x(x+1)}{x+4}$

www.algconcepts.com/vocabulary_review

Resource Manager

Reproducible Masters
Chapter 15 Resource Masters
• *Assessment*, pp. 693–701, 704–706, 707–710, 711–714

Technology/Multimedia
MindJogger Videoquizzes
ExamView® Pro

Objectives and Examples

- **Lesson 15–2** Multiply and divide rational expressions.

$$\frac{6x}{x^2 + x - 2} \cdot \frac{x^2 - 4}{2x^2}$$

$$= \frac{\overset{3}{\cancel{6x}}}{(\cancel{x+2})(x-1)} \cdot \frac{(x-2)(\cancel{x+2})}{\cancel{2x^2}_{x}}$$

$$= \frac{3(x-2)}{x(x-1)}$$

$$\frac{a^3}{2c} \div \frac{a^2}{c^3} = \frac{\cancel{a^3}^{a}}{2\cancel{c}} \cdot \frac{\cancel{c^3}^{c^2}}{\cancel{a^2}}$$ *The reciprocal of* $\frac{a^2}{c^3}$ *is* $\frac{c^3}{a^2}$.

$$= \frac{ac^2}{2}$$

Review Exercises

Find each product or quotient.

15. $\frac{4x^2y}{y^2z} \cdot \frac{z}{6y}$ $\frac{2x^2}{3y^2}$
16. $\frac{4a - 4b}{a} \cdot \frac{a^3}{8a - 8b}$ $\frac{a^2}{2}$
17. $\frac{x + 2}{x + 3} \cdot \frac{x}{x^2 - x - 6}$ $\frac{x}{(x + 3)(x - 3)}$
18. $\frac{6m^2n}{10p} \div 3m$ $\frac{mn}{5p}$
19. $\frac{y^2 - 4y + 4}{3} \div \frac{y^2 - 4}{9}$ $\frac{3(y - 2)}{y + 2}$
20. $\frac{x^2 - 9}{2x} \div \frac{6 - 2x}{8x^2}$ $-2x(x + 3)$

- **Lesson 15–3** Divide polynomials by binomials.

$$\begin{array}{r} 2x - 3 \\ x + 2\overline{)2x^2 + x - 6} \\ (-)\ 2x^2 + 4x \\ \hline -3x - 6 \\ (-)\ -3x - 6 \\ \hline 0 \end{array}$$

$2x^2 \div x = 2x$
Multiply $2x$ and $x + 2$.
Subtract.
$-3x \div x = -3$
Subtract.

Find each quotient.

21. $(a^2 + 6a) \div (a + 6)$ a
22. $(x^2 - 5x - 6) \div (x + 1)$ $x - 6$
23. $(4y^2 + 8y + 5) \div (2y + 3)$ $2y + 1 + \frac{2}{2y + 3}$
24. $(x^3 - 1) \div (x - 1)$ $x^2 + x + 1$

- **Lesson 15–4** Add and subtract rational expressions with like denominators.

$$\frac{x^2}{x + 3} - \frac{9}{x + 3} = \frac{x^2 - 9}{x + 3}$$

$$= \frac{(\cancel{x+3})(x-3)}{\cancel{x+3}}$$

$$= x - 3$$

Find each sum or difference. Write in simplest form.

25. $\frac{2x}{8} + \frac{4x}{8}$ $\frac{3x}{4}$
26. $\frac{11}{15m} - \frac{2}{15m}$ $\frac{3}{5m}$
27. $\frac{3x}{x + 4} + \frac{2x}{x + 4}$ $\frac{5x}{x + 4}$
28. $\frac{8x}{2x - 3} - \frac{2x + 9}{2x - 3}$ 3

Skills and Concepts

The **Objectives and Examples** section reviews the skills and concepts of the chapter and shows completely worked examples.

The **Review Exercises** provide practice for the corresponding objectives.

ExamView® Pro

Use the networkable ***ExamView® Pro Testmaker CD-ROM*** to:

- Create **multiple versions** of tests.
- Create **modified** tests for **inclusion** students with one mouse click.
- **Edit** existing questions and **add** your own questions.
- Build tests aligned with state standards using built-in **state curriculum correlations**.
- Change **English** tests to **Spanish** with one mouse click and vice versa.

Mixed Problem Solving
See pages 724-731.

Applications and Problem Solving

This section provides additional practice in solving real-world problems that involve the concepts of this chapter.

Objectives and Examples

- **Lesson 15–5** Add and subtract rational expressions with unlike denominators.

$$\frac{y}{y+2} - \frac{2}{y} \quad \textit{The LCD is } y(y+2).$$
$$= \frac{y}{y+2} \cdot \frac{y}{y} - \frac{2}{y} \cdot \frac{y+2}{y+2}$$
$$= \frac{y^2}{y(y+2)} - \frac{2(y+2)}{y(y+2)}$$
$$= \frac{y^2 - 2(y+2)}{y(y+2)}$$
$$= \frac{y^2 - 2y - 4}{y(y+2)}$$

Review Exercises

Find each sum or difference. Write in simplest form.

29. $\frac{3}{4x} + \frac{5}{8x^2}$ **$\frac{6x+5}{8x^2}$**
30. $\frac{5}{x} - \frac{x+1}{3x}$ **$\frac{14-x}{3x}$**
31. $\frac{4}{a} + \frac{3}{a-2}$ **$\frac{7a-8}{a(a-2)}$**
32. $\frac{y+2}{y^2-4} - \frac{2}{y+2}$ **$\frac{-y+6}{y^2-4}$**

Objectives and Examples

- **Lesson 15–6** Solve rational equations.

$$\frac{x}{4} - \frac{x}{6} = \frac{1}{4}$$
$$12\left(\frac{x}{4} - \frac{x}{6}\right) = 12\left(\frac{1}{4}\right)$$
$$\overset{3}{\cancel{12}}\left(\frac{x}{\underset{1}{\cancel{4}}}\right) - \overset{2}{\cancel{12}}\left(\frac{x}{\underset{1}{\cancel{6}}}\right) = \overset{3}{\cancel{12}}\left(\frac{1}{\underset{1}{\cancel{4}}}\right)$$
$$3x - 2x = 3$$
$$x = 3$$

Review Exercises

Solve each equation. Check your solution.

33. $\frac{x}{3} - \frac{3x}{4} = \frac{1}{12}$ **$-\frac{1}{5}$**
34. $\frac{6}{3x} - \frac{3}{x} = 1$ **-1**
35. $\frac{x+2}{x} + \frac{2}{3x} = \frac{1}{3}$ **-4**
36. $\frac{m}{m+1} + \frac{5}{m-1} = 1$ **$-\frac{3}{2}$**

Applications and Problem Solving

37. **Carpentry** Determine the number of pieces of $\frac{3}{8}$-inch plywood that are in a stack 30 inches high. *(Lesson 15–2)* **80 pieces**

38. **Boating** The top speed of a boat in still water is 5 miles per hour. At this speed, a 21-mile trip downstream takes the same amount of time as a 9-mile trip upstream. Find the rate of the current. *(Lesson 15–6)* **2 mph**

39. **Geometry** The volume of a rectangular prism is $4x^3 + 14x^2 + 10x$ cubic inches. The height of the prism is $2x + 5$ inches, and the length of the prism is $2x$ inches. *(Lesson 15–3)*
    a. Find the polynomial that represents the width. **$x + 1$**
    b. If $x = 2$, find the length, width, height, and volume of the prism. **4 in., 3 in., 9 in., 108 in^3**

Assessment, pp. 695–696

15 NAME ______ DATE ______ PERIOD ______

Chapter 15 Test, Form 1B

Write the letter for the correct answer in the blank at the right of each problem.

Find the excluded value(s) for each rational expression.

1. $\frac{a}{a+7}$
   A. -7 B. 0 C. 1 D. 7 — 1. A
2. $\frac{x+4}{(x+4)(x+3)}$
   A. 0 B. $\frac{1}{x+3}$ C. $-3, -4$ D. 3, 4 — 2. C

Simplify each rational expression.

3. $\frac{8}{20}$
   A. $\frac{1}{8}$ B. $\frac{1}{3}$ C. $\frac{2}{5}$ D. $\frac{4}{5}$ — 3. C
4. $\frac{(x+2)(x-5)}{(x-5)(x+3)}$
   A. $\frac{2}{3}$ B. $x + \frac{2}{3}$ C. $x + 2$ D. $\frac{x+2}{x+3}$ — 4. D

Find each product.

5. $\frac{gh}{gm} \cdot \frac{m}{n}$
   A. h B. $\frac{h}{n}$ C. 1 D. $\frac{hm}{nh}$ — 5. B
6. $\frac{2a+2}{a^2-4} \cdot \frac{a+2}{a^2+a}$
   A. $\frac{2}{a(a-2)}$ B. $\frac{(a+2)^2}{a^2+a-4}$ C. $\frac{a+2}{a-2}$ D. $\frac{2}{a-2}$ — 6. A

Find each quotient.

7. $\frac{x^2}{y} \div \frac{x}{y^2}$
   A. $\frac{x}{y}$ B. $\frac{x^3}{y^3}$ C. xy D. $\frac{x^2}{y^2}$ — 7. C
8. $\frac{4c^2}{b+3} \div \frac{4c}{b+3}$
   A. $\frac{4c^2}{(b+3)^2}$ B. $\frac{c}{b+3}$ C. c D. 1 — 8. C
9. $(12a - 3) \div (4a - 1)$
   A. $\frac{1}{3a}$ B. 3 C. 4 D. $\frac{1}{3}$ — 9. B
10. $(8b^2 + 20b) \div (2b + 5)$
    A. $\frac{b}{5}$ B. $\frac{b}{b+5}$ C. $b + 4$ D. $4b$ — 10. D

Assessment

Four forms of Chapter 15 Test are available in the *Chapter 15 Resource Masters.*

Chapter 15 Test, Form 1B, is shown at the left. Chapter 15 Test, Form 2B, is shown on the next page.

	Form of Test		Level
1A	Multiple Choice	pp. 693–694	Average
1B	Multiple Choice	pp. 695–696	Basic
2A	Free Response	pp. 697–698	Average
2B	Free Response	pp. 699–700	Basic

CHAPTER 15 Test

1. **Explain** why $x = -3$ is an excluded value for $\frac{2x}{x+3}$. **The denominator is 0 when $x = -3$.**

2. **Identify** and correct the error that was made while finding the difference.

$$\frac{4}{x+1} - \frac{x-3}{x+1} = \frac{4-x-3}{x+1}$$
$$= \frac{1-x}{x+1}$$

The numerator should be $4 - (x - 3)$ or $4 - x + 3$.

3. **Find** the least common denominator of $\frac{5}{6a}$ and $\frac{2}{3a^2}$. **$6a^2$**

Find the excluded value(s) for each rational expression.

4. $\frac{8}{x}$ **0**

5. $\frac{6m}{m(m-2)}$ **0, 2**

6. $\frac{2n}{n^2-9}$ **3, −3**

Simplify each expression.

7. $\frac{9a^3b^2}{15ab^5}$ **$\frac{3a^2}{5b^3}$**

8. $\frac{x^3-x^2}{x-1}$ **x^2**

9. $\frac{x-2}{x^2-5x+6}$ **$\frac{1}{x-3}$**

Find each sum, difference, product, or quotient. Write in simplest form.

10. $\frac{z^3}{8} \cdot \frac{10x^2}{z^3}$ **$\frac{5x^2}{4}$**

11. $\frac{x^2-1}{3x} \div \frac{1-x}{9x}$ **$-3(x+1)$**

12. $\frac{x^2-x-2}{x^2-4} \cdot \frac{x+2}{x^2+4x+3}$ **$\frac{1}{x+3}$**

13. $\frac{4a+4b}{a} \div \frac{10a+10b}{a^2}$ **$\frac{2a}{5}$**

14. $\frac{5}{8x} + \frac{11}{8x}$ **$\frac{2}{x}$**

15. $\frac{t}{t+5} - \frac{t-6}{t+5}$ **$\frac{6}{t+5}$**

16. $\frac{6}{5y} + \frac{7}{10y^2}$ **$\frac{12y+7}{10y^2}$**

17. $\frac{2}{x+4} - \frac{x}{x^2-16}$ **$\frac{x-8}{(x+4)(x-4)}$**

Find each quotient.

18. $(y^2 + 10y + 16) \div (y + 2)$ **$y + 8$**

19. $(x^3 + x^2 - x - 1) \div (x - 1)$ **$x^2 + 2x + 1$**

20. $(a^2 - 10) \div (a + 3)$ **$a - 3 - \frac{1}{a+3}$**

21. $(x^2 - 5x + 8) \div (x - 2)$ **$x - 3 + \frac{2}{x-2}$**

Solve each equation. Check your solution.

22. $\frac{n+2}{3} + \frac{n}{2} = \frac{1}{2}$ **$-\frac{1}{5}$**

23. $\frac{5}{x+2} - \frac{1}{4x+8} = \frac{1}{4}$ **17**

24. **Geometry** Find the measure of the area of the rectangle in simplest form. **$\frac{x+5}{x^2-12x+35}$**

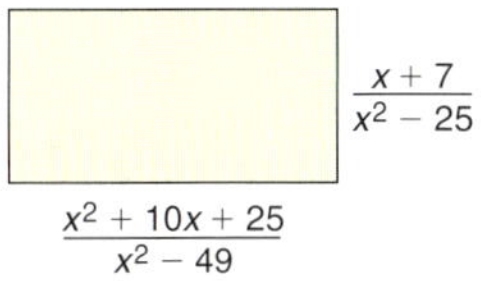

25. **Transportation** A tugboat pushing a barge up the Mississippi River takes 1 hour longer to travel 36 miles up the river than to travel the same distance down the river. If the rate of the current is 3 miles per hour, find the speed of the tugboat and barge in still water. **15 mph**

www.algconcepts.com/chapter_test

Chapter Test

End-of-Year Tests

A Second Semester Test for Chapters 9–15 and a Final Test for Chapters 1–15 are available in the *Chapter 15 Resource Masters*, pp. 707–714.

***Assessment*, pp. 699–700**

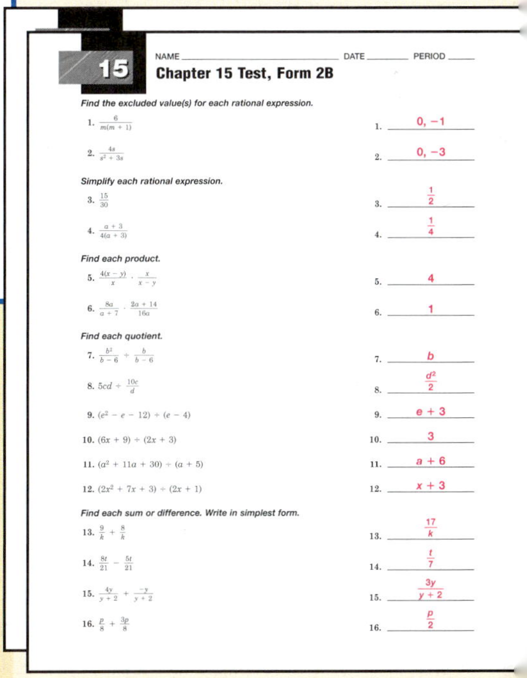

NAME ______ DATE ______ PERIOD ______

15 Chapter 15 Test, Form 2B

Find the excluded value(s) for each rational expression.

1. $\frac{6}{m(m+1)}$ 1. **0, −1**
2. $\frac{4s}{s^2+3s}$ 2. **0, −3**

Simplify each rational expression.

3. $\frac{15}{30}$ 3. **$\frac{1}{2}$**
4. $\frac{a+3}{4(a+3)}$ 4. **$\frac{1}{4}$**

Find each product.

5. $\frac{4(x-y)}{x} \cdot \frac{x}{x-y}$ 5. **4**
6. $\frac{8a}{a+7} \cdot \frac{2a+14}{16a}$ 6. **1**

Find each quotient.

7. $\frac{b^2}{b-6} \div \frac{b}{b-6}$ 7. **b**
8. $5cd \div \frac{10c}{d}$ 8. **$\frac{d^2}{2}$**
9. $(e^2 - e - 12) \div (e - 4)$ 9. **$e + 3$**
10. $(6x + 9) \div (2x + 3)$ 10. **3**
11. $(a^2 + 11a + 30) \div (a + 5)$ 11. **$a + 6$**
12. $(2x^2 + 7x + 3) \div (2x + 1)$ 12. **$x + 3$**

Find each sum or difference. Write in simplest form.

13. $\frac{9}{k} + \frac{8}{k}$ 13. **$\frac{17}{k}$**
14. $\frac{8t}{21} - \frac{5t}{21}$ 14. **$\frac{t}{7}$**
15. $\frac{4y}{y+2} + \frac{-y}{y+2}$ 15. **$\frac{3y}{y+2}$**
16. $\frac{p}{8} + \frac{3p}{8}$ 16. **$\frac{p}{2}$**

Chapter Test Bonus Question

Write an expression in simplest form for the mean of the reciprocals of any two integers x and y. **$\frac{x+y}{2xy}$**

Preparing for Standardized Tests

Pages 680–681 are part of a complete test preparation course that is described in detail on page T9 of the Teacher's Handbook. The test items on these pages were written in the same style as those in state proficiency tests and standardized tests like ACT and SAT.

Diagnosis and Prescription

Each of the 10 test questions on page 681 is cross-referenced to the lesson where that SAT or ACT skill is covered. If students miss a particular type of problem, you can have them study that skill.

(See chart at the bottom of page 681. Note that SPT = State Proficiency Test, SAT = Scholastic Assessment Test, and ACT = American College Test.)

Assessment, p. 704

NAME ______________ DATE ________ PERIOD ______

15 Chapter 15 Cumulative Review

1. Write an equation for the sentence below. *(Lesson 1–1)*
   The product of a number k and 17 is equal to 68. — 1. $17k = 68$
2. Find $-15 + (-3)$. *(Lesson 2–3)* — 2. -18
3. Write $\frac{3}{8}$, 4.2, 0.95, 1.7, and -3 in order from least to greatest. *(Lesson 3–1)* — 3. $-3, \frac{3}{8}, 0.95, 1.7, 4.2$
4. Find $-\frac{3}{8}\left(-\frac{1}{3}\right)$. *(Lesson 4–1)* — 4. $\frac{1}{8}$
5. Use the percent proportion to find 8% of 250. *(Lesson 5–3)* — 5. 20
6. If y varies inversely as x and $y = 4$ when $x = 9$, find y when $x = 2$. *(Lesson 6–6)* — 6. 18
7. Write an equation in point-slope form of the line with slope $-\frac{1}{3}$ and y-intercept 7. *(Lesson 7–2)* — 7. $y - 7 = -\frac{1}{3}x$
8. Simplify $-\sqrt{\frac{25}{49}}$. *(Lesson 8–5)* — 8. $-\frac{5}{7}$
9. Find $(x + 2y)(2x + y)$. *(Lesson 9–4)* — 9. $2x^2 + 5xy + 2y^2$
10. Factor $7k^3 - 7k$. *(Lesson 10–5)* — 10. $7k(k - 1)(k + 1)$
11. Describe how the graph of $y = \frac{1}{2}x^2$ changes from its parent graph of $y = x^2$. Then name the vertex. *(Lesson 11–2)* — 11. widens; (0, 0)
12. Solve $-\frac{1}{2} \leq a + \frac{1}{3}$. *(Lesson 12–2)* — 12. $\{a | a \geq -\frac{5}{6}\}$
13. What is the solution of the system $y = x + 5$ and $3x = -2y$? *(Lesson 13–3)* — 13. $(-2, 3)$

For Questions 14–15, solve each equation. (Lesson 14–5)

14. $\sqrt{\frac{x}{3}} + 2 = 7$ — 14. 75
15. $\frac{\sqrt{y-3}}{y} = \frac{1}{4}$ — 15. 4 or 12
16. Simplify $\frac{m^2 - 25}{m^2 - 3m - 10}$. *(Lesson 15–1)* — 16. $\frac{m+5}{m+2}$

CHAPTER 15 Preparing for Standardized Tests

Right Triangle Problems

You'll need to know how to identify right triangles and how to apply the Pythagorean Theorem.

> **Test-Taking Tip**
> Look for special right triangles, such as 3-4-5 triangles.

If a and b are the lengths of the legs of a triangle and c is the length of the hypotenuse, then $a^2 + b^2 = c^2$.

Example 1

A 32-foot telephone pole is braced with a cable from the top of the pole to a point 7 feet from the base. What is the length of the cable, rounded to the nearest tenth?

Ⓐ 31.2 ft Ⓑ 32.8 ft Ⓒ 34.3 ft Ⓓ 36.2 ft

Hint Since telephone poles are vertical and the ground is horizontal, the pole forms a right angle at the base.

Solution Draw a figure.

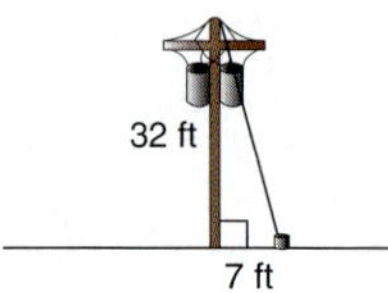

The cable forms the hypotenuse of a right triangle. Use the Pythagorean Theorem to find the length of the hypotenuse.

$a^2 + b^2 = c^2$	*Pythagorean Theorem*
$32^2 + 7^2 = c^2$	*Replace a with 32 and b with 7.*
$1024 + 49 = c^2$	*Multiply.*
$1073 = c^2$	*Add.*
$\sqrt{1073} = \sqrt{c^2}$	*Take the square root of each side.*
$32.8 \approx c$	*Use a calculator.*

To the nearest tenth, the length of the cable is 32.8 feet.

The answer is B.

Example 2

In the figure, $\overline{MO}$ is perpendicular to $\overline{LN}$, LO is equal to 4, MO is equal to ON, and LM is equal to 6. What is MN?

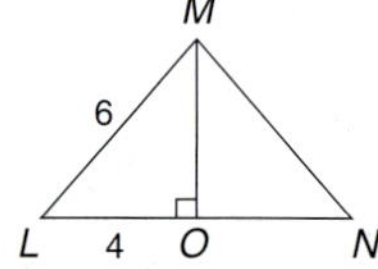

Ⓐ $2\sqrt{10}$ Ⓑ $3\sqrt{5}$ Ⓒ $4\sqrt{5}$ Ⓓ $3\sqrt{10}$ Ⓔ $6\sqrt{4}$

Hint Identify the types of triangles shown in the figure.

Solution $\triangle LMO$ is a right triangle. Use the Pythagorean Theorem to find MO.

$a^2 + b^2 = c^2$	*Pythagorean Theorem*
$a^2 + 4^2 = 6^2$	*Replace b with 4 and c with 6.*
$a^2 + 16 = 36$	
$a^2 + 16 - 16 = 36 - 16$	*Subtract 16.*
$a^2 = 20$	*Simplify.*
$\sqrt{a^2} = \sqrt{20}$	*Take the square root.*
$a = \sqrt{4(5)}$ or $2\sqrt{5}$	

$\triangle MON$ is an isosceles triangle. Use the Pythagorean Theorem to find MN.

$a^2 + b^2 = c^2$	*Pythagorean Theorem*
$(2\sqrt{5})^2 + (2\sqrt{5})^2 = c^2$	*$a = 2\sqrt{5}$, $b = 2\sqrt{5}$*
$20 + 20 = c^2$	*Simplify.*
$\sqrt{40} = \sqrt{c^2}$	*Take the square root of each side.*
$2\sqrt{10} = c$	

The answer is A.

Resource Manager

Reproducible Masters

Chapter 15 Resource Masters

- *Assessment*, pp. 704–706

Preparing for Standardized Tests

For test-taking strategies and more practice, see pages 732–749.

After you work each problem, record your answer on the answer sheet provided or on a sheet of paper.

Multiple Choice

1. Quadrilateral $ABCD$ is a rectangle. $\overline{AD}$ is 5 centimeters long, and $\overline{CD}$ is 12 centimeters long. What is length of $\overline{AC}$? *(Lesson 8–7)* **A**

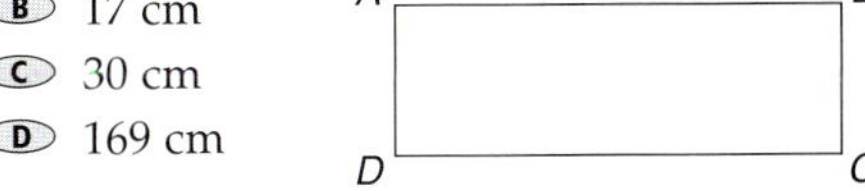

   Ⓐ 13 cm
   Ⓑ 17 cm
   Ⓒ 30 cm
   Ⓓ 169 cm

2. What is the distance between points G and H in the graph? *(Lesson 14–2)* **B**
   Ⓐ 4
   Ⓑ $4\sqrt{2}$
   Ⓒ 8
   Ⓓ $8\sqrt{2}$

3. The hypotenuse of an isosceles right triangle has a length of 20 units. What is the length of one of the legs of the triangle? *(Lesson 8–7)* **B**
   Ⓐ 10   Ⓑ $10\sqrt{2}$ Ⓒ $10\sqrt{3}$
   Ⓓ 20   Ⓔ $20\sqrt{2}$

4. The streets on a jogging route form a right triangle. How long is the jogging route? *(Lesson 8–7)* **C**
   Ⓐ 10 km
   Ⓑ 14 km
   Ⓒ 24 km
   Ⓓ 100 km

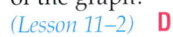

5. What is the equation of the graph? *(Lesson 11–2)* **D**

   Ⓐ $y = -x^2 + 2$
   Ⓑ $y = x^2 - 2$
   Ⓒ $y = (x + 2)^2$
   Ⓓ $y = x^2 + 2$

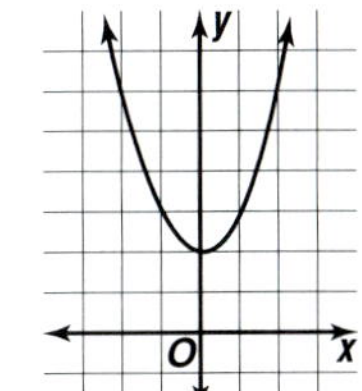

6. What is the length of $\overline{BC}$? *(Lesson 8–7)* **B**
   Ⓐ 6
   Ⓑ $4\sqrt{3}$
   Ⓒ $2\sqrt{13}$
   Ⓓ 8
   Ⓔ $2\sqrt{38}$

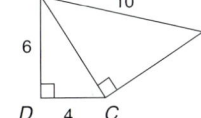

7. The base of a 25-foot ladder is placed 10 feet from the side of a house. How high does the ladder reach? *(Lesson 8–7)* **C**
   Ⓐ 12.5 ft   Ⓑ 15 ft
   Ⓒ 22.9 ft   Ⓓ 35 f

8. $\sqrt{3} + \sqrt{4}$ is between which pair of numbers? *(Lesson 14–4)* **A**
   Ⓐ 3 and 4
   Ⓑ 4 and 5
   Ⓒ 5 and 6
   Ⓓ 6 and 7

Grid In

9. Triangles ABC and XYZ are similar right triangles. If $\angle B$ measures 55°, what is the degree measure of $\angle Z$? *(Lesson 3–6)* **35**

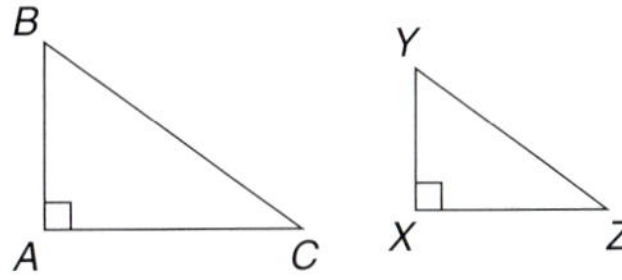

Extended Response

10. You want to fit a square cake inside a circular cake carrier that is 12 inches in diameter. *(Lesson 8–7)* **See margin.**

   Part A Draw a diagram showing side s of the *largest* square cake that will fit inside the carrier.

   Part B Use your diagram to write an equation and then find the value of s.

www.algconcepts.com/standardized_test

A bubble-in answer sheet for these practice problems is available on page A1 of the *Chapter 15 Resource Masters.*

Additional Practice

Additional test practice questions are available in the *Chapter 15 Resource Masters,* pp. 705–706.

Answers

10A.

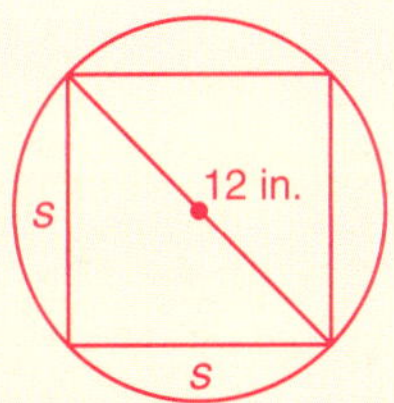

10B. $s^2 + s^2 = 12^2$; $6\sqrt{2}$ in.

***Assessment,* pp. 705–706**

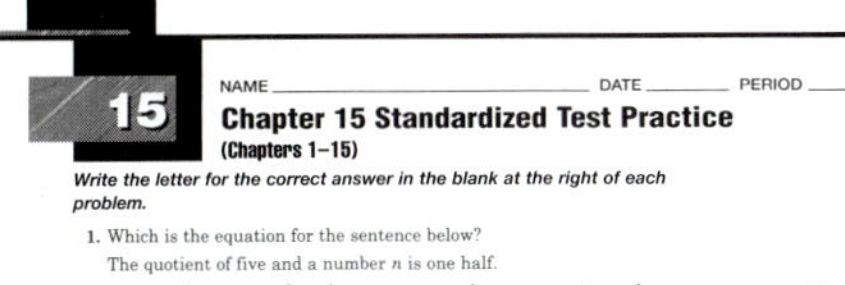

15 NAME ______ DATE ______ PERIOD ______

Chapter 15 Standardized Test Practice
(Chapters 1–15)

Write the letter for the correct answer in the blank at the right of each problem.

1. Which is the equation for the sentence below?
   The quotient of five and a number n is one half.
   A. $5n = \frac{1}{2}$ B. $\frac{5}{n} = \frac{1}{2}$ C. $5 = \frac{1}{2} \cdot n$ D. $\frac{n}{5} = 1\frac{1}{2}$ 1. B

2. Simplify $-5x + (-2x)$.
   A. $-7x$ B. $-3x$ C. $7x$ D. 3   2. A

3. Which of the following is *not* a true statement about the set of data below?
   5, 8, 12, 12, 12, 14, 18, 20
   A. The mode is 12.
   B. The median is 12.
   C. The range is 25.
   D. The mean is larger than either the mode or the median.   3. C

4. The product of two negative integers is
   A. undefined.   B. zero.
   C. always positive.   D. always negative.   4. C

5. One letter from the word CANADA is chosen at random. What is the probability that a letter A is chosen?
   A. 5   B. 3   C. $\frac{1}{2}$ D. $\frac{1}{6}$ 5. C

6. Which of the following is a linear equation?
   A. $y = \frac{1}{x}$ B. $y - x = 17$ C. $y = x^2$ D. $xy = 12$ 6. B

7. Determine the slope of the line passing through points at (4, 1) and (0, −2).
   A. $-\frac{5}{2}$ B. $-\frac{4}{3}$ C. $\frac{3}{4}$ D. undefined   7. C

8. Which of the following is equivalent to $\frac{a^9}{a^3}$?
   A. $\frac{a^7}{a}$ B. a^3 C. 6   D. $\frac{1}{a^6}$ 8. A

9. Which expression is equivalent to $z^2 - 25$?
   A. $(z - 5)(z - 5)$ B. $(z - 5)^2$
   C. $(z + 25)(z - 25)$ D. $(z + 5)(z - 5)$ 9. D

10. The expression $5x$ is the GCF of which pair of monomials?
   A. 5 and x B. x and x^5
   C. $5xy$ and $15xz$ D. $10xy$ and $12wx$ 10. C

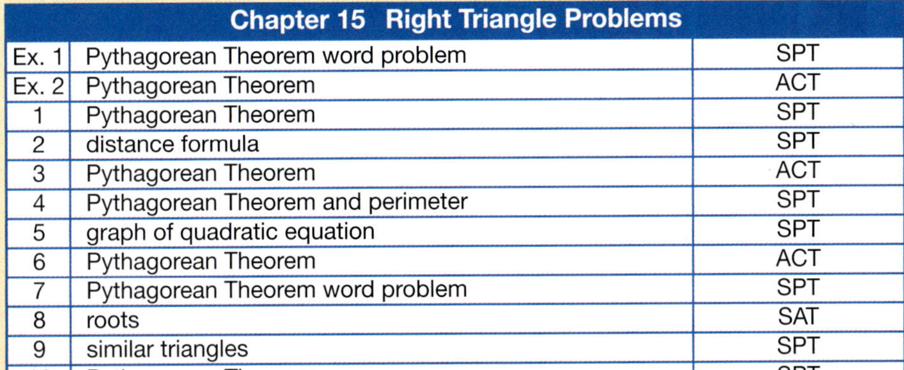

Chapter 15 Right Triangle Problems		
Ex. 1	Pythagorean Theorem word problem	SPT
Ex. 2	Pythagorean Theorem	ACT
1	Pythagorean Theorem	SPT
2	distance formula	SPT
3	Pythagorean Theorem	ACT
4	Pythagorean Theorem and perimeter	SPT
5	graph of quadratic equation	SPT
6	Pythagorean Theorem	ACT
7	Pythagorean Theorem word problem	SPT
8	roots	SAT
9	similar triangles	SPT
10	Pythagorean Theorem	SPT

Student Handbook

Skills Handbook

Reference Handbook

PREREQUISITE SKILLS REVIEW

Prerequisite Skills Review

Operations with Decimals

- To add or subtract decimals, line up the decimal points. You may want to *annex*, or place zeros at the end of the decimals, to help align the columns. Then add or subtract.

Examples

1 **Find 18.39 + 6.4.**

$$\begin{array}{r} 18.39 \\ +\ 6.4 \\ \hline \end{array} \quad \rightarrow \quad \begin{array}{r} 18.39 \\ +\ 6.40 \\ \hline 24.79 \end{array}$$

Annex a zero to align the columns.

2 **Find 5 − 2.17.**

$$\begin{array}{r} 5 \\ -\ 2.17 \\ \hline \end{array} \quad \rightarrow \quad \begin{array}{r} 5.00 \\ -\ 2.17 \\ \hline 2.83 \end{array}$$

Annex two zeros to align the columns.

- To multiply decimals, multiply as with whole numbers. Then add the total number of decimal places in the factors. Place the same number of decimal places in the product, counting from right to left.
- To divide decimals, move the decimal point in the divisor to the right, and then move the decimal point in the dividend the same number of places. Align the decimal point in the quotient with the decimal point in the dividend.

Examples

3 **Find 7.3(0.61).**

$$\begin{array}{r} 7.3 \\ \times\ 0.61 \\ \hline 73 \\ 438 \\ \hline 4.453 \end{array} \quad \begin{array}{l} \leftarrow \text{1 decimal place} \\ \leftarrow \text{2 decimal places} \\ \\ \\ \leftarrow \text{3 decimal places} \end{array}$$

4 **Find 8.12 ÷ 5.8.**

$$\begin{array}{r} 1.4 \\ 5.8\overline{)8.12} \\ 5\ 8 \\ \hline 2\ 32 \\ 2\ 32 \\ \hline 0 \end{array}$$

Move each decimal point right 1 place.

Find each sum or difference.

1. 4.8 + 1.1 **5.9**
2. 10.25 + 0.16 **10.41**
3. 36.5 − 11.2 **25.3**
4. 13.6 + 20.41 **34.01**
5. 42.99 − 12.63 **30.36**
6. 5.37 − 2.47 **2.90**
7. 24 + 7.3 **31.3**
8. 7.28 − 1.1 **6.18**
9. 0.7 + 1.682 **2.382**
10. 6.2 − 4.01 **2.19**
11. 4.02 + 5.9 + 0.03 **9.95**
12. 18 − 16.39 **1.61**

Find each product or quotient.

13. 8(0.4) **3.2**
14. 5.72 ÷ 2 **2.86**
15. 1.6(1.9) **3.04**
16. 7.2(3.5) **25.20**
17. 24(0.86) **20.64**
18. 4.86 ÷ 0.3 **16.2**
19. 11.5 ÷ 4.6 **2.5**
20. 9.3(30.23) **281.139**
21. 0.5(2.41)(6.7) **8.0735**
22. $\frac{50.4}{36}$ **1.4**
23. $\frac{6.46}{0.68}$ **9.5**
24. $\frac{9.264}{0.24}$ **38.6**

Simplifying Fractions

- A fraction is in simplest form when the greatest common factor (GCF) of the numerator and the denominator is 1.
- To write a fraction in simplest form, divide both the numerator and the denominator by the GCF.

Examples **Write each fraction in simplest form.**

 $\frac{6}{9}$

The GCF of 6 and 9 is 3.

$$\frac{6}{9} = \frac{2}{3}$$ (÷ 3 numerator and denominator) *Divide 6 and 9 by 3.*

2 $\frac{4}{20}$

The GCF of 4 and 20 is 4.

$$\frac{4}{20} = \frac{1}{5}$$ (÷ 4 numerator and denominator) *Divide 4 and 20 by 4.*

- To change an improper fraction to a mixed number, divide the numerator by the denominator. Write the remainder as a fraction in simplest form.
- To change a mixed number to an improper fraction, multiply the whole number by the denominator and add the numerator. Write this sum over the denominator.

Examples

3 **Write $\frac{8}{5}$ as a mixed number.**

$\frac{8}{5}$ means 8 ÷ 5.

$$\begin{array}{r} 1 \\ 5\overline{)8} \\ -5 \\ \hline 3 \end{array} \rightarrow 1\text{ R3 or } 1\frac{3}{5}$$

4 **Write $2\frac{3}{4}$ as an improper fraction.**

$$2\frac{3}{4} = \frac{(2 \times 4) + 3}{4} = \frac{8 + 3}{4} = \frac{11}{4}$$

Write each fraction in simplest form.

1. $\frac{4}{8}$ $\frac{1}{2}$	2. $\frac{3}{9}$ $\frac{1}{3}$	3. $\frac{2}{12}$ $\frac{1}{6}$	4. $\frac{9}{15}$ $\frac{3}{5}$	5. $\frac{8}{20}$ $\frac{2}{5}$
6. $\frac{6}{21}$ $\frac{2}{7}$	7. $\frac{25}{30}$ $\frac{5}{6}$	8. $\frac{30}{40}$ $\frac{3}{4}$	9. $\frac{12}{16}$ $\frac{3}{4}$	10. $\frac{16}{36}$ $\frac{4}{9}$

Write each improper fraction as a mixed number.

11. $\frac{7}{6}$ $1\frac{1}{6}$	12. $\frac{5}{3}$ $1\frac{2}{3}$	13. $\frac{9}{2}$ $4\frac{1}{2}$	14. $\frac{12}{5}$ $2\frac{2}{5}$	15. $\frac{10}{4}$ $2\frac{1}{2}$
16. $\frac{11}{2}$ $5\frac{1}{2}$	17. $\frac{21}{10}$ $2\frac{1}{10}$	18. $\frac{16}{6}$ $2\frac{2}{3}$	19. $\frac{24}{14}$ $1\frac{5}{7}$	20. $\frac{30}{8}$ $3\frac{3}{4}$

Write each mixed number as an improper fraction.

21. $2\frac{1}{3}$ $\frac{7}{3}$	22. $6\frac{1}{2}$ $\frac{13}{2}$	23. $2\frac{4}{5}$ $\frac{14}{5}$	24. $4\frac{2}{5}$ $\frac{22}{5}$	25. $6\frac{1}{7}$ $\frac{43}{7}$
26. $3\frac{5}{8}$ $\frac{29}{8}$	27. $2\frac{7}{10}$ $\frac{27}{10}$	28. $3\frac{5}{12}$ $\frac{41}{12}$	29. $1\frac{10}{11}$ $\frac{21}{11}$	30. $5\frac{8}{9}$ $\frac{53}{9}$

Multiplying and Dividing Fractions

- To multiply fractions, multiply the numerators and multiply the denominators.
- When a numerator and denominator have a common factor, you can simplify before multiplying.

Examples **Find each product.**

1 $\frac{1}{2} \cdot \frac{3}{5}$

$$\frac{1}{2} \cdot \frac{3}{5} = \frac{1 \cdot 3}{2 \cdot 5} = \frac{3}{10}$$

2 $\frac{4}{9} \cdot \frac{3}{5}$

$$\frac{4}{9} \cdot \frac{3}{5} = \frac{4}{\cancel{9}_3} \cdot \frac{\cancel{3}^1}{5}$$ *Divide by the GCF, 3.*

$$= \frac{4 \cdot 1}{3 \cdot 5} = \frac{4}{15}$$

- To divide fractions, multiply by the reciprocal of the divisor.

Examples **Find each quotient.**

3 $\frac{5}{6} \div \frac{6}{7}$

$$\frac{5}{6} \div \frac{6}{7} = \frac{5}{6} \cdot \frac{7}{6}$$ *Multiply by the reciprocal.*

$$= \frac{5 \cdot 7}{6 \cdot 6} = \frac{35}{36}$$

4 $\frac{2}{5} \div 4$

$$\frac{2}{5} \div 4 = \frac{2}{5} \div \frac{4}{1}$$

$$= \frac{2}{5} \cdot \frac{1}{4}$$ *Multiply by the reciprocal.*

$$= \frac{\cancel{2}^1}{5} \cdot \frac{1}{\cancel{4}_2}$$ *Divide by the GCF, 2.*

$$= \frac{1 \cdot 1}{5 \cdot 2} = \frac{1}{10}$$

Find each product or quotient.

1. $\frac{1}{3} \cdot \frac{1}{2}$ $\frac{1}{6}$
2. $\frac{2}{5} \div \frac{1}{2}$ $\frac{4}{5}$
3. $\frac{1}{4} \div \frac{4}{5}$ $\frac{5}{16}$
4. $\frac{2}{7} \cdot \frac{1}{3}$ $\frac{2}{21}$
5. $\frac{1}{8} \div \frac{1}{5}$ $\frac{5}{8}$
6. $\frac{4}{9} \cdot \frac{2}{3}$ $\frac{8}{27}$
7. $\frac{2}{7} \div 7$ $\frac{2}{49}$
8. $\frac{6}{7} \cdot \frac{1}{5}$ $\frac{6}{35}$
9. $\frac{1}{2} \cdot \frac{2}{7}$ $\frac{1}{7}$
10. $\frac{2}{3} \cdot \frac{1}{6}$ $\frac{1}{9}$
11. $\frac{5}{9} \div \frac{2}{3}$ $\frac{5}{6}$
12. $\frac{1}{8} \div \frac{1}{2}$ $\frac{1}{4}$
13. $\frac{2}{9} \div \frac{4}{7}$ $\frac{7}{18}$
14. $\frac{3}{4} \cdot \frac{5}{6}$ $\frac{5}{8}$
15. $\frac{5}{12} \cdot \frac{3}{7}$ $\frac{5}{28}$
16. $\frac{3}{4} \div \frac{5}{6}$ $\frac{9}{10}$
17. $\frac{2}{9} \div \frac{2}{3}$ $\frac{1}{3}$
18. $\frac{2}{3} \div 4$ $\frac{1}{6}$
19. $\frac{7}{9} \cdot \frac{6}{7}$ $\frac{2}{3}$
20. $\frac{3}{10} \cdot \frac{5}{18}$ $\frac{1}{12}$
21. $\frac{5}{12} \cdot \frac{8}{25}$ $\frac{2}{15}$
22. $\frac{3}{8} \div 12$ $\frac{1}{32}$
23. $\frac{9}{20} \cdot \frac{4}{15}$ $\frac{3}{25}$
24. $\frac{8}{21} \div \frac{12}{15}$ $\frac{10}{21}$

Adding and Subtracting Fractions

- To add or subtract fractions with like denominators, add or subtract the numerators. Simplify if necessary.

Examples **Find each sum or difference.**

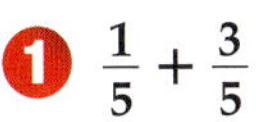

1 $\frac{1}{5} + \frac{3}{5}$

$\frac{1}{5} + \frac{3}{5} = \frac{1+3}{5}$ *Add the numerators.*

$= \frac{4}{5}$

2 $\frac{7}{12} - \frac{5}{12}$

$\frac{7}{12} - \frac{5}{12} = \frac{7-5}{12}$ *Subtract the numerators.*

$= \frac{2}{12}$

$= \frac{1}{6}$ *Simplify.*

- To add or subtract fractions with unlike denominators, first find the least common denominator (LCD). Rewrite each fraction with the LCD, and then add or subtract the numerators. Simplify if necessary.

Examples **Find each sum or difference.**

3 $\frac{1}{3} - \frac{2}{9}$

The LCD of 3 and 9 is 9.

$\frac{1}{3} - \frac{2}{9} = \frac{3}{9} - \frac{2}{9}$ *Rewrite $\frac{1}{3}$ as $\frac{3}{9}$.*

$= \frac{3-2}{9}$ *Subtract the numerators.*

$= \frac{1}{9}$

4 $\frac{3}{10} + \frac{1}{4}$

The LCD of 10 and 4 is 20.

$\frac{3}{10} + \frac{1}{4} = \frac{6}{20} + \frac{5}{20}$ *Rewrite the fractions using the LCD, 20.*

$= \frac{6+5}{20}$ *Add the numerators.*

$= \frac{11}{20}$

Find each sum or difference.

1. $\frac{1}{5} + \frac{2}{5}$ $\frac{3}{5}$
2. $\frac{7}{8} - \frac{2}{8}$ $\frac{5}{8}$
3. $\frac{1}{9} + \frac{4}{9}$ $\frac{5}{9}$
4. $\frac{6}{7} - \frac{2}{7}$ $\frac{4}{7}$

5. $\frac{1}{15} + \frac{3}{15}$ $\frac{4}{15}$
6. $\frac{4}{13} + \frac{1}{13}$ $\frac{5}{13}$
7. $\frac{5}{6} - \frac{4}{6}$ $\frac{1}{6}$
8. $\frac{9}{11} - \frac{3}{11}$ $\frac{6}{11}$

9. $\frac{3}{8} + \frac{1}{8}$ $\frac{1}{2}$
10. $\frac{5}{6} - \frac{1}{6}$ $\frac{2}{3}$
11. $\frac{7}{10} - \frac{3}{10}$ $\frac{2}{5}$
12. $\frac{1}{12} + \frac{7}{12}$ $\frac{2}{3}$

13. $\frac{1}{3} + \frac{1}{5}$ $\frac{8}{15}$
14. $\frac{5}{8} - \frac{1}{4}$ $\frac{3}{8}$
15. $\frac{4}{9} + \frac{1}{3}$ $\frac{7}{9}$
16. $\frac{9}{10} - \frac{3}{5}$ $\frac{3}{10}$

17. $\frac{3}{7} + \frac{1}{2}$ $\frac{13}{14}$
18. $\frac{3}{4} - \frac{2}{5}$ $\frac{7}{20}$
19. $\frac{2}{9} + \frac{2}{3}$ $\frac{8}{9}$
20. $\frac{9}{10} - \frac{3}{4}$ $\frac{3}{20}$

21. $\frac{2}{15} + \frac{7}{15}$ $\frac{3}{5}$
22. $\frac{1}{2} + \frac{1}{6}$ $\frac{2}{3}$
23. $\frac{5}{6} - \frac{3}{4}$ $\frac{1}{12}$
24. $\frac{3}{4} + \frac{1}{12}$ $\frac{5}{6}$

25. $\frac{4}{7} - \frac{1}{6}$ $\frac{17}{42}$
26. $\frac{2}{15} + \frac{2}{3}$ $\frac{4}{5}$
27. $\frac{5}{6} + \frac{1}{8}$ $\frac{23}{24}$
28. $\frac{7}{8} - \frac{3}{12}$ $\frac{5}{8}$

Fractions and Decimals

- To change a fraction to a decimal, divide the numerator by the denominator.
- A **terminating decimal** is a decimal like 0.75, in which the division ends, or terminates, when the remainder is zero.
- A **repeating decimal** is a decimal like 0.545454 . . . whose digits do not end. Since it is impossible to write all the digits, you can use bar notation to show that 54 repeats. We can write 0.545454 . . . as $0.\overline{54}$.

Examples **Write each fraction as a decimal.**

1 $\frac{2}{5}$

$$5\overline{)2.0} = 0.4,\quad 2\,0,\quad 0$$

$\frac{2}{5} = 0.4$

2 $\frac{4}{9}$

$$9\overline{)4.00} = 0.44,\quad 36,\quad 40,\quad 36,\quad 4$$

The pattern is repeating.

$\frac{4}{9} = 0.\overline{4}$

- Every terminating decimal can be expressed as a fraction with a denominator of 10, 100, and so on.
- Every repeating decimal can be expressed as a fraction.

Examples **Write each decimal as a fraction.**

3 **0.8**

$$0.8 = \frac{8}{10} = \frac{4}{5}$$

4 $0.\overline{1}$

Let $N = 0.\overline{1}$ or 0.111 . . . .

Then $10N = 1.\overline{1}$ or 1.111 . . . .

$$\begin{aligned} 10N &= 1.111\ldots \\ -\ 1N &= 0.111\ldots \\ \hline 9N &= 1 \\ N &= \frac{1}{9} \end{aligned}$$

Subtract. *N = 1N*

So, $0.\overline{1} = \frac{1}{9}$.

Write each fraction as a decimal.

1. $\frac{3}{4}$ **0.75** **2.** $\frac{1}{5}$ **0.2** **3.** $\frac{3}{8}$ **0.375** **4.** $\frac{2}{9}$ **$0.\overline{2}$**

5. $\frac{4}{11}$ **$0.\overline{36}$** **6.** $\frac{7}{10}$ **0.7** **7.** $\frac{2}{15}$ **$0.1\overline{3}$** **8.** $\frac{1}{6}$ **$0.1\overline{6}$**

9. $\frac{4}{15}$ **$0.2\overline{6}$** **10.** $\frac{3}{20}$ **0.15** **11.** $\frac{5}{6}$ **$0.8\overline{3}$** **12.** $\frac{5}{8}$ **0.625**

Write each decimal as a fraction in simplest form.

13. 0.9 $\frac{9}{10}$ **14.** 0.6 $\frac{3}{5}$ **15.** 0.25 $\frac{1}{4}$ **16.** $0.\overline{3}$ $\frac{1}{3}$

17. $0.\overline{5}$ $\frac{5}{9}$ **18.** 0.4 $\frac{2}{5}$ **19.** 0.16 $\frac{4}{25}$ **20.** $0.\overline{6}$ $\frac{2}{3}$

21. 0.125 $\frac{1}{8}$ **22.** 0.35 $\frac{7}{20}$ **23.** $0.\overline{8}$ $\frac{8}{9}$ **24.** $0.\overline{7}$ $\frac{7}{9}$

Decimals and Percents

- To express a decimal as a percent, first express the decimal as a fraction with a denominator of 100. Then express the fraction as a percent.

Examples **Write each decimal as a percent.**

1 **0.38**

$$0.38 = \frac{38}{100} = 38\%$$

2 **0.4**

$$0.4 = 0.40 = \frac{40}{100} = 40\%$$

3 **0.07**

$$0.07 = \frac{7}{100} = 7\%$$

4 **2.55**

$$2.55 = \frac{255}{100} = 255\%$$

Shortcut

To write a decimal as a percent, multiply by 100 and add the % symbol.
0.62 = 0.62 = 62%

- To express a percent as a decimal, rewrite the percent as a fraction with a denominator of 100. Then express the fraction as a decimal.

Examples **Write each percent as a decimal.**

5 **26%**

$$26\% = \frac{26}{100} = 0.26$$

6 **9%**

$$9\% = \frac{9}{100} = 0.09$$

7 **87.5%**

$$87.5\% = \frac{87.5}{100} = 0.875$$

8 **125%**

$$125\% = \frac{125}{100} = 1.25$$

Shortcut

To write a percent as a decimal, divide by 100 and remove the % symbol.
62% = 62% = 0.62

Write each decimal as a percent.

1. 0.62 **62%**
2. 0.99 **99%**
3. 0.14 **14%**
4. 0.20 **20%**
5. 0.15 **15%**
6. 0.8 **80%**
7. 0.5 **50%**
8. 0.06 **6%**
9. 0.42 **42%**
10. 0.03 **3%**
11. 0.1 **10%**
12. 1.76 **176%**
13. 0.08 **8%**
14. 3.10 **310%**
15. 1.05 **105%**
16. 2.6 **260%**

Write each percent as a decimal.

17. 12% **0.12**
18. 37% **0.37**
19. 86% **0.86**
20. 51% **0.51**
21. 30% **0.3**
22. 90% **0.9**
23. 2% **0.02**
24. 55% **0.55**
25. 5% **0.05**
26. 12.5% **0.125**
27. 73.6% **0.736**
28. 134% **1.34**
29. 208% **2.08**
30. 120% **1.2**
31. 60.5% **0.605**
32. 200% **2**

Fractions and Percents

- To express a percent as a fraction, write the percent as a fraction with a denominator of 100 and simplify.

Example 1 **Express 24% as a fraction in simplest form.**

$24\% = \frac{24}{100}$ *Write as a fraction with a denominator of 100.*

$= \frac{\overset{6}{\cancel{24}}}{\underset{25}{\cancel{100}}}$ *Divide by the GCF, 4.*

$= \frac{6}{25}$

- To express a fraction as a percent, write a proportion and solve.

Example 2 **Express $\frac{3}{5}$ as a percent.**

$\frac{3}{5} = \frac{n}{100}$ *Write a proportion.*

$3 \times 100 = 5 \times n$ *Find the cross products.*

$300 = 5n$

$\frac{300}{5} = \frac{5n}{5}$ *Divide each side by 5.*

$60 = n$

So, $\frac{3}{5} = \frac{60}{100}$ or 60%.

Write each percent as a fraction in simplest form.

1. 25% $\frac{1}{4}$
2. 80% $\frac{4}{5}$
3. 70% $\frac{7}{10}$
4. 15% $\frac{3}{20}$
5. 6% $\frac{3}{50}$
6. 56% $\frac{14}{25}$
7. 40% $\frac{2}{5}$
8. 98% $\frac{49}{50}$
9. 19% $\frac{19}{100}$
10. 35% $\frac{7}{20}$
11. 22% $\frac{11}{50}$
12. 64% $\frac{16}{25}$

Write each fraction as a percent.

13. $\frac{3}{10}$ **30%**
14. $\frac{1}{5}$ **20%**
15. $\frac{3}{4}$ **75%**
16. $\frac{13}{25}$ **52%**
17. $\frac{37}{50}$ **74%**
18. $\frac{1}{20}$ **5%**
19. $\frac{19}{20}$ **95%**
20. $\frac{42}{50}$ **84%**
21. $\frac{2}{25}$ **8%**
22. $\frac{9}{10}$ **90%**
23. $\frac{8}{32}$ **25%**
24. $\frac{22}{40}$ **55%**

Comparing and Ordering Rational Numbers

- To compare rational numbers, it is usually easier and faster if you write the numbers as decimals. In some cases, it may be easier if you write the numbers as fractions having the same denominator or as percents.

Examples **Replace each ● with <, >, or = to make a true sentence.**

1 $\frac{3}{5}$ ● 0.65

$\frac{3}{5} = 0.60$

Since $0.60 < 0.65$, $\frac{3}{5} < 0.65$.

2 0.18 ● 2%

$2\% = 0.02$

Since $0.18 > 0.02$, $0.18 > 2\%$.

- To order rational numbers, first write the numbers as decimals. Then write the decimals in order from least to greatest and write the corresponding rational numbers in the same order.

Example **3** **Write $\frac{1}{2}$, 55%, and 0.20 in order from least to greatest.**

Write each number as a decimal.

$\frac{1}{2} = 0.50$ $\quad 55\% = 0.55 \quad 0.20 = 0.20$

Write the decimals in order from least to greatest. $\quad 0.20 < 0.50 < 0.55$

↓ ↓ ↓

Write the corresponding rational numbers in the same order. $\quad 0.20 < \frac{1}{2} < 55\%$

The numbers in order from least to greatest are 0.20, $\frac{1}{2}$, and 55%.

Replace each ● with <, >, or = to make a true sentence.

1. 0.35 ● $\frac{1}{4}$ **>**
2. 65% ● 0.7 **<**
3. 80% ● $\frac{4}{5}$ **=**
4. $\frac{1}{10}$ ● 1% **>**
5. 12.5% ● $\frac{1}{8}$ **=**
6. 36% ● 3.6 **<**
7. $\frac{2}{3}$ ● 0.7 **<**
8. 0.2 ● 20% **=**
9. $\frac{5}{6}$ ● $\frac{5}{7}$ **>**
10. $\frac{3}{8}$ ● 0.375 **=**
11. 0.9 ● 10% **>**
12. 30% ● 3.9% **>**
13. 0.9 ● $0.\overline{9}$ **<**
14. 0.15 ● $\frac{2}{15}$ **>**
15. 51% ● 51 **<**
16. 78% ● $\frac{7}{8}$ **<**
17. $\frac{6}{11}$ ● 50% **>**
18. $0.\overline{4}$ ● $\frac{4}{9}$ **=**

Write the numbers in each set in order from least to greatest.

19. 0.52, 5%, $\frac{1}{2}$ **5%, $\frac{1}{2}$, 0.52**
20. $\frac{1}{3}$, 40%, $\frac{1}{4}$ **$\frac{1}{4}$, $\frac{1}{3}$, 40%**
21. 20%, 0.1, $\frac{3}{10}$ **0.1, 20%, $\frac{3}{10}$**
22. 0.19, $\frac{1}{5}$, 15% **15%, 0.19, $\frac{1}{5}$**
23. 63%, $\frac{2}{3}$, 0.06 **0.06, 63%, $\frac{2}{3}$**
24. $\frac{1}{9}$, $0.\overline{2}$, 19% **$\frac{1}{9}$, 19%, $0.\overline{2}$**

PREREQUISITE SKILLS REVIEW

EXTRA PRACTICE

Answers

Lesson 1-1

8. 5 minus 2 times k
9. 3 divided by x
10. 7 minus the quotient of m and x
11. the quotient of 6 and m plus 7
12. the quotient of 7 and 2 plus r

Lesson 1-2

13. $6 + 5 \cdot 2 - 1$
    - $= 6 + 10 - 1$ *Substitution*
    - $= 16 - 1$ *Substitution*
    - $= 15$ *Substitution*
14. $9(8 \div 2 - 3)$
    - $= 9(4 - 3)$ *Substitution*
    - $= 9(1)$ *Substitution*
    - $= 9$ *Multiplicative Identity*
15. $24 - 1(3 + 15)$
    - $= 24 - 1(18)$ *Substitution*
    - $= 24 - 18$ *Multiplicative Identity*
    - $= 6$ *Substitution*
16. $5(6) + 6(14 - 14)$
    - $= 30 + 6(0)$ *Substitution*
    - $= 30 + 0$ *Multiplicative Property of Zero*
    - $= 30$ *Additive Identity*
17. $11 + (7 - 28 \div 4)$
    - $= 11 + (7 - 7)$ *Substitution*
    - $= 11 + 0$ *Substitution*
    - $= 11$ *Additive Identity*
18. $\frac{7 + 21 \div 3}{3(2 - 1) + 4}$
    - $= \frac{7 + 7}{3(1) + 4}$ *Substitution*
    - $= \frac{7 + 7}{3 + 4}$ *Multiplicative Identity*
    - $= \frac{14}{7}$ *Substitution*
    - $= 2$ *Substitution*

Lesson 1-3

1. Assoc. (+) 2. Assoc. (×)
3. Comm. (×) 4. Comm. (×)
5. Assoc. (+) 6. Comm. (+)
7. Assoc. (×) 8. Assoc. (+)
9. Comm. (×) 10. Assoc. (×)
11. Comm. (+) 12. Comm. (+)

Extra Practice

Lesson 1-1 *(Pages 4–7)* **Write an algebraic expression for each verbal expression.**

1. 4 less than x **$x - 4$**
2. the difference of 3 and m **$3 - m$**
3. 6 more than the quotient of b and 5 **$\frac{b}{5} + 6$**
4. 2 less than 7 times k **$7k - 2$**

Write a verbal expression for each algebraic expression. **8–12. See margin.**

5. $b + 4$ **b plus 4**
6. $12x$ **12 times x**
7. $6 - k$ **6 minus k**
8. $5 - 2k$
9. $\frac{3}{x}$
10. $7 - \frac{m}{x}$
11. $\frac{6}{m} + 7$
12. $\frac{7}{2} + r$

Write an equation for each sentence.

13. Six times s plus seven equals fourteen. **$6s + 7 = 14$**
14. Three minus k equals twelve plus two times j. **$3 - k = 12 + 2j$**
15. Five divided by m is equal to two times n minus three. **$\frac{5}{m} = 2n - 3$**

Lesson 1-2 *(Pages 8–13)* **Find the value of each expression.**

1. $32 + 4 \cdot 2$ **40**
2. $6 - 3 \cdot 4 + 7$ **1**
3. $16 \div 4 + 3$ **7**
4. $6 \cdot 2 \cdot 4 - 1$ **47**
5. $25 \div 5 - (3 + 2)$ **0**
6. $\frac{3(7 + 1)}{3 \cdot 4}$ **2**
7. $[2(6 - 1)] \div 2$ **5**
8. $2(5 + 3) - 6$ **10**

Name the property of equality shown by each statement.

9. If $3y + 4 = 5x$ and $5x = 10$, then $3y + 4 = 10$. **Transitive**
10. If $13 + y = 18$, then $18 = 13 + y$. **Symmetric**
11. $9 - 7x = 9 - 7x$ **Reflexive**
12. $2(6 + 4) = 2(10)$ **Substitution**

13–18. See margin for properties.

Find the value of each expression. Identify the property used in each step.

13. $6 + 5 \cdot 2 - 1$ **15**
14. $9(8 \div 2 - 3)$ **9**
15. $24 - 1(3 + 15)$ **6**
16. $5(6) + 6(14 - 14)$ **30**
17. $11 + (7 - 28 \div 4)$ **11**
18. $\frac{7 + 21 \div 3}{3(2 - 1) + 4}$ **2**

Evaluate each algebraic expression if $m = 6$ and $b = 3$.

19. $8m + 2b$ **54**
20. $m(b - 3)$ **0**
21. $m - 3b + 4$ **1**
22. $\frac{6 \cdot 2 + b \cdot m}{m - 4}$ **15**
23. $\frac{m(6 - 2)}{b + 1}$ **6**
24. $(2m + 1)(6 - b)$ **39**

Lesson 1-3 *(Pages 14–18)* **Name the property shown by each statement.** **1–6. See margin.**

1. $(7 + 7) + 2 = 7 + (7 + 2)$
2. $(3 \cdot m) \cdot 6 = 3 \cdot (m \cdot 6)$
3. $9 \cdot 13 = 13 \cdot 9$
4. $6 \cdot 3 \cdot k = k \cdot 6 \cdot 3$
5. $k + (p + q) = (k + p) + q$
6. $m + 5 = 5 + m$

7–12. See margin for properties.

Simplify each expression. Identify the properties used in each step.

7. $5(6r)$ **$30r$**
8. $(k + 1) + 4$ **$k + 5$**
9. $11 \cdot p \cdot 2$ **$22p$**
10. $(s \cdot 4) \cdot 3$ **$12s$**
11. $7 + y + 1$ **$8 + y$**
12. $13 + w + 12$ **$25 + w$**

Answer

Page 693

Lesson 1-4

13. $6m + (3k + m) + 2k + 4(m + 1)$
    - $= 6m + 3k + m + 2k + 4m + 4$ *Distributive Property*
    - $= 6m + m + 4m + 3k + 2k + 4$ *Commutative (+)*
    - $= (6m + m + 4m) + (3k + 2k) + 4$ *Associative (+)*
    - $= 11m + 5k + 4$ *Substitution*

Lesson 1–4 *(Pages 19–23)* Simplify each expression. 3. $12a + 7b$ 8. $11m + 24n$

1. $6b + 3b$ $9b$
2. $4(3k - 2k)$ $4k$
3. $4a + 2b + 8a + 5b$
4. $9 + 12s - 2s$ $9 + 10s$
5. $3 + 4p - 2p$ $3 + 2p$
6. $8 + k - 7 + 6k$ $7k + 1$
7. $ab - b + 3ab$ $4ab - b$
8. $8(m + 3n) + 3m$
9. $6st - st + t$ $5st + t$
10. $6(3 + 2m)$ $18 + 12m$
11. $5(7 + 2r) + 2(r - 3)$ $29 + 12r$
12. $4 + (7 + 5x)3$ $25 + 15x$
13. Write $6m + (3k + m) + 2k + 4(m + 1)$ in simplest form. Indicate the property that justifies each step. $11m + 5k + 4$; See margin for properties.
14. What is the value of $4x$ multiplied by the quantity $2x$ minus 3 if x equals 4? 80

Lesson 1–5 *(Pages 24–29)* Solve each problem. Use any strategy.

1. Simone has 16 players on her soccer team. There are 6 more boys than girls. How many girls and how many boys are there? 5 girls, 11 boys
2. Dillon put $350 into a bank account that pays 6% simple interest. How much money will he have in four years? $434
3. How many ways can you make 50¢ using dimes, nickels, and pennies? 36 ways
4. Kate lives 24 miles from school. Jake lives 6 miles closer to school than Kate does. How far does Jake live from school? 18 miles
5. Nine paintings are for sale at an art gallery. Tara must choose two of the paintings.
   a. Make a chart or diagram to represent the problem. See margin.
   b. How many different combinations are there in all? 36
   c. How many different combinations would there be if there were eight paintings? 28

Lesson 1–6 *(Pages 32–37)* Determine whether each is a good sample. Describe what caused the bias in each poor sample. Explain. 1–4. See margin.

1. Every other student in physical education class is surveyed to determine how many hours per week students exercise.
2. Every student at a school is asked to name their favorite subject in school.
3. Surveys are placed on every fifth car in a mall parking lot concerning the construction of a new lot.
4. Six people at a music concert are randomly chosen to find out their opinion of the concert.

Refer to the chart at the right.

5. Make a frequency table to organize the data. See margin.
6. What number of hours is most frequent? 4 hours
7. How many students use the Internet for less than two hours per week? 3 students
8. How many more students use the Internet six or more hours than those who use it less than four hours? 2 students

Time on the Internet per Week (h)		
4	9	4
2	5	2
6	1	10
10	8	4
3	2	9
10	4	6
1	7	1

Answers
Lesson 1–6

1. No, the sample is not random enough.
2. Yes, the sample is large enough.
3. Yes, a large enough sample is used and the parking would concern the people already parked in the lot.
4. No, the sample would not be large enough.
5.

Time on the Internet per Week (h)		
Hours	Tally	Frequency
1	III	3
2	III	3
3	I	1
4	IIII	4
5	I	1
6	II	2
7	I	1
8	I	1
9	II	2
10	III	3

EXTRA PRACTICE

Answer *Lesson 1–5*

5a.

Painting #1	Painting #2	Total Combinations
1	2 3 4 5 6 7 8 9	8
2	3 4 5 6 7 8 9	7
3	4 5 6 7 8 9	6
4	5 6 7 8 9	5
5	6 7 8 9	4
6	7 8 9	3
7	8 9	2
8	9	1
9		0
		36

Answers

Lesson 1–7

1.

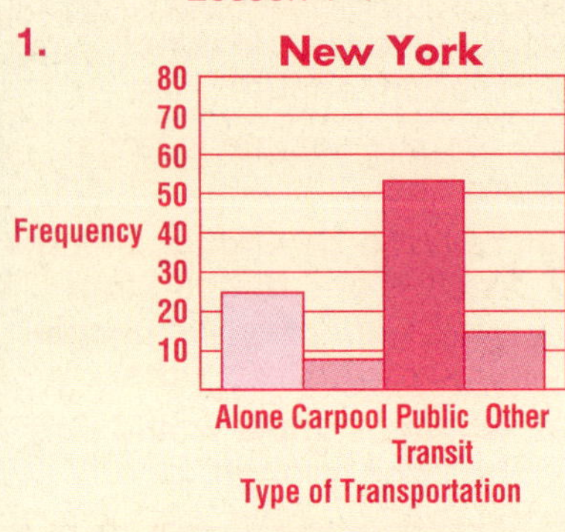

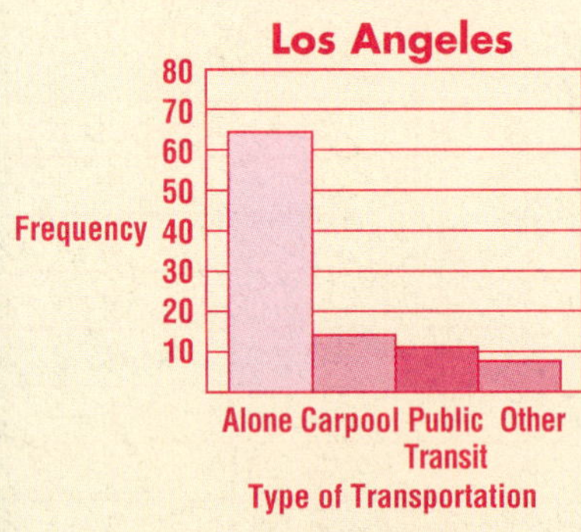

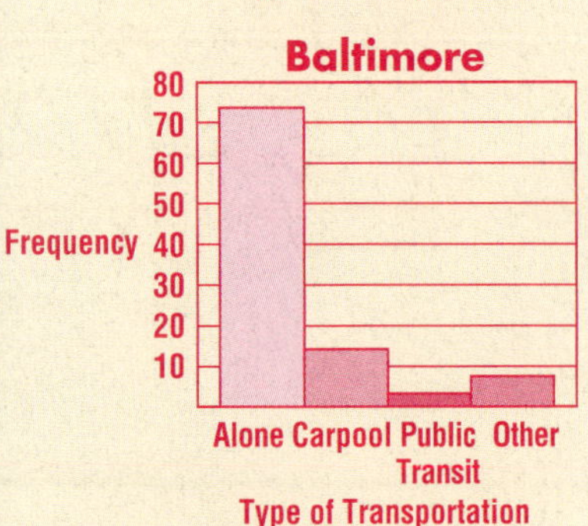

2. The public transit system in New York is used much more than it is in Baltimore and Los Angeles. In Los Angeles and Baltimore, people tend to drive alone.
3. Sample answer: The public transit system in New York may be more developed than the other two cities.

Lesson 2–1

7. −5 −4 −3 −2 −1 0 1 2 3 4 5
8. −5 −4 −3 −2 −1 0 1 2 3 4 5
9. −5 −4 −3 −2 −1 0 1 2 3 4 5
10. −5 −4 −3 −2 −1 0 1 2 3 4 5
11. −5 −4 −3 −2 −1 0 1 2 3 4 5
12. −5 −4 −3 −2 −1 0 1 2 3 4 5

Lesson 1–7 *(Pages 38–43)* **People in three major cities were surveyed to see what transportation they used to get to work. The table shows the results.**

1. Make a histogram for each city that shows the results of the survey.
2. How do the histograms compare?
3. Make some assumptions to support your findings. **1–3. See margin.**

Type	New York, NY	Los Angeles, CA	Baltimore, MD
drive alone	24.0%	65.2%	75.5%
carpool	8.5%	15.4%	14.2%
public transit	53.4%	10.5%	2.7%
other	14.0%	8.9%	7.6%

Source: *Time Almanac 2000*

The record low temperatures for select states are shown.

4. How many states are represented? **27 states**
5. What is the lowest record low temperature? **−62°C**
6. Which temperature occurs most frequently? **−40°C**
7. How many states have record low temperatures of less than −50°C? **5 states**

Stem	Leaf
−6	2
−5	7 2 1 1
−4	6 4 4 3 0 0 0 0
−3	8 8 8 7 6 4 3
−2	8 7 7 7 6
−1	9 4 $-1 \mid 4 = -14°C$

Lesson 2–1 *(Pages 52–57)* **Name the coordinate of each point.**

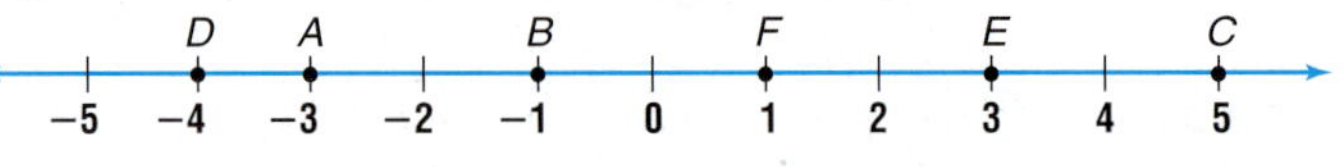

1. *A* **−3**   2. *B* **−1**   3. *C* **5**   4. *D* **−4**   5. *E* **3**   6. *F* **1**

Graph each set of numbers on a number line. **7–12. See margin.**

7. {−1, 0, 4}   8. {−3, −2, −1}   9. {0, 3, 5}
10. {−4, −2, 1, 2}   11. {−1, 0, 1, 2}   12. {−4, 2, 3, 4}

Replace each ● with $<$ or $>$ to make a true sentence.

13. $-3 \bullet 0$ **$<$**   14. $-4 \bullet -3$ **$<$**   15. $-8 \bullet 2$ **$<$**   16. $0 \bullet -1$ **$>$**
17. $|6| \bullet -2$ **$>$**   18. $-2 \bullet |-3|$ **$<$**   19. $|-7| \bullet |-4|$ **$>$**   20. $|16| \bullet -12$ **$>$**

Evaluate each expression.

21. $|7|$ **7**   22. $|-5|$ **5**   23. $|-7| - 5$ **2**   24. $-11 + |4|$ **−7**
25. $|4| - |-12|$ **−8**   26. $|-8| + |-3|$ **11**   27. $|5| - 11$ **−6**   28. $|18| + |-21|$ **39**

Lesson 2–2 *(Pages 58–63)* **Write the ordered pair that names each point.**

1. *G* **(2, 2)**   2. *H* **(−4, 0)**   3. *I* **(2, −3)**
4. *J* **(−1, −1)**   5. *K* **(0, 5)**   6. *L* **(−3, 4)**

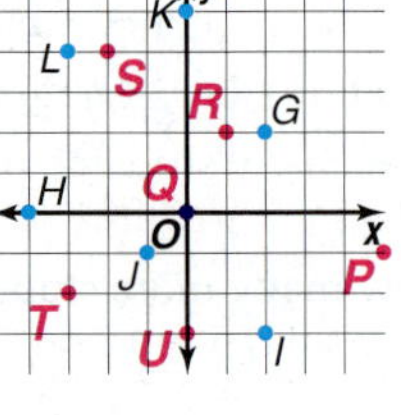

Graph each point on a coordinate plane.

7. *P*(5, −1)   8. *Q*(0, 0)   9. *R*(1, 2)
10. *S*(−2, 4)   11. *T*(−3, −2)   12. *U*(0, −3)

Name the quadrant in which each point is located.

13. (6, 4) **I**   14. (−2, 1) **II**   15. (−3, −3) **III**   16. (0, 0) **none**

Lesson 2–3 *(Pages 64–69)* Find each sum.

1. $2 + 14$ **16**
2. $8 + 4$ **12**
3. $-16 + (-7)$ **−23**
4. $-13 + (-26)$ **−39**
5. $(-21) + 14$ **−7**
6. $-19 + 17$ **−2**
7. $-5 + (-13) + (-24)$ **−42**
8. $2 + 14 + (-12)$ **4**
9. $-16 + (-4) + (-11) + 5$ **−26**
10. $12 + (-7) + 6 + (-18)$ **−7**

Simplify each expression.

11. $6b + 2b$ **8*b***
12. $-8y + 23y$ **15*y***
13. $17k + (-18k)$ **−1*k***
14. $7m - 7m$ **0**
15. $-8r + (-4r) + 12r$ **0**
16. $-6p + (-3p) + (-1p)$ **−10*p***
17. $7c + 11c + (-12c)$ **6*c***
18. $-5x - 6x$ **−11*x***

Lesson 2–4 *(Pages 70–74)* Find each difference.

1. $16 - 7$ **9**
2. $16 - 21$ **−5**
3. $-10 - (-9)$ **−1**
4. $-4 - 6$ **−10**
5. $3 - 9$ **−6**
6. $7 - (-12)$ **19**
7. $-12 - (-4)$ **−8**
8. $0 - 11$ **−11**

Evaluate each expression if $x = 6$, $y = -8$, $z = -3$, and $w = 4$.

9. $y - z$ **−5**
10. $3 - z$ **6**
11. $6 + y$ **−2**
12. $w - y$ **12**
13. $14 - y - x$ **16**
14. $6 + x - z$ **15**
15. $y + z + w$ **−7**
16. $w - z + 11$ **18**

Simplify each expression.

17. $7k - 4k$ **3*k***
18. $6m + (-3m)$ **3*m***
19. $3y - 7y$ **−4*y***
20. $-14n - 17n$ **−31*n***
21. $13s - (-2s)$ **15*s***
22. $9r - (-2r) - 4r$ **7*r***
23. $8c - 12c + 2c$ **−2*c***
24. $b + 2b - 3b$ **0**

25. Corey spends \$112 on movies each month. He spends \$80 less on books. How much does he spend on books each month? **\$32**

Lesson 2–5 *(Pages 75–79)* Find each product.

1. $2(4)$ **8**
2. $3(-12)$ **−36**
3. $-4(-13)$ **52**
4. $-5(8)$ **−40**
5. $10(-3)(2)$ **−60**
6. $-4(2)(-3)1$ **24**
7. $6(-1)(-1)(-2)$ **−12**
8. $-8(-4)2$ **64**

Evaluate each expression if $x = -2$, $y = 4$, and $z = -6$.

9. $6x$ **−12**
10. $2yz$ **−48**
11. xyz **48**
12. $-8xy$ **64**
13. $4x - 3z$ **10**
14. $6y - z$ **30**
15. $2x + z$ **−10**
16. $10x + 5y$ **0**

Simplify each expression.

17. $-6(2b)$ **−12*b***
18. $7(-4m)$ **−28*m***
19. $-10(-3k)$ **30*k***
20. $4(5c)$ **20*c***
21. $(6g)(9h)$ **54*gh***
22. $8m(-6n)$ **−48*mn***
23. $(-2r)(-3s)$ **6*rs***
24. $(-5p)2q$ **−10*pq***

Extra Practice 695

Answer

Lesson 3-1

16. 

0.7 is to the right of 0.6. So, 0.7 > 0.6.

EXTRA PRACTICE

Lesson 2–6 *(Pages 82–85)* Find each quotient.

1. $49 \div 7$ **7**
2. $-12 \div 6$ **−2**
3. $-36 \div (-12)$ **3**
4. $21 \div (-3)$ **−7**
5. $36 \div (-6)$ **−6**
6. $-20 \div 4$ **−5**
7. $-16 \div (-4)$ **4**
8. $-32 \div (-4)$ **8**
9. $\frac{8}{2}$ **4**
10. $\frac{-54}{-9}$ **6**
11. $\frac{63}{-7}$ **−9**
12. $\frac{-64}{8}$ **−8**

Evaluate each expression if $x = -5$, $y = -3$, $z = 2$, and $w = 7$.

13. $25 \div x$ **−5**
14. $-42 \div w$ **−6**
15. $3 \div y$ **−1**
16. $2x \div z$ **−5**
17. $-3x \div y$ **−5**
18. $x \div (-1)$ **5**
19. $xyz \div 10$ **3**
20. $yzw \div 2w$ **−3**
21. $\frac{3y}{-3}$ **3**
22. $\frac{6-y}{y}$ **−3**
23. $\frac{w}{-7}$ **−1**
24. $\frac{w-x}{y}$ **−4**

Lesson 3–1 *(Pages 94–99)* Replace each ● with <, >, or = to make a true sentence. **4. >**

1. -2 ● -5 **>**
2. -3.4 ● -4 **>**
3. -0.32 ● -0.3 **<**
4. $4(-2)(-1)$ ● $2(-3)$
5. 0.6 ● $\frac{3}{5}$ **=**
6. $\frac{2}{3}$ ● 0.7 **<**
7. $\frac{7}{9}$ ● $\frac{7}{8}$ **<**
8. $-\frac{3}{18}$ ● $-\frac{2}{9}$ **>**

Write the numbers in each set from least to greatest.

9. $\frac{5}{6}, 0.3, \frac{1}{3}$ **$0.3, \frac{1}{3}, \frac{5}{6}$**
10. $0.4, -\frac{2}{3}, \frac{3}{4}$ **$-\frac{2}{3}, 0.4, \frac{3}{4}$**
11. $-\frac{4}{5}, -\frac{6}{7}, -0.7$ **$-\frac{6}{7}, -\frac{4}{5}, -0.7$**
12. $\frac{1}{2}, -\frac{7}{10}, -\frac{2}{3}$ **$-\frac{7}{10}, -\frac{2}{3}, \frac{1}{2}$**
13. $0.\overline{4}, \frac{3}{8}, \frac{4}{6}$ **$\frac{3}{8}, 0.\overline{4}, \frac{4}{6}$**
14. $-\frac{1}{8}, \frac{2}{9}, -\frac{3}{12}$ **$-\frac{3}{12}, -\frac{1}{8}, \frac{2}{9}$**

15. Compare the numbers $\frac{6}{7}$ and $\frac{7}{8}$ using an inequality. **$\frac{6}{7} < \frac{7}{8}$**
16. Using a number line, explain why $0.7 > 0.6$. **See margin.**

Lesson 3–2 *(Pages 100–103)* Find each sum or difference.

1. $-13.1 + (-21.4)$ **−34.5**
2. $-8.14 - 0.13 + 1.11$ **−7.16**
3. $6.2 - (-3.4)$ **9.6**
4. $-11.12 - 2.15 + 5.28 - 3.12$ **−11.11**
5. $15.9 + 6.25 - 3.48 - 2.13$ **16.54**
6. $-17.6 + 0.3 - 3.7$ **−21**
7. $-\frac{4}{5} + \frac{2}{3}$ **$-\frac{2}{15}$**
8. $\frac{6}{7} - \frac{1}{8}$ **$\frac{41}{56}$**
9. $-\frac{3}{4} + \frac{2}{3}$ **$-\frac{1}{12}$**
10. $3\frac{6}{7} - \left(-4\frac{5}{8}\right)$ **$8\frac{27}{56}$**
11. $-6\frac{1}{6} + 6\frac{5}{6}$ **$\frac{2}{3}$**
12. $2\frac{8}{9} - 1\frac{1}{3}$ **$1\frac{5}{9}$**
13. Find the value of $z - 0.25$ if $z = 0.5$. **0.25**
14. Evaluate $m + 5\frac{3}{4}$ if $m = -3\frac{1}{2}$. **$2\frac{1}{4}$**

Lesson 3–3 *(Pages 104–109)* **Find the mean, median, mode, and range for each set of data.** **1–8. See margin.**

1. 8, 3, 2, 8, 1, 5, 8, 5
2. 11, 5, 18, 10, 14, 14, 18, 18, 14, 17, 15
3. 1.2, 5.4, 2.3, 3.2, 1.2, 5.3
4. 123, 153, 123, 114, 148, 114, 135
5. 47, 29, 77, 99, 50 47, 29
6. 6.5, 7.8, 5.7, 3.9, 9.9, 3.8, 5.5, 5.7
7.

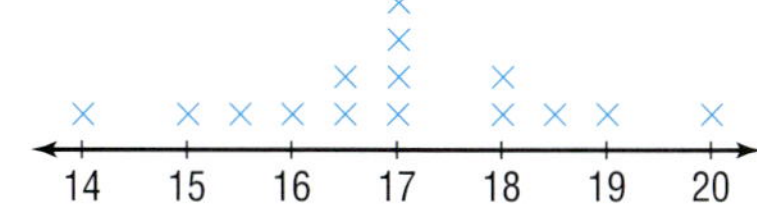

8.

Stem	Leaf
11	0 1 7
12	3 4 4
13	5 5 6 7
14	2 2

11|0 = 110

Write a set of data with six numbers that satisfies each set of conditions.

9. The median is equal to the mean. **Sample answer: 4, 5, 6, 6, 7, 8**
10. The median is less than the mean. **Sample answer: 2, 4, 6, 8, 10, 30**

Lesson 3–4 *(Pages 112–116)* **Find the solution of each equation if the replacement sets are $x = \{-2, -1, 0, 1\}$, $m = \{-1, 0, 1, 2\}$, and $d = \{6, 7, 8, 9\}$.**

1. $x + 3 = 2$ **−1**
2. $2d = 16$ **8**
3. $6m + 15 = 21$ **1**
4. $7m - 11 = \frac{3m}{2}$ **2**
5. $\frac{14 - 2}{x} = -6$ **−2**
6. $\frac{2}{3(1 + 5)} = \frac{6 - 5}{d}$ **9**

Solve each equation. **13. 10.5**

7. $-17.8 - 12.2 = p$ **−30**
8. $6.5 - 13.2 = y$ **−6.7**
9. $x = 7 \cdot 2 + 1$ **15**
10. $j = 16 - 6 \cdot 2 \div 3$ **12**
11. $7 \cdot 4 - 4 \div 2 = r$ **26**
12. $w = 6.1 - 3.6 \cdot 4$ **−8.3**
13. $n = 10.3 + 4.2 \div 3 - 1.2$
14. $[12 - (4 + 1)] = z$ **7**
15. $t = -4.2 - 3.3 \div 3$ **−5.3**
16. $\frac{7 \cdot 3 + 3}{(3 - 1) \cdot 2} = k$ **6**
17. $\frac{-13 - 5}{3 \cdot 2} = s$ **−3**
18. $\frac{6 - 3 \cdot 2}{12 \div 3 + 1} = m$ **0**
19. $x = \frac{5 \div (2 + 3)}{2 \cdot 2 - 3}$ **1**
20. $\frac{36 \div 9 + 1}{2 \cdot 5} = a$ **$\frac{1}{2}$**
21. $h = \frac{15 + 10 \div 2}{-3 \cdot 1 + 2}$ **−20**
22. Find the solution of $6 - 3(8) = r$. **−18**
23. What is the value of k if $k = \frac{16 + 2}{-9}$? **−2**

Lesson 3–5 *(Pages 117–121)* **Solve each equation. Use algebra tiles if necessary.**

1. $g + 13 = 7$ **−6**
2. $-8 = w + -6$ **−2**
3. $m - 8 = -23$ **−15**
4. $-3 + q = 9$ **12**
5. $-21 = a + 15$ **−36**
6. $t - 12 = -3$ **9**
7. $d - (-6) = 14$ **8**
8. $s - 4 = 12$ **16**
9. $k - 9 = 18$ **27**
10. $c - (-3) = -8$ **−11**
11. $13 + y = 4$ **−9**
12. $h - 11 = 5$ **16**
13. $-24 = x + 7$ **−31**
14. $16 = 9 + p$ **7**
15. $6 = -12 + b$ **18**
16. $-5 = -12 + b$ **7**
17. $m + 15 = 9$ **−6**
18. $-8 = w + (-6)$ **−2**
19. What is the value of y if $y - 4 = -15$? **−11**
20. When x is divided by 8, the result is 7. Find the value of x. **56**

Answers

Lesson 3–3

1. 5; 5; 8; 7
2. 14; 14; 14, 18; 13
3. 3.1; 2.75; 1.2; 4.2
4. 130; 123; 114, 123; 39
5. 54; 47; 29, 47; 70
6. 6.1; 5.7; 5.7; 6.1
7. 17; 17; 17; 6
8. 128; 129.5; 124, 135, 142; 32

Extra Practice 697

EXTRA PRACTICE

Lesson 3–6 *(Pages 122–127)* **Solve each equation. Check your solution.**

1. $-18 = w + 3$ **−21**
2. $m + (-5) = 6$ **11**
3. $-16.4 = r - 7.4$ **−9**
4. $m - 16 = -13$ **3**
5. $5 = x - (-3)$ **2**
6. $x + (-8) = -7$ **1**
7. $45 + j = -27$ **−72**
8. $-56 = g - 32$ **−24**
9. $n - (-26) = 41$ **15**
10. $6.4 + k = -3.2$ **−9.6**
11. $-2.8 + k = 3.1$ **5.9**
12. $s + 2.5 = -1.3$ **−3.8**
13. $13.4 + p = -2.4$ **−15.8**
14. $17.3 = -2.7 + d$ **20**
15. $y + 2.17 = 5.67$ **3.5**
16. $\frac{5}{7} = z - \frac{6}{21}$ **1**
17. $n - \frac{5}{8} = -\frac{3}{4}$ $\mathbf{-\frac{1}{8}}$
18. $-\frac{2}{3} = j + \frac{7}{12}$ $\mathbf{-\frac{5}{4}}$
19. Solve for m if $-20 + m = -7$. **13**
20. What is the solution of $h - 4 = 14$? **18**

Lesson 3–7 *(Pages 128–131)* **Solve each equation. Check your solution.**

1. $|m| = -3$ **∅**
2. $4 = |x|$ **{−4, 4}**
3. $|r| + 4 = 8$ **{−4, 4}**
4. $|y| - 8 = 11$ **{−19, 19}**
5. $|x + 5| = 10$ **{−15, 5}**
6. $|t + 8| = -2$ **∅**
7. $|d - 4| = 10$ **{−6, 14}**
8. $6 = |-3 + p|$ **{−3, 9}**
9. $-4 = |5 + s|$ **∅**
10. $|n - 11| = 15$ **{−4, 26}**
11. $-4 = |-2 + j|$ **∅**
12. $|-2 + p| = 0$ **{2}**
13. $|k - (-2)| = 5$ **{−7, 3}**
14. $12 = |n + (-3)|$ **{−9, 15}**
15. $2 + |s + 3| = 8$ **{−9, 3}**
16. $5 = |-4 + g| + 17$ **∅**
17. $16 = |z - 4| + 3$ **{−9, 17}**
18. $|s + 4| - 3 = 7$ **{−14, 6}**
19. How many solutions exist for $-3 = |4 + d|$? **none**
20. How many solutions exist for $8 = |2 + k|$? **2**

Lesson 4–1 *(Pages 140–145)* **Find each product.** **16.** $\mathbf{-\frac{25}{16}}$ **or** $\mathbf{-1\frac{9}{16}}$

1. $3(8.2)$ **24.6**
2. $-7.3(3)$ **−21.9**
3. $-6.2(3.5)$ **−21.7**
4. $(2.1)(-1)$ **−2.1**
5. $-3.2(-0.5)$ **1.6**
6. $-2.1(3)(-2)$ **12.6**
7. $16.2(0)$ **0**
8. $8.5(-1)(-2.2)$ **18.7**
9. $\frac{2}{5}\left(\frac{1}{9}\right)$ $\mathbf{\frac{2}{45}}$
10. $\left(-\frac{1}{2}\right)\left(\frac{3}{8}\right)$ $\mathbf{-\frac{3}{16}}$
11. $-\frac{1}{4}\left(-\frac{6}{5}\right)$ $\mathbf{\frac{3}{10}}$
12. $-1 \cdot \frac{1}{8}$ $\mathbf{-\frac{1}{8}}$
13. $-\frac{6}{7}(0)$ **0**
14. $4\left(-\frac{3}{4}\right)$ **−3**
15. $\left(-\frac{3}{4}\right)\left(\frac{5}{6}\right)$ $\mathbf{-\frac{5}{8}}$
16. $\frac{1}{2}\left(-3\frac{1}{8}\right)$

Simplify each expression.

17. $-7.2(2p)$ **−14.4*p***
18. $(-9y)(-4.1)$ **36.9*y***
19. $(6.2k)(-3)$ **−18.6*k***
20. $-5(0.5j)$ **−2.5*j***
21. $2x(-3.4y)$ **−6.8*xy***
22. $5.4m(-2n)$ **−10.8*mn***
23. $\frac{7}{13}p(26)$ **14*p***
24. $\left(\frac{3}{5}r\right)\left(-\frac{1}{3}\right)$ $\mathbf{-\frac{1}{5}r}$
25. $\left(\frac{1}{2}x\right)\left(\frac{2}{3}y\right)$ $\mathbf{\frac{1}{3}xy}$
26. $\frac{5}{6}k(6)$ **5*k***
27. $\left(-\frac{2}{3}s\right)\left(\frac{4}{5}r\right)$ $\mathbf{-\frac{8}{15}rs}$
28. $-\frac{1}{4}w(-3t)$ $\mathbf{\frac{3}{4}tw}$

698 Extra Practice

Lesson 4–2 *(Pages 146–151)* Find the number of possible outcomes by drawing a tree diagram.

Men	Women
Troy	Ann
Malik	Lorena
Ben	Ellen
Aaron	

Exercise 2

1. tossing a coin twice **4**
2. choosing different teams of one man and one woman **12**
3. At a dinner party, you can choose either chicken or beef for your main dish, soda or juice for your drink, and pie, ice cream, or cake for your dessert. How many different meals are possible? **12 meals**
4. When choosing auto insurance, you have many choices, as shown in the table. How many different types of coverage are possible? Find the number of possible outcomes by using the Fundamental Counting Principle. **108**

Liability	Underinsured Drivers	Collision	Comprehensive
\$50,000	\$50,000	\$100	\$100
\$100,000	\$100,000	\$500	\$500
\$150,000	\$150,000	\$1000	\$1000
	none		

5. choosing four cards from a standard 52-card deck **270,725**
6. There are 5 multiple-choice problems on a quiz (with possible answers of a, b, c, and d). How many different combinations of answers are possible? **1024**

Lesson 4–3 *(Pages 154–159)* Find each quotient. **16. $-\frac{8}{5}$ or $-1\frac{3}{5}$**

1. $12 \div 1.5$ **8**
2. $3.1 \div (-3.1)$ **−1**
3. $-8.2 \div (-4.1)$ **2**
4. $-18.6 \div 6.2$ **−3**
5. $0 \div 7.3$ **0**
6. $-9.3 \div (-3.1)$ **3**
7. $1.6 \div (-16)$ **−0.1**
8. $-0.4 \div 0.2$ **−2**
9. $-\frac{1}{6} \div \left(-\frac{1}{5}\right)$ **$\frac{5}{6}$**
10. $\frac{1}{3} \div (-5)$ **$-\frac{1}{15}$**
11. $-4 \div \frac{2}{3}$ **−6**
12. $8 \div \left(-\frac{1}{4}\right)$ **−32**
13. $6 \div \left(-\frac{3}{4}\right)$ **−8**
14. $-\frac{2}{5} \div \left(-\frac{7}{8}\right)$ **$\frac{16}{35}$**
15. $4\frac{5}{8} \div \frac{5}{8}$ **$\frac{37}{5}$ or $7\frac{2}{5}$**
16. $-\frac{4}{3} \div \frac{5}{6}$

Evaluate each expression if $j = -\frac{1}{8}$, $k = \frac{3}{4}$, and $m = \frac{1}{6}$.

17. $\frac{8}{j}$ **−64**
18. $\frac{k}{2}$ **$\frac{3}{8}$**
19. $\frac{m}{5}$ **$\frac{1}{30}$**
20. $\frac{4}{m}$ **24**
21. $\frac{j}{k}$ **$-\frac{1}{6}$**
22. $\frac{2m}{k}$ **$\frac{4}{9}$**
23. $\frac{j}{8}$ **$-\frac{1}{64}$**
24. $\frac{km}{j}$ **−1**

Lesson 4–4 *(Pages 160–164)* Solve each equation. **10. $\frac{24}{5}$ or $4\frac{4}{5}$**

1. $6k = -54$ **−9**
2. $0 = 8p$ **0**
3. $0.5y = 12$ **24**
4. $3.2m = -6.4$ **−2**
5. $7.5 = -1.5w$ **−5**
6. $2.1t = -6.3$ **−3**
7. $-3n = 2$ **$-\frac{2}{3}$**
8. $-7 = -3s$ **$\frac{7}{3}$**
9. $\frac{1}{3}x = -4$ **−12**
10. $-3 = -\frac{5}{8}v$
11. $\frac{4}{5} = \frac{4}{9}r$ **$\frac{9}{5}$ or $1\frac{4}{5}$**
12. $-\frac{2}{3}z = \frac{1}{8}$ **$-\frac{3}{16}$**
13. $-\frac{1}{2}a = 3$ **−6**
14. $10 = \frac{c}{4}$ **40**
15. $\frac{21}{13} = -7j$ **$-\frac{3}{13}$**
16. $-\frac{1}{18} = \frac{17}{18}b$ **$-\frac{1}{17}$**
17. Solve $-1.5x = 18$. Then check your solution. **−12**
18. What is the solution of $\frac{s}{16} = \frac{1}{4}$? **4**

Answers

Lesson 4–2

1.
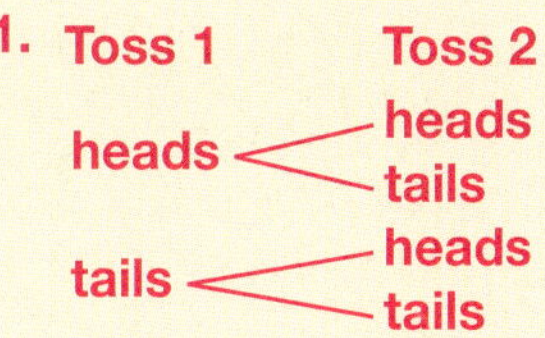

2.
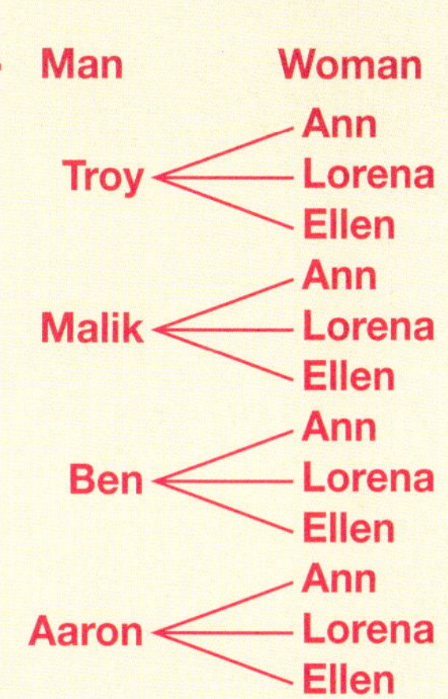

3.
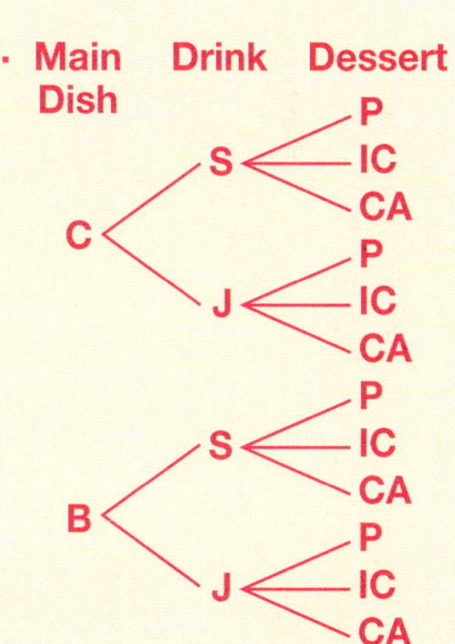

Extra Practice 699

Lesson 4-5 *(Pages 165–170)* **Solve each equation. Check your solution.**

1. $6 + 2y = 8$ **1**
2. $3 - 3r = 4$ **$-\frac{1}{3}$**
3. $4 + (-0.5y) = 8$ **−8**
4. $-7s - 3 = 18$ **−3**
5. $0 = 1.5a + 3$ **−2**
6. $3.1 = 0.7w + 1$ **3**
7. $6.3 = 0.4t - 2.1$ **21**
8. $6c + 1.5 = 3$ **0.25**
9. $\frac{r}{5} - 1 = 5$ **30**
10. $6 = -\frac{y}{3} + 3$ **−9**
11. $10 = -\frac{x}{2} + 6$ **−8**
12. $8 - \frac{b}{4} = 2$ **24**
13. $8 = -\frac{x}{3} + 2$ **−18**
14. $\frac{6 - p}{2} = -3$ **12**
15. $-\frac{7}{8}c + 1 = 4$ **$-\frac{24}{7}$ or $-3\frac{3}{7}$**
16. $\frac{8 + m}{3} = -8$ **−32**
17. What is the solution of $3 = \frac{r + 10}{-3}$? **−19**
18. Find the value of w in the equation $6 + \frac{2}{3}w = 12$. **9**

Write an equation and solve each problem.

19. Three less than half a number x is 12. Find the number. **$\frac{1}{2}x - 3 = 12$; 30**
20. Start with a number k. If you add 3, multiply by 7 and subtract 4, you get 18. What is the value of k? **$7(k + 3) - 4 = 18$; $\frac{1}{7}$**

Lesson 4-6 *(Pages 171–175)* **Solve each equation. Check your solution.** **7. no solution**

1. $6a = 3a + 3$ **1**
2. $16 + 5r = 3r$ **−8**
3. $8x + 6 = 12 - 3x$ **$\frac{6}{11}$**
4. $7k + 3 = 2k$ **$-\frac{3}{5}$**
5. $1.4y = -0.6y - 2$ **−1**
6. $4 + 0.8p = -3.2p$ **−1**
7. $6.4 + m = m - 2.3$
8. $\frac{1}{3}f = \frac{2}{3}f + 4$ **−12**
9. $\frac{1}{8}t + 3 = -4 - \frac{1}{6}t$ **−24**
10. $\frac{1}{2}w - 2 = \frac{2}{3}w$ **−12**
11. $1 - \frac{7}{8}h = \frac{1}{4}h$ **$\frac{8}{9}$**
12. $\frac{2}{5}j + 3 = -\frac{3}{5}j$ **−3**
13. Find the solution of $1.7x + 11 = -6.3x - 13$. **−3**
14. Six times a number k is 17 less than 2 times the number.
    a. Write an equation to represent the problem. **$6k = 2k - 17$**
    b. Find the number. **$-\frac{17}{4}$ or $-4\frac{1}{4}$**

Lesson 4-7 *(Pages 176–179)* **Solve each equation. Check your solution.**

1. $16 = -2(p - 1)$ **−7**
2. $6(x + 5) = 12$ **−3**
3. $8x + 6 = 3(4 - x)$ **$\frac{6}{11}$**
4. $8(6 - q) - 5 = -21$ **8**
5. $6t = 3(2t + 5)$ **no solution**
6. $-7 = 3k - 2(2k)$ **7**
7. $7(z + 3) = 2z + 4$ **$-\frac{17}{5}$ or $-3\frac{2}{5}$**
8. $-1(r + 1) - 2 = 2 + (r - 1)$ **−2**
9. $5x = 3(1.5x - 3)$ **−18**
10. $-4(m - 2.2) = -2(m + 1.4)$ **5.8**
11. $6 + \frac{1}{3}(9x + 3) = 6x + 2$ **$\frac{5}{3}$ or $1\frac{2}{3}$**
12. $4\left(\frac{1}{2}x + \frac{1}{2}\right) = 2x + 2$ **identity**
13. Find the solution of $4[8 + 2(m - 2)] = 8$. **−1**
14. What is the value of r in $7(8 - r) + 3(2 + 2r) = 9$? **53**

700 Extra Practice

Lesson 5-1 (*Pages 188–193*) Solve each proportion.

1. $\frac{9}{x} = \frac{3}{10}$ **30**
2. $\frac{6}{4} = \frac{y}{10}$ **15**
3. $\frac{b}{12} = \frac{32}{8}$ **48**
4. $\frac{1}{3} = \frac{5}{r}$ **15**
5. $\frac{21}{28} = \frac{h}{7}$ **5.25**
6. $\frac{4}{5} = \frac{a}{100}$ **80**
7. $\frac{25}{3} = \frac{500}{x}$ **60**
8. $\frac{p}{24} = \frac{8}{4}$ **48**
9. $\frac{z}{6} = \frac{9}{2}$ **27**
10. $\frac{6.2}{s} = \frac{3.1}{2}$ **4**
11. $\frac{0.3}{0.4} = \frac{x}{8}$ **6**
12. $\frac{6+m}{m-3} = \frac{1}{2}$ **−15**
13. $\frac{7+b}{4} = \frac{3}{5}$ $-\frac{23}{5}$
14. $\frac{7-d}{8+d} = \frac{4}{9}$ $\frac{31}{13}$
15. $\frac{6}{d-2} = \frac{7}{d}$ **14**
16. $\frac{k+1}{3} = \frac{k}{6}$ **−2**
17. Are $\frac{18}{4}$ and $\frac{9}{2}$ equivalent ratios? Explain your reasoning. **Yes;** $\frac{18 \div 2}{4 \div 2} = \frac{9}{2}$.
18. Find the value of x that makes $\frac{17}{3} = \frac{x}{9}$ a proportion. **51**

Convert each measurement as indicated.

19. 7200 pounds to tons **3.6 T**
20. 4.5 feet to inches **54 in.**
21. 2.1 quarts to pints **4.2 pt**
22. 460 meters to kilometers **0.46 km**
23. 13 grams to milligrams **13,000 mg**
24. 16 milliliters to liters **0.016 L**

Lesson 5-2 (*Pages 194–197*) On a map, the scale is 1.5 inches = 100 miles. Find the actual distance for each map distance.

	From	To	Map Distance	
1.	Paris, France	Brussels, Belgium	3.15 inches	**210 mi**
2.	Amsterdam, Holland	Frankfurt, Germany	4.5 inches	**300 mi**
3.	Luxembourg, Luxembourg	Frankfurt, Germany	$2\frac{1}{4}$ inches	**150 mi**
4.	Luxembourg, Luxembourg	Paris, France	$3\frac{3}{5}$ inches	**240 mi**

5. **Puzzles** The picture on a puzzle box shows the puzzle to be 4 inches tall. The actual height is 12 inches. What is the scale of the picture? **1 in. = 3 in.**

Lesson 5-3 (*Pages 198–203*) Express each fraction or ratio as a percent.

1. 11 to 44 **25%**
2. 36 out of 18 **200%**
3. 8 out of 32 **25%**
4. 12 to 24 **50%**
5. $\frac{12}{20}$ **60%**
6. $\frac{27}{9}$ **300%**
7. $\frac{10}{8}$ **125%**
8. $\frac{7}{21}$ $33\frac{1}{3}\%$
9. 4 to 5 **80%**
10. 3 out of 10 **30%**
11. 6 to 8 **75%**
12. 2 out of 20 **10%**
13. $\frac{18}{27}$ $66\frac{2}{3}\%$
14. 5 out of 25 **20%**
15. 3 to 8 $37\frac{1}{2}\%$
16. $\frac{13}{100}$ **13%**
17. 12 to 3 **400%**
18. 6 out of 12 **50%**
19. 10 to 5 **200%**
20. 5 out of 6 $83\frac{1}{3}\%$
21. One out of three people at the conference agreed with the speaker. $33\frac{1}{3}\%$
22. Two fifths of the students passed the test. **40%**

Use the percent proportion to find each number.

23. What number is 30% of 120? **36**
24. 125% of what number is 15? **12**
25. What percent of 21 is 7? $33\frac{1}{3}\%$
26. Find 10% of 12. **1.2**
27. 50 is what percent of 200? **25%**
28. 16 is 25% of what number? **64**

Extra Practice 701

Lesson 5-4 *(Pages 204–209)* **Use the percent equation to find each number.**

1. What number is 110% of 36? **39.6**
2. 30 is 300% of what number? **10**
3. Find 25% of 120. **30**
4. What number is 30% of 200? **60**
5. What number is 10% of 25? **2.5**
6. Find 200% of 18. **36**
7. 15 is 75% of what number? **20**
8. 16 is 40% of what number? **40**
9. Find 5% of 120. **6**
10. What number is 500% of 5? **25**
11. 35 is 20% of what number? **175**
12. Find 150% of 80. **120**
13. 45 is 80% of what number? **56.25**
14. What number is 18% of 324? **58.32**
15. **Banking** How long will it take Kristin to earn $180 if she invests $4000 at a rate of 6%? **9 months**
16. **Banking** How much interest will Michael earn if he invests $575 at a rate of 7% for 3 years? **$120.75**

Lesson 5-5 *(Pages 212–217)* **Find the percent of increase or decrease. Round to the nearest percent.**

1. original: 20 new: 22 **10% inc.**
2. original: 125 new: 100 **20% dec.**
3. original: 18 new: 9 **50% dec.**
4. original: 600 new: 750 **25% inc.**
5. original: 30 new: 10 **$66\frac{2}{3}$% dec.**
6. original: 28 new: 21 **25% dec.**
7. original: 12 new: 15 **25% inc.**
8. original: 50 new: 70 **40% inc.**

The cost of an item and a sales tax rate are given. Find the total price of each item to the nearest cent.

9. painting: $600; 6.5% **$639**
10. shoes: $85; 4% **$88.40**
11. book: $24.95; 3.5% **$25.82**
12. shirt: $29.99; 6% **$31.79**
13. piano: $1600; 5% **$1680**
14. sweater: $45.99; 7% **$49.21**

The original cost of an item and a discount rate are given. Find the sale price of each item to the nearest cent.

15. jacket: $120; 40% **$72**
16. hat: $23.99; 30% **$16.79**
17. basketball: $25; 10% **$22.50**
18. video game: $49.99; 5% **$47.49**
19. bicycle: $425; 15% **$361.25**
20. snow skis: $225; 25% **$168.75**
21. What number is 20% less than $75? **$60**
22. Find the percent of increase from $75 to $85. **$13\frac{1}{3}$%**

Lesson 5-6 *(Pages 219–223)* **Refer to the application on page 219. Find the probability of each outcome if a pair of dice are rolled.** **1–6. See margin.**

1. 2 odd numbers
2. a sum of 7
3. a sum of 4
4. even number on the first die
5. a sum greater than 11
6. a sum of less than 10

Find the odds of each outcome if the spinner at the right is spun.

7. greater than 10 **3:5**
8. blue **1:3**
9. blue or yellow **1:1**
10. not red **1:1**
11. an even number **8:0**
12. not a two **7:1**
13. an odd number **0:8**
14. less than 4 **1:7**
15. What is the probability that you select a seven at random from a standard deck of cards? **$\frac{1}{13}$**

EXTRA PRACTICE

Answers

Lesson 5–6

1. $\frac{1}{4}$
2. $\frac{1}{6}$
3. $\frac{1}{12}$
4. $\frac{1}{2}$
5. $\frac{1}{36}$
6. $\frac{5}{6}$

Lesson 5–7

1. $\frac{1}{100}$
2. $\frac{1}{100}$
3. $\frac{1}{20}$
4. $\frac{1}{4}$
5. $\frac{1}{20}$
6. $\frac{1}{4}$

Lesson 6–1

7. {(1, −0.5), (2, 0), (3, 1), (4, 3)}
8. {(−3, 4), (−2, 3), (−1, 2), (0, 1)}
9. {(1, −3), (2, −2), (−3, −1), (−1, 0)}

Lesson 6–2

5. {(−2, −12), (−1, −6), (0, 0), (1, 6), (2, 12), (3, 18)}

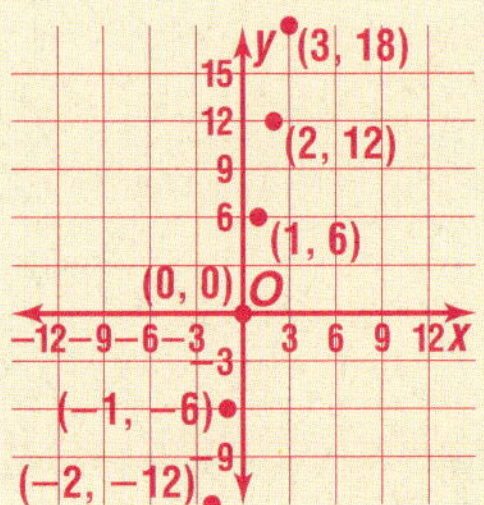

6. {(−2, 5), (−1, 3), (0, 1), (1, −1), (2, −3), (3, −5)}

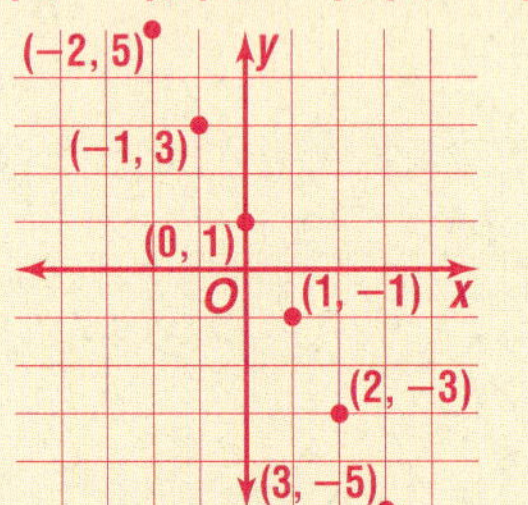

7. {(−2, −10), (−1, −9), (0, −8), (1, −7), (2, −6), (3, −5)}

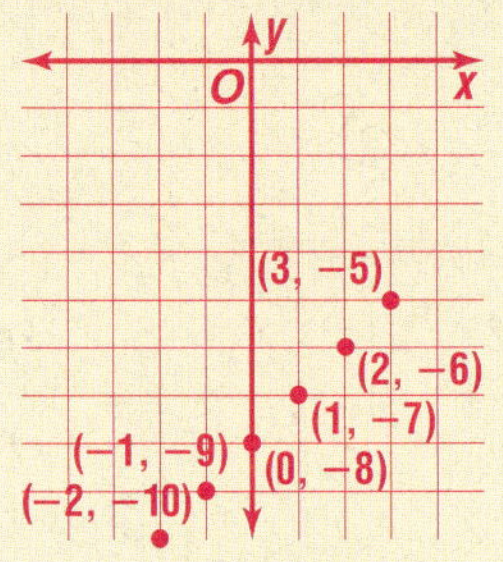

702 Extra Practice

Answers

8. {(−2, −7), (−1, −6), (0, −5), (1, −4), (2, −3), (3, −2)}

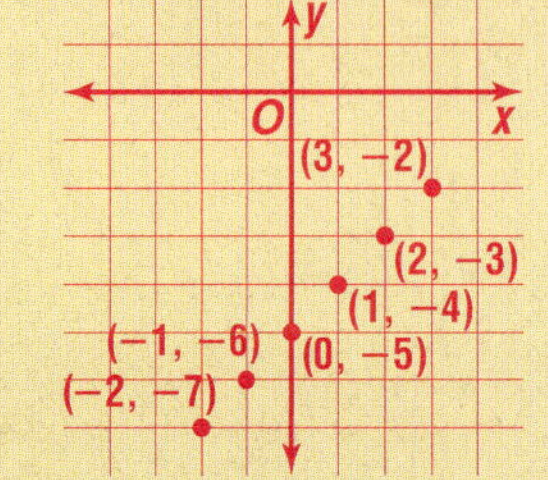

9. {(−2, 2), (−1, −1), (0, −4), (1, −7), (2, −10), (3, −13)}

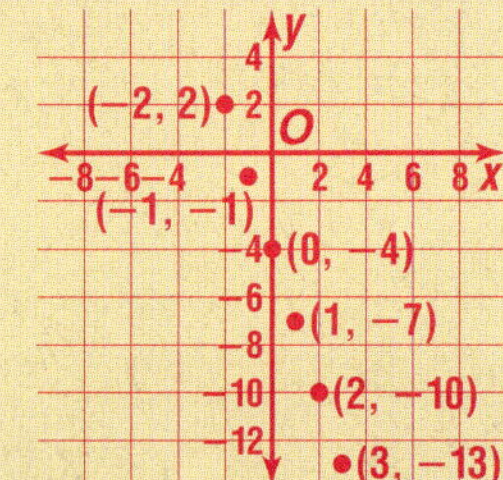

Lesson 5–7 *(Pages 224–229)* **A card is drawn from a deck of ten cards numbered 1 through 10. The card is replaced in the deck and another card is drawn. Find the probability of each outcome.** **1–6. See margin.**

1. P(9 and then a 7) 2. P(6 and a 4) 3. P(an even and then a 2)
4. P(two numbers less than 6) 5. P(8 and then an odd number) 6. P(two odd numbers)

7. What is the probability of tossing a coin three times and getting two heads and a tail? $\frac{3}{8}$
8. What is the probability of tossing a coin three times and getting a head, then a tail, then a head? $\frac{1}{8}$

Determine whether each event is *mutually exclusive* or *inclusive*. Then find each probability.

9. There are 18 cars in a lot. There are 6 red cars, 8 blue cars, 3 white cars, and 1 purple car. What is the probability of randomly choosing a white or blue car? M, $\frac{11}{18}$
10. In rolling a die, what is the probability that it is either an even or a four? I, $\frac{1}{2}$

Lesson 6–1 *(Pages 238–243)* **Express each relation as a table and as a graph. Then determine the domain and the range.** **1–6. See Solutions Manual.**

1. $\{(6, 3), (2, 4), (3, 2), (5, 5)\}$
2. $\{(5.9, -3), (-2, 3.1), (0, 0), (-1, 7)\}$
3. $\{(-1, 6), (2, -5), (-2, 9), (0, 4)\}$
4. $\{(1, -3), (2.4, 6), (0, -5.1), (-4, 4)\}$
5. $\left\{\left(\frac{1}{4}, \frac{1}{6}\right), \left(-\frac{2}{3}, \frac{4}{5}\right), \left(-1, \frac{1}{3}\right)\right\}$
6. $\left\{(3, 1), \left(\frac{3}{4}, 2\right), (2, 0), \left(-\frac{2}{3}, -3\right)\right\}$

Express each relation as a set of ordered pairs. **7–9. See margin.**

7.

x	y
1	−0.5
2	0
3	1
4	3

8.

x	y
−3	4
−2	3
−1	2
0	1

9.

x	y
1	−3
2	−2
−3	−1
−1	0

Lesson 6–2 *(Pages 244–249)* **Which ordered pairs are solutions of each equation?**

1. $6x + 2y = 12$ **a and c** a. (1, 3) b. (1, 1) c. (2, 0) d. (2, 2)
2. $3a + 6 = 7b$ **d** a. (4, 3) b. (1, 0) c. (2, 3) d. (5, 3)
3. $2m + n = 7$ **b** a. (1, 6) b. (3, 1) c. (0, 2) d. (0, 6)
4. $3r - 15 = -6s$ **a and b** a. $\left(\frac{1}{3}, \frac{7}{3}\right)$ b. (1, 2) c. (2, 1) d. $\left(\frac{1}{3}, \frac{1}{2}\right)$

Solve each equation if the domain is $\{-2, -1, 0, 1, 2, 3\}$. Graph the solution set. **5–12. See margin.**

5. $y = 6x$ 6. $y = -2x + 1$ 7. $y = x - 8$ 8. $x - y = 5$
9. $y = -3x - 4$ 10. $3x - 2y = 6$ 11. $4 - x = y$ 12. $7 = 6x + 4y$

Find the domain of each equation if the range is $\{-1, 0, 1, 2\}$. **13–20. See margin.**

13. $y = 3x + 3$ 14. $7 - 4x = y$ 15. $6y = 2x$ 16. $x = 5y - 2$
17. $2y = 3x + 1$ 18. $5x - 15 = y$ 19. $2x = y - 4$ 20. $2y - 2x = 4$

Answers

Lesson 6–2 (continued)

10. $\left\{(-2, -6), \left(-1, -4\frac{1}{2}\right), (0, -3), \left(1, -1\frac{1}{2}\right), (2, 0), \left(3, 1\frac{1}{2}\right)\right\}$

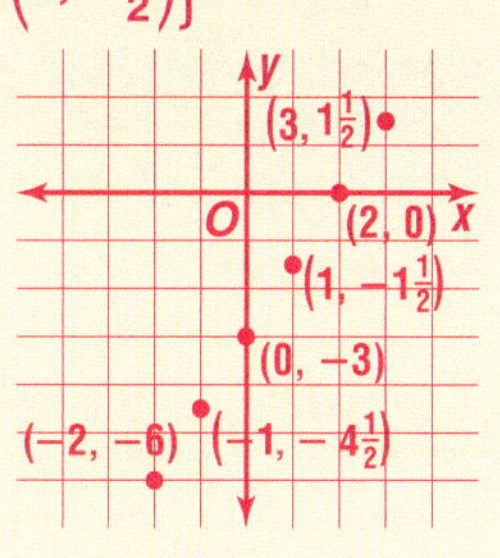

11. $\{(-2, 6), (-1, 5), (0, 4), (1, 3), (2, 2), (3, 1)\}$

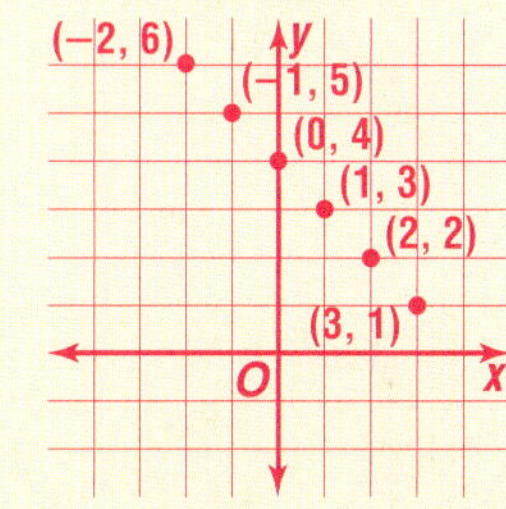

12. $\left\{\left(-2, 4\frac{3}{4}\right), \left(-1, 3\frac{1}{4}\right), \left(0, 1\frac{3}{4}\right), \left(1, \frac{1}{4}\right), \left(2, -1\frac{1}{4}\right), \left(3, -2\frac{3}{4}\right)\right\}$

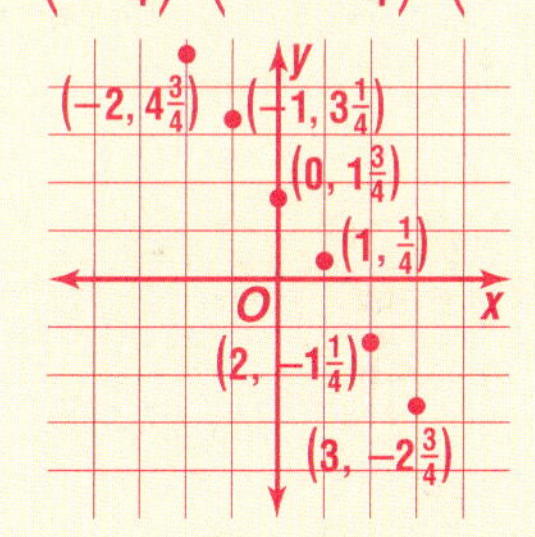

EXTRA PRACTICE

Answers

13. $\left\{\left(-1\frac{1}{3}, -1\right), (-1, 0), \left(-\frac{2}{3}, 1\right), \left(-\frac{1}{3}, 2\right)\right\}$
14. $\left\{(2, -1), \left(1\frac{3}{4}, 0\right), \left(1\frac{1}{2}, 1\right), \left(1\frac{1}{4}, 2\right)\right\}$
15. $\{(-3, -1), (0, 0), (3, 1), (6, 2)\}$
16. $\{(-7, -1), (-2, 0), (3, 1), (8, 2)\}$
17. $\left\{(-1, -1), \left(-\frac{1}{3}, 0\right), \left(\frac{1}{3}, 1\right), (1, 2)\right\}$
18. $\left\{\left(2\frac{4}{5}, -1\right), (3, 0), \left(3\frac{1}{5}, 1\right), \left(3\frac{2}{5}, 2\right)\right\}$
19. $\left\{\left(-2\frac{1}{2}, -1\right), (-2, 0), \left(-1\frac{1}{2}, 1\right), (-1, 2)\right\}$
20. $\{(-3, -1), (-2, 0), (-1, 1), (0, 2)\}$

Answers

Lesson 6–3

1. no
2. yes; $A = 7, B = -2, C = 3$
3. no
4. yes; $A = -4, B = 16, C = 0$
5. yes; $A = 1, B = -2, C = 0$
6. yes; $A = -1, B = 3, C = 4$
7. yes; $A = 1, B = -1, C = -2$
8. yes; $A = 1, B = 3, C = 9$
9. yes; $A = 6, B = -1, C = -3$
10. no
11. yes; $A = 8, B = 4, C = 0$
12. no

13.

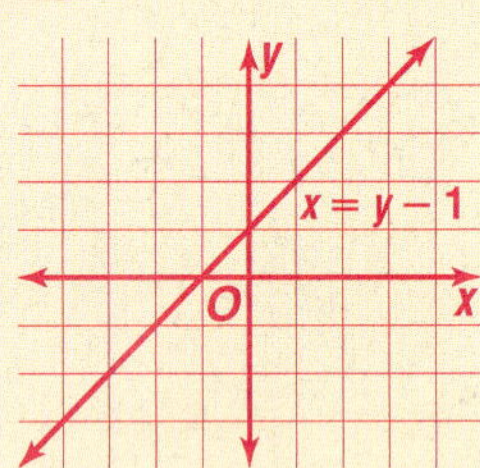

14.

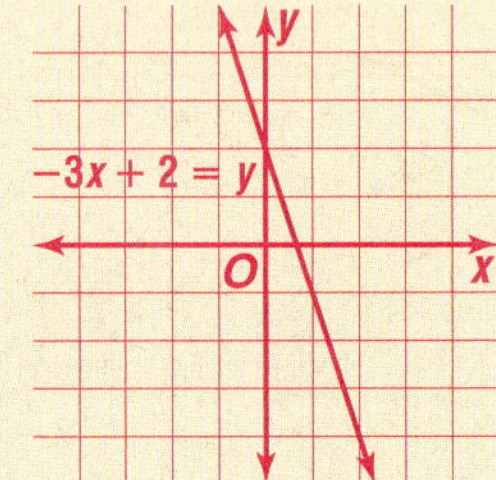

15.

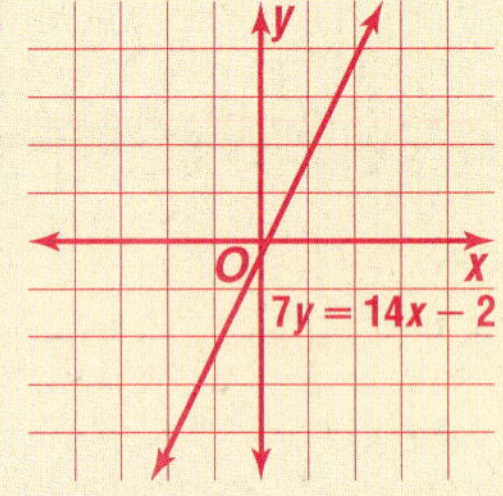

16.

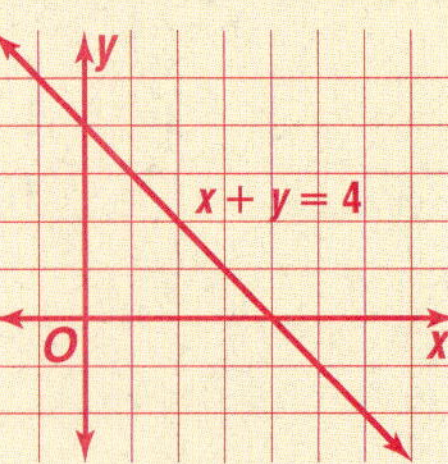

17.

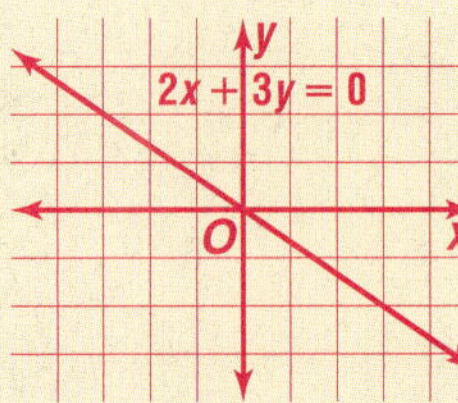

18.

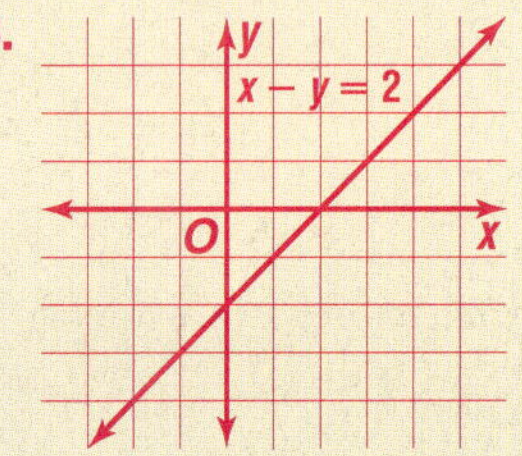

EXTRA PRACTICE

Lesson 6–3 *(Pages 250–255)* **Determine whether each equation is a linear equation. Explain. If an equation is linear, identify *A*, *B*, and *C*.** 1–12. See margin.

1. $4xy = 3$	2. $7x - 2y = 3$	3. $2xy + x = y$	4. $16y = 4x$
5. $x = 2y$	6. $3y = x + 4$	7. $y - 2 = x$	8. $x + 3y = 9$
9. $6x + 3 = y$	10. $2x^2 + y^2 = 8$	11. $4y + 8x = 0$	12. $y + xy = 1$

Graph each equation. 13–20. See margin.

13. $x = y - 1$	14. $-3x + 2 = y$	15. $7y = 14x - 2$	16. $x + y = 4$
17. $2x + 3y = 0$	18. $x - y = 2$	19. $16 = 2x + 4y$	20. $5x - y = 15$

Lesson 6–4 *(Pages 256–261)* **Determine whether each relation is a function.**

1. $\{(3, 6), (4, -2), (-1, -3), (0, 6)\}$ yes
2. $\{(-1, 8), (0, 3), (3, 3), (-2, 2)\}$ yes
3. $\{(-1, 5), (3, 1), (5, 4), (-1, 3)\}$ no
4. $\{(-2, 4), (0, 0), (4, 2), (3, 0)\}$ yes
5. $\{(0, 3), (1, 1), (1, -1), (2, 4)\}$ no
6. $\{(3, 3), (-3, 1), (4, 2), (-2, 3)\}$ yes

7. no

x	y
−4	2
−3	3
−2	2
−4	3
−5	4

8. yes

x	y
1	1
3	3
−1	1
2	2
0	0

9. no

x	y
−1	−3
2	−2
−3	−3
0	−1
−1	0

Use the vertical line test to determine whether each relation is a function.

10. yes

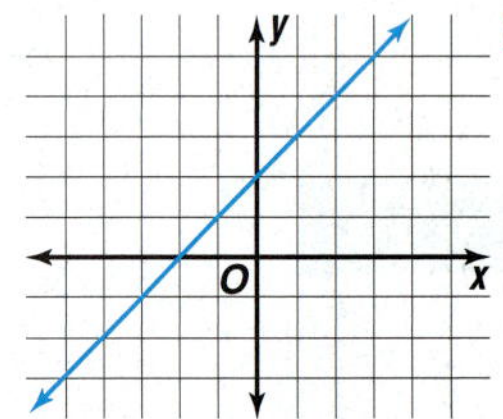

11. yes

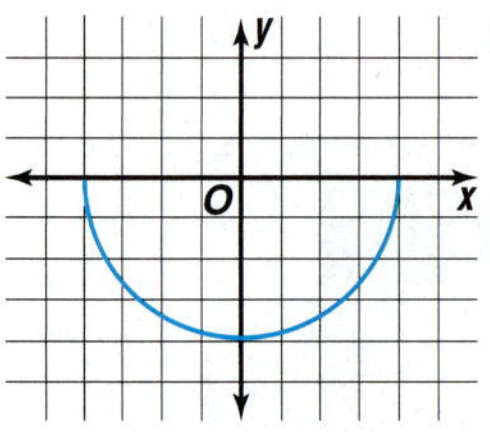

12. no

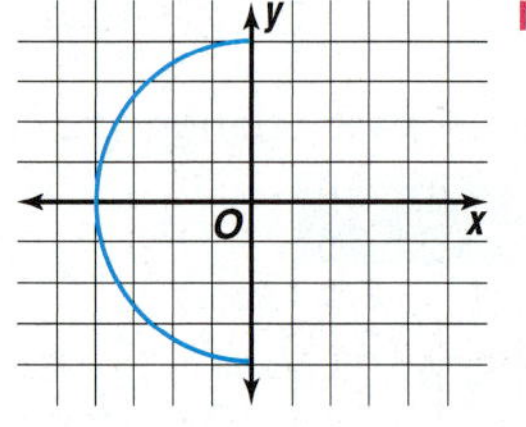

If $f(x) = -3x - 1$ and $g(x) = 4x + 5$, find each value. 19. −2.5 21. 8.3

13. $f(3)$ −10	14. $g(2)$ 13	15. $g(-1)$ 1	16. $f(-4)$ 11	17. $g(-1.25)$ 0	18. $f(5)$ −16
19. $f(0.5)$	20. $g(0.2)$ 5.8	21. $f(-3.1)$	22. $g(-1.5)$ −1	23. $f(1.2)$ −4.6	24. $g(2.1)$ 13.4
25. $f\left(\frac{1}{8}\right)$ $-1\frac{3}{8}$	26. $g\left(\frac{2}{3}\right)$ $7\frac{2}{3}$	27. $f\left(\frac{1}{2}\right)$ $-2\frac{1}{2}$	28. $g\left(-\frac{3}{5}\right)$ $\frac{13}{5}$	29. $f\left(-\frac{5}{8}\right)$ $\frac{7}{8}$	30. $g\left(\frac{1}{4}\right)$ 6

Lesson 6–5 *(Pages 264–269)* **Determine whether each equation is a direct variation. Verify the answer with a graph.** 1–6. See margin for graphs. 1. no 3. no 4. no

1. $x = y + 2$ 2. $y = 3x$ yes 3. $y = 4x + 2$ 4. $y = 2x + 2$ 5. $y = 2x$ yes 6. $x = -2$ no

Solve. Assume that *y* varies directly as *x*.

7. If $y = -3$ when $x = -6$, find x when $y = 16$. 32
8. If $y = 9$ when $x = 5$, find y when $x = 10$. 18
9. Find x when $y = 24$ if $y = 16$ when $x = 6$. 9
10. Find x when $y = 3$ if $y = 15$ when $x = 4$. $\frac{4}{5}$
11. If $x = 13$ when $y = 7$, find y when $x = 26$. 14
12. Find y when $x = 2.5$ if $y = 10$ when $x = 6$. $4\frac{1}{6}$

Answers

19.

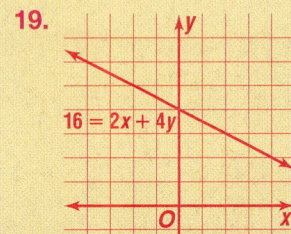

16 = 2x + 4y

20.

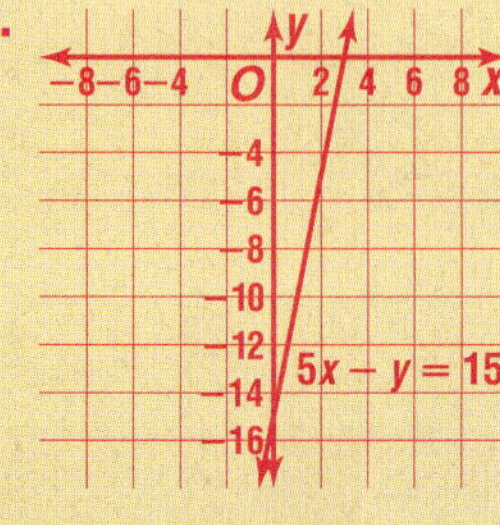

Lesson 6–5

1.

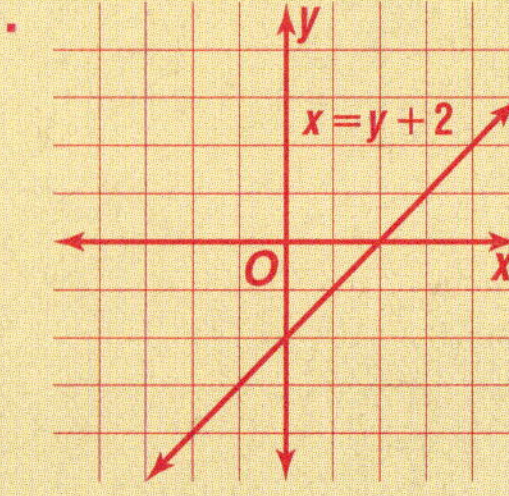

Lesson 6–6 *(Pages 270–275)* **Solve. Assume that y varies inversely as x.**

1. Suppose $y = 14$ when $x = 7$. Find x when $y = 18$. $5\frac{4}{9}$
2. Find y when $x = 8$ if $y = 15$ when $x = 12$. 22.5
3. If $y = 2.5$ when $x = 6.5$, find y when $x = 13$. 1.25
4. Suppose $y = \frac{5}{8}$ when $x = \frac{1}{3}$. Find y when $x = -\frac{1}{2}$. $-\frac{5}{12}$

Find the constant of variation. Then write an equation for each statement.

5. y varies inversely as x, and $y = 7$ when $x = 4$. 28; $xy = 28$
6. y varies directly as x, and $y = 8$ when $x = -1$. -8; $y = -8x$
7. y varies inversely as x, and $y = -3.5$ when $x = 7$. -24.5; $xy = -24.5$
8. y varies directly as x, and $y = 4$ when $x = -12$. $-\frac{1}{3}$; $y = -\frac{1}{3}x$

Lesson 7–1 *(Pages 284–289)* **Determine the slope of each line.**

1. 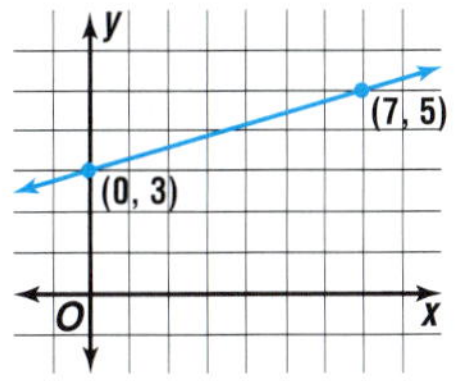

$\frac{2}{7}$

2. 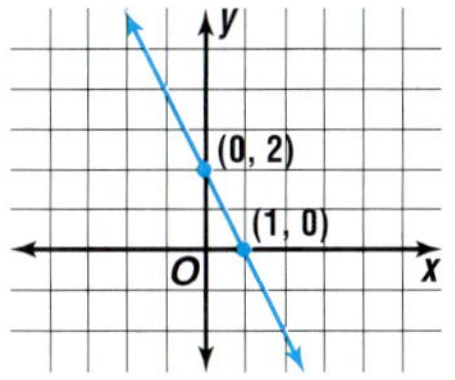

-2

3. 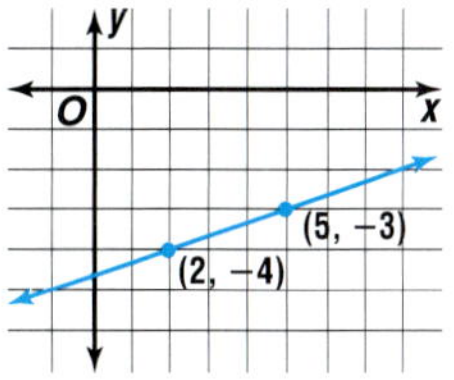

$\frac{1}{3}$

Determine the slope of the line passing through the points whose coordinates are listed in each table.

4. 1

x	y
−4	2
−2	4
0	6
2	8

5. −6

x	y
−1	12
0	6
1	0
2	−6

6. $\frac{1}{5}$

x	y
0	1
5	2
10	3
15	4

Lesson 7–2 *(Pages 290–295)* **Write the point-slope form of an equation for each line passing through the given point and having the given slope.** 1–12. See margin.

1. $(6, 4)$, $m = -3$
2. $(-7, 3)$, $m = 2$
3. $(2, 2)$, $m = -1$
4. $(3, -2)$, $m = 6$
5. $(1, 5)$, $m = -5$
6. $(1, 2)$, $m = -4$
7. $(-7, 2)$, $m =$ none
8. $(-3, -3)$, $m = \frac{1}{2}$
9. $(-4, -2)$, $m = \frac{4}{5}$
10. $(4, 0)$, $m = \frac{1}{3}$
11. $\left(-\frac{1}{2}, -1\right)$, $m = 4$
12. $\left(\frac{3}{4}, 4\right)$, $m = -2$

Write the point-slope form of an equation for each line. 13–18. See margin.

13. the line through points at $(8, 4)$ and $(7, 6)$
14. the line through points at $(0, -2)$ and $(-4, 2)$
15. the line through points at $(4, 7)$and $(-1, 0)$
16. the line through points at $(2, 1)$ and $(1, 2)$
17. the line through points at $(4, -8)$ and $(6, -2)$
18. the line through points at $(-3, 1)$ and $(2, -3)$
19. Write an equation in point-slope form of a line that has a slope of 1.5 and passes through the point $(16, -5)$. $y + 5 = 1.5(x - 16)$

Answers

Page 704

Lesson 6–5 (continued)

2. 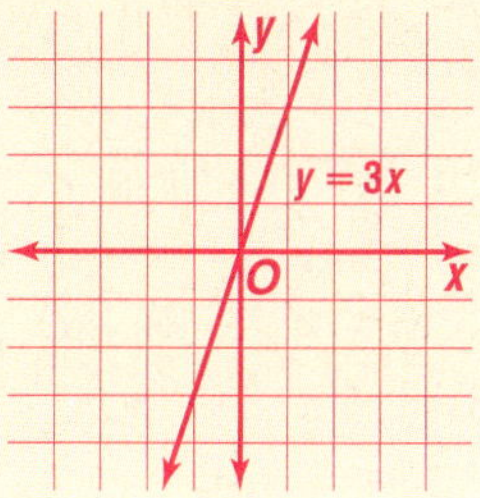

3. 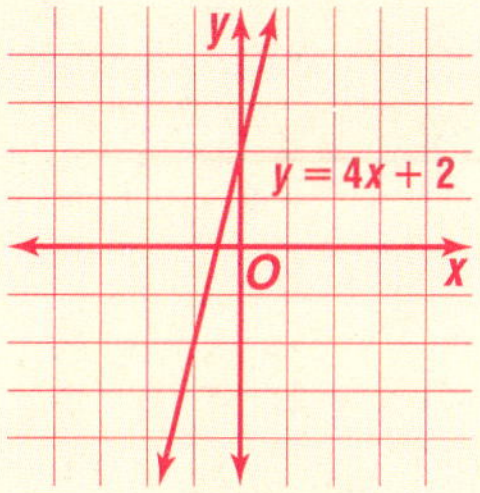

4. 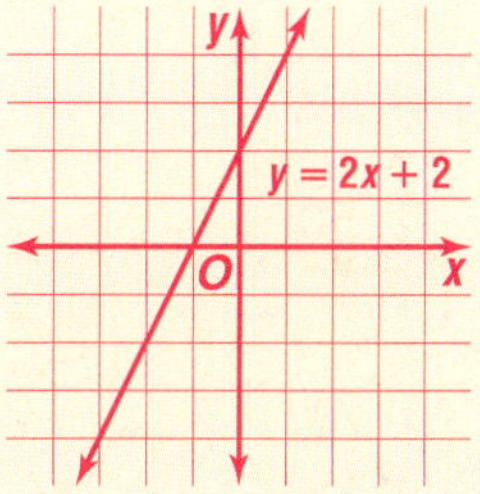

5. 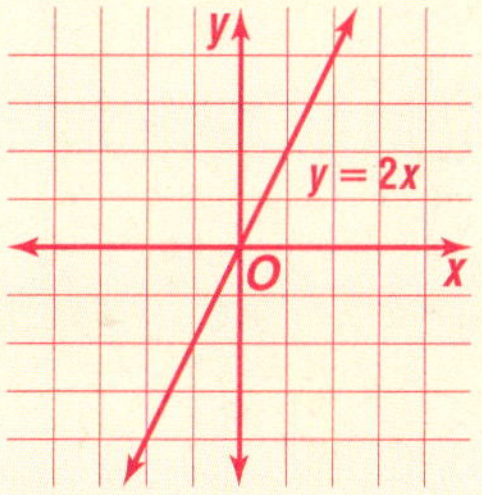

6. 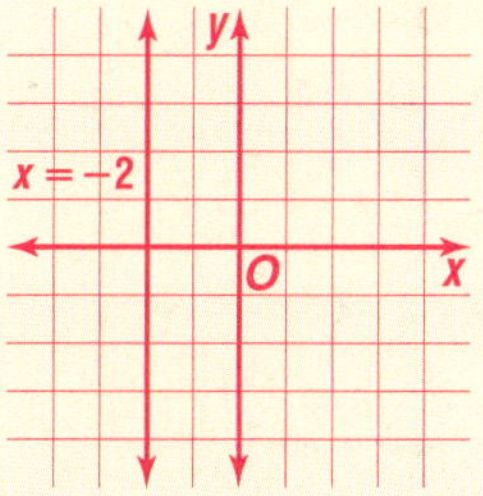

Lesson 7–2

1. $y - 4 = -3(x - 6)$
2. $y - 3 = 2(x + 7)$
3. $y - 2 = -1(x - 2)$
4. $y + 2 = 6(x - 3)$
5. $y - 5 = -5(x - 1)$
6. $y - 2 = -4(x - 1)$
7. $x = -7$
8. $y + 3 = \frac{1}{2}(x + 3)$
9. $y + 2 = \frac{4}{5}(x + 4)$

Answers

10. $y - 0 = \frac{1}{3}(x - 4)$
11. $y + 1 = 4\left(x + \frac{1}{2}\right)$
12. $y - 4 = -2\left(x - \frac{3}{4}\right)$
13. $y - 4 = -2(x - 8)$ or $y - 6 = -2(x - 7)$
14. $y + 2 = -1(x - 0)$ or $y + 4 = -1(x - 2)$
15. $y - 7 = \frac{7}{5}(x - 4)$ or $y - 0 = \frac{7}{5}(x + 1)$
16. $y - 1 = -1(x - 2)$ or $y - 2 = -1(x - 1)$
17. $y + 8 = 3(x - 4)$ or $y + 2 = 3(x - 6)$
18. $y - 1 = \frac{4}{5}(x + 3)$ or $y + 3 = \frac{4}{5}(x - 2)$

Extra Practice 705

Answers

Lesson 7-3

1. $y = -3x + 8$
2. $y = 2x - 3$
3. $y = -x - 6$
4. $y = 3x + 2$
5. $y = 4$
6. $y = -2x + 1$
7. $y = 6x + 4$
8. $y = -1.4x + 7$
9. $y = \frac{5}{7}x - 7$
10. $y = -\frac{2}{3}x + 9$
11. $y = -\frac{4}{3}x - 3$
12. $y = -\frac{1}{4}x + 6$
13. $y = 4x - 6$
14. $y = 3x + 12$
15. $y = -2x - 6$
16. $y = -4x - 2$
17. $y = -x + 4$
18. $y = 8$
19. $y = 6x - 78$
20. $y = 5x + 43$
21. $y = -\frac{4}{5}x + \frac{42}{5}$
22. $y = \frac{3}{8}x + \frac{11}{8}$
23. $y = \frac{1}{3}x + 5$
24. $y = -\frac{2}{3}x - 3$
25. $y = -3x + 10$
26. $y = -\frac{1}{3}x + 1$
27. $y = \frac{2}{3}x - \frac{1}{3}$
28. $y = -6x + 34$
29. $y = -\frac{7}{11}x + \frac{19}{11}$
30. $y = 2x - 4$
31. $y = -11x + 58$
32. $y = -\frac{2}{5}x + \frac{6}{5}$
33. $y = -x + 11$
34. $y = -\frac{1}{5}x - \frac{12}{5}$
35. $y = \frac{1}{4}x$
36. $y = 2x - 10$

EXTRA PRACTICE

Lesson 7-3 *(Pages 296–301)* **Write an equation in slope-intercept form of the line with each slope and y-intercept. 1–12. See margin.**

1. $m = -3, b = 8$
2. $m = 2, b = -3$
3. $m = -1, b = -6$
4. $m = 3, b = 2$
5. $m = 0, b = 4$
6. $m = -2, b = 1$
7. $m = 6, b = 4$
8. $m = -1.4, b = 7$
9. $m = \frac{5}{7}, b = -7$
10. $m = -\frac{2}{3}, b = 9$
11. $m = -\frac{4}{3}, b = -3$
12. $m = -\frac{1}{4}, b = 6$

Write an equation in slope-intercept form of the line having the given slope and passing through the given point. 13–24. See margin.

13. $m = 4, (1, -2)$
14. $m = 3, (-5, -3)$
15. $m = -2, (-3, 0)$
16. $m = -4, (-2, 6)$
17. $m = -1, (0, 4)$
18. $m = 0, (7, 8)$
19. $m = 6, (12, -6)$
20. $m = 5, (-8, 3)$
21. $m = -\frac{4}{5}, (3, 6)$
22. $m = \frac{3}{8}, (-1, 1)$
23. $m = \frac{1}{3}, (0, 5)$
24. $m = -\frac{2}{3}, (-6, 1)$

Write an equation in slope-intercept form of the line passing through each pair of points. 25–36. See margin.

25. (2, 4) and (3, 1)
26. (0, 1) and (3, 0)
27. (−4, −3) and (2, 1)
28. (5, 4) and (7, −8)
29. (−2, 3) and (9, −4)
30. (8, 12) and (−2, −8)
31. (5, 3) and (6, −8)
32. (3, 0) and (8, −2)
33. (7, 4) and (4, 7)
34. (−2, −2) and (3, −3)
35. (8, 2) and (−4, −1)
36. (6, 2) and (3, −4)

Lesson 7-4 *(Pages 302–307)* **Determine whether each scatter plot has a *positive* relationship, *negative* relationship, or *no* relationship. If there is a relationship, describe it.** **1. no relationship 2. positive 3. negative**

1.
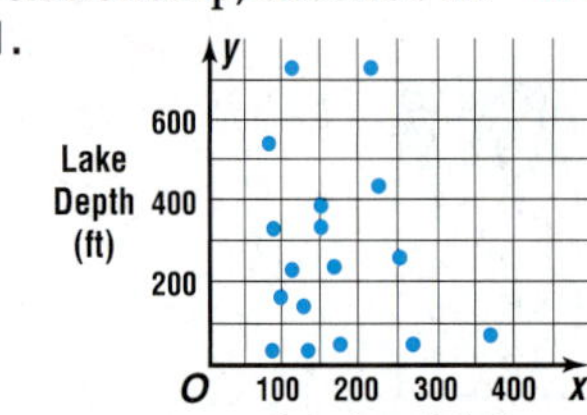

2.
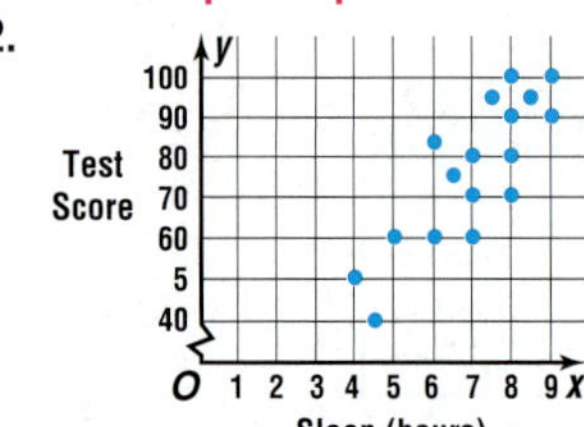

3.
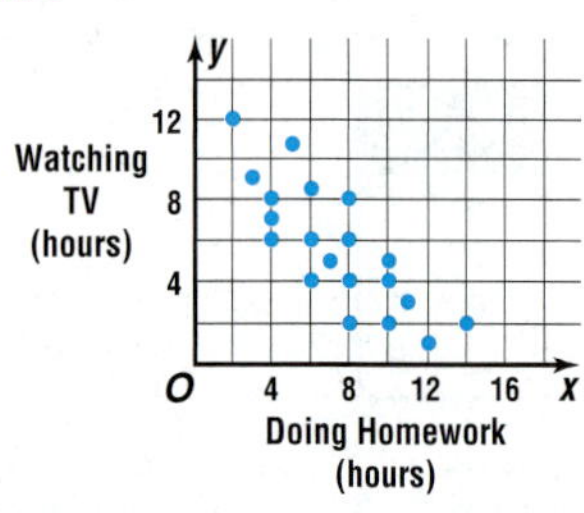

Determine whether a scatter plot of the data for the following would show *positive*, *negative*, or *no* relationship between the variables.

4. study time and score on a test **positive**
5. your shoe size and your age **no relationship**

Lesson 7-5 *(Pages 310–315)* **Determine the x-intercept and y-intercept of the graph of each equation. Then graph the equation. 1–12. See Solutions Manual for graphs.**

1. $2x + 8y = 16$ **8, 2**
2. $x + 2y = 2$ **2, 1**
3. $x + y = 3$ **3, 3**
4. $2x - 9y = 18$ **9, −2**
5. $6x - y = 4$ **$\frac{2}{3}, -4$**
6. $7y - x = -3$ **$3, -\frac{3}{7}$**
7. $6x + y = -2$ **$-\frac{1}{3}, -2$**
8. $4y + 3x = -4$ **$-\frac{4}{3}, -1$**
9. $x + \frac{2}{3}y = 6$ **6, 9**
10. $\frac{1}{2}x + 4y = 1$ **$2, \frac{1}{4}$**
11. $\frac{2}{3}x + \frac{1}{3}y = \frac{4}{3}$ **2, 4**
12. $3y - 2x = -4$ **$2, -\frac{4}{3}$**

Determine the slope and y-intercept of the graph of each equation. Then graph the equation. 13–24. See margin for graphs. **17. $-\frac{1}{2}, -3$ 21. 2, −4 22. $-\frac{7}{8}, -7$**

13. $y = 8 - x$ **−1, 8**
14. $y = 5x - 2$ **5, −2**
15. $y = 3x + 6$ **3, 6**
16. $y = 4x + 8$ **4, 8**
17. $y = -\frac{1}{2}x - 3$
18. $y = \frac{4}{3}x - 2$ **$\frac{4}{3}, -2$**
19. $y = \frac{1}{6}x + 5$ **$\frac{1}{6}, 5$**
20. $y = -\frac{3}{4}x + 6$ **$-\frac{3}{4}, 6$**
21. $-2y + 4x = 8$
22. $7x + 8y = -56$
23. $9y + 3x = 6$ **$-\frac{1}{3}, \frac{2}{3}$**
24. $-5x + y = 6$ **5, 6**

706 Extra Practice

Lesson 7–6 *(Pages 316–321)* Graph each pair of equations. Describe any similarities or differences and explain why they are a family of graphs. **1–8. See margin.**

1. $y = 3x + 2$; $y = 3x$
2. $y = -2x + 1$; $y = 2x + 1$
3. $y = 3x$; $y = 4x$
4. $y = x + 4$; $y = 5x + 4$
5. $y = -4x + 1$; $y = -4x - 2$
6. $-\frac{3}{4}x + 2 = y$; $-x + 2 = y$
7. $y = \frac{1}{3}x$; $y = 3x$
8. $y = \frac{1}{2}x + \frac{1}{2}$; $y = 2x + 2$

9–16. See margin.
Compare and contrast the graphs of each pair of equations. Verify by graphing the equations.

9. $y = 2x$; $y = 2x + 1$
10. $y = 0.25x + 1$; $y = 4x + 1$
11. $y = -3x - 2$; $y = 3x - 2$
12. $y = x - 4$; $y = x + 6$
13. $y = \frac{1}{2}x + 2$; $y = x + 2$
14. $y = -\frac{2}{3}x + 2$; $y = -\frac{2}{3}x$
15. $y = -5x + 4$; $y = -\frac{1}{2}x + 4$
16. $y = \frac{1}{3}x$; $5 + x = 3y$

17–20. Sample answers are given.
Change $y = 0.5x - 5$ so that the graph of the new equation fits each description.

17. y-intercept is 2, same slope **$y = 0.5x + 2$**
18. negative slope, y-intercept is -3 **$y = -0.5x - 3$**
19. shifted up 2 units, slope is 2 **$y = 2x - 3$**
20. less steep positive slope, same y-intercept **$y = 0.4x - 5$**

Lesson 7–7 *(Pages 322–327)* Determine whether the graphs of each pair of equations are *parallel*, *perpendicular*, or *neither*.

1. $y = 3x + 2$; $y = -\frac{1}{3}x + 5$ **perp.**
2. $y = 3x + 4$; $y = 3x + 2$ **parallel**
3. $y = 3x + 2$; $y = \frac{1}{5}x - \frac{7}{5}$ **neither**
4. $y = 2x + 4$; $3y = 6x + 6$ **parallel**
5. $y = \frac{1}{2}x + 4$; $4x - 3y = 12$ **neither**
6. $2y = 10x - \frac{2}{5}$; $\frac{1}{5}x + y = 3$ **perp.**
7. $y = \frac{2}{3}x - 6$; $2x - 4 = 3y$ **parallel**
8. $2x + y = 6$; $y = \frac{1}{2}x + 4$ **perp.**

Write an equation in slope-intercept form of the line that is parallel to the graph of each equation and passes through the given point.

9. $y = 8x + 5$; $(0, 4)$ **$y = 8x + 4$**
10. $y = 3$; $(-1, 2)$ **$y = 2$**
11. $x = 2$; $(4, 3)$ **$x = 4$**

Write an equation in slope-intercept form of the line that is perpendicular to the graph of each equation and passes through the given point.

12. $y = x + 5$; $(-2, 1)$ **$y = -x - 1$**
13. $y = 4x$; $(0, 0)$ **$y = -\frac{1}{4}x$**
14. $6x - 2y = 3$; $(0, 1)$ **$y = -\frac{1}{3}x + 1$**

Lesson 8–1 *(Pages 336–340)* Write each expression using exponents. **3. $4^2 5^3 6$**

1. $7 \cdot 7 \cdot 7$ **7^3**
2. $(-3)(-3)(-3)$ **$(-3)^3$**
3. $5 \cdot 5 \cdot 5 \cdot 6 \cdot 4 \cdot 4$
4. 6 squared **6^2**
5. $m \cdot m \cdot n \cdot n \cdot n$ **m^2n^3**
6. 2 cubed **2^3**
7. $4 \cdot r \cdot r \cdot r \cdot s \cdot s$ **$4r^3s^2$**
8. $(-2)(k)(k)(j)(j)$ **$-2k^2j^2$**

Write each power as a multiplication expression. **10. $(-4)(-4)(-4)(-4)$ 11. $8 \cdot 8 \cdot 8 \cdot 8 \cdot 2 \cdot 2 \cdot 2$**

9. 2^3 **$2 \cdot 2 \cdot 2$**
10. $(-4)^4$
11. $8^4 2^3$
12. k^5 **$k \cdot k \cdot k \cdot k \cdot k$**
13. w^2z^2 **$w \cdot w \cdot z \cdot z$**
14. $-6xy^2$ **$-6 \cdot x \cdot y \cdot y$**

Evaluate each expression if $x = 4$, $y = -1$, $z = -2$, and $w = 1.5$.

15. y^3 **-1**
16. $w(yz + 4)$ **9**
17. $z^3 + 2xy$ **-16**
18. $2x^2 - z^5$ **64**
19. $-2(y^3 + w)$ **-1**
20. wxy **-6**

Answers

Lesson 7–6

1.

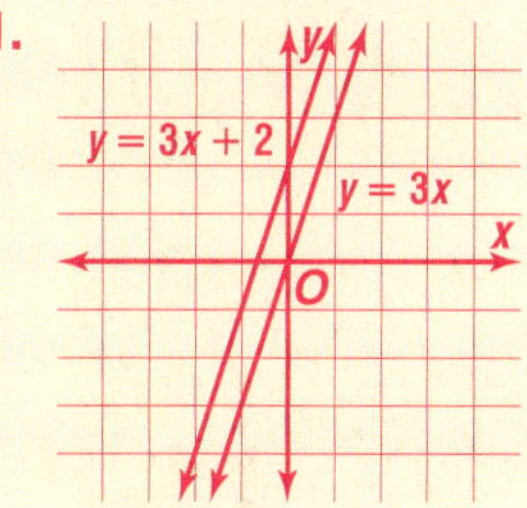

same slope, different y-intercepts

2.

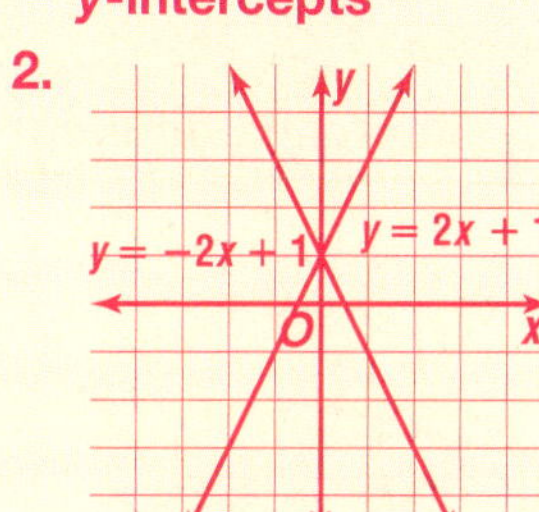

same y-intercept, different slopes

3.

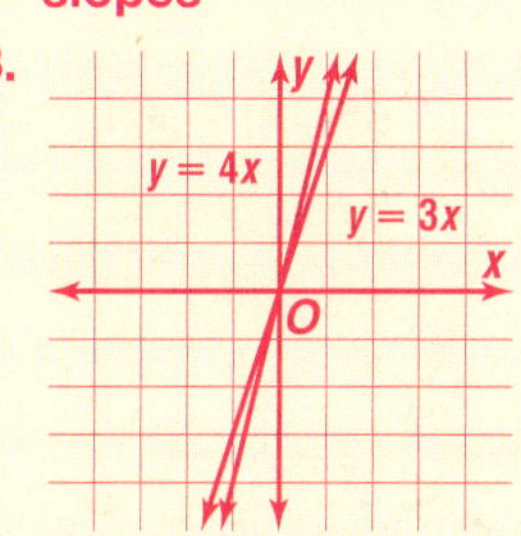

same x- and y-intercepts, different slopes

4.

$y = x + 4$

$y = 5x + 4$

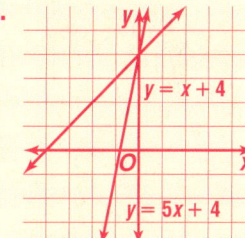

same y-intercept, different slopes

5.

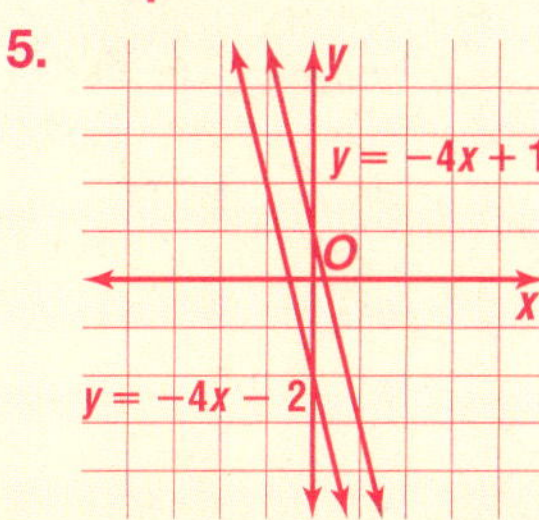

same slope, different y-intercepts

EXTRA PRACTICE

Answers

6.

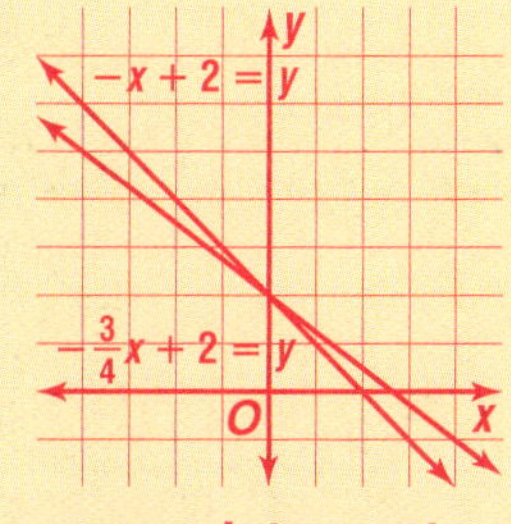

same y-intercept, different slopes

7.

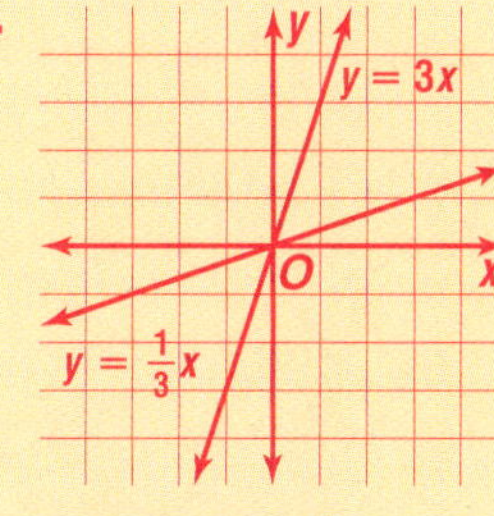

same x- and y-intercepts, different slopes

8.

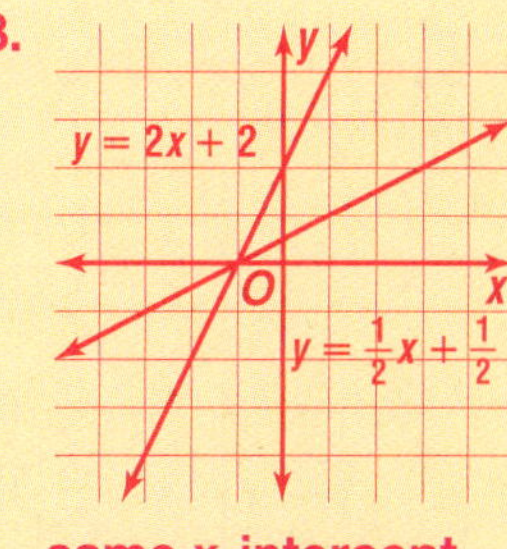

same x-intercept, different slopes

Answers

Page 707

Lesson 7–6 (continued)

9. same slope, different y-intercept;

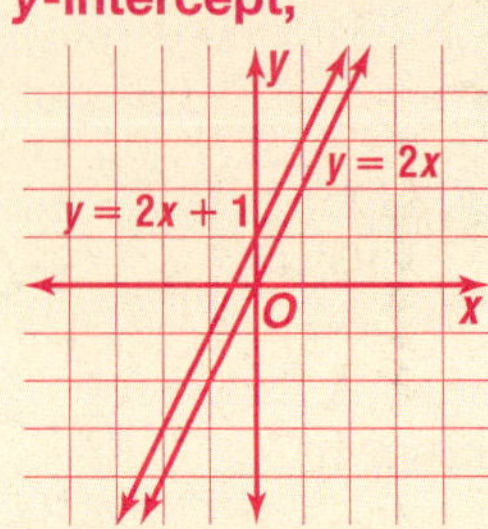

10. same y-intercept, different slopes;

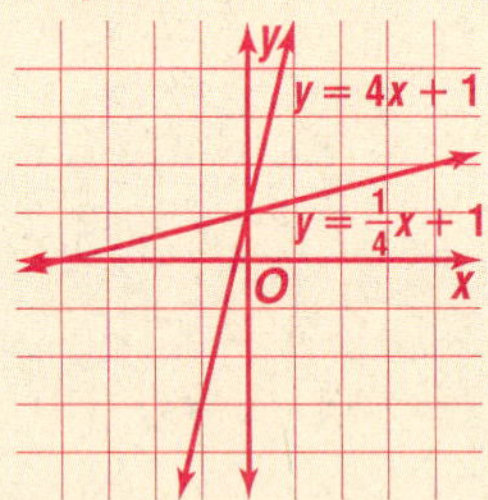

11. same y-intercept, different slopes;

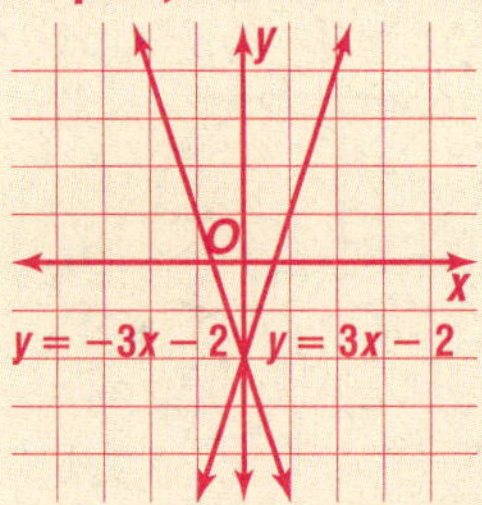

12. same slope, different y-intercepts;

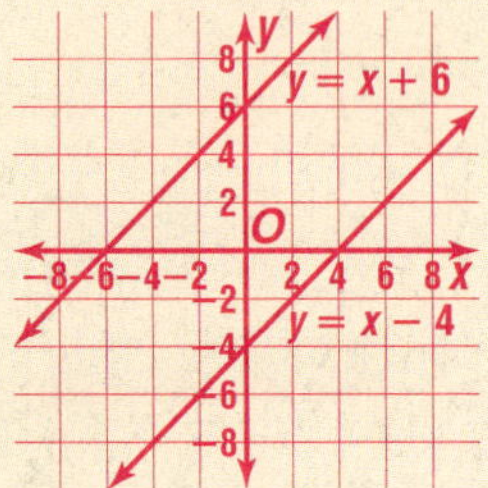

13. same y-intercept; different slopes;

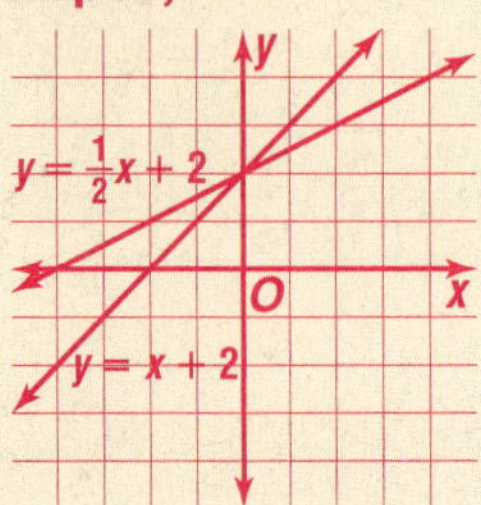

Lesson 8–2 *(Pages 341–346)* **Simplify each expression.** **13. $3p^6q^9$** **14. $-12x^3y^4z$**

1. $2^2 \cdot 2^5$ **2^7**
2. $6^3 \cdot 6^4$ **6^7**
3. $x^4 \cdot x^4$ **x^8**
4. $y \cdot y^5$ **y^6**
5. $(a^3)(a^4)$ **a^7**
6. $(g^6h^3)(g^4h^3)$ **$g^{10}h^6$**
7. $(r^2t^2)(rt^3)$ **r^3t^5**
8. $(2k^2j)(-3kj)$ **$-6k^3j^2$**
9. $(6m^3)(7m^2)$ **$42m^5$**
10. $(2a^2b)(9ab^5)$ **$18a^3b^6$**
11. $(6a^3b)(4ac)$ **$24a^4bc$**
12. $(8y^4)(3y^4)$ **$24y^8$**
13. $(-3p^5q^2)(-pq^7)$
14. $(3xyz)(-4x^2y^3)$
15. $\frac{4^5}{4^2}$ **4^3**
16. $\frac{8^8}{8^7}$ **8**
17. $\frac{15m^{12}}{3m^5}$ **$5m^7$**
18. $\frac{-4r^4s^5}{-rs^4}$ **$4r^3s$**
19. $\frac{-16m^5n^9p^{10}}{2m^2n^3p^6}$ **$-8m^3n^6p^4$**
20. $\left(\frac{2}{3}x^5y\right)(-12x^3y^2)$ **$-8x^8y^3$**
21. Evaluate m^0. **1**
22. Find the product of mn and $-3m^2n$. **$-3m^3n^2$**

1. $\frac{1}{8^3} = \frac{1}{512}$ **2. $\frac{1}{2^4} = \frac{1}{16}$** **3. $\frac{1}{10^3} = \frac{1}{1000}$**

Lesson 8–3 *(Pages 347–351)* **Write each expression using positive exponents. Then evaluate the expression.**

1. 8^{-3}
2. 2^{-4}
3. 10^{-3}
4. 5^{-1} **$\frac{1}{5^1} = \frac{1}{5}$**
5. 9^{-3} **$\frac{1}{9^3} = \frac{1}{729}$**
6. 7^{-2} **$\frac{1}{7^2} = \frac{1}{49}$**
7. 3^{-5} **$\frac{1}{3^5} = \frac{1}{243}$**
8. 2^{-6} **$\frac{1}{2^6} = \frac{1}{64}$**

Simplify each expression.

9. $(m^6)(m^{-2})$ **m^4**
10. $r^0s^1t^4$ **st^4**
11. $6x^{-2}y^{-4}z$ **$\frac{6z}{x^2y^4}$**
12. $(a^{-3})(a^2)$ **$\frac{1}{a}$**
13. $\frac{1}{k^{-4}}$ **k^4**
14. $\frac{m}{m^{-5}}$ **m^6**
15. $\frac{6b^4}{3b^{-2}}$ **$2b^6$**
16. $(r^{-3})(s^4)$ **$\frac{s^4}{r^3}$**
17. $-\frac{r^2s^5}{rs^6}$ **$-\frac{r}{s}$**
18. $\frac{2a^2b^5}{ab^7}$ **$\frac{2a}{b^2}$**
19. $\frac{-7cd^0}{14c}$ **$-\frac{1}{2}$**
20. $-\frac{10a^7b^4}{2ab}$ **$-5a^6b^3$**
21. $-\frac{4w^5z^4}{10z^7}$ **$-\frac{2w^5}{5z^3}$**
22. $-\frac{2p^2q^7}{8p^4q}$ **$-\frac{q^6}{4p^2}$**
23. $\frac{56yz^4}{14y^5z^4}$ **$\frac{4}{y^4}$**
24. $\frac{x^{-3}y^4z^{-2}}{xy^2z^{-2}}$ **$\frac{y^2}{x^4}$**
25. Evaluate $7m^{-2}n^{-3}$ if $m = 4$ and $n = 2$. **$\frac{7}{128}$**
26. Find the value of $5a^3b^{-1}$ if $a = -1$ and $b = 3$. **$-\frac{5}{3}$**

Lesson 8–4 *(Pages 352–356)* **Express each measure in standard form.** **1–8. See margin.**

1. 1.5 megaohms
2. 168 billion dollars
3. 2 megahertz
4. 76 milliamperes
5. 400 nanoseconds
6. 1.2 million dollars
7. 18 kilobytes
8. 93 micrograms

Express each number in scientific notation. **14. 2.693×10^2** **15. 8.3×10^{-6}** **17. 1.6×10^{-5}**

9. 178 **1.78×10^2**
10. 0.0098 **9.8×10^{-3}**
11. 0.032 **3.2×10^{-2}**
12. 106,000 **1.06×10^5**
13. 13.8 **1.38×10**
14. 269.3
15. 0.0000083
16. 100 **1×10^2**
17. 0.000016
18. 1.2 **1.2×10^0**
19. 0.3 **3×10^{-1}**
20. 400,300 **4.003×10^5**
21. 17 **1.7×10**
22. 1852 **1.852×10^3**
23. 1900 **1.9×10^3**
24. 0.000000103 **1.03×10^{-7}**

25–30. See margin.

Evaluate each expression. Express each result in scientific notation and standard form.

25. $(6 \times 10^3)(4 \times 10^5)$
26. $(7 \times 10)(3.5 \times 10^2)$
27. $(2.1 \times 10^3)(1 \times 10^4)$
28. $(4 \times 10^{-3})(7 \times 10^4)$
29. $(1.5 \times 10^5)(6 \times 10^{-3})$
30. $(5 \times 10^{-1})(2.5 \times 10^{-4})$

Answers

14. same slope, different y-intercepts;

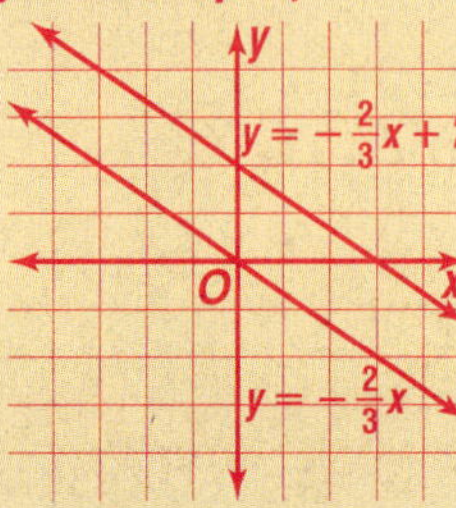

15. same y-intercept, different slopes;

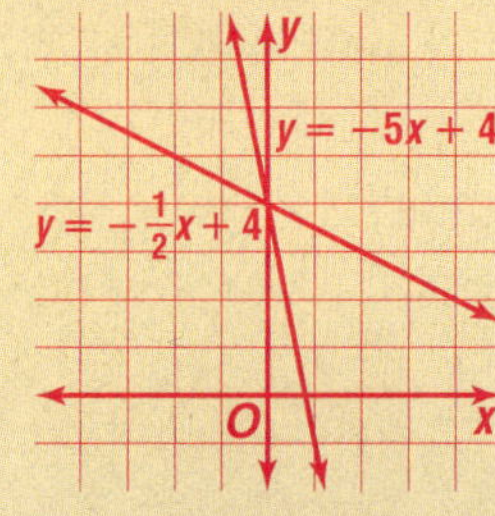

16. same slope, different y-intercepts;

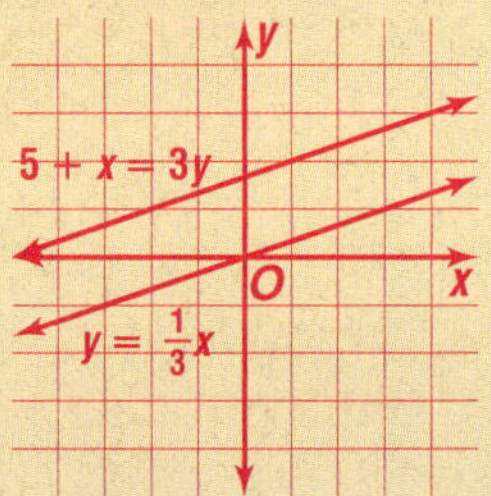

708 Extra Practice

Lesson 8–5 *(Pages 357–361)* **Simplify.**

1. $\sqrt{64}$ **8**
2. $-\sqrt{9}$ **−3**
3. $\sqrt{121}$ **11**
4. $-\sqrt{225}$ **−15**
5. $\sqrt{\frac{144}{81}}$ $\frac{4}{3}$
6. $-\sqrt{\frac{225}{100}}$ $-\frac{3}{2}$
7. $\sqrt{\frac{256}{16}}$ **4**
8. $\sqrt{\frac{36}{64}}$ $\frac{3}{4}$
9. $-\sqrt{\frac{225}{441}}$ $-\frac{5}{7}$
10. $-\sqrt{\frac{121}{289}}$ $-\frac{11}{17}$
11. $-\sqrt{0.36}$ **−0.6**
12. $\sqrt{0.81}$ **0.9**
13. $\sqrt{0.0025}$ **0.05**
14. $-\sqrt{0.0049}$ **−0.07**
15. $\sqrt{0.0289}$ **0.17**
16. $-\sqrt{0.000196}$ **−0.014**
17. Find the negative square root of 25. **−5**
18. If $x = \sqrt{1024}$, what is the value of x? **32**

Lesson 8–6 *(Pages 362–365)* **Estimate each square root to the nearest whole number.**

1. $\sqrt{5}$ **2**
2. $\sqrt{10}$ **3**
3. $\sqrt{11}$ **3**
4. $\sqrt{15}$ **4**
5. $\sqrt{18}$ **4**
6. $\sqrt{24}$ **5**
7. $\sqrt{61}$ **8**
8. $\sqrt{126}$ **11**
9. $\sqrt{153}$ **12**
10. $\sqrt{412}$ **20**
11. $\sqrt{483}$ **22**
12. $\sqrt{504}$ **22**
13. $\sqrt{555}$ **24**
14. $\sqrt{621}$ **25**
15. $\sqrt{709}$ **27**
16. $\sqrt{981}$ **31**
17. $\sqrt{70.3}$ **8**
18. $\sqrt{81.4}$ **9**
19. $\sqrt{121.6}$ **11**
20. $\sqrt{153.2}$ **12**
21. $\sqrt{9.35}$ **3**
22. $\sqrt{13.6}$ **4**
23. $\sqrt{0.021}$ **0**
24. $\sqrt{0.29}$ **1**
25. Tell whether 11 is closer to $\sqrt{119}$ or $\sqrt{125}$. $\sqrt{119}$
26. Which is closer to $\sqrt{285}$, 16 or 17? **17**

Lesson 8–7 *(Pages 366–371)* **If c is the measure of the hypotenuse and a and b are the measures of the legs, find each missing measure. Round to the nearest tenth if necessary.**

1. 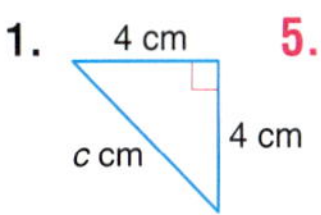

5.7

2. 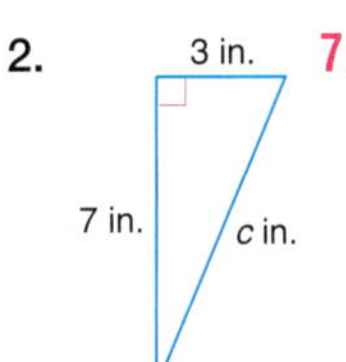

7.6

3. 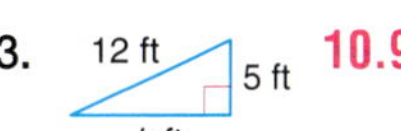

10.9

4. 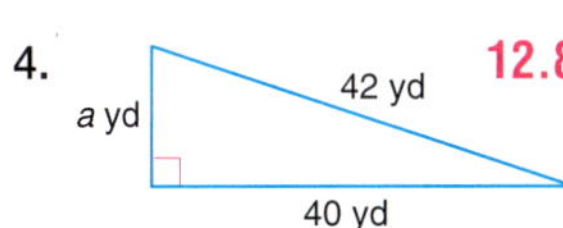

12.8

5. 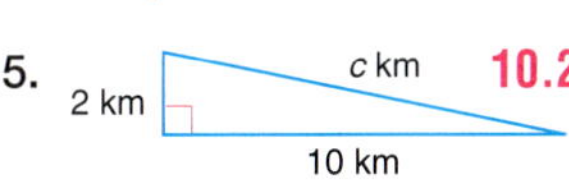

10.2

6. 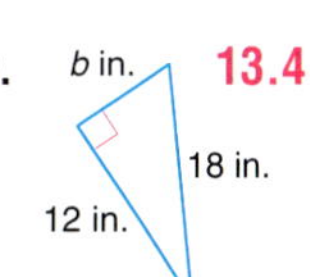

13.4

7. $a = 15, b = 17, c = ?$ **22.7**
8. $b = 12, c = 38, a = ?$ **36.1**
9. $a = 4, b = 5, c = ?$ **6.4**
10. $a = 10, c = 21, b = ?$ **18.5**
11. $a = 3, c = 8, b = ?$ **7.4**
12. $b = 20, c = 40, a = ?$ **34.6**

The lengths of three sides of a triangle are given. Determine whether each triangle is a right triangle.

13. 5 ft, 12 ft, 13 ft **yes**
14. 4 mi, 5 mi, 7 mi **no**
15. 7 cm, 11 cm, 17 cm **no**
16. 56 in., 70 in., 84 in. **no**
17. 19 m, 24 m, 28 m **no**
18. 31 mm, 37 mm, 49 mm **no**
19. 3 ft, 6 ft, 12 ft **no**
20. 2 in., 8 in., 10 in. **no**
21. 6 mi, 8 mi, 10 mi **yes**
22. 10 cm, 24 cm, 26 cm **yes**
23. 20 m, 21 m, 29 m **yes**
24. 18 in., 26 in., 32 in. **no**

Answers

Page 708
Lesson 8–4

1. 1,500,000
2. 168,000,000,000
3. 2,000,000
4. 0.076
5. 0.0000004
6. 1,200,000
7. 18,000
8. 0.000093
25. $2.4 \times 10^9 = 2{,}400{,}000{,}000$
26. $2.45 \times 10^4 = 24{,}500$
27. $2.1 \times 10^7 = 21{,}000{,}000$
28. $2.8 \times 10^2 = 280$
29. $9 \times 10^2 = 900$
30. $1.25 \times 10^{-4} = 0.000125$

Extra Practice 709

EXTRA PRACTICE

Answers

Lesson 9–1

1. yes; a product of a number and a variable
2. no; includes division
3. no; includes addition
4. no; includes a negative exponent
5. yes; a variable with a positive exponent
6. no; includes subtraction
7. yes; a variable with a positive exponent
8. yes; a product of a number and variables with positive exponents

Lesson 9–1 *(Pages 382–387)* **Determine whether each expression is a monomial. Explain why or why not.** **1–8. See margin for explanations.**

1. $6x$ **yes**
2. $\frac{4}{k}$ **no**
3. $13b + 2a$ **no**
4. $7a^2b^{-1}$ **no**
5. m^2 **yes**
6. $3x - y$ **no**
7. s^4 **yes**
8. $-3x^2yz^3$ **yes**

State whether each expression is a polynomial. If it is a polynomial, identify it as a *monomial*, *binomial*, or *trinomial*. **10. trinomial**

9. $11y$ **monomial**
10. $9xy^2 - y^3 + x^2$
11. $r^2 - r$ **binomial**
12. $\frac{1}{4}x - 2$ **binomial**
13. $3x - 11 + y - 2$ **trinomial**
14. $6j^3k^2\ell - 7j^2$ **binomial**
15. $2.3mn^2$ **monomial**
16. $7r^2s^3t$ **monomial**

Find the degree of each polynomial.

17. y^2 **2**
18. $6x^3$ **3**
19. 4 **0**
20. $3j^2 - 2k + m$ **2**
21. $16m^2n + 14m^5n^3$ **8**
22. $11p^5q^2 - 6pq^7$ **8**
23. $x^2 + 2x + 3x^6$ **6**
24. $7a^2b^3 - 2ab$ **5**

25. $x^3 + 4x^2 - x$ **26. $-6x^5 + 2x - 3$** **27. $5x^5 - 4x^2 + 3x - 2$**

Arrange the terms of each polynomial so that the powers of x are in descending order.

25. $4x^2 + x^3 - x$
26. $2x - 6x^5 - 3$
27. $5x^5 - 4x^2 + 3x - 2$
28. $6mx^5 - 4m^2x^6 + 4mx$ **$-4m^2x^6 + 6mx^5 + 4mx$**
29. $3x^4y^5 + 2x^2y^3 + 6x^5y$ **$6x^5y + 3x^4y^5 + 2x^2y^3$**
30. $2x - x^2 + 2$ **$-x^2 + 2x + 2$**

Lesson 9–2 *(Pages 388–393)* **Find each sum.**

1. $\begin{array}{r} 7x - 2 \\ (+)\ x + 4 \\ \hline \mathbf{8x + 2} \end{array}$
2. $\begin{array}{r} 6x^2 - 2x - 1 \\ (+)\ 3x^2 - 4x - 7 \\ \hline \mathbf{9x^2 - 6x - 8} \end{array}$
3. $\begin{array}{r} 2xy - 3x + 2 \\ (+)\ 4xy \quad\ \ - 7 \\ \hline \mathbf{6xy - 3x - 5} \end{array}$
4. $(-3y + 7) + (4y - 2)$ **$y + 5$**
5. $(7ab - 2a) + (2ab + 3b)$ **$9ab - 2a + 3b$**
6. $(4x^2 - 2xy + y^2) + (xy + 3y^2)$ **$4x^2 - xy + 4y^2$**

Find each difference.

7. $\begin{array}{r} 3x - 4 \\ (-)\ 2x + 3 \\ \hline \mathbf{x - 7} \end{array}$
8. $\begin{array}{r} 5m^2 - m + 2 \\ (-)\ 2m^2 + 2m + 5 \\ \hline \mathbf{3m^2 - 3m - 3} \end{array}$
9. $\begin{array}{r} 11a^2 - 2a - 1 \\ (-)\ 5a^2 + 6a + 2 \\ \hline \mathbf{6a^2 - 8a - 3} \end{array}$
10. $(13x - 2) - (9x + 4)$ **$4x - 6$**
11. $(7m + 4n) - (2m - n)$ **$5m + 5n$**
12. $(6a + b) - (2b + 4)$ **$6a - b - 4$**

Find each sum or difference. **15. $2y^3 + y^2 + y - 6$** **16. $x^2y + 5xy - 3x^2 - 7y^2$**

13. $(2x + 3) + (4x^2 + x - 7)$ **$4x^2 + 3x - 4$**
14. $(mn + 2pn - mp) - (2pn - mp)$ **mn**
15. $(y^2 + 2y - 3) + (2y^3 - y - 3)$
16. $(x^2y + 5xy - 9y^2) - (3x^2 - 2y^2)$
17. $(3x^2 + 2x - 4) + (5x^2 - 9)$ **$8x^2 + 2x - 13$**
18. $(x^2y + 9xy - y^2) - (2x^2y - y^2)$ **$-x^2y + 9xy$**

Lesson 9–3 *(Pages 394–398)* **Find each product.** **2. $-21b + 12$** **5. $-10y^2 + 5y$**

1. $4(3x + 2)$ **$12x + 8$**
2. $-3(7b - 4)$
3. $a(2a - 4)$ **$2a^2 - 4a$**
4. $2n(n - 5)$ **$2n^2 - 10n$**
5. $-y(10y - 5)$
6. $-3x(5x + 8)$
7. $7p(p^2 - 3)$ **$7p^3 - 21p$**
8. $-5m(3m^3 - 2m)$
9. $2r^3(7 - r + r^2)$
10. $-6x^3(x^2 - 4x)$
11. $0.75k(8k^3 - k^2)$
12. $2.4z^2(2z^2 - 3)$

6. $-15x^2 - 24x$ **8. $-15m^4 + 10m^2$** **9. $14r^3 - 2r^4 + 2r^5$** **10. $-6x^5 + 24x^4$**

Solve each equation. **11. $6k^4 - 0.75k^3$** **12. $4.8z^4 - 7.2z^2$**

13. $-3(4 - x) = 18$ **10**
14. $21 = 7(y - 11)$ **14**
15. $10x - 9 = 4(x - 1) + 1$ **1**
16. $7(r + 8) - 4 = -4(-7 - r)$ **-8**
17. $6(s - 7) = 3(4s + 4)$ **-9**
18. $-a + 6(a + 4) = 3(a + 5) + 1$ **-4**

710 Extra Practice

Lesson 9–4 (Pages 399–404) Find each product. Use the Distributive Property or the FOIL method.

7. $28p^2 + 2p - 6$ 9. $3d^2 + 16d - 35$ 10. $32x^2 + 32xy + 6y^2$

1. $(m + 3)(m + 2)$ $m^2 + 5m + 6$
2. $(x - 4)(x + 6)$ $x^2 + 2x - 24$
3. $(y - 5)(y - 7)$ $y^2 - 12y + 35$
4. $(r + 9)(r - 3)$ $r^2 + 6r - 27$
5. $(2s - 3)(s + 5)$ $2s^2 + 7s - 15$
6. $(a - 2)(4a + 8)$ $4a^2 - 16$
7. $(7p - 3)(4p + 2)$
8. $(2k + 2)(2k + 1)$ $4k^2 + 6k + 2$
9. $(d + 7)(3d - 5)$
10. $(8x + 2y)(4x + 3y)$
11. $(3m + n)(m - 3)$
12. $(7r + 8s)(5r - 3s)$
13. $(6x - 1)(x - 2)$ $6x^2 - 13x + 2$
14. $(7p - 2)(3p + n)$
15. $(2p - 1)(p - 3)$ $2p^2 - 7p + 3$
16. $(b + a)(a - b)$ $a^2 - b^2$
17. $(5c + d)(3c - 2d)$
18. $(p + 1)(p + m)$
19. $(j^2 - 2)(j + 5)$
20. $(z^2 + r)(z^2 - r)$ $z^4 - r^2$
21. $(n^2 + 1)(2n^2 - 3)$

11. $3m^2 + mn - 3n - 9m$ 12. $35r^2 + 19rs - 24s^2$ 14. $21p^2 - 6p + 7np - 2n$
17. $15c^2 - 7cd - 2d^2$ 18. $p^2 + p + m + mp$ 19. $j^3 + 5j^2 - 2j - 10$ 21. $2n^4 - n^2 - 3$

Lesson 9–5 (Pages 405–409) Find each product.

12. $3r^2 - 4rs - 4s^2$ 13. $4x^2 - 28xy + 49y^2$

1. $(m + 5)^2$ $m^2 + 10m + 25$
2. $(n - 3)^2$ $n^2 - 6n + 9$
3. $(2x + 3)^2$ $4x^2 + 12x + 9$
4. $(3y - x)^2$ $9y^2 - 6xy + x^2$
5. $(4a + b)(4a - b)$ $16a^2 - b^2$
6. $(2k + 2p)^2$ $4k^2 + 8kp + 4p^2$
7. $(4x - 2)(4x + 2)$ $16x^2 - 4$
8. $(1 + p)^2$ $1 + 2p + p^2$
9. $(a - 2b)(a + 2b)$ $a^2 - 4b^2$
10. $(6 + 3m)^2$ $36 + 36m + 9m^2$
11. $(2 - 4t)^2$ $4 - 16t + 16t^2$
12. $(r - 2s)(3r + 2s)$
13. $(2x - 7y)^2$
14. $(m + 3n)^2$ $m^2 + 6mn + 9n^2$
15. $2(p - q)^2$ $2p^2 - 4pq + 2q^2$
16. $4(r + s)^2$ $4r^2 + 8rs + 4s^2$
17. $k(2 + k)^2$ $4k + 4k^2 + k^3$
18. $4s(s - 1)^2$ $4s^3 - 8s^2 + 4s$
19. $3r(r - 2)^2$ $3r^3 - 12r^2 + 12r$
20. $y(y + 1)(y - 2)$ $y^3 - y^2 - 2y$
21. $p(p + 2)(2p - 3)$
22. $2(j + 3)(j - 3)$ $2j^2 - 18$
23. $(x + 3)(x - 1)(x + 2)$ $x^3 + 4x^2 + x - 6$
24. $6(m + 5)(m - 1)$ $6m^2 + 24m - 30$

21. $2p^3 + p^2 - 6p$

25. The area of a circle is given by the formula $A = \pi r^2$, where r is the radius of the circle. Suppose a circle has a radius of $k - 4$ inches.
    a. Write an equation to find the area of the circle. $A = \pi(k - 4)^2$
    b. Find the area to the nearest hundredth if $k = 6$. 12.56 in^2

Lesson 10–1 (Pages 420–425) Find the factors of each number. Then classify each number as *prime* or *composite*.

5. 1, 2, 4, 8, 13, 26, 52, 104; C

1. 57 1, 3, 19, 57; C
2. 22 1, 2, 11, 22; C
3. 65 1, 5, 13, 65; C
4. 17 1, 17; P
5. 104
6. 18 1, 2, 3, 6, 9, 18; C
7. 81 1, 3, 9, 27, 81; C
8. 73 1, 73; P

Factor each monomial.

9. $12x^2$ $2 \cdot 2 \cdot 3 \cdot x \cdot x$
10. $28m^2n$ $2 \cdot 2 \cdot 7 \cdot m \cdot m \cdot n$
11. $-33j^2k^2$ $-1 \cdot 3 \cdot 11 \cdot j \cdot j \cdot k \cdot k$
12. $54p^3$ $2 \cdot 3 \cdot 3 \cdot 3 \cdot p \cdot p \cdot p$
13. $81ab^2$ $3 \cdot 3 \cdot 3 \cdot 3 \cdot a \cdot b \cdot b$
14. $75xy$ $3 \cdot 5 \cdot 5 \cdot x \cdot y$
15. $-13p^2q^2$ $-1 \cdot 13 \cdot p \cdot p \cdot q \cdot q$
16. $105r^4$ $3 \cdot 5 \cdot 7 \cdot r \cdot r \cdot r \cdot r$

Find the GCF of each set of numbers or monomials. 28. $-x^2y$

17. 13, 33 1
18. 50, 75 25
19. 32, 84, 144 4
20. 32, 64, 96 32
21. $-21, 15xy$ 3
22. $4x^2, 2x, 8x$ $2x$
23. $-3r^2s^2, -17rs$ $-1rs$
24. $14rs, 12rst, 6t$ 2
25. $7kr, 21k^2, 2kr$ k
26. $-16c^2d, -4cd, -8cd^2$ $-4cd$
27. $24m^2n, 51m, 63m^2n^2$ $3m$
28. $-1x^3yz, -7x^2y^2, -2x^3yz$

EXTRA PRACTICE

Extra Practice 711

Answers

Page 713

Lesson 11-1

1. 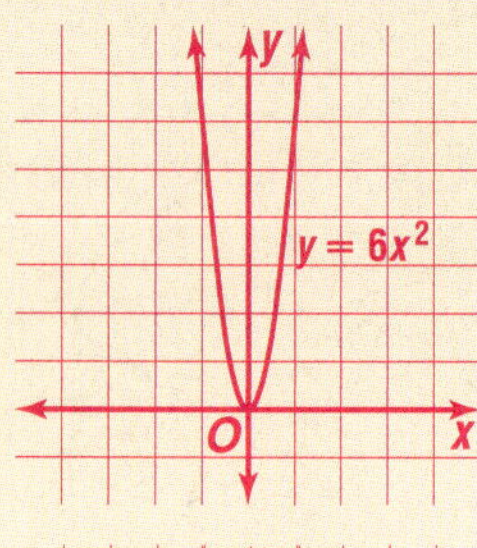

2. 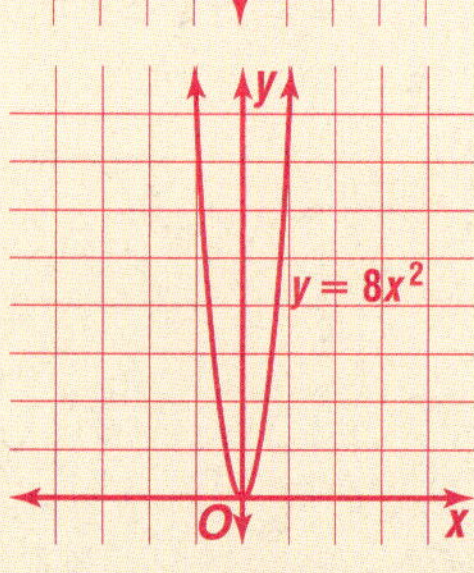

3. 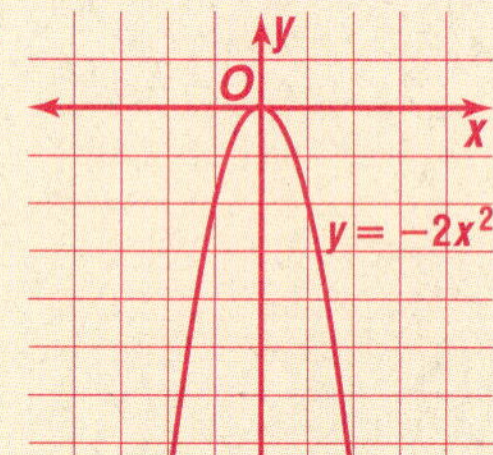

4. 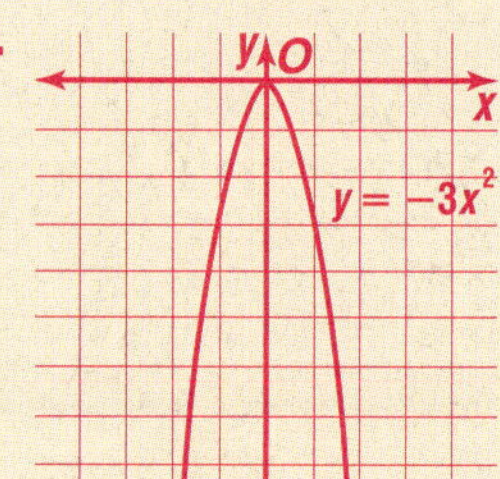

5. 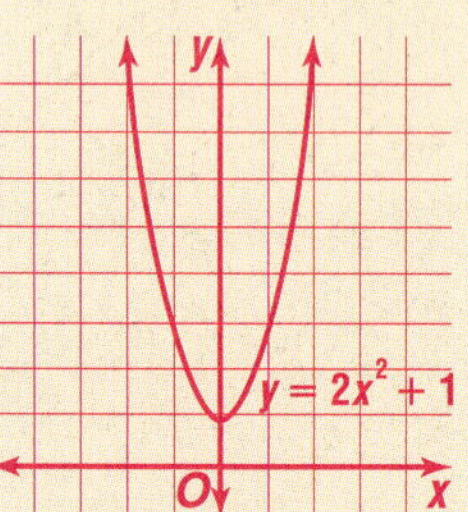

6. 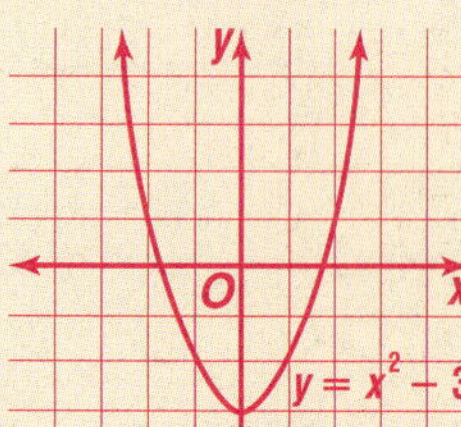

7. 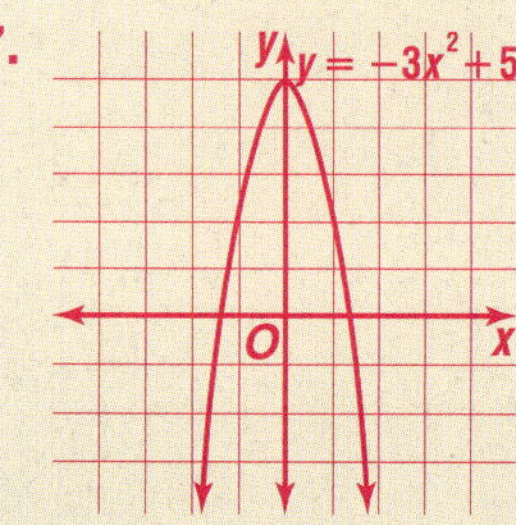

EXTRA PRACTICE

Lesson 10-2 *(Pages 428–433)* **Factor each polynomial. If the polynomial cannot be factored, write *prime*.** 11. $2c(8b^2 - a + 2b)$ 12. $2(6mn - 7m^2 + 8n)$

1. $4m + 12$ $4(m + 3)$
2. $13n + n^2$ $n(13 + n)$
3. $2k^2 + 6k$ $2k(k + 3)$
4. $7s^2t + 3$ prime
5. $9pq^3 - 21pq^2$ $3pq^2(3q - 7)$
6. $x^2y + 7y$ $y(x^2 + 7)$
7. $3k - 4j$ prime
8. $14m^2n - 18m$ $2m(7mn - 9)$
9. $17cd - 14mn$ prime
10. $16a^2b^2 - 3ab$ $ab(16ab - 3)$
11. $16b^2c - 2ac + 4bc$
12. $12mn - 14m^2 + 16n$
13. $6ab - 7bc + 12ac$ prime
14. $m^2n^3p + mn - mp$ $m(mn^3p + n - p)$
15. $x^2y + 15xy + 5x$ $x(xy + 15y + 5)$

Find each quotient. 22. $3b - 2$ 26. $5s + 4rt$ 27. $y^2 + 2x$

16. $(16x + 4y^2) \div 4$ $4x + y^2$
17. $(2rs + r) \div r$ $2s + 1$
18. $(7xy - 3x) \div x$ $7y - 3$
19. $(15a + 3b^2) \div 3$ $5a + b^2$
20. $(6m^2n - 9m) \div 3m$ $2mn - 3$
21. $(18cd^2 - 9cd) \div 9cd$ $2d - 1$
22. $(21a^2b - 14a^2) \div 7a^2$
23. $(16xy - 12xy) \div 4xy$ 1
24. $(12a^2b + 4b) \div 4b$ $3a^2 + 1$
25. $(36st^2 - 9st) \div 9st$ $4t - 1$
26. $(15rs^2 + 12r^2st) \div 3rs$
27. $(5xy^2z + 10x^2z) \div 5xz$
28. $(13r^2s - 26r^2) \div 13r^2$ $s - 2$
29. $(32np + m^2np^2) \div np$ $32 + m^2p$
30. $(20r^2st + 15rs) \div 5rs$ $4rt + 3$

Lesson 10-3 *(Pages 434–439)* **Factor each trinomial. If the trinomial cannot be factored, write *prime*.** 8. $(d + 10)(d + 11)$ 12. $3(2c + 1)(c - 1)$

1. $x^2 + 4x + 4$ $(x + 2)(x + 2)$
2. $n^2 + 6n + 9$ $(n + 3)(n + 3)$
3. $t^2 - 8t + 16$ $(t - 4)(t - 4)$
4. $w^2 + 5w - 2$ prime
5. $r^2 + r - 12$ $(r + 4)(r - 3)$
6. $p^2 - 8p + 4$ prime
7. $s^2 - 7s - 8$ $(s - 8)(s + 1)$
8. $d^2 + 21d + 110$
9. $y^2 - 10y + 3$ prime
10. $q^2 - 3q - 28$ $(q - 7)(q + 4)$
11. $4z^2 - 16z - 18$ prime
12. $6c^2 - 3c - 3$
13. $2m^2 - 3m - 20$ $(2m + 5)(m - 4)$
14. $8y^2 + 6y - 2$ $2(y + 1)(4y - 1)$
15. $3r^2 - 12r - 15$ $3(r + 1)(r - 5)$

Lesson 10-4 *(Pages 440–444)* **Factor each trinomial. If the trinomial cannot be factored, write *prime*.**

1. $5x^2 + 13x + 6$ $(5x + 3)(x + 2)$
2. $4w^2 + 7w + 3$ $(4w + 3)(w + 1)$
3. $2a^2 - 9a + 4$ $(2a - 1)(a - 4)$
4. $12t^2 + 18t + 2$ $2(6t^2 + 9t + 1)$
5. $4c^2 - 8c - 3$ prime
6. $3r^2 + 15r + 12$ $3(r + 1)(r + 4)$
7. $8y^2 - 16y + 6$ $2(2y - 1)(2y - 3)$
8. $20n^2 - 40n + 20$ $20(n - 1)(n - 1)$
9. $10m + 3 + 3m^2$ $(3m + 1)(m + 3)$
10. $13 + 2m^2 + 14m$ prime
11. $3q^2 - 4 - 4q$ $(3q + 2)(q - 2)$
12. $12d + 8 - 8d^2$ $-4(2d + 1)(d - 2)$
13. $2m^2 - 3mn - 2n^2$ $(2m + n)(m - 2n)$
14. $2a^2 + 4ab + 2b^2$ $2(a + b)(a + b)$
15. $12p^2 + 10mp + 2m^2$ $2(3p + m)(2p + m)$
16. $18x^2 + 15xy + 3y^2$ $3(3x + y)(2x + y)$

17. A rectangle has dimensions of $(y + 5)$ inches and $(y - 4)$ inches.
    a. Express the area as a trinomial. $y^2 + y - 20$
    b. If y units are removed from the length, express the new area. $5y - 20$

Answers

8. 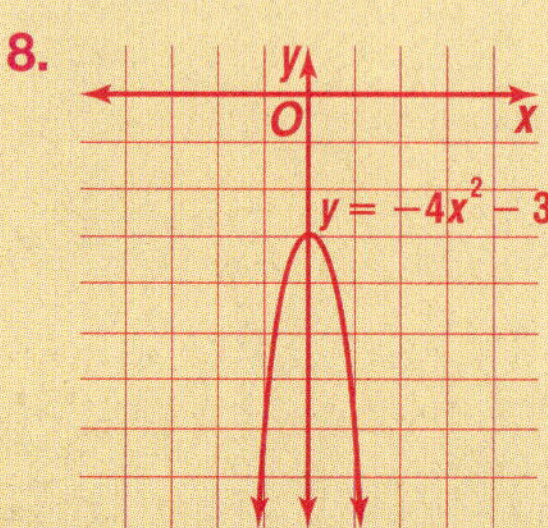

9. 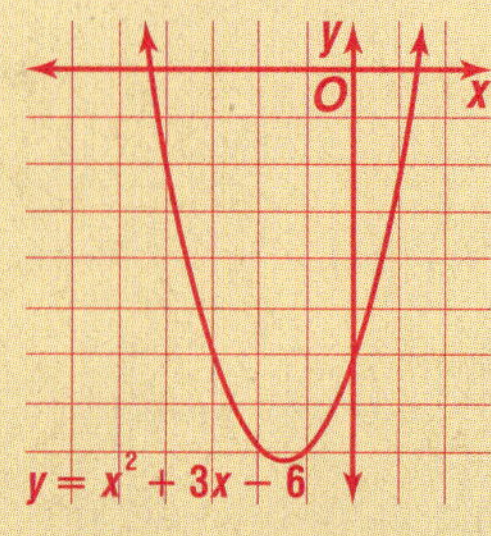

10. 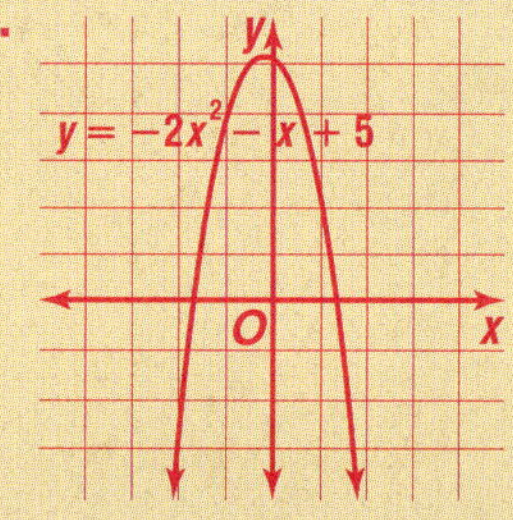

Lesson 10–5 *(Pages 445–449)* Determine whether each trinomial is a perfect square trinomial. If so, factor it. **1. $(x - 7)^2$ 4. $(d + 5)^2$ 5. $(r + 12)^2$ 6. $(3c + 2)^2$**

1. $x^2 - 14x + 49$
2. $a^2 - 4a + 4$ **$(a - 2)^2$**
3. $y^2 - 16y + 3$ **no**
4. $d^2 + 10d + 25$
5. $r^2 + 24r + 144$
6. $9c^2 + 12c + 4$
7. $4k^2 - 28k + 49$
8. $16m^2 - 8m + 1$

7. $(2k - 7)^2$ 8. $(4m - 1)^2$ 9. $(x + 3)(x - 3)$ 10. $(2 + 7z)(2 - 7z)$ 11. $4(3x + 1)(3x - 1)$

Determine whether each trinomial is the difference of squares. If so, factor it.

9. $x^2 - 9$
10. $4 - 49z^2$
11. $36x^2 - 4$
12. $17 - 3p^2$ **no**
13. $s^2 - 16r^2$ **$(s + 4r)(s - 4r)$**
14. $9 - m^2n^2$ **$(3 + mn)(3 - mn)$**
15. $32k^2 - 50$ **$2(4k + 5)(4k - 5)$**
16. $12b^2 - 48$ **$12(b + 2)(b - 2)$**

Factor each polynomial. If the polynomial cannot be factored, write *prime*. **23. $3b(ab + 3)(ab - 3)$**

17. $a^2 - a - 12$
18. $2s^2 - rs - r^2$
19. $2r^2 - 3$ **prime**
20. $4m^2 + 48m + 144$
21. $6m^2 - 16m - 6$
22. $2xy - 8x$ **$2x(y - 4)$**
23. $3a^2b^3 - 27b$
24. $16x^2 - 8xy + y^2$

17. $(a - 4)(a + 3)$ 18. $(2s + r)(s - r)$ 20. $4(m + 6)^2$ 21. $2(3m + 1)(m - 3)$ 24. $(4x - y)^2$

Lesson 11–1 *(Pages 458–463)* Graph each quadratic function by making a table of values.

1. $y = 6x^2$
2. $y = 8x^2$
3. $y = -2x^2$
4. $y = -3x^2$
5. $y = 2x^2 + 1$
6. $y = x^2 - 3$
7. $y = -3x^2 + 5$
8. $y = -4x^2 - 3$
9. $y = x^2 + 3x - 6$
10. $y = -2x^2 - x + 5$
11. $y = 3x^2 - 2x$
12. $y = 0.25x^2 - 2x - 4$

1–12. See margin.

Write the equation of the axis of symmetry and the coordinates of the vertex of the graph of each quadratic function. Then graph the function. **13–24. See Solutions Manual.**

13. $y = 2x^2$
14. $y = 3x^2$
15. $y = 8x^2$
16. $y = x^2 + 2x$
17. $y = 3x^2 - 6x$
18. $y = -2x^2 + 1$
19. $y = 4x^2 - x + 2$
20. $y = -x^2 + 6x - 4$
21. $y = \frac{1}{2}x^2 - x + 1$
22. $y = -\frac{1}{2}x^2 - 2x + 2$
23. $y = 2x^2 + 3x - 1$
24. $y = -x^2 - 2x - 1$

Match each function with its graph.

25. $y = (x + 2)^2 - 2$ **C**
26. $y = -2x^2 - x - 1$ **A**
27. $y = -2x^2 + x + 2$ **B**

A. 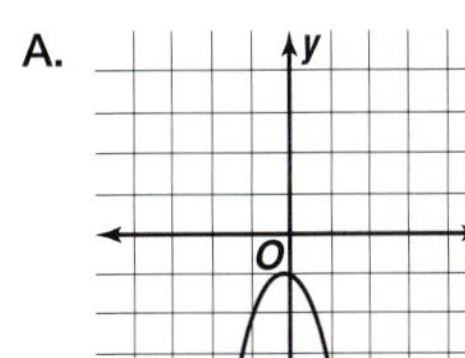

B. 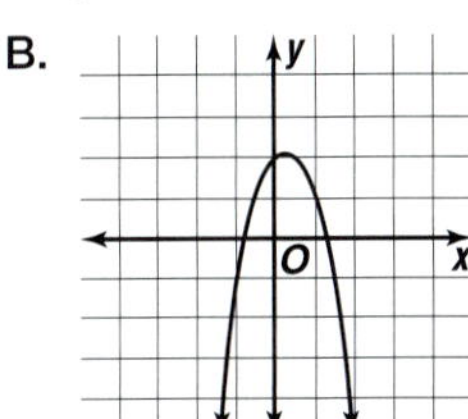

C. 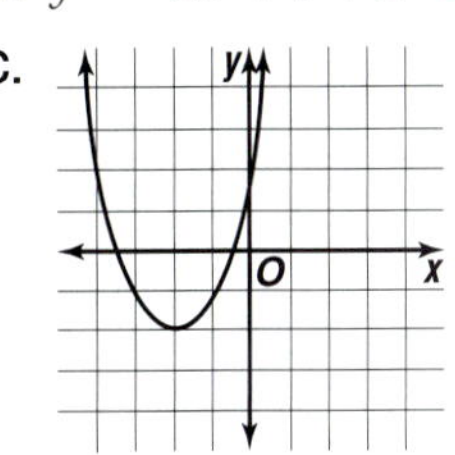

Lesson 11–2 *(Pages 464–467)* Graph each group of equations on the same screen. Compare and contrast the graphs. **1–3. See margin.**

1. $y = x^2$, $y = 2x^2$, $y = 4x^2$
2. $y = x^2 + 1$, $y = x^2 - 4$, $y = x^2 - 8$
3. $y = (2x - 1)^2$, $y = (2x - 2)^2$, $y = (2x - 3)^2$

Describe how each graph changes from the parent graph of $y = x^2$. Then name the vertex of each graph. **4–15. See margin.**

4. $y = 8x^2$
5. $y = -2x^2$
6. $y = \frac{1}{2}x^2$
7. $y = (x - 1)^2$
8. $y = (x + 3)^2$
9. $y = (3x + 2)^2$
10. $y = -(x + 3)^2$
11. $y = -2x^2 - 3$
12. $y = 0.25x^2 + 0.5$
13. $y = -3x^2 + 8$
14. $y = (4x - 1)^2 + 3$
15. $y = (x + 2)^2 - 2$

Answers

Lesson 11–1 (continued)

11.  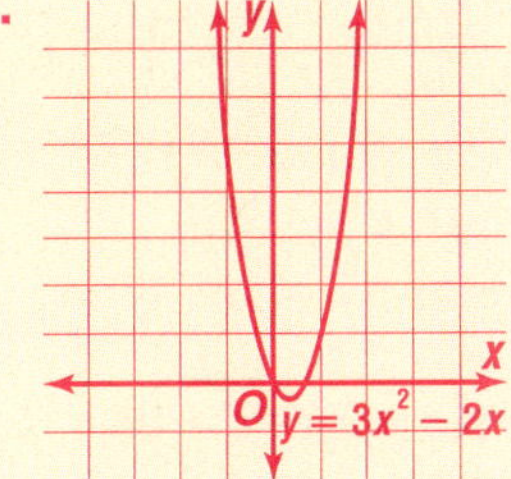

12. 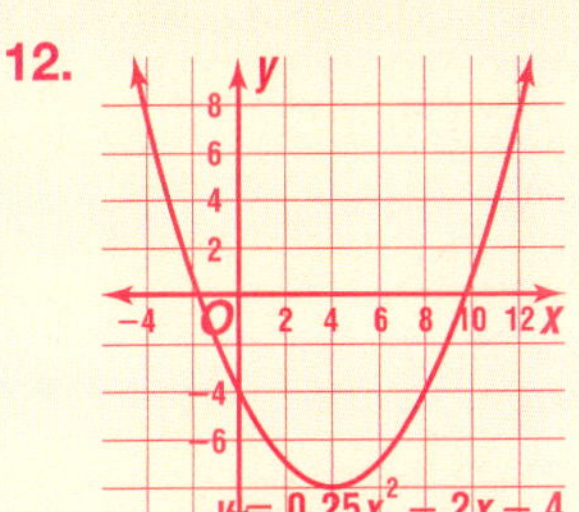

Lesson 11–2

1. 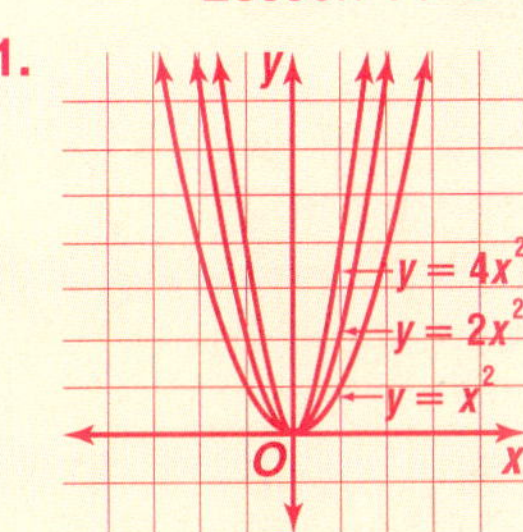

narrows, opens up

2. 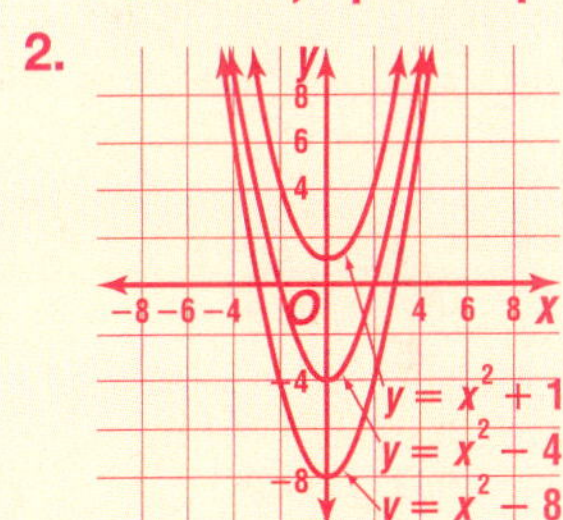

opens up, shifts down

3. 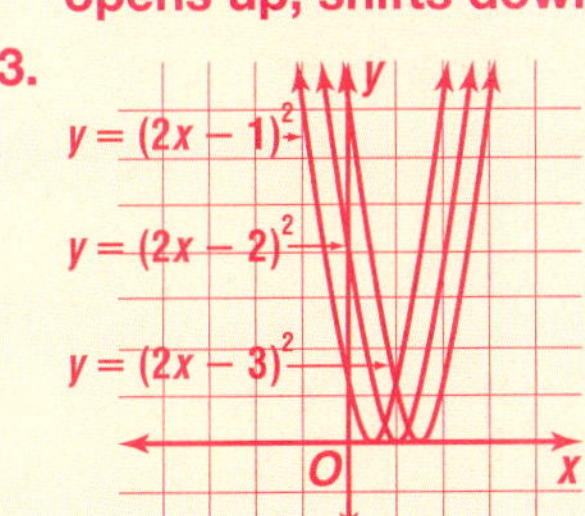

opens up, shifts right

4. narrows; (0, 0)
5. opens down, narrows; (0, 0)
6. widens; (0, 0)
7. shifts right; (1, 0)
8. shifts left; (−3, 0)
9. shifts left, narrows; $\left(-\frac{2}{3}, 0\right)$
10. shifts left, opens down; (−3, 0)

Answers *Lesson 11–2 (continued)*

11. opens down, shifts down, narrows; (0, −3)
12. widens, shifts up; $\left(0, \frac{1}{2}\right)$
13. narrows, opens down, shifts up; (0, 8)
14. narrows, opens up, shifts right, shifts up; $\left(\frac{1}{4}, 3\right)$
15. shifts left, shifts down; (−2, −2)

EXTRA PRACTICE

Answers

Lesson 11-3

1. 4, −6
2. −5, 1
3. between 0 and −1; between 4 and 5
4. between −7 and −6; between −1 and 0
5. between −7 and −6; between −1 and 0
6. between 1 and 2; between 4 and 5
7. −2, −1
8. none
9. between −5 and −4; between −2 and −1
10. between 0 and 1; between 10 and 11
11. between −3 and −2; between 1 and 2
12. −10, −2
13. between −12 and −11; between −1 and 0
14. between −2 and −1; between 1 and 2
15. none
16. between −7 and −6; between 1 and 2

Page 715
Lesson 11-6

1. −4, 1
2. −5, −4
3. 3, 4
4. −6, 1
5. −5, 3
6. 2
7. −3, 6
8. $-4+\sqrt{30}, -4-\sqrt{30}$
9. −3, 8
10. −3, −2
11. −3
12. −2, 4
13. −2, 3
14. −7, 3
15. −2, −1
16. 2, 9
17. $1+\sqrt{2}, 1-\sqrt{2}$
18. −6, 1
19. $-1, -\frac{2}{3}$
20. $-2, \frac{2}{3}$
21. $-3, \frac{1}{2}$

Lesson 11-3 *(Pages 468–473)* Solve each equation by graphing the related function. If exact roots cannot be found, state the consecutive integers between which the roots are located.

1. $-x^2 - 2x + 24 = 0$
2. $x^2 + 4x - 5 = 0$
3. $x^2 - 4x - 2 = 0$
4. $x^2 + 7x + 3 = 0$
5. $x^2 + 7x + 5 = 0$
6. $x^2 - 6x + 6 = 0$
7. $-x^2 - 3x - 2 = 0$
8. $4x^2 + 2x + 1 = 0$
9. $-x^2 - 6x - 6 = 0$
10. $x^2 - 11x + 4 = 0$
11. $-2x^2 - 2x + 10 = 0$
12. $x^2 + 12x + 20 = 0$
13. $-x^2 - 12x - 3 = 0$
14. $-3x^2 - x + 8 = 0$
15. $x^2 + 3x + 4 = 0$
16. $x^2 + 5x - 9 = 0$

1–16. See margin.

Use a quadratic equation to determine the two numbers that satisfy each situation.

17. Their sum is 21 and their product is 104. **13, 8**
18. Their difference is 8 and their product is 20. **10, 2 or −10, −2**
19. Their sum is 13 and their product is 22. **2, 11**
20. Their sum is 32 and their product is 135. **5, 27**

Lesson 11-4 *(Pages 474–477)* Solve each equation. Check your solution.

1. $2r(r - 4) = 0$ **0, 4**
2. $4k(k + 5) = 0$ **0, −5**
3. $(s + 4)(s - 3) = 0$ **−4, 3**
4. $(m - 4)(m - 5) = 0$ **4, 5**
5. $(3x - 4)(x - 2) = 0$ **$\frac{4}{3}$, 2**
6. $(2y + 2)(2y - 4) = 0$ **−1, 2**
7. $(t + 2)(6t + 1) = 0$ **$-2, -\frac{1}{6}$**
8. $n^2 + n - 6 = 0$ **−3, 2**
9. $k^2 + 4k + 4 = 0$ **−2**
10. $p^2 - 5p + 6 = 0$ **2, 3**
11. $q^2 - 2q - 15 = 0$ **−3, 5**
12. $x^2 + 2x - 3 = 0$ **−3, 1**
13. $j^2 + 9j + 20 = 0$ **−5, −4**
14. $r^2 - 16r = 0$ **0, 16**
15. $z^3 - 25z = 0$ **0, −5, 5**

For each problem, define a variable. Then use an equation to solve the problem.

16. Find two integers whose sum is 15 and whose product is 36. **12 and 3**
17. The length of a swimming pool is 15 feet longer than it is wide. The area in square feet is 1350. Find the dimensions of the pool. **45 ft by 30 ft**
18. Find two integers whose difference is 12 and whose product is 13. **13 and 1 or −13 and −1**

Lesson 11-5 *(Pages 478–482)* Find the value of *c* that makes each trinomial a perfect square.

1. $r^2 - 16r + c$ **64**
2. $k^2 + 12k + c$ **36**
3. $p^2 - 4p + c$ **4**
4. $n^2 + 2n + c$ **1**
5. $f^2 - 8f + c$ **16**
6. $s^2 + 18s + c$ **81**
7. $x^2 + 20x + c$ **100**
8. $r^2 + 14r + c$ **49**
9. $w^2 + 30w + c$ **225**
10. $h^2 + 10h + c$ **25**
11. $z^2 - 2z + c$ **1**
12. $m^2 - 6m + c$ **9**
13. $q^2 + 26q + c$ **169**
14. $t^2 + 28t + c$ **196**
15. $y^2 + 22y + c$ **121**
16. $z^2 + 24z + c$ **144**

Solve each equation by completing the square. **20. $-\frac{15}{2} \pm \frac{7\sqrt{5}}{2}$**

17. $z^2 + 10z + 12 = 0$ **$-5 \pm \sqrt{13}$**
18. $h^2 - 8h - 15 = 0$ **$4 \pm \sqrt{31}$**
19. $y^2 + 3y + 1 = 0$ **$-\frac{3}{2} \pm \frac{\sqrt{5}}{2}$**
20. $w^2 + 15w = 5$
21. $m^2 + 2m = 0$ **0, −2**
22. $t^2 + 2t = 18$ **$-1 \pm \sqrt{19}$**
23. $r^2 - 20r + 24 = 0$ **$10 \pm 2\sqrt{19}$**
24. $p^2 - 2p = 32$ **$1 \pm \sqrt{33}$**
25. $q^2 - 7q + 12 = 0$ **4, 3**
26. $n^2 - 4n - 16 = 0$ **$2 \pm 2\sqrt{5}$**
27. $x^2 + 10x = 12$ **$-5 \pm \sqrt{37}$**
28. $r^2 + 12r = 0$ **0, −12**

714 Extra Practice

Answers

22. $-1, \frac{1}{2}$
23. −1
24. $-1, -\frac{1}{4}$

Lesson 11-6 *(Pages 483–488)* Use the Quadratic Formula to solve each equation.

1. $j^2 + 3j - 4 = 0$
2. $w^2 + 9w + 20 = 0$
3. $m^2 - 7m + 12 = 0$
4. $n^2 + 5n - 6 = 0$
5. $k^2 + 2k - 15 = 0$
6. $z^2 - 4z + 4 = 0$
7. $d^2 - 3d - 18 = 0$
8. $s^2 + 8s - 14 = 0$
9. $x^2 - 5x - 24 = 0$
10. $t^2 + 5t + 6 = 0$
11. $r^2 + 6r + 9 = 0$
12. $y^2 - 2y - 8 = 0$
13. $y^2 - y - 6 = 0$
14. $d^2 + 4d - 21 = 0$
15. $s^2 + 3s + 2 = 0$
16. $m^2 - 11m = -18$
17. $4k^2 - 4 = 8k$
18. $-3r^2 - 15r = -18$
19. $3p^2 + 5p + 2 = 0$
20. $6x^2 + 8x = 8$
21. $2p^2 + 5p = 3$
22. $4q^2 - 2 = -2q$
23. $-2c^2 - 4c - 2 = 0$
24. $-8m^2 - 10m = 2$

1–24. See margin.

Lesson 11-7 *(Pages 489–493)* Graph each exponential function. Then state the y-intercept. **1–16. See Solutions Manual.**

1. $y = 2^x$
2. $y = 5^x$
3. $y = 3^x - 1$
4. $y = 4^x + 4$
5. $y = 2^x - 3$
6. $y = 2^x + 3$
7. $y = 4^x + 3$
8. $y = 2^x - 6$
9. $y = 3^x - 3$
10. $y = 2^{4x} + 1$
11. $y = 4^{0.5x}$
12. $y = 5^{3x} + 1$
13. $y = 4^{3x} - 2$
14. $y = 3^{3x} - 2$
15. $y = 2^{0.5x} - 1$
16. $y = 2^{3x} - 4$

Find the amount of money in a bank account given the following conditions.

17. initial deposit = \$6000, annual rate = 6.5%, time = 3 years **\$7247.70**
18. initial deposit = \$1000, annual rate = 12%, time = 15 years **\$5473.57**
19. initial deposit = \$2500, annual rate = 2%, time = 6 years **\$2815.41**
20. initial deposit = \$5100, annual rate = 9%, time = 4 years **\$7199.07**

Lesson 12-1 *(Pages 504–508)* Write an inequality to describe each number. **1–12. See margin.**

1. a number less than 10
2. a number that is at least −4
3. a number greater than −2
4. a number less than or equal to 5
5. a number greater than 3
6. a number more than 12
7. a minimum number of 7
8. a number less than −1
9. a maximum number of 8
10. a number greater than −6
11. a minimum number of −8
12. a number more than −11

Graph each inequality on a number line. **13–28. See margin.**

13. $m < 8$
14. $n \leq -4$
15. $x > 2$
16. $z < -7$
17. $r \geq 15$
18. $m \leq -5$
19. $s > 8.4$
20. $y > 6.2$
21. $w \leq 1.3$
22. $\ell \geq -2.4$
23. $p < -3.2$
24. $j > 4.3$
25. $t < \frac{1}{3}$
26. $r \leq -2\frac{1}{4}$
27. $y \leq \frac{1}{2}$
28. $q \geq 3\frac{5}{8}$

Write an inequality for each graph.

29. 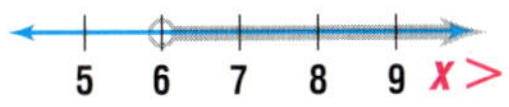

$x > 6$

30. 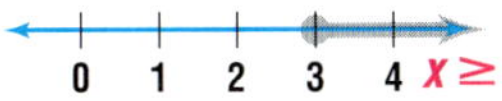

$x \geq 3$

31. 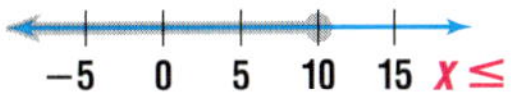

$x \leq 10$

32. 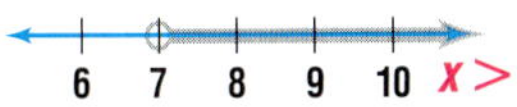

$x > 7$

33. 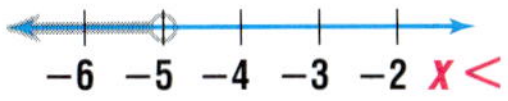

$x < -5$

34. 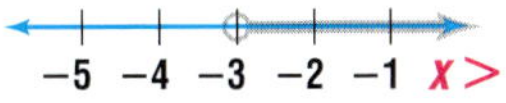

$x > -3$

35. 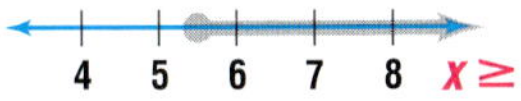

$x \geq 5.5$

36. 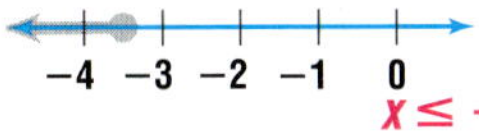

$x \leq -3.5$

37. 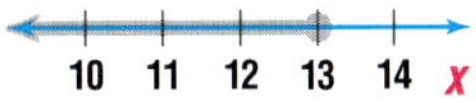

$x \leq 13$

Answers

Lesson 12-1

1. $x < 10$
2. $x \geq -4$
3. $x > -2$
4. $x \leq 5$
5. $x > 3$
6. $x > 12$
7. $x \geq 7$
8. $x < -1$
9. $x \leq 8$
10. $x > -6$
11. $x \geq -8$
12. $x > -11$

13. 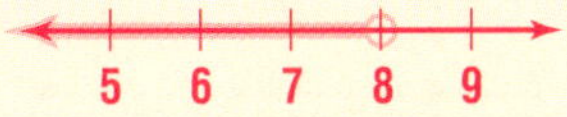

14. 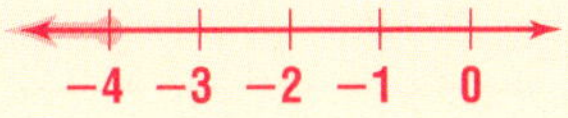

15. 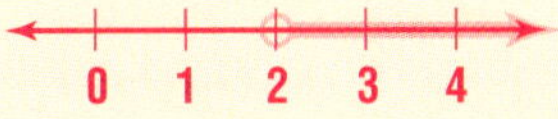

16. 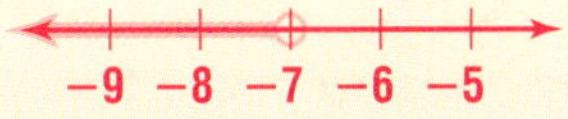

17. 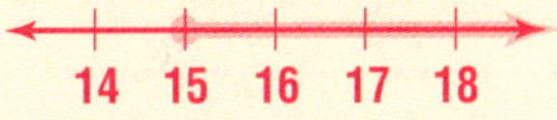

18. 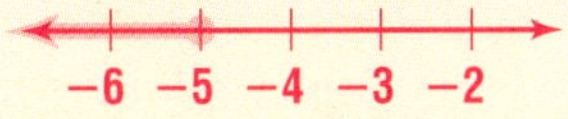

19. 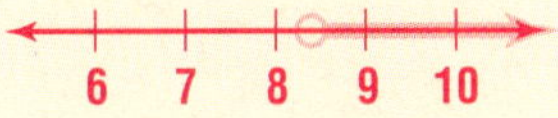

20. 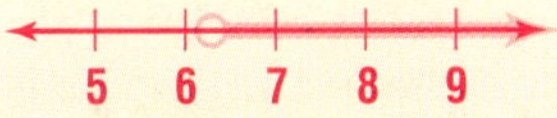

21. 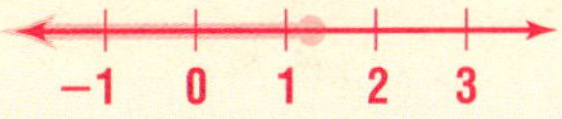

22. 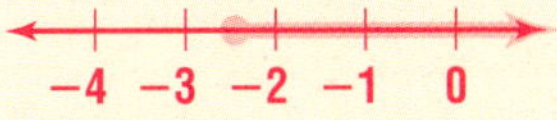

23. 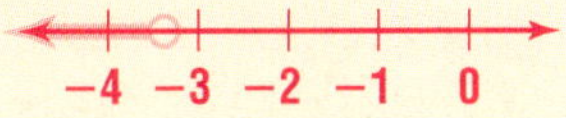

24. 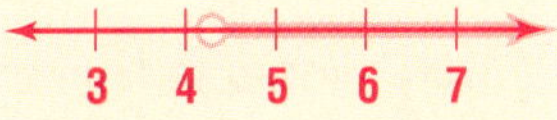

25. 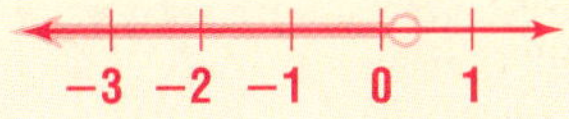

26. 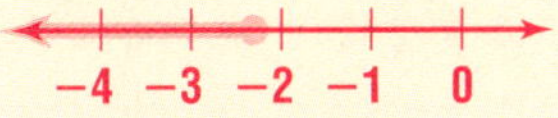

27. 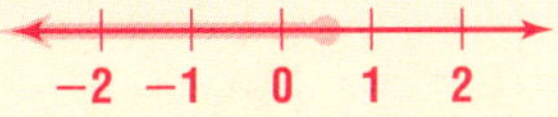

28. 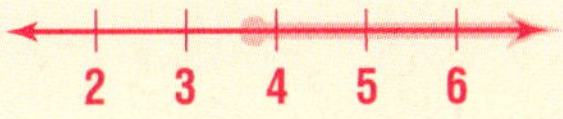

Extra Practice 715

Lesson 12–2 *(Pages 509–513)* Solve each inequality. Check your solution.

1. $r + 3 < 8$ $\{r \mid r < 5\}$
2. $m + 4 > -2$ $\{m \mid m > -6\}$
3. $j - 3 > 5$ $\{j \mid j > 8\}$
4. $k - 6 < -13$ $\{k \mid k < -7\}$
5. $-12 + w < 15$ $\{w \mid w < 27\}$
6. $p + 11 \geq 5$ $\{p \mid p \geq -6\}$
7. $0.4 + p \geq 1.2$ $\{p \mid p \geq 0.8\}$
8. $x - 6.2 < 4$ $\{x \mid x < 10.2\}$
9. $4.3 < 2.1 + y$ $\{y \mid y > 2.2\}$
10. $\frac{1}{4} + r \leq 3\frac{3}{4}$ $\left\{r \mid r \leq 3\frac{1}{2}\right\}$
11. $k + 10.6 \geq -3.4$ $\{k \mid k \geq -14\}$
12. $\frac{1}{6} + s \leq \frac{1}{3}$ $\left\{s \mid s \leq \frac{1}{6}\right\}$

Solve each inequality. Graph the solution. **13–24. See margin for graphs.**

13. $2n > n + 3$ $\{n \mid n > 3\}$
14. $8 + 5x > 6x$ $\{x \mid x < 8\}$
15. $3y - 2 < 4y$ $\{y \mid y > -2\}$
16. $5d < 8 + 4d$ $\{d \mid d < 8\}$
17. $9 \geq 3t - 4 - 2t$ $\{t \mid t \leq 13\}$
18. $-2a < -3(a - 2)$ $\{a \mid a < 6\}$
19. $6u \leq 5(u + 1)$ $\{u \mid u \leq 5\}$
20. $11p \leq 2(5p + 4)$ $\{p \mid p \leq 8\}$
21. $7s < 3(2s - 3)$ $\{s \mid s < -9\}$
22. $3(x - 3) \leq 4x$ $\{x \mid x \geq -9\}$
23. $7b < 4(2b - 3)$ $\{b \mid b > 12\}$
24. $2c + 3 \leq 3(c - 5)$ $\{c \mid c \geq 18\}$

Lesson 12–3 *(Pages 514–518)* Solve each inequality. Check your solution.

1. $-3k < 15$ $\{k \mid k > -5\}$
2. $2r > -10$ $\{r \mid r > -5\}$
3. $-4d \geq -12$ $\{d \mid d \leq 3\}$
4. $\frac{x}{5} \geq 6$ $\{x \mid x \geq 30\}$
5. $-\frac{y}{4} < 8$ $\{y \mid y > -32\}$
6. $\frac{p}{7} \leq -3$ $\{p \mid p \leq -21\}$
7. $-9\ell < 27$ $\{\ell \mid \ell > -3\}$
8. $2s > 20$ $\{s \mid s > 10\}$
9. $-t > 11$ $\{t \mid t < -11\}$
10. $-\frac{d}{2} > 12$ $\{d \mid d < -24\}$
11. $\frac{m}{6} \leq -5$ $\{m \mid m \leq -30\}$
12. $-\frac{b}{4} < -9$ $\{b \mid b > 36\}$
13. $-3p < 2$ $\left\{p \mid p > -\frac{2}{3}\right\}$
14. $-8y \geq -4$ $\left\{y \mid y \leq \frac{1}{2}\right\}$
15. $7v \leq 3$ $\left\{v \mid v \leq \frac{3}{7}\right\}$
16. $2 < \frac{2}{3}j$ $\{j \mid j > 3\}$
17. $\frac{5}{8}t \geq -5$ $\{t \mid t \geq -8\}$
18. $-\frac{1}{4}w < 6$ $\{w \mid w > -24\}$
19. $0.01r \leq 8$ $\{r \mid r \leq 800\}$
20. $2.1m > 6.3$ $\{m \mid m > 3\}$
21. $-2.25j \geq 9$ $\{j \mid j \leq -4\}$
22. $\frac{k}{7.2} < -3$ $\{k \mid k < -21.6\}$
23. $-\frac{r}{2.5} > 4$ $\{r \mid r < -10\}$
24. $\frac{y}{0.4} \geq 2$ $\{y \mid y \geq 0.8\}$

Lesson 12–4 *(Pages 519–523)* Solve each inequality. Check your solution. 8. $\{x \mid x < -1\}$

1. $2a + 7 \leq 11$ $\{a \mid a \leq 2\}$
2. $5r - 3 > 27$ $\{r \mid r > 6\}$
3. $6 - 4m > 10$ $\{m \mid m < -1\}$
4. $1 - 3s \geq 13$ $\{s \mid s \leq -4\}$
5. $3 + 5d \leq -12$ $\{d \mid d \leq -3\}$
6. $-8x - 1 < 15$ $\{x \mid x > -2\}$
7. $6 + 2b \geq -2.4$ $\{b \mid b \geq -4.2\}$
8. $5.1x - 2.4 < -7.5$
9. $3.3 + 4k > -8.7$ $\{k \mid k > -3\}$
10. $\frac{6 - r}{8} < 7$ $\{r \mid r > -50\}$
11. $\frac{j}{3} + 7 > 9$ $\{j \mid j > 6\}$
12. $\frac{4 + 3m}{7} \leq -5$ $\{m \mid m \leq -13\}$
13. $2x - 1 < 8x + 2$
14. $11 \leq -(t + 4)$ $\{t \mid t \leq -15\}$
15. $4(3 - 6r) > 18$ $\left\{r \mid r < -\frac{1}{4}\right\}$
16. $\frac{2}{3}(k - 6) > 4$ $\{k \mid k > 12\}$
17. $\frac{1}{3}(y + 3) < \frac{1}{2}(y - 2)$ $\{y \mid y > 12\}$
18. $\frac{1}{8}(m + 3) \geq \frac{1}{4}(m - 3)$ $\{m \mid m \leq 9\}$

13. $\left\{x \mid x > -\frac{1}{2}\right\}$

Write and solve an inequality for each situation.

19. Three fifths times the sum of a number and 5 is greater than 15. $\frac{3}{5}(n + 5) > 15$; $\{n \mid n > 20\}$
20. Four times the difference of a number and 3 is less than 24. $4(n - 3) < 24$; $\{n \mid n < 9\}$

EXTRA PRACTICE

Answers

Lesson 12–2

13. 2 3 4 5 6
14. 5 6 7 8 9
15. −3 −2 −1 0 1
16. 5 6 7 8 9
17. 11 12 13 14 15
18. 4 5 6 7 8
19. 2 3 4 5 6
20. 5 6 7 8 9
21. −11 −10 −9 −8 −7
22. −11 −10 −9 −8 −7
23. 11 12 13 14 15
24. 16 17 18 19 20

Page 717

Lesson 12–5

1. $\{m \mid 2 < m < 6\}$
2. $\{j \mid -12 < j < 4\}$
3. $\{r \mid -1 \leq r < 6\}$
4. $\{x \mid 2 \leq x < 3\}$
5. $\{y \mid -3 \leq y \leq 5\}$
6. $\{s \mid -7 \leq s < -4\}$
7. −2 −1 0 1 2 3 4 5 6 7 8
8. −5 −4 −3 −2 −1 0 1 2 3 4 5
9. 7 8 9 10 11 12 13 14 15 16 17
10. −20 −10 0 10 20 30
11. −10 −9 −8 −7 −6 −5 −4 −3 −2 −1 0
12. 1 2 3 4 5 6 7 8 9 10 11
13. $\{n \mid 2 < n < 5\}$; −1 0 1 2 3 4 5 6 7 8 9
14. $\{x \mid -11 \leq x < 2\}$; −14 −12 −10 −8 −6 −4 −2 0 2 4 6
15. $\{v \mid 4 < v < 18\}$; 0 2 4 6 8 10 12 14 16 18 20
16. $\{q \mid 3 \leq q \leq 7\}$; −1 0 1 2 3 4 5 6 7 8 9

716 Extra Practice

Answers

17. $\{b \mid 3 < b < 29\}$;

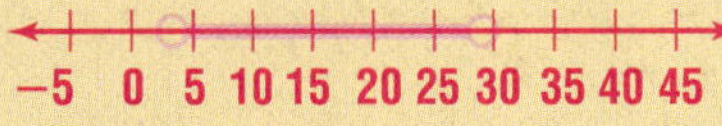

18. $\{c \mid 1 \leq c < 4\}$;

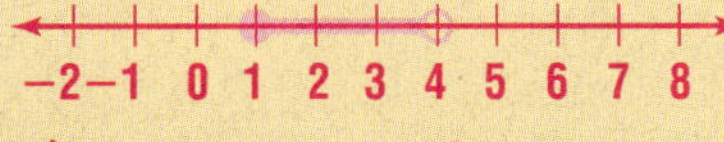

19. $\{t \mid t > 12 \text{ or } t < -7\}$; −12 −9 −6 −3 0 3 6 9 12 14 16
20. $\{d \mid -3 < d \leq 4\}$; −4 −3 −2 −1 0 1 2 3 4 5 6
21. $\{k \mid 27 < k < 30\}$; 22 23 24 25 26 27 28 29 30 31 32
22. $\{u \mid u > 9 \text{ or } u < -9\}$; −15 −12 −9 −6 −3 0 3 6 9 12 15

Lesson 12–5 *(Pages 524–529)* **Write each compound inequality without using *and*.**

1. $m > 2$ and $m < 6$
2. $j > -12$ and $j < 4$
3. $r < 6$ and $r \geq -1$
4. $x < 3$ and $x \geq 2$
5. $y \leq 5$ and $y \geq -3$
6. $s \geq -7$ and $s < -4$

1–6. See margin.

Graph the solution of each compound inequality. **7–12. See margin.**

7. $w > 6$ or $w < 2$
8. $a < -4$ or $a \geq 4$
9. $z > 12$ and $z \leq 15$
10. $h \leq 20$ and $h \geq -3$
11. $s < -7$ or $s > -5$
12. $f \leq 8$ or $f > 9$

Solve each compound inequality. Graph the solution. **13–24. See margin.**

13. $4 < 2n < 10$
14. $-4 \leq x + 7 < 9$
15. $12 > v - 6 > -2$
16. $9 \leq 3q \leq 21$
17. $24 > b - 5 > -2$
18. $8 > c + 4 \geq 5$
19. $t + 3 > 15$ or $t - 5 < -12$
20. $-12 \leq -3d < 9$
21. $5.4 < 0.2k < 6$
22. $u - 12 > -3$ or $u + 11 < 2$
23. $-4 \leq g - \frac{1}{3} < 1$
24. $\frac{p}{3} > 3$ or $\frac{p}{3} \leq -2$

Lesson 12–6 *(Pages 530–534)* **Solve each inequality. Graph the solution.**

1. $|d + 7| > 15$
2. $|5c| > 30$
3. $|a - 2| < 17$
4. $|z + 3| > 12$
5. $|j - 12| < 10$
6. $|\ell + 8| \leq -14$
7. $|t - 6| < 9$
8. $|m - 4| \leq 3$
9. $|4x| < -20$
10. $|s + 5| < 8$
11. $|n + 1| \geq 7$
12. $|-2v| > 14$
13. $|y - 3| > 16$
14. $|r - 7| \leq -11$
15. $|-6b| > 18$
16. $|w - 4| \leq 1.5$

1–16. See margin.

For each graph, write an inequality involving absolute value. **17. $|x + 3| \leq 2$**

17. –7 –6 –5 –4 –3 –2 –1 0 1 2 3

18. –3 –2 –1 0 1 2 3 4 5 6 7 **$|x - 2| < 4$**

19. –1 0 1 2 3 4 5 6 7 8 9 **$|x - 5| \leq 1$**

Write an inequality involving absolute value for each statement. Do not solve.

20. Alli's quiz score was within 5 points of her average of 85. **$|s - 85| < 5$**
21. The 5-inch-wide picture frame was made with an accuracy of 0.1 inch. **$|w - 5| \leq 0.1$**

Lesson 12–7 *(Pages 535–539)* **Graph each inequality.** **1–20. See Solutions Manual.**

1. $x < -3$
2. $y > -6$
3. $y \geq x - 3$
4. $y < x + 2$
5. $y \geq -4x$
6. $6 \geq 2x + 4y$
7. $y > -2x + 6$
8. $x + 5y \geq 15$
9. $x + y \leq -2$
10. $-3x + 2 > y$
11. $y \leq -2(x - 1)$
12. $8 \geq y - 4x$
13. $y > -x - 8$
14. $7 + 3y \leq x$
15. $2x - y \geq -6$
16. $x - y \leq 5$
17. $3(6x + y) < 4$
18. $y \leq 4(x - 2)$
19. $-(4x - 3) \geq 2y$
20. $-2(3x + y) < 1$

For Exercises 21–24, write an inequality and graph the solution. **21–24. See margin for graphs.**

21. The difference of a number and three is less than or equal to eight. **$x - 3 \leq 8$**
22. Three times a number is greater than negative six. **$3x > -6$**
23. One half the sum of a number and eight is greater than or equal to twelve. **$\frac{1}{2}(x + 8) \geq 12$**
24. A number minus three is less than another number. **$x - 3 < y$**

Answers

Lesson 12–7

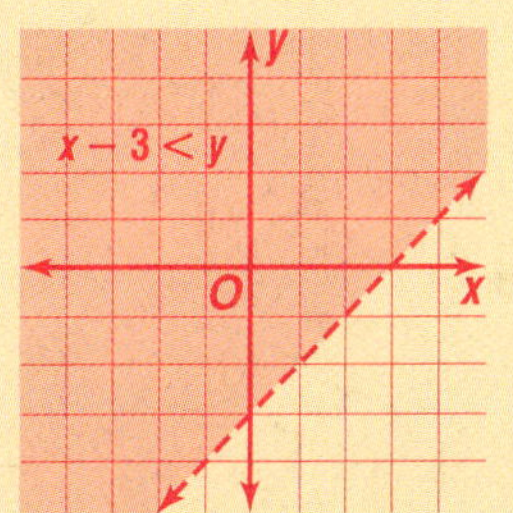

Answers

Lesson 12–5 (continued)

23. $\{g \mid -3\frac{2}{3} \leq g < 1\frac{1}{3}\}$;

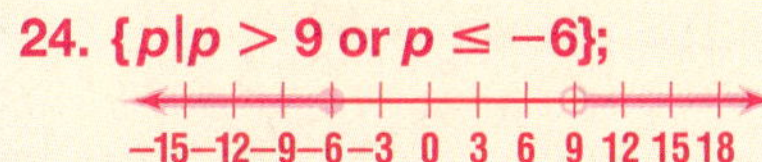

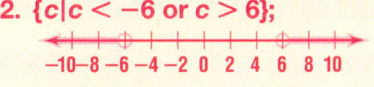

24. $\{p \mid p > 9$ or $p \leq -6\}$;
–15 –12 –9 –6 –3 0 3 6 9 12 15 18

Lesson 12–6

1. $\{d \mid d < -22$ or $d > 8\}$;
–24 –20 –16 –12 –8 –4 0 4 8 12 16
2. $\{c \mid c < -6$ or $c > 6\}$;
–10 –8 –6 –4 –2 0 2 4 6 8 10
3. $\{a \mid -15 < a < 19\}$;
–25 –20 –15 –10 –5 0 5 10 15 20 25
4. $\{z \mid z < -15$ or $z > 9\}$;
–18 –15 –12 –9 –6 –3 0 3 6 9 12
5. $\{j \mid 2 < j < 22\}$;
–8 –4 0 4 8 12 16 20 24 28 32
6. no solution
7. $\{t \mid -3 < t < 15\}$;
–12 –9 –6 –3 0 3 6 9 12 15 18
8. $\{m \mid 1 \leq m \leq 7\}$;
–1 0 1 2 3 4 5 6 7 8 9
9. no solution
10. $\{s \mid -13 < s < 3\}$;
–18 –15 –12 –9 –6 –3 0 3 6 9 12
11. $\{n \mid n \leq -8$ or $n \geq 6\}$;
–10 –8 –6 –4 –2 0 2 4 6 8 10
12. $\{v \mid v < -7$ or $v > 7\}$;
–10 –8 –6 –4 –2 0 2 4 6 8 10
13. $\{y \mid y < -13$ or $y > 19\}$;
–25 –20 –15 –10 –5 0 5 10 15 20 25
14. no solution
15. $\{b \mid b < -3$ or $b > 3\}$;
–5 –4 –3 –2 –1 0 1 2 3 4 5
16. $\{w \mid 2.5 \leq w \leq 5.5\}$;
–1 0 1 2 3 4 5 6 7 8 9

Answers

Lesson 13-2

1. inconsistent
2. consistent, independent
3. consistent, independent

Page 719
Lesson 13-6

1.
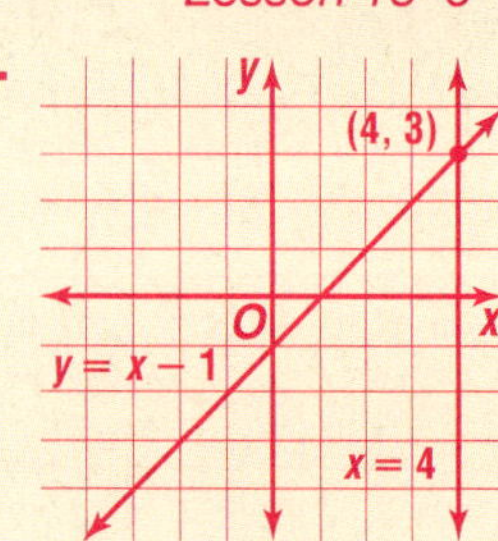

2.
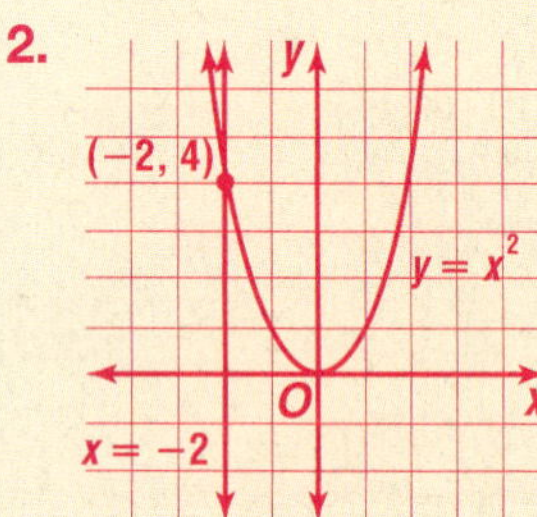

3.
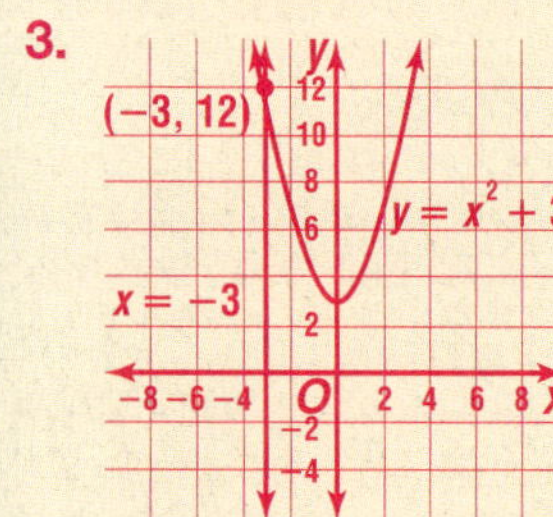

4.
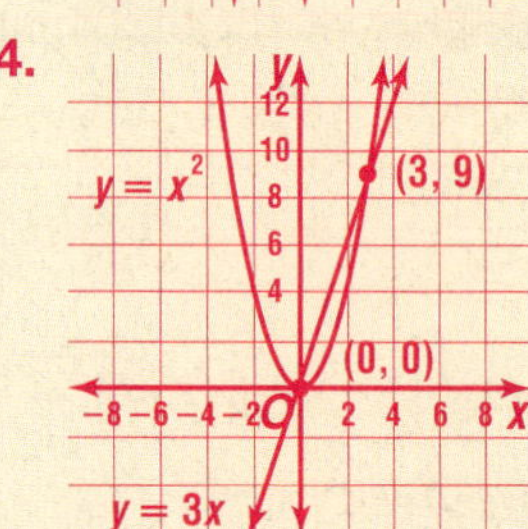

5.
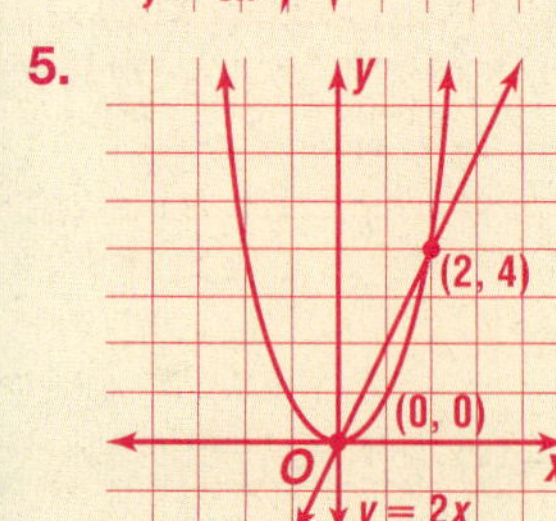

6.
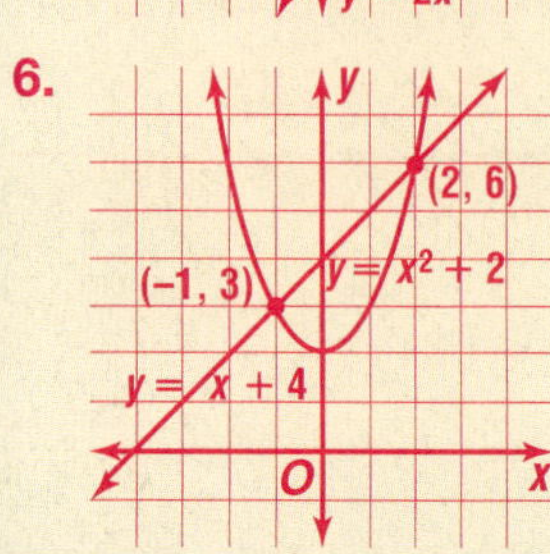

EXTRA PRACTICE

Lesson 13-1 *(Pages 550–553)* **Solve each system of equations by graphing.**

1. $x = -2$, $y = 3$ **(−2, 3)**
2. $x = 5$, $y = x$ **(5, 5)**
3. $x = -1$, $y = x - 2$ **(−1, −3)**
4. $x = -3$, $y = x + 4$ **(−3, 1)**
5. $y = -x - 4$, $y = x + 4$ **(−4, 0)**
6. $y = x + 2$, $y = 2x - 1$ **(3, 5)**
7. $y = -x - 1$, $y = x - 1$ **(0, −1)**
8. $x = 2$, $x + 2y = 4$ **(2, 1)**
9. $y = 4x + 1$, $y = 3x$ **(−1, −3)**
10. $y = -3x - 2$, $2x - y = 2$ **(0, −2)**
11. $y = 2$, $x - y = 3$ **(5, 2)**
12. $x - y = 6$, $2x + y = 3$ **(3, −3)**
13. $y = -\frac{1}{2}x$, $y = 3x + 7$ **(−2, 1)**
14. $y = \frac{1}{2}x - 3$, $x - y = 6$ **(6, 0)**
15. $\frac{1}{4}x - 2 = y$, $x = -4y$ **(4, −1)**
16. $2x + 3y = 12$, $4x + y = 4$ **(0, 4)**

1–16. See Solutions Manual for graphs.

Lesson 13-2 *(Pages 554–559)* **State whether each system is** *consistent and independent, consistent and dependent,* **or** *inconsistent.* **1–3. See margin.**

1.
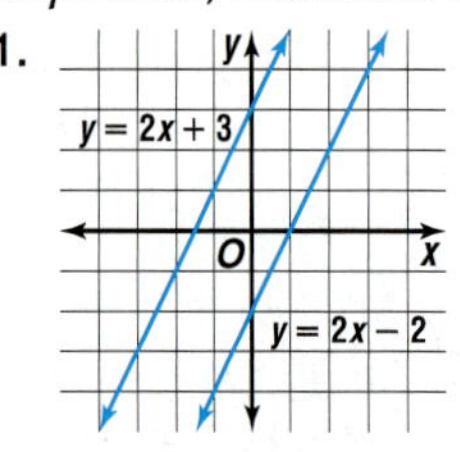

2.
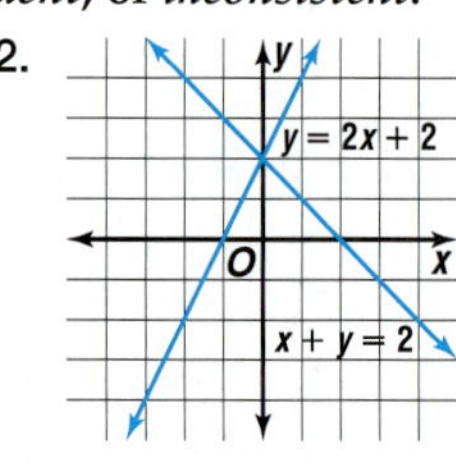

3.
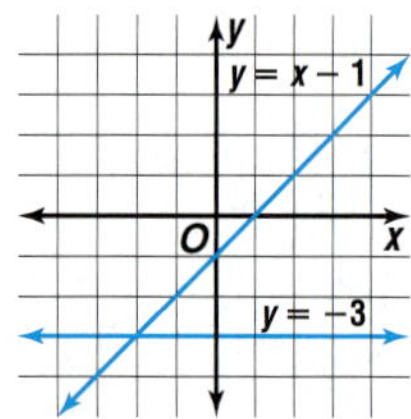

4–19. See margin for graphs.

Determine whether each system of equations has *one* **solution,** *no* **solution, or** *infinitely many* **solutions by graphing. If the system has one solution, name it.**

4. $y = 2x + 1$, $y = -x - 2$ **(−1, −1)**
5. $x = 5$, $x - 4y = 1$ **(5, 1)**
6. $y = 3$, $x - y = 2$ **(5, 3)**
7. $y = x + 7$, $x = 7 - y$ **(0, 7)**
8. $y = 2x - 6$, $y = 2x + 4$
9. $y = 10x - 16$, $y = 4x - 4$ **(2, 4)**
10. $x + 2y = 5$, $x + y = 4$ **(3, 1)**
11. $x + 4y = 5$, $2x + 6y = 6$ **(−3, 2)**
12. $y = \frac{1}{4}x - 1$, $y = -x + 4$ **(4, 0)**
13. $\frac{1}{3}y = -x - \frac{1}{3}$, $y = -3x - 1$
14. $\frac{1}{3}x - 6y = 9$, $x - 18y = 27$
15. $y = \frac{3}{2}x$, $y = -\frac{2}{3}x + 13$ **(6, 9)**
16. $x + y = 12$, $2x - y = 3$ **(5, 7)**
17. $x - 2y = -3$, $-2x + 4y = 8$
18. $4x + 6y = 12$, $2x - 6 = -3y$
19. $3x + 3y = 6$, $4x - y = 3$ **(1, 1)**

8, 17. no solution 13, 14, 18. infinitely many
9. $\left(\frac{1}{2}, -4\right)$ **10.** $\left(\frac{13}{3}, \frac{5}{3}\right)$ **11.** $\left(-\frac{8}{3}, -\frac{4}{3}\right)$

Lesson 13-3 *(Pages 560–565)* **Use substitution to solve each system of equations.**

1. $y = -x$, $x - y = -12$ **(−6, 6)**
2. $x = 5 - y$, $-3x + y = -3$ **(2, 3)**
3. $y = x + 1$, $-x + 3y = -15$ **(−9, −8)**
4. $y = x + 2$, $2x - y = 1$ **(3, 5)**
5. $y = 4x$, $x + y = 3$ $\left(\frac{3}{5}, \frac{12}{5}\right)$
6. $x = 3y$, $x + 3y = 4$ $\left(2, \frac{2}{3}\right)$
7. $y = 2x$, $2x - y = -1$ **no solution**
8. $y = 3x - 1$, $x + y = 4$ $\left(\frac{5}{4}, \frac{11}{4}\right)$
9. $y = 6x - 7$, $-2x - y = 3$
10. $x = 6 - y$, $x - 2y = 1$
11. $x + 4y = -8$, $3x - 6y = 0$
12. $x = 3 + y$, $3y + 2 = x$ $\left(\frac{7}{2}, \frac{1}{2}\right)$
13. $x - 7y = 0$, $2x + y = 0$ **(0, 0)**
14. $x - 4y = 8$, $6y + 8 = 2x$ **(−8, −4)**
15. $6x - 3y = -2$, $y + 1 = x$ $\left(-\frac{5}{3}, -\frac{8}{3}\right)$
16. $x - \frac{1}{2}y = 14$, $x + \frac{1}{2}y = 2$ **(8, −12)**

Lesson 13–4 (Pages 566–571) Use elimination to solve each system of equations.

1. $x + y = 4$, $x - y = 16$ **(10, −6)**
2. $x + 2y = 4$, $x + 3y = 6$ **(0, 2)**
3. $x = 13 - y$, $x - y = -3$ **(5, 8)**
4. $x - 6y = -1$, $x + 3y = -10$ **(−7, −1)**
5. $x - 3y = 7$, $x + 2y = 8$ **$\left(\frac{38}{5}, \frac{1}{5}\right)$**
6. $x - 4y = 5$, $-5x + 10y = -5$
7. $x - y = -2$, $2x + 2y = 16$ **(3, 5)**
8. $x - y = 14$, $3x - 6y = 15$ **(23, 9)**
9. $x - 13y = -2$, $7y - x = 5$
10. $x - 6y = 9$, $4x - 8y = 12$
11. $x + y = 8$, $16 + 3x = y$ **(−2, 10)**
12. $x - y = 9$, $4x - 2y = -8$
13. $x = 5 - y$, $6x - 4y = 0$ **(2, 3)**
14. $x = 6 - y$, $2x - 8y = 2$ **(5, 1)**
15. $4x + 3y = 12$, $8x - 4y = -12$
16. $6x - 3y = 18$, $6x + 3y = 12$ **$\left(\frac{5}{2}, -1\right)$**

6. (−3, −2) 9. $\left(-\frac{17}{2}, -\frac{1}{2}\right)$ 10. $\left(0, -\frac{3}{2}\right)$ 12. (−13, −22) 15. $\left(\frac{3}{10}, \frac{18}{5}\right)$

Lesson 13–5 (Pages 572–577) Use elimination to solve each system of equations.

1. $x + 5y = 10$, $x + y = -6$ **(−10, 4)**
2. $x + 4y = 5$, $x - 2y = -7$ **(−3, 2)**
3. $x - 6y = 9$, $x + 3y = -9$ **(−3, −2)**
4. $-x - y = 4$, $x + y = -4$
5. $2x + y = -2$, $2x - y = 4$
6. $6x + y = -9$, $3x - y = 6$
7. $3x - y = 0$, $4x + 2y = 10$ **(1, 3)**
8. $2x + 12y = 24$, $x + 7y = 18$ **(−24, 6)**
9. $4x - 8y = 0$, $x + 3y = -10$
10. $2x - 8y = 0$, $-x + 7y = -3$
11. $x - 8y = 5$, $-2x + 8y = -2$
12. $15x - y = 4$, $-6x + y = 5$ **(1, 11)**
13. $4x - 16y = 24$, $2x + 2y = 12$ **(6, 0)**
14. $x - 13y = 4$, $-2x + 10y = -4$
15. $-7x - y = 6$, $-3x + y = 4$ **(−1, 1)**
16. $6x - 3y = 18$, $10x + y = -12$

4. infinitely many 5. $\left(\frac{1}{2}, -3\right)$ 6. $\left(-\frac{1}{3}, -7\right)$ 9. (−4, −2) 10. (−4, −1) 11. (−3, −1) 14. $\left(\frac{3}{4}, -\frac{1}{4}\right)$ 16. $\left(-\frac{1}{2}, -7\right)$

Lesson 13–6 (Pages 580–585) Solve each system of equations by graphing.

1. $x = 4$, $y = x - 1$ **(4, 3)**
2. $x = -2$, $y = x^2$ **(−2, 4)**
3. $x = -3$, $y = x^2 + 3$ **(−3, 12)**
4. $y = x^2$, $y = 3x$ **(0, 0); (3, 9)**
5. $y = x^2$, $y = 2x$ **(0, 0); (2, 4)**
6. $y = x^2 + 2$, $y = x + 4$ **(−1, 3); (2, 6)**
7. $y = x^2 + 1$, $y = -x + 1$ **(−1, 2); (0, 1)**
8. $y = x^2 + 3$, $y = 5x - 3$ **(2, 7); (3, 12)**
9. $y = 4x^2$, $y = 8x$ **(0, 0); (2, 16)**
10. $y = x^2 + 12$, $y = -x - 8$
11. $y = x^2 + 4$, $y = 3x + 2$ **(1, 5); (2, 8)**
12. $y = -x^2 + 3$, $y = 3x + 5$ **(−2, −1); (−1, 2)**

1–12. See margin for graphs. 10. no solution

Use substitution to solve each system of equations. **18. $\left(-2, -\frac{1}{2}\right)$; $\left(2, -\frac{1}{2}\right)$**

13. $y = 2x^2$, $x = -2$ **(−2, 8)**
14. $y = -x^2$, $y = 5x + 6$ **(−2, −4); (−3, −9)**
15. $y = -2x^2$, $y = 4x - 6$ **(−3, −18); (1, −2)**
16. $y = -3x^2$, $x = -5$ **(−5, −75)**
17. $y = \frac{1}{2}x^2 - 1$, $x = 4$ **(4, 7)**
18. $y = -\frac{1}{2}$, $y = \frac{1}{2}x^2 - \frac{5}{2}$
19. $y = -\frac{1}{2}x^2 + \frac{7}{2}$, $x = 2x + 1$ **(1, 3); (−5, −9)**
20. $y = \frac{1}{4}x^2 - 2$, $y = 2$ **(4, 2); (−4, 2)**
21. $y = -x + 4$, $y = x^2 + 2$ **(−2, 6); (1, 3)**
22. $y = x^2 + 3$, $y = -7x^2 + 5$ **$\left(\frac{1}{2}, \frac{13}{4}\right)$; $\left(-\frac{1}{2}, \frac{13}{4}\right)$**
23. $y = x^2 + x + 4$, $y = 6$ **(−2, 6); (1, 6)**
24. $y = 3x^2 + 6x - 9$, $y = -12$ **(−1, −12)**

Answers
Lesson 13–6 (continued)

7.
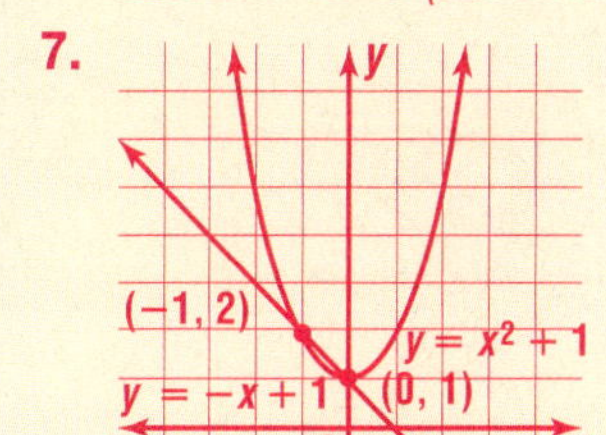

8.
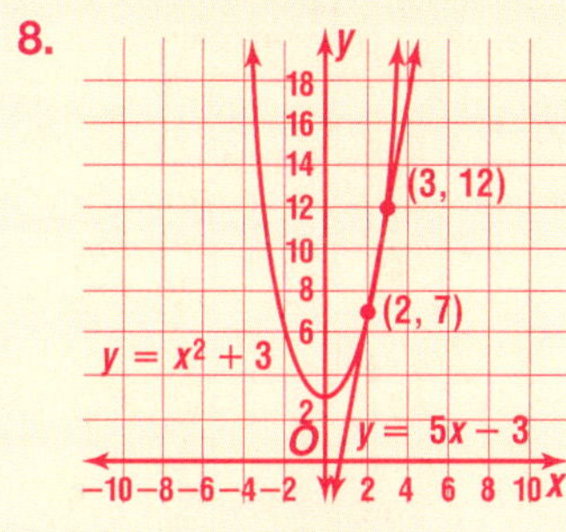

9.
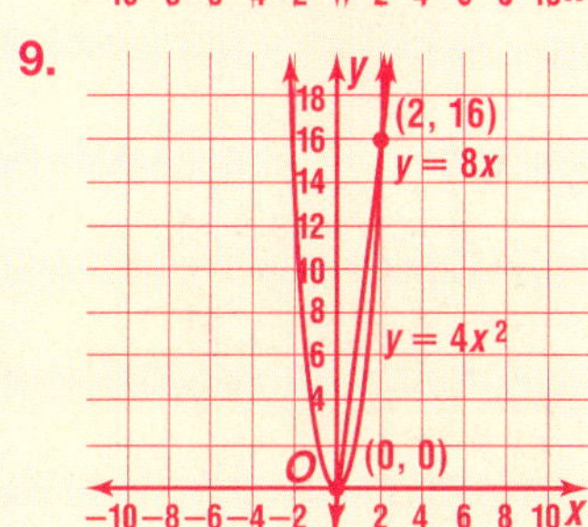

10.
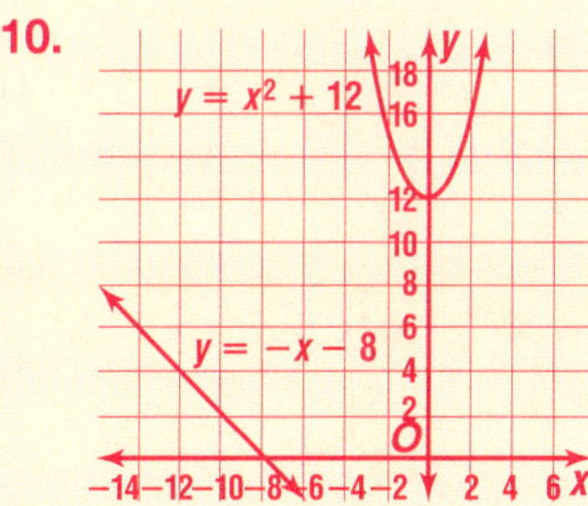

11.
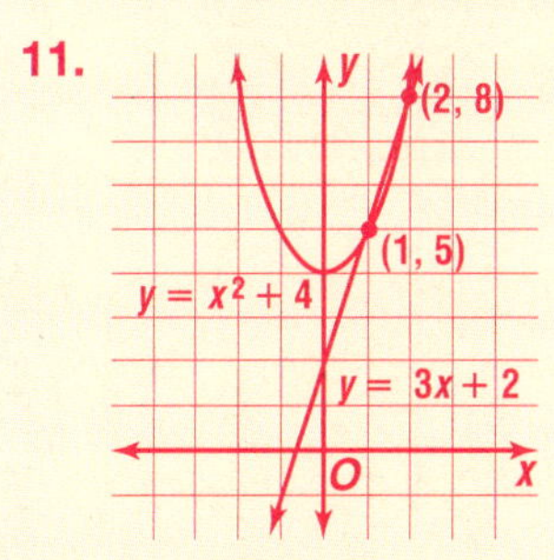

12.
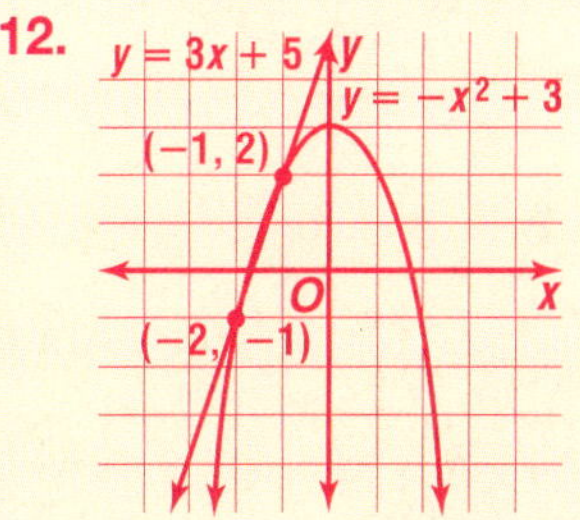

EXTRA PRACTICE

EXTRA PRACTICE

Answers

Lesson 13–7

1. 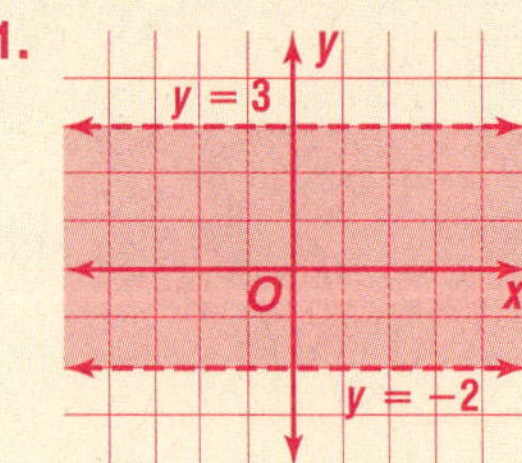

2. 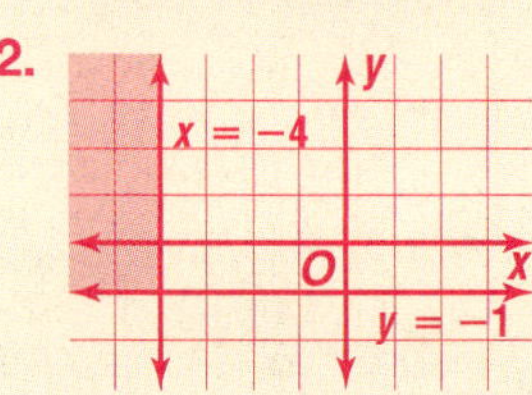

3. 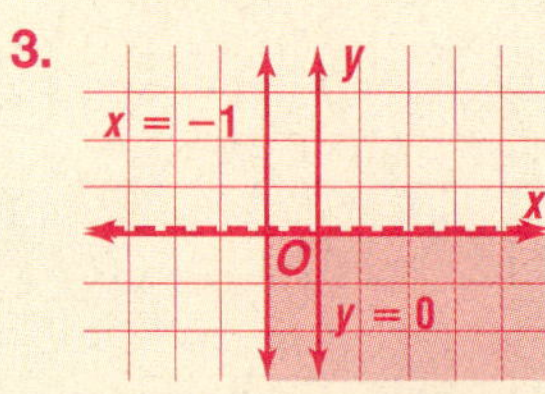

4. 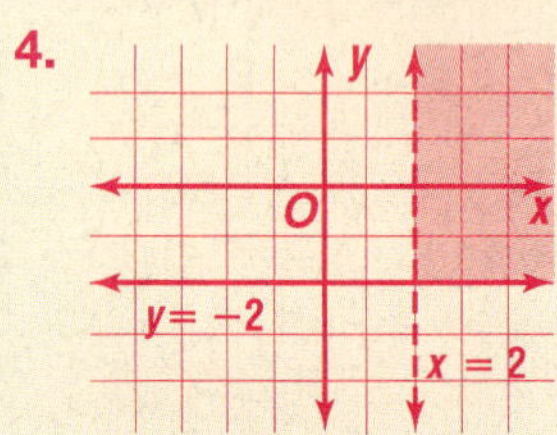

5. 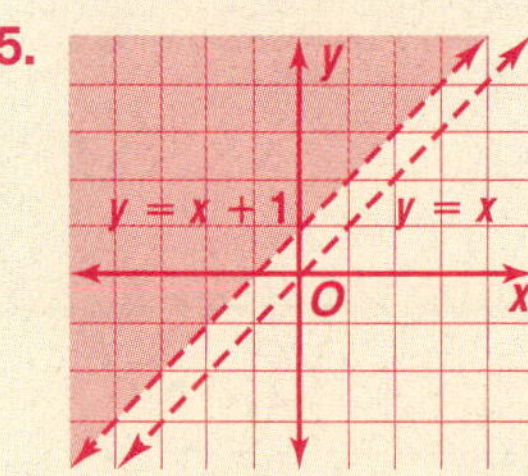

6. 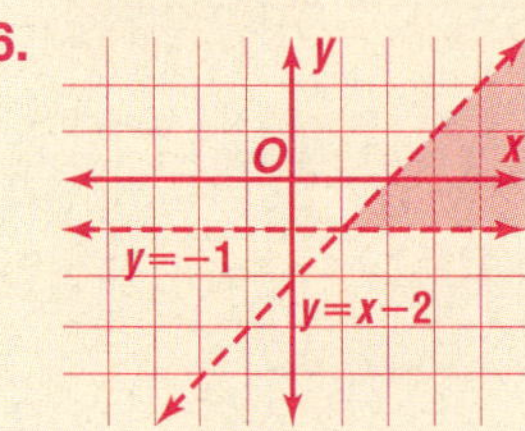

7. 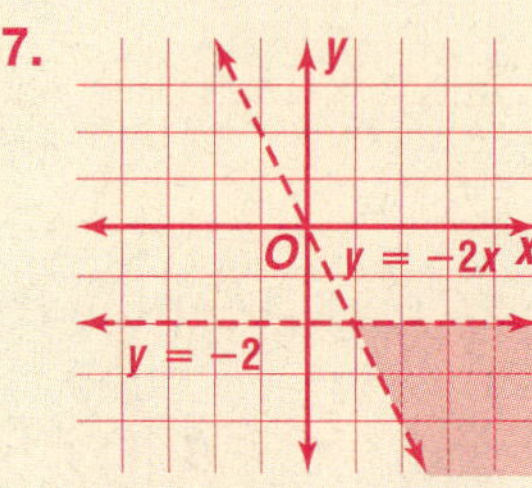

8. 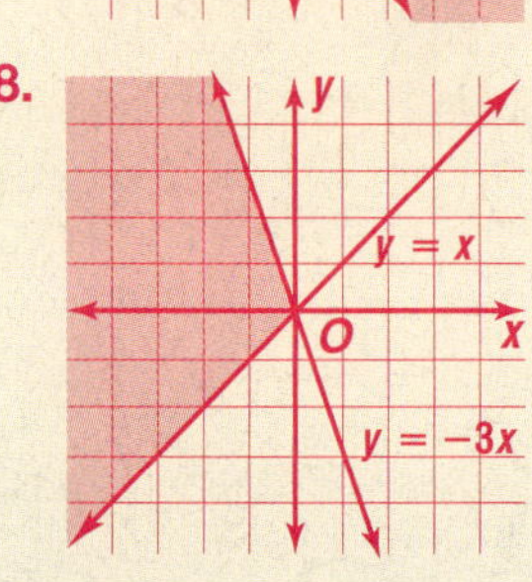

Lesson 13–7 *(Pages 586–591)* **Solve each system of inequalities by graphing.**

1. $y > -2$, $y < 3$
2. $x \le -4$, $y \ge -1$
3. $y < 0$, $x \ge -1$
4. $y \ge -2$, $x > 2$
5. $y > x$, $y > x + 1$
6. $y < x - 2$, $y > -1$
7. $y < -2$, $y > -2x$
8. $y \ge x$, $y \le -3x$
9. $x < 3$, $y < x + 1$
10. $y \ge -2$, $y \le -2x - 1$
11. $x + y > 0$, $2x + y \le 1$
12. $y \ge x + 1$, $y < 2x + 2$
13. $y - 2 \le x$, $y + 4 \ge x$
14. $y \le 3x - 1$, $2x + 4 < y$
15. $2x + 4 > y$, $y - 2 \le x$
16. $-3y - x > 2$, $2y + x < 0$

1–16. See margin.

Lesson 14–1 *(Pages 600–605)* **Name the set or sets of numbers to which each real number belongs. Let N = natural numbers, W = whole numbers, Z = integers, Q = rational numbers, and I = irrational numbers.**

1. $\frac{21}{7}$ **N, W, Z, Q**
2. $\sqrt{13}$ **I**
3. $\sqrt{121}$ **N, W, Z, Q**
4. 47.13013001 . . . **I**
5. $-\sqrt{38}$ **I**
6. 0.631 **Q**
7. $\frac{1}{4}$ **Q**
8. $-\frac{8}{2}$ **Z, Q**
9. −3 **Z, Q**
10. $-\frac{1}{6}$ **Q**
11. 0.949949994 . . . **I**
12. $-\sqrt{64}$ **Z, Q**

Find an approximation, to the nearest tenth, for each square root. Then graph the square root on a number line. **13–24. See margin for graphs.**

13. $-\sqrt{5}$ **−2.2**
14. $\sqrt{14}$ **3.7**
15. $\sqrt{18}$ **4.2**
16. $\sqrt{29}$ **5.4**
17. $\sqrt{63}$ **7.9**
18. $-\sqrt{71}$ **−8.4**
19. $\sqrt{82}$ **9.1**
20. $-\sqrt{93}$ **−9.6**
21. $\sqrt{102}$ **10.1**
22. $\sqrt{145}$ **12.0**
23. $-\sqrt{201}$ **−14.2**
24. $\sqrt{305}$ **17.5**

Determine whether each number is *rational* or *irrational*. If it is irrational, find two consecutive integers between which its graph lies on the number line.

25. $\sqrt{4}$ **R**
26. $-\sqrt{19}$
27. $-\sqrt{62}$
28. $\sqrt{33}$ **I; 5 and 6**
29. $-\sqrt{54}$
30. $\sqrt{49}$ **R**
31. $-\sqrt{225}$ **R**
32. $-\sqrt{15}$ **I; −3 and −4**
33. $\sqrt{196}$ **R**
34. $-\sqrt{152}$
35. $-\sqrt{181}$
36. $\sqrt{8}$ **I; 2 and 3**

26. I; −4 and −5 **27. I; −7 and −8** **29. I; −7 and −8** **34. I; −12 and −13** **35. I; −13 and −14**

Lesson 14–2 *(Pages 606–611)* **Find the distance between each pair of points. Round to the nearest tenth, if necessary.** **1–9. See margin.**

1. $Q(2, 16)$, $R(3, 15)$
2. $C(-4, -2)$, $D(-6, -5)$
3. $P(-8, 1)$, $Q(10, -7)$
4. $A(3, -9)$, $B(1, -2)$
5. $G(-6, -14)$, $H(-7, 2)$
6. $X(0, 0)$, $Y(9, 9)$
7. $E(12, -2)$, $F(-3, -4)$
8. $M(15, 3)$, $N(8, -11)$
9. $V(-4, 4)$, $W(6, -6)$

Find the value of *a* if the points are the indicated distance apart. **10–18. See margin.**

10. $A(a, 3)$, $B(6, 5)$; $d = 2$
11. $G(-1, 5)$, $H(-8, a)$; $d = \sqrt{85}$
12. $X(9, a)$, $Y(5, -2)$; $d = 4$
13. $P(6, 1)$, $Q(a, -7)$; $d = \sqrt{113}$
14. $C(-9, -2)$, $D(0, a)$; $d = \sqrt{90}$
15. $Q(a, -1)$, $R(4, 5)$; $d = 10$
16. $E(7, a)$, $F(-2, 4)$; $d = \sqrt{90}$
17. $M(a, 3)$, $N(-1, 5)$; $d = \sqrt{8}$
18. $V(-3, -3)$, $W(a, 4)$; $d = \sqrt{50}$

Answers

9. 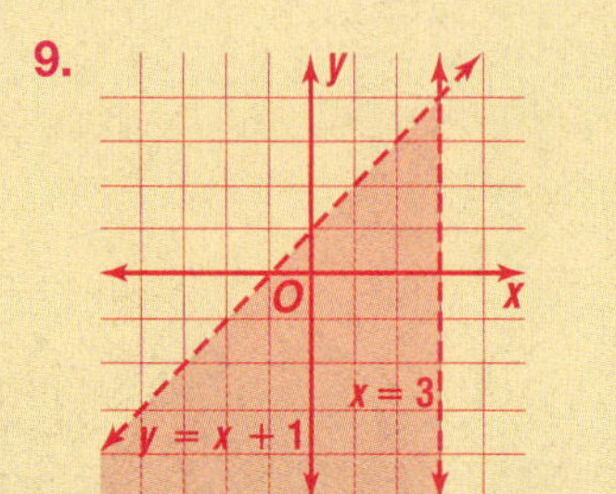

10. 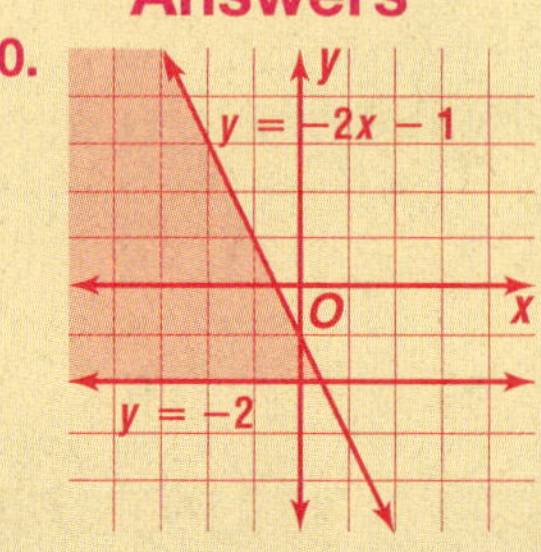

11. 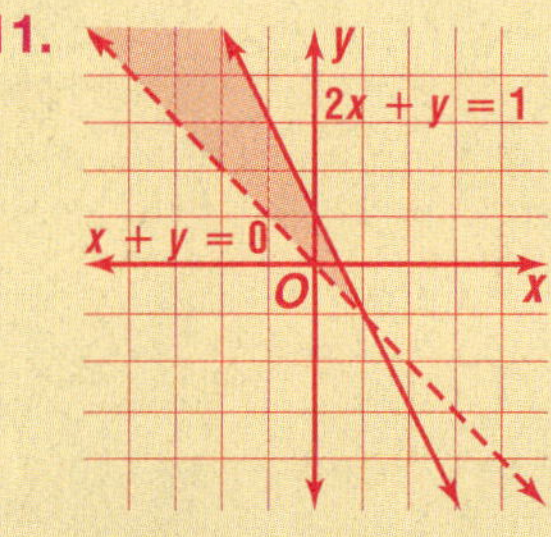

Lesson 14–3 *(Pages 614–619)* **Simplify each expression. Leave in radical form.**

1. $\sqrt{54}$ $3\sqrt{6}$ 2. $\sqrt{80}$ $4\sqrt{5}$ 3. $\sqrt{75}$ $5\sqrt{3}$ 4. $\sqrt{300}$ $10\sqrt{3}$
5. $\sqrt{3}\cdot\sqrt{12}$ 6 6. $\sqrt{4}\cdot\sqrt{8}$ $4\sqrt{2}$ 7. $2\sqrt{6}\cdot\sqrt{18}$ $12\sqrt{3}$ 8. $\sqrt{10}\cdot\sqrt{15}$ $5\sqrt{6}$
9. $\frac{\sqrt{18}}{\sqrt{9}}$ $\sqrt{2}$ 10. $\frac{\sqrt{12}}{\sqrt{4}}$ $\sqrt{3}$ 11. $\frac{\sqrt{20}}{\sqrt{8}}$ $\frac{\sqrt{10}}{2}$ 12. $\frac{\sqrt{75}}{\sqrt{5}}$ $\sqrt{15}$
13. $\frac{4}{2+\sqrt{12}}$ $-1+\sqrt{3}$ 14. $\frac{6}{3-\sqrt{8}}$ $18+12\sqrt{2}$ 15. $\frac{3}{5+\sqrt{3}}$ $\frac{15-3\sqrt{3}}{22}$ 16. $\frac{5}{4+\sqrt{10}}$ $\frac{20-5\sqrt{10}}{6}$

Simplify each expression. Use absolute value symbols if necessary.

17. $\sqrt{18a^2b}$ 18. $\sqrt{21m^8}$ $m^4\sqrt{21}$ 19. $\sqrt{120x^2y}$ 20. $\sqrt{50jk}$ $5\sqrt{2jk}$
21. $\sqrt{16r^2s^3}$ 22. $\sqrt{17n^3}$ $n\sqrt{17n}$ 23. $\sqrt{32c^4d^3}$ 24. $\sqrt{12xy^2}$ $2|y|\sqrt{3x}$
25. $\sqrt{53m^2n^4}$ 26. $\sqrt{51g^2h^2}$ 27. $\sqrt{66mn^4}$ $n^2\sqrt{66m}$ 28. $\sqrt{48j^6k^2}$
29. $\sqrt{8a^2bc^2}$ 30. $\sqrt{75x^6y^6}$ 31. $\sqrt{30rs^2t^4}$ 32. $\sqrt{196a^6}$ $14|a^3|$

17. $3|a|\sqrt{2b}$ 19. $2|x|\sqrt{30y}$ 21. $4|r|s\sqrt{s}$ 23. $4c^2d\sqrt{2d}$ 25. $|m|n^2\sqrt{53}$
26. $|gh|\sqrt{51}$ 28. $4|j^3k|\sqrt{3}$ 29. $2|ac|\sqrt{2b}$ 30. $5|x^3y^3|\sqrt{3}$ 31. $t^2|s|\sqrt{30r}$

Lesson 14–4 *(Pages 620–623)* **Simplify each expression.** 5. simplest form 8. $-9\sqrt{3}$

1. $5\sqrt{3}+7\sqrt{3}$ $12\sqrt{3}$ 2. $4\sqrt{6}-2\sqrt{6}$ $2\sqrt{6}$ 3. $5\sqrt{7}+2\sqrt{7}$ $7\sqrt{7}$
4. $13\sqrt{5}-5\sqrt{5}$ $8\sqrt{5}$ 5. $-11\sqrt{3}+6\sqrt{2}$ 6. $-6\sqrt{5}-5\sqrt{5}$ $-11\sqrt{5}$
7. $4\sqrt{2}+3\sqrt{2}-5\sqrt{2}$ $2\sqrt{2}$ 8. $-3\sqrt{3}-2\sqrt{3}-4\sqrt{3}$ 9. $2\sqrt{7}+8\sqrt{7}-14\sqrt{7}$
10. $4\sqrt{12}-7\sqrt{3}$ $\sqrt{3}$ 11. $-2\sqrt{24}+3\sqrt{6}$ $-\sqrt{6}$ 12. $-6\sqrt{32}+4\sqrt{8}$ $-16\sqrt{2}$
13. $4\sqrt{2}-3\sqrt{3}+6\sqrt{2}$ 14. $-7\sqrt{5}+6\sqrt{3}-2\sqrt{3}$ 15. $\sqrt{105}-\sqrt{12}-\sqrt{18}$
16. $-8\sqrt{24}+6\sqrt{12}-3\sqrt{2}$ 17. $4\sqrt{27}-2\sqrt{48}+3\sqrt{20}$ 18. $\sqrt{52}-\sqrt{18}+\sqrt{120}$

19. If an equilateral triangle has a side of length $5\sqrt{6}$, what is the measure of the perimeter? $15\sqrt{6}$

9. $-4\sqrt{7}$ 13. $10\sqrt{2}-3\sqrt{3}$ 14. $-7\sqrt{5}+4\sqrt{3}$ 15. $\sqrt{105}-2\sqrt{3}-3\sqrt{2}$
16. $-16\sqrt{6}+12\sqrt{3}-3\sqrt{2}$ 17. $4\sqrt{3}+6\sqrt{5}$ 18. $2\sqrt{13}-3\sqrt{2}+2\sqrt{30}$

Lesson 14–5 *(Pages 624–629)* **Solve each equation. Check your solution.** 12. no solution

1. $\sqrt{m}=4$ 16 2. $\sqrt{y}=-2$ no solution 3. $\sqrt{2d}=4$ 8
4. $-\sqrt{5k}=15$ no solution 5. $\sqrt{8t}=4$ 2 6. $\sqrt{r-9}=9$ 90
7. $\sqrt{t+5}=7$ 44 8. $\sqrt{n-3}=2$ 7 9. $\sqrt{h+5}=0$ -5
10. $\sqrt{2x+6}=6$ 15 11. $\sqrt{4m+5}-6=2$ $\frac{59}{4}$ 12. $\sqrt{3r-2}+5=4$
13. $r=\sqrt{r+12}$ 4 14. $k=\sqrt{3k+10}$ 5 15. $x-10=\sqrt{x+2}$ 14
16. $m=\sqrt{9m+4}-2$ 0, 5 17. $2+\sqrt{4d+4}=d$ 8 18. $5-\sqrt{5r-6}=9-r$ 11

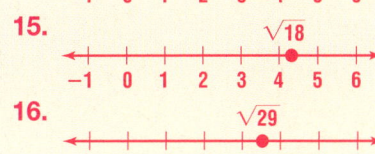

Answers

Lesson 14–2

21.

22.

23. $-\sqrt{201}$ −15 −14 −13 −12 −11 −10 −9 −8

24. $\sqrt{305}$ 13 14 15 16 17 18 19 20

1. $\sqrt{2}$ or 1.4
2. $\sqrt{13}$ or 3.6
3. $\sqrt{388}$ or 19.7
4. $\sqrt{53}$ or 7.3
5. $\sqrt{257}$ or 16.0
6. $\sqrt{162}$ or 12.7
7. $\sqrt{229}$ or 15.1
8. $\sqrt{245}$ or 15.7
9. $\sqrt{200}$ or 14.1
10. 6
11. −1 or 11
12. −2
13. −1 or 13
14. −5 or 1
15. 12 or −4
16. 1 or 7
17. −3 or 1
18. −4 or −2

Answers

Lesson 13–7 (continued)

12.
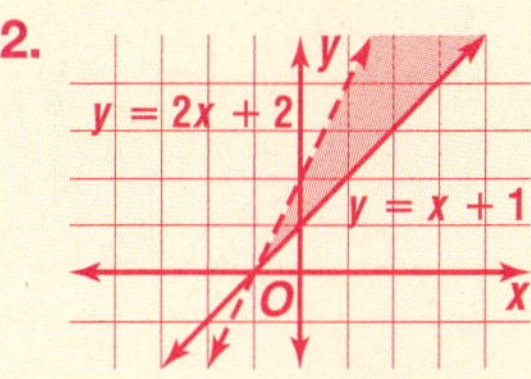

13.
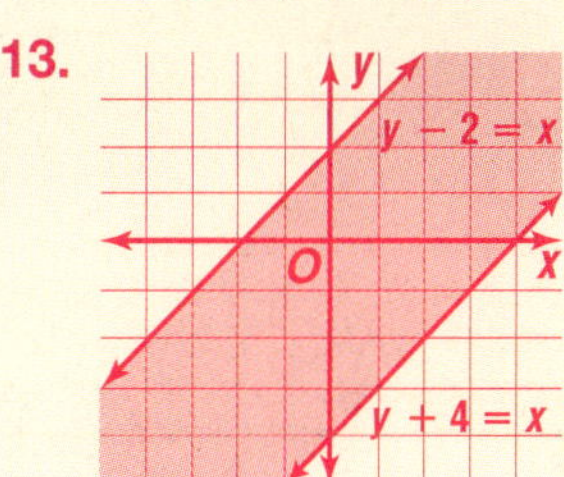

14.
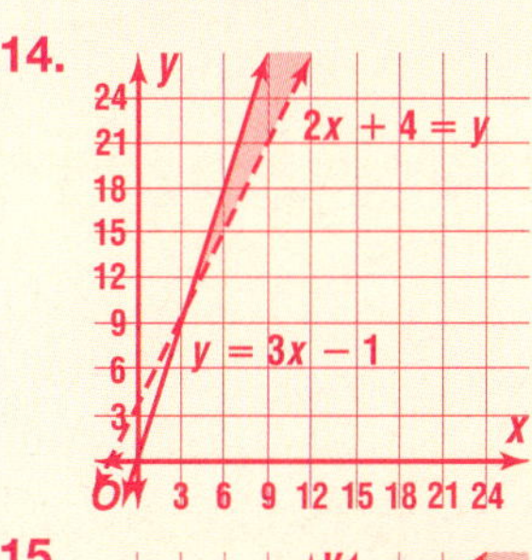

15.
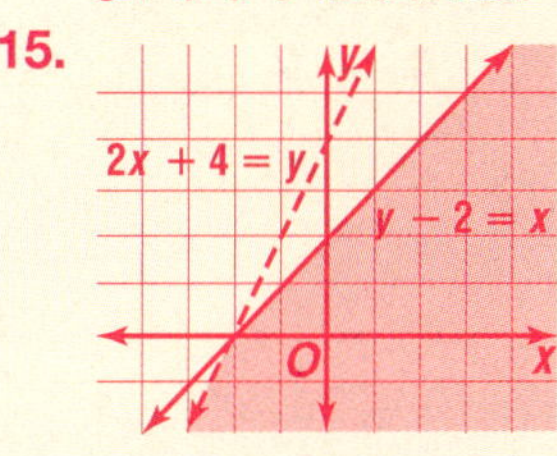

16.
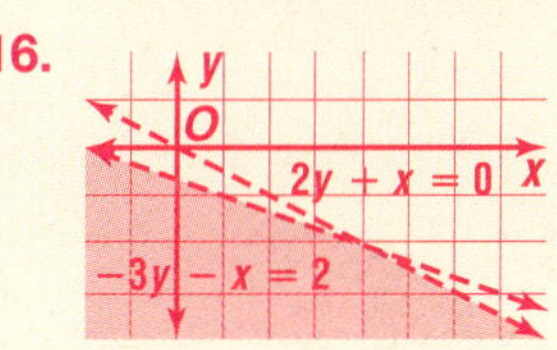

Lesson 14–1

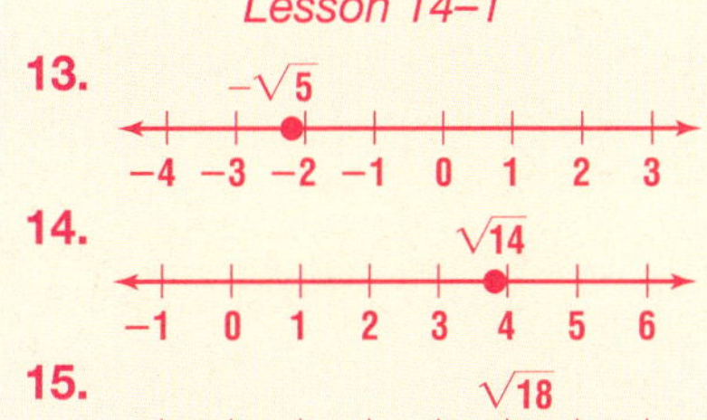

16.
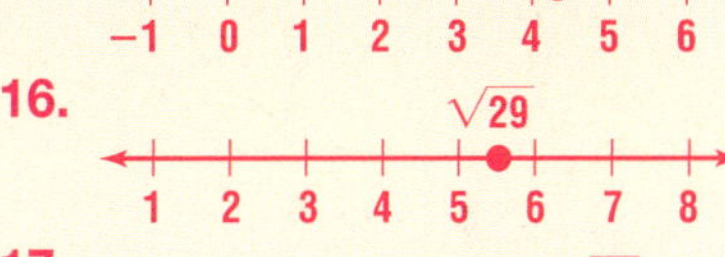

17.
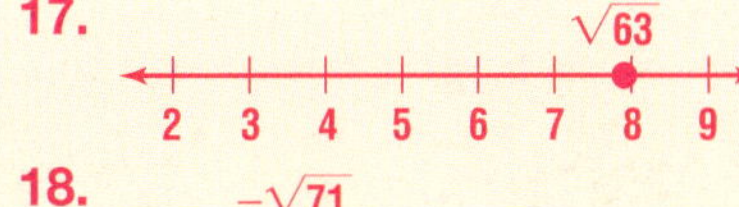

18.
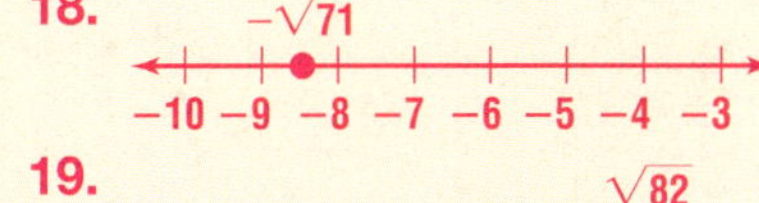

19.
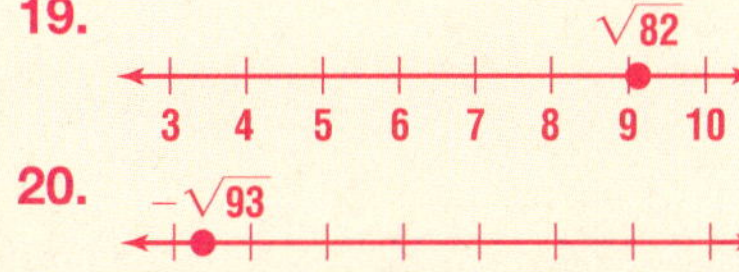

EXTRA PRACTICE

EXTRA PRACTICE

Lesson 15-1 (Pages 638–643) Find the excluded value(s) for each rational expression.

1. $\frac{y}{y-6}$ **6**
2. $\frac{2m}{3m+6}$ **−2**
3. $\frac{-n}{-8+4n}$ **2**
4. $\frac{j+1}{2j-6}$ **3**
5. $\frac{6}{r(r-2)}$ **0, 2**
6. $\frac{8k}{k(k+1)}$ **0, −1**
7. $\frac{2x}{(x-3)(x+1)}$ **3, −1**
8. $\frac{7m}{(m+1)(m+1)}$ **−1**
9. $\frac{2s}{s^2+2s}$ **0, −2**
10. $\frac{q-2}{q^2-25}$ **5, −5**
11. $\frac{3+p}{p^2-p-2}$ **2, −1**
12. $\frac{x^2-2x+3}{x^2+x-6}$ **2, −3**

Simplify each rational expression. **18. $\frac{m-1}{m+3}$** **21. $\frac{n-3}{n-2}$**

13. $\frac{8}{22}$ **$\frac{4}{11}$**
14. $\frac{6}{24}$ **$\frac{1}{4}$**
15. $\frac{30a}{36b}$ **$\frac{5a}{6b}$**
16. $\frac{2y^2}{4y}$ **$\frac{y}{2}$**
17. $\frac{2(x+1)}{x(x+1)}$ **$\frac{2}{x}$**
18. $\frac{(m-1)(m+2)}{(m+2)(m+3)}$
19. $\frac{2d^2-2d}{d^2-1}$ **$\frac{2d}{d+1}$**
20. $\frac{-4j+12}{j^2-9}$ **$-\frac{4}{j+3}$**
21. $\frac{n^2+2n-15}{n^2+3n-10}$
22. $\frac{2y^2+2y}{y^2-y-2}$ **$\frac{2y}{y-2}$**
23. $\frac{x^2+2x-8}{x^2-7x+10}$ **$\frac{x+4}{x-5}$**
24. $\frac{r^2-r-6}{r^3-6r^2+9r}$ **$\frac{r+2}{r(r-3)}$**

Lesson 15-2 (Pages 644–649) Find each product.

1. $\frac{xy}{x}\cdot\frac{z}{y}$ **z**
2. $\frac{2s}{3t}\cdot\frac{6}{4s^2}$ **$\frac{1}{st}$**
3. $\frac{7b^2}{c^2}\cdot\frac{3c}{b^2}$ **$\frac{21}{c}$**
4. $\frac{2p+1}{p^2}\cdot\frac{2p^2}{4p+2}$ **1**
5. $\frac{x(x+5)}{3}\cdot\frac{3}{x(2x+10)}$ **$\frac{1}{2}$**
6. $\frac{6r}{r+2}\cdot\frac{4r+8}{18}$ **$\frac{4r}{3}$**
7. $\frac{3n+6}{n}\cdot\frac{n^2}{n^2+4n+4}$ **$\frac{3n}{n+2}$**
8. $\frac{d^2+8d+16}{d^3}\cdot\frac{d^2}{d+4}$ **$\frac{d+4}{d}$**
9. $\frac{m}{m^2+4m+3}\cdot\frac{m+1}{m}$ **$\frac{1}{m+3}$**

Find each quotient. **17. $\frac{d-2}{6(d-4)}$**

10. $\frac{m}{n^2}\div\frac{2}{mn}$ **$\frac{m^2}{2n}$**
11. $\frac{cd}{3a^2b}\div\frac{c^2}{3a^2}$ **$\frac{d}{bc}$**
12. $6rs\div\frac{3r^2}{s}$ **$\frac{2s^2}{r}$**
13. $\frac{2}{t-1}\div\frac{t}{2t-2}$ **$\frac{4}{t}$**
14. $\frac{y^2-2y+1}{2y}\div(y-1)$ **$\frac{y-1}{2y}$**
15. $\frac{c^2-4}{3}\div\frac{c-2}{c+2}$ **$\frac{(c+2)^2}{3}$**
16. $\frac{x^2-9}{x^2}\div\frac{x+3}{x}$ **$\frac{x-3}{x}$**
17. $\frac{d}{d^2-16}\div\frac{6d}{d^2+2d-8}$
18. $\frac{b^2-4}{16b^2}\div\frac{2b+4}{4b}$ **$\frac{b-2}{8b}$**

6. $y+3$ **7. $m-3$** **9. $s-5$** **12. $x+3+\frac{21}{x-3}$** **13. $t+2+\frac{12}{t-2}$** **14. $y-4+\frac{32}{y+4}$**

Lesson 15-3 (Pages 650–655) Find each quotient. **15. $2p^2-3p+2-\frac{9}{2p+3}$**

1. $(6x-3)\div(2x-1)$ **3**
2. $(m^2+4m)\div(m+4)$ **m**
3. $(16a^2+8a)\div(4a+2)$ **$4a$**
4. $(k^2-5k-6)\div(k+1)$ **$k-6$**
5. $(s^2-5s+6)\div(s-3)$ **$s-2$**
6. $(2y^2-2y-24)\div(2y-8)$
7. $(m^2-6m+9)\div(m-3)$
8. $(r^2+3r-18)\div(r-3)$ **$r+6$**
9. $(s^2-3s-10)\div(s+2)$
10. $(4r^3-12r^2)\div(r-3)$ **$4r^2$**
11. $(6p^3-10p^2)\div(3p-5)$ **$2p^2$**
12. $(x^2+12)\div(x-3)$
13. $(t^2+8)\div(t-2)$
14. $(y^2+16)\div(y+4)$
15. $(4p^3-5p-3)\div(2p+3)$

722 Extra Practice

Lesson 15-4 (Pages 656–661) Find each sum or difference. Write in simplest form.

1. $\frac{9}{r} + \frac{6}{r}$ $\frac{15}{r}$
2. $\frac{2a}{3} + \frac{5a}{3}$ $\frac{7a}{3}$
3. $\frac{4}{j} - \frac{3}{j}$ $\frac{1}{j}$
4. $\frac{2b}{4} - \frac{4b}{4}$ $-\frac{b}{2}$
5. $\frac{3}{2s} + \frac{1}{2s}$ $\frac{2}{s}$
6. $\frac{6}{3b} - \frac{9}{3b}$ $-\frac{1}{b}$
7. $\frac{2a}{a} + \frac{7a}{a}$ 9
8. $\frac{y}{16y} - \frac{y}{16y}$ 0
9. $\frac{15}{8x} + \frac{1}{8x}$ $\frac{2}{x}$
10. $\frac{2}{9k} + \frac{5}{9k}$ $\frac{7}{9k}$
11. $\frac{4}{6p} + \frac{10}{6p}$ $\frac{7}{3p}$
12. $\frac{14}{25r} - \frac{19}{25r}$ $-\frac{1}{5r}$
13. $\frac{11}{10s} - \frac{12}{10s}$ $-\frac{1}{10s}$
14. $\frac{9}{2y} - \frac{3}{2y}$ $\frac{3}{y}$
15. $\frac{6}{6+m} + \frac{m}{6+m}$ 1
16. $\frac{6s}{2-s} + \frac{3}{2-s}$ $\frac{6s+3}{2-s}$
17. $\frac{2p}{3+j} - \frac{p}{3+j}$
18. $\frac{16}{m-5} - \frac{10}{m-5}$
19. $\frac{3r}{r+3} + \frac{9}{r+3}$ 3
20. $\frac{-2x-4}{x+7} + \frac{2x+11}{x+7}$
21. $\frac{8}{6m-3} - \frac{4}{6m-3}$
22. $\frac{4x+5}{2x+6} + \frac{2-4x}{2x+6}$
23. $\frac{8-6r}{5r-2} + \frac{11r-10}{5r-2}$ 1
24. $\frac{12t+1}{t+1} + \frac{3-8t}{t+1}$ 4

17. $\frac{p}{3+j}$ 18. $\frac{6}{m-5}$ 20. $\frac{7}{x+7}$ 21. $\frac{4}{6m-3}$ 22. $\frac{7}{2x+6}$

Lesson 15-5 (Pages 662–667) Find the LCM for each pair of expressions.

1. $4mn, 6m^2$ $12m^2n$
2. $8x^2y, 20xy$ $40x^2y$
3. $4cd, 18d$ $36cd$
4. $d^2-9, d+1$ $(d-3)(d+3)(d+1)$
5. a^2+a-2, a^2+5a-6 $(a+2)(a-1)(a+6)$
6. x^2+x-6, x^2-x-12 $(x-2)(x+3)(x-4)$

Write each pair of rational expressions with the same LCD.

7. $\frac{2}{6mn}, \frac{1}{30}$ $\frac{10}{30mn}, \frac{mn}{30mn}$
8. $\frac{5}{4p}, \frac{3}{2p}$ $\frac{5}{4p}, \frac{6}{4p}$
9. $-\frac{1}{6rs}, \frac{3}{4st}$ $-\frac{2t}{12rst}, \frac{9r}{12rst}$
10. $\frac{5a}{3a^2b}, -\frac{4}{2a}$ $\frac{10a}{6a^2b}, -\frac{12ab}{6a^2b}$
11. $\frac{1}{y-2}, -\frac{3}{y^2-4}$
12. $\frac{m^2}{2m+4}, \frac{6m}{4m+8}$

11. $\frac{y+2}{y^2-4}, -\frac{3}{y^2-4}$ 12. $\frac{2m^2}{4m+8}, \frac{6m}{4m+8}$

Find each sum or difference in simplest form.

13. $\frac{r}{8} - \frac{r}{3}$ $-\frac{5r}{24}$
14. $\frac{2p}{6} - \frac{p}{5}$ $\frac{2p}{15}$
15. $-\frac{r}{4} + \frac{3r}{7}$ $\frac{5r}{28}$
16. $\frac{6t}{9} + \frac{t}{6}$ $\frac{5t}{6}$
17. $\frac{2m}{5p^2} + \frac{2}{6p}$ $\frac{6m+5p}{15p^2}$
18. $\frac{5x}{x-1} + \frac{7}{x}$
19. $\frac{6m}{j^2} + \frac{7m}{3j}$
20. $\frac{2t}{3+r} - \frac{9}{r^2}$
21. $\frac{4}{b^3} + \frac{3}{b}$ $\frac{3b^2+4}{b^3}$
22. $\frac{6y+1}{2y+1} - \frac{1}{4y+2}$ $\frac{12y+1}{4y+2}$
23. $\frac{1}{x^2+x-2} + \frac{3}{x+2}$ $\frac{3x-2}{(x+2)(x-1)}$
24. $\frac{4}{b^2-16} + \frac{8}{b+4}$ $\frac{8b-28}{(b-4)(b+4)}$

18. $\frac{5x^2+7x-7}{x(x-1)}$ 19. $\frac{18m+7mj}{3j^2}$ 20. $\frac{2r^2t-9r-27}{r^3+3r^2}$

Lesson 15-6 (Pages 668–673) Solve each equation. Check your solution.

1. $2 = \frac{2x}{5} + \frac{x}{10}$ 4
2. $\frac{y}{5} + \frac{3y}{5} = \frac{4}{5}$ 1
3. $\frac{m}{8} - \frac{3m}{4} = \frac{1}{2}$ $-\frac{4}{5}$
4. $\frac{d}{5} = \frac{6d}{5} - \frac{3}{10}$ $\frac{3}{10}$
5. $\frac{8}{7v} + \frac{3}{4v} = -2$ $-\frac{53}{56}$
6. $\frac{s+1}{2} + \frac{s}{4} = -2$ $-\frac{10}{3}$
7. $\frac{2b-1}{2} = \frac{b}{3} - \frac{1}{4}$ $\frac{3}{8}$
8. $\frac{m-1}{3} + \frac{m+4}{3} = -5$ -9
9. $\frac{4}{c+1} + \frac{2}{c+1} = -3$ -3
10. $\frac{7}{3d} - \frac{5}{9d} = -3$ $-\frac{16}{27}$
11. $\frac{1}{2m} + \frac{3}{m} = \frac{1}{4}$ 14
12. $\frac{1}{r} + \frac{2}{r-5} = \frac{3}{r}$
13. $\frac{1}{3a+5} - \frac{4a}{6a+10} = \frac{1}{2}$ $-\frac{3}{7}$
14. $\frac{1}{k-3} + \frac{k}{2} = \frac{k-2}{2}$ 2
15. $\frac{r+2}{r^2-4} + \frac{1}{r-2} = \frac{1}{r}$

12, 15. no solution

EXTRA PRACTICE

Extra Practice 723

Mixed Problem Solving

Chapter 1 Writing Expressions and Equations

1. **Car Rental** A rental company charges \$120 dollars for a weekly car rental plus \$0.25 per mile. *(Lesson 1-1)*
   a. Write an expression for the total cost of driving 250 miles. **120 + 0.25(250)**
   b. Write an expression for the total cost of driving m miles. **120 + 0.25*m***
2. **Entertainment** The cost to go to the movies for a adults and c children is $8.50a + 5.50c$. If Mr. Washington takes his three children, what is the total cost? *(Lesson 1-2)* **\$25**
3. **Fencing** Emma wants to fence her horse pasture. Write and evaluate an expression for the amount of fence that she will need. *(Lesson 1-3)* **50 + 71 + 109 + 64; 294 ft**

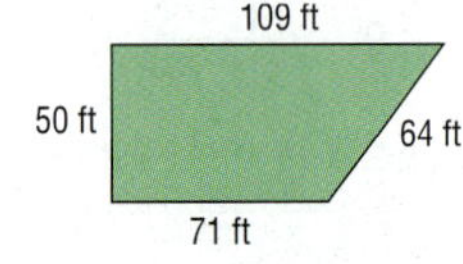

4. **Money** Hats cost \$12 each, and T-shirts cost \$8 each. Use the Distributive Property to write an expression for the cost of 11 hats and shirts for a softball team. Then find the total cost. *(Lesson 1-4)* **11(12 + 8); \$220**
5. **Savings** Marta deposited \$300 into a savings account that pays 5% simple interest. How much more money will she have after 5 years than after 2 years? *(Lesson 1-5)* **\$45**
6. **Education** The table shows the number of hours that each student in the orchestra practiced in a week.

Hours of Practice							
14	3	2	15	4	14	13	3
13	17	5	4	12	3	13	15
16	21	12	13	3	14	5	23

   a. Make a frequency table of the data. *(Lesson 1-6)* **See margin.**
   b. Make a stem-and-leaf plot of the data. *(Lesson 1-7)* **See margin.**

Chapter 2 Integers

1. **Astronomy** The table gives the surface temperature of the nine planets in our solar system. Order the temperatures from least to greatest. *(Lesson 2-1)* **See margin.**

Planet Temperatures (°F)				
333	867	59	−81	−162
−323	−330	−369	−218	

Source: *The World Almanac For Kids*

2. \$6, \$18, \$36; See margin for graph.

2. **Babysitting** Ellie earns \$6 per hour for babysitting. Find her total earnings for 1, 3, and 6 hours. Graph the ordered pairs (number of hours, earnings). *(Lesson 2-2)*
3. **Money** Nantan borrowed \$30 from his parents. In the next two weeks, he paid them \$40 and then borrowed \$15 and \$25. How much does he owe them? *(Lesson 2-3)* **\$30**
4. **Geography** The highest elevation in the world is Mt. Everest in Tibet, Nepal. It is 29,035 feet above sea level. The lowest elevation in the world is the Dead Sea in Israel and Jordan. It is 30,384 feet lower than Mr. Everest. What is the elevation of the lowest point in the world? *(Lesson 2-4)* **1349 ft below sea level**
5. **Farming** From 2000 to 2002, the number of farms in Missouri decreased by about 1000 per year. If there were 109,000 farms in 2000, how many farms were there in Missouri in 2002? *(Lesson 2-5)* **107,000**
6. **Education** The table shows the cost of higher education in public institutions for several years. Find the average yearly change from 1999 to 2002. *(Lesson 2-6)* **+\$313 per year**

Year	Cost per year
1999	\$7107
2000	\$7310
2001	\$7586
2002	\$8046

Source: U.S. Dept. of Education

Answers

Chapter 1

6a. See below.

6b.

Stem	Leaf
0	2 3 3 3 3 4 4 5 5
1	2 2 3 3 3 3 4 4 4 5 5 6 7
2	1 3

$1|2 = 12$

Chapter 2

1. −369, −330, −323, −218, −162, −81, 59, 333, 867
2. \$6, \$18, \$36

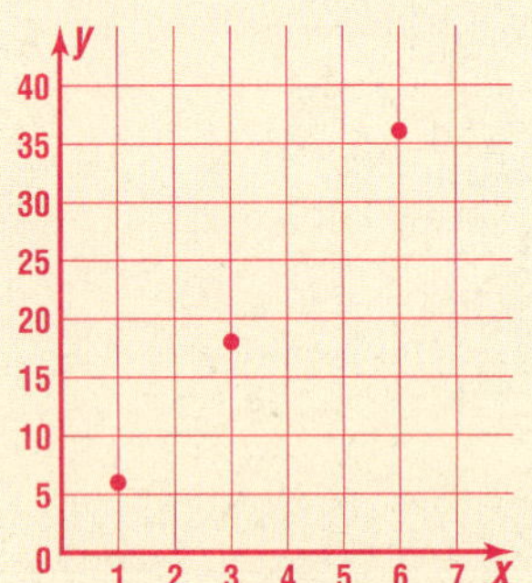

6a. Sample answer:

Number of hours practiced	Tally	Frequency
0–5	𝍸 ||||	9
6–10		0
11–15	𝍸 𝍸 |	11
16–20	||	2
21–25	||	2

Chapter 3 Addition and Subtraction Equations

1. **Shopping** B's Market sells oranges 10 for \$3.99. QuickMart sells 8 oranges for \$2.59. Which is the better buy? *(Lesson 3-1)* **8 oranges for \$2.59**

2. **Population** The table shows the population for certain years in Pennsylvania. What is the change in population from 1999 to 2002? *(Lesson 3-2)* **+1,837,079**

Year	Population
1999	10,498,012
2000	11,881,643
2001	12,281,054
2002	12,335,091

Source: U.S. Bureau of the Census

3. **Parks** The table shows the acreage of five National Seashores. What is the mean acreage? *(Lesson 3-3)* **39,912.8 acres**

Seashore	Acreage
Assateague Island	39,733
Canaveral	57,662
Cape Cod	43,605
Cape Hatteras	30,321
Cape Lookout	28,243

Source: *Time Almanac*

4. **Health** The body mass index (BMI) is found by dividing weight in kilograms by height in meters squared or $BMI = \frac{w}{h^2}$. For a person that is 1.7 meters tall, what are the weights that correspond to a BMI of 18.5 and a BMI of 25? *(Lesson 3-4)* **53.465 kg, 72.25 kg**

5. **History** The *Mayflower* sailed for America in 1620. The Jamestown settlers had sailed for America 14 years earlier. Write an equation that represents these years. In what year did the Jamestown settlers sail for America? *(Lesson 3-6)* **$x + 14 = 1620$; 1606**

6. **Games** To win a radio contest, Sam guesses the price of a car as \$25,000. To win, he must be within \$1000 of the actual price. Write and solve an absolute value equation to find the greatest and least values that the price could be so that Sam wins the car. *(Lesson 3-7)* **7. $|s - 25{,}000| = 1000$; 26,000 and 24,000**

Chapter 4 Multiplication and Division Equations

1. **Track** For his training, Marcus runs 55 miles each week. If he ran 8 miles every day from Monday through Saturday, how many miles will he run on Sunday? *(Lesson 4-1)* **7**

2. **School** If the morning consists of four class periods and Soledad needs to take Biology, Algebra, American History, and English 9, how many different morning schedules are possible? *(Lesson 4-2)* **24**

3. **Food** A recipe for a fruit slush drink calls for $2\frac{1}{4}$ cups of sugar for each batch. How many batches can be made with 9 cups of sugar? *(Lesson 4-3)* **4**

4. **Geometry** Find the height of the trapezoid if its area is 120 square inches. Use the formula $A = \frac{1}{2}h(b_1 + b_2)$. *(Lesson 4-4)* **6 in.**

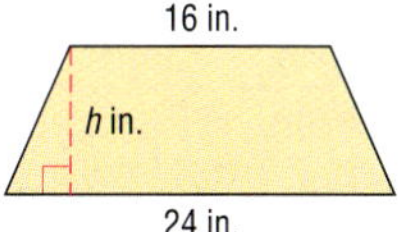

5. $23.99 + 0.10m = 43.99$; $m = 200$

5. **Money** Jessica received a cell phone bill for \$43.99. If her phone plan charges \$23.99 each month plus \$0.10 per minute, write an equation for the number of minutes that Jessica talked. How many minutes did she talk? *(Lesson 4-5)*

6. **Car Rental** Able Car Rental charges \$50 a day to rent a car and \$0.25 per mile. Century Car Rental charges \$30 a day and \$0.35 per mile. The equations that represent the total amount A charged by each company for m miles driven are given. For how many miles will the companies charge the same amount? *(Lesson 4-6)* **200 mi**

   Able: $A = 50 + 0.25m$
   Century: $A = 30 + 0.35m$

7. **Gardening** Liam has 60 feet of rock to put a walkway around his garden. Write an equation to find the width w of the garden. Find the garden's dimensions. *(Lesson 4-7)* **$2w + 2(3w + 2) = 60$; 7 ft by 23 ft**

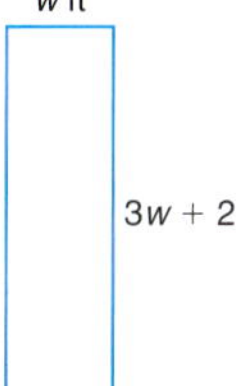

MIXED PROBLEM SOLVING

Mixed Problem Solving 725

Answers

Chapter 6

1a. D: {1984, 1988, 1992, 1996, 2000, 2004}; R: {21.81, 21.34, 21.81, 22.12, 21.84, 22.05}

1b.

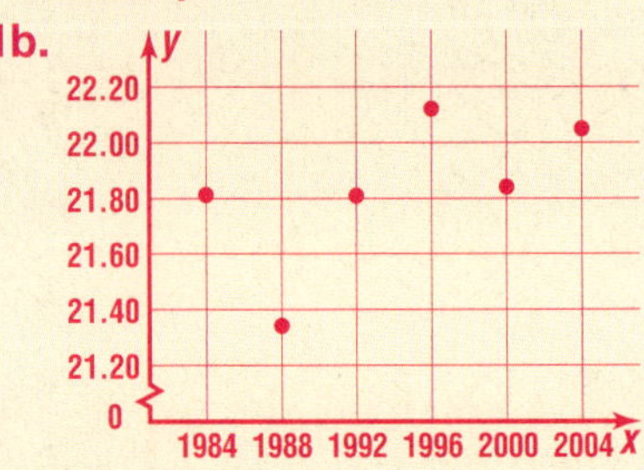

3.

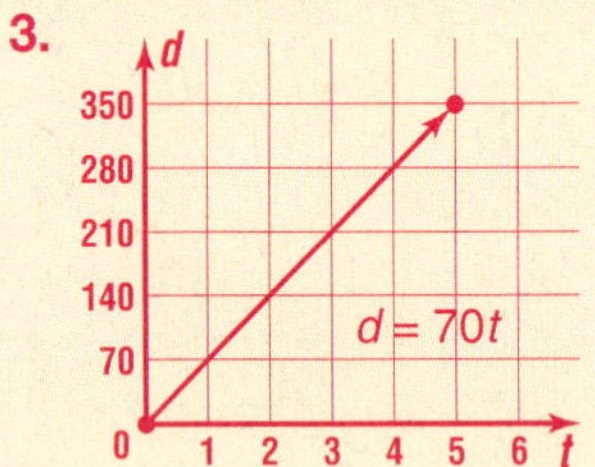

MIXED PROBLEM SOLVING

Chapter 5 Proportional Reasoning and Probability

1. **Crafts** Lanelle makes stuffed bears out of fabric for gifts. It takes $\frac{3}{4}$ of a yard for 1 bear. How many yards are needed for 8 bears? *(Lesson 5-1)* **6 yd**

2. **Models** Conor built a model of a stegosaurus with a length of 18 inches for his science project. The scale of the model is 3 inches = 5 feet. What is the actual length of a stegosaurus? *(Lesson 5-2)* **30 ft**

3. **Money** The table shows the number of each type of pizza sold at Pizza Palace over a lunch hour. What percent of each type of pizza were sold that hour? *(Lesson 5-3)*

Type of Pizza	Number Sold
Pepperoni	15
Cheese	24
Other	11

3. 30% pepperoni, 48% cheese, 22% other

4. **Fund-Raisers** A charity sponsored a race to raise funds for cancer research. Adults paid \$25 to participate and youths paid \$10. There were 45 more youth participants than adults. They earned a total of \$19,000. How many youth and adults participated? *(Lesson 5-4)* **530 adults and 575 youth**

5. **Money** Yuan bought a pair of shoes for \$80 that were originally \$95. Find the percent of increase or decrease in price. Round to the nearest percent. *(Lesson 5-5)* **16% decrease**

6. **Games** To win a prize at a carnival, you must spin the spinner and land on a number greater than 5. What are the odds of spinning a number greater than 5? *(Lesson 5-6)* **3:5**

7. **Music** Of the 16 CDs in Mariah's car, 5 are jazz, 3 are classical, 6 are soundtracks, and 2 are pop music. If Mariah chooses a CD at random, what is the probability that it is a soundtrack or a pop CD? *(Lesson 5-7)* $\frac{1}{2}$

Chapter 6 Functions and Graphs

1. **Olympics** The table shows the winning times for the women's Olympics 200-meter dash since 1984. *(Lesson 6-1)*

Year	Time (seconds)
1984	21.81
1988	21.34
1992	21.81
1996	22.12
2000	21.84
2004	22.05

Source: *Time Almanac*

   a. Determine the domain and range of the relation. **See margin.**

   b. Graph the relation. **See margin.**

   c. Between which consecutive games was the decrease in time the greatest? **between 1984 and 1988**

2. **Mail** The equation $C = 1.42 + 0.42b$ gives the cost C to send a book in the mail for each pound b over 1 pound. Find the total cost for mailing books that weigh 3, 5, and 7 pounds. *(Lesson 6-2)* **\$2.26, \$3.10, \$3.94**

3. **Driving** Raina drove at a constant rate of 70 miles per hour. The equation that can be used to find the distance d traveled in t hours is given by $d = 70t$. Graph the equation. *(Lesson 6-3)* **See margin.**

4. **Money** Michael's monthly cell phone bill is given by $C(x) = 24.99 + 0.15x$, where x is the number of minutes he talks. Suppose Michael talks for 184 minutes one month. What will be his phone bill? *(Lesson 6-4)* **\$52.59**

5. **Landscaping** Ben charged Mr. Diaz \$18 to landscape his yard for 3 hours. At this rate, what would Ben charge Ms. Lewis for 7 hours of landscaping? *(Lesson 6-5)* **\$42**

6. **Geometry** The volume of a rectangular prism is 100 cubic meters. The area A of its base varies inversely as its height h. Write an equation that represents this inverse variation. *(Lesson 6-6)* $Ah = 100$

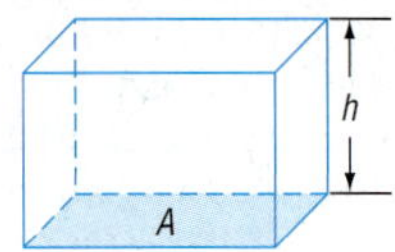

726 Mixed Problem Solving

Chapter 7 Linear Equations

1. **Entertainment** The linear function $C(x) = 19.50x$ represents the total cost of x tickets to an amusement park. Explain how changes in x affect $C(x)$. What is the slope of the graph of the equation? *(Lesson 7-1)* **See margin.**

2. **Elections** The table shows the national voter turnout in federal elections in 1996 and 2000. Write the point-slope form of an equation for the line passing through these points. *(Lesson 7-2)* **See margin.**

Year	Voters
1996	96,456,345
2000	105,586,274

Source: Federal Election Commission

3. **Business** Jamie holds scrapbooking parties. She charges \$50 for supplies and \$12 per person. Write an equation in slope-intercept form for the function if y is the total cost for x people. *(Lesson 7-3)* $y = 12x + 50$

4. **Driving** Determine whether a scatter plot for the distance traveled in a car for the time elapsed would show a *positive*, *negative*, or *no* relationship between the variables. Explain. *(Lesson 7-4)* **See margin.**

5. **Business** Denzel charges \$400 for a landscape plan and \$25 per hour for installation. Write a linear function to represent the total cost $f(x)$ as a function of the number of hours worked x. *(Lesson 7-5)* $f(x) = 400 + 25x$

6. **Track** Jason is running at a pace of 10 miles per hour while Cruz is running at a pace of 6 miles per hour. Make a conjecture about the similarities and differences in the graphs of the equations that represent the distance traveled by each runner d in time t. Verify by graphing the equations. *(Lesson 7-6)* **See margin.**

7. **Geometry** Determine whether $\triangle ABC$ is a right triangle if the vertices are $A(2, 1)$, $B(10, 15)$, and $C(3, 19)$. Justify your answer. *(Lesson 7-7)* **See margin.**

Chapter 8 Powers and Roots

1. **Sales** The Band Boosters will sell food at the basketball game. To determine how much popcorn to order, Monisha uses the formula $V = \pi r^2 h$ to find the volume of a cylindrical popcorn container. In the formula, $\pi \approx 3.14$, r is the radius of the base of the cylinder, and h is the height. Find the volume of the cylinder. *(Lesson 8-1)* **283 in³**

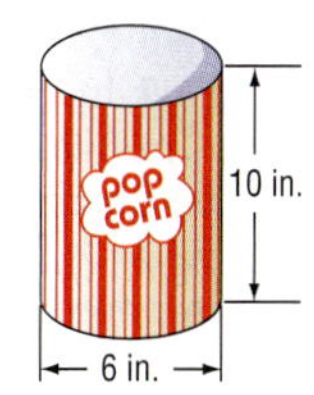

2. **Gardening** Hana is planting a vegetable garden. She wants it to be square as shown. Hana decides that the garden is not large enough so she doubles the length and width. Write an expression for the area of the new garden. *(Lesson 8-2)*

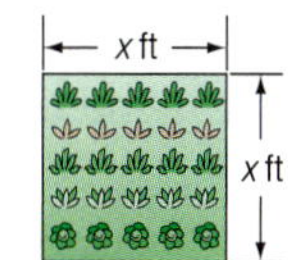

2. $A = (2x)(2x) = 4x^2$ 3. $\frac{1}{10^3} = 0.001$

3. **Measurement** There are 10^{-3} grams in a milligram. Express 10^{-3} using positive exponents and evaluate. *(Lesson 8-3)*

4. **Astronomy** Mercury is 36 million miles from the Sun. Express the distance in scientific notation. *(Lesson 8-4)* 3.6×10^7

5. **Geometry** The area of a circle is given by the formula $A = \pi r^2$. If the area of a circle is 121π square feet, what is the radius? *(Lesson 8-5)* **11 ft**

6. **Decorating** Fiona's square room has an area of 154 square feet. What is a whole-number estimate of the length of a side? *(Lesson 8-6)* **12 ft**

7. **Skateboarding** How far from the base of the platform does the skateboarding ramp reach? *(Lesson 8-7)* **12 ft**

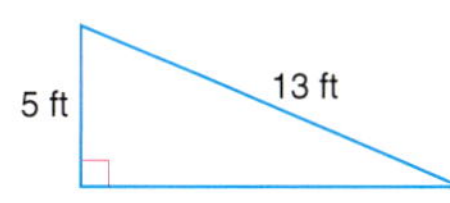

Answers

Chapter 7

1. As x increases, $C(x)$ also increases; the slope is 19.50.
2. $y - 105{,}586{,}274 = \frac{9{,}129{,}929}{4}(x - 2000)$ **or** $y - 96{,}456{,345} = \frac{9{,}129{,}929}{4}(x - 1996)$
4. positive; the longer the car is driving, the further it will travel
6. The two graphs have the same y-intercept (0, 0), but different slopes. See students' graphs.
7. Yes, the slope of $\overline{AB} = \frac{7}{4}$ and the slope of $\overline{BC} = -\frac{4}{7}$.

Chapter 9 Polynomials

1. **History** Manassas National Battlefield Park in Virginia has an area of 5067 acres. A numeral in base 10 can be written in polynomial form such as $256 = 2(10)^2 + 5(10)^1 + 6(10)^0$. Write the number of acres in polynomial form. *(Lesson 9-1)* **$5067 = 5(10)^3 + 0(10)^2 + 6(10) + 7(10)^0$**

2. **Framing** José designs picture frames in nonstandard sizes. The width of one rectangular frame is $2x + 5$ and the length is $5x - 9$. *(Lesson 9-2)*
   a. What is the perimeter of this type of frame written as a polynomial? **$14x - 8$**
   b. What is the perimeter of the frame if $x = 7$ inches? **90 in.**

3. **Gardening** A garden located on the grounds of a new museum has width a and length $3a - 7$. *(Lesson 9-3)*
   a. Write an expression in simplest form for the area of the garden. **$3a^2 - 7a$**
   b. If a is 12 feet, find the area of the garden. **348 ft^2**

4. **Shipping** Packages and More sells a variety of mailing boxes. *(Lesson 9-4)*

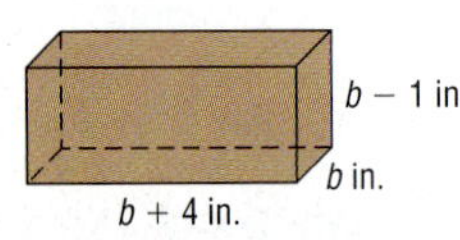

   a. Find the length of box when $b = 10$ inches. **14 inches**
   b. Express the volume of the box shown as a polynomial. **$b^3 + 3b^2 - 4b$**
   c. What is the volume of the box in cubic inches when $b = 1$ foot? **2112 in^3**

5. **Quilting** Deidra is designing a quilt to use as a wall hanging in her room. She made the center portion of the quilt as a square with sides x inches long. She will surround the square with rectangles that have a width 2 inches smaller than the square and a length 2 inches longer than the square. Express the area of one rectangle as a polynomial in simplest form. *(Lesson 9-5)* **$x^2 - 4$**

Chapter 10 Factoring

1. **Parties** Latisha has 90 candies and 54 cookies. If she divided the candies and cookies evenly among the bags, what is the greatest number of gift bags she can make so that there are no candies or cookies left over? *(Lesson 10-1)* **18**

2. **Home Improvement** Carmen is installing hardwood floors in part of her house. Write an expression in factored form that represents the area that will be hardwood. *(Lesson 10-2)* **$6x(x + 5)$**

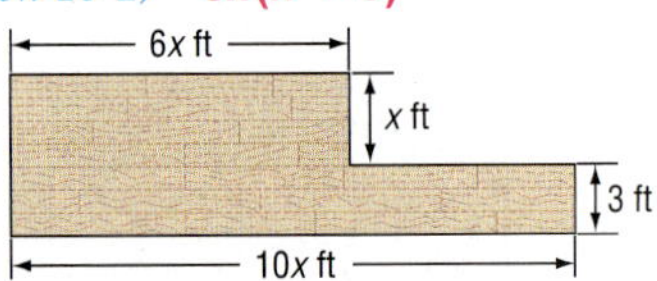

3. **Construction** Kevin is going to put a hot tub that measures 8 feet by 6 feet within his new deck so that no decking will be installed below it. How many square feet of decking will Kevin need? Express the answer in factored form. *(Lesson 10-3)* **$(w + 12)(w - 4)$**

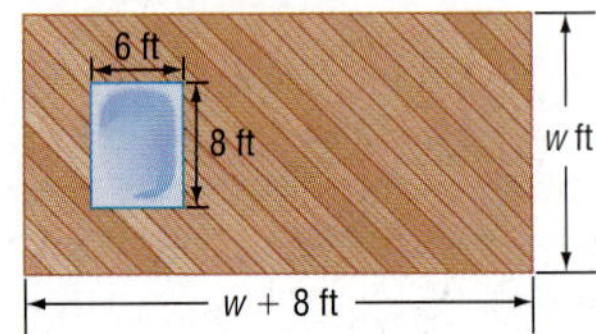

4. **Decorating** Cheyenne is painting a mural on one wall of her living room. The wall that she will paint has a window as shown. Write an expression for the area that will be painted in factored form. *(Lesson 10-4)* **$2(x + 3)(x - 8)$**

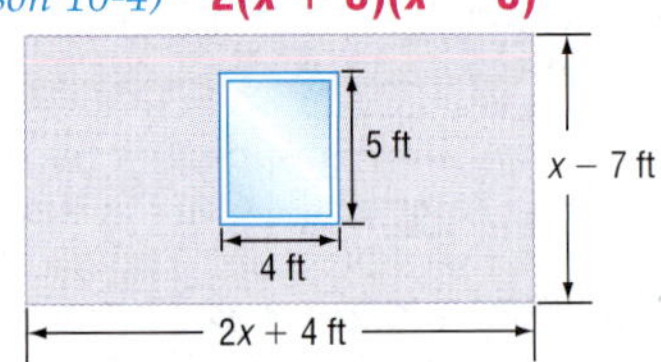

5. **Construction** Tavon drew plans for a square shed to put in his backyard. He then decided that he didn't want the shed to be square, so he reduced one dimension by a number and increased the other dimension by that same number. The new area of the shed floor is $x^2 - 16$. Factor this expression. *(Lesson 10-5)* **$(x - 4)(x + 4)$**

728 Mixed Problem Solving

Chapter 11 Quadratic and Exponential Functions

1. **Sports** The height of a ball in feet is given by $h = -16t^2 + 64t + 5$ if the initial velocity is 64 feet per second and the ball is released 5 feet off the ground. What is the maximum height the ball reaches? *(Lesson 11-1)* **69 ft**

2. **Families of Graphs** Suppose the parent graph $y = x^2 + 7$ is shifted 4 units to the right. Write the new equation and name the vertex of the new graph. *(Lesson 11-2)* $y = (x - 4)^2 + 7$; **(4, 7)**

3. **Construction** Lexi will surround a rectangular sitting area with rose bushes. She has enough bushes to extend 44 feet. *(Lesson 11-3)* **3b. See margin.**
   a. Let w represent the width of the area. Write a quadratic equation for the area A of the sitting area. (*Hint:* Write an expression for the length in terms of the width.) $A = -w^2 + 22w$
   b. Graph the equation you wrote. Can the width be 20 feet? Explain.

4. **Quilting** The length of a quilt is 3 feet more than the width. The area of the quilt is 40 square feet. Find the length and width of the quilt. *(Lesson 11-4)* **5 ft by 8 ft**

5. **Construction** The area of a daycare center's sandbox is 56 square feet. The walkway around it has equal width all the way around. What are the dimensions of the sandbox? *(Lesson 11-5)* **7 ft by 8 ft**

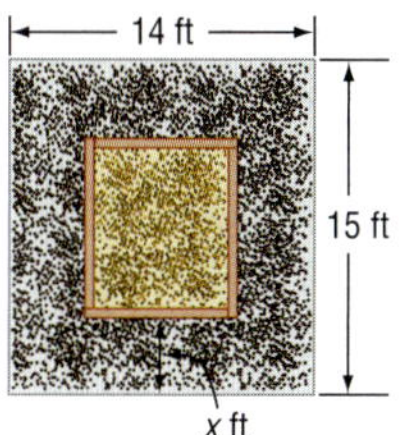

6. **Models** A model rocket is launched from a height of 2 feet off the ground. Use the function $h = -16t^2 + 155t + 2$ to find the approximate time that the rocket hits the ground after it is launched. *(Lesson 11-6)* **about 9.7 seconds**

7. **Savings** Piper deposits \$200 in a compound interest savings account with an annual interest rate of 6%. Use $A(t) = 200(1.06)^t$, where t is the number of years, to find how much will she have after 6 years. *(Lesson 11-7)* **\$283.70**

Chapter 12 Inequalities

1. **Travel** On most airlines, a piece of luggage cannot weigh over 50 pounds. Write an inequality to represent this situation and graph it on a number line. *(Lesson 12-1)* $x \le 50$; **see margin for graph.**

2. **Driving** Christopher drove from Bozeman, Montana, to Minneapolis, Minnesota, to visit his brother. He estimated that the trip was more than 1100 miles. On the first day he drove 540 miles. On the second day he had driven 280 miles when he stopped for lunch. How many more miles did he have to drive? *(Lesson 12-2)* **more than 280 mi**

3. **School** Dewayne is preparing for a final exam. He can study 25 pages of his textbook per day. If the textbook has fewer than 310 pages, how many days will he need to study? *(Lesson 12-3)* **fewer than 12.4 days**

4. **Money** Ms. Parker is taking the French Club to a movie. The tickets cost \$7.50 each, and there are 25 students in the club. If she wants to spend no more than \$250, how much can she spend on refreshments? *(Lesson 12-4)* **\$62.50 or less**

5. **Track** If Jackson runs at a brisk pace, he needs to run between 50 and 60 minutes. If he runs at a slower pace, he needs to run between 55 and 75 minutes. Write an inequality that represents his possible running times. Graph the inequality. *(Lesson 12-5)* $50 \le x \le 75$; **see margin for graph.**

6. **Sewing** Mira's job is to cut fabric at a fabric store. The fabric must be within $\frac{1}{8}$ yard of the desired length. The customer wants 2 yards of fabric. Write and solve an absolute value inequality that represents the possible lengths of the fabric. *(Lesson 12-6)* **See margin.**

7. **Fund-Raising** The band is selling small boxes of candy for \$6 and large boxes for \$10. They must raise at least \$1170 for uniforms. Write and graph an inequality that describes the number of each size of box that could be sold to meet the goal. *(Lesson 12-7)* $6s + 10\ell \ge 1170$; **See margin for graph.**

Answers

Chapter 11

3b.

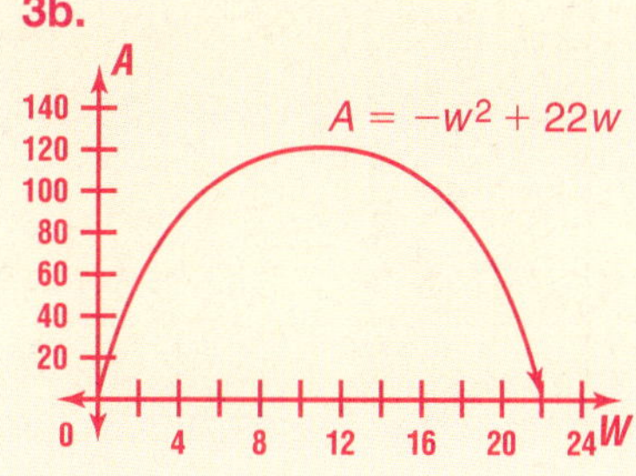

Chapter 12

1.

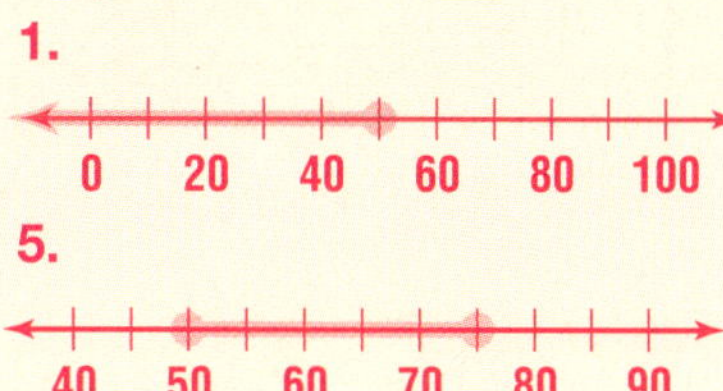

5.

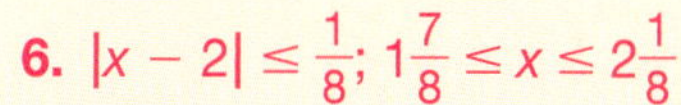

6. $|x - 2| \le \frac{1}{8}$; $1\frac{7}{8} \le x \le 2\frac{1}{8}$

7.

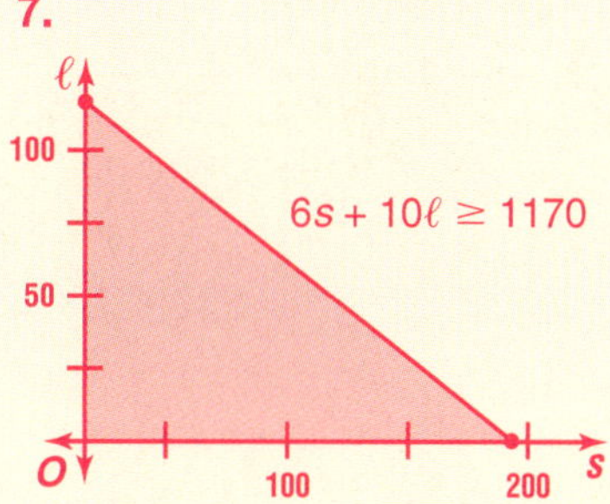

MIXED PROBLEM SOLVING

Answers

Chapter 13

1.

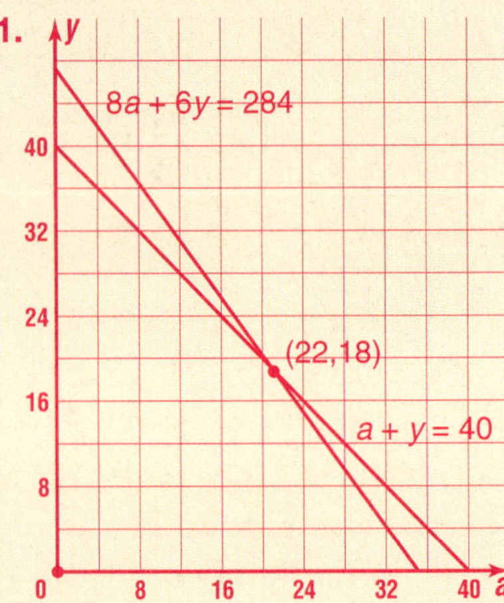

Chapter 14

2a.

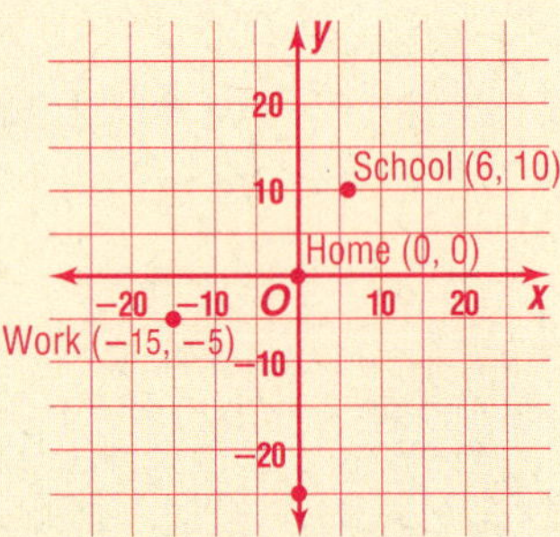

Chapter 13 Systems of Equations and Inequalities

1. **Money** A soccer team is celebrating the end of the season by going to a professional soccer game. Kara buys 40 tickets and spends $284. Youth tickets are $6 each and adult tickets are $8 each. Write and graph two equations that represent this situation. How many of each type of ticket did Kara buy? *(Lesson 13-1)* **See margin for graph; 18 youth and 22 adult.**

2. **Sports** The 97th World Series was played in 2003. By 2003, an American League team had won the World Series 17 times more than a National League team. Write a system of equations to represent the situation. Find the solution of the system and explain what it means. *(Lesson 13-3)* **(57, 40); AL 57 wins and NL 40 wins**

3. **Food** Inez is making a fruit punch by mixing a juice that is 25% water with a juice that is 40% water. How much of each juice should be used to make 5 gallons of a juice that is 35.5% water? *(Lesson 13-3)* **1.5 gal of 25% and 3.5 gal of 40%**

4. **Movies** Jamar bought 2 movie tickets and 1 bag of popcorn for $20. The next weekend, he bought 1 movie ticket and 1 bag of popcorn for $12.50. How much does a ticket cost? *(Lesson 13-4)* **$7.50**

5. **Money** The cost of 5 hamburgers and 4 hot dogs at Burger Joe's is $16. It costs $19 for 8 hamburgers and 2 hot dogs. Write a system of equations to represent this situation. What is the cost of a hamburger? a hot dog? *(Lesson 13-5)* **hamburger: $2, hot dog: $1.50**

6. **Geometry** The area a of a rectangle that has a perimeter of 56 feet and a width of w feet is represented by $a = -w^2 + 28w$. Use a system of equations to find the width of a rectangle that has a perimeter of 56 feet and an area of 187 square feet. *(Lesson 13-6)* **11 ft or 17 ft**

7. **Health** Alvaro walks or jogs at least 3 miles a day. He walks at a rate of 4 miles per hour and jogs 8 miles per hour. He has a half hour to exercise today. How much time can he spend at both? *(Lesson 13-7)* **Sample answer: walk 10 min, jog 20 min**

Chapter 14 Radical Expressions

1. **Geometry** The radius of a cylinder is given by $r = \sqrt{\frac{V}{\pi h}}$, where V is the volume of the cylinder, h is the height of the cylinder, and $\pi \approx 3.14$. What is the radius of a cylinder with a volume of 200 cubic feet and a height of 6 feet? *(Lesson 14-1)* **about 3.3 ft**

2. **Walking** Jariah's school is 10 miles north and 6 miles east of her house. Her workplace is 5 miles south and 15 miles west of her house. *(Lesson 14-2)*
   a. Draw a diagram on a coordinate grid to represent this situation. **See margin.**
   b. How far is Jariah's school from her work? Round to the nearest tenth. **about 25.8 mi**

3. **Roller Coasters** The first hill of a small roller coaster is 50 feet tall. If the coaster travels a horizontal distance of 90 feet as it climbs the hill, how long is the track up the hill? *(Lesson 14-3)* **$10\sqrt{106}$ ft**

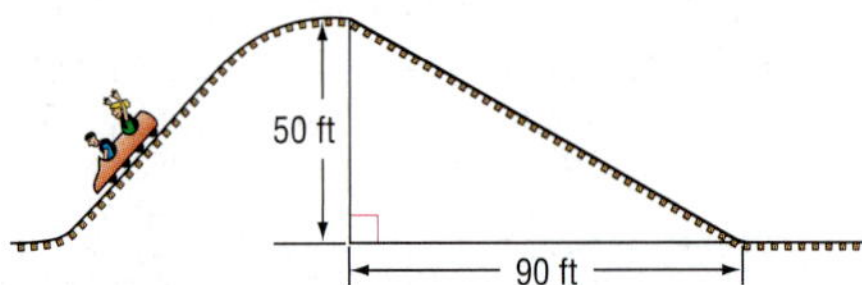

4. **Coordinate Geometry** What is the perimeter of $\triangle ABC$? *(Lesson 14-4)* **$4\sqrt{2} + 2\sqrt{5}$ units**

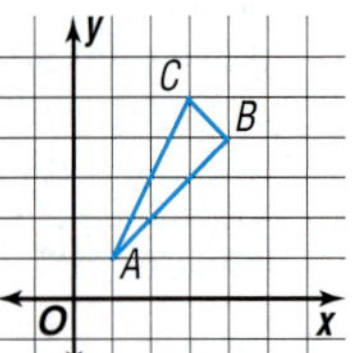

5a. $r = \sqrt{\frac{A}{\pi}}$

5. **Geometry** The formula $A = \pi r^2$ gives the area A of a circle with radius r. *(Lesson 14-5)*
   a. Write an equation for r in terms of A.
   b. What is the radius of a circle with an area of 52 square centimeters? Round to the nearest tenth. **about 4.1 cm**
   c. If the area of a circle is doubled, what is the change in the radius? **It increases by a factor of $\sqrt{2}$.**

730 Mixed Problem Solving

Chapter 15 Rational Expressions and Equations

1. **Elections** Chloe has a 75% chance of winning her student council election. Express 75% as a fraction in simplest form. *(Lesson 15-1)* $\frac{3}{4}$

2. **Gift Wrapping** It takes $1\frac{2}{3}$ yards of ribbon to wrap a gift. How many gifts can be wrapped with a roll of ribbon that is 15 yards long? *(Lesson 15-2)* **9**

3. **Home Design** Find the length of the kitchen if the area is $x^2 + 10x + 24$ square feet. *(Lesson 15-3)* **$x + 6$ feet**

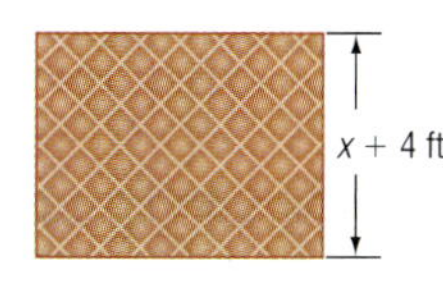

4. **Sports** The time that it takes Mario to paddle his kayak 3 miles down a stream with a current of 15 miles per hour is given by $\frac{3}{x + 15}$, where x is Mario's paddling speed. The time that it would take Mario to paddle 5 miles down the same stream is given by $\frac{5}{x + 15}$. To find the time that it would take Mario to paddle 8 miles down the stream, add the two expressions. *(Lesson 15-4)* $\frac{8}{x + 15}$

5. **Geometry** Find an expression for the perimeter of the isosceles triangle. *(Lesson 15-5)* $\frac{12x^2 + 27x - 3}{(x - 1)(x + 2)}$

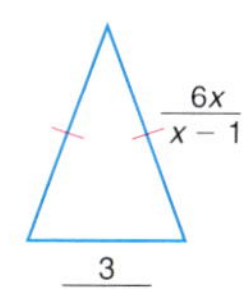

6. **School** Mandy's algebra teacher gives a 10-point quiz each week. After five weeks, Mandy has a total score of 36 for an average of 7.2 points on the quizzes. She would like to raise her average to 9 points. On how many quizzes must Mandy earn a 10 to reach her goal? *(Lesson 15-6)* **9**

MIXED PROBLEM SOLVING

Mixed Problem Solving 731

Preparing for Standardized Tests

Becoming a Better Test-Taker

At some time in your life, you will probably have to take a standardized test. Sometimes this test may determine if you go on to the next grade or course, or even if you will graduate from high school. This section of your textbook is dedicated to making you a better test-taker.

TYPES OF TEST QUESTIONS In the following pages, you will see examples of four types of questions commonly seen on standardized tests. A description of each type of question is shown in the table below.

Type of Question	Description	See Pages
multiple choice	4 or 5 possible answer choices are given from which you choose the best answer.	733–736
gridded response	You solve the problem. Then you enter the answer in aspecial grid and shade in the corresponding circles.	737–740
short response	You solve the problem, showing your work and/or explaning your reasoning.	741–744
extended response	You solve a multi-part problem, showing your work and/orexplaining your reasoning.	745–749

PRACTICE After being introduced to each type of question, you can practice that type of question. Each set of practice questions is divided into five sections that represent the concepts most commonly assessed on standardized tests.

- Number and Operations
- Algebra
- Geometry
- Data Analysis and Probability
- Measurement

USING A CALCULATOR On some tests, you are permitted to use a calculator. You should check with your teacher to determine if calculator use is permitted on the test you will be taking, and if so, what type of calculator can be used.

Test-Taking Tip

If you are allowed to use a calculator, make sure you are familiar with how it works so that you won't waste time trying to figure out the calculator when taking the test.

TEST-TAKING TIPS In addition to Test-Taking Tips like the one shown at the right, here are some additional thoughts that might help you.

- Get a good night's rest before the test. Cramming the night before does not improve your results.
- Budget your time when taking a test. Don't dwell on problems that you cannot solve. Just leave that question blank on your answer sheet.
- Watch for key words like NOT and EXCEPT. Also look for order words like LEAST, GREATEST, FIRST, and LAST.

732 Preparing for Standardized Tests

Multiple-Choice Questions

Multiple-choice questions are the most common type of questions on standardized tests. These questions are sometimes called *selected-response questions*. You are asked to choose the best answer from four or five possible answers.

To record a multiple-choice answer, you may be asked to shade in a bubble that is a circle or an oval, or to just write the letter of your choice. Always make sure that your shading is dark enough and completely covers the bubble.

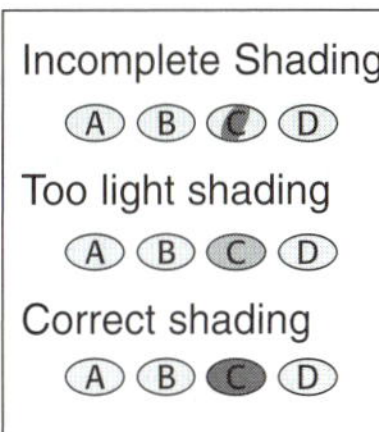

The answer to a multiple-choice question is usually not immediately obvious from the choices, but you may be able to eliminate some of the possibilities by using your knowledge of mathematics. Another answer choice might be that the correct answer is not given.

Example 1

A storm signal flag is used to warn small craft of wind speeds that are greater than 38 miles per hour. The length of the square flag is always three times the length of the side of the black square. If y is the area of the black square and x is the length of the side of the flag, which equation describes the relationship between x and y?

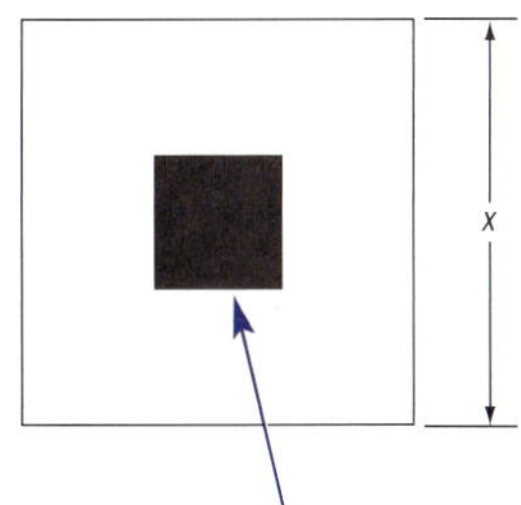

If x is 3 times the length of this side, then this side length is $\frac{1}{3}x$.

Strategy

Elimination
You can eliminate any obvious wrong answers.

(A) $y = \frac{1}{3}x^2$

(B) $y = \frac{1}{9}x^2$

(C) $y = x^2 - 1$

(D) $y = 3x$

(E) $y = 9x$

For the area of a square, $A = s^2$. So, $A = x \cdot x$ or x^2.

The area of the black square is part of the area of the flag, which is x^2. Eliminate choices D and E because they do not include x^2.

$A = \left(\frac{1}{3}x\right)^2$ or $\frac{1}{9}x^2$ square units

So, $y = \frac{1}{9}x^2$. This is choice B.

Use some random numbers to check your choice.

Multiples of 3 make calculations easier.

Length of Flag (x)	Length of Black Square	Area of Black Square	Area = $\frac{1}{9}x^2$
12	4	16	$16 \stackrel{?}{=} \frac{1}{9}(12^2)$ ✓
27	9	81	$81 \stackrel{?}{=} \frac{1}{9}(27^2)$ ✓
60	20	400	$400 \stackrel{?}{=} \frac{1}{9}(60^2)$ ✓

Preparing for Standardized Tests 733

Many multiple-choice questions are actually two- or three-step problems. If you do not read the question carefully, you may select a choice that is an intermediate step instead of the correct final answer.

Example 2 **Barrington can skateboard down a hill five times as fast as he can walk up the hill. If it takes 9 minutes to walk up the hill and skateboard back down, how many minutes does it take him to walk up the hill?**

Strategy
Reread the Problem
Read the problem carefully to find what the question is asking.

Ⓕ **1.5 min** Ⓖ **4.5 min** Ⓗ **7.2 min** Ⓙ **7.5 min**

Before involving any algebra, let's think about the problem using random numbers.

Skating is five times as fast as walking, so walking time equals 5 times the skate time. Use a table to find a pattern

Skate Time	Skate Time × 5 = Walk Time
6 min	$6 \cdot 5 = 30$ min
3 min	$3 \cdot 5 = 15$ min
2 min	$2 \cdot 5 = 10$ min
x min	$x \cdot 5 = 5x$ min

Use the pattern to find a general expression for walk time given any skate time.

The problem states that the walk time and the skate time total 9 minutes.

Use the expression to write an equation for the problem.

$x + 5x = 9$ skate time + walk time = 9 minutes
$6x = 9$ Add like terms.
$x = 1.5$ Divide each side by 6.

Looking at the choices, you might think that choice F is the correct answer. But what does x represent, and what is the problem asking?

The problem asks for the time it takes to walk up the hill, but the value of x is the time it takes to skateboard. So, the actual answer is found using $5x$ or $5(1.5)$, which is 7.5 minutes.

The correct choice is J.

Example 3 **The Band Boosters are making ice cream to sell at an Open House. Each batch of ice cream calls for 5 cups of milk. They plan to make 20 batches. How many gallons of milk do they need?**

Strategy
Units of Measure
Make certain your answer reflects the correct unit of measure.

Ⓐ **800** Ⓑ **100** Ⓒ **25** Ⓓ **12.5** Ⓔ **6.25**

The Band Boosters need 5×20 or 100 cups of milk. However, choice B is not the correct answer. The question asks for *gallons* of milk.

4 cups = 1 quart and 4 quarts = 1 gallon, so 1 gallon = 4×4 or 16 cups.

$100 \text{ }\cancel{\text{cups}} \times \frac{1 \text{ gallon}}{16 \text{ }\cancel{\text{cups}}} = 6.25$ gallons, which is choice E.

734 Preparing for Standardized Tests

Multiple Choice Practice

Choose the best answer.

Number and Operations

1. One mile on land is 5280 feet, while one nautical mile is 6076 feet. What is the ratio of the length of a nautical mile to the length of a land mile as a decimal rounded to the nearest hundredth? C

(A) 0.87 (B) 1.01 (C) 1.15 (D) 5.68

2. The star Proxima Centauri is 24,792,500 million miles from Earth. The star Epsilon Eridani is 6.345×10^{13} miles from Earth. In scientific notation, how much farther from Earth is Epsilon Eridani than Proxima Centauri? B

(A) 0.697×10^{14} mi (B) 3.866×10^{13} mi

(C) 6.097×10^{13} mi (D) 38.658×10^{12} mi

3. In 1976, the cost per gallon for regular unleaded gasoline was 61 cents. In 2002, the cost was $1.29 per gallon. To the nearest percent, what was the percent of increase in the cost per gallon of gas from 1976 to 2002? D

(A) 1% (B) 53% (C) 95% (D) 111%

4. The serial numbers on a particular model of personal data assistant (PDA) consist of two letters followed by five digits. How many serial numbers are possible if any letter of the alphabet and any digit 0–9 can be used in any position in the serial number? B

(A) 676,000,000 (B) 67,600,000

(C) 6,760,000 (D) 676,000

Algebra

5. The graph shows the approximate relationship between the latitude of a location in the Northern Hemisphere and its distance in miles from the equator. If y represents the distance of a location from the equator and x represents the measure of latitude, which equation describes the relationship between x and y? C

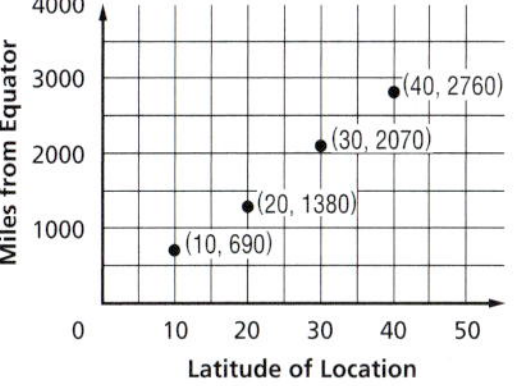

(A) $y = x + 69$ (B) $y = x + 690$

(C) $y = 69x$ (D) $y = 10x$

6. A particular prepaid phone card can be used from a pay phone. The charge is 30 cents to connect and then 4.5 cents per minute. If y is the total cost of a call in cents where x is the number of minutes, which equation describes the relation between x and y? A

(A) $y = 4.5x + 30$ (B) $y = 30x + 4.5$

(C) $y = 0.45x + 0.30$ (D) $y = 0.30x + 0.45$

7. Katie drove to the lake for a weekend outing. The lake is 100 miles from her home. On the trip back, she drove for an hour, stopped for lunch for an hour, and then finished the trip home. Which graph best represents her trip home and the distance from her home at various times? B

(A)

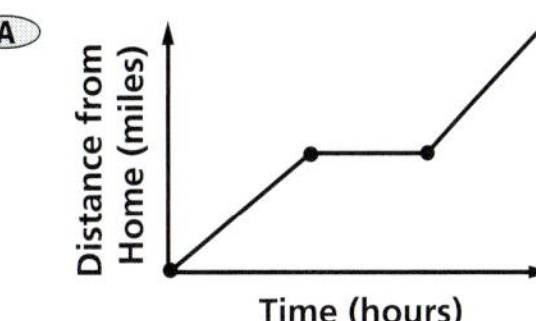

(B)

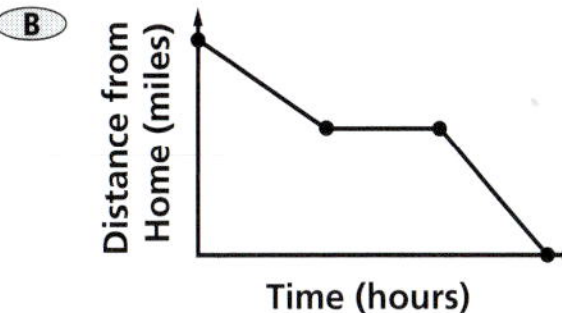

(C)

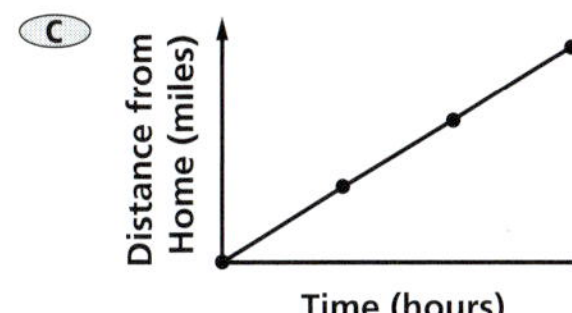

(D)

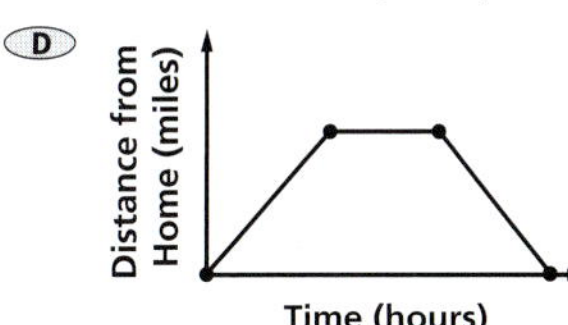

8. Temperature can be given in degrees Fahrenheit or degrees Celsius. The formula $F = \frac{9}{5}C + 32$ can be used to change any temperature given in degrees Celsius to degrees Fahrenheit. Solve the formula for C. A

(A) $C = \frac{5}{9}(F - 32)$ (B) $C = F + 32 - \frac{9}{5}$

(C) $C = \frac{5}{9}F - 32$ (D) $C = \frac{9}{5}(F - 32)$

Preparing for Standardized Tests 735

Geometry

9. Which of the following statements are true about the 4-inch quilt square? E

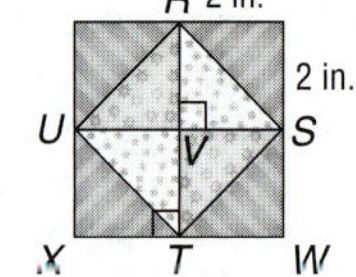

(A) *VSWT* is a square.

(B) $UVTX \cong VSWT$

(C) Four right angles are formed at *V*.

(D) Only A and B are true.

(E) A, B, and C are true.

10. At the Daniels County Fair, the carnival rides are positioned as shown. What is the value of *x*? A

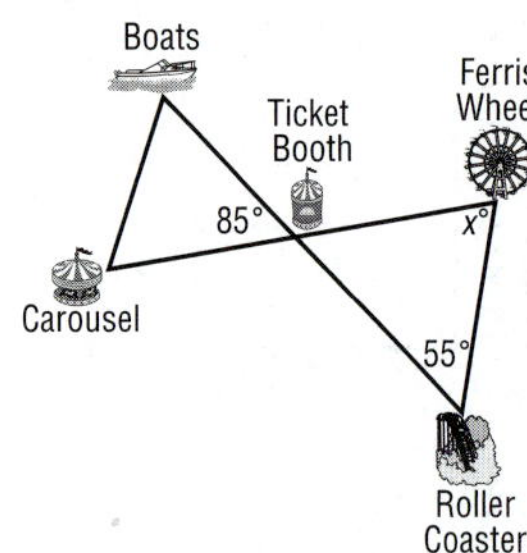

(A) 40 (B) 47.5

(C) 55 (D) 70

(E) 85

11. The diagram shows a map of the Clearwater Wilderness hiking area. To the nearest tenth of a mile, what is the distance from the Parking Lot to Bear Ridge using the most direct route? A

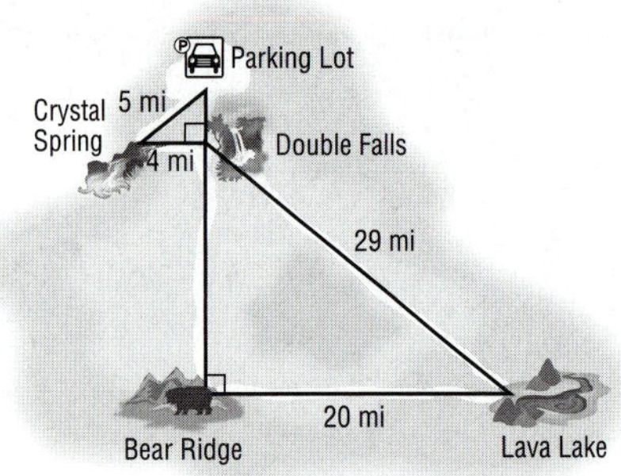

(A) 24 mi (B) 25.5 mi

(C) 26 mi (D) 30.4 mi

Measurement

12. Laura expects about 60 people to attend a party. She estimates that she will need one quart of punch for every two people. How many gallons of punch should she prepare? A

(A) 7.5

(B) 15

(C) 30

(D) 34

13. Stone Mountain Manufacturers are designing two sizes of cylindrical cans below. What is the ratio of the volume of can A to the volume of can B? D

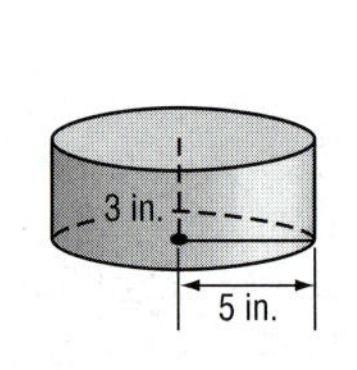

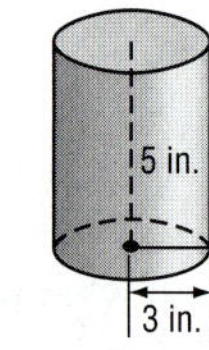

(A) 9 to 25 (B) 25 to 3

(C) 3 to 5 (D) 5 to 3

Data Analysis and Probability

14. The 2000 populations of the five least-populated U.S. states are shown in the table. Which statement is true about this set of data? E

State	Population
Alaska	626,932
North Dakota	642,200
South Dakota	754,844
Vermont	608,827
Wyoming	493,782

Source: U.S. Census Bureau

(A) The mode of the data set is 642,200.

(B) The median of the data set is 626,932.

(C) The mean of the data set is 625,317.

(D) A and C are true.

(E) B and C are true.

736 Preparing for Standardized Tests

Gridded-Response Questions

Gridded-response questions are other types of questions on standardized tests. These questions are sometimes called *student-produced responses* or *grid-ins*, because you must create the answer yourself, not just choose from four or five possible answers.

For gridded response, you must mark your answer on a grid printed on an answer sheet. The grid contains a row of four or five boxes at the top, two rows of ovals or circles with decimal and fraction symbols, and four or five columns of ovals, numbered 0–9. Since there is no negative symbol on the grid, answers are never negative. At the right is an example of a grid from an answer sheet.

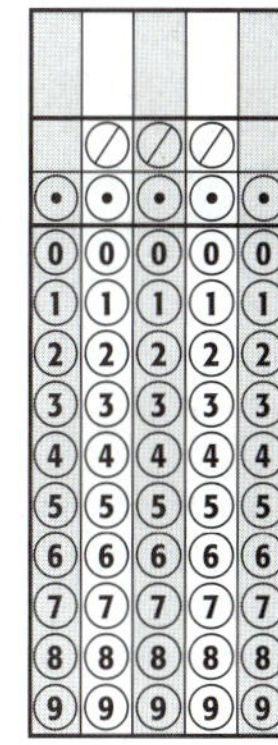

How do you correctly fill in the grid?

Example 

Diego drove 174 miles to his grandmother's house. He made the drive in 3 hours without any stops. At this rate, how far in miles can Diego drive in 5 hours?

What value do you need to find?

You need to find the number of miles Diego can drive in 5 hours.

Write a proportion for the problem. Let m represent the number of miles.

$$\begin{array}{r} \text{miles} \rightarrow \\ \text{hours} \rightarrow \end{array} \frac{174}{3} = \frac{m}{5} \begin{array}{l} \leftarrow \text{miles} \\ \leftarrow \text{hours} \end{array}$$

Solve the proportion.

$\frac{174}{3} = \frac{m}{5}$ Original proportion

$870 = 3m$ Find the cross products.

$290 = m$ Divide each side by 3.

How do you fill in the grid for the answer?

- Write your answer in the answer boxes.
- Write only one digit or symbol in each answer box.
- Do not write any digits or symbols outside the answer boxes.
- You may write your answer with the first digit in the left answer box, or with the last digit in the right answer box. You may leave blank any boxes you do not need on the right or the left side of your answer.
- Fill in only one bubble for every answer box that you have written in. Be sure not to fill in a bubble under a blank answer box

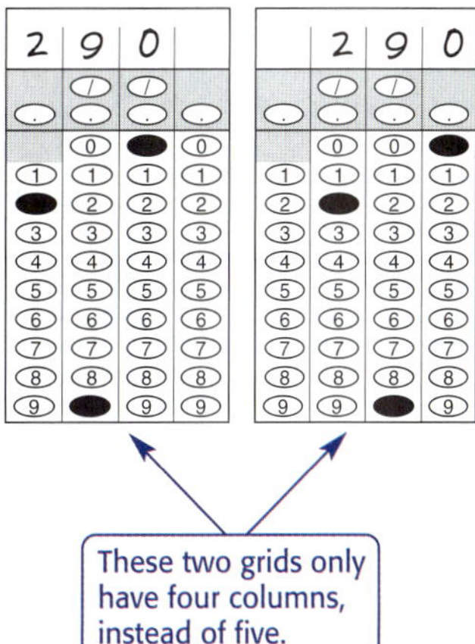

These two grids only have four columns, instead of five.

Preparing for Standardized Tests 737

Many gridded response questions result in an answer that is a fraction or a decimal. These values can also be filled in on the grid.

How do you grid decimals and fractions?

Example 2 **What is the slope of the line that passes through (−2, 3) and (2, 4)?**

Strategy
Decimals and Fractions
Fill in the grid with decimal and fraction answers.

Let $(-2, 3) = (x_1, y_1)$ and $(2., 4) = (x_2, y_2)$.

$m = \frac{y_2 - y_1}{x_2 - x_1}$ Slope formula

$= \frac{4 - 3}{2 - (-2)}$ or $\frac{1}{4}$ Substitute and simplify.

How do you grid the answer?

You can either grid the fraction $\frac{1}{4}$, or rewrite it as 0.25 and grid the decimal. Be sure to write the decimal point or fraction bar in the answer box. The following are acceptable answer responses that represent $\frac{1}{4}$ and 0.25.

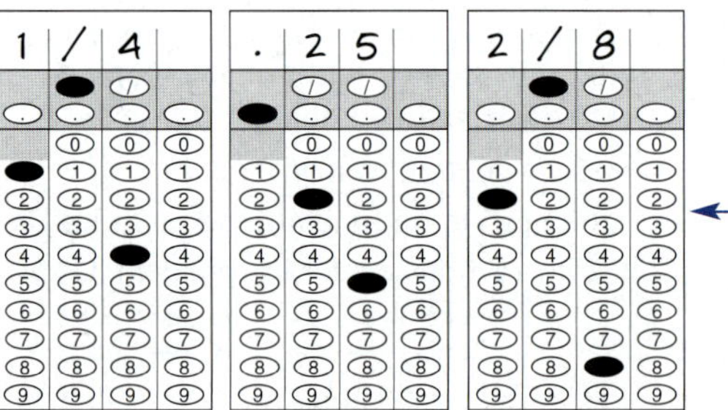

Fractions do not have to be written in lowest terms. Any equivalent fraction that fits the grid will be counted as correct.

Some problems may result in an answer that is a mixed number. Before filling in the grid, change the mixed number to an equivalent improper fraction or decimal. For example, if the answer is $1\frac{1}{2}$, do not enter 1 1/2 as this will be interpreted as $\frac{11}{2}$. Instead, either enter 3/2 or 1.5.

How do you grid mixed numbers?

Example 3 **Amber's cookie recipe calls for $1\frac{1}{3}$ cups of coconut. If Amber plans to make 4 batches of cookies, how much coconut does she need?**

Find the amount of coconut needed using a proportion.

$$\frac{1\frac{1}{3}}{1} = \frac{x}{4}$$

$$4\left(1\frac{1}{3}\right) = 1x$$

$$4\left(\frac{4}{3}\right) = x$$

$$\frac{16}{3} = x$$

Leave the answer as the improper fraction $\frac{16}{3}$, as you cannot correctly grid $5\frac{1}{3}$.

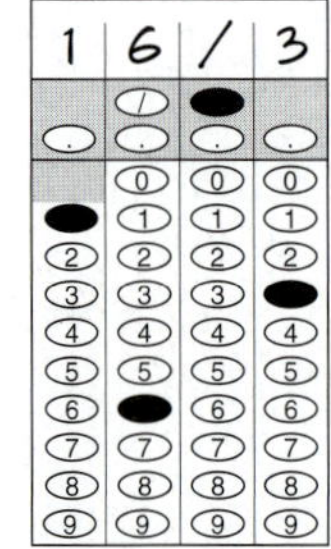

738 Preparing for Standardized Tests

Gridded-Response Practice

Solve each problem. Then copy and complete a grid like the one shown on page 738.

Number and Operations

1. China has the most days of school per year for children with 251 days. If there are 365 days in a year, what percent of the days of the year do Chinese students spend in school? Round to the nearest tenth of a percent. **68.8**

2. Charles is building a deck and wants to buy some long boards that he can cut into various lengths without wasting any lumber. He would like to cut any board into all lengths of 24 inches, 48 inches, or 60 inches. In feet, what is the shortest length of boards that he can buy? **20**

3. At a sale, an item was discounted 20%. After several weeks, the sale price was discounted an additional 25%. What was the total percent discount from the original price of the item? **20**

4. The Andromeda Spiral galaxy is 2.2×10^6 light-years from Earth. The Ursa Minor dwarf is 2.5×10^5 light-years from Earth. How many times as far is the Andromeda Spiral as Ursa Minor dwarf from Earth? **8.8**

5. Twenty students want to attend the World Language Convention. The school budget will only allow for four students to attend. In how many ways can five students be chosen from the twenty students to attend the convention? **4845**

Algebra

6. Find the y-intercept of the graph of the equation $3x + 4y - 5 = 0$. **5/4 or 1.25**

7. Name the x-coordinate of the solution of the system of equations $2x - y = 7$ and $3x + 2y = 7$. **3**

8. Solve $2b - 2(3b - 5) = 8(b - 7)$ for b. **11/2 or 5.5**

9. Kersi read 36 pages of a novel in 2 hours. Find the number of hours it will take him to read the remaining 135 pages if he reads them at the same rate? **15/2 or 7.5**

10. If $f(x) = x + 3$ and $g(x) = x^2 - 2x + 5$, find $6[f(g(1))]$. **42**

11. The Donaldsons have a fish pond in their yard that measures 8 feet by 15 feet. They want to put a walkway around the pond that measures the same width on all sides of the pond, as shown in the diagram. They want the total area of the pond and walkway to be 294 square feet. What will be the width of the walkway in feet? **3**

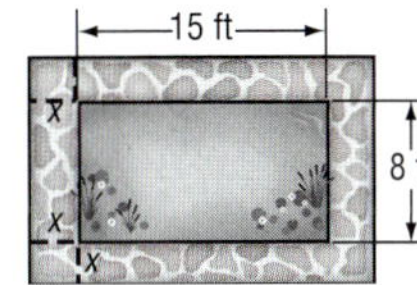

Geometry

12. $\triangle MNP$ is reflected over the x-axis. What is the x-coordinate of the image of point N? **2**

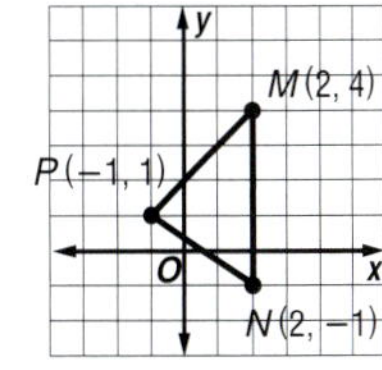

13. The pattern for the square tile shown in the diagram is to be enlarged so that it will measure 15 inches on a side. By what scale factor must the pattern be enlarged? **24**

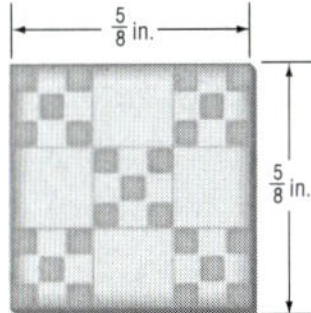

14. Find the measure of $\angle A$ to the nearest degree. **56**

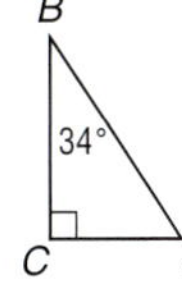

Test-Taking Tip

Question 14

Remember that the hypotenuse of a right triangle is always opposite the right angle.

Preparing for Standardized Tests 739

15. Use the diagram for $\triangle ABO$ and $\triangle XBY$. Find the length of $\overline{BX}$. **9**

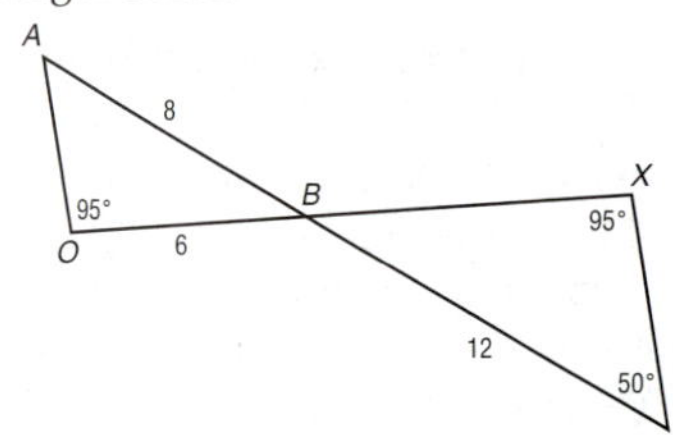

16. A triangle has a perimeter of 96 centimeters. The ratio of measures of its three sides is 6:8:10. Find the length of the longest side in centimeters. **40**

17. The scale on a map of Texas is 0.75 inch = 5 miles. The distance on the map from San Antonio to Dallas is 8.25 inches. What is the actual distance from San Antonio to Dallas in miles? **275**

Measurement

18. Pluto is the farthest planet from the Sun in this solar system at 2756 million miles. If light travels at 186,000 miles per second, how many minutes does it take for a particular ray of light to reach Pluto from the Sun? Round to the nearest minute. **247**

19. Noah drove 342 miles and used 12 gallons of gas. At this same rate, how many gallons of gas will he use on his entire trip of 1140 miles? **40**

20. The jumping surface of a trampoline is shaped like a circle with a diameter of 14 feet. Find the area of the jumping surface. Use 3.14 for π and round to the nearest square foot. **154**

21. A cone is drilled out of a cylinder of wood. If the cone and cylinder have the same base and height, find the volume of the remaining wood. Use 3.14 for π and round to the nearest cubic inch. **1055**

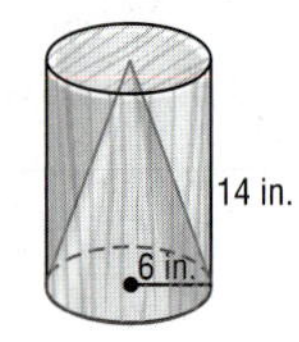

22. The front of a storage building is shaped like a trapezoid as shown. Find the area in square feet of the front of the storage building. **176**

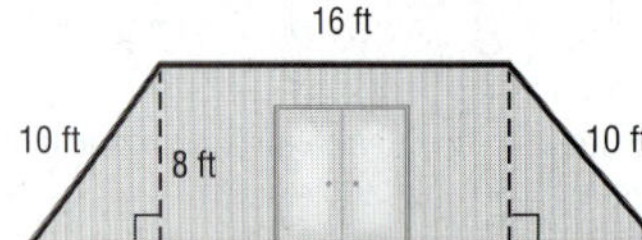

Data Analysis and Probability

23. The table shows the average size in acres of farms in the six states with the largest farms. Find the median of the farm data in acres.
2600

Average Size of Farms for 2001	
State	**Acres per Farm**
Alaska	1586
Arizona	3644
Montana	2124
Nevada	2267
New Mexico	2933
Wyoming	3761

24. The Lindley Park Pavilion is available to rent for parties. There is a fee to rent the pavilion and then a charge per hour. The graph shows the total amount you would pay to rent the pavilion for various numbers of hours. If a function is written to model the charge to rent the pavilion, where x is the number of hours and y is the total charge, what is the rate of change of the function? **10**

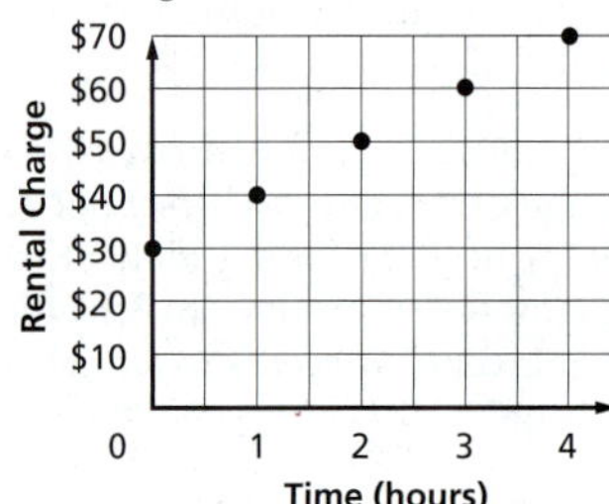

25. A particular game is played by rolling three tetrahedral (4-sided) dice. The faces of each die are numbered with the digits 1–4. How many outcomes are in the sample space for the event of rolling the three dice once? **64**

26. In a carnival game, the blindfolded contestant draws two toy ducks from a pond without replacement. The pond contains 2 yellow ducks, 10 black ducks, 22 white ducks, and 8 red ducks. The best prize is won by drawing two yellow ducks. What is the probability of drawing two yellow ducks? Write your answer as a percent rounded to the nearest tenth of a percent. **0.1**

740 Preparing for Standardized Tests

Short-Response Questions

Short-response questions require you to provide a solution to the problem, as well as any method, explanation, and/or justification you used to arrive at the solution. These are sometimes called *constructed-response, open-response, open-ended, free-response,* or *student-produced questions.* The following is a sample rubric, or scoring guide, for scoring short-response questions.

Credit	Score	Criteria
Full	2	Full credit: The answer is correct and a full explanation is provided that shows each step in arriving at the final answer.
Partial	1	Partial credit: There are two different ways to receive partial credit. • The answer is correct, but the explanation provided is incomplete or incorrect. • The answer is incorrect, but the explanation and method of solving the problem is correct.
None	0	No credit: Either an answer is not provided or the answer does not make sense.

On some standardized tests, no credit is given for a correct answer if your work is not shown.

Example 

Susana is painting two large rooms at her art studio. She has calculated that each room has 4000 square feet to be painted. It says on the can of paint that one gallon covers 300 square feet of smooth surface for one coat and that two coats should be applied for best results. What is the minimum number of 5-gallon cans of paint Susana needs to buy to apply two coats in the two rooms of her studio?

Full Credit Solution

First find the total number of square feet to be painted.

$$4000 \times 2 = 8000 \text{ ft}^2$$

Since 1 gallon covers 300 ft^2, multiply 8000 ft^2 by the unit rate $\frac{1 \text{ gal}}{300 \text{ ft}^2}$.

$$8000 \text{ ft}^2 \times \frac{1 \text{ gal}}{300 \text{ ft}^2} = \frac{8000}{300} \text{ gal}$$

$$= 26\frac{2}{3} \text{ gal.}$$

Each can of paint contains 5 gallons, so divide $26\frac{2}{3}$ gallons by 5 gallons

$$26\frac{2}{3} \div 5 = \frac{80}{3} \div 5 = \frac{\overset{16}{\cancel{80}}}{3} \times \frac{1}{\underset{1}{\cancel{5}}} = \frac{16}{3} = 5\frac{1}{3}$$

Since Susana cannot buy a fraction of a can of paint, she needs to but 6 cans of paint.

The steps, calculations, and reasoning are clearly stated.

The solution of the problem is clearly stated.

Preparing for Standardized Tests 741

Partial Credit Solution

In this sample solution, the answer is correct; however there is no justification for any of the calculations.

There is no explanation of how $26\frac{2}{3}$ was obtained.

$$26\frac{2}{3} \div 5 = \frac{80}{3} \div 5$$

$$= \frac{\overset{16}{\cancel{80}}}{3} \times \frac{1}{\underset{1}{\cancel{5}}}$$

$$= \frac{16}{3}$$

$$= 5\frac{1}{3}$$

Susana will need to buy 6 cans of paint.

Partial Credit Solution

In this sample solution, the answer is incorrect. However, after the first statement, all of the calculations and reasoning are correct.

There are 4000 ft² to be painted and one gallon of paint covers 300 ft².

$$4000\text{ft}^2 \times \frac{1\text{ gal}}{300\text{ ft}^2} = \frac{4000}{300}\text{ gal}$$

$$= 13\frac{1}{3}\text{ gal}$$

Each can of paint contains five gallons. So 2 cans would contain 10 gallons, which is not enough. Three cans of paint would contain 15 gallons which is enough.

Therefore, Susana will need to buy 3 cans of paint.

The first step of doubling the square footage for painting the second room was left out.

No Credit Solution

The wrong operations are used, so the answer is incorrect. Also, there are no units of measure given with any of the calculations.

$300 \times 2 = 600$

$600 \div 5 = 120$

$4000 \div 120 = 33\frac{1}{3}$

Susana will need 34 cans of paint.

742 Preparing for Standardized Tests

Short-Response Practice

Solve each problem. Show all your work.

Number and Operations

1. The world's slowest fish is the sea horse. The average speed of a sea horse is 0.001 mile per hour. What is the rate of speed of a sea horse in feet per minute? **0.088 ft/min**

2. Two buses arrive at the Central Avenue bus stop at 8 A.M. The route for the City Loop bus takes 35 minutes, while the route for the By-Pass bus takes 20 minutes. What is the next time that the two buses will both be at the Central Avenue bus stop? **10:20 A.M.**

3. Toya's Clothing World purchased some denim jackets for \$35. The jackets are marked up 40%. Later in the season, the jackets are discounted 25%. How much does the store lose or gain on the sale of one jacket at the discounted price? **gain of \$1.75**

4. A femtosecond is 10^{-15} second, and a millisecond is 10^{-3} second. How many times faster is a millisecond than a femtosecond? **10^{12} times greater**

5. Find the next three terms in the sequence.

   1, 3, 9, 27, . . . **81, 243, 729**

Algebra

6. Find the slope of the graph of $5x - 2y + 1 = 0$.

7. Simplify $5 + x(1 - x) + 3x$. Write the result in the form $ax^2 + bx + c$. **$-x^2 + 4x + 5$**

8. Solve $17 - 3x \geq 23$. **$x \leq -2$**

9. The table shows what Gerardo charges in dollars for his consulting services for various numbers of hours. Write an equation that can be used to find the charge for any amount of time, where y is the total charge in dollars and x is the number of hours. **$y = 25 + 15x$**

Hours	Charge	Hours	Charge
0	\$25	2	\$55
1	\$40	3	\$70

10. The population of Clark County, Nevada, was 1,375,765 in 2000 and 1,464,653 in 2001. Let x represent the years since 2000 and y represent the total population of Clark County. Suppose the county continues to increase at the same rate. Write an equation that represents the population of the county for any year after 2000. **$y = 88{,}888x + 1{,}375{,}765$**

Geometry

11. $\triangle ABC$ is dilated with scale factor 2.5. Find the coordinates of dilated $\triangle A'B'C'$? **$A'(2.5, 2.5)$, $B'(5, -5)$, $C'(-5, -2.5)$**

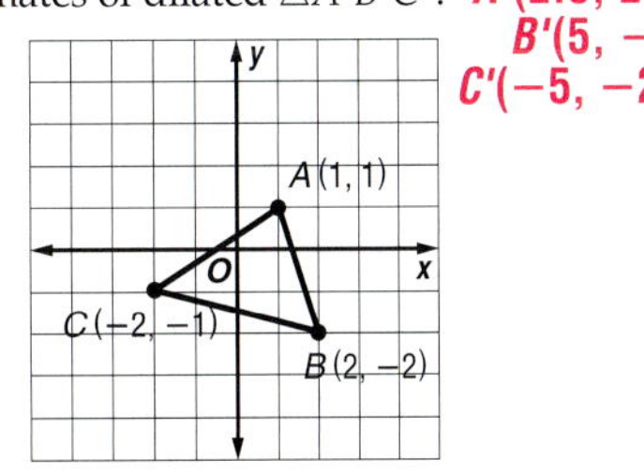

12. At a particular time in its flight, a plane is 10,000 feet above a lake. The distance from the lake to the airport is 5 miles. Find the distance in feet from the plane to the airport. **about 28,230 ft**

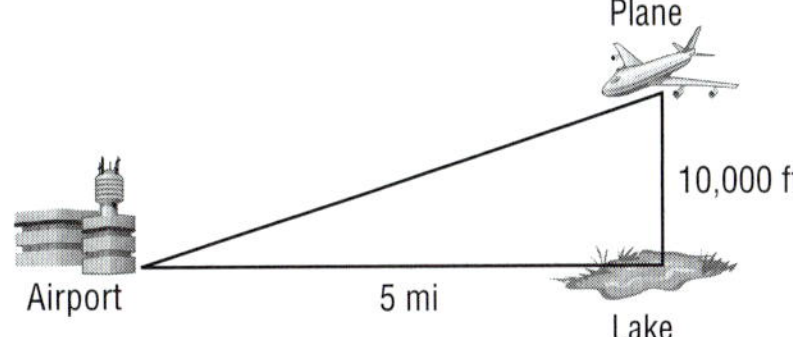

13. Refer to the diagram of the two similar triangles below. Find the value of a. **61.5**

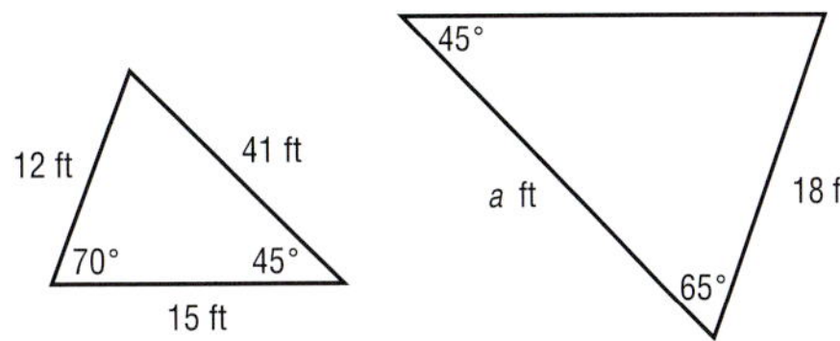

14. The vertices of two triangles are $P(3, 3)$, $Q(7, 3)$, $R(3, 10)$, and $S(-1, 4)$, $T(3, 4)$, $U(-1, 11)$. Which transformation moves $\triangle PQR$ to $\triangle STU$? **translation $(x - 4)$, $(y + 1)$**

Preparing for Standardized Tests 743

15. Aaron has a square garden in his front yard. The length of a side of the garden is 9 feet. Casey wants to plant 6 flowering bushes evenly spaced out along the diagonal. Approximately how many inches apart should he plant the bushes? **30.5 in. apart**

Measurement

16. One inch is equivalent to approximately 2.54 centimeters. Nikki is 61 inches tall. What is her height in centimeters? **154.94 cm**

17. During the holidays, Evan works at Cheese Haus. He packages gift baskets containing a variety of cheeses and sausages. During one four-hour shift, he packaged 20 baskets. At this rate, how many baskets will he package if he works 26 hours in one week? **130 baskets**

18. Ms. Ortega built a box for her garden and placed a round barrel inside to be used for a fountain in the center of the box. The barrel touches the box at its sides as shown. She wants to put potting soil in the shaded corners of the box at a depth of 6 inches. How many cubic feet of soil will she need? Use 3.14 for π and round to the nearest tenth of a cubic foot. **1.7 ft^3**

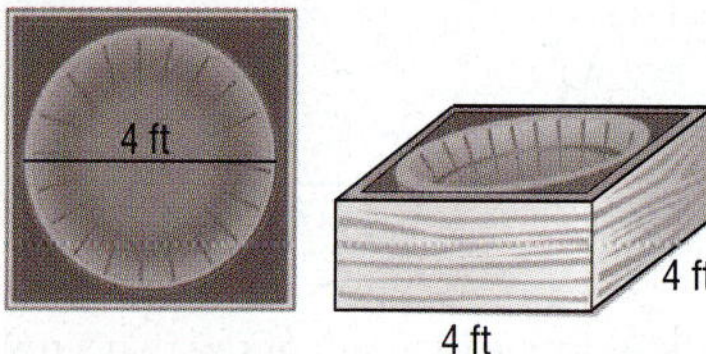

19. A line segment has its midpoint located at $(1, -5)$ and one endpoint at $(-2, -7)$. Find the length of the line to the nearest tenth. **7.2**

Test-Taking Tip

Question 20

Most standardized tests will include any commonly used formulas at the front of the test booklet. Quickly review the list before you begin so that you know what formulas are available.

20. A child's portable swimming pool is 6 feet across and is filled to a depth of 8 inches. One gallon of water is 231 cubic inches. What is the volume of water in the pool in gallons? Use 3.14 as an approximation for π and round to the nearest gallon. **141 gal**

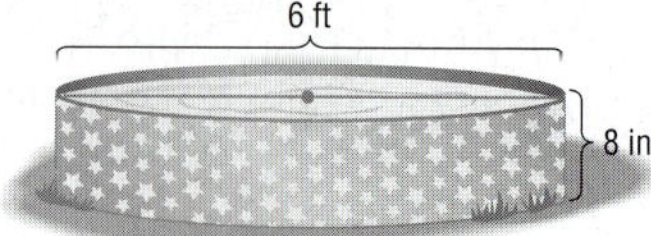

Data Analysis and Probability

21. The table shows the five lowest recorded temperatures on Earth. Find the mean of the temperatures. **−108.16°F**

Location	Temperature (°F)
Vostok, Antarctica	−138.6
Plateau Station, Antarctica	−129.2
Oymyakon, Russia	−96.0
Verkhoyansk, Russia	−90.0
Northice, Greenland	−87.0

Source: *The World Almanac*

22. Two six-sided dice are rolled. The sum of the numbers of dots on the faces of the two dice is recorded. What is the probability that the sum is 10? $\frac{1}{12}$

23. The table shows the amount of a particular chemical that is needed to treat various sizes of swimming pools. Write the equation for a line to model the data. Let x represent the capacity of the pool in gallons and y represent the amount of the chemical in ounces. $y = 0.003x$

Pool Capacity (gal)	Amount of Chemical (oz)
5000	15
10,000	30
15,000	45
20,000	60
25,000	75

24. Fifty balls are labeled from 1 through 50. Two balls are drawn without replacement. What is the probability that both balls show an even number? $\frac{12}{49}$

744 Preparing for Standardized Tests

Extended-Response Questions

Extended-response questions are often called *open-ended* or *constructed-response questions.* Most extended-response questions have multiple parts. You must answer all parts correctly to receive full credit.

Extended-response questions are similar to short-response questions in that you must show all of your work in solving the problem, and a rubric is used to determine whether you receive full, partial, or no credit. The following is a sample rubric for scoring extended-response questions.

Credit	Score	Criteria
Full	4	A correct solution is given that is supported by well-developed, accurate explanations
Partial	3, 2, 1	A generally correct solution is given that may contain minor flaws in reasoning or computation or an incomplete solution. The more correct the solution, the greater the score.
None	0	An incorrect solution is given indicating no mathematical understanding of the concept, or no solution is given.

On some standardized tests, no credit is given for a correct answer if your work is not shown.

Make sure that when the problem says to *Show your work,* show every aspect of your solution including figures, sketches of graphing calculator screens, or reasoning behind computations.

Example 1

The table shows the population density in the United States on April 1 in each decade of the 20th century.

a. Make a scatter plot of the data.

b. Alaska and Hawaii became states in the same year. Between what two census dates do you think this happened. Why did you choose those years?

c. Use the data and your graph to predict the population density in 2010. Explain your reasoning

U.S. Population Density

Year	People Per Square Mile
1910	31.0
1920	35.6
1930	41.2
1940	44.2
1950	50.7
1960	50.6
1970	57.4
1980	64.0
1990	70.3
2000	79.6

Source: *U.S. Census Bureau*

Full Credit Solution

Part a A complete scatter plot includes a title for the graph, appropriate scales and labels for the axes, and correctly graphed points.

- The student should determine that the year data should go on the x-axis while the people per square mile data should go on the y-axis.
- On the x-axis, each square should represent 10 years.
- The y-axis could start at 0, or it could show data starting at 30 with a broken line to indicate that some of the scale is missing.

PREPARING FOR STANDARDIZED TESTS

Preparing for Standardized Tests 745

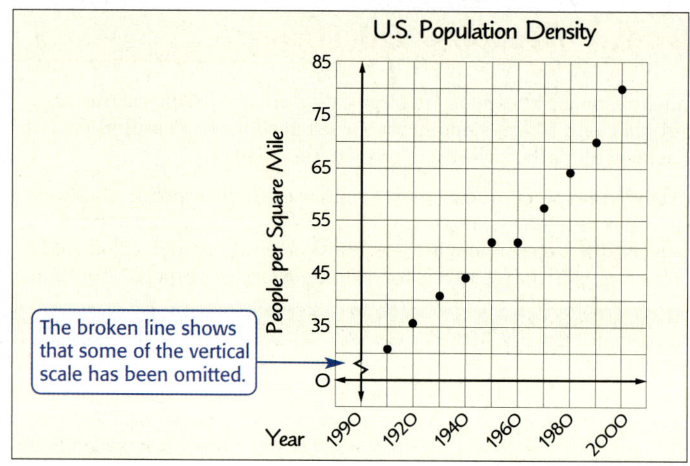

Part b

1950–1960 because when Alaska became a state it added little population but a lot of land, which made the people per square mile ratio less.

Part c

About 85.0, because the population per square mile would probably get larger so I connected the first point and the last point. The rate of change for each year was $\frac{79.6 - 31.0}{2000 - 1910}$ or about 0.54. I added 10×0.54, or 5.4 to 79.6 to get the next 10-year point.

Actually, any estimate from 84 to 86 might be acceptable. You could also use different points to find the equation for a line of best fit for the data, and then find the corresponding y value for $x = 2010$.

Partial Credit Solution

Part a This sample answer includes no labels for the graph or the axes and one of the points is not graphed correctly.

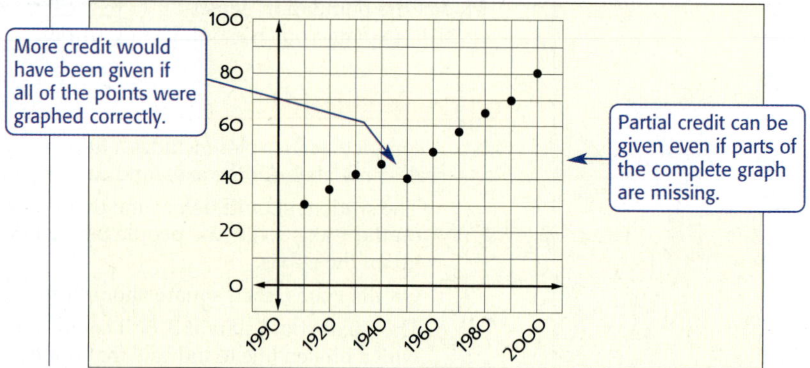

746 Preparing for Standardized Tests

Part b Partial credit is given because the reasoning is correct, but the reasoning was based on the incorrect graph in Part a.

> 1940–1950, because when Alaska became a state it added little population but a lot of land, which made the people per square mile ratio less.

Part c Full credit is given for Part c.

> Suppose I draw a line of best fit through points (1910, 31.0) and (1990, 70.3). The slope would be $\frac{70.3 - 31.0}{1990 - 1910}$ or about 0.49. Now use the slope and one of the points to find the y-intercept.
>
> $y = mx + b$
>
> $70.3 = 0.49(1990) + b$
>
> $-904.8 = b$
>
> So an equation of my line of best fit is $y = 0.49x - 904.8$.
>
> If $x = 2010$, then $y = 0.49(2010) - 904.8$ or about 80.1 people per square mile in the year 2010.

This sample answer might have received a score of 2 or 1. Had the student graphed all points correctly and gotten Part B correct, the score would probably have been a 3.

No Credit Solution

Part a

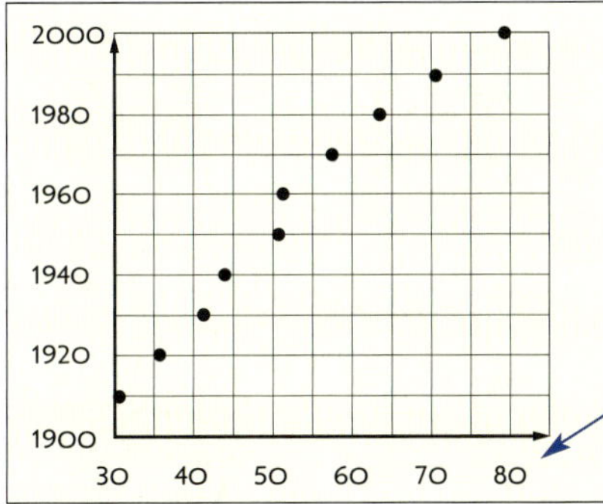

The people per square mile data is dependent on the year, so the years should have been graphed on the *x*-axis. No labels have been included to identify what the graph represents.

Part b

> I have no idea.

Part c

> 85, because it is the next grid line.

In this sample answer, the student does not understand how to represent data on a graph or how to interpret the data after the points are graphed.

Preparing for Standardized Tests 747

PREPARING FOR STANDARDIZED TESTS

Answers

1a. Italy had a 227% increase.

1b. 1970; you could get more marks for every U.S. dollar and buy more merchandise.

1c. There were about 16.7 yen for 1 franc.

2a. Sample answer: In each case, mass ÷ volume is approximately equal to the density.

2b. Sample answer: The greater the radius, the greater the gravity. However, the relationship does not look like it is strictly linear.

2c. Sample answer: No, you would need to examine the similar statistics for all the planets.

3a. The glider was released when $x = 0$. The altitude of 10,000 feet is the altitude of the glider upon release.

3b. During 10 to 20 minutes after release, the rate of descent was greatest as the slope of the line is greatest and the line is steepest.

3c. It appears to have taken about 65 minutes. It is the x-intercept, where the altitude y is 0.

3d. $y = -200x + 13{,}000$

3e. The slope is the rate of descent of the glider, which is 200 feet per minute.

4a. $a_n = 5 + 15n$; If x represents the days and y represents the number of minutes, Anna will drive 20 minutes on the first day (1, 20) and 35 minutes on the second day (2, 35). Using these points, I found the point-slope form of the equation to be $y = 15x + 5$. Then, I changed the equation into the formula with variables a_n and n.

4b. $a_{14} = 5 + 15(14)$ or 215 minutes; sample answer: I substituted $n = 14$ days into the formula.

4c. No; the total of all the minutes for 14 days is only 1645 minutes while 30 hours is 1800 minutes. He will be short 155 minutes.

Extended Response Practice

Solve each problem. Show all your work.

Number and Operations

a–c. See margin.

1. The table shows what one dollar in U.S. money was worth in five countries in 1970 and in 2001.

Money Equivalent to One U.S. Dollar		
Country	1970 Value	2001 Value
France	5.5 francs	7 francs
Germany	3.6 marks	2 marks
Great Britain	0.4 pounds	0.67 pounds
Italy	623 lire	2040 lire
Japan	358 yen	117 yen

   a. For which country was the percent of increase or decrease in the number of units of currency that was equivalent to $1 the greatest from 1970 to 2001?

   b. Suppose a U.S. citizen traveled to Germany in 1970 and in 2001. In which year would the traveler receive a better value for their money? Explain.

   c. In 2001, what was the value of one franc in yen?

2. The table shows some data about the planets and the Sun. The radius is given in miles and the volume, mass, and gravity quantities are related to the volume and mass of Earth, which has a value of 1. a–c. See margin.

	Volume	Mass	Density	Radius	Gravity
Sun	1,304,000	332,950	0.26	434,474	28
Mercury	0.056	0.0553	0.98	1516	0.38
Venus	0.857	0.815	0.95	3760	0.91
Moon	0.0203	0.0123	0.61	1079	0.17
Mars	0.151	0.107	0.71	2106	0.38
Jupiter	1321	317.83	0.24	43,441	2.36
Saturn	764	95.16	0.12	36,184	0.92
Uranus	63	14.54	0.23	15,759	0.89
Neptune	58	17.15	0.30	15,301	1.12
Pluto	0.007	0.0021	0.32	743	0.06

   a. Make and test a conjecture relating volume, mass, and density.

   b. Describe the relationship between radius and gravity.

   c. Can you be sure that the relationship in part b holds true for all planets? Explain.

Algebra

3. The graph shows the altitude of a glider during various times of his flight after being released from a tow plane. a–e. See margin.

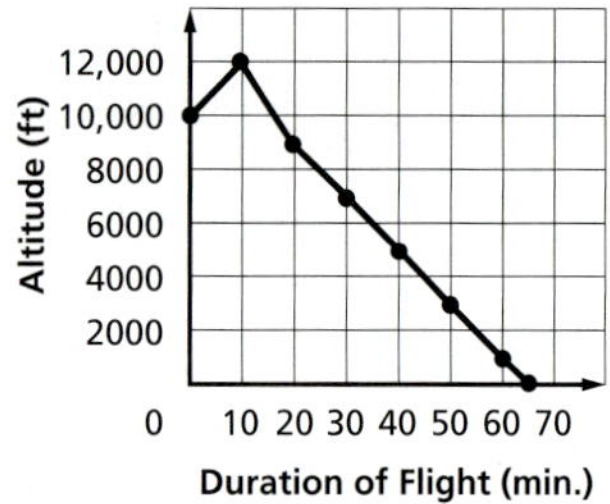

   a. What point on the graph represents the moment the glider was released from the tow plane? Explain the meaning of this point in terms of altitude.

   b. During which time period did the greatest rate of descent of the glider take place? Explain your reasoning.

   c. How long did it take the glider to reach an altitude of 0 feet? Where is this point on the graph?

   d. What is the equation of a line that represents the glider's altitude y as the time increased from 20 to 60 minutes?

   e. Explain what the slope of the line in part d represents?

4. John has just received his learner's permit which allows him to practice driving with a licensed driver. His mother has agreed to take him driving every day for two weeks. On the first day, John will drive for 20 minutes. Each day after that, John's mother has agreed he can drive 15 minutes more than the day before. a–c. See margin.

   a. Write a formula for the nth term of the sequence. Explain how you found the formula.

   b. For how many minutes will John drive on the last day? Show how you found the number of minutes.

   c. John's driver's education teacher requires that each student drive for 30 hours with an adult outside of class. Will John fulfill this requirement? Explain.

Geometry

5. Polygon $QUAD$ is shown on a coordinate plane. **a–c. See margin.**

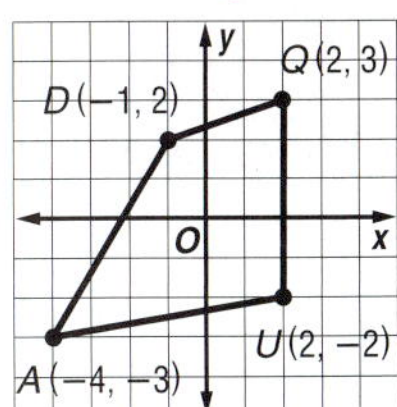

a. Find the coordinates of the vertices of $Q'U'A'D'$, which is the image of $QUAD$ after a reflection over the y-axis. Explain.

b. Suppose point $M(a, b)$ is reflected over the y-axis. What will be the coordinates of the image M'?

c. Describe a reflection that will make $QUAD$ look "upside down." What will be the coordinates of the vertices of the image?

6. The diagram shows a sphere with a radius of a and a cylinder with a radius and a height of a. **a–c. See margin.**

a. What is the ratio of the volume of the sphere to the volume of the cylinder?

b. What is the ratio of the surface area of the sphere to the surface area of the cylinder?

c. The ratio of the volume of another cylinder is 3 times the volume of the sphere shown. Give one possible set of measures for the radius and height of the cylinder in terms of a.

Measurement

7. Alexis is using a map of the province of Saskatchewan in Canada. The scale for the map shows that 2 centimeters on the map is 30 kilometers in actual distance. **a–c. See margin.**

a. The distance on the map between two cities measures 7 centimeters. What is the actual distance between the two cities in kilometers? Show how you found the distance.

b. Alexis is more familiar with distances in miles. The distance between two other cities is 8 centimeters. If one kilometer is about 0.62 mile, what is the distance in miles?

c. Alexis' entire trip measures 54 centimeters. If her car averages 25 miles per gallon of gasoline, how many gallons will she need to complete the trip? Round to the nearest gallon. Explain.

Test-Taking Tip

Question 5

In a reflection over the x-axis, the x-coordinate remains the same, and the y-coordinate changes its sign. In a reflection over the y-axis, the y-coordinate remains the same, and the x-coordinate changes its sign.

8. The diagram shows a pattern for a quilt square called Colorful Fan. **a–c. See margin.**

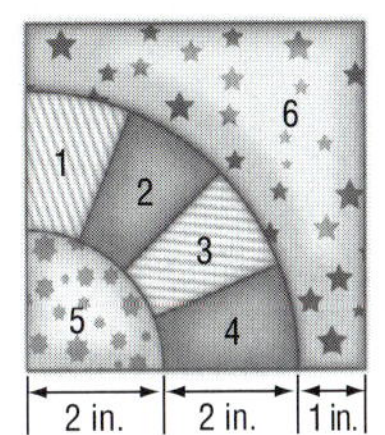

a. What is the area of region 6? Explain.

b. What is the area of region 1? Explain.

c. What is the ratio of the area of region 5 to the area of the entire square? Show how you found the ratio. Leave the ratio in terms of π.

Data Analysis and Probability

9. The table shows the Olympics winning times in the women's 1500-meter speed skating event. The times are to the nearest second.

Year	Time (s)	Year	Time (s)
1960	172	1984	124
1964	143	1988	121
1968	142	1992	126
1972	141	1994	122
1976	137	1998	118
1980	131	2002	114

a. Make a scatter plot of the data.

b. Use the data and your graph to predict the winning time in 2010. **a–b. See margin.**

10. There are 1320 ways for three students to win first, second, and third place during a debate.

a. How many students are on the debate team?

b. What is the probability that a student will come in one of the first three places if each student has an equal chance of succeeding?

c. If the teacher announces the third place winner, what is the probability that one particular other student will win first or second place? **a–c. See margin.**

Answers

5a. $Q'(-2, 3)$, $U'(-2, -2)$, $A'(4, -3)$, $D'(1, 2)$; sample answer: Since the reflection occurred over the y-axis, I found the opposite of the x-coordinate for each point.

5b. $(-a, b)$

5c. A reflection over the x-axis will make the polygon look "upside down." The coordinates of the vertices will be $Q'(2, -3)$, $U'(2, 2)$, $A'(-4, 3)$, $D'(-1, -2)$.

6a. The ratio is 4 to 3.

6b. The ratio is 1 to 1.

6c. Sample answer: $r = a$; $h = 4a$

7a. The distance is 105 kilometers. Sample answer: I set up the proportion $\frac{2 \text{ cm}}{30 \text{ km}} = \frac{7 \text{ cm}}{x \text{ km}}$; $x = 105$.

7b. The distance is 74.4 miles.

7c. She will need about 20 gallons. Sample answer: I found 54 cm = 810 km = 502.2 mi. Then I divided 25 mi/gal into 502.2 mi to get 20.088 gal.

8a. The area of region 6 is about 12.4 in^2.

8b. The area of region 1 is about 2.4 in^2.

8c. $\frac{\pi}{25}$; sample answer: The area of region 5 is $\frac{1}{4}\pi 2^2$, or π in^2. The area of the square is 5^2 or 25. So, the ratio is π to 25 or $\frac{\pi}{25}$.

9a.

1500-Meter Speed Skating Winning Times

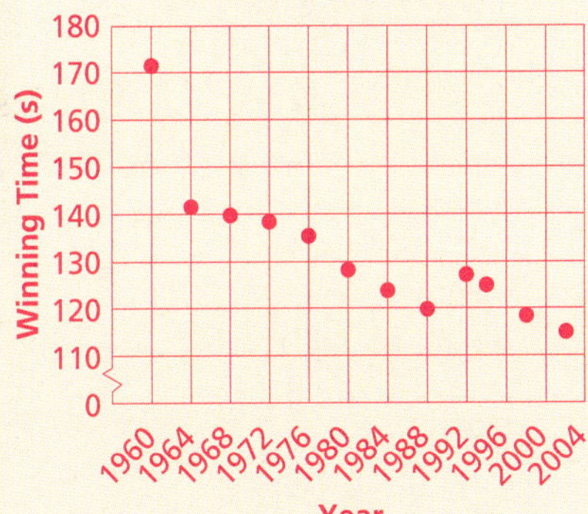

9b. Sample answer: In recent years, the times have decreased by 4 seconds every 4 years. Since 2010 is 8 years from 2002, the winning time in 2010 will be about 114 – 2(4) or 106 seconds.

10a. There are 12 students on the debate team because there are 12 · 11 · 10 or 1320 ways for three students to win the top three places during the debate.

10b. $\frac{\text{ways to come in the first 3 places}}{\text{number of students}} = \frac{3}{20}$ or 15%

10c. $\frac{2}{19}$ or about 10.5%

PREPARING FOR STANDARDIZED TESTS

Preparing for Standardized Tests 749

Graphing Calculator Tutorial

General Information

- On a T1-84 Plus, any blue commands written above the calculator keys are accessed with the 2nd key, which is also blue. Similarly, the yellow functions on a T1-83 Plus are accessed with the yellow 2nd key. On both the T1-84 Plus and the T1-83 Plus, any green characters or commands above the keys are accessed with the ALPHA key which is also green. In this text, commands that are accessed by the 2nd and ALPHA keys are shown in brackets. For example, 2nd [QUIT] means to press the 2nd key followed by the key below the yellow QUIT command.
- 2nd [ENTRY] copies the previous calculation so it can be edited or reused.
- 2nd [ANS] copies the previous answer so it can be used in another calculation.
- 2nd [QUIT] will return you to the home (or text) screen.
- 2nd [A-LOCK] allows you to use the green characters above the keys without pressing ALPHA before typing each letter. (This is handy for programming.)
- Negative numbers are entered using the (–) key, not the minus sign, –.
- The variable x can be entered using the X,T,θ,n key, rather than using ALPHA [X].
- 2nd [OFF] turns the calculator off.

Basic Keystrokes

Some commonly used mathematical functions are shown in the table below. As with any scientific calculator, graphing calculators observe the order of operations.

Mathematical Operation	Example	Keys	Display
evaluate expressions	Evaluate 2 + 5.	2 [+] 5 [ENTER]	2 + 5 7
multiplication	Evaluate 3(9.1 + 0.8).	3 [(] 9.1 [+] .8 [)] [ENTER]	3(9.1 + .8) 29.7
division	Evaluate $\frac{8-5}{4}$.	[(] 8 [−] 5 [)] [÷] 4 [ENTER]	(8 − 5)/4 .75
exponents	Find 3^5.	3 [^] 5 [ENTER]	3^5 243
roots	Find $\sqrt{14}$.	[2nd] [√] 14 [ENTER]	√(14 3.741657387
opposites	Enter −3.	[(−)] 3	−3
variable expressions	Enter $x^2 + 4x - 3$.	[X,T,θ,n] [x^2] [+] 4 [X,T,θ,n] [−] 3	X^2 + 4X − 3

Key Skills

Each Graphing Calculator Exploration in the Student Edition requires the use of certain key skills. Use this section as a reference for further instruction.

A: Entering and Graphing Equations

Press [Y=]. Use the [X,T,θ,n] key to enter *any* variable for your equation. To see a graph of the equation, press [GRAPH].

B: Setting Your Viewing Window

Press [WINDOW]. Use the arrow or [ENTER] keys to move the cursor and edit the window settings. Xmin and Xmax represent the minimum and maximum values along the *x*-axis. Similarly, Ymin and Ymax represent the minimum and maximum values along the *y*-axis. Xscl and Yscl refer to the spacing between tick marks placed on the *x*- and *y*-axes. Suppose Xscl = 1. Then the numbers along the *x*-axis progress by 1 unit. Set Xres to 1.

C: The Standard Viewing Window

A good window to start with to graph an equation is the **standard viewing window.** It appears in the [WINDOW] screen as follows.

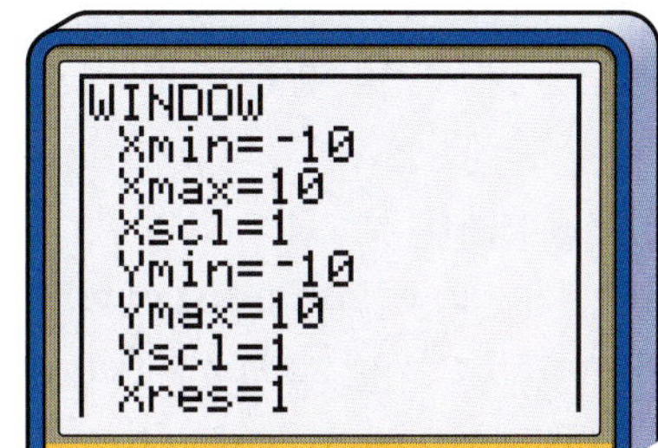

To easily set the values for the standard viewing window, press [ZOOM] 6.

D: Zoom Features

To easily access a viewing window that shows only integer coordinates, press [ZOOM] 8 [ENTER].

To easily access a viewing window for statistical graphs of data you have entered, press [ZOOM] 9.

E: Using the Trace Feature

To trace a graph, press TRACE. A flashing cursor appears on a point of your graph. At the bottom of the screen, x- and y-coordinates for the point are shown. At the top left of the screen, the equation of the graph is shown. Use the left and right arrow keys to move the cursor along the graph. Notice how the coordinates change as the cursor moves from one point to the next. If more than one equation is graphed, use the up and down arrow keys to move from one graph to another.

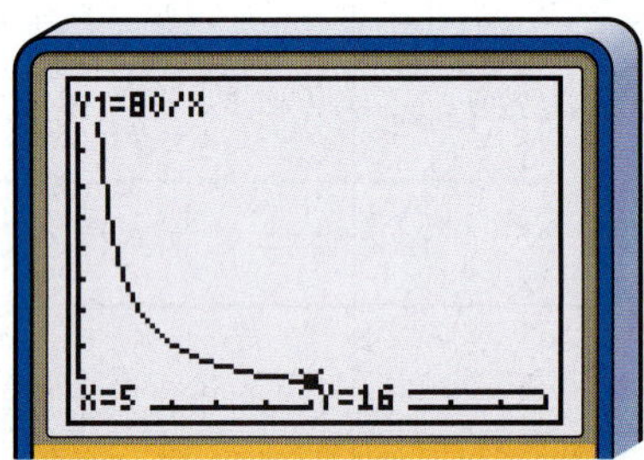

F: Setting or Making a Table

Press 2nd [TBLSET]. Use the arrow or ENTER keys to move the cursor and edit the table settings. Indpnt represents the x-variable in your equation. Set Indpnt to *Ask* so that you may enter any value for x into your table. Depend represents the y-variable in your equation. Set Depend to *Auto* so that the calculator will find y for any value of x.

G: Using the Table

Before using the table, you must enter at least one equation in the Y= screen. Then press 2nd [TABLE]. Enter any value for x as shown at the bottom of the screen. The function entered as Y_1 will be evaluated at this value for x. In the two columns labeled X and Y_1, you will see the values for x that you entered and the resulting y-values.

H: Entering Inequalities

Press 2nd [TEST]. From this menu, you can enter the $=$, $\neq$, $>$, $\geq$, $<$, and $\leq$ symbols.

Chapter	Page(s)	Key Skills
2	61	B
3	106	I
5	214	K
6	272	A, B, E
7	317	A, D, E
8	307	A, B, E
10	422	K
11	471, 491	A, D, E, F, G, I, J
12	521	A, D, E, H
13	551	A, D
14	625	A, B
15	638–639	A, B, E

I: Entering and Deleting Lists

Press STAT ENTER. Under L_1, enter your list of numerical data. To delete the data in the list, use your arrow keys to highlight L_1. Press CLEAR ENTER. Remember to clear all lists before entering a new set of data.

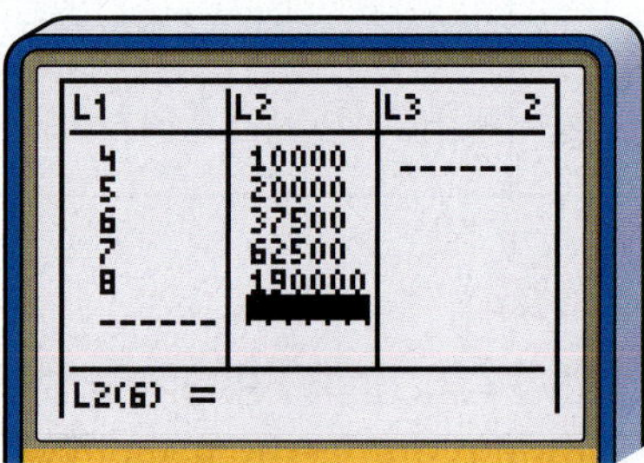

J: Plotting Statistical Data in Lists

Press Y=. If appropriate, clear equations. Use the arrow keys until Plot1 is highlighted. Plot1 represents a Stat Plot, which enables you to graph the numerical data in the lists. Press ENTER to turn the Stat Plot on and off. You may need to display different types of statistical graphs. To set the details of a Stat Plot, press 2nd [STAT PLOT] ENTER. A screen like the one below appears.

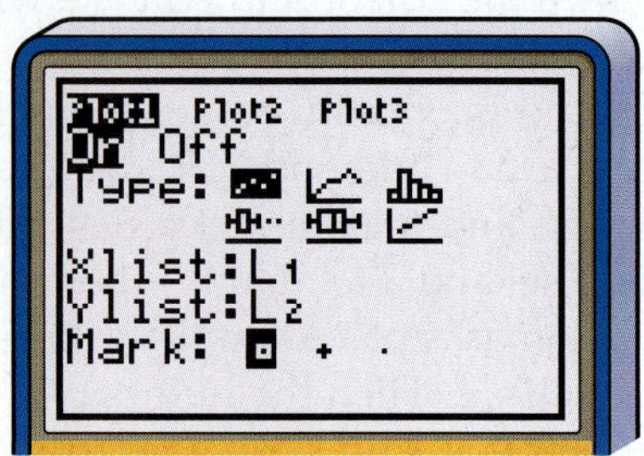

At the top of the screen, you can choose from one of three plots to store settings. The second line allows you to turn a Stat Plot on and off. Then you may select the type of plot: scatter plot, line plot, histogram, two types of box-and-whisker plots, or a normal probability plot. For this text, you will mainly use the scatter plot, line plot, and histogram. Next, choose which lists of data you would like to display along the x- and y-axes. Finally, choose the symbol that will represent each data point. To see a graph of the statistical data, press [ZOOM] 9.

K: Programming on the TI–83 Plus/TI–84 Plus

The TI–83 Plus and TI–84 Plus have programming features that allow you to write and execute a series of commands to perform tasks that may be too complex to perform otherwise. Each program is given a name. Commands begin with a colon (:), followed by an expression or an instruction. Most calculator features are accessible from the program mode.

When you press [PRGM], you see three menus: EXEC, EDIT, and NEW. EXEC allows you to execute a stored program by selecting the name of the program from the menu. EDIT allows you to edit or change an existing program. NEW allows you to create a new program.

The following example illustrates how to create and execute a new program that stores an expression as Y and evaluates the expression for the designated value of X.

1. Press [PRGM] [▶] [▶] [ENTER] to create a new program.
2. Type EVAL [ENTER] to name the program. (Be sure that the A-LOCK is on.) You are now in the program editor, which allows you to enter commands. The colon (:) in the first column of the line indicates that it is the beginning of a command line.
3. The first command line will ask the user to choose a value for x. Press [PRGM] [▶] 3 [2nd] [A-LOCK] "ENTER X" [ENTER]. (To enter a space between words, press the 0 key when the A-LOCK is on.)
4. The second command line will allow the user to enter any value for x into the calculator. Press [PRGM] [▶] 1 [X,T,θ,n] [ENTER].
5. The expression to be evaluated for the value of x is $x - 7$. To store the expression as Y, press [X,T,θ,n] [−] 7 [STO▸] [ALPHA] [Y] [ENTER].
6. Finally, we want to display the value for the expression. Press [PRGM] [▶] 3 [ALPHA] [Y] [ENTER]. At this point, you have completed writing the program. It should appear on your calculator like the screen shown below.

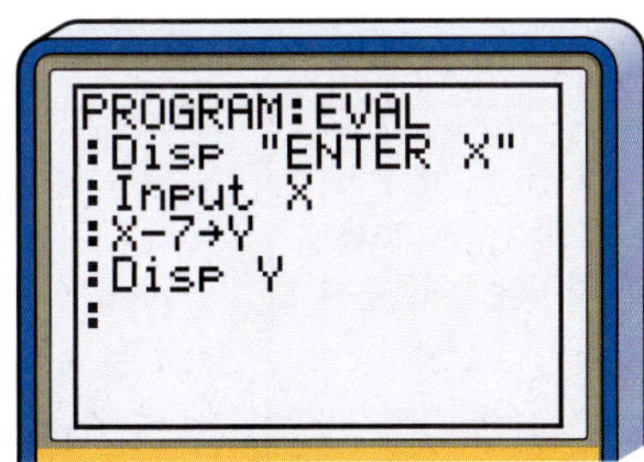

7. Now press [2nd] [QUIT] to return to the home screen.
8. To execute the program, press [PRGM]. Then press the down arrow to locate the program name and press [ENTER] twice. The program asks for a value for x. Input any value for which the expression is defined and press [ENTER]. To immediately re-execute the program, simply press [ENTER] when the word *Done* appears on the screen. To break during program execution, press [ON].
9. To delete a program, press [2nd] [MEM] 2 7. Use the arrow and [ENTER] keys to select a program you wish to delete. Then press [DEL] 2.

While a graphing calculator cannot do everything, it can make some tasks easier. To prepare for whatever lies ahead, you should try to learn as much as you can. The future will definitely involve technology. Using a graphing calculator is a good start toward becoming familiar with technology.

Problem-Solving Strategy Workshop

Objectives Students should:
- write numerical and algebraic expressions in a spreadsheet,
- write an article summarizing their findings, and
- research the retail costs of videos, books, and CDs.

How to Use the Workshop

You may want to introduce the workshop at the beginning of Chapter 1, with the intent that it be completed by the end of the chapter. This should motivate students to learn about translating words into expressions and about organizing data.

▶ **Problem-Solving Pointer**
Stress that the shipping and handling applies to each video ordered. Suggest that students find the least and greatest average costs per movie by listing the cost of 12 movies from the same category plus their shipping costs. Then have students write an expression that summarizes each average cost.

As students begin researching the costs of joining a buying club, point out that some clubs may not clearly state their shipping charges.

Students should add their spreadsheets and articles to their portfolios at this time.

Problem-Solving Strategy Workshops

Use a Table

Project

As a new DVD club member, you can choose seven movies. Each movie costs 1¢ plus $1.69 for shipping and handling. Within three years, you must order at least five more movies, each at the regular club price, plus the same shipping fee.

Type of Movie	Regular Club Price
Children's	$12.99
New Release	$24.99
All-Time Favorite	$16.99
Classic	$8.99

Suppose you buy a total of 12 movies. What is the lowest possible average cost per movie? the highest possible average cost per movie?

Working on the Project

Work with a partner and choose a strategy to help analyze and solve the problem. Here are some questions to help you get started.

- How much do you pay for the first shipment of seven 1¢ movies?
- What are the least and greatest amounts you can spend on the five required regular-priced movies?

Technology Tools

- Use a **spreadsheet** to calculate the average cost of the movies.
- Use **word processing software** to write your newspaper article.

Research For more information about CD clubs, visit: www.algconcepts.com

Presenting the Project

Write an article for the school newspaper discussing the advantages and disadvantages of joining a DVD, book, or CD club.

- Research prices at retail or online stores. What would you pay for the same number of DVDs, books, or CDs required by the club?
- Show how the average cost per DVD, book, or CD changes as you buy more at the regular club price.

754 Problem-Solving Strategy Workshops

Draw a Diagram

Project

Sears Tower, the Magnificent Mile, Lake Shore Drive, Wrigley Field—you're in Chicago to see the sights! Suppose you are at the intersection of Illinois Street and Wells Street. You need to meet your group at the intersection of Ohio Street and Dearborn Street. How many ways can you walk there if you can only cross at intersections and can't backtrack?

Working on the Project

Work with a partner and choose a strategy. Develop a plan. Here are some suggestions to help you get started.

- Which intersections could you walk to from Illinois and Wells if you can't backtrack?
- Draw the map on grid paper and label the intersections with letters or numbers.
- Draw the different routes you can take on the grid paper.

Technology Tools

- Using **word processing software** to write an explanation of your solution.
- Use **drawing software** to draw your routes.
- Use **presentation software** to present your project.

Research For more information about Chicago, visit: www.algconcepts.com

Presenting the Project

Make a poster showing all of the routes. Include an explanation of your strategy for solving this problem. Make sure your explanation includes the following:

- a discussion of the number of possible routes if you have to meet your group one block past Ohio and Dearborn at Ontario and Dearborn, and
- a conjecture about the number of routes between any two intersections on the map.

Problem-Solving Strategy Workshop

Objectives Students should:

- use a diagram,
- make an organized list, and
- discuss how many different routes can be identified between two points on a grid.

How to Use the Workshop

You may want to introduce the workshop at the beginning of Chapter 2, with the intent that it be completed by the end of the chapter. The project should motivate students to learn about coordinate planes.

▶ Problem-Solving Pointer

Provide students with grid paper for drawing the routes. Review the advantages of making an organized list as opposed to just randomly identifying routes. Suggest that students label all the possible intersections that can be crossed in order to simplify the listing of routes.

Students should add their posters and explanations to their portfolios at this time.

Problem-Solving Strategy Workshops 755

Problem-Solving Strategy Workshop

Objectives Students should:
- solve equations,
- make an organized list, and
- use guess and check to solve a problem.

How to Use the Workshop

You may want to introduce the workshop at the beginning of Chapter 3, with the intent that it is to be completed by the end of the chapter. Make sure students understand that they are to find several combinations of rides that will require the use of a total of exactly 50 tickets.

▶ **Problem Solving Pointer** You might consider providing each pair of students with 50 paper "tickets" that they can use to model each combination of rides they are considering.

The students should add their spreadsheets and ad campaigns to their portfolios at this time.

PROBLEM-SOLVING STRATEGY WORKSHOP

Use an Equation

Project

At Funland, an amusement park, you can buy a 50-ticket pass to use at any time. The table shows the number of tickets needed for each ride.

You can use all 50 tickets by riding one 5-ticket ride 10 times. In what other ways can you use exactly 50 tickets?

Ride	Tickets Needed
The Mighty Axe	7
Log Chute	6
Kite-Eating Tree	5
Mystery Mine Ride	7
Ripsaw Roller Coaster	7
Screaming Yellow Eagle	6
Skyscraper Ferris Wheel	5
Tumbler	6

Working on the Project

Work with a partner to solve the problem.

- If you ride each ride once, can you use exactly 50 tickets?
- Suppose you ride six 5-ticket rides. Can you use all of the remaining tickets with none left over if you do not ride any more 5-ticket rides?

Technology Tools

- Use a **spreadsheet** to find the number of tickets a person can use for different combinations of rides.
- Use **word processing software** to design an ad campaign.
- Use **presentation software** to present your project.

Research For more information about amusement parks, visit: www.algconcepts.com

Presenting the Project 

Suppose the 50-ticket pass costs \$25 and an all-day pass for one person costs \$18.95. Design an ad campaign that answers the questions below.

- Suppose the amusement park is open from 10:00 A.M. to 9 P.M. About how many tickets could you use during that time period?
- What are the advantages and disadvantages of buying a 50-ticket pass? an all-day pass?

756 Problem-Solving Strategy Workshops

Look for a Pattern

Project

The United States government began issuing nine-digit Social Security numbers in 1935. An example is 123-45-6789. Each person can have only one number and the numbers are never reused. When do you think the government will run out of new numbers?

Year	Population	Births
1930	123,202,624	2,618,000
1940	132,164,569	2,559,000
1950	151,325,798	3,632,000
1960	179,323,175	4,258,000
1970	203,302,031	3,731,000
1980	226,542,203	3,612,000
1990	248,709,873	4,158,000
2000	281,421,906	4,059,000

Working on the Project

Work with a partner and choose a strategy to solve the problem.

- How many Social Security numbers are possible if each of the nine digits can be 0 through 9?
- Estimate the maximum number of Social Security numbers that could have been issued in 1935.
- Estimate the total number of people who had Social Security numbers by the year 2010 if every U.S. citizen had a Social Security number.

Technology Tools

- Use a **calculator** to find the total number of unique Social Security numbers.
- Use **spreadsheet software** to estimate the total number of Social Security numbers that will have been issued by the year 2010.

Research For more information about Social Security numbers, visit: www.algconcepts.com

Presenting the Project

Write a one-page paper describing your results. Include the following:

- the total number of Social Security numbers possible,
- estimates for the total number of Social Security numbers issued by the year 2010 and the number issued each year after the year 2010, and
- the year you think that the government will run out of new Social Security numbers.

Objectives Students should:

- use the Fundamental Counting Principle to find the total number of Social Security numbers possible,
- estimate population growth using a table and problem-solving techniques,
- use an equation to estimate the year when the U.S. will run out of Social Security numbers, and
- write a description of their problem-solving method for this project.

How to Use the Workshop

You may want to introduce the workshop at the beginning of Chapter 4, with the intent that it be completed by the end of the chapter. Students will probably employ a variety of strategies to estimate the number of Social Security numbers issued by the year 2010. One strategy is to estimate the number of Social Security numbers issued in 1935 and then add the estimated number of people born each year from 1935 to 2010.

Problem Solving Pointer Consider having students compare their estimates to the actual government figures found by searching the Internet. You may also want to have a discussion about whether all possible combinations of numerals are actually used. For example, was the number 000-00-0000 ever issued? Another discussion point might be whether every U.S. citizen actually has a Social Security number.

Problem-Solving Strategy Workshop

Objectives Students should:
- make a pattern for the surface area of a rectangular prism and find its surface area,
- write a polynomial expression for the surface area of a rectangular prism,
- use guess-and-check or another problem-solving strategy to make patterns for three rectangular prisms with surface areas of 150 square inches, and
- write a paper promoting the design they feel is best for the new product, including a product name and logo or design.

How to Use the Workshop

You may want to introduce the workshop at the beginning of Chapter 9, with the intent that it be completed by the end of the chapter. This should motivate students to learn about using addition and multiplication with polynomial expressions so they can calculate the dimensions of the boxes.

▶ Problem-Solving Pointer
Drawing patterns on grid paper will make it easy for students to find the surface area of a prism. You may want to have extra grid paper on hand for students.

Students should add their calculations, papers, and designs to their portfolios at this time.

Guess and Check

Project

Congratulations! You have been selected to design the box for a new caramel corn snack. The manufacturers would like you to present three possible box designs. Each box must be a rectangular prism with a surface area of 150 square inches. Make a pattern for each of the three box designs. In a one-page paper, explain which box design you think will be best for the new product snack.

Working on the Project

Work with a partner and choose a strategy. Here are some suggestions to help you get started.

- Cut apart an empty cereal box along its edges to make a pattern for the surface area of a rectangular prism.
- Find the area of each of the six faces and the surface area of the prism.
- Suppose you don't know the measurements of the prism. Let ℓ represent the length, w represent the width, and h represent the height of the prism. Write a polynomial expression for the surface area of the prism.

Technology Tools

- Use a **calculator** to find the surface area of your patterns.
- Use **drawing software** to make patterns for your boxes.

Research For more information about packaging, visit: www.algconcepts.com

Presenting the Project

Prepare a portfolio of your box designs. Make sure your portfolio contains the following information:

- your calculations of the surface area of each box design,
- a one-page paper promoting the box design you think will be best, and
- a name for the new product and a logo or design for the box.

758 Problem-Solving Strategy Workshops

Use a Graph

Project

Do you want to be a millionaire? In this project, you will make a plan for saving money. Your goal is to reach $1,000,000 at the end of a 40-year period, following the guidelines listed below.

1. Your initial deposit is any amount of your choice.
2. The money is invested for exactly 40 years.
3. The annual growth rate of your investment is 5%.
4. Interest is calculated at the end of each year.
5. Choose an additional amount of money to deposit after interest is calculated each year. This amount is fixed, or the same, for all 40 years.
6. After 40 years, you should have as close to $1,000,000 as possible.

Working on the Project

Work with a partner and choose a strategy to help analyze and solve the problem. Here are some questions to help you get started.

- At the beginning of year 1, you decide to invest $5000. How much money will you have at the end of year 1, including interest?
- At the end of year 1, you invest an additional $2000. How much money will you have at the end of year 2, including interest?

Technology Tools

- Use a **spreadsheet** to calculate each year-end balance.
- Use **graphing software** to make a graph of your investment over 40 years.
- Use **presentation software** to present your project.

Research For more information about investing, visit: www.algconcepts.com

Presenting the Project

Write a one-page paper describing your investment plan. Include:

- successful and unsuccessful strategies used to plan,
- a table or spreadsheet showing each year-end balance, and
- a graph of your savings over the 40-year period.

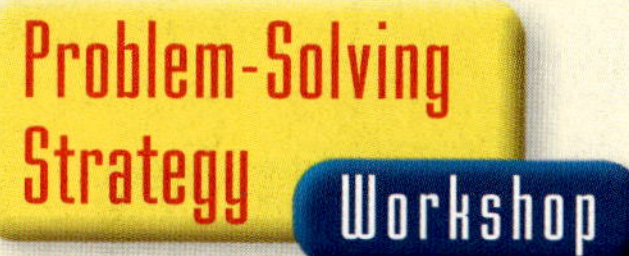

Objectives Students should:

- use a table or spreadsheet to record and examine the amount of money in an investment plan,
- graph a function which represents the growth of the balance, and
- use a variety of problem-solving strategies to find the appropriate amount of money to save each year.

How to Use the Workshop

You may want to introduce the workshop at the beginning of Chapter 11, with the intent that it be completed by the end of the chapter. If you prefer, students could change the amount of money that they want to have at the end of 40 years. You could also have students compare investing just one amount for 40 years to investing an initial amount and then adding money each year.

▶ **Problem Solving Pointer** Be sure students understand that they are to start with an initial amount of money for one year, calculate the yearly interest on that amount, and then add some other amount before the next year begins. The next year, they will find the interest on the new amount and then add more money to the account.

The students should add their table or spreadsheet, graph, and papers to their portfolios at this time.

Problem-Solving Strategy Workshop

Objectives Students should:
- plan a one-week food menu that meets the given guidelines,
- write inequalities to represent the guidelines, and
- determine whether a sample menu meets the guidelines.

How to Use the Workshop

You may want to introduce the workshop at the beginning of Chapter 12, with the intent that it be completed by the end of the chapter. This should motivate students to learn how inequalities apply to the fields of nutrition and diet.

▶ **Problem Solving Pointer** As students plan their menus, ask them to consider other criteria they might use in their selections, such as what foods they like, what foods go well together, and what kind of preparation the foods need.

You may want to have students record what they actually eat for one week and calculate the nutritional value for comparison to the menus they plan.

The students should add their menus, nutritional information, inequality lists, and any tables or spreadsheets to their portfolios at this time.

Check for Reasonableness

Project

Plan a one-week menu of foods that require little or no preparation.
1. Total Calories must be between 1200 and 1800 per day.
2. Calories from fat must be at most 30% of the total Calories. Note that 1 gram of fat has 9 Calories.
3. Total sodium must be less than 2400 milligrams.
4. Total protein must be at least 46 grams.

Working on the Project

Work with a partner to solve the problem.

- Write inequalities to represent the Calories, fat (g), sodium (mg), and protein (g) allowed per day.
- Determine whether the sample menu in the table below satisfies the nutritional guidelines.

Technology Tools

- Use a **spreadsheet** to help write a menu that meets all guidelines.
- Use **publishing software** to illustrate your menu.

Food (one serving)	Calories	Fat (g)	Sodium (mg)	Protein (g)
canned chicken noodle soup	90	2	940	6
granola cereal and skim milk	270	9	20	5
canned tuna in water (2 ounces)	60	0.5	250	13
frozen dinner	450	20	1530	10
white bread (3 slices)	240	3	480	6

Research For more information about nutrition, visit: www.algconcepts.com

Presenting the Project

Compile a portfolio of your work for this project. Include the menu for each of seven days, the nutritional value of each food, and a list of inequalities to show that all guidelines were met.

760 Problem-Solving Strategy Workshops

Use a Formula

Project

Although penguins waddle slowly on land, the gentoo penguin of Antarctica is thought to be the fastest swimming bird in the world. Use the formula $d = rt$ to model the distance d, rate r, and time t of the penguin. Suppose the penguin swims a distance of 20 miles. The equation that relates t and r to d is $20 = rt$.

In this project, you will graph $20 = rt$ and describe the graph. You will then choose four other values for d and graph their equations.

Working on the Project

Work with a partner and choose a strategy. Here are some suggestions to help you get started.

- Choose several values of r and solve the equation for t. Write the results as ordered pairs (r, t).
- Graph the ordered pairs on a coordinate plane.
- Research the average distance that four other animals could travel in one hour.

Technology Tools

- Use a **spreadsheet** or a **graphing calculator** to prepare your graphs.
- Use **presentation software** to prepare and give your presentation.

Research For more information about animal speeds, visit: www.algconcepts.com

Presenting the Project

Prepare a presentation of your findings for your animals. Make sure that your presentation includes:

- a table of ordered pairs and a graph for each value of d,
- an explanation of your findings, including a comparison of the graphs, and
- a discussion of how these graphs differ from the graph of an equation such as $d = 20t$. Use the terms *direct variation* and *inverse variation* in your discussion.

Objectives Students should:

- find ordered pairs that satisfy an equation by substituting and solving equations by division,
- graph ordered pairs in a coordinate plane,
- research how fast various animals can travel,
- graph inverse variations for different values of the constant of variation, and
- compare graphs of direct and inverse variations.

How to Use the Workshop

You may want to introduce the workshop at the beginning of Chapter 15, with the intent that it be completed by the end of the chapter. This should motivate students to learn about operating with rational expressions and solving rational equations.

▶ Problem-Solving Pointer
Students who use graphing calculators or graphing software may notice that the graphs have branches both in quadrants I and III. Have them substitute negative values of r in the original equation to see that the graphs make mathematical sense, but point out that in this real-world application time and rate cannot be negative. As students graph equations of the form $d = rt$ for fixed values of d for different animals, have them observe how the shape of the graph changes. They should notice that the smaller the value of d, the more the graph bows in toward the origin.

Students should add their materials to their portfolios at this time.

Glossary/Glosario

A mathematics multilingual glossary is available at www.math.glencoe.com.

The glossary is available in the following languages.

Arabic	Haitian Creole	Russian	Urdu	Bengali	Hmong
Spanish	Vietnamese	Cantonese	Korean	Tagalog	

En www.math.glencoe.com encontrarás un glosario matemático plurilingüe.

Disponible en los siguientes idiomas:

Árabe	Criollo haitiano	Ruso	Urdu	Bengalí	Hmong
Español	Vietnamita	Cantonés	Coreano	Tagalo	

A

absolute value The absolute value of a number a is its distance from zero on a number line and is represented by $|a|$. (p. 55)

The *absolute value* of -2 is 2, or $|-2| = 2$.

valor absoluto El valor absoluto de un número a es la distancia que a dista de cero en la recta numérica, la que se denota por $|a|$. (pág. 55)

El *valor absoluto* de -2 es 2 ó $|-2| = 2$.

additive inverse A number that when added to a given number results in a sum of zero. (p. 65)

The *additive inverse* of 4 is -4 because $4 + (-4) = 0$.

inverso aditivo Número que sumado a un número dado da cero. (pág. 65)

El *inverso aditivo* de 4 es -4 porque $4 + (-4) = 0$.

algebraic expression An expression consisting of one or more numbers and variables along with one or more arithmetic operations. (p. 4)

$4x$, ab, and $5x^2 - 7$ are *algebraic expressions.*

expresión algebraica Expresión que consta de uno o más números y variables, junto con una o más operaciones matemáticas. (pág. 4)

$4x$, ab y $5x^2 - 7$ son todas *expresiones algebraicas.*

arithmetic sequence A numerical pattern that increases or decreases at a constant rate or value. The difference between successive terms of the sequence is constant. (p. 110)

2, 5, 8, 11, 14, ... is an *arithmetic sequence.*

sucesión aritmética Patrón numérico que aumenta o disminuye a una tasa o valor constante. La diferencia entre términos consecutivos de la sucesión es constante. (pág. 110)

2, 5, 8, 11, 14, ... es una *sucesión aritmética.*

B

base in a percent proportion The number to which the percentage is compared. (p. 199)

$$\frac{\textit{percentage} \rightarrow 1}{\textit{base} \rightarrow 5} = \frac{20}{100} \leftarrow \textit{percent}$$

base de una proporción porcentual Número al que se compara la parte. (pág. 199)

$$\frac{\textit{parte} \rightarrow 1}{\textit{base} \rightarrow 5} = \frac{20}{100} \leftarrow \textit{porcentaje}$$

base of an expression In an expression of the form x^n, the base is x. (p. 336)

In 5^6, the base is 5.

base de una expresión En una expresión de la forma x^n, la base es x. (pág. 336)

La base de 5^6 es 5.

best-fit line The line that most closely approximates the data in a scatter plot. (p. 308)

recta de óptimo ajuste Recta que aproxima más estrechamente los datos de una gráfica de dispersión. (pág. 308)

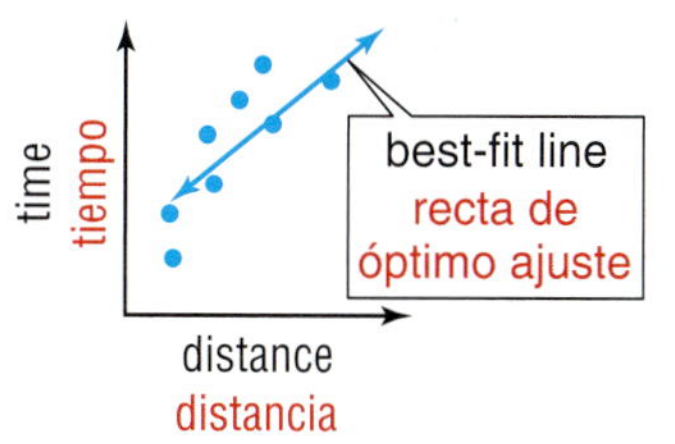

---

biased sample A sample in which one or more parts of the population are favored over others. (p. 32)

muestra sesgada Muestra en que se ha favorecido una o más partes de una población en lugar de otras. (pág. 32)

---

binomial An algebraic expression that has two terms. (p. 383) — $x + 1$, $x^2 + y$, and $a - 2b$

binomio Expresión algebraica de dos términos. (pág. 383) — $x + 1$, $x^2 + y$, y $a - 2b$

---

boundary A line or curve that separates the coordinate plane into regions. (p. 535)

frontera Recta o curva que divide el plano coordenado en regiones. (pág. 535)

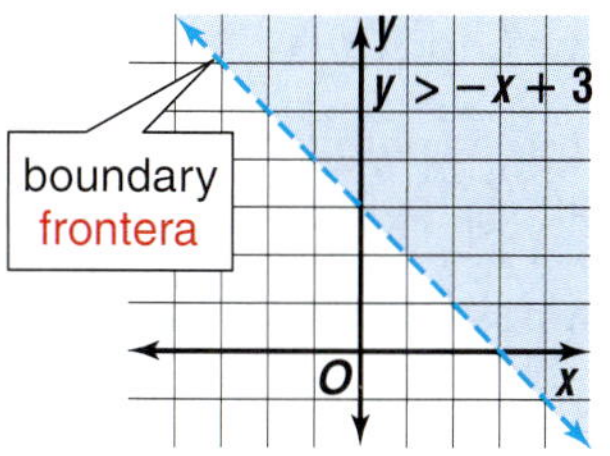

---

box-and-whisker plot A diagram that divides a set of data into four parts using the median and quartiles. (p. 210)

diagrama de caja y patillas Diagrama que divide un conjunto de datos en cuatro partes usando la mediana y los cuartiles. (pág. 210)

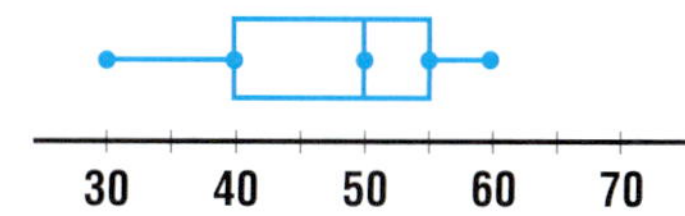

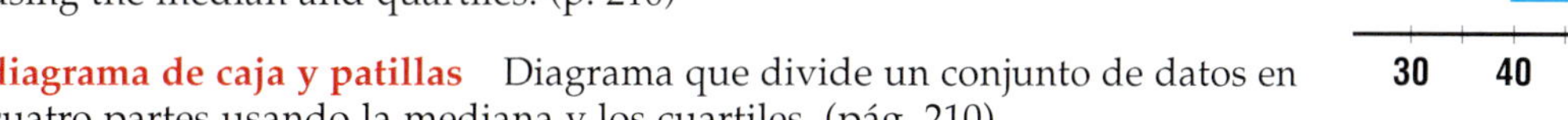

C

circle graph A graph that compares parts of a set of data as a percent of the whole set. (p. 200)

gráfica circular Gráfica en que se comparan partes de un todo como porcentaje del todo. (pág. 200)

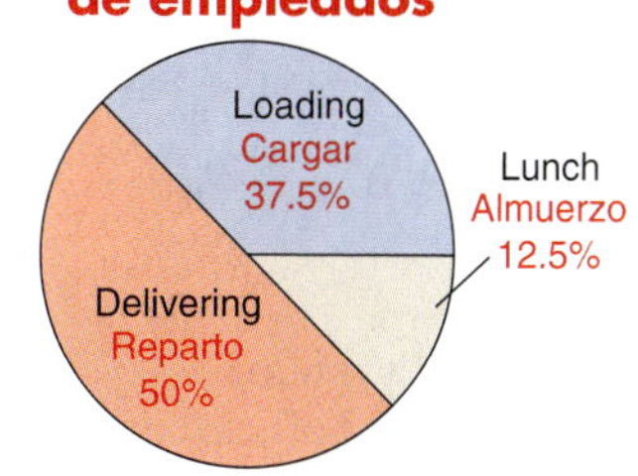

---

coefficient The numerical factor of a term. (p. 20) — The *coefficient* of $3x$ is 3.

coeficiente El factor numérico de un término. (pág. 20) — El *coeficiente* de $3x$ es 3.

---

combination An arrangement or listing in which order is not important. (p. 152) — Choosing toppings for a pizza is a *combination*.

combinación Arreglo o lista en que el orden no es importante. (pág. 152) — Elegir los ingredientes para una pizza es una *combinación*.

GLOSSARY

common factor A whole number that is a factor of each number in a set of numbers. (p. 422)

5 is a *common factor* of 10, 15, 25, and 100.

factor común Un número que es un factor de cada número en un conjunto de números. (pág. 422)

5 es un *factor común* de 10, 15, 25 y 100.

---

common multiples Multiples that are shared by two or more numbers. (p. 662)

Some *common multiples* of 2 and 3 are 6, 12, and 18.

múltiplos comunes Múltiplos compartidos por dos o más números. (pág. 662)

Algunos *múltiplos comunes* de 2 y 3 son 6, 12 y 18.

---

complements Two events are complements if the sum of their probabilities is 1. (p. 223)

When tossing a coin, heads and tails are *complements*.

complementarios Dos eventos son complementarios si sus probabilidades suman 1. (pág. 223)

Al lanzar una moneda, cara y escudo son eventos *complementarios*.

---

composite number A whole number greater than 1 that has more than two factors. (p. 420)

4 is a *composite number*. It has three factors: 1, 2, and 4.

número compuesto Número entero mayor que 1 que posee más de dos factores. (pág. 420)

4 es un *número compuesto*. Tiene tres factores: 1, 2 y 4.

---

compound event An event which consists of two or more simple events. (p. 224)

Drawing a card and then drawing a second card is a *compound event*.

evento compuesto Evento que consta de dos o más eventos simples. (pág. 224)

Sacar una baraja y luego sacar una segunda baraja es un *evento compuesto*.

---

compound inequality Two or more inequalities that are connected by the words *and* or *or*. (p. 524)

$x > 3$ and $x < 5$
or
$x < -3$ or $x > 7$

desigualdad compuesta Dos o más desigualdades unidas por las palabras *y* u *o*. (pág. 524)

$x > 3$ y $x < 5$
ó
$x < -3$ ó $x > 7$

---

conjugate of a binomial The conjugate of the binomial $a\sqrt{b} + c\sqrt{d}$ is $a\sqrt{b} - c\sqrt{d}$. (p. 616)

$6\sqrt{2} + 4\sqrt{3}$ and $6\sqrt{2} - 4\sqrt{3}$ are *conjugates*.

conjugado de un binomio El conjugado de $a\sqrt{b} + c\sqrt{d}$ es $a\sqrt{b} - c\sqrt{d}$. (pág. 616)

$6\sqrt{2} + 4\sqrt{3}$ y $6\sqrt{2} - 4\sqrt{3}$ son *conjugados*.

---

consistent system A system of equations that has at least one ordered pair that satisfies both equations. (p. 554)

sistema consistente Sistema de ecuaciones para el que hay por lo menos un par ordenado que satisface ambas ecuaciones. (pág. 554)

---

constant of variation The number k in an equation of the form $y = kx$. (p. 264)

In the equation of $y = 4x$, the *constant of variation* is 4.

constante de variación El número k en una ecuación de la forma $y = kx$. (pág. 264)

En la equación, $y = 4x$, la *constante de variación* es 4.

coordinate The number that corresponds to a point on a number line. (p. 53) **coordinate grid (see coordinate plane)**

coordenada Número asociado con un punto de una recta numérica. (pág. 53) **cuadriculado coordenado (véase plano coordenado)**

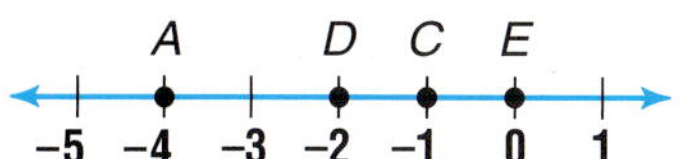

The *coordinate* of point A is -4.
La *coordenada* de punto A es -4.

coordinate plane A plane, also called a coordinate grid or coordinate system, in which a horizontal number line and a vertical number line intersect at their zero points. The two number lines are called *axes*. Their intersection point is called the *origin*. (p. 58)

coordinate system *See coordinate plane.*

plano coordenado Plano, también llamado cuadriculado coordenado o sistema coordenado, en el que se han trazado una recta numérica horizontal y una vertical que se intersecan en sus puntos cero. Las rectas se llaman *ejes* y su punto de intersección se llama *origin*. (pág. 58)

sistema coordenado *Véase plano coordenado.*

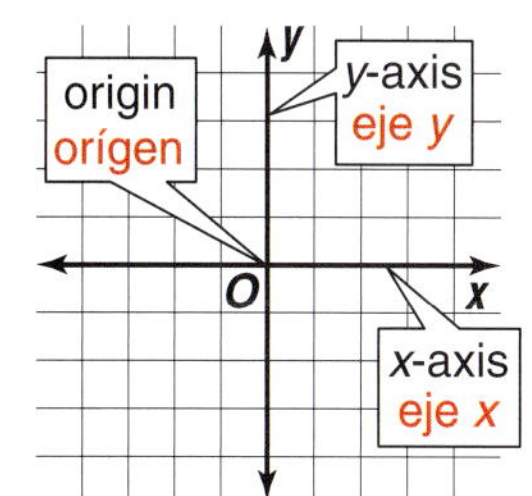

counterexample A specific case which proves a statement false. (p. 16)

contraejemplo Ejemplo que prueba la falsedad de un enunciado. (pág. 16)

$1 \div 2 = \frac{1}{2}$ is a *counterexample* to "The quotient of two integers is always an integer."

$1 \div 2 = \frac{1}{2}$ es un *contraejemplo* a "El cociente de dos enteros es siempre un entero."

cross products The product of the terms on the diagonals when two ratios are compared. (p. 95)

productos cruzados Producto de los términos en diagonal de dos razones que se están comparando. (pág. 95)

In the proportion $\frac{2}{3} = \frac{8}{12}$, the *cross products* are 2×12 and 3×8.

En la proporción $\frac{2}{3} = \frac{8}{12}$, los *productos cruzados* son 2×12 y 3×8.

D

data Numerical information gathered for statistical purposes. (p. 32)

datos Información numérica que se reúne con fines estadísticos. (pág. 32)

deductive reasoning The process of using facts, rules, definitions, or properties to reach a valid conclusion. (p. 30)

razonamiento deductivo Proceso de usar hechos, reglas, definiciones o propiedades para sacar conclusiones válidas. (pág. 30)

degree of a monomial The sum of the exponents of all its variables. (p. 384)

grado de un monomio Suma de los exponentes de todas sus variables. (pág. 384)

The *degree* of the monomial $4x^3$ is 3.

El *grado* de un monomio $4x^3$ es 3.

degree of a polynomial The greatest degree of any term in the polynomial. (p. 384)

The *degree* of the polynomial $8x^4 + 2x^3 - 1$ is 4.

grado de un polinomio Grado mayor de los términos del polinomio. (pág. 384)

El *grado* de un polinomio $8x^4 + 2x^3 - 1$ es 4.

---

dependent system A consistent system of equations that has an infinite number of solutions. (p. 554)

sistema dependiente Sistema consistente de ecuaciones que posee un número infinito de soluciones. (pág. 554)

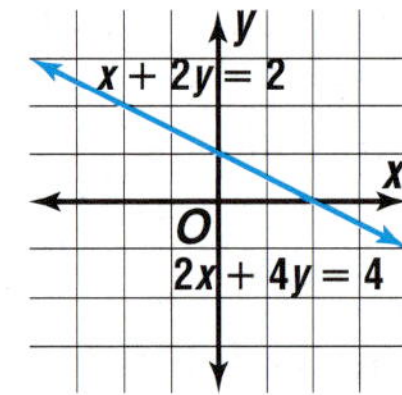

---

dependent variable The variable in a relation whose value depends on the value of the independent variable. (p. 264)

variable dependiente Variable de una relación cuyo valor depende del valor de la variable independiente. (pág. 264)

---

difference of squares Two perfect squares separated by a subtraction sign. (p. 447)

$a^2 - b^2 = (a + b)(a - b)$

diferencia de cuadrados Cuadrados perfectos separados por el signo de sustracción. (pág. 447)

---

dimensional analysis The process of carrying units throughout a computation. (p. 190)

$\frac{55 \text{ miles}}{1 \text{ hour}} \cdot 3 \text{ hours} = 165 \text{ miles}$

análisis dimensional En un cálculo, proceso de tomar en cuenta las unidades. (pág. 190)

$\frac{55 \text{ millas}}{1 \text{ hora}} \cdot 3 \text{ horas} = 165 \text{ millas}$

---

direct variation An equation of the form $y = kx$, where $k = 0$. (p. 264)

$y = 4x$
or
$y = \frac{1}{3}x$

variación directa Ecuación de la forma $y = kx$, con $k = 0$. (pág. 264)

$y = 4x$
o
$y = \frac{1}{3}x$

Distance Formula The distance between two points, with coordinates (x_1, y_1) and (x_2, y_2), is given by $d = \sqrt{(x_2 - x_1)^2 + (y_2 - y_1)^2}$. (p. 607)

Distance between (2, −1) and (5, 3), is $d = \sqrt{(5 - 2)^2 + (3 - (-1))^2}$ or 5.

fórmula de la distancia La distancia entre los puntos (x_1, y_1) y (x_2, y_2), viene dada por $d = \sqrt{(x_2 - x_1)^2 + (y_2 - y_1)^2}$. (pág. 607)

La distancia entre los puntos (2, −1) y (5, 3), es $d = \sqrt{(5 - 2)^2 + (3 - (-1))^2}$ ó 5.

---

domain The set of the first numbers or abscissas of the ordered pairs in a relation. (p. 238)

In the relation {(1, 2), (3, 5), (4, 0)}, the *domain* is {1, 3, 4}.

dominio Conjunto de los primeros números o abscisas de los pares ordenados de una relación. (pág. 238)

En la relación {(1, 2), (3, 5), (4, 0)}, el *dominio* es {1, 3, 4}.

GLOSSARY

empty set A set with no elements shown by the symbol { } or ∅. A solution set with no members. Also called a null set. (p. 130)

conjunto vacío Conjunto sin elementos, denotado por { } o ∅. Conjunto-solución sin elementos. (pág. 130)

---

equation A mathematical sentence stating that two expressions are equal. (p. 5)

$3 \times (7 + 8) = 9 \times 5$

ecuación Enunciado matemático que afirma la igualdad de dos expresiones. (pág. 5)

---

equivalent equations Equations that have the same solution. (p. 122)

$2x + 3 = 9$ and $2x = 6$

ecuaciones equivalentes Ecuaciones que tienen la misma solución. (pág. 122)

---

equivalent expressions Expressions that have the same value or that have the same mathematical meaning for all replacement values of their variables. (p. 20)

$3 + 2 = 10 - 5$ and $2x + 3x = 5x$

expresiones equivalentes Expresiones que tienen el mismo valor o el mismo significado matemático para todos los valores de sustitución de sus variables. (pág. 20)

$3 + 2 = 10 - 5$ y $2x + 3x = 5x$

---

evaluate Find the value of an expression. (p. 10)

evaluar Calcular el valor de una expresión. (pág. 10)

---

event A specific outcome or type of outcome. (p. 147)

evento Resultado específico o tipo de resultado. (pág. 147)

---

excluded values Any values of a variable that result in a denominator of 0 must be excluded from the domain of that variable. (p. 638)

The *excluded value* of $y = \frac{1}{x-1}$ is 1.

valores excluidos Cualquier valor de una variable que dé un denominador nulo debe excluirse del dominio de dicha variable. (pág. 638)

El *valor excluido* de $y = \frac{1}{x-1}$ es 1.

---

experimental probability The ratio of the number of positive outcomes to the total number of events or trials in a probability experiment. (p. 220)

probabilidad experimental En un experimento probabilístico, razón del número de resultados positivos al número total de eventos o pruebas. (pág. 220)

---

exponent A number that indicates how many times a number or expression is to be multiplied by itself. (p. 336)

In the expression 5^3, the *exponent* is 3.

exponente Número que indica el número de veces que un número o expresión se multiplica por sí misma. (pág. 336)

En la expresión 5^3, el *exponente* es 3.

---

exponential function A function that can be described by an equation of the form $y = a^x$, where $a > 0$ and $a \neq 1$. (p. 489)

$y = 2^x$

función exponencial Función de la forma $y = a^x$, con $a > 0$ y $a \neq 1$. (pág. 489)

expression A mathematical combination of numbers, variables, and operations. (p. 4)

expresión Combinación de números, variables y operaciones. (pág. 4)

$5 + 2^3$, $4x$, $6x + y^2$

F

factor In an algebraic or numerical expression, the quantities being multiplied are called factors. (p. 4)

factor En una expresión numérica o algebraica, se llaman factores las cantidades que se multiplican. (pág. 4)

3 and 11 are *factors* of 33.

3 y 11 son *factores* de 33.

---

factor a polynomial Express a polynomial as the product of monomials and polynomials. (p. 428)

factorizar un polinomio Escribir un polinomio como producto de monomios y polinomios. (pág. 428)

$8y^2 + 10y$ in factored form is $2y(4y = 5)$.

$8y^2 + 10y$ en forma factorizada es $2y(4y = 5)$.

---

factorial The expression of $n!$, read n factorial, where n is greater than zero, is the product of all positive integers beginning with n and counting backward to 1. (p. 153)

factorial La expresión $n!$, léase n factorial, con n un número natural, es el producto de todos los números naturales, empezando en n y contando hacia atrás hasta llegar a 1. (pág. 153)

$5! = 5 \times 4 \times 3 \times 2 \times 1 = 120$

---

family of graphs Graphs and equations of graphs that have at least one characteristic in common. (p. 316)

familia de gráficas Gráficas y sus ecuaciones que poseen por lo menos un rasgo común. (pág. 316)

---

formula An equation that states a rule for the relationship between certain quantities. (p. 24)

fórmula Ecuación que establece una relación entre ciertas cantidades. (pág. 24)

$A = \pi r^2$ is the *formula* for finding the area of a circle.

$A = \pi r^2$ es la *fórmula* para el área de un círculo.

---

frequency table A table of tally marks used to record and display how often events occur. (p. 33)

tabla de frecuencias Tabla en que se lleva la cuenta de la frecuencia de eventos. (pág. 33)

Age Edad	Tally Cuenta	Frequency Frecuencia
1-10	~~||||~~ |||	8
11-20	~~||||~~ ||||	9
21-30	~~||||~~	5
31-40	~~||||~~ ~~||||~~ ||	12
41-50	~~||||~~ |||	8

---

function A relation in which exactly one element of the range is paired with each element of the domain. (p. 256)

función Relación en que a cada elemento del dominio le corresponde un único elemento del rango. (pág. 256)

$\{(5, -2), (3, 2), (4, -1), (-2, 2)\}$

---

function notation A way to name a function that is defined by an equation. (p. 258)

notación funcional Forma de indicar una función definida por una ecuación. (pág. 258)

In *function notation*, the equation $y = 3x - 8$ is written as $f(x) = 3x - 8$.

En *notación funcional*, la ecuación $y = 3x - 8$ se escribe $f(x) = 3x - 8$.

functional value The element in the range that corresponds to a specific element in the domain. (p. 258)

If $f(x) = 3x - 8$, then $f(2) = 3(2) - 8$ or -2.

valor funcional Elemento del rango que corresponde a un elemento específico del dominio. (pág. 258)

Si $f(x) = 3x - 8$, entonces $f(2) = 3(2) - 8$ ó -2.

G

geometric sequence A sequence in which each term after the nonzero first term is found by multiplying the previous term by a constant called the common ratio r, where $r \neq 0$ or 1. (p. 494)

1, 4, 16, 64, 256, …

sucesión geométrica Sucesión en que cada término después del primero se obtiene del anterior multiplicándolo por una constante, la llamada razón común r, con $r \neq 0$ ó 1. (pág. 494)

---

graph To draw, or plot, the points named by certain numbers or ordered pairs on a number line or coordinate plane. (p. 53)

graficar Marcar en una recta numérica o plano coordenado los puntos dados por ciertos números o pares ordenados. (pág. 53)

---

greatest common factor (GCF) The greatest number that is a factor of two or more numbers. (p. 422)

The *greatest common factor* of 30, 60, and 15 is 15.

máximo común divisor (MCD) Número que es el factor común mayor de dos o más números. (pág. 422)

El *máximo común divisor* de 30, 60 y 15 es 15.

H

half-plane The two regions of the coordinate plane separated by the graph of a linear equation. (p. 535)

semiplano Cualquiera de las dos regiones de un plano coordenado que divide la gráfica de una ecuación lineal. (pág. 535)

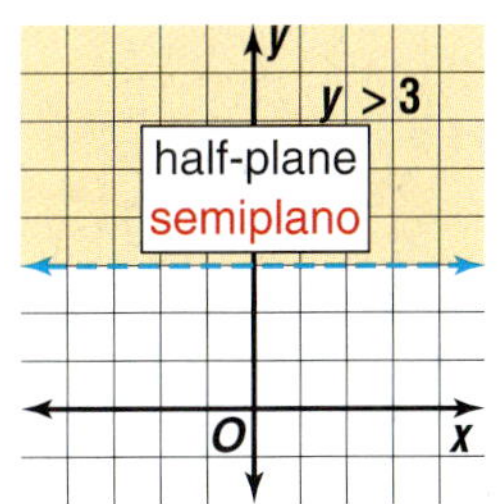

---

histogram A bar graph in which the data are organized into equal intervals. The height of the bar represents the frequency in that interval. (p. 39)

histograma Gráfica de barras en que los datos están organizados en intervalos iguales. La altura de cada barra viene dada por la frecuencia del intervalo correspondiente. (pág. 39)

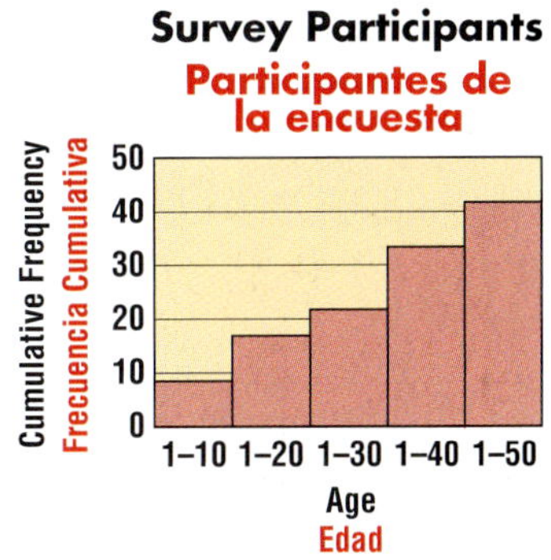

---

hypotenuse The side opposite the right angle in a right triangle. (p. 366)

hipotenusa En un triángulo rectángulo, el lado opuesto al ángulo recto. (pág. 366)

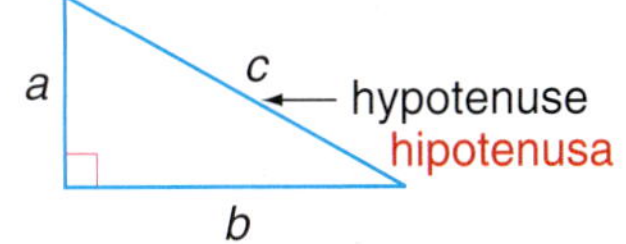

identity An equation that is true for every value of the variable. (p. 172)

$2x = 2x$

identidad Ecuación que se cumple para todo valor de la variable. (pág. 172)

---

inclusive events Two events that can occur at the same time and whose outcomes may be the same. (p. 227)

Rain on Saturday and rain on Sunday are *inclusive events*.

eventos inclusivos Dos eventos que pueden ocurrir simultáneamente y cuyos resultados pueden ser los mismos. (pág. 227)

Lluvia el sábado y lluvia el domingo son *eventos inclusivos*.

---

inconsistent system A system of equations with no ordered pair that satisfies both equations. (p. 555)

sistema inconsistente Sistema de ecuaciones para el cual no existe un par ordenado que satisfaga ambas ecuaciones. (pág. 555)

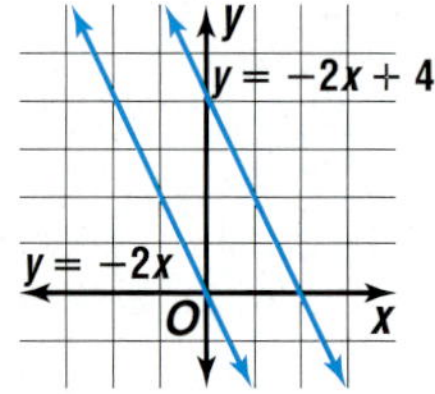

---

independent events Two or more events in which the outcome of one event does not affect the outcome of the other events. (p. 224)

Rolling a number cube and tossing a coin are *independent events*.

eventos independientes Dos o más eventos en los que el resultado de un evento no afecta el resultado del otro. (pág. 224)

Lanzar un dado y lanzar una moneda son *eventos independientes*.

---

independent system A system of equations with exactly one solution. (p. 554)

sistema independiente Sistema de ecuaciones con una única solución. (pág. 554)

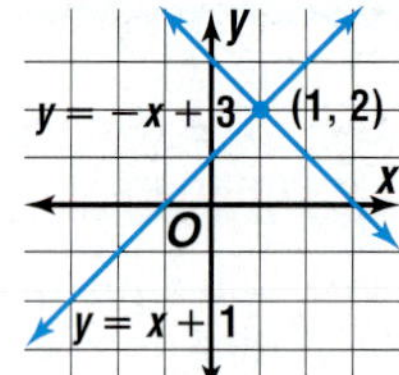

---

independent variable The variable in a function whose value is subject to choice. (p. 264)

variable independiente Variable de una función cuyo valor se puede elegir. (pág. 264)

---

inductive reasoning A conclusion based on a pattern of examples. (p. 30)

razonamiento inductivo Conclusión que se basa en un patrón de ejemplos. (pág. 30)

---

inequality An open sentence that uses the symbol $<$, $\leq$, $>$, or $\geq$ to compare two quantities. (p. 95)

$x < 9$, $y \geq -3$

desigualdad Enunciado abierto que usa uno de los símbolos $<$, $\leq$, $>$ o $\geq$ para comparar dos cantidades. (pág. 95)

---

integers The set of whole numbers and their opposites $\{\ldots, -2, -1, 0, 1, 2, \ldots\}$. (p. 52)

enteros Conjunto de los números enteros junto con sus opuestos, $\{\ldots, -2, -1, 0, 1, 2, \ldots\}$. (pág. 52)

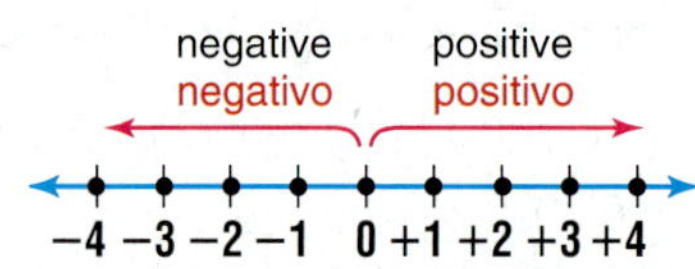

intersection The set of elements common to two or more sets as in compound inequalities and Venn diagrams. (p. 524)

intersección Conjunto de elementos comunes a dos o más conjuntos, como en el caso de desigualdades compuestas o diagramas de Venn. (pág. 524)

---

inverse variation An equation of the form $xy = k$, where $k \neq 0$. (p. 270)

variación inversa Ecuación de la forma $xy = k$, con $k \neq 0$. (pág. 270)

$xy = 9$

---

irrational numbers Numbers that cannot be expressed as fractions, terminating decimals, or repeating decimals. (p. 362)

números irracionales Números que no pueden escribirse como fracciones, decimales exactos o decimales periódicos. (pág. 362)

$\pi, \sqrt{2}, 0.10110111\ldots$

L

least common denominator (LCD) The least common multiple of the denominators of two or more fractions. (p. 663)

2 is the *least common denominator* of $\frac{1}{3}, \frac{3}{4}$, and $\frac{5}{6}$.

mínimo común denominador (mcd) El múltiplo común menor de los denominadores de dos o más fracciones. (pág. 663)

2 es el *mínimo común denominador* de $\frac{1}{3}, \frac{3}{4}$ y $\frac{5}{6}$.

---

least common multiple (LCM) The least nonzero number that is a common multiple of two or more numbers. (p. 662)

36 is the *least common multiple* of 3, 9, and 12.

mínimo común múltiplo (MCM) Número no nulo que es el múltiplo común menor de dos o más números. (pág. 662)

36 es el *mínimo común múltiplo* de 3, 9 y 12.

---

legs The sides of a right triangle that form the right angle. (p. 366)

catetos En un triángulo rectángulo, los lados que forman el ángulo. (pág. 366)

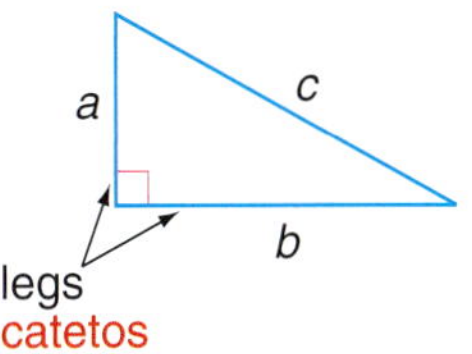

---

like terms Terms that contain the same variables raised to the same power. (p. 20)

$5x^2$ and $6x^2$ are *like terms.*

términos semejantes Términos que tienen las mismas variables elevadas a los mismos exponentes. (pág. 20)

$5x^2$ y $6x^2$ son *términos semejantes.*

---

line graph Numerical data displayed to show trends or changes over time. (p. 38)

gráfica lineal Datos numéricos exhibidos de modo que muestren tendencias o cambios en el tiempo. (pág. 38)

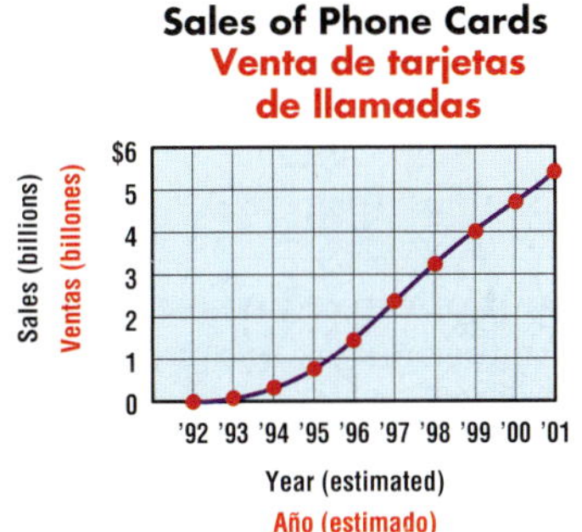

linear equation An equation that can be written in the form $Ax + By = C$, where $A \neq 0$ and $B \neq 0$, whose graph is a straight line. (p. 250)

ecuación lineal Ecuación de la forma $Ax + By = C$, con $A \neq 0$ y $B \neq 0$. Su gráfica es una recta. (pág. 250)

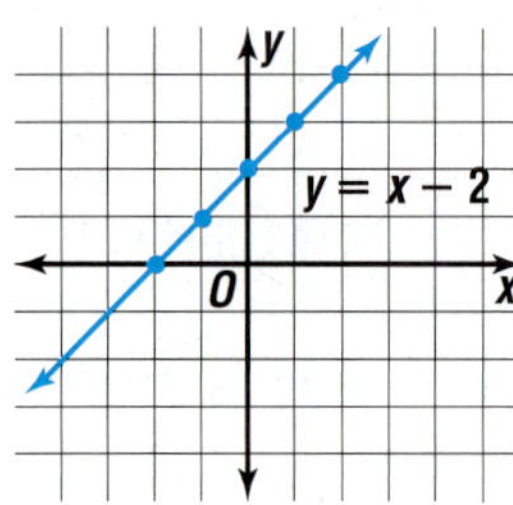

M

mean The sum of the numbers in a set of data divided by the number of pieces of data in the set. (p. 104)

For {1, 2, 2, 6, 9}

$mean = \dfrac{1 + 2 + 2 + 6 + 9}{5}$ or 4

media Suma de los números de un conjunto de datos dividida entre el número de datos. (pág. 104)

Para {1, 2, 2, 6, 9}

$media = \dfrac{1 + 2 + 2 + 6 + 9}{5}$ ó 4

---

measure of central tendency A number or piece of data that represents the whole set of data. (p. 104)

medida de tendencia central Número o dato que representa todo un conjunto de datos. (pág. 104)

---

measure of variation A measure that describes the spread of the values in a set of data. (p. 106)

medida de variación Medida que describe la dispersión de los números de un conjunto de datos. (pág. 106)

---

median The middle number in a set of data when the data are arranged in numerical order. If the data has an even number, the median is the mean of the two middle numbers. (p. 104)

For {1, 2, 2, 6, 9}
↑ (under the middle 2)
The *median* is 2.

mediana Número central de un conjunto de datos, cuando éstos se han ordenado numéricamente. Si el número de datos es par, la mediana es la media de los dos valores centrales. (pág. 104)

Para {1, 2, 2, 6, 9}
↑ (under the middle 2)
La *mediana* es 2.

---

mode The number(s) or item(s) that appear most often in a set of data. (p. 104)

For {1, 2, 2, 6, 9}
↑ (under the second 2)
The *mode* is 2.

moda Dato(s) más frecuente(s) de un conjunto de datos. (pág. 104)

Para {1, 2, 2, 6, 9}
↑ (under the second 2)
la *moda* es 2.

---

monomial A number, a variable, or a product of a number and one or more variables. (p. 382)

$3, y, 2x, 5xy^2$

monomio Número, variable o producto de un número por una o más variables. (pág. 382)

---

multiplicative inverse The number that when multiplied by a given number results in a product of one. (p. 154)

The *multiplicative inverse* of 4 is $\frac{1}{4}$ because $4 \times \frac{1}{4} = 1$.

inverso multiplicativo Número que multiplicado por un número dado da uno. (pág. 154)

El *inverso multiplicativo* de 4 es $\frac{1}{4}$ porque $4 \times \frac{1}{4} = 1$.

multi-step equations Equations with more than one operation. (p. 165)

$3x - 4 = 11, \frac{x}{3} + 5 = 14$

ecuaciones de varios pasos Ecuaciones con más de una operación. (pág. 165)

---

mutually exclusive events Two or more events whose outcomes can never be the same. (p. 226)

Rolling a 2 and rolling a 4 on a dice are *mutually exclusive events.*

eventos mutuamente excluyentes Dos o más eventos cuyos resultados no son jamás los mismos. (pág. 226)

Sacar un 2 y sacar un 4 en un dado son *eventos mutuamente excluyentes.*

natural numbers The set {1, 2, 3, … }. (p. 52)

números naturales El conjunto {1, 2, 3, … }. (pág. 52)

---

negative number Any number that is less than zero. (p. 52)

$-4, -\frac{1}{2}$

número negativo Cualquier número menor que cero. (pág. 52)

---

number line A line with equal distances marked off to represent numbers. (p. 52)

recta numérica Recta en que se han marcado distancias a espacios iguales para representar números. (pág. 52)

negative / negativo ← → positive / positivo

−4 −3 −2 −1 0 +1 +2 +3 +4

---

numerical expression A mathematical expression that has a combination of numbers and at least one operation. (p. 4)

$4 + 2 \cdot 3$ is a *numerical expression.*

expresión numérica Combinación de números y por lo menos menos una operación. (pág. 4)

$4 + 2 \cdot 3$ es un *expresión numérica.*

odds The ratio that compares the number of ways an event can occur (successes) to the number of ways the event cannot occur (failures). (p. 221)

posibilidades Razón del número de maneras en que un evento puede ocurrir (éxitos) al número de maneras en que el evento no puede ocurrir (fracasos). (pág. 221)

---

open sentence A mathematical statement with one or more variables. (p. 112)

$n - 7 = 14$
$15 = m + 2$

enunciado abierto Enunciado matemático con una o más variables. (pág. 112)

---

opposites Every positive rational number and its negative pair. (p. 65)

-1 and 1, $\frac{1}{4}$ and $-\frac{1}{4}$

opuestos Cada número racional positivo y su par negativo. (pág. 65)

-1 y 1, $\frac{1}{4}$ y $-\frac{1}{4}$

ordered pair A pair of numbers used to locate a point in the coordinate plane or the solution of an equation in two variables. An ordered pair is written in the form (x-coordinate, y-coordinate). (p. 58)

par ordenado Par de números que se usa para ubicar un punto en el plano coordenado o la solución de una ecuación de dos variables. Un par ordenado se escribe de la forma (coordenada x, coordenada y). (pág. 58)

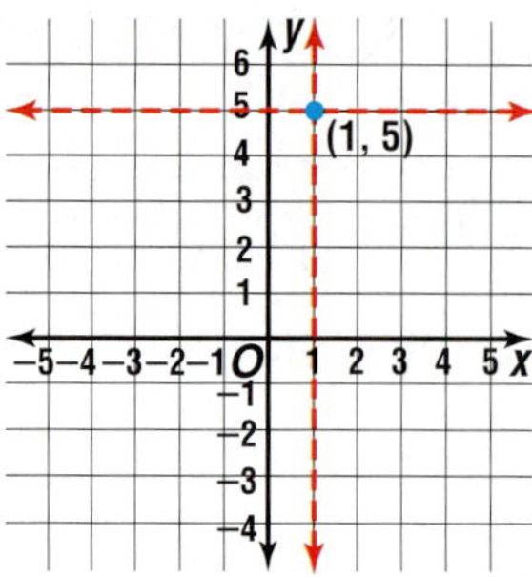

---

origin *See coordinate plane.*

origen *Ver plano coordenado.*

---

outcome One possible result of a probability event. (p. 146)

4 is an *outcome* when a number cube is rolled.

resultado Resultado posible de un evento probabilístico. (pág. 146)

4 es un *resultado* cuando se lanza un cubo numérico.

parabola The graph of a quadratic function. (p. 458)

parábola Gráfica de una ecuación cuadrática. (pág. 458)

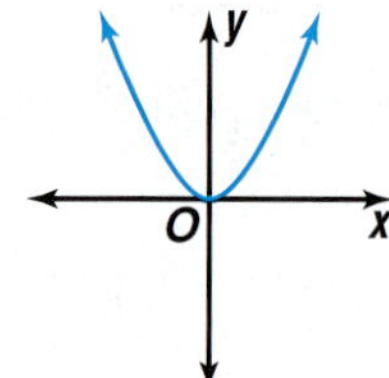

---

parallel lines Lines in the same plane that never intersect and have the same slope. (p. 322)

rectas paralelas Rectas en el plano que no se intersecan y que tienen la misma pendiente. (pág. 322)

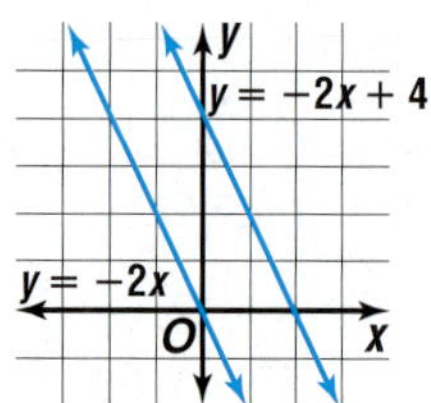

---

parent graph The simplest of the graphs or the anchor graph in a family of graphs. (p. 318)

gráfica madre La más simple de las gráficas o gráfica áncora de una familia de gráficas. (pág. 318)

---

percent A ratio that compares a number to 100. (p. 198)

76 out of 100 is 76 *percent* or 76%.

porcentaje Razón que compara un número con 100. (pág. 198)

76 de 100 es 76 *por ciento* ó 76%.

---

percent of change A ratio that compares the change in a quantity to the original amount. (p. 212)

porcentaje de cambio Razón que compara el cambio de una cantidad con la cantidad original. (pág. 212)

perfect square A number whose square root is a rational number. (p. 336)

25 is a *perfect square* since $25 = 5^2$.

cuadrado perfecto Número cuya raíz cuadrada es un número racional. (pág. 336)

25 es un *cuadrado perfecto* porque $25 = 5^2$.

---

perfect square trinomial Trinomials that are the square of a binomial. (p. 445)

$x^2 + 2x + 1$ is a *perfect square trinomial* since $x^2 + 2x + 1 = (x + 1)^2$.

trinomio cuadrado perfecto Trinomio que es el cuadrado de un binomio. (pág. 445)

$x^2 + 2x + 1$ es un *trinomio cuadrado perfecto* porque $x^2 + 2x + 1 = (x + 1)^2$.

---

permutation An arrangement or listing of a group of objects in which order is important. (p. 152)

Choosing a chairperson and a co-chair is a *permutation*.

permutación Arreglo o lista en la que el orden es importante. (pág. 152)

Elegir a un director y a un subdirector es una *permutación*.

---

perpendicular lines Lines which meet to form right angles and whose slopes have a product of -1. (p. 324)

rectas perpendiculares Rectas que al intersecarse forman un ángulo recto y si sus pendientes existen, su producto es -1. (pág. 324)

---

point-slope form An equation of the form $y - y_1 = m(x - x_1)$, where m is the slope and (x_1, y_1) is a given point on a nonvertical line. (p. 290)

In $y - 4 = 2(x - 3)$, the slope is 2 and (3, 4) is a point on the line.

forma punto-pendiente Ecuación de la forma $y - y_1 = m(x - x_1)$, donde m es la pendiente y (x_1, y_1) un punto dado de una recta no vertical. (pág. 290)

En la ecuación $y - 4 = 2(x - 3)$, el pendiente es 2 y (3, 4) es un punto en la recta.

---

polynomial A monomial or sum of monomials. (p. 383)

$5x$, $3x^2 + 2$

polinomio Un monomio o una suma de monomios. (pág. 383)

---

power An expression of the form x^n, read *x to the nth power*. (p. 336)

7^4 is 7 raised to the fourth *power*, or $7 \times 7 \times 7 \times 7$.

potencia Expresión de la forma x^n, que se lee *x a la potencia enésima*. (pág. 336)

7^4 es 7 elevado a la cuarta *potencia*, ó $7 \times 7 \times 7 \times 7$.

---

prime factorization Expressing a composite number as a product of its prime factors. (pp. 358, 421)

The *prime factorization* of 63 is $3 \times 3 \times 7$.

factorización prima Número compuesto escrito como producto de sus factores primos. (ppág. 358, 421)

El *factorización prima* de 63 es $3 \times 3 \times 7$.

---

prime number A whole number, greater than 1, whose only factors are 1 and itself. (p. 420)

2 is a prime number. It has two factors: 1 and 2.

número primo Número entero mayor que 1 cuyos únicos factores son 1 y sí mismo. (pág. 420)

2 es un número primo. Tiene dos factores: 1 y 2.

prime polynomial A polynomial that cannot be written as a product of two polynomials with integral coefficients. (p. 430)

$2x + 5y + 4$

polinomio primo Polinomio que no puede escribirse como producto de dos polinomios con coeficientes enteros. (pág. 430)

---

probability The ratio of the number of favorable outcomes for an event to the number of possible outcomes of the event. (p. 219)

$P(a) = \dfrac{\text{number of favorable outcomes}}{\text{total number of possible outcomes}}$

probabilidad Razón del número de resultados favorables de un evento al número total de resultados posibles. (pág. 219)

$P(a) = \dfrac{\text{número de resultados favorables}}{\text{número total de resultados posibles}}$

---

product The result obtained by multiplying two or more numbers or variables. (p. 4)

33 is the *product* of 3 and 11.

producto El resultado de la multiplicación de dos o más números o variables. (pág. 4)

33 es el *producto* de 3 y 11.

---

proportion An equation of the form $\frac{a}{b} = \frac{c}{d}$ stating that two ratios are equivalent. (p. 188)

$\frac{3}{12} = \frac{1}{4}$ or $\frac{x}{15} = \frac{2}{5}$

proporción Ecuación de la forma $\frac{a}{b} = \frac{c}{d}$ en que se establece la equivalencia de dos razones. (pág. 188)

Pythagorean Theorem If a and b are the measures of the legs of a right triangle and c is the measure of the hypotenuse, then $c^2 = a^2 + b^2$. (p. 366)

teorema de Pitágoras Si a y b son las longitudes de los catetos de un triángulo rectángulo y c es la longitud de su hipotenusa, entonces $c^2 = a^2 + b^2$. (pág. 366)

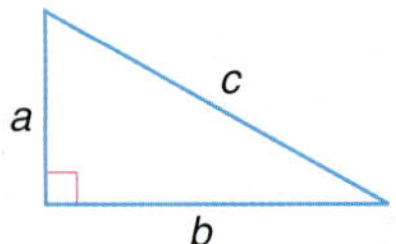

Q

quadrant One of four regions into which the x- and y-axes separate the coordinate plane. (p. 60)

cuadrante Cada una de las cuatro regiones en que los ejes x y y dividen el plano coordenado. (pág. 60)

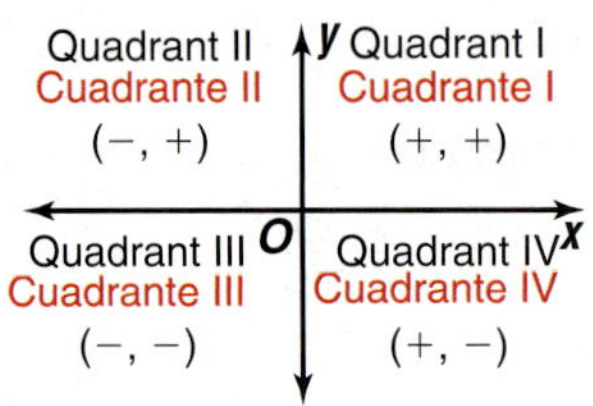

---

quadratic equation An equation of the form $ax^2 + bx + c = 0$, where $a \neq 0$. (p. 468)

$2x^2 + 5x + 18 = 0$

ecuación cuadrática Ecuación de la forma $ax^2 + bx + c = 0$, donde $a \neq 0$. (pág. 468)

Quadratic Formula The solutions of a quadratic equation in the form $ax^2 + bx + c = 0$, where $a \neq 0$, are given by the formula

$x = \dfrac{-b \pm \sqrt{b^2 - 4ac}}{2a}$. (p. 484)

fórmula cuadrática Las soluciones de la ecuación cuadrática $ax^2 + bx + c = 0$, donde $a \neq 0$, vienen dadas por la fórmula

$x = \dfrac{-b \pm \sqrt{b^2 - 4ac}}{2a}$. (pág. 484)

quadratic function A function described by an equation of the form $f(x) = ax^2 + bx + c$, where $a \neq 0$. (p. 458)

función cuadrática Función de la forma $f(x) = ax^2 + bx + c$, donde $a \neq 0$. (pág. 458)

quadratic-linear system of equations A system of equations involving a linear and a quadratic function. (p. 580)

sistema cuadrático-lineal de ecuaciones Sistema con una ecuación lineal y una cuadrática. (pág. 580)

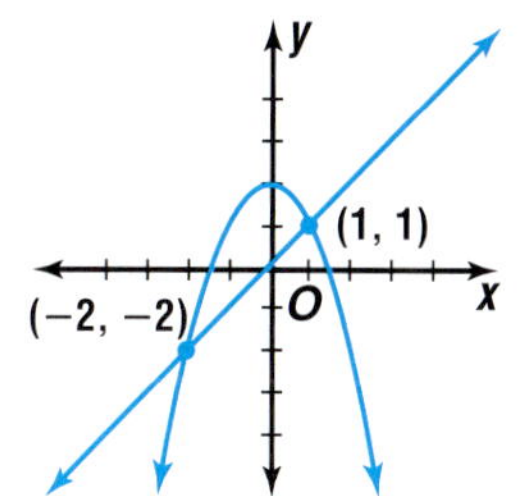

R

radical equation An equation that contains a radical expression with the variable in the radicand. (p. 624)

ecuación radical Ecuación que contiene una expresión radical con la variable en el radicando. (pág. 624)

$y = 2\sqrt{x} + 1$

radical expression An expression that contains a square root. (p. 358)

expresión radical Expresión que contiene una raíz cuadrada. (pág. 358)

$\sqrt{3}$, $\sqrt[4]{14}$, $2\sqrt{x} + 1$

radical sign The symbol $\sqrt{\ }$, used to indicate a root. (p. 357)

signo radical El símbolo $\sqrt{\ }$, que indica una raíz. (pág. 357)

random Outcomes occur at random if each outcome is equally likely to occur. (p. 220)

Tossing a coin and getting heads or tails is *random.*

aleatoriamente Los resultados ocurren aleatoriamente si cada uno tiene la misma probabilidad de ocurrir. (pág. 220)

Lanzar una moneda y sacar cara o escudo es un evento *aleatorio.*

range of a data set The difference between the greatest number and the least number in a set of data. (p. 106)

{1, 2, 2, 6, 9}
↑ ↑
range = 9 − 1 or 8

rango de un conjunto de datos Diferencia entre el dato mayor y el menor. (pág. 106)

rango = 9 − 1 ó 8

range of a relation The set of second numbers in the ordered pairs of a relation. (p. 238)

In the relation {(1, 2), (3, 5), (4, 0)}, the *range* is {0, 2, 5}.

rango de una relación Conjunto de los segundos números de los pares ordenados de la relación. (pág. 238)

En la relación {(1, 2), (3, 5), (4, 0)}, el *rango* es {0, 2, 5}.

rate The ratio of two measurements having different units of measure. (p. 190)

60 miles per hour

tasa Razón de dos medidas de unidades distintas. (pág. 190)

60 millas por hora

rate of change The change in a quantity over time. (p. 285)

tasa de cambio El cambio de una cantidad en el tiempo. (pág. 285)

ratio A comparison of two numbers by division. (p. 188) — 2 to 3; 1:2, $\frac{3}{5}$

razón Comparación de dos números mediante división. (pág. 188)

---

rational equation An equation that contains rational expressions. (p. 668) — $\frac{2x}{x+3} = 0$

ecuación racional Ecuación que contiene expresiones racionales. (pág. 668)

---

rational expression An algebraic fraction whose numerator and denominator are polynomials. (p. 638) — $\frac{2x}{x+3}$

expresión racional Fracción cuyo numerador y denominador son polinomios. (pág. 638)

---

rational function A function that contains rational expressions represented as the quotient of two polynomials in the form $f(x) = \frac{g(x)}{h(x)}$, where $h(x) \neq 0$. (p. 638) — $f(x) = \frac{2x}{x+3}$

función racional Función que es el cociente de dos polinomios, de la forma $f(x) = \frac{g(x)}{h(x)}$, donde $h(x) \neq 0$. (pág. 638)

---

rational numbers The set of numbers that can be written in the form $\frac{a}{b}$, where a and b are integers and $b \neq 0$. (p. 94) — $1 = \frac{1}{1}, \frac{2}{9}, -2.3 = -2\frac{3}{10}$

números racionales Conjunto de los números de la forma $\frac{a}{b}$, donde a y b son enteros y $b \neq 0$. (pág. 94)

---

real numbers The set of rational numbers and irrational numbers together. (p. 600) — $-1, 0, \sqrt{2}$

números reales Conjunto de los números racionales junto con el de los irracionales. (pág. 600)

---

reciprocal The multiplicative inverse of a number. (p. 154) — $\frac{2}{3}$ and $\frac{3}{2}$

recíproco El inverso multiplicativo de un número. (pág. 154)

---

relation A set of ordered pairs. (p. 238) — $\{(9, 3), (-1, 4), (4, 4), (-2, 0)\}$

relación Conjunto de pares ordenados. (pág. 238)

---

rise The vertical change between any two points on a line on the coordinate plane. (p. 284)

elevación Cambio vertical entre dos puntos cualesquiera de una recta en el plano coordenado. (pág. 284)

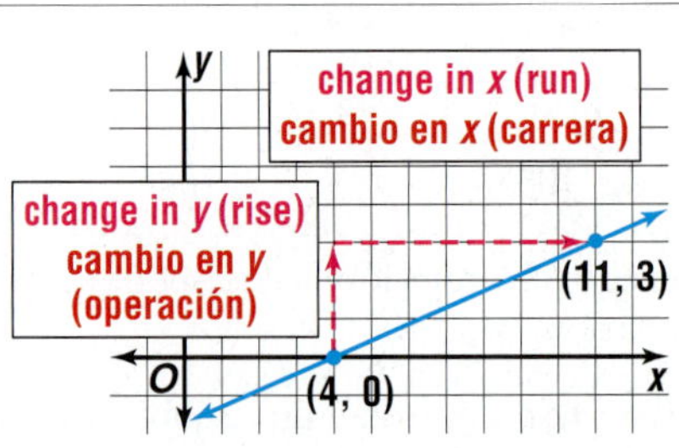

---

root The solution of a quadratic equation. (p. 468)

raíz Solución de una ecuación cuadrática. (pág. 468)

---

run The horizontal change between any two points on a line on the coordinate plane. (p. 284)

carrera Cambio horizontal entre dos puntos cualesquiera de una recta en el plano coordenado. (pág. 284)

sample Some portion of a larger group selected to represent that group. (p. 32)

muestra Parte de un grupo más grande seleccionada para representarlo. (pág. 32)

---

sample space The list of all possible outcomes for an event. (p. 146)

The *sample space* for rolling a number cube is {1, 2, 3, 4, 5, 6}.

espacio muestral Lista de todos los resultados posibles de un evento. (pág. 146)

El *espacio muestral* de lanzar un dado es s {1, 2, 3, 4, 5, 6}.

---

scale A ratio or rate between the actual size of an object and the size of its model. (p. 194)

escala Razón o tasa entre el tamaño natural de un objeto y el de su modelo. (pág. 194)

---

scale drawing A drawing that is similar to an actual object, but is either enlarged or reduced. (p. 194)

dibujo a escala Dibujo semejante a un objeto real, pero más grande o más pequeño. (pág. 194)

---

scale model A replica of an original object that is too large or too small to be built at actual size. (p. 194)

maqueta Reproducción de un objeto real demasiado grande o demasiado pequeño como para ser construido de tamaño natural. (pág. 194)

---

scatter plot A graph that shows the general relationship between two sets of data. (p. 302)

gráfica de dispersión Gráfica que muestra la relación general entre dos conjuntos de datos. (pág. 302)

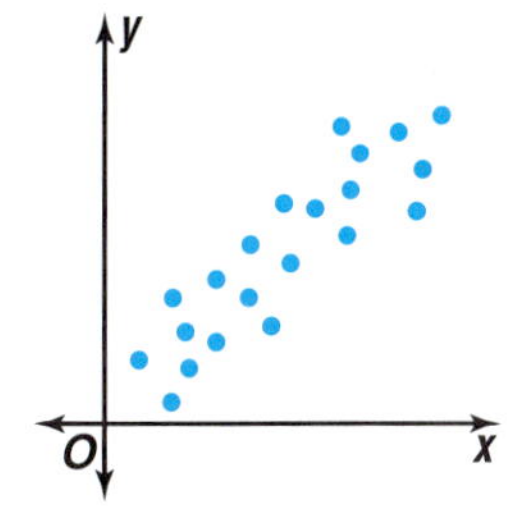

---

scientific notation A system of writing numbers using multiplication and powers of ten. A number of the form $a \times 10^n$, where $1 < a < 10$ and n is an integer. (p. 353)

$95{,}700 = 9.57 \times 10^4$
$0.000024 = 2.4 \times 10^{-5}$

notación científica Forma de escribir números usando multiplicación y potencias de diez. Un número de la forma $a \times 10^n$, donde $1 < a < 10$ y n un entero. (pág. 353)

---

set-builder notation A concise way of writing a solution set. (p. 510)

$\{t|t < 17\}$ represents the set of all numbers t such that t is less than 17.

notación de conjuntos Forma concisa de escribir un conjunto-solución. (pág. 510)

$\{t|t < 17\}$ representa el conjunto de todos los números t tal que t es menos que 17.

simplest form of an expression An expression is in simplest form when it is replaced by an equivalent expression having no like terms or parentheses. (p. 20)

forma reducida de una expresión Una expresión está reducida si no tiene términos semejantes o paréntesis. (pág. 20)

---

simplify Write an expression in simplest form. (p. 15)

reducir Simplificar una expresión. (pág. 15)

---

slope The ratio of the change in the y-coordinates (rise) to the corresponding change in the x-coordinates (run) as you move from one point to another along a line. (p. 284)

pendiente Razón del cambio en las coordenadas y (elevación) al cambio correspondiente en las coordenadas x (carrera) a medida que uno se desplaza de izquierda a derecha. (pág. 284)

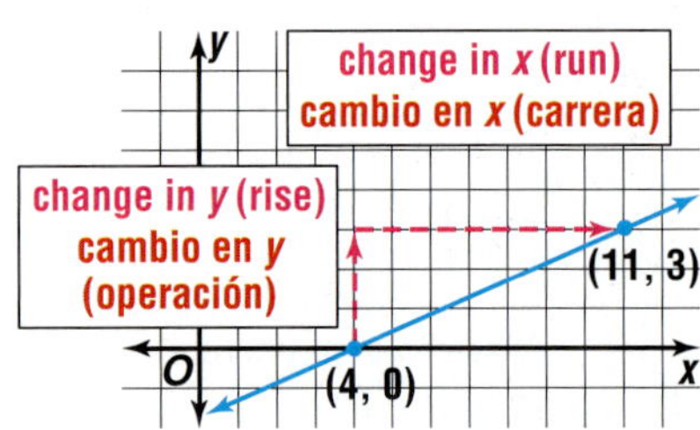

---

slope formula The slope m of a line passing through two points is the ratio of the difference in the y-coordinates to the corresponding difference in the x-coordinates. (p. 284)

$$m = \frac{y_2 - y_1}{x_2 - x_1}$$

fórmula de la pendiente La pendiente m de una recta que pasa por dos puntos es la razón de la diferencia en las coordenadas y a la diferencia correspondiente en las coordenadas x. (pág. 284)

---

slope-intercept form An equation of the form $y = mx + b$, where m is the slope and b is the y-intercept. (p. 296)

forma pendiente-intersección Ecuación de la forma $y = mx + b$, donde m es la pendiente y b es la intersección y. (pág. 296)

---

solution A replacement value for the variable in an open sentence. A value for the variable that makes an equation true. (p. 112)

The *solution* of $12 = x + 7$ is 5.

solución Valor de sustitución de la variable de un enunciado abierto. Valor de la variable que hace verdadera una ecuación. (pág. 112)

La *solución* de $12 = x + 7$ es 5.

---

solution of a system of equations The ordered pair representing the solution common to both equations in a system of equations. (p. 550)

solución de un sistema de ecuaciones Pares ordenados cuyas coordenadas satisfacen ambas ecuaciones del sistema. (pág. 550)

---

solve an equation The process of finding all values of the variable that make an equation a true statement. (p. 112)

resolver una ecuación Proceso de hallar todos los valores de la variable que hacen verdadera una ecuación. (pág. 112)

---

square root One of the two equal factors of a number. If $a^2 = b$, then a is the square root of b. (p. 357)

12 is a *square root* of 144 since $12^2 = 144$.

raíz cuadrada Uno de dos factores iguales de un número. Si $a^2 = b$, a es una raíz cuadrada de b. (pág. 357)

12 es una *raíz cuadrada* de 144 porque $12^2 = 144$.

standard form of a linear equation A linear equation of the form $Ax + By = C$, where $A \geq 0$, and A and B are real numbers and not both zero. (p. 250)

forma estándar de una ecuación lineal Ecuación lineal de la forma $Ax + By = C$, donde $A \geq 0$ y A y B son números reales que no son ambos nulos. (pág. 250)

---

statement A sentence that is either true or false, but not both. (p. 112)

enunciado Frase que es verdadera o falsa, pero no ambas. (pág. 112)

---

stem-and-leaf plot A display of data in which each piece of the data is separated into two numbers that are used to form a stem and a leaf. (p. 40)

diagrama de tallo y hojas Presentación de datos en que cada uno está dividido en dos números que se usan para formar un tallo y una hoja. (pág. 40)

Stem Tallo	Leaf Hojas
0	7
1	4 7
2	4 7 7 7 9
3	5 7 8
4	
5	9

1|4 = 14

---

system of equations A set of two or more equations with the same variables. (p. 550)

sistema de ecuaciones Conjunto de dos o más ecuaciones con las mismas variables. (pág. 550)

$2x - 3 = y$
$3x + 5y = 7$

---

system of inequalities A set of two or more inequalities with the same variables. (p. 586)

sistema de desigualdades Conjunto de dos o más desigualdades con las mismas variables. (pág. 586)

$y > 2x + 6$
$4x - 3y < 12$

T

term in an expression A number, a variable, or a product or quotient of numbers and variables. (p. 20)

término en una expresión Número, variable, producto o cociente de números y variables. (pág. 20)

---

theoretical probability The ratio of the number of favorable outcomes to the total number of possible outcomes. (p. 220)

The *theoretical probability* of rolling a 3 on a number cube is $\frac{1}{6}$.

probabilidad teórica Razón del número de resultados favorables al número total de resultados posibles. (pág. 220)

La *probabilidad teórica* de sacar un 3 en un cubo numerado es $\frac{1}{6}$.

---

tree diagram A diagram used to show the total number of possible outcomes of an event. (p. 146)

bread — **filling**
white — turkey, cheese
wheat — turkey, cheese

diagrama de árbol Diagrama que se usa para mostrar todos los resultados posibles de un evento. (pág. 146)

pan — **relleno**
blanco — pavo, queso
intebral — pavo, queso

trinomial An algebraic expression that has three terms. (p. 383)

trinomio Expresión algebraica de tres términos. (pág. 383)

$x^2 + 2x + 1$
$a + b - c$

union The graph of a compound inequality containing the word *or*; the solution is a solution of either inequality, not necessarily both. (p. 525)

unión La gráfica de una desigualdad compuesta que contiene la palabra *o*; la solución es una solución de cualquiera de las desigualdades, no necesariamente de ambas. (pág. 525)

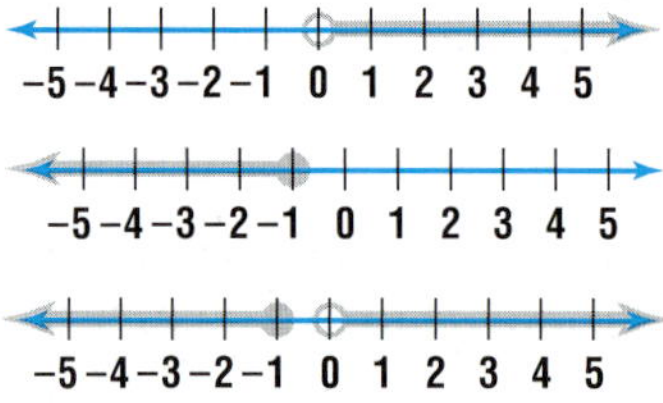

unit rate A rate with a denominator of 1. (p. 190)

tasa unitaria Tasa de denominador 1. (pág. 190)

55 miles per hour
99¢ per pound

55 millas por hora
99¢ por libra

variable A letter or other symbol used to represent an unspecified number or value. (p. 4)

variable Letra u otro símbolo que se usa para representar un número o valor general. (pág. 4)

Venn diagram A diagram that uses circles to show relationships among sets of numbers or objects. (p. 52)

diagrama de Venn Diagrama que usa círculos para mostrar relaciones entre conjuntos de números u objetos. (pág. 52)

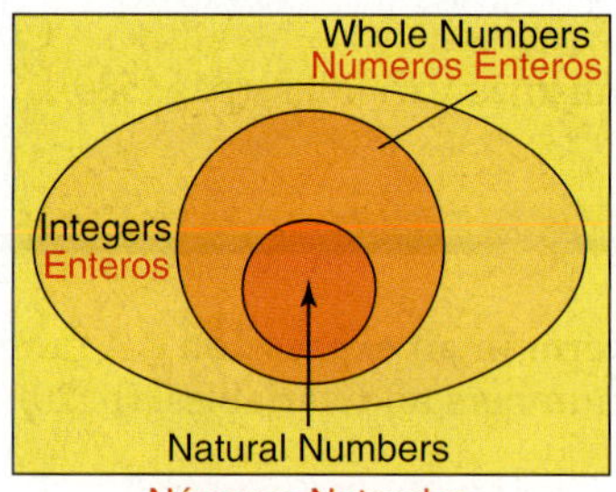

vertex of a parabola The maximum or minimum point of a parabola. (p. 459)

vértice de una parábola Su punto máximo o mínimo. (pág. 459)

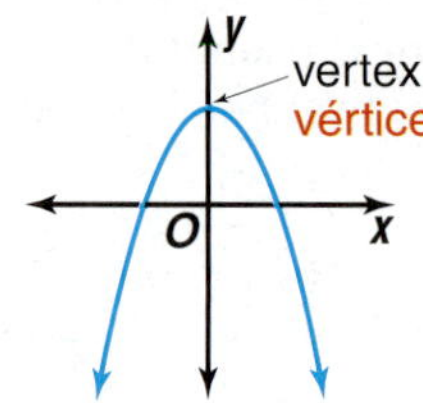

vertical line test A test used to determine if a relation is a function. (p. 257)

prueba de la recta vertical Prueba que se usa para determinar si una relación es una función. (pág. 257)

whole numbers The set {0, 1, 2, 3, ...}. (p. 16)

números enteros El conjunto {0, 1, 2, 3, ...}. (pág. 16)

x-axis The horizontal number line on a coordinate plane. (p. 58)

eje x La recta numérica horizontal de un plano coordenado. (pág. 58)

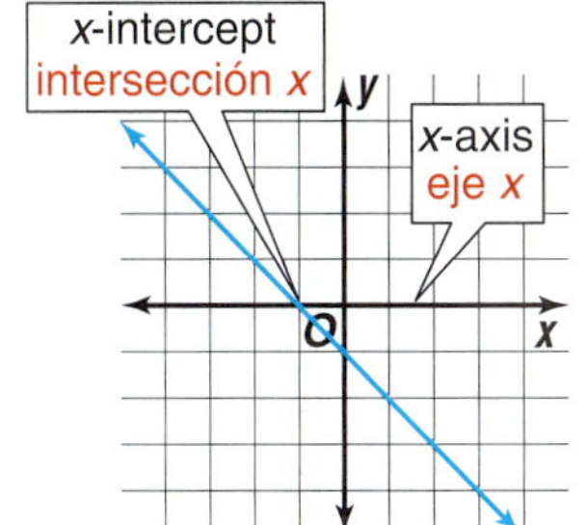

x-coordinate The first number in an ordered pair. (pp. 59, 238)

coordenada x Primer número de un par ordenado. (págs. 59, 238)

x-intercept The x-coordinate at which a graph intersects the x-axis. (p. 296)

intersección x La coordenada x donde una gráfica interseca el eje x. (pág. 296)

y-axis The vertical number line on a coordinate plane. (p. 58)

eje y La recta numérica vertical de un plano coordenado. (pág. 58)

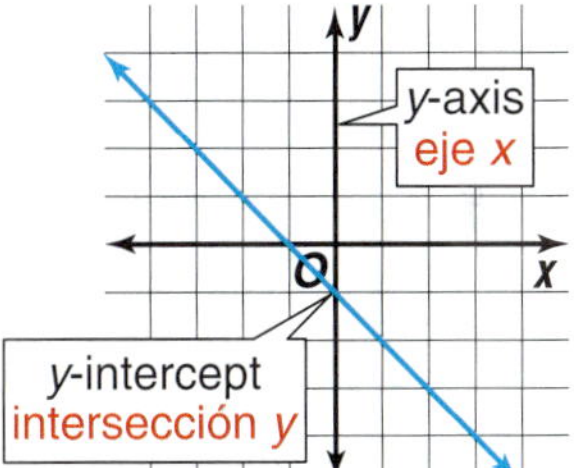

y-coordinate The second number in an ordered pair. (pp. 59, 253)

coordenada y Segundo número de un par ordenado. (págs. 59, 253)

y-intercept The y-coordinate at which a graph intersects the y-axis. (p. 296)

intersección y La coordenada y donde una gráfica interseca el eje y. (pág. 296)

Z

zero pair The result of a positive counter paired with a negative counter. (p. 65)

par nulo Ficha positiva junto con su ficha negativa. (pág. 65)

zeros The roots, or x-intercepts, of a function. (p. 468)

ceros Las raíces, o intersecciones x, de una función. (pág. 468)

SELECTED ANSWERS

Selected Answers

Chapter 1 The Language of Algebra

Page 3 Check Your Readiness
1. 18 **3.** 0 **5.** 64 **7.** 6 **9.** 120 **11.** 7 **13.** 6 **15.** 6 **17.** 0.99 **19.** 0.15 **21.** 0.55 **23.** 1 **25.** 17.5, 18, 18.5, 19 **27.** 14, 18, 20, 21, 26, 29, 30, 32, 37

Pages 6–7 Lesson 1–1
1. Sample answer: $3 + 7$, $10(6)$, $4 - 1$; $2x$, $5 + a$, gh **5.** Sample answer: $7m$ **7.** Sample answer: 7 divided by q **9.** $6 + m = 17$ **11.** Sample answer: 5 increased by r is 15. **13.** $9w + 3$ **15.** rs **17.** $5a + 3$ **19.** $1 - n$ **21.** $10 + h \cdot 1$ or $10 + h$ **23.** Sample answer: the product of 9 and x **25.** Sample answer: the difference of 6 and y **27.** Sample answer: 8 less than the quotient of 3 and r **29.** $3 + w = 15$ **31.** $2 = \frac{7}{x}$ **33.** $3 - 5y = 2z$ **35.** Sample answer: 3 times r is 18. **37.** Sample answer: h equals 10 minus i. **39.** Sample answer: The quotient of t and 4 is 16. **41.** Sample answer: Let x = the variable; $4x - 7 = 15 + c + 2x$ **43a.** $20 + 10(15 - 1)$ **43b.** $20 + 10(m - 1)$

Pages 11–13 Lesson 1–2
1. parentheses, brackets, fraction bar **5.** Multiply 6 and 2. **7.** Subtract 4 from 10. **9.** 3 **11.** Transitive
13. $8(4 - 8 \div 2)$
$= 8(4 - 4)$ *Substitution*
$= 8(0)$ *Substitution*
$= 0$ *Multiplicative Property of 0*
15. 8 **17.** 13 **19.** 14 **21.** 26 **23.** 40 **25.** 21 **27.** 8 **29.** Reflexive **31.** Substitution **33.** Transitive
35. $8(9 - 3 \cdot 2) = 8(9 - 6)$ *Substitution*
$= 8(3)$ *Substitution*
$= 24$ *Substitution*
37. $10(6 - 5) - (20 \div 2)$
$= 10(1) - 10$ *Substitution*
$= 10 - 10$ *Multiplicative Identity*
$= 0$ *Substitution*
39. $6(12 - 48 \div 4) + 7 \cdot 1$
$= 6(12 - 12) + 7$ *Substitution*
$= 6(0) + 7$ *Substitution*
$= 0 + 7$ *Multiplicative Property of 0*
$= 7$ *Additive Identity*
41. 50 **43.** 23 **45.** 40 **47.** 5 **49.** 26 **51.** $311,392 **53.** 44 feet **55.** Sample answer: Eight more than x is 12. **57.** Sample answer: The quotient of 25 and n is 5. **59.** $3c = 27$ **61.** $b + 10 - 1 = 18$ **63a.** $4(20) + 7$ **63b.** 87

Page 13 Quiz 1
1. $2v - 5$ **3.** 13 **5.** 56 ft

Pages 17–18 Lesson 1–3
1. For any whole numbers that are multiplied, the product is a whole number. **3.** Jessie; the Associative Property can be applied to addition or to multiplication, but not to both at the same time. **5.** Associative (+) **7.** $(3 + 47 + p)(7 - 6)$; $50 + p$; Commutative (+) **9a.** 188 **9b.** Use the Commutative Property of Addition to change the order of the numbers to $69 + 31 + 80 + 8$. Then use the Substitution Property, adding 69 and 31 first. **11.** Commutative (+) **13.** Commutative (×) **15.** Associative (×) **17.** $r \cdot (30 \cdot 5)$; $150r$; Associative (×) **19.** $(6 + 3) + y$; $9 + y$; Associative (+) **21.** $2 \cdot 7 \cdot j$; $14j$; Commutative (×) **23.** false; sample counterexample:
$(4 \div 2) \div 2 \stackrel{?}{=} 4 \div (2 \div 2)$
$2 \div 2 \stackrel{?}{=} 4 \div 1$
$1 \neq 4$
25. 100 board feet
27. Sample answer:
$(8 - 5) - 3 \stackrel{?}{=} 8 - (5 - 3)$
$3 - 3 \stackrel{?}{=} 8 - 2$
$0 \neq 6$
29. 6 **31.** 7 **33.** A

Pages 21–23 Lesson 1–4
1. Sample answer: $1 + 2x + 3x + ab + 5ab$ **3.** b, e; Write all the expressions in simplest form and find two that are the same. **5.** $8p$, $9p$ **7.** $6bc$, bc **9.** $y + 2$ **11.** $12 - 18m$ **13.** $7a + 7t$ **15.** $21f$ **17.** $3r$ **19.** $2g + 6$ **21.** $6x + 5y$ **23.** $15am - 12$ **25.** $7a + 5b$ **27.** $4y$ **29.** $46xy$ **31.** $3r + 4s$
33. $5(2n + 3r) + 4n + 3(r + 2)$
$= 10n + 15r + 4n + 3r + 6$ *Dist. Prop.*
$= 10n + 4n + 15r + 3r + 6$ *Commutative (+)*
$= (10 + 4)n + (15 + 3)r + 6$ *Dist. Prop.*
$= 14n + 18r + 6$ *Substitution*
35. 11 **37.** $52 **39a.** 179.8 lb **39b.** 31 lb **41a.** 96 ft^2 **41b.** $4(8 + 10 + 6)$ or 96 **43.** Commutative (×) **45.** Symmetric **47.** 6 **49a.** $d + 30$ **49b.** 170

Pages 27–29 Lesson 1–5
1. Sample answer: You should look back to determine whether the answer fits the problem, whether there are other possible answers, or whether there might be a better way to solve the problem. **3.** no **5.** $n + 2$ **7a.** 25 $5 bills, 32 $1 bills **7b.** Sample answer: Use an equation. Let x represent the number of $1 bills. Then $x - 7$ represents the number of $5 bills. Write and solve the equation $11(10) + 1x + 5(x - 7) = 267$.

9. Craig, 16; mother 40; sample explanation: Let m represent the age of the mother. Then $m - 24$ represents Craig's age. Solve the equation $m + (m - 24) = 56$. **11.** 7 craft books, 13 cookbooks

13a. Sample answer:

Person	1	2	3	4	5	6
Number of Handshakes	5	4	3	2	1	0

13b. 15 **13c.** 66 **15.** 9:00 P.M. **17a.** $d = rt$ **17b.** 703 mi **19.** 20 people **21.** $8b$ **23.** $24a + 45$ **25.** A

Pages 34–37 Lesson 1–6

1. A frequency table shows the number of times a single event occurs. A cumulative frequency table shows the number of times an event occurs plus the previous events. **3.** No, the sample is not large enough.

5.

Number of Goals	Tally	Frequency
1	III	3
2	~~IIII~~	5
3	I	1
4	IIII	4
5	III	3
6	III	3
7	I	1
8	I	1

7. 1 time **9a.** 100 s

9b.

Cycle (s)	Frequency	Cumulative Frequency
80	33	33
90	42	75
100	60	135
110	25	160

9c. 75 times **11.** Yes; the sample is random and representative of drivers in the area. **13.** No; the sample is not representative of the entire country. **15.** No; people might prefer a golfer who is from the same area.

17a.

Quiz Score	Tally	Frequency
6	II	2
7	IIII	4
8	~~IIII~~ ~~IIII~~ I	11
9	~~IIII~~ IIII	9
10	IIII	4

17b. 8 **17c.** Rather than adding all of the numbers, this formula uses multiplication to find the sum. Then the sum is divided by the number of scores, 30. **17d.** 8.3 **19.** That is a typical breakfast time and they want the business. **21.** Sample answers: a stadium, restaurant, school campus **25.** 275 miles **27.** nine more than x

Page 37 Quiz 2

1. $11 + 6 + 2a$, Commutative (+); $17 + 2a$

5.

Eye Color		
Color	Tally	Frequency
brown	~~IIII~~ ~~IIII~~	10
blue	IIII	4
green	~~IIII~~ I	6
hazel	~~IIII~~	5

Blue occurs the least.

Pages 41–43 Lesson 1–7

1. Line graphs usually show change over time, while histograms compare quantities of similar nature. **2.** A correct graph has a title, horizontal and vertical axes labeled, and equally spaced units on both axes. **3.** Manuel; line graphs are usually used to show trends over time. Histograms are usually used to how frequency of items in sets of data. **5.** 1997–1998 **7.** 14–15 cm

9.

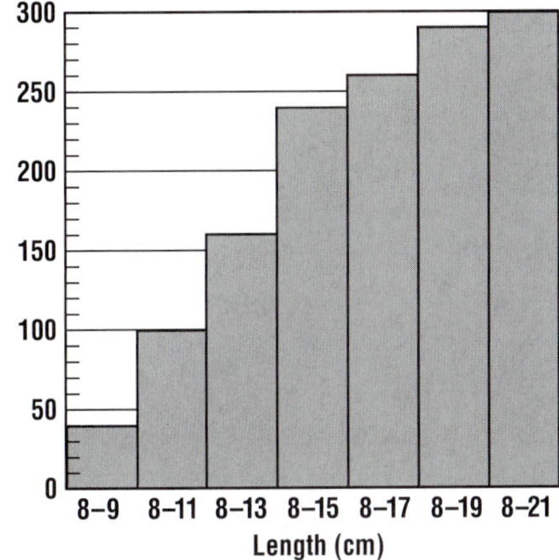

11a. 81°F **11b.** 12 days **11c.** 63 **13.** 1993

15.

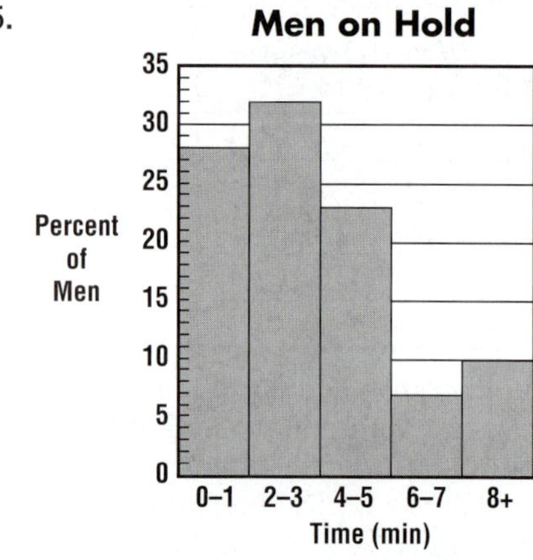

17. Sample answer: The histograms show that a greater percent of men than women are willing stay on hold for 0–1 minutes **19.** 123 **21.** 71, 73, 92, and 100 occur three times. **23.** Sample answer: More students in second period scored 80 or more, so second period did better **25.** Yes; it is a random sample. **27.** $8x - 3y$

Pages 44–46 Chapter 1 Study Guide and Assessment

1. term **3.** algebraic expression **5.** counterexample **7.** sample **9.** frequency tables **11.** $5n$ **13.** $2y - 6 = 14$ **15.** 7 **17.** 14 **19.** 9 **21.** Commutative (×), $20c$ **23.** Commutative (×), $9x$ **25.** Associative (+), $g + 3$ **27.** $7v - 7$ **29.** $2h + ah$ **31.** $22 - 3d$ **33a.** 1200 **33b.** 80 **35.** 0

37.

Number	Cumulative Frequency
0	6
1	9
2	13
3	15
4	19

39. 10 **41.** 10–11

43.

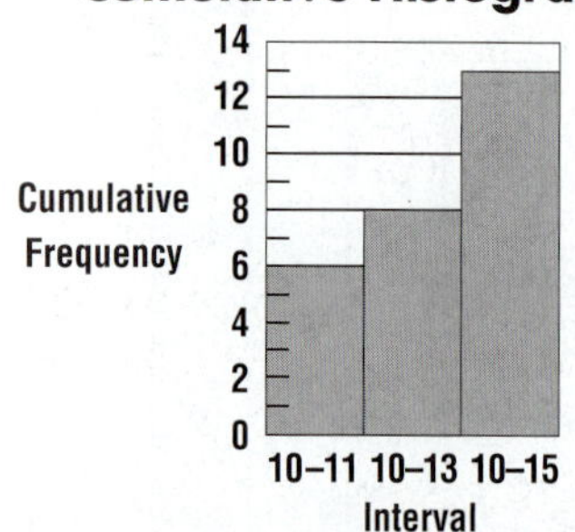

45. 8 **47a.** 100; 62 **47b.** 88 **47c.** 10

Page 49 Preparing for Standardized Tests

1. C **3.** D **5.** C **7.** A **9.** 234

Chapter 2 Integers

Page 51 Check Your Readiness

1. $10x$ **3.** $19d$ **5.** $6y$ **7.** $6r$ **9.** $25p$ **11.** 19 **13.** 1 **15.** 54 **17.** 2 **19.** 42 **21.** 6 **23.** $32q$ **25.** $36x$ **27.** $26c$ **29.** $36st$

Pages 55–57 Lesson 2-1

1. Sample answer: temperature, elevation
3. Neither; 0 is neither negative nor positive.
5. −6 **7.** +150 **9.** 1

11.

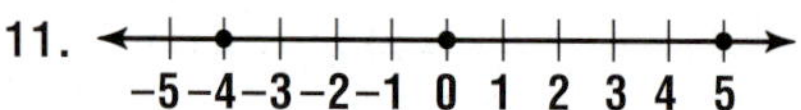

13. < **15.** 10 **17.** −22, −17, −14, −7, −1, 9, 10, 13, 22, 26, 35 **19.** 2 **21.** 0 **23.** −4

25. −5 −4 −3 −2 −1 0 1 2 3 4 5

27. −5 −4 −3 −2 −1 0 1 2 3 4 5

29. −5 −4 −3 −2 −1 0 1 2 3 4 5

31. > **33.** > **35.** < **37.** > **39.** 6 **41.** 2 **43.** 9 **45.** Aways; $|5| = 5$ and $|-5| = 5$, so $|5| = |-5|$. **47.** 78, 14, −14, −25, −36 **49a.** −36° **49b.** −26° **49c.** −36°

51.

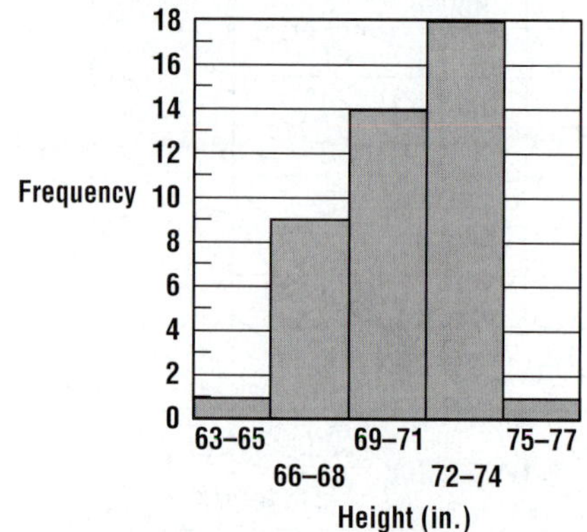

53. 19 presidents **55.** $6a$ **57.** $7m + 2n$ **59.** $7x + 11y$

Pages 61–63 Lesson 2-2

1. Start at the origin. Move 5 units to the left. Then move 1 unit up and draw a dot.

3.

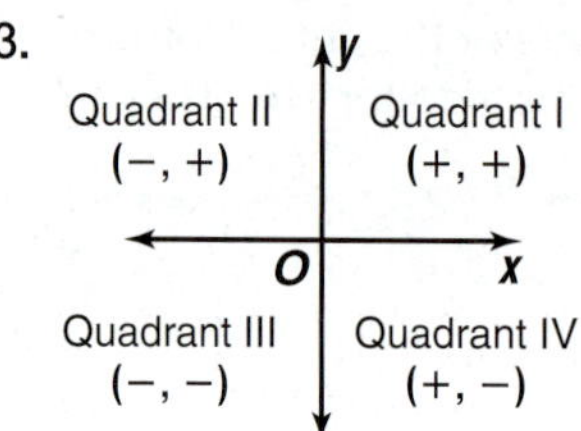

5. (3, −2)

7. 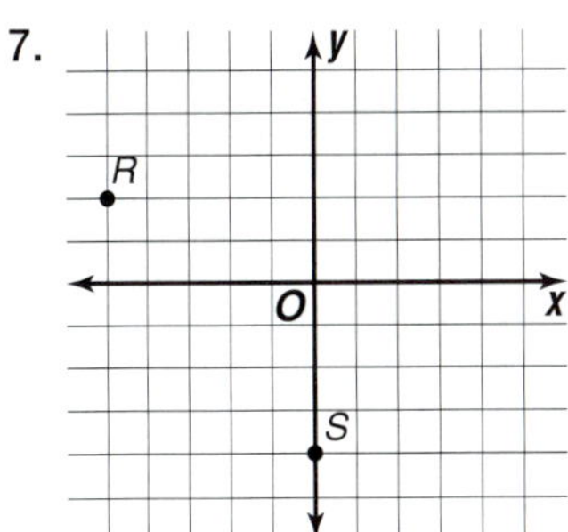

9. none 11. (1, −1) 13. (0, −4) 15. (−3, −3)

17–22. 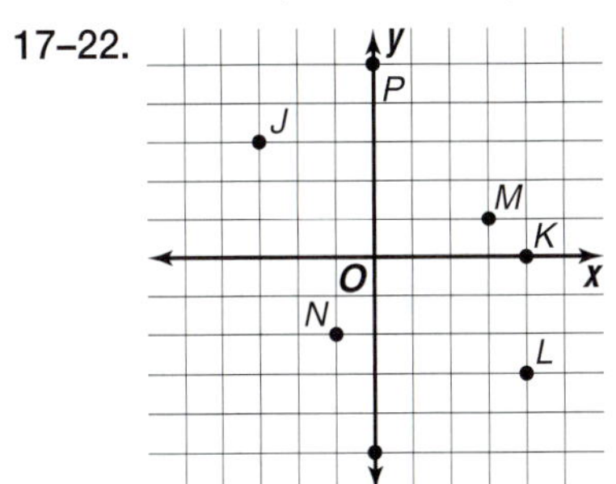

23. III 25. none 27. II 29. origin 31. I 33. IV
35a. \$3, \$9, \$15

35b–c. 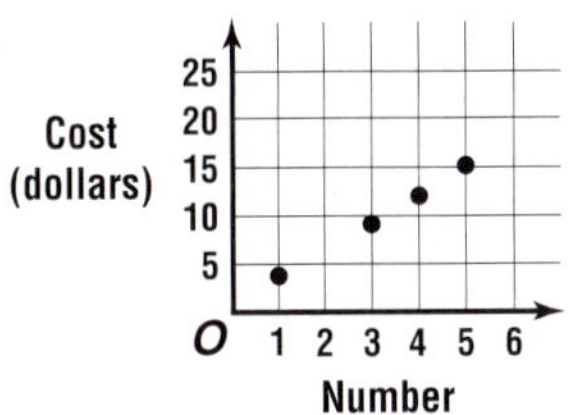

37a. $A(2, 1)$, $B(6, 1)$, $C(6, 5)$

37b.

37c. The triangles have the same shape, but the second is larger. 39. −8685 41. Commutative (+) 43. Closure (+) 45. A

Page 63 Quiz 1
1. > 3. < 5. 11 7. III 9. II

Pages 68–69 Lesson 2–3
1. 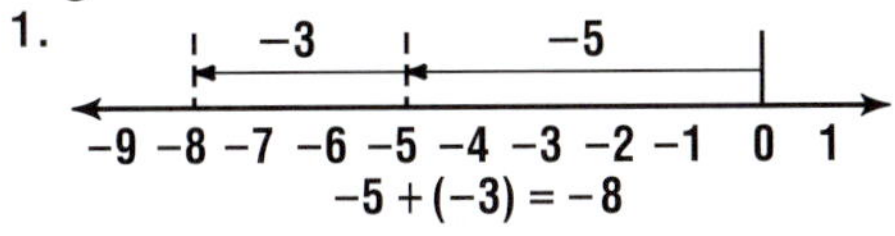

3. 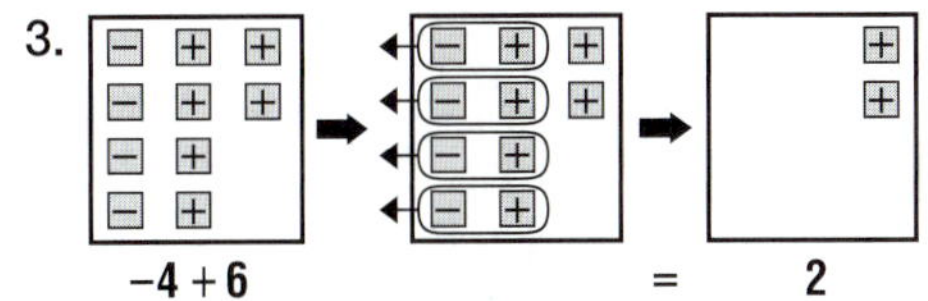

5. + 7. − 9. − 11. 16 13. −1 15. −5 17. $2x$
19. $2a$ 21. 12 23. 21 25. −11 27. −8 29. 9
31. −13 33. −8 35. −21 37. −55 39. 13
41. −22 43. −4 45. −1 47. −3 49. $-15x$
51. $3m$ 53. 0 55. $8y$ 57. −9 59. −1

61. 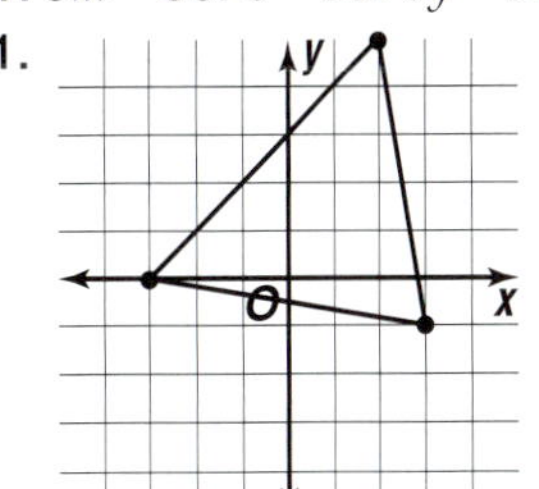

63. IV 65. II 67. −5 69. −10 71. 0 73. C

Pages 73–74 Lesson 2–4
1. To subtract an integer, add its additive inverse.
3. 10 + (−3) 5. −4 + (−8) 7. −2 9. 11 11. 5
13. 8 15. 13 17. 7 19. −8 21. −13 23. −11
25. 16 27. −6 29. −9 31. 12 33. 7 35. −17
37. 3 39. −10 41. −12 43. $25n$ 45. 50
47. −34°F 49a. false; $2 - 5 \neq 5 - 2$ 49b. false; $(6 - 2) - 3 \neq 6 - (2 - 3)$ 49c. true 51. −20
53. −27 55. −8 57. > 59. <

Pages 77–79 Lesson 2–5
1. 3(−3) = −9 3. −12 5. 80 7. 36 9. 28 11. $-18x$

13a. 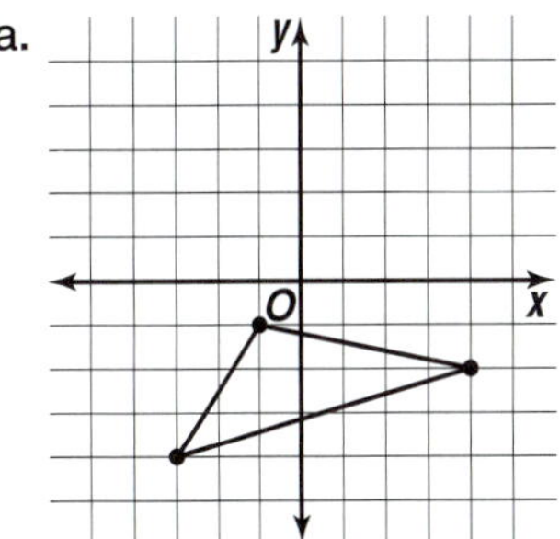

13b. It was reflected over the x-axis. 15. −48
17. −9 19. 45 21. 0 23. −24 25. −39 27. 27
29. −84 31. −1 33. 99 35. −8 37. 30 39. 13
41. $-8a$ 43. $32mn$ 45. −22 47. −100(5) = −500

49a.

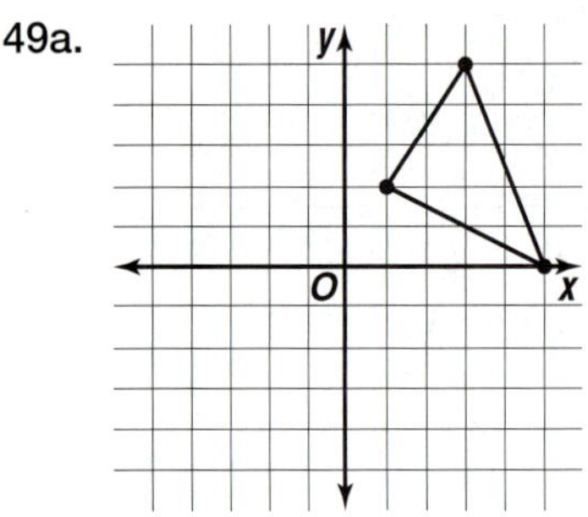

49b. It was reflected over both the x- and y-axis. **51.** 4 **53.** −5 **55.** −8 **57.** −2 **59.** 0 **61.** C

Page 79 Quiz 2
1. −7 **3.** 32 **5.** 13 **7.** −45 **9.** −9

Pages 84–85 Lesson 2–6
1. $-10 \div 2 = -5, -10 \div (-5) = 2$ **3.** −5 **5.** −5 **7.** 4 **9.** −8 **11.** 12 **13.** 1 **15.** 6 **17.** −6 **19.** 5 **21.** −9 **23.** 10 **25.** −8 **27.** −5 **29.** −12 **31.** −6 **33.** −5 **35.** −3 **37.** 6 **39.** −7 **41.** −2 **43a.** −7000 farms **43b.** 10 to 49 and 2000 acres or more **45.** 74 **47.** −54 **49.** 16 **51.** 80 **53.** −2 **55.** 4 **57.** B

Pages 86–88 Chapter 2 Study Guide and Assessment
1. negative numbers **3.** opposites, additive inverses, or zero pairs **5.** absolute value **7.** integers **9.** ordered pair **11.** > **13.** < **15.** −4, −3, −2, 0, 4, 7 **17.** (3, 2) **19.** (−4, −3) **21.** I **23.** none **25.** −6 **27.** −10 **29.** 5 **31.** −5 **33.** $2x$ **35.** $4m$ **37.** −8 **39.** 9 **41.** −4 **43.** −1 **45.** −1 **47.** 35 **49.** 36 **51.** 27 **53.** $-42m$ **55.** −7 **57.** −6 **59.** −8 **61.** $213 **63.** $-40(2) = -80$

Page 91 Preparing for Standardized Tests
1. A **3.** B **5.** B **7.** E **9.** −1

Chapter 3 Addition and Subtraction Equations

Page 93 Check Your Readiness

1. $\frac{1}{3}$ **3.** $\frac{1}{9}$ **5.** $\frac{2}{5}$ **7.** 5.6 **9.** 23.7 **11.** 33.1 **13.** 17.2 **15.** 5.2 **17.** 4.05 **19.** 37.6 **21.** $\frac{1}{2}$ **23.** $1\frac{3}{10}$ **25.** $\frac{1}{6}$

Pages 97–99 Lesson 3–1

1. Sample answer: $-\frac{1}{2}, \frac{4}{3}$, and 0.9 **5.** 0.75 **7.** 0.625

9. −5 −4 −3 −2 −1 0 1 2 3 4 5

Since −3 is to the left of $-2.\overline{6}$ on the number line, −3 is less than $-2.\overline{6}$. **11.** < **13.** = **15.** $\frac{7}{10}, \frac{6}{8}, \frac{4}{5}$ **17.** < **19.** < **21.** < **23.** > **25.** < **27.** > **29.** $-2.002 > -2.02$

31. $-\frac{5}{6}$ −1 $-\frac{8}{10}$ 0

Since $-\frac{5}{6}$ is to the left of $-\frac{8}{10}$ on the number line, $-\frac{8}{10} > -\frac{5}{6}$. **33.** $\frac{1}{2}$, or 0.6, $\frac{5}{8}$ **35.** $-\frac{2}{6}, -\frac{1}{4}, -\frac{1}{8}$ **37.** $\frac{3}{8}, \frac{3}{5}, 0.\overline{6}$ **39.** a dozen eggs for $1.59; $0.13 < $0.14 **41.** a 25-yd by 12-in. roll for $2.99; $0.0004 and $0.0003 are both > $0.0002 **43a.** Firefly, Japanese Beetle, Mealworm **43b.** Elm Leaf Beetle **45.** −7 **47.** −4 **49.** 48 **51.** 3822

Pages 102–103 Lesson 3–2
1. Add and subtract in the order in which they are given; or group the positive numbers together and the negative numbers together and then add.

3. 0.9 **5.** −15.05 **7.** 0 **9.** $\frac{1}{3}$ **11.** −13.3 **13.** −5.82 **15.** −1.26 **17.** −29.48 **19.** $-\frac{3}{8}$ **21.** $-5\frac{3}{10}$ **23.** $\frac{19}{35}$ **25.** $3\frac{1}{8}$ 27. $-2\frac{3}{20}$ **29.** 6.9 **31.** $-6\frac{1}{6}$ **33.** −$153.58 or a decrease of $153.58 **35.** $\frac{1}{4}$ foot above normal **37.** −20 **39.** $-12bc$ **41.** $11y + 14$

Pages 107–109 Lesson 3–3
1. When a set of data has one or more extremely high or low values, the mean does not accurately describe the values in the set of data. Consider the set of data 3, 10, 5, 16, 148. The mean is 36.4. **3.** Sonia; a few really high or really low values could influence the mean so that it is not representative of Eric's scores. **5.** 42 **7.** 0.151 **9.** 47.25 **11.** 15.5; 12.45; 12.5; 21.1 **13.** 62.3; 60; 72; 17 **15a.** Ms. Diaz: 201.625; 200.6; 210.0; 19.1; Mr. Cruz: 201.625; 200.6; 210.0; 59.1 **15b.** Ms. Diaz; her sales are more consistent **17.** 9; 8; none; 19 **19.** 0.25; 0.3; 0.3; 0.2 **21.** 60.8; 24; 20; 180 **23.** 10.375; 10.1; 10.2; 3.8 **25.** $10.8\overline{3}$; 10.5; 10; 2.5 **27.** 19.5; 19.2; 18.3, 19.0, 21.0; 2.7 **29.** Sample answer: 3.1, 1.2, 1.0, 1.5 **31.** Sample answer: 1, 7, 7, 7, 7, 7 **33.** 96 **35a.** $33,425; $29,050; none **35b.** mean; highest value **35c.** median; lowest value **37.** −9.9 **39.** −1

41. −5 −4 −3 −2 −1 0 1 2 3 4 5

Pages 114–116 Lesson 3–4
1. An open sentence is neither true nor false; a statement is either true or false. **3.** Replace n with each element of the replacement set to see which replacement results in a true statement. **5.** 6

7. 11.7 **9.** $\frac{4}{5}$ **11.** 4 **13.** −3 **15.** 3 **17.** 4 **19.** −5 **21.** 2 **23.** 42 **25.** 11 **27.** 6 **29.** −1 **31.** 18 **33.** $5\frac{1}{4}$ **35.** 2 **37.** 6 h **39.** 65 days **41.** 41.2; 40; 40; 8 **43.** $2\frac{1}{3}$ **45.** $-4\frac{1}{12}$ **47.** D

Page 116 Quiz 1

1. < **3.** = **5.** 1.69 **7.** 2.8; 2.85; 3.0; 3.6 **9.** −2

Pages 120–121 Lesson 3–5

1.

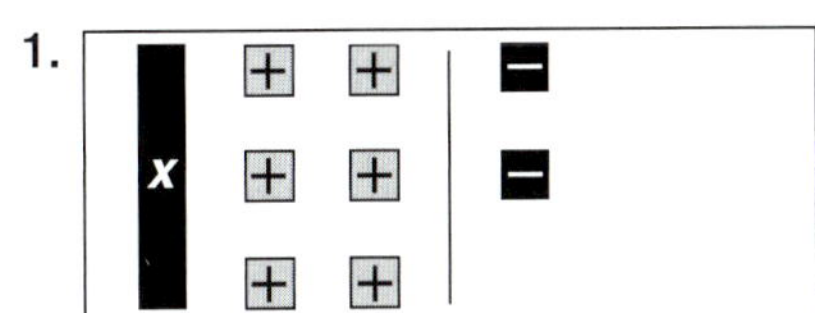

3. $x + (-2) = 12$ **5.** $7 = -6 + x$ **7.** −4 **9.** −4 **11.** 6 **13.** 3 **15.** −6 **17.** −7 **19.** 8 **21.** 0 **23.** −3 **25.** 1 **27.** 2 **29.** 16 **31.** −7 **33.** −21 **35.** 11 **37a.** $t - 8 = -2$ **37b.** 6°F **39.** 2 **41.** 1 **43.** 54 **45.** Commutative (+)

Pages 125–127 Lesson 3–6

1. Replace m with −4 to see if $12 + m$ is equal to 8 **3.** 15; In the first equation, add 5 to each side to get $x = 17$. By replacing x with 17 in the second equation, the result is 17 − 2 or 15. **5.** Sample answer: −21 **7.** Sample answer: +9 **9.** Sample answer: −9.3 **11.** −15 **13.** 1.1 **15.** $-\frac{1}{6}$ **17a.** Sample answer: $x - 1788 = 124$ **17b.** 1912 **19.** −4 **21.** 3 **23.** 13 **25.** −13 **27.** −48 **29.** 34 **31.** −3.8 **33.** $\frac{2}{3}$ **35.** $\frac{10}{9}$ or $1\frac{1}{9}$ **37.** 13 **39.** $m - (-15) = 20$; $m = 5$ **41a.** $c + 27.16 = 32.50$ **41b.** \$5.34 **43.** 0 **45.** 3.25 **47.** 6 **49.** \$6.76 **51.** D

Page 127 Quiz 2

1. −4 **3.** 23 **5a.** Let c = cost of Chicago meal; $3.89 = c - 0.62$. **5b.** \$4.51

Pages 130–131 Lesson 3–7

1. Sample answer: $|a - 2| = -5$ **3.** Sample answer: The value inside the absolute value sign can either be positive or negative. You have to consider both cases, so there may be up to 2 solutions. **5.** {−3, 3} **7.** ∅ **9.** {−7, 13} **11.** $|s - 145| = 6$; {139, 151} **13.** {−2, 2} **15.** {−6, 6} **17.** ∅ **19.** {1} **21.** ∅ **23.** {−21, 1} **25.** {−19, 7} **27.** {−9, 13} **29.** ∅ **31.** 0 **33.** $|p - 165{,}000| = 10{,}000$; {155,000, 175,000} **35.** A **37.** 48 **39.** −14 **41.** B

Pages 132–134 Chapter 3 Study Guide and Assessment

1. median **3.** range **5.** rational number **7.** inequality **9.** rational number **11.** > **13.** > **15.** < **17.** $-1.1, 0, \frac{1}{8}, 0.25$ **19.** 10.5 **21.** −1 **23.** $5.\overline{72}$; 6; 5; 7 **25.** 12.92; 10; 10; 28 **27.** 0 **29.** −1 **31.** 45 **33.** −2 **35.** −4 **37.** 0 **39.** $x + 4 = 11$; $x = 7$ **41.** 12 **43.** 17 **45.** 9 **47.** $\frac{1}{2}$ **49.** $x + 12 = -108$; −120 **51.** {−8, 8} **53.** {−4, 8} **55.** ∅ **57.** \$70.625; \$72.5; \$50, \$75; \$45

Page 137 Preparing for Standardized Tests

1. E **3.** C **5.** B **7.** A **9.** 0.6

Chapter 4 Multiplication and Division Equations

Page 139 Check Your Readiness

1. 54 **3.** 0 **5.** 60 **7.** 20 **9.** 18 **11.** −7 **13.** −7 **15.** −7 **17.** −8 **19.** $\frac{1}{4}$ **21.** $\frac{2}{3}$ **23.** $\frac{3}{5}$ **25.** $7a$ **27.** t **29.** $4x - 8$ **31.** $2b - 3.2$ **33.** $32m - 12$

Pages 143–145 Lesson 4–1

1. Always, because rational numbers are closed under multiplication. **3.** $\frac{1}{4} \cdot \frac{5}{6} = \frac{5}{24}$ **5.** −9 **7.** −14.7 **9.** −2.3 **11.** $-\frac{9}{4}$ or $-2\frac{1}{4}$ **13.** $\frac{3}{5}y$ **15.** $6cd$ **17.** 24.4 **19.** −7.1 **21.** −51 **23.** −29 **25.** 57.6 **27.** $\frac{5}{8}$ **29.** 0 **31.** −1 **33.** $\frac{21}{8}$ or $2\frac{5}{8}$ **35.** $-3x$ **37.** $8.8rs$ **39.** $\frac{6}{25}ab$ **41.** $\frac{5}{4}s$ or $1\frac{1}{4}s$ **43.** −4 **45.** \$135 **47.** $-\frac{2401}{40}$ m or $-60\frac{1}{4}$ m **49.** ∅ **51.** {−12, −6} **53.** 1.4 **55.** $-\frac{5}{8}$ **57a.** 73 in. **57b.** 64 in.

Pages 148–151 Lesson 4–2

1. Sample answer: the number of ways two or more events can occur is the product of the number of ways each event can occur. **3.** Ling; three outcomes have 2 heads: HHT, HTH, THH. Three outcomes have 2 tails: TTH, THT, HTT. **5.** sample space **7.** outcome **9.** 36 **11.** 8 outcomes:

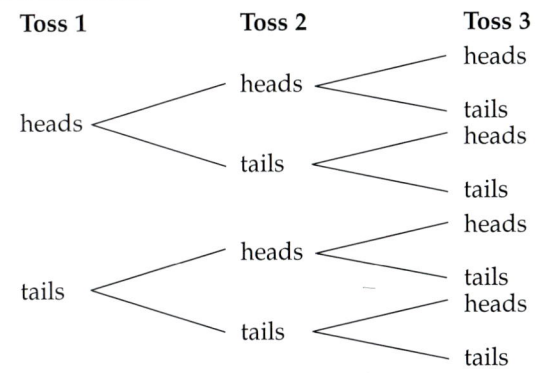

Toss 1	Toss 2	Toss 3
heads	heads	heads
		tails
	tails	heads
		tails
tails	heads	heads
		tails
	tails	heads
		tails

13. 15 outcomes:

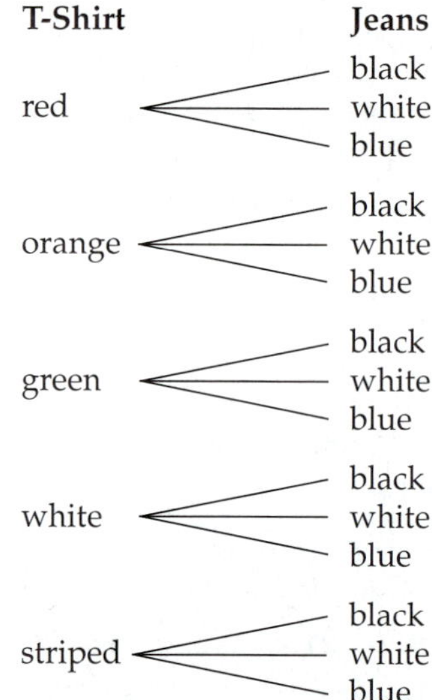

15. 32 **17.** 30 **19.** Sample answer: choosing a red, green, or white car with or without a sunroof

21. 20 **23a.** 64 **23b.** 729 **25.** −16.2 **27.** $\frac{6}{35}$

29. $-\frac{5}{12}$ **31.** B

Pages 156–159 Lesson 4-3

1. No; the product of a negative number and a positive number can never equal 1.

5. $-\frac{3}{2}$ **7.** $-x$ **9.** 0.2 **11.** $-\frac{3}{14}$ **13.** $\frac{36}{5}$ or $7\frac{1}{5}$

15. $-\frac{15}{2}$ or $-7\frac{1}{2}$ **17.** 16 students **19.** −2.5

21. −0.1 **23.** 12 **25.** 0 **27.** −1 **29.** $\frac{14}{15}$ **31.** 45

33. $-\frac{3}{32}$ **35.** −21 **37.** 12 **39.** $\frac{1}{10}$ **41.** $-\frac{2}{5}$

43. $-\frac{4}{5}$ **45.** 18 **47.** 2 min **49.** $12.50

51. 48 **53.** 6; 6; 8; 5 **55.** 20.5; 20; 20; 18 **57.** =

59. $\frac{1}{2}$-pound bag of cashews for $3.15; $6.30 < $6.92

61. 22-ounce bottle for $1.09; $0.06 > $0.05 **63.** A

Page 159 Quiz 1

1. −18.8 **3.** $\frac{20}{8}$ or $2\frac{1}{2}$ **5.** 27 **7.** 2.7 **9.** $\frac{28}{9}$ or $3\frac{1}{9}$

Pages 163–164 Lesson 4-4

1. Divide each side by 4. **3.** Sample answer: $-8a = 56$ **5.** 8 **7.** 45 **9.** 4 **11.** 6 **13.** 15 **15.** −7 **17.** 28 **19.** 5 **21.** 64 **23.** −70 **25.** 27

27. $-\frac{5}{2}$ or $-2\frac{1}{2}$ **29.** 2 **31.** −8 **33.** $8x = 112$; 14

35. $\frac{3}{5}$ yr = −9; −15 **37a.** $6n = 288$ **37b.** 48

39. 110 mph **41.** 20 **43.** $\frac{7}{12}$ **45.** 0 **47.** 57 ft

49. D

Pages 168–170 Lesson 4-5

1. Sample answer: If necessary, first undo addition and/or subtraction of constants, then undo multiplication or division of variables. **3.** Jean; Soto did not multiply each side by 4 in the second step. **5.** Add 5 to each side. **7.** 2 **9.** −5 **11.** −20

13a. $y = 6(x - 1) + 21$ **13b.** $\frac{1}{3}$ yr or 4 months

15. −1 **17.** −7 **19.** 0 **21.** 6 **23.** 9.6 **25.** 8.1

27. 30 **29.** −14 **31.** 68 **33.** −7.2 **35.** $\frac{n-2}{-3} = \frac{1}{3}$; 1;

−3 −2 −1 0 1 2 3 4

37. 3

39a. $8c + 15 = 143$ **39b.** $16 **41a.** $3x + 12 = 39$ **41b.** 9 **43.** 3 **45.** 6 **47.** 4 **49.** −81

Pages 173–175 Lesson 4-6

1. Add $2x$ to each side, subtract 5 from each side, divide each side by 5. **3.** An equation that is an identity is true for every value of the variable; an equation with no solution is never true. **5.** Subtract $0.5p$ from each side. **7.** 2 **9.** identity **11.** −5

13a. $3n = 5n - 24$ **13b.** 12 **15.** 19 **17.** 2 **19.** $\frac{5}{2}$ or $2\frac{1}{2}$ **21.** no solution **23.** 0.7 **25.** −1 **27.** 3.4

29. −20 **31.** 60 **33.** 4.2 **35a.** $3w - 18 = 2w$ **35b.** 18 cm by 36 cm **37.** Sample answer: $2x = x + 6$ **39.** −9 **41.** 12 **43.** 264 numbers **45.** 31 **47.** 3.3

Page 175 Quiz 2

1. −6 **3.** 63 **5.** 72 **7.** 12 **9.** no solution

Pages 178–179 Lesson 4-7

1. Use the Distributive Property to remove the parentheses. **5.** identity **7.** −4 **9.** 8 **11.** 6

13. 5 **15.** −10 **17.** $\frac{1}{3}$ **19.** $\frac{7}{2}$ or $3\frac{1}{2}$ **21.** 2.5

23. no solution **25.** 42 **27.** identity **29.** −2 **31.** base 1 = 20 in.; base 2 = 8 in. **33.** 17, 19 **35.** 1

37a. $\frac{1}{2}x - 4 = 7$ **37b.** 22 **39.** −382, −369, −364, −292, −229, 867

Pages 180–182 Chapter 4 Study Guide and Assessment

1. event **3.** tree diagram **5.** identity **7.** outcomes **9.** outcomes **11.** 22.8

13. $-\frac{3}{4}$ **15.** $-10.8y$ **17.** $-\frac{6}{7}s$

19. 6 outcomes:

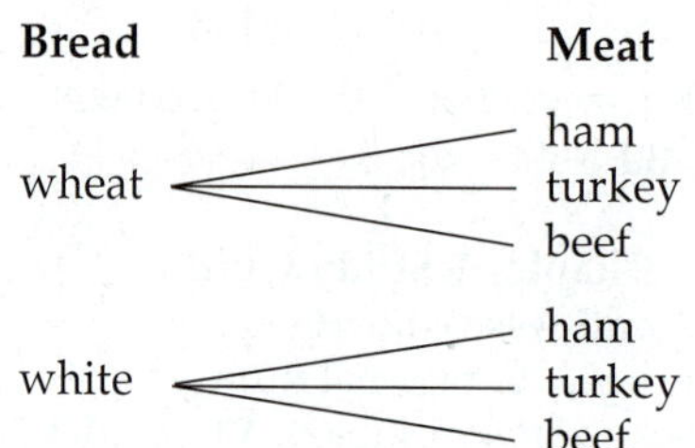

21. −2 **23.** $\frac{8}{15}$ **25.** $-\frac{1}{8}$ **27.** $-\frac{3}{4}$ **29.** 8 **31.** −4 **33.** 12.3 **35.** 9 **37.** −2 **39.** 0 **41.** 2 **43.** −25 **45.** 6 **47.** 7.3 **49.** 1 **51.** 2 **53.** identity **55.** 5 times **57.** 6 in.

Page 185 Preparing for Standardized Tests
1. A **3.** C **5.** E **7.** D **9.** 2

Chapter 5 Proportional Reasoning and Probability

Page 187 Check Your Readiness
1. 3 **3.** 9 **5.** 15 **7.** 0.25 **9.** 0.45 **11.** 1.25 **13.** $\frac{1}{3}$ **15.** $\frac{9}{10}$ **17.** $\frac{5}{6}$ **19.** $\frac{1}{100}$ **21.** $\frac{3}{20}$ **23.** $\frac{7}{25}$ **25.** 0.77 **27.** 0.62 **29.** 0.24

Pages 191–193 Lesson 5-1
1. Multiply 12 and 5. Then divide by 6. **3.** 75(4) = 100(3) **5.** 8(15) = 12(10) **7.** 60 **9.** 3.8 **11.** 5 **13.** 2 kg **15.** 4.5 gal **17.** 15 **19.** 4 **21.** 75 **23.** 9 **25.** 10.5 **27.** 3.5 **29.** 2.1 **31.** 5 **33.** 6 **35.** 37 **37.** Yes; the cross products are equal. **39.** 60 in. **41.** 3000 mL **43.** 4 gal **45.** 0.52 km **47.** 9 pt **49.** no **51.** 750 lb, 1050 lb **53.** It decreases. **55.** −6 **57.** −2 **59.** $-\frac{1}{2}$ **61.** 12 **63.** $-\frac{2}{5}$

Pages 196–197 Lesson 5-2
1. Sample answer:

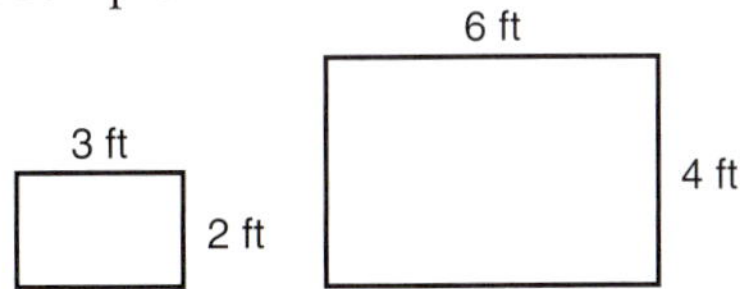

3. 150 mi **5.** 1 m = 35.2 m **7.** 120 mi **9.** 100 mi **11.** 30 mi **13.** 1 in. = 15 in. **15.** 0.5 c **17.** 1 in. = 30.2 mi **19.** $\frac{4}{3}$ **21.** −4

Pages 201–203 Lesson 5-3
1. P is the percentage, B is the base, and r is the percent or the rate per hundred. **3.** $\frac{12}{24} = \frac{r}{100}$ **5.** $\frac{P}{60} = \frac{45}{100}$ **7.** 60% **9.** 70% **11.** 25% **13.** 16.4 **15a.** 48%, 51%, 1%
15b. **Senators in 108th Congress**

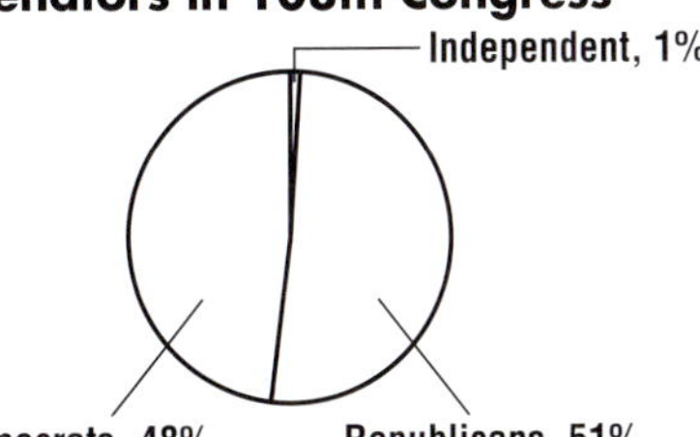

17. 62.5% **19.** 80% **21.** 75% **23.** 150% **25.** 60% **27.** 6% **29.** 36% **31.** 190 **33.** 12.5 **35.** 60 **37.** 0.5% **39.** 9.3 **41.** 50%

43. **Mother's Day Flowers**

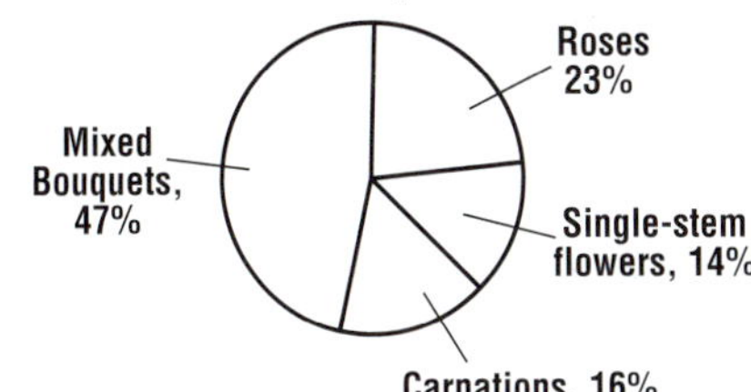

45. 1 in. = 70 mi **47.** 2500 m **49.** −5.2 **51.** $\frac{2}{5}$

Page 203 Quiz 1
1. 27 **3.** 100 ft **5.** 950

Pages 207–209 Lesson 5-4
1. r represents the percent rate; R is the rate written in decimal form. **3.** Nikki; in the percent equation, the R is the decimal form of the percent. **5.** 41.8 **7.** 550 **9.** $75 **11.** 112 **13.** 117 **15.** 24 **17.** 625 **19.** 88 **21.** 64 **23.** 62.5 **25.** $50 **27a.** $P = 0.29(150)$ **27b.** 44 **29.** 7 adults **31.** $x = y$ **33a.** 180 mi **33b.** 210 mi **33c.** 135 mi **35.** −31 **37.** 5

Pages 215–217 Lesson 5-5
1. First method: find the amount of the tax, then add the tax to the cost of the item. Second method: add the tax rate to 100%, then find the total cost. **5.** $\frac{4}{14} = \frac{r}{100}$ **7.** 20% decrease **9.** $94.50 **11.** $1080 **13.** 83% **15.** 12% increase **17.** 16% decrease **19.** 133% increase **21.** $42.80 **23.** $39.52 **25.** $26.64 **27.** $266 **29.** $29.75 **31.** $14.21 **33.** 100% **35.** 1999 and 2001; The amount of increase is larger and the base is smaller between these years. **37.** less than **39.** 62.5% **41.** 35% **43.** 1.5 kg

Page 217 Quiz 2
1. 25 **3.** 20 **5.** $140.81

Pages 222–223 Lesson 5-6
1. Both theoretical and experimental probability are ratios that compare the number of favorable outcomes to the number of possible outcomes. Theoretical probability is based on known characteristics. It tells what should happen. Experimental probability is the result of an experiment or simulation. **3.** Yes; when the number of favorable outcomes is greater than the number of unfavorable outcomes, the odds of the outcome are greater than 1. **5.** $\frac{1}{36}$ **7.** 1:1 **9.** $\frac{1}{2}$ **11.** $\frac{1}{9}$ **13.** 0 **15.** 1:1 **17.** 7:1 **19.** 5:3 **21.** $\frac{1}{4}$ **23a.** 9:00, $\frac{1}{20}$; 1:00, $\frac{1}{40}$; 4:00, $\frac{1}{8}$ **23b.** 4:00 **25a.** $\frac{1}{2}$ **25b.** 9:11 **27.** $36 **29.** $180 **31.** A

Pages 227–229 Lesson 5–7

1. *P*(*A* and *B*) is the probability that both *A* and *B* occur; *P*(*A* or *B*) is the probability that either *A* or *B* occur. **3.** $\frac{1}{24}$ **5.** mutually exclusive, $\frac{2}{13}$ **7.** inclusive, $\frac{4}{13}$ **9.** 20.1% **11.** $\frac{1}{4}$ **13.** $\frac{1}{20}$ **15.** $\frac{3}{20}$ **17.** mutually exclusive, $\frac{2}{3}$ **19.** 0.013 **21a.** 34.2% **21b.** 28% **21c.** 88.3% **23.** $\frac{1}{3}$ **25.** $\frac{3}{5}$ **27.** D

Pages 230–232 Chapter 5 Study Guide and Assessment

1. proportion **3.** impossible **5.** odds **7.** unit rate **9.** Discount **11.** 45 **13.** 4 **15.** 10 mi **17.** 30 mi **19.** 1 in. = 30.5 ft **21.** 60 **23.** 150% **25.** 50 **27.** 8.1 **29.** 150 **31.** 25% increase **33.** $33.\overline{3}$% decrease **35.** 100% **37.** $\frac{1}{11}$ **39.** $\frac{7}{11}$ **41.** $\frac{1}{2}$ **43.** mutually exclusive, $\frac{2}{3}$ **45.** $339.15

Page 235 Preparing for Standardized Tests

1. B **3.** C **5.** A **7.** C **9.** 2

Chapter 6 Functions and Graphs

Page 237 Check Your Readiness

1, 3, 5, 7, 9.

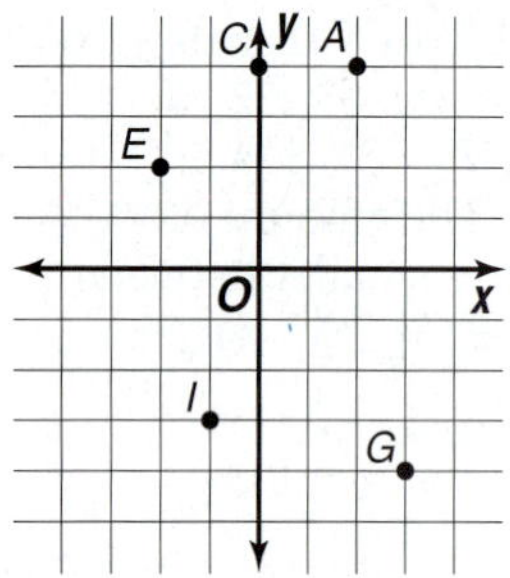

11. −6 **13.** 6 **15.** 10 **17.** −6 **19.** 10 **21.** 4 **23.** −15 **25.** 30 **27.** 4

Pages 241–243 Lesson 6–1

1. ordered pairs, table, graph **3.** The domain is the set of all first coordinates from the ordered pairs. The range is the set of all second coordinates from the ordered pairs.

5.

x	y
−3	−3.5
−1	3.9
0	2
5	2.5

domain: {−3, −1, 0, 5},
range: {−3.5, 3.9, 2, 2.5}

7. {(−3, −2), (1, 2), (−4, −2)}

x	y
−3	−2
1	2
−4	−2

domain: {−3, 1, −4},
range: {−2, 2}

9.

x	y
4	3
−2	3
−2	4
4	4

domain: {4, −2},
range: {3, 4}

11.

x	y
−2	0
3	−7
2	−5
−6	3
1	5

domain: {−2, 3, 2, −6, 1},
range: {0, −7, −5, 3, 5}

13.

x	y
$-\frac{1}{2}$	$\frac{1}{2}$
1	0
$-\frac{1}{2}$	−5

domain: $\left\{-\frac{1}{2}, 1\right\}$,

range: $\left\{\frac{1}{2}, 0, -5\right\}$

15. {(−3, 2), (−1, 1), (1, −2), (1, 3)}

x	y
−3	2
−1	1
1	−2
1	3

domain domain: {−3, −1, 1},
range: {2, 1, −2, 3}

17. $\{(-3, 2), (-2, 2), (-1, 4), (2, -2), (3, -2)\}$

x	y
−3	2
−2	2
−1	4
2	−2
3	−2

domain: $\{-3, -2, -1, 2, 3\}$,
range: $\{2, 4, -2\}$

19. $\{(-3, -3), (-3, -1), (-1, 0), (2, -2), (3, 4)\}$

x	y
−3	−3
−3	−1
−1	0
2	−2
3	4

domain: $\{-3, -1, 2, 3\}$,
range: $\{-3, -1, 0, -2, 4\}$

21. $\{(-1, 1), (0, 2), (1, 3), (2, 4)\}$

23. $\{(0, 0), (-1, -0.5), (1, 0.5), (-2, -1)\}$

25a. $\{(8, 4), (9, 3), (10, 2), (11, 1), (12, 0)\}$

25b.

x	y
8	4
9	3
10	2
11	1
12	0

27a. 3, 4, 5, 6, 7, 8, 9, 10, 11, 12, 13, 14, 15

27b. $\{(3, 1), (4, 1), (5, 2), (6, 2), (7, 3), (8, 3), (9, 4), (10, 3), (11, 3), (12, 2), (13, 2), (14, 1), (15, 1)\}$

29. 4:6 or 2:3 **31.** −24 **33.** B

Pages 247–249 Lesson 6–2

1. When the values are substituted in the equation, the equation is true. **3.** Dan; when using variables other than x and y, it can be assumed that the domain goes with the variable that comes first alphabetically. Thus, m is the domain. **5.** c, d

7. $\{(-2, 7), (-1, 2), (0, -3), (1, -8), (2, -13)\}$

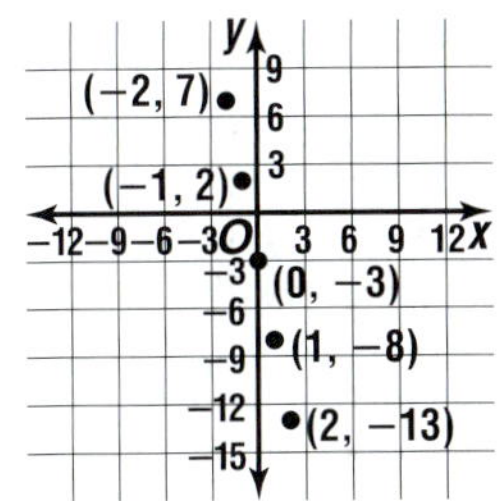

9. $\left\{(-2, -5.5), (-1, -4), (0, -2.5), (1, -1), \left(2, \frac{1}{2}\right)\right\}$

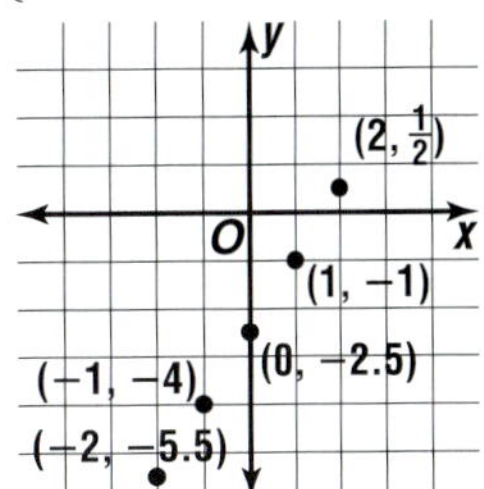

11. $\{(3, 25), (4, 24), (8, 20), (12, 16)\}$

13. d **15.** b **17.** c

19. $\{(-1, 3), (0, 0), (1, -3), (2, -6), (3, -9)\}$

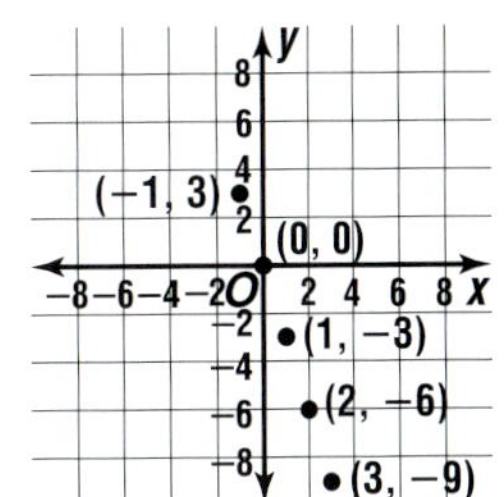

21. $\{(-1, -7), (0, -6), (1, -5), (2, -4), (3, -3)\}$

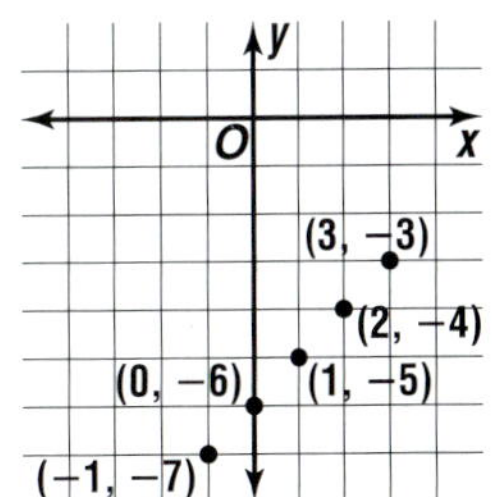

23. $\{(-1, 10), (0, 5), (1, 0), (2, -5), (3, -10)\}$

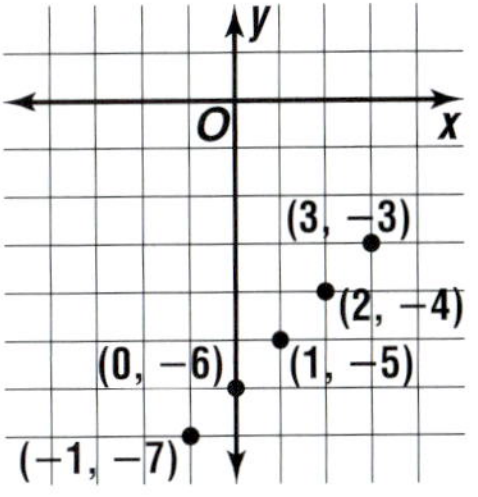

25. $\{(-1, -3), (0, -2), (1, -1), (2, 0), (3, 1)\}$

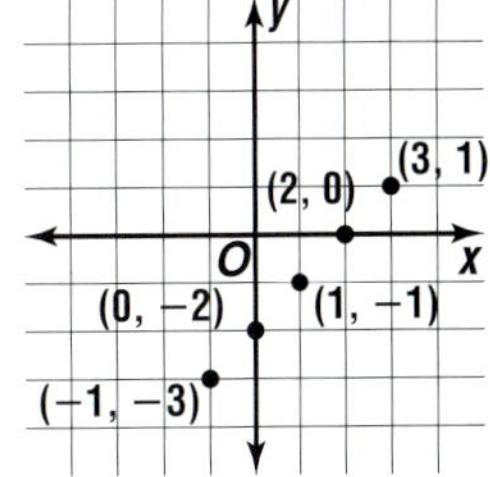

SELECTED ANSWERS

27. {(−1, 6), (0, 4), (1, 2), (2, 0), (3, −2)}

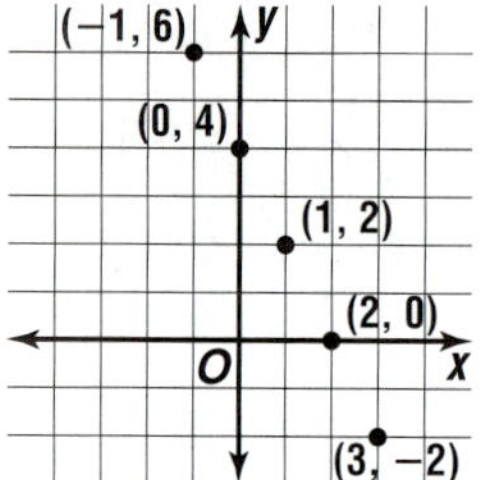

29. {(−1, −7), (0, −4), (1, −1), (2, 2), (3, 5)}

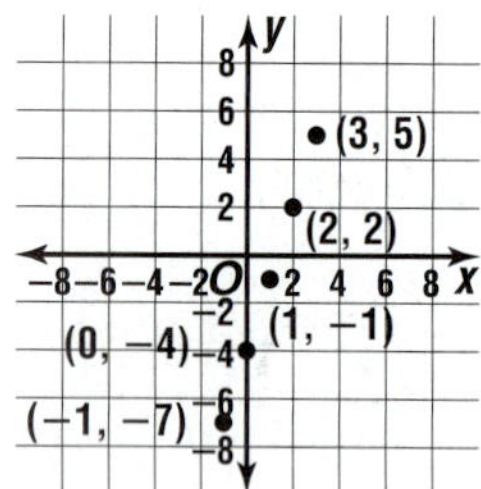

31. {0, 1, 2, 3} **33.** {−1, 0, 1, 2} **35.** 3

37a.

w	*y*
6.0	0.15
6.8	0.17
7.2	0.18
8.0	0.2
8.8	0.22

37b. {(6.0, 0.15), (6.8, 0.17), (7.2, 0.18), (8.0, 0.2), (8.8, 0.22)}

37c.

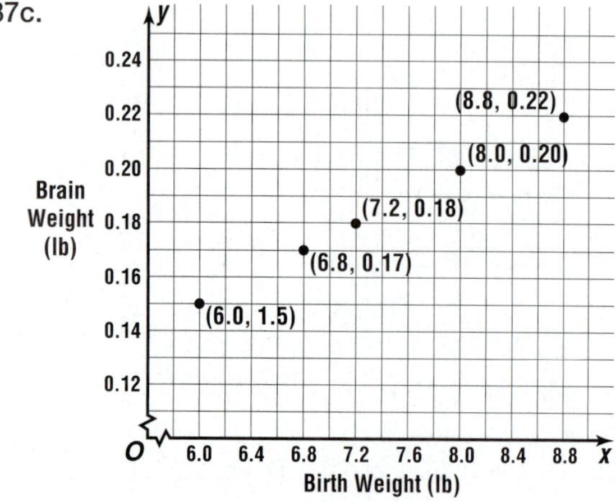

39a. $x + y = 90$; $x = 2y$ **39b.** (60, 30)
41. \$10.99 **43.** 5 **45.** −1 **47.** C

Pages 254–255 Lesson 6–3

1. The graph of $x = -8$ is a straight line parallel to the y-axis. For every value of y, $x = -8$. **3.** no
5. yes; $A = 2$, $B = 0$, $C = 8$

7. **9.**

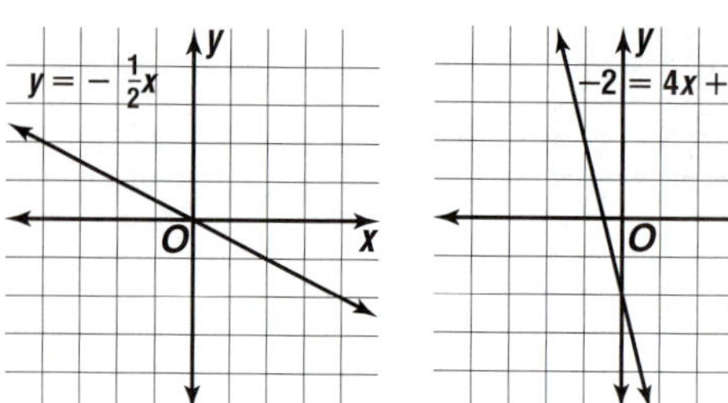

11.

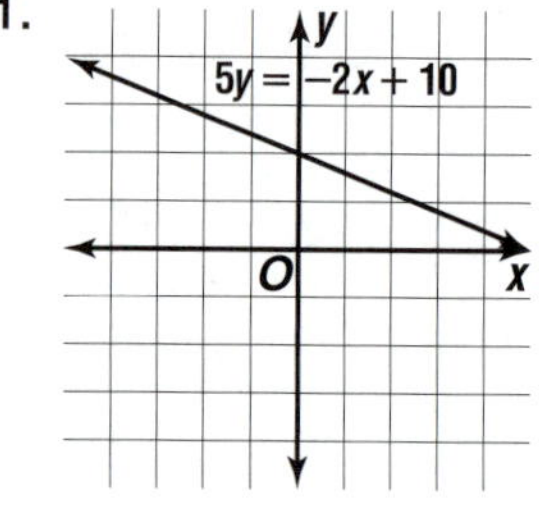

13. no **15.** yes; $A = -6$, $B = 2$, $C = 0$
17. yes; $A = -1$, $B = 1$, $C = 0$ **19.** no
21. yes; $A = 5$, $B = 1$, $C = 9$

23. **25.**

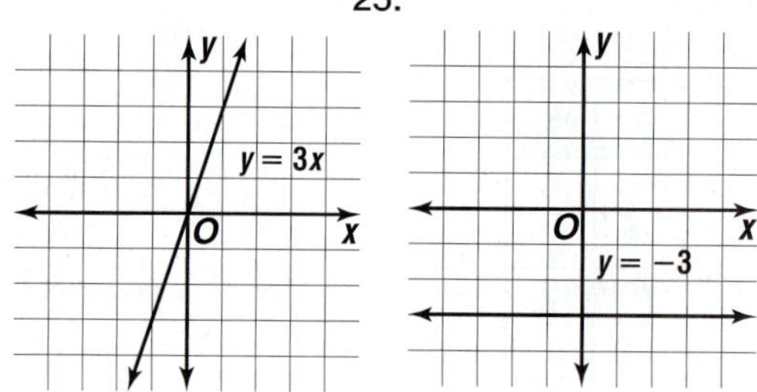

27.

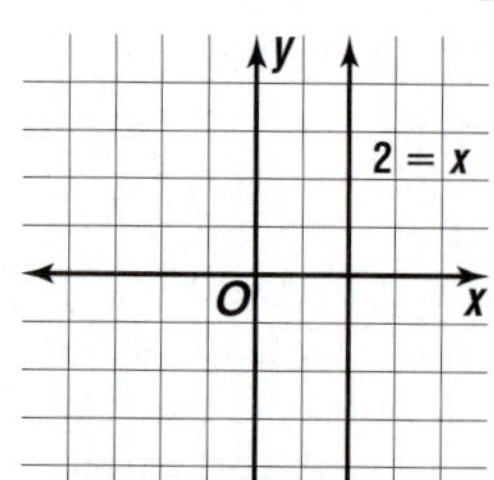

29.

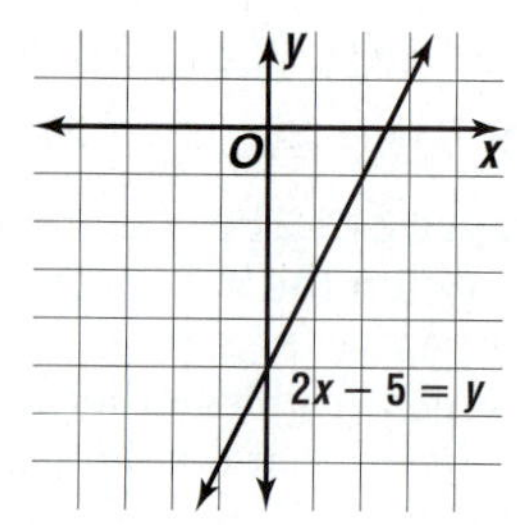

31.

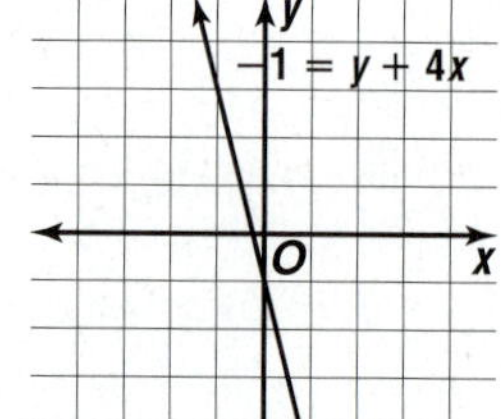

33.

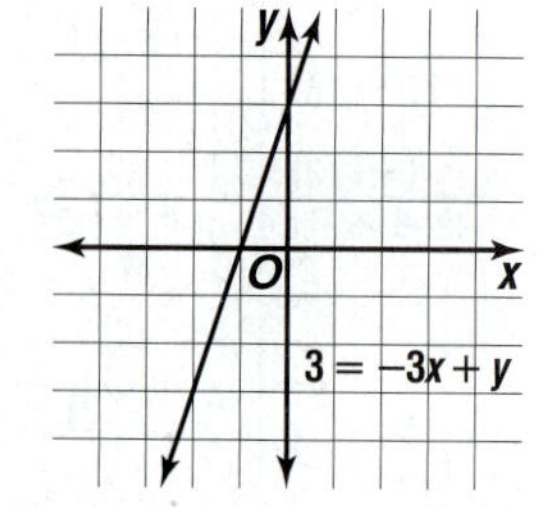

35.

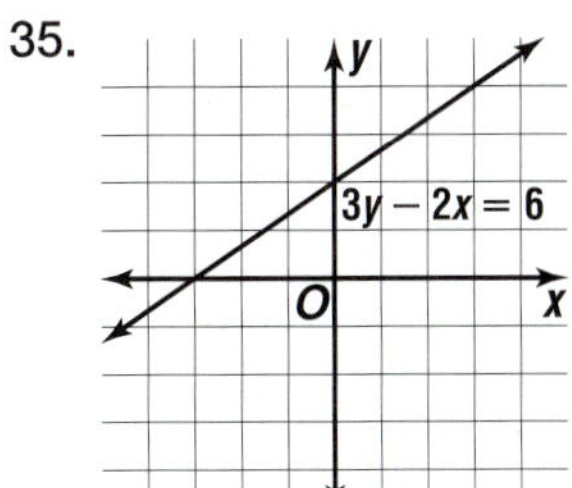

37.

x	y
0	−2
1	−1
2	0
3	1
4	2

39.

x	y
−2	−4
−1	−1
0	2
1	5
2	8

41a. $3x + 6y = 90$ **41b.** Sample answer: $\{(2, 14), (6, 12), (12, 9), (18, 6), (24, 3)\}$

41c.

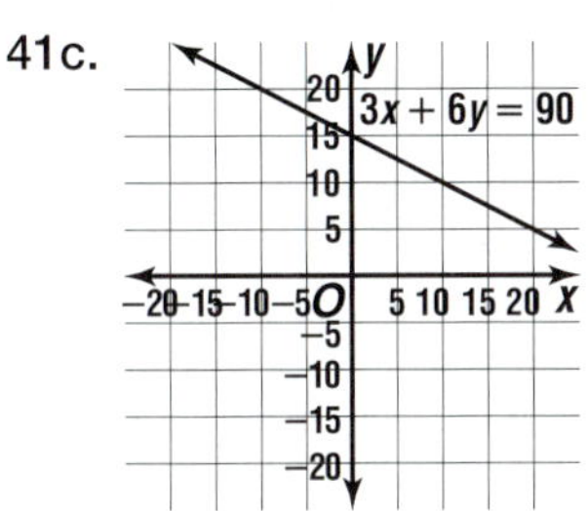

43. $\{-8, -1, 0, 2, 4\}$ **45.** 6 **47.** 4 **49.** $67 + 2h = 79$

Page 255 Quiz 1

1.

x	y
1	−1
3	5
0	−2
3	3

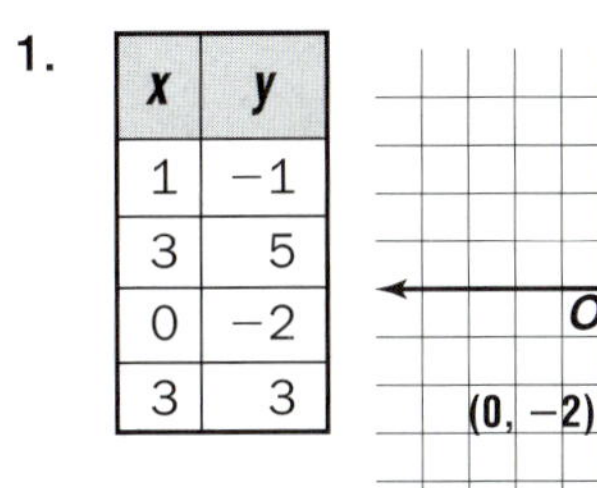

3. b, c

5a.

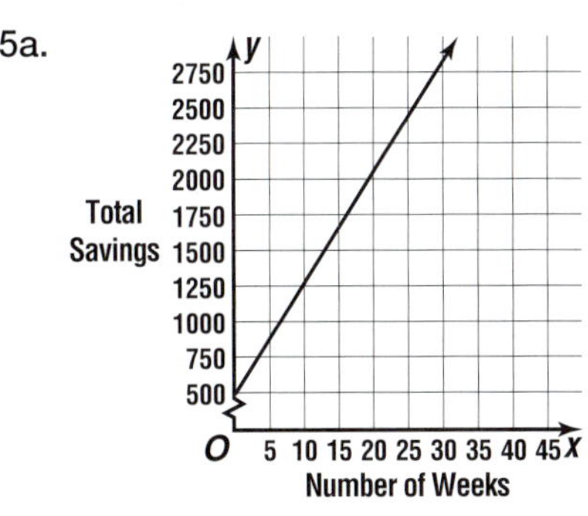

5b. 25 weeks **5c.** $y = 500 + 80x$

$y = 500 + 80(25)$

$y = 500 + 2000$

$y = 2500$

Pages 259–261 Lesson 6–4

1. Sample answer: A function is a relation. However, in a relation, each member of the domain can be paired with more than one member of the range. In a function, each member of the domain is paired with exactly one member of the range. **3.** Zina is correct. There are some graphs of straight lines that do not represent a function. The graph of $x = 2$ is a straight line. However, it is not a function since there are many y values for a single x value. **5.** yes **7.** no **9.** −15 **11.** $4b - 3$ **13.** 3.25 s **15.** yes **17.** no **19.** yes **21.** no **23.** no **25.** yes **27.** yes **29.** −1 **31.** 13 **33.** 3.4 **35.** −1.7 **37.** 3 **39.** $-3\frac{1}{2}$ **41.** $-3h + 4$ **43.** $6a + 4$ **45.** −12

47a. $p(a) = 0.5a + 110$ **47b.** 128

47c.

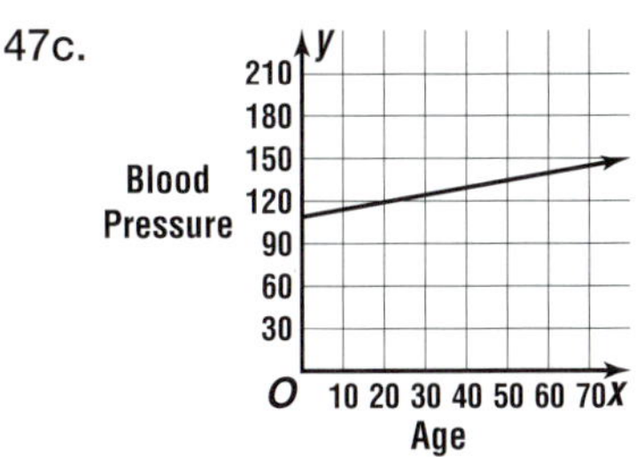

49a. −7

49b.

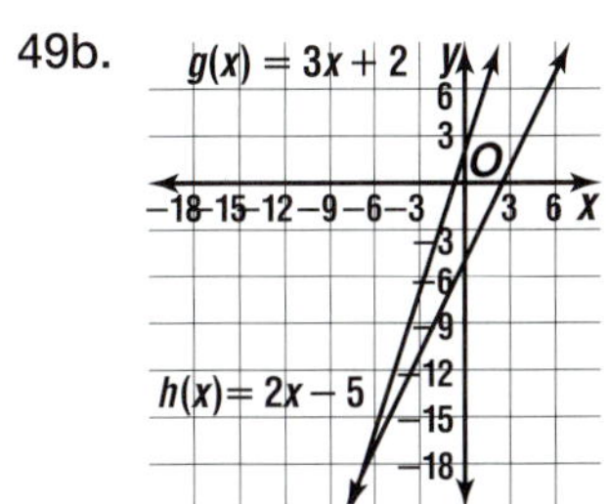

The lines intersect at $x = -7$.

51. $\{(2, 28), (4, 14), (7, 8)\}$ **53.** 30% **55.** $\frac{3}{20}$ **57.** 1.2

Pages 267–269 Lesson 6–5

1. Divide each side of $y = kx$ by x, where $x \neq 0$

3. Sample answer: The length of the trip depends upon the amount of gasoline used. So, the length of the trip represents the dependent variable and the amount of gasoline used represents the independent variable. **5.** 5 **7.** $\frac{2}{5}$

9. yes

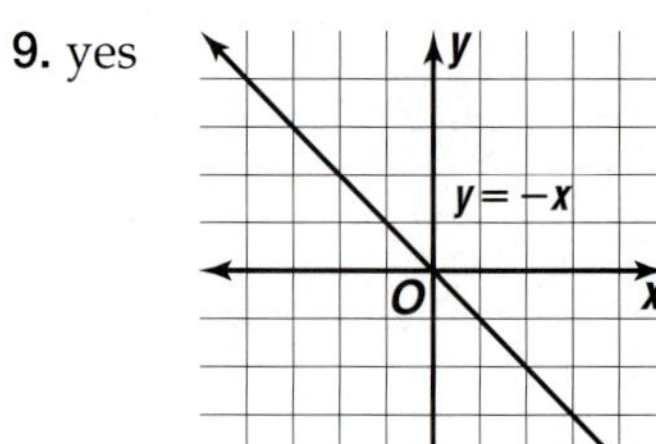

11. 13 **13.** 4.7 yd

15. no

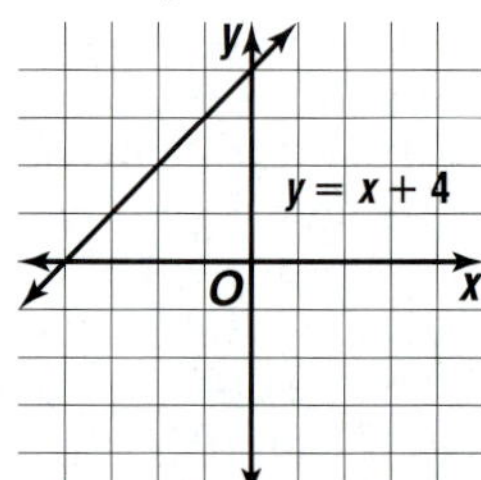

17. no

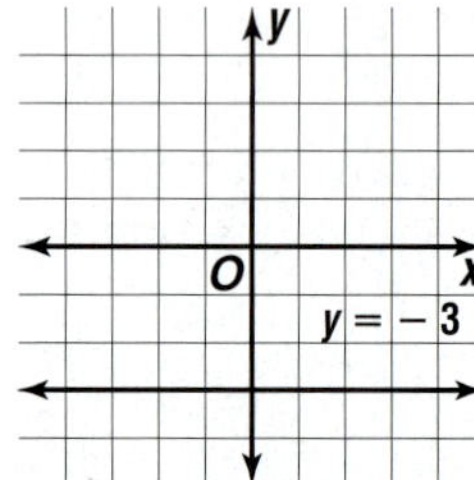

19. no

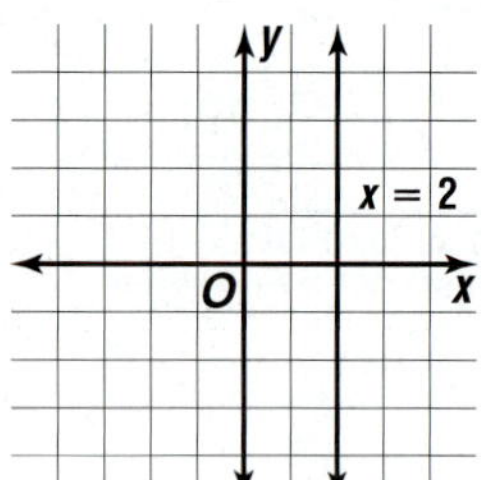

21. 25 **23.** 20 **25.** $26\frac{1}{4}$ **27.** $2\frac{2}{3}$ **29.** 0.6 bu

31. \$3.15 **33a.** $h = 7f$

33b.

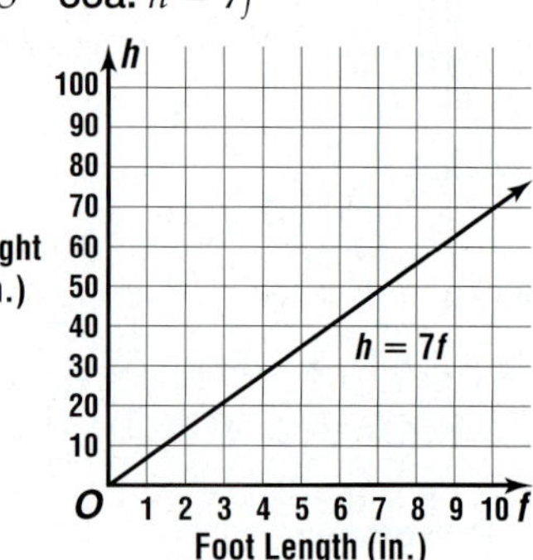

33c. about 5.5 ft **35.** 11 **37a.** $x - 50 = 550$
37b. about 600 pounds **39.** 8 **41.** B

Page 269 Quiz 2
1. yes **3.** −7.4 **5.** 8 gal

Pages 273–275 Lesson 6–6
1. Multiply the value of x by the value of y.
3. Sample answer: In a direct variation, as x increases, y increases. In an inverse variation, as x increases, y decreases. **5.** direct; 5 **7.** direct; $\frac{1}{3}$
9. direct; $\frac{1}{4}$ **11.** 6 **13.** 3.25 hours **15.** 10
17. 0.09 **19.** $\frac{5}{8}$ **21.** $-\frac{1}{6}$; $y = -\frac{1}{6}x$ **23.** 0.3; $y = 0.3x$
25. a; inverse **27.** c; direct **29a.** 36 **29b.** 2.4 m
29c. $\ell w = 36$
29d.

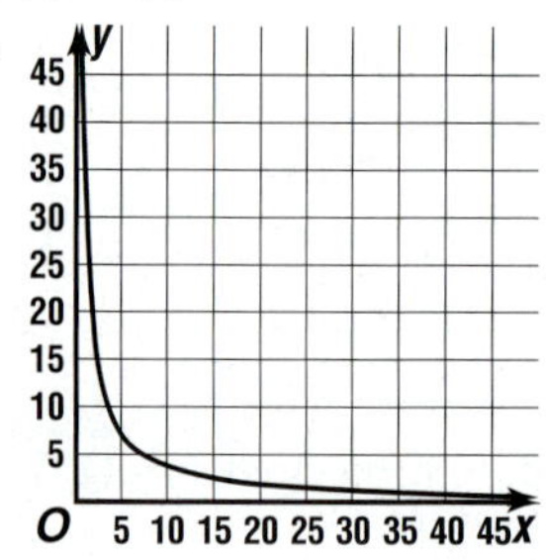

31. 27 **33.** −14.7 **35.** 457

Pages 276–278 Chapter 6 Study Guide and Assessment
1. linear equation **3.** functional notation **5.** inverse
7. $y = kx$ **9.** $y = 2x + 3$
11.

x	y
−1	0
0	0
2	3
4	1

(2, 3)
(4, 1)
(0, 0)
(−1, 0)

domain: {−1, 0, 2, 4},
range: {0, 3, 1}

13.

x	y
−1.5	2
1	3.5
4.5	−5.5

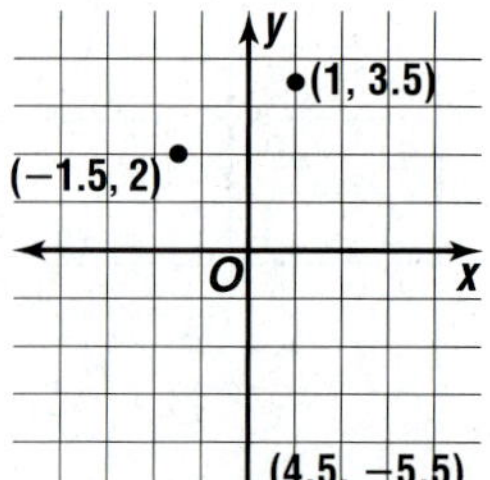

domain: {−1.5, 1, 4.5},
range: {2, 3.5, −5.5}

15. {(−4, 2), (−2, 1), (−2, −3), (0, −2), (1, 3)}

x	y
−4	2
−2	1
−2	−3
0	−2
1	3

domain: {−4, −2, 0, 1},
range: {2, 1, −3, −2, 3}

17. $\left\{\left(-1, \frac{1}{2}\right), (0, 0), (2, 1), (-4, 2)\right\}$

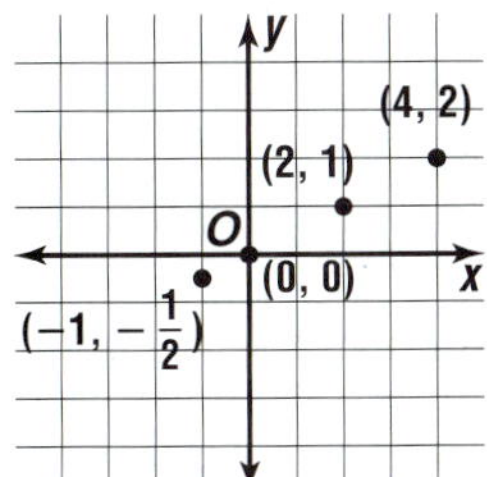

19. $\{(-1, 6.5), (0, 5), (2, 2), (4, -1)\}$

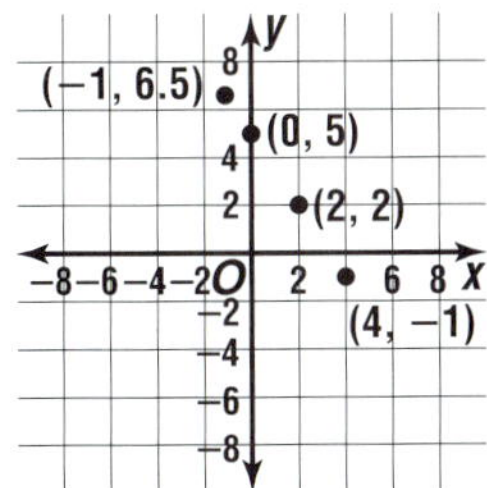

21. $\{-2, -1, 0, 1\}$ **23.** yes; $A = -1, B = -1, C = 0$

25. no

27.

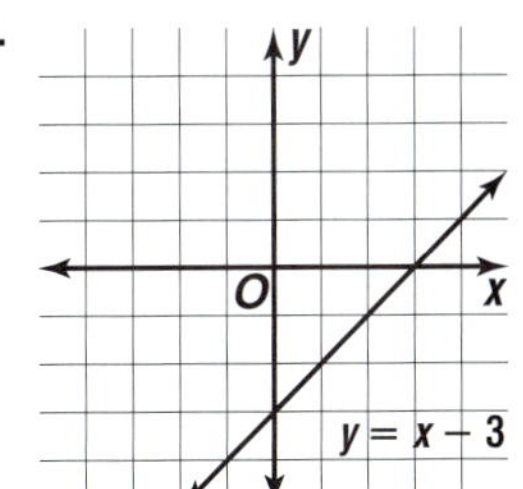

29.

30.

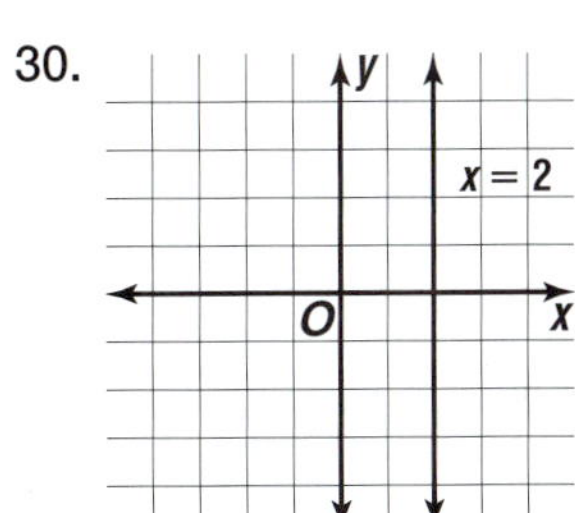

33. no **35.** yes **37.** 4.5 **39.** 21 **41.** 12.5 **43.** 5 **45.** 8; $xy = 8$ **47.** $\{(5, 1), (10, 2), (15, 3), (20, 4), (25, 5)\}$ domain: $\{5, 10, 15, 20, 25\}$, range: $\{1, 2, 3, 4, 5\}$ **49.** 132 volts

Page 281 Preparing for Standardized Tests

1. A **3.** E **5.** E **7.** B **9.** 3/4

Chapter 7 Linear Equations

Page 283 Check Your Readiness

1. 4 **3.** -5 **5.** -1 **7.** -6 **9.** 18 **11.** 3 **13.** $-\frac{1}{2}$ **15.** -1 **17.** $\frac{6}{5}$ **19.** (3, 4) **21.** (3, 0) **23.** (0, −3)

Pages 287–289 Lesson 7–1

1a. Sample answer:

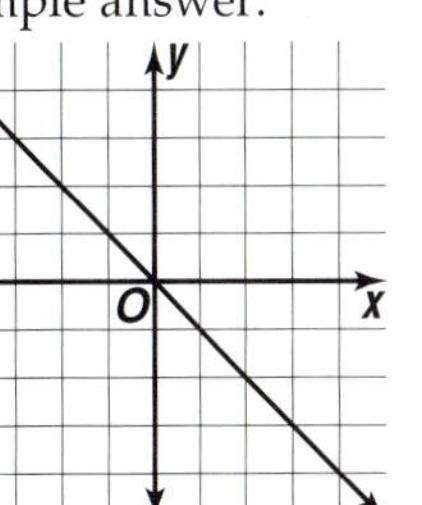

1b. Sample answer:

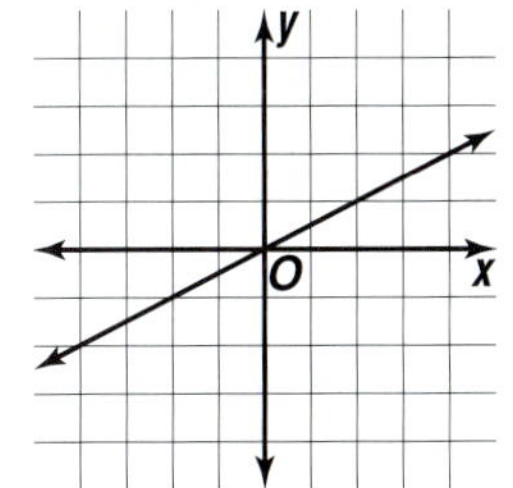

1c. Sample answer:

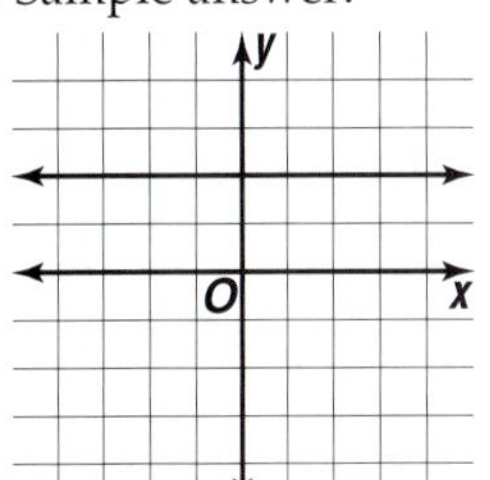

1d. Sample answer:

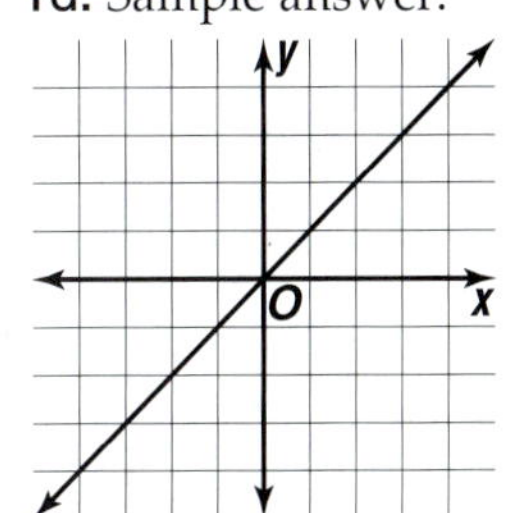

3. Percy; any two points on the line can be used to find the slope. **5.** 2 **7.** $\frac{3}{4}$ **9.** 4 **11.** $-\frac{6}{5}$ **13.** $-\frac{5}{3}$ **15.** $\frac{1}{3}$ **17.** -4 **19.** -2 **21.** $\frac{1}{3}$ **23.** $-\frac{2}{5}$ **25.** $-\frac{3}{8}$ **27.** $\frac{7}{100}$ **29a.** As x increases by 1, $f(x)$ increases by 0.25. The slope is 0.25. **29b.** Sample answer: domain: positive integers; range: positive multiples of 0.25 **31.** 12 **33.** no **35.** -2

Pages 293–295 Lesson 7–2

1. the coordinates of a point on the line and the slope of the line, or the coordinates of two points on the line **3.** $m = 2, (1, 6)$ **5.** $m = -3, (-9, -5)$ **7.** $y - 4 = \frac{3}{4}(x + 2)$ **9.** $y - 1 = 0$ or $y = 1$ **11.** $y - 3 = -\frac{5}{6}(x)$ or $y + 2 = -\frac{5}{6}(x - 6)$ **13.** $y - 5 = -2(x + 2)$ or $y + 7 = -2(x - 4)$ **15.** $y - 4 = -3(x + 1)$ **17.** $y - 1 = 6(x - 9)$ **19.** $y + 2 = \frac{4}{5}(x + 2)$ **21.** $x = 4$ **23.** $y + 4 = -\frac{5}{2}\left(x - \frac{1}{2}\right)$ **25.** $y - 4 = \frac{1}{4}(x + 2)$ or $y - 5 = \frac{1}{4}(x - 2)$

27. $y = \frac{3}{4}(x + 2)$ or $y - 3 = \frac{3}{4}(x - 2)$ **29.** $y - 3 = -\frac{5}{2}(x + 3)$ or $y + 2 = -\frac{5}{2}(x + 1)$ **31.** $y - 5 = \frac{1}{2}(x - 4)$ or $y - 3 = \frac{1}{2}(x)$ **33.** $y + 4 = 3(x - 1)$ or $y - 5 = 3(x - 4)$ **35.** $y - 7 = -\frac{5}{8}(x + 3)$ or $y - 2 = -\frac{5}{8}(x - 5)$ **37.** $y - 6 = -\frac{1}{7}(x - 3)$ or $y - 7 = -\frac{1}{7}(x + 4)$ **39a.** Sample answer: $y - 5 = 5(x - 1)$ **39b.** Sample answer: domain: positive integers; range: positive multiples of 5 **41.** (4, 8) **43.** {0, 1, 2} **45.** {1, 2, 3} **47.** 300% increase **49.** 8.4

Pages 299–301 Lesson 7–3

1. Sample answer:

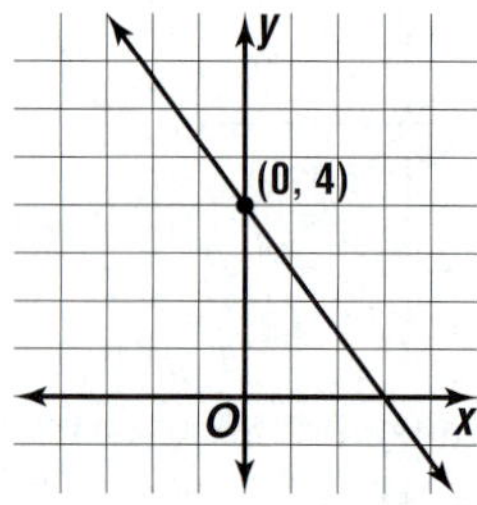

3. Jacquie; the line $y = 2$ does not intersect the x-axis. **5.** $y = \frac{1}{4}x - 3$ **7.** $y = 4x - 7$ **9.** $y = -\frac{1}{2}x - 3$ **11.** $y = 2x - 8$ **13a.** $y = 25x + 100$ **13b.** the pounds of cans he plans to collect each week **13c.** 400 lb **15.** $y = -2x + 5$ **17.** $y = \frac{3}{2}x + 7$ **19.** $y = -\frac{2}{5}x + 3$ **21.** $y = 4$ **23.** $y = 3x - 17$ **25.** $y = 5x$ **27.** $y = -2x + 13$ **29.** $y = -5x + 26$ **31.** $y = \frac{1}{2}x - \frac{9}{2}$ **33.** $y = -\frac{5}{4}x + 5$ **35.** $y = 4x - 12$ **37.** $y = -3$ **39.** $y = -\frac{4}{5}x + \frac{4}{5}$ **41.** $y = -\frac{3}{2}x + 6$ **43.** $y = -\frac{6}{5}x + \frac{8}{5}$ **45.** $y = -3x + 11$ **47a.** $y = 8x$ **47b.** cost per person **47c.** $320 **49.** $y - 5 = -2(x - 4)$

51.

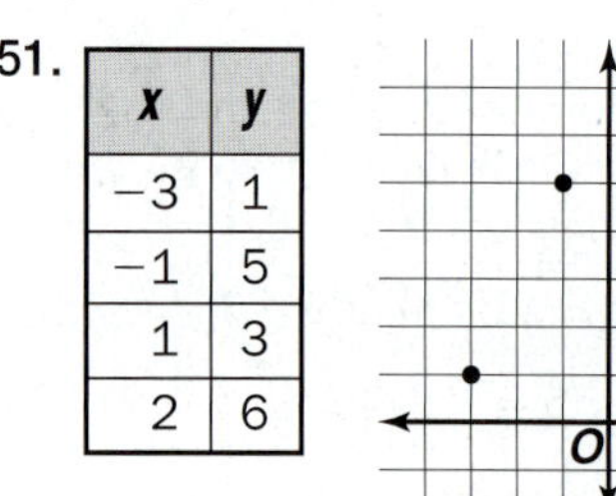

x	y
−3	1
−1	5
1	3
2	6

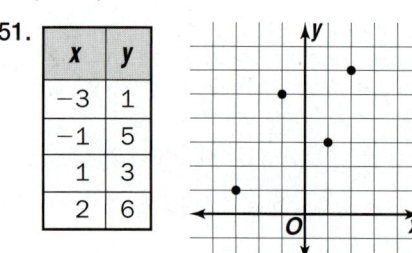

domain: {−3, −1, 1, 2}, range: {1, 5, 3, 6}
53. 9.144 km **55.** −5 **57.** A

Page 301 Quiz 1

1. $-\frac{4}{3}$ **3.** $y - 4 = -2(x - 5)$ **5a.** 90

5b. The slope represents how much each problem is worth and the y-intercept represents the extra points students get with the curve.

Pages 305–307 Lesson 7–4

1. Graph the two sets of data as ordered pairs, where the x-coordinate is a member of one data set and the y-coordinate is a member of the other data set. The set of x-coordinates is the domain and the independent variable. The set of y-coordinates is the range and the dependent variable.

5a.

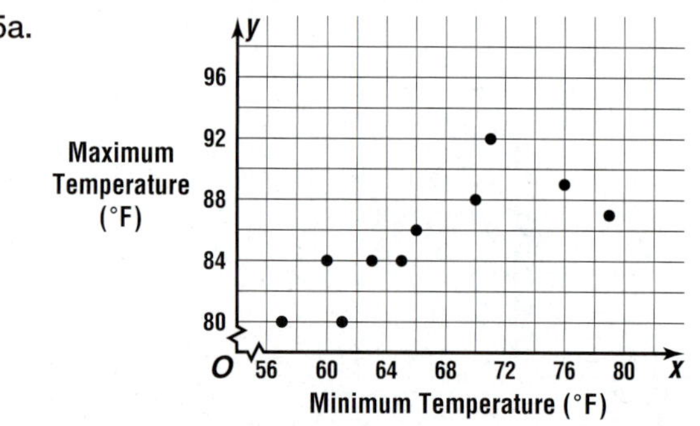

5b. Yes; as maximum temperature increases, so does the minimum temperature. There is a positive relationship. **7.** Negative relationship; as speed increases, CO_2 decreases.

9.

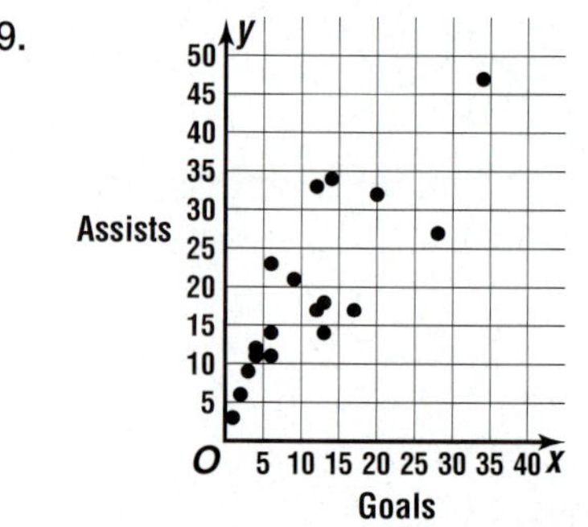

There is a positive relationship; as the number of goals increases, the number of assists increases **11.** Sample answer: Have higher maximum speed limits because CO_2 emissions decrease as speed increases. **13.** no **15.** negative

17a.

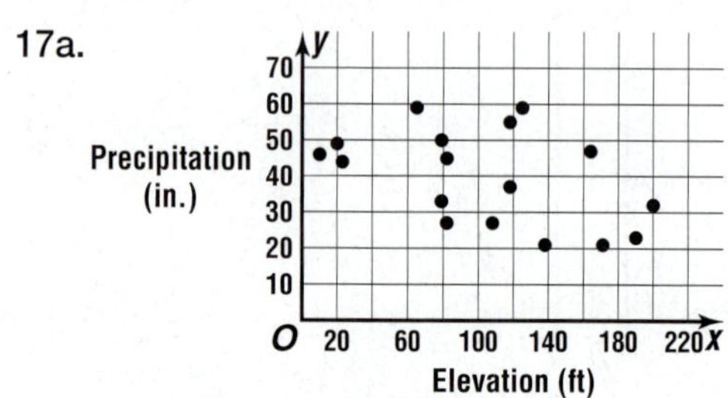

17b. No; there does not appear to be a relationship between elevation and precipitation. **17c.** zero **19.** $y = 5x + 2$ **21.** 1 cm = 5 m or 1:500 **23.** D

Pages 314–315 Lesson 7–5

1. two points, then draw a line through them

3. $-3, -3$

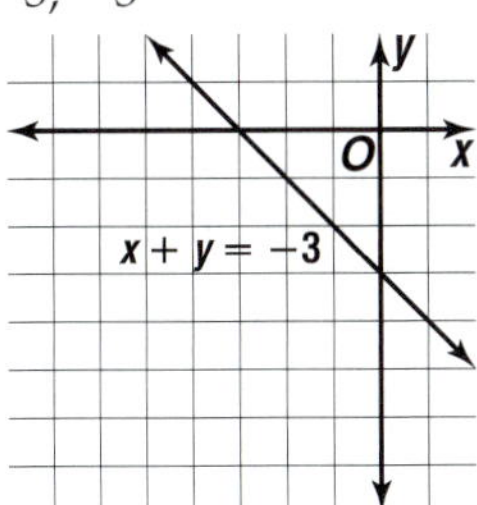

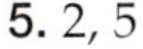

5. $2, 5$

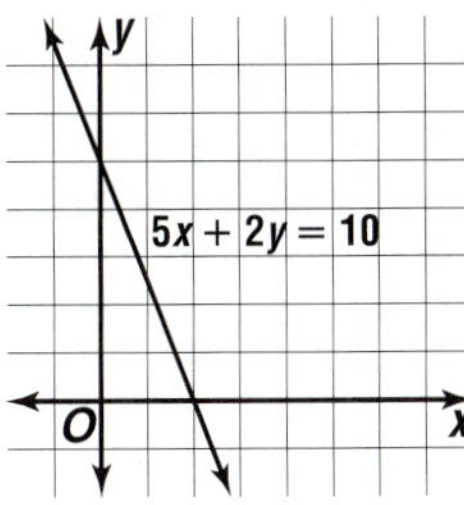

7. $m = 1, b = 1$

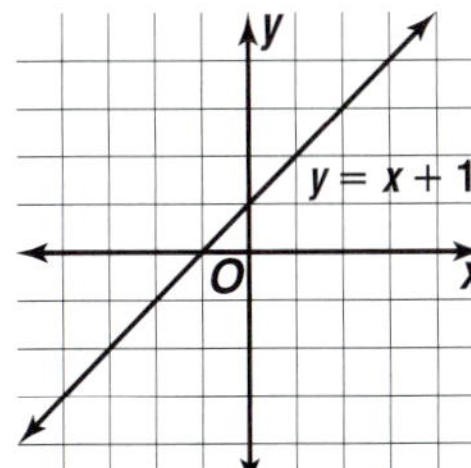

9. $m = \frac{1}{4}, b = -3$

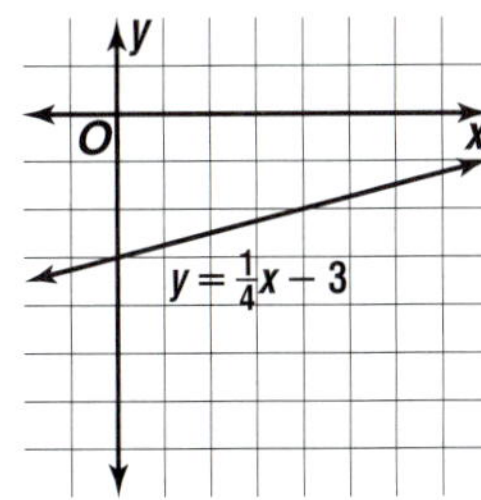

11. $m = \frac{9}{5}, b = 32$

13. $-5, -5$

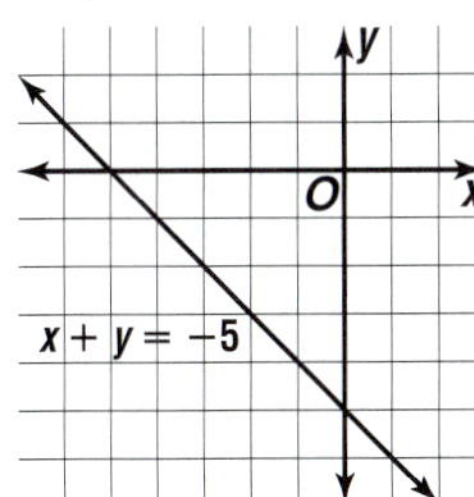

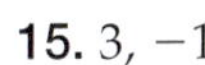

15. $3, -1$

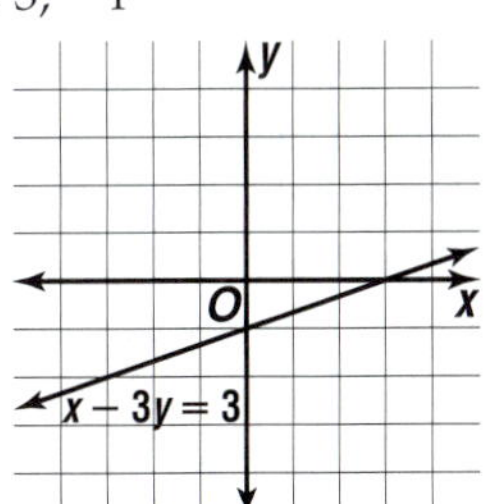

17. $-3, -6$

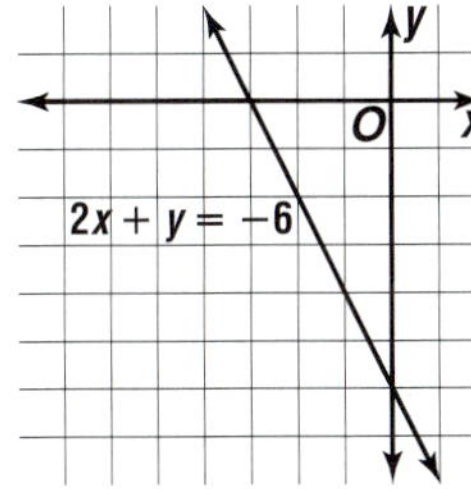

19. $-4, 3$

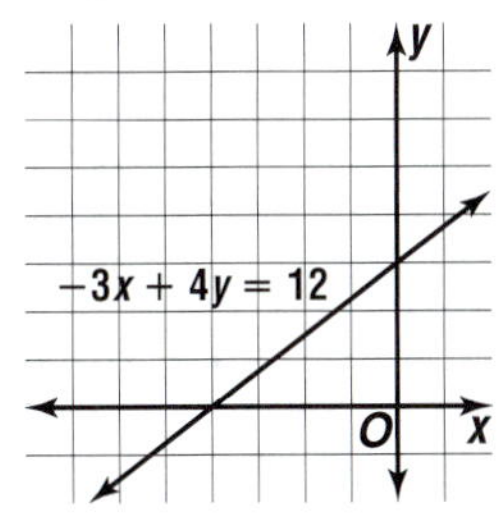

21. $2, -7$

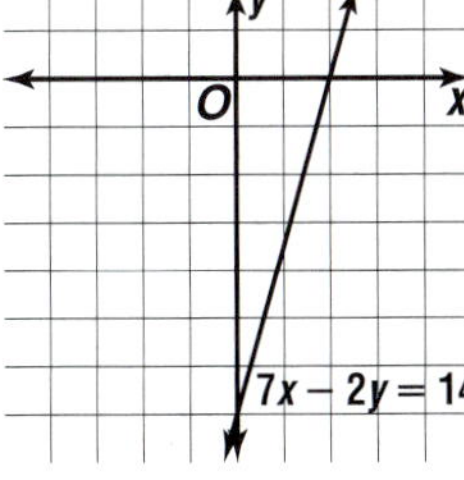

23. $4, 8$

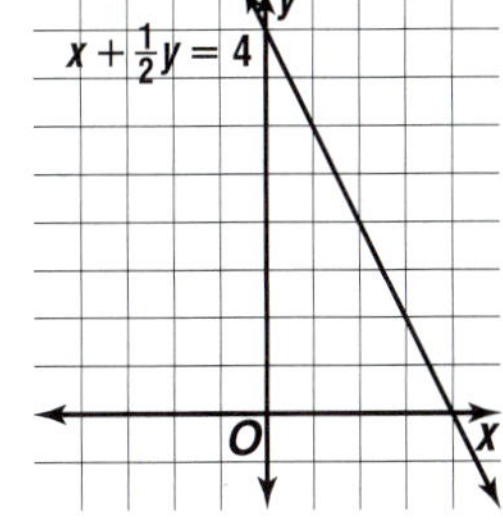

25. $m = -1, b = 2$

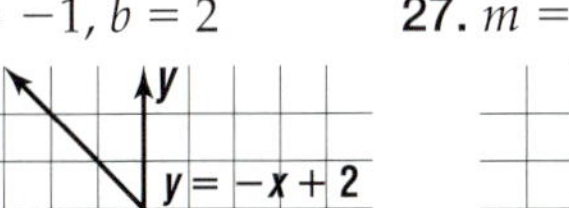

27. $m = 3, b = -1$

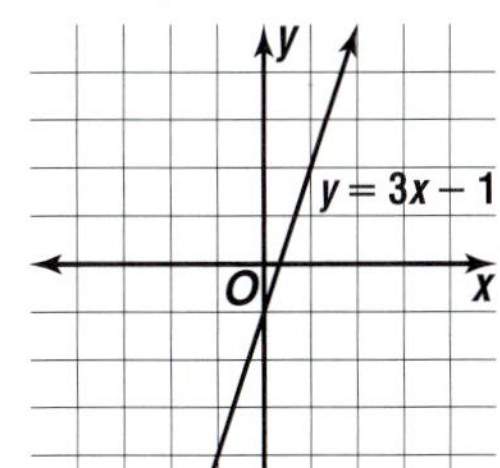

29. $m = 0, b = 4$

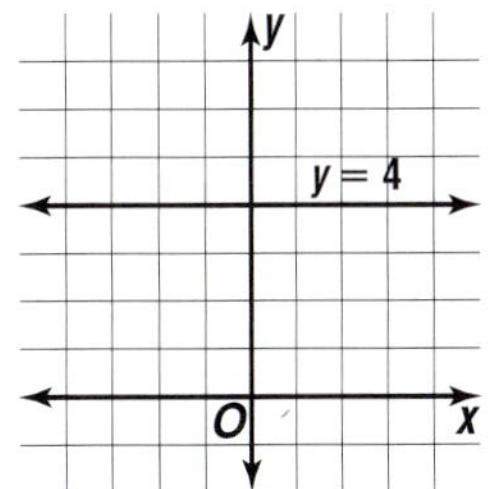

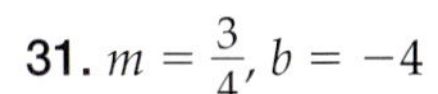

31. $m = \frac{3}{4}, b = -4$

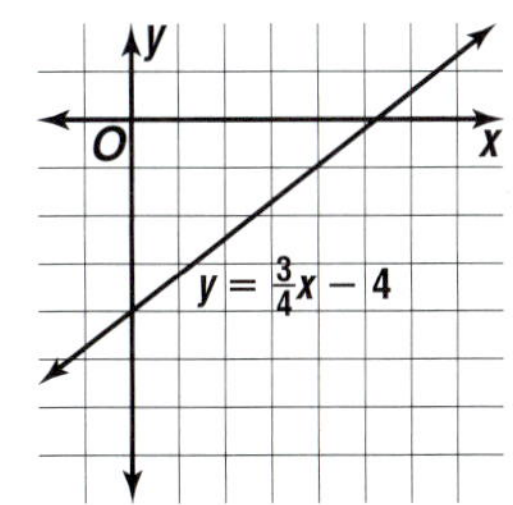

33. $m = 2, b = 3$

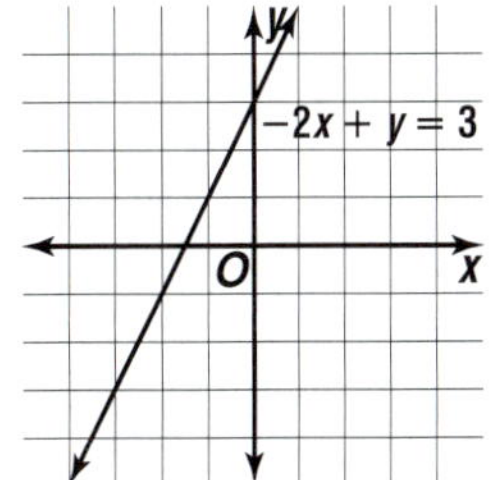

35. $m = -\frac{3}{5}, b = 2$

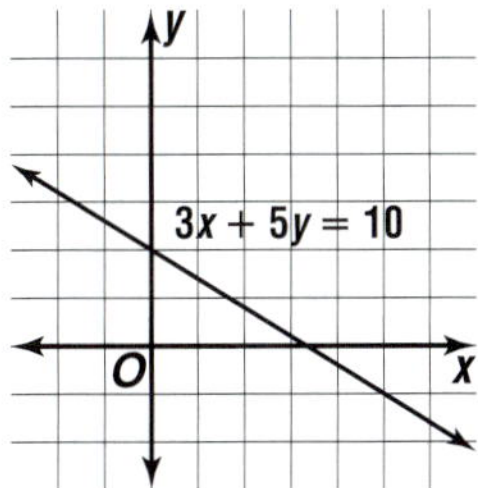

37a.

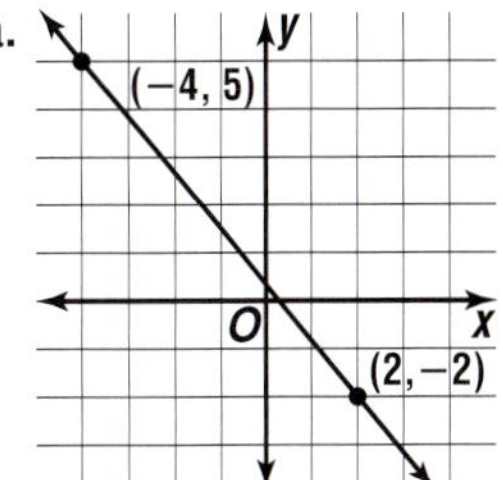

37b. $y = -\frac{7}{6}x + \frac{1}{3}$

37c. $m = -\frac{7}{6}, b = \frac{1}{3}$

39a. $f(x) = \frac{1}{4}x + 1$

39b.

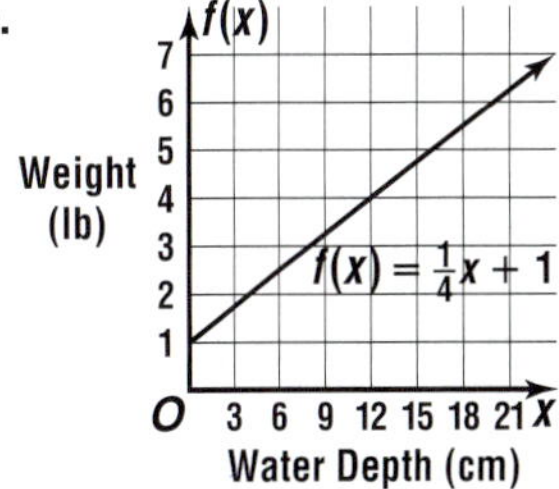

39c. about $5\frac{1}{2}$ lb

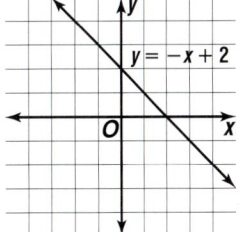

41. Sample answer:

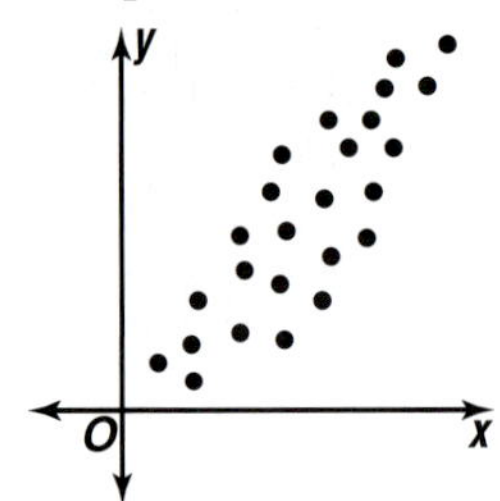

43. $\frac{1}{6}$ **45.** 0 **47.** 19 **49.** A

Pages 319–321 Lesson 7–6

1a. shifted down 4 units **1b.** less steep **1c.** steeper **1d.** shifted up 3 units **3a.** Slope represents rate. **3b.** A greater rate results in a steeper positive slope.

5.

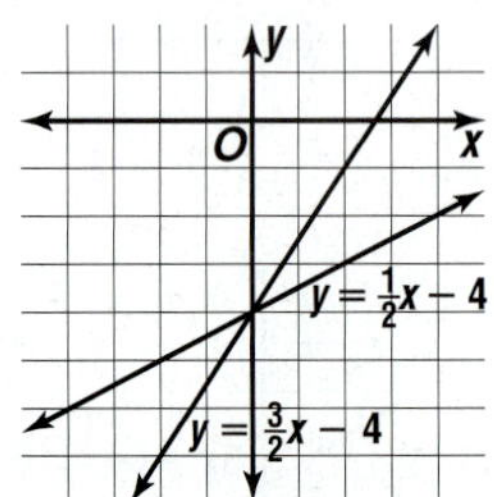

same y-intercept, different slopes

7. same y-intercept, different slopes **9.** $y = x$

11.

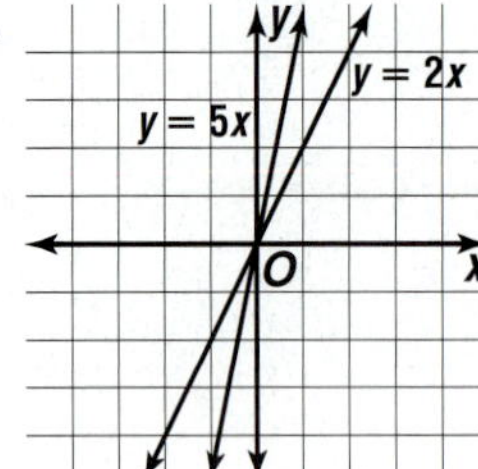

same y-intercept, different slopes

13. same slope, different y-intercepts

15.

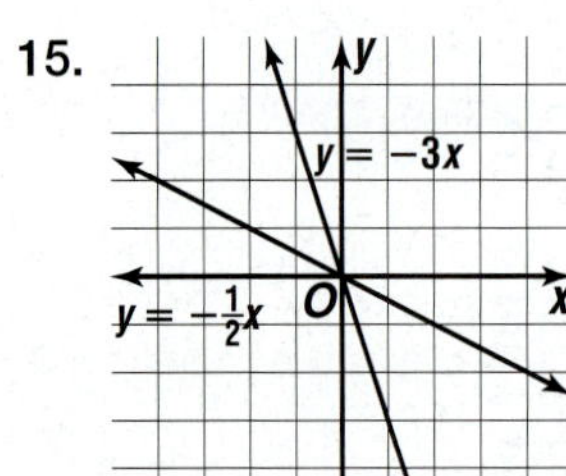

same y-intercept, different slopes

17. same y-intercept, different slopes **19.** same slope, different y-intercepts **21.** same y-intercept, different slopes **23.** $y = -\frac{4}{5}x$ **25.** $y = -\frac{4}{5}x + 2$ **27.** Sample answer: $y = -x + 6$ **29.** $y = 2x + 1$; Sample answer: $y = 2x$ **31a.** Yes; they have the same y-intercept, 0. **31b.** miles per gallon **31c.** A barge gets 500 miles per gallon of fuel; its graph has the steepest slope. **33.** Yes; they have the same y-intercept, 7. **35.** $m = -2, b = 0$ **37.** 48 **39.** C

Page 321 Quiz 2

1. Positive; as air temperature increases, the temperature of a glass of water would increase.
3. 4, 1

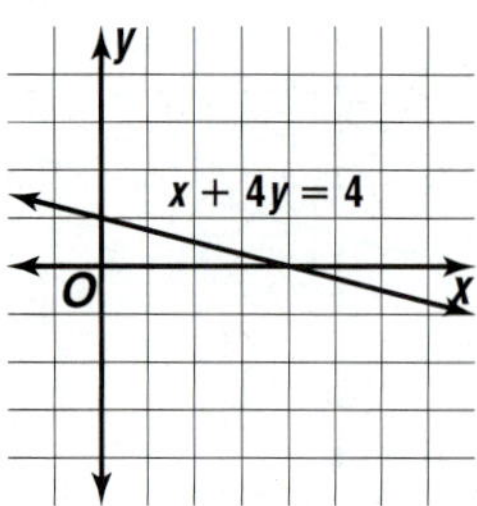

5a. Different slopes; they are a family because they have the same y-intercept, 0. **5b.** cost per movie **5c.** Store B

Pages 325–327 Lesson 7–7

1. Parallel lines have equal slopes. Perpendicular lines have slopes that are negative reciprocals. **3.** $4, -\frac{1}{4}$ **5.** $\frac{2}{3}, -\frac{3}{2}$ **7.** neither **9.** The lines are perpendicular. **11.** $y = \frac{3}{4}x - \frac{3}{2}$ **13.** $y = \frac{8}{3}x + 4$ **15.** perpendicular **17.** neither **19.** perpendicular **21.** parallel **23.** perpendicular **25.** $y = -\frac{1}{2}x$ **27.** $y = -4x - 7$ **29.** $x = 2$ **31.** $y = \frac{1}{4}x + \frac{3}{4}$ **33.** $y = -\frac{1}{2}x + 1$ **35.** $y = -\frac{5}{3}x - \frac{5}{3}$ **37.** neither **39.** Yes; the product of their slopes is −1. **41.** 4 **43.** same y-intercept, different slopes **45.** mutually exclusive **47.** mutually exclusive **49.** 12.85

Pages 328–330 Chapter 7 Study Guide and Assessment

1. false; y-intercept **3.** false; parallel **5.** true **7.** false; horizontal **9.** true **11.** $\frac{2}{3}$ **13.** −2 **15.** $y - 5 = 3(x - 2)$ **17.** $y - 3 = -5(x)$ **19.** $y - 4 = -\frac{1}{6}(x - 10)$ or $y - 7 = -\frac{1}{6}(x + 8)$ **21.** $y = 3x - 9$ **23.** $y = 2x$ **25.** $y = 11x - 6$ **27.** no relationship

29. 2, −4

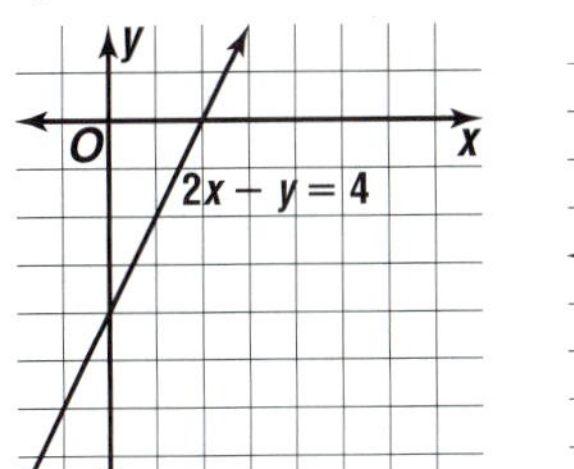

31. $m = -2, b = 3$

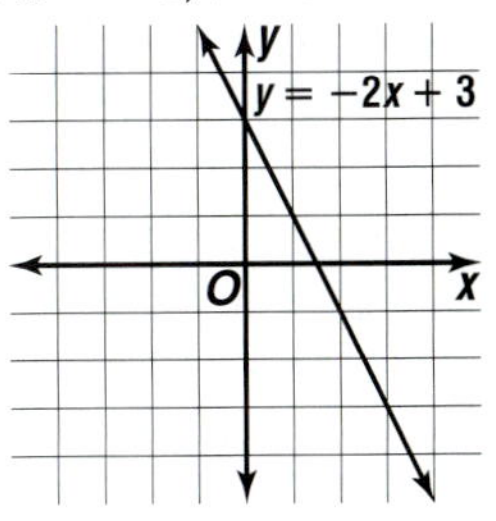

33. same y-intercept, different slope **35.** same slope, different y-intercepts **37.** neither **39.** parallel

41a.

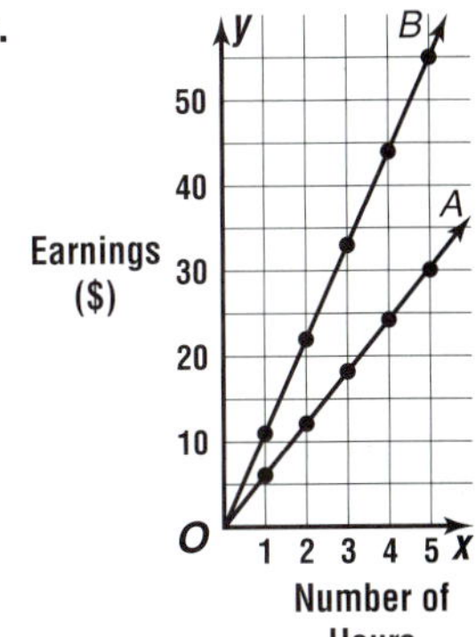

41b. It would be steeper than graph A and less steep than graph B.

Page 333 Preparing for Standardized Tests
1. B **3.** E **5.** C **7.** A **9.** 4

Chapter 8 Powers and Roots

Page 335 Check Your Readiness
1. 36 **3.** 9 **5.** 625 **7.** 225 **9.** −48 **11.** $\frac{1}{16}$ **13.** 6 **15.** $\frac{1}{8}$ **17.** 3 **19.** 7.5 **21.** 6 **23.** 30.08 **25.** 4.5 **27.** 4.6

Pages 339–340 Lesson 8–1
1. A perfect square is the product of a number and itself. **3.** Becky; $(6n)(6n)(6n) = 6 \cdot 6 \cdot 6 \cdot n \cdot n \cdot n$, not $6n^3$. **5.** a^5 **7.** $12 \cdot 12 \cdot 12 \cdot 12$ **9.** $m \cdot m \cdot m \cdot m \cdot n \cdot n \cdot n$ **11.** 162 **13.** $2^3 3^2 5$ **15.** $(-2)^4$ **17.** 7^3 **19.** $2^2 3^3 5$ **21.** $3x^2y^2$ **23.** $3 \cdot 3 \cdot 3 \cdot 3 \cdot 3$ **25.** $2 \cdot 2 \cdot 2 \cdot 2 \cdot 3 \cdot 3$ **27.** $y \cdot y \cdot y$ **29.** $6 \cdot a \cdot b \cdot b \cdot b \cdot b$ **31.** −32 **33.** −128 **35.** 24 **37.** 0.5 **39.** 4 **41.** 486 **43a.** 200.96 ft^2 **43b.** 21 bags **45.** $y = -2x + 6$

47.

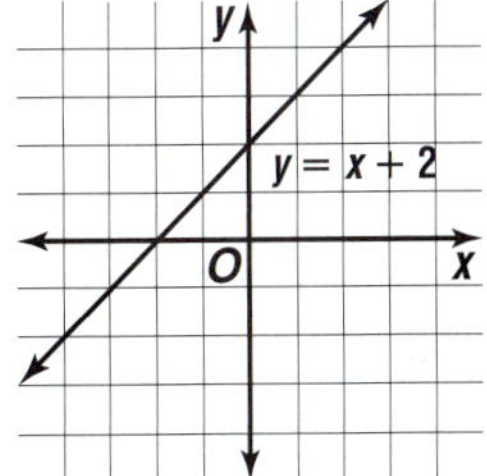

49. 31.2 **51.** 50% **53.** B

Pages 344–345 Lesson 8–2
1. In the first case, the bases are the same, but in the second case, the bases are different. **3.** y^7 **5.** t^5 **7.** $12a^5$ **9.** n^3 **11.** xy^3 **13.** b^5 **15.** $10^3 = 1000$ **17.** 5^4 **19.** d^6 **21.** a^3b^5 **23.** r^7t^8 **25.** $15a^2$ **27.** $8x^3y^5$ **29.** m^7b^2 **31.** $m^3n^3a^2$ **33.** 9 **35.** k^5 **37.** $-12ab^4$ **39.** 4 **41.** $15a^3c$ **43.** $2x^3$ **45.** $-16x^2$ **47.** 10^6 or 1,000,000 times **49a.** 10^{12} **49b.** 4^{15} **49c.** x^8 **49d.** Multiply the exponents; $(x^a)^b = x^{ab}$. **51.** 3 **53.** 15 **55.** 9.5 **57.** 3 **59.** D

Pages 349–351 Lesson 8–3
1. $\frac{1}{5^2} = \frac{1}{25}$ **3a.** Booker **3b.** Antonio **5.** $\frac{1}{3^3} = \frac{1}{27}$ **7.** $\frac{1}{n^3}$ **9.** $\frac{r^2}{pq^2}$ **11.** a^7 **13.** $\frac{c^3}{9a}$ **15.** $\frac{1}{2^5} = \frac{1}{32}$ **17.** $\frac{1}{4}$ **19.** $\frac{1}{r^{10}}$ **21.** a^2 **23.** $\frac{t^4}{s^2}$ **25.** $\frac{15r}{s^2}$ **27.** m^6 **29.** $\frac{1}{k^8}$ **31.** $\frac{a}{n^2}$ **33.** $\frac{1}{y^9}$ **35.** $3b^9$ **37.** $\frac{x^3}{4}$ **39.** $\frac{x^2}{7}$ **41.** $\frac{5}{8b^5c}$ **43.** 18 **45.** 2^{-4} **47.** radar microwaves **49.** $-10a^7$ **51.** n^6 **53.** z^4 **55.** 9^1 **57.** B

Page 351 Quiz 1
1. 6^3 **3.** 10^1 **5.** $2^5 3$ **7.** x^5 **9.** $-\frac{1}{3b^3}$

Pages 355–356 Lesson 8–4
1. no, 2.35×10^4 **5.** 68,000 **7.** 0.64 **9.** 0.0003 **11.** 0.0065 **13.** 5.69×10 **15.** 2.3×10^{-5} **17.** $2.6 \times 10^3 = 2600$ **19.** 5,800,000,000 **21.** 0.0000000039 **23.** 0.009 **25.** 5.28×10^3 **27.** 2.683×10^2 **29.** 3.2×10^{-4} **31.** 4.296×10^{-3} **33.** 1.2×10^{-2} **35.** 2.2×10^{-8} **37.** $6 \times 10^6 =$ 6,000,000 **39.** $7.5 \times 10^2 = 750$ **41.** $2 \times 10^3 = 2000$ **43.** $5.5 \times 10^{-5} = 0.000055$ **45.** 5,000,000 **47.** Earth, Venus, Mars **49.** between 0.3 and 2 mm **51.** y^3 **53.** $-6c^6$ **55.** x^6 **57.** a^3b^3 **59.** $-6x^2y^6$ **61.** C

Pages 360–361 Lesson 8–5
1. 1, 4, 9, 16, 25, 36, 49, 64, 81, 100 **3.** $3^2 = 9$ and $\sqrt{9} = 3$ **5.** 11^2 **7.** $2^2 \cdot 7^2$ **9.** 2 **11.** −11 **13.** $\frac{1}{2}$ **15.** 52.5 mi **17.** 10 **19.** −14 **21.** 23 **23.** −26 **25.** −22 **27.** 17 **29.** $-\frac{3}{10}$ **31.** $-\frac{1}{4}$ **33.** $\frac{6}{7}$ **35.** $\frac{3}{4}$ **37.** −0.05 **39.** 0.03 **41.** 6 **43.** 52 m **45.** false; $-6(-6) \neq -36$ **47.** 6.3×10^4 **49.** 7.6×10^{-4} **51.** $\frac{1}{x^4}$ **53.** $\frac{1}{x^2y^3}$ **55.** Sample answer: $2x + 6 = x + 1$

Page 361 Quiz 2
1. 7×10^5 **3.** 21 **5.** square house, 30 ft × 30 ft

Pages 364–365 Lesson 8–6
1. Its decimal value does not terminate or repeat. **3.** Sample answer: The area of a square garden is 200 square feet. Estimate the length of one side of the garden. **5.** 81, 100 **7.** 484, 529 **9.** 8 **11.** 16

13. 2 **15.** 4 **17.** 6 **19.** 11 **21.** 20 **23.** 24 **25.** 8 **27.** 11 **29.** 12 **31.** 0 **33.** $\sqrt{34}$ **35.** 17 h **37.** 31 mi^2 **39.** 10, 11, 12, 13, 14, 15 **41.** 16 **43.** $\frac{13}{11}$ **45.** 0.000000004 **47.** $y = 5x - 17$ **49.** $y = -5x + 29$ **51.** C

Pages 369–371 Lesson 8–7

1. Sample answer:

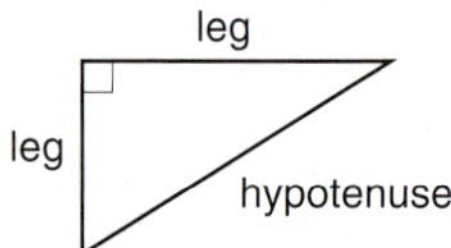

3. false **5.** false **7.** 14.1 **9.** 8.1 **11.** yes **13.** 13 **15.** 8.2 **17.** 4.2 **19.** 6.7 **21.** 21 **23.** 11.4 **25.** no **27.** no **29.** yes **31.** yes **33.** 9.4 m **35.** 6.9 ft **37.** 8 **39.** 14 **41.** 15 **43.** 10 **45a.** \$100 **45b.** \$5 **47.** C

Pages 374–376 Chapter 8 Study Guide and Assessment

1. prime factorization **3.** irrational numbers **5.** exponent **7.** hypotenuse **9.** perfect squares **11.** 9^5 **13.** $2^2 3^3$ **15.** $12 \cdot 12 \cdot 12$ **17.** $y \cdot y$ **19.** $5 \cdot x \cdot x \cdot y \cdot y \cdot y \cdot z \cdot z$ **21.** b^7 **23.** a^4b^3 **25.** y^4 **27.** $3x^3$ **29.** $\frac{1}{2^3} = \frac{1}{8}$ **31.** $\frac{1}{x^4}$ **33.** y^7 **35.** $-5b^3$ **37.** 0.0000000015 **39.** 2.4×10^5 **41.** 4.88×10^9 **43.** 6×10^{11} **45.** 11 **47.** $\frac{2}{9}$ **49.** 4 **51.** 14 **53.** 7 **55.** 11.2 **57.** 10.4 **59.** 9 m

Page 379 Preparing for Standardized Tests

1. C **3.** B **5.** C **7.** D **9.** 25

Chapter 9 Polynomials

Page 381 Check Your Readiness

1. $7a$ **3.** $16g$ **5.** $8t$ **7.** $3m + 6$ **9.** $3p + 3q$ **11.** $ab + ac$ **13.** $6a + b$ **15.** $3y + 3$ **17.** n^6 **19.** $15d^2$ **21.** $12a^3$ **23.** $12jk^2$ **25.** $48nm^2$ **27.** $24g^4h^4$

Pages 385–387 Lesson 9–1

1. Sample answer: $\frac{2}{x}$; it includes division. **3.** $2y^2 + 5xy + 3x^2$ **5.** yes; product of numbers and variables **7.** yes; binomial **9.** no **11.** 1 **13.** 2 **15a.** 20 **15b.** 9 **17.** no; includes addition **19.** yes; product of numbers and variables **21.** yes; product of numbers and variables **23.** yes; trinomial **25.** no **27.** yes; binomial **29.** yes; binomial **31.** yes; monomial **33.** yes; monomial **35.** 2 **37.** 2 **39.** 1 **41.** 3 **43.** 6 **45.** 5 **47.** $-x^5 + x^2 - x + 25$ **49.** $5x^7 - 10x^6 + 3wx^2 + 6w^3x$ **51.** $7 + 2x - x^2 + 5x^3$ **53.** $y^6 + 5xy^4 - x^2y^3 + 3x^3y$ **55.** false **57a.** $-0.006t^4 + 0.14t^3 - 0.53t^2 + 1.79t$ **57b.** 4 **59.** yes **61.** yes **63.** 7 **65.** 11 **67.** $\frac{x}{y^3}$ **69.** B

Pages 392–393 Lesson 9–2

1. Arrange like terms in column form. **3.** $-2a - 9b$ **5.** $-x^2 - 8x - 5$ **7.** $-4xy^2 - 6x^2y + y^3$ **9.** $8y + 2$ **11.** $3x^2 - x + 2$ **13.** $2x + 2$ **15.** $x^2 + 3x - 3$ **17.** $x^2 - x + 5$ **19.** $4x^2 + 3x + 3$ **21.** $14x + y$ **23.** $3n^2 + 13n + 11$ **25.** $7n^2 - 8n + 5$ **27.** $5x + 8$ **29.** $-x^2 + 3x - 3$ **31.** $-3x + y$ **33.** $a^2 + 3a - 1$ **35.** $5x^2 - xy - 2y^2$ **37.** $2pq + pr$ **39.** $-2x^2y - 7x^2y^2$ **41.** $-2x^2 + 6x - 7$ **43a.** $2w + 4$ in. **43b.** 34 in. **45.** 2 **47.** 3 **49.** x^5y^3 **51.** $10xy^2$

Pages 396–398 Lesson 9–3

1. Distributive Property **3.** Consuelo; Shawn forgot to multiply $2x$ and 4. **5.** $x^2 + 2x$ **7.** $-4x + 8$ **9.** $4z^3 - 8z^2$ **11.** $24y^2 + 9y - 15$ **13.** 4 **15.** 2 **17.** $3x^2$ **19.** $-3y - 9$ **21.** $x^2 - 5x$ **23.** $3z^2 - 2z$ **25.** $8x^2 - 12x$ **27.** $-2a^2 + 4a$ **29.** $-30y + 10y^2$ **31.** $5d^3 + 15d$ **33.** $-10a^3 + 14a^2 - 4a$ **35.** $40n^4 + 35n^3 - 15n^2$ **37.** $-14a^2 + 35a - 77$ **39.** $1.2c^2 - 12$ **41.** $x^2 - 2x$ **43.** $4y^2 - 6y + 2$ **45.** 5 **47.** -2 **49.** -1 **51.** 2 **53.** -3 **55.** 17 **57.** $4x^3 - 8x^2 + 4x$ **59.** $-4a^3 - 10a^2 - 10a + 66$ **61.** $21n^3 - 6n^2 - 46n + 28$ **63.** $8t^2 + t$ **65a.** $7y^2 + 35y$ units^2 **65b.** 1050 in^2 **67.** $7x - 2$ **69.** $3x^2 - 2x - 3$ **71.** 5 **73.** -11 **75.** $\frac{3}{10}$ **77.** 1:1

Page 398 Quiz 1

1. 4 **3.** $-2x^2 + 8x - 2$ **5a.** x ft, $2x + 40$ ft **5b.** $2x^2 + 40x$ ft^2

Pages 402–404 Lesson 9–4

1.

x	3
x^2	$3x$
$-2x$	-6

$(x - 2)(x + 3) = x^2 + x - 6$

3. The Distributive Property is used to multiply both two binomials and a binomial and a monomial. But when you multiply two binomials, there are four multiplications to perform. With a binomial and a monomial there are only two multiplications to perform. **5.** $6x$ **7.** $13m$ **9.** $14y$ **11.** $w^2 - 2w - 35$ **13.** $2y^2 + y - 3$ **15.** $6y^2 + 11y - 35$ **17.** $2a^2 - ab - 10b^2$ **19.** $m^4 + 3m^3 - 2m^2 - 6m$ **21.** $x^2 + 12x + 32$ **23.** $a^2 + 4a - 21$ **25.** $n^2 - 16n + 55$ **27.** $z^2 + 2z - 24$ **29.** $3x^2 - 10x - 8$ **31.** $3z^2 - 17z + 20$ **33.** $5n^2 - 13n - 6$ **35.** $9a^2 + 6a + 1$ **37.** $12h^2 - h - 6$ **39.** $2x^2 + x - 21$ **41.** $6y^2 - 24yz + 24z^2$ **43.** $6x^2 + 13xy - 5y^2$ **45.** $x^3 - 2x^2 + x - 2$ **47.** $3x^4 + 8x^2 - 3$ **49.** $x^3 - x^2 - 6x$ **51.** $x^2 - 9$ **53a.** $3x + 2$ **53b.** $3x^2 + 4x - 1$ **55a.** $4x^3 + 2x^2 - 20x$ in. **55b.** 450 in^3 **57a.** $10 - 2t$ **57b.** 6 in. **59.** 2^3x^3 **61.** $3^2(-2)^3$

63. 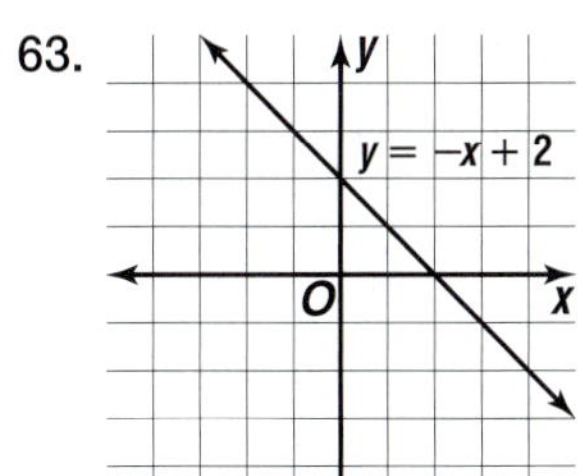

65. 84 students

Pages 408–409 Lesson 9–5
1a. ii **1b.** iii **1c.** i **3.** $y^2 + 4y + 4$ **5.** $m^2 - 18m + 81$ **7.** $x^2 - 49$ **9.** 90.25% **11.** $x^2 + 10x + 25$ **13.** $w^2 - 16w + 64$ **15.** $m^2 - 6mn + 9n^2$ **17.** $x^2 - 16y^2$ **19.** $25 + 10k + k^2$ **21.** $y^2 - 9z^2$ **23.** $9x^2 - 25$ **25.** $16 + 16x + 4x^2$ **27.** $a^2 - 6ab + 9b^2$ **29.** $x^3 + 2x^2 + x$ **31.** $x^2 - 4y^2$ **33.** $(40 - 1)(40 + 1) = 1600 - 1$ or 1599 **35a.** $\frac{1}{2}x^2 + \left(-\frac{9}{2}\right)$ **35b.** 8 square units **35c.** about 9.5 units **37.** $3x^2 - 4x - 3$ **39.** $y = 3x + 4$ **41.** $y = \frac{2}{5}x - 3$ **43.** B

Page 409 Quiz 2
1. $c^2 + 10c + 16$ **3.** $c^2 - 2c + 1$ **5.** $12x$ cm^2

Pages 412–414 Chapter 9 Study Guide and Assessment
1. i **3.** j **5.** c **7.** a **9.** e **11.** yes; monomial **13.** yes; trinomial **15.** 0 **17.** 4 **19.** $7cd - d^2 + 3d^3$ **21.** $9x^2 - 3x + 6$ **23.** $7s^2 - 6s - 2$ **25.** $4g^2 + 8g$ **27.** $-10s^2t^2 + 2st^2$ **29.** $3x - 15$ **31.** $2m^3 + 8m^2$ **33.** $10x^2 + 15x - 10$ **35.** −3 **37.** $y^2 + 4y - 12$ **39.** $x^2 - 4x + 3$ **41.** $2m^2 - 9m + 4$ **43.** $12y^2 + 10y + 2$ **45.** $10x^2 - 7x - 12$ **47.** $4x^2 + 12x + 9$ **49.** $x^2 - 4x + 4$ **51.** $25m^2 - 30mn + 9n^2$ **53.** $4a^2 - 9$ **55.** $25a^2 + 10a + 1$ **57.** $10\frac{1}{2}x^2 + 22x - 16$ **59a.** $64x^2 - 32x + 4$ **59b.** $(48x - 12)(48x - 12)$

Page 417 Preparing for Standardized Tests
1. C **3.** D **5.** B **7.** A **9.** 265

Chapter 10 Factoring

Page 419 Check Your Readiness
1. 3^4 **3.** 10^5 **5.** $2^2 5^4$ **7.** p^3q^3 **9.** 5^2 **11.** k^9 **13.** $6m^2n$ **15.** $2x + 2y$ **17.** $48b + 16$ **19.** $14f - 8$

Pages 424–425 Lesson 10–1
1. 23, 29, 31, 37, 41, 43, 47 **3.** Write $8x^2$ and $16x$ as the product of prime numbers and variables where no variable has an exponent. Then multiply the common prime factors and variables. The GCF is $8x$. **5.** $3 \cdot 7$ **7.** $3 \cdot 17$ **9.** $2^2 \cdot 3^3$ **11.** 1, 2, 3, 6, 7, 14, 21, 42; composite **13.** $2 \cdot 2 \cdot 2 \cdot 3 \cdot x \cdot x \cdot y$ **15.** 5 **17.** 1 **19.** $7y^2$ **21.** 72 cm, 1 cm **23.** 1, 2, 4, 5, 10, 20; composite **25.** 1, 3, 5, 9, 15, 45; composite **27.** 1, 7, 13, 91; composite **29.** $-1 \cdot 3 \cdot 5 \cdot a \cdot a \cdot b$ **31.** $2 \cdot 5 \cdot 5 \cdot m \cdot m \cdot n \cdot n$ **33.** $2 \cdot 3 \cdot 3 \cdot 5 \cdot y \cdot z \cdot z$ **35.** 4 **37.** 18 **39.** 9 **41.** $3y$ **43.** 9 **45.** 1 **47.** 1 **49.** 5 **51.** 5 **53.** 5 **55.** 12 in. by 12 in. **57.** Every even number has a factor of 2. So, any even number greater than 2 has at least 3 factors—the number itself, 1, and 2—and is therefore composite. **59.** $4y^2 + 4y + 1$ **61.** $z^2 + 7z + 12$ **63.** $2n^2 + 9n + 4$ **65.** $6a - 2a^3$ **67.** 8.17×10^8

Pages 431–433 Lesson 10–2
1. Sample answer:

$x + 2$

x	x^2	x	x

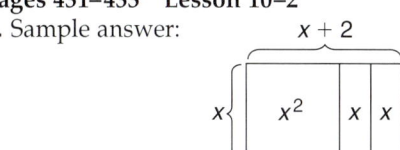

3. The Distributive Property is used to factor out a common factor from the terms of an expression. Examples: $5xy + 10x^2y^2 = 5xy(1 + 2xy)$, $2a^2 + 8a + 10 = 2(a^2 + 4a + 5)$ **5.** x **7.** $6y$ **9.** $5m^2n$ **11.** $2x(x + 2)$ **13.** prime **15.** $2ab(a^2b + 4 + 8ab^2)$ **17.** $c + 3$ **19.** $3(3x + 5)$ **21.** $2x(4 + xy)$ **23.** $3c^2d(1 - 2d)$ **25.** $mn(36 - 11n)$ **27.** prime **29.** $y^3(12x + y)$ **31.** $3y(8x + 6xy - 1)$ **33.** $x(1 + xy^3 + x^2y^2)$ **35.** $2a(6xy - 7y + 10x)$ **37.** $9x^2 - 7y^2$ **39.** $a + 2$ **41.** $2xy^2 + 3z$ **43.** $2x^2 + 3$ **45.** $(4x + y)$ ft **47a.** $8(a + b + 8)$ **47b.** $3(4a + b + 12)$ **49.** prime **51.** composite **53.** composite **55.** $3x^2 - 2x - 12$ **57.** Sample answer: $x^2 + 3x - 4$

Page 433 Quiz 1
1. $2^3 \cdot 3$ **3.** x^3; $x^3(a + 7b + 11c)$ **5.** $8(x + 1)$

Pages 438–439 Lesson 10–3
1.

	x	6
x	x^2	$6x$
1	$1x$	6

$(x + 1)(x + 6)$

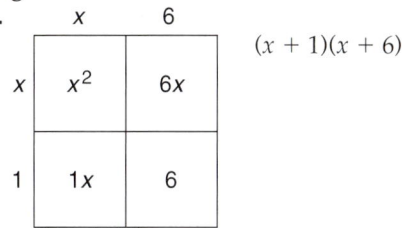

3. −10, −3 **5.** −4, −3 **7.** −6, 1 **9.** $(x + 3)(x + 2)$ **11.** $(a - 1)(a - 4)$ **13.** $(x + 5)(x - 2)$ **15.** prime **17.** $2(c - 7)(c + 1)$ **19.** $(b + 4)(b + 1)$ **21.** $(a + 3)(a + 4)$ **23.** $(y + 9)(y + 3)$ **25.** $(x - 5)(x - 3)$ **27.** $(c - 9)(c - 4)$ **29.** prime **31.** $(c + 3)(c - 1)$ **33.** $(r - 6)(r + 3)$ **35.** $(n + 15)(m - 2)$ **37.** $(x - 9)(x - 8)$ **39.** $(r + 24)(r - 2)$ **41.** $3(y - 4)(y - 3)$ **43.** $m(m + 1)(m + 2)$ **45.** $2a(a + 8)(a - 1)$ **47.** Sample answer: $x^2 + 7x - 9$ **49a.** $x^2 - 5x - 24$ **49b.** $(x - 8)(x + 3)$

51a.

	C	c
C	CC	Cc
c	Cc	cc

51b.

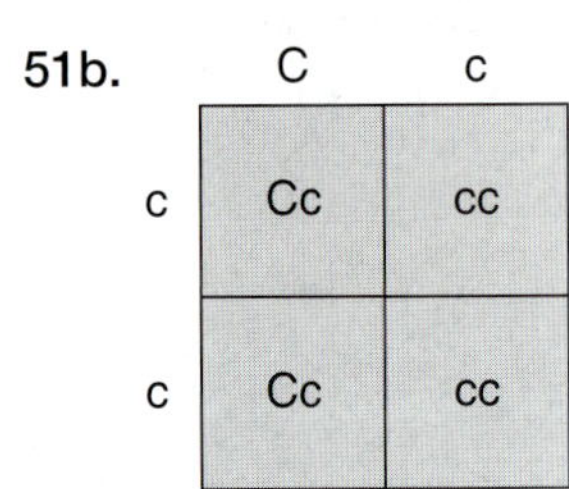

53. $2x^2 + 5y^2$ **55.** $3a^2 + 4ab - 3b^2$ **57.** 4 **59.** 1 **61a.** $y = 0.2x - 398$ **61b.** the average increase each year **61c.** \$3 billion

Pages 443–444 Lesson 10–4

1a. $2x^2 - 5x - 12 = (x - 4)(2x + 3)$ **1b.** $6x^2 - 11x + 3 = (2x - 3)(3x - 1)$ **1c.** $5x^2 + 28x + 15 = (5x + 3)(x + 5)$ **3.** $(2a + 3)(a + 1)$ **5.** $(5x + 3)(x + 2)$ **7.** $(2x + 7)(x - 3)$ **9.** $(5a - 2)(2a - 1)$ **11.** $2(3x + 5)(x + 1)$ **13.** $(2y + 1)(y + 3)$ **15.** $(2a + 1)(2a + 3)$ **17.** $(2q + 3)(q - 6)$ **19.** $(7a + 1)(a + 3)$ **21.** $(3x + 2)(x + 4)$ **23.** $(3x + 5)(x + 3)$ **25.** prime **27.** $(7x - 1)(2x + 5)$ **29.** $(2m - 1)(4m - 3)$ **31.** $3(2x + 5)(x - 2)$ **33.** $x(2x - 3)(x + 4)$ **35.** $(6y - 1)(y + 2)$ **37.** $(2a - b)(a + 3b)$ **39.** $(3k + 5m)(3k + 5m)$ **41.** $3x$, $2x - 1$, $x + 3$ **43a.** $2(y^2 - 7y - 2)$ **43b.** $2(y - 9)(y - 1)$ **43c.** 56 in^2, 18 in^3 **45.** $(x + 16)(x - 2)$ **47.** $3a(a^2 - 5a + 2)$ **49.** 25 ft **51.** 10

Page 444 Quiz 2

1. $(x + 5)(x - 2)$ **3.** $(2x + 7)(x + 1)$ **5.** $(x + 3)(x - 2)$

Pages 448–449 Lesson 10–5

1. No; 7 is not a perfect square. **3.** $(y + 7)^2$ **5.** $(a - 5)^2$ **7.** $2(2x - 5y)(2x + 5y)$ **9.** $3(x^2 + 5)$ **11.** $3(y - 1)(y + 8)$ **13.** $(r + 4)^2$ **15.** $(a + 1)^2$ **17.** $(2z - 5)^2$ **19.** $(3a + 4)^2$ **21.** $(7 + z)^2$ **23.** $(a - 6)(a + 6)$ **25.** $(1 - 3m)(1 + 3m)$ **27.** no **29.** $2(z - 7)(z + 7)$ **31.** Sample answer: $25x^2 - 4$; $(5x - 4)(5x + 4)$ **33.** $5(x^2 + 5)$ **35.** $(y - 3)(y - 2)$ **37.** $2(x - 6)(x + 6)$ **39.** $xy(8y - 13x)$ **41.** prime **43.** $2(5n + 1)(2n + 3)$ **45.** $(4w - 3)(2w + 5)$ **47.** $5(x + 1)(x + 2)$ **49.** $2x(x - 4)(x + 4)$ **51.** 2, 4 **53.** 4 **55.** $(4x - 1)(x + 3)$ **57.** $(m - 7)(m + 2)$ **59.** a^2 **61.** $\frac{d}{3c^2}$ **63.** C

Pages 450–452 Chapter 10 Study Guide and Assessment

1. false, $2^2 \cdot 3$ **3.** true **5.** true **7.** false, $x^2 - 9$ **9.** true **11.** 5 **13.** 1 **15.** $9x$ **17.** $5(x + 6y)$ **19.** $6a(2b - 3a)$ **21.** $3xy(1 + 4xy)$ **23.** $5ab - 1$ **25.** $(x - 5)(x - 3)$ **27.** $(x - 4)(x + 2)$ **29.** $(x + 7)(x - 5)$ **31.** $2(n - 6)(n + 2)$ **33.** $(3x + 5)(x + 1)$ **35.** $(3a - 2)(2a + 1)$ **37.** $(2y + 3)(y - 6)$ **39.** $5(3a - 1)(a - 1)$ **41.** $(a - 6)^2$ **43.** $(5x + 2)^2$ **45.** $(2x + 3)(2x - 3)$ **47.** $12(c + 1)(c - 1)$

49.

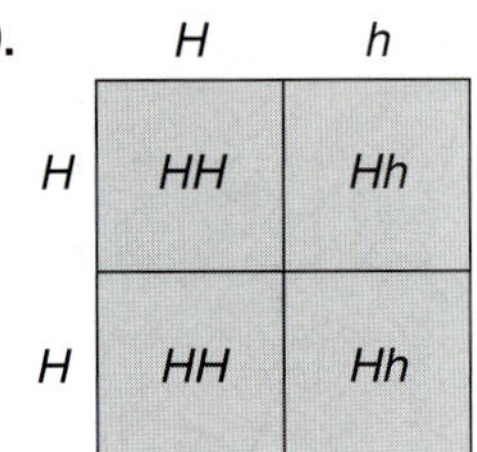

Page 455 Preparing for Standardized Tests

1. C **3.** D **5.** D **7.** B **9.** 0.5

Chapter 11 Quadratic and Exponential Functions

Page 457 Check Your Readiness

1.

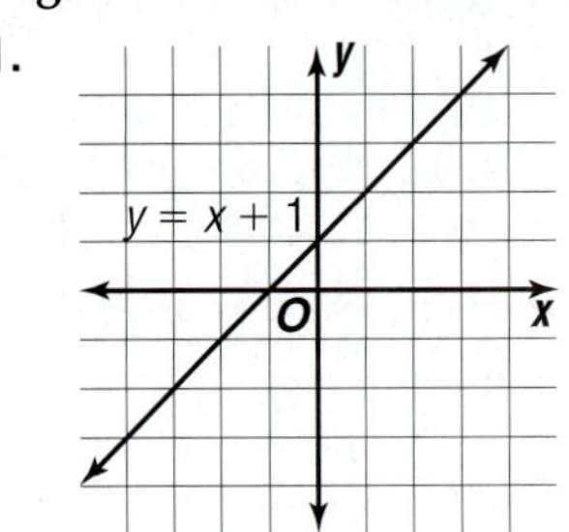

3.

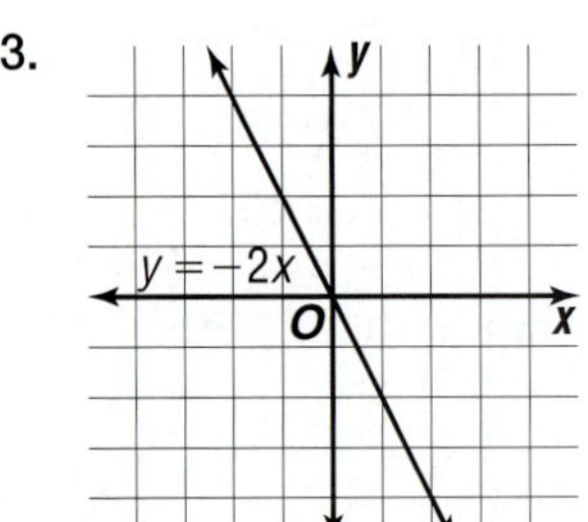

5.

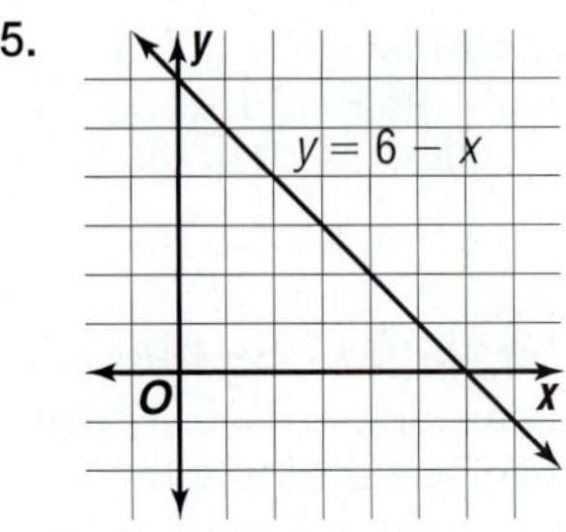

7. 18 **9.** 4 **11.** 2 **13.** $(x + 2)(x + 8)$ **15.** $(b - 5)(b + 3)$ **17.** $(2c - 1)(c + 3)$

Pages 461–463 Lesson 11–1

1. Sample answer: Quadratic functions have a degree of 2, but linear functions have a degree of 1. The graphs of quadratic functions are not straight lines as with linear functions. **3.** the vertex **5.** 1, 0, 4 **7.** 3, 4, 0

9.

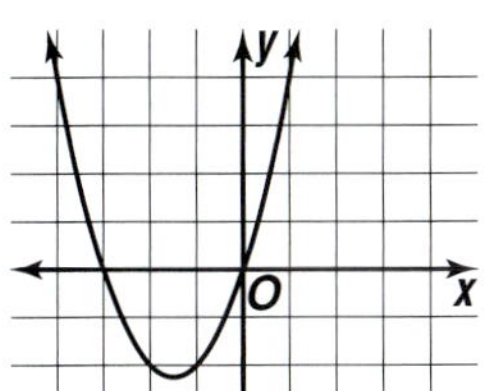

11. $x = 0$; (0, 2)

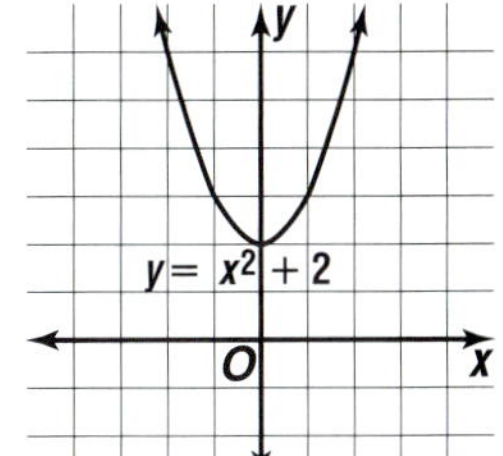

13. $x = 2$; (2, 7)

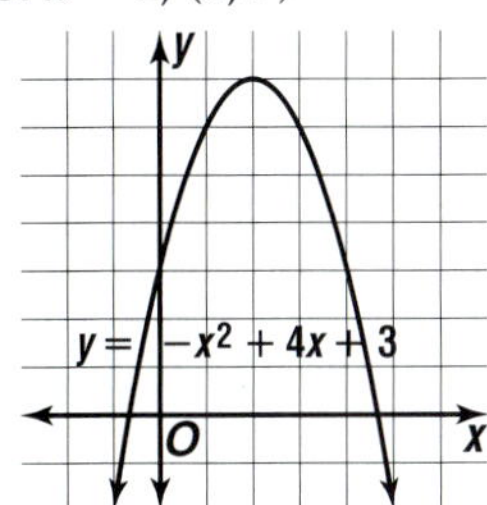

15a.

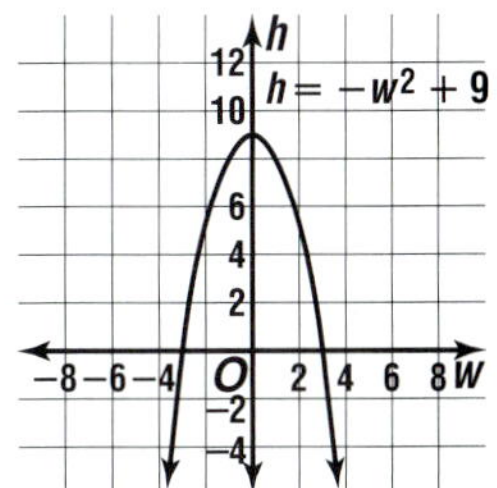

15b. 9 ft **15c.** 6 ft

17.

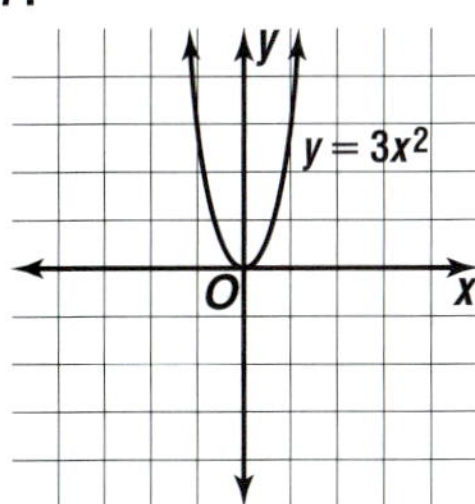

19.

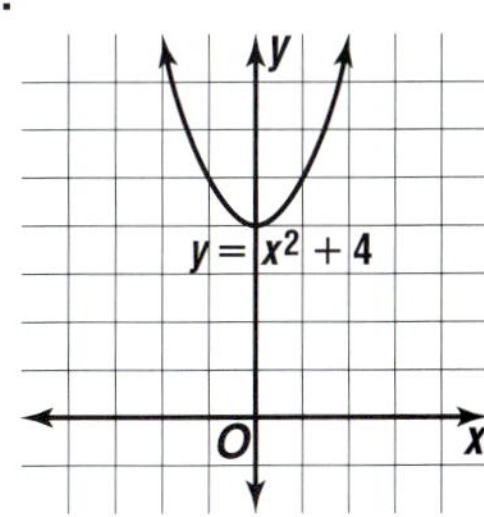

21.

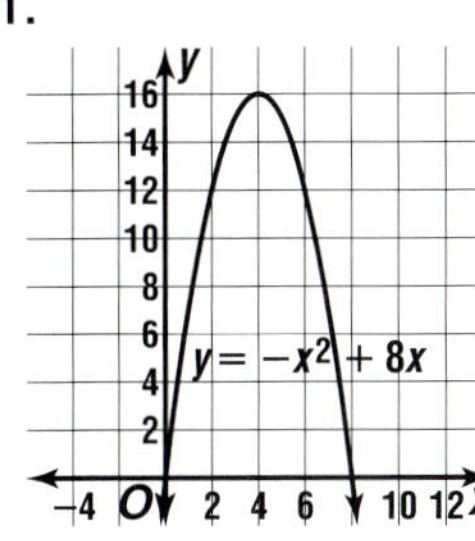

23.

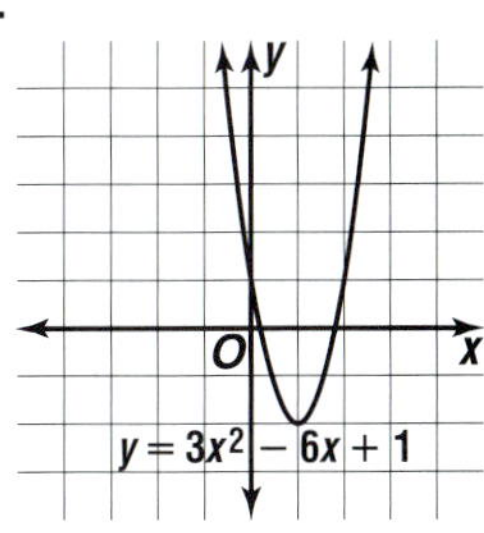

25.

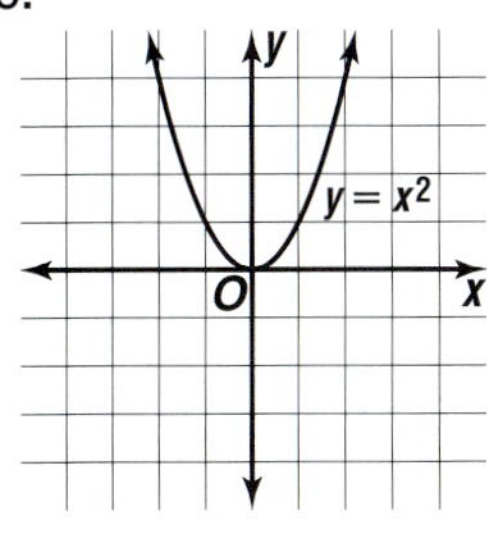

27.

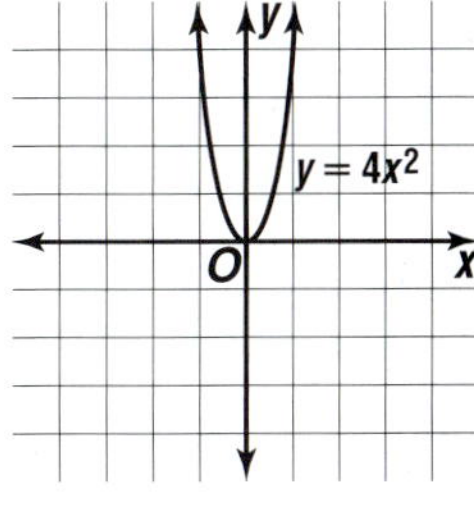

29.

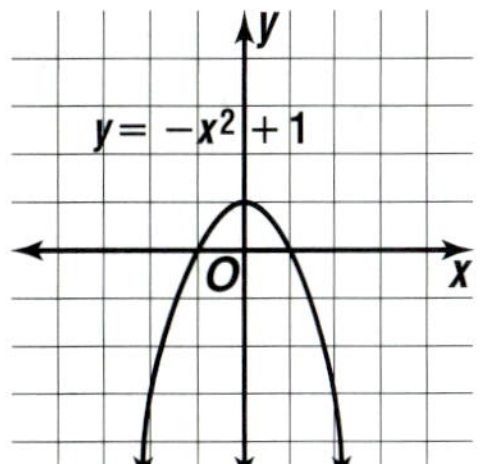

31.

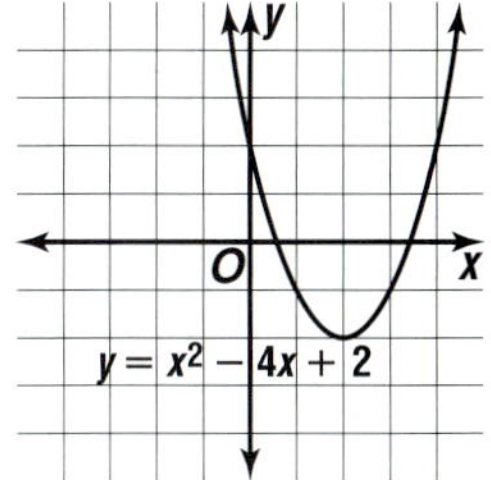

33.

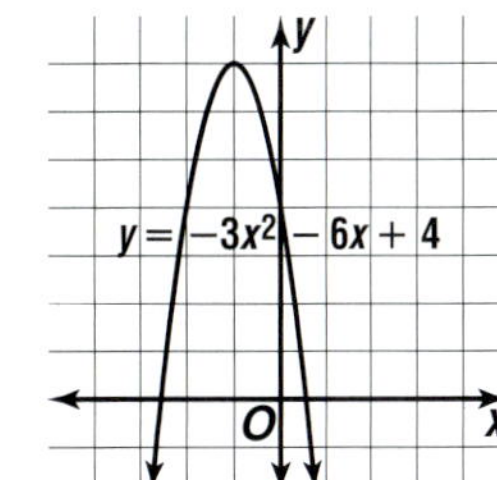

35.

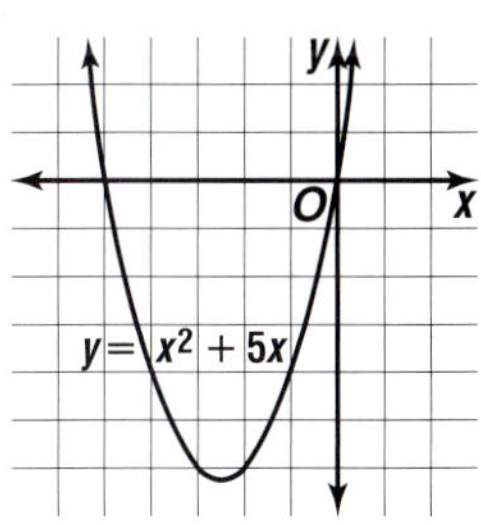

37. C **39.** B **41.** $x = -1$ **43a.** 150 **43b.** $22,500 **45.** $(x - 7)(x + 7)$ **47.** $2(g - 4)^2$ **49.** $-x^3 - 4x^2 - 8x + 5$; 3 **51.** 1.4 mi

Pages 466–467 Lesson 11–2

1. extremely narrow with vertex at (0, 0), opening up **3.** B, A, C **5.** widens; (0, 0) **7.** left 4 units; (−4, 0) **9.** $y = 0.5(x - 4)^2 + 3$ **11.** shifts left, opens up **12.** same axis of symmetry, shifts down, opens down **13.** narrows; (0, 0) **15.** right 7 units; (7, 0) **17.** opens down, narrows; (0, 0) **19.** narrows, up 1 unit; (0, 1) **21.** widens, down 8 units; (0, −8) **23.** left 1 unit, down 5 units; (−1, −5) **25.** $y = (x + 8)^2$ **27.** $h(t) = -4.9(t - 34)^2 + 80$ **29.** $y = (x - 2)^2 - 4$ **31.** (−1.5, −1.5) **33.** $a^2 + 3a - 28$ **35.** B

Pages 471–473 Lesson 11–3

1. Sample answer: The x-intercepts are the values for x where $f(x)$ equals 0. Before solving a quadratic function, you always set it equal to zero.
3. Sample answer:

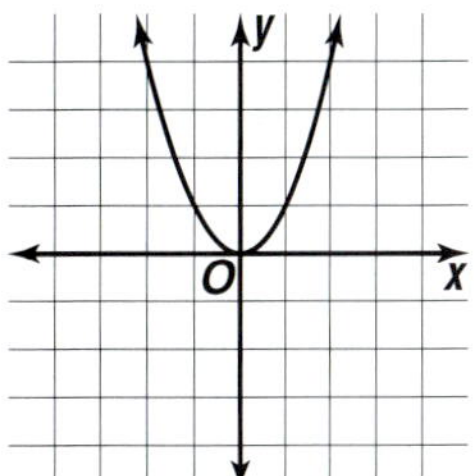

5. −1, 4 **7.** 2, 3 **9.** between −2 and −1; between 2 and 3 **11.** −1 **13.** −2, 7 **15.** 5 **17.** between −5 and −4; between 0 and 1 **19.** between −1 and 0; between 2 and 3 **21.** −4, −8 or 4, 8

23.

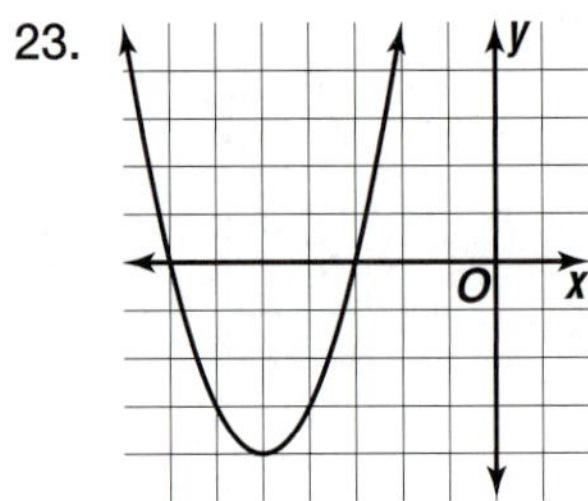

25. between 46 and 47 s **27.** The value of the function changes from negative when $x = 1$ to positive when $x = 2$. To do this, the value of the function would have to be 0 for some value of x between 1 and 2. Thus, the value of x represents the x-intercept of the function. This is the root of the related equation. **29.** up 4 units; (0, 4) **31.** 4

Page 473 Quiz 1

1. $x = -3$; $(-3, -14)$ **3.** $x = -1$; $(-1, 7)$

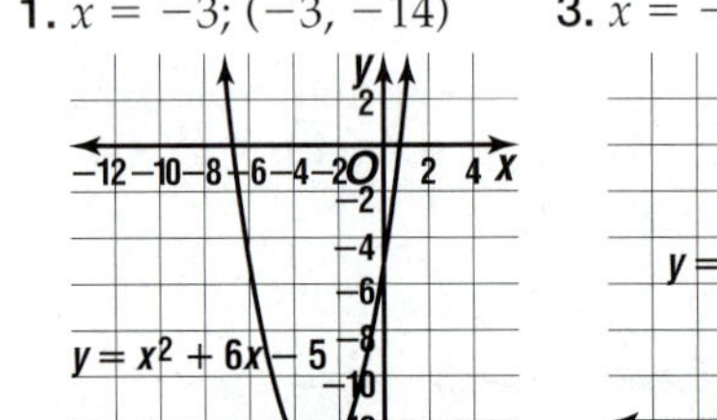

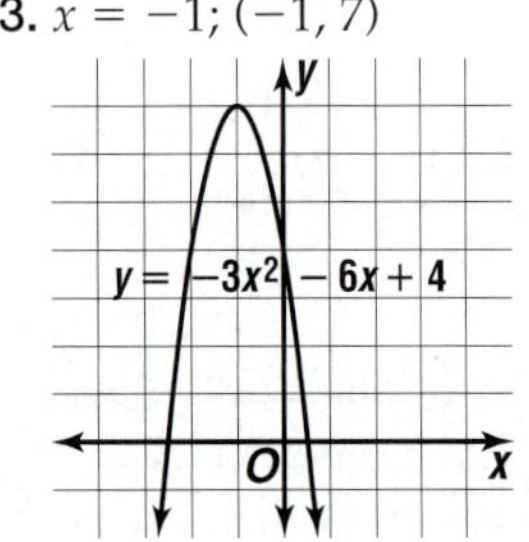

5. right 6 units; (6, 0) **7.** 2, 3 **9.** between −1 and 0; between 2 and 3

Pages 476–477 Lesson 11–4

1. Any number times 0 is 0. Therefore, if two or more numbers are multiplied and the result is 0, at least one of those numbers has to equal 0.
3. They are both correct, but factoring can only be used when the equation is not prime. **5.** 0, 5 **7.** 3 **9.** −2, 4 **11.** 0, −2 **13.** −7, 6 **15.** −1, 4 **17.** −8, −2 **19.** 3, 8 **21.** −3, 4 **23.** 0, 3 **25.** 2 **27.** −1, 5 **29.** 25 ft by 35 ft **31.** 8, 11; −11, −8 **33.** 3 ft by 1 ft by 1 ft **35a.** −5, 2 **35b.** Sample answer: $y = (x + 5)(x - 2)$ **37.** no real solution **39.** The graph of $y = 0.25x^2$ is wider than the graph of $y = x^2$. The graph of $y = 4x^2$ is narrower than the graph of $y = x^2$.

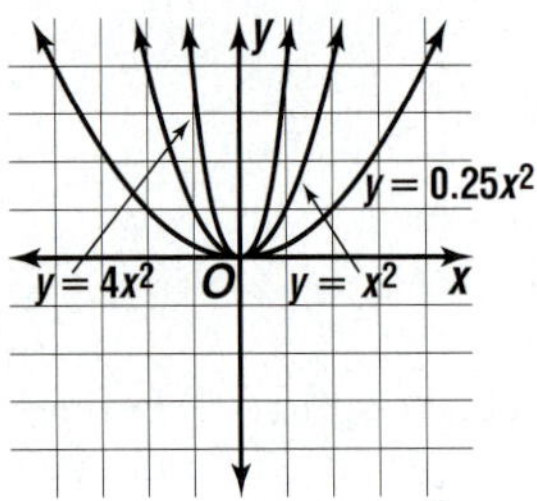

41. $(b + 7)(b - 6)$

Pages 481–482 Lesson 11–5

1. The equation is not factorable with integers and the coefficient of x^2 is 1. **3.** yes **5.** no **7.** 1 **9.** $\frac{9}{4}$ **11.** −3, −4 **13.** 9 ft by 11 ft **15.** 36 **17.** 1 **19.** 144 **21.** $\frac{25}{4}$ **23.** −3, 1 **25.** −5, −1 **27.** 0, 14 **29.** $1 \pm \sqrt{31}$ **31.** $3 \pm \sqrt{19}$ **33.** −5, 2 **35.** 3 s or 5 s **37.** −18 or 18 **39.** 2, 5 **41.** −3 **43a.** $2b(a + 4b)$ **43b.** $2a + 12b$ **43c.** 9 in. by 2 in., 18 in^2, 22 in.

Pages 486–487 Lesson 11–6

1. Sometimes; quadratic equations can have no real solutions. **3.** 48 **5.** 192 **7.** ±5 **9.** −3, −1 **11.** −6, −1 **13.** 2, 3 **15.** no real solutions **17.** 0, 8 **19.** $\frac{-1 \pm \sqrt{29}}{2}$ **21.** $\frac{-2 \pm \sqrt{7}}{2}$ **23.** $\frac{-7 \pm \sqrt{69}}{2}$ **25.** $\frac{5 \pm \sqrt{249}}{16}$ **27a.** 6.25 s **27b.** about 1.02 s **29.** 4 **31.** 81 **33.** B

Page 487 Quiz 2

1. −6, 2 **3.** 3, 7 **5.** $3 \pm \sqrt{2}$ **7.** no real solutions **9.** $\frac{-5 \pm \sqrt{13}}{2}$

Pages 491–493 Lesson 11–7

1. Sample answer: The graph begins almost flat, but then for increasing x-values, it becomes more and more steep. **3.** 256 **5.** 1.26 **7.** −5

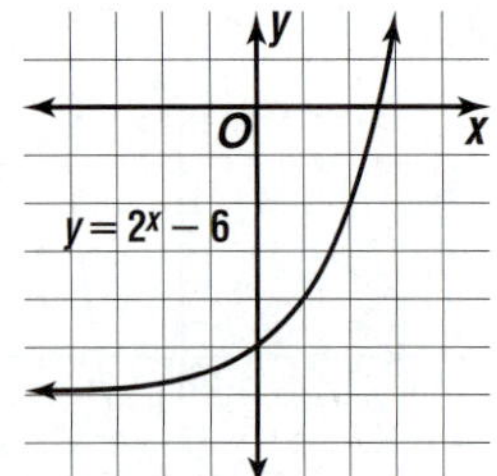

9a.

Year	Balance	Pattern
0	500	$500(1.02)^0$
1	500(1.02)1.02 = 510	$500(1.02)^1$
2	(500 · 1.02)1.02 = 520.20	$500(1.02)^2$
3	(500 · 1.02 · 102)1.02 = 530.604	$500(1.02)^3$

9b. $T(x) = 500(1.02)^x$ **9c.** \$714.12 **9d.** \$1428.25

11. 1

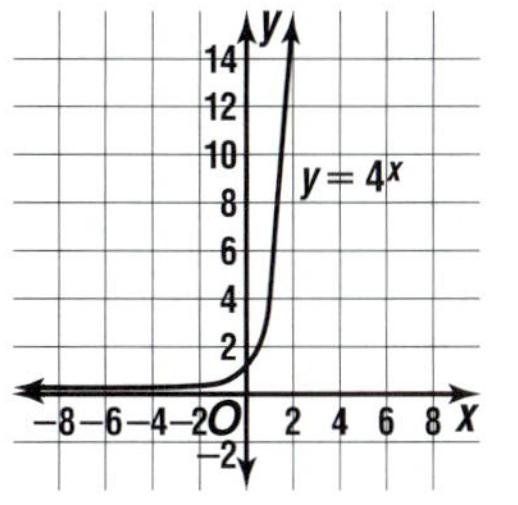

13. 2

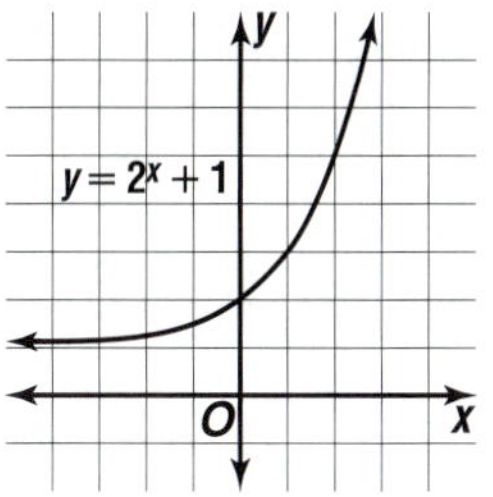

15. −2

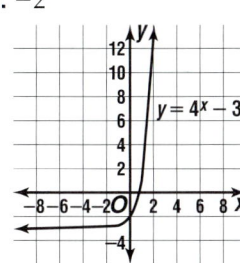

17. 0

19. \$5304.50 **21.** \$3920.88 **23.** \$7986.70

25a.

Day	Cents	Pattern
1	1	2^0
2	$2 \cdot 1$	2^1
3	$2 \cdot 2 \cdot 1$	2^2
4	$2 \cdot 2 \cdot 2 \cdot 1$	2^3
5	$2 \cdot 2 \cdot 2 \cdot 2 \cdot 1$	2^4

25b. $y = 2^{x-1}$ **25c.** January 17 **27.** −2, 2
29a. $(2x + 4)x = 30$ **29b.** 3 yd by 10 yd
31. $y - 6 = 0$

Pages 496–498 Chapter 11 Study Guide and Assessment

1. f **3.** c **5.** j **7.** a **9.** d **11.** $x = 0$; (0, −3)
13. $x = -1$; (−1, 20)

15.

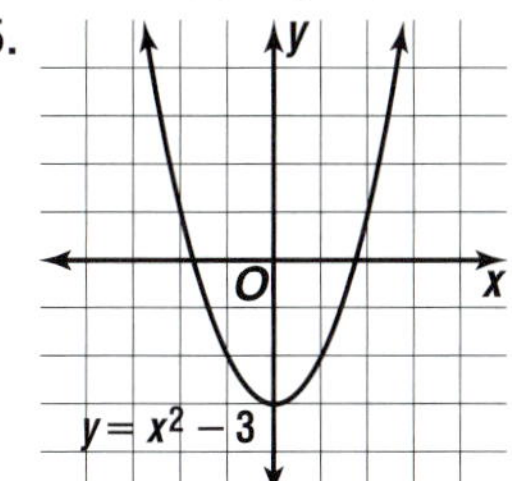

17.

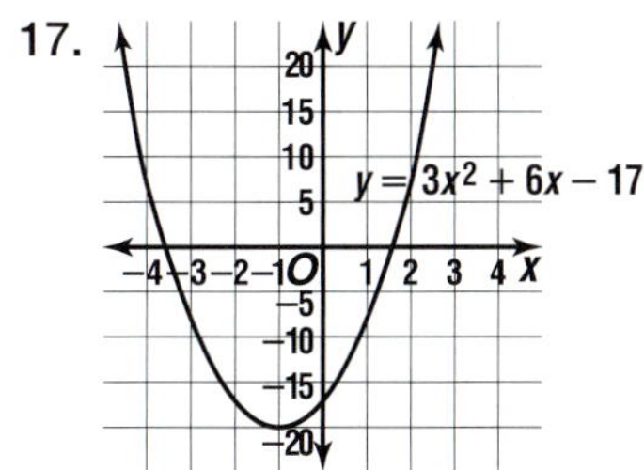

19. shifts right; (8, 0) **21.** widens; (0, 0)
23. shifts right and up; (1, 7) **25.** −3, 4
27. no real solutions **29.** $-1 < x < 0$, 3
31. −4, 5 **33.** −8 **35.** −1, 1 **37.** 4 **39.** −3, 7
41. −1, −9 **43.** $-\frac{1}{3}$, 1 **45.** $\frac{-5 \pm \sqrt{13}}{2}$

47. 1 **49.** 4

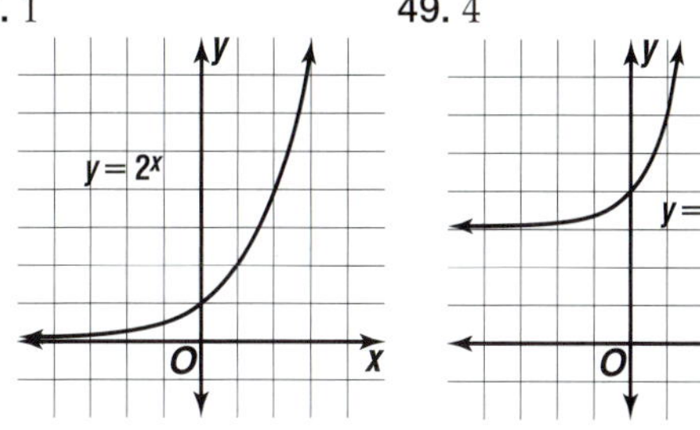

51. \$5.50

Page 501 Preparing for Standardized Tests

1. D **3.** A **5.** C **7.** C **9.** 2

Chapter 12 Inequalities

Page 503 Check Your Readiness

1. < **3.** > **5.** <

7.

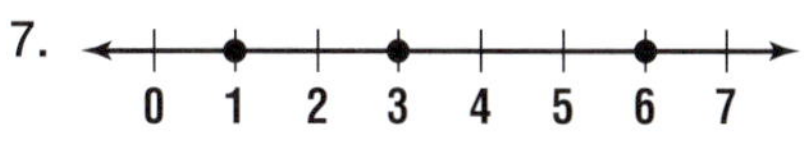

9.

11. 5 **13.** 10 **15.** 39 **17.** −6, 6 **19.** −12, 16
21. −10, 10 **23.** 9 **25.** 15 **27.** 7 **29.** 4 **31.** 21

Pages 506–508 Lesson 12–1

1. i. b; ii. a; iii. d; iv. c **3.** The graph of $x \leq 4$ has a bullet at 4. The graph of $x < 4$ has a circle at 4.
5. $x < 14$

7. 9.

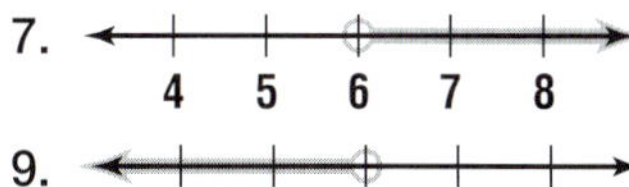

11. $x < -1$ **13.** $x > 4$ **15.** $x < 9$ **17.** $x \geq -8$

19.

21.

23.

25.

27.

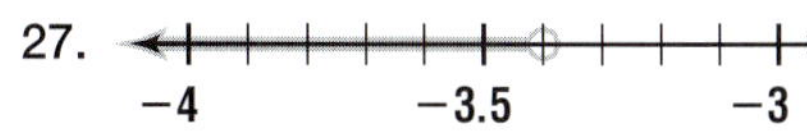

29.

31. $x > -6$ 33. $x \leq 3$ 35. $x \leq 5$ 37. $x \geq 2.5$

39. $f < 3$;

1 2 3 4 5

41. $w \leq 3600$;

3400 3600 3800

43. 7 yr 45. no real solutions 47. $2^3 \cdot 3$; two

49. $-6x^2y$

Pages 511–513 Lesson 12–2

1. They are the same. 5. $8 + n \leq 12$ 7. $\{v \mid v \geq 17\}$

9. $\{h \mid h \leq 21\}$ 11. $\left\{y \mid y < 1\frac{5}{6}\right\}$

13. $\{x \mid x < -1\}$;

−3 −2 −1 0 1

15. $\{n \mid n > 3\}$ 17. $\{b \mid b < -5\}$ 19. $\{r \mid r < 19\}$

21. $\{w \mid w > 22\}$ 23. $\{t \mid t \leq 0\}$ 25. $\{c \mid c > -4\}$

27. $\{d \mid d < 5.4\}$ 29. $\{z \mid z \leq -0.5\}$ 31. $\left\{v \mid v \leq 2\frac{1}{2}\right\}$

33. $\{g \mid g < 13\}$;

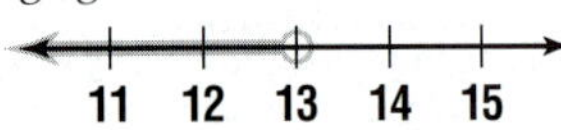

35. $\{x \mid x \geq -16\}$;

−18 −17 −16 −15 −14

37. $\{t \mid t \geq 9\}$;

7 8 9 10 11

39. $x - 8 \leq 12$; $x \leq 20$ 41. $t \geq 99$ 43. $p \geq 99$

45. Sometimes; examples are $x > x$ and $2x + 1 \leq 2x$.

47. $x < -8$ 49. $x \geq 5$ 51. $(3x + 2)(x - 5)$ 53. 32

Pages 516–518 Lesson 12–3

1. For both, reverse the sign when dividing or multiplying by a negative number.

3.

2 3 4 5 6

5. $\{x \mid x \geq -9\}$ 7. $\{x \mid x \geq 20\}$ 9. $\{a \mid a \leq -30\}$

11. $\{h \mid h \geq -2\}$ 13. $\{d \mid d \leq -3\}$ 15. $\{g \mid g > -36\}$

17. $\{x \mid x \geq 7\}$ 19. $\{z \mid z > -3\}$ 21. $\{c \mid c > -54\}$

23. $\left\{k \mid k > \frac{5}{3}\right\}$ 25. $\left\{w \mid w > \frac{1}{2}\right\}$ 27. $\{y \mid y \geq 6\}$

29. $\{v \mid v \geq 2\}$ 31. $\{s \mid s \leq 280\}$ 33. $\{n \mid n < 105\}$

35. $x < 45$ 37. at least 17 lawns 39. $\{z \mid z \leq 4\}$

41. $p > 600$ 43. $4v^2 - 1$ 45. $\frac{1}{400}$

Page 518 Quiz 1

1. $x < -1$

3. $\{x \mid x > 5\}$;

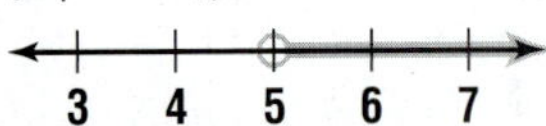

5. $\{z \mid z < -3\}$;

−5 −4 −3 −2 −1

7. $\{x \mid x > -5\}$ 9. $\left\{v \mid v \leq \frac{15}{2}\right\}$

Pages 522–523 Lesson 12–4

1. subtraction, then division 3. −7 5. −3.1

7. $\{y \mid y > 4\}$ 9. $\{x \mid x < -1\}$ 11. $\{w \mid w > -2\}$

13. $s \geq 92$ 15. $\{b \mid b < 12\}$ 17. $\{x \mid x \geq 4\}$

19. $\{g \mid g > 2\}$ 21. $\{v \mid v \leq 0.5\}$ 23. $\{m \mid m \leq -3\}$

25. $\{d \mid d < 1\}$ 27. $\{x \mid x \leq 2\}$ 29. $\{y \mid y < 1\}$

31. $\{b \mid b < 11\}$ 33. $\{x \mid x < 1\}$ 35. at least \$260

37. $x > \$1484$ 39. $\{p \mid p > 7\}$ 41. $\{x \mid x \leq -40\}$

43. $m \leq \$19.50$ 45. A

Pages 526–529 Lesson 12–5

1. an inequality made from two inequalities

3. intersection 5. union 7. $5 < \ell < 10$

9.

−2 −1 0 1 2 3 4 5 6 7 8

11. $\{c \mid 3 < c < 8\}$;

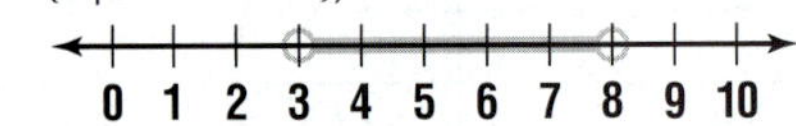

13. $\{v \mid 0 \leq v \leq 4\}$;

−3 −2 −1 0 1 2 3 4 5 6 7

15. $\{j \mid j \leq -1 \text{ or } j > 0\}$;

−5 −4 −3 −2 −1 0 1 2 3 4 5

17. $9 \leq t \leq 11.5$;

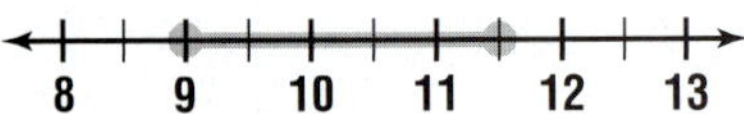

19. $-8 < h < 8$ 21. $2 \leq g \leq 5$ 23. $6 < x \leq 8$

25.

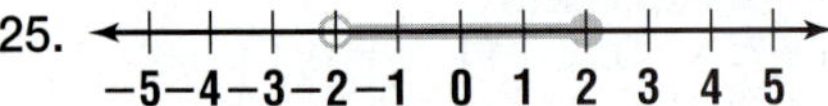

27.

13 14 15 16 17 18 19 20 21 22 23

29.

2 3 4 5 6 7 8 9 10 11 12

31. $\{x \mid 4 \leq x \leq 8\}$;

0 1 2 3 4 5 6 7 8 9 10

33. $\{w \mid 0 < w \leq 3\}$;

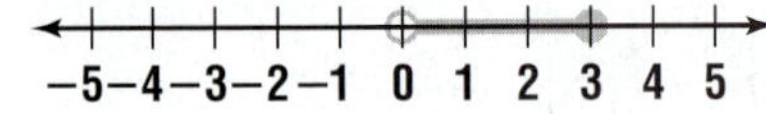

35. $\{h \mid -3 \leq h \leq 3\}$;

37. $\{c \mid -5 < c < -4\}$;

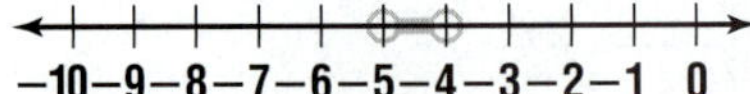

39. $\{z \mid z \leq -7 \text{ or } z > 4\}$;

−12 −10 −8 −6 −4 −2 0 2 4 6 8

41. $\{r \mid r \leq -1 \text{ or } r > 0\}$;

−5 −4 −3 −2 −1 0 1 2 3 4 5

43. $\{p \mid p \geq -1\}$;

−5 −4 −3 −2 −1 0 1 2 3 4 5

45. $\{d \mid d > 1.5\}$;

0 1 2 3 4 5

47. $\{w \mid w \leq -6 \text{ or } w > -2\}$;

−7 −6 −5 −4 −3 −2 −1 0 1 2 3

49. $x < -3$ or $x > 3$ 51. $-3 \leq x < 5$ 53. $\{x \mid x \leq 3$ or $x > 9\}$ 55. $\{k \mid 11 \leq k \leq 13\}$ 57a. Sample answer: $11 \leq h \leq 14$

57b. 8 9 10 11 12 13 14 15 16 17 18

59. $6.6 < p < 9.4$ 61. $\{y \mid y < 0\}$ 63. $\{x \mid x \geq 5\}$
65. $\{b \mid b > 6\}$

67a.

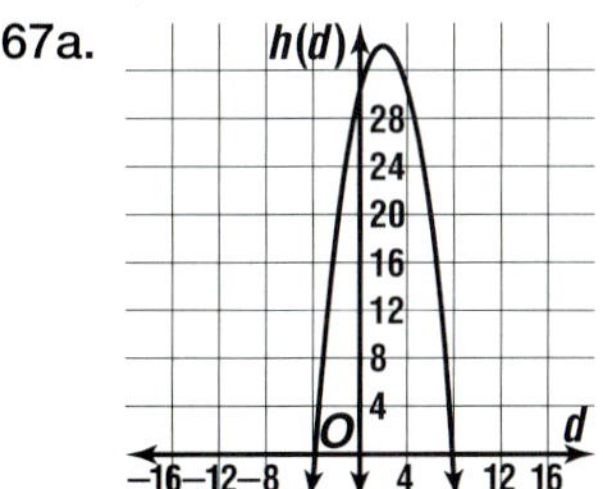

67b. 34 in. 69. $3b - 18$ 71. C

Pages 532–534 Lesson 12-6

1. $|x| < 7$ includes all numbers between −7 and 7. $|x| > 7$ includes all numbers to the left of −7 and to the right of 7. 3. Mia; $|x| \leq 0$ includes only 0. $|x| \geq 0$ includes all real numbers. 5. $x \leq 3$ and $x \geq -3$ 7. $x > 8$ or $x < -8$

9. $\{x \mid -4 < x < 8\}$;

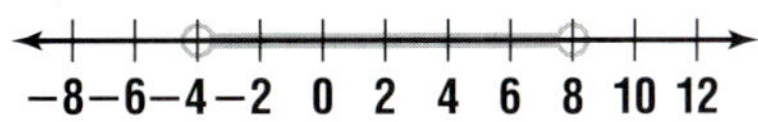

11. $\{t \mid t \geq 8 \text{ or } t \leq 2\}$;

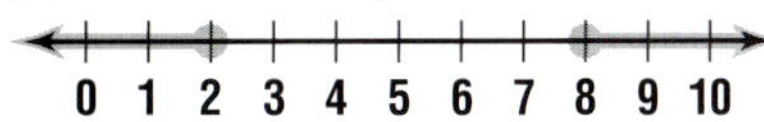

13. $\{s \mid s \geq -1 \text{ or } s \leq -7\}$;

−9 −8 −7 −6 −5 −4 −3 −2 −1 0 1

15. $\{m \mid -6 < m < 4\}$;

−6 −5 −4 −3 −2 −1 0 1 2 3 4

17. $\{z \mid -9 \leq z \leq -5\}$;

−12 −11 −10 −9 −8 −7 −6 −5 −4 −3 −2

19. $\{p \mid -1 < p < 3\}$;

−5 −4 −3 −2 −1 0 1 2 3 4 5

21. $\{t \mid -2 \leq t \leq 2\}$;

−5 −4 −3 −2 −1 0 1 2 3 4 5

23. $\{k \mid -5.5 < k < 1.5\}$;

−6 −5 −4 −3 −2 −1 0 1 2 3 4

25. $\{y \mid y \leq -9 \text{ or } y \geq 3\}$;

−14 −12 −10 −8 −6 −4 −2 0 2 4 6

27. $\{x \mid x < -2 \text{ or } x > 2\}$;

−5 −4 −3 −2 −1 0 1 2 3 4 5

29. $\{r \mid r \leq 3 \text{ or } r \geq 5\}$;

−2 −1 0 1 2 3 4 5 6 7 8

31. $\{h \mid h < -6 \text{ or } h > 12\}$;

−6 −4 −2 0 2 4 6 8 10 12 14

33. $|s - 90| < 4$ 35. $|s - 65| \leq 3$
37. $|x - 1| \geq 1$

39. $\{x \mid 0 < x < 8\}$;

−1 0 1 2 3 4 5 6 7 8 9

41. $\{x \mid x < -10 \text{ or } x > 8\}$;

−10 −8 −6 −4 −2 0 2 4 6 8 10

43. $|m - 3.25| \leq 0.05$; $3.20 \leq m \leq 3.30$

45. −8 −7 −6 −5 −4 −3 −2 −1 0 1 2

47. −7 −6 −5 −4 −3 −2 −1 0 1 2 3

49. $\{y \mid y \leq -1\}$ 51. −2, 6

Page 534 Quiz 2

1. $\{x \mid x < 1\}$ 3. $\{d \mid d \geq 1.2\}$
5. $\{n \mid -5 \leq n < 1\}$;

−7 −6 −5 −4 −3 −2 −1 0 1 2 3

7. $\{y \mid 5 < y < 9\}$;

2 3 4 5 6 7 8 9 10 11 12

9. $\{x \mid -4 < x < 4\}$;

−5 −4 −3 −2 −1 0 1 2 3 4 5

Pages 538–539 Lesson 12–7

1. Graph a solid line for ≤ and ≥. Graph a dashed line for < and >.

3.

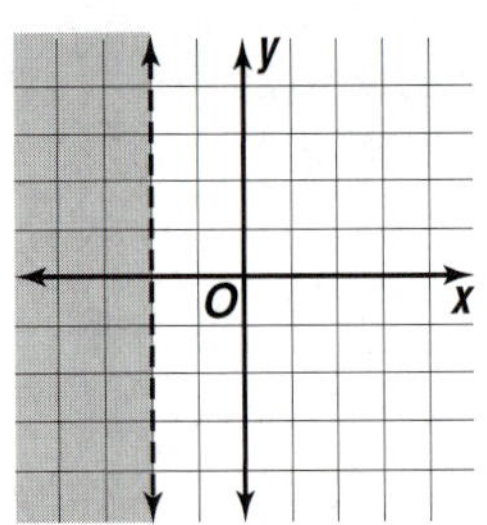

5.

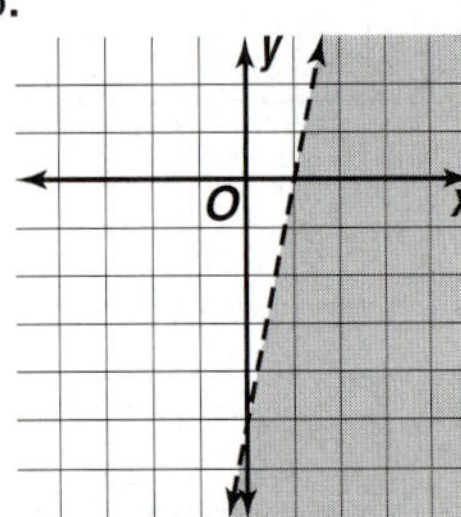

7.

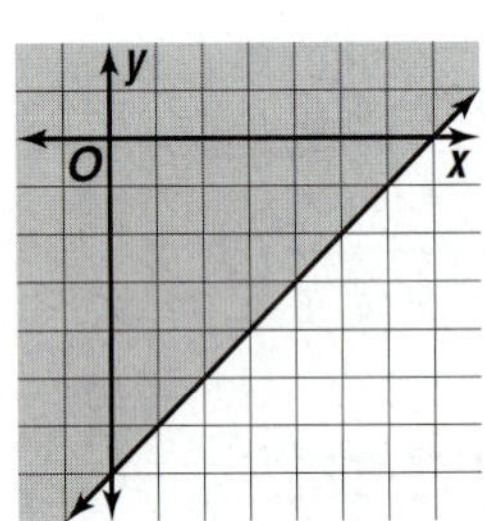

9.

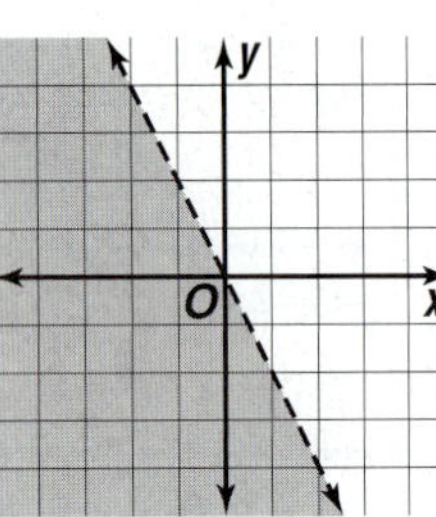

11.

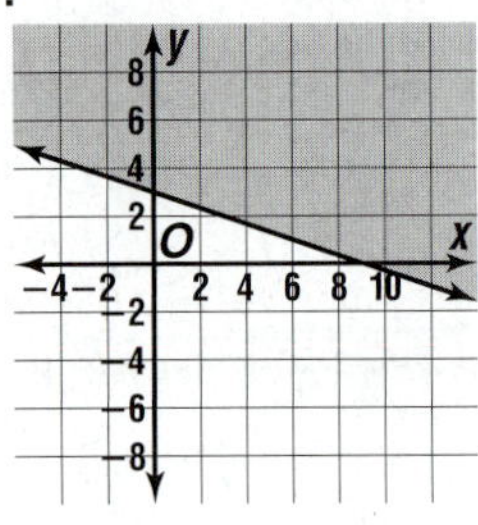

13.

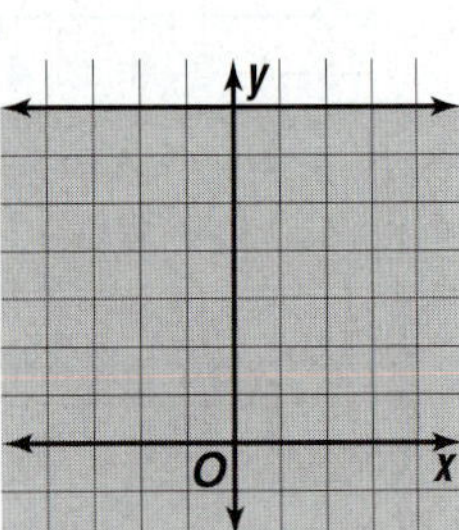

15.

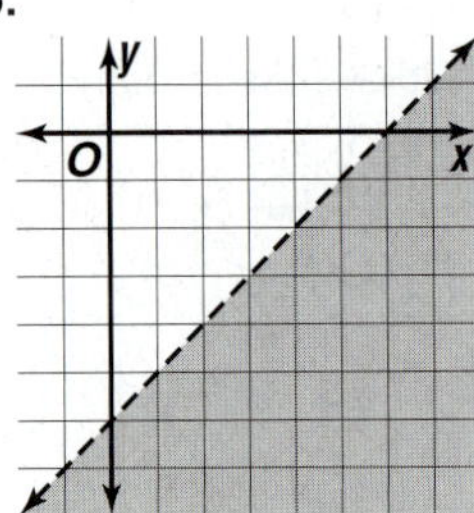

17.

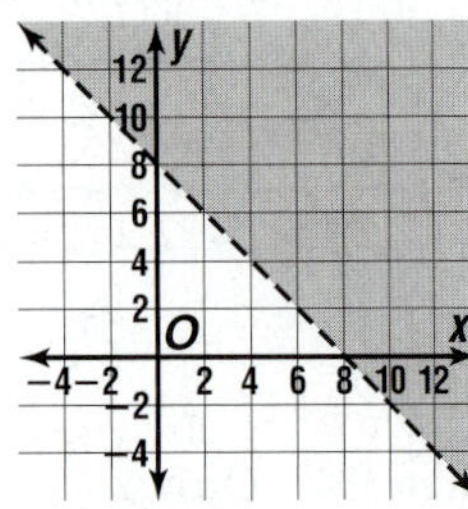

19.

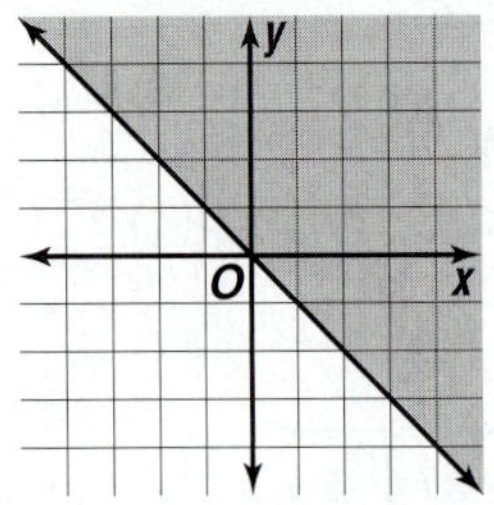

21.

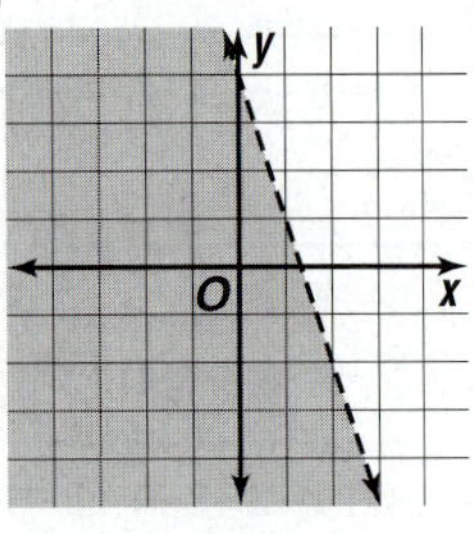

23.

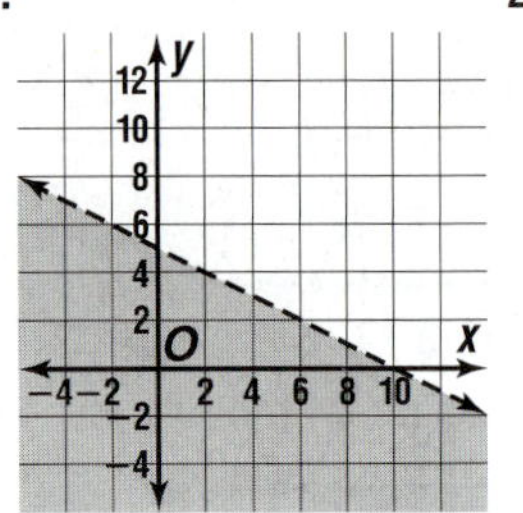

25.

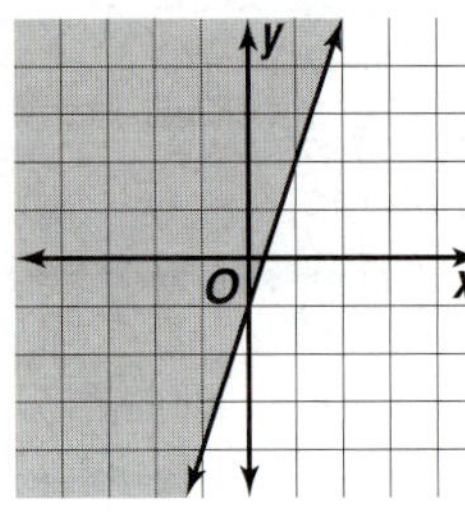

27.

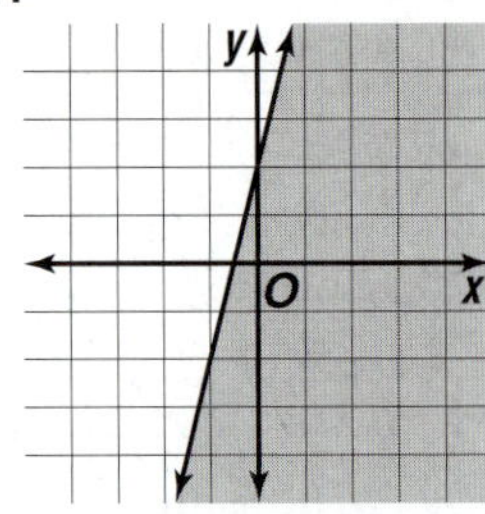

29.

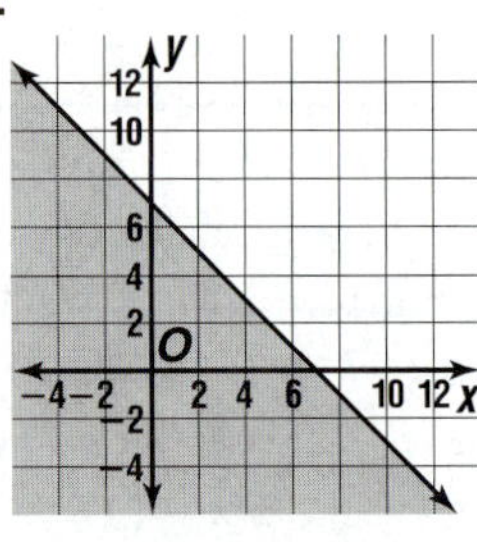

31.

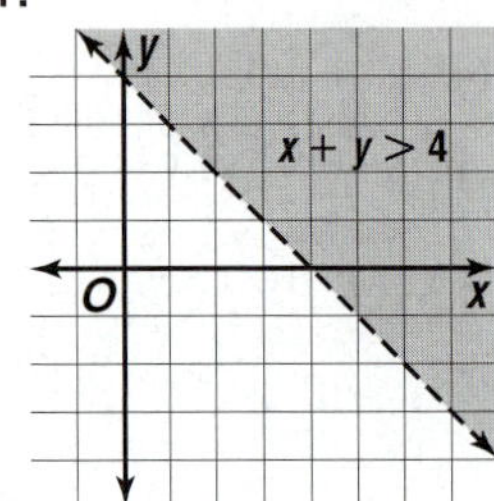

33a.

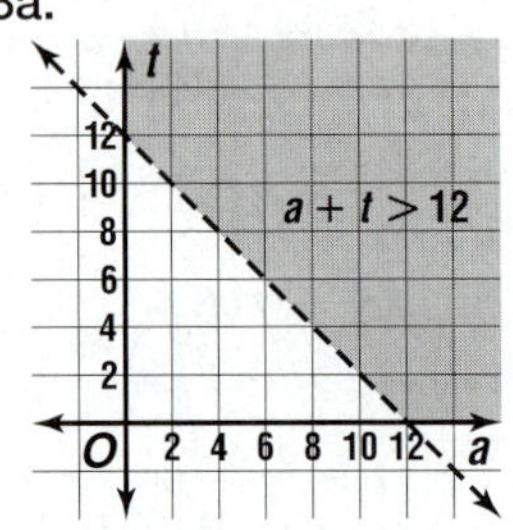

33b. Sample answer: {(4, 9), (2, 11), (3, 16)}

33c. positive, whole

35.

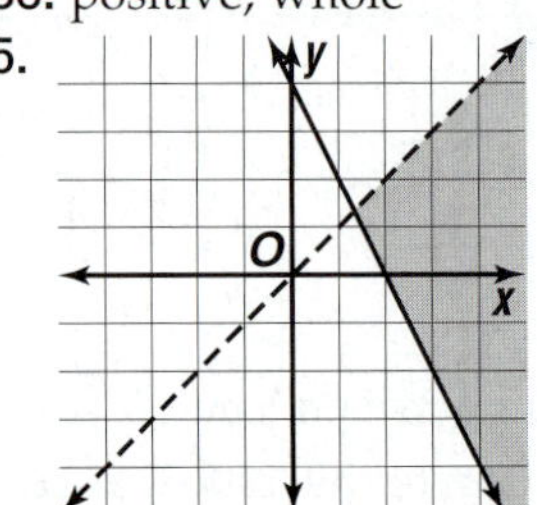

37. $\{h \mid 5 < h < 9\}$;

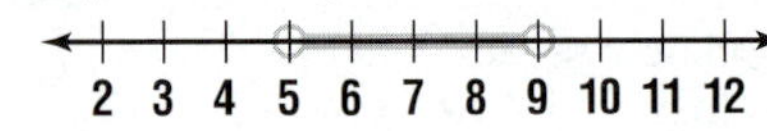

39. $\{b \mid b < -2 \text{ or } b > 1\}$;

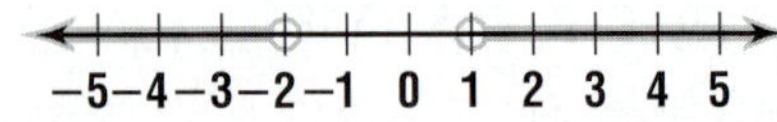

41. 8.483×10^2

Pages 542–544 Chapter 12 Study Guide and Assessment

1. union **3.** half-plane **5.** boundary **7.** half-plane **9.** set-builder notation

11.

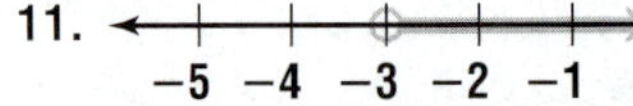

13. 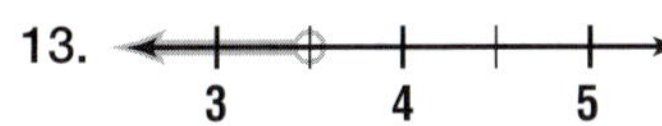

15. $x \geq -5$ **17.** $\{x \mid x > 4\}$ **19.** $\left\{y \mid y \geq 1\frac{1}{6}\right\}$
21. $\{x \mid x \leq 3\}$ **23.** $\{n \mid n < -5\}$ **25.** $\{x \mid x > 15\}$
27. $\{w \mid w \leq 8\}$ **29.** $\{x \mid x \geq 4\}$ **31.** $\{t \mid t \leq 9\}$
33. $4x - 3 > 25$; $\{x \mid x > 7\}$

35. $\{y \mid -7 \leq y < -1\}$;

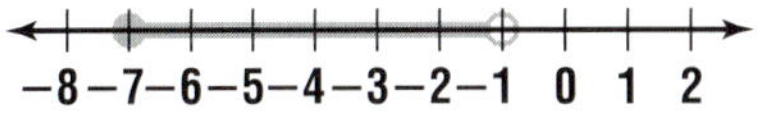

−8 −7 −6 −5 −4 −3 −2 −1 0 1 2

37. $\{a \mid a \leq 1 \text{ or } a \geq 2\}$;

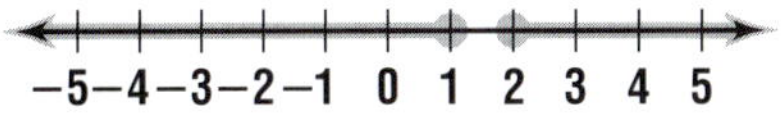

−5 −4 −3 −2 −1 0 1 2 3 4 5

39. $\{t \mid -4 \leq t \leq 6\}$;

−4 −3 −2 −1 0 1 2 3 4 5 6

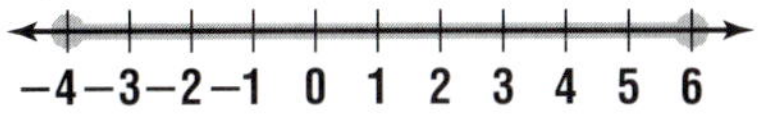

41. $\{x \mid -4 < x < -2\}$;

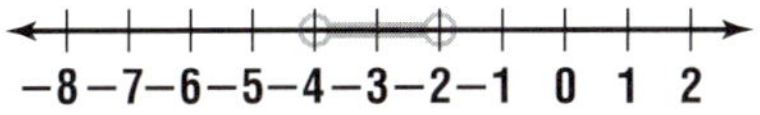

−8 −7 −6 −5 −4 −3 −2 −1 0 1 2

43. $\{y \mid y \neq 2\}$;

−5 −4 −3 −2 −1 0 1 2 3 4 5

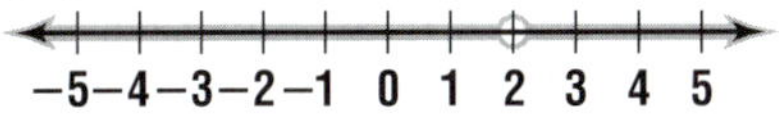

45. $|s - 80| > 5$

47.

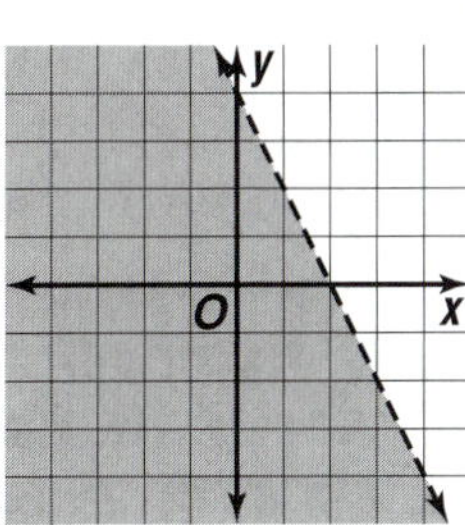

49.

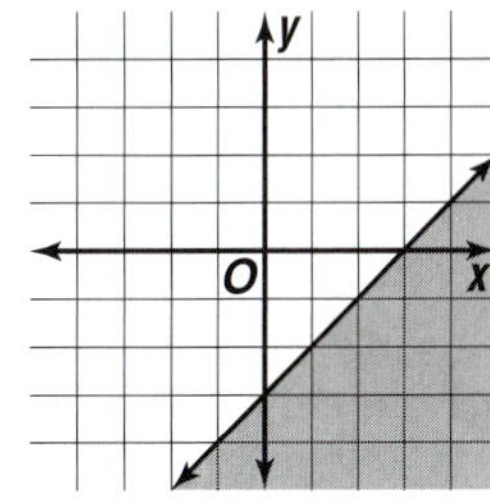

51. $\{x \mid 30 < x < 60\}$

Page 547 Preparing for Standardized Tests
1. A **3.** C **5.** A **7.** B **9.** 270

Chapter 13 Systems of Equations and Inequalities

Page 549 Check Your Readiness

1.

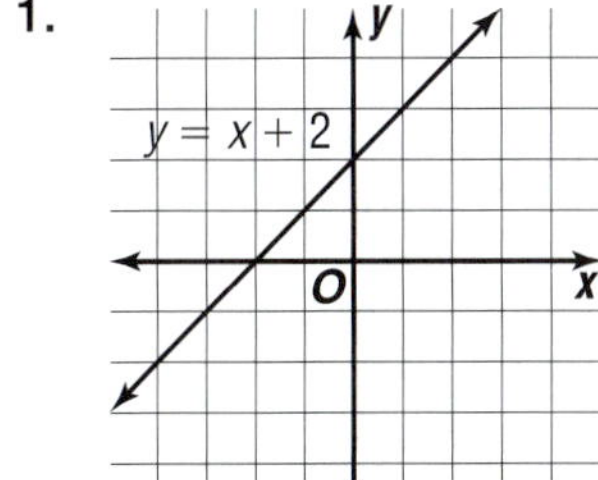

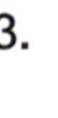
3.

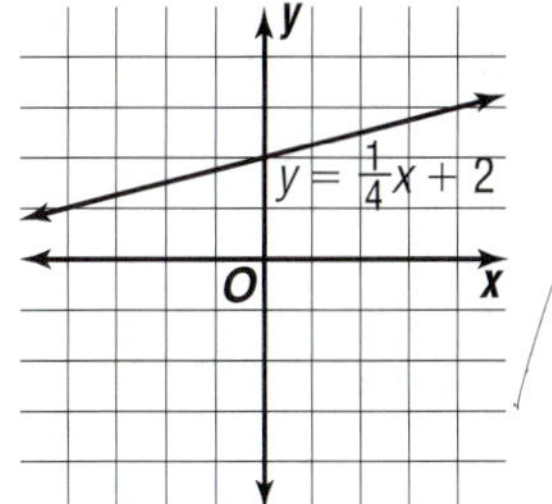

5.

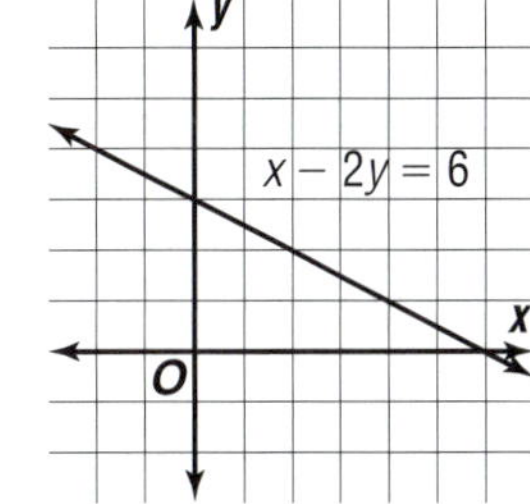

7. 3 **9.** 4 **11.** 2

13.

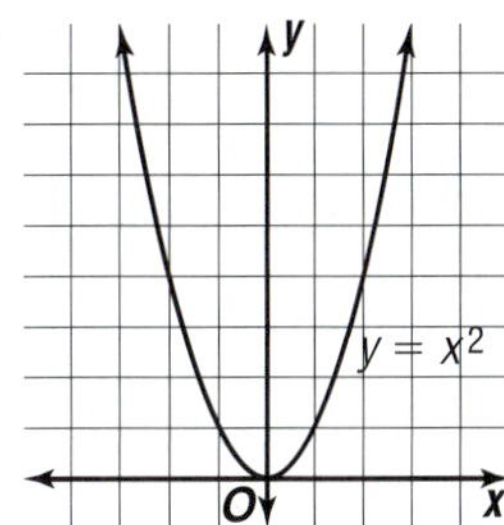

15.

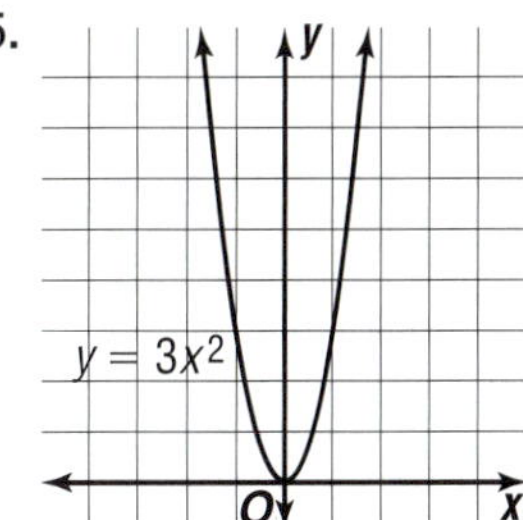

17.

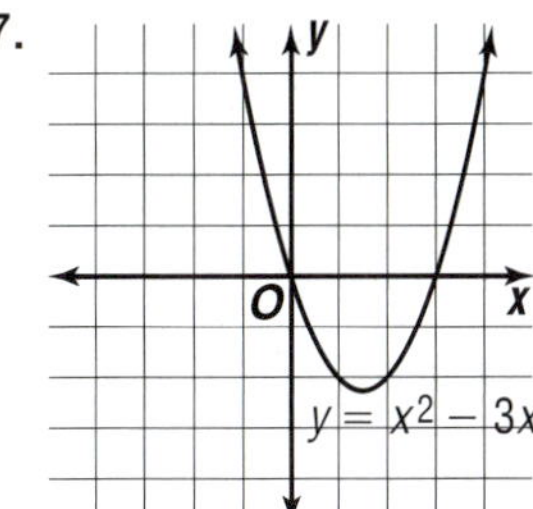

19.

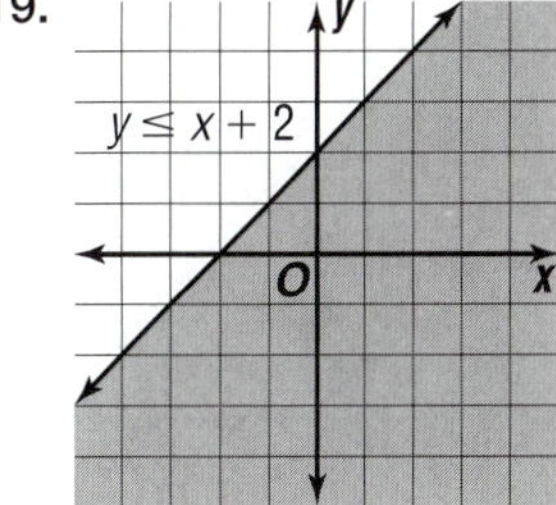

21.

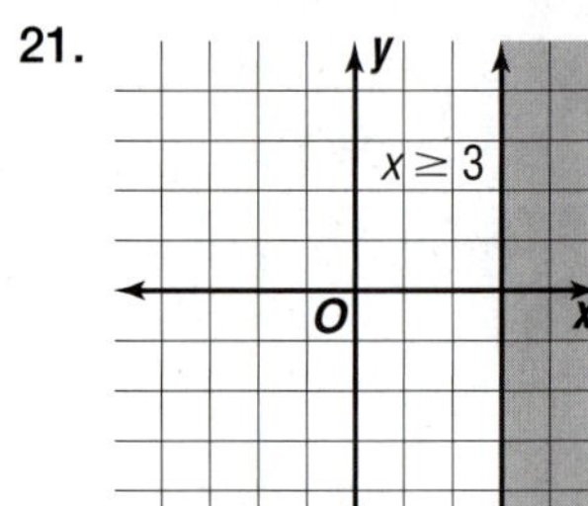

23.

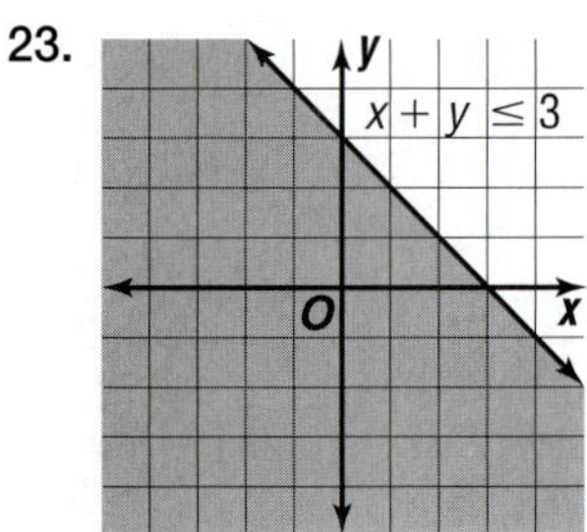

Pages 552–553 Lesson 13–1

1. the ordered pair for the point at which the graphs of the equations intersect **3a.** $(-4, 2)$ **3b.** $(2, -2)$ **3c.** $(2, 4)$

5. $(-4, 1)$

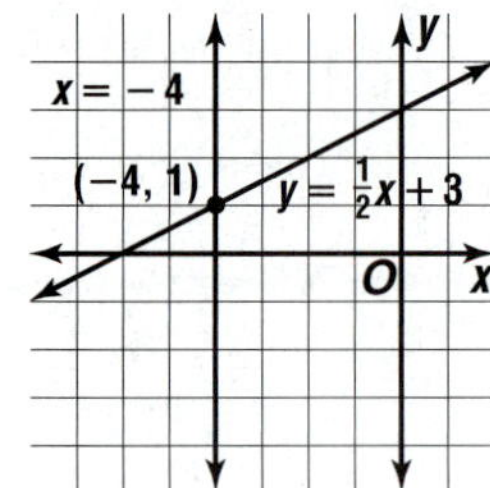

7. $(1, -3)$

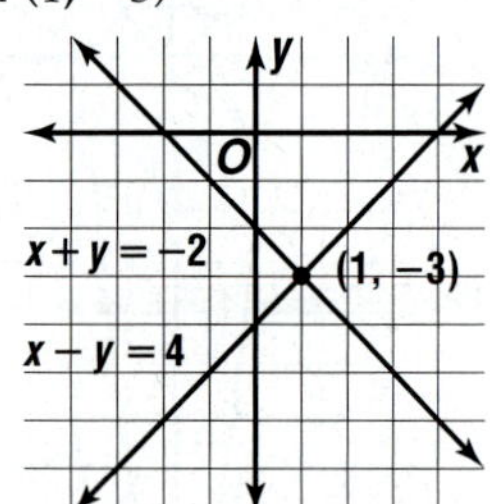

9. $(5, -4)$

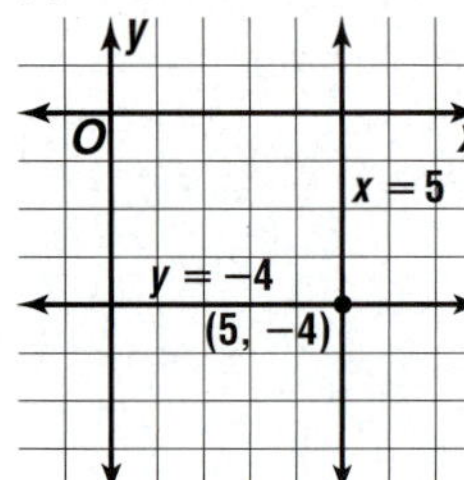

11. $(0, 2)$

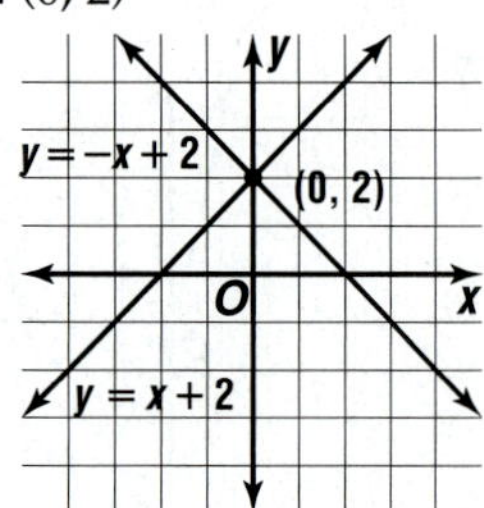

13. $(3, 6)$

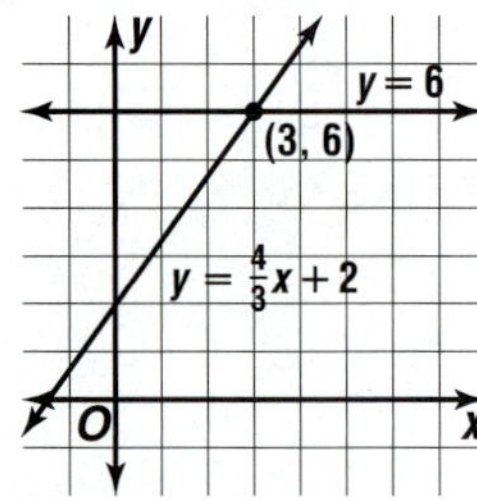

15. $(5, 2)$

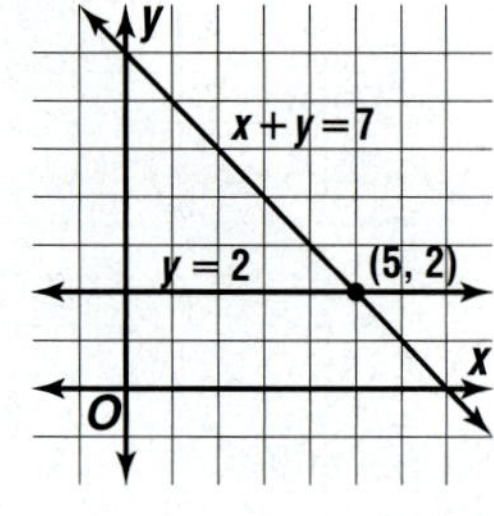

17. $(-2, -3)$

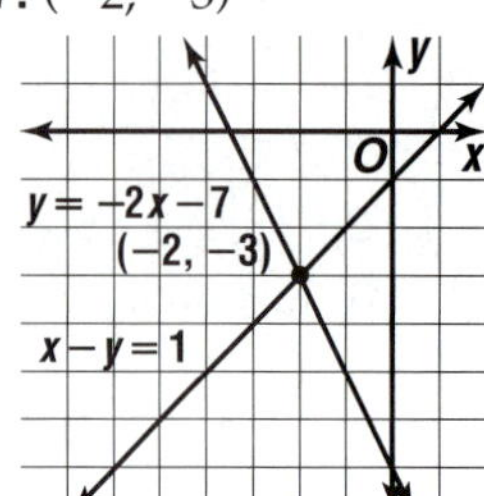

19. $(-6, 3)$

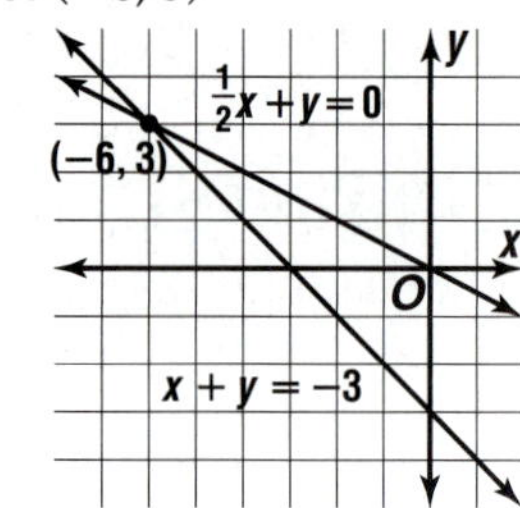

21. $(2, 6)$ **23.** $(10, -6)$

25. Sample answer: $y = x + 4$ and $y = -x + 4$

27. $\{x \mid 1 < x < 5\}$

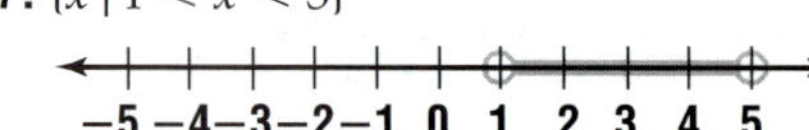

29. $\{m \mid m > 2 \text{ or } m < -4\}$

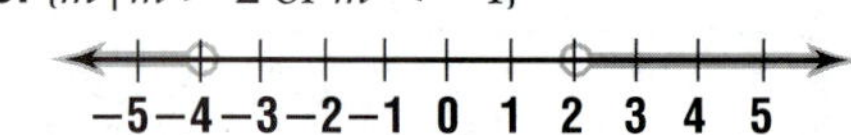

31. $\frac{3}{2}, 2$

Pages 557–559 Lesson 13–2

1. two parallel lines, two intersecting lines, the same line **5.** consistent and dependent

7. $(1, -1)$

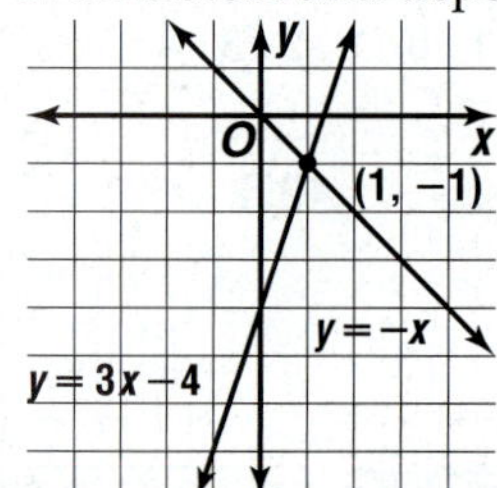

9a. Consistent and independent; there is one solution, (3, 150). **9b.** After 3 seconds, both the cat and dog will be 150 feet from the dog's original starting position. **11.** inconsistent **13.** consistent and dependent **15.** consistent and independent

17. infinitely many

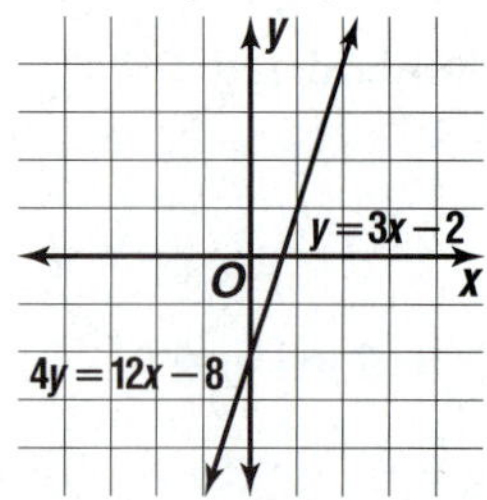

19. $(5, -2)$

y = −x + 3
(5, −2)
$y = \frac{1}{5}x - 3$

21. no solution

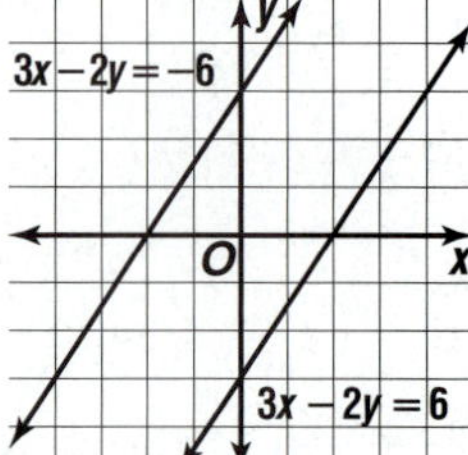

23. $(-1, 3)$

(−1, 3)
−x + y = 4
6x + y = −3

25. one; (5, 1) **27.** There is no solution because the graphs of the equations are parallel lines. The second dog never catches up with the first.

29. Sample answer: $y = -\frac{1}{3}x$

31.

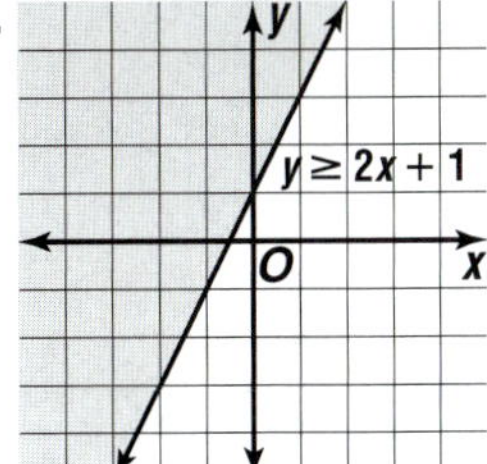

33. $4(x - 2)$ **35.** $3a(ab^2 + 2b - 3)$

Pages 564–565 Lesson 13-3

1. when the exact coordinates are not easily determined from the graph **3.** Both are correct because you can substitute for either variable. **5.** (4, 12) **7.** (4, 0) **9.** no solution **11.** (1, 1) **13.** (−3, −9) **15.** no solution **17.** (−4, 1) **19.** (3, 4) **21.** (2, 2) **23.** (2, −10) **25.** $\left(4, \frac{1}{3}\right)$ **27.** infinitely many **29.** (13, 30) **31a.** 5.5 hr **31b.** 357.5 mi **33.** No; the graphs of equations *A*, *B*, and *C* could form a triangle. Any two of the lines would be intersecting, but there is no point where all three lines intersect.

35. (−1, 4)

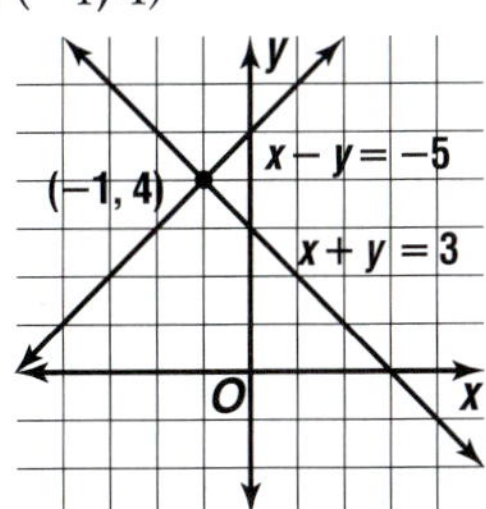

37. $-3 < x \leq 2$ **39.** 460

Page 565 Quiz 1

1. (4, −1) **3.** no solution

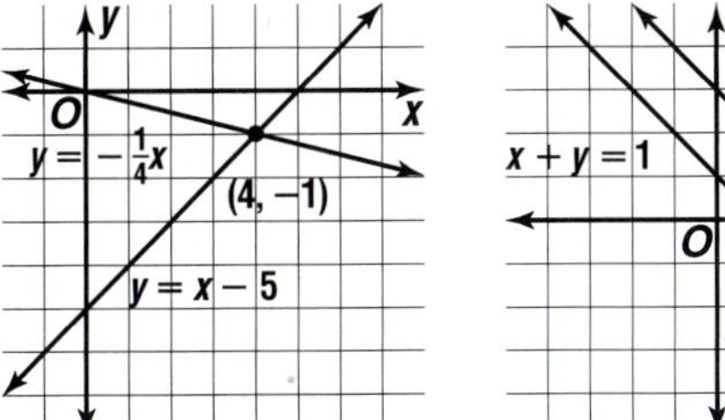

5. 12.5 lb of the \$3.90, 37.5 lb of the \$4.30

Pages 569–571 Lesson 13-4

1. when the coefficients of terms with the same variable are the same or additive inverses **3.** addition **5.** subtraction **7.** (8, −1) **9.** (2, 2) **11.** $\left(\frac{16}{3}, 3\right)$ **13a.** $x + y = 42$ and $x = 2y + 3$ **13b.** 29, 13 **15.** (14, 5) **17.** (−5, 15) **19.** (0, 6) **21.** (8, −1) **23.** (−15, −22) **25.** $\left(5, \frac{5}{2}\right)$ **27.** infinitely many **29.** (−11, −4) **31.** (2, 8) **33.** $\left(2, \frac{1}{2}\right)$ **35a.** $13x + 2y = 137.50$ and $9x + 2y = 103.50$ **35b.** \$8.50 child, \$13.50 adult **37a.** $x + 2y = 29$ and $y = 3x - 10$ **37b.** 7 and 11 **39.** (11, −38) **41.** consistent and dependent **43.** $y < 4$ **45.** $x < -2$ **47.** $a^2 - 6a + 9$ **49.** $y - 1 = -5(x - 6)$ or $y + 4 = -5(x - 7)$

Pages 575–577 Lesson 13-5

1. when neither of the variables in a system of equations can be eliminated by simply adding or subtracting the equations **3a.** substitution **3b.** elimination (×) **3c.** elimination (+) **3d.** elimination (−) **5.** Multiply second equation by 2. Then add. **7.** (2, 2) **9.** (−3, 2) **11.** no solution **13a.** 2 mph **13b.** 14 mph **15.** (2, 0) **17.** (−1, 3) **19.** $\left(\frac{1}{2}, \frac{3}{4}\right)$ **21.** $\left(\frac{5}{2}, 1\right)$ **23.** (−6, −3) **25.** infinitely many **27.** (−2, −1) **29.** $\left(\frac{7}{9}, 0\right)$ **31.** Sample answer: substitution or multiplication; (−2, 4) **33.** Sample answer: graphing; no solution **35a.** $6a + 10b = 108$ and $4a + 12b = 104$ **35b.** \$8, \$6 **37.** (7, −1) **39.** $\left(\frac{3}{2}, 1\right)$ **41.** $m < -4$ **43.** $a \geq -4$ **45.** 5

Pages 583–585 Lesson 13-6

1. (−1, −7), (2, 0) **3.** Sample answer: The following methods can be used to solve linear systems of equations: graphing, substitution, elimination using addition or subtraction, and elimination using multiplication. Graphing and substitution can be used to solve quadratic-linear systems of equations. Graphing is useful when you want to estimate the solution. The other methods are useful when you want to find the exact solution.

5. (0, 0), (2, 4)

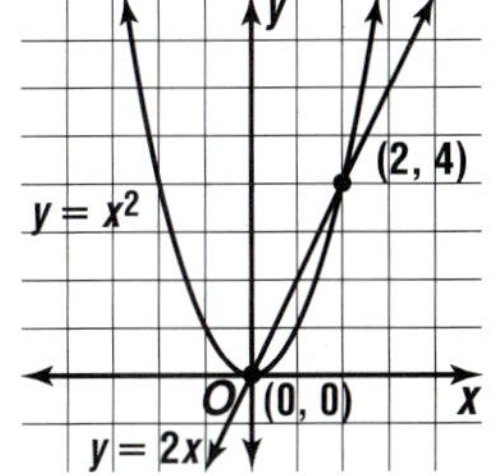

7. no solution

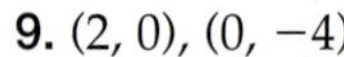

9. (2, 0), (0, −4)

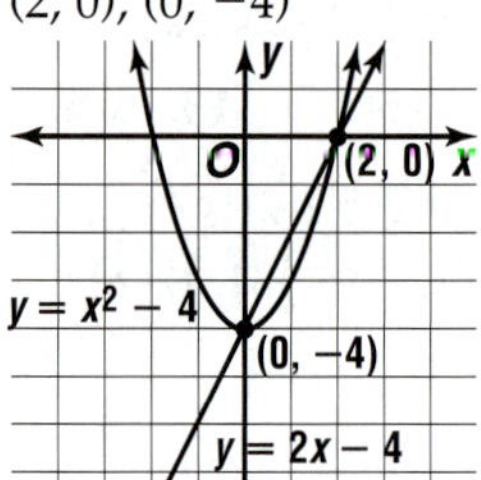

11. no solution

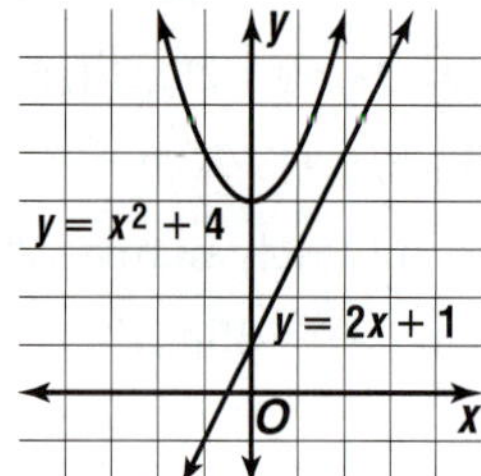

13. (−1, 4), (2, 1)

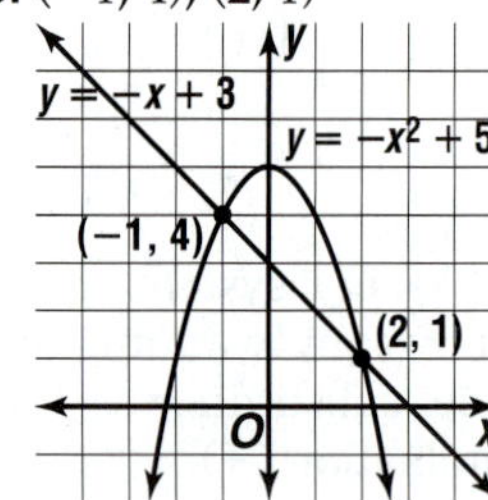

15. (−4, 7) **17.** (8, 32), (−2, 2) **19.** no solution **21.** (3, 0), (0, −9) **23a.** $A = 16$ and $A = 4x^2 - 28x + 40$ **23b.** $\ell = 8$ ft, $w = 2$ ft, $h = 1$ ft **23c.** 16 ft^3 **25.** Side length 4 units; solve the quadratic-linear system of equations $p = s^2$ and $p = 4s$, where p represents perimeter and s represents side length. The solutions are (0, 0) and (4, 16). **27.** (2, −1) **29.** {−7, 7} **31.** {4, −8}

Page 585 Quiz 2

1. (4, 2) **3.** (6, 2) **5a.** $x - y = 38$, $x = 3y - 2$ **5b.** 58, 20

Pages 589–590 Lesson 13–7

1. A system of linear equations may have at most one solution if the equations are distinct. A system of inequalities may have an infinite number of solutions. **3.** Kyle; all of the points in region B satisfy both inequalities. **5.** no

7.

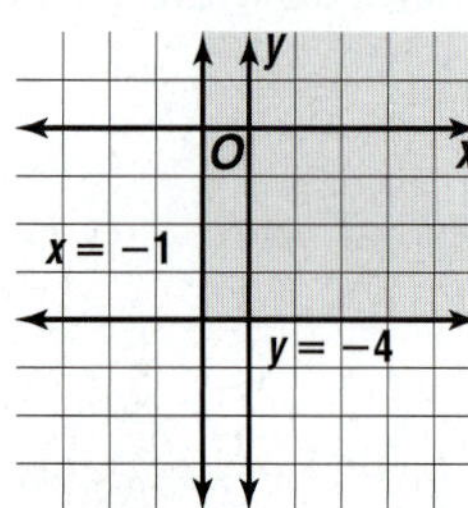

9.

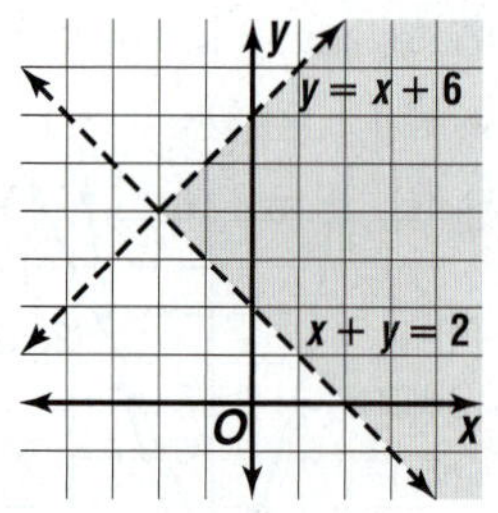

11. Sample answer: 2 dozen sugar, 4 dozen choc. chip; 4 dozen sugar, 3 dozen choc. chip; 5 dozen sugar, 2 dozen choc. chip

13.

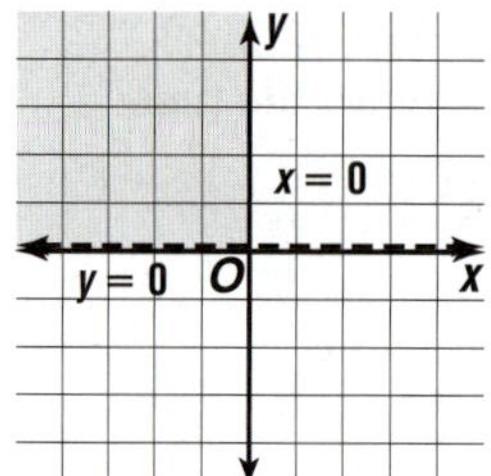

15.

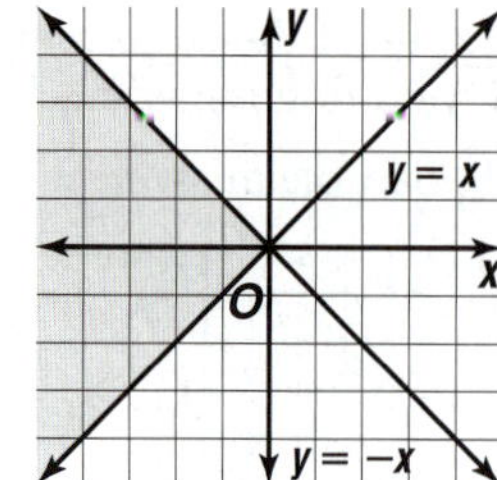

17.

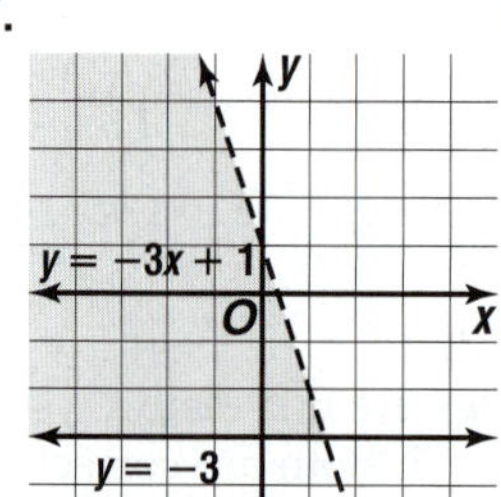

19. no solution

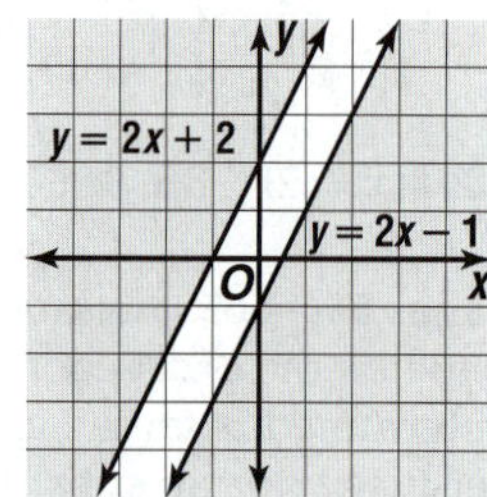

21.

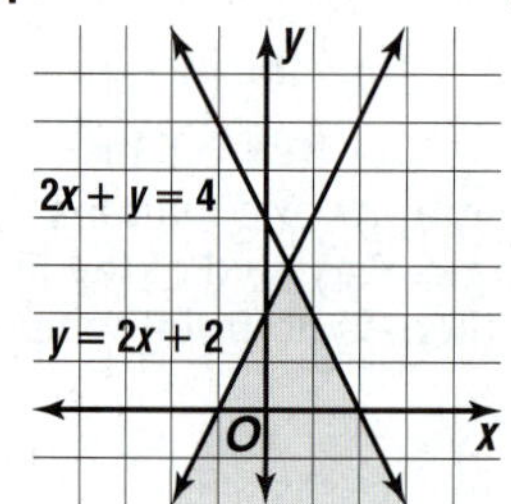

23.

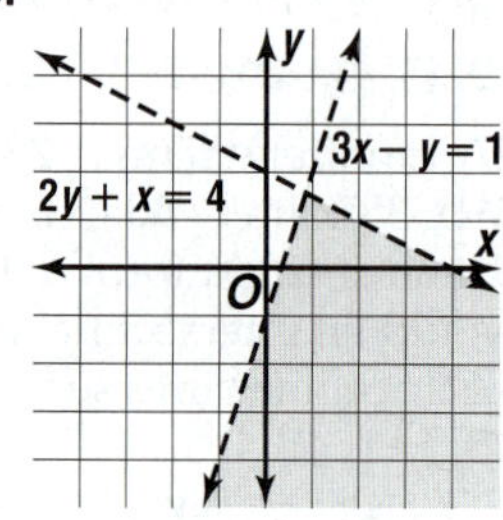

25.

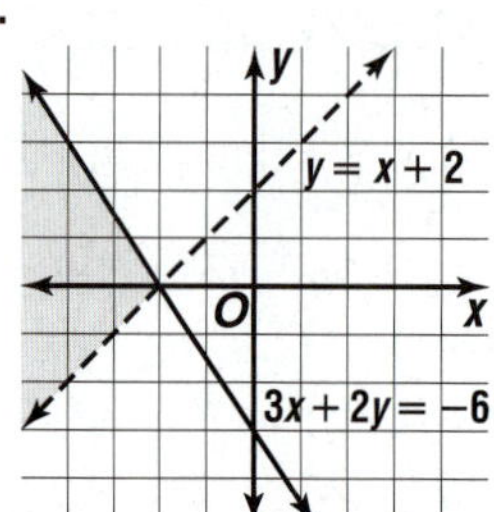

27. People who use the phone less than 75 minutes a month should select the $0.15 per minute plan. People who use the phone more than 75 minutes a month should select the $0.05 per minute plan. The middle plan is never the cheapest.

29. (2, 4), (−2, 4)

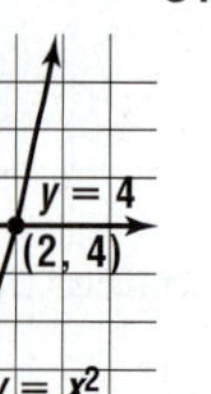

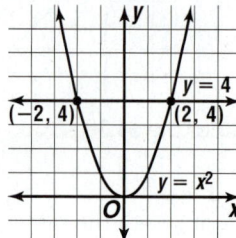

31. no solution

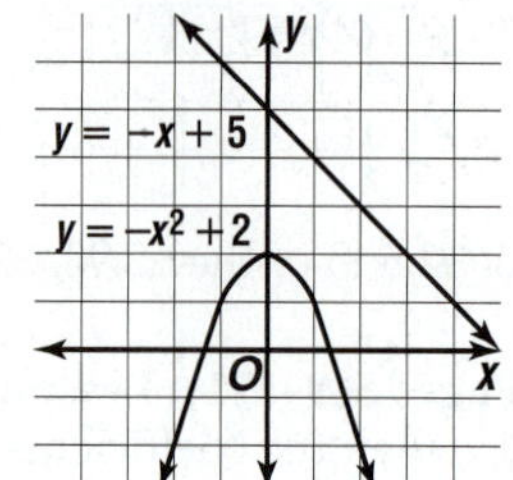

33. 4 lb of the $5.25, 1 lb of the $6.50 **35.** 5 **37.** A

Pages 592–594 Chapter 13 Study Guide and Assessment

1. true **3.** false; all **5.** true **7.** true **9.** false; elimination

11. (1, 4) **13.** (−2, 2)

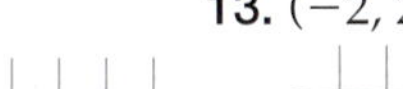

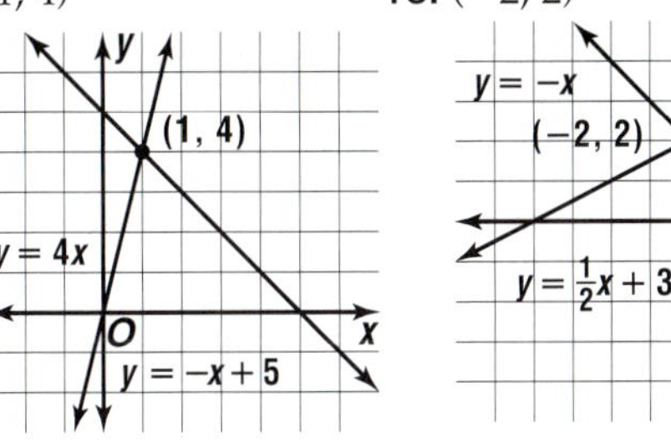

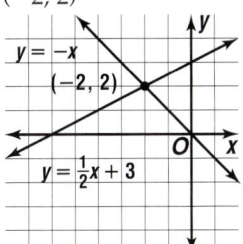

15. (−6, 0) 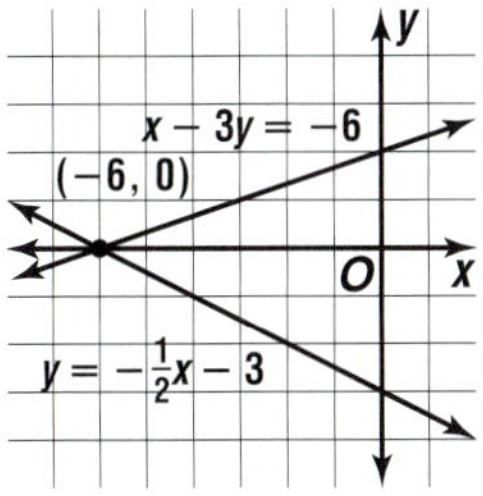 **17.** (3, −2)

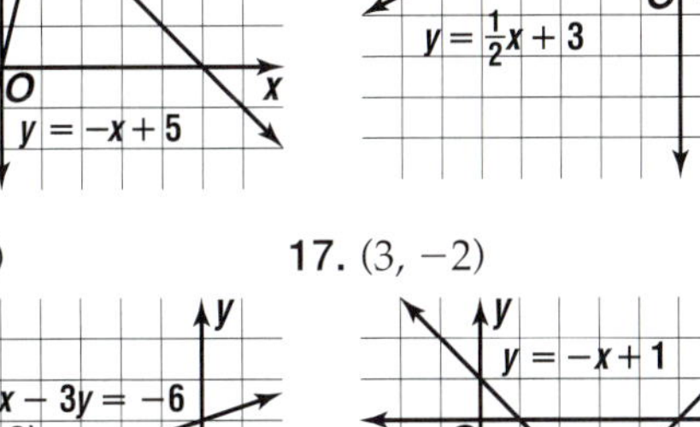

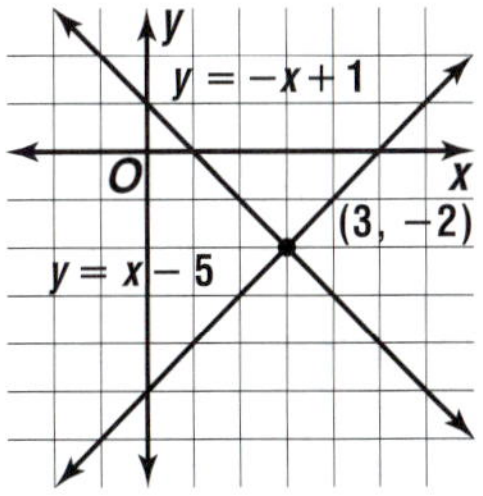

19. infinitely many

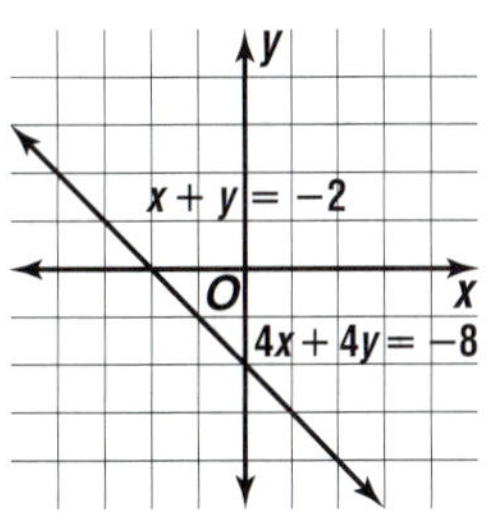

21. $\left(\frac{1}{2}, 1\right)$ **23.** no solution **25.** infinitely many **27.** (4, −6) **29.** no solution **31.** infinitely many **33.** (2, 0) **35.** $\left(\frac{6}{11}, \frac{1}{11}\right)$ **37.** (22, 31)

39. (−2, −4), (1, −1)

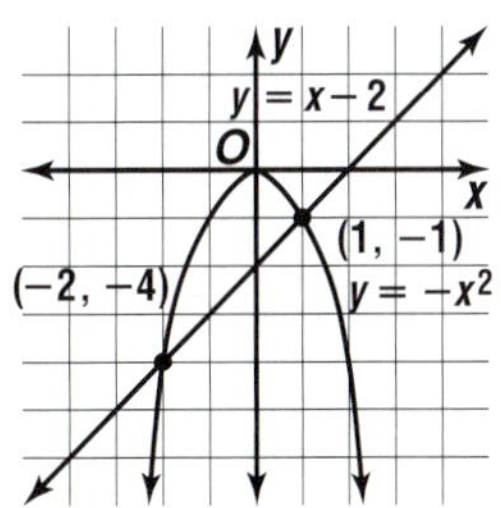

41. no solution

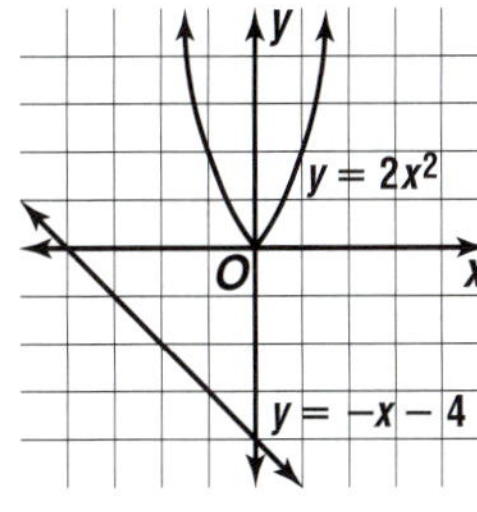

43. (4, 18), (−1, 3) **45.** (3, 3)

47. **49.**

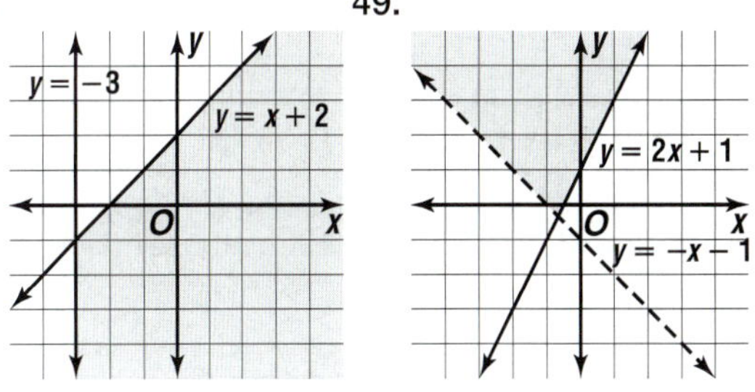

51. 8 lb of almonds, 12 lb of cashews **53.** 16 ft by 9 ft

Page 597 Preparing for Standardized Tests

1. D **3.** A **5.** B **7.** D **9.** 934

Chapter 14 Radical Expressions

Page 599 Check Your Readiness

1. $\frac{1}{2}$ **3.** 2 **5.** $-\frac{1}{11}$ **7.** 1 **9.** 4 **11.** 7 **13.** $13p$ **15.** $4v$ **17.** $6r + 3$ **19.** $y^2 - 9$ **21.** $z^2 - 25$ **23.** $9c^2 - 25$

Pages 603–605 Lesson 14–1

1. Sample answer: −4 **3.** Sample answer: $\sqrt{13}$ **5.** I **7.** Q

9. −4.5

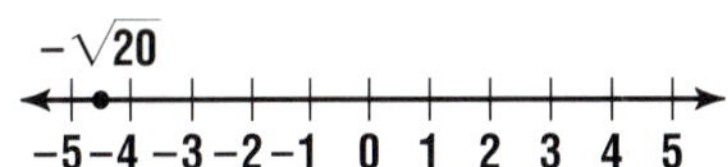

11. 7.3

13. rational **15.** rational **17.** Q **19.** W, Z, Q **21.** Z, Q **23.** N, W, Z, Q **25.** Z, Q **27.** Q

29. 1.7

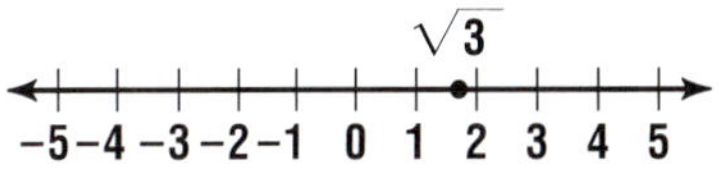

$\sqrt{3}$

−5 −4 −3 −2 −1 0 1 2 3 4 5

31. −3.2

$-\sqrt{10}$

−5 −4 −3 −2 −1 0 1 2 3 4 5

33. −6.3

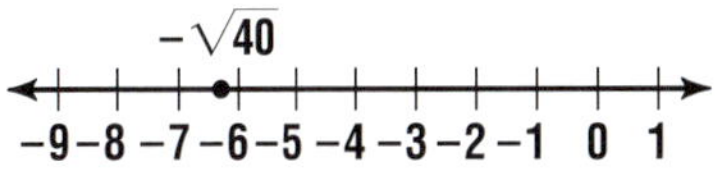

$-\sqrt{40}$

−9 −8 −7 −6 −5 −4 −3 −2 −1 0 1

35. 7.2

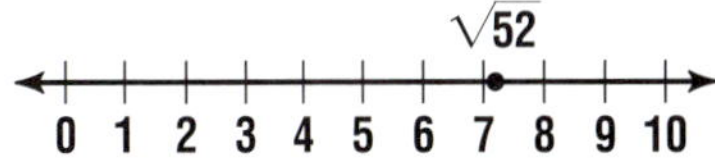

37. 10.4

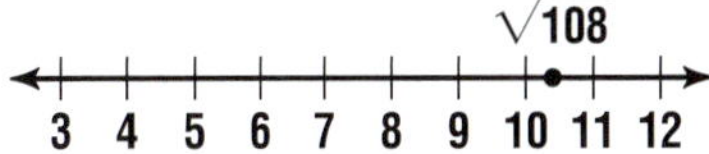

SELECTED ANSWERS

39. 15.8

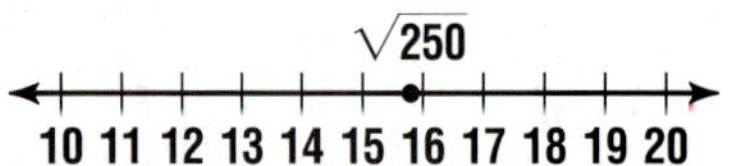

41. rational **43.** irrational; 6 and 7 **45.** rational **47.** irrational; −2 and −3 **49.** irrational; 11 and 12 **51.** rational

53. $-\sqrt{5}$, $\sqrt{2}$, π

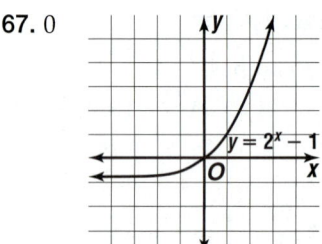

55. about 8.5 s **57.** 5.2 ohms and 1200 watts or 4.6 ohms and 1500 watts **59a.** $\frac{1}{2}$ **59b.** $\frac{1}{3}$ **59c.** $\frac{1}{4}$

61. (3, 11), (1, 3) **63.** $\{x \mid x < -7\}$ **65.** $\{m \mid m \leq -1.8\}$

67. 0 ($y = 2^x - 1$)

69. $4(a + 3)(a - 3)$ **71.** 4 **73.** A

Pages 608–611 Lesson 14–2

1. Pythagorean Theorem **3.** Both are correct since it does not matter which ordered pair is first when using the Distance Formula as long as the coordinates are used in the same order. **5.** 5 **7.** 9 or −3 **9.** 9.5 mi **11.** 12 **13.** $\sqrt{58}$ or 7.6 **15.** $\sqrt{185}$ or 13.6 **17.** $\sqrt{18}$ or 4.2 **19.** 17 or −13 **21.** −12 or 6 **23.** 2 or −14 **25.** $\sqrt{194}$ or 13.9 **27.** 8 or 32 **29.** No; no two sides have the same measure. **31.** 16.3 mi

33. 3.5 ($\sqrt{12}$)

35. 11.6 ($\sqrt{135}$)

37. $\{k \mid k \geq 6\}$ **39.** $\{b \mid b \leq -8\}$ **41.** $\varnothing$

Page 611 Quiz 1

1. W, Z, Q **3.** I **5.** 1.22, $1.\overline{2}$, $\sqrt{2}$ **7.** $\sqrt{98}$ or 9.9 **9.** 0 or 6

Pages 617–619 Lesson 14–3

1. to ensure nonnegative results **3.** Greg is correct. In its simplest form, the expression is written as $\frac{3 - \sqrt{5}}{2}$ since the numbers 6, 2, and 4 in $\frac{6 - 2\sqrt{5}}{4}$ are all divisible by 2. **5.** $2 - \sqrt{8}$; −4 **7.** $5\sqrt{3}$ **9.** $3\sqrt{10}$ **11.** $\frac{\sqrt{21}}{7}$ **13.** $6|x|\sqrt{y}$ **15.** $4\sqrt{2}$ units **17.** $7\sqrt{2}$ **19.** $10\sqrt{5}$ **21.** 4 **23.** 15 **25.** 4 **27.** $2\sqrt{10}$ **29.** $\frac{2\sqrt{5}}{5}$ **31.** $\frac{8 + 2\sqrt{3}}{13}$ **33.** $\frac{24 + 4\sqrt{7}}{29}$ **35.** $2|m|\sqrt{10}$ **37.** $3|a|b\sqrt{6b}$ **39.** $6|x|y^2z^2\sqrt{xz}$ **41a.** $2\sqrt{3}$ mph **41b.** 10 in. **43.** 2 or −4 **45.** Z, Q **47.** no solution **49.** $\sqrt{79}$

Pages 621–623 Lesson 14–4

1. Sample answer: Add like terms. **3.** The Distributive Property allows you to add like terms. Radicals with like radicands can be added or subtracted. **5.** none **7.** $8\sqrt{3}, -18\sqrt{3}$ **9.** $2\sqrt{5}$ **11.** $-5\sqrt{7}$ **13.** $-6\sqrt{2}$ **15.** $6\sqrt{3}$ **17.** $14\sqrt{5}$ **19.** $4\sqrt{7}$ **21.** $-2\sqrt{5}$ **23.** $-9\sqrt{2} + 8\sqrt{5}$ **25.** $7\sqrt{6}$ **27.** $14\sqrt{3}$ **29.** $-4\sqrt{2}$ **31.** $8\sqrt{2} + 6\sqrt{3}$ **33.** $-2\sqrt{5} - 6\sqrt{6}$ **35.** $5\sqrt{7}$ **37a.** $4\sqrt{3}$ **37b.** $6\sqrt{3} + 6$ **39.** $3n|p|\sqrt{5mn}$ **41.** $\sqrt{101}$ or 10.0 **43.** $-1 \cdot 2 \cdot 13 \cdot a \cdot a \cdot a \cdot b \cdot b$ **44.** $2 \cdot 2 \cdot 3 \cdot 3 \cdot x \cdot y \cdot y \cdot y \cdot y \cdot z$ **45.** $a^2 - 4c^2$ **46.** A

Pages 627–629 Lesson 14–5

1. Isolate the radical. **5.** $a + 5 = 4$ **7.** 121 **9.** 4 **11.** 35.5 ft **13.** no solution **15.** −12 **17.** 2 **19.** 75 **21.** 63 **23.** no solution **25.** 2 and 3 **27.** 6 **29.** 10 **31.** 140 **33a.** 1.6 m/s^2 **33b.** 9.8 m/s^2 **35.** $-5\sqrt{5}$ **37.** $-\sqrt{6} - 3\sqrt{2}$ **39.** $\sqrt{5}$

41.

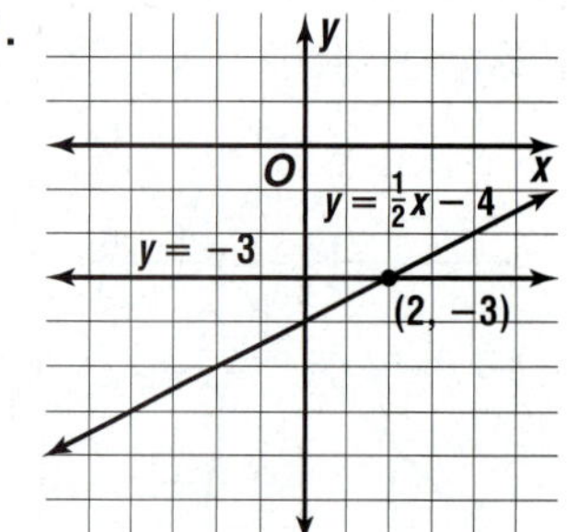

43. $x > 1$ **45.** 22.5

Page 629 Quiz 2

1. $3\sqrt{10}$ **3.** $\frac{7 - \sqrt{5}}{44}$ **5.** $12\sqrt{5}$ **7.** $7\sqrt{2}$ **9.** no solution

Pages 630–632 Study Guide and Assessment

1. $\sqrt{(x_2 - x_1)^2 + (y_2 - y_1)^2}$ **3.** radicals **5.** $\sqrt{7ab}$ **7.** whole **9.** no solution **11.** Q **13.** Z, Q

15. 2.2

$\sqrt{5}$

−5 −4 −3 −2 −1 0 1 2 3 4 5

17. 10.5

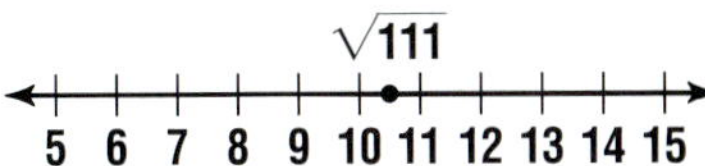

19. 5 **21.** $\sqrt{72}$ or 8.5 **23.** −1 or −5 **25.** $5\sqrt{2}$
27. $3\sqrt{3}$ **29.** $\frac{4+\sqrt{5}}{11}$ **31.** $10|a|c^2\sqrt{3b}$
33. $7\sqrt{5}$ **35.** 0 **37.** $22\sqrt{2}$ **39.** $-14\sqrt{5}$ **41.** 25
43. 9 **45.** no solution **47.** 6 **49.** 150
51. integers, rational numbers **53.** $10\sqrt{2}$ lb
55a. 3844 m **55b.** 2601 m

Page 635 Preparing for Standardized Tests

1. A **3.** A **5.** A **7.** C **9.** 50

Chapter 15 Rational Expressions and Equations

Page 637 Check Your Readiness

1. 0, 1 **3.** 1, −2 **5.** 6, 2 **7.** k^7 **9.** $12d^5$ **11.** $30m^2n^2$
13. $w^2 + 4w$ **15.** $15c - 20$ **17.** $r^3 + 3r^2 - r$
19. $y^2 + 2y - 3$ **21.** $p^2 - p - 30$ **23.** $2s^2 + s - 10$

Pages 641–643 Lesson 15–1

1. The denominator is 0 when $x = 2$. **3.** Darnell; x is not a factor of $x + 3$ and $x + 4$. Therefore, it cannot be divided out. **5.** $4x(4x - 5)$ **7.** $(a - 5)(a + 4)$
9. $(x - 6)(x - 4)$ **11.** 0, −6 **13.** $\frac{2}{5}$ **15.** $\frac{2y}{3z^3}$ **17.** $\frac{3}{4}$
19. $\frac{x-2}{x+1}$ **21.** $\frac{-(x+5)}{x+6}$ **23.** −5 **25.** 10 **27.** −2, 3
29. 0, −5 **31.** 4, −4 **33.** $\frac{1}{3}$ **35.** $\frac{2c}{3d}$ **37.** $-4bc$
39. $\frac{1}{5}$ **41.** $\frac{x+3}{x+1}$ **43.** $\frac{x}{2}$ **45.** $\frac{r+2}{3}$ **47.** $\frac{y+3}{y-4}$
49. $\frac{y+2}{y-1}$ **51.** $\frac{z-5}{z+3}$ **53.** $\frac{8m}{m-2}$ **55.** $\frac{c-5}{c(c+6)}$
57. $\frac{-x}{4+x}$ **59.** $\frac{4}{3+a}$ **61.** 9 yr **63.** 8 lumens/m^2
65. 16 **67.** 19 **69.** $7\sqrt{3}$ **71.** $6\sqrt{3}$ **73.** 5 **75.** D

Pages 647–649 Lesson 15–2

1. The reciprocal of $\frac{a^2-25}{3a}$ was used instead of the reciprocal of $\frac{a+5}{15a^2}$. **3.** $\frac{4}{x^2}$ **5.** $\frac{x+5}{x-4}$ **7.** $\frac{a^2}{bd}$
9. $\frac{2}{x+4}$ **11.** $4a$ **13.** $\frac{3x^2}{2}$ **15.** 12 flags **17.** $\frac{xy}{4}$
19. $3a$ **21.** $\frac{2s}{3}$ **23.** $\frac{x}{(x+4)^2}$ **25.** $\frac{2(x-3)}{x+3}$ **27.** $\frac{n}{n+2}$
29. $6a^2b$ **31.** $2a$ **33.** $\frac{y+4}{y^2}$ **35.** $\frac{(m+1)(m-1)}{2}$
37. $\frac{-(x+2)}{x-3}$ **39.** $\frac{-3a^2(a+3)}{2}$ **41.** $\frac{1}{x-y}$ **43a.** $\frac{x^4}{625}$
43b. $\frac{81}{625}$ or about 13% **45.** $\frac{7b^3}{a}$ **47.** $\frac{-4}{x-2}$
49. (1, 3) **51.** (−3, −4) **53a.** (5, −75), (8, 0), (10, 50)
53b.

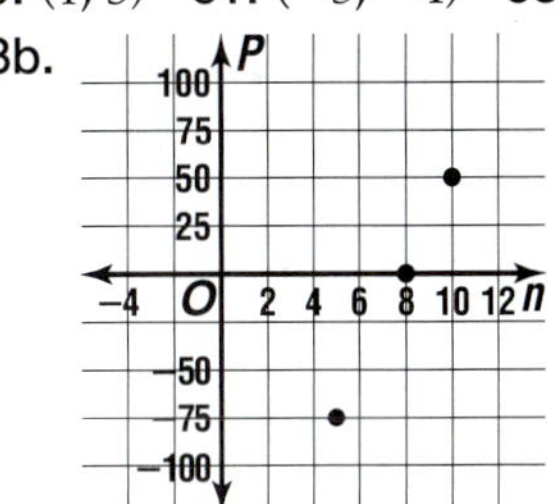

Pages 653–655 Lesson 15–3

1. dividend: $3k^2 - 7k - 5$; divisor: $3k + 2$; quotient: $k - 3$; remainder: $\frac{1}{3k+2}$ **3.** The quotient is a factor of the dividend if the remainder is 0. **5.** $3a$ **7.** x
9. $x + 2 - \frac{1}{2x-1}$ **11.** $2x + 3$ *in.* **13.** x **15.** $3r^2$
17. $x + 4$ **19.** $a + 5$ **21.** $t - 2 - \frac{32}{3t-4}$
23. $b + 2 - \frac{4}{2b-1}$ **25.** $x^2 - 3x + 9$
27. $4x^3 + 4x^2 + 2x + 3 + \frac{4}{x-1}$ **29.** $x - 3$
31a. 936 h **31b.** 140 h **31c.** 51 h **31d.** 5 h
33. 27 **35.** 1 **37.** one solution **39.** infinitely many **41.** $y \geq -6$

Page 655 Quiz 1

1. 2; m **3.** $\frac{1}{2}$ **5.** $\frac{-2(a+1)}{a-1}$ **7.** y **9.** $x^2 + x + 1$

Pages 659–661 Lesson 15–4

1. They are additive inverses.

3.

a	b	$a + b$	$a - b$	$a \cdot b$	$a \div b$
$\frac{2}{t}$	$\frac{1}{t}$	$\frac{3}{t}$	$\frac{1}{t}$	$\frac{2}{t^2}$	2
$\frac{12}{y}$	$\frac{4}{y}$	$\frac{16}{y}$	$\frac{8}{y}$	$\frac{48}{y^2}$	3
$\frac{x}{x-1}$	$\frac{1}{x-1}$	$\frac{x+1}{x-1}$	1	$\frac{x}{x^2-2x+1}$	x

5. $\frac{3}{4}$ **7.** $\frac{7}{x}$ **9.** −1 **11.** 2 **13.** $\frac{10y}{7x-2y}$ in. **15.** $\frac{7x}{8}$
17. $\frac{t}{2}$ **19.** $\frac{7}{6r}$ **21.** n **23.** $\frac{x}{12}$ **25.** $-\frac{1}{z}$ **27.** 0
29. 2 **31.** 3 **33.** $\frac{2y}{y-2}$ **35a.** Sample answer: $\frac{15{,}000}{5} - \frac{0.3(15{,}000)}{5}$ **35b.** $2100 **37.** $a + 4$
39. $y - 5$ **41.** $\frac{3axy}{10}$ **43.** $9x^2 + 6xy + y^2$

SELECTED ANSWERS

Pages 665–667 Lesson 15–5

1. Sample answer: $\frac{1}{2x}, \frac{1}{4x}$ **3.** Malik; he found the LCD and then added the terms. Ashley incorrectly added the numerators and the denominators.
5. 72 **7.** $30xy^2$ **9.** $\frac{4}{a^2}, \frac{5a}{a^2}$ **11.** $\frac{5t}{12}$ **13.** $\frac{2 + 3b}{ab^2}$
15. $\frac{-3x + 8}{x^2 - 4}$ **17.** 2042 **19.** $12a^2b^2$
21. $(y + 2)(y - 2)$ **23.** $(x + 3)(x - 3)(3x + 1)$
25. $\frac{35}{10t}, \frac{8}{10t}$ **27.** $\frac{5z}{xyz}, \frac{6x}{xyz}$ **29.** $\frac{30k}{3(3k + 1)}, \frac{k}{3(3k + 1)}$
31. $\frac{3d}{10}$ **33.** $\frac{15}{4x}$ **35.** $\frac{m + 3}{3t}$ **37.** $\frac{14x - 1}{6x^2}$
39. $\frac{2y + x^2 + 3x}{xy}$ **41.** $\frac{5x + 2y^2}{20x^2y}$ **43.** $\frac{2a + 3}{a + 3}$
45. $\frac{1}{(y - 1)^2}$ **47.** $\frac{22x + 5}{4(2x - 1)}$ **49.** $\frac{x^2 + 6x - 11}{(x + 1)(x - 3)}$
51. $\frac{m^2 - 5m + 5n}{m(m - n)}$ **53.** 17.5 min **55.** $\frac{16}{a}$ **57.** 2
59. $9\sqrt{19}$ **61.** in simplest form **63.** (2, 5)

Page 667 Quiz 2

1. $x + 8$ **3.** $\frac{8a}{9}$ **5.** 168

Pages 671–673 Lesson 15–6

1. Sample answer: In a linear equation, the variable cannot appear in the denominator and it cannot be raised to a power higher than 1. **3.** $-\frac{7}{3}$ **5.** $-\frac{10}{3}$
7. 3 **9.** 30 mph **11.** $-\frac{2}{3}$ **13.** 7 **15.** −3 **17.** 9
19. $\frac{4}{3}$ **21.** 7 **23.** $\frac{1}{12}$ **25.** 3 **27.** $\frac{1}{6}$ **29.** 6 **31.** −13
33. $-\frac{3}{2}$ **35.** $\frac{7}{5}$ **37.** 5 **39.** 2 **41.** 7 **43.** $\frac{9}{2n^2}$
45. Z, Q **47a.** $c + s \leq 16, 10c + 15s \leq 200$

47b.

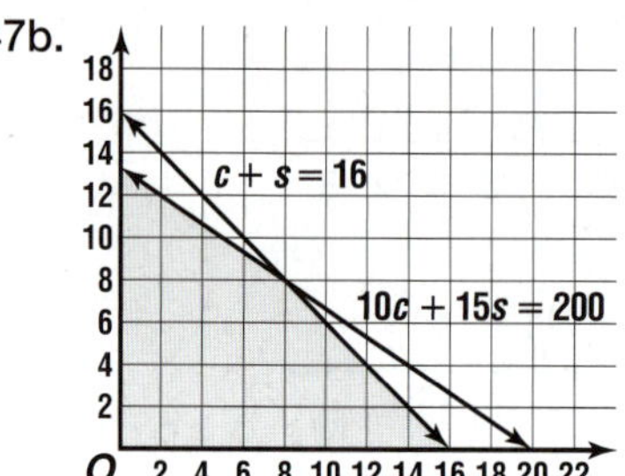

47c. Sample answer: (7, 8), he could plant corn for 7 days and soybeans for 8 days; (2, 12), he could plant corn for 2 days and soybeans for 12 days.
49. (4, −1);

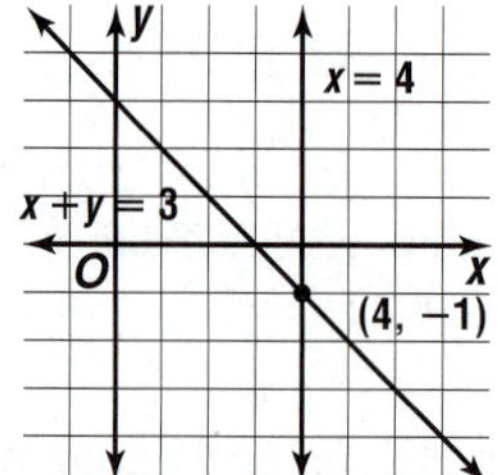

Pages 676–678 Study Guide and Assessment

1. denominator **3.** $\frac{x}{3x + 6y}$ **5.** greatest common factor **7.** $\frac{1}{a + b}$ **9.** $2x^2$ **11.** $\frac{xz}{4y^2}$ **13.** $\frac{n - 4}{n - 2}$
15. $\frac{2x^2}{3y^2}$ **17.** $\frac{x}{(x + 3)(x - 3)}$ **19.** $\frac{3(y - 2)}{y + 2}$ **21.** a
23. $2y + 1 + \frac{2}{2y + 3}$ **25.** $\frac{3x}{4}$ **27.** $\frac{5x}{x + 4}$ **29.** $\frac{6x + 5}{8x^2}$
31. $\frac{7a - 8}{a(a - 2)}$ **33.** $-\frac{1}{5}$ **35.** −4 **37.** 80 pieces
39a. $x + 1$ **39b.** 4 in., 3 in., 9 in., 108 in^3

Page 681 Preparing for Standardized Tests

1. A **3.** B **5.** D **7.** C **9.** 35

Photo Credits

Cover: (l)Getty Images, (r)Johnny Crawford/The Image Works; **v** file photo; **ix** Elizabeth Simpson/Getty Images; **x** Chip Simons/Getty Images; **xi** Grafton Marshall Smith/CORBIS; **xii** KS Studio; **xiii** Alan Carey; **xiv** Rich Brommer; **xv** Larry Tackett/Masterfile; **xvi** Rob Atkins/Getty Images; **xvii** SuperStock; **xviii** Lynn M. Stone; **xix** Garry Black/Masterfile; **xx** Jeri Gleiter/Getty Images; **xxi** K. L. Gay/©Arizona State Parks; **xxii** Mark Reinstein/Getty Images; **xxiii** NASA; **2** Amos Nachoum/Getty Images; **4** Mark Burnett; **7** Elizabeth Simpson/Getty Images; **9** Darrell Eager Photography; **11** Kenji Kerins Photo; **12** StudiOhio; **17** NPS photo/Cecil W. Stoughton/U.S. Department of Interior; **21** Robert Huntzinger/CORBIS; **28** Ed Shay; **31** George Diebold/CORBIS; **34** Mark Burnett; **35** Doug Martin; **36** Elaine Shay; **38** Rudi VonBriel; **50** Daryl Pederson/Alaska Stock; **54** SuperStock; **61** Chip Simons/Getty Images; **92** ThinkStock; **97 103 110** Mark Burnett; **113** StudiOhio; **115** Grafton Marshall Smith/CORBIS; **116** Phillip Colla/Innerspace Visions; **119** (t)Darren Bennett/Animals Animals, (b)Tom McHugh/Photo Researchers; **124** Giraudon/Art Resource, NY; **126** Mark Burnett; **127** Andrew Redington/Getty Images; **138** Brand X Pictures; **144** Mark Burnett; **146** (t)Janice Sheldon/Lonely Planet Images, (b) Carolina Biological Supply/Visuals Unlimited; **147** Mark Burnett; **150** KS Studio; **153** Aaron Haupt Photography; **157** Mark Burnett; **158** Steve Lissau; **161** Frank Balthis; **166** Alan Oddie/PhotoEdit; **170** Tony Freeman/PhotoEdit; **175** Michael Newman/PhotoEdit; **177** Andre Jenny/Lonely Planet Images; **179** Digital Vision, Ltd.; **186** Neal & Molly Jansen/SuperStock; **192** Mark Burnett; **193** Doug Martin; **194** AP/Wide World Photos; **195** Getty Images; **197** David R. Frazier Photolibrary/Photo Researchers; **209 210** SuperStock; **217** Alan Carey; **222** Mark Burnett; **229** Lucy Nicholson/Reuters/CORBIS; **236** RubberBall; **240** Peter Gridley/Getty Images; **249** Ariel Skelley/CORBIS; **253** Gary Randall/Getty Images; **254** Rich Brommer; **261 263** SuperStock; **265** NASA; **271** SuperStock; **274** Aaron Haupt Photography; **282** Vic Bider/Index Stock Imagery; **288** Paul L. Ruben; **299** Courtesy Aluminum Can Company of America; **302** Larry Tackett/Masterfile; **304** Aaron Haupt Photography; **307** Creative Focus/Masterfile; **315** Ex Rouchon/Explorer/Photo Researchers; **317** Renee Lynn/CORBIS; **326** Roberto Soncin Gerometta/Lonely Planet Images; **334** Michael Heller/911 Pictures; **337** (l)SSEC/UW-Madison, (r)file photo; **340** SuperStock; **345** Rob Atkins/Getty Images; **346** SuperStock; **349** Alfred Pasieka/Peter Arnold, Inc.; **351** SuperStock; **354** Albert Normandin/Masterfile; **356** Hal Horwitz/CORBIS; **357** ©Tribune Media Services, Inc. All Rights Reserved. Reprinted with permission; **359** Bill Brooks/Masterfile; **364** NASA/CORBIS; **368** SuperStock; **371** Mark Tomalty/Masterfile; **372–373** Icon Images; **380** SuperStock; **385** Syd Greenberg/Photo Researchers; **393** Courtesy Carol McAdoo; **410** J. A. Kraulis/Masterfile; **411** SuperStock; **418** Michael Boys/CORBIS; **427** Mark Burnett; **431** Aaron Haupt Photography; **434** Lynn M. Stone; **456** NASA; **460** Aidan O'Rourke; **467** Garry Black/Masterfile; **468** G. Randall/Getty Images; **480 485** Aaron Haupt Photography; **488** Aidan O'Rourke; **493** E. Alan McGee/Getty Images; **494–495** Mark Burnett; **502** Doug Pensinger/Getty Images; **509** Telegraph Colour Library/Getty Images; **517** Jeri Gleiter/Getty Images; **523** Ron Holt/Aristock, Inc.; **529** Peter Christopher/Masterfile; **532 534** Mark Burnett; **535** Richard Pasley/Stock Boston; **539** Mark Burnett; **540–541** Aaron Haupt Photography; **548** Michael Gilday/Alamy Images; **557** Joseph DiChello; **563** Lonely Planet Images; **567** K. L. Gay/©Arizona State Parks; **571** Mak-1; **575** David R. Frazier Photolibrary/Photo Researchers; **578–579** Sarah Jones

(Debut Art)/Getty Images; **585** VCG/Getty Images; **591** Mug Shots/CORBIS; **598** Gary Crabbe/Earth Scenes; **604** David Parker/ Science Photo Library/Photo Researchers; **608** Mark Reinstein/Getty Images; **610** Gordon R. Gainer/CORBIS, **611** Miles Ertman/Masterfile; **612** Mark Burnett; **614** FOX TROT ©1998 Bill Amend. Reprinted with permission of UNIVERSAL PRESS SYNDICATE. All rights reserved.; **623** CLOSE TO HOME ©1994 John McPherson. Reprinted with permission of UNIVERSAL PRESS SYNDICATE. All rights reserved.; **628** Peter Brock/Getty Images News Services; **636** Digital Vision, Ltd.; **640** Aaron Haupt Photography; **646** John Zimmerman/Getty Images; **666** NASA; **668** Aaron Haupt Photography; **671** Roberto Soncin Gerometta/ Lonely Planet Images; **673** Science Photo Library/Photo Researchers; **674–675** David Lissey/Getty Images; **683** (t br)Aaron Haupt Photography, (bl)Patrix Ravaux/Masterfile, (bc)Gary Randall/Getty Images; **724** Aaron Haupt Photography.

Index

Red type denotes items only in the Teacher's Wraparound Edition.

INDEX

D

E

H

I

INDEX

K

L

M

N

Q

R

S

T

U

V

W

X

Z

Properties

Substitution (=)	If $a = b$, then a may be replaced by b.
Reflexive (=)	$a = a$
Symmetric (=)	If $a = b$, then $b = a$.
Transitive (=)	If $a = b$ and $b = c$, then $a = c$.
Additive Identity	For any number a, $a + 0 = 0 + a = a$.
Multiplicative Identity	For any number a, $a \cdot 1 = 1 \cdot a = a$.
Multiplicative (0)	For any number a, $a \cdot 0 = 0 \cdot a = 0$.
Multiplicative (−1)	For any number a, $-1 \cdot a = -a$.
Additive Inverse	For any number a, there is exactly one number $-a$ such that $a + (-a) = 0$.
Multiplicative Inverse	For any number $\frac{a}{b}$, where $a, b \neq 0$, there is exactly one number $\frac{b}{a}$ such that $\frac{a}{b} \cdot \frac{b}{a} = 1$.
Commutative (+)	For any numbers a and b, $a + b = b + a$.
Commutative (×)	For any numbers a and b, $a \cdot b = b \cdot a$.
Associative (+)	For any numbers a, b, and c, $(a + b) + c = a + (b + c)$.
Associative (×)	For any numbers a, b, and c, $(a \cdot b) \cdot c = a \cdot (b \cdot c)$.
Distributive	For any numbers a, b, and c, $a(b + c) = ab + ac$ and $a(b - c) = ab - ac$.
Comparison	For any numbers a and b, exactly one of the following sentences is true: $a < b$, $a > b$, or $a = b$.
Addition (=)	For any numbers a, b, and c, if $a = b$, then $a + c = b + c$.
Subtraction (=)	For any numbers a, b, and c, if $a = b$, then $a - c = b - c$.
Division and Multiplication (=)	For any numbers a, b, and c, with $c \neq 0$, if $a = b$, then $ac = bc$ and $\frac{a}{c} = \frac{b}{c}$.
Product Property of Square Roots	For any numbers a and b, with $a, b \geq 0$, $\sqrt{ab} = \sqrt{a} \cdot \sqrt{b}$.
Quotient Property of Square Roots	For any numbers a and b, with $a \geq 0$ and $b > 0$, $\sqrt{\frac{a}{b}} = \frac{\sqrt{a}}{\sqrt{b}}$.
Zero Product	For any numbers a and b, if $ab = 0$, then $a = 0$, $b = 0$, or both a and b equal 0.
Addition (>)*	For any numbers a, b, and c, if $a > b$, then $a + c > b + c$.
Subtraction (>)*	For any numbers a, b, and c, if $a > b$, then $a - c > b - c$.
Division and Multiplication (>)*	For any numbers a, b, and c, **1.** if $a > b$ and $c > 0$, then $ac > bc$ and $\frac{a}{c} > \frac{b}{c}$. **2.** if $a > b$ and $c < 0$, then $ac < bc$ and $\frac{a}{c} < \frac{b}{c}$.

* *These properties are also true for $<$, $\geq$, and $\leq$.*